Passive Solar Design Handbook

Passive Solar Design Handbook

Part One: Total Environmental Action, Inc.
Revised by Bruce Anderson

Part Two: Los Alamos Scientific Laboratory

Part Three: Los Alamos National Laboratory

VNR VAN NOSTRAND REINHOLD COMPANY
New York

Portions of this report were prepared for the
U.S. Department of Energy
Assistant Secretary for Conservation and Solar Energy
Office of Solar Applications
Passive and Hybrid Solar Building Program
Washington, D.C. 20585

Copyright © 1984 by Van Nostrand Reinhold Company Inc.
Library of Congress Catalog Card Number 83-21805
ISBN 0-442-20810-3

All rights reserved. No part of this work covered by the
copyright hereon may be reproduced or used in any form or
by any means—graphic, electronic, or mechanical, including
photocopying, recording, taping, or information storage and
retrieval systems—without written permission of the
publisher.

Printed in the United States
Designed by Loudan Enterprises

Published by Van Nostrand Reinhold Company Inc.
115 Fifth Avenue
New York, New York 10003

Van Nostrand Reinhold Company Limited
Molly Millars Lane
Wokingham, Berkshire RG11 2PY, England

Van Nostrand Reinhold
480 La Trobe Street
Melbourne, Victoria 3000, Australia

Macmillan of Canada
Division of Canada Publishing Corporation
164 Commander Boulevard
Agincourt, Ontario M1S 3C7, Canada

16　15　14　13　12　11　10　9　8　7　6　5　4　3　2

Library of Congress Cataloging in Publication Data
Main entry under title:
Passive solar design handbook.

　Includes bibliographies and index.
　1. Solar houses—Design and construction.　2. Solar
energy—Passive systems.　I. Total Environmental Action,
inc.　II. Los Alamos National Laboratory.
TH7414.P26　　1984　　　690'.869　　　83-21805
ISBN 0-442-20810-3

Contents

Preface 10

ONE: PASSIVE SOLAR DESIGN CONCEPTS 11

A

Background to Passive Solar Building Design 12

A.1	Definition of Passive Solar Energy Systems 12
A.2	Five Basic System Types 12
A.2.a	*Direct-Gain Systems 13*
A.2.b	*Thermosiphoning Systems 13*
A.2.c	*Thermal Storage Walls 13*
A.2.d	*Thermal Storage Roofs 14*
A.2.e	*Attached Sunspaces 15*
A.3	Advantages and Disadvantages of Passive Solar Energy Systems 15

B

Basics of Solar Building Design 17

B.1	Energy Conservation 17
B.2	Comfort 17
B.3	Solar Position 18
B.4	Siting 20
B.5	Daylighting 21

C

Direct Gain 22

C.1	Introduction 22
C.2	Solar Gain 22
C.3	Design Fundamentals 23
C.3.a	*Thermal Mass 23*
C.3.b	*Movable Insulation 27*
C.3.c	*Thermal Performance 28*
C.4	Advantages and Disadvantages of Direct-Gain Systems 29
C.5	Examples of Direct-Gain Buildings 29
C.5.a	*The David Wright House; Santa Fe, New Mexico 29*
C.5.b	*St. George's School; Wallasey, England 30*
C.5.c	*MIT Solar Building V; Cambridge, Massachusetts 30*

D

Thermosiphoning Collectors 32

D.1	Introduction 32
D.2	Design Fundamentals 32
D.2.a	*Basic System Design 32*
D.2.b	*Airflow 33*
D.2.c	*Applications 36*
D.2.d	*Costs 36*
D.2.e	*Thermal Performance 36*
D.3	A Thermosiphoning Collector 36
D.4	Heat Collection and Storage Systems 38
D.5	Advantages and Disadvantages of Thermosiphoning Collectors 38
D.6	Examples of Thermosiphoning Systems 38
D.6.a	*The Paul Davis House; Albuquerque, New Mexico 38*
D.6.b	*The Mark Jones House; Santa Fe, New Mexico 39*
D.6.c	*A Thermocirculation Water Heater 40*

E

Thermal Storage Walls 41

E.1	Introduction 41
E.2	Thermal Storage Wall Systems 41
E.2.a	*Basic Designs 41*
E.2.b	*Design Variations 42*
E.3	Design Fundamentals 44
E.3.a	*Building Integration 44*
E.3.b	*Costs 44*
E.3.c	*Thermal Performance 44*
E.4	A Basic Trombe-Wall Configuration 46
E.4.a	*Materials 46*
E.4.b	*Design 47*
E.4.c	*Construction and Installation 47*

E.5	Advantages and Disadvantages of Thermal Storage Walls 47	
E.6	Examples of Thermal Storage Walls 49	
E.6.a	*Benedictine Monastery; Pecos, New Mexico 49*	
E.6.b	*The Doug Kelbaugh House; Princeton, New Jersey 50*	

F

Thermal Storage Roofs 52

F.1	Introduction 52
F.2	Basic System Configuration 52
F.3	Advantages and Disadvantages of Thermal Storage Roofs 52
F.4	Examples of Thermal Storage Roofs 52
F.4.a	*The Atascadero House; Atascadero, California 52*
F.4.b	*The Winters House; Winters, California 55*

G

Attached Sunspaces 56

G.1	Introduction 56
G.2	Design Fundamentals 57
G.2.a	*Glazing Systems and Performance 57*
G.2.b	*Building Integration 58*
G.2.c	*Ventilation 58*
G.2.d	*Thermal Storage 59*
G.3	Methods of Heat Transfer from Sunspace to Building 60
G.3.a	*Direct Solar Transmission 60*
G.3.b	*Direct Air Exchange 60*
G.3.c	*Conduction through Common Walls 62*
G.3.d	*Storage in and Transfer from Rockbeds 62*
G.4	A Compromise Sunspace Design 62
G.5	Advantages and Disadvantages of Attached Sunspaces 62
G.6	Examples of Attached Sunspaces 64
G.6.a	*Unit I; First Village, New Mexico 64*
G.6.b	*The Brookhaven House; Upton, New York 67*

H

Passive Solar Cooling 68

H.1	Introduction 68
H.2	Cooling Techniques 68
H.2.a	*Solar Control 68*
H.2.b	*Convective Cooling 74*
H.2.c	*Evaporative Cooling 78*
H.2.d	*Radiative Cooling 78*
H.2.e	*Ground Cooling 82*
H.3	Peak Load Reduction 83
H.4	Applications to Large Buildings 83

TWO: PASSIVE SOLAR DESIGN ANALYSIS 85

I

Sensitivity Analysis of Thermal Storage Walls 86

I.1	Effects of Thermal Storage Mass 86
I.2	Effects of Thermocirculation Vents 89
I.3	Properties of Trombe Walls 90
I.4	Effects of Multiple Glazings and Selective Surfaces 90
I.5	Wall Absorptance and Emittance 93
I.6	Properties of Glass 93
I.7	Effects of Overhangs 94
I.8	Effects of Glazing Tilt and Orientation 97
I.9	Effect of Wind Velocity 98
I.10	Effects of Wall-to-Room Heat Transfer Coefficient 98
I.11	Effect of Ground Reflectance 99
I.12	Effects of Changing Upper or Lower Control Temperature Settings 101
I.13	Effects of Additional Internal Mass 103
I.14	Effects of Side Shading 103
I.15	Mixed Systems 105

J

Estimating Temperature Swings in Direct-Gain Buildings 106

J.1	General Considerations 106
J.1.a	*Materials 106*
J.1.b	*Thickness 107*
J.1.c	*Location 107*
J.1.d	*Orientation 108*
J.2	Estimation Procedure 109
J.3	Example 109

K

Designing Fan-Forced Rockbeds 114

K.1	Hybrid Applications with Passive Solar Components 114
K.1.a	*Single-Zone Situations 114*
K.1.b	*The Two-Zone Approach 115*
K.2	Rockbed Characteristics 116
K.3	Design Procedure 118
K.4	Example 118
K.5	Options 120
K.6	Other Design Pointers 121

L

Economic Analysis 122

L.1	Introduction 122
L.2	Life-Cycle Cost Equation 122
L.3	Cost Evaluation 123
L.3.a	*Categorization of Costs 123*

L.3.b	*Passive Cost Estimation* 124		O.1.b.7	*Number of Glazings* 162
L.3.c	*Envelope Cost Estimation* 124		O.1.b.8	*Temperature Bounds* 163
L.3.d	*Conventional Cost Estimation* 124		O.1.b.9	*Nighttime Thermostat Setback* 164
L.3.e	*Final Passive Add-on Cost Estimation* 124		O.1.c	*Overheating Sensitivity Data* 164
L.4	Cost Optimization Concepts 124		O.1.c.1	*Thermal Storage Mass and Thermal Comfort* 164
L.5	Cost Optimization Procedure 129		O.1.c.2	*Lightweight Surfaces and Thermal Comfort* 166

THREE: PASSIVE SOLAR DESIGN PERFORMANCE 137

M

Review of Performance Analysis Methods 138

M.1	Terminology 138
M.2	Solar Load Ratio (*SLR*) Correlations 139
M.3	Annual Calculation— The *LCR* Method 140
M.4	Sensitivity Data 140
M.5	Units 141
M.6	Cooling Considerations 141

N

Balancing Conservation and Solar 142

N.1	Conservation Versus Solar 142
N.2	Conservation Formulas 143
N.3	Outline of Design Procedure 144
N.4	Step-by-Step Discussion of Design Procedure 145
N.5	Choosing Final Design Parameters 147
N.6	Example: The Method in Action 148
N.7	Life-Cycle Costs— A Different Approach 150
N.7.a	*Backup Heat* 150
N.7.b	*Review of the Example* 150
N.8	Cooling Issues 151
N.9	Other Advantages to a Balanced Approach 152

O

Direct Gain 153

O.1	High-Mass Direct-Gain Buildings 155
O.1.a	*The Reference Designs* 155
O.1.b	*Heating Performance Sensitivity Data* 156
O.1.b.1	*Thermal Storage Mass* 157
O.1.b.2	*Properties of the Thermal Storage Medium* 158
O.1.b.3	*Solar Absorptance of Mass Surfaces* 159
O.1.b.4	*Solar Absorptance of Lightweight Surfaces* 159
O.1.b.5	*Mass Distribution* 160
O.1.b.6	*Mass Coverings* 161
O.1.d	*Solar Load Ratio (SLR) Correlations* 166
O.2	Low-Mass Sun-Tempered Buildings 167
O.2.a	*The Concept of Sun-Tempering* 167
O.2.b	*The Reference Design* 168
O.2.c	*Performance of the Reference Design* 168
O.2.d	*Sensitivity Data* 171
O.2.d.1	*Mass Surface Area* 171
O.2.d.2	*Number of Glazings* 171
O.2.d.3	*Solar Absorptance of Gypsum-Board Surface* 171
O.2.e	*Example* 171

P

Sunspaces 174

P.1	The Reference Designs 174
P.2	Sensitivity Data 179
P.2.a	*Thermal Storage Volume* 179
P.2.b	*Water Container Shape* 180
P.2.c	*Orientation* 180
P.2.d	*Glazing Tilt* 181
P.2.e	*Sunspace Width* 184
P.2.f	*Number of Glazings* 184
P.2.g	*Glazing Air Gap* 185
P.2.h	*Glazing Thickness* 185
P.2.i	*Solar Absorptances* 186
P.2.j	*Lightweight Objects* 188
P.2.k	*Vent Area* 189
P.2.l	*Sunspace Infiltration Rate* 190
P.2.m	*Common-Wall Thermal Resistance* 190
P.2.n	*Thermostat Setpoints* 191
P.3	Solar Load Ratio Correlations 193
P.3.a	*Solar Load Ratio (SLR)* 193
P.3.b	*Absorbed Solar Radiation (S)* 193
P.3.c	*Effective Sunspace Load (L)* 193
P.3.d	*Degree-Days (DD)* 194
P.3.e	*SLR Correlations* 194
P.3.f	*Correlations for S* 194
P.4	Monthly Calculation— The *SLR* Method 195
P.4.a	*Calculating LCR$_s$* 195
P.4.b	*Calculating S* 196
P.4.c	*Calculating DD* 197
P.4.d	*Off-Reference Nighttime Insulation* 197

Q

Monthly Calculation—The *SLR* Method 198

Q.1	A Step-by-Step Procedure 198

Q.2	Calculating *S* 198		10.4.b	*Simple Payback, Definition #2 (SPBK$_2$)* 234
Q.3	Calculating *DD* 198		10.4.c	*Discounted Payback (DPBK)* 239
Q.4	Off-Reference Nighttime Insulation 199			
Q.5	The Correlation Equations 199			
Q.6	Mixed Systems 200			

APPENDIXES 201

Appendix 1	Properties of Glazing Materials 202		**Appendix 11**	Interest Tables 240
1.1	Glass 202		11.1	Compound Interest Tables 240
1.2	Fiberglass-Reinforced Polyester 202		11.2	Differential Series Present Worth Factor (*DSPWF*) Tables 247
1.3	Films 203		11.3	Remaining Loan Balance Factor 253
1.4	Clear Thermoplastics 204		**Appendix 12**	Tables of *LCR* and D Versus Versus *SSF* 256
1.5	Insulating Panels 204		**Appendix 13**	Fixed Charge Rate (*FCR*) Tables 285
Appendix 2	Specific Heats and Heat Capacities of Materials 205		**Appendix 14**	Fuel Cost Leveling Factors (*FF*) at Different Escalation Rates 294
Appendix 3	Chart of Mean Monthly and Annual Hours of Sunshine Throughout the United States 206		**Appendix 15**	Fixed Charge Rate (*FCR*) Plotted Against Various Other Rates and Factors 299
Appendix 4	Maps of Mean Monthly Hours of Sunshine in the United States 207		**Appendix 16**	Calculation of Diurnal Heat Capacity 303
Appendix 5	Sun-Path Diagrams for Various N Latitudes 214		**Appendix 17**	Conservation Factor Tables 307
Appendix 6	Conversion Factors 216		**Appendix 18**	Solar Load Ratio (*SLR*) Correlations 391
Appendix 7	Cost of Energy 217		**Appendix 19**	Weather Data 402
Appendix 8	Thermal Performance Values for Materials, Air Spaces, and Windows 218		**Appendix 20**	Solar Radiation Approximations 431
Appendix 9	Definitions of Economic Terms and Formulas 223		20.1	Incident Radiation 431
9.1	Time Value of Money and Discounting 223		20.2	Transmitted Radiation 431
9.2	Capital Recovery Factor 224		20.3	Transmittance Corrections 431
9.3	Differential-Series Present-Worth Factors and Alternative Fuel Costs, Operation and Maintenance Expenses, and Property Taxes 225		20.3.a	*Extinction Coefficient–Thickness Product 431*
			20.3.b	*Ground Reflectance 442*
			20.4	Absorbed Radiation 442
			20.5	Calculation Procedure 446
9.4	Remaining Loan Balance and Mortgage Interest Tax Deductions 227		**Appendix 21**	*LCR* Tables 461
9.5	Real Versus Nominal Terms 227		**Appendix 22**	Sensitivity Data 627
Appendix 10	Generalized Techniques of Economic Analysis 228		22.1	High-Mass Direct-Gain Systems 627
10.1	Cost Curves 228		22.1.a	*Heating Performance 627*
10.2	Dollar per Million Btu ($/MMBtu) Cost 229		22.1.a.1	*Mass Thickness 628*
			22.1.a.2	*ρck Product 639*
10.3	Annual Cash-Flow Analysis 232		22.1.a.3	*Mass Absorptance 640*
10.4	Payback Analysis 234		22.1.a.4	*Lightweight Absorptance Factor 641*
10.4.a	*Simple Payback, Definition #1 (SPBK$_1$)* 234		22.1.a.5	*Mass Distribution 642*
			22.1.a.6	*R-Value of the Mass Covering 643*
			22.1.a.7	*Number of Glazings 645*
			22.1.a.8	*Temperature Bounds 647*
			22.1.a.9	*Nighttime Setback 649*
			22.1.b	*Overheating 650*
			22.1.b.1	*Mass-Area-to-Glazing-Area Ratio 650*
			22.1.b.2	*Lightweight Absorption Factor 653*

22.2		Sun-Tempered Direct-Gain Systems 654
22.2.a		Heating Performance 654
22.2.a.1		Mass-Area-to-Glazing-Area Ratio 654
22.2.a.2		Number of Glazings 656
22.2.a.3		Mass Absorptance 658
22.3		Sunspaces 660
22.3.a		Heating Performance 660
22.3.a.1		Storage-Volume-to-Projected-Area Ratio 660
22.3.a.2		Common-Wall Thickness 662
22.3.a.3		Water-Volume-to-Projected-Area Ratio 664
22.3.a.4		Sunspace Orientation 666
22.3.a.5		Glazing Tilt—Fixed-Length, Single-Plane Glazing 668
22.3.a.6		Glazing Tilt—Variable-Length, Single-Plane Glazing 670
22.3.a.7		Vertical Glazing Height 674
22.3.a.8		Upper Glazing Tilt 676
22.3.a.9		Sunspace Width 678
22.3.a.10		Number of Glazings 684
22.3.a.11		Air-Gap Width Between Glazings 691
22.3.a.12		Glazing Thickness 693
22.3.a.13		Common-Wall Absorptance 695
22.3.a.14		Floor Absorptance 697
22.3.a.15		Floor and Wall Absorptance 699
22.3.a.16		Water-Container Absorptance 703
22.3.a.17		Lightweight Absorption Fraction 705
22.3.a.18		Vent Area Fraction 709
22.3.a.19		Sunspace Infiltration Rate 717
22.3.a.20		Common-Wall R-Value 719
22.3.a.21		Minimum Room Setpoint 723
22.3.a.22		Minimum Sunspace Setpoint 727

Glossary 731

References 734
Coded References 734
Additional References 736

Index 739

Preface

The *Passive Solar Design Handbook* is the product of over five years of writing and revising. The original Volume One was published by the Department of Energy in 1980, as was Volume Two. Volume Three followed in 1982. The present edition combines all three volumes into a single book that has been revised and updated to supply new information, remove inconsistencies, and correct earlier errata. Portions of the original Volume Two that were superseded by the methods and data in Volume Three have been omitted.

Part One of the present edition, newly revised by Bruce Anderson, is primarily qualitative. It provides a complete introduction to the concepts and terminology of passive solar energy design and is written as a guide for the uninitiated. Parts Two and Three are primarily quantitative. They present fast, simple analytical methods for gauging the relative performance of various types of passive solar heating systems and are intended as guides for architects, engineers, and researchers.

Passive solar design has matured from a curiosity to a commonplace. Combined with energy conservation, active-solar methods, and other energy options, passive solar has gained mainstream respectability as a major means of reducing energy consumption in a wide variety of building types. The *Passive Solar Design Handbook* is dedicated to enlarging the role of passive solar energy in the building designs of the future.

Part One

Passive Solar Design Concepts

Total Environmental Action, Inc.

Revised by Bruce Anderson

A Background to Passive Solar Building Design

A.1 DEFINITION OF PASSIVE SOLAR ENERGY SYSTEMS

Passive solar energy systems are methods for heating or cooling buildings or for heating water in which thermal energy flows by natural means (i.e., without pumps or fans). This natural thermal energy transfer may involve flows of thermal energy into and out of buildings, into and out of thermal energy storage, or around and through a conditioned space.

The above definition contradicts the common supposition that using solar energy requires an array of collectors, a thermal energy storage system, thermal energy transport systems that use pumps or fans, and the attachment to or installation in a building of the various components without greatly affecting the building's architectural fabric (roof, walls, floor, etc.). Solar energy systems that *do* include these various elements are termed "active" because of the moving parts and the power requirements of the fans and pumps.

Simple definitions of passive and active energy systems, however, tend to exclude techniques that combine natural thermal energy flow with mechanically powered energy flow. For example, a fan added to a passive system may improve the energy transfer or may provide an additional level of control over the amount and time of such transfer. Because combined-flow systems fit neither simple definition, they are sometimes called "hybrid."

The term "passive" may also overlap with some energy conservation definitions. Energy conservation in buildings is usually conceived in terms of reducing energy consumption, whether the conserved energy is renewable (e.g., solar) or nonrenewable (e.g., fossil fuel). Although solar energy systems can reduce consumption of fossil fuels, they are not generally regarded as energy conservation systems because they don't necessarily reduce a building's total energy use. Consequently, many simple techniques, such as combining south-facing glass and thermal mass, are usually not considered energy conservation measures.

Conceding a degree of overlap, a range of techniques for reducing a building's consumption of nonrenewable energy might look like this:

Energy Conservation Techniques • Passive Solar Design • Hybrid Solar Systems • Active Solar Systems

Energy conservation techniques tend to be the simplest and least costly of the four categories, while active systems tend to be the most complex and most expensive. In cost and complexity, passive systems generally fall somewhere between energy conservation techniques and active systems.

The proper semantic cataloging of these various general techniques is unimportant in comparison to the concepts they try to communicate. With good cause, the above terms may not be adopted universally; but as our understanding and use of these techniques evolve, so will our vocabulary for communicating about them.

The cataloging of specific passive solar techniques is also fraught with ambiguity. Dozens of variations occur within the five basic types of passive heating systems described in these pages. Sometimes they fit neatly, and sometimes they fit only through the use of a giant shoehorn. The great variety and diversity of passive system types, although complicating their classification, can also be an advantage: there are many bases on which to determine the one that is most appropriate for a particular climate, site, and building type.

Wise decision-making requires more than just a passing interest in the subject. Analyzing, predicting, and evaluating a passive system's thermal and economic performance is a complex task. Fortunately, however, passive solar designs usually apply simple concepts and are easy to build.

A.2 FIVE BASIC SYSTEM TYPES

While passive solar energy systems can be grouped in various ways, the approach used here is to describe them according to five physically identifiable methods of operation:

1. Direct-Gain Systems
2. Thermosiphoning Systems
3. Thermal Storage Walls
4. Thermal Storage Roofs
5. Attached Sunspaces

The physical images that these five methods evoke can help to simplify the task of introducing passive solar energy systems to millions of people. At the same time, this classification system is sufficiently flexible to permit innovation.

The following brief descriptions simply introduce the basic components of these five system types. Each in turn is discussed in detail in Chapters C through G.

A.2.a DIRECT-GAIN SYSTEMS

Direct-gain systems use solar radiation that enters through glass or plastic glazing directly into the space that is to be heated. See figure A-1. Nearly all of the solar radiation entering the room is immediately converted into heat. Thermal mass for storing excess solar heat is most effective when located so that it receives direct exposure to sunlight, as in a concrete floor. To reduce heat loss and thereby to increase overall thermal performance, insulation may be applied at night to the glass—either inside or outside. During the heating season, south-facing glass takes advantage of the sun's low position in the sky; in the summer, when the sun is high in the sky, the glass is shaded by overhangs, awnings, or trees.

A.2.b THERMOSIPHONING SYSTEMS

Figure A-2 illustrates the simplest form of a thermosiphoning air collector. As the air in the space between the glass and the blackened absorber surface is heated, it expands and becomes lighter, rises through the collector, and flows into the room through a vent at the top. Cooled room air, drawn into the collector through another vent at the base of the wall, replaces the warmed air leaving the collector. It, too, is heated and subsequently expelled from the top of the collector into the room. This process continues as long as there is enough solar radiation to raise the temperature of the collector above the temperature of the room.

A.2.c THERMAL STORAGE WALLS

In many applications of passive solar heating, thermal energy storage is located between a wall of glass or plastic and the space to be heated. See figure A-3. There are two general types of thermal storage walls. The first type uses heavy masonry materials, one foot or so thick. The outside surface of the wall, which is painted a dark color, heats up as the sun irradiates it; the heat is conducted through the wall and is then transmitted to the interior spaces several hours after the sun's energy strikes the wall. The second type uses water instead of masonry. Tubes, 55-gallon drums, and specially fabricated water-walls are commonly used for thermal storage.

In masonry thermal storage walls, vent holes frequently are placed at the top and bottom in order to allow room air to enter, rise in the warm space between the storage wall and the glass, and reenter the room. This combines the thermosiphoning process with a thermal storage wall. Such systems are usually called "Trombe walls," after Felix Trombe of Odeillo, France, who gave a substantial boost to their development.

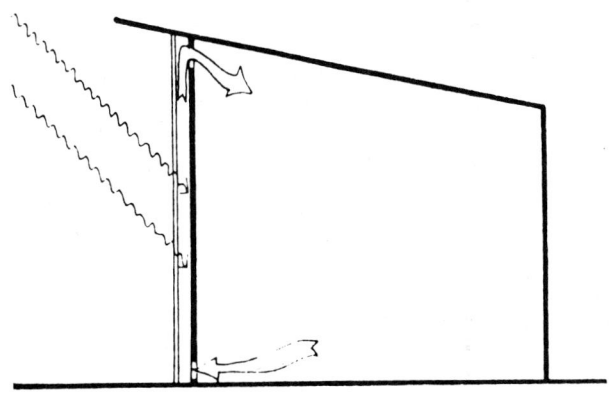

A-2: A thermosiphoning air heater [AND-4].

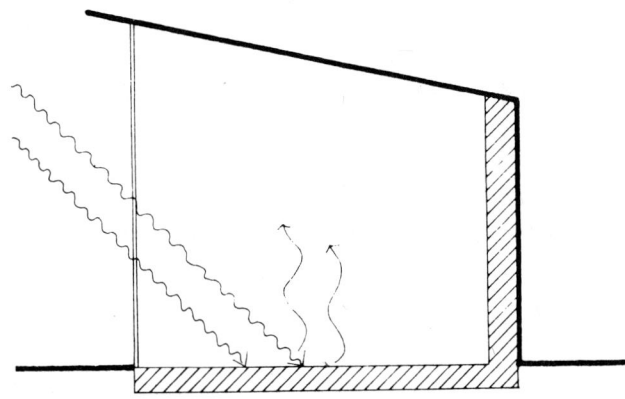

A-1: Direct gain [AND-4].

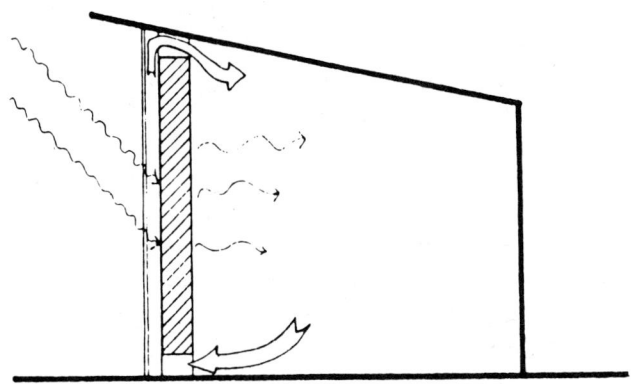

A-3: Thermal storage wall with thermocirculation vents [AND-4].

Fans can be used to increase and control airflow. Even when vents are used, however, the majority of the heat is absorbed by and conducted through the masonry material. In many systems, manual or automatic dampers prevent the nighttime reverse flow of air that would cool the space. As with direct-gain systems, movable insulation may be used to cover the glass at night to reduce heat loss and thereby increase overall thermal performance, especially in cold climates.

A.2.d THERMAL STORAGE ROOFS

Some passive designs locate the thermal storage on the roof. The most widely known system of this type, developed by Harold Hay, is known by its trademark name, Skytherm. See figure A-4. Skytherm uses roof ponds—water stored in large, clear vinyl bags that are supported on a black, waterproof liner. Solar radiation is absorbed by both the water and the black liners. This heat is then

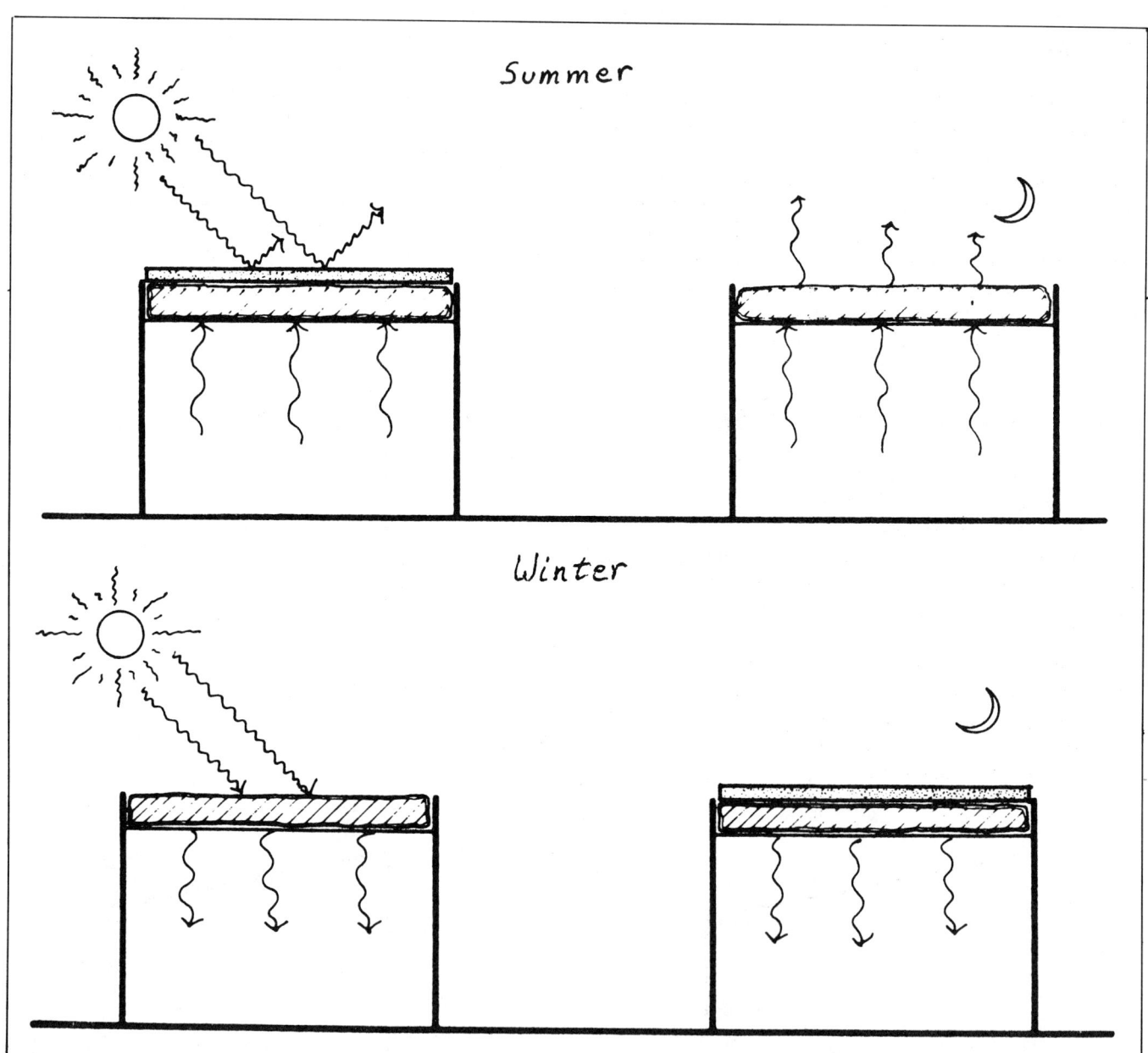

A-4: Skytherm thermal storage roofs—summer and winter operation [AND-4].

conducted through the ceiling, in contact with the bags, to the space below. Insulating panels cover the roof ponds at night to minimize heat loss.

In many climates, this system can also be used for cooling during the summer. The water absorbs heat from the space below and radiates the heat to the cold night sky. By flooding the tops of the sealed water bags, occupants can effectively take advantage of evaporative cooling. The insulating panels cover and shade the roof ponds during the day to minimize solar heat gains and are removed at night to permit cooling. This system, installed on flat roofs, is most applicable to southern latitudes.

Other variations, developed for heating in colder climates at higher latitudes, use glazing incorporated into a south-sloping roof pitch and water containers or "thermoponds" supported by the ceiling.

A.2.e ATTACHED SUNSPACES

Greenhouses and other "solar rooms" can be attached to new or existing buildings. See figure A-5. Overheated sunspace air may be delivered directly to the building that is to be heated, or the building and the sunspace can have a common thermal storage wall or rockbed. The heat stored in the thermal storage wall is shared by both the sunspace and the building. Some of the heat from the wall warms the greenhouse, thereby extending the growing season inside. At the same time, the wall helps to moderate greenhouse temperatures on sunny days. The sunspace provides substantial quantities of excess energy to heat the building, while simultaneously acting as a buffer zone to reduce heat loss from the building to the outdoors.

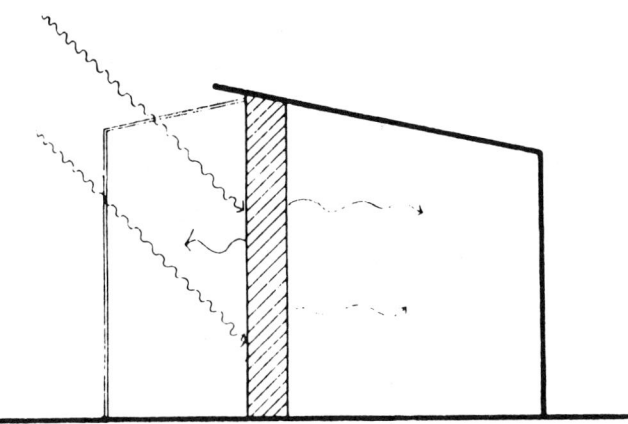

A-5: Solar greenhouse [AND-4].

A.3 ADVANTAGES AND DISADVANTAGES OF PASSIVE SOLAR ENERGY SYSTEMS

When compared with active systems, passive solar energy systems have both advantages and disadvantages. Most advantages stem from their inherent simplicity, which results in greater reliability, lower costs, and longer system life. Most of the disadvantages are related to "market acceptability" by homebuyers and the building industry, and to lower efficiency of operation.

Since passive systems have few or no moving parts and usually employ conventional building materials, their performance is reliable. Passive solar energy systems perform their solar task effortlessly and quietly, without mechanical or electrical commands or requirements. Generally, there are few maintenance requirements. If any external control is required (e.g., shutters), operating it is usually a simple task that can be performed easily by the occupant. Using conventional building materials, such as glass, concrete, and brick, involves well-known construction techniques, and these materials are long-lasting.

Simplicity, low initial costs of conventional materials, low maintenance costs, and long system life all contribute to the cost-effectiveness of well-designed passive systems. But perhaps the most significant reason for passive's cost-effectiveness is the long system life. Throughout the life of the building, a passive system should retain, if not improve, its value in at least equal proportion to the rest of the building. It should require no more maintenance than any other wall or roof. This is particularly true when the passive system fulfills the dual role of admitting solar energy and forming an integral part of the building's surface and structure.

While passive's simplicity lowers the capital cost (because it involves no motorized dampers, automatic valves, sophisticated control systems, or high-technology details or materials), a further consequence of this simplicity is the relative lack of legal barriers and certification requirements. Although there may be some new materials developed for passive that will require certification, certification procedures have already been established for the standard construction materials used in normal passive solar applications.

From the viewpoint of society as a whole, passive solar offers many benefits. The most obvious is the savings in fossil fuels, which helps the economy and preserves these resources for their optimum applications. The economy profits both by reducing the balance of payments deficit and by freeing capital from fossil fuel expenditures for better uses. The decrease in fossil fuel use also has a beneficial environmental impact. After installation, solar heating and cooling systems depend on few, if any, transmission lines, pipelines, or strip mines for their continued operation; they produce no dangerous radioactive wastes and no polluted air or water. Solar energy systems have few negative societal impacts, since they use materials that are renewable and can be recycled.

For optimum performance, some passive systems require daily moving of shutters or vents. To some people this is an imposition; others find it not only tolerable but a pleasant way of growing closer to their environment.

Many people feel that the radiant heat from large surfaces—a characteristic of most passive systems—is more comfortable than heat produced by conventional methods, which usually heat the air first. A frequent problem with some passive designs is large temperature swings; however, in well-designed systems, these swings are small—generally on the order of 5 Fahrenheit degrees. Some people, in any case, prefer warmer room temperatures during the cold winter months than can be maintained by a simple passive system.

It is difficult to incorporate passive solar designs without significantly affecting building appearance. This, in turn, may affect the willingness of builders to build it or buyers to purchase it.

The primary disadvantage of passive design is that, in most cases, it increases construction cost. For a system with a large solar fraction (i.e., a large decrease in conventional fuel bills) the installation cost can approach that of active systems. However, the financial outlay can be limited by building only a small system. Because the chances of system failure are so small, this makes the decision to use passive designs relatively risk-free.

B Basics of Solar Building Design

B.1 ENERGY CONSERVATION

Passive solar heating and cooling methods extend our understanding of energy conservation and of designing with, rather than against, nature. Countless pages have been written on the philosophy and the applications of energy conservation and of climate-responsive and site-sensitive design. Although these subjects are not discussed in detail here, a few observations on the relationship between passive design and conservation are in order.

Energy conservation is the best first step in the thermal design of buildings. This makes sense from both an economic and a practical engineering standpoint. The procedure for determining the optimum level of energy conservation to accommodate in a design is no different for a passive or active solar house than for any other building. Insulation is added until a point of diminishing returns is reached—that is, until the cost of additional insulation begins to exceed the life-cycle cost of the fuel that the added insulation will save. The same procedure is used for determining the optimum size of the solar energy system: its size is increased until the cost of any additional collection area (including all of the associated system costs) would exceed the life-cycle cost of the fuel to be saved from the additional collector. The optimum conservation mix has been achieved when there is a three-way equality: the added cost of further energy conservation equals the added cost of additional solar collection equals the life-cycle cost of the fuel to be saved by either.

By reducing energy demands through energy conservation, conventional energy systems can be reduced in size and complexity. For example, the traditionally elaborate heat-distribution system may be eliminated. Instead, a central space heater (through-the-wall type, beneath-the-floor type, or even a wood stove) can supply the remaining heat requirements for a full-size house.

Many designers have found that, when energy conservation and solar energy systems (whether active or passive) are analyzed according to the same economic criteria, climate determines the appropriate mix. In cold, cloudy climates, energy conservation measures are usually more cost-effective than solar energy systems until the buildings' heating loads are reduced to a very small fraction of conventional loads. On the other hand, in sunny climates, solar heating may be more cost-effective than energy conservation for reducing both fuel bills and fossil fuel use.

B.2 COMFORT

The human body uses three basic mechanisms to maintain comfort: convection, evaporation/respiration, and radiation. Factors affecting comfort are air temperature, humidity, air speed, and mean radiant temperature. Mean radiant temperature is particularly important to passive design.

Mean radiant temperature (MRT) is the temperature that the surfaces of an imaginary black enclosure (equal in size and shape to the actual room involved) would have to have if they were to exchange an amount of heat (by radiation) with a person equal to the amount of heat that the person is actually exchanging with all of the surfaces of the room he or she is in. Different combinations of MRT and air temperature can produce the same comfort sensation. Under typical conditions, a change of plus or minus 1.5 Fahrenheit degrees MRT is balanced by a change of plus or minus 2 Fahrenheit degrees air temperature. The following pairs of combined numbers (in degress Fahrenheit) produce the equivalent sensation of 70°F [HAG-2]:

Air Temperature	49°	56°	63°	70°	77°	84°	91°
Mean Radiant Temperature (MRT)	85°	80°	75°	70°	65°	60°	55°

The first pair of numbers, for example, indicates that if the air temperature is 49°F and the temperatures of the wall surfaces and other surrounding surfaces average 85°F, then the comfort sensation will be of a temperature of 70°F. The last pair of numbers reveals that if the air temperature is 91°F and the MRT of the surrounding surfaces is 55°F, the comfort sensation experienced by most lightly-clad occupants will be similar to the one they would experience in a room with an air and surface temperature of 70°F.

This can have important design implications. For example, the better a wall is insulated, the higher will be its interior surface temperature during cold weather. This warmer surface temperature allows interior air temperatures to remain lower and still provide comfort; in turn, heat loss to the outside is reduced because of the smaller difference between indoor and outdoor air temperatures. This lower heat loss reduces the size of the heating system required and reduces the need to deliver heat to the building's perimeter.

The effect of MRT is crucial for providing comfort in passive solar design because many passive systems rely on warm (or cool) surfaces to exchange energy with the air. For heating, the often-higher MRTs provide comfort at relatively low air temperatures.

B.3 SOLAR POSITION

The sun's position is designated by the two angles shown in figure B-1. Solar altitude (β) is the angle measured upward from the horizontal to the sun. It equals 0° when the sun is on the horizon and 90° when the sun is directly overhead. Solar azimuth (ϕ) is the angle measured in a horizontal plane from true south to the projection of the sun on the horizon. It equals 0° at south, 90° at east and west, and 180° at north [ASH-1]. (Some references instead measure the azimuth from true north.)

The sun's path across the sky varies with the time of year, but it always follows a circular arc across the sky dome. Figure B-2 shows three diurnal paths: the summer solstice (21 June), the vernal and autumnal equinoxes (21 March and 21 September), and the winter solstice (21 December). The numbers in circles represent times of day.

Solar altitude and azimuth can be determined for the twenty-first day of each month and for any hour of the day by using sun-path diagrams. A different diagram is required for each latitude, although interpolation between graphs is reasonably accurate. Reprinted in figure B-3 is a representative diagram for 40°N latitude. Other diagrams are given in Appendix 5 and in references [RAM], [BEN], and [MAZ-2].

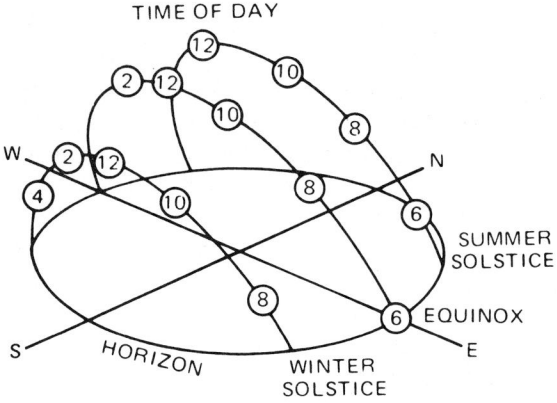

B-2: The sun's daily path across the sky: the sun is higher in the sky during the summer than it is during the winter, due to the tilt of the earth's axis [AND-1].

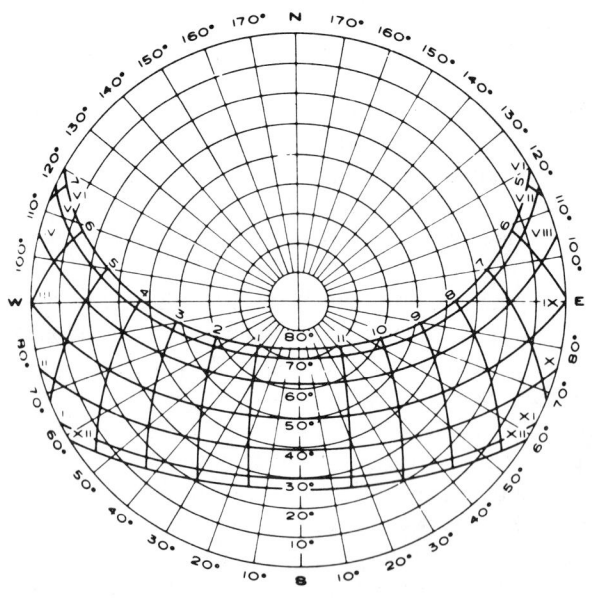

B-1: Measuring the sun's position: the solar altitude (β) is the angle between the sun and the horizon; the azimuth (ϕ) is measured from true south [AND-1].

B-3: Sun-path diagram for 40°N latitude [RAM].

By using the diagram in figure B-3, you can determine (for example) the solar altitude and azimuth at 4:00 P.M. solar time on April 21 in New York City (40°N). Locate the April line—the dark line running left-to-right that is numbered "IV" (April is the fourth month). Next, locate the 4:00 P.M. line—the dark up-and-down line that is numbered "4." The intersection of these lines indicates the solar position. Solar altitude is read from the concentric circles; in this case it is 30°. The solar azimuth is read from the radial lines, 80°W in this case.

A primary application of solar angle information is in the determining of shading angles for windows and collector surfaces. Shading should protect a surface from excessive sun but not from useful solar energy. The above references are excellent guides for designing shading devices. In addition, reference [LAU] contains an excellent discussion of this subject.

The quantity of solar radiation that penetrates a south-facing window on an average sunny day in the winter is greater than that through the same window on an average sunny day in the summer. There are a number of reasons for this.

First, although more daylight hours occur during the summer than during the winter, there are more hours of possible sunshine on a south-facing window in winter than in summer. For example, at 35°N latitude, there are fourteen hours of sunshine on June 21; but, since the sun remains north of east until after 8:30 A.M. and moves to the north of west before 3:30 P.M., direct sunshine occurs for only seven hours on the south-facing wall. On December 21, however, the sun is on the south wall for the full ten hours that it is above the horizon.

Second, since the sun is closer to the horizon during the winter, the rays strike the windows at more nearly perpendicular angles than they do in the summer, when the sun is at a higher altitude. At 35°N, 150 units of energy may strike a square foot of window during an average winter hour; during the summer this number would be 100 units.

Third, the more nearly the angles at which the sun's rays hit the windows approach right angles, the greater the transmittance (fraction of radiant energy that, having entered a layer of absorbing material, reaches the farther boundary) of the sunlight through the glazing becomes.

Fourth, a small roof overhang, such as is usually to be found above a window, will shield it from most direct summer irradiation.

About twice as much solar radiation is transmitted through unshaded south-facing windows in winter as in summer. If the windows are shaded in the summer, the difference is significantly greater. In contrast, east- and west-facing vertical surfaces have greater solar gain during the summer than during the winter. Figure B-4 is a plot of the average solar radiation values on vertical walls of various orientations in New York City [AND-2].

Vertical surfaces are more adaptable to passive systems than tilted surfaces, such as roofs, are. The amount of solar energy striking a south-facing vertical surface in northern latitudes during the winter is almost identical to that striking a steeply tilted surface. With reflective surfaces—such as snow—on the ground, a south-facing vertical surface may actually receive more incident energy during the middle of the winter than a south-facing tilted one does. Figure B-5 compares incident energy on south-facing surfaces at tilts ranging from vertical to

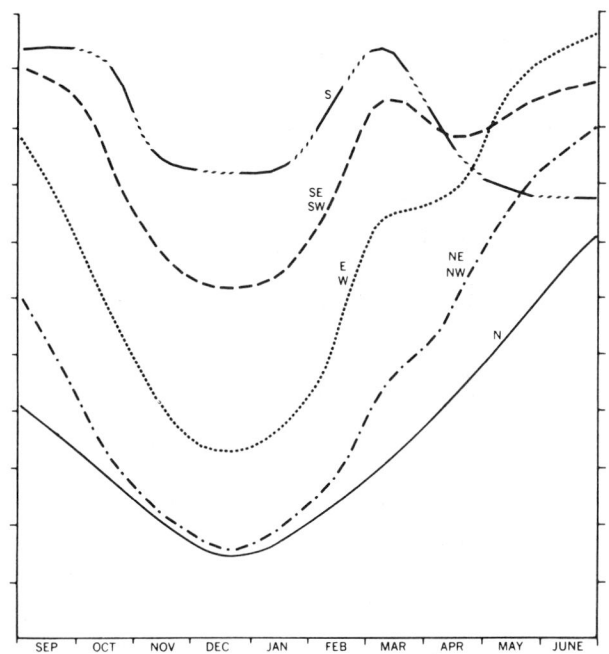

B-4: Relative average solar radiation on vertical walls in New York City [AND-2].

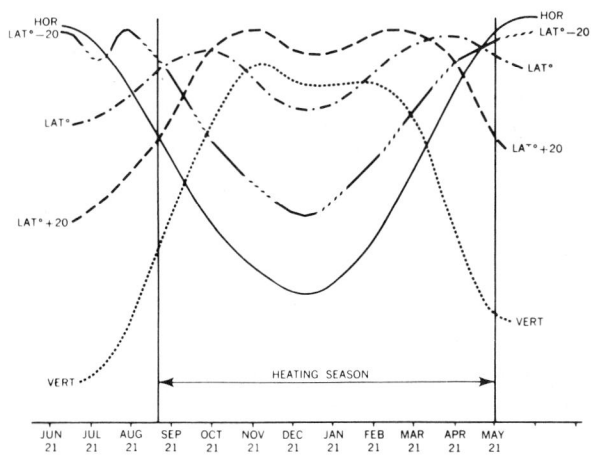

B-5: Effect of tilt on the direct radiation striking south-facing surfaces at 40°N latitude [JOR].

horizontal. The curves represent energy levels for buildings at 40°N latitude. Notice that during the primary heating months, from October 1 to March 15, tilted surfaces receive only a little more direct radiation than vertical surfaces receive. The difference becomes even less significant in more northern latitudes [AND-2]. Keep in mind also that tilted surfaces receive substantially more solar irradiation in the summer than vertical surfaces do and are more difficult to shade.

Figure B-6 compares the daily total sky radiation on clear days, both on earth and outside the atmosphere, at 42°N latitude on south-facing vertical surfaces. Notice that the amount of solar radiation on earth at this latitude from October 1 to March 15 fluctuates very little, enabling south-facing vertical surfaces to perform well throughout the heating season. In the same figure are three curves based on actually recorded daily radiation incident upon a south-facing vertical surface; included in the figure are weekly means and the minimum and maximum records. These data, recorded at Blue Hill, Massachusetts, are representative of daily radiation figures recorded at other localities at about 42°N latitude that have reasonably dust-free atmospheres and about 50% sunshine during daylight hours in December and January. Until fairly recently, Blue Hill Observatory, which began collecting data in 1945, was the only weather station in the country that maintained recorded data for radiation incident upon south-facing vertical surfaces.

B.4 SITING

Siting issues must be addressed early in the design process. If the site does not have proper solar exposure (if, for example, it is sloped sharply north or is darkly shaded by evergreens or large buildings), the building designed for the site also has less chance for good solar exposure. In general, south-facing facades (including those facing east and west of south) should be exposed to sunlight during the winter months and shaded during the summer.

There are four solar land-planning considerations particularly appropriate to housing subdivisions:

Lot Orientation. South-facing houses assure lower energy consumption during both summer and winter. Figure B-7 compares percentages of solar radiation in New York City on vertical surfaces deviating from true south. Surfaces that are oriented up to 30° east or west of south receive nearly the same amount of solar radiation as surfaces facing true south. Thus, most engineers recommend that solar apertures should be within 30° of

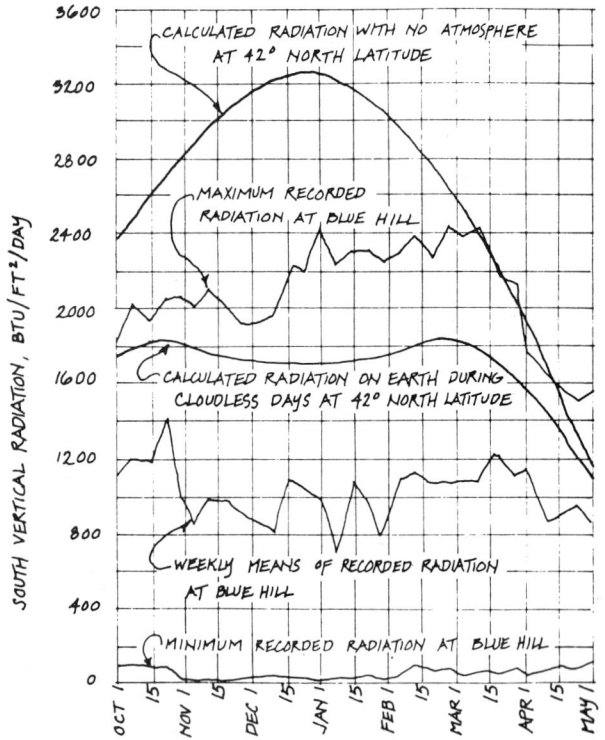

B-6: Calculated and recorded daily radiation incident upon a south-facing vertical surface at Blue Hill, Massachusetts (near Boston) [JOR].

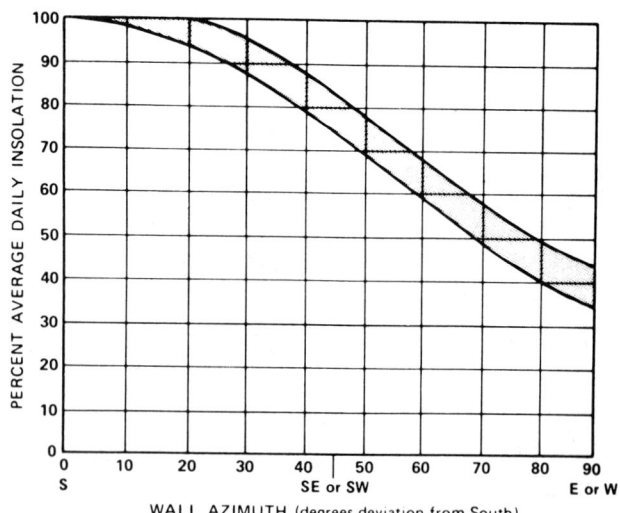

B-7: The percentage of solar radiation on vertical walls for orientations away from true south [AND-2].

true south, with true south preferred. True south may not always be optimal, but it will almost always be better than 30° east or west of it [OLG-1]. Except in complex terrain, this land planning change in subdivisions can be accomplished with little or no increase in development costs.

Solar Access. In most construction situations, long-term, shade-free, direct access to sunlight is necessary to guarantee effective solar use over the life of the building. Vegetation must be properly planned to permit access to breezes during the summer and to sunlight during the winter.

Street Width. Nonporous, heat-absorbing pavements dramatically affect the developed environment. Although proper landscaping and land planning can reduce the impact of such surfaces, street design is also important. Narrow streets, for example, save valuable land and can be shaded more easily than wide streets can. Parking bays rather than on-street parking can promote shading, too. Pedestrian and bicycle systems are readily integrated into such a plan.

Landscaping. Proper vegetation and landscaping can provide environmental beauty, enhance comfort, and save energy. Large deciduous trees, such as oak, hackberry, sycamore, ash, and maple, provide shade, promote evapotranspiration, and shed their leaves in the winter to admit the sun for heating.

Proper landscaping can have subtle but multiplying effects on many energy-consuming activities. For example, shading and landscaping may often be decisive factors when people consider walking or bicycling. Glaring, unshaded asphalt creates desertlike microclimates in the summer. Landscape designs that encourage home gardening can save energy in the food sector.

B.5 DAYLIGHTING

The bonus provided by natural daylighting in passive solar designs is impressive. In many cases, solar glazing may save more energy and money by reducing the need for artificial lighting than it saves by reducing fuel bills.

J. W. Griffith, past president of the Illuminating Engineering Society, reported on lighting studies at Southern Methodist University [GRI-1]:

> There is a considerable loss of contrast from overhead lighting when compared to . . . sidewall lighting systems. If it takes 10 to 15 percent more illumination to make up for each 1 percent loss of contrast, most of these tasks would require two to three times as much illumination . . . from overhead sources [as from] sidewall lighting.

In a 1977 ASHRAE report, Mr. Griffith said [GRI-2]:

> Daylight illumination from windows has been shown to be three to four times as effective in increasing visual performance as equal illumination from conventional electric lighting if properly utilized.

Years ago, classroom seating in many elementary schools was arranged so that daylight from sidewall windows would come in over the right shoulders of left-handed people and over the left shoulders of right-handed people. This kept glaring light sources out of what illuminating engineers call the offending or glare zone. As lighting contrast increases, visual performance improves rapidly.

Natural lighting systems, however, are relatively difficult to design and engineer; the Illuminating Engineering Society's *IES Lighting Handbook* and other industry sources contain valuable daylighting techniques and guidelines [IES], [IDC].

C Direct Gain

C.1 INTRODUCTION

By far the most common way solar energy is used to heat buildings is by means of its penetration through the windows of our country's 70,000,000 buildings. Without this direct-penetration energy, the United States would use 2% to 3% more nonrenewable energy than it does now. Windows vary considerably in their thermal performance, however. Loose-fitting, single-glazed windows, for example, usually lose more heat than they admit in the form of solar heat gain. On the other hand, a properly designed window facing south, with perhaps a reflective surface on the ground (such as snow or an aluminized, mirrored surface) and with means of reducing heat loss significantly at night (such as movable insulation), can supply almost as much solar energy to a building as a good active solar collector of the same surface area.

In colder and cloudier climates, windows provide fewer than 40,000 Btu; and in warmer and sunnier locations, as many as 70,000 Btu. Another important factor that helps to determine the amount of energy provided is the size of the window area in comparison to the heating load. The larger the window area in comparison to the heating load, the greater the quantity of excess, wasted solar heat and the lower the contribution of useful heat by each square foot of window.

Little is understood of the way in which light energy is reflected, transformed, reradiated, and ultimately distributed once it enters a building. This lack of understanding has resulted in, and continues to result in, significant errors in the integration of glass into the fabric of buildings. Even today, many buildings that rely for their heating primarily on the direct gain of sunlight through south-facing glass and are designed by knowledgeable passive-solar building designers do a poor job of satisfying human comfort and other needs of the building's occupants. Although a thermally elegant direct-gain system can perform like a finely tuned machine and satisfy human comfort to a high degree, an improperly designed direct-gain system can create discomfort.

Design considerations for properly designed direct-gain systems include the following:

1. The timing must be right—that is, the sun must enter the building at the right time of year. (Corollary: the sun must be kept out at the right times of day and year—that is, the solar heat gain must correspond to the comfort needs of the building's occupants.)
2. The amount of solar heat gain must correspond to the heating load of the building. Too much glazing should be avoided because it might result in large temperature swings. While some occupants may accept large fluctuations in temperature, the designer should not count on it as an excuse for a poorly engineered direct-gain system.
3. The right type of glazing should be used. Clear glass has its place, but so do clear and translucent materials such as diffusing glass, plastic films, fiberglass-based glazings, and acrylics. Reflective glazings may be appropriate for reducing unwanted solar gains. See Appendix 1 for a discussion of various glazings.
4. Heat loss back through the glazing should be reduced to as low a level as possible.
5. The manner in which sunlight enters the building must be compatible with the program needs of the building. One factor to consider is that many activities are not amenable to direct solar gain. Many people, for example, simply do not like working in direct sunlight; in fact, north light is often preferred, and it is generally recognized that lighting from the side (in most cases) assists in proper distribution of light throughout a space, reducing glare. Another factor is that too much glass can infringe on privacy inside a building space by creating a "fishbowl" effect.

C.2 SOLAR GAIN

There is growing recognition that direct-gain systems are easier to design if they are sized to provide only a limited portion of the heating load. For systems designed to satisfy only daytime heating requirements (usually around 10% of the total heating burden), necessary glass areas can be easily incorporated into most building facades. Although heat storage requirements in these instances should not be ignored, they are not the overwhelming considerations that they are in systems with high (greater than 20%) solar fractions. Most of the other problems associated with direct-gain systems are also reduced.

The glass should be placed facing as close to south as possible. Since shading is more difficult during the summer, deviations from true south should be avoided beyond southeast and southwest (45°). However, even in cold climates, east and west glass areas can admit somewhat more solar energy during the winter than they lose, assuming thermal shades cover the windows at night to reduce heat loss.

Glass placed vertically admits nearly as much heat during the winter as does glass placed at the tilted angles that are optimum for many active-system collectors. Tilted glass presents designers with a number of problems. It is relatively difficult to shade during the summer; left unshaded, it admits unwanted heat. In addition, large areas of tilted glazing may be difficult to integrate into a building's overall design; and tilted glazing is noticeably more difficult to keep clean. Many codes require the use of tempered glass in all tilted configurations located above occupied space.

Window areas should be incorporated into the walls and roof of a given building in such a way as to distribute the heat to as much of the building as possible. To reach northern rooms, vertical clerestory windows for easier summer shading are preferred over sloping skylights. Sawtooth roofs distribute south sun throughout interior portions of large-roofed buildings.

A principal consideration in locating windows is to enhance the exposure of the thermal mass to direct sunlight. See section C.3. For example, if the building has heavy floors and walls, windows should be placed so that as much of the floor and wall surface as possible is bathed in light. Translucent glass with good transmittance will scatter the light through the space, distributing the heat to many surfaces at once. This can both keep interior temperatures down and improve lighting quality. Translucent surfaces produce more glare, and this must be considered in their design.

In the final design of a building, these many thermal considerations must be blended with aesthetics and with the overall purposes of the building.

C.3 DESIGN FUNDAMENTALS

C.3.a THERMAL MASS

Using mass heat storage in the design of residential buildings is as old as the concept of shelter itself. For centuries, massive adobe Indian dwellings in the southwestern United States have tempered the effects of large fluctuations in daily outdoor temperatures on interior comfort conditions. The centercore fireplace and hearth mass of colonial houses stored heat from the fire during the day and slowly released that heat at night to keep the houses warm as the fire died.

Thermal mass is necessary to absorb solar radiation and store it for later use, usually in the evening. Thermal mass also serves to dampen temperature swings in a direct-gain system, keeping temperatures within acceptable limits.

The amount of mass required depends on the type of mass and its location in the building. Types of thermal storage mass include concrete, brick, gypsum, oak, rockbeds, and other heavy materials.

Commonly, the thermal mass is in the floor or in vertical walls or partitions. The best location for the mass is in direct sunlight; however, this is not always possible. Some of the mass may be in indirect sunlight—that is, on the walls and ceiling of a sunlit room. This is the second best location. Thermal mass may also be located on the floors, walls, and ceilings of rooms remote from sunlight.

Since mass is heavy and since floors are adjacent to windows, a common method for storing heat is to locate the thermal mass in the floor. Yet even the best floors designed for solar exposure are often covered or shaded by rugs and furniture, and rarely can be counted on for heat storage.

The floor mass need not be thicker than four inches. Additional thickness saves only marginally more heat. Dramatic thicknesses—for example, several feet of earth topped by a floor slab—will have a tempering effect on interior temperatures during long, cold, cloudy spells, but the extra building cost rarely can be justified solely on the basis of reduced fuel bills. Nevertheless, in combination with extremely tight construction, these extra-thick floors can enable buildings to coast through bad weather with little or no backup heat while thermostat temperatures are kept low.

Thermal mass is often costly, and its inclusion can inconvenience the construction process. For example, extra floor area may be needed to accommodate thick masonry walls. If the use of thermal mass requires a different type of construction than normal, this may jeopardize the building's acceptance by the building industry.

Remote thermal mass, in the form of rockbeds, has been used in an attempt to counteract these problems. Rockbeds are often two feet deep and are usually located beneath the floor slab of the building. Overheated room air is blown horizontally through the bed. The heat rises through the floor slab directly into the space. Unfortunately, because of the moderate temperatures of overheated rooms, large volumes of air must be moved. This requires large fans and considerable fan energy. Also, the resultant increase in the temperature of the rockbed is so slight that little heat is actually stored. Rockbeds are not advisable for direct-gain systems, unless room air temperatures in the 80s and 90s can be stored. They can, however, be considered for attached sunspaces, and they are discussed in this connection in more detail in Chapter G.

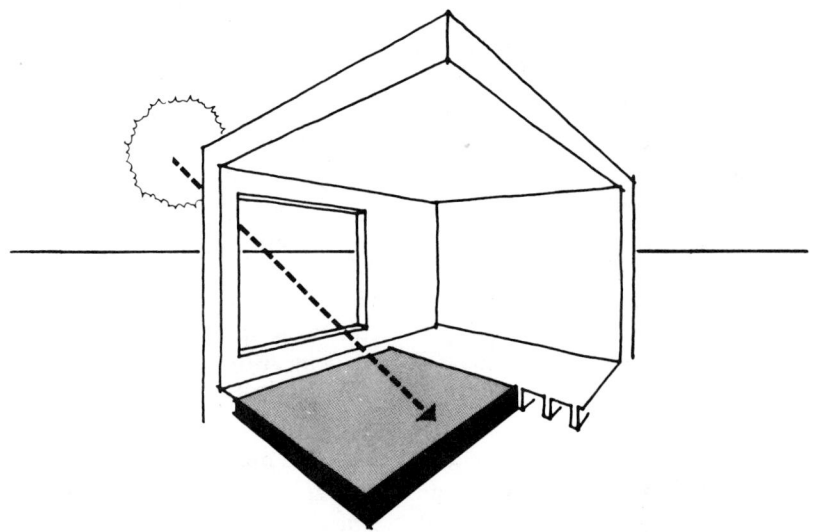

This pattern is defined as thermal mass that has one surface exposed to the living space and a 'back' surface that is insulated. The exposed surface is further defined as being in direct sunlight for at least six hours a day. Architecturally, this pattern combined with Pattern 2 is useful for direct gain passive rooms. The mass can be either a directly irradiated floor slab as shown, or a directly irradiated exterior wall. As with Patterns 2 and 3, the mass element is one-sided, that is, heat moves into and out of the mass from the same surface.

Mass Sizing Table

Material Thickness	Directly Irradiated Mass Surface Area to Passive Aperture Area Ratio				
	Concrete	Brick	Gypsum	Oak	Pine
½"	-	-	76	-	-
1"	14	17	38	17	21
1½"	-	-	26	-	-
2"	7	8	20	10	12
3"	5	6	-	10	12
4"	4	5	-	11	12
6"	3	5	-	11	13
8"	3	5	-	11	13

C-1: Pattern 1: floor or wall in direct sunlight [TEA-3].

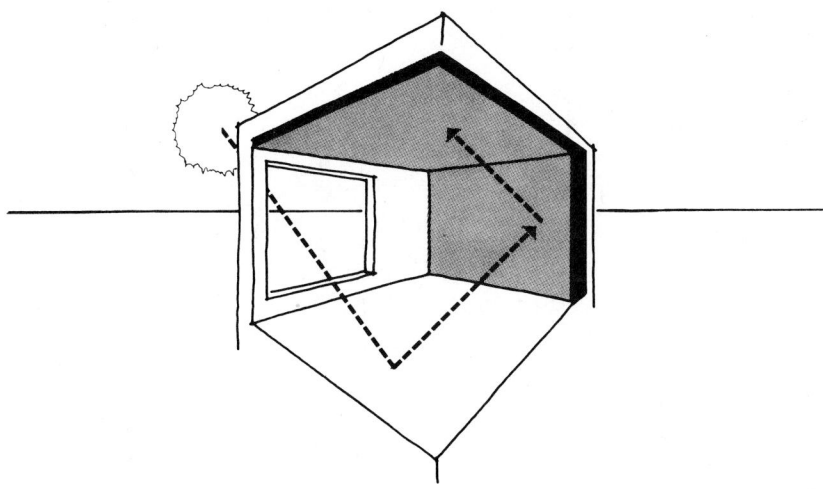

The mass in this pattern is defined in the same way as in Pattern 1, that is, the mass is one-sided and insulated on the back side. The distinction here is that the mass is not receiving direct beam radiation, but rather reflected sunlight. In a simple direct gain space, some of the mass will be a Pattern 1 case (perhaps a floor slab near the aperture, for example), and some mass will be a Pattern 2 case (perhaps the ceiling, for example). Much of the mass in such a space will be directly irradiated some of the time and indirectly sunlit for the rest of the day. In these cases an interpolation between Patterns 1 and 2 must be carried out.

Mass Sizing Table

Material Thickness	Indirectly Irradiated Mass Surface Area to Passive Aperture Area Ratio				
	Concrete	Brick	Gypsum	Oak	Pine
½"	-	-	114	-	-
1"	25	30	57	28	36
1½"	-	-	39	-	-
2"	12	15	31	17	21
3"	8	11	-	17	20
4"	7	9	-	19	21
6"	5	9	-	19	22
8"	5	10	-	19	22

C-2: Pattern 2: floor, wall, or ceiling in indirect sunlight [TEA-3].

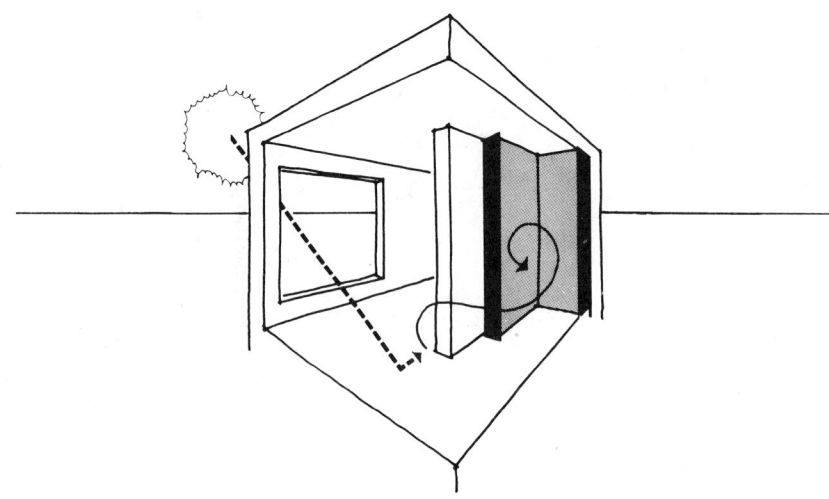

As in Patterns 1 and 2, the mass in this pattern is one-sided. The distinction here is that the mass material is charged neither by direct radiation nor by reflected radiation, but rather is heated by the room air that is warmed as a result of solar gains elsewhere in the building. Architecturally, this pattern is useful for mass materials in spaces deeper within a passive building, away from the rooms receiving solar gains. However, the solar-heated air must reach the remote mass either by natural or forced convection. Reasonable judgement is required here—a hallway open to a south room could be included; a back room totally closed off from the solar heated spaces could not be included.

Mass Sizing Table

Material Thickness	Remote Mass Surface Area to Passive Aperture Area Ratio				
	Concrete	Brick	Gypsum	Oak	Pine
½"	-	-	114	-	-
1"	27	32	57	32	39
1½"	-	-	42	-	-
2"	17	20	35	24	27
3"	15	17	-	26	28
4"	14	17	-	24	30
6"	14	18	-	28	31
8"	15	19	-	28	31

C-3: Pattern 3: floor, wall, or ceiling remote from sunlight [TEA-3].

On-site, indigenous building materials should not be underestimated as a means of easily and inexpensively adding large quantities of mass to buildings. Local materials also require relatively little energy to transport. Adobe buildings in the Southwest demonstrate some of the possibilities. This type of regional approach to building design can help to relax the demand on other, more depletable resources, such as wood, metals, and plastics.

Several procedures have been developed for estimating the amount of thermal mass necessary for a given building area. Two methods are presented in Part Two of this book. Another method is presented in *The Thermal Mass Pattern Book* [TEA-3], parts of which are excerpted here. Pattern 1 (figure C-1) shows the thermal mass in the floor or wall in direct sunlight. Pattern 2 (figure C-2) shows the mass in the floor, wall, or ceiling in indirect sunlight. Pattern 3 (figure C-3) shows the mass in the floor, wall, or ceiling remote from sunlight. The accompanying tables show the required ratio of mass surface area to passive aperture area for the various thicknesses and types of thermal mass shown.

For example, for 2 inches of concrete in direct sunlight (Pattern 1), 7 square feet of mass must be used per square foot of glazing. For the same concrete mass in a room remote from sun (Pattern 3), 17 square feet of mass are necessary per square foot of glazing.

For small passive solar areas, extra thermal mass is not necessary because the solar gains can be used directly to offset daytime heating loads. For a tightly constructed house, thermal mass is not necessary until the passive solar area exceeds 6% or 7% of the floor area [TEA-3].

C.3.b MOVABLE INSULATION

An inescapable reality affecting a building's thermal performance is the high heat-loss rate of glass compared to the low rate for well-insulated walls. In fact, a single sheet of glass has a heat-loss rate that is nearly thirty times that of a high-quality wall. Insulating glass has one-half the heat-loss rate of single glass.

Reducing heat loss through glass is one of the best ways to improve the performance of direct-gain systems. Adding layers of glazing is one of the best alternatives, especially in cold climates. In climates of more than 5,000 degree-days,* triple glazing is economically justifiable. Although each layer of glass also blocks 10% to 20% of the solar gain, using more than three layers is sometimes cost-effective in cold climates, especially if they are installed inexpensively. Multilayered glazing systems (of four and even five layers) that use high transmittance glazings (up to 9% transmittance) are becoming available. Low heat-loss glazings are also being developed.

Insulating the glass at night is another alternative. The essence of the concept is the transformation of a high heat-loss window that admits large quantities of solar heat during the day into a low heat-loss wall at night. The range of choices is large [SHU-2], [LAN-4]. Examples include:

- Sheets of rigid insulation manually inserted at night and removed in the morning;
- Roller-shade devices using wood or plastic slats such as those commonly used in many European countries;
- Framed and hinged insulation panels;
- Roller-like shade devices of one or more sheets of aluminized Mylar, sometimes in combination with cloth and other materials;
- Sun-powered louvers, such as Skylid, that automatically open when the sun shines and close when it does not; and
- Mechanically powered systems, such as Beadwall, that use blowers to fill the air space between two layers of glazing with insulating beads at night.

The R-values** of most movable insulating devices range from 4 to 10. Figure C-4 shows the effect on the U-value** of different glazing systems when movable in-

Solar Transmission Values	Single† Glass	Double Glass	Triple Glass
Nominal Solar "Transmittance"	0.87	0.76	0.66
Approximate Seasonal Transmittance	0.80–0.85	0.64–0.72	0.51–0.61
U-values (winter)‡ (in Btu/hr/ft^2/°F)			
Nominal U-value	1.15	0.55	0.35
With R4 Insulating Cover	0.21	0.17	0.14
Average U-value with R4 Cover in Place, 16 hr/day (3/4 of the degree-days)	0.45	0.27	0.20
Average U-value with R4 Cover in Place, 12 hr/day (2/3 of the degree-days)	0.52	0.29	0.21
With R10 Insulating Cover	0.09	0.085	0.078
Average U-value with R10 Cover in Place, 16 hr/day (3/4 of the degree-days)	0.36	0.20	0.15
Average U-value with R10 Cover in Place, 12 hr/day (2/3 of the degree-days)	0.44	0.24	0.17

*Degree-days are explained in the Glossary.
**The terms R-value and U-value are explained in the Glossary under the headings *Resistance* and *Coefficient of heat transmission*, respectively.

†for 1/8-inch grade B window glass only.
‡values are slightly different in summer.

C-4: U-values for glazing with movable insulation.

sulation is added. Also listed are U-values that are averages for glazings covered by insulation for different lengths of time. More degree-days occur at night than during the day; approximately two-thirds of the 24-hour daily degree-days occur while the movable insulation is in place if it is set in place for 12 hours at night. About three-fourths of the daily degree-days occur while it is in place if it is set in place for 16 out of the full 24 hours (that is, from late afternoon to early morning).

To find the annual reduction in conduction heat loss that is accomplished by using movable insulation, multiply the resulting difference in the U-value by the number of degree-days, and multiply the resulting figure by 24 (hours per day). The equation is:

$$E = (U_{gl} - U_a) \times \left(\frac{\text{degree-days}}{1 \text{ year}}\right) \times \left(\frac{24 \text{ hours}}{1 \text{ day}}\right)$$

or:

$$E = U_\Delta \times \text{degree-days} \times 24$$

where:
- E is the annual energy savings in Btu/ft²/yr;
- U_{gl} is the U-value of the glass without insulation, in Btu/hr/ft²/°F;
- U_a is the average U-value of the glass with the insulation in place part of the time; and
- U_Δ is the difference in U-values.

Suppose that an R10 panel is considered for double glazing in Minneapolis, and assume that the insulation will be in place two-thirds of the time and that the house will be kept at approximately 72°F. Figure C–4 indicates that the U-value of the glass (U_{gl}) is 0.55 Btu/hr/ft²/°F. U_a is 0.24. Minneapolis averages 8,382 degree-days per year. The above equation becomes:

$$E = (0.55 - 0.24) \times 8{,}382 \times 24 = 62{,}362 \text{ Btu/ft}^2/\text{yr}$$

This annual energy savings is roughly equivalent to the heating energy obtained from 18 kWh of electric resistance heating, 1 gallon of oil burned in a poorly adjusted furnace, or 1 square foot of an active solar collector of average design. A tight-fitting shutter also reduces air infiltration losses and raises the mean radiant temperature and corresponding feeling of comfort. If the window is particularly drafty when uninsulated, the shutter can save considerably more energy than is suggested by the above calculations—in some cases even double the normal amount.

Glass loses heat at a rate directly related to the temperature of the air space between it and the shutter. Many insulating shutters do not save as much energy as anticipated because they do not seal tightly when closed. As a result, room air finds its way between the glass and the shutter, and the temperature of the air space increases. Subsequent heat loss is greater, and the insulating effect of the shutter is diminished. For tightly sealed windows, this is usually not an important concern because room air moves very little through air spaces of up to ¼ inch in width. In general, the more loosely fitting the window is, the more tightly sealing the shutter needs to be.

C.3.c THERMAL PERFORMANCE

The thermal performance of direct-gain systems varies substantially, depending on the design. The factors that most affect performance are the glazing system and the amount, type, and location of the thermal mass.

To provide good performance and to avoid overheating, thermal storage mass should be designed in accordance with the methods described previously. The performance methods developed at Los Alamos Scientific Laboratory are based on these thermal mass configurations. The appropriate glazing treatment depends considerably on the climate. In all climates in the United States, a minimum of double glazing is usually recommended. In sunny climates, such as those in the Southwest and in other warm climates, double glazing can provide respectable performance.

Figure C–5 shows the solar savings fraction for direct-gain systems in several cities. Figures in each case are for a passive system with a load collector ratio (*LCR*) of 24. The *LCR* is the building heat-loss coefficient (*BLC*) divided by the direct-gain area (*A*). A detailed explanation of these terms is given in Parts Two and Three of this book. Three glazing treatments are shown in figure C–5: double glazing, double glazing with R9 night insulation, and triple glazing. The calculations are for a direct-gain system with thermal storage mass of 30 Btu/°F/ft² glass area (12 inches of concrete).

City:	Type of Glazing:		
	Double-Glazed	Double-Glazed With R9 Nighttime Insulation	Triple-Glazed
Boston	<0.1	0.38	0.26
Denver	0.38	0.62	0.49
Madison	<0.1	0.34	0.21
Nashville	<0.1	0.49	0.36
Seattle	<0.1	0.42	0.30

C–5: Effect on solar savings fraction of different forms of insulation, applied to a direct-gain building with *LCR* = 24 in five cities.

It is apparent from the table that double glazing alone cannot provide solar savings fractions greater than 0.10 in northern climates. Only in Denver (among the cities represented) is double glazing satisfactory. The addition of night insulation or triple glazing results in dramatic increases in performance in all climates.

A few words of caution are in order. Insulating material placed on the inside surface of one or more layers of glass is similar in design to a flat-plate collector. It will get hot when the sun hits it, and the resultant thermal stresses can cause glass breakage. The more tightly the shutter fits, the hotter the glass will become. A highly reflective surface on the material facing the glass is the best solution to this potential problem, but even it is not foolproof.

Designers and owners must also beware of condensation. If the insulation is located on the indoors side of the glazing, then the inside glass surface will be only slightly above outside air temperature. Condensation of moisture out of the air between the insulation and the glazing is probable under these circumstances, especially if the inside humidity is high and the outside air temperature is low. Water droplets or ice will form on the inside glass, sometimes accummulating on the sill and causing damage.

Insulation may also be located outside the glazing (or in rare cases between the glasses in double-glazing, as in Beadwall) with resulting advantages and disadvantages. If the insulation is outside the glazing, tight sealing becomes extremely important because of wind and the infiltration of outside air into the space between the insulation and the glazing. Operation of the movable insulation is also more difficult with outside placement than with inside placement. However, because the insulation is outside, the inner glass surface remains warm, and condensation is not a problem. In addition, the insulation is much more effective as a shade in the summer if it is located outside. The overheating problem described above does not occur.

C.4 ADVANTAGES AND DISADVANTAGES OF DIRECT-GAIN SYSTEMS

Advantages of using a direct-gain system are:

- Glazing is a relatively inexpensive form of solar collector, is widely available, and has been thoroughly tested;
- The overall system can be one of the least expensive means of solar heating;
- Direct gain is the simplest solar energy system to conceptualize and can be the easiest to build (in many instances, it is achieved by simply relocating windows);
- The glazing serves multiple functions, allowing solar radiation to enter the building while also admitting natural daylight and providing visual access to the outside; and
- When called upon to provide only a small fraction of the heating needs of a building, direct-gain systems do not necessarily require thermal storage mass.

Disadvantages of this type of system are:

- Large expanses of glass can result in too much glare during the day and in loss of privacy at night;
- The ultraviolet radiation in sunlight will fade fabrics and photographs;
- If the design is to achieve large energy savings, relatively large glazing areas and concomitantly large amounts of thermal mass are required to decrease temperature swings;
- Thermal mass adds substantially to a building's overall construction cost, particularly if it serves no structural purpose;
- Internal daytime temperature swings of 15 to 20 Fahrenheit degrees are common; and
- Providing for reduced heat loss at night through the glazing can be expensive and awkward.

C.5 EXAMPLES OF DIRECT-GAIN BUILDINGS

C.5.a THE DAVID WRIGHT HOUSE; SANTA FE, NEW MEXICO

See figure C-6. Direct gain is through the south side of this semicylindrical house, which has 384 square feet of insulating glass. There are few windows elsewhere. The 17-inch-thick adobe walls and 2-foot-thick adobe floor are insulated on the outside by 2 inches of polyurethane foam. The house loses about 13,000 Btu per degree-day. On a clear January day, as many as 500,000 Btu enter the house through the south windows. The temperature of the house is permitted to fluctuate between 60° and 80°F. This fluctuation, in combination with the large expanse of glass and the large volume of thermal mass, permits 90% of the house's heating needs to be satisfied by solar energy in the 6,200-degree-day climate of Santa Fe.

At night, heat loss through the south-facing glass is reduced by folding, insulating shutters made of 2-inch-thick foam insulation covered with canvas. Since the shutters have no side seal, they are not as effective as they would be if tightly sealed.

C-6: The David Wright House; Santa Fe, New Mexico [AND-1].

C.5.b ST. GEORGE'S SCHOOL; WALLASEY, ENGLAND

See figure C-7. St. George's School is located on the west coast of England at 53.4°N latitude. Despite rather poor wintertime solar conditions, the school's annex has operated with nearly no backup heat since 1962. More than half of its heat comes from the double-glazed south wall, whose dimensions are 230 feet by 27 feet. Approximately one-third of the heat comes from the electric lighting, and the balance comes from the metabolic heat of the students. The two layers of glass are spaced 24 inches apart, permitting entry between them for repairs and other operations. The inner glass has a rippled surface, diffusing the penetrating light to the floor, walls, ceilings, students, and furniture. Several operable windows with clear glass provide view and ventilation. There is no movable insulation.

The building has considerable thermal mass. The floors are 9- to 10-inch-thick concrete. The partitions are 9-inch-thick brick, as is the upper portion of the north wall. The sloping roof is 7-inch-thick concrete. Five inches of expanded polystyrene have been positioned outside the sloping ceiling.

The lights are operated on a timer, typically from 6:00 A.M. to 8:00 P.M., and are used as little as possible in the summer.

There is no mechanical ventilation of fresh air. Windows and vents are relied upon entirely, with mixed success. The limitation on this system is noise from students in the corridors, which prevents the opening of interior vents for cross-ventilation from various classrooms. Students and teachers use different amounts of clothing as necessary. The adjustment in clothing occurs almost automatically.

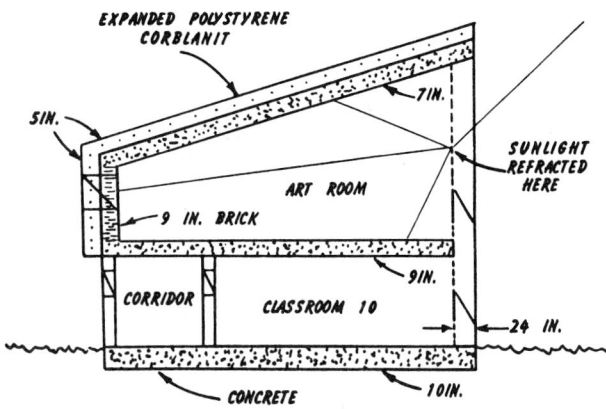

C-7: Schematic diagram of the St. George's School annex; Wallasey, England [PER].

C.5.c MIT SOLAR BUILDING V; CAMBRIDGE, MASSACHUSETTS

See figure C-8. Avoiding large interior temperature swings, accommodating and efficiently locating adequate amounts of thermal mass, and limiting heat loss through large expanses of south-facing glass pose the greatest challenges to designers using direct-gain systems. In 1978, the Massachusetts Institute of Technology (MIT) completed its Solar Building V in Cambridge, Massachusetts, to demonstrate new approaches to these difficult areas.

The 866-square-foot building uses three materials that have been under development for several years. The first, Heat Mirror, is a transparent insulation that, when applied to the building's 180 square feet of south-facing glass, admits solar heat and greatly reduces heat loss by reflecting heat that would otherwise exit through the glass back into the building.

The second material is phase-change ceiling tiles. The tiles, 1.25 inches thick, contain two layers of a Glauber's salt mixture, each ⅜-inch thick. The tiles absorb heat, melting the salts at 73°F. The salts solidify at night, slowly releasing their stored heat into the building.

The third product is a mirrored venetian blind–type louver that reflects the solar irradiation up to the ceiling tiles. It also reduces problems of glare and sends natural lighting deep into the room by reflecting it off the ceiling. Figure C-9 shows the operation of these three components.

C-8: The south facade of MIT Solar Building V; Cambridge, Massachusetts [MAH].

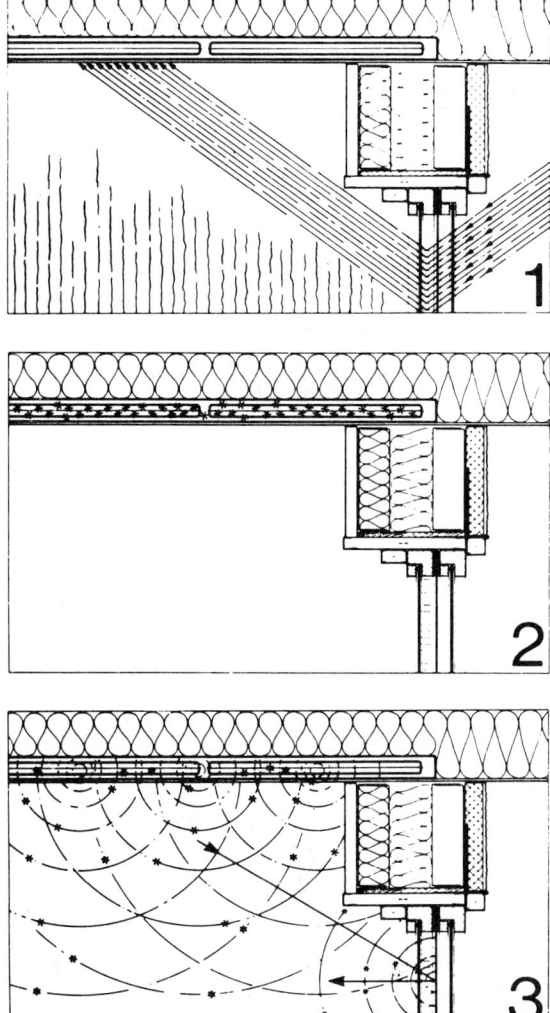

C-9: Operation of Heat Mirror insulation, phase-change ceiling tiles, and special louvers. The half-inch-wide solar louvers are sandwiched between the double glazings just inside the Heat Mirror. 1. The louvers' concave shape directs incoming sunlight toward the ceiling, where rising room heat is also collected. 2. The tiles hold heat when temperatures are above 73°F. 3. Heat is released when the room cools, as phase-change salts recrystallize [MAH].

Direct Gain / 31

D Thermosiphoning Collectors

D.1 INTRODUCTION

Of the five major types of passive heating systems, thermosiphoning collectors constitute the only type that does not lose heat when the sun is not shining (provided that the design includes reverse-flow prevention devices, as discussed in this chapter). Thermosiphoning collectors are also the passive-system type most similar to conventional active systems.

Second to direct gain, thermosiphoning collectors are the most widespread building application of solar heating in the world. Hundreds of thousands of thermosiphoning water heaters have been used for decades. In a domestic hot water system, the solar collector is at an elevation below that of the hot water tank. See figure D-1. Warmed liquid rises through the collector and up to the water tank, where it is stored.

In some systems, an electric heating element is located in the top third of the tank to boost the temperature of the water when cloud cover prevents the sun from providing the required amount of heat. In other systems, the solar tank preheats the water prior to final heating in a conventional water heater. Thermosiphoning systems can provide most of the domestic water heating needs year-round for houses in the southern states. In the north, freeze protection is a major concern.

As thermosiphoning collectors continue to become better understood, they are being used with increasing frequency as wall panels to complement south-facing windows in supplying solar heat directly to buildings. The amount of supplemental thermal mass required in each case depends on the correlation between the heating needs of the building and the amount of heat supplied by the panels. For example, many existing buildings have high heat-loss rates compared to available solar gains on south-facing wall surfaces. These surfaces can be converted to thermosiphoning collectors that supply solar heat only when the sun shines and when the building needs heating. In conventional, wood-framed, residential construction, up to 25% of the heating load can be supplied in this manner, without supplemental thermal storage.

D.2 DESIGN FUNDAMENTALS

D.2.a BASIC SYSTEM DESIGN

A thermosiphoning collector is similar to a flat-plate collector in an active system. A layer or two of glass or plastic covers a black absorber. Depending on the design, the air may flow in a channel in front of or behind the absorber. The air may also flow *through* the absorber if the absorber is perforated. The collector is backed by insulation.

In figure D-2, the thermosiphoning collector is mounted on or integrated into the wall. Openings at the bottom and top of the wall permit convective airflow to take place from the building, through the heated collector, and back to the building.

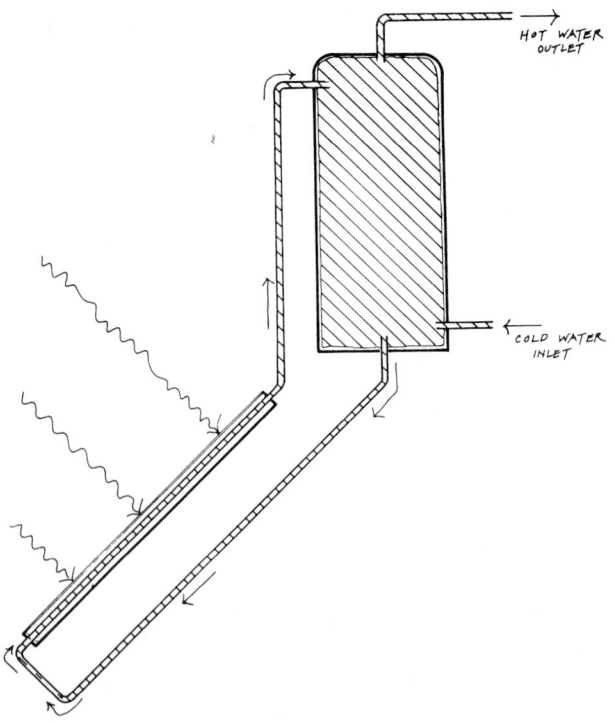

D-1: Thermosiphoning collector used to heat water.

The slowly moving air must come in contact with as much absorber surface area as possible, but the flow of air must not die impeded by the configuration of the absorber surface, which is designed to maximize heat transfer. The amount of heat transferred from the absorber to the flowing air is in direct proportion to the heat transfer coefficient (h).* Typical values for h in thermosiphoning systems range from 1 to 3 Btu/hr/ft^2/°F.

Steve Baer and others have used up to six layers of expanded metal lath in thermosiphoning systems. The air rises in front of the lath, passes through it, and leaves the collector through a channel behind the lath. See figure D-3. The value for h in this design is about 3. A flat or corrugated metal surface has a heat transfer coefficient of less than 2 in thermosiphoning panels.

Thermosiphoning collectors should be constructed as carefully as active collectors are. Exposed, unglazed portions should be well-insulated—particularly the upper areas, which are likely to be hottest. Avoid the use of polystyrene insulation in the collector, as it has a melting temperature of about 180°F. If the collector stagnates during a sunny day, its temperature can reach over 300°F. Wood construction should be avoided inside the collector.

Plans and design details for thermosiphoning collectors have been published [MOR-2].

D.2.b AIRFLOW

Designing a [thermosiphoning] system is a somewhat tricky and difficult task. If you aren't very respectful of the will of the air, the system won't work.

[BAE-1]

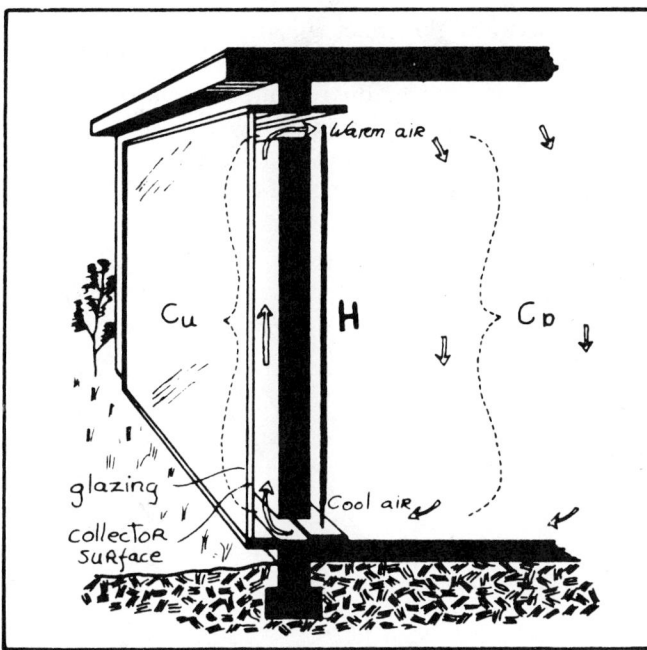

D-2: A wall-mounted thermosiphoning collector. (The length H is the distance from the midpoint of the opening at the bottom of the wall to the midpoint of the opening at the top of the wall.) [MOR-1].

As in active collectors, the hotter the absorber in a thermosiphoning collector becomes, the greater the heat loss is and the lower the collector's efficiency becomes. Good airflow keeps the absorber cool and transports the maximum possible amount of heat into the building. Flow channels should be as large as possible, and bends and turns in the ducts should be minimized.

In contrast to conventional air heating collectors, in which the airflow is powered by a fan and in which the air channel is only ½ inch to 1 inch deep, the air channel past the absorbers in a thermosiphoning collector is usually from 3 to 6 inches deep. In fact, Steve Baer's rule-of-thumb is that it be one-fifteenth of the vertical length of the collector. Others suggest one-twentieth. This means that, if the collector is 15 feet long, the air channel should be 8 to 12 inches deep. Baer also recommends that the air ducts to and from the collector have a cross-sectional area that is about 5% of the collector area. For example, if the area of the collector is 150 square feet, the cross-sectional area of the air duct to and from the rockbed should be about 7½ square feet [BAE-2].

Convective airflow is created by a difference in temperature between the two sides of the loop (most commonly, the difference between the average temperature in the collector and the average temperature of the adjacent room). It is also affected by the height of the loop.

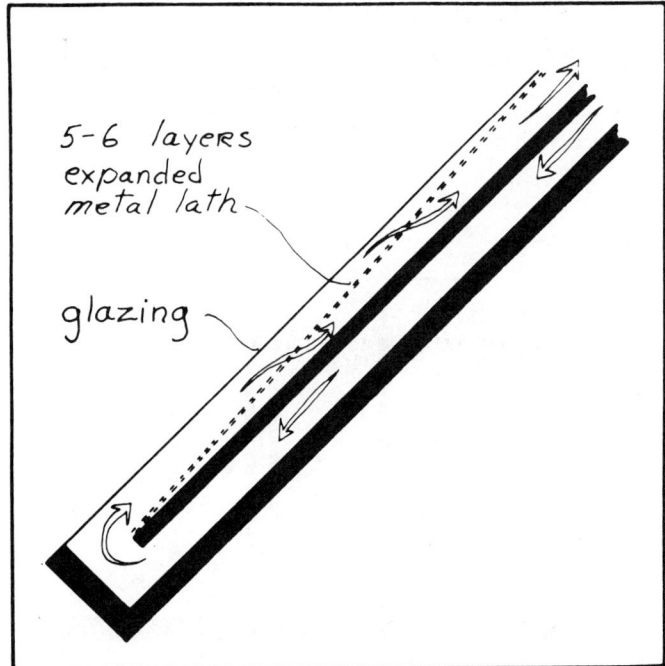

D-3: Typical airflow through a thermosiphoning collector using expanded metal lath as the absorber [MOR-1].

---

*Heat transfer coefficient is explained in the Glossary under the heading *Conductance*.

Thermosiphoning Collectors / 33

To obtain the best airflow of the system in figure D-2, it is necessary to:

- Increase the temperature of the collector (C_U);
- Decrease the temperature of the room (C_D); and
- Increase the height of the collector.

The pressure driving the airflow is obtained from the following equation:

$$P_C = 7.6H \times \left(\frac{1}{t_U} - \frac{1}{t_D}\right)$$

where:

t_U ("temperature up") is the average temperature at U, expressed on the Rankine scale (°F + 460);

t_D ("temperature down") is average temperature at D, expressed on the Rankine scale (°F + 460);

H is the height of the air column, in feet; and

P_C ("collector pressure") is the resulting driving pressure, in inches of water.

The resulting pressure driving the airflow in a thermosiphoning collector like the one shown in Figure D-2 can be calculated as follows:

Suppose the collector is 8 feet tall, its average temperature is 120°F, and the average room temperature is 70°F. The pressure driving the air is:

$$7.6 \times 8 \times \left(\frac{1}{[70 + 460]} - \frac{1}{[120 + 460]}\right)$$
$$= P_C = .005 \text{ inches of water}$$

In a properly designed thermosiphoning collector with unrestricted vent openings, flow rates can be as high as 5 cubic feet of air per minute per square foot of collector, on clear sunny days. This is about twice the flow rate of most active air systems. Figure D-4 is a graph that plots flow rate against the difference in air temperature (Δt) between hot and cold sides of the loop. The larger the Δt, the greater the flow rate. This relationship, in fact, helps to regulate the airflow. The greater the solar radiation, the hotter the collector and the faster the airflow. During nonsunny hours, when the temperature drops, the airflow stops completely. Figure D-4 also plots the effect of restricted airflow caused by reduction in vent size to one-half and one-fourth the sizes recommended above.

Given two identical half-wall collectors, positioned as shown in figure D-5, the collector in B will operate the more efficiently because the entire C_U side is hot, whereas the lower half of the C_U side in A contains cool air. In consequence, the average Δt in B is greater than the average Δt in A, and the flow rate is better in B, as well.

The vertical height of the collector (measured vertically from the middle of the bottom opening to the middle of the top opening) should be at least 6 feet, if it is to achieve the necessary stack (chimney) effect. The collector should be tilted at an angle of not less than 45° with the ground, both to receive optimum solar exposure and to allow the air to flow freely upward.

An alternative flow path for the air through a thermosiphoning collector is shown in figure D-6. Here, in effect, the collector is double-glazed. The incoming cool air flows down the front channel between the exterior and interior glazings and rises by convection up the back channel between the inner glazing and the absorber plate. The primary advantage of this reversal in design is that the hot air can be ducted under a masonry floor before it enters the building—a system that obviates complicated duct work. The floor then acts as a radiant heating element. The efficiency of this collector is nearly

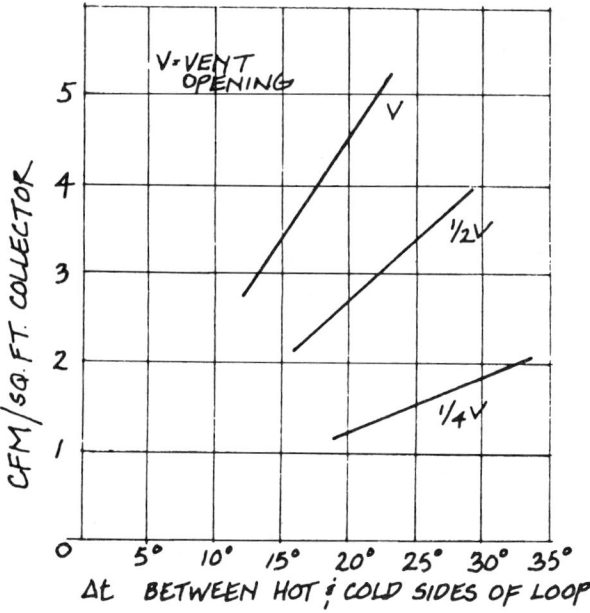

D-4: Airflow rates through thermosiphoning collectors equipped with vent openings of various sizes [MOR-3].

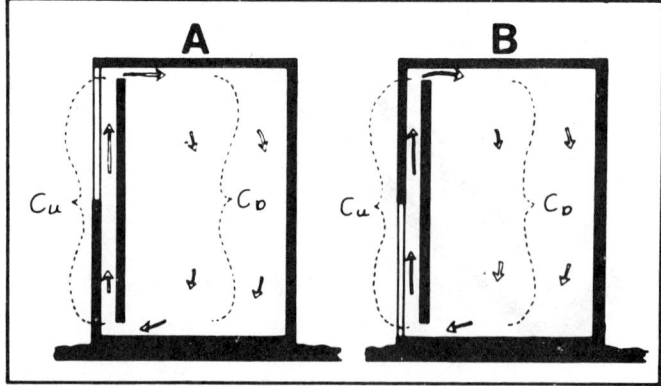

D-5: Alternative configurations for half-wall collectors [MOR-1].

identical to that of the conventional thermosiphoning collector [MOR-3], [SAN].

In an improperly designed system, reverse convection can occur. In such a case, a cool collector can draw heat from the house or from storage. Up to 20% of the heat gained during a sunny day can be lost through this process overnight [MOR-3].

There are three primary methods for automatically preventing reverse convection. One is to build the collector in a location below the heat storage area and below the house. A second is to install backdraft dampers that automatically close when air flows in the wrong direction. A collector designed for preexisting walls (illustrated in figure D-7) uses lightweight, thin plastic film. Warm airflow gently pushes the dampers open. Reverse, cool airflow causes the plastic to fall back against the screen opening, stopping airflow. For best results, both top and bottom vents should be equipped with such dampers.

Figure D-8 shows the third method of minimizing reverse convection. Both the intake and outlet vents are near the top of the panel. The back of the absorber is insulated to about R4 and is centered between the glazing and the walls. Inlet cool air drops into the channel behind the absorber. The solar heated air rises in the front channel. During no-sun periods, the air in both channels cools and settles to the bottom of the "U-tube." Only minor reverse convection occurs.

The main office building of the National Scientific Research Center in Odeillo, France (now more than fifteen years old) uses this type of thermosiphoning wall panel in combination with windows. Together, the windows and wall panels contribute about 50% of the building's heat. See figure D-9. No provision has been made to store the heat elsewhere than in the building's thermal

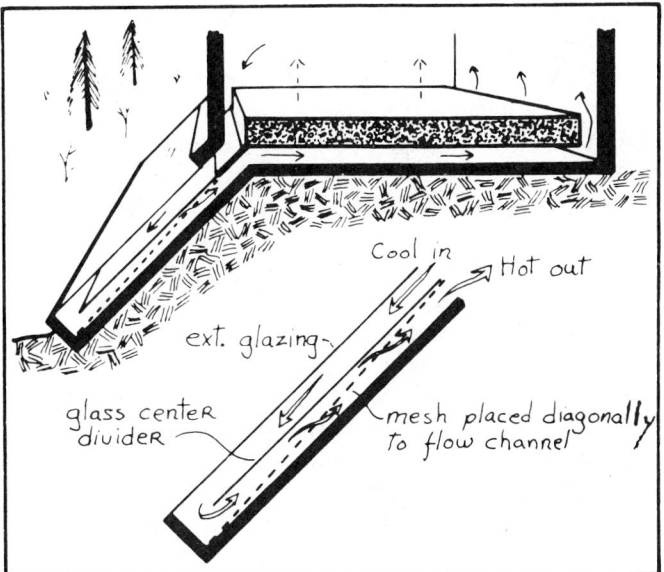

D-6: Center-glazed collector with masonry floor heat storage [MOR-1].

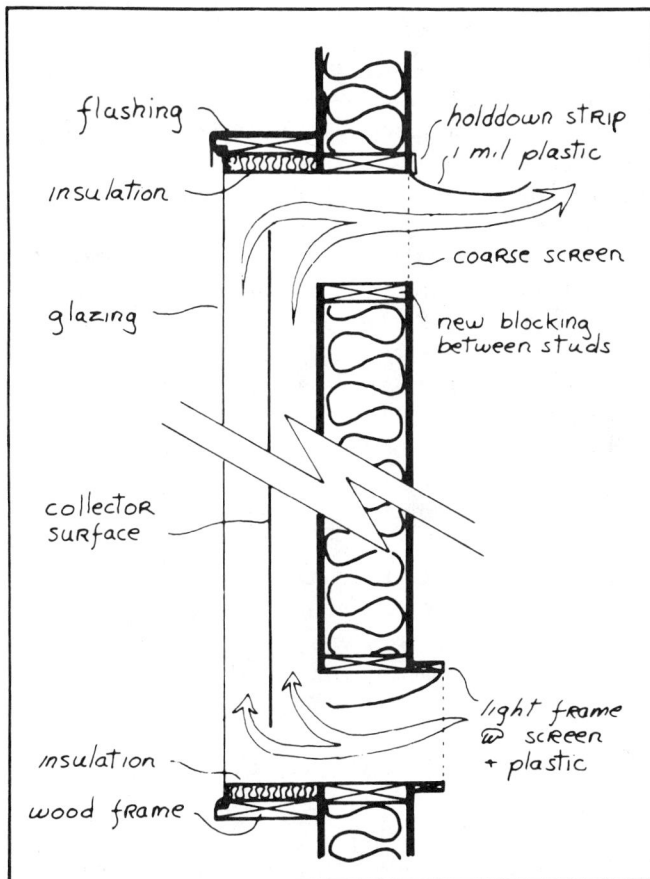

D-7: Simplified construction details for a thermosiphoning collector built into a preexisting wall. Notice the plastic backdraft damper [MOR-1].

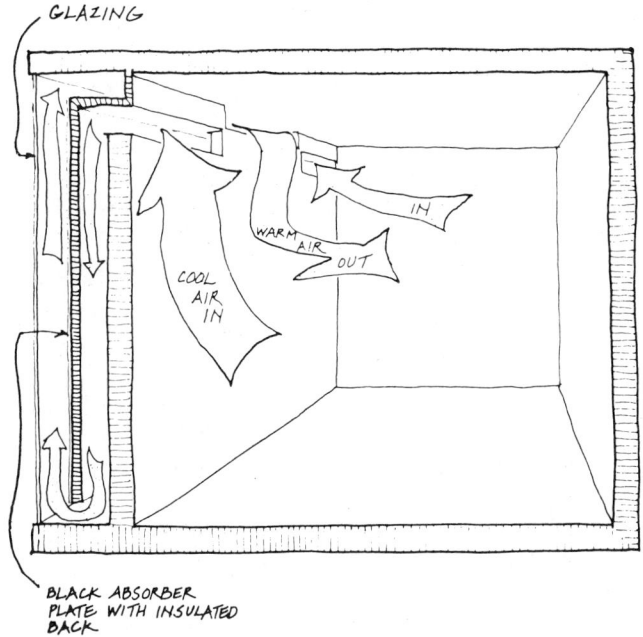

D-8: A "U-tube" collector designed to prevent reverse convection effectively.

mass, which is composed of reinforced, poured concrete. The panels are designed to be closed easily and "turned off" during warm weather. Their design allows cool air to settle to the bottom of the air passages, inhibiting reverse thermocirculation of cold air through the panels at night and back into the building.

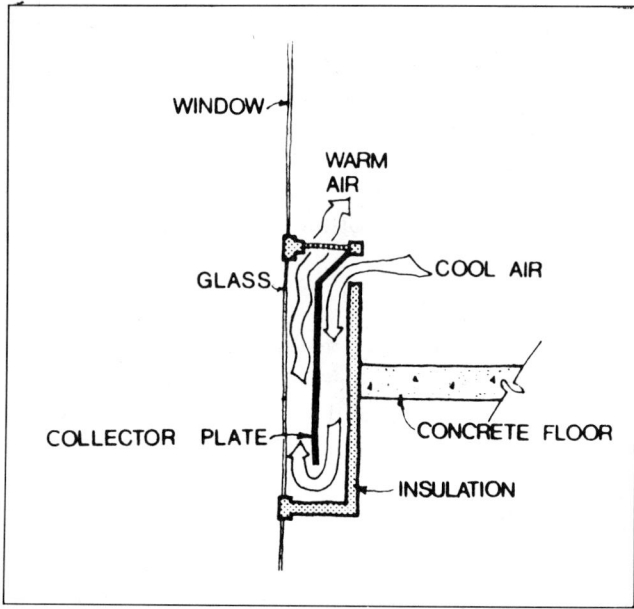

D-9: Schematic diagram of a thermosiphoning wall panel in the main office building of the National Scientific Research Center; Odeillo, France [AND-4].

D.2.c APPLICATIONS

Thermosiphoning collectors are best suited to commercial structures in which the heating load is large in comparison to the collector area. If all of the heat can be used immediately by the building, additional thermal storage—which can increase the total system cost—is unnecessary. If any thermal storage *is* added, it will necessarily be indirect (and, therefore, not as effective as directly irradiated mass).

The collectors can also be used in both single-family and multifamily housing. Additional thermal mass may be required; without it, the low heating loads (especially in well-insulated housing) translate into a smaller fraction of the building's total heat that the collectors can supply.

Architecturally, thermosiphoning collectors—with their conventional wall construction and glass-surfaced appearance—are well-suited for integration into most new commercial construction. With minor design modifications, the system is suitable for cost-effective retrofit application over existing exterior walls.

D.2.d COSTS

First costs depend primarily on labor. Most materials are available from a number of competitive manufacturers. In the case of manufactured panels, initial shop fabrication costs may be high, given the unconventionality of the product, but the contractor will value their convenience. Estimates of installed costs range from $10 to $25 per square foot, depending on construction methods and building types. Operating costs are nonexistent, and maintenance costs should be very low.

D.2.e THERMAL PERFORMANCE

Performance depends largely on the delicate, natural convection currents set up in the system. Airflow is low to nonexistent at times of little or no sun, but it increases rapidly under sunny conditions. Flow rates are generally higher than in thermal storage walls: as noted earlier, rates of up to 5 cubic feet per minute per square foot of collector are possible. The resulting output of 90 Btu/hr/ft^2 represents a collection efficiency during periods of good, sunny conditions similar to that of low-temperature, flat-plate collectors used in standard active-system designs.

The chief determinant of overall performance is the ratio of the building's heating load to the collector area. For a given heating load, effective performance deteriorates rapidly after the optimum point with increasing collector size. Estimates of useful delivered energy (using simplified analyses) range from 20,000 to 80,000 Btu per square foot per heating season. The high numbers in this range are typically associated with low solar heating fractions and sunny climates, and the low numbers with moderate solar heating fractions or very cloudy climates. 40,000 Btu per square foot per heating season is a typical figure in moderate climates such as that of Boston, Massachusetts, when the solar contribution is 10% of the heating load.

To increase solar performance, increase the thermal mass so that it can store larger quantities of solar heat effectively. To maximize system performance, especially in well-insulated (low-load) buildings, providing additional heat storage (by such means as doubling the thickness of exposed gypsum board) should be considered. Overheating, with resulting poorer performance and lower comfort levels, may result whenever systems are designed to provide over 10% of the seasonal load.

D.3 A THERMOSIPHONING COLLECTOR

The air heater in figure D-10 has an exterior glazing that covers a composite wall consisting of the absorber plate, rigid insulation, and interior finish. Trim, air grilles, and backdraft dampers complete the design. Its insulated, low-mass construction will cause it to undergo greater temperature fluctuations than thermal walls undergo. Consequently, its sealants, glazings, and other materials must withstand greater thermal ranges.

This design uses tempered insulating glass. Economy of installation and availability of replacement material is assured by using stock sizes manufactured for sliding glass doors (normally 34 by 76, 34 by 90, 46 by 76, or 46 by 90 inches). The glazing supports should minimize thermal contact with the absorber behind the glazing. This particular design doubles as a wood-frame structural wall; the glazing details are also made of wood, a natural insulating material. The core of the wall is highly insulated. The interior surface can be coated with a conventional interior finish.

The absorber plate is made of corrugated metal siding, a readily available building material that can be purchased complete with compatible fasteners and preformed EDPM or neoprene closure strips. This simplifies construction of an airtight, durable heater. The ribs provide structural stability. The plate must be spaced the proper distance from the wall. For an 8-foot-high panel, the rule-of-thumb calls for a space of 5 to 6 inches.

The absorber plate is made of mill-finished aluminum and must be prepared for painting with an etching cleaner (prepainted aluminum is available from some suppliers). The recommended finish is a thin coat of flat black enamel, such as the high-temperature spray paints commonly used for wood stoves, barbecue grills, and engine blocks.

The panel is vented to the space to be heated through a full-width register that provides manual control of the airflow. Continuous linear diffusers, normally available for commercial HVAC systems, can be used.

Backdrafts at night or during cold, cloudy weather can be prevented by using one-way dampers made of very lightweight, very thin (1 milliliter) plastic film. The collector shown in figure D–11 must be specially fabricated. The plastic film is attached with double-sided adhesive tape to a punched or die-cut 24-gauge galvanized sheet.

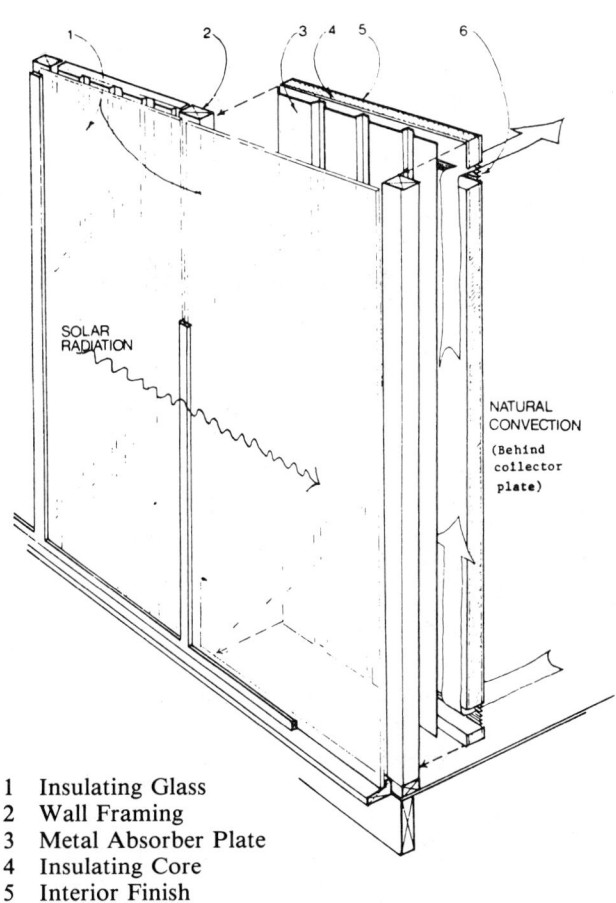

1 Insulating Glass
2 Wall Framing
3 Metal Absorber Plate
4 Insulating Core
5 Interior Finish
6 Continuous Air Vents
(Backdraft dampers required. See figure D–11.)

D–10: A typical thermosiphoning air collector (backdraft dampers not shown) [AND-4].

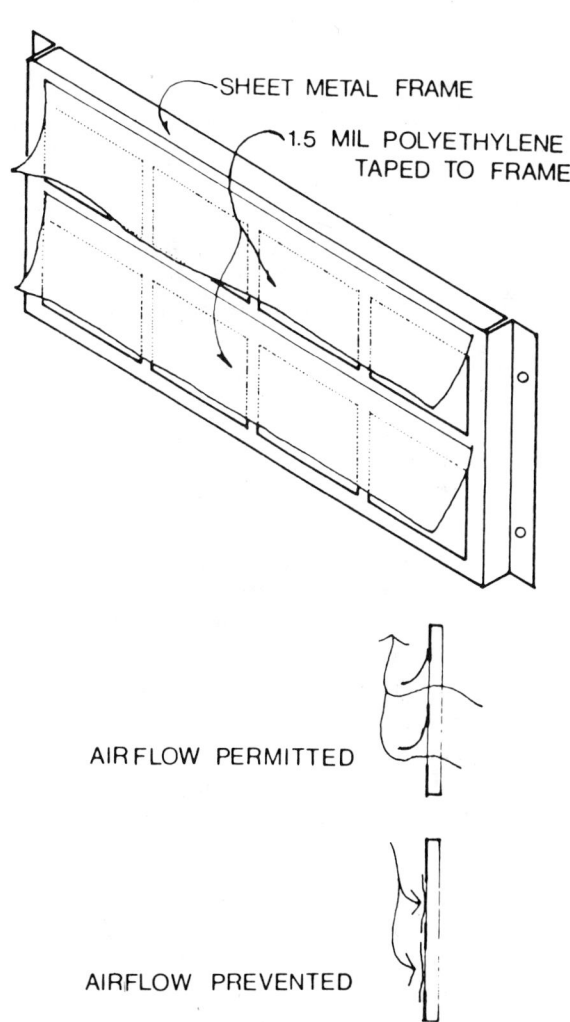

D–11: A backdraft damper designed for a thermosiphoning collector [AND-4].

D.4 HEAT COLLECTION AND STORAGE SYSTEMS

Thermosiphoning systems should include heat storage if they are large enough to supply more than 25% to 30% of a building's heat. Some heat can be stored within the enclosing walls of the building, but this option is limited because heat transfer is fairly poor from warmed air to the interior wall surfaces and extremely poor from the warmed air to the floor. The amount of heat successfully stored depends directly on the size of the room air temperature swing; so the occupants' tolerance to temperature fluctuations must be considered. In a residence for an elderly couple, for example, a swing of 5 to 8 Fahrenheit degrees may be the tolerance limit, whereas a fluctuation of 20 Fahrenheit degrees or more may be acceptable in a warehouse.

The thermal mass needed for a thermosiphoning system is determined in the same way as for a direct gain system, assuming Pattern 3. See section C.3.

Another option for heat storage is to use a rockbed (as in the case of the Paul Davis House, discussed in section D.6.a). In this case, the storage is separate from the building. The storage should contain at least 200 pounds of rock per square foot of collector. It should be located as high above the collector as possible, but below the house [MOR-2]. This type of thermosiphoning system will collect and deliver 30% of the solar irradiation that strikes it in cold climates, and 50% in mild climates.

The rocks used for storage should be close to uniform in size (i.e., you should not mix 1-inch rock with 4-inch rock). The cross-sectional area of the rockbed receiving air from the collector should be from 50% to 75% of the surface area of the collector. The warm air from the collector should flow down through the rocks, and the supply air to the house should flow in the reverse direction. Optimum rock size depends on the rockbed's depth. Baer recommends gravel as small as 1 inch in diameter for rockbeds that are 2 feet deep and up to 6 inches in diameter for depths of 4 feet [BAE-1]. In active systems, the best level of heat transfer is obtained at bed depths of at least 20 rock diameters. That is, if the rocks are 4 inches in diameter, the bed should be at least 6½ feet deep if it is to remove most of the heat from the air before the air penetrates completely through it and returns to the collector. This depth should be considered a maximum for thermosiphoning collector rockbeds.

D.5 ADVANTAGES AND DISADVANTAGES OF THERMOSIPHONING COLLECTORS

Advantages of thermosiphoning collectors are:

- They provide one of the least expensive ways to utilize solar heat;
- Glare and ultraviolet degradation of fabrics are not problems;
- If the collectors are going to be used to provide only a small fraction of the heating needs of a building, thermal storage is not necessarily needed;
- They are easily incorporated into south facades;
- They are readily adaptable to existing buildings; and
- Because the collectors can be thermally isolated from the building interior, nighttime heat losses can be lower than for any other passive design.

Disadvantages of this type of system are:

- The collector is an add-on device to the building (a possible advantage in retrofitting);
- Careful engineering and construction are required to ensure proper airflows and adequate thermal isolation at night;
- Because the thermal energy is delivered as warmed air, the heat is difficult to store for later retrieval (air has poor mass-to-mass heat-transfer characteristics in comparison to mass directly irradiated by the sun); and
- When thermal storage is used with thermosiphoning collectors, the system works best when the collectors are located below both the building and the storage, but this configuration is difficult to achieve with conventional construction.

D.6 EXAMPLES OF THERMOSIPHONING SYSTEMS

D.6.a THE PAUL DAVIS HOUSE; ALBUQUERQUE, NEW MEXICO

The Paul Davis House has used an air-heating thermosiphoning collector, in combination with a rockbed, since 1972. The system was designed by Steve Baer. Airflow is shown in figure D-12. The collector (36 feet wide, 12 feet long) is incorporated into the support structure of the porch, at an elevation below that of the house.

A single layer of glass covers the absorber, which consists of six layers of expanded metal lath. Warm air rises through the collector, becoming heated in the process. From there it travels through the rockbed (10 feet wide, 4 feet deep) located below the porch; the rockbed contains 330 pounds of fist-size rock per square foot of collector. As the air moves through the rockbed, it loses its heat and falls back to the collector inlet.

At night, a damper between the collector and the thermal storage is closed to prevent convection. Floor registers allow a thermosiphoning loop to heat the house from the storage. If the house needs heat during the day, the floor registers admit hot air from the collector, bypassing storage.

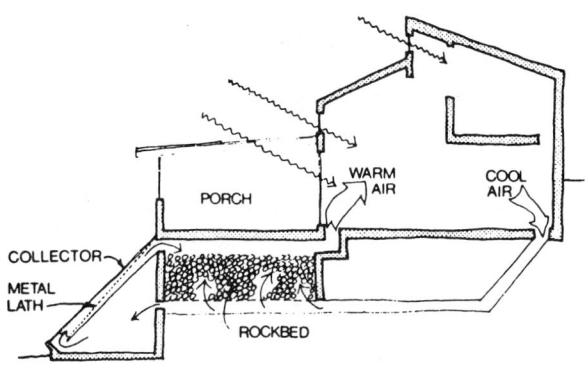

D-12: Schematic diagram of the Paul Davis House; Albuquerque, New Mexico [AND-4].

This system supplies over half of the heating load of this 1,000-square-foot house. The thermal storage is sufficient to last through two gray winter days. The backup system is wood heat.

D.6.b THE MARK JONES HOUSE; SANTA FE, NEW MEXICO

See figures D-13 and D-14. This 2,650-square-foot house combines a thermosiphoning system for collecting and storing solar heat with a forced-air distribution system [HUN-1]. The collector is 34 feet wide and 18 feet long. The absorber consists of three layers of ⅜-inch mesh wire lath set on top of black galvanized sheet-metal pans. A single layer of fiber-reinforced plastic covers the absorber. Thirty tons of ½- to 3-inch-diameter washed gravel—a quantity equivalent to 100 pounds per square foot of collector—fill the 4-foot-deep rockbed.

Behind the collector and next to the rockbed is a greenhouse. Portions of the absorber and of the collector backing were intentionally omitted during construction to permit light to enter. The greenhouse acts as the return air plenum, since the air—which moves slowly, but in large volumes—must pass through it en route from storage to the collector.

The house's performance results, monitored by Los Alamos Scientific Laboratory during a representative two-week period from December 26, 1978, to January 8, 1979, are shown in figure D-15. From December 29 to

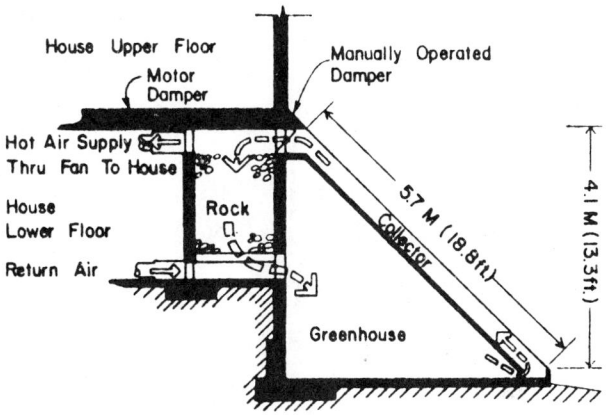

D-14: Schematic diagram of the solar heating system of the Mark Jones House [HUN-1].

D-13: The Mark Jones House; Santa Fe, New Mexico [HUN-1].

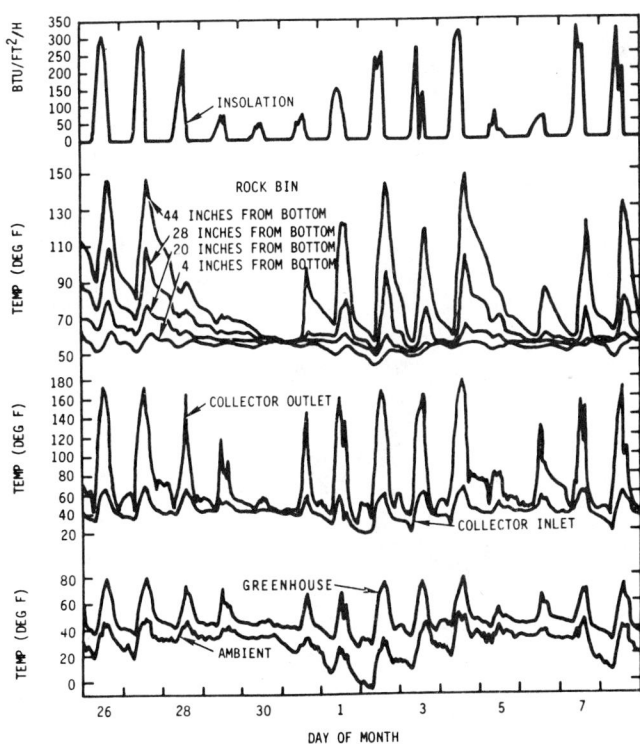

D-15: Representative performance of the Mark Jones House's thermosiphoning system, December 26 through January 8, 1978-79 [SAN].

Thermosiphoning Collectors / 39

December 31, the weather was cloudy (top curve), and all of the heat from the rockbed was depleted (upper center set of four curves). The outlet temperatures of the collector (lower center set of two curves) soared to more than 170°F during sunny weather and heated the top of the rockbed to nearly 150°F.

The rockbed is nearly devoid of stored heat each morning during winter. The greenhouse temperature remains (except on rare occasions) within the range for supporting indoor plant growth.

A 1,000-cfm (cubic foot per minute) fan, activated by an indoor thermostat, supplies solar heat by drawing air from the top of the rockbed. Return air from the house enters the rockbed at the bottom. Air distribution to the house takes place through a conventional duct system. A 15-kilowatt electric strip heater provides backup, but it was not needed in the first winter of operation, which began in February of 1978. The total installed cost of the solar energy system was $12.22 per square foot of collector—a total of about $6,500 in 1978.

D.6.c A THERMOCIRCULATION WATER HEATER

Zomeworks, Inc., of Albuquerque, New Mexico, manufactures a passive-solar domestic hot water system that operates by thermocirculation. See figure D-16. It consists of a 66-gallon, glass-lined storage tank with integral heat exchanger, two 17-square-foot collectors with low-iron single glazing, and an expansion tank and relief valve. The heat-transfer fluid in the collector loop is a nontoxic antifreeze.

The storage tank is positioned above the collectors. When sunshine heats the solar collectors, a thermal gradiant develops, and the antifreeze rises and gives up its heat to the water in the storage tank. After doing so, it cools and sinks to the collectors again. This system requires no pump to move the fluid through the collector loop. Some manufacturers make similar systems that use Freon rather than water or antifreeze as the heat-transfer fluid.

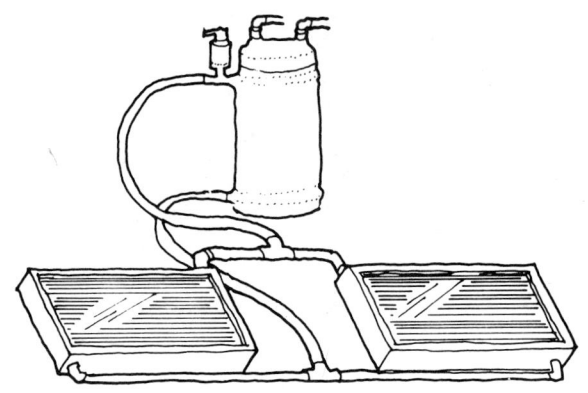

D-16: Thermocirculation water heater designed by Zomeworks.

E Thermal Storage Walls

E.1 INTRODUCTION

In a number of ways, thermal storage walls combine features of direct-gain systems and thermosiphoning collectors. As in direct-gain systems, large amounts of thermal mass are placed in direct sunlight; and as in thermosiphoning systems, heat flow to the room can be controlled.

Thermal storage walls, however, compensate for some of the disadvantages of the other two systems. Mass is used very efficiently because it is directly in the sun, in contrast to thermosiphoning systems which must rely on mass that is remote from the sun. Because the mass is placed between the glass and the space to be heated, the large fluctuations in room temperature sometimes associated with direct-gain systems are eliminated.

Per unit of thermal storage mass used, the thermal storage wall makes best use of the material because the temperature swing in the material is greatest. Even so, temperature swings in the heated space can be relatively small.

E.2 THERMAL STORAGE WALL SYSTEMS

E.2.a BASIC DESIGNS

There are two types of thermal storage walls. One uses foot-thick, heavy masonry material (concrete, adobe, brick, etc.). The wall is painted a dark color and heats as the sunlight passes through the glazing and strikes it. Usually, but not necessarily, vents are placed at the bottom and top. I vents are used, cool room air is drawn in at the bottom, rises in the warm space between the mass and the glazing, and enters the room through the top vents. Such systems are usually called "Trombe walls," after Felix Trombe of Odeillo, France, who—with architect Jacques Michel—substantially boosted their development in the 1960s by building several homes in the Pyrenees that incorporated this design. The concept was originated and patented in the 1880s by E. L. Morse of Salem, Massachusetts. His walls, complete with top and bottom dampers, used glass-covered slate.

The second general type of thermal storage wall uses water. The waterwall represented in figure E–1 uses

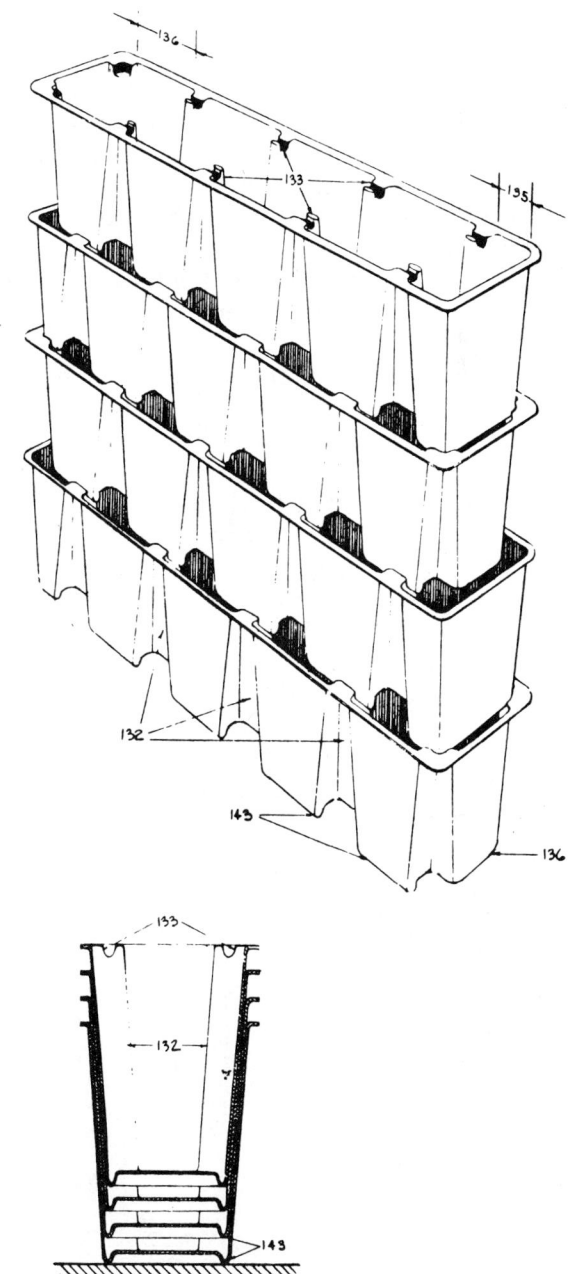

E-1: Waterwall modules designed by One Design [MAL].

modules of cast fiberglass—reinforced polyester. The black modules are about 8 feet long, 2 feet high, and 16 to 20 inches wide. They nest inside one another during transport from factory to site [EEE].

In some instances, waterwalls are more convenient than concrete thermal walls. Because water maintains a more uniform temperature throughout the thickness of the wall, its absorption surface remains at a lower temperature than the absorption surfaces of Trombe walls. This is the primary reason that waterwalls are slightly more efficient than Trombe walls.

E.2.b DESIGN VARIATIONS

Figure E-2 shows movable insulation combined with a concrete thermal storage wall. Beadwall, reflective Mylar roller shades, and hinged or sliding insulating shutters have been used. The economic value of movable insulation in passive systems increases as climatic extremes become more severe. However, most concrete storage wall systems to date have not used movable insulation because it increases first costs and is inconvenient. Triple glazing is an attractive alternative.

Another variation is to use thermal storage walls to induce ventilation. Dampers are positioned as shown in figure E-3. The solar-heated air from between the glass and the warm concrete is exhausted through vents in other exterior walls. This system should not be considered unless the dampers can be closed tightly during the winter; otherwise, air leaks will be a major source of heat loss.

An alternative to a solid concrete wall facing south is vertical solar louvers, a set of rectangular, masonry columns situated directly behind south-facing glazing and oriented in the southeast-northwest direction. See figure

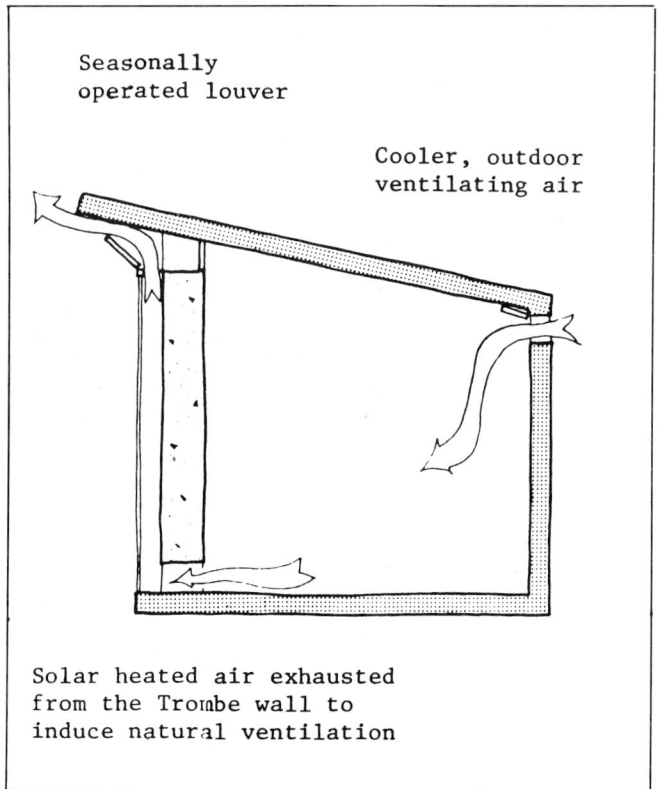

E-3: Thermal storage wall—cooling [AND-4].

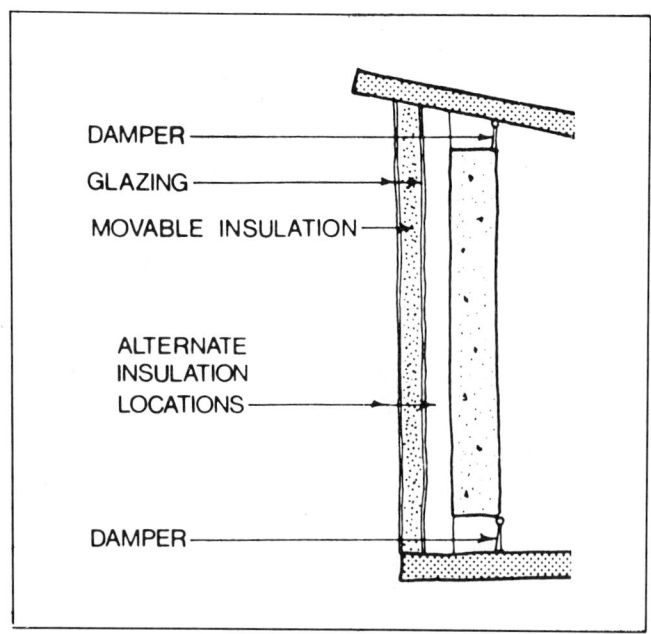

E-2: Movable insulation with a concrete thermal storage wall (set for nighttime operation) [AND-4].

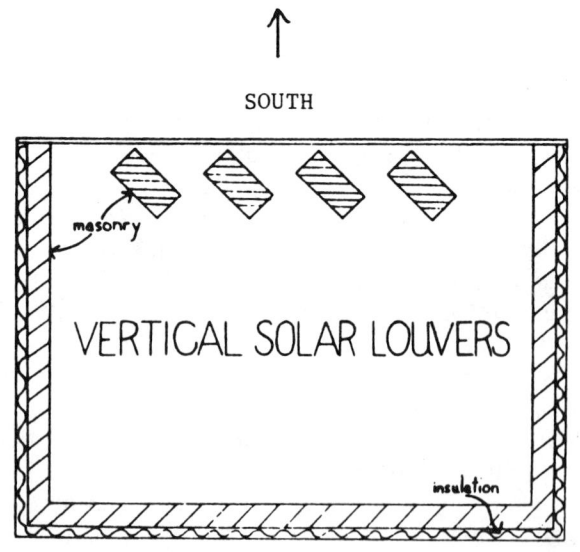

E-4: Vertical solar louvers [BIE].

42 / CHAPTER E

E-4. This system combines features of direct gain and Trombe walls. It can quickly heat the building in the early morning through direct gain, and the wall columns can absorb afternoon heat from the sun. The system also permits access to the glazing system for cleaning and maintenance and readily allows use of movable insulation between the glazing and the walls. Notice that the heat wave through the concrete louvers will be delayed by several hours because of their westerly orientation. The main disadvantage of this system is that the louvers are situated in the living space [BIE].

Figure E-5 is an example of one type of waterwall. This system, whose trade name is Drumwall, was first developed by Steve Baer of Zomeworks, Inc. It uses 55-gallon drums filled with water. Insulating panels, hinged at the base of each wall, cover the single layer of glass at night to reduce heat loss. When the panels are open and lying flat on the ground, the aluminum surface reflects additional solar irradiation onto the drums. During the summer, the panels in their closed position shade the glass.

The waterwall in figure E-6 uses water-filled vertical tubes. As an added means of comfort control, the tubes are separated from the living space by a wall through which room air can pass and sweep past the warm tubes. A fan controls the flow of air. A thermal curtain closes between the tubes and the glass at night to reduce heat loss.

Figure E-7 gives the heat-storage capacity for cylindrical tubes of various diameters. Corrugated, galvanized culverts and fiberglass-reinforced polyester are the most commonly used cylinders.

To overcome the high cost of concrete—and at the same time store large amounts of heat—Wayne and Susan Nichols, designer/builders in Santa Fe, New Mexico, developed a "water-loaded Trombe wall." It consists of cast-concrete tanks, 4 feet by 8 feet by 10 inches (outside dimensions); the tank wall is 2 inches thick, leaving a 6-inch cavity. After the wall is installed, a plastic bag, filled with water and then sealed, is placed in the cavity. At night, the single-glazed wall is covered outside by a Steve Baer–style hinged, insulating, reflecting shutter.

Data taken on the wall indicate that the thermal resistance of the outside 2-inch-thick concrete wall is too great during the charging mode. Temperature differences of 40 Fahrenheit degrees are observed across this wall. Temperature differences across the water and inner wall are small (5 Fahrenheit degrees or less). The Nichols have concluded that walls of this type should be thicker, for more heat storage, and that the outer wall should be made of metal to reduce its resistance to heat flow.

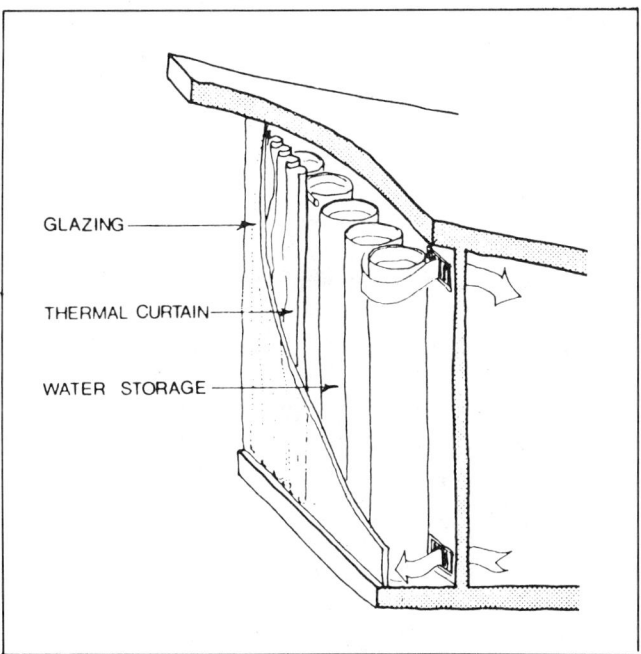

E-6: Thermal curtain with water-tube storage designed by Kalwall [AND-4].

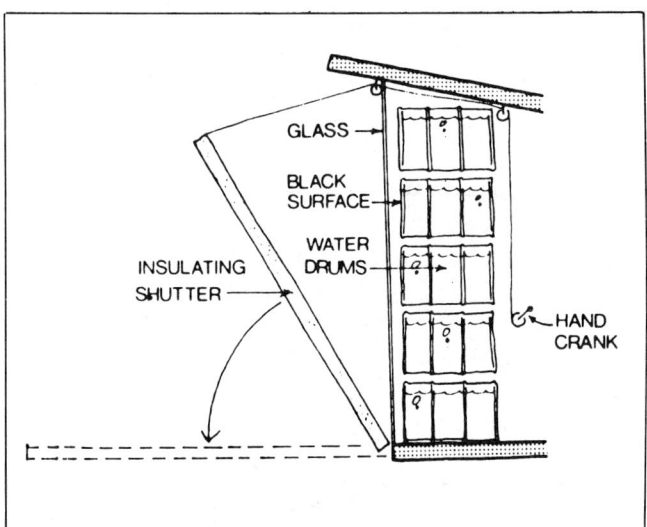

E-5: Hand-operated insulating shutter with water-drum storage designed by Zomeworks [AND-4].

Tube Diameter	8 in.	12 in.	18 in.	24 in.
Volume/linear ft	0.34 ft³	0.788 ft³	1.77 ft³	3.14 ft³
Weight of contained water /linear ft	21.7 lbs	49 lbs	110 lbs	196 lbs
Heat storage capacity/linear ft	21.7 Btu/°F	49 Btu/°F	110 Btu/°F	196 Btu/°F

E-7: Heat-storage capacity per linear foot possessed by cylindrical containers of water with different tube diameters.

E.3 DESIGN FUNDAMENTALS

E.3.a BUILDING INTEGRATION

Thermal storage walls provide temperature stability in passive buildings and are appropriate for a variety of building types. The air vents for thermocirculation somewhat control the timespan over which heat is delivered to the space. Since the wall is opaque, it eliminates the excessive glare associated with direct sunlight. Ultraviolet damage to goods and furnishings is avoided—an especially significant advantage in retail stores and commercial buildings.

Trombe walls are fire-resistant. They provide structural security for warehouses and manufacturing plants and structural ability in high-rise constructions. Finish details can be very rough, to suit manufacturing and industrial applications, or more polished to fit residential designs. Windows placed at suitable intervals supply daylighting and offer occupants views of their surroundings.

E.3.b COSTS

First costs vary according to variations in the construction and detailing of the thermal storage wall and in the cost of the exterior glazing. In localities where constructions using above-grade poured concrete and masonry block are common, building a Trombe wall is generally inexpensive. If an experienced subcontractor is available, or if materials can be obtained cheaply through local suppliers, the cost of the exterior glazing will be low. Other types of thermal storage walls, including waterwalls, are comparable in price.

Cost estimates prepared for the Trombe wall design described in section E.4 (with plaster interior finish) vary from a low of $11 per square foot to a high of $27 per square foot, in 1980 dollars. The lower cost is applicable to retrofit situations, where a mass wall already exists and inexpensive glazing is used. To obtain a true net additional cost for this type of passive solar heating, the cost of conventional construction that is replaced by the thermal wall should be subtracted. Since the most expensive conventional residential exterior wall, including insulation and interior finish, usually runs between $2.50 and $4 per square foot, the true first cost of the Trombe wall may be estimated at $9 to $25 per square foot (all dollar amounts in 1980 dollars).

Operating costs for these walls are zero, and little or no maintenance is required. In many climates, maintenance for thermal walls is comparable to that for vinyl siding: occasional washing (every 2 to 4 years) of the exterior glazing is advised. Harsh industrial environments may degrade plastic glazing; "refinishing" coatings are available from leading manufacturers and (in such situations) may be applied on a 3- to 5-year basis.

E.3.c THERMAL PERFORMANCE

Thermally, these walls perform reliably. Heat losses, even under the worst conditions, are not very different from those permitted by conventionally constructed walls. Their overall U-value of 0.23 (reverse thermocirculation prevented) enables them to meet ASHRAE's energy performance standards for single-family residences located in climates that do not exceed 5,200 degree days. If solar gains are considered, the thermal walls are net heat producers.

Solar energy collection takes place at low to moderate temperatures (generally not exceeding 150°F for the outside surface of Trombe walls, and even lower for waterwalls). This provides a high level of instantaneous efficiency (generally comparable to that of active-system flat-plate collectors). Except in the Deep South, the ver-

E-8: Fluctuations over a 1-week period in the room-side and sun-side surface temperatures of ventless Trombe walls of different thicknesses [BAL-2].

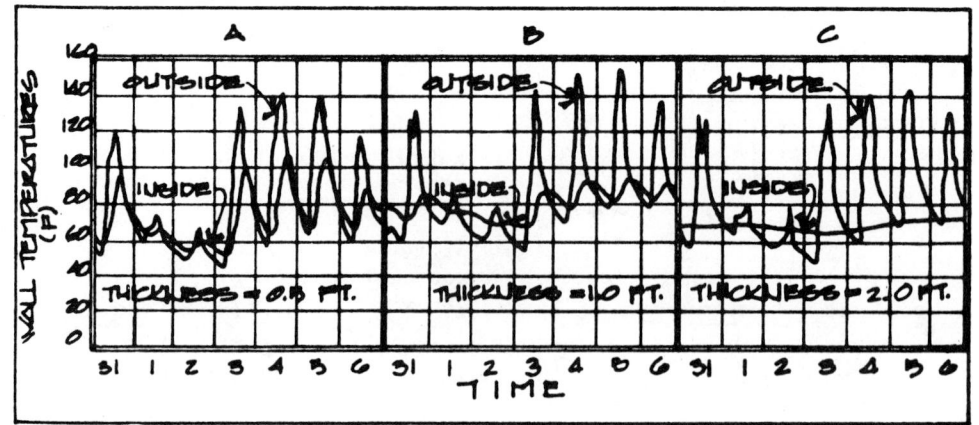

tical, south-wall orientation results in good winter heating performance and minimal summer overheating. Air is delivered through the vents and to the room at moderate temperatures (generally not exceeding 90°F—or 20 to 30 Fahrenheit degrees higher than the room air entering the space between the glazing and the wall). Normal airflow is approximately 1 cubic foot per minute per square foot of wall.

Although there are many ways to assess the relative benefits of different thicknesses of thermal storage walls, 10 to 16 inches is the optimum range of thickness for both Trombe walls and waterwalls.

Figure E-8 shows the calculated 7-day temperature fluctuations of the room-side and sun-side surfaces of three ventless Trombe walls in Los Alamos, New Mexico. The ratio of the building load to the glass area is 0.5 Btu/hr/ft²/°F. One wall is 6 inches thick, another 12 inches thick, and the third 24 inches thick. The daily fluctuations on the inside wall surface are markedly different for the three cases. They are very pronounced (45 Fahrenheit degrees) for the thin wall and nonexistent for the thick wall. The long-term effect of two days of cloudy weather is observed on the inside of the thick wall as a 10-Fahrenheit-degree variation.

The amount of thermal energy contributed annually by walls of different thicknesses is not markedly different. See figure E-9. The 1-foot-thick wall is the best of the three, giving a net annual solar-heating contribution of 68%. Although the net annual contributions of thin and thick walls are nearly the same, the fluctuations in room temperature are smaller as wall thickness increases. Notice that the solar fraction continues to increase (although by progressively minute amounts) for waterwalls of increasing thickness. (The solar fraction for waterwalls is represented by the curve for wall materials of infinite conductivity.)

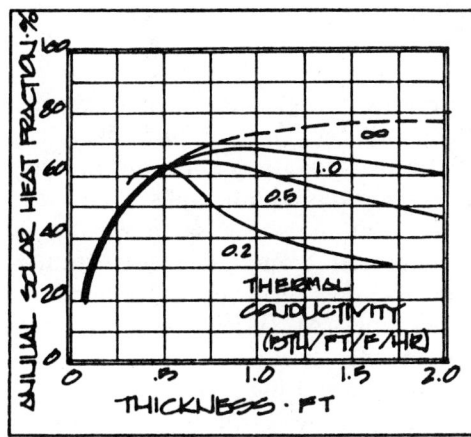

E-9: Annual thermal contribution provided by thermal storage walls of different conductivities and thicknesses [BAL-2]. (In this figure, thermal conductivity of 1.0 represents that of the Trombe wall, ∞ [infinity] that of water, and 0.5 (probably) that of a common concrete block wall.)

Trombe walls can be made different thicknesses to produce different time delays in the arrival of the heat wave from the hot, sunny side to the room side. In general, a building can be heated by direct gain or by a thermosiphoning system during the day and then by the thermal storage wall at night. Figure E-10 shows variations in inside surface temperature swing and in time delay between the irradiation and the occurrence of peak temperatures on the inside surface, based on differences in wall thickness. These representations are for a double-glazed solid concrete wall during a normal sunny winter day.

Figure E-11 shows the solar savings fraction (in various cities) for Trombe walls in passive systems with a load collector ratio (LCR) of 24. (The LCR in this case is the building heat loss coefficient (BLC) divided by the Trombe wall area (A).) Three glazing treatments are shown: double glazing, double glazing with R9 night insulation, and triple glazing. The calculations are for a Trombe wall with thermal storage mass of 30 Btu/°F/ft² glass area (12 inches of concrete).

Nighttime insulation dramatically improves performance of thermal storage walls, especially in cold climates. The R-value, and therefore the panel construction, need not be substantial. Insulation with an R-value of 4 to 6 will provide about 75% of the solar savings fraction of R9 insulation. Only rarely will insulation with an R-value higher than 9 be economically warranted. A simple timer is the only control necessary for opening and closing the insulation.

Wall Thickness (in inches)	Inside Surface Temperature Swing (in Fahrenheit degrees)	Time of Temperature Peak at Inside Surface
8	27	6:00 P.M.
12	13	8:00 P.M.
16	6.5	10:30 P.M.
20	3.0	1:30 A.M.
24	1.3	4:30 A.M.

E-10: Characteristics of solid concrete walls equipped with outside double glazing (figures are for sunny days).

City:	Double-Glazed	Double-Glazed With R9 Nighttime Insulation	Triple-Glazed
Boston	0.25	0.41	0.30
Denver	0.50	0.67	0.56
Madison	0.19	0.37	0.26
Nashville	0.35	0.46	0.41
Seattle	0.29	0.45	0.35

E-11: Effect on solar savings fraction of different forms of insulation applied to a Trombe wall with $LCR = 24$ in five cities.

In nearly all climates, thermal walls should have at least two glazings. An alternative to movable insulation in cold climates is an additional glazing layer. Movable insulation with more than two glazings is rarely economical. Multiple glazings have a greater effect on performance at relatively high solar load fractions than they have at lower ones.

E.4 A BASIC TROMBE-WALL CONFIGURATION

Although a variety of Trombe walls have been built, and although the design can be adjusted for specific climates, the design in figure E-12 is basic and cost-effective for heating in most of the United States. The modular dimensions and particular construction details used here should help to simplify the tasks of designers and builders. Although this drawing has been prepared to show many details and specific dimensions, the exact configuration of a Trombe wall can vary considerably from this model without adversely affecting its performance.

E.4.a MATERIALS

The basic design shown in figure E-12 consists of an outer glazing system, an inner thermal-energy storage wall, backdraft dampers for airflow control, and various optional trim and structural integration details.

The outer curtain-wall/window-wall system consists of aluminum framing in combination with two layers of glass or translucent or semitransparent low-cost plastic. Unlike in a direct-gain passive system, views *out* are not possible; views *in,* showing the rough concrete wall surface, may be undesirable. Maximum system temperatures, even under stagnation conditions, range from 150° to 180°F, well below the stagnation temperatures of metal flat-plate collectors. Heat-resistant plastics easily withstand these lower-range maximum temperatures to which Trombe wall glazing is subjected.

The thermal storage wall is concrete—either cast in place or laid with solid concrete masonry units and concrete mortar. The concrete used should be regular stone concrete (about 140 lbs/ft^3); lightweight aggregates should not be used. When (as in most cases) the Trombe wall serves as both a heat-storage wall and a structural wall, the necessary reinforcing wire or bar and any structural anchors can be added without altering the wall's solar performance characteristics. In general, the junctions between the inner storage wall and the foundation, floors, adjacent side walls, and roof should be treated as normal construction situations. A primary exception to this is the extreme importance of eliminating or changing details that would permit direct conduction of heat to masonry and metal that are exposed to the weather. For this reason, the concrete wall is thermally isolated from

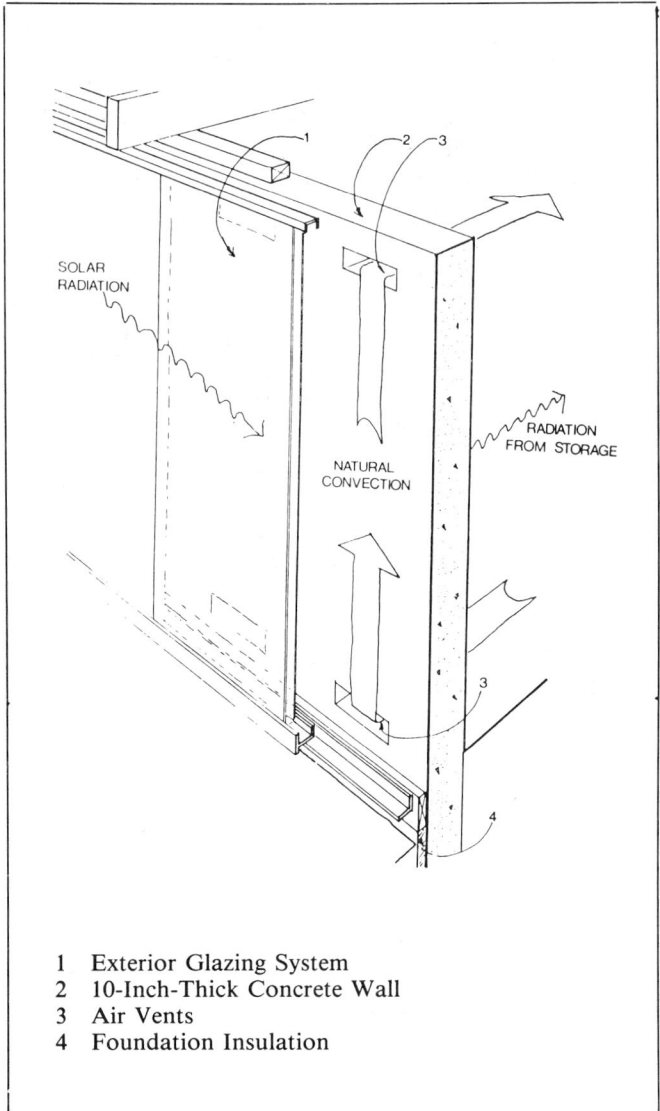

1 Exterior Glazing System
2 10-Inch-Thick Concrete Wall
3 Air Vents
4 Foundation Insulation

E-12: Trombe wall design [AND-4].

the metal frame of the glazing system by wooden blocking and from adjacent conventional concrete construction by preformed vinyl or rubber control joints. Foundations directly below Trombe walls should be protected with rigid insulation in the same way as perimeter heating systems in slab-on-grade construction are.

Backdraft dampers serve the same function as backdraft dampers in HVAC systems—that of preventing air circulation in the "wrong" direction. However, in Trombe walls, slowly rising solar-heated air in the cavity between the concrete wall and the glazing exerts a slight pressure to open them, while falling cool air exerts a slight reverse pressure that forces them to close. Dampers are commercially available or can be custom-fabricated. In many cases, as discussed earlier, vents need not be used.

The interior finish on the Trombe wall (if any) must not prevent the wall's heat from reaching the room. A conventional architectural concrete finish, such as exposed aggregate, or a sandblasted or brushed surface may be used. The surface may be sealed and painted any color. A plastic skim coat or plaster may be used. Sheet materials, however, such as plywood or hardwood paneling, should not be used. Gypsum board can be used only if excellent, continuous contact between the board and the wall is maintained—a difficult task indeed.

The exterior surface should be cleaned with a masonry cleaner prior to painting. Although any dark color may be used on particularly rough-textured walls, flat black paint is preferred.

E.4.b DESIGN

The concrete storage wall in this basic design is 10 inches thick and nominally 8 feet high. A 7-foot, 10-inch height is suitable for cast-in-place construction. Walls can be of any length. Vent holes, if used, should be provided at intervals along the entire length. Vent holes in concrete block walls are nominally $3\frac{5}{8}$ inches by $15\frac{5}{8}$ inches. Single blocks are left out of lower and upper courses. In poured concrete walls, 4-by-15 inch openings are preferred. The total cross-sectional area of the vents (upper plus lower) need not be greater than 1% of the total wall area. The upper and lower vents are placed as close to the ceiling and floor as is practical; in no case should the vertical distance between vents be less than 6 feet. Decorative grilles or registers are installed over these openings on the interior face. The lower grille includes the backdraft damper.

The exterior glazing system is mounted 3 to 4 inches away from the outer, darkened concrete surface. At the place where the aluminum glazing supports are attached to the wall, wood or other insulating material is used as a thermal separator. The glazing is extended above and below the face of the storage wall, fully exposing it to the sun. Since glazing is the weatherskin of the building, it must be airtight and water-resistant.

Trombe walls without vents are easier to build if windows are incorporated into the wall. The direct gain through these windows will heat the building during the day. Simultaneously, the Trombe wall will store solar heat for use during the night. Figure E-13 shows an example of such a wall. This wall is used in the Brookhaven House, designed by Total Environmental Action, Inc., for Brookhaven National Laboratories under a Department of Energy contract. The wall consists of two layers of paving brick covered by triple-glazed, float-glass panels mounted in milled wood strips. In the summer, when the sun is high in the sky, the wall can be shaded by a retractable canvas awning.

E.4.c CONSTRUCTION AND INSTALLATION

Building the Trombe wall described in figure E-12 normally requires only general contracting skills. Depending on contractor preference, the installation of the glazing system usually can be handled by the manufacturer's representative. This enables the building's owner to obtain a better warranty on its weather-tightness. The storage wall should be constructed at the lowest cost possible, given the thermal, structural, and interior finish requirements outlined above. If the contractor or subcontractors normally use poured concrete only in foundation work or if multistory installations are planned, the solid masonry-unit wall is preferable.

Work scheduling presents no problem if the contractor carefully reviews construction requirements in advance. The glazing system is usually fabricated to site dimensions; therefore, to avoid delays in closing the building, these dimensions should be established early in construction, and orders should be placed early for the glazing. Concrete finishing work may require having the appropriate trades on the job site at other than the normal times.

E.5 ADVANTAGES AND DISADVANTAGES OF THERMAL STORAGE WALLS

Advantages of thermal storage walls are:

- Glare and ultraviolet degradation of fabrics are not problems;
- Temperature swings in the living space are lower than with direct-gain or thermosiphoning systems; and
- The time delay between the absorption of radiant energy by the wall surface and delivery of the resulting heat to the interior space provides warmth in the evening when most houses need it.

Disadvantages of this type of system are:

- Two south walls—a glazed wall and a mass wall—are needed;
- Massive walls tend to be costly and are not generally used in modern residential construction (although thermal storage walls may be the least expensive way to achieve the required thermal storage, since they are compactly located behind the glass);
- The mass wall occupies valuable space within the building; and
- In cold climates, considerable heat is lost at night to the outside, from the warm wall through the glazing, unless the glazing is kept insulated at night—and movable insulation tends to be expensive and awkward.

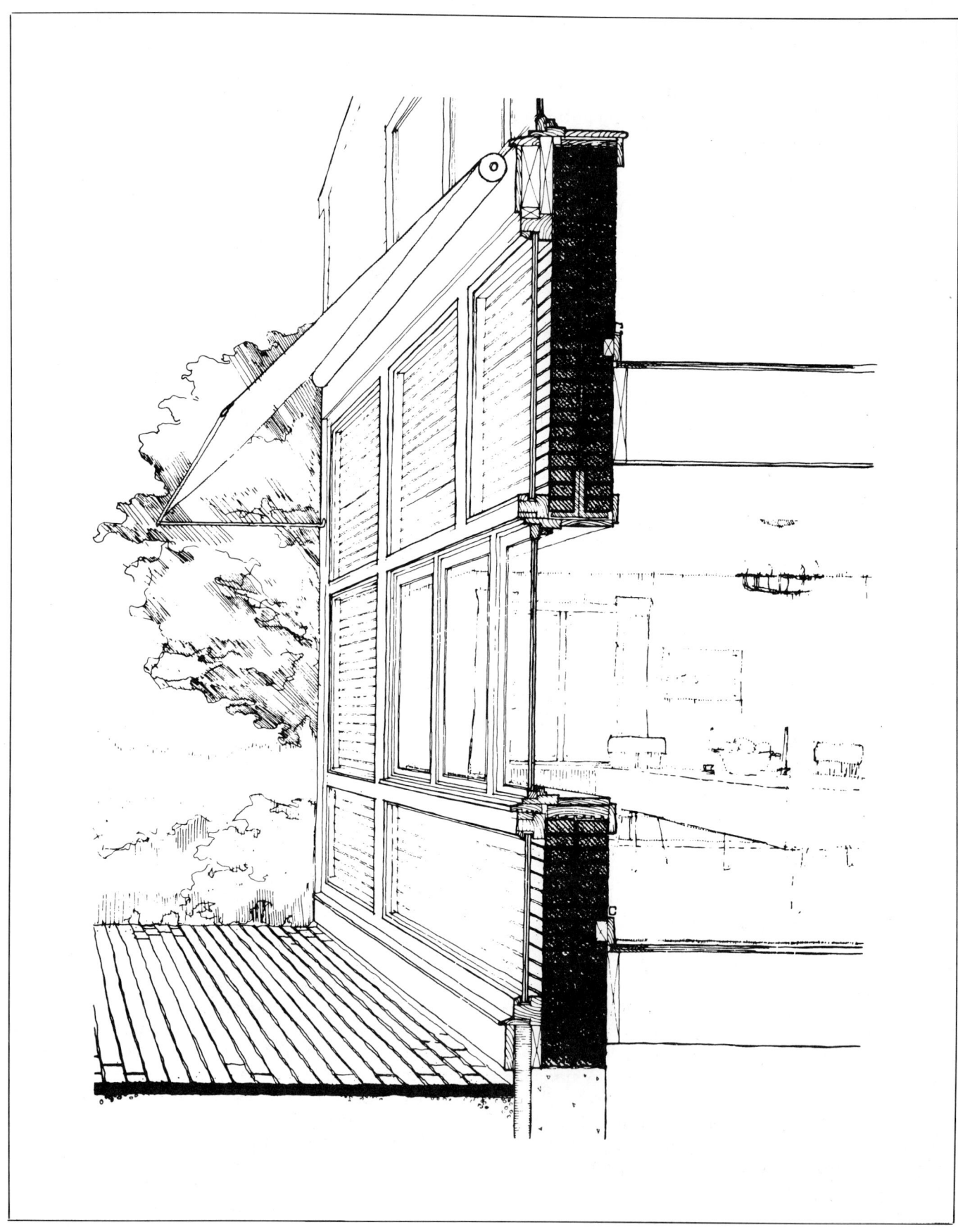

E-13: Trombe wall of the Brookhaven House; Upton, New York [TEA-2].

E.6 EXAMPLES OF THERMAL STORAGE WALLS

E.6.a BENEDICTINE MONASTERY; PECOS, NEW MEXICO

Ten miles south of Santa Fe, New Mexico, a 9,320-square-foot office/warehouse building for the book publishing operations of a Benedictine monastery combines direct gain with a Drumwall. See figure E-14.

The south surface is almost entirely glass; the window area is 1,356 square feet, with 440 square feet of Drumwall. The Drumwall consists of 138 water-filled oil drums enclosed in an insulated cabinet. The top of the cabinet is a counter-top work surface in the 2,660-square-foot offices. Heat passes by natural convection through vents in the cabinet, eliminating the need for fans.

Insulating panels are hinged to the exterior base of the wall. In the panels' louvered, horizontal position, they reflect additional solar radiation onto the Drumwall. In their raised position, they reduce heat loss. During the summer, they shade the wall.

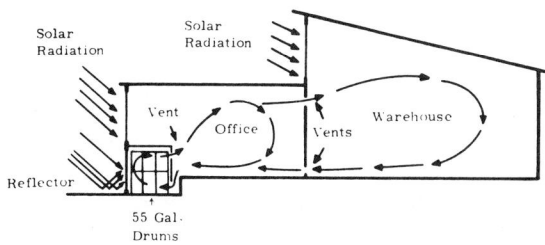

E-15: Schematic diagram of the Benedictine Monastery's office building/warehouse, showing the building's solar heating systems [SAN].

E-14: The Benedictine Monastery's office building/warehouse; Pecos, New Mexico [STR].

The 4,900-square-foot warehouse is heated by direct gain through clerestory windows. Excess warm air from the offices is occasionally vented into the warehouse. See figure E-15. The building is masonry with rigid foam insulation applied to the exterior surface.

The sun provides 90% of the building's heat. The office has a temperature swing of 15 Fahrenheit degrees, with an average wintertime low of 63°F. The warehouse has a temperature swing of 10 Fahrenheit degrees, with an average wintertime low of 48°F.

E.6.b THE DOUG KELBAUGH HOUSE; PRINCETON, NEW JERSEY

Architect Doug Kelbaugh of Princeton, New Jersey, designed his own two-story Trombe-wall house. See figure E-16. North, east, and west walls are standard wood frame constructions and have minimal window areas. Windows on the south side are incorporated into the Trombe wall. The Trombe wall also incorporates a standard, commercially available greenhouse. The total south wall collection area, including the greenhouse and a two-story Trombe wall, is 600 square feet. See figure E-17.

As computed using conventional heat-loss analyses, the design load is 65,000 Btu/hr. The empirically determined load is 56,300. Estimated consumption of gas in the backup heating system was 121 ccf (hundred cubic feet) during a 4,500 degree-day winter. Actual consumption during its first winter of operation (with 4,500 degree-days) was 338 ccf; actual consumption during its second year (with 5,556 degree days), was 246 ccf.

Indoor temperature swings were 3 to 6 Fahrenheit degrees during a 24-hour cycle. The seasonal low and high temperatures were 58° and 68°F downstairs and 62° and 72°F upstairs. The estimated averages were 63°F downstairs and 67°F upstairs. (Actual comfort levels were somewhat higher because of the radiating warmth from the Trombe wall.)

In addition to the vents at the top of the Trombe wall, four fans are used to ventilate the wall during the summer. The wall, in turn, ventilates the entire house by pulling air across the rooms from windows on the north wall [KEL-1], [KEL-2], [KEL-3].

E-16: South elevation of the Doug Kelbaugh House; Princeton, New Jersey [KEL-1].

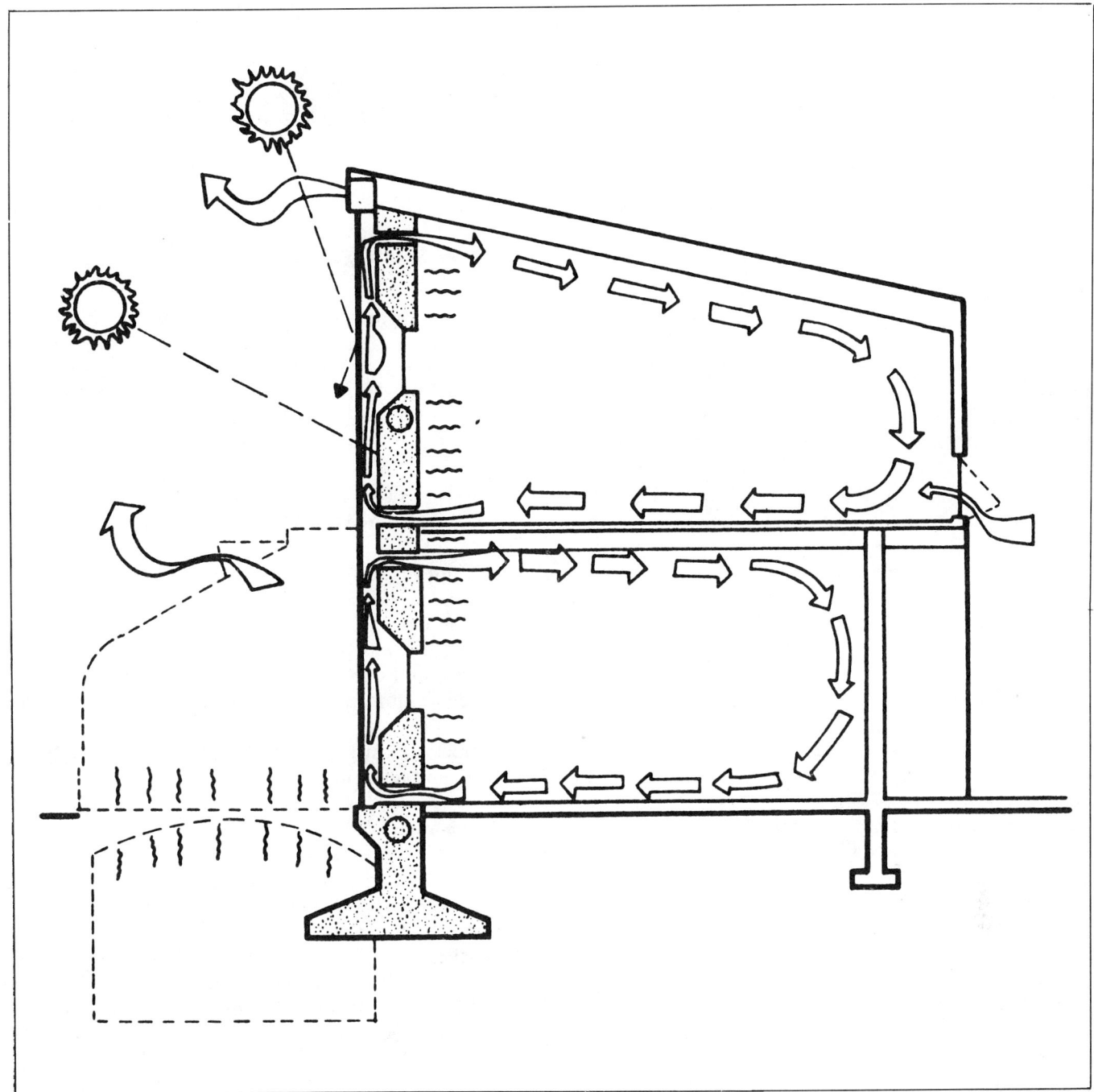

E-17: Schematic diagram of the Doug Kelbaugh House, showing heat flows [SAN].

Thermal Storage Walls / 51

F Thermal Storage Roofs

F.1 INTRODUCTION

In many respects, thermal storage roofs—or roof ponds—are similar to thermal walls: the collector and heat-storage mass are part of the same unit.

Roof ponds consist of waterbed-like transparent bags filled with water that, when exposed to solar irradiation, collect, store, and distribute heat. This heat passes downard from the supporting metal ceiling to the living space, gently warming it. In the summer, heat passes upward to the ceiling and into the water-filled "thermoponds," cooling the house. Then during the night, the water gives up its heat to the sky by thermal radiation, convection, and evaporation. Movable insulation is used to enhance the roof pond's performance. Figure F-1 shows this system's seasonal adaptability.

Water can store more energy per unit weight than other common building materials can. Roof ponds typically have water depths of 8 to 12 inches. Because of the free convection of water within a water bag, the bags operate isothermally and quickly transfer any temperature gain or loss to the building space. This is quite different from concrete thermal storage walls, which exhibit a time-lag effect between the time the outer surface changes temperature and the time the inner surface changes temperature.

Initially, the lack of a time-delay in temperature transference may seem a serious disadvantage of roof ponds, since—without other controls—it would eliminate the natural diurnal cycling. However, the movable insulation largely controls the roof pond's temperature. The temperature drops as low as 65°F in the summer and is permitted to climb as high as 85°F in the winter. Typical daylong heat-collection efficiency is 45% [YEL-3]. Less than half of the heat is transferred downward into the house; the remainder is lost through and around the closed insulation panels.

F.2 BASIC SYSTEM CONFIGURATION

Thermal storage roofs have been built in two basic configurations: flat and south-sloping. Generally, the flat-roof system is used in the lower latitudes, where the sun rises high enough in winter to bathe the ponds with substantial solar irradiation.

In more northerly latitudes, south-sloping glazing admits the low-angled solar irradiation and sheds snow. The roof ponds themselves, however, lie flat above the ceiling of the house. The other surfaces of the space formed by the extension of the glazing are well-insulated and are faced with reflective foil.

F.3 ADVANTAGES AND DISADVANTAGES OF THERMAL STORAGE ROOFS

Advantages of thermal storage roofs are:

- This system can provide both heating and cooling;
- Compared with those of many passive systems, the heating and cooling effects of thermal storage roofs are more uniformly distributed throughout the building;
- Temperature swings in the building may be small; and
- Glare and ultraviolet degradation are not problems.

Disadvantages of this type of system are:

- The heavy weight of the thermal mass above the ceiling might be psychologically unacceptable (especially in an earthquake-prone area);
- The thermal storage roof area needs to be at least 50% of the total floor area to produce a significant fraction of the thermal energy needs of the building; and
- Structural support for the heavy thermal mass can be costly.

F.4 EXAMPLES OF THERMAL STORAGE ROOFS

F.4.a THE ATASCADERO HOUSE; ATASCADERO, CALIFORNIA

Figure F-2 is a cross-sectional view of the roof pond and movable insulation system used on a house designed by Harold Hay in Atascadero, California [NIL]. The house was designed according to earthquake codes. The ceiling above the approximately 1,100 square feet of living space is completely covered with 8 inches of water, sealed in clear, ultraviolet-inhibited, 20-millimeter-thick

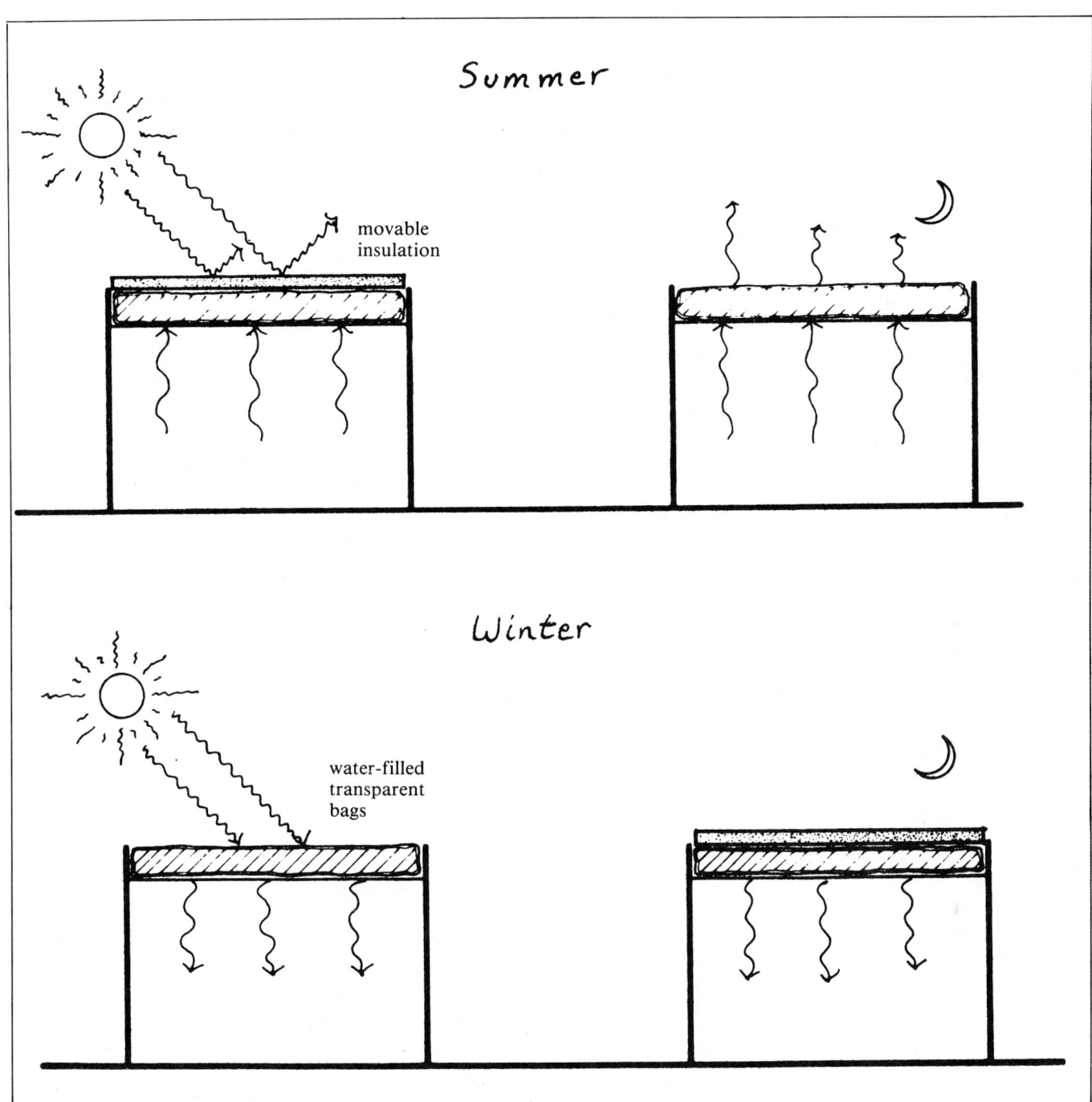

F-1: Skytherm thermal storage roofs—summer and winter operation [AND-4].

polyvinyl-chloride water bags as shown. Underneath these 53,600 pounds of water is a layer of black polyethylene to help absorb solar irradiation at the bottom of the bags.

An air cell of inflated clear sheet-plastic above the water bags enhances the greenhouse effect during the heating season. This air cell is deflated in the summer months to enhance radiative cooling. A 40-millimeter-thick steel deck-roof/ceiling supports the water bags and provides good heat transfer to and from the living space. Above the roof ponds, a system of movable insulating panels is mounted on horizontal steel tracks. The insulation consists of 2-inch-thick rigid polyurethane, faced on both sides with aluminum foil. The panels are moved by

Thermal Storage Roofs / 53

a ⅙-horsepower motor, operating for about 10 minutes per day.

Thermal storage roofs are unique in that they are the only passive solar heating system that can also provide substantial cooling effects. Within the building, comfort is accomplished by radiation of interior heat to the cool ceiling, with subsequent combinations of night-sky radiation, night-air convection, and—when the water bags are covered with additional water—evaporative cooling.

Figure F-3 shows the system's monthly performance over a 9-month period after construction. As the graph indicates, only small variations occurred in indoor temperature. During the winter and summer, temperatures typically fluctuated between 66° and 73°F, while the outdoor average daily temperature fluctuated between 47° and 82°F throughout the entire year. This house is 100% solar heated and cooled and has no backup system. Occupants have found the heating and cooling to be "superior" to conventional systems previously experienced [NIL].

Cloud cover and high relative humidity will reduce radiative heat transfer to the sky; clear skies and low relative humidity permit the best heat-transfer rates. Cooling works best if the water bags "see" the entire night sky—the flat configuration in the Atascadero House allows this. A south-sloping glazing configuration does not see the entire night sky, nor does the glazing easily transmit heat; therefore, radiative heat transfer to the sky in such a case is poor. However, the south-sloping design is intended for use in northern latitudes, where cooling is not essential.

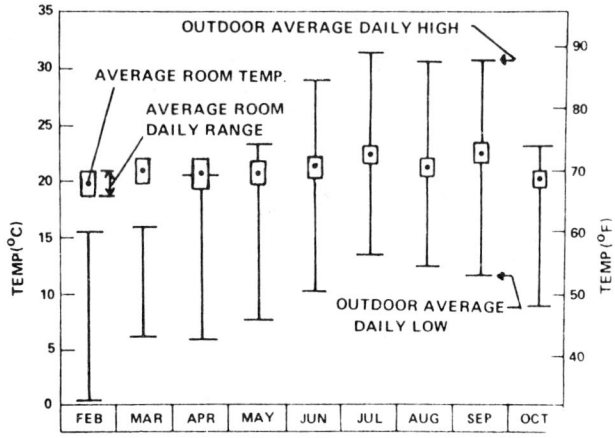

F-3: Monthly average temperatures of the Atascadero House; Atascadero, California [NIL].

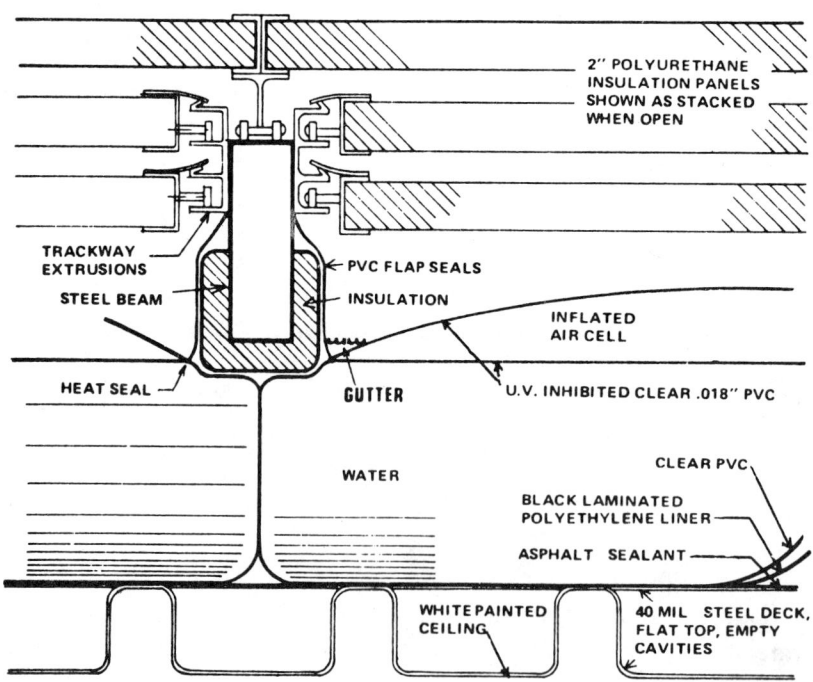

F-2: Schematic diagram of a water-bag and panel system [NIL].

54 / CHAPTER F

F.4.b THE WINTERS HOUSE; WINTERS, CALIFORNIA

A variation on the house at Atascadero was designed by John Hammond and built in Winters, California. See figure F-4. Winters has approximately 2,600 heating degree-days and 1,300 cooling degree-days per year. The average temperature in the hottest summer month (July) is 78°F, with an average daily maximum of 97°F and an average daily minimum of 59°F. In the coldest winter month (January), the average temperature is 45°F, with an average daily maximum of 54.5°F and an average daily minimum of 35°F. The 1,250-square-foot house is wood-framed and rests on a concrete slab. Walls and ceilings are R11 and R19, respectively. Windows are single-paned with movable insulating shutters. There are approximately 120 square feet of south-facing windows and approximately 100 square feet of windows on the other three walls.

On the roof, running the length of the center of the house, is a series of 6-by-8-foot 12-inch-deep galvanized steel pans, coated inside with tar. They cover one-third of the roof area and hold approximately 13,200 pounds of water. The bottoms of the pans form the ceiling of the house. The pans have vinyl sheet covers to prevent evaporation and to reduce heat loss. Movable 8-foot-square insulated lids cover the sealed pans. Conventional hydraulic pistons open and close the lids.

The house was completed in 1975. Even during the hottest summer weather, the maximum interior temperature has not exceeded 78°F. Cooling is obtained both by ventilating the house at night and by opening the lids so that the ponds can radiate to the cool night sky. As in the case of the south-sloping glazing configuration described earlier, nighttime radiative cooling in this system is somewhat inhibited by the angled lids.

During the winter of 1975-76, the performance of the house was closely monitored. No auxiliary heating was used, except for occasional fires in a Franklin stove. The natural-gas space heater was never used. During a typical day, the temperature fluctuation was between 3 and 7 Fahrenheit degrees. The average daily maximum was above 70°F and the average daily minimum was above 65°F [HAM].

F-4: The Winters House; Winters, California [HAM].

G Attached Sunspaces

G.1 INTRODUCTION

Perhaps the greatest disadvantage of direct-gain systems is the potential in poorly designed systems for wide fluctuations in indoor temperatures. If wide fluctuations *are* permissible, however, reducing the thermal mass in the system will substantially reduce its cost. Conversely, the thermal mass required to temper wide temperature fluctuations in commercial greenhouses (rising over 100°F on a cold, sunny day and dropping dramatically at night) is usually expensive, as is a mechanical system to circulate the overheated air to a rockbed for later use. Fortunately, some spaces in buildings can easily tolerate wider-than-normal temperature fluctuations, and extra efforts to moderate them may be unnecessary. Examples of such spaces are greenhouses, atriums, sunporches, and garages. South-facing, glazed corridors are an example in larger buildings.

Although building costs are high even for such very simple and unrefined spaces, these spaces can take significant advantage of the sun's energy. The term "attached sunspace" designates a space whose temperature fluctuates widely because of direct gain from the sun. The overheated air can be used immediately to help heat the adjacent building, or it can be stored for use later when the sun is no longer shining. At virtually all times of the day, the attached sunspace has an indoor temperature that is higher than the outdoor temperature. This normally higher temperature tends to lessen the heat loss from the rest of the building. Because the sunspace both supplies solar heat and reduces heat loss by acting as a buffer zone, the portion of the building adjacent to the sunspace "sees" a milder outdoor climate.

An ordinary south-facing buffer zone can be changed into a sunspace by adding a south-facing transparent surface. This space can, in turn, provide solar heat to the attached building. Buffer zones in other orientations can have a similar, but not so dramatic, effect.

A north-facing buffer zone, though it will not provide solar heat, will maintain temperatures between those of the indoors and those of the outdoors, resulting in a reduction in heat loss from the adjacent building. Placing auxiliary spaces—such as garages, corridors, restrooms, and other areas not requiring constant temperatures—around the building is becoming an increasingly important option for conserving energy.

East- and west-facing buffer zones can be regarded as sunspaces, but they will not provide as much energy as south-facing buffer zones, and they may be a source of serious overheating problems in summer. An east-facing greenhouse can provide early day sunlight, heat, and humidity. Augmented with moderate amounts of thermal mass to slow the cooling speed of the space, it can continue to act as a buffer zone throughout the remainder of the day. The east side of the building adjacent to this sunspace may be the kind of space whose use necessitates higher temperatures early in the day and permits cooler temperatures at night. In many homes, kitchens are such spaces. On the other hand, living rooms and bedrooms may well be allowed to remain cool during the day and only begin to warm in the afternoon from the heat gained from west-facing sunspaces. In any case, the living patterns of the occupants should be considered. An east- or west-facing buffer zone should not use very much east or west glass, since—if it did—the high solar gains in the summer would cause severe overheating.

Although sunspaces can be relatively simple to build, it may be expensive to bring their construction to the same level of quality and durability as that of the rest of the building. For example, it is possible to build a simple, lightweight frame onto a house to support thin-film plastics. The resulting enclosure will be an excellent sunspace and will provide considerable heat to the building. The temporary nature of the structure (compared with the construction of the rest of the building), its lack of permanent and firm foundations, and the tenuous connections of the structure to the building, detract little from its thermal performance.

On the other hand, commercial-quality construction, coupled with commercially-available greenhouse structures, can be expensive. In general, permanent sunspaces are most economical when they have purposes in addition to providing heat, and when they are built to a standard of quality that will enhance the appraised value of the building. In many urban and suburban locations, zoning or aesthetic regulations prohibit the construction of relatively inexpensive sunspaces.

An important function of sunspaces can be as a place for growing garden and house plants. If plant growth is important, design considerations become more complex. Glazing types, temperature fluctuations, humidity levels, and auxiliary heat must all be viewed differently in such a case.

Cold-weather plants, such as lettuce and cabbage, can tolerate cold, sometimes mildly freezing temperatures. Few house plants can tolerate freezing temperatures, but many can endure rather cool temperatures. Still other plants require stable or high temperatures. When such conditions are needed in the sunspace itself, it is difficult for the sunspace to provide excess amounts of energy to the adjacent building. Although there frequently is excess solar heat during sunny weather, greenhouses in most climates require other sources of heat in order to maintain high or stable temperatures during long, cold, cloudy spells. Multiple-layer (up to quadruple) glazing has been used to reduce the need for auxiliary energy. Alternatively, movable insulation may be used to cover single or double glazing at night or during cloudy weather.

Warm ground temperatures and proper light levels are the two most critical elements for successful plant growth. Multiple-layer glazing can reduce light levels. This is a crucial issue in cloudy environments, particularly those receiving less than 50% of possible sunshine. Circulating overheated sunspace air through rockbeds situated under the plants can raise plant-bed temperatures, increasing the growth rate of most plants.

Evaporation of water from plant beds and transpiration by the plants are two factors known to influence the microclimate of a sunspace significantly. Humidity resulting from evaporation and transpiration can seriously affect thermal performance because large amounts of energy are required to evaporate water (approximately 1,000 Btu per pound of water). In fact, peak temperatures can be significantly reduced by the diversion of solar heat to this activity. Excessively humid air, coupled with potentially unpleasant odors, in some cases makes it undesirable to circulate overheated greenhouse air through the building. On the other hand, house plants' functions in odor-cleansing and in producing moderately moist air are often recognized as benefits.

Greenhouse environments are rather complex ecological systems. Unexpected (often undesirable) plant and animal growth is sure to occur. Many greenhouses built onto existing houses to provide both solar heat and vegetable production have suffered from significant insect and plant disease problems. Although much remains unknown about the ecological causes of such problems, preliminary findings indicate that the more complex the environment within a sunspace is, the more likely it is that a natural balance can eventually be achieved.

G.2 DESIGN FUNDAMENTALS

G.2.a GLAZING SYSTEMS AND PERFORMANCE

Single Glazing. For maximum light transmission, whether for plant growth or for solar-heating buildings, single-layer glass or plastic is preferred as the glazing for sunspaces. Single-layer glazing loses large amounts of heat. Nonetheless, a properly designed, single-glazed sunspace with adequate thermal mass, when attached to a building, will regularly provide heat to the building—in addition to maintaining adequate sunspace temperatures—in climates where temperatures rarely drop below zero and where the sun shines at least half the time. In colder climates, with the same 50% possible sunshine, the sunspace will provide heat to the building during sunny weather, but auxiliary heat will be needed to keep the sunspace from freezing during long, cold, cloudy spells.

Double Glazing. Most sunspaces are built with double glazing, without incorporation of additional methods for reducing heat loss. Heat loss through double glazing is still high, however. For the same conditions described above, a properly designed, double-glazed, attached sunspace in a 50% possible sunshine climate will remain above freezing the entire winter, in all but the coldest climates of the continental United States; but if heat conservation and production are desired beyond merely keeping the sunspace above freezing (for example, to assist plant growth), additional layers of glazing are necessary in most climates of greater than 6,000 degree-days per year.

Triple and Quadruple Glazing. Each additional layer of glazing increases the likelihood that more complex design decisions will be required. In order to maintain sufficiently high levels of light transmission, the third and fourth layers must be composed of very clear film or very clear glass. Particular care must be taken to prevent potential structural damage resulting from condensation of moisture between glazings. Light transmission through the composite layers should be in excess of 65%. U-values for triple and quadruple glazing can be as low as 0.35 and 0.24 Btu/hr/ft^2/°F, respectively.

Movable Insulation. Movable insulation is necessary to permit maximum levels of sunshine and to allow a minimum of nighttime heat loss. As with direct-gain systems in which the insulation is applied to ordinary building glazing, movable insulation for sunspaces presents numerous design constraints. The most critical of these involves the placement of the insulation during the day when it is not covering the glazing. A second significant issue is avoiding potential interference by plants with the movement of the insulation. A third problem is obtaining a tight fit when the insulation is covering the glazing. a fourth major consideration is cost. These four issues are similar to those faced in designing movable insulation for direct-gain systems.

Additional considerations include the sunspace's ecological vulnerability to mold and other unwanted plant growths and to unwelcome insect life, and the entire system's susceptibility to long-term decay in both thermal performance and physical integrity because of high moisture levels.

City:	Double-Glazed	Double-Glazed With R9 Night-time Insulation
Boston	0.34	0.48
Denver	0.58	0.72
Madison	0.29	0.42
Nashville	0.45	0.60
Seattle	0.37	0.50

G-1: Effect on solar savings fraction of different forms of insulation applied to an attached sunspace with $LCR = 24$ in five cities.

Figure G-1 shows the solar savings fraction in several cities for sunspaces utilizing a passive system with a load collector ratio (LCR) of 24. The LCR is the building heat loss coefficient (BLC) divided by the direct-gain area (A). Three glazing treatments are shown: double glazing, double glazing with R9 night insulation, and triple glazing. The calculations are for a sunspace with thermal storage mass of 30 Btu/°F/ft² glass area (12 inches of concrete).

G.2.b BUILDING INTEGRATION

Designers are increasingly aware of the advantages of "embedding" the sunspace in the building. Wrapping the building around the sunspace in this manner produces many advantages:

1. Heat loss from both the sunspace and the building is significantly reduced.
2. Heat is easily transferred directly from the sunspace to a large portion of the adjoining building.
3. Large amounts of natural light can penetrate deeply into a building that might otherwise rely entirely on artificial light. During daylight hours, heat loss back through the glass and into the sunspace is negligible.
4. The sunspace is easily heated by the building (when the sun is not shining) through the large amount of wall surface common to both.
5. In such a location, the sunspace is more likely to be functionally incorporated as part of an expanded living space.
6. The building itself can be built more compactly, while the wall area that it shares with the sunspace can provide the impression of a large exterior surface area.
7. Building costs are somewhat reduced compared to those for sunspaces attached to the south sides of buildings. For example, the common walls between the sunspace and the building are less costly than those exposed to the outdoors. In addition, their foundations do not need protection from frost and, therefore, do not need to be as deep. The compact building design results in less perimeter foundation work and a smaller total exterior area.

Attic space is often a practical location for a sunspace. The roof is framed in a conventional manner. The south slope is glazed. End walls and north-sloping surfaces, as well as the floor, are well-insulated. The surfaces are then covered with a dark surface material, such as black-painted plywood; less durable materials will also do for this purpose. When the house calls for heat, a thermostat triggers a fan to circulate solar-heated sunspace air from the attic to the house.

If the only purpose of the solar attic is to provide heating to the house, the choice of glazing is relatively unimportant—a double layer of glass or plastic will be sufficient. However, if the sunspace is to have other functions, such as growing plants, measures to keep the temperatures from dropping too low at night must be introduced.

G.2.c VENTILATION

Even the most well-designed sunspace will require ventilation during periods of intense sunshine and hot weather. Even during winter heating conditions, some controlled ventilation may be required to reduce humidity and maintain normal carbon dioxide levels. If mechanical ventilation is used, it should be capable of producing up to six complete air changes per hour if it is to prevent extreme overheating of the sunspace.

Natural ventilation is preferred to energy-consuming mechanical ventilation. Exhaust vents should be as close to the roof ridge as possible, and intake vents should be as low as possible. Air volume flow rates and, in turn, necessary vent sizes can be estimated. The flow of the air (Q), in cubic feet per minute (cfm), is approximately:

$$Q = 9.4 A \sqrt{H(t_u - t_d)}$$

where:
- H is the vertical distance between the intake vent and the exhaust vent, in feet;
- t_u is the average temperature in the sunspace, in Fahrenheit degrees;
- t_d is the average outdoor temperature, in Fahrenheit degrees; and
- A is the total area of the inlet *or* outlet vents (assumed equal).

For example, if the outdoor temperature is 80°F, the desired average temperature of the sunspace is no higher than 90°F, the vertical distance between the intake and exhaust vents is 10 feet, and the vent area is 1 square foot, then:

$$Q = 9.4 \, (1) \sqrt{10 \, (90 - 80)} = 94 \text{ cfm}$$

Each square foot of vent, therefore, will permit an airflow of 94 cubic feet per minute. The heat capacity of air is 0.018 Btu/ft³/°F. Therefore, the amount of heat exhausted through one square foot of vent per hour is:

$$(94 \text{ ft}^3/\text{min}) \times (10°F) \times (0.018 \text{ Btu/ft}^3/°F) \times (60 \text{ min})$$
$$= 1,015 \text{ Btu/hr/ft}^2$$

A representative value for heat gain through glass is 200 Btu/hr/ft². Therefore, each square foot of inlet and exhaust vent can accommodate the solar energy transmitted by 5 square feet of glass. Heat storage capacity will temper the amount of heat that must be vented. Most sunspaces perform well with 1 square foot of inlet and exhaust vent for each 5 to 15 square feet of glass.

G.2.d THERMAL STORAGE

The floor is the easiest and most obvious place to locate thermal mass in sunspaces. Whether composed of earth or of man-made materials—such as concrete or tile (laid directly onto the earth)—the floor has a vast storage capacity and thus moderates temperature fluctuations. Foundation walls should be insulated down to the footers—to at least R12 in cold climates.

Walls between the sunspace and the building are by far the most effective place to locate thermal mass in new construction. These walls receive full sunshine during the winter months and conduct some of their heat into the house; the remaining heat warms the sunspace. A single-glazing cover over the wall will trap more heat for the building, keeping the sunspace cooler. Walls between the sunspace and the building are easily shaded during the summer. Most of the design guidelines for conventional thermal storage walls apply to those used in sunspaces.

Containers of water can also serve as thermal mass. A storage volume of 4 to 6 gallons of water per square foot of south-facing glass is adequate, depending on the tolerable temperature swings.

Systems for circulating warm (and often humid) sunspace air through rockbeds are still in the exploratory stage, and little quantifiable information is available. As with remote rockbeds, discussed in section C.3.a, the airflow is in one direction only. Air circulation from the building through the rockbed is unnecessary. In fact, it often may be undesirable because the air is too humid. If the rockbed is located underneath an uninsulated floor slab, the heat will conduct through the slab and radiate and convect into the building. In cold climates with less than 50% possible sunshine, a greenhouse will probably need to regain the lost excess heat at night. If the rockbeds are located beneath the plant beds, the warmed rock will heat the soil, aiding plant growth.

Ordinary washed rock, 1 to 2 inches in diameter, is appropriate for most situations. If the rock is the only thermal mass in the sunspace (besides the floor), approximately 2 cubic feet per square foot of south-facing aperture is required. The rock need not be more than 2 to 3 feet deep. The air should flow as shown in figure G-2. For sunspaces having no other thermal mass, an air flow of approximately 6 cubic feet per minute per square foot

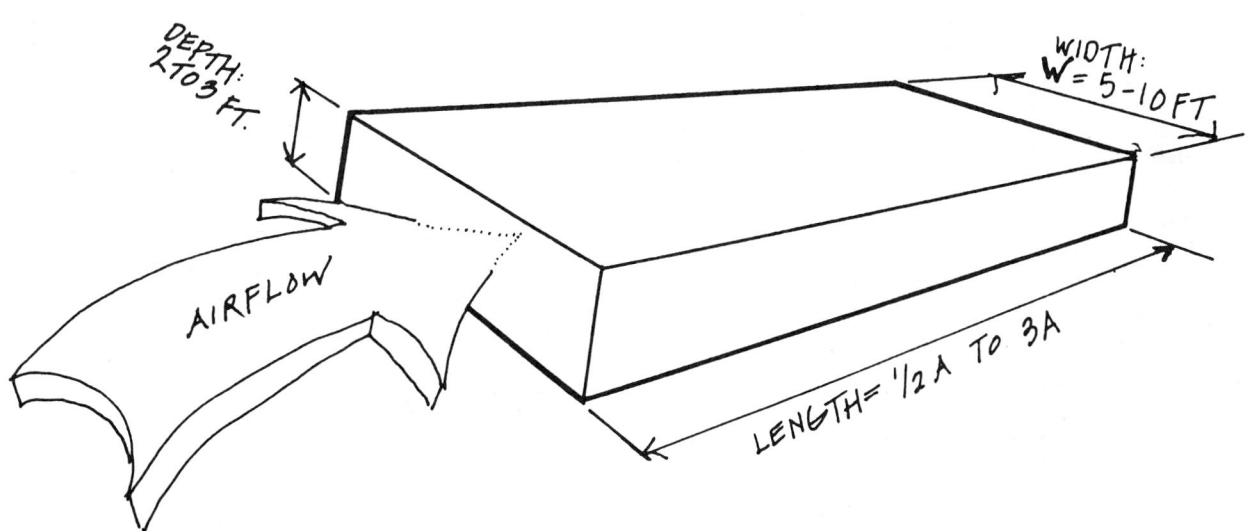

G-2: Preferred configuration for horizontal gravel beds that are to be placed under floor slabs or planting beds.

of south-facing aperture will be required to keep temperatures from rising above 85°F. The right fan size for this amount of airflow through rockbeds of the configuration shown in figure G-2 is in the range of ¼-horsepower per 500 cubic feet per minute.

In many designs, the sunspace is only large enough to supply the daytime heating needs of the building, which usually account for 10% of the total heating needs of the building. In such cases, the exchange of air between the sunspace and the building is not considered undesirable; also, no thermal mass is needed in the sunspace. Instead, excess heat is circulated to the building during the day, and at night building heat can be circulated to the sunspace.

G.3 METHODS OF HEAT TRANSFER FROM SUNSPACE TO BUILDING

The four basic methods for transferring thermal energy from sunspaces into buildings are:

1. Direct Solar Transmission
2. Direct Air Exchange
3. Conduction through Common Walls
    a. Massive Walls
    b. Frame Construction
4. Storage in and Transfer from Rockbeds

Cf. figure G-3.

Although these basic themes are discussed individually here, they can also be used in combinations. For example, in addition to installing a common heat-storage wall to conduct heat from the sunspace to the building, forced or natural airflow (direct air exchange) can be used to supplement heat transfer.

G.3.a DIRECT SOLAR TRANSMISSION

Frequently, some portion of the common wall between the sunspace and the building is made of glass. Depending on the design, a significant percentage of the light that penetrates the sunspace can enter the building directly through this common glass—especially when the sun is low in the sky, during the winter months. Both thermal mass and plants can intercept sufficient amounts of energy to maintain moderate sunspace temperatures and sustain healthy plant growth. The resulting environment acts as a buffer zone, reducing heat loss through the glass. A properly designed sunspace also helps shade the glass during the summer to help keep the building cool. Warm greenhouse air can be vented by natural convection to the outside. The vented air can induce natural ventilation through the building and into the sunspace.

Whether the common glass is single- or double-glazed depends on the strategy for maintaining air temperatures in the sunspace. If, for example, the sunspace is expected to remain above 45°F most of the time, heat loss from the building through the glass to the temperate environment of a sunspace will be small, and single glazing will suffice. In fact, what heat loss there is will help keep the sunspace above the desired 45°F. Double glazing is recommended when air temperatures in the sunspace are likely to drop below 45°F for long periods of time. For example, if the space is permitted to freeze with such frequency that only cold-weather vegetables (such as those of the cabbage family) are able to grow, double glazing is a logical choice for limiting heat loss from the building to the sunspace.

Whether single- or double-glazed, this common glass wall is protected from outdoor weather by the sunspace, making highest-quality construction unnecessary and reducing building costs.

G.3.b DIRECT AIR EXCHANGE

Sunspace heat can be transferred directly to the building in two basic ways: by natural air convection or with fans.

Often there is no common wall between the sunspace and the building. Instead, devices as simple as curtains can be used as desired to separate the sunspace thermally and physically from the rest of the building. If common walls are used, large windows and doors that open and close automatically or manually can be used to permit natural air convection. The greater the vertical distance between the intake and exhaust vents and the warmer the sunspace, the greater the airflow.

Heat transfer rates can be computed approximately as follows:

$$\text{Btu/hr} = 1.08 \, \text{cfm} \times (t_s - t_b)$$
$$Q = 9.4 \, A \sqrt{H(t_s - t_b)}$$

where:

- Q is the airflow (in cubic feet per minute) between the building and the sunspace;
- t_s is the average temperature in the sunspace, in Fahrenheit degrees;
- t_b is the average temperature in the building in Fahrenheit degrees;
- A is the total area of the inlet or outlet vent openings in square feet (assumed equal); and
- H is the vertical distance between the vent openings, in feet.

Various methods can be used to operate a fan for transferring warm air to the building.

1. Manually: The fan can be turned on or off according to the observations of the user.
2. Electric Clock Switch: At a certain time every day, the fan automatically comes on and at another time, automatically switches off.

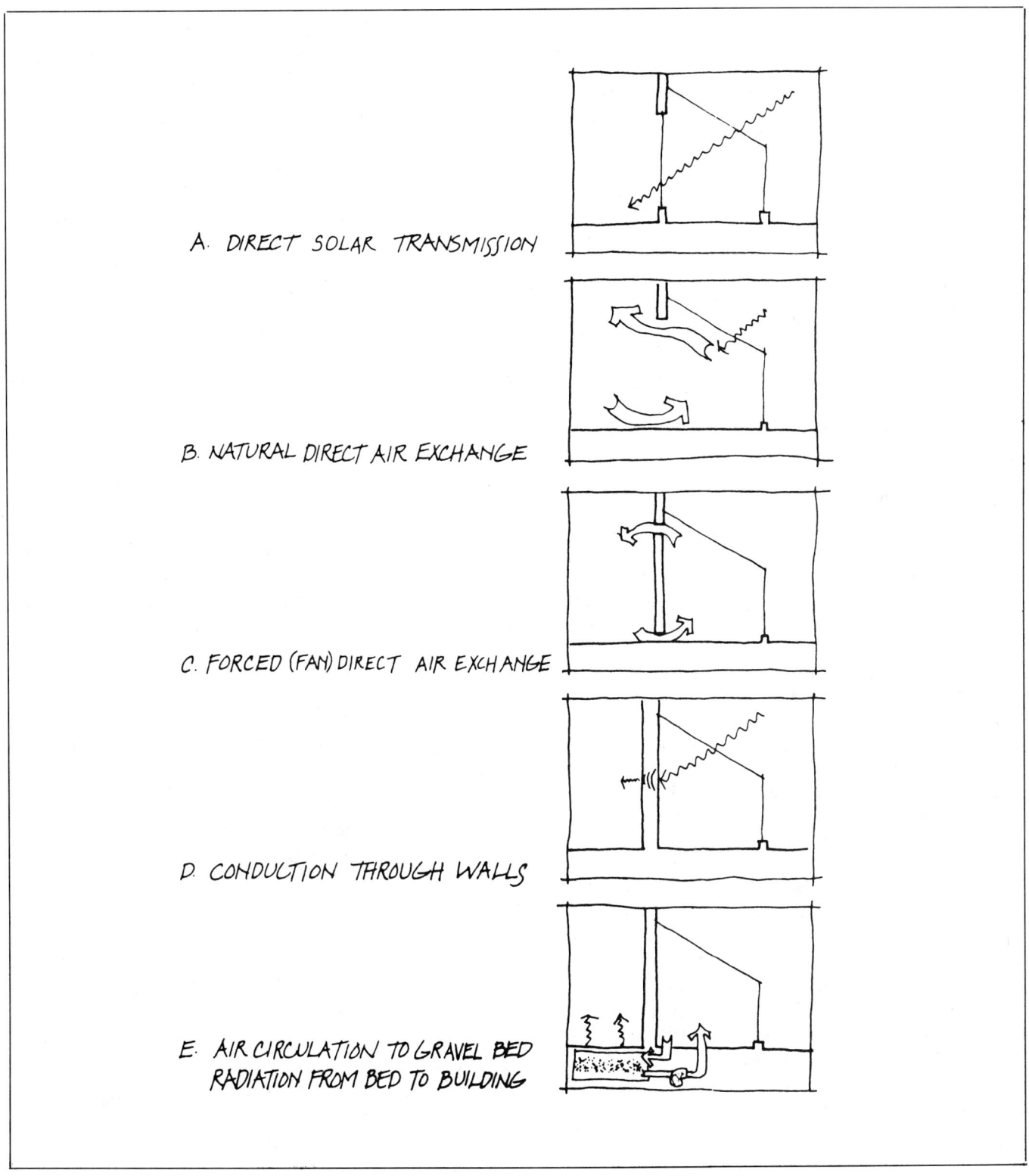

G-3: Heat-transfer methods—sunspace to building.

3. Temperature Sensor Control: When the sunspace temperature reaches a certain level, the fan switches on; when it drops below a certain level, the fan switches off. A thermostat in the building can, if necessary, override the sunspace sensor to keep the building from becoming overheated. Automatically dumping heat to the outside may be required. Differential thermostats can also be used.

Fans can direct the air to locations in the building, such as the north side, that would not otherwise receive solar heat. In large buildings, sunspace air can be used as a source of warm fresh air for ventilation or as makeup air for exhaust fans.

Moisture content, odors, and insects must be considered in evaluating this method of transferring sunspace heat to the building. Although many sunspaces will be relatively dry, others—in which large amounts of water are used for plants—will be very moist.

G.3.c CONDUCTION THROUGH COMMON WALLS

Massive Walls. A massive wall effectively connects a sunspace to the adjacent building. The wall should not be insulated. In effect, the wall functions very much like a Trombe wall: the sun's heat is absorbed on the sunspace-facing surface and is conducted to the inside where it is radiated and convected to the interior building space. The mass of the wall buffers the interior of the building from the temperature extremes of the sunspace. This mass effect works equally well in the summer to buffer the building from the daytime highs in the sunspace. Shading will prevent the wall from delivering heat to the interior of the building. Design considerations for the mass wall—thickness and material choices—are similar to those for Trombe walls. The sunspace effectively represents an expansion of the vertical air space in the Trombe-wall design in a way that allows it to function as a useful space.

Frame Construction. In general, common walls between the building and the sunspace need little or no insulation. Significant exceptions to this rule are walls exposed to the sun during the summer and walls through which heat gain is undesirable. During the winter, only small amounts of heat will be conducted into the house through wood-frame walls, even if they are poorly insulated.

G.3.d STORAGE IN AND TRANSFER FROM ROCKBEDS

As described in section G.2.d, overheated sunspace air can be blown by fans through rockbeds. These rockbeds may be located in the building under uninsulated floor slabs. Heat radiates up through the floor slab and directly into the room to be heated. Due to the relatively low (65°F to 75°F) temperatures in the rockbeds, the rooms will not overheat, and controlling the radiation through the slab from the rockbed is unnecessary. The fans can be left off during mild weather. Using fans to circulate air from rockbeds to the building—rather than letting it radiate through the floor—is another method of transferring heat to the house, but radiant floor heating is a more effective and more comfortable method of distributing the heat to the building when it is possible.

Moisture must be kept out of rockbeds, just as it must be kept away from floor slabs. Since temperatures in rockbeds are similar to those in the building, insulation levels there need to be only slightly greater than they would be for floor slabs in good energy-conserving construction.

G.4 A COMPROMISE SUNSPACE DESIGN

It is difficult to sort through the confusing multitude of design options for attached sunspaces. Few engineering details have been analyzed in sufficient depth to develop sound rules of thumb. The thermodynamics are so complicated that they preclude easy analysis for purposes of inclusion in the architectural design process. Figure G-4 is a "compromise sunspace design." It is applicable in most climates in the United States and allows for many possible sunspace uses. Although its net energy contribution to the building will vary depending on climate and use, it is a good compromise.

G.5 ADVANTAGES AND DISADVANTAGES OF ATTACHED SUNSPACES

Advantages of attached sunspaces are:

- Temperature swings in adjacent living spaces are small;
- Sunspaces provide space for growing vegetables and other plants;
- Heat loss from buildings is reduced because sunspaces act as buffer zones;
- Sunspaces are readily adaptable to existing buildings; and
- Since the sunspace serves more than one function, it is a natural and integrated part of the building design.

Disadvantages of this type of system are:

- Thermal performance varies from one design to another, which makes the performance difficult to predict; and
- Sunspaces are a relatively expensive means of solar heating.

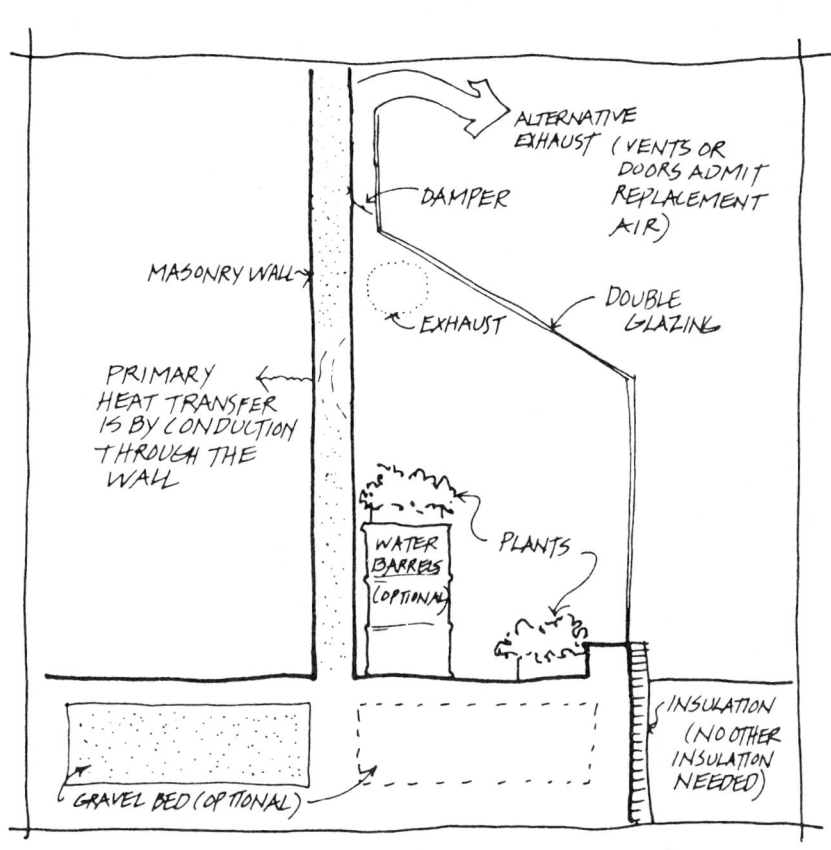

Summer temperatures can be kept close to outdoor temperatures with adequate ventilation. Mechanical ventilation and/or shading will be needed in hot, humid climates.

Winter temperatures are likely to be as follows:

Up to 8,000 degree-days and more than 70% possible sunshine:

Up to 8,000 degree-days and less than 70% possible sunshine: } 45°F – 85°F

More than 8,000 degree-days and more than 70% possible sunshine: 35°F – 85°F with occasional need for auxiliary heat

More than 8,000 degree-days and less than 70% possible sunshine: Up to 85°F with frequent need for auxiliary heat

G-4: A compromise sunspace design applicable to many climates.

Attached Sunspaces / 63

G.6 EXAMPLES OF ATTACHED SUNSPACES

G.6.a UNIT I; FIRST VILLAGE, NEW MEXICO

Unit I is located in First Village, a small, planned environmental community designed and built by Susan and Wayne Nichols and located six miles south of Santa Fe, New Mexico. See figure G–5. The basic floor plan of the 2,300-square-foot, two-story home (figure G–6) wraps the living space around a triangular shaped, 20-foot-high greenhouse located on the south side of the building. The south wall and the roof of the greenhouse are composed of two layers of glass and cover 409 square feet of area.

The roof is mounted at a 50° angle. The walls in common with the house are adobe. The wall is 14 inches thick at the first floor level and 10 inches thick at the upper level.

Solar heat is absorbed by the wall during the day and works its way through the wall and into the living spaces at night. The wall tends to flatten out the fluctuations between the surface temperature on the darkened adobe mass-wall during the day and the temperature in the unshuttered greenhouse at night. On a sunny winter day, the outside surface temperature of the wall can be as high as 110°F; it can drop to 45°F on a very cold (0°F) winter night. The average surface temperature is about 80°F.

G–5: Unit I, First Village; near Santa Fe, New Mexico [STR].

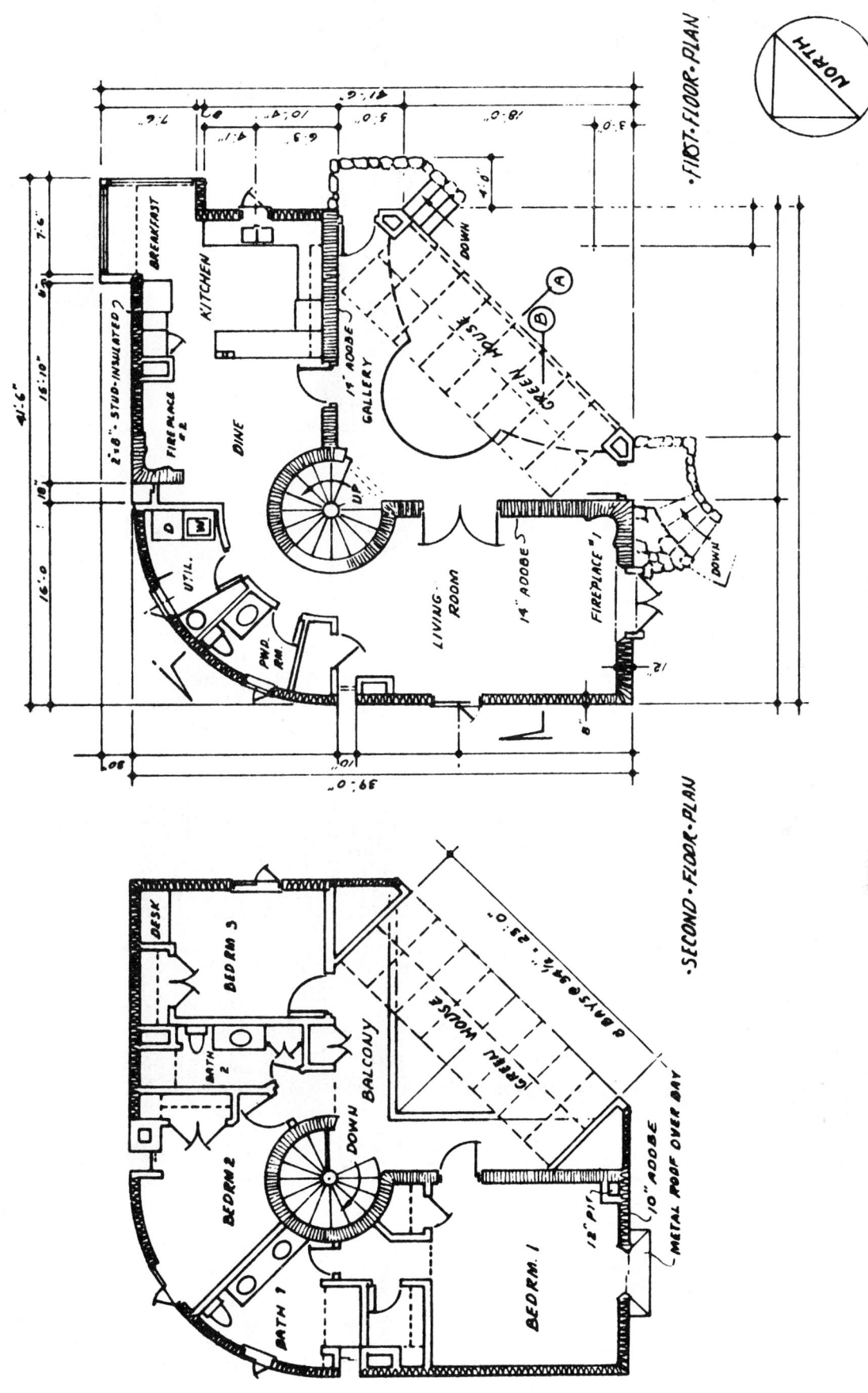

G-6: Floor plan of Unit I [SAN].

Excess heat from the greenhouse can be circulated by two ⅓-horsepower fans through two rockbeds located beneath the house. See figure G-7. The air, at a rate of approximately 2½ cubic feet per minute per square foot of glass, is then circulated back to the greenhouse. The heat stored in the rock conducts through the 7-inch-thick floor slab, which is covered with quarry tile. One horizontal rockbed is located underneath the living room, and the other underneath the dining room. The rockbeds are 2 feet deep and 10 feet wide. One is 19 feet long; the other is 15 feet long. The rockbeds contain a total of 24 cubic yards of 3- to 6-inch-diameter round riverbed rock. Floor temperatures range between 75°F during sunny weather and about 65°F after a cloudy spell. Baseboard electric heaters with individual thermostats provide backup heat for each room.

Cool indoor temperatures are maintained during the summer in several ways. First the adobe mass-wall is almost totally shaded by the balcony over the first floor and by the roof over the second. The large thermal mass of the house, along with the mild average summer temperature in Santa Fe (about 75°F) and the large day/night fluctuations (35 Fahrenheit degrees, on the average) in outdoor temperature, keeps indoor temperatures comfortable. Although the greenhouse temperatures vary greatly (65° to 95°F), air vents, windows, and doors can be opened near the base of the greenhouse, and a large vent window at the highest point can be opened to allow the warm air to exhaust. Greenhouse interior temperatures rarely exceed exterior temperatures.

Figure G-8 shows a plot of data gathered during the period from December 26, 1978, to January 8, 1979. These figures are generally representative of the lowest outside temperatures normally experienced in the Santa Fe area and illustrate the thermal stability of the house, the temperature fluctuations of the sunspace, and the consumption of electric heat. During most of the winter, internal temperatures both upstairs and downstairs normally hold in the upper 60s and the rockbeds (which supply heat through the floors) normally maintain a temperature of 68° to 72°F on sunny days.

The peak temperature for the lower level in the living room of the house during the summer was 76°F, despite peak outdoor temperatures of 95°F. Peak afternoon temperatures of 85°F have been recorded in the upstairs bedrooms, but they quickly drop after sunset to 70°F or less.

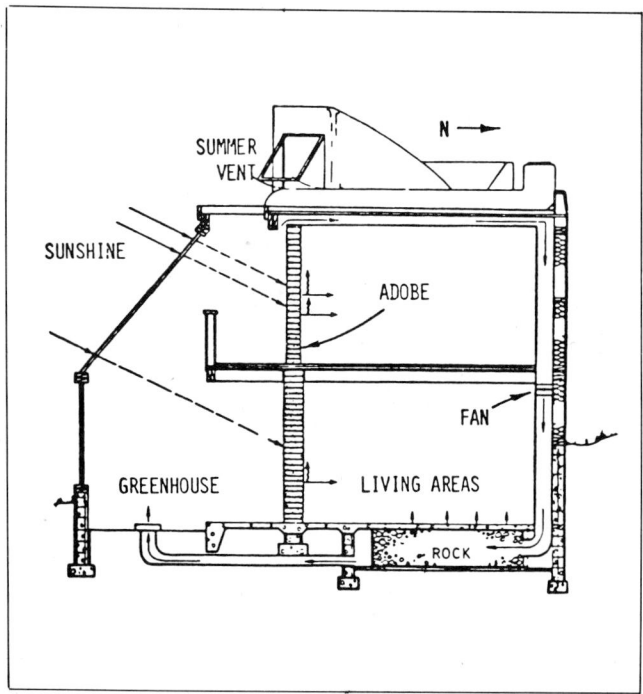

G-7: Schematic diagram of Unit I, showing its solar heating system [BAL-3].

G-8: Representative performance of Unit I's solar heating system, December 26 through January 8, 1978-79 [SAN].

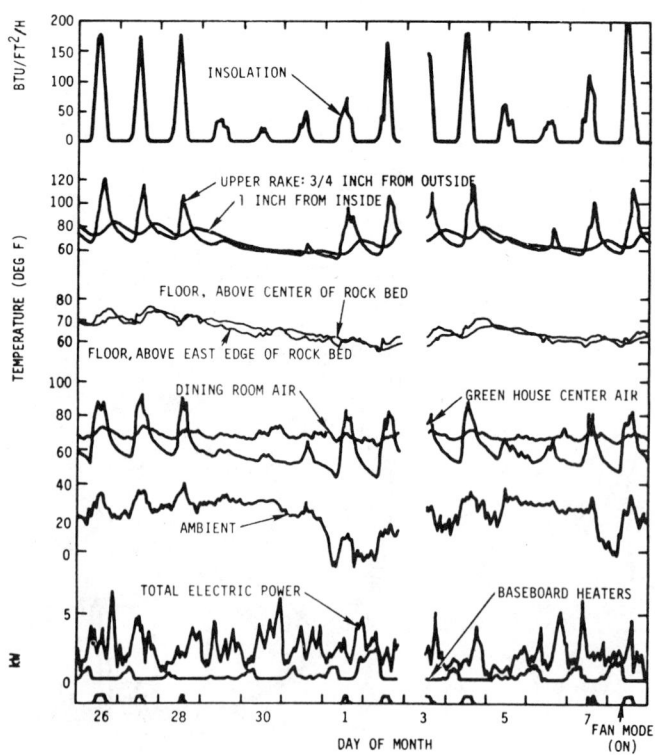

G.6.b THE BROOKHAVEN HOUSE; UPTON, NEW YORK

The Brookhaven House was designed by Total Environmental Action, Inc. and is located at Brookhaven National Laboratories in Upton, New York. See figure G–9. Particular care was taken to make the house energy-conserving by installing triple-glazed windows throughout and by using various techniques for preventing heat loss and air infiltration. The Trombe wall consists of triple-glazed, float-glass panels mounted in milled wood strips attached to a brick wall.

The Brookhaven House uses a greenhouse kit. Tight-fitting doors separate the sunspace from the rest of the house, allowing the occupants to control the flow of heat from the sunspace into the rest of the house. The balcony in the sunspace offers an additional sunny area for the second floor.

The wall between the sunspace and the kitchen/family room is a thermal mass wall consisting of standard paving brick. An exhaust fan at the top of the two-story greenhouse automatically exhausts air when there is any overheating.

G–9: The Brookhaven House; Upton, New York.

H Passive Solar Cooling

H.1 INTRODUCTION

Just as there is a distinction between active and passive solar heating, there is one between active and passive solar cooling. Although many passive solar cooling techniques are not strictly "solar," they are included here to represent cooling methods that require little or no mechanical power.

Fortunately, in virtually all climates, buildings can be designed and constructed to remain cool during hot weather. In most cases, it is also possible to eliminate completely the need for mechanical cooling methods by using passive methods, although this sometimes requires extraordinary measures.

By far the most important step is solar control—keeping the sun's energy from hitting and entering the building. Other methods discussed here include:

- Convective cooling;
- Evaporative cooling;
- Radiative cooling; and
- Ground cooling.

The state-of-the-art for passive cooling lags behind that for solar heating. The climatic data needed for making wise decisions is quite sparse. Analytical codes are nearly nonexistent. Little hardware is available. The relatively short length of this chapter is indicative of the situation.

H.2 COOLING TECHNIQUES

H.2.a SOLAR CONTROL

The least costly, yet most effective means of "solar cooling" is to keep the sun's energy out of the building. This is most effectively accomplished by shading the building's windows, walls, and roof. In fact, in locales where monthly mean temperature averages are below 70°F, controlling solar heat gain can virtually eliminate the need for other forms of cooling. Figures H-1 and H-2 are maps showing the normal daily average temperatures in the United States for July and August. As an approximation, the band of the United States along the 70°F line indicates the geographical limit of where the use of solar control can eliminate the need for other forms of cooling.

Most techniques for reducing heat loss from residences during the winter are also effective for reducing unwanted heat gain during the summer. For example, heavily insulated walls permit very little heat to penetrate during the summer. Multiple-layer windows likewise reduce heat flow into the building during hot weather. Weatherstripping restricts uncontrolled hot airflow into the building. Proper orientation of windows—especially minimization of east- and west-facing glazing in favor of winter-heat-gaining south-facing glass—is most important in reducing summer solar heat gain. Where east- or west-facing glazing is used, it is especially important to shade it, a difficult job because of the low and variable sun angles; vertical, movable devices may work the best.

Shading walls and roofs is critical in many hot climates. Heavily insulated walls and roofs need less shading than poorly insulated ones.

Light-colored surfaces can also be used to reduce heat gain. A dark-colored sunlit roof may be 60 to 80 Fahrenheit degrees hotter than a light-colored roof. Again, heavy insulation reduces the need for such considerations.

The shape of a surface can also affect heat gain. Much of the escaping of heat from a hot surface is caused by the flow of air across the surface. Many surfaces can induce their own natural convection currents due to their shape. Exposure to breezes helps cooling considerably.

Although shading the building is important, shading windows is far more so. Information on controlling solar heat gain is abundant. The most significant sources include the following references: [ASH-3], [OLG-2], [RAM].

The most effective shading arrangements prevent the sun from striking the buildings and employ devices such as overhangs or awnings on the buildings' exteriors. Unfortunately, the amount of shading that fixed overhangs provide varies according to the seasons of the sun rather than with the climatic seasons. The middle of the sun's summer is June 21, the solstice, but the hottest weather in the earth's northern hemisphere occurs from the end of July to the end of August, when the sun is lower in the

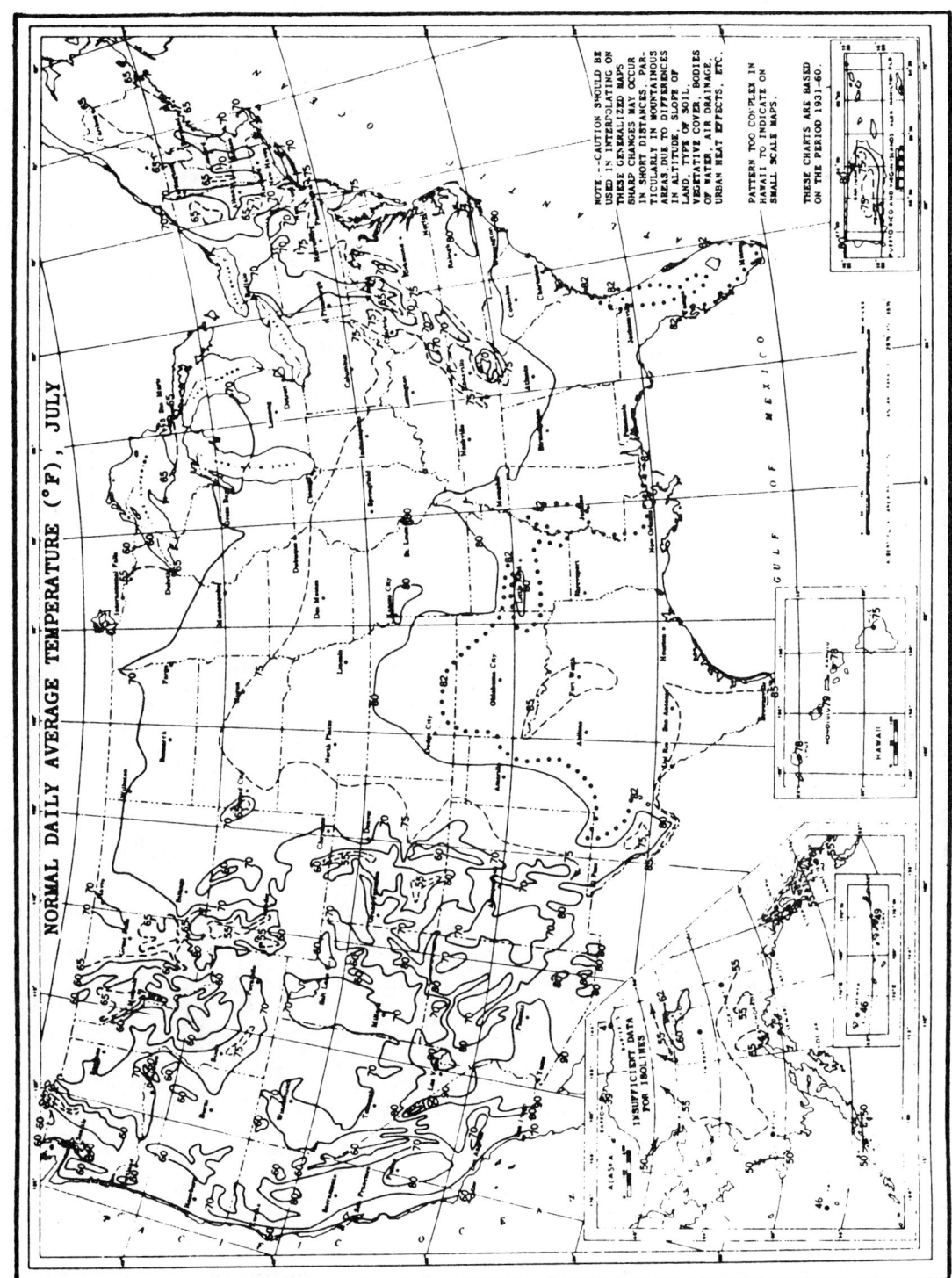

H-1: Normal daily average temperature for the month of July (in degrees Fahrenheit) [COM].

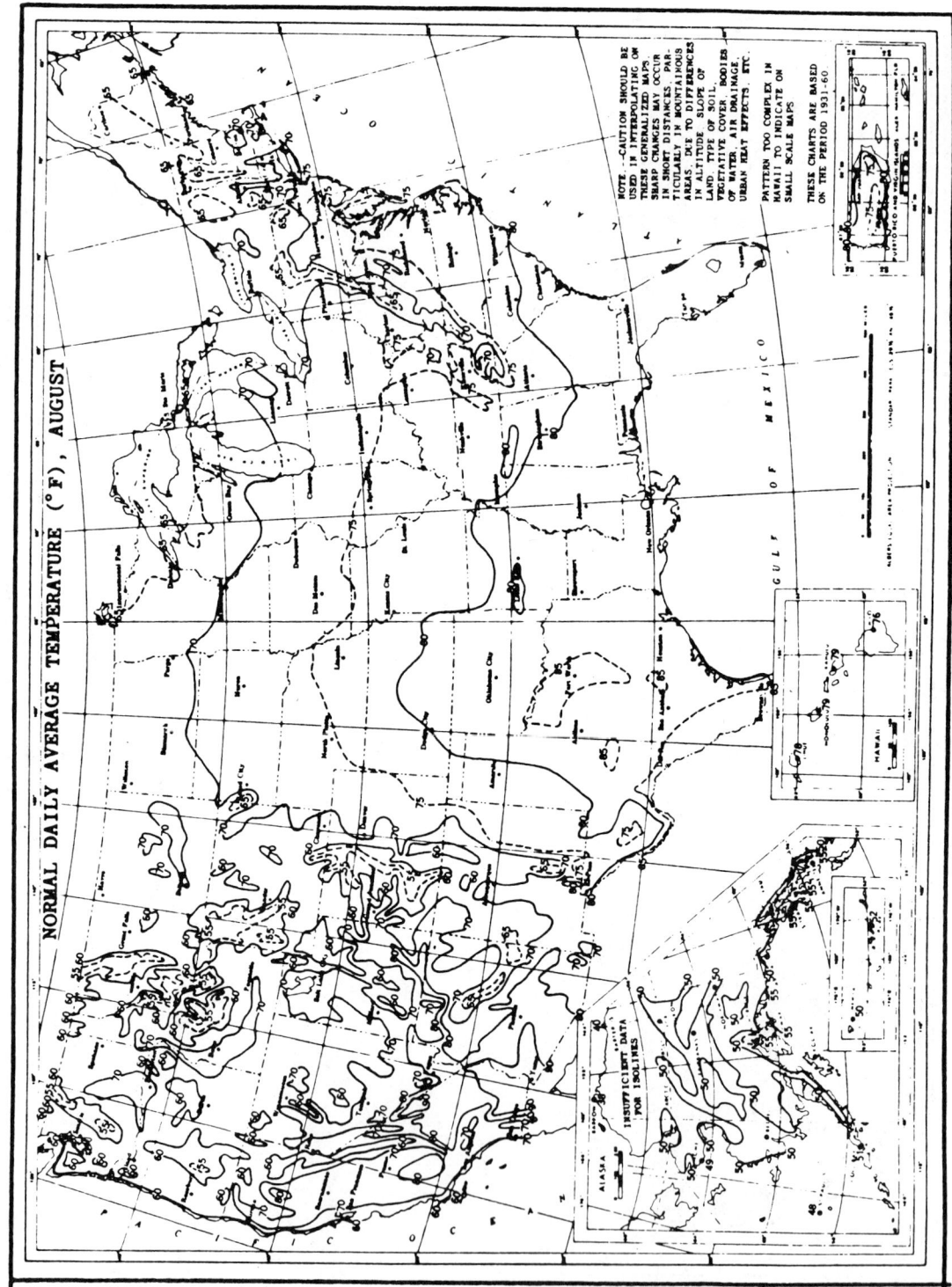

H-2: Normal daily average temperature for the month of August (in degrees Fahrenheit) [COM].

sky. A fixed overhang designed for optimal shading on August 10 creates the same size shadow on May 1. The overhang designed for optimal shading on September 21, when the weather is still somewhat warm and solar heat gain is unwelcome, produces the same shading on March 21, when the weather is cooler and the solar heat gain would be welcome.

Shading from deciduous vegetation more closely follows the climatic seasons and, therefore, the energy needs of buildings. On March 21, for example, most trees are bare and sunlight will pass readily around their branches. On September 21, however, the trees are still in full leaf and provide necessary shading. Deciduous trees in front of south-facing windows can provide shade from the intense midday summer sun. An overhanging trellis with a climbing vine that sheds its leaves in winter is an excellent alternative. It should be recognized, however, that a deciduous tree completely bare of leaves still blocks from 20% to 40% of the sun's direct radiation, reducing solar gain proportionately. This may be too severe a penalty to bear in a cold climate.

Operable shading devices are more versatile and adaptable to human comfort than fixed devices and vegetation are, but when attached to the outsides of buildings they are difficult to maintain; in addition, most designs deteriorate rapidly. Efforts to make them more durable are usually unsuccessful; however, with rising fuel prices and greater emphasis on shading, increased efforts have been made to produce better operable shading devices. Awnings are perhaps the simplest and most reliable operable devices, but their aesthetic appeal is limited. The fact that most of these devices do not operate automatically is not necessarily a drawback.

Shading east- and west-facing glass is difficult because, in both summer and winter, when the sun reaches the eastern and western skies, it is at a low altitude. See figure H-3. Overhangs do not prevent the penetration of the sun during the summer any more than they do during the winter. Vertical louvers or other vertical extensions of the building are probably the best means of shading such glass.

Where view and light are important on east and west facades, window area can be minimized by using eye-level, shallow, horizontal windows under deep overhangs. The time during which the sun penetrates the windows is short, and the solar impact is greatly reduced.

H-3: Sun angles for various dates, wall faces, and times, at 40°N latitude [AND-2].

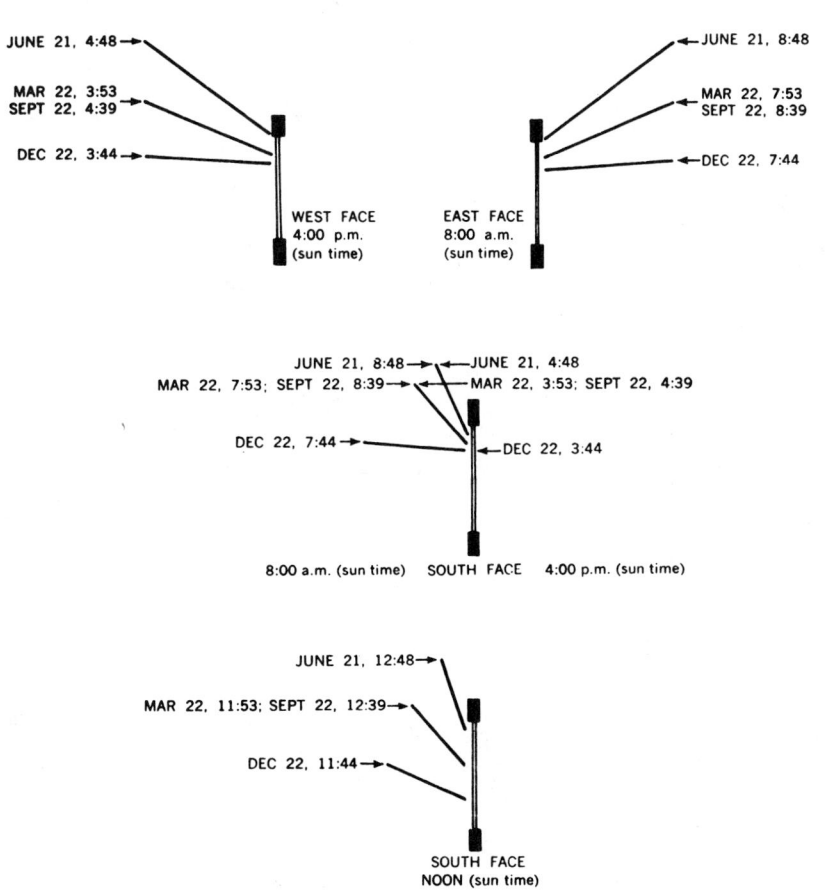

Another method of shading glass on east and west walls is to orient the glass to face either north or south. By facing the glass north, only indirect irradiation is admitted, a favorable lighting effect for many tasks. By facing the glass south, solar heat is admitted during the winter. Figure H-4 shows a method for orienting the glass toward the south to provide full shading during the summer.

Shading coefficients are important in comparing the relative effectiveness of various shading devices. By definition, a single layer of clear, double-strength glass has a shading coefficient of 1.00. The shading coefficient for any other glazing system in combination with shading devices is the ratio of the solar heat gain through that system to the solar heat gain that would occur through the standard single-layer double-strength glass under the same solar conditions. Solar gain through a glazing system is thus the product of its shading coefficient times the solar gain for clear, double-strength glass. Figure H-5 shows some typical shading coefficients obtained under various shading conditions.

Using different types of glass for different sun orientations is one method of sun control. If reducing heat gain is critical, heat-absorbing and heat-reflecting glass can help, especially on east and west facades. The important factors to consider in their use include the following:

1. Such glass reduces solar heat gain. Although this can be an advantage in the summer, it is a disadvantage in the winter.
2. Except for glare control, heat-absorbing and heat-reflecting glass are almost always unnecessary in north, north-northeast, and north-northwest orientations. Little solar heat is gained on these facades except in the latitudes south of 30°N.

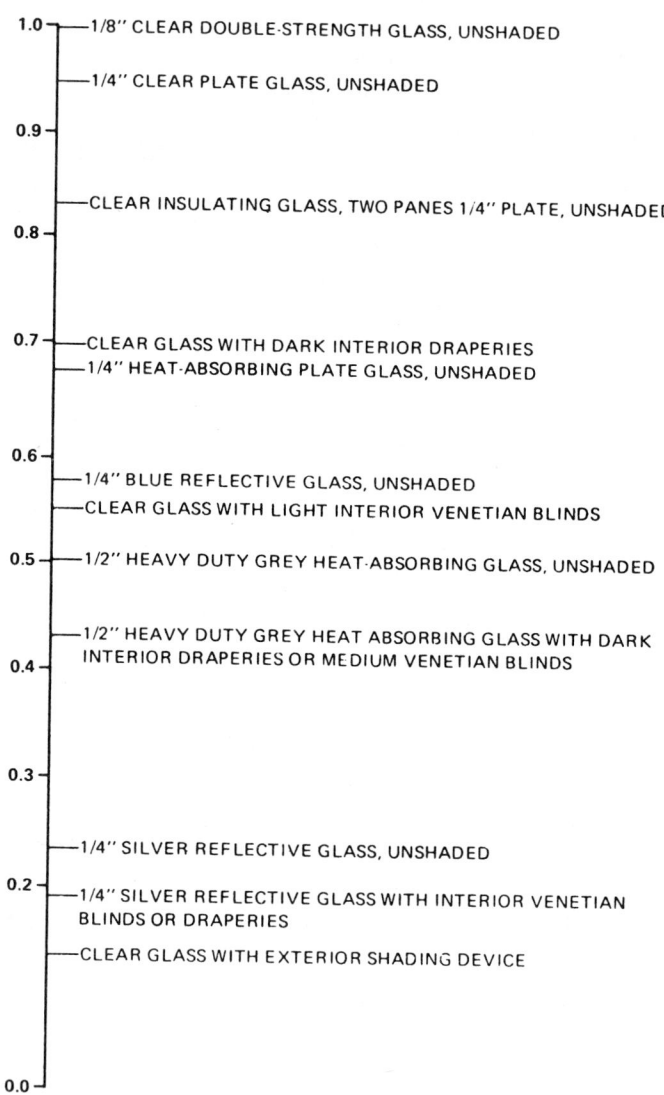

H-5: Shading coefficients obtained under various shading conditions [AND-1].

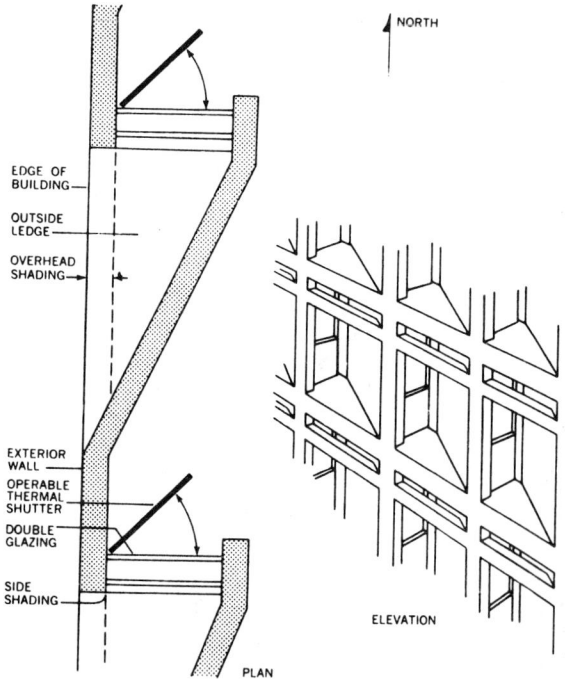

H-4: Sawtooth arrangement of windows on the west facade of a building, allowing solar heat gain during the winter but disallowing it during the summer [AND-2].

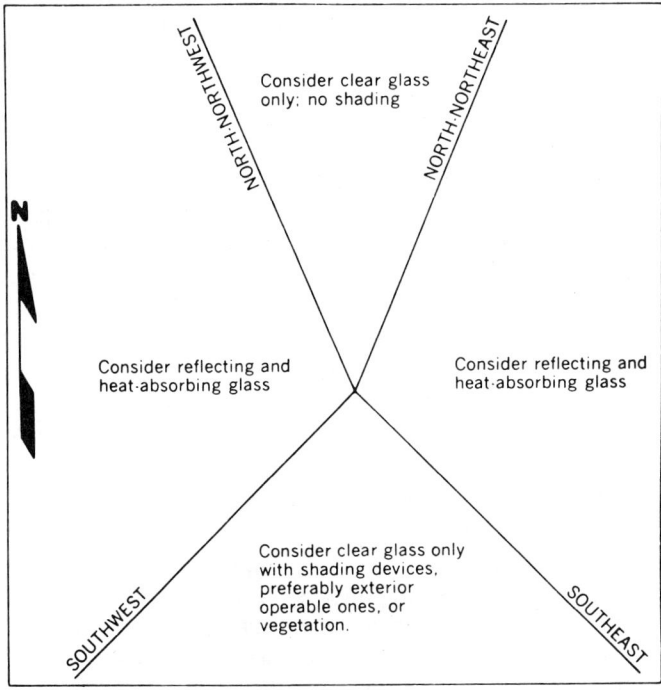

H-6: Appropriate glass types for various orientations of windows (approximations for the continental United States) [AND-2].

3. In latitudes south of 40°N, heat-absorbing and heat-reflecting glass usually should not be considered for south-facing windows (except as a necessary means of controlling glare or of excluding winter solar heat—e.g., in large office buildings). The heat gain through south-facing glass is relatively small in the summer. See figure H-6.

4. Vegetation and operable shading devices are more sensible solutions than heat-absorbing or heat-reflecting glass for south, southeast, and southwest orientations. Shading devices on the exterior of the building are the most effective; devices (such as venetian blinds) situated between two layers of glass are the next most effective; interior devices (such as blinds, shades, and draperies) are the least effective because they only stop the sun's rays after these have penetrated the building. See figure H-7. Even so, highly reflective devices are only slightly less effective on the inside than they are between the two layers of glass; and because a highly reflective device is effective anywhere it is placed, its location is not as important as the location of a shading device that is not reflective.

H-7: Possible locations for shading devices, in order of effectiveness [AND-2].

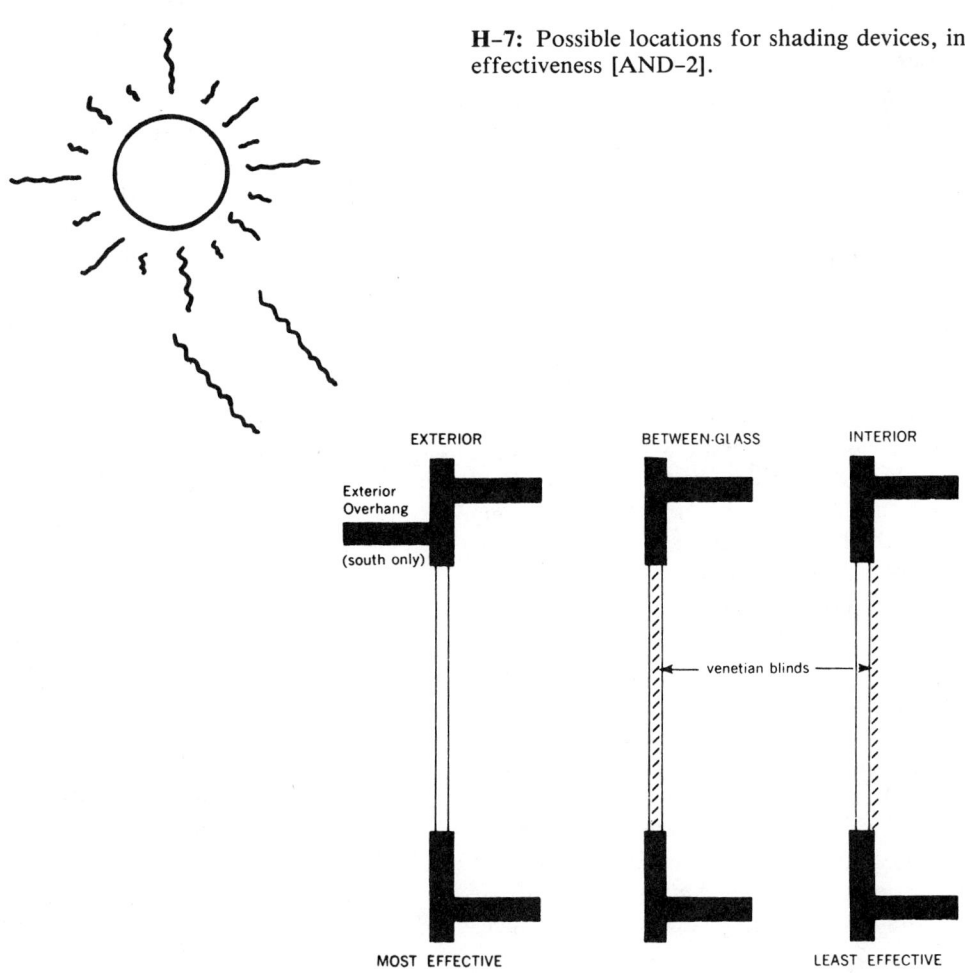

Passive Solar Cooling / 73

5. Some advertisements for heat-absorbing and heat-reflecting glass suggest that these products will reduce both the initial cost of air-conditioning equipment and the cost of its operation, especially in terms of energy consumption. The savings, however, are usually obtained by comparing costs with those for all-glass buildings (buildings entirely enclosed by glass of any type), rather than with those for buildings already designed to conserve energy. Rarely is it mentioned that substantial savings could be achieved by beginning with opaque, well-insulated walls that have reduced glass areas on the north, east, and west walls and with a well-designed building that allows the sun to penetrate through the south-facing glass during the winter but that provides shading during the summer.

All four (or more) facades of a building need not—and, in fact, should not—be of identical appearance. This is particularly true of buildings that have large areas of glass. Although there may be economic, social, and personal reaons for building glass boxes, glass has certainly been misused as a design element.

Large buildings with comparatively small exterior wall and roof areas and comparatively large interior floor areas often require air-conditioning year-round. This is because tremendous amounts of internal heat are generated from the activities of people, from burning lights, and from operation of equipment. Using glass that is shaded twelve months a year instead of only during the summer provides the most successful solution in these buildings. Every effort should be made to reduce the amount of heat that is produced by lights and machines. Energy-conscious designs for such buildings now commonly employ waste-heat recovery systems that remove heat from overheated areas and deliver it to areas that need it. Heat not immediately required may be stored for later retrieval. Dependence on artificial lighting should be reduced by using more natural lighting (through windows), by lowering lighting levels, or by placing light fixtures directly where light is needed ("task lighting").

Designers should also consider the shading effects of buildings on one another and on the surrounding environment—i.e., whether the shading occurs on buildings that directly or indirectly use the sun's heat or light—as well as their effects on wild vegetation or gardens.

H.2.b CONVECTIVE COOLING

At temperatures below 91°F, the movement of air across human skin creates a cooling sensation caused by heat leaving the skin through convection and by the evaporation of perspiration. Air movement at up to 50 feet per minute goes unnoticed. At speeds over 200 feet per minute, it becomes annoying. Air movement above 300 feet per minute adversely affects health and reduces indoor work efficiency [OLG-1].

The most common way to create air movement without using mechanical power is to open windows and allow breezes to blow into a building. This simple concept is often forgotten. Although an open window can admit dust, pollen, and, in many cases, warm air, its cooling effect should not be underestimated. Proper window location can aid natural ventilation. Air inlet locations govern the airflow patterns in the building. An inlet window high in the wall will create an airflow above the living area. Lower openings will direct the air through the occupied area. Outlet locations have comparatively little effect on airflow patterns, but should be as large as possible and located higher up than the inlets. "Deep" plans (e.g., two to three rooms stacked "in series," and very large, deep offices) are notoriously difficult to ventilate using windows only.

Land planning also influences natural ventilation through buildings. Natural breezes should not be blocked from entering buildings. Building shape, proper clustering of buildings, and landscaping features such as vegetation and fences can be used to enhance natural wind flow patterns [GEI].

The "stack effect" in buildings can induce ventilation even when there is no breeze: ventilation occurs when warm air rises to the top of a tall space. An opening at the top naturally exhausts the warm air while openings at floor level admit cooler outdoor air to replace it. Natural ventilation can be further induced by using belvederes, wind vanes, and wind scoops.

In figure H-8, a solar collector exhausts its hot air to the outdoors and pulls house air through it, creating natural ventilation. Many variations of this "solar chimney" have been used widely in the past, and many are being developed again today. In some active solar systems in which air serves as the heat-transfer medium, the collectors are vented to the outside during hot summer weather, pulling building air through them to induce ventilation.

In designing for stack-effect ventilation, it should be kept in mind that the greatest airflow is achieved by maximizing both the height of the stack and the temperature of air in the stack. As noted earlier, the airflow is determined by the inlet area and the square root of the height times the average temperature difference:

$$Q = 9.4 A \sqrt{h(t_u - t_d)}$$

where:

- Q is cubic feet per minute of airflow;
- A is the area of the inlets in square feet;
- h is the height in feet between inlets and outlets;
- t_u is the average temperature of the air in the "chimney," in Fahrenheit degrees; and
- t_d is the average temperature of the return air (normally just the ambient outside temperature) in Fahrenheit degrees.

It is better to add heat (presumably using a passive air-heating collector) at the bottom of the chimney or stack than at the top. In this way the entire column of air in the chimney is hot, creating the buoyancy required to cause the air to flow.

A common element in Iranian architecture is the wind tower (or wind catcher), which harnesses the prevailing summer winds to cool the inside of a building. Wind towers resemble chimneys, with one end in the basement and the other end rising well above the roof. Doors open into the basement at the lower part of the tower and into a central hall on the main floor. By opening and closing these and other doors, airflow through different parts of the building is controlled. Reference [BAH] provides an excellent discussion of these structures.

Thermal mass in buildings can be advantageous in areas where nighttime temperatures drop below comfort levels. Cool late-night and early-morning air can be circulated through a building and the coolness stored in thermal mass for use during the day. An alternative to using thermal mass as an integral part of the building is to use a rockbed. A fan circulates cool air through it at night; then, during the day, room air can be circulated through the cool rockbed, cooling the building.

Figure H-9 shows the fluctuation in inside air temperature of a night-air-cooled office building near Davis, California, over a 4-day period in the summer. The indoor temperature fluctuated from a low of 65°F (the outdoor nighttime low temperature) to a high of nearly 80°F. Outdoor daytime temperatures during the same period soared well over 100°F.

In locales where mean daily temperatures drop to less than 65°F, nighttime ventilation in combination with thermal mass should be seriously considered. Figures H-10 and H-11 are maps of normal daily minimum temperatures in the United States for the months of July and August. As an approximation, the band along the 65°F line indicates the geographical limit of practicable night-air cooling.

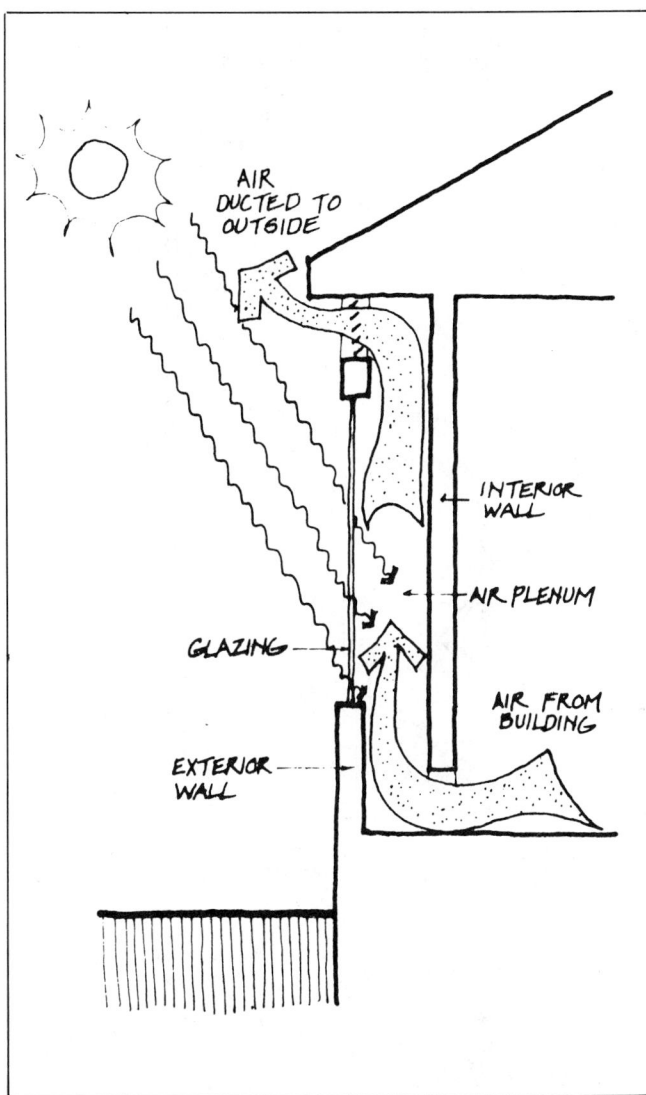

H-8: A wall designed as a solar collector to induce ventilation through a building [CRO].

H-9: Indoor temperature fluctuations over a 4-day period in an office building in Davis, California, that is cooled by the circulation of nighttime outdoor air [BAI].

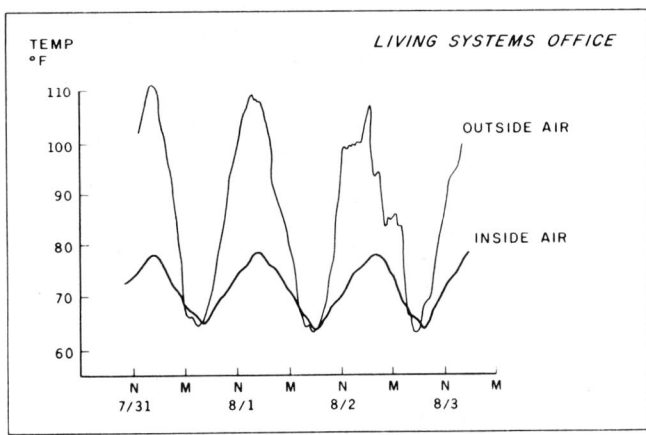

Passive Solar Cooling / 75

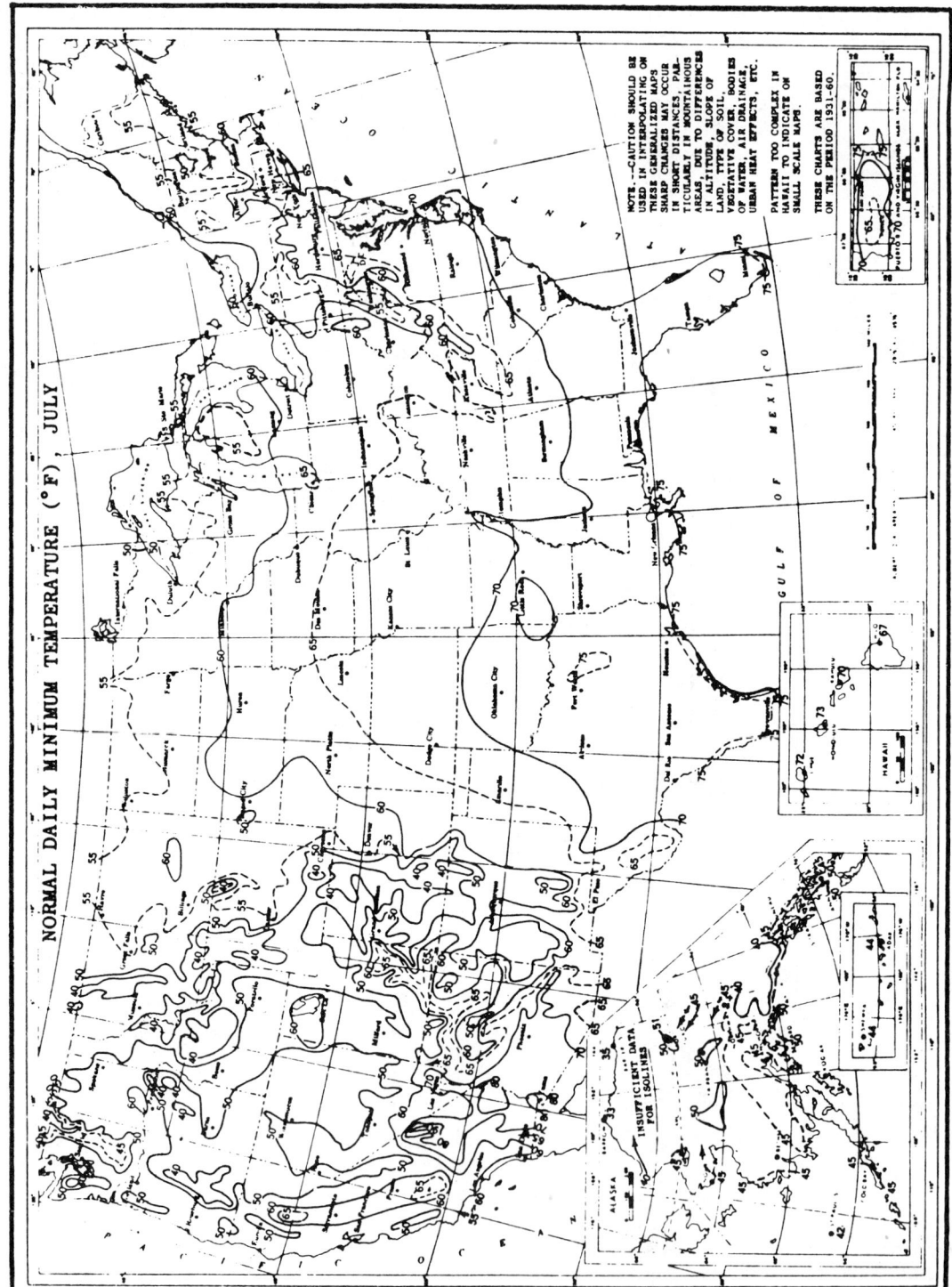

H-10: Normal daily minimum temperatures for the month of July (in degrees Fahrenheit) [COM].

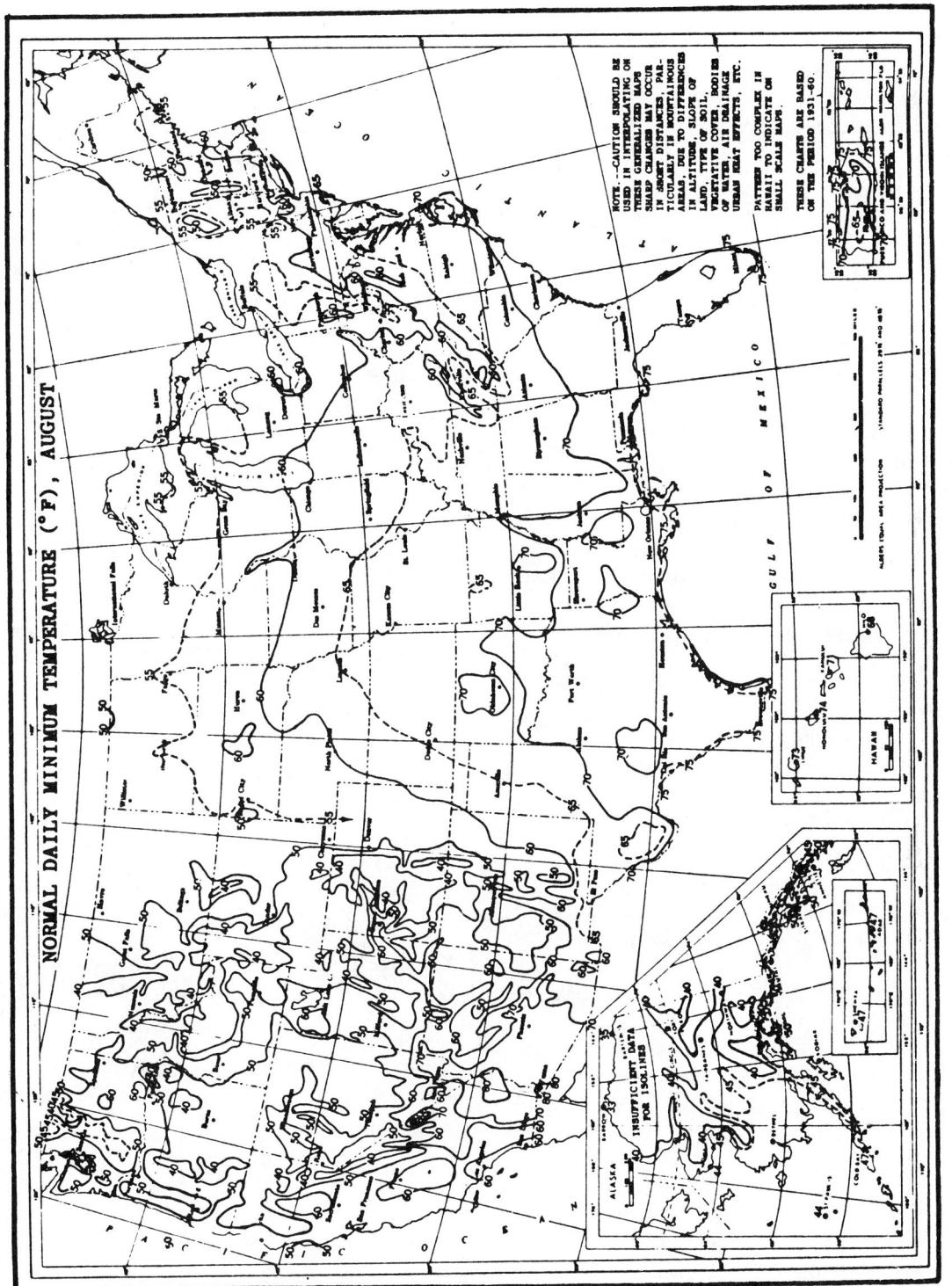

H-11: Normal daily minimum temperatures for the month of August (in degrees Fahrenheit) [COM].

H.2.c EVAPORATIVE COOLING

When the dry bulb temperature of air is higher than the wet bulb temperature, humidifying the air will cool it. The cooling effect of moving air is enhanced if it contacts water prior to coming into contact with people. Although the evaporation of water into the air occurs under many circumstances, the most important means are to have a large water-to-air surface contact area and to increase the turbulence or movement of the water that is in contact with the air. For example, large shallow ponds create large surface contact areas. Moving streams and sprays from water fountains increase turbulence and surface exposure. Air blow across the surface of a large, shallow pond provides both effects.

Mechanical evaporative coolers can increase this evaporative cooling effect, usually with an energy requirement substantially below that of conventional compression and absorption air-conditioning.

Evaporative cooling can also keep roofs of buildings cool. Flooded roofs, roof sprays, and roof ponds have been used successfully in severely hot climates such as those in Arizona and Florida.

In fact, water performs two functions: it protects roof materials from ultraviolet damage, and (more importantly) it cools the roof as it evaporates. At an evaporation rate of 10 to 15 inches per month, water can produce a cooling rate averaging between 115 and 175 Btu per square foot per day [SEL].

Wind is a major factor in the rate of evaporation, and windflows across roof ponds should be enhanced as much as possible. Usually, only crude calculations of evaporative cooling potential are possible because of limited data on wind speed; however, where wind speed is known, useful approximations of evaporative cooling effects can be made.

Problems associated with evaporative cooling include insect and plant growth in the wet environment and a build-up of minerals such as lime on evaporative surfaces.

Water sprays are effective in lessening solar heat absorption by roofs on hot summer days [YEL-2]. For example, during the intense sunshine of Arizona summers, directing a fine water spray on horizontal roofs for approximately 40 seconds of each minute keeps the roof surfaces within 5°F of the ambient dry bulb temperatures during the middle of the day. During the early morning and afternoon hours, the same level of spraying reduces roof temperatures below the ambient dry bulb temperature and within a few degrees of the ambient wet bulb temperature. Approximately 0.3 pounds of water per square foot of roof surface is needed to accomplish this. The power required to pump the spray at the required water pressure (10 psi) is approximately 0.01 hp (7.46 watts) for every 1,000 square feet of roof area. The cooling effect is approximately 300 Btu/hr/ft² (88,000 watts) [YEL-2].

The nighttime cooling capacity of sprayed roofs has not yet been explored in detail. Evaporation from roof ponds, however, has been found to produce a significant cooling effect. Evaporation of 11.4 pounds of water an hour at 75°F produces the equivalent cooling of 1 ton of refrigeration [YEL-2], [YEL-3].

H.2.d RADIATIVE COOLING

A constant exchange of thermal energy occurs between objects that can "see" each other. A net transfer of energy occurs from the warmer object to the cooler object. The earth, for example, radiates heat to clear night skies, which would otherwise be quite cold even during hot weather. In the North, the sky can also be very cool during the day. The earth's radiation to the sky is primarily at wavelengths between 6 and 15 microns [BUD].

To be effective, a radiating surface must be carefully protected from the warming effects of breezes. One of the best sources of engineering data and analyses of radiative cooling is [BLI]. Figure H-12 shows the net radiative loss rate (R) from an exposed, thermally black,* horizontal surface at air temperature to the sky. Three parameters—air temperature, humidity level, and dewpoint—are involved in the computations; two of these must be known. The graph shows that the heat loss rate (cooling effect) is usually between 15 and 30 Btu/hr/ft². During a 10-hour night, the total cooling effect is between 150 and 300 Btu/ft².** This range holds in many

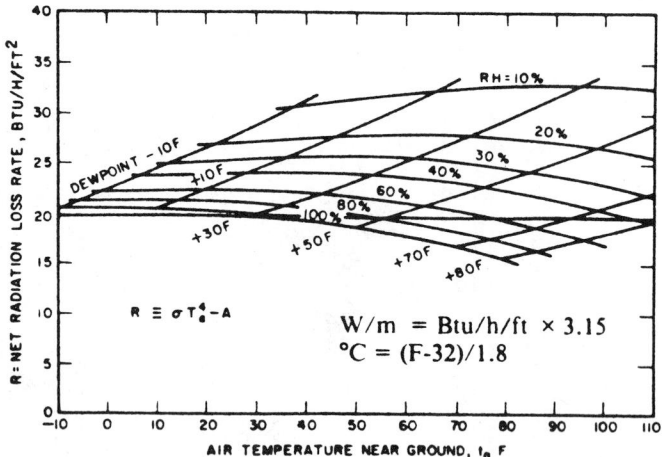

H-12: Typical values of radiative heat loss from horizontal surfaces to the sky, at sea level [BLI].

---

*A "thermally black" surface is one that absorbs all heat that strikes it.

**In comparison, a good solar heat collection system produces up to 1,000 Btu/ft² during a sunny day and an average of 200 to 500 Btu/ft² a day during a heating season.

climates, making nighttime radiative cooling widely applicable; nevertheless, it is far less effective in humid climates with high dewpoints than elsewhere. Precise radiative cooling effects in humid, high-dewpoint climates have yet to be established.

Most radiation from the surface of the earth to the sky occurs at night. The percentage of radiation varies greatly at different angles to the surface—from a maximum of 100% possible perpendicular to the surface (towards the zenith) to virtually none at the horizon. The most effective radiant cooling surface is a horizontal one. Obstructions such as trees and walls reduce nighttime radiative cooling [BAI].

A vertical surface with no obstructions yields about 40% of the radiant cooling of a horizontal surface. For additional information on radiative cooling, the following references are useful: [ABR], [BAI], [BAR], [BRO], [DUB], [GEI], [KNO], [NEU], [PIT], [PLE], [REI], [SEL].

Probably the classic radiative cooling concept is the one first developed by John Yellot and Harold Hay. Hay continued its development and called the system Skytherm. Skytherm uses roof ponds in which water is contained in black plastic bags. The water cools by nighttime radiation. During the day, the ponds are shaded by sliding panels that fit above them.

The cooled water in turn cools the house by absorbing room heat through the house's metal ceiling. The tops of the bags can be flooded with water to provide additional cooling by evaporation. With only slight additions of mechanical power, 600 to 1,300 Btu/ft^2 can be dissipated to the sky during a summer night, through a combination of radiation and evaporation [YEL-3]. This system has been used successfully to take care of the entire cooling needs of a house in Atascadero, California. See figure H–13. Annual temperature profiles, inside and outside, are shown in figure H–14 for an earlier version of the system with similar performance characteristics.

H-13: The Atascadero House; Atascadero, California [SAN].

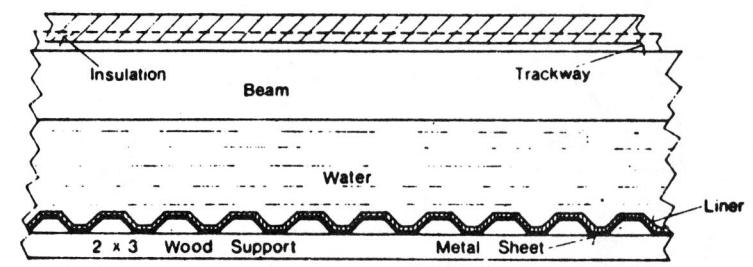
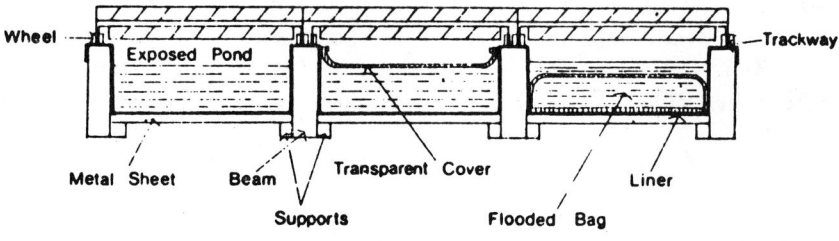

Cross-section of ceiling ponds.

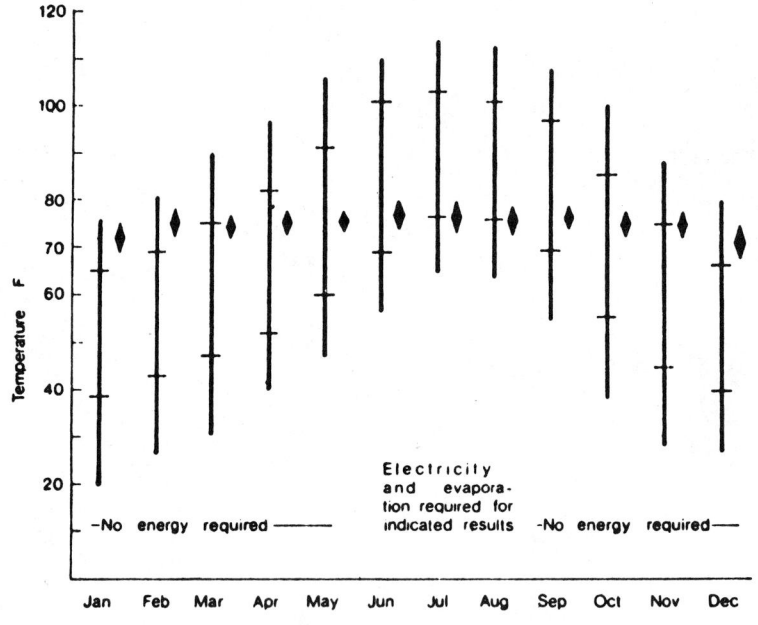

H-14: Schematic diagram of a ceiling-pond system (top), and outdoor and indoor temperatures, by month, at the Atascadero House (bottom) [YEL-3].

Figures H-15, H-16, and H-17 present three different designs that use radiative cooling. Figure H-15 shows a house developed in 1975 by John Hammond in Winters, California, that incorporates a form of the roof-pond system. The percentage of roof covered by the system is smaller than that in later versions of the system, and the function of the sliding insulation panels is served by a hinged panel powered by a hydraulic ram. Even during the hottest weather, the air temperature in this house has not exceeded 78°F.

The Cool Pool concept (figure H-16) was developed and tested by Living Systems in Davis, Indio, and Sacramento, California. The roof pond radiates its heat to the sky, and the coolest water settles nearest to the radiant wall panel, cooling the house.

David Bainbridge used a thermosiphon cooling slab in one of his designs. See figure H-17. The placement of the tank at ground level makes maintenance easier. A modification of this concept is a nighttime-air cooling system developed for integration into the structure of a

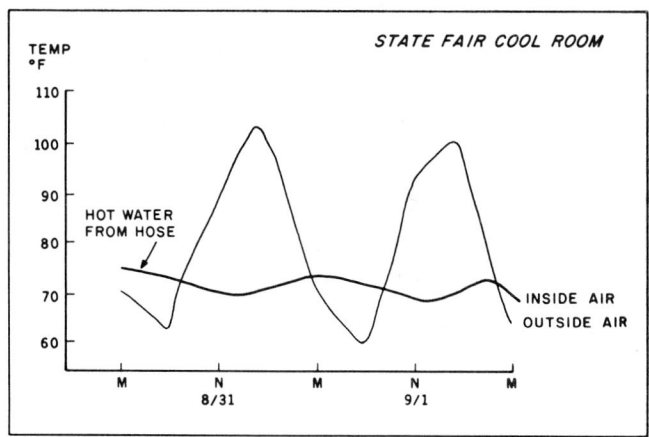

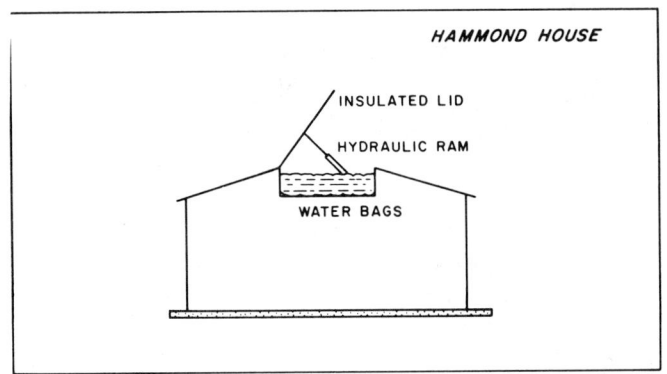

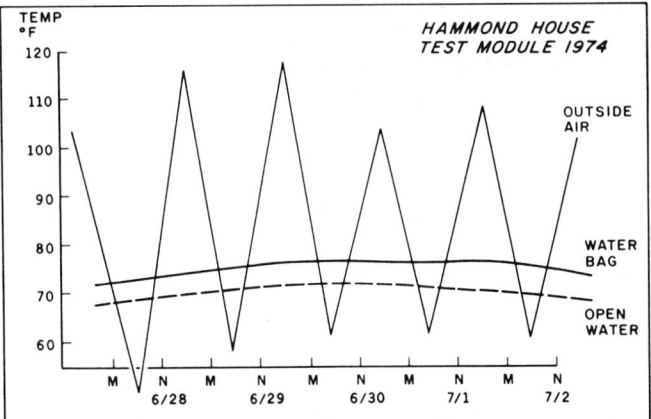

H-15: Diagram (top) and performance records (bottom) of the Winters House. The design is a modification of the Skytherm radiative cooling system [BAI].

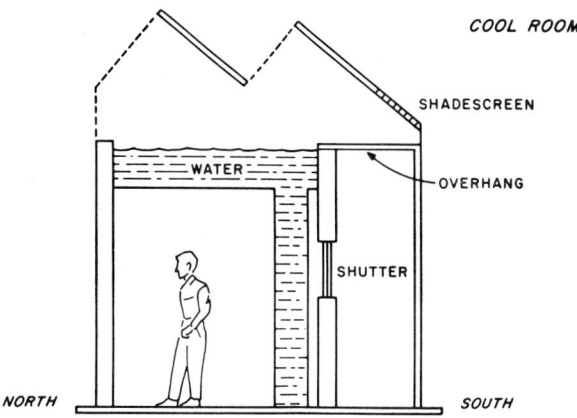

H-16: Diagram (top) and performance records (bottom) of the Cool Pool designed by Living Systems [BAI].

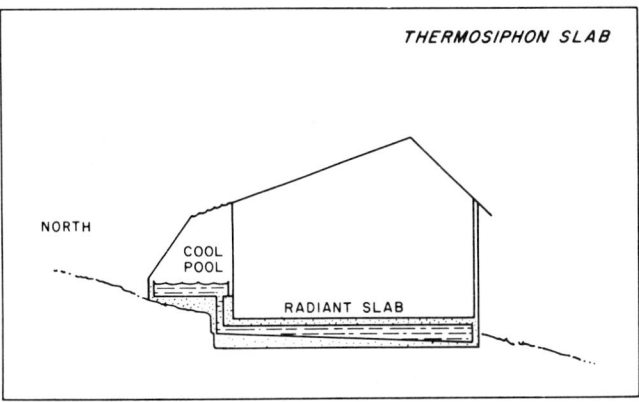

H-17: A thermosiphon cooling slab designed by Living Systems [BAI].

building. The heat dissipator is a sill-wall inclined 45° to the sky, with a ducted metal plate and a selective cold-body surface coating. See figure H-18. The warm room air is ducted through the heat dissipator, where it is cooled. From there, it passes through ducts in a concrete floor slab that stores the coolness. Airflow can be enhanced by using fans. This system can be easily integrated into multistory constructions; with modifications, solar heating can also be accommodated.

Other radiative cooling concepts, as well as the Atascadero House and the Winters House, are discussed in Chapter F.

H.2.e GROUND COOLING

Since the ground is nearly always cooler than the air at times when cooling is required, the greater the portion of a building that is in contact with the ground, the cooler the building will be. Situating a building below ground level or in the side of a hill is the easiest way to obtain more ground contact; berming earth as high as possible or covering buildings with earth are other ways of achieving the same result. High levels of comfort and quiet are usually attained in well-designed underground housing.

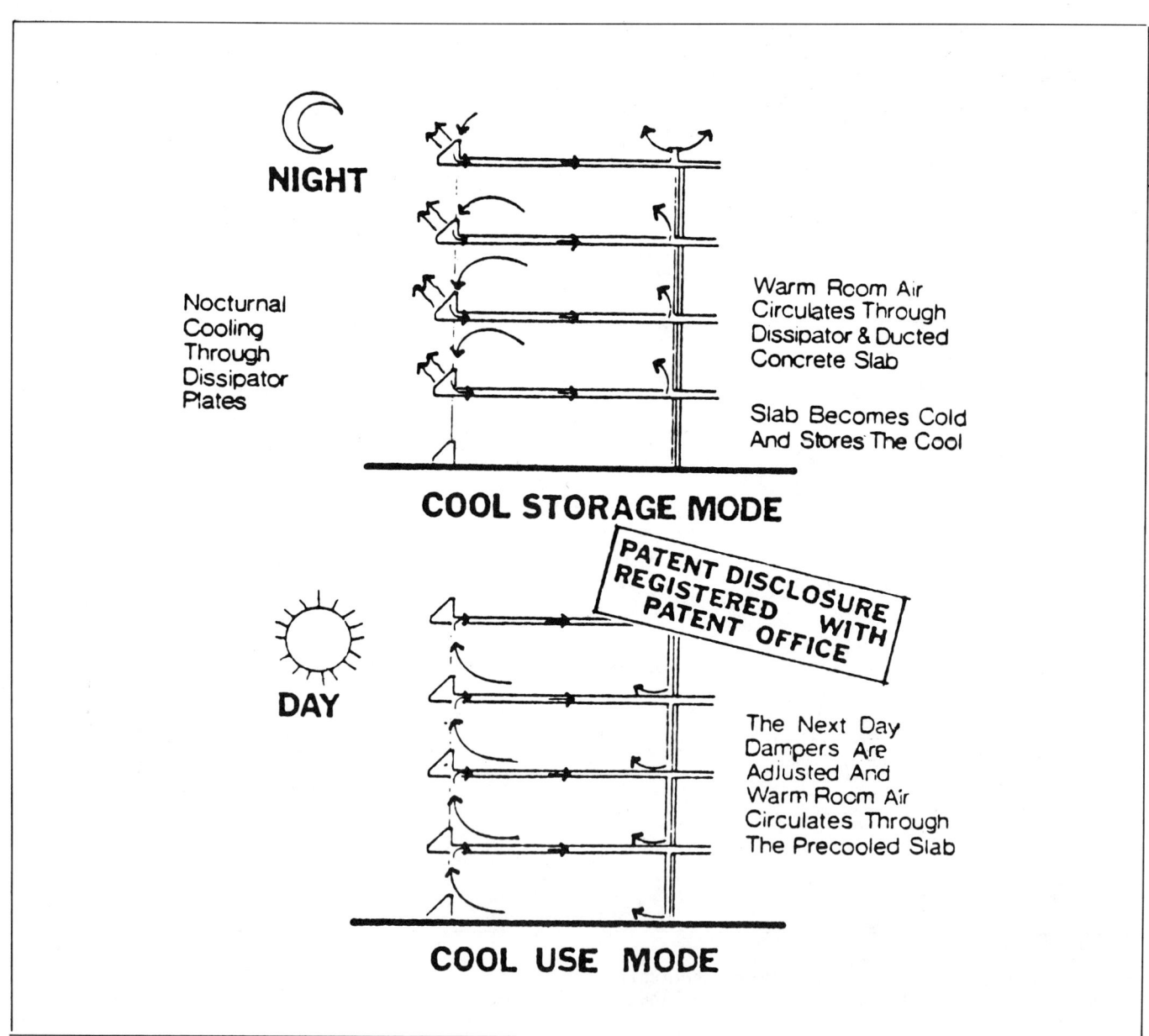

H-18: A nighttime cooling system for multistory buildings [PIN].

82 / *CHAPTER H*

Underground housing in cold climates requires good insulation of all walls that are in contact with the ground. The insulation keeps the interior surface of the walls warm and dry. In hot climates, the walls in contact with the earth should remain uninsulated. The resulting cool interior surface will tend to be damp from condensation, especially in humid climates. Particular care should be taken, therefore, to waterproof the outside surface of any wall that is in contact with the ground.

H.3 PEAK LOAD REDUCTION

Various means of reducing a building's cooling loads also have the effect of reducing the local public utility's peak loads. Frequently, a building's electrical cooling requirements can be eliminated entirely. In other cases, passive cooling systems incorporating thermal mass (roof ponds, for example) can be integrated with off-peak active cooling systems. In a typical situation, the thermal mass would absorb the excessive daytime solar heat, delaying the building's need for cooling until the hours of peak demand had passed; mechanical cooling would then be supplied in the succeeding hours at a lower level of power (consonant with the gradual dissipation of stored heat from the thermal mass), reducing the size and quantity of mechanical equipment needed to meet the building's cooling needs.

In general, if mechanical equipment is operated at night to store ambient coolness for use during the day, the equipment will operate with relatively high efficiency and will avoid placing an additional burden on the local public utility during the peak hours of local electrical consumption.

H.4 APPLICATIONS TO LARGE BUILDINGS

As with passive solar heating, passive solar cooling need not be limited to residential-scale structures. Perhaps the first step in making a large building compatible with passive cooling is to decrease the cooling load of the building. Major savings can be realized (through natural lighting and natural ventilation) by increasing the exterior surface area of the building in proportion to its overall volume. Wise thermal design of the increased external wall area (specifying high levels of insulation and thermally efficient glazing systems) can compensate for the losses incurred by increased surface exposure. The potential annual energy savings resulting from the combination of natural lighting, winter solar heat gain, and summer natural ventilation should not be underestimated.

Passive cooling methods present several challenges when applied to large buildings. Foremost is the difficulty of providing the needed quantity of airflow through the building during nighttime ventilation. Given the relatively warm water produced in radiative and evaporative cooling, the challenge is to cool effectively. Most mechanical cooling systems are designed to use water cooled below 65°F.

The larger the building, the greater the likelihood that mechanically driven fans will be required to circulate cool night air. For the system to be desirable, the electrical consumption of these large fans must not exceed that of conventional cooling systems. Analytical methods to determine the net cooling effect of nighttime-air ventilation on a seasonal basis still need to be developed. However, estimates can be made.

As a general guideline for buildings in climates such as that of Sacramento, California, 1 cfm per square foot of floor area can provide about 35 Btu of cooling per square foot of floor area per day. Building mass should be about 40 pounds per square foot of floor area, with about 1¾ square feet of exposed surface area per square foot of floor area. Larger airflows are desirable. Airflows of 3 (or more) cfm per square foot, in combination with an interior building mass of 75 to 150 pounds per square foot of floor area, can provide about 100 Btu of cooling per day per square foot of floor area. Similar cooling effects of building mass will occur in other areas of the country where the minimum daily temperature is 65°F or below during most of the cooling season. In structures exposed to this same temperature range, the building mass will be cooled to between 65° and 70°F.

Part Two
Passive Solar Design Analysis

Los Alamos Scientific Laboratory
University of California

J. Douglas Balcomb
Dennis Barley
Robert McFarland
Joseph Perry, Jr.
William Wray
Scott Noll

I Sensitivity Analysis of Thermal Storage Walls

I.1 EFFECTS OF THERMAL STORAGE MASS

Sizing a thermal storage wall is an especially important design decision because it has major cost implications. The wall itself is expensive, and the space it takes up in the building is valuable.

Thermal wall performance improves markedly with increasing thermal storage mass, up to the point at which the mass is able to store sufficient daytime solar heat to handle the building's nighttime heating load. Making provision for several-day heat storage is comparatively ineffective as a means of increasing the year-round heating performance of the storage wall. Thermal wall performance is greatly improved at high values of solar saving fraction (*SSF*) (corresponding to low values of *LCR*) and is significantly worsened at low solar savings fraction values.

Performance results for thermal walls in six cities—Boston, Albuquerque, Madison, Nashville, Medford (Oregon), and Santa Maria (California)—are shown in figure I-1. The calculations represented are for Trombe walls and waterwalls without nighttime insulation. Several effects can be determined from the graphs.

In the case of waterwalls, performance improves continuously with increases in thermal storage mass, although the growth of the *SSF* very soon becomes negligible. The "knee" of each curve reflects a solar savings fraction "height" that is relatively great when the corresponding *LCR* value is low and the wall is thick enough to allow the *SSF* to reach its tapering-off point. *SSF* (when attended by a low *LCR* value) grows rapidly between the *M* values of 10 and 30 in most building environments. See section I.3. This is the primary reason for the rule of thumb that more thermal storage mass should be provided at higher values of solar savings fraction.

In case of Trombe walls (masonry), there is normally an optimum wall thickness. Performance declines after a certain point because the increased thermal storage capacity of a thicker wall is outweighed by the increased difficulty the heat has in passing through the wall.

Study of the large body of calculations that have been done on this subject leads to the following conclusions:

1. A change in the amount of thermal mass in a waterwall has approximately 35% less effect on the wall's heating performance when R9 nighttime insulation is regularly used than when no nighttime insulation is used.
2. The optimum thickness of a Trombe wall is slightly greater—and is less readily identifiable—when R9 nighttime insulation is regularly used than when no nighttime insulation is used.
3. The greater the thermal conductivity of a Trombe wall's material, the greater the wall's optimum thickness will be.
4. The presence of vents in a Trombe wall becomes important in a case involving either a thick wall or a wall of low thermal conductivity. This is because the convective loop effect of thermocirculation is a major mechanism for heating the building in such a case. In one study, thermocirculation vents increased the optimum thickness of the walls being tested from approximately 12 inches to approximately 18 inches.
5. The presence of vents in a Trombe wall is more important at low values of solar savings fraction than at high values.

The curves in figure I-1 indicate that the effect of thermal storage varies significantly from one locality to another. Presumably, this is caused by differences in local weather patterns, but no systematic study has yet been completed to determine the exact causes. An attempt to compact the data on the curves shown in figure I-1 into a single universal curve proved unsuccessful.

The cities compared in figure I-1 represent a wide range of geographical locations and climates. The designer should choose from among them the location whose climate is most similar to his or her own, and then use its graph as a basis for estimating local performance variations.

Up to this point, discussion has focused on the effect changes in mass have on the annual solar savings performance of the building. An equally important consideration involves the temperature swings observed within the heated space. These temperature swings are greatly affected by the amount and type of thermal storage mass.

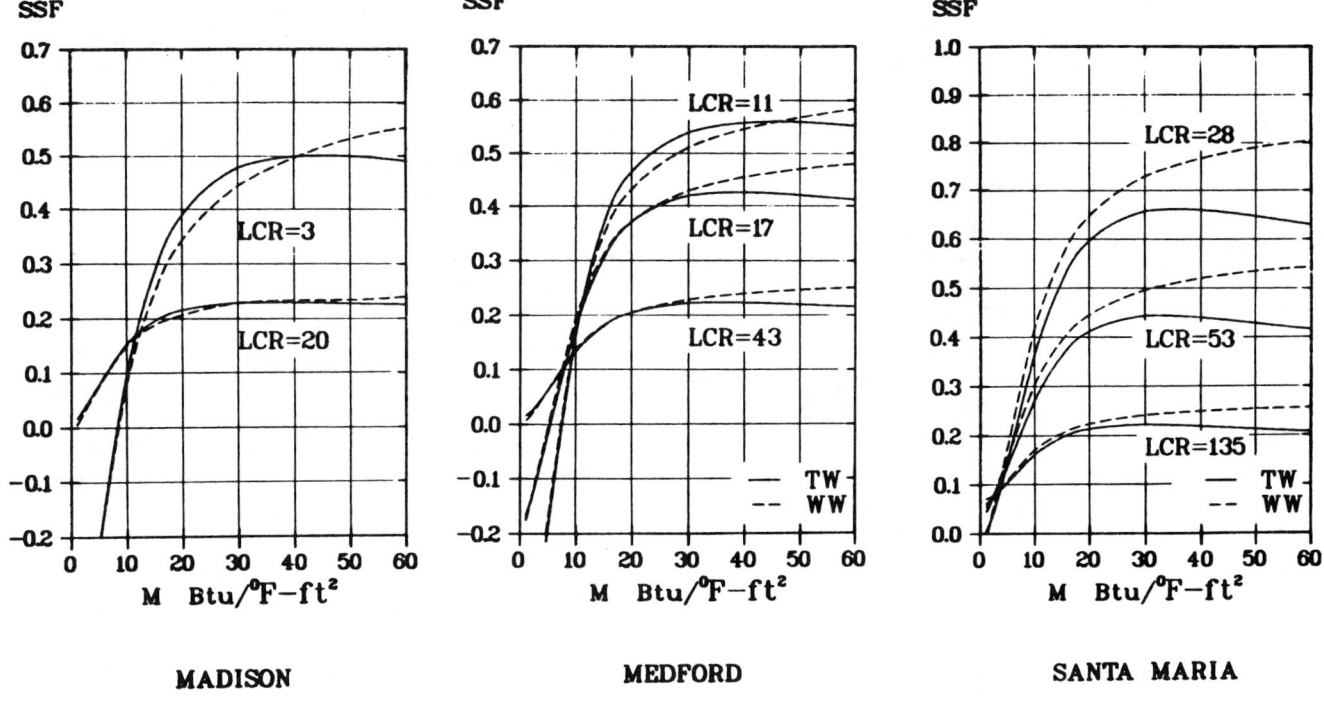

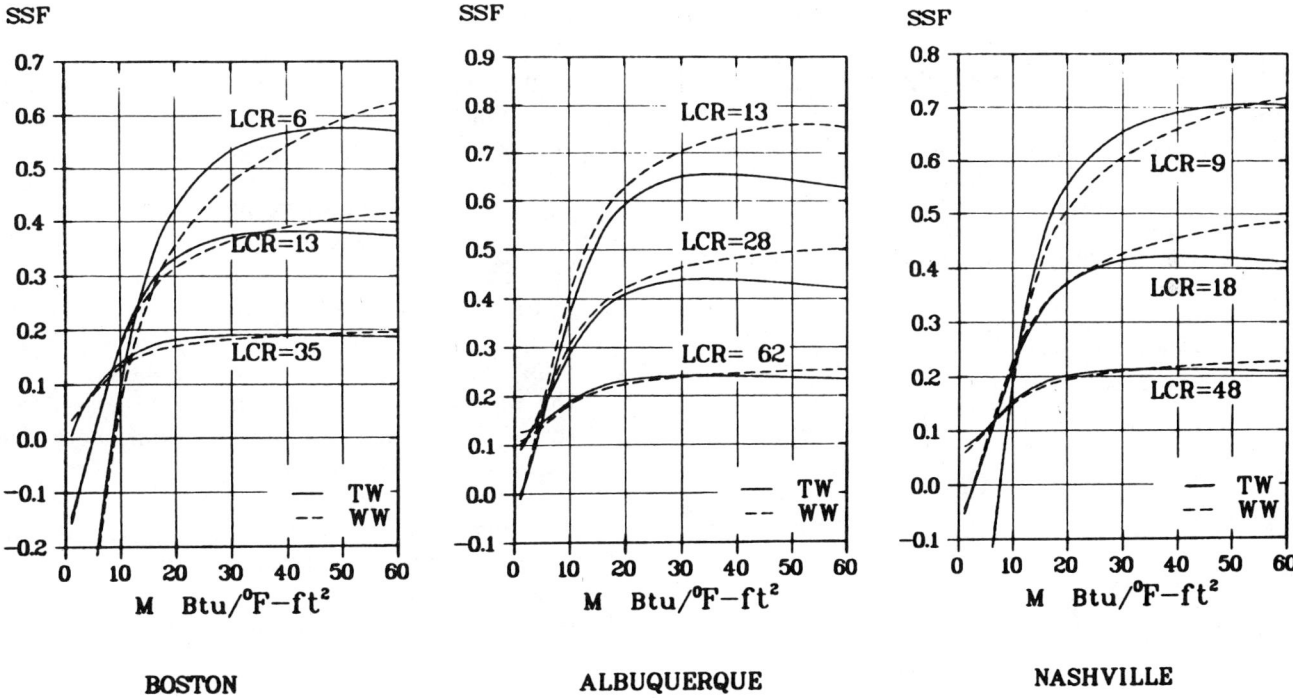

I-1: Effect of different quantities of thermal storage wall mass on the solar savings fraction (*SSF*) of Trombe walls and waterwalls in six cities. The reference value is 45 Btu/°F/ft².

SSF = solar savings fraction
LCR = load collector ratio
TW = Trombe wall
WW = waterwall
M = thermal storage capacity of wall per unit area

Figure I–2 presents data on temperature swings recorded in test rooms at the Los Alamos Scientific Laboratory. These test rooms are small, 40-square-foot frame structures where passive solar heating elements can be tested under carefully controlled conditions. During the time of these tests, the rooms were operated without any auxiliary heat—that is, in a "free-running" mode. The entire 10-by-5-foot south side of each test room is glazed with two sheets of ⅛-inch-thick Plexiglas to provide for solar collection. Except for the passive thermal storage element itself, there is very little mass in the room. The overall building load coefficient (BLC) of each test room, excluding the south facade, is approximately 300 Btu/DD (degree-day), and the LCR is approximately 6 [BAL-6].

The curves in figure I–2 show the room temperatures inside three test rooms and the corresponding outside temperatures during a sunny day (February 24) that followed a sequence of clear, sunny days, and during two days (March 1 and 2) that followed a sequence of cold and stormy weather. On March 1 the sky was totally overcast, and on March 2 there was some sunshine. Each of the three rooms involved in the testing had one of the special features below:

- A 16-inch-thick unvented Trombe wall built of solid concrete blocks;
- A wall made up of three 19-inch-diameter fiberglass tubes, filled with water, which stood immediately behind the glazing and which contained 73 pounds of water per square foot of glazing; or
- A wall made up of four 12-inch-diameter water-filled fiberglass tubes, containing 32 pounds of water per square foot of glazing.

In each of the rooms containing water-filled tubes, the spaces between and above the tubes were blocked with expanded polystyrene sections to prevent convective exchange between the glazing side of the tubes and the room.

The following conclusions can be drawn directly from the test observations:

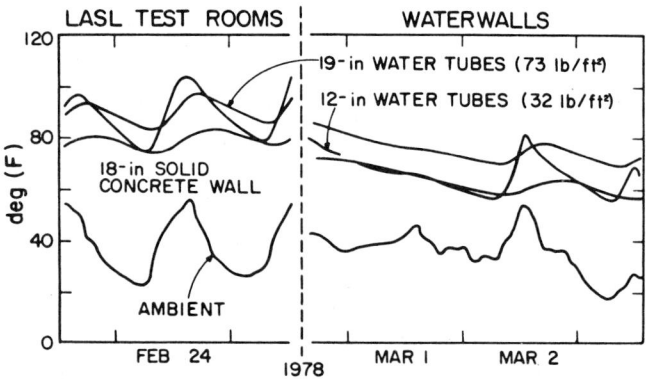

I–2: Test results obtained from three test rooms of virtually identical dimensions—one equipped with an unvented Trombe wall, and the other two with different-sized waterwalls. The room temperatures plotted are globe temperature measurements made in the center of the room.

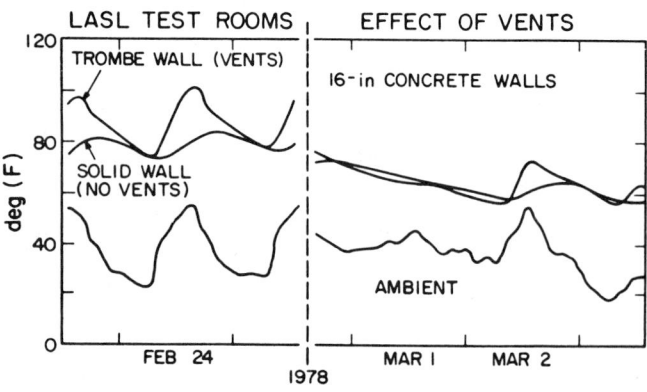

I–3: Test results obtained from two test rooms of virtually identical dimensions—one equipped with an unvented Trombe wall, and the other with a vented Trombe wall (the upper vent and the lower vent each taking up approximately 3% of the wall's total surface area). The room temperatures plotted are globe temperature measurements made in the center of the room.

1. Despite the two rooms having comparable net heat capacities (approximately 30 Btu/°F/ft^2 of glazing), the temperature fluctuation in the unvented Trombe-wall room (10 Fahrenheit degrees) was much smaller than that measured in the waterwall room that had the 12-inch-diameter tubes (24 Fahrenheit degrees). This was caused by the smoothing effect of the diffusion of heat through the wall, an effect that did not occur in the water-filled tubes because of the rapid transportation of heat by convection across the tubes.
2. The minimum temperature observed each night in the Trombe-wall room was approximately equal to the minimum temperature observed on the same night in the waterwall room that had the 12-inch-diameter tubes. This result held true, during both sunny weather and cloudy weather, through the entire winter.
3. The temperature swing in the waterwall room equipped with 19-inch-diameter tubes was smaller than that in the room equipped with 12-inch-diameter tubes, roughly in proportion to the ratio of thermal mass in the two rooms.

Another test made during the same time period compared a room containing a Trombe wall without vents (the same room as before) with a room in which vents at the bottom and top of an otherwise identical Trombe wall were left open. See figure I–3.

The marked difference in temperature ranges between the two rooms shows the effect of allowing heated air from the thermocirculation vents to be dumped into the room during the day. The temperature swing in the room without vents was approximately 10 Fahrenheit degrees; in the room with vents, it was approximately 27 Fahrenheit degrees. Since the room had no mass other than the Trombe wall, there was no way for the thermocirculation heat to be stored within the space; but even with additional mass inside the space, the temperature swing in this room would have been substantially larger than in the ventless Trombe-wall room because of the difficulty of transferring heat from warm air to a mass surface.

General conclusions that can be drawn from the curves in figures I-2 and I-3 are:

1. An unvented Trombe wall is much more effective than waterwalls of the same heat capacity in reducing temperature swings within a space. This advantage is largely lost, however, if the wall is vented.
2. Venting a Trombe wall would be an advantage in situations where warm air could be utilized effectively during the day. This might be the case in very cold climates, in a building that experiences a large amount of infiltration, or with a Trombe wall that has a relatively low value of solar savings fraction.
3. A waterwall's primary advantage is its ability to obtain a very high thermal storage value within a reasonable volume of space. It is impractical to attempt to achieve a thermal storage value of 90 Btu/°F/ft² with a Trombe wall because to do so would require a 3-foot-thick wall. This would be not only awkward and uneconomical, but also ineffective, since a large amount of heat would be unable to pass through such a thick wall. On the other hand, a waterwall having this heat storage capacity could be made that was only 18 inches thick. It would perform efficiently and have a small temperature swing.

Trombe walls and waterwalls perform very similarly with waterwalls having some advantage in efficiency for the same thermal storage mass (e.g., 45 Btu/°F/ft²). The difference between the two is less when nighttime insulation is used regularly, also when low solar savings fractions are obtained. The reason waterwalls perform more efficiently is that the surface temperatures of waterwalls are lower than those of Trombe walls. Overall, however, this effect is not as prominent as one might think. Although heat losses are higher for Trombe walls during the peak operating time (around noon), there is a converse effect associated with the fact that Trombe walls' surface temperatures are slightly lower than those of waterwalls during most of the remaining time—including the entire night. Therefore, although Trombe wall heat losses are considerably higher for a short period of time, they are lower for a long period, and the two effects tend to cancel one another so that the net heat losses for both types of wall are very nearly the same.

I.2 EFFECTS OF THERMOCIRCULATION VENTS

Thermocirculation vents are commonly thought an almost indispensible feature of masonry thermal storage walls. This is partly due to the fact that such vents were used by Felix Trombe and Jacques Michel in the original experimental house at Odeillo (figure I-4), and other designers have simply followed along. As indicated above, however, vents are not always desirable, and some designers prefer not to use them.

In the original Trombe house, the thermocirculation vents were designed for summer and winter use. During the winter, warm air in the space between the wall and the glazing would flow out the top vent and into the room; it would be replaced in the space by cold air drawn into the lower vent from the room. During the summer, hot air would rise and flow out an upper external vent (located on the south side of the house) to the outside; this would reduce the pressure in the house slightly, causing cool air to be drawn into the house at the north side of the building. Notice that the summer mode would be defeated by a southerly breeze, which would flow into the upper external south vent and force warm air into the house. This could be prevented by using a backdraft damper or a wind turbine.

As a general rule, vents provide heat to the building during the day in the form of heated air; if this air is useful, the vents may well be desired. In other circumstances, they would simply lead to overheating of the building.

A thorough study of the effect of vent area in different climates (with different values of *LCR*) indicated that there is normally an optimum vent size for a given wall, depending primarily on the solar savings fraction. See figure I-5.

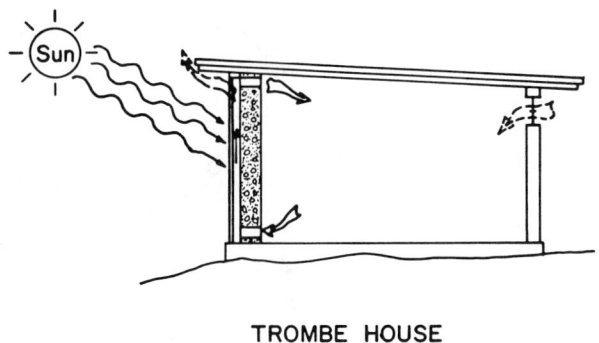

TROMBE HOUSE

I-4: Schematic diagram of the original Trombe house in the French Pyrenees, showing the thermocirculation vents. The solid arrows show winter operation; the dotted arrows show summer operation.

"Vent area" in figure I-5 is the area of the lower vents (which is equal to the area of the upper vents), measured as a percentage of the total Trombe wall area. If vents are to be used, they should be accompanied by some mechanism for preventing backflow at night—such as a passive backdraft damper—or performance will be severely impaired at values of LCR below 24 (unless nighttime insulation is used).

A fixed 3% vent area slightly underestimates the performance that would be achieved at the larger values of solar savings fraction for which a 3% vent area is much too large. Figure I-6 shows the improvement obtained with the optimum vent area, compared to the 3% vent area.

Solar Savings Fraction	Recommended Vent Area	Comment
0.25	3%	Performance levels off above 3%
0.50	1%	Performance levels off above 1%
0.75	0.5%	Performance decreases above 1%

I-5: Recommended vent areas (expressed as a percentage of the entire wall area) for Trombe walls having different values of solar savings fraction.

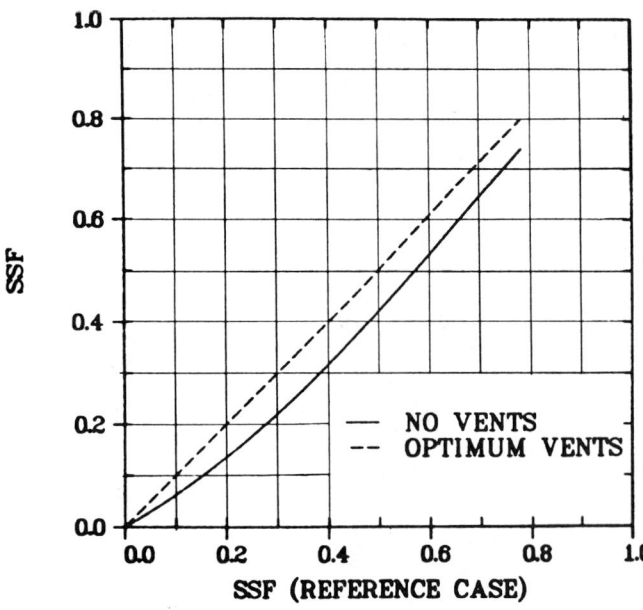

I-6: Comparison of the performance (SSF) of a ventless Trombe wall and that of a Trombe wall with optimum-sized vents. Performance in the reference case (a Trombe wall with a 3% vent area) would be represented by a straight diagonal line from the lower left corner of the graph to the upper right corner. The curves plotted on this graph can be applied at any location.

Unvented Trombe walls constitute an attractive design possibility because of their reduced tendency to overheat. However, the performance of the walls will be decreased somewhat in the absence of vents, as indicated in figure I-6. For example, a building in Albuquerque that would have given 70% solar savings without vents will give 76% solar savings when equipped with upper and lower vents that are each equal to 0.5% of the wall area.

Many designers utilize direct-gain windows for daytime building heating, instead of placing vents in Trombe walls. If vents are to be used, provision should be made to block them during the summer, or, better yet, to close the indoors-opening upper vent and open an upper vent to the outside, while leaving the lower vent open. This will purge much of the heat collected during the summer and will help ventilate the house.

I.3 PROPERTIES OF TROMBE WALLS

The effect of changes in density (ϱ), heat capacity (c), thermal conductivity (k), and wall thickness (L) can be described in two groupings of these properties: ϱcL and ϱck. The grouping ϱcL is also expressed as M. M is simply the thermal storage capacity of the wall per unit area.

The reference value of M is 45 Btu/°F/ft^2; and the reference value of ϱck is 30 Btu^2/°F^2/hr/ft^4. The effect of changing M at constant values of ϱck (i.e., the effect of changing wall thickness (L)) is discussed in section I.2. The effect of changing ϱck at constant values of M is shown in figure I-7, for Trombe walls in the cities of Albuquerque, Boston, and Madison.

In the case of thick walls ($M = 45$), performance (measured as SSF) increases steadily as thermal conductivity (k) increases. However, in the case of thin walls ($M = 15$), there exists an optimum level of thermal conductivity (k).

I.4 EFFECTS OF MULTIPLE GLAZINGS AND SELECTIVE SURFACES

Instead of using movable insulation, a designer may decide to use more than two sheets of glass (multiple glazings) as a covering for a thermal storage wall or may decide to place a selective surface on the outside of the thermal storage wall. Both strategies improve performance markedly. Performance continues to increase when a fourth or (in some instances) a fifth sheet of glass is added. At some point, however, the decrease in transmittance associated with reflection and absorption in the added glass layers more than offsets the reduction in heat-loss rate.

A selective surface is used to reduce the infrared radiation from the wall to the glass, a major source of energy loss through glazing. A selective surface performs more efficiently than an infrared reflector situated on the glass surface because a reflective coating would necessarily involve a reduction in solar transmittance. In these studies the absorptance of the selective surface is 0.9 and the emittance is 0.1.

The effects of changing the number of glazings and using either a selective surface or night insulation are highly interactive and nonlinear. The effects produced by various combinations of glazings and selective surfaces on walls in six cities are shown in figure I-8.

The effectiveness of a selective surface used on thermal storage walls has been verified by data collected at the Los Alamos Scientific Laboratory in four thermal storage wall test rooms. Temperatures in the test rooms were held relatively constant by means of electric auxiliary heaters within each space. The load was increased by artificially maintaining forced ventilation at a level of approximately four air changes per hour over a 12-day period. The average inside temperature was held relatively constant at an average value of 71°F, the average outside temperature was 43°F. The calculated average heat load of the building, not including the south-facing glazing, is approximately 280 watts under these conditions.

Of the four test rooms, two were equipped with a selective surface and two were not. One of the rooms without the selective surface incorporated a Beadwall for nighttime insulation. One of the rooms with the selective surface and the two rooms without it utilized 16-inch-thick solid concrete unvented Trombe walls. The other room equipped with the selective surface utilized four water-filled 12-inch-diameter fiberglass tubes standing vertically behind the glazing. The selective surface consisted of black chrome on a thin copper foil substrate (wall surface absorptance (α) = 0.93; infrared emittance (ϵ) = 0.08), which was glued to the outer surface of the Trombe wall and of the water-filled tubes.

The performance of the rooms with the selective surface was notably superior to that of the rooms without, even in the case of the room with Beadwall nighttime insulation. The power consumption in each room is shown in figure I-9.

The weather during this 12-day period was relatively mild, with large diurnal swings. The time of year was near the vernal equinox, so sun angles were moderately high. There was a mild storm during the middle of the period, accompanied by two days of cloudy, overcast weather; otherwise, the weather was clear and sunny.

The data from this test are obscured somewhat by the fact that the four test rooms were not exactly comparable. The effective aperture area of the rooms without the selective surface was approximately 13% less than that of the rooms with the selective surface. This difference in load collector ratio (LCR), when corrected, would be expected to lower the performance level in the third and fourth rooms by about 4% in solar savings, as noted in the last column.

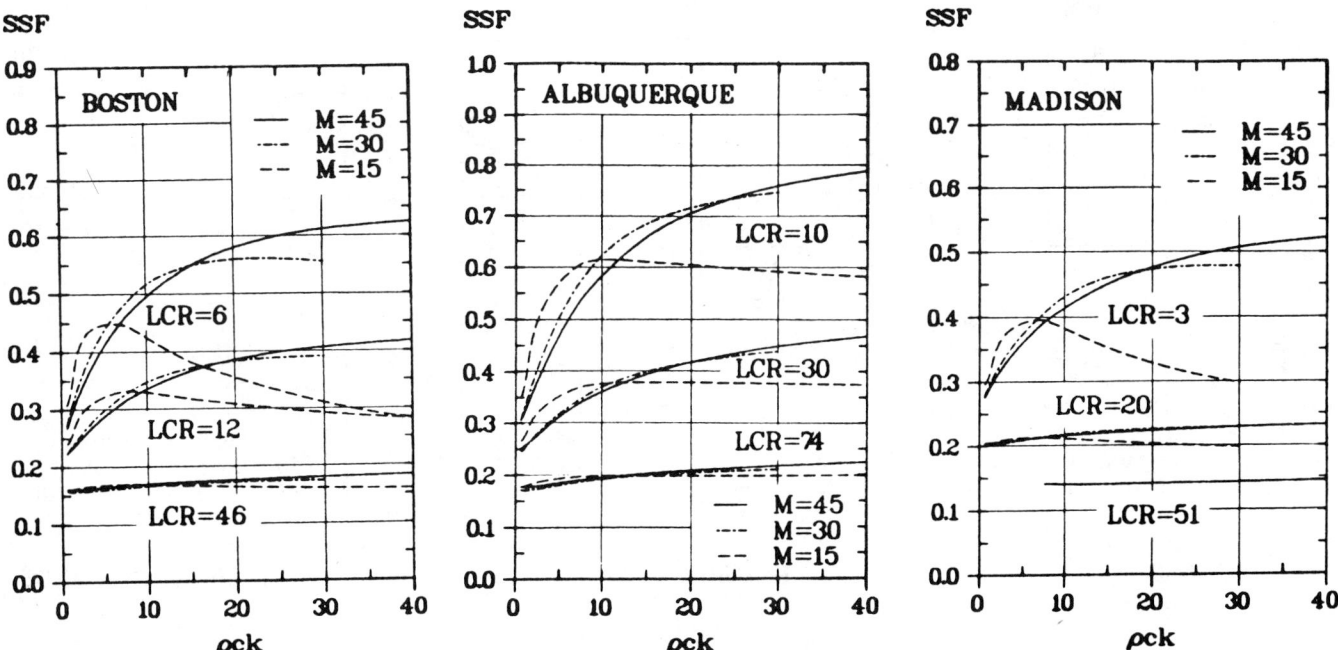

I-7: Effect of different values of wall property parameter ρck on Trombe wall performance in three cities, when the property $M (= \rho cL)$ is held constant.

The curious thing about these results is that they show a remarkable improvement in performance when the selective surface, rather than a normal black surface, is issued—an improvement even larger than that obtained with nighttime insulation. The improvement with nighttime insulation is about what one would expect; but the improvement with the selective surface is larger than had been anticipated. Subsequent results indicate that the Beadwall and selective surface rooms are quite close in performance.

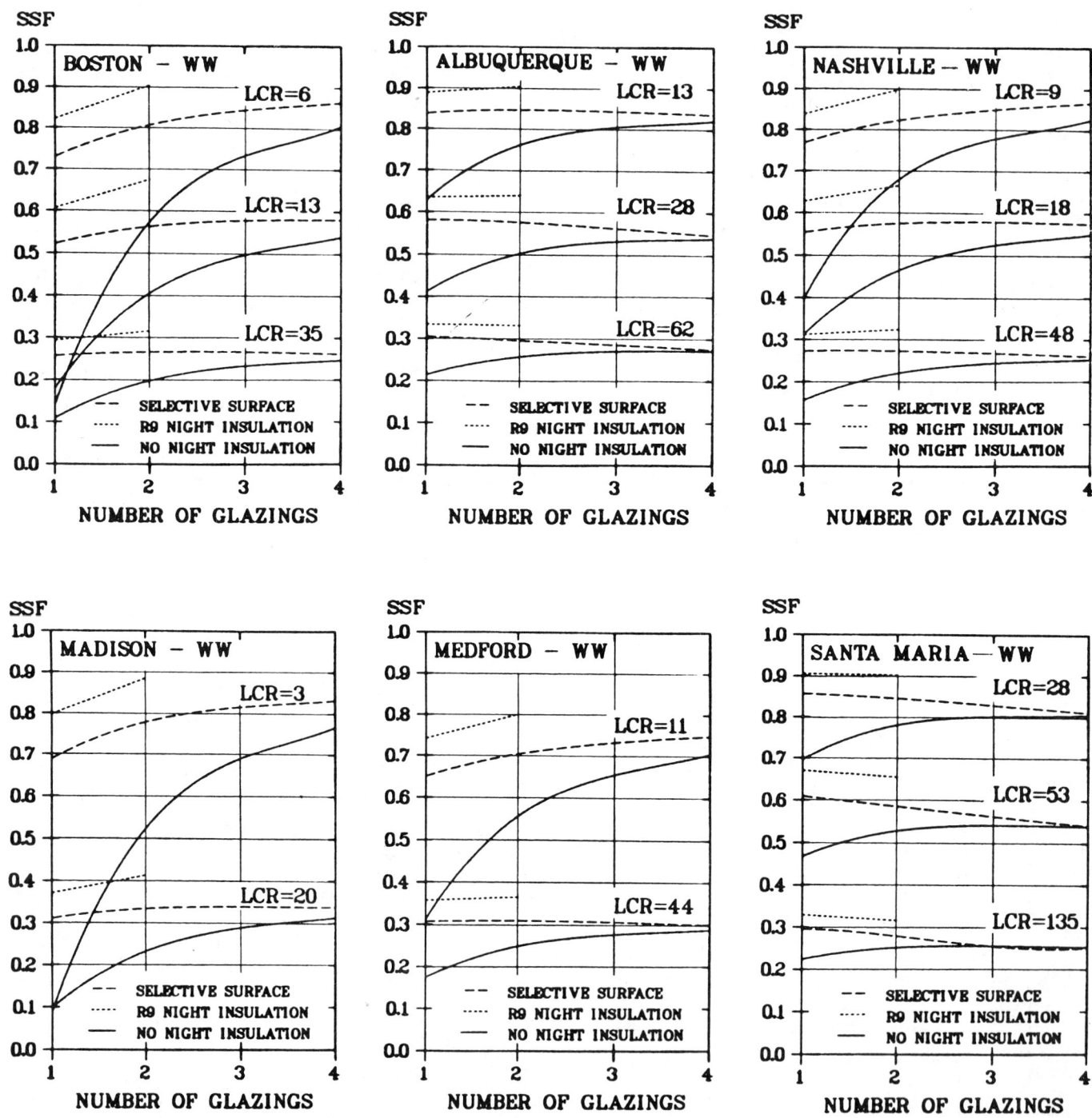

I-8: Effect of different numbers of glazings, alone or combined with a selective surface or R9 nighttime insulation, on the performance of waterwalls in six cities. Continuous curves have been drawn to show trends; however, the calculations apply only to integer numbers of glazings.

	Selective Surface	Average Power (in Watts)	Solar Savings Fraction	LCR	Corrected SSF
Trombe Wall	No	148	0.47	20.0	0.47
Trombe Wall with Beadwall	No	106	0.62	20.0	0.62
Trombe Wall	Yes	68	0.76	17.4	0.71
Waterwall	Yes	67	0.76	17.6	0.72

I-9: Test data for rooms with and without selective surfaces. In each case, *SSF* (observed) = 1 − calculated auxiliary load. A first-order correction was applied to adjust the third and fourth cases to the same *LCR* value (20 Btu/degree-day/ft^2) possessed by the first and second cases.

The test results indicate that a selective surface is a very effective design strategy, and that the performance increase it produces may be larger than one might expect on the basis of the graphs in figure I-8.

I.5 WALL ABSORPTANCE AND EMITTANCE

Although many early Trombe-wall buildings featured flat black paint on the Trombe wall, a few designers have begun to use other dark colors—such as brown, blue, and green—for aesthetic reasons. The lower set of three curves in figure I-10 shows the effect of changing wall surface absorptance on Trombe walls in Los Alamos while infrared emittance is held constant at 0.9. The upper set of three curves in figure I-10 shows the effect of changing infrared emittance while holding absorptance constant at 0.95.

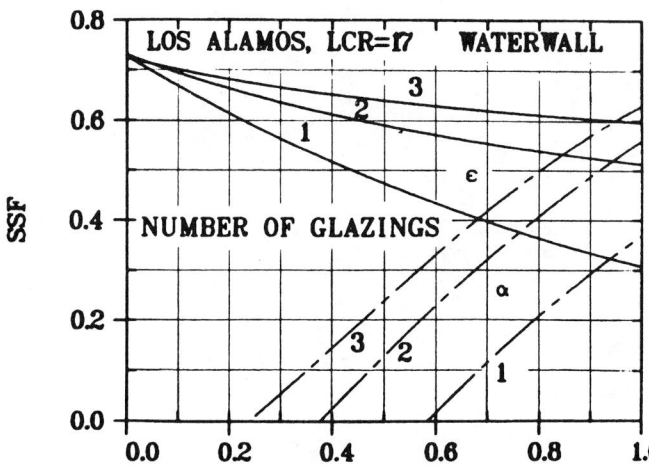

I-10: Effect of different values of effective solar absorptance (α) at a constant value of infrared emittance (= 0.9) and effect of different values of infrared emittance (ϵ) at a constant value of effective solar absorptance (= 0.95) on the solar savings fraction (*SSF*) of a waterwall in Los Alamos, New Mexico.

I.6 PROPERTIES OF GLASS

Two possible ways to improve performance are to use low-iron glass and to increase the glass spacing.

Low-iron glass provides relatively slight improvement. The major reason for this is that the energy absorbed in glass of any type is not lost by the system but retained as increased glass temperature, which in turn reduces heat loss from the wall to the glass. The heat retained by the glass is not as useful as it would have been had it penetrated the glass to the absorber surface, but neither is it useless. The increase in solar savings fraction caused solely by using perfectly clear glass rather than ordinary ⅛-inch-thick glass (extinction coefficient = 0.5 per inch) is shown in figure I-11.

Performance can also be improved by increasing the glass spacing to reduce the heat conduction of the air between the glass sheets. The optimum spacing is about ¾ inch, rather than the ¼ inch normally used in manufactured, sealed units. The increase in solar savings fraction caused solely by this increase in spacing is shown in figure I-12.

City	LCR	Increase in SSF	
		2 glazings	4 glazings
Albuquerque	28	0.03	0.06
Los Alamos	17	0.06	0.09
Madison	3	0.07	0.07
Medford	17	0.04	0.06
Boston	13	0.05	0.08
Santa Maria	53	0.04	0.07
Nashville	18	0.04	0.07

I-11: Increase in solar savings fraction, at different values of *LCR*, attributable to use of low-iron glass (results for double and quadruple glazings, measured in seven cities).

City	LCR	Increase in SSF	
		2 glazings	4 glazings
Albuquerque	28	0.03	0.04
Los Alamos	17	0.06	0.07
Madison	3	0.10	0.09
Medford	17	0.05	0.06
Boston	13	0.06	0.07
Santa Maria	53	0.03	0.03
Nashville	18	0.05	0.05

I-12: Increase in solar savings fraction, at different values of *LCR*, attributable to an increase in spacing—from ½ inch to ¾ inch—between layers of glazing (results for double and quadruple glazings, measured in seven cities).

One effect that has not been studied is the effect reducing glass reflectance. Undoubtedly, this would greatly enhance performance, especially if multiple glazings were being used. Solar heat lost to reflection is about 16% for normal double glazing. This could be reduced to approximately 4% by means of various surface-etching techniques or antireflective coatings.

I.7 EFFECTS OF OVERHANGS

The normal use of overhangs in passive solar applications is to reduce summertime overheating. If the overhang is properly designed, there is no blockage of the midwinter sun and almost entire blockage of the midsummer sun. Figure I-13 shows a simple, convenient scheme for determining the noon sun angles on the summer solstice, the winter solstice, and the equinoxes.

It is good design practice to allow about 5° of leeway at the window top in locating the overhang. See figure I-14. This way, full illumination of the window will occur from mid-November until February 1. It is also good design practice to allow for about 5° of leeway at the window bottom for the summer design condition. This will provide complete noontime blockage of the window from about May 15 until August 1.

Notice that the glazing should not extend to the base of the overhang; if it did, the top portion of the window would never receive direct sun, but it would still lose as much heat as any other part of the window.

If the overhang is in place year-round, the design of the angles becomes a trade-off between a sacrifice of solar heating during the spring months—when the sun angles are high, but the weather is still cold—and overheating during the late summer and fall months—when the sun angles are low, but temperatures are still warm. This effect is borne out in figure I-15, which shows heating and cooling conditions for Los Alamos during each month of the year. The calculations show that, in the case of no shading, solar heating performance remains good right into the spring, but a significant cooling load is imposed during the summer and fall months. The heating and cooling loads cannot be subtracted from one another because they occur at different times during the month.

In the case of an overhang, heating performance is unaffected during the midwinter months because the overhang causes no shading of the glass. However the heating performance is seriously reduced during the spring—especially during May. The benefit of the overhang is that it greatly decreases cooling loads throughout the summer. However, the overhang is relatively ineffective at reducing cooling loads in the fall. (The overhang used for this example had a separation (Y/H) of 0.25 and an overhang (X/H) of 0.25. Figure I-16 defines these terms diagrammatically.)

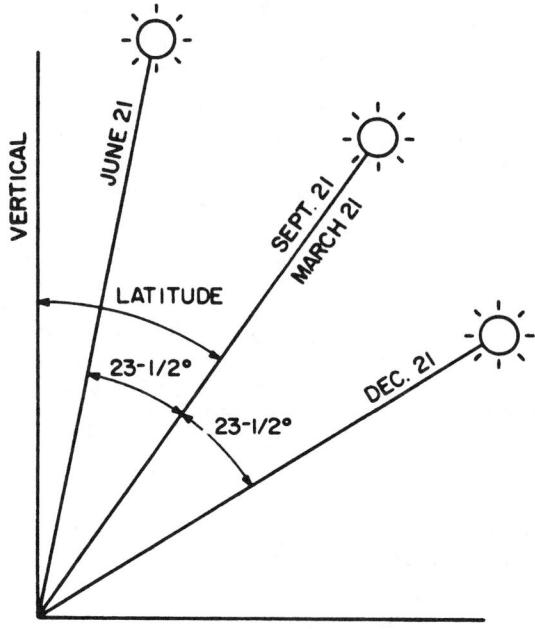

I-13: Simple method for calculating noontime sun angles: on March 21 and September 21, the sun angle measured from the zenith is identical to the latitude of the site; on December 21 and June 21, this angle is either increased or decreased by 23.5°.

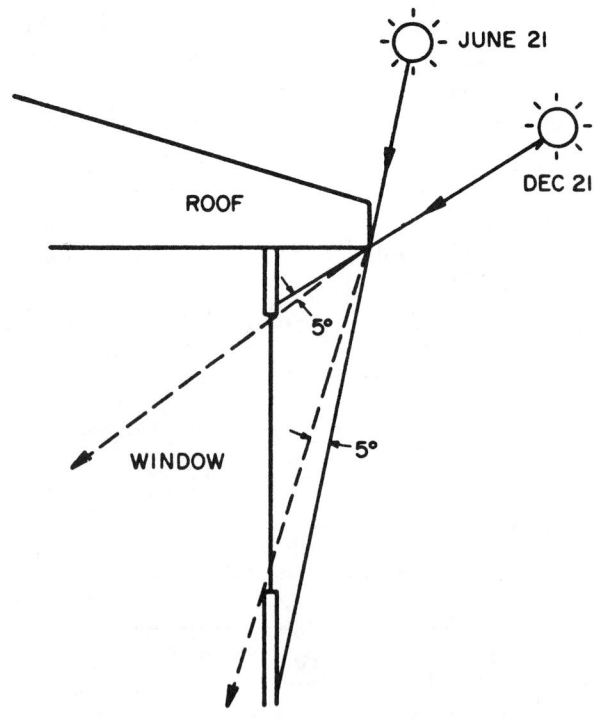

I-14: Modifying roof overhang geometry to provide desired summer shading and winter solar gain. The 5° allowances shown extend the desired effects over a ten-week season.

A number of similar calculations have been done for a waterwall in Albuquerque; these are summarized in figure I-17. The calculations were all performed for a diffuse ground reflectance equal to 0.3. No end effects of the overhang were taken into account. Thus the results are applicable either to a very long overhang and window combination or to an overhang which extends for some distance beyond the ends of the window. The curve summarizes the effect on both heating loads and cooling loads of a number of geometrical configurations. The following observations can be made on the basis of the data in figure I-17:

1. In the case of a window without an overhang, the cooling load amounts to 32% of the heating load. The *SSF* for the case being calculated is 0.49.
2. In the case of a south-facing glazing accompanied by an overhang with the geometrical configuration of $Y/H = 0.28$ and $X/H = 0.38$, the *SSF* is 0.47—a minor reduction of 0.02 from the *SSF* for the preceding case. The cooling load is 26% of the total winter load—a reduction of 6.0%.
3. Any overhang that significantly reduces a glazing's cooling load also significantly reduces the glazing's solar heating contribution.

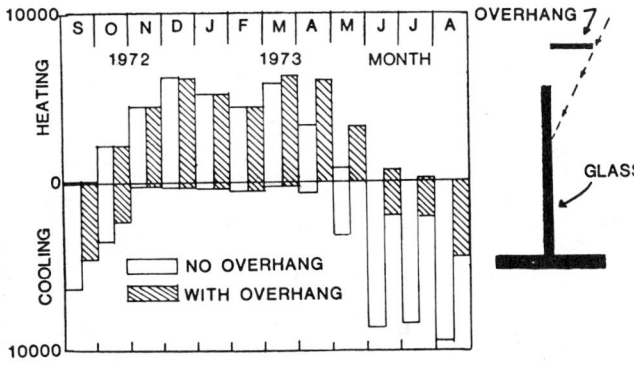

I-15: Effect on monthly heating and cooling loads of a Trombe-wall building in Los Alamos, New Mexico, attributable to shading by an overhang. The overhang's length is 25% of the glass's height, and the overhang is situated a distance above the top of the glass that is equal to its length.

An alternative to fixed shading is movable shading. Movable shading is awkward and not much favored by designers, but it is quite effective. Left on until late in the fall, the shade can reduce instances of overheating substantially. It can then be taken off and stored until late in the spring, when the heating season is over.

Another option is to use nighttime insulation as shading. This is a particularly attractive approach, since the shading also serves as daytime insulation in the summer. Nighttime insulation markedly improves performance during the winter, and it is especially effective at reduc-

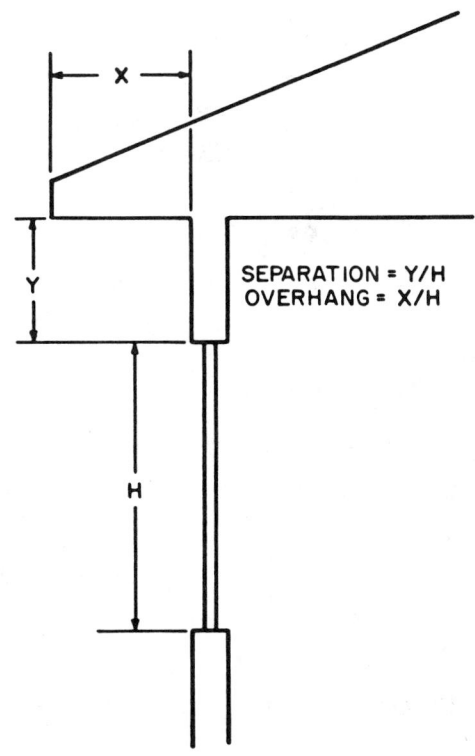

I-16: Overhang geometry.

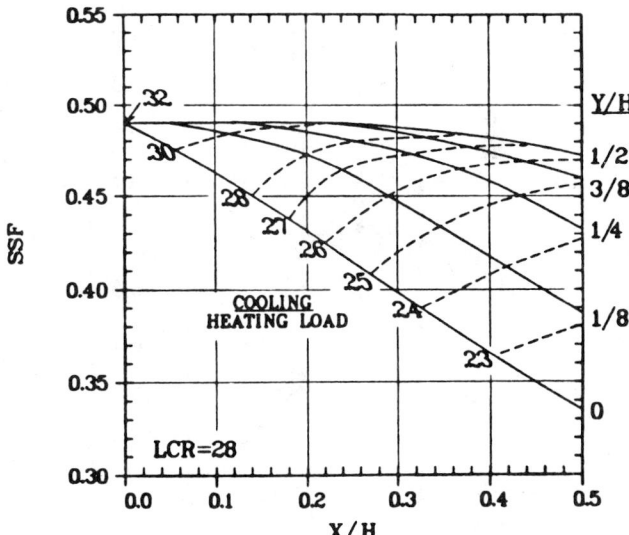

I-17: Effect on heating and cooling loads of a waterwall building in Albuquerque, New Mexico, with $LCR = 28$, attributable to overhang geometry. The X and Y parameters refer to the length of the overhang and to the separation between the top of the window and the bottom of the overhang, respectively; H refers to the height of the window. The cooling load is given as a fraction of the heating load.

ing summer overheating; it also provides a very simple and effective means of accommodating changes in the weather. Nighttime insulation designed for placement outside the window is particularly effective for summertime shading. If it is located inside the window, the designer must be particularly careful to avoid material damage associated with the buildup of heat between the glazing and the insulation. This can be accomplished by

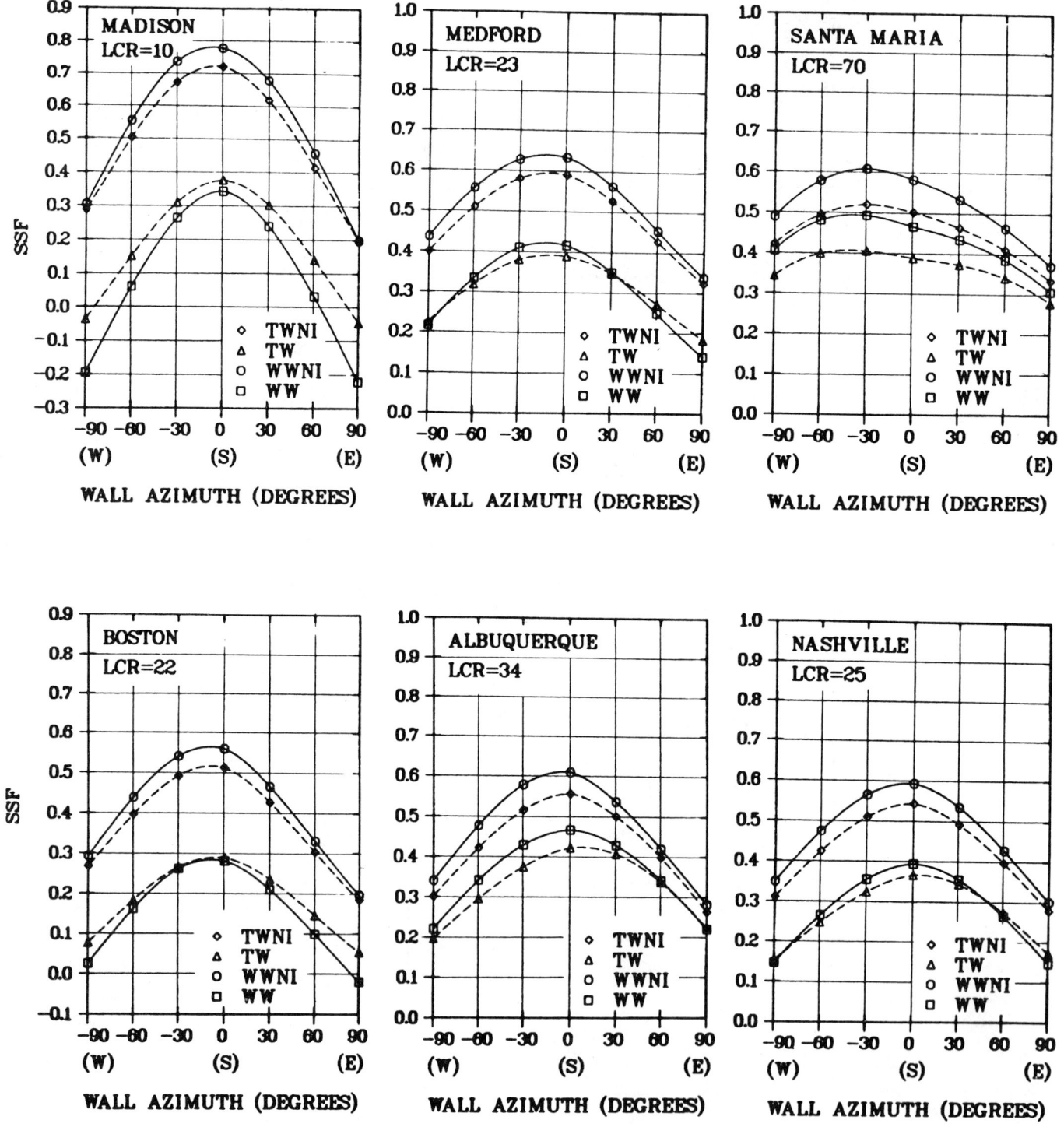

I-18: Effect on solar savings fraction of orientation of thermal storage walls in six cities.

using a light-colored or reflective outer surface. Thermal stress breakage of glazing can also be a problem. Use of tempered glass reduces the likelihood of this occurring.

Perhaps the simplest solution to the summertime overheating problem is to put any of a variety of old-fashioned shading devices outside the windows. Rolldown bamboo or canvas shades would be particularly effective. Very low-cost solutions, such as greenhouse shading materials or whitewash, could be used in situations where aesthetics are not of great concern.

I.8 EFFECTS OF GLAZING TILT AND ORIENTATION

Although vertical south-facing glazing is normally used in passive applications, there are likely to be instances in which the designer prefers or is constrained to use some other tilt or orientation. The effects of different orientations are shown in figure I-18.

The effect of changes in wall orientation (azimuth) is relatively minor for small deviations from true south. The optimum orientation averages about 6° west of true south for localities studied. Within the range of 20° east to 32° west of true south, the performance decrease (measured as SSF) is less than 0.10 on an annual basis.

The results for Santa Maria, California, show an exception to the general trends noted above. The optimum wall orientation is far to the west of south. This is presumably due to morning fog along the California coast.

Figure I-19 shows that a vertical wall (tilt = 90°) is far from the best and that a tilt of approximately 55° is near the optimum.* The reason that the optimum orientation is sloped is that, during the winter heating season, there is more sun received per unit of glazing area on a tilted surface than on a perpendicular (vertical) surface. The choice of vertical glazing by most designers is made in consideration of the inconvenience and lost space associated with a tilted surface, especially when the surface is serving a dual function as collector and as the south wall of the building. For clerestory windows, these considerations do not apply with the same force, and tilted surfaces are often used. On the other hand, it is *much* harder to shade a tilted surface than a vertical surface during the summer.

The calculations were all done for a ground reflectance of 0.3, corresponding to a normal foreground. If the ground were snow-covered (reflectance = 0.7), then the effect of tilt would be less because the performance of a vertical collector would be enhanced more than that of a sloped glazing.

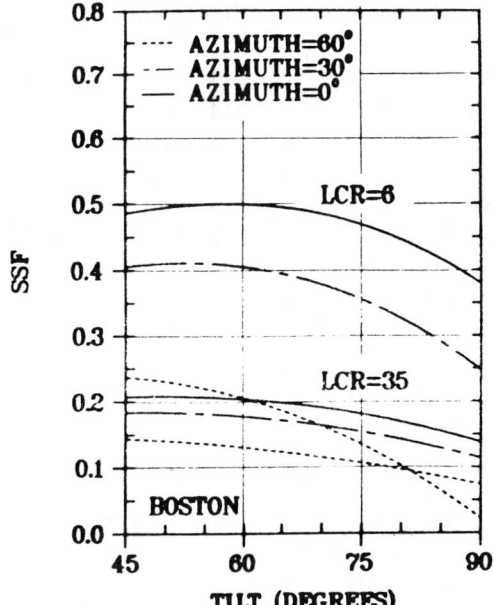

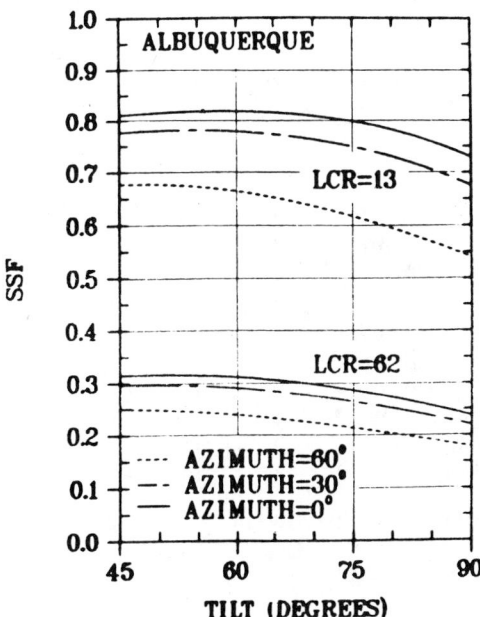

I-19: Effect on solar savings fraction of tilt (measured from the horizontal) of thermal storage walls in two cities (values are calculated for walls with three different orientations and two different LCR values for each city).

---

*Optimum slope is latitude-dependent. As a rule of thumb, a slope equal to the degrees of latitude + 15° is near-optimum for winter heating with south-facing collector surfaces.

I.9 EFFECT OF WIND VELOCITY

The effect of wind, as measured in hour-by-hour calculations, is to increase the heat loss coefficient of the outside glass surface. For the double-glazed case, this is not very significant. The reference calculation was done with the window-surface wind velocity equal to the hourly value of wind velocity given on the weather tape (wind velocity is usually measured at airports).

Several calculations were done with the window-surface wind velocity reduced to half the free-stream wind velocity, and others with the value set to zero. The resulting solar savings fractions are shown in figure I-20.

Of course, another significant effect of wind is that it increases the infiltration of air into the building. Although measurement of this effect was *not* included in the hour-by-hour calculations, accounting for the effect would be appropriate. Reasonable models, based on data, do exist.

I.10 EFFECTS OF WALL-TO-ROOM HEAT TRANSFER COEFFICIENT

The heat transfer coefficient between a vertical surface and a room is normally about 1.5 Btu/hr/ft²/°F. The convective effect contributes about one-third of this quantity; and the radiative effect the remaining two-

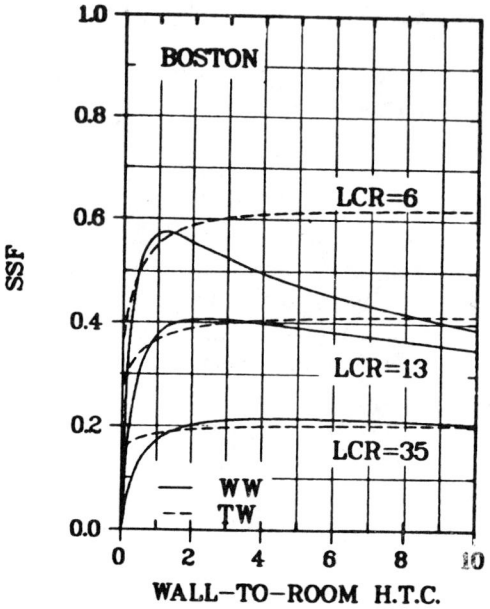

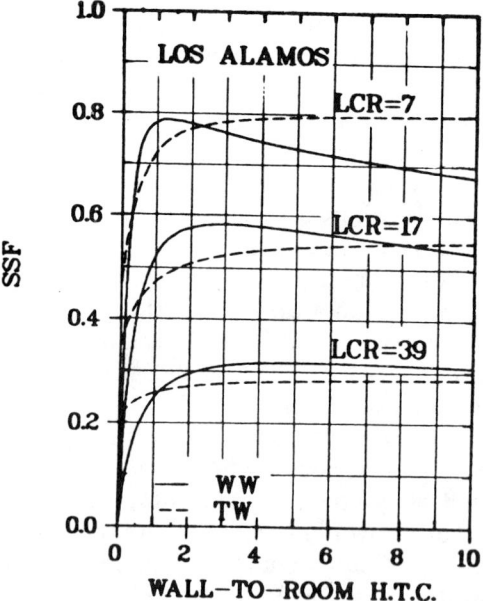

City	Wall	No. of Glazings	LCR	SSF at Different Ratios of Surface Velocity/Wind Velocity		
				1	1/2	0
Boston	WW	1	8	.133	.232	.443
Boston	WW	1	18	.171	.224	.339
Boston	WW	1	53	.103	.125	.174
Boston	WW	2	18	.399	.424	.483
Boston	WW	3	18	.490	.506	.544
Boston	WWNI	1	18	.606	.633	.691
Boston	TW	1	8	.276	.349	.498
Boston	TW	1	18	.202	.244	.335
Los Alamos	WW	1	17	.374	.404	.452
Los Alamos	WW	2	17	.561	.574	.599
Los Alamos	WW	3	17	.635	.644	.659

Average Wind Velocity: Boston: 12 mph
Los Alamos: 4 mph

I-20: Solar savings fractions of thermal storage walls with different compositions, *LCR* values, and numbers of glazings, that result when the walls are exposed to different wind velocities. (WW = waterwall; TW = Trombe wall; NI = nighttime insulation.)

I-21: Effect on solar savings fraction of different values of wall-to-room heat transfer coefficient (*HTC*) applied to thermal storage walls in two cities. The normal value for *HTC* is 1.5 Btu/°F/hr/ft²; values above 1.5 would presumably require auxiliary machinery, such as a fan to blow air past the wall.

thirds. Fortunately, this value is near optimum for passive thermal storage wall applications. Figure I-21 shows the effect of changes in this coefficient in the cities of Los Alamos and Boston. The optimum value of heat transfer coefficient for a waterwall is between 1 and 4, depending on solar savings fraction; performance falls off markedly for *SSF* values less than 0.5. Heat transfer coefficient values above 1.5 are perhaps of academic interest only, since they could be obtained only by forced ventilation.

At values of heat transfer coefficient greater than the optimum, the waterwall becomes too closely coupled to the room. In the limiting case of perfect coupling, the room and water-wall temperatures would be identical. In this situation, the waterwall's temperature cannot increase above that of the room, and so the waterwall cannot store as much heat. If the heat transfer coefficient is too low, then the waterwall cannot heat the room effectively. In the limiting case, in which the coupling is zero, the waterwall is completely ineffective. Between these two extremes, then, there exists an optimum thermal coupling level.

In the case of a Trombe wall, there is no optimum value because of the resistance of the wall itself, which adds to the surface's heat transfer coefficient.

One design option that has been used is to enclose a waterwall in a small room of its own—for example, by providing a layer of interior wallboard between the waterwall and the room. This has the advantages of giving the room a normal-looking interior wall surface, hiding the water containers, and protecting the containers somewhat. If there is no convective connection between the room and the water containers, the effective overall heat transfer coefficient is approximately 0.5 Btu/hr/ft²/°F. This is well below the optimum value for all locations. In Los Alamos, the performance penalty would be moderately large, reducing the solar savings fraction from 0.56 to 0.38 (for *LCR* = 17). In Boston, the effect would be to reduce the annual solar savings fraction from 0.40 to 0.30 (for *LCR* = 13). The effect is smaller because in Boston the load is smaller and the sun is less intense, and therefore the total amount of heat to be transported is smaller.

Two variations to this approach have also been used. Each involves providing a duct opening at the bottom and top of the partition wall that separates the room from the waterwall. In one case, the convection path is natural, with a convective loop set up between the two spaces. In the other, a small fan—which can be run on a thermostat as the building needs heat—is installed. The advantage to the natural convective approach is that it is passive and quite simple. Manual dampers can be installed to allow the room's occupants to close off the convection whenever they wish.

The advantages of the fan-forced approach are that it is more controllable and that it is automatic, providing heat on demand. In both variations, insulating the wall between the water tubes and the room is a poor idea. To do so would block a major natural flow-path. The radiant heating effect provided by the large warm surface is generally more beneficial and comfortable than heat provided in the form of warmed air, irrespective of whether the convection is forced or natural

I.11 EFFECT OF GROUND REFLECTANCE

The effect of changing ground reflectance is shown as the dotted curves in figure I-22. A ground reflectance of 0.3, which was assumed for the reference calculation and is reasonably representative of an average foreground, improves the performance from 0.44—the annual *SSF* for a completely black foreground (with reflectance = 0)—to 0.56. For this calculation, the foreground is assumed to be completely diffuse—that is, the reflected sunlight is assumed to be scattered equally in all directions.

The presence of snow on the ground can significantly improve performance, increasing reflectance to a value of approximately 0.7. The snow is generally present when it is needed and melts in the spring, watering the grass and other plants. Healthy plants provide a darker foreground in the summer and reduce the effects of overheating due to reflection during warm weather.

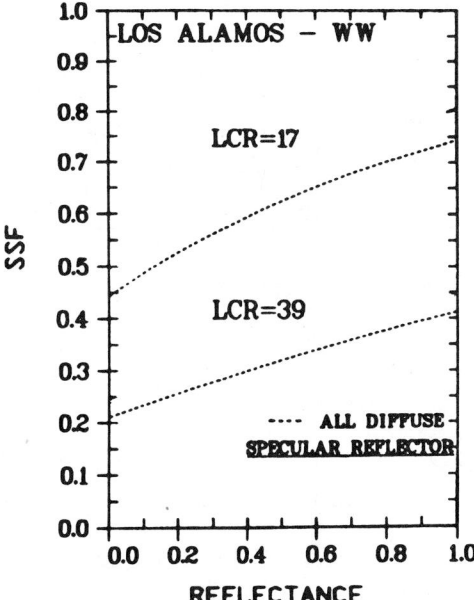

I-22: Effect on solar savings fraction of different values of ground reflectance applied to a waterwall in Los Alamos, New Mexico, at two different *LCR* values.

A more generalized plot, useful in any location, is shown in figure I-23. This graph shows the effect of changing diffuse reflectance either to 0.9 or to zero. The X marks on the plot are the actual results for the numerous cases run in several locations, and the solid lines are least-squares polynomial function fits to these points. The polynomials are as follows:

$X = SSF$ for cases in which reflectance $= 0.3$

$SSF = 0.8241X - 0.1293X^2 + 0.2763X^3$ for cases in which reflectance $= 0.0$

$SSF = 1.405X - 0.2165X^2 - 0.1868X^3$ for cases in which reflectance $= 0.9$

Reference [HUN-2] contains a table of ground reflectance estimates that is useful in conjunction with figure I-23. See figure I-24.

The four curves in figure I-25 pertain to specular reflectors. "Specular" means shiny; the surface of a specular reflector reflects a collimated beam as a collimated beam, just as a mirror would. Specular reflectors are more effective than diffuse ones because they are more efficient at reflecting all of the light that hits them onto the collecting surface—provided that they are properly oriented and located with respect to the collecting surface. It was assumed in all of the calculations that the reflector was of a size equal to the solar collection area and that it was positioned in front of the vertical south-

Rural Areas	Reflectance
Fields with snow cover	
1. Field with wooded area in background	0.66, 0.73
2. Open field (soil and dry grass) near road	0.61, 0.70
3. Trees dispersed in field	0.62
Wooded areas	
4. Conifer forest (with heavy snow cover)	0.61
5. Deciduous forest (with heavy snow cover)	0.72
Water	
6. Open water	0.16
7. Ice/snow-covered water	0.68
8. Partially open waterway (trees and houses in background)	0.43, 0.66
Urban Areas	
9. Urban area (commercial, institutional)	0.16, 0.38
10. Residential area (dwellings and roadway)	0.21, 0.35, 0.45
11. Educational institution	0.36, 0.42
12. Recreational area (park)	0.49

I-24: Average reflectance values for twelve representative winter landscapes.

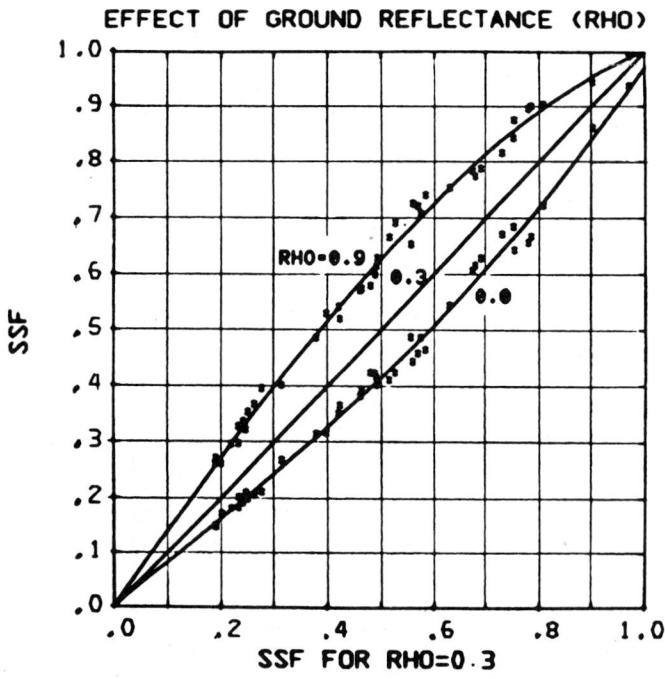

I-23: Comparison of solar savings fractions obtained at ground reflectance (*RHO*) values of 0.0 and 0.9 with those obtained at the reference value (*RHO* = 0.3). The small black squares pinpoint results of simulations done at a variety of locations; the plotted lines are applicable to thermal storage walls in any location.

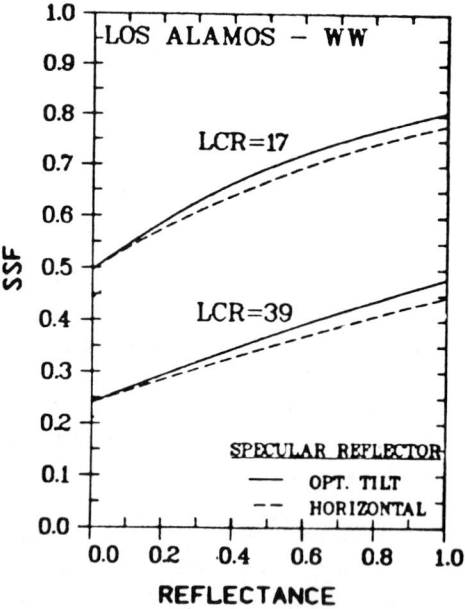

I-25: Effect on solar savings fraction of specular reflectors applied horizontally (dashed curves) or adjusted monthly to achieve optimum reflectance (solid curves) to a waterwall in Los Alamos, New Mexico.

facing collection wall. Calculations of end effects assumed the reflector to be five times as long as it was wide. Beyond a length ratio of 2:1, end effects are very small. The diffuse reflectance of the ground beyond the reflector area was assumed to be 0.3.

The curves marked "HORIZONTAL" pertain to a reflector that is positioned horizontally on the south side of the solar collection wall, with the north edge of the reflector set against the bottom of the solar collection wall. This configuration is convenient, especially when the flat roof of a building can be used as the base for the reflector in front of a second story collection wall—a situation that occurs frequently in the architecture of the Southwest.

Reflector performance can be increased at more northerly latitudes by tilting the reflector down slightly from horizontal. For Los Alamos (latitude = 36°N), the optimum fixed tilt was found to be horizontal. The change in optimum tilt is probably about one to one, so that at 46°N the optimum tilt is probably about 10 degrees downward.

Another design approach is to adjust the reflector tilt to account for variations of sun angles at different seasons. This is rather easily done if the reflector is hinged at the bottom of the collection wall. The curve marked "OPT. TILT" has been calculated for the case in which the reflector is periodically adjusted to enable the noon direct-beam sunlight striking the extreme outer edge of the reflector to reflect onto the extreme top edge of the collection wall. This angle is easily determined by reference to sun charts,* or it can be determined experimentally on the site by observing actual reflections. An example of a sun chart is shown in figure I-26.

Any actual reflector will fall somewhere between the limiting cases of "perfectly specular" and "perfectly diffuse." Data taken on a painted white surface shows that, even there, significant forward scattering occurs and that the reflected energy is predominantly in the same direction as it would be if a specular reflector were used.

I.12 EFFECTS OF CHANGING UPPER OR LOWER CONTROL TEMPERATURE SETTINGS

Changing the upper control temperature setpoint (reference value = 75°F) has little effect on performance; however, changing the lower setpoint (the thermostat setting; reference value = 65°F) has a major effect. Figure I-27 shows these effects for Albuquerque and Madison in the case of a waterwall. The upper, solid

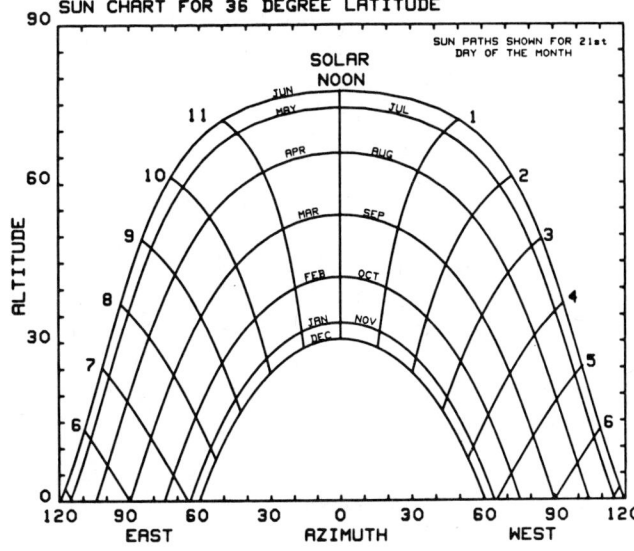

I-26: A sun chart expressed in rectangular coordinates for latitude 36°N. Solar azimuth (relative to true south) and altitude (elevation) angles are given for different hours of the day (relative to solar noon) for the 21st day of each month.

lines correspond to an arrangement with no upper temperature limit; the lower, dashed lines correspond to a fixed room temperature—that is, to an arrangement in which the upper temperature limit is equal to the thermostat setpoint.

The plots in figure I-27 may be confusing because of the definition of SSF, which is:

$$SSF = 1 - \left(\frac{\text{Auxiliary}}{\text{Base 65°F Heating Load}}\right)$$

This definition has been used on all other plots involving solar savings fraction. The auxiliary increases as the thermostat setpoint is increased, and this reduces the value of SSF because the base 65°F heating load is unchanged. Basically, the curves show the relative change in auxiliary heat as the thermostat setting is changed. An alternate definition of solar savings uses a reference thermal load defined in terms of the base T heating load, where T is the thermostat setting. Results based on this definition are shown in figure I-28.

The corresponding heating loads for the particular years chosen for the analysis are as shown in figure I-29.

---

*An excellent description of sun charts and their use is given in [MAZ-2] which contains charts for latitudes 28°N to 56°N at 4° intervals. More detailed sun charts are given in [BEN], which covers all latitudes with a spacing of 2° over the range of principal interest.

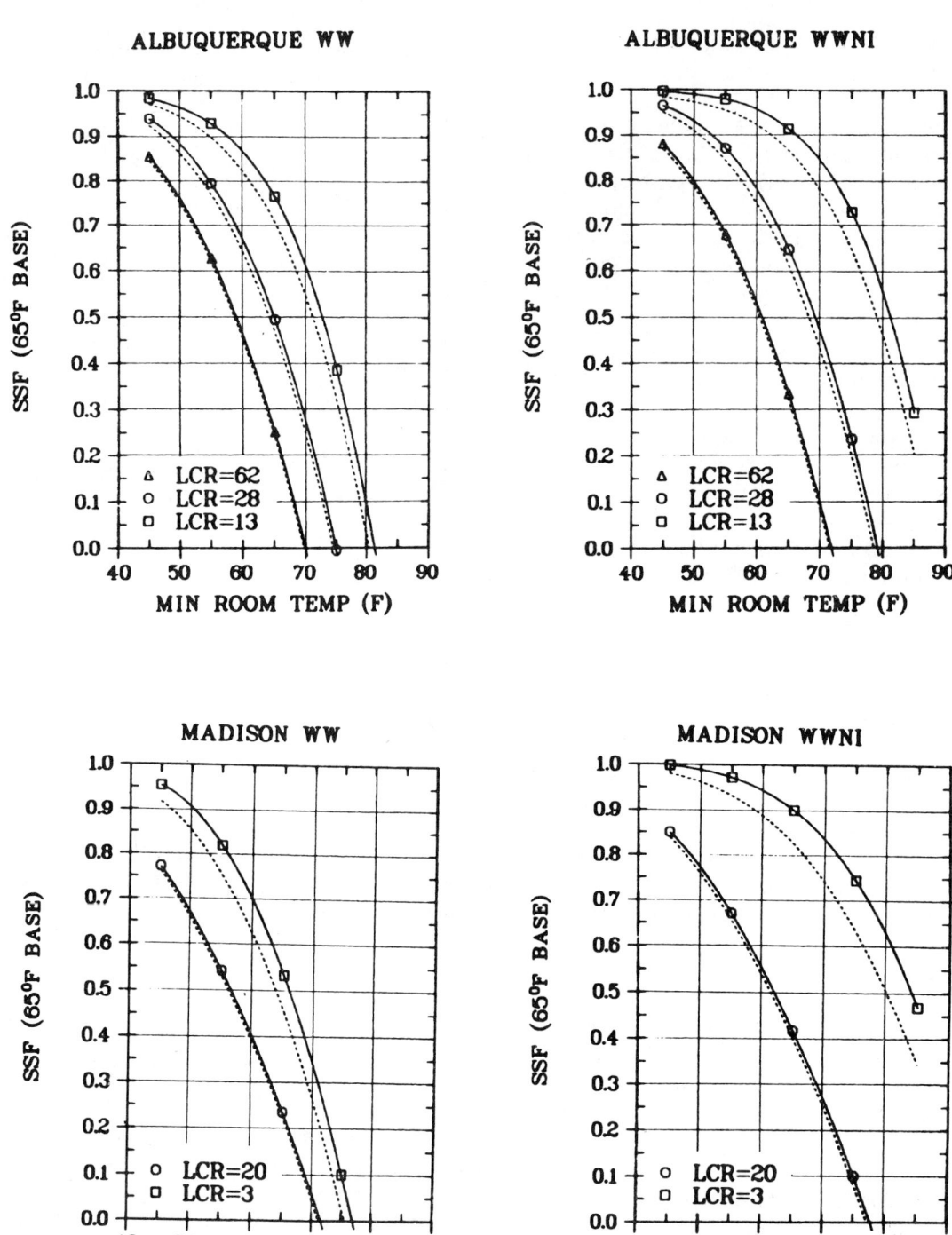

I-27: Effect on solar savings fraction of different minimum air temperature thermostat settings applied to insulated and uninsulated waterwalls in two cities. The lower (dotted) curve in each pair of curves corresponds to a situation in which room temperature is held constant throughout; the upper (solid) curve in each pair corresponds to a situation in which there is no maximum temperature limit. Solar savings fraction is calculated relative to a 69°F base temperature load.

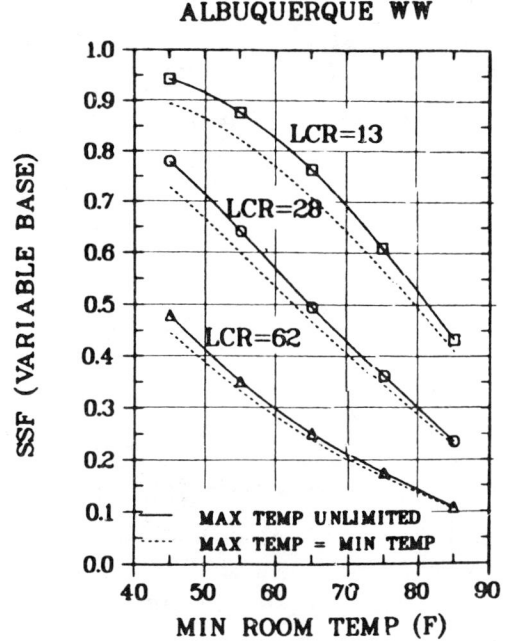
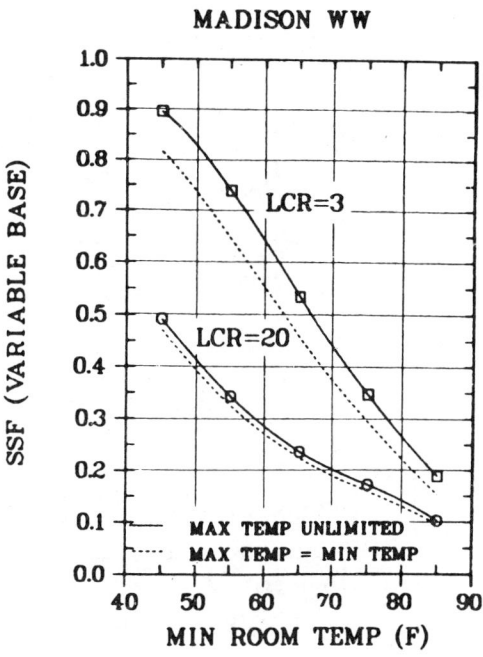

I-28: Effect on solar savings fraction of different minimum air temperature thermostat settings applied to waterwalls in two cities. These curves match those in figure I-27, except that here solar savings fraction is computed for a load where base temperature is taken as equal to the minimum air temperature thermostat setting.

Base	Albuquerque	Madison
45°F	1,297	3,573
55°F	2,670	5,541
65°F	4,649	7,963
75°F	7,320	10,956
85°F	10,536	14,416

I-29: Annual heating loads (in degree-days) at different base thermostat settings for thermal storage walls in two cities.

I.13 EFFECTS OF ADDITIONAL INTERNAL MASS

Additional mass in the building, like the mass in the thermal storage wall itself, increases performance by storing heat that would otherwise be vented. The magnitude of this effect is shown in figure I-30. The effect is more pronounced at high values of SSF than at low values.

In order to be able to plot the results in terms of a ratio of mass surface area to floor area, it was assumed that the building load coefficient (*BLC*) was 6 Btu/DD/ft² of building floor area, exclusive of the solar wall. (The abscissa could be adjusted to account for a different building load coefficient; for example, if the *BLC* were 9 Btu/DD/ft², then the abscissa value of 1.0 would be relabeled 1.5, etc.)

I.14 EFFECTS OF SIDE SHADING

Side shading reduces solar gains and, consequently, performance. The base case is a side shade that has a configuration with the following ratios:

$WC/HC = 1$
$XS/HC = 0$
$YS/HC = 1.5$
$ZS/HC = 2.0$

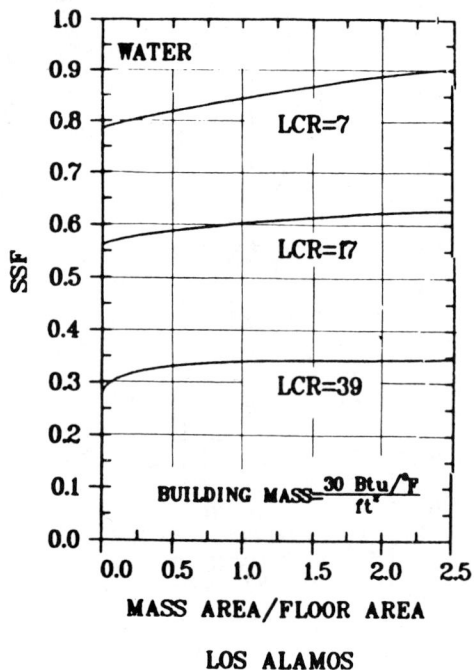

LOS ALAMOS

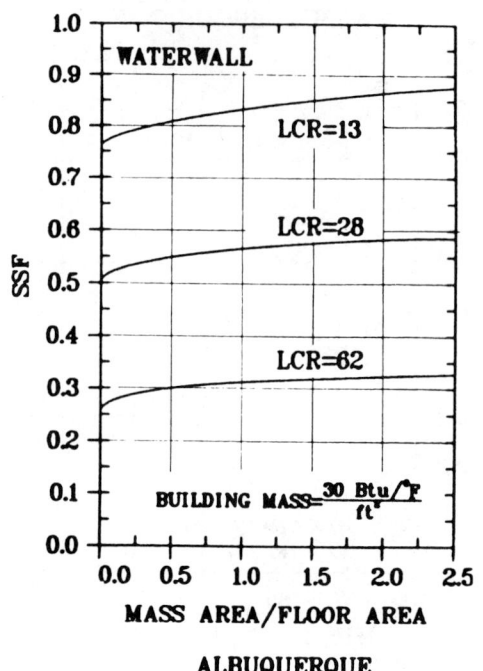

ALBUQUERQUE

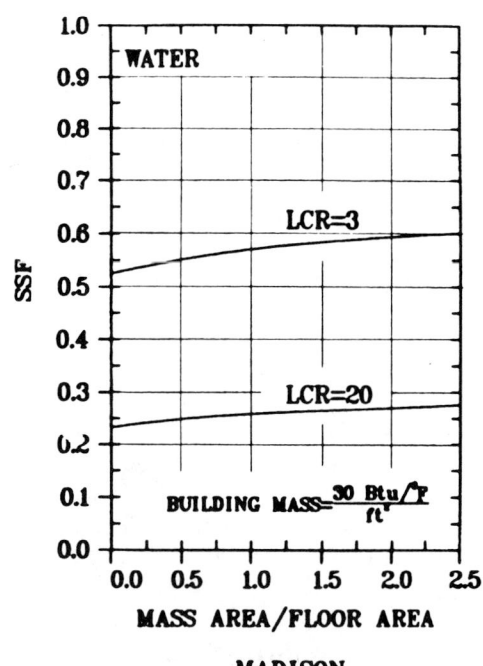

MADISON

I-30: Effect on solar savings fraction of additional mass applied to waterwalls in three cities. Results are expressed in terms of the ratio of mass surface area to building floor area, assuming a building load coefficient of 6 Btu/DD/ft² of interior building floor area. The additional mass is assumed to be so thick that thickness is not a factor. The reference value of the ratio of mass area to floor area is 0.

Figure I-31 defines these terms diagramatically. The shade results in a solar savings fraction obtainable through the equation:

$$SSF = SSF_o \times [1 + (RATIO \times D)]$$

where:
  SSF_o is SSF with no side shade
  $RATIO$ is a parameter determinable from figure I-32.
    ($RATIO = -1$, for the base case)

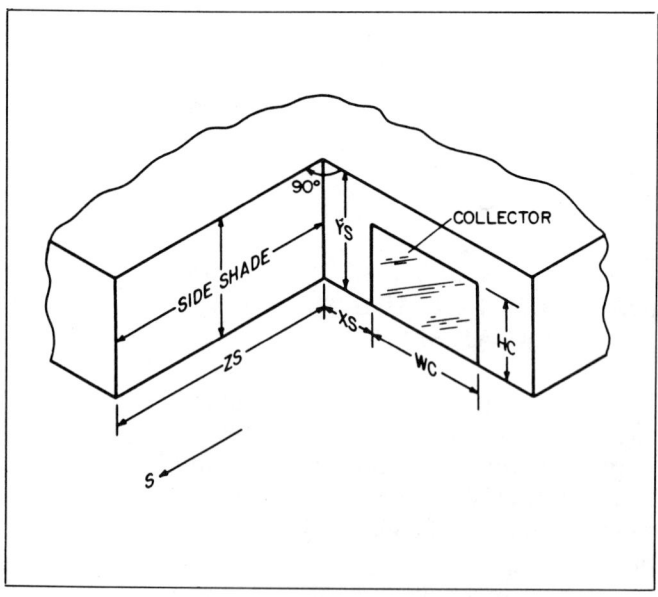

I-31: Illustration of the parameters WC, XS, YS, ZS, and HC, which are used in specifying the geometry of side shading.

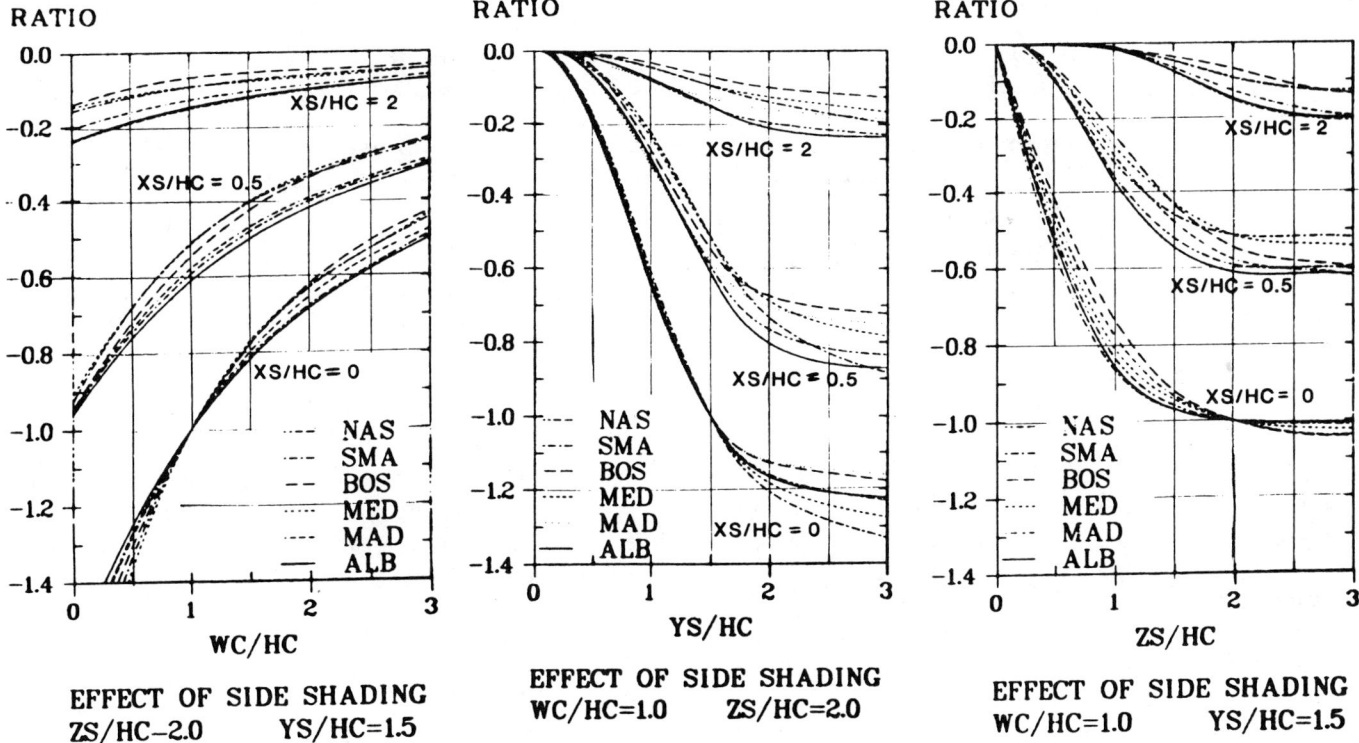

I-32: Effect on the value of RATIO of different side-shading geometries applied in six cities.

Values for D calculated from data recorded in six different cities are listed in figure I-33. In this figure, the word "fin" refers to any southerly projection on the east or west side of the collection area.

Simulation results for other values of XS, YS, ZS, and WC are shown in figure I-32 in terms of the single parameters, RATIO.

I.15 MIXED SYSTEMS

Almost all buildings need windows—for aesthetic reasons, to supply natural daylighting, and to serve as emergency exits. Typical window areas are in the range of 10% to 20% of the building floor area. Ten percent is a normal minimum specified by building code; architects frequently employ 20% or even more. It is appropriate to use these windows as direct-gain solar collection elements as much as possible, locating them on the south side of the building.

An effective design strategy is to combine direct gain and a Trombe wall in the same building. A normal procedure in cold climates is to base the window area on the nonsolar considerations mentioned above, but to locate the windows on the building's south side and in clerestories as much as possible. Additional solar gain is then made possible by using Trombe walls between pairs of south windows. This approach has superior comfort and performance characteristics in comparison to a pure direct-gain or Trombe-wall approach. It is better than the pure Trombe wall because fewer energy-losing windows have to be located in the nonsouth walls to achieve the 10% window area minimum. It is better than a pure direct-gain building because the timing of energy delivery is more uniform, there is a much less sudden drop in mean radiant temperature at nightfall, and the large temperature swing associated with a large-area direct-gain building is avoided.

City	LCR	D East Fin	D West Fin
Albuquerque (ALB)	28	0.280	0.274
Madison (MAD)	3	0.683	0.722
Medford (MED)	17	0.188	0.306
Boston (BOS)	6	0.501	0.711
Nashville (NAS)	18	0.276	0.275
Santa Maria (SMA)	53	0.153	0.191

I-33: Values for D (east and west fins) obtained from waterwalls in six cities.

J Estimating Temperature Swings in Direct-Gain Buildings

During the final phase of designing a building, it is desirable to do a detailed estimate of the building's interior temperature swing, in cases in which this may be a problem. When the direct-gain approach is used, the building's interior temperature swing can be fairly large if the *LCR* is small or if thermal storage is inadequate.

Estimation of the temperature swing to be expected in a direct-gain structure is complicated because of changing sun angles and the difficulties of determining internal reflections and absorptances. Complete and rigorous analysis is impractical. What is needed is an estimating procedure that yields appropriate answers and accounts quantitatively for the important physical effects: thickness and material properties of the thermal storage materials, location of these materials, and orientation of the surfaces. Such a method is described in this chapter.

J.1 GENERAL CONSIDERATIONS

The amount of heat stored in a wall, floor, or other material of a building is given simply by:

Heat Stored (in Btu) = Mass (in lb) × Specific Heat (in Btu/lb/°F) × Average ΔT (in Fahrenheit degrees)

The difficulty in using this formula lies in determining the average ΔT. If the temperature change took place so slowly that the whole mass of material changed temperature at the same time and in the same amount throughout the day, and if the temperature swing of the mass were equal to the temperature swing of the room, then applying the formula would be straightforward; but these assumptions do not apply in practice. The thermal coupling of the various elements in a room or throughout a building is relatively weak, so the storage mass temperature does not follow the room temperature very closely. In addition, the thermal conductivity of most building materials (such as masonry) that are commonly used for thermal storage is low enough that there will be an appreciable lag in temperature throughout the body of the material, even if it is only a few inches thick.

The most important factors in estimating interior temperature swings are the materials used for thermal storage, the thickness and location of those materials, and their orientation.

J.1.a MATERIALS

To be effective for heat storage, a material should have the following characteristics:

1. It should have a high thermal heat capacity. This means that, for any given storage mass thickness, the product of density (ϱ) and specific heat (c) should be large.
2. It should have a high thermal conductivity (k). The deeper portions of the wall cannot participate in the charging and discharging cycle if they are isolated from the room by a layer of material having a low thermal conductivity.

As a basis for ranking various materials, the grouping of properties $\varrho c k$ provides a good measure of the relative thermal storage effectiveness of a thick wall.

Since materials of high density also usually have a high degree of thermal conductivity, it follows that wall materials that are good insulators are poor for thermal storage. Materials such as Styrofoam and fiberboard are nearly worthless for heat storage; wood is moderately poor; and concrete, rock, brick, and adobe are relatively good. The very properties that enable a wall to perform well as an interior thermal storage element make it perform poorly as an exterior insulating element. Unfortunately, it is a common construction practice to make interior partitions of lightweight frame construction and to place the more massive construction (if any is used at all) in exterior walls. Frequently, the most massive elements are placed outside the thermal insulation (for example, in the case of a brick facade used on the exterior of an insulated, frame wall), rendering the mass ineffective for heat storage.

J.1.b THICKNESS

The thickness of the storage mass is very important because material deep within the surface is so isolated from the room as to be largely ineffective in the daily charging and discharging cycle. Furthermore, there is a time delay associated with conduction through the storage mass's material. Heat making a round trip several inches into the material may reappear at the surface 24 hours later when the wall is again charging. Beyond the thickness at which this out-of-phase effect begins to be created, added material actually *decreases* performance.

Figure J-1 shows the relative effectiveness of walls made of various materials in storing heat during a periodic sequence of similar days. Numerical values are provided in figures J-2 and J-3. The wall in each case is assumed to be insulated on the side away from the room. Using the appropriate curve in figure J-1, it is possible to assess the value of the one-inch difference between a six-inch-thick concrete wall and a five-inch-thick concrete wall. The extra (sixth) inch is shown to be only 35% as effective as the first inch (the one at the surface) was, so the designer might seriously consider placing the extra inch somewhere else. Notice that the effect of thicknesses greater than eight inches becomes negative because of the time-phasing considerations discussed above.

J.1.c LOCATION

Surface location is important. The most effective heat storage is accomplished when materials are directly or indirectly irradiated by the sun. Surface temperatures will generally be higher than the corresponding room temperature during the time when they are "in the sun." Lightweight materials exposed to the sun can store little heat themselves. As a result, they become quite hot, transferring absorbed solar radiation to the room air by convection and to other building surfaces by infrared radiation, in roughly equal proportion. Light-colored surfaces reflect the sun in a diffuse pattern to other regions of the room, which are thereby illuminated and heated. The energy exchange mechanisms are multistage and complex, defying rigorous analysis.

The most important distinction to be made in storage surface location is based on whether the surface receives energy by radiation (that is, by direct or scattered solar radiation or by infrared radiation from objects heated directly by the sun) or only by convection from the room air. In this book, these two ways of situating surfaces will henceforth be identified as "direct" and "indirect," respectively. "Direct" surfaces are much more effective than "indirect" surfaces for heat storage. If the room air must be heated before the surface receives the heat, then large temperature swings are inevitable.

Surfaces that enclose direct-gain spaces are classified as "direct" because they may receive energy by either a primary or secondary radiative process, not involving heating of the room air. The temperature of these surfaces will usually be quite close to the room air temperature—some surfaces slightly warmer and some slightly cooler—except for those in direct sun, which temporarily will be significantly warmer. Surfaces that enclose spaces having no direct gain can receive solar heat only

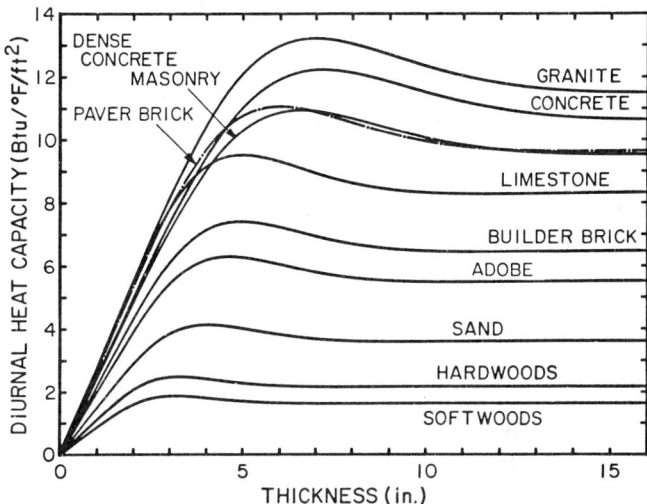

J-1: Diurnal heat capacities of various materials, expressed as a function of thickness. The material properties used to generate these curves are given in figure J-2; numerical values are given in figure J-3. For interior partition walls, one-half the total wall thickness should be used for determining the diurnal heat capacity for each of the two exposed surfaces.

Material	Density (ρ) lb/ft^3	Specific Heat (c) Btu/lb/°F	Thermal Conductivity (k) Btu/°F/hr/ft
Granite	167	0.20	1.050
Concrete	143	0.21	1.000
Limestone	153	0.22	0.540
Common Brick	120	0.22	0.417
Paver Brick	135	0.24	0.758
Adobe	120	0.20	0.332
Sand	95	0.19	0.190
Softwood	32	0.33	0.067
Hardwood	45	0.30	0.092
Concrete Masonry	140	0.21	0.822

J-2: Characteristic properties of various materials—density (ρ), specific heat (*c*), and thermal conductivity (*k*).

Estimating Temperature Swings / 107

from air heated in other rooms and are classified as "indirect." If there is a large opening (such as an open doorway) between a direct-gain space and another space (such as a back hallway or bedroom), then the heated air will convect very strongly from one room to the other, carrying the heat to the enclosing surfaces of the second room. If the door is closed and the path is blocked, however, then the enclosing surfaces will be isolated and ineffective for heat storage.

J.1.d ORIENTATION

Surface orientation is important because natural convective heat transfer always causes the heated air to rise.

Downward convection to the floor is essentially zero—the measured heat transfer coefficients show only a small effect due to conduction down through a 1-inch surface layer of air. Energy flow to the floor is primarily due to direct solar radiation or to infrared radiation from the ceiling and walls. Thus the floor tends to be cooler than the room and to be ineffective for heat storage unless it is directly sunlit.

Conversely, convection to the ceiling is very pronounced—so much so that the ceiling temperature is frequently warmer than the average room air temperature, even in back rooms. Walls represent an intermediate situation, having a moderately strong convective exchange with the room air.

Thickness	Granite	Concrete	Concrete Masonry (140 lb.)	Paver Brick	Limestone	Builder Brick	Adobe	Sand	Hardwood	Softwood
Direct dhc (Classes 1 and 2)										
1	2.78	2.50	2.44	2.70	2.80	2.20	2.00	1.50	1.12	.87
2	5.54	4.99	4.85	5.36	5.52	4.33	3.92	2.90	2.08	1.61
3	8.18	7.37	7.13	7.81	7.81	6.11	5.44	3.86	2.48	1.88
4	10.46	9.47	9.04	9.71	9.17	7.16	6.20	4.14	2.42	1.81
6	13.00	11.94	10.88	11.07	9.30	7.23	6.05	3.82	2.20	1.65
8	13.05	12.14	10.69	10.55	8.63	6.70	5.62	3.62	2.17	1.64
12	11.81	10.99	9.68	9.68	8.29	6.45	5.49	3.61	2.18	1.64
16	11.51	10.65	9.53	9.65	8.33	6.48	5.52	3.62	2.18	1.64
Indirect dhc (Class 3)										
1	2.48	2.28	2.23	2.42	2.48	2.03	1.86	1.43	1.08	.85
2	3.84	3.63	3.56	3.73	3.71	3.23	3.00	2.39	1.77	1.42
3	4.36	4.21	4.10	4.19	4.05	3.64	3.39	2.70	1.91	1.54
4	4.51	4.40	4.26	4.29	4.06	3.68	3.41	2.67	1.82	1.46
6	4.45	4.37	4.19	4.16	3.84	3.48	3.19	2.46	1.68	1.35
8	4.30	4.22	4.03	3.99	3.69	3.32	3.05	2.38	1.67	1.34
12	4.11	4.02	3.86	3.86	3.65	3.28	3.04	2.39	1.68	1.35
16	4.10	4.00	3.85	3.87	3.66	3.29	3.04	2.39	1.68	1.35

J-3: Diurnal heat capacities (in Btu/°F/ft²) of various materials, Classes 1, 2, and 3. Classifications for materials in various locations are given in figure J-4.

J.2 ESTIMATION PROCEDURE

A simple technique has been devised for estimating the diurnal temperature swing to be expected in a direct-gain building. The method is based on computing a "diurnal heat capacity" of the entire building, determined from the thermal properties of each enclosing surface. The method can be described in a seven-step procedure, as follows:

Step 1. Calculate the surface area (A) of all of the building's interior mass walls, floors, and ceilings, and assign a classification to each in accordance with the criteria in figure J-4 and the previous discussion.

Location	Classification
The surface of any massive material that receives some direct sun, except a covered floor.	Class 1
Covered floor (or any covered surface)	Class 4
Walls that enclose a direct-gain room for which the solar gains exceed the room daytime losses.	Class 2
All ceilings, except in sealed rooms (such as closets).	Class 2
Walls that enclose other rooms which communicate by convection with direct gain rooms.	Class 3
Uncovered floor in direct-gain rooms (not directly sunlit)	Class 3
Floors other than in direct-gain rooms.	Class 4
All surfaces in closed-off rooms	Class 4

J-4: Classifications for surfaces in different locations, to be used in calculating effective surface areas. In this listing, include gypsum-board surfaces and include wood surfaces that are more than ½ inch thick. Do not include insulating materials—such as fiberglass ceiling panels—or surfaces that are covered with heavy fabric, rugs, or other insulation.

Latitude	Qtran (in Btu/day/ft²)
20	1,390
25	1,443
30	1,460
35	1,445
40	1,390
45	1,287
50	1,128
55	899
60	584

J-5: *Qtran* values for ⅛-inch-thick double glass at various N latitudes.

Step 2. For Class 1 surfaces, estimate the fraction of the solar day (f) during which the surface is sunlit, and estimate the absorptance of the surface (α). The estimate need not be very precise. Absorptance values can be estimated visually using the following guide:

For very dark surfaces, α = 0.8 to 0.9.
For most surfaces, α = 0.5 to 0.6.
For light-colored surfaces, α = 0.3 to 0.4.

Also refer to figure I-11.

Step 3. Look up in figure J-3 the diurnal heat capacity (*dhc*) for the material that composes each Class 1, Class 2, and Class 3 surface in the building. Notice that there are two *dhc* tables in figure J-3—one for Class 1 and Class 2 surfaces (direct), and another for Class 3 surfaces (indirect). For interior mass walls, the effective wall thickness is one-half the total thickness; both surfaces should be counted. (For a method of calculating the diurnal heat capacity of a material—whatever its properties and thickness—and for a fuller explanation, consult Appendix 10).

Step 4. Calculate the diurnal heat capacity (*DHC*) of each surface, using the formula:

$DHC = A \times (dhc) \times (1 + \alpha f)$ [for Class 1]
$DHC = A \times (dhc)$ [for Classes 2 and 3]
$DHC = 0$ [for Class 4]

Step 5. Estimate the daytime heat capacity of the building's furnishings. In lieu of other information, use the figure 2 Btu/°F for each square foot of floor space:

$$DHC = 2 \times A_{floor}$$

Step 6. Sum the *DHC* values to obtain the total diurnal heat capacity (Total *DHC*) of the building's interior.

Step 7. Use the following formula to estimate the January, clear-day temperature swing (ΔT):

$$\Delta T = 0.61 \times \text{(Collection Area)} \times \frac{Qtran}{\text{Total } DHC}$$

Qtran for ⅛-inch-thick double glass can be calculated for latitudes between 20°N and 60°N using the data in figure J-5.

J.3 EXAMPLE

A numerical example will serve to illustrate the above method and to answer most questions. The house to be analyzed is at 35°N latitude. The construction is massive (adobe), insulated on the outsides of the exterior walls with 2 inches of polyurethane foam, with extensive south-facing glazing and a clerestory window across part of the roof. The floor is concrete in the bedrooms, and brick on sand in the remaining areas. A floor plan and a section through the house are shown in figure J-6.

Estimating Temperature Swings

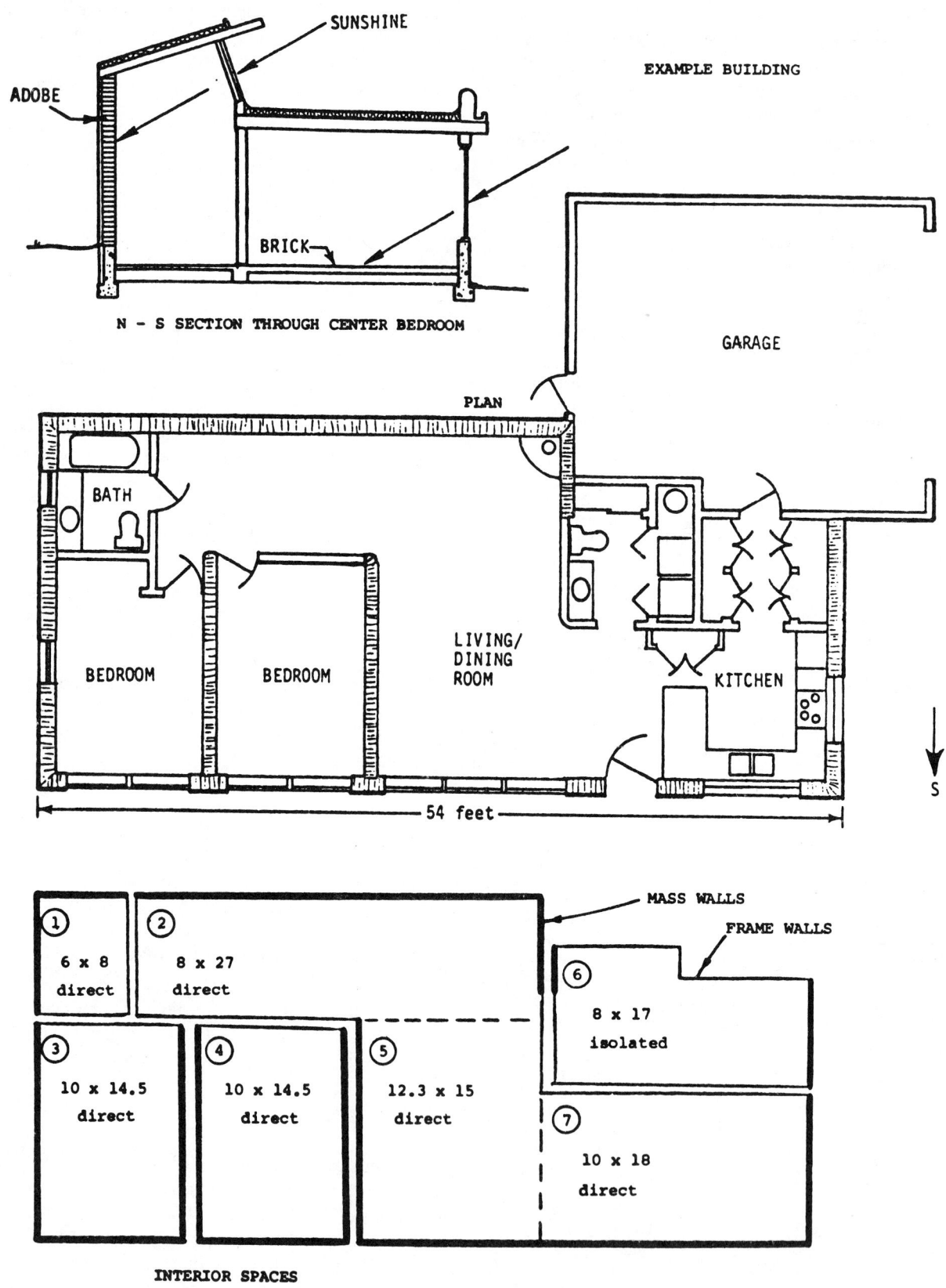

J-6: Floor plan and schematic diagrams of an example house, illustrating how to measure and classify a building's interior space for purposes of calculating total diurnal heat capacity.

110 / *CHAPTER J*

Surface type: _____

SPACE	Area, in square feet							(Class 1)×(1+αf)
	GROSS	NET	Class 4	Class 3	Class 2	Class 1	α	f

J-7: Blank worksheet.

WORKSHEET — Surface type: **Mass Walls**
Area, in square feet

SPACE	GROSS	NET	Class 4	Class 3	Class 2	Class 1	α	f	(Class 1)×(1+α·ε·f)
1	192	192	78	76	24		.7	.6	35
2	429	429	86	243	100		.7	.8	164
3	357	305	72	81	152		.7	.5	213
4	351	311	72	87	152		.7	.5	213
5	251	174	50	53	76		.7	.5	106
6	72	92	50						
7	257	215	111	84	20		.7	.5	28
TOTALS	1929	1709	561	624	524				759

WORKSHEET — Surface type: **Frame Walls**
Area, in square feet

SPACE	GROSS	NET	Class 4	Class 3	Class 2	Class 1	α	f	(Class 1)×(1+α·ε·f)
1	192	167	58		131				
2	325	240	65		175				
3	92	80	18		62				
4	92	80	18		62				
5	46	46	9		37				
6	368	368	368						
7	166	139	79		60				
TOTALS	1291	1120	593		527				

WORKSHEET — Surface type: **Ceiling**
Area, in square feet

SPACE	GROSS	NET	Class 4	Class 3	Class 2	Class 1	α	f	(Class 1)×(1+α·ε·f)
1	92	92			92				
2	216	324			324				
3	145	217			217				
4	145	217			217				
5	184	276			276				
6	136	136	136						
7	170	270			270				
TOTALS	1054	1512	136		1376				

WORKSHEET — Surface type: **Floor**
Area, in square feet

SPACE	GROSS	NET	Class 4	Class 3	Class 2	Class 1	α	f	(Class 1)×(1+α·ε·f)
1	B	48	24	24					
2	B	216	72	144					
3	C	145	72	23		50	.7	.5	67
4	C	145	72	23		50	.7	.5	67
5	B	184	61	41		72	.7	.6	116
6	B	136	136						
7	B	170	50			130	.7	.5	157
BRICK	B	764	343	209		212			273
CONCRETE	C	290	144	46		100			134

J-8: Examples of filled-in worksheets.

(*Step 1.*) The first step of the calculation process is facilitated by dividing the house up into different internal spaces. This is shown in the bottom diagram of figure J-6. In this particular house, all rooms are directly heated by the sun, except a small pantry and bathroom adjacent to the kitchen area. The northwest bathroom receives direct sun through the clerestory window. The bottom diagram shows the inside dimensions of each of the spaces, as well as the nature of the heat exchange (direct, indirect, or isolated) in each. There are no indirect spaces in this particular house. The main heat-storing walls in the house are shown on the bottom diagram by heavy lines. These are adobe walls. The other walls are of frame construction.

By means of this interior-space diagram, it is a relatively quick procedure to estimate the surface area of all of the enclosing planes that surround each interior space. A convenient way of doing this is to fill in a separate worksheet for each distinct type of surface. See figure J-7. In the example house, the distinct types are adobe mass walls, frame walls, ceiling (which is composed entirely of exposed wood), brick floor, and concrete floor.

In figure J-8, worksheets for each surface type are shown filled out for the example house. The gross internal surface area is calculated first and listed in the first column. The additional piece of information required for this calculation is the ceiling height, which in the example is 9.2 feet except in the clerestory area, where it is 13 feet. A net area for each surface is then determined by substracting estimates of the window or door opening areas within the walls—which do not contribute significantly to heat storage—from the gross area. For the floor, the net area is equal to the gross area. For the ceiling, the net area exceeds the gross area because of the deep wood ceiling beams employed there; the result is a 50% increase in the net surface area of the ceiling compared to the projected area. To save space, the calculations for brick floors and concrete floors were put on a single worksheet.

The net surface area must then be allocated according to the four classes of surfaces previously identified. The allocation is relatively straightforward, but some assumptions need to be made. For this example, the following assumptions were made:

1. Wall surfaces hidden behind cabinets in the northwest bathroom and in the kitchen are ineffective and therefore belong in Class 4.
2. The doors to the small bathroom and pantry next to the kitchen are kept closed; therefore, internal surfaces in those areas are Class 4.
3. Of the gross wall surface area, 20% is covered by objects (such as chairs pushed against the wall, or pictures hung on it, rendering that portion Class 4.
4. One-half of the floor area in the bathroom and bedrooms is covered, and one-third of the floor area in the remaining rooms is covered; these covered areas are assigned to Class 4.

(*Step 2.*) The enhancement factor $(1 + \alpha f)$ is computed for each of the Class 1 surfaces. The worksheets show the α and f values that were used. The areas of the Class 1 surfaces are then multiplied by the $(1 + \alpha f)$ values, and each column on the worksheet is totaled.

(*Step 3.*) The totals from the worksheets are transferred to the table shown in figure J-9. Values of *dhc* are obtained from figure J-3 and listed.

(*Step 4.*) The (modified) areas are multiplied by the *dhc* values to determine the *DHC* values.

(*Step 5.*) The *DHC* of furnishings is estimated as the floor area (1,054 square feet) times 2 Btu/°F/ft².

(*Step 6.*) The *DHC* values are summed to obtain the total diurnal heat capacity (18,686 Btu/°F).

(*Step 7.*) The glazing area is 310 square feet. The latitude is 35°N. Therefore the expected inside air temperature swing (ΔT) is:

$$\Delta T = 0.61 \times 310 \times \frac{1{,}445}{18{,}686} = 15 \text{ Fahrenheit degrees}$$

This is in good agreement with the actual, observed value.

More recent work has been completed on the calculation and use of diurnal heat capacity. It is presented in reference [BAL-5].

Category	Class	Modified Area	dhc	DHC
Mass Walls	1	759	5.84	4,428
	2	624	5.84	3,644
Frame Walls	2	527	1.25	658
Brick Floor	1	273	5.73	1,564
	3	209	3.54	740
Concrete Floor	1	134	9.47	1,269
	3	46	4.40	202
Ceiling	2	1,376	2.96	4,073
Furniture	2	1,054	2	2,108
Total Diurnal Heat Capacity				18,686 Btu/°F

J-9: Numerical totals from the worksheets, assembled for purposes of calculating total diurnal heat capacity.

K Designing Fan-Forced Rockbeds

K.1 HYBRID APPLICATIONS WITH PASSIVE SOLAR COMPONENTS

An effective and favorite hybrid application involves using a rockbed in conjunction with a passive solar building. In many applications, the rockbed is located beneath the source of hot air; as a result, natural convection cannot be used to transfer the heat. In such a case, a fan is normally employed, which technically makes that part of the system an active element.

Rockbeds can be used effectively in situations in which a passive system creates an excess of thermal energy, in the form of air that is heated above the comfort level. It is desirable to remove this overheated air for three reasons:

- To reduce the air temperature in the space and improve thermal comfort;
- To store the heat thus removed for later retrieval; and
- To redistribute heat from the upper south part of the building, where hot air tends to accumulate, to the lower north part of the building, which normally tends to be colder. Imbalances in temperature in the building, which might be created by passive solar elements operating alone, can be corrected.

In most passive solar applications, the temperature of the available air is in the range of 85° to 95°F, although in some instances it may be appreciably higher. For example, air that is removed from the space between a Trombe wall and its glazing may be at a temperature of from 100° to 130°F. In either case, however, the situation is not analogous to one involving an air-heating solar collector because the temperatures are appreciably lower, which means that the design of the rockbed and the distribution of the heat from the rockbed must be treated quite differently.

Very special, careful design is needed to arrange for the removal of heat from the rockbed in the form of warm air, which is then blown into the space to be heated. The air temperatures that can be achieved are low, and the necessary flow rates are therefore high. The effect of high air-velocity at low temperatures may be cold and unpleasant.

A much-preferred approach is to remove the stored heat by means of radiation and convection from the surface of the rockbed container. In this case, the rockbed is thermally coupled to the space that is to be heated, rather than being thermally isolated from it. A convenient approach that is often used is to place the rockbed underneath the floor of the building—although it would also be possible to place it behind one of the walls. Distribution of heat from the rockbed to the space is entirely passive. The floor or wall-surface temperature will only be a few degrees above the room temperature. If the arrangement is properly designed, the net result will be a very comfortable interior space, heating slowly from a large radiant panel.

A small amount of insulation between the rockbed and the living space is not harmful. For example, a rug on the floor over an underfloor rockbed is quite acceptable.

Owner experience with underfloor rockbeds has produced very favorable reports. Comfort is greatly improved by keeping the floor temperatures 5 to 10 Fahrenheit degrees above what they normally would be. By increasing surface temperatures (and thus increasing the mean radiant temperature within the space), air temperatures can be reduced, and energy savings can be realized that are even greater than might be expected from the actual amount of heat the rockbed releases.

Both horizontal-flow and vertical-flow rockbeds have been used. The horizontal-flow situation tends to be easier to build and seems to have no significant disadvantage, although the problem of "channeling" air above a horizontal-flow rockbed must be carefully avoided.

K.1.a SINGLE-ZONE SITUATIONS

Fan-forced rockbeds can be used effectively in situations in which the building to be heated is essentially a single thermal zone, provided that large temperature swings are permissible in that zone. Heat is stored during the high-temperature peak in the afternoon and is released at night to prevent temperatures from dropping too radically. This fairly accurately describes the situation in a greenhouse environment, in which high temperatures of 70° to 90°F may be observed in the afternoon and the main concern at night is to prevent freezing conditions. In this instance, the allowable temperature swing may be sufficient to make fan-forced distribution of heat from the rockbed practicable.

This type of application was employed in the greenhouse of the Doug Balcomb House in Los Alamos, New Mexico. (This is an earlier residence, not to be confused with the Unit 1 house in First Village near Santa Fe.) The rockbed was located within a planting bench at the side of the greenhouse. The 11-ton rockbed was typically cycled between 50° and 65°F, alternately storing and releasing about 60,000 Btu of heat. Both charging and distribution were by means of a 1,300-cfm fan that blew air horizontally one way through the rockbed. It was found that no ducting was required, since at this high air exchange rate there was no temperature stratification in the room.

A slightly different configuration was used in the greenhouse of the Kenneth Balcomb House in Glenwood Springs, Colorado. Here, the horizontal, one-way rockbed was underneath the floor of the greenhouse, with the major part of the heat distribution occurring by conduction up through the floor.

Designers have also used rockbeds in single-zone residential situations. However, large volumes of air have to be moved through the system, and special attention is required to size the air-supply grilles properly.

K.1.b THE TWO-ZONE APPROACH

Another technique, which is generally more suitable for residential applications, is to divide the building into two thermal zones and accept fairly large temperature swings in one zone in order to stabilize temperatures in the other zone. This design approach is shown schematically in figure K-1. It is basically the attached sunspace approach.

In Zone 1—the direct-gain space—large temperature swings can be expected because there is a large quantity of excess heat. Heat storage takes place in the mass separating the zones and in the floor of Zone 1. Depending on the size of Zone 1, the surface area of its enclosing mass, and the glazing area, temperature swings of 25 to 35 Fahrenheit degrees can be anticipated. Such swings can be completely acceptable (and perhaps even advantageous), if the space is used as a greenhouse, a sun room, an atrium, a conservatory, a transit area, or a vestibule or airlock entry.

A principal advantage of this approach is reduced temperature swings in Zone 2—a buffered space protected from the extremes of Zone 1 by the time-delay and heat-capacity effects of the mass wall. With a little care in design, one can phase the time of heat arrival into Zone 2 in a way that maintains a nearly constant temperature in the zone.

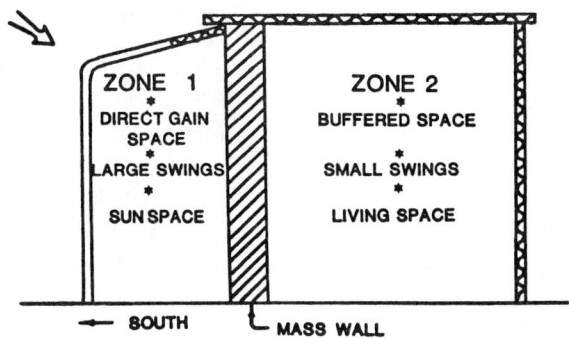

K-1: Two-zone approach to passive solar heating.

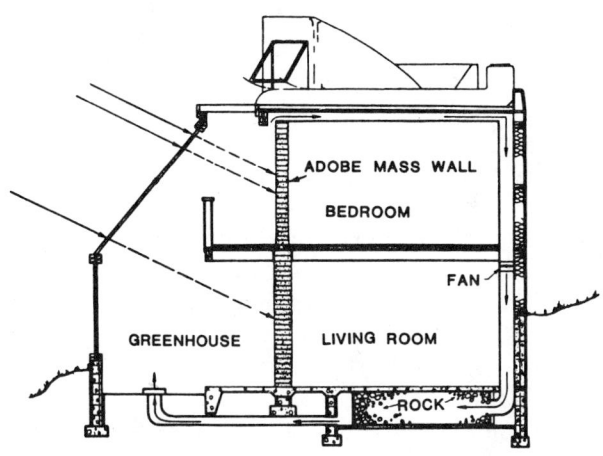

K-2: Schematic diagram of Unit 1, First Village; near Santa Fe, New Mexico.

An example of the effective use of the two-zone approach is Doug Balcomb's Santa Fe house (Unit 1, First Village). A cross section of the house is shown in figure K-2. The house was designed and built by Susan and Wayne Nichols. The rockbed system was designed by Herman Barkmann.*

Temperatures that are experienced in the two-story greenhouse typically range from 52° to 85°F in winter and from 65° to 87°F in summer, with a wintertime low of 43°F and a summertime high of 95°F. Ambient outside temperatures at the site range from a wintertime low of −14°F to a summertime high of 95°F. There is no auxiliary heat in the greenhouse. In addition to accommodating houseplants and garden vegetables, the space serves as the only house entry, as a major transit area, and as the access-way to the bedrooms upstairs. The

---

*For a complete description, see reference [SAN].

temperature swings do not interfere with any of these functions.

The living space is separated from the greenhouse by an adobe mass wall (14 inches thick downstairs and 10 inches thick upstairs). Temperatures in the downstairs living room range from 63° to 75°F in the winter—with a typical daily swing of 6 to 7 Fahrenheit degrees—and from 70° to 78°F in the summer—with a typical daily swing of 4 to 5 Fahrenheit degrees. The small wintertime swing is partly due to the stabilizing effect of underfloor rockbeds, which are heated using 85°F greenhouse air. During sunny winter days, 60,000 to 160,000 Btu of heat are removed from the greenhouse and stored in the underfloor rockbeds. This heat is removed from the greenhouse during its peak-temperature hours, when the heat would otherwise largely be lost. The heat removal also results in a decrease in maximum wintertime greenhouse temperatures from 95° to 85°F, thus improving the comfort in the greenhouse.

Airflow through the underfloor rockbed in Unit 1 is from the north side of the house horizontally toward the south side. Air is delivered to the north plenum through a duct leading from the top of the greenhouse; it returns from the south plenum through a duct to the greenhouse, completing the circuit. There are two rockbeds and two fans; the combined airflow is 1,750 cfm. The fan operated on 134 of the 176 days in the 1978-79 winter analysis period, transporting a total of 9.4 MMBtu to the rockbed—an average of 70,000 Btu per operating day. The fan operated an average of 3.5 hours per day, drawing a total of 531 watts, for an overall electrical consumption of 850,000 Btu (249 kWh). Therefore, the coefficient of performance (*COP*) of the system is 11.0.

Overall, the house was 89% solar heated, and the fan-forced rockbed system accounted for 30% of the solar heat utilized in the living areas of the house; the other 70% was supplied by passive means.*

The weight of the 7-inch-thick concrete slab covering the rockbed is borne by the rocks. Settling has not been a problem. There is no evidence (such as odors or mildew) to indicate that condensation in the rockbed is a problem. One would not expect condensation under the conditions observed. Wintertime floor temperatures above the rockbed are normally in the range of 60° to 70°F, with typical daily swings above the rockbed entrance of 5 Fahrenheit degrees during sunny days.

The two-zone approach works effectively in conjunction with a fan-forced rockbed only because the temperatures achieved in Zone 1 are appreciably higher than the desired temperature level of Zone 2 and because there is excess heat in Zone 1. If both of these conditions are not present, then there is no point in building a fan-forced rockbed. Whether these conditions will prevail depends on the configuration of the particular design and on the climate of the locality.

K.2 ROCKBED CHARACTERISTICS

The terminology used for rockbed airflow is shown in figure K-3, which depicts airflow into one face of the rockbed (which might be either a vertical face or a horizontal face, depending on the direction of airflow) and through the bed. The face of the rockbed at which the flow enters is designated the front face, and the face at which it exits is designated the back face.

The face velocity is the velocity of the air just before entering the front face—not the actual velocity of the air between rocks, which would be much higher. The following formulas characterize most rockbeds containing typical-shaped rocks:

Rockbed Volume = Face Area × Rockbed Length
Face Velocity = Air Flow/Face Area

For most rock (independent of rock size):

Density = 165 lb/ft³ (solid)
Void = 42%
Solid = 58%
Heat Capacity = 0.21 Btu/lb/°F

Therefore:

Volumetric Heat Capacity = 165 × 0.58 × 0.21
= 20 Btu/ft³/°F
Heat Storage = 20 × (Working Rockbed ΔT)
× (Rockbed Volume)

The figure for density indicated above is appropriate for most types of hard rock used by gravel companies and others in the formulation of concrete, etc. Some types of porous volcanic rock have a lower density and are probably inappropriate for use in a rockbed if other rock is available.

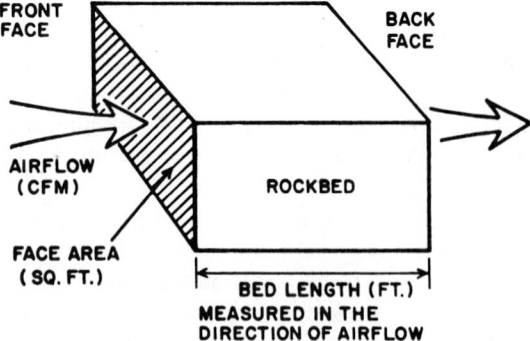

K-3: Terminology for rockbed airflow.

---

*A more comprehensive discussion of these data is presented in reference [BAL-3].

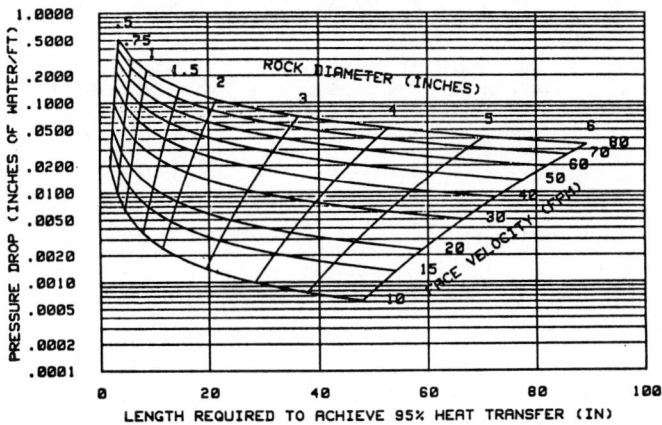

K-4: Performance map showing the pressure drop per foot of rockbed length for different values of rock diameter and rockbed face velocity. The rockbed length required to achieve 95% heat exchange from the air to the rock can also be determined from the map [DUN].

K-5: Values of pressure drop ($\Delta P/L$), expressed in inches of water per foot of rockbed length.

The pressure drop and heat transfer characteristics of a rockbed are charted in figure K-4. Heat transfer in rockbeds is very good and seldom imposes a constraint on design, since a few inches is often sufficient length (or thickness) within which to obtain adequate heat exchange. Pressure drop is usually the variable of greater concern. The independent variables in figure K-4 are rock diameter (in inches) and face velocity (in feet per minute). Based on these two inputs, it is possible to determine directly the pressure drop per unit of rockbed length (inches of water per foot of rockbed length) and the length required to achieve 95% heat transfer (in inches). The 95% figure will be used subsequently in the design procedure. Tabular values of the pressure drop per unit of rockbed length are given in figure K-5. Pressure drop is proportional to air density. For altitudes other than sea level, ΔP should be corrected as follows: Pressure Drop = Sea Level Pressure Drop × Air Density Ratio.

Another requirement for the calculation is some knowledge of the properties of air. The relevant properties are listed in figure K-6.

Face Velocity (in ft/min)	Rock Size (in inches)								
	.50	.75	1.00	1.50	2.00	3.00	4.00	5.00	6.00
10	.0193	.0100	.0064	.0035	.0024	.0014	.0010	.0008	.0006
15	.0337	.0181	.0119	.0069	.0048	.0029	.0021	.0016	.0013
20	.0512	.0283	.0191	.0113	.0079	.0049	.0035	.0028	.0023
30	.0956	.0550	.0380	.0232	.0166	.0105	.0077	.0060	.0050
40	.1525	.0901	.0632	.0392	.0283	.0182	.0134	.0106	.0087
50	.2220	.1335	.0947	.0595	.0433	.0279	.0206	.0163	.0135
60	.3040	.1853	.1324	.0839	.0613	.0398	.0294	.0234	.0194
70	.3986	.2454	.1765	.1126	.0825	.0537	.0398	.0316	.0262
80	.5057	.3139	.2268	.1454	.1069	.0698	.0518	.0412	.0342

Altitude in ft	Density in lb/ft³	Heat Capacity		Air Density Ratio (ADR)
		in Btu/°F/ft³	in Btu/hr/°F/CFM	
0	.0750	.0180	1.08	1.00
1,000	.0724	.0174	1.04	0.96
2,000	.0698	.0167	1.01	0.93
3,000	.0672	.0161	0.97	0.90
4,000	.0648	.0155	0.93	0.86
5,000	.0625	.0150	0.90	0.83
6,000	.0601	.0144	0.86	0.80
8,000	.0559	.0134	0.81	0.74
10,000	.0516	.0124	0.74	0.69

K-6: Properties of air at different altitudes above sea level.

K.3 DESIGN PROCEDURE

As a rule of thumb, it is not advisable to design for the transfer of more than approximately one-third of the net heat out of a space to a rockbed. "Net heat" means the solar energy input minus the daytime losses. Exceeding one-third results in either excessive airflow rates or excessive sunspace temperatures. The bulk of the solar heat should be passively stored within the enclosing surfaces of the sunspace. The active part of the system then becomes a minor element and the solar heating is not totally dependent on external power or on the reliable functioning of mechanical equipment.

The design procedure described below involves finding a good match between the pressure-rise characteristics of the fan and the pressure-drop characteristics of the rockbed and associated ducting.

In order to begin the procedure, it is necessary to determine two values from the building: the working air ΔT and the working rockbed ΔT.

The working air ΔT is the temperature of the air flowing into the rockbed minus the temperature of the air returning from the rockbed. The working rockbed ΔT is the temperature swing of the rocks; it determines the effective heat storage capacity of a given amount of rock. As a rule of thumb, the working rockbed ΔT can be taken as one-half of the working air ΔT.

The working air ΔT is determined by the air temperature available in the sunspace and the average temperature at which the rockbed will tend to become thermally stable. In the applications previously described, the sunspace temperature is typically between 80° and 90°F, and the rockbed temperature is typically 65°F to 70°F. Since the air exits the rockbed at something less than the average temperature of the rockbed, a slightly lower value can be used for the return air temperature. Thus a typical working air ΔT is 15 to 20 Fahrenheit degrees and a typical working bed ΔT is 7 to 10 Fahrenheit degrees.

With these two values in hand, you can begin a sizing procedure as follows:

Step 1. Establish rockbed size and airflow requirements:

$$\text{Rockbed Volume (in ft}^3\text{)} = \frac{\text{Desired Heat Storage (in Btu)}}{(20 \text{ Btu/°F/ft}^3) \times (\text{Working Bed } \Delta T \text{ [in °F]})}$$

$$\text{Airflow (in cfm)} = \frac{\text{Heat Transport Rate (in Btu/hr)}}{(\text{Working Air } \Delta T) \times (1.08 \text{ Btu/hr/cfm/°F})^*}$$

Step 2. Select rockbed length (in ft), and calculate face area (in ft²):

$$\text{Face Area (in ft}^2\text{)} = \frac{\text{Bed Volume (in ft}^3\text{)}}{\text{Bed Length (in ft)}}$$

Step 3. Calculate face velocity in feet per minute (fpm):

$$\text{Face Velocity (in fpm)} = \frac{\text{Airflow (in cfm)}}{\text{Face Area (in ft}^2\text{)}}$$

Step 4. Select rock size. Look up $\Delta P/L$ in figure K-5 or figure K-4. Calculate ΔP by multiplying this value by rockbed length:

$$\Delta P = \frac{\Delta P}{L} \times \text{Bed Length}$$

Check to see if the selected rockbed length is greater than the length recommended in figure K-4 for good heat transfer. (Return to Step 2 if necessary.)

Step 5. Add ducting ΔP. Calculate duct velocity level (not over 700 fpm for low noise level):

$$\text{Duct Velocity (in fpm)} = \frac{\text{Airflow (in cfm)}}{\text{Duct Area (in ft}^2\text{)}}$$

Find $\Delta P/L$ in figure K-7. Determine equivalent length of elbows ($\sim 15\ L/D$ per turn.)**

Step 6. Consult a catalog of fan sizes, and select a fan that matches the airflow and total ΔP values calculated above.

You can return to Step 2 or Step 4, or both, to obtain a good match.

K.4 EXAMPLE

It is desired to remove excess heat from a sunspace to an underfloor rockbed. The maximum sunspace temperature is to be limited to 85°F, by the removal of 15,000 Btuhr. The rockbed temperature is normally about 70°F, so the working air ΔT is 15 Fahrenheit degrees. The duct has a diameter of 18 inches, a length of 60 feet, and a total of 5 turns.

Given:

Heat Transport Rate = 15,000 Btu/hr (peak) at 15 Fahrenheit degrees working air ΔT

Desired Heat Storage = 2 days' accumulation (or 8 hours', at peak rate)

Step 1: Desired Heat Storage = 15,000 × 8
= 120,000 Btu

$$\text{Rockbed Volume} = \left(\frac{120,000}{20 \times 7.5}\right) = 800 \text{ ft}^3$$

$$\text{Airflow} = \left(\frac{15,000}{15 \times 0.84}\right) = 1,190 \text{ cfm}$$

Step 2: Desired Rockbed Length = 10 ft

$$\text{Face Area} = \left(\frac{800}{10}\right) = 80 \text{ ft}^2$$

---

*See figure K-6 for air properties at altitudes other than sea level.

**Consult Chapter 31 of reference [ASH-2] for more details.

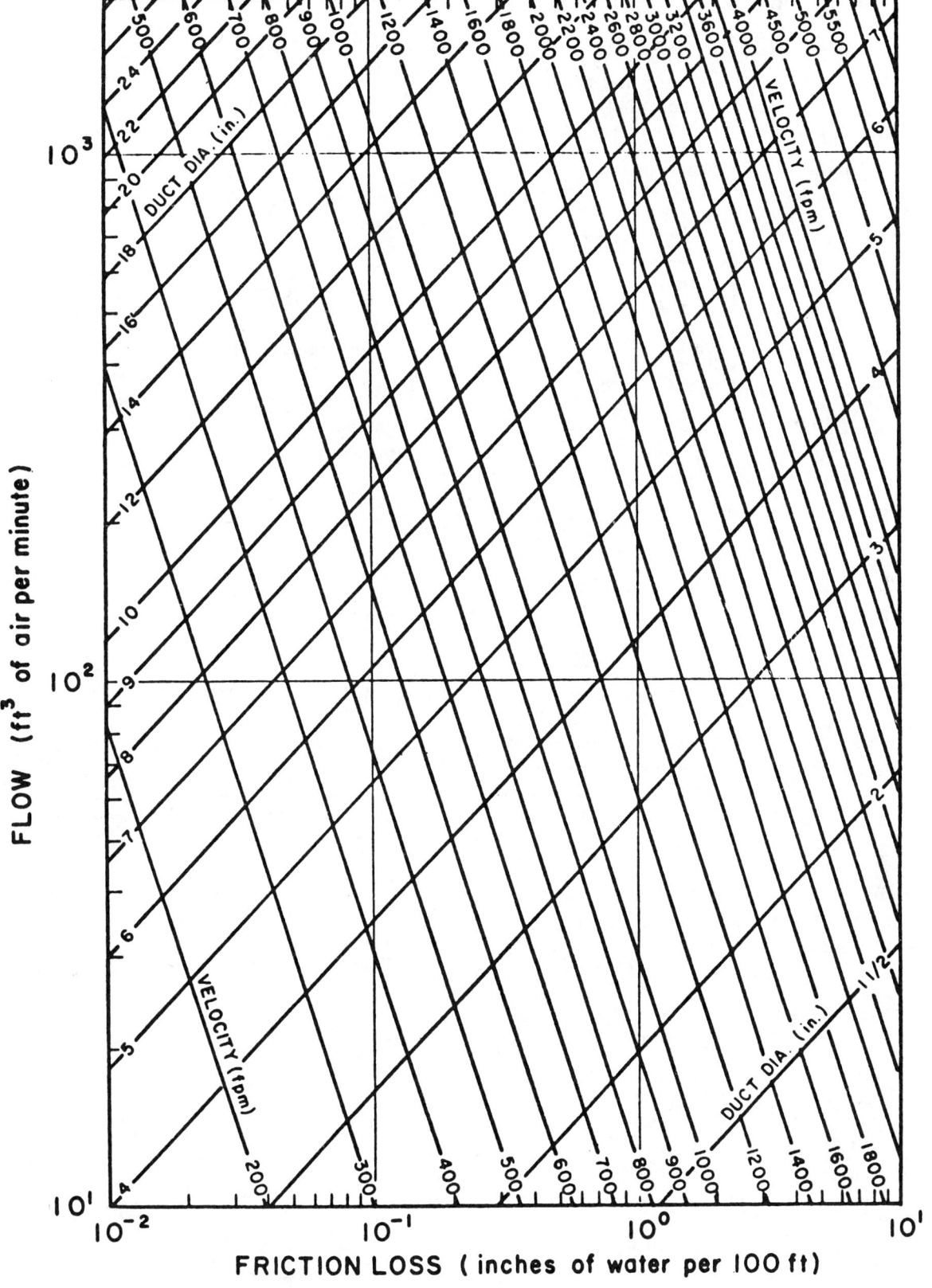

K-7: Design chart for airflow in ducts [SEA].

Designing Fan-Forced Rockbeds / 119

Step 3: Face Velocity $= \left(\dfrac{1{,}190}{80}\right) = 14.9$ fpm

Step 4: Desired Rock Size $= 2$ inches in diameter
$\Delta P/L = 0.0048$ inches of water per foot
(found in figure K-4)
$\Delta P = 0.0048 \times 10 = 0.048$ inches of water

Minimum length for good heat transfer is 13 inches (found in figure K-4).

Step 5: Duct Area $= \left(\dfrac{\pi}{4}\right) \times \left(\dfrac{18^2}{144}\right) = 1.77$ ft^2

Duct Velocity $= \left(\dfrac{1{,}190}{1.77}\right) = 672$ fpm

$\Delta P/L = \left(\dfrac{0.04 \text{ inches of water}}{100 \text{ feet}}\right)$
(found in figure K-7)

$\Delta P = \left(\dfrac{0.04 \times [60 + (5 \times 15 \times 18 \div 12)]}{100}\right)$
$= 0.07$ inches of water

Step 6: Total $\Delta P = 0.048 + 0.07 = 0.118$ inches of water

One might, therefore, choose a fan that will deliver 1,200 cfm at 0.12 inches of water (with no need to change length or rock size).

K.5 OPTIONS

Rockbed design is very flexible. The configuration of the bed is at least as important as the rock size, so the designer can reconfigure the bed to reduce the pressure drop, if necessary. The flow can be horizontal or vertical, as has been discussed. Another option is the shape of the bed.

Consider, for example, an underfloor rockbed configured as shown in figure K-8. If the pressure drop were too large with that configuration, the designer could use the layout shown in figure K-9 instead. If two-inch-diameter rock were being used, the pressure drop would be seven times less with the second configuration than with the first. Another way to achieve the same effect would be to use the configuration shown in figure K-10.

Many other options have been tried, including the following:

- Heat can be distributed through the rockbed by means of perforated pipe of the type used for septic field drainage;
- Concrete blocks can be used to form flow channels to distribute warm air through the rockbed (in this case making the mass of the concrete blocks an effective heat storage mass, as well); and
- The foundation walls of the building can be used as plenum walls, with wire mesh used to construct the front and rear faces of a horizontal-flow rockbed in much the same way as it would be used to build a fence.

The last-named construction option is particularly simple and is often used.

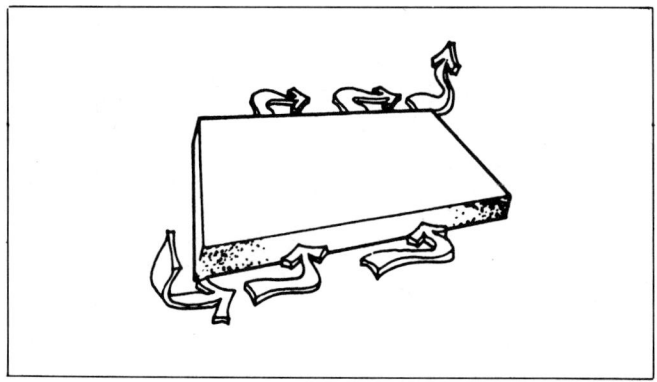

K-9: Rockbed designed for side-to-side airflow.

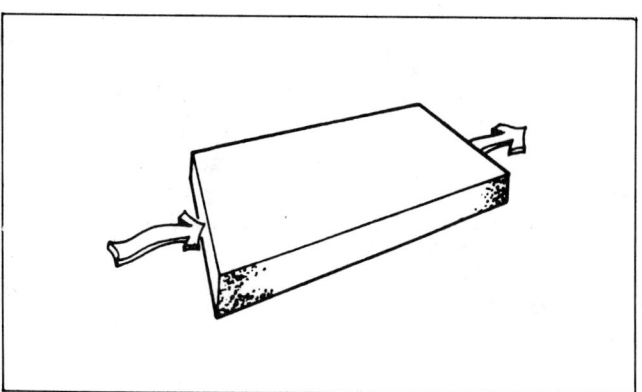

K-8: Rockbed designed for end-to-end airflow.

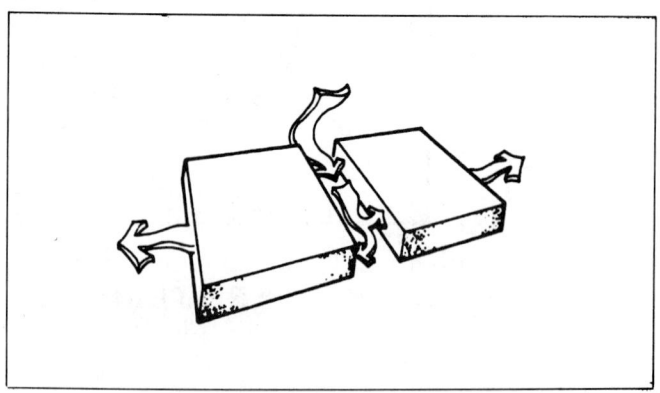

K-10 Split rockbed designed for divided airflow.

K.6 OTHER DESIGN POINTERS

Key points to keep in mind during design are:

- Use a high-flow, low-ΔP squirrel-cage blower to obtain high efficiency and quiet operation;
- Lay out the rockbeds with their inlet and exit plenums arranged in such a way that all of the rock has air flowing through it;
- Complete the airflow circuit by returning the air to the sunspace from which it was taken; and
- Use backdraft dampers to prevent reverse thermocirculation at night.

Correct mounting of the fan is essential to quiet operation. One very successful approach is to hang the blower on spring mounts inside the air duct. This will mechanically isolate it, thus preventing the fan's vibration from coupling to the building. In addition, the air duct can be covered with an acoustic absorber (such as a thin layer of fiberglass bat insulation) to deaden the noise of the blower, belt, and motor. A correctly installed 1,000-cfm blower should make no more noise than an electric refrigerator.

Another important design point is to avoid poor airflow distribution through the rockbed. This can be done by sizing the rockbed inlet plenum such that the pressure drop through the rockbed is several times greater than the pressure head of the air in the plenum. The formula for the pressure head is:

$$\Delta P = 6.2 \times 10^{-8} \times v^2$$

The velocity (v) is in feet per minute, and the pressure head is in inches of water.

In the previous example, the rockbed ΔP is .048 inches of water. To keep the pressure head five times lower than this would require a plenum velocity of:

$$\sqrt{\frac{0.048}{5 \times 6.2 \times 10^{-8}}} = 393 \text{ fpm}$$

This in turn would require an inlet plenum cross section of:

$$\frac{1,190}{393} = 3 \text{ ft}^2$$

L Economic Analysis

L.1 INTRODUCTION

The subject of economics arises when someone who thought solar energy was going to be free is discouraged from building a solar house by the high initial expense. There are two basic components of the cost of maintaining comfortable temperatures in any house: heat supply costs (fuel, for example) and energy conservation costs (insulation, double-glazed windows, etc.).

Solar houses generally have two sources of heat supply: the solar heating system and a back-up (or auxiliary) heating system. In such cases, the three principal elements of the total cost are the investment for the solar heating system, the investment for energy conservation, and the recurring cost of auxiliary heating. These three cost elements are interrelated in such a way that any one of them can be decreased through an increase in one or both of the other two. The purpose of this chapter is to show how a designer can take the fullest possible advantage of this interrelationship.

Two aspects of this process are discussed: *cost evaluation* and *cost optimization*. Cost evaluation consists of determining the overall economic merit of a given design. In this chapter, life-cycle cost (or life-cycle savings) is used as an evaluating criterion. Optimization consists of manipulating the design variables so as to minimize the total cost.

Most people tend to associate the term "economics" with money. For an individual on the housing market, this may well be the predominant concern. Basically, however, "economics" pertains to the way in which resources are utilized. From a world-wide perspective, the nature of the energy problem is physical—it is petroleum that is in limited supply, not dollar bills! Therefore, the resources that are consumed in constructing a solar heating system must be considered: if a particular system consumes more energy in construction than it subsequently saves in heating, no financial subsidy can alter the impact of the design on resource reserves. Consequently, the methods presented in this chapter are discussed in terms of both monetary economics and physical economics.

L.2 LIFE-CYCLE COST EQUATION

The total cost of supplying and conserving heat in a home can be expressed in the following equation, one side of which has three terms corresponding to the three elements of cost listed above (solar, conservation, and auxiliary):

$$LCC = \bigl(\{[FC + (A \times VC)] \times E_1\} + (CC \times E_1) + [L \times (1 - F) \times DD \times FP \times E_2]\bigr) \div 10^6$$

where:

- LCC = uniform annual life-cycle cost
- FC = fixed cost of the solar-heating system (the costs independent of collection area)
- A = solar collection area (in ft²)
- VC = variable cost of the solar heating system per square foot of collection area
- CC = conservation cost (insulation, etc.)
- L = building load coefficient (BLC) (in Btu/DD)
- F = solar savings fraction (SSF)
- DD = annual heating degree-days
- FP = auxiliary fuel price in the first year of operation
- E_1 and E_2 = economic parameters for converting initial costs and recurring costs to a common basis

	Monetary	Physical
LCC	$/yr	MMBtu^p/yr
FC	$	MMBtu^p
VC	$/ft²	MMBtu^p/ft²
CC	$	MMBtu^p
FP	$/MMBtu	MMBtu^p/MMBtu
E_1	FCR(yr^{-1})	1/SL (yr^{-1})
E_2	FF	1

(Note: SL = expected system life, in years.)

L-1: Units for economic parameters—monetary and physical.

In terms of monetary economics, *LCC* is expressed as a "uniform annual cost." This is the amount of a uniform expenditure which, if repeated every year for a given number of years, would be equivalent to a different non-uniform series of expenditures over the same time period. Units for the various economic parameters in the above equation are listed in figure L–1. (In figure L–1 "*SL*" stands for "[expected] system life" and "MMBtu^P" for "million Btu of primary source energy.") As explained in Appendix 10, E_1 is called the fixed charge rate (*FCR*), and E_2 is called the fuel-cost leveling factor (*FF*).

The methodology used in considering physical economics is one presented by the Center for Advanced Computation at the University of Illinois. *LCC* has units of energy that correspond to the energy resources embodied in an item. For example, about 4 Btu of energy resources are embodied in each Btu of electrical energy delivered to a heating load, taking into account power-plant inefficiency, construction of transmission lines, etc. So $FP = 4$ Btu^P/Btu, with Btu^P denoting primary energy resources. Again, see figure L–1 for units for the economic parameters in the *LCC* equation.

L.3 COST EVALUATION

One component of a proper economic analysis is accurate accounting or estimation of the initial costs of the design to be evaluated. Where the physical components are easily separable and definable, standard cost estimating procedures can be used with a fair degree of accuracy. With passive designs, separability is more difficult because many of the passive features are integral elements within the design of the building. In addition, collection and storage elements often serve dual functions, leading to confusion as to the precise charges to be made and credits to be allowed.

To deal with these issues, a simplified cost-evaluating procedure is outlined below.

L.3.a CATEGORIZATION OF COSTS

Costing is essentially a method of accounting, and the use of a consistent categorization scheme can help the designer/architect keep things organized. One such scheme, called the *functional elements approach,* assigns each passive-design-related item to one of four functional categories: collection, storage, distribution, or controls. Two additional categories are auxiliary heating equipment and building envelope construction (insulation features). Figure L–2 shows how various design elements fall into one another of these categories.

L–2: Categorization of design options by functional element.

Collection	Storage	Distribution	Controls	Auxiliary Heating Equipment	Envelope
Glazing	Containment	Ducting	Overhang	Ducted Warm Air: Heat Pump	Walls
Framing	Material	Piping	Movable Insulation	Resistance Wire Furnace	Ceiling
Absorption	Support	Vents & Dampers	Reflectors/ Glare Control		Windows
Reflectors		Blowers, Pumps, and Fans	Mechanical/Electrical: Thermostats Timers Wiring	Hydronic Radiators: Boiler	Doors
Support				Direct: Zone Electric Resistance or Radiant Panels	Infiltration
		Ancillary Equipment		Combustion Stove (Wood, Coal, Oil, Other): Fireplace Window Unit	

L.3.b PASSIVE COST ESTIMATION

Once the design has been conceptualized and the desired or available design options have been categorized by functional element—as shown in figure L-2—cost estimates must be made or obtained for each of the items. Common architectural practice is to state variable costs in unit terms—for example, in dollars per square foot of glazing or of residence floor area ($/ft$_g^2$ or $/ft$_R^2$). Some costs will be on a per-item basis and will not vary with the size of the system. These are called fixed costs, and are stated in total dollar terms.

Unit or variable costs (*VC*) and fixed costs (*FC*) have both a materials and an installation or labor component. Depending upon the circumstances, owner-builders may or may not want to account for their contributed labor.

Figure L-3 shows a sample worksheet that can be used for cost evaluating purposes. The worksheet includes columns for the name of the item under consideration, a description of the item, the cost-estimating unit (e.g., $/ft^2), elements of the cost, the amount desired, the total cost, and additional notes. Figure L-4 shows some typical entries, using the collection function as an example.

Cost evaluating worksheets can be completed for each of the functional elements in the design. From these, the total costs of a variety of designs can be estimated and broken down into collector-area-dependent (variable) and collector-area-independent (fixed) cost components. In most passive applications, the majority of costs will depend upon collector area. This implies that one can design with a small, moderate, or large collector area (taken as a percentage of floor area) and not be subject to substantial fixed or "sunk" costs.

L.3.c ENVELOPE COST ESTIMATION

Increasing the level of a building's insulation has the dual effect of increasing the building envelope construction costs and decreasing the annual heating requirements. In order to determine appropriate conservation construction measures, information is needed regarding the cost and thermal effectiveness of the various design options available. For example, options for wall insulation might include 3½-inch-thick (R11), 5½-inch-thick (R19), and 9½-inch-thick (R30) fiberglass batt insulation, at costs of $0.57, $0.70, and $0.98, respectively, per square foot of wall area (figures are in 1982 dollars). Elements of the cost that are common to all conservation options (for example, gypsum board, exterior sheathing, and paint, in the above example for walls) need not be included in this accounting procedure. Cost information can be obtained from construction material suppliers, insulation contractors, and so forth.

L.3.d CONVENTIONAL COST ESTIMATION

In many instances, passive solar design elements (glazing, storage, etc.) replace or augment construction items that otherwise would have been installed. To arrive at the *add-on* costs attributable to the passive design, credit must be given for the avoided costs of those items that were replaced or augmented. For instance, if "normal" construction practice called for a 4-inch-thick slab on grade, and a 6-inch-thick slab were poured for direct-gain thermal mass storage, only the additional 2-inch thickness of the larger slab should be counted as an add-on cost.

In the case of passive solar collection area, an allowance for the cost of the insulated wall that is displaced by the solar aperture should be deducted from the square-foot cost of the passive element.

Figure L-5 lists some common replacement credits that should be taken when estimating passive costs.

L.3.e FINAL PASSIVE ADD-ON COST ESTIMATION

Once the add-on costs and credits are specified (figure L-3 can be used as a sample format), a particular design can be chosen and assigned cost parameters as follows:

VC = variable costs, in dollars per square foot of glazing ($/ft$_g^2$), for passive solar add-on costs, after credit allowance for replacement of conventional construction items.

FC = fixed costs, in dollars, after credit allowance for replacement of conventional construction items.

The net add-on construction cost (CC_n) for the passive design calculated as

$$CC_n = FC + A(VC)$$

The right-hand cost term appears in the original *LCC* equation in section L.2.

L.4 COST OPTIMIZATION CONCEPTS

In the above discussion, the building load (*L*) the solar collection area (*A*), and the solar savings fraction (*SSF* or *F*) were treated as fixed numbers. The question now is how to manipulate these parameters to achieve the best possible economic advantage.

For a fixed level of building load (say, L_1), the annual cost of heat delivered to a passive home is the sum of the annual backup or auxiliary system heating costs and the annual charges associated with the add-on costs of the passive design. As the fraction of the annual heating load provided by the passive solar *SSF* increases, the annual cost of auxiliary energy decreases. However, because of the timing relationships between heating energy demand and passive solar heat supply, increasing glazing

L-3: Blank cost-evaluating worksheet.

Project Number: 16   Functional Element: COLLECTION   Sheet # 1 of 4
Design Description: TROMBE WALL   Date: 4/15/78
Location: ALBUQ., N.M.
Calculated by: SEYMOUR SUNSHINE

ITEM	DESCRIPTION	COST UNIT	COST				AMOUNT	TOTAL COST	NOTES
			MAT.	LABOR	O&P	TOTAL			
GLAZING	Patio Glass 76"x46" 5/8" Tempered Double	$/ft²	2.50	.75	.33	3.58	12 units (292 ft²)	$1045.36	
FRAMING	Aluminum Framing	$/ft	3.00	.50	.35	3.85	180'	693.03	
ABSORB-TION	Black Heat-Res. Paint	$/ft²	.05	.15	.02	.22	350 ft²	77.00	
TOTAL								$1815.39	

L–4: Example of filled-in cost-evaluating worksheet.

FUNCTIONAL ELEMENT	PASSIVE SOLAR FEATURE	DISPLACED CONSTRUCTION FEATURES			
		STORAGE WALL	STORAGE ROOF	DIRECT GAIN	ATTACHED SUNSPACE
Collection	Glazing Framing Absorption	Normal Wood Frame, Concrete, or Masonry Wall with Insulation	None None None	Normal Wood Frame, Concrete, or Masonry Wall with Insulation	None None None
Storage	Containment Material Support	Normal Foundation	Roof Structure Replaced Interior & Exterior Walls Replaced with Load Bearing Walls Normal Foundation	Conventional Slab on Grade if Augmented Interior Walls Replaced with Mass Normal Foundation	Adjoining Exterior Wall if Made Massive to Provide Storage
Distribution	Ducting Vents & Dampers Blowers, Pumps, Fans	None None None	None None None	None None None	None None None
Controls	Overhang Movable Insulation Reflectors Mechanical/Electrical	Replaced Trim Drapes, etc. None None	None None None None	Replaced Trim Drapes, etc. None None	None None None None
Auxiliary	Changes in the auxiliary heating system may allow changes in the conventional distribution and control items in which case the extra costs or credits should be accounted for.				

L-5: Conventional construction items that are commonly replaced with passive design elements.

Economic Analysis / 127

area additions become necessary to satisfy each additional increment in solar savings fraction (F).

This can be shown by taking any of the load collector ratio (LCR) curves, solving for collector area requirements versus solar savings fraction (using a specific value for the load (L)), and plotting the results as shown in figure L-6. Figure L-7 illustrates the annual dollar charges that are due to auxiliary and passive solar heating, as a function of SSF. As solar savings fraction increases, less auxiliary energy is needed because home heating requirements are offset by passive solar heat gains; however, the attendant increase in collector area implies increased construction costs. When converted into annual figures, the auxiliary and solar heat costs are summed to give the total annual cost of heat. System size is optimized when the total annual cost is minimized. This corresponds to SSF_o in figure L-7. Notice that an optimal heat supply cost (HSC_1) for this building's thermal load (L_1) occurs at SSF_o. At a different load (say, L_2), a different minimal cost (HSC_2) is obtained, but at the same solar savings fraction, SSF_o.

Figure L-8 illustrates the annual primary energy charges that are due to auxiliary and passive solar heating, as a function of SSF. Notice that optimal solar savings fraction in this figure (SSF_p) is higher than the one in figure L-7 (SSF_o); this demonstrates that the primary energy net effectiveness of passive solar collection area is typically greater than the monetary cost effectiveness of such area.

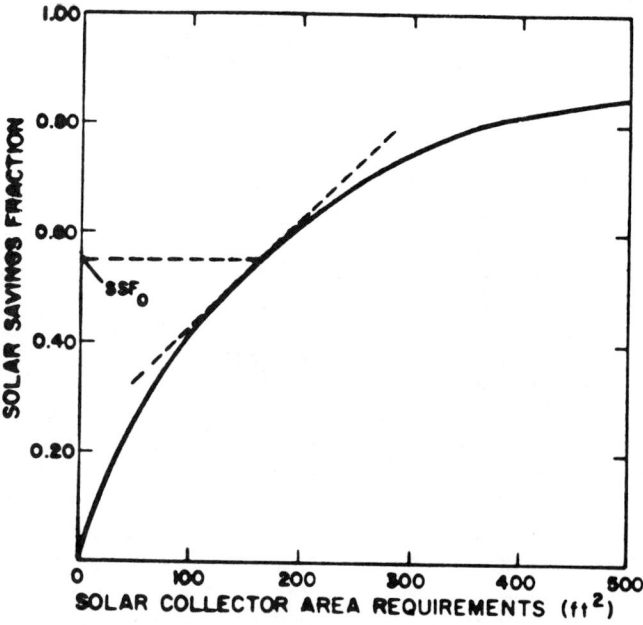

L-6: Typical solar collector areas required to achieve different values of solar savings fraction.

Figures L-7 and L-8 show only *heat supply* costs. A designer's overall objective, though, is to minimize the combined cost of heat supply and conservation. Figure L-9 shows the effect of variations in building load (L)* on these costs. Decreasing a building's thermal load has the dual effect of increasing insulation expenses and decreasing the heat supply cost (HSC). The combined cost has a minimum value at a point corresponding to the building's optimum thermal load. Although figure L-9 shows the building's thermal load as a single continuous variable, it is actually the sum of component loads, including walls, ceiling, windows, doors, and infiltration. In the step-by-step procedure presented in the next section, insulation levels are optimized separately for each portion of the building shell.

In an optimally designed solar heating system, the collection area is set such that an additional square foot of area costs exactly as much as the additional life-cycle fuel savings associated with it. This condition corresponds to a particular slope on the performance curve, as shown in figure L-6. The slope that determines this optimal design point is governed by the ratio of the system's cost per unit of collection area to the unit cost of fuel. Mathematically, this condition is satisfied when:

$$D = EP$$

where:

$$D = \frac{\Delta F}{\Delta (A/L)}$$

and:

$$EP = \frac{VC \times E_1 \times 10^6}{FP \times DD \times E_2}$$

D is related to the slope of the performance curve in figure L-6, and EP is an economic parameter equal to the value of D and corresponding to minimum heat supply cost (e.g., HSC_1 in figure L-7).

This formula is convenient because all performance related factors are included in D, which is a numerical value that relates a change in solar savings fraction to a change in the inverse of the load collector ratio. D values are listed with LCR values in Appendix 12. Conversely, all monetary or physical economic considerations are contained in the economic parameter EP; any changes in assumed economic conditions affect only the term EP. With this simple separation of terms, the optimal system size can be determined by finding the value of SSF at the point at which D equals EP. A method for determining optimum values of SSF and LCR is included in the procedure outlined below.

---

*L is the same as the building load coefficient (BLC), used earlier in this book.

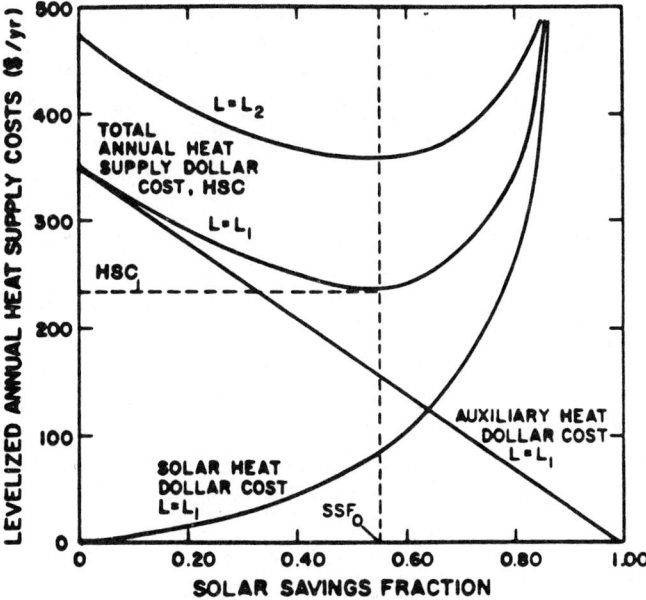

L-7: Annual heat supply costs for a typical building. The values on the curve marked "total annual heat supply dollar cost" are the sums of the corresponding values on the curves marked "solar heat dollar cost" and "auxiliary heat dollar cost." The three curves just named all apply to the load $L = L_1$. The uppermost curve, marked $L = L_2$, shows the situation for a larger load. The minimum dollar cost for each load (only HSC_1—the minimum dollar cost for $L = L_1$—is drawn in) is obtained at the intersection of the total dollar cost curve for the load and the dollar-optimum solar savings fraction (SSF_o) line (the vertical dashed line).

L.5 COST OPTIMIZATION PROCEDURE

The process of jointly optimizing building insulation levels, the size of the passive solar energy system, and auxiliary energy usage follows a straightforward step-by-step procedure. The steps are summarized below; detailed instructions follow.

Step 1. Plot D vs. LCR.
Step 2. Calculate the economic parameter (EP).
Step 3. Determine optimal values of LCR and SSF.
Step 4. Calculate the unit cost of the load (ICH).
Step 5. Select optimum conservation levels for building components (walls, ceiling, windows, doors, and infiltration levels).
Step 6. Determine the building's thermal load (L) and the associated passive solar collection area (A).
Step 7. Determine LCC of the conventional and optimized designs, and calculate the net present value of savings (NPV).

Step 1. Plot D vs. LCR.

Values of D are tabulated in Appendix 12 for the six simple (unmixed) system types considered in this part of the handbook. When applicable, these values can be plotted directly. For mixed systems, or systems with other-than-R9 nighttime insulation, D functions must be plotted for several of the listed system types, and the designer must then interpolate among the curves. This interpolation procedure is directly analogous to that illustrated for SSF in figures L-10 and L-11.

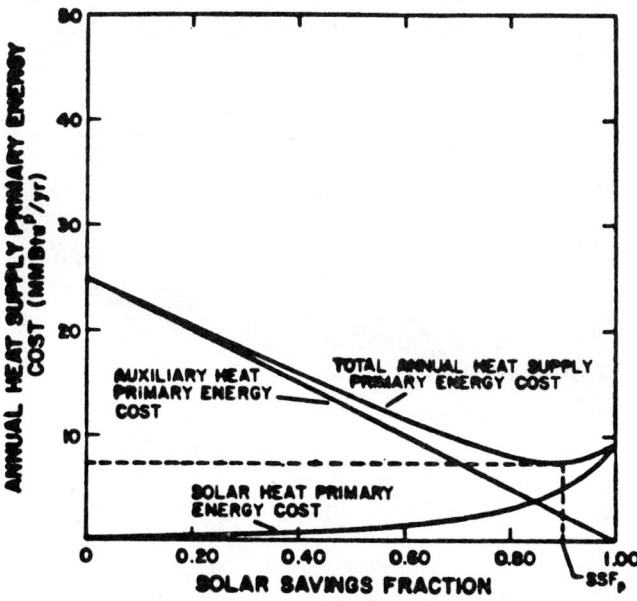

L-8: The annual cost of primary energy for heat supply, expressed as a function of solar savings fraction. The energy optimum is at a much higher value of SSF than the dollar optimum.

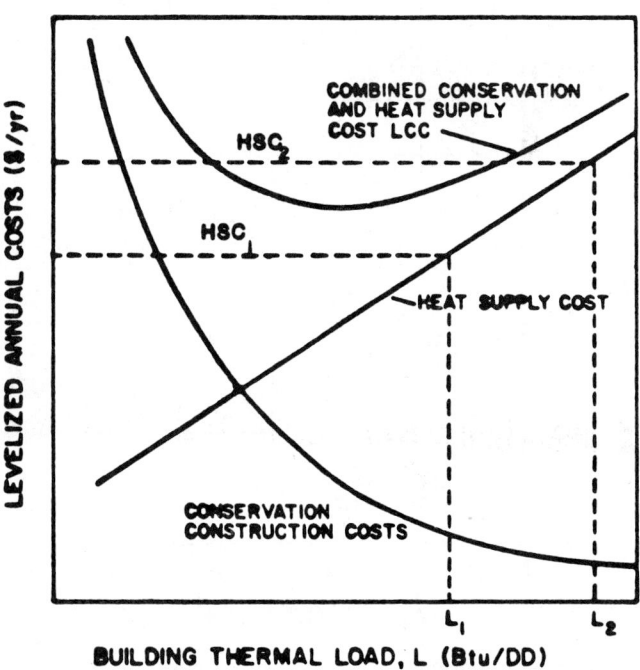

L-9: Annual cost as a function of a building's thermal load. The combined conservation and heat supply cost is the sum of the two components; its minimum value is reached at an optimum load that is somewhat less than L_1.

Economic Analysis / 129

To proceed with the Dodge City example (a 60% waterwall, 40% direct-gain system, with R4 nighttime insulation), first plot *D* vs. *LCR* for the waterwall and direct-gain systems, with and without night insulation. This is illustrated in figure L–12. For each system type, values for *D* and *LCR* that correspond to the same solar savings fraction (*SSF*) are used for the plots. For example, Appendix 12 indicates that a waterwall system with *SSF* = 0.40 requires an *LCR* of 27 (see page 264). The value of *D* for a waterwall with *SSF* = 0.40 is 8.1; the point (*LCR* = 27, *D* = 8.1) is then plotted in figure L–12.

After the four curves have been plotted, a curve must be interpolated between the waterwall and direct-gain curves to account for the 60%–40% system mix. Figure L–13 indicates that R4 nighttime insulation provides 78% of the effect of R9 nighttime insulation. In generating the final *D* curve, you must interpolate between the two existing mixed system curves in order to account for the change in nighttime insulation R-value to a value of R4. This is shown in figure L–14 (cf. figure L–11).

Step 2. Calculate *EP*.

To calculate *EP*, you must define the parameter values of the formula:

$$EP = \frac{VC \times E_1 \times 10^6}{FP \times DD \times E_2}$$

a. Determine *VC* from the cost-evaluation section.

b. Specify *FP*, corrected for the furnace efficiency or coefficient of performance (*COP*). Common use conversions are listed in figure L–15. To get $/MMBtu or MMBtu^p/MMBtu corrected for the furnace effi-

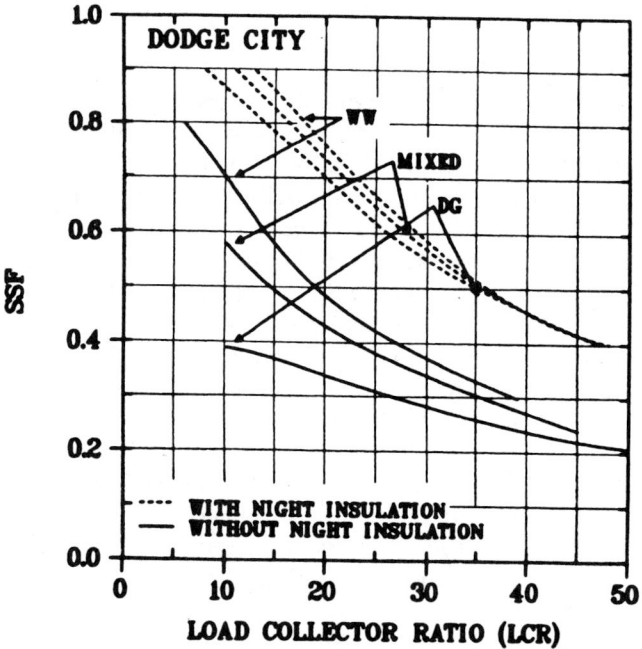

L–10: Generation of a performance curve for a mixed system—here, 40% direct gain and 60% waterwall, located in Dodge City, Kansas.

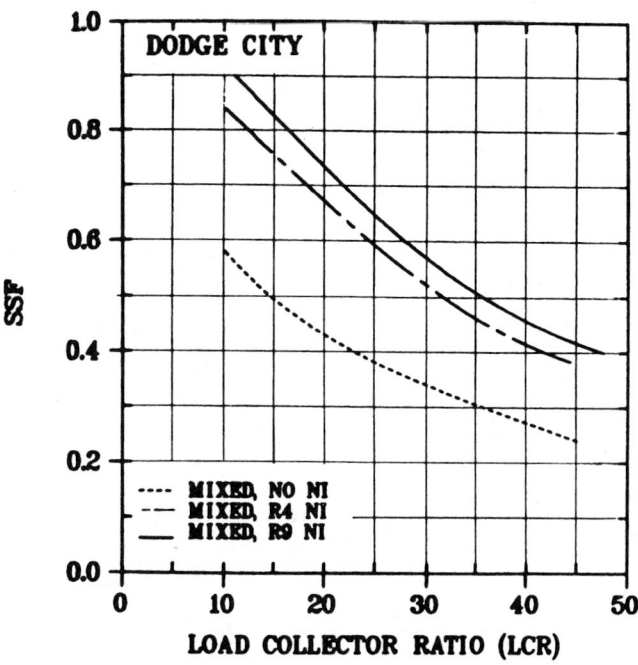

L–11: Performance curves for mixed systems in Dodge City, Kansas. The upper and lower curves are identical to those labeled "mixed" in figure L–10; the curve for R4 nighttime insulation was interpolated, based on a *Y* value of 0.78.

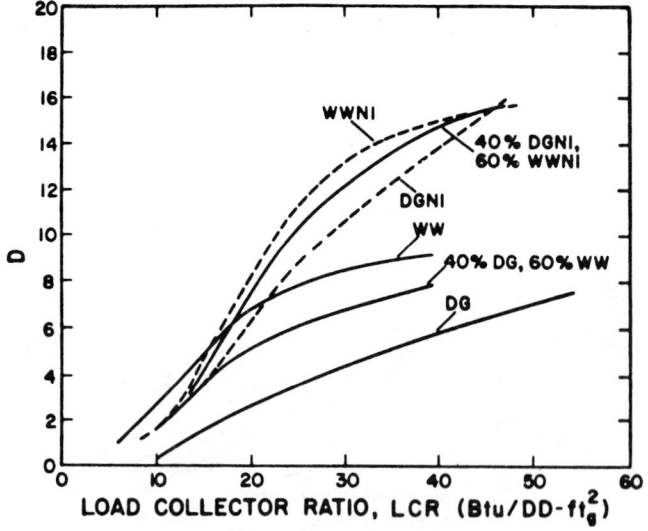

L–12: The derivative function *D* plotted as a function of the load collector ratio. This plot shows how to determine *D* in a mixed system—here, 40% direct gain and 60% waterwall. The procedure is analogous to that used in figure L–10 to determine *SSF* in a mixed system. Notice that the interpolation should be done vertically—that is, along the lines that have constant load collector ratio.

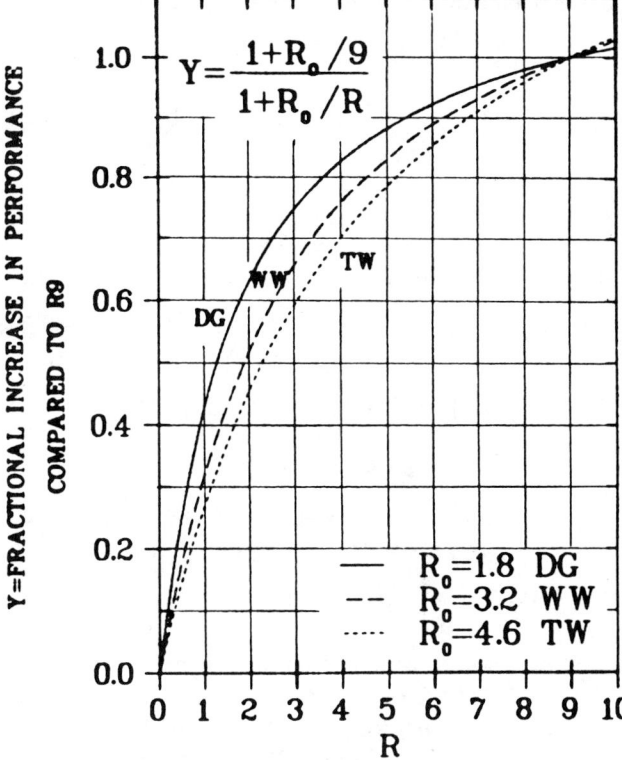

L-13: Effectiveness of various nighttime insulations in comparison to R9 nighttime insulation. The improvement factor (Y) is given by an equation in which R_0 (a constant) is given for three different passive types: direct gain (DG), waterwall (WW), and Trombe wall (TW).

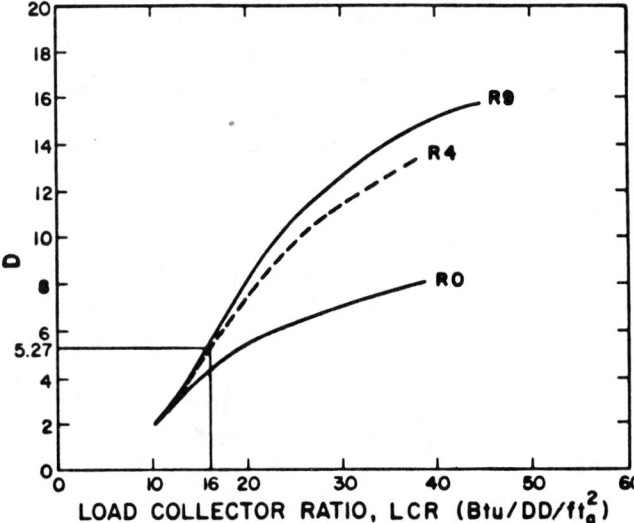

L-14: Determining the derivative function D for a system equipped with nighttime insulation other than R9. Again, the interpolation is done along lines of constant load collector ratio. The procedure is analogous to that used in figure L-11 to determine SSF in an R4 system.

Caution: See the explanation accompanying Appendix 13 of the limiting assumptions underlying the FCR values. More detailed calculating procedures for FCR are explained in Appendix 10.

ciency, or coefficient of performance (COP), divide the above dollar or primary energy unit cost by the corresponding COP.

c. Specify DD—the 65°F-base heating degree-days listed by location in the LCR tables in Appendix 12.

d. Specify E_1. For monetary analysis, FCR must be determined. Determine values for the mortgage or loan interest rate (i), loan term (T) in years, and period of financial analysis or ownership period (PT) in years. Specific values for FCR are contained in Appendix 13 at the end of this section. A series of tables are given for two cases: first, the passive solar heating system has a resale of one-half its original cost. Value by the end of the analysis period ($V = 0.5$); and second, all of the passive system cost is recovered in the resale value by the end of the analysis period ($V = 1.0$). FCR is extremely sensitive to this parameter, especially when the ownership period is short. For either case, each subset of tables corresponds to a value of i that is between 0.06 and 0.15, in increments of 0.01. Choose the table that corresponds to the specified i; then find the value of FCR corresponding to the matched pair of T and PT. For example, if a passive solar home is financed with a 30-year ($T = 30$), 13% ($i = 0.13$) mortgage, and the ownership period is expected to be 8 years ($PT = 8$)—at which point the passive components will be appraised at full cost ($V = 1.00$)—Appendix 13 indicates an FCR value of 0.075.* For energy analysis, use $E_1 = 1/SL$.

e. Specify E_2. In monetary analysis, $E_2 = FF$. FF is a fuel cost leveling factor that accounts for escalating fuel costs over the analysis period. Define the annual fuel escalation rate (e). Find the appropriate list for e in Appendix 14; then, within that table, find the value for FF that corresponds to the matched pair of i and PT specified above. For energy analysis, use $E_2 = 1$.

f. Calculate $EP = \dfrac{VC \times E_1 \times 10^6}{FP \times E_2 \times DD}$

To get $/MMBtu for	multiply	by
Gas	$/therm	10.00
Oil	$/gallon	7.21
Electric Resistance	$/kWh	293.00
To get MMBtu^p/ MMBtu for	use the following reference values	
Gas	1.1 MMBtu^p/MMBtu	
Oil	1.2 MMBtu^p/MMBtu	
Electric Resistance	4.07 MMBtu^p/MMBtu	

(Note: MMBtu^p denotes MMBtu of primary source energy.)

L-15: Factors used in determining fuel price (FP).

Economic Analysis / 131

Example Case for Monetary Analysis

Returning now to the Dodge City example,

a. $VC = \$15.00/\text{ft}^2$
b. $FP = \$0.05/\text{kWh} \times \left(\dfrac{293 \text{ kWh}}{\text{MMBtu}}\right) \div 1.0 \,[\text{the } COP]$
   $= \$14.65/\text{MMBtu}$
c. $DD = 5{,}046$
d. $i = 0.09$, $T = 30$, $PT = 8$, $V = 1.0$
   From Appendix 13, $FCR = 0.0360$
e. $e = 0.08$, $i = 0.09$, $PT = 8$
   From Appendix 14, $FF = 1.387$
f. $EP = \dfrac{VC \times FCR \times 10^6}{FP \times FF \times DD}$

$EP = \dfrac{15 \times 0.0360 \times 10^6}{14.65 \times 1.387 \times 5{,}046}$

$EP = 5.27 \text{ Btu}/DD/\text{ft}_g^2$

Example Case for Primary Energy Analysis

Based on data recorded at the Center for Advanced Computation (CAC) in Champaign-Urbana, Illinois, primary energy requirements per square foot of passive solar collection area were determined for the 60% water-wall, 40% direct-gain design. On a separate issue, CAC analysis indicates that for every 1 Btu of electrical energy delivered to the home, an average of 4.07 Btu of primary energy were consumed in the process. Following the procedure as before to calculate EP:

a. $VC = 0.224 \text{ MMBtu}^p/\text{ft}_g^2$
b. for electrical resistance heating,
   $FP = 4.07 \text{ Btu}^p/\text{Btu} / 1.0 \,[\text{the } COP]$
   $= 4.07 \text{ MMBtu}^p/\text{MMBtu}$
c. $DD = 5{,}046$
d. $E_1 = 1/30 = 0.0333$
e. $E_2 = 1$
f. $EP = \left(\dfrac{0.224 \times 0.0333 \times 10^6}{4.07 \times 1 \times 5{,}046}\right) = 0.3632 \text{ Btu}/DD/\text{ft}_g^2$

Step 3. Determine LCR_o and SSF_o.

Applying the condition given in the equation $D = EP$ in section L.4, locate the value $EP = 5.27$ (monetary example) on the D scale of figure L-14. The corresponding LCR value is shown to be approximately 16. Figure L-11 indicates a corresponding solar savings fraction of $SSF_o = 0.74$.

For the net energy example, EP was calculated to be $EP = 0.36$. Since the portion of the $D(F)$ versus LCR curve in figure L-14 (and the SSF versus LCR curve in figure L-11) corresponding to $EP = 0.36$ is lacking, consider selecting $LCR = 10$ as the design curve end point. Figure L-11 indicates a corresponding $SSF = 0.84$. Since values of LCR less than 10 are difficult to obtain physically, architectural constraints will determine the collection area in this case.

Step 4. Determine the incremental cost of heat supply (ICH) per unit of load.

Use the following formula:

$$ICH = \left(\dfrac{VC}{LCR_o}\right) + \left(\dfrac{FP \times E_2 \times DD \times (1 - SSF_o)}{E_1 \times 10^6}\right)$$

Monetary Example

Using the results from Step 3 with $LCR_o = 16$ and $SSF_o = 0.74$ for the R4 mixed design in Dodge City:

$$ICH = \left(\dfrac{15.0}{16}\right) + \left(\dfrac{14.65 \times 1.387 \times 5{,}046 \times (1 - 0.74)}{10^6 \times 0.036}\right)$$

$ICH = 0.9375 + 0.740 = \$1.678/\text{Btu}/DD$

Primary Energy Example

$$ICH = \left(\dfrac{0.224}{10}\right) + \left(\dfrac{4.07 \times 1.0 \times 5{,}046 \times (1 - 0.84)}{0.0333 \times 10^6}\right)$$

$ICH = 0.0224 + 0.0986 = 0.121 \text{ MMBtu}^p/\text{Btu}/DD$

Step 5. Select optimum conservation levels.

Compile a list of conservation options, categorized by building part (with gross areas), type of construction, U-value, and construction cost ($/\text{ft}^2$ for monetary economic analysis, and $\text{MMBtu}^p/\text{ft}^2$ for physical economic analysis). Determine the net unit costs for each option and choose the lowest cost options. Refer to figure L-16 for an example of this procedure.

a. Determine the surface areas of walls, ceilings, windows, and doors other than south exposures, and the total heated volume of the home.
b. Through local contractors, building component manuals, cost manuals, etc., determine the U-value (in $\text{Btu}/\text{hr}/\text{ft}^2/°\text{F}$) and the construction cost (in $/\text{ft}^2$) of conservation options. For net energy analysis, refer to manuals on energy cost of materials. The most extensive listing has been compiled by the CAC. For each conservation option, determine the primary energy input per square foot ($\text{MMBtu}^p/\text{ft}^2$).
c. Determine net unit costs for each conservation measure, and choose the lowest cost option for each building component. Net square foot costs (NC) are determined by summing the square-foot conservation construction cost (CC) with the corresponding square-foot cost of heat supply (ICH), determined in Step 4. Use the formula:

$$NC = CC + (24 \times U \times ICH)$$

d. For each building component, choose the conservation option with the lowest net unit cost.

Building Shell Component	Surface Area	Type	U-Value	Conservation Cost (CC)		Net Unit Cost (NC)	
				Monetary (in \$/ft^2)	Energetic (in MMBtu^p/ft^2)	Monetary (in \$/ft^2)	Energetic (in MMBtu^p/ft^2)
Windows	66	Single Casement	1.64	20.0	.0154	86.05	4.78
		Double	1.09	24.3	.0309	68.20	3.20
		Double + Storm	0.82	29.20	.0417	[62.20]	[2.42]
Doors	42	Solid Core	1.42				
Opaque Walls	772	3½-inch-thick Batt	0.085	0.571	.0323	3.99	0.279
		5½-inch-thick	0.051	0.696	.0347	2.75	0.183
		9½-inch-thick	0.032	0.983	.0402	[2.27]	[.133]
Ceiling	1512	3½-inch-thick wool (minimum)	0.068	0.152	.0681	2.89	0.266
		5½-inch-thick	0.044	0.263	.0700	2.04	0.198
		9½-inch-thick	0.026	0.484	.0739	1.53	0.149
		14-inch-thick	0.018	0.735	.0702	[1.46]	0.131
		18-inch-thick	0.014	0.957	.0821	1.52	[.123]

U-values for windows and doors include losses due to infiltration.

L-16: Conservation options and net unit costs. (U-values for windows and doors include losses due to infiltration.)

Optimization Example

The reference house is a single-story building with an east-west dimension of 54 feet and a north-south dimension of 28 feet (a total area of 1,512 square feet). The ceiling height is 8 feet and the nonsouth-facing window area constitutes 7.5% of the nonsouth-facing wall area. The house has two doors, each measuring 3 feet by 7 feet. Suppose that conventional construction practice is to use single-pane casement windows, 3½-inch-thick batt R11 wall insulation, R20 mineral wool ceiling insulation, and R10 perimeter insulation. Optimization of conservation levels will indicate whether or not conventional construction practice provides the lowest possible value of combined heating and energy conservation costs. Parts a through c of Step 5 are summarized below; they are compiled in figure L-16.

a. Surface Areas

The various dimensions of the reference house are computed in figure L-17.

b. U-Values and Cost

In order to determine optimum window and door constructions, one must include infiltration losses in the calculated U-values. These losses can be determined by using the crack-length method. Figure L-16 summarizes U-values and costs for various construction options.

```
Nonsouth
Windows (28 + 28 + 54) × 8 × 0.075    = 66 ft²
Doors         3 × 7 × 2                 = 42 ft²
Nonsouth
Opaque
Walls  {[(28 + 28 + 54) × 8] − 66} − 42 = 772 ft²
Ceiling       28 × 54                   = 1,512 ft²
Volume        28 × 54 × 8               = 12,096 ft³
Perimeter  (28 + 54) × 2                = 164 ft
```

L-17: Surface, perimeter, and volume dimensions of the reference house.

c. Determine Net Unit Costs (NC)

As an illustrative example, consider nonsouth opaque walls constructed with 5½-inch-thick fiberglass batt insulation. As shown in figure L-16, the corresponding U-value is 0.051 Btu/hr/ft^2/°F, with monetary costs of \$0.696/ft^2 and calculated primary energy costs of 34,670 Btu^p/ft^2. Net unit cost (NC) is calculated as:

Monetary: $NC = 0.696 + (24 \times 0.051 \times 1.678)$
$= \$2.75/\text{ft}^2$

Energetic: $NC = 0.0347 + (24 \times 0.051 \times 0.121)$
$= 0.183 \text{ MMBtu}^p/\text{ft}^2$

The results for the other conservation options are listed in figure L-16. Only one option for doors (solid core wood) is assumed. A base infiltration rate of three-quarters of an air change per hour (ACH) is assumed, and no weatherstripping or caulking options are analyzed. One-third of the air change rate is taken into account through the U-values listed for windows and doors. The remaining one-half ACH is due to other infiltration leaks in the home.

The monetary optimization results suggest a home design with double glazing plus storm windows, R30 walls (9½-inch-thick insulation), and R56 ceiling (14-inch-thick insulation). The net energy optimization results suggest an increase in the thickness of the ceiling insulation to 18 inches, or R70. Triple glazing and thicker wall insulation may have been warranted from the standpoint of energy optimization, but these options were not analyzed here.

Step 6. Calculate building load and collector area.

With the results from Step 5, one now can determine the final building load (in Btu/DD) by summing conductance losses and infiltration losses.

Conductance Losses

Conductance losses are equal to the losses through windows (L_g), doors (L_d), walls (L_w), roof (L_r), and perimeter (L_p). For all surfaces but the perimeter, use the general formula:

$$L \text{ (in Btu/DD)} = 24(°F/hr/DD) \times U \text{ (in Btu/hr/ft}^2/°F) \times A \text{ (in ft}^2)$$

For the perimeter, use the following formula:

$$L_p \text{ (in Btu/DD)} = \frac{100 \times \text{perimeter}}{(\text{number of stories}) \times (\text{R-value of perimeter insulation} + 5)}$$

Therefore, conductance losses for the monetary and primary energy analysis are calculated as:

$L_g = 24 \times 0.82 \times 66 = 1,299$
$L_d = 24 \times 1.42 \times 42 = 1,431$
$L_w = 24 \times 0.032 \times 772 = 593$
$L_r = 24 \times 0.018 \times 1,512 = 653$ for monetary analysis
$L_r = 24 \times 0.014 \times 1,512 = 508$ for primary energy analysis
$L_p = \left(\frac{100 \times 165}{1 \times (10+5)}\right) = 965$

Total conductance losses = 4,941 Btu/DD for monetary analysis
= 4,796 Btu/DD for primary energy analysis

L-18: Values of air density ratio (ADR) for different elevations. Sea level air density is 0.075 lb/ft³.

Infiltration Losses

Infiltration losses through windows and doors were included in the conductance loss U-values in figure L-16. However, additional infiltration losses can occur through wall outlets, soleplate, duct systems, fireplaces, vents, and so forth. These losses can be estimated by the air change method.

To calculate infiltration losses (L_i), use the formula:

L_i = Heated volume (in ft³) × ACH (air changes/hr) × 0.432 × ADR (air density ratio)

The ADR can be found by using figure L-18.

$$L_i = 12,096 \times 0.50 \times 0.432 \times 0.91 = 2,378 \text{ Btu/DD}$$

Calculate the total building load (L) and the collector area (A) for the optimized design.

For monetary analysis, total losses (L) sum to 7,319 Btu/DD (4,941 conductance + 2,378 infiltration). For the primary energy analysis total losses (L) sum to 7,174 Btu/DD (4,796 conductance + 2,378 infiltration). Collector area can be determined by dividing L by the optimum LCR value found in Step 3. That is:

$$A = \frac{L \text{ (in Btu/DD)}}{LCR \text{ (in Btu/DD/ft}^2)}$$

In the monetary example:

$$A = \left(\frac{7,319}{16}\right) = 457 \text{ ft}^2$$

In the primary energy example:

$$A = \left(\frac{7,174}{10}\right) = 717 \text{ ft}^2$$

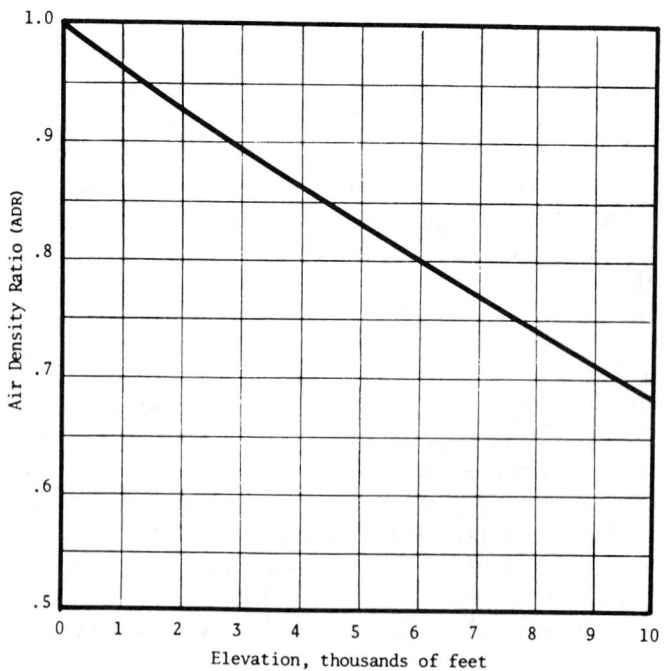

INTEREST RATE : R

YEAR	0.04	0.05	0.06	0.07	0.08	0.09	0.10	0.11	0.12	0.13	0.14	0.15
1	0.962	0.952	0.943	0.935	0.926	0.917	0.909	0.901	0.893	0.885	0.877	0.870
2	1.886	1.859	1.833	1.808	1.783	1.759	1.736	1.713	1.690	1.668	1.647	1.626
3	2.775	2.723	2.673	2.624	2.577	2.531	2.487	2.444	2.402	2.361	2.322	2.283
4	3.630	3.546	3.465	3.387	3.312	3.240	3.170	3.102	3.037	2.974	2.914	2.855
5	4.452	4.329	4.212	4.100	3.993	3.890	3.791	3.696	3.605	3.517	3.433	3.352
6	5.242	5.076	4.917	4.767	4.623	4.486	4.355	4.231	4.111	3.998	3.889	3.784
7	6.002	5.786	5.582	5.389	5.206	5.033	4.868	4.712	4.564	4.423	4.288	4.160
8	6.733	6.463	6.210	5.971	5.747	5.535	5.335	5.146	4.968	4.799	4.639	4.487
9	7.435	7.108	6.802	6.515	6.247	5.995	5.759	5.537	5.328	5.132	4.946	4.772
10	8.111	7.722	7.360	7.024	6.710	6.418	6.145	5.889	5.650	5.426	5.216	5.019
11	8.760	8.306	7.887	7.499	7.139	6.805	6.495	6.207	5.938	5.687	5.453	5.234
12	9.385	8.863	8.384	7.943	7.536	7.161	6.814	6.492	6.194	5.918	5.660	5.421
13	9.986	9.394	8.853	8.358	7.904	7.487	7.103	6.750	6.424	6.122	5.842	5.583
14	10.563	9.899	9.295	8.745	8.244	7.786	7.367	6.982	6.628	6.303	6.002	5.724
15	11.118	10.380	9.712	9.108	8.559	8.061	7.606	7.191	6.811	6.462	6.142	5.847
16	11.652	10.838	10.106	9.447	8.851	8.313	7.824	7.379	6.974	6.604	6.265	5.954
17	12.166	11.274	10.477	9.763	9.122	8.544	8.022	7.549	7.120	6.729	6.373	6.047
18	12.659	11.690	10.828	10.059	9.372	8.756	8.201	7.702	7.250	6.840	6.467	6.128
19	13.134	12.085	11.158	10.336	9.604	8.950	8.365	7.839	7.366	6.938	6.550	6.198
20	13.590	12.462	11.470	10.594	9.818	9.129	8.514	7.963	7.469	7.025	6.623	6.259
21	14.029	12.821	11.764	10.836	10.017	9.292	8.649	8.075	7.562	7.102	6.687	6.312
22	14.451	13.163	12.042	11.061	10.201	9.442	8.772	8.176	7.645	7.170	6.743	6.359
23	14.857	13.489	12.303	11.272	10.371	9.580	8.883	8.266	7.718	7.230	6.792	6.399
24	15.247	13.799	12.550	11.469	10.529	9.707	8.985	8.348	7.784	7.283	6.835	6.434
25	15.622	14.094	12.783	11.654	10.675	9.823	9.077	8.422	7.843	7.330	6.873	6.464
26	15.983	14.375	13.003	11.826	10.810	9.929	9.161	8.488	7.896	7.372	6.906	6.491
27	16.330	14.643	13.211	11.987	10.935	10.027	9.237	8.548	7.943	7.409	6.935	6.514
28	16.663	14.898	13.406	12.137	11.051	10.116	9.307	8.602	7.984	7.441	6.961	6.534
29	16.984	15.141	13.591	12.278	11.158	10.198	9.370	8.650	8.022	7.470	6.983	6.551
30	17.292	15.373	13.765	12.409	11.258	10.274	9.427	8.694	8.055	7.496	7.003	6.566

L-19: Series present worth factor (*SPWF*) table. This table is based on the variables *PT* (years of ownership) and *R* (interest rate).

Step 7. Determine the *LCC* of the conventional and optimized designs, and calculate the net present value of savings (*NPV*).

First, calculate the total building load (*L*) of the conventional reference home:

$$L_g = 24 \times 1.64 \times 66 = 2{,}598$$
$$L_d = 24 \times 1.42 \times 42 = 1{,}431$$
$$L_w = 24 \times 0.085 \times 772 = 1{,}575$$
$$L_r = 24 \times 0.044 \times 1{,}512 = 1{,}597$$
$$L_p = \left(\frac{100 \times 164}{1 \times (10 + 5)}\right) = 965$$
$$L_i = 12{,}096 \times 0.50 \times 0.432 \times 0.91 = 2{,}378$$

Total = 10,543 Btu/DD

Second, calculate the conservation costs (*CC*) of the conventional home. Conservation elements that are similar for both the conventional homes and the optimized homes (i.e., the doors and perimeter insulation, in this example) are *not* included in *CC*. *CC* is determined by multiplying each component surface area by the corresponding $/ft² unit cost shown in figure L-16, and then summing the products:

$$CC = \underset{\text{windows}}{(66 \text{ ft}^2 \times \$20/\text{ft}^2)} + \underset{\text{walls}}{(772 \text{ ft}^2 \times \$0.571/\text{ft}^2)}$$
$$+ \underset{\text{ceiling}}{(1{,}512 \text{ ft}^2 \times \$0.263/\text{ft}^2)}$$
$$CC = 1{,}320 + 441 + 398 = \$2{,}159$$

Third, calculate the conservation costs (*CC*) and solar costs (*FC* + (*VC* × *A*)) of the optimized home. Refer to figure L-16 for unit conservation costs of the optimized design (monetary):

$$CC = \underset{\text{windows}}{(66 \text{ ft}^2 \times \$29.20/\text{ft}^2)} + \underset{\text{walls}}{(772 \text{ ft}^2 \times \$0.983/\text{ft}^2)}$$
$$+ \underset{\text{ceiling}}{(1{,}512 \text{ ft}^2 \times \$0.735/\text{ft}^2)}$$
$$CC = 1{,}927 + 759 + 1{,}111 = \$3{,}797$$

With *FC* = 0 and *VC* = $15/ft² for the 40% direct-gain, 60% waterwall design, passive solar add-on costs are:

$$FC + (VC \times A) = 0 + (\$15/\text{ft}^2 \times 457 \text{ ft}^2) = \$6{,}855$$

Fourth, calculate *LCC* for both the conventional (*LCC_c*) and optimized passive solar (*LCC_s*) designs.

Restating the original equation for *LCC*:

$$LCC = \{[FC + (VC \times A)] \times E_1\} + [L \times (1 - F) \times DD \times FP \times E_2 \div 10^6]$$

The two calculations, then, are as follows:

Conventional (LCC_c):
$$LCC_c = \{[0 + (15 \times 0)] \times 0.036\} + (2{,}159 \times 0.036)$$
$$+ [10{,}543 \times (1 - 0) \times 5{,}046 \times 14.65$$
$$\times 1.387 \div 10^6]$$
$$LCC = 0 + 78 + 1{,}081 = \$1{,}159/\text{yr}$$

Optimized Solar (LCC_s):
$$LCC_s = \{[0 + (15 \times 457)] \times 0.036\} + (3{,}797 \times 0.036)$$
$$+ [7{,}319 \times (1 - 0.74) \times 5{,}046 \times 14.65$$
$$\times 1.387 \div 10^6]$$
$$LCC_s = 247 + 137 + 195 = \$579/\text{yr}$$

Comparing the values for *LCC*, one can see that the uniform annual cost of the optimized passive solar home design is about half that of the conventional home design.

Fifth, calculate the net present value of savings (*NPV*).

NPV is an indication of the dollar savings that are due to passive solar and upgraded conservation levels over the ownership period. *NPV* includes fuel savings, tax savings, and the profit realized at the time of home resale that are due to the passive and conservation additions. The formula for *NPV* is:

$$NPV = (LCC_c - LCC_s) \times SPWF$$

SPWF, a series present worth factor, can be found in figure L-19. The *SPWF* value is obtained by using the variables *PT* and *i*, specified above. In the Dodge City example, for *i* = 0.09 and *PT* = 8, *SPWF* from figure L-19 equals 5.535. Therefore:

$$NPV = (1{,}159 - 579) \times 5.535 = \$3{,}210$$

This indicates that over the eight-year ownership period, the discounted value of savings will be $3,210.

Part Three

Passive Solar Design Performance

*Los Alamos National Laboratory
University of California*

*Robert W. Jones
J. Douglas Balcomb
Claudia Kosiewicz
Gloria Lazarus
Robert McFarland
William Wray*

Review of Performance Analysis Methods

M.1 TERMINOLOGY

A thorough survey of the terminology used in passive solar design analysis is provided in Part Two of this handbook; the present section reviews key terms. For further explanation of terms, refer to the Glossary.

Two design parameters of primary importance in determining the solar heating performance of a building are the solar projected area and the building load coefficient. Implicit in the definitions (given below) of these is a subdivision of the building's vertical wall areas according to their solar or nonsolar functions—that is, into the solar wall, and the other walls.

The term "solar wall" refers collectively to all glazed building walls for which solar gains are to be accounted. In the northern hemisphere, solar walls should be oriented to face due south or in a direction within 30 degrees east or west of due south. The solar wall includes all major solar features such as thermal storage walls, direct-gain windows, attached sunspaces, or a combination of these features. There is some ambiguity regarding features that are less obviously "solar" in nature—especially ordinary windows—but as a general rule, windows should be included as direct gain elements of the solar wall unless they are oriented more than 90 degrees east or west of true south. The solar aperture is the portion of the solar wall that is actually covered by glazing. Its area—the net glazing area—is smaller than the area of the solar wall by an amount equal to the area of the window mullions and other framing members that shade the wall.

The solar projected area (A_p) is the principal net glazing area projected onto a vertical plane. For vertical-glazing systems, this is simply the net glazing area, or the collection area (A_c) as defined in Part Two of this handbook. For tilted-glazing systems, which account for most sunspace systems, the solar projected area is the principal net glazing area projected onto a vertical plane—that is, the net glazing area multiplied by the sine of the tilt angle. The term "principal" signals that any east- or west-facing glazings that exist in sunspace end walls are excluded from the projected area. A separate procedure is used to account for these glazings.

Defining projected area, rather than actual area, as the primary area parameter is an arbitrary convention; however, the level of performance provided by typical sunspace configurations is more dependent on their projected area than on their actual area. Projected area is also convenient for comparing alternative systems on the same solar wall. For example, a sunspace and a Trombe wall that occupy the same area of solar wall may have different glazing areas but the same projected area. Actual area can be used with equal success, as long as the convention is applied consistently throughout an analysis. In order to maintain the actual area convention in the calculation of quantities that were previously referenced to projected area (including load collector ratio and solar radiation absorbed per unit of projected area), modifications of tabulated quantities—such as those in figures 18-7 and 18-10, and in Appendix 21—are required.

The building load coefficient (BLC) assesses the nonsolar parts of the building—that is, everything but the solar wall. It is the net load (or heat loss) of the nonsolar portions of the building per day per degree of indoor-outdoor temperature difference. In practice, BLC is usually calculated as the total steady-state building loss coefficient minus the solar wall conduction loss coefficient. The unit of measurement for BLC is the Btu/DD (Btu per degree-day). The word "net" indicates that heat losses through the solar wall are excluded. In some other publications, the term "net load coefficient" (NLC) has the same meaning as BLC.

The building load over any given time period depends on BLC and the actual indoor and outdoor temperatures experienced over the time period. The buiding load used in design analysis is the net reference load. The term "reference" indicates that the indoor temperature is taken to be a constant reference temperature—normally 65°F.

The effect of the outdoor temperature on the building load is expressed in terms of daily mean temperatures, each computed as the mean of the daily minimum and daily maximum temperatures. The sum of the differences between the fixed base temperature and the daily mean temperatures over an entire month or year equals the monthly or annual heating degree-days (DD). (Temperature differences are counted only when the outdoor mean temperature is less than the base temperature.) Thus:

$$\text{net reference load} = BLC \times DD$$

where the reference temperature in the net reference load and the base temperature in DD are the same.

The base temperature in the definition of DD is traditionally 65°F; it is to this base that degree-day records are customarily available. The base 65°F corresponds to the 65°F reference temperature in the usual definition of the net reference load. Other base temperatures are sometimes employed in design analysis, however, so DD should be recognized as a quantity whose value is defined relative to an arbitrary base.

The passive solar building is characterized by its load collector ratio (LCR), the ratio of the building load coefficient to the projected area:

$$LCR = \frac{BLC}{A_p}$$

The subject of Part Three of this handbook is the performance of passive solar heating systems in terms of the partial displacement of conventional space heat requirements by solar heat. A design analysis that is limited to this aspect of performance seeks to estimate the auxiliary heat requirements, either monthly or yearly. One aim of passive solar design is to minimize this quantity without impairing comfort or convenience.

The term "solar savings" is used to refer to the calculated extent to which a solar design has reduced the auxiliary heat requirement of the solar building relative to a particular reference building. The reference building is defined as one that is identical in all aspects to the solar building except that it has an energy-neutral wall in place of the solar wall possessed by the solar building. The energy-neutral wall has neither solar gains nor heat losses. The reference building has a constant indoor temperature equal to the heating thermostat setpoint in the solar building (normally 65°F). The total heating energy requirement of the reference building is the net reference load; therefore, the solar savings figure equals the net reference load minus the auxiliary heat requirement of the solar building.

Caution should be exercised in interpreting the solar savings beyond the terms of this definition. For example, the figure for solar savings is *not* equivalent to the solar energy contribution to the actual building's space heating consumption. The solar savings figure is, in fact, smaller than the actual solar contribution because credit is taken for neither solar gains that offset losses through the solar wall nor solar gains that raise the room temperature above the thermostat setpoint. Another common misinterpretation of the term "solar savings" is to associate it with the savings in space-heating requirements achieved by the solar building relative to some "standard" building. This association is correct only if the "standard" building is the same as the reference building defined above—that is, one with an energy-neutral wall in place of the solar wall. Otherwise, it may be preferable to define relative solar savings in terms of a particular definition of a standard building. See section O.2.e for an example.

The solar savings fraction (SSF) is the ratio of the solar savings to the net reference load:

$$SSF = \frac{\text{net reference load} - \text{auxiliary heat}}{\text{net reference load}}$$

or:

$$SSF = 1 - \left(\frac{\text{auxiliary heat}}{\text{net reference load}}\right)$$

Although SSF is the most commonly used performance measure of a solar heating or cooling system, it is subject to misinterpretation as an expression of a system's effectiveness. For example, the cautions mentioned above in connection with solar savings apply to SSF also. Another possible misapplication of SSF is as a measure of comparison between two solar buildings. Such a comparison is only valid when the two buildings have the same building load coefficient. The quantity SSF is primarily useful as an intermediate step in the calculation of the auxiliary heat requirement of the solar building, which is always a valid performance measure:

$$\text{auxiliary heat} = (1 - SSF) \times BLC \times DD$$

M.2 SOLAR LOAD RATIO (SLR) CORRELATIONS

The basis of performance evaluation is computer simulation in which the thermal behavior of a building is represented by a numerical model which has been validated by comparing its predictions with the measured behavior of experimental buildings. Computer simulation is an accurate and versatile tool in design analysis, but it requires a level of technical support (computing equipment and engineering staff) that is not always available or justifiable, given a particular design budget or potential energy savings. Therefore, simplified analytical methods amenable to "hand" calculation have been developed. The basis of these simplified methods is a set of monthly solar load ratio (SLR) correlations.

The quantity SLR is a variable that incorporates sufficient building and site information to allow prediction of an annual auxiliary heat requirement of acceptable accuracy for a particular building type based on building load coefficient (BLC) and the twelve monthly values of SLR. A general definition of SLR is:

$$SLR = Q_s/Q_{load}$$

where:

Q_s = monthly solar radiation input to the solar aperture

Q_{load} = a monthly building load

The specific definitions of Q_s and Q_{load} differ somewhat from one passive system type to another. Refer to Chapters O and P on direct-gain and sunspace systems or Appendix 18 for details.

SLR correlations are correlations of the monthly *SSF*s with the monthly *SLR*s. In order to develop the correlations, an assortment of reference designs is first defined for each building type. Each reference design is a detailed specification of the passive solar features of a building; the one design parameter that remains variable is *LCR*. A large quantity of performance data is then generated by computer simulation for each reference design. The data consists of monthly *SSF*s for several different *LCR* values and for several different cities. Finally, correlations are established between the monthly *SSF*s and the corresponding monthly *SLR*s. These correlations take the form of a specific formula for the monthly *SSF* in terms of the monthly *SLR*. The criterion used to determine the formula is that the mean square deviation in the *annual SSF* is a minimum. Refer to Chapters O and P for the details of the correlations for direct-gain and sunspace systems, or to Appendix 18 for a summary of the correlations.

The monthly *SLR* correlations are used in the calculation of the monthly *SSF*s, and hence also in the calculation of the monthly and annual auxiliary heat. The monthly calculation, or "*SLR* method," is summarized in Chapter Q.

M.3 ANNUAL CALCULATION —THE *LCR* METHOD

Application of the monthly calculation method produces figures that can be used to calculate the annual auxiliary heat and annual *SSF*. The monthly calculation has been performed for 219 locations in the United States and Canada for each reference design, using a variety of *LCR* values. Refer to Chapters O and P for a description of the particular design assumptions made for direct-gain and sunspace systems. The design parameters for all of the system types are summarized in Appendix 21, which also contains the results of the calculations of annual *SSF* versus *LCR*.

The annual calculation, or "*LCR* method," makes use of the *LCR* tables reproduced in Appendix 21 to produce a useful first estimate of annual building performance. The method consists of the following steps:

1. Obtain building information:
   a. Building load coefficient (*BLC*)
   b. Projected area (A_p)
   c. Load collector ratio (*LCR*) (= BLC/A_p)
2. Enter the *LCR* tables (Appendix 21):
   a. Find the desired city
   b. Find the desired reference design
   c. Determine the annual *SSF* by interpolation
   d. Note the annual heating degree-days (*DD*)
3. Calculate the annual auxiliary heat:

$$Q = (1 - SSF) \times BLC \times DD$$

If the building has a mixture of two or more system types, then the following procedure may be used. The projected area (Step 1b) equals the combined projected areas of all of the systems. Steps 2a–2c are performed for each system type; this amounts to assuming that the entire solar wall is first of one type, then the next type, and so forth. The final *SSF* is the sum of each *SSF* determined in Step 2c, weighted according to the respective projected areas of the individual systems, and divided by the weighting factor.* The auxiliary heat is calculated in the normal way (Step 3) using the average *SSF*.

M.4 SENSITIVITY DATA

The monthly and annual performance calculations—the *SLR* and *LCR* methods—are based on specific reference designs. The monthly calculation allows some flexibility in the choice of certain design parameters, but probably not enough to encompass all design variability that will be encountered. The annual method allows no departure at all from the reference designs. To treat off-reference designs, the sensitivity data reproduced in Appendix 22 may be used.

The sensitivity data express the effect on the annual *SSF* of varying one or two design parameters. The data are usually expressed as plots of the annual *SSF* versus one design parameter while the other design parameters are held fixed at their reference values. Dots are plotted on the sensitivity curves to represent the reference conditions. In most cases, each set of sensitivity data is presented for six cities in the United States, chosen to represent wide geographical and climatological ranges. The reference city that is closest in climate and geographical characteristics to the location of interest should be chosen to represent the actual location.

Two applications of sensitivity data are important. One is quantitative, as a final step in a design analysis. In this application, an analysis is performed on the basis of the reference design which most nearly resembles the design of interest. Then, the annual *SSF* is corrected by use of sensitivity data.

---

*For example, to calculate the final *SSF* of a 70% sunspace, 30% direct-gain system, the following equation could be used:

$$\text{Final } SSF = \frac{(7 \times \text{sunspace } SSF) + (3 \times \text{direct-gain } SSF)}{10}$$

The second important application of sensitivity data is qualitative, wherein the data serve as a guide, perhaps very early in the design process, to the relative significance of various design parameters and their preferred values. The discussions of the sensitivity data for direct-gain and sunspace systems in Chapters O and P are in this qualitative vein.

Selected sensitivity data are presented in Chapters O and P to illustrate the discussion. A much more complete body of data is presented in Appendix 21 for routine reference.

M.5 UNITS

English units are used in this volume to be consistent with customary design and engineering practice in the United States. Selected quantities are listed in figure M–1 with their metric (SI) system equivalents.

Quantity	English Unit	Equivalent SI Unit
Energy	Btu	1054 J
Power	Btu/hr	0.2928 W
Energy flux	Btu/hr/ft²	3.152 W/m²
Length	ft	0.3048 m
Area	ft²	0.09290 m²
Volume	ft³	0.02832 m³
Mass	lbm	0.4536 kg
Density	lbm/ft³	16.02 kg/m³
Heat capacity	Btu/°F	1897 J/°K
Specific heat	Btu/lbm/°F	4182 J/kg/°K
Thermal conductivity (k)	Btu/hr/°F/ft	1.729 W/°K/m
Conductance (U)	Btu/hr/°F ft²	5.673 W/°K/m²
Degree days (DD)	°F/day	5/9°K/day
Building load coefficient (BLC)	Btu/°F/day	0.0220 W/°K
Load collector ratio (LCR)	Btu/°F/day/ft²	0.236 W/°K/m²

M–1: Selected quantities, with their English system units and approximate equivalents in SI (Système International) units.

Quantity	English Unit	"SI" Hybrid Equivalents
BLC	Btu/°F/day	1.90 kJ/°C/day
		0.527 kJ/°C/day
LCR	Btu/°F/day/ft²	20.4 kJ/°C/day/m²
		5.67 Wh/°C/day/m²

Note: The prefix k means 1,000. The abbreviation Wh means Watt-hour. To convert BLC in W/°C to kJ/°C/day, multiply by 86.4; and to Wh/°C/day, multiply by 24.

M–2: English system units, with equivalents in "SI" hybrid units, for the quantities BLC and LCR.

The SI unit of temperature, the Kelvin (K), is not as commonly used in design and engineering practice as the degree Celsius (C). To convert a temperature in degrees C to K, add 273.15; no conversion is needed to convert a temperature increment (which is what occurs in combination with other units such as J/C) from C to K. To convert a temperature in degrees F to degrees C, subtract 32 and multiply by ⅝; to convert a temperature increment (which is what occurs in combination with other units such as Btu/°F) in degrees F to degrees C, multiply by ⅝ only.

The English mass unit is most properly the "slug," but the unit "pound-mass" (lbm) is often used instead; it refers to the mass of an object that weighs 1 pound. The distinction between weight and mass is not very important in solar design analysis, so the abbreviation for "pound-mass" is shortened to "lb" in the following chapters.

Insulation levels are usually quoted as R-values, which are thermal resistances in units hr/°F/ft²/Btu. The SI equivalent to an R-value is usually expressed as a U-value, or conductance, in units W/K/m². The relationship is $U = 5.673/R$ where U is the conductance in W/K/m² and R is the R-value in hr/°F/ft²/Btu.

The most customary SI unit for a building heat-loss coefficient is W/°C; therefore, it is natural for BLC to have the unit W/°C. Similarly, a natural unit for LCR is W/°C/m². But it is more convenient in solar design analysis to adopt units that contain days as the time unit because of the frequent occurrence of combinations involving degree-days. The resulting units will not be SI units, but they will be good compromises for application to solar design analysis. See figure M–2.

M.6 COOLING CONSIDERATIONS

The important issues of summertime comfort levels and cooling loads have not been addressed thus far in Part Three; nevertheless *year-round* comfort and energy efficiency is the goal of a balanced passive solar design. A passive solar heating system may aggravate summertime discomfort or increase the summertime cooling load because of excessive solar gains through the solar aperture, but suitable seasonal solar controls can largely alleviate this potential problem. Indeed, some of the elements of a good passive solar design for heating also enhance summertime comfort—namely, mass for thermal storage, and conservation measures in the building envelope. Other important controls are shading and ventilation. It is a significant design challenge to integrate the necessary controls tastefully and effectively with the passive solar heating system and the rest of the building, so that the heating system's performance is not unnecessarily compromised and yet does not significantly aggravate discomfort or cooling loads during the summer.

Performance Analysis Methods / 141

N Balancing Conservation and Solar

Optimizing a building's thermal design is somewhat like climbing a mountain. Seen from a distance, the mountain, covered with trees, seems fairly smooth, and it's easy to make out the overall shape and locate the vicinity of the summit. A hiker on the mountain, though, has an entirely different perspective. He or she can't tell much about the overall shape of the mountain but is absorbed in the detailed surrounding structure, which was invisible to and ignored by the faraway viewer. The hiker may attain the top of a local outcropping and think that the summit has been reached, when it is really still quite a distance higher up.

So it is with the designer trying to find the best thermal design strategy. He or she may waste a lot of time looking for an optimum in the wrong place altogether. All the designer really wants to do is get near the top, knowing that attaining an exact optimum is rather meaningless because costs are not well known and because there are various other uncertainties. The contours of constant life-cycle costs define a surface that is comparable to a rather flat-topped mountain with steep sides. Once the designer is in the right vicinity, only small changes in life-cycle costs can be made, even by relatively large shifts in the level of conservation or passive solar. This is because the performance changes are roughly offset by the cost changes. At this point, the designer will be far more concerned with other factors that determine the design, such as convenience of framing, availability of materials, and aesthetics.

Experience shows that, in most instances, energy economics are not the main factor in the design process. A simple procedure that gives answers in the right ballpark will receive much wider use and thus have a much greater impact on design than a complex procedure that produces more precise answers but will rarely be used.

N.1 CONSERVATION VERSUS SOLAR

An old adage of the solar designer is "First insulate, then use solar." This is good advice, but how much conservation is appropriate before starting to use solar? Should the wall be R19, R25, or R40? How does the answer vary with climate and with type of solar energy system? How does the answer vary with the degree of solar contribution? These questions can all be lumped into a general question, "What is the optimum mix of solar energy and energy conservation?" Although this question has been answered in complex mathematical terms [BAL-7], what has been lacking is a simple method of determining the proper insulation levels and the proper degree of airtightness to use in conjunction with a passive solar design in a particular desired location. Such a method has now been developed and is presented in this chapter.

Energy conservation and passive solar are both strategies for saving energy; they compete with one another for builders' investment dollars. Both strategies can be characterized as behaving according to a law of diminishing returns. This is easily seen in the case of wall insulation. If the designer increases the overall wall resistance to heat flow from R5 to R10, this will decrease the wall's heat loss by a factor of 2. Suppose this has a monetary value of $100 per year for a 400 ft^2 exterior wall in a particular climate because it reduces the dollar value of annual energy flow through the wall from $200 to $100. Suppose also that the additional cost of this improvement is $100, so that the simple payback time is one year.

Pleased with this result, the designer might consider an additional increase from R10 to R20. The cost of this improvement might be an additional $200 if the cost per R remains constant, which is normally a reasonable assumption. This reduces the annual energy cost for the wall from $100 to $50, or a $50 per year reduction. Notice that the payback time for this second increment is four years.

The designer, still pleased with the economics, decides on a further increase in wall insulation from R20 to R40, at an additional investment of $400. The annual energy cost for the wall drops from $50 to $25, for a savings of $25 per year. The simple payback time for this last increment is sixteen years.

It's easy to see that the incremental return associated with each incremental investment is decreasing and that the payback time is increasing rapidly. The designer may decide to stop at R40. The overall cost of going from R5 to R40 has been $700 and the annual energy savings is $175 so that the overall payback time is four years. If much smaller increments had been used, the payback time for this example would have increased from half a year for the first small increment to thirty-two years for the last increment.*

Note that it is the *incremental* savings compared to the *incremental* cost—not the average—that is important in determining when to stop.

This characteristic of decreasing annual savings associated with each increase in investment is what is meant by reference to a law of diminishing returns. It characterizes virtually every conservation strategy, from additional wall insulation to additional ceiling insulation to increased perimeter insulation to reduced air infiltration. It also characterizes investments in passive solar heating.

The first added passive solar collection area is very effective because it saves auxiliary heat during the entire heating season. Consider, however, adding additional solar collection area to a building that already has a very high solar contribution. Since the building only requires backup heat during periods of either extremely cold weather or prolonged cloudiness, the added collection area is beneficial only during these rare periods. Although the cost of the last collection area that is added may be the same as the first area that was added, its cost effectiveness is much less because the energy savings are much less. This is another example of diminishing returns, which also underscores the necessity of calculating on an incremental basis rather than on an average basis.

Because of these characteristics, it is usually true that a mix of conservation and passive solar strategies will produce the maximum energy savings for a given initial investment—an arrangement that allows both energy conservation and passive solar to work in their most cost-effective ranges.

It is clear from the above example that the best strategy depends on the starting point and also on the climate. At each point in the design, it is possible to weigh the options and to determine the best strategy for increasing performance. The decision always involves a trade-off between the cost of the improvement versus the increased performance. An optimum mix is achieved when the incremental cost/benefit of each conservation option is just equal to the incremental cost/benefit of the passive solar strategy being used.

The mathematics of this balancing process have been reduced to a set of formulas that indicate appropriate conservation levels to use with a particular value of solar savings fraction (*SSF*). The formulas incorporate incremental cost information for both solar and conservation improvements, as well as a conservation factor (*CF*) that accounts for climate and for various system-performance considerations. The procedure to be described begins with these formulas, carries through a building load calculation and an annual solar performance calculation, and concludes with a summary of the first-cost and performance implications of the initial assumption.

What about cooling issues? These should be of major concern to designers who are working either in locations that have hot summer climates or with buildings that have large internal gains. The method in this chapter only accounts for winter performance. The designer should appreciate that passive solar heating features can increase cooling loads and that appropriate shading strategies should be used to minimize this effect. The designer should also appreciate that building mass added to make the passive solar heating work may also help passive summer cooling. Overall, however, it is recommended that the designer lean toward more conservation and less solar than the formulas indicate in any situations in which cooling is a significant issue. This is discussed further in section N.8.

The method to be described provides a technique for keeping the solar and conservation levels in proper balance. The other side of the decision involves determining how far to go with the combined strategies. Here, there are two possibilities. In the first and most likely situation, the designer faces an economic limit to the total combined investment allowable in conservation and passive solar. In the second situation, life-cycle costing alone determines the degree of energy savings to incorporate—weighing the incremental present value of future fuel savings against the cost of the additional investment. A simple procedure for determining the life-cycle optimum solar savings fraction (*SSF*) is presented in Chapter L of this handbook; it is reviewed briefly in section N.7.

N.2 CONSERVATION FORMULAS

The following formulas can be used to determine the optimum conservation levels to employ in conjunction with different passive solar strategies.

For walls, ceilings, floors, and east, west, and north windows:

$$R = CF \times \sqrt{\frac{\text{Passive System Cost (in \$/ft}^2\text{)}}{\text{Insulation Cost (in \$/R/ft}^2\text{)}}}$$

For the perimeter:

$$R = 2.04 \times CF \times \sqrt{\frac{\text{Passive System Cost (in \$/ft}^2\text{)}}{\text{Insulation Cost (in \$/R/linear ft)}}} - 5$$

For a heated basement:

$$R = 3.26 \times CF \times \sqrt{\frac{\text{Passive System Cost (in \$/ft}^2\text{)}}{\text{Insulation Cost (in \$/linear ft)}}} - 8$$

---

*The payback time for an incremental improvement varies as R^2.

For infiltration:

$$ACH = \cfrac{7.5 \times \sqrt{\cfrac{\text{Cost to Increase } 1/ACH \text{ by } 1}{\text{(in \$/ft}^3\text{)}}}}{\sqrt{\cfrac{\text{Passive System Cost}}{\text{(in \$/R/ ft}^2\text{)}}}}{CF}$$

The quantity R is the "R-value"—the thermal resistance, in hr/°F/ft²/Btu. The passive-system cost in these formulas is the cost per square foot of projected area. (Projected area is the principal net glazing area projected on a vertical plane.) The quantity ACH is the infiltration rate, in air changes/hour. The cost of reducing ACH is expressed as the cost of increasing the quantity $1/ACH$ by 1 per cubic foot heated volume.

The conservation factor (CF) can be determined for any desired value of SSF, for any passive-system type from the tables in Appendix 17.

N.3 OUTLINE OF DESIGN PROCEDURE

The process consists of choosing a value of SSF (perhaps somewhat arbitrarily) and then computing the implications of this choice. The process can then be repeated if necessary. The steps are as follows:

Step 1. Choose a starting-point solar savings fraction (SSF). One way to pick the SSF is by using the map in figure N-1 for guidance.

Step 2. Find the appropriate value of CF from Appendix 17 for the passive system selected.

Step 3. Determine the appropriate incremental costs for conservation measures and for the selected passive solar energy system.

Step 4. Calculate R-values and other conservation levels, using the formulas given earlier and appropriate incremental costs. Select the closest practical, buildable values.

Step 5. Calculate the building load coefficient (BLC).

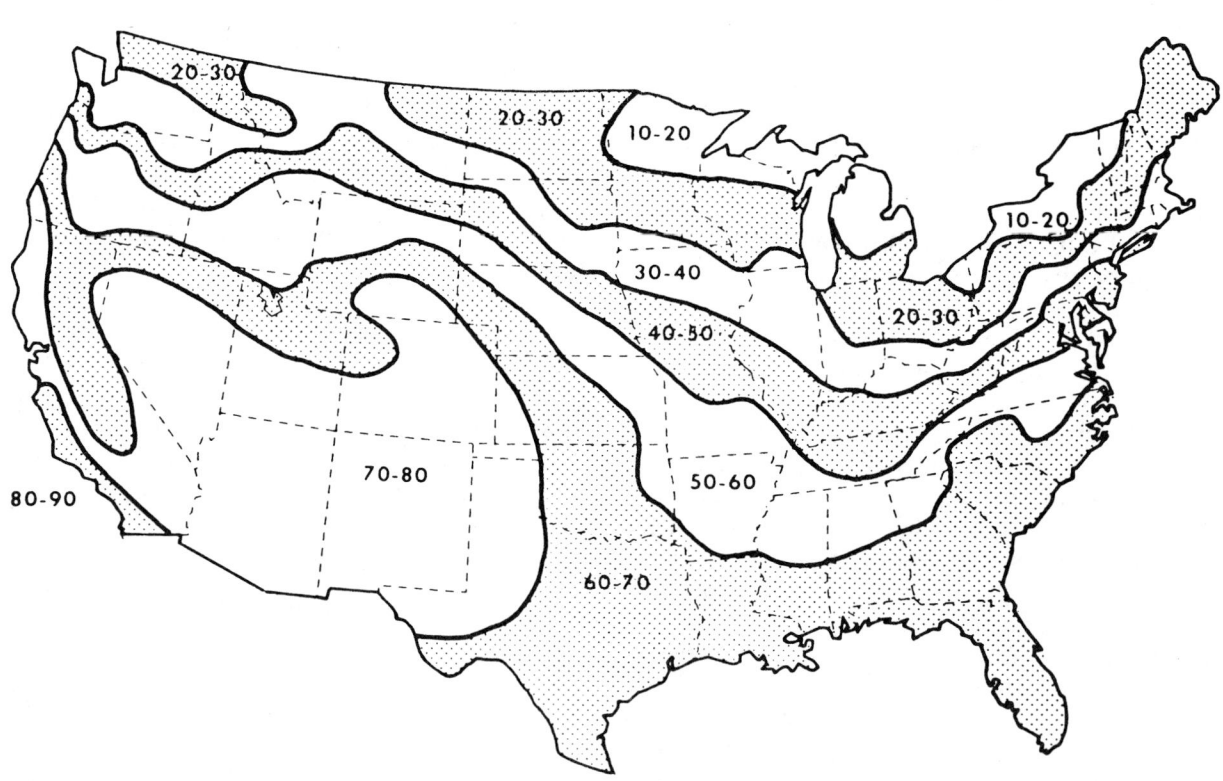

N-1: Starting-point values of solar savings fraction (suggested).

Step 6. Look up the load collector ratio (*LCR*) in Appendix 21 that corresponds to the selected value of *SSF*. Compute the required net passive solar projected area:

$$A_p = \frac{BLC}{LCR}$$

Step 7. Estimate the total add-on cost of the passive-solar and conservation levels selected.

Step 8. Estimate the annual backup heating using the formula:

auxiliary heat = $(1 - SSF) \times BLC \times$ annual DD

If the performance from Step 8 and costs from Step 7 are satisfactory, then you can stop here. Otherwise, repeat Steps 1 through 8, adjusting *SSF* upward to reduce auxiliary heating or downward to reduce costs. Each case calculated is an optimum mix. One can calculate several cases, if desired, and then choose from among them.

N.4 STEP-BY-STEP DISCUSSION OF DESIGN PROCEDURE

The following discussion should assist in correctly doing each step the first time through.

Step 1. Getting Started

In most situations, the building program will have already determined much about the design and the passive-system choice. Considerations of space layout, daylighting, lot shape and orientation, function, aesthetics, and preference will have already provided a starting point. Life-cycle costs may be the dominant consideration for some designers and total dollar constraints the dominant consideration for others. You can save time by making an intelligent choice for the initial value of *SSF*. There is probably no universal best way of choosing *SSF*, but several ways of getting started are described below.

One way of getting started is simply to choose one of the three *SSF* values listed in Appendix 17. This avoids any need to interpolate in Appendix 17 or Appendix 21. The three different values of *SSF* listed in the *CF* table—0.10, 0.50, and 0.80—describe different levels of passive solar use: a threshold level, a level appropriate for a typical solar building, and a level appropriate for a major solar building.

The threshold level is characterized by the values for $SSF = 0.10$. That is to say, the building should be insulated to the levels indicated by the *CF* for $SSF = 0.10$ before solar should be considered at all. This essentially quantifies what is meant by the phrase, "conservation first."

During the schematic design phase of the design process, one can select either the $SSF = 0.50$ level (if the objective is to obtain a typical passive solar building) or the $SSF = 0.80$ level (if the goal is to design a building that has very heavy reliance on passive solar).

A second possible way of getting started is to use the map shown in figure N–1. This map shows the optimum value of *SSF* for a particular set of conservation and passive-solar costs. It shows how climate affects the *SSF* choice if the ratio of costs is held constant. The designer might wish to use this map to obtain a "feel" for how optimum *SSF* values vary across the country. As a starting point, the designer might lean ± 0.10 with respect to the *SSF* figures given in this map, depending on whether solar heating or minimum first cost is to be emphasized. Other reasons for leaning to the high or low side may be that local energy costs are particularly high or particularly low or that serious concerns exist over added summertime cooling loads imposed by the solar elements.

A third possible way of getting started is to choose an *LCR* value. It frequently turns out that a good choice is $LCR = 20$ Btu/DD/ft^2 in cold climates and $LCR = 30$ Btu/DD/ft^2 in warm climates. Then you simply look up the corresponding value of *SSF* from the tables in Appendix 21.

A fourth possible way of getting started is to find the life-cycle optimum *SSF*. This will then lead to a three-way optimum involving future fuel costs and initial conservation and solar costs. To determine the life-cycle optimum point, first, calculate a/h, where:

$a =$ cost per ft^2 of projected area for the passive solar system

$h = \dfrac{(\text{cost of backup heat, leveled over the accounting period}) \times (\text{degree-days})}{\text{fixed charge rate}}$

The cost of backup heat is expressed in dollars per Btu. For other than electric heat, the furnace efficiency should be taken into account in the following manner:

$$\text{cost of backup heat} = \frac{\text{cost of heating fuel}}{\text{furnace efficiency}}$$

The fixed charge rate is an overall factor that converts an initial investment into an equivalent annual cash-flow, accounting for interest, taxes, depreciation, maintenance, discount rate, and resale value. For a detailed discussion of level fuel cost and of fixed charge rate, see Chapter L.

Now calculate *D* for various values of *SSF*,* and determine the *SSF* for which $D = a/h$ for the particular

---

*Determining *D*: *D* is the derivative of *SSF* with respect to $1/LCR$. See section L.4. In principle, *D* could be determined by differencing in the Appendix 21 tables, where *LCR* is given for values of *SSF* at 0.1 spacing. This is not satisfactory in practice because the roundoff in *LCR* results in significant error. A more satisfactory approach is to use the values of *CF* given in Appendix 17 and of *LCR* given in Appendix 21 at $SSF = 0.1, 0.5,$ and 0.8, and then to calculate *D* from the formula: $D = (1 - SSF)/(CF^2/24 - 1/LCR)$. Values of *D* at intermediate points can then be obtained by plotting *D* vs *SSF* and drawing a smooth curve.

passive solar system under consideration. This is the maximum *SSF* that is cost effective for the particular fuel rates used. This procedure, although straightforward, does require the estimation of future fuel costs and is, therefore, of limited value. An advantage, however, is that no iteration is required.

Step 2. Looking up *CF*

For this step, it will be necessary to interpolate in the tables for *CF* if the value of *SSF* is other than 0.10, 0.50, or 0.80. If a mix of more than one passive system is used, calculate *CF* based on a weighted average, according to the relative areas of the various passive system types being used.

Step 3. Determining Unit Costs

In computing costs, use the full installed cost, including materials, installation labor, overhead, and profits. Incremental costs for both the conservation and solar strategies are needed. These are the costs to be inserted into the formulas for the R-values and the *ACH* value. Costs should be computed for relatively small changes in either conservation or solar, which is why they are referred to as *incremental* costs.

In determining the incremental cost of wall insulation, therefore, it is not appropriate to simply divide the cost per square foot of the wall by the R-value of the wall. Instead, it is desirable to determine the costs and R-values of two different wall sections and to calculate the incremental cost by dividing the increase in cost per square foot by the increase in R-value. For example, if the cost of an R12.7 wall is $2.05/ft² and the cost of an R28 wall is $2.69/ft², then the incremental cost is $(2.69 - 2.05)/(28 - 12.7) = 0.042$ \$/R/ft². Incremental costs of 3.5¢/R/ft² in the range from R12 to R20 and 5¢/R/ft² in the range from R20 to R28 are typical for frame construction.

In order to determine unit costs for windows, it is necessary to use the same technique as above, in which the value for $R(1/U)$ is determined for different integer numbers of glazings and the incremental cost is determined in the same fashion.

For perimeter and basement insulation, the incremental cost needed is per R-value per linear foot of perimeter.

The cost of reducing infiltration presents a special case in which it is very difficult to make any meaningful estimates. If the designer is fortunate enough to have information on actual costs and the *ACH* values corresponding to these costs, then these should be used. More commonly, rough estimates must be made instead.

Down to an infiltration value of half an air change per hour, reductions can be achieved simply by decreasing the natural infiltration characteristics of the house. This is done by installing a plastic vapor-barrier, which is recommended anyway for all insulated construction.

Further reductions can be achieved by carefully caulking all cracks, caulking overlaps in the vapor-barrier, and sealing other cracks such as those that may occur around utility outlet boxes, along sill plates, and through ceiling vents. As in any conservation strategy, the first gains are the least expensive and the most cost-effective. As more and more care is taken in the process, costs escalate rapidly. A significant portion of the cost will be the labor cost associated with installing a very effective vapor-barrier with caulked overlaps covering all surfaces. Details for installation of these vapor-barriers have been developed in the Canadian plains provinces [ERD]. They are neither simple nor inexpensive.

In order to achieve lower and lower values of infiltration safely, a series of strategies is normally used. Little information is available on costs, but the scenario presented in figure N-2 is a reasonable possibility for a 1,500 ft² house.

It is strongly recommended by a number of different groups that a heat recovery unit (HRU) be used in any building in which the natural infiltration rate is less than one-half air change per hour.

The use of one-half air change per hour as a lower limit is somewhat arbitrary, but it is supported by an accumulating body of evidence from several researchers.

Effective ACH*	Strategy	Cost Increase, $	Cost $
1.5			0
	add plastic vapor barrier	125	
1.0			125
	seal joints, foam cracks, more care	375	
0.5			500
	add window heat recovery unit (HRU), more care	250	
0.375			750
	add another HRU, more tightening	250	
0.300			1,000
	install large HRU instead, plus distribution system, 70% effective, more tightening	250	
0.250			1,250
	larger HRU, 80% effective, more tightening	250	
0.214			1,500
	larger HRU, 90% effective, more tightening	250	
0.187			1,750

*The effective ACH is simply the sum of the natural ACH plus the forced ACH times (1 − effectiveness factor).

N-2: Different strategies for lowering infiltration in a 1,500-ft² house, and their costs.

For example, the value of ½ ACH is recommended for adoption as a standard in Denmark [FAN]. It is felt that, by maintaining an outside air change rate equal to at least ½ ACH, the various hazards that are associated with low air makeup rates can be avoided reasonably well. These hazards include accumulation of formaldehydes (which emanate from the resins in plywood, chipboard, and other building materials), radon gas (which leaks up under the house through cracks in the rock and soil), oxides of nitrogen and carbon monoxide (from combustion), cigarette smoke, high humidity, and odors [ROS]. By installing a heat recovery unit, the outside air dilution rate can be maintained at a safe level at the same time as the heat losses associated with this air exchange are reduced by the effectiveness factor of the unit.

A simple window-box heat recovery unit may be acceptable in a situation in which the natural infiltration rate is reasonably close to ½ ACH and one can assume sufficient mixing of the air within the building so that the room that has the heat recovery unit exchanges air with the other rooms. As the building is made progressively tighter, this may not be adequate, and several heat recovery units may be required for different rooms or an air distribution system may be required in conjunction with one main heat recovery unit. Many passive-system buildings are not equipped with a central heating system, relying instead on several small baseboard electric or other unit heaters scattered throughout the building. Therefore the cost of the distribution system must be accounted for in determining the incremental cost of the reduced ACH.

In the above table, the incremental cost of increasing $1/ACH$ is constant and is equal to \$375 per unit of $1/ACH$. This inverse relationship seems intuitively reasonable and corresponds roughly to the assumed constancy of the ratio of cost per R that is used in the other portions of the method. The \$375 cost of increasing $1/ACH$ by one unit corresponds to \$0.0312/ft^3 of house volume for a 12,000-ft^3 house.

Step 4. Selecting Conservation Levels

Typically, the formulas for selecting conservation levels give an R-value or other conservation level that is between practical choices. Select the closest practical value. Notice that the incremental cost numbers used earlier should apply within the range finally selected.

Step 5. Calculating BLC

Simple formulas for calculating the BLC are given in figure N-3. Wall area, perimeter length, ceiling area, building volume, and window area are needed. The formulas are based on ASHRAE procedures. Alternatively, one can use the following formula:

$$BLC = 24 \times \left(\frac{\text{design heat loss} - \text{design heat loss of solar aperture (in Btu/hr)}}{\text{design temperature difference (in F degrees)}}\right)$$

Step 6. Looking up LCR and Calculating A_p

Tables in Appendix 21 give values of LCR corresponding to nine different levels of SSF for 94 passive system types and for 219 locations. Simply look up the needed LCR under the appropriate SSF. Interpolation may be needed. For a mix of passive-system types, calculate LCR based on a weighted average, according to the relative areas of the various passive-system types being used.

Step 7. Estimating Add-on Cost

The building is now completely specified. Use standard cost estimating procedures. The unit-cost values determined earlier will be applicable within their valid range.

Step 8. Calculating Auxiliary Heat Requirements

The objective of the balancing process should be to minimize auxiliary heat within the cost constraints.

N.5 CHOOSING FINAL DESIGN PARAMETERS

The process described above is useful during the initial phases of the design process for "getting in the right ballpark." The natures of both the construction process and the optimization curves are such that homing in on an exact, optimum final answer is not worth the effort.

For one thing, the assumption that the costs per R are constant is not a very good approximation, considering that discrete wall construction details must be individually analyzed. Different wall construction techniques may lead to the same R-values but different costs. Wall-framing materials come only in particular sizes. Nonetheless, the constant cost per R does characterize the general trend and therefore is useful in locating the *region* of the optimum.

Another consideration is that the nature of the cost-effectiveness curve at optimum is very flat, meaning that, as one moves away from the optimum mix, the overall cost effectiveness of the design changes very slowly. For mixes within about 20% of the optimum mix, the changes are very nearly insignificant. This is a fortunate result because it allows the designer considerable latitude in selecting wall sections and construction techniques and in sizing the solar aperture, without having to worry that the cost effectiveness of the design may be seriously compromised.

The implication of the foregoing comments is that decisions made during schematic design will probably carry through to final design unless some problems of construction-detailing change the final R-values or the final solar collection areas.

N.6 EXAMPLE: THE METHOD IN ACTION

Consider a proposed building to be designed for Mason City, Iowa. The designer has developed a rough-sketch plan for a building with approximately 1,500 ft² of floor area plus 200 ft² of semienclosed sunspace. Tentatively, a 40%–60% combination of sunspace (system SSD1) and nighttime-insulated direct gain (system DGC3) has been selected based on client requirements. The designer now wants to estimate the appropriate conservation levels, size the glazing, predict annual performance, and estimate costs.

Step 1. From among the various possible ways of getting started, the designer opts to use the map shown on figure N-1. Figure N-1 recommends $SSF = 0.36$ for northern Iowa. The designer wants an explicitly passive solar building and so opts to increase SSF to 0.45 as a starting point.

Step 2. Values of CF given in Appendix 17 are as follows:

System	$SSF = 0.10$	$SSF = 0.50$	$SSF = 0.45$ (interpolated)
SSD1 (40%)	1.4	3.1	2.9
DGC3 (60%)	1.4	1.8	1.7
Area-weighted average		$[(0.4 \times 2.9) + (0.6 \times 1.7)]$	2.2

Step 3. The designer then obtains cost numbers for two wall sections:

R24 wall cost = $3.00/ft²
R49 wall cost = $4.25/ft²

Therefore, the incremental cost of wall insulation within this range is:

$$(4.25 - 3.00)/(49 - 24) = \$0.05/R/ft^2$$

Other conservation costs are calculated in similar fashion. Assume that the following incremental costs are determined:

Ceiling Insulation = $0.03/R/ft²
Perimeter Insulation = $0.25/R/linear ft (3-ft depth)
Window Glazing = $4/ft² for one added glazing (this leads to $4.40/R/ft²)
Infiltration Decrease = $0.0312/(1/ACH)/ft³

The incremental costs of passive solar are determined by the designer as follows:

sunspace: $12/ft² of projected area
direct gain: $15/ft² of projected area

In this determination, the designer first calculates the full cost of the sunspace, $39.20/ft² of glazing, and then subtracts a reasonable estimate of the amenity (appraisal) value, $30/ft², to obtain the incremental cost, $9.20/ft², that is chargeable solely to the sunspace heating potential. The cost is to be expressed in $/ft² of projected area, so that the cost per square foot of tilted glazing must be converted to the cost per square foot of vertical projection of the glazing. Since 1 ft² of 50°-tilted glazing projects onto 0.77 ft², the converted cost is $9.20 /0.77 = $12/ft² of projected area. For direct gain, the designer determines the cost of the double-glazed window, $9/ft², subtracts the cost of the replaced opaque wall, $3/ft², and then adds the cost of the movable insulation, $9/ft², to obtain the cost per square foot used, $15/ft². The area-weighted passive solar cost is:

$$(12 \times 0.4) + (\$15 \times 0.6) = \$13.80/ft^2$$

assuming that the designer intends to scale the size of the sunspace and the size of the direct-gain area incrementally according to a 40/60 proportion.

Step 4. The designer can use the formulas as follows:

$$R_{wall} = CF \times \sqrt{\frac{\text{Passive System Cost (in \$/ft)}}{\text{Wall Insulation Cost (in \$/R/ft}^2\text{)}}}$$

$$R_{wall} = 2.2 \times \sqrt{13.8/0.5} = R37$$

Similarly:

$$R_{ceiling} = 2.2 \times \sqrt{13.8/0.03} = R47$$

$$R_{perimeter} = 2.04 \times 2.2 \times \sqrt{13.8/0.25} - 5 = R28$$

$$R_{windows} = 2.2 \times \sqrt{13.8/4.4} = R3.9$$

$$ACH = 7.5 \times \sqrt{0.0312/13.8} /2.2 = 0.16 \text{ ACH}$$

The designer must now take this guidance and make intelligent choices of actual, buildable values to use. Assume that the following choices are made:

Wall = R32
Ceiling = R50
Perimeter = R20
Windows = R2.7 (triple glazing)
Infiltration = 0.187 ACH

Remember that the windows here are facing east, west, and north. The south-facing solar windows are double glazed with night insulation.

Step 5. The designer now calculates the BLC, using the formulas given in figure N-3 and the building dimensions listed in the table at the top of the facing page.

Step 6. From Appendix 21, the designer obtains values of LCR. The system types are SSD1 (semienclosed sunspace, 50° tilted glazing, mass wall) and DGC3 (direct gain, double glazing, R9 nighttime insulation). The LCR values are as given in the second table on the facing page.

Dimensions	Formula	BLC
Wall area = 865 ft²	24 × 865/32	649
Ceiling area = 1,500 ft²	24 × 1,500/50	720
Perimeter length = 165 ft²	100 × 165/(20 + 5)	660
E,W,N windows = 65 ft²	26 × 65/3	563
Volume = 12,000 ft³	0.432 × 12,000 × 0.96 × 0.187	931
		BLC = 3,523 Btu/DD

System	SSF = 0.40	SSF = 0.50	SSF = 0.45 (by interpolation)
SSD1 (40%)	17	9	13
DGC3 (60%)	26	19	23

Walls

$$L_w = 24 \times \left(\frac{\text{wall area}}{\text{R-value of walls}}\right)$$

where wall area = (perimeter × ceiling height) − (nonsouth window area + south window area)

East-, West-, and North-Facing Windows

$$L_g = 26 \times \left(\frac{\text{nonsouth window area}}{\text{number of glazings}}\right)$$

Perimeter (slab on grade)

$$L_p = 100 \times \left(\frac{\text{length of perimeter foundation}}{(\text{R-value of perimeter insulation} + 5)}\right)$$

Floor (over vented crawl space)

$$L_f = 24 \times \left(\frac{\text{area of ground floor}}{\text{R-value of floor}}\right)$$

Basement (heated basement or other fully-bermed wall, including floor losses)

$$L_b = 256 \times \left(\frac{\text{length of wall}}{\text{R-value of wall insulation} + 8}\right)$$

Roof

$$L_r = 24 \times \left(\frac{\text{roof area}}{\text{R-value of roof}}\right)$$

Infiltration:

L_i = (0.432) × average air changes per hour × air density ratio × ceiling height × combined area of all floors

Final BLC

$$BLC = L_w + L_g + L_r + L_p + L_i + L_b + L_f$$

N-3: Simple formulas for calculating the *BLC* of various surfaces.

Area-weighted average SSF = [(0.4 × 13) + (0.6 × 23)] = 19

A_p = BLC/LCR = 3,523/19 = 185 ft²
Sunspace = 74 ft² (40%)
Direct gain = 111 ft² (60%).

Step 7. The solar add-on cost is 185 × 13.80 = $2,553. To calculate the conservation add-on cost, the designer needs to know the starting point or reference level—that is, the base-case values.* These are as follows:

Walls	− assumed base case is R11, overall; add-on cost is (32 − 11) × 0.05 × 865 =	$ 908
Ceiling	− assumed base case is R19, overall; add-on cost is (50 − 19) × 0.03 × 1,500 =	$1,395
Perimeter	− assumed base case is R0; add-on cost is 20 × 0.25 × 165 =	$ 825
Windows facing E,W, N	− assumed base case is single glazing; add-on cost is (3 − 1) × 4 × 65 =	$ 520
Infiltration	− assumed base case is *ACH* = 1; add-on cost = 1750 − 125 =	$1,625
Total conservation add-on cost =		$5,273
Total add-on cost =		$7,826

Step 8.
Auxiliary heat = (1 − 0.45) × 3,523 × 7,901
= 15.3 MMBtu

*The optimum mix is independent of the base-case assumed, but the add-on conservation cost is not.

Balancing Conservation & Solar / 149

For comparison purposes, the base-case *BLC* can be calculated, using the same formulas as in Step 5, to obtain 13,749 Btu/DD. The base-case auxiliary heat is therefore:

$$\text{Auxiliary heat} = (1 - 0) \times 13,749 \times 7,901$$
$$= 108.6 \text{ MMBtu}$$

assuming that the net energy effect of the base-case south-facing wall is exactly neutral. Therefore, the energy saved by the combined conservation and solar strategies is

$$108.6 - 15.3 = 93.3 \text{ MMBtu/year}$$

Therefore, the designer would have spent an extra $7,826 to save 93.3 MMBtu annually—an investment of $83.88 per million Btu saved annually.

N.7 LIFE-CYCLE COSTS —A DIFFERENT APPROACH

The straightforward approach to using life-cycle costs is to calculate a/h, as described earlier, and then to find a value of *SSF* that results in $D = a/h$. Another, somewhat backward approach is described below.

The global optimum is reached when $D = a/h$. By turning this equation around, it is possible to find a present-day cost of fuel that will justify the initial cost of the system. This occurs when $h = a/D$. The formula used to estimate *D* is:

$$D = (1 - SSF)/[(CF^2/24) - (1/LCR)]$$

Continuing with the example (*SSF* = 0.45, *CF* = 2.2, and *LCR* = 19):

$$D = (1 - 0.45)/[(2.2^2/24) - (1/19)]$$
$$D = 3.7 \text{ Btu/DD/ft}^2$$

Therefore:

$$h = a/D = 13.80/3.7 = 3.7$$

But:

$$h = \left(\frac{\text{(cost of backup heat, leveled over the accounting period)} \times \text{(degree-days)}}{\text{fixed charge rate } (FCR)} \right)$$

If the fixed charge rate is 0.075 (see section L.5, Step 2) then the cost of backup heat, leveled over the accounting period, is equal to:

$$h \times FCR/DD$$
$$3.7 \times 0.075/7,901$$
$$\$0.0000351/\text{Btu}$$
$$\$35.10/\text{MMBtu}$$

If the cost of fuel leveled over the accounting period is 1.5 times the current cost of fuel, then:

Present cost of fuel = $23.40/MMBtu

This is the present cost of fuel that justifies the added initial cost of $7,826. This fuel cost corresponds to the cost of electric heat at 7¢/kWh and an efficiency of 100% (electric resistance).

N.7.a BACKUP HEAT

It is worth noting at this point that resistance electric heat may be a reasonable choice for this building. The following factors should be considered:

1. The annual dollar amount of auxiliary heating is small, corresponding to only $313/year at 7¢/kWh.
2. Consequently, it is doubtful that the initial cost of a heat pump would be warranted, especially since the backup heat is required generally during extreme weather when a heat pump would be operating purely in a resistance mode. However, if a heat pump is to be installed anyway for summertime cooling, then it may also be used for wintertime heating at little additional cost.
3. The size of the backup is small—only about 4.2 kW capacity.*
4. It is inexpensive to obtain final zone control of resistance electric heat in small capacities. Experience shows that good zone control is essential in passive-solar buildings.
5. It is well-documented that passive solar buildings intrinsically tend to demand heat during off-peak hours of the electric utility—that is, from 11:00 P.M. to 7:00 A.M. [BAK], [PYD]. Thus, the building has a good load pattern for the utility and could qualify for off-peak rates where these are available. With minor control changes, the load could be made to be entirely off-peak [PYD].

It might also be reasonable to augment the backup heat with a small, efficient wood-burning stove in the main area of the building.

N.7.b REVIEW OF THE EXAMPLE

This building is a superinsulated passive-solar structure that uses less than 15% of the auxiliary heating energy of a typical, contemporary structure in the same location. It will be cost-effective and comfortable.

---

*The *BLC* of the sunspace is about 639 and of the direct gain about 391, giving an overall building *BLC* = 4,553 Btu/DD. The design temperature is −11°F, so that the backup is 4,553 × (65 − 11) = 346,000 Btu/day = 14,420 Btu/hr = 4.2 kW.

As in the previous example, the following costs are assumed:

System Cost: $13.80/ft²
Wall Insulation Cost: $ 0.05/R/ft²

The optimum wall insulation is given by:

$$R_{wall} = CF \times \sqrt{13.8/0.05}$$
$$R_{wall} = 16.6 \times CF$$

the value 16.6 × *CF* can then be mapped for a particular passive system to see the nature of the results. An example is shown plotted in figure N-4, showing contours of optimum-mix wall-insulation levels.

Because the cost values appear under the square root sign, the contours will move slowly as cost values change. Notice also that it is only the ratio of costs that is important. This means that both inflation and region-to-region variations in materials and labor costs will not change the optimum R-values much because the ratio of costs should remain relatively stable.

N.8 COOLING ISSUES

How do considerations of summer cooling effect the conservation-solar balancing process described in this chapter? Since the analytical procedure is based solely on considerations of the heating season, and since simplified methods are not yet developed for cooling-load estimation, it is necessary to rely for guidance on experience and on the results of a few case studies that have been made.

Advice from those working on passive solar projects in the Southeast is to focus on the range of *SSF* from 0.2 to 0.5 rather than on the range from 0.5 to 0.7 as indicated in figure N-1 [HIL]. This will reduce the magnitude of cooling loads created by passive-solar heating elements, and can be done without changing the amount of annual auxiliary heating if conservation levels are increased. In effect, the balance is simply shifted toward an increased reliance on conservation and away from passive solar. As previously indicated, the initial cost will not increase much because the changes are taking

N-4: Optimum wall insulation R-values, based on The *SSF* values given in figure N-1 and on an assumed incremental insulation cost of 5¢/R/ft² and an assumed incremental passive system cost of $13.80/ft². The passive system is a semienclosed sunspace with 50° sloped glazing (SSD1).

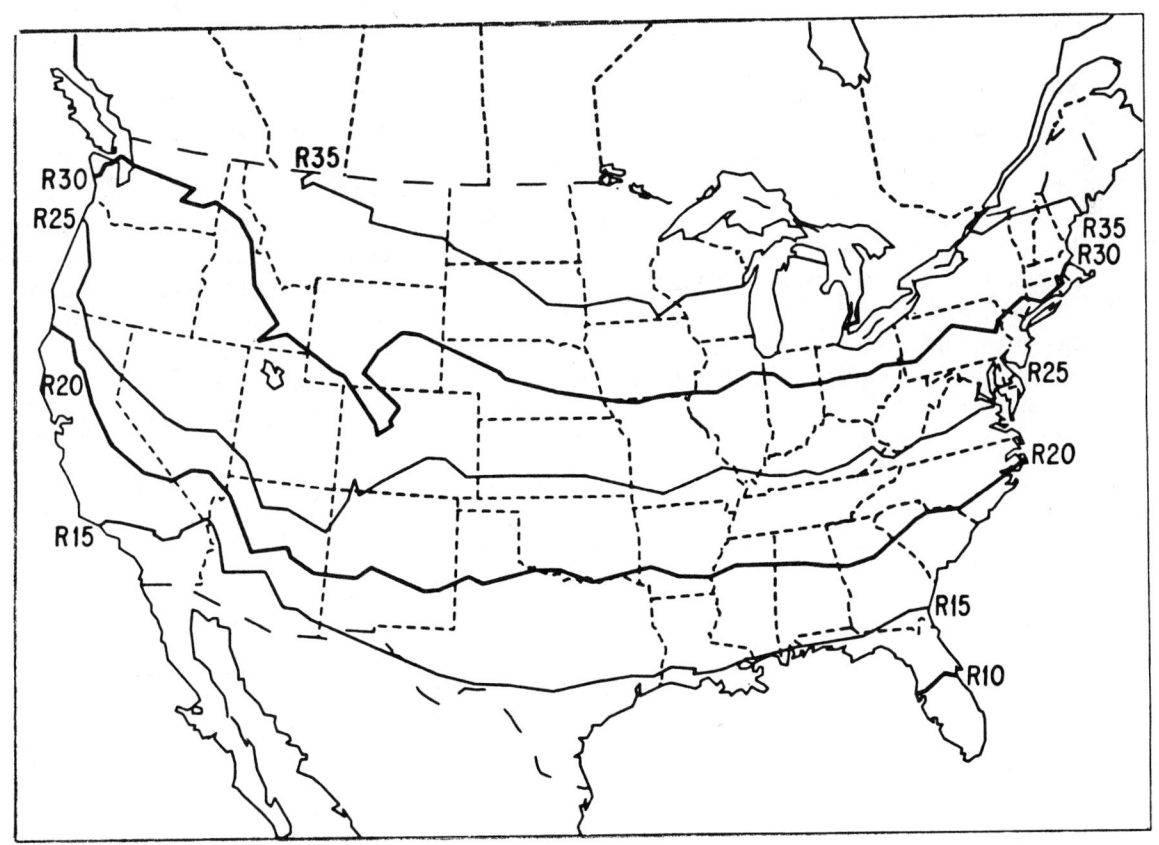

Balancing Conservation & Solar / 151

place near an optimum point, at which the initial cost curve (keeping annual auxiliary heat constant) is reasonably flat. The added insulation will also help in reducing summertime cooling loads.

As a general rule, *SSF* should be reduced by 0.10 or 0.20 below the values attained through the procedures in this chapter in situations in which cooling loads caused by passive-solar elements will increase summertime cooling requirements. Another possibility that should be considered in such situations is providing for complete sun protection over the passive-solar elements. Movable insulation can often be used for this purpose and is especially effective if it is located outside the window.

N.9 OTHER ADVANTAGES TO A BALANCED APPROACH

Other considerations that favor a mix of conservation and solar can be seen by investigating problems that arise from either extreme taken alone. These problems can be largely avoided with a balanced design; the resulting building will be more comfortable, more inhabitable more resilient, and more salable, less backup heat will be needed, and the initial cost will be less.

The extreme of energy conservation attempts to isolate the inside of a building from the outside. A very cloistered and seemingly artificial environment is created. One has to go outside to see the sky, clouds, birds, weather, and trees bending in the wind. This type of indoor environment may be exactly what some people desire, but it is certainly not the ideal of others.

The other extreme is an overdone solar building, without adequate insulation. In this case, virtually every south-facing surface is a solar collector. Storage, control, and distribution of huge solar gains become a problem, increasing cost and forcing unwelcome compromises in space allocation. Large temperature swings may become unavoidable. A traditional, oversized backup heating plant is needed, with careful attention to distribution and zone control, and this adds to the cost. Cold wall and window surfaces require a higher inside air temperature to maintain comfort.

With a balanced approach, most of these problems simply go away. A friendly and livable indoor environment is achieved, maintaining good visual connection with the outdoors but with comfortable shelter from its extremes. The atmosphere seems natural rather than artificial—working with nature and not against it. Since the solar collection area is less than the entire south facade of the house, there is more architectural freedom in the design. Thermal storage requirements are met easily by interior mass walls serving a dual function and thus not seeming out of place. Good insulation and other conservation strategies reduce the size and cost of the backup heating plant. Elaborate distribution of heat is not particularly critical. Icy drafts from cold wall and window surfaces disappear.

Another advantage of balanced design is resilience. Suppose backup heating fuel or electricity were simply not available for an extended period. Suppose a total disruption of all utility service occurred. Suppose the average outside temperature were 10°F and cloudy. The balanced design building fares better under these conditions than either of the extremes. Because of high internal heat capacity and good insulation, the building temperature drops slowly (over a period of 3 to 5 days) and eventually levels off. Because even cloudy-day solar gains are significant, this stable temperature condition is well above the outside temperature and certainly well above freezing (at least in the 48 contiguous United States). The extreme energy-conservation house may cool quickly or more slowly, depending on its internal heat capacity, but will eventually reach low temperatures because of the absence of any heat source. The extreme solar house will cool quickly because of poor insulation, and its north rooms may be in danger of freezing.

Marketing is another consideration. Energy performance is of increasing concern to those who purchase buildings, both residential and commercial. Solar is visible, well-publicized, and has demonstrated market appeal. Good insulation and other conservation strategies are less evident but are well-understood to be essential to both energy savings and comfort. Together, they make a consistent and very compelling argument to an increasingly knowledgeable buying public.

Direct Gain

A direct-gain building is a type of passive system in which solar radiation enters the living space directly through windows, as illustrated schematically in figure O-1. These solar gains either serve to meet part of the current heating needs of the structure or are stored in the building mass to meet heating needs that arise later. Space-heating needs not satisfied by the solar gains are met by conventional backup systems that generally use nonrenewable sources of energy. Other incidental sources of internal heat displace part of the thermal load, but for many residential applications these sources represent only a small part of the total requirement.

As indicated in figure O-1, thermal insulation is placed on the outside of massive elements of a direct-gain building's shell in order to avoid isolating the mass from the living space and therby neutralizing its effectiveness as a thermal storage medium. Floor slabs can contribute to the thermal storage capacity of a structure, provided they are not isolated by carpets and cushioning pads. It is advisable to limit losses from floor slabs by providing perimeter insulation on the foundation walls, but placing insulation beneath slabs is not generally con-

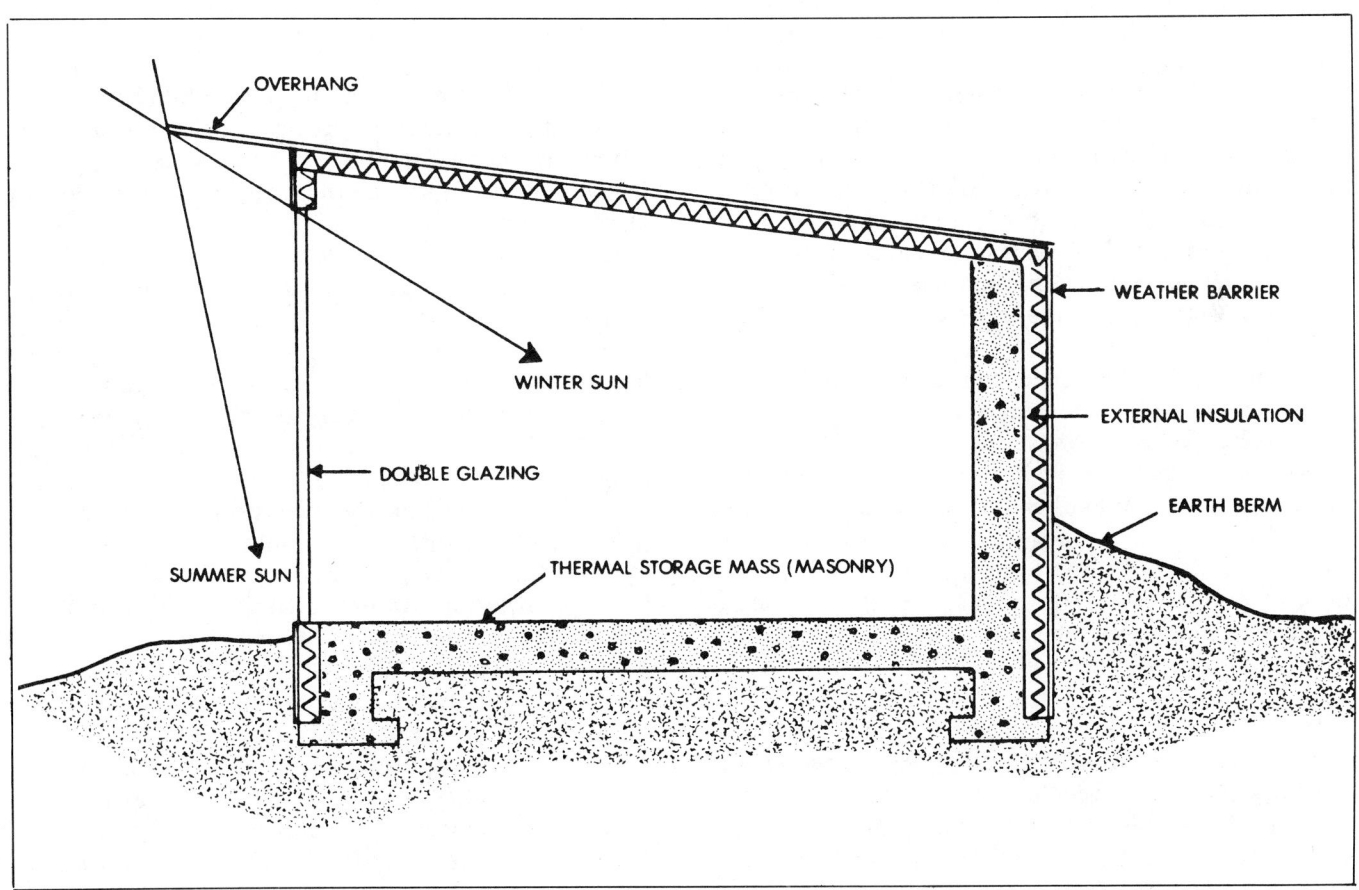

O-1: Direct-gain building.

sidered cost-effective. An earth berm is frequently placed on the north side of passive solar buildings to provide extra protection from cold winter winds.

Figure O-1 also illustrates an overhang employed to shade the solar aperture from the high summer sun, while permitting rays from the low winter sun to penetrate and warm the living space. Usually, two glazing layers are mounted in direct-gain apertures. A single glazing layer is undesirable because of the large heat losses that single glazing allows. In very cold climates, triple (or even quadruple) glazing is advisable as a means of controlling losses, but it has not yet been widely accepted in the marketplace. Another method of controlling losses is to use movable opaque insulation that covers the apertures at night and can be removed during the daylight hours.

Direct-gain buildings have a relatively high level of market appeal because they involve less departure from conventional construction than other types of passive solar heating systems. Their actual effectiveness, however, is limited to areas with mild or moderate winter climates, unless triple glazing or nighttime insulation is employed to limit heat losses through the windows. Improperly designed direct-gain systems may overheat during the day, particularly if inadequate effective thermal storage mass has been provided. In fact, failure to provide sufficient effective thermal storage mass within the building shell is the most common error observed in direct-gain structures. This design error often leads to uncomfortable temperature swings, and the resulting structure yields little energy savings. Finally, direct-gain buildings can be susceptible to glare and to fabric degradation caused by admission of solar radiation into the living space. The best procedure for minimizing these problems is to distribute the transmitted solar radiation as uniformly as possible through appropriate window placement and through the use of diffusive blinds or glazings.

On the positive side, direct-gain buildings are attractive and take excellent advantage of views to the south. The incremental cost relative to an otherwise nonsolar building is small, especially if the structure is inherently massive, as are the adobe buildings of the Southwest and the block wall on slab-on-grade buildings of the Southeast. When designed appropriately for their location, direct-gain buildings can provide a cost-effective means of reducing energy consumption for space-heating without requiring any sacrifice of comfort or aesthetic values.

Thermal storage mass plays a vital role in the effective performance of direct-gain buildings. As a general rule, buildings that contain very little thermal mass are unable to store heat for nighttime use; thus, only the daytime portion of the heat load can be met by solar gains, and overheating can be a serious problem if the solar gains are excessive. Furthermore, heat losses through an aperture whose effectiveness is impaired by insufficient thermal mass can exceed the useful solar gain through it. Lightly constructed stud-wall buildings with no floor slab (or with a carpeted floor slab) are the worst offenders in this category.

The effectiveness of thermal storage mass in direct-gain buildings depends on the mass's thickness, surface area, and thermal properties (volumetric heat capacity and thermal conductivity). The best materials to use are those that are capable of storing large quantities of heat (high volumetric heat capacity) and that can readily transport heat from the mass surface to the mass interior for storage and back to the surface again to meet the building heat load (high thermal conductivity). As a general rule, mass surfaces should be relatively dark in color compared to nonmassive surfaces, in order to promote preferential absorption of solar radiation by the thermal storage medium; for optimal effectiveness, mass surfaces should be located in building zones that experience direct solar gains. The question of appropriate colors in direct-gain buildings is complex and is addressed in more detail in sections O.1.b.3, O.1.b.4, and O.1.b.5. Mass located in zones that are not directly illuminated by solar radiation entering through south-facing windows will be ineffective unless care is taken to insure adequate free convective exchanges with directly heated zones or unless a forced-air distribution system is employed. In either case, remote thermal storage mass will be less effective than that located in direct-gain zones.

For analtyical purposes, direct-gain buildings are divided into two broad categories, distinguished by the amount of thermal storage mass within the solar-heated living space per unit of collection area. Buildings having large amounts of thermal mass, called *high-mass, direct-gain* buildings, and those having little mass, called *low-mass, sun-tempered* buildings, are considered separately in this chapter.

The design procedures discussed in this chapter are based on analyses performed with the SUNSPOT thermal network computer program [WRA-1], [WRA-2]. SUNSPOT is a direct-gain adaptation of the PASOLE program that was developed by R. D. McFarland and originally applied to thermal storage wall systems [MCF-1]. Both programs employ hourly historical weather data to operate a system of nodes and connections (the network), whose behavior is governed by the first law of thermodynamics (conservation of energy). Temperatures are calculated hourly at each node by determining the correct balance of heat fluxes through the connections. SUNSPOT was validated for direct-gain buildings by comparison with experimental data from test cells at the Los Alamos National Laboratory.

Much of the information presented in this chapter is available in reference [WRA-3]. This report describes a greatly simplified procedure for applying the monthly solar load ratio method, which may be of interest to some readers.

O.1 HIGH-MASS DIRECT-GAIN BUILDINGS

O.1.a THE REFERENCE DESIGNS

The high-mass reference designs comprise nine different direct-gain systems. They are described below in terms of a set of specified parameters. The projected area (A_p) and the building load coefficient (BLC) remain variable.

There are configurations with heat capacities of 30, 45, and 60 Btu/°F/ft² of glazing. The heat capacity should generally equal or exceed the lower limit of this range to prevent excessive temperature swings. The range corresponds to 150–300 pounds of high-density masonry or concrete per square foot of glazing area. Thermal storage mass in the reference designs is located on the floor and north wall of the illuminated zones. The three levels of thermal storage mass are achieved with the following mass configurations:

1. A mass thickness of 2 inches and a mass surface area of six times the glazing area;
2. A mass thickness of 6 inches and a mass surface area of three times the glazing area; and
3. A mass thickness of 4 inches and a mass surface area of six times the glazing area.

Designation	Thermal Storage Capacity* (Btu/ft²/°F)	Mass Thickness** (inches)	Mass-Area-to-Glazing-Area Ratio	No. of Glazings	Nighttime Insulation
A1	30	2	6	2	no
A2	30	2	6	3	no
A3	30	2	6	2	yes
B1	45	6	3	2	no
B2	45	6	3	3	no
B3	45	6	3	2	yes
C1	60	4	6	2	no
C2	60	4	6	3	no
C3	60	4	6	2	yes

*per unit of projected area
**Assuming volumetric heat capacity of 30 Btu/ft³/°F.

O-2: Direct-gain reference designs.

For each of these mass configurations, three glazing configurations are included: double glazing, triple glazing, and double glazing with R9 nighttime insulation. Altogether, these variations comprise nine different references designs, which are summarized in figure O-2. A selection of the high-mass reference design characteristics is given in figure O-3.

Thermal Storage

Heat capacity	thickness	mass area to glazing area
30 Btu/°F/ft² of glazing	2 inches	6
45 Btu/°F/ft² of glazing	6 inches	3
60 Btu/°F/ft² of glazing	4 inches	6

Masonry properties	
Thermal conductivity (k)	1.0 Btu/hr/ft/°F
Density (ρ)	150 lb/ft³
Specific heat (c)	0.2 Btu/lb/°F
Infrared emittance of surface	0.9

Glazing properties	
Transmission characteristics	diffuse
Infrared emittance of surface	0.9

Control range	
Room air	65° to 75°F

Lightweight absorption fraction	0.2
Simulates effect of solar radiation absorption on lightweight walls or objects by transferring given fraction of transmitted and reflected solar radiation directly to room air	

Additional assumptions for annual method	
Solar absorptance of mass floor and walls	0.8
Ground reflectance	0.3
Internal heat generation	0
Glazing properties:	
Orientation	due south
Shading	none
Number of panes	2 or 3
Index of refraction	1.526
Extinction coefficient	0.5 inch^{-1}
Thickness of each pane	1/8 inch
Air gap between panes	0.5 inch
Nighttime insulation thermal resistance, when used (only with double glazing)	R9

O-3: Properties of high-mass reference designs.

The reference design characteristics are not necessarily recommended for all direct-gain applications. For example, 45 Btu/°F/ft² may be more thermal storage than is necessary in buildings with low solar fractions—that is, in buildings in which the solar gains are able to meet no more than daytime heating needs. On the other hand, 45 Btu/°F/ft² may be insufficient to prevent occasional overheating in cases where the solar aperture is sized to provide a large fraction of the building's heating needs. In short, the design of thermal storage mass in a direct-gain building is an issue that involves a great deal more than simply the selection of a particular overall heat capacity. There are more details of this issue in sections O.1.2.1, O.1.2.2, O.1.2.5, and O.1.2.6.

Direct Gain / 155

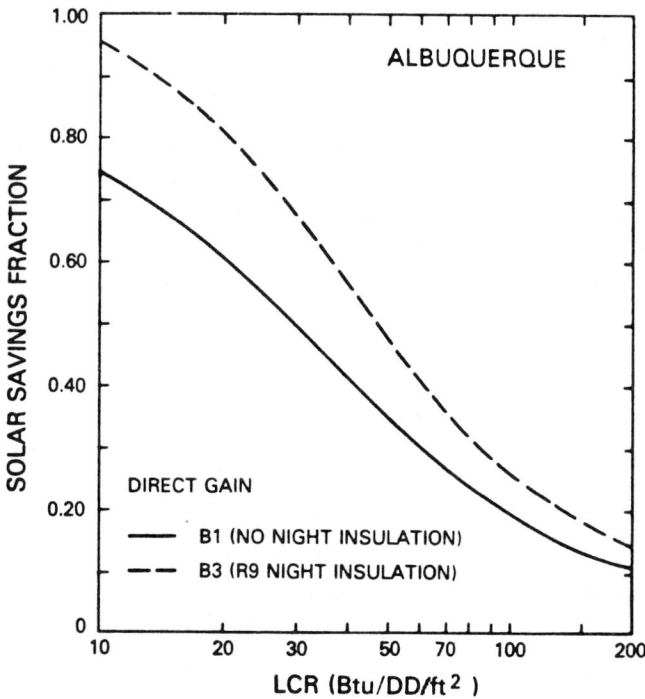

O-4: Effect on solar savings fraction of different values of *LCR*, applied to two direct-gain systems in Albuquerque, New Mexico.

These reference designs, in combination with the effect of parameter variations that are represented by the sensitivity data presented below, cover most direct-gain configurations of practical importance. The concept of a sun-tempered building has led to designs that have too little thermal mass to be amenable to analysis by the techniques presented here. These low-mass buildings are discussed in section O.2.

The parameters excluded from the definitions of the reference designs—the projected area (A_p) and the building load coefficient (*BLC*)—are combined into the load collector ratio ($LCR = BLC/A_p$). The quantity *LCR* then becomes the principal building variable for a given reference design. Performance estimates for a given location are functions of *LCR*. There are two options. The annual calculation, or *LCR* method, is the easiest and fastest to perform, but it has limited generality. Refer to section M.3 and to Appendix 21 for details. The monthly calculation, or *SLR* method, is lengthier but more general. Refer to section O.1.d, Chapter Q, and Appendix 18 for details. As an example of a performance estimate achieved by the *LCR* method, figure O-4 is a plot of the annual solar savings fraction (*SSF*) versus *LCR* for reference designs B1 and B3 (see figure O-2) in Albuquerque, New Mexico. The plot is obtained by simply drawing smooth curves through points obtained from the *LCR* table for Albuquerque in Appendix 21. A logarithmic scale is used for *LCR*, but this is not necessary.

O.1.b HEATING PERFORMANCE SENSITIVITY DATA

Data presented in this section express the sensitivity of the annual *SSF* to selected design parameters. In any given set of data, one or two parameters are varied while the others are held fixed at their reference values. The reference values are indicated by dots on the curves.

In order to avoid cluttering the text, only sample sensitivity curves are presented in this chapter. The sample plots illustrate the effect of various parameter variations on performance in Albuquerque, which has a moderate winter climate. In several instances, plots for other cities are also included. The complete set of sensitivity curves is contained in Appendix 22, which includes results for six reference cities: Albuquerque, New Mexico; Boston, Massachusetts; Madison, Wisconsin; Medford, Oregon; Nashville, Tennessee; and Santa Maria, California.

The above cities were chosen to represent wide geographical and climatological ranges. To apply the data, choose one of the reference cities (on the basis of climate similarity) to represent the city of interest. Two gross measures are of use in making the selection: the 65°F base heating degree-days and the solar radiation incident on a vertical surface. These data are available as monthly averages for numerous locations in the United States and Canada in Appendix 19, where they are tabulated as *D65* and *VS* respectively. Figure O-5 shows the January values of *D65* and *VS* for each of the six reference cities.

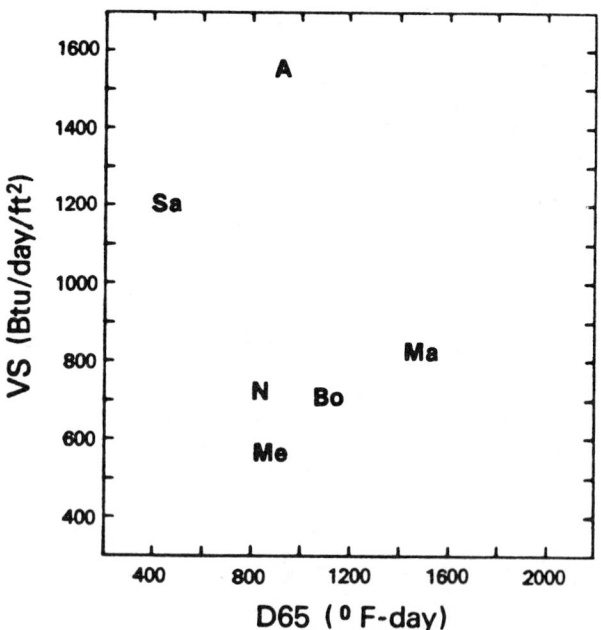

O-5: Climates of six cities mapped according to daily average solar radiation incident upon a vertical surface (*VS*) in each city and the January 65°F-base heating degree-days (*D65*) in each city. The cities are Albuquerque (A), Boston (Bo), Madison (Ma), Medford (Me), Nashville (N), and Santa Maria (Sa).

O.1.b.1 Thermal Storage Mass

Figure O-6 illustrates the sensitivity of annual heating performance to thermal storage mass thickness for a double-glazed configuration. The *LCR* in figure O-6 was held constant at 12. An additional parameter that has an important effect on mass sensitivity is the ratio of the mass surface area to the net glazing area (A_m/A_g). Four mass-area-to-glazing-area ratios (A_m/A_g = 2, 3, 6, and 10) are included on each graph, and the three reference design conditions are indicated by solid points.

It is assumed in the reference designs that the thermal storage mass is uniformly illuminated by the transmitted solar radiation. In practice, the mass can be considered sufficiently well-coupled to the solar gains if it is in the same zone as the illuminated mass, so that the various mass surfaces can exchange long-wave infrared and reflected short-wave radiation with each other. See section O.1.b.5 for details of the effect of the distribution of mass and solar radiation. Mass that is only convectively coupled to the direct-gain zone is only partially effective. A conservative analytical practice would be to disregard mass that is remote from the direct-gain zone unless special measures, such as forced air distribution, were taken to enhance the coupling.

Upon inspection of the complete set of mass sensitivity data, the following patterns emerge for high density (150 lb/ft³) concrete with a fixed A_m/A_g ratio:

1. Performance variations for mass thicknesses between 4 and 8 inches are small. The thickness may be reduced to 4 inches without incurring significant performance penalties. This generalization is independent of location, configuration, and mass surface area.
2. The range of mass thicknesses between 2 and 4 inches can be considered a transition region. In this region, performance penalties for reduced thicknesses become significant, but in some cases may be considered acceptable as design cost trade-offs.
3. For mass thicknesses below 2 inches, performance falls off much more rapidly than in the transition region. Under most conditions, it is not advisable to employ mass thicknesses of less than 2 inches particularly if a high *SSF* is sought.
4. The above rules apply to high-density masonry, of roughly 130–150 lb/ft³. Lower-density masonry has a lower thermal conductivity and, therefore, a smaller effective thickness for diurnal heat storage. The same heat storage capacity must, therefore, be achieved with material spread over a larger area.

These same patterns are observed whether or not the system includes nighttime insulation. It would be a mistake to assume that all points on the sensitivity curves represent designs that can be achieved in practice. In par-

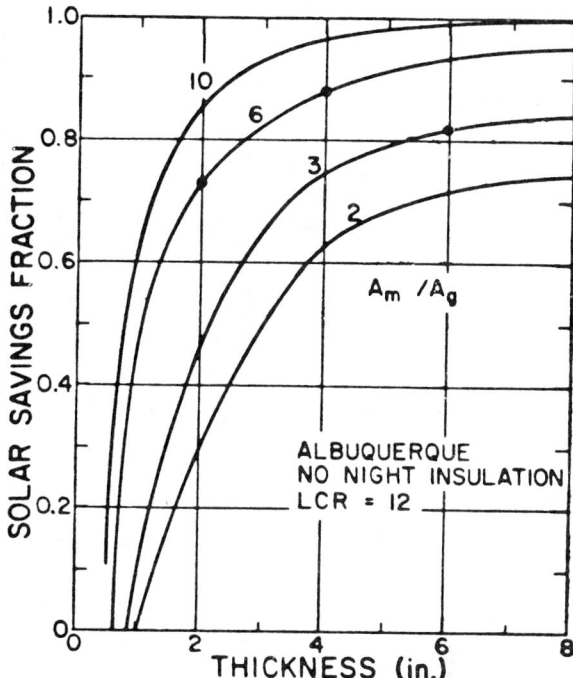

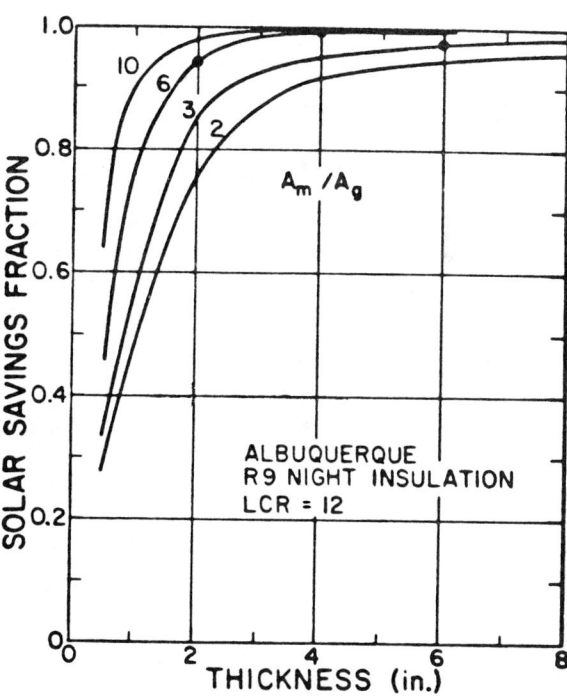

O-6: Effect on solar savings fraction of different mass thicknesses applied to insulated and uninsulated double-glazed configurations. Reference conditions are indicated by dots. Each graph shows results for configurations with four different mass-area-to-glazing-area (A_m/A_g) ratios.

ticular, the high A_m/A_g ratios may not be obtainable for large glazing areas that cause the *LCR* to be small: there may simply not be enough surface available on which to place the mass. This problem is aggravated by the fact that only mass that is located in building zones that experience direct solar gains can be credited to the system unless special measures have been taken to enhance the convective coupling with illuminated zones. Convective coupling can be enhanced either by a forced-air distribution system or by properly sizing the doorways and other apertures between adjacent zones [WRA-4], [WRA-5], [WEB]. Clerestory windows are often used to provide direct solar gains to the northern zones in a building, thereby reducing or eliminating the need for strong convective coupling to southern zones.

O.1.b.2 Properties of the Thermal Storage Medium

The effect of changes in specific heat (c), density (ϱ), thermal conductivity (k), and storage mass thickness (L), can be represented by two groupings of these properties. The first grouping, ϱcL, gives the thermal storage capacity of the mass layer per unit of surface area. The effect of changes in L was discussed in the previous section. That entire discussion can be converted to an analysis of performance sensitivity due to changes in the grouping ϱcL simply by multiplying all values of L (in feet) by 30 Btu/ft³/°F, the reference value of ϱc used for concrete.

The effect of varying the magnitude of the second grouping, ϱck, is illustrated in figure O-7 for Albuquerque. *LCR*s of 12, 48, and 120 are included for configurations with and without nighttime insulation. In order to facilitate interpretation of these plots, the ϱck products for various building materials are given in figure O-8. Performance is insensitive to variations in ϱck between 20 and 40 and only moderately sensitive to variations between 10 and 20. However, if the value of ϱck drops much below 10, performance begins to erode. The performance penalty is greatest for small values of *LCR* (high *SSF*) because the need for effective heat storage in these cases is most critical.

This observation does not necessarily imply that materials with ϱck products less than 10 should not be used for thermal storage mass in direct-gain buildings. Adobe, for example, has a ϱck product of approximately 6 (see figure O-4), but has, nevertheless, been successfully used as the thermal storage medium in many passive-solar homes in the Southwest. However, the

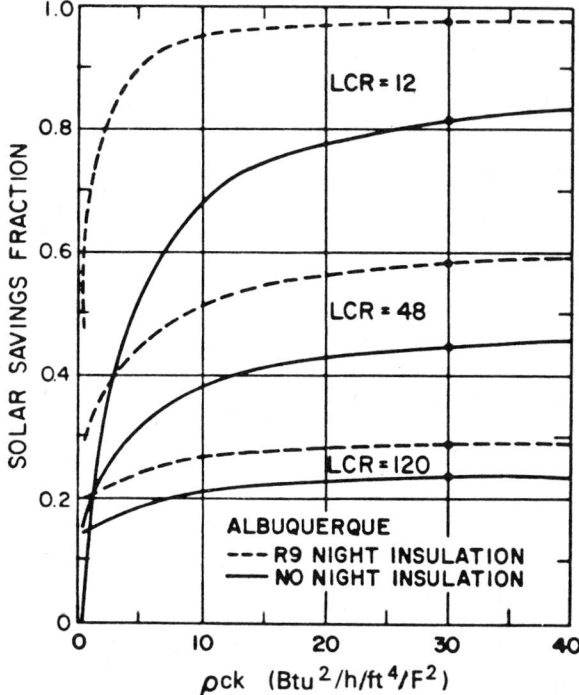

O-7: Effect on solar savings fraction of changes in the parameter ϱck, applied to 6-inch-thick double-glazed configurations with an A_m/A_g ratio of 3. Reference conditions are indicated by dots.

Material	Density (ρ) (in lb/ft³)	Specific Heat (c) (in Btu/lb/°F)	Thermal Conductivity (k) (in Btu/hr/ft/°F)	ρck
Magnesite Brick	158	0.22	2.2	76.5
Marble	162	0.21	1.5	51.0
Concrete (High Density Reference)	150	0.20	1.00	30.0
Plaster	132	0.43	0.42	23.8
Chrome Brick	200	0.17	0.67	22.8
Fireclay Brick	112	0.20	0.58	13.0
Concrete (Stone)	144	0.16	0.54	12.4
Concrete (Lightweight Aggravate)	120	0.21	0.43	10.84
Brick, Building	123	0.20	0.40	9.80
Adobe	80	0.20	0.38	6.08
Sand	95	0.19	0.19	3.43
Gypsum Board	50	0.26	0.10	1.30

O-8: Properties of building materials [ASH-2].

mass-area-to-glazing-area ratios in these buildings may be larger than the reference design values. The thick adobe walls are traditional and appear frequently in the Southwest, irrespective of the application of other passive solar design concepts. In a sense, therefore, the abundant thermal storage mass in adobe direct-gain buildings is free; this is also the case for any building that is inherently massive for structural reasons—provided that the structural mass is not thermally isolated from the living space.

In conclusion, the designer is advised to adjust the amount of thermal storage mass included in direct-gain buildings on the basis of the material properties. As a general rule, materials having relatively small values of $\varrho c k$ will necessitate the use of larger mass surface areas.

O.1.b.3 Solar Absorptance of Mass Surfaces

The effect on performance of the average solar absorptance of the thermal storage mass surface is illustrated in figure O-9 for Albuquerque. Although other design criteria relate to the extent of the performance penalty that is acceptable as a consequence of using light colors on the surface of thermal storage mass, it seems unwise to employ a color scheme that yields an average mass surface absorptance of less than 0.5 for direct-gain buildings resembling the reference designs. Below that value, the performance of buildings with small *LCR*s and no nighttime insulation begins to fall rapidly. This constraint is not at all severe, as indicated by the solar absorptances listed in figure O-10. A medium paint in any color and most unpainted masonry materials will exceed the suggested average absorptance of 0.5.

The results discussed above are for a direct-gain building with a mass-area-to-glazing-area ratio of 3. As the mass surface area increases relative to the glazing area, it becomes less important to employ dark colors on the mass surfaces. Indeed, for very large values of A_m/A_g (about 6 or greater), it may actually be desirable to employ light colors on at least some of the mass surfaces as a means of distributing solar radiation more uniformly about the living space.

O.1.b.4 Solar Absorption of Lightweight Surfaces

Lightweight objects, such as plants, rugs, wall hangings, furniture, or nonmassive partition walls, can impair the performance of direct-gain buildings, especially if they are dark-colored and are exposed to direct sunlight. When solar radiation is absorbed on the surface of a lightweight object that has little heat capacity, the surface temperature rises rapidly, causing heat to be convected rapidly to the room air. This process can cause

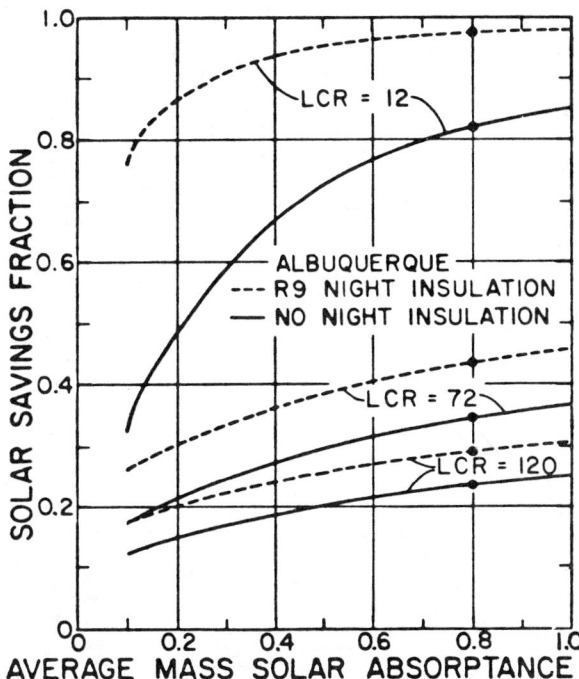

O-9: Effect on solar savings fraction of different average mass solar absorptance, applied to 6-inch-thick double-glazed configurations with an A_m/A_g ratio of 3. Reference conditions are indicated by dots.

Flat black paint	0.95
Black lacquer	0.92
Dark gray paint	0.91
Black concrete	0.91
Dark blue lacquer	0.91
Black oil paint	0.90
Gray slate	0.90
Stafford blue bricks	0.89
Dark olive drab paint	0.89
Dark brown paint	0.88
Dark blue-gray paint	0.88
Azure blue or dark green lacquer	0.88
Brown concrete	0.85
Medium brown paint	0.84
Medium light brown paint	0.80
Silver-gray slate	0.80
Brown or green lacquer	0.79
Medium rust paint	0.78
Light gray oil paint	0.75
Red oil paint	0.74
Red bricks	0.70
Uncolored concrete	0.65
Moderately light buff bricks	0.60
Medium dull green paint	0.59
Medium orange paint	0.58
Medium yellow paint	0.57
Medium blue paint	0.51
Medium Kelly green paint	0.51
Light green paint	0.47
White semigloss paint	0.30
White gloss paint	0.25
White lacquer	0.21

O-10: Solar absorptance of various materials.

overheating, which necessitates ventilation-cooling and thereby reduces the net solar gain of the structure. The extent to which performance is degraded is illustrated in figure O-11, where the annual *SSF* is plotted against the fraction of transmitted and reflected radiation absorbed on massless surfaces.

The lightweight absorption fraction is the fraction of solar radiation that directly heats the room air when it is transmitted through the glazing and when it is reflected from interior mass surfaces. The lightweight absorption fraction simulates lightweight walls, floors, and ceilings, and also lightweight furnishings placed in the direct-gain space by the occupant. (Notice that a lightweight absorption fraction is also employed in the analysis of sunspace performance, but in that case it simulates only the presence of plants and lightweight furnishings. Lightweight walls and ceiling are explicitly accounted for in the sunspace models. See section P.2.j. Determination of the lightweight absorption fraction in a given direct-gain space can only be done in practice as a rough estimate. Prediction of the lightweight absorption fraction of a space under design may be impossible because it depends on the way in which the space is later furnished. But the following procedure may be useful, at least as a way of visualizing in an approximate way the meaning of the lightweight absorption fraction:

1. Stand near the south-facing glazing, looking north. Estimate the fraction of your view that consists of lightweight floor, walls, ceiling, and other lightweight objects. If the lightweight objects have substantially different solar absorptances, do Step 1 separately for each different absorptance.
2. Estimate the solar absorptances of the lightweight objects.
3. Multiply the fractions from Step 1 by the absorptances from Step 2 and add the products. The result is an estimate of the lightweight absorption fraction.

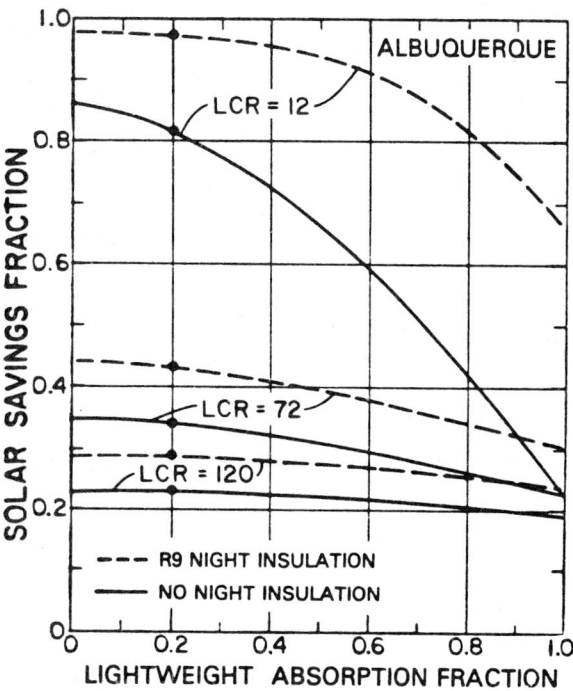

O-11: Effect on solar savings fraction of different lightweight absorption fractions, applied to 6-inch-thick double-glazed configurations with an A_m/A_g ratio of 3. Reference conditions are indicated by dots.

Both designer and occupant should try to minimize the occurrence of lightweight absorption in order to get the best possible performance from their passive solar heating system. Lightweight objects, especially those that are dark-colored, should be kept out of direct sunlight, and an effort should be made to avoid excessive solar-absorbing clutter in direct-gain zones. Whenever possible, use dark colors on high-mass surfaces and light colors on low-mass surfaces.

O.1.b.5 Mass Distribution

The effect of mass distribution in direct-gain buildings for systems with transparent or diffusing glazing is presented in figure O-12. These results were obtained with

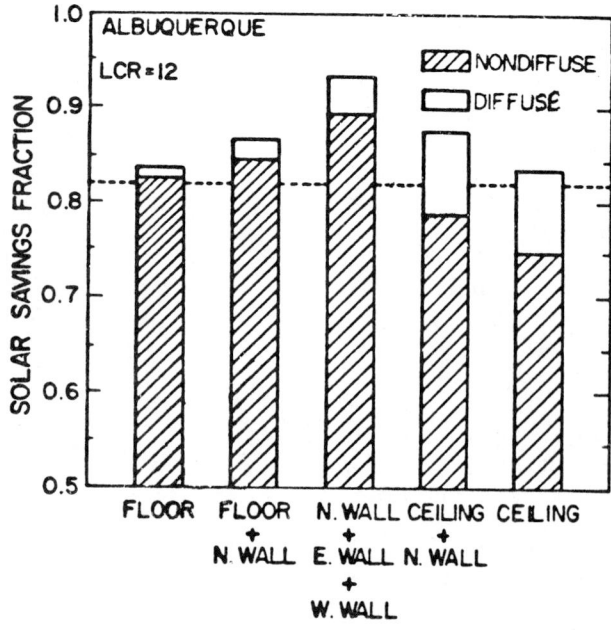

O-12: Effect on solar savings fraction of various distributions of thermal mass in a direct-gain space with a 6-inch-thick double-glazed configuration, an A_m/A_g ratio of 3, and no nighttime insulation.

the three-dimensional SUNSPOT Level II computer model. The cross-hatched bars show the annual SSF achieved by a reference direct-gain design (mass thickness = 6 inches, $A_m/A_g = 3$) with nondiffusing double glazing and without nighttime insulation, as the location of the thermal storage mass is progressively raised from the floor to the ceiling. The blank bars show the performance increments that result from the introduction of diffusing glazing that is otherwise assumed to have the same transmission characteristics as that of the original reference design. The dashed line, which traverses the entire graph, shows the performance predicted by the two-dimensional (SUNSPOT Level I) model, which was used in the other sensitivity studies and in the solar load ratio correlations. The Level I model does not recognize variations in mass location or diffusive characteristics of the glazing. From the results presented in figure O–12, it is evident that optimum performance is achieved by locating thermal storage mass on vertical surfaces and by employing diffusing glazing (provided that the diffusing glazing does not significantly reduce solar transmittance). This observation is a consequence of the low conductances for heat-flow from the room air to the floor and from the ceiling to the room air, which reduce the effectiveness of horizontal surfaces. Vertical surfaces, on the other hand, easily receive and release heat by convection. The following are suggestions for placement and coloring of massive and lightweight surfaces:

1. The lightweight surfaces in a direct-gain building should be relatively light in color, while the massive surfaces should be relatively dark. This precaution should assure adequate distribution of transmitted radiation, as well as preferential absorption on massive surfaces.
2. The first surface struck by transmitted solar radiation should be of a light to moderate color. This measure will help promote uniform heating of the direct-gain zone, but should only be invoked for a massive surface if large areas of other relatively dark massive surfaces are exposed in the same zone.
3. Massive surfaces should be distributed about a direct-gain zone as uniformly as possible. The use of masonry for interior partitions is especially effective because both sides of the partition can face direct-gain zones, thereby doubling the effective surface area, and because vertical surfaces are most favorably oriented for free convective heat exchange with the room air.

In connection with the above recommendations, it is suggested that the reader review the solar absorptances listed in figure O–10. Moderately light-colored buff bricks and concrete blocks painted a light beige color both have a solar absorptance of about 0.6, which is considerably higher than most people would expect.

O.1.b.6 Mass Coverings

The question of the extent to which various types of floor or wall coverings might hurt the performance of direct-gain buildings if they are placed over thermal storage mass surfaces arises repeatedly. Concrete floor slabs, for example, provide the most significant storage medium in many buildings but generally require a covering of some sort for aesthetic reasons. There are many other circumstances that might lead a builder or occupant to cover mass surfaces, and the effect of this action can be assessed by studying figure O–13 and the related figures in Appendix 22. Performance sensitivity for decorative coverings with R-values of from 0.0 to 2.0 is given in figure O–13 for Albuquerque. The covering is assumed to blanket the entire mass surface. The solar absorptance ($ALFC$) of the covering is 0.8—the same value used for the mass surface itself in the reference design. In Appendix 22, plots for Albuquerque, Madison, and Santa Maria are presented for $ALFC = 0.8$ and (to show the effect of employing light-colored covering materials) for $ALFC = 0.3$.

Inspection of figure O–13 shows that the effect of mass coverings on performance can be quite drastic, especially for buildings with small LCRs and no nighttime insulation. The performance of buildings in this

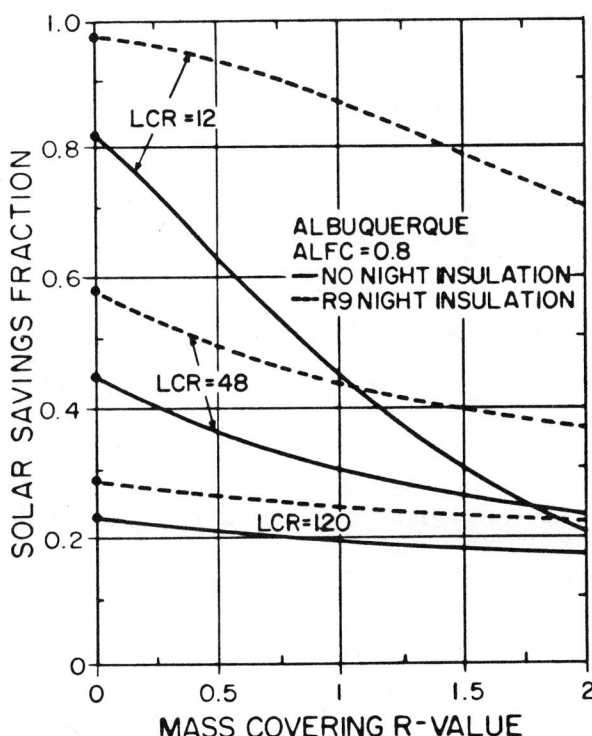

O–13: Effect on solar savings fraction of coverings with different R-values, applied to a 6-inch-thick double-glazed configuration with an A_m/A_g ratio of 3. Reference conditions are indicated by dots.

Direct Gain / 161

Material	R-Value
Tile, asphalt, linoleum, vinyl, rubber (1/16 inch)	0.05
Terrazzo tile (1.0 inch)	0.08
Tile, asphalt, linoleum, vinyl, rubber (1/8 inch)	0.10
Cork tile (1/8 inch)	0.28
Plywood, Douglas fir (1/4 inch)	0.31
Gypsum or plasterboard (3/8 inch)	0.32
Gypsum or plasterboard (1/2 inch)	0.45
Plywood, Douglas fir (1/2 inch)	0.62
Wood, hardwood finish (3/4 inch)	0.68
Carpet and rubber pad	1.23
Gypsum or plasterboard (3/8 inch over 1/2-inch furring strips	1.68
Gypsum or plasterboard (1/2 inch over 1/2-inch furring strips)	1.81
Carpet and fibrous pad	2.08

O-14: R-values of various wall and floor coverings [ASH-2].

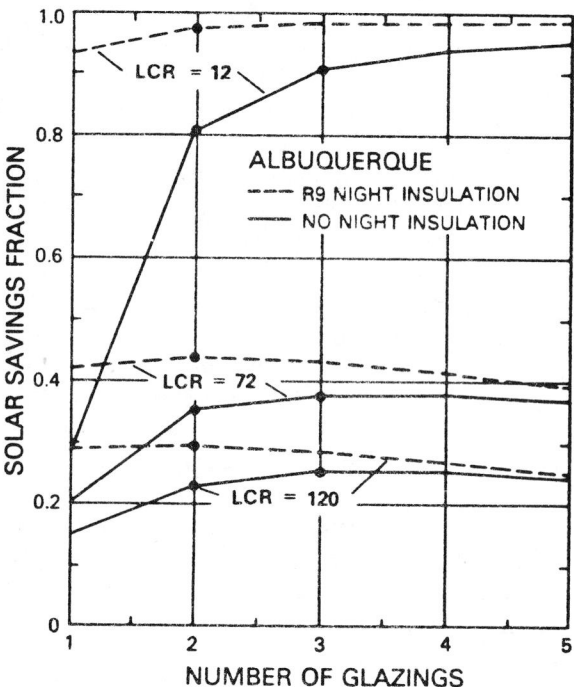

category, as has been indicated by earlier results, is very dependent on the effective use of thermal storage mass. Penalties exceed 10% for coverings with R-values of 0.25 or greater that are situated in buildings with *LCR*s less than 24 and with no nighttime insulation. It is, therefore, suggested that 0.25 be considered a maximum acceptable R-value except in buildings that have nighttime insulation and/or *LCR*s of 48 or above—in which case, 0.50 might be taken as the upper limit. If, as is most likely, the entire mass surface is not to be covered, simply scale the penalty indicated by the appropriate R-value sensitivity curve by the ratio of the covered mass area to the total mass area to obtain an estimate of performance.

The R-values of various wall and floor coverings are given in figure O-14, which shows that many attractive alternatives to bare, unconditioned concrete surfaces are available that will not significantly impair performance. Concrete surfaces that have been dyed and/or scored to resemble tile or other decorative materials are also attractive and have the advantage of causing no reduction in performance whatsoever.

O.1.b.7 Number of Glazings

The number of glazing layers required on direct-gain buildings depends largely on climate and on whether or not nighttime insulation is employed. In figure O-15, the sensitivity of performance to the number of glazings is given for Albuquerque, which has a moderate climate, and for Madison, which has a severe climate. Notice that double glazing yields no significant increase in the performance of nighttime insulated systems in Albuquerque, but is very effective in combination with nighttime insulation in the harsh Madison climate. For systems lacking nighttime insulation, double glazing is a must in

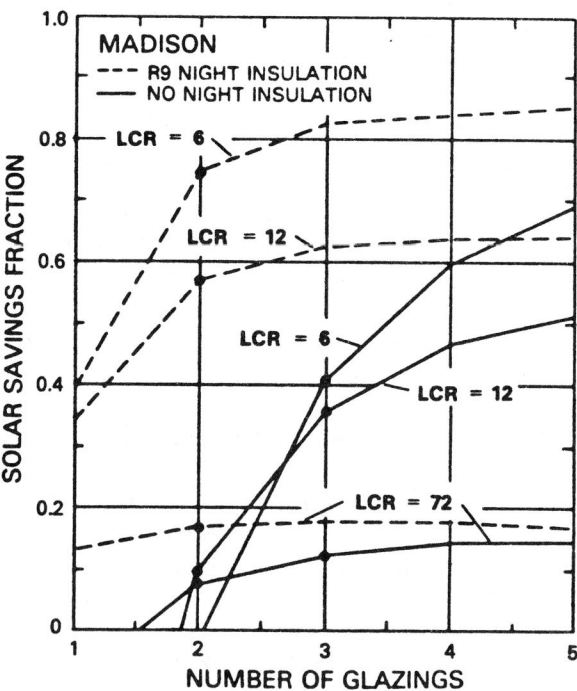

O-15: Effect on solar savings fraction of different numbers of glazings, applied to a 6-inch-thick configuration with an A_m/A_g ratio of 3, in two cities. Reference conditions are indicated by dots.

Albuquerque, and triple or quadruple glazing is advisable in Madison. The number of glazings recommended on direct-gain buildings is summarized in figure O-16 for three representative cities.

O.1.b.8 Temperature Bounds

The effect of interior temperature bounds on the performance of a direct-gain building with no night insulation located in Albuquerque is shown in figure O-17. The minimum room temperature, which appears on the horizontal axis, is the thermostat setpoint. These data were calculated in the absence of internal sources. If internal sources were included, enough heat would be introduced to the structure to increase the inside air temperature by 5 to 7 Fahrenheit degrees in a typical residence without requiring any additional auxiliary heat. (The temperature rise could be considerably more in a building sealed and insulated to current standards.) The 65°F thermostat setpoint employed for the reference design would then represent a setpoint of 70° to 72°F in the presence of internal sources.

The *SSF* on the vertical axis of figure O-17 is calculated relative to a net reference load with a reference temperature equal to the reference thermostat setting of 65°F. The sensitivity curves, therefore, show how much energy is saved *relative to the reference setpoint* when the actual setpoint is varied. The lower boundary of the sensitivity bands gives the performance when the maximum room temperature is limited to 75°F by ventilation cooling, while the upper boundary represents performance at a maximum room temperature of 85°F. For a given site, nighttime insulation status, and *LCR*, this set of sensitivity plots enables one to determine the effect of simultaneous variations of the minimum room air temperature from 55° to 75°F and the maximum room air temperature from 75° to 85°F. The two temperature-swing extremes that may therefore be assessed are a constant room air temperature of 75°F (no temperature swing) and room air temperature variations from 55° to 85°F (a 30 Fahrenheit-degree temperature swing).

Energy-efficient performance is, of course, very sensitive to the minimum thermostat setting under all conditions. Notice, however, that allowing a high maximum temperature inside a building (with its associated discomfort) improves performance only for small *LCR*s, and then primarily when the thermostat for auxiliary heat is set relatively high. Buildings with small *LCR*s often have much more transmitted solar radiation than can be utilized during the day, and a higher allowed maximum room temperature permits a larger part of the excess heat to be stored for nighttime use.

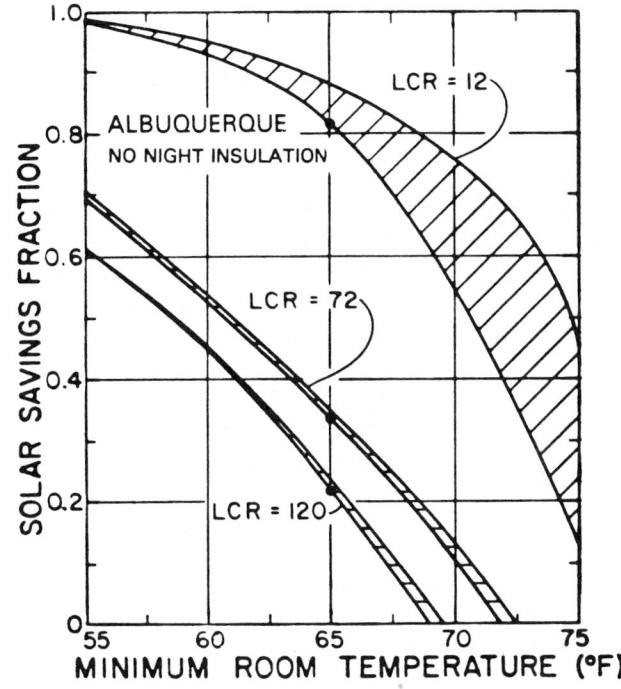

O-17: Effect on solar savings fraction of different base thermostat settings, applied to a 6-inch-thick double-glazed configuration with an A_m/A_g ratio of 3. The cross-hatched bands correspond to ranges of maximum room temperature—the lower edge of each band corresponding to 75°F and the upper edge to 85°F. The reference value is 75°F; reference conditions are indicated by dots.

O-16: Recommended number of glazings for direct-gain buildings in three cities with different types of winter climate.

Location	No Nighttime Insulation	R9 Nighttime Insulation
Santa Maria (Mild Climate)	1 or 2	1
Albuquerque (Moderate Climate)	2	1 or 2
Madison (Severe Climate)	3 or 4	2

O.1.b.9 Nighttime Thermostat Setback

The benefits obtainable from setting the thermostat back at night are not as large in a passive-solar home as they would be in a home of conventional design. For much of the night, a high-solar-fraction direct-gain building with a small *LCR* will hold the inside air temperature above the thermostat setting without the aid of auxiliary heat. Only in the early morning hours when heat stored in the mass of the structure has been depleted is there generally a demand for heat from the backup system. The thermostat setback, therefore, goes unnoticed much of the time. At larger values of *LCR*, less solar heat is stored for nighttime use and, under these conditions, setting the thermostat back can yield significant savings.

The effect of nighttime setback on direct-gain buildings in Albuquerque is illustrated in figure O–18 for systems with and without nighttime insulation. The setback period begins at 11:00 P.M. each evening and ends at 6:00 A.M. the next morning. A 10-Fahrenheit-degree setback —from the reference setting of 65°F to 55°F—can yield up to a 0.15 increase in annual *SSF*. Flat regions on these sensitivity curves indicate low temperature ranges that are seldom, if ever, reached in practice, regardless of the thermostat setpoint.

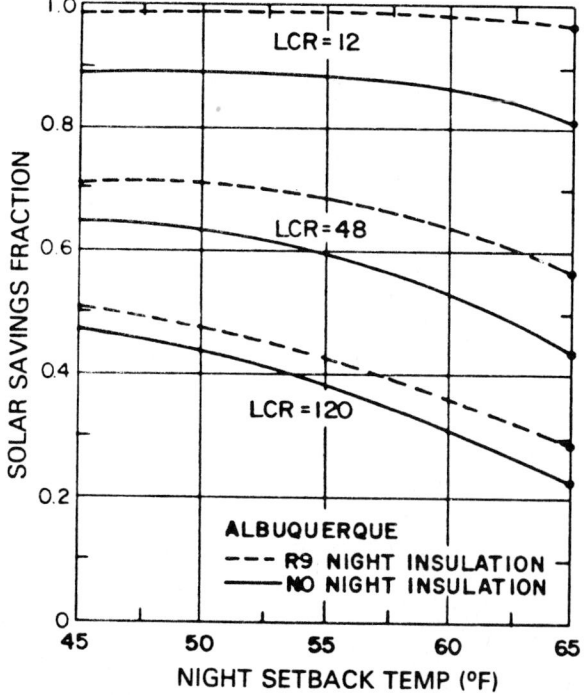

O–18: Effect on solar savings fraction of different nighttime thermostat settings, applied to a 6-inch-thick double-glazed configuration with an A_m/A_g ratio of 3. In this graph, nighttime was taken to mean the hours of 11:00 P.M. to 6:00 A.M. The reference thermostat setting is 65°F; reference conditions are indicated by dots.

O.1.c OVERHEATING SENSITIVITY DATA

The inside air temperature has been assumed to be limited to 75°F in all performance data reported so far. It is assumed that the excess heat is instantaneously vented to the outside air, provided that the ambient temperature is less than 75°F. If the ambient temperature is 75°F or above, an idealized auxiliary cooling system is assumed to remove the excess heat. In practice, ventilation cooling must generally be provided by occupant intervention —that is, by opening windows, doors, or special vents whenever the temperature becomes uncomfortably warm in the living space. An alternative form of intervention would be for the occupant to reduce the solar gains by partially closing drapes or other shading devices. Since the inside air temperature was limited to 75°F in the data reported so far, the effect of occupant intervention on performance is included. However, no measure of occupant discomfort or inconvenience has been provided. In this section a relative measure is given of the effect of thermal storage mass and lightweight solar absorbing surfaces on thermal comfort in direct-gain buildings.

O.1.c.1 Thermal Storage Mass and Thermal Comfort

In order to quantify the tendency of a given direct-gain building to overheat, a series of data is presented for the month of January in which the maximum inside air temperature is not limited by any form of ventilation or shading. The maximum air temperature occurring within the building under these conditions provides a relative basis for determining the effect of various parameters on the thermal comfort of the living space.

The maximum room temperature reached in Albuquerque is plotted as a function of the mass-area-to-glazing-area ratio for mass thicknesses of 2, 4, and 8 inches in figure O–19. Each figure includes sensitivity curves for three *LCR*s. The reference design mass-area-to-glazing-area ratios are indicated on each curve by a solid print, except in the case of the graph for the mass thickness of 8 inches, in which the thickness does not correspond to any of the reference designs. Remember that the temperatures indicated on these figures may only be reached once during the one-month test period and even then only if no occupant intervention is allowed. The very high temperatures indicated for some configurations are, therefore, not as serious a problem as one might at first suppose; they do indicate, however, on a relative basis, the extent to which the occupant may be kept busy adjusting ventilation rates and shading in order to keep the structure comfortable. To illustrate this point, a complete temperature histogram is presented in figure O–20 for a building in Albuquerque during January. The building has a thermal storage mass thickness of 4 inches, an A_m/A_g ratio of 3, and an *LCR* of 12. The his-

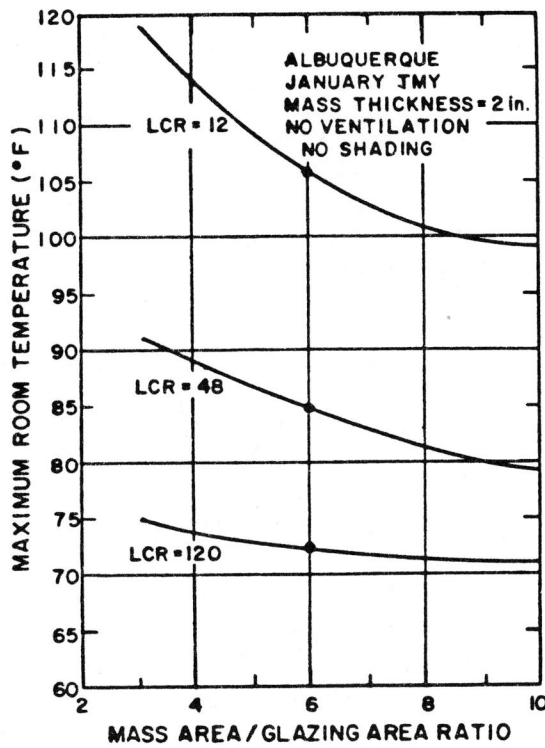

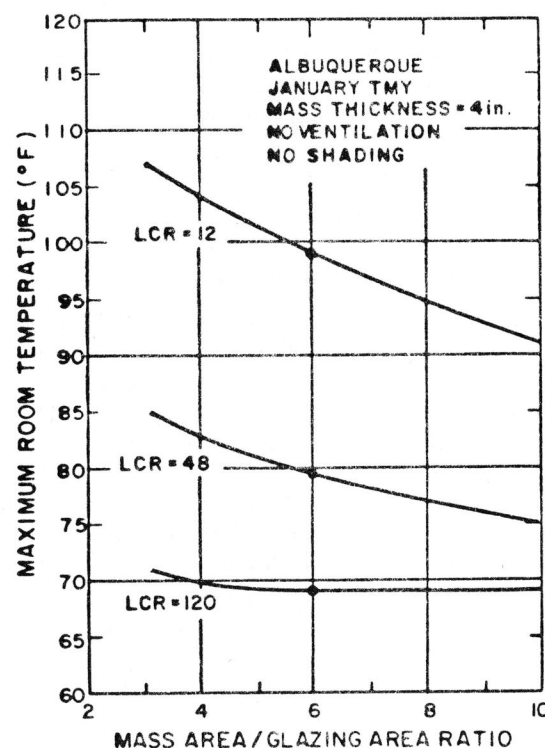

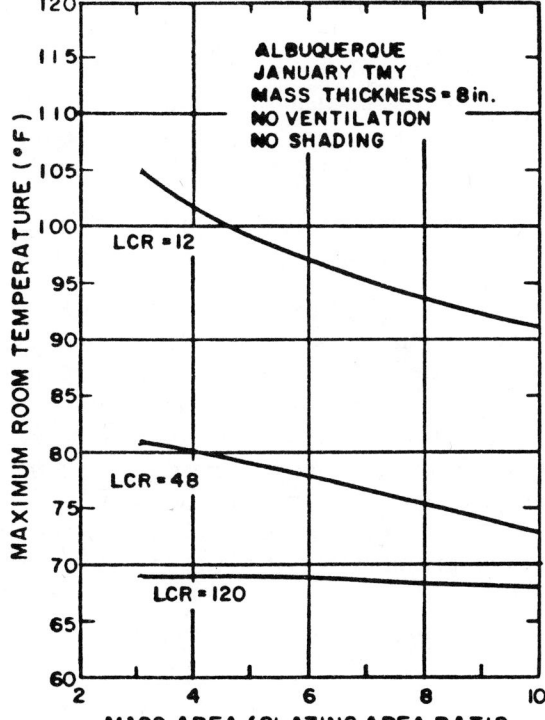

O-19: Maximum room temperatures (in degrees Fahrenheit) obtained in January at different mass-area-to-glazing-area (A_m/A_g) ratios, applied to double-glazed configurations of three different thicknesses. Reference conditions are indicated by dots.

O-20: January room-temperature histogram for a 4-inch-thick double-glazed configuration with an A_m/A_g ratio of 3 and with no nighttime insulation.

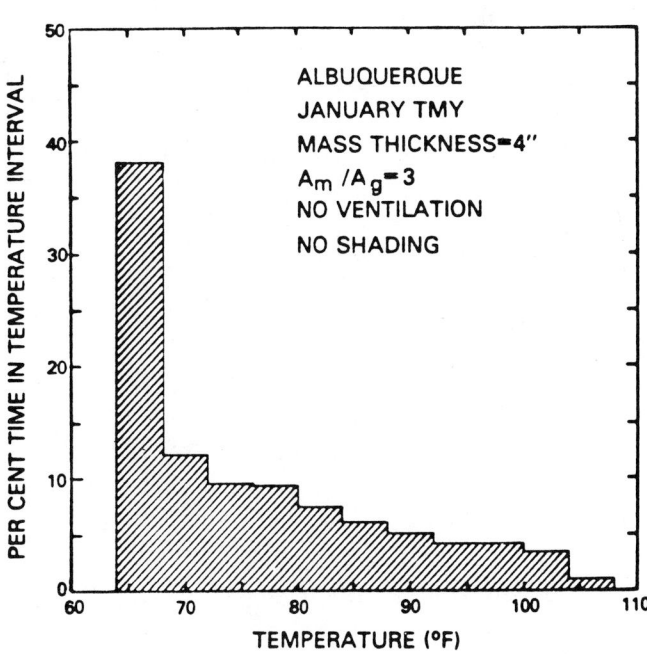

Direct Gain / 165

togram corresponds to the condition represented by a maximum temperature of 108°F on the left end of the upper curve of the graph for mass thickness of 4 inches in figure O–19. Notice that the histogram indicates that the building will have an air temperature in the range of 104° to 108°F only 1% of the time during January, and then only if no shading or ventilation is allowed.

As expected, buildings with small *LCR*s (which imply large projected areas relative to the loads) are most sensitive to the amount of thermal storage mass. Notice that comfort continues to improve as the mass-area-to-glazing-area ratio approaches 10 and that increasing the mass thickness from 2 to 4 inches yields a significant reduction in the maximum room temperature. However, increasing the mass thickness from 4 to 8 inches does not result in a significant improvement in building comfort. These results indicate that one can best provide for thermal comfort in a direct-gain building by employing thermal storage mass with a thickness of about 4 inches, spread over as large an area as is practicable.

O.1.c.2 Lightweight Surfaces and Thermal Comfort

The discomfort caused by solar absorption on non-massive surfaces can be demonstrated by the same type of illustration as is used to demonstrate the beneficial effect of thermal storage mass. Figure O–21 shows that the maximum room temperature in Albuquerque decreases rapidly as the percent of transmitted solar radiation absorbed on lightweight surfaces decreases. This is a graphic illustration of the importance of avoiding excessive solar-absorbing clutter in direct gain zones.

O.1.d SOLAR LOAD RATIO (*SLR*) CORRELATIONS

A general definition of the monthly solar load ratio (*SLR*) is:

$$SLR = Q_s/Q_{load}$$

where:

Q_s is a monthly solar radiation input to the solar aperture

Q_{load} is a monthly building heat load

For direct-gain systems, Q_s is the solar radiation absorbed in the direct-gain space and Q_{load} is the net reference load plus the effective load of the solar wall. Expressing these quantities per unit of projected area, the direct-gain *SLR* may be written as:

$$SLR = (S/DD)/(LCR + G)$$

where:

S is the monthly absorbed solar radiation per unit of projected area (in Btu/ft²)

DD is the monthly Fahrenheit heating degree-days

LCR is the building load collector ratio (in Btu/DD/ft²)

G is the effective solar wall load coefficient per unit of projected area (in Btu/DD/ft²)

The quantity G is an adjustable parameter that is determined implicitly in the correlation procedure; its values are listed below in figure O–22. The values of G are determined so as to minimize the squared deviations in the annual *SSF*; they are not equal to the solar wall load coefficients, but might instead be called the *effective* solar wall load coefficients.

The *SLR* correlations are correlations of the monthly *SSF*s with the monthly *SLR*s. The correlations have the form:

$$SSF = 1 - K[1 - F(SLR)]$$

where:

$$F(SLR) = \begin{cases} A \times SLR, & SLR < R \\ B - C\exp(-D \times SLR), & SLR \geq R \end{cases}$$

and:

$$K = 1 + G/LCR$$

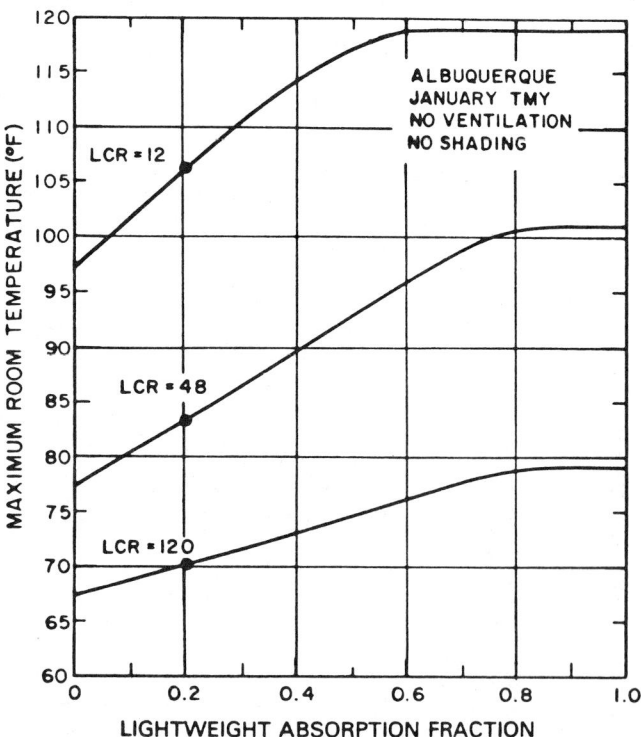

O–21: Maximum room temperatures (in degrees Fahrenheit) obtained in January at different lightweight absorption fractions, applied to a 6-inch-thick double-glazed configuration with an A_m/A_g ratio of 3. Reference conditions are indicated by dots.

The coefficients *A, B, C, D, R,* and *G* (which also appears in the definition of *SLR*) are adjustable and are determined through the correlation procedure for each reference design. Their values are listed in figure O-22 for each of the reference designs characterized in figure O-2. These correlation coefficients are also tabulated in Appendix 18 along with the coefficients for thermal storage wall and sunspace systems. The standard deviations are listed also as an expression of the precision with which the annual *SSF* may be estimated by use of the *SLR* correlations.

For a given reference design, the coefficient *G* is fixed; therefore, one parameter, *LCR,* and one monthly variable, *S/DD,* determine the monthly *SLR* and hence the monthly *SSF*. This relationship is expressed for each of the reference designs in graphical form in figure 18-11 of Appendix 18, of which figure O-23 is a sample. These figures show a family of monthly *SSF*-versus-*S/DD* curves, in which the various curves in the family correspond to various *LCR* values.

O.2 LOW-MASS SUN-TEMPERED BUILDINGS

O.2.a THE CONCEPT OF SUN-TEMPERING

The concept of a sun-tempered building has emerged from attempts to induce the generally conservative community of mainstream builders to construct passive-solar homes that derive part of their space-heating needs from the sun. The mainstream builder, as opposed to the innovator or early adopter of new shelter concepts, serves the mass-housing market and is characterized by a reluctance to make radical changes in the design of houses that have been successfully built and marketed in the past. Such a builder generally sees himself or herself as a businessperson whose primary job is to be responsive to the demands of customers.

The concept of sun-tempering is best presented as a series of minor modifications to home designs that have already been proven in the marketplace. A higher standard of insulation and sealing may also be incorporated at this time. The first step is one that must be taken during the initial planning of a new development. It is necessary to orient the long axis of each house that is to be sun-tempered in the east-west direction. The second step is to allot a substantial portion (say, 80%) of the window space that would ordinarily be used throughout the facade to the south side of the building. The remaining window area is usually placed mostly on the east and west sides, with very little or no window area to the north. The third and final step is to double- or triple-glaze all windows. The cost of these modifications to the builder and to the builder's customers is small; nevertheless, under appropriate circumstances (to be addressed shortly), a significant part of the space-heating load will be displayed by direct solar gains. Notice that sun-tempering also reduces cooling loads; further reductions in this area can be achieved by moving the east-facing and west-facing windows to the north side of the building (unless they can be well-shaded during the cooling season).

To get an idea of the glazing areas involved in sun-tempering, consider a small, single-family, detached residence with a 1,500-square-foot floorspace. Typical houses of this type have a total window area that equals about 15% of the floorspace—that is, $0.15 \times 1,500$ ft^2 = 225 ft^2. Placing 80% of the window area on the south side of the building yields a solar collection area of 180 square feet. At this point, sun-tempered buildings can be divided into two very different categories. Suppose, in the first case, that the home in question is built on a concrete floor slab but is otherwise of frame construction. The floor slab will be at least 4 inches thick, which is

Reference Design Designation (see figure O-2)	A	B	C	D	R	G	stdv
A1	0.5650	1.009	1.044	0.712	0.3931	9.36	0.046
A2	0.5906	1.006	1.065	0.81	0.4681	5.28	0.039
A3	0.5442	0.972	1.130	0.927	0.7068	2.64	0.036
B1	0.5739	0.995	1.251	1.061	0.7905	9.60	0.042
B2	0.6180	1.000	1.276	1.156	0.7528	5.52	0.035
B3	0.5601	0.984	1.352	1.151	0.8879	2.38	0.032
C1	0.6344	0.989	1.527	1.438	0.8632	9.60	0.039
C2	0.6763	0.999	1.400	1.394	0.7604	5.28	0.033
C3	0.6182	0.986	1.566	1.437	0.8990	2.40	0.031

O-22: Direct-gain solar load ratio correlation coefficients.

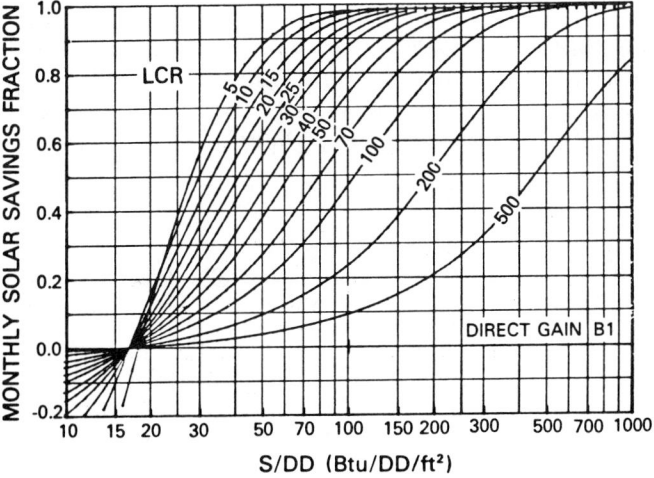

O-23: Monthly solar savings fraction plotted against monthly solar radiation absorbed per degree-day (*S/DD*), for direct-gain reference design designation B1.

within the recommended mass thickness range, and will have a gross area of 1,500 square feet. Even if only half of the gross area of the slab is available for thermal storage, there still exists a mass-area-to-glazing-area ratio of:

$$\left(\frac{A_m}{A_g}\right) = \left(\frac{750}{180}\right) = 4.2$$

This figure is in the range of the reference values used for the high-mass buildings discussed in the previous section. Therefore, although the building may be considered a lightweight structure, it is thermally massive relative to the small solar collection area. Sun-tempered buildings built on thermally accessible floor slabs or having other high-mass elements within the insulation of the building envelope are thus amenable to analysis by the methods presented in the preceding section. The mass sensitivity curves provide a means of correcting departures from the reference values of the thickness and area ratios that appear in performance estimates obtained from the SLR correlation.

If, however, the sun-tempered building has a wood frame floor over a crawl space, and there are no massive elements within the insulated shell of the structure, then the building is a low-mass case that represents the second category. These buildings have very little thermal storage relative to the size of the solar collection area. The most significant thermal storage medium within the structure is the gypsum board that lines the walls and the ceiling. The correlation method of the preceding section is not applicable to the analysis of these buildings. The characteristics of low-mass, sun-tempered reference designs that are considered representative of this building type are presented in the following section.

O.2.b THE REFERENCE DESIGN

All of the previously specified characteristics of the reference designs for high-mass direct-gain buildings (see figure O–3) apply to the present case except those pertaining to the thermal storage mass, which are replaced by the values given in figure O–24. Night insulation is not included and only double-glazed windows are considered.

A gypsum-board thickness of ½ inch is used in most wood-frame houses. The properties given in figure O–24 were obtained from reference [ASH-2], and the A_m/A_g ratio of 20 was calculated for a 1,500-square-foot structure with a collection area of 180 square feet, assuming that only the south zones were available for thermal storage. Sensitivity data presented below allow performance estimates based on other assumptions concerning availability of gypsum board for thermal storage.

O.2.c PERFORMANCE OF THE REFERENCE DESIGN

Plots of SSF versus LCR for low-mass sun-tempered buildings, with and without nighttime insulation, are presented in figure O–25 for seven cities. The graphs are ordered according to increasing severity of the winter climate of each city, as measured by the ratio of the solar radiation incident on a vertical, south-facing surface to the 65°F-base heating degree-days. This ratio is tabulated in figure O–26. The solar radiation and degree-days sums are for those months in which there are 50 heating degree-days or more. The data are from Appendix 19, where they are tabulated as VS and D65.

The highest solar savings fraction observed for the case with no nighttime insulation is 0.18, which occurs in Albuquerque at LCR of about 72. The SSF decreases for LCRs greater than 72 because the solar gain decreases relative to the building load. The SSF also decreases for LCRs less than 72 because further increases in solar gain relative to the building load are not fully utilizable. It is important to note at this point that design decisions should be based on auxiliary heat requirements, not on SSF. The quantity SSF is a ratio that allows presentation of performance information in a compact form; however, as seen in the following example, it is necessary to calculate the auxiliary heat to determine the effect of design decisions on energy consumption.

The SSF of the sun-tempered building with no night-time insulation decreases as the location is shifted from Albuquerque to the increasingly severe climates of Charleston, Nashville, Medford, and Boston, finally dropping to a local maximum SSF of 0.03 in Madison, Wisconsin. This decrease of SSF in harsh winter climates was expected. However, the decrease of the maximum SSF in the milder climate (relative to Albuquerque's) of Santa Maria was unforeseen. Apparently, the low-mass sun-tempered buildings can, at best, meet only the daytime portion of the heat load, and have insufficient thermal storage for nighttime carryover. As a result, Albuquerque, in which a relatively high fraction of the annual heat load occurs during the daylight hours, is a preferred location.

Thermal storage capacity (per square foot of glazing)	10.8	Btu/°F/ft²
Gypsum board properties		
Thermal conductivity (k)	0.0923	Btu/hr/ft/°F
Density (ρ)	50	lb/ft³
Specific heat (c)	0.26	Btu/lb/°F
Thickness	0.5	inch
Mass-area-to-glazing-area ratio	20	

O–24: Characteristics of thermal storage mass in the reference low-mass sun-tempered building design.

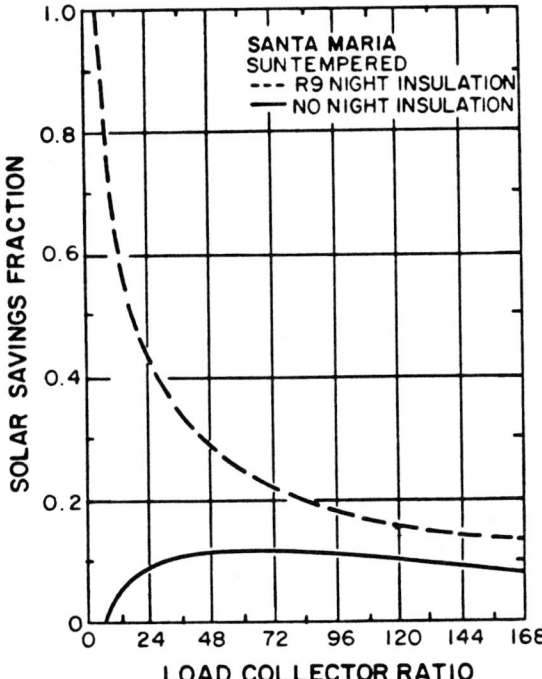
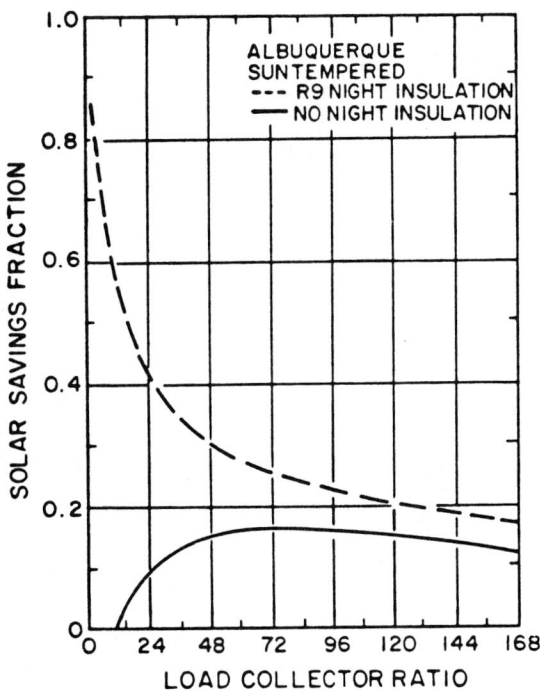
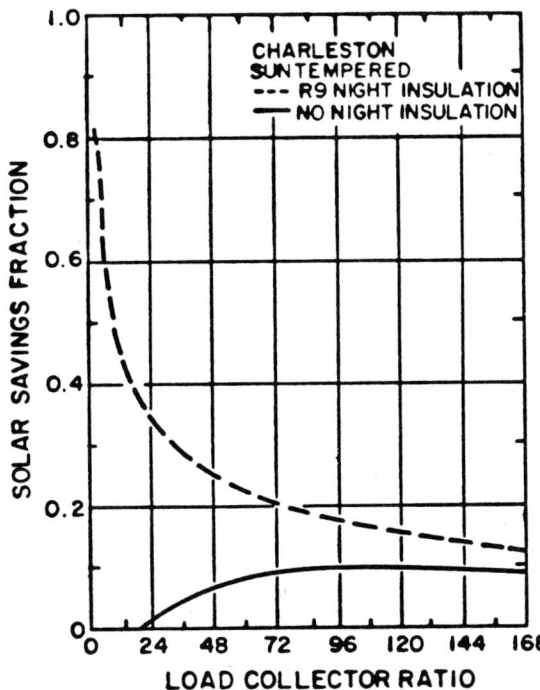
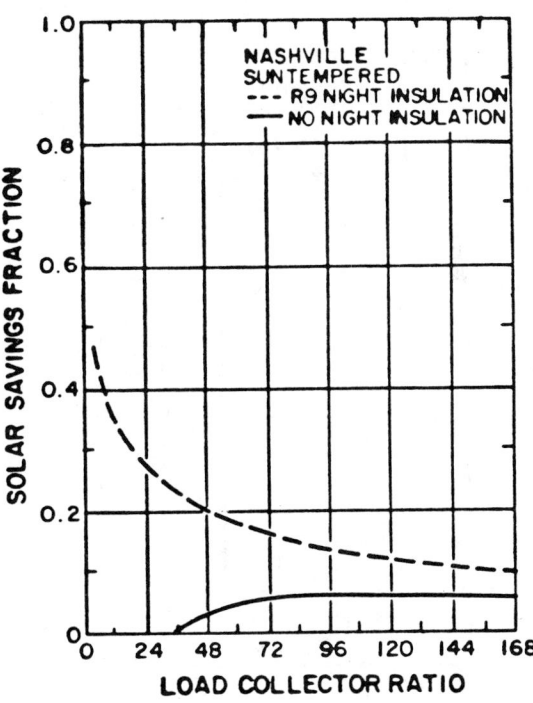

O-25: Effect on solar savings fraction of different load collector ratios, applied to low-mass sun-tempered buildings in seven cities with progressively more severe winter climates.

Direct Gain / 169

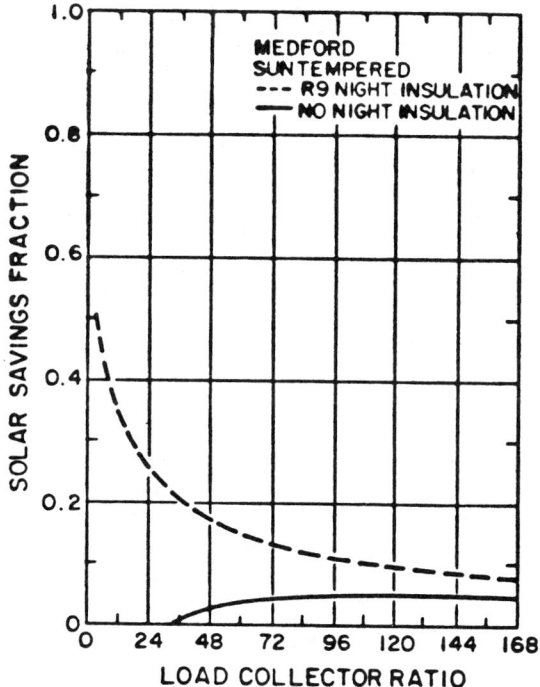

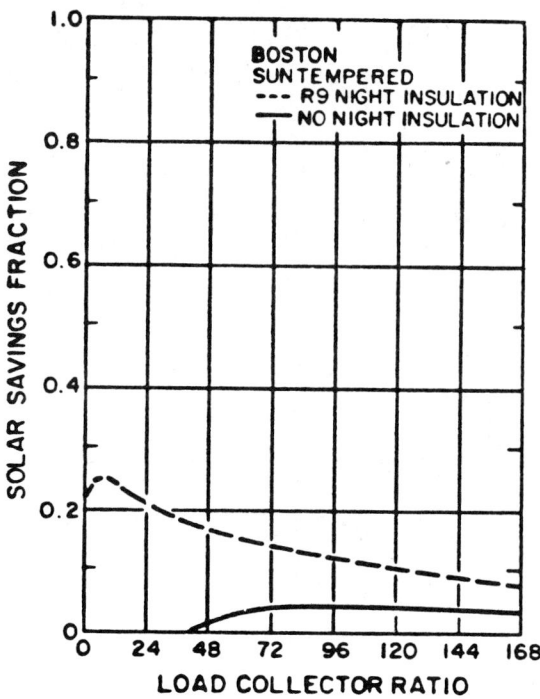

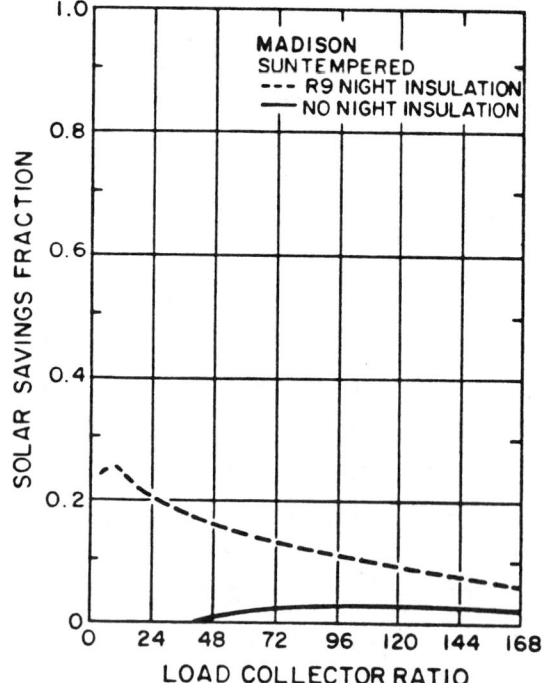

O-25: Continued.

City	VS/D65 (Btu/ft^2/°F/day)
Santa Maria	140
Albuquerque	117
Charleston	100
Nashville	88
Medford	76
Boston	57
Madison	46

O-26: Heating season VS/D65 ratio.

The addition of nighttime insulation will, of course, improve the performance of sun-tempered buildings by reducing nighttime heat losses that would otherwise offset solar gains. Solar fractions are still far below those obtainable in higher-mass structures. For sun-tempered buildings with R9 nighttime insulation at *LCR* of 24, an *SSF* of about 0.40 is realized in Santa Maria, Charleston, and Albuquerque, about 0.30 in Nashville and Medford, and about 0.20 in Boston and Madison.

Monthly *SLR* correlations for low-mass sun-tempered buildings are not presented because the sensitivity of these structures to short-period weather patterns renders the correlations too poor to be useful.

O.2.d SENSITIVITY DATA

O.2.d.1 Mass Surface Area

The effect of varying the mass-area-to-glazing-area ratio about the reference value of 20 is shown in figure O-27 for Albuquerque. Notice that the performance at low *LCR*s can be significantly improved by adding more ½-inch-thick gypsum board. An area ratio of 40 would correspond to a 1,500-square-foot frame house with 180 square feet of south-facing glazing for which the gypsum board in northern, as well as southern, zones is available for thermal storage. The northern zones are accessible for thermal storage if a forced-air distribution system is employed or if wide, full-ceiling-height doorways are used to enhance free convective heat exchanges with the southern zones that experience direct solar gains [WRA-4], [WRA-5], [WEB]. Figure O-27 and the related graphs in Appendix 22 can also be used to account for thermal storage in furniture and other objects. Each increase of 10 in the A_m/A_g ratio is equivalent to an increase of 5.4 Btu/°F/ft² of glazing in the heat capacity of the structure.

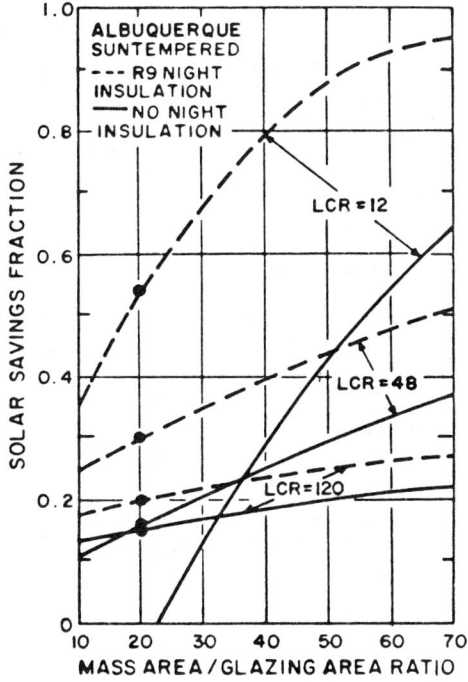

O-27: Effect on solar savings fraction of different A_m/A_g ratios, applied to a low-mass sun-tempered building in Albuquerque. The *LCR* value of 12 is not characteristic of sun-tempered buildings, but is shown for comparison. Reference conditions are indicated by dots.

O.2.d.2 Number of Glazings

The effect that varying the number of glazings has on low-mass sun-tempered buildings is illustrated in figure O-28 for Santa Maria, Albuquerque, and Madison. Notice that triple glazing appears desirable at all locations for systems without nighttime insulation and that, even with nighttime insulation, double glazing is advisable in Albuquerque and Madison.

O.2.d.3 Solar Absorptance of Gypsum-Board Surface

The solar absorptance of the gypsum-board surface has almost no effect on performance, as illustrated in figure O-29 for Albuquerque. The same lack of sensitivity is noted for the other cities in Appendix 22. In the case of a sun-tempered building, the ratio of mass surface area to glazing area is so large that multiple reflections within the living space cause most of the transmitted radiation to be absorbed in the storage mass, regardless of the building's interior color.

O.2.e EXAMPLE

An illustrative example might help clarify the concept of sun-tempering and the application of the above data to a performance analysis. Consider a typical 1,500-square-foot, single-family, detached residence in Boston, Massachusetts. Assume that the windows are double-glazed, are uniformly distributed about the perimeter of the building, and have a total area equal to 15% of the floorspace (225 ft²). Assume also that the building is 34 feet deep by 44 feet wide and that the long axis is oriented in a north-south direction. Under these conditions, the solar collection area is:

$$½ \times \left(\frac{34}{34 + 44}\right) \times 225 = 49 \text{ ft}^2$$

Now, if the building is of wood-frame construction with a crawl space (no floor slab) and the surface area of gypsum board available for thermal storage is 3,600 square feet, then the mass-area-to-glazing-area ratio is:

$$\left(\frac{A_m}{A_g}\right) = \left(\frac{3,600}{49}\right) = 73$$

A typical modern, wood-frame, single-family, detached residence will have a building heat load of about 7 Btu/DD/ft² of floorspace. The 1,500-square-foot home in the example will, therefore, lose a total of about 10,500 Btu/DD. Since the south-facing windows are considered solar collectors, their contribution to the static load must be subtracted from the above total. A U-value of 0.65 Btu/hr/ft²/°F times the window area of 49 ft² yields a loss of 31.9 Btu/hr/°F or 764 Btu/DD. The *BLC*, therefore, becomes 9,736 Btu/DD. Dividing by the collection area of 49 ft² yields an *LCR* of 199. Ex-

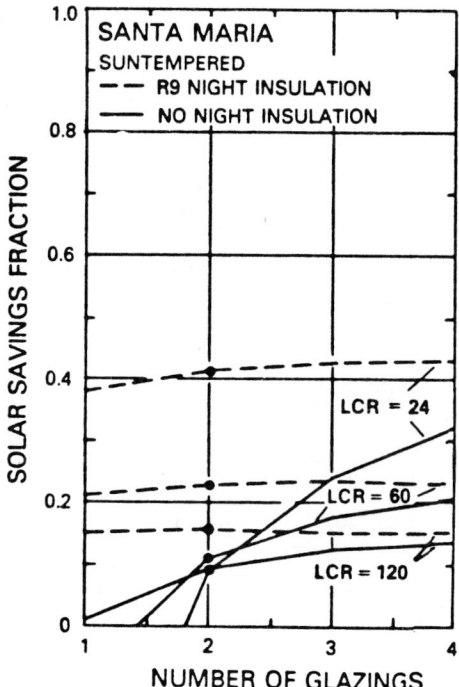
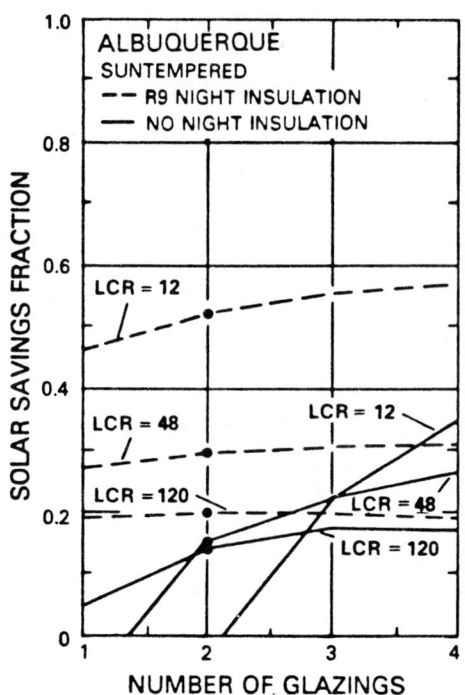
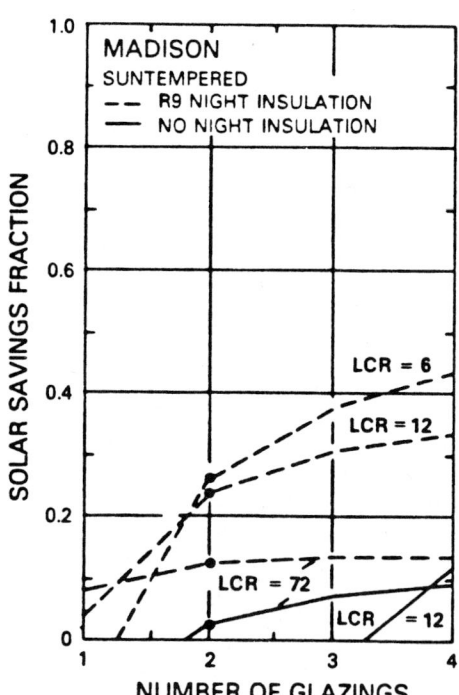

O-28: Effect on solar savings fraction of different numbers of glazings, applied to low-mass sun-tempered buildings in three cities. Reference conditions are indicated by dots.

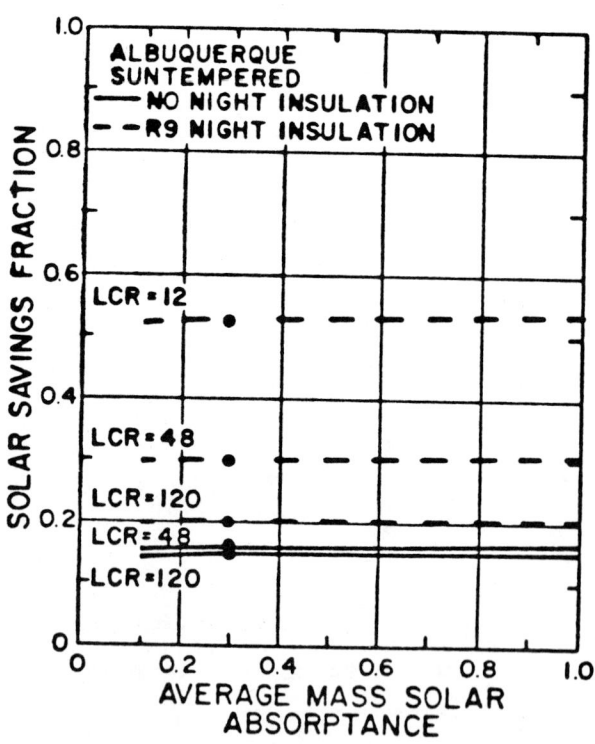

O-29: Effect on solar savings fraction of different values of average mass solar absorptance, applied to a low-mass sun-tempered building in Albuquerque. Reference conditions are indicated by dots.

trapolating beyond the scale on the horizontal axis of the graph for Boston in figure O-25, one obtains an *SSF* of about 0.035. Since the building used in this example has an area ratio of 73, rather than the reference value of 20 represented by the graph for Madison in figure O-28, the mass sensitivity curves must be used to scale this result. However, inspection of the graph for Boston in section 22.2.a.1 of Appendix 22 shows that a sun-tempered building in Boston with no nighttime insulation is not very sensitive to mass surface area ratios between 20 and 70 for an *LCR* of 72 and that the sensitivity is decreasing as the *LCR* increases. Since the building addressed in this example has an *LCR* of 199, mass sensitivity can be neglected altogether for the range of surface area ratios involved. Therefore, even though this building would not ordinarily be thought of as incorporating a passive solar energy system, 3.5% of the heat load experienced by all elements of the shell except the south-facing windows will be met by direct solar gains. The annual auxiliary energy required, therefore, is:

$$Q_{aux} = (1 - SSF) \times BLC \times DD$$
$$Q_{aux} = 0.965 \times 9,763 \times 5,621$$
$$Q_{aux} = 53.0 \text{ MMBtu}$$

This typical wood-frame building in Boston can be used as an example of a sun-tempered building by orienting the long axis in the east-west direction and by placing 80% of the 225 square feet of window space on the south side. The solar collection area becomes 180 square feet. The steady-state load through this collection area is the U-value (0.65 Btu/hr/ft²/°F) times the area (180 ft²) times 24 hours (to yield units in Btu/DD). The result is 2,808 Btu/DD. Subtraction of this quantity from the gross heat load of 10,500 Btu/DD yields 7,692 as the *BLC* for the sun-tempered building. Dividing by 180 ft² yields an *LCR* of 43. Inspection of the Boston graph in figure O-25 reveals an *SSF* of no more than 0.05. Therefore, the annual auxiliary heat required is:

$$Q'_{aux} = (1 - SSF) \times BLC \times DD$$
$$Q'_{aux} = 0.995 \times 7,692 \times 5,621$$
$$Q'_{aux} = 43.0 \text{ MMBtu}$$

Notice that, although the sun-tempered building has a lower *SSF* than the original structure, it requires less auxiliary heat. *The bottom line on performance is the auxiliary heat required.* The *SSF* should not be used to compare the performance of one building with another unless both have the same building load coefficient.

The performance of the sun-tempered building can be compared with that of the typical residential structure it is intended to replace by introducing a relative *SSF* defined as follows:

$$RSSF = \frac{(Q_{aux} - Q'_{aux})}{Q_{aux}}$$

where:

Q_{aux} is the annual auxiliary heat required by the original structure

Q'_{aux} is the same quantity for the modified structure

In the present example:

$$RSSF = \left(\frac{53 - 43}{53}\right) = 0.19$$

The sun-tempered building is thus shown to save almost 20% of the energy that would ordinarily be required for space-heating by its untempered counterpart, and these savings are achieved simply be reorienting the building and rearranging the windows.

Clearly, even the low-mass, sun-tempered building represents a step in the right direction that should not be discouraged. In fact, in mild climates where the heating loads are small, any further steps may not be economically justifiable. However, it would be a gross oversight not to point out that it is possible to improve both performance and comfort greatly by placing recommended levels of thermal storage mass within the building. The use of three or more glazings, or nighttime insulation, can further improve performance. Multiple glazings and nighttime insulation are, of course, desirable on all the windows of a building—not just those that face south—because they reduce heat losses through the windows. One should also consider the merits of combining direct gain with other types of passive solar systems, such as thermal storage walls, that are better suited to harsh winter climates.

P Sunspaces

An attached sunspace is a solar collector that is also a useful space, capable of serving other building functions. Other terms for it are "solarium" and "sun room." The term "attached greenhouse" is also often used, but can imply a plant-growing function as well as a solar heating function. Sunspace solar heating is compatible with limited plant raising, but no account has been taken in the data presented here of the energy transfer due to the evaporation of water from plants and planters.

The term "attached" generally means that the sunspace shares a common wall or walls with other building spaces. It can also have a more restricted meaning, denoting a sunspace with only one common wall, as if it were literally "attached" to a building. "Attached" is used in this restricted sense in this handbook in order to distinguish it from a more fully integrated sunspace with three common walls, which is a called here a "semienclosed" sunspace.

The sunspace performs its passive solar heating function by transmitting solar radiation through its glazing and absorbing it on its interior surfaces. The solar radiation is converted into heat upon absorption. Some of this heat is rapidly transferred by natural convection to the sunspace air, and some of it flows into massive elements in the sunspace (floor, walls, and water containers) to be returned later. The sunspace is thus a direct-gain space in which heat is used directly to maintain a temperature suitable for its intended secondary function, such as to serve as occasional living space. The primary purpose of the sunspace as a solar heating system, however, is to deliver heat to the adjacent building spaces. This may occur by conduction through a masonry common wall and by natural convection through openings (doors, windows, or special vents) in the common wall. The sunspace system then resembles an indirect-gain, thermal storage wall. Lastly, the common wall may be insulated so that heat transfer occurs only by natural convection through openings; then the sunspace system is in the isolated-gain category.

Some sunspaces are operated as hybrid systems in which a fan is used to transfer heated air from the sunspace to other building spaces or to storage.

Sunspaces have become very popular building features, both for the attractive space they provide and for their ability to deliver solar heat to adjacent spaces. Their solar heating performance can indeed be very effective, often exceeding the performance of any other passive solar system occupying the same area of south wall. The purpose of this chapter is to provide the design-analysis information needed to assure that a sunspace design will be as effective as possible.

P.1 THE REFERENCE DESIGNS

Numerous sunspace designs have been built and are of interest to designers. Any design-analysis method must, therefore, be flexible enough to address a variety of designs. Some of the design features of interest are the glazing geometry, the common-wall configuration, the type of thermal storage, and the type of movable nighttime insulation used. Twenty-eight reference designs have been chosen that reflect various combinations of typical choices of these features. These designs are described below and summarized in figure P-1.

The principal sunspace glazing is assumed to face true south. Thus wall locations are referred to by the compass directions: the principal glazing is the south wall, the principal common wall is the north wall, and the end walls are the east and west walls.

Two types of sunspace are defined according to the degree to which the sunspace has been integrated with the rest of the building. One type is the *attached* sunspace, whose north wall is common with the rest of the building and is 30 feet wide in the east-west direction. The other is the *semienclosed* sunspace that has three common walls: the north, east, and west. The semienclosed sunspaces discussed here are all 24 feet wide (east-west) and 12 feet deep (north-south). The north common wall in each case is 9 feet high.

Two geometrical shapes of the attached sunspace and three of the semienclosed sunspace are treated. The two attached sunspace geometries are: a single plane of glazing on the south wall, tilted up from the horizontal by 50°; and two planes of south-facing glazing, the lower one 6 feet high and vertical, the upper one tilted at 30°. The three semienclosed geometries are: a single, vertical plane of glazing on the south wall; a single, 50°-tilted

Designation	Type	Tilt (degrees)	Common Wall	End Walls	Nighttime Insulation
A1	attached	50	masonry	opaque	no
A2	attached	50	masonry	opaque	yes
A3	attached	50	masonry	glazed	no
A4	attached	50	masonry	glazed	yes
A5	attached	50	insulated	opaque	no
A6	attached	50	insulated	opaque	yes
A7	attached	50	insulated	glazed	no
A8	attached	50	insulated	glazed	yes
B1	attached	90/30	masonry	opaque	no
B2	attached	90/30	masonry	opaque	yes
B3	attached	90/30	masonry	glazed	no
B4	attached	90/30	masonry	glazed	yes
B5	attached	90/30	insulated	opaque	no
B6	attached	90/30	insulated	opaque	yes
B7	attached	90/30	insulated	glazed	no
B8	attached	90/30	insulated	glazed	yes
C1	semienclosed	90	masonry	common	no
C2	semienclosed	90	masonry	common	yes
C3	semienclosed	90	insulated	common	no
C4	semienclosed	90	insulated	common	yes
D1	semienclosed	50	masonry	common	no
D2	semienclosed	50	masonry	common	yes
D3	semienclosed	50	insulated	common	no
D4	semienclosed	50	insulated	common	yes
E1	semienclosed	90/30	masonry	common	no
E2	semienclosed	90/30	masonry	common	yes
E3	semienclosed	90/30	insulated	common	no
E4	semienclosed	90/30	insulated	common	yes

P-1: Sunspace reference designs.

plane of glazing on the south wall; and two planes of south-facing glazing, the lower one 6 feet high and vertical, the upper one tilted at 30°. Altogether, these comprise five geometrical configurations, designated A through E. They are illustrated in figure P-2.

The reference designs include two types of common wall between the sunspace and the adjacent building. One is lightweight and insulated, corresponding to a frame wall, with a thermal resistance of R20 (20 hr/°F/ft²/Btu); and one is uninsulated, 12-inch-thick, high-density masonry with a thermal conductivity of 1.0 Btu/hr/ft/°F and a volumetric heat capacity of 30 Btu/ft³/°F. In the lightweight-wall configuration, the design includes a row of water containers in the sunspace for thermal storage. The row extends the full east-west width of the sunspace. The containers are twice as high as they are deep. The water volume is one cubic foot per square foot of common-wall area. The containers are immediately adjacent to, but detached from, the sunspace floor and wall; they are thermally coupled to the wall and floor directly by radiation and indirectly by convection through the sunspace air.

Both wall configurations include thermocirculation vents in the common wall whose areas total 6% of the north wall area (3% for the top vents and 3% for the bottom vents). The vent centers are separated by a height of 8 feet. There is no reverse thermocirculation.

For each geometry and wall configuration, movable insulation may or may not be applied at night to the sunspace glazing. When used, the nighttime insulation has a thermal resistance of R9 (9 hr/°F/ft²/Btu), and is in place from 5:30 P.M. to 7:30 A.M. solar time.

The sunspace's east and west end walls may or may not be glazed. When not glazed, the end walls are insulated to a thermal resistance of R20 (20 hr/°F/ft²/Btu). In either case, the glazing system has optical and thermal properties equivalent to two panes of ordinary, eighth-inch-thick window glass with a half-inch air gap between them, with the exception that the transmitted solar radiation is assumed to be perfectly diffuse.

The sunspace floor is a 6-inch-thick slab of masonry material with a thermal conductivity of 0.5 Btu/hr/ft/°F and a volumetric heat capacity of 30 Btu/ft³/°F. There is conduction through underlying soil to a fixed temperature deep in the earth and through perimeter insulation to the ambient air.

The floor, the common wall in the case of the masonry-wall configuration, and the water containers in the case of the lightweight-wall configuration are capable of heat storage. The other sunspace surfaces are assumed to have no mass.

The sunspace-side surfaces of the common wall have solar absorptances of 0.7 if they are lightweight and of 0.8 if they are masonry. The water containers have a solar absorptance of 0.9. The sunspace floor has a solar absorptance of 0.8. The other surfaces (ceiling and end walls, if not glazed) have solar absorptances of 0.3. The solar radiation not absorbed is assumed to be diffusely reflected. The glazed surfaces are assumed to reflect 24% of any internally reflected solar radiation that is incident on them; the other 76% of incident radiation is transmitted back outside. Radiation absorbed at nonmassive surfaces is assumed to produce heat that transfers immediately to sunspace air. It is assumed that the absorption of solar radiation on other lightweight objects such as furniture and plants results in 20% of transmitted and reflected radiation being transferred rapidly to the sunspace air.

Sunspace heat losses occur primarily through the glazing. There are also losses through the R20 opaque surfaces (ceiling and end walls, if not glazed). A sunspace infiltration rate of 0.5 air changes per hour is assumed. No account is taken of heat transfer due to water evaporation that would occur if plants were present. (The effect of evaporation from plants can be a significant cause of heat-loss from the sunspace).

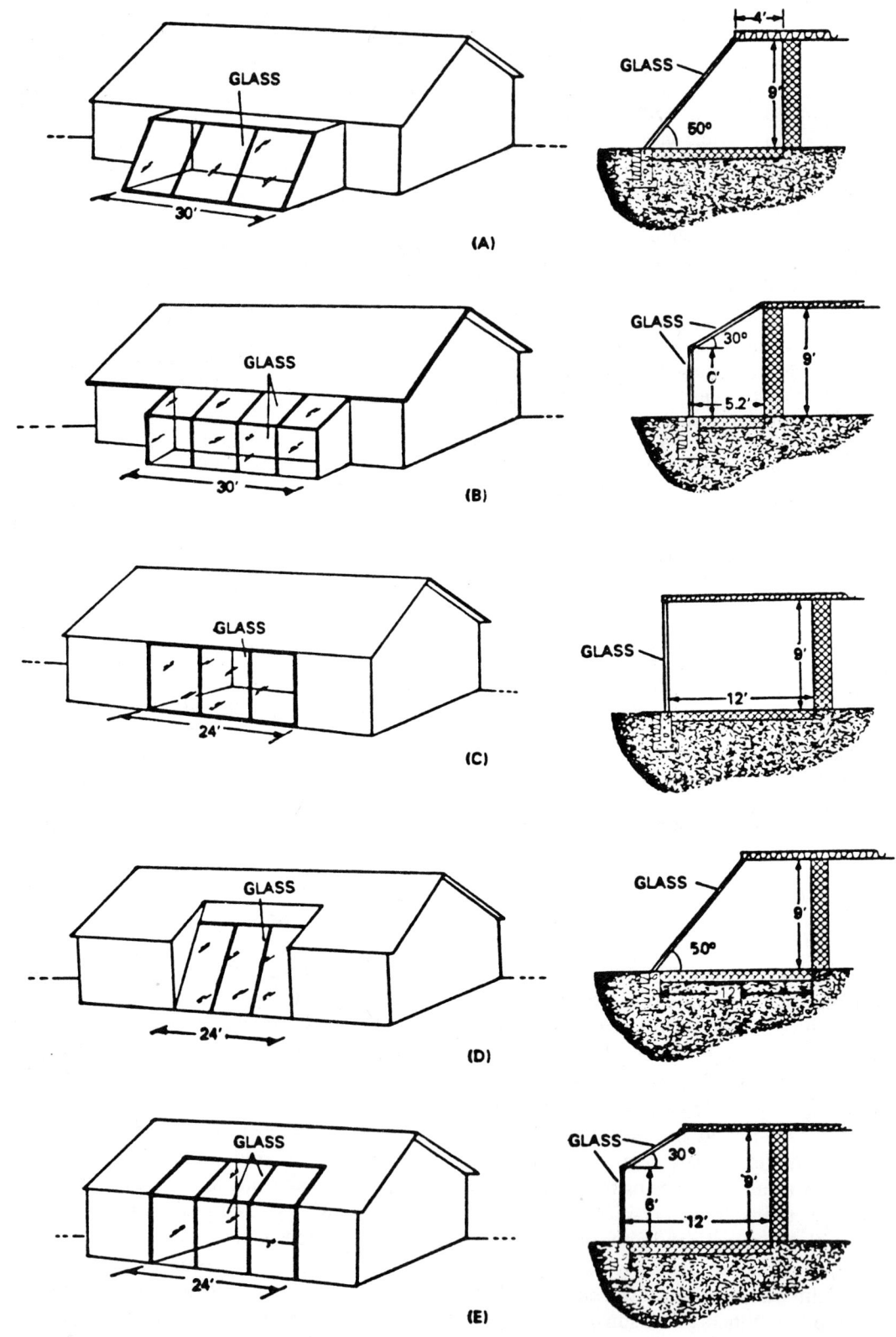

P-2: Sunspace geometries (not drawn to scale). The architectural detail beside the sunspace in drawing (D) is not significant; no shading was accounted for.

176 / *CHAPTER P*

Thermal storage capacity	
masonry common wall, per square foot of common wall area	30.0 Btu/F ft^2
water containers, per square foot of common wall area	62.4 Btu/F ft^2
floor, per square foot of floor area	15.0 Btu/F ft^2
Masonry properties	
thermal conductivity, wall	1.0 Btu/h ft F
thermal conductivity, floor	0.5 Btu/h ft F
density	150 lb/ft^3
specific heat	0.2 Btu/lb F
infrared emittance of surface	0.9
Glazing properties	
transmission characteristics	diffuse
infrared emittance of surface	0.9
Control range	
sunspace, heating thermostat setpoint	45 F
sunspace, cooling thermostat setpoint	95 F
room dead band, difference between heating and cooling thermostat setpoints	10 F
Thermocirculation vents	
vent area/projected area (sum of both upper and lower vents)	0.06
height between vents	8 ft
reverse flow	none
Lightweight absorption fraction	0.2
simulates effect of solar radiation absorption on lightweight objects by transferring given fraction of transmitted and reflected solar radiation directly to sunspace air	
Additional assumptions for annual method	
sunspace solar absorptances:	
common wall, lightweight	0.7
common wall, masonry	0.8
water containers	0.9
floor	0.8
other surfaces	0.3
ground reflectance	0.3
internal heat generation	0
room lower thermostat setpoint	65 F
sunspace opaque wall thermal resistance	R20
sunspace infiltration rate	0.5 air changes/hour
sunspace glazing properties:	
orientation	due south
shading	none
number of panes	2
index of refraction	1.526
extinction coefficient	0.5 in.$^{-1}$
thickness of each pane	1/8 in.
air gap between panes	0.5 in.
night insulation thermal resistance, when used	R9

P-3: Reference design characteristics.

Auxiliary heating is assumed to maintain the sunspace temperature at 45°F. Ventilation cooling is assumed to limit the sunspace temperature to 95°F if possible. The heating thermostat setpoint in the adjacent building is 65°F. Ventilation or auxiliary cooling are assumed to limit the room temperature to 75°F. There are no nonauxiliary internal heat sources in either the sunspace or the adjacent building.

It is important to realize that a design analysis is not necessarily restricted to the exact specifications of one of the above-described reference designs. This is because of certain intrinsic flexibilities in the analytical method. For one thing, the sunspace dimensions that define the five geometrical configurations are specified only to establish the *relative* sunspace dimensions. It is the sunspace's *shape,* not its absolute dimensions, that determines its performance per unit of projected area. For example, configuration A is nominally 30 feet wide by 9 feet high, with a 4-foot ceiling and a glazing tilt of 50°, but a sunspace 40 feet wide by 12 feet high with a 5-foot, 4-inch ceiling and a glazing tilt of 50° is also configuration A, being just one-third larger in all its dimensions than the first sunspace but identical in shape. The effect of the absolute size of the sunspace is accounted for by the projected area, an independent parameter.

Another aspect of design flexibility involves the use of the monthly (or *SLR*) rather than the annual (or *LCR*) analysis method. The essential idea of the *SLR* method is that the analyst determines the amount of solar radiation absorbed in the sunspace, the sunspace's heat loss coefficient, and the degree-days base temperature according to the properties of the sunspace and building. Thus, there is flexibility in the choice of such things as the sunspace's glazing geometry and orientation, the sunspace's glazing and opaque-wall thermal resistances, the sunspace's infiltration rate, the building thermostat setpoint, and the building internal heat generation rate.

A summary of reference design characteristics is presented in figure P-3. Figure P-3 shows the parameters that are common to both the monthly and the annual method and shows the additional restrictions that characterize the annual method.

Some of these parameters have been revised since the previous reports on sunspace performance [MCF-2], [JON]. The revised parameters are: the constant for thermal conductivity of masonry in the masonry common walls has been changed from 0.8 to 1.0 Btu/hr/ft/°F to be consistent with the value used in the analysis of thermal storage wall and direct-gain systems; the constants for solar absorptance of the masonry and insulated common walls have been changed from 0.9 to 0.8 and from 0.9 to 0.7, respectively, to conform more nearly to typical practice; the value for sunspace infiltration rate has been changed from 0.2 to 0.5 air changes per hour to conform more nearly to readily achievable practical rates; and a "lightweight absorption fraction"

of 0.2 was added to account (to some extent) for the presence of lightweight objects such as furniture and plants in the sunspace. See section P.2.j for a more detailed discussion of the lightweight absorption fraction. The enlarged value for masonry thermal conductivity tends to increase the calculated value for system heating performance; the reduced absorptances, the increased sunspace infiltration rate, and the lightweight absorption fraction tend to reduce the calculated value for heating performance. These changes, plus other technical improvements in the simulation model, have produced small variations in the performance predictions here relative to previously published reports [MCF-2], [JON].

The parameters excluded from the definitions of the reference designs—the projected area (A_p) and the building load coefficient (*BLC*)—are combined into the load collector ratio ($LCR = BLC/A_p$). The quantity *LCR* then becomes the principal building variable for a given reference design. Performance estimates for a given location are functions of *LCR*. There are two options. The annual calculation, or *LCR* method, is the easiest and fastest to perform, but it has limited applicability. Refer to section M.3 and Appendix 21 for details. The monthly calculation, or *SLR* method, is lengthier, but more general. Refer to section P.4 and Appendix 18 for details.

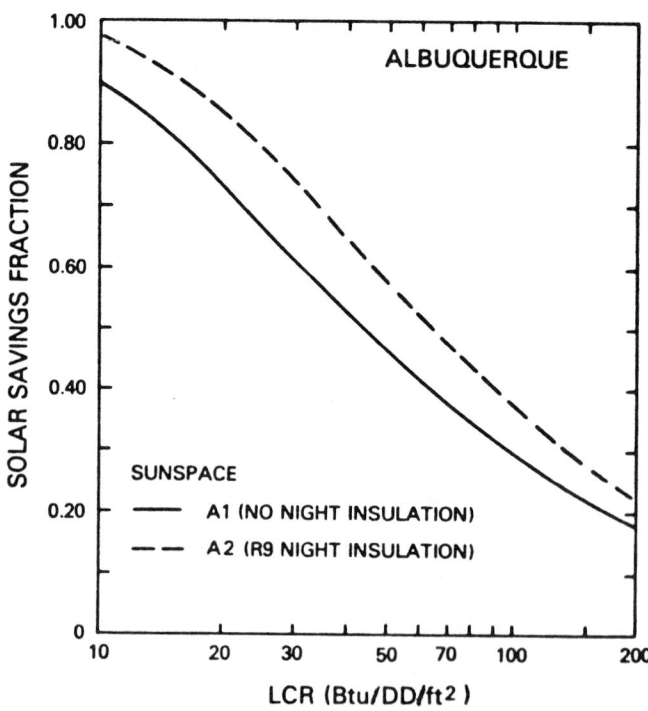

P-4: Effect on solar savings fraction of different *LCR* values, applied to sunspaces with and without nighttime insulation in Albuquerque, New Mexico.

As an example of a performance estimate by the *LCR* method, figure P-4 is a plot of the annual solar savings fraction (*SSF*) versus *LCR* reference designs A1 and A2 (see figure P-1) in Albuquerque, New Mexico. The plot is obtained by drawing smooth curves through points obtained from the *LCR* table for Albuquerque in Appendix 21. A logarithmic scale is used for *LCR*, but this isn't necessary.

P.2 SENSITIVITY DATA

Data are presented in this section that express the sensitivity of the annual solar savings fraction (*SSF*) to selected design parameters. In any given set of data, one or two parameters are varied while the others are held fixed at their reference values. The reference values are indicated by dots on the curves.

Each set of data is presented for six different reference cities in the United States: Albuquerque, New Mexico; Bismarck, North Dakota; Boston, Massachusetts; Ely, Nevada; Nashville, Tennessee; and Seattle, Washington. The cities were chosen to represent wide geographical and climatological ranges. To apply the data, choose one of the reference cities to represent the location of interest, on the basis of climate similarity. Two gross measures are of use in making the choice: the 65°F base heating degree-days and the solar radiation incident on a vertical surface. These data are available as monthly averages for numerous locations in the United States and Canada in Appendix 19, where they are tabulated as *D65* and *VS*, respectively. Figure P-5 shows the January values of *D65* and *VS* for each of the six reference cities.

Two applications of sensitivity data are important. One is quantitative, as a final step in a design analysis. In this application, an analysis is performed on the basis of the reference design that most nearly resembles the design of interest. Then the annual *SSF* is corrected by use of sensitivity data.

Another application of sensitivity data is qualitative, wherein the data serve as a guide, perhaps very early in the design process, to the relative significance of various design parameters and to their preferred values. The discussion of the sensitivity data that follows is in this qualitative vein. Selected data illustrate the discussion, but readers are urged to make use of the full set of sensitivity data in Appendix 22 for routine reference.

P.2.a THERMAL STORAGE VOLUME

A primary element in any passive solar heating system is the thermal storage medium. In the sunspace reference designs, this is masonry in the floor and either masonry in the common wall or water in containers. A significant design parameter in these designs is the ratio of the volume of thermal storage material (exclusive of the floor) to the projected area—that is, the number of cubic feet of thermal storage material per square foot of projected area. (Recall that the projected area is the sunspace's principal net glazing area projected onto a vertical plane. See the definition in Chapter M.) In the cases involving attached sunspaces with masonry walls, the thermal storage ratio is just the wall thickness: a ratio of 1 foot implies a 1-foot-thick wall. In the cases involving lightweight, insulated walls, the thermal storage ratio is the volume ratio of the water storage containers: a ratio of 1 foot implies 1 cubic foot of water per square foot of projected area. In both types of cases (geometries A and B), the reference design value is 1 foot for the attached sunspaces.

Figure P-6 illustrates the sensitivity of the annual solar savings fraction to the thermal storage volume ratio for geometry A. For the masonry-wall cases, there are curves for several different masonry densities. The thermal conductivity is strongly correlated with the density and is varied as a function of the density. The relationship by which this is done is $k = 0.049 \exp(0.02\varrho)$, where k is the thermal conductivity (in Btu/hr/ft/°F) and ϱ is the density (lb/ft³). This is a least-squares fit to data for many types of concrete and other masonry materials. For 150 lb/ft³ masonry, the performance generally reaches a maximum in the vicinity of a storage volume ratio of 1 foot—that is, a wall thicknesss of 1 foot. For lower-density materials, optimum wall thickness is a little lower. For the insulated-wall cases, there is

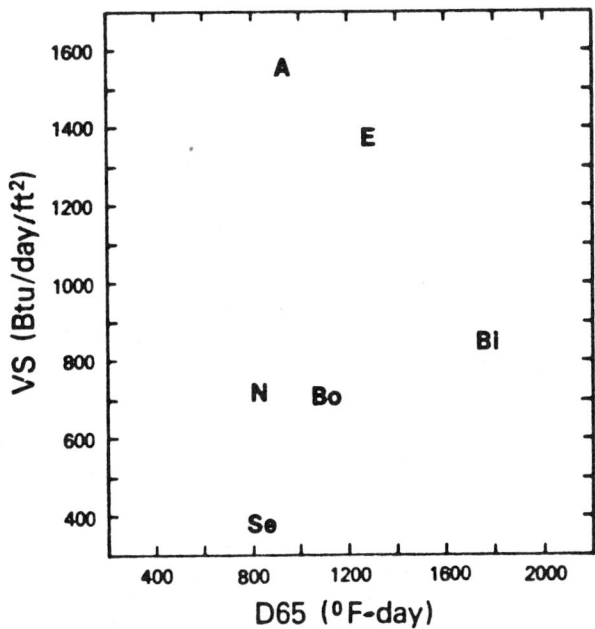

P-5: Climates of six cities mapped according to daily average solar radiation incident upon a vertical surface (*VS*) in each city and the January 65°F-base heating degree-days (*D65*) in each city. The cities are Albuquerque (A), Bismarck (Bi), Boston (Bo), Ely (E), Nashville (N), and Seattle (Se).

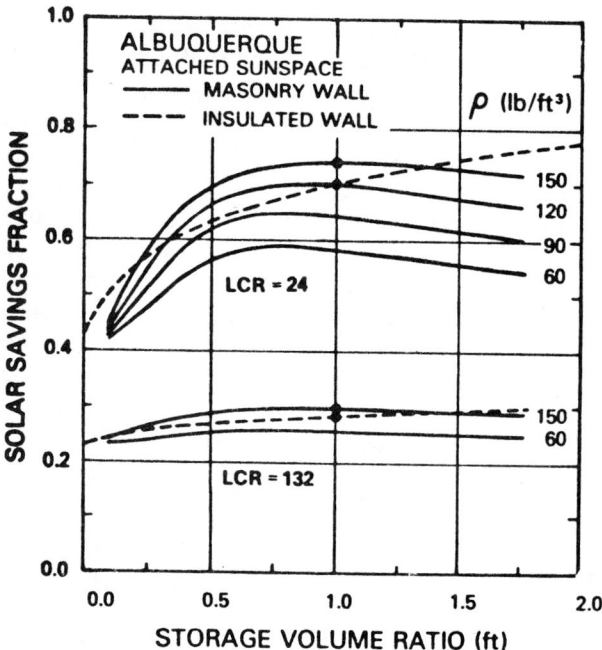

P-6: Effect on solar savings fraction of different ratios of thermal storage volume (in ft³) to projected area (in ft²), applied to geometry A with insulated end walls, no nighttime insulation, and both masonry and insulated walls (A1 and A5). In the masonry-wall case, the storage volume/projected area ratio is just the wall thickness; in the insulated wall case, the storage volume/projected area ratio is the ratio of the volume of the water containers in the sunspace to the projected area. Reference conditions are indicated by dots.

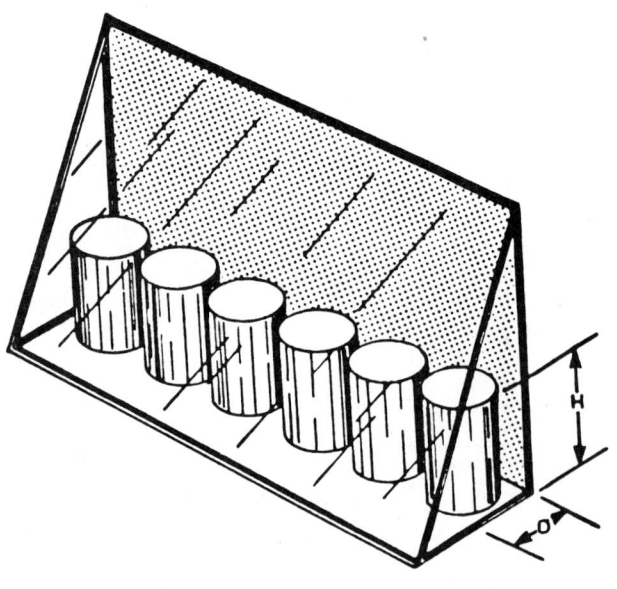

P-7: Water container shape, defined by height (H), north-south depth (D), and shape parameter (H/D). The reference value of H/D is 2.

no optimum storage volume ratio: the performance increases indefinitely with increasing water volume, although the rate of increase diminishes noticeably for water volumes above 1.0 cubic foot per square foot of projected area. A recommended minimum is 0.5, but as much water should be used as is compatible with other design constraints.

In the cases involving semienclosed sunspaces with masonry walls, all three common walls (east, north, and west) are masonry, so the volume of thermal storage mass per unit of projected area is greater than in the attached geometries. In fact, it is twice as great for geometry C, 1.69 times as great for geometry D, and 1.93 times for geometry E. The larger masonry wall area accounts to a large extent for the performance advantage of the semienclosed over the attached geometries. The sensitivities of performance in the semienclosed geometries to the masonry thickness and density are nearly the same as in the attached geometries. See Appendix 22 for relevant data. The reference thickness is 1 foot, and the reference density is 150 lb/ft³, as in the attached geometries.

To be consistent with the cases involving semienclosed sunspaces with masonry-walls, the reference design value of the water storage volume in the cases involving semienclosed sunspaces with lightweight, insulated walls has also been made proportional to the common-wall area: 2 cubic feet per square foot of projected area for geometry C, 1.69 for geometry D, and 1.93 for geometry E. Data on sensitivity to water storage volume for the geometry-C, lightweight wall configuration are available in Appendix 22.

P.2.b WATER CONTAINER SHAPE

In the insulated-wall cases, the *shape* of the water-storage containers has a small influence on the average annual heating performance. One shape parameter is the overall height-to-depth ratio of the row of containers (H/D in figure P-7). The reference value of this parameter is 2, but slightly improved performance can be obtained with very small or very large values (always keeping the total volume constant)—corresponding to tall, thin containers covering the whole wall, or to low, flat containers covering the whole floor. A performance improvement of approximately 0.05 can be achieved by adopting one of these extreme shapes. The improved performance occurs because these extreme shapes have a higher surface-to-volume ratio and, therefore, absorb more solar radiation and release more heat by convection.

P.2.c ORIENTATION

The performance of a sunspace depends on the orientation of its principal glazing, relative to true south.

Typical performance sensitivity to orientation is illustrated by figure P-8, which shows plots of annual *SSF* versus orientation for Albuquerque. The optimum orientation is always close to true south, although the performance is relatively insensitive to orientation within a few degrees of true south. Taking an average over all six reference cities, the *SSF* is reduced by about 0.05 when the orientation is either 30 degrees west or 30 degrees east of true south. The reduction in *SSF* is about 0.10 at 40 degrees, and about 0.20 at 60 degrees. This estimate pertains to both glazed-end-wall and insulated-end-wall cases. More extreme cases are represented by Seattle and Bismarck. For Seattle, the *SSF* is reduced on the average by about 0.05 for orientations either 35 degrees west or east of south, by about 0.10 at 50 degrees, and by about 0.20 at 70 degrees. The greater latitude for error in the orientation in Seattle is due primarily to the cloudy climate, in which a high percentage of solar radiation is diffuse. Since the diffuse radiation is not strongly directional, optimally oriented glazing is not as important there as it is in sunnier climates. For Bismarck, the *SSF* is reduced on the average by about 0.05 for orientations of 20 degrees west or east of true south, by about 0.10 at 25 degrees, and by about 0.20 at 40 degrees. The smaller latitude for error in the orientation in Bismarck is due primarily to the extremely cold climate: glazings at off-optimum orientations suffer above-average heat losses relative to solar heat gains. As a general rule of thumb, the sunspace glazing may be oriented in a range within 25 to 35 degrees east or west of south, with a solar savings reduction relative to a true-south orientation of about 0.05 or less. The smaller range of acceptable orientation pertains to extremely cold climates; the larger range to mild, cloudy climates.

P.2.d GLAZING TILT

Figure P-9 illustrates the dependence of annual heating performance on the glazing tilt. In these sensitivity data, the glazing area is held constant while the tilt is varied. This is done by adjusting the height of an opaque knee wall at the base of the glazing. See figure P-10. Notice that the presence of the opaque knee wall causes the projected area—that is, the vertical projection

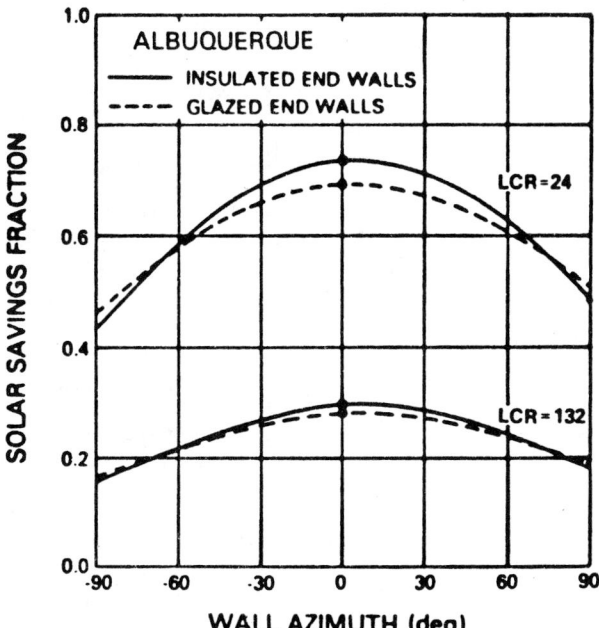

P-8: Effect on solar savings fraction of different sunspace orientations (azimuth = 0 when the principal glazing faces true south; it is positive when the glazing faces east of true south and negative when the glazing faces west of true south), applied to geometry A with a masonry common wall, no nighttime insulation, and both insulated and glazed end walls (A1 and A3). Reference conditions are indicated by dots.

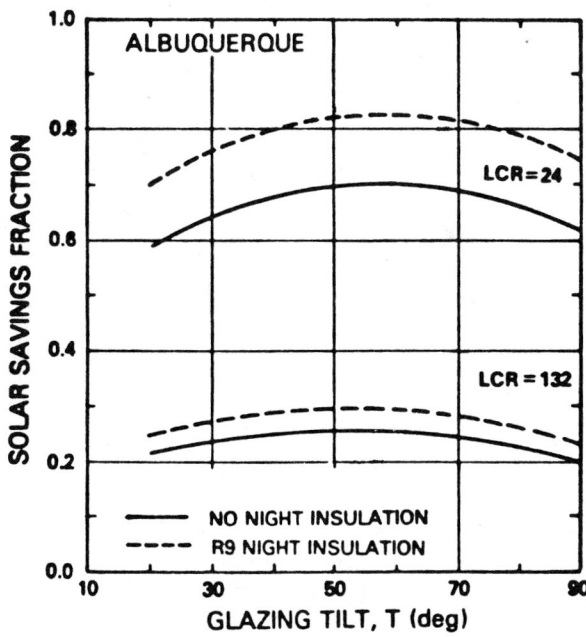

P-9: Effect on solar savings fraction of changes in glazing tilt (T), applied to a sunspace with a masonry common wall, insulated end walls, and a single tilted glazing plane. The glazing area remains constant; tilt is varied by means of a variable-height knee wall.

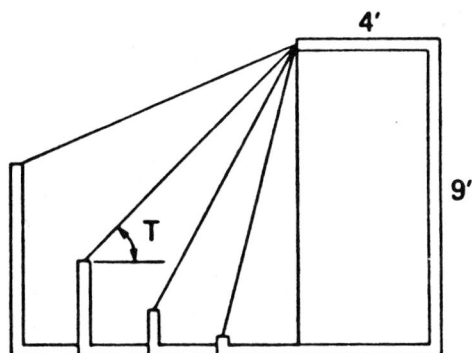

P-10: Geometry of the sunspace tested in figure P-9, showing different positions of the variable-height knee wall.

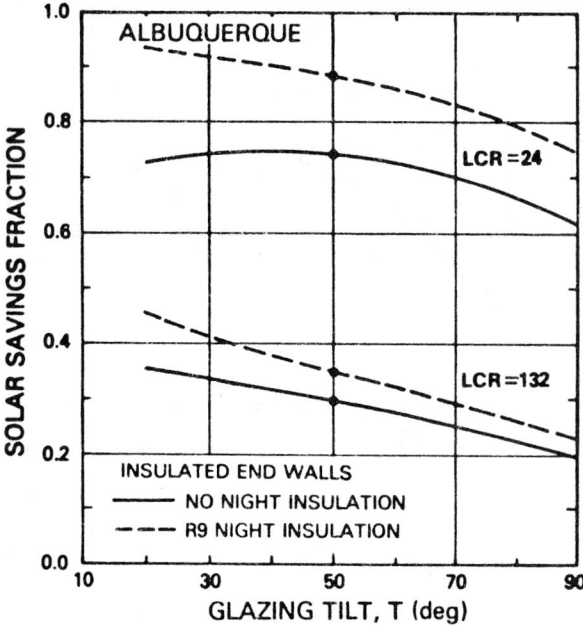

of the glazing area—to vary with the tilt angle. Thus, technically, *LCR* varies with the tilt angle in this case because *BLC* is held constant. However, the *LCR* values that label figure P-9 and the corresponding graphs in Appendix 22 refer to the vertical-glazing configuration in which the projected area and glazing area are equal.

The optimum glazing tilt depends on latitude, but also on climate and on the building *LCR*. Therefore, there is no simple rule for tilt that depends on latitude alone. In any case, however, performance is very insensitive to tilt within several degrees of the optimum. As a general rule, a tilt in the range 50° to 65° is very close to the optimum. Within this range, for every case among those in the sensitivity data, the solar savings are at least 95% of the optimum values.

A related design issue also concerns the glazing tilt, but does not carry with it the constraint of holding glazing area constant. The sensitivity of annual heating performance to glazing tilt in this sense is illustrated by figure P-11. Because the glazing area increases as the tilt decreases, the optimum tilt is shifted toward smaller angles. See figure P-12. Indeed, in many cases there is no optimum at all, rather the heating performance continues to increase without limit as the glazing plane approaches horizontal. This does not mean, however, that very shallow glazing angles necessarily lead to good designs. On the contrary, there are important disadvantages associated with shallow glazings, including higher cost, potential snow buildup, poor water drainage, low head room, and summer overheating problems. A reasonable compromise is to stay within the 50 to 65 degree tilt range suggested in the previous paragraph. In fact, some designers would emphasize the advantages of steep glazing angles to the point of choosing tilts above the optimum range, perhaps even vertical.

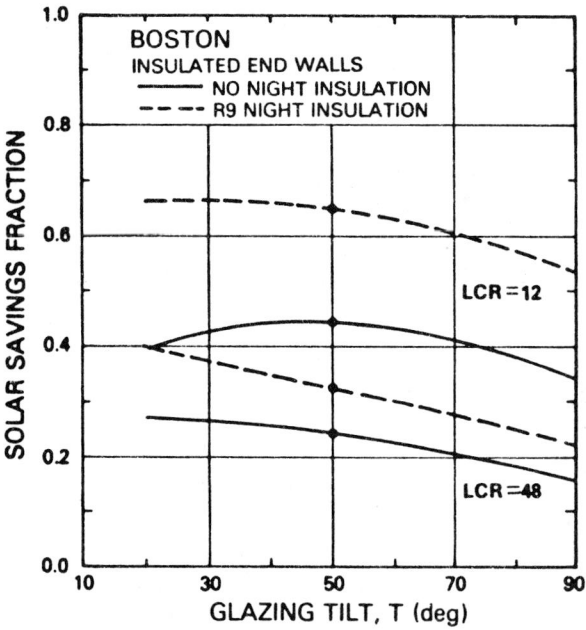

P-11: Effect on solar savings fraction of changes in glazing tilt (*T*), applied to sunspaces resembling geometry A—with a masonry common wall, insulated end walls, and a single tilted glazing plane—in two cities. The glazing area varies as the tilt is varied. Reference conditions are indicated by dots.

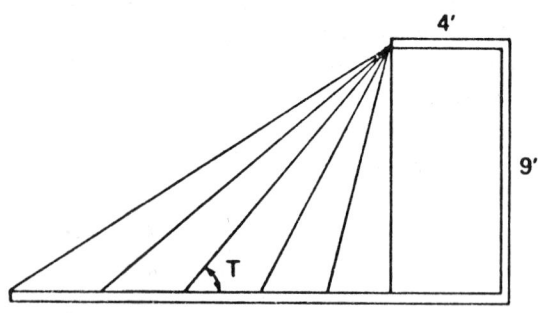

P-12: Geometry of the sunspaces tested in figure P-11, showing different glazing areas at different tilts.

Sensitivity data have been developed for cases in which some portion of the sunspace glazing is vertical. Figure P-13 illustrates the effect of incorporating increasing amounts of vertical glazing. The geometry of this design is shown in figure P-14. There is a heating performance penalty associated with a vertical glazing segment in a sunspace, but the configuration may still be chosen on the basis of other criteria, as discussed in the previous paragraph.

Once a vertical glazing segment is chosen, another design decision arises, concerning whether or not to place a tilted glazing segment above the vertical one, and, if so, at what tilt. The consequences of this decision are illustrated in figure P-15, which assumes a fixed, six-foot-high vertical glazing and shows the effect of varying the tilt of the upper glazing segment. See figure P-16. The best heating performance can usually be obtained with rather low tilts. Such tilts can also offer other advantages—such as large floor area and good light distribution—but the disadvantages of low tilts mentioned above apply in this case, too. Fortunately, the system's sensitivity to variation in this parameter is not very great, which leaves the designer relatively free to act on criteria other than heating performance.

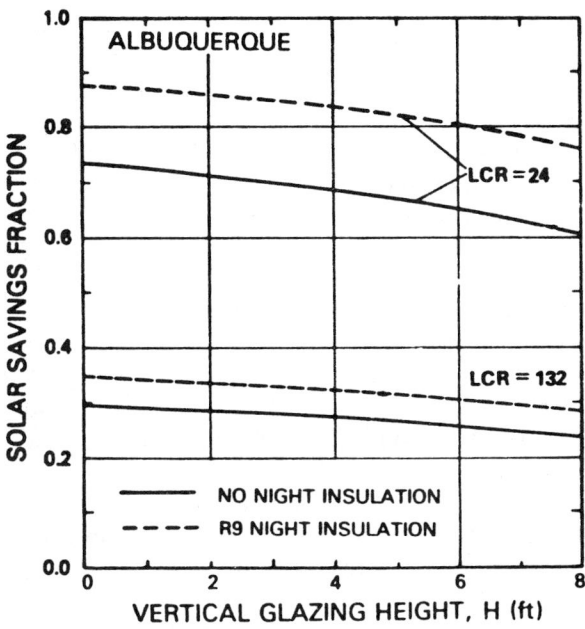

P-13: Effect on solar savings fraction of different vertical glazing heights (H), applied to a sunspace with a masonry common wall and insulated end walls.

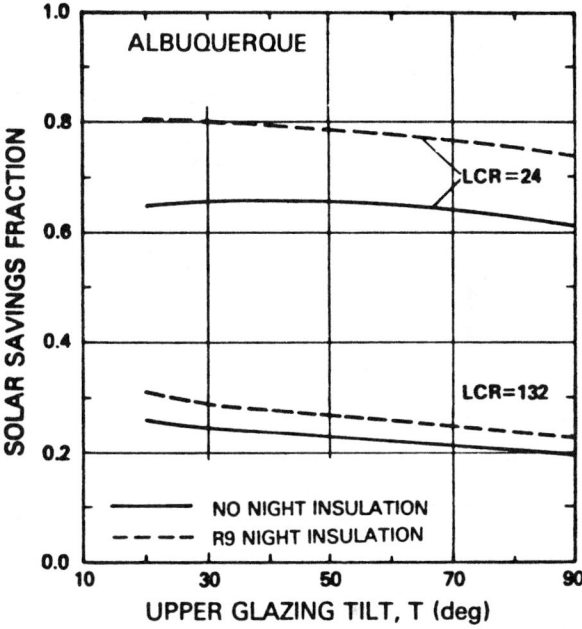

P-15: Effect on solar savings fraction of different upper glazing tilts (T), applied to a sunspace resembling geometry B with insulated end walls and a 6-foot-high vertical glazing.

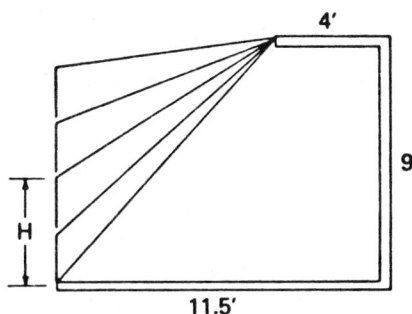

P-14: Geometry of the sunspace tested in figure P-13, showing different vertical glazing heights.

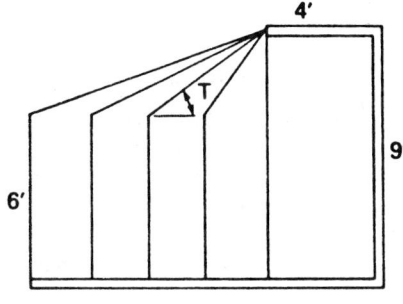

P-16: Geometry of the sunspace tested in figure P-15, showing different upper glazing tilts. When the upper glazing tilt is 30°, the sunspace's geometry is very close to geometry B, except that geometry B does not include a 4-foot-long ceiling.

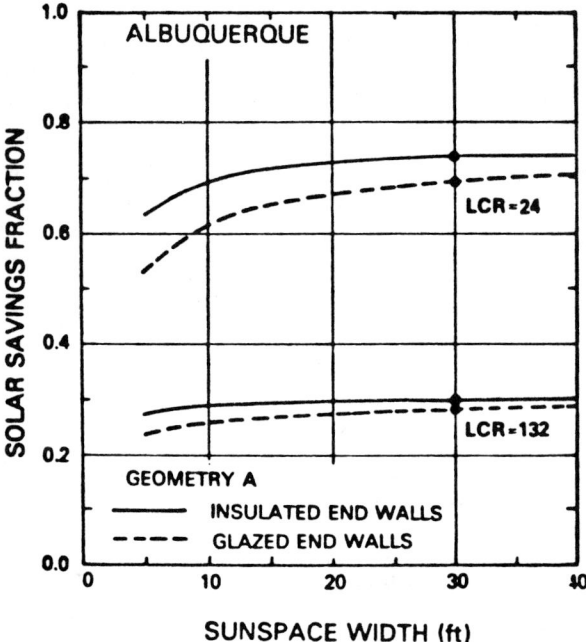

P-17: Effect on solar savings fraction of different east-west sunspace widths, applied to geometry A with a masonry common wall, no nighttime insulation, and both insulated and glazed end walls (A1 and A3). Reference conditions are indicated by dots.

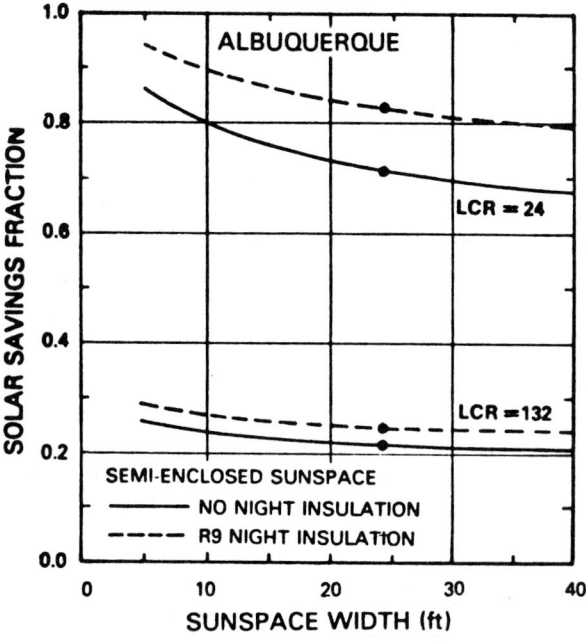

P-18: Effect on solar savings fraction of different east-west sunspace widths, applied to geometry C with a masonry common wall and both with and without R9 nighttime insulation (C1 and C2). Reference conditions are indicated by dots.

P.2.e SUNSPACE WIDTH

The east-west width of a sunspace influences its heating performance primarily because the projected area is proportional to the width. A more subtle question relates to the effectiveness of a *unit* of projected area as a function of sunspace width. This relationship for attached sunspaces is illustrated by figure P-17. Notice that the curves in this figure are plotted for fixed values of the *LCR*. Therefore, as the width varies, the building load is implicitly varied also, so that the *LCR* value remains constant. It is seen that the performance generally improves with sunspace width. This is because the end-wall losses become relatively less important with width. The effect in cases involving insulated end walls is very small for sunspace widths greater than 20 feet. The effect is more significant in cases involving glazed end walls, where the solar radiation and heat losses through the end walls are more severe.

The situation is different for semienclosed sunspaces. In fact, performance in most cases actually decreases as the sunspace width increases, as illustrated by figure P-18. Three reasons for this can be identified:

1. The end-wall "losses" are to the adjacent room, and, therefore, not losses at all.
2. The thermal storage mass per unit of projected area decreases as the sunspace width increases because of the fixed amount of mass associated with the end walls.
3. The fraction of solar radiation transmitted through the glazing that is absorbed in the sunspace decreases as the sunspace width increases.

The third reason is due to the reduced effectiveness of the sunspace as a cavity absorber as its width increases.

P.2.f NUMBER OF GLAZINGS

Figure P-19 illustrates the effect recorded in the sensitivity data of varying the number of panes of glazing in the sunspace. The number of glazings affects the performance in two ways. Each glazing pane reflects and absorbs a fraction of the incident solar radiation and thereby reduces the solar gain to the sunspace. But each glazing pane increases the thermal resistance of the glazing system and thereby decreases the heat loss from the sunspace. There is, therefore, a tradeoff between these two effects—one tending to decrease heating performance, and the other tending to increase it. Each added pane of glazing reduces the solar gain by an approximately constant fraction, but reduces the heat loss by a diminishing fraction. The benefit of adding glazings thus diminishes with number to a point at which the overall performance begins to fall. The optimum number of glazings depends on climate and on the particular system

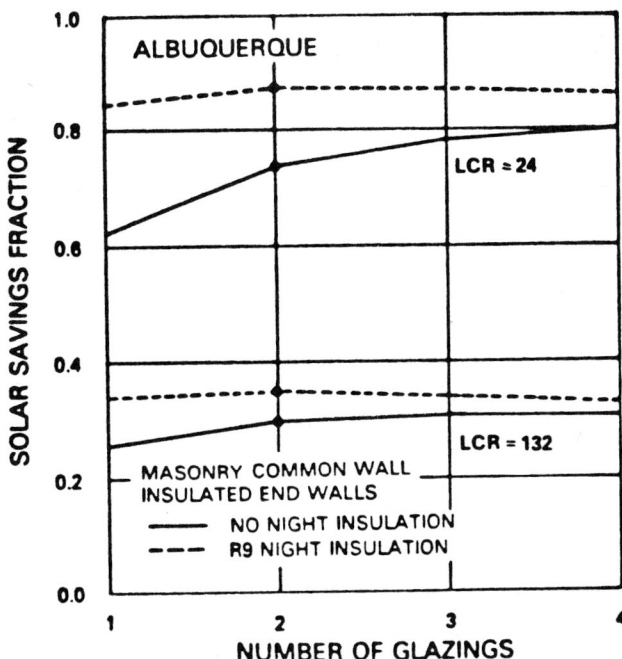

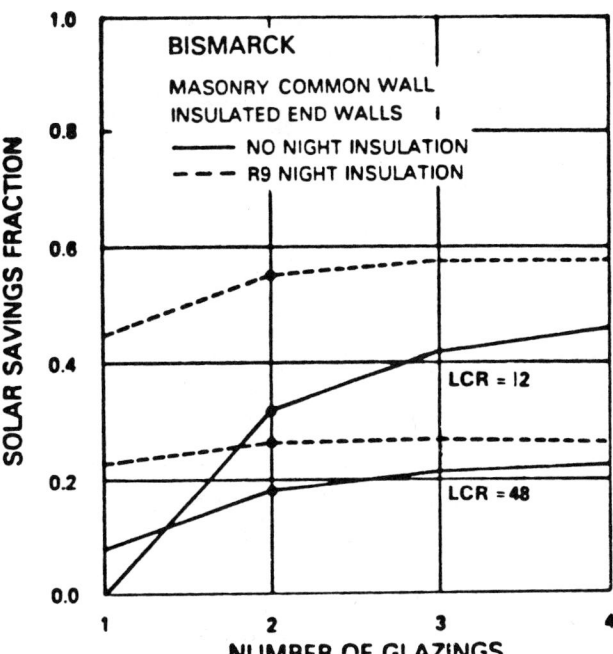

design—especially on the presence or absence of nighttime insulation. In sunny, mild climates even two glazings may be too many, if nighttime insulation is used. Without nighttime insulation, two glazings would almost always be advisable. In cold climates, three or even four glazings can improve performance, but nighttime insulation on double glazing may be an even better choice.

P.2.g GLAZING AIR GAP

The thermal resistance of double glazing depends on the width of the air gap between the panes. Therefore, the annual heating performance also depends on the air gap, as is shown in figure P-20. The performance is sensitive to the air gap primarily in the range below 0.5 inch, but in very cold climates an air gap of up to 1 inch may be worthwhile. The reference designs specify a 0.5-inch air gap between glazings.

P.2.h GLAZING THICKNESS

The amount of solar radiation absorbed in the glazing depends on the number and thickness of the panes and on the transparency of the glazing material. The transparency of glazing material is expressed by the "extinction coefficient," which is the fraction of solar radiation absorbed per unit length of travel in the material. For ordinary window glass the extinction coefficient is about 0.5 per inch. Figure P-21 illustrates the sensitivity of annual heating performance to glazing thickness, assuming an extinction coefficient of 0.5 per inch. The reference designs have a glazing thickness of 0.125 inch (⅛ inch).

P-19: Effect on solar savings fraction of different numbers of glazings, applied to sunspaces with geometry A—with a masonry common wall, insulated end walls, and both with and without R9 nighttime insulation (A1 and A2)—in two cities. Reference conditions are indicated by dots.

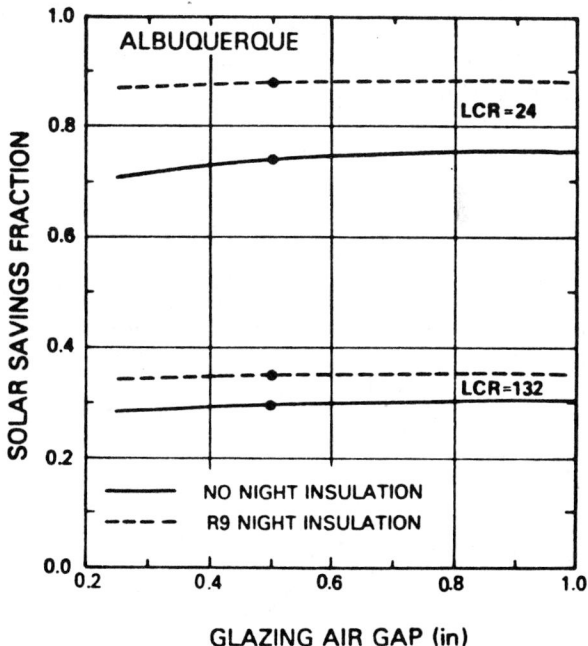

P-20: Effect on solar savings fraction of different distances of air gap between glazings; applied to a sunspace of geometry A with a masonry common wall, insulated end walls, and both with and without R9 nighttime insulation (A1 and A2). Reference conditions are indicated by dots.

Sunspaces / 185

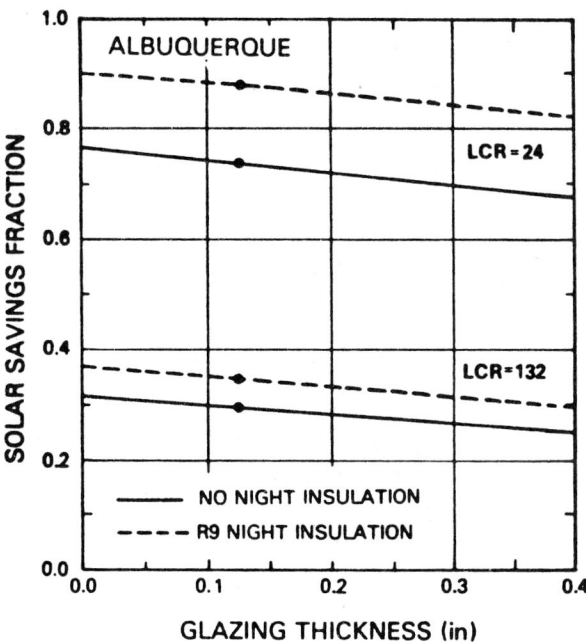

P-21: Effect on solar savings fraction of different glazing thicknesses, applied to a sunspace of geometry A with a masonry common wall, insulated end walls, and both with and without R9 nighttime insulation (A1 and A2). Reference conditions are indicated by dots.

The glazing thickness sensitivity data may be used for other extinction coefficients by using an equivalent thickness (equal to the actual thickness multiplied by the ratio of the actual extinction coefficient to the reference extinction coefficient). A short table of approximate extinction coefficients and thickness multipliers is given in figure P-22.

Material	Extinction coefficient (per inch)	Multiply thickness by
low-iron glass	0.1	0.2
common glass	0.5	1.0
fiberglass reinforced polyester	2.0	4.0
acrylic	0.2	0.4
polycarbonate	0.6	1.2

P-22: Extinction coefficients and thickness multipliers for obtaining equivalent thicknesses of various materials.

P.2.i SOLAR ABSORPTANCES

The solar absorptance of a surface is the fraction of incident solar energy that is absorbed upon striking that surface. The absorptances of a few interior surface finishes are listed in figure P-23.

Optical flat black paint	0.98
Flat black paint	0.95
Black lacquer	0.92
Dark gray paint	0.91
Black concrete	0.91
Dark blue lacquer	0.91
Black oil paint	0.90
Stafford blue bricks	0.89
Dark olive drab paint	0.89
Dark brown paint	0.88
Dark blue-gray paint	0.88
Azure blue or dark green lacquer	0.88
Brown concrete	0.85
Medium brown paint	0.84
Medium light brown paint	0.80
Brown or green lacquer	0.79
Medium rust paint	0.78
Light gray oil paint	0.75
Red oil paint	0.74
Red bricks	0.70
Uncolored concrete	0.65
Moderately light buff bricks	0.60
Medium dull green paint	0.59
Medium orange paint	0.58
Medium yellow paint	0.57
Medium blue paint	0.51
Medium Kelly green paint	0.51
Light green paint	0.47
White semigloss paint	0.30
White gloss paint	0.25
Silver paint	0.25
White lacquer	0.21
Polished aluminum reflector sheet	0.12
Aluminized mylar film	0.10
Laboratory vapor deposited coatings	0.02

P-23: Solar absorptance of various materials [GUB]. Variations in texture, tone, overcoats, pigments, binders, etc., can affect these values.

Figure P-24 illustrates the sensitivity of annual heating performance to the solar absorptances of various sunspace interior surfaces. In the first graph, the absorptance of the sunspace's masonry common wall is varied while the absorptances of the other surfaces are held constant. The reference designs have a masonry common-wall absorptance of 0.8. In the second graph, absorptance of the sunspace floor is varied. The floor also has a reference absorptance of 0.8. In the third graph, the absorptances of both the floor and wall are varied simultaneously. The fourth graph illustrates the sensitivity of the annual heating performance of the lightweight common-wall reference designs to the absorptance of the water containers. The reference absorptance is 0.9. Dark-colored (high-absorptance) sunspace surfaces tend to perform better than light-colored ones because less solar radiation is reflected back out through the glazing, but dark *mass* surfaces (wall, floor, and water containers) are especially important because solar

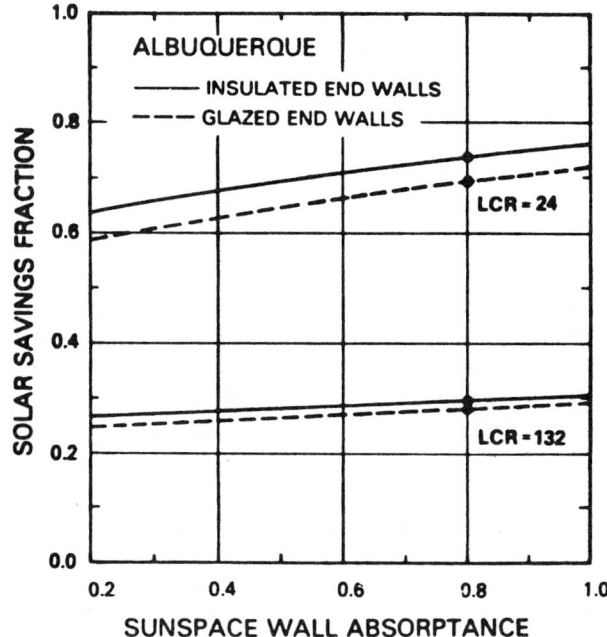
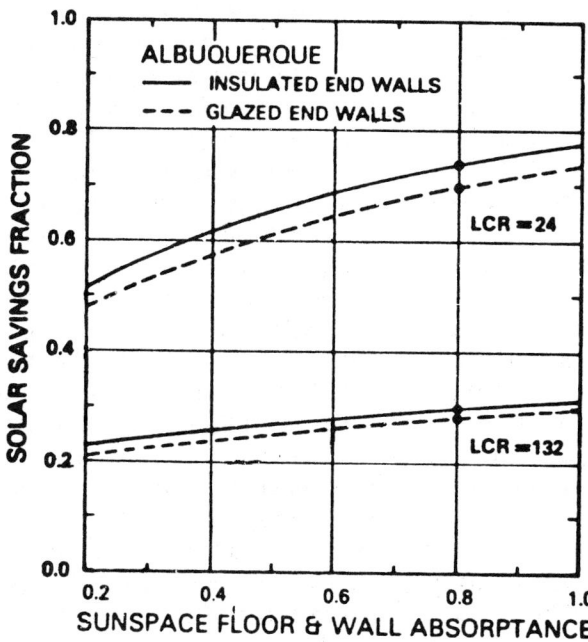
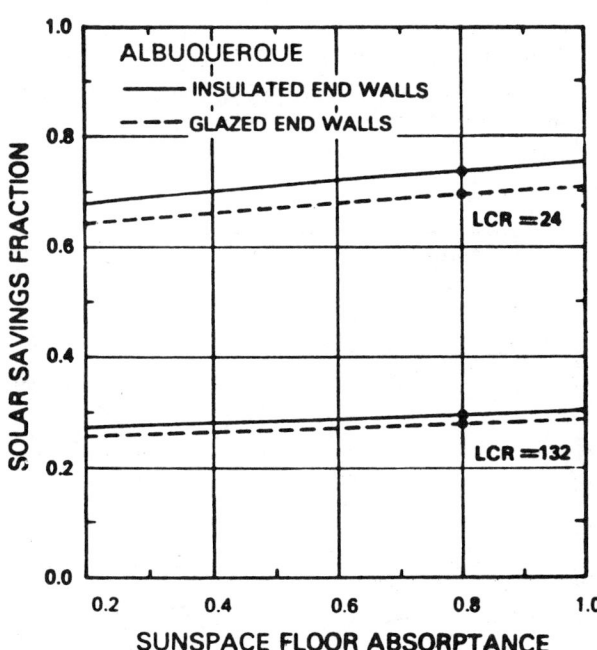
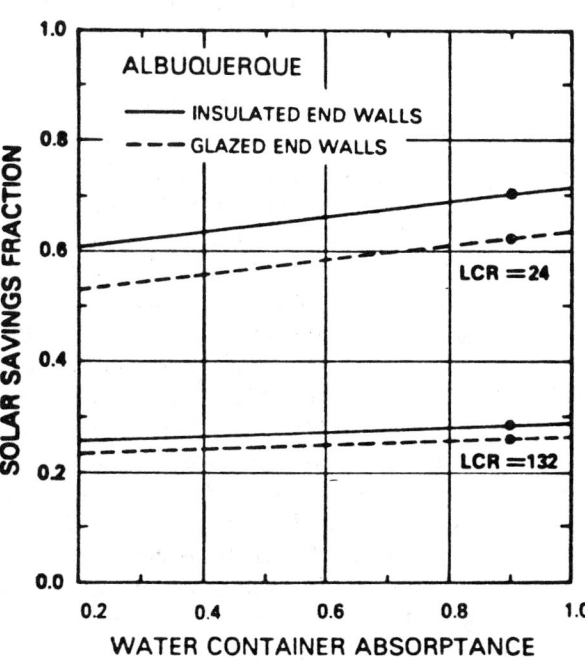

P-24: Effect on solar savings fraction of differences in the solar absorptance of various surfaces of a sunspace: the masonry common wall, the floor, the wall and floor in combination, and the water containers. All changes were applied to a sunspace of geometry A with a masonry common wall, no nighttime insulation, and both insulated and glazed end walls (A1 and A3 in the first three graphs, A5 and A7 in the fourth). Reference conditions are indicated by dots.

Sunspaces / 187

radiation absorbed by massive elements is partially retained as stored heat to be released later when it is needed. The performance sensitivity to mass absorptance is greatest for high-*SSF* (low-*LCR*) designs in which heat storage is essential if auxiliary heat is to be displaced during nighttime or cloudy periods. For low-*SSF* (high-*LCR*) designs, high mass absorptance is not as important because most of the solar energy collected is needed immediately for daytime heating requirements.

The sensitivity of the annual heating performance to the absorptance of lightweight ceiling and end walls is negligible. (No data are presented.) The reason for this is the near cancellation of opposing effects: diffuse reflections from lightweight surfaces are partly lost from the sunspace by transmission outward through glazings and are partly absorbed by massive elements in the sunspace. It is possible that *specular* reflections from lightweight surfaces would be predominantly absorbed by massive elements and would, therefore, lead to improved performance, but this speculation has not been tested. Because the effects cancel each other out, criteria other than heating performance may be used to select the finishes for these surfaces. It is probably good design practice to use light colors on these surfaces to improve the light distribution in the sunspace and to offset the visual effect of the dark colors on the mass surfaces.

P.2.j LIGHTWEIGHT OBJECTS

Lightweight objects such as furniture and plants in the sunspace have the effect of rapidly putting some solar energy into the form of heated air. This is because the solar radiation absorbed by lightweight objects causes a rapid surface temperature increase and consequently a rapid transfer of heat by convection to the surrounding air. This effect has been simulated by assuming that a certain fraction, the "lightweight absorption fraction," of solar radiation directly heats the sunspace air when it is transmitted through the glazing, as well as each time it is reflected from interior surfaces. The lightweight absorption fraction is 0.2 in all of the sunspace reference designs. This is in addition to heat transferred rapidly to the sunspace air due to absorption on the lightweight sunspace surfaces, namely the ceiling and (when they are lightweight) the end walls and common wall.

Notice that the lightweight absorption fraction simulates only one aspect of plants in the sunspace. The other aspect, which may have a very important effect on sunspace performance, is the transfer of latent heat due to water evaporation. The effect of evaporation is not explicitly accounted for in the analytical methods presented here.

The sensitivity of the annual heating performance to the lightweight absorption fraction is illustrated in figure P-25. It can be seen that the annual heating performance risks being seriously degraded by the presence of lightweight objects in high-*SSF* (low-*LCR*) designs because a high *SSF* requires substantial daytime heat storage with which lightweight objects interfere. For low-*SSF* (high-*LCR*) designs, on the other hand, the performance is not very sensitive to the lightweight absorption fraction because the solar gains are mainly needed for daytime air heating. In fact, the performance in these cases can even be improved slightly by a moderate lightweight absorption fraction.

Determination of the lightweight absorption fraction in a given sunspace can only be done in practice as a rough estimate. Prediction of the lightweight absorption fraction of a sunspace under design may be impossible because it depends on the way in which the sunspace is later furnished. But the following procedure may be useful, at least as a way of visualizing in an approximate way the meaning of the lightweight absorption fraction:

1. Stand near the south-facing glazing, looking north. Estimate the fraction of your view of the sunspace floor, north wall, and ceiling that is intercepted by lightweight objects. If the lightweight objects have substantially different solar absorptances, do Step 1 separately for each different absorptance.
2. Estimate the solar absorptances of the lightweight objects.
3. Multiply the fractions from Step 1 by the absorptances from Step 2 and add up the products. The result is an estimate of the lightweight absorption fraction.

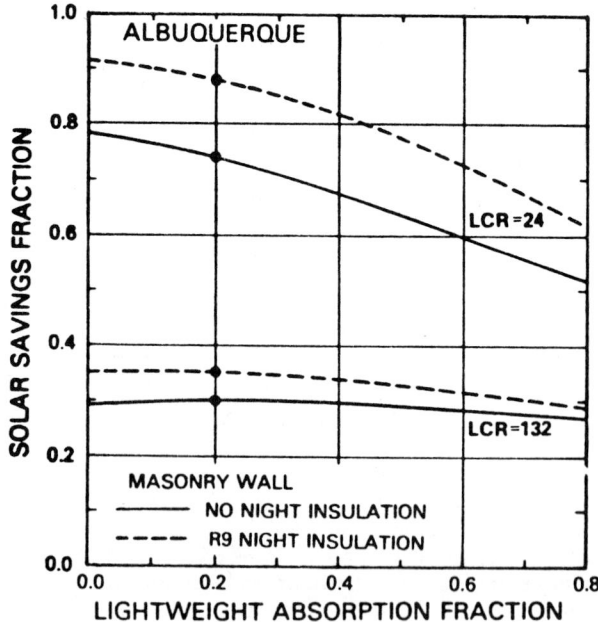

P-25: Effect on solar savings fraction of different values of lightweight absorption fraction, applied to a sunspace of geometry A with a masonry common wall, insulated end walls, and both with and without R9 nighttime insulation (A1 and A2). Reference conditions are indicated by dots.

P.2.k VENT AREA

It is assumed in all of the sunspace reference designs that convective heat transfer occurs from the sunspace to the adjacent room through openings in the sunspace common wall. This convection occurs freely because of the temperature difference between the two spaces. Reverse flow from the room to the sunspace is assumed in the reference designs to be prevented by backflow dampers. The amount of convective flow through the openings depends in part on the area of the openings. The effectiveness of that area depends on the configuration of the openings—in particular, on their vertical dimensions, or on the vertical distance between their centers when they are in pairs, one above the other. In the reference designs, it is assumed that the openings are in the form of Trombe-like thermocirculation vents—pairs of vents, with one near the floor and one near the ceiling. The reference height between the vent centers is 8 feet. The reference designs all have a combined vent area of 6% of the projected area. See figure P-26. Figure P-27 illustrates the sensitivity of the annual heating performance to the combined vent area expressed as a fraction of the projected area.

If the vertical height between the vent centers is not the 8-foot reference height, an equivalent area can be calculated to use with the sensitivity data. The equivalent area can be obtained by multiplying the actual area by a factor from figure P-28. The relationship is that the parameter $A \sqrt{H}$ characterizes the vents with respect to their performance; therefore, $A' = A \sqrt{H/H'}$, where A' is the equivalent area, A is the actual area, H' is the 8-foot reference height, and H is the actual height.

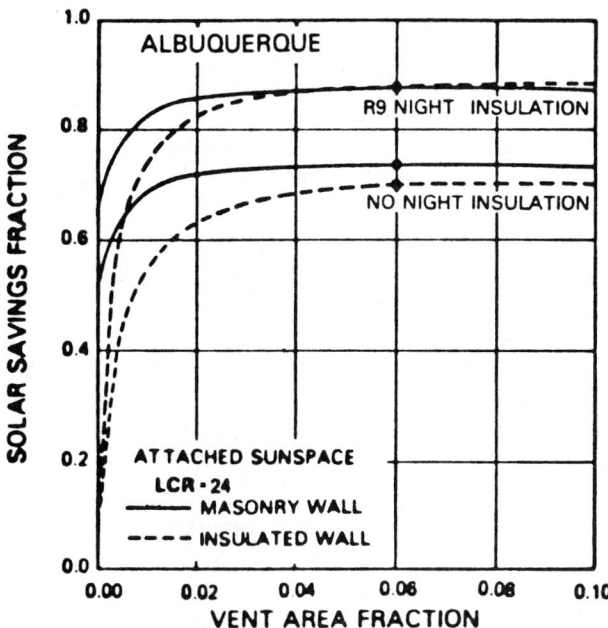

P-27: Effect on solar savings fraction of different vent area fractions, applied to a sunspace of geometry A with both a masonry and an insulated common wall, insulated end walls, and both with and without R9 nighttime insulation (A1, A2, A5, and A6). The vent area is the combined area of the top and bottom vents. Reference conditions are indicated by dots.

Vertical height between vent centers (in feet)	Multiply area by
4	0.71
6	0.87
8	1.00
10	1.12
12	1.22
14	1.32
16	1.41
18	1.50
20	1.58

P-28: Multiplication factors for obtaining thermocirculation vent areas equivalent to that of the reference case.

Height of opening (in feet)	Multiply area by
1	0.25
2	0.35
3	0.43
4	0.50
6	0.61
8	0.71
10	0.79

P-29: Multiplication factors for obtaining single-opening areas equivalent to the thermocirculation vent area of the reference case.

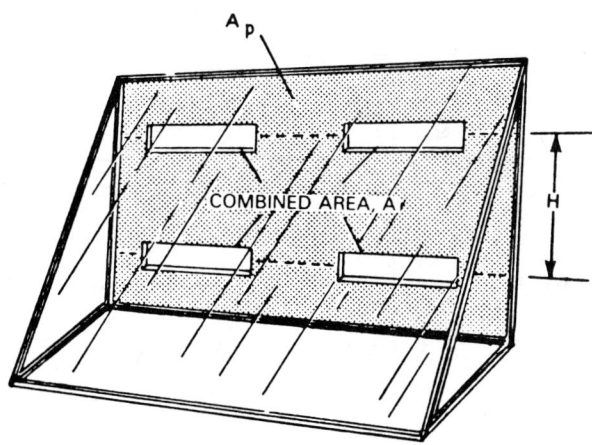

P-26: Common wall vents. The projected area (A_p) is equal to the north common-wall area in the sunspace reference designs. The combined vent area is labeled A. The vent area fraction (A/A_p) is equal to 0.06 for the reference designs. The vertical distance (H) between upper and lower vent centers is equal to 8 feet for the reference designs.

If the common-wall openings are single openings (windows and doors) rather than vent pairs, then the height of the opening determines the effectiveness of the area. (The "height" in this case means the vertical distance between the top and bottom of the opening, not the distance from the floor to the opening.) The equivalent area multiplier is found from figure P-29.

P.2.1 SUNSPACE INFILTRATION RATE

Figure P-30 illustrates the sensitivity of the annual heating performance to the sunspace infiltration rate. There is not as much performance sensitivity to the sunspace infiltration rate as might have been thought because the dominant sunspace heat losses are the noninfiltrative glazing losses. See figure P-38 for the size of sunspace infiltration losses compared to the total sunspace losses. Nevertheless, the infiltration component is significant and should be minimized by use of good caulking and weatherstripping practices, especially in very cold climates.

P.2.m COMMON-WALL THERMAL RESISTANCE

The common-wall thermal resistance pertains to the lightweight, insulated-wall cases. Figure P-31 illustrates

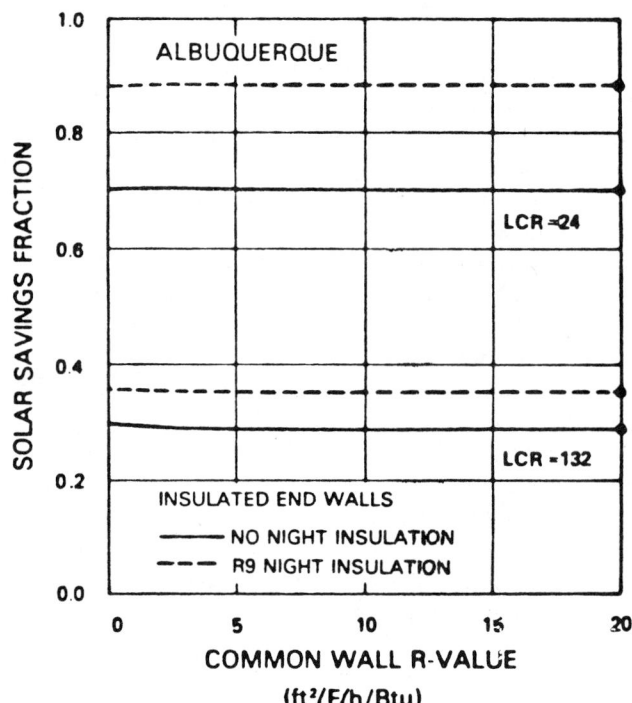

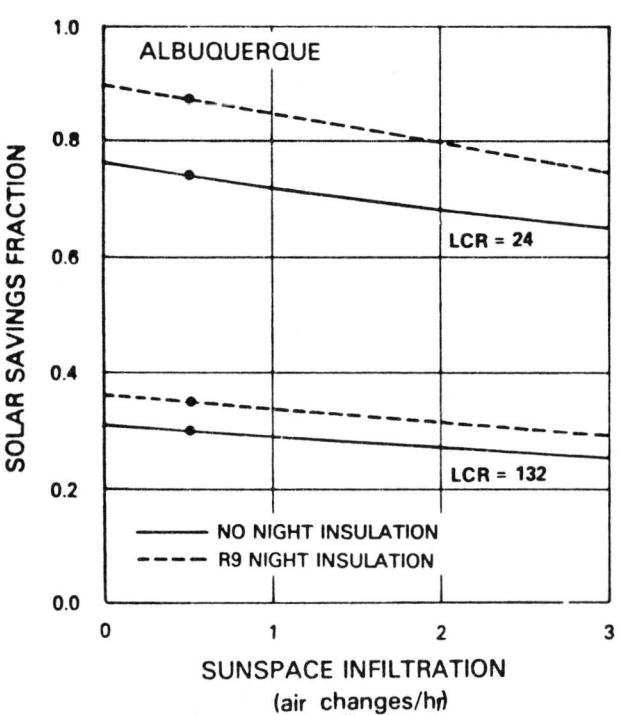

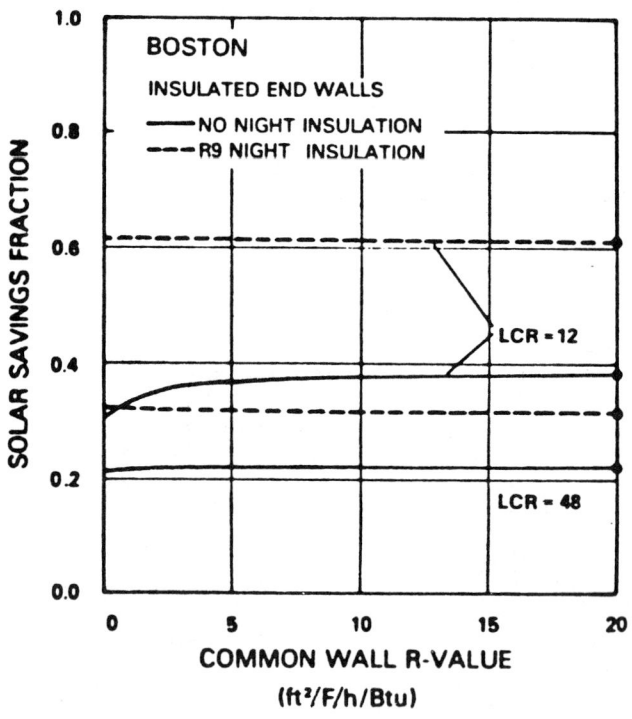

P-30: Effect on solar savings fraction of different rates of sunspace infiltration, applied to a sunspace of geometry A with a masonry common wall, insulated end walls, and both with and without R9 nighttime insulation (A1 and A2). Reference conditions are indicated by dots.

P-31: Effect on solar savings fraction of different common-wall R-values applied to sunspaces of geometry A—with a lightweight insulated common wall, insulated end walls, and both with and without R9 nighttime insulation (A5 and A6)—in two cities. Wall-surface-to-air resistances are not included. Reference conditions are indicated by dots.

the sensitivity of the annual solar savings to the common-wall thermal resistance. The R-value plotted in these figures includes only the conductive resistance; the wall-surface-to-air resistances are accounted for separately. For example, an R-value of zero refers to a wall that consists of an opaque film with no conductive resistance.

It is seen that the only important sensitivity of annual heating performance to the common-wall thermal resistance is below an R-value of about 5, and even then only in the relatively cold, cloudy climates. This is because daytime heat transfer from the sunspace to the adjacent room is predominantly by convection through openings, and nighttime losses by conduction through the wall are small because of the typically small temperature difference. Caution should be used, however, in concluding on the basis of these data that insulation need not be used in a lightweight common wall. In sunny, mild climates, the heating performance will not be much affected by the amount of common-wall insulation, but insulation may be very important in these cases to minimize summer heat gain through the common wall. In cloudy, cold climates, some common-wall insulation should be used, especially if the design is not similar to any of the reference designs. For example, less thermal storage in the sunspace or less thermal resistance in the sunspace to the exterior can cause the sunspace's temperatures to drop to lower minimum temperatures at night. There may then be significant heat loss through the common wall if it is not well-insulated.

P.2.n THERMOSTAT SETPOINTS

Although the thermostat setpoints are not design parameters, but rather features of occupant behavior, it is important to realize that they may have a significant effect on the building heating performance. In particular, the annual solar savings fraction is very sensitive to the room minimum thermostat setpoint (that is, the auxiliary heat thermostat for the building adjacent to the sunspace). This is illustrated in figures P-32 and P-33. Notice that the solar savings fraction in figure P-32 is in relation to a reference building with the variable minimum setpoint. Therefore the solar savings sensitivity to the setpoint does not reflect the sensitivity of the building load to the setpoint. On the other hand, the *SSF* in figure P-33 is defined relative to a reference building with a fixed 65°F minimum setpoint. The data, therefore, reflect the sensitivity of the total energy savings to the setpoint.

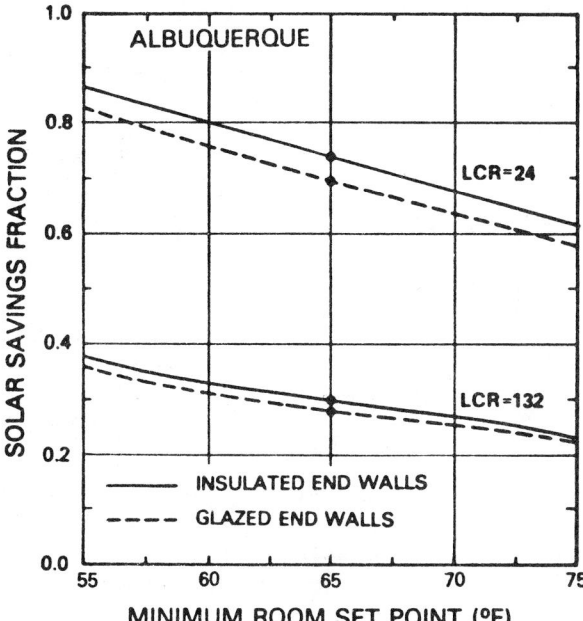

P-32: Effect on solar savings fraction of different minimum room thermostat settings, applied to a sunspace of geometry A with a masonry common wall, both insulated and glazed end walls, and no nighttime insulation (A1 and A3). The solar savings fraction in this case is defined relative to a net reference load evaluated at the variable minimum thermostat setting. Reference conditions are indicated by dots.

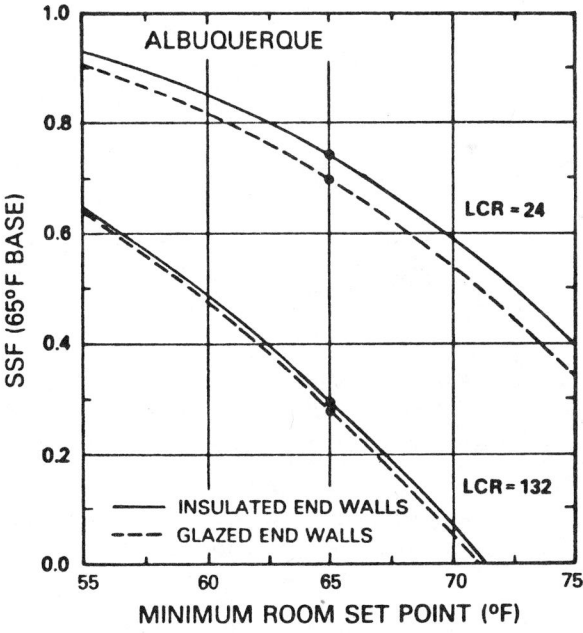

P-33: Effect on solar savings fraction of different minimum room thermostat settings, applied to a sunspace of geometry A with a masonry common wall, both insulated and glazed end walls, and no nighttime insulation (A1 and A3). The solar savings fraction in this case is defined relative to a 65°F reference load. Reference conditions are indicated by dots.

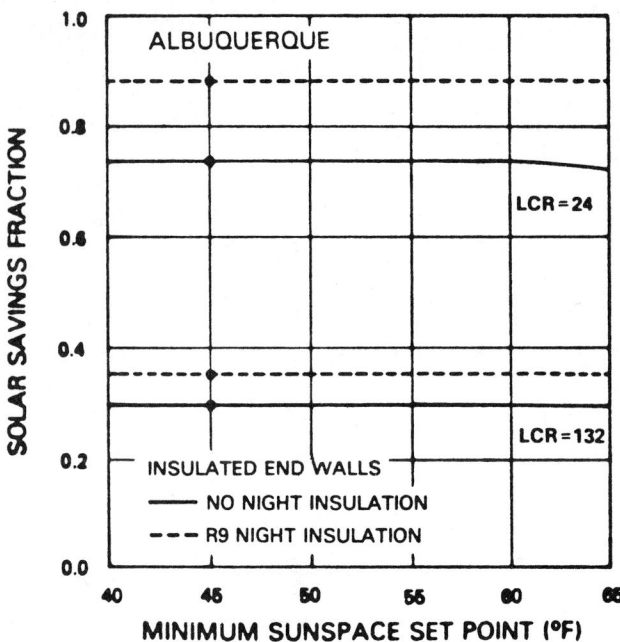

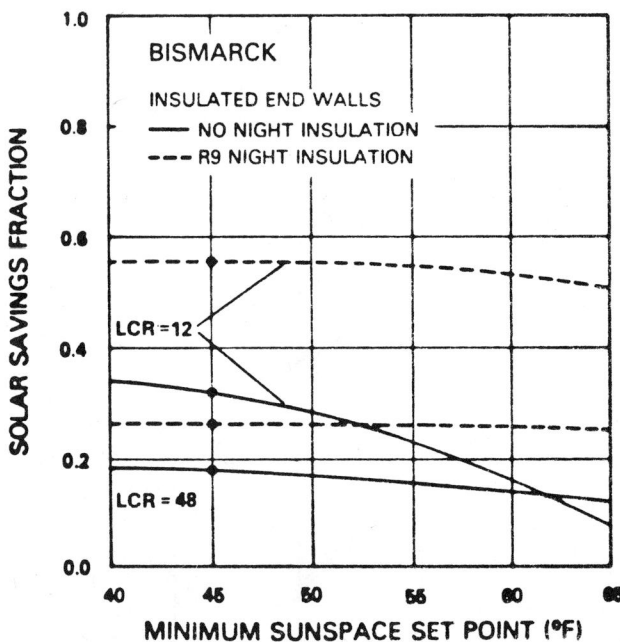

P-34: Effect on solar savings fraction of different minimum sunspace thermostat settings, applied to sunspaces of geometry A—with a masonry common wall, insulated end walls, and both with and without R9 nighttime insulation (A1 and A2)—in two cities. The definition of solar savings fraction includes the sunspace auxiliary heat. Reference conditions are indicated by dots.

No account was taken of incidental internal heat generation (appliances, people, etc.) in these data. With a normal amount of internal heat and a normal building heat loss coefficient, internal sources can be expected to raise the room temperature by 5 to 7 Fahrenheit degrees. In buildings with lower-than-average heat-loss characteristics, the effect on room temperature of these incidental heat-generating sources will be greater. This means that the room setpoint sensitivity data may be corrected to account for internal heat generation by modifying the setpoint temperatures on the horizontal axis. To make this correction, increase the temperatures by the quantity:

$$\Delta T(F) = \frac{\text{internal heat generation rate (in Btu/day)}}{\text{total load coefficient (in Btu/°F/day)}}$$

The total load coefficient is the building load coefficient plus the solar wall load coefficient.

The sunspace in the reference designs is heated as necessary to maintain its air temperature at 45°F. The sunspace auxiliary heat is included in the definition of *SSF*:

$$SSF = 1 - \left(\frac{\text{auxiliary heat}}{\text{net reference load}}\right)$$

Therefore, *SSF* is sensitive to the sunspace's minimum thermostat setpoint to the extent that sunspace heating is needed. This sensitivity is illustrated for both Albuquerque and Bismarck in figure P-34. The figure shows that there is little sensitivity to a minimum sunspace thermostat setpoint in the range 40°–50°F except in extremely cold climates, such as Bismarck's. This means that an insignificant amount of auxiliary heat is required by the sunspace for setpoints up to about 50°F except in a place like Bismarck.

With R9 nighttime insulation, virtually no sunspace auxiliary heat is needed for setpoints up to about 55°F, and very little is needed even up to 65°F. This does not mean, however, that a sunspace temperature would never fall below the indicated levels. Occasional extended periods of cold and cloudiness can be expected to occur, during which some auxiliary heat needs to be provided—for example, to protect temperature sensitive plants. This can normally be done by simply keeping a door or window open to the sunspace from the building, so that some auxiliary heat supplied to the adjacent spaces can flow into the sunspace. However, the total auxiliary heat required by the sunspace over a long term will be small in comparison to the total building auxiliary heat. All this is conditioned, of course, on the sunspace's having a heat storage capacity and a thermal integrity comparable to those of the reference designs—namely, one cubic foot of water or high-density masonry per square foot of projected area, double glazing, R20 insulation in opaque exterior walls, and an infiltration rate of 0.5 air changes per hour.

P.3 SOLAR LOAD RATIO CORRELATIONS

P.3.a SOLAR LOAD RATIO (SLR)

A general definition of the monthly solar load ratio (*SLR*) is

$$SLR = Q_s/Q_{load}$$

For sunspaces, Q_s is the monthly solar radiation absorbed in the sunspace minus the monthly effective heat loss from the sunspace to ambient, and Q_{load} is the monthly net reference load. Expressing these quantities per unit of projected area, the sunspace *SLR* may be written as:

$$SLR' = (S - L)/(LCR \times DD)$$

where:
- *S* is the monthly absorbed solar radiation per unit of projected area (in Btu/ft²)
- *L* is the effective monthly sunspace load per unit of projected area (in Btu/ft²)
- *DD* is the monthly Fahrenheit heating degree-days
- *LCR* is the building load collector ratio (in Btu/DD/ft²)

The quantities *S*, *L*, and *DD* are discussed in more detail below.

P.3.b ABSORBED SOLAR RADIATION

The quantity *S* is the monthly solar radiation absorbed in the sunspace per unit of projected area. Included in *S* is solar radiation absorbed in the floor, in the common wall, in lightweight objects represented by the lightweight absorption fraction, and (if present) in the ceiling, in water storage containers, and in opaque end walls. Some solar radiation is also absorbed in the glazing, and this is taken into account in the sunspace models, but the definition of *S* excludes this type of radiation.

As in a direct-gain room, the geometry of the sunspace, as well as its solar absorptances, determines how much transmitted radiation is ultimately absorbed. In addition, radiation may enter and leave the sunspace through several glazings with various orientations. The values of *S* used in the *SLR* correlations were determined from a particular correlation between *S* and the monthly solar radiation incident on a horizontal surface. Refer to section P.3.b for details. For information on determining *S* for off-reference sunspace configurations, refer to Appendix 20.

P.3.c EFFECTIVE SUNSPACE LOAD (L)

The quantity *L* is the effective sunspace load per unit of projected area. The term "effective" means that *L* is not the actual sunspace load, but rather a quantity that simulates the effect of the sunspace load on the ability of the sunspace's *SLR* to be an accurate predictor of the building's auxiliary heat requirement. The effective sunspace load is written:

$$L = LCR_s \times H \times DD$$

where:
- LCR_s is the sunspace load collector ratio (in Btu/DD/ft²)
- *H* is an adjustable parameter that is evaluated in the correlation procedure described below
- *DD* is the Fahrenheit heating degree-days

The sunspace load collector ratio (*LCR*) is the sunspace load coefficient (in Btu/DD) divided by the projected area (in ft²). The sunspace load coefficients for the reference designs were determined by ASHRAE-type methods; the values are listed in figure P-35. There is more information on evaluating LCR_s in section P.4.

Reference Design Designation (see Figure P-1)	C	D	Coefficients LCR$_s$	H	stdv
A1	0.9587	0.4770	18.6	0.83	0.027
A2	0.9982	0.6614	10.4	0.77	0.026
A3	0.9552	0.4230	23.6	0.83	0.030
A4	0.9956	0.6277	12.4	0.80	0.026
A5	0.9300	0.4041	18.6	0.96	0.031
A6	0.9981	0.6660	10.4	0.86	0.028
A7	0.9219	0.3225	23.6	0.96	0.035
A8	0.9922	0.6173	12.4	0.90	0.028
B1	0.9683	0.4954	16.3	0.84	0.028
B2	1.0029	0.6802	8.5	0.74	0.026
B3	0.9689	0.4685	19.3	0.82	0.029
B4	1.0029	0.6641	9.7	0.76	0.026
B5	0.9408	0.3866	16.3	0.97	0.030
B6	1.0068	0.6778	8.5	0.84	0.028
B7	0.9395	0.3363	19.3	0.95	0.032
B8	1.0047	0.6469	9.7	0.87	0.027
C1	1.0087	0.7683	16.3	0.76	0.025
C2	1.0412	0.9281	10.0	0.78	0.027
C3	0.9699	0.5106	16.3	0.79	0.024
C4	1.0152	0.7523	10.0	0.81	0.025
D1	0.9889	0.6643	17.8	0.84	0.028
D2	1.0493	0.8753	9.9	0.70	0.028
D3	0.9570	0.5285	17.8	0.90	0.029
D4	1.0356	0.8142	9.9	0.73	0.028
E1	0.9968	0.7004	19.6	0.77	0.027
E2	1.0468	0.9054	10.8	0.76	0.027
E3	0.9565	0.4827	19.6	0.81	0.028
E4	1.0214	0.7694	10.8	0.79	0.027

P-35: Sunspace solar load ratio (*SLR*) correlation coefficients.

P.3.d DEGREE-DAYS

The quantity DD is the monthly heating degree-days, evaluated from the means of the daily minimum and maximum temperatures. It is defined relative to a base temperature of 65°F in the *development* of the SLR correlations; but in the *application* of the SLR correlations to a monthly calculation, the base temperature for DD is a variable, as explained in sections P.4 and Q.3.

A potential point of misunderstanding involves the use of DD in the effective sunspace load (L). It might be thought that this amounts to an assumption that the average sunspace temperature is the temperature-base in the definition of DD. This is not the case, and two comments should help to clarify this point. First, the adjustable parameter H, which is evaluated in the correlation procedure, has the effect of correcting L for the choice of the base temperature in DD. Second, the average sunspace temperature is accounted for implicitly in the correlation because it is a function of the climate as represented by the variable S/DD, used below.

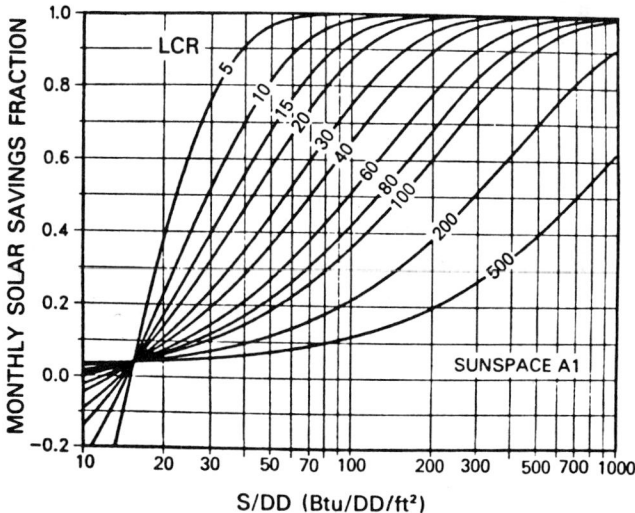

P-36: Monthly solar savings fraction plotted against monthly solar radiation absorbed per degree-day (S/DD), for sunspace design designation A1.

P.3.e SLR CORRELATIONS

It is convenient to rewrite the expression for SLR' by substituting the L equation in section P.3.c into the SLR' equation in section P.3.a. The result is:

$$SLR' = (S/DD - LCR_s \times H)/LCR$$

The SLR correlations have the form:

$$SSF = 1 - C \exp(-D \times SLR')$$

where C and D (and H, which appears in the definition of SLR') are adjustable coefficients determined in the correlation procedure for each reference design. In this form of the equation, there is no portion linear in SLR', and the constant term is fixed at 1 rather than being adjustable. The correlation coefficients for all of the sunspace reference designs are listed in figure P-35. The standard deviations, as indications of the precisions of the correlations, are also listed. These correlation coefficients are also tabulated in Appendix 18, along with the coefficients for thermal storage wall and direct-gain systems.

A graph of the SLR correlations is useful. Referring to the SLR equation in this section for a given reference design, the parameters LCR_s and H are fixed. Therefore, one parameter, LCR, and one monthly variable, S/DD, determine the monthly SLR' and hence the monthly SSF. This relationship is plotted for each of the reference designs in figure 18–12 of Appendix 18. A representative example is given in figure P–36. The figure shows a family of monthly SSF-versus-S/DD curves, in which the various curves in the family correspond to various LCR values.

P.3.f CORRELATIONS FOR S

Implicit in the SLR correlations are correlations for S, the monthly solar radiation absorbed in the sunspace per unit of projected area. These correlations have the form:

$$S/Q_h = B_1 + B_2 Y + B_3 Y^2 + K_T(B_4 + B_5 Y + B_6 Y^2)$$

where:

Q_h is monthly solar radiation on a horizontal surface (in Btu/ft²)

P-37: Coefficients for solar radiation correlations.

Reference Design								
Geometry	End walls	B_1	B_2	B_3	B_4	B_5	B_6	stdv
A	insulated	0.72008	−0.15181	0.49973	−0.15039	0.14384	3.6374	0.054
A	glazed	0.81554	−0.23988	0.60252	−0.16445	0.33730	3.1695	0.048
B	insulated	0.58932	−0.09693	0.38955	−0.14699	−0.39149	3.9171	0.044
B	glazed	0.62569	−0.13941	0.43331	−0.14982	−0.26401	3.5685	0.040
C	----	0.39436	−0.21103	0.58815	−0.24083	−0.60746	4.6546	0.046
D	----	0.73147	−0.15418	0.50763	−0.15276	0.14608	3.6950	0.055
E	----	0.61661	−0.10127	0.40733	−0.15367	−0.40940	4.0969	0.046

Y is $(LAT - DEC)/100$
LAT is the latitude (degrees)
DEC is the mid-month solar declination (degrees)
K_T is the average monthly clearness ratio
B_1 through B_6 are constants that depend on the reference design and are listed in figure P-37.

The required monthly variables are HS, $LAT - DEC$, and K_T, tabulated for numerous locations in Appendix 19. The quantity Q_h is obtained by multiplying HS by the number of days in the month.

P.4 MONTHLY CALCULATION —THE SLR METHOD

The purpose of the *SLR* correlations is their application to the calculation of the monthly auxiliary heat requirement. The monthly calculation, or *SLR* method, is summarized in Chapter Q. Some specifics that relate to sunspaces are discussed below.

The monthly calculation procedure applies specifically to the 28 sunspace reference designs listed in figure P-1; therefore, the monthly method begins with the analyst choosing the reference design that is most similar to the actual design to be analyzed. The applicability of the method is not strictly limited to one of the reference designs—an important degree of flexibility may be invoked in the choices of certain design parameters. The areas of flexibility concern the sunspace configuration and orientation, the thermostat setpoint of the building's auxiliary heating system, and nonauxiliary internal heat generation in the building. The possibilities are summarized below.

Virtually any sunspace configuration and orientation can be analyzed by means of the monthly calculation method, provided that the common wall and thermal storage configuration correspond to one of the reference designs. The sunspace configuration and orientation enter the monthly calculation through the quantities LCR_s and S.

P-38: Values of LCR_s for different reference designs and with different forms of insulation.

P.4.a CALCULATING LCR_s

One aspect of accounting for the sunspace configuration is through LCR_s, the sunspace load collector ratio. If the sunspace design has heat-loss characteristics that are substantially the same as the chosen reference design, then the reference value of LCR_s may be used. The reference values of LCR_s are listed in figure P-35. If the heat-loss characteristics of the design differ primarily in infiltration rate and in the thermal resistance of the nighttime insulation, then the value of the LCR_s may be found in figure P-38 or P-39.

In figure P-39, the formula used for calculating LCR_s is:

$$LCR_s = \left(\frac{B_0 + (B_1 \times R)}{B_2 + R}\right) + (B_3 \times ACH) \text{ (in Btu/DD/ft}^2\text{)}$$

where:
R = thermal resistance of night insulation (in hr/°F/ft²/Btu)
ACH = sunspace infiltration rate (in air changes/hour)

If the heat-loss characteristics differ from the reference design with respect to the areas and thermal resistances (R-values) of opaque exterior walls, or with respect to the area, number of panes, and interpane air gap of the glazed surfaces, then a simplified ASHRAE-type method may be used to estimate an off-reference value of LCR_s. First, the sunspace load coefficient (in Btu/DD) is estimated as a sum of the load coefficients that

Reference Design		Coefficients			
Geometry	End walls	B_0	B_1	B_2	B_3
A	insulated	34.9	6.80	2.06	3.4
A	glazed	46.1	8.06	2.10	3.4
B	insulated	33.7	5.63	2.20	1.9
B	glazed	39.9	6.43	2.17	1.9
C	---	28.3	5.92	2.07	5.2
D	---	30.4	6.20	1.93	4.1
E	---	35.0	6.84	1.99	4.1

P-39: Coefficients for the LCR_s formula.

Reference Design			no NI			R4 NI			R9 NI		
Geom	End walls	ACH =	0.5	1.0	2.0	0.5	1.0	2.0	0.5	1.0	2.0
A	insulated		18.6	20.3	23.7	11.9	13.6	17.0	10.4	12.1	15.5
A	glazed		23.7	25.4	28.8	14.5	16.2	19.6	12.4	14.1	17.5
B	insulated		16.3	17.2	19.1	10.0	11.0	12.9	8.5	9.4	11.3
B	glazed		19.3	20.3	22.2	11.6	12.5	14.4	9.7	10.7	12.6
C	—		16.3	18.9	24.1	11.2	13.8	19.0	10.0	12.6	17.8
D	—		17.8	19.9	24.0	11.4	13.4	17.5	9.9	12.0	16.0
E	—		19.6	21.7	25.8	12.5	14.5	18.6	10.8	12.9	17.0

NI = nighttime insulation, in place 14 hours per day; the reference value of the nighttime insulation thermal resistance is R9.
ACH = sunspace infiltration rate (air changes/hour); the reference value of the sunspace infiltration rate is 0.5 air changes/hour.

are due to losses through the glazings, through infiltration, through the floor-slab perimeter, and through opaque exterior walls and ceiling. The sunspace common wall is not included in this estimate. Then, the sunspace load coefficient is divided by the projected area (in ft²). The result is LCR_s (in Btu/DD/ft²). If a nonreference value of LCR_s is used, then the monthly SSF-versus-S/DD curves (figure 18–12) cannot be used to find the monthly SSFs. Instead, SLR' and SSF must be calculated from the two equations in section P.3.e.

If the recommended load coefficients below are used, the calculation will be consistent with those done to determine the reference values of LCR_s.

The recommended glazing load coefficient is generalized to take account of the air gap between glazings, as well as the number of glazing layers. The additional precision is desirable in the case of the sunspace load coefficient because the glazing load is the dominant contribution to the total. If there is no nighttime insulation, the glazing load coefficient (in Btu/DD) is:

$$L_g = \frac{26 \times \text{glazing area (in ft}^2)}{(N + 1.517t - 0.4295)}$$

where:

N = number of glazing layers, and
t = thickness of the glazing air gap in inches.*

If there is movable nighttime insulation:

$$L_g = 26 \times \text{glazing area (in ft}^2) \times \left[\tfrac{1}{3} \left(\frac{1}{N + 1.517t - 0.4295} \right) + \tfrac{2}{3} \left(\frac{1}{1.083\,R + N + 1.517t - 0.4295} \right) \right]$$

where R is the R-value of nighttime insulation (in hr/°F/ft²/Btu). It is assumed in this formula that the nighttime insulation is removed 8 hours out of 24 and is in place 16 hours out of 24; thus the fractions ⅓ and ⅔. These proportions can be changed if desired.

The recommended infiltration load coefficient is:

$$L_i = 0.432 \times \text{sunspace volume (in ft}^3) \times \text{air changes per hour}$$

The recommended perimeter load coefficient is:

$$L_p = \frac{100 \times \text{length of sunspace exterior perimeter (in ft)}}{\text{R-value of perimeter insulation} + 5}$$

The recommended load coefficient for opaque walls or ceiling is:

$$L_{w,c} = \frac{24 \times \text{wall or ceiling area (in ft}^2)}{\text{R-value of wall or ceiling}}$$

To illustrate a typical LCR_s calculation, the value for sunspace reference design A1 is calculated below. The glazing configuration is: two glazing layers, 0.5-inch air gap, no nighttime insulation, 353 ft² glazing area. Thus:

$$L_g = \left(\frac{26 \times 353}{2 + (1.517 \times 0.5) - 0.4295} \right) = 3{,}941 \text{ Btu/DD}$$

The infiltration rate is 0.5 air changes per hour, and the sunspace volume is 2,099 ft³. Thus:

$$L_i = 0.432 \times 2{,}099 \times 0.5 = 453 \text{ Btu/DD}$$

The sunspace perimeter is insulated to R12 and is 53.1 feet long. Thus:

$$L_p = \left(\frac{100 \times 53.1}{(12 + 5)} \right) = 312 \text{ Btu/DD}$$

The ceiling is insulated to R20 and is 120 ft² in area. Thus:

$$L_c = \left(\frac{24 \times 120}{20} \right) = 144 \text{ Btu/DD}$$

The end walls are insulated to R20 and are 140 ft² in total area. Thus:

$$L_w = \left(\frac{24 \times 140}{20} \right) = 168 \text{ Btu/DD}$$

The total sunspace load is:

$$L = L_g + L_i + L_p + L_c + L_w = 5{,}018 \text{ Btu/DD}$$

The sunspace load collector ratio is the above load coefficient divided by the 270-ft² projected area:

$$LCR_s = 5{,}018/270 = 18.6 \text{ Btu/DD/ft}^2$$

This value agrees with that used in the SLR correlations for configuration A1. See figure P–35. The method is, however, simplified relative to the one used to calculate the reference LCR_s values. It will not always give an exact agreement, but it will be within 5%—sufficiently accurate for use in a monthly SLR calculation.

P.4.b CALCULATING S

The other aspect of accounting for the sunspace configuration is through S, the monthly solar radiation absorbed in the sunspace per unit of projected area. It is through S that the sunspace glazing geometry and orientation are accounted for. If the sunspace is one of the reference designs, then S may be determined from the correlation expressed by the equation in section P.3.f. Otherwise, if a sunspace conforms to one of the reference geometries but is not oriented to true south, or if a sunspace geometry differs from the reference designs in its relative dimensions, glazing tilt angles, number of glazing layers, etc., then values of S specific to the design in question should be determined. The data in Appendix 20 provide information that is needed to do this. The method consists of determining the solar radiation incident on each glazed surface, the fraction of the incident radiation that is transmitted, and the fraction of the transmitted radiation that is absorbed.

---

*The formula is valid for t in the range of 0 to 0.5 inches. For values of L greater than 0.5, use 0.5. If $N = 1$, use $t = 0.35$ inch.

P.4.c CALCULATING DD

The reference designs have an auxiliary heating system thermostat setpoint of 65°F and contain no nonauxiliary internal heat generation. But internal heat generation and an off-reference thermostat setpoint may be accounted for in the monthly calculation by the choice of the base temperature for the monthly heating degree-days (DD). This procedure is reviewed in Chapter Q. The needed parameters are:

T_{set}—the heating thermostat setpoint (in °F)
Q_{int}—the internal heat generation rate (in Btu/day)
BLC—the building load coefficient (in Btu/DD)

Ref. Design	U_c (in Btu/hr/ft²/°F)								
	no NI			R4 NI			R9 NI		
ACH =	0.5	1.0	2.0	0.5	1.0	2.0	0.5	1.0	2.0
A1, 2	0.27	0.28	0.29	0.23	0.24	0.26	0.21	0.23	0.25
A3, 4	0.29	0.30	0.31	0.25	0.26	0.28	0.23	0.24	0.27
A5, 6	0.05	0.05	0.05	0.05	0.05	0.05	0.04	0.05	0.05
A7, 8	0.05	0.05	0.05	0.05	0.05	0.05	0.05	0.05	0.05
B1, 2	0.26	0.26	0.28	0.21	0.22	0.24	0.19	0.20	0.22
B3, 4	0.28	0.28	0.29	0.22	0.23	0.25	0.21	0.22	0.23
B5, 6	0.05	0.05	0.05	0.04	0.05	0.05	0.04	0.04	0.05
B7, 8	0.05	0.05	0.05	0.05	0.05	0.05	0.04	0.04	0.05
C1, 2	0.38	0.41	0.46	0.30	0.34	0.41	0.28	0.32	0.39
C3, 4	0.09	0.09	0.09	0.08	0.09	0.09	0.08	0.08	0.09
D1, 2	0.36	0.38	0.41	0.28	0.31	0.36	0.26	0.29	0.34
D3, 4	0.08	0.08	0.08	0.07	0.07	0.08	0.07	0.07	0.08
E1, 2	0.41	0.43	0.46	0.32	0.35	0.40	0.29	0.32	0.38
E3, 4	0.09	0.09	0.09	0.08	0.08	0.09	0.08	0.08	0.08

NI = nighttime insulation, in place 14 hours per day; the reference value of the nighttime insulation thermal resistance is R9.
ACH = sunspace infiltration rate (air changes/hour); the reference value of the sunspace infiltration rate is 0.5 air changes/hour.

P-40: Values for solar wall steady-state conductance (U_c) for different reference designs and with different forms of insulation. The reference designs are illustrated in figure P-1.

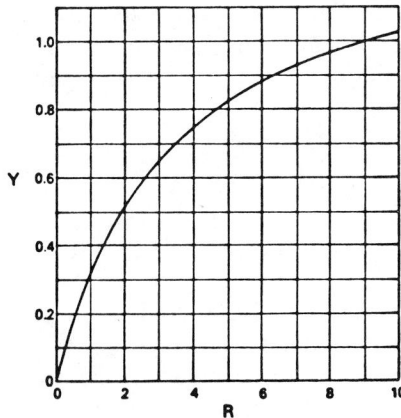

P-42: Plot of values computed from the equation for fractional performance improvement (Y), obtained by adding **nighttime** insulation with thermal resistance of numerical value R.

A_p—the projected area (in ft²)
U_c—the steady-state conductance of the solar wall (in Btu/hr/ft²/°F)

Values of U_c for the sunspace reference designs are listed in figure P-40.

The U_c value for nonreference designs may be found from the formula:

$$U_c = \frac{1}{\left(\frac{1}{A_w U_w}\right) + \left(\frac{24}{LCR_s}\right)}$$

where:

LCR_s is the sunspace's load collector ratio
U_w is the common-wall steady-state conductance (in Btu/hr/ft²/°F)
A_w is the common-wall area, expressed in units of the projected area (for example, A_w is 1 for the attached geometries A and B because the common-wall area is the same as the projected area)

Values of U_w and A_w for the reference designs are tabulated in figure P-41.

P.4.d OFF-REFERENCE NIGHTTIME INSULATION

For values of nighttime insulation thermal resistance less than the reference value of R9 (9 hr/°F/ft²/Btu), a nonlinear interpolation between no-nighttime-insulation and R9-nighttime-insulation results can be done by means of the quantity Y, which is the fractional performance improvement obtained by adding nighttime insulation of some thermal resistance R-value between 0 and 9. In other words, if SSF_0 is the monthly solar savings fraction for no nighttime insulation and SSF_9 is the monthly solar savings fraction for R9 nighttime insulation, then the monthly solar savings fraction for an intermediate thermal resistance R is:

$$SSF_R = SSF_0 + Y(SSF_9 - SSF_0)$$

A simple formula for Y that describes the average performance of night insulation is:

$$Y = \frac{(1 + R_o/9)}{(1 + R_o/R)}$$

where $R_o = 3.4$ for sunspaces. This relationship is also expressed graphically in figure P-42.

Wall Type	U_w	Geometry	A_w
masonry (12")	0.42	A, B	1
insulated (R20)	0.05	C	2
		D	1.69
		E	1.93

P-41: Values of U_w and A_w for different wall types and geometries.

Q Monthly Calculation — The SLR Method

A central tool in passive-solar design analysis is the monthly calculation, or "*SLR* method," based on the monthly *SLR* correlations. A summary of the method is presented below.

For ease in application to the monthly calculation, the *SLR* is expressed in terms of two quantities: *LCR*, which is a building parameter; and a monthly variable, *S/DD*, in which *S* is the monthly solar radiation absorbed in the building per unit of projected area and *DD* is the monthly heating degree-days. Refer to Chapters O and P or to Appendix 18 for the specific formulas for *SLR* in terms of *LCR* and *S/DD*.

Q.1 A STEP-BY-STEP PROCEDURE

The steps of a simplified monthly calculation are listed below to review the basic elements of the method:

1. Obtain building information:
   a. Building load coefficient (*BLC*)
   b. Projected area (A_p)
   c. Load collector ratio (*LCR* = *BLC*/A_p)
2. Obtain site and climate information:
   a. Latitude
   b. Latitude minus midmonth declination, calculated monthly (Appendix 19)
   c. Clearness ratio (K_T), calculated monthly (Appendix 19)
   d. Incident solar radiation on a horizontal surface, calculated monthly (Appendix 19)
   e. Heating degree-days (*DD*), calculated monthly (Appendix 19)
3. Obtain absorbed solar radiation (*S*), calculated monthly (Appendix 18 or 20)
4. Obtain monthly solar savings fractions (*SSF*):
   a. Calculate *S/DD*, monthly
   b. Determine *SSF*, monthly (two options) (Appendix 18):
      i) Graphical, *SSF*-versus-*S/DD* curves
      ii) Analytical, equations for *SLR* and *SSF*
5. Calculate auxiliary heat requirement (*Q*):
   a. Monthly *Q* = (1 − *SSF*) × *BLC* × *DD*
   b. Annual *Q* = sum of monthly *Q* values
6. Calculate annual solar savings fraction:
   SSF = (1 − *Q*)/(*BLC* × *DD*)

Q.2 CALCULATING *S*

The average monthly clearness ratio (K_T) is defined in Appendix 19 and is tabulated there along with the monthly solar radiation, temperature, and degree-days data. The role of K_T in the calculation procedure is as a second parameter, along with the latitude-minus-midmonth-declination (*LAT* − *DEC*) parameter, in the solar radiation data in Appendix 20.

Figures for absorbed solar radiation as a fraction of transmitted radiation are given in figure 20–13 and in tables at the end of Appendix 20. Figure 20–13 is for thermal storage walls, and the tables at the end of the appendix are for direct-gain and sunspace systems.

Q.3 CALCULATING *DD*

The other major variable in monthly *SLR* is *DD*, the monthly heating degree-days. This is nominally the 65°F-base heating degree-days that pertain to the reference case of a heating thermostat setpoint of 65°F and no internal heat generation. (A distinction is made between auxiliary heat deliberately generated by a furnace, space heaters, wood stoves, etc., and internal heat-generation incidental to other activities, such as that produced by appliances, hot water, and people.) The effect of the thermostat setpoint and the rate of internal heat-generation are taken into account in the monthly calculation method by the choice of the base temperature for *DD* (Step 2e). The base temperature needed for the monthly calculation is the "solar base temperature" defined by:

$$T_{bs} = (T_{set} - Q_{int})/[BLC + (24 \times A_p \times U_c)]$$

where:

T_{set} is the heating thermostat setpoint (in °F)
Q_{int} is the internal heat generation rate (in Btu/day)
BLC is the building load coefficient (in Btu/°F/day)
A_p is the projected area (in ft²)
U_c is the steady-state conductance of the solar wall (in Btu/hr/ft²/°F)

Values of U_c for the sunspace reference designs are tabulated in Chapter P. The quantity *BLC* + (24 × A_p × U_c) is the total load coefficient.

The nonsolar base temperature is:

$$T_{bns} = (T_{set} - Q_{int})/BLC$$

It is used in calculating the annual SSF (Step 6).

An estimate of Q_{int}, the internal heat generation rate, is a key part of the above procedure. This quantity may vary substantially from building to building, and it may have a large effect on the building's energy performance. Estimates of Q_{int} can be made by itemizing the anticipated sources of internal heat generation, each source's rate of heat generation, and the number of hours per day that each source generates heat at this rate. If this procedure is too detailed for the level of accuracy desired, more general estimates may be satisfactory. A rate of 20,000 Btu/day/person may be used as typical of residential buildings.

Judgment should be exercised in accounting for internal heat generation in cases in which a substantial fraction of the heat is generated during daytime hours, and in which the average monthly internal heat generation is comparable to the net reference load. In such cases, the daytime internal sources may not be entirely useful, but may lead instead to overheating. A conservative analytical practice would be to neglect the daytime internal heat generation in such cases to the extent that it exceeds the nighttime heat generation.

Q.4 OFF-REFERENCE NIGHTTIME INSULATION

For values of nighttime insulation thermal resistance that are less than the reference value of R9 (9 hr/°F/ft²/Btu), a non-linear interpolation between no-nighttime-insulation and R9-nighttime-insulation results can be done. The interpolation is done by use of a quantity Y that is the fractional performance improvement obtained by adding nighttime insulation of an R-value between 0 and 9. In other words:

$$SSF_R = SSF_0 + Y(SSF_9 - SSF_0)$$

where SSF_R, SSF_0, and SSF_9 are the monthly SSFs for systems with nighttime insulation thermal resistances of R, 0, and 9, respectively.

The value of Y depends on R and on the system type. The following function is a useful approximation:

$$Y = \frac{(1 + R_o/9)}{(1 + R_o/R)}$$

System Type	R_o
Waterwall	3.2
Trombe wall	4.6
Direct gain	1.8
Sunspace	3.4

Q-1: R_o values for various system types.

where R is the thermal resistance of the night insulation, and R_o is a constant that depends on the system configuration, both in units hr/°F/ft²/Btu. Values of R_o are given in figure Q-1. Remember that the values for the waterwall, Trombe-wall, and direct-gain cases apply just to the particular solar aperture configurations for which they were derived—namely, to double glazing and a Trombe wall with an 18-inch thickness and a thermal conductivity of 1.0 Btu/hr/ft/°F. For other numbers of glazings and wall configurations, the value of R_o should be corected. An approximate way to do this is to let $R_o = 1/U_c$, where U_c is the steady-state conductance of the solar wall.

Q.5 THE CORRELATION EQUATIONS

For sunspace systems and for the thermal storage wall correlations in Appendix 18:

$$SLR' = (S/DD - LCR_s \times H)/LCR$$

The quantity SLR' contains the sunspace load coefficient (LCR_s), which may be set to an off-reference value according to the properties of the particular sunspace being evaluated. Refer to Chapter P for details. If an off-reference value of LCR_s is used, Step 4b above must be performed by its analytical option, that is, the evaluation of SLR' and SSF from the formulas for these quantities in section P.3.e.

When the analytical option of Step 4b is used, the equations are as follows:

$$SSF = 1 - K(1 - F(X))$$
$$F(X) = AX, \quad (X \leq R)$$
$$F(X) = B - C \exp(-DX), \quad (X > R)$$
$$K = 1 + G/LCR$$
$$X = (S/DD - LCR_s \times H)/(K \times LCR)$$

where (for thermal storage wall and sunspace systems):

$$G = 0$$
$$R = -9$$
$$B = 1$$

and (for direct gain):

$$LCR_s = H = 0.$$

The significance of $R = -9$ for thermal storage wall and sunspace systems is that the linear portion of $F(X)$ is not used; R is therefore set sufficiently small that X is always greater than R. Any value of R less than -9 would also be correct. The quantity $\exp(-DX)$ above is the exponential function with argument $-DX$—that is, e^{-DX}, where $e = 2.718\ldots$ is the base of natural logarithms.

Q.6 MIXED SYSTEMS

Designs that combine two or more passive system types may be analyzed by treating *BLC* as if it were divided into portions in the same ratios as the relative projected areas of the various passive system types. Then, each system is analyzed independently and the results are combined. This amounts to an assumption that each system independently serves a portion of the load—as if the building were divided into zones with no heat-flow across imaginary boundaries between zones. This might be a realistic assumption in an actual multi-zone case with equal thermostat settings. It is less realistic in cases where two or more systems serve the same zone. In either case, one would expect that heat-flow between zones would be beneficial; and therefore, a mixed system calculation based on this assumption will tend to underestimate the combined performance.

The procedure is similar to the step-by-step procedure listed above for single types but with certain steps modified as follows:

Step 1. BLC for the whole building is calculated in the normal way. None of the solar wall heat losses are included. A_p for the whole building is the sum of the individual values of A_p. An additional step is to calculate individual values of *BLC* for each system as the entire building *BLC* times the ratio of the individual A_p to the whole building A_p. Calculate one *LCR* for the whole building; notice that *LCR* is the same for each individual system and for the whole building because of the manner of apportioning *BLC*.

Step 2. There is no change in Steps 2a–2d. Step 2e requires calculating the entire building's solar base temperature with the equation:

$$T_{bs} = (T_{set} - Q_{int})/\{BLC + [24 \times \Sigma(A_p \times U_c)]\}$$

where T_{set}, Q_{int}, and *BLC* refer to the whole building as before, Σ means a sum over all of the individual systems, and A_p and U_c refer to each individual system in the sum.

Step 3. S is calculated for each individual system.
Step 4. The monthly solar savings fractions are obtained for each individual system using the correlation that is appropriate to each system type.

There are two alternatives for Step 5:

Step 5A. Calculate the whole building *SSF* for each month as a weighted average of the *SSF*s for the individual systems. The weighting factors are the ratios of the individual system A_p to the whole building A_p. Then apply Step 5 as for a single-system type to obtain the annual auxiliary heat.

or:

Step 5B. Apply Step 5 to each individual system to obtain the monthly and annual auxiliary heats. Individual values of *BLC* for each system are used. The whole building auxiliary heat is the sum of the values for each individual system.
Step 6. Calculate the annual *SSF* as before, using the whole building auxiliary heat and *BLC*.

Alternative Step 5B treats the individual passive solar systems explicitly as independent systems—as if each served a separate zone, each with its own auxiliary heater. Mathematically, however, Steps 5A and 5B are equivalent under the assumptions used in the mixed-system analysis to this point—namely, that individual-system values of *BLC* be used in proportion to the projected areas, and that the same degree-day base temperature for each system be used. The advantage of Step 5B over 5A appears in cases in which the solar systems serve distinct zones. The two assumptions mentioned above may then be relaxed if desired: the actual *BLC* for each zone may be used; and the degree-day base temperature may be different in each zone. If the base temperature is calculated independently for each zone, then the T_{bs} equation from section Q.3 is applied to each individual zone, with Q_{int}, *BLC*, A_p, and U_c, the values that pertain to each zone.

Appendixes

Appendix 1 Properties of Glazing Materials

In choosing materials for a passive system, the most important considerations are appearance, durability, performance, and cost. Since the glazing is the face of a system, the glazing material has a large overall impact. Whether the glazing is clear or cloudy, shiny or dull, flat or bowed, dramatically affects the appearance of the system. Durability is a critical factor, since the glazing provides the outermost barrier to water, cold air, ultraviolet radiation, and long-term weathering. Finally, the transmittance (both short- and long-wave) of the glazing directly affects the overall efficiency of the system. These and other characteristic properties of various glazing materials are given in figure 1-1.

Although many factors influence the selection of a proper glazing for a passive system, the choice is not as crucial as for active collectors because the conditions are not as demanding. The maximum temperatures reached by a Trombe wall, for example, are not as high as those reached by a stagnating active collector because the Trombe wall's mass has a moderating effect on peak solar gains. On the other hand, the glazing for a thermosiphoning collector should be able to meet the standards set for active collectors because the thermal mass of the absorber plate is usually similar to that of active collectors. A stagnating thermosiphoning collector can reach peak temperatures close to those reached by stagnating active collectors.

The most desirable qualities in a glazing material are:

1. resistance to degradation from heat, light, and weather;
2. high transmittance of solar radiation and low transmittance of infrared or thermal radiation;
3. low cost;
4. ease of handling and fabrication; and
5. attractive appearance.

1.1 GLASS

Glass is often more expensive than other glazing materials, but when all factors are taken into account, it is a popular choice. Glass can be purchased either tempered or nontempered. Although nontempered glass is less expensive, fully tempered glass is usually preferred because of its greater resistance to breakage; in addition, when it does break, it produces only tiny fragments instead of dangerous shards.

For high transmittance the recommended glass is fully tempered "water white" sheet glass, which has a very low iron oxide content (0.01%) and thus the highest transmittance (91% for all thicknesses). Tempered float glass is less expensive but has a higher iron oxide content (0.12%) and a transmittance of between 79% and 86%, depending on its thickness (for this example, between ¼ inch and ⅛ inch). Various low-iron tempered float glasses are manufactured; one, for example, has an iron content of 0.05% and a transmittance that varies between 88% and 89% for thicknesses of 3/16 inch to ⅛ inch. As the iron content decreases, the dependence of transmittance on thickness also decreases. Window glass (nontempered) has a low iron oxide content and a transmittance of 91%; however, its use should be limited to low stress applications, such as vertical window glazing.

Glass is rigid, attractive, highly durable, and resistant to weathering and to chemical and light deterioration. Unfortunately, its high density (weight) makes it difficult to handle.

1.2 FIBERGLASS-REINFORCED POLYESTER

Several manufacturers have developed fiberglass-reinforced polyester (FRP) glazing materials that are formulated to resist ultraviolet and thermal degradation. Although FRP glazings appear cloudy, their solar transmittance (84% to 90%) is only slightly less than that of low-iron glass.

Some FRP glazings are available in flat sheets or in 4- and 5-foot-wide rolls and in thicknesses of 0.024, 0.040 and 0.060 inches. Most FRPs are easy to cut, drill, and install. However, two problems are often apparent. One is a wavy appearance due to buckling from expansion. These materials have a relatively large coefficient of thermal expansion. The problem may become progressively worse as the material continues to undergo thermal cycling. In order to minimize this problem, at least one manufacturer has developed double-glazed panels where the FRP is "stretched" onto an aluminum frame. The theory is that since aluminum and the FRP have nearly the same coefficient of thermal expansion, the glazing will be tight regardless of the temperature. These panels reduce but do not entirely eliminate the wavy effect. Some FRPs are available in a corrugated form that also reduces buckling.

The second potential problem relates to thermal degradation at high temperatures. Some FRPs experience losses in transmittance of 1%, 3%, and 11%, when exposed to temperatures of 150°F, 200°F, and 300°F, respectively, for 300 hours. For most passive applications this will not be a problem, but for some—such as nonvented thermal storage walls—it must be seriously considered.

FRP glazings are particularly suitable for greenhouse and sunspace applications in which the operating temperatures are lower than (for instance) a stagnating Trombe wall. Do-it-yourselfers will appreciate FRP glazings' ease of handling.

1.3 FILMS

Plastic films offer high transmittance and are relatively inexpensive. Although some plastic films have excellent resistance to temperature, a high coefficient of expansion makes them tend to sag at high temperatures. They can also be difficult to seal and handle, bowing between supports and sticking to certain surfaces due to their tendency to acquire electrostatic charge. In addition, they may be relatively transparent to infrared radiation, thus reducing solar collection efficiency.

Embrittlement in some plastic films is caused by direct exposure to ultraviolet radiation, and this effect is tremendously accelerated at higher temperatures. If the films are used at low temperatures, 4 to 5 years may elapse before embrittlement is likely to occur. Any hot spot (e.g., one near a hot metal support) may cause plastic film to embrittle much more rapidly. Recently devel-

	Thickness (in.)	Cost ($/ft.²)	Transmittance	Weight/Area (lb/ft.²)	Thermal Expansion (°F⁻¹ x 10⁻⁴)	Ease in Handling	Strength	Sheet Size (ft.)	Remarks
Water white glass "Solatex" (ASG)	0.125	0.99	0.90	1.60	0.47	Poor	Good (tempered)	2, 3, or 4x8	Very durable—no degradation
Float glass	0.125	2.35	0.84	1.60	0.47	Poor	Good (tempered)	4x8	Very durable—no degradation
Window glass (ASG SS Lustra-glass)	0.090	1.80	0.91	1.20	0.47	Poor	Poor (non-tempered)	4x7	Fragile
Sunlite Premium II (Kalwall)	0.040	0.60	0.88	0.29	2.00	Excellent	Very good	4 or 5 width rolls	Maximum temperature 300°F
Filon w/Tedlar (Vistron Corp.)	--	1.00	0.86	0.25	2.30	Very good	Very good	4.25x16	Maximum temperature 300°F
Flexiguard 7410 (3M)	7 mil	0.38	0.89	0.053	--	Fair	Good	4x150 roll	Maximum temperature 275°F
Tedlar (Dupont)	4 mil	0.05	0.95	0.029	2.80	Fair	Good, some embrittlement	up to 5.33 width roll (64 in.)	4-5 yr. lifetime at 150°F
Teflon FEP 100A (Dupont)	1 mil	0.58	0.96	0.02	5.85	Poor	Fair, not for exterior glazing	4.83 width roll (58 in.)	Maximum temperature 300°F
Swedcast 300 Acrylic (Swedlow Inc.)	0.125	0.81	0.93	0.77	4	Excellent	Very good	9 wide	Maximum service temperature 200°F
Lucite Acrylic (Dupont)	0.125	1.14	0.92	0.73	4	Very good	Very good	4x8	Maximum temperature 200°F
Tuffak-Twinwall (Rhom & Haas)	--	1.25 (2 layers)	Equiv. to 0.89 for 1 layer	0.25	3.3	Very good	High impact strength fatigue cracking	4x8	5% reduction in transmittance over 5 years
Acrylite SDP (Cyro)	--	2.15 (2 layers)	Equiv. to 0.93 for 1 layer	1.00	4	Very good	Good	6x8	Maximum temperature 230°F
Sun-lite Insulated Panels (Kalwall)	--	2.50 (2 layers)	Equiv. to 0.88 for 1 layer	0.7	--	Good	Good	4x8 4x10 4x12 4x14	Maximum temperature 300°F
Solar Glass Panels (ASG)	--	2.99 (2 layers)	Equiv. to 0.90 for 1 layer	4.5	0.47	Poor	Good	3 or 4x6 3 or 4x8	Very durable

1-1: Characteristic properties of various glazing materials [TEM].

oped films are expected to be less susceptible to ultraviolet degradation.

One type of plastic film is often heat-shrunk in order to make it taut, thereby improving its appearance and its useful life. As a result of heating, two effects occur. First, upon heating, temporary expansion of the plastic takes place at a level determined by the coefficient of thermal expansion. Second, as the plastic cools, it shrinks permanently. Because the two effects are opposite, one cannot observe the extent of the shrinkage until the material has cooled. As a result, the shrinkage is often too great, and the tension in the film warps the frame. Perhaps the safest procedure is not to shrink the material artificially, but to install it and let the shrinkage occur slowly as a result of exposure to normal operating temperatures.

One major disadvantage of all thin plastic films is their significant transmittance of long-wave thermal radiation, between 4 and 2 μm. This reduces efficiency by increasing the heat loss through the glazing. Glass has a transmittance in this region of less than 1%, but the transmittance for plastic film ranges from 17% to 57%.

1.4 CLEAR THERMOPLASTICS

Rigid plastic glazings have high impact and fracture resistance, ease of handling and fabrication, and attractive appearance. Most of these products can be categorized as either acrylics or polycarbonates. Acrylics have a slightly higher transmittance then tempered water-white glass and exhibit good resistance to ultraviolet light and weathering. They are usually clear and, as long as they remain unscratched, are as attractive as glass.

Acrylics do suffer problems at higher temperatures, but, except in the case of stagnating Trombe walls or convective loop collectors, this should not be a major concern for passive applications. These plastics have a large coefficient of thermal expansion, tending to bow in on the hot side, which in turn imposes severe stresses on the glazing supports. The bowing effect can be reduced by employing a mounting detail that allows freedom of movement, but this creates a more critical sealing detail.

Polycarbonates are stronger and can operate at higher temperatures than acrylics, but they have lower transmittance and suffer from ultraviolet degradation (yellowing upon prolonged exposure to the sun). Polycarbonates also have a large coefficient of thermal expansion and tend to bow inward at high temperatures.

1.5 INSULATING PANELS

Some glazing materials are manufactured in double-layer "insulating" panels that consist of a rigid sandwich of two glazing layers with an air space between. The relatively high initial cost of these components may be offset by the substantial labor savings during installation.

Although a panel using polycarbonate material can be relatively inexpensive, it shares the serious disadvantages of all polycarbonates: ultraviolet degradation, low transmittance, and a large coefficient of thermal expansion. Likewise, panels using acrylics have the disadvantages associated with acrylics: a low melting point and a very large coefficient of thermal expansion.

A serious consideration when using any plastic glazing material in either passive or active systems is the possibility of fire or fume inhalation. This consideration is particularly important in systems in which air is moved from behind the plastic glazing and distributed to the living area.

Appendix 2

Specific Heats and Heat Capacities of Materials

Materials in the chart below are rated on an equal volume basis.

Material	Specific Heat (Btu/lb/°F)	Density (lb/ft^3)	Heat Capacity (Btu/ft^3/°F)
Copper	0.092	556	51.2
Aluminum	0.214	171	36.6
Asphalt	0.22	132	29.0
Glass	0.18	154	27.7
White Oak	0.57	47	26.8
Limestone	0.217	103	22.4
Gypsum	0.26	78	20.3
Sand	0.191	94.6	18.1
White Pine	0.67	27	18.1
White Fir	0.65	27	17.6
Clay	0.22	63	13.9
Air (75°F)	0.24	0.075	0.018
Water	1.00	62.5	62.5
Iron (Scrap)	0.112	489	55
Concrete	0.27	140	38
Brick	0.20	140	28
Marble	0.21	180	38

Appendix 3

Chart of Mean Monthly and Annual Hours of Sunshine Throughout the United States

Figure 3-1 presents averages over as many as thirty years of monthly and annual hours of sunshine received at various localities in the United States.

STATE AND STATION	YEARS	JAN.	FEB.	MAR.	APR.	MAY	JUNE	JULY	AUG.	SEPT.	OCT.	NOV.	DEC.	ANNUAL
ALA. BIRMINGHAM	30	138	152	207	248	293	294	269	265	244	234	182	136	2662
MOBILE	22	157	158	212	253	301	289	249	259	235	254	195	146	2708
MONTGOMERY	30	160	168	227	267	317	311	288	290	260	250	200	156	2894
ALASKA ANCHORAGE	19	78	114	210	254	268	288	255	184	128	96	68	49	1992
FAIRBANKS	20	54	120	224	302	319	334	274	164	122	85	71	36	2105
JUNEAU	29	71	102	171	200	230	251	193	161	123	67	60	51	1680
NOME	27	72	109	193	226	285	297	204	146	142	101	67	42	1884
ARIZ. PHOENIX	30	248	244	314	346	404	404	377	351	334	307	267	236	3832
PRESCOTT	14	222	230	293	323	378	392	323	305	315	286	254	228	3549
TUCSON	13	255	266	317	350	399	394	329	329	335	317	280	258	3829
YUMA	30	258	266	337	365	419	420	404	380	351	330	285	262	4077
ARK. FT. SMITH	30	146	156	202	234	268	303	321	305	261	230	174	147	2747
LITTLE ROCK	30	143	158	2:3	243	291	316	321	316	265	251	181	142	2840
CALIF. EUREKA	30	120	138	180	209	247	261	244	205	195	164	127	108	2198
FRESNO	29	153	192	283	330	389	418	435	406	355	306	221	144	3632
LOS ANGELES	30	224	217	273	264	292	299	352	336	295	263	249	220	3284
RED BLUFF	15	156	186	246	302	366	396	438	407	341	277	199	154	3468
SACRAMENTO	30	134	169	255	300	367	405	437	406	347	283	197	122	3422
SAN DIEGO	30	216	212	262	242	261	253	293	277	255	234	236	217	2958
SAN FRANCISCO	30	165	182	251	281	314	330	300	272	267	243	198	156	2959
COLO. DENVER	30	207	205	247	252	281	311	321	297	274	246	200	192	3033
GRAND JUNCTION	30	169	182	243	265	314	350	349	311	291	255	198	168	3095
PUEBLO	30	224	217	261	271	299	340	349	318	290	265	225	211	3270
CONN. HARTFORD	30	141	166	206	223	267	285	299	268	220	193	137	136	2541
NEW HAVEN	30	155	178	215	234	274	291	309	284	238	215	157	154	2704
D. C. WASHINGTON	30	138	160	205	226	267	288	291	264	233	207	162	135	2576
FLA. APALACHICOLA	26	193	195	233	274	328	296	273	259	236	263	216	175	2941
JACKSONVILLE	30	192	189	241	267	296	260	255	248	199	205	191	170	2713
KEY WEST	30	229	238	285	296	307	273	277	269	236	237	226	225	3098
LAKELAND	7	204	186	222	251	285	268	252	242	203	209	212	198	2732
MIAMI	30	222	227	266	275	280	251	267	263	216	215	212	209	2903
PENSACOLA	30	175	180	232	270	311	302	278	284	249	265	206	166	2918
TAMPA	30	223	220	260	283	320	276	257	252	232	243	227	209	3001
GA. ATLANTA	25	154	165	218	266	309	304	284	285	247	241	188	160	2821
MACON	30	177	178	235	279	321	314	292	295	253	236	202	168	2950
SAVANNAH	30	175	173	229	274	307	279	267	256	212	216	197	167	2752
HAWAII HILO	7	153	135	161	112	106	158	184	134	137	153	106	131	1670
HONOLULU	30	227	202	250	255	276	280	293	290	279	257	221	211	3041
LIHUE	10	171	162	176	176	211	246	246	236	246	210	170	161	2411
IDAHO BOISE	30	116	144	218	274	322	352	412	378	311	232	143	104	3006
POCATELLO	30	111	143	211	255	300	338	380	347	296	230	145	108	2864
ILL. CAIRO	15	124	160	218	254	298	324	345	336	279	254	181	145	2918
CHICAGO	30	126	142	199	221	274	300	333	299	247	216	136	118	2611
MOLINE	18	132	139	189	214	255	279	337	300	251	214	130	123	2563
PEORIA	30	134	149	198	229	273	303	336	299	259	222	149	122	2673
SPRINGFIELD	30	127	149	193	224	282	304	346	312	266	225	152	122	2702
IND. EVANSVILLE	30	123	145	199	237	294	322	342	318	274	236	156	120	2766
FT. WAYNE	30	113	136	191	217	281	310	342	306	242	210	120	102	2570
INDIANAPOLIS	30	118	140	193	227	278	313	342	313	265	222	139	118	2668
TERRE HAUTE	24	125	148	189	231	274	302	341	305	253	235	150	122	2675
IOWA BURLINGTON	19	148	165	217	241	284	315	353	327	270	243	175	147	2885
CHARLES CITY	22	137	157	190	226	258	285	336	290	241	207	130	115	2572
DES MOINES	30	155	170	203	236	276	303	346	299	263	227	156	136	2770
SIOUX CITY	30	164	177	216	254	300	320	363	320	270	236	160	146	2926
KAN. CONCORDIA	30	180	172	214	243	281	315	348	308	249	245	189	172	2916
DODGE CITY	30	205	191	249	268	305	335	359	335	290	266	218	198	3219
TOPEKA	18	159	160	193	215	260	287	310	304	263	229	173	149	2702
WICHITA	30	187	186	233	254	291	321	350	325	277	245	206	182	3057
KY. LOUISVILLE	30	115	135	188	221	283	303	324	295	256	219	148	114	2601
LA. NEW ORLEANS	30	160	158	213	247	292	287	260	269	241	260	200	157	2744
SHREVEPORT	19	151	172	214	240	298	332	339	322	289	273	208	177	3015
MAINE EASTPORT	22	133	151	196	201	245	248	275	260	205	175	105	115	2309
PORTLAND	30	155	174	213	226	268	286	312	294	229	202	146	148	2653
MD. BALTIMORE	30	148	170	211	229	270	295	299	272	238	212	164	145	2653
MASS. BLUE HILL OBS.	10	125	136	165	182	233	248	266	241	211	181	134	135	2257
BOSTON	30	148	168	212	222	263	283	300	280	232	207	152	148	2615
NANTUCKET	22	128	156	214	227	278	284	291	279	242	208	149	129	2585
MICH. ALPENA	24	86	124	198	228	261	303	339	285	204	159	70	67	2324
DETROIT	30	90	128	180	212	263	295	321	284	226	189	98	89	2375
LANSING	30	84	119	175	215	272	305	344	294	228	182	87	73	2378
ESCANABA	30	112	148	204	226	266	283	316	267	198	162	90	94	2366
GRAND RAPIDS	30	74	117	178	218	277	308	349	304	231	188	92	70	2406
MARQUETTE	30	78	113	172	207	248	268	305	251	186	142	68	66	2104
SAULT STE. MARIE	30	83	123	187	217	252	269	309	256	165	133	61	62	2117
MINN. DULUTH	30	125	163	221	235	268	282	328	277	203	166	100	107	2475
MINNEAPOLIS	30	140	166	200	231	272	302	343	296	237	193	115	112	2607
MISS. JACKSON	12	130	147	199	244	280	287	279	287	235	223	185	150	2646
VICKSBURG	30	136	141	199	232	284	304	291	297	254	244	183	140	2705
MO. COLUMBIA	30	147	164	207	232	281	296	341	298	262	225	166	138	2757
KANSAS CITY	30	154	170	211	235	278	313	347	308	266	235	178	151	2846
ST. JOSEPH	23	154	165	211	231	274	301	347	287	260	224	168	144	2766
ST. LOUIS	30	137	152	202	235	283	301	325	289	256	223	166	125	2694
SPRINGFIELD	30	145	164	213	238	278	305	342	310	269	233	183	140	2820
MONT. BILLINGS	21	140	154	208	236	283	301	372	332	258	213	136	129	2762
GREAT FALLS	19	154	176	245	261	299	299	381	342	256	206	132	133	2884
HAVRE	30	136	174	234	268	311	312	384	339	260	202	132	122	2874
HELENA	30	138	168	215	241	292	292	342	336	258	202	137	121	2742
MISSOULA	25	85	109	167	209	261	260	378	328	246	178	90	66	2377
NEBR. LINCOLN	30	173	172	213	244	287	316	356	309	266	237	174	160	2907
NORTH PLATTE	30	181	179	221	246	282	310	343	304	264	242	184	169	2925
OMAHA	30	172	188	222	259	305	332	379	311	270	248	166	145	2997
VALENTINE	30	185	194	229	252	296	323	369	326	275	242	174	172	3037
NEV. ELY	22	186	197	262	260	300	354	359	344	303	255	204	187	3211
LAS VEGAS	8	239	251	314	336	386	411	383	364	345	301	258	250	3838
RENO	30	185	199	267	306	354	376	414	391	336	273	212	170	3483
WINNEMUCCA	30	142	155	207	255	312	346	395	375	316	242	177	139	3061
N. H. CONCORD	23	136	153	192	196	229	261	286	260	214	179	122	126	2354
MT. WASHINGTON OBS.	18	94	98	133	141	162	145	150	143	139	159	89	87	1540
N. J. ATLANTIC CITY	30	151	173	210	233	273	287	298	271	239	218	177	153	2683
TRENTON	30	145	168	203	235	277	294	309	273	239	208	160	142	2653
N. MEX. ALBUQUERQUE	30	221	218	273	299	343	365	340	317	299	279	245	219	3418
ROSWELL	21	218	223	286	306	330	333	341	313	266	266	242	216	3340
N. Y. ALBANY	30	125	151	194	213	266	301	317	286	224	192	115	112	2496
BINGHAMTON	30	94	119	151	170	226	256	266	230	184	158	92	79	2025
BUFFALO	30	110	125	180	212	274	319	338	297	239	183	97	84	2458
NEW YORK	30	154	171	213	237	268	289	302	271	235	213	169	155	2677
ROCHESTER	30	93	123	172	209	274	314	333	294	224	173	97	86	2392
SYRACUSE	30	87	115	165	197	261	295	316	276	211	163	81	74	2241
N. C. ASHEVILLE	30	146	161	211	247	289	292	268	250	235	222	179	146	2646
CAPE HATTERAS	9	152	168	206	259	293	301	286	265	214	202	169	154	2669
CHARLOTTE	30	165	177	230	267	313	316	291	277	247	243	198	167	2891
GREENSBORO	30	157	171	217	231	298	302	287	272	243	236	190	163	2767
RALEIGH	29	154	168	220	255	290	284	277	253	224	215	184	156	2680
WILMINGTON	30	179	180	237	279	314	312	286	273	237	238	206	178	2919
N. DAK. BISMARCK	30	141	170	205	236	279	294	358	307	243	198	130	125	2686
DEVILS LAKE	30	150	177	220	250	291	297	352	302	230	198	123	124	2714
FARGO	30	132	170	210	232	283	288	343	293	222	187	112	114	2586
WILLISTON	29	141	168	215	260	305	312	377	328	247	206	131	129	2819
OHIO CINCINNATI (ABBE)	30	115	137	186	222	273	309	323	295	253	205	138	118	2574
CLEVELAND	30	79	111	167	209	274	301	325	288	235	187	99	77	2352
COLUMBUS	30	112	132	177	215	270	296	323	291	250	210	131	101	2508
DAYTON	10	114	136	195	222	281	313	323	307	268	229	152	124	2664
SANDUSKY	30	100	128	183	229	285	312	343	302	248	201	111	91	2533
TOLEDO	30	93	120	170	203	263	296	331	298	241	196	106	92	2409
OKLA. OKLAHOMA CITY	29	175	182	235	253	290	329	352	331	282	243	201	175	3048
TULSA	18	152	164	200	213	244	287	314	308	281	241	207	172	2783
OREG. BAKER	22	118	143	198	251	302	313	406	368	289	215	132	100	2835
PORTLAND	30	77	97	142	203	246	249	329	275	218	134	87	65	2122
ROSEBURG	30	69	96	148	205	257	278	369	329	255	146	81	50	2283
PA. HARRISBURG	30	132	160	203	230	277	297	319	282	233	200	140	131	2604
PHILADELPHIA	30	142	166	203	231	270	281	288	253	225	205	158	142	2564
PITTSBURGH	25	89	114	163	200	239	260	283	250	234	180	114	76	2202
READING	30	133	151	195	220	259	275	293	259	219	198	144	127	2473
SCRANTON	30	108	138	178	199	251	269	290	249	213	183	120	105	2303
R. I. PROVIDENCE	30	145	168	211	221	271	285	292	267	226	207	153	143	2589
S. C. CHARLESTON	30	188	189	243	284	323	308	297	281	244	239	210	187	2993
COLUMBIA	30	173	183	233	274	312	312	291	283	243	242	202	166	2914
GREENVILLE	26	166	176	227	274	307	300	278	274	239	232	192	157	2822
S. DAK. HURON	30	153	177	213	250	295	321	367	320	260	212	142	134	2844
RAPID CITY	30	164	182	222	245	278	300	348	317	266	228	164	144	2858
TENN. CHATTANOOGA	30	126	146	187	239	290	295	278	266	247	220	169	128	2591
KNOXVILLE	30	124	144	189	237	281	288	277	248	237	213	157	120	2515
MEMPHIS	30	135	152	204	244	296	321	319	314	261	243	180	139	2808
NASHVILLE	30	123	142	196	241	285	308	292	279	250	224	168	126	2634
TEX. ABILENE	13	190	199	250	259	29C	347	335	322	276	245	223	201	3137
AMARILLO	30	207	199	258	276	305	338	350	328	288	260	229	205	3243
AUSTIN	30	148	152	207	221	266	302	331	320	261	242	180	160	2790
BROWNSVILLE	30	147	152	187	210	272	297	326	311	246	252	165	151	2716
CORPUS CHRISTI	24	160	165	212	237	295	329	366	341	276	264	194	164	3003
DALLAS	30	155	159	220	238	279	326	341	325	274	240	191	163	2911
DEL RIO	27	173	173	230	237	259	279	331	319	252	240	195	178	2866
EL PASO	30	234	236	299	329	373	369	336	327	300	287	257	236	3583
GALVESTON	30	151	149	203	230	288	322	305	292	257	264	199	151	2811
HOUSTON	30	144	141	193	212	266	298	294	281	238	239	181	146	2633
PORT ARTHUR	30	153	149	209	235	292	317	285	281	252	256	191	148	2768
SAN ANTONIO	30	148	153	213	224	258	292	325	307	261	241	183	160	2765
UTAH SALT LAKE CITY	30	137	155	227	269	329	358	377	346	306	249	171	135	3059
VT. BURLINGTON	30	103	127	184	185	244	270	291	266	199	152	77	80	2178
VA. LYNCHBURG	26	153	169	216	243	288	297	288	264	235	217	177	158	2705
NORFOLK	30	156	174	223	257	304	311	296	282	237	220	182	161	2803
RICHMOND	30	144	166	211	248	280	296	286	263	230	211	176	152	2663
WASH. NORTH HEAD	22	76	97	135	182	221	214	226	186	170	123	87	66	1783
SEATTLE	30	74	99	154	201	247	234	304	248	197	122	77	62	2019
SPOKANE	30	78	120	197	262	308	309	397	350	264	177	86	57	2605
TATOOSH ISLAND	30	70	100	135	182	229	217	235	190	175	129	71	60	1793
WALLA WALLA	30	72	106	194	262	317	335	411	367	280	198	92	51	2685
W. VA. ELKINS	24	110	119	158	198	227	256	225	236	211	186	131	103	2160
PARKERSBURG	30	91	111	155	200	252	277	286	264	230	189	117	93	2265
WIS. GREEN BAY	30	121	148	194	210	251	279	314	266	213	176	110	106	2388
MADISON	30	126	147	196	214	258	285	336	288	230	198	116	108	2502
MILWAUKEE	30	116	134	191	218	267	293	340	292	235	193	125	106	2510
WYO. CHEYENNE	30	191	197	243	237	259	304	318	286	265	242	188	170	2900
LANDER	30	200	208	260	264	301	340	361	326	280	233	186	185	3144
SHERIDAN	30	160	179	226	245	286	303	367	333	266	221	153	145	2884
P. R. SAN JUAN	30	231	229	273	252	240	245	264	257	219	229	217	222	2878

3-1: Average monthly and annual hours of sunshine received at 155 different American localities [AND-1].

Appendix 4: Maps of Mean Monthly Hours of Sunshine in the United States

Figures 4-1 through 4-12 present maps of the United States showing mean hours of sunshine received during each month of the year.

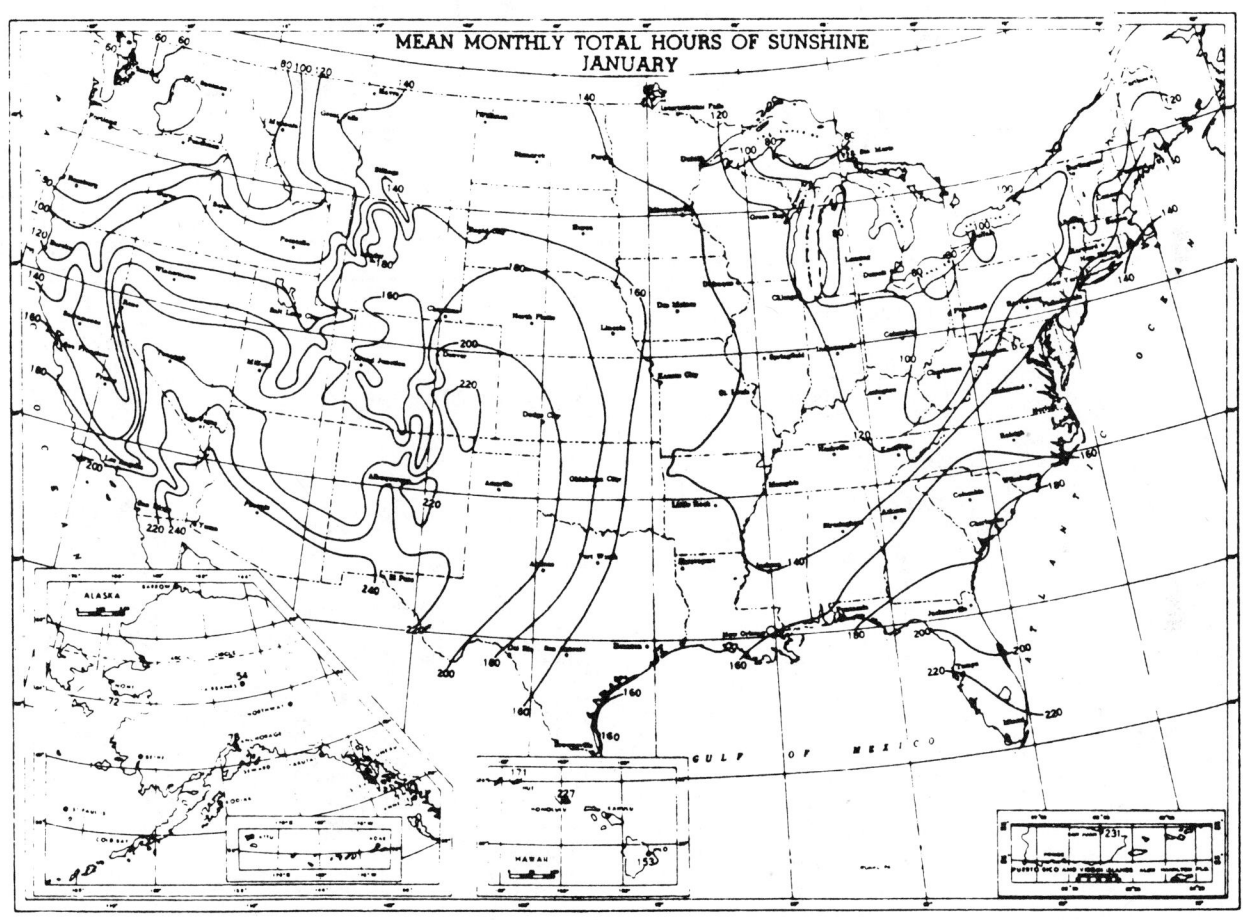

4-1: Mean monthly hours of sunshine—January [COM].

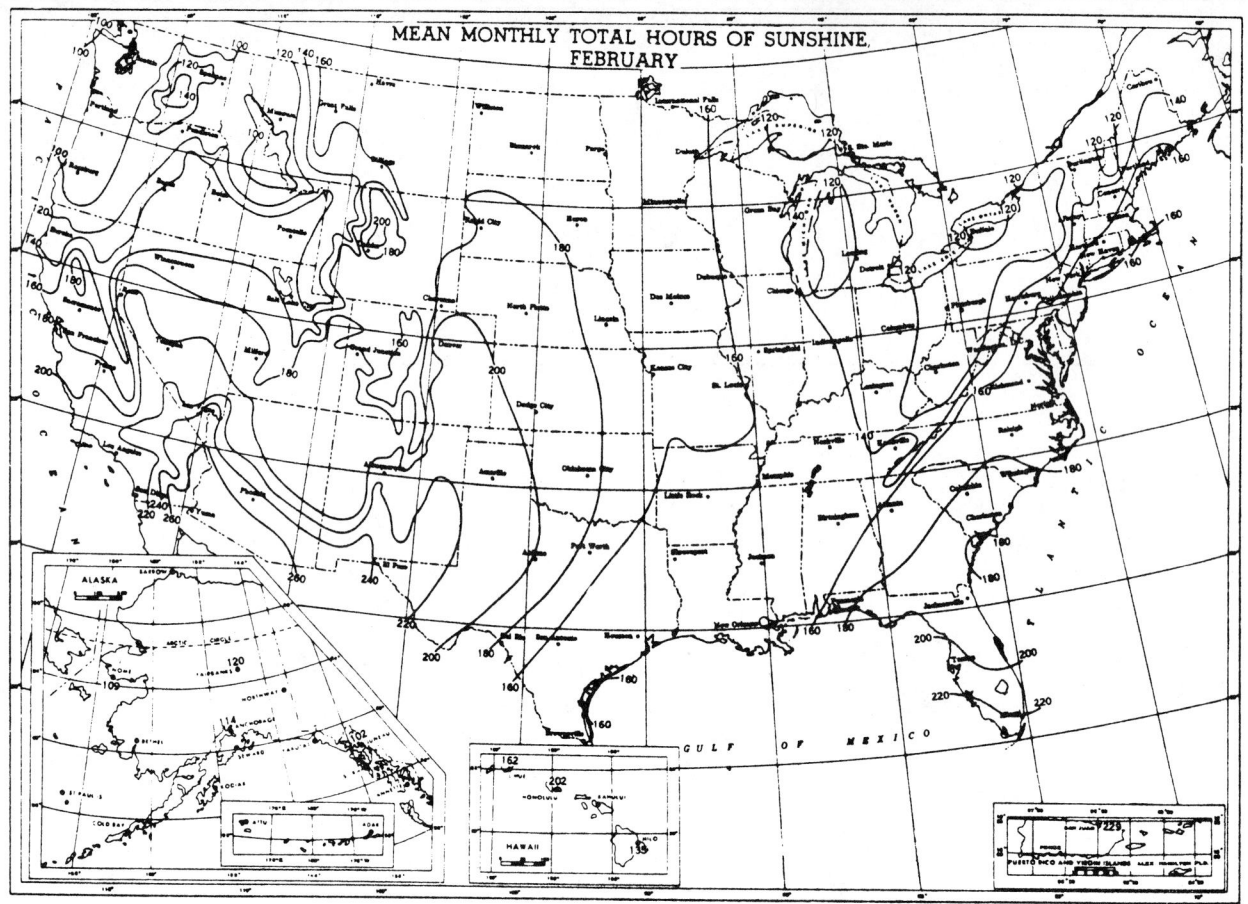

4-2: Mean monthly hours of sunshine—February [COM].

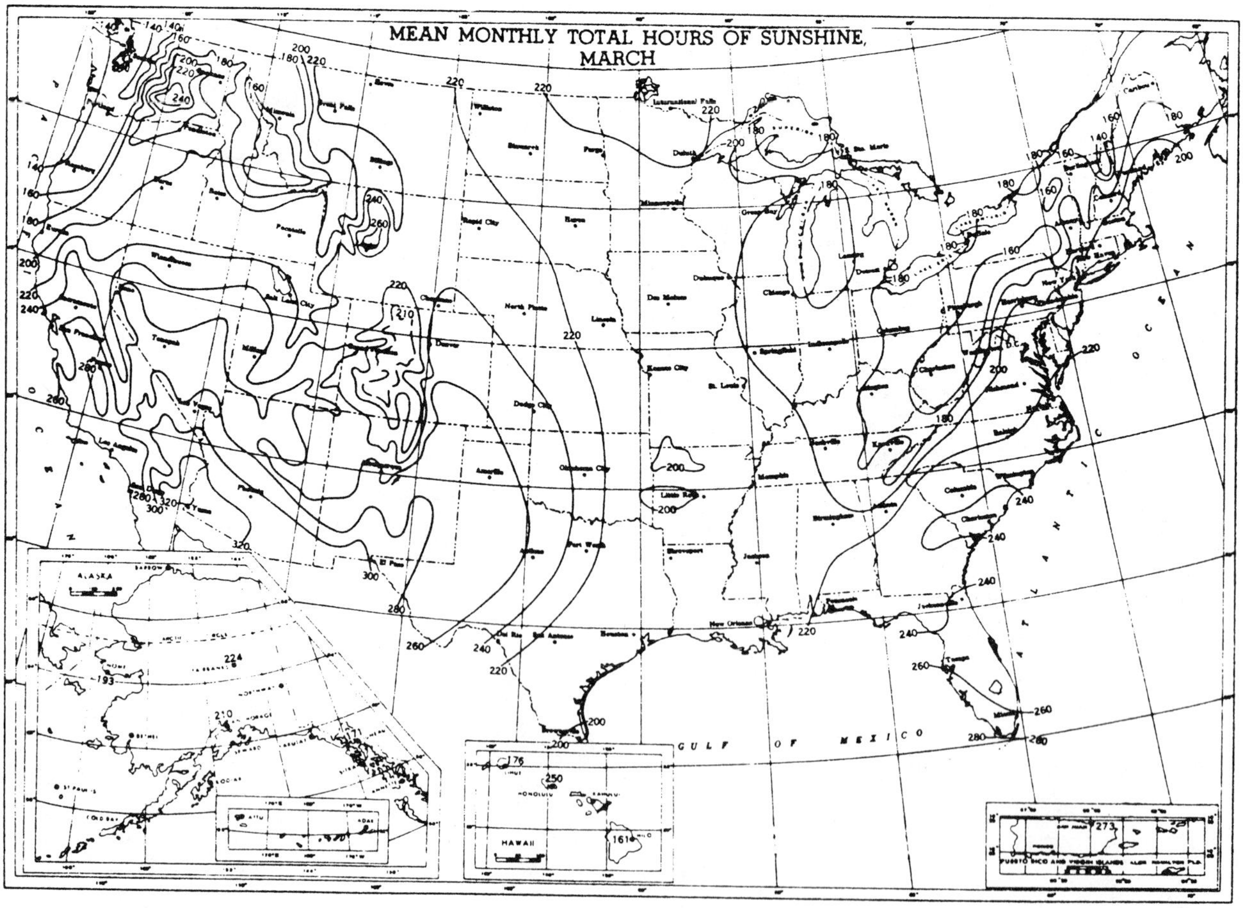

4-3: Mean monthly hours of sunshine—March [COM].

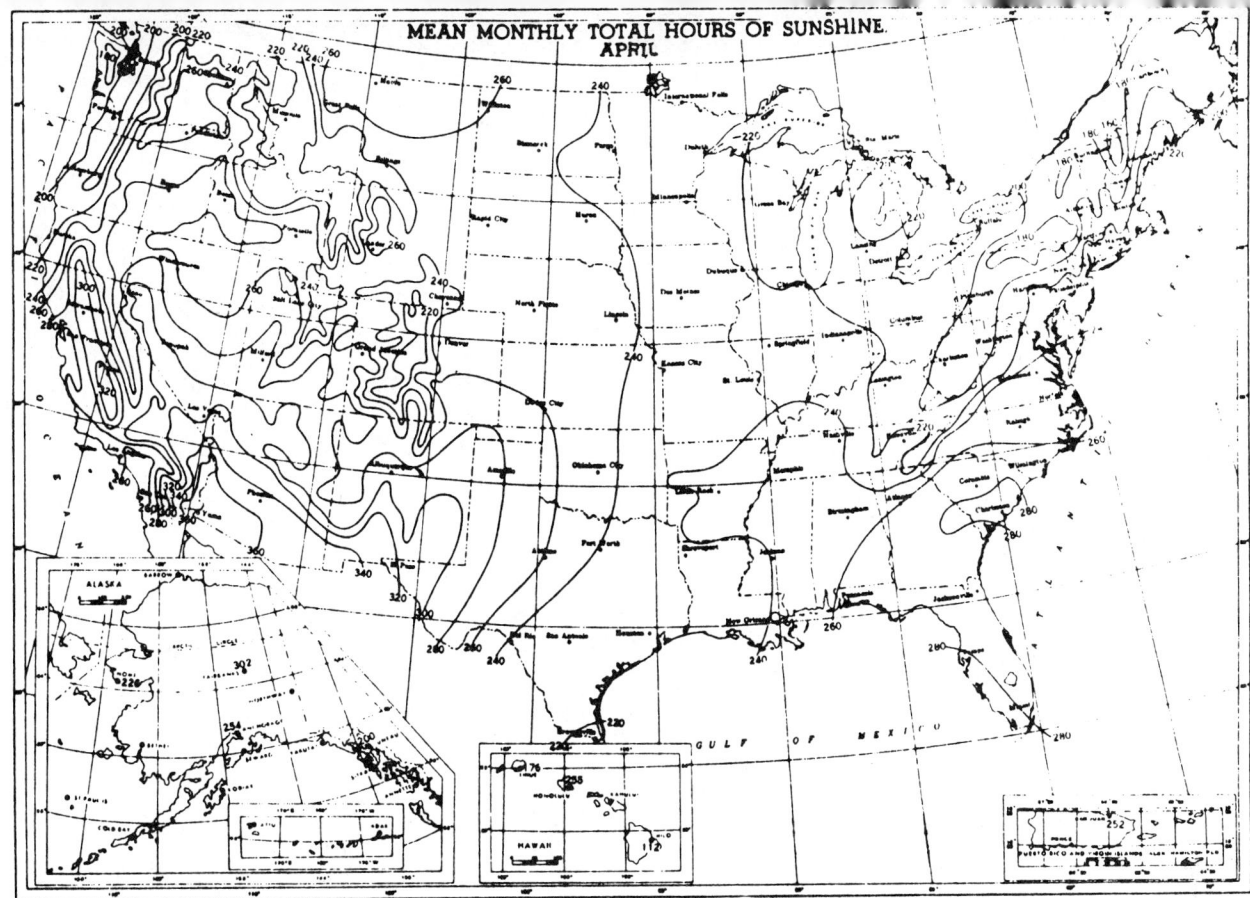

4-4: Mean monthly hours of sunshine—April [COM].

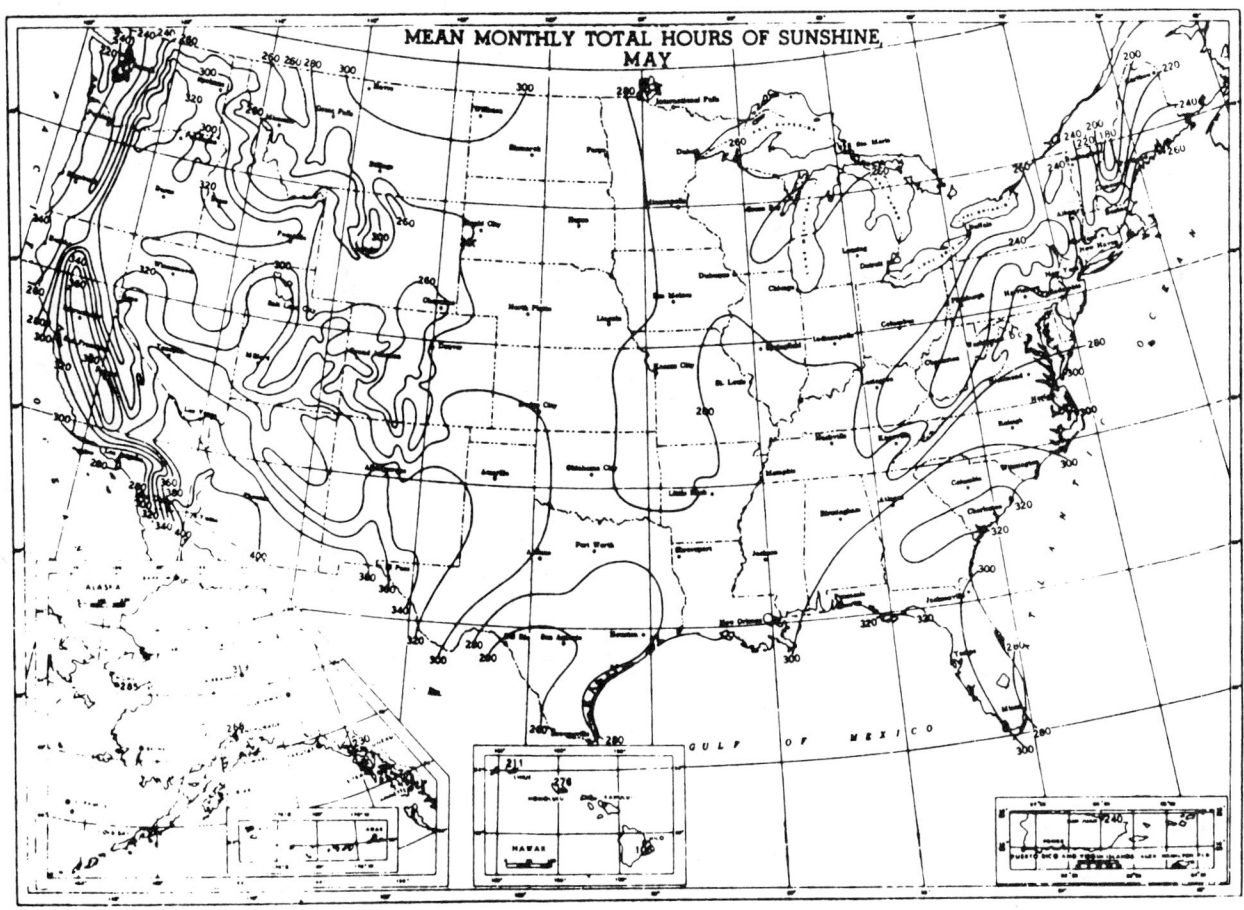

4-5: Mean monthly hours of sunshine—May [COM].

Maps of Monthly & Annual Sunshine / 209

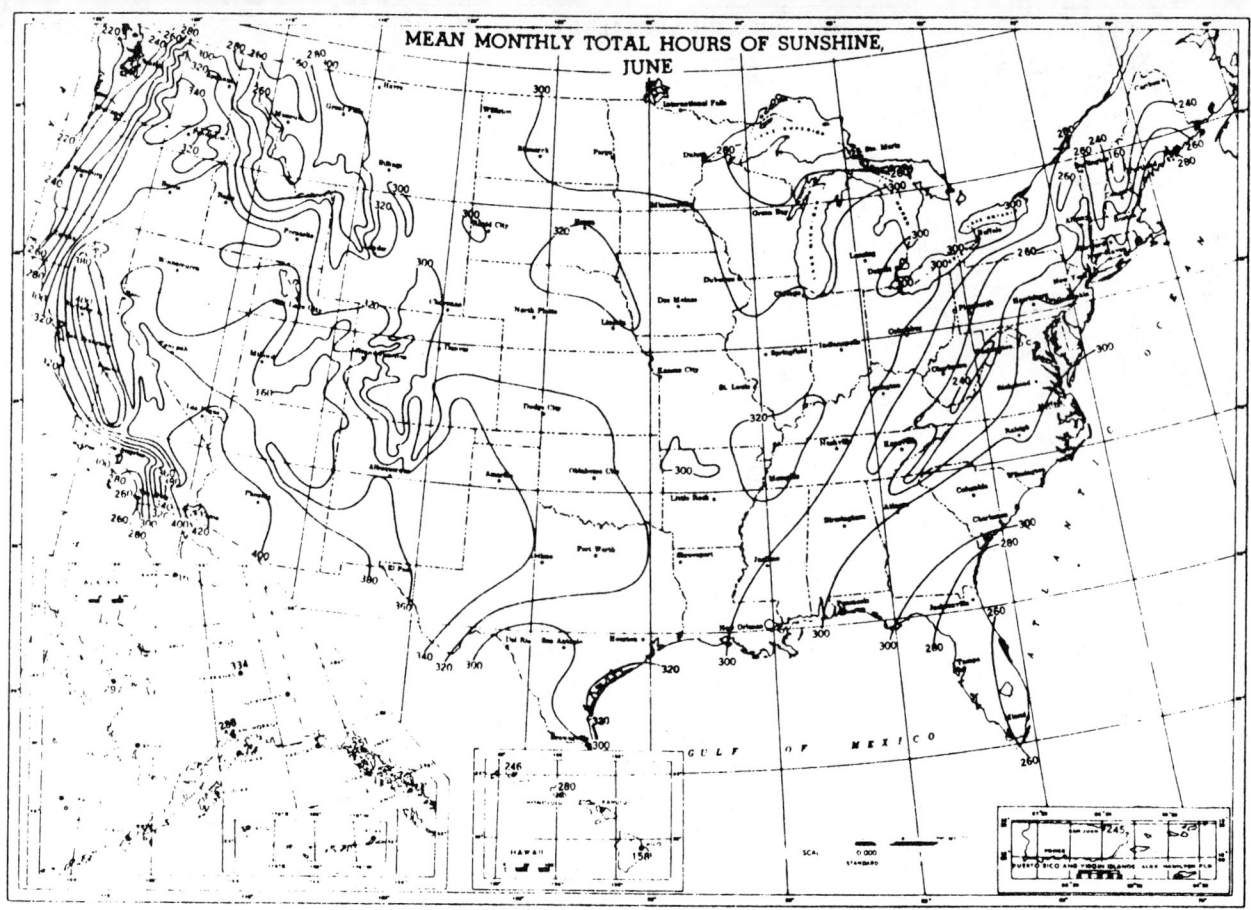

4-6: Mean monthly hours of sunshine—June [COM].

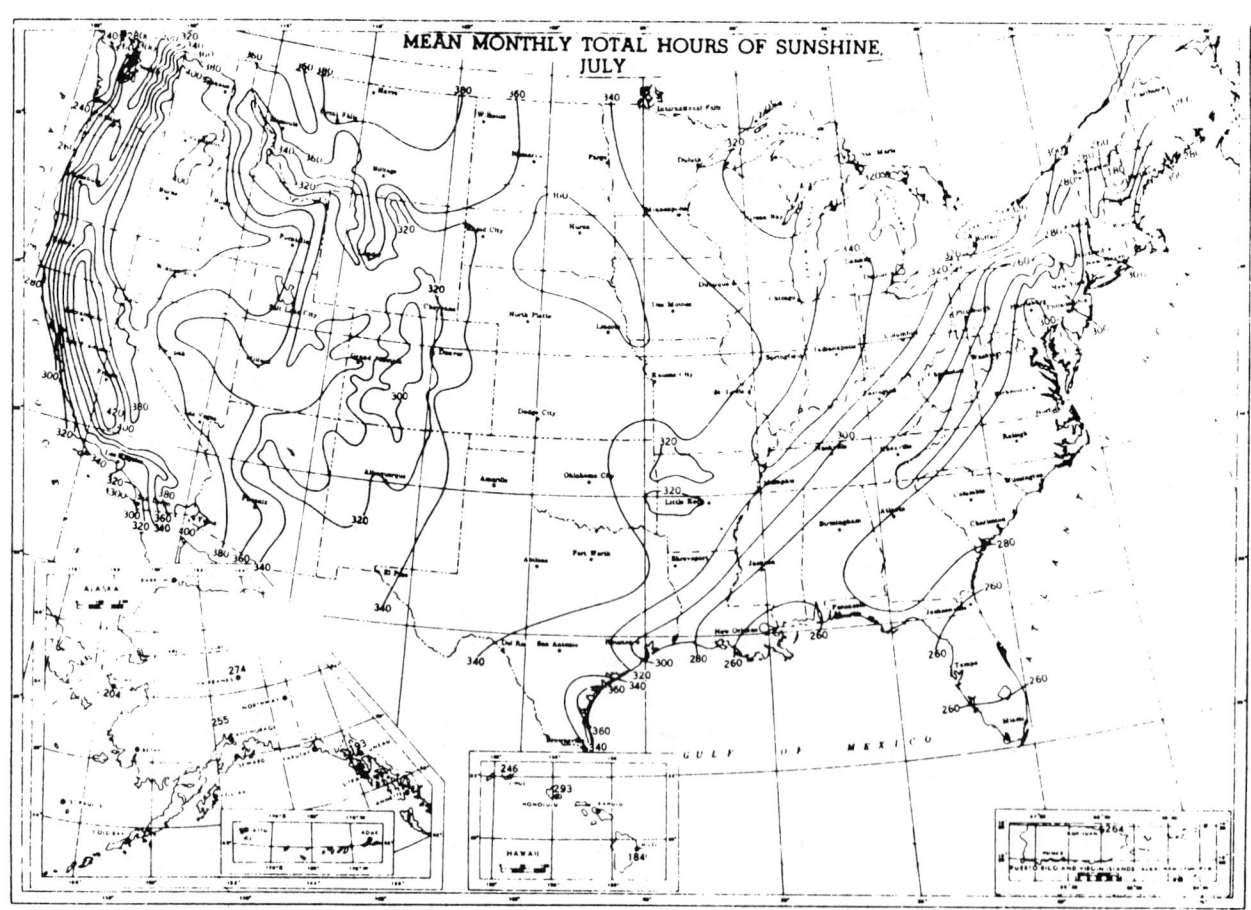

4-7: Mean monthly hours of sunshine—July [COM].

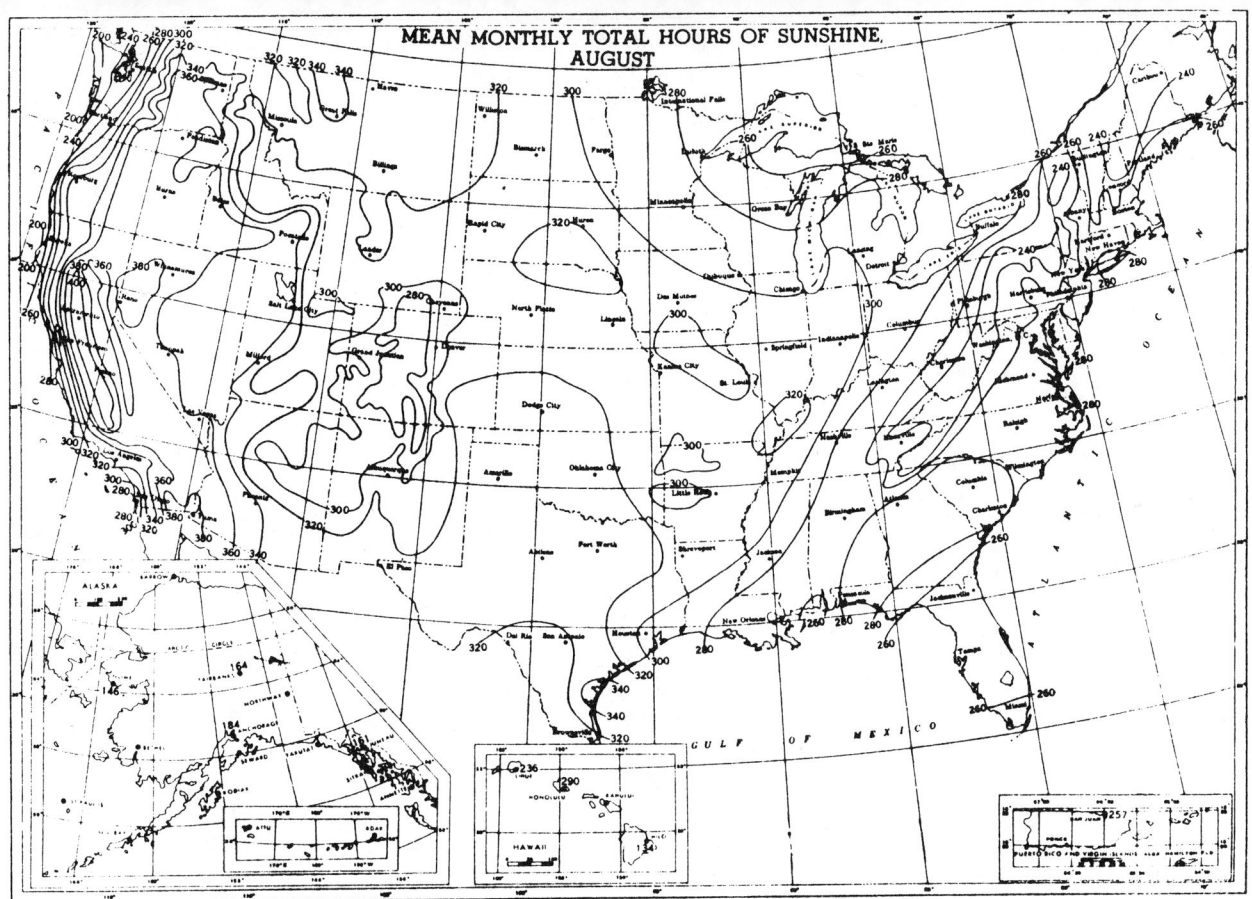

4-8: Mean monthly hours of sunshine—August [COM].

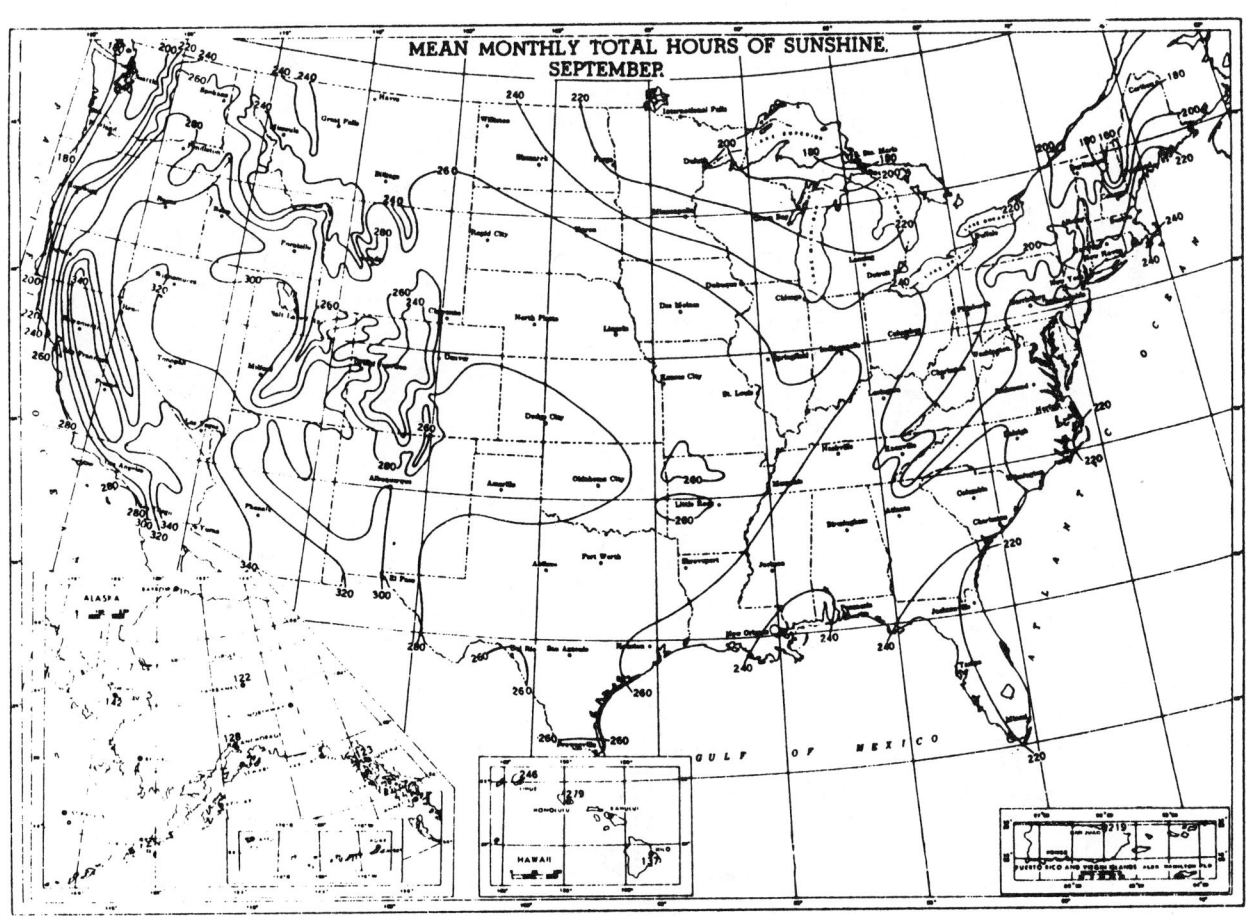

4-9: Mean monthly hours of sunshine—September [COM].

Maps of Monthly & Annual Sunshine / 211

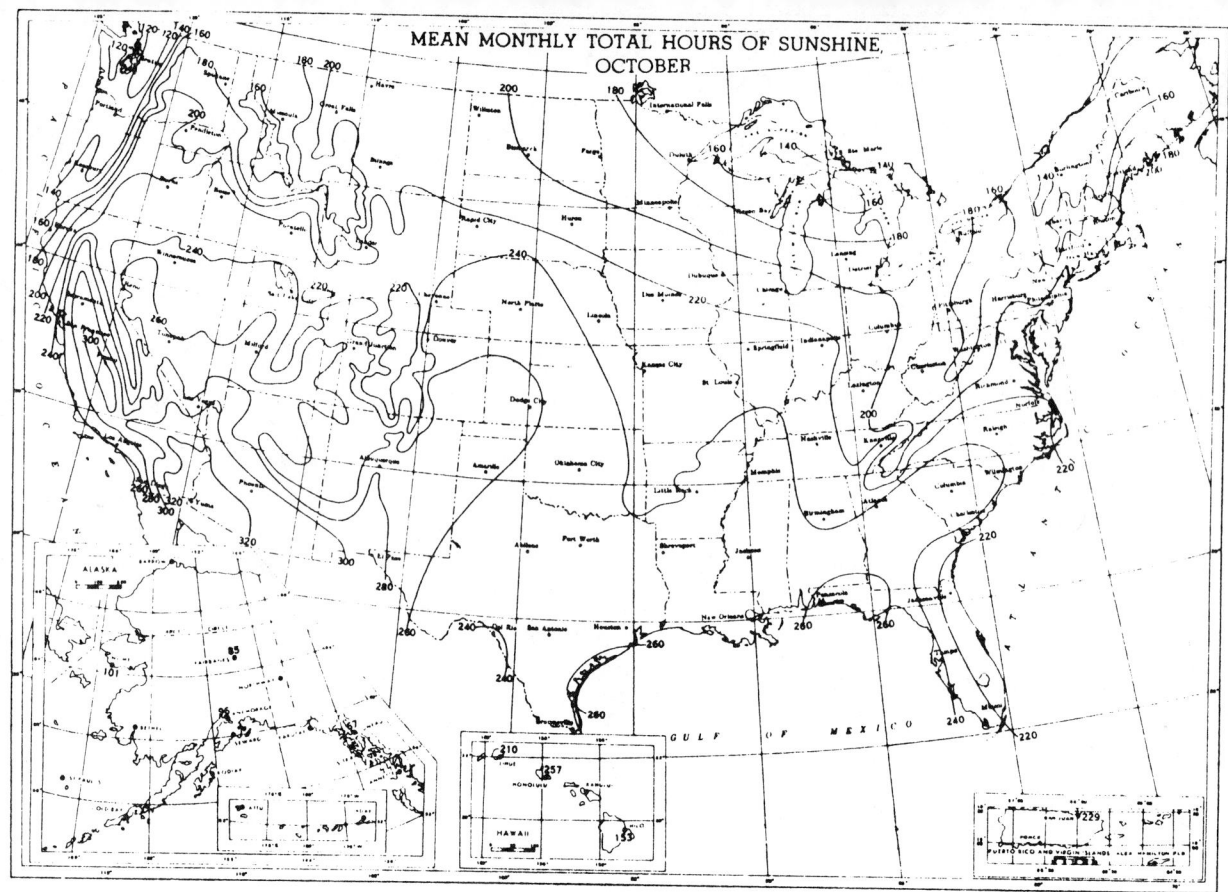

4-10: Mean monthly hours of sunshine—October [COM].

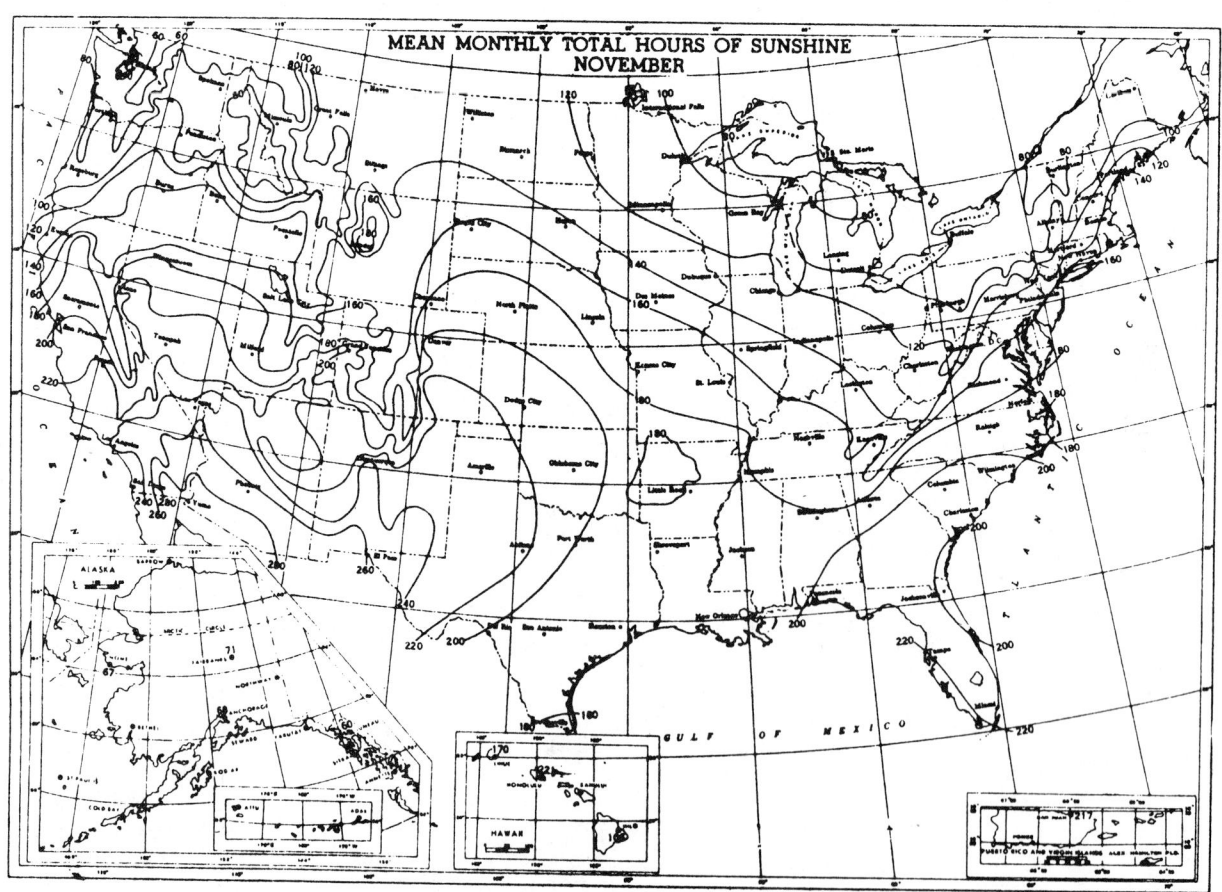

4-11: Mean monthly hours of sunshine—November [COM].

212 / *APPENDIX 4*

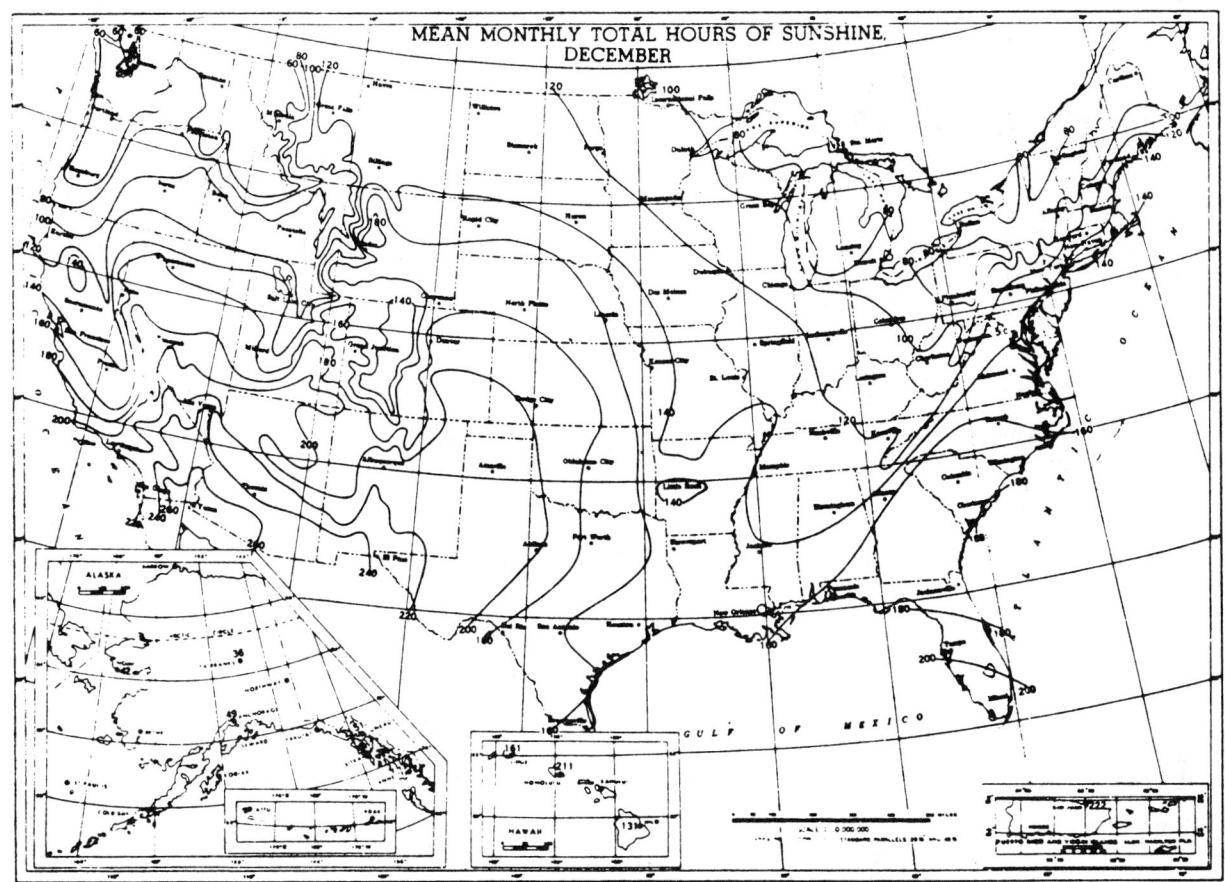

4-12: Mean monthly hours of sunshine—December [COM].

Appendix 5   Sun-Path Diagrams for Various N Latitudes

Figures 5-1 through 5-7 show sun-path diagrams for seven different N latitudes. The sun-path diagram for 40°N is shown in figure B-3.

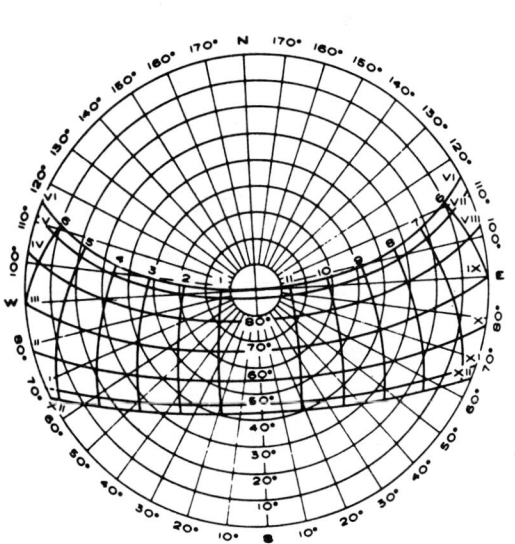

5-1: Sun-path diagram for 24°N latitude [RAM].

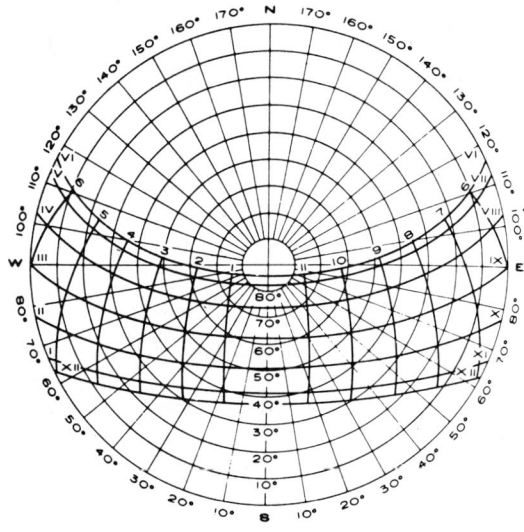

5-2: Sun-path diagram for 28°N latitude [RAM].

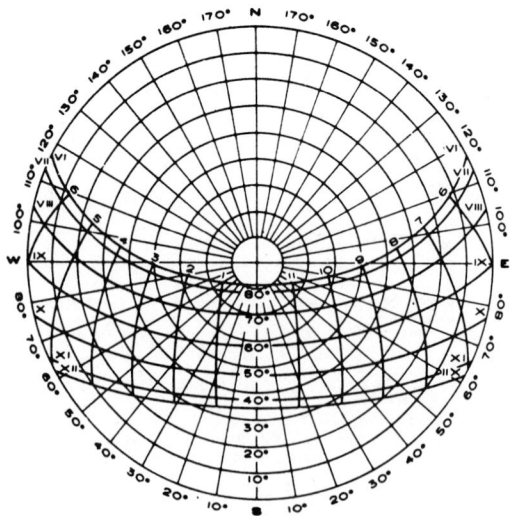

5-3: Sun-path diagram for 32°N latitude [RAM].

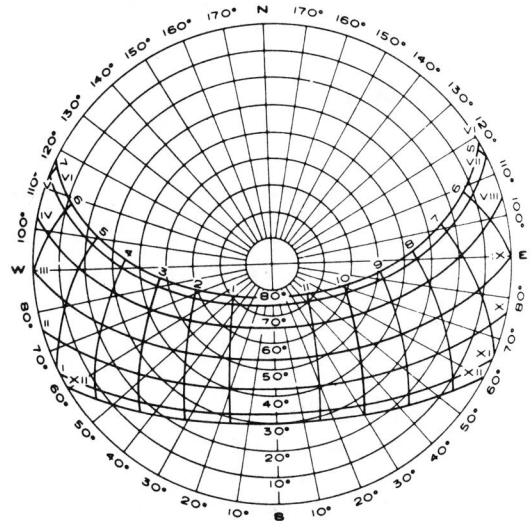

5-4: Sun-path diagram for 36°N latitude [RAM].

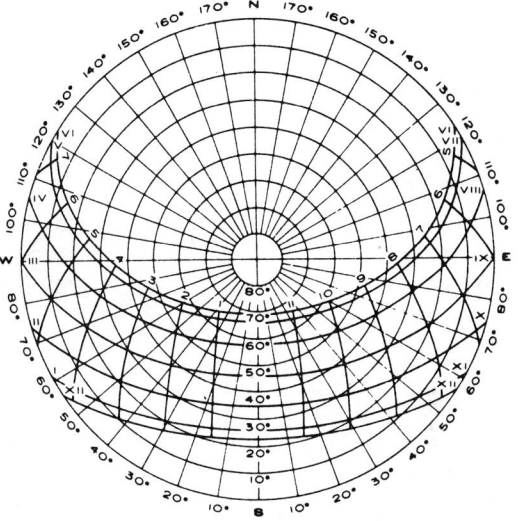

5-5: Sun-path diagram for 44°N latitude [RAM].

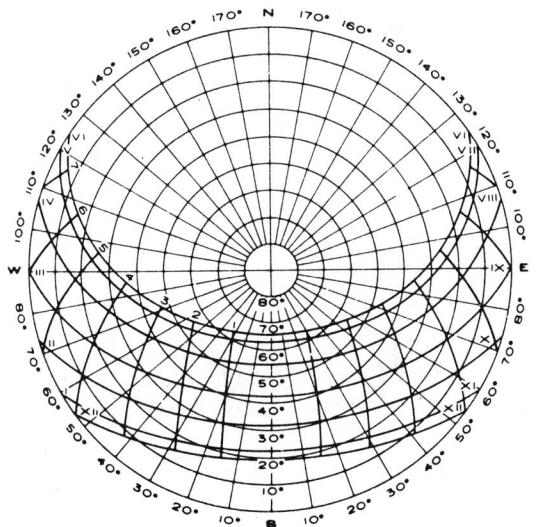

5-6: Sun-path diagram for 48°N latitude [RAM].

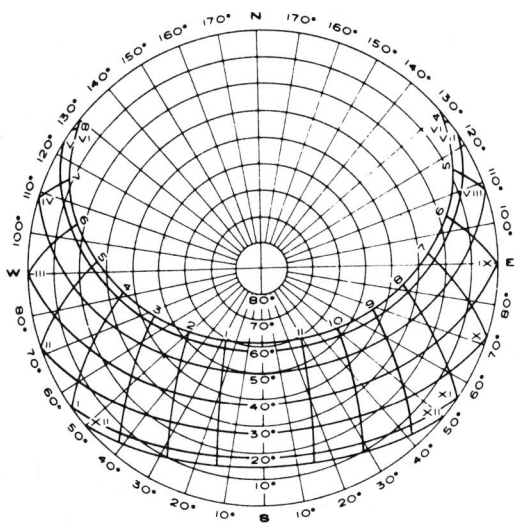

5-7: Sun-path diagram for 52°N latitude [RAM].

Sun-Path Diagrams / 215

Appendix 6 Conversion Factors

Length	1 foot	0.3048 meter
Area	1 square foot	0.0929 square meter
Volume	1 cubic foot	0.0283 cubic meter
Volume	1 cubic foot	7.48 gallons (U.S.)
Volume	1 gallon (U.S.)	3.7854 liters
Weight	1 pound	0.4356 kilograms
Weight	1 gallon of water	8.3453 pounds
Density	1 pound/cubic foot	16.02 kilograms/cubic meter
Pressure	1 pound/square inch	703.07 kilograms/square meter
Pressure	1 foot of water	0.4335 pounds/square inch
Barometric Pressure	1 foot of water	0.8826 inch of mercury
Speed (Linear)	1 foot/minute	0.00051 meter/second
Speed (Linear)	1 mile/hour	0.44704 meter/second
Speed (Volumetric)	1 cubic foot/minute	0.47195 liter/second
Energy	1 Btu	1,054 joules
Energy	1 therm	100,000 Btu
Energy	1 ton of refrigeration	12,000 Btu
Energy Flux	1 Btu/hour/square foot	3.152 watts/square meter
Power	1 Btu/hour	0.2928 watt
Power	1 horsepower	0.7457 kilowatt
Heat Capacity	1 Btu/°F	1,897 joules/°C
Specific Heat	1 Btu/pound/°F	4,182 joules/kilogram/°C
Thermal Conductivity	1 Btu/hour/foot/°F	1.729 watts/meter/°C
Conductance	1 Btu/hour/square foot/°F	5.673 watts/square meter/°C
Degree-Days	1 °F-day	⅝ °C-day
Building Load Coefficient	1 Btu/°F-day	0.022 watt/°C-day
Load Collector Ratio	1 Btu/°F-day/square foot	0.236 watt/°C-day/square meter

Appendix 7 Cost of Energy

The chart shown in figure 7-1 helps you to calculate the actual cost of providing heat. It converts the unit price of the energy source—whether gas, oil, or electricity—into the cost per million Btu produced inside the building.

Typically, only 50% to 75% of the heat content of gas or oil is delivered inside the building. The rest goes up the chimney. Electrical resistance heating is 100% efficient.

Here is an example of how the chart works:

1. Find the point on the appropriate vertical scale that corresponds to the retail price of the fuel—for example, oil at $1.00/gallon (or equivalently, electricity at $0.025/kWh and gas at $0.74/therm).
2. Move right to find the retail cost for 1 million Btu of that fuel, or $7.40.
3. Continue right to intersect the oblique line representing the efficiency of heat delivery, or 60% in this case.
4. Drop down to find the actual cost per million Btu of heat produced, or $12.35.

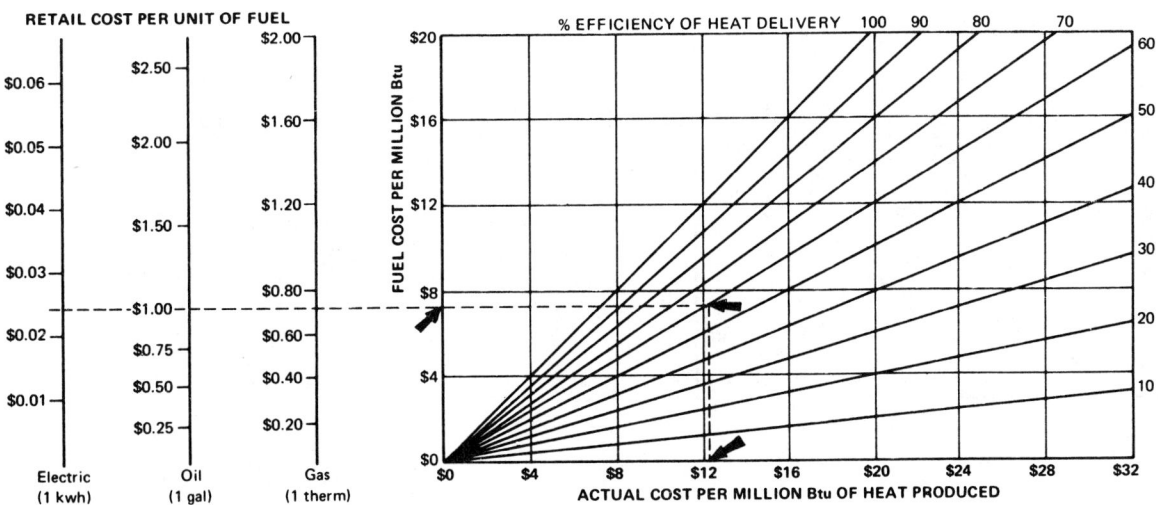

7-1: Cost-of-energy chart [AND-1].

Appendix 8

Thermal Performance Values for Materials, Air Spaces, and Windows

The tables in figures 8–1, 8–2, and 8–3 provide R-values and other data relating to the thermal performance of various building materials, air films, and air spaces. The table in figure 8–4 provides winter and summer U-values for various windows and skylights.

R-VALUES OF BUILDING MATERIALS

Material and Description		Density (lb/ft³)	R-value* per inch thickness	R-value* for listed thickness
Building Boards, Panels, Flooring				
Asbestos-cement board	1/8"	120	0.25	—
Asbestos-cement board		120	—	0.03
Gypsum or plaster board	3/8"	50	—	0.32
Gypsum or plaster board	1/2"	50	—	0.45
Plywood (see Siding Materials)		34	1.25	—
Sheathing, wood fiber (impregnated or coated)	25/32"	20	—	2.06
Wood fiber board (laminated or homogenous)		26	2.38	—
Wood fiber, hardboard type		65	0.72	—
Wood fiber, hardboard type	1/4"	65	—	0.18
Wood subfloor	25/32"	—	—	0.98
Wood, hardwood finish	3/4"	—	—	0.68
Building Paper				
Vapor-permeable felt		—	—	0.06
Vapor-seal, 2 layers of mopped 15 lb felt		—	—	0.12
Vapor-seal plastic film		—	—	negl.
Finish Materials				
Carpet and fibrous pad		—	—	2.08
Carpet and rubber pad		—	—	1.23
Cork tile	1/8"	—	—	0.28
Terrazzo	1"	—	—	0.08
Tile (asphalt, linoleum, vinyl, rubber)		—	—	0.05
Gypsumboard	1/2"	—	—	0.45
Gypsumboard	5/8"	—	—	0.56
Hardwood flooring	25/32"	—	—	0.68
Insulating Materials				
Blankets and Batts:				
Mineral wool, fibrous form (from rock, slag or glass)		0.5 1.5–4.0 3.2–3.6	3.12 3.70 4.00	— — —
Wood fiber		9	2.44	—
Boards and Slabs:				
Cellular glass	90°F			—

8–1: R-values of building materials [ASH-1].

R-VALUES OF BUILDING MATERIALS

Material and Description		Density (lb/ft³)	R-value* per inch thickness	R-value* for listed thickness
	60°F		2.56	—
	30°F		2.70	—
	0°F		2.86	—
Corkboard	90°F	6.5-8.0	3.57	—
	60°F		3.70	—
	30°F		3.85	—
	0°F		4.00	—
	90°F		3.22	—
	60°F	12	3.33	—
	30°F		3.45	—
	0°F		3.57	—
Glass fiber	90°F		3.85	—
	60°F	4.0-9.0	4.17	—
	30°F		4.55	—
	0°F		4.76	—
Expanded rubber (rigid)	75°F	4.5	4.55	—
Expanded polyurethane (R-11 blown; 1" thickness or more)	100°F		5.56	—
	75°F		5.88	—
	50°F	1.5-2.5	6.25	—
	25°F		5.88	—
	0°F		5.88	—
Expanded polystyrene, extruded	75°F		3.85	—
	60°F	1.9	4.00	—
	30°F		4.17	—
	0°F		4.55	—
Expanded polystyrene, molded beads	75°F		3.57	—
	30°F	1.0	3.85	—
	0°F		4.17	—
Mineral fiberboard, felted core or roof insulation		16-17	2.94	—
acoustical tile[1]		18	2.86	—
acoustical tile[1]		21	2.73	—
Mineral fiberboard, molded acoustical tile[1]		23	2.38	—
Wood or cane fiberboard				
acoustical tile	1/2"	—	—	1.19
acoustical tile	3/4"	—	—	1.78
interior finish		15	2.86	—

R-VALUES OF BUILDING MATERIALS

Material and Description		Density (lb/ft³)	R-value* per inch thickness	R-value* for listed thickness
Insulating roof deck[2]	1"	—	—	2.78
	2"	—	—	5.56
	3"	—	—	8.33
Shredded wood (cemented, preformed slabs)		22	1.67	—
Loose Fills:				
Macerated paper or pulp		2.5-3.5	3.57	—
Mineral wool	90°F		3.33	—
	60°F	2.0-5.0	3.70	—
	30°F		4.00	—
	0°F		4.35	—
Perlite (expanded)	90°F		2.63	—
	60°F	5.0-8.0	2.78	—
	30°F		2.94	—
	0°F		3.12	—
Vermiculite (expanded)	90°F		2.08	—
	60°F	7.0-8.2	2.18	—
	30°F		2.27	—
	0°F		2.38	—
Sawdust or shavings		0.8-15	2.22	—
Masonry Materials—Concretes				
Cement mortar		116	0.20	—
Gypsum-fiber concrete (87½% gypsum, 12½% concrete)		51	0.60	—
Lightweight aggregates (expanded shale, clay or slate; expanded slags, or cinders; pumice; perlite or vermiculite; cellular concretes)		120	0.19	—
		100	0.28	—
		80	0.40	—
		60	0.59	—
		40	0.86	—
		20	1.43	—
Sand and gravel or stone aggregate (oven dried)		140	0.11	—
Sand and gravel or stone aggregate (not dried)		140	0.08	—
Stucco		116	0.20	—
Masonry Units				
Brick, common[3]		120	0.20	—
Brick, face[3]		130	0.11	—

8-1: Continued.

R-VALUES OF BUILDING MATERIALS

Material and Description		Density (lb/ft³)	R-value* per inch thickness	R-value* for listed thickness
Clay tile, hollow				
1 cell deep	3"	—	—	0.80
1 cell deep	4"	—	—	1.11
2 cells deep	6"	—	—	1.52
2 cells deep	8"	—	—	1.85
3 cells deep	10"	—	—	2.22
3 cells deep	12"	—	—	2.50
Concrete block, 3 oval core				
Sand and gravel aggregate	4"	—	—	0.71
	8"	—	—	1.11
	12"	—	—	1.28
Cinder aggregate	3"	—	—	0.86
	4"	—	—	1.11
	8"	—	—	1.72
	12"	—	—	1.89
Lightweight aggregate (expanded shale, clay slate or slag; pumice)	3"	—	—	1.27
	4"	—	—	1.50
	8"	—	—	2.00
	12"	—	—	2.72
Concrete blocks, rectangular core				
Sand and gravel aggregate				
2 core, 36 lb[4]	8"	—	—	1.04
same, filled cores[5]		—	—	1.93
Lightweight aggregates				
3 core, 19 lb[4]	6"	—	—	1.65
same, filled cores[5]		—	—	2.99
2 core, 24 lb[4]	8"	—	—	2.18
same, filled cores[5]		—	—	5.03
3 core, 38 lb[4]	12"	—	—	2.48
same, filled cores[5]		—	—	5.82
Stone, lime or sand		150–175	0.08	—
Granite, marble			0.05	—
Plastering Materials				
Cement plaster, sand aggregate		116	0.20	—
Gypsum plaster				
Lightweight aggregate	1/2"	45	—	0.32
Lightweight aggregate	3/8"	45	—	0.39
Same, on metal lath	3/4"	—	—	0.47
Perlite aggregate		45	0.67	—
Sand aggregate		105	0.18	—

R-VALUES OF BUILDING MATERIALS

Material and Description		Density (lb/ft³)	R-value* per inch thickness	R-value* for listed thickness
Same, on metal lath	3/4"	—	—	0.10
Same, on wood lath	3/4"	—	—	0.40
Vermiculite aggregate		45	0.59	—
Roofing Materials				
Asbestos-cement shingles		120	—	0.21
Asphalt roll roofing		70	—	0.15
Built-up roofing	3/8"	70	—	0.44
Slate roofing	1/2"	—	—	0.05
Wood shingles		—	—	0.94
Siding Materials				
Shingles				
Asbestos-cement		120	—	0.21
Wood, 16" with 7½" exposure		—	—	0.80
Wood, double 16" with 12" exposure		—	—	1.19
Wood, plus insulating backer board	5/16"	—	—	1.40
Siding				
Asbestos-cement lapped	1/4"	—	—	0.21
Asphalt roll siding		—	—	0.15
Asphalt insulating siding	1/2"	—	—	1.46
Wood, drop (1" × 8")		—	—	0.79
Wood, drop (½" × 8" lapped)		—	—	0.81
Wood, bevel (¾" × 10", lapped)		—	—	1.05
Plywood, lapped	3/8"	—	—	0.59
Plywood	1/4"	—	—	0.31
	3/8"	—	—	0.47
	1/2"	—	—	0.62
	5/8"	—	—	0.78
	3/4"	—	—	0.94
Stucco	1/2"	116	0.20	—
Sheathing, insulating board (regular density)	25/32"	—	—	1.32
		—	—	2.04
Woods				
Hardwoods (maple, oak)		45	0.91	—
Softwoods (fir, pine)		32	1.25	—

8-1: Continued.

R-VALUES OF BUILDING MATERIALS

Material and Description		Density (lb/ft³)	R-value* per inch thickness	R-value* for listed thickness
	25/32"	32	—	0.98
	1–5/8"	32	—	2.03
	2–5/8"	32	—	3.28
	3–5/8"	32	—	4.55
Wood Doors				
Solid core	1"	—	—	1.56
	1–1/4"	—	—	1.82
	1–1/2"	—	—	2.04
	2"	—	—	2.33

* Representative values intended for use as design values of dry building materials in normal use.
1 R-values of acoustical tile depend upon the board and the type, size and depth of perforations; these are average values.
2 Roof deck insulation is made in thicknesses to meet these standards; thickness may vary somewhat with manufacturer.
3 Face brick and common brick do not always have these densities and R-values.
4 Weights of blocks approximately 7–5/8" high by 15–3/8" long.
5 Vermiculite, perlite, or mineral wool insulation.

8-1: Continued.

R-VALUES OF AIR SPACES

Orientation & Thickness of Air Space		Direction of Heat Flow	R-value for Air Space Facing:‡ Non-reflective surface	Fairly reflective surface	Highly reflective surface
Horizontal	¾"	up*	0.87	1.71	2.23
	4"	up*	0.94	1.99	2.73
	¾"	up†	0.76	1.63	2.26
	4"	up†	0.80	1.87	2.75
	¾"	down*	1.02	2.39	3.55
	1½"	down*	1.14	3.21	5.74
	4"	down*	1.23	4.02	8.94
	¾	down†	0.84	2.08	3.25
	1½"	down†	0.93	2.76	5.24
	4"	down†	0.99	3.38	8.03
45° slope	¾"	up*	0.94	2.02	2.78
	4"	up*	0.96	2.13	3.00
	¾"	up†	0.81	1.90	2.81
	4"	up†	0.82	1.98	3.00
	¾"	down*	1.02	2.40	3.57
	4"	down*	1.08	2.75	4.41
	¾"	down†	0.84	2.09	3.34
	4"	down†	0.90	2.50	4.36
Vertical	¾"	across*	1.01	2.36	3.48
	4"	across*	1.01	2.34	3.45
	¾"	across†	0.84	2.10	3.28
	4"	across†	0.91	2.16	3.44

‡ One side of the air space is a non-reflective surface.
* Winter conditions.
† Summer conditions.

8-3: R-values of air spaces [ASH-1].

R-VALUES OF AIR FILMS

Type and Orientation of Air Film		Direction of Heat Flow	R-value for Air Film On: Non-reflective surface	Fairly reflective surface	Highly reflective surface
Still air:					
Horizontal		up	0.61	1.10	1.32
Horizontal		down	0.92	2.70	4.55
45° slope		up	0.62	1.14	1.37
45° slope		down	0.76	1.67	2.22
Vertical		across	0.68	1.35	1.70
Moving air:					
15 mph wind		any*	0.17	—	—
7½ mph wind		any†	0.25	—	—

* Winter conditions.
† Summer conditions.

8-2: R-values of air films [ASH-1].

U-VALUES OF WINDOWS AND SKYLIGHTS		
Description	U-values[1]	
	Winter	Summer
Vertical panels:		
Single pane flat glass	1.13	1.06
Insulating glass—double[2]		
3/16" air space	0.69	0.64
1/4" air space	0.65	0.61
1/2" air space	0.58	0.56
Insulating glass—triple[2]		
1/4" air spaces	0.47	0.45
1/2" air spaces	0.36	0.35
Storm windows		
1-4" air space	0.56	0.54
Glass blocks[3]		
6 × 6 × 4" thick	0.60	0.57
8 × 8 × 4" thick	0.56	0.54
same, with cavity divider	0.48	0.46
Single plastic sheet	1.09	1.00
Horizontal panels:[4]		
Single pane flat glass	1.22	0.83
Insulating glass—double[2]		
3/16" air space	0.75	0.49
1/4" air space	0.70	0.46
1/2" air space	0.66	0.44
Glass blocks[3]		
11 × 11 × 3" thick, with cavity divider	0.53	0.35
12 × 12 × 4" thick, with cavity divider	0.51	0.34
Plastic bubbles[5]		
single-walled	1.15	0.80
double-walled	0.70	0.46

[1] in units of $Btu/hr/ft^2/°F$
[2] double and triple refer to the number of lights of glass.
[3] nominal dimensions.
[4] U-values for horizontal panels are for heat flow *up* in winter and *down* in summer.
[5] based on area of opening, not surface.

8-4: U-values of windows and skylights [ASH-1].

Appendix 9

Definition of Economic Terms and Formulas

9.1 TIME VALUE OF MONEY AND DISCOUNTING

The purpose of investment analysis is to compare the initial cost of a capital asset with an expected stream of benefits (and related costs) that accrue as a direct result of that investment. In the case of solar investments, you must compare the initial dollar outlay for components, materials, and labor against the projected value of fuel savings over a predetermined period of time. There is always a cost associated with any investment because—once the money is spent—the opportunity to invest it elsewhere is lost. One measure of this opportunity cost is the current rate of interest on bank savings or certificates of deposit.

If you invest $1,000 at 8% for three years, the expected return at the end of the third year can be measured in one of two ways depending on whether the compounding of interest is discrete or continuous. See figure 9-1.

Alternatively, you can calculate the present value of investment required to generate a future worth of $1,000 by the end of the third year, by discounting the future worth at the 8% rate of interest. This is done by using a single-payment present worth factor (PWF), which is the inverse of the single-payment future worth factor (FWF). See figure 9-2.

Therefore, the following formulas apply to the discrete compounding convention:

$$FW = (1 + r)^t \times PW$$

$$PW = \left(\frac{1}{(1 + r)^t}\right) \times FW$$

where:

FW = future worth of an investment
PW = present worth of an investment

$$\left(\frac{1}{(1 + r)^t}\right) = FWF(t,r)^* = \text{future worth factor (single payment)}$$
$$= PWF(t,r)^* = \text{present worth factor (single payment)}$$

r = opportunity cost, interest rate, or discount rate on capital investments
t = number of compounding or discounting periods

The present-worth convention allows comparison of alternative investment, cost, and benefit streams on an equal footing because all dollar amounts are measured in terms of *today's* dollar value. Suppose that the option exists to make one of two alternative energy conservation investments for the same initial dollar cost, but that the expected stream of future benefits differs in terms of timing—not in terms of total amount. The two hypothetical options are shown in figure 9-3.

	Future Worth of Investment	Single-Payment Future Worth Factor	Present Worth of Investment
Discrete Compounding:	$1,260	= (1 + 0.08)^3 \times	$1,000
Continuous Compounding:	$1,271	= $_{\exp}(0.08 \times 3) \times$	$1,000

9-1: Calculating returns on an investment with a present worth of $1,000.

	Present Worth of Investment	Single-Payment Present Worth Factor	Future Worth of Investment
Discrete Discounting:	$794	$= (1 + 0.08)^{-3} \times$	$1,000
Continuous Discounting:	$786	$= {}_{\exp}-(0.08 \times 3)\times$	$1,000

9-2: Calculating present worth of investments with a future worth of $1,000.

---

*The expression "(t,r)" after FWF and after PWF is used to show that values for FWF and PWF are determined by the values of t and r according to the formula shown. This can be read as, "FWF (or PWF), a function of t and r."

Year (t)	Option 1	Option 2
1	$300	$150
2	$250	$175
3	$200	$200
4	$100	$225
5	$ 50	$150
	$900	$900

9-3: Two hypothetical options for investments in alternative energy conservation, varying in timing but not in total amount.

Using a discount rate of 8%, the discounted present worth of these benefit streams can be calculated using the following formula:

$$PW_j = \sum_{t=1}^{T} \left(\frac{1}{(1+r)^t}\right) \times B_{jt}$$

where:
 Σ denotes summation
 T = the total period of analysis (in years)
 t = an index pertaining to year
 j = an index pertaining to the option
 B_{jt} = dollar benefits of option j in year t

Thus:

$$PW_1 = \left[\left(\frac{1}{1.08^1}\right) \times 300\right] + \left[\left(\frac{1}{1.08^2}\right) \times 250\right]$$
$$+ \left[\left(\frac{1}{1.08^3}\right) \times 200\right] + \left[\left(\frac{1}{1.08^4}\right) \times 100\right]$$
$$+ \left[\left(\frac{1}{1.08^5}\right) \times 50\right] = \$758.41$$

And:

$$PW_2 = \left[\left(\frac{1}{1.08^1}\right) \times 150\right] + \left[\left(\frac{1}{1.08^2}\right) \times 175\right]$$
$$+ \left[\left(\frac{1}{1.08^3}\right) \times 200\right] + \left[\left(\frac{1}{1.08^4}\right) \times 225\right]$$
$$+ \left[\left(\frac{1}{1.08^5}\right) \times 150\right] = \$715.16$$

Although both options yield equivalent undiscounted benefits, Option 1 produces larger discounted benefits because the larger annual benefits are weighted toward the earlier years and hence discounted at a lower relative weight. The discounted present-worth technique is especially useful for evaluating alternatives that have different investment lives, capital requirements, and so forth.

The rate at which you choose to discount future benefits is a matter of subjective choice. Lower discount rates assign a relatively high value to longer-term savings; higher discount rates give preference to shorter-term savings. In the above examples, an 8% discount rate was used for illustrative purposes only, and should not be interpreted as a suggested value. The discount rates that have been used in energy analyses usually fall in the range between 6% and 20%.

The sensitivity of this parameter is illustrated by the following simple example. Suppose an initial investment of $1,500 is made with the expectation of receiving a benefit of $200 per year through electric-resistance fuel savings for the next 20 years. The discounted present worth of savings is calculated as:

$$PW = \sum_{t=1}^{20} \left(\frac{1}{(1+r)^t}\right) \times 200$$

$$PW = 200 \sum_{t=1}^{20} \left(\frac{1}{(1+r)^t}\right)$$

The $200 can be separated from the summation because in this example the nominal value of fuel savings does not vary with t (instead, it is the same every year). The remaining term, $\sum_{t=1}^{20} \left(\frac{1}{(1+r)^t}\right)$ is called the uniform series present worth factor, $SPWF(T,r)$, and varies according to T (20 years, in the above example) and r (interest). The present worth calculation is repeated for various discount rates over the assumed 20-year period and is shown in figure 9–4.

r	Uniform Series PWF $SPWF(20,r)$	×	Uniform Annual Benefit B	=	Discounted Present Worth PW
5%	12.462	×	$200	=	$2,492.40
10%	8.514	×	$200	=	$1,702.80
15%	6.259	×	$200	=	$1,251.80
20%	4.870	×	$200	=	$ 974.00

9-4: Present worth calculations for various discount rates over a 20-year period.

As figure 9–4 indicates, the present worth of the electricity savings benefits can be greater or less than the $1,500 initial outlay, depending upon the rate at which one discounts future benefits.

9.2 CAPITAL RECOVERY FACTOR

The added materials and labor cost of features of the passive-solar design can be paid for in one of several ways:

- 100% out-of-pocket cash;
- Down-payment and partial loan or mortgage financing; or
- No down-payment and 100% loan or mortgage financing.

In new construction, the additional cost is usually incorporated into the first mortgage. For a retrofit application, a separate home improvement loan might be arranged. Regardless of the financing method used, certain techniques of financial analysis allow you to convert an initial investment cost into an equivalent annual cost. One such technique is to use a capital recovery factor (*CRF*). The *CRF*, when multiplied by the initial value of the loan, gives an equivalent payment to be paid each period for the entire term of the loan such that the discounted present worth of these equivalent periodic payments equals the initial loan amount plus interest charges. This is described by the following formula:

$$LP = \sum_{t=1}^{T} \left(\frac{1}{(1+i)^t} \right) \times AMP_t$$

where:
- LP = loan principal amount
- i = mortgage or loan interest rate
- T = term of the loan
- AMP = equivalent annual payment for the mortgage or loan

Rearranging this formula, you can see that:

$$AMP = \left(\frac{1}{\sum_{t=1}^{T} \left(\frac{1}{(1+i)^t} \right)} \right) \times LP$$

By definition:

$$\frac{1}{\sum_{t=1}^{T}\left[\left(\frac{1}{1+i}\right)^t\right]} = \frac{i}{\left[1 - \left(\frac{1}{1+i}\right)^T\right]} = CRF(T,i)$$

Therefore:

$$AMP = CRF(T,i) \times LP$$

Thus, the *CRF* is seen to be equivalent to the inverse of the uniform-series (equivalent period payment) present worth factor:

$$CRF(T,i) = \frac{1}{SPWF(T,i)}$$

Example: Mr. Jones finances a 20-year, 10% mortgage for his $60,000 home, $5,000 of which is attributable to the passive-solar addition. An arrangement for 100% financing has been made. What is the annual payment associated with the passive addition?

$$AMP = CRF(20, 0.10) \times 5,000 = \frac{0.10}{\left[1 - \left(\frac{1}{1.1}\right)^{20}\right]}$$

$$\times 5,000 = 0.11746 \times 5,000 = \$587.30$$

Example: Mr. Jones again finances the cost of his passive-solar addition through the mortgage, but the terms have changed. He makes a 20% down payment and finances the remainder with a 25-year, 9.75% mortgage. What is the annual mortgage payment associated with the passive addition?

$$AMP = (1 - 0.2) \times 5,000 \times CRF(25, 0.0975) = 4,000$$

$$\times \frac{0.0975}{\left[1 - \left(\frac{1}{1.0975}\right)^{25}\right]} = 4,000 \times 0.108057$$

$$= \$432.23$$

Explanation: In this second example, $432.23 represents the annual mortgage payment associated with the loan portion (80%) of the passive-solar addition. The equivalent-payment counterpart of the $1,000 down payment (*ADP*) can also be calculated with a *CRF* based upon either the investment opportunity cost of the out-of-pocket cash or the cost of down-payment financing. Assuming Mr. Jones pays the $1,000 down payment in cash and by doing so has foregone the opportunity to invest that money at 8%, the *ADP* of the down payment equals:

$$ADP = 0.2 \times 5,000 \times CRF(25, 0.08) = 1,000$$

$$\times \frac{0.08}{\left[1 - \left(\frac{1}{1.08}\right)^{25}\right]} = 1,000 \times 0.09368$$

$$= \$93.68$$

Therefore, the total equivalent annual cost (*EAC*) equals the sum of the two equivalent payments (*AMP* + *ADP*):

$$EAC = AMP + ADP = \$432.23 + \$93.68 = \$525.91$$

If the down-payment opportunity or financing cost equals the mortgage interest rate, we have the simple calculation:

$$EAC = CRF \times TC$$

where:
- TC = total add-on cost of the passive solar addition.

This is because *CRF* is identical when calculating both *AMP* and *ADP*.

9.3 DIFFERENTIAL-SERIES PRESENT-WORTH FACTORS AND ALTERNATIVE FUEL COSTS, OPERATION AND MAINTENANCE EXPENSES, AND PROPERTY TAXES

The present-worth and capital recovery calculations also can be combined to estimate the equivalent annual cost of the auxiliary heating fuel source. The procedural steps are outlined below.

Step. 1. Identify the fuel source, acquire unit cost data (P_j), convert to $/MMBtu using the appropriate conversion factor (CF_j), and correct for the heat conversion coefficient of performance (COP). This gives the initial fuel price (P_o) in $/MMBtu:

$$P_o = \frac{P_j}{COP_j \times CF_j}$$

Step 2. Specify the discount rate (r), specify the expected annual fuel escalation rate (e), and determine the relevant period of analysis in years (PT).

Step 3. The following formula can be used to calculate the equivalent annual cost of fuel (P):

$$P = \sum_{t=1}^{PT}\left[\left(\frac{1}{1+r}\right)^t \times P_o \times (1+e)^t\right] \times CRF(PT,r)$$

$$\overline{P} = \sum_{t=1}^{PT}\left[\left(\frac{1+e}{1+r}\right)^t\right] \times P_o \times CRF(PT,r)$$

The term $\sum_{t=1}^{PT}\left[\left(\frac{1+e}{1+r}\right)^t\right]$ is called a differential series present worth factor (DSPWF) because the numerator contains an escalation rate (e) and the denominator a discount rate (r). Through geometric series definitions:

$$DSPWF(PT,e,r) = \frac{(1+e) \times \left[1 - \left(\frac{1+e}{1+r}\right)^{PT}\right]}{(r-e)}$$

$$= \sum_{t=1}^{PT}\left[\left(\frac{1+e}{1+r}\right)^t\right]$$

Therefore:

$$P = DSPWF(PT,e,r) \times P_o \times CRF(PT,r)$$

Example: Mr. Jones wants to compare the additional annual cost of his $5,000 passive-solar investment with the expected fuel savings over a 20-year period. He has installed a full-sized central heating system that burns natural gas with combustion efficiency of 60%, and he has found out from his local utility that current gas prices are $.25/Therm. He expects the price of natural gas to escalate at a rate of about 11% per year and feels his discount rate is about 9%. What is the equivalent annual cost of fuel (P)?

Step 1.

$$P_o = \frac{\$.25/\text{Therm}}{0.60 \times 0.1 \text{ MMBtu/Therm}}$$

$$P_o = \$4.17/\text{MMBtu}$$

Step 2.

$$r = 9\%$$
$$e = 11\%$$
$$PT = 20 \text{ years}$$

Step 3.

$$P = DSPWF(20,.11,.09) \times 4.17 \times CRF(20,.09)$$

$$P = \frac{1.11 \times \left[1 - \left(\frac{1.11}{1.09}\right)^{20}\right]}{(0.09 - 0.11)} \times 4.17 \times \frac{0.09}{\left[1 - \left(\frac{1}{1.09}\right)^{20}\right]}$$

$$P = 24.34 \times 4.17 \times 0.10955 = \$11.12/\text{MMBtu}$$

Explanation: When the discount rate (r) equals the fuel escalation rate (e) the first term in brackets is undefined. However, it can be shown that under this condition the first term exactly equals PT, the period of analysis. In the above example, if $e = 9\%$ had been assumed, then:

$$r = e = 9\%$$

and:

$$P = 20 \times 4.17 \times 0.10955 = \$9.14/\text{MMBtu}$$

Annualized operation, maintenance, and insurance charges and property taxes also can be calculated using DSPWF. A simple and frequently used method of estimate operation, maintenance, and insurance charges is to express the first year's annual charge as a percentage (OMI) of the total system cost. For example, if $OMI = 0.01$ and $TC = \$5,000$:

$$OMI_1 = OMI \times TC = 0.01 \times 5,000 = \$50$$

In subsequent years, OMI_t can be expected to increase at an undetermined annual escalation rate (γ). The present worth of the expenditures over PT years is therefore:

$$PW(OMI) = \sum_{t=1}^{PT}\left[\left(\frac{1+\gamma}{1+r}\right)^t\right] \times OMI \times TC$$

$$= DSPWF(PT,\gamma,r) \times OMI \times TC$$

With $PT = 20$, $\gamma = 0.07$, and $r = 0.09$, it follows that:

$$PW(\overline{OMI}) = 16.56 \times 0.01 \times 5,000 = \$828$$

The annualized value $\overline{OMI}$ is given by:

$$\overline{OMI} = CRF(PT,r) \times PW(OMI) = 0.10955 \times 828 = \$90.71$$

Property taxes can be figured in the same manner. Given a first year effective property tax rate (PTXR) of 0.02, and an annual escalation rate in property value of $\delta = 0.10$, it follows that:

$$\begin{aligned}
\overline{PTXR} &= CRF(PT,r) \times PW(PTXR) \\
&= CRF(PT,r) \times DSPWF(PT,\delta,r) \times PTXR \\
&\quad \times (1 - FSLTX) \times TC \\
&= CRF(20,0.09) \times DSPWF(20,0.10,0.09) \\
&\quad \times 0.02 \times 0.70 \times 5,000 \\
&= 0.10955 \times 22.04 \times 0.02 \times 0.70 \times 5,000 \\
&= \$169.02
\end{aligned}$$

FSLTX is the federal and state marginal income tax rate. Property taxes are deductible from gross income, so $1 - FSLTX$ is the actual portion of property tax paid.

9.4 REMAINING LOAN BALANCE AND MORTGAGE INTEREST TAX DEDUCTIONS

Section 9.2 discussed the calculation of annual mortgage payments. Although each annual payment is identical in current-year dollars, the amount apportioned to payment of principal or of interest varies. Interest charges constitute the majority of the payment in the early years of the mortgage, but decline over time. The remaining principal to be paid in year t—as a fraction of the initial loan financed for T years at rate i—is equal to:

$$RBF(T,i,t) = \left\{ \frac{CRFN(T,i)}{i} \times \left[1 - (1 + i)^t\right] \right\} + (1 + i)^t$$

It can be shown that $0 \leq RBF \leq 1$; and that $RBF = 1$ when $t = 0$; and that $RBF = 0$ when $t \geq T$.

Mortgage interest deductions (ID_t) are determined by calculating interest costs in any one year based upon the remaining loan balance, and by using the marginal income tax rate (*FSLTX*) to value the tax deduction:

$$ID_t = (1 - DPR) \times TC \times RBF(T,i,t) \times i \times FSLTX$$

Example: Given a total add-on passive cost of $2,000 financed with a loan of 30 years at $i = 0.10$ and a downpayment ratio = 20%, what is the value of interest tax deductions in the 10th year if the marginal tax rate equals 30%?

$$ID_{10} = (1 - 0.2) \times 2,000 \times RBF(30, 0.10, 10) \times 0.10 \times 0.30$$

$$ID_{10} = 1,600 \times 0.90 \times 0.10 \times 0.30 = \$43.20$$

9.5 REAL VERSUS NOMINAL TERMS

Throughout the previous discussion, discount, interest, and fuel escalation rates have been stated in *nominal terms*—that is to say, they include the effects of inflation. Sometimes it is convenient to divide these parameters into two components: a real (inflation-free) component and the expected inflation rate. This can be done by defining the discount rate by the following simplistic expression:

$$r = R + \theta$$

where:
  r = the nominal rate of discount
  R = the real rate of discount
  θ = the rate of inflation

We could do the same with fuel escalation rates (e):

$$e = E + \theta$$

where:
  e = the nominal fuel escalation rate
  E = the real fuel escalation rate
  θ = the rate of inflation

Note: If the economy were to enter into a period of deflation (a period in which the general price level declined), θ would be negative during that period.

Unless otherwise specified, discount, interest, and fuel escalation rates are always stated in *nominal* terms throughout Chapter L and the Appendixes.

Appendix 10

Generalized Techniques of Economic Analysis

This appendix is an extension of the economic analysis presented in Chapter L. All economic and financial parameters can be treated as variables that take on values assigned by the user. Methods are demonstrated for the calculation of cost curves, fixed charge rates, delivered cost of heat by system size, and net present value. In addition, cash flow analysis techniques are presented and a section on definitions and calculation of payback period are included. Whereas Chapter L dealt with the joint optimization of passive collection area and building thermal load, this appendix deals with optimization of collector size only in cases in which building thermal load is predetermined.

Suppose the design under consideration is a 60% double-glazed 12-inch-thick waterwall design and 40% double-glazed direct-gain system with R4 night insulation to be built in Dodge City, Kansas. The generalized performance characteristics for single-option systems can be found in Appendix 12, where load collector ratios (in Btu/DD/ft$_g^2$) for particular solar savings fractions are listed. The method for determining load collector ratios for mixed systems is given in Chapter I.

10.1 COST CURVES

Step 1. Estimate the building heating load coefficient (L) for all surfaces other than the south wall. The Dodge City reference example's single-story house will be used for illustrative purposes. In that example, $L = 7,420$ Btu/DD. See Chapter I.
Total annual heating load (*HLOAD*) can be calculated as:

$$HLOAD = L \times DD = 7,420 \text{ Btu/DD} \times 5,046 \text{ DD}$$
$$= 37.44 \text{ MMBtu}$$

Step 2. Using the *LCR* ratios, calculate a solar performance curve that relates collector area (A) to annual solar savings fraction (F). For instance, for the point at which $F = 0.5$, $LCR = 30.3$, and:

$$A \times 0.5 = \frac{L}{LCR} = \frac{7,420}{30.3} = 245 \text{ ft}_g^2$$

Repeating this calculation for all values of F, the $A(F)$ curve shown in figure 10–1 is defined.

Step 3. Identify the maximum southern exposure area available for glazing. Suppose the length of the home will be 54 feet and the glazing unit height will be 92 inches. Without consideration of framing, this implies that $54 \times 92/12$ ($\cong 414$ ft$_g^2$) is the maximum possible exposure area for this reference design. By linear interpolation, this implies a maximum possible F of approximately 0.71.* Although economic considerations alone may dictate large glazing areas, physical design constraints may limit one's options.

Step 4. Using the cost estimation method outlined in Chapter L, derive the estimated add-on passive solar construction cost in $/ft$_g^2$ for collector-area-*dependent* or variable costs and in $ for collector-area-*independent* or fixed costs. In the formulas that follow, these are denoted as:

VC = variable cost ($/ft$_g^2$)
FC = fixed cost ($)

Step 5. Calculate the total cost (*TC*) and average cost (*AC*) curves. The $A(F)$ curve is the backbone of the economic analysis, since it provides the link between physical design (A) and expected performance (F). The first connection with economics is made through the estimation of *FC* and *VC*, which enables one to calculate the total cost (*TC*) and average cost (*AC*) curves for passive

F	0.10	0.20	0.30	0.40	0.50	0.60	0.70	0.80	0.90
LCR	210.4	97.6	60.1	40.7	30.3	23.5	18.3	13.6	8.6
A	35	76	123	182	245	316	405	546	863

10–1: *A(F)* values for $L = 7,420$ Btu/DD.

---

*Given a collector area A that is not included in figure 10–1, the solar savings fraction F that corresponds to A can be estimated by linear interpolation between the surrounding solar fractions F_1 and $F_2(F_2, F_1)$ by the following formula:

$$F = \frac{(F_2 - F_1)}{(A_2 - A_1)} \times (A - A_1) + F_1.$$

With A equal to 414 in the above example, this implies:

$$F = \frac{(0.80 - 0.70)}{(546 - 405)} \times (414 - 405) + 0.70 = 0.71$$

solar. Suppose that VC is estimated to be \$15/ft$_g^2$ after allowance is made for replacement credits. Since the majority of costs in most passive designs are dependent upon the glazing area, FC is assumed to be zero in the following example. It should be emphasized that not all passive designs are characterized by zero fixed costs; care should be taken when analyzing a specific design to ascertain the fixed-cost level.

The following equations should be used to calculate TC and AC for each solar savings fraction (F).

$$TC(F) = FC + (VC \times A(F))$$

and

$$AC(F) = \left(\frac{FC + (VC \times A(F))}{HLOAD \times F}\right) = \left(\frac{TC(F)}{HLOAD \times F}\right)$$

$TC(F)$ indicates the initial total cost of a design that has been sized to provide a given level of solar savings fraction F. $AC(F)$ is defined as the per-unit cost (in \$/MMBtu/yr of providing a given solar savings fraction F of the total annual heating load (i.e., $HLOAD \times F$). Figure 10–2 shows the calculation of TC and AC by solar savings fraction (F). At this point, $TC(F)$ can be used to get an indication of the respective first costs associated with systems of various sizes and annual thermal yields. If a budgeted maximum—say, TC—is available for the first cost, a maximum system size can be determined by solving for A in the original $TC(F)$ equation. That is:

$$A = (TC - FC) \div VC$$

For example, if the owner is willing to add a maximum of \$3,500 to the first cost of the home, the maximum possible system size would be:

$$A = (3,500 - 0) \div 15 = 233 \text{ ft}_g^2$$

In this example, the budget allows for 233 ft$_g^2$, whereas the physical sizing constraint allows for 414 ft$_g^2$; the budget, therefore, is the binding constraint on design size.

Calculation of these cost curves alone may provide sufficient information about economic performance for some designers. Those who want information on equivalent annual costs, optimal system size, cash flow, or payback analysis should proceed through the ensuing sections of this appendix.

F	0.20	0.30	0.40	0.50	0.60	0.70	0.80
A	76	123	182	245	316	405	546
TC	1,140	1,845	2,730	3,675	4,740	6,075	8,190
HLOAD × F	7.49	11.23	14.98	18.72	22.46	26.21	29.95
AC	152.20	164.30	182.20	196.30	211.00	231.80	273.50

10–2: Total cost (TC) and average cost (AC) calculations, based on different solar savings fractions (F).

10.2 DOLLAR PER MILLION Btu (\$/MMBtu) COST

The purpose of this section is to illustrate a relatively simple method for calculating the delivered cost of heat (DCH) to the home, taking into account all factors that enter into thorough cash-flow analysis. In the method to be described, DCH is made level over the specified ownership or life-cycle period. Usually, an auxiliary heat source is necessary to provide a portion of the annual heating load, with the passive-solar system satisfying the remaining load. For this reason, the dollar per million Btu cost of heat to the home is made up of two components: solar and auxiliary. DCH takes into consideration both components.

Reference tables contained in Appendix 11 have been generated to minimize time spent with formulas. Even so, the calculation of DCH takes some time and patience; it becomes easier with practice.

Step 1. Assign values to the following financial and economic parameters:

- PT = period of financial analysis in years. PT does not necessarily have to equal the expected life of the passive solar design; it may equal an expected ownership period, with PT equal to the number of years between time of home purchase or system installation and time of resale.
- DPR = down-payment fraction of a mortgage or home improvement loan, if used to finance the passive solar addition (0 DPR 1).
- T = term of the mortgage or home improvement loan, in years. (The financial analysis assumes payment on an annual rather than on a monthly basis, to simplify the analysis.)
- i = annual interest rate on the mortgage or home improvement loan (nominal rate).*
- r = annual discount rate or opportunity cost of money (nominal rate).*
- θ = expected annual inflation rate over the PT years of analysis (assumed to be constant).
- $FSLTX$ = federal, state, and local marginal income tax rate.
- $PTXR$ = effective property tax rate, expressed as a percentage of TC.
- OMI = passive system operation, maintenance, and insurance charge, expressed as a percentage of TC.
- P_o = current cost of competing backup fuel alternative, corrected for the coefficient of performance (COP). P_o is stated in MMBtu.

---
*For a discussion of the distinction between nominal and real interest rates, see section 9.5.

e = annual escalation rate of competing fuel over the PT years (assumed to be constant).

V = an assessment valuation factor that is meant to take into account the value of the passive addition assigned by the property assessor in the PTth year. If none of the additional cost is reflected in the resale price, then $V = 0$; if all of the additional cost is included in the appraisal of market value, then $V = 1.0$. It is conceivable in the future that, with increasing home heating costs, $V > 1.0$ may be possible.

BSC = backup system capital cost ($).

OMB = backup system operation (exclusive of heating fuel), maintenance and insurance charge expressed as a percentage of BSC.

Continuing with the sample case for Dodge City, the following values are assigned:

$PT = 10$ years
$DPR = 0.15$ (15%)
$T = 30$ years
$i = 0.10$ (10%)
$r = 0.07$ (7%)
$\theta = 0.06$ (6%)
$FSLTX = 0.25$ (25%)
$PTXR = 0.02$ (2%)
$OMI = 0.01$ (1%)
$P_o = 14.65$ (in \$/MMBtu for electric resistance)
$e = 0.08$ (8%)
$V = 0.75$ (75%)
$BSC = \$500$
$OMB = 0.005$ (0.5%)

Step 2. Calculate a fixed charge rate (FCR) based upon the following formula:

$$FCR = CRFN(PT,r) \times (X_1 - X_2 - X_3 - X_3 + X_4)$$

First, find the appropriate value for $CRFN(PT,i)$ in the Compound Interest Tables in Appendix 11. $CRFN(PT,r)$ is the nominal capital recovery factor based on PT years at discount rate r.

$$CRFN(10,.07) = 0.14238$$

Second, calculate X_1. X_1 is a factor that accounts for the mortgage or loan down payment and for the principal and interest costs over the years of the loan term (T):

$$X_1 = DPR + \left[(1 - DPR) \times \left(\frac{CRFN(T,i)}{CRFN(PT,r)}\right)\right]$$

$$X_1 = 0.15 + \left[(1 - 0.15) \times \left(\frac{0.10608}{0.14238}\right)\right]$$

$$X_1 = 0.78329$$

Third, calculate X_2. X_2 is a factor that accounts for mortgage interest tax deductions over the years of ownership (PT):

$$X_2 = (1 - DPR) \times FSLTX \times \left[\left(\frac{CRFN(T,i)}{CRFN(PT,r)}\right) + [(i - CRFN(T,i)) \times DSPWF(PT,i,r)]\right]$$

$DSPWF(PT,i,r)$ is a differential series present worth factor, whose value under different conditions can be found in the Differential Series Present-Worth Factor Tables in Appendix 11.

$$DSPWF(10,0.10,0.07) = 11.68$$

$$X_2 = (1 - 0.15) \times 0.25 \times \left\{\left(\frac{0.10608}{0.14238}\right) + [(0.10 - 0.10608) \times 11.68]\right\}$$

$$X_2 = 0.85 \times 0.25 \times 0.67403 = 0.14323$$

Fourth, calculate X_3. X_3 is a factor that accounts for the profit that is realized from the sale of the passive home at the end of the PTth year. The X_3 calculation is based on the difference between the assessed market value of the passive elements in the home and the unpaid principal balance of the mortgage or home improvement loan.

$$X_3 = \left(FWF(PT,\delta) \times PWF(PT,r) \times V\right) - [(1 - DPR) \times RBF(T,i,PT) \times PWF(PT,r)]$$

$FWF(PT,\delta)$ is a future-worth factor that converts the initial total cost (TC) of the passive addition into market value at the end of the PTth year. δ is the annual escalation rate of property values over the PT years (not necessarily equal to θ, the rate of inflation, but assumed to be equal to θ in this example). $FWF(PT,\delta)$ can be found in the Compound Interest Tables of Appendix 11. $PWF(PT,r)$ is a present-worth factor that restates the market value at the end of the PTth year as a first-year discounted value. $PWF(PT,r)$ also can be found in the Compound Interest Tables.

$RBF(T,i,PT)$ is simply a fraction that identifies the portion of the initial loan principal that remains unpaid at the time of sale; it is a function of PT and of the loan term factors T and i ($0 \leq RBF \leq 1$). $RBF(T,i,PT)$ can be found in the Remaining Loan Balance Factor Tables in Appendix 11. It follows that:

$FWF(10,0.06) = 1.79083$
$PWF(10,0.07) = 0.50835$
$V = 0.75$
$RBF(30,0.10,10) = 0.90$

$X_3 = 1.79083 \times 0.50835 \times 0.75 -$
$[(1 - 0.15) \times 0.90 \times 0.50835]$
$X_3 = 0.68277 - 0.38889 = 0.29388$

Fifth, calculate X_4. X_4 is a factor that accounts for property taxes and operation, maintenance, and insurance expenditures over the PT years.

$$X_4 = DSPWF(PT,\theta,r) \times \{[PTXR \times (1 - FSLTX)] + OMI\}$$

Notice that the property tax rate ($PTXR$) is adjusted to reflect federal, state, and local income tax deductions. $DSPWF(PT,\theta,r)$ is used as a correction factor for $PTXR$ and OMI to account for the inflationary increases in these expenditures. If the expenditure escalation rates are expected to deviate from θ, the corresponding factors δ (for $PTXR$) and γ (for OMI) should replace θ, in order to allow calculation of two separate $DSPWF$ factors. For example, if $\delta = 0.08$ and $\gamma = 0.07$:

$DSPWF(PT,\delta,r) = DSPWF(10,0.08,0.07) = 10.53$
$DSPWF(PT,\gamma,r) = DSPWF(10,0.07,0.07) = 10.00$

and:

$X_4 = [DSPWF(PT,\delta,r) \times PTXR \times (1 - FSLTX)]$
$\quad\quad + (DSPWF(PT,\gamma,r) \times OMI)$
$X_4 = (10.53 \times 0.02 \times 0.75) + (10.00 \times 0.01) = 0.25795$

In the sample case, $\theta = \delta = \gamma$, so:

$X_4 = DSPWF(PT,\theta,r) \times \{[PTXR \times (1 - FSLTX)] + OMI\}$
$X_4 = DSPWF(10,0.06,0.07) \times \{[0.02 \times (1 - 0.25)] + 0.01\}$
$X_4 = 9.50 \times 0.025 = 0.2375$

With X_1, X_2, X_3, and X_4 now determined, the final calculation for FCR can be obtained by using the FCR equation given at the beginning of Step 2:

$FCR = CRFN(PT,r) \times (X_1 - X_2 - X_3 + X_4)$
$FCR = 0.14238 \times (0.78329 - 0.14323$
$\quad\quad - 0.29388 + 0.2375)$
$FCR = 0.14238 \times (0.58368)$
$FCR = 0.08310$

It is important to remember that FCR is quite sensitive to the parameter values that enter into the calculation of X_1, X_2, X_3, and X_4. Care should be taken to analyze the parameters that are subject to the most uncertainty (e.g., V and r) so that potential deviations in FCR may be determined. The graphs in Appendix 15 illustrate the sensitivity of FCR to variations in the parameter values.

Step 3. Calculate $AAC(F)$, the average annual cost of passive solar heat (in $/MMBtu). Use the following formula:

$$AAC(F) = AC(F) \times FCR$$

Figure 10-3 shows calculations of $AAC(F)$ at different F values.

Step 4. Calculate the average annual cost of the backup system over PT years. Three components need to be determined: average annual capital costs, operation and maintenance costs, and backup fuel costs.

Average annual capital, operation, and maintenance costs ($ABSC$) are given by:

$$ABSC = \frac{BSC \times X_5 \times CRFN(PT,r)}{HLOAD}$$

where:

$X_5 = X_1 + (DSPWF(PT,\theta,r) \times OMB)$

with X_1 at its previously defined value. Therefore:

$X_5 = 0.78329 + (9.50 \times 0.005) = 0.83079$

and:

$$ABSC = \left(\frac{500 \times 0.83079 \times 0.14238}{37.44}\right) = \$1.58/MMBtu$$

Average annual cost of backup fuel ($\overline{P}$) is defined by:

$\overline{P} = CRFN(PT,r) \times DSPWF(PT,e,r) \times P_o$

In the sample case:

$\overline{P} = CRFN(10,.07) \times DSPWF(10,0.08,0.07) \times 14.65$

After the Compound Interest Tables and the Differential Series Present-Worth Factor Tables in Appendix 11 are used to determine values for $CRFN$ and $DSPWF$, the equation becomes:

$\overline{P} = 0.14238 \times 10.53 \times 14.65$
$\overline{P} = \$21.96/MMBtu$

Step 5. Calculate $DCH(F)$, the average annual value of the delivered cost of heat to the home, passive solar and backup combined, using the following equation:

$DCH(F) = (AAC(F) \times F) + [\overline{P} \times (1 - F)] + ABSC$

Figure 10-4 shows calculations of the various components of this equation for the sample case, at different F values. The optimum system-size occurs where $DCH(F)$ is at a minimum. This corresponds to $DCH(.50) = \$20.72/MMBtu$, where:

$A = 245\ ft_g^2$
$TC = \$3,675$
$LOAD \times F = 18.72\ MMBtu/yr$
$AC = \$196.30/MMBtu/yr$

F	0.20	0.30	0.40	0.50	0.60	0.70	0.80
AAC(F) × F	2.53	4.10	6.06	8.16	10.52	13.48	18.18
P × (1 − F)	17.57	15.37	13.18	10.98	8.78	6.59	4.39
ABSC	1.58	1.58	1.58	1.58	1.58	1.58	1.58
DCH(F)	21.68	21.05	20.82	20.72	20.88	21.65	24.15

F	0.20	0.30	0.40	0.50	0.60	0.70	0.80
AC(F)	152.20	164.30	182.20	196.30	211.00	231.80	273.50
AAC(F)	12.65	13.65	15.14	16.31	17.53	19.26	22.73

10-3: Calculations of average annual cost of passive solar heat ($AAC(F)$), in $/MMBtu, at different values of F.

10-4: Calculations of the various components in the $DCH(F)$ equation given in Step 5.

Techniques of Economic Analysis / 231

Building the system at its optimum size may be impossible due to design and budget constraints; however, knowledge of optimum system-size provides additional information for the designer, architect, or builder.

Step 6. Calculate $NPV(F)$, the net present value of annual passive system savings over the PT years, discounted at r. $SPWF(PT,r)$ can be found in the Compound Interest Tables in Appendix 11. The equation for $NPV(F)$ is:

$$NPV(F) = [(\overline{P} + ABSC) - DCH(\hat{F})] \times HLOAD \\ \times SPWF(PT,r)$$
$$NPV(F) = [(21.96 + 1.58) - 20.72] \times 37.44 \\ \times 7.02359$$
$$NPV(F) = \$741$$

$NPV(F)$ will be maximized at $\hat{F}$, which is always the point at which $DCH(F)$ is minimized. $NPV(F)$ values are shown in figure 10–5.

F	0.20	0.30	0.40	0.50	0.60	0.70	0.80
DCH(F)	21.68	21.05	20.82	20.72	20.88	21.65	24.15
NPV(F)	489	654	715	741	699	497	−160

10–5: Net present worth of annual passive system savings over PT years, discounted at r, at different values of F. These figures for $NPV(F)$ apply to the example case, in which $A = 245$ ft$_g^2$.

With a budget constraint of \$3,500, the maximum collector area is restricted to 233 ft$_g^2$, which—from the $A(F)$ relationship—implies an annual solar heating fraction of approximately 0.48. The optimum 0.50 system-size meets the physical sizing constraint in this case, but exceeds the available budget of \$3,500 by \$175.

The table in figure 10–6 is included as a worksheet for this section.

10.3 ANNUAL CASH-FLOW ANALYSIS

Simplified cash-flow analysis allows year-by-year accounting of expected dollar inflows and outflows. The inflows occur through fuel savings and interest deductions, while the outflows occur through annual mortgage payments, operation and maintenance costs, property taxes, and insurance premiums. Resale value appears in the final year (the PTth year) of the cash-flow analysis. The simplified cash-flow analysis disregards the time value of money, hence, discounting of costs and benefits to present-worth dollars is not done.

To make the cash-flow analysis correspond to the \$/MMBtu cost derivations of section 10.2, all inflows and outflows must be discounted to arrive at an equivalent NPV calculation after PT years. Both the simple and the discounted approaches are shown below. In the sample problem, the cash-flow analysis is conducted for the 245 ft$_g^2$, 50% design, with $TC = \$3,675$.

Step 1. Calculate net annual cash flow (NCF_t). NCF_t is defined as the difference between annual conventional system costs and annual passive-solar costs. In the cash-flow analysis, an end-of-year convention is used. When dollar cash flows are discounted, all dollars are discounted to $t = 0$, which is the beginning of the series that ends at $t = 1$. At $t = 0$, the only cash flow relates to the initial out-of-pocket expense for the passive and auxiliary systems' down payment (DP):

$$DP = DPR \times (TC(F) + BSC)$$

In the sample problem:

$$DP = 0.15 \times (3{,}675 + 500) = \$626.25$$

The difference between the conventional (0.15×500) and the passive/auxiliary system cost in the first year (NCF_o), is therefore defined as:

$$NCF_o = (0.15 \times 500) - 626.25 = -\$551.25.$$

In this cash-flow analysis, negative values for NCF indicate a net outflow of money (negative passive-solar system savings), while positive values for NCF indicate a net inflow of money (positive passive-solar system savings).

NCF_t in subsequent years ($t = 1, \ldots, PT$) is given by the following formula:

$$NCF_t = \{AMPB_t + [OMB \times BSC \\ \times (1 + \theta)^t] + [HLOAD \times P_o \\ \times (1 + e)^t]\} - [AMP_t \\ + AMPB_t + [(1 - F) \times HLOAD \\ \times P_o \times (1 + e)^t] + (TC(F) \\ \times \{OMI + [PTXR \times (1 - \\ FSLTX)]\} \times (1 + \theta)^t) + \\ [OMB \times BSC \times (1 + \theta)^t] \\ - (i \times RB_t \times FSLTX)]$$

Resale value RV of the solar home should be accounted for at the time of sale (at the end of the PTth year) with the following formula:

$$RV(PT) = \{(FWF(PT,\delta) \times V) - [(1 - DPR) \\ \times RBF(T,i,PT)]\} \times TC$$

where:
$$AMPB_t = CRFN(T,r) \times (1 - DPR) \times BSC$$
$$AMP_t = CRFN(T,r) \times (1 - DPR) \times TC(F)$$
$$RB_t = RBF(T,i,t) \times (1 - DPR) \times TC(F)$$
$$t = 1, 2, \ldots, PT\text{—an index of years.}$$

The original NCF_t equation above can be simplified by cancellation of the common terms. Therefore:

$$NCF_t = [HLOAD \times P_o \times (1 + e)^t] - [AMP_t \\ + [(1 - F) \times HLOAD \times P_o \times (1 + e)^t] \\ + (TC(F) \times \{OMI + [PTXR \\ \times (1 - FSLTX)]\} \times (1 + \theta)^t) - (i \times RB_t \\ \times FSLTX)]$$

AMP_t is the constant annual mortgage payment and RB_t

F	.10	.20	.30	.40	.50	.60	.70	.80	.90
LCR									
A									
TC									
HLOAD * F									
AC									
AAC									
AAC*F									
P*(1-F)									
ABSC									
DCH									
NPV									
SPBK1									

10-6: Worksheet for generalized economic analysis.

is the remaining dollar balance on the mortgage or loan principal. The equation immediately above shows that the net annual cash flow is the difference between the yearly cost of the conventional system (first term in brackets) and the yearly cost of the passive solar system (second—and extremely long-term in brackets). Extending the sample case, the following first year ($t = 1$) values appear:

$HLOAD \times P_o \times (1 + e)^t = 37.44 \times 14.65 \times (1.08)^1$
$= \$592$
$AMP_1 = 0.10608 \times (1 - 0.15) \times 3{,}675 = \$331$
$RB_1 = 0.99 \times (1 - 0.15) \times 3{,}675 = \$3{,}092$
$i \times RB_1 \times FSLTX = 0.10 \times 3{,}092 \times 0.25 = \$78$
$(1 - F) \times HLOAD \times P_o \times (1 + e)^t = (1 - 0.50)$
$\times 37.44 \times 14.65 \times (1.08)^1 = \$297$
$TC(F) \times \{OMI + [PTXR \times (1 - FSLTX)]\}$
$\times (1 + \theta)^t = 3{,}675 \times [0.01 + (0.02 \times 0.75)]$
$\times (1.06)^1 = \$97$

Therefore, the first year net cash flow is:

$NCF_1 = 592 - 331 + 296 + 97 - 78 = \$-55$

When put in tabular form, as shown in figures 10–7 and 10–8, the annual cash-flow calculation becomes a simple matter of filling in the numbers. Clean copies of these forms are supplied in figures 10–9 and 10–10. Energy costs and savings should be escalated at the rate $(1 + e)$ each year; operation and maintenance costs, insurance premiums, and property taxes should be escalated at $(1 + \theta)$ each year; the annual mortgage payment remains constant; and interest deductions will vary according to the yearly value of RBF_t.

The down payment is assumed to be made in cash at the beginning of the first period ($t = 0$) and is not discounted. Net annual cash flow (NCF_t) in figure 10–8 is shown for years $t = 0, \ldots, PT$. Resale value only enters the calculations in the last year. In this example:

$RV = [(1.79 \times 0.75) - (0.85 \times 0.90)] \times 3{,}675 = \$2{,}122$

When each yearly entry is multiplied by a present worth factor ($PWF(t,r)$), NCF_t is converted to a discounted net cash flow ($DNCF_t$).

Step 2. Interpret the cash flow analysis. Column 12 of figure 10–8 gives the sum of the discounted net annual cash flows, which—when accumulated over PT years—should coincide with the net present value of savings (NPV) calculated in Step 6 of section 10.2. Therefore:

$$\sum_{t=1}^{PT} DNCF_t = NPV$$

For $F = 50\%$, $NPV = \$741$, which closely conforms to the cash-flow result of $\$746$. Differences due to rounding error are to be expected.

Without consideration of the down payment, a positive cash flow is generated by the fourth year. This is reflected by the positive cash flow shown in the column 9 entry of figure 10–8 at $t = 4$. If all the entries in column 9 are summed, one can see that the cost of the initial down payment is not recovered until the year of resale.

Year	(1) Conventional Energy Cost	(2) Down Payment	(3) Annual Mortgage Payment	(4) Passive Backup Energy Cost	(5) Operation Maintenance Insurance Taxes	(6) Mortgage Interest Tax Deductions	(7) Resale Value
t	LOAD * Po * (1+e)^t	DPR * TC	CRFN(T,r) * (1-DPR) * TC	(1-F) * HLOAD * Po * (1+e)^t	[PTXR*(1-FSLTX) + OMI] * TC * (1+θ)^t	(1-DPR) * TC* * RBF(T,i,t) * i * FSLTX	[FWF(PT,δ) * V -(1-DPR) * RBF(T,i,PT)] * TC
0	0	551	0	0	0	0	0
1	592	0	331	297	97	78	0
2	640	0	331	320	103	77	0
3	691	0	331	345	109	76	0
4	746	0	331	373	116	76	0
5	806	0	331	403	123	75	0
6	870	0	331	435	130	74	0
7	940	0	331	470	138	74	0
8	1015	0	331	508	146	73	0
9	1096	0	331	548	155	72	0
10	1184	0	331	592	164	70	2122

10–7: Filled-in table for cash-flow analysis—annual calculations.

Year	(8) Annual Solar Cost $(2)+(3)+(4)+(5)-(6)-(7)$	(9) Net Annual Cash Flow $NCF_t = (1)-(8)$	(10) Present Worth Factor $PWF(t,r)$	(11) Discounted Net Annual Cash Flow $DNCF_t = NCF_t * PWF(t,r)$	(12) Cumulative Discounted Net Annual Cash Flow $DNCF_t = \Sigma NCF_t * PWF(t,r)$
0	551	-551	1.00	-551	-551
1	647	-55	.93	-51	-602
2	677	-37	.87	-33	-635
3	710	-19	.82	-15	-650
4	744	2	.76	1	-649
5	782	24	.71	17	-632
6	822	48	.67	33	-599
7	866	74	.62	46	-553
8	913	103	.58	60	-493
9	963	133	.54	72	-421
10	-1105	2289	.51	1167	746

10-8: Filled-in table for cash-flow analysis—annual summations.

CASH FLOW ANALYSIS ~ ANNUAL CALCULATIONS

Year	(1) Conventional Energy Cost	(2) Down Payment	(3) Annual Mortgage Payment	(4) Passive Backup Energy Cost	(5) Operation Maintenance Insurance Taxes	(6) Mortgage Interest Tax Deductions	(7) Resale Value
t	$LOAD * P_o * (1+e)^t$	$DPR * TC$	$CRFN(T,r) * (1-DPR) * TC$	$(1-F) * HLOAD * P_o * (1+e)^t$	$[PTXR*(1-FSLTX) + OMI] * TC * (1+\theta)^t$	$(1-DPR) * TC * RBF(T,i,t) * 1 * FSLTX$	$[FWF(PT,\delta) * V - (1-DPR) * RBF(T,i,PT)] * TC$
0							
1							
2							
3							
4							
5							
6							
7							
8							
9							
10							

10-9: Blank table for cash-flow analysis—annual calculations.

Year	(8) Annual Solar Cost	(9) Net Annual Cash Flow	(10) Present Worth Factor	(11) Discounted Net Annual Cash Flow	(12) Cumulative Discounted Net Annual Cash Flow
t	(2)+(3)+(4)+(5)−(6)−(7)	$NCF_t = (1)-(8)$	$PWF(t,r)$	$DNCF_t = NCF_t * PWF(t,r)$	$DNCF_t = \sum_t NCF_t * PWF(t,r)$
0					
1					
2					
3					
4					
5					
6					
7					
8					
9					
10					

10-10: Blank table for cash-flow analysis—annual summations.

Techniques of Economic Analysis / 237

10.4 PAYBACK ANALYSIS

Another commonly used economic indicator is the payback period: the time required for the passive-solar system to yield savings sufficient to recoup its initial cost. A proper definition of payback will account for all costs and benefits associated with the passive-solar investment, as well as for the time value of accrued savings. However, because of their ease of computation, other approximate definitions are more frequently used. Several definitions, with a sample calculation of each, are given in the sections that follow. It should be noted that these definitions ignore the impact of resale value upon the payback period. As a result, the payback period may exceed the ownership period (PT).

10.4.a SIMPLE PAYBACK, DEFINITION #1 ($SPBK_1$)

The most basic definition of simple payback is calculated by dividing the total initial cost (TC) by the value of fuel savings in the first year ($P_o \times HLOAD \times F$). Using the system described in section 10.3 for cash flow analysis:

$$SPBK_1 = \left(\frac{TC}{P_o \times HLOAD \times F}\right)$$

$$= \left(\frac{3,675}{14.65 \times 37.44 \times 0.50}\right)$$

$$= 13.4 \text{ years}$$

This simple definition does have certain drawbacks. $SPBK_1$ ignores other system costs, including operation and maintenance costs, insurance premiums, property taxes, and loan interest costs. Fuel and other expense escalations over time are also ignored, and savings are not discounted to account for the opportunity cost of expended money.

10.4.b SIMPLE PAYBACK, DEFINITION #2 ($SPBK_2$)

A second definition of simple payback includes the additional costs due to operation and maintenance costs, insurance premiums, and taxes and accounts for the stream of fuel savings and system costs over time. $SPBK_2$ is defined as the year N for which:

$$\sum_{t=1}^{N}\left[[P_o \times (1 + e)^t \times HLOAD \times F] - (TC \right.$$
$$\left. \times \{OMI + [PTXR \times (1 - FSLTX)]\} \right.$$
$$\left. \times (1 + \theta)^t)\right] \geq TC$$

From the cash-flow analysis, this formula implies that $DPBK_1$ is defined by N such that:

$$\sum_{t=1}^{N} [(1) - (4) - (5)]_t \geq TC$$

will define $SPBK2$. The bracketed numbers (1), (4), and (5) refer to column entries in figure 10–7.

t	$[(1) - (4) - (5)]_t$	$\sum_{t=1}^{N} [(1) - (4) - (5)]_t$
1	$200	$200
2	$217	$417
3	$236	$653
4	$257	$910
5	$280	$1,190
6	$305	$1,495
7	$332	$1,827
8	$362	$2,189
9	$393	$2,582
10	$428	$3,010
11	$466	$3,476
12	$506	$3,982

10–11: Cumulative savings defined in terms of $SPBK_2$—the second definition of simple payback.

In figure 10–11, $SPBK_2$ is shown to be approximately 11.5 years, at which point cumulative savings exceed the initial cost of $3,675. This definition, although more complex, also has drawbacks. $SPBK_2$ does not account for loan interest costs and interest tax deductions; in addition, savings are not discounted to account for the opportunity cost of expended money.

10.4.c DISCOUNTED PAYBACK ($DPBK$)

The definition of discounted payback ($DPBK$) accounts for the time value of money, implying that a savings of X dollars in year one is *not* equivalent to a savings of X dollars in a later year. $DPBK$ is defined as the year N for which:

$$\sum_{t=1}^{N}\left[\left(\frac{1}{(1 + r)}\right)^t\right] [[P_o \times (1 + e)^t \times HLOAD \times F]$$
$$- [TC \times \{OMI + [PTXR \times (1 - FSLTX)]\}$$
$$\times (1 + \theta)^t)] \geq TC$$

From the cash-flow analysis, this formula implies that $DPBK$ is defined by N such that:

$$\left[\left(\sum_{t=1}^{N} \frac{1}{(1 + r)}\right)^t\right] [(1) - (4) - (5)]_t \geq TC$$

If calculated, *DPBK* is shown to be approximately 19 years.

Another method for calculating *DPBK,* which can be done independently of the cash flow method shown above, is as follows. *DPBK* can be defined as the year N in which:

$$\left(1 + \frac{\{OMI + [PTXR \times (1 - FSLTX)]\} \times DSPWF(N,\theta,r)}{DSPWF(N,e,r)}\right)$$

$$\leq \left(\frac{P_o \times HLOAD \times F}{TC}\right)$$

Filling in the values from the example case:

$$\left(1 + \frac{\{[0.01 + (0.02 \times 0.75)] \times DSPWF(N,0.06,0.07)\}}{DSPWF(N,0.08,0.07)}\right)$$

$$\leq \left(\frac{14.65 \times 37.44 \times 0.50}{3,675}\right)$$

or:

$$\left(\frac{1 + (0.025 \times DSPWF(N,0.06,0.07))}{DSPWF(N,0.08,0.07)}\right) \leq .075$$

DSPWF values can be found in the Differential Series Present-Worth Factor Tables in Appendix 11 for the appropriate values of θ, e, and r.

In figure 10-12, *DPBK* is shown to be a little more than 17 years.

DPBK corrects for the shortcoming of simple payback definitions, with the exception that it ignores the interest costs and resultant tax deductions associated with a mortgage or loan.

t	DSPWF (t,0.06,0.07)	DSPWF (t,0.08,0.07)	$\frac{(1 + 0.025) \times DSPWF(N,0.06,0.07)}{DSPWF(N,0.08,0.07)}$
1	0.99	1.01	1.015
2	1.97	2.03	0.517
3	2.94	3.06	0.351
4	3.91	4.09	0.268
5	4.86	5.14	0.218
6	5.81	6.20	0.184
7	6.74	7.27	0.161
8	7.67	8.34	0.143
9	8.59	9.43	0.129
10	9.50	10.53	0.118
11	10.40	11.64	0.108
12	11.30	12.75	0.101
13	12.18	13.88	0.094
14	13.06	15.02	0.088
15	13.93	16.17	0.083
16	14.79	17.33	0.079
17	15.64	18.50	0.0751
18	16.48	19.69	0.072
19	17.32	20.88	0.068
20	18.15	22.08	0.066

10-12: Values of present worth factor (*DSPWF*—at different interest rates) and of discounted payback (*DPBK*) for different years (*t*).

Appendix 11 Interest Tables

The following tables provide numerical values for future worth factors (*FWF*), present worth factors (*PWF*), series present worth factors (*SPWF*), nominal capital recovery factors (*CRFN*), differential series present worth factors (*DSPWF*), and remaining loan balance factors (*RBF*).

Definitions are provided below for each of these factors.

Compound Interest Tables:

$$FWF(t,r) = (1 + r)^t$$

$$PWF(t,r) = \left(\frac{1}{(1 + r)^t}\right) = (1 + r)^{-t}$$

$$SPWF(T,r) = \sum_{t=1}^{T}\left[\left(\frac{1}{1 + r}\right)^t\right] = \left(\frac{1 - \left(\frac{1}{1 + r}\right)^T}{r}\right) = \left(\frac{1}{CRFN(T,r)}\right)$$

$$CRFN(T,r) = \left(\frac{1}{\sum_{t=1}^{T}\left[\left(\frac{1}{1+r}\right)^T\right]}\right) = \left(\frac{r}{1 - \left(\frac{1}{1+r}\right)^T}\right) = \left(\frac{1}{SPWF(T,r)}\right)$$

Differential Series Present-Worth Factor Tables:

$$DSPWF(T,e,r) = \sum_{t=1}^{T}\left[\left(\frac{1 + e}{1 + r}\right)^t\right] = \left\{\frac{r - e}{\left[(1 + e) \times 1 - \left(\frac{1 + e}{1 + r}\right)^T\right]}\right\}^{-1}$$

Note: if $e = r$, then $DSPWF(T,e,r) = T$.

Remaining Loan Balance Factor Tables:

$$RBF(T,i,t) = \left(\frac{CRFN(T,i)}{i}\right) \times [1 - (1 + i)^t] + (1 + i)^t$$

Note: When $t = 0$, $RBF(T,i,0) = 1$; when $t = T$, $RBF(T,i,T) = 0$.

11.1 COMPOUND INTEREST TABLES

The various factors presented in these tables are all functions of the variables Year and Interest Rate (*I* or *R*).

COMPOUND INTEREST TABLES

INTEREST RATE (I OR R) = .04

YEAR	FWF(YEAR,RATE) FUTURE WORTH FACTOR	PWF(YEAR,RATE) PRESENT WORTH FACTOR	SPWF(YEAR,RATE) SERIES PRESENT WORTH FACTOR	CRFN(YEAR,RATE) NOMINAL CAPITAL RECOVERY FACTOR
1	1.04000	0.96154	0.96154	1.04000
2	1.08160	0.92456	1.88609	0.53020
3	1.12486	0.88900	2.77509	0.36035
4	1.16986	0.85480	3.62989	0.27549
5	1.21665	0.82193	4.45182	0.22463
6	1.26532	0.79031	5.24214	0.19076
7	1.31593	0.75992	6.00205	0.16661
8	1.36856	0.73069	6.73274	0.14853
9	1.42330	0.70259	7.43533	0.13449
10	1.48024	0.67556	8.11089	0.12329
11	1.53944	0.64958	8.76047	0.11415
12	1.60102	0.62460	9.38507	0.1065_
13	1.66506	0.60057	9.98564	0.10014
14	1.73166	0.57748	10.56312	0.09467
15	1.80093	0.55526	11.11838	0.08994
16	1.87296	0.53391	11.65229	0.08582
17	1.94788	0.51337	12.16566	0.08220
18	2.02579	0.49363	12.65929	0.07899
19	2.10683	0.47464	13.13393	0.07614
20	2.19110	0.45639	13.59032	0.07358
21	2.27874	0.43883	14.02915	0.07128
22	2.36989	0.42196	14.45111	0.06920
23	2.46468	0.40573	14.85683	0.06731
24	2.56327	0.39012	15.24695	0.06559
25	2.66580	0.37512	15.62207	0.06401
26	2.77243	0.36069	15.98276	0.06257
27	2.88332	0.34682	16.32957	0.06124
28	2.99865	0.33348	16.66304	0.06001
29	3.11860	0.32065	16.98369	0.05888
30	3.24334	0.30832	17.29199	0.05783

COMPOUND INTEREST TABLES

INTEREST RATE (I OR R) = .05

YEAR	FWF(YEAR,RATE) FUTURE WORTH FACTOR	PWF(YEAR,RATE) PRESENT WORTH FACTOR	SPWF(YEAR,RATE) SERIES PRESENT WORTH FACTOR	CRFN(YEAR,RATE) NOMINAL CAPITAL RECOVERY FACTOR
1	1.05000	0.95238	0.95238	1.05001
2	1.10250	0.90703	1.85941	0.53781
3	1.15762	0.86384	2.72325	0.36722
4	1.21550	0.82270	3.54596	0.28202
5	1.27628	0.78353	4.32948	0.23098
6	1.34009	0.74622	5.07570	0.19702
7	1.40709	0.71068	5.78639	0.17282
8	1.47744	0.67684	6.46323	0.15472
9	1.55131	0.64461	7.10784	0.14069
10	1.62888	0.61392	7.72176	0.12951
11	1.71032	0.58468	8.30644	0.12039
12	1.79583	0.55684	8.86328	0.11283
13	1.88562	0.53033	9.39361	0.10646
14	1.97990	0.50507	9.89868	0.10103
15	2.07890	0.48102	10.37970	0.09634
16	2.18284	0.45812	10.83782	0.09227
17	2.29198	0.43630	11.27412	0.08870
18	2.40658	0.41553	11.68964	0.08555
19	2.52690	0.39574	12.08538	0.08275
20	2.65324	0.37689	12.46228	0.08024
21	2.78590	0.35895	12.82122	0.07800
22	2.92520	0.34185	13.16308	0.07597
23	3.07145	0.32558	13.48865	0.07414
24	3.22502	0.31007	13.79872	0.07247
25	3.38627	0.29531	14.09403	0.07095
26	3.55558	0.28125	14.37528	0.06956
27	3.73335	0.26785	14.64313	0.06829
28	3.92002	0.25510	14.89823	0.06712
29	4.11602	0.24295	15.14118	0.06605
30	4.32181	0.23138	15.37256	0.06505

Interest Tables / 241

COMPOUND INTEREST TABLES

INTEREST RATE (I OR R) = .06

YEAR	FWF(YEAR,RATE) FUTURE WORTH FACTOR	PWF(YEAR,RATE) PRESENT WORTH FACTOR	SPWF(YEAR,RATE) SERIES PRESENT WORTH FACTOR	CRFN(YEAR,RATE) NOMINAL CAPITAL RECOVERY FACTOR
1	1.06000	0.94340	0.94340	1.06001
2	1.12360	0.89000	1.83339	0.54544
3	1.19101	0.83962	2.67301	0.37411
4	1.26247	0.79210	3.46511	0.28860
5	1.33822	0.74726	4.21237	0.23740
6	1.41851	0.70496	4.91733	0.20337
7	1.50362	0.66506	5.58239	0.17914
8	1.59384	0.62741	6.20980	0.16104
9	1.68946	0.59190	6.80170	0.14702
10	1.79083	0.55840	7.36010	0.13587
11	1.89828	0.52679	7.88689	0.12679
12	2.01217	0.49697	8.38386	0.11928
13	2.13290	0.46884	8.85270	0.11296
14	2.26087	0.44230	9.29501	0.10759
15	2.39652	0.41727	9.71228	0.10296
16	2.54031	0.39365	10.10592	0.09895
17	2.69273	0.37137	10.47729	0.09545
18	2.85429	0.35035	10.82764	0.09236
19	3.02555	0.33052	11.15815	0.08962
20	3.20707	0.31181	11.46996	0.08719
21	3.39950	0.29416	11.76412	0.08501
22	3.60346	0.27751	12.04163	0.08305
23	3.81967	0.26180	12.30343	0.08128
24	4.04884	0.24698	12.55041	0.07968
25	4.29177	0.23300	12.78341	0.07823
26	4.54927	0.21981	13.00322	0.07690
27	4.82222	0.20737	13.21059	0.07570
28	5.11155	0.19563	13.40622	0.07459
29	5.41824	0.18456	13.59078	0.07358
30	5.74332	0.17411	13.76489	0.07265

INTEREST RATE (I OR R) = .07

YEAR	FWF(YEAR,RATE) FUTURE WORTH FACTOR	PWF(YEAR,RATE) PRESENT WORTH FACTOR	SPWF(YEAR,RATE) SERIES PRESENT WORTH FACTOR	CRFN(YEAR,RATE) NOMINAL CAPITAL RECOVERY FACTOR
1	1.07000	0.93458	0.93458	1.07000
2	1.14490	0.87344	1.80802	0.55310
3	1.22504	0.81630	2.62432	0.38106
4	1.31079	0.76290	3.38721	0.29523
5	1.40255	0.71299	4.10020	0.24389
6	1.50072	0.66634	4.76654	0.20980
7	1.60577	0.62275	5.38929	0.18555
8	1.71817	0.58201	5.97130	0.16747
9	1.83845	0.54393	6.51523	0.15349
10	1.96714	0.50835	7.02359	0.14238
11	2.10483	0.47509	7.49868	0.13336
12	2.25217	0.44401	7.94269	0.12590
13	2.40982	0.41497	8.35766	0.11965
14	2.57850	0.38782	8.74547	0.11435
15	2.75900	0.36245	9.10792	0.10980
16	2.95212	0.33874	9.44666	0.10586
17	3.15877	0.31658	9.76323	0.10243
18	3.37988	0.29586	10.05910	0.09941
19	3.61647	0.27651	10.33561	0.09675
20	3.86962	0.25842	10.59403	0.09439
21	4.14049	0.24151	10.83554	0.09229
22	4.43032	0.22571	11.06125	0.09041
23	4.74044	0.21095	11.27220	0.08871
24	5.07227	0.19715	11.46935	0.08719
25	5.42732	0.18425	11.65360	0.08581
26	5.80723	0.17220	11.82579	0.08456
27	6.21373	0.16093	11.98672	0.08343
28	6.64868	0.15040	12.13713	0.08239
29	7.11409	0.14056	12.27769	0.08145
30	7.61206	0.13137	12.40906	0.08059

COMPOUND INTEREST TABLES

INTEREST RATE (I OR R) = .08

YEAR	FWF(YEAR,RATE) FUTURE WORTH FACTOR	PWF(YEAR,RATE) PRESENT WORTH FACTOR	SPWF(YEAR,RATE) SERIES PRESENT WORTH FACTOR	CRFN(YEAR,RATE) NOMINAL CAPITAL RECOVERY FACTOR
1	1.08000	0.92593	0.92593	1.08000
2	1.16640	0.85734	1.78326	0.56077
3	1.25971	0.79383	2.57710	0.38804
4	1.36049	0.73503	3.31213	0.30192
5	1.46932	0.68058	3.99271	0.25046
6	1.58687	0.63017	4.62288	0.21632
7	1.71382	0.58349	5.20637	0.19207
8	1.85092	0.54027	5.74664	0.17402
9	1.99899	0.50025	6.24689	0.16008
10	2.15891	0.46319	6.71008	0.14903
11	2.33162	0.42888	7.13896	0.14008
12	2.51815	0.39711	7.53607	0.13270
13	2.71960	0.36770	7.90377	0.12652
14	2.93717	0.34046	8.24423	0.12130
15	3.17214	0.31524	8.55947	0.11683
16	3.42591	0.29189	8.85136	0.11298
17	3.69998	0.27027	9.12163	0.10963
18	3.99598	0.25025	9.37188	0.10670
19	4.31565	0.23171	9.60359	0.10413
20	4.66090	0.21455	9.81814	0.10185
21	5.03377	0.19866	10.01680	0.09983
22	5.43647	0.18394	10.20074	0.09803
23	5.87138	0.17032	10.37105	0.09642
24	6.34109	0.15770	10.52875	0.09498
25	6.84838	0.14602	10.67477	0.09368
26	7.39624	0.13520	10.80997	0.09251
27	7.98794	0.12519	10.93515	0.09145
28	8.62696	0.11591	11.05107	0.09049
29	9.31712	0.10733	11.15839	0.08962
30	10.06248	0.09938	11.25777	0.08883

COMPOUND INTEREST TABLES

INTEREST RATE (I OR R) = .09

YEAR	FWF(YEAR,RATE) FUTURE WORTH FACTOR	PWF(YEAR,RATE) PRESENT WORTH FACTOR	SPWF(YEAR,RATE) SERIES PRESENT WORTH FACTOR	CRFN(YEAR,RATE) NOMINAL CAPITAL RECOVERY FACTOR
1	1.09000	0.91743	0.91743	1.09001
2	1.18810	0.84168	1.75911	0.56847
3	1.29502	0.77219	2.53130	0.39506
4	1.41158	0.70843	3.23972	0.30867
5	1.53862	0.64993	3.88966	0.25709
6	1.67709	0.59627	4.48593	0.22292
7	1.82803	0.54704	5.03296	0.19869
8	1.99255	0.50187	5.53483	0.18068
9	2.17187	0.46043	5.99526	0.16680
10	2.36734	0.42241	6.41767	0.15582
11	2.58040	0.38754	6.80521	0.14695
12	2.81263	0.35554	7.16075	0.13965
13	3.06576	0.32618	7.48693	0.13357
14	3.34168	0.29925	7.78618	0.12843
15	3.64243	0.27454	8.06072	0.12406
16	3.97024	0.25187	8.31259	0.12030
17	4.32756	0.23108	8.54366	0.11705
18	4.71704	0.21200	8.75566	0.11421
19	5.14156	0.19449	8.95015	0.11173
20	5.60430	0.17843	9.12858	0.10955
21	6.10868	0.16370	9.29228	0.10762
22	6.65846	0.15018	9.44247	0.10591
23	7.25771	0.13778	9.58025	0.10438
24	7.91090	0.12641	9.70666	0.10302
25	8.62287	0.11597	9.82263	0.10181
26	9.39892	0.10639	9.92902	0.10072
27	10.24481	0.09761	10.02663	0.09974
28	11.16683	0.08955	10.11618	0.09885
29	12.17183	0.08216	10.19833	0.09806
30	13.26729	0.07537	10.27371	0.09734

Interest Tables / 243

COMPOUND INTEREST TABLES

INTEREST RATE (I OR R) = .10

YEAR	FWF(YEAR,RATE) FUTURE WORTH FACTOR	PWF(YEAR,RATE) PRESENT WORTH FACTOR	SPWF(YEAR,RATE) SERIES PRESENT WORTH FACTOR	CRFN(YEAR,RATE) NOMINAL CAPITAL RECOVERY FACTOR
1	1.10000	0.90909	0.90909	1.10000
2	1.21000	0.82645	1.73554	0.57619
3	1.33100	0.75132	2.48685	0.40212
4	1.46410	0.68301	3.16987	0.31547
5	1.61050	0.62092	3.79079	0.26380
6	1.77155	0.56448	4.35526	0.22961
7	1.94870	0.51316	4.86842	0.20541
8	2.14357	0.46651	5.33493	0.18745
9	2.35793	0.42410	5.75903	0.17364
10	2.59372	0.38555	6.14458	0.16275
11	2.85309	0.35050	6.49507	0.15396
12	3.13840	0.31863	6.81370	0.14676
13	3.45223	0.28967	7.10337	0.14078
14	3.79745	0.26333	7.36670	0.13575
15	4.17719	0.23939	7.60609	0.13147
16	4.59491	0.21763	7.82372	0.12782
17	5.05440	0.19785	8.02157	0.12466
18	5.55983	0.17986	8.20143	0.12193
19	6.11581	0.16351	8.36494	0.11955
20	6.72738	0.14864	8.51358	0.11746
21	7.40011	0.13513	8.64872	0.11562
22	8.14012	0.12285	8.77156	0.11401
23	8.95412	0.11168	8.88324	0.11257
24	9.84952	0.10153	8.98477	0.11130
25	10.83447	0.09230	9.07706	0.11017
26	11.91790	0.08391	9.16097	0.10916
27	13.10968	0.07628	9.23725	0.10826
28	14.42064	0.06934	9.30659	0.10745
29	15.86269	0.06304	9.36963	0.10673
30	17.44893	0.05731	9.42694	0.10608

COMPOUND INTEREST TABLES

INTEREST RATE (I OR R) = .11

YEAR	FWF(YEAR,RATE) FUTURE WORTH FACTOR	PWF(YEAR,RATE) PRESENT WORTH FACTOR	SPWF(YEAR,RATE) SERIES PRESENT WORTH FACTOR	CRFN(YEAR,RATE) NOMINAL CAPITAL RECOVERY FACTOR
1	1.11000	0.90090	0.90090	1.11000
2	1.23210	0.81162	1.71252	0.58394
3	1.36763	0.73119	2.44372	0.40922
4	1.51807	0.65873	3.10245	0.32233
5	1.68505	0.59345	3.69590	0.27057
6	1.87041	0.53464	4.23054	0.23638
7	2.07615	0.48166	4.71220	0.21222
8	2.30453	0.43393	5.14612	0.19432
9	2.55802	0.39093	5.53705	0.18060
10	2.83940	0.35219	5.88924	0.16980
11	3.15173	0.31728	6.20652	0.16112
12	3.49842	0.28584	6.49236	0.15403
13	3.88325	0.25752	6.74988	0.14815
14	4.31040	0.23200	6.98187	0.14323
15	4.78454	0.20901	7.19088	0.13907
16	5.31084	0.18829	7.37917	0.13552
17	5.89503	0.16963	7.54880	0.13247
18	6.54347	0.15282	7.70162	0.12984
19	7.26325	0.13768	7.83930	0.12756
20	8.06220	0.12403	7.96334	0.12558
21	8.94904	0.11174	8.07508	0.12384
22	9.93343	0.10067	8.17575	0.12231
23	11.02609	0.09069	8.26644	0.12097
24	12.23896	0.08171	8.34814	0.11979
25	13.58523	0.07361	8.42175	0.11874
26	15.07960	0.06631	8.48807	0.11781
27	16.73833	0.05974	8.54781	0.11699
28	18.57954	0.05382	8.60163	0.11626
29	20.62328	0.04849	8.65012	0.11561
30	22.89183	0.04368	8.69380	0.11502

COMPOUND INTEREST TABLES

INTEREST RATE (I OR R) = .12

YEAR	FWF(YEAR,RATE) FUTURE WORTH FACTOR	PWF(YEAR,RATE) PRESENT WORTH FACTOR	SPWF(YEAR,RATE) SERIES PRESENT WORTH FACTOR	CRFN(YEAR,RATE) NOMINAL CAPITAL RECOVERY FACTOR
1	1.12000	0.89286	0.89286	1.12000
2	1.25440	0.79719	1.69005	0.59170
3	1.40493	0.71178	2.40183	0.41635
4	1.57352	0.63552	3.03735	0.32924
5	1.76234	0.56743	3.60478	0.27741
6	1.97382	0.50663	4.11141	0.24323
7	2.21067	0.45235	4.56376	0.21912
8	2.47596	0.40388	4.96764	0.20130
9	2.77307	0.36061	5.32825	0.18768
10	3.10584	0.32197	5.65022	0.17698
11	3.47853	0.28748	5.93770	0.16842
12	3.89596	0.25668	6.19437	0.16144
13	4.36347	0.22917	6.42354	0.15568
14	4.88708	0.20462	6.62816	0.15087
15	5.47353	0.18270	6.81086	0.14682
16	6.13035	0.16312	6.97398	0.14339
17	6.86599	0.14564	7.11963	0.14046
18	7.68991	0.13004	7.24966	0.13794
19	8.61269	0.11611	7.36577	0.13576
20	9.64621	0.10367	7.46944	0.13388
21	10.80376	0.09256	7.56200	0.13224
22	12.10020	0.08264	7.64464	0.13081
23	13.55221	0.07379	7.71842	0.12956
24	15.17848	0.06588	7.78431	0.12846
25	16.99988	0.05882	7.84313	0.12750
26	19.03986	0.05252	7.89565	0.12665
27	21.32463	0.04689	7.94254	0.12590
28	23.88358	0.04187	7.98441	0.12524
29	26.74960	0.03738	8.02180	0.12466
30	29.95955	0.03338	8.05517	0.12414

INTEREST RATE (I OR R) = .13

YEAR	FWF(YEAR,RATE) FUTURE WORTH FACTOR	PWF(YEAR,RATE) PRESENT WORTH FACTOR	SPWF(YEAR,RATE) SERIES PRESENT WORTH FACTOR	CRFN(YEAR,RATE) NOMINAL CAPITAL RECOVERY FACTOR
1	1.13000	0.88496	0.88496	1.13001
2	1.27690	0.78315	1.66810	0.59949
3	1.44289	0.69305	2.36115	0.42353
4	1.63047	0.61332	2.97447	0.33620
5	1.84243	0.54276	3.51724	0.28432
6	2.08194	0.48032	3.99756	0.25015
7	2.35259	0.42506	4.42262	0.22611
8	2.65842	0.37616	4.79878	0.20839
9	3.00401	0.33289	5.13167	0.19487
10	3.39453	0.29459	5.42626	0.18429
11	3.83581	0.26070	5.68696	0.17584
12	4.33447	0.23071	5.91766	0.16899
13	4.89794	0.20417	6.12183	0.16335
14	5.53467	0.18068	6.30251	0.15867
15	6.25416	0.15989	6.46240	0.15474
16	7.06720	0.14150	6.60390	0.15143
17	7.98593	0.12522	6.72911	0.14861
18	9.02409	0.11081	6.83993	0.14620
19	10.19721	0.09807	6.93799	0.14413
20	11.52283	0.08678	7.02477	0.14235
21	13.02079	0.07680	7.10157	0.14081
22	14.71347	0.06796	7.16954	0.13948
23	16.62619	0.06015	7.22968	0.13832
24	18.78758	0.05323	7.28291	0.13731
25	21.22995	0.04710	7.33001	0.13643
26	23.98982	0.04168	7.37169	0.13565
27	27.10846	0.03689	7.40858	0.13498
28	30.63252	0.03264	7.44122	0.13439
29	34.61472	0.02889	7.47011	0.13387
30	39.11458	0.02557	7.49568	0.13341

Interest Tables / 245

COMPOUND INTEREST TABLES

INTEREST RATE (I OR R) = .14

YEAR	FWF(YEAR,RATE) FUTURE WORTH FACTOR	PWF(YEAR,RATE) PRESENT WORTH FACTOR	SPWF(YEAR,RATE) SERIES PRESENT WORTH FACTOR	CRFN(YEAR,RATE) NOMINAL CAPITAL RECOVERY FACTOR
1	1.14000	0.87719	0.87719	1.14000
2	1.29960	0.76947	1.64666	0.60729
3	1.48154	0.67497	2.32163	0.43073
4	1.68895	0.59208	2.91372	0.34321
5	1.92541	0.51937	3.43308	0.29128
6	2.19496	0.45559	3.88867	0.25716
7	2.50225	0.39964	4.28831	0.23319
8	2.85256	0.35056	4.63887	0.21557
9	3.25192	0.30751	4.94638	0.20217
10	3.70718	0.26975	5.21612	0.19171
11	4.22618	0.23662	5.45274	0.18339
12	4.81785	0.20756	5.66030	0.17667
13	5.49234	0.18207	5.84237	0.17116
14	6.26126	0.15971	6.00208	0.16661
15	7.13783	0.14010	6.14218	0.16281
16	8.13711	0.12289	6.26507	0.15962
17	9.27631	0.10780	6.37287	0.15692
18	10.57498	0.09456	6.46743	0.15462
19	12.05546	0.08295	6.55038	0.15266
20	13.74321	0.07276	6.62314	0.15099
21	15.66724	0.06383	6.68697	0.14955
22	17.86063	0.05599	6.74296	0.14830
23	20.36110	0.04911	6.79207	0.14723
24	23.21162	0.04308	6.83515	0.14630
25	26.46124	0.03779	6.87294	0.14550
26	30.16577	0.03315	6.90609	0.14480
27	34.38895	0.02908	6.93517	0.14419
28	39.20335	0.02551	6.96068	0.14366
29	44.69179	0.02238	6.98305	0.14320
30	50.94858	0.01963	7.00268	0.14280

COMPOUND INTEREST TABLES

INTEREST RATE (I OR R) = .15

YEAR	FWF(YEAR,RATE) FUTURE WORTH FACTOR	PWF(YEAR,RATE) PRESENT WORTH FACTOR	SPWF(YEAR,RATE) SERIES PRESENT WORTH FACTOR	CRFN(YEAR,RATE) NOMINAL CAPITAL RECOVERY FACTOR
1	1.15000	0.86957	0.86957	1.15000
2	1.32250	0.75614	1.62571	0.61512
3	1.52087	0.65752	2.28323	0.43798
4	1.74900	0.57175	2.85498	0.35027
5	2.01135	0.49718	3.35216	0.29832
6	2.31305	0.43233	3.78448	0.26424
7	2.66001	0.37594	4.16042	0.24036
8	3.05901	0.32690	4.48732	0.22285
9	3.51785	0.28426	4.77159	0.20957
10	4.04553	0.24719	5.01877	0.19925
11	4.65236	0.21494	5.23372	0.19107
12	5.35021	0.18691	5.42062	0.18448
13	6.15273	0.16253	5.58315	0.17911
14	7.07564	0.14133	5.72448	0.17469
15	8.13698	0.12289	5.84737	0.17102
16	9.35752	0.10687	5.95424	0.16795
17	10.76114	0.09293	6.04716	0.16537
18	12.37530	0.08081	6.12797	0.16319
19	14.23158	0.07027	6.19823	0.16134
20	16.36630	0.06110	6.25933	0.15976
21	18.82124	0.05313	6.31246	0.15842
22	21.64441	0.04620	6.35866	0.15727
23	24.89105	0.04017	6.39884	0.15628
24	28.62468	0.03493	6.43377	0.15543
25	32.91837	0.03038	6.46415	0.15470
26	37.85609	0.02642	6.49057	0.15407
27	43.53448	0.02297	6.51353	0.15353
28	50.06462	0.01997	6.53351	0.15306
29	57.57428	0.01737	6.55088	0.15265
30	66.21037	0.01510	6.56598	0.15230

11.2 DIFFERENTIAL SERIES PRESENT WORTH FACTOR (*DSPWF*) TABLES

DSPWF is a function of the variables Year, *R* (annual discount rate), and *E*. *E* represents a general annual escalation rate; values used for *E* include inflation (θ), fuel escalation (e), mortgage rate (i), property appreciation (δ), and operation and maintenance escalation (γ).

DIFFERENTIAL SERIES PRESENT WORTH FACTOR : DSPWF(YEAR,E,R)

NOMINAL ESCALATION RATE : E =.04

NOMINAL DISCOUNT RATE : R

YEAR	0.04	0.05	0.06	0.07	0.08	0.09	0.10	0.11	0.12	0.13	0.14	0.15
1	1.00	0.99	0.98	0.97	0.96	0.95	0.95	0.94	0.93	0.92	0.91	0.90
2	2.00	1.97	1.94	1.92	1.89	1.86	1.84	1.81	1.79	1.77	1.74	1.72
3	3.00	2.94	2.89	2.83	2.78	2.73	2.68	2.64	2.59	2.55	2.50	2.46
4	4.00	3.91	3.81	3.73	3.64	3.56	3.48	3.41	3.33	3.26	3.20	3.13
5	5.00	4.86	4.72	4.59	4.47	4.35	4.24	4.13	4.03	3.92	3.83	3.74
6	6.00	5.80	5.62	5.44	5.27	5.11	4.95	4.81	4.67	4.53	4.40	4.28
7	7.00	6.74	6.49	6.26	6.04	5.83	5.63	5.44	5.26	5.09	4.93	4.78
8	8.00	7.66	7.35	7.05	6.78	6.51	6.27	6.03	5.81	5.61	5.41	5.22
9	9.00	8.58	8.19	7.83	7.49	7.17	6.87	6.59	6.33	6.08	5.85	5.63
10	10.00	9.49	9.02	8.58	8.17	7.79	7.44	7.11	6.80	6.52	6.25	6.00
11	11.00	10.39	9.83	9.31	8.83	8.39	7.98	7.60	7.25	6.92	6.61	6.33
12	12.00	11.28	10.63	10.02	9.47	8.96	8.49	8.06	7.66	7.29	6.94	6.63
13	13.00	12.17	11.41	10.71	10.08	9.50	8.97	8.49	8.04	7.63	7.25	6.90
14	14.00	13.04	12.17	11.39	10.67	10.02	9.43	8.89	8.39	7.94	7.52	7.14
15	15.00	13.91	12.92	12.04	11.24	10.52	9.86	9.26	8.72	8.23	7.78	7.36
16	16.00	14.76	13.66	12.67	11.79	10.99	10.27	9.62	9.03	8.49	8.01	7.56
17	17.00	15.61	14.38	13.29	12.31	11.44	10.65	9.95	9.31	8.74	8.22	7.74
18	18.00	16.46	15.09	13.89	12.82	11.87	11.02	10.26	9.58	8.96	8.41	7.91
19	19.00	17.29	15.79	14.47	13.31	12.28	11.36	10.55	9.82	9.17	8.58	8.05
20	20.00	18.12	16.47	15.04	13.78	12.67	11.69	10.82	10.05	9.36	8.74	8.19
21	21.00	18.93	17.14	15.59	14.23	13.04	12.00	11.07	10.26	9.53	8.89	8.31
22	22.00	19.74	17.80	16.12	14.67	13.40	12.29	11.31	10.45	9.69	9.02	8.42
23	23.00	20.55	18.45	16.64	15.09	13.74	12.56	11.54	10.64	9.84	9.14	8.52
24	24.00	21.34	19.08	17.15	15.49	14.06	12.82	11.75	10.80	9.98	9.25	8.61
25	25.00	22.13	19.70	17.64	15.88	14.37	13.07	11.94	10.96	10.10	9.35	8.69
26	26.00	22.91	20.31	18.12	16.25	14.66	13.30	12.13	11.11	10.22	9.44	8.76
27	27.00	23.68	20.91	18.58	16.62	14.95	13.52	12.30	11.24	10.33	9.53	8.83
28	28.00	24.45	21.49	19.03	16.96	15.21	13.73	12.46	11.37	10.42	9.60	8.89
29	29.00	25.20	22.07	19.47	17.30	15.47	13.93	12.61	11.48	10.51	9.67	8.94
30	30.00	25.95	22.64	19.90	17.62	15.72	14.11	12.75	11.59	10.60	9.74	8.99

DIFFERENTIAL SERIES PRESENT WORTH FACTOR : DSPWF(YEAR,E,R)

NOMINAL ESCALATION RATE : E =.05

NOMINAL DISCOUNT RATE : R

YEAR	0.04	0.05	0.06	0.07	0.08	0.09	0.10	0.11	0.12	0.13	0.14	0.15
1	1.01	1.00	0.99	0.98	0.97	0.96	0.95	0.95	0.94	0.93	0.92	0.91
2	2.03	2.00	1.97	1.94	1.92	1.89	1.87	1.84	1.82	1.79	1.77	1.75
3	3.06	3.00	2.94	2.89	2.84	2.79	2.74	2.69	2.64	2.59	2.55	2.51
4	4.10	4.00	3.91	3.82	3.73	3.65	3.57	3.49	3.41	3.34	3.27	3.20
5	5.15	5.00	4.86	4.73	4.60	4.48	4.36	4.25	4.14	4.03	3.93	3.84
6	6.21	6.00	5.80	5.62	5.44	5.27	5.11	4.96	4.82	4.68	4.54	4.42
7	7.27	7.00	6.74	6.50	6.26	6.04	5.84	5.64	5.45	5.27	5.11	4.95
8	8.35	8.00	7.67	7.36	7.06	6.79	6.53	6.28	6.05	5.83	5.62	5.43
9	9.44	9.00	8.59	8.20	7.84	7.50	7.18	6.89	6.61	6.35	6.10	5.87
10	10.54	10.00	9.50	9.03	8.59	8.19	7.81	7.46	7.13	6.83	6.54	6.27
11	11.66	11.00	10.40	9.84	9.33	8.85	8.41	8.00	7.62	7.27	6.95	6.64
12	12.78	12.00	11.29	10.64	10.04	9.49	8.98	8.52	8.09	7.69	7.32	6.98
13	13.91	13.00	12.17	11.42	10.73	10.10	9.53	9.00	8.52	8.07	7.66	7.28
14	15.05	14.00	13.05	12.19	11.41	10.70	10.05	9.46	8.92	8.43	7.98	7.56
15	16.21	15.00	13.92	12.94	12.06	11.27	10.55	9.90	9.30	8.76	8.27	7.82
16	17.37	16.00	14.78	13.68	12.70	11.82	11.02	10.31	9.66	9.07	8.54	8.05
17	18.55	17.00	15.63	14.41	13.32	12.35	11.48	10.70	9.99	9.36	8.78	8.26
18	19.74	18.00	16.47	15.12	13.92	12.86	11.91	11.06	10.31	9.62	9.01	8.46
19	20.94	19.00	17.30	15.82	14.51	13.35	12.32	11.41	10.60	9.87	9.22	8.64
20	22.15	20.00	18.13	16.50	15.08	13.82	12.72	11.74	10.87	10.10	9.41	8.80
21	23.37	21.00	18.95	17.18	15.63	14.28	13.09	12.05	11.13	10.32	9.59	8.95
22	24.60	22.00	19.76	17.84	16.17	14.72	13.45	12.35	11.37	10.52	9.76	9.08
23	25.85	23.00	20.57	18.48	16.69	15.14	13.80	12.63	11.60	10.70	9.91	9.20
24	27.11	24.00	21.36	19.12	17.20	15.55	14.12	12.89	11.81	10.87	10.05	9.32
25	28.38	25.00	22.15	19.74	17.69	15.94	14.44	13.14	12.01	11.03	10.17	9.42
26	29.66	26.00	22.93	20.36	18.17	16.32	14.73	13.37	12.20	11.18	10.29	9.51
27	30.96	27.00	23.71	20.96	18.64	16.68	15.02	13.60	12.37	11.32	10.40	9.60
28	32.26	28.00	24.48	21.55	19.10	17.04	15.29	13.81	12.54	11.45	10.50	9.68
29	33.58	29.00	25.24	22.12	19.54	17.37	15.55	14.01	12.69	11.56	10.59	9.75
30	34.91	30.00	25.99	22.69	19.97	17.70	15.80	14.20	12.84	11.67	10.68	9.81

Interest Tables / 247

DIFFERENTIAL SERIES PRESENT WORTH FACTOR : DSPWF(YEAR,E,R)

NOMINAL ESCALATION RATE : E =.06

NOMINAL DISCOUNT RATE : R

YEAR	0.04	0.05	0.06	0.07	0.08	0.09	0.10	0.11	0.12	0.13	0.14	0.15
1	1.02	1.01	1.00	0.99	0.98	0.97	0.96	0.95	0.95	0.94	0.93	0.92
2	2.06	2.03	2.00	1.97	1.94	1.92	1.89	1.87	1.84	1.82	1.79	1.77
3	3.12	3.06	3.00	2.94	2.89	2.84	2.79	2.74	2.69	2.64	2.60	2.55
4	4.20	4.10	4.00	3.91	3.82	3.73	3.65	3.57	3.49	3.42	3.35	3.28
5	5.30	5.14	5.00	4.86	4.73	4.60	4.48	4.36	4.25	4.14	4.04	3.94
6	6.42	6.20	6.00	5.81	5.62	5.45	5.28	5.12	4.97	4.83	4.69	4.55
7	7.56	7.27	7.00	6.74	6.50	6.27	6.05	5.85	5.65	5.46	5.29	5.12
8	8.72	8.35	8.00	7.67	7.36	7.07	6.80	6.54	6.29	6.06	5.85	5.64
9	9.91	9.44	9.00	8.59	8.21	7.85	7.51	7.20	6.90	6.63	6.37	6.12
10	11.12	10.54	10.00	9.50	9.04	8.60	8.20	7.83	7.48	7.15	6.85	6.56
11	12.35	11.65	11.00	10.40	9.85	9.34	8.87	8.43	8.03	7.65	7.30	6.97
12	13.61	12.77	12.00	11.30	10.65	10.06	9.51	9.01	8.54	8.11	7.72	7.35
13	14.89	13.90	13.00	12.18	11.43	10.75	10.13	9.56	9.03	8.55	8.10	7.69
14	16.20	15.04	14.00	13.06	12.20	11.43	10.72	10.08	9.49	8.96	8.47	8.01
15	17.53	16.20	15.00	13.93	12.96	12.09	11.30	10.58	9.93	9.34	8.80	8.31
16	18.88	17.36	16.00	14.79	13.70	12.73	11.85	11.06	10.35	9.70	9.11	8.58
17	20.27	18.53	17.00	15.64	14.43	13.35	12.38	11.52	10.74	10.04	9.40	8.83
18	21.68	19.72	18.00	16.48	15.14	13.95	12.90	11.95	11.11	10.35	9.67	9.06
19	23.11	20.92	19.00	17.32	15.84	14.54	13.39	12.37	11.46	10.65	9.92	9.27
20	24.58	22.13	20.00	18.15	16.53	15.11	13.87	12.77	11.79	10.93	10.16	9.47
21	26.07	23.35	21.00	18.97	17.21	15.67	14.33	13.15	12.11	11.19	10.37	9.65
22	27.59	24.58	22.00	19.78	17.87	16.21	14.77	13.51	12.41	11.43	10.58	9.82
23	29.14	25.82	23.00	20.59	18.52	16.74	15.20	13.86	12.69	11.66	10.76	9.97
24	30.72	27.08	24.00	21.39	19.16	17.25	15.61	14.19	12.95	11.88	10.94	10.11
25	32.33	28.34	25.00	22.18	19.79	17.75	16.00	14.50	13.21	12.08	11.10	10.24
26	33.97	29.62	26.00	22.96	20.40	18.23	16.38	14.80	13.45	12.27	11.25	10.36
27	35.64	30.92	27.00	23.74	21.00	18.70	16.75	15.09	13.67	12.45	11.39	10.47
28	37.34	32.22	28.00	24.51	21.60	19.16	17.11	15.37	13.89	12.62	11.52	10.58
29	39.08	33.54	29.00	25.27	22.18	19.60	17.45	15.63	14.09	12.77	11.64	10.67
30	40.85	34.86	30.00	26.02	22.75	20.04	17.78	15.88	14.28	12.92	11.76	10.76

DIFFERENTIAL SERIES PRESENT WORTH FACTOR : DSPWF(YEAR,E,R)

NOMINAL ESCALATION RATE : E =.07

NOMINAL DISCOUNT RATE : R

YEAR	0.04	0.05	0.06	0.07	0.08	0.09	0.10	0.11	0.12	0.13	0.14	0.15
1	1.03	1.02	1.01	1.00	0.99	0.98	0.97	0.96	0.96	0.95	0.94	0.93
2	2.09	2.06	2.03	2.00	1.97	1.95	1.92	1.89	1.87	1.84	1.82	1.80
3	3.18	3.12	3.06	3.00	2.94	2.89	2.84	2.79	2.74	2.69	2.65	2.60
4	4.30	4.19	4.10	4.00	3.91	3.82	3.73	3.65	3.57	3.50	3.42	3.35
5	5.45	5.29	5.14	5.00	4.86	4.73	4.61	4.48	4.37	4.26	4.15	4.05
6	6.64	6.41	6.20	6.00	5.81	5.63	5.45	5.29	5.13	4.98	4.83	4.70
7	7.86	7.55	7.27	7.00	6.75	6.50	6.28	6.06	5.86	5.66	5.48	5.30
8	9.11	8.72	8.35	8.00	7.67	7.37	7.08	6.81	6.55	6.31	6.08	5.86
9	10.40	9.90	9.44	9.00	8.59	8.21	7.86	7.52	7.21	6.92	6.64	6.39
10	11.73	11.11	10.53	10.00	9.50	9.04	8.62	8.22	7.85	7.50	7.17	6.87
11	13.10	12.34	11.64	11.00	10.41	9.86	9.35	8.89	8.45	8.05	7.67	7.32
12	14.51	13.59	12.76	12.00	11.30	10.66	10.07	9.53	9.03	8.57	8.14	7.74
13	15.95	14.87	13.89	13.00	12.19	11.45	10.77	10.15	9.58	9.06	8.58	8.14
14	17.44	16.17	15.03	14.00	13.07	12.22	11.45	10.75	10.11	9.53	8.99	8.50
15	18.97	17.50	16.18	15.00	13.94	12.98	12.11	11.32	10.61	9.97	9.38	8.84
16	20.55	18.85	17.35	16.00	14.80	13.72	12.75	11.88	11.09	10.38	9.74	9.16
17	22.17	20.23	18.52	17.00	15.65	14.45	13.38	12.42	11.55	10.78	10.08	9.45
18	23.84	21.64	19.70	18.00	16.50	15.17	13.98	12.93	11.99	11.15	10.40	9.72
19	25.56	23.07	20.90	19.00	17.33	15.87	14.58	13.43	12.41	11.51	10.70	9.98
20	27.32	24.53	22.10	20.00	18.16	16.56	15.15	13.91	12.82	11.84	10.98	10.21
21	29.14	26.01	23.32	21.00	18.99	17.24	15.71	14.37	13.20	12.16	11.25	10.43
22	31.01	27.53	24.55	22.00	19.80	17.90	16.25	14.82	13.56	12.46	11.49	10.64
23	32.93	29.07	25.79	23.00	20.61	18.56	16.78	15.25	13.91	12.75	11.73	10.83
24	34.91	30.64	27.05	24.00	21.41	19.20	17.30	15.66	14.25	13.02	11.95	11.00
25	36.95	32.25	28.31	25.00	22.20	19.83	17.80	16.06	14.57	13.27	12.15	11.17
26	39.04	33.88	29.59	26.00	22.99	20.44	18.29	16.45	14.87	13.52	12.34	11.32
27	41.20	35.54	30.88	27.00	23.77	21.05	18.76	16.82	15.16	13.75	12.52	11.47
28	43.41	37.24	32.18	28.00	24.54	21.65	19.22	17.18	15.44	13.96	12.69	11.60
29	45.70	38.97	33.49	29.00	25.30	22.23	19.67	17.52	15.71	14.17	12.85	11.72
30	48.04	40.73	34.81	30.00	26.06	22.80	20.11	17.85	15.96	14.36	13.00	11.84

DIFFERENTIAL SERIES PRESENT WORTH FACTOR : DSPWF(YEAR,E,R)

NOMINAL ESCALATION RATE : E =.08

NOMINAL DISCOUNT RATE : R

YEAR	0.04	0.05	0.06	0.07	0.08	0.09	0.10	0.11	0.12	0.13	0.14	0.15
1	1.04	1.03	1.02	1.01	1.00	0.99	0.98	0.97	0.96	0.96	0.95	0.94
2	2.12	2.09	2.06	2.03	2.00	1.97	1.95	1.92	1.89	1.87	1.84	1.82
3	3.24	3.17	3.11	3.06	3.00	2.95	2.89	2.84	2.79	2.74	2.70	2.65
4	4.40	4.29	4.19	4.09	4.00	3.91	3.82	3.74	3.66	3.58	3.50	3.43
5	5.61	5.45	5.29	5.14	5.00	4.86	4.73	4.61	4.49	4.37	4.26	4.16
6	6.86	6.63	6.41	6.20	6.00	5.81	5.63	5.46	5.29	5.14	4.99	4.84
7	8.16	7.85	7.55	7.27	7.00	6.75	6.51	6.28	6.07	5.86	5.67	5.49
8	9.52	9.10	8.71	8.34	8.00	7.68	7.37	7.09	6.82	6.56	6.32	6.09
9	10.92	10.39	9.89	9.43	9.00	8.60	8.22	7.87	7.54	7.23	6.94	6.66
10	12.38	11.71	11.10	10.53	10.00	9.51	9.05	8.63	8.23	7.86	7.52	7.20
11	13.89	13.08	12.33	11.64	11.00	10.41	9.87	9.37	8.90	8.47	8.07	7.70
12	15.47	14.48	13.58	12.75	12.00	11.31	10.67	10.09	9.55	9.05	8.59	8.17
13	17.10	15.92	14.85	13.88	13.00	12.20	11.46	10.79	10.17	9.61	9.09	8.61
14	18.80	17.41	16.15	15.02	14.00	13.07	12.23	11.47	10.77	10.14	9.56	9.02
15	20.56	18.93	17.48	16.17	15.00	13.94	12.99	12.13	11.35	10.64	10.00	9.41
16	22.39	20.50	18.82	17.33	16.00	14.81	13.74	12.78	11.91	11.13	10.42	9.78
17	24.29	22.12	20.20	18.50	17.00	15.66	14.47	13.40	12.45	11.59	10.82	10.12
18	26.26	23.78	21.60	19.69	18.00	16.51	15.19	14.02	12.97	12.04	11.20	10.45
19	28.31	25.48	23.03	20.88	19.00	17.35	15.89	14.61	13.47	12.46	11.56	10.75
20	30.43	27.24	24.48	22.08	20.00	18.18	16.59	15.19	13.95	12.86	11.90	11.03
21	32.64	29.05	25.96	23.30	21.00	19.00	17.27	15.75	14.42	13.25	12.22	11.30
22	34.94	30.91	27.47	24.53	22.00	19.82	17.94	16.30	14.87	13.62	12.52	11.55
23	37.32	32.82	29.00	25.77	23.00	20.63	18.59	16.83	15.30	13.97	12.81	11.79
24	39.79	34.78	30.57	27.02	24.00	21.43	19.24	17.35	15.72	14.31	13.08	12.01
25	42.36	36.81	32.17	28.28	25.00	22.23	19.87	17.85	16.12	14.63	13.34	12.22
26	45.03	38.89	33.79	29.55	26.00	23.01	20.49	18.34	16.51	14.94	13.59	12.41
27	47.80	41.03	35.45	30.84	27.00	23.79	21.10	18.82	16.89	15.24	13.82	12.60
28	50.68	43.23	37.14	32.13	28.00	24.57	21.70	19.28	17.25	15.52	14.04	12.77
29	53.66	45.49	38.86	33.44	29.00	25.33	22.28	19.74	17.60	15.79	14.25	12.93
30	56.77	47.82	40.61	34.77	30.00	26.09	22.86	20.18	17.93	16.04	14.45	13.08

DIFFERENTIAL SERIES PRESENT WORTH FACTOR : DSPWF(YEAR,E,R)

NOMINAL ESCALATION RATE : E =.09

NOMINAL DISCOUNT RATE : R

YEAR	0.04	0.05	0.06	0.07	0.08	0.09	0.10	0.11	0.12	0.13	0.14	0.15
1	1.05	1.04	1.03	1.02	1.01	1.00	0.99	0.98	0.97	0.96	0.96	0.95
2	2.15	2.12	2.09	2.06	2.03	2.00	1.97	1.95	1.92	1.90	1.87	1.85
3	3.30	3.23	3.17	3.11	3.06	3.00	2.95	2.89	2.84	2.79	2.74	2.70
4	4.50	4.40	4.29	4.19	4.09	4.00	3.91	3.82	3.74	3.66	3.58	3.50
5	5.77	5.60	5.44	5.29	5.14	5.00	4.87	4.74	4.61	4.49	4.38	4.27
6	7.09	6.85	6.62	6.40	6.20	6.00	5.81	5.63	5.46	5.30	5.14	4.99
7	8.48	8.15	7.84	7.54	7.26	7.00	6.75	6.51	6.29	6.08	5.87	5.68
8	9.94	9.50	9.09	8.70	8.34	8.00	7.68	7.38	7.09	6.83	6.57	6.33
9	11.47	10.90	10.37	9.88	9.43	9.00	8.60	8.23	7.88	7.55	7.24	6.95
10	13.06	12.35	11.70	11.09	10.52	10.00	9.51	9.06	8.64	8.25	7.88	7.54
11	14.74	13.86	13.06	12.31	11.63	11.00	10.42	9.88	9.38	8.92	8.49	8.09
12	16.50	15.43	14.45	13.56	12.75	12.00	11.31	10.68	10.10	9.57	9.07	8.62
13	18.34	17.05	15.89	14.83	13.87	13.00	12.20	11.47	10.81	10.19	9.63	9.11
14	20.27	18.74	17.37	16.13	15.01	14.00	13.08	12.25	11.49	10.80	10.17	9.59
15	22.29	20.49	18.89	17.45	16.16	15.00	13.95	13.01	12.15	11.38	10.68	10.03
16	24.41	22.31	20.45	18.80	17.32	16.00	14.82	13.76	12.80	11.94	11.16	10.46
17	26.63	24.20	22.06	20.17	18.49	17.00	15.67	14.49	13.43	12.48	11.63	10.86
18	28.96	26.16	23.71	21.56	19.67	18.00	16.52	15.21	14.05	13.01	12.08	11.24
19	31.40	28.20	25.41	22.98	20.86	19.00	17.36	15.92	14.64	13.51	12.50	11.60
20	33.96	30.31	27.16	24.43	22.06	20.00	18.20	16.61	15.22	14.00	12.91	11.95
21	36.64	32.50	28.96	25.91	23.28	21.00	19.02	17.30	15.79	14.47	13.30	12.27
22	39.45	34.78	30.80	27.41	24.50	22.00	19.84	17.97	16.34	14.92	13.67	12.58
23	42.39	37.14	32.70	28.94	25.74	23.00	20.65	18.63	16.88	15.35	14.03	12.87
24	45.48	39.59	34.66	30.50	26.98	24.00	21.45	19.27	17.40	15.78	14.37	13.15
25	48.71	42.14	36.67	32.09	28.24	25.00	22.25	19.91	17.90	16.18	14.70	13.41
26	52.10	44.78	38.73	33.71	29.51	26.00	23.04	20.53	18.40	16.57	15.01	13.66
27	55.66	47.53	40.86	35.36	30.80	27.00	23.82	21.14	18.88	16.95	15.31	13.89
28	59.38	50.38	43.04	37.04	32.09	28.00	24.59	21.74	19.35	17.32	15.59	14.11
29	63.28	53.33	45.29	38.75	33.40	29.00	25.36	22.33	19.80	17.67	15.86	14.33
30	67.37	56.40	47.60	40.49	34.72	30.00	26.12	22.91	20.24	18.01	16.12	14.53

Interest Tables / 249

DIFFERENTIAL SERIES PRESENT WORTH FACTOR : DSPWF(YEAR,E,R)

NOMINAL ESCALATION RATE : E = .10

NOMINAL DISCOUNT RATE : R

YEAR	0.04	0.05	0.06	0.07	0.08	0.09	0.10	0.11	0.12	0.13	0.14	0.15
1	1.06	1.05	1.04	1.03	1.02	1.01	1.00	0.99	0.98	0.97	0.96	0.96
2	2.18	2.15	2.11	2.08	2.06	2.03	2.00	1.97	1.95	1.92	1.90	1.87
3	3.36	3.29	3.23	3.17	3.11	3.06	3.00	2.95	2.89	2.84	2.79	2.75
4	4.61	4.50	4.39	4.29	4.19	4.09	4.00	3.91	3.82	3.74	3.66	3.58
5	5.93	5.76	5.60	5.44	5.28	5.14	5.00	4.87	4.74	4.62	4.50	4.38
6	7.33	7.08	6.84	6.62	6.40	6.20	6.00	5.81	5.64	5.47	5.30	5.15
7	8.82	8.47	8.14	7.83	7.54	7.26	7.00	6.75	6.52	6.29	6.08	5.88
8	10.38	9.92	9.49	9.08	8.70	8.34	8.00	7.68	7.38	7.10	6.84	6.58
9	12.04	11.44	10.88	10.36	9.88	9.42	9.00	8.60	8.23	7.89	7.56	7.25
10	13.79	13.03	12.33	11.68	11.08	10.52	10.00	9.52	9.07	8.65	8.26	7.90
11	15.64	14.70	13.83	13.03	12.30	11.62	11.00	10.42	9.89	9.39	8.93	8.51
12	17.60	16.45	15.39	14.43	13.55	12.74	12.00	11.32	10.69	10.12	9.59	9.09
13	19.68	18.28	17.01	15.86	14.82	13.87	13.00	12.21	11.49	10.82	10.21	9.66
14	21.87	20.20	18.69	17.33	16.11	15.00	14.00	13.09	12.26	11.51	10.82	10.19
15	24.19	22.21	20.43	18.85	17.43	16.15	15.00	13.96	13.03	12.18	11.41	10.71
16	26.64	24.31	22.24	20.40	18.77	17.31	16.00	14.83	13.78	12.83	11.97	11.20
17	29.24	26.52	24.12	22.00	20.13	18.47	17.00	15.69	14.51	13.46	12.52	11.67
18	31.98	28.83	26.07	23.65	21.52	19.65	18.00	16.54	15.23	14.08	13.04	12.12
19	34.89	31.25	28.09	25.34	22.94	20.84	19.00	17.38	15.94	14.68	13.55	12.55
20	37.96	33.78	30.19	27.08	24.39	22.04	20.00	18.21	16.64	15.26	14.04	12.96
21	41.20	36.44	32.36	28.87	25.86	23.25	21.00	19.04	17.33	15.83	14.51	13.35
22	44.64	39.22	34.62	30.70	27.35	24.48	22.00	19.86	18.00	16.38	14.97	13.73
23	48.27	42.14	36.97	32.59	28.88	25.71	23.00	20.67	18.66	16.92	15.41	14.09
24	52.11	45.19	39.40	34.53	30.43	26.96	24.00	21.47	19.31	17.44	15.83	14.43
25	56.18	48.39	41.92	36.53	32.01	28.21	25.00	22.27	19.95	17.95	16.24	14.76
26	60.48	51.74	44.54	38.58	33.62	29.48	26.00	23.06	20.57	18.45	16.64	15.07
27	65.02	55.25	47.26	40.69	35.26	30.76	27.00	23.85	21.19	18.93	17.02	15.37
28	69.83	58.93	50.08	42.86	36.94	32.05	28.00	24.62	21.79	19.41	17.38	15.66
29	74.92	62.78	53.01	45.09	38.64	33.35	29.00	25.39	22.38	19.86	17.74	15.94
30	80.30	66.82	56.05	47.38	40.37	34.67	30.00	26.15	22.97	20.31	18.08	16.20

DIFFERENTIAL SERIES PRESENT WORTH FACTOR : DSPWF(YEAR,E,R)

NOMINAL ESCALATION RATE : E = .11

NOMINAL DISCOUNT RATE : R

YEAR	0.04	0.05	0.06	0.07	0.08	0.09	0.10	0.11	0.12	0.13	0.14	0.15
1	1.07	1.06	1.05	1.04	1.03	1.02	1.01	1.00	0.99	0.98	0.97	0.97
2	2.21	2.17	2.14	2.11	2.08	2.06	2.03	2.00	1.97	1.95	1.92	1.90
3	3.42	3.36	3.29	3.23	3.17	3.11	3.05	3.00	2.95	2.90	2.84	2.80
4	4.72	4.61	4.49	4.39	4.29	4.19	4.09	4.00	3.91	3.83	3.74	3.66
5	6.10	5.93	5.75	5.59	5.43	5.28	5.14	5.00	4.87	4.74	4.62	4.50
6	7.58	7.32	7.07	6.84	6.61	6.40	6.19	6.00	5.82	5.64	5.47	5.31
7	9.16	8.80	8.45	8.13	7.82	7.53	7.26	7.00	6.75	6.52	6.30	6.09
8	10.84	10.36	9.90	9.47	9.07	8.69	8.33	8.00	7.69	7.39	7.11	6.84
9	12.64	12.01	11.41	10.86	10.35	9.87	9.42	9.00	8.61	8.24	7.90	7.57
10	14.56	13.75	13.00	12.30	11.66	11.07	10.51	10.00	9.52	9.08	8.66	8.27
11	16.61	15.59	14.66	13.80	13.01	12.29	11.62	11.00	10.43	9.90	9.41	8.95
12	18.79	17.54	16.40	15.36	14.40	13.53	12.73	12.00	11.33	10.71	10.13	9.60
13	21.12	19.60	18.22	16.97	15.83	14.80	13.86	13.00	12.22	11.50	10.84	10.24
14	23.61	21.78	20.12	18.64	17.30	16.09	14.99	14.00	13.10	12.28	11.53	10.85
15	26.27	24.08	22.12	20.37	18.81	17.40	16.14	15.00	13.97	13.04	12.20	11.43
16	29.11	26.51	24.21	22.17	20.36	18.74	17.29	16.00	14.84	13.79	12.85	12.00
17	32.13	29.08	26.40	24.04	21.95	20.10	18.46	17.00	15.70	14.53	13.49	12.55
18	35.36	31.80	28.69	25.97	23.59	21.49	19.64	18.00	16.55	15.26	14.11	13.08
19	38.81	34.68	31.09	27.98	25.27	22.90	20.82	19.00	17.39	15.97	14.71	13.59
20	42.49	37.71	33.61	30.07	27.00	24.34	22.02	20.00	18.23	16.67	15.29	14.08
21	46.42	40.93	36.24	32.23	28.78	25.81	23.23	21.00	19.06	17.36	15.87	14.56
22	50.61	44.32	39.00	34.47	30.61	27.30	24.45	22.00	19.88	18.03	16.42	15.01
23	55.08	47.91	41.88	36.79	32.48	28.82	25.68	23.00	20.69	18.69	16.96	15.46
24	59.86	51.71	44.91	39.21	34.41	30.36	26.93	24.00	21.50	19.35	17.49	15.89
25	64.95	55.72	48.07	41.71	36.40	31.94	28.18	25.00	22.30	19.99	18.00	16.30
26	70.39	59.96	51.39	44.31	38.44	33.54	29.45	26.00	23.09	20.61	18.50	16.70
27	76.20	64.44	54.86	47.00	40.53	35.18	30.72	27.00	23.87	21.23	18.99	17.08
28	82.39	69.18	58.49	49.80	42.68	36.84	32.01	28.00	24.65	21.84	19.46	17.45
29	89.00	74.19	62.30	52.69	44.90	38.54	33.31	29.00	25.42	22.43	19.93	17.81
30	96.06	79.49	66.28	55.70	47.17	40.26	34.62	30.00	26.18	23.02	20.38	18.16

DIFFERENTIAL SERIES PRESENT WORTH FACTOR : DSPWF(YEAR,E,R)

NOMINAL ESCALATION RATE : E = .12

NOMINAL DISCOUNT RATE : R

YEAR	0.04	0.05	0.06	0.07	0.08	0.09	0.10	0.11	0.12	0.13	0.14	0.15
1	1.08	1.07	1.06	1.05	1.04	1.03	1.02	1.01	1.00	0.99	0.98	0.97
2	2.24	2.20	2.17	2.14	2.11	2.08	2.05	2.03	2.00	1.97	1.95	1.92
3	3.49	3.42	3.35	3.29	3.23	3.17	3.11	3.05	3.00	2.95	2.90	2.85
4	4.83	4.71	4.60	4.49	4.38	4.28	4.19	4.09	4.00	3.91	3.83	3.75
5	6.28	6.09	5.92	5.75	5.58	5.43	5.28	5.14	5.00	4.87	4.74	4.62
6	7.84	7.57	7.31	7.06	6.83	6.61	6.39	6.19	6.00	5.82	5.64	5.48
7	9.52	9.14	8.78	8.44	8.12	7.81	7.53	7.26	7.00	6.76	6.53	6.31
8	11.33	10.81	10.33	9.88	9.46	9.06	8.68	8.33	8.00	7.69	7.39	7.12
9	13.28	12.60	11.97	11.39	10.84	10.33	9.86	9.42	9.00	8.61	8.25	7.90
10	15.37	14.51	13.71	12.97	12.28	11.65	11.06	10.51	10.00	9.53	9.08	8.67
11	17.63	16.54	15.54	14.62	13.77	12.99	12.28	11.61	11.00	10.43	9.91	9.42
12	20.07	18.71	17.48	16.35	15.32	14.38	13.52	12.73	12.00	11.33	10.72	10.15
13	22.69	21.02	19.52	18.16	16.92	15.80	14.78	13.85	13.00	12.22	11.51	10.86
14	25.51	23.49	21.68	20.05	18.59	17.26	16.07	14.98	14.00	13.11	12.29	11.55
15	28.55	26.13	23.97	22.04	20.31	18.77	17.38	16.13	15.00	13.98	13.06	12.22
16	31.82	28.93	26.38	24.12	22.10	20.31	18.71	17.28	16.00	14.85	13.81	12.88
17	35.35	31.93	28.93	26.29	23.96	21.90	20.07	18.45	17.00	15.71	14.55	13.51
18	39.14	35.13	31.62	28.56	25.88	23.53	21.45	19.62	18.00	16.56	15.28	14.13
19	43.23	38.53	34.47	30.95	27.88	25.20	22.86	20.81	19.00	17.40	15.99	14.74
20	47.63	42.17	37.48	33.44	29.95	26.92	24.30	22.00	20.00	18.24	16.69	15.33
21	52.37	46.05	40.66	36.05	32.09	28.69	25.76	23.21	21.00	19.07	17.38	15.90
22	57.48	50.18	44.01	38.78	34.32	30.51	27.24	24.43	22.00	19.89	18.06	16.46
23	62.98	54.60	47.56	41.64	36.63	32.38	28.76	25.66	23.00	20.71	18.73	17.01
24	68.90	59.30	51.31	44.63	39.02	34.30	30.30	26.90	24.00	21.52	19.38	17.54
25	75.28	64.32	55.27	47.76	41.50	36.27	31.87	28.15	25.00	22.32	20.02	18.05
26	82.14	69.68	59.46	51.04	44.08	38.29	33.46	29.41	26.00	23.11	20.65	18.56
27	89.54	75.39	63.88	54.47	46.75	40.37	35.09	30.69	27.00	23.90	21.27	19.05
28	97.50	81.48	68.55	58.06	49.52	42.51	36.75	31.97	28.00	24.68	21.88	19.52
29	*****	87.98	73.49	61.82	52.39	44.71	38.43	33.27	29.00	25.45	22.48	19.99
30	*****	94.91	78.70	65.76	55.36	46.97	40.15	34.58	30.00	26.22	23.07	20.44

DIFFERENTIAL SERIES PRESENT WORTH FACTOR : DSPWF(YEAR,E,R)

NOMINAL ESCALATION RATE : E = .13

NOMINAL DISCOUNT RATE : R

YEAR	0.04	0.05	0.06	0.07	0.08	0.09	0.10	0.11	0.12	0.13	0.14	0.15
1	1.09	1.08	1.07	1.06	1.05	1.04	1.03	1.02	1.01	1.00	0.99	0.98
2	2.27	2.23	2.20	2.17	2.14	2.11	2.08	2.05	2.03	2.00	1.97	1.95
3	3.55	3.48	3.41	3.35	3.29	3.23	3.17	3.11	3.05	3.00	2.95	2.90
4	4.94	4.82	4.71	4.59	4.48	4.38	4.28	4.18	4.09	4.00	3.91	3.83
5	6.46	6.27	6.08	5.91	5.74	5.58	5.42	5.28	5.14	5.00	4.87	4.75
6	8.10	7.82	7.55	7.29	7.05	6.82	6.60	6.39	6.19	6.00	5.82	5.65
7	9.89	9.49	9.11	8.76	8.42	8.11	7.81	7.52	7.25	7.00	6.76	6.53
8	11.83	11.29	10.78	10.31	9.86	9.44	9.05	8.68	8.33	8.00	7.69	7.40
9	13.94	13.23	12.56	11.94	11.36	10.82	10.32	9.85	9.41	9.00	8.61	8.25
10	16.24	15.31	14.46	13.67	12.93	12.26	11.63	11.05	10.50	10.00	9.53	9.09
11	18.73	17.55	16.48	15.49	14.58	13.74	12.97	12.26	11.61	11.00	10.44	9.92
12	21.44	19.97	18.63	17.41	16.30	15.29	14.36	13.50	12.72	12.00	11.34	10.73
13	24.38	22.56	20.93	19.45	18.10	16.88	15.77	14.76	13.84	13.00	12.23	11.52
14	27.57	25.36	23.38	21.59	19.99	18.54	17.23	16.05	14.97	14.00	13.11	12.30
15	31.05	28.37	25.98	23.86	21.96	20.26	18.73	17.35	16.12	15.00	13.99	13.07
16	34.82	31.61	28.77	26.25	24.02	22.04	20.27	18.69	17.27	16.00	14.86	13.83
17	38.92	35.09	31.73	28.78	26.18	23.88	21.85	20.04	18.43	17.00	15.72	14.57
18	43.37	38.84	34.89	31.45	28.44	25.79	23.47	21.42	19.61	18.00	16.57	15.30
19	48.21	42.88	38.26	34.27	30.80	27.78	25.14	22.82	20.79	19.00	17.42	16.02
20	53.47	47.22	41.86	37.25	33.27	29.83	26.85	24.25	21.99	20.00	18.26	16.72
21	59.19	51.89	45.69	40.39	35.86	31.97	28.61	25.71	23.19	21.00	19.09	17.41
22	65.40	56.92	49.77	43.71	38.57	34.17	30.42	27.19	24.41	22.00	19.91	18.09
23	72.14	62.33	54.12	47.22	41.40	36.47	32.27	28.70	25.63	23.00	20.73	18.76
24	79.47	68.16	58.76	50.92	44.36	38.84	34.18	30.23	26.87	24.00	21.54	19.42
25	87.43	74.43	63.71	54.84	47.46	41.30	36.14	31.79	28.12	25.00	22.34	20.06
26	96.09	81.18	68.98	58.97	50.70	43.85	38.15	33.38	29.38	26.00	23.13	20.69
27	*****	88.44	74.60	63.33	54.10	46.50	40.22	35.00	30.65	27.00	23.92	21.32
28	*****	96.25	80.60	67.94	57.65	49.24	42.35	36.65	31.93	28.00	24.70	21.93
29	*****	*****	86.99	72.80	61.36	52.09	44.53	38.33	33.23	29.00	25.48	22.53
30	*****	*****	93.80	77.94	65.25	55.04	46.77	40.04	34.53	30.00	26.25	23.12

Interest Tables / 251

DIFFERENTIAL SERIES PRESENT WORTH FACTOR : DSPWF(YEAR,E,R)

NOMINAL ESCALATION RATE : E = .14

NOMINAL DISCOUNT RATE : R

YEAR	0.04	0.05	0.06	0.07	0.08	0.09	0.10	0.11	0.12	0.13	0.14	0.15
1	1.10	1.09	1.08	1.07	1.06	1.05	1.04	1.03	1.02	1.01	1.00	0.99
2	2.30	2.26	2.23	2.20	2.17	2.14	2.11	2.08	2.05	2.03	2.00	1.97
3	3.61	3.54	3.48	3.41	3.35	3.28	3.22	3.17	3.11	3.05	3.00	2.95
4	5.06	4.93	4.81	4.70	4.59	4.48	4.38	4.28	4.18	4.09	4.00	3.91
5	6.64	6.44	6.25	6.07	5.90	5.73	5.57	5.42	5.27	5.13	5.00	4.87
6	8.38	8.08	7.80	7.53	7.28	7.04	6.81	6.59	6.39	6.19	6.00	5.82
7	10.28	9.86	9.46	9.09	8.74	8.41	8.10	7.80	7.52	7.25	7.00	6.76
8	12.36	11.79	11.25	10.75	10.28	9.84	9.43	9.04	8.67	8.33	8.00	7.69
9	14.65	13.89	13.18	12.52	11.91	11.34	10.81	10.31	9.84	9.41	9.00	8.62
10	17.15	16.16	15.25	14.41	13.63	12.90	12.23	11.61	11.04	10.50	10.00	9.53
11	19.90	18.63	17.48	16.41	15.44	14.54	13.72	12.95	12.25	11.60	11.00	10.44
12	22.91	21.32	19.87	18.55	17.35	16.25	15.25	14.33	13.49	12.71	12.00	11.34
13	26.20	24.23	22.44	20.83	19.37	18.05	16.84	15.75	14.75	13.83	13.00	12.24
14	29.82	27.39	25.21	23.26	21.50	19.92	18.49	17.20	16.03	14.97	14.00	13.12
15	33.78	30.82	28.19	25.85	23.75	21.88	20.20	18.69	17.33	16.11	15.00	14.00
16	38.13	34.55	31.40	28.60	26.13	23.93	21.97	20.22	18.66	17.26	16.00	14.87
17	42.89	38.60	34.84	31.54	28.64	26.07	23.81	21.80	20.01	18.42	17.00	15.73
18	48.11	42.99	38.54	34.67	31.28	28.31	25.71	23.41	21.39	19.59	18.00	16.58
19	53.83	47.76	42.53	38.00	34.07	30.66	27.68	25.07	22.79	20.77	19.00	17.43
20	60.10	52.94	46.81	41.55	37.02	33.11	29.72	26.78	24.21	21.97	20.00	18.27
21	66.98	58.57	51.42	45.34	40.14	35.68	31.84	28.53	25.66	23.17	21.00	19.10
22	74.52	64.67	56.38	49.37	43.42	38.36	34.03	30.33	27.14	24.38	22.00	19.93
23	82.78	71.30	61.71	53.66	46.89	41.16	36.31	32.17	28.64	25.61	23.00	20.75
24	91.83	78.50	67.44	58.24	50.55	44.10	38.66	34.07	30.17	26.84	24.00	21.56
25	*****	86.31	73.61	63.12	54.41	47.17	41.11	36.02	31.72	28.09	25.00	22.36
26	*****	94.80	80.24	68.31	58.49	50.38	43.64	38.02	33.31	29.35	26.00	23.16
27	*****	*****	87.37	73.84	62.80	53.73	46.26	40.07	34.92	30.62	27.00	23.95
28	*****	*****	95.04	79.74	67.34	57.24	48.98	42.18	36.56	31.90	28.00	24.73
29	*****	*****	*****	86.02	72.14	60.92	51.80	44.35	38.23	33.19	29.00	25.51
30	*****	*****	*****	92.72	77.20	64.76	54.72	46.57	39.93	34.49	30.00	26.28

DIFFERENTIAL SERIES PRESENT WORTH FACTOR : DSPWF(YEAR,E,R)

NOMINAL ESCALATION RATE : E = .15

NOMINAL DISCOUNT RATE : R

YEAR	0.04	0.05	0.06	0.07	0.08	0.09	0.10	0.11	0.12	0.13	0.14	0.15
1	1.11	1.10	1.08	1.07	1.06	1.06	1.05	1.04	1.03	1.02	1.01	1.00
2	2.33	2.29	2.26	2.23	2.20	2.17	2.14	2.11	2.08	2.05	2.03	2.00
3	3.68	3.61	3.54	3.47	3.41	3.34	3.28	3.22	3.16	3.11	3.05	3.00
4	5.18	5.05	4.92	4.81	4.69	4.58	4.48	4.37	4.28	4.18	4.09	4.00
5	6.83	6.62	6.43	6.24	6.06	5.89	5.72	5.57	5.42	5.27	5.13	5.00
6	8.66	8.35	8.06	7.78	7.52	7.27	7.03	6.80	6.59	6.38	6.19	6.00
7	10.68	10.24	9.83	9.44	9.07	8.72	8.40	8.09	7.79	7.51	7.25	7.00
8	12.91	12.31	11.75	11.22	10.72	10.26	9.82	9.41	9.03	8.66	8.32	8.00
9	15.38	14.58	13.83	13.13	12.48	11.88	11.31	10.79	10.30	9.84	9.40	9.00
10	18.12	17.06	16.09	15.19	14.36	13.59	12.87	12.21	11.60	11.03	10.50	10.00
11	21.14	19.78	18.54	17.40	16.35	15.39	14.50	13.69	12.94	12.24	11.60	11.00
12	24.48	22.76	21.20	19.77	18.48	17.29	16.21	15.22	14.31	13.47	12.71	12.00
13	28.18	26.02	24.08	22.33	20.74	19.30	17.99	16.80	15.72	14.73	13.83	13.00
14	32.26	29.60	27.21	25.07	23.15	21.42	19.85	18.44	17.17	16.01	14.96	14.00
15	36.78	33.51	30.61	28.02	25.71	23.65	21.80	20.14	18.65	17.31	16.10	15.00
16	41.78	37.80	34.29	31.19	28.44	26.01	23.84	21.91	20.18	18.63	17.25	16.00
17	47.30	42.49	38.29	34.60	31.35	28.49	25.97	23.73	21.75	19.98	18.41	17.00
18	53.41	47.64	42.62	38.26	34.45	31.12	28.19	25.62	23.36	21.35	19.58	18.00
19	60.17	53.27	47.33	42.19	37.75	33.89	30.52	27.58	25.01	22.75	20.76	19.00
20	67.63	59.44	52.43	46.42	41.26	36.81	32.95	29.61	26.71	24.17	21.95	20.00
21	75.89	66.19	57.97	50.97	45.00	39.89	35.50	31.72	28.45	25.61	23.15	21.00
22	85.03	73.59	63.97	55.85	48.98	43.14	38.16	33.89	30.24	27.09	24.36	22.00
23	95.13	81.69	70.49	61.10	53.22	46.57	40.94	36.15	32.07	28.58	25.58	23.00
24	*****	90.57	77.56	66.75	57.73	50.19	43.84	38.49	33.96	30.11	26.82	24.00
25	*****	*****	85.23	72.81	62.54	54.00	46.88	40.91	35.89	31.66	28.06	25.00
26	*****	*****	93.55	79.33	67.66	58.03	50.06	43.42	37.88	33.23	29.32	26.00
27	*****	*****	*****	86.34	73.11	62.28	53.38	46.02	39.92	34.84	30.58	27.00
28	*****	*****	*****	93.87	78.91	66.76	56.85	48.72	42.02	36.47	31.86	28.00
29	*****	*****	*****	*****	85.09	71.49	60.48	51.51	44.17	38.14	33.15	29.00
30	*****	*****	*****	*****	91.67	76.48	64.27	54.40	46.38	39.83	34.45	30.00

11.3 REMAINING LOAN BALANCE FACTOR

The remaining loan balance factor is a function of the variables Year, T (term of the loan), and I (loan interest rate).

REMAINING LOAN BALANCE FACTOR : RBF(T,I,YEAR)

LOAN TERM : T = 5

LOAN INTEREST RATE : I

YEAR	0.04	0.05	0.06	0.07	0.08	0.09	0.10	0.11	0.12	0.13	0.14	0.15
1	0.82	0.82	0.82	0.83	0.83	0.83	0.84	0.84	0.84	0.85	0.85	0.85
2	0.62	0.63	0.63	0.64	0.65	0.65	0.66	0.66	0.67	0.67	0.68	0.68
3	0.42	0.43	0.44	0.44	0.45	0.45	0.46	0.46	0.47	0.47	0.48	0.48
4	0.22	0.22	0.22	0.23	0.23	0.24	0.24	0.24	0.25	0.25	0.26	0.26
5	0.00	0.00	0.00	0.00	0.00	0.00	0.00	0.00	0.00	0.00	0.00	0.00

REMAINING LOAN BALANCE FACTOR : RBF(T,I,YEAR)

LOAN TERM : T =10

LOAN INTEREST RATE : I

YEAR	0.04	0.05	0.06	0.07	0.08	0.09	0.10	0.11	0.12	0.13	0.14	0.15
1	0.92	0.92	0.92	0.93	0.93	0.93	0.94	0.94	0.94	0.95	0.95	0.95
2	0.83	0.84	0.84	0.85	0.86	0.86	0.87	0.87	0.88	0.88	0.89	0.89
3	0.74	0.75	0.76	0.77	0.78	0.78	0.79	0.80	0.81	0.82	0.82	0.83
4	0.65	0.66	0.67	0.68	0.69	0.70	0.71	0.72	0.73	0.74	0.75	0.75
5	0.55	0.56	0.57	0.58	0.60	0.61	0.62	0.63	0.64	0.65	0.66	0.67
6	0.45	0.46	0.47	0.48	0.49	0.50	0.52	0.53	0.54	0.55	0.56	0.57
7	0.34	0.35	0.36	0.37	0.38	0.39	0.40	0.41	0.43	0.44	0.45	0.45
8	0.23	0.24	0.25	0.26	0.27	0.27	0.28	0.29	0.30	0.31	0.32	0.32
9	0.12	0.12	0.13	0.13	0.14	0.14	0.15	0.15	0.16	0.16	0.17	0.17
10	0.00	0.00	0.00	0.00	0.00	0.00	0.00	0.00	0.00	0.00	0.00	0.00

REMAINING LOAN BALANCE FACTOR : RBF(T,I,YEAR)

LOAN TERM : T =15

LOAN INTEREST RATE : I

YEAR	0.04	0.05	0.06	0.07	0.08	0.09	0.10	0.11	0.12	0.13	0.14	0.15
1	0.95	0.95	0.96	0.96	0.96	0.97	0.97	0.97	0.97	0.98	0.98	0.98
2	0.90	0.90	0.91	0.92	0.92	0.93	0.93	0.94	0.94	0.95	0.95	0.95
3	0.84	0.85	0.86	0.87	0.88	0.89	0.90	0.90	0.91	0.92	0.92	0.93
4	0.79	0.80	0.81	0.82	0.83	0.84	0.85	0.86	0.87	0.88	0.89	0.90
5	0.73	0.74	0.76	0.77	0.78	0.80	0.81	0.82	0.83	0.84	0.85	0.86
6	0.67	0.68	0.70	0.72	0.73	0.74	0.76	0.77	0.78	0.79	0.81	0.82
7	0.61	0.62	0.64	0.66	0.67	0.69	0.70	0.72	0.73	0.74	0.76	0.77
8	0.54	0.56	0.57	0.59	0.61	0.62	0.64	0.66	0.67	0.68	0.70	0.71
9	0.47	0.49	0.51	0.52	0.54	0.56	0.57	0.59	0.60	0.62	0.63	0.65
10	0.40	0.42	0.43	0.45	0.47	0.48	0.50	0.51	0.53	0.54	0.56	0.57
11	0.33	0.34	0.36	0.37	0.39	0.40	0.42	0.43	0.45	0.46	0.47	0.49
12	0.25	0.26	0.28	0.29	0.30	0.31	0.33	0.34	0.35	0.37	0.38	0.39
13	0.17	0.18	0.19	0.20	0.21	0.22	0.23	0.24	0.25	0.26	0.27	0.28
14	0.09	0.09	0.10	0.10	0.11	0.11	0.12	0.13	0.13	0.14	0.14	0.15
15	0.00	0.00	0.00	0.00	0.00	0.00	0.00	0.00	0.00	0.00	0.00	0.00

REMAINING LOAN BALANCE FACTOR : RBF(T,I,YEAR)

LOAN TERM : T =20

LOAN INTEREST RATE : I

YEAR	0.04	0.05	0.06	0.07	0.08	0.09	0.10	0.11	0.12	0.13	0.14	0.15
1	0.97	0.97	0.97	0.98	0.98	0.98	0.98	0.98	0.99	0.99	0.99	0.99
2	0.93	0.94	0.94	0.95	0.95	0.96	0.96	0.97	0.97	0.97	0.98	0.98
3	0.90	0.90	0.91	0.92	0.93	0.94	0.94	0.95	0.95	0.96	0.96	0.97
4	0.86	0.87	0.88	0.89	0.90	0.91	0.92	0.93	0.93	0.94	0.95	0.95
5	0.82	0.83	0.85	0.86	0.87	0.88	0.89	0.90	0.91	0.92	0.93	0.93
6	0.78	0.79	0.81	0.83	0.84	0.85	0.87	0.88	0.89	0.90	0.91	0.91
7	0.73	0.75	0.77	0.79	0.81	0.82	0.83	0.85	0.86	0.87	0.88	0.89
8	0.69	0.71	0.73	0.75	0.77	0.78	0.80	0.82	0.83	0.84	0.85	0.87
9	0.64	0.67	0.69	0.71	0.73	0.75	0.76	0.78	0.79	0.81	0.82	0.84
10	0.60	0.62	0.64	0.66	0.68	0.70	0.72	0.74	0.76	0.77	0.79	0.80
11	0.55	0.57	0.59	0.61	0.64	0.66	0.68	0.70	0.71	0.73	0.75	0.76
12	0.50	0.52	0.54	0.56	0.59	0.61	0.63	0.65	0.67	0.68	0.70	0.72
13	0.44	0.46	0.49	0.51	0.53	0.55	0.57	0.59	0.61	0.63	0.65	0.66
14	0.39	0.41	0.43	0.45	0.47	0.49	0.51	0.53	0.55	0.57	0.59	0.60
15	0.33	0.35	0.37	0.39	0.41	0.43	0.45	0.46	0.48	0.50	0.52	0.54
16	0.27	0.28	0.30	0.32	0.34	0.35	0.37	0.39	0.41	0.42	0.44	0.46
17	0.20	0.22	0.23	0.25	0.26	0.28	0.29	0.31	0.32	0.34	0.35	0.36
18	0.14	0.15	0.16	0.17	0.18	0.19	0.20	0.22	0.23	0.24	0.25	0.26
19	0.07	0.08	0.08	0.09	0.09	0.10	0.11	0.11	0.12	0.13	0.13	0.14
20	0.00	0.00	0.00	0.00	0.00	0.00	0.00	0.00	0.00	0.00	0.00	0.00

REMAINING LOAN BALANCE FACTOR : RBF(T,I,YEAR)

LOAN TERM : T =25

LOAN INTEREST RATE : I

YEAR	0.04	0.05	0.06	0.07	0.08	0.09	0.10	0.11	0.12	0.13	0.14	0.15
1	0.98	0.98	0.98	0.98	0.99	0.99	0.99	0.99	0.99	0.99	0.99	1.00
2	0.95	0.96	0.96	0.97	0.97	0.98	0.98	0.98	0.98	0.99	0.99	0.99
3	0.93	0.93	0.94	0.95	0.96	0.96	0.97	0.97	0.97	0.98	0.98	0.98
4	0.90	0.91	0.92	0.93	0.94	0.95	0.95	0.96	0.96	0.97	0.97	0.98
5	0.87	0.88	0.90	0.91	0.92	0.93	0.94	0.95	0.95	0.96	0.96	0.97
6	0.84	0.86	0.87	0.89	0.90	0.91	0.92	0.93	0.94	0.95	0.95	0.96
7	0.81	0.83	0.85	0.86	0.88	0.89	0.90	0.91	0.92	0.93	0.94	0.95
8	0.78	0.80	0.82	0.84	0.85	0.87	0.88	0.90	0.91	0.92	0.93	0.94
9	0.75	0.77	0.79	0.81	0.83	0.85	0.86	0.88	0.89	0.90	0.91	0.92
10	0.71	0.74	0.76	0.78	0.80	0.82	0.84	0.85	0.87	0.88	0.89	0.90
11	0.68	0.70	0.73	0.75	0.77	0.79	0.81	0.83	0.85	0.86	0.87	0.89
12	0.64	0.67	0.69	0.72	0.74	0.76	0.78	0.80	0.82	0.84	0.85	0.86
13	0.60	0.63	0.66	0.68	0.71	0.73	0.75	0.77	0.79	0.81	0.82	0.84
14	0.56	0.59	0.62	0.64	0.67	0.69	0.72	0.74	0.76	0.78	0.79	0.81
15	0.52	0.55	0.58	0.60	0.63	0.65	0.68	0.70	0.72	0.74	0.76	0.78
16	0.48	0.50	0.53	0.56	0.59	0.61	0.63	0.66	0.68	0.70	0.72	0.74
17	0.43	0.46	0.49	0.51	0.54	0.56	0.59	0.61	0.63	0.65	0.67	0.69
18	0.38	0.41	0.44	0.46	0.49	0.51	0.54	0.56	0.58	0.60	0.62	0.64
19	0.34	0.36	0.38	0.41	0.43	0.46	0.48	0.50	0.52	0.55	0.57	0.59
20	0.28	0.31	0.33	0.35	0.37	0.40	0.42	0.44	0.46	0.48	0.50	0.52
21	0.23	0.25	0.27	0.29	0.31	0.33	0.35	0.37	0.39	0.41	0.42	0.44
22	0.18	0.19	0.21	0.23	0.24	0.26	0.27	0.29	0.31	0.32	0.34	0.35
23	0.12	0.13	0.14	0.16	0.17	0.18	0.19	0.20	0.22	0.23	0.24	0.25
24	0.06	0.07	0.07	0.08	0.09	0.09	0.10	0.11	0.11	0.12	0.13	0.13
25	0.00	0.00	0.00	0.00	0.00	0.00	0.00	0.00	0.00	0.00	0.00	0.00

REMAINING LOAN BALANCE FACTOR : RBF(T,I,YEAR)

LOAN TERM : T =30

LOAN INTEREST RATE : I

YEAR	0.04	0.05	0.06	0.07	0.08	0.09	0.10	0.11	0.12	0.13	0.14	0.15
1	0.98	0.98	0.99	0.99	0.99	0.99	0.99	0.99	1.00	1.00	1.00	1.00
2	0.96	0.97	0.97	0.98	0.98	0.98	0.99	0.99	0.99	0.99	0.99	1.00
3	0.94	0.95	0.96	0.97	0.97	0.98	0.98	0.98	0.99	0.99	0.99	0.99
4	0.92	0.94	0.94	0.95	0.96	0.97	0.97	0.98	0.98	0.98	0.99	0.99
5	0.90	0.92	0.93	0.94	0.95	0.96	0.96	0.97	0.97	0.98	0.98	0.98
6	0.88	0.90	0.91	0.92	0.94	0.94	0.95	0.96	0.97	0.97	0.98	0.98
7	0.86	0.88	0.89	0.91	0.92	0.93	0.94	0.95	0.96	0.96	0.97	0.97
8	0.84	0.86	0.87	0.89	0.91	0.92	0.93	0.94	0.95	0.96	0.96	0.97
9	0.81	0.83	0.85	0.87	0.89	0.90	0.92	0.93	0.94	0.95	0.95	0.96
10	0.79	0.81	0.83	0.85	0.87	0.89	0.90	0.92	0.93	0.94	0.95	0.95
11	0.76	0.79	0.81	0.83	0.85	0.87	0.89	0.90	0.91	0.93	0.94	0.94
12	0.73	0.76	0.79	0.81	0.83	0.85	0.87	0.89	0.90	0.91	0.92	0.93
13	0.70	0.73	0.76	0.79	0.81	0.83	0.85	0.87	0.88	0.90	0.91	0.92
14	0.67	0.71	0.73	0.76	0.79	0.81	0.83	0.85	0.87	0.88	0.89	0.91
15	0.64	0.68	0.71	0.73	0.76	0.78	0.81	0.83	0.85	0.86	0.88	0.89
16	0.61	0.64	0.68	0.70	0.73	0.76	0.78	0.80	0.82	0.84	0.86	0.87
17	0.58	0.61	0.64	0.67	0.70	0.73	0.75	0.78	0.80	0.82	0.83	0.85
18	0.54	0.58	0.61	0.64	0.67	0.70	0.72	0.75	0.77	0.79	0.81	0.83
19	0.51	0.54	0.57	0.60	0.63	0.66	0.69	0.71	0.74	0.76	0.78	0.80
20	0.47	0.50	0.53	0.57	0.60	0.62	0.65	0.68	0.70	0.72	0.74	0.76
21	0.43	0.46	0.49	0.53	0.55	0.58	0.61	0.64	0.66	0.68	0.71	0.73
22	0.39	0.42	0.45	0.48	0.51	0.54	0.57	0.59	0.62	0.64	0.66	0.68
23	0.35	0.38	0.41	0.43	0.46	0.49	0.52	0.54	0.57	0.59	0.61	0.63
24	0.30	0.33	0.36	0.38	0.41	0.44	0.46	0.49	0.51	0.53	0.56	0.58
25	0.26	0.28	0.31	0.33	0.35	0.38	0.40	0.43	0.45	0.47	0.49	0.51
26	0.21	0.23	0.25	0.27	0.29	0.32	0.34	0.36	0.38	0.40	0.42	0.43
27	0.16	0.18	0.19	0.21	0.23	0.25	0.26	0.28	0.30	0.32	0.33	0.35
28	0.11	0.12	0.13	0.15	0.16	0.17	0.18	0.20	0.21	0.22	0.24	0.25
29	0.06	0.06	0.07	0.08	0.08	0.09	0.10	0.10	0.11	0.12	0.13	0.13
30	0.00	0.00	0.00	0.00	0.00	0.00	0.00	0.00	0.00	0.00	0.00	0.00

Appendix 12
Tables of LCR and D Versus SSF

These tables are included in the revised edition of this handbook for a very limited purpose: to help the reader work through the cost optimization procedure described in section L.5. In any instance in which LCR values *alone* are sought, Appendix 21—which contains revised LCR values reflecting improvements in wall design and measuring techniques—should be used in preference to this appendix.

For each location identified in this appendix, two sets of numbers are tabulated for each of the six system types measured and for values of SSF ranging from 0.1 to 0.9. In the upper set, values of LCR (in Btu/DD/ft²) are tabulated for use with the load collector ratio method. This method is explained in section M.3. In the lower set of numbers, values of D are tabulated for use in the economic optimization procedure. D is the derivative of SSF with respect to $1/LCR$ and is expressed in Btu/DD/ft². The physical significance of D is that it represents the equivalent additional load (in Btu/DD) that can be satisfied fully by one additional square foot of solar collection area.

The six system types are abbreviated as follows:

WW = Waterwall
$WWNI$ = Waterwall with nighttime insulation (R9)
TW = Trombe wall
$TWNI$ = Trombe wall with nighttime insulation (R9)
DG = Direct gain
$DGNI$ = Direct gain with nighttime insulation (R9)

BIRMINGHAM ALABAMA
33.6 N
2844 DD
T(JAN)=44

		SSF = .1	.2	.3	.4	.5	.6	.7	.8	.9
L	WW	186	87	53	37	27	21	15	10	5
C	WWNI	281	134	85	60	45	36	29	23	16
R	TW	179	82	50	33	23	16	11	7	4
	TWNI	265	126	79	56	42	32	25	18	12
	DG	192	83	47	28	15	6	-	-	-
	DGNI	296	139	86	61	45	34	25	18	11
D	WW	17.1	14.8	12.6	11.4	9.5	7.4	4.6	2.2	.4
	WWNI	26.1	25.0	21.2	19.4	18.0	15.6	12.9	9.3	3.4
	TW	16.3	13.9	11.6	8.9	6.6	4.5	2.0	1.6	.9
	TWNI	24.9	22.5	20.0	16.8	15.8	12.7	9.0	5.5	3.6
	DG	16.6	12.9	8.9	5.1	2.0	-	-	-	2.2
	DGNI	27.7	24.3	21.8	19.0	15.9	11.9	8.2	4.8	1.8

MOBILE ALABAMA
30.7 N
1684 DD
T(JAN)=51

		SSF = .1	.2	.3	.4	.5	.6	.7	.8	.9
L	WW	318	144	89	64	48	37	29	21	13
C	WWNI	441	205	127	92	71	56	45	36	25
R	TW	293	135	83	57	40	30	22	15	10
	TWNI	412	191	120	86	65	51	39	29	19
	DG	354	159	96	63	43	29	18	10	-
	DGNI	469	216	135	96	72	55	42	30	20
D	WW	29.8	24.2	23.0	21.1	18.3	15.1	10.5	5.9	2.0
	WWNI	42.6	34.9	32.1	28.4	25.3	20.3	13.4	7.5	5.6
	TW	27.0	23.4	19.7	16.1	12.6	9.5	6.7	4.2	1.8
	TWNI	39.1	33.2	31.4	28.4	25.2	20.3	14.5	8.9	3.7
	DG	31.7	26.4	21.5	15.9	11.0	7.0	3.6	1.2	-
	DGNI	43.8	37.6	34.9	31.0	26.5	20.2	14.3	8.6	3.5

MONTGOMERY ALABAMA
32.3 N
2269 DD
T(JAN)=48

		SSF = .1	.2	.3	.4	.5	.6	.7	.8	.9
L	WW	236	111	68	48	36	27	21	15	8
C	WWNI	339	165	103	73	56	44	35	28	20
R	TW	224	103	63	43	30	22	15	11	7
	TWNI	320	153	96	68	51	40	30	23	15
	DG	257	115	67	42	27	16	9	-	-
	DGNI	362	171	107	75	56	42	32	23	15
D	WW	22.7	18.9	16.6	15.2	12.8	10.3	6.8	3.6	1.0
	WWNI	32.7	29.9	25.2	22.5	19.3	15.9	4.4	10.2	4.3
	TW	20.6	17.2	15.0	11.8	8.9	6.4	4.4	2.6	1.0
	TWNI	31.0	27.2	24.5	22.2	19.4	15.6	11.1	6.8	2.8
	DG	22.9	18.4	14.1	9.4	5.7	1.2	-	-	-
	DGNI	34.9	30.0	27.2	23.7	20.3	15.1	10.5	6.3	2.4

PHOENIX ARIZONA
33.4 N
1552 DD
T(JAN)=51

		SSF = .1	.2	.3	.4	.5	.6	.7	.8	.9
L	WW	467	219	139	100	75	58	45	34	22
C	WWNI	620	293	188	136	104	82	65	51	36
R	TW	436	202	126	87	63	46	34	25	16
	TWNI	583	275	176	126	95	73	56	42	28
	DG	555	256	157	107	75	53	37	24	12
	DGNI	673	316	201	143	107	82	62	45	29
D	WW	42.5	39.7	37.2	32.9	28.2	23.2	17.0	9.9	3.7
	WWNI	56.5	54.3	50.3	40.8	35.1	28.6	18.8	7.9	7.9
	TW	40.1	35.6	33.1	25.6	20.3	15.2	10.9	7.0	3.1
	TWNI	53.4	47.3	41.3	35.4	29.5	24.0	17.3	12.6	5.1
	DG	50.4	44.2	37.5	29.5	21.9	15.1	9.4	4.8	1.4
	DGNI	62.2	57.3	52.7	46.2	39.2	30.0	21.1	12.8	5.3

PRESCOTT ARIZONA
34.6 N
4456 DD
T(JAN)=37

		SSF = .1	.2	.3	.4	.5	.6	.7	.8	.9
L	WW	189	89	56	40	30	23	17	12	7
C	WWNI	286	135	88	63	48	39	31	25	18
R	TW	183	85	53	36	25	18	13	9	5
	TWNI	269	128	82	59	45	35	27	20	14
	DG	198	89	52	32	20	10	-	-	-
	DGNI	300	142	90	64	49	37	28	20	13
D	WW	17.1	16.5	14.3	13.1	11.3	8.9	5.7	2.9	.7
	WWNI	26.2	25.4	23.7	21.5	20.3	18.0	14.6	9.3	3.9
	TW	16.9	14.9	12.7	10.0	7.6	5.5	3.5	2.0	.9
	TWNI	24.9	23.8	21.7	20.2	17.7	14.1	10.0	6.2	2.5
	DG	17.6	14.6	10.6	6.8	3.6	1.0	-	-	-
	DGNI	27.7	25.8	23.7	21.4	17.8	13.4	9.3	5.5	2.1

TUCSON ARIZONA
32.1 N
1752 DD
T(JAN)=51

		SSF = .1	.2	.3	.4	.5	.6	.7	.8	.9
L	WW	425	199	127	92	70	55	43	32	21
C	WWNI	571	268	172	125	97	77	62	49	36
R	TW	394	185	116	81	59	43	32	23	16
	TWNI	533	252	161	116	89	69	54	40	27
	DG	498	231	143	98	69	49	34	22	12
	DGNI	613	289	185	132	100	77	59	43	29
D	WW	39.0	36.4	33.8	31.6	27.6	23.3	16.7	9.8	3.7
	WWNI	53.0	49.5	46.6	44.5	41.0	35.4	29.5	18.8	8.0
	TW	36.9	33.2	29.0	23.9	19.1	14.7	10.4	6.7	3.3
	TWNI	49.8	46.1	43.6	39.9	35.1	28.2	20.2	12.5	5.2
	DG	46.0	40.7	34.7	27.2	20.1	14.2	8.8	4.5	1.4
	DGNI	57.3	52.6	49.4	44.2	37.9	29.0	20.6	12.7	5.3

WINSLOW ARIZONA
35.0 N
4733 DD
T(JAN)=33

		SSF = .1	.2	.3	.4	.5	.6	.7	.8	.9
L	WW	165	78	49	34	25	19	14	9	4
C	WWNI	255	123	79	57	43	34	27	21	15
R	TW	162	75	46	31	22	15	10	7	4
	TWNI	242	116	74	53	40	31	24	18	12
	DG	168	74	42	24	12	-	-	-	-
	DGNI	268	127	81	57	43	32	24	17	11
D	WW	15.3	14.0	12.2	10.8	8.9	6.8	4.1	1.9	.2
	WWNI	23.8	23.4	20.8	19.2	17.6	15.3	11.9	7.8	3.2
	TW	14.6	13.1	10.8	8.4	6.2	4.3	2.7	1.5	.6
	TWNI	22.7	21.6	19.5	17.2	14.2	12.2	8.6	5.3	2.1
	DG	14.7	11.6	7.7	4.2	1.2	-	-	-	-
	DGNI	25.1	23.1	20.9	18.4	15.2	11.3	7.7	4.5	1.6

YUMA ARIZONA
32.7 N
1010 DD
T(JAN)=55

		SSF = .1	.2	.3	.4	.5	.6	.7	.8	.9
L	WW	728	342	219	155	116	89	70	52	35
C	WWNI	932	442	283	205	154	119	94	74	53
R	TW	670	313	194	135	97	72	53	38	25
	TWNI	877	413	264	187	140	108	82	61	40
	DG	887	412	254	174	124	89	64	44	26
	DGNI	1021	481	305	215	160	122	92	67	44
D	WW	68.9	61.9	58.9	49.2	42.7	35.4	26.9	16.1	6.6
	WWNI	88.7	80.0	77.5	67.6	57.4	49.4	41.5	27.2	11.6
	TW	61.7	53.7	47.8	39.7	31.2	24.0	17.3	10.4	5.4
	TWNI	81.2	75.7	68.5	59.7	51.2	41.9	29.6	18.1	7.5
	DG	81.2	71.5	61.0	49.4	37.6	27.3	18.2	10.4	4.0
	DGNI	94.8	88.2	77.6	67.9	57.7	44.5	31.3	19.2	8.2

Tables of LCR & D Versus SSF / 257

	SSF =	.1	.2	.3	.4	.5	.6	.7	.8	.9
FORT SMITH	WW	158	74	45	31	23	17	13	8	-
ARKANSAS	WWNI	246	119	75	53	40	32	26	20	14
	TW	153	70	43	29	20	14	9	6	3
35.3 N	TWNI	233	111	70	50	37	29	22	17	11
3336 DD	DG	156	67	36	20	-	-	-	-	-
T(JAN)=39	DGNI	258	121	76	53	40	30	22	16	10
	WW	15.0	12.4	10.8	9.7	8.0	6.1	3.6	1.6	-
	WWNI	23.5	21.8	18.8	17.6	16.5	14.2	11.7	7.4	3.1
	TW	14.0	12.1	9.9	7.5	5.3	3.7	2.1	1.2	.4
D	TWNI	22.3	19.9	17.9	16.4	14.2	11.4	8.1	5.0	2.0
	DG	13.4	10.1	6.2	2.8	-	-	-	-	-
	DGNI	24.6	21.3	19.3	16.9	14.0	10.4	7.1	4.1	1.5

	SSF =	.1	.2	.3	.4	.5	.6	.7	.8	.9
LITTLE ROCK	WW	157	72	44	31	22	17	12	8	-
ARKANSAS	WWNI	246	117	74	52	40	31	25	20	14
	TW	153	69	42	28	19	13	9	6	3
34.7 N	TWNI	232	110	69	49	37	29	22	16	11
3354 DD	DG	154	65	35	18	-	-	-	-	-
T(JAN)=39	DGNI	257	120	75	53	39	29	22	15	10
	WW	14.6	12.1	10.5	9.4	7.7	5.8	3.4	1.5	-
	WWNI	23.0	21.4	18.5	17.2	16.1	13.9	11.4	7.3	3.0
	TW	13.8	11.7	9.6	7.3	5.2	3.5	2.2	1.2	.4
D	TWNI	21.9	19.6	17.6	16.0	13.9	11.2	7.9	4.9	2.3
	DG	13.0	9.6	5.8	2.4	-	-	-	-	-
	DGNI	24.2	21.0	18.9	16.5	13.7	10.2	7.0	4.0	1.5

	SSF =	.1	.2	.3	.4	.5	.6	.7	.8	.9
BAKERSFIELD	WW	323	143	89	63	46	35	26	18	10
CALIFORNIA	WWNI	449	203	127	91	69	54	42	32	22
	TW	297	135	81	55	39	28	20	14	8
35.4 N	TWNI	418	191	119	85	63	48	36	27	17
2185 DD	DG	359	158	93	60	40	26	15	-	-
T(JAN)=48	DGNI	475	216	135	94	69	52	39	27	17
	WW	28.6	22.6	19.7	15.7	11.9	8.5	4.5	1.3	-
	WWNI	41.9	34.7	33.5	26.6	21.1	16.2	11.1	4.6	-
	TW	26.9	22.7	18.8	15.1	11.6	8.5	5.7	3.3	1.2
D	TWNI	38.0	33.2	30.7	26.9	22.3	18.1	12.7	7.6	3.0
	DG	31.5	25.4	19.8	14.3	9.5	4.8	2.3	-	-
	DGNI	42.9	37.5	34.0	28.8	23.6	18.2	12.4	7.2	2.8

	SSF =	.1	.2	.3	.4	.5	.6	.7	.8	.9
DAGGETT	WW	360	165	103	73	55	42	32	24	15
CALIFORNIA	WWNI	491	228	144	103	79	62	49	39	27
	TW	334	153	94	65	46	34	25	17	11
34.9 N	TWNI	461	213	135	96	72	56	43	32	21
2203 DD	DG	411	185	112	74	51	35	23	13	-
T(JAN)=47	DGNI	526	244	153	108	81	61	46	33	22
	WW	33.0	28.9	24.2	20.3	16.2	11.6	6.5	2.2	-
	WWNI	46.6	40.2	37.7	32.4	26.8	21.3	14.2	5.9	-
	TW	30.6	26.6	22.6	18.3	14.4	10.8	7.4	4.6	1.9
D	TWNI	42.1	38.2	35.1	31.6	27.0	22.0	15.6	9.5	3.9
	DG	36.9	31.1	25.1	19.0	13.5	8.7	4.8	1.9	-
	DGNI	48.3	43.0	39.6	34.5	29.0	22.3	15.5	9.3	3.8

	SSF =	.1	.2	.3	.4	.5	.6	.7	.8	.9
FRESNO	WW	272	122	74	51	37	27	20	13	7
CALIFORNIA	WWNI	383	179	110	77	58	45	34	26	18
	TW	252	114	68	45	31	22	15	10	5
36.8 N	TWNI	361	166	102	72	53	40	30	22	14
2650 DD	DG	295	128	72	44	27	15	-	-	-
T(JAN)=45	DGNI	408	186	114	79	57	42	31	22	13
	WW	25.8	19.8	17.2	15.1	11.7	8.5	5.8	2.8	.6
	WWNI	37.0	29.9	27.4	24.9	21.6	17.1	12.7	8.8	3.6
	TW	22.6	18.8	15.1	11.6	8.7	6.2	3.9	2.2	.7
D	TWNI	34.1	28.4	25.3	22.0	18.0	14.4	10.2	6.0	2.3
	DG	25.6	19.6	14.1	9.3	5.4	2.2	-	-	-
	DGNI	37.7	31.8	27.9	23.2	18.6	14.3	9.6	5.5	2.0

	SSF =	.1	.2	.3	.4	.5	.6	.7	.8	.9
LONG BEACH	WW	527	249	156	112	85	67	53	40	27
CALIFORNIA	WWNI	689	331	208	150	115	91	74	59	42
	TW	488	228	142	99	72	53	39	29	19
33.8 N	TWNI	650	307	195	139	106	83	64	48	32
1606 DD	DG	630	291	180	123	87	63	44	30	17
T(JAN)=54	DGNI	753	354	224	159	120	92	70	52	34
	WW	50.5	43.9	40.6	37.1	33.3	28.8	20.7	12.4	4.9
	WWNI	66.1	59.1	54.6	51.7	47.0	42.1	34.8	22.4	9.6
	TW	45.3	40.4	35.5	29.2	24.5	17.5	13.0	8.3	4.2
D	TWNI	62.3	55.2	51.2	46.5	41.5	33.3	23.9	14.8	6.2
	DG	57.9	50.8	43.8	34.6	26.2	18.8	12.3	6.8	2.4
	DGNI	71.4	63.6	58.1	52.3	45.0	34.6	24.7	15.3	6.5

	SSF =	.1	.2	.3	.4	.5	.6	.7	.8	.9
LOS ANGELES	WW	563	259	161	115	87	68	54	41	27
CALIFORNIA	WWNI	737	342	215	153	117	93	75	60	43
	TW	513	238	147	101	73	54	40	29	20
33.9 N	TWNI	687	320	200	143	108	84	65	49	33
1819 DD	DG	665	304	187	127	90	65	46	31	18
T(JAN)=54	DGNI	793	369	231	163	123	94	72	53	35
	WW	52.5	45.3	41.1	37.7	33.7	29.6	21.2	12.8	5.1
	WWNI	71.1	61.4	55.3	52.1	47.4	42.5	35.7	22.9	9.8
	TW	47.7	41.4	36.1	29.7	23.7	18.3	13.0	8.7	4.3
D	TWNI	64.3	56.3	51.7	47.0	42.6	33.9	24.3	15.1	6.3
	DG	61.0	52.0	44.7	35.3	27.0	19.4	12.7	7.1	2.7
	DGNI	74.0	64.7	58.9	53.0	45.9	35.3	25.2	15.7	6.7

	SSF =	.1	.2	.3	.4	.5	.6	.7	.8	.9
MOUNT SHASTA	WW	142	62	37	25	18	13	9	5	-
CALIFORNIA	WWNI	230	106	66	46	35	27	21	16	11
	TW	137	61	36	23	15	10	7	4	-
41.3 N	TWNI	216	99	62	43	32	25	19	14	9
5890 DD	DG	130	50	23	-	-	-	-	-	-
T(JAN)=34	DGNI	237	107	66	46	33	25	18	12	7
	WW	12.4	10.1	8.5	7.0	5.4	3.6	1.9	.5	-
	WWNI	21.0	18.3	16.6	15.0	13.2	11.4	8.5	5.7	-
	TW	12.1	9.7	7.6	5.6	3.9	2.4	1.3	.5	-
D	TWNI	20.0	17.2	15.4	13.6	11.6	9.2	6.5	3.9	1.5
	DG	10.4	6.5	2.7	-	-	-	-	-	-
	DGNI	21.4	18.3	15.9	13.7	11.0	8.1	5.4	2.9	.9

SAN DIEGO CALIFORNIA 32.7 N 1507 DD T(JAN)=55		SSF =	.1	.2	.3	.4	.5	.6	.7	.8	.9
	L	WW	612	284	179	129	98	77	61	46	31
	C	WWNI	796	373	235	170	131	104	84	67	48
	R	TW	562	261	163	113	82	61	46	33	23
		TWNI	744	348	220	158	120	94	73	54	37
		DG	734	338	210	144	103	75	53	37	22
		DGNI	864	403	255	181	137	105	80	59	40
	D	WW	56.7	49.9	47.0	43.3	39.0	33.5	24.4	14.7	6.0
		WWNI	74.2	66.1	62.3	59.7	53.5	48.3	39.3	25.5	10.9
		TW	51.9	46.0	40.5	33.6	26.8	20.9	15.3	9.9	4.9
		TWNI	69.7	62.0	58.3	52.9	47.6	38.0	27.3	16.9	7.1
		DG	67.0	58.8	51.2	41.1	31.7	23.1	15.5	8.9	3.5
		DGNI	80.7	71.9	66.4	60.1	51.4	39.9	28.5	17.7	7.6

SAN FRANCISCO CALIFORNIA 37.6 N 3042 DD T(JAN)=48		SSF =	.1	.2	.3	.4	.5	.6	.7	.8	.9
	L	WW	332	163	103	72	53	40	31	22	13
	C	WWNI	453	225	145	103	77	60	47	37	26
	R	TW	313	149	92	63	45	33	23	16	10
		TWNI	428	210	134	94	71	54	41	30	20
		DG	385	180	109	72	49	32	20	11	-
		DGNI	491	239	151	106	79	59	44	32	20
	D	WW	33.2	30.6	25.5	22.0	19.0	14.7	10.5	5.8	1.9
		WWNI	45.3	43.2	37.9	32.5	29.4	25.5	19.3	13.1	5.4
		TW	40.8	34.6	28.3	23.3	17.9	13.9	9.9	4.0	1.3
		TWNI	42.8	39.6	34.2	30.4	25.7	20.4	14.6	8.8	3.6
		DG	37.0	30.9	24.5	18.1	12.4	7.7	3.9	1.3	-
		DGNI	48.7	44.0	38.2	33.4	27.1	20.8	14.4	8.6	3.4

SANTA MARIA CALIFORNIA 34.9 N 3053 DD T(JAN)=50		SSF =	.1	.2	.3	.4	.5	.6	.7	.8	.9
	L	WW	327	164	107	77	58	46	35	26	17
	C	WWNI	448	224	148	108	83	66	53	42	30
	R	TW	311	152	97	67	49	36	27	19	12
		TWNI	423	211	138	100	76	59	46	34	23
		DG	383	185	116	79	55	38	26	16	7
		DGNI	485	241	156	112	85	65	50	36	24
	D	WW	32.7	29.1	25.5	25.7	22.4	19.1	13.2	7.6	2.8
		WWNI	44.8	42.8	42.6	37.2	33.8	29.0	24.9	15.9	6.8
		TW	30.9	28.2	24.8	20.2	15.9	12.0	8.4	5.1	2.6
		TWNI	42.3	41.6	37.8	34.0	30.3	23.9	17.1	10.5	4.4
		DG	37.7	33.6	28.3	21.5	15.5	10.3	6.0	2.7	-.6
		DGNI	48.5	46.9	42.3	37.7	32.0	24.3	17.1	10.4	4.3

COLORADO SPRINGS COLORADO 38.8 N 6473 DD T(JAN)=29		SSF =	.1	.2	.3	.4	.5	.6	.7	.8	.9
	L	WW	130	60	38	26	19	15	11	7	-
	C	WWNI	211	102	65	47	36	29	23	18	13
	R	TW	128	59	36	24	17	11	8	5	-
		TWNI	201	96	61	44	33	26	20	15	10
		DG	120	52	26	10	-	-	-	-	-
		DGNI	220	104	66	47	35	26	20	14	9
	D	WW	12.5	10.5	9.2	8.1	6.9	4.9	2.8	1.1	-
		WWNI	20.1	18.7	17.4	16.0	14.6	13.2	10.8	6.8	2.8
		TW	11.8	10.2	8.3	6.3	4.5	3.0	1.8	.8	.3
		TWNI	19.4	17.6	16.2	14.7	13.1	10.4	7.4	4.5	1.8
		DG	17.5	-	-	-	-	-	-	-	-
		DGNI	20.9	19.0	16.9	15.2	12.5	9.2	6.3	3.6	1.3

NEEDLES CALIFORNIA 34.8 N 1428 DD T(JAN)=52		SSF =	.1	.2	.3	.4	.5	.6	.7	.8	.9
	L	WW	508	239	152	108	81	62	47	35	23
	C	WWNI	668	317	202	147	111	86	68	53	37
	R	TW	474	220	136	94	67	49	36	26	17
		TWNI	630	297	189	134	100	77	59	44	29
		DG	608	280	171	116	81	57	40	26	13
		DGNI	729	342	217	153	114	86	65	47	30
	D	WW	47.4	42.8	40.9	34.3	29.0	23.6	17.5	10.2	3.9
		WWNI	62.3	57.4	54.6	49.8	41.8	34.9	28.8	19.1	8.1
		TW	43.3	38.5	33.1	27.3	21.3	16.1	11.4	6.9	3.3
		TWNI	59.0	53.9	49.2	43.2	36.7	29.9	21.1	12.8	5.2
		DG	55.1	49.8	40.2	31.7	23.3	16.2	10.1	5.2	1.5
		DGNI	67.6	62.2	55.5	48.0	40.6	31.2	21.7	13.1	5.4

OAKLAND CALIFORNIA 37.7 N 2909 DD T(JAN)=49		SSF =	.1	.2	.3	.4	.5	.6	.7	.8	.9
	L	WW	348	170	107	75	56	42	32	23	14
	C	WWNI	472	234	153	106	80	62	49	38	27
	R	TW	327	155	96	66	47	34	24	17	11
		TWNI	446	219	139	98	73	56	43	31	21
		DG	405	189	115	76	51	35	22	12	-
		DGNI	512	249	157	110	82	61	46	33	21
	D	WW	34.8	31.7	26.6	23.1	20.0	15.3	11.0	6.1	2.0
		WWNI	47.2	44.8	39.1	33.8	30.6	26.4	19.8	13.5	5.6
		TW	31.7	27.6	23.2	18.7	14.3	10.4	7.0	4.2	2.0
		TWNI	41.4	40.8	35.4	31.5	26.6	21.0	15.0	9.1	3.6
		DG	38.8	32.5	25.8	19.2	13.3	8.3	4.4	1.6	-
		DGNI	50.6	45.6	39.7	34.7	28.1	21.6	15.0	8.9	3.5

RED BLUFF CALIFORNIA 40.1 N 2688 DD T(JAN)=45		SSF =	.1	.2	.3	.4	.5	.6	.7	.8	.9
	L	WW	258	115	69	48	35	26	19	13	6
	C	WWNI	365	170	104	74	56	43	33	26	18
	R	TW	239	108	64	43	30	21	15	10	5
		TWNI	345	158	98	69	51	39	29	21	14
		DG	276	119	68	41	25	14	-	-	-
		DGNI	389	177	109	76	55	41	30	21	13
	D	WW	24.4	18.5	16.7	14.8	11.4	8.3	5.7	2.7	.6
		WWNI	35.5	28.4	26.0	24.1	21.6	17.1	12.6	8.7	3.5
		TW	21.4	17.0	14.1	11.1	8.4	6.0	3.1	2.1	3.7
		TWNI	32.6	27.0	24.3	21.6	17.7	14.2	10.1	6.0	2.3
		DG	23.9	18.4	13.2	8.6	4.9	1.8	-	-	-
		DGNI	35.9	30.1	26.7	22.6	18.2	14.0	9.4	5.4	1.9

SACRAMENTO CALIFORNIA 38.5 N 2843 DD T(JAN)=45		SSF =	.1	.2	.3	.4	.5	.6	.7	.8	.9
	L	WW	258	119	73	50	36	26	19	13	6
	C	WWNI	368	174	108	76	57	44	34	26	18
	R	TW	244	110	66	44	31	22	15	10	5
		TWNI	347	162	101	71	52	39	30	21	14
		DG	283	124	71	44	27	15	-	-	-
		DGNI	393	182	112	78	57	42	31	22	13
	D	WW	23.9	20.2	17.4	15.0	11.6	8.3	5.7	2.7	.5
		WWNI	34.2	30.7	27.1	24.4	21.7	17.1	12.4	8.7	3.5
		TW	22.0	18.5	15.0	11.8	8.6	6.0	3.9	2.0	3.6
		TWNI	32.6	28.4	25.2	22.1	17.9	14.1	10.1	6.0	2.3
		DG	24.7	19.4	13.9	9.2	5.1	2.0	-	-	-
		DGNI	36.3	31.4	27.6	23.2	18.4	14.1	9.4	5.4	1.9

Tables of LCR & D Versus SSF / 259

DENVER COLORADO 39.7 N 6016 DD T(JAN)=30		SSF =	.1	.2	.3	.4	.5	.6	.7	.8	.9
	L	WW	136	63	39	27	20	15	11	7	-
	C	WWNI	218	105	67	48	37	29	24	19	13
	R	TW	132	61	38	25	17	12	8	5	1
		TWNI	207	99	63	45	34	27	21	15	10
		DG	127	54	28	13	-	-	-	-	-
		DGNI	227	108	68	48	36	27	20	14	9
	D	WW	12.9	11.1	9.6	8.4	7.1	5.1	3.0	1.2	-
		WWNI	21.8	19.1	18.2	16.2	15.0	13.6	10.8	6.9	2.9
		TW	12.5	10.6	8.7	6.5	4.7	3.2	1.9	.9	.3
		TWNI	20.2	18.2	16.5	15.1	13.3	10.6	7.5	4.6	1.9
		DG	11.2	8.0	4.3	1.1	-	-	-	-	-
		DGNI	21.8	19.6	17.4	15.6	12.8	9.5	6.5	3.7	1.3

EAGLE COLORADO 39.6 N 8426 DD T(JAN)=18		SSF =	.1	.2	.3	.4	.5	.6	.7	.8	.9
	L	WW	95	43	26	17	12	8	5	-	-
	C	WWNI	172	81	52	37	28	22	17	13	9
	R	TW	96	43	25	16	10	7	4	1	-
		TWNI	163	77	49	34	26	20	15	11	7
		DG	72	22	-	-	-	-	-	-	-
		DGNI	174	81	51	35	26	19	14	9	5
	D	WW	8.5	7.1	5.8	4.7	3.3	1.9	.6	-	-
		WWNI	16.2	15.2	13.4	12.2	10.8	9.3	7.1	4.7	1.9
		TW	8.7	7.0	5.4	3.7	2.3	1.1	.5	-	-
		TWNI	15.3	13.8	12.5	11.1	9.5	7.6	5.3	3.2	1.2
		DG	5.4	1.5	-	-	-	-	-	-	-
		DGNI	16.2	14.3	12.6	10.8	8.6	6.2	4.0	2.1	.6

GRAND JUNCTION COLORADO 39.1 N 5605 DD T(JAN)=27		SSF =	.1	.2	.3	.4	.5	.6	.7	.8	.9
	L	WW	135	61	37	25	18	13	9	5	-
	C	WWNI	222	104	66	46	35	28	22	17	12
	R	TW	133	60	36	23	16	11	7	4	1
		TWNI	210	98	61	43	33	25	19	14	-
		DG	125	50	24	-	-	-	-	-	-
		DGNI	229	106	66	46	34	25	18	13	8
	D	WW	12.0	10.3	8.7	7.5	5.7	4.1	2.2	.7	-
		WWNI	20.1	18.7	16.9	15.2	13.9	11.8	9.1	6.0	2.4
		TW	11.8	9.8	7.9	5.9	4.1	2.6	1.5	.6	1.1
		TWNI	19.2	17.4	15.6	14.1	11.9	9.6	6.8	4.1	1.6
		DG	10.2	6.7	3.0	-	-	-	-	-	-
		DGNI	20.9	18.4	16.3	14.1	11.5	8.5	5.7	3.2	1.1

PUEBLO COLORADO 38.3 N 5394 DD T(JAN)=30		SSF =	.1	.2	.3	.4	.5	.6	.7	.8	.9
	L	WW	145	67	41	29	21	16	12	8	-
	C	WWNI	234	109	70	50	38	30	24	19	14
	R	TW	143	65	39	26	18	13	9	5	3
		TWNI	220	103	66	47	36	28	21	16	11
		DG	140	59	32	16	-	-	-	-	-
		DGNI	242	113	71	50	38	28	21	15	9
	D	WW	12.8	11.8	10.2	9.0	7.5	5.5	3.2	1.4	-
		WWNI	21.2	20.3	18.6	16.9	15.4	13.9	11.2	7.1	3.0
		TW	12.9	11.0	9.1	7.0	5.1	3.4	2.1	1.1	.4
		TWNI	20.2	18.9	17.2	15.7	13.7	11.0	7.8	4.8	1.9
		DG	11.8	8.8	5.1	1.9	-	-	-	-	-
		DGNI	22.1	20.1	18.1	16.2	13.4	9.9	6.8	3.9	1.4

HARTFORD CONNECTICUT 41.9 N 6350 DD T(JAN)=25		SSF =	.1	.2	.3	.4	.5	.6	.7	.8	.9
	L	WW	62	26	14	9	5	-	-	-	-
	C	WWNI	134	62	39	28	21	16	13	10	7
	R	TW	66	28	16	9	5	-	-	-	-
		TWNI	127	59	37	26	19	15	11	8	5
		DG	-	-	-	-	-	-	-	-	-
		DGNI	131	60	37	25	18	13	9	6	3
	D	WW	5.2	3.8	2.8	1.7	.6	-	-	-	-
		WWNI	12.4	11.3	10.0	8.9	7.9	6.6	5.1	3.3	1.3
		TW	5.6	4.2	2.8	1.6	.6	-	-	-	-
		TWNI	11.6	10.6	9.3	8.2	6.9	5.5	3.8	2.3	.8
		DG	-	-	-	-	-	-	-	-	-
		DGNI	11.8	10.2	8.8	7.3	5.7	3.9	2.4	1.1	.2

WILMINGTON DELAWARE 39.7 N 4940 DD T(JAN)=32		SSF =	.1	.2	.3	.4	.5	.6	.7	.8	.9
	L	WW	95	44	26	18	12	9	5	-	-
	C	WWNI	171	83	52	37	28	22	18	14	9
	R	TW	97	44	26	17	11	7	4	1	1
		TWNI	163	78	49	35	26	20	15	11	7
		DG	74	24	-	-	-	-	-	-	-
		DGNI	174	82	51	36	26	19	14	10	5
	D	WW	8.9	7.2	6.0	4.8	3.5	2.1	.7	-	-
		WWNI	16.1	15.5	13.4	12.3	11.1	9.3	7.4	4.8	1.9
		TW	8.7	7.2	5.5	4.1	2.7	1.4	.3	-	-
		TWNI	15.4	14.1	12.7	11.3	9.6	7.7	5.4	3.3	1.3
		DG	5.7	1.9	-	-	-	-	-	-	-
		DGNI	16.5	14.5	12.8	10.9	8.8	6.4	4.1	2.2	.7

WASHINGTON DC 38.9 N 5010 DD T(JAN)=32		SSF =	.1	.2	.3	.4	.5	.6	.7	.8	.9
	L	WW	92	42	25	17	11	8	4	-	-
	C	WWNI	169	80	51	36	27	21	17	13	9
	R	TW	94	42	25	16	10	6	3	-	-
		TWNI	160	76	48	34	25	19	15	11	7
		DG	69	19	-	-	-	-	-	-	-
		DGNI	171	80	50	35	25	19	13	9	5
	D	WW	8.4	6.9	5.5	4.4	3.1	1.8	.5	-	-
		WWNI	15.6	14.9	13.1	11.7	10.7	8.9	7.1	4.6	1.8
		TW	8.4	6.8	5.2	3.6	2.3	1.2	.4	-	-
		TWNI	14.9	13.7	12.2	10.9	9.3	7.4	5.2	3.1	1.2
		DG	5.0	.9	-	-	-	-	-	-	-
		DGNI	15.9	14.0	12.3	10.4	8.4	6.0	3.9	2.1	.6

APALACHICOLA FLORIDA 29.7 N 1361 DD T(JAN)=54		SSF =	.1	.2	.3	.4	.5	.6	.7	.8	.9
	L	WW	405	182	114	81	61	48	37	27	18
	C	WWNI	547	249	156	113	87	68	55	44	31
	R	TW	371	170	104	72	51	38	28	20	13
		TWNI	511	233	147	105	79	62	48	36	24
		DG	464	208	126	84	59	41	28	17	8
		DGNI	585	267	168	119	89	68	52	38	25
	D	WW	36.6	31.0	29.9	26.7	23.1	19.7	15.3	8.0	2.9
		WWNI	50.2	42.8	41.4	39.2	34.7	34.5	25.8	16.3	6.9
		TW	33.7	29.1	25.2	20.5	16.3	12.4	8.8	5.6	2.5
		TWNI	46.7	40.7	38.4	34.5	31.0	24.7	17.6	10.9	4.5
		DG	41.1	34.7	29.2	22.5	16.4	11.1	6.6	3.1	-
		DGNI	53.0	46.7	43.0	38.3	32.9	25.2	17.8	10.8	4.5

DAYTONA BEACH, FLORIDA
29.2 N, 902 DD, T(JAN)=58

	SSF = .1	.2	.3	.4	.5	.6	.7	.8	.9
L WW	603	281	180	130	100	79	62	48	32
C WWNI	784	367	235	172	133	106	86	69	49
R TW	556	260	164	114	83	62	47	34	23
TWNI	735	344	221	159	122	96	74	56	38
DG	725	337	211	146	105	76	55	38	23
DGNI	852	399	256	183	139	107	82	61	41
D WW	55.5	50.8	49.4	45.2	39.8	35.4	25.3	15.3	6.3
WWNI	73.8	66.4	65.2	61.3	55.2	49.0	41.9	26.4	11.4
TW	50.7	46.6	41.7	34.3	27.9	21.7	15.9	10.5	5.4
TWNI	67.6	63.0	59.8	54.4	49.3	39.1	28.1	17.5	7.3
DG	65.9	59.8	52.8	42.3	32.8	24.1	16.2	9.4	3.7
DGNI	78.1	73.2	68.4	61.6	53.3	41.1	29.4	18.3	7.9

JACKSONVILLE, FLORIDA
30.5 N, 1327 DD, T(JAN)=55

	SSF = .1	.2	.3	.4	.5	.6	.7	.8	.9
L WW	425	191	120	87	66	52	41	31	20
C WWNI	573	260	164	119	92	74	60	47	35
R TW	390	179	111	77	55	41	30	22	15
TWNI	533	244	154	111	85	66	52	39	26
DG	489	221	136	92	65	46	32	21	11
DGNI	611	279	177	126	96	73	56	41	27
D WW	37.8	32.9	32.3	29.8	26.0	22.6	15.8	9.3	3.5
WWNI	52.8	44.9	44.0	43.2	38.5	33.9	28.1	18.4	7.7
TW	35.4	31.2	27.1	22.4	18.0	13.9	10.0	6.4	3.1
TWNI	48.4	42.9	41.4	37.7	33.9	27.1	19.4	12.0	5.0
DG	43.5	37.7	32.3	25.2	18.8	13.1	8.1	4.1	1.2
DGNI	55.2	49.2	46.6	41.9	36.1	27.8	19.7	12.1	5.1

MIAMI, FLORIDA
25.8 N, 206 DD, T(JAN)=67

	SSF = .1	.2	.3	.4	.5	.6	.7	.8	.9
L WW	2285	1138	731	516	389	307	242	186	130
C WWNI	2804	1401	906	641	483	382	307	242	174
R TW	2102	1013	641	449	329	245	184	135	94
TWNI	2638	1304	832	589	446	344	265	198	133
DG	2912	1403	886	620	452	336	249	181	121
DGNI	3119	1534	975	692	520	398	303	224	151
D WW	229.	218.	189.	166.	153.	131.	101.	64.	28.
WWNI	280.	274.	240.	205.	188.	178.	136.	92.	39.
TW	206.	185.	163.	137.	110.	84.	63.	42.	23.
TWNI	264.	247.	214.	192.	169.	136.	97.	61.	25.
DG	286.	255.	226.	188.	149.	114.	81.	52.	25.
DGNI	312.	287.	251.	228.	190.	149.	107.	68.	30.

ORLANDO, FLORIDA
28.5 N, 733 DD, T(JAN)=60

	SSF = .1	.2	.3	.4	.5	.6	.7	.8	.9
L WW	723	346	224	163	124	98	78	60	41
C WWNI	927	446	289	211	163	130	105	84	60
R TW	663	319	202	142	104	78	58	43	30
TWNI	867	418	270	195	149	117	91	68	46
DG	879	421	265	185	134	99	72	51	32
DGNI	1010	487	314	226	171	132	101	75	50
D WW	69.8	64.7	61.8	56.4	49.8	43.6	31.5	19.1	8.2
WWNI	90.9	83.1	81.9	75.0	67.1	60.0	50.3	32.1	13.8
TW	64.0	58.4	52.1	43.2	35.1	27.4	20.1	12.8	7.0
TWNI	83.7	78.4	73.7	66.7	59.7	47.7	34.2	21.2	8.9
DG	84.3	76.8	67.6	54.9	43.0	32.2	22.0	13.1	5.6
DGNI	97.3	92.0	84.6	76.2	65.8	50.7	36.3	22.6	9.8

TALLAHASSEE, FLORIDA
30.4 N, 1563 DD, T(JAN)=53

	SSF = .1	.2	.3	.4	.5	.6	.7	.8	.9
L WW	359	162	101	72	55	43	33	24	16
C WWNI	491	226	141	102	79	62	50	40	28
R TW	330	152	93	64	46	34	25	18	12
TWNI	458	211	133	95	72	56	43	32	22
DG	405	183	110	74	51	35	23	14	5
DGNI	523	240	151	107	80	62	47	34	22
D WW	33.2	27.3	26.3	24.0	20.6	17.8	12.3	7.0	2.5
WWNI	46.5	38.5	36.9	36.6	32.0	28.0	23.7	15.0	6.4
TW	30.2	26.2	22.5	18.2	14.2	11.0	7.8	4.9	2.4
TWNI	42.7	36.7	35.1	31.5	28.2	22.6	16.2	10.0	4.1
DG	36.2	30.5	25.3	19.2	13.8	9.1	5.2	2.2	.3
DGNI	48.2	41.9	39.0	34.7	29.9	22.9	16.1	9.8	4.0

TAMPA, FLORIDA
28.0 N, 718 DD, T(JAN)=60

	SSF = .1	.2	.3	.4	.5	.6	.7	.8	.9
L WW	720	343	223	162	124	98	78	59	41
C WWNI	926	441	288	211	162	129	105	83	60
R TW	663	317	201	141	104	78	58	43	30
TWNI	866	415	269	195	149	117	90	68	46
DG	878	418	264	184	134	98	72	50	32
DGNI	1008	484	312	225	171	132	101	75	50
D WW	67.7	64.8	61.7	56.0	49.6	43.5	31.8	19.5	8.2
WWNI	87.2	83.2	82.5	74.9	66.5	59.8	50.9	32.0	13.8
TW	63.4	57.8	51.8	43.1	35.0	27.4	21.0	12.8	6.5
TWNI	82.1	78.7	73.7	66.5	59.6	47.6	34.1	21.2	8.9
DG	83.5	75.8	67.2	54.9	42.9	31.9	21.9	13.1	5.6
DGNI	95.7	91.5	84.2	76.1	65.6	50.6	36.2	22.5	9.8

WEST PALM BEACH, FLORIDA
26.7 N, 299 DD, T(JAN)=65

	SSF = .1	.2	.3	.4	.5	.6	.7	.8	.9
L WW	1523	761	494	350	264	208	164	126	88
C WWNI	1886	943	619	441	331	261	211	167	120
R TW	1407	682	433	304	223	167	125	91	63
TWNI	1775	884	568	403	305	237	183	136	92
DG	1930	935	592	415	302	224	165	119	79
DGNI	2093	1038	663	471	355	272	207	153	103
D WW	152.	151.	130.	112.	103.	90.	68.	43.	19.
WWNI	189.	189.	168.	141.	128.	121.	96.	64.	28.
TW	137.	125.	111.	93.	75.	58.	42.	28.	15.
TWNI	178.	171.	150.	132.	118.	94.	67.	42.	18.
DG	192.	171.	152.	125.	99.	75.	53.	34.	16.
DGNI	209.	197.	172.	155.	132.	102.	73.	46.	21.

ATLANTA, GEORGIA
33.6 N, 3095 DD, T(JAN)=42

	SSF = .1	.2	.3	.4	.5	.6	.7	.8	.9
L WW	172	79	48	33	25	19	14	9	4
C WWNI	264	126	79	56	42	33	27	21	15
R TW	166	75	46	31	21	15	10	7	4
TWNI	249	118	74	52	39	30	23	17	11
DG	173	74	41	23	10	-	-	-	-
DGNI	276	129	80	56	42	32	23	17	10
D WW	15.5	13.5	11.5	10.2	8.5	6.6	4.0	1.8	.2
WWNI	24.6	22.9	20.0	18.3	16.9	14.6	12.1	7.7	3.2
TW	15.0	12.8	10.5	8.1	5.8	4.1	2.5	1.4	.5
TWNI	23.3	21.2	18.8	17.0	14.8	11.9	8.4	5.2	2.1
DG	14.8	11.2	7.3	4.1	1.0	-	-	-	-
DGNI	25.9	22.6	20.3	17.6	14.8	11.0	7.5	4.4	1.6

AUGUSTA
GEORGIA

33.4 N
2547 DD
T(JAN)=46

		SSF =	.1	.2	.3	.4	.5	.6	.7	.8	.9
L		WW	213	101	62	43	32	25	19	13	7
C		WWNI	312	152	95	67	51	41	33	26	18
R		TW	204	94	58	39	27	20	14	9	6
		TWNI	295	141	89	63	48	37	28	21	14
		DG	228	101	59	37	22	12	–	–	–
		DGNI	332	157	98	69	52	39	29	21	13
D		WW	20.5	15.0	13.8	11.6	9.4	6.0	4.8
		WWNI	30.0	23.8	22.4	21.0	18.0	15.0	13.8	3.1
		TW	18.7	16.3	13.7	10.7	8.7	5.7	3.8	2.2	4.0
		TWNI	28.5	25.2	22.7	20.7	18.0	14.5	10.4	6.4	2.9
		DG	20.3	16.3	12.0	7.8	4.3	1.6	.9	–	
		DGNI	32.0	27.7	25.1	21.9	18.5	13.9	9.7	5.7	2.6 2.2

MACON
GEORGIA

32.7 N
2240 DD
T(JAN)=48

		SSF =	.1	.2	.3	.4	.5	.6	.7	.8	.9
L		WW	244	114	70	49	37	28	21	15	9
C		WWNI	349	169	105	74	57	45	36	28	20
R		TW	230	106	65	44	31	22	16	11	7
		TWNI	330	156	98	69	52	41	31	23	16
		DG	266	119	70	44	28	17	9	–	–
		DGNI	373	175	109	77	57	43	33	24	15
D		WW	23.8	17.1	15.0		13.4	11.0	7.2	3.8	1.1
		WWNI	34.0	29.7	25.0		19.9	16.7	10.6	4.4
		TW	21.2	18.4	15.5	12.2	9.1	6.7	4.6	2.8	1.1
		TWNI	32.1	27.5	25.0	22.7	19.9	16.1	11.5	7.1	2.9
		DG	23.7	19.2	14.7	10.2	6.1	3.1	.9	–	
		DGNI	35.9	30.5	27.8	24.3	20.7	15.6	10.9	6.5	2.6

SAVANNAH
GEORGIA

32.1 N
1952 DD
T(JAN)=50

		SSF =	.1	.2	.3	.4	.5	.6	.7	.8	.9
L		WW	278	131	81	57	43	33	26	19	11
C		WWNI	389	189	118	84	65	51	41	32	23
R		TW	260	122	75	51	37	26	19	13	9
		TWNI	367	175	110	79	59	46	36	27	18
		DG	309	141	84	55	37	24	14	6	–
		DGNI	417	197	124	88	66	50	38	27	18
D		WW	27.5	20.4	18.7	15.7	15.9	13.3	9.0	4.9	1.6
		WWNI	38.9	33.7	28.9	26.2	22.6	19.1	12.1	5.1
		TW	24.5	21.4	18.2	14.3	11.1	8.2	5.7	3.5	1.5
		TWNI	36.1	31.2	28.7	25.8	22.8	18.4	13.1	8.1	3.3
		DG	28.4	23.5	18.7	13.4	8.9	4.1	2.4	.9	
		DGNI	40.5	35.0	32.0	28.0	23.9	18.1	12.7	7.7	3.1

BOISE
IDAHO

43.6 N
5833 DD
T(JAN)=29

		SSF =	.1	.2	.3	.4	.5	.6	.7	.8	.9
L		WW	132	58	34	22	15	10	5	–	–
C		WWNI	215	101	63	44	32	25	19	14	9
R		TW	127	56	32	20	13	8	4	–	–
		TWNI	203	95	59	41	30	22	16	12	7
		DG	117	42	–	–	–	–	–	–	–
		DGNI	221	102	62	42	30	22	15	10	5
D		WW	11.8	7.2	5.4	3.6	2.1	.7	–		
		WWNI	21.1	15.8	13.6	11.3	9.2	6.5	4.4	1.7
		TW	11.4	6.5	4.6	3.0	1.6	.6	–		
		TWNI	19.2	14.3	12.1	9.8	7.5	5.3	3.1	1.1
		DG	9.3	4.6	–	–	–	–	–	–	
		DGNI	20.5	17.4	14.5	11.9	9.0	6.6	4.1	2.1	.6

LEWISTON
IDAHO

46.4 N
5464 DD
T(JAN)=31

		SSF =	.1	.2	.3	.4	.5	.6	.7	.8	.9
L		WW	110	46	26	15	9	4	–	–	–
C		WWNI	192	88	54	37	27	20	15	11	7
R		TW	108	46	25	15	9	4	–	–	–
		TWNI	181	82	50	34	25	18	13	9	6
		DG	85	–	–	–	–	–	–	–	–
		DGNI	194	87	52	34	24	17	11	7	3
D		WW	9.2	7.1	4.9	3.0	1.5	–			
		WWNI	18.5	14.8	13.3	10.9	8.7	6.3	4.8	3.3	1.3
		TW	9.3	6.7	4.6	2.9	1.5	.4	–		
		TWNI	16.5	14.0	11.9	9.7	7.7	5.8	4.1	2.3	.8
		DG	5.8	–	–	–	–	–	–	–	
		DGNI	17.4	14.4	11.6	9.0	6.7	4.7	2.8	1.3	.2

POCATELLO
IDAHO

42.9 N
7063 DD
T(JAN)=23

		SSF =	.1	.2	.3	.4	.5	.6	.7	.8	.9
L		WW	105	47	28	18	12	8	3	–	–
C		WWNI	182	88	55	39	29	22	17	13	9
R		TW	104	47	27	17	11	6	3	–	–
		TWNI	173	82	51	36	26	20	15	11	7
		DG	84	27	–	–	–	–	–	–	–
		DGNI	186	87	54	37	26	19	14	9	5
D		WW	9.9	7.8	5.9	4.4	2.8	1.5	–		
		WWNI	17.7	15.6	14.2	12.1	10.4	8.5	6.2	4.1	1.6
		TW	9.5	7.5	5.5	3.7	2.3	1.1	–		
		TWNI	16.8	14.7	12.8	11.0	9.1	7.0	4.9	2.9	1.1
		DG	6.5	2.1	–	–	–	–	–	–	
		DGNI	17.6	15.3	12.8	10.5	8.1	5.8	3.7	1.8	.5

CHICAGO
ILLINOIS

41.8 N
6127 DD
T(JAN)=24

		SSF =	.1	.2	.3	.4	.5	.6	.7	.8	.9
L		WW	73	31	17	11	6	–	–	–	–
C		WWNI	149	69	43	30	22	17	14	10	7
R		TW	78	33	18	11	6	2	–	–	–
		TWNI	141	65	40	28	21	16	12	9	5
		DG	34	–	–	–	–	–	–	–	–
		DGNI	148	67	41	28	20	14	10	7	3
D		WW	6.3	4.6	3.4	2.2	1.1	–			
		WWNI	13.1	12.7	10.6	9.3	8.4	6.9	5.3	3.5	1.3
		TW	6.5	4.9	3.4	2.0	.9	–			
		TWNI	12.6	11.4	10.0	8.8	7.3	5.8	4.0	2.4	.9
		DG	1.3	–	–	–	–	–	–	–	
		DGNI	13.1	11.3	9.6	7.9	6.2	4.3	2.6	1.3	.2

MOLINE
ILLINOIS

41.4 N
6395 DD
T(JAN)=21

		SSF =	.1	.2	.3	.4	.5	.6	.7	.8	.9
L		WW	72	30	17	10	6	–	–	–	–
C		WWNI	147	68	42	30	22	17	13	10	7
R		TW	76	32	18	10	6	2	–	–	–
		TWNI	139	64	40	28	21	16	12	9	5
		DG	29	–	–	–	–	–	–	–	–
		DGNI	145	66	40	27	19	14	10	6	3
D		WW	6.2	4.4	3.2	2.1	1.0	–			
		WWNI	13.1	12.3	10.4	9.1	8.3	6.9	5.3	3.4	1.3
		TW	6.4	4.7	3.3	1.9	.8	–			
		TWNI	12.6	11.2	9.7	8.7	7.2	5.7	4.0	2.4	.9
		DG	.8	–	–	–	–	–	–	–	
		DGNI	13.0	11.0	9.4	7.8	6.1	4.2	2.6	1.2	.2

262 / APPENDIX 12

		SSF =	.1	.2	.3	.4	.5	.6	.7	.8	.9
SPRINGFIELD ILLINOIS 39.8 N 5558 DD T(JAN)=27	L	WW	90	39	22	14	10	6	-	-	-
		WWNI	168	78	49	34	25	20	16	12	8
	C	TW	92	39	23	14	11	5	2	-	-
	R	TWNI	159	74	46	32	24	18	14	10	6
		DG	61	-	-	-	-	-	-	-	-
		DGNI	169	77	47	32	23	17	12	8	4
	D	WW	7.9	5.9	4.7	3.6	2.4	1.1	-	-	-
		WWNI	15.0	14.1	12.0	10.6	9.8	8.2	6.4	4.2	1.7
		TW	7.8	6.1	4.5	2.9	1.7	.8	.1	-	-
		TWNI	14.4	12.8	11.2	10.0	8.5	6.8	4.8	2.9	1.1
		DG	4.0	-	-	-	-	-	-	-	-
		DGNI	15.2	13.0	11.2	9.4	7.6	5.4	3.4	1.8	.5

		SSF =	.1	.2	.3	.4	.5	.6	.7	.8	.9
EVANSVILLE INDIANA 38.0 N 4629 DD T(JAN)=33	L	WW	104	46	26	17	12	8	5	-	-
		WWNI	182	86	54	37	28	22	17	13	9
	C	TW	104	45	26	17	11	7	2	-	-
	R	TWNI	173	81	50	35	26	20	15	11	7
		DG	81	25	-	-	-	-	-	-	-
		DGNI	186	85	52	36	26	19	14	9	5
	D	WW	9.3	7.2	5.6	4.5	3.3	1.9	.6	-	-
		WWNI	16.9	15.6	13.3	11.6	10.9	9.1	7.2	4.7	1.9
		TW	9.1	7.2	5.5	3.7	2.3	1.3	.5	-	-
		TWNI	16.1	14.2	12.3	11.1	9.4	7.5	5.3	3.2	1.2
		DG	5.9	1.9	-	-	-	-	-	-	-
		DGNI	17.1	14.5	12.5	10.6	8.6	6.2	4.0	2.1	.6

		SSF =	.1	.2	.3	.4	.5	.6	.7	.8	.9
FORT WAYNE INDIANA 41.0 N 6209 DD T(JAN)=25	L	WW	62	25	13	8	7	-	-	-	-
		WWNI	136	62	39	27	20	16	12	9	6
	C	TW	67	28	15	8	4	5	-	-	-
	R	TWNI	128	59	37	25	19	14	11	8	5
		DG	-	-	-	-	-	-	-	-	-
		DGNI	132	60	36	24	17	12	9	5	2
	D	WW	5.1	3.6	2.4	1.3	7.4	6.2	4.7	3.1	1.2
		WWNI	12.1	11.3	9.6	8.3	7.4	6.2	4.7	3.1	1.2
		TW	5.5	4.0	2.9	1.8	6.5	5.2	3.6	2.1	.8
		TWNI	11.5	10.3	8.9	7.9	6.5	5.2	3.6	2.1	.8
		DG	-	-	-	-	-	-	-	-	-
		DGNI	11.7	9.9	8.4	6.9	5.3	3.6	2.2	1.0	.1

		SSF =	.1	.2	.3	.4	.5	.6	.7	.8	.9
INDIANAPOLIS INDIANA 39.7 N 5577 DD T(JAN)=28	L	WW	74	31	17	11	7	-	-	-	-
		WWNI	148	69	43	30	23	17	14	11	7
	C	TW	77	33	18	11	6	2	-	-	-
	R	TWNI	141	65	41	28	21	16	12	9	6
		DG	35	-	-	-	-	-	-	-	-
		DGNI	147	67	41	28	20	14	10	7	3
	D	WW	6.5	4.6	3.4	2.3	1.3	-	-	-	-
		WWNI	13.0	12.6	10.6	9.7	8.5	7.1	5.5	3.6	1.4
		TW	6.6	4.9	3.4	2.1	1.2	-	-	-	-
		TWNI	12.4	11.4	9.9	8.9	7.4	5.9	4.1	2.4	.9
		DG	-	-	-	-	-	-	-	-	-
		DGNI	13.4	11.3	9.6	8.0	6.3	4.4	2.7	1.3	.3

		SSF =	.1	.2	.3	.4	.5	.6	.7	.8	.9
SOUTH BEND INDIANA 41.7 N 6462 DD T(JAN)=24	L	WW	60	24	12	6	-	-	-	-	-
		WWNI	137	61	38	26	19	15	11	9	6
	C	TW	66	26	14	7	18	13	10	7	5
	R	TWNI	129	58	36	25	18	13	10	7	5
		DG	-	-	-	-	-	-	-	-	-
		DGNI	132	58	35	23	16	12	8	5	1
	D	WW	4.8	3.2	1.9	7.8	6.9	5.6	4.2	2.8	1.0
		WWNI	11.8	10.8	9.1	7.8	6.9	5.6	4.2	2.8	1.0
		TW	5.3	3.6	2.2	1.0	-	-	-	-	-
		TWNI	11.2	9.9	8.5	7.4	6.1	4.8	3.3	1.9	.7
		DG	-	-	-	-	-	-	-	-	-
		DGNI	11.3	9.4	7.8	6.3	4.8	3.2	1.9	.8	-

		SSF =	.1	.2	.3	.4	.5	.6	.7	.8	.9
BURLINGTON IOWA 40.8 N 6149 DD T(JAN)=23	L	WW	83	36	20	13	8	5	-	-	-
		WWNI	159	74	46	32	24	19	15	11	8
	C	TW	85	36	21	13	7	4	-	-	-
	R	TWNI	151	70	43	30	22	17	13	9	6
		DG	50	-	-	-	-	-	-	-	-
		DGNI	159	72	44	30	22	16	11	7	4
	D	WW	7.2	5.4	4.1	3.0	1.9	.6	-	-	-
		WWNI	14.3	13.4	11.4	10.0	9.2	7.6	5.9	3.9	1.5
		TW	7.3	5.6	4.0	2.6	1.3	6.4	4.4	2.6	1.0
		TWNI	13.7	12.2	10.6	9.5	7.9	6.4	4.4	2.6	1.0
		DG	3.0	-	-	-	-	-	-	-	-
		DGNI	14.4	12.2	10.5	8.8	6.9	4.9	3.1	1.5	.4

		SSF =	.1	.2	.3	.4	.5	.6	.7	.8	.9
DES MOINES IOWA 41.5 N 6710 DD T(JAN)=19	L	WW	78	33	18	11	7	-	-	-	-
		WWNI	155	71	44	31	23	18	14	11	7
	C	TW	81	34	19	11	6	3	-	-	-
	R	TWNI	146	67	41	29	21	16	12	9	6
		DG	41	-	-	-	-	-	-	-	-
		DGNI	153	69	42	29	20	15	11	7	3
	D	WW	6.6	4.8	3.6	2.5	1.5	.2	-	-	-
		WWNI	13.9	12.7	10.8	9.5	8.6	7.3	5.6	3.7	1.4
		TW	6.8	5.1	3.6	2.3	1.0	-	-	-	-
		TWNI	13.2	11.6	10.1	9.0	7.5	6.1	4.2	2.5	1.0
		DG	2.0	-	-	-	-	-	-	-	-
		DGNI	13.7	11.5	9.8	8.2	6.5	4.5	2.8	1.4	.3

		SSF =	.1	.2	.3	.4	.5	.6	.7	.8	.9
MASON CITY IOWA 43.1 N 7901 DD T(JAN)=14	L	WW	68	27	14	8	4	-	-	-	-
		WWNI	143	65	40	28	21	16	12	9	6
	C	TW	72	29	16	9	4	1	-	-	-
	R	TWNI	136	62	38	26	19	15	11	8	5
		DG	-	-	-	-	-	-	-	-	-
		DGNI	141	63	37	25	18	13	9	6	2
	D	WW	5.5	3.7	2.6	1.5	.3	-	-	-	-
		WWNI	13.0	11.3	9.7	8.5	7.5	6.3	4.8	3.2	1.2
		TW	5.9	4.2	2.7	1.4	.4	-	-	-	-
		TWNI	12.2	10.6	9.1	8.0	6.7	5.3	3.7	2.2	.8
		DG	-	-	-	-	-	-	-	-	-
		DGNI	12.3	10.2	8.5	7.1	5.4	3.7	2.3	1.0	.1

Tables of LCR & D Versus SSF / 263

SIOUX CITY, IOWA
42.4 N, 6953 DD, T(JAN)=18

	SSF=	.1	.2	.3	.4	.5	.6	.7	.8	.9
L	WW	77	32	17	11	7	-	-	-	-
C	WWNI	153	70	44	30	22	17	14	10	7
R	TW	80	33	18	11	6	-	-	-	-
	TWNI	144	66	41	28	21	16	12	9	6
	DG	37	-	-	-	-	-	-	-	-
	DGNI	151	68	41	28	20	14	10	7	3
D	WW	6.5	4.6	3.3	2.3	1.2	-	-	-	-
	WWNI	14.0	12.4	10.6	9.2	8.4	7.0	5.4	3.5	1.4
	TW	6.7	5.0	3.4	2.0	.9	-	-	-	-
	TWNI	13.2	11.4	9.9	8.8	7.3	5.9	4.1	2.4	.9
	DG	1.6	-	-	-	-	-	-	-	-
	DGNI	13.6	11.3	9.5	7.9	6.2	4.3	2.7	1.3	.3

DODGE CITY, KANSAS
37.8 N, 5046 DD, T(JAN)=31

	SSF=	.1	.2	.3	.4	.5	.6	.7	.8	.9
L	WW	140	64	39	27	19	15	10	6	-
C	WWNI	228	107	67	48	36	29	23	18	13
R	TW	138	62	37	25	17	11	8	5	2
	TWNI	215	100	63	45	34	26	20	15	10
	DG	132	54	27	10	-	-	-	-	-
	DGNI	236	109	68	47	35	26	20	14	8
D	WW	12.5	10.6	9.1	8.1	6.6	4.7	2.7	1.0	-
	WWNI	20.6	19.6	17.1	15.6	14.4	12.8	10.3	6.6	2.7
	TW	12.2	10.3	8.4	6.3	4.5	2.9	1.7	.8	.2
	TWNI	19.7	17.8	16.0	14.6	12.7	10.2	7.2	4.4	1.8
	DG	10.8	7.5	3.9	.4	-	-	-	-	-
	DGNI	21.6	18.9	16.8	14.9	12.3	9.1	6.2	3.5	1.2

GOODLAND, KANSAS
39.4 N, 6119 DD, T(JAN)=28

	SSF=	.1	.2	.3	.4	.5	.6	.7	.8	.9
L	WW	120	55	33	23	17	12	8	5	-
C	WWNI	202	95	61	43	33	26	21	16	12
R	TW	118	54	32	21	14	10	6	4	1
	TWNI	190	90	57	40	31	24	18	13	9
	DG	105	42	18	-	-	-	-	-	-
	DGNI	206	97	60	42	31	24	17	12	7
D	WW	10.8	9.4	7.8	6.7	5.4	3.7	1.9	.6	-
	WWNI	19.7	17.2	16.0	14.2	12.9	11.7	9.3	5.9	2.5
	TW	18.1	9.0	7.2	5.3	3.7	2.4	1.3	.5	.1
	TWNI	18.1	16.4	14.6	13.2	11.6	9.2	6.5	4.0	1.6
	DG	8.7	5.4	1.7	-	-	-	-	-	-
	DGNI	19.3	17.1	15.2	13.4	10.9	8.0	5.4	3.0	1.0

TOPEKA, KANSAS
39.1 N, 5243 DD, T(JAN)=28

	SSF=	.1	.2	.3	.4	.5	.6	.7	.8	.9
L	WW	112	50	29	20	14	10	6	-	-
C	WWNI	194	91	57	40	30	23	19	14	10
R	TW	112	49	29	18	12	8	4	2	-
	TWNI	183	85	53	37	28	21	16	12	8
	DG	93	33	-	-	-	-	-	-	-
	DGNI	198	91	56	39	28	21	15	10	6
D	WW	10.0	7.9	6.4	5.4	4.0	2.6	1.1	-	-
	WWNI	17.7	16.5	14.2	12.6	11.6	9.9	7.9	5.1	2.1
	TW	16.9	14.9	6.1	4.3	2.8	1.7	.8	.2	-
	TWNI	16.8	14.9	13.2	11.9	10.1	8.1	5.7	3.5	1.4
	DG	7.1	3.4	-	-	-	-	-	-	-
	DGNI	18.2	15.6	13.5	11.6	9.4	6.8	4.5	2.4	.8

WICHITA, KANSAS
37.6 N, 4687 DD, T(JAN)=31

	SSF=	.1	.2	.3	.4	.5	.6	.7	.8	.9
L	WW	138	63	38	26	19	14	10	6	-
C	WWNI	224	106	66	47	35	28	22	17	12
R	TW	135	61	36	24	16	11	8	-	-
	TWNI	212	99	62	44	33	25	19	14	9
	DG	128	52	25	-	-	-	-	-	-
	DGNI	232	107	66	46	34	26	19	13	8
D	WW	12.6	10.3	8.7	7.6	6.1	4.4	2.4	.8	-
	WWNI	20.7	19.1	16.8	15.1	14.1	12.2	9.8	6.3	2.6
	TW	12.0	10.1	8.1	5.9	4.1	2.7	1.5	.7	.1
	TWNI	19.8	17.5	15.6	14.2	12.2	9.8	7.0	4.3	1.7
	DG	10.5	7.1	3.3	-	-	-	-	-	-
	DGNI	21.4	18.7	16.4	14.3	11.8	8.7	5.9	3.3	1.1

LEXINGTON, KENTUCKY
38.0 N, 4729 DD, T(JAN)=33

	SSF=	.1	.2	.3	.4	.5	.6	.7	.8	.9
L	WW	96	42	24	16	11	7	4	-	-
C	WWNI	174	82	51	36	27	21	17	13	9
R	TW	98	42	25	15	10	6	3	-	-
	TWNI	165	77	48	34	25	19	15	11	7
	DG	71	17	-	-	-	-	-	-	-
	DGNI	177	81	50	34	25	18	13	9	5
D	WW	8.5	6.6	5.2	4.1	2.9	1.6	.4	-	-
	WWNI	16.0	14.9	12.7	11.2	10.4	8.7	7.0	4.5	1.8
	TW	8.5	6.7	5.0	3.2	2.0	1.1	.4	-	-
	TWNI	15.3	13.5	11.8	10.6	9.0	7.2	5.1	3.1	1.2
	DG	4.9	.4	-	-	-	-	-	-	-
	DGNI	16.2	13.8	11.9	10.1	8.2	5.8	3.8	2.0	.6

LOUISVILLE, KENTUCKY
38.2 N, 4645 DD, T(JAN)=33

	SSF=	.1	.2	.3	.4	.5	.6	.7	.8	.9
L	WW	98	44	25	17	12	8	4	-	-
C	WWNI	176	84	52	37	27	21	17	13	9
R	TW	99	44	25	16	10	6	3	-	-
	TWNI	167	78	49	34	25	19	15	11	7
	DG	74	21	-	-	-	-	-	-	-
	DGNI	179	83	51	35	25	19	14	9	5
D	WW	8.9	6.9	5.4	4.3	3.1	1.8	.5	-	-
	WWNI	16.2	15.3	13.1	11.4	10.6	8.9	7.1	4.6	1.8
	TW	8.7	6.9	5.2	3.5	2.2	1.2	.4	-	-
	TWNI	15.6	13.8	12.1	10.8	9.2	7.4	5.2	3.1	1.2
	DG	5.3	1.2	-	-	-	-	-	-	-
	DGNI	16.5	14.2	12.2	10.3	8.4	6.0	3.9	2.1	.6

BATON ROUGE, LOUISIANA
30.5 N, 1670 DD, T(JAN)=51

	SSF=	.1	.2	.3	.4	.5	.6	.7	.8	.9
L	WW	312	142	87	62	46	36	28	20	12
C	WWNI	430	203	125	90	69	54	43	34	24
R	TW	287	133	81	55	39	28	21	15	9
	TWNI	405	188	118	84	63	49	38	28	19
	DG	346	156	93	61	41	27	17	8	-
	DGNI	460	213	133	94	70	53	40	29	19
D	WW	30.3	23.6	22.1	20.0	17.2	14.1	9.7	5.4	1.7
	WWNI	43.0	34.4	31.8	27.2	23.9	19.6	12.7	7.9	5.3
	TW	26.6	22.8	19.3	15.2	12.0	9.0	6.2	3.9	1.6
	TWNI	39.0	32.7	30.5	24.1	19.4	13.8	8.5	3.5	-
	DG	31.2	25.5	20.4	14.8	10.2	6.3	-	-	-
	DGNI	43.5	37.0	33.8	29.7	25.2	19.3	13.5	8.1	3.3

		SSF =	.1	.2	.3	.4	.5	.6	.7	.8	.9
LAKE CHARLES LOUISIANA 30.1 N 1498 DD T(JAN)=52	L	WW	330	144	89	63	48	37	28	21	13
	C	WWNI	457	205	127	92	70	55	44	35	25
	R	TW	301	136	83	56	40	29	21	15	10
		TWNI	425	192	120	85	64	50	39	29	19
		DG	364	160	95	62	42	28	18	9	
		DGNI	482	218	136	96	71	54	41	30	19
	D	WW	29.3	23.9	23.1	20.6	17.8	14.4	10.0	5.6	1.8
		WWNI	43.2	34.2	33.1	31.7	28.2	24.7	19.7	12.0	5.4
		TW	27.1	23.0	22.0	19.4	15.8	12.4	8.3	4.0	1.7
		TWNI	38.8	32.9	30.9	27.9	24.8	19.9	14.1	8.7	3.5
		DG	31.8	25.9	20.9	15.5	10.7	6.7	3.4	1.0	
		DGNI	43.6	37.3	34.3	30.6	25.8	19.8	13.9	8.3	3.3
NEW ORLEANS LOUISIANA 30.0 N 1465 DD T(JAN)=53	L	WW	374	169	104	74	56	44	34	25	16
	C	WWNI	506	235	146	105	80	64	51	40	29
	R	TW	342	158	97	66	47	35	25	18	12
		TWNI	474	219	137	98	74	57	44	33	22
		DG	423	191	115	76	52	36	24	14	6
		DGNI	542	249	156	110	83	63	48	35	23
	D	WW	35.5	28.2	26.8	24.6	21.3	17.7	12.5	7.1	2.5
		WWNI	50.6	39.6	37.4	36.6	32.7	28.6	23.2	15.1	6.4
		TW	31.5	27.9	27.2	23.2	18.6	14.8	10.1	5.9	2.2
		TWNI	45.0	37.9	35.6	32.1	28.6	22.9	16.4	10.1	4.1
		DG	38.1	31.8	26.1	19.7	14.2	9.5	5.4	2.3	.3
		DGNI	50.6	43.2	39.8	35.4	30.2	23.2	16.4	10.0	4.1
SHREVEPORT LOUISIANA 32.5 N 2167 DD T(JAN)=47	L	WW	250	115	70	50	37	29	22	15	9
	C	WWNI	356	170	105	75	57	45	36	29	20
	R	TW	234	108	66	45	32	23	16	11	7
		TWNI	336	158	99	70	53	41	32	23	16
		DG	271	120	71	45	29	18	9		
		DGNI	380	177	110	78	58	44	33	24	15
	D	WW	23.9	19.2	17.5	15.9	13.6	10.9	7.3	3.9	1.1
		WWNI	34.8	29.7	26.5	25.6	23.3	20.1	16.5	10.8	4.5
		TW	21.5	19.5	17.1	12.2	10.7	7.9	5.4	4.7	1.2
		TWNI	32.2	27.6	25.4	22.9	20.1	16.2	11.6	7.1	2.9
		DG	24.0	19.5	14.9	10.2	6.4	3.3	.9		
		DGNI	35.9	30.8	27.9	24.6	20.8	15.8	11.0	6.6	2.6
CARIBOU MAINE 46.9 N 9632 DD T(JAN)=11	L	WW	48	18	8						
	C	WWNI	119	55	34	24	18	13	10	7	5
	R	TW	54	21	10	3					
		TWNI	113	52	32	22	16	12	9	6	4
		DG									
		DGNI	114	51	31	20	14	10	6	3	1
	D	WW	3.8	2.3	.9						
		WWNI	11.1	9.5	8.4	7.4	6.2	4.9	3.5	2.2	.8
		TW	4.5	2.8	1.4	.1					
		TWNI	10.5	8.9	7.9	6.7	5.4	4.1	2.9	1.6	.6
		DG									
		DGNI	10.3	8.5	7.0	5.4	3.9	2.6	1.4	.4	

		SSF =	.1	.2	.3	.4	.5	.6	.7	.8	.9
PORTLAND MAINE 43.6 N 7498 DD T(JAN)=21	L	WW	60	24	13	7					
	C	WWNI	133	61	38	27	20	15	12	9	6
	R	TW	65	27	14	8	3				
		TWNI	126	58	36	25	19	14	11	8	5
		DG									
		DGNI	130	58	35	24	17	12	8	5	2
	D	WW	4.8	3.4	2.3	1.2					
		WWNI	12.2	10.5	9.7	8.4	7.3	6.1	4.7	3.0	1.2
		TW	5.4	3.8	2.4	1.2					
		TWNI	11.6	10.0	8.8	7.7	6.5	5.1	3.6	2.1	.8
		DG									
		DGNI	11.5	9.8	8.2	6.8	5.2	3.6	2.1	.9	.1
BALTIMORE MARYLAND 39.2 N 4729 DD T(JAN)=33	L	WW	101	46	28	19	13	9	6		
	C	WWNI	178	86	55	39	29	23	18	14	10
	R	TW	102	46	27	18	12	8	4	2	
		TWNI	169	81	51	36	27	21	16	12	8
		DG	81	29							
		DGNI	182	86	54	37	27	20	15	10	6
	D	WW	9.5	7.7	6.4	5.2	3.9	2.4	1.0		
		WWNI	16.8	16.1	14.0	12.7	11.6	9.7	7.8	5.0	2.0
		TW	9.1	7.6	5.9	4.2	2.7	1.6	5.6	3.4	
		TWNI	16.1	14.7	13.1	11.8	10.0	8.0	5.6	3.4	1.3
		DG	6.4	2.8							
		DGNI	17.1	15.2	13.4	11.4	9.3	6.7	4.4	2.4	.7
BOSTON MASSACHUSETTS 42.4 N 5621 DD T(JAN)=29	L	WW	81	35	20	13	9	5			
	C	WWNI	159	73	46	32	24	19	15	11	8
	R	TW	85	36	21	13	8	4			
		TWNI	150	69	43	30	23	17	13	10	6
		DG	49								
		DGNI	158	72	44	30	22	16	11	8	4
	D	WW	6.8	5.3	4.2	3.1	2.0	.8			
		WWNI	14.1	13.2	11.5	10.3	9.3	7.7	6.1	4.0	1.6
		TW	7.2	5.5	4.1	2.8	1.4	.5	4.5	2.7	1.0
		TWNI	13.5	12.1	10.8	9.6	8.1	6.5	4.5	2.7	1.0
		DG	2.9								
		DGNI	14.0	12.1	10.6	8.9	7.1	5.0	3.2	1.6	.4
ALPENA MICHIGAN 45.1 N 8518 DD T(JAN)=18	L	WW	53	19	8						
	C	WWNI	126	57	35	24	17	13	10	7	5
	R	TW	58	22	10						
		TWNI	119	54	33	22	16	12	9	6	4
		DG									
		DGNI	121	53	31	20	14	10	6	3	1
	D	WW	4.0	2.2	.7						
		WWNI	11.6	9.5	8.5	7.1	5.8	4.6	3.3	2.2	.8
		TW	4.7	2.8	1.4						
		TWNI	11.0	9.0	7.7	6.4	5.2	4.0	2.7	1.6	.5
		DG									
		DGNI	10.8	8.5	6.7	5.2	3.7	2.4	1.3	.4	.1

		SSF =	.1	.2	.3	.4	.5	.6	.7	.8	.9
DETROIT MICHIGAN	L	WW		62	26	14	8					
	C	WWNI		137	63	39	27	20	16	12	9	6
42.4 N	R	TW		68	28	15	8	3				
6228 DD		TWNI		130	60	37	26	19	14	11	8	5
T(JAN)=25		DG		134								
		DGNI			60	36	25	17	12	9	5	2
	D	WW		5.2	3.6	2.5	1.2					
		WWNI		12.0	11.4	9.6	8.4	7.4	6.1	4.6	3.0	1.1
		TW		5.6	4.0	2.6	1.3					
		TWNI		11.5	10.3	9.0	7.9	6.5	5.1	3.6	2.1	.8
		DG										
		DGNI		11.8	10.0	8.4	6.8	5.2	3.6	2.1	.9	.1

		SSF =	.1	.2	.3	.4	.5	.6	.7	.8	.9
FLINT MICHIGAN	L	WW		54	21	10						
	C	WWNI		129	58	36	25	18	14	11	8	5
43.0 N	R	TW		60	24	12	5					
7041 DD		TWNI		122	55	34	23	17	13	9	7	4
T(JAN)=22		DG		124								
		DGNI			54	32	21	15	10	7		
	D	WW		4.1	2.6	1.3						
		WWNI		11.6	9.8	8.7	7.3	6.3	5.2	3.8	2.5	.9
		TW		4.8	3.1	1.7	.5					
		TWNI		10.8	9.4	8.0	6.9	5.6	4.4	3.0	1.8	.6
		DG										
		DGNI		10.6	8.8	7.1	5.7	4.2	2.8	1.6	.6	

		SSF =	.1	.2	.3	.4	.5	.6	.7	.8	.9
GRAND RAPIDS MICHIGAN	L	WW		57	22	10						
	C	WWNI		133	59	37	25	19	14	11	8	5
42.9 N	R	TW		63	25	12	6					
6801 DD		TWNI		125	56	34	24	17	13	10	7	4
T(JAN)=23		DG		128								
		DGNI			56	33	22	15	11	7		
	D	WW		4.4	2.8	1.5						
		WWNI		11.5	10.3	8.8	7.5	6.5	5.2	3.8	2.5	.9
		TW		5.0	3.2	1.9	.6					
		TWNI		10.9	9.6	8.2	7.0	5.7	4.4	3.1	1.8	.6
		DG										
		DGNI		10.9	9.0	7.4	5.9	4.3	2.9	1.6	.6	

		SSF =	.1	.2	.3	.4	.5	.6	.7	.8	.9
SAULT STE. MARIE MICHIGAN	L	WW.		45	15							
	C	WWNI		116	53	33	22	16	12	9	6	4
46.5 N	R	TW		51	19	8						
9193 DD		TWNI		110	50	31	21	15	11	8	6	3
T(JAN)=14		DG		110								
		DGNI			49	29	19	12	8	5	2	
	D	WW		3.4	1.5							
		WWNI		10.8	9.0	7.9	6.6	5.3	4.0	2.8	1.8	.7
		TW		4.1	2.2	.8						
		TWNI		10.2	8.5	7.2	5.9	4.7	3.5	2.4	1.3	.4
		DG										
		DGNI		10.0	7.9	6.1	4.5	3.1	1.9	.9	.2	

		SSF =	.1	.2	.3	.4	.5	.6	.7	.8	.9
TRAVERSE CITY MICHIGAN	L	WW		51	18							
	C	WWNI		127	56	34	23	17	13	9	7	4
44.7 N	R	TW		58	21	9						
7698 DD		TWNI		120	53	32	22	16	11	8	6	4
T(JAN)=21		DG										
		DGNI		121	52	30	20	13	9	6	3	
	D	WW		3.6	1.9							
		WWNI		11.3	9.1	8.1	6.7	5.5	4.3	3.1	2.0	.7
		TW		4.4	2.6	1.1						
		TWNI		10.6	8.9	7.4	6.2	4.9	3.7	2.6	1.4	.5
		DG										
		DGNI		10.2	8.1	6.4	4.9	3.4	2.2	1.1	.3	

		SSF =	.1	.2	.3	.4	.5	.6	.7	.8	.9
DULUTH MINNESOTA	L	WW		42	15							
	C	WWNI		113	52	32	22	16	12	9	7	4
46.8 N	R	TW		49	18	8						
9756 DD		TWNI		107	49	30	21	15	11	8	6	3
T(JAN)=9		DG										
		DGNI		107	48	28	18	12	8	5	3	
	D	WW		3.2	1.5							
		WWNI		10.2	9.0	7.9	6.6	5.4	4.2	3.0	2.0	.9
		TW		3.9	2.2	.8						
		TWNI		9.8	8.4	7.2	6.0	4.8	3.6	2.5	1.4	.5
		DG										
		DGNI		9.6	7.8	6.1	4.6	3.3	2.0	1.0	.2	

		SSF =	.1	.2	.3	.4	.5	.6	.7	.8	.9
INTER. FALLS MINNESOTA	L	WW		38	10							
	C	WWNI		109	49	30	21	15	11	8	6	4
48.6 N	R	TW		45	16							
10547 DD		TWNI		104	47	28	19	14	10	7	5	
T(JAN)=2		DG										
		DGNI		103	45	26	16	11	7	4		
	D	WW		2.7	.4							
		WWNI		9.8	8.4	7.2	5.7	4.5	3.5	2.5	1.6	.5
		TW		3.5	1.6							
		TWNI		9.3	7.9	6.6	5.3	4.1	3.1	2.1	1.2	.4
		DG										
		DGNI		9.0	7.1	5.3	3.8	2.5	1.4	.6		

		SSF =	.1	.2	.3	.4	.5	.6	.7	.8	.9
MINNEAPOLIS MINNESOTA	L	WW		54	21	10						
	C	WWNI		126	58	36	25	18	14	11	8	5
44.9 N	R	TW		59	24	12	5					
8159 DD		TWNI		120	55	34	23	17	13	9	7	4
T(JAN)=12		DG										
		DGNI		122	54	32	22	15	10	7	4	
	D	WW		4.3	2.5	1.3						
		WWNI		11.6	10.1	8.6	7.3	6.4	5.3	3.8	2.5	.9
		TW		4.8	3.2	1.7	.5					
		TWNI		10.9	9.4	8.0	6.9	5.7	4.4	3.1	1.8	.6
		DG										
		DGNI		10.8	8.8	7.2	5.7	4.3	2.8	1.6	.6	

ROCHESTER, MINNESOTA
43.9 N, 8227 DD, T(JAN)=13

	SSF =	.1	.2	.3	.4	.5	.6	.7	.8	.9
L	WW	53	20	9	-	18	14	11	8	-
C	WWNI	126	57	35	24	18	14	11	8	5
R	TW	59	23	12	5	-	-	-	-	-
	TWNI	120	54	33	23	17	13	9	7	4
	DG	-	-	-	-	-	-	-	-	-
	DGNI	121	54	32	21	15	10	7	4	1
D	WW	4.1	2.5	1.3	-	-	-	-	-	-
	WWNI	11.3	9.8	8.5	7.3	6.4	5.4	4.0	2.6	1.0
	TW	4.7	3.1	1.7	-	-	-	-	-	-
	TWNI	10.7	9.3	7.9	6.5	5.7	4.5	3.1	1.8	.7
	DG	-	-	-	-	-	-	-	-	-
	DGNI	10.6	8.7	7.1	5.8	4.3	2.9	1.6	.6	-

JACKSON, MISSISSIPPI
32.3 N, 2300 DD, T(JAN)=47

	SSF =	.1	.2	.3	.4	.5	.6	.7	.8	.9
L	WW	230	108	66	46	34	26	20	14	8
C	WWNI	331	161	100	71	54	43	34	27	19
R	TW	218	101	62	42	29	21	15	10	6
	TWNI	313	149	94	66	50	39	30	22	15
	DG	248	111	65	41	26	15	5	-	-
	DGNI	354	167	104	73	55	41	31	22	14
D	WW	22.4	18.2	16.1	14.7	12.5	10.0	6.6	3.4	.9
	WWNI	32.3	29.0	25.0	23.8	22.0	18.9	15.6	10.0	4.2
	TW	20.5	17.5	14.6	11.5	8.6	4.2	2.5	1.0	-
	TWNI	30.5	26.5	24.0	21.7	19.0	15.3	10.9	6.7	2.7
	DG	22.2	17.9	13.4	8.9	5.2	2.4	.2	-	-
	DGNI	34.1	29.3	26.5	23.1	19.5	14.8	10.3	6.1	2.4

MERIDIAN, MISSISSIPPI
32.3 N, 2388 DD, T(JAN)=47

	SSF =	.1	.2	.3	.4	.5	.6	.7	.8	.9
L	WW	216	103	63	44	33	25	19	13	7
C	WWNI	315	154	97	69	53	41	33	26	19
R	TW	206	96	59	40	28	20	14	10	6
	TWNI	298	144	90	64	48	38	29	21	14
	DG	233	104	61	38	24	13	-	-	-
	DGNI	335	160	100	71	53	40	30	21	14
D	WW	21.0	17.8	15.4	14.2	11.9	9.5	6.2	3.2	.8
	WWNI	30.6	28.8	24.4	22.8	20.7	18.3	15.1	9.7	4.1
	TW	19.2	16.8	14.1	11.5	8.2	5.9	4.0	2.3	-
	TWNI	28.9	25.9	23.3	21.1	18.3	14.8	10.5	6.5	2.6
	DG	20.9	17.0	12.6	8.2	4.6	1.9	-	-	-
	DGNI	32.6	28.4	25.7	22.4	18.8	14.2	9.8	5.8	2.3

COLUMBIA, MISSOURI
38.8 N, 5083 DD, T(JAN)=29

	SSF =	.1	.2	.3	.4	.5	.6	.7	.8	.9
L	WW	102	45	26	17	12	8	5	-	-
C	WWNI	182	85	53	37	28	22	17	13	9
R	TW	103	45	26	16	10	6	3	-	-
	TWNI	173	80	50	35	26	20	15	11	7
	DG	79	23	-	-	-	-	-	-	-
	DGNI	185	84	51	36	26	19	14	9	5
D	WW	9.1	7.0	5.5	4.5	3.2	1.9	.6	-	-
	WWNI	16.5	15.2	13.1	11.6	10.7	9.1	7.1	4.6	1.9
	TW	8.9	7.1	5.3	3.6	2.2	1.0	-	-	-
	TWNI	15.8	13.9	12.2	10.9	9.3	7.5	5.2	3.2	1.2
	DG	5.6	1.6	-	-	-	-	-	-	-
	DGNI	16.8	14.3	12.4	10.5	8.5	6.1	4.0	2.1	.6

KANSAS CITY, MISSOURI
39.3 N, 5357 DD, T(JAN)=27

	SSF =	.1	.2	.3	.4	.5	.6	.7	.8	.9
L	WW	102	45	26	17	12	8	5	-	-
C	WWNI	182	85	53	37	28	22	17	14	9
R	TW	103	45	26	17	11	7	4	-	-
	TWNI	172	80	50	35	26	20	15	11	7
	DG	79	25	-	-	-	-	-	-	-
	DGNI	184	84	52	36	26	19	14	10	5
D	WW	9.1	7.1	5.7	4.6	3.4	2.0	.7	-	-
	WWNI	16.5	15.5	13.3	11.8	10.9	9.2	7.4	4.8	1.9
	TW	8.9	7.2	5.5	3.8	2.3	1.3	.5	-	-
	TWNI	15.8	14.0	12.4	11.1	9.5	7.6	5.3	3.2	1.3
	DG	5.7	1.9	-	-	-	-	-	-	-
	DGNI	16.9	14.5	12.5	10.7	8.7	6.2	4.1	2.2	.6

SAINT LOUIS, MISSOURI
38.7 N, 4750 DD, T(JAN)=31

	SSF =	.1	.2	.3	.4	.5	.6	.7	.8	.9
L	WW	113	50	29	19	14	10	6	-	-
C	WWNI	194	91	57	40	30	23	18	14	10
R	TW	113	49	29	18	12	8	4	-	-
	TWNI	184	86	53	37	28	21	16	12	8
	DG	94	32	-	-	-	-	-	-	-
	DGNI	199	91	56	38	28	21	15	10	6
D	WW	10.2	8.0	6.3	5.2	3.9	2.5	1.0	-	-
	WWNI	17.8	16.4	14.1	12.4	11.5	9.7	7.8	5.0	2.0
	TW	9.8	7.9	6.1	4.2	2.7	1.6	.7	-	-
	TWNI	17.0	15.0	13.1	11.7	10.0	8.0	5.7	3.4	1.3
	DG	7.0	3.3	-	-	-	-	-	-	-
	DGNI	18.2	15.5	13.4	11.4	9.3	6.7	4.4	2.4	.7

SPRINGFIELD, MISSOURI
37.2 N, 4570 DD, T(JAN)=33

	SSF =	.1	.2	.3	.4	.5	.6	.7	.8	.9
L	WW	123	55	33	22	16	11	8	4	-
C	WWNI	207	97	61	43	32	25	20	16	11
R	TW	122	54	32	21	14	9	6	-	-
	TWNI	196	91	57	40	30	23	18	13	8
	DG	108	41	16	-	-	-	-	-	-
	DGNI	213	97	60	42	31	23	17	12	7
D	WW	11.0	8.9	7.3	6.3	4.9	3.3	1.7	.4	-
	WWNI	18.7	17.6	15.2	13.6	12.6	10.9	8.8	5.7	2.3
	TW	10.6	8.8	6.9	5.0	3.4	2.1	1.1	-	-
	TWNI	17.5	15.3	13.1	12.8	11.0	8.9	6.3	3.8	1.5
	DG	8.4	5.0	1.0	-	-	-	-	-	-
	DGNI	19.2	16.8	14.7	12.7	10.4	7.6	5.1	2.8	.9

BILLINGS, MONTANA
45.8 N, 7265 DD, T(JAN)=22

	SSF =	.1	.2	.3	.4	.5	.6	.7	.8	.9
L	WW	91	41	23	15	9	6	-	-	-
C	WWNI	165	80	50	35	26	20	15	12	8
R	TW	92	40	23	14	9	5	-	-	-
	TWNI	158	75	47	32	24	18	14	10	6
	DG	64	-	-	-	-	-	-	-	-
	DGNI	168	78	48	33	23	17	12	8	4
D	WW	8.5	6.5	4.7	3.4	2.0	.8	-	-	-
	WWNI	16.0	14.2	12.8	10.8	9.3	7.8	5.8	3.9	1.5
	TW	8.2	6.7	4.6	2.9	1.7	.6	-	-	-
	TWNI	15.2	13.4	11.5	9.9	8.3	6.4	4.5	2.7	1.0
	DG	4.4	-	-	-	-	-	-	-	-
	DGNI	15.8	13.6	11.3	9.4	7.2	5.1	3.2	1.6	.4

Tables of LCR & D Versus SSF / 267

CUT BANK, MONTANA
48.6 N
9033 DD
T(JAN)=16

	SSF =	.1	.2	.3	.4	.5	.6	.7	.8	.9
L	WW	77	33	18	10	5	-	-	-	-
C	WWNI	153	71	44	31	23	17	13	10	6
R	TW	80	34	18	11	5	15	11	9	5
	TWNI	144	67	42	29	21	16	12	8	5
	DG	40	-	-	-	-	-	-	-	-
	DGNI	151	69	42	28	20	14	9	6	-
D	WW	6.5	5.0	3.2	1.8	.4	-	-	-	-
	WWNI	14.1	12.7	11.1	9.7	7.7	6.1	4.3	2.9	1.1
	TW	7.0	5.0	3.3	1.8	.6	-	-	-	-
	TWNI	13.3	11.7	10.1	8.5	6.9	5.1	3.6	2.0	.7
	DG	1.7	-	-	-	-	-	-	-	-
	DGNI	13.8	11.7	9.5	7.5	5.6	3.8	2.2	.9	-

DILLON, MONTANA
45.2 N
8354 DD
T(JAN)=20

	SSF =	.1	.2	.3	.4	.5	.6	.7	.8	.9
L	WW	90	40	24	15	10	6	-	-	-
C	WWNI	168	78	50	35	26	20	16	12	8
R	TW	92	40	23	15	9	5	-	-	-
	TWNI	158	74	47	33	24	18	14	10	6
	DG	64	-	-	-	-	-	-	-	-
	DGNI	168	78	48	33	24	17	12	8	4
D	WW	7.8	6.6	5.1	3.7	2.3	1.0	-	-	-
	WWNI	15.2	14.4	12.9	11.4	9.7	8.1	5.9	4.0	1.5
	TW	8.1	6.3	4.7	3.1	1.8	.8	-	-	-
	TWNI	14.6	13.3	11.9	10.3	8.6	6.6	4.7	2.7	1.0
	DG	4.5	-	-	-	-	-	-	-	-
	DGNI	15.3	13.6	11.7	9.7	7.5	5.4	3.4	1.7	.4

GLASGOW, MONTANA
48.2 N
8969 DD
T(JAN)= 9

	SSF =	.1	.2	.3	.4	.5	.6	.7	.8	.9
L	WW	60	23	10	-	-	-	-	-	-
C	WWNI	130	61	38	26	19	14	10	8	5
R	TW	64	25	13	5	-	-	-	-	-
	TWNI	124	58	35	24	17	13	9	7	4
	DG	-	-	-	-	-	-	-	-	-
	DGNI	128	58	34	22	15	10	7	4	-
D	WW	5.0	2.9	1.2	-	-	-	-	-	-
	WWNI	12.4	10.6	9.0	7.2	6.1	4.9	3.5	2.3	.8
	TW	5.3	3.4	1.8	.4	-	-	-	-	-
	TWNI	11.7	10.0	8.2	6.8	5.5	4.1	2.9	1.6	.6
	DG	-	-	-	-	-	-	-	-	-
	DGNI	11.6	9.4	7.3	5.6	4.0	2.6	1.4	.5	-

GREAT FALLS, MONTANA
47.5 N
7652 DD
T(JAN)=20

	SSF =	.1	.2	.3	.4	.5	.6	.7	.8	.9
L	WW	86	38	21	13	7	-	-	-	-
C	WWNI	161	77	48	33	25	19	14	10	7
R	TW	88	38	21	13	7	3	-	-	-
	TWNI	153	72	45	31	23	17	12	9	5
	DG	56	-	-	-	-	-	-	-	-
	DGNI	163	75	46	31	22	15	11	7	3
D	WW	7.8	5.9	4.0	2.6	1.1	-	-	-	-
	WWNI	14.9	13.7	12.1	10.1	8.5	6.9	4.8	3.3	1.2
	TW	8.1	5.8	4.0	2.4	1.2	.2	-	-	-
	TWNI	14.3	12.7	10.9	9.2	7.5	5.6	4.0	2.3	.8
	DG	3.4	-	-	-	-	-	-	-	-
	DGNI	15.0	12.9	10.5	8.4	6.3	4.4	2.6	1.2	.2

HELENA, MONTANA
46.6 N
8190 DD
T(JAN)=18

	SSF =	.1	.2	.3	.4	.5	.6	.7	.8	.9
L	WW	77	34	19	11	6	-	-	-	-
C	WWNI	151	71	45	31	23	17	13	10	6
R	TW	80	34	19	11	6	16	12	8	5
	TWNI	144	67	42	29	21	14	10	6	2
	DG	41	-	-	-	-	-	-	-	-
	DGNI	151	69	42	29	20				
D	WW	6.9	5.2	3.5	2.1	.7	6.4	4.6	3.0	1.1
	WWNI	13.7	13.0	11.4	9.7	8.0	5.3	3.8	2.1	.8
	TW	6.9	5.1	3.5	2.0	.8	4.0	2.4	1.0	-
	TWNI	13.1	11.9	10.4	8.7	7.1				
	DG	2.0	-	-	-	-				
	DGNI	13.8	11.9	9.8	7.8	5.8				

LEWISTOWN, MONTANA
47.0 N
8586 DD
T(JAN)=19

	SSF =	.1	.2	.3	.4	.5	.6	.7	.8	.9
L	WW	76	33	19	11	6	-	-	-	-
C	WWNI	153	71	45	31	23	18	13	10	7
R	TW	80	34	19	11	6	16	12	9	5
	TWNI	144	67	42	29	21	14	10	6	3
	DG	41	-	-	-	-				
	DGNI	151	69	42	29	20				
D	WW	6.5	5.2	3.5	2.2	.9	6.6	4.8	3.2	1.2
	WWNI	13.8	12.8	11.3	9.7	8.1	5.5	3.9	2.2	.8
	TW	6.9	5.1	3.5	2.0	.8	4.2	2.5	1.1	.2
	TWNI	13.2	11.8	10.3	8.8	7.3				
	DG	1.9	-	-	-	-				
	DGNI	13.6	11.8	9.9	8.0	6.0				

MILES CITY, MONTANA
46.4 N
7889 DD
T(JAN)=15

	SSF =	.1	.2	.3	.4	.5	.6	.7	.8	.9
L	WW	78	33	18	10	5	-	-	-	-
C	WWNI	151	72	44	31	22	17	13	10	6
R	TW	80	34	18	10	5	15	11	8	5
	TWNI	144	67	41	28	21	14	9	6	2
	DG	39	-	-	-	-				
	DGNI	151	69	41	28	19				
D	WW	6.8	4.7	3.1	1.8	.6	6.3	4.6	3.1	1.2
	WWNI	14.7	12.5	10.8	9.2	7.7	5.3	3.7	2.2	.8
	TW	6.9	5.0	3.2	1.8	.7	3.9	2.3	1.0	.1
	TWNI	13.7	11.7	9.9	8.3	6.9				
	DG	1.6	-	-	-	-				
	DGNI	14.0	11.5	9.3	7.5	5.6				

MISSOULA, MONTANA
46.9 N
7931 DD
T(JAN)=21

	SSF =	.1	.2	.3	.4	.5	.6	.7	.8	.9
L	WW	66	27	13	-	-	-	-	-	-
C	WWNI	140	65	40	28	20	15	11	8	5
R	TW	71	28	15	7	-	13	10	7	4
	TWNI	133	61	38	26	18	11	7	4	1
	DG	-	-	-	-	-				
	DGNI	138	62	37	24	16				
D	WW	5.7	3.7	1.8	-	-	4.9	3.4	2.2	.8
	WWNI	12.4	11.4	9.7	8.2	6.4	4.2	2.9	1.6	.5
	TW	5.8	3.9	2.2	.8	-	2.8	1.5	.5	-
	TWNI	12.0	10.5	9.0	7.3	5.7				
	DG	-	-	-	-	-				
	DGNI	12.4	10.2	8.1	6.1	4.3				

268 / APPENDIX 12

GRAND ISLAND, NEBRASKA
41.0 N, 6425 DD, T(JAN)=22

	SSF =	.1	.2	.3	.4	.5	.6	.7	.8	.9
L	WW	98	43	25	16	11	7	4	–	–
C	WWNI	178	82	52	36	27	21	17	13	9
R	TW	99	43	25	10	6	6	3	1	1
	TWNI	168	78	48	34	25	19	15	11	7
	DG	73	18	–	–	–	–	–	–	–
	DGNI	179	82	50	34	25	18	13	9	5
D	WW	8.4	6.7	5.2	4.2	3.0	1.6	.4	–	–
	WWNI	16.4	14.7	12.8	11.3	10.2	8.8	7.0	4.5	1.8
	TW	8.6	6.8	5.0	3.5	2.1	1.3	–	–	–
	TWNI	16.6	13.6	11.9	10.6	9.1	7.3	5.3	3.1	1.2
	DG	5.1	–	–	–	–	–	–	–	–
	DGNI	16.2	13.8	11.9	10.2	8.2	5.9	3.8	2.0	.6

NORTH OMAHA, NEBRASKA
41.4 N, 6601 DD, T(JAN)=20

	SSF =	.1	.2	.3	.4	.5	.6	.7	.8	.9
L	WW	85	36	20	13	–	5	–	–	–
C	WWNI	163	75	47	33	24	19	15	12	8
R	TW	87	37	21	13	8	4	2	1	–
	TWNI	154	71	44	30	23	17	13	10	6
	DG	52	–	–	–	–	–	–	–	–
	DGNI	162	73	44	31	22	16	12	8	4
D	WW	7.2	5.4	4.1	3.2	2.1	.8	–	–	–
	WWNI	14.8	13.3	11.4	10.1	9.2	7.9	6.2	4.0	1.6
	TW	7.4	5.5	4.1	2.6	1.4	.6	–	–	–
	TWNI	13.9	12.2	10.7	9.6	8.1	6.5	4.6	2.7	1.1
	DG	3.1	–	–	–	–	–	–	–	–
	DGNI	14.5	12.2	10.5	8.9	7.1	5.0	3.2	1.6	.4

NORTH PLATTE, NEBRASKA
41.1 N, 6743 DD, T(JAN)=23

	SSF =	.1	.2	.3	.4	.5	.6	.7	.8	.9
L	WW	103	46	27	18	13	9	5	–	–
C	WWNI	181	85	54	38	29	22	18	14	10
R	TW	103	46	27	17	11	7	4	1	–
	TWNI	172	80	50	35	27	20	16	11	7
	DG	81	27	–	–	–	–	–	–	–
	DGNI	184	85	53	37	27	20	14	10	6
D	WW	9.1	7.5	6.0	5.0	3.6	2.2	.9	–	–
	WWNI	17.3	15.1	13.9	12.3	10.9	9.7	7.6	4.9	2.0
	TW	9.2	7.4	5.7	4.1	2.6	1.5	.4	–	–
	TWNI	16.1	14.4	12.7	11.3	9.8	7.8	5.5	3.3	1.3
	DG	6.1	2.3	–	–	–	–	–	–	–
	DGNI	17.0	14.8	12.8	11.1	9.0	6.5	4.2	2.3	.7

SCOTTSBLUFF, NEBRASKA
41.9 N, 6774 DD, T(JAN)=25

	SSF =	.1	.2	.3	.4	.5	.6	.7	.8	.9
L	WW	103	47	28	19	13	9	6	–	–
C	WWNI	180	86	55	39	29	23	18	14	10
R	TW	103	46	28	18	12	7	4	2	–
	TWNI	171	81	51	36	27	21	16	12	8
	DG	82	29	–	–	–	–	–	–	–
	DGNI	183	86	54	37	27	20	15	10	6
D	WW	9.5	7.9	6.3	5.3	3.9	2.4	1.0	–	–
	WWNI	17.8	15.4	14.4	12.6	11.3	10.1	7.8	5.1	2.0
	TW	9.4	7.7	5.9	4.1	2.7	1.6	.7	.1	–
	TWNI	16.5	14.7	13.0	11.7	10.1	8.1	5.7	3.4	1.3
	DG	6.5	2.9	–	–	–	–	–	–	–
	DGNI	17.4	15.3	13.3	11.5	9.3	6.7	4.4	2.4	.7

ELKO, NEVADA
40.8 N, 7483 DD, T(JAN)=23

	SSF =	.1	.2	.3	.4	.5	.6	.7	.8	.9
L	WW	113	52	31	21	15	10	7	–	–
C	WWNI	195	92	58	42	31	24	19	15	10
R	TW	114	51	30	20	14	9	5	2	1
	TWNI	184	87	55	39	29	22	17	12	8
	DG	97	36	–	–	–	–	–	–	–
	DGNI	199	93	58	40	29	22	16	11	6
D	WW	10.2	8.7	7.2	5.8	4.2	2.7	1.2	–	–
	WWNI	17.9	17.0	15.2	13.6	12.0	10.3	7.6	5.1	2.0
	TW	10.1	8.3	6.5	4.7	3.1	1.8	.5	–	–
	TWNI	17.1	15.6	14.1	12.4	10.5	8.3	5.8	3.5	1.3
	DG	7.8	4.2	–	–	–	–	–	–	–
	DGNI	18.4	16.4	14.3	12.3	9.7	7.1	4.6	2.5	.7

ELY, NEVADA
39.3 N, 7814 DD, T(JAN)=24

	SSF =	.1	.2	.3	.4	.5	.6	.7	.8	.9
L	WW	111	52	32	22	16	12	8	4	–
C	WWNI	191	91	59	42	32	26	20	16	11
R	TW	111	51	31	20	14	9	6	3	1
	TWNI	180	86	55	39	30	23	18	13	9
	DG	96	38	16	–	–	–	–	–	–
	DGNI	195	92	58	41	31	23	17	12	7
D	WW	10.1	9.1	7.8	6.7	5.2	3.5	1.8	.4	–
	WWNI	17.8	17.1	15.5	14.4	13.0	11.5	8.9	5.8	2.3
	TW	10.2	8.7	7.0	5.1	3.6	2.4	.9	–	–
	TWNI	16.9	15.9	14.6	13.2	11.4	9.1	6.4	3.9	1.5
	DG	8.0	4.9	1.2	–	–	–	–	–	–
	DGNI	18.3	16.8	15.1	13.2	10.7	7.9	5.2	2.9	1.0

LAS VEGAS, NEVADA
36.1 N, 2601 DD, T(JAN)=44

	SSF =	.1	.2	.3	.4	.5	.6	.7	.8	.9
L	WW	311	144	90	64	48	37	28	20	12
C	WWNI	430	204	129	92	70	56	44	35	24
R	TW	291	134	83	56	40	29	21	15	9
	TWNI	405	191	120	86	64	50	38	28	19
	DG	350	158	95	62	42	28	17	9	–
	DGNI	461	216	136	96	71	54	41	29	19
D	WW	29.2	25.4	22.7	20.9	17.5	14.0	9.8	5.4	1.7
	WWNI	41.4	36.4	33.2	31.6	28.6	24.0	19.1	12.7	5.3
	TW	26.6	23.4	19.8	15.8	12.3	9.2	6.3	3.9	1.7
	TWNI	38.2	34.1	31.2	28.2	24.0	19.6	13.9	8.5	3.4
	DG	31.5	26.5	21.2	15.4	10.6	6.5	3.2	.9	–
	DGNI	43.0	38.2	35.0	30.5	25.7	19.6	13.6	8.2	3.2

LOVELOCK, NEVADA
40.1 N, 5990 DD, T(JAN)=29

	SSF =	.1	.2	.3	.4	.5	.6	.7	.8	.9
L	WW	160	73	45	31	22	16	12	7	–
C	WWNI	248	118	74	53	40	31	25	19	13
R	TW	153	70	42	28	19	13	9	5	3
	TWNI	235	110	70	49	37	28	22	16	10
	DG	156	66	35	18	–	–	–	–	–
	DGNI	259	121	75	53	39	29	21	15	9
D	WW	14.8	12.5	10.7	9.1	7.3	5.2	3.0	1.2	–
	WWNI	24.1	20.8	19.7	17.3	15.4	13.4	10.1	6.7	2.7
	TW	14.0	11.8	9.5	7.1	5.2	3.5	2.0	1.0	.3
	TWNI	22.3	19.7	17.8	15.8	13.5	10.8	7.6	4.6	1.8
	DG	13.5	9.6	5.7	2.2	–	–	–	–	–
	DGNI	24.2	21.5	18.8	16.3	13.2	9.9	6.6	3.7	1.3

Tables of LCR & D Versus SSF / 269

RENO NEVADA		SSF =	.1	.2	.3	.4	.5	.6	.7	.8	.9
39.5 N	L	WW									
6022 DD	C	WWNI	166	77	47	33	24	18	13	8	-
T(JAN)=32	R	TW	258	122	77	55	42	33	26	20	14
		TWNI	161	74	45	30	21	14	10	6	3
		DG	243	115	73	51	39	30	23	17	11
		DGNI	166	71	39	22	-	-	-	-	-
	D		269	126	79	56	41	31	23	16	10
		WW	15.3	13.2	11.7	10.0	8.2	5.9	3.6	1.5	-
		WWNI	24.6	22.3	20.1	18.5	16.5	14.4	10.9	7.2	2.9
		TW	14.6	12.5	10.2	7.8	5.7	3.9	-	-	-
		TWNI	22.8	20.6	18.8	16.8	14.4	11.5	8.1	4.9	2.4
		DG	14.4	10.8	6.9	3.3	-	-	-	-	-
		DGNI	25.2	22.4	19.9	17.5	14.1	10.6	7.2	4.1	1.4

TONOPAH NEVADA		SSF =	.1	.2	.3	.4	.5	.6	.7	.8	.9
38.1 N	L	WW	163	75	47	33	24	18	13	9	4
5900 DD	C	WWNI	251	120	76	55	42	33	26	21	15
T(JAN)=30	R	TW	156	72	44	30	21	14	10	6	3
		TWNI	238	113	72	51	39	30	23	17	11
		DG	161	70	39	24	-	-	-	-	-
	D	DGNI	263	124	78	55	41	31	23	16	10
		WW	15.3	13.1	11.5	10.1	8.5	6.3	3.8	1.7	-
		WWNI	24.6	21.5	18.4	16.9	14.8	11.8	8.8	7.3	3.1
		TW	14.5	12.5	10.3	7.8	5.8	4.0	2.5	1.5	-
		TWNI	22.8	20.5	18.6	16.9	14.7	11.8	8.3	5.1	3.4
		DG	14.2	10.8	6.9	3.4	-	-	-	-	-
		DGNI	24.2	22.4	19.9	17.6	14.5	10.8	7.4	4.3	1.5

WINNEMUCCA NEVADA		SSF =	.1	.2	.3	.4	.5	.6	.7	.8	.9
40.9 N	L	WW	131	60	37	25	18	13	9	5	-
6629 DD	C	WWNI	214	103	65	47	35	28	22	17	12
T(JAN)=28	R	TW	129	59	35	23	16	11	7	4	1
		TWNI	203	96	61	43	32	25	19	14	9
		DG	121	49	24	-	-	-	-	-	-
	D	DGNI	222	104	65	46	34	25	18	13	7
		WW	12.2	10.3	8.8	7.3	5.7	3.9	2.1	-	-
		WWNI	20.2	18.8	16.9	15.4	13.7	11.8	8.8	5.9	2.4
		TW	11.7	9.9	7.9	5.8	4.0	2.6	1.5	-	-
		TWNI	19.2	17.4	15.8	14.0	11.9	9.4	6.7	4.0	1.6
		DG	10.2	6.7	2.9	-	-	-	-	-	-
		DGNI	21.0	18.6	16.4	14.2	11.3	8.4	5.6	3.1	1.0

CONCORD NEW HAMPSHIRE		SSF =	.1	.2	.3	.4	.5	.6	.7	.8	.9
43.2 N	L	WW	56	23	12	6	-	-	-	-	-
7360 DD	C	WWNI	129	59	37	26	19	15	12	9	6
T(JAN)=21	R	TW	62	25	13	7	-	-	-	-	-
		TWNI	123	56	35	24	18	14	10	7	5
		DG									
	D	DGNI	125	56	34	23	16	12	8	5	-
		WW	4.4	3.1	2.0	.9	-	-	-	-	-
		WWNI	11.5	10.2	9.4	8.1	7.1	5.9	4.5	2.9	1.1
		TW	5.1	3.6	2.2	1.0	-	-	-	-	-
		TWNI	11.1	9.8	8.6	7.5	6.3	5.0	3.4	2.0	.7
		DG									
		DGNI	11.0	9.4	7.9	6.5	5.0	3.4	2.0	.9	-

NEWARK NEW JERSEY		SSF =	.1	.2	.3	.4	.5	.6	.7	.8	.9
40.7 N	L	WW	91	42	25	16	11	8	4	-	-
5034 DD	C	WWNI	167	80	51	36	27	21	17	13	9
T(JAN)=31	R	TW	93	42	25	16	10	6	3	1	-
		TWNI	158	76	48	34	25	19	15	11	7
		DG	67	18	-	-	-	-	-	-	-
	D	DGNI	169	79	49	34	25	19	13	9	5
		WW	8.5	6.7	5.5	4.3	3.1	1.7	.5	-	-
		WWNI	15.7	15.1	12.9	11.8	10.6	8.9	7.1	4.6	1.8
		TW	8.3	6.8	5.2	3.5	2.1	1.0	-	-	-
		TWNI	15.0	13.6	12.2	10.5	9.3	7.4	5.2	3.1	1.2
		DG	5.0	.8	-	-	-	-	-	-	-
		DGNI	16.0	13.9	12.3	10.4	8.4	6.0	3.9	2.1	.6

ALBUQUERQUE NEW MEXICO		SSF =	.1	.2	.3	.4	.5	.6	.7	.8	.9
35.0 N	L	WW	178	84	53	37	28	21	16	11	6
4292 DD	C	WWNI	270	130	83	60	46	37	30	23	17
T(JAN)=35	R	TW	171	80	50	34	24	17	12	8	5
		TWNI	255	122	78	56	43	33	26	19	13
		DG	183	82	47	29	16	-	-	-	-
	D	DGNI	283	135	86	61	46	35	26	19	12
		WW	16.8	14.9	13.3	12.1	10.3	8.0	5.0	2.5	.5
		WWNI	26.3	24.2	22.1	20.5	19.0	16.9	13.6	8.7	3.6
		TW	16.0	14.2	11.9	10.1	7.9	4.9	3.2	1.8	-
		TWNI	24.5	22.5	20.6	19.1	16.6	13.3	9.5	5.8	3.8
		DG	16.5	13.3	9.3	5.6	2.6	-	-	-	-
		DGNI	27.1	24.6	22.3	20.0	16.6	12.5	8.6	5.1	1.9

CLAYTON NEW MEXICO		SSF =	.1	.2	.3	.4	.5	.6	.7	.8	.9
36.4 N	L	WW	160	73	46	32	24	18	14	9	4
5212 DD	C	WWNI	251	117	75	54	41	33	27	21	15
T(JAN)=33	R	TW	154	71	43	29	20	14	10	6	4
		TWNI	235	111	70	50	38	30	23	17	11
		DG	157	68	38	21	-	-	-	-	-
	D	DGNI	260	122	77	54	41	31	23	17	10
		WW	14.3	13.1	11.3	10.2	8.8	6.6	4.0	1.9	-
		WWNI	23.6	21.3	20.1	18.1	16.9	15.3	12.5	7.9	3.3
		TW	14.2	12.2	10.1	7.8	5.8	4.1	2.5	1.4	-
		TWNI	22.0	20.3	18.3	16.9	15.1	11.9	8.5	5.2	2.1
		DG	13.6	10.6	6.8	3.4	-	-	-	-	-
		DGNI	24.1	21.8	19.7	17.8	14.7	11.0	7.6	4.4	1.7

FARMINGTON NEW MEXICO		SSF =	.1	.2	.3	.4	.5	.6	.7	.8	.9
36.7 N	L	WW	145	66	41	29	21	16	11	7	-
5713 DD	C	WWNI	234	108	70	50	38	30	24	19	13
T(JAN)=29	R	TW	142	64	39	26	18	13	8	5	3
		TWNI	220	103	65	47	35	27	21	16	10
		DG	138	58	31	15	-	-	-	-	-
	D	DGNI	241	112	70	50	37	28	21	15	9
		WW	12.7	11.7	10.1	9.0	7.3	5.4	3.1	1.3	-
		WWNI	21.3	19.8	18.7	16.8	15.5	13.7	10.6	6.9	2.9
		TW	12.8	10.9	9.0	6.9	5.0	3.4	2.0	1.0	.3
		TWNI	20.1	18.7	17.0	15.6	13.5	10.8	7.7	4.7	1.9
		DG	11.6	8.6	4.9	1.7	-	-	-	-	-
		DGNI	21.9	20.0	18.1	16.1	13.2	9.8	6.6	3.8	1.3

LOS ALAMOS NEW MEXICO		SSF =	.1	.2	.3	.4	.5	.6	.7	.8	.9
		WW	128	58	36	26	19	14	10	7	–
	L	WWNI	210	100	64	46	35	28	23	18	13
35.9 N	C	TW	125	58	35	24	16	11	8	5	3
6359 DD	R	TWNI	199	94	60	43	33	26	20	15	10
T(JAN)=29		DG	116	49	25	–	–	–	–	–	–
		DGNI	217	102	64	46	34	26	19	14	9
		WW	11.9	10.0	9.2	8.2	7.0	4.9	2.8	1.1	–
		WWNI	19.9	18.1	16.8	16.1	14.8	13.4	10.9	6.8	2.9
	D	TW	11.5	10.0	8.1	6.2	4.4	3.1	1.8	.9	.3
		TWNI	18.9	17.0	16.0	14.7	13.1	11.4	7.4	4.6	1.9
		DG	9.9	7.1	3.5	–	–	–	–	–	–
		DGNI	20.5	18.2	16.7	15.1	12.4	9.2	6.3	3.6	1.3

ROSWELL NEW MEXICO		SSF =	.1	.2	.3	.4	.5	.6	.7	.8	.9
		WW	200	95	59	42	32	24	18	13	7
	L	WWNI	299	142	92	66	50	40	32	25	18
33.4 N	C	TW	194	89	55	38	27	19	14	9	6
3697 DD	R	TWNI	283	134	86	61	47	36	28	21	14
T(JAN)=38		DG	214	95	56	35	22	12	–	–	–
		DGNI	316	149	95	67	51	38	29	21	13
		WW	18.2	17.3	14.9	13.7	11.7	9.3	5.9	3.0	.7
		WWNI	27.3	27.0	24.6	22.1	20.9	18.3	14.8	9.5	4.0
	D	TW	17.6	15.7	13.4	10.5	8.0	4.5	3.7	1.0	1.0
		TWNI	26.0	22.5	20.8	18.0	14.5	10.3	6.3	2.6
		DG	18.8	15.7	11.6	7.5	4.2	1.5	–	–	–
		DGNI	29.1	27.1	24.6	22.0	18.4	13.8	9.6	5.7	2.2

TRUTH OR CONSEQ. NEW MEXICO		SSF =	.1	.2	.3	.4	.5	.6	.7	.8	.9
		WW	229	109	69	49	37	29	22	16	9
	L	WWNI	332	160	103	74	57	45	37	29	21
33.2 N	C	TW	220	102	64	44	31	23	16	11	7
3392 DD	R	TWNI	314	150	96	69	53	41	32	24	16
T(JAN)=40		DG	251	114	69	44	29	18	9	–	–
		DGNI	354	168	107	76	58	44	33	24	15
		WW	21.5	19.7	17.6	16.3	14.1	11.3	7.5	4.0	1.2
		WWNI	31.2	30.1	27.7	25.2	24.0	21.0	16.8	10.9	4.6
	D	TW	20.4	18.3	15.6	12.9	10.4	8.0	4.8	2.8	1.3
		TWNI	29.7	27.7	25.4	23.7	20.4	16.4	11.7	7.2	3.0
		DG	22.8	19.4	15.0	10.4	6.5	3.4	1.0	–	–
		DGNI	33.5	30.8	28.1	25.2	21.2	16.0	11.2	6.7	2.6

TUCUMCARI NEW MEXICO		SSF =	.1	.2	.3	.4	.5	.6	.7	.8	.9
		WW	194	90	56	40	30	23	17	12	7
	L	WWNI	291	138	88	63	48	39	31	25	18
35.2 N	C	TW	187	86	53	36	25	18	13	9	5
4047 DD	R	TWNI	274	130	82	59	45	35	27	20	13
T(JAN)=37		DG	203	90	53	32	20	10	–	–	–
		DGNI	306	144	91	65	49	37	28	20	13
		WW	17.7	16.0	14.3	12.9	11.2	8.8	5.6	2.9	.7
		WWNI	27.2	25.4	23.3	21.5	20.0	17.9	14.6	9.3	3.9
	D	TW	17.0	15.0	12.7	9.9	7.6	5.5	3.6	2.0	.9
		TWNI	25.4	23.6	21.7	20.0	17.6	14.0	10.0	6.2	2.5
		DG	17.8	14.7	10.6	6.7	3.5	1.0	–	–	–
		DGNI	28.2	25.8	23.5	21.2	17.7	13.3	9.2	5.5	2.1

ZUNI NEW MEXICO		SSF =	.1	.2	.3	.4	.5	.6	.7	.8	.9
		WW	144	66	41	29	22	16	12	8	–
	L	WWNI	231	109	69	50	38	31	25	19	14
35.1 N	C	TW	140	64	39	26	18	13	9	6	3
5815 DD	R	TWNI	218	103	65	47	36	28	21	16	11
T(JAN)=30		DG	137	59	32	16	–	–	–	–	–
		DGNI	238	112	71	50	38	28	21	15	9
		WW	13.1	11.6	10.2	9.2	7.7	5.7	3.3	1.4	–
		WWNI	22.5	19.5	18.9	17.1	15.8	14.2	11.4	7.2	3.0
	D	TW	12.9	11.1	9.1	6.9	5.1	3.5	2.2	1.1	.4
		TWNI	20.5	18.6	17.2	15.8	13.9	11.1	7.9	4.8	2.0
		DG	11.9	8.8	5.1	2.0	–	–	–	–	–
		DGNI	22.2	20.2	18.2	16.4	13.5	10.1	6.9	4.0	1.4

ALBANY NEW YORK		SSF =	.1	.2	.3	.4	.5	.6	.7	.8	.9
		WW	56	23	12	6	–	–	–	–	–
	L	WWNI	129	59	37	26	19	15	11	9	6
42.7 N	C	TW	62	25	13	7	–	–	–	–	–
6888 DD	R	TWNI	123	56	35	24	18	14	10	7	5
T(JAN)=21		DG	–	–	–	–	–	–	–	–	–
		DGNI	125	56	34	23	16	12	8	5	–
		WW	4.5	3.1	2.0	.8	–	–	–	–	–
		WWNI	11.8	10.4	9.2	8.0	7.0	5.8	4.4	2.9	1.1
	D	TW	5.0	3.6	2.2	.9	–	–	–	–	–
		TWNI	11.0	9.8	8.5	7.5	6.2	4.9	3.4	2.0	.7
		DG	–	–	–	–	–	–	–	–	–
		DGNI	11.0	9.3	7.8	6.4	4.9	3.3	1.9	.8	–

BINGHAMTON NEW YORK		SSF =	.1	.2	.3	.4	.5	.6	.7	.8	.9
		WW	42	15	–	–	–	–	–	–	–
	L	WWNI	115	52	32	22	16	12	10	7	5
42.2 N	C	TW	50	19	8	–	–	–	–	–	–
7285 DD	R	TWNI	109	49	30	21	15	11	8	6	4
T(JAN)=22		DG	–	–	–	–	–	–	–	–	–
		DGNI	109	48	28	19	13	9	6	3	–
		WW	3.1	1.5	–	–	–	–	–	–	–
		WWNI	10.3	8.7	7.7	6.6	5.7	4.7	3.6	2.3	.8
	D	TW	3.8	2.3	.9	–	–	–	–	–	–
		TWNI	9.7	8.4	7.2	6.1	5.1	4.0	2.8	1.6	.6
		DG	–	–	–	–	–	–	–	–	–
		DGNI	9.3	7.6	6.2	4.9	3.6	2.3	1.2	.4	–

BUFFALO NEW YORK		SSF =	.1	.2	.3	.4	.5	.6	.7	.8	.9
		WW	46	17	6	–	–	–	–	–	–
	L	WWNI	121	54	33	23	17	13	10	7	5
42.9 N	C	TW	54	20	9	–	–	–	–	–	–
6927 DD	R	TWNI	115	51	31	21	16	12	9	6	4
T(JAN)=24		DG	–	–	–	–	–	–	–	–	–
		DGNI	115	50	29	20	13	9	6	3	–
		WW	3.4	1.8	.4	–	–	–	–	–	–
		WWNI	10.7	9.2	8.0	6.8	5.8	4.7	3.5	2.3	.8
	D	TW	4.2	2.6	1.2	–	–	–	–	–	–
		TWNI	10.0	8.7	7.4	6.4	5.2	4.0	2.8	1.6	.6
		DG	–	–	–	–	–	–	–	–	–
		DGNI	9.7	8.0	6.5	5.1	3.8	2.4	1.3	.4	–

			SSF =	.1	.2	.3	.4	.5	.6	.7	.8	.9
ASHEVILLE			WW	138	64	39	27	20	15	11	7	-
NORTH CAROLINA	L		WWNI	225	106	68	49	37	29	23	18	13
	C		TW	136	62	38	25	17	12	8	5	2
35.4 N	R		TWNI	212	100	64	45	34	26	20	15	10
4237 DD			DG	131	55	28	12	-	-	-	-	-
T(JAN)=38			DGNI	232	109	68	48	36	27	20	14	9
			WW	12.5	11.2	9.5	8.2	6.7	4.9	2.8	1.1	-
			WWNI	21.0	19.5	18.0	16.2	14.6	12.7	10.4	6.6	2.7
	D		TW	12.4	10.5	8.5	6.5	4.6	3.1	1.8	.8	.2
			TWNI	19.8	18.3	16.5	14.9	12.9	10.4	7.3	4.5	1.8
			DG	11.1	7.9	4.2	-	-	-	-	-	-
			DGNI	21.6	19.4	17.3	15.2	12.6	9.3	6.3	3.6	1.3

			SSF =	.1	.2	.3	.4	.5	.6	.7	.8	.9
CAPE HATTERAS			WW	214	97	60	42	31	23	18	12	7
NORTH CAROLINA	L		WWNI	312	148	93	66	50	39	31	25	18
	C		TW	200	92	56	38	26	19	13	9	5
35.3 N	R		TWNI	294	138	87	61	46	36	27	20	13
2731 DD			DG	223	98	56	34	21	10	-	-	-
T(JAN)=45			DGNI	329	154	96	67	50	38	28	20	13
			WW	20.4	16.5	14.9	12.5	10.9	8.7	5.6	2.8	.6
			WWNI	31.2	25.7	21.8	18.5	19.1	17.2	14.3	9.1	3.8
	D		TW	18.6	15.7	13.1	10.0	7.6	4.8	3.6	1.9	.5
			TWNI	28.5	24.5	22.2	19.6	17.6	14.0	9.9	6.1	2.5
			DG	20.0	15.4	11.2	7.0	3.7	1.1	-	-	-
			DGNI	31.3	27.2	23.8	21.0	17.7	13.3	9.2	5.4	2.1

			SSF =	.1	.2	.3	.4	.5	.6	.7	.8	.9
CHARLOTTE			WW	177	82	50	35	25	19	14	9	4
NORTH CAROLINA	L		WWNI	270	129	81	57	43	34	27	21	15
	C		TW	170	78	47	31	22	15	10	7	4
35.2 N	R		TWNI	255	121	76	53	40	31	24	18	12
3218 DD			DG	180	77	43	24	12	-	-	-	-
T(JAN)=42			DGNI	283	133	82	58	43	32	24	17	11
			WW	16.2	14.0	11.8	10.5	8.7	6.8	4.1	1.9	-
			WWNI	25.3	23.7	20.4	18.5	17.2	14.8	12.3	7.9	3.3
	D		TW	15.5	13.2	10.8	8.3	6.1	4.2	2.6	1.4	.5
			TWNI	23.9	21.6	19.1	17.3	15.1	12.1	8.6	5.3	2.1
			DG	15.5	11.7	7.8	4.1	1.2	-	-	-	-
			DGNI	26.5	23.2	20.7	18.0	15.1	11.2	7.7	4.5	1.7

			SSF =	.1	.2	.3	.4	.5	.6	.7	.8	.9
GREENSBORO			WW	151	70	43	29	22	16	12	7	-
NORTH CAROLINA	L		WWNI	239	113	72	51	39	31	24	19	14
	C		TW	147	67	41	27	19	13	9	5	3
36.1 N	R		TWNI	226	107	67	48	36	28	21	16	10
3825 DD			DG	147	62	33	16	-	-	-	-	-
T(JAN)=39			DGNI	249	117	73	51	38	28	21	15	9
			WW	13.8	12.2	10.1	8.9	7.2	5.4	3.1	1.3	-
			WWNI	22.6	20.9	18.8	16.6	15.4	13.2	10.9	7.0	2.9
	D		TW	13.5	11.4	9.3	7.0	5.1	3.4	2.0	1.0	.3
			TWNI	21.2	19.1	17.2	15.6	13.5	10.8	7.7	4.7	1.9
			DG	12.5	9.1	5.2	1.9	-	-	-	-	-
			DGNI	23.1	20.7	18.3	15.9	13.3	9.8	6.6	3.8	1.4

			SSF =	.1	.2	.3	.4	.5	.6	.7	.8	.9
MASSENA			WW	42	14	-	-	-	-	-	-	-
NEW YORK	L		WWNI	115	51	32	22	16	12	9	7	5
	C		TW	50	18	8	-	-	-	-	-	-
44.9 N	R		TWNI	109	49	30	21	15	11	8	6	4
8237 DD			DG	-	-	-	-	-	-	-	-	-
T(JAN)=14			DGNI	109	47	28	18	13	9	6	3	-
			WW	3.0	1.4	-	-	-	-	-	-	-
			WWNI	10.1	8.7	7.8	6.5	5.6	4.4	3.3	2.1	-
	D		TW	3.8	2.2	.8	-	-	-	-	-	-
			TWNI	9.7	8.3	7.1	6.1	5.0	3.8	2.6	1.5	.5
			DG	-	-	-	-	-	-	-	-	-
			DGNI	9.3	7.6	6.1	4.8	3.4	2.2	1.1	.3	-

			SSF =	.1	.2	.3	.4	.5	.6	.7	.8	.9
NYC (CNTRL. PARK)			WW	85	37	22	14	10	6	-	-	-
NEW YORK	L		WWNI	159	76	48	34	25	20	16	12	8
	C		TW	87	38	22	14	8	5	2	-	-
40.8 N	R		TWNI	151	72	45	32	24	18	14	10	6
4848 DD			DG	56	-	-	-	-	-	-	-	-
T(JAN)=32			DGNI	161	74	46	32	23	17	12	8	4
			WW	8.0	5.8	4.7	3.6	2.4	1.1	-	-	-
			WWNI	15.0	13.8	11.8	10.9	9.3	8.8	6.5	4.2	1.7
	D		TW	7.6	6.1	4.5	3.0	1.7	.8	.1	-	-
			TWNI	14.3	12.6	11.3	10.1	8.5	6.8	4.8	2.9	1.1
			DG	3.8	-	-	-	-	-	-	-	-
			DGNI	15.2	12.9	11.2	9.4	7.6	5.4	3.5	1.8	.5

			SSF =	.1	.2	.3	.4	.5	.6	.7	.8	.9
ROCHESTER			WW	50	18	8	-	-	-	-	-	-
NEW YORK	L		WWNI	124	55	34	24	17	13	10	8	5
	C		TW	56	22	11	4	-	-	-	-	-
43.1 N	R		TWNI	117	53	32	22	16	12	9	6	4
6719 DD			DG	-	-	-	-	-	-	-	-	-
T(JAN)=24			DGNI	118	52	31	20	14	10	7	4	-
			WW	3.8	2.1	.8	-	-	-	-	-	-
			WWNI	10.9	9.6	8.3	7.0	6.0	4.8	3.7	2.4	.9
	D		TW	4.4	2.8	1.4	-	-	-	-	-	-
			TWNI	10.2	9.0	7.6	6.6	5.3	4.2	2.9	1.7	.6
			DG	-	-	-	-	-	-	-	-	-
			DGNI	10.1	8.3	6.8	5.3	3.9	2.6	1.4	.5	-

			SSF =	.1	.2	.3	.4	.5	.6	.7	.8	.9
SYRACUSE			WW	50	18	8	-	-	-	-	-	-
NEW YORK	L		WWNI	124	55	34	24	17	13	10	8	5
	C		TW	56	22	11	4	-	-	-	-	-
43.1 N	R		TWNI	117	53	32	22	16	12	9	6	4
6678 DD			DG	-	-	-	-	-	-	-	-	-
T(JAN)=24			DGNI	118	52	31	20	14	10	7	4	-
			WW	3.8	2.1	.9	-	-	-	-	-	-
			WWNI	10.9	9.7	8.3	7.0	6.1	5.0	3.8	2.4	.9
	D		TW	4.4	2.8	1.4	-	-	-	-	-	-
			TWNI	10.3	9.0	7.7	6.6	5.4	4.3	2.9	1.7	.6
			DG	-	-	-	-	-	-	-	-	-
			DGNI	10.1	8.4	6.8	5.4	4.0	2.6	1.4	.5	-

RALEIGH-DURHAM NORTH CAROLINA		SSF =	.1	.2	.3	.4	.5	.6	.7	.8	.9
	L	WW	155	73	45	31	23	17	12	8	–
35.9 N	C	WWNI	244	117	75	53	40	31	25	20	14
3514 DD	R	TW	151	69	42	28	19	13	9	6	3
T(JAN)=40		TWNI	230	110	69	49	37	29	22	16	11
		DG	153	65	36	19	–	–	–	–	–
	D	DGNI	254	120	75	53	39	29	22	15	9
		WW	14.4	12.8	10.6	9.3	7.6	5.8	3.4	1.4	–
		WWNI	23.4	21.8	19.3	17.3	15.8	13.6	11.1	7.2	3.0
		TW	13.7	11.9	9.7	7.3	5.4	3.6	2.2	1.1	1.3
		TWNI	21.8	20.1	17.8	16.1	13.9	11.2	7.9	4.8	1.9
		DG	13.3	9.8	5.9	2.5	–	–	–	–	–
	D	DGNI	23.9	21.4	19.0	16.5	13.7	10.2	6.9	4.0	1.4

BISMARCK NORTH DAKOTA		SSF =	.1	.2	.3	.4	.5	.6	.7	.8	.9
	L	WW	60	24	11	5	–	–	–	–	–
46.8 N	C	WWNI	132	61	38	26	19	14	11	8	5
9044 DD	R	TW	64	26	13	7	–	–	–	–	–
T(JAN)= 8		TWNI	125	58	36	24	18	13	10	7	4
		DG	–	–	–	–	–	–	–	–	–
	D	DGNI	129	58	34	23	16	11	7	4	–
		WW	4.9	3.1	1.6	–	–	–	–	–	–
		WWNI	12.3	10.6	9.2	7.7	6.6	5.4	3.9	2.6	1.0
		TW	5.3	3.6	2.1	.8	–	–	–	–	–
		TWNI	11.6	10.0	8.4	7.1	5.8	4.5	3.1	1.8	.7
		DG	3.0	–	–	–	–	–	–	–	–
	D	DGNI	11.6	9.5	7.7	6.1	4.5	3.0	1.7	.7	–

FARGO NORTH DAKOTA		SSF =	.1	.2	.3	.4	.5	.6	.7	.8	.9
	L	WW	48	17	–	–	–	–	–	–	–
46.9 N	C	WWNI	119	55	34	23	17	12	9	7	5
9271 DD	R	TW	54	20	9	–	–	–	–	–	–
T(JAN)= 6		TWNI	114	52	31	21	15	11	8	6	4
		DG	–	–	–	–	–	–	–	–	–
	D	DGNI	114	51	29	19	13	9	6	3	–
		WW	3.7	1.6	–	–	–	–	–	–	–
		WWNI	10.9	9.4	7.9	6.5	5.5	4.5	3.2	2.1	.8
		TW	4.3	2.5	1.0	–	–	–	–	–	–
		TWNI	10.3	8.8	7.3	6.1	4.9	3.8	2.6	1.5	.5
		DG	–	–	–	–	–	–	–	–	–
	D	DGNI	10.1	8.1	6.3	4.8	3.4	2.2	1.1	.3	–

MINOT NORTH DAKOTA		SSF =	.1	.2	.3	.4	.5	.6	.7	.8	.9
	L	WW	54	19	7	–	–	–	–	–	–
48.3 N	C	WWNI	126	58	35	24	17	13	10	7	5
9407 DD	R	TW	59	22	10	–	–	–	–	–	–
T(JAN)= 8		TWNI	120	55	33	22	16	12	9	6	4
		DG	–	–	–	–	–	–	–	–	–
	D	DGNI	122	54	31	20	14	9	6	3	–
		WW	4.2	2.1	.4	–	–	–	–	–	–
		WWNI	11.5	9.9	8.3	6.6	5.5	4.5	3.2	2.1	.8
		TW	4.7	2.8	1.2	–	–	–	–	–	–
		TWNI	10.9	9.3	7.6	6.2	5.0	3.8	2.6	1.5	.5
		DG	–	–	–	–	–	–	–	–	–
	D	DGNI	10.8	8.6	6.6	5.0	3.5	2.3	1.2	.3	–

AKRON-CANTON OHIO		SSF =	.1	.2	.3	.4	.5	.6	.7	.8	.9
	L	WW	61	25	13	7	–	–	–	–	–
40.9 N	C	WWNI	137	62	39	27	20	15	12	9	6
6224 DD	R	TW	67	27	14	8	–	–	–	–	–
T(JAN)=26		TWNI	130	59	36	25	18	14	10	7	5
		DG	–	–	–	–	–	–	–	–	–
	D	DGNI	133	59	35	24	17	12	8	5	2
		WW	4.9	3.3	2.2	1.0	–	–	–	–	–
		WWNI	12.0	10.9	9.3	8.1	7.2	5.9	4.5	2.9	1.1
		TW	5.4	3.8	2.4	1.1	–	–	–	–	–
		TWNI	11.4	10.1	8.7	7.6	6.3	5.0	3.5	2.0	.8
		DG	–	–	–	–	–	–	–	–	–
	D	DGNI	11.5	9.7	8.1	6.6	5.1	3.4	2.0	.9	.1

CINCINNATI OHIO		SSF =	.1	.2	.3	.4	.5	.6	.7	.8	.9
	L	WW	81	36	20	13	8	5	–	–	–
39.1 N	C	WWNI	156	74	46	32	24	19	15	11	8
5070 DD	R	TW	84	36	21	13	8	4	–	–	–
T(JAN)=31		TWNI	148	70	43	30	22	17	13	9	6
		DG	49	–	–	–	–	–	–	–	–
	D	DGNI	157	72	44	30	22	16	11	8	4
		WW	7.3	5.4	4.1	3.0	1.9	–	–	–	–
		WWNI	14.3	13.6	11.5	10.0	9.2	7.7	6.1	3.9	1.6
		TW	7.3	5.6	4.1	2.6	1.4	.5	–	–	–
		TWNI	13.7	12.2	10.7	9.5	8.0	6.4	4.5	2.7	1.0
		DG	3.0	–	–	–	–	–	–	–	–
	D	DGNI	14.4	12.3	10.5	8.8	7.0	4.9	3.1	1.6	.4

CLEVELAND OHIO		SSF =	.1	.2	.3	.4	.5	.6	.7	.8	.9
	L	WW	59	23	11	5	–	–	–	–	–
41.4 N	C	WWNI	135	60	37	26	19	14	11	8	6
6154 DD	R	TW	65	26	13	7	–	–	–	–	–
T(JAN)=27		TWNI	128	57	35	24	18	13	10	7	4
		DG	–	–	–	–	–	–	–	–	–
	D	DGNI	130	57	34	23	16	11	8	5	1
		WW	4.6	3.0	1.7	.5	–	–	–	–	–
		WWNI	11.7	10.6	8.9	7.7	6.7	5.4	4.2	2.7	1.0
		TW	5.1	3.4	2.1	.8	–	–	–	–	–
		TWNI	11.1	9.7	8.3	7.2	5.9	4.7	3.2	1.9	.7
		DG	–	–	–	–	–	–	–	–	–
	D	DGNI	11.2	9.2	7.6	6.1	4.6	3.1	1.8	.7	–

COLUMBUS OHIO		SSF =	.1	.2	.3	.4	.5	.6	.7	.8	.9
	L	WW	67	28	15	9	5	–	–	–	–
40.0 N	C	WWNI	141	65	41	29	21	16	13	10	7
5702 DD	R	TW	71	30	16	9	5	–	–	–	–
T(JAN)=28		TWNI	133	62	38	27	20	15	11	8	5
		DG	–	–	–	–	–	–	–	–	–
	D	DGNI	138	63	38	26	19	13	9	6	3
		WW	5.7	4.1	2.9	1.8	.7	–	–	–	–
		WWNI	12.7	11.8	10.0	8.8	7.9	6.5	5.1	3.3	1.3
		TW	6.0	4.4	3.0	1.7	.6	–	–	–	–
		TWNI	12.1	10.8	9.4	8.3	6.9	5.5	3.8	2.3	.9
		DG	–	–	–	–	–	–	–	–	–
	D	DGNI	12.5	10.6	8.9	7.3	5.7	4.0	2.4	1.1	.2

Tables of LCR & D Versus SSF / 273

DAYTON, OHIO (39.9 N, 5641 DD, T(JAN)=28)

	SSF =	.1	.2	.3	.4	.5	.6	.7	.8	.9
L	WW	72	30	17	10	6	-	-	-	-
	WWNI	146	68	43	30	22	17	13	10	7
C	TW	76	32	18	10	6	-	-	-	-
	TWNI	139	64	40	28	21	16	12	9	6
R	DG	30	-	-	-	-	-	-	-	-
	DGNI	145	66	40	27	20	14	10	7	3
D	WW	6.2	4.4	3.3	2.2	1.1	-	-	-	-
	WWNI	13.1	12.4	10.4	9.2	8.4	6.9	5.4	3.5	1.4
	TW	6.4	4.8	3.3	2.0	.8	-	-	-	-
	TWNI	12.6	11.2	9.8	8.7	7.3	5.8	4.1	2.4	.9
	DG	.9	-	-	-	-	-	-	-	-
	DGNI	13.1	11.1	9.4	7.8	6.2	4.3	2.7	1.3	.3

TOLEDO, OHIO (41.6 N, 6381 DD, T(JAN)=25)

	SSF =	.1	.2	.3	.4	.5	.6	.7	.8	.9
L	WW	62	25	13	7	-	-	-	-	-
	WWNI	138	62	39	27	20	15	12	9	6
C	TW	68	27	15	8	3	-	-	-	-
	TWNI	130	59	36	25	19	14	10	8	5
R	DG	-	-	-	-	-	-	-	-	-
	DGNI	134	60	36	24	17	12	8	5	2
D	WW	5.0	3.4	2.3	1.0	-	-	-	-	-
	WWNI	12.1	11.1	9.4	8.2	7.2	5.9	4.5	2.9	1.1
	TW	5.5	3.8	2.4	1.2	-	-	-	-	-
	TWNI	11.5	10.2	8.8	7.7	6.4	5.0	3.5	2.0	.8
	DG	-	-	-	-	-	-	-	-	-
	DGNI	11.6	9.8	8.2	6.7	5.1	3.5	2.0	.9	.1

YOUNGSTOWN, OHIO (41.3 N, 6426 DD, T(JAN)=26)

	SSF =	.1	.2	.3	.4	.5	.6	.7	.8	.9
L	WW	52	20	9	-	-	-	-	-	-
	WWNI	126	57	35	24	18	14	10	8	5
C	TW	58	23	11	5	-	-	-	-	-
	TWNI	120	54	33	23	17	12	9	7	4
R	DG	-	-	-	-	-	-	-	-	-
	DGNI	121	53	32	21	15	10	7	4	1
D	WW	4.0	2.4	1.1	-	-	-	-	-	-
	WWNI	11.2	9.9	8.4	7.2	6.3	5.1	3.9	2.6	1.0
	TW	4.6	3.0	1.6	-	-	-	-	-	-
	TWNI	10.5	9.2	7.9	6.8	5.6	4.4	3.1	1.8	.6
	DG	-	-	-	-	-	-	-	-	-
	DGNI	10.4	8.6	7.0	5.6	4.2	2.8	1.6	.6	-

OKLAHOMA CITY, OKLAHOMA (35.4 N, 3695 DD, T(JAN)=37)

	SSF =	.1	.2	.3	.4	.5	.6	.7	.8	.9
L	WW	161	74	45	31	23	17	13	8	-
	WWNI	251	119	75	53	40	32	26	20	14
C	TW	156	71	43	28	20	14	9	6	3
	TWNI	237	111	70	50	37	29	22	16	11
R	DG	158	67	36	19	-	-	-	-	-
	DGNI	262	122	76	53	40	30	22	16	10
D	WW	14.7	12.5	10.7	9.6	8.0	6.0	3.6	1.6	-
	WWNI	23.5	21.3	19.0	17.5	16.3	14.7	11.8	7.7	3.1
	TW	14.0	11.9	9.8	7.5	5.4	3.7	2.3	1.2	.4
	TWNI	22.2	19.9	17.8	16.3	14.2	11.4	8.1	5.0	2.0
	DG	13.1	10.0	6.1	2.8	-	-	-	-	-
	DGNI	24.4	21.1	19.1	16.8	14.0	10.4	7.1	4.1	1.5

TULSA, OKLAHOMA (36.2 N, 3680 DD, T(JAN)=37)

	SSF =	.1	.2	.3	.4	.5	.6	.7	.8	.9
L	WW	147	67	41	28	21	15	11	7	-
	WWNI	235	111	70	49	38	30	24	19	13
C	TW	144	65	39	26	18	12	8	-	-
	TWNI	222	104	65	46	35	27	21	15	10
R	DG	141	58	30	14	-	-	-	-	-
	DGNI	245	113	70	49	37	28	20	14	9
D	WW	13.5	11.2	9.7	8.6	7.0	5.2	3.0	1.2	-
	WWNI	21.8	19.8	17.7	16.4	15.2	13.1	10.7	6.8	2.8
	TW	12.9	10.8	8.8	6.7	4.8	3.2	1.9	1.0	.2
	TWNI	20.8	18.5	16.7	15.2	13.2	10.6	7.5	4.6	1.8
	DG	11.7	8.4	4.7	1.3	-	-	-	-	-
	DGNI	22.7	19.7	17.8	15.5	12.9	9.5	6.5	3.7	1.3

ASTORIA, OREGON (46.1 N, 5295 DD, T(JAN)=41)

	SSF =	.1	.2	.3	.4	.5	.6	.7	.8	.9
L	WW	122	59	36	24	17	11	7	-	-
	WWNI	201	100	64	46	34	26	20	15	10
C	TW	121	56	34	22	14	9	6	2	-
	TWNI	191	94	60	42	31	23	17	13	8
R	DG	112	46	19	-	-	-	-	-	-
	DGNI	208	101	64	44	32	23	17	11	6
D	WW	12.2	10.2	8.3	6.4	4.5	2.6	1.1	-	-
	WWNI	20.1	19.7	16.6	14.7	12.4	10.1	7.2	4.6	1.8
	TW	11.5	9.6	7.3	5.3	3.5	2.0	.9	.2	-
	TWNI	19.1	17.5	15.5	13.2	10.8	8.2	5.8	3.4	1.2
	DG	9.9	5.8	1.7	-	-	-	-	-	-
	DGNI	20.8	18.4	15.8	13.0	10.0	7.3	4.6	2.4	.7

BURNS, OREGON (43.6 N, 7212 DD, T(JAN)=25)

	SSF =	.1	.2	.3	.4	.5	.6	.7	.8	.9
L	WW	100	44	26	17	11	7	-	-	-
	WWNI	179	84	52	37	28	21	16	12	8
C	TW	101	44	25	16	10	6	2	-	-
	TWNI	170	79	49	34	25	19	14	10	6
R	DG	76	19	-	-	-	-	-	-	-
	DGNI	182	83	51	35	25	18	13	8	4
D	WW	8.9	7.0	5.5	4.0	2.4	1.2	-	-	-
	WWNI	16.3	15.0	13.3	11.7	9.9	8.2	5.9	4.0	1.5
	TW	8.8	6.9	5.1	3.3	1.7	.6	-	-	-
	TWNI	15.6	13.8	12.3	10.5	8.8	6.7	4.8	2.8	1.0
	DG	5.5	.8	-	-	-	-	-	-	-
	DGNI	16.6	14.3	12.2	10.0	7.7	5.6	3.5	1.7	.4

MEDFORD, OREGON (42.4 N, 4930 DD, T(JAN)=37)

	SSF =	.1	.2	.3	.4	.5	.6	.7	.8	.9
L	WW	134	57	34	22	15	9	5	-	-
	WWNI	222	100	62	44	32	25	19	14	9
C	TW	130	56	32	20	13	8	4	-	-
	TWNI	209	94	58	40	29	22	16	12	7
R	DG	118	42	-	-	-	-	-	-	-
	DGNI	227	102	62	42	30	21	15	10	5
D	WW	11.3	9.4	7.4	5.4	3.5	1.9	.6	-	-
	WWNI	20.2	16.9	15.9	13.6	11.4	9.0	6.3	4.2	1.1
	TW	11.2	8.7	6.4	4.5	2.9	1.6	.6	-	-
	TWNI	18.7	16.2	14.2	12.1	9.8	7.4	5.2	3.0	1.1
	DG	9.0	4.5	-	-	-	-	-	-	-
	DGNI	19.9	17.2	14.1	11.8	8.9	6.5	4.1	2.0	.5

NORTH BEND, OREGON
43.4 N, 4688 DD, T(JAN)=45

SSF =	.1	.2	.3	.4	.5	.6	.7	.8	.9
L WW	172	86	55	39	29	21	15	10	5
WWNI	260	130	86	63	48	37	29	23	16
C TW	168	81	51	34	24	17	12	8	4
TWNI	247	123	80	58	44	33	25	19	12
R DG	181	84	49	30	17	–	–	–	–
DGNI	275	137	88	63	47	35	26	18	11
D WW	17.2	16.8	14.4	11.9	9.9	7.0	4.5	2.0	.3
WWNI	26.0	24.4	21.4	18.3	15.9	11.8	7.9	3.2	
TW	16.6	14.7	12.1	9.5	7.0	4.9	2.9	1.5	.5
TWNI	24.7	21.9	19.1	16.3	12.6	9.0	5.4	2.1	
DG	17.7	17.0	9.7	5.7	2.5	–	–	–	–
DGNI	27.5	26.3	23.2	20.2	16.1	12.1	8.1	4.6	1.7

PENDLETON, OREGON
45.7 N, 5240 DD, T(JAN)=32

SSF =	.1	.2	.3	.4	.5	.6	.7	.8	.9
L WW	117	49	28	17	10	5	–	–	–
WWNI	204	90	56	39	28	21	15	11	7
C TW	115	48	26	16	8	–	–	–	–
TWNI	190	85	52	35	25	19	14	10	6
R DG	94	22	–	–	–	–	–	–	–
DGNI	205	91	54	36	25	17	12	8	3
D WW	9.5	7.5	5.3	3.3	1.7	.5	–	–	–
WWNI	18.4	15.3	13.7	11.2	8.9	7.1	5.0	3.4	1.3
TW	9.7	7.1	4.9	3.2	1.7	.6	–	–	–
TWNI	16.9	14.4	12.2	10.0	8.0	6.0	4.2	2.4	.9
DG	6.5	–	–	–	–	–	–	–	–
DGNI	17.8	14.8	12.0	9.5	7.0	4.9	2.9	1.4	.3

PORTLAND, OREGON
45.6 N, 4792 DD, T(JAN)=38

SSF =	.1	.2	.3	.4	.5	.6	.7	.8	.9
L WW	135	56	32	20	13	8	–	–	–
WWNI	223	100	61	42	31	23	17	13	8
C TW	130	55	31	19	12	7	3	–	–
TWNI	209	94	57	39	28	21	15	11	7
R DG	117	38	–	–	–	–	–	–	–
DGNI	228	100	60	40	28	20	14	9	4
D WW	11.4	8.5	6.6	4.6	2.8	1.4	–	–	–
WWNI	20.2	16.6	14.8	12.7	10.5	8.2	5.8	3.8	1.5
TW	11.2	8.1	6.0	3.2	1.2	–	–	–	–
TWNI	19.0	15.6	13.5	11.3	9.1	6.8	4.8	2.8	1.0
DG	8.6	3.5	–	–	–	–	–	–	–
DGNI	20.3	16.4	13.5	10.8	8.2	5.9	3.6	1.8	.4

REDMOND, OREGON
44.3 N, 6643 DD, T(JAN)=30

SSF =	.1	.2	.3	.4	.5	.6	.7	.8	.9
L WW	125	56	33	22	15	10	6	–	–
WWNI	209	98	61	43	32	25	19	15	10
C TW	122	54	32	20	13	9	5	2	–
TWNI	196	92	57	40	30	22	17	12	8
R DG	110	41	12	–	–	–	–	–	–
DGNI	213	98	61	42	30	22	16	11	6
D WW	11.1	9.0	7.5	5.9	4.2	2.5	1.1	–	–
WWNI	19.8	17.4	15.4	14.0	12.0	9.9	7.3	4.9	1.9
TW	11.0	8.7	6.6	4.8	3.1	1.8	.9	.1	–
TWNI	18.4	16.1	14.5	12.5	10.4	8.1	5.7	3.4	1.3
DG	8.8	4.7	1.2	–	–	–	–	–	–
DGNI	19.8	16.9	14.7	12.3	9.6	7.1	4.5	2.4	.7

SALEM, OREGON
44.9 N, 4852 DD, T(JAN)=39

SSF =	.1	.2	.3	.4	.5	.6	.7	.8	.9
L WW	137	57	34	22	15	9	5	–	–
WWNI	226	101	62	43	32	24	18	14	9
C TW	133	56	32	20	13	8	4	1	–
TWNI	212	95	58	40	29	22	16	11	7
R DG	121	41	–	–	–	–	–	–	–
DGNI	231	102	62	42	30	21	15	10	5
D WW	11.5	8.8	7.2	5.3	3.5	1.9	.6	–	–
WWNI	20.3	16.9	15.1	13.4	11.3	8.9	6.3	4.2	1.6
TW	11.3	8.5	6.6	4.5	2.8	1.5	.6	–	–
TWNI	19.1	15.8	14.0	12.0	9.7	7.4	5.2	3.0	1.1
DG	9.1	4.3	–	–	–	–	–	–	–
DGNI	20.4	16.8	14.2	11.6	8.9	6.4	4.0	2.0	.5

ALLENTOWN, PENNSYLVANIA
40.6 N, 5827 DD, T(JAN)=28

SSF =	.1	.2	.3	.4	.5	.6	.7	.8	.9
L WW	76	33	19	12	8	5	–	–	–
WWNI	152	71	45	32	24	18	15	11	8
C TW	80	35	20	12	7	4	–	–	–
TWNI	143	67	42	30	22	17	13	9	6
R DG	43	–	–	–	–	–	–	–	–
DGNI	150	69	43	29	21	16	11	7	4
D WW	6.6	5.2	4.0	2.9	1.8	.6	–	–	–
WWNI	13.9	12.9	11.4	10.1	9.1	7.6	5.9	3.9	1.5
TW	6.9	5.4	3.9	2.5	1.3	–	–	–	–
TWNI	13.1	12.0	10.6	9.4	8.0	6.3	4.4	2.6	1.0
DG	2.4	–	–	–	–	–	–	–	–
DGNI	13.6	11.9	10.3	8.7	6.9	4.8	3.0	1.5	.4

ERIE, PENNSYLVANIA
42.1 N, 6851 DD, T(JAN)=25

SSF =	.1	.2	.3	.4	.5	.6	.7	.8	.9
L WW	53	19	8	–	–	–	–	–	–
WWNI	129	57	35	24	17	13	10	7	5
C TW	59	22	11	4	–	–	–	–	–
TWNI	122	54	33	22	16	12	9	6	4
R DG	–	–	–	–	–	–	–	–	–
DGNI	124	53	31	21	14	10	6	4	1
D WW	3.8	2.2	.8	–	–	–	–	–	–
WWNI	11.5	9.4	8.3	7.1	6.0	4.8	3.6	2.3	.9
TW	4.6	2.8	1.4	.1	–	–	–	–	–
TWNI	10.7	9.1	7.7	6.6	5.3	4.1	2.9	1.6	.6
DG	–	–	–	–	–	–	–	–	–
DGNI	10.4	8.4	6.8	5.3	3.9	2.6	1.4	.5	–

HARRISBURG, PENNSYLVANIA
40.2 N, 5224 DD, T(JAN)=30

SSF =	.1	.2	.3	.4	.5	.6	.7	.8	.9
L WW	83	37	22	14	10	6	–	–	–
WWNI	158	75	47	34	25	20	16	12	8
C TW	86	38	22	14	9	5	2	–	–
TWNI	150	71	44	31	23	18	14	10	6
R DG	55	–	–	–	–	–	–	–	–
DGNI	158	74	46	32	23	17	12	8	4
D WW	7.5	5.9	4.6	3.5	2.4	1.1	–	–	–
WWNI	14.7	13.8	12.1	10.7	9.8	8.2	6.5	4.2	1.7
TW	7.6	6.0	4.4	2.9	1.7	.8	.1	–	–
TWNI	14.1	12.7	11.2	10.1	8.5	6.8	4.8	2.8	1.1
DG	3.6	–	–	–	–	–	–	–	–
DGNI	14.8	12.8	11.2	9.4	7.5	5.3	3.4	1.8	.5

PHILADELPHIA PENNSYLVANIA		SSF =	.1	.2	.3	.4	.5	.6	.7	.8	.9
39.9 N	L	WW		95	43	26	17	12	8	5	-	-
4865 DD	C	WWNI		171	82	52	37	28	22	17	13	9
T(JAN)=32	R	TW		96	43	25	16	6	6	4	-	-
		TWNI		162	77	49	34	26	20	15	11	7
		DG		72	22	-	-	-	-	-	-	-
		DGNI		174	81	51	35	26	19	14	9	5
	D	WW		8.8	6.9	5.7	4.5	3.3	1.9	.6	-	-
		WWNI		16.0	15.3	13.2	10.8	10.8	9.1	7.3	4.7	1.9
		TW		8.6	7.0	5.3	3.7	2.3	1.3	.5	-	-
		TWNI		15.3	13.8	12.4	11.1	9.4	7.6	5.3	3.2	1.2
		DG		5.4	1.4	-	-	-	-	-	-	-
		DGNI		16.3	14.2	12.5	10.6	8.6	6.2	4.0	2.1	.6

PITTSBURGH PENNSYLVANIA		SSF =	.1	.2	.3	.4	.5	.6	.7	.8	.9
40.5 N	L	WW		62	25	13	7	-	-	-	-	-
5930 DD	C	WWNI		137	62	39	27	20	15	12	9	6
T(JAN)=28	R	TW		68	28	15	8	3	-	-	-	-
		TWNI		130	59	37	25	19	14	11	8	5
		DG		-	-	-	-	-	-	-	-	-
		DGNI		134	60	36	24	17	12	8	5	2
	D	WW		5.1	3.5	2.3	1.1	-	-	-	-	-
		WWNI		12.2	11.1	9.6	8.2	7.2	5.9	4.6	3.0	1.1
		TW		5.5	3.9	2.5	1.2	.3	-	-	-	-
		TWNI		11.5	10.3	8.9	7.7	6.4	5.0	3.5	2.1	.8
		DG		-	-	-	-	-	-	-	-	-
		DGNI		11.7	9.9	8.2	6.7	5.1	3.5	2.1	.9	.1

SCRANTON PENNSYLVANIA		SSF =	.1	.2	.3	.4	.5	.6	.7	.8	.9
41.3 N	L	WW		63	26	14	8	4	-	-	-	-
6277 DD	C	WWNI		137	63	40	28	20	16	12	9	6
T(JAN)=26	R	TW		68	28	15	9	3	-	-	-	-
		TWNI		130	60	37	26	19	14	11	8	5
		DG		-	-	-	-	-	-	-	-	-
		DGNI		134	61	37	25	18	13	9	6	2
	D	WW		5.2	3.7	2.5	1.4	.3	-	-	-	-
		WWNI		12.5	11.2	9.8	8.5	7.5	6.3	4.9	3.1	1.2
		TW		5.6	4.1	2.7	1.4	.4	-	-	-	-
		TWNI		11.7	10.5	9.1	8.0	6.7	5.3	3.7	2.2	.8
		DG		-	-	-	-	-	-	-	-	-
		DGNI		11.9	10.1	8.5	7.0	5.4	3.7	2.2	1.0	.1

PROVIDENCE RHODE ISLAND		SSF =	.1	.2	.3	.4	.5	.6	.7	.8	.9
41.7 N	L	WW		77	34	20	13	8	5	-	-	-
5972 DD	C	WWNI		153	71	45	32	24	19	15	11	8
T(JAN)=28	R	TW		81	35	20	12	7	4	-	-	-
		TWNI		145	67	42	30	22	17	13	9	6
		DG		44	-	-	-	-	-	-	-	-
		DGNI		152	69	43	30	22	16	11	8	4
	D	WW		6.5	5.3	4.2	3.1	1.9	.7	-	-	-
		WWNI		14.0	12.7	11.6	10.3	9.2	7.8	6.1	3.9	1.6
		TW		7.0	5.5	3.9	2.5	1.4	.5	-	-	-
		TWNI		13.2	12.1	10.7	9.6	8.1	6.5	4.5	2.7	1.0
		DG		2.5	-	-	-	-	-	-	-	-
		DGNI		13.6	11.9	10.5	8.8	7.0	5.0	3.1	1.6	.4

CHARLESTON SOUTH CAROLINA		SSF =	.1	.2	.3	.4	.5	.6	.7	.8	.9
32.9 N	L	WW		252	118	72	51	38	29	22	16	9
2146 DD	C	WWNI		358	173	108	76	58	46	37	29	21
T(JAN)=49	R	TW		238	110	67	46	32	23	17	11	7
		TWNI		339	161	101	71	54	42	32	24	16
		DG		276	124	73	47	30	19	10	-	-
		DGNI		384	180	113	79	59	45	34	24	16
	D	WW		24.6	19.9	17.9	16.1	13.7	11.3	7.5	4.0	1.2
		WWNI		35.0	30.8	27.0	25.9	23.5	20.3	17.0	10.9	4.6
		TW		21.9	19.0	15.9	12.6	9.5	7.0	4.8	2.9	1.3
		TWNI		33.0	28.4	25.9	23.2	20.5	16.5	11.8	7.2	3.0
		DG		24.8	20.1	15.5	10.7	6.7	3.5	1.1	-	-
		DGNI		36.8	31.6	28.4	24.9	21.2	16.0	11.2	6.7	2.7

COLUMBIA SOUTH CAROLINA		SSF =	.1	.2	.3	.4	.5	.6	.7	.8	.9
33.9 N	L	WW		215	102	62	43	32	25	19	13	7
2598 DD	C	WWNI		315	153	96	68	52	41	33	26	18
T(JAN)=45	R	TW		206	95	58	39	28	20	14	9	6
		TWNI		298	142	89	63	48	37	28	21	14
		DG		231	102	60	37	23	12	-	-	-
		DGNI		335	158	99	69	52	39	29	21	13
	D	WW		20.6	17.2	15.1	13.7	11.6	9.4	6.1	3.1	.8
		WWNI		30.2	27.9	23.8	22.4	20.9	18.0	15.9	9.6	4.0
		TW		18.9	16.4	13.7	10.7	7.9	5.7	3.9	2.3	.9
		TWNI		28.6	25.3	22.8	20.6	18.1	14.6	10.4	6.4	2.6
		DG		20.5	16.4	12.1	7.8	4.3	1.7	-	-	-
		DGNI		32.2	27.8	25.1	21.9	18.5	13.9	9.7	5.7	2.2

GREENVILLE SOUTH CAROLINA		SSF =	.1	.2	.3	.4	.5	.6	.7	.8	.9
34.9 N	L	WW		178	83	51	35	26	20	14	10	4
3163 DD	C	WWNI		272	129	82	58	44	35	28	22	15
T(JAN)=42	R	TW		172	78	48	32	22	16	11	7	4
		TWNI		257	121	76	54	41	31	24	18	12
		DG		182	78	44	25	13	-	-	-	-
		DGNI		285	134	83	58	43	33	24	17	11
	D	WW		16.4	14.2	12.0	10.8	8.9	7.0	4.3	2.0	.3
		WWNI		25.4	23.8	20.7	18.8	17.5	15.1	12.5	8.0	3.3
		TW		15.6	13.4	11.0	8.5	6.3	4.3	2.7	1.5	.5
		TWNI		24.0	21.8	19.4	17.6	15.3	12.3	8.7	5.4	2.2
		DG		15.7	12.0	8.1	4.4	1.4	-	-	-	-
		DGNI		26.7	23.4	21.0	18.3	15.3	11.4	7.8	4.6	1.7

HURON SOUTH DAKOTA		SSF =	.1	.2	.3	.4	.5	.6	.7	.8	.9
44.4 N	L	WW		63	25	12	6	-	-	-	-	-
8054 DD	C	WWNI		136	62	39	27	19	15	11	9	6
T(JAN)=13	R	TW		67	27	14	7	-	-	-	-	-
		TWNI		129	59	36	25	18	14	10	7	5
		DG		-	-	-	-	-	-	-	-	-
		DGNI		133	59	35	24	16	12	8	5	1
	D	WW		5.0	3.2	1.9	.7	-	-	-	-	-
		WWNI		12.7	10.9	9.3	7.8	6.9	5.6	4.2	2.8	1.1
		TW		5.5	3.7	2.2	1.0	-	-	-	-	-
		TWNI		11.8	10.1	8.5	7.4	6.1	4.8	3.3	1.9	.7
		DG		-	-	-	-	-	-	-	-	-
		DGNI		11.8	9.7	7.9	6.3	4.8	3.2	1.9	.8	-

		SSF =	.1	.2	.3	.4	.5	.6	.7	.8	.9
PIERRE SOUTH DAKOTA 44.4 N 7677 DD T(JAN)=16	L C R	WW WWNI TW TWNI DG DGNI	77 153 80 144 36 151	32 70 33 67 — 68	17 44 18 41 — 41	10 30 10 28 — 28	6 22 6 21 — 20	— 17 — 16 — 14	— 13 — 12 — 10	— 10 — 8 — 6	— 7 — 5 — 3
	D	WW WWNI TW TWNI DG DGNI	6.4 14.4 6.8 13.3 1.3 13.6	4.5 12.3 4.8 11.5 — 11.2	3.1 10.5 3.2 9.8 — 9.3	2.0 9.0 1.9 8.5 — 7.6	.9 8.0 7.0 — 5.9	6.6 5.6 4.1	5.0 3.9 2.5	3.3 2.3 1.2	1.3 .9 .2
RAPID CITY SOUTH DAKOTA 44.0 N 7324 DD T(JAN)=22	L C R	WW WWNI TW TWNI DG DGNI	91 166 92 158 64 168	40 79 40 75 — 78	23 50 23 47 — 48	15 35 14 33 — 33	10 26 9 24 — 24	6 20 5 18 — 17	— 16 — 14 — 12	— 12 — 10 — 8	— 8 — 6 — 4
	D	WW WWNI TW TWNI DG DGNI	8.3 15.9 8.2 15.1 4.4 15.7	6.4 14.1 6.4 13.4 — 13.5	4.8 12.7 4.7 11.5 — 11.4	3.7 10.9 3.3 10.1 — 9.6	2.3 9.5 1.8 8.5 — 7.5	1.1 8.2 .8 6.7 — 5.4	6.2 4.7 3.4	4.1 2.8 1.7	1.6 1.1 .4
SIOUX FALLS SOUTH DAKOTA 43.6 N 7838 DD T(JAN)=14	L C R	WW WWNI TW TWNI DG DGNI	68 143 72 135 — 140	28 66 30 62 — 63	15 41 16 38 — 38	8 28 9 26 — 25	4 21 4 19 — 18	— 16 — 15 — 13	— 12 — 11 — 9	— 9 — 8 — 6	— 6 — 5 — 2
	D	WW WWNI TW TWNI DG DGNI	5.6 13.2 6.0 12.3 — 12.5	3.8 11.5 4.3 10.7 — 10.4	2.6 9.9 2.7 9.1 — 8.6	1.4 8.5 1.5 8.0 — 7.0	.3 7.5 .4 6.6 — 5.4	6.2 5.3 3.7	4.7 3.7 2.2	3.1 2.1 1.0	1.2 .8 .1
CHATTANOOGA TENNESSEE 35.0 N 3505 DD T(JAN)=40	L C R	WW- WWNI TW TWNI DG DGNI	136 220 133 208 127 228	63 106 61 99 52 108	38 67 36 62 25 67	26 47 24 44 — 47	19 35 16 33 — 34	14 28 11 25 — 26	10 22 7 19 — 19	6 17 4 14 — 13	— 12 — 9 — 8
	D	WW WWNI TW TWNI DG DGNI	12.7 21.4 12.2 19.9 10.6 21.6	10.6 19.7 10.2 17.9 7.1 18.8	8.7 16.8 8.1 15.6 3.3 16.5	7.6 15.0 6.0 14.2 — 14.3	6.0 14.1 4.2 12.2 — 11.8	4.3 11.9 2.7 9.8 — 8.7	2.4 9.8 1.5 6.9 — 5.8	.8 6.3 .7 4.2 — 3.3	— 2.6 — 1.7 — 1.1

		SSF =	.1	.2	.3	.4	.5	.6	.7	.8	.9
KNOXVILLE TENNESSEE 35.8 N 3478 DD T(JAN)=41	L C R	WW WWNI TW TWNI DG DGNI	143 230 140 217 136 238	66 110 63 103 56 111	39 69 38 64 28 69	27 48 25 45 10 48	20 37 17 34 — 35	14 29 11 26 — 26	10 23 8 20 — 19	6 18 — 15 — 14	— 13 — 12 — 10
	D	WW WWNI TW TWNI DG DGNI	13.2 21.9 12.7 20.5 11.3 22.4	11.0 20.1 10.6 18.4 7.7 19.4	9.0 17.2 8.4 16.0 3.9 17.0	7.9 15.4 6.3 14.5 1.3 14.7	6.4 14.5 4.4 12.5 — 12.2	4.6 12.2 2.9 10.1 — 9.0	2.6 10.0 1.7 7.1 — 6.0	.9 6.5 .8 4.3 — 3.4	— 2.7 — .2 — 1.7 1.2 .9
MEMPHIS TENNESSEE 35.0 N 3227 DD T(JAN)=40	L C R	WW WWNI TW TWNI DG DGNI	161 251 156 237 159 263	74 119 70 112 66 122	44 75 42 70 35 75	31 52 28 49 18 53	22 40 19 37 — 39	17 31 13 28 — 29	12 25 9 22 — 22	8 20 6 16 — 15	— 14 — 11 — 9
	D	WW WWNI TW TWNI DG DGNI	14.8 23.3 14.0 22.2 13.3 24.6	12.1 21.7 11.8 19.6 9.7 21.0	10.4 18.4 9.6 17.5 5.8 18.8	9.2 17.0 7.3 15.9 2.4 16.3	7.6 15.9 5.2 13.7 — 13.6	5.7 13.6 3.5 11.1 — 10.1	3.3 11.1 2.2 7.9 — 6.9	1.7 7.2 1.1 4.8 — 3.9	— 3.0 — 1.3 — 1.9 1.4 .9
NASHVILLE TENNESSEE 36.1 N 3696 DD T(JAN)=38	L C R	WW WWNI TW TWNI DG DGNI	124 207 122 195 108 212	56 98 54 92 41 98	32 61 32 57 15 60	22 42 21 40 — 42	16 32 14 30 — 31	11 25 9 23 — 23	8 20 6 17 — 17	4 16 3 13 — 11	— 11 — 8 — 7
	D	WW WWNI TW TWNI DG DGNI	11.2 19.4 10.8 18.3 8.5 19.8	8.8 17.6 8.8 16.0 4.9 16.7	7.2 14.8 6.8 14.0 — 14.6	6.1 13.4 4.9 12.7 — 12.5	4.8 12.6 3.3 10.9 — 10.3	3.2 10.7 2.0 8.8 — 7.5	1.6 8.6 1.1 6.2 — 5.0	.3 5.6 .4 3.8 — 2.8	— 2.3 — 1.5 — .9
ABILENE TEXAS 32.4 N 2610 DD T(JAN)=44	L C R	WW WWNI TW TWNI DG- DGNI	236 339 224 321 257 362	111 164 104 153 116 171	69 104 64 97 69 108	49 74 44 69 44 77	37 57 31 52 29 57	28 45 22 41 18 44	22 36 16 31 9 33	15 29 11 23 — 24	9 20 7 16 — 15
	D	WW WWNI TW TWNI DG DGNI	22.6 32.6 20.8 30.9 23.2 34.5	19.5 30.0 18.3 27.7 19.3 30.6	17.3 26.7 15.5 25.3 17.4 28.1	16.0 25.1 12.3 23.1 13.3 24.7	13.7 23.6 9.3 20.2 11.8 20.9	11.0 20.3 6.9 16.2 8.7 15.8	7.3 16.6 4.6 11.6 4.3 11.0	3.9 10.7 2.8 7.2 1.6 6.6	1.1 4.5 1.2 2.9 — 2.6

		SSF =	.1	.2	.3	.4	.5	.6	.7	.8	.9
AMARILLO TEXAS		WW	178	83	52	36	27	21	16	11	9
	L	WWNI	271	130	83	59	45	36	29	23	20
35.2 N	C	TW	173	79	49	33	23	16	11	8	7
4183 DD	R	TWNI	256	122	77	55	42	33	25	19	15
T(JAN)=36		DG	183	80	46	27	15	–	–	–	–
		DGNI	285	134	85	60	45	34	25	18	15
		WW	16.5	14.5	12.8	11.6	9.9	7.7	4.8	2.4	.5
		WWNI	25.2	24.2	21.5	19.8	16.4	16.5	13.5	8.6	3.6
	D	TW	15.6	13.8	11.5	9.0	6.7	4.7	3.1	1.7	.7
		TWNI	24.0	22.1	20.1	18.5	16.3	13.0	9.3	5.7	2.3
		DG	15.2	12.8	8.9	5.2	2.2	–	–	–	–
		DGNI	26.6	24.0	21.7	19.4	16.2	12.1	8.4	5.0	1.9

		SSF =	.1	.2	.3	.4	.5	.6	.7	.8	.9
AUSTIN TEXAS		WW	311	143	89	63	48	37	28	21	13
	L	WWNI	432	203	127	91	70	55	45	35	25
30.3 N	C	TW	288	134	82	56	40	29	21	15	10
1737 DD	R	TWNI	404	189	119	85	64	50	39	29	19
T(JAN)=50		DG	348	157	95	62	42	28	18	–	–
		DGNI	460	214	135	95	71	54	41	30	19
		WW	29.0	24.1	23.0	20.8	18.0	14.7	10.2	5.7	1.9
		WWNI	41.6	35.3	33.0	31.8	28.5	25.0	20.1	13.1	5.5
	D	TW	26.7	23.2	19.6	15.9	12.4	9.3	6.5	4.1	1.7
		TWNI	38.3	33.3	31.2	28.1	25.0	19.1	14.3	8.8	3.6
		DG	33.7	26.2	19.6	15.6	10.8	6.7	3.5	1.1	–
		DGNI	43.4	37.7	34.7	30.8	26.0	20.0	14.0	8.4	3.4

		SSF =	.1	.2	.3	.4	.5	.6	.7	.8	.9
BROWNSVILLE TEXAS		WW	656	308	192	137	105	82	65	49	33
	L	WWNI	846	402	252	180	138	111	89	70	50
25.9 N	C	TW	609	280	175	121	88	65	48	35	24
650 DD	R	TWNI	797	373	235	168	128	99	77	57	38
T(JAN)=60		DG	799	365	226	155	111	80	58	40	24
		DGNI	929	432	272	194	146	112	85	63	42
		WW	60.7	53.8	49.0	46.4	41.9	34.5	25.8	15.6	6.3
		WWNI	78.3	71.7	64.3	62.0	57.7	50.8	40.0	26.6	11.3
	D	TW	54.4	49.4	43.2	35.8	28.2	22.9	16.2	10.7	5.3
		TWNI	74.1	66.0	61.2	56.6	44.2	34.0	28.6	17.7	7.3
		DG	71.5	63.6	54.7	44.2	34.0	24.8	16.7	9.7	3.9
		DGNI	86.3	76.2	70.7	63.6	53.8	42.0	30.0	18.6	7.9

		SSF =	.1	.2	.3	.4	.5	.6	.7	.8	.9
CORPUS CHRISTI TEXAS		WW	469	225	146	105	80	62	49	37	24
	L	WWNI	621	299	195	141	109	86	69	55	39
27.8 N	C	TW	442	208	131	92	67	50	37	26	18
930 DD	R	TWNI	586	282	182	131	100	78	60	45	30
T(JAN)=56		DG	564	264	165	113	81	58	41	27	15
		DGNI	677	325	208	149	113	87	66	48	32
		WW	43.4	43.3	39.0	35.3	31.5	25.6	18.8	11.1	4.3
		WWNI	57.6	57.5	53.8	49.2	44.2	40.3	30.9	20.5	8.7
	D	TW	40.8	37.6	33.2	27.5	21.9	16.8	11.6	7.6	3.7
		TWNI	54.5	53.4	49.2	44.5	38.8	31.2	22.2	13.7	5.6
		DG	52.3	47.2	40.5	32.1	24.1	17.1	10.9	5.7	1.9
		DGNI	63.0	60.5	55.8	50.3	41.9	32.4	22.9	14.1	5.9

		SSF =	.1	.2	.3	.4	.5	.6	.7	.8	.9
DALLAS TEXAS		WW	237	109	68	48	36	28	21	15	9
	L	WWNI	343	162	102	73	56	45	36	28	20
32.8 N	C	TW	223	103	64	43	31	22	16	11	7
2290 DD	R	TWNI	321	151	95	68	52	40	31	23	15
T(JAN)=45		DG	255	114	68	43	28	17	8	–	–
		DGNI	362	169	106	75	57	43	32	23	15
		WW	22.2	18.6	17.4	15.8	13.5	10.8	7.2	3.8	1.1
		WWNI	33.4	28.7	26.3	25.4	23.2	20.2	16.3	10.5	4.4
	D	TW	20.7	17.9	15.2	12.0	9.2	6.7	4.6	2.8	1.1
		TWNI	30.6	26.8	25.1	22.8	20.1	16.1	11.5	7.1	2.9
		DG	22.9	18.7	14.4	9.9	6.1	3.1	.8	–	–
		DGNI	34.2	29.9	27.5	24.4	20.6	15.6	10.9	6.5	2.5

		SSF =	.1	.2	.3	.4	.5	.6	.7	.8	.9
DEL RIO TEXAS		WW	375	167	105	75	57	44	34	25	16
	L	WWNI	512	232	146	106	81	64	51	40	29
29.4 N	C	TW	343	158	97	66	48	35	26	18	12
1523 DD	R	TWNI	476	218	138	98	74	58	44	33	22
T(JAN)=51		DG	424	191	115	77	53	36	24	14	5
		DGNI	543	248	157	111	83	63	48	35	23
		WW	33.4	28.3	28.3	24.5	21.3	17.6	12.4	7.0	2.5
		WWNI	47.5	39.9	39.2	37.8	32.2	28.1	22.9	14.9	6.3
	D	TW	31.3	27.0	23.0	19.0	15.0	11.3	8.0	5.0	2.2
		TWNI	43.4	38.2	36.6	30.3	28.3	22.9	16.3	10.0	4.1
		DG	37.7	31.8	26.4	20.1	14.4	9.5	5.5	2.3	.3
		DGNI	49.3	43.7	40.2	35.8	30.4	23.3	16.3	9.8	4.0

		SSF =	.1	.2	.3	.4	.5	.6	.7	.8	.9
EL PASO TEXAS		WW	274	131	82	58	44	34	26	19	12
	L	WWNI	384	189	119	85	65	52	42	33	24
31.8 N	C	TW	258	122	76	52	37	27	19	14	9
2678 DD	R	TWNI	363	175	111	79	60	47	36	27	18
T(JAN)=44		DG	307	141	85	56	37	25	15	7	–
		DGNI	413	197	125	89	66	51	38	28	18
		WW	27.4	22.9	20.5	19.4	16.7	13.6	9.3	5.1	1.6
		WWNI	38.4	34.5	30.2	28.9	27.6	23.7	19.2	12.4	5.2
	D	TW	24.5	21.8	18.5	14.7	11.3	8.4	5.9	3.6	1.6
		TWNI	36.2	31.4	29.1	26.7	23.1	18.7	13.4	8.3	3.4
		DG	28.5	24.1	19.1	14.3	9.2	5.5	2.6	.3	–
		DGNI	40.6	35.4	32.5	28.8	24.4	18.5	13.0	7.9	3.2

		SSF =	.1	.2	.3	.4	.5	.6	.7	.8	.9
FORT WORTH TEXAS		WW	225	104	65	46	34	26	20	14	8
	L	WWNI	328	155	98	71	54	43	34	27	19
32.8 N	C	TW	212	99	61	41	29	21	15	10	6
2382 DD	R	TWNI	307	146	92	66	50	39	30	22	15
T(JAN)=45		DG	240	108	64	40	25	15	6	–	–
		DGNI	345	162	102	73	54	41	31	22	14
		WW	21.0	18.1	16.6	14.9	12.7	10.0	6.7	3.5	.9
		WWNI	32.1	28.1	25.4	24.2	22.2	19.2	15.5	10.1	4.2
	D	TW	19.4	17.2	14.5	11.9	8.8	6.3	4.4	2.5	1.1
		TWNI	29.4	26.1	24.3	21.9	19.3	15.4	11.0	6.7	2.7
		DG	21.6	17.6	13.4	9.0	5.4	2.4	–	–	–
		DGNI	32.7	28.9	26.4	23.4	19.6	14.9	10.3	6.1	2.4

278 / APPENDIX 12

		SSF =	.1	.2	.3	.4	.5	.6	.7	.8	.9
MIDLAND-ODESSA TEXAS 31.9 N 2621 DD T(JAN)=44	L C R	WW WWNI TW TWNI DG DGNI	265 373 249 352 294 400	127 182 118 170 136 191	79 115 74 108 82 121	56 83 50 77 54 86	43 64 36 59 36 65	33 51 28 46 24 50	26 41 19 35 14 37	19 32 13 27 6 27	11 23 9 18 - 18
	D	WW WWNI TW TWNI DG DGNI	26.4 37.3 23.7 34.8 27.4 39.0	22.4 33.7 21.1 30.9 22.4 34.6	20.1 29.9 18.0 28.4 18.4 31.9	19.0 28.5 16.1 26.1 13.3 28.3	16.3 26.9 13.4 23.0 8.9 24.0	13.4 23.3 8.2 18.4 5.2 18.2	9.1 19.1 5.7 13.2 2.4 12.8	5.0 12.3 3.5 8.2 .5 7.7	1.6 5.2 1.7 3.4 - 3.1
PORT ARTHUR TEXAS 29.9 N 1518 DD T(JAN)=52	L C R	WW WWNI TW TWNI DG DGNI	339 466 311 435 380 496	153 215 143 201 170 228	95 134 88 126 102 143	67 96 60 90 68 101	51 74 43 68 46 76	40 59 31 53 31 58	31 47 23 41 20 44	22 37 16 30 11 32	14 27 11 20 - 21
	D	WW WWNI TW TWNI DG DGNI	31.5 44.7 28.5 40.9 33.9 46.0	25.6 36.5 21.4 34.9 28.3 39.6	24.5 34.7 20.8 32.8 23.1 36.5	22.4 33.2 17.0 29.7 17.2 32.6	19.4 30.3 13.3 26.4 12.1 27.6	15.8 26.9 10.1 21.2 7.8 21.3	11.1 21.3 7.1 15.1 4.2 15.0	6.3 13.9 4.4 9.3 1.6 9.0	2.1 5.8 2.1 3.8 - 3.7
SAN ANGELO TEXAS 31.4 N 2240 DD T(JAN)=46	L C R	WW WWNI TW TWNI DG DGNI	269 377 253 357 300 405	129 185 120 172 138 194	80 117 75 109 84 123	57 84 51 78 55 87	43 65 36 59 37 66	33 51 26 46 24 50	26 41 19 36 15 38	19 33 13 30 6 27	11 23 9 18 - 18
	D	WW WWNI TW TWNI DG DGNI	26.9 37.7 24.1 35.5 27.4 39.7	22.6 34.2 21.4 31.3 23.6 35.0	20.4 30.2 18.2 28.8 18.8 32.3	19.0 28.8 14.6 26.2 13.5 28.4	16.3 26.8 11.2 23.1 9.9 24.1	13.4 23.2 8.3 18.5 5.3 18.3	9.1 19.2 5.8 13.5 2.5 12.9	5.0 12.3 3.6 9.3 .5 7.8	1.6 5.2 1.6 3.1 - 3.1
SAN ANTONIO TEXAS 29.5 N 1570 DD T(JAN)=51	L C R	WW WWNI TW TWNI DG DGNI	345 474 318 443 389 505	156 219 146 204 175 232	97 137 90 129 105 146	70 99 62 92 70 103	52 76 44 70 48 78	41 60 32 54 33 59	31 48 24 42 21 45	23 38 17 31 12 33	14 27 11 21 - 21
	D	WW WWNI TW TWNI DG DGNI	31.9 44.9 29.2 41.4 34.8 46.6	26.6 37.5 25.2 35.8 29.1 40.8	25.6 36.0 21.5 33.7 23.8 37.6	22.8 34.5 17.7 30.4 18.0 33.4	19.9 30.6 13.8 27.0 12.7 28.4	16.3 27.0 10.4 21.6 8.2 21.8	11.4 21.7 7.3 15.4 4.5 15.3	6.5 14.6 4.6 9.5 1.8 9.2	2.2 6.0 2.0 3.9 - 3.8

		SSF =	.1	.2	.3	.4	.5	.6	.7	.8	.9
HOUSTON TEXAS 30.0 N 1434 DD T(JAN)=52	L C R	WW WWNI TW TWNI DG DGNI	336 466 309 433 375 492	148 209 140 196 166 223	93 131 86 124 99 140	66 95 59 88 66 99	50 73 42 67 45 74	39 58 31 52 30 57	30 46 22 40 19 43	22 36 16 30 11 31	14 26 10 20 - 20
	D	WW WWNI TW TWNI DG DGNI	29.1 42.2 27.7 38.8 32.7 43.9	25.4 35.9 23.8 34.3 27.3 39.0	24.1 35.3 20.5 32.2 26.1 35.9	21.7 32.7 16.6 29.2 16.7 32.1	18.9 29.3 13.1 25.9 11.7 27.1	15.3 25.9 9.8 20.7 7.4 20.8	10.7 20.6 6.9 14.7 3.9 14.5	6.0 13.5 4.1 9.1 1.3 8.8	2.0 5.7 1.8 3.7 - 3.5
LAREDO TEXAS 27.5 N 876 DD T(JAN)=56	L C R	WW WWNI TW TWNI DG DGNI	528 693 495 652 639 756	252 331 231 312 297 360	163 216 145 200 184 230	115 156 101 143 126 163	87 117 73 108 89 122	67 92 54 83 64 93	52 74 44 64 45 70	39 58 28 48 30 51	26 41 19 32 17 34
	D	WW WWNI TW TWNI DG DGNI	48.3 63.4 46.0 60.2 59.0 69.8	48.1 63.1 41.0 58.9 51.9 67.0	43.3 59.6 36.2 52.7 44.7 59.5	36.9 51.5 29.2 46.8 35.4 53.1	33.0 45.2 23.7 41.0 26.6 44.5	27.1 40.5 18.0 32.8 18.8 34.3	20.0 32.4 12.5 23.3 12.1 24.1	11.9 21.5 7.7 14.2 6.4 14.8	4.7 9.2 4.0 5.9 2.2 6.2
LUBBOCK TEXAS 33.6 N 3545 DD T(JAN)=39	L C R	WW WWNI TW TWNI DG DGNI	206 306 198 288 220 323	97 146 92 137 135 153	61 94 57 87 58 97	43 67 39 63 37 69	32 51 28 48 23 52	25 41 20 37 13 39	19 33 14 29 - 30	13 26 10 21 - 21	8 19 6 14 - 14
	D	WW WWNI TW TWNI DG DGNI	19.1 28.5 18.3 27.1 19.7 30.3	17.4 25.7 16.2 25.2 16.4 27.7	15.3 29.1 13.7 22.9 12.2 25.1	14.1 22.6 10.8 21.2 8.0 22.5	12.2 21.5 8.3 18.6 4.6 18.9	9.7 18.8 5.9 14.9 1.9 14.3	6.3 15.8 4.0 10.6 - 9.9	3.3 9.9 2.3 6.6 - 5.9	.9 4.2 1.1 2.7 - 2.3
LUFKIN TEXAS 31.2 N 1940 DD T(JAN)=49	L C R	WW WWNI TW TWNI DG DGNI	273 385 254 361 299 409	125 182 118 170 135 191	78 113 73 107 80 120	55 81 49 76 52 85	42 63 35 58 34 63	32 50 25 45 22 48	25 40 18 35 13 36	18 32 13 26 - 26	11 22 8 17 - 17
	D	WW WWNI TW TWNI DG DGNI	25.8 38.0 23.6 34.6 27.0 38.8	21.2 31.8 20.5 32.2 22.3 33.5	19.7 29.1 17.3 29.8 17.6 30.8	18.2 28.4 13.8 27.8 12.5 27.3	15.6 25.8 10.6 22.3 8.3 23.1	12.6 22.4 7.9 17.9 4.8 17.6	8.6 18.1 5.5 12.1 2.1 12.3	4.7 11.8 3.4 7.9 - 7.4	1.4 4.9 1.5 3.2 - 2.9

SHERMAN, TEXAS
33.7 N, 2864 DD, T(JAN)=42

	SSF = .1	.2	.3	.4	.5	.6	.7	.8	.9
WW	191	88	54	38	28	21	16	11	6
WWNI	286	136	85	61	47	37	30	23	17
TW	182	83	51	34	24	17	12	8	5
TWNI	271	127	80	57	43	33	26	19	13
DG	197	86	49	29	17	–	–	–	–
DGNI	302	141	88	62	46	35	26	19	12

D	.1	.2	.3	.4	.5	.6	.7	.8	.9
WW	17.8	14.9	11.9	–	10.2	7.9	5.0	2.4	–
WWNI	27.5	24.2	21.8	20.7	18.9	16.5	13.4	8.6	3.6
TW	16.6	14.3	11.9	–	16.5	13.2	9.4	4.8	–.7
TWNI	25.6	22.6	20.7	18.8	16.5	13.2	9.4	5.8	2.3
DG	17.1	13.5	9.5	5.6	2.6	–	–	–	–
DGNI	28.1	24.7	22.3	19.8	16.6	12.4	8.6	5.0	1.9

WACO, TEXAS
31.6 N, 2058 DD, T(JAN)=47

	SSF = .1	.2	.3	.4	.5	.6	.7	.8	.9
WW	263	121	75	53	40	31	24	17	10
WWNI	373	176	110	79	61	48	39	30	22
TW	245	114	70	47	34	24	18	12	8
TWNI	350	164	103	74	56	43	33	25	17
DG	287	129	77	49	32	21	12	–	–
DGNI	396	185	116	82	61	47	35	25	16

D	.1	.2	.3	.4	.5	.6	.7	.8	.9
WW	24.7	20.5	19.2	17.3	14.9	11.9	8.1	4.4	1.3
WWNI	36.6	30.8	27.7	24.8	21.6	17.7	11.3	4.7
TW	22.7	19.7	16.7	13.2	10.2	7.5	5.2	3.2	1.4
TWNI	33.5	29.0	27.0	21.6	21.6	17.3	12.3	7.6	3.1
DG	25.2	21.2	16.6	11.7	7.6	4.2	1.1	–	–
DGNI	37.6	32.5	29.7	26.4	22.2	17.0	11.8	7.1	2.8

WICHITA FALLS, TEXAS
34.0 N, 2904 DD, T(JAN)=41

	SSF = .1	.2	.3	.4	.5	.6	.7	.8	.9
WW	207	96	59	41	31	24	18	13	7
WWNI	304	145	91	65	50	40	32	25	18
TW	197	90	55	37	26	19	13	9	5
TWNI	288	136	85	61	46	36	28	21	14
DG	217	96	56	34	21	11	–	–	–
DGNI	322	151	94	67	50	38	28	20	13

D	.1	.2	.3	.4	.5	.6	.7	.8	.9
WW	19.6	16.3	14.6	13.4	11.4	9.0	5.8	2.9	–
WWNI	29.2	26.1	23.2	20.1	20.6	17.9	14.6	9.4	3.7
TW	18.0	15.6	13.1	10.2	7.7	5.5	3.7	2.3	3.9
TWNI	27.1	24.1	22.1	17.7	17.7	14.3	10.2	6.3	2.9
DG	19.1	15.4	11.2	7.1	3.9	1.1	–	–	–
DGNI	30.3	26.5	24.2	21.4	18.0	13.6	9.4	5.6	2.1

BRYCE CANYON, UTAH
37.7 N, 9133 DD, T(JAN)=20

	SSF = .1	.2	.3	.4	.5	.6	.7	.8	.9
WW	98	47	29	20	15	11	7	3	–
WWNI	171	85	55	40	30	24	19	15	11
TW	99	46	28	19	12	8	5	3	–
TWNI	163	80	52	37	28	22	17	12	8
DG	80	32	–	–	–	–	–	–	–
DGNI	175	85	54	39	29	22	16	11	7

D	.1	.2	.3	.4	.5	.6	.7	.8	.9
WW	9.7	8.3	7.1	6.0	4.7	3.0	1.4	.2	–
WWNI	17.1	16.2	15.0	13.7	12.2	11.0	8.6	5.5	2.3
TW	9.5	8.1	6.4	4.6	3.1	2.0	1.0	.3	–
TWNI	16.3	15.2	13.9	12.5	10.9	8.6	6.1	3.7	1.5
DG	6.9	3.7	–	–	–	–	–	–	–
DGNI	17.4	15.9	14.1	12.5	10.1	7.3	4.9	2.7	.9

CEDAR CITY, UTAH
37.7 N, 6137 DD, T(JAN)=29

	SSF = .1	.2	.3	.4	.5	.6	.7	.8	.9
WW	139	64	39	27	20	15	11	7	1
WWNI	223	106	68	49	37	29	23	18	13
TW	135	62	38	25	17	12	9	5	2
TWNI	211	100	64	45	34	27	20	15	10
DG	130	55	29	12	–	–	–	–	–
DGNI	231	109	68	48	36	27	20	14	9

D	.1	.2	.3	.4	.5	.6	.7	.8	.9
WW	12.9	11.1	9.6	8.4	6.9	5.0	2.8	1.1	–
WWNI	22.1	19.0	18.3	16.3	14.9	13.2	10.4	6.7	2.8
TW	12.5	10.6	8.6	6.4	4.7	3.2	1.9	.9	1.3
TWNI	20.3	18.2	16.5	15.0	13.1	10.5	7.4	4.5	1.8
DG	11.3	8.0	4.3	1.0	–	–	–	–	–
DGNI	21.8	19.7	17.4	15.4	12.7	9.4	6.4	3.6	1.3

SALT LAKE CITY, UTAH
40.8 N, 5983 DD, T(JAN)=28

	SSF = .1	.2	.3	.4	.5	.6	.7	.8	.9
WW	131	59	35	23	16	11	7	3	–
WWNI	215	101	64	45	33	26	20	15	11
TW	127	57	34	22	14	9	6	3	–
TWNI	203	95	59	42	31	23	18	13	8
DG	118	45	18	–	–	–	–	–	–
DGNI	221	102	63	44	32	23	17	11	6

D	.1	.2	.3	.4	.5	.6	.7	.8	.9
WW	11.7	9.9	7.9	6.4	4.6	3.1	1.4	–	–
WWNI	20.9	17.6	16.2	14.2	12.4	10.5	7.8	5.2	2.1
TW	11.6	9.3	7.1	5.2	3.5	2.1	1.0	–	–
TWNI	19.1	16.9	14.8	13.0	10.8	8.6	6.0	3.6	1.4
DG	9.5	5.5	1.5	–	–	–	–	–	–
DGNI	20.5	17.7	15.2	12.9	10.2	7.5	4.9	2.6	.8

BURLINGTON, VERMONT
44.5 N, 7876 DD, T(JAN)=17

	SSF = .1	.2	.3	.4	.5	.6	.7	.8	.9
WW	42	14	–	–	–	–	–	–	–
WWNI	115	51	32	22	16	12	9	7	5
TW	50	18	8	–	–	–	–	–	–
TWNI	109	49	30	21	15	11	8	6	4
DG	–	–	–	–	–	–	–	–	–
DGNI	109	47	28	19	13	9	6	3	–

D	.1	.2	.3	.4	.5	.6	.7	.8	.9
WW	3.1	1.5	–	–	–	–	–	–	–
WWNI	10.0	8.8	7.8	6.6	5.6	4.5	3.3	2.2	.8
TW	3.8	2.3	.9	–	–	–	–	–	–
TWNI	9.6	8.4	7.2	6.1	5.0	3.9	2.7	1.5	.5
DG	–	–	–	–	–	–	–	–	–
DGNI	9.2	7.6	6.2	4.8	3.5	2.2	1.1	.3	–

NORFOLK, VIRGINIA
36.9 N, 3488 DD, T(JAN)=40

	SSF = .1	.2	.3	.4	.5	.6	.7	.8	.9
WW	162	75	46	32	23	17	13	8	–
WWNI	251	120	76	54	41	32	26	20	14
TW	156	72	44	29	20	14	9	6	3
TWNI	237	113	71	50	38	29	22	17	11
DG	160	68	37	20	–	–	–	–	–
DGNI	263	123	77	54	40	30	22	16	10

D	.1	.2	.3	.4	.5	.6	.7	.8	.9
WW	15.2	12.6	11.1	9.5	7.9	6.0	3.6	1.6	–
WWNI	24.5	21.6	19.5	17.9	15.9	13.6	11.6	7.3	3.4
TW	14.3	12.1	10.0	8.1	5.4	3.8	2.3	1.2	–
TWNI	22.7	20.1	18.2	16.2	14.2	11.4	8.1	4.9	1.3
DG	13.9	10.2	6.3	2.9	–	–	–	–	–
DGNI	24.9	21.9	19.3	16.8	14.1	10.4	7.1	4.1	1.5

280 / APPENDIX 12

		SSF =	.1	.2	.3	.4	.5	.6	.7	.8	.9
RICHMOND VIRGINIA		WW	132	61	37	25	18	13	9	5	-
37.5 N	L	WWNI	216	103	65	46	35	27	22	17	12
3939 DD	C	TW	130	59	35	23	16	11	7	4	2
T(JAN)=37	R	TWNI	205	97	61	43	32	25	19	14	9
		DG	122	49	24	-	-	-	-	-	-
		DGNI	223	105	65	45	34	25	18	13	8
		WW	12.1	10.3	8.5	7.3	5.7	4.1	2.2	.7	-
		WWNI	20.5	19.0	16.9	14.9	13.6	11.5	9.4	6.1	2.5
	D	TW	11.7	9.8	7.9	5.8	4.1	2.6	1.4	.6	.1
		TWNI	19.2	17.5	15.5	13.9	11.9	9.6	6.7	4.1	1.6
		DG	10.1	6.7	2.9	-	-	-	-	-	-
		DGNI	20.8	18.5	16.3	13.9	11.5	8.4	5.6	3.2	1.1

		SSF =	.1	.2	.3	.4	.5	.6	.7	.8	.9
ROANOKE VIRGINIA		WW	124	58	35	24	18	13	9	5	-
37.3 N	L	WWNI	206	99	63	45	34	27	21	17	12
4307 DD	C	TW	123	56	34	22	15	10	7	4	1
T(JAN)=36	R	TWNI	195	93	59	42	32	24	19	14	9
		DG	113	46	21	-	-	-	-	-	-
		DGNI	212	100	63	44	33	24	18	13	7
		WW	11.6	9.9	8.3	7.1	5.6	3.9	2.1	.6	-
		WWNI	19.3	18.5	16.4	14.7	13.5	11.5	9.4	6.0	2.5
	D	TW	11.3	9.5	7.6	5.6	3.9	2.5	1.4	.6	.1
		TWNI	18.5	17.0	15.3	13.7	11.8	9.5	6.7	4.1	1.6
		DG	9.5	6.2	2.4	-	-	-	-	-	-
		DGNI	20.1	18.0	15.9	13.7	11.3	8.3	5.5	3.1	1.0

		SSF =	.1	.2	.3	.4	.5	.6	.7	.8	.9
OLYMPIA WASHINGTON		WW	112	48	28	17	10	5	-	-	-
47.0 N	L	WWNI	191	90	55	38	28	21	16	11	7
5530 DD	C	TW	109	48	27	16	10	5	-	-	-
T(JAN)=37	R	TWNI	181	84	52	36	26	19	14	10	6
		DG	90	23	-	-	-	-	-	-	-
		DGNI	195	89	54	36	25	18	12	8	3
		WW	10.1	7.3	5.6	3.6	1.9	.5	-	-	-
		WWNI	19.1	15.6	13.3	11.7	9.5	7.2	5.0	3.2	1.2
	D	TW	9.8	7.3	4.9	3.2	1.8	.7	-	-	-
		TWNI	17.3	14.5	12.4	10.3	8.2	6.1	4.2	2.4	.8
		DG	6.6	1.1	-	-	-	-	-	-	-
		DGNI	18.3	14.9	12.3	9.6	7.1	5.0	3.0	1.4	.2

		SSF =	.1	.2	.3	.4	.5	.6	.7	.8	.9
SEATTLE-TACOMA WASHINGTON		WW	128	54	30	19	11	6	-	-	-
47.4 N	L	WWNI	213	98	59	41	30	22	16	11	7
5185 DD	C	TW	123	53	29	18	11	6	-	-	-
T(JAN)=38	R	TWNI	199	91	55	38	27	20	14	10	6
		DG	108	33	-	-	-	-	-	-	-
		DGNI	217	97	58	38	27	19	13	8	3
		WW	11.1	8.0	6.0	3.9	2.1	.7	-	-	-
		WWNI	20.3	16.5	14.0	12.1	9.7	7.3	5.1	3.2	1.2
	D	TW	10.8	7.9	5.4	3.6	2.0	.8	-	-	-
		TWNI	18.6	15.3	13.0	10.7	8.4	6.2	4.3	2.4	.8
		DG	8.0	2.5	-	-	-	-	-	-	-
		DGNI	19.9	15.9	12.9	10.0	7.4	5.1	3.1	1.4	.2

		SSF =	.1	.2	.3	.4	.5	.6	.7	.8	.9
SPOKANE WASHINGTON		WW	91	37	20	10	-	-	-	-	-
47.6 N	L	WWNI	169	78	47	32	23	17	12	9	6
6835 DD	C	TW	92	38	20	11	5	-	-	-	-
T(JAN)=25	R	TWNI	161	73	44	30	21	15	11	8	4
		DG	56	-	-	-	-	-	-	-	-
		DGNI	171	75	45	29	20	13	9	5	-
		WW	7.6	5.3	3.3	1.4	-	-	-	-	-
		WWNI	15.4	13.0	11.2	9.3	7.2	5.5	3.7	2.5	.9
	D	TW	7.6	5.3	3.3	1.7	.5	-	-	-	-
		TWNI	14.7	12.3	10.3	8.2	6.4	4.7	3.3	1.8	.6
		DG	2.9	-	-	-	-	-	-	-	-
		DGNI	15.0	12.3	9.7	7.3	5.2	3.5	1.9	.7	-

		SSF =	.1	.2	.3	.4	.5	.6	.7	.8	.9
YAKIMA WASHINGTON		WW	117	48	26	15	8	-	-	-	-
46.6 N	L	WWNI	201	90	55	38	27	20	14	10	7
6009 DD	C	TW	114	47	25	15	8	4	-	-	-
T(JAN)=28	R	TWNI	190	85	51	35	25	18	13	9	5
		DG	92	-	-	-	-	-	-	-	-
		DGNI	205	89	53	35	24	16	11	7	3
		WW	9.6	7.0	4.9	2.8	1.2	-	-	-	-
		WWNI	18.3	14.9	13.0	10.8	8.4	6.5	4.5	3.1	1.2
	D	TW	9.5	6.8	4.5	2.8	1.4	.2	-	-	-
		TWNI	16.9	14.0	11.8	9.5	7.5	5.5	3.9	2.2	.8
		DG	6.1	-	-	-	-	-	-	-	-
		DGNI	17.7	14.4	11.5	8.8	6.4	4.5	2.6	1.1	.2

		SSF =	.1	.2	.3	.4	.5	.6	.7	.8	.9
CHARLESTON WEST VIRGINIA		WW	93	41	23	15	10	7	3	-	-
38.4 N	L	WWNI	170	80	50	35	26	20	16	12	9
4590 DD	C	TW	94	41	24	15	9	5	2	-	-
T(JAN)=34	R	TWNI	161	75	47	33	24	19	14	10	7
		DG	66	-	-	-	-	-	-	-	-
		DGNI	172	79	48	33	24	18	13	8	5
		WW	8.2	6.4	4.9	3.7	2.6	1.3	.1	-	-
		WWNI	15.5	14.5	12.6	10.9	9.9	8.2	6.6	4.3	1.7
	D	TW	8.2	6.4	4.8	3.1	1.9	.9	.2	-	-
		TWNI	14.9	13.3	11.6	10.3	8.6	7.0	4.9	2.9	1.1
		DG	4.5	-	-	-	-	-	-	-	-
		DGNI	15.7	13.5	11.5	9.7	7.8	5.5	3.5	1.8	.5

		SSF =	.1	.2	.3	.4	.5	.6	.7	.8	.9
HUNTINGTON WEST VIRGINIA		WW	99	44	26	17	12	8	4	-	-
38.4 N	L	WWNI	178	84	53	37	28	21	17	13	9
4624 DD	C	TW	101	44	26	16	10	6	3	-	-
T(JAN)=34	R	TWNI	169	79	49	34	26	20	15	11	7
		DG	76	22	-	-	-	-	-	-	-
		DGNI	181	83	51	35	26	19	14	9	5
		WW	8.9	7.1	5.4	4.3	3.1	1.7	.5	-	-
		WWNI	16.3	15.2	13.2	11.5	10.5	8.7	7.0	4.5	1.8
	D	TW	8.8	7.0	5.3	3.6	2.2	1.1	.4	-	-
		TWNI	15.6	14.0	12.2	10.8	9.1	7.3	5.1	3.1	1.2
		DG	5.5	1.3	-	-	-	-	-	-	-
		DGNI	16.6	14.3	12.3	10.3	8.4	6.0	3.9	2.0	.6

EAU CLAIRE WISCONSIN
44.9 N
8388 DD
T(JAN)=12

		SSF =	.1	.2	.3	.4	.5	.6	.7	.8	.9
L		WW	50	19	8	-	-	-	-	-	-
C		WWNI	121	56	34	24	17	13	10	8	5
R		TW	56	22	11	8	6	5	4	3	1
		TWNI	115	53	32	22	16	12	9	6	4
		DG	-	-	-	-	-	-	-	-	-
		DGNI	117	52	31	20	14	10	7	4	1
D		WW	3.9	2.2	.9	-	-	-	-	-	-
		WWNI	10.4	9.7	8.4	7.0	6.1	5.1	3.7	2.4	.9
		TW	4.5	2.9	1.4	-	-	-	-	-	-
		TWNI	10.4	9.1	7.7	6.6	5.4	4.2	2.9	1.7	.6
		DG	-	-	-	-	-	-	-	-	-
		DGNI	10.2	8.5	6.9	5.4	4.0	2.6	1.4	.5	-

GREEN BAY WISCONSIN
44.5 N
8098 DD
T(JAN)=15

		SSF =	.1	.2	.3	.4	.5	.6	.7	.8	.9
L		WW	57	22	11	5	-	-	-	-	-
C		WWNI	132	59	37	26	19	14	11	8	5
R		TW	63	25	13	6	-	-	-	-	-
		TWNI	125	56	35	24	17	13	10	7	4
		DG	-	-	-	-	-	-	-	-	-
		DGNI	127	56	34	22	16	11	7	4	-
D		WW	4.4	2.9	1.6	.3	-	-	-	-	-
		WWNI	11.8	10.1	9.1	7.7	6.6	5.4	4.0	2.6	1.0
		TW	5.1	3.4	1.9	.7	-	-	-	-	-
		TWNI	11.2	9.7	8.3	7.1	5.9	4.6	3.2	1.8	.7
		DG	-	-	-	-	-	-	-	-	-
		DGNI	11.1	9.2	7.5	6.0	4.5	3.0	1.7	.7	-

LA CROSSE WISCONSIN
43.9 N
7417 DD
T(JAN)=16

		SSF =	.1	.2	.3	.4	.5	.6	.7	.8	.9
L		WW	59	23	12	6	-	-	-	-	-
C		WWNI	133	61	37	26	19	15	11	9	6
R		TW	64	26	13	7	-	-	-	-	-
		TWNI	126	57	35	24	18	13	10	7	5
		DG	-	-	-	-	-	-	-	-	-
		DGNI	129	57	34	23	16	11	8	5	1
D		WW	4.8	3.0	1.9	.7	-	-	-	-	-
		WWNI	12.0	10.5	9.1	7.8	6.9	5.8	4.3	2.8	1.1
		TW	5.2	3.6	2.1	1.2	-	-	-	-	-
		TWNI	11.2	9.9	8.4	7.3	6.1	4.8	3.3	1.9	.7
		DG	-	-	-	-	-	-	-	-	-
		DGNI	11.3	9.4	7.8	6.3	4.8	3.2	1.9	.8	-

MADISON WISCONSIN
43.1 N
7730 DD
T(JAN)=17

		SSF =	.1	.2	.3	.4	.5	.6	.7	.8	.9
L		WW	62	25	13	7	-	-	-	-	-
C		WWNI	137	62	39	27	20	15	12	9	6
R		TW	67	27	15	8	3	-	-	-	-
		TWNI	130	59	36	25	19	14	11	8	5
		DG	-	-	-	-	-	-	-	-	-
		DGNI	133	59	36	24	17	12	8	5	2
D		WW	4.9	3.4	2.3	1.1	-	-	-	-	-
		WWNI	12.3	10.7	9.6	8.3	7.3	6.2	4.6	3.0	1.1
		TW	5.5	3.9	2.4	1.2	.1	-	-	-	-
		TWNI	11.7	10.2	8.9	7.7	6.5	5.1	3.5	2.1	.8
		DG	-	-	-	-	-	-	-	-	-
		DGNI	11.7	9.8	8.2	6.6	5.2	3.5	2.1	.9	.1

MILWAUKEE WISCONSIN
42.9 N
7444 DD
T(JAN)=19

		SSF =	.1	.2	.3	.4	.5	.6	.7	.8	.9
L		WW	64	26	14	8	-	-	-	-	-
C		WWNI	140	63	39	27	20	16	12	9	6
R		TW	69	28	15	8	3	-	-	-	-
		TWNI	132	60	37	26	19	14	11	8	5
		DG	-	-	-	-	-	-	-	-	-
		DGNI	136	60	36	25	17	12	9	5	2
D		WW	5.0	3.6	2.4	1.2	-	-	-	-	-
		WWNI	12.9	10.7	9.8	8.5	7.4	6.1	4.6	3.0	1.1
		TW	5.6	4.0	2.5	1.3	.2	-	-	-	-
		TWNI	11.8	10.4	9.0	7.8	6.5	5.1	3.6	2.1	.8
		DG	-	-	-	-	-	-	-	-	-
		DGNI	11.8	10.0	8.3	6.8	5.2	3.6	2.1	.9	.1

CASPER WYOMING
42.9 N
7555 DD
T(JAN)=23

		SSF =	.1	.2	.3	.4	.5	.6	.7	.8	.9
L		WW	107	49	30	20	14	10	7	5	1
C		WWNI	184	90	57	41	31	24	19	15	10
R		TW	107	49	29	19	12	8	5	2	-
		TWNI	175	84	53	38	28	22	17	12	8
		DG	90	34	16	5	-	-	-	-	-
		DGNI	190	90	56	39	29	21	16	11	6
D		WW	10.3	8.4	6.9	5.7	4.3	2.8	1.2	-	-
		WWNI	17.7	16.2	14.4	13.3	11.8	10.4	7.9	5.2	2.1
		TW	9.8	8.2	6.4	4.6	3.0	1.8	3.6	2.2	-
		TWNI	16.9	15.3	13.7	12.2	10.5	8.4	5.9	3.5	1.4
		DG	7.3	3.8	1.4	-	-	-	-	-	-
		DGNI	18.1	16.1	14.0	12.1	9.7	7.1	4.7	2.5	.8

CHEYENNE WYOMING
41.1 N
7255 DD
T(JAN)=27

		SSF =	.1	.2	.3	.4	.5	.6	.7	.8	.9
L		WW	111	52	32	22	16	12	8	5	1
C		WWNI	190	91	58	42	32	25	21	16	11
R		TW	112	51	31	20	14	9	6	3	1
		TWNI	181	86	55	39	30	23	18	13	9
		DG	96	38	16	-	-	-	-	-	-
		DGNI	195	92	58	41	31	23	17	12	7
D		WW	10.3	8.9	7.7	6.6	5.4	3.6	1.9	-	-
		WWNI	17.7	17.2	15.5	14.3	12.9	11.7	9.3	5.9	2.4
		TW	10.1	8.7	7.0	5.1	3.5	2.3	1.3	5.5	1.6
		TWNI	16.9	15.8	14.5	13.1	11.6	9.2	6.5	4.0	1.6
		DG	7.9	5.0	1.2	-	-	-	-	-	-
		DGNI	18.3	16.7	14.9	13.3	10.8	7.9	5.3	3.0	1.0

ROCK SPRINGS WYOMING
41.6 N
8410 DD
T(JAN)=19

		SSF =	.1	.2	.3	.4	.5	.6	.7	.8	.9
L		WW	98	45	28	19	13	9	6	-	-
C		WWNI	176	84	54	38	29	23	18	14	10
R		TW	100	45	27	17	11	7	4	2	-
		TWNI	167	79	50	36	27	21	16	12	8
		DG	78	28	-	-	-	-	-	-	-
		DGNI	179	84	53	37	27	20	15	10	6
D		WW	9.1	7.7	6.3	5.2	3.8	2.4	.9	-	-
		WWNI	16.4	15.7	14.1	12.8	11.4	9.9	7.6	5.0	2.0
		TW	9.0	7.5	5.8	4.1	2.6	1.5	.7	3.4	1.3
		TWNI	15.7	14.4	13.1	11.7	10.0	8.0	5.6	3.4	1.3
		DG	6.2	2.6	-	-	-	-	-	-	-
		DGNI	16.8	15.1	13.3	11.5	9.2	6.7	4.4	2.3	.7

SHERIDAN
WYOMING

44.8 N
7708 DD
T(JAN)=21

		SSF =	.1	.2	.3	.4	.5	.6	.7	.8	.9
L		WW	86	38	22	14	9	5	-	-	-
C		WWNI	161	77	48	34	25	19	15	12	8
R		TW	88	38	22	14	8	4	-	-	-
		TWNI	153	72	45	32	23	18	13	10	6
		DG	57	-	-	-	-	-	-	-	-
		DGNI	163	75	46	32	23	17	12	8	4
D		WW	8.0	6.2	4.5	3.3	1.9	.8	-	-	-
		WWNI	14.9	13.8	12.4	10.6	9.1	7.8	5.8	3.9	1.5
		TW	7.7	6.0	4.4	2.8	1.5	-	-	-	-
		TWNI	14.2	12.8	11.2	9.7	8.2	6.6	4.5	2.7	1.0
		DG	3.7	-	-	-	-	-	-	-	-
		DGNI	15.1	13.1	11.0	9.2	7.1	5.1	3.2	1.6	.4

CANADA

EDMONTON
ALBERTA

53.6 N
10268 DD
T(JAN)= 7

		SSF =	.1	.2	.3	.4	.5	.6	.7	.8	.9
L		WW	66	26	10	-	-	-	-	-	-
C		WWNI	139	65	39	27	19	13	10	7	4
R		TW	69	27	13	-	-	-	-	-	-
		TWNI	132	61	37	25	17	12	9	6	3
		DG	-	-	-	-	-	-	-	-	-
		DGNI	136	61	35	23	15	10	6	3	1
D		WW	5.5	3.1	.9	-	-	-	-	-	-
		WWNI	13.1	11.3	9.0	7.2	5.5	4.2	2.8	1.8	.6
		TW	5.8	4.8	1.7	-	-	-	-	-	-
		TWNI	12.3	10.3	8.3	6.5	5.0	3.6	2.5	1.4	.4
		DG	-	-	-	-	-	-	-	-	-
		DGNI	12.5	9.8	7.3	5.2	3.5	2.2	1.0	.2	-

SUFFIELD
ALBERTA

50.3 N
8644 DD
T(JAN)=17

		SSF =	.1	.2	.3	.4	.5	.6	.7	.8	.9
L		WW	92	40	23	14	8	4	-	-	-
C		WWNI	170	79	50	35	26	19	15	11	7
R		TW	93	40	23	14	8	4	-	-	-
		TWNI	160	75	46	32	23	17	13	9	6
		DG	64	-	-	-	-	-	-	-	-
		DGNI	170	78	48	32	23	16	11	7	3
D		WW	7.9	6.3	4.5	3.0	1.4	.2	-	-	-
		WWNI	15.6	14.3	12.5	10.8	8.8	7.1	5.0	3.3	1.3
		TW	8.2	6.1	4.3	2.7	1.4	-	-	-	-
		TWNI	14.8	13.2	11.5	9.6	7.8	5.8	4.1	2.4	.8
		DG	4.2	-	-	-	-	-	-	-	-
		DGNI	15.6	13.3	11.1	8.9	6.6	4.7	2.8	1.3	.2

NANAIMO
BRITISH COLUMBIA

49.2 N
5515 DD
T(JAN)=37

		SSF =	.1	.2	.3	.4	.5	.6	.7	.8	.9
L		WW	140	61	35	21	13	7	-	-	-
C		WWNI	229	105	65	44	32	23	17	12	8
R		TW	135	58	32	20	12	7	-	-	-
		TWNI	214	98	60	40	29	21	15	10	6
		DG	126	43	-	-	-	-	-	-	-
		DGNI	233	106	63	42	29	20	14	9	4
D		WW	12.2	9.5	6.9	4.4	2.6	1.1	-	-	-
		WWNI	21.5	18.2	15.4	12.9	9.9	7.4	5.4	3.4	1.3
		TW	12.0	8.8	6.1	4.1	2.3	1.0	-	-	-
		TWNI	19.8	16.7	14.1	11.2	8.7	6.5	4.5	2.5	.9
		DG	9.7	4.1	-	-	-	-	-	-	-
		DGNI	21.3	17.4	13.9	10.7	8.0	5.5	3.3	1.6	.3

VANCOUVER
BRITISH COLUMBIA

49.3 N
5515 DD
T(JAN)=37

		SSF =	.1	.2	.3	.4	.5	.6	.7	.8	.9
L		WW	111	47	26	15	9	4	-	-	-
C		WWNI	193	89	54	37	27	20	15	11	7
R		TW	109	46	25	15	8	4	-	-	-
		TWNI	181	83	50	34	24	18	13	9	6
		DG	86	-	-	-	-	-	-	-	-
		DGNI	195	88	52	34	24	17	11	7	3
D		WW	9.5	6.9	4.7	2.8	1.4	.1	-	-	-
		WWNI	18.3	15.2	12.7	10.8	8.4	6.6	4.9	3.2	1.2
		TW	9.5	6.7	4.5	2.2	1.4	-	-	-	-
		TWNI	16.9	14.1	11.7	9.5	7.6	5.7	4.0	2.3	.8
		DG	5.9	-	-	-	-	-	-	-	-
		DGNI	17.8	14.3	11.4	8.8	6.6	4.6	2.7	1.3	.2

WINNIPEG
MANITOBA

49.9 N
10679 DD
T(JAN)= -

		SSF =	.1	.2	.3	.4	.5	.6	.7	.8	.9
L		WW	53	21	9	-	-	-	-	-	-
C		WWNI	125	58	36	25	18	13	10	7	5
R		TW	59	23	11	4	-	-	-	-	-
		TWNI	119	55	34	23	17	12	9	6	4
		DG	-	-	-	-	-	-	-	-	-
		DGNI	121	54	32	21	15	10	6	3	1
D		WW	4.3	2.7	1.0	-	-	-	-	-	-
		WWNI	11.4	10.1	8.8	7.4	5.9	4.7	3.4	2.2	.8
		TW	4.8	3.1	1.6	.2	-	-	-	-	-
		TWNI	10.7	9.4	8.1	6.6	5.3	4.1	2.8	1.6	.5
		DG	-	-	-	-	-	-	-	-	-
		DGNI	10.8	9.0	7.1	5.4	3.9	2.5	1.3	.4	-

DARTMOUTH
NOVA SCOTIA

44.6 N
7361 DD
T(JAN)=26

		SSF =	.1	.2	.3	.4	.5	.6	.7	.8	.9
L		WW	73	31	18	11	7	4	-	-	-
C		WWNI	148	68	43	30	23	18	14	11	7
R		TW	76	33	18	11	6	3	-	-	-
		TWNI	140	64	40	28	20	16	12	9	6
		DG	34	-	-	-	-	-	-	-	-
		DGNI	146	66	41	28	20	15	11	7	3
D		WW	6.0	4.8	3.7	2.7	1.5	.3	-	-	-
		WWNI	13.5	11.9	11.1	9.8	8.8	7.5	5.7	3.7	1.4
		TW	6.6	5.0	3.5	2.2	1.1	.3	-	-	-
		TWNI	12.9	11.4	10.2	9.1	7.7	6.1	4.3	2.5	1.0
		DG	1.5	-	-	-	-	-	-	-	-
		DGNI	13.1	11.3	9.8	8.3	6.5	4.6	2.9	1.4	.3

MOOSONEE ONTARIO			SSF =	.1	.2	.3	.4	.5	.6	.7	.8	.9
			WW	34	11	-	-	-	-	-	-	-
51.3 N	L		WWNI	102	48	30	21	15	11	8	6	4
11572 DD	C		TW	42	15	-	-	-	-	-	-	-
T(JAN)=-1	R		TWNI	98	45	28	19	14	10	7	5	3
			DG	-	-	-	-	-	-	-	-	-
			DGNI	96	43	26	17	11	7	4	-	-
			WW	2.5	.9	-	-	-	-	-	-	-
			WWNI	9.7	8.3	7.3	6.4	4.9	3.6	2.5	1.6	.5
		D	TW	3.4	1.7	-	-	-	-	-	-	-
			TWNI	9.2	7.8	6.9	5.6	4.3	3.2	2.2	1.2	.4
			DG	-	-	-	-	-	-	-	-	-
			DGNI	8.7	7.1	5.6	4.0	2.7	1.6	.6	-	-
OTTAWA ONTARIO			SSF =	.1	.2	.3	.4	.5	.6	.7	.8	.9
			WW	65	26	14	8	-	-	-	-	-
45.5 N	L		WWNI	140	63	39	27	20	16	12	9	6
8735 DD	C		TW	69	28	15	8	3	-	-	-	-
T(JAN)=13	R		TWNI	133	60	37	26	19	14	11	8	5
			DG	-	-	-	-	-	-	-	-	-
			DGNI	137	60	36	25	17	12	9	5	2
			WW	5.1	3.6	2.4	1.2	-	-	-	-	-
			WWNI	12.6	10.7	9.7	8.6	7.4	6.0	4.6	3.0	1.1
		D	TW	5.7	4.0	2.5	1.3	.2	-	-	-	-
			TWNI	12.0	10.2	9.0	7.8	6.5	5.1	3.5	2.1	.8
			DG	-	-	-	-	-	-	-	-	-
			DGNI	12.0	10.0	8.4	6.8	5.2	3.6	2.1	.9	.1
TORONTO ONTARIO			SSF =	.1	.2	.3	.4	.5	.6	.7	.8	.9
			WW	73	31	17	11	7	-	-	-	-
43.7 N	L		WWNI	148	68	43	30	22	17	14	10	7
6827 DD	C		TW	76	32	18	11	6	2	-	-	-
T(JAN)=25	R		TWNI	140	65	40	28	21	16	12	9	6
			DG	32	-	-	-	-	-	-	-	-
			DGNI	146	66	40	28	20	14	10	7	3
			WW	6.0	4.6	3.4	2.4	1.2	-	-	-	-
			WWNI	13.7	11.9	10.9	9.5	8.3	7.1	5.5	3.6	1.4
		D	TW	6.5	4.9	3.3	2.0	1.0	.1	-	-	-
			TWNI	12.8	11.3	10.0	8.8	7.5	5.9	4.1	2.4	.9
			DG	1.2	-	-	-	-	-	-	-	-
			DGNI	13.0	11.2	9.5	8.0	6.3	4.4	2.7	1.3	.3
NORMANDIN QUEBEC			SSF =	.1	.2	.3	.4	.5	.6	.7	.8	.9
			WW	49	19	9	-	-	-	-	-	-
48.8 N	L		WWNI	118	56	35	24	18	13	10	7	5
10528 DD	C		TW	54	22	11	4	-	-	-	-	-
T(JAN)= 4	R		TWNI	112	52	33	23	16	12	9	6	4
			DG	-	-	-	-	-	-	-	-	-
			DGNI	113	52	31	21	14	10	6	4	-
			WW	4.1	2.5	1.0	-	-	-	-	-	-
			WWNI	11.4	9.9	8.6	7.7	6.2	4.8	3.4	2.2	.8
		D	TW	4.6	2.9	1.5	.2	-	-	-	-	-
			TWNI	10.6	9.2	8.1	6.7	5.4	4.1	2.8	1.6	.5
			DG	-	-	-	-	-	-	-	-	-
			DGNI	10.6	8.7	7.1	5.5	3.9	2.6	1.4	.4	-

Appendix 13 — Fixed Charge Rate (FCR) Tables

The fixed charge rate (*FCR*) converts an initial dollar outlay and subsequent dollar cash flows into an equivalent annual payment series. As defined in this context, *FCR* takes into account the mortgage or loan down payment, the principal and interest costs over the loan term, the mortgage interest tax deductions over the ownership years, property taxes, insurance, operation and maintenance expenditures, and the resale profit realized at the time of sale. In order to present the *FCR* values in tabular form, a number of limiting assumptions were required. These are as follows:

Down payment ratio = 0.20
Loan term = T
Ownership period = PT
Mortgage interest rate = discount rate = i
Property tax, insurance, operation and maintenance expense escalation rates = 0.07
Resale valuation factor (V)* = 1.0 or 0.50
Federal and state income tax rates = 0.30
First year property tax rate (stated as fraction of add-on cost) = 0.02
First year operation and maintenance cost rate (stated as fraction of add-on cost) = 0.01

Values for *FCR* based upon these assumptions are contained in the tables below. The actual *FCR* calculation procedure is contained in Appendix 10.

Appendix 15, in a series of graphs, indicates the sensitivity of *FCR* to variations in the above parameters. The importance of these parameters to the *FCR* calculation can be readily seen in the graphs.

FIXED CHARGE RATE TABLES : FCR(PT,T,I)

INTEREST RATE (I) = 0.04   RESALE FACTOR, V = 1.00

OWNERSHIP PERIOD : PT

LOAN TERM T	2	4	6	8	10	12	14	16	18	20	22	24	26	28	30
2	-.007	-.006	-.006	-.006	-.006	-.007	-.007	-.008	-.008	-.008	-.009	-.010	-.010	-.011	-.011
4	-.011	-.009	-.008	-.008	-.008	-.008	-.008	-.008	-.009	-.009	-.010	-.010	-.011	-.011	-.012
6	-.012	-.011	-.009	-.009	-.009	-.009	-.009	-.009	-.009	-.010	-.010	-.011	-.011	-.012	-.013
8	-.012	-.012	-.011	-.010	-.010	-.010	-.010	-.010	-.010	-.010	-.011	-.011	-.012	-.012	-.013
10	-.013	-.012	-.012	-.011	-.011	-.011	-.010	-.011	-.011	-.011	-.011	-.012	-.012	-.013	-.014
12	-.013	-.013	-.012	-.012	-.012	-.011	-.011	-.011	-.011	-.012	-.012	-.012	-.013	-.013	-.014
14	-.013	-.013	-.013	-.013	-.012	-.012	-.012	-.012	-.012	-.012	-.013	-.013	-.013	-.014	-.014
16	-.013	-.013	-.013	-.013	-.013	-.013	-.013	-.013	-.013	-.013	-.013	-.013	-.014	-.014	-.015
18	-.014	-.013	-.013	-.013	-.013	-.013	-.013	-.013	-.013	-.013	-.014	-.014	-.014	-.015	-.015
20	-.014	-.014	-.014	-.014	-.014	-.014	-.014	-.014	-.014	-.014	-.014	-.014	-.015	-.015	-.016
22	-.014	-.014	-.014	-.014	-.014	-.014	-.014	-.014	-.014	-.014	-.015	-.015	-.015	-.016	-.016
24	-.014	-.014	-.014	-.014	-.014	-.014	-.014	-.014	-.015	-.015	-.015	-.015	-.016	-.016	-.017
26	-.014	-.014	-.014	-.014	-.014	-.014	-.015	-.015	-.015	-.015	-.015	-.016	-.016	-.016	-.017
28	-.014	-.014	-.014	-.014	-.014	-.015	-.015	-.015	-.015	-.015	-.016	-.016	-.016	-.017	-.017
30	-.014	-.014	-.014	-.014	-.014	-.015	-.015	-.015	-.015	-.016	-.016	-.016	-.017	-.017	-.018

---

*V is an assessment valuation factor that is meant to take into account the value of the passive addition assigned by the property assessor in year *PT*. If none of the additional cost is reflected in the resale price, then $V = 0$; if all of the additional cost is appraised in the market value, then $V = 1.0$. It is conceivable in the future that, with increasing home heating costs, $V = 1.0$ may be possible. A series of *FCR* values is given in this appendix for $V = 1.0$ and $V = 0.50$.

FIXED CHARGE RATE TABLES : FCR(PT,T,I)

INTEREST RATE (I) = 0.05   RESALE FACTOR, V = 1.00

OWNERSHIP PERIOD : PT

LOAN TERM T	2	4	6	8	10	12	14	16	18	20	22	24	26	28	30
2	0.003	0.005	0.006	0.006	0.007	0.007	0.008	0.009	0.009	0.010	0.011	0.011	0.012	0.013	0.014
4	-.002	0.001	0.003	0.004	0.005	0.006	0.007	0.008	0.008	0.009	0.010	0.011	0.011	0.012	0.013
6	-.003	-.001	0.001	0.003	0.004	0.005	0.006	0.007	0.007	0.008	0.009	0.010	0.011	0.011	0.012
8	-.004	-.003	-.001	0.001	0.003	0.004	0.005	0.006	0.006	0.007	0.008	0.009	0.010	0.011	0.012
10	-.005	-.003	-.002	-.000	0.001	0.003	0.004	0.005	0.006	0.007	0.007	0.008	0.009	0.010	0.011
12	-.005	-.004	-.003	-.001	0.000	0.001	0.003	0.004	0.005	0.006	0.007	0.008	0.008	0.009	0.010
14	-.005	-.004	-.003	-.002	-.001	0.000	0.002	0.003	0.004	0.005	0.006	0.007	0.008	0.009	0.010
16	-.005	-.004	-.003	-.002	-.001	-.000	0.001	0.002	0.003	0.004	0.005	0.006	0.007	0.008	0.009
18	-.005	-.005	-.004	-.003	-.002	-.001	0.000	0.001	0.002	0.004	0.005	0.006	0.006	0.007	0.008
20	-.006	-.005	-.004	-.003	-.002	-.001	-.000	0.001	0.002	0.003	0.004	0.005	0.006	0.007	0.008
22	-.006	-.005	-.004	-.003	-.003	-.002	-.001	0.000	0.001	0.002	0.003	0.004	0.005	0.006	0.007
24	-.006	-.005	-.004	-.004	-.003	-.002	-.001	-.000	0.001	0.002	0.003	0.004	0.005	0.006	0.007
26	-.006	-.005	-.004	-.004	-.003	-.002	-.001	-.001	0.000	0.001	0.002	0.003	0.004	0.005	0.006
28	-.006	-.005	-.005	-.004	-.003	-.002	-.002	-.001	0.000	0.001	0.002	0.003	0.004	0.005	0.006
30	-.006	-.005	-.005	-.004	-.003	-.003	-.002	-.001	-.000	0.001	0.002	0.002	0.003	0.004	0.006

FIXED CHARGE RATE TABLES : FCR(PT,T,I)

INTEREST RATE (I) = 0.06   RESALE FACTOR, V = 1.00

OWNERSHIP PERIOD : PT

LOAN TERM T	2	4	6	8	10	12	14	16	18	20	22	24	26	28	30
2	0.012	0.015	0.017	0.019	0.020	0.021	0.023	0.024	0.026	0.028	0.029	0.031	0.033	0.035	0.037
4	0.007	0.011	0.014	0.016	0.018	0.020	0.021	0.023	0.025	0.026	0.028	0.030	0.032	0.034	0.036
6	0.005	0.008	0.012	0.014	0.016	0.018	0.020	0.022	0.024	0.025	0.027	0.029	0.031	0.033	0.035
8	0.004	0.007	0.009	0.012	0.015	0.017	0.019	0.021	0.022	0.024	0.026	0.028	0.030	0.032	0.034
10	0.003	0.006	0.008	0.011	0.013	0.015	0.017	0.019	0.021	0.023	0.025	0.027	0.029	0.031	0.033
12	0.003	0.005	0.007	0.009	0.012	0.014	0.016	0.018	0.020	0.022	0.024	0.026	0.028	0.030	0.032
14	0.003	0.005	0.007	0.009	0.011	0.013	0.015	0.017	0.019	0.021	0.023	0.025	0.027	0.029	0.031
16	0.003	0.004	0.006	0.008	0.010	0.012	0.014	0.016	0.018	0.020	0.022	0.024	0.026	0.028	0.030
18	0.003	0.004	0.006	0.008	0.009	0.011	0.013	0.015	0.017	0.020	0.022	0.024	0.026	0.028	0.030
20	0.002	0.004	0.006	0.007	0.009	0.011	0.013	0.015	0.017	0.019	0.021	0.023	0.025	0.027	0.029
22	0.002	0.004	0.005	0.007	0.009	0.010	0.012	0.014	0.016	0.018	0.020	0.022	0.024	0.026	0.028
24	0.002	0.004	0.005	0.007	0.008	0.010	0.012	0.014	0.015	0.017	0.019	0.021	0.024	0.026	0.028
26	0.002	0.004	0.005	0.006	0.008	0.010	0.011	0.013	0.015	0.017	0.019	0.021	0.023	0.025	0.027
28	0.002	0.003	0.005	0.006	0.008	0.009	0.011	0.013	0.015	0.016	0.018	0.020	0.022	0.025	0.027
30	0.002	0.003	0.005	0.006	0.008	0.009	0.011	0.013	0.014	0.016	0.018	0.020	0.022	0.024	0.026

FIXED CHARGE RATE TABLES : FCR(PT,T,I)

INTEREST RATE (I) = 0.07   RESALE FACTOR, V = 1.00

OWNERSHIP PERIOD : PT

LOAN TERM T	2	4	6	8	10	12	14	16	18	20	22	24	26	28	30
2	0.022	0.026	0.029	0.031	0.033	0.035	0.037	0.040	0.042	0.045	0.047	0.050	0.052	0.055	0.057
4	0.015	0.021	0.025	0.028	0.031	0.033	0.036	0.038	0.041	0.043	0.046	0.048	0.051	0.053	0.056
6	0.013	0.018	0.022	0.026	0.029	0.031	0.034	0.037	0.039	0.042	0.044	0.047	0.049	0.052	0.055
8	0.012	0.016	0.019	0.023	0.027	0.030	0.032	0.035	0.038	0.040	0.043	0.046	0.048	0.051	0.054
10	0.012	0.015	0.018	0.021	0.025	0.028	0.031	0.034	0.036	0.039	0.042	0.044	0.047	0.050	0.053
12	0.011	0.014	0.017	0.020	0.023	0.026	0.029	0.032	0.035	0.038	0.041	0.043	0.046	0.049	0.052
14	0.011	0.013	0.016	0.019	0.022	0.025	0.028	0.031	0.034	0.037	0.040	0.042	0.045	0.048	0.051
16	0.011	0.013	0.016	0.018	0.021	0.024	0.027	0.030	0.033	0.036	0.038	0.041	0.044	0.047	0.050
18	0.010	0.013	0.015	0.018	0.020	0.023	0.026	0.029	0.032	0.035	0.038	0.040	0.043	0.046	0.049
20	0.010	0.013	0.015	0.017	0.020	0.022	0.025	0.028	0.031	0.034	0.037	0.040	0.042	0.045	0.048
22	0.010	0.012	0.015	0.017	0.019	0.022	0.025	0.027	0.030	0.033	0.036	0.039	0.042	0.044	0.047
24	0.010	0.012	0.014	0.017	0.019	0.022	0.024	0.027	0.029	0.032	0.035	0.038	0.041	0.044	0.047
26	0.010	0.012	0.014	0.017	0.019	0.021	0.024	0.026	0.029	0.032	0.034	0.037	0.040	0.043	0.046
28	0.010	0.012	0.014	0.016	0.019	0.021	0.023	0.026	0.028	0.031	0.034	0.037	0.040	0.042	0.045
30	0.010	0.012	0.014	0.016	0.018	0.021	0.023	0.026	0.028	0.031	0.033	0.036	0.039	0.042	0.045

FIXED CHARGE RATE TABLES : FCR(PT,T,I)

INTEREST RATE (I) = 0.08   RESALE FACTOR, V = 1.00

OWNERSHIP PERIOD : PT

LOAN TERM T	2	4	6	8	10	12	14	16	18	20	22	24	26	28	30
2	0.032	0.037	0.040	0.043	0.046	0.049	0.052	0.055	0.058	0.061	0.064	0.067	0.070	0.073	0.076
4	0.024	0.031	0.036	0.040	0.043	0.046	0.050	0.053	0.056	0.059	0.062	0.065	0.068	0.072	0.075
6	0.022	0.027	0.032	0.037	0.041	0.044	0.048	0.051	0.054	0.057	0.060	0.064	0.067	0.070	0.073
8	0.020	0.025	0.029	0.034	0.038	0.042	0.046	0.049	0.052	0.056	0.059	0.062	0.065	0.069	0.072
10	0.020	0.024	0.028	0.032	0.036	0.040	0.044	0.047	0.051	0.054	0.057	0.061	0.064	0.067	0.071
12	0.019	0.023	0.026	0.030	0.034	0.038	0.042	0.046	0.049	0.053	0.056	0.059	0.063	0.066	0.069
14	0.019	0.022	0.026	0.029	0.033	0.037	0.041	0.044	0.048	0.051	0.055	0.058	0.062	0.065	0.068
16	0.019	0.022	0.025	0.028	0.032	0.035	0.039	0.043	0.047	0.050	0.054	0.057	0.060	0.064	0.067
18	0.018	0.021	0.025	0.028	0.031	0.035	0.038	0.042	0.045	0.049	0.053	0.056	0.059	0.063	0.066
20	0.018	0.021	0.024	0.027	0.031	0.034	0.037	0.041	0.044	0.048	0.051	0.055	0.058	0.062	0.065
22	0.018	0.021	0.024	0.027	0.030	0.033	0.037	0.040	0.043	0.047	0.051	0.054	0.058	0.061	0.064
24	0.018	0.021	0.024	0.027	0.030	0.033	0.036	0.039	0.043	0.046	0.050	0.053	0.057	0.060	0.064
26	0.018	0.021	0.024	0.027	0.030	0.033	0.036	0.039	0.042	0.046	0.049	0.052	0.056	0.059	0.063
28	0.018	0.021	0.023	0.026	0.029	0.032	0.035	0.039	0.042	0.045	0.048	0.052	0.055	0.059	0.062
30	0.018	0.021	0.023	0.026	0.029	0.032	0.035	0.038	0.041	0.045	0.048	0.051	0.055	0.058	0.062

FIXED CHARGE RATE TABLES : FCR(PT,T,I)

INTEREST RATE (I) = 0.09   RESALE FACTOR, V = 1.00

OWNERSHIP PERIOD : PT

LOAN TERM T	2	4	6	8	10	12	14	16	18	20	22	24	26	28	30
2	0.041	0.047	0.051	0.055	0.059	0.062	0.066	0.069	0.073	0.076	0.080	0.083	0.087	0.090	0.094
4	0.033	0.041	0.047	0.051	0.055	0.059	0.063	0.067	0.071	0.074	0.078	0.081	0.085	0.088	0.092
6	0.030	0.036	0.043	0.048	0.053	0.057	0.061	0.065	0.068	0.072	0.076	0.079	0.083	0.087	0.090
8	0.029	0.034	0.039	0.045	0.050	0.054	0.059	0.063	0.066	0.070	0.074	0.078	0.081	0.085	0.088
10	0.028	0.032	0.037	0.042	0.047	0.052	0.056	0.061	0.065	0.068	0.072	0.076	0.080	0.083	0.087
12	0.027	0.032	0.036	0.041	0.045	0.050	0.055	0.059	0.063	0.067	0.071	0.074	0.078	0.082	0.085
14	0.027	0.031	0.035	0.039	0.044	0.048	0.053	0.057	0.061	0.065	0.069	0.073	0.077	0.080	0.084
16	0.027	0.030	0.034	0.038	0.043	0.047	0.051	0.056	0.060	0.064	0.068	0.072	0.075	0.079	0.083
18	0.026	0.030	0.034	0.038	0.042	0.046	0.050	0.054	0.058	0.063	0.067	0.070	0.074	0.078	0.082
20	0.026	0.030	0.034	0.037	0.041	0.045	0.049	0.053	0.057	0.061	0.065	0.069	0.073	0.077	0.081
22	0.026	0.030	0.033	0.037	0.041	0.045	0.048	0.052	0.056	0.060	0.064	0.068	0.072	0.076	0.080
24	0.026	0.029	0.033	0.037	0.040	0.044	0.048	0.052	0.056	0.060	0.064	0.067	0.071	0.075	0.079
26	0.026	0.029	0.033	0.036	0.040	0.044	0.047	0.051	0.055	0.059	0.063	0.067	0.070	0.074	0.078
28	0.026	0.029	0.033	0.036	0.040	0.043	0.047	0.051	0.055	0.058	0.062	0.066	0.070	0.074	0.077
30	0.026	0.029	0.033	0.036	0.040	0.043	0.047	0.050	0.054	0.058	0.062	0.065	0.069	0.073	0.077

FIXED CHARGE RATE TABLES : FCR(PT,T,I)

INTEREST RATE (I) = 0.10   RESALE FACTOR, V = 1.00

OWNERSHIP PERIOD : PT

LOAN TERM T	2	4	6	8	10	12	14	16	18	20	22	24	26	28	30
2	0.051	0.058	0.063	0.067	0.071	0.075	0.079	0.083	0.087	0.091	0.095	0.099	0.103	0.106	0.110
4	0.041	0.051	0.058	0.063	0.068	0.072	0.076	0.081	0.085	0.089	0.093	0.097	0.100	0.104	0.107
6	0.038	0.045	0.053	0.059	0.064	0.069	0.074	0.078	0.082	0.086	0.090	0.094	0.098	0.102	0.105
8	0.037	0.043	0.049	0.056	0.061	0.066	0.071	0.076	0.080	0.084	0.088	0.092	0.096	0.100	0.103
10	0.036	0.041	0.047	0.053	0.058	0.064	0.069	0.073	0.078	0.082	0.086	0.090	0.094	0.098	0.102
12	0.035	0.040	0.045	0.051	0.056	0.061	0.067	0.071	0.076	0.080	0.084	0.088	0.092	0.096	0.100
14	0.035	0.040	0.044	0.049	0.054	0.059	0.065	0.069	0.074	0.079	0.083	0.087	0.091	0.095	0.098
16	0.035	0.039	0.044	0.048	0.053	0.058	0.063	0.068	0.072	0.077	0.081	0.085	0.089	0.093	0.097
18	0.034	0.039	0.043	0.048	0.052	0.057	0.062	0.066	0.071	0.076	0.080	0.084	0.088	0.092	0.096
20	0.034	0.038	0.043	0.047	0.052	0.056	0.061	0.065	0.070	0.074	0.079	0.083	0.087	0.091	0.094
22	0.034	0.038	0.042	0.047	0.051	0.055	0.060	0.064	0.069	0.073	0.078	0.082	0.086	0.090	0.093
24	0.034	0.038	0.042	0.046	0.051	0.055	0.059	0.064	0.068	0.072	0.077	0.081	0.085	0.089	0.092
26	0.034	0.038	0.042	0.046	0.050	0.055	0.059	0.063	0.067	0.072	0.076	0.080	0.084	0.088	0.092
28	0.034	0.038	0.042	0.046	0.050	0.054	0.058	0.063	0.067	0.071	0.075	0.079	0.083	0.087	0.091
30	0.034	0.038	0.042	0.046	0.050	0.054	0.058	0.062	0.066	0.071	0.075	0.079	0.083	0.086	0.090

FIXED CHARGE RATE TABLES : FCR(PT,T,I)

INTEREST RATE (I) = 0.11   RESALE FACTOR, V = 1.00

OWNERSHIP PERIOD : PT

LOAN TERM T	2	4	6	8	10	12	14	16	18	20	22	24	26	28	30
2	0.061	0.068	0.074	0.079	0.084	0.088	0.093	0.097	0.101	0.106	0.110	0.114	0.117	0.121	0.125
4	0.050	0.061	0.068	0.074	0.080	0.085	0.089	0.094	0.098	0.103	0.107	0.111	0.115	0.118	0.122
6	0.047	0.055	0.063	0.070	0.076	0.081	0.086	0.091	0.096	0.100	0.104	0.108	0.112	0.116	0.119
8	0.045	0.052	0.059	0.066	0.072	0.078	0.083	0.088	0.093	0.097	0.102	0.106	0.110	0.114	0.117
10	0.044	0.050	0.056	0.063	0.069	0.075	0.081	0.086	0.091	0.095	0.099	0.104	0.108	0.111	0.115
12	0.043	0.049	0.055	0.061	0.067	0.073	0.078	0.083	0.088	0.093	0.097	0.102	0.106	0.109	0.113
14	0.043	0.048	0.054	0.059	0.065	0.070	0.076	0.081	0.086	0.091	0.096	0.100	0.104	0.108	0.111
16	0.042	0.048	0.053	0.058	0.064	0.069	0.074	0.080	0.085	0.089	0.094	0.098	0.102	0.106	0.110
18	0.042	0.047	0.052	0.057	0.063	0.068	0.073	0.078	0.083	0.088	0.092	0.097	0.101	0.105	0.108
20	0.042	0.047	0.052	0.057	0.062	0.067	0.072	0.077	0.082	0.086	0.091	0.095	0.099	0.103	0.107
22	0.042	0.047	0.052	0.056	0.061	0.066	0.071	0.076	0.081	0.085	0.090	0.094	0.098	0.102	0.106
24	0.042	0.047	0.051	0.056	0.061	0.066	0.070	0.075	0.080	0.084	0.089	0.093	0.097	0.101	0.105
26	0.042	0.046	0.051	0.056	0.061	0.065	0.070	0.075	0.079	0.084	0.088	0.092	0.096	0.100	0.104
28	0.042	0.046	0.051	0.056	0.060	0.065	0.070	0.074	0.079	0.083	0.087	0.092	0.096	0.100	0.103
30	0.042	0.046	0.051	0.055	0.060	0.065	0.069	0.074	0.078	0.083	0.087	0.091	0.095	0.099	0.103

FIXED CHARGE RATE TABLES : FCR(PT,T,I)

INTEREST RATE (I) = 0.12   RESALE FACTOR, V = 1.00

OWNERSHIP PERIOD : PT

LOAN TERM T	2	4	6	8	10	12	14	16	18	20	22	24	26	28	30
2	0.070	0.079	0.085	0.091	0.096	0.101	0.106	0.111	0.115	0.120	0.124	0.128	0.132	0.135	0.138
4	0.059	0.070	0.079	0.085	0.091	0.097	0.102	0.107	0.112	0.116	0.120	0.124	0.128	0.132	0.135
6	0.055	0.064	0.073	0.081	0.087	0.093	0.098	0.104	0.108	0.113	0.117	0.121	0.125	0.129	0.132
8	0.053	0.061	0.068	0.076	0.083	0.090	0.095	0.100	0.105	0.110	0.115	0.119	0.123	0.126	0.130
10	0.052	0.059	0.066	0.073	0.080	0.086	0.092	0.098	0.103	0.108	0.112	0.116	0.120	0.124	0.127
12	0.051	0.058	0.064	0.071	0.077	0.084	0.090	0.095	0.100	0.105	0.110	0.114	0.118	0.122	0.125
14	0.051	0.057	0.063	0.069	0.075	0.081	0.087	0.093	0.098	0.103	0.108	0.112	0.116	0.120	0.123
16	0.050	0.056	0.062	0.068	0.074	0.080	0.085	0.091	0.096	0.101	0.106	0.110	0.114	0.118	0.122
18	0.050	0.056	0.061	0.067	0.073	0.078	0.084	0.089	0.095	0.100	0.104	0.109	0.113	0.116	0.120
20	0.050	0.055	0.061	0.066	0.072	0.077	0.083	0.088	0.093	0.098	0.103	0.107	0.111	0.115	0.119
22	0.050	0.055	0.061	0.066	0.071	0.077	0.082	0.087	0.092	0.097	0.102	0.106	0.110	0.114	0.117
24	0.050	0.055	0.060	0.066	0.071	0.076	0.081	0.086	0.091	0.096	0.101	0.105	0.109	0.113	0.116
26	0.050	0.055	0.060	0.065	0.071	0.076	0.081	0.086	0.091	0.095	0.100	0.104	0.108	0.112	0.116
28	0.050	0.055	0.060	0.065	0.070	0.075	0.080	0.085	0.090	0.095	0.099	0.103	0.107	0.111	0.115
30	0.050	0.055	0.060	0.065	0.070	0.075	0.080	0.085	0.090	0.094	0.099	0.103	0.107	0.110	0.114

FIXED CHARGE RATE TABLES : FCR(PT,T,I)

INTEREST RATE (I) = 0.13   RESALE FACTOR, V = 1.00

OWNERSHIP PERIOD : PT

LOAN TERM T	2	4	6	8	10	12	14	16	18	20	22	24	26	28	30
2	0.080	0.089	0.096	0.102	0.108	0.113	0.119	0.124	0.128	0.133	0.137	0.141	0.145	0.148	0.151
4	0.067	0.080	0.089	0.096	0.103	0.109	0.114	0.120	0.124	0.129	0.133	0.137	0.141	0.145	0.148
6	0.063	0.073	0.083	0.091	0.098	0.105	0.110	0.116	0.121	0.125	0.130	0.134	0.138	0.141	0.145
8	0.061	0.069	0.078	0.087	0.094	0.101	0.107	0.112	0.117	0.122	0.127	0.131	0.135	0.138	0.141
10	0.060	0.067	0.075	0.083	0.090	0.097	0.104	0.109	0.115	0.119	0.124	0.128	0.132	0.136	0.139
12	0.059	0.066	0.073	0.080	0.087	0.094	0.101	0.107	0.112	0.117	0.121	0.126	0.129	0.133	0.136
14	0.059	0.065	0.072	0.079	0.085	0.092	0.098	0.104	0.110	0.115	0.119	0.123	0.127	0.131	0.134
16	0.058	0.065	0.071	0.077	0.084	0.090	0.096	0.102	0.108	0.113	0.117	0.121	0.125	0.129	0.132
18	0.058	0.064	0.070	0.077	0.083	0.089	0.095	0.100	0.106	0.111	0.115	0.120	0.124	0.127	0.131
20	0.058	0.064	0.070	0.076	0.082	0.088	0.093	0.099	0.104	0.109	0.114	0.118	0.122	0.126	0.129
22	0.058	0.064	0.070	0.076	0.081	0.087	0.093	0.098	0.103	0.108	0.113	0.117	0.121	0.125	0.128
24	0.058	0.064	0.069	0.075	0.081	0.087	0.092	0.097	0.102	0.107	0.112	0.116	0.120	0.124	0.127
26	0.058	0.063	0.069	0.075	0.081	0.086	0.091	0.097	0.102	0.106	0.111	0.115	0.119	0.123	0.126
28	0.058	0.063	0.069	0.075	0.080	0.086	0.091	0.096	0.101	0.106	0.110	0.114	0.118	0.122	0.125
30	0.057	0.063	0.069	0.075	0.080	0.086	0.091	0.096	0.101	0.105	0.110	0.114	0.118	0.121	0.125

FIXED CHARGE RATE TABLES : FCR(PT,T,I)

INTEREST RATE (I) = 0.14   RESALE FACTOR, V = 1.00

OWNERSHIP PERIOD : PT

LOAN TERM T	2	4	6	8	10	12	14	16	18	20	22	24	26	28	30
2	0.089	0.100	0.107	0.114	0.120	0.126	0.131	0.137	0.141	0.146	0.150	0.154	0.158	0.161	0.164
4	0.076	0.090	0.100	0.107	0.114	0.121	0.126	0.132	0.137	0.141	0.146	0.150	0.153	0.157	0.160
6	0.071	0.082	0.093	0.102	0.109	0.116	0.122	0.128	0.133	0.138	0.142	0.146	0.149	0.153	0.156
8	0.069	0.078	0.087	0.097	0.105	0.112	0.118	0.124	0.129	0.134	0.138	0.142	0.146	0.149	0.152
10	0.068	0.076	0.084	0.093	0.101	0.108	0.115	0.121	0.126	0.131	0.135	0.139	0.143	0.146	0.149
12	0.067	0.075	0.082	0.090	0.097	0.105	0.112	0.118	0.123	0.128	0.132	0.137	0.140	0.144	0.147
14	0.067	0.074	0.081	0.088	0.095	0.102	0.109	0.115	0.121	0.126	0.130	0.134	0.138	0.141	0.145
16	0.066	0.073	0.080	0.087	0.094	0.100	0.107	0.113	0.118	0.123	0.128	0.132	0.136	0.139	0.143
18	0.066	0.073	0.079	0.086	0.093	0.099	0.105	0.111	0.116	0.122	0.126	0.130	0.134	0.138	0.141
20	0.066	0.072	0.079	0.085	0.092	0.098	0.104	0.110	0.115	0.120	0.125	0.129	0.133	0.136	0.139
22	0.066	0.072	0.079	0.085	0.091	0.097	0.103	0.109	0.114	0.119	0.123	0.128	0.131	0.135	0.138
24	0.066	0.072	0.078	0.085	0.091	0.097	0.102	0.108	0.113	0.118	0.122	0.126	0.130	0.134	0.137
26	0.065	0.072	0.078	0.084	0.090	0.096	0.102	0.107	0.112	0.117	0.121	0.126	0.129	0.133	0.136
28	0.065	0.072	0.078	0.084	0.090	0.096	0.102	0.107	0.112	0.116	0.121	0.125	0.129	0.132	0.135
30	0.065	0.072	0.078	0.084	0.090	0.096	0.101	0.106	0.111	0.116	0.120	0.124	0.128	0.131	0.135

FIXED CHARGE RATE TABLES : FCR(PT,T,I)

INTEREST RATE (I) = 0.15   RESALE FACTOR, V = 1.00

OWNERSHIP PERIOD : PT

LOAN TERM T	2	4	6	8	10	12	14	16	18	20	22	24	26	28	30
2	0.099	0.110	0.118	0.125	0.132	0.138	0.144	0.149	0.154	0.158	0.163	0.166	0.170	0.173	0.176
4	0.084	0.099	0.110	0.118	0.126	0.132	0.138	0.144	0.149	0.154	0.158	0.162	0.165	0.168	0.171
6	0.079	0.091	0.103	0.112	0.120	0.127	0.133	0.139	0.144	0.149	0.153	0.157	0.161	0.164	0.167
8	0.077	0.087	0.097	0.107	0.115	0.123	0.129	0.135	0.140	0.145	0.149	0.153	0.157	0.160	0.163
10	0.076	0.085	0.094	0.102	0.111	0.119	0.125	0.131	0.137	0.142	0.146	0.150	0.154	0.157	0.160
12	0.075	0.083	0.091	0.100	0.107	0.115	0.122	0.128	0.134	0.139	0.143	0.147	0.151	0.154	0.157
14	0.074	0.082	0.090	0.098	0.105	0.112	0.119	0.125	0.131	0.136	0.141	0.145	0.148	0.151	0.154
16	0.074	0.082	0.089	0.096	0.104	0.110	0.117	0.123	0.129	0.134	0.138	0.142	0.146	0.149	0.152
18	0.074	0.081	0.088	0.096	0.102	0.109	0.115	0.121	0.127	0.132	0.136	0.140	0.144	0.147	0.150
20	0.074	0.081	0.088	0.095	0.102	0.108	0.114	0.120	0.125	0.130	0.135	0.139	0.143	0.146	0.149
22	0.074	0.081	0.088	0.094	0.101	0.107	0.113	0.119	0.124	0.129	0.134	0.138	0.141	0.144	0.147
24	0.073	0.080	0.087	0.094	0.101	0.107	0.113	0.118	0.123	0.128	0.132	0.136	0.140	0.143	0.146
26	0.073	0.080	0.087	0.094	0.100	0.106	0.112	0.118	0.123	0.127	0.132	0.136	0.139	0.142	0.145
28	0.073	0.080	0.087	0.094	0.100	0.106	0.112	0.117	0.122	0.127	0.131	0.135	0.139	0.142	0.145
30	0.073	0.080	0.087	0.093	0.100	0.106	0.111	0.117	0.122	0.126	0.131	0.134	0.138	0.141	0.144

FIXED CHARGE RATE TABLES : FCR(PT,T,I)

INTEREST RATE (I) = 0.04   RESALE FACTOR, V = 0.50

OWNERSHIP PERIOD : PT

LOAN TERM T	2	4	6	8	10	12	14	16	18	20	22	24	26	28	30
2	0.274	0.148	0.107	0.087	0.076	0.068	0.063	0.060	0.058	0.057	0.056	0.055	0.055	0.056	0.056
4	0.270	0.146	0.105	0.086	0.074	0.067	0.062	0.059	0.057	0.056	0.055	0.055	0.055	0.055	0.056
6	0.269	0.144	0.104	0.084	0.073	0.066	0.062	0.059	0.056	0.055	0.054	0.054	0.054	0.055	0.055
8	0.268	0.143	0.102	0.083	0.072	0.065	0.061	0.058	0.056	0.055	0.054	0.054	0.054	0.054	0.055
10	0.268	0.142	0.101	0.082	0.071	0.064	0.060	0.057	0.055	0.054	0.053	0.053	0.053	0.053	0.054
12	0.268	0.142	0.101	0.081	0.070	0.064	0.059	0.056	0.054	0.053	0.053	0.052	0.053	0.053	0.054
14	0.267	0.141	0.100	0.081	0.070	0.063	0.058	0.056	0.054	0.053	0.052	0.052	0.052	0.053	0.053
16	0.267	0.141	0.100	0.080	0.069	0.062	0.058	0.055	0.053	0.052	0.052	0.051	0.052	0.052	0.053
18	0.267	0.141	0.100	0.080	0.069	0.062	0.057	0.054	0.053	0.052	0.051	0.051	0.051	0.052	0.053
20	0.267	0.141	0.100	0.080	0.068	0.061	0.057	0.054	0.052	0.051	0.051	0.050	0.051	0.051	0.052
22	0.267	0.141	0.099	0.079	0.068	0.061	0.056	0.054	0.052	0.051	0.050	0.050	0.050	0.051	0.052
24	0.267	0.141	0.099	0.079	0.068	0.061	0.056	0.053	0.051	0.050	0.050	0.050	0.050	0.050	0.051
26	0.267	0.141	0.099	0.079	0.068	0.061	0.056	0.053	0.051	0.050	0.049	0.049	0.049	0.050	0.051
28	0.267	0.140	0.099	0.079	0.068	0.060	0.056	0.053	0.051	0.050	0.049	0.049	0.049	0.050	0.051
30	0.267	0.140	0.099	0.079	0.067	0.060	0.056	0.053	0.051	0.049	0.049	0.049	0.049	0.049	0.050

FIXED CHARGE RATE TABLES : FCR(PT,T,I)

INTEREST RATE (I) = 0.05   RESALE FACTOR, V = 0.50

OWNERSHIP PERIOD : PT

LOAN TERM T	2	4	6	8	10	12	14	16	18	20	22	24	26	28	30
2	0.282	0.157	0.116	0.096	0.085	0.078	0.074	0.071	0.069	0.068	0.068	0.068	0.069	0.070	0.071
4	0.277	0.153	0.114	0.094	0.084	0.077	0.073	0.070	0.068	0.068	0.067	0.068	0.068	0.069	0.070
6	0.276	0.151	0.111	0.093	0.082	0.076	0.072	0.069	0.067	0.067	0.066	0.067	0.067	0.068	0.070
8	0.275	0.150	0.110	0.091	0.081	0.074	0.071	0.068	0.067	0.066	0.066	0.066	0.067	0.068	0.069
10	0.275	0.149	0.109	0.090	0.079	0.073	0.069	0.067	0.066	0.065	0.065	0.065	0.066	0.067	0.068
12	0.274	0.148	0.108	0.089	0.078	0.072	0.068	0.066	0.065	0.064	0.064	0.064	0.065	0.066	0.068
14	0.274	0.148	0.107	0.088	0.077	0.071	0.068	0.065	0.064	0.063	0.063	0.064	0.065	0.066	0.067
16	0.274	0.148	0.107	0.088	0.077	0.070	0.067	0.064	0.063	0.063	0.063	0.063	0.064	0.065	0.066
18	0.274	0.147	0.107	0.087	0.076	0.070	0.066	0.064	0.063	0.062	0.062	0.063	0.063	0.064	0.066
20	0.274	0.147	0.106	0.087	0.076	0.069	0.065	0.063	0.062	0.061	0.061	0.062	0.063	0.064	0.065
22	0.274	0.147	0.106	0.087	0.076	0.069	0.065	0.063	0.061	0.061	0.061	0.061	0.062	0.063	0.065
24	0.274	0.147	0.106	0.086	0.075	0.069	0.065	0.062	0.061	0.060	0.060	0.061	0.062	0.063	0.064
26	0.273	0.147	0.106	0.086	0.075	0.069	0.064	0.062	0.060	0.060	0.060	0.060	0.061	0.062	0.064
28	0.273	0.147	0.106	0.086	0.075	0.068	0.064	0.062	0.060	0.059	0.059	0.060	0.061	0.062	0.063
30	0.273	0.147	0.106	0.086	0.075	0.068	0.064	0.061	0.060	0.059	0.059	0.059	0.060	0.061	0.063

FIXED CHARGE RATE TABLES : FCR(PT,T,I)

INTEREST RATE (I) = 0.06   RESALE FACTOR, V = 0.50

OWNERSHIP PERIOD : PT

LOAN TERM T	2	4	6	8	10	12	14	16	18	20	22	24	26	28	30
2	0.290	0.165	0.125	0.105	0.095	0.088	0.084	0.082	0.081	0.080	0.080	0.081	0.082	0.083	0.085
4	0.285	0.161	0.122	0.103	0.093	0.087	0.083	0.081	0.079	0.079	0.079	0.080	0.081	0.082	0.084
6	0.283	0.158	0.119	0.101	0.091	0.085	0.081	0.079	0.078	0.078	0.078	0.079	0.080	0.081	0.083
8	0.282	0.156	0.117	0.099	0.089	0.084	0.080	0.078	0.077	0.077	0.077	0.078	0.079	0.080	0.082
10	0.281	0.156	0.116	0.097	0.088	0.082	0.079	0.077	0.076	0.076	0.076	0.077	0.078	0.079	0.081
12	0.281	0.155	0.115	0.096	0.086	0.081	0.078	0.076	0.075	0.075	0.075	0.076	0.077	0.079	0.080
14	0.281	0.155	0.114	0.095	0.085	0.080	0.076	0.075	0.074	0.074	0.074	0.075	0.076	0.078	0.079
16	0.281	0.154	0.114	0.095	0.085	0.079	0.075	0.074	0.073	0.073	0.073	0.074	0.075	0.077	0.079
18	0.280	0.154	0.113	0.094	0.084	0.078	0.075	0.073	0.072	0.072	0.073	0.074	0.075	0.076	0.078
20	0.280	0.154	0.113	0.094	0.084	0.077	0.074	0.072	0.071	0.071	0.072	0.073	0.074	0.076	0.077
22	0.280	0.154	0.113	0.094	0.083	0.077	0.073	0.071	0.071	0.071	0.071	0.072	0.073	0.075	0.077
24	0.280	0.153	0.113	0.093	0.083	0.077	0.073	0.071	0.070	0.070	0.070	0.071	0.073	0.074	0.076
26	0.280	0.153	0.113	0.093	0.083	0.076	0.073	0.071	0.070	0.069	0.070	0.071	0.072	0.074	0.075
28	0.280	0.153	0.112	0.093	0.082	0.076	0.072	0.070	0.069	0.069	0.069	0.070	0.072	0.073	0.075
30	0.280	0.153	0.112	0.093	0.082	0.076	0.072	0.070	0.069	0.069	0.069	0.070	0.071	0.073	0.074

FIXED CHARGE RATE TABLES : FCR(PT,T,I)

INTEREST RATE (I) = 0.07   RESALE FACTOR, V = 0.50

OWNERSHIP PERIOD : PT

LOAN TERM T	2	4	6	8	10	12	14	16	18	20	22	24	26	28	30
2	0.299	0.174	0.133	0.115	0.104	0.098	0.095	0.093	0.092	0.092	0.092	0.093	0.094	0.096	0.098
4	0.292	0.169	0.130	0.112	0.102	0.096	0.093	0.091	0.090	0.090	0.091	0.092	0.093	0.095	0.096
6	0.290	0.165	0.127	0.109	0.100	0.094	0.091	0.089	0.089	0.089	0.089	0.090	0.092	0.093	0.095
8	0.289	0.163	0.124	0.107	0.098	0.093	0.090	0.088	0.087	0.088	0.088	0.089	0.091	0.092	0.094
10	0.288	0.162	0.123	0.105	0.096	0.091	0.088	0.087	0.086	0.086	0.087	0.088	0.089	0.091	0.093
12	0.288	0.162	0.122	0.104	0.094	0.089	0.087	0.085	0.085	0.085	0.086	0.087	0.088	0.090	0.092
14	0.287	0.161	0.121	0.103	0.093	0.088	0.085	0.084	0.084	0.084	0.085	0.086	0.087	0.089	0.091
16	0.287	0.161	0.121	0.102	0.092	0.087	0.084	0.083	0.082	0.083	0.084	0.085	0.086	0.088	0.090
18	0.287	0.160	0.120	0.101	0.092	0.086	0.083	0.082	0.081	0.082	0.083	0.084	0.085	0.087	0.089
20	0.287	0.160	0.120	0.101	0.091	0.085	0.082	0.081	0.080	0.081	0.082	0.083	0.085	0.087	0.089
22	0.287	0.160	0.120	0.101	0.091	0.085	0.082	0.080	0.080	0.080	0.081	0.082	0.084	0.086	0.088
24	0.287	0.160	0.119	0.100	0.090	0.085	0.081	0.080	0.079	0.079	0.080	0.082	0.083	0.085	0.087
26	0.287	0.160	0.119	0.100	0.090	0.084	0.081	0.079	0.079	0.079	0.080	0.081	0.082	0.084	0.086
28	0.287	0.160	0.119	0.100	0.090	0.084	0.081	0.079	0.078	0.078	0.079	0.080	0.082	0.084	0.086
30	0.287	0.160	0.119	0.100	0.090	0.084	0.080	0.079	0.078	0.078	0.079	0.080	0.081	0.083	0.085

FIXED CHARGE RATE TABLES : FCR(PT,T,I)

INTEREST RATE (I) = 0.08   RESALE FACTOR, V = 0.50

OWNERSHIP PERIOD : PT

LOAN TERM T	2	4	6	8	10	12	14	16	18	20	22	24	26	28	30
2	0.307	0.182	0.142	0.124	0.114	0.108	0.105	0.103	0.103	0.103	0.104	0.105	0.106	0.108	0.110
4	0.299	0.177	0.138	0.121	0.111	0.106	0.103	0.101	0.101	0.101	0.102	0.103	0.105	0.106	0.108
6	0.297	0.172	0.135	0.118	0.109	0.104	0.101	0.099	0.099	0.100	0.100	0.102	0.103	0.105	0.107
8	0.296	0.170	0.132	0.115	0.106	0.101	0.099	0.098	0.097	0.098	0.099	0.100	0.102	0.104	0.105
10	0.295	0.169	0.130	0.113	0.104	0.099	0.097	0.096	0.096	0.096	0.097	0.099	0.100	0.102	0.104
12	0.294	0.168	0.129	0.111	0.102	0.098	0.095	0.094	0.094	0.095	0.096	0.097	0.099	0.101	0.103
14	0.294	0.168	0.128	0.110	0.101	0.096	0.094	0.093	0.093	0.094	0.095	0.096	0.098	0.100	0.102
16	0.294	0.167	0.127	0.109	0.100	0.095	0.092	0.092	0.092	0.092	0.094	0.095	0.097	0.099	0.101
18	0.294	0.167	0.127	0.109	0.099	0.094	0.091	0.090	0.091	0.091	0.092	0.094	0.096	0.098	0.100
20	0.294	0.167	0.127	0.108	0.099	0.093	0.091	0.089	0.089	0.090	0.091	0.093	0.095	0.097	0.099
22	0.293	0.166	0.126	0.108	0.098	0.093	0.090	0.089	0.089	0.089	0.090	0.092	0.094	0.096	0.098
24	0.293	0.166	0.126	0.108	0.098	0.092	0.089	0.088	0.088	0.089	0.090	0.091	0.093	0.095	0.097
26	0.293	0.166	0.126	0.107	0.097	0.092	0.089	0.088	0.087	0.088	0.089	0.090	0.092	0.094	0.096
28	0.293	0.166	0.126	0.107	0.097	0.092	0.089	0.087	0.087	0.087	0.088	0.090	0.092	0.094	0.096
30	0.293	0.166	0.126	0.107	0.097	0.091	0.088	0.087	0.087	0.087	0.088	0.089	0.091	0.093	0.095

FIXED CHARGE RATE TABLES : FCR(PT,T,I)

INTEREST RATE (I) = 0.09   RESALE FACTOR, V = 0.50

OWNERSHIP PERIOD : PT

LOAN TERM T	2	4	6	8	10	12	14	16	18	20	22	24	26	28	30
2	0.315	0.190	0.151	0.133	0.123	0.118	0.115	0.114	0.114	0.114	0.115	0.116	0.118	0.120	0.122
4	0.307	0.184	0.147	0.129	0.120	0.115	0.113	0.112	0.111	0.112	0.113	0.114	0.116	0.118	0.120
6	0.304	0.180	0.142	0.126	0.117	0.113	0.110	0.109	0.109	0.110	0.111	0.112	0.114	0.116	0.118
8	0.302	0.177	0.139	0.123	0.115	0.110	0.108	0.107	0.107	0.108	0.109	0.111	0.112	0.114	0.116
10	0.302	0.176	0.137	0.120	0.112	0.108	0.106	0.105	0.105	0.106	0.107	0.109	0.111	0.113	0.115
12	0.301	0.175	0.136	0.118	0.110	0.106	0.104	0.104	0.104	0.105	0.106	0.107	0.109	0.111	0.113
14	0.301	0.174	0.135	0.117	0.108	0.104	0.102	0.102	0.102	0.103	0.104	0.106	0.108	0.110	0.112
16	0.301	0.174	0.134	0.116	0.107	0.103	0.101	0.100	0.101	0.102	0.103	0.105	0.107	0.109	0.111
18	0.300	0.173	0.134	0.116	0.107	0.102	0.100	0.099	0.099	0.100	0.102	0.104	0.105	0.107	0.110
20	0.300	0.173	0.133	0.115	0.106	0.101	0.099	0.098	0.098	0.099	0.101	0.102	0.104	0.106	0.108
22	0.300	0.173	0.133	0.115	0.105	0.100	0.098	0.097	0.097	0.098	0.100	0.101	0.103	0.105	0.108
24	0.300	0.173	0.133	0.115	0.105	0.100	0.097	0.096	0.097	0.097	0.099	0.100	0.102	0.104	0.107
26	0.300	0.173	0.133	0.114	0.105	0.100	0.097	0.096	0.096	0.097	0.098	0.100	0.102	0.104	0.106
28	0.300	0.173	0.132	0.114	0.104	0.099	0.097	0.096	0.096	0.096	0.097	0.099	0.101	0.103	0.105
30	0.300	0.172	0.132	0.114	0.104	0.099	0.096	0.095	0.095	0.096	0.097	0.098	0.100	0.102	0.104

FIXED CHARGE RATE TABLES : FCR(PT,T,I)

INTEREST RATE (I) = 0.10   RESALE FACTOR, V = 0.50

OWNERSHIP PERIOD : PT

LOAN TERM T	2	4	6	8	10	12	14	16	18	20	22	24	26	28	30
2	0.324	0.199	0.160	0.142	0.133	0.128	0.125	0.124	0.124	0.125	0.126	0.128	0.129	0.131	0.133
4	0.314	0.192	0.155	0.138	0.129	0.125	0.122	0.122	0.122	0.122	0.124	0.125	0.127	0.129	0.131
6	0.311	0.187	0.150	0.134	0.126	0.122	0.120	0.119	0.119	0.120	0.121	0.123	0.125	0.127	0.128
8	0.309	0.184	0.146	0.131	0.123	0.119	0.117	0.117	0.117	0.118	0.119	0.121	0.123	0.125	0.126
10	0.308	0.182	0.144	0.128	0.120	0.116	0.115	0.114	0.115	0.116	0.117	0.119	0.121	0.123	0.125
12	0.308	0.181	0.143	0.126	0.118	0.114	0.113	0.112	0.113	0.114	0.115	0.117	0.119	0.121	0.123
14	0.307	0.181	0.142	0.124	0.116	0.112	0.111	0.111	0.111	0.112	0.114	0.116	0.117	0.119	0.121
16	0.307	0.180	0.141	0.123	0.115	0.111	0.109	0.109	0.110	0.111	0.112	0.114	0.116	0.118	0.120
18	0.307	0.180	0.140	0.123	0.114	0.110	0.108	0.107	0.108	0.109	0.111	0.113	0.115	0.117	0.119
20	0.307	0.180	0.140	0.122	0.113	0.109	0.107	0.106	0.107	0.108	0.110	0.111	0.113	0.115	0.118
22	0.307	0.179	0.140	0.122	0.113	0.108	0.106	0.105	0.106	0.107	0.109	0.110	0.112	0.114	0.117
24	0.307	0.179	0.139	0.122	0.112	0.108	0.105	0.105	0.105	0.106	0.108	0.109	0.111	0.113	0.116
26	0.306	0.179	0.139	0.121	0.112	0.107	0.105	0.104	0.104	0.105	0.107	0.109	0.111	0.113	0.115
28	0.306	0.179	0.139	0.121	0.112	0.107	0.105	0.104	0.104	0.105	0.106	0.108	0.110	0.112	0.114
30	0.306	0.179	0.139	0.121	0.112	0.107	0.104	0.103	0.104	0.104	0.106	0.107	0.109	0.111	0.113

FIXED CHARGE RATE TABLES : FCR(PT,T,I)

INTEREST RATE (I) = 0.11 RESALE FACTOR, V = 0.50

LOAN TERM T \ OWNERSHIP PERIOD : PT

T	2	4	6	8	10	12	14	16	18	20	22	24	26	28	30
2	0.332	0.207	0.169	0.151	0.142	0.138	0.136	0.135	0.135	0.136	0.137	0.138	0.140	0.142	0.144
4	0.321	0.200	0.163	0.147	0.138	0.134	0.132	0.132	0.132	0.133	0.134	0.136	0.137	0.139	0.141
6	0.318	0.194	0.158	0.142	0.135	0.131	0.129	0.129	0.129	0.130	0.131	0.133	0.135	0.137	0.139
8	0.316	0.191	0.154	0.138	0.131	0.128	0.126	0.126	0.126	0.128	0.129	0.131	0.132	0.134	0.136
10	0.315	0.189	0.151	0.135	0.128	0.125	0.124	0.123	0.124	0.125	0.127	0.128	0.130	0.132	0.134
12	0.315	0.188	0.150	0.133	0.125	0.122	0.121	0.121	0.122	0.123	0.125	0.126	0.128	0.130	0.132
14	0.314	0.187	0.148	0.132	0.124	0.120	0.119	0.119	0.120	0.121	0.123	0.125	0.127	0.128	0.130
16	0.314	0.187	0.148	0.131	0.122	0.118	0.117	0.117	0.118	0.120	0.121	0.123	0.125	0.127	0.129
18	0.314	0.186	0.147	0.130	0.121	0.117	0.116	0.116	0.117	0.118	0.120	0.122	0.123	0.125	0.127
20	0.313	0.186	0.147	0.129	0.121	0.116	0.115	0.114	0.115	0.117	0.118	0.120	0.122	0.124	0.126
22	0.313	0.186	0.146	0.129	0.120	0.116	0.114	0.114	0.114	0.115	0.117	0.119	0.121	0.123	0.125
24	0.313	0.186	0.146	0.129	0.120	0.115	0.113	0.113	0.113	0.115	0.116	0.118	0.120	0.122	0.124
26	0.313	0.186	0.146	0.128	0.119	0.115	0.113	0.112	0.113	0.114	0.115	0.117	0.119	0.121	0.123
28	0.313	0.185	0.146	0.128	0.119	0.115	0.112	0.112	0.112	0.113	0.115	0.116	0.118	0.120	0.122
30	0.313	0.185	0.146	0.128	0.119	0.114	0.112	0.112	0.112	0.113	0.114	0.116	0.118	0.120	0.122

FIXED CHARGE RATE TABLES : FCR(PT,T,I)

INTEREST RATE (I) = 0.12 RESALE FACTOR, V = 0.50

LOAN TERM T \ OWNERSHIP PERIOD : PT

T	2	4	6	8	10	12	14	16	18	20	22	24	26	28	30
2	0.340	0.216	0.177	0.160	0.152	0.148	0.146	0.145	0.145	0.146	0.148	0.149	0.151	0.153	0.154
4	0.329	0.207	0.171	0.155	0.147	0.143	0.142	0.141	0.142	0.143	0.144	0.146	0.148	0.149	0.151
6	0.325	0.201	0.165	0.151	0.143	0.140	0.138	0.138	0.139	0.140	0.141	0.143	0.145	0.146	0.148
8	0.323	0.198	0.161	0.146	0.139	0.136	0.135	0.135	0.136	0.137	0.138	0.140	0.142	0.144	0.146
10	0.322	0.196	0.158	0.143	0.136	0.133	0.132	0.132	0.133	0.134	0.136	0.138	0.139	0.141	0.143
12	0.321	0.195	0.156	0.140	0.133	0.130	0.129	0.130	0.131	0.132	0.134	0.135	0.137	0.139	0.141
14	0.321	0.194	0.155	0.139	0.131	0.128	0.127	0.128	0.129	0.130	0.132	0.133	0.135	0.137	0.139
16	0.320	0.193	0.154	0.138	0.130	0.126	0.125	0.126	0.127	0.128	0.130	0.132	0.134	0.135	0.137
18	0.320	0.193	0.154	0.137	0.129	0.125	0.124	0.124	0.125	0.126	0.128	0.130	0.132	0.134	0.136
20	0.320	0.193	0.153	0.136	0.128	0.124	0.123	0.123	0.123	0.125	0.127	0.129	0.131	0.133	0.134
22	0.320	0.192	0.153	0.136	0.127	0.123	0.122	0.122	0.122	0.124	0.126	0.127	0.129	0.131	0.133
24	0.320	0.192	0.153	0.136	0.127	0.123	0.121	0.121	0.122	0.123	0.124	0.126	0.128	0.130	0.132
26	0.320	0.192	0.153	0.135	0.127	0.122	0.121	0.120	0.121	0.122	0.124	0.125	0.127	0.129	0.131
28	0.320	0.192	0.152	0.135	0.126	0.122	0.120	0.120	0.120	0.122	0.123	0.125	0.127	0.129	0.131
30	0.320	0.192	0.152	0.135	0.126	0.122	0.120	0.120	0.120	0.121	0.123	0.124	0.126	0.128	0.130

FIXED CHARGE RATE TABLES : FCR(PT,T,I)

INTEREST RATE (I) = 0.13 RESALE FACTOR, V = 0.50

LOAN TERM T \ OWNERSHIP PERIOD : PT

T	2	4	6	8	10	12	14	16	18	20	22	24	26	28	30
2	0.348	0.224	0.186	0.170	0.161	0.157	0.156	0.155	0.156	0.157	0.158	0.160	0.161	0.163	0.164
4	0.336	0.215	0.179	0.164	0.156	0.153	0.151	0.151	0.152	0.153	0.154	0.156	0.158	0.159	0.161
6	0.332	0.208	0.173	0.159	0.152	0.149	0.147	0.147	0.148	0.149	0.151	0.152	0.154	0.156	0.157
8	0.330	0.205	0.168	0.154	0.148	0.145	0.144	0.144	0.145	0.146	0.148	0.149	0.151	0.153	0.154
10	0.329	0.203	0.165	0.150	0.144	0.141	0.141	0.141	0.142	0.143	0.145	0.147	0.148	0.150	0.152
12	0.328	0.201	0.163	0.148	0.141	0.138	0.138	0.138	0.139	0.141	0.142	0.144	0.146	0.148	0.149
14	0.327	0.200	0.162	0.146	0.139	0.136	0.135	0.136	0.137	0.138	0.140	0.142	0.144	0.146	0.147
16	0.327	0.200	0.161	0.145	0.137	0.134	0.133	0.134	0.135	0.136	0.138	0.140	0.142	0.144	0.145
18	0.327	0.199	0.161	0.144	0.136	0.133	0.132	0.132	0.133	0.135	0.136	0.138	0.140	0.142	0.144
20	0.327	0.199	0.160	0.143	0.135	0.132	0.130	0.131	0.132	0.133	0.135	0.137	0.139	0.141	0.142
22	0.327	0.199	0.160	0.143	0.135	0.131	0.130	0.130	0.130	0.132	0.134	0.136	0.137	0.139	0.141
24	0.326	0.199	0.160	0.143	0.134	0.130	0.129	0.129	0.130	0.131	0.133	0.135	0.136	0.138	0.140
26	0.326	0.199	0.159	0.142	0.134	0.130	0.128	0.128	0.129	0.130	0.132	0.134	0.135	0.137	0.139
28	0.326	0.198	0.159	0.142	0.134	0.130	0.128	0.128	0.130	0.131	0.133	0.135	0.137	0.138	
30	0.326	0.198	0.159	0.142	0.134	0.129	0.128	0.128	0.128	0.129	0.131	0.132	0.134	0.136	0.138

FIXED CHARGE RATE TABLES : FCR(PT,T,I)

INTEREST RATE (I) = 0.14   RESALE FACTOR, V = 0.50

OWNERSHIP PERIOD : PT

LOAN TERM T	2	4	6	8	10	12	14	16	18	20	22	24	26	28	30
2	0.357	0.233	0.195	0.179	0.171	0.167	0.166	0.165	0.166	0.167	0.168	0.170	0.171	0.173	0.174
4	0.343	0.223	0.188	0.172	0.165	0.162	0.161	0.161	0.162	0.163	0.164	0.166	0.167	0.169	0.170
6	0.339	0.215	0.181	0.167	0.160	0.157	0.156	0.157	0.157	0.159	0.160	0.162	0.163	0.165	0.167
8	0.337	0.211	0.175	0.162	0.156	0.153	0.152	0.153	0.154	0.155	0.157	0.158	0.160	0.162	0.163
10	0.335	0.209	0.172	0.157	0.152	0.149	0.149	0.149	0.151	0.152	0.154	0.155	0.157	0.159	0.160
12	0.335	0.208	0.170	0.155	0.148	0.146	0.146	0.147	0.148	0.149	0.151	0.153	0.154	0.156	0.158
14	0.334	0.207	0.169	0.153	0.146	0.143	0.143	0.144	0.145	0.147	0.148	0.150	0.152	0.154	0.155
16	0.334	0.206	0.168	0.152	0.145	0.142	0.141	0.142	0.143	0.145	0.146	0.148	0.150	0.152	0.153
18	0.333	0.206	0.167	0.151	0.143	0.140	0.139	0.140	0.141	0.143	0.145	0.146	0.148	0.150	0.151
20	0.333	0.206	0.167	0.150	0.143	0.139	0.138	0.138	0.140	0.141	0.143	0.145	0.147	0.148	0.150
22	0.333	0.205	0.167	0.150	0.142	0.139	0.137	0.137	0.138	0.140	0.142	0.144	0.145	0.147	0.149
24	0.333	0.205	0.166	0.150	0.142	0.138	0.137	0.137	0.138	0.139	0.141	0.142	0.144	0.146	0.148
26	0.333	0.205	0.166	0.149	0.141	0.138	0.136	0.136	0.137	0.138	0.140	0.142	0.143	0.145	0.147
28	0.333	0.205	0.166	0.149	0.141	0.137	0.136	0.136	0.136	0.138	0.139	0.141	0.143	0.144	0.146
30	0.333	0.205	0.166	0.149	0.141	0.137	0.136	0.135	0.136	0.137	0.139	0.140	0.142	0.144	0.145

FIXED CHARGE RATE TABLES : FCR(PT,T,I)

INTEREST RATE (I) = 0.15   RESALE FACTOR, V = 0.50

OWNERSHIP PERIOD : PT

LOAN TERM T	2	4	6	8	10	12	14	16	18	20	22	24	26	28	30
2	0.365	0.241	0.204	0.188	0.180	0.177	0.176	0.176	0.176	0.177	0.179	0.180	0.182	0.183	0.184
4	0.350	0.230	0.196	0.181	0.174	0.171	0.170	0.170	0.171	0.172	0.174	0.175	0.177	0.178	0.180
6	0.346	0.222	0.188	0.175	0.169	0.166	0.165	0.166	0.167	0.168	0.169	0.171	0.172	0.174	0.175
8	0.343	0.218	0.183	0.169	0.164	0.161	0.161	0.162	0.163	0.164	0.166	0.167	0.169	0.170	0.172
10	0.342	0.216	0.179	0.165	0.159	0.157	0.157	0.158	0.159	0.161	0.162	0.164	0.165	0.167	0.168
12	0.341	0.215	0.177	0.162	0.156	0.154	0.154	0.155	0.156	0.158	0.159	0.161	0.162	0.164	0.165
14	0.341	0.214	0.176	0.160	0.154	0.151	0.151	0.152	0.153	0.155	0.157	0.158	0.160	0.161	0.163
16	0.340	0.213	0.175	0.159	0.152	0.149	0.149	0.150	0.151	0.153	0.154	0.156	0.158	0.159	0.161
18	0.340	0.212	0.174	0.158	0.151	0.148	0.147	0.148	0.149	0.151	0.153	0.154	0.156	0.157	0.159
20	0.340	0.212	0.174	0.157	0.150	0.147	0.146	0.146	0.148	0.149	0.151	0.153	0.154	0.156	0.157
22	0.340	0.212	0.173	0.157	0.149	0.146	0.145	0.145	0.146	0.148	0.150	0.151	0.153	0.155	0.156
24	0.340	0.212	0.173	0.157	0.149	0.146	0.144	0.145	0.146	0.147	0.149	0.150	0.152	0.154	0.155
26	0.340	0.212	0.173	0.156	0.149	0.145	0.144	0.144	0.145	0.146	0.148	0.149	0.151	0.153	0.154
28	0.340	0.211	0.173	0.156	0.148	0.145	0.144	0.144	0.144	0.146	0.147	0.149	0.150	0.152	0.153
30	0.340	0.211	0.173	0.156	0.148	0.145	0.143	0.143	0.144	0.145	0.147	0.148	0.150	0.151	0.153

Appendix 14: Fuel Cost Leveling Factors (FF) at Different Escalation Rates

Throughout this appendix, the fuel cost leveling factors are determined by the parameters E (escalation rate—constant for each table), PT (ownership period, in years—the far left column in each table), and R (interest rate—the top row in each table). The values are general in that no limiting assumptions were used to generate them.

FUEL COST LEVELIZATION FACTORS : FF(PT,E,R)

ESCALATION RATE : E = .04

INTEREST RATE : R

(PT)	0.04	0.05	0.06	0.07	0.08	0.09	0.10	0.11	0.12	0.13	0.14	0.15
2	1.060	1.060	1.060	1.060	1.060	1.060	1.060	1.060	1.060	1.060	1.059	1.059
4	1.102	1.101	1.101	1.100	1.100	1.099	1.099	1.098	1.098	1.098	1.097	1.097
6	1.145	1.143	1.142	1.141	1.140	1.139	1.137	1.136	1.135	1.134	1.133	1.132
8	1.188	1.186	1.184	1.181	1.179	1.177	1.175	1.173	1.170	1.168	1.166	1.164
10	1.233	1.229	1.225	1.222	1.218	1.215	1.211	1.208	1.204	1.201	1.198	1.195
12	1.279	1.273	1.267	1.262	1.257	1.251	1.246	1.241	1.236	1.231	1.227	1.222
14	1.325	1.317	1.310	1.302	1.294	1.287	1.280	1.273	1.266	1.260	1.254	1.247
16	1.373	1.362	1.352	1.342	1.332	1.322	1.312	1.303	1.295	1.286	1.278	1.270
18	1.422	1.408	1.394	1.381	1.368	1.355	1.343	1.332	1.321	1.310	1.300	1.290
20	1.472	1.454	1.436	1.419	1.403	1.388	1.373	1.359	1.345	1.332	1.320	1.308
22	1.522	1.500	1.478	1.458	1.438	1.419	1.401	1.384	1.367	1.352	1.338	1.324
24	1.574	1.547	1.520	1.495	1.471	1.449	1.427	1.407	1.388	1.370	1.354	1.338
26	1.627	1.594	1.562	1.532	1.504	1.477	1.452	1.429	1.407	1.386	1.368	1.350
28	1.680	1.641	1.603	1.568	1.535	1.504	1.475	1.448	1.424	1.401	1.380	1.360
30	1.735	1.688	1.644	1.603	1.565	1.530	1.497	1.467	1.439	1.414	1.391	1.369

FUEL COST LEVELIZATION FACTORS : FF(PT,E,R)

ESCALATION RATE : E = .05

INTEREST RATE : R

(PT)	0.04	0.05	0.06	0.07	0.08	0.09	0.10	0.11	0.12	0.13	0.14	0.15
2	1.076	1.076	1.075	1.075	1.075	1.075	1.075	1.075	1.075	1.075	1.075	1.074
4	1.129	1.128	1.127	1.127	1.126	1.125	1.125	1.124	1.124	1.123	1.122	1.122
6	1.184	1.182	1.181	1.179	1.177	1.176	1.174	1.173	1.171	1.170	1.168	1.167
8	1.241	1.238	1.235	1.232	1.229	1.226	1.223	1.220	1.218	1.215	1.212	1.210
10	1.300	1.295	1.290	1.285	1.281	1.276	1.271	1.267	1.262	1.258	1.254	1.250
12	1.361	1.354	1.347	1.339	1.332	1.325	1.318	1.312	1.305	1.299	1.293	1.287
14	1.425	1.414	1.404	1.394	1.384	1.374	1.364	1.355	1.346	1.338	1.329	1.321
16	1.491	1.476	1.462	1.448	1.435	1.422	1.409	1.397	1.385	1.374	1.363	1.352
18	1.559	1.540	1.521	1.503	1.485	1.468	1.452	1.437	1.422	1.407	1.393	1.380
20	1.630	1.605	1.581	1.558	1.536	1.514	1.494	1.474	1.456	1.438	1.421	1.406
22	1.703	1.671	1.641	1.612	1.585	1.559	1.534	1.510	1.488	1.467	1.447	1.428
24	1.778	1.739	1.702	1.667	1.634	1.602	1.572	1.544	1.518	1.493	1.470	1.448
26	1.856	1.809	1.764	1.721	1.681	1.644	1.608	1.576	1.545	1.517	1.490	1.466
28	1.936	1.879	1.826	1.775	1.728	1.684	1.643	1.605	1.570	1.538	1.508	1.481
30	2.019	1.952	1.888	1.829	1.774	1.723	1.676	1.633	1.594	1.558	1.525	1.495

FUEL COST LEVELIZATION FACTORS : FF(PT,E,R)

ESCALATION RATE : E = .06

INTEREST RATE : R

(PT)	0.04	0.05	0.06	0.07	0.08	0.09	0.10	0.11	0.12	0.13	0.14	0.15
2	1.091	1.091	1.091	1.091	1.091	1.090	1.090	1.090	1.090	1.090	1.090	1.090
4	1.156	1.155	1.154	1.154	1.153	1.152	1.151	1.151	1.150	1.149	1.148	1.148
6	1.224	1.222	1.220	1.218	1.216	1.214	1.213	1.211	1.209	1.207	1.205	1.204
8	1.296	1.292	1.288	1.285	1.281	1.277	1.274	1.270	1.267	1.264	1.260	1.257
10	1.371	1.365	1.359	1.353	1.347	1.341	1.335	1.329	1.324	1.318	1.313	1.308
12	1.450	1.441	1.431	1.422	1.413	1.404	1.396	1.387	1.379	1.371	1.363	1.356
14	1.533	1.520	1.506	1.493	1.480	1.468	1.456	1.444	1.432	1.421	1.410	1.400
16	1.621	1.602	1.583	1.565	1.548	1.531	1.515	1.499	1.483	1.469	1.455	1.441
18	1.712	1.687	1.662	1.639	1.616	1.594	1.572	1.552	1.532	1.514	1.496	1.479
20	1.808	1.775	1.744	1.713	1.684	1.656	1.629	1.603	1.579	1.556	1.534	1.513
22	1.909	1.867	1.827	1.789	1.752	1.717	1.684	1.652	1.623	1.595	1.569	1.544
24	2.015	1.962	1.912	1.865	1.820	1.777	1.737	1.699	1.664	1.631	1.600	1.572
26	2.125	2.061	2.000	1.942	1.887	1.836	1.789	1.744	1.703	1.665	1.629	1.597
28	2.241	2.163	2.089	2.019	1.954	1.894	1.838	1.787	1.739	1.695	1.655	1.619
30	2.363	2.268	2.179	2.097	2.021	1.950	1.886	1.827	1.773	1.724	1.679	1.638

FUEL COST LEVELIZATION FACTORS : FF(PT,E,R)

ESCALATION RATE : E = .07

INTEREST RATE : R

(PT)	0.04	0.05	0.06	0.07	0.08	0.09	0.10	0.11	0.12	0.13	0.14	0.15
2	1.107	1.107	1.106	1.106	1.106	1.106	1.106	1.106	1.105	1.105	1.105	1.105
4	1.184	1.183	1.182	1.181	1.180	1.179	1.178	1.177	1.176	1.176	1.175	1.174
6	1.266	1.263	1.261	1.259	1.256	1.254	1.252	1.250	1.248	1.245	1.243	1.241
8	1.353	1.349	1.344	1.340	1.335	1.331	1.327	1.323	1.318	1.314	1.310	1.306
10	1.446	1.439	1.431	1.424	1.416	1.409	1.402	1.395	1.389	1.382	1.375	1.369
12	1.546	1.534	1.522	1.511	1.500	1.489	1.478	1.468	1.458	1.448	1.438	1.429
14	1.651	1.634	1.617	1.601	1.585	1.569	1.554	1.539	1.525	1.511	1.498	1.485
16	1.764	1.740	1.716	1.694	1.672	1.650	1.630	1.610	1.591	1.572	1.555	1.538
18	1.883	1.851	1.820	1.789	1.760	1.732	1.705	1.679	1.654	1.631	1.608	1.587
20	2.010	1.968	1.927	1.888	1.850	1.814	1.780	1.747	1.716	1.686	1.658	1.632
22	2.146	2.091	2.039	1.989	1.941	1.896	1.853	1.813	1.774	1.738	1.705	1.673
24	2.290	2.221	2.155	2.093	2.033	1.978	1.925	1.876	1.830	1.788	1.748	1.710
26	2.443	2.357	2.275	2.199	2.126	2.059	1.996	1.938	1.884	1.834	1.787	1.745
28	2.605	2.500	2.400	2.307	2.220	2.140	2.065	1.997	1.934	1.876	1.824	1.775
30	2.778	2.650	2.529	2.418	2.314	2.220	2.133	2.054	1.982	1.916	1.857	1.803

FUEL COST LEVELIZATION FACTORS : FF(PT,E,R)

ESCALATION RATE : E = .08

INTEREST RATE : R

(PT)	0.04	0.05	0.06	0.07	0.08	0.09	0.10	0.11	0.12	0.13	0.14	0.15
2	1.122	1.122	1.122	1.122	1.122	1.121	1.121	1.121	1.121	1.121	1.120	1.120
4	1.212	1.211	1.210	1.209	1.208	1.207	1.206	1.205	1.203	1.202	1.201	1.200
6	1.309	1.306	1.303	1.301	1.298	1.295	1.293	1.290	1.287	1.285	1.282	1.280
8	1.413	1.408	1.403	1.397	1.392	1.387	1.382	1.377	1.372	1.367	1.363	1.358
10	1.526	1.517	1.508	1.499	1.490	1.482	1.473	1.465	1.457	1.449	1.441	1.434
12	1.648	1.634	1.620	1.606	1.592	1.579	1.566	1.554	1.541	1.530	1.518	1.507
14	1.779	1.758	1.738	1.718	1.698	1.679	1.661	1.643	1.625	1.608	1.592	1.576
16	1.921	1.892	1.863	1.835	1.808	1.781	1.756	1.732	1.708	1.685	1.663	1.643
18	2.074	2.034	1.995	1.957	1.921	1.886	1.852	1.820	1.789	1.760	1.732	1.705
20	2.239	2.186	2.134	2.085	2.037	1.992	1.948	1.907	1.868	1.831	1.796	1.763
22	2.418	2.348	2.281	2.217	2.157	2.099	2.045	1.993	1.945	1.900	1.857	1.817
24	2.610	2.521	2.436	2.355	2.279	2.208	2.141	2.078	2.019	1.965	1.914	1.867
26	2.817	2.705	2.599	2.499	2.405	2.318	2.236	2.161	2.091	2.027	1.967	1.913
28	3.041	2.901	2.770	2.648	2.534	2.428	2.331	2.242	2.160	2.085	2.017	1.955
30	3.283	3.111	2.950	2.802	2.665	2.539	2.425	2.321	2.226	2.140	2.063	1.993

FUEL COST LEVELIZATION FACTORS : FF(PT,E,R)

ESCALATION RATE : E = .09

INTEREST RATE : R

(PT)	0.04	0.05	0.06	0.07	0.08	0.09	0.10	0.11	0.12	0.13	0.14	0.15
2	1.138	1.138	1.138	1.137	1.137	1.137	1.137	1.136	1.136	1.136	1.136	1.136
4	1.241	1.240	1.238	1.237	1.236	1.235	1.233	1.232	1.231	1.230	1.229	1.228
6	1.353	1.350	1.347	1.344	1.341	1.338	1.334	1.331	1.328	1.326	1.323	1.320
8	1.476	1.470	1.464	1.457	1.451	1.445	1.439	1.434	1.428	1.422	1.417	1.411
10	1.611	1.600	1.589	1.579	1.568	1.558	1.548	1.539	1.529	1.520	1.510	1.502
12	1.758	1.741	1.724	1.708	1.691	1.676	1.660	1.646	1.631	1.617	1.603	1.590
14	1.919	1.893	1.869	1.844	1.821	1.798	1.776	1.754	1.733	1.713	1.694	1.675
16	2.095	2.059	2.024	1.990	1.957	1.925	1.894	1.864	1.836	1.808	1.782	1.757
18	2.288	2.238	2.190	2.143	2.099	2.056	2.015	1.975	1.937	1.901	1.867	1.835
20	2.499	2.432	2.368	2.306	2.247	2.191	2.137	2.086	2.038	1.992	1.949	1.908
22	2.730	2.642	2.558	2.478	2.402	2.330	2.262	2.198	2.137	2.081	2.028	1.978
24	2.983	2.869	2.761	2.659	2.563	2.473	2.388	2.309	2.235	2.166	2.102	2.043
26	3.260	3.115	2.979	2.850	2.730	2.619	2.515	2.419	2.330	2.248	2.173	2.104
28	3.564	3.381	3.211	3.051	2.904	2.768	2.643	2.528	2.423	2.327	2.240	2.160
30	3.896	3.669	3.458	3.263	3.084	2.920	2.771	2.636	2.513	2.402	2.302	2.212

FUEL COST LEVELIZATION FACTORS : FF(PT,E,R)

ESCALATION RATE : E = .10

INTEREST RATE : R

(PT)	0.04	0.05	0.06	0.07	0.08	0.09	0.10	0.11	0.12	0.13	0.14	0.15
2	1.154	1.154	1.153	1.153	1.153	1.153	1.152	1.152	1.152	1.152	1.151	1.151
4	1.270	1.269	1.267	1.266	1.265	1.263	1.262	1.261	1.259	1.258	1.257	1.255
6	1.399	1.396	1.392	1.388	1.385	1.381	1.378	1.374	1.371	1.367	1.364	1.361
8	1.542	1.535	1.527	1.520	1.513	1.506	1.500	1.493	1.486	1.480	1.473	1.467
10	1.700	1.688	1.675	1.663	1.651	1.639	1.627	1.616	1.605	1.594	1.584	1.573
12	1.876	1.856	1.836	1.817	1.798	1.779	1.761	1.744	1.726	1.710	1.694	1.678
14	2.070	2.040	2.011	1.982	1.954	1.927	1.900	1.875	1.850	1.826	1.803	1.781
16	2.287	2.243	2.201	2.160	2.120	2.082	2.045	2.009	1.975	1.942	1.911	1.881
18	2.526	2.466	2.407	2.351	2.297	2.245	2.195	2.147	2.101	2.058	2.017	1.977
20	2.793	2.711	2.632	2.556	2.484	2.415	2.349	2.287	2.228	2.172	2.120	2.070
22	3.089	2.980	2.875	2.776	2.681	2.592	2.508	2.429	2.355	2.285	2.220	2.159
24	3.418	3.275	3.139	3.011	2.890	2.777	2.671	2.572	2.481	2.395	2.316	2.243
26	3.784	3.599	3.425	3.263	3.110	2.969	2.838	2.717	2.606	2.503	2.409	2.322
28	4.191	3.956	3.736	3.531	3.342	3.168	3.009	2.862	2.729	2.608	2.498	2.397
30	4.644	4.347	4.072	3.818	3.586	3.375	3.182	3.008	2.851	2.710	2.582	2.468

FUEL COST LEVELIZATION FACTORS : FF(PT,E,R)

ESCALATION RATE : E = .11

INTEREST RATE : R

(PT)	0.04	0.05	0.06	0.07	0.08	0.09	0.10	0.11	0.12	0.13	0.14	0.15
2	1.170	1.170	1.169	1.169	1.169	1.168	1.168	1.168	1.168	1.167	1.167	1.167
4	1.300	1.299	1.297	1.295	1.294	1.292	1.291	1.289	1.288	1.286	1.285	1.283
6	1.447	1.442	1.438	1.434	1.430	1.426	1.422	1.418	1.414	1.411	1.407	1.403
8	1.611	1.602	1.594	1.586	1.578	1.570	1.562	1.555	1.547	1.540	1.532	1.525
10	1.795	1.780	1.766	1.752	1.738	1.724	1.711	1.698	1.685	1.673	1.660	1.648
12	2.002	1.979	1.956	1.933	1.911	1.890	1.869	1.848	1.828	1.809	1.790	1.772
14	2.236	2.200	2.165	2.131	2.098	2.066	2.035	2.005	1.976	1.948	1.921	1.894
16	2.498	2.446	2.396	2.347	2.300	2.254	2.210	2.168	2.128	2.089	2.051	2.015
18	2.793	2.720	2.650	2.582	2.517	2.454	2.394	2.337	2.283	2.231	2.181	2.134
20	3.126	3.026	2.930	2.838	2.750	2.666	2.587	2.512	2.440	2.373	2.309	2.249
22	3.502	3.367	3.239	3.116	3.000	2.891	2.788	2.691	2.600	2.515	2.435	2.361
24	3.926	3.747	3.578	3.419	3.269	3.128	2.997	2.875	2.761	2.656	2.559	2.469
26	4.404	4.171	3.952	3.747	3.556	3.378	3.214	3.063	2.924	2.796	2.679	2.572
28	4.945	4.644	4.363	4.103	3.862	3.642	3.440	3.255	3.087	2.935	2.796	2.671
30	5.555	5.171	4.815	4.489	4.190	3.919	3.673	3.451	3.251	3.071	2.910	2.765

FUEL COST LEVELIZATION FACTORS : FF(PT,E,R)

ESCALATION RATE : E = .12

INTEREST RATE : R

(PT)	0.04	0.05	0.06	0.07	0.08	0.09	0.10	0.11	0.12	0.13	0.14	0.15
2	1.186	1.186	1.185	1.185	1.185	1.184	1.184	1.184	1.183	1.183	1.183	1.183
4	1.331	1.329	1.327	1.325	1.324	1.322	1.320	1.319	1.317	1.315	1.314	1.312
6	1.495	1.491	1.486	1.481	1.477	1.472	1.468	1.464	1.459	1.455	1.451	1.447
8	1.683	1.673	1.664	1.654	1.645	1.636	1.628	1.619	1.610	1.602	1.594	1.586
10	1.896	1.879	1.862	1.846	1.830	1.815	1.799	1.784	1.770	1.756	1.742	1.728
12	2.138	2.111	2.084	2.058	2.033	2.008	1.984	1.960	1.937	1.915	1.893	1.872
14	2.415	2.373	2.333	2.293	2.255	2.217	2.181	2.146	2.112	2.079	2.048	2.017
16	2.731	2.670	2.610	2.553	2.497	2.443	2.392	2.342	2.294	2.248	2.204	2.162
18	3.092	3.005	2.921	2.840	2.762	2.687	2.616	2.548	2.483	2.421	2.362	2.307
20	3.505	3.384	3.267	3.156	3.050	2.950	2.854	2.763	2.678	2.597	2.521	2.449
22	3.978	3.813	3.655	3.506	3.365	3.231	3.106	2.988	2.878	2.775	2.679	2.589
24	4.519	4.298	4.088	3.891	3.706	3.533	3.372	3.222	3.083	2.954	2.836	2.726
26	5.140	4.847	4.572	4.316	4.078	3.857	3.653	3.465	3.293	3.135	2.991	2.859
28	5.852	5.469	5.113	4.784	4.481	4.203	3.948	3.717	3.507	3.316	3.144	2.988
30	6.669	6.174	5.718	5.299	4.918	4.572	4.259	3.977	3.724	3.498	3.295	3.113

FUEL COST LEVELIZATION FACTORS : FF(PT,E,R)

ESCALATION RATE : E = .13

INTEREST RATE : R

(PT)	0.04	0.05	0.06	0.07	0.08	0.09	0.10	0.11	0.12	0.13	0.14	0.15
2	1.202	1.202	1.201	1.201	1.201	1.200	1.200	1.200	1.199	1.199	1.199	1.198
4	1.362	1.360	1.358	1.356	1.354	1.352	1.350	1.348	1.347	1.345	1.343	1.341
6	1.546	1.541	1.535	1.530	1.525	1.520	1.515	1.510	1.506	1.501	1.496	1.492
8	1.758	1.747	1.736	1.726	1.716	1.706	1.696	1.686	1.676	1.667	1.658	1.649
10	2.002	1.983	1.964	1.946	1.928	1.910	1.893	1.876	1.859	1.843	1.827	1.812
12	2.284	2.253	2.222	2.192	2.163	2.135	2.107	2.080	2.053	2.028	2.003	1.979
14	2.610	2.562	2.515	2.469	2.424	2.381	2.339	2.298	2.259	2.221	2.185	2.149
16	2.988	2.916	2.847	2.779	2.714	2.651	2.590	2.532	2.476	2.423	2.372	2.322
18	3.426	3.323	3.223	3.127	3.034	2.946	2.862	2.781	2.704	2.632	2.562	2.497
20	3.935	3.789	3.649	3.516	3.389	3.268	3.154	3.046	2.943	2.847	2.756	2.671
22	4.525	4.324	4.133	3.952	3.781	3.619	3.468	3.326	3.193	3.069	2.953	2.845
24	5.212	4.940	4.682	4.440	4.213	4.001	3.804	3.621	3.452	3.295	3.151	3.018
26	6.012	5.647	5.305	4.986	4.690	4.417	4.165	3.933	3.721	3.527	3.350	3.188
28	6.944	6.461	6.012	5.598	5.216	4.868	4.550	4.261	3.999	3.763	3.549	3.356
30	8.030	7.397	6.814	6.281	5.796	5.357	4.961	4.606	4.287	4.002	3.748	3.521

FUEL COST LEVELIZATION FACTORS : FF(PT,E,R)

ESCALATION RATE : E = .14

INTEREST RATE : R

(PT)	0.04	0.05	0.06	0.07	0.08	0.09	0.10	0.11	0.12	0.13	0.14	0.15
2	1.218	1.218	1.217	1.217	1.217	1.216	1.216	1.216	1.215	1.215	1.215	1.214
4	1.394	1.391	1.389	1.387	1.385	1.383	1.381	1.379	1.377	1.375	1.373	1.371
6	1.598	1.592	1.586	1.581	1.575	1.569	1.564	1.559	1.553	1.548	1.543	1.538
8	1.836	1.824	1.812	1.801	1.789	1.778	1.767	1.756	1.745	1.735	1.725	1.714
10	2.115	2.093	2.072	2.051	2.031	2.011	1.991	1.972	1.953	1.935	1.917	1.900
12	2.441	2.405	2.370	2.336	2.302	2.270	2.238	2.207	2.177	2.148	2.120	2.093
14	2.823	2.767	2.713	2.660	2.608	2.558	2.510	2.463	2.418	2.375	2.333	2.292
16	3.272	3.188	3.107	3.028	2.952	2.879	2.808	2.741	2.676	2.613	2.554	2.497
18	3.800	3.678	3.560	3.447	3.338	3.234	3.135	3.040	2.950	2.864	2.783	2.706
20	4.423	4.248	4.082	3.922	3.771	3.627	3.491	3.362	3.241	3.127	3.020	2.919
22	5.156	4.913	4.682	4.463	4.257	4.062	3.880	3.709	3.550	3.401	3.263	3.134
24	6.023	5.689	5.374	5.078	4.801	4.543	4.303	4.081	3.875	3.686	3.511	3.351
26	7.048	6.595	6.171	5.776	5.411	5.074	4.763	4.479	4.219	3.981	3.765	3.568
28	8.260	7.653	7.089	6.570	6.094	5.659	5.263	4.904	4.579	4.286	4.023	3.785
30	9.697	8.890	8.148	7.472	6.857	6.303	5.804	5.357	4.958	4.601	4.284	4.002

FUEL COST LEVELIZATION FACTORS : FF(PT,E,R)

ESCALATION RATE : E =.15

INTEREST RATE : R

(PT)	0.04	0.05	0.06	0.07	0.08	0.09	0.10	0.11	0.12	0.13	0.14	0.15
2	1.235	1.234	1.234	1.233	1.233	1.233	1.232	1.232	1.231	1.231	1.231	1.230
4	1.426	1.423	1.421	1.419	1.416	1.414	1.412	1.410	1.408	1.405	1.403	1.401
6	1.651	1.645	1.639	1.632	1.626	1.620	1.614	1.608	1.602	1.597	1.591	1.585
8	1.918	1.905	1.892	1.879	1.866	1.853	1.841	1.829	1.817	1.806	1.794	1.783
10	2.234	2.210	2.186	2.162	2.140	2.117	2.095	2.074	2.053	2.032	2.012	1.993
12	2.609	2.568	2.528	2.490	2.452	2.415	2.379	2.344	2.310	2.277	2.245	2.214
14	3.054	2.990	2.928	2.867	2.808	2.751	2.695	2.642	2.590	2.540	2.492	2.446
16	3.585	3.488	3.393	3.302	3.214	3.129	3.047	2.969	2.894	2.822	2.753	2.687
18	4.219	4.075	3.936	3.803	3.676	3.554	3.438	3.327	3.222	3.122	3.027	2.937
20	4.977	4.769	4.571	4.382	4.202	4.032	3.871	3.719	3.575	3.440	3.314	3.195
22	5.884	5.591	5.313	5.050	4.801	4.568	4.350	4.146	3.955	3.778	3.613	3.460
24	6.971	6.564	6.180	5.820	5.483	5.170	4.880	4.611	4.362	4.134	3.923	3.730
26	8.277	7.717	7.194	6.708	6.259	5.845	5.464	5.116	4.798	4.508	4.245	4.006
28	9.847	9.086	8.382	7.734	7.140	6.600	6.108	5.664	5.263	4.902	4.577	4.286
30	11.737	10.712	9.773	8.918	8.143	7.445	6.818	6.258	5.758	5.314	4.919	4.569

Appendix 15
Fixed Charge Rate (FCR) Plotted Against Various Other Rates and Factors

Figures 15–1 through 15–11 present graphs showing fixed charge rate measured against various other rates and factors. The quantity V is an assessment valuation factor that is meant to take into account the value of the passive addition assigned by the property assessor in year PT. If none of the additional cost were reflected in the resale price, V would equal 0; if all of the additional cost were appraised in the market value, V would equal 1. For figures 15–1 through 15–10 curves are given for $V = 1.0$ and $V = 0.5$. Figure 15–11 plots fixed charge rate against V itself.

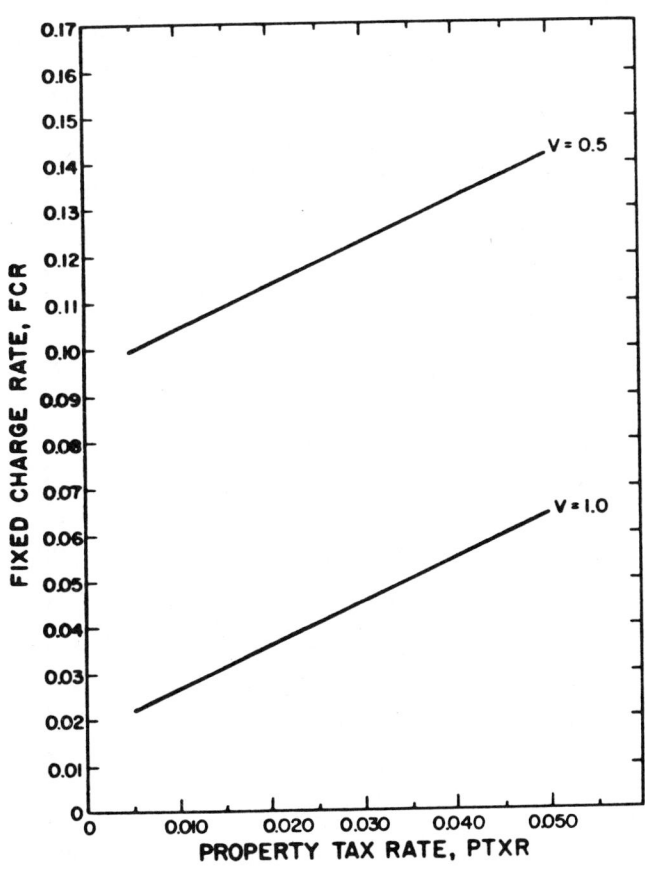

15-1: Fixed charge rate versus property tax rate ($PTXR$).

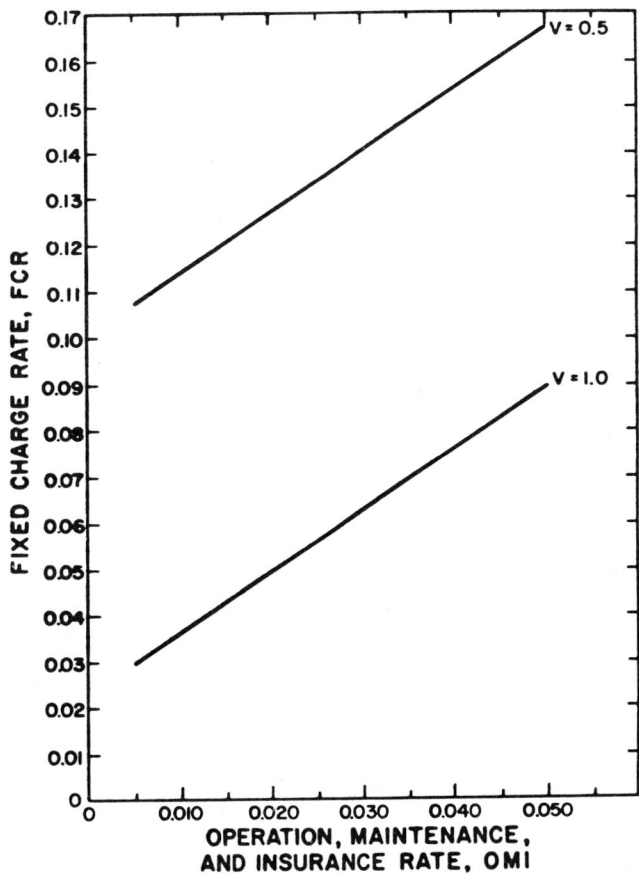

15-2: Fixed charge rate versus operation, maintenance, and insurance rate (OMI).

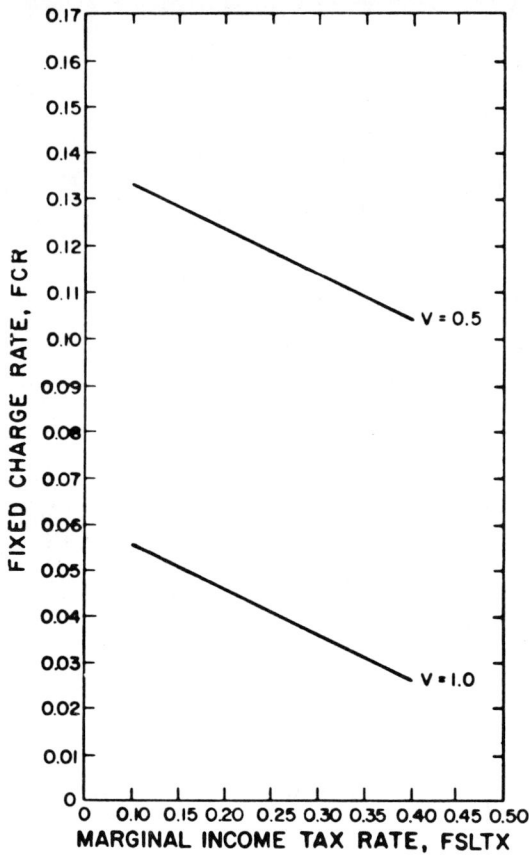

15-3: Fixed charge rate versus marginal state and federal income tax rate (*FSLTX*).

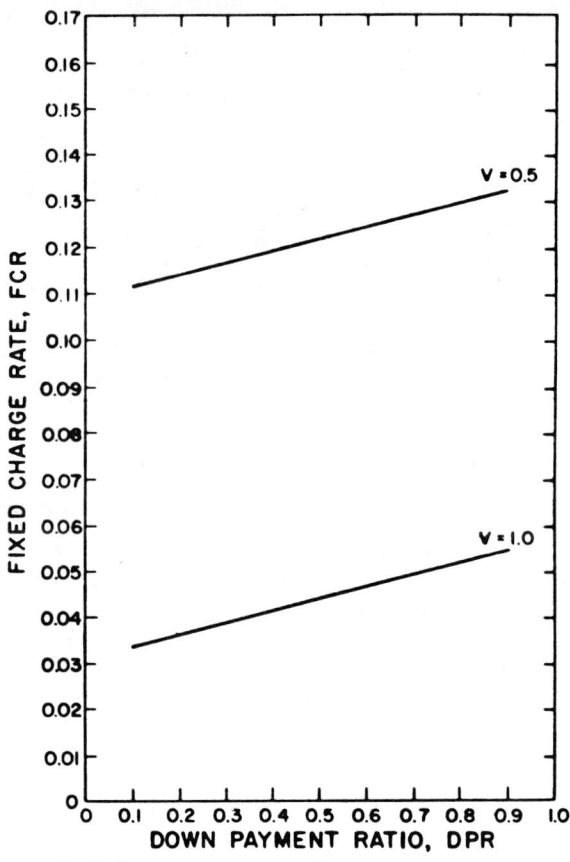

15-4: Fixed charge rate versus down payment ratio (*DPR*).

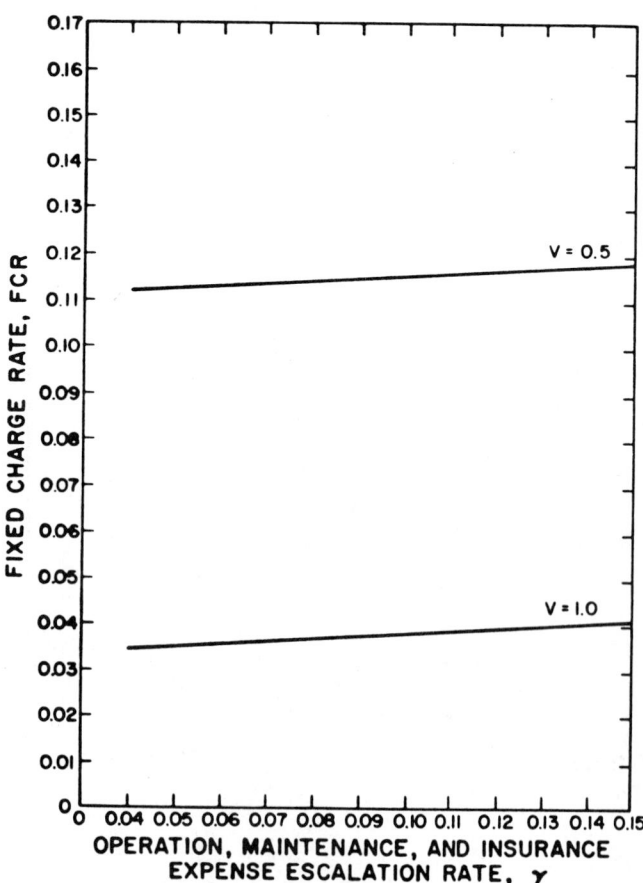

15-5: Fixed charge rate versus operation, maintenance, and insurance expense escalation rate (γ).

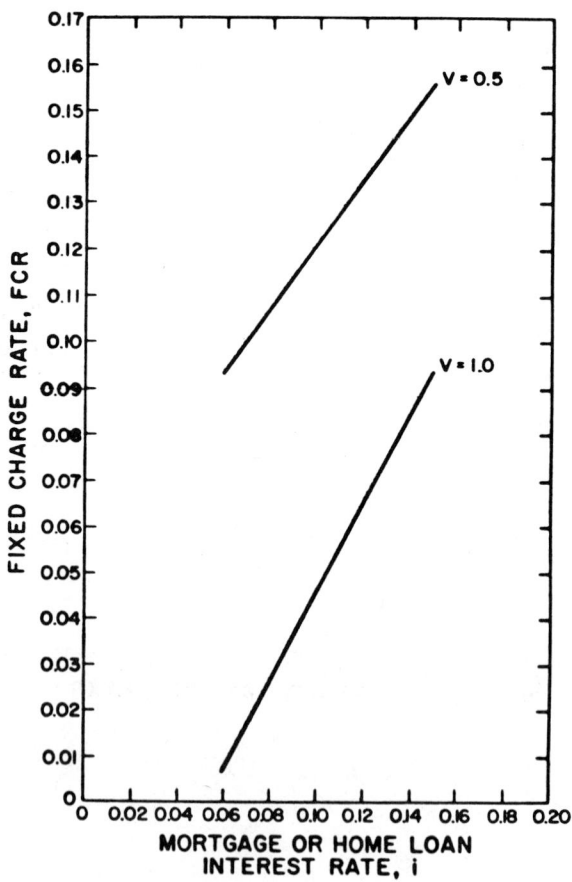

15-6: Fixed charge rate versus mortgage or home-loan interest rate (i).

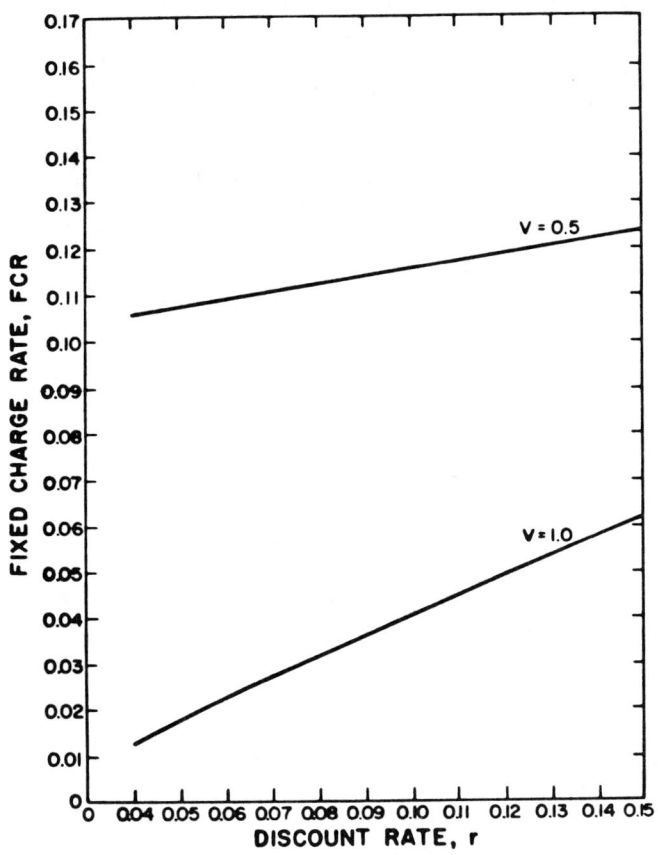

15-7: Fixed charge rate versus discount rate (r).

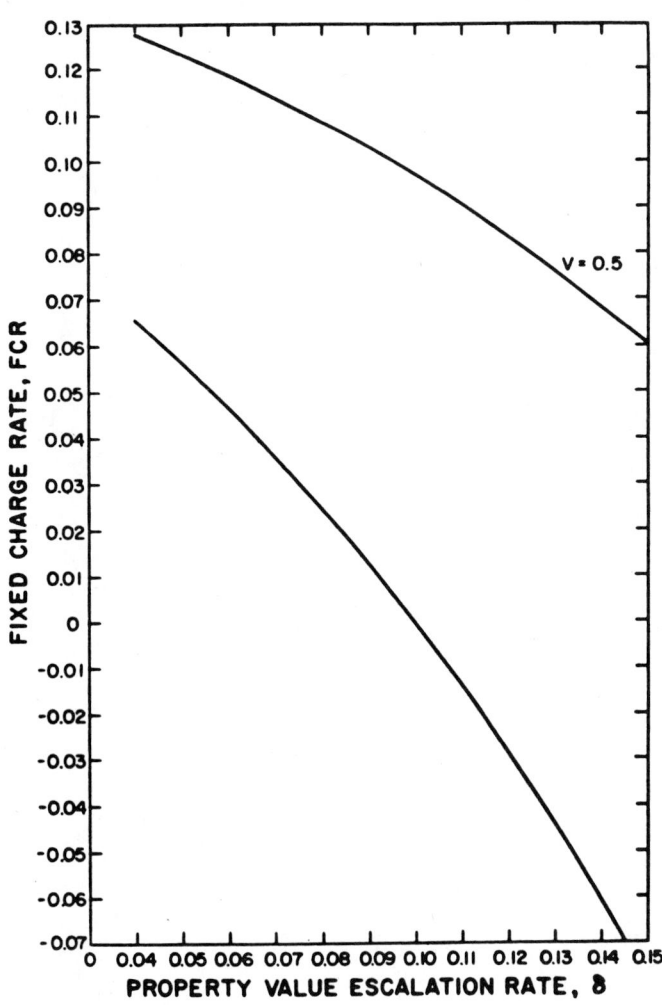

15-8: Fixed charge rate versus property value escalation rate (δ).

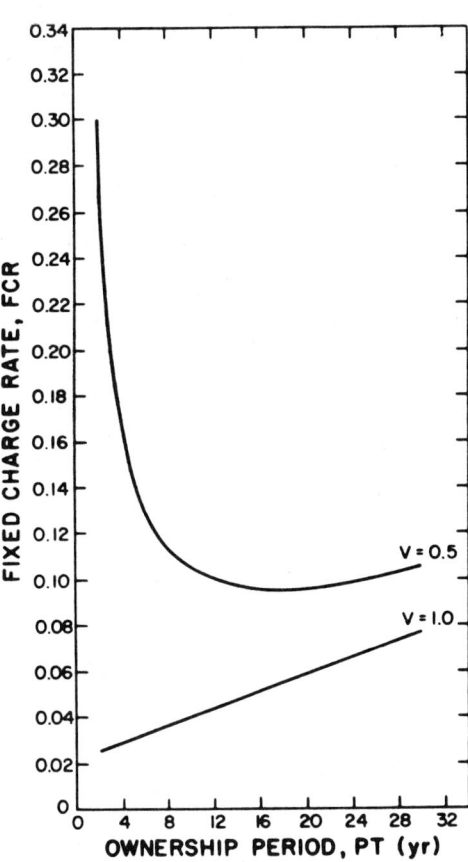

15-9: Fixed charge rate versus ownership period (PT).

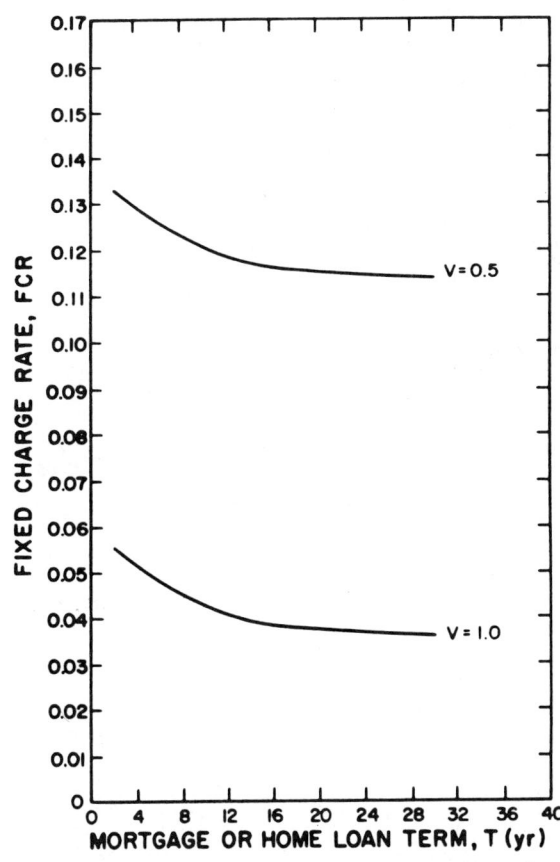

15-10: Fixed charge rate versus mortgage or home-loan term (T).

Fixed Charge Rate (FCR) *Graphs* / 301

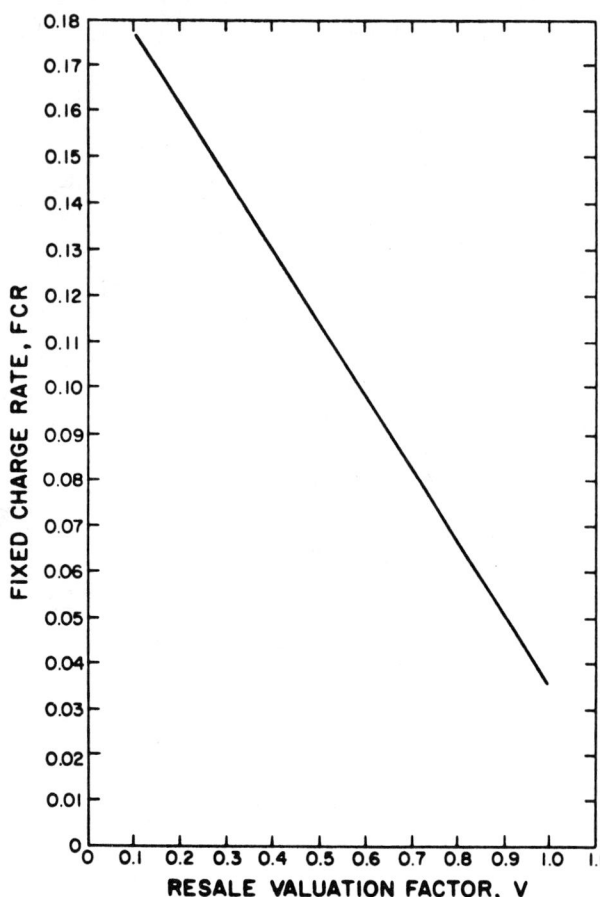

15-11: Fixed charge rate versus resale valuation factor (*V*).

Appendix 16   Calculation of Diurnal Heat Capacity

This appendix consists of equations and tables that are useful in calculating diurnal heat capacity (*dhc*) values given in figure J-3 for any wall thickness and wall properties. The diurnal heat capacity of a wall surface is the daily amount of heat per unit of surface area that is stored and then given back per unit of temperature swing. The variation is assumed to be sinusoidal, with one-day periodicity; the heat flow is assumed to be one-dimensional; the properties of the wall's compositional material are assumed to remain constant. Given these assumptions, the heat diffusion equation can be solved in closed form, and mathematical expressions can be given for the diurnal heat capacity.

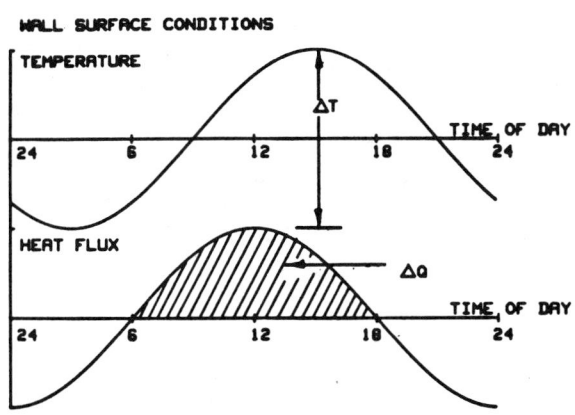

16-1: Variations in wall surface temperature and heat flux over a 24-hour period.

As shown in figure 16-1 the heat flux at the wall surface is sinusoidal for a sinusoidal variation in surface temperature, but it is shifted in phase. The diurnal heat capacity is the integral under one-half of the sine wave, ΔQ (expressed in Btu/ft²), divided by the magnitude of temperature swing, ΔT (expressed in Fahrenheit degrees).*

---

*Diurnal heat capacity is related to the "thermal admittance" parameter discussed by some authors, especially European authors. It is equal to $P/2\pi$ times the thermal admittance, where P is the periodicity.

Although the actual temperature wave forms that are observed in heat exchanges in passive-solar walls are not purely sinusoidal, the great bulk of heat stored and retrieved each day during a series of sunny days is due to the fundamental, 24-hour harmonic. The higher harmonics of the wave form do not penetrate the wall very far; therefore, the temperature variation within the wall becomes quite sinusoidal at fairly shallow depths. Since harmonics do not affect net diurnal heat capacity, they can be ignored in the analysis. The only significant odd harmonic is the third, which can be dealt with by means of a correction factor applied to the fundamental harmonic. (This factor enters the analysis in the 0.733 constant used in Step 7 of the calculation in Chapter J.)

Base Case: Wall of Infinite Thickness

If the wall is very thick, then the mathematical solution is particularly compact. The diurnal heat capacity (*dhc*) is simply:

$$dhc = s = \sqrt{k \times \varrho \times c \times \left(\frac{P}{2\pi}\right)}$$

where:
- k = thermal conductivity (in Btu/ft/hr/°F)
- ϱ = density (in lb/ft³)
- c = heat capacity (in Btu/lb/°F)
- P = periodicity (in hr) = 24 hr

Since this expression involves only material properties (assuming P = constant = 24 hours in all cases), it is also a property and is designated by the symbol s. It is expressed in Btu/°F/ft².

Class 1 Procedure: Wall of Finite Thickness, Insulated on the Back Side

In this case, the diurnal heat capacity is modified by a factor, F_1:

$$dhc = s \times F_1$$

where:
- F_1 = function of x
- $x = L\sqrt{\pi\varrho c/Pk}$
- L = wall thickness (in ft)

The parameter x is a dimensionless thickness. If x is greater than about 3, then the factor F_1 is very close to unity, and the wall is effectively infinite.

x	F_1	x	F_1	x	F_1	x	F_1	x	F_1	x	F_1
		.30	.423	.60	.816	.90	1.076	1.40	1.122	2.00	1.024
.01	.014	.31	.437	.61	.828	.91	1.081	1.42	1.118	2.05	1.019
.02	.028	.32	.451	.62	.839	.92	1.086	1.44	1.115	2.10	1.015
.03	.042	.33	.465	.63	.850	.93	1.091	1.46	1.111	2.15	1.011
.04	.057	.34	.479	.64	.861	.94	1.095	1.48	1.107	2.20	1.008
.05	.071	.35	.493	.65	.872	.95	1.100	1.50	1.104	2.25	1.005
.06	.085	.36	.506	.66	.883	.96	1.104	1.52	1.100	2.30	1.002
.07	.099	.37	.520	.67	.894	.97	1.107	1.54	1.096	2.35	1.000
.08	.113	.38	.534	.68	.904	.98	1.111	1.56	1.092	2.40	.999
.09	.127	.39	.548	.69	.914	.99	1.114	1.58	1.089	2.45	.997
.10	.141	.40	.561	.70	.924	1.00	1.117	1.60	1.085	2.50	.996
.11	.156	.41	.575	.71	.934	1.02	1.123	1.62	1.081	2.55	.995
.12	.170	.42	.588	.72	.943	1.04	1.128	1.64	1.077	2.60	.995
.13	.184	.43	.602	.73	.952	1.06	1.132	1.66	1.074	2.65	.994
.14	.198	.44	.615	.74	.961	1.08	1.136	1.68	1.070	2.70	.994
.15	.212	.45	.628	.75	.970	1.10	1.138	1.70	1.067	2.75	.994
.16	.226	.46	.642	.76	.979	1.12	1.140	1.72	1.063	2.80	.994
.17	.240	.47	.655	.77	.987	1.14	1.142	1.74	1.060	2.85	.994
.18	.254	.48	.668	.78	.996	1.16	1.143	1.76	1.057	2.90	.995
.19	.269	.49	.681	.79	1.004	1.18	1.143	1.78	1.053	2.95	.995
.20	.283	.50	.694	.80	1.011	1.20	1.143	1.80	1.050	3.00	.995
.21	.297	.51	.707	.81	1.019	1.22	1.142	1.82	1.047	3.25	.997
.22	.311	.52	.719	.82	1.026	1.24	1.141	1.84	1.044	3.50	.999
.23	.325	.53	.732	.83	1.033	1.26	1.140	1.86	1.041	3.75	1.000
.24	.339	.54	.744	.84	1.040	1.28	1.138	1.88	1.039	4.00	1.000
.25	.353	.55	.757	.85	1.047	1.30	1.136	1.90	1.036	4.25	1.000
.26	.367	.56	.769	.86	1.053	1.32	1.133	1.92	1.033	4.50	1.000
.27	.381	.57	.781	.87	1.059	1.34	1.131	1.94	1.031	4.75	1.000
.28	.395	.58	.793	.88	1.065	1.36	1.128	1.96	1.029	5.00	1.000
.29	.409	.59	.805	.89	1.071	1.38	1.125	1.98	1.026		

16-2: Function F_1 table. The function F_1 is the ratio of the diurnal heat capacity of a wall of finite thickness to that of a wall of infinite thickness. The dimensionless thickness is expressed in the equation:

$$x = L\sqrt{\pi \rho c / Pk}$$

$$F_1 = \sqrt{\frac{\cosh 2x - \cos 2x}{\cosh 2x + \cos 2x}}$$

Values of F_1 for different values of x, as well as the general equation, are given in figure 16-2. The function is also plotted in figure 16-3.

The Class 1 situation also applies to a wall that receives equal amounts of heat from both sides—for example, an internal partition wall. Since the situation is symmetric, there is no heat flow through a plane at the center of the wall; therefore, this plane is mathematically equivalent to an insulated surface. For an internal partition, the thickness L to be used in the above equation should be equal to one-half the partition thickness, and the resulting diurnal heat capacity applies to both outside surfaces.

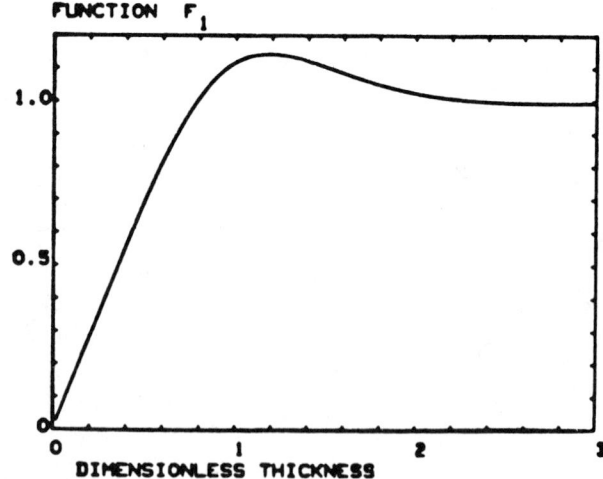

16-3: Graph of function F_1 values.

r	\multicolumn{11}{c}{dimensionless thickness x}

r	0.0	.2	.4	.6	.8	1.0	1.2	1.4	1.6	1.8	2.0	3.0
.00	1.000	1.000	1.000	1.000	1.000	.999	.999	.999	.999	.999	.999	.999
.05	.999	.997	.994	.987	.980	.974	.969	.967	.965	.965	.964	.965
.10	.995	.992	.985	.973	.960	.948	.939	.934	.932	.931	.930	.932
.15	.989	.985	.974	.957	.938	.922	.910	.903	.900	.898	.898	.900
.20	.981	.976	.961	.940	.916	.896	.882	.873	.869	.867	.867	.869
.25	.970	.964	.947	.922	.894	.871	.854	.845	.840	.838	.837	.840
.30	.958	.951	.931	.902	.872	.846	.828	.817	.812	.810	.810	.812
.35	.944	.936	.914	.883	.849	.821	.802	.791	.786	.783	.783	.786
.40	.928	.920	.896	.863	.827	.798	.778	.766	.761	.758	.758	.761
.45	.912	.903	.878	.842	.805	.775	.755	.743	.737	.734	.734	.737
.50	.894	.885	.859	.822	.784	.753	.732	.720	.714	.712	.711	.715
.60	.857	.848	.820	.782	.743	.711	.690	.678	.672	.670	.669	.673
.70	.819	.809	.781	.743	.704	.673	.652	.640	.634	.632	.632	.635
.80	.781	.771	.743	.706	.668	.637	.617	.606	.600	.598	.597	.601
.90	.743	.734	.707	.671	.634	.605	.585	.575	.569	.567	.566	.569
1.00	.707	.698	.672	.638	.602	.575	.556	.546	.541	.539	.538	.541
1.10	.673	.664	.640	.607	.573	.547	.530	.520	.515	.513	.512	.515
1.20	.640	.632	.609	.578	.547	.522	.505	.496	.491	.489	.489	.492
1.30	.610	.602	.581	.551	.522	.498	.483	.474	.469	.468	.467	.470
1.40	.581	.574	.554	.527	.499	.477	.462	.454	.449	.448	.447	.450
1.50	.555	.548	.529	.504	.478	.457	.443	.435	.431	.429	.429	.431
1.60	.530	.524	.506	.482	.458	.438	.425	.418	.414	.413	.412	.414
1.70	.507	.501	.485	.463	.440	.421	.409	.402	.398	.397	.397	.398
1.80	.486	.480	.465	.444	.423	.405	.394	.387	.384	.382	.382	.384
1.90	.466	.461	.447	.427	.407	.390	.379	.373	.370	.369	.368	.370
2.00	.447	.443	.429	.411	.392	.376	.366	.360	.357	.356	.356	.357
2.20	.414	.410	.398	.382	.365	.351	.342	.337	.334	.333	.333	.334
2.40	.385	.381	.371	.357	.342	.329	.321	.316	.314	.313	.312	.314
2.60	.359	.356	.347	.334	.321	.310	.302	.298	.295	.295	.294	.296
2.80	.336	.334	.326	.314	.302	.292	.285	.281	.279	.278	.278	.279
3.00	.316	.314	.307	.296	.286	.276	.270	.267	.265	.264	.264	.265
3.20	.298	.296	.290	.280	.271	.262	.257	.253	.252	.251	.251	.252
3.40	.282	.280	.274	.266	.257	.250	.244	.241	.240	.239	.239	.240
3.60	.268	.266	.261	.253	.245	.238	.233	.230	.229	.228	.228	.229
3.80	.254	.253	.248	.241	.234	.227	.223	.220	.219	.219	.218	.219
4.00	.243	.241	.237	.230	.223	.218	.214	.211	.210	.209	.209	.210
4.50	.217	.216	.212	.207	.201	.197	.193	.191	.190	.190	.190	.190
5.00	.196	.195	.192	.188	.183	.179	.176	.175	.174	.173	.173	.174

16-4: Function F_2 table. The function F_2 is the ratio of diurnal heat capacity based on room air temperature fluctuation to that based on wall-surface temperature fluctuation.

$$F_2 = \frac{1}{\sqrt{(r + \cos \phi)^2 + \sin^2 \phi}}$$

$$\phi = \frac{\pi}{4} + \arctan\left(\frac{\sin 2x}{\sinh 2x}\right)$$

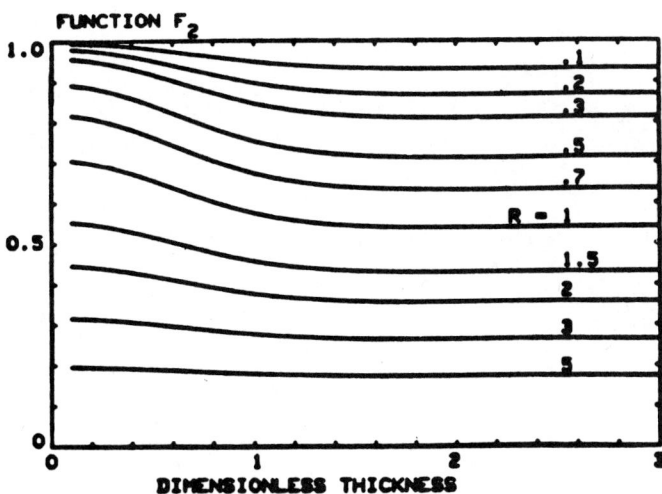

16-5: Graph of function F_2 values.

Calculating Diurnal Heat Capacity / 305

Class 2

Example: 8-inch-thick external concrete wall, insulated on the outside:

$$k = 1.0 \text{ Btu/ft/hr/}°F$$
$$\varrho = 143 \text{ lb/ft}^3$$
$$c = 0.21 \text{ Btu/lb/}°F$$
$$s = 10.71 \text{ Btu/}°F/\text{ft}^2$$
$$x = 1.32, F_1 = 1.133$$
$$dhc = 12.14 \text{ Btu/}°F/\text{ft}^2$$

Example B: 8-inch-thick internal concrete partition, heated on both sides:

$$x = 0.66, F_1 = 0.88$$
$$dhc = 2sF_1 = 18.94 \text{ Btu/}°F/\text{ft}^2$$

Notice that the internal wall is 56% more effective than the external wall because of the increased amount of exposed surface.

The cases given so far apply to a wall heated radiantly by the sun; they all give the diurnal heat storage in terms of the wall *surface* temperature swing:

$$dhc = \frac{\Delta Q}{\Delta T_s}$$

These are the Class 2 situations discussed in Chapter J.

If the wall is heated indirectly, by the room air, then the situation is Class 3. In these cases the primary resistance to heat flow into the wall is due to the film coefficient between the room air and the surface. This is typically about 1.5 Btu/°F/hr/ft². (This value accounts for both convection from the room air and thermal radiation from objects in the room and from other walls also heated by the room air.) For Class 3 situations, the diurnal heat capacity is given in terms of room-air temperature swing instead of wall-surface temperature swing:

$$dhc = \frac{\Delta Q}{\Delta T_r}$$

Class 3 Procedure

Calculate the diurnal heat capacity in terms of wall surface temperature, $dhc = s \times F_1$, as for Class 2. This is then modified by another factor F_2 as follows

$$dhc = s \times F_1 \times F_2$$

where:

F_2 = function of r and x
$r = 2s\pi F_1/hP$
x = same as before

Values of F_2 are given in figure 16-4 for different values of r and x; plots of these values are shown in figure 16-5.

Imagine, for example, an 8-inch-thick concrete wall, insulated on the back side, with $h = 1.5$:

$$sF_1 = 12.14$$
$$\left. \begin{array}{l} r = 2.12 \\ x = 1.32 \end{array} \right\} F_x = 0.348$$
$$dhc = 4.22 \text{ Btu/}°F/\text{ft}^2$$

The derivation of these formulas and a procedure for analyzing the daytime heat capacity of a layered wall is given in reference [DAV].

Appendix 17 Conservation Factor Tables

The conservation factor (CF) used to determine optimum conservation levels is given in this appendix for various localities and system types. The equation for CF is as follows:

$$CF = \sqrt{24 \times \left(\frac{1}{LCR} + \frac{(1 - SSF)}{D}\right)}$$

where:
- LCR = load collector ratio = BLC/A_p
- SSF = solar savings fraction
- $D = d(SSF)/d(1/LCR)$

Values of SSF were calculated using the same correlations and correlation constants given in Appendix 21, which also contains descriptions of the various systems and the reference designs.

BIRMINGHAM, AL				BIRMINGHAM, AL				MOBILE, AL				MONTGOMERY, AL			
	SOLAR SAVINGS				SOLAR SAVINGS				SOLAR SAVINGS				SOLAR SAVINGS		
	10%	50%	80%		10%	50%	80%		10%	50%	80%		10%	50%	80%
WW A1	1.2	1.9	2.6	DG C2	1.0	1.3	1.8	TW D4	.7	.9	1.1	WW A1	1.0	1.6	2.2
WW A2	1.0	1.5	1.9	DG C3	.9	1.1	1.4	TW D5	.7	.9	1.1	WW A2	.9	1.3	1.7
WW A3	1.0	1.3	1.8	SS A1	.9	1.4	1.9	TW E1	.7	.9	1.1	WW A3	.8	1.2	1.5
WW A4	.9	1.3	1.7	SS A2	.8	1.1	1.4	TW E2	.7	.9	1.1	WW A4	.8	1.1	1.5
WW A5	.9	1.2	1.6	SS A3	1.0	1.5	2.1	TW E3	.6	.8	1.0	WW A5	.8	1.1	1.4
WW A6	.9	1.2	1.6	SS A4	.8	1.1	1.5	TW E4	.6	.8	1.0	WW A6	.8	1.1	1.4
WW B1	1.2	1.8	2.5	SS A5	.9	1.5	2.1	TW F1	.9	1.4	1.8	WW B1	1.0	1.5	2.0
WW B2	.9	1.2	1.6	SS A6	.8	1.1	1.5	TW F2	.9	1.2	1.5	WW B2	.8	1.1	1.4
WW B3	.8	1.1	1.4	SS A7	1.0	1.8	2.5	TW F3	.8	1.2	1.5	WW B3	.7	1.0	1.2
WW B4	.8	1.1	1.3	SS A8	.8	1.1	1.5	TW F4	.9	1.2	1.5	WW B4	.7	.9	1.2
WW B5	.8	1.1	1.4	SS B1	1.0	1.6	2.1	TW G1	1.0	1.4	1.8	WW B5	.7	1.0	1.2
WW C1	.8	1.1	1.4	SS B2	.8	1.2	1.6	TW G2	.9	1.3	1.6	WW C1	.7	.9	1.2
WW C2	.8	1.1	1.4	SS B3	1.1	1.6	2.2	TW G3	.9	1.3	1.6	WW C2	.7	1.0	1.2
WW C3	.7	.9	1.2	SS B4	.9	1.2	1.6	TW G4	1.0	1.4	1.8	WW C3	.6	.8	1.1
WW C4	.8	1.0	1.2	SS B5	1.1	1.8	2.5	TW H1	1.0	1.4	1.8	WW C4	.7	.9	1.1
TW A1	1.2	2.0	2.6	SS B6	.9	1.2	1.6	TW H2	1.1	1.4	1.9	TW A1	1.1	1.6	2.2
TW A2	1.1	1.6	2.1	SS B7	1.2	2.0	2.8	TW H3	1.1	1.6	2.0	TW A2	1.0	1.4	1.8
TW A3	1.0	1.5	2.0	SS B8	.9	1.3	1.7	TW H4	1.3	1.8	2.3	TW A3	.9	1.3	1.7
TW A4	1.0	1.5	1.9	SS C1	1.0	1.5	2.0	TW I1	1.0	1.4	1.9	TW A4	.9	1.3	1.6
TW B1	1.2	1.9	2.5	SS C2	.9	1.2	1.6	TW I2	.8	1.1	1.4	TW B1	1.1	1.6	2.1
TW B2	1.1	1.6	2.1	SS C3	1.2	1.8	2.4	TW I3	.7	1.0	1.3	TW B2	1.0	1.4	1.8
TW B3	1.1	1.6	2.1	SS C4	1.0	1.4	1.8	TW I4	.7	.9	1.2	TW B3	1.0	1.4	1.8
TW B4	1.2	1.6	2.1	SS D1	.8	1.2	1.6	TW I5	.7	.9	1.2	TW B4	1.0	1.4	1.9
TW C1	1.2	1.8	2.4	SS D2	.7	1.0	1.2	TW J1	.7	.9	1.2	TW C1	1.1	1.6	2.1
TW C2	1.2	1.7	2.3	SS D3	.9	1.3	1.8	TW J2	.7	.9	1.2	TW C2	1.1	1.5	2.0
TW C3	1.3	1.8	2.4	SS D4	.7	1.0	1.3	TW J3	.6	.8	1.0	TW C3	1.1	1.6	2.1
TW C4	1.4	2.0	2.6	SS E1	.9	1.3	1.8	TW J4	.7	.8	1.1	TW C4	1.2	1.7	2.3
TW D1	1.3	2.0	2.7	SS E2	.8	1.1	1.4	DG A1	.9	1.4	2.8	TW D1	1.1	1.7	2.2
TW D2	1.0	1.4	1.8	SS E3	1.0	1.6	2.1	DG A2	.9	1.2	1.7	TW D2	.9	1.2	1.5
TW D3	.9	1.3	1.6	SS E4	.8	1.1	1.5	DG A3	.8	1.0	1.3	TW D3	.8	1.1	1.4
TW D4	.9	1.2	1.5					DG B1	.9	1.1	1.8	TW D4	.8	1.0	1.3
TW D5	.9	1.1	1.5	MOBILE, AL				DG B2	.8	1.0	1.4	TW D5	.8	1.0	1.3
TW E1	.8	1.2	1.5		SOLAR SAVINGS			DG B3	.8	.9	1.2	TW E1	.7	1.0	1.3
TW E2	.8	1.1	1.5		10%	50%	80%	DG C1	.8	1.0	1.4	TW E2	.8	1.0	1.3
TW E3	.7	1.0	1.3	WW A1	.9	1.3	1.8	DG C2	.8	.9	1.2	TW E3	.7	.9	1.1
TW E4	.8	1.0	1.3	WW A2	.8	1.1	1.4	DG C3	.7	.9	1.0	TW E4	.7	.9	1.2
TW F1	1.3	1.9	2.6	WW A3	.7	1.0	1.3	SS A1	.7	1.0	1.3	TW F1	1.1	1.6	2.2
TW F2	1.1	1.7	2.2	WW A4	.7	.9	1.2	SS A2	.6	.8	1.1	TW F2	1.0	1.4	1.9
TW F3	1.1	1.6	2.1	WW A5	.7	.9	1.2	SS A3	.7	1.1	1.4	TW F3	1.0	1.4	1.8
TW F4	1.2	1.6	2.1	WW A6	.7	.9	1.1	SS A4	.6	.8	1.1	TW F4	1.0	1.4	1.8
TW G1	1.3	1.9	2.5	WW B1	.9	1.2	1.6	SS A5	.7	1.1	1.5	TW G1	1.1	1.6	2.1
TW G2	1.2	1.7	2.3	WW B2	.7	.9	1.2	SS A6	.6	.8	1.1	TW G2	1.1	1.5	1.9
TW G3	1.3	1.7	2.3	WW B3	.6	.8	1.1	SS A7	.8	1.2	1.7	TW G3	1.1	1.5	2.0
TW G4	1.4	1.9	2.5	WW B4	.6	.8	1.0	SS A8	.6	.8	1.1	TW G4	1.2	1.6	2.1
TW H1	1.4	2.0	2.6	WW B5	.6	.8	1.1	SS B1	.8	1.1	1.5	TW H1	1.2	1.7	2.2
TW H2	1.4	2.0	2.6	WW C1	.6	.8	1.0	SS B2	.7	.9	1.2	TW H2	1.2	1.7	2.2
TW H3	1.5	2.1	2.8	WW C2	.6	.8	1.1	SS B3	.8	1.2	1.5	TW H3	1.3	1.8	2.4
TW H4	1.7	2.4	3.2	WW C3	.6	.7	.9	SS B4	.7	.9	1.2	TW H4	1.5	2.1	2.7
TW I1	1.4	2.2	2.9	WW C4	.6	.8	1.0	SS B5	.9	1.3	1.7	TW I1	1.2	1.8	2.3
TW I2	1.0	1.4	1.8	TW A1	.9	1.4	1.8	SS B6	.7	.9	1.2	TW I2	.9	1.2	1.6
TW I3	1.0	1.3	1.7	TW A2	.8	1.2	1.5	SS B7	.9	1.4	1.9	TW I3	.9	1.2	1.5
TW I4	.9	1.2	1.6	TW A3	.8	1.1	1.4	SS B8	.7	.9	1.2	TW I4	.8	1.1	1.4
TW I5	.9	1.2	1.5	TW A4	.8	1.1	1.4	SS C1	.8	1.1	1.4	TW I5	.8	1.1	1.4
TW J1	.9	1.2	1.5	TW B1	.9	1.3	1.7	SS C2	.7	.9	1.2	TW J1	.8	1.1	1.4
TW J2	.9	1.2	1.5	TW B2	.8	1.2	1.5	SS C3	.9	1.3	1.7	TW J2	.8	1.1	1.4
TW J3	.8	1.0	1.3	TW B3	.8	1.2	1.5	SS C4	.8	1.0	1.3	TW J3	.7	.9	1.2
TW J4	.8	1.1	1.4	TW B4	.9	1.2	1.6	SS D1	.6	.9	1.1	TW J4	.7	1.0	1.2
DG A1	1.2	3.2	---	TW C1	.9	1.3	1.7	SS D2	.5	.7	.9	DG A1	1.0	1.9	---
DG A2	1.1	1.7	3.2	TW C2	.9	1.3	1.7	SS D3	.6	1.0	1.3	DG A2	1.0	1.4	2.2
DG A3	1.0	1.3	1.8	TW C3	1.0	1.3	1.7	SS D4	.5	.7	1.0	DG A3	.9	1.1	1.6
DG B1	1.2	1.9	---	TW C4	1.0	1.5	1.9	SS E1	.7	.9	1.2	DG B1	1.0	1.4	4.3
DG B2	1.1	1.5	2.2	TW D1	1.0	1.4	1.8	SS E2	.6	.8	1.0	DG B2	1.0	1.2	1.8
DG B3	1.0	1.2	1.6	TW D2	.7	1.0	1.3	SS E3	.8	1.1	1.5	DG B3	.9	1.1	1.4
DG C1	1.1	1.5	3.6	TW D3	.7	.9	1.2	SS E4	.6	.9	1.1	DG C1	.9	1.2	1.9

MONTGOMERY, AL

SOLAR SAVINGS

	10%	50%	80%
DG C2	.9	1.1	1.5
DG C3	.8	1.0	1.2
SS A1	.8	1.2	1.6
SS A2	.7	1.0	1.3
SS A3	.8	1.3	1.7
SS A4	.7	1.0	1.3
SS A5	.8	1.3	1.8
SS A6	.7	1.0	1.3
SS A7	.9	1.5	2.0
SS A8	.7	1.0	1.3
SS B1	.9	1.3	1.8
SS B2	.8	1.1	1.4
SS B3	.9	1.4	1.9
SS B4	.8	1.1	1.4
SS B5	1.0	1.5	2.1
SS B6	.8	1.1	1.4
SS B7	1.1	1.7	2.3
SS B8	.8	1.1	1.4
SS C1	.9	1.3	1.7
SS C2	.8	1.1	1.4
SS C3	1.1	1.6	2.1
SS C4	.9	1.2	1.6
SS D1	.7	1.0	1.3
SS D2	.6	.8	1.1
SS D3	.8	1.1	1.5
SS D4	.6	.9	1.1
SS E1	.8	1.1	1.5
SS E2	.7	.9	1.2
SS E3	.9	1.3	1.8
SS E4	.7	1.0	1.3

PHOENIX, AZ

SOLAR SAVINGS

	10%	50%	80%
WW A1	.7	1.0	1.3
WW A2	.6	.8	1.0
WW A3	.6	.7	1.0
WW A4	.5	.7	.9
WW A5	.5	.7	.9
WW A6	.5	.7	.9
WW B1	.6	.9	1.2
WW B2	.5	.7	.9
WW B3	.5	.6	.8
WW B4	.5	.6	.8
WW B5	.5	.6	.8
WW C1	.5	.6	.8
WW C2	.5	.6	.8
WW C3	.5	.6	.7
WW C4	.5	.6	.8
TW A1	.7	1.0	1.3
TW A2	.6	.9	1.1
TW A3	.6	.8	1.1
TW A4	.6	.8	1.0
TW B1	.7	1.0	1.3
TW B2	.7	.9	1.1
TW B3	.6	.9	1.1
TW B4	.7	.9	1.2
TW C1	.7	1.0	1.3
TW C2	.7	1.0	1.3
TW C3	.7	1.0	1.3
TW C4	.8	1.1	1.5
TW D1	.7	1.0	1.3
TW D2	.6	.8	1.0
TW D3	.5	.7	.9
TW D4	.5	.7	.9
TW D5	.5	.7	.9
TW E1	.5	.7	.9
TW E2	.5	.7	.9
TW E3	.5	.6	.8
TW E4	.5	.6	.8
TW F1	.7	1.0	1.3
TW F2	.7	.9	1.2
TW F3	.6	.9	1.1
TW F4	.7	.9	1.1
TW G1	.7	1.0	1.3
TW G2	.7	.9	1.2
TW G3	.7	1.0	1.2
TW G4	.8	1.0	1.3
TW H1	.8	1.0	1.4
TW H2	.8	1.1	1.4
TW H3	.9	1.2	1.5
TW H4	1.0	1.3	1.7
TW I1	.8	1.0	1.4
TW I2	.6	.8	1.0
TW I3	.6	.8	1.0
TW I4	.6	.7	.9
TW I5	.6	.7	.9
TW J1	.5	.7	.9
TW J2	.5	.7	.9
TW J3	.5	.6	.8
TW J4	.5	.7	.8
DG A1	.7	.9	1.3
DG A2	.7	.8	1.2
DG A3	.6	.7	1.0
DG B1	.7	.8	1.1
DG B2	.7	.8	1.0
DG B3	.6	.7	.9
DG C1	.6	.7	.9
DG C2	.6	.7	.9
DG C3	.6	.7	.8
SS A1	.5	.8	1.0
SS A2	.5	.7	.9
SS A3	.6	.8	1.1
SS A4	.5	.7	.9
SS A5	.6	.9	1.1
SS A6	.5	.7	.9
SS A7	.6	.9	1.3
SS A8	.5	.7	.9
SS B1	.6	.9	1.1
SS B2	.5	.7	.9
SS B3	.6	.9	1.2
SS B4	.5	.7	1.0
SS B5	.7	1.0	1.3
SS B6	.5	.7	1.0
SS B7	.7	1.1	1.4
SS B8	.5	.7	1.0
SS C1	.6	.8	1.0
SS C2	.5	.7	.9
SS C3	.7	1.0	1.3
SS C4	.6	.8	1.0
SS D1	.5	.7	.9
SS D2	.4	.6	.8
SS D3	.5	.7	1.0
SS D4	.4	.6	.8
SS E1	.5	.7	1.0
SS E2	.5	.6	.8
SS E3	.6	.9	1.1
SS E4	.5	.7	.9

PRESCOTT, AZ

SOLAR SAVINGS

	10%	50%	80%
WW A1	1.1	1.6	2.1
WW A2	.9	1.2	1.6
WW A3	.9	1.1	1.5
WW A4	.8	1.1	1.4
WW A5	.8	1.1	1.3
WW A6	.8	1.0	1.3
WW B1	1.1	1.5	1.9
WW B2	.8	1.1	1.4
WW B3	.7	.9	1.2
WW B4	.7	.9	1.2
WW B5	.7	.9	1.2
WW C1	.7	.9	1.2
WW C2	.7	.9	1.2
WW C3	.7	.8	1.0
WW C4	.7	.9	1.1
TW A1	1.1	1.6	2.1
TW A2	1.0	1.3	1.7
TW A3	1.0	1.3	1.6
TW A4	.9	1.2	1.6
TW B1	1.1	1.6	2.0
TW B2	1.0	1.4	1.8
TW B3	1.0	1.3	1.7
TW B4	1.1	1.4	1.8
TW C1	1.1	1.5	2.0
TW C2	1.1	1.5	1.9
TW C3	1.2	1.5	2.0
TW C4	1.3	1.7	2.2
TW D1	1.2	1.6	2.1
TW D2	.9	1.2	1.5
TW D3	.8	1.1	1.4
TW D4	.8	1.0	1.3
TW D5	.8	1.0	1.3
TW E1	.8	1.0	1.3
TW E2	.8	1.0	1.3
TW E3	.7	.9	1.1
TW E4	.7	.9	1.1
TW F1	1.1	1.6	2.1
TW F2	1.0	1.4	1.8
TW F3	1.0	1.3	1.7
TW F4	1.0	1.4	1.7
TW G1	1.2	1.6	2.1
TW G2	1.1	1.5	1.9
TW G3	1.1	1.5	1.9
TW G4	1.2	1.6	2.0
TW H1	1.2	1.6	2.1
TW H2	1.3	1.7	2.1
TW H3	1.4	1.8	2.3
TW H4	1.6	2.1	2.6
TW I1	1.3	1.7	2.2
TW I2	.9	1.2	1.6
TW I3	.9	1.1	1.5
TW I4	.8	1.1	1.3
TW I5	.8	1.0	1.3
TW J1	.8	1.0	1.3
TW J2	.8	1.0	1.3
TW J3	.7	.9	1.1
TW J4	.7	.9	1.2
DG A1	1.1	1.8	---
DG A2	1.0	1.4	2.1
DG A3	1.0	1.1	1.5
DG B1	1.1	1.4	3.0
DG B2	1.0	1.2	1.7
DG B3	.9	1.1	1.3
DG C1	1.0	1.2	1.8

PRESCOTT, AZ

SOLAR SAVINGS

	10%	50%	80%
DG C2	1.0	1.1	1.4
DG C3	.9	1.0	1.2
SS A1	.8	1.2	1.6
SS A2	.7	.9	1.2
SS A3	.8	1.3	1.7
SS A4	.7	1.0	1.3
SS A5	.8	1.3	1.7
SS A6	.7	1.0	1.2
SS A7	.9	1.5	2.0
SS A8	.7	1.0	1.3
SS B1	.9	1.3	1.7
SS B2	.8	1.0	1.4
SS B3	1.0	1.3	1.8
SS B4	.8	1.1	1.4
SS B5	1.0	1.5	2.0
SS B6	.8	1.1	1.4
SS B7	1.1	1.7	2.2
SS B8	.8	1.1	1.4
SS C1	.9	1.3	1.6
SS C2	.8	1.1	1.4
SS C3	1.1	1.5	2.0
SS C4	.9	1.2	1.5
SS D1	.7	1.0	1.3
SS D2	.6	.8	1.1
SS D3	.8	1.1	1.5
SS D4	.6	.9	1.1
SS E1	.8	1.1	1.4
SS E2	.7	.9	1.2
SS E3	.9	1.3	1.7
SS E4	.7	1.0	1.3

TUCSON, AZ

SOLAR SAVINGS

	10%	50%	80%
WW A1	.7	1.0	1.3
WW A2	.6	.8	1.0
WW A3	.6	.8	1.0
WW A4	.6	.7	.9
WW A5	.5	.7	.9
WW A6	.5	.7	.9
WW B1	.7	.9	1.2
WW B2	.6	.7	.9
WW B3	.5	.6	.8
WW B4	.5	.6	.8
WW B5	.5	.7	.8
WW C1	.5	.6	.8
WW C2	.5	.7	.8
WW C3	.5	.6	.7
WW C4	.5	.6	.8
TW A1	.7	1.0	1.3
TW A2	.7	.9	1.1
TW A3	.6	.8	1.1
TW A4	.6	.8	1.1
TW B1	.7	1.0	1.3
TW B2	.7	.9	1.2
TW B3	.7	.9	1.1
TW B4	.7	.9	1.2
TW C1	.8	1.0	1.3
TW C2	.7	1.0	1.3
TW C3	.8	1.0	1.3
TW C4	.8	1.1	1.5
TW D1	.8	1.0	1.3
TW D2	.6	.8	1.0
TW D3	.6	.7	.9

Conservation Factor Tables / 309

TUCSON, AZ

SOLAR SAVINGS

	10%	50%	80%
WW A1	.7	1.0	1.3
WW A2	.6	.8	1.0
WW A3	.6	.8	1.0
WW A4	.6	.7	.9
WW A5	.5	.7	.9
WW A6	.5	.7	.9
WW B1	.7	.9	1.2
WW B2	.6	.7	.9
WW B3	.5	.6	.8
WW B4	.5	.6	.8
WW B5	.5	.7	.8
WW C1	.5	.6	.8
WW C2	.5	.7	.8
WW C3	.5	.6	.7
WW C4	.5	.6	.8
TW A1	.7	1.0	1.3
TW A2	.7	.9	1.1
TW A3	.6	.8	1.1
TW A4	.6	.8	1.1
TW B1	.7	1.0	1.3
TW B2	.7	.9	1.2
TW B3	.7	.9	1.1
TW B4	.7	.9	1.2
TW C1	.8	1.0	1.3
TW C2	.7	1.0	1.3
TW C3	.8	1.0	1.3
TW C4	.8	1.1	1.5
TW D1	.8	1.0	1.3
TW D2	.6	.8	1.0
TW D3	.6	.7	.9
DG B1	.7	.8	1.1
DG B2	.7	.8	1.0
DG B3	.7	.7	.9
DG C1	.6	.7	.9
DG C2	.6	.7	.9
DG C3	.6	.7	.8
SS A1	.6	.8	1.0
SS A2	.5	.7	.9
SS A3	.6	.8	1.1
SS A4	.5	.7	.9
SS A5	.6	.9	1.1
SS A6	.5	.7	.9
SS A7	.6	1.0	1.3
SS A8	.5	.7	.9
SS B1	.6	.9	1.1
SS B2	.5	.7	.9
SS B3	.6	.9	1.2
SS B4	.5	.7	1.0
SS B5	.7	1.0	1.3
SS B6	.5	.7	.9
SS B7	.7	1.1	1.4
SS B8	.6	.8	1.0
SS C1	.6	.8	1.0
SS C2	.6	.7	.9
SS C3	.7	1.0	1.3
SS C4	.6	.8	1.0
SS D1	.5	.7	.9
SS D2	.4	.6	.7
SS D3	.5	.7	1.0
SS D4	.4	.6	.8
SS E1	.5	.7	.9
SS E2	.5	.6	.8
SS E3	.6	.9	1.1
SS E4	.5	.7	.9

WINSLOW, AZ

SOLAR SAVINGS

	10%	50%	80%
WW A1	1.2	1.8	2.3
WW A2	1.0	1.3	1.8
WW A3	.9	1.2	1.6
WW A4	.9	1.2	1.5
WW A5	.9	1.1	1.5
WW A6	.9	1.1	1.4
WW B1	1.1	1.6	2.2
WW B2	.9	1.2	1.5
WW B3	.8	1.0	1.3
WW B4	.8	1.0	1.3
WW B5	.8	1.0	1.3
WW C1	.8	1.0	1.3
WW C2	.8	1.0	1.3
WW C3	.7	.9	1.1
WW C4	.7	.9	1.2
TW A1	1.2	1.8	2.4
TW A2	1.1	1.5	1.9
TW A3	1.0	1.4	1.8
TW A4	1.0	1.3	1.7
TW B1	1.2	1.7	2.3
TW B2	1.1	1.5	1.9
TW B3	1.1	1.5	1.9
TW B4	1.1	1.5	2.0
TW C1	1.2	1.7	2.2
TW C2	1.2	1.6	2.1
TW C3	1.2	1.7	2.2
TW C4	1.3	1.8	2.4
TW D1	1.3	1.8	2.4
TW D2	.9	1.3	1.6
TW D3	.9	1.2	1.5
TW D4	.8	1.1	1.4
TW D5	.8	1.1	1.4
TW E1	.8	1.1	1.4
TW E2	.8	1.1	1.4
TW E3	.7	.9	1.2
TW E4	.8	1.0	1.2
TW F1	1.2	1.8	2.3
TW F2	1.1	1.5	2.0
TW F3	1.1	1.5	1.9
TW F4	1.1	1.5	1.9
TW G1	1.3	1.7	2.3
TW G2	1.2	1.6	2.1
TW G3	1.2	1.6	2.1
TW G4	1.3	1.7	2.3
TW H1	1.3	1.8	2.3
TW H2	1.4	1.8	2.3
TW H3	1.5	1.9	2.5
TW H4	1.7	2.2	2.9
TW I1	1.4	1.9	2.6
TW I2	1.0	1.3	1.7
TW I3	.9	1.2	1.6
TW I4	.9	1.1	1.4
TW I5	.9	1.1	1.4
TW J1	.8	1.1	1.4
TW J2	.9	1.1	1.4
TW J3	.8	1.0	1.2
TW J4	.8	1.0	1.3
DG A1	1.2	2.2	---
DG A2	1.1	1.5	2.5
DG A3	1.0	1.2	1.7
DG B1	1.2	1.6	---
DG B2	1.1	1.3	1.9
DG B3	1.0	1.1	1.5
DG C1	1.1	1.3	2.3

WINSLOW, AZ

SOLAR SAVINGS

	10%	50%	80%
DG C2	1.0	1.2	1.6
DG C3	.9	1.0	1.3
SS A1	.9	1.3	1.8
SS A2	.7	1.0	1.4
SS A3	.9	1.4	1.9
SS A4	.7	1.1	1.4
SS A5	.9	1.4	2.0
SS A6	.7	1.0	1.4
SS A7	1.0	1.7	2.3
SS A8	.8	1.1	1.4
SS B1	1.0	1.5	2.0
SS B2	.8	1.1	1.5
SS B3	1.0	1.5	2.1
SS B4	.8	1.2	1.5
SS B5	1.1	1.7	2.3
SS B6	.8	1.2	1.5
SS B7	1.2	1.9	2.6
SS B8	.9	1.2	1.6
SS C1	1.0	1.4	1.8
SS C2	.9	1.2	1.5
SS C3	1.2	1.7	2.2
SS C4	1.0	1.3	1.7
SS D1	.8	1.1	1.5
SS D2	.7	.9	1.2
SS D3	.8	1.2	1.7
SS D4	.7	.9	1.2
SS E1	.9	1.2	1.6
SS E2	.7	1.0	1.3
SS E3	1.0	1.5	2.0
SS E4	.8	1.1	1.4

YUMA, AZ

SOLAR SAVINGS

	10%	50%	80%
WW A1	.5	.8	1.0
WW A2	.5	.6	.8
WW A3	.4	.6	.8
WW A4	.4	.6	.7
WW A5	.4	.5	.7
WW A6	.4	.5	.7
WW B1	.5	.7	.9
WW B2	.4	.6	.7
WW B3	.4	.5	.7
WW B4	.4	.5	.7
WW B5	.4	.5	.7
WW C1	.4	.5	.6
WW C2	.4	.5	.7
WW C3	.4	.5	.6
WW C4	.4	.5	.6
TW A1	.6	.8	1.1
TW A2	.5	.7	.9
TW A3	.5	.6	.8
TW A4	.5	.6	.8
TW B1	.6	.8	1.0
TW B2	.5	.7	.9
TW B3	.5	.7	.9
TW B4	.5	.7	1.0
TW C1	.6	.8	1.0
TW C2	.6	.8	1.0
TW C3	.6	.8	1.1
TW C4	.7	.9	1.2
TW D1	.6	.8	1.0
TW D2	.5	.6	.8
TW D3	.4	.6	.7

YUMA, AZ

SOLAR SAVINGS

	10%	50%	80%
TW D4	.4	.6	.7
TW D5	.4	.6	.7
TW E1	.4	.5	.7
TW E2	.4	.5	.7
TW E3	.4	.5	.6
TW E4	.4	.5	.7
TW F1	.6	.8	1.0
TW F2	.5	.7	.9
TW F3	.5	.7	.9
TW F4	.5	.7	.9
TW G1	.6	.8	1.0
TW G2	.6	.7	1.0
TW G3	.6	.8	1.0
TW G4	.6	.8	1.1
TW H1	.6	.8	1.1
TW H2	.6	.9	1.1
TW H3	.7	.9	1.2
TW H4	.8	1.1	1.4
TW I1	.6	.8	1.1
TW I2	.5	.6	.8
TW I3	.5	.6	.8
TW I4	.4	.6	.7
TW I5	.4	.6	.7
TW J1	.4	.6	.7
TW J2	.4	.6	.7
TW J3	.4	.5	.7
TW J4	.4	.5	.7
DG A1	.5	.7	1.0
DG A2	.5	.7	.9
DG A3	.5	.6	.8
DG B1	.5	.6	.8
DG B2	.5	.6	.8
DG B3	.5	.6	.7
DG C1	.5	.6	.7
DG C2	.5	.6	.7
DG C3	.5	.5	.7
SS A1	.4	.6	.8
SS A2	.4	.5	.7
SS A3	.5	.7	.9
SS A4	.4	.5	.7
SS A5	.5	.7	.9
SS A6	.4	.5	.7
SS A7	.5	.8	1.0
SS A8	.4	.5	.7
SS B1	.5	.7	.9
SS B2	.4	.6	.8
SS B3	.5	.7	1.0
SS B4	.4	.6	.8
SS B5	.5	.8	1.0
SS B6	.4	.6	.8
SS B7	.6	.8	1.1
SS B8	.4	.6	.8
SS C1	.5	.6	.8
SS C2	.4	.6	.7
SS C3	.6	.8	1.0
SS C4	.5	.6	.8
SS D1	.4	.5	.7
SS D2	.3	.5	.6
SS D3	.4	.6	.8
SS D4	.4	.5	.6
SS E1	.4	.6	.8
SS E2	.4	.5	.7
SS E3	.5	.7	.9
SS E4	.4	.5	.7

FORT SMITH, AR
SOLAR SAVINGS

	10%	50%	80%
WW A1	1.3	2.1	2.8
WW A2	1.1	1.6	2.0
WW A3	1.0	1.4	1.9
WW A4	1.0	1.4	1.8
WW A5	1.0	1.3	1.7
WW A6	.9	1.3	1.6
WW B1	1.3	2.0	2.7
WW B2	1.0	1.3	1.7
WW B3	.9	1.1	1.5
WW B4	.8	1.1	1.4
WW B5	.8	1.1	1.4
WW C1	.8	1.1	1.4
WW C2	.8	1.1	1.4
WW C3	.8	1.0	1.2
WW C4	.8	1.0	1.3
TW A1	1.3	2.1	2.8
TW A2	1.2	1.7	2.2
TW A3	1.1	1.6	2.1
TW A4	1.1	1.5	2.0
TW B1	1.3	2.0	2.6
TW B2	1.2	1.7	2.2
TW B3	1.2	1.7	2.2
TW B4	1.2	1.7	2.3
TW C1	1.3	1.9	2.5
TW C2	1.3	1.8	2.4
TW C3	1.3	1.9	2.5
TW C4	1.5	2.1	2.8
TW D1	1.4	2.2	2.9
TW D2	1.0	1.4	1.8
TW D3	1.0	1.3	1.7
TW D4	.9	1.2	1.5
TW D5	.9	1.2	1.5
TW E1	.9	1.2	1.6
TW E2	.9	1.2	1.5
TW E3	.8	1.0	1.3
TW E4	.8	1.1	1.4
TW F1	1.4	2.1	2.8
TW F2	1.2	1.8	2.3
TW F3	1.2	1.7	2.2
TW F4	1.2	1.7	2.2
TW G1	1.4	2.0	2.7
TW G2	1.3	1.8	2.4
TW G3	1.3	1.8	2.4
TW G4	1.4	2.0	2.6
TW H1	1.5	2.1	2.7
TW H2	1.5	2.1	2.7
TW H3	1.6	2.2	2.9
TW H4	1.9	2.6	3.3
TW I1	1.5	2.3	3.1
TW I2	1.1	1.5	1.9
TW I3	1.0	1.4	1.8
TW I4	1.0	1.3	1.6
TW I5	.9	1.2	1.6
TW J1	.9	1.3	1.6
TW J2	.9	1.2	1.6
TW J3	.8	1.1	1.4
TW J4	.9	1.1	1.4
DG A1	1.3	---	---
DG A2	1.2	1.8	3.8
DG A3	1.1	1.3	1.9
DG B1	1.3	2.2	---
DG B2	1.2	1.5	2.5
DG B3	1.1	1.3	1.6
DG C1	1.1	1.6	---

FORT SMITH, AR
SOLAR SAVINGS

	10%	50%	80%
DG C2	1.1	1.4	1.9
DG C3	1.0	1.2	1.4
SS A1	1.0	1.5	2.0
SS A2	.8	1.2	1.5
SS A3	1.1	1.6	2.2
SS A4	.8	1.2	1.6
SS A5	1.0	1.7	2.3
SS A6	.8	1.2	1.5
SS A7	1.1	2.0	2.7
SS A8	.8	1.2	1.6
SS B1	1.1	1.7	2.2
SS B2	.9	1.3	1.7
SS B3	1.2	1.8	2.4
SS B4	.9	1.3	1.7
SS B5	1.2	2.0	2.7
SS B6	.9	1.3	1.7
SS B7	1.4	2.2	3.1
SS B8	.9	1.3	1.8
SS C1	1.1	1.6	2.1
SS C2	1.0	1.3	1.7
SS C3	1.3	1.9	2.6
SS C4	1.0	1.4	1.9
SS D1	.9	1.3	1.7
SS D2	.7	1.0	1.3
SS D3	.9	1.4	1.9
SS D4	.8	1.0	1.4
SS E1	1.0	1.4	1.9
SS E2	.8	1.1	1.5
SS E3	1.1	1.7	2.3
SS E4	.9	1.2	1.6

LITTLE ROCK, AR
SOLAR SAVINGS

	10%	50%	80%
WW A1	1.3	2.1	2.9
WW A2	1.1	1.6	2.1
WW A3	1.0	1.5	1.9
WW A4	1.0	1.4	1.8
WW A5	1.0	1.3	1.7
WW A6	1.0	1.3	1.7
WW B1	1.3	2.1	2.8
WW B2	1.0	1.3	1.7
WW B3	.9	1.2	1.5
WW B4	.8	1.1	1.4
WW B5	.9	1.1	1.4
WW C1	.8	1.1	1.5
WW C2	.9	1.1	1.5
WW C3	.8	1.0	1.3
WW C4	.8	1.0	1.3
TW A1	1.3	2.1	2.9
TW A2	1.2	1.7	2.3
TW A3	1.1	1.6	2.1
TW A4	1.1	1.6	2.1
TW B1	1.3	2.0	2.7
TW B2	1.2	1.7	2.3
TW B3	1.2	1.7	2.2
TW B4	1.2	1.8	2.3
TW C1	1.3	2.0	2.6
TW C2	1.3	1.9	2.5
TW C3	1.4	1.9	2.6
TW C4	1.5	2.1	2.8
TW D1	1.5	2.3	3.1
TW D2	1.0	1.5	1.9
TW D3	1.0	1.3	1.7

LITTLE ROCK, AR
SOLAR SAVINGS

	10%	50%	80%
TW D4	.9	1.2	1.6
TW D5	.9	1.2	1.6
TW E1	.9	1.2	1.6
TW E2	.9	1.2	1.6
TW E3	.8	1.0	1.3
TW E4	.8	1.1	1.4
TW F1	1.4	2.1	2.8
TW F2	1.2	1.8	2.4
TW F3	1.2	1.7	2.2
TW F4	1.2	1.7	2.2
TW G1	1.4	2.1	2.8
TW G2	1.3	1.9	2.5
TW G3	1.4	1.9	2.5
TW G4	1.5	2.0	2.6
TW H1	1.5	2.1	2.8
TW H2	1.5	2.1	2.8
TW H3	1.6	2.3	3.0
TW H4	1.9	2.6	3.4
TW I1	1.6	2.4	3.3
TW I2	1.1	1.5	2.0
TW I3	1.0	1.4	1.8
TW I4	1.0	1.3	1.6
TW I5	1.0	1.3	1.6
TW J1	.9	1.3	1.6
TW J2	.9	1.3	1.6
TW J3	.8	1.1	1.4
TW J4	.9	1.1	1.4
DG A1	1.4	---	---
DG A2	1.2	1.9	4.4
DG A3	1.1	1.4	2.0
DG B1	1.3	2.4	---
DG B2	1.2	1.6	2.6
DG B3	1.1	1.3	1.7
DG C1	1.2	1.7	---
DG C2	1.1	1.4	2.0
DG C3	1.0	1.2	1.5
SS A1	1.0	1.5	2.0
SS A2	.8	1.2	1.5
SS A3	1.1	1.7	2.3
SS A4	.8	1.2	1.6
SS A5	1.0	1.7	2.3
SS A6	.8	1.2	1.6
SS A7	1.1	2.0	2.8
SS A8	.8	1.2	1.6
SS B1	1.1	1.7	2.3
SS B2	.9	1.3	1.7
SS B3	1.2	1.8	2.5
SS B4	.9	1.3	1.7
SS B5	1.2	2.0	2.7
SS B6	.9	1.3	1.7
SS B7	1.4	2.3	3.1
SS B8	.9	1.4	1.8
SS C1	1.1	1.6	2.1
SS C2	1.0	1.3	1.7
SS C3	1.3	2.0	2.6
SS C4	1.1	1.5	1.9
SS D1	.9	1.3	1.7
SS D2	.7	1.0	1.3
SS D3	.9	1.4	1.9
SS D4	.8	1.1	1.4
SS E1	1.0	1.4	1.9
SS E2	.8	1.1	1.5
SS E3	1.1	1.7	2.3
SS E4	.9	1.2	1.6

BAKERSFIELD, CA
SOLAR SAVINGS

	10%	50%	80%
WW A1	.9	1.4	1.9
WW A2	.7	1.1	1.5
WW A3	.7	1.0	1.4
WW A4	.7	1.0	1.3
WW A5	.7	.9	1.2
WW A6	.7	.9	1.2
WW B1	.8	1.3	1.8
WW B2	.7	.9	1.3
WW B3	.6	.8	1.1
WW B4	.6	.8	1.1
WW B5	.6	.8	1.1
WW C1	.6	.8	1.1
WW C2	.6	.8	1.1
WW C3	.6	.8	1.0
WW C4	.6	.8	1.0
TW A1	.9	1.4	1.9
TW A2	.8	1.2	1.6
TW A3	.8	1.1	1.5
TW A4	.8	1.1	1.5
TW B1	.9	1.4	1.9
TW B2	.8	1.2	1.6
TW B3	.8	1.2	1.6
TW B4	.9	1.2	1.7
TW C1	.9	1.3	1.8
TW C2	.9	1.3	1.8
TW C3	.9	1.4	1.8
TW C4	1.0	1.5	2.0
TW D1	.9	1.4	2.0
TW D2	.7	1.0	1.4
TW D3	.7	1.0	1.3
TW D4	.7	.9	1.2
TW D5	.7	.9	1.2
TW E1	.6	.9	1.2
TW E2	.7	.9	1.2
TW E3	.6	.8	1.0
TW E4	.6	.8	1.1
TW F1	.9	1.4	1.9
TW F2	.8	1.2	1.7
TW F3	.8	1.2	1.6
TW F4	.9	1.2	1.6
TW G1	.9	1.4	1.9
TW G2	.9	1.3	1.7
TW G3	.9	1.3	1.8
TW G4	1.0	1.4	1.9
TW H1	1.0	1.4	2.0
TW H2	1.0	1.5	2.0
TW H3	1.1	1.6	2.1
TW H4	1.3	1.8	2.5
TW I1	1.0	1.5	2.1
TW I2	.8	1.1	1.4
TW I3	.7	1.0	1.4
TW I4	.7	.9	1.2
TW I5	.7	.9	1.2
TW J1	.7	.9	1.2
TW J2	.7	.9	1.2
TW J3	.6	.8	1.1
TW J4	.6	.9	1.1
DG A1	.9	1.4	8.4
DG A2	.9	1.2	1.8
DG A3	.8	1.0	1.4
DG B1	.9	1.2	2.2
DG B2	.8	1.1	1.5
DG B3	.8	1.0	1.2
DG C1	.8	1.0	1.5

BAKERSFIELD, CA

	SOLAR SAVINGS		
	10%	50%	80%
DG C2	.8	1.0	1.3
DG C3	.7	.9	1.1
SS A1	.7	1.1	1.5
SS A2	.6	.9	1.2
SS A3	.7	1.1	1.6
SS A4	.6	.9	1.2
SS A5	.7	1.2	1.7
SS A6	.6	.9	1.2
SS A7	.8	1.3	1.9
SS A8	.6	.9	1.2
SS B1	.8	1.2	1.6
SS B2	.7	1.0	1.3
SS B3	.8	1.2	1.7
SS B4	.7	1.0	1.3
SS B5	.8	1.4	1.9
SS B6	.7	1.0	1.3
SS B7	.9	1.5	2.1
SS B8	.7	1.0	1.4
SS C1	.8	1.1	1.5
SS C2	.7	1.0	1.3
SS C3	.9	1.3	1.8
SS C4	.8	1.1	1.4
SS D1	.6	.9	1.3
SS D2	.5	.8	1.0
SS D3	.6	1.0	1.4
SS D4	.5	.8	1.1
SS E1	.7	1.0	1.4
SS E2	.6	.8	1.1
SS E3	.8	1.2	1.7
SS E4	.6	.9	1.2

DAGGETT, CA

	SOLAR SAVINGS		
	10%	50%	80%
WW A1	.8	1.2	1.5
WW A2	.7	.9	1.2
WW A3	.6	.9	1.1
WW A4	.6	.8	1.1
WW A5	.6	.8	1.0
WW A6	.6	.8	1.0
WW B1	.7	1.1	1.4
WW B2	.6	.8	1.1
WW B3	.6	.7	.9
WW B4	.6	.7	.9
WW B5	.6	.7	.9
WW C1	.5	.7	.9
WW C2	.6	.7	.9
WW C3	.5	.7	.8
WW C4	.5	.7	.9
TW A1	.8	1.2	1.6
TW A2	.7	1.0	1.3
TW A3	.7	1.0	1.2
TW A4	.7	.9	1.2
TW B1	.8	1.2	1.5
TW B2	.7	1.0	1.3
TW B3	.7	1.0	1.3
TW B4	.8	1.1	1.4
TW C1	.8	1.2	1.5
TW C2	.8	1.1	1.5
TW C3	.9	1.2	1.5
TW C4	.9	1.3	1.7
TW D1	.8	1.2	1.5
TW D2	.7	.9	1.2
TW D3	.6	.8	1.1

DAGGETT, CA

	SOLAR SAVINGS		
	10%	50%	80%
TW D4	.6	.8	1.0
TW D5	.6	.8	1.0
TW E1	.6	.8	1.0
TW E2	.6	.8	1.0
TW E3	.5	.7	.9
TW E4	.6	.7	.9
TW F1	.8	1.2	1.6
TW F2	.8	1.0	1.4
TW F3	.7	1.0	1.3
TW F4	.8	1.0	1.3
TW G1	.8	1.2	1.5
TW G2	.8	1.1	1.4
TW G3	.8	1.1	1.4
TW G4	.9	1.2	1.6
TW H1	.9	1.2	1.6
TW H2	.9	1.3	1.6
TW H3	1.0	1.4	1.8
TW H4	1.2	1.6	2.0
TW I1	.9	1.2	1.6
TW I2	.7	.9	1.2
TW I3	.7	.9	1.1
TW I4	.6	.8	1.1
TW I5	.6	.8	1.0
TW J1	.6	.8	1.0
TW J2	.6	.8	1.0
TW J3	.6	.7	.9
TW J4	.6	.8	1.0
DG A1	.8	1.1	1.8
DG A2	.8	1.0	1.4
DG A3	.7	.9	1.2
DG B1	.8	1.0	1.4
DG B2	.8	.9	1.2
DG B3	.7	.8	1.0
DG C1	.7	.8	1.1
DG C2	.7	.8	1.1
DG C3	.7	.8	.9
SS A1	.6	.9	1.2
SS A2	.5	.8	1.0
SS A3	.7	1.0	1.3
SS A4	.5	.8	1.0
SS A5	.7	1.0	1.4
SS A6	.5	.8	1.0
SS A7	.7	1.1	1.5
SS A8	.5	.8	1.0
SS B1	.7	1.0	1.4
SS B2	.6	.8	1.1
SS B3	.7	1.1	1.4
SS B4	.6	.9	1.1
SS B5	.8	1.2	1.6
SS B6	.6	.8	1.1
SS B7	.8	1.3	1.7
SS B8	.6	.9	1.1
SS C1	.7	.9	1.2
SS C2	.6	.8	1.1
SS C3	.8	1.1	1.5
SS C4	.7	.9	1.2
SS D1	.5	.8	1.0
SS D2	.5	.7	.9
SS D3	.6	.9	1.2
SS D4	.5	.7	.9
SS E1	.6	.9	1.1
SS E2	.5	.7	1.0
SS E3	.7	1.0	1.4
SS E4	.6	.8	1.0

FRESNO, CA

	SOLAR SAVINGS		
	10%	50%	80%
WW A1	1.0	1.6	2.4
WW A2	.8	1.3	1.8
WW A3	.8	1.2	1.6
WW A4	.8	1.1	1.6
WW A5	.8	1.1	1.5
WW A6	.7	1.1	1.5
WW B1	.9	1.5	2.3
WW B2	.8	1.1	1.5
WW B3	.7	1.0	1.3
WW B4	.7	.9	1.3
WW B5	.7	.9	1.3
WW C1	.7	.9	1.3
WW C2	.7	1.0	1.3
WW C3	.6	.8	1.1
WW C4	.7	.9	1.2
TW A1	1.0	1.6	2.4
TW A2	.9	1.4	2.0
TW A3	.9	1.3	1.8
TW A4	.9	1.3	1.8
TW B1	1.0	1.6	2.3
TW B2	.9	1.4	2.0
TW B3	.9	1.4	1.9
TW B4	1.0	1.4	2.0
TW C1	1.0	1.6	2.2
TW C2	1.0	1.5	2.1
TW C3	1.0	1.6	2.2
TW C4	1.1	1.7	2.4
TW D1	1.1	1.7	2.5
TW D2	.8	1.2	1.6
TW D3	.8	1.1	1.5
TW D4	.7	1.0	1.4
TW D5	.7	1.0	1.4
TW E1	.7	1.0	1.4
TW E2	.7	1.0	1.4
TW E3	.6	.9	1.2
TW E4	.7	.9	1.2
TW F1	1.0	1.6	2.4
TW F2	.9	1.4	2.0
TW F3	.9	1.4	1.9
TW F4	1.0	1.4	1.9
TW G1	1.0	1.6	2.3
TW G2	1.0	1.5	2.1
TW G3	1.0	1.5	2.1
TW G4	1.1	1.6	2.3
TW H1	1.1	1.7	2.4
TW H2	1.2	1.7	2.4
TW H3	1.3	1.8	2.5
TW H4	1.4	2.1	2.9
TW I1	1.1	1.8	2.7
TW I2	.9	1.2	1.7
TW I3	.8	1.2	1.6
TW I4	.8	1.1	1.4
TW I5	.8	1.1	1.4
TW J1	.7	1.1	1.4
TW J2	.8	1.1	1.4
TW J3	.7	.9	1.2
TW J4	.7	1.0	1.3
DG A1	1.0	1.9	---
DG A2	.9	1.4	2.5
DG A3	.9	1.1	1.7
DG B1	1.0	1.5	---
DG B2	.9	1.3	1.9
DG B3	.8	1.1	1.5
DG C1	.9	1.3	2.4

FRESNO, CA

	SOLAR SAVINGS		
	10%	50%	80%
DG C2	.8	1.1	1.6
DG C3	.8	1.0	1.3
SS A1	.7	1.2	1.8
SS A2	.6	1.0	1.4
SS A3	.8	1.3	2.0
SS A4	.6	1.0	1.4
SS A5	.8	1.4	2.0
SS A6	.6	1.0	1.4
SS A7	.8	1.6	2.4
SS A8	.6	1.0	1.5
SS B1	.8	1.4	2.0
SS B2	.7	1.1	1.5
SS B3	.9	1.5	2.2
SS B4	.7	1.1	1.5
SS B5	.9	1.6	2.4
SS B6	.7	1.1	1.5
SS B7	1.0	1.8	2.7
SS B8	.7	1.1	1.6
SS C1	.9	1.3	1.8
SS C2	.8	1.1	1.5
SS C3	1.0	1.6	2.2
SS C4	.8	1.2	1.7
SS D1	.7	1.1	1.5
SS D2	.6	.9	1.2
SS D3	.7	1.2	1.7
SS D4	.6	.9	1.2
SS E1	.7	1.2	1.7
SS E2	.6	1.0	1.3
SS E3	.8	1.4	2.0
SS E4	.7	1.0	1.4

LONG BEACH, CA

	SOLAR SAVINGS		
	10%	50%	80%
WW A1	.7	.9	1.2
WW A2	.6	.8	1.0
WW A3	.5	.7	.9
WW A4	.5	.7	.9
WW A5	.5	.7	.8
WW A6	.5	.6	.8
WW B1	.6	.8	1.1
WW B2	.5	.7	.9
WW B3	.5	.6	.8
WW B4	.5	.6	.8
WW B5	.5	.6	.8
WW C1	.5	.6	.8
WW C2	.5	.6	.8
WW C3	.4	.6	.7
WW C4	.5	.6	.7
TW A1	.7	1.0	1.3
TW A2	.6	.8	1.1
TW A3	.6	.8	1.0
TW A4	.6	.8	1.0
TW B1	.7	.9	1.2
TW B2	.6	.8	1.1
TW B3	.6	.8	1.1
TW B4	.7	.9	1.1
TW C1	.7	1.0	1.2
TW C2	.7	.9	1.2
TW C3	.7	1.0	1.2
TW C4	.8	1.1	1.4
TW D1	.7	.9	1.2
TW D2	.6	.7	.9
TW D3	.5	.7	.9

LONG BEACH, CA					LOS ANGELES, CA					LOS ANGELES, CA					MOUNT SHASTA, CA			
	SOLAR SAVINGS					SOLAR SAVINGS					SOLAR SAVINGS					SOLAR SAVINGS		
	10%	50%	80%			10%	50%	80%			10%	50%	80%			10%	50%	80%
WW A1	.7	.9	1.2		WW A1	.6	.9	1.2		DG C2	.6	.7	.8		TW D4	1.0	1.3	1.8
WW A2	.6	.8	1.0		WW A2	.5	.7	1.0		DG C3	.5	.6	.8		TW D5	1.0	1.3	1.7
WW A3	.5	.7	.9		WW A3	.5	.7	.9		SS A1	.5	.7	.9		TW E1	1.0	1.4	1.8
WW A4	.5	.7	.9		WW A4	.5	.7	.9		SS A2	.4	.6	.8		TW E2	1.0	1.3	1.8
WW A5	.5	.7	.8		WW A5	.5	.7	.8		SS A3	.5	.7	1.0		TW E3	.8	1.1	1.5
WW A6	.5	.6	.8		WW A6	.5	.6	.8		SS A4	.4	.6	.8		TW E4	.9	1.2	1.5
WW B1	.6	.8	1.1		WW B1	.6	.8	1.1		SS A5	.5	.8	1.0		TW F1	1.4	2.5	3.8
WW B2	.5	.7	.9		WW B2	.5	.7	.9		SS A6	.4	.6	.8		TW F2	1.3	2.1	2.9
WW B3	.5	.6	.8		WW B3	.5	.6	.8		SS A7	.5	.8	1.1		TW F3	1.3	1.9	2.7
WW B4	.5	.6	.8		WW B4	.5	.6	.8		SS A8	.4	.6	.8		TW F4	1.3	1.9	2.7
WW B5	.5	.6	.8		WW B5	.5	.6	.8		SS B1	.5	.8	1.0		TW G1	1.5	2.4	3.5
WW C1	.5	.6	.8		WW C1	.4	.6	.7		SS B2	.5	.6	.8		TW G2	1.4	2.1	3.0
WW C2	.5	.6	.8		WW C2	.5	.6	.8		SS B3	.5	.8	1.0		TW G3	1.4	2.1	3.0
WW C3	.4	.6	.7		WW C3	.4	.5	.7		SS B4	.5	.7	.9		TW G4	1.6	2.3	3.1
WW C4	.5	.6	.7		WW C4	.5	.6	.7		SS B5	.6	.9	1.2		TW H1	1.6	2.4	3.5
TW A1	.7	1.0	1.3		TW A1	.7	1.0	1.2		SS B6	.5	.7	.8		TW H2	1.6	2.4	3.3
TW A2	.6	.8	1.1		TW A2	.6	.8	1.1		SS B7	.6	.9	1.2		TW H3	1.8	2.6	3.5
TW A3	.6	.8	1.0		TW A3	.6	.8	1.0		SS B8	.5	.7	.9		TW H4	2.0	2.9	4.0
TW A4	.6	.8	1.0		TW A4	.6	.8	1.0		SS C1	.6	.8	1.0		TW I1	1.7	3.2	6.4
TW B1	.7	.9	1.2		TW B1	.7	.9	1.2		SS C2	.5	.7	.9		TW I2	1.2	1.7	2.3
TW B2	.6	.8	1.1		TW B2	.6	.8	1.1		SS C3	.7	.9	1.2		TW I3	1.1	1.6	2.1
TW B3	.6	.8	1.1		TW B3	.6	.8	1.1		SS C4	.6	.7	1.0		TW I4	1.0	1.4	1.8
TW B4	.7	.9	1.1		TW B4	.6	.9	1.1		SS D1	.4	.6	.8		TW I5	1.0	1.4	1.8
TW C1	.7	1.0	1.2		TW C1	.7	.9	1.2		SS D2	.4	.5	.7		TW J1	1.0	1.4	1.9
TW C2	.7	.9	1.2		TW C2	.7	.9	1.2		SS D3	.4	.7	.9		TW J2	1.0	1.4	1.8
TW C3	.7	1.0	1.2		TW C3	.7	1.0	1.2		SS D4	.4	.5	.7		TW J3	.9	1.2	1.5
TW C4	.8	1.1	1.4		TW C4	.8	1.1	1.4		SS E1	.5	.6	.8		TW J4	.9	1.2	1.6
TW D1	.7	.9	1.2		TW D1	.7	.9	1.2		SS E2	.4	.6	.7		DG A1	1.5	---	---
TW D2	.6	.7	.9		TW D2	.5	.7	.9		SS E3	.5	.8	1.0		DG A2	1.3	2.2	---
TW D3	.5	.7	.9		TW D3	.5	.7	.9		SS E4	.4	.6	.8		DG A3	1.1	1.5	2.3
DG B1	.6	.7	1.0		TW D4	.5	.7	.8							DG B1	1.5	---	---
DG B2	.6	.7	.9		TW D5	.5	.7	.8		MOUNT SHASTA, CA					DG B2	1.3	1.8	4.4
DG B3	.6	.7	.9		TW E1	.5	.6	.8			SOLAR SAVINGS				DG B3	1.1	1.4	1.9
DG C1	.6	.7	.9		TW E2	.5	.6	.8			10%	50%	80%		DG C1	1.3	2.2	---
DG C2	.6	.7	.8		TW E3	.4	.6	.7		WW A1	1.4	2.6	4.1		DG C2	1.2	1.6	2.5
DG C3	.6	.6	.8		TW E4	.5	.6	.8		WW A2	1.2	1.8	2.6		DG C3	1.0	1.3	1.7
SS A1	.5	.7	1.0		TW F1	.7	.9	1.2		WW A3	1.1	1.7	2.3		SS A1	1.0	1.7	2.6
SS A2	.4	.6	.8		TW F2	.6	.8	1.1		WW A4	1.1	1.6	2.2		SS A2	.8	1.3	1.8
SS A3	.5	.8	1.0		TW F3	.6	.8	1.0		WW A5	1.0	1.5	2.1		SS A3	1.0	2.0	3.3
SS A4	.4	.6	.8		TW F4	.6	.8	1.1		WW A6	1.0	1.5	2.0		SS A4	.8	1.3	1.9
SS A5	.5	.8	1.0		TW G1	.7	1.0	1.2		WW B1	1.4	2.7	5.8		SS A5	1.0	2.0	3.2
SS A6	.4	.6	.8		TW G2	.7	.9	1.1		WW B2	1.0	1.5	2.0		SS A6	.8	1.3	1.8
SS A7	.6	.9	1.2		TW G3	.7	.9	1.2		WW B3	.9	1.3	1.7		SS A7	1.1	2.5	4.8
SS A8	.4	.6	.8		TW G4	.7	1.0	1.3		WW B4	.9	1.2	1.6		SS A8	.8	1.4	2.0
SS B1	.6	.8	1.1		TW H1	.7	1.0	1.3		WW B5	.9	1.2	1.6		SS B1	1.1	2.0	3.0
SS B2	.5	.7	.9		TW H2	.8	1.0	1.3		WW C1	.9	1.2	1.7		SS B2	.9	1.4	1.9
SS B3	.6	.8	1.1		TW H3	.8	1.1	1.4		WW C2	.9	1.2	1.6		SS B3	1.2	2.2	3.5
SS B4	.5	.7	.9		TW H4	1.0	1.3	1.6		WW C3	.8	1.1	1.4		SS B4	.9	1.4	2.0
SS B5	.6	.9	1.2		TW I1	.7	1.0	1.3		WW C4	.8	1.1	1.4		SS B5	1.2	2.4	3.8
SS B6	.5	.7	.9		TW I2	.6	.8	1.0		TW A1	1.4	2.5	3.9		SS B6	.9	1.4	2.0
SS B7	.7	1.0	1.3		TW I3	.5	.7	.9		TW A2	1.2	2.0	2.8		SS B7	1.3	2.8	5.3
SS B8	.5	.7	.9		TW I4	.5	.7	.9		TW A3	1.2	1.8	2.6		SS B8	1.0	1.5	2.1
SS C1	.6	.8	1.0		TW I5	.5	.7	.9		TW A4	1.2	1.8	2.4		SS C1	1.2	1.9	2.7
SS C2	.5	.7	.9		TW J1	.5	.7	.8		TW B1	1.4	2.4	3.5		SS C2	1.0	1.5	2.0
SS C3	.7	.9	1.2		TW J2	.5	.7	.9		TW B2	1.3	2.0	2.8		SS C3	1.4	2.3	3.4
SS C4	.6	.8	1.0		TW J3	.5	.6	.8		TW B3	1.3	1.9	2.7		SS C4	1.1	1.6	2.2
SS D1	.4	.6	.8		TW J4	.5	.6	.8		TW B4	1.3	2.0	2.7		SS D1	.9	1.5	2.2
SS D2	.4	.5	.7		DG A1	.6	.8	1.2		TW C1	1.4	2.3	3.2		SS D2	.8	1.1	1.5
SS D3	.5	.7	.9		DG A2	.6	.8	1.1		TW C2	1.4	2.1	2.9		SS D3	.9	1.7	2.5
SS D4	.4	.5	.7		DG A3	.6	.7	.9		TW C3	1.4	2.2	3.0		SS D4	.8	1.2	1.6
SS E1	.5	.7	.9		DG B1	.6	.7	1.0		TW C4	1.5	2.4	3.3		SS E1	1.0	1.7	2.6
SS E2	.4	.6	.8		DG B2	.6	.7	.9		TW D1	1.6	2.9	5.4		SS E2	.8	1.3	1.7
SS E3	.6	.8	1.0		DG B3	.6	.7	.8		TW D2	1.1	1.6	2.2		SS E3	1.1	2.0	3.2
SS E4	.5	.6	.8		DG C1	.6	.7	.8		TW D3	1.0	1.5	2.0		SS E4	.9	1.4	1.9

Conservation Factor Tables / 313

NEEDLES, CA				NEEDLES, CA				OAKLAND, CA				RED BLUFF, CA			
	SOLAR SAVINGS				SOLAR SAVINGS				SOLAR SAVINGS				SOLAR SAVINGS		
	10%	50%	80%		10%	50%	80%		10%	50%	80%		10%	50%	80%
WW A1	.6	.9	1.3	DG C2	.6	.7	.9	TW D4	.6	.8	1.1	WW A1	1.0	1.7	2.5
WW A2	.6	.8	1.0	DG C3	.6	.6	.8	TW D5	.6	.8	1.1	WW A2	.9	1.3	1.8
WW A3	.5	.7	.9	SS A1	.5	.8	1.0	TW E1	.6	.8	1.1	WW A3	.8	1.2	1.7
WW A4	.5	.7	.9	SS A2	.5	.6	.9	TW E2	.6	.8	1.1	WW A4	.8	1.1	1.6
WW A5	.5	.7	.9	SS A3	.6	.8	1.1	TW E3	.5	.7	.9	WW A5	.8	1.1	1.5
WW A6	.5	.6	.8	SS A4	.5	.7	.9	TW E4	.6	.8	1.0	WW A6	.8	1.1	1.5
WW B1	.6	.8	1.1	SS A5	.5	.8	1.1	TW F1	.9	1.2	1.7	WW B1	1.0	1.6	2.4
WW B2	.5	.7	.9	SS A6	.5	.6	.9	TW F2	.8	1.1	1.5	WW B2	.8	1.1	1.5
WW B3	.5	.6	.8	SS A7	.6	.9	1.3	TW F3	.8	1.1	1.4	WW B3	.7	1.0	1.3
WW B4	.5	.6	.8	SS A8	.5	.7	.9	TW F4	.8	1.1	1.4	WW B4	.7	1.0	1.3
WW B5	.5	.6	.8	SS B1	.6	.8	1.1	TW G1	.9	1.2	1.7	WW B5	.7	1.0	1.3
WW C1	.5	.6	.8	SS B2	.5	.7	.9	TW G2	.8	1.2	1.5	WW C1	.7	1.0	1.3
WW C2	.5	.6	.8	SS B3	.6	.9	1.2	TW G3	.9	1.2	1.6	WW C2	.7	1.0	1.3
WW C3	.4	.6	.7	SS B4	.5	.7	1.0	TW G4	.9	1.3	1.7	WW C3	.6	.9	1.1
WW C4	.5	.6	.8	SS B5	.6	1.0	1.3	TW H1	.9	1.3	1.7	WW C4	.7	.9	1.2
TW A1	.7	1.0	1.3	SS B6	.5	.7	.9	TW H2	1.0	1.3	1.8	TW A1	1.0	1.7	2.5
TW A2	.6	.8	1.1	SS B7	.7	1.0	1.4	TW H3	1.0	1.4	1.9	TW A2	.9	1.4	2.0
TW A3	.6	.8	1.0	SS B8	.5	.7	1.0	TW H4	1.2	1.7	2.2	TW A3	.9	1.3	1.8
TW A4	.6	.8	1.0	SS C1	.6	.8	1.0	TW I1	.9	1.3	1.8	TW A4	.9	1.3	1.8
TW B1	.7	.9	1.3	SS C2	.5	.7	.9	TW I2	.7	1.0	1.3	TW B1	1.0	1.6	2.3
TW B2	.6	.8	1.1	SS C3	.7	.9	1.2	TW I3	.7	.9	1.2	TW B2	1.0	1.4	2.0
TW B3	.6	.8	1.1	SS C4	.6	.8	1.0	TW I4	.7	.9	1.1	TW B3	.9	1.4	1.9
TW B4	.6	.9	1.1	SS D1	.5	.7	.9	TW I5	.6	.9	1.1	TW B4	1.0	1.5	2.0
TW C1	.7	.9	1.3	SS D2	.4	.6	.7	TW J1	.6	.8	1.1	TW C1	1.0	1.6	2.3
TW C2	.7	.9	1.2	SS D3	.5	.7	1.0	TW J2	.6	.9	1.1	TW C2	1.0	1.5	2.1
TW C3	.7	1.0	1.3	SS D4	.4	.6	.8	TW J3	.6	.8	1.0	TW C3	1.1	1.6	2.2
TW C4	.8	1.1	1.4	SS E1	.5	.7	.9	TW J4	.6	.8	1.0	TW C4	1.2	1.8	2.4
TW D1	.7	.9	1.3	SS E2	.5	.6	.8	DG A1	.8	1.2	2.2	TW D1	1.1	1.7	2.6
TW D2	.6	.7	1.0	SS E3	.6	.8	1.1	DG A2	.8	1.1	1.5	TW D2	.8	1.2	1.6
TW D3	.5	.7	.9	SS E4	.5	.7	.9	DG A3	.7	.9	1.3	TW D3	.8	1.1	1.5
TW D4	.5	.7	.9					DG B1	.8	1.0	1.6	TW D4	.8	1.0	1.4
TW D5	.5	.7	.9	OAKLAND, CA				DG B2	.8	1.0	1.3	TW D5	.8	1.0	1.4
TW E1	.5	.6	.8		SOLAR SAVINGS			DG B3	.7	.9	1.1	TW E1	.7	1.0	1.4
TW E2	.5	.7	.8		10%	50%	80%	DG C1	.7	.9	1.3	TW E2	.7	1.0	1.4
TW E3	.4	.6	.7	WW A1	.8	1.2	1.7	DG C2	.7	.9	1.1	TW E3	.7	.9	1.2
TW E4	.5	.6	.8	WW A2	.7	1.0	1.3	DG C3	.7	.8	1.0	TW E4	.7	.9	1.2
TW F1	.7	1.0	1.3	WW A3	.7	.9	1.2	SS A1	.6	.9	1.3	TW F1	1.0	1.7	2.4
TW F2	.6	.9	1.1	WW A4	.6	.9	1.2	SS A2	.5	.8	1.1	TW F2	1.0	1.5	2.0
TW F3	.6	.8	1.1	WW A5	.6	.8	1.1	SS A3	.6	1.0	1.4	TW F3	1.0	1.4	1.9
TW F4	.6	.8	1.1	WW A6	.6	.8	1.1	SS A4	.5	.8	1.1	TW F4	1.0	1.4	1.9
TW G1	.7	1.0	1.3	WW B1	.8	1.1	1.5	SS A5	.6	1.0	1.4	TW G1	1.1	1.7	2.4
TW G2	.7	.9	1.2	WW B2	.6	.9	1.1	SS A6	.5	.8	1.1	TW G2	1.0	1.5	2.1
TW G3	.7	.9	1.2	WW B3	.6	.8	1.0	SS A7	.7	1.1	1.6	TW G3	1.1	1.5	2.1
TW G4	.8	1.0	1.3	WW B4	.6	.8	1.0	SS A8	.5	.8	1.1	TW G4	1.2	1.7	2.3
TW H1	.7	1.0	1.3	WW B5	.6	.8	1.0	SS B1	.7	1.0	1.4	TW H1	1.2	1.7	2.4
TW H2	.8	1.0	1.4	WW C1	.6	.7	1.0	SS B2	.6	.8	1.1	TW H2	1.2	1.7	2.4
TW H3	.8	1.1	1.5	WW C2	.6	.8	1.0	SS B3	.7	1.1	1.5	TW H3	1.3	1.9	2.6
TW H4	1.0	1.3	1.7	WW C3	.5	.7	.9	SS B4	.6	.8	1.2	TW H4	1.5	2.2	3.0
TW I1	.7	1.0	1.3	WW C4	.6	.7	.9	SS B5	.8	1.2	1.6	TW I1	1.2	1.9	2.8
TW I2	.6	.8	1.0	TW A1	.9	1.3	1.7	SS B6	.6	.8	1.2	TW I2	.9	1.3	1.7
TW I3	.6	.7	1.0	TW A2	.8	1.1	1.4	SS B7	.8	1.3	1.8	TW I3	.8	1.2	1.6
TW I4	.5	.7	.9	TW A3	.7	1.0	1.3	SS B8	.6	.9	1.2	TW I4	.8	1.1	1.5
TW I5	.5	.7	.9	TW A4	.7	1.0	1.3	SS C1	.7	1.0	1.3	TW I5	.8	1.1	1.4
TW J1	.5	.7	.9	TW B1	.9	1.2	1.6	SS C2	.6	.9	1.1	TW J1	.8	1.1	1.5
TW J2	.5	.7	.9	TW B2	.8	1.1	1.4	SS C3	.8	1.2	1.6	TW J2	.8	1.1	1.4
TW J3	.5	.6	.8	TW B3	.8	1.1	1.4	SS C4	.7	1.0	1.3	TW J3	.7	.9	1.2
TW J4	.5	.6	.8	TW B4	.8	1.1	1.5	SS D1	.5	.8	1.1	TW J4	.7	1.0	1.3
DG A1	.6	.9	1.3	TW C1	.9	1.2	1.6	SS D2	.5	.7	.9	DG A1	1.0	2.0	---
DG A2	.6	.8	1.1	TW C2	.8	1.2	1.6	SS D3	.6	.9	1.2	DG A2	1.0	1.4	2.6
DG A3	.6	.7	1.0	TW C3	.9	1.2	1.6	SS D4	.5	.7	.9	DG A3	.9	1.2	1.7
DG B1	.6	.8	1.1	TW C4	1.0	1.4	1.8	SS E1	.6	.9	1.2	DG B1	1.0	1.6	---
DG B2	.6	.7	1.0	TW D1	.9	1.2	1.7	SS E2	.5	.7	1.0	DG B2	.9	1.3	2.0
DG B3	.6	.7	.9	TW D2	.7	.9	1.2	SS E3	.7	1.0	1.4	DG B3	.9	1.1	1.5
DG C1	.6	.7	.9	TW D3	.6	.9	1.1	SS E4	.5	.8	1.1	DG C1	.9	1.3	2.5

RED BLUFF, CA
SOLAR SAVINGS
	10%	50%	80%
DG C2	.9	1.1	1.6
DG C3	.8	1.0	1.3
SS A1	.8	1.3	1.9
SS A2	.7	1.0	1.4
SS A3	.8	1.4	2.1
SS A4	.7	1.1	1.5
SS A5	.8	1.4	2.1
SS A6	.7	1.0	1.4
SS A7	.9	1.7	2.6
SS A8	.7	1.1	1.5
SS B1	.9	1.4	2.1
SS B2	.7	1.1	1.5
SS B3	.9	1.5	2.2
SS B4	.8	1.1	1.6
SS B5	1.0	1.7	2.4
SS B6	.8	1.1	1.6
SS B7	1.0	1.9	2.8
SS B8	.8	1.2	1.6
SS C1	.9	1.3	1.8
SS C2	.8	1.1	1.5
SS C3	1.0	1.6	2.3
SS C4	.9	1.2	1.7
SS D1	.7	1.1	1.6
SS D2	.6	.9	1.2
SS D3	.7	1.2	1.8
SS D4	.6	.9	1.3
SS E1	.8	1.2	1.7
SS E2	.7	1.0	1.4
SS E3	.9	1.5	2.1
SS E4	.7	1.1	1.5

SACRAMENTO, CA
SOLAR SAVINGS
	10%	50%	80%
WW A1	1.0	1.7	2.5
WW A2	.8	1.3	1.8
WW A3	.8	1.2	1.7
WW A4	.8	1.1	1.6
WW A5	.8	1.1	1.5
WW A6	.8	1.1	1.5
WW B1	1.0	1.6	2.4
WW B2	.8	1.1	1.5
WW B3	.7	1.0	1.3
WW B4	.7	.9	1.3
WW B5	.7	1.0	1.3
WW C1	.7	.9	1.3
WW C2	.7	1.0	1.3
WW C3	.6	.8	1.1
WW C4	.7	.9	1.2
TW A1	1.0	1.7	2.5
TW A2	.9	1.4	2.0
TW A3	.9	1.3	1.8
TW A4	.9	1.3	1.8
TW B1	1.0	1.6	2.3
TW B2	.9	1.4	2.0
TW B3	.9	1.4	1.9
TW B4	1.0	1.4	2.0
TW C1	1.0	1.6	2.3
TW C2	1.0	1.5	2.1
TW C3	1.1	1.6	2.2
TW C4	1.2	1.7	2.5
TW D1	1.1	1.7	2.6
TW D2	.8	1.2	1.7
TW D3	.8	1.1	1.5

SACRAMENTO, CA (cont.)
SOLAR SAVINGS
	10%	50%	80%
TW D4	.7	1.0	1.4
TW D5	.8	1.0	1.4
TW E1	.7	1.0	1.4
TW E2	.7	1.0	1.4
TW E3	.7	.9	1.2
TW E4	.7	.9	1.2
TW F1	1.0	1.7	2.4
TW F2	1.0	1.4	2.1
TW F3	.9	1.4	1.9
TW F4	1.0	1.4	2.0
TW G1	1.1	1.6	2.4
TW G2	1.0	1.5	2.1
TW G3	1.1	1.5	2.1
TW G4	1.2	1.7	2.3
TW H1	1.1	1.7	2.4
TW H2	1.2	1.7	2.4
TW H3	1.3	1.9	2.6
TW H4	1.5	2.1	3.0
TW I1	1.2	1.8	2.8
TW I2	.9	1.3	1.7
TW I3	.8	1.2	1.6
TW I4	.8	1.1	1.5
TW I5	.8	1.1	1.4
TW J1	.8	1.1	1.5
TW J2	.8	1.1	1.4
TW J3	.7	.9	1.2
TW J4	.7	1.0	1.3
DG A1	1.0	1.9	---
DG A2	1.0	1.4	2.6
DG A3	.9	1.2	1.7
DG B1	1.0	1.5	---
DG B2	.9	1.3	2.0
DG B3	.9	1.1	1.5
DG C1	.9	1.3	2.6
DG C2	.9	1.1	1.6
DG C3	.8	1.0	1.3
SS A1	.8	1.3	1.9
SS A2	.6	1.0	1.4
SS A3	.8	1.4	2.1
SS A4	.6	1.0	1.5
SS A5	.8	1.4	2.1
SS A6	.6	1.0	1.4
SS A7	.8	1.6	2.5
SS A8	.7	1.0	1.5
SS B1	.9	1.4	2.1
SS B2	.7	1.1	1.5
SS B3	.9	1.5	2.2
SS B4	.7	1.1	1.6
SS B5	.9	1.6	2.4
SS B6	.7	1.1	1.6
SS B7	1.0	1.8	2.8
SS B8	.7	1.1	1.6
SS C1	.9	1.3	1.9
SS C2	.8	1.1	1.5
SS C3	1.0	1.6	2.3
SS C4	.8	1.2	1.7
SS D1	.7	1.1	1.6
SS D2	.6	.9	1.2
SS D3	.7	1.2	1.8
SS D4	.6	.9	1.3
SS E1	.7	1.2	1.7
SS E2	.6	1.0	1.3
SS E3	.8	1.4	2.1
SS E4	.7	1.0	1.5

SAN DIEGO, CA
SOLAR SAVINGS
	10%	50%	80%
WW A1	.6	.9	1.1
WW A2	.5	.7	.9
WW A3	.5	.7	.8
WW A4	.5	.6	.8
WW A5	.5	.6	.8
WW A6	.5	.6	.8
WW B1	.6	.8	1.0
WW B2	.5	.6	.8
WW B3	.4	.6	.7
WW B4	.4	.6	.7
WW B5	.5	.6	.7
WW C1	.4	.5	.7
WW C2	.4	.6	.7
WW C3	.4	.5	.7
WW C4	.4	.5	.7
TW A1	.6	.9	1.2
TW A2	.6	.8	1.0
TW A3	.5	.7	.9
TW A4	.5	.7	.9
TW B1	.6	.9	1.1
TW B2	.6	.8	1.0
TW B3	.6	.8	1.0
TW B4	.6	.8	1.0
TW C1	.6	.9	1.1
TW C2	.6	.9	1.1
TW C3	.7	.9	1.2
TW C4	.7	1.0	1.3
TW D1	.6	.9	1.1
TW D2	.5	.7	.9
TW D3	.5	.6	.8
TW D4	.5	.6	.8
TW D5	.5	.6	.8
TW E1	.5	.6	.8
TW E2	.5	.6	.8
TW E3	.4	.5	.7
TW E4	.4	.6	.7
TW F1	.6	.9	1.1
TW F2	.6	.8	1.0
TW F3	.6	.8	1.0
TW F4	.6	.8	1.0
TW G1	.7	.9	1.1
TW G2	.6	.8	1.1
TW G3	.6	.9	1.1
TW G4	.7	.9	1.2
TW H1	.7	.9	1.2
TW H2	.7	1.0	1.2
TW H3	.8	1.0	1.3
TW H4	.9	1.2	1.5
TW I1	.7	.9	1.2
TW I2	.5	.7	.9
TW I3	.5	.7	.9
TW I4	.5	.6	.8
TW I5	.5	.6	.8
TW J1	.5	.6	.8
TW J2	.5	.6	.8
TW J3	.4	.6	.7
TW J4	.5	.6	.7
DG A1	.6	.8	1.1
DG A2	.6	.8	1.0
DG A3	.6	.7	.9
DG B1	.6	.7	.9
DG B2	.6	.7	.9
DG B3	.6	.6	.8
DG C1	.5	.6	.8

SAN DIEGO, CA (cont.)
SOLAR SAVINGS
	10%	50%	80%
DG C2	.5	.6	.8
DG C3	.5	.6	.7
SS A1	.5	.7	.9
SS A2	.4	.6	.7
SS A3	.5	.7	.9
SS A4	.4	.6	.7
SS A5	.5	.7	.9
SS A6	.4	.6	.7
SS A7	.5	.8	1.0
SS A8	.4	.6	.7
SS B1	.5	.7	1.0
SS B2	.4	.6	.8
SS B3	.5	.7	1.0
SS B4	.4	.6	.8
SS B5	.6	.8	1.1
SS B6	.4	.6	.8
SS B7	.6	.9	1.2
SS B8	.5	.6	.8
SS C1	.5	.7	.9
SS C2	.5	.6	.8
SS C3	.6	.9	1.1
SS C4	.5	.7	.9
SS D1	.4	.6	.7
SS D2	.4	.5	.6
SS D3	.4	.6	.8
SS D4	.4	.5	.7
SS E1	.4	.6	.8
SS E2	.4	.5	.7
SS E3	.5	.7	1.0
SS E4	.4	.6	.7

SAN FRANCISCO, CA
SOLAR SAVINGS
	10%	50%	80%
WW A1	.9	1.3	1.7
WW A2	.7	1.0	1.3
WW A3	.7	.9	1.2
WW A4	.7	.9	1.2
WW A5	.6	.9	1.1
WW A6	.6	.8	1.1
WW B1	.8	1.1	1.5
WW B2	.7	.9	1.2
WW B3	.6	.8	1.0
WW B4	.6	.8	1.0
WW B5	.6	.8	1.0
WW C1	.6	.8	1.0
WW C2	.6	.8	1.0
WW C3	.5	.7	.9
WW C4	.6	.7	1.0
TW A1	.9	1.3	1.7
TW A2	.8	1.1	1.5
TW A3	.7	1.0	1.4
TW A4	.7	1.0	1.3
TW B1	.9	1.2	1.7
TW B2	.8	1.1	1.5
TW B3	.8	1.1	1.5
TW B4	.8	1.1	1.5
TW C1	.9	1.2	1.7
TW C2	.9	1.2	1.6
TW C3	.9	1.3	1.7
TW C4	1.0	1.4	1.9
TW D1	.9	1.3	1.7
TW D2	.7	1.0	1.3
TW D3	.6	.9	1.2

Conservation Factor Tables / 315

SAN FRANCISCO, CA				SANTA MARIA, CA				SANTA MARIA, CA				COLORADO SPRINGS, CO			
	SOLAR SAVINGS				SOLAR SAVINGS				SOLAR SAVINGS				SOLAR SAVINGS		
	10%	50%	80%		10%	50%	80%		10%	50%	80%		10%	50%	80%
WW A1	.9	1.3	1.7	WW A1	.9	1.2	1.5	DG C2	.7	.8	1.0	TW D4	.9	1.2	1.5
WW A2	.7	1.0	1.3	WW A2	.7	.9	1.2	DG C3	.7	.8	.9	TW D5	.9	1.2	1.5
WW A3	.7	.9	1.2	WW A3	.7	.9	1.1	SS A1	.6	.8	1.1	TW E1	.9	1.2	1.5
WW A4	.7	.9	1.2	WW A4	.6	.8	1.1	SS A2	.5	.7	.9	TW E2	.9	1.2	1.5
WW A5	.6	.9	1.1	WW A5	.6	.8	1.0	SS A3	.6	.9	1.2	TW E3	.8	1.0	1.3
WW A6	.6	.8	1.1	WW A6	.6	.8	1.0	SS A4	.5	.7	.9	TW E4	.8	1.1	1.3
WW B1	.8	1.1	1.5	WW B1	.8	1.0	1.3	SS A5	.6	.9	1.2	TW F1	1.4	2.0	2.6
WW B2	.7	.9	1.2	WW B2	.6	.8	1.0	SS A6	.5	.7	.9	TW F2	1.3	1.7	2.2
WW B3	.6	.8	1.0	WW B3	.6	.7	.9	SS A7	.7	1.0	1.4	TW F3	1.2	1.6	2.1
WW B4	.6	.8	1.0	WW B4	.6	.7	.9	SS A8	.5	.7	.9	TW F4	1.2	1.7	2.1
WW B5	.6	.8	1.0	WW B5	.6	.7	.9	SS B1	.7	.9	1.2	TW G1	1.4	2.0	2.6
WW C1	.6	.8	1.0	WW C1	.5	.7	.9	SS B2	.6	.8	1.0	TW G2	1.3	1.8	2.3
WW C2	.6	.8	1.0	WW C2	.6	.7	.9	SS B3	.7	.9	1.3	TW G3	1.4	1.8	2.3
WW C3	.5	.7	.9	WW C3	.5	.7	.8	SS B4	.6	.8	1.0	TW G4	1.5	2.0	2.5
WW C4	.6	.7	1.0	WW C4	.6	.7	.9	SS B5	.7	1.0	1.4	TW H1	1.5	2.0	2.6
TW A1	.9	1.3	1.7	TW A1	.9	1.2	1.5	SS B6	.6	.8	1.0	TW H2	1.5	2.0	2.6
TW A2	.8	1.1	1.5	TW A2	.8	1.0	1.3	SS B7	.8	1.1	1.5	TW H3	1.6	2.2	2.8
TW A3	.7	1.0	1.4	TW A3	.7	.9	1.2	SS B8	.6	.8	1.0	TW H4	1.9	2.5	3.2
TW A4	.7	1.0	1.3	TW A4	.7	.9	1.2	SS C1	.7	.9	1.2	TW I1	1.6	2.2	2.9
TW B1	.9	1.2	1.7	TW B1	.9	1.2	1.5	SS C2	.6	.8	1.0	TW I2	1.1	1.5	1.9
TW B2	.8	1.1	1.5	TW B2	.8	1.0	1.3	SS C3	.8	1.1	1.5	TW I3	1.0	1.4	1.7
TW B3	.8	1.1	1.5	TW B3	.8	1.0	1.3	SS C4	.7	.9	1.2	TW I4	1.0	1.2	1.6
TW B4	.8	1.1	1.5	TW B4	.8	1.1	1.4	SS D1	.5	.7	.9	TW I5	.9	1.2	1.5
TW C1	.9	1.2	1.7	TW C1	.9	1.2	1.5	SS D2	.4	.6	.8	TW J1	.9	1.2	1.6
TW C2	.9	1.2	1.6	TW C2	.9	1.1	1.4	SS D3	.6	.8	1.0	TW J2	.9	1.2	1.6
TW C3	.9	1.3	1.7	TW C3	.9	1.2	1.5	SS D4	.5	.6	.8	TW J3	.8	1.1	1.3
TW C4	1.0	1.4	1.9	TW C4	1.0	1.3	1.7	SS E1	.6	.8	1.0	TW J4	.9	1.1	1.4
TW D1	.9	1.3	1.7	TW D1	.9	1.1	1.5	SS E2	.5	.7	.9	DG A1	1.4	4.7	---
TW D2	.7	1.0	1.3	TW D2	.7	.9	1.1	SS E3	.7	.9	1.2	DG A2	1.2	1.8	3.4
TW D3	.6	.9	1.2	TW D3	.6	.8	1.0	SS E4	.5	.7	.9	DG A3	1.1	1.3	1.9
DG B1	.8	1.1	1.7	TW D4	.6	.8	1.0					DG B1	1.4	2.1	---
DG B2	.8	1.0	1.3	TW D5	.6	.8	1.0	COLORADO SPRINGS, CO				DG B2	1.2	1.5	2.3
DG B3	.7	.9	1.1	TW E1	.6	.8	1.0		SOLAR SAVINGS			DG B3	1.1	1.2	1.6
DG C1	.7	.9	1.3	TW E2	.6	.8	1.0		10%	50%	80%	DG C1	1.2	1.5	4.1
DG C2	.7	.9	1.2	TW E3	.5	.7	.9	WW A1	1.3	2.0	2.7	DG C2	1.1	1.3	1.8
DG C3	.7	.8	1.0	TW E4	.6	.7	.9	WW A2	1.1	1.5	2.0	DG C3	1.0	1.1	1.4
SS A1	.6	.9	1.3	TW F1	.9	1.2	1.5	WW A3	1.0	1.4	1.8	SS A1	1.0	1.5	2.0
SS A2	.5	.8	1.1	TW F2	.8	1.0	1.3	WW A4	1.0	1.3	1.7	SS A2	.8	1.1	1.5
SS A3	.6	1.0	1.4	TW F3	.8	1.0	1.3	WW A5	1.0	1.3	1.6	SS A3	1.1	1.6	2.2
SS A4	.5	.8	1.1	TW F4	.8	1.0	1.3	WW A6	1.0	1.3	1.6	SS A4	.8	1.2	1.6
SS A5	.6	1.0	1.5	TW G1	.9	1.2	1.5	WW B1	1.3	1.9	2.5	SS A5	1.1	1.6	2.2
SS A6	.5	.8	1.1	TW G2	.8	1.1	1.4	WW B2	1.0	1.3	1.6	SS A6	.8	1.2	1.5
SS A7	.7	1.1	1.7	TW G3	.9	1.1	1.4	WW B3	.9	1.1	1.4	SS A7	1.2	1.9	2.7
SS A8	.5	.8	1.1	TW G4	.9	1.2	1.6	WW B4	.8	1.1	1.4	SS A8	.8	1.2	1.6
SS B1	.7	1.0	1.5	TW H1	.9	1.2	1.6	WW B5	.9	1.1	1.4	SS B1	1.1	1.6	2.2
SS B2	.6	.9	1.2	TW H2	1.0	1.3	1.6	WW C1	.8	1.1	1.4	SS B2	.9	1.3	1.6
SS B3	.7	1.1	1.5	TW H3	1.0	1.4	1.7	WW C2	.9	1.1	1.4	SS B3	1.2	1.8	2.4
SS B4	.6	.9	1.2	TW H4	1.2	1.6	2.0	WW C3	.8	1.0	1.2	SS B4	.9	1.3	1.7
SS B5	.8	1.2	1.7	TW I1	.9	1.2	1.6	WW C4	.8	1.0	1.3	SS B5	1.3	1.9	2.6
SS B6	.6	.9	1.2	TW I2	.7	.9	1.2	TW A1	1.4	2.0	2.7	SS B6	.9	1.3	1.7
SS B7	.8	1.3	1.8	TW I3	.7	.9	1.1	TW A2	1.2	1.7	2.1	SS B7	1.4	2.2	3.0
SS B8	.6	.9	1.2	TW I4	.6	.8	1.0	TW A3	1.1	1.5	2.0	SS B8	.9	1.3	1.7
SS C1	.7	1.0	1.3	TW I5	.6	.8	1.0	TW A4	1.1	1.5	1.9	SS C1	1.1	1.5	2.0
SS C2	.7	.9	1.2	TW J1	.6	.8	1.0	TW B1	1.4	1.9	2.5	SS C2	1.0	1.3	1.6
SS C3	.9	1.2	1.6	TW J2	.6	.8	1.0	TW B2	1.2	1.7	2.2	SS C3	1.4	1.9	2.5
SS C4	.7	1.0	1.3	TW J3	.6	.7	.9	TW B3	1.2	1.6	2.1	SS C4	1.1	1.4	1.8
SS D1	.5	.8	1.1	TW J4	.6	.7	.9	TW B4	1.3	1.7	2.2	SS D1	.9	1.2	1.6
SS D2	.5	.7	.9	DG A1	.8	1.1	1.8	TW C1	1.4	1.9	2.5	SS D2	.7	1.0	1.3
SS D3	.6	.9	1.3	DG A2	.8	1.0	1.4	TW C2	1.3	1.8	2.3	SS D3	.9	1.4	1.9
SS D4	.5	.7	1.0	DG A3	.7	.9	1.2	TW C3	1.4	1.9	2.4	SS D4	.7	1.0	1.4
SS E1	.6	.9	1.2	DG B1	.8	.9	1.4	TW C4	1.5	2.1	2.7	SS E1	1.0	1.4	1.8
SS E2	.5	.7	1.0	DG B2	.8	.9	1.2	TW D1	1.5	2.1	2.8	SS E2	.8	1.1	1.4
SS E3	.7	1.0	1.5	DG B3	.7	.8	1.0	TW D2	1.1	1.4	1.8	SS E3	1.1	1.7	2.2
SS E4	.6	.8	1.1	DG C1	.7	.8	1.1	TW D3	1.0	1.3	1.7	SS E4	.9	1.2	1.6

DENVER, CO				DENVER, CO				EAGLE, CO				GRAND JUNCTION, CO			
	SOLAR SAVINGS				SOLAR SAVINGS				SOLAR SAVINGS				SOLAR SAVINGS		
	10%	50%	80%		10%	50%	80%		10%	50%	80%		10%	50%	80%
WW A1	1.3	2.0	2.6	DG C2	1.1	1.3	1.8	TW D4	1.1	1.4	1.9	WW A1	1.3	2.2	3.0
WW A2	1.1	1.5	1.9	DG C3	1.0	1.1	1.4	TW D5	1.1	1.4	1.8	WW A2	1.1	1.6	2.2
WW A3	1.0	1.4	1.8	SS A1	1.0	1.5	2.0	TW E1	1.1	1.5	1.9	WW A3	1.1	1.5	2.0
WW A4	1.0	1.3	1.7	SS A2	.8	1.1	1.5	TW E2	1.1	1.4	1.9	WW A4	1.0	1.4	1.9
WW A5	1.0	1.3	1.6	SS A3	1.0	1.6	2.2	TW E3	.9	1.2	1.6	WW A5	1.0	1.4	1.8
WW A6	.9	1.2	1.6	SS A4	.8	1.2	1.6	TW E4	1.0	1.2	1.6	WW A6	1.0	1.3	1.7
WW B1	1.3	1.9	2.5	SS A5	1.0	1.6	2.2	TW F1	1.7	2.8	4.2	WW B1	1.3	2.1	3.0
WW B2	1.0	1.3	1.6	SS A6	.8	1.2	1.5	TW F2	1.5	2.3	3.2	WW B2	1.0	1.3	1.8
WW B3	.8	1.1	1.4	SS A7	1.2	1.9	2.7	TW F3	1.5	2.1	2.9	WW B3	.9	1.2	1.5
WW B4	.8	1.1	1.4	SS A8	.8	1.2	1.6	TW F4	1.5	2.1	2.9	WW B4	.9	1.1	1.4
WW B5	.8	1.1	1.4	SS B1	1.1	1.6	2.2	TW G1	1.7	2.7	3.8	WW B5	.9	1.1	1.4
WW C1	.8	1.1	1.4	SS B2	.9	1.3	1.6	TW G2	1.6	2.4	3.2	WW C1	.8	1.1	1.5
WW C2	.8	1.1	1.4	SS B3	1.2	1.8	2.4	TW G3	1.6	2.3	3.2	WW C2	.9	1.2	1.5
WW C3	.7	1.0	1.2	SS B4	.9	1.3	1.7	TW G4	1.8	2.5	3.4	WW C3	.8	1.0	1.3
WW C4	.8	1.0	1.3	SS B5	1.2	1.9	2.6	TW H1	1.8	2.7	3.8	WW C4	.8	1.0	1.3
TW A1	1.4	2.0	2.6	SS B6	.9	1.3	1.7	TW H2	1.9	2.7	3.6	TW A1	1.3	2.2	3.0
TW A2	1.2	1.6	2.1	SS B7	1.4	2.2	3.0	TW H3	2.0	2.8	3.8	TW A2	1.2	1.8	2.4
TW A3	1.1	1.5	2.0	SS B8	.9	1.3	1.7	TW H4	2.3	3.2	4.3	TW A3	1.1	1.6	2.2
TW A4	1.1	1.5	1.9	SS C1	1.1	1.5	2.0	TW I1	2.1	3.9	8.0	TW A4	1.1	1.6	2.1
TW B1	1.3	1.9	2.5	SS C2	1.0	1.3	1.6	TW I2	1.3	1.8	2.4	TW B1	1.3	2.1	2.8
TW B2	1.2	1.7	2.1	SS C3	1.3	1.9	2.4	TW I3	1.2	1.7	2.3	TW B2	1.2	1.8	2.4
TW B3	1.2	1.6	2.1	SS C4	1.0	1.4	1.8	TW I4	1.1	1.5	1.9	TW B3	1.2	1.7	2.3
TW B4	1.2	1.7	2.2	SS D1	.9	1.2	1.7	TW I5	1.1	1.4	1.9	TW B4	1.3	1.8	2.4
TW C1	1.3	1.9	2.4	SS D2	.7	1.0	1.3	TW J1	1.1	1.5	2.0	TW C1	1.3	2.0	2.7
TW C2	1.3	1.8	2.3	SS D3	.9	1.4	1.9	TW J2	1.1	1.5	1.9	TW C2	1.3	1.9	2.5
TW C3	1.3	1.9	2.4	SS D4	.7	1.0	1.3	TW J3	1.0	1.2	1.6	TW C3	1.4	2.0	2.6
TW C4	1.5	2.0	2.6	SS E1	1.0	1.4	1.8	TW J4	1.0	1.3	1.6	TW C4	1.5	2.2	2.9
TW D1	1.4	2.1	2.7	SS E2	.8	1.1	1.4	DG A1	1.8	---	---	TW D1	1.5	2.3	3.3
TW D2	1.0	1.4	1.8	SS E3	1.1	1.7	2.2	DG A2	1.5	2.6	---	TW D2	1.1	1.5	1.9
TW D3	1.0	1.3	1.6	SS E4	.9	1.2	1.6	DG A3	1.3	1.6	2.5	TW D3	1.0	1.4	1.8
TW D4	.9	1.2	1.5					DG B1	1.8	---	---	TW D4	.9	1.2	1.6
TW D5	.9	1.2	1.5	EAGLE, CO				DG B2	1.5	2.1	---	TW D5	.9	1.2	1.6
TW E1	.9	1.2	1.5		SOLAR SAVINGS			DG B3	1.3	1.5	2.1	TW E1	.9	1.2	1.6
TW E2	.9	1.2	1.5		10%	50%	80%	DG C1	1.5	2.8	---	TW E2	.9	1.2	1.6
TW E3	.8	1.0	1.3	WW A1	1.7	3.0	4.6	DG C2	1.3	1.7	2.9	TW E3	.8	1.1	1.4
TW E4	.8	1.0	1.3	WW A2	1.4	2.0	2.8	DG C3	1.2	1.4	1.8	TW E4	.8	1.1	1.4
TW F1	1.4	2.0	2.6	WW A3	1.3	1.8	2.5	SS A1	1.1	1.9	2.8	TW F1	1.4	2.1	3.0
TW F2	1.2	1.7	2.2	WW A4	1.2	1.7	2.3	SS A2	.9	1.4	1.9	TW F2	1.3	1.8	2.4
TW F3	1.2	1.6	2.1	WW A5	1.2	1.7	2.2	SS A3	1.2	2.3	3.6	TW F3	1.2	1.7	2.3
TW F4	1.2	1.6	2.1	WW A6	1.2	1.6	2.1	SS A4	.9	1.4	2.0	TW F4	1.3	1.7	2.3
TW G1	1.4	2.0	2.5	WW B1	1.7	3.4	7.5	SS A5	1.2	2.2	3.4	TW G1	1.4	2.1	2.9
TW G2	1.3	1.8	2.3	WW B2	1.2	1.6	2.1	SS A6	.9	1.4	1.9	TW G2	1.3	1.9	2.5
TW G3	1.3	1.8	2.3	WW B3	1.0	1.4	1.8	SS A7	1.3	2.9	5.7	TW G3	1.4	1.9	2.5
TW G4	1.4	1.9	2.5	WW B4	1.0	1.3	1.7	SS A8	1.0	1.5	2.1	TW G4	1.5	2.1	2.7
TW H1	1.5	2.0	2.6	WW B5	1.0	1.3	1.7	SS B1	1.3	2.2	3.2	TW H1	1.5	2.2	2.9
TW H2	1.5	2.0	2.6	WW C1	1.0	1.3	1.7	SS B2	1.0	1.5	2.1	TW H2	1.5	2.1	2.8
TW H3	1.6	2.2	2.8	WW C2	1.0	1.3	1.7	SS B3	1.4	2.5	3.9	TW H3	1.7	2.3	3.0
TW H4	1.8	2.5	3.2	WW C3	.9	1.1	1.4	SS B4	1.1	1.6	2.1	TW H4	1.9	2.6	3.5
TW I1	1.5	2.2	2.9	WW C4	.9	1.2	1.5	SS B5	1.5	2.7	4.2	TW I1	1.6	2.5	3.5
TW I2	1.1	1.5	1.9	TW A1	1.7	2.9	4.3	SS B6	1.1	1.6	2.1	TW I2	1.1	1.5	2.0
TW I3	1.0	1.4	1.7	TW A2	1.5	2.2	3.1	SS B7	1.6	3.4	6.3	TW I3	1.1	1.4	1.9
TW I4	.9	1.2	1.6	TW A3	1.4	2.0	2.8	SS B8	1.1	1.6	2.3	TW I4	1.0	1.3	1.7
TW I5	.9	1.2	1.5	TW A4	1.4	1.9	2.6	SS C1	1.4	2.1	2.9	TW I5	1.0	1.3	1.6
TW J1	.9	1.2	1.6	TW B1	1.6	2.7	3.8	SS C2	1.2	1.6	2.1	TW J1	1.0	1.3	1.7
TW J2	.9	1.2	1.5	TW B2	1.5	2.2	3.0	SS C3	1.7	2.6	3.7	TW J2	1.0	1.3	1.7
TW J3	.8	1.0	1.3	TW B3	1.5	2.1	2.9	SS C4	1.3	1.8	2.4	TW J3	.8	1.1	1.4
TW J4	.8	1.1	1.4	TW B4	1.5	2.2	2.9	SS D1	1.0	1.6	2.3	TW J4	.9	1.1	1.4
DG A1	1.3	3.9	---	TW C1	1.6	2.5	3.4	SS D2	.8	1.2	1.6	DG A1	1.4	---	---
DG A2	1.2	1.7	3.3	TW C2	1.6	2.3	3.2	SS D3	1.1	1.8	2.7	DG A2	1.2	1.9	5.0
DG A3	1.1	1.3	1.8	TW C3	1.6	2.4	3.2	SS D4	.9	1.2	1.7	DG A3	1.1	1.4	2.0
DG B1	1.3	2.0	---	TW C4	1.8	2.6	3.5	SS E1	1.2	1.9	2.8	DG B1	1.4	2.5	---
DG B2	1.2	1.5	2.3	TW D1	1.9	3.5	6.5	SS E2	.9	1.4	1.8	DG B2	1.2	1.6	2.7
DG B3	1.1	1.2	1.6	TW D2	1.3	1.8	2.3	SS E3	1.3	2.3	3.5	DG B3	1.1	1.3	1.7
DG C1	1.2	1.5	3.7	TW D3	1.2	1.6	2.2	SS E4	1.0	1.5	2.0	DG C1	1.2	1.7	---

GRAND JUNCTION, CO

	SOLAR SAVINGS		
	10%	50%	80%
DG C2	1.1	1.4	2.0
DG C3	1.0	1.2	1.5
SS A1	1.0	1.6	2.2
SS A2	.8	1.2	1.6
SS A3	1.1	1.8	2.6
SS A4	.9	1.3	1.7
SS A5	1.1	1.8	2.6
SS A6	.8	1.2	1.7
SS A7	1.2	2.2	3.3
SS A8	.9	1.3	1.8
SS B1	1.1	1.8	2.5
SS B2	.9	1.3	1.8
SS B3	1.2	2.0	2.8
SS B4	.9	1.4	1.8
SS B5	1.3	2.1	3.0
SS B6	.9	1.4	1.8
SS B7	1.4	2.5	3.6
SS B8	1.0	1.4	1.9
SS C1	1.2	1.6	2.2
SS C2	1.0	1.3	1.8
SS C3	1.4	2.0	2.7
SS C4	1.1	1.5	2.0
SS D1	.9	1.3	1.9
SS D2	.8	1.1	1.4
SS D3	1.0	1.5	2.1
SS D4	.8	1.1	1.5
SS E1	1.0	1.5	2.1
SS E2	.8	1.2	1.6
SS E3	1.1	1.8	2.6
SS E4	.9	1.3	1.7

PUEBLO, CO

	SOLAR SAVINGS		
	10%	50%	80%
WW A1	1.3	1.9	2.5
WW A2	1.1	1.5	1.9
WW A3	1.0	1.3	1.7
WW A4	1.0	1.3	1.6
WW A5	.9	1.2	1.6
WW A6	.9	1.2	1.5
WW B1	1.2	1.8	2.4
WW B2	.9	1.2	1.6
WW B3	.8	1.1	1.4
WW B4	.8	1.0	1.3
WW B5	.8	1.1	1.3
WW C1	.8	1.1	1.3
WW C2	.8	1.1	1.4
WW C3	.7	.9	1.2
WW C4	.8	1.0	1.2
TW A1	1.3	1.9	2.6
TW A2	1.1	1.6	2.1
TW A3	1.1	1.5	1.9
TW A4	1.1	1.4	1.9
TW B1	1.3	1.9	2.4
TW B2	1.2	1.6	2.1
TW B3	1.2	1.6	2.0
TW B4	1.2	1.6	2.1
TW C1	1.3	1.8	2.4
TW C2	1.3	1.7	2.2
TW C3	1.3	1.8	2.3
TW C4	1.4	2.0	2.6
TW D1	1.4	2.0	2.6
TW D2	1.0	1.3	1.7
TW D3	.9	1.3	1.6

PUEBLO, CO

	SOLAR SAVINGS		
	10%	50%	80%
TW D4	.9	1.2	1.5
TW D5	.9	1.1	1.5
TW E1	.9	1.2	1.5
TW E2	.9	1.1	1.5
TW E3	.8	1.0	1.3
TW E4	.8	1.0	1.3
TW F1	1.3	1.9	2.5
TW F2	1.2	1.6	2.1
TW F3	1.2	1.6	2.0
TW F4	1.2	1.6	2.0
TW G1	1.3	1.9	2.5
TW G2	1.3	1.7	2.2
TW G3	1.3	1.7	2.2
TW G4	1.4	1.9	2.4
TW H1	1.4	1.9	2.5
TW H2	1.5	2.0	2.5
TW H3	1.6	2.1	2.7
TW H4	1.8	2.4	3.1
TW I1	1.5	2.1	2.8
TW I2	1.1	1.4	1.8
TW I3	1.0	1.3	1.7
TW I4	.9	1.2	1.5
TW I5	.9	1.2	1.5
TW J1	.9	1.2	1.5
TW J2	.9	1.2	1.5
TW J3	.8	1.0	1.3
TW J4	.8	1.1	1.3
DG A1	1.3	3.1	---
DG A2	1.2	1.7	3.0
DG A3	1.1	1.3	1.8
DG B1	1.3	1.9	---
DG B2	1.2	1.4	2.2
DG B3	1.1	1.2	1.6
DG C1	1.2	1.4	3.0
DG C2	1.1	1.3	1.7
DG C3	1.0	1.1	1.4
SS A1	1.0	1.4	1.9
SS A2	.8	1.1	1.5
SS A3	1.0	1.6	2.1
SS A4	.8	1.2	1.5
SS A5	1.0	1.6	2.2
SS A6	.8	1.1	1.5
SS A7	1.1	1.9	2.6
SS A8	.8	1.2	1.6
SS B1	1.1	1.6	2.1
SS B2	.9	1.2	1.6
SS B3	1.1	1.7	2.3
SS B4	.9	1.3	1.6
SS B5	1.2	1.9	2.5
SS B6	.9	1.2	1.6
SS B7	1.3	2.1	2.9
SS B8	.9	1.3	1.7
SS C1	1.1	1.5	1.9
SS C2	.9	1.2	1.6
SS C3	1.3	1.8	2.4
SS C4	1.0	1.4	1.8
SS D1	.8	1.2	1.6
SS D2	.7	1.0	1.3
SS D3	.9	1.4	1.8
SS D4	.7	1.0	1.3
SS E1	.9	1.4	1.8
SS E2	.8	1.1	1.4
SS E3	1.1	1.6	2.2
SS E4	.8	1.2	1.5

HARTFORD, CT

	SOLAR SAVINGS		
	10%	50%	80%
WW A1	2.6	---	---
WW A2	2.1	---	---
WW A3	2.0	4.8	---
WW A4	1.9	3.8	---
WW A5	1.8	3.3	8.7
WW A6	1.8	3.1	6.3
WW B1	---	0.0	0.0
WW B2	1.6	2.6	3.8
WW B3	1.4	2.1	3.0
WW B4	1.3	1.8	2.4
WW B5	1.3	1.7	2.2
WW C1	1.3	1.9	2.7
WW C2	1.3	1.9	2.5
WW C3	1.1	1.5	1.9
WW C4	1.1	1.5	2.0
TW A1	2.6	---	---
TW A2	2.2	---	---
TW A3	2.1	5.2	---
TW A4	2.0	3.9	15.0
TW B1	2.5	---	---
TW B2	2.2	5.5	---
TW B3	2.2	4.3	---
TW B4	2.2	4.1	8.3
TW C1	2.4	10.3	---
TW C2	2.3	4.6	16.2
TW C3	2.3	4.5	9.1
TW C4	2.5	4.6	8.2
TW D1	---	0.0	0.0
TW D2	1.7	2.8	4.3
TW D3	1.6	2.6	3.9
TW D4	1.4	2.0	2.7
TW D5	1.4	1.9	2.5
TW E1	1.4	2.2	3.1
TW E2	1.4	2.0	2.8
TW E3	1.2	1.6	2.1
TW E4	1.2	1.6	2.1
TW F1	2.9	---	---
TW F2	2.4	---	---
TW F3	2.3	6.3	---
TW F4	2.3	4.5	---
TW G1	2.9	---	---
TW G2	2.6	11.7	---
TW G3	2.5	5.3	---
TW G4	2.7	4.9	10.9
TW H1	3.0	---	---
TW H2	2.9	6.3	---
TW H3	3.0	5.6	14.5
TW H4	3.3	5.8	10.7
TW I1	---	0.0	0.0
TW I2	1.9	3.0	4.5
TW I3	1.7	2.7	3.8
TW I4	1.5	2.1	2.8
TW I5	1.4	1.9	2.5
TW J1	1.5	2.2	3.1
TW J2	1.5	2.1	2.8
TW J3	1.2	1.6	2.1
TW J4	1.2	1.7	2.2
DG A1	---	0.0	0.0
DG A2	2.2	---	---
DG A3	1.6	2.4	10.7
DG B1	---	0.0	0.0
DG B2	2.1	---	---
DG B3	1.6	2.1	3.4
DG C1	3.1	---	---

HARTFORD, CT

	SOLAR SAVINGS		
	10%	50%	80%
DG C2	1.8	3.5	---
DG C3	1.5	1.9	2.6
SS A1	1.7	4.7	---
SS A2	1.3	2.0	2.9
SS A3	1.9	---	---
SS A4	1.3	2.2	3.2
SS A5	1.8	---	---
SS A6	1.3	2.1	3.0
SS A7	2.1	---	---
SS A8	1.3	2.3	3.6
SS B1	2.0	---	---
SS B2	1.4	2.1	3.0
SS B3	2.2	---	---
SS B4	1.4	2.3	3.3
SS B5	2.3	---	---
SS B6	1.4	2.3	3.2
SS B7	2.7	---	---
SS B8	1.5	2.5	3.7
SS C1	2.4	---	---
SS C2	1.6	2.5	3.6
SS C3	2.7	---	---
SS C4	1.8	2.8	4.2
SS D1	1.5	3.5	---
SS D2	1.1	1.7	2.3
SS D3	1.6	5.0	---
SS D4	1.1	1.8	2.5
SS E1	1.8	---	---
SS E2	1.3	2.0	2.9
SS E3	2.0	---	---
SS E4	1.4	2.2	3.2

WILMINGTON, DE

	SOLAR SAVINGS		
	10%	50%	80%
WW A1	1.8	3.9	7.8
WW A2	1.5	2.3	3.4
WW A3	1.4	2.1	2.9
WW A4	1.3	2.0	2.7
WW A5	1.3	1.9	2.5
WW A6	1.3	1.8	2.5
WW B1	2.0	---	---
WW B2	1.3	1.8	2.4
WW B3	1.1	1.5	2.0
WW B4	1.0	1.4	1.8
WW B5	1.1	1.4	1.8
WW C1	1.1	1.5	1.9
WW C2	1.1	1.4	1.9
WW C3	.9	1.2	1.5
WW C4	1.0	1.3	1.6
TW A1	1.8	3.6	6.2
TW A2	1.6	2.6	3.7
TW A3	1.5	2.3	3.2
TW A4	1.5	2.2	3.0
TW B1	1.8	3.1	4.8
TW B2	1.6	2.5	3.5
TW B3	1.6	2.4	3.3
TW B4	1.6	2.4	3.3
TW C1	1.8	2.9	4.1
TW C2	1.7	2.6	3.6
TW C3	1.8	2.7	3.7
TW C4	1.9	2.9	4.0
TW D1	2.2	7.3	---
TW D2	1.3	1.9	2.6
TW D3	1.3	1.8	2.4

WILMINGTON, DE

	SOLAR SAVINGS		
	10%	50%	80%
WW A1	1.8	3.9	7.8
WW A2	1.5	2.3	3.4
WW A3	1.4	2.1	2.9
WW A4	1.3	2.0	2.7
WW A5	1.3	1.9	2.5
WW A6	1.3	1.8	2.5
WW B1	2.0	---	---
WW B2	1.3	1.8	2.4
WW B3	1.1	1.5	2.0
WW B4	1.0	1.4	1.8
WW B5	1.1	1.4	1.8
WW C1	1.1	1.5	1.9
WW C2	1.1	1.4	1.9
WW C3	.9	1.2	1.5
WW C4	1.0	1.3	1.6
TW A1	1.8	3.6	6.2
TW A2	1.6	2.6	3.7
TW A3	1.5	2.3	3.2
TW A4	1.5	2.2	3.0
TW B1	1.8	3.1	4.8
TW B2	1.6	2.5	3.5
TW B3	1.6	2.4	3.3
TW B4	1.6	2.4	3.3
TW C1	1.8	2.9	4.1
TW C2	1.7	2.6	3.6
TW C3	1.8	2.7	3.7
TW C4	1.9	2.9	4.0
TW D1	2.2	7.3	---
TW D2	1.3	1.9	2.6
TW D3	1.3	1.8	2.4
DG B1	2.0	---	---
DG B2	1.6	2.4	---
DG B3	1.3	1.6	2.3
DG C1	1.7	---	---
DG C2	1.4	1.9	4.1
DG C3	1.2	1.5	1.9
SS A1	1.3	2.2	3.2
SS A2	1.0	1.5	2.1
SS A3	1.4	2.7	4.5
SS A4	1.0	1.6	2.2
SS A5	1.4	2.6	4.2
SS A6	1.0	1.6	2.1
SS A7	1.5	3.9	10.9
SS A8	1.1	1.7	2.3
SS B1	1.5	2.5	3.8
SS B2	1.1	1.6	2.2
SS B3	1.6	3.0	4.8
SS B4	1.2	1.7	2.3
SS B5	1.7	3.2	5.3
SS B6	1.2	1.7	2.3
SS B7	1.9	4.5	12.9
SS B8	1.2	1.8	2.5
SS C1	1.5	2.5	3.6
SS C2	1.2	1.8	2.3
SS C3	1.8	3.1	4.6
SS C4	1.4	2.0	2.6
SS D1	1.1	1.8	2.6
SS D2	.9	1.3	1.8
SS D3	1.2	2.1	3.1
SS D4	.9	1.4	1.8
SS E1	1.3	2.2	3.3
SS E2	1.0	1.5	2.0
SS E3	1.5	2.7	4.3
SS E4	1.1	1.6	2.2

WASHINGTON, DC

	SOLAR SAVINGS		
	10%	50%	80%
WW A1	1.9	4.8	---
WW A2	1.5	2.5	3.8
WW A3	1.4	2.2	3.2
WW A4	1.4	2.1	2.9
WW A5	1.3	2.0	2.7
WW A6	1.3	1.9	2.6
WW B1	2.2	---	---
WW B2	1.3	1.9	2.5
WW B3	1.1	1.6	2.1
WW B4	1.1	1.4	1.9
WW B5	1.1	1.4	1.8
WW C1	1.1	1.5	2.0
WW C2	1.1	1.5	2.0
WW C3	.9	1.2	1.6
WW C4	1.0	1.3	1.7
TW A1	1.9	4.1	10.1
TW A2	1.6	2.8	4.1
TW A3	1.6	2.5	3.5
TW A4	1.5	2.3	3.3
TW B1	1.8	3.5	5.7
TW B2	1.7	2.7	3.8
TW B3	1.6	2.5	3.6
TW B4	1.7	2.6	3.6
TW C1	1.8	3.1	4.5
TW C2	1.8	2.8	3.9
TW C3	1.8	2.8	4.0
TW C4	2.0	3.1	4.3
TW D1	2.4	---	---
TW D2	1.4	2.0	2.8
TW D3	1.3	1.9	2.5
TW D4	1.2	1.6	2.1
TW D5	1.1	1.5	2.0
TW E1	1.2	1.7	2.2
TW E2	1.2	1.6	2.1
TW E3	1.0	1.3	1.7
TW E4	1.0	1.4	1.8
TW F1	2.0	4.0	8.3
TW F2	1.7	2.9	4.3
TW F3	1.7	2.6	3.8
TW F4	1.7	2.6	3.6
TW G1	2.0	3.6	5.8
TW G2	1.8	2.9	4.2
TW G3	1.9	2.8	4.0
TW G4	2.0	3.0	4.1
TW H1	2.1	3.4	5.1
TW H2	2.1	3.2	4.6
TW H3	2.2	3.3	4.7
TW H4	2.5	3.8	5.2
TW I1	2.6	---	---
TW I2	1.5	2.1	2.9
TW I3	1.4	2.0	2.6
TW I4	1.2	1.7	2.2
TW I5	1.2	1.6	2.1
TW J1	1.2	1.7	2.3
TW J2	1.2	1.7	2.2
TW J3	1.0	1.4	1.8
TW J4	1.1	1.4	1.8
DG A1	2.2	---	---
DG A2	1.7	4.1	---
DG A3	1.4	1.8	3.1
DG B1	2.2	---	---
DG B2	1.6	2.7	---
DG B3	1.3	1.7	2.4
DG C1	1.7	---	---

WASHINGTON, DC

	SOLAR SAVINGS		
	10%	50%	80%
DG C2	1.4	2.1	6.1
DG C3	1.2	1.5	2.0
SS A1	1.3	2.3	3.5
SS A2	1.0	1.6	2.1
SS A3	1.4	2.9	5.3
SS A4	1.1	1.6	2.3
SS A5	1.4	2.8	4.7
SS A6	1.1	1.6	2.2
SS A7	1.6	4.5	---
SS A8	1.1	1.7	2.4
SS B1	1.5	2.7	4.1
SS B2	1.1	1.7	2.3
SS B3	1.6	3.2	5.8
SS B4	1.2	1.8	2.4
SS B5	1.7	3.5	6.4
SS B6	1.2	1.7	2.4
SS B7	1.9	5.5	---
SS B8	1.2	1.9	2.6
SS C1	1.6	2.7	4.1
SS C2	1.3	1.8	2.5
SS C3	1.9	3.4	5.4
SS C4	1.4	2.0	2.8
SS D1	1.1	1.9	2.8
SS D2	.9	1.3	1.8
SS D3	1.2	2.2	3.4
SS D4	1.0	1.4	1.9
SS E1	1.3	2.4	3.7
SS E2	1.1	1.5	2.1
SS E3	1.5	2.9	5.0
SS E4	1.1	1.7	2.3

APALACHICOLA, FL

	SOLAR SAVINGS		
	10%	50%	80%
WW A1	.8	1.2	1.5
WW A2	.7	.9	1.2
WW A3	.6	.9	1.1
WW A4	.6	.8	1.1
WW A5	.6	.8	1.0
WW A6	.6	.8	1.0
WW B1	.7	1.1	1.4
WW B2	.6	.8	1.1
WW B3	.6	.7	.9
WW B4	.6	.7	.9
WW B5	.6	.7	.9
WW C1	.5	.7	.9
WW C2	.6	.7	.9
WW C3	.5	.7	.8
WW C4	.5	.7	.9
TW A1	.8	1.2	1.6
TW A2	.7	1.0	1.3
TW A3	.7	1.0	1.2
TW A4	.7	.9	1.2
TW B1	.8	1.2	1.5
TW B2	.7	1.0	1.3
TW B3	.7	1.0	1.3
TW B4	.8	1.1	1.4
TW C1	.8	1.2	1.5
TW C2	.8	1.1	1.5
TW C3	.8	1.2	1.5
TW C4	.9	1.3	1.7
TW D1	.8	1.2	1.5
TW D2	.7	.9	1.2
TW D3	.6	.8	1.1

APALACHICOLA, FL

	SOLAR SAVINGS		
	10%	50%	80%
TW D4	.6	.8	1.0
TW D5	.6	.8	1.0
TW E1	.6	.8	1.0
TW E2	.6	.8	1.0
TW E3	.5	.7	.9
TW E4	.6	.7	.9
TW F1	.8	1.2	1.6
TW F2	.8	1.1	1.4
TW F3	.7	1.0	1.3
TW F4	.8	1.0	1.3
TW G1	.8	1.2	1.6
TW G2	.8	1.1	1.4
TW G3	.8	1.1	1.5
TW G4	.9	1.2	1.6
TW H1	.9	1.2	1.6
TW H2	.9	1.3	1.6
TW H3	1.0	1.4	1.8
TW H4	1.2	1.6	2.0
TW I1	.9	1.2	1.6
TW I2	.7	.9	1.2
TW I3	.7	.9	1.1
TW I4	.6	.8	1.1
TW I5	.6	.8	1.1
TW J1	.6	.8	1.0
TW J2	.6	.8	1.0
TW J3	.6	.7	.9
TW J4	.6	.8	1.0
DG A1	.8	1.1	1.8
DG A2	.8	1.0	1.4
DG A3	.7	.9	1.2
DG B1	.8	1.0	1.4
DG B2	.8	.9	1.2
DG B3	.7	.8	1.0
DG C1	.7	.9	1.1
DG C2	.7	.8	1.1
DG C3	.7	.8	.9
SS A1	.6	.9	1.2
SS A2	.5	.7	1.0
SS A3	.6	.9	1.2
SS A4	.5	.7	1.0
SS A5	.6	1.0	1.3
SS A6	.5	.7	.9
SS A7	.7	1.1	1.4
SS A8	.5	.8	1.0
SS B1	.7	1.0	1.3
SS B2	.6	.8	1.1
SS B3	.7	1.0	1.3
SS B4	.6	.8	1.1
SS B5	.7	1.1	1.5
SS B6	.6	.8	1.1
SS B7	.8	1.2	1.6
SS B8	.6	.8	1.1
SS C1	.7	.9	1.2
SS C2	.6	.8	1.1
SS C3	.8	1.1	1.5
SS C4	.7	.9	1.2
SS D1	.5	.8	1.0
SS D2	.5	.6	.9
SS D3	.6	.8	1.1
SS D4	.5	.7	.9
SS E1	.6	.8	1.1
SS E2	.5	.7	.9
SS E3	.7	1.0	1.3
SS E4	.6	.8	1.0

DAYTONA BEACH, FL
SOLAR SAVINGS

	10%	50%	80%
WW A1	.6	.9	1.1
WW A2	.5	.7	.9
WW A3	.5	.7	.8
WW A4	.5	.6	.8
WW A5	.5	.6	.8
WW A6	.5	.6	.8
WW B1	.6	.8	1.0
WW B2	.5	.6	.8
WW B3	.4	.6	.7
WW B4	.5	.6	.7
WW B5	.5	.6	.7
WW C1	.4	.6	.7
WW C2	.5	.6	.7
WW C3	.4	.5	.7
WW C4	.4	.6	.7
TW A1	.6	.9	1.2
TW A2	.6	.8	1.0
TW A3	.6	.7	.9
TW A4	.6	.7	.9
TW B1	.6	.9	1.1
TW B2	.6	.8	1.0
TW B3	.6	.8	1.0
TW B4	.6	.8	1.1
TW C1	.7	.9	1.2
TW C2	.7	.9	1.1
TW C3	.7	.9	1.2
TW C4	.7	1.0	1.3
TW D1	.7	.9	1.1
TW D2	.5	.7	.9
TW D3	.5	.6	.8
TW D4	.5	.6	.8
TW D5	.5	.6	.8
TW E1	.5	.6	.8
TW E2	.5	.6	.8
TW E3	.4	.5	.7
TW E4	.5	.6	.7
TW F1	.7	.9	1.2
TW F2	.6	.8	1.0
TW F3	.6	.8	1.0
TW F4	.6	.8	1.0
TW G1	.7	.9	1.2
TW G2	.6	.8	1.1
TW G3	.7	.9	1.1
TW G4	.7	.9	1.2
TW H1	.7	.9	1.2
TW H2	.7	1.0	1.2
TW H3	.8	1.1	1.3
TW H4	.9	1.2	1.6
TW I1	.7	.9	1.2
TW I2	.6	.7	.9
TW I3	.5	.7	.9
TW I4	.5	.7	.8
TW I5	.5	.7	.8
TW J1	.5	.6	.8
TW J2	.5	.6	.8
TW J3	.5	.6	.7
TW J4	.5	.6	.8
DG A1	.6	.8	1.1
DG A2	.6	.8	1.0
DG A3	.6	.7	.9
DG B1	.6	.7	.9
DG B2	.6	.7	.9
DG B3	.6	.7	.8
DG C1	.6	.6	.8

DAYTONA BEACH, FL
SOLAR SAVINGS

	10%	50%	80%
DG C2	.6	.6	.8
DG C3	.5	.6	.7
SS A1	.5	.7	.9
SS A2	.4	.6	.7
SS A3	.5	.7	.9
SS A4	.4	.6	.7
SS A5	.5	.7	1.0
SS A6	.4	.6	.7
SS A7	.5	.8	1.1
SS A8	.4	.6	.8
SS B1	.6	.8	1.0
SS B2	.5	.6	.8
SS B3	.6	.8	1.0
SS B4	.5	.6	.8
SS B5	.6	.9	1.1
SS B6	.5	.6	.8
SS B7	.6	.9	1.2
SS B8	.5	.7	.8
SS C1	.5	.7	.9
SS C2	.5	.6	.8
SS C3	.6	.9	1.1
SS C4	.5	.7	.9
SS D1	.4	.6	.8
SS D2	.4	.5	.6
SS D3	.5	.6	.8
SS D4	.4	.5	.7
SS E1	.5	.6	.8
SS E2	.4	.5	.7
SS E3	.5	.7	1.0
SS E4	.4	.6	.8

JACKSONVILLE, FL
SOLAR SAVINGS

	10%	50%	80%
WW A1	.8	1.1	1.4
WW A2	.6	.9	1.1
WW A3	.6	.8	1.1
WW A4	.6	.8	1.0
WW A5	.6	.8	1.0
WW A6	.6	.8	1.0
WW B1	.7	1.0	1.3
WW B2	.6	.8	1.0
WW B3	.5	.7	.9
WW B4	.5	.7	.9
WW B5	.6	.7	.9
WW C1	.5	.7	.9
WW C2	.5	.7	.9
WW C3	.5	.6	.8
WW C4	.5	.7	.8
TW A1	.8	1.1	1.5
TW A2	.7	1.0	1.2
TW A3	.7	.9	1.2
TW A4	.7	.9	1.1
TW B1	.8	1.1	1.4
TW B2	.7	1.0	1.3
TW B3	.7	1.0	1.2
TW B4	.8	1.0	1.3
TW C1	.8	1.1	1.4
TW C2	.8	1.1	1.4
TW C3	.8	1.1	1.4
TW C4	.9	1.2	1.6
TW D1	.8	1.1	1.4
TW D2	.6	.9	1.1
TW D3	.6	.8	1.0

JACKSONVILLE, FL
SOLAR SAVINGS

	10%	50%	80%
TW D4	.6	.8	1.0
TW D5	.6	.8	1.0
TW E1	.6	.7	.9
TW E2	.6	.7	1.0
TW E3	.5	.7	.8
TW E4	.5	.7	.9
TW F1	.8	1.1	1.5
TW F2	.7	1.0	1.3
TW F3	.7	1.0	1.2
TW F4	.7	1.0	1.3
TW G1	.8	1.1	1.5
TW G2	.8	1.0	1.3
TW G3	.8	1.1	1.4
TW G4	.9	1.2	1.5
TW H1	.9	1.2	1.5
TW H2	.9	1.2	1.5
TW H3	1.0	1.3	1.7
TW H4	1.1	1.5	1.9
TW I1	.9	1.2	1.5
TW I2	.7	.9	1.1
TW I3	.6	.8	1.1
TW I4	.6	.8	1.0
TW I5	.6	.8	1.0
TW J1	.6	.8	1.0
TW J2	.6	.8	1.0
TW J3	.5	.7	.9
TW J4	.6	.7	.9
DG A1	.7	1.0	1.6
DG A2	.8	1.0	1.3
DG A3	.7	.8	1.1
DG B1	.8	.9	1.3
DG B2	.7	.9	1.1
DG B3	.7	.8	1.0
DG C1	.7	.8	1.0
DG C2	.7	.8	1.0
DG C3	.6	.7	.9
SS A1	.6	.8	1.1
SS A2	.5	.7	.9
SS A3	.6	.9	1.2
SS A4	.5	.7	.9
SS A5	.6	.9	1.2
SS A6	.5	.7	.9
SS A7	.7	1.0	1.4
SS A8	.5	.7	.9
SS B1	.7	.9	1.2
SS B2	.6	.8	1.0
SS B3	.7	1.0	1.3
SS B4	.6	.8	1.0
SS B5	.7	1.1	1.4
SS B6	.6	.8	1.0
SS B7	.8	1.2	1.5
SS B8	.6	.8	1.0
SS C1	.7	.9	1.1
SS C2	.6	.8	1.0
SS C3	.8	1.1	1.4
SS C4	.7	.9	1.1
SS D1	.5	.7	.9
SS D2	.5	.6	.8
SS D3	.6	.8	1.0
SS D4	.5	.6	.8
SS E1	.6	.8	1.0
SS E2	.5	.7	.9
SS E3	.7	.9	1.2
SS E4	.5	.7	.9

MIAMI, FL
SOLAR SAVINGS

	10%	50%	80%
WW A1	---	.4	.6
WW A2	.3	.4	.5
WW A3	.3	.3	.4
WW A4	.2	.3	.4
WW A5	.2	.3	.4
WW A6	.2	.3	.4
WW B1	.3	.4	.5
WW B2	.2	.3	.4
WW B3	.2	.3	.4
WW B4	.2	.3	.4
WW B5	.2	.3	.4
WW C1	.2	.3	.4
WW C2	.2	.3	.4
WW C3	.2	.3	.3
WW C4	.2	.3	.4
TW A1	---	.5	.6
TW A2	.3	.4	.5
TW A3	.3	.4	.5
TW A4	.3	.4	.5
TW B1	.3	.4	.6
TW B2	.3	.4	.5
TW B3	.3	.4	.5
TW B4	.3	.4	.5
TW C1	.3	.5	.6
TW C2	.3	.4	.6
TW C3	.3	.5	.6
TW C4	.4	.5	.7
TW D1	.3	.4	.6
TW D2	.3	.4	.5
TW D3	.2	.3	.4
TW D4	.2	.3	.4
TW D5	.3	.3	.4
TW E1	.2	.3	.4
TW E2	.2	.3	.4
TW E3	.2	.3	.4
TW E4	.2	.3	.4
TW F1	.3	.4	.6
TW F2	.3	.4	.5
TW F3	.3	.4	.5
TW F4	.3	.4	.5
TW G1	.3	.5	.6
TW G2	.3	.4	.6
TW G3	.3	.4	.6
TW G4	.4	.5	.6
TW H1	.4	.5	.6
TW H2	.4	.5	.6
TW H3	.4	.5	.7
TW H4	.5	.6	.8
TW I1	.3	.5	.6
TW I2	.3	.4	.5
TW I3	.3	.4	.5
TW I4	.3	.3	.4
TW I5	.3	.4	.4
TW J1	.2	.3	.4
TW J2	.3	.3	.4
TW J3	.2	.3	.4
TW J4	.2	.3	.4
DG A1	.3	.4	.5
DG A2	.3	.4	.5
DG A3	.3	.4	.5
DG B1	.3	.3	.4
DG B2	.3	.4	.4
DG B3	.3	.3	.4
DG C1	.3	.3	.4

MIAMI, FL

	SOLAR SAVINGS		
	10%	50%	80%
DG C2	.3	.3	.4
DG C3	.3	.3	.4
SS A1	---	.3	.4
SS A2	.2	.3	.4
SS A3	---	.4	.5
SS A4	.2	.3	.4
SS A5	---	.4	.5
SS A6	.2	.3	.4
SS A7	---	.4	.5
SS A8	.2	.3	.4
SS B1	.3	.4	.5
SS B2	.2	.3	.4
SS B3	.3	.4	.5
SS B4	.2	.3	.4
SS B5	.3	.4	.6
SS B6	.2	.3	.4
SS B7	.3	.5	.6
SS B8	.2	.3	.4
SS C1	.3	.4	.5
SS C2	.3	.3	.4
SS C3	.3	.4	.6
SS C4	.3	.4	.5
SS D1	.2	.3	.4
SS D2	.2	.3	.3
SS D3	---	.3	.4
SS D4	.2	.3	.3
SS E1	.2	.3	.4
SS E2	.2	.3	.4
SS E3	.3	.4	.5
SS E4	.2	.3	.4

ORLANDO, FL

	SOLAR SAVINGS		
	10%	50%	80%
WW A1	.6	.8	1.0
WW A2	.5	.6	.8
WW A3	.5	.6	.8
WW A4	.4	.6	.7
WW A5	.4	.6	.7
WW A6	.4	.5	.7
WW B1	.5	.7	.9
WW B2	.4	.6	.7
WW B3	.4	.5	.6
WW B4	.4	.5	.6
WW B5	.4	.5	.7
WW C1	.4	.5	.6
WW C2	.4	.5	.7
WW C3	.4	.5	.6
WW C4	.4	.5	.6
TW A1	.6	.8	1.0
TW A2	.5	.7	.9
TW A3	.5	.7	.8
TW A4	.5	.7	.8
TW B1	.6	.8	1.0
TW B2	.5	.7	.9
TW B3	.5	.7	.9
TW B4	.6	.7	.9
TW C1	.6	.8	1.0
TW C2	.6	.8	1.0
TW C3	.6	.8	1.0
TW C4	.7	.9	1.2
TW D1	.6	.8	1.0
TW D2	.5	.6	.8
TW D3	.4	.6	.7
TW D4	.4	.6	.7
TW D5	.4	.6	.7
TW E1	.4	.5	.7
TW E2	.4	.6	.7
TW E3	.4	.5	.6
TW E4	.4	.5	.7
TW F1	.6	.8	1.0
TW F2	.5	.7	.9
TW F3	.5	.7	.9
TW F4	.5	.7	.9
TW G1	.6	.8	1.0
TW G2	.6	.8	1.0
TW G3	.6	.8	1.0
TW G4	.7	.8	1.1
TW H1	.6	.8	1.1
TW H2	.7	.9	1.1
TW H3	.7	.9	1.2
TW H4	.8	1.1	1.4
TW I1	.6	.8	1.1
TW I2	.5	.7	.8
TW I3	.5	.6	.8
TW I4	.5	.6	.7
TW I5	.5	.6	.7
TW J1	.4	.6	.7
TW J2	.4	.6	.7
TW J3	.4	.5	.6
TW J4	.4	.5	.7
DG A1	.5	.7	.9
DG A2	.6	.7	.9
DG A3	.5	.6	.8
DG B1	.5	.6	.8
DG B2	.5	.6	.8
DG B3	.5	.6	.8
DG C1	.5	.6	.7
DG C2	.5	.6	.7
DG C3	.5	.5	.6
SS A1	.4	.6	.8
SS A2	.4	.5	.7
SS A3	.5	.6	.8
SS A4	.4	.5	.7
SS A5	.5	.7	.9
SS A6	.4	.5	.7
SS A7	.5	.7	.9
SS A8	.4	.5	.7
SS B1	.5	.7	.9
SS B2	.4	.6	.7
SS B3	.5	.7	.9
SS B4	.4	.6	.7
SS B5	.5	.8	1.0
SS B6	.4	.6	.7
SS B7	.6	.8	1.1
SS B8	.4	.6	.7
SS C1	.5	.6	.8
SS C2	.4	.6	.7
SS C3	.6	.8	1.0
SS C4	.5	.6	.8
SS D1	.4	.5	.7
SS D2	.3	.4	.6
SS D3	.4	.6	.7
SS D4	.4	.5	.6
SS E1	.4	.6	.7
SS E2	.4	.5	.6
SS E3	.5	.7	.9
SS E4	.4	.5	.7

TALLAHASSEE, FL

	SOLAR SAVINGS		
	10%	50%	80%
WW A1	.8	1.2	1.6
WW A2	.7	1.0	1.3
WW A3	.7	.9	1.2
WW A4	.7	.9	1.1
WW A5	.6	.8	1.1
WW A6	.6	.8	1.1
WW B1	.8	1.1	1.5
WW B2	.7	.9	1.1
WW B3	.6	.8	1.0
WW B4	.6	.8	1.0
WW B5	.6	.8	1.0
WW C1	.6	.8	1.0
WW C2	.6	.8	1.0
WW C3	.5	.7	.9
WW C4	.6	.7	.9
TW A1	.9	1.3	1.7
TW A2	.8	1.1	1.4
TW A3	.7	1.0	1.3
TW A4	.7	1.0	1.3
TW B1	.9	1.2	1.6
TW B2	.8	1.1	1.4
TW B3	.8	1.1	1.4
TW B4	.8	1.1	1.5
TW C1	.9	1.2	1.6
TW C2	.9	1.2	1.5
TW C3	.9	1.2	1.6
TW C4	1.0	1.4	1.8
TW D1	.9	1.2	1.6
TW D2	.7	.9	1.2
TW D3	.7	.9	1.1
TW D4	.6	.8	1.1
TW D5	.6	.8	1.1
TW E1	.6	.8	1.0
TW E2	.6	.8	1.1
TW E3	.6	.7	.9
TW E4	.6	.8	1.0
TW F1	.9	1.3	1.6
TW F2	.8	1.1	1.4
TW F3	.8	1.1	1.4
TW F4	.8	1.1	1.4
TW G1	.9	1.3	1.6
TW G2	.9	1.2	1.5
TW G3	.9	1.2	1.5
TW G4	1.0	1.3	1.7
TW H1	1.0	1.3	1.7
TW H2	1.0	1.3	1.7
TW H3	1.1	1.4	1.8
TW H4	1.2	1.7	2.1
TW I1	.9	1.3	1.7
TW I2	.7	1.0	1.3
TW I3	.7	.9	1.2
TW I4	.7	.9	1.1
TW I5	.7	.9	1.1
TW J1	.6	.8	1.1
TW J2	.7	.9	1.1
TW J3	.6	.8	1.0
TW J4	.6	.8	1.0
DG A1	.8	1.2	2.1
DG A2	.8	1.1	1.5
DG A3	.8	.9	1.2
DG B1	.8	1.0	1.5
DG B2	.8	1.0	1.3
DG B3	.7	.9	1.1
DG C1	.7	.9	1.2

TALLAHASSEE, FL

	SOLAR SAVINGS		
	10%	50%	80%
DG C2	.7	.9	1.1
DG C3	.7	.8	1.0
SS A1	.6	.9	1.2
SS A2	.6	.8	1.0
SS A3	.7	1.0	1.3
SS A4	.6	.8	1.0
SS A5	.7	1.0	1.4
SS A6	.6	.8	1.0
SS A7	.7	1.1	1.5
SS A8	.6	.8	1.0
SS B1	.7	1.0	1.4
SS B2	.6	.9	1.1
SS B3	.7	1.1	1.4
SS B4	.6	.9	1.1
SS B5	.8	1.2	1.6
SS B6	.6	.9	1.1
SS B7	.8	1.3	1.7
SS B8	.6	.9	1.1
SS C1	.7	1.0	1.3
SS C2	.7	.9	1.1
SS C3	.9	1.2	1.6
SS C4	.7	1.0	1.2
SS D1	.6	.8	1.0
SS D2	.5	.7	.9
SS D3	.6	.9	1.2
SS D4	.5	.7	.9
SS E1	.6	.9	1.1
SS E2	.6	.7	1.0
SS E3	.7	1.0	1.4
SS E4	.6	.8	1.0

TAMPA, FL

	SOLAR SAVINGS		
	10%	50%	80%
WW A1	.6	.8	1.0
WW A2	.5	.6	.8
WW A3	.5	.6	.8
WW A4	.4	.6	.7
WW A5	.4	.6	.7
WW A6	.4	.5	.7
WW B1	.5	.7	.9
WW B2	.4	.6	.7
WW B3	.4	.5	.7
WW B4	.4	.5	.6
WW B5	.4	.5	.7
WW C1	.4	.5	.6
WW C2	.4	.5	.7
WW C3	.4	.5	.6
WW C4	.4	.5	.6
TW A1	.6	.8	1.0
TW A2	.5	.7	.9
TW A3	.5	.7	.8
TW A4	.5	.7	.8
TW B1	.6	.8	1.0
TW B2	.5	.7	.9
TW B3	.5	.7	.9
TW B4	.6	.7	.9
TW C1	.6	.8	1.0
TW C2	.6	.8	1.0
TW C3	.6	.8	1.0
TW C4	.7	.9	1.2
TW D1	.6	.8	1.0
TW D2	.5	.6	.8
TW D3	.4	.6	.7

Conservation Factor Tables / 321

TAMPA, FL
SOLAR SAVINGS

	10%	50%	80%
WW A1	.6	.8	1.0
WW A2	.5	.6	.8
WW A3	.5	.6	.8
WW A4	.4	.6	.7
WW A5	.4	.6	.7
WW A6	.4	.5	.7
WW B1	.5	.7	.9
WW B2	.4	.6	.7
WW B3	.4	.5	.7
WW B4	.4	.5	.6
WW B5	.4	.5	.7
WW C1	.4	.5	.6
WW C2	.4	.5	.7
WW C3	.4	.5	.6
WW C4	.4	.5	.6
TW A1	.6	.8	1.0
TW A2	.5	.7	.9
TW A3	.5	.7	.8
TW A4	.5	.7	.8
TW B1	.6	.8	1.0
TW B2	.5	.7	.9
TW B3	.5	.7	.9
TW B4	.6	.7	.9
TW C1	.6	.8	1.0
TW C2	.6	.8	1.0
TW C3	.6	.8	1.0
TW C4	.7	.9	1.2
TW D1	.6	.8	1.0
TW D2	.5	.6	.8
TW D3	.4	.6	.7
DG B1	.5	.6	.8
DG B2	.6	.6	.8
DG B3	.5	.6	.7
DG C1	.5	.6	.7
DG C2	.5	.6	.7
DG C3	.5	.5	.6
SS A1	.4	.6	.8
SS A2	.4	.5	.7
SS A3	.5	.6	.8
SS A4	.4	.5	.7
SS A5	.5	.7	.9
SS A6	.4	.5	.7
SS A7	.5	.7	.9
SS A8	.4	.5	.7
SS B1	.5	.7	.9
SS B2	.4	.6	.7
SS B3	.5	.7	.9
SS B4	.4	.6	.7
SS B5	.5	.8	1.0
SS B6	.4	.6	.7
SS B7	.6	.8	1.1
SS B8	.4	.6	.7
SS C1	.5	.6	.8
SS C2	.4	.6	.7
SS C3	.6	.8	1.0
SS C4	.5	.6	.8
SS D1	.4	.5	.7
SS D2	.3	.4	.6
SS D3	.4	.6	.7
SS D4	.4	.5	.6
SS E1	.4	.6	.7
SS E2	.4	.5	.6
SS E3	.5	.7	.9
SS E4	.4	.5	.7

WEST PALM BEACH, FL
SOLAR SAVINGS

	10%	50%	80%
WW A1	---	.5	.7
WW A2	.3	.4	.6
WW A3	.3	.4	.5
WW A4	.3	.4	.5
WW A5	.3	.4	.5
WW A6	.3	.4	.5
WW B1	.4	.5	.6
WW B2	.3	.4	.5
WW B3	.3	.4	.5
WW B4	.3	.4	.5
WW B5	.3	.4	.5
WW C1	.3	.4	.4
WW C2	.3	.4	.5
WW C3	.3	.3	.4
WW C4	.3	.4	.4
TW A1	.4	.6	.7
TW A2	.4	.5	.6
TW A3	.3	.5	.6
TW A4	.3	.5	.6
TW B1	.4	.5	.7
TW B2	.4	.5	.6
TW B3	.4	.5	.6
TW B4	.4	.5	.7
TW C1	.4	.6	.7
TW C2	.4	.5	.7
TW C3	.4	.6	.7
TW C4	.5	.6	.8
TW D1	.4	.5	.7
TW D2	.3	.4	.6
TW D3	.3	.4	.5
TW D4	.3	.4	.5
TW D5	.3	.4	.5
TW E1	.3	.4	.5
TW E2	.3	.4	.5
TW E3	.3	.3	.4
TW E4	.3	.4	.5
TW F1	.4	.5	.7
TW F2	.4	.5	.6
TW F3	.4	.5	.6
TW F4	.4	.5	.6
TW G1	.4	.6	.7
TW G2	.4	.5	.7
TW G3	.4	.5	.7
TW G4	.4	.6	.8
TW H1	.4	.6	.8
TW H2	.5	.6	.8
TW H3	.5	.7	.8
TW H4	.6	.8	1.0
TW I1	.4	.6	.7
TW I2	.3	.5	.6
TW I3	.3	.4	.6
TW I4	.3	.4	.5
TW I5	.3	.4	.5
TW J1	.3	.4	.5
TW J2	.3	.4	.5
TW J3	.3	.4	.5
TW J4	.3	.4	.5
DG A1	.4	.5	.6
DG A2	.4	.5	.6
DG A3	.4	.4	.6
DG B1	.4	.4	.5
DG B2	.4	.4	.5
DG B3	.4	.4	.5
DG C1	.3	.4	.5

WEST PALM BEACH, FL
SOLAR SAVINGS

	10%	50%	80%
DG C2	.3	.4	.5
DG C3	.3	.4	.5
SS A1	.3	.4	.5
SS A2	.3	.4	.5
SS A3	.3	.4	.6
SS A4	.3	.4	.5
SS A5	---	.5	.6
SS A6	.3	.4	.5
SS A7	---	.5	.6
SS A8	.3	.4	.5
SS B1	.3	.5	.6
SS B2	.3	.4	.5
SS B3	.3	.5	.6
SS B4	.3	.4	.5
SS B5	.4	.5	.7
SS B6	.3	.4	.5
SS B7	.4	.6	.7
SS B8	.3	.4	.5
SS C1	.3	.4	.6
SS C2	.3	.4	.5
SS C3	.4	.5	.7
SS C4	.3	.4	.6
SS D1	.3	.4	.5
SS D2	.2	.3	.4
SS D3	.3	.4	.5
SS D4	.2	.3	.4
SS E1	.3	.4	.5
SS E2	.3	.3	.4
SS E3	.3	.5	.6
SS E4	.3	.4	.5

ATLANTA, GA
SOLAR SAVINGS

	10%	50%	80%
WW A1	1.2	2.1	2.8
WW A2	1.1	1.5	2.0
WW A3	1.0	1.4	1.8
WW A4	1.0	1.3	1.8
WW A5	.9	1.3	1.7
WW A6	.9	1.3	1.6
WW B1	1.2	2.0	2.7
WW B2	.9	1.3	1.7
WW B3	.8	1.1	1.5
WW B4	.8	1.1	1.4
WW B5	.8	1.1	1.4
WW C1	.8	1.1	1.4
WW C2	.8	1.1	1.4
WW C3	.7	1.0	1.2
WW C4	.8	1.0	1.3
TW A1	1.3	2.1	2.8
TW A2	1.1	1.7	2.2
TW A3	1.1	1.6	2.1
TW A4	1.1	1.5	2.0
TW B1	1.3	2.0	2.6
TW B2	1.2	1.7	2.2
TW B3	1.2	1.6	2.2
TW B4	1.2	1.7	2.2
TW C1	1.3	1.9	2.5
TW C2	1.3	1.8	2.4
TW C3	1.3	1.9	2.5
TW C4	1.4	2.1	2.7
TW D1	1.4	2.1	2.9
TW D2	1.0	1.4	1.8
TW D3	.9	1.3	1.7

ATLANTA, GA
SOLAR SAVINGS

	10%	50%	80%
TW D4	.9	1.2	1.5
TW D5	.9	1.2	1.5
TW E1	.9	1.2	1.6
TW E2	.9	1.2	1.5
TW E3	.8	1.0	1.3
TW E4	.8	1.1	1.4
TW F1	1.3	2.0	2.7
TW F2	1.2	1.7	2.3
TW F3	1.2	1.7	2.2
TW F4	1.2	1.7	2.2
TW G1	1.3	2.0	2.7
TW G2	1.3	1.8	2.4
TW G3	1.3	1.8	2.4
TW G4	1.4	2.0	2.6
TW H1	1.4	2.0	2.7
TW H2	1.5	2.1	2.7
TW H3	1.6	2.2	2.9
TW H4	1.8	2.5	3.3
TW I1	1.5	2.3	3.1
TW I2	1.1	1.5	1.9
TW I3	1.0	1.4	1.8
TW I4	.9	1.2	1.6
TW I5	.9	1.2	1.6
TW J1	.9	1.2	1.6
TW J2	.9	1.2	1.6
TW J3	.8	1.1	1.4
TW J4	.8	1.1	1.4
DG A1	1.3	5.5	---
DG A2	1.2	1.8	3.7
DG A3	1.0	1.3	1.9
DG B1	1.3	2.1	---
DG B2	1.2	1.5	2.4
DG B3	1.0	1.3	1.6
DG C1	1.1	1.6	---
DG C2	1.1	1.3	1.9
DG C3	1.0	1.2	1.4
SS A1	.9	1.4	2.0
SS A2	.8	1.1	1.5
SS A3	1.0	1.6	2.2
SS A4	.8	1.2	1.5
SS A5	1.0	1.6	2.2
SS A6	.8	1.1	1.5
SS A7	1.1	1.9	2.6
SS A8	.8	1.2	1.6
SS B1	1.1	1.6	2.2
SS B2	.9	1.2	1.6
SS B3	1.1	1.7	2.3
SS B4	.9	1.3	1.7
SS B5	1.2	1.9	2.6
SS B6	.9	1.3	1.7
SS B7	1.3	2.1	2.9
SS B8	.9	1.3	1.7
SS C1	1.1	1.6	2.1
SS C2	.9	1.3	1.7
SS C3	1.3	1.9	2.6
SS C4	1.0	1.4	1.9
SS D1	.8	1.2	1.6
SS D2	.7	1.0	1.3
SS D3	.9	1.4	1.9
SS D4	.7	1.0	1.3
SS E1	.9	1.4	1.8
SS E2	.8	1.1	1.4
SS E3	1.1	1.6	2.2
SS E4	.8	1.2	1.5

AUGUSTA, GA

	SOLAR SAVINGS		
	10%	50%	80%
WW A1	1.1	1.7	2.3
WW A2	.9	1.3	1.7
WW A3	.9	1.2	1.6
WW A4	.8	1.2	1.5
WW A5	.8	1.1	1.4
WW A6	.8	1.1	1.4
WW B1	1.1	1.6	2.1
WW B2	.8	1.1	1.5
WW B3	.7	1.0	1.3
WW B4	.7	1.0	1.2
WW B5	.8	1.0	1.2
WW C1	.7	1.0	1.2
WW C2	.7	1.0	1.3
WW C3	.7	.9	1.1
WW C4	.7	.9	1.2
TW A1	1.1	1.7	2.3
TW A2	1.0	1.4	1.9
TW A3	1.0	1.3	1.7
TW A4	1.0	1.3	1.7
TW B1	1.1	1.7	2.2
TW B2	1.0	1.4	1.9
TW B3	1.0	1.4	1.8
TW B4	1.1	1.5	1.9
TW C1	1.1	1.6	2.1
TW C2	1.1	1.6	2.1
TW C3	1.2	1.6	2.1
TW C4	1.3	1.8	2.4
TW D1	1.2	1.8	2.3
TW D2	.9	1.2	1.6
TW D3	.8	1.1	1.5
TW D4	.8	1.1	1.4
TW D5	.8	1.1	1.4
TW E1	.8	1.1	1.4
TW E2	.8	1.1	1.4
TW E3	.7	.9	1.2
TW E4	.7	1.0	1.2
TW F1	1.1	1.7	2.3
TW F2	1.0	1.5	1.9
TW F3	1.0	1.4	1.8
TW F4	1.1	1.4	1.9
TW G1	1.2	1.7	2.2
TW G2	1.1	1.6	2.0
TW G3	1.1	1.6	2.0
TW G4	1.2	1.7	2.2
TW H1	1.2	1.7	2.3
TW H2	1.3	1.8	2.3
TW H3	1.4	1.9	2.5
TW H4	1.6	2.2	2.8
TW I1	1.3	1.9	2.5
TW I2	1.0	1.3	1.7
TW I3	.9	1.2	1.6
TW I4	.8	1.1	1.4
TW I5	.8	1.1	1.4
TW J1	.8	1.1	1.4
TW J2	.8	1.1	1.4
TW J3	.7	1.0	1.2
TW J4	.8	1.0	1.3
DG A1	1.1	2.1	---
DG A2	1.0	1.5	2.4
DG A3	.9	1.2	1.6
DG B1	1.1	1.5	---
DG B2	1.0	1.3	1.9
DG B3	.9	1.1	1.4
DG C1	1.0	1.3	2.1

AUGUSTA, GA

	SOLAR SAVINGS		
	10%	50%	80%
DG C2	.9	1.2	1.5
DG C3	.9	1.0	1.3
SS A1	.8	1.3	1.7
SS A2	.7	1.0	1.3
SS A3	.9	1.4	1.8
SS A4	.7	1.0	1.3
SS A5	.9	1.4	1.9
SS A6	.7	1.0	1.3
SS A7	1.0	1.6	2.2
SS A8	.7	1.0	1.4
SS B1	.9	1.4	1.9
SS B2	.8	1.1	1.4
SS B3	1.0	1.5	2.0
SS B4	.8	1.1	1.5
SS B5	1.1	1.6	2.2
SS B6	.8	1.1	1.5
SS B7	1.1	1.8	2.4
SS B8	.8	1.2	1.5
SS C1	1.0	1.3	1.7
SS C2	.8	1.1	1.5
SS C3	1.1	1.6	2.1
SS C4	.9	1.3	1.6
SS D1	.7	1.1	1.4
SS D2	.6	.9	1.1
SS D3	.8	1.2	1.6
SS D4	.7	.9	1.2
SS E1	.8	1.2	1.6
SS E2	.7	1.0	1.3
SS E3	.9	1.4	1.9
SS E4	.7	1.0	1.4

MACON, GA

	SOLAR SAVINGS		
	10%	50%	80%
WW A1	1.0	1.6	2.1
WW A2	.9	1.2	1.6
WW A3	.8	1.1	1.5
WW A4	.8	1.1	1.4
WW A5	.8	1.0	1.4
WW A6	.8	1.0	1.3
WW B1	1.0	1.5	1.9
WW B2	.8	1.1	1.4
WW B3	.7	.9	1.2
WW B4	.7	.9	1.2
WW B5	.7	.9	1.2
WW C1	.7	.9	1.2
WW C2	.7	.9	1.2
WW C3	.6	.8	1.0
WW C4	.7	.9	1.1
TW A1	1.0	1.6	2.1
TW A2	.9	1.3	1.8
TW A3	.9	1.3	1.6
TW A4	.9	1.2	1.6
TW B1	1.0	1.5	2.0
TW B2	1.0	1.4	1.8
TW B3	1.0	1.3	1.7
TW B4	1.0	1.4	1.8
TW C1	1.1	1.5	2.0
TW C2	1.0	1.5	1.9
TW C3	1.1	1.5	2.0
TW C4	1.2	1.7	2.2
TW D1	1.1	1.6	2.1
TW D2	.8	1.2	1.5
TW D3	.8	1.1	1.4

MACON, GA

	SOLAR SAVINGS		
	10%	50%	80%
TW D4	.8	1.0	1.3
TW D5	.8	1.0	1.3
TW E1	.7	1.0	1.3
TW E2	.7	1.0	1.3
TW E3	.7	.9	1.1
TW E4	.7	.9	1.2
TW F1	1.1	1.6	2.1
TW F2	1.0	1.4	1.8
TW F3	1.0	1.3	1.7
TW F4	1.0	1.3	1.7
TW G1	1.1	1.6	2.1
TW G2	1.0	1.5	1.9
TW G3	1.1	1.5	1.9
TW G4	1.2	1.6	2.1
TW H1	1.2	1.6	2.1
TW H2	1.2	1.7	2.1
TW H3	1.3	1.8	2.3
TW H4	1.5	2.1	2.7
TW I1	1.2	1.7	2.3
TW I2	.9	1.2	1.6
TW I3	.8	1.1	1.5
TW I4	.8	1.1	1.3
TW I5	.8	1.0	1.3
TW J1	.8	1.0	1.3
TW J2	.8	1.0	1.3
TW J3	.7	.9	1.1
TW J4	.7	.9	1.2
DG A1	1.0	1.8	---
DG A2	1.0	1.4	2.1
DG A3	.9	1.1	1.5
DG B1	1.0	1.4	3.2
DG B2	.9	1.2	1.7
DG B3	.9	1.1	1.3
DG C1	.9	1.2	1.8
DG C2	.9	1.1	1.4
DG C3	.8	1.0	1.2
SS A1	.8	1.2	1.6
SS A2	.7	1.0	1.2
SS A3	.8	1.3	1.7
SS A4	.7	1.0	1.3
SS A5	.8	1.3	1.7
SS A6	.7	1.0	1.2
SS A7	.9	1.5	2.0
SS A8	.7	1.0	1.3
SS B1	.9	1.3	1.7
SS B2	.7	1.0	1.4
SS B3	.9	1.4	1.8
SS B4	.7	1.1	1.4
SS B5	1.0	1.5	2.0
SS B6	.7	1.1	1.4
SS B7	1.1	1.7	2.2
SS B8	.8	1.1	1.4
SS C1	.9	1.2	1.6
SS C2	.8	1.1	1.4
SS C3	1.1	1.5	2.0
SS C4	.9	1.2	1.5
SS D1	.7	1.0	1.3
SS D2	.6	.8	1.1
SS D3	.7	1.1	1.5
SS D4	.6	.9	1.1
SS E1	.8	1.1	1.4
SS E2	.7	.9	1.2
SS E3	.9	1.3	1.7
SS E4	.7	1.0	1.3

SAVANNAH, GA

	SOLAR SAVINGS		
	10%	50%	80%
WW A1	1.0	1.4	1.9
WW A2	.8	1.1	1.5
WW A3	.8	1.0	1.3
WW A4	.7	1.0	1.3
WW A5	.7	1.0	1.2
WW A6	.7	.9	1.2
WW B1	.9	1.3	1.7
WW B2	.7	1.0	1.3
WW B3	.7	.9	1.1
WW B4	.7	.9	1.1
WW B5	.7	.9	1.1
WW C1	.6	.8	1.1
WW C2	.7	.9	1.1
WW C3	.6	.8	1.0
WW C4	.6	.8	1.0
TW A1	1.0	1.5	1.9
TW A2	.9	1.2	1.6
TW A3	.8	1.2	1.5
TW A4	.8	1.1	1.5
TW B1	1.0	1.4	1.9
TW B2	.9	1.2	1.6
TW B3	.9	1.2	1.6
TW B4	.9	1.3	1.7
TW C1	1.0	1.4	1.8
TW C2	1.0	1.4	1.8
TW C3	1.0	1.4	1.8
TW C4	1.1	1.6	2.0
TW D1	1.0	1.5	1.9
TW D2	.8	1.1	1.4
TW D3	.7	1.0	1.3
TW D4	.7	.9	1.2
TW D5	.7	.9	1.2
TW E1	.7	.9	1.2
TW E2	.7	.9	1.2
TW E3	.6	.8	1.0
TW E4	.6	.8	1.1
TW F1	1.0	1.4	1.9
TW F2	.9	1.3	1.6
TW F3	.9	1.2	1.6
TW F4	.9	1.2	1.6
TW G1	1.0	1.4	1.9
TW G2	1.0	1.3	1.7
TW G3	1.0	1.4	1.7
TW G4	1.1	1.5	1.9
TW H1	1.1	1.5	1.9
TW H2	1.1	1.5	2.0
TW H3	1.2	1.6	2.1
TW H4	1.4	1.9	2.4
TW I1	1.1	1.5	2.0
TW I2	.8	1.1	1.4
TW I3	.8	1.1	1.4
TW I4	.7	1.0	1.2
TW I5	.7	1.0	1.2
TW J1	.7	1.0	1.2
TW J2	.7	1.0	1.2
TW J3	.6	.8	1.1
TW J4	.7	.9	1.1
DG A1	.9	1.5	4.9
DG A2	.9	1.2	1.8
DG A3	.8	1.0	1.4
DG B1	.9	1.2	2.1
DG B2	.9	1.1	1.5
DG B3	.8	1.0	1.2
DG C1	.8	1.1	1.5

SAVANNAH, GA				BOISE, ID				LEWISTON, ID				LEWISTON, ID			
	SOLAR SAVINGS				SOLAR SAVINGS				SOLAR SAVINGS				SOLAR SAVINGS		
	10%	50%	80%		10%	50%	80%		10%	50%	80%		10%	50%	80%
DG C2	.8	1.0	1.3	TW D4	1.0	1.4	2.0	WW A1	1.7	---	---	DG C2	1.4	2.9	---
DG C3	.8	.9	1.1	TW D5	1.0	1.4	1.9	WW A2	1.5	---	---	DG C3	1.2	1.7	2.6
SS A1	.7	1.1	1.4	TW E1	1.0	1.5	2.2	WW A3	1.4	5.1	---	SS A1	1.3	6.3	---
SS A2	.6	.9	1.1	TW E2	1.0	1.5	2.0	WW A4	1.4	3.7	---	SS A2	1.0	1.9	3.0
SS A3	.8	1.1	1.5	TW E3	.9	1.2	1.7	WW A5	1.3	3.2	15.9	SS A3	1.4	---	---
SS A4	.6	.9	1.2	TW E4	.9	1.2	1.7	WW A6	1.3	3.0	8.6	SS A4	1.0	2.1	3.5
SS A5	.8	1.2	1.6	TW F1	1.5	3.2	9.6	WW B1	2.1	---	---	SS A5	1.3	---	---
SS A6	.6	.9	1.2	TW F2	1.4	2.4	4.5	WW B2	1.3	2.4	4.2	SS A6	1.0	2.0	3.3
SS A7	.8	1.3	1.8	TW F3	1.3	2.3	3.8	WW B3	1.1	2.0	3.2	SS A7	1.4	---	---
SS A8	.6	.9	1.2	TW F4	1.4	2.2	3.6	WW B4	1.1	1.6	2.4	SS A8	1.1	2.3	4.2
SS B1	.8	1.2	1.6	TW G1	1.5	3.0	6.3	WW B5	1.1	1.6	2.2	SS B1	1.5	---	---
SS B2	.7	1.0	1.3	TW G2	1.5	2.5	4.3	WW C1	1.1	1.8	2.8	SS B2	1.1	2.0	3.2
SS B3	.8	1.2	1.7	TW G3	1.5	2.5	4.0	WW C2	1.1	1.7	2.5	SS B3	1.6	---	---
SS B4	.7	1.0	1.3	TW G4	1.6	2.6	4.1	WW C3	.9	1.4	1.9	SS B4	1.2	2.2	3.5
SS B5	.9	1.4	1.8	TW H1	1.6	2.9	5.3	WW C4	1.0	1.4	1.9	SS B5	1.6	---	---
SS B6	.7	1.0	1.3	TW H2	1.7	2.8	4.5	TW A1	1.7	---	---	SS B6	1.2	2.1	3.4
SS B7	1.0	1.5	2.0	TW H3	1.8	2.9	4.6	TW A2	1.6	---	---	SS B7	1.8	---	---
SS B8	.7	1.0	1.3	TW H4	2.1	3.3	5.1	TW A3	1.5	5.1	---	SS B8	1.2	2.4	4.1
SS C1	.8	1.1	1.5	TW I1	1.8	---	---	TW A4	1.5	3.7	---	SS C1	1.6	---	---
SS C2	.7	1.0	1.3	TW I2	1.2	1.9	2.8	TW B1	1.7	---	---	SS C2	1.3	2.3	3.9
SS C3	1.0	1.4	1.8	TW I3	1.1	1.7	2.6	TW B2	1.6	5.2	---	SS C3	1.8	---	---
SS C4	.8	1.1	1.4	TW I4	1.0	1.5	2.1	TW B3	1.6	4.0	---	SS C4	1.4	2.6	4.6
SS D1	.6	.9	1.2	TW I5	1.0	1.5	2.0	TW B4	1.6	3.7	11.5	SS D1	1.1	4.0	---
SS D2	.6	.8	1.0	TW J1	1.0	1.5	2.2	TW C1	1.7	10.9	---	SS D2	.9	1.6	2.5
SS D3	.7	1.0	1.4	TW J2	1.0	1.5	2.1	TW C2	1.7	4.2	---	SS D3	1.2	---	---
SS D4	.6	.8	1.0	TW J3	.9	1.3	1.7	TW C3	1.7	4.0	12.7	SS D4	1.0	1.7	2.6
SS E1	.7	1.0	1.3	TW J4	.9	1.3	1.7	TW C4	1.9	4.1	10.4	SS E1	1.3	---	---
SS E2	.6	.8	1.1	DG A1	1.5	---	---	TW D1	2.3	---	---	SS E2	1.1	1.9	3.2
SS E3	.8	1.2	1.6	DG A2	1.3	2.7	---	TW D2	1.4	2.6	4.8	SS E3	1.4	---	---
SS E4	.7	.9	1.2	DG A3	1.2	1.7	2.9	TW D3	1.3	2.4	4.3	SS E4	1.1	2.1	3.5
				DG B1	1.5	---	---	TW D4	1.2	1.8	2.7				
BOISE, ID				DG B2	1.3	2.2	---	TW D5	1.1	1.7	2.4	POCATELLO, ID			
	SOLAR SAVINGS			DG B3	1.1	1.5	2.3	TW E1	1.2	2.0	3.2		SOLAR SAVINGS		
	10%	50%	80%	DG C1	1.3	4.5	---	TW E2	1.1	1.9	2.8		10%	50%	80%
WW A1	1.4	3.5	---	DG C2	1.2	1.8	4.0	TW E3	1.0	1.5	2.1	WW A1	1.6	4.1	---
WW A2	1.2	2.2	3.9	DG C3	1.0	1.4	1.9	TW E4	1.0	1.5	2.1	WW A2	1.3	2.3	4.2
WW A3	1.1	1.9	3.2	SS A1	1.1	2.1	3.9	TW F1	1.8	---	---	WW A3	1.3	2.1	3.4
WW A4	1.1	1.8	2.9	SS A2	.9	1.5	2.2	TW F2	1.7	---	---	WW A4	1.2	1.9	3.1
WW A5	1.1	1.7	2.7	SS A3	1.1	2.8	8.7	TW F3	1.7	7.3	---	WW A5	1.2	1.8	2.8
WW A6	1.1	1.7	2.6	SS A4	.9	1.5	2.3	TW F4	1.7	4.4	---	WW A6	1.2	1.8	2.7
WW B1	1.5	---	---	SS A5	1.1	2.6	6.1	TW G1	1.9	---	---	WW B1	1.7	---	---
WW B2	1.1	1.6	2.4	SS A6	.9	1.5	2.2	TW G2	1.8	---	---	WW B2	1.2	1.7	2.5
WW B3	.9	1.4	2.1	SS A7	1.2	4.5	---	TW G3	1.8	5.2	---	WW B3	1.0	1.5	2.1
WW B4	.9	1.3	1.8	SS A8	.9	1.6	2.5	TW G4	2.0	4.6	15.9	WW B4	1.0	1.4	1.8
WW B5	.9	1.3	1.8	SS B1	1.2	2.4	4.7	TW H1	2.0	---	---	WW B5	1.0	1.3	1.8
WW C1	.9	1.4	1.9	SS B2	1.0	1.6	2.3	TW H2	2.1	6.2	---	WW C1	1.0	1.4	2.0
WW C2	.9	1.4	1.9	SS B3	1.3	3.0	8.8	TW H3	2.2	5.3	26.4	WW C2	1.0	1.4	1.9
WW C3	.8	1.1	1.5	SS B4	1.0	1.6	2.4	TW H4	2.5	5.4	13.7	WW C3	.9	1.2	1.6
WW C4	.9	1.2	1.6	SS B5	1.3	3.1	9.3	TW I1	2.7	---	---	WW C4	.9	1.2	1.6
TW A1	1.5	3.2	11.7	SS B6	1.0	1.6	2.4	TW I2	1.5	2.7	5.0	TW A1	1.6	3.6	34.2
TW A2	1.3	2.3	4.2	SS B7	1.5	5.3	---	TW I3	1.4	2.4	4.1	TW A2	1.4	2.5	4.5
TW A3	1.2	2.1	3.5	SS B8	1.0	1.7	2.6	TW I4	1.2	1.9	2.8	TW A3	1.4	2.3	3.7
TW A4	1.2	2.0	3.2	SS C1	1.3	2.3	4.3	TW I5	1.2	1.8	2.5	TW A4	1.3	2.1	3.4
TW B1	1.4	2.9	6.2	SS C2	1.1	1.6	2.4	TW J1	1.2	2.0	3.2	TW B1	1.6	3.1	7.2
TW B2	1.3	2.3	3.8	SS C3	1.5	2.8	5.8	TW J2	1.2	1.9	2.9	TW B2	1.5	2.4	4.0
TW B3	1.3	2.2	3.5	SS C4	1.2	1.8	2.7	TW J3	1.0	1.5	2.1	TW B3	1.4	2.3	3.7
TW B4	1.4	2.2	3.5	SS D1	.9	1.8	3.1	TW J4	1.1	1.5	2.1	TW B4	1.5	2.4	3.7
TW C1	1.5	2.6	4.6	SS D2	.8	1.3	1.8	DG A1	2.2	---	---	TW C1	1.6	2.8	4.9
TW C2	1.4	2.4	3.8	SS D3	1.0	2.0	3.8	DG A2	1.6	---	---	TW C2	1.6	2.5	4.0
TW C3	1.5	2.5	3.9	SS D4	.8	1.3	1.9	DG A3	1.3	2.1	8.4	TW C3	1.6	2.6	4.0
TW C4	1.6	2.7	4.2	SS E1	1.1	2.2	4.5	DG B1	2.1	---	---	TW C4	1.8	2.8	4.3
TW D1	1.7	---	---	SS E2	.9	1.4	2.1	DG B2	1.6	9.0	---	TW D1	1.9	---	---
TW D2	1.1	1.8	2.7	SS E3	1.2	2.7	6.4	DG B3	1.3	1.9	3.3	TW D2	1.2	1.9	2.7
TW D3	1.1	1.7	2.5	SS E4	.9	1.5	2.3	DG C1	1.7	---	---	TW D3	1.2	1.7	2.5

POCATELLO, ID

	SOLAR SAVINGS		
	10%	50%	80%
WW A1	1.6	4.1	---
WW A2	1.3	2.3	4.2
WW A3	1.3	2.1	3.4
WW A4	1.2	1.9	3.1
WW A5	1.2	1.8	2.8
WW A6	1.2	1.8	2.7
WW B1	1.7	---	---
WW B2	1.2	1.7	2.5
WW B3	1.0	1.5	2.1
WW B4	1.0	1.4	1.8
WW B5	1.0	1.3	1.8
WW C1	1.0	1.4	2.0
WW C2	1.0	1.4	1.9
WW C3	.9	1.2	1.6
WW C4	.9	1.2	1.6
TW A1	1.6	3.6	34.2
TW A2	1.4	2.5	4.5
TW A3	1.4	2.3	3.7
TW A4	1.3	2.1	3.4
TW B1	1.6	3.1	7.2
TW B2	1.5	2.4	4.0
TW B3	1.4	2.3	3.7
TW B4	1.5	2.4	3.7
TW C1	1.6	2.8	4.9
TW C2	1.6	2.5	4.0
TW C3	1.6	2.6	4.0
TW C4	1.8	2.8	4.3
TW D1	1.9	---	---
TW D2	1.2	1.9	2.7
TW D3	1.2	1.7	2.5
DG B1	1.7	---	---
DG B2	1.4	2.4	---
DG B3	1.2	1.6	2.3
DG C1	1.5	---	---
DG C2	1.3	1.9	4.9
DG C3	1.1	1.4	2.0
SS A1	1.2	2.2	4.1
SS A2	.9	1.5	2.2
SS A3	1.2	3.0	15.1
SS A4	1.0	1.6	2.4
SS A5	1.2	2.7	7.3
SS A6	.9	1.5	2.3
SS A7	1.4	6.6	---
SS A8	1.0	1.7	2.6
SS B1	1.3	2.6	5.1
SS B2	1.0	1.6	2.3
SS B3	1.4	3.3	13.7
SS B4	1.1	1.7	2.5
SS B5	1.5	3.4	13.7
SS B6	1.1	1.7	2.4
SS B7	1.6	10.4	---
SS B8	1.1	1.8	2.7
SS C1	1.4	2.5	4.7
SS C2	1.1	1.7	2.5
SS C3	1.6	3.1	6.6
SS C4	1.3	1.9	2.8
SS D1	1.0	1.9	3.3
SS D2	.8	1.3	1.9
SS D3	1.1	2.2	4.1
SS D4	.9	1.3	1.9
SS E1	1.2	2.3	5.1
SS E2	1.0	1.5	2.2
SS E3	1.3	2.9	7.9
SS E4	1.0	1.6	2.4

CHICAGO, IL

	SOLAR SAVINGS		
	10%	50%	80%
WW A1	2.1	---	---
WW A2	1.8	5.0	---
WW A3	1.7	3.5	---
WW A4	1.7	3.0	6.4
WW A5	1.6	2.8	5.0
WW A6	1.6	2.6	4.5
WW B1	3.7	---	---
WW B2	1.5	2.3	3.4
WW B3	1.3	1.9	2.8
WW B4	1.2	1.7	2.2
WW B5	1.2	1.6	2.1
WW C1	1.2	1.8	2.5
WW C2	1.2	1.8	2.4
WW C3	1.0	1.4	1.8
WW C4	1.1	1.4	1.9
TW A1	2.1	---	---
TW A2	1.9	5.3	---
TW A3	1.8	3.8	---
TW A4	1.8	3.3	6.2
TW B1	2.1	---	---
TW B2	2.0	4.0	---
TW B3	1.9	3.6	7.2
TW B4	2.0	3.5	6.1
TW C1	2.1	5.1	---
TW C2	2.0	3.9	7.4
TW C3	2.1	3.8	6.7
TW C4	2.2	4.0	6.7
TW D1	3.8	---	---
TW D2	1.6	2.6	3.8
TW D3	1.5	2.4	3.5
TW D4	1.3	1.9	2.5
TW D5	1.3	1.8	2.3
TW E1	1.3	2.0	2.9
TW E2	1.3	1.9	2.6
TW E3	1.1	1.5	2.0
TW E4	1.1	1.5	2.0
TW F1	2.4	---	---
TW F2	2.1	6.4	---
TW F3	2.0	4.2	---
TW F4	2.0	3.7	7.3
TW G1	2.4	---	---
TW G2	2.2	4.9	---
TW G3	2.2	4.2	9.7
TW G4	2.4	4.1	7.3
TW H1	2.5	7.9	---
TW H2	2.5	4.8	12.8
TW H3	2.6	4.7	8.5
TW H4	3.0	5.1	8.4
TW I1	5.6	---	---
TW I2	1.7	2.7	4.0
TW I3	1.6	2.4	3.5
TW I4	1.4	1.9	2.6
TW I5	1.3	1.8	2.4
TW J1	1.4	2.1	2.9
TW J2	1.4	2.0	2.7
TW J3	1.1	1.5	2.0
TW J4	1.2	1.6	2.1
DG A1	4.5	---	---
DG A2	1.9	---	---
DG A3	1.5	2.2	5.2
DG B1	3.7	---	---
DG B2	1.9	---	---
DG B3	1.5	2.0	3.1
DG C1	2.2	---	---

CHICAGO, IL

	SOLAR SAVINGS		
	10%	50%	80%
DG C2	1.6	2.9	---
DG C3	1.4	1.8	2.5
SS A1	1.5	3.7	---
SS A2	1.2	1.9	2.7
SS A3	1.7	---	---
SS A4	1.2	2.0	3.0
SS A5	1.6	---	---
SS A6	1.2	2.0	2.8
SS A7	1.8	---	---
SS A8	1.2	2.2	3.3
SS B1	1.8	5.0	---
SS B2	1.3	2.0	2.8
SS B3	2.0	---	---
SS B4	1.3	2.1	3.1
SS B5	2.0	---	---
SS B6	1.3	2.1	3.0
SS B7	2.3	---	---
SS B8	1.4	2.3	3.4
SS C1	2.0	---	---
SS C2	1.5	2.3	3.3
SS C3	2.3	---	---
SS C4	1.6	2.6	3.8
SS D1	1.3	2.9	7.2
SS D2	1.0	1.6	2.2
SS D3	1.4	3.6	---
SS D4	1.1	1.7	2.3
SS E1	1.6	---	---
SS E2	1.2	1.9	2.7
SS E3	1.8	---	---
SS E4	1.3	2.1	3.0

MOLINE, IL

	SOLAR SAVINGS		
	10%	50%	80%
WW A1	2.1	---	---
WW A2	1.8	4.5	---
WW A3	1.7	3.3	7.3
WW A4	1.7	2.9	5.3
WW A5	1.6	2.7	4.5
WW A6	1.6	2.6	4.1
WW B1	3.7	---	---
WW B2	1.5	2.3	3.3
WW B3	1.3	1.9	2.7
WW B4	1.2	1.7	2.2
WW B5	1.2	1.6	2.1
WW C1	1.2	1.8	2.5
WW C2	1.2	1.7	2.3
WW C3	1.0	1.4	1.8
WW C4	1.1	1.4	1.9
TW A1	2.1	---	---
TW A2	1.9	4.8	---
TW A3	1.8	3.6	8.1
TW A4	1.8	3.2	5.5
TW B1	2.1	---	---
TW B2	1.9	3.9	8.7
TW B3	1.9	3.5	6.2
TW B4	2.0	3.4	5.6
TW C1	2.1	4.8	---
TW C2	2.0	3.8	6.6
TW C3	2.1	3.8	6.2
TW C4	2.2	4.0	6.3
TW D1	3.8	---	---
TW D2	1.6	2.5	3.7
TW D3	1.5	2.3	3.4
TW D4	1.3	1.9	2.5
TW D5	1.3	1.8	2.3
TW E1	1.3	2.0	2.8
TW E2	1.3	1.9	2.6
TW E3	1.1	1.5	2.0
TW E4	1.1	1.5	2.0
TW F1	2.4	---	---
TW F2	2.1	5.4	---
TW F3	2.0	4.0	9.6
TW F4	2.0	3.6	6.3
TW G1	2.4	---	---
TW G2	2.2	4.6	20.3
TW G3	2.2	4.0	7.6
TW G4	2.4	4.0	6.6
TW H1	2.5	6.5	---
TW H2	2.5	4.6	9.1
TW H3	2.6	4.6	7.7
TW H4	3.0	5.0	7.8
TW I1	5.7	---	---
TW I2	1.7	2.7	3.9
TW I3	1.6	2.4	3.4
TW I4	1.4	1.9	2.6
TW I5	1.3	1.8	2.4
TW J1	1.4	2.1	2.8
TW J2	1.4	1.9	2.6
TW J3	1.1	1.5	2.0
TW J4	1.2	1.6	2.1
DG A1	4.6	---	---
DG A2	1.9	---	---
DG A3	1.5	2.2	5.0
DG B1	3.8	---	---
DG B2	1.9	7.9	---
DG B3	1.5	2.0	3.0
DG C1	2.3	---	---
DG C2	1.6	2.8	---
DG C3	1.4	1.8	2.4
SS A1	1.5	3.5	9.7
SS A2	1.2	1.9	2.6
SS A3	1.7	---	---
SS A4	1.2	2.0	2.9
SS A5	1.6	---	---
SS A6	1.2	1.9	2.8
SS A7	1.8	---	---
SS A8	1.2	2.1	3.2
SS B1	1.8	4.6	---
SS B2	1.3	2.0	2.8
SS B3	2.0	---	---
SS B4	1.3	2.1	3.0
SS B5	2.0	---	---
SS B6	1.3	2.1	3.0
SS B7	2.3	---	---
SS B8	1.4	2.3	3.3
SS C1	2.0	8.1	---
SS C2	1.5	2.3	3.2
SS C3	2.3	---	---
SS C4	1.6	2.5	3.6
SS D1	1.3	2.8	5.7
SS D2	1.1	1.6	2.2
SS D3	1.4	3.5	19.2
SS D4	1.1	1.7	2.3
SS E1	1.6	6.9	---
SS E2	1.2	1.9	2.7
SS E3	1.8	---	---
SS E4	1.3	2.1	2.9

SPRINGFIELD, IL				SPRINGFIELD, IL				EVANSVILLE, IN				FORT WAYNE, IN			
	SOLAR SAVINGS				SOLAR SAVINGS				SOLAR SAVINGS				SOLAR SAVINGS		
	10%	50%	80%		10%	50%	80%		10%	50%	80%		10%	50%	80%
WW A1	1.8	7.2	---	DG C2	1.5	2.2	---	TW D4	1.1	1.6	2.1	WW A1	2.6	---	---
WW A2	1.6	2.7	4.3	DG C3	1.3	1.6	2.1	TW D5	1.1	1.5	2.0	WW A2	2.2	---	---
WW A3	1.5	2.4	3.5	SS A1	1.3	2.5	3.9	TW E1	1.1	1.7	2.2	WW A3	2.1	---	---
WW A4	1.4	2.2	3.2	SS A2	1.1	1.6	2.2	TW E2	1.1	1.6	2.1	WW A4	2.0	---	---
WW A5	1.4	2.1	2.9	SS A3	1.5	3.4	8.0	TW E3	1.0	1.3	1.7	WW A5	1.9	5.4	---
WW A6	1.4	2.0	2.8	SS A4	1.1	1.7	2.4	TW E4	1.0	1.4	1.8	WW A6	1.9	4.4	---
WW B1	2.3	---	---	SS A5	1.4	3.1	5.9	TW F1	1.8	3.9	7.0	WW B1	---	---	---
WW B2	1.3	1.9	2.6	SS A6	1.1	1.7	2.3	TW F2	1.7	2.8	4.2	WW B2	1.7	2.9	4.8
WW B3	1.2	1.7	2.2	SS A7	1.6	---	---	TW F3	1.6	2.6	3.7	WW B3	1.5	2.3	3.5
WW B4	1.1	1.5	1.9	SS A8	1.1	1.8	2.5	TW F4	1.6	2.5	3.5	WW B4	1.3	1.9	2.5
WW B5	1.1	1.5	1.9	SS B1	1.5	2.9	4.8	TW G1	1.9	3.5	5.5	WW B5	1.3	1.8	2.4
WW C1	1.1	1.6	2.1	SS B2	1.2	1.8	2.4	TW G2	1.8	2.9	4.1	WW C1	1.4	2.1	3.0
WW C2	1.1	1.5	2.0	SS B3	1.7	3.7	8.7	TW G3	1.8	2.8	3.9	WW C2	1.4	2.0	2.7
WW C3	1.0	1.3	1.6	SS B4	1.2	1.9	2.5	TW G4	1.9	2.9	4.1	WW C3	1.1	1.5	2.0
WW C4	1.0	1.3	1.7	SS B5	1.7	4.0	9.5	TW H1	2.0	3.4	5.0	WW C4	1.2	1.6	2.0
TW A1	1.8	4.9	---	SS B6	1.2	1.8	2.5	TW H2	2.0	3.2	4.5	TW A1	2.6	---	---
TW A2	1.7	3.0	4.7	SS B7	1.9	---	---	TW H3	2.1	3.3	4.6	TW A2	2.3	---	---
TW A3	1.6	2.6	3.9	SS B8	1.3	2.0	2.7	TW H4	2.4	3.7	5.1	TW A3	2.2	---	---
TW A4	1.6	2.5	3.5	SS C1	1.7	2.9	4.8	TW I1	2.5	---	---	TW A4	2.1	6.6	---
TW B1	1.9	3.8	7.1	SS C2	1.3	1.9	2.6	TW I2	1.4	2.1	2.9	TW B1	2.6	---	---
TW B2	1.7	2.8	4.2	SS C3	1.9	3.7	6.6	TW I3	1.3	1.9	2.6	TW B2	2.3	---	---
TW B3	1.7	2.7	3.8	SS C4	1.4	2.1	2.9	TW I4	1.2	1.7	2.2	TW B3	2.2	8.1	---
TW B4	1.7	2.7	3.8	SS D1	1.2	2.1	3.1	TW I5	1.2	1.6	2.1	TW B4	2.3	5.4	---
TW C1	1.9	3.3	5.0	SS D2	1.0	1.4	1.9	TW J1	1.2	1.7	2.3	TW C1	2.5	---	---
TW C2	1.8	2.9	4.2	SS D3	1.3	2.4	3.8	TW J2	1.2	1.7	2.2	TW C2	2.4	7.0	---
TW C3	1.9	3.0	4.2	SS D4	1.0	1.5	2.0	TW J3	1.0	1.4	1.8	TW C3	2.4	5.8	---
TW C4	2.0	3.2	4.6	SS E1	1.4	2.6	4.5	TW J4	1.0	1.4	1.8	TW C4	2.5	5.6	---
TW D1	2.5	---	---	SS E2	1.1	1.6	2.2	DG A1	2.1	---	---	TW D1	---	---	---
TW D2	1.4	2.1	2.9	SS E3	1.6	3.3	6.4	DG A2	1.6	3.9	---	TW D2	1.8	3.2	5.6
TW D3	1.3	2.0	2.7	SS E4	1.2	1.8	2.4	DG A3	1.3	1.8	3.1	TW D3	1.7	2.9	4.9
TW D4	1.2	1.7	2.2					DG B1	2.0	---	---	TW D4	1.4	2.1	2.9
TW D5	1.2	1.6	2.1	EVANSVILLE, IN				DG B2	1.5	2.6	---	TW D5	1.4	1.9	2.6
TW E1	1.2	1.7	2.3		SOLAR SAVINGS			DG B3	1.3	1.7	2.4	TW E1	1.5	2.4	3.5
TW E2	1.2	1.7	2.2		10%	50%	80%	DG C1	1.6	---	---	TW E2	1.4	2.2	3.1
TW E3	1.0	1.4	1.8	WW A1	1.7	4.5	12.2	DG C2	1.4	2.0	5.3	TW E3	1.2	1.7	2.2
TW E4	1.0	1.4	1.8	WW A2	1.5	2.5	3.7	DG C3	1.2	1.5	2.0	TW E4	1.2	1.7	2.2
TW F1	2.0	4.7	26.1	WW A3	1.4	2.2	3.1	SS A1	1.2	2.3	3.4	TW F1	3.0	---	---
TW F2	1.8	3.1	4.9	WW A4	1.3	2.1	2.9	SS A2	1.0	1.6	2.1	TW F2	2.6	---	---
TW F3	1.7	2.8	4.1	WW A5	1.3	2.0	2.7	SS A3	1.3	2.8	4.8	TW F3	2.4	---	---
TW F4	1.8	2.7	3.9	WW A6	1.3	1.9	2.6	SS A4	1.0	1.6	2.2	TW F4	2.4	---	---
TW G1	2.0	4.0	7.3	WW B1	2.0	---	---	SS A5	1.3	2.7	4.4	TW G1	3.0	---	---
TW G2	1.9	3.1	4.7	WW B2	1.3	1.8	2.5	SS A6	1.0	1.6	2.2	TW G2	2.7	---	---
TW G3	1.9	3.0	4.3	WW B3	1.1	1.6	2.1	SS A7	1.4	4.2	13.5	TW G3	2.7	---	---
TW G4	2.1	3.2	4.4	WW B4	1.0	1.4	1.9	SS A8	1.0	1.7	2.3	TW G4	2.8	7.0	---
TW H1	2.1	3.7	5.8	WW B5	1.1	1.4	1.8	SS B1	1.4	2.6	3.9	TW H1	3.1	---	---
TW H2	2.2	3.4	5.0	WW C1	1.1	1.5	2.0	SS B2	1.1	1.7	2.3	TW H2	3.0	---	---
TW H3	2.3	3.5	5.0	WW C2	1.1	1.5	1.9	SS B3	1.5	3.1	5.2	TW H3	3.1	8.5	---
TW H4	2.6	4.0	5.6	WW C3	.9	1.2	1.6	SS B4	1.1	1.7	2.4	TW H4	3.4	7.4	---
TW I1	2.8	---	---	WW C4	1.0	1.3	1.6	SS B5	1.6	3.4	5.8	TW I1	---	---	---
TW I2	1.5	2.2	3.0	TW A1	1.7	4.0	7.7	SS B6	1.1	1.7	2.3	TW I2	1.9	3.4	5.9
TW I3	1.4	2.0	2.8	TW A2	1.6	2.7	4.0	SS B7	1.8	5.0	20.9	TW I3	1.8	2.9	4.6
TW I4	1.3	1.7	2.3	TW A3	1.5	2.4	3.5	SS B8	1.2	1.8	2.5	TW I4	1.5	2.2	3.0
TW I5	1.2	1.6	2.1	TW A4	1.5	2.3	3.2	SS C1	1.5	2.6	3.9	TW I5	1.4	2.0	2.7
TW J1	1.2	1.8	2.4	TW B1	1.7	3.4	5.4	SS C2	1.2	1.8	2.4	TW J1	1.5	2.4	3.5
TW J2	1.2	1.7	2.3	TW B2	1.6	2.6	3.7	SS C3	1.8	3.3	5.1	TW J2	1.5	2.2	3.1
TW J3	1.1	1.4	1.8	TW B3	1.6	2.5	3.5	SS C4	1.4	2.0	2.7	TW J3	1.2	1.7	2.2
TW J4	1.1	1.4	1.9	TW B4	1.6	2.5	3.5	SS D1	1.1	1.9	2.7	TW J4	1.3	1.7	2.3
DG A1	2.4	---	---	TW C1	1.7	3.0	4.4	SS D2	.9	1.3	1.8	DG A1	---	---	---
DG A2	1.7	5.7	---	TW C2	1.7	2.7	3.8	SS D3	1.2	2.2	3.2	DG A2	2.2	---	---
DG A3	1.4	1.9	3.3	TW C3	1.7	2.8	3.9	SS D4	.9	1.4	1.9	DG A3	1.6	2.5	---
DG B1	2.3	---	---	TW C4	1.9	3.0	4.2	SS E1	1.3	2.3	3.5	DG B1	---	---	---
DG B2	1.6	2.9	---	TW D1	2.2	---	---	SS E2	1.0	1.5	2.1	DG B2	2.1	---	---
DG B3	1.4	1.7	2.5	TW D2	1.3	2.0	2.7	SS E3	1.4	2.8	4.6	DG B3	1.6	2.2	3.8
DG C1	1.8	---	---	TW D3	1.2	1.9	2.5	SS E4	1.1	1.7	2.2	DG C1	3.5	---	---

326 / APPENDIX 17

FORT WAYNE, IN			
	\multicolumn{3}{c}{SOLAR SAVINGS}		
	10%	50%	80%
DG C2	1.8	4.7	---
DG C3	1.5	2.0	2.9
SS A1	1.7	---	---
SS A2	1.3	2.1	3.1
SS A3	1.9	---	---
SS A4	1.3	2.3	3.5
SS A5	1.8	---	---
SS A6	1.3	2.2	3.3
SS A7	2.0	---	---
SS A8	1.3	2.5	4.1
SS B1	2.0	---	---
SS B2	1.4	2.3	3.2
SS B3	2.3	---	---
SS B4	1.4	2.4	3.6
SS B5	2.3	---	---
SS B6	1.4	2.4	3.5
SS B7	2.7	---	---
SS B8	1.5	2.7	4.1
SS C1	2.5	---	---
SS C2	1.7	2.8	4.4
SS C3	2.8	---	---
SS C4	1.8	3.2	5.2
SS D1	1.5	5.5	---
SS D2	1.1	1.8	2.5
SS D3	1.6	---	---
SS D4	1.2	1.9	2.6
SS E1	1.9	---	---
SS E2	1.3	2.2	3.2
SS E3	2.0	---	---
SS E4	1.4	2.4	3.6

INDIANAPOLIS, IN			
	\multicolumn{3}{c}{SOLAR SAVINGS}		
	10%	50%	80%
WW A1	2.2	---	---
WW A2	1.9	5.7	---
WW A3	1.8	3.6	9.7
WW A4	1.7	3.2	6.1
WW A5	1.6	2.9	5.0
WW A6	1.6	2.7	4.5
WW B1	4.5	---	---
WW B2	1.5	2.4	3.5
WW B3	1.3	2.0	2.8
WW B4	1.2	1.7	2.3
WW B5	1.2	1.6	2.2
WW C1	1.2	1.8	2.5
WW C2	1.3	1.8	2.4
WW C3	1.0	1.4	1.8
WW C4	1.1	1.5	1.9
TW A1	2.2	---	---
TW A2	2.0	6.0	---
TW A3	1.9	4.0	10.6
TW A4	1.8	3.4	6.1
TW B1	2.2	---	---
TW B2	2.0	4.3	11.5
TW B3	2.0	3.7	7.0
TW B4	2.0	3.6	6.1
TW C1	2.2	5.5	---
TW C2	2.1	4.0	7.3
TW C3	2.1	4.0	6.7
TW C4	2.3	4.2	6.7
TW D1	4.4	---	---
TW D2	1.6	2.6	3.9
TW D3	1.5	2.4	3.5

INDIANAPOLIS, IN			
	\multicolumn{3}{c}{SOLAR SAVINGS}		
	10%	50%	80%
TW D4	1.3	1.9	2.6
TW D5	1.3	1.8	2.4
TW E1	1.3	2.1	2.9
TW E2	1.3	1.9	2.6
TW E3	1.1	1.5	2.0
TW E4	1.1	1.6	2.0
TW F1	2.4	---	---
TW F2	2.2	8.0	---
TW F3	2.1	4.4	15.6
TW F4	2.1	3.8	7.1
TW G1	2.5	---	---
TW G2	2.3	5.3	---
TW G3	2.3	4.4	8.8
TW G4	2.4	4.3	7.3
TW H1	2.6	10.1	---
TW H2	2.6	5.0	10.8
TW H3	2.7	4.9	8.4
TW H4	3.0	5.2	8.4
TW I1	---	---	---
TW I2	1.7	2.8	4.1
TW I3	1.6	2.5	3.5
TW I4	1.4	2.0	2.6
TW I5	1.3	1.9	2.5
TW J1	1.4	2.1	2.9
TW J2	1.4	2.0	2.7
TW J3	1.1	1.6	2.0
TW J4	1.2	1.6	2.1
DG A1	---	---	---
DG A2	2.0	---	---
DG A3	1.5	2.2	5.6
DG B1	4.8	---	---
DG B2	1.9	---	---
DG B3	1.5	2.0	3.1
DG C1	2.4	---	---
DG C2	1.7	3.0	---
DG C3	1.4	1.8	2.5
SS A1	1.5	3.6	9.1
SS A2	1.2	1.9	2.7
SS A3	1.7	---	---
SS A4	1.2	2.0	2.9
SS A5	1.6	---	---
SS A6	1.2	2.0	2.8
SS A7	1.8	---	---
SS A8	1.2	2.1	3.2
SS B1	1.8	4.8	---
SS B2	1.3	2.0	2.8
SS B3	1.9	---	---
SS B4	1.3	2.1	3.0
SS B5	2.0	---	---
SS B6	1.3	2.1	3.0
SS B7	2.3	---	---
SS B8	1.4	2.3	3.3
SS C1	2.1	---	---
SS C2	1.5	2.3	3.3
SS C3	2.4	---	---
SS C4	1.6	2.6	3.8
SS D1	1.3	2.8	5.6
SS D2	1.0	1.6	2.2
SS D3	1.4	3.5	12.5
SS D4	1.1	1.7	2.3
SS E1	1.6	8.7	---
SS E2	1.2	1.9	2.7
SS E3	1.8	---	---
SS E4	1.3	2.1	3.0

SOUTH BEND, IN			
	\multicolumn{3}{c}{SOLAR SAVINGS}		
	10%	50%	80%
WW A1	2.6	---	---
WW A2	2.4	---	---
WW A3	2.2	---	---
WW A4	2.1	---	---
WW A5	2.0	---	---
WW A6	2.0	---	---
WW B1	---	---	---
WW B2	1.8	3.5	8.5
WW B3	1.5	2.6	4.4
WW B4	1.4	2.0	2.7
WW B5	1.3	1.9	2.5
WW C1	1.4	2.3	3.5
WW C2	1.4	2.1	3.0
WW C3	1.1	1.6	2.1
WW C4	1.2	1.6	2.1
TW A1	2.5	---	---
TW A2	2.4	---	---
TW A3	2.3	---	---
TW A4	2.2	---	---
TW B1	2.6	---	---
TW B2	2.4	---	---
TW B3	2.3	---	---
TW B4	2.4	---	---
TW C1	2.6	---	---
TW C2	2.5	---	---
TW C3	2.5	---	---
TW C4	2.6	10.0	---
TW D1	---	---	---
TW D2	1.9	4.0	13.9
TW D3	1.7	3.5	8.6
TW D4	1.5	2.2	3.2
TW D5	1.4	2.0	2.8
TW E1	1.5	2.7	4.3
TW E2	1.5	2.4	3.5
TW E3	1.2	1.7	2.4
TW E4	1.2	1.8	2.3
TW F1	3.1	---	---
TW F2	2.8	---	---
TW F3	2.6	---	---
TW F4	2.6	---	---
TW G1	3.3	---	---
TW G2	2.9	---	---
TW G3	2.9	---	---
TW G4	3.0	---	---
TW H1	3.4	---	---
TW H2	3.2	---	---
TW H3	3.3	---	---
TW H4	3.6	---	---
TW I1	---	---	---
TW I2	2.0	4.2	14.9
TW I3	1.8	3.4	6.3
TW I4	1.5	2.3	3.3
TW I5	1.5	2.1	2.9
TW J1	1.6	2.7	4.2
TW J2	1.5	2.4	3.4
TW J3	1.3	1.8	2.4
TW J4	1.3	1.8	2.4
DG A1	---	---	---
DG A2	2.3	---	---
DG A3	1.7	2.7	---
DG B1	---	---	---
DG B2	2.2	---	---
DG B3	1.6	2.4	4.6
DG C1	4.8	---	---

SOUTH BEND, IN			
	\multicolumn{3}{c}{SOLAR SAVINGS}		
	10%	50%	80%
DG C2	1.9	---	---
DG C3	1.5	2.1	3.3
SS A1	1.7	---	---
SS A2	1.3	2.3	3.5
SS A3	2.0	---	---
SS A4	1.3	2.6	4.3
SS A5	1.7	---	---
SS A6	1.3	2.4	3.9
SS A7	2.0	---	---
SS A8	1.4	2.9	5.6
SS B1	2.1	---	---
SS B2	1.4	2.4	3.6
SS B3	2.4	---	---
SS B4	1.5	2.7	4.2
SS B5	2.3	---	---
SS B6	1.5	2.6	4.0
SS B7	2.8	---	---
SS B8	1.5	3.0	5.2
SS C1	2.8	---	---
SS C2	1.7	3.2	6.3
SS C3	3.1	---	---
SS C4	1.9	3.8	8.6
SS D1	1.5	---	---
SS D2	1.2	1.9	2.8
SS D3	1.6	---	---
SS D4	1.2	2.0	2.9
SS E1	2.0	---	---
SS E2	1.3	2.4	3.9
SS E3	2.1	---	---
SS E4	1.4	2.7	4.5

BURLINGTON, IA			
	\multicolumn{3}{c}{SOLAR SAVINGS}		
	10%	50%	80%
WW A1	1.9	---	---
WW A2	1.6	3.0	5.1
WW A3	1.5	2.6	3.9
WW A4	1.5	2.4	3.5
WW A5	1.4	2.2	3.2
WW A6	1.4	2.1	3.1
WW B1	2.5	---	---
WW B2	1.4	2.0	2.8
WW B3	1.2	1.7	2.3
WW B4	1.1	1.5	2.0
WW B5	1.1	1.5	2.0
WW C1	1.1	1.6	2.2
WW C2	1.1	1.6	2.1
WW C3	1.0	1.3	1.7
WW C4	1.0	1.4	1.7
TW A1	1.9	7.7	---
TW A2	1.7	3.2	5.6
TW A3	1.6	2.8	4.4
TW A4	1.6	2.6	3.8
TW B1	1.9	4.4	11.7
TW B2	1.8	3.0	4.7
TW B3	1.7	2.8	4.2
TW B4	1.8	2.9	4.2
TW C1	1.9	3.6	5.8
TW C2	1.9	3.1	4.6
TW C3	1.9	3.2	4.6
TW C4	2.0	3.4	4.9
TW D1	2.7	---	---
TW D2	1.5	2.2	3.1
TW D3	1.4	2.0	2.8

Conservation Factor Tables / 327

BURLINGTON, IA				DES MOINES, IA				DES MOINES, IA				MASON CITY, IA			
	SOLAR SAVINGS				SOLAR SAVINGS				SOLAR SAVINGS				SOLAR SAVINGS		
	10%	50%	80%		10%	50%	80%		10%	50%	80%		10%	50%	80%
WW A1	1.9	---	---	WW A1	2.0	---	---	DG C2	1.5	2.4	---	TW D4	1.3	1.9	2.5
WW A2	1.6	3.0	5.1	WW A2	1.7	3.2	5.6	DG C3	1.3	1.7	2.2	TW D5	1.3	1.8	2.3
WW A3	1.5	2.6	3.9	WW A3	1.6	2.7	4.2	SS A1	1.4	2.9	5.1	TW E1	1.3	2.0	2.9
WW A4	1.5	2.4	3.5	WW A4	1.5	2.5	3.7	SS A2	1.1	1.8	2.4	TW E2	1.3	1.9	2.6
WW A5	1.4	2.2	3.2	WW A5	1.5	2.3	3.4	SS A3	1.6	5.6	---	TW E3	1.1	1.5	2.0
WW A6	1.4	2.1	3.1	WW A6	1.4	2.2	3.2	SS A4	1.1	1.9	2.6	TW E4	1.1	1.5	2.0
WW B1	2.5	---	---	WW B1	2.7	---	---	SS A5	1.5	4.1	16.5	TW F1	2.4	---	---
WW B2	1.4	2.0	2.8	WW B2	1.4	2.1	2.9	SS A6	1.1	1.8	2.5	TW F2	2.1	6.2	---
WW B3	1.2	1.7	2.3	WW B3	1.2	1.8	2.4	SS A7	1.7	---	---	TW F3	2.0	4.2	---
WW B4	1.1	1.5	2.0	WW B4	1.1	1.6	2.1	SS A8	1.2	2.0	2.8	TW F4	2.1	3.7	6.9
WW B5	1.1	1.5	2.0	WW B5	1.1	1.5	2.0	SS B1	1.7	3.5	6.7	TW G1	2.4	---	---
WW C1	1.1	1.6	2.2	WW C1	1.2	1.7	2.2	SS B2	1.2	1.9	2.6	TW G2	2.3	4.9	---
WW C2	1.1	1.6	2.1	WW C2	1.2	1.6	2.2	SS B3	1.8	6.4	---	TW G3	2.3	4.2	8.8
WW C3	1.0	1.3	1.7	WW C3	1.0	1.3	1.7	SS B4	1.3	2.0	2.8	TW G4	2.4	4.2	7.0
WW C4	1.0	1.4	1.7	WW C4	1.0	1.4	1.8	SS B5	1.9	6.3	---	TW H1	2.6	7.6	---
TW A1	1.9	7.7	---	TW A1	2.0	---	---	SS B6	1.3	2.0	2.7	TW H2	2.5	4.8	11.0
TW A2	1.7	3.2	5.6	TW A2	1.8	3.5	6.1	SS B7	2.1	---	---	TW H3	2.7	4.7	8.2
TW A3	1.6	2.8	4.4	TW A3	1.7	3.0	4.6	SS B8	1.3	2.1	3.0	TW H4	3.0	5.1	8.2
TW A4	1.6	2.6	3.8	TW A4	1.7	2.7	4.0	SS C1	1.8	3.5	7.0	TW I1	10.1	---	---
TW B1	1.9	4.4	11.7	TW B1	2.0	4.9	17.4	SS C2	1.4	2.1	2.8	TW I2	1.7	2.7	3.9
TW B2	1.8	3.0	4.7	TW B2	1.8	3.2	5.0	SS C3	2.1	4.7	12.3	TW I3	1.6	2.4	3.5
TW B3	1.7	2.8	4.2	TW B3	1.8	3.0	4.4	SS C4	1.5	2.3	3.2	TW I4	1.4	2.0	2.6
TW B4	1.8	2.9	4.2	TW B4	1.8	3.0	4.3	SS D1	1.3	2.4	3.9	TW I5	1.3	1.8	2.4
TW C1	1.9	3.6	5.8	TW C1	2.0	3.8	6.3	SS D2	1.0	1.5	2.0	TW J1	1.4	2.1	2.9
TW C2	1.9	3.1	4.6	TW C2	1.9	3.2	4.8	SS D3	1.4	2.8	5.1	TW J2	1.4	2.0	2.7
TW C3	1.9	3.2	4.6	TW C3	2.0	3.3	4.8	SS D4	1.0	1.6	2.1	TW J3	1.1	1.6	2.0
TW C4	2.0	3.4	4.9	TW C4	2.1	3.5	5.1	SS E1	1.5	3.3	7.7	TW J4	1.2	1.6	2.1
TW D1	2.7	---	---	TW D1	2.9	---	---	SS E2	1.1	1.8	2.4	DG A1	6.0	---	---
TW D2	1.5	2.2	3.1	TW D2	1.5	2.3	3.2	SS E3	1.7	4.5	---	DG A2	2.0	---	---
TW D3	1.4	2.0	2.8	TW D3	1.4	2.1	2.9	SS E4	1.2	1.9	2.6	DG A3	1.5	2.2	5.3
DG B1	2.5	---	---	TW D4	1.2	1.7	2.3					DG B1	4.2	---	---
DG B2	1.7	3.3	---	TW D5	1.2	1.7	2.2	MASON CITY, IA				DG B2	1.9	---	---
DG B3	1.4	1.8	2.6	TW E1	1.3	1.9	2.5		SOLAR SAVINGS			DG B3	1.5	2.0	3.1
DG C1	1.9	---	---	TW E2	1.2	1.8	2.4		10%	50%	80%	DG C1	2.3	---	---
DG C2	1.5	2.3	---	TW E3	1.1	1.4	1.9	WW A1	2.1	---	---	DG C2	1.7	2.9	---
DG C3	1.3	1.6	2.2	TW E4	1.1	1.5	1.9	WW A2	1.8	5.0	---	DG C3	1.4	1.8	2.5
SS A1	1.4	2.7	4.6	TW F1	2.1	16.8	---	WW A3	1.7	3.5	11.8	SS A1	1.5	4.0	---
SS A2	1.1	1.7	2.4	TW F2	1.9	3.7	6.6	WW A4	1.7	3.1	6.0	SS A2	1.2	1.9	2.7
SS A3	1.5	4.3	---	TW F3	1.9	3.2	5.0	WW A5	1.6	2.8	4.8	SS A3	1.7	---	---
SS A4	1.1	1.8	2.4	TW F4	1.9	3.0	4.5	WW A6	1.6	2.7	4.4	SS A4	1.2	2.1	3.1
SS A5	1.5	3.6	9.5	TW G1	2.2	5.1	18.0	WW B1	4.2	---	---	SS A5	1.6	---	---
SS A6	1.1	1.8	2.5	TW G2	2.0	3.6	5.8	WW B2	1.5	2.3	3.4	SS A6	1.2	2.0	2.9
SS A7	1.6	---	---	TW G3	2.0	3.4	5.1	WW B3	1.3	2.0	2.7	SS A7	1.8	---	---
SS A8	1.1	1.9	2.7	TW G4	2.2	3.5	5.0	WW B4	1.2	1.7	2.2	SS A8	1.3	2.3	3.4
SS B1	1.6	3.3	5.9	TW H1	2.3	4.4	7.9	WW B5	1.2	1.6	2.1	SS B1	1.8	5.8	---
SS B2	1.2	1.8	2.5	TW H2	2.3	3.8	5.8	WW C1	1.2	1.8	2.5	SS B2	1.3	2.1	2.9
SS B3	1.7	4.8	---	TW H3	2.4	3.9	5.7	WW C2	1.2	1.8	2.4	SS B3	2.0	---	---
SS B4	1.2	1.9	2.7	TW H4	2.8	4.4	6.2	WW C3	1.0	1.4	1.8	SS B4	1.3	2.2	3.2
SS B5	1.8	5.0	---	TW I1	3.5	---	---	WW C4	1.1	1.5	1.9	SS B5	2.0	---	---
SS B6	1.2	1.9	2.6	TW I2	1.6	2.4	3.3	TW A1	2.1	---	---	SS B6	1.3	2.2	3.1
SS B7	2.0	---	---	TW I3	1.5	2.2	3.0	TW A2	2.0	5.3	---	SS B7	2.4	---	---
SS B8	1.3	2.1	2.9	TW I4	1.3	1.8	2.4	TW A3	1.9	3.8	12.7	SS B8	1.4	2.4	3.5
SS C1	1.7	3.3	6.1	TW I5	1.3	1.7	2.3	TW A4	1.8	3.3	5.9	SS C1	2.0	---	---
SS C2	1.3	2.0	2.7	TW J1	1.3	1.9	2.6	TW B1	2.2	---	---	SS C2	1.5	2.3	3.3
SS C3	2.0	4.2	9.7	TW J2	1.3	1.8	2.4	TW B2	2.0	4.1	13.4	SS C3	2.3	---	---
SS C4	1.5	2.2	3.1	TW J3	1.1	1.5	1.9	TW B3	1.9	3.6	6.8	SS C4	1.6	2.6	3.7
SS D1	1.2	2.3	3.6	TW J4	1.1	1.5	1.9	TW B4	2.0	3.5	5.9	SS D1	1.4	3.1	9.4
SS D2	1.0	1.5	2.0	DG A1	2.9	---	---	TW C1	2.1	5.1	---	SS D2	1.1	1.6	2.3
SS D3	1.3	2.6	4.6	DG A2	1.8	---	---	TW C2	2.1	3.9	7.1	SS D3	1.5	4.0	---
SS D4	1.0	1.5	2.1	DG A3	1.4	2.0	3.8	TW C3	2.1	3.9	6.5	SS D4	1.1	1.7	2.4
SS E1	1.4	3.0	6.2	DG B1	2.8	---	---	TW C4	2.2	4.1	6.5	SS E1	1.7	---	---
SS E2	1.1	1.7	2.3	DG B2	1.7	3.6	---	TW D1	4.2	---	---	SS E2	1.2	1.9	2.8
SS E3	1.6	3.9	12.5	DG B3	1.4	1.8	2.7	TW D2	1.6	2.6	3.8	SS E3	1.8	---	---
SS E4	1.2	1.8	2.6	DG C1	2.0	---	---	TW D3	1.5	2.4	3.5	SS E4	1.3	2.1	3.1

328 / APPENDIX 17

SIOUX CITY, IA			
	\% SOLAR SAVINGS		
	10%	50%	80%
WW A1	2.0	---	---
WW A2	1.7	3.4	6.9
WW A3	1.6	2.8	4.6
WW A4	1.5	2.6	4.0
WW A5	1.5	2.4	3.6
WW A6	1.5	2.3	3.4
WW B1	2.8	---	---
WW B2	1.4	2.1	3.0
WW B3	1.2	1.8	2.5
WW B4	1.1	1.6	2.1
WW B5	1.1	1.6	2.0
WW C1	1.2	1.7	2.3
WW C2	1.2	1.7	2.2
WW C3	1.0	1.3	1.7
WW C4	1.0	1.4	1.8
TW A1	2.0	---	---
TW A2	1.8	3.7	7.5
TW A3	1.7	3.1	5.1
TW A4	1.7	2.8	4.3
TW B1	2.0	5.5	---
TW B2	1.8	3.3	5.5
TW B3	1.8	3.1	4.7
TW B4	1.8	3.1	4.6
TW C1	2.0	4.0	7.2
TW C2	1.9	3.4	5.1
TW C3	2.0	3.4	5.1
TW C4	2.1	3.6	5.3
TW D1	3.0	---	---
TW D2	1.5	2.3	3.3
TW D3	1.4	2.2	3.0
TW D4	1.2	1.8	2.3
TW D5	1.2	1.7	2.2
TW E1	1.3	1.9	2.6
TW E2	1.2	1.8	2.4
TW E3	1.1	1.5	1.9
TW E4	1.1	1.5	1.9
TW F1	2.2	---	---
TW F2	1.9	3.9	8.4
TW F3	1.9	3.3	5.6
TW F4	1.9	3.1	4.8
TW G1	2.2	5.9	---
TW G2	2.1	3.8	6.6
TW G3	2.1	3.5	5.5
TW G4	2.2	3.6	5.3
TW H1	2.3	4.7	9.9
TW H2	2.3	4.0	6.3
TW H3	2.5	4.1	6.1
TW H4	2.8	4.5	6.6
TW I1	3.7	---	---
TW I2	1.6	2.5	3.4
TW I3	1.5	2.2	3.1
TW I4	1.3	1.8	2.4
TW I5	1.3	1.7	2.3
TW J1	1.3	1.9	2.6
TW J2	1.3	1.9	2.5
TW J3	1.1	1.5	1.9
TW J4	1.1	1.5	2.0
DG A1	3.0	---	---
DG A2	1.8	---	---
DG A3	1.4	2.0	4.0
DG B1	2.8	---	---
DG B2	1.8	3.9	---
DG B3	1.4	1.9	2.8
DG C1	2.0	---	---

SIOUX CITY, IA			
	\% SOLAR SAVINGS		
	10%	50%	80%
DG C2	1.6	2.5	---
DG C3	1.3	1.7	2.3
SS A1	1.4	3.1	6.0
SS A2	1.1	1.8	2.5
SS A3	1.6	---	---
SS A4	1.2	1.9	2.8
SS A5	1.5	4.7	---
SS A6	1.1	1.9	2.6
SS A7	1.7	---	---
SS A8	1.2	2.1	3.0
SS B1	1.7	3.8	8.9
SS B2	1.2	2.1	2.9
SS B3	1.9	---	---
SS B4	1.3	2.1	2.9
SS B5	1.9	---	---
SS B6	1.3	2.0	2.8
SS B7	2.2	---	---
SS B8	1.3	2.2	3.1
SS C1	1.8	3.8	10.9
SS C2	1.4	2.1	2.9
SS C3	2.1	5.2	---
SS C4	1.5	2.3	3.3
SS D1	1.3	2.5	4.3
SS D2	1.0	1.5	2.1
SS D3	1.4	3.0	6.2
SS D4	1.0	1.6	2.2
SS E1	1.5	3.8	---
SS E2	1.2	1.8	2.5
SS E3	1.7	5.6	---
SS E4	1.2	2.0	2.7

DODGE CITY, KS			
	\% SOLAR SAVINGS		
	10%	50%	80%
WW A1	1.3	2.1	2.8
WW A2	1.1	1.6	2.1
WW A3	1.0	1.5	1.9
WW A4	1.0	1.4	1.8
WW A5	1.0	1.3	1.7
WW A6	1.0	1.3	1.7
WW B1	1.3	2.0	2.7
WW B2	1.0	1.3	1.7
WW B3	.9	1.2	1.5
WW B4	.9	1.1	1.4
WW B5	.9	1.1	1.4
WW C1	.8	1.1	1.4
WW C2	.9	1.1	1.5
WW C3	.8	1.0	1.2
WW C4	.8	1.0	1.3
TW A1	1.3	2.1	2.8
TW A2	1.2	1.7	2.3
TW A3	1.1	1.6	2.0
TW A4	1.1	1.6	2.0
TW B1	1.3	2.0	2.7
TW B2	1.2	1.7	2.3
TW B3	1.2	1.7	2.2
TW B4	1.3	1.8	2.3
TW C1	1.3	2.0	2.6
TW C2	1.3	1.9	2.4
TW C3	1.4	1.9	2.5
TW C4	1.5	2.1	2.8
TW D1	1.5	2.2	3.0
TW D2	1.1	1.4	1.9
TW D3	1.0	1.3	1.7

DODGE CITY, KS			
	\% SOLAR SAVINGS		
	10%	50%	80%
TW D4	.9	1.2	1.6
TW D5	.9	1.2	1.5
TW E1	.9	1.2	1.6
TW E2	.9	1.2	1.6
TW E3	.8	1.0	1.3
TW E4	.8	1.1	1.4
TW F1	1.4	2.1	2.8
TW F2	1.3	1.8	2.3
TW F3	1.2	1.7	2.2
TW F4	1.3	1.7	2.2
TW G1	1.4	2.1	2.7
TW G2	1.3	1.9	2.4
TW G3	1.4	1.9	2.4
TW G4	1.5	2.0	2.6
TW H1	1.5	2.1	2.8
TW H2	1.5	2.1	2.7
TW H3	1.7	2.3	2.9
TW H4	1.9	2.6	3.4
TW I1	1.6	2.4	3.2
TW I2	1.1	1.5	2.0
TW I3	1.1	1.4	1.8
TW I4	1.0	1.3	1.6
TW I5	1.0	1.3	1.6
TW J1	.9	1.3	1.6
TW J2	1.0	1.3	1.6
TW J3	.8	1.1	1.4
TW J4	.9	1.1	1.4
DG A1	1.4	---	---
DG A2	1.2	1.8	4.1
DG A3	1.1	1.4	2.0
DG B1	1.4	2.3	---
DG B2	1.2	1.6	2.5
DG B3	1.1	1.3	1.7
DG C1	1.2	1.6	---
DG C2	1.1	1.4	1.9
DG C3	1.0	1.2	1.5
SS A1	1.0	1.6	2.1
SS A2	.8	1.2	1.6
SS A3	1.1	1.7	2.4
SS A4	.8	1.2	1.6
SS A5	1.0	1.7	2.4
SS A6	.8	1.2	1.6
SS A7	1.1	2.1	2.9
SS A8	.9	1.3	1.7
SS B1	1.1	1.7	2.3
SS B2	.9	1.3	1.7
SS B3	1.2	1.9	2.5
SS B4	.9	1.3	1.8
SS B5	1.3	2.0	2.8
SS B6	.9	1.3	1.7
SS B7	1.4	2.3	3.2
SS B8	1.0	1.4	1.8
SS C1	1.1	1.6	2.1
SS C2	1.0	1.3	1.7
SS C3	1.4	2.0	2.6
SS C4	1.1	1.5	1.9
SS D1	.9	1.3	1.7
SS D2	.7	1.0	1.3
SS D3	.9	1.5	2.0
SS D4	.8	1.1	1.4
SS E1	1.0	1.5	2.0
SS E2	.8	1.2	1.5
SS E3	1.1	1.8	2.4
SS E4	.9	1.2	1.6

GOODLAND, KS			
	\% SOLAR SAVINGS		
	10%	50%	80%
WW A1	1.5	2.3	3.1
WW A2	1.2	1.7	2.2
WW A3	1.1	1.6	2.0
WW A4	1.1	1.5	1.9
WW A5	1.0	1.4	1.8
WW A6	1.0	1.4	1.8
WW B1	1.5	2.3	3.1
WW B2	1.0	1.4	1.8
WW B3	.9	1.2	1.6
WW B4	.9	1.2	1.5
WW B5	.9	1.2	1.5
WW C1	.9	1.2	1.5
WW C2	.9	1.2	1.5
WW C3	.8	1.0	1.3
WW C4	.8	1.1	1.4
TW A1	1.5	2.3	3.1
TW A2	1.3	1.9	2.4
TW A3	1.2	1.7	2.2
TW A4	1.2	1.7	2.2
TW B1	1.4	2.2	2.9
TW B2	1.3	1.9	2.4
TW B3	1.3	1.8	2.4
TW B4	1.3	1.9	2.4
TW C1	1.4	2.1	2.8
TW C2	1.4	2.0	2.6
TW C3	1.5	2.1	2.7
TW C4	1.6	2.3	3.0
TW D1	1.6	2.5	3.4
TW D2	1.1	1.5	2.0
TW D3	1.0	1.4	1.8
TW D4	1.0	1.3	1.6
TW D5	1.0	1.3	1.6
TW E1	1.0	1.3	1.7
TW E2	1.0	1.3	1.7
TW E3	.8	1.1	1.4
TW E4	.9	1.1	1.4
TW F1	1.5	2.3	3.1
TW F2	1.3	1.9	2.5
TW F3	1.3	1.8	2.4
TW F4	1.3	1.8	2.4
TW G1	1.5	2.2	3.0
TW G2	1.4	2.0	2.6
TW G3	1.5	2.0	2.6
TW G4	1.6	2.2	2.8
TW H1	1.6	2.3	3.0
TW H2	1.6	2.3	2.9
TW H3	1.8	2.4	3.1
TW H4	2.0	2.8	3.6
TW I1	1.7	2.7	3.6
TW I2	1.2	1.6	2.1
TW I3	1.1	1.5	1.9
TW I4	1.0	1.3	1.7
TW I5	1.0	1.3	1.7
TW J1	1.0	1.3	1.7
TW J2	1.0	1.3	1.7
TW J3	.9	1.1	1.4
TW J4	.9	1.2	1.5
DG A1	1.5	---	---
DG A2	1.3	2.0	---
DG A3	1.2	1.4	2.1
DG B1	1.5	3.2	---
DG B2	1.3	1.7	3.0
DG B3	1.1	1.3	1.8
DG C1	1.3	1.8	---

Conservation Factor Tables

GOODLAND, KS

	SOLAR SAVINGS		
	10%	50%	80%
DG C2	1.2	1.5	2.1
DG C3	1.1	1.2	1.5
SS A1	1.1	1.7	2.3
SS A2	.9	1.3	1.7
SS A3	1.1	1.9	2.6
SS A4	.9	1.3	1.7
SS A5	1.1	1.9	2.6
SS A6	.9	1.3	1.7
SS A7	1.3	2.3	3.3
SS A8	.9	1.3	1.8
SS B1	1.2	1.9	2.5
SS B2	1.0	1.4	1.8
SS B3	1.3	2.0	2.8
SS B4	1.0	1.4	1.9
SS B5	1.4	2.2	3.0
SS B6	1.0	1.4	1.8
SS B7	1.5	2.6	3.6
SS B8	1.0	1.5	1.9
SS C1	1.2	1.7	2.3
SS C2	1.0	1.4	1.8
SS C3	1.5	2.1	2.8
SS C4	1.1	1.6	2.0
SS D1	.9	1.4	1.9
SS D2	.8	1.1	1.4
SS D3	1.0	1.6	2.2
SS D4	.8	1.1	1.5
SS E1	1.1	1.6	2.1
SS E2	.9	1.2	1.6
SS E3	1.2	1.9	2.6
SS E4	.9	1.3	1.7

TOPEKA, KS

	SOLAR SAVINGS		
	10%	50%	80%
WW A1	1.5	3.0	4.6
WW A2	1.3	2.0	2.8
WW A3	1.2	1.8	2.5
WW A4	1.2	1.7	2.3
WW A5	1.2	1.7	2.2
WW A6	1.1	1.6	2.1
WW B1	1.7	3.4	6.4
WW B2	1.1	1.6	2.1
WW B3	1.0	1.4	1.8
WW B4	1.0	1.3	1.7
WW B5	1.0	1.3	1.7
WW C1	1.0	1.3	1.8
WW C2	1.0	1.3	1.7
WW C3	.9	1.1	1.4
WW C4	.9	1.2	1.5
TW A1	1.6	2.9	4.3
TW A2	1.4	2.2	3.1
TW A3	1.3	2.0	2.8
TW A4	1.3	2.0	2.6
TW B1	1.6	2.7	3.8
TW B2	1.4	2.2	3.0
TW B3	1.4	2.1	2.9
TW B4	1.5	2.2	2.9
TW C1	1.6	2.5	3.5
TW C2	1.5	2.3	3.2
TW C3	1.6	2.4	3.2
TW C4	1.7	2.6	3.6
TW D1	1.9	3.6	6.0
TW D2	1.2	1.8	2.3
TW D3	1.1	1.6	2.2
TW D4	1.1	1.4	1.9
TW D5	1.0	1.4	1.8
TW E1	1.0	1.5	1.9
TW E2	1.0	1.4	1.9
TW E3	.9	1.2	1.6
TW E4	.9	1.2	1.6
TW F1	1.6	2.9	4.2
TW F2	1.5	2.3	3.2
TW F3	1.5	2.2	2.9
TW F4	1.5	2.2	2.9
TW G1	1.7	2.7	3.9
TW G2	1.6	2.4	3.2
TW G3	1.6	2.4	3.2
TW G4	1.7	2.5	3.4
TW H1	1.8	2.7	3.8
TW H2	1.8	2.7	3.6
TW H3	1.9	2.8	3.8
TW H4	2.2	3.2	4.3
TW I1	2.0	4.0	7.1
TW I2	1.3	1.8	2.4
TW I3	1.2	1.7	2.3
TW I4	1.1	1.5	1.9
TW I5	1.1	1.5	1.9
TW J1	1.1	1.5	2.0
TW J2	1.1	1.5	1.9
TW J3	.9	1.3	1.6
TW J4	1.0	1.3	1.7
DG A1	1.7	---	---
DG A2	1.5	2.6	---
DG A3	1.2	1.6	2.5
DG B1	1.7	---	---
DG B2	1.4	2.1	---
DG B3	1.2	1.5	2.1
DG C1	1.5	2.9	---
DG C2	1.3	1.7	2.9
DG C3	1.1	1.4	1.8
SS A1	1.2	2.0	2.8
SS A2	.9	1.4	1.9
SS A3	1.3	2.3	3.5
SS A4	1.0	1.5	2.0
SS A5	1.2	2.3	3.4
SS A6	1.0	1.4	1.9
SS A7	1.3	3.0	5.1
SS A8	1.0	1.5	2.1
SS B1	1.3	2.2	3.2
SS B2	1.0	1.5	2.1
SS B3	1.4	2.5	3.7
SS B4	1.1	1.6	2.1
SS B5	1.5	2.7	4.1
SS B6	1.1	1.6	2.1
SS B7	1.6	3.4	5.7
SS B8	1.1	1.7	2.2
SS C1	1.4	2.1	2.9
SS C2	1.1	1.6	2.1
SS C3	1.6	2.6	3.7
SS C4	1.2	1.8	2.4
SS D1	1.0	1.7	2.3
SS D2	.9	1.2	1.6
SS D3	1.1	1.9	2.7
SS D4	.9	1.3	1.7
SS E1	1.2	1.9	2.7
SS E2	1.0	1.4	1.8
SS E3	1.3	2.3	3.4
SS E4	1.0	1.5	2.0

WICHITA, KS

	SOLAR SAVINGS		
	10%	50%	80%
WW A1	1.3	2.3	3.1
WW A2	1.1	1.7	2.2
WW A3	1.1	1.5	2.0
WW A4	1.0	1.5	1.9
WW A5	1.0	1.4	1.8
WW A6	1.0	1.4	1.8
WW B1	1.4	2.2	3.1
WW B2	1.0	1.4	1.8
WW B3	.9	1.2	1.6
WW B4	.9	1.1	1.5
WW B5	.9	1.2	1.5
WW C1	.9	1.2	1.5
WW C2	.9	1.2	1.5
WW C3	.8	1.0	1.3
WW C4	.8	1.1	1.3
TW A1	1.4	2.3	3.1
TW A2	1.2	1.8	2.4
TW A3	1.2	1.7	2.2
TW A4	1.2	1.6	2.1
TW B1	1.4	2.1	2.9
TW B2	1.3	1.8	2.4
TW B3	1.2	1.8	2.3
TW B4	1.3	1.8	2.4
TW C1	1.4	2.1	2.7
TW C2	1.4	2.0	2.6
TW C3	1.4	2.0	2.7
TW C4	1.5	2.2	2.9
TW D1	1.5	2.4	3.3
TW D2	1.1	1.5	2.0
TW D3	1.0	1.4	1.8
TW D4	.9	1.3	1.6
TW D5	.9	1.3	1.6
TW E1	.9	1.3	1.7
TW E2	.9	1.3	1.6
TW E3	.8	1.1	1.4
TW E4	.9	1.1	1.4
TW F1	1.4	2.2	3.0
TW F2	1.3	1.9	2.5
TW F3	1.3	1.8	2.3
TW F4	1.3	1.8	2.3
TW G1	1.5	2.2	2.9
TW G2	1.4	2.0	2.6
TW G3	1.4	2.0	2.6
TW G4	1.5	2.1	2.7
TW H1	1.5	2.2	2.9
TW H2	1.6	2.2	2.9
TW H3	1.7	2.4	3.1
TW H4	1.9	2.7	3.5
TW I1	1.7	2.6	3.6
TW I2	1.1	1.6	2.0
TW I3	1.1	1.5	1.9
TW I4	1.0	1.3	1.7
TW I5	1.0	1.3	1.6
TW J1	1.0	1.3	1.7
TW J2	1.0	1.3	1.7
TW J3	.9	1.1	1.4
TW J4	.9	1.2	1.5
DG A1	1.4	---	---
DG A2	1.3	2.0	6.3
DG A3	1.1	1.4	2.1
DG B1	1.4	2.8	---
DG B2	1.2	1.6	2.8
DG B3	1.1	1.3	1.7
DG C1	1.2	1.8	---

WICHITA, KS

	SOLAR SAVINGS		
	10%	50%	80%
DG C2	1.1	1.4	2.1
DG C3	1.0	1.2	1.5
SS A1	1.0	1.6	2.2
SS A2	.9	1.2	1.6
SS A3	1.1	1.8	2.5
SS A4	.9	1.3	1.7
SS A5	1.1	1.8	2.5
SS A6	.9	1.3	1.7
SS A7	1.2	2.2	3.2
SS A8	.9	1.3	1.7
SS B1	1.2	1.8	2.5
SS B2	.9	1.3	1.8
SS B3	1.2	2.0	2.7
SS B4	1.0	1.4	1.8
SS B5	1.3	2.2	3.0
SS B6	1.0	1.4	1.8
SS B7	1.4	2.5	3.5
SS B8	1.0	1.4	1.9
SS C1	1.2	1.7	2.2
SS C2	1.0	1.4	1.8
SS C3	1.4	2.1	2.8
SS C4	1.1	1.5	2.0
SS D1	.9	1.4	1.8
SS D2	.8	1.1	1.4
SS D3	1.0	1.5	2.1
SS D4	.8	1.1	1.5
SS E1	1.0	1.5	2.1
SS E2	.9	1.2	1.6
SS E3	1.2	1.9	2.5
SS E4	.9	1.3	1.7

LEXINGTON, KY

	SOLAR SAVINGS		
	10%	50%	80%
WW A1	1.8	6.3	---
WW A2	1.5	2.7	4.1
WW A3	1.5	2.4	3.4
WW A4	1.4	2.2	3.1
WW A5	1.4	2.1	2.9
WW A6	1.3	2.0	2.8
WW B1	2.2	---	---
WW B2	1.3	1.9	2.6
WW B3	1.1	1.7	2.2
WW B4	1.1	1.5	1.9
WW B5	1.1	1.5	1.9
WW C1	1.1	1.6	2.1
WW C2	1.1	1.5	2.0
WW C3	1.0	1.3	1.6
WW C4	1.0	1.3	1.7
TW A1	1.8	4.8	11.5
TW A2	1.6	3.0	4.4
TW A3	1.6	2.6	3.8
TW A4	1.5	2.5	3.4
TW B1	1.8	3.8	6.3
TW B2	1.7	2.8	4.1
TW B3	1.6	2.7	3.8
TW B4	1.7	2.7	3.8
TW C1	1.8	3.3	4.9
TW C2	1.8	2.9	4.1
TW C3	1.8	3.0	4.2
TW C4	2.0	3.2	4.5
TW D1	2.4	---	---
TW D2	1.4	2.1	2.9
TW D3	1.3	1.9	2.6

LEXINGTON, KY				LOUISVILLE, KY				LOUISVILLE, KY				BATON ROUGE, LA			
	SOLAR SAVINGS				SOLAR SAVINGS				SOLAR SAVINGS				SOLAR SAVINGS		
	10%	50%	80%		10%	50%	80%		10%	50%	80%		10%	50%	80%
WW A1	1.8	6.3	---	WW A1	1.8	5.3	22.9	DG C2	1.4	2.1	9.0	TW D4	.7	.9	1.2
WW A2	1.5	2.7	4.1	WW A2	1.5	2.6	3.9	DG C3	1.2	1.6	2.0	TW D5	.7	.9	1.2
WW A3	1.5	2.4	3.4	WW A3	1.4	2.3	3.3	SS A1	1.3	2.4	3.5	TW E1	.7	.9	1.2
WW A4	1.4	2.2	3.1	WW A4	1.4	2.1	3.0	SS A2	1.0	1.6	2.2	TW E2	.7	.9	1.2
WW A5	1.4	2.1	2.9	WW A5	1.3	2.0	2.8	SS A3	1.4	3.0	5.2	TW E3	.6	.8	1.0
WW A6	1.3	2.0	2.8	WW A6	1.3	2.0	2.7	SS A4	1.0	1.7	2.3	TW E4	.6	.8	1.1
WW B1	2.2	---	---	WW B1	2.1	---	---	SS A5	1.3	2.9	4.7	TW F1	1.0	1.4	1.9
WW B2	1.3	1.9	2.6	WW B2	1.3	1.9	2.5	SS A6	1.0	1.6	2.2	TW F2	.9	1.2	1.6
WW B3	1.1	1.7	2.2	WW B3	1.1	1.6	2.2	SS A7	1.5	4.7	---	TW F3	.9	1.2	1.5
WW B4	1.1	1.5	1.9	WW B4	1.1	1.5	1.9	SS A8	1.1	1.7	2.4	TW F4	.9	1.2	1.6
WW B5	1.1	1.5	1.9	WW B5	1.1	1.4	1.9	SS B1	1.5	2.7	4.1	TW G1	1.0	1.4	1.8
WW C1	1.1	1.6	2.1	WW C1	1.1	1.5	2.0	SS B2	1.1	1.7	2.3	TW G2	.9	1.3	1.7
WW C2	1.1	1.5	2.0	WW C2	1.1	1.5	2.0	SS B3	1.6	3.3	5.7	TW G3	1.0	1.3	1.7
WW C3	1.0	1.3	1.6	WW C3	.9	1.2	1.6	SS B4	1.2	1.8	2.4	TW G4	1.1	1.4	1.9
WW C4	1.0	1.3	1.7	WW C4	1.0	1.3	1.7	SS B5	1.6	3.6	6.3	TW H1	1.0	1.5	1.9
TW A1	1.8	4.8	11.5	TW A1	1.8	4.4	9.3	SS B6	1.2	1.8	2.4	TW H2	1.1	1.5	1.9
TW A2	1.6	3.0	4.4	TW A2	1.6	2.8	4.2	SS B7	1.8	5.8	---	TW H3	1.2	1.6	2.1
TW A3	1.6	2.6	3.8	TW A3	1.5	2.5	3.6	SS B8	1.2	1.9	2.6	TW H4	1.4	1.9	2.4
TW A4	1.5	2.5	3.4	TW A4	1.5	2.4	3.3	SS C1	1.6	2.8	4.2	TW I1	1.0	1.5	2.0
TW B1	1.8	3.8	6.3	TW B1	1.8	3.6	5.9	SS C2	1.3	1.9	2.5	TW I2	.8	1.1	1.4
TW B2	1.7	2.8	4.1	TW B2	1.6	2.7	3.9	SS C3	1.9	3.5	5.6	TW I3	.8	1.0	1.3
TW B3	1.6	2.7	3.8	TW B3	1.6	2.6	3.6	SS C4	1.4	2.1	2.8	TW I4	.7	1.0	1.2
TW B4	1.7	2.7	3.8	TW B4	1.7	2.6	3.7	SS D1	1.1	2.0	2.8	TW I5	.7	1.0	1.2
TW C1	1.8	3.3	4.9	TW C1	1.8	3.2	4.7	SS D2	.9	1.4	1.8	TW J1	.7	.9	1.2
TW C2	1.8	2.9	4.1	TW C2	1.7	2.8	4.0	SS D3	1.2	2.3	3.4	TW J2	.7	.9	1.2
TW C3	1.8	3.0	4.2	TW C3	1.8	2.9	4.1	SS D4	1.0	1.4	1.9	TW J3	.6	.8	1.1
TW C4	2.0	3.2	4.5	TW C4	1.9	3.1	4.4	SS E1	1.3	2.4	3.7	TW J4	.7	.9	1.1
TW D1	2.4	---	---	TW D1	2.3	---	---	SS E2	1.0	1.6	2.1	DG A1	.9	1.4	3.8
TW D2	1.4	2.1	2.9	TW D2	1.4	2.1	2.8	SS E3	1.5	3.0	4.9	DG A2	.9	1.2	1.8
TW D3	1.3	1.9	2.6	TW D3	1.3	1.9	2.6	SS E4	1.1	1.7	2.3	DG A3	.8	1.0	1.4
DG B1	2.3	---	---	TW D4	1.2	1.6	2.1					DG B1	.9	1.2	2.0
DG B2	1.6	2.9	---	TW D5	1.1	1.6	2.0	BATON ROUGE, LA				DG B2	.9	1.1	1.5
DG B3	1.3	1.7	2.5	TW E1	1.2	1.7	2.3		SOLAR SAVINGS			DG B3	.8	1.0	1.2
DG C1	1.8	---	---	TW E2	1.2	1.6	2.2		10%	50%	80%	DG C1	.8	1.0	1.5
DG C2	1.4	2.2	---	TW E3	1.0	1.3	1.7	WW A1	.9	1.4	1.9	DG C2	.8	1.0	1.3
DG C3	1.2	1.6	2.1	TW E4	1.0	1.4	1.8	WW A2	.8	1.1	1.4	DG C3	.7	.9	1.1
SS A1	1.3	2.4	3.6	TW F1	1.9	4.2	8.2	WW A3	.7	1.0	1.3	SS A1	.7	1.0	1.4
SS A2	1.0	1.6	2.2	TW F2	1.7	3.0	4.4	WW A4	.7	1.0	1.3	SS A2	.6	.9	1.1
SS A3	1.4	3.1	5.5	TW F3	1.7	2.7	3.9	WW A5	.7	.9	1.2	SS A3	.7	1.1	1.5
SS A4	1.1	1.7	2.3	TW F4	1.7	2.6	3.7	WW A6	.7	.9	1.2	SS A4	.6	.9	1.1
SS A5	1.4	3.0	4.9	TW G1	2.0	3.7	6.0	WW B1	.9	1.3	1.7	SS A5	.7	1.1	1.5
SS A6	1.0	1.6	2.2	TW G2	1.8	3.0	4.3	WW B2	.7	1.0	1.2	SS A6	.6	.9	1.1
SS A7	1.5	5.2	---	TW G3	1.9	2.9	4.1	WW B3	.6	.9	1.1	SS A7	.8	1.3	1.7
SS A8	1.1	1.8	2.4	TW G4	2.0	3.1	4.2	WW B4	.6	.8	1.1	SS A8	.6	.9	1.1
SS B1	1.5	2.8	4.3	TW H1	2.1	3.5	5.3	WW B5	.7	.9	1.1	SS B1	.8	1.2	1.5
SS B2	1.1	1.7	2.3	TW H2	2.1	3.3	4.7	WW C1	.6	.8	1.1	SS B2	.7	.9	1.2
SS B3	1.6	3.4	6.1	TW H3	2.2	3.4	4.8	WW C2	.6	.9	1.1	SS B3	.8	1.2	1.6
SS B4	1.2	1.8	2.5	TW H4	2.5	3.9	5.3	WW C3	.6	.8	1.0	SS B4	.7	.9	1.2
SS B5	1.7	3.8	6.8	TW I1	2.6	---	---	WW C4	.6	.8	1.0	SS B5	.9	1.3	1.8
SS B6	1.2	1.8	2.4	TW I2	1.5	2.2	2.9	TW A1	.9	1.4	1.9	SS B6	.7	.9	1.2
SS B7	1.9	6.8	---	TW I3	1.4	2.0	2.7	TW A2	.8	1.2	1.6	SS B7	.9	1.4	1.9
SS B8	1.2	1.9	2.6	TW I4	1.2	1.7	2.2	TW A3	.8	1.1	1.5	SS B8	.7	1.0	1.3
SS C1	1.6	2.9	4.4	TW I5	1.2	1.6	2.1	TW A4	.8	1.1	1.4	SS C1	.8	1.1	1.4
SS C2	1.3	1.9	2.6	TW J1	1.2	1.7	2.3	TW B1	.9	1.4	1.8	SS C2	.7	1.0	1.2
SS C3	1.9	3.7	5.9	TW J2	1.2	1.7	2.2	TW B2	.9	1.2	1.6	SS C3	.9	1.4	1.8
SS C4	1.4	2.1	2.9	TW J3	1.0	1.4	1.8	TW B3	.9	1.2	1.6	SS C4	.8	1.1	1.4
SS D1	1.1	2.0	2.9	TW J4	1.1	1.4	1.8	TW B4	.9	1.3	1.6	SS D1	.6	.9	1.2
SS D2	.9	1.4	1.8	DG A1	2.2	---	---	TW C1	.9	1.4	1.8	SS D2	.5	.7	1.0
SS D3	1.2	2.3	3.5	DG A2	1.6	4.6	---	TW C2	.9	1.3	1.7	SS D3	.7	1.0	1.3
SS D4	1.0	1.4	1.9	DG A3	1.3	1.9	3.2	TW C3	1.0	1.4	1.8	SS D4	.6	.8	1.0
SS E1	1.3	2.5	3.9	DG B1	2.2	---	---	TW C4	1.1	1.5	2.0	SS E1	.7	1.0	1.3
SS E2	1.1	1.6	2.1	DG B2	1.6	2.8	---	TW D1	1.0	1.4	1.9	SS E2	.6	.8	1.1
SS E3	1.5	3.1	5.2	DG B3	1.3	1.7	2.4	TW D2	.8	1.0	1.4	SS E3	.8	1.2	1.5
SS E4	1.1	1.7	2.3	DG C1	1.7	---	---	TW D3	.7	1.0	1.3	SS E4	.6	.9	1.2

Conservation Factor Tables / 331

LAKE CHARLES, LA

	SOLAR SAVINGS		
	10%	50%	80%
WW A1	.9	1.4	1.9
WW A2	.8	1.1	1.4
WW A3	.7	1.0	1.3
WW A4	.7	1.0	1.3
WW A5	.7	.9	1.2
WW A6	.7	.9	1.2
WW B1	.9	1.3	1.7
WW B2	.7	1.0	1.2
WW B3	.6	.9	1.1
WW B4	.6	.8	1.1
WW B5	.7	.9	1.1
WW C1	.6	.8	1.1
WW C2	.6	.9	1.1
WW C3	.6	.8	1.0
WW C4	.6	.8	1.0
TW A1	.9	1.4	1.9
TW A2	.8	1.2	1.6
TW A3	.8	1.1	1.5
TW A4	.8	1.1	1.4
TW B1	.9	1.4	1.8
TW B2	.9	1.2	1.6
TW B3	.9	1.2	1.6
TW B4	.9	1.3	1.6
TW C1	.9	1.4	1.8
TW C2	.9	1.3	1.7
TW C3	1.0	1.4	1.8
TW C4	1.1	1.5	2.0
TW D1	1.0	1.4	1.9
TW D2	.8	1.0	1.4
TW D3	.7	1.0	1.3
TW D4	.7	.9	1.2
TW D5	.7	.9	1.2
TW E1	.7	.9	1.2
TW E2	.7	.9	1.2
TW E3	.6	.8	1.0
TW E4	.6	.8	1.1
TW F1	.9	1.4	1.9
TW F2	.9	1.2	1.6
TW F3	.9	1.2	1.6
TW F4	.9	1.2	1.6
TW G1	1.0	1.4	1.9
TW G2	.9	1.3	1.7
TW G3	1.0	1.3	1.7
TW G4	1.1	1.4	1.9
TW H1	1.0	1.5	1.9
TW H2	1.1	1.5	1.9
TW H3	1.2	1.6	2.1
TW H4	1.4	1.9	2.4
TW I1	1.0	1.5	2.0
TW I2	.8	1.1	1.4
TW I3	.8	1.0	1.3
TW I4	.7	1.0	1.2
TW I5	.7	.9	1.2
TW J1	.7	.9	1.2
TW J2	.7	.9	1.2
TW J3	.6	.8	1.1
TW J4	.7	.9	1.1
DG A1	.9	1.4	4.0
DG A2	.9	1.2	1.8
DG A3	.8	1.0	1.4
DG B1	.9	1.2	2.1
DG B2	.9	1.1	1.5
DG B3	.8	1.0	1.2
DG C1	.8	1.0	1.5

LAKE CHARLES, LA

	SOLAR SAVINGS		
	10%	50%	80%
DG C2	.8	1.0	1.3
DG C3	.7	.9	1.1
SS A1	.7	1.0	1.4
SS A2	.6	.8	1.1
SS A3	.7	1.1	1.5
SS A4	.6	.9	1.1
SS A5	.7	1.1	1.5
SS A6	.6	.8	1.1
SS A7	.8	1.3	1.7
SS A8	.6	.9	1.1
SS B1	.8	1.1	1.5
SS B2	.7	.9	1.2
SS B3	.8	1.2	1.6
SS B4	.7	.9	1.2
SS B5	.8	1.3	1.8
SS B6	.7	.9	1.2
SS B7	.9	1.4	1.9
SS B8	.7	1.0	1.3
SS C1	.8	1.1	1.5
SS C2	.7	1.0	1.2
SS C3	.9	1.3	1.8
SS C4	.8	1.1	1.4
SS D1	.6	.9	1.2
SS D2	.5	.7	1.0
SS D3	.6	1.0	1.3
SS D4	.6	.8	1.0
SS E1	.7	1.0	1.3
SS E2	.6	.8	1.1
SS E3	.8	1.1	1.5
SS E4	.6	.9	1.1

NEW ORLEANS, LA

	SOLAR SAVINGS		
	10%	50%	80%
WW A1	.8	1.2	1.6
WW A2	.7	1.0	1.3
WW A3	.7	.9	1.2
WW A4	.7	.9	1.1
WW A5	.6	.8	1.1
WW A6	.6	.8	1.1
WW B1	.8	1.1	1.5
WW B2	.6	.9	1.1
WW B3	.6	.8	1.0
WW B4	.6	.8	1.0
WW B5	.6	.8	1.0
WW C1	.6	.8	1.0
WW C2	.6	.8	1.0
WW C3	.5	.7	.9
WW C4	.6	.7	.9
TW A1	.8	1.3	1.7
TW A2	.8	1.1	1.4
TW A3	.7	1.0	1.3
TW A4	.7	1.0	1.3
TW B1	.8	1.2	1.6
TW B2	.8	1.1	1.4
TW B3	.8	1.1	1.4
TW B4	.8	1.1	1.5
TW C1	.9	1.2	1.6
TW C2	.9	1.2	1.5
TW C3	.9	1.2	1.6
TW C4	1.0	1.4	1.8
TW D1	.9	1.2	1.6
TW D2	.7	.9	1.2
TW D3	.6	.9	1.1

NEW ORLEANS, LA

	SOLAR SAVINGS		
	10%	50%	80%
TW D4	.6	.8	1.1
TW D5	.6	.8	1.1
TW E1	.6	.8	1.0
TW E2	.6	.8	1.1
TW E3	.6	.7	.9
TW E4	.6	.8	1.0
TW F1	.9	1.3	1.6
TW F2	.8	1.1	1.4
TW F3	.8	1.1	1.4
TW F4	.8	1.1	1.4
TW G1	.9	1.3	1.6
TW G2	.8	1.2	1.5
TW G3	.9	1.2	1.5
TW G4	1.0	1.3	1.7
TW H1	.9	1.3	1.7
TW H2	1.0	1.3	1.7
TW H3	1.1	1.4	1.9
TW H4	1.2	1.7	2.1
TW I1	.9	1.3	1.7
TW I2	.7	1.0	1.3
TW I3	.7	.9	1.2
TW I4	.7	.9	1.1
TW I5	.7	.9	1.1
TW J1	.6	.8	1.1
TW J2	.6	.9	1.1
TW J3	.6	.8	1.1
TW J4	.6	.8	1.0
DG A1	.8	1.2	2.1
DG A2	.8	1.1	1.5
DG A3	.7	.9	1.2
DG B1	.8	1.0	1.6
DG B2	.8	1.0	1.3
DG B3	.7	.9	1.1
DG C1	.7	.9	1.2
DG C2	.7	.9	1.1
DG C3	.7	.8	1.0
SS A1	.6	.9	1.2
SS A2	.5	.8	1.0
SS A3	.7	1.0	1.3
SS A4	.5	.8	1.0
SS A5	.7	1.0	1.4
SS A6	.5	.8	1.0
SS A7	.7	1.1	1.5
SS A8	.6	.8	1.0
SS B1	.7	1.0	1.4
SS B2	.6	.8	1.1
SS B3	.7	1.1	1.4
SS B4	.6	.9	1.1
SS B5	.8	1.2	1.6
SS B6	.6	.9	1.1
SS B7	.8	1.3	1.7
SS B8	.6	.9	1.1
SS C1	.7	1.0	1.3
SS C2	.6	.9	1.1
SS C3	.8	1.2	1.6
SS C4	.7	1.0	1.2
SS D1	.6	.8	1.0
SS D2	.5	.7	.9
SS D3	.6	.9	1.2
SS D4	.5	.7	.9
SS E1	.6	.9	1.1
SS E2	.5	.7	1.0
SS E3	.7	1.0	1.4
SS E4	.6	.8	1.0

SHREVEPORT, LA

	SOLAR SAVINGS		
	10%	50%	80%
WW A1	1.0	1.6	2.1
WW A2	.9	1.2	1.6
WW A3	.8	1.1	1.5
WW A4	.8	1.1	1.4
WW A5	.8	1.0	1.3
WW A6	.8	1.0	1.3
WW B1	1.0	1.4	1.9
WW B2	.8	1.1	1.4
WW B3	.7	.9	1.2
WW B4	.7	.9	1.2
WW B5	.7	.9	1.2
WW C1	.7	.9	1.2
WW C2	.7	.9	1.2
WW C3	.6	.8	1.0
WW C4	.7	.9	1.1
TW A1	1.0	1.6	2.1
TW A2	.9	1.3	1.7
TW A3	.9	1.3	1.6
TW A4	.9	1.2	1.6
TW B1	1.0	1.5	2.0
TW B2	1.0	1.4	1.8
TW B3	.9	1.3	1.7
TW B4	1.0	1.4	1.8
TW C1	1.1	1.5	2.0
TW C2	1.0	1.5	1.9
TW C3	1.1	1.5	2.0
TW C4	1.2	1.7	2.2
TW D1	1.1	1.6	2.1
TW D2	.8	1.2	1.5
TW D3	.8	1.1	1.4
TW D4	.7	1.0	1.3
TW D5	.8	1.0	1.3
TW E1	.7	1.0	1.3
TW E2	.7	1.0	1.3
TW E3	.7	.9	1.1
TW E4	.7	.9	1.1
TW F1	1.1	1.6	2.1
TW F2	1.0	1.4	1.8
TW F3	1.0	1.3	1.7
TW F4	1.0	1.3	1.7
TW G1	1.1	1.6	2.1
TW G2	1.0	1.4	1.9
TW G3	1.1	1.5	1.9
TW G4	1.2	1.6	2.1
TW H1	1.2	1.6	2.1
TW H2	1.2	1.7	2.1
TW H3	1.3	1.8	2.3
TW H4	1.5	2.0	2.7
TW I1	1.2	1.7	2.3
TW I2	.9	1.2	1.6
TW I3	.8	1.1	1.5
TW I4	.8	1.0	1.3
TW I5	.8	1.0	1.3
TW J1	.8	1.0	1.3
TW J2	.8	1.0	1.3
TW J3	.7	.9	1.1
TW J4	.7	.9	1.2
DG A1	1.0	1.7	---
DG A2	1.0	1.4	2.1
DG A3	.9	1.1	1.5
DG B1	1.0	1.4	3.1
DG B2	.9	1.2	1.7
DG B3	.9	1.1	1.3
DG C1	.9	1.2	1.8

SHREVEPORT, LA

	SOLAR SAVINGS		
	10%	50%	80%
DG C2	.9	1.1	1.4
DG C3	.8	1.0	1.2
SS A1	.8	1.2	1.6
SS A2	.7	.9	1.2
SS A3	.8	1.2	1.7
SS A4	.7	1.0	1.3
SS A5	.8	1.3	1.7
SS A6	.7	.9	1.2
SS A7	.9	1.5	2.0
SS A8	.7	1.0	1.3
SS B1	.9	1.3	1.7
SS B2	.7	1.0	1.4
SS B3	.9	1.4	1.8
SS B4	.7	1.1	1.4
SS B5	1.0	1.5	2.0
SS B6	.7	1.0	1.4
SS B7	1.1	1.6	2.2
SS B8	.8	1.1	1.4
SS C1	.9	1.2	1.6
SS C2	.8	1.1	1.4
SS C3	1.0	1.5	2.0
SS C4	.9	1.2	1.5
SS D1	.7	1.0	1.3
SS D2	.6	.8	1.1
SS D3	.7	1.1	1.5
SS D4	.6	.9	1.1
SS E1	.8	1.1	1.4
SS E2	.7	.9	1.2
SS E3	.9	1.3	1.7
SS E4	.7	1.0	1.3

CARIBOU, ME

	SOLAR SAVINGS		
	10%	50%	80%
WW A1	3.1	---	---
WW A2	2.5	---	---
WW A3	2.3	---	---
WW A4	2.2	---	---
WW A5	2.1	---	---
WW A6	2.1	---	---
WW B1	---	---	---
WW B2	1.8	4.1	---
WW B3	1.6	2.8	7.8
WW B4	1.4	2.0	2.9
WW B5	1.3	1.9	2.6
WW C1	1.4	2.4	4.2
WW C2	1.4	2.2	3.3
WW C3	1.2	1.6	2.1
WW C4	1.2	1.6	2.2
TW A1	3.0	---	---
TW A2	2.6	---	---
TW A3	2.4	---	---
TW A4	2.3	---	---
TW B1	2.9	---	---
TW B2	2.5	---	---
TW B3	2.4	---	---
TW B4	2.4	---	---
TW C1	2.7	---	---
TW C2	2.6	---	---
TW C3	2.6	---	---
TW C4	2.7	---	---
TW D1	---	---	---
TW D2	1.9	4.7	---
TW D3	1.8	4.2	---

CARIBOU, ME

	SOLAR SAVINGS		
	10%	50%	80%
TW D4	1.5	2.3	3.4
TW D5	1.4	2.1	2.9
TW E1	1.6	2.8	5.9
TW E2	1.5	2.4	3.9
TW E3	1.2	1.8	2.5
TW E4	1.3	1.8	2.4
TW F1	3.5	---	---
TW F2	3.0	---	---
TW F3	2.8	---	---
TW F4	2.7	---	---
TW G1	3.6	---	---
TW G2	3.1	---	---
TW G3	3.0	---	---
TW G4	3.1	---	---
TW H1	3.7	---	---
TW H2	3.4	---	---
TW H3	3.4	---	---
TW H4	3.8	---	---
TW I1	---	---	---
TW I2	2.1	5.1	---
TW I3	1.9	3.7	---
TW I4	1.6	2.4	3.5
TW I5	1.5	2.1	3.0
TW J1	1.6	2.8	5.2
TW J2	1.6	2.4	3.8
TW J3	1.3	1.8	2.4
TW J4	1.3	1.8	2.5
DG A1	---	---	---
DG A2	2.4	---	---
DG A3	1.7	2.8	---
DG B1	---	---	---
DG B2	2.4	---	---
DG B3	1.7	2.4	5.5
DG C1	---	---	---
DG C2	2.0	---	---
DG C3	1.5	2.1	3.5
SS A1	1.8	---	---
SS A2	1.3	2.5	4.4
SS A3	2.2	---	---
SS A4	1.4	3.0	9.7
SS A5	1.9	---	---
SS A6	1.4	2.7	5.5
SS A7	2.3	---	---
SS A8	1.4	3.7	---
SS B1	2.2	---	---
SS B2	1.5	2.6	4.3
SS B3	2.7	---	---
SS B4	1.5	2.9	5.7
SS B5	2.5	---	---
SS B6	1.5	2.8	5.2
SS B7	3.3	---	---
SS B8	1.6	3.5	---
SS C1	3.1	---	---
SS C2	1.8	3.6	---
SS C3	3.3	---	---
SS C4	1.9	4.2	---
SS D1	1.7	---	---
SS D2	1.2	2.0	3.2
SS D3	1.7	---	---
SS D4	1.2	2.1	3.5
SS E1	2.2	---	---
SS E2	1.4	2.7	6.2
SS E3	2.3	---	---
SS E4	1.5	3.0	10.4

PORTLAND, ME

	SOLAR SAVINGS		
	10%	50%	80%
WW A1	2.6	---	---
WW A2	2.2	---	---
WW A3	2.0	---	---
WW A4	2.0	5.3	---
WW A5	1.9	4.1	---
WW A6	1.8	3.7	---
WW B1	---	---	---
WW B2	1.7	2.8	4.3
WW B3	1.4	2.2	3.3
WW B4	1.3	1.8	2.5
WW B5	1.3	1.8	2.3
WW C1	1.4	2.0	2.9
WW C2	1.4	1.9	2.6
WW C3	1.1	1.5	1.9
WW C4	1.2	1.5	2.0
TW A1	2.5	---	---
TW A2	2.3	---	---
TW A3	2.2	---	---
TW A4	2.1	4.9	---
TW B1	2.5	---	---
TW B2	2.3	---	---
TW B3	2.2	5.5	---
TW B4	2.2	4.7	---
TW C1	2.5	---	---
TW C2	2.3	5.6	---
TW C3	2.4	5.1	---
TW C4	2.5	5.1	12.7
TW D1	---	---	---
TW D2	1.8	3.1	4.9
TW D3	1.7	2.8	4.4
TW D4	1.4	2.1	2.8
TW D5	1.4	1.9	2.5
TW E1	1.5	2.3	3.3
TW E2	1.4	2.1	2.9
TW E3	1.2	1.6	2.2
TW E4	1.2	1.7	2.2
TW F1	2.9	---	---
TW F2	2.5	---	---
TW F3	2.4	---	---
TW F4	2.4	5.9	---
TW G1	3.0	---	---
TW G2	2.7	---	---
TW G3	2.6	8.3	---
TW G4	2.8	5.8	---
TW H1	3.1	---	---
TW H2	3.0	14.2	---
TW H3	3.1	6.8	---
TW H4	3.4	6.6	23.1
TW I1	---	---	---
TW I2	1.9	3.2	5.2
TW I3	1.8	2.8	4.2
TW I4	1.5	2.1	2.9
TW I5	1.4	2.0	2.6
TW J1	1.5	2.3	3.3
TW J2	1.5	2.2	3.0
TW J3	1.2	1.7	2.2
TW J4	1.3	1.7	2.2
DG A1	---	---	---
DG A2	2.2	---	---
DG A3	1.6	2.5	---
DG B1	---	---	---
DG B2	2.2	---	---
DG B3	1.6	2.2	3.6
DG C1	3.3	---	---

PORTLAND, ME

	SOLAR SAVINGS		
	10%	50%	80%
DG C2	1.9	4.1	---
DG C3	1.5	1.9	2.8
SS A1	1.7	---	---
SS A2	1.3	2.1	3.1
SS A3	1.9	---	---
SS A4	1.3	2.3	3.5
SS A5	1.7	---	---
SS A6	1.3	2.2	3.3
SS A7	2.0	---	---
SS A8	1.3	2.5	4.1
SS B1	2.0	---	---
SS B2	1.4	2.2	3.2
SS B3	2.3	---	---
SS B4	1.4	2.4	3.6
SS B5	2.2	---	---
SS B6	1.4	2.4	3.5
SS B7	2.7	---	---
SS B8	1.5	2.6	4.1
SS C1	2.5	---	---
SS C2	1.6	2.7	4.0
SS C3	2.8	---	---
SS C4	1.8	3.0	4.7
SS D1	1.5	5.1	---
SS D2	1.1	1.8	2.5
SS D3	1.6	---	---
SS D4	1.2	1.8	2.6
SS E1	1.9	---	---
SS E2	1.3	2.1	3.2
SS E3	2.0	---	---
SS E4	1.4	2.3	3.6

BALTIMORE, MD

	SOLAR SAVINGS		
	10%	50%	80%
WW A1	1.7	3.5	6.3
WW A2	1.4	2.2	3.2
WW A3	1.3	2.0	2.8
WW A4	1.3	1.9	2.6
WW A5	1.2	1.8	2.4
WW A6	1.2	1.7	2.4
WW B1	1.9	5.7	---
WW B2	1.2	1.7	2.3
WW B3	1.1	1.5	2.0
WW B4	1.0	1.4	1.8
WW B5	1.0	1.4	1.7
WW C1	1.0	1.4	1.9
WW C2	1.0	1.4	1.8
WW C3	.9	1.2	1.5
WW C4	.9	1.2	1.6
TW A1	1.7	3.3	5.4
TW A2	1.5	2.4	3.5
TW A3	1.5	2.2	3.1
TW A4	1.4	2.1	2.9
TW B1	1.7	3.0	4.4
TW B2	1.6	2.4	3.3
TW B3	1.5	2.3	3.2
TW B4	1.6	2.3	3.2
TW C1	1.7	2.7	3.9
TW C2	1.7	2.5	3.5
TW C3	1.7	2.6	3.6
TW C4	1.9	2.8	3.9
TW D1	2.1	4.9	---
TW D2	1.3	1.9	2.5
TW D3	1.2	1.7	2.3

BALTIMORE, MD					BOSTON, MA					BOSTON, MA					ALPENA, MI			
	SOLAR SAVINGS					SOLAR SAVINGS					SOLAR SAVINGS					SOLAR SAVINGS		
	10%	50%	80%			10%	50%	80%			10%	50%	80%			10%	50%	80%
WW A1	1.7	3.5	6.3		WW A1	2.0	---	---		DG C2	1.6	2.5	---		TW D4	1.5	2.6	4.1
WW A2	1.4	2.2	3.2		WW A2	1.7	3.3	6.0		DG C3	1.3	1.7	2.3		TW D5	1.5	2.2	3.2
WW A3	1.3	2.0	2.8		WW A3	1.6	2.8	4.3		SS A1	1.4	2.9	5.2		TW E1	1.6	3.6	---
WW A4	1.3	1.9	2.6		WW A4	1.6	2.5	3.8		SS A2	1.1	1.8	2.5		TW E2	1.6	2.8	5.2
WW A5	1.2	1.8	2.4		WW A5	1.5	2.4	3.5		SS A3	1.6	5.7	---		TW E3	1.3	1.9	2.7
WW A6	1.2	1.7	2.4		WW A6	1.5	2.3	3.3		SS A4	1.2	1.9	2.7		TW E4	1.3	1.9	2.6
WW B1	1.9	5.7	---		WW B1	2.9	---	---		SS A5	1.5	4.1	---		TW F1	3.7	---	---
WW B2	1.2	1.7	2.3		WW B2	1.4	2.1	2.9		SS A6	1.1	1.8	2.5		TW F2	3.3	---	---
WW B3	1.1	1.5	2.0		WW B3	1.2	1.8	2.4		SS A7	1.7	---	---		TW F3	3.1	---	---
WW B4	1.0	1.4	1.8		WW B4	1.2	1.6	2.1		SS A8	1.2	2.0	2.9		TW F4	3.0	---	---
WW B5	1.0	1.4	1.7		WW B5	1.2	1.5	2.0		SS B1	1.7	3.5	6.9		TW G1	4.0	---	---
WW C1	1.0	1.4	1.9		WW C1	1.2	1.7	2.3		SS B2	1.2	1.9	2.6		TW G2	3.5	---	---
WW C2	1.0	1.4	1.8		WW C2	1.2	1.7	2.2		SS B3	1.8	6.6	---		TW G3	3.3	---	---
WW C3	.9	1.2	1.5		WW C3	1.0	1.3	1.7		SS B4	1.3	2.0	2.8		TW G4	3.4	---	---
WW C4	.9	1.2	1.6		WW C4	1.1	1.4	1.8		SS B5	1.9	6.6	---		TW H1	4.2	---	---
TW A1	1.7	3.3	5.4		TW A1	2.0	---	---		SS B6	1.3	2.0	2.7		TW H2	3.8	---	---
TW A2	1.5	2.4	3.5		TW A2	1.8	3.6	6.5		SS B7	2.2	---	---		TW H3	3.7	---	---
TW A3	1.5	2.2	3.1		TW A3	1.7	3.0	4.8		SS B8	1.3	2.1	3.0		TW H4	4.0	---	---
TW A4	1.4	2.1	2.9		TW A4	1.7	2.8	4.1		SS C1	1.9	3.7	7.9		TW I1	---	---	---
TW B1	1.7	3.0	4.4		TW B1	2.0	5.2	---		SS C2	1.4	2.1	2.9		TW I2	2.2	---	---
TW B2	1.6	2.4	3.3		TW B2	1.9	3.3	5.2		SS C3	2.2	5.0	19.7		TW I3	2.0	9.7	---
TW B3	1.5	2.3	3.2		TW B3	1.8	3.0	4.6		SS C4	1.5	2.3	3.2		TW I4	1.6	2.6	4.2
TW B4	1.6	2.3	3.2		TW B4	1.9	3.0	4.4		SS D1	1.3	2.4	3.9		TW I5	1.5	2.3	3.4
TW C1	1.7	2.7	3.9		TW C1	2.0	3.9	6.6		SS D2	1.0	1.5	2.0		TW J1	1.7	3.5	---
TW C2	1.7	2.5	3.5		TW C2	2.0	3.3	5.0		SS D3	1.4	2.9	5.2		TW J2	1.6	2.8	4.8
TW C3	1.7	2.6	3.6		TW C3	2.0	3.3	4.9		SS D4	1.0	1.6	2.1		TW J3	1.3	1.9	2.6
TW C4	1.9	2.8	3.9		TW C4	2.1	3.6	5.2		SS E1	1.5	3.4	8.4		TW J4	1.3	1.9	2.6
TW D1	2.1	4.9	---		TW D1	3.1	---	---		SS E2	1.2	1.8	2.4		DG A1	---	---	---
TW D2	1.3	1.9	2.5		TW D2	1.5	2.3	3.2		SS E3	1.7	4.6	---		DG A2	2.5	---	---
TW D3	1.2	1.7	2.3		TW D3	1.4	2.1	3.0		SS E4	1.2	1.9	2.7		DG A3	1.7	3.2	---
DG B1	1.9	---	---		TW D4	1.3	1.8	2.3							DG B1	---	---	---
DG B2	1.5	2.3	---		TW D5	1.2	1.7	2.2		ALPENA, MI					DG B2	2.5	---	---
DG B3	1.3	1.6	2.2		TW E1	1.3	1.9	2.6			SOLAR SAVINGS				DG B3	1.7	2.7	---
DG C1	1.6	6.2	---		TW E2	1.3	1.8	2.4			10%	50%	80%		DG C1	---	---	---
DG C2	1.4	1.9	3.6		TW E3	1.1	1.4	1.9		WW A1	3.0	---	---		DG C2	2.0	---	---
DG C3	1.2	1.4	1.9		TW E4	1.1	1.5	1.9		WW A2	2.7	---	---		DG C3	1.5	2.3	4.4
SS A1	1.3	2.1	3.1		TW F1	2.2	---	---		WW A3	2.6	---	---		SS A1	1.8	---	---
SS A2	1.0	1.5	2.0		TW F2	2.0	3.8	7.1		WW A4	2.5	---	---		SS A2	1.3	2.7	5.7
SS A3	1.4	2.5	4.0		TW F3	1.9	3.3	5.2		WW A5	2.4	---	---		SS A3	2.1	---	---
SS A4	1.0	1.5	2.1		TW F4	1.9	3.1	4.6		WW A6	2.3	---	---		SS A4	1.4	3.5	---
SS A5	1.3	2.5	3.8		TW G1	2.3	5.5	---		WW B1	---	---	---		SS A5	1.8	---	---
SS A6	1.0	1.5	2.1		TW G2	2.1	3.7	6.0		WW B2	1.9	---	---		SS A6	1.3	3.1	---
SS A7	1.5	3.4	7.3		TW G3	2.1	3.4	5.2		WW B3	1.6	4.0	---		SS A7	2.1	---	---
SS A8	1.0	1.6	2.2		TW G4	2.2	3.6	5.2		WW B4	1.4	2.2	3.3		SS A8	1.4	5.9	---
SS B1	1.4	2.4	3.5		TW H1	2.4	4.5	8.4		WW B5	1.4	2.0	2.9		SS B1	2.2	---	---
SS B2	1.1	1.6	2.2		TW H2	2.4	3.9	6.0		WW C1	1.5	2.9	11.3		SS B2	1.5	2.8	5.2
SS B3	1.5	2.8	4.4		TW H3	2.5	4.0	5.9		WW C2	1.5	2.5	4.0		SS B3	2.6	---	---
SS B4	1.1	1.7	2.3		TW H4	2.8	4.4	6.4		WW C3	1.2	1.7	2.3		SS B4	1.5	3.3	13.6
SS B5	1.6	3.0	4.8		TW I1	3.8	---	---		WW C4	1.2	1.7	2.4		SS B5	2.4	---	---
SS B6	1.1	1.6	2.2		TW I2	1.6	2.4	3.4		TW A1	2.9	---	---		SS B6	1.5	3.2	8.4
SS B7	1.8	4.0	8.4		TW I3	1.5	2.2	3.0		TW A2	2.7	---	---		SS B7	3.0	---	---
SS B8	1.2	1.7	2.4		TW I4	1.3	1.8	2.4		TW A3	2.6	---	---		SS B8	1.6	4.5	---
SS C1	1.5	2.3	3.3		TW I5	1.3	1.7	2.3		TW A4	2.5	---	---		SS C1	4.0	---	---
SS C2	1.2	1.7	2.3		TW J1	1.3	1.9	2.6		TW B1	2.9	---	---		SS C2	1.8	---	---
SS C3	1.8	2.9	4.3		TW J2	1.3	1.8	2.4		TW B2	2.7	---	---		SS C3	3.7	---	---
SS C4	1.3	1.9	2.5		TW J3	1.1	1.5	1.9		TW B3	2.6	---	---		SS C4	2.0	---	---
SS D1	1.1	1.8	2.5		TW J4	1.1	1.5	2.0		TW B4	2.6	---	---		SS D1	1.6	---	---
SS D2	.9	1.3	1.7		DG A1	3.2	---	---		TW C1	2.8	---	---		SS D2	1.2	2.2	3.7
SS D3	1.2	2.0	2.9		DG A2	1.9	---	---		TW C2	2.6	---	---		SS D3	1.7	---	---
SS D4	.9	1.3	1.8		DG A3	1.5	2.0	4.0		TW C3	2.6	---	---		SS D4	1.2	2.3	4.0
SS E1	1.3	2.1	3.1		DG B1	3.0	---	---		TW C4	2.7	---	---		SS E1	2.0	---	---
SS E2	1.0	1.4	1.9		DG B2	1.8	3.9	---		TW D1	---	---	---		SS E2	1.4	3.3	---
SS E3	1.4	2.6	3.9		DG B3	1.5	1.9	2.7		TW D2	2.0	---	---		SS E3	2.2	---	---
SS E4	1.1	1.6	2.1		DG C1	2.0	---	---		TW D3	1.9	---	---		SS E4	1.5	3.8	---

DETROIT, MI	10%	50%	80%	DETROIT, MI	10%	50%	80%	FLINT, MI	10%	50%	80%	GRAND RAPIDS, MI	10%	50%	80%
WW A1	2.5	---	---	DG C2	1.8	5.1	---	TW D4	1.5	2.4	3.5	WW A1	2.7	---	---
WW A2	2.2	---	---	DG C3	1.5	2.0	3.0	TW D5	1.5	2.2	3.0	WW A2	2.6	---	---
WW A3	2.0	---	---	SS A1	1.7	---	---	TW E1	1.6	3.1	6.0	WW A3	2.4	---	---
WW A4	2.0	---	---	SS A2	1.3	2.2	3.2	TW E2	1.6	2.6	4.0	WW A4	2.3	---	---
WW A5	1.9	---	---	SS A3	1.9	---	---	TW E3	1.3	1.8	2.5	WW A5	2.2	---	---
WW A6	1.8	5.4	---	SS A4	1.3	2.4	3.8	TW E4	1.3	1.8	2.5	WW A6	2.2	---	---
WW B1	---	---	---	SS A5	1.7	---	---	TW F1	3.7	---	---	WW B1	---	---	---
WW B2	1.7	3.0	5.3	SS A6	1.3	2.3	3.5	TW F2	3.2	---	---	WW B2	1.9	5.2	---
WW B3	1.4	2.4	3.6	SS A7	2.0	---	---	TW F3	3.0	---	---	WW B3	1.6	3.1	6.8
WW B4	1.3	1.9	2.6	SS A8	1.3	2.6	4.5	TW F4	2.9	---	---	WW B4	1.4	2.1	3.0
WW B5	1.3	1.8	2.4	SS B1	2.0	---	---	TW G1	4.0	---	---	WW B5	1.4	2.0	2.7
WW C1	1.4	2.1	3.1	SS B2	1.4	2.3	3.3	TW G2	3.4	---	---	WW C1	1.5	2.5	4.3
WW C2	1.4	2.0	2.8	SS B3	2.3	---	---	TW G3	3.2	---	---	WW C2	1.4	2.3	3.4
WW C3	1.1	1.5	2.0	SS B4	1.4	2.5	3.8	TW G4	3.3	---	---	WW C3	1.2	1.6	2.2
WW C4	1.2	1.6	2.1	SS B5	2.3	---	---	TW H1	4.1	---	---	WW C4	1.2	1.7	2.2
TW A1	2.5	---	---	SS B6	1.4	2.4	3.7	TW H2	3.7	---	---	TW A1	2.7	---	---
TW A2	2.3	---	---	SS B7	2.7	---	---	TW H3	3.7	---	---	TW A2	2.6	---	---
TW A3	2.2	---	---	SS B8	1.5	2.7	4.4	TW H4	4.0	---	---	TW A3	2.5	---	---
TW A4	2.1	---	---	SS C1	2.5	---	---	TW I1	---	---	---	TW A4	2.4	---	---
TW B1	2.5	---	---	SS C2	1.6	2.8	4.7	TW I2	2.2	7.6	---	TW B1	2.8	---	---
TW B2	2.3	---	---	SS C3	2.8	---	---	TW I3	2.0	4.2	---	TW B2	2.6	---	---
TW B3	2.2	---	---	SS C4	1.8	3.2	5.7	TW I4	1.6	2.5	3.6	TW B3	2.5	---	---
TW B4	2.2	6.0	---	SS D1	1.5	---	---	TW I5	1.5	2.3	3.1	TW B4	2.5	---	---
TW C1	2.5	---	---	SS D2	1.1	1.8	2.6	TW J1	1.7	3.0	5.4	TW C1	2.7	---	---
TW C2	2.3	10.1	---	SS D3	1.6	---	---	TW J2	1.6	2.6	3.9	TW C2	2.6	---	---
TW C3	2.4	6.4	---	SS D4	1.2	1.9	2.7	TW J3	1.3	1.8	2.5	TW C3	2.6	---	---
TW C4	2.5	5.9	---	SS E1	1.9	---	---	TW J4	1.3	1.9	2.5	TW C4	2.7	---	---
TW D1	---	---	---	SS E2	1.3	2.2	3.4	DG A1	---	---	---	TW D1	---	---	---
TW D2	1.8	3.3	6.3	SS E3	2.1	---	---	DG A2	2.5	---	---	TW D2	2.0	6.5	---
TW D3	1.7	3.0	5.4	SS E4	1.4	2.4	3.8	DG A3	1.7	3.0	---	TW D3	1.8	5.1	---
TW D4	1.4	2.1	2.9					DG B1	---	---	---	TW D4	1.5	2.4	3.5
TW D5	1.4	2.0	2.6	FLINT, MI	10%	50%	80%	DG B2	2.4	---	---	TW D5	1.4	2.1	3.0
TW E1	1.5	2.4	3.7					DG B3	1.7	2.6	6.4	TW E1	1.6	3.0	6.1
TW E2	1.4	2.2	3.1	WW A1	2.9	---	---	DG C1	---	---	---	TW E2	1.5	2.6	4.0
TW E3	1.2	1.7	2.2	WW A2	2.7	---	---	DG C2	2.1	---	---	TW E3	1.2	1.8	2.5
TW E4	1.2	1.7	2.2	WW A3	2.5	---	---	DG C3	1.6	2.2	3.7	TW E4	1.3	1.8	2.5
TW F1	2.9	---	---	WW A4	2.4	---	---	SS A1	1.8	---	---	TW F1	3.4	---	---
TW F2	2.6	---	---	WW A5	2.3	---	---	SS A2	1.3	2.5	4.1	TW F2	3.0	---	---
TW F3	2.4	---	---	WW A6	2.3	---	---	SS A3	2.1	---	---	TW F3	2.9	---	---
TW F4	2.4	---	---	WW B1	---	---	---	SS A4	1.4	2.9	5.8	TW F4	2.8	---	---
TW G1	3.0	---	---	WW B2	1.9	5.2	---	SS A5	1.8	---	---	TW G1	3.7	---	---
TW G2	2.7	---	---	WW B3	1.6	3.1	6.9	SS A6	1.4	2.7	4.9	TW G2	3.2	---	---
TW G3	2.6	---	---	WW B4	1.4	2.1	3.0	SS A7	2.1	---	---	TW G3	3.1	---	---
TW G4	2.8	9.0	---	WW B5	1.4	2.0	2.7	SS A8	1.4	3.5	---	TW G4	3.2	---	---
TW H1	3.1	---	---	WW C1	1.5	2.6	4.3	SS B1	2.2	---	---	TW H1	3.8	---	---
TW H2	3.0	---	---	WW C2	1.5	2.3	3.4	SS B2	1.5	2.6	4.1	TW H2	3.5	---	---
TW H3	3.1	14.1	---	WW C3	1.2	1.6	2.2	SS B3	2.6	---	---	TW H3	3.5	---	---
TW H4	3.4	8.1	---	WW C4	1.2	1.7	2.3	SS B4	1.5	3.0	5.1	TW H4	3.8	---	---
TW I1	---	---	---	TW A1	2.8	---	---	SS B5	2.5	---	---	TW I1	---	---	---
TW I2	1.9	3.5	6.6	TW A2	2.7	---	---	SS B6	1.5	2.9	4.8	TW I2	2.1	7.9	---
TW I3	1.8	3.0	4.9	TW A3	2.6	---	---	SS B7	3.1	---	---	TW I3	1.9	4.2	18.2
TW I4	1.5	2.2	3.0	TW A4	2.5	---	---	SS B8	1.6	3.5	8.2	TW I4	1.6	2.5	3.6
TW I5	1.4	2.0	2.7	TW B1	2.9	---	---	SS C1	3.7	---	---	TW I5	1.5	2.2	3.1
TW J1	1.5	2.4	3.6	TW B2	2.7	---	---	SS C2	1.8	4.1	---	TW J1	1.6	3.0	5.5
TW J2	1.5	2.2	3.2	TW B3	2.6	---	---	SS C3	3.6	---	---	TW J2	1.6	2.6	3.9
TW J3	1.2	1.7	2.2	TW B4	2.6	---	---	SS C4	2.0	5.1	---	TW J3	1.3	1.8	2.5
TW J4	1.3	1.7	2.3	TW C1	2.8	---	---	SS D1	1.7	---	---	TW J4	1.3	1.9	2.5
DG A1	---	---	---	TW C2	2.7	---	---	SS D2	1.2	2.1	3.1	DG A1	---	---	---
DG A2	2.2	---	---	TW C3	2.7	---	---	SS D3	1.7	---	---	DG A2	2.5	---	---
DG A3	1.6	2.5	---	TW C4	2.7	---	---	SS D4	1.2	2.2	3.3	DG A3	1.7	3.0	---
DG B1	---	---	---	TW D1	---	---	---	SS E1	2.2	---	---	DG B1	---	---	---
DG B2	2.1	---	---	TW D2	2.0	6.5	---	SS E2	1.4	2.8	5.2	DG B2	2.4	---	---
DG B3	1.6	2.3	3.9	TW D3	1.9	5.1	---	SS E3	2.3	---	---	DG B3	1.7	2.5	6.4
DG C1	3.4	---	---					SS E4	1.5	3.1	6.4	DG C1	---	---	---

GRAND RAPIDS, MI

	SOLAR SAVINGS		
	10%	50%	80%
DG C2	2.0	---	---
DG C3	1.5	2.2	3.7
SS A1	1.8	---	---
SS A2	1.3	2.5	4.1
SS A3	2.1	---	---
SS A4	1.4	2.9	5.9
SS A5	1.8	---	---
SS A6	1.3	2.7	4.9
SS A7	2.1	---	---
SS A8	1.4	3.5	16.8
SS B1	2.2	---	---
SS B2	1.4	2.6	4.2
SS B3	2.5	---	---
SS B4	1.5	2.9	5.2
SS B5	2.4	---	---
SS B6	1.5	2.9	4.9
SS B7	3.0	---	---
SS B8	1.6	3.5	8.1
SS C1	3.3	---	---
SS C2	1.8	4.1	---
SS C3	3.4	---	---
SS C4	1.9	5.0	---
SS D1	1.6	---	---
SS D2	1.2	2.0	3.1
SS D3	1.7	---	---
SS D4	1.2	2.1	3.3
SS E1	2.1	---	---
SS E2	1.4	2.7	5.2
SS E3	2.2	---	---
SS E4	1.5	3.0	6.4

SAULT STE. MARIE, MI

	SOLAR SAVINGS		
	10%	50%	80%
WW A1	3.8	---	---
WW A2	3.5	---	---
WW A3	3.3	---	---
WW A4	3.1	---	---
WW A5	3.0	---	---
WW A6	2.9	---	---
WW B1	---	---	---
WW B2	2.1	---	---
WW B3	1.7	---	---
WW B4	1.5	2.5	4.0
WW B5	1.4	2.2	3.2
WW C1	1.6	4.1	---
WW C2	1.6	2.8	5.4
WW C3	1.2	1.8	2.5
WW C4	1.3	1.8	2.5
TW A1	3.5	---	---
TW A2	3.3	---	---
TW A3	3.1	---	---
TW A4	2.9	---	---
TW B1	3.4	---	---
TW B2	3.1	---	---
TW B3	3.0	---	---
TW B4	2.9	---	---
TW C1	3.2	---	---
TW C2	3.0	---	---
TW C3	2.9	---	---
TW C4	3.0	---	---
TW D1	---	---	---
TW D2	2.2	---	---
TW D3	2.0	---	---

SAULT STE. MARIE, MI

	SOLAR SAVINGS		
	10%	50%	80%
TW D4	1.6	2.9	5.3
TW D5	1.5	2.4	3.6
TW E1	1.7	---	---
TW E2	1.6	3.4	12.9
TW E3	1.3	2.0	3.0
TW E4	1.3	2.0	2.9
TW F1	6.0	---	---
TW F2	4.7	---	---
TW F3	4.1	---	---
TW F4	3.7	---	---
TW G1	8.8	---	---
TW G2	4.9	---	---
TW G3	4.2	---	---
TW G4	4.0	---	---
TW H1	7.7	---	---
TW H2	4.8	---	---
TW H3	4.4	---	---
TW H4	4.6	---	---
TW I1	---	---	---
TW I2	2.4	---	---
TW I3	2.1	---	---
TW I4	1.7	3.0	5.2
TW I5	1.6	2.5	3.8
TW J1	1.8	6.3	---
TW J2	1.7	3.3	7.2
TW J3	1.4	2.0	2.9
TW J4	1.4	2.0	2.9
DG A1	---	---	---
DG A2	2.8	---	---
DG A3	1.8	3.8	---
DG B1	---	---	---
DG B2	2.7	---	---
DG B3	1.7	3.0	---
DG C1	---	---	---
DG C2	2.2	---	---
DG C3	1.6	2.6	7.1
SS A1	1.9	---	---
SS A2	1.4	3.4	---
SS A3	2.3	---	---
SS A4	1.4	12.0	---
SS A5	2.0	---	---
SS A6	1.4	4.5	---
SS A7	2.4	---	---
SS A8	1.5	---	---
SS B1	2.4	---	---
SS B2	1.5	3.4	8.8
SS B3	3.2	---	---
SS B4	1.6	4.6	---
SS B5	2.7	---	---
SS B6	1.6	4.2	---
SS B7	3.7	---	---
SS B8	1.7	---	---
SS C1	---	---	---
SS C2	2.0	---	---
SS C3	5.8	---	---
SS C4	2.2	---	---
SS D1	1.8	---	---
SS D2	1.3	2.6	5.1
SS D3	1.8	---	---
SS D4	1.3	2.7	5.9
SS E1	2.8	---	---
SS E2	1.5	---	---
SS E3	2.6	---	---
SS E4	1.6	---	---

TRAVERSE CITY, MI

	SOLAR SAVINGS		
	10%	50%	80%
WW A1	3.1	---	---
WW A2	3.3	---	---
WW A3	3.1	---	---
WW A4	3.0	---	---
WW A5	2.9	---	---
WW A6	2.8	---	---
WW B1	---	---	---
WW B2	2.1	---	---
WW B3	1.7	---	---
WW B4	1.5	2.4	3.8
WW B5	1.4	2.2	3.1
WW C1	1.6	3.8	---
WW C2	1.6	2.8	5.0
WW C3	1.2	1.8	2.4
WW C4	1.3	1.8	2.5
TW A1	3.0	---	---
TW A2	3.1	---	---
TW A3	3.0	---	---
TW A4	2.9	---	---
TW B1	3.1	---	---
TW B2	3.0	---	---
TW B3	2.9	---	---
TW B4	2.8	---	---
TW C1	3.0	---	---
TW C2	2.9	---	---
TW C3	2.8	---	---
TW C4	2.9	---	---
TW D1	---	---	---
TW D2	2.1	---	---
TW D3	2.0	---	---
TW D4	1.6	2.9	5.0
TW D5	1.5	2.4	3.5
TW E1	1.7	8.0	---
TW E2	1.6	3.3	7.7
TW E3	1.3	2.0	2.9
TW E4	1.3	2.0	2.8
TW F1	4.6	---	---
TW F2	4.2	---	---
TW F3	3.9	---	---
TW F4	3.6	---	---
TW G1	5.8	---	---
TW G2	4.5	---	---
TW G3	4.0	---	---
TW G4	3.9	---	---
TW H1	6.2	---	---
TW H2	4.6	---	---
TW H3	4.3	---	---
TW H4	4.5	---	---
TW I1	---	---	---
TW I2	2.4	---	---
TW I3	2.1	---	---
TW I4	1.7	2.9	4.9
TW I5	1.6	2.5	3.7
TW J1	1.8	5.1	---
TW J2	1.7	3.2	6.2
TW J3	1.4	2.0	2.8
TW J4	1.4	2.0	2.8
DG A1	---	---	---
DG A2	2.8	---	---
DG A3	1.8	3.7	---
DG B1	---	---	---
DG B2	2.7	---	---
DG B3	1.7	3.0	---
DG C1	---	---	---

TRAVERSE CITY, MI

	SOLAR SAVINGS		
	10%	50%	80%
DG C2	2.2	---	---
DG C3	1.6	2.5	6.2
SS A1	1.9	---	---
SS A2	1.4	3.2	8.3
SS A3	2.2	---	---
SS A4	1.4	4.8	---
SS A5	1.9	---	---
SS A6	1.4	3.8	---
SS A7	2.1	---	---
SS A8	1.5	---	---
SS B1	2.4	---	---
SS B2	1.5	3.2	6.7
SS B3	2.9	---	---
SS B4	1.6	4.0	---
SS B5	2.6	---	---
SS B6	1.6	3.8	25.2
SS B7	3.3	---	---
SS B8	1.7	---	---
SS C1	---	---	---
SS C2	2.0	---	---
SS C3	4.9	---	---
SS C4	2.1	---	---
SS D1	1.7	---	---
SS D2	1.3	2.5	4.4
SS D3	1.8	---	---
SS D4	1.3	2.6	4.9
SS E1	2.6	---	---
SS E2	1.5	4.8	---
SS E3	2.4	---	---
SS E4	1.6	6.9	---

DULUTH, MN

	SOLAR SAVINGS		
	10%	50%	80%
WW A1	3.8	---	---
WW A2	3.2	---	---
WW A3	2.9	---	---
WW A4	2.8	---	---
WW A5	2.6	---	---
WW A6	2.5	---	---
WW B1	---	---	---
WW B2	2.0	---	---
WW B3	1.7	4.4	---
WW B4	1.5	2.3	3.4
WW B5	1.4	2.1	2.9
WW C1	1.6	3.0	9.7
WW C2	1.5	2.5	4.0
WW C3	1.2	1.7	2.3
WW C4	1.3	1.8	2.4
TW A1	3.5	---	---
TW A2	3.1	---	---
TW A3	2.9	---	---
TW A4	2.8	---	---
TW B1	3.4	---	---
TW B2	3.0	---	---
TW B3	2.8	---	---
TW B4	2.8	---	---
TW C1	3.2	---	---
TW C2	2.9	---	---
TW C3	2.9	---	---
TW C4	2.9	---	---
TW D1	---	---	---
TW D2	2.1	---	---
TW D3	2.0	---	---

DULUTH, MN	10%	50%	80%
WW A1	3.8	---	---
WW A2	3.2	---	---
WW A3	2.9	---	---
WW A4	2.8	---	---
WW A5	2.6	---	---
WW A6	2.5	---	---
WW B1	---	---	---
WW B2	2.0	---	---
WW B3	1.7	4.4	---
WW B4	1.5	2.3	3.4
WW B5	1.4	2.1	2.9
WW C1	1.6	3.0	9.7
WW C2	1.5	2.5	4.0
WW C3	1.2	1.7	2.3
WW C4	1.3	1.8	2.4
TW A1	3.5	---	---
TW A2	3.1	---	---
TW A3	2.9	---	---
TW A4	2.8	---	---
TW B1	3.4	---	---
TW B2	3.0	---	---
TW B3	2.8	---	---
TW B4	2.8	---	---
TW C1	3.2	---	---
TW C2	2.9	---	---
TW C3	2.9	---	---
TW C4	2.9	---	---
TW D1	---	---	---
TW D2	2.1	---	---
TW D3	2.0	---	---
DG B1	---	---	---
DG B2	2.7	---	---
DG B3	1.8	2.8	---
DG C1	---	---	---
DG C2	2.2	---	---
DG C3	1.6	2.4	4.5
SS A1	2.0	---	---
SS A2	1.4	3.0	6.4
SS A3	2.5	---	---
SS A4	1.5	4.2	---
SS A5	2.1	---	---
SS A6	1.5	3.4	---
SS A7	2.5	---	---
SS A8	1.5	---	---
SS B1	2.5	---	---
SS B2	1.6	3.0	5.6
SS B3	3.4	---	---
SS B4	1.6	3.7	---
SS B5	2.9	---	---
SS B6	1.6	3.5	10.9
SS B7	4.2	---	---
SS B8	1.7	5.8	---
SS C1	---	---	---
SS C2	2.0	---	---
SS C3	4.8	---	---
SS C4	2.1	---	---
SS D1	1.9	---	---
SS D2	1.3	2.3	4.0
SS D3	1.9	---	---
SS D4	1.3	2.5	4.3
SS E1	2.8	---	---
SS E2	1.5	3.8	---
SS E3	2.7	---	---
SS E4	1.6	4.5	---

INTL. FALLS, MN	10%	50%	80%
WW A1	4.3	---	---
WW A2	4.3	---	---
WW A3	3.9	---	---
WW A4	3.6	---	---
WW A5	3.4	---	---
WW A6	3.3	---	---
WW B1	---	---	---
WW B2	2.2	---	---
WW B3	1.8	---	---
WW B4	1.5	2.5	4.0
WW B5	1.5	2.2	3.2
WW C1	1.7	4.7	---
WW C2	1.6	2.9	5.4
WW C3	1.3	1.8	2.5
WW C4	1.3	1.9	2.6
TW A1	3.8	---	---
TW A2	3.7	---	---
TW A3	3.5	---	---
TW A4	3.2	---	---
TW B1	3.8	---	---
TW B2	3.4	---	---
TW B3	3.2	---	---
TW B4	3.1	---	---
TW C1	3.5	---	---
TW C2	3.2	---	---
TW C3	3.1	---	---
TW C4	3.1	---	---
TW D1	---	---	---
TW D2	2.3	---	---
TW D3	2.1	---	---
TW D4	1.7	3.0	5.3
TW D5	1.6	2.4	3.6
TW E1	1.8	---	---
TW E2	1.7	3.6	12.7
TW E3	1.4	2.1	3.1
TW E4	1.4	2.0	2.9
TW F1	---	---	---
TW F2	7.9	---	---
TW F3	5.2	---	---
TW F4	4.3	---	---
TW G1	---	---	---
TW G2	7.0	---	---
TW G3	4.8	---	---
TW G4	4.3	---	---
TW H1	---	---	---
TW H2	5.6	---	---
TW H3	4.8	---	---
TW H4	4.9	---	---
TW I1	---	---	---
TW I2	2.5	---	---
TW I3	2.2	---	---
TW I4	1.8	3.1	5.3
TW I5	1.6	2.6	3.8
TW J1	1.9	11.7	---
TW J2	1.8	3.4	7.1
TW J3	1.4	2.1	2.9
TW J4	1.4	2.1	2.9
DG A1	---	---	---
DG A2	3.0	---	---
DG A3	1.8	4.1	---
DG B1	---	---	---
DG B2	2.8	---	---
DG B3	1.8	3.2	---
DG C1	---	---	---

INTL. FALLS, MN (cont.)	10%	50%	80%
DG C2	2.2	---	---
DG C3	1.6	2.7	8.1
SS A1	2.2	---	---
SS A2	1.5	4.0	---
SS A3	3.1	---	---
SS A4	1.6	---	---
SS A5	2.2	---	---
SS A6	1.5	8.4	---
SS A7	2.8	---	---
SS A8	1.7	---	---
SS B1	2.9	---	---
SS B2	1.6	3.8	12.3
SS B3	5.7	---	---
SS B4	1.7	6.8	---
SS B5	3.3	---	---
SS B6	1.7	5.3	---
SS B7	---	---	---
SS B8	1.9	---	---
SS C1	---	---	---
SS C2	2.1	---	---
SS C3	---	---	---
SS C4	2.3	---	---
SS D1	2.1	---	---
SS D2	1.4	2.8	5.7
SS D3	2.1	---	---
SS D4	1.4	3.1	7.0
SS E1	5.0	---	---
SS E2	1.6	---	---
SS E3	3.3	---	---
SS E4	1.7	---	---

MINNEAPOLIS, MN	10%	50%	80%
WW A1	2.7	---	---
WW A2	2.3	---	---
WW A3	2.2	---	---
WW A4	2.1	---	---
WW A5	2.0	---	---
WW A6	1.9	---	---
WW B1	---	---	---
WW B2	1.7	3.3	8.3
WW B3	1.5	2.5	4.2
WW B4	1.3	1.9	2.7
WW B5	1.3	1.8	2.5
WW C1	1.4	2.2	3.4
WW C2	1.4	2.1	2.9
WW C3	1.1	1.5	2.0
WW C4	1.2	1.6	2.1
TW A1	2.7	---	---
TW A2	2.4	---	---
TW A3	2.3	---	---
TW A4	2.2	---	---
TW B1	2.7	---	---
TW B2	2.4	---	---
TW B3	2.3	---	---
TW B4	2.3	---	---
TW C1	2.6	---	---
TW C2	2.4	---	---
TW C3	2.5	---	---
TW C4	2.6	7.4	---
TW D1	---	---	---
TW D2	1.8	3.6	---
TW D3	1.7	3.3	9.1
TW D4	1.5	2.2	3.1
TW D5	1.4	2.0	2.7
TW E1	1.5	2.6	4.2
TW E2	1.5	2.3	3.3
TW E3	1.2	1.7	2.3
TW E4	1.2	1.7	2.3
TW F1	3.1	---	---
TW F2	2.7	---	---
TW F3	2.6	---	---
TW F4	2.5	---	---
TW G1	3.2	---	---
TW G2	2.8	---	---
TW G3	2.8	---	---
TW G4	2.9	---	---
TW H1	3.3	---	---
TW H2	3.1	---	---
TW H3	3.2	---	---
TW H4	3.6	16.0	---
TW I1	---	---	---
TW I2	2.0	3.9	---
TW I3	1.8	3.2	6.0
TW I4	1.5	2.3	3.2
TW I5	1.5	2.1	2.8
TW J1	1.6	2.6	4.0
TW J2	1.5	2.3	3.3
TW J3	1.2	1.7	2.3
TW J4	1.3	1.8	2.3
DG A1	---	---	---
DG A2	2.3	---	---
DG A3	1.7	2.7	---
DG B1	---	---	---
DG B2	2.2	---	---
DG B3	1.6	2.3	4.4
DG C1	4.2	---	---
DG C2	1.9	---	---
DG C3	1.5	2.0	3.1
SS A1	1.8	---	---
SS A2	1.3	2.3	3.6
SS A3	2.1	---	---
SS A4	1.4	2.7	4.7
SS A5	1.9	---	---
SS A6	1.4	2.5	4.1
SS A7	2.2	---	---
SS A8	1.4	3.1	7.5
SS B1	2.2	---	---
SS B2	1.4	2.5	3.7
SS B3	2.6	---	---
SS B4	1.5	2.7	4.4
SS B5	2.5	---	---
SS B6	1.5	2.7	4.2
SS B7	3.1	---	---
SS B8	1.6	3.1	5.7
SS C1	2.7	---	---
SS C2	1.7	3.1	6.0
SS C3	3.0	---	---
SS C4	1.8	3.5	8.3
SS D1	1.6	---	---
SS D2	1.2	1.9	2.8
SS D3	1.7	---	---
SS D4	1.2	2.0	3.0
SS E1	2.1	---	---
SS E2	1.4	2.5	4.1
SS E3	2.3	---	---
SS E4	1.5	2.7	4.7

Conservation Factor Tables

ROCHESTER, MN				ROCHESTER, MN				JACKSON, MS				MERIDIAN, MS			
	SOLAR SAVINGS				SOLAR SAVINGS				SOLAR SAVINGS				SOLAR SAVINGS		
	10%	50%	80%		10%	50%	80%		10%	50%	80%		10%	50%	80%
WW A1	2.8	---	---	DG C2	1.9	---	---	TW D4	.8	1.0	1.3	WW A1	1.1	1.7	2.3
WW A2	2.4	---	---	DG C3	1.5	2.1	3.1	TW D5	.8	1.0	1.3	WW A2	.9	1.3	1.7
WW A3	2.2	---	---	SS A1	1.8	---	---	TW E1	.8	1.0	1.3	WW A3	.9	1.2	1.6
WW A4	2.1	---	---	SS A2	1.3	2.3	3.5	TW E2	.8	1.0	1.3	WW A4	.8	1.2	1.5
WW A5	2.0	---	---	SS A3	2.1	---	---	TW E3	.7	.9	1.1	WW A5	.8	1.1	1.4
WW A6	2.0	---	---	SS A4	1.4	2.6	4.4	TW E4	.7	.9	1.2	WW A6	.8	1.1	1.4
WW B1	---	---	---	SS A5	1.9	---	---	TW F1	1.1	1.7	2.2	WW B1	1.1	1.6	2.1
WW B2	1.8	3.2	6.7	SS A6	1.4	2.5	3.9	TW F2	1.0	1.4	1.9	WW B2	.8	1.1	1.5
WW B3	1.5	2.5	4.0	SS A7	2.2	---	---	TW F3	1.0	1.4	1.8	WW B3	.7	1.0	1.3
WW B4	1.4	1.9	2.7	SS A8	1.4	3.0	6.1	TW F4	1.0	1.4	1.8	WW B4	.7	1.0	1.2
WW B5	1.3	1.8	2.5	SS B1	2.2	---	---	TW G1	1.1	1.6	2.2	WW B5	.7	1.0	1.2
WW C1	1.4	2.2	3.3	SS B2	1.5	2.5	3.6	TW G2	1.1	1.5	2.0	WW C1	.7	1.0	1.2
WW C2	1.4	2.1	2.9	SS B3	2.6	---	---	TW G3	1.1	1.5	2.0	WW C2	.7	1.0	1.3
WW C3	1.1	1.6	2.0	SS B4	1.5	2.7	4.2	TW G4	1.2	1.7	2.1	WW C3	.7	.9	1.1
WW C4	1.2	1.6	2.1	SS B5	2.5	---	---	TW H1	1.2	1.7	2.2	WW C4	.7	.9	1.2
TW A1	2.7	---	---	SS B6	1.5	2.6	4.0	TW H2	1.2	1.7	2.2	TW A1	1.1	1.7	2.3
TW A2	2.5	---	---	SS B7	3.1	---	---	TW H3	1.3	1.9	2.4	TW A2	1.0	1.4	1.9
TW A3	2.3	---	---	SS B8	1.6	3.0	5.2	TW H4	1.6	2.1	2.8	TW A3	1.0	1.3	1.8
TW A4	2.3	---	---	SS C1	2.9	---	---	TW I1	1.2	1.8	2.4	TW A4	.9	1.3	1.7
TW B1	2.7	---	---	SS C2	1.7	3.1	5.4	TW I2	.9	1.3	1.6	TW B1	1.1	1.7	2.2
TW B2	2.4	---	---	SS C3	3.1	---	---	TW I3	.9	1.2	1.5	TW B2	1.0	1.4	1.9
TW B3	2.4	---	---	SS C4	1.9	3.5	7.1	TW I4	.8	1.1	1.4	TW B3	1.0	1.4	1.9
TW B4	2.4	---	---	SS D1	1.6	---	---	TW I5	.8	1.1	1.4	TW B4	1.1	1.5	1.9
TW C1	2.6	---	---	SS D2	1.2	1.9	2.8	TW J1	.8	1.1	1.4	TW C1	1.1	1.6	2.2
TW C2	2.5	---	---	SS D3	1.7	---	---	TW J2	.8	1.1	1.4	TW C2	1.1	1.6	2.1
TW C3	2.5	12.1	---	SS D4	1.2	2.0	2.9	TW J3	.7	.9	1.2	TW C3	1.2	1.6	2.1
TW C4	2.6	7.1	---	SS E1	2.1	---	---	TW J4	.7	1.0	1.2	TW C4	1.3	1.8	2.4
TW D1	---	---	---	SS E2	1.4	2.4	3.9	DG A1	1.1	1.9	---	TW D1	1.2	1.8	2.3
TW D2	1.9	3.6	9.1	SS E3	2.3	---	---	DG A2	1.0	1.4	2.3	TW D2	.9	1.2	1.6
TW D3	1.8	3.3	7.1	SS E4	1.5	2.7	4.4	DG A3	.9	1.1	1.6	TW D3	.8	1.1	1.5
TW D4	1.5	2.2	3.1					DG B1	1.0	1.5	6.0	TW D4	.8	1.1	1.4
TW D5	1.4	2.0	2.7	JACKSON, MS				DG B2	1.0	1.3	1.8	TW D5	.8	1.1	1.4
TW E1	1.5	2.6	4.0		SOLAR SAVINGS			DG B3	.9	1.1	1.4	TW E1	.8	1.1	1.4
TW E2	1.5	2.3	3.3		10%	50%	80%	DG C1	.9	1.2	2.0	TW E2	.8	1.1	1.4
TW E3	1.2	1.7	2.3	WW A1	1.1	1.7	2.2	DG C2	.9	1.1	1.5	TW E3	.7	.9	1.2
TW E4	1.3	1.7	2.3	WW A2	.9	1.3	1.7	DG C3	.8	1.0	1.2	TW E4	.7	1.0	1.2
TW F1	3.3	---	---	WW A3	.8	1.2	1.5	SS A1	.8	1.2	1.6	TW F1	1.2	1.7	2.3
TW F2	2.8	---	---	WW A4	.8	1.1	1.5	SS A2	.7	1.0	1.3	TW F2	1.0	1.5	1.9
TW F3	2.7	---	---	WW A5	.8	1.1	1.4	SS A3	.8	1.3	1.8	TW F3	1.0	1.4	1.8
TW F4	2.6	---	---	WW A6	.8	1.1	1.4	SS A4	.7	1.0	1.3	TW F4	1.0	1.4	1.9
TW G1	3.4	---	---	WW B1	1.0	1.5	2.0	SS A5	.9	1.3	1.8	TW G1	1.2	1.7	2.2
TW G2	2.9	---	---	WW B2	.8	1.1	1.4	SS A6	.7	1.0	1.3	TW G2	1.1	1.6	2.0
TW G3	2.9	---	---	WW B3	.7	1.0	1.3	SS A7	.9	1.5	2.1	TW G3	1.1	1.6	2.0
TW G4	3.0	---	---	WW B4	.7	.9	1.2	SS A8	.7	1.0	1.3	TW G4	1.2	1.7	2.2
TW H1	3.4	---	---	WW B5	.7	1.0	1.2	SS B1	.9	1.4	1.8	TW H1	1.2	1.7	2.3
TW H2	3.2	---	---	WW C1	.7	.9	1.2	SS B2	.8	1.1	1.4	TW H2	1.3	1.8	2.3
TW H3	3.3	---	---	WW C2	.7	1.0	1.2	SS B3	.9	1.4	1.9	TW H3	1.4	1.9	2.5
TW H4	3.7	11.6	---	WW C3	.7	.9	1.1	SS B4	.8	1.1	1.4	TW H4	1.6	2.2	2.8
TW I1	---	---	---	WW C4	.7	.9	1.1	SS B5	1.0	1.6	2.1	TW I1	1.3	1.9	2.5
TW I2	2.0	3.8	9.7	TW A1	1.1	1.7	2.2	SS B6	.8	1.1	1.4	TW I2	.9	1.3	1.7
TW I3	1.9	3.2	5.6	TW A2	1.0	1.4	1.8	SS B7	1.1	1.7	2.3	TW I3	.9	1.2	1.6
TW I4	1.6	2.3	3.2	TW A3	.9	1.3	1.7	SS B8	.8	1.1	1.5	TW I4	.8	1.1	1.4
TW I5	1.5	2.1	2.8	TW A4	.9	1.3	1.7	SS C1	.9	1.3	1.7	TW I5	.8	1.1	1.4
TW J1	1.6	2.6	3.9	TW B1	1.1	1.6	2.1	SS C2	.8	1.1	1.4	TW J1	.8	1.1	1.4
TW J2	1.6	2.3	3.3	TW B2	1.0	1.4	1.8	SS C3	1.1	1.6	2.1	TW J2	.8	1.1	1.4
TW J3	1.3	1.7	2.3	TW B3	1.0	1.4	1.8	SS C4	.9	1.2	1.6	TW J3	.7	1.0	1.2
TW J4	1.3	1.8	2.3	TW B4	1.0	1.4	1.9	SS D1	.7	1.0	1.4	TW J4	.8	1.0	1.3
DG A1	---	---	---	TW C1	1.1	1.6	2.1	SS D2	.6	.8	1.1	DG A1	1.1	2.1	---
DG A2	2.4	---	---	TW C2	1.1	1.5	2.0	SS D3	.8	1.1	1.5	DG A2	1.0	1.5	2.4
DG A3	1.7	2.7	---	TW C3	1.1	1.6	2.1	SS D4	.6	.9	1.1	DG A3	.9	1.2	1.6
DG B1	---	---	---	TW C4	1.2	1.8	2.3	SS E1	.8	1.1	1.5	DG B1	1.1	1.5	---
DG B2	2.3	---	---	TW D1	1.2	1.7	2.3	SS E2	.7	.9	1.2	DG B2	1.0	1.3	1.9
DG B3	1.6	2.3	4.3	TW D2	.9	1.2	1.6	SS E3	.9	1.4	1.8	DG B3	.9	1.1	1.4
DG C1	6.0	---	---	TW D3	.8	1.1	1.4	SS E4	.7	1.0	1.3	DG C1	1.0	1.3	2.1

338 / APPENDIX 17

MERIDIAN, MS

	SOLAR SAVINGS		
	10%	50%	80%
DG C2	.9	1.2	1.5
DG C3	.9	1.0	1.3
SS A1	.8	1.2	1.7
SS A2	.7	1.0	1.3
SS A3	.9	1.3	1.8
SS A4	.7	1.0	1.3
SS A5	.9	1.4	1.9
SS A6	.7	1.0	1.3
SS A7	1.0	1.6	2.1
SS A8	.7	1.0	1.4
SS B1	.9	1.4	1.8
SS B2	.8	1.1	1.4
SS B3	1.0	1.5	2.0
SS B4	.8	1.1	1.5
SS B5	1.0	1.6	2.2
SS B6	.8	1.1	1.4
SS B7	1.1	1.8	2.4
SS B8	.8	1.1	1.5
SS C1	.9	1.3	1.7
SS C2	.8	1.1	1.5
SS C3	1.1	1.6	2.1
SS C4	.9	1.3	1.6
SS D1	.7	1.1	1.4
SS D2	.6	.9	1.1
SS D3	.8	1.2	1.6
SS D4	.6	.9	1.2
SS E1	.8	1.2	1.6
SS E2	.7	1.0	1.2
SS E3	.9	1.4	1.9
SS E4	.7	1.0	1.4

COLUMBIA, MO

	SOLAR SAVINGS		
	10%	50%	80%
WW A1	1.7	4.1	8.8
WW A2	1.4	2.4	3.5
WW A3	1.4	2.1	3.0
WW A4	1.3	2.0	2.8
WW A5	1.3	1.9	2.6
WW A6	1.3	1.8	2.5
WW B1	2.0	---	---
WW B2	1.2	1.8	2.4
WW B3	1.1	1.5	2.1
WW B4	1.0	1.4	1.8
WW B5	1.0	1.4	1.8
WW C1	1.0	1.5	1.9
WW C2	1.1	1.5	1.9
WW C3	.9	1.2	1.6
WW C4	1.0	1.3	1.6
TW A1	1.7	3.7	6.6
TW A2	1.5	2.6	3.8
TW A3	1.5	2.3	3.3
TW A4	1.5	2.2	3.1
TW B1	1.7	3.2	5.0
TW B2	1.6	2.5	3.6
TW B3	1.5	2.4	3.4
TW B4	1.6	2.5	3.4
TW C1	1.7	2.9	4.2
TW C2	1.7	2.7	3.7
TW C3	1.7	2.7	3.8
TW C4	1.9	3.0	4.1
TW D1	2.2	---	---
TW D2	1.3	2.0	2.6
TW D3	1.2	1.8	2.4

COLUMBIA, MO (continued)

	SOLAR SAVINGS		
	10%	50%	80%
TW D4	1.1	1.6	2.0
TW D5	1.1	1.5	2.0
TW E1	1.1	1.6	2.2
TW E2	1.1	1.6	2.1
TW E3	1.0	1.3	1.7
TW E4	1.0	1.3	1.7
TW F1	1.8	3.6	6.1
TW F2	1.6	2.7	3.9
TW F3	1.6	2.5	3.5
TW F4	1.6	2.5	3.4
TW G1	1.9	3.3	5.1
TW G2	1.7	2.8	3.9
TW G3	1.8	2.7	3.8
TW G4	1.9	2.9	3.9
TW H1	2.0	3.2	4.7
TW H2	2.0	3.1	4.3
TW H3	2.1	3.2	4.4
TW H4	2.4	3.6	4.9
TW I1	2.4	---	---
TW I2	1.4	2.1	2.8
TW I3	1.3	1.9	2.5
TW I4	1.2	1.6	2.1
TW I5	1.2	1.6	2.0
TW J1	1.2	1.7	2.2
TW J2	1.2	1.6	2.1
TW J3	1.0	1.3	1.7
TW J4	1.0	1.4	1.8
DG A1	2.0	---	---
DG A2	1.6	3.5	---
DG A3	1.3	1.8	3.0
DG B1	2.0	---	---
DG B2	1.5	2.5	---
DG B3	1.3	1.6	2.3
DG C1	1.6	---	---
DG C2	1.4	2.0	4.4
DG C3	1.2	1.5	1.9
SS A1	1.2	2.2	3.3
SS A2	1.0	1.5	2.1
SS A3	1.3	2.8	4.7
SS A4	1.0	1.6	2.2
SS A5	1.3	2.7	4.3
SS A6	1.0	1.6	2.1
SS A7	1.4	4.0	11.7
SS A8	1.0	1.7	2.3
SS B1	1.4	2.6	3.8
SS B2	1.1	1.7	2.2
SS B3	1.5	3.0	5.0
SS B4	1.1	1.7	2.3
SS B5	1.6	3.3	5.5
SS B6	1.1	1.7	2.3
SS B7	1.8	4.6	14.6
SS B8	1.2	1.8	2.5
SS C1	1.5	2.5	3.7
SS C2	1.2	1.8	2.4
SS C3	1.8	3.1	4.8
SS C4	1.3	2.0	2.7
SS D1	1.1	1.9	2.7
SS D2	.9	1.3	1.8
SS D3	1.2	2.1	3.2
SS D4	.9	1.4	1.8
SS E1	.3	2.2	3.4
SS E2	1.0	1.5	2.0
SS E3	1.4	2.8	4.4
SS E4	1.1	1.6	2.2

KANSAS CITY, MO

	SOLAR SAVINGS		
	10%	50%	80%
WW A1	1.7	3.5	5.8
WW A2	1.4	2.2	3.1
WW A3	1.3	2.0	2.7
WW A4	1.3	1.9	2.6
WW A5	1.2	1.8	2.4
WW A6	1.2	1.7	2.3
WW B1	1.9	5.1	---
WW B2	1.2	1.7	2.3
WW B3	1.1	1.5	2.0
WW B4	1.0	1.4	1.8
WW B5	1.0	1.4	1.7
WW C1	1.0	1.4	1.9
WW C2	1.0	1.4	1.8
WW C3	.9	1.2	1.5
WW C4	.9	1.2	1.6
TW A1	1.7	3.3	5.2
TW A2	1.5	2.4	3.4
TW A3	1.4	2.2	3.0
TW A4	1.4	2.1	2.9
TW B1	1.7	3.0	4.3
TW B2	1.5	2.4	3.3
TW B3	1.5	2.3	3.1
TW B4	1.6	2.4	3.2
TW C1	1.7	2.7	3.8
TW C2	1.6	2.5	3.4
TW C3	1.7	2.6	3.5
TW C4	1.8	2.8	3.8
TW D1	2.1	4.7	10.8
TW D2	1.3	1.9	2.5
TW D3	1.2	1.7	2.3
TW D4	1.1	1.5	2.0
TW D5	1.1	1.5	1.9
TW E1	1.1	1.6	2.1
TW E2	1.1	1.5	2.0
TW E3	.9	1.3	1.6
TW E4	1.0	1.3	1.7
TW F1	1.8	3.3	4.9
TW F2	1.6	2.5	3.5
TW F3	1.6	2.4	3.2
TW F4	1.6	2.3	3.1
TW G1	1.8	3.1	4.4
TW G2	1.7	2.6	3.6
TW G3	1.7	2.6	3.5
TW G4	1.9	2.7	3.7
TW H1	1.9	3.0	4.2
TW H2	1.9	2.9	3.9
TW H3	2.1	3.1	4.1
TW H4	2.4	3.5	4.6
TW I1	2.3	5.7	17.4
TW I2	1.4	2.0	2.6
TW I3	1.3	1.8	2.4
TW I4	1.2	1.6	2.1
TW I5	1.1	1.5	2.0
TW J1	1.2	1.6	2.1
TW J2	1.2	1.6	2.1
TW J3	1.0	1.3	1.7
TW J4	1.0	1.3	1.7
DG A1	1.9	---	---
DG A2	1.5	3.1	---
DG A3	1.3	1.7	2.8
DG B1	1.9	---	---
DG B2	1.5	2.3	---
DG B3	1.3	1.6	2.2
DG C1	1.6	6.2	---

KANSAS CITY, MO (continued)

	SOLAR SAVINGS		
	10%	50%	80%
DG C2	1.4	1.9	3.5
DG C3	1.2	1.5	1.9
SS A1	1.2	2.1	3.1
SS A2	1.0	1.5	2.0
SS A3	1.3	2.6	4.1
SS A4	1.0	1.6	2.1
SS A5	1.3	2.5	3.8
SS A6	1.0	1.5	2.1
SS A7	1.4	3.6	7.0
SS A8	1.0	1.6	2.2
SS B1	1.4	2.4	3.5
SS B2	1.1	1.6	2.2
SS B3	1.5	2.8	4.3
SS B4	1.1	1.7	2.3
SS B5	1.6	3.1	4.7
SS B6	1.1	1.7	2.2
SS B7	1.8	4.1	7.9
SS B8	1.2	1.8	2.4
SS C1	1.5	2.3	3.3
SS C2	1.2	1.7	2.3
SS C3	1.7	2.9	4.2
SS C4	1.3	1.9	2.5
SS D1	1.1	1.8	2.5
SS D2	.9	1.3	1.7
SS D3	1.2	2.0	2.9
SS D4	.9	1.3	1.8
SS E1	1.2	2.1	3.1
SS E2	1.0	1.5	2.0
SS E3	1.4	2.6	3.9
SS E4	1.1	1.6	2.1

SAINT LOUIS, MO

	SOLAR SAVINGS		
	10%	50%	80%
WW A1	1.6	3.3	5.5
WW A2	1.4	2.2	3.1
WW A3	1.3	1.9	2.7
WW A4	1.2	1.8	2.5
WW A5	1.2	1.8	2.4
WW A6	1.2	1.7	2.3
WW B1	1.8	4.4	---
WW B2	1.2	1.7	2.2
WW B3	1.0	1.5	1.9
WW B4	1.0	1.3	1.7
WW B5	1.0	1.3	1.7
WW C1	1.0	1.4	1.8
WW C2	1.0	1.4	1.8
WW C3	.9	1.2	1.5
WW C4	.9	1.2	1.6
TW A1	1.6	3.2	5.0
TW A2	1.4	2.4	3.3
TW A3	1.4	2.2	3.0
TW A4	1.4	2.1	2.8
TW B1	1.6	2.9	4.2
TW B2	1.5	2.3	3.2
TW B3	1.5	2.2	3.1
TW B4	1.5	2.3	3.1
TW C1	1.6	2.7	3.7
TW C2	1.6	2.5	3.4
TW C3	1.6	2.5	3.4
TW C4	1.8	2.8	3.8
TW D1	2.0	4.3	10.0
TW D2	1.3	1.8	2.5
TW D3	1.2	1.7	2.3

SAINT LOUIS, MO				SPRINGFIELD, MO				SPRINGFIELD, MO				BILLINGS, MT			
	SOLAR SAVINGS				SOLAR SAVINGS				SOLAR SAVINGS				SOLAR SAVINGS		
	10%	50%	80%		10%	50%	80%		10%	50%	80%		10%	50%	80%
WW A1	1.6	3.3	5.5	WW A1	1.5	2.8	4.0	DG C2	1.3	1.7	2.6	TW D4	1.1	1.6	2.2
WW A2	1.4	2.2	3.1	WW A2	1.3	1.9	2.6	DG C3	1.1	1.3	1.7	TW D5	1.1	1.5	2.1
WW A3	1.3	1.9	2.7	WW A3	1.2	1.8	2.3	SS A1	1.1	1.8	2.5	TW E1	1.1	1.7	2.3
WW A4	1.2	1.8	2.5	WW A4	1.2	1.7	2.2	SS A2	.9	1.4	1.8	TW E2	1.1	1.6	2.2
WW A5	1.2	1.8	2.4	WW A5	1.1	1.6	2.1	SS A3	1.2	2.1	3.0	TW E3	1.0	1.3	1.8
WW A6	1.2	1.7	2.3	WW A6	1.1	1.6	2.0	SS A4	.9	1.4	1.9	TW E4	1.0	1.4	1.8
WW B1	1.8	4.4	---	WW B1	1.6	2.9	4.5	SS A5	1.2	2.1	3.0	TW F1	1.9	4.5	---
WW B2	1.2	1.7	2.2	WW B2	1.1	1.6	2.0	SS A6	.9	1.4	1.8	TW F2	1.7	3.0	5.3
WW B3	1.0	1.5	1.9	WW B3	1.0	1.4	1.8	SS A7	1.3	2.6	4.0	TW F3	1.6	2.7	4.3
WW B4	1.0	1.3	1.7	WW B4	1.0	1.3	1.6	SS A8	.9	1.4	1.9	TW F4	1.6	2.6	3.9
WW B5	1.0	1.3	1.7	WW B5	1.0	1.3	1.6	SS B1	1.3	2.1	2.9	TW G1	1.9	3.8	8.8
WW C1	1.0	1.4	1.8	WW C1	.9	1.3	1.7	SS B2	1.0	1.5	2.0	TW G2	1.8	3.0	4.9
WW C2	1.0	1.4	1.8	WW C2	1.0	1.3	1.7	SS B3	1.3	2.3	3.3	TW G3	1.8	2.9	4.4
WW C3	.9	1.2	1.5	WW C3	.8	1.1	1.4	SS B4	1.0	1.5	2.0	TW G4	1.9	3.0	4.5
WW C4	.9	1.2	1.6	WW C4	.9	1.2	1.5	SS B5	1.4	2.5	3.6	TW H1	2.0	3.6	6.2
TW A1	1.6	3.2	5.0	TW A1	1.5	2.7	3.8	SS B6	1.0	1.5	2.0	TW H2	2.0	3.3	5.1
TW A2	1.4	2.4	3.3	TW A2	1.4	2.1	2.9	SS B7	1.5	3.0	4.5	TW H3	2.1	3.4	5.1
TW A3	1.4	2.2	3.0	TW A3	1.3	1.9	2.6	SS B8	1.1	1.6	2.1	TW H4	2.4	3.8	5.6
TW A4	1.4	2.1	2.8	TW A4	1.3	1.9	2.5	SS C1	1.3	2.0	2.7	TW I1	2.5	---	---
TW B1	1.6	2.9	4.2	TW B1	1.5	2.5	3.5	SS C2	1.1	1.5	2.0	TW I2	1.4	2.1	3.0
TW B2	1.5	2.3	3.2	TW B2	1.4	2.1	2.8	SS C3	1.6	2.5	3.4	TW I3	1.3	2.0	2.8
TW B3	1.5	2.2	3.1	TW B3	1.4	2.0	2.7	SS C4	1.2	1.7	2.3	TW I4	1.2	1.7	2.2
TW B4	1.5	2.3	3.1	TW B4	1.4	2.1	2.8	SS D1	1.0	1.5	2.1	TW I5	1.2	1.6	2.1
TW C1	1.6	2.7	3.7	TW C1	1.5	2.4	3.2	SS D2	.8	1.2	1.5	TW J1	1.2	1.7	2.4
TW C2	1.6	2.5	3.4	TW C2	1.5	2.2	3.0	SS D3	1.1	1.8	2.4	TW J2	1.2	1.7	2.3
TW C3	1.6	2.5	3.4	TW C3	1.6	2.3	3.1	SS D4	.8	1.2	1.6	TW J3	1.0	1.4	1.8
TW C4	1.8	2.8	3.8	TW C4	1.7	2.5	3.4	SS E1	1.1	1.8	2.5	TW J4	1.0	1.4	1.8
TW D1	2.0	4.3	10.0	TW D1	1.8	3.1	4.6	SS E2	.9	1.3	1.7	DG A1	2.1	---	---
TW D2	1.3	1.8	2.5	TW D2	1.2	1.7	2.2	SS E3	1.3	2.2	3.0	DG A2	1.6	4.1	---
TW D3	1.2	1.7	2.3	TW D3	1.1	1.6	2.1	SS E4	1.0	1.4	1.9	DG A3	1.3	1.8	3.3
DG B1	1.8	---	---	TW D4	1.0	1.4	1.8					DG B1	2.0	---	---
DG B2	1.4	2.2	---	TW D5	1.0	1.4	1.8	BILLINGS, MT				DG B2	1.5	2.7	---
DG B3	1.2	1.6	2.2	TW E1	1.0	1.4	1.9		SOLAR SAVINGS			DG B3	1.3	1.7	2.5
DG C1	1.5	3.9	---	TW E2	1.0	1.4	1.8		10%	50%	80%	DG C1	1.6	---	---
DG C2	1.3	1.8	3.3	TW E3	.9	1.2	1.5	WW A1	1.8	6.9	---	DG C2	1.4	2.1	---
DG C3	1.1	1.4	1.8	TW E4	.9	1.2	1.6	WW A2	1.5	2.6	4.5	DG C3	1.2	1.5	2.1
SS A1	1.2	2.1	3.0	TW F1	1.6	2.7	3.7	WW A3	1.4	2.3	3.6	SS A1	1.3	2.6	4.8
SS A2	1.0	1.5	2.0	TW F2	1.4	2.2	3.0	WW A4	1.3	2.1	3.2	SS A2	1.0	1.7	2.4
SS A3	1.3	2.5	3.8	TW F3	1.4	2.1	2.7	WW A5	1.3	2.0	3.0	SS A3	1.4	4.4	---
SS A4	1.0	1.5	2.1	TW F4	1.4	2.1	2.7	WW A6	1.3	2.0	2.9	SS A4	1.1	1.8	2.6
SS A5	1.2	2.4	3.6	TW G1	1.6	2.6	3.5	WW B1	2.0	---	---	SS A5	1.4	3.4	13.4
SS A6	1.0	1.5	2.0	TW G2	1.5	2.3	3.0	WW B2	1.3	1.9	2.6	SS A6	1.0	1.7	2.5
SS A7	1.4	3.3	6.2	TW G3	1.6	2.3	3.0	WW B3	1.1	1.6	2.2	SS A7	1.6	---	---
SS A8	1.0	1.6	2.2	TW G4	1.7	2.4	3.2	WW B4	1.0	1.4	1.9	SS A8	1.1	1.9	2.8
SS B1	1.4	2.3	3.4	TW H1	1.7	2.6	3.5	WW B5	1.0	1.4	1.9	SS B1	1.5	3.1	6.1
SS B2	1.1	1.6	2.1	TW H2	1.8	2.5	3.4	WW C1	1.1	1.5	2.1	SS B2	1.1	1.8	2.5
SS B3	1.4	2.7	4.1	TW H3	1.9	2.7	3.6	WW C2	1.1	1.5	2.0	SS B3	1.6	4.5	---
SS B4	1.1	1.6	2.2	TW H4	2.2	3.1	4.1	WW C3	.9	1.2	1.6	SS B4	1.2	1.9	2.7
SS B5	1.5	2.9	4.5	TW I1	1.9	3.4	5.1	WW C4	1.0	1.3	1.7	SS B5	1.7	4.6	---
SS B6	1.1	1.6	2.2	TW I2	1.3	1.8	2.3	TW A1	1.8	4.7	---	SS B6	1.2	1.8	2.6
SS B7	1.7	3.7	7.0	TW I3	1.2	1.7	2.2	TW A2	1.6	2.9	4.9	SS B7	1.9	---	---
SS B8	1.1	1.7	2.3	TW I4	1.1	1.5	1.9	TW A3	1.5	2.5	4.0	SS B8	1.2	2.0	2.9
SS C1	1.4	2.3	3.2	TW I5	1.1	1.4	1.8	TW A4	1.5	2.4	3.6	SS C1	1.5	2.8	5.2
SS C2	1.2	1.7	2.2	TW J1	1.1	1.5	1.9	TW B1	1.8	3.7	8.6	SS C2	1.2	1.9	2.6
SS C3	1.7	2.8	4.0	TW J2	1.1	1.5	1.9	TW B2	1.6	2.7	4.3	SS C3	1.8	3.6	7.7
SS C4	1.3	1.9	2.5	TW J3	.9	1.2	1.6	TW B3	1.6	2.6	3.9	SS C4	1.4	2.1	2.9
SS D1	1.0	1.7	2.4	TW J4	1.0	1.3	1.6	TW B4	1.6	2.6	3.9	SS D1	1.1	2.2	3.7
SS D2	.9	1.3	1.7	DG A1	1.7	---	---	TW C1	1.8	3.2	5.2	SS D2	.9	1.4	2.0
SS D3	1.1	2.0	2.8	DG A2	1.4	2.4	---	TW C2	1.7	2.8	4.3	SS D3	1.2	2.5	4.8
SS D4	.9	1.3	1.7	DG A3	1.2	1.6	2.4	TW C3	1.8	2.9	4.3	SS D4	1.0	1.5	2.1
SS E1	1.2	2.0	3.0	DG B1	1.6	---	---	TW C4	1.9	3.1	4.6	SS E1	1.3	2.8	6.8
SS E2	1.0	1.4	1.9	DG B2	1.4	1.9	5.4	TW D1	2.2	---	---	SS E2	1.0	1.6	2.3
SS E3	1.3	2.5	3.7	DG B3	1.2	1.5	2.0	TW D2	1.3	2.0	2.9	SS E3	1.5	3.6	15.5
SS E4	1.0	1.5	2.1	DG C1	1.4	2.4	---	TW D3	1.2	1.9	2.7	SS E4	1.1	1.8	2.5

CUT BANK, MT				CUT BANK, MT				DILLON, MT				GLASGOW, MT			
	SOLAR SAVINGS				SOLAR SAVINGS				SOLAR SAVINGS				SOLAR SAVINGS		
	10%	50%	80%		10%	50%	80%		10%	50%	80%		10%	50%	80%
WW A1	2.0	---	---	DG C2	1.5	2.7	---	TW D4	1.1	1.5	2.1	WW A1	2.4	---	---
WW A2	1.6	---	---	DG C3	1.3	1.7	2.5	TW D5	1.1	1.5	2.0	WW A2	2.0	---	---
WW A3	1.5	3.7	---	SS A1	1.4	5.8	---	TW E1	1.1	1.6	2.2	WW A3	1.9	---	---
WW A4	1.5	3.1	---	SS A2	1.1	1.9	3.0	TW E2	1.1	1.6	2.1	WW A4	1.9	---	---
WW A5	1.5	2.8	10.7	SS A3	1.5	---	---	TW E3	1.0	1.3	1.7	WW A5	1.8	---	---
WW A6	1.4	2.7	7.3	SS A4	1.1	2.1	3.6	TW E4	1.0	1.3	1.7	WW A6	1.8	---	---
WW B1	2.6	---	---	SS A5	1.5	---	---	TW F1	1.8	3.8	---	WW B1	---	---	---
WW B2	1.4	2.3	3.8	SS A6	1.1	2.0	3.3	TW F2	1.6	2.7	4.7	WW B2	1.6	3.3	7.5
WW B3	1.2	1.9	3.0	SS A7	1.7	---	---	TW F3	1.6	2.5	3.9	WW B3	1.4	2.5	4.4
WW B4	1.1	1.6	2.3	SS A8	1.2	2.3	4.4	TW F4	1.6	2.5	3.7	WW B4	1.3	1.9	2.7
WW B5	1.1	1.6	2.2	SS B1	1.6	---	---	TW G1	1.9	3.4	7.2	WW B5	1.3	1.8	2.5
WW C1	1.1	1.8	2.6	SS B2	1.2	2.0	3.1	TW G2	1.7	2.8	4.5	WW C1	1.3	2.2	3.5
WW C2	1.2	1.7	2.4	SS B3	1.8	---	---	TW G3	1.8	2.7	4.1	WW C2	1.3	2.0	3.0
WW C3	1.0	1.4	1.8	SS B4	1.3	2.2	3.5	TW G4	1.9	2.9	4.2	WW C3	1.1	1.5	2.0
WW C4	1.0	1.4	1.9	SS B5	1.8	---	---	TW H1	2.0	3.3	5.6	WW C4	1.1	1.6	2.1
TW A1	2.0	---	---	SS B6	1.2	2.1	3.4	TW H2	2.0	3.1	4.7	TW A1	2.4	---	---
TW A2	1.7	8.9	---	SS B7	2.1	---	---	TW H3	2.1	3.2	4.8	TW A2	2.1	---	---
TW A3	1.7	4.0	---	SS B8	1.3	2.4	4.1	TW H4	2.4	3.6	5.3	TW A3	2.0	---	---
TW A4	1.6	3.3	---	SS C1	1.8	---	---	TW I1	2.4	---	---	TW A4	2.0	---	---
TW B1	1.9	---	---	SS C2	1.4	2.2	3.6	TW I2	1.4	2.0	2.9	TW B1	2.4	---	---
TW B2	1.8	4.2	---	SS C3	2.0	---	---	TW I3	1.3	1.9	2.6	TW B2	2.1	---	---
TW B3	1.7	3.6	---	SS C4	1.5	2.5	4.2	TW I4	1.2	1.6	2.2	TW B3	2.1	---	---
TW B4	1.8	3.4	9.7	SS D1	1.2	3.8	---	TW I5	1.2	1.6	2.1	TW B4	2.1	---	---
TW C1	1.9	5.5	---	SS D2	1.0	1.6	2.4	TW J1	1.2	1.7	2.3	TW C1	2.3	---	---
TW C2	1.9	3.8	29.4	SS D3	1.3	---	---	TW J2	1.2	1.6	2.2	TW C2	2.2	---	---
TW C3	1.9	3.8	10.6	SS D4	1.0	1.7	2.6	TW J3	1.0	1.3	1.8	TW C3	2.3	---	---
TW C4	2.0	3.9	8.9	SS E1	1.5	---	---	TW J4	1.0	1.4	1.8	TW C4	2.4	7.7	---
TW D1	2.8	---	---	SS E2	1.1	1.9	3.1	DG A1	2.0	---	---	TW D1	---	---	---
TW D2	1.5	2.5	4.3	SS E3	1.6	---	---	DG A2	1.6	3.4	---	TW D2	1.7	3.6	9.7
TW D3	1.4	2.3	3.9	SS E4	1.2	2.1	3.5	DG A3	1.3	1.8	3.1	TW D3	1.6	3.4	8.2
TW D4	1.2	1.8	2.6					DG B1	2.0	---	---	TW D4	1.4	2.2	3.1
TW D5	1.2	1.7	2.4	DILLON, MT				DG B2	1.5	2.5	---	TW D5	1.3	2.0	2.7
TW E1	1.2	2.0	3.1		SOLAR SAVINGS			DG B3	1.3	1.6	2.4	TW E1	1.4	2.6	4.3
TW E2	1.2	1.8	2.7		10%	50%	80%	DG C1	1.6	---	---	TW E2	1.4	2.3	3.4
TW E3	1.0	1.5	2.0	WW A1	1.7	4.5	---	DG C2	1.4	2.0	5.3	TW E3	1.2	1.7	2.3
TW E4	1.1	1.5	2.0	WW A2	1.4	2.4	4.1	DG C3	1.2	1.5	2.0	TW E4	1.2	1.7	2.3
TW F1	2.1	---	---	WW A3	1.4	2.1	3.3	SS A1	1.2	2.4	4.3	TW F1	2.7	---	---
TW F2	1.9	---	---	WW A4	1.3	2.0	3.0	SS A2	1.0	1.6	2.3	TW F2	2.4	---	---
TW F3	1.8	4.6	---	WW A5	1.3	1.9	2.8	SS A3	1.4	3.4	---	TW F3	2.3	---	---
TW F4	1.8	3.8	---	WW A6	1.2	1.8	2.7	SS A4	1.0	1.7	2.5	TW F4	2.3	---	---
TW G1	2.1	---	---	WW B1	1.9	---	---	SS A5	1.3	3.0	9.7	TW G1	2.7	---	---
TW G2	2.0	5.7	---	WW B2	1.2	1.8	2.5	SS A6	1.0	1.6	2.3	TW G2	2.5	---	---
TW G3	2.0	4.3	---	WW B3	1.1	1.5	2.1	SS A7	1.5	---	---	TW G3	2.5	---	---
TW G4	2.2	4.2	13.0	WW B4	1.0	1.4	1.9	SS A8	1.0	1.7	2.6	TW G4	2.6	---	---
TW H1	2.3	---	---	WW B5	1.0	1.4	1.8	SS B1	1.4	2.8	5.3	TW H1	2.9	---	---
TW H2	2.3	5.0	---	WW C1	1.0	1.5	2.0	SS B2	1.1	1.7	2.4	TW H2	2.8	---	---
TW H3	2.4	4.7	18.0	WW C2	1.1	1.5	2.0	SS B3	1.5	3.6	---	TW H3	2.9	---	---
TW H4	2.7	5.0	11.8	WW C3	.9	1.2	1.6	SS B4	1.1	1.8	2.6	TW H4	3.3	---	---
TW I1	3.4	---	---	WW C4	1.0	1.3	1.6	SS B5	1.6	3.8	---	TW I1	---	---	---
TW I2	1.6	2.6	4.5	TW A1	1.7	3.9	---	SS B6	1.1	1.7	2.5	TW I2	1.8	3.9	10.2
TW I3	1.5	2.4	3.8	TW A2	1.5	2.6	4.4	SS B7	1.8	---	---	TW I3	1.7	3.2	6.3
TW I4	1.3	1.9	2.7	TW A3	1.5	2.3	3.7	SS B8	1.2	1.9	2.7	TW I4	1.5	2.2	3.2
TW I5	1.2	1.8	2.4	TW A4	1.4	2.2	3.3	SS C1	1.5	2.6	4.6	TW I5	1.4	2.0	2.8
TW J1	1.3	2.0	3.1	TW B1	1.7	3.3	6.9	SS C2	1.2	1.8	2.5	TW J1	1.5	2.6	4.2
TW J2	1.3	1.9	2.7	TW B2	1.6	2.5	4.0	SS C3	1.8	3.2	6.4	TW J2	1.4	2.3	3.4
TW J3	1.1	1.5	2.0	TW B3	1.5	2.4	3.7	SS C4	1.3	2.0	2.8	TW J3	1.2	1.7	2.3
TW J4	1.1	1.5	2.1	TW B4	1.6	2.5	3.6	SS D1	1.1	2.0	3.4	TW J4	1.2	1.7	2.4
DG A1	2.7	---	---	TW C1	1.7	2.9	4.8	SS D2	.9	1.3	1.9	DG A1	---	---	---
DG A2	1.7	---	---	TW C2	1.7	2.6	4.0	SS D3	1.2	2.3	4.3	DG A2	2.1	---	---
DG A3	1.4	2.1	5.8	TW C3	1.7	2.7	4.0	SS D4	.9	1.4	2.0	DG A3	1.6	2.6	---
DG B1	2.5	---	---	TW C4	1.9	2.9	4.3	SS E1	1.3	2.5	5.6	DG B1	---	---	---
DG B2	1.7	5.5	---	TW D1	2.1	---	---	SS E2	1.0	1.5	2.2	DG B2	2.0	---	---
DG B3	1.4	1.9	3.1	TW D2	1.3	1.9	2.8	SS E3	1.4	3.1	10.9	DG B3	1.5	2.3	4.4
DG C1	1.9	---	---	TW D3	1.2	1.8	2.6	SS E4	1.1	1.7	2.4	DG C1	2.7	---	---

Conservation Factor Tables / 341

GLASGOW, MT

	SOLAR SAVINGS		
	10%	50%	80%
DG C2	1.7	7.8	---
DG C3	1.4	2.0	3.2
SS A1	1.7	---	---
SS A2	1.3	2.4	3.9
SS A3	2.0	---	---
SS A4	1.3	2.8	5.6
SS A5	1.8	---	---
SS A6	1.3	2.6	4.5
SS A7	2.1	---	---
SS A8	1.4	3.4	10.4
SS B1	2.0	---	---
SS B2	1.4	2.5	3.9
SS B3	2.3	---	---
SS B4	1.4	2.8	4.8
SS B5	2.3	---	---
SS B6	1.4	2.7	4.5
SS B7	2.8	---	---
SS B8	1.5	3.3	6.8
SS C1	2.3	---	---
SS C2	1.6	3.1	6.1
SS C3	2.6	---	---
SS C4	1.7	3.5	8.0
SS D1	1.5	---	---
SS D2	1.1	2.0	3.0
SS D3	1.6	---	---
SS D4	1.2	2.1	3.2
SS E1	1.9	---	---
SS E2	1.3	2.6	4.5
SS E3	2.1	---	---
SS E4	1.4	2.8	5.3

GREAT FALLS, MT

	SOLAR SAVINGS		
	10%	50%	80%
WW A1	1.8	---	---
WW A2	1.5	3.7	---
WW A3	1.5	2.9	---
WW A4	1.4	2.6	7.0
WW A5	1.4	2.4	5.3
WW A6	1.3	2.3	4.7
WW B1	2.2	---	---
WW B2	1.3	2.1	3.3
WW B3	1.1	1.8	2.7
WW B4	1.1	1.5	2.2
WW B5	1.1	1.5	2.1
WW C1	1.1	1.7	2.4
WW C2	1.1	1.6	2.3
WW C3	1.0	1.3	1.7
WW C4	1.0	1.4	1.8
TW A1	1.8	---	---
TW A2	1.6	4.0	---
TW A3	1.6	3.2	---
TW A4	1.6	2.9	6.7
TW B1	1.8	---	---
TW B2	1.7	3.4	---
TW B3	1.6	3.1	7.7
TW B4	1.7	3.1	6.3
TW C1	1.8	4.1	---
TW C2	1.8	3.3	7.9
TW C3	1.8	3.3	6.9
TW C4	2.0	3.6	6.7
TW D1	2.5	---	---
TW D2	1.4	2.3	3.7
TW D3	1.3	2.1	3.4

GREAT FALLS, MT

	SOLAR SAVINGS		
	10%	50%	80%
TW D4	1.2	1.7	2.4
TW D5	1.2	1.6	2.3
TW E1	1.2	1.9	2.8
TW E2	1.2	1.8	2.5
TW E3	1.0	1.4	1.9
TW E4	1.0	1.4	1.9
TW F1	2.0	---	---
TW F2	1.8	4.5	---
TW F3	1.7	3.5	---
TW F4	1.7	3.2	7.9
TW G1	2.0	---	---
TW G2	1.9	4.0	---
TW G3	1.9	3.6	10.7
TW G4	2.0	3.6	7.6
TW H1	2.1	5.4	---
TW H2	2.1	4.1	14.5
TW H3	2.3	4.1	9.0
TW H4	2.6	4.5	8.6
TW I1	2.8	---	---
TW I2	1.5	2.4	3.9
TW I3	1.4	2.2	3.4
TW I4	1.2	1.8	2.5
TW I5	1.2	1.7	2.3
TW J1	1.2	1.9	2.8
TW J2	1.2	1.8	2.6
TW J3	1.0	1.4	2.0
TW J4	1.1	1.5	2.0
DG A1	2.3	---	---
DG A2	1.7	---	---
DG A3	1.4	2.0	4.5
DG B1	2.2	---	---
DG B2	1.6	3.7	---
DG B3	1.3	1.8	2.9
DG C1	1.7	---	---
DG C2	1.4	2.4	---
DG C3	1.2	1.6	2.4
SS A1	1.4	3.6	---
SS A2	1.1	1.8	2.8
SS A3	1.5	---	---
SS A4	1.1	2.0	3.2
SS A5	1.4	---	---
SS A6	1.1	1.9	3.0
SS A7	1.6	---	---
SS A8	1.1	2.1	3.7
SS B1	1.6	4.9	---
SS B2	1.2	1.9	2.9
SS B3	1.7	---	---
SS B4	1.2	2.1	3.2
SS B5	1.8	---	---
SS B6	1.2	2.0	3.1
SS B7	2.0	---	---
SS B8	1.3	2.2	3.6
SS C1	1.6	5.5	---
SS C2	1.3	2.1	3.2
SS C3	1.9	---	---
SS C4	1.4	2.3	3.7
SS D1	1.2	2.8	---
SS D2	1.0	1.5	2.3
SS D3	1.3	3.6	---
SS D4	1.0	1.6	2.4
SS E1	1.4	---	---
SS E2	1.1	1.8	2.8
SS E3	1.6	---	---
SS E4	1.2	2.0	3.1

HELENA, MT

	SOLAR SAVINGS		
	10%	50%	80%
WW A1	2.0	---	---
WW A2	1.7	10.5	---
WW A3	1.6	3.6	---
WW A4	1.5	3.1	13.2
WW A5	1.5	2.8	7.7
WW A6	1.4	2.7	6.3
WW B1	2.7	---	---
WW B2	1.4	2.3	3.7
WW B3	1.2	1.9	2.9
WW B4	1.1	1.6	2.3
WW B5	1.1	1.6	2.1
WW C1	1.2	1.8	2.6
WW C2	1.2	1.7	2.4
WW C3	1.0	1.4	1.8
WW C4	1.0	1.4	1.9
TW A1	2.0	---	---
TW A2	1.8	7.5	---
TW A3	1.7	3.9	---
TW A4	1.7	3.3	10.2
TW B1	2.0	---	---
TW B2	1.8	4.1	---
TW B3	1.8	3.6	12.9
TW B4	1.8	3.4	8.4
TW C1	2.0	5.4	---
TW C2	1.9	3.8	12.2
TW C3	1.9	3.7	9.3
TW C4	2.1	3.9	8.4
TW D1	2.9	---	---
TW D2	1.5	2.5	4.2
TW D3	1.4	2.3	3.9
TW D4	1.2	1.8	2.6
TW D5	1.2	1.7	2.4
TW E1	1.3	2.0	3.0
TW E2	1.2	1.8	2.7
TW E3	1.0	1.5	2.0
TW E4	1.1	1.5	2.0
TW F1	2.1	---	---
TW F2	1.9	---	---
TW F3	1.8	4.5	---
TW F4	1.9	3.8	13.0
TW G1	2.2	---	---
TW G2	2.0	5.6	---
TW G3	2.0	4.3	27.4
TW G4	2.2	4.1	10.6
TW H1	2.3	---	---
TW H2	2.3	5.0	---
TW H3	2.4	4.7	13.1
TW H4	2.7	5.0	10.9
TW I1	3.4	---	---
TW I2	1.6	2.6	4.4
TW I3	1.5	2.4	3.8
TW I4	1.3	1.9	2.7
TW I5	1.3	1.8	2.4
TW J1	1.3	2.0	3.0
TW J2	1.3	1.9	2.7
TW J3	1.1	1.5	2.0
TW J4	1.1	1.5	2.1
DG A1	2.8	---	---
DG A2	1.8	---	---
DG A3	1.4	2.1	5.7
DG B1	2.5	---	---
DG B2	1.7	5.6	---
DG B3	1.4	1.9	3.1
DG C1	1.9	---	---

HELENA, MT

	SOLAR SAVINGS		
	10%	50%	80%
DG C2	1.5	2.7	---
DG C3	1.3	1.7	2.5
SS A1	1.4	4.7	---
SS A2	1.1	1.9	2.9
SS A3	1.6	---	---
SS A4	1.1	2.1	3.4
SS A5	1.5	---	---
SS A6	1.1	2.0	3.2
SS A7	1.7	---	---
SS A8	1.2	2.2	4.0
SS B1	1.7	---	---
SS B2	1.2	2.0	3.0
SS B3	1.8	---	---
SS B4	1.3	2.2	3.4
SS B5	1.8	---	---
SS B6	1.3	2.1	3.3
SS B7	2.1	---	---
SS B8	1.3	2.3	3.9
SS C1	1.8	---	---
SS C2	1.4	2.2	3.6
SS C3	2.1	---	---
SS C4	1.5	2.5	4.1
SS D1	1.3	3.4	---
SS D2	1.0	1.6	2.4
SS D3	1.3	5.3	---
SS D4	1.0	1.7	2.5
SS E1	1.5	---	---
SS E2	1.1	1.9	3.0
SS E3	1.7	---	---
SS E4	1.2	2.1	3.4

LEWISTOWN, MT

	SOLAR SAVINGS		
	10%	50%	80%
WW A1	2.0	---	---
WW A2	1.7	4.2	---
WW A3	1.6	3.1	12.7
WW A4	1.5	2.8	6.8
WW A5	1.5	2.6	5.2
WW A6	1.4	2.5	4.6
WW B1	2.6	---	---
WW B2	1.4	2.2	3.4
WW B3	1.2	1.9	2.7
WW B4	1.1	1.6	2.2
WW B5	1.1	1.6	2.1
WW C1	1.2	1.7	2.5
WW C2	1.2	1.7	2.3
WW C3	1.0	1.3	1.8
WW C4	1.0	1.4	1.8
TW A1	2.0	---	---
TW A2	1.8	4.5	---
TW A3	1.7	3.4	14.1
TW A4	1.7	3.0	6.5
TW B1	2.0	---	---
TW B2	1.8	3.7	15.2
TW B3	1.8	3.3	7.5
TW B4	1.8	3.2	6.2
TW C1	1.9	4.5	---
TW C2	1.9	3.6	7.7
TW C3	1.9	3.5	6.8
TW C4	2.1	3.7	6.7
TW D1	2.9	---	---
TW D2	1.5	2.4	3.8
TW D3	1.4	2.2	3.5

LEWISTOWN, MT			
	\multicolumn{3}{c}{SOLAR SAVINGS}
	10%	50%	80%
WW A1	2.0	---	---
WW A2	1.7	4.2	---
WW A3	1.6	3.1	12.7
WW A4	1.5	2.8	6.8
WW A5	1.5	2.6	5.2
WW A6	1.4	2.5	4.6
WW B1	2.6	---	---
WW B2	1.4	2.2	3.4
WW B3	1.2	1.9	2.7
WW B4	1.1	1.6	2.2
WW B5	1.1	1.6	2.1
WW C1	1.2	1.7	2.5
WW C2	1.2	1.7	2.3
WW C3	1.0	1.3	1.8
WW C4	1.0	1.4	1.8
TW A1	2.0	---	---
TW A2	1.8	4.5	---
TW A3	1.7	3.4	14.1
TW A4	1.7	3.0	6.5
TW B1	2.0	---	---
TW B2	1.8	3.7	15.2
TW B3	1.8	3.3	7.5
TW B4	1.8	3.2	6.2
TW C1	1.9	4.5	---
TW C2	1.9	3.6	7.7
TW C3	1.9	3.5	6.8
TW C4	2.1	3.7	6.7
TW D1	2.9	---	---
TW D2	1.5	2.4	3.8
TW D3	1.4	2.2	3.5
DG B1	2.6	---	---
DG B2	1.7	4.4	---
DG B3	1.4	1.9	3.0
DG C1	1.9	---	---
DG C2	1.6	2.6	---
DG C3	1.3	1.7	2.4
SS A1	1.4	3.8	---
SS A2	1.1	1.8	2.8
SS A3	1.5	---	---
SS A4	1.1	2.0	3.2
SS A5	1.5	---	---
SS A6	1.1	1.9	3.0
SS A7	1.6	---	---
SS A8	1.2	2.1	3.6
SS B1	1.6	5.5	---
SS B2	1.2	2.0	2.9
SS B3	1.8	---	---
SS B4	1.2	2.1	3.2
SS B5	1.8	---	---
SS B6	1.2	2.1	3.1
SS B7	2.1	---	---
SS B8	1.3	2.3	3.6
SS C1	1.8	10.3	---
SS C2	1.4	2.1	3.2
SS C3	2.1	---	---
SS C4	1.5	2.4	3.7
SS D1	1.2	2.9	---
SS D2	1.0	1.6	2.3
SS D3	1.3	3.8	---
SS D4	1.0	1.6	2.4
SS E1	1.5	---	---
SS E2	1.1	1.9	2.8
SS E3	1.6	---	---
SS E4	1.2	2.0	3.1

MILES CITY, MT			
	\multicolumn{3}{c}{SOLAR SAVINGS}
	10%	50%	80%
WW A1	2.0	---	---
WW A2	1.6	4.5	---
WW A3	1.5	3.3	8.5
WW A4	1.5	2.9	5.9
WW A5	1.5	2.7	4.8
WW A6	1.4	2.5	4.4
WW B1	2.6	---	---
WW B2	1.4	2.2	3.3
WW B3	1.2	1.9	2.7
WW B4	1.1	1.6	2.2
WW B5	1.1	1.6	2.1
WW C1	1.1	1.8	2.5
WW C2	1.2	1.7	2.3
WW C3	1.0	1.4	1.8
WW C4	1.0	1.4	1.9
TW A1	2.0	---	---
TW A2	1.7	4.7	---
TW A3	1.7	3.5	9.4
TW A4	1.6	3.1	5.9
TW B1	2.0	---	---
TW B2	1.8	3.8	10.1
TW B3	1.7	3.4	6.8
TW B4	1.8	3.3	5.9
TW C1	1.9	4.7	---
TW C2	1.9	3.7	7.1
TW C3	1.9	3.6	6.5
TW C4	2.1	3.8	6.5
TW D1	2.8	---	---
TW D2	1.5	2.4	3.7
TW D3	1.4	2.3	3.5
TW D4	1.2	1.8	2.5
TW D5	1.2	1.7	2.3
TW E1	1.2	2.0	2.8
TW E2	1.2	1.8	2.6
TW E3	1.0	1.5	2.0
TW E4	1.1	1.5	2.0
TW F1	2.1	---	---
TW F2	1.9	5.5	---
TW F3	1.8	3.9	11.3
TW F4	1.8	3.5	6.9
TW G1	2.1	---	---
TW G2	2.0	4.6	21.5
TW G3	2.0	4.0	8.4
TW G4	2.2	3.9	7.1
TW H1	2.3	6.7	---
TW H2	2.3	4.6	10.2
TW H3	2.4	4.5	8.2
TW H4	2.7	4.8	8.2
TW I1	3.4	---	---
TW I2	1.6	2.6	3.9
TW I3	1.5	2.3	3.5
TW I4	1.3	1.9	2.6
TW I5	1.2	1.8	2.4
TW J1	1.3	2.0	2.9
TW J2	1.3	1.9	2.6
TW J3	1.1	1.5	2.0
TW J4	1.1	1.5	2.0
DG A1	2.8	---	---
DG A2	1.7	---	---
DG A3	1.4	2.1	4.9
DG B1	2.5	---	---
DG B2	1.7	4.8	---
DG B3	1.4	1.9	3.0
DG C1	1.9	---	---

MILES CITY, MT			
	\multicolumn{3}{c}{SOLAR SAVINGS}
	10%	50%	80%
DG C2	1.5	2.7	---
DG C3	1.3	1.7	2.4
SS A1	1.4	4.2	---
SS A2	1.1	1.9	2.8
SS A3	1.6	---	---
SS A4	1.2	2.1	3.2
SS A5	1.5	---	---
SS A6	1.1	2.0	3.0
SS A7	1.7	---	---
SS A8	1.2	2.3	3.6
SS B1	1.7	6.7	---
SS B2	1.2	2.0	2.9
SS B3	1.8	---	---
SS B4	1.3	2.2	3.2
SS B5	1.9	---	---
SS B6	1.3	2.1	3.1
SS B7	2.2	---	---
SS B8	1.3	2.4	3.6
SS C1	1.8	---	---
SS C2	1.4	2.2	3.2
SS C3	2.0	---	---
SS C4	1.5	2.5	3.7
SS D1	1.3	3.2	11.2
SS D2	1.0	1.6	2.3
SS D3	1.4	4.3	---
SS D4	1.0	1.7	2.4
SS E1	1.5	---	---
SS E2	1.2	1.9	2.8
SS E3	1.7	---	---
SS E4	1.2	2.1	3.1

MISSOULA, MT			
	\multicolumn{3}{c}{SOLAR SAVINGS}
	10%	50%	80%
WW A1	2.3	---	---
WW A2	2.1	---	---
WW A3	2.0	---	---
WW A4	1.9	---	---
WW A5	1.9	---	---
WW A6	1.8	---	---
WW B1	---	---	---
WW B2	1.7	---	---
WW B3	1.4	3.3	---
WW B4	1.3	2.1	3.2
WW B5	1.3	1.9	2.8
WW C1	1.3	2.6	5.9
WW C2	1.3	2.2	3.8
WW C3	1.1	1.6	2.2
WW C4	1.1	1.6	2.3
TW A1	2.3	---	---
TW A2	2.2	---	---
TW A3	2.1	---	---
TW A4	2.0	---	---
TW B1	2.3	---	---
TW B2	2.2	---	---
TW B3	2.1	---	---
TW B4	2.2	---	---
TW C1	2.3	---	---
TW C2	2.2	---	---
TW C3	2.3	---	---
TW C4	2.4	---	---
TW D1	---	---	---
TW D2	1.7	---	---
TW D3	1.6	---	---

MISSOULA, MT			
	\multicolumn{3}{c}{SOLAR SAVINGS}
	10%	50%	80%
TW D4	1.4	2.3	3.9
TW D5	1.3	2.1	3.1
TW E1	1.4	3.1	14.1
TW E2	1.4	2.5	4.8
TW E3	1.2	1.8	2.6
TW E4	1.2	1.8	2.5
TW F1	2.7	---	---
TW F2	2.4	---	---
TW F3	2.3	---	---
TW F4	2.4	---	---
TW G1	2.8	---	---
TW G2	2.6	---	---
TW G3	2.6	---	---
TW G4	2.7	---	---
TW H1	3.0	---	---
TW H2	2.9	---	---
TW H3	3.0	---	---
TW H4	3.3	---	---
TW I1	---	---	---
TW I2	1.9	---	---
TW I3	1.7	4.9	---
TW I4	1.5	2.4	4.0
TW I5	1.4	2.2	3.3
TW J1	1.5	3.0	8.6
TW J2	1.5	2.5	4.5
TW J3	1.2	1.8	2.6
TW J4	1.2	1.8	2.6
DG A1	---	---	---
DG A2	2.1	---	---
DG A3	1.6	2.9	---
DG B1	---	---	---
DG B2	2.0	---	---
DG B3	1.5	2.5	10.8
DG C1	2.8	---	---
DG C2	1.7	---	---
DG C3	1.4	2.2	4.0
SS A1	1.6	---	---
SS A2	1.2	2.6	5.4
SS A3	1.8	---	---
SS A4	1.3	3.2	---
SS A5	1.6	---	---
SS A6	1.2	2.8	8.6
SS A7	1.8	---	---
SS A8	1.3	4.5	---
SS B1	1.9	---	---
SS B2	1.3	2.6	5.1
SS B3	2.1	---	---
SS B4	1.4	3.1	8.2
SS B5	2.1	---	---
SS B6	1.4	2.9	7.0
SS B7	2.5	---	---
SS B8	1.5	3.9	---
SS C1	2.4	---	---
SS C2	1.6	4.9	---
SS C3	2.6	---	---
SS C4	1.7	7.6	---
SS D1	1.4	---	---
SS D2	1.1	2.1	3.7
SS D3	1.5	---	---
SS D4	1.1	2.2	4.0
SS E1	1.8	---	---
SS E2	1.3	3.0	12.4
SS E3	1.9	---	---
SS E4	1.4	3.3	---

GRAND ISLAND, NB SOLAR SAVINGS	10%	50%	80%	GRAND ISLAND, NB SOLAR SAVINGS	10%	50%	80%	NORTH OMAHA, NB SOLAR SAVINGS	10%	50%	80%	NORTH PLATTE, NB SOLAR SAVINGS	10%	50%	80%
WW A1	1.7	3.5	5.7	DG C2	1.4	1.9	3.5	TW D4	1.2	1.6	2.1	WW A1	1.6	2.9	4.3
WW A2	1.4	2.2	3.1	DG C3	1.2	1.4	1.9	TW D5	1.2	1.6	2.1	WW A2	1.3	2.0	2.7
WW A3	1.3	2.0	2.7	SS A1	1.2	2.2	3.1	TW E1	1.2	1.7	2.3	WW A3	1.2	1.8	2.4
WW A4	1.3	1.9	2.5	SS A2	1.0	1.5	2.0	TW E2	1.2	1.7	2.2	WW A4	1.2	1.7	2.3
WW A5	1.2	1.8	2.4	SS A3	1.3	2.7	4.3	TW E3	1.0	1.4	1.8	WW A5	1.2	1.6	2.2
WW A6	1.2	1.7	2.3	SS A4	1.0	1.6	2.2	TW E4	1.0	1.4	1.8	WW A6	1.1	1.6	2.1
WW B1	1.9	5.0	---	SS A5	1.3	2.6	4.0	TW F1	2.0	4.4	10.4	WW B1	1.7	3.2	5.4
WW B2	1.2	1.7	2.3	SS A6	1.0	1.5	2.1	TW F2	1.8	3.0	4.5	WW B2	1.2	1.6	2.1
WW B3	1.1	1.5	2.0	SS A7	1.5	3.8	8.7	TW F3	1.7	2.8	3.9	WW B3	1.0	1.4	1.8
WW B4	1.0	1.4	1.8	SS A8	1.0	1.7	2.3	TW F4	1.8	2.7	3.7	WW B4	1.0	1.3	1.7
WW B5	1.0	1.4	1.7	SS B1	1.4	2.5	3.6	TW G1	2.0	3.8	6.3	WW B5	1.0	1.3	1.6
WW C1	1.0	1.4	1.9	SS B2	1.1	1.6	2.2	TW G2	1.9	3.1	4.4	WW C1	1.0	1.3	1.7
WW C2	1.0	1.4	1.8	SS B3	1.5	2.9	4.6	TW G3	1.9	3.0	4.1	WW C2	1.0	1.3	1.7
WW C3	.9	1.2	1.5	SS B4	1.1	1.7	2.3	TW G4	2.1	3.1	4.3	WW C3	.9	1.1	1.4
WW C4	.9	1.2	1.6	SS B5	1.6	3.1	4.9	TW H1	2.1	3.6	5.3	WW C4	.9	1.2	1.5
TW A1	1.7	3.3	5.1	SS B6	1.1	1.7	2.3	TW H2	2.2	3.4	4.7	TW A1	1.6	2.9	4.1
TW A2	1.5	2.4	3.4	SS B7	1.8	4.3	9.4	TW H3	2.3	3.5	4.8	TW A2	1.4	2.2	3.0
TW A3	1.4	2.2	3.0	SS B8	1.2	1.8	2.4	TW H4	2.6	3.9	5.4	TW A3	1.4	2.0	2.7
TW A4	1.4	2.1	2.9	SS C1	1.5	2.3	3.3	TW I1	2.9	---	---	TW A4	1.3	1.9	2.6
TW B1	1.7	3.0	4.3	SS C2	1.2	1.7	2.3	TW I2	1.5	2.2	3.0	TW B1	1.6	2.6	3.7
TW B2	1.5	2.4	3.3	SS C3	1.7	2.9	4.1	TW I3	1.4	2.0	2.7	TW B2	1.5	2.2	2.9
TW B3	1.5	2.3	3.1	SS C4	1.3	1.9	2.5	TW I4	1.3	1.7	2.2	TW B3	1.4	2.1	2.8
TW B4	1.6	2.3	3.2	SS D1	1.1	1.8	2.6	TW I5	1.2	1.6	2.1	TW B4	1.5	2.2	2.9
TW C1	1.7	2.7	3.8	SS D2	.9	1.3	1.7	TW J1	1.2	1.8	2.3	TW C1	1.6	2.5	3.4
TW C2	1.6	2.5	3.4	SS D3	1.2	2.1	3.0	TW J2	1.2	1.7	2.2	TW C2	1.6	2.3	3.1
TW C3	1.7	2.6	3.5	SS D4	.9	1.4	1.8	TW J3	1.1	1.4	1.8	TW C3	1.6	2.4	3.2
TW C4	1.8	2.8	3.8	SS E1	1.3	2.2	3.2	TW J4	1.1	1.4	1.8	TW C4	1.8	2.6	3.5
TW D1	2.1	4.7	11.7	SS E2	1.0	1.5	2.0	DG A1	2.4	---	---	TW D1	1.9	3.4	5.3
TW D2	1.3	1.9	2.5	SS E3	1.4	2.7	4.0	DG A2	1.7	5.2	---	TW D2	1.2	1.7	2.3
TW D3	1.2	1.7	2.3	SS E4	1.1	1.6	2.2	DG A3	1.4	1.9	3.3	TW D3	1.1	1.6	2.1
TW D4	1.1	1.5	2.0					DG B1	2.4	---	---	TW D4	1.1	1.4	1.8
TW D5	1.1	1.5	1.9	NORTH OMAHA, NB SOLAR SAVINGS				DG B2	1.7	2.9	---	TW D5	1.0	1.4	1.8
TW E1	1.1	1.6	2.1		10%	50%	80%	DG B3	1.4	1.7	2.5	TW E1	1.1	1.5	1.9
TW E2	1.1	1.5	2.0	WW A1	1.8	5.5	---	DG C1	1.8	---	---	TW E2	1.1	1.4	1.9
TW E3	.9	1.3	1.6	WW A2	1.6	2.7	3.9	DG C2	1.5	2.1	---	TW E3	.9	1.2	1.5
TW E4	1.0	1.3	1.7	WW A3	1.5	2.3	3.3	DG C3	1.3	1.6	2.1	TW E4	.9	1.2	1.6
TW F1	1.8	3.3	4.9	WW A4	1.4	2.2	3.0	SS A1	1.4	2.6	3.9	TW F1	1.7	2.8	4.0
TW F2	1.6	2.5	3.5	WW A5	1.4	2.1	2.8	SS A2	1.1	1.7	2.3	TW F2	1.5	2.3	3.1
TW F3	1.5	2.4	3.2	WW A6	1.4	2.0	2.7	SS A3	1.5	3.5	8.6	TW F3	1.5	2.1	2.9
TW F4	1.6	2.3	3.1	WW B1	2.3	---	---	SS A4	1.1	1.8	2.4	TW F4	1.5	2.1	2.8
TW G1	1.8	3.0	4.4	WW B2	1.3	1.9	2.6	SS A5	1.4	3.2	5.8	TW G1	1.7	2.7	3.7
TW G2	1.7	2.6	3.6	WW B3	1.2	1.6	2.2	SS A6	1.1	1.7	2.3	TW G2	1.6	2.4	3.2
TW G3	1.7	2.6	3.5	WW B4	1.1	1.5	1.9	SS A7	1.6	---	---	TW G3	1.6	2.3	3.1
TW G4	1.9	2.7	3.6	WW B5	1.1	1.5	1.9	SS A8	1.1	1.8	2.6	TW G4	1.8	2.5	3.3
TW H1	1.9	3.0	4.2	WW C1	1.1	1.6	2.1	SS B1	1.6	3.0	4.6	TW H1	1.8	2.7	3.6
TW H2	1.9	2.9	3.9	WW C2	1.1	1.5	2.0	SS B2	1.2	1.8	2.4	TW H2	1.8	2.6	3.5
TW H3	2.1	3.0	4.1	WW C3	1.0	1.3	1.6	SS B3	1.7	3.8	8.8	TW H3	2.0	2.8	3.7
TW H4	2.4	3.5	4.6	WW C4	1.0	1.3	1.7	SS B4	1.2	1.9	2.6	TW H4	2.2	3.2	4.2
TW I1	2.3	5.6	34.6	TW A1	1.9	4.5	19.5	SS B5	1.8	4.1	9.1	TW I1	2.1	3.8	6.1
TW I2	1.4	2.0	2.6	TW A2	1.7	2.9	4.3	SS B6	1.2	1.9	2.5	TW I2	1.3	1.8	2.4
TW I3	1.3	1.8	2.4	TW A3	1.6	2.6	3.6	SS B7	2.0	---	---	TW I3	1.2	1.7	2.2
TW I4	1.2	1.6	2.0	TW A4	1.6	2.4	3.4	SS B8	1.3	2.0	2.7	TW I4	1.1	1.5	1.9
TW I5	1.1	1.5	2.0	TW B1	1.9	3.7	6.1	SS C1	1.7	2.8	4.3	TW I5	1.1	1.4	1.9
TW J1	1.1	1.6	2.1	TW B2	1.7	2.8	4.0	SS C2	1.3	1.9	2.5	TW J1	1.1	1.5	2.0
TW J2	1.1	1.6	2.0	TW B3	1.7	2.6	3.7	SS C3	1.9	3.6	5.7	TW J2	1.1	1.5	1.9
TW J3	1.0	1.3	1.7	TW B4	1.7	2.7	3.7	SS C4	1.4	2.1	2.8	TW J3	.9	1.2	1.6
TW J4	1.0	1.3	1.7	TW C1	1.9	3.2	4.7	SS D1	1.2	2.1	3.1	TW J4	1.0	1.3	1.6
DG A1	1.9	---	---	TW C2	1.8	2.9	4.0	SS D2	1.0	1.4	1.9	DG A1	1.8	---	---
DG A2	1.5	3.1	---	TW C3	1.9	3.0	4.1	SS D3	1.3	2.5	3.8	DG A2	1.5	2.6	---
DG A3	1.3	1.7	2.8	TW C4	2.0	3.2	4.4	SS D4	1.0	1.5	2.0	DG A3	1.2	1.6	2.5
DG B1	1.9	---	---	TW D1	2.5	---	---	SS E1	1.4	2.7	4.3	DG B1	1.7	---	---
DG B2	1.5	2.3	---	TW D2	1.4	2.1	2.8	SS E2	1.1	1.6	2.2	DG B2	1.4	2.0	---
DG B3	1.3	1.6	2.2	TW D3	1.3	1.9	2.6	SS E3	1.6	3.4	6.2	DG B3	1.2	1.5	2.0
DG C1	1.6	5.6	---					SS E4	1.2	1.8	2.4	DG C1	1.5	2.7	---

NORTH PLATTE, NB
SOLAR SAVINGS

	10%	50%	80%
DG C2	1.3	1.7	2.8
DG C3	1.1	1.4	1.7
SS A1	1.2	2.0	2.8
SS A2	1.0	1.4	1.9
SS A3	1.3	2.4	3.5
SS A4	1.0	1.5	2.0
SS A5	1.3	2.3	3.4
SS A6	1.0	1.4	2.0
SS A7	1.4	3.1	5.2
SS A8	1.0	1.5	2.1
SS B1	1.4	2.2	3.1
SS B2	1.1	1.5	2.1
SS B3	1.4	2.5	3.7
SS B4	1.1	1.6	2.1
SS B5	1.5	2.7	4.0
SS B6	1.1	1.6	2.1
SS B7	1.7	3.4	5.6
SS B8	1.1	1.7	2.3
SS C1	1.4	2.1	2.8
SS C2	1.1	1.6	2.1
SS C3	1.6	2.6	3.5
SS C4	1.3	1.8	2.3
SS D1	1.0	1.7	2.3
SS D2	.9	1.2	1.6
SS D3	1.1	1.9	2.7
SS D4	.9	1.3	1.7
SS E1	1.2	1.9	2.7
SS E2	1.0	1.4	1.8
SS E3	1.3	2.3	3.4
SS E4	1.0	1.5	2.0

SCOTTSBLUFF, NB
SOLAR SAVINGS

	10%	50%	80%
WW A1	1.6	2.8	4.1
WW A2	1.3	2.0	2.7
WW A3	1.2	1.8	2.4
WW A4	1.2	1.7	2.2
WW A5	1.2	1.6	2.1
WW A6	1.1	1.6	2.1
WW B1	1.7	3.0	5.1
WW B2	1.1	1.6	2.0
WW B3	1.0	1.4	1.8
WW B4	1.0	1.3	1.6
WW B5	1.0	1.3	1.6
WW C1	1.0	1.3	1.7
WW C2	1.0	1.3	1.7
WW C3	.9	1.1	1.4
WW C4	.9	1.2	1.5
TW A1	1.6	2.8	3.9
TW A2	1.4	2.1	2.9
TW A3	1.3	2.0	2.6
TW A4	1.3	1.9	2.5
TW B1	1.6	2.6	3.5
TW B2	1.4	2.1	2.8
TW B3	1.4	2.1	2.7
TW B4	1.5	2.1	2.8
TW C1	1.6	2.4	3.3
TW C2	1.5	2.3	3.0
TW C3	1.6	2.3	3.1
TW C4	1.7	2.6	3.4
TW D1	1.9	3.2	5.0
TW D2	1.2	1.7	2.2
TW D3	1.1	1.6	2.1

SCOTTSBLUFF, NB
SOLAR SAVINGS

	10%	50%	80%
TW D4	1.0	1.4	1.8
TW D5	1.0	1.4	1.8
TW E1	1.0	1.4	1.9
TW E2	1.0	1.4	1.8
TW E3	.9	1.2	1.5
TW E4	.9	1.2	1.6
TW F1	1.7	2.7	3.8
TW F2	1.5	2.2	3.0
TW F3	1.4	2.1	2.8
TW F4	1.5	2.1	2.7
TW G1	1.7	2.6	3.6
TW G2	1.6	2.3	3.1
TW G3	1.6	2.3	3.0
TW G4	1.7	2.4	3.2
TW H1	1.8	2.6	3.5
TW H2	1.8	2.6	3.4
TW H3	1.9	2.7	3.6
TW H4	2.2	3.1	4.1
TW I1	2.0	3.5	5.7
TW I2	1.3	1.8	2.4
TW I3	1.2	1.7	2.2
TW I4	1.1	1.5	1.9
TW I5	1.1	1.4	1.8
TW J1	1.1	1.5	1.9
TW J2	1.1	1.5	1.9
TW J3	.9	1.2	1.6
TW J4	1.0	1.3	1.6
DG A1	1.7	---	---
DG A2	1.4	2.5	---
DG A3	1.2	1.6	2.4
DG B1	1.7	---	---
DG B2	1.4	2.0	6.0
DG B3	1.2	1.5	2.0
DG C1	1.5	2.5	---
DG C2	1.3	1.7	2.7
DG C3	1.1	1.3	1.7
SS A1	1.2	1.9	2.7
SS A2	.9	1.4	1.9
SS A3	1.3	2.3	3.4
SS A4	1.0	1.5	2.0
SS A5	1.2	2.2	3.3
SS A6	1.0	1.4	1.9
SS A7	1.4	2.9	5.0
SS A8	1.0	1.5	2.1
SS B1	1.3	2.2	3.1
SS B2	1.0	1.5	2.0
SS B3	1.4	2.5	3.6
SS B4	1.1	1.6	2.1
SS B5	1.5	2.7	3.9
SS B6	1.1	1.6	2.1
SS B7	1.7	3.3	5.4
SS B8	1.1	1.6	2.2
SS C1	1.4	2.0	2.7
SS C2	1.1	1.6	2.0
SS C3	1.6	2.5	3.4
SS C4	1.2	1.7	2.3
SS D1	1.0	1.6	2.3
SS D2	.8	1.2	1.6
SS D3	1.1	1.8	2.6
SS D4	.9	1.3	1.7
SS E1	1.2	1.9	2.7
SS E2	1.0	1.4	1.8
SS E3	1.3	2.3	3.3
SS E4	1.0	1.5	2.0

ELKO, NV
SOLAR SAVINGS

	10%	50%	80%
WW A1	1.5	2.6	4.2
WW A2	1.2	1.9	2.6
WW A3	1.2	1.7	2.3
WW A4	1.1	1.6	2.2
WW A5	1.1	1.5	2.1
WW A6	1.1	1.5	2.0
WW B1	1.5	2.8	5.8
WW B2	1.1	1.5	2.0
WW B3	1.0	1.3	1.7
WW B4	.9	1.2	1.6
WW B5	.9	1.2	1.6
WW C1	.9	1.3	1.7
WW C2	.9	1.3	1.7
WW C3	.8	1.1	1.4
WW C4	.9	1.1	1.5
TW A1	1.5	2.6	3.9
TW A2	1.3	2.0	2.9
TW A3	1.3	1.9	2.6
TW A4	1.3	1.8	2.5
TW B1	1.5	2.4	3.5
TW B2	1.4	2.0	2.8
TW B3	1.3	2.0	2.7
TW B4	1.4	2.0	2.8
TW C1	1.5	2.3	3.2
TW C2	1.5	2.2	3.0
TW C3	1.5	2.2	3.0
TW C4	1.6	2.4	3.3
TW D1	1.7	3.0	5.5
TW D2	1.2	1.6	2.2
TW D3	1.1	1.5	2.0
TW D4	1.0	1.4	1.8
TW D5	1.0	1.3	1.7
TW E1	1.0	1.4	1.8
TW E2	1.0	1.4	1.8
TW E3	.9	1.1	1.5
TW E4	.9	1.2	1.5
TW F1	1.6	2.6	3.8
TW F2	1.4	2.1	3.0
TW F3	1.4	2.0	2.7
TW F4	1.4	2.0	2.7
TW G1	1.6	2.5	3.6
TW G2	1.5	2.2	3.0
TW G3	1.5	2.2	3.0
TW G4	1.6	2.3	3.2
TW H1	1.7	2.5	3.5
TW H2	1.7	2.5	3.4
TW H3	1.8	2.6	3.5
TW H4	2.1	3.0	4.0
TW I1	1.8	3.3	6.5
TW I2	1.2	1.7	2.3
TW I3	1.2	1.6	2.1
TW I4	1.1	1.4	1.9
TW I5	1.0	1.4	1.8
TW J1	1.0	1.4	1.9
TW J2	1.0	1.4	1.8
TW J3	.9	1.2	1.5
TW J4	.9	1.2	1.6
DG A1	1.6	---	---
DG A2	1.4	2.3	---
DG A3	1.2	1.5	2.3
DG B1	1.6	---	---
DG B2	1.3	1.9	4.7
DG B3	1.2	1.4	1.9
DG C1	1.4	2.3	---

ELKO, NV
SOLAR SAVINGS

	10%	50%	80%
DG C2	1.2	1.6	2.5
DG C3	1.1	1.3	1.7
SS A1	1.1	1.8	2.7
SS A2	.9	1.3	1.8
SS A3	1.1	2.1	3.4
SS A4	.9	1.4	1.9
SS A5	1.1	2.0	3.2
SS A6	.9	1.3	1.9
SS A7	1.2	2.6	5.2
SS A8	.9	1.4	2.0
SS B1	1.2	2.0	3.0
SS B2	1.0	1.4	2.0
SS B3	1.3	2.3	3.6
SS B4	1.0	1.5	2.1
SS B5	1.4	2.4	3.9
SS B6	1.0	1.5	2.0
SS B7	1.5	3.0	5.7
SS B8	1.0	1.5	2.2
SS C1	1.3	1.9	2.7
SS C2	1.1	1.5	2.0
SS C3	1.5	2.4	3.4
SS C4	1.2	1.7	2.2
SS D1	.9	1.5	2.2
SS D2	.8	1.1	1.6
SS D3	1.0	1.7	2.6
SS D4	.8	1.2	1.6
SS E1	1.1	1.7	2.6
SS E2	.9	1.3	1.8
SS E3	1.2	2.1	3.3
SS E4	.9	1.4	1.9

ELY, NV
SOLAR SAVINGS

	10%	50%	80%
WW A1	1.5	2.3	3.2
WW A2	1.2	1.7	2.2
WW A3	1.1	1.6	2.0
WW A4	1.1	1.5	1.9
WW A5	1.1	1.4	1.8
WW A6	1.0	1.4	1.8
WW B1	1.5	2.3	3.2
WW B2	1.1	1.4	1.8
WW B3	.9	1.2	1.6
WW B4	.9	1.2	1.5
WW B5	.9	1.2	1.5
WW C1	.9	1.2	1.5
WW C2	.9	1.2	1.5
WW C3	.8	1.0	1.3
WW C4	.9	1.1	1.4
TW A1	1.5	2.3	3.1
TW A2	1.3	1.9	2.5
TW A3	1.2	1.7	2.3
TW A4	1.2	1.7	2.2
TW B1	1.5	2.2	2.9
TW B2	1.3	1.9	2.4
TW B3	1.3	1.8	2.4
TW B4	1.4	1.9	2.4
TW C1	1.5	2.1	2.8
TW C2	1.4	2.0	2.6
TW C3	1.5	2.1	2.7
TW C4	1.6	2.3	3.0
TW D1	1.7	2.5	3.4
TW D2	1.1	1.5	2.0
TW D3	1.1	1.4	1.8

ELY, NV

SOLAR SAVINGS

	10%	50%	80%
WW A1	1.5	2.3	3.2
WW A2	1.2	1.7	2.2
WW A3	1.1	1.6	2.0
WW A4	1.1	1.5	1.9
WW A5	1.1	1.4	1.8
WW A6	1.0	1.4	1.8
WW B1	1.5	2.3	3.2
WW B2	1.1	1.4	1.8
WW B3	.9	1.2	1.6
WW B4	.9	1.2	1.5
WW B5	.9	1.2	1.5
WW C1	.9	1.2	1.5
WW C2	.9	1.2	1.5
WW C3	.8	1.0	1.3
WW C4	.9	1.1	1.4
TW A1	1.5	2.3	3.1
TW A2	1.3	1.9	2.5
TW A3	1.2	1.7	2.3
TW A4	1.2	1.7	2.2
TW B1	1.5	2.2	2.9
TW B2	1.3	1.9	2.4
TW B3	1.3	1.8	2.4
TW B4	1.4	1.9	2.4
TW C1	1.5	2.1	2.8
TW C2	1.4	2.0	2.6
TW C3	1.5	2.1	2.7
TW C4	1.6	2.3	3.0
TW D1	1.7	2.5	3.4
TW D2	1.1	1.5	2.0
TW D3	1.1	1.4	1.8
DG B1	1.5	3.1	---
DG B2	1.3	1.7	3.0
DG B3	1.2	1.3	1.8
DG C1	1.3	1.9	---
DG C2	1.2	1.5	2.1
DG C3	1.1	1.2	1.5
SS A1	1.0	1.6	2.2
SS A2	.9	1.2	1.6
SS A3	1.1	1.8	2.6
SS A4	.9	1.3	1.7
SS A5	1.1	1.8	2.6
SS A6	.9	1.2	1.7
SS A7	1.2	2.2	3.3
SS A8	.9	1.3	1.8
SS B1	1.2	1.8	2.5
SS B2	1.0	1.3	1.8
SS B3	1.3	2.0	2.8
SS B4	1.0	1.4	1.8
SS B5	1.3	2.1	3.0
SS B6	1.0	1.4	1.8
SS B7	1.5	2.5	3.7
SS B8	1.0	1.4	1.9
SS C1	1.3	1.7	2.3
SS C2	1.1	1.4	1.8
SS C3	1.5	2.1	2.8
SS C4	1.2	1.6	2.0
SS D1	.9	1.3	1.9
SS D2	.8	1.1	1.4
SS D3	1.0	1.5	2.1
SS D4	.8	1.1	1.5
SS E1	1.0	1.5	2.1
SS E2	.9	1.2	1.6
SS E3	1.2	1.9	2.6
SS E4	.9	1.3	1.7

LAS VEGAS, NV

SOLAR SAVINGS

	10%	50%	80%
WW A1	.8	1.2	1.6
WW A2	.7	1.0	1.3
WW A3	.7	.9	1.2
WW A4	.7	.9	1.1
WW A5	.6	.8	1.1
WW A6	.6	.8	1.1
WW B1	.8	1.1	1.5
WW B2	.7	.9	1.1
WW B3	.6	.8	1.0
WW B4	.6	.8	1.0
WW B5	.6	.8	1.0
WW C1	.6	.7	1.0
WW C2	.6	.8	1.0
WW C3	.5	.7	.9
WW C4	.6	.7	.9
TW A1	.9	1.2	1.7
TW A2	.8	1.1	1.4
TW A3	.7	1.0	1.3
TW A4	.7	1.0	1.3
TW B1	.9	1.2	1.6
TW B2	.8	1.1	1.4
TW B3	.8	1.1	1.4
TW B4	.8	1.1	1.4
TW C1	.9	1.2	1.6
TW C2	.9	1.2	1.5
TW C3	.9	1.2	1.6
TW C4	1.0	1.4	1.8
TW D1	.9	1.2	1.6
TW D2	.7	.9	1.2
TW D3	.7	.9	1.1
TW D4	.6	.8	1.1
TW D5	.6	.8	1.1
TW E1	.6	.8	1.0
TW E2	.6	.8	1.0
TW E3	.6	.7	.9
TW E4	.6	.8	1.0
TW F1	.9	1.2	1.6
TW F2	.8	1.1	1.4
TW F3	.8	1.1	1.4
TW F4	.8	1.1	1.4
TW G1	.9	1.2	1.6
TW G2	.9	1.2	1.5
TW G3	.9	1.2	1.5
TW G4	1.0	1.3	1.7
TW H1	1.0	1.3	1.7
TW H2	1.0	1.3	1.7
TW H3	1.1	1.4	1.8
TW H4	1.2	1.7	2.1
TW I1	.9	1.3	1.7
TW I2	.7	1.0	1.3
TW I3	.7	.9	1.2
TW I4	.7	.9	1.1
TW I5	.7	.9	1.1
TW J1	.6	.8	1.1
TW J2	.6	.9	1.1
TW J3	.6	.8	1.0
TW J4	.6	.8	1.0
DG A1	.8	1.2	2.1
DG A2	.8	1.1	1.5
DG A3	.8	.9	1.2
DG B1	.8	1.0	1.5
DG B2	.8	1.0	1.3
DG B3	.7	.9	1.1
DG C1	.7	.9	1.2

LAS VEGAS, NV

SOLAR SAVINGS

	10%	50%	80%
DG C2	.7	.9	1.1
DG C3	.7	.8	1.0
SS A1	.7	1.0	1.3
SS A2	.6	.8	1.1
SS A3	.7	1.1	1.4
SS A4	.6	.8	1.1
SS A5	.7	1.1	1.4
SS A6	.6	.8	1.1
SS A7	.8	1.2	1.6
SS A8	.6	.8	1.1
SS B1	.8	1.1	1.4
SS B2	.6	.9	1.2
SS B3	.8	1.1	1.5
SS B4	.6	.9	1.2
SS B5	.8	1.2	1.7
SS B6	.6	.9	1.2
SS B7	.9	1.4	1.8
SS B8	.7	.9	1.2
SS C1	.7	1.0	1.3
SS C2	.7	.9	1.1
SS C3	.9	1.2	1.6
SS C4	.7	1.0	1.2
SS D1	.6	.8	1.1
SS D2	.5	.7	.9
SS D3	.6	.9	1.2
SS D4	.5	.7	1.0
SS E1	.6	.9	1.2
SS E2	.6	.8	1.0
SS E3	.7	1.1	1.4
SS E4	.6	.8	1.1

LOVELOCK, NV

SOLAR SAVINGS

	10%	50%	80%
WW A1	1.2	1.9	2.6
WW A2	1.0	1.4	1.9
WW A3	1.0	1.3	1.8
WW A4	.9	1.3	1.7
WW A5	.9	1.2	1.6
WW A6	.9	1.2	1.6
WW B1	1.2	1.8	2.5
WW B2	.9	1.2	1.6
WW B3	.8	1.1	1.4
WW B4	.8	1.0	1.3
WW B5	.8	1.0	1.3
WW C1	.8	1.0	1.4
WW C2	.8	1.1	1.4
WW C3	.7	.9	1.2
WW C4	.8	1.0	1.2
TW A1	1.2	1.9	2.6
TW A2	1.1	1.6	2.1
TW A3	1.0	1.5	1.9
TW A4	1.0	1.4	1.9
TW B1	1.2	1.8	2.5
TW B2	1.1	1.6	2.1
TW B3	1.1	1.5	2.1
TW B4	1.2	1.6	2.1
TW C1	1.2	1.8	2.4
TW C2	1.2	1.7	2.3
TW C3	1.3	1.8	2.4
TW C4	1.4	2.0	2.6
TW D1	1.3	2.0	2.8
TW D2	1.0	1.3	1.7
TW D3	.9	1.2	1.6

LOVELOCK, NV

SOLAR SAVINGS

	10%	50%	80%
TW D4	.9	1.1	1.5
TW D5	.9	1.1	1.5
TW E1	.8	1.1	1.5
TW E2	.8	1.1	1.5
TW E3	.7	1.0	1.3
TW E4	.8	1.0	1.3
TW F1	1.3	1.9	2.6
TW F2	1.1	1.6	2.2
TW F3	1.1	1.5	2.1
TW F4	1.1	1.6	2.1
TW G1	1.3	1.9	2.5
TW G2	1.2	1.7	2.3
TW G3	1.2	1.7	2.3
TW G4	1.4	1.8	2.4
TW H1	1.4	1.9	2.6
TW H2	1.4	1.9	2.6
TW H3	1.5	2.1	2.7
TW H4	1.7	2.4	3.1
TW I1	1.4	2.1	3.0
TW I2	1.0	1.4	1.8
TW I3	1.0	1.3	1.7
TW I4	.9	1.2	1.5
TW I5	.9	1.2	1.5
TW J1	.9	1.2	1.5
TW J2	.9	1.2	1.5
TW J3	.8	1.0	1.3
TW J4	.8	1.1	1.3
DG A1	1.2	2.8	---
DG A2	1.1	1.6	3.1
DG A3	1.0	1.3	1.8
DG B1	1.2	1.8	---
DG B2	1.1	1.4	2.2
DG B3	1.0	1.2	1.6
DG C1	1.1	1.5	3.4
DG C2	1.0	1.3	1.8
DG C3	.9	1.1	1.4
SS A1	.9	1.4	2.0
SS A2	.8	1.1	1.5
SS A3	1.0	1.6	2.3
SS A4	.8	1.1	1.6
SS A5	.9	1.6	2.3
SS A6	.8	1.1	1.5
SS A7	1.0	1.9	2.8
SS A8	.8	1.2	1.6
SS B1	1.0	1.6	2.2
SS B2	.8	1.2	1.6
SS B3	1.1	1.7	2.4
SS B4	.9	1.2	1.7
SS B5	1.1	1.9	2.6
SS B6	.9	1.2	1.7
SS B7	1.2	2.1	3.1
SS B8	.9	1.3	1.7
SS C1	1.0	1.5	2.0
SS C2	.9	1.2	1.6
SS C3	1.2	1.8	2.4
SS C4	1.0	1.4	1.8
SS D1	.8	1.2	1.7
SS D2	.7	1.0	1.3
SS D3	.9	1.4	1.9
SS D4	.7	1.0	1.3
SS E1	.9	1.3	1.9
SS E2	.8	1.1	1.4
SS E3	1.0	1.6	2.3
SS E4	.8	1.2	1.6

RENO, NV

	SOLAR SAVINGS		
	10%	50%	80%
WW A1	1.2	1.8	2.5
WW A2	1.0	1.4	1.9
WW A3	.9	1.3	1.7
WW A4	.9	1.2	1.6
WW A5	.9	1.2	1.6
WW A6	.9	1.2	1.5
WW B1	1.2	1.7	2.4
WW B2	.9	1.2	1.6
WW B3	.8	1.1	1.4
WW B4	.8	1.0	1.3
WW B5	.8	1.0	1.3
WW C1	.8	1.0	1.3
WW C2	.8	1.0	1.3
WW C3	.7	.9	1.2
WW C4	.7	1.0	1.2
TW A1	1.2	1.9	2.5
TW A2	1.1	1.5	2.0
TW A3	1.0	1.4	1.9
TW A4	1.0	1.4	1.8
TW B1	1.2	1.8	2.4
TW B2	1.1	1.5	2.1
TW B3	1.1	1.5	2.0
TW B4	1.1	1.6	2.1
TW C1	1.2	1.7	2.3
TW C2	1.2	1.7	2.2
TW C3	1.2	1.7	2.3
TW C4	1.4	1.9	2.5
TW D1	1.3	1.9	2.6
TW D2	1.0	1.3	1.7
TW D3	.9	1.2	1.6
TW D4	.8	1.1	1.4
TW D5	.9	1.1	1.4
TW E1	.8	1.1	1.5
TW E2	.8	1.1	1.4
TW E3	.7	1.0	1.2
TW E4	.8	1.0	1.3
TW F1	1.2	1.8	2.5
TW F2	1.1	1.6	2.1
TW F3	1.1	1.5	2.0
TW F4	1.1	1.5	2.0
TW G1	1.3	1.8	2.4
TW G2	1.2	1.7	2.2
TW G3	1.2	1.7	2.2
TW G4	1.3	1.8	2.4
TW H1	1.3	1.9	2.5
TW H2	1.4	1.9	2.5
TW H3	1.5	2.0	2.7
TW H4	1.7	2.3	3.1
TW I1	1.4	2.0	2.8
TW I2	1.0	1.4	1.8
TW I3	1.0	1.3	1.7
TW I4	.9	1.2	1.5
TW I5	.9	1.1	1.5
TW J1	.9	1.2	1.5
TW J2	.9	1.2	1.5
TW J3	.8	1.0	1.3
TW J4	.8	1.0	1.3
DG A1	1.2	2.6	---
DG A2	1.1	1.6	2.8
DG A3	1.0	1.2	1.8
DG B1	1.2	1.8	---
DG B2	1.1	1.4	2.1
DG B3	1.0	1.2	1.5
DG C1	1.1	1.4	2.8
DG C2	1.0	1.2	1.7
DG C3	.9	1.1	1.3
SS A1	.9	1.4	1.9
SS A2	.7	1.1	1.5
SS A3	.9	1.5	2.2
SS A4	.7	1.1	1.5
SS A5	.9	1.5	2.2
SS A6	.7	1.1	1.5
SS A7	1.0	1.8	2.6
SS A8	.7	1.1	1.5
SS B1	1.0	1.5	2.1
SS B2	.8	1.2	1.6
SS B3	1.0	1.6	2.3
SS B4	.8	1.2	1.6
SS B5	1.1	1.8	2.5
SS B6	.8	1.2	1.6
SS B7	1.2	2.0	2.9
SS B8	.9	1.2	1.7
SS C1	1.0	1.4	1.9
SS C2	.9	1.2	1.6
SS C3	1.2	1.7	2.3
SS C4	1.0	1.3	1.7
SS D1	.8	1.1	1.6
SS D2	.7	.9	1.2
SS D3	.8	1.3	1.8
SS D4	.7	1.0	1.3
SS E1	.9	1.3	1.8
SS E2	.7	1.0	1.4
SS E3	1.0	1.5	2.2
SS E4	.8	1.1	1.5

TONOPAH, NV

	SOLAR SAVINGS		
	10%	50%	80%
WW A1	1.2	1.8	2.4
WW A2	1.0	1.4	1.8
WW A3	.9	1.3	1.6
WW A4	.9	1.2	1.6
WW A5	.9	1.2	1.5
WW A6	.9	1.1	1.4
WW B1	1.1	1.6	2.2
WW B2	.9	1.2	1.5
WW B3	.8	1.0	1.3
WW B4	.8	1.0	1.3
WW B5	.8	1.0	1.3
WW C1	.8	1.0	1.3
WW C2	.8	1.0	1.3
WW C3	.7	.9	1.1
WW C4	.7	.9	1.2
TW A1	1.2	1.8	2.4
TW A2	1.1	1.5	1.9
TW A3	1.0	1.4	1.8
TW A4	1.0	1.4	1.8
TW B1	1.2	1.7	2.3
TW B2	1.1	1.5	2.0
TW B3	1.1	1.5	1.9
TW B4	1.1	1.5	2.0
TW C1	1.2	1.7	2.2
TW C2	1.2	1.6	2.1
TW C3	1.2	1.7	2.2
TW C4	1.4	1.9	2.4
TW D1	1.3	1.8	2.4
TW D2	.9	1.3	1.6
TW D3	.9	1.2	1.5
TW D4	.8	1.1	1.4
TW D5	.8	1.1	1.4
TW E1	.8	1.1	1.4
TW E2	.8	1.1	1.4
TW E3	.7	.9	1.2
TW E4	.8	1.0	1.2
TW F1	1.2	1.8	2.3
TW F2	1.1	1.5	2.0
TW F3	1.1	1.5	1.9
TW F4	1.1	1.5	1.9
TW G1	1.3	1.8	2.3
TW G2	1.2	1.6	2.1
TW G3	1.2	1.6	2.1
TW G4	1.3	1.8	2.3
TW H1	1.3	1.8	2.4
TW H2	1.4	1.8	2.4
TW H3	1.5	2.0	2.5
TW H4	1.7	2.3	2.9
TW I1	1.4	1.9	2.6
TW I2	1.0	1.3	1.7
TW I3	.9	1.2	1.6
TW I4	.9	1.1	1.5
TW I5	.9	1.1	1.4
TW J1	.9	1.1	1.4
TW J2	.9	1.1	1.4
TW J3	.8	1.0	1.2
TW J4	.8	1.0	1.3
DG A1	1.2	2.3	---
DG A2	1.1	1.5	2.6
DG A3	1.0	1.2	1.7
DG B1	1.2	1.6	---
DG B2	1.1	1.3	2.0
DG B3	1.0	1.1	1.5
DG C1	1.1	1.3	2.3
DG C2	1.0	1.2	1.6
DG C3	.9	1.0	1.3
SS A1	.9	1.3	1.8
SS A2	.7	1.0	1.4
SS A3	.9	1.4	2.0
SS A4	.7	1.1	1.4
SS A5	.9	1.5	2.0
SS A6	.7	1.1	1.4
SS A7	1.0	1.7	2.4
SS A8	.8	1.1	1.5
SS B1	1.0	1.5	2.0
SS B2	.8	1.1	1.5
SS B3	1.0	1.6	2.1
SS B4	.8	1.2	1.6
SS B5	1.1	1.7	2.3
SS B6	.8	1.2	1.5
SS B7	1.2	1.9	2.7
SS B8	.9	1.2	1.6
SS C1	1.0	1.4	1.8
SS C2	.9	1.2	1.5
SS C3	1.2	1.7	2.2
SS C4	1.0	1.3	1.7
SS D1	.8	1.1	1.5
SS D2	.7	.9	1.2
SS D3	.8	1.2	1.7
SS D4	.7	.9	1.2
SS E1	.9	1.2	1.7
SS E2	.7	1.0	1.3
SS E3	1.0	1.5	2.0
SS E4	.8	1.1	1.4

WINNEMUCCA, NV

	SOLAR SAVINGS		
	10%	50%	80%
WW A1	1.4	2.3	3.3
WW A2	1.1	1.7	2.3
WW A3	1.1	1.5	2.1
WW A4	1.0	1.5	1.9
WW A5	1.0	1.4	1.9
WW A6	1.0	1.4	1.8
WW B1	1.4	2.2	3.5
WW B2	1.0	1.4	1.8
WW B3	.9	1.2	1.6
WW B4	.9	1.1	1.5
WW B5	.9	1.2	1.5
WW C1	.9	1.2	1.5
WW C2	.9	1.2	1.5
WW C3	.8	1.0	1.3
WW C4	.8	1.1	1.4
TW A1	1.4	2.3	3.2
TW A2	1.2	1.8	2.5
TW A3	1.2	1.7	2.3
TW A4	1.2	1.6	2.2
TW B1	1.4	2.1	3.0
TW B2	1.3	1.8	2.5
TW B3	1.2	1.8	2.4
TW B4	1.3	1.8	2.5
TW C1	1.4	2.1	2.8
TW C2	1.4	2.0	2.6
TW C3	1.4	2.0	2.7
TW C4	1.5	2.2	3.0
TW D1	1.5	2.4	3.7
TW D2	1.1	1.5	2.0
TW D3	1.0	1.4	1.8
TW D4	.9	1.3	1.6
TW D5	.9	1.2	1.6
TW E1	.9	1.3	1.7
TW E2	.9	1.3	1.6
TW E3	.8	1.1	1.4
TW E4	.8	1.1	1.4
TW F1	1.4	2.2	3.2
TW F2	1.3	1.9	2.6
TW F3	1.3	1.8	2.4
TW F4	1.3	1.8	2.4
TW G1	1.5	2.2	3.0
TW G2	1.4	2.0	2.7
TW G3	1.4	2.0	2.6
TW G4	1.5	2.1	2.8
TW H1	1.5	2.2	3.0
TW H2	1.6	2.2	3.0
TW H3	1.7	2.4	3.2
TW H4	1.9	2.7	3.6
TW I1	1.6	2.6	4.0
TW I2	1.1	1.6	2.1
TW I3	1.1	1.5	1.9
TW I4	1.0	1.3	1.7
TW I5	1.0	1.3	1.7
TW J1	1.0	1.3	1.7
TW J2	1.0	1.3	1.7
TW J3	.9	1.1	1.4
TW J4	.9	1.2	1.5
DG A1	1.4	---	---
DG A2	1.3	2.0	---
DG A3	1.1	1.4	2.1
DG B1	1.4	2.8	---
DG B2	1.2	1.7	3.0
DG B3	1.1	1.3	1.8
DG C1	1.2	1.8	---

WINNEMUCCA, NV

	SOLAR SAVINGS		
	10%	50%	80%
DG C2	1.1	1.4	2.1
DG C3	1.0	1.2	1.5
SS A1	1.0	1.6	2.3
SS A2	.8	1.2	1.7
SS A3	1.1	1.8	2.8
SS A4	.8	1.3	1.8
SS A5	1.1	1.8	2.7
SS A6	.8	1.2	1.7
SS A7	1.2	2.2	3.7
SS A8	.8	1.3	1.8
SS B1	1.1	1.8	2.6
SS B2	.9	1.3	1.8
SS B3	1.2	2.0	3.0
SS B4	.9	1.4	1.9
SS B5	1.3	2.1	3.2
SS B6	.9	1.4	1.9
SS B7	1.4	2.5	4.0
SS B8	1.0	1.4	2.0
SS C1	1.2	1.7	2.3
SS C2	1.0	1.4	1.8
SS C3	1.4	2.1	2.9
SS C4	1.1	1.5	2.0
SS D1	.9	1.4	2.0
SS D2	.7	1.1	1.4
SS D3	.9	1.5	2.2
SS D4	.8	1.1	1.5
SS E1	1.0	1.5	2.2
SS E2	.8	1.2	1.6
SS E3	1.1	1.9	2.7
SS E4	.9	1.3	1.8

CONCORD, NH

	SOLAR SAVINGS		
	10%	50%	80%
WW A1	2.8	---	---
WW A2	2.3	---	---
WW A3	2.2	---	---
WW A4	2.1	---	---
WW A5	2.0	5.6	---
WW A6	1.9	4.4	---
WW B1	---	---	---
WW B2	1.8	2.9	4.8
WW B3	1.5	2.3	3.5
WW B4	1.3	1.9	2.5
WW B5	1.3	1.8	2.4
WW C1	1.4	2.1	3.0
WW C2	1.4	2.0	2.7
WW C3	1.1	1.5	2.0
WW C4	1.2	1.6	2.0
TW A1	2.7	---	---
TW A2	2.4	---	---
TW A3	2.3	---	---
TW A4	2.2	6.9	---
TW B1	2.7	---	---
TW B2	2.4	---	---
TW B3	2.3	8.7	---
TW B4	2.4	5.5	---
TW C1	2.6	---	---
TW C2	2.5	7.2	---
TW C3	2.5	5.9	---
TW C4	2.6	5.7	---
TW D1	---	---	---
TW D2	1.9	3.3	5.6
TW D3	1.7	3.0	4.9

CONCORD, NH

	SOLAR SAVINGS		
	10%	50%	80%
TW D4	1.5	2.1	2.9
TW D5	1.4	2.0	2.6
TW E1	1.5	2.4	3.5
TW E2	1.5	2.2	3.1
TW E3	1.2	1.7	2.2
TW E4	1.2	1.7	2.2
TW F1	3.2	---	---
TW F2	2.7	---	---
TW F3	2.6	---	---
TW F4	2.5	---	---
TW G1	3.3	---	---
TW G2	2.8	---	---
TW G3	2.8	---	---
TW G4	2.9	7.2	---
TW H1	3.3	---	---
TW H2	3.1	---	---
TW H3	3.2	8.8	---
TW H4	3.6	7.5	---
TW I1	---	---	---
TW I2	2.0	3.4	5.9
TW I3	1.8	3.0	4.6
TW I4	1.5	2.2	3.0
TW I5	1.5	2.0	2.7
TW J1	1.6	2.4	3.5
TW J2	1.5	2.2	3.1
TW J3	1.3	1.7	2.2
TW J4	1.3	1.7	2.3
DG A1	---	---	---
DG A2	2.3	---	---
DG A3	1.7	2.5	---
DG B1	---	---	---
DG B2	2.3	---	---
DG B3	1.6	2.3	3.9
DG C1	4.4	---	---
DG C2	1.9	5.0	---
DG C3	1.5	2.0	2.9
SS A1	1.8	---	---
SS A2	1.3	2.2	3.2
SS A3	2.1	---	---
SS A4	1.3	2.4	3.7
SS A5	1.8	---	---
SS A6	1.3	2.3	3.4
SS A7	2.1	---	---
SS A8	1.4	2.6	4.5
SS B1	2.1	---	---
SS B2	1.4	2.3	3.3
SS B3	2.5	---	---
SS B4	1.5	2.5	3.7
SS B5	2.4	---	---
SS B6	1.5	2.4	3.6
SS B7	3.0	---	---
SS B8	1.6	2.8	4.4
SS C1	2.7	---	---
SS C2	1.7	2.8	4.4
SS C3	3.1	---	---
SS C4	1.9	3.2	5.2
SS D1	1.6	---	---
SS D2	1.2	1.8	2.6
SS D3	1.7	---	---
SS D4	1.2	1.9	2.7
SS E1	2.0	---	---
SS E2	1.4	2.2	3.4
SS E3	2.2	---	---
SS E4	1.4	2.5	3.8

NEWARK, NJ

	SOLAR SAVINGS		
	10%	50%	80%
WW A1	1.9	4.5	---
WW A2	1.5	2.5	3.7
WW A3	1.4	2.2	3.1
WW A4	1.4	2.1	2.9
WW A5	1.3	2.0	2.7
WW A6	1.3	1.9	2.6
WW B1	2.2	---	---
WW B2	1.3	1.8	2.5
WW B3	1.1	1.6	2.1
WW B4	1.1	1.4	1.9
WW B5	1.1	1.4	1.8
WW C1	1.1	1.5	2.0
WW C2	1.1	1.5	2.0
WW C3	.9	1.2	1.6
WW C4	1.0	1.3	1.6
TW A1	1.9	4.0	8.3
TW A2	1.6	2.7	4.0
TW A3	1.6	2.4	3.5
TW A4	1.5	2.3	3.2
TW B1	1.8	3.4	5.4
TW B2	1.7	2.6	3.7
TW B3	1.6	2.5	3.5
TW B4	1.7	2.5	3.5
TW C1	1.8	3.0	4.4
TW C2	1.8	2.7	3.8
TW C3	1.8	2.8	3.9
TW C4	2.0	3.1	4.2
TW D1	2.4	---	---
TW D2	1.4	2.0	2.7
TW D3	1.3	1.9	2.5
TW D4	1.2	1.6	2.1
TW D5	1.1	1.5	2.0
TW E1	1.2	1.7	2.2
TW E2	1.2	1.6	2.1
TW E3	1.0	1.3	1.7
TW E4	1.0	1.4	1.8
TW F1	2.0	3.9	7.3
TW F2	1.7	2.8	4.2
TW F3	1.7	2.6	3.7
TW F4	1.7	2.5	3.5
TW G1	2.0	3.5	5.6
TW G2	1.8	2.9	4.1
TW G3	1.9	2.8	3.9
TW G4	2.0	3.0	4.1
TW H1	2.1	3.4	5.0
TW H2	2.1	3.2	4.5
TW H3	2.2	3.3	4.6
TW H4	2.5	3.7	5.1
TW I1	2.6	---	---
TW I2	1.5	2.1	2.9
TW I3	1.4	1.9	2.6
TW I4	1.2	1.7	2.2
TW I5	1.2	1.6	2.1
TW J1	1.2	1.7	2.3
TW J2	1.2	1.7	2.2
TW J3	1.0	1.4	1.8
TW J4	1.1	1.4	1.8
DG A1	2.2	---	---
DG A2	1.7	4.0	---
DG A3	1.4	1.8	3.1
DG B1	2.2	---	---
DG B2	1.6	2.6	---
DG B3	1.3	1.7	2.4
DG C1	1.7	---	---

NEWARK, NJ

	SOLAR SAVINGS		
	10%	50%	80%
DG C2	1.4	2.0	5.4
DG C3	1.2	1.5	2.0
SS A1	1.3	2.3	3.5
SS A2	1.1	1.6	2.2
SS A3	1.5	3.0	5.5
SS A4	1.1	1.7	2.3
SS A5	1.4	2.8	4.8
SS A6	1.1	1.6	2.2
SS A7	1.6	4.8	---
SS A8	1.1	1.7	2.4
SS B1	1.5	2.7	4.1
SS B2	1.2	1.7	2.3
SS B3	1.6	3.3	5.9
SS B4	1.2	1.8	2.4
SS B5	1.7	3.5	6.4
SS B6	1.2	1.8	2.4
SS B7	1.9	5.7	---
SS B8	1.2	1.9	2.6
SS C1	1.6	2.6	3.9
SS C2	1.3	1.8	2.4
SS C3	1.9	3.3	5.2
SS C4	1.4	2.0	2.7
SS D1	1.2	1.9	2.9
SS D2	.9	1.4	1.8
SS D3	1.3	2.2	3.4
SS D4	1.0	1.4	1.9
SS E1	1.3	2.4	3.7
SS E2	1.1	1.6	2.1
SS E3	1.5	3.0	5.0
SS E4	1.1	1.7	2.3

ALBUQUERQUE, NM

	SOLAR SAVINGS		
	10%	50%	80%
WW A1	1.1	1.6	2.1
WW A2	.9	1.3	1.6
WW A3	.9	1.2	1.5
WW A4	.9	1.1	1.4
WW A5	.8	1.1	1.4
WW A6	.8	1.1	1.3
WW B1	1.1	1.5	2.0
WW B2	.8	1.1	1.4
WW B3	.7	1.0	1.2
WW B4	.7	.9	1.2
WW B5	.8	1.0	1.2
WW C1	.7	.9	1.2
WW C2	.7	1.0	1.2
WW C3	.7	.8	1.1
WW C4	.7	.9	1.1
TW A1	1.2	1.7	2.2
TW A2	1.0	1.4	1.8
TW A3	1.0	1.3	1.7
TW A4	1.0	1.3	1.6
TW B1	1.1	1.6	2.1
TW B2	1.0	1.4	1.8
TW B3	1.0	1.4	1.8
TW B4	1.1	1.4	1.8
TW C1	1.2	1.6	2.0
TW C2	1.1	1.5	2.0
TW C3	1.2	1.6	2.0
TW C4	1.3	1.8	2.3
TW D1	1.2	1.7	2.2
TW D2	.9	1.2	1.5
TW D3	.8	1.1	1.4

ALBUQUERQUE, NM
SOLAR SAVINGS

	10%	50%	80%
WW A1	1.1	1.6	2.1
WW A2	.9	1.3	1.6
WW A3	.9	1.2	1.5
WW A4	.9	1.1	1.4
WW A5	.8	1.1	1.4
WW A6	.8	1.1	1.3
WW B1	1.1	1.5	2.0
WW B2	.8	1.1	1.4
WW B3	.7	1.0	1.2
WW B4	.7	.9	1.2
WW B5	.8	1.0	1.2
WW C1	.7	.9	1.2
WW C2	.7	1.0	1.2
WW C3	.7	.8	1.1
WW C4	.7	.9	1.1
TW A1	1.2	1.7	2.2
TW A2	1.0	1.4	1.8
TW A3	1.0	1.3	1.7
TW A4	1.0	1.3	1.6
TW B1	1.1	1.6	2.1
TW B2	1.0	1.4	1.8
TW B3	1.0	1.4	1.8
TW B4	1.1	1.4	1.8
TW C1	1.2	1.6	2.0
TW C2	1.1	1.5	2.0
TW C3	1.2	1.6	2.0
TW C4	1.3	1.8	2.3
TW D1	1.2	1.7	2.2
TW D2	.9	1.2	1.5
TW D3	.8	1.1	1.4
DG B1	1.1	1.5	3.7
DG B2	1.0	1.2	1.7
DG B3	.9	1.1	1.4
DG C1	1.0	1.2	1.9
DG C2	1.0	1.1	1.5
DG C3	.9	1.0	1.2
SS A1	.9	1.2	1.6
SS A2	.7	1.0	1.3
SS A3	.9	1.3	1.8
SS A4	.7	1.0	1.3
SS A5	.9	1.4	1.8
SS A6	.7	1.0	1.3
SS A7	1.0	1.6	2.1
SS A8	.7	1.0	1.4
SS B1	1.0	1.4	1.8
SS B2	.8	1.1	1.4
SS B3	1.0	1.5	1.9
SS B4	.8	1.1	1.4
SS B5	1.1	1.6	2.1
SS B6	.8	1.1	1.4
SS B7	1.2	1.8	2.4
SS B8	.8	1.1	1.5
SS C1	1.0	1.3	1.7
SS C2	.8	1.1	1.4
SS C3	1.1	1.6	2.0
SS C4	.9	1.2	1.6
SS D1	.7	1.0	1.4
SS D2	.6	.9	1.1
SS D3	.8	1.2	1.6
SS D4	.7	.9	1.2
SS E1	.8	1.2	1.5
SS E2	.7	1.0	1.2
SS E3	1.0	1.4	1.8
SS E4	.8	1.0	1.3

CLAYTON, NM
SOLAR SAVINGS

	10%	50%	80%
WW A1	1.2	1.8	2.4
WW A2	1.0	1.4	1.8
WW A3	.9	1.3	1.6
WW A4	.9	1.2	1.6
WW A5	.9	1.2	1.5
WW A6	.9	1.1	1.5
WW B1	1.2	1.7	2.2
WW B2	.9	1.2	1.5
WW B3	.8	1.0	1.3
WW B4	.8	1.0	1.3
WW B5	.8	1.0	1.3
WW C1	.8	1.0	1.3
WW C2	.8	1.0	1.3
WW C3	.7	.9	1.1
WW C4	.7	.9	1.2
TW A1	1.2	1.8	2.4
TW A2	1.1	1.5	1.9
TW A3	1.0	1.4	1.8
TW A4	1.0	1.4	1.8
TW B1	1.2	1.7	2.3
TW B2	1.1	1.5	2.0
TW B3	1.1	1.5	1.9
TW B4	1.1	1.6	2.0
TW C1	1.2	1.7	2.2
TW C2	1.2	1.6	2.1
TW C3	1.3	1.7	2.2
TW C4	1.4	1.9	2.4
TW D1	1.3	1.9	2.4
TW D2	1.0	1.3	1.6
TW D3	.9	1.2	1.5
TW D4	.8	1.1	1.4
TW D5	.9	1.1	1.4
TW E1	.8	1.1	1.4
TW E2	.8	1.1	1.4
TW E3	.7	.9	1.2
TW E4	.8	1.0	1.2
TW F1	1.3	1.8	2.3
TW F2	1.1	1.6	2.0
TW F3	1.1	1.5	1.9
TW F4	1.1	1.5	1.9
TW G1	1.3	1.8	2.3
TW G2	1.2	1.6	2.1
TW G3	1.2	1.6	2.1
TW G4	1.3	1.8	2.3
TW H1	1.4	1.8	2.4
TW H2	1.4	1.9	2.4
TW H3	1.5	2.0	2.5
TW H4	1.7	2.3	2.9
TW I1	1.4	2.0	2.5
TW I2	1.0	1.3	1.7
TW I3	1.0	1.3	1.6
TW I4	.9	1.2	1.5
TW I5	.9	1.1	1.4
TW J1	.9	1.1	1.4
TW J2	.9	1.1	1.4
TW J3	.8	1.0	1.2
TW J4	.8	1.0	1.3
DG A1	1.2	2.4	---
DG A2	1.1	1.6	2.6
DG A3	1.0	1.2	1.7
DG B1	1.2	1.7	---
DG B2	1.1	1.3	2.0
DG B3	1.0	1.1	1.5
DG C1	1.1	1.3	2.3

CLAYTON, NM
SOLAR SAVINGS

	10%	50%	80%
DG C2	1.0	1.2	1.6
DG C3	.9	1.1	1.3
SS A1	.9	1.3	1.8
SS A2	.8	1.1	1.4
SS A3	.9	1.4	1.9
SS A4	.8	1.1	1.4
SS A5	.9	1.5	2.0
SS A6	.8	1.1	1.4
SS A7	1.0	1.7	2.3
SS A8	.8	1.1	1.4
SS B1	1.0	1.5	2.0
SS B2	.8	1.2	1.5
SS B3	1.1	1.6	2.1
SS B4	.8	1.2	1.5
SS B5	1.1	1.7	2.3
SS B6	.8	1.2	1.5
SS B7	1.2	1.9	2.6
SS B8	.9	1.2	1.6
SS C1	1.0	1.4	1.8
SS C2	.9	1.2	1.5
SS C3	1.2	1.7	2.2
SS C4	1.0	1.3	1.7
SS D1	.8	1.1	1.5
SS D2	.7	.9	1.2
SS D3	.8	1.3	1.7
SS D4	.7	.9	1.2
SS E1	.9	1.2	1.6
SS E2	.8	1.0	1.3
SS E3	1.0	1.5	2.0
SS E4	.8	1.1	1.4

FARMINGTON, NM
SOLAR SAVINGS

	10%	50%	80%
WW A1	1.3	1.9	2.6
WW A2	1.1	1.5	1.9
WW A3	1.0	1.3	1.8
WW A4	1.0	1.3	1.7
WW A5	.9	1.2	1.6
WW A6	.9	1.2	1.6
WW B1	1.3	1.8	2.5
WW B2	.9	1.2	1.6
WW B3	.8	1.1	1.4
WW B4	.8	1.1	1.3
WW B5	.8	1.1	1.3
WW C1	.8	1.1	1.4
WW C2	.8	1.1	1.4
WW C3	.7	.9	1.2
WW C4	.8	1.0	1.2
TW A1	1.3	1.9	2.6
TW A2	1.1	1.6	2.1
TW A3	1.1	1.5	1.9
TW A4	1.1	1.5	1.9
TW B1	1.3	1.9	2.5
TW B2	1.2	1.6	2.1
TW B3	1.2	1.6	2.1
TW B4	1.2	1.6	2.1
TW C1	1.3	1.8	2.4
TW C2	1.3	1.7	2.3
TW C3	1.3	1.8	2.4
TW C4	1.4	2.0	2.6
TW D1	1.4	2.0	2.7
TW D2	1.0	1.4	1.7
TW D3	.9	1.3	1.6

FARMINGTON, NM
SOLAR SAVINGS

	10%	50%	80%
TW D4	.9	1.2	1.5
TW D5	.9	1.2	1.5
TW E1	.9	1.2	1.5
TW E2	.9	1.2	1.5
TW E3	.8	1.0	1.3
TW E4	.8	1.0	1.3
TW F1	1.3	1.9	2.6
TW F2	1.2	1.7	2.2
TW F3	1.2	1.6	2.0
TW F4	1.2	1.6	2.1
TW G1	1.3	1.9	2.5
TW G2	1.3	1.7	2.2
TW G3	1.3	1.7	2.3
TW G4	1.4	1.9	2.4
TW H1	1.4	2.0	2.6
TW H2	1.5	2.0	2.5
TW H3	1.6	2.1	2.7
TW H4	1.8	2.4	3.1
TW I1	1.5	2.1	2.9
TW I2	1.1	1.4	1.8
TW I3	1.0	1.3	1.7
TW I4	.9	1.2	1.5
TW I5	.9	1.2	1.5
TW J1	.9	1.2	1.5
TW J2	.9	1.2	1.5
TW J3	.8	1.0	1.3
TW J4	.8	1.1	1.4
DG A1	1.3	3.2	---
DG A2	1.2	1.7	3.1
DG A3	1.1	1.3	1.8
DG B1	1.3	1.9	---
DG B2	1.2	1.4	2.2
DG B3	1.1	1.2	1.6
DG C1	1.2	1.5	3.3
DG C2	1.1	1.3	1.8
DG C3	1.0	1.1	1.4
SS A1	.9	1.4	1.9
SS A2	.8	1.1	1.5
SS A3	1.0	1.6	2.2
SS A4	.8	1.1	1.5
SS A5	1.0	1.6	2.2
SS A6	.8	1.1	1.5
SS A7	1.1	1.8	2.6
SS A8	.8	1.2	1.6
SS B1	1.1	1.6	2.1
SS B2	.9	1.2	1.6
SS B3	1.1	1.7	2.3
SS B4	.9	1.2	1.6
SS B5	1.2	1.8	2.5
SS B6	.9	1.2	1.6
SS B7	1.3	2.1	2.9
SS B8	.9	1.3	1.7
SS C1	1.1	1.5	1.9
SS C2	.9	1.2	1.6
SS C3	1.3	1.8	2.4
SS C4	1.0	1.4	1.8
SS D1	.8	1.2	1.6
SS D2	.7	1.0	1.3
SS D3	.9	1.3	1.8
SS D4	.7	1.0	1.3
SS E1	.9	1.3	1.8
SS E2	.8	1.1	1.4
SS E3	1.1	1.6	2.2
SS E4	.8	1.2	1.5

LOS ALAMOS, NM SOLAR SAVINGS				LOS ALAMOS, NM SOLAR SAVINGS				ROSWELL, NM SOLAR SAVINGS				TRUTH OR CONSEQ., NM SOLAR SAVINGS			
	10%	50%	80%		10%	50%	80%		10%	50%	80%		10%	50%	80%
WW A1	1.4	2.1	2.7	DG C2	1.1	1.3	1.9	TW D4	.8	1.0	1.3	WW A1	1.0	1.4	1.8
WW A2	1.1	1.5	2.0	DG C3	1.0	1.2	1.4	TW D5	.8	1.0	1.3	WW A2	.8	1.1	1.4
WW A3	1.1	1.4	1.8	SS A1	1.0	1.4	1.9	TW E1	.7	1.0	1.3	WW A3	.8	1.0	1.3
WW A4	1.0	1.4	1.7	SS A2	.8	1.1	1.5	TW E2	.8	1.0	1.3	WW A4	.7	1.0	1.2
WW A5	1.0	1.3	1.7	SS A3	1.0	1.6	2.1	TW E3	.7	.9	1.1	WW A5	.7	.9	1.2
WW A6	1.0	1.3	1.6	SS A4	.8	1.2	1.5	TW E4	.7	.9	1.1	WW A6	.7	.9	1.2
WW B1	1.4	1.9	2.5	SS A5	1.0	1.6	2.2	TW F1	1.1	1.6	2.0	WW B1	.9	1.3	1.7
WW B2	1.0	1.3	1.7	SS A6	.8	1.1	1.5	TW F2	1.0	1.4	1.8	WW B2	.7	1.0	1.2
WW B3	.9	1.1	1.4	SS A7	1.1	1.9	2.6	TW F3	1.0	1.3	1.7	WW B3	.7	.9	1.1
WW B4	.9	1.1	1.4	SS A8	.8	1.2	1.6	TW F4	1.0	1.3	1.7	WW B4	.7	.8	1.1
WW B5	.9	1.1	1.4	SS B1	1.1	1.6	2.1	TW G1	1.1	1.6	2.0	WW B5	.7	.9	1.1
WW C1	.9	1.1	1.4	SS B2	.9	1.2	1.6	TW G2	1.1	1.4	1.8	WW C1	.6	.8	1.1
WW C2	.9	1.1	1.4	SS B3	1.2	1.7	2.3	TW G3	1.1	1.4	1.9	WW C2	.7	.9	1.1
WW C3	.8	1.0	1.2	SS B4	.9	1.3	1.7	TW G4	1.2	1.6	2.0	WW C3	.6	.8	1.0
WW C4	.8	1.0	1.3	SS B5	1.2	1.9	2.5	TW H1	1.2	1.6	2.1	WW C4	.6	.8	1.0
TW A1	1.4	2.1	2.7	SS B6	.9	1.3	1.6	TW H2	1.2	1.6	2.1	TW A1	1.0	1.4	1.9
TW A2	1.2	1.7	2.2	SS B7	1.4	2.1	2.9	TW H3	1.3	1.8	2.3	TW A2	.9	1.2	1.5
TW A3	1.2	1.6	2.0	SS B8	.9	1.3	1.7	TW H4	1.5	2.0	2.6	TW A3	.8	1.1	1.4
TW A4	1.2	1.5	2.0	SS C1	1.2	1.6	2.0	TW I1	1.2	1.7	2.2	TW A4	.8	1.1	1.4
TW B1	1.4	2.0	2.6	SS C2	1.0	1.3	1.7	TW I2	.9	1.2	1.5	TW B1	1.0	1.4	1.8
TW B2	1.2	1.7	2.2	SS C3	1.4	1.9	2.5	TW I3	.9	1.1	1.4	TW B2	.9	1.2	1.6
TW B3	1.2	1.7	2.1	SS C4	1.1	1.4	1.8	TW I4	.8	1.0	1.3	TW B3	.9	1.2	1.5
TW B4	1.3	1.7	2.2	SS D1	.9	1.2	1.6	TW I5	.8	1.0	1.3	TW B4	.9	1.3	1.6
TW C1	1.4	1.9	2.5	SS D2	.7	1.0	1.3	TW J1	.8	1.0	1.3	TW C1	1.0	1.4	1.8
TW C2	1.3	1.8	2.4	SS D3	.9	1.4	1.8	TW J2	.8	1.0	1.3	TW C2	1.0	1.3	1.7
TW C3	1.4	1.9	2.4	SS D4	.7	1.0	1.3	TW J3	.7	.9	1.1	TW C3	1.0	1.4	1.8
TW C4	1.5	2.1	2.7	SS E1	1.0	1.4	1.8	TW J4	.7	.9	1.2	TW C4	1.1	1.5	2.0
TW D1	1.5	2.1	2.8	SS E2	.8	1.1	1.4	DG A1	1.1	1.7	---	TW D1	1.0	1.4	1.8
TW D2	1.1	1.4	1.8	SS E3	1.1	1.7	2.2	DG A2	1.0	1.3	2.0	TW D2	.8	1.0	1.3
TW D3	1.0	1.3	1.7	SS E4	.9	1.2	1.5	DG A3	.9	1.1	1.5	TW D3	.7	1.0	1.2
TW D4	.9	1.2	1.5					DG B1	1.1	1.4	2.7	TW D4	.7	.9	1.2
TW D5	.9	1.2	1.5	ROSWELL, NM SOLAR SAVINGS				DG B2	1.0	1.2	1.6	TW D5	.7	.9	1.2
TW E1	.9	1.2	1.5		10%	50%	80%	DG B3	.9	1.0	1.3	TW E1	.7	.9	1.2
TW E2	.9	1.2	1.5	WW A1	1.0	1.6	2.0	DG C1	1.0	1.1	1.7	TW E2	.7	.9	1.2
TW E3	.8	1.0	1.3	WW A2	.9	1.2	1.6	DG C2	.9	1.1	1.4	TW E3	.6	.8	1.0
TW E4	.8	1.1	1.3	WW A3	.8	1.1	1.4	DG C3	.9	.9	1.2	TW E4	.7	.8	1.0
TW F1	1.4	2.0	2.7	WW A4	.8	1.1	1.4	SS A1	.8	1.2	1.6	TW F1	1.0	1.4	1.8
TW F2	1.3	1.7	2.2	WW A5	.8	1.0	1.3	SS A2	.7	.9	1.2	TW F2	.9	1.2	1.6
TW F3	1.2	1.7	2.1	WW A6	.8	1.0	1.3	SS A3	.9	1.3	1.7	TW F3	.9	1.2	1.5
TW F4	1.3	1.7	2.1	WW B1	1.0	1.4	1.9	SS A4	.7	1.0	1.3	TW F4	.9	1.2	1.5
TW G1	1.5	2.0	2.6	WW B2	.8	1.0	1.3	SS A5	.8	1.3	1.7	TW G1	1.0	1.4	1.8
TW G2	1.4	1.8	2.3	WW B3	.7	.9	1.2	SS A6	.7	1.0	1.2	TW G2	1.0	1.3	1.7
TW G3	1.4	1.8	2.3	WW B4	.7	.9	1.1	SS A7	.9	1.5	2.0	TW G3	1.0	1.3	1.7
TW G4	1.5	2.0	2.5	WW B5	.7	.9	1.2	SS A8	.7	1.0	1.3	TW G4	1.1	1.4	1.8
TW H1	1.5	2.1	2.6	WW C1	.7	.9	1.1	SS B1	.9	1.3	1.7	TW H1	1.1	1.5	1.9
TW H2	1.6	2.1	2.6	WW C2	.7	.9	1.2	SS B2	.8	1.0	1.4	TW H2	1.1	1.5	1.9
TW H3	1.7	2.2	2.8	WW C3	.6	.8	1.0	SS B3	1.0	1.4	1.8	TW H3	1.2	1.6	2.0
TW H4	1.9	2.5	3.2	WW C4	.7	.9	1.1	SS B4	.8	1.1	1.4	TW H4	1.4	1.9	2.4
TW I1	1.6	2.3	3.0	TW A1	1.1	1.6	2.1	SS B5	1.0	1.5	2.0	TW I1	1.1	1.5	1.9
TW I2	1.1	1.5	1.9	TW A2	1.0	1.3	1.7	SS B6	.8	1.1	1.4	TW I2	.8	1.1	1.4
TW I3	1.1	1.4	1.8	TW A3	.9	1.2	1.6	SS B7	1.1	1.7	2.2	TW I3	.8	1.0	1.3
TW I4	1.0	1.3	1.6	TW A4	.9	1.2	1.6	SS B8	.8	1.1	1.4	TW I4	.7	1.0	1.2
TW I5	1.0	1.2	1.6	TW B1	1.1	1.5	2.0	SS C1	.9	1.2	1.6	TW I5	.7	1.0	1.2
TW J1	1.0	1.2	1.6	TW B2	1.0	1.3	1.7	SS C2	.8	1.1	1.3	TW J1	.7	.9	1.2
TW J2	1.0	1.3	1.6	TW B3	1.0	1.3	1.7	SS C3	1.1	1.5	1.9	TW J2	.7	.9	1.2
TW J3	.8	1.1	1.3	TW B4	1.0	1.4	1.8	SS C4	.9	1.2	1.5	TW J3	.7	.8	1.0
TW J4	.9	1.1	1.4	TW C1	1.1	1.5	2.0	SS D1	.7	1.0	1.3	TW J4	.7	.9	1.1
DG A1	1.4	---	---	TW C2	1.1	1.5	1.9	SS D2	.6	.8	1.1	DG A1	1.0	1.4	3.3
DG A2	1.3	1.8	3.6	TW C3	1.1	1.5	2.0	SS D3	.8	1.1	1.5	DG A2	.9	1.2	1.7
DG A3	1.1	1.3	1.9	TW C4	1.2	1.7	2.2	SS D4	.6	.9	1.1	DG A3	.9	1.0	1.4
DG B1	1.4	2.2	---	TW D1	1.1	1.6	2.1	SS E1	.8	1.1	1.4	DG B1	.9	1.2	1.9
DG B2	1.2	1.5	2.4	TW D2	.9	1.1	1.5	SS E2	.7	.9	1.2	DG B2	.9	1.1	1.4
DG B3	1.1	1.3	1.6	TW D3	.8	1.1	1.4	SS E3	.9	1.3	1.7	DG B3	.9	1.0	1.2
DG C1	1.2	1.6	5.0					SS E4	.7	1.0	1.3	DG C1	.9	1.0	1.4

TRUTH OR CONSEQ., NM				TUCUMCARI, NM				ZUNI, NM				ZUNI, NM			
	SOLAR SAVINGS				SOLAR SAVINGS				SOLAR SAVINGS				SOLAR SAVINGS		
	10%	50%	80%		10%	50%	80%		10%	50%	80%		10%	50%	80%
DG C2	.9	1.0	1.2	TW D4	.8	1.0	1.3	WW A1	1.3	1.9	2.5	DG C2	1.1	1.3	1.7
DG C3	.8	.9	1.1	TW D5	.8	1.0	1.3	WW A2	1.1	1.5	1.9	DG C3	1.0	1.1	1.4
SS A1	.8	1.1	1.4	TW E1	.8	1.0	1.3	WW A3	1.0	1.3	1.7	SS A1	.9	1.4	1.8
SS A2	.6	.9	1.1	TW E2	.8	1.0	1.3	WW A4	1.0	1.3	1.7	SS A2	.8	1.1	1.4
SS A3	.8	1.1	1.5	TW E3	.7	.9	1.1	WW A5	.9	1.2	1.6	SS A3	1.0	1.5	2.0
SS A4	.6	.9	1.2	TW E4	.7	.9	1.1	WW A6	.9	1.2	1.5	SS A4	.8	1.1	1.5
SS A5	.8	1.2	1.6	TW F1	1.1	1.6	2.1	WW B1	1.3	1.8	2.4	SS A5	1.0	1.5	2.1
SS A6	.6	.9	1.1	TW F2	1.0	1.4	1.8	WW B2	.9	1.2	1.6	SS A6	.8	1.1	1.4
SS A7	.9	1.3	1.8	TW F3	1.0	1.3	1.7	WW B3	.8	1.1	1.4	SS A7	1.1	1.8	2.5
SS A8	.7	.9	1.2	TW F4	1.0	1.3	1.7	WW B4	.8	1.1	1.3	SS A8	.8	1.1	1.5
SS B1	.9	1.2	1.6	TW G1	1.2	1.6	2.0	WW B5	.8	1.1	1.3	SS B1	1.1	1.5	2.1
SS B2	.7	1.0	1.2	TW G2	1.1	1.5	1.9	WW C1	.8	1.1	1.3	SS B2	.9	1.2	1.6
SS B3	.9	1.2	1.6	TW G3	1.1	1.5	1.9	WW C2	.8	1.1	1.4	SS B3	1.1	1.6	2.2
SS B4	.7	1.0	1.3	TW G4	1.2	1.6	2.0	WW C3	.7	.9	1.2	SS B4	.9	1.2	1.6
SS B5	.9	1.4	1.8	TW H1	1.2	1.6	2.1	WW C4	.8	1.0	1.2	SS B5	1.2	1.8	2.4
SS B6	.7	1.0	1.3	TW H2	1.3	1.7	2.1	TW A1	1.3	1.9	2.6	SS B6	.9	1.2	1.6
SS B7	1.0	1.5	2.0	TW H3	1.4	1.8	2.3	TW A2	1.2	1.6	2.1	SS B7	1.3	2.0	2.7
SS B8	.7	1.0	1.3	TW H4	1.6	2.1	2.6	TW A3	1.1	1.5	1.9	SS B8	.9	1.3	1.6
SS C1	.8	1.1	1.4	TW I1	1.2	1.7	2.2	TW A4	1.1	1.5	1.9	SS C1	1.1	1.5	1.9
SS C2	.7	1.0	1.2	TW I2	.9	1.2	1.5	TW B1	1.3	1.9	2.4	SS C2	.9	1.2	1.6
SS C3	1.0	1.4	1.7	TW I3	.9	1.1	1.5	TW B2	1.2	1.6	2.1	SS C3	1.3	1.8	2.4
SS C4	.8	1.1	1.4	TW I4	.8	1.1	1.3	TW B3	1.2	1.6	2.0	SS C4	1.0	1.4	1.8
SS D1	.7	.9	1.2	TW I5	.8	1.0	1.3	TW B4	1.2	1.6	2.1	SS D1	.8	1.2	1.5
SS D2	.6	.8	1.0	TW J1	.8	1.0	1.3	TW C1	1.3	1.8	2.4	SS D2	.7	.9	1.2
SS D3	.7	1.0	1.3	TW J2	.8	1.0	1.3	TW C2	1.3	1.7	2.3	SS D3	.9	1.3	1.7
SS D4	.6	.8	1.0	TW J3	.7	.9	1.1	TW C3	1.3	1.8	2.3	SS D4	.7	1.0	1.3
SS E1	.7	1.0	1.3	TW J4	.7	.9	1.2	TW C4	1.5	2.0	2.6	SS E1	.9	1.3	1.7
SS E2	.6	.8	1.1	DG A1	1.1	1.8	---	TW D1	1.4	2.0	2.6	SS E2	.8	1.0	1.4
SS E3	.8	1.2	1.6	DG A2	1.0	1.4	2.1	TW D2	1.0	1.4	1.7	SS E3	1.0	1.6	2.1
SS E4	.7	.9	1.2	DG A3	.9	1.1	1.5	TW D3	.9	1.3	1.6	SS E4	.8	1.1	1.5
				DG B1	1.1	1.4	2.9	TW D4	.9	1.2	1.5				
TUCUMCARI, NM				DG B2	1.0	1.2	1.7	TW D5	.9	1.2	1.5	ALBANY, NY			
	SOLAR SAVINGS			DG B3	.9	1.0	1.3	TW E1	.9	1.2	1.5		SOLAR SAVINGS		
	10%	50%	80%	DG C1	1.0	1.2	1.7	TW E2	.9	1.2	1.5		10%	50%	80%
WW A1	1.1	1.6	2.1	DG C2	.9	1.1	1.4	TW E3	.8	1.0	1.3	WW A1	2.8	---	---
WW A2	.9	1.2	1.6	DG C3	.9	1.0	1.2	TW E4	.8	1.0	1.3	WW A2	2.4	---	---
WW A3	.9	1.1	1.5	SS A1	.8	1.2	1.6	TW F1	1.3	1.9	2.5	WW A3	2.2	---	---
WW A4	.8	1.1	1.4	SS A2	.7	1.0	1.3	TW F2	1.2	1.7	2.1	WW A4	2.1	---	---
WW A5	.8	1.0	1.3	SS A3	.9	1.3	1.7	TW F3	1.2	1.6	2.0	WW A5	2.0	---	---
WW A6	.8	1.0	1.3	SS A4	.7	1.0	1.3	TW F4	1.2	1.6	2.0	WW A6	2.0	5.7	---
WW B1	1.0	1.5	1.9	SS A5	.9	1.3	1.8	TW G1	1.4	1.9	2.5	WW B1	---	---	---
WW B2	.8	1.1	1.4	SS A6	.7	1.0	1.3	TW G2	1.3	1.7	2.2	WW B2	1.8	3.1	5.3
WW B3	.7	.9	1.2	SS A7	.9	1.5	2.0	TW G3	1.3	1.7	2.2	WW B3	1.5	2.4	3.6
WW B4	.7	.9	1.2	SS A8	.7	1.0	1.3	TW G4	1.4	1.9	2.4	WW B4	1.3	1.9	2.6
WW B5	.7	.9	1.2	SS B1	.9	1.3	1.8	TW H1	1.4	2.0	2.5	WW B5	1.3	1.8	2.4
WW C1	.7	.9	1.2	SS B2	.8	1.1	1.4	TW H2	1.5	2.0	2.5	WW C1	1.4	2.2	3.1
WW C2	.7	.9	1.2	SS B3	1.0	1.4	1.9	TW H3	1.6	2.1	2.7	WW C2	1.4	2.0	2.8
WW C3	.7	.8	1.0	SS B4	.8	1.1	1.4	TW H4	1.8	2.4	3.1	WW C3	1.1	1.5	2.0
WW C4	.7	.9	1.1	SS B5	1.0	1.5	2.0	TW I1	1.5	2.1	2.8	WW C4	1.2	1.6	2.1
TW A1	1.1	1.6	2.1	SS B6	.8	1.1	1.4	TW I2	1.1	1.4	1.8	TW A1	2.8	---	---
TW A2	1.0	1.3	1.7	SS B7	1.1	1.7	2.3	TW I3	1.0	1.3	1.7	TW A2	2.5	---	---
TW A3	.9	1.3	1.6	SS B8	.8	1.1	1.4	TW I4	.9	1.2	1.5	TW A3	2.3	---	---
TW A4	.9	1.2	1.6	SS C1	.9	1.2	1.6	TW I5	.9	1.2	1.5	TW A4	2.2	---	---
TW B1	1.1	1.6	2.0	SS C2	.8	1.1	1.4	TW J1	.9	1.2	1.5	TW B1	2.8	---	---
TW B2	1.0	1.4	1.7	SS C3	1.1	1.5	2.0	TW J2	.9	1.2	1.5	TW B2	2.4	---	---
TW B3	1.0	1.3	1.7	SS C4	.9	1.2	1.5	TW J3	.8	1.0	1.3	TW B3	2.4	---	---
TW B4	1.0	1.4	1.8	SS D1	.7	1.0	1.3	TW J4	.8	1.1	1.3	TW B4	2.4	6.4	---
TW C1	1.1	1.5	2.0	SS D2	.6	.8	1.1	DG A1	1.3	3.2	---	TW C1	2.6	---	---
TW C2	1.1	1.5	1.9	SS D3	.8	1.1	1.5	DG A2	1.2	1.7	3.0	TW C2	2.5	14.9	---
TW C3	1.1	1.5	2.0	SS D4	.6	.9	1.1	DG A3	1.1	1.3	1.8	TW C3	2.5	6.8	---
TW C4	1.2	1.7	2.2	SS E1	.8	1.1	1.5	DG B1	1.3	1.9	---	TW C4	2.6	6.1	---
TW D1	1.2	1.6	2.1	SS E2	.7	.9	1.2	DG B2	1.2	1.4	2.2	TW D1	---	---	---
TW D2	.9	1.2	1.5	SS E3	.9	1.3	1.8	DG B3	1.1	1.2	1.6	TW D2	1.9	3.4	6.3
TW D3	.8	1.1	1.4	SS E4	.7	1.0	1.3	DG C1	1.2	1.5	3.0	TW D3	1.7	3.1	5.4

Conservation Factor Tables / 351

ALBANY, NY	SOLAR SAVINGS		
	10%	50%	80%
WW A1	2.8	---	---
WW A2	2.4	---	---
WW A3	2.2	---	---
WW A4	2.1	---	---
WW A5	2.0	---	---
WW A6	2.0	5.7	---
WW B1	---	---	---
WW B2	1.8	3.1	5.3
WW B3	1.5	2.4	3.6
WW B4	1.3	1.9	2.6
WW B5	1.3	1.8	2.4
WW C1	1.4	2.2	3.1
WW C2	1.4	2.0	2.8
WW C3	1.1	1.5	2.0
WW C4	1.2	1.6	2.1
TW A1	2.8	---	---
TW A2	2.5	---	---
TW A3	2.3	---	---
TW A4	2.2	---	---
TW B1	2.8	---	---
TW B2	2.4	---	---
TW B3	2.4	---	---
TW B4	2.4	6.4	---
TW C1	2.6	---	---
TW C2	2.5	14.9	---
TW C3	2.5	6.8	---
TW C4	2.6	6.1	---
TW D1	---	---	---
TW D2	1.9	3.4	6.3
TW D3	1.7	3.1	5.4
DG B1	---	---	---
DG B2	2.3	---	---
DG B3	1.7	2.3	4.0
DG C1	5.0	---	---
DG C2	1.9	6.5	---
DG C3	1.5	2.0	3.0
SS A1	1.8	---	---
SS A2	1.3	2.2	3.3
SS A3	2.1	---	---
SS A4	1.4	2.5	3.8
SS A5	1.9	---	---
SS A6	1.3	2.3	3.5
SS A7	2.2	---	---
SS A8	1.4	2.7	4.7
SS B1	2.1	---	---
SS B2	1.4	2.4	3.4
SS B3	2.5	---	---
SS B4	1.5	2.6	3.8
SS B5	2.4	---	---
SS B6	1.5	2.5	3.7
SS B7	3.0	---	---
SS B8	1.6	2.8	4.5
SS C1	2.8	---	---
SS C2	1.7	2.9	4.7
SS C3	3.1	---	---
SS C4	1.9	3.3	5.7
SS D1	1.6	---	---
SS D2	1.2	1.8	2.6
SS D3	1.7	---	---
SS D4	1.2	1.9	2.8
SS E1	2.0	---	---
SS E2	1.4	2.3	3.5
SS E3	2.2	---	---
SS E4	1.5	2.5	3.9

BINGHAMTON, NY	SOLAR SAVINGS		
	10%	50%	80%
WW A1	4.8	---	---
WW A2	5.3	---	---
WW A3	4.3	---	---
WW A4	3.8	---	---
WW A5	3.5	---	---
WW A6	3.3	---	---
WW B1	---	---	---
WW B2	2.3	---	---
WW B3	1.9	4.3	---
WW B4	1.6	2.4	3.3
WW B5	1.5	2.1	2.9
WW C1	1.7	3.1	6.2
WW C2	1.6	2.6	3.9
WW C3	1.3	1.8	2.3
WW C4	1.3	1.8	2.4
TW A1	4.1	---	---
TW A2	4.1	---	---
TW A3	3.8	---	---
TW A4	3.4	---	---
TW B1	4.2	---	---
TW B2	3.6	---	---
TW B3	3.4	---	---
TW B4	3.2	---	---
TW C1	3.7	---	---
TW C2	3.3	---	---
TW C3	3.2	---	---
TW C4	3.2	---	---
TW D1	---	---	---
TW D2	2.4	---	---
TW D3	2.2	---	---
TW D4	1.7	2.7	4.0
TW D5	1.6	2.4	3.2
TW E1	1.8	3.9	---
TW E2	1.7	3.0	4.8
TW E3	1.4	2.0	2.7
TW E4	1.4	2.0	2.7
TW F1	---	---	---
TW F2	---	---	---
TW F3	5.9	---	---
TW F4	4.5	---	---
TW G1	---	---	---
TW G2	9.7	---	---
TW G3	5.1	---	---
TW G4	4.5	---	---
TW H1	---	---	---
TW H2	6.0	---	---
TW H3	5.0	---	---
TW H4	5.1	---	---
TW I1	---	---	---
TW I2	2.6	---	---
TW I3	2.3	10.2	---
TW I4	1.8	2.8	4.1
TW I5	1.7	2.5	3.4
TW J1	1.9	3.7	16.0
TW J2	1.8	3.0	4.6
TW J3	1.4	2.0	2.7
TW J4	1.4	2.0	2.7
DG A1	---	---	---
DG A2	3.3	---	---
DG A3	1.9	3.6	---
DG B1	---	---	---
DG B2	3.0	---	---
DG B3	1.9	2.9	---
DG C1	---	---	---

BINGHAMTON, NY	SOLAR SAVINGS		
	10%	50%	80%
DG C2	2.4	---	---
DG C3	1.7	2.5	4.5
SS A1	2.1	---	---
SS A2	1.5	2.8	4.8
SS A3	2.6	---	---
SS A4	1.5	3.5	---
SS A5	2.1	---	---
SS A6	1.5	3.2	6.7
SS A7	2.5	---	---
SS A8	1.6	4.9	---
SS B1	2.7	---	---
SS B2	1.6	3.0	4.7
SS B3	3.7	---	---
SS B4	1.7	3.4	6.6
SS B5	3.1	---	---
SS B6	1.7	3.3	6.1
SS B7	4.7	---	---
SS B8	1.8	4.4	---
SS C1	---	---	---
SS C2	2.1	---	---
SS C3	---	---	---
SS C4	2.3	---	---
SS D1	1.9	---	---
SS D2	1.3	2.3	3.4
SS D3	2.0	---	---
SS D4	1.3	2.4	3.7
SS E1	3.1	---	---
SS E2	1.6	3.3	13.5
SS E3	2.9	---	---
SS E4	1.7	3.8	---

BUFFALO, NY	SOLAR SAVINGS		
	10%	50%	80%
WW A1	3.9	---	---
WW A2	4.0	---	---
WW A3	3.6	---	---
WW A4	3.3	---	---
WW A5	3.1	---	---
WW A6	3.0	---	---
WW B1	---	---	---
WW B2	2.2	---	---
WW B3	1.8	4.5	---
WW B4	1.5	2.3	3.4
WW B5	1.5	2.1	2.9
WW C1	1.6	3.1	6.1
WW C2	1.6	2.6	4.0
WW C3	1.2	1.7	2.3
WW C4	1.3	1.8	2.4
TW A1	3.5	---	---
TW A2	3.6	---	---
TW A3	3.3	---	---
TW A4	3.1	---	---
TW B1	3.7	---	---
TW B2	3.3	---	---
TW B3	3.1	---	---
TW B4	3.0	---	---
TW C1	3.4	---	---
TW C2	3.1	---	---
TW C3	3.1	---	---
TW C4	3.1	---	---
TW D1	---	---	---
TW D2	2.3	---	---
TW D3	2.1	---	---
TW D4	1.7	2.7	4.1
TW D5	1.6	2.3	3.3
TW E1	1.8	4.0	---
TW E2	1.7	3.0	4.9
TW E3	1.3	2.0	2.7
TW E4	1.4	2.0	2.7
TW F1	---	---	---
TW F2	5.7	---	---
TW F3	4.5	---	---
TW F4	3.9	---	---
TW G1	---	---	---
TW G2	5.5	---	---
TW G3	4.4	---	---
TW G4	4.1	---	---
TW H1	---	---	---
TW H2	5.1	---	---
TW H3	4.6	---	---
TW H4	4.8	---	---
TW I1	---	---	---
TW I2	2.5	---	---
TW I3	2.2	---	---
TW I4	1.8	2.8	4.1
TW I5	1.6	2.5	3.4
TW J1	1.9	3.8	9.4
TW J2	1.8	3.0	4.6
TW J3	1.4	2.0	2.7
TW J4	1.4	2.0	2.7
DG A1	---	---	---
DG A2	3.0	---	---
DG A3	1.9	3.5	---
DG B1	---	---	---
DG B2	2.8	---	---
DG B3	1.8	2.9	---
DG C1	---	---	---
DG C2	2.3	---	---
DG C3	1.7	2.5	4.6
SS A1	2.0	---	---
SS A2	1.4	2.9	4.9
SS A3	2.5	---	---
SS A4	1.5	3.5	11.3
SS A5	2.0	---	---
SS A6	1.5	3.2	6.8
SS A7	2.4	---	---
SS A8	1.5	5.3	---
SS B1	2.5	---	---
SS B2	1.6	3.0	4.8
SS B3	3.3	---	---
SS B4	1.6	3.5	6.8
SS B5	2.9	---	---
SS B6	1.6	3.4	6.2
SS B7	4.0	---	---
SS B8	1.7	4.5	---
SS C1	---	---	---
SS C2	2.1	---	---
SS C3	10.6	---	---
SS C4	2.2	---	---
SS D1	1.9	---	---
SS D2	1.3	2.3	3.5
SS D3	1.9	---	---
SS D4	1.3	2.4	3.8
SS E1	2.8	---	---
SS E2	1.5	3.4	9.0
SS E3	2.7	---	---
SS E4	1.6	3.9	---

MASSENA, NY			
	SOLAR SAVINGS		
	10%	50%	80%
WW A1	4.1	---	---
WW A2	4.0	---	---
WW A3	3.5	---	---
WW A4	3.2	---	---
WW A5	3.0	---	---
WW A6	2.9	---	---
WW B1	---	---	---
WW B2	2.2	---	---
WW B3	1.8	4.3	---
WW B4	1.5	2.3	3.3
WW B5	1.5	2.1	2.9
WW C1	1.6	3.1	6.9
WW C2	1.6	2.6	3.9
WW C3	1.2	1.7	2.3
WW C4	1.3	1.8	2.4
TW A1	3.7	---	---
TW A2	3.6	---	---
TW A3	3.4	---	---
TW A4	3.1	---	---
TW B1	3.8	---	---
TW B2	3.3	---	---
TW B3	3.1	---	---
TW B4	3.0	---	---
TW C1	3.5	---	---
TW C2	3.2	---	---
TW C3	3.1	---	---
TW C4	3.1	---	---
TW D1	---	---	---
TW D2	2.3	---	---
TW D3	2.1	---	---
TW D4	1.7	2.7	4.0
TW D5	1.6	2.3	3.2
TW E1	1.8	3.9	---
TW E2	1.7	2.9	4.9
TW E3	1.3	2.0	2.7
TW E4	1.4	1.9	2.6
TW F1	---	---	---
TW F2	5.6	---	---
TW F3	4.4	---	---
TW F4	3.9	---	---
TW G1	---	---	---
TW G2	5.3	---	---
TW G3	4.4	---	---
TW G4	4.1	---	---
TW H1	16.3	---	---
TW H2	5.0	---	---
TW H3	4.6	---	---
TW H4	4.8	---	---
TW I1	---	---	---
TW I2	2.5	---	---
TW I3	2.2	---	---
TW I4	1.8	2.8	4.1
TW I5	1.6	2.4	3.4
TW J1	1.9	3.7	---
TW J2	1.8	2.9	4.6
TW J3	1.4	2.0	2.7
TW J4	1.4	2.0	2.7
DG A1	---	---	---
DG A2	3.0	---	---
DG A3	1.9	3.5	---
DG B1	---	---	---
DG B2	2.9	---	---
DG B3	1.8	2.8	---
DG C1	---	---	---

MASSENA, NY			
	SOLAR SAVINGS		
	10%	50%	80%
DG C2	2.3	---	---
DG C3	1.7	2.4	4.5
SS A1	2.1	---	---
SS A2	1.5	2.9	5.4
SS A3	2.8	---	---
SS A4	1.5	3.8	---
SS A5	2.2	---	---
SS A6	1.5	3.3	11.9
SS A7	2.6	---	---
SS A8	1.6	11.7	---
SS B1	2.7	---	---
SS B2	1.6	3.0	5.1
SS B3	3.9	---	---
SS B4	1.7	3.6	9.1
SS B5	3.1	---	---
SS B6	1.7	3.4	7.3
SS B7	5.2	---	---
SS B8	1.8	5.0	---
SS C1	---	---	---
SS C2	2.1	---	---
SS C3	15.6	---	---
SS C4	2.2	---	---
SS D1	2.0	---	---
SS D2	1.3	2.3	3.7
SS D3	2.0	---	---
SS D4	1.4	2.5	4.0
SS E1	3.3	---	---
SS E2	1.6	3.6	---
SS E3	3.0	---	---
SS E4	1.7	4.1	---

NEW YORK CITY, NY			
	SOLAR SAVINGS		
	10%	50%	80%
WW A1	2.0	---	---
WW A2	1.7	3.1	5.3
WW A3	1.6	2.6	4.0
WW A4	1.5	2.4	3.6
WW A5	1.5	2.3	3.3
WW A6	1.4	2.2	3.1
WW B1	2.7	---	---
WW B2	1.4	2.1	2.8
WW B3	1.2	1.7	2.4
WW B4	1.1	1.6	2.0
WW B5	1.1	1.5	2.0
WW C1	1.2	1.6	2.2
WW C2	1.2	1.6	2.1
WW C3	1.0	1.3	1.7
WW C4	1.0	1.4	1.8
TW A1	2.0	10.4	---
TW A2	1.8	3.3	5.7
TW A3	1.7	2.9	4.4
TW A4	1.7	2.7	3.9
TW B1	2.0	4.6	---
TW B2	1.8	3.1	4.8
TW B3	1.8	2.9	4.3
TW B4	1.8	2.9	4.2
TW C1	2.0	3.6	5.9
TW C2	1.9	3.2	4.7
TW C3	2.0	3.2	4.7
TW C4	2.1	3.5	5.0
TW D1	2.9	---	---
TW D2	1.5	2.3	3.1
TW D3	1.4	2.1	2.9

NEW YORK CITY, NY			
	SOLAR SAVINGS		
	10%	50%	80%
TW D4	1.2	1.7	2.3
TW D5	1.2	1.7	2.2
TW E1	1.3	1.8	2.5
TW E2	1.2	1.8	2.3
TW E3	1.0	1.4	1.8
TW E4	1.1	1.5	1.9
TW F1	2.2	7.4	---
TW F2	1.9	3.5	6.2
TW F3	1.8	3.1	4.8
TW F4	1.9	3.0	4.3
TW G1	2.2	4.8	---
TW G2	2.0	3.5	5.5
TW G3	2.0	3.3	4.9
TW G4	2.2	3.4	4.9
TW H1	2.3	4.2	7.3
TW H2	2.3	3.7	5.6
TW H3	2.4	3.8	5.6
TW H4	2.7	4.3	6.1
TW I1	3.4	---	---
TW I2	1.6	2.4	3.3
TW I3	1.5	2.2	2.9
TW I4	1.3	1.8	2.4
TW I5	1.3	1.7	2.2
TW J1	1.3	1.9	2.5
TW J2	1.3	1.8	2.4
TW J3	1.1	1.5	1.9
TW J4	1.1	1.5	1.9
DG A1	2.8	---	---
DG A2	1.8	---	---
DG A3	1.4	2.0	3.7
DG B1	2.7	---	---
DG B2	1.7	3.5	---
DG B3	1.4	1.8	2.7
DG C1	2.0	---	---
DG C2	1.5	2.4	---
DG C3	1.3	1.6	2.2
SS A1	1.4	2.7	4.5
SS A2	1.1	1.7	2.4
SS A3	1.6	4.1	---
SS A4	1.1	1.8	2.5
SS A5	1.5	3.6	9.1
SS A6	1.1	1.8	2.4
SS A7	1.7	---	---
SS A8	1.2	1.9	2.7
SS B1	1.6	3.2	5.6
SS B2	1.2	1.8	2.5
SS B3	1.8	4.6	---
SS B4	1.2	1.9	2.7
SS B5	1.8	4.9	---
SS B6	1.2	1.9	2.6
SS B7	2.1	---	---
SS B8	1.3	2.0	2.9
SS C1	1.8	3.4	6.5
SS C2	1.4	2.0	2.8
SS C3	2.1	4.4	12.5
SS C4	1.5	2.3	3.1
SS D1	1.2	2.2	3.5
SS D2	1.0	1.5	2.0
SS D3	1.3	2.6	4.4
SS D4	1.0	1.5	2.1
SS E1	1.5	3.0	5.9
SS E2	1.1	1.7	2.3
SS E3	1.7	3.8	14.6
SS E4	1.2	1.8	2.5

ROCHESTER, NY			
	SOLAR SAVINGS		
	10%	50%	80%
WW A1	3.5	---	---
WW A2	3.3	---	---
WW A3	3.1	---	---
WW A4	2.9	---	---
WW A5	2.8	---	---
WW A6	2.7	---	---
WW B1	---	---	---
WW B2	2.1	---	---
WW B3	1.7	3.8	---
WW B4	1.5	2.3	3.2
WW B5	1.4	2.1	2.9
WW C1	1.6	2.9	5.4
WW C2	1.6	2.5	3.7
WW C3	1.2	1.7	2.3
WW C4	1.3	1.8	2.3
TW A1	3.3	---	---
TW A2	3.2	---	---
TW A3	3.0	---	---
TW A4	2.8	---	---
TW B1	3.4	---	---
TW B2	3.0	---	---
TW B3	2.9	---	---
TW B4	2.8	---	---
TW C1	3.2	---	---
TW C2	3.0	---	---
TW C3	2.9	---	---
TW C4	3.0	---	---
TW D1	---	---	---
TW D2	2.2	---	---
TW D3	2.0	---	---
TW D4	1.6	2.6	3.9
TW D5	1.5	2.3	3.2
TW E1	1.7	3.6	---
TW E2	1.6	2.8	4.6
TW E3	1.3	1.9	2.7
TW E4	1.3	1.9	2.6
TW F1	5.4	---	---
TW F2	4.2	---	---
TW F3	3.8	---	---
TW F4	3.5	---	---
TW G1	6.2	---	---
TW G2	4.3	---	---
TW G3	3.9	---	---
TW G4	3.8	---	---
TW H1	5.8	---	---
TW H2	4.4	---	---
TW H3	4.2	---	---
TW H4	4.5	---	---
TW I1	---	---	---
TW I2	2.4	---	---
TW I3	2.1	6.2	---
TW I4	1.7	2.7	4.0
TW I5	1.6	2.4	3.3
TW J1	1.8	3.5	7.7
TW J2	1.7	2.9	4.4
TW J3	1.4	1.9	2.6
TW J4	1.4	2.0	2.6
DG A1	---	---	---
DG A2	2.8	---	---
DG A3	1.8	3.4	---
DG B1	---	---	---
DG B2	2.7	---	---
DG B3	1.8	2.8	---
DG C1	---	---	---

ROCHESTER, NY

	SOLAR SAVINGS		
	10%	50%	80%
DG C2	2.2	---	---
DG C3	1.6	2.4	4.3
SS A1	2.0	---	---
SS A2	1.4	2.7	4.6
SS A3	2.4	---	---
SS A4	1.4	3.3	8.9
SS A5	2.0	---	---
SS A6	1.4	3.0	6.0
SS A7	2.3	---	---
SS A8	1.5	4.4	---
SS B1	2.4	---	---
SS B2	1.5	2.9	4.6
SS B3	3.1	---	---
SS B4	1.6	3.3	6.1
SS B5	2.8	---	---
SS B6	1.6	3.2	5.7
SS B7	3.7	---	---
SS B8	1.7	4.1	---
SS C1	---	---	---
SS C2	2.0	7.2	---
SS C3	5.1	---	---
SS C4	2.1	---	---
SS D1	1.8	---	---
SS D2	1.3	2.2	3.4
SS D3	1.9	---	---
SS D4	1.3	2.3	3.6
SS E1	2.6	---	---
SS E2	1.5	3.2	7.4
SS E3	2.6	---	---
SS E4	1.6	3.6	---

SYRACUSE, NY

	SOLAR SAVINGS		
	10%	50%	80%
WW A1	3.5	---	---
WW A2	3.3	---	---
WW A3	3.0	---	---
WW A4	2.8	---	---
WW A5	2.7	---	---
WW A6	2.6	---	---
WW B1	---	---	---
WW B2	2.1	---	---
WW B3	1.7	3.5	---
WW B4	1.5	2.2	3.1
WW B5	1.4	2.1	2.8
WW C1	1.6	2.8	4.8
WW C2	1.5	2.4	3.6
WW C3	1.2	1.7	2.2
WW C4	1.3	1.8	2.3
TW A1	3.3	---	---
TW A2	3.2	---	---
TW A3	3.0	---	---
TW A4	2.8	---	---
TW B1	3.4	---	---
TW B2	3.0	---	---
TW B3	2.9	---	---
TW B4	2.8	---	---
TW C1	3.2	---	---
TW C2	2.9	---	---
TW C3	2.9	---	---
TW C4	3.0	---	---
TW D1	---	---	---
TW D2	2.2	---	---
TW D3	2.0	---	---

SYRACUSE, NY

	SOLAR SAVINGS		
	10%	50%	80%
TW D4	1.6	2.6	3.7
TW D5	1.5	2.2	3.1
TW E1	1.7	3.4	9.9
TW E2	1.6	2.8	4.3
TW E3	1.3	1.9	2.6
TW E4	1.3	1.9	2.6
TW F1	5.3	---	---
TW F2	4.1	---	---
TW F3	3.7	---	---
TW F4	3.4	---	---
TW G1	5.8	---	---
TW G2	4.2	---	---
TW G3	3.8	---	---
TW G4	3.7	---	---
TW H1	5.5	---	---
TW H2	4.3	---	---
TW H3	4.2	---	---
TW H4	4.4	---	---
TW I1	---	---	---
TW I2	2.4	---	---
TW I3	2.1	5.0	---
TW I4	1.7	2.6	3.8
TW I5	1.6	2.3	3.2
TW J1	1.8	3.3	6.5
TW J2	1.7	2.8	4.1
TW J3	1.4	1.9	2.6
TW J4	1.4	1.9	2.6
DG A1	---	---	---
DG A2	2.8	---	---
DG A3	1.8	3.3	---
DG B1	---	---	---
DG B2	2.7	---	---
DG B3	1.8	2.7	9.9
DG C1	---	---	---
DG C2	2.2	---	---
DG C3	1.6	2.4	4.0
SS A1	2.0	---	---
SS A2	1.4	2.7	4.3
SS A3	2.4	---	---
SS A4	1.4	3.2	7.2
SS A5	2.0	---	---
SS A6	1.4	2.9	5.4
SS A7	2.4	---	---
SS A8	1.5	4.0	---
SS B1	2.4	---	---
SS B2	1.5	2.8	4.3
SS B3	3.1	---	---
SS B4	1.6	3.2	5.6
SS B5	2.8	---	---
SS B6	1.6	3.1	5.2
SS B7	3.8	---	---
SS B8	1.7	3.9	---
SS C1	---	---	---
SS C2	2.0	5.1	---
SS C3	4.9	---	---
SS C4	2.1	7.5	---
SS D1	1.8	---	---
SS D2	1.3	2.2	3.2
SS D3	1.9	---	---
SS D4	1.3	2.3	3.5
SS E1	2.6	---	---
SS E2	1.5	3.0	6.2
SS E3	2.6	---	---
SS E4	1.6	3.4	10.0

ASHEVILLE, NC

	SOLAR SAVINGS		
	10%	50%	80%
WW A1	1.4	2.3	3.2
WW A2	1.2	1.7	2.2
WW A3	1.1	1.5	2.0
WW A4	1.1	1.5	1.9
WW A5	1.0	1.4	1.8
WW A6	1.0	1.4	1.8
WW B1	1.4	2.3	3.2
WW B2	1.0	1.4	1.8
WW B3	.9	1.2	1.6
WW B4	.9	1.2	1.5
WW B5	.9	1.2	1.5
WW C1	.9	1.2	1.5
WW C2	.9	1.2	1.5
WW C3	.8	1.0	1.3
WW C4	.8	1.1	1.4
TW A1	1.4	2.3	3.1
TW A2	1.3	1.8	2.5
TW A3	1.2	1.7	2.3
TW A4	1.2	1.7	2.2
TW B1	1.4	2.2	2.9
TW B2	1.3	1.8	2.4
TW B3	1.3	1.8	2.4
TW B4	1.3	1.9	2.4
TW C1	1.4	2.1	2.8
TW C2	1.4	2.0	2.6
TW C3	1.4	2.0	2.7
TW C4	1.6	2.3	3.0
TW D1	1.6	2.5	3.4
TW D2	1.1	1.5	2.0
TW D3	1.0	1.4	1.8
TW D4	1.0	1.3	1.6
TW D5	1.0	1.3	1.6
TW E1	.9	1.3	1.7
TW E2	.9	1.3	1.6
TW E3	.8	1.1	1.4
TW E4	.9	1.1	1.4
TW F1	1.5	2.3	3.1
TW F2	1.3	1.9	2.5
TW F3	1.3	1.8	2.4
TW F4	1.3	1.8	2.4
TW G1	1.5	2.2	3.0
TW G2	1.4	2.0	2.6
TW G3	1.4	2.0	2.6
TW G4	1.5	2.1	2.8
TW H1	1.6	2.3	3.0
TW H2	1.6	2.2	3.0
TW H3	1.7	2.4	3.1
TW H4	2.0	2.7	3.6
TW I1	1.7	2.6	3.7
TW I2	1.2	1.6	2.1
TW I3	1.1	1.5	1.9
TW I4	1.0	1.3	1.7
TW I5	1.0	1.3	1.7
TW J1	1.0	1.3	1.7
TW J2	1.0	1.3	1.7
TW J3	.9	1.1	1.4
TW J4	.9	1.2	1.5
DG A1	1.5	---	---
DG A2	1.3	2.0	---
DG A3	1.1	1.4	2.1
DG B1	1.4	3.0	---
DG B2	1.3	1.7	3.0
DG B3	1.1	1.3	1.8
DG C1	1.3	1.8	---

ASHEVILLE, NC

	SOLAR SAVINGS		
	10%	50%	80%
DG C2	1.2	1.5	2.1
DG C3	1.0	1.2	1.5
SS A1	1.0	1.6	2.2
SS A2	.8	1.2	1.6
SS A3	1.1	1.8	2.5
SS A4	.8	1.2	1.7
SS A5	1.1	1.8	2.5
SS A6	.8	1.2	1.6
SS A7	1.2	2.1	3.0
SS A8	.9	1.3	1.7
SS B1	1.2	1.8	2.4
SS B2	.9	1.3	1.8
SS B3	1.2	1.9	2.7
SS B4	.9	1.4	1.8
SS B5	1.3	2.1	2.9
SS B6	.9	1.4	1.8
SS B7	1.4	2.4	3.4
SS B8	1.0	1.4	1.9
SS C1	1.2	1.7	2.3
SS C2	1.0	1.4	1.8
SS C3	1.4	2.1	2.8
SS C4	1.1	1.5	2.0
SS D1	.9	1.3	1.8
SS D2	.8	1.1	1.4
SS D3	1.0	1.5	2.1
SS D4	.8	1.1	1.4
SS E1	1.0	1.5	2.1
SS E2	.8	1.2	1.6
SS E3	1.1	1.8	2.5
SS E4	.9	1.3	1.7

CAPE HATTERAS, NC

	SOLAR SAVINGS		
	10%	50%	80%
WW A1	1.1	1.8	2.4
WW A2	.9	1.3	1.8
WW A3	.9	1.2	1.6
WW A4	.9	1.2	1.6
WW A5	.8	1.1	1.5
WW A6	.8	1.1	1.5
WW B1	1.1	1.6	2.2
WW B2	.8	1.2	1.5
WW B3	.8	1.0	1.3
WW B4	.7	1.0	1.3
WW B5	.8	1.0	1.3
WW C1	.7	1.0	1.3
WW C2	.8	1.0	1.3
WW C3	.7	.9	1.1
WW C4	.7	.9	1.2
TW A1	1.1	1.8	2.4
TW A2	1.0	1.5	1.9
TW A3	1.0	1.4	1.8
TW A4	1.0	1.3	1.8
TW B1	1.1	1.7	2.3
TW B2	1.0	1.5	2.0
TW B3	1.0	1.5	1.9
TW B4	1.1	1.5	2.0
TW C1	1.1	1.7	2.2
TW C2	1.1	1.6	2.1
TW C3	1.2	1.7	2.2
TW C4	1.3	1.8	2.4
TW D1	1.2	1.8	2.4
TW D2	.9	1.3	1.6
TW D3	.8	1.2	1.5

354 / APPENDIX 17

CAPE HATTERAS, NC

	SOLAR SAVINGS		
	10%	50%	80%
WW A1	1.1	1.8	2.4
WW A2	.9	1.3	1.8
WW A3	.9	1.2	1.6
WW A4	.9	1.2	1.6
WW A5	.8	1.1	1.5
WW A6	.8	1.1	1.5
WW B1	1.1	1.6	2.2
WW B2	.8	1.2	1.5
WW B3	.8	1.0	1.3
WW B4	.7	1.0	1.3
WW B5	.8	1.0	1.3
WW C1	.7	1.0	1.3
WW C2	.8	1.0	1.3
WW C3	.7	.9	1.1
WW C4	.7	.9	1.2
TW A1	1.1	1.8	2.4
TW A2	1.0	1.5	1.9
TW A3	1.0	1.4	1.8
TW A4	1.0	1.3	1.8
TW B1	1.1	1.7	2.3
TW B2	1.0	1.5	2.0
TW B3	1.0	1.5	1.9
TW B4	1.1	1.5	2.0
TW C1	1.1	1.7	2.2
TW C2	1.1	1.6	2.1
TW C3	1.2	1.7	2.2
TW C4	1.3	1.8	2.4
TW D1	1.2	1.8	2.4
TW D2	.9	1.3	1.6
TW D3	.8	1.2	1.5
DG B1	1.1	1.6	---
DG B2	1.0	1.3	2.0
DG B3	.9	1.1	1.5
DG C1	1.0	1.3	2.3
DG C2	.9	1.2	1.6
DG C3	.9	1.1	1.3
SS A1	.8	1.3	1.7
SS A2	.7	1.0	1.4
SS A3	.9	1.4	1.9
SS A4	.7	1.0	1.4
SS A5	.9	1.4	1.9
SS A6	.7	1.0	1.4
SS A7	1.0	1.6	2.3
SS A8	.7	1.1	1.4
SS B1	.9	1.4	1.9
SS B2	.8	1.1	1.5
SS B3	1.0	1.5	2.0
SS B4	.8	1.1	1.4
SS B5	1.0	1.7	2.3
SS B6	.8	1.1	1.5
SS B7	1.1	1.8	2.5
SS B8	.8	1.2	1.6
SS C1	1.0	1.4	1.8
SS C2	.8	1.2	1.5
SS C3	1.1	1.7	2.2
SS C4	.9	1.3	1.7
SS D1	.7	1.1	1.5
SS D2	.6	.9	1.2
SS D3	.8	1.2	1.6
SS D4	.7	.9	1.2
SS E1	.8	1.2	1.6
SS E2	.7	1.0	1.3
SS E3	.9	1.4	2.0
SS E4	.7	1.1	1.4

CHARLOTTE, NC

	SOLAR SAVINGS		
	10%	50%	80%
WW A1	1.2	2.0	2.7
WW A2	1.0	1.5	2.0
WW A3	1.0	1.4	1.8
WW A4	.9	1.3	1.7
WW A5	.9	1.3	1.6
WW A6	.9	1.2	1.6
WW B1	1.2	1.9	2.5
WW B2	.9	1.3	1.6
WW B3	.8	1.1	1.4
WW B4	.8	1.1	1.4
WW B5	.8	1.1	1.4
WW C1	.8	1.1	1.4
WW C2	.8	1.1	1.4
WW C3	.7	.9	1.2
WW C4	.8	1.0	1.3
TW A1	1.2	2.0	2.7
TW A2	1.1	1.6	2.1
TW A3	1.1	1.5	2.0
TW A4	1.1	1.5	1.9
TW B1	1.2	1.9	2.5
TW B2	1.1	1.6	2.2
TW B3	1.1	1.6	2.1
TW B4	1.2	1.7	2.2
TW C1	1.3	1.9	2.5
TW C2	1.2	1.8	2.3
TW C3	1.3	1.8	2.4
TW C4	1.4	2.0	2.7
TW D1	1.4	2.1	2.8
TW D2	1.0	1.4	1.8
TW D3	.9	1.3	1.7
TW D4	.9	1.2	1.5
TW D5	.9	1.2	1.5
TW E1	.9	1.2	1.5
TW E2	.9	1.2	1.5
TW E3	.8	1.0	1.3
TW E4	.8	1.0	1.3
TW F1	1.3	2.0	2.6
TW F2	1.2	1.7	2.2
TW F3	1.1	1.6	2.1
TW F4	1.2	1.6	2.1
TW G1	1.3	1.9	2.6
TW G2	1.2	1.8	2.3
TW G3	1.3	1.8	2.3
TW G4	1.4	1.9	2.5
TW H1	1.4	2.0	2.6
TW H2	1.4	2.0	2.6
TW H3	1.5	2.1	2.8
TW H4	1.8	2.5	3.2
TW I1	1.5	2.2	3.0
TW I2	1.0	1.4	1.9
TW I3	1.0	1.3	1.7
TW I4	.9	1.2	1.6
TW I5	.9	1.2	1.5
TW J1	.9	1.2	1.6
TW J2	.9	1.2	1.6
TW J3	.8	1.0	1.3
TW J4	.8	1.1	1.4
DG A1	1.3	3.5	---
DG A2	1.2	1.7	3.3
DG A3	1.0	1.3	1.9
DG B1	1.2	2.0	---
DG B2	1.1	1.5	2.3
DG B3	1.0	1.2	1.6
DG C1	1.1	1.5	4.1

CHARLOTTE, NC

	SOLAR SAVINGS		
	10%	50%	80%
DG C2	1.0	1.3	1.8
DG C3	.9	1.1	1.4
SS A1	.9	1.4	1.9
SS A2	.8	1.1	1.5
SS A3	1.0	1.6	2.1
SS A4	.8	1.1	1.5
SS A5	1.0	1.6	2.2
SS A6	.8	1.1	1.5
SS A7	1.1	1.9	2.6
SS A8	.8	1.2	1.5
SS B1	1.1	1.6	2.1
SS B2	.9	1.2	1.6
SS B3	1.1	1.7	2.3
SS B4	.9	1.3	1.6
SS B5	1.2	1.9	2.5
SS B6	.9	1.2	1.6
SS B7	1.3	2.1	2.9
SS B8	.9	1.3	1.7
SS C1	1.1	1.5	2.0
SS C2	.9	1.3	1.6
SS C3	1.3	1.9	2.5
SS C4	1.0	1.4	1.8
SS D1	.8	1.2	1.6
SS D2	.7	1.0	1.3
SS D3	.9	1.4	1.8
SS D4	.7	1.0	1.3
SS E1	.9	1.4	1.8
SS E2	.8	1.1	1.4
SS E3	1.0	1.6	2.2
SS E4	.8	1.2	1.5

GREENSBORO, NC

	SOLAR SAVINGS		
	10%	50%	80%
WW A1	1.3	2.2	3.0
WW A2	1.1	1.6	2.1
WW A3	1.0	1.5	1.9
WW A4	1.0	1.4	1.8
WW A5	1.0	1.4	1.8
WW A6	1.0	1.3	1.7
WW B1	1.3	2.1	2.9
WW B2	1.0	1.3	1.7
WW B3	.9	1.2	1.5
WW B4	.9	1.1	1.4
WW B5	.9	1.1	1.4
WW C1	.8	1.1	1.5
WW C2	.9	1.2	1.5
WW C3	.8	1.0	1.3
WW C4	.8	1.0	1.3
TW A1	1.3	2.2	3.0
TW A2	1.2	1.8	2.3
TW A3	1.1	1.6	2.2
TW A4	1.1	1.6	2.1
TW B1	1.3	2.1	2.8
TW B2	1.2	1.8	2.3
TW B3	1.2	1.7	2.3
TW B4	1.3	1.8	2.3
TW C1	1.3	2.0	2.7
TW C2	1.3	1.9	2.5
TW C3	1.4	2.0	2.6
TW C4	1.5	2.2	2.9
TW D1	1.5	2.3	3.2
TW D2	1.1	1.5	1.9
TW D3	1.0	1.4	1.8

GREENSBORO, NC

	SOLAR SAVINGS		
	10%	50%	80%
TW D4	.9	1.2	1.6
TW D5	.9	1.2	1.6
TW E1	.9	1.2	1.6
TW E2	.9	1.2	1.6
TW E3	.8	1.1	1.4
TW E4	.8	1.1	1.4
TW F1	1.4	2.2	2.9
TW F2	1.3	1.8	2.4
TW F3	1.2	1.7	2.3
TW F4	1.3	1.7	2.3
TW G1	1.4	2.1	2.8
TW G2	1.3	1.9	2.5
TW G3	1.4	1.9	2.5
TW G4	1.5	2.1	2.7
TW H1	1.5	2.2	2.9
TW H2	1.5	2.2	2.8
TW H3	1.7	2.3	3.0
TW H4	1.9	2.6	3.4
TW I1	1.6	2.5	3.4
TW I2	1.1	1.5	2.0
TW I3	1.1	1.4	1.9
TW I4	1.0	1.3	1.7
TW I5	1.0	1.3	1.6
TW J1	.9	1.3	1.7
TW J2	1.0	1.3	1.7
TW J3	.8	1.1	1.4
TW J4	.9	1.1	1.4
DG A1	1.4	---	---
DG A2	1.2	1.9	4.9
DG A3	1.1	1.4	2.0
DG B1	1.4	2.5	---
DG B2	1.2	1.6	2.7
DG B3	1.1	1.3	1.7
DG C1	1.2	1.7	---
DG C2	1.1	1.4	2.0
DG C3	1.0	1.2	1.5
SS A1	1.0	1.6	2.1
SS A2	.8	1.2	1.6
SS A3	1.1	1.7	2.4
SS A4	.8	1.2	1.6
SS A5	1.0	1.7	2.4
SS A6	.8	1.2	1.6
SS A7	1.1	2.1	2.9
SS A8	.8	1.3	1.7
SS B1	1.1	1.7	2.3
SS B2	.9	1.3	1.7
SS B3	1.2	1.9	2.6
SS B4	.9	1.3	1.8
SS B5	1.2	2.0	2.8
SS B6	.9	1.3	1.7
SS B7	1.4	2.3	3.3
SS B8	1.0	1.4	1.8
SS C1	1.1	1.6	2.2
SS C2	1.0	1.3	1.7
SS C3	1.4	2.0	2.7
SS C4	1.1	1.5	1.9
SS D1	.9	1.3	1.8
SS D2	.7	1.0	1.4
SS D3	.9	1.5	2.0
SS D4	.8	1.1	1.4
SS E1	1.0	1.5	2.0
SS E2	.8	1.2	1.5
SS E3	1.1	1.8	2.4
SS E4	.9	1.2	1.6

RALEIGH-DURHAM, NC SOLAR SAVINGS	10%	50%	80%
WW A1	1.3	2.2	3.0
WW A2	1.1	1.6	2.1
WW A3	1.0	1.5	1.9
WW A4	1.0	1.4	1.8
WW A5	1.0	1.3	1.7
WW A6	1.0	1.3	1.7
WW B1	1.3	2.1	2.9
WW B2	1.0	1.3	1.7
WW B3	.9	1.2	1.5
WW B4	.8	1.1	1.4
WW B5	.9	1.1	1.4
WW C1	.8	1.1	1.5
WW C2	.9	1.1	1.5
WW C3	.8	1.0	1.3
WW C4	.8	1.0	1.3
TW A1	1.3	2.2	2.9
TW A2	1.2	1.7	2.3
TW A3	1.1	1.6	2.1
TW A4	1.1	1.6	2.1
TW B1	1.3	2.0	2.8
TW B2	1.2	1.7	2.3
TW B3	1.2	1.7	2.3
TW B4	1.3	1.8	2.3
TW C1	1.3	2.0	2.6
TW C2	1.3	1.9	2.5
TW C3	1.4	2.0	2.6
TW C4	1.5	2.1	2.8
TW D1	1.5	2.3	3.1
TW D2	1.0	1.5	1.9
TW D3	1.0	1.3	1.8
TW D4	.9	1.2	1.6
TW D5	.9	1.2	1.6
TW E1	.9	1.2	1.6
TW E2	.9	1.2	1.6
TW E3	.8	1.0	1.3
TW E4	.8	1.1	1.4
TW F1	1.4	2.1	2.9
TW F2	1.2	1.8	2.4
TW F3	1.2	1.7	2.3
TW F4	1.2	1.7	2.3
TW G1	1.4	2.1	2.8
TW G2	1.3	1.9	2.5
TW G3	1.4	1.9	2.5
TW G4	1.5	2.0	2.7
TW H1	1.5	2.1	2.8
TW H2	1.5	2.1	2.8
TW H3	1.6	2.3	3.0
TW H4	1.9	2.6	3.4
TW I1	1.6	2.4	3.4
TW I2	1.1	1.5	2.0
TW I3	1.0	1.4	1.9
TW I4	1.0	1.3	1.7
TW I5	1.0	1.3	1.6
TW J1	.9	1.3	1.7
TW J2	1.0	1.3	1.6
TW J3	.8	1.1	1.4
TW J4	.9	1.1	1.4
DG A1	1.4	---	---
DG A2	1.2	1.9	4.6
DG A3	1.1	1.4	2.0
DG B1	1.3	2.4	---
DG B2	1.2	1.6	2.6
DG B3	1.1	1.3	1.7
DG C1	1.2	1.7	---
DG C2	1.1	1.4	2.0
DG C3	1.0	1.2	1.5
SS A1	1.0	1.5	2.1
SS A2	.8	1.2	1.6
SS A3	1.0	1.7	2.3
SS A4	.8	1.2	1.6
SS A5	1.0	1.7	2.4
SS A6	.8	1.2	1.6
SS A7	1.1	2.0	2.9
SS A8	.8	1.2	1.7
SS B1	1.1	1.7	2.3
SS B2	.9	1.3	1.7
SS B3	1.2	1.8	2.5
SS B4	.9	1.3	1.7
SS B5	1.2	2.0	2.8
SS B6	.9	1.3	1.7
SS B7	1.4	2.3	3.2
SS B8	.9	1.4	1.8
SS C1	1.1	1.6	2.2
SS C2	1.0	1.3	1.7
SS C3	1.4	2.0	2.7
SS C4	1.1	1.5	1.9
SS D1	.9	1.3	1.7
SS D2	.7	1.0	1.3
SS D3	.9	1.5	2.0
SS D4	.8	1.1	1.4
SS E1	1.0	1.5	2.0
SS E2	.8	1.1	1.5
SS E3	1.1	1.7	2.4
SS E4	.9	1.2	1.6

BISMARCK, ND SOLAR SAVINGS	10%	50%	80%
WW A1	2.3	---	---
WW A2	2.0	---	---
WW A3	1.8	---	---
WW A4	1.8	6.2	---
WW A5	1.7	4.3	---
WW A6	1.7	3.7	---
WW B1	---	---	---
WW B2	1.6	2.7	4.5
WW B3	1.4	2.2	3.4
WW B4	1.2	1.8	2.5
WW B5	1.2	1.7	2.3
WW C1	1.3	2.0	2.9
WW C2	1.3	1.9	2.7
WW C3	1.1	1.5	1.9
WW C4	1.1	1.5	2.0
TW A1	2.3	---	---
TW A2	2.1	---	---
TW A3	2.0	---	---
TW A4	1.9	5.0	---
TW B1	2.3	---	---
TW B2	2.1	---	---
TW B3	2.0	5.7	---
TW B4	2.1	4.7	---
TW C1	2.3	---	---
TW C2	2.2	5.6	---
TW C3	2.2	5.1	---
TW C4	2.3	5.0	16.2
TW D1	---	---	---
TW D2	1.7	3.0	5.2
TW D3	1.6	2.8	4.7
TW D4	1.4	2.0	2.8
TW D5	1.3	1.9	2.5
TW E1	1.4	2.3	3.4
TW E2	1.4	2.1	3.0
TW E3	1.1	1.6	2.2
TW E4	1.2	1.6	2.2
TW F1	2.6	---	---
TW F2	2.3	---	---
TW F3	2.2	---	---
TW F4	2.2	6.4	---
TW G1	2.6	---	---
TW G2	2.4	---	---
TW G3	2.4	14.0	---
TW G4	2.5	5.9	---
TW H1	2.8	---	---
TW H2	2.7	---	---
TW H3	2.8	6.9	---
TW H4	3.2	6.6	---
TW I1	---	---	---
TW I2	1.8	3.2	5.5
TW I3	1.7	2.8	4.4
TW I4	1.4	2.1	2.9
TW I5	1.4	1.9	2.6
TW J1	1.5	2.3	3.4
TW J2	1.4	2.1	3.0
TW J3	1.2	1.6	2.2
TW J4	1.2	1.7	2.2
DG A1	---	---	---
DG A2	2.0	---	---
DG A3	1.6	2.4	---
DG B1	---	---	---
DG B2	2.0	---	---
DG B3	1.5	2.1	3.6
DG C1	2.6	---	---
DG C2	1.7	3.8	---
DG C3	1.4	1.9	2.8
SS A1	1.7	---	---
SS A2	1.3	2.2	3.3
SS A3	1.9	---	---
SS A4	1.3	2.5	4.1
SS A5	1.7	---	---
SS A6	1.3	2.3	3.6
SS A7	2.0	---	---
SS A8	1.3	2.8	5.2
SS B1	2.0	---	---
SS B2	1.4	2.3	3.4
SS B3	2.2	---	---
SS B4	1.4	2.5	3.9
SS B5	2.2	---	---
SS B6	1.4	2.4	3.7
SS B7	2.7	---	---
SS B8	1.5	2.8	4.7
SS C1	2.2	---	---
SS C2	1.5	2.6	4.2
SS C3	2.5	---	---
SS C4	1.7	3.0	5.0
SS D1	1.5	---	---
SS D2	1.1	1.8	2.6
SS D3	1.6	---	---
SS D4	1.2	1.9	2.8
SS E1	1.8	---	---
SS E2	1.3	2.2	3.5
SS E3	2.0	---	---
SS E4	1.4	2.5	3.9

FARGO, ND SOLAR SAVINGS	10%	50%	80%
WW A1	3.0	---	---
WW A2	2.6	---	---
WW A3	2.4	---	---
WW A4	2.3	---	---
WW A5	2.2	---	---
WW A6	2.2	---	---
WW B1	---	---	---
WW B2	1.9	4.8	---
WW B3	1.6	3.1	6.3
WW B4	1.4	2.1	3.0
WW B5	1.4	2.0	2.7
WW C1	1.5	2.6	4.2
WW C2	1.4	2.3	3.3
WW C3	1.2	1.6	2.2
WW C4	1.2	1.7	2.2
TW A1	2.9	---	---
TW A2	2.6	---	---
TW A3	2.5	---	---
TW A4	2.4	---	---
TW B1	2.9	---	---
TW B2	2.6	---	---
TW B3	2.5	---	---
TW B4	2.5	---	---
TW C1	2.7	---	---
TW C2	2.6	---	---
TW C3	2.6	---	---
TW C4	2.7	---	---
TW D1	---	---	---
TW D2	1.9	5.8	---
TW D3	1.8	5.0	---
TW D4	1.5	2.4	3.5
TW D5	1.4	2.1	3.0
TW E1	1.6	3.0	5.7
TW E2	1.5	2.6	3.9
TW E3	1.2	1.8	2.5
TW E4	1.3	1.8	2.5
TW F1	3.5	---	---
TW F2	3.0	---	---
TW F3	2.9	---	---
TW F4	2.8	---	---
TW G1	3.7	---	---
TW G2	3.2	---	---
TW G3	3.1	---	---
TW G4	3.2	---	---
TW H1	3.8	---	---
TW H2	3.5	---	---
TW H3	3.5	---	---
TW H4	3.8	---	---
TW I1	---	---	---
TW I2	2.1	6.5	---
TW I3	1.9	4.2	13.9
TW I4	1.6	2.5	3.6
TW I5	1.5	2.2	3.1
TW J1	1.7	3.0	5.2
TW J2	1.6	2.6	3.9
TW J3	1.3	1.8	2.5
TW J4	1.3	1.9	2.5
DG A1	---	---	---
DG A2	2.4	---	---
DG A3	1.7	3.0	---
DG B1	---	---	---
DG B2	2.3	---	---
DG B3	1.7	2.6	6.2
DG C1	---	---	---

FARGO, ND

	SOLAR SAVINGS		
	10%	50%	80%
DG C2	2.0	---	---
DG C3	1.5	2.2	3.7
SS A1	1.9	---	---
SS A2	1.4	2.7	4.5
SS A3	2.4	---	---
SS A4	1.4	3.4	8.1
SS A5	2.0	---	---
SS A6	1.4	3.0	5.7
SS A7	2.4	---	---
SS A8	1.5	4.6	---
SS B1	2.4	---	---
SS B2	1.5	2.8	4.4
SS B3	3.0	---	---
SS B4	1.6	3.2	5.9
SS B5	2.7	---	---
SS B6	1.6	3.1	5.4
SS B7	3.7	---	---
SS B8	1.7	4.0	12.7
SS C1	3.3	---	---
SS C2	1.8	4.0	---
SS C3	3.4	---	---
SS C4	1.9	4.9	---
SS D1	1.8	---	---
SS D2	1.2	2.2	3.3
SS D3	1.8	---	---
SS D4	1.3	2.3	3.6
SS E1	2.5	---	---
SS E2	1.5	3.0	6.0
SS E3	2.5	---	---
SS E4	1.6	3.4	7.8

MINOT, ND

	SOLAR SAVINGS		
	10%	50%	80%
WW A1	2.6	---	---
WW A2	2.3	---	---
WW A3	2.2	---	---
WW A4	2.1	---	---
WW A5	2.1	---	---
WW A6	2.0	---	---
WW B1	---	---	---
WW B2	1.8	4.5	---
WW B3	1.5	3.0	6.5
WW B4	1.3	2.1	2.9
WW B5	1.3	1.9	2.7
WW C1	1.4	2.5	4.2
WW C2	1.4	2.2	3.3
WW C3	1.1	1.6	2.2
WW C4	1.2	1.7	2.2
TW A1	2.6	---	---
TW A2	2.4	---	---
TW A3	2.3	---	---
TW A4	2.2	---	---
TW B1	2.6	---	---
TW B2	2.4	---	---
TW B3	2.3	---	---
TW B4	2.3	---	---
TW C1	2.5	---	---
TW C2	2.4	---	---
TW C3	2.4	---	---
TW C4	2.5	---	---
TW D1	---	---	---
TW D2	1.8	5.3	---
TW D3	1.7	4.7	---

MINOT, ND

	SOLAR SAVINGS		
	10%	50%	80%
TW D4	1.5	2.4	3.5
TW D5	1.4	2.1	2.9
TW E1	1.5	3.0	5.8
TW E2	1.5	2.5	3.9
TW E3	1.2	1.8	2.5
TW E4	1.2	1.8	2.5
TW F1	3.1	---	---
TW F2	2.7	---	---
TW F3	2.6	---	---
TW F4	2.6	---	---
TW G1	3.2	---	---
TW G2	2.9	---	---
TW G3	2.8	---	---
TW G4	3.0	---	---
TW H1	3.4	---	---
TW H2	3.2	---	---
TW H3	3.3	---	---
TW H4	3.6	---	---
TW I1	---	---	---
TW I2	2.0	5.9	---
TW I3	1.8	4.1	22.0
TW I4	1.5	2.4	3.6
TW I5	1.5	2.2	3.1
TW J1	1.6	2.9	5.2
TW J2	1.5	2.5	3.8
TW J3	1.3	1.8	2.5
TW J4	1.3	1.8	2.5
DG A1	---	---	---
DG A2	2.3	---	---
DG A3	1.6	2.9	---
DG B1	---	---	---
DG B2	2.2	---	---
DG B3	1.6	2.5	5.9
DG C1	4.3	---	---
DG C2	1.9	---	---
DG C3	1.5	2.2	3.6
SS A1	1.8	---	---
SS A2	1.3	2.7	4.6
SS A3	2.2	---	---
SS A4	1.4	3.3	9.5
SS A5	1.9	---	---
SS A6	1.4	3.0	5.9
SS A7	2.2	---	---
SS A8	1.4	4.7	---
SS B1	2.2	---	---
SS B2	1.5	2.8	4.5
SS B3	2.6	---	---
SS B4	1.5	3.2	6.1
SS B5	2.4	---	---
SS B6	1.5	3.1	5.5
SS B7	3.1	---	---
SS B8	1.6	4.0	---
SS C1	2.8	---	---
SS C2	1.7	3.9	---
SS C3	3.0	---	---
SS C4	1.9	4.7	---
SS D1	1.6	---	---
SS D2	1.2	2.2	3.4
SS D3	1.7	---	---
SS D4	1.2	2.3	3.6
SS E1	2.2	---	---
SS E2	1.4	3.0	6.4
SS E3	2.3	---	---
SS E4	1.5	3.4	8.7

AKRON-CANTON, OH

	SOLAR SAVINGS		
	10%	50%	80%
WW A1	2.6	---	---
WW A2	2.3	---	---
WW A3	2.2	---	---
WW A4	2.1	---	---
WW A5	2.0	---	---
WW A6	2.0	---	---
WW B1	---	---	---
WW B2	1.8	3.2	5.9
WW B3	1.5	2.5	3.8
WW B4	1.3	1.9	2.6
WW B5	1.3	1.8	2.5
WW C1	1.4	2.2	3.2
WW C2	1.4	2.1	2.9
WW C3	1.1	1.5	2.0
WW C4	1.2	1.6	2.1
TW A1	2.6	---	---
TW A2	2.4	---	---
TW A3	2.3	---	---
TW A4	2.2	---	---
TW B1	2.6	---	---
TW B2	2.4	---	---
TW B3	2.3	---	---
TW B4	2.3	9.6	---
TW C1	2.6	---	---
TW C2	2.4	---	---
TW C3	2.5	9.1	---
TW C4	2.6	6.8	---
TW D1	---	---	---
TW D2	1.8	3.6	7.2
TW D3	1.7	3.2	6.0
TW D4	1.5	2.2	3.0
TW D5	1.4	2.0	2.7
TW E1	1.5	2.5	3.9
TW E2	1.5	2.3	3.3
TW E3	1.2	1.7	2.3
TW E4	1.2	1.7	2.3
TW F1	3.1	---	---
TW F2	2.7	---	---
TW F3	2.6	---	---
TW F4	2.5	---	---
TW G1	3.2	---	---
TW G2	2.8	---	---
TW G3	2.8	---	---
TW G4	2.9	---	---
TW H1	3.3	---	---
TW H2	3.2	---	---
TW H3	3.2	---	---
TW H4	3.6	10.3	---
TW I1	---	---	---
TW I2	2.0	3.8	7.6
TW I3	1.8	3.2	5.2
TW I4	1.5	2.3	3.1
TW I5	1.5	2.1	2.8
TW J1	1.6	2.6	3.8
TW J2	1.5	2.3	3.3
TW J3	1.2	1.7	2.3
TW J4	1.3	1.8	2.3
DG A1	---	---	---
DG A2	2.3	---	---
DG A3	1.7	2.6	---
DG B1	---	---	---
DG B2	2.2	---	---
DG B3	1.6	2.3	4.2
DG C1	4.4	---	---

AKRON-CANTON, OH

	SOLAR SAVINGS		
	10%	50%	80%
DG C2	1.9	---	---
DG C3	1.5	2.1	3.1
SS A1	1.7	---	---
SS A2	1.3	2.2	3.3
SS A3	1.9	---	---
SS A4	1.3	2.4	3.8
SS A5	1.7	---	---
SS A6	1.3	2.3	3.5
SS A7	2.0	---	---
SS A8	1.3	2.7	4.6
SS B1	2.0	---	---
SS B2	1.4	2.3	3.4
SS B3	2.3	---	---
SS B4	1.4	2.5	3.8
SS B5	2.3	---	---
SS B6	1.4	2.5	3.7
SS B7	2.7	---	---
SS B8	1.5	2.8	4.5
SS C1	2.7	---	---
SS C2	1.7	3.0	5.1
SS C3	3.0	---	---
SS C4	1.8	3.5	6.3
SS D1	1.5	---	---
SS D2	1.1	1.8	2.6
SS D3	1.6	---	---
SS D4	1.2	1.9	2.8
SS E1	1.9	---	---
SS E2	1.3	2.3	3.5
SS E3	2.1	---	---
SS E4	1.4	2.5	3.9

CINCINNATI, OH

	SOLAR SAVINGS		
	10%	50%	80%
WW A1	2.1	---	---
WW A2	1.7	3.7	7.6
WW A3	1.6	3.0	4.9
WW A4	1.6	2.7	4.2
WW A5	1.5	2.5	3.8
WW A6	1.5	2.4	3.5
WW B1	3.0	---	---
WW B2	1.4	2.2	3.1
WW B3	1.3	1.9	2.5
WW B4	1.2	1.6	2.1
WW B5	1.2	1.6	2.1
WW C1	1.2	1.7	2.3
WW C2	1.2	1.7	2.2
WW C3	1.0	1.4	1.7
WW C4	1.1	1.4	1.8
TW A1	2.1	---	---
TW A2	1.9	4.0	8.3
TW A3	1.8	3.3	5.4
TW A4	1.7	3.0	4.5
TW B1	2.1	6.8	---
TW B2	1.9	3.5	5.9
TW B3	1.8	3.2	5.0
TW B4	1.9	3.2	4.8
TW C1	2.0	4.2	7.8
TW C2	2.0	3.5	5.4
TW C3	2.0	3.5	5.3
TW C4	2.2	3.8	5.6
TW D1	3.2	---	---
TW D2	1.5	2.4	3.4
TW D3	1.4	2.2	3.1

CINCINNATI, OH

	SOLAR SAVINGS		
	10%	50%	80%
WW A1	2.1	---	---
WW A2	1.7	3.7	7.6
WW A3	1.6	3.0	4.9
WW A4	1.6	2.7	4.2
WW A5	1.5	2.5	3.8
WW A6	1.5	2.4	3.5
WW B1	3.0	---	---
WW B2	1.4	2.2	3.1
WW B3	1.3	1.9	2.5
WW B4	1.2	1.6	2.1
WW B5	1.2	1.6	2.1
WW C1	1.2	1.7	2.3
WW C2	1.2	1.7	2.2
WW C3	1.0	1.4	1.7
WW C4	1.1	1.4	1.8
TW A1	2.1	---	---
TW A2	1.9	4.0	8.3
TW A3	1.8	3.3	5.4
TW A4	1.7	3.0	4.5
TW B1	2.1	6.8	---
TW B2	1.9	3.5	5.9
TW B3	1.8	3.2	5.0
TW B4	1.9	3.2	4.8
TW C1	2.0	4.2	7.8
TW C2	2.0	3.5	5.4
TW C3	2.0	3.5	5.3
TW C4	2.2	3.8	5.6
TW D1	3.2	---	---
TW D2	1.5	2.4	3.4
TW D3	1.4	2.2	3.1
DG B1	3.1	---	---
DG B2	1.8	4.7	---
DG B3	1.4	1.9	2.8
DG C1	2.1	---	---
DG C2	1.6	2.6	---
DG C3	1.3	1.7	2.3
SS A1	1.4	3.0	5.2
SS A2	1.1	1.8	2.5
SS A3	1.6	5.5	---
SS A4	1.1	1.9	2.6
SS A5	1.5	4.2	---
SS A6	1.1	1.8	2.6
SS A7	1.7	---	---
SS A8	1.2	2.0	2.8
SS B1	1.7	3.6	7.1
SS B2	1.2	1.9	2.6
SS B3	1.8	7.0	---
SS B4	1.3	2.0	2.8
SS B5	1.9	7.4	---
SS B6	1.3	2.0	2.7
SS B7	2.1	---	---
SS B8	1.3	2.1	3.0
SS C1	1.9	4.3	14.0
SS C2	1.4	2.2	3.0
SS C3	2.2	6.3	---
SS C4	1.5	2.4	3.4
SS D1	1.3	2.4	3.9
SS D2	1.0	1.5	2.0
SS D3	1.4	2.9	5.2
SS D4	1.0	1.6	2.1
SS E1	1.5	3.5	8.7
SS E2	1.1	1.8	2.4
SS E3	1.7	4.8	---
SS E4	1.2	1.9	2.7

CLEVELAND, OH

	SOLAR SAVINGS		
	10%	50%	80%
WW A1	2.7	---	---
WW A2	2.6	---	---
WW A3	2.4	---	---
WW A4	2.3	---	---
WW A5	2.2	---	---
WW A6	2.2	---	---
WW B1	---	---	---
WW B2	1.9	4.2	---
WW B3	1.6	2.9	5.1
WW B4	1.4	2.1	2.9
WW B5	1.4	1.9	2.6
WW C1	1.5	2.5	3.8
WW C2	1.4	2.2	3.2
WW C3	1.2	1.6	2.1
WW C4	1.2	1.7	2.2
TW A1	2.7	---	---
TW A2	2.6	---	---
TW A3	2.5	---	---
TW A4	2.4	---	---
TW B1	2.8	---	---
TW B2	2.6	---	---
TW B3	2.5	---	---
TW B4	2.5	---	---
TW C1	2.7	---	---
TW C2	2.6	---	---
TW C3	2.6	---	---
TW C4	2.7	---	---
TW D1	---	---	---
TW D2	1.9	4.8	---
TW D3	1.8	4.1	---
TW D4	1.5	2.4	3.4
TW D5	1.4	2.1	2.9
TW E1	1.6	2.9	4.9
TW E2	1.5	2.5	3.7
TW E3	1.2	1.8	2.4
TW E4	1.3	1.8	2.4
TW F1	3.4	---	---
TW F2	3.0	---	---
TW F3	2.9	---	---
TW F4	2.8	---	---
TW G1	3.6	---	---
TW G2	3.2	---	---
TW G3	3.1	---	---
TW G4	3.2	---	---
TW H1	3.8	---	---
TW H2	3.5	---	---
TW H3	3.5	---	---
TW H4	3.8	---	---
TW I1	---	---	---
TW I2	2.1	5.2	---
TW I3	1.9	3.8	8.0
TW I4	1.6	2.4	3.4
TW I5	1.5	2.2	3.0
TW J1	1.6	2.9	4.6
TW J2	1.6	2.5	3.7
TW J3	1.3	1.8	2.4
TW J4	1.3	1.9	2.5
DG A1	---	---	---
DG A2	2.5	---	---
DG A3	1.7	2.9	---
DG B1	---	---	---
DG B2	2.3	---	---
DG B3	1.7	2.5	5.4
DG C1	---	---	---

CLEVELAND, OH

	SOLAR SAVINGS		
	10%	50%	80%
DG C2	2.0	---	---
DG C3	1.5	2.2	3.5
SS A1	1.7	---	---
SS A2	1.3	2.4	3.7
SS A3	2.0	---	---
SS A4	1.3	2.7	4.6
SS A5	1.8	---	---
SS A6	1.3	2.6	4.1
SS A7	2.0	---	---
SS A8	1.4	3.1	6.5
SS B1	2.1	---	---
SS B2	1.4	2.5	3.8
SS B3	2.5	---	---
SS B4	1.5	2.8	4.4
SS B5	2.4	---	---
SS B6	1.5	2.7	4.3
SS B7	2.9	---	---
SS B8	1.6	3.2	5.7
SS C1	3.3	---	---
SS C2	1.8	3.7	8.8
SS C3	3.4	---	---
SS C4	1.9	4.3	---
SS D1	1.6	---	---
SS D2	1.2	2.0	2.9
SS D3	1.7	---	---
SS D4	1.2	2.1	3.0
SS E1	2.1	---	---
SS E2	1.4	2.6	4.2
SS E3	2.2	---	---
SS E4	1.5	2.8	4.9

COLUMBUS, OH

	SOLAR SAVINGS		
	10%	50%	80%
WW A1	2.5	---	---
WW A2	2.1	---	---
WW A3	1.9	12.2	---
WW A4	1.9	4.6	---
WW A5	1.8	3.8	15.9
WW A6	1.8	3.5	7.8
WW B1	---	---	---
WW B2	1.6	2.7	4.1
WW B3	1.4	2.2	3.2
WW B4	1.3	1.8	2.4
WW B5	1.3	1.7	2.3
WW C1	1.3	2.0	2.8
WW C2	1.3	1.9	2.6
WW C3	1.1	1.5	1.9
WW C4	1.1	1.5	2.0
TW A1	2.4	---	---
TW A2	2.2	---	---
TW A3	2.1	9.0	---
TW A4	2.0	4.5	---
TW B1	2.4	---	---
TW B2	2.2	8.6	---
TW B3	2.1	5.0	---
TW B4	2.2	4.5	10.3
TW C1	2.4	---	---
TW C2	2.3	5.2	---
TW C3	2.3	4.9	11.3
TW C4	2.4	5.0	9.5
TW D1	---	---	---
TW D2	1.7	3.0	4.7
TW D3	1.6	2.7	4.2

COLUMBUS, OH

	SOLAR SAVINGS		
	10%	50%	80%
TW D4	1.4	2.0	2.8
TW D5	1.3	1.9	2.5
TW E1	1.4	2.3	3.2
TW E2	1.4	2.1	2.9
TW E3	1.2	1.6	2.1
TW E4	1.2	1.6	2.2
TW F1	2.7	---	---
TW F2	2.4	---	---
TW F3	2.3	---	---
TW F4	2.3	5.3	---
TW G1	2.8	---	---
TW G2	2.5	---	---
TW G3	2.5	6.7	---
TW G4	2.6	5.5	14.4
TW H1	2.9	---	---
TW H2	2.8	8.4	---
TW H3	2.9	6.3	25.4
TW H4	3.3	6.4	12.4
TW I1	---	---	---
TW I2	1.9	3.2	4.9
TW I3	1.7	2.8	4.1
TW I4	1.5	2.1	2.9
TW I5	1.4	2.0	2.6
TW J1	1.5	2.3	3.2
TW J2	1.5	2.1	2.9
TW J3	1.2	1.6	2.2
TW J4	1.2	1.7	2.2
DG A1	---	---	---
DG A2	2.1	---	---
DG A3	1.6	2.4	---
DG B1	---	---	---
DG B2	2.1	---	---
DG B3	1.6	2.2	3.5
DG C1	2.9	---	---
DG C2	1.8	3.8	---
DG C3	1.4	1.9	2.7
SS A1	1.6	5.1	---
SS A2	1.2	2.0	2.9
SS A3	1.8	---	---
SS A4	1.2	2.2	3.2
SS A5	1.7	---	---
SS A6	1.2	2.1	3.1
SS A7	1.9	---	---
SS A8	1.3	2.4	3.6
SS B1	1.9	---	---
SS B2	1.3	2.2	3.1
SS B3	2.1	---	---
SS B4	1.4	2.3	3.3
SS B5	2.1	---	---
SS B6	1.4	2.3	3.3
SS B7	2.5	---	---
SS B8	1.4	2.5	3.7
SS C1	2.3	---	---
SS C2	1.6	2.6	3.9
SS C3	2.6	---	---
SS C4	1.7	3.0	4.5
SS D1	1.4	3.6	---
SS D2	1.1	1.7	2.4
SS D3	1.5	5.6	---
SS D4	1.1	1.8	2.5
SS E1	1.8	---	---
SS E2	1.3	2.1	3.0
SS E3	1.9	---	---
SS E4	1.3	2.3	3.3

DAYTON, OH

	SOLAR SAVINGS		
	10%	50%	80%
WW A1	2.3	---	---
WW A2	1.9	8.4	---
WW A3	1.8	3.9	---
WW A4	1.7	3.3	7.0
WW A5	1.7	3.0	5.4
WW A6	1.6	2.8	4.8
WW B1	7.2	---	---
WW B2	1.5	2.5	3.6
WW B3	1.3	2.0	2.8
WW B4	1.2	1.7	2.3
WW B5	1.2	1.7	2.2
WW C1	1.3	1.9	2.6
WW C2	1.3	1.8	2.4
WW C3	1.1	1.4	1.8
WW C4	1.1	1.5	1.9
TW A1	2.3	---	---
TW A2	2.0	8.0	---
TW A3	1.9	4.2	---
TW A4	1.9	3.6	6.7
TW B1	2.3	---	---
TW B2	2.0	4.5	---
TW B3	2.0	3.9	7.8
TW B4	2.0	3.7	6.5
TW C1	2.2	6.0	---
TW C2	2.1	4.2	8.0
TW C3	2.2	4.1	7.1
TW C4	2.3	4.3	7.1
TW D1	5.4	---	---
TW D2	1.6	2.7	4.0
TW D3	1.5	2.5	3.6
TW D4	1.3	1.9	2.6
TW D5	1.3	1.8	2.4
TW E1	1.4	2.1	2.9
TW E2	1.3	2.0	2.7
TW E3	1.1	1.6	2.0
TW E4	1.2	1.6	2.1
TW F1	2.5	---	---
TW F2	2.2	---	---
TW F3	2.1	4.8	---
TW F4	2.1	4.0	7.9
TW G1	2.6	---	---
TW G2	2.3	5.8	---
TW G3	2.3	4.6	10.6
TW G4	2.5	4.5	7.8
TW H1	2.7	---	---
TW H2	2.6	5.3	14.3
TW H3	2.8	5.1	9.2
TW H4	3.1	5.4	8.9
TW I1	---	---	---
TW I2	1.8	2.8	4.2
TW I3	1.6	2.5	3.6
TW I4	1.4	2.0	2.7
TW I5	1.4	1.9	2.5
TW J1	1.4	2.1	3.0
TW J2	1.4	2.0	2.7
TW J3	1.2	1.6	2.1
TW J4	1.2	1.6	2.1
DG A1	---	---	---
DG A2	2.0	---	---
DG A3	1.5	2.3	6.3
DG B1	9.3	---	---
DG B2	1.9	---	---
DG B3	1.5	2.0	3.2
DG C1	2.5	---	---
DG C2	1.7	3.1	---
DG C3	1.4	1.8	2.5
SS A1	1.5	3.7	16.0
SS A2	1.2	1.9	2.7
SS A3	1.7	---	---
SS A4	1.2	2.0	3.0
SS A5	1.6	---	---
SS A6	1.2	2.0	2.8
SS A7	1.8	---	---
SS A8	1.2	2.2	3.2
SS B1	1.8	5.2	---
SS B2	1.3	2.1	2.9
SS B3	2.0	---	---
SS B4	1.3	2.2	3.1
SS B5	2.0	---	---
SS B6	1.3	2.2	3.0
SS B7	2.3	---	---
SS B8	1.4	2.3	3.4
SS C1	2.1	---	---
SS C2	1.5	2.4	3.4
SS C3	2.4	---	---
SS C4	1.7	2.7	3.9
SS D1	1.4	2.9	6.3
SS D2	1.1	1.6	2.2
SS D3	1.4	3.7	---
SS D4	1.1	1.7	2.3
SS E1	1.6	---	---
SS E2	1.2	1.9	2.7
SS E3	1.8	---	---
SS E4	1.3	2.1	3.0

TOLEDO, OH

	SOLAR SAVINGS		
	10%	50%	80%
WW A1	2.5	---	---
WW A2	2.2	---	---
WW A3	2.1	---	---
WW A4	2.0	---	---
WW A5	1.9	---	---
WW A6	1.9	10.3	---
WW B1	---	---	---
WW B2	1.7	3.1	5.7
WW B3	1.5	2.4	3.8
WW B4	1.3	1.9	2.6
WW B5	1.3	1.8	2.4
WW C1	1.4	2.2	3.2
WW C2	1.4	2.0	2.8
WW C3	1.1	1.5	2.0
WW C4	1.2	1.6	2.1
TW A1	2.5	---	---
TW A2	2.3	---	---
TW A3	2.2	---	---
TW A4	2.1	---	---
TW B1	2.5	---	---
TW B2	2.3	---	---
TW B3	2.3	---	---
TW B4	2.3	7.1	---
TW C1	2.5	---	---
TW C2	2.4	---	---
TW C3	2.4	7.4	---
TW C4	2.5	6.3	---
TW D1	---	---	---
TW D2	1.8	3.4	7.0
TW D3	1.7	3.1	5.8
TW D4	1.4	2.2	3.0
TW D5	1.4	2.0	2.7
TW E1	1.5	2.5	3.8
TW E2	1.4	2.2	3.2
TW E3	1.2	1.7	2.3
TW E4	1.2	1.7	2.3
TW F1	3.0	---	---
TW F2	2.6	---	---
TW F3	2.5	---	---
TW F4	2.5	---	---
TW G1	3.1	---	---
TW G2	2.7	---	---
TW G3	2.7	---	---
TW G4	2.8	---	---
TW H1	3.2	---	---
TW H2	3.1	---	---
TW H3	3.2	---	---
TW H4	3.5	9.0	---
TW I1	---	---	---
TW I2	2.0	3.7	7.4
TW I3	1.8	3.1	5.1
TW I4	1.5	2.2	3.1
TW I5	1.4	2.1	2.8
TW J1	1.5	2.5	3.7
TW J2	1.5	2.3	3.2
TW J3	1.2	1.7	2.3
TW J4	1.3	1.7	2.3
DG A1	---	---	---
DG A2	2.3	---	---
DG A3	1.6	2.6	---
DG B1	---	---	---
DG B2	2.2	---	---
DG B3	1.6	2.3	4.1
DG C1	3.7	---	---
DG C2	1.9	6.4	---
DG C3	1.5	2.0	3.0
SS A1	1.7	---	---
SS A2	1.3	2.2	3.3
SS A3	1.9	---	---
SS A4	1.3	2.4	3.8
SS A5	1.7	---	---
SS A6	1.3	2.3	3.5
SS A7	2.0	---	---
SS A8	1.3	2.7	4.7
SS B1	2.0	---	---
SS B2	1.4	2.3	3.4
SS B3	2.3	---	---
SS B4	1.4	2.5	3.8
SS B5	2.2	---	---
SS B6	1.4	2.5	3.7
SS B7	2.7	---	---
SS B8	1.5	2.8	4.5
SS C1	2.6	---	---
SS C2	1.7	2.9	5.0
SS C3	2.9	---	---
SS C4	1.8	3.3	6.1
SS D1	1.5	---	---
SS D2	1.1	1.8	2.6
SS D3	1.6	---	---
SS D4	1.2	1.9	2.7
SS E1	1.9	---	---
SS E2	1.3	2.3	3.5
SS E3	2.1	---	---
SS E4	1.4	2.5	3.9

YOUNGSTOWN, OH

	SOLAR SAVINGS		
	10%	50%	80%
WW A1	3.3	---	---
WW A2	3.0	---	---
WW A3	2.8	---	---
WW A4	2.7	---	---
WW A5	2.5	---	---
WW A6	2.5	---	---
WW B1	---	---	---
WW B2	2.0	5.9	---
WW B3	1.7	3.2	6.6
WW B4	1.5	2.2	3.0
WW B5	1.4	2.0	2.7
WW C1	1.5	2.7	4.3
WW C2	1.5	2.4	3.4
WW C3	1.2	1.7	2.3
WW C4	1.3	1.7	2.3
TW A1	3.1	---	---
TW A2	3.0	---	---
TW A3	2.8	---	---
TW A4	2.7	---	---
TW B1	3.2	---	---
TW B2	2.9	---	---
TW B3	2.8	---	---
TW B4	2.7	---	---
TW C1	3.0	---	---
TW C2	2.8	---	---
TW C3	2.8	---	---
TW C4	2.9	---	---
TW D1	---	---	---
TW D2	2.1	8.4	---
TW D3	2.0	5.6	---
TW D4	1.6	2.5	3.6
TW D5	1.5	2.2	3.0
TW E1	1.7	3.2	5.9
TW E2	1.6	2.7	4.0
TW E3	1.3	1.9	2.6
TW E4	1.3	1.9	2.5
TW F1	4.5	---	---
TW F2	3.7	---	---
TW F3	3.4	---	---
TW F4	3.2	---	---
TW G1	4.8	---	---
TW G2	3.8	---	---
TW G3	3.6	---	---
TW G4	3.6	---	---
TW H1	4.8	---	---
TW H2	4.1	---	---
TW H3	4.0	---	---
TW H4	4.2	---	---
TW I1	---	---	---
TW I2	2.3	12.0	---
TW I3	2.0	4.5	---
TW I4	1.7	2.6	3.7
TW I5	1.6	2.3	3.2
TW J1	1.7	3.1	5.3
TW J2	1.7	2.7	3.9
TW J3	1.3	1.9	2.5
TW J4	1.4	1.9	2.6
DG A1	---	---	---
DG A2	2.7	---	---
DG A3	1.8	3.1	---
DG B1	---	---	---
DG B2	2.6	---	---
DG B3	1.7	2.7	7.1
DG C1	---	---	---

YOUNGSTOWN, OH

	SOLAR SAVINGS		
	10%	50%	80%
DG C2	2.1	---	---
DG C3	1.6	2.3	3.8
SS A1	1.9	---	---
SS A2	1.4	2.5	4.0
SS A3	2.2	---	---
SS A4	1.4	2.9	5.3
SS A5	1.9	---	---
SS A6	1.4	2.7	4.6
SS A7	2.2	---	---
SS A8	1.5	3.5	10.7
SS B1	2.3	---	---
SS B2	1.5	2.7	4.1
SS B3	2.8	---	---
SS B4	1.5	3.0	4.9
SS B5	2.6	---	---
SS B6	1.6	2.9	4.7
SS B7	3.3	---	---
SS B8	1.6	3.5	7.0
SS C1	5.2	---	---
SS C2	1.9	4.4	---
SS C3	4.3	---	---
SS C4	2.1	5.6	---
SS D1	1.7	---	---
SS D2	1.2	2.1	3.0
SS D3	1.8	---	---
SS D4	1.3	2.2	3.2
SS E1	2.3	---	---
SS E2	1.5	2.8	4.8
SS E3	2.4	---	---
SS E4	1.5	3.1	5.8

OKLAHOMA CITY, OK

	SOLAR SAVINGS		
	10%	50%	80%
WW A1	1.2	2.0	2.7
WW A2	1.1	1.5	2.0
WW A3	1.0	1.4	1.8
WW A4	1.0	1.3	1.7
WW A5	.9	1.3	1.7
WW A6	.9	1.3	1.6
WW B1	1.3	1.9	2.6
WW B2	.9	1.3	1.6
WW B3	.8	1.1	1.4
WW B4	.8	1.1	1.4
WW B5	.8	1.1	1.4
WW C1	.8	1.1	1.4
WW C2	.8	1.1	1.4
WW C3	.7	1.0	1.2
WW C4	.8	1.0	1.3
TW A1	1.3	2.0	2.7
TW A2	1.1	1.7	2.2
TW A3	1.1	1.5	2.0
TW A4	1.1	1.5	2.0
TW B1	1.3	1.9	2.6
TW B2	1.2	1.7	2.2
TW B3	1.2	1.6	2.1
TW B4	1.2	1.7	2.2
TW C1	1.3	1.9	2.5
TW C2	1.3	1.8	2.4
TW C3	1.3	1.9	2.4
TW C4	1.4	2.1	2.7
TW D1	1.4	2.1	2.8
TW D2	1.0	1.4	1.8
TW D3	.9	1.3	1.7
TW D4	.9	1.2	1.5
TW D5	.9	1.2	1.5
TW E1	.9	1.2	1.5
TW E2	.9	1.2	1.5
TW E3	.8	1.0	1.3
TW E4	.8	1.1	1.3
TW F1	1.3	2.0	2.7
TW F2	1.2	1.7	2.2
TW F3	1.2	1.6	2.1
TW F4	1.2	1.6	2.1
TW G1	1.4	2.0	2.6
TW G2	1.3	1.8	2.3
TW G3	1.3	1.8	2.3
TW G4	1.4	1.9	2.5
TW H1	1.4	2.0	2.7
TW H2	1.5	2.0	2.6
TW H3	1.6	2.2	2.8
TW H4	1.8	2.5	3.2
TW I1	1.5	2.3	3.0
TW I2	1.1	1.5	1.9
TW I3	1.0	1.4	1.8
TW I4	.9	1.2	1.6
TW I5	.9	1.2	1.5
TW J1	.9	1.2	1.6
TW J2	.9	1.2	1.6
TW J3	.8	1.1	1.3
TW J4	.8	1.1	1.4
DG A1	1.3	4.4	---
DG A2	1.2	1.8	3.5
DG A3	1.1	1.3	1.9
DG B1	1.3	2.1	---
DG B2	1.2	1.5	2.4
DG B3	1.0	1.2	1.6
DG C1	1.1	1.5	5.2
DG C2	1.1	1.3	1.8
DG C3	1.0	1.1	1.4
SS A1	1.0	1.5	2.0
SS A2	.8	1.1	1.5
SS A3	1.0	1.6	2.2
SS A4	.8	1.2	1.5
SS A5	1.0	1.6	2.2
SS A6	.8	1.2	1.5
SS A7	1.1	1.9	2.7
SS A8	.8	1.2	1.6
SS B1	1.1	1.6	2.2
SS B2	.9	1.2	1.6
SS B3	1.1	1.8	2.4
SS B4	.9	1.3	1.7
SS B5	1.2	1.9	2.6
SS B6	.9	1.3	1.7
SS B7	1.3	2.2	3.0
SS B8	.9	1.3	1.7
SS C1	1.1	1.5	2.0
SS C2	.9	1.3	1.7
SS C3	1.3	1.9	2.5
SS C4	1.0	1.4	1.8
SS D1	.8	1.2	1.6
SS D2	.7	1.0	1.3
SS D3	.9	1.4	1.9
SS D4	.7	1.0	1.3
SS E1	.9	1.4	1.8
SS E2	.8	1.1	1.4
SS E3	1.1	1.7	2.2
SS E4	.9	1.2	1.6

TULSA, OK

	SOLAR SAVINGS		
	10%	50%	80%
WW A1	1.3	2.2	3.0
WW A2	1.1	1.6	2.2
WW A3	1.1	1.5	2.0
WW A4	1.0	1.4	1.9
WW A5	1.0	1.4	1.8
WW A6	1.0	1.3	1.7
WW B1	1.4	2.2	3.0
WW B2	1.0	1.4	1.8
WW B3	.9	1.2	1.5
WW B4	.9	1.1	1.5
WW B5	.9	1.1	1.5
WW C1	.9	1.2	1.5
WW C2	.9	1.2	1.5
WW C3	.8	1.0	1.3
WW C4	.8	1.1	1.3
TW A1	1.4	2.2	3.0
TW A2	1.2	1.8	2.4
TW A3	1.2	1.7	2.2
TW A4	1.2	1.6	2.1
TW B1	1.4	2.1	2.8
TW B2	1.2	1.8	2.4
TW B3	1.2	1.8	2.3
TW B4	1.3	1.8	2.4
TW C1	1.4	2.0	2.7
TW C2	1.3	1.9	2.5
TW C3	1.4	2.0	2.6
TW C4	1.5	2.2	2.9
TW D1	1.5	2.4	3.2
TW D2	1.1	1.5	1.9
TW D3	1.0	1.4	1.8
TW D4	.9	1.3	1.6
TW D5	.9	1.2	1.6
TW E1	.9	1.3	1.6
TW E2	.9	1.3	1.6
TW E3	.8	1.1	1.4
TW E4	.8	1.1	1.4
TW F1	1.4	2.2	3.0
TW F2	1.3	1.9	2.5
TW F3	1.3	1.8	2.3
TW F4	1.3	1.8	2.3
TW G1	1.4	2.2	2.9
TW G2	1.4	1.9	2.5
TW G3	1.4	1.9	2.5
TW G4	1.5	2.1	2.7
TW H1	1.5	2.2	2.9
TW H2	1.6	2.2	2.9
TW H3	1.7	2.3	3.0
TW H4	1.9	2.7	3.5
TW I1	1.7	2.5	3.5
TW I2	1.1	1.6	2.0
TW I3	1.1	1.5	1.9
TW I4	1.0	1.3	1.7
TW I5	1.0	1.3	1.6
TW J1	1.0	1.3	1.7
TW J2	1.0	1.3	1.7
TW J3	.9	1.1	1.4
TW J4	.9	1.1	1.5
DG A1	1.4	---	---
DG A2	1.3	1.9	5.6
DG A3	1.1	1.4	2.0
DG B1	1.4	2.6	---
DG B2	1.2	1.6	2.8
DG B3	1.1	1.3	1.7
DG C1	1.2	1.8	---

TULSA, OK

	SOLAR SAVINGS		
	10%	50%	80%
DG C2	1.1	1.4	2.0
DG C3	1.0	1.2	1.5
SS A1	1.0	1.6	2.1
SS A2	.9	1.2	1.6
SS A3	1.1	1.8	2.4
SS A4	.9	1.2	1.7
SS A5	1.1	1.8	2.4
SS A6	.9	1.2	1.6
SS A7	1.2	2.1	3.0
SS A8	.9	1.3	1.7
SS B1	1.2	1.8	2.4
SS B2	.9	1.3	1.7
SS B3	1.2	1.9	2.6
SS B4	1.0	1.4	1.8
SS B5	1.3	2.1	2.9
SS B6	1.0	1.4	1.8
SS B7	1.4	2.4	3.4
SS B8	1.0	1.4	1.9
SS C1	1.2	1.7	2.2
SS C2	1.0	1.4	1.8
SS C3	1.4	2.1	2.7
SS C4	1.1	1.5	2.0
SS D1	.9	1.3	1.8
SS D2	.8	1.1	1.4
SS D3	1.0	1.5	2.0
SS D4	.8	1.1	1.4
SS E1	1.0	1.5	2.0
SS E2	.8	1.2	1.5
SS E3	1.2	1.8	2.5
SS E4	.9	1.3	1.7

ASTORIA, OR

	SOLAR SAVINGS		
	10%	50%	80%
WW A1	1.6	5.1	---
WW A2	1.3	2.4	10.9
WW A3	1.2	2.1	4.6
WW A4	1.2	2.0	3.8
WW A5	1.1	1.9	3.4
WW A6	1.1	1.8	3.2
WW B1	1.6	---	---
WW B2	1.1	1.7	2.7
WW B3	1.0	1.5	2.3
WW B4	1.0	1.4	1.9
WW B5	1.0	1.3	1.9
WW C1	1.0	1.4	2.1
WW C2	1.0	1.4	2.0
WW C3	.9	1.2	1.6
WW C4	.9	1.2	1.6
TW A1	1.6	4.0	---
TW A2	1.4	2.6	11.6
TW A3	1.3	2.3	5.1
TW A4	1.3	2.2	4.1
TW B1	1.6	3.2	---
TW B2	1.4	2.5	5.5
TW B3	1.4	2.4	4.5
TW B4	1.4	2.4	4.3
TW C1	1.6	2.8	7.9
TW C2	1.5	2.6	4.8
TW C3	1.6	2.6	4.7
TW C4	1.7	2.8	4.9
TW D1	1.8	---	---
TW D2	1.2	1.9	3.0
TW D3	1.1	1.7	2.8

ASTORIA, OR				BURNS, OR				BURNS, OR				MEDFORD, OR			
	\multicolumn{3}{c}{SOLAR SAVINGS}			\multicolumn{3}{c}{SOLAR SAVINGS}			\multicolumn{3}{c}{SOLAR SAVINGS}			\multicolumn{3}{c}{SOLAR SAVINGS}
	10%	50%	80%		10%	50%	80%		10%	50%	80%		10%	50%	80%
WW A1	1.6	5.1	---	WW A1	1.7	---	---	DG C2	1.4	2.1	---	TW D4	1.0	1.6	2.3
WW A2	1.3	2.4	10.9	WW A2	1.4	2.6	5.4	DG C3	1.2	1.5	2.1	TW D5	1.0	1.5	2.1
WW A3	1.2	2.1	4.6	WW A3	1.3	2.3	3.9	SS A1	1.2	2.5	5.1	TW E1	1.0	1.6	2.5
WW A4	1.2	2.0	3.8	WW A4	1.3	2.1	3.5	SS A2	1.0	1.6	2.3	TW E2	1.0	1.6	2.3
WW A5	1.1	1.9	3.4	WW A5	1.3	2.0	3.2	SS A3	1.3	3.7	---	TW E3	.9	1.3	1.8
WW A6	1.1	1.8	3.2	WW A6	1.2	1.9	3.0	SS A4	1.0	1.7	2.5	TW E4	.9	1.3	1.8
WW B1	1.6	---	---	WW B1	1.9	---	---	SS A5	1.2	3.1	---	TW F1	1.6	5.9	---
WW B2	1.1	1.7	2.7	WW B2	1.2	1.8	2.7	SS A6	1.0	1.6	2.4	TW F2	1.5	3.1	---
WW B3	1.0	1.5	2.3	WW B3	1.1	1.6	2.3	SS A7	1.4	---	---	TW F3	1.4	2.7	7.0
WW B4	1.0	1.4	1.9	WW B4	1.0	1.4	1.9	SS A8	1.0	1.8	2.7	TW F4	1.5	2.6	5.3
WW B5	1.0	1.3	1.9	WW B5	1.0	1.4	1.9	SS B1	1.4	2.9	6.9	TW G1	1.6	4.1	---
WW C1	1.0	1.4	2.1	WW C1	1.0	1.5	2.1	SS B2	1.1	1.7	2.5	TW G2	1.6	3.0	9.2
WW C2	1.0	1.4	2.0	WW C2	1.1	1.5	2.0	SS B3	1.5	4.1	---	TW G3	1.6	2.9	6.2
WW C3	.9	1.2	1.6	WW C3	.9	1.2	1.6	SS B4	1.1	1.8	2.7	TW G4	1.7	3.0	5.7
WW C4	.9	1.2	1.6	WW C4	1.0	1.3	1.7	SS B5	1.5	4.2	---	TW H1	1.8	3.7	---
TW A1	1.6	4.0	---	TW A1	1.7	4.8	---	SS B6	1.1	1.8	2.6	TW H2	1.8	3.3	7.3
TW A2	1.4	2.6	11.6	TW A2	1.5	2.8	5.9	SS B7	1.7	---	---	TW H3	1.9	3.4	6.5
TW A3	1.3	2.3	5.1	TW A3	1.5	2.5	4.4	SS B8	1.1	1.9	2.9	TW H4	2.2	3.8	6.8
TW A4	1.3	2.2	4.1	TW A4	1.4	2.3	3.8	SS C1	1.5	2.8	6.9	TW I1	2.0	---	---
TW B1	1.6	3.2	---	TW B1	1.7	3.6	18.7	SS C2	1.2	1.8	2.6	TW I2	1.3	2.1	3.4
TW B2	1.4	2.5	5.5	TW B2	1.5	2.7	4.7	SS C3	1.8	3.6	12.7	TW I3	1.2	1.9	3.0
TW B3	1.4	2.4	4.5	TW B3	1.5	2.5	4.2	SS C4	1.3	2.0	3.0	TW I4	1.1	1.6	2.3
TW B4	1.4	2.4	4.3	TW B4	1.6	2.6	4.1	SS D1	1.1	2.0	3.8	TW I5	1.1	1.6	2.2
TW C1	1.6	2.8	7.9	TW C1	1.7	3.1	5.9	SS D2	.9	1.4	2.0	TW J1	1.1	1.7	2.5
TW C2	1.5	2.6	4.8	TW C2	1.7	2.8	4.5	SS D3	1.1	2.4	5.1	TW J2	1.1	1.6	2.4
TW C3	1.6	2.6	4.7	TW C3	1.7	2.8	4.5	SS D4	.9	1.4	2.0	TW J3	.9	1.3	1.8
TW C4	1.7	2.8	4.9	TW C4	1.8	3.1	4.8	SS E1	1.2	2.7	8.9	TW J4	1.0	1.4	1.9
TW D1	1.8	---	---	TW D1	2.1	---	---	SS E2	1.0	1.6	2.3	DG A1	1.7	---	---
TW D2	1.2	1.9	3.0	TW D2	1.3	2.0	3.0	SS E3	1.4	3.4	---	DG A2	1.4	3.6	---
TW D3	1.1	1.7	2.8	TW D3	1.2	1.9	2.8	SS E4	1.1	1.7	2.5	DG A3	1.2	1.8	3.6
DG B1	1.6	---	---	TW D4	1.1	1.6	2.2					DG B1	1.7	---	---
DG B2	1.4	2.4	---	TW D5	1.1	1.5	2.1	MEDFORD, OR				DG B2	1.4	2.7	---
DG B3	1.2	1.6	2.4	TW E1	1.1	1.7	2.4		\multicolumn{3}{c}{SOLAR SAVINGS}		DG B3	1.2	1.7	2.6
DG C1	1.4	---	---	TW E2	1.1	1.6	2.2		10%	50%	80%	DG C1	1.5	---	---
DG C2	1.2	1.9	---	TW E3	1.0	1.3	1.8	WW A1	1.4	---	---	DG C2	1.3	2.0	---
DG C3	1.1	1.4	2.1	TW E4	1.0	1.3	1.8	WW A2	1.3	2.7	---	DG C3	1.1	1.5	2.2
SS A1	1.1	2.2	6.1	TW F1	1.8	4.6	---	WW A3	1.2	2.3	5.6	SS A1	1.1	2.4	6.4
SS A2	.9	1.5	2.3	TW F2	1.6	3.0	6.4	WW A4	1.2	2.1	4.4	SS A2	.9	1.5	2.4
SS A3	1.2	2.8	---	TW F3	1.6	2.7	4.7	WW A5	1.2	2.0	3.9	SS A3	1.1	3.6	---
SS A4	.9	1.5	2.5	TW F4	1.6	2.6	4.2	WW A6	1.2	2.0	3.6	SS A4	.9	1.6	2.6
SS A5	1.2	2.6	---	TW G1	1.8	3.8	19.4	WW B1	1.6	---	---	SS A5	1.1	3.0	---
SS A6	.9	1.5	2.4	TW G2	1.7	3.0	5.5	WW B2	1.1	1.8	3.0	SS A6	.9	1.6	2.5
SS A7	1.3	5.8	---	TW G3	1.7	2.9	4.8	WW B3	1.0	1.6	2.4	SS A7	1.2	---	---
SS A8	.9	1.6	2.7	TW G4	1.9	3.0	4.8	WW B4	1.0	1.4	2.0	SS A8	.9	1.7	2.8
SS B1	1.2	2.5	15.9	TW H1	1.9	3.5	7.6	WW B5	1.0	1.4	1.9	SS B1	1.2	2.8	---
SS B2	1.0	1.6	2.4	TW H2	2.0	3.3	5.5	WW C1	1.0	1.5	2.2	SS B2	1.0	1.7	2.5
SS B3	1.3	3.1	---	TW H3	2.1	3.4	5.4	WW C2	1.0	1.5	2.1	SS B3	1.3	4.0	---
SS B4	1.0	1.6	2.6	TW H4	2.4	3.8	5.9	WW C3	.9	1.2	1.6	SS B4	1.0	1.7	2.7
SS B5	1.4	3.3	---	TW I1	2.3	---	---	WW C4	.9	1.3	1.7	SS B5	1.4	4.2	---
SS B6	1.0	1.6	2.5	TW I2	1.4	2.1	3.1	TW A1	1.5	6.4	---	SS B6	1.0	1.7	2.7
SS B7	1.5	---	---	TW I3	1.3	2.0	2.8	TW A2	1.4	2.9	---	SS B7	1.5	---	---
SS B8	1.0	1.7	2.8	TW I4	1.2	1.7	2.3	TW A3	1.3	2.5	6.1	SS B8	1.0	1.8	3.0
SS C1	1.3	2.6	---	TW I5	1.1	1.6	2.1	TW A4	1.3	2.3	4.7	SS C1	1.4	3.0	---
SS C2	1.1	1.7	2.7	TW J1	1.2	1.7	2.4	TW B1	1.5	3.8	---	SS C2	1.1	1.8	2.9
SS C3	1.6	3.2	---	TW J2	1.2	1.7	2.3	TW B2	1.4	2.7	6.6	SS C3	1.6	3.8	---
SS C4	1.2	1.9	3.0	TW J3	1.0	1.4	1.8	TW B3	1.4	2.5	5.2	SS C4	1.2	2.0	3.3
SS D1	.9	1.8	4.2	TW J4	1.0	1.4	1.8	TW B4	1.4	2.6	4.8	SS D1	1.0	2.0	4.4
SS D2	.8	1.3	1.9	DG A1	2.0	---	---	TW C1	1.5	3.1	10.8	SS D2	.8	1.3	2.0
SS D3	1.0	2.1	6.9	DG A2	1.6	3.9	---	TW C2	1.5	2.8	5.5	SS D3	1.0	2.3	7.0
SS D4	.8	1.3	2.0	DG A3	1.3	1.8	3.3	TW C3	1.6	2.8	5.3	SS D4	.8	1.4	2.1
SS E1	1.1	2.3	---	DG B1	1.9	---	---	TW C4	1.7	3.0	5.5	SS E1	1.1	2.6	---
SS E2	.9	1.4	2.3	DG B2	1.5	2.7	---	TW D1	1.8	---	---	SS E2	.9	1.5	2.4
SS E3	1.2	2.8	---	DG B3	1.3	1.7	2.5	TW D2	1.2	2.0	3.3	SS E3	1.2	3.3	---
SS E4	1.0	1.6	2.5	DG C1	1.6	---	---	TW D3	1.1	1.8	3.0	SS E4	1.0	1.7	2.6

Conservation Factor Tables

NORTH BEND, OR
SOLAR SAVINGS

	10%	50%	80%
WW A1	1.3	1.9	2.9
WW A2	1.0	1.4	2.0
WW A3	1.0	1.3	1.9
WW A4	.9	1.3	1.8
WW A5	.9	1.2	1.7
WW A6	.9	1.2	1.6
WW B1	1.2	1.8	2.9
WW B2	.9	1.2	1.7
WW B3	.8	1.1	1.5
WW B4	.8	1.0	1.4
WW B5	.8	1.1	1.4
WW C1	.8	1.0	1.4
WW C2	.8	1.1	1.4
WW C3	.7	.9	1.2
WW C4	.7	1.0	1.3
TW A1	1.3	1.9	2.8
TW A2	1.1	1.6	2.2
TW A3	1.0	1.5	2.1
TW A4	1.0	1.4	2.0
TW B1	1.3	1.8	2.7
TW B2	1.1	1.6	2.2
TW B3	1.1	1.6	2.2
TW B4	1.2	1.6	2.2
TW C1	1.3	1.8	2.5
TW C2	1.2	1.7	2.4
TW C3	1.3	1.8	2.5
TW C4	1.4	2.0	2.7
TW D1	1.3	2.0	3.1
TW D2	1.0	1.3	1.8
TW D3	.9	1.2	1.7
TW D4	.9	1.1	1.5
TW D5	.9	1.1	1.5
TW E1	.8	1.1	1.5
TW E2	.8	1.1	1.5
TW E3	.7	1.0	1.3
TW E4	.8	1.0	1.3
TW F1	1.3	1.9	2.8
TW F2	1.1	1.6	2.3
TW F3	1.1	1.6	2.2
TW F4	1.1	1.6	2.2
TW G1	1.3	1.9	2.7
TW G2	1.2	1.7	2.4
TW G3	1.2	1.7	2.4
TW G4	1.3	1.9	2.6
TW H1	1.4	1.9	2.7
TW H2	1.4	1.9	2.7
TW H3	1.5	2.1	2.9
TW H4	1.7	2.4	3.3
TW I1	1.4	2.2	3.4
TW I2	1.0	1.4	1.9
TW I3	1.0	1.3	1.8
TW I4	.9	1.2	1.6
TW I5	.9	1.2	1.5
TW J1	.9	1.2	1.6
TW J2	.9	1.2	1.6
TW J3	.8	1.0	1.3
TW J4	.8	1.1	1.4
DG A1	1.2	3.0	---
DG A2	1.1	1.6	3.5
DG A3	1.0	1.3	1.9
DG B1	1.2	1.9	---
DG B2	1.1	1.4	2.4
DG B3	1.0	1.2	1.6
DG C1	1.1	1.5	---

NORTH BEND, OR
SOLAR SAVINGS

	10%	50%	80%
DG C2	1.0	1.3	1.9
DG C3	.9	1.1	1.4
SS A1	.9	1.4	2.1
SS A2	.7	1.1	1.5
SS A3	.9	1.5	2.4
SS A4	.7	1.1	1.6
SS A5	.9	1.5	2.4
SS A6	.7	1.1	1.5
SS A7	1.0	1.7	3.0
SS A8	.7	1.1	1.6
SS B1	1.0	1.5	2.3
SS B2	.8	1.2	1.6
SS B3	1.0	1.6	2.5
SS B4	.8	1.2	1.7
SS B5	1.1	1.8	2.8
SS B6	.8	1.2	1.7
SS B7	1.2	2.0	3.3
SS B8	.8	1.2	1.8
SS C1	1.0	1.5	2.1
SS C2	.9	1.2	1.7
SS C3	1.2	1.8	2.6
SS C4	1.0	1.4	1.9
SS D1	.8	1.1	1.7
SS D2	.6	.9	1.3
SS D3	.8	1.3	2.0
SS D4	.7	1.0	1.4
SS E1	.8	1.3	2.0
SS E2	.7	1.0	1.5
SS E3	1.0	1.5	2.4
SS E4	.8	1.1	1.6

PENDLETON, OR
SOLAR SAVINGS

	10%	50%	80%
WW A1	1.6	---	---
WW A2	1.4	---	---
WW A3	1.4	4.0	---
WW A4	1.3	3.3	---
WW A5	1.3	2.9	8.6
WW A6	1.3	2.8	6.8
WW B1	2.0	---	---
WW B2	1.3	2.3	4.0
WW B3	1.1	1.9	3.1
WW B4	1.1	1.6	2.3
WW B5	1.1	1.6	2.2
WW C1	1.1	1.7	2.7
WW C2	1.1	1.7	2.5
WW C3	.9	1.3	1.8
WW C4	1.0	1.4	1.9
TW A1	1.6	---	---
TW A2	1.5	---	---
TW A3	1.5	4.2	---
TW A4	1.5	3.4	11.9
TW B1	1.6	---	---
TW B2	1.5	4.4	---
TW B3	1.5	3.2	17.1
TW B4	1.6	3.5	9.2
TW C1	1.7	6.1	---
TW C2	1.6	3.9	14.2
TW C3	1.7	3.8	10.2
TW C4	1.8	3.9	9.1
TW D1	2.2	---	---
TW D2	1.3	2.5	4.5
TW D3	1.2	2.3	4.1

PENDLETON, OR
SOLAR SAVINGS

	10%	50%	80%
TW D4	1.1	1.8	2.6
TW D5	1.1	1.7	2.4
TW E1	1.1	2.0	3.1
TW E2	1.1	1.8	2.8
TW E3	1.0	1.4	2.0
TW E4	1.0	1.5	2.0
TW F1	1.7	---	---
TW F2	1.6	---	---
TW F3	1.6	5.1	---
TW F4	1.7	3.9	17.6
TW G1	1.8	---	---
TW G2	1.8	7.0	---
TW G3	1.8	4.6	---
TW G4	1.9	4.3	11.6
TW H1	2.0	---	---
TW H2	2.0	5.4	---
TW H3	2.1	4.9	14.6
TW H4	2.4	5.1	11.8
TW I1	2.5	---	---
TW I2	1.4	2.6	4.7
TW I3	1.3	2.4	3.9
TW I4	1.2	1.9	2.7
TW I5	1.2	1.7	2.5
TW J1	1.2	2.0	3.1
TW J2	1.2	1.9	2.8
TW J3	1.0	1.5	2.1
TW J4	1.1	1.5	2.1
DG A1	2.0	---	---
DG A2	1.5	---	---
DG A3	1.3	2.1	6.4
DG B1	2.0	---	---
DG B2	1.5	5.1	---
DG B3	1.3	1.9	3.2
DG C1	1.6	---	---
DG C2	1.4	2.7	---
DG C3	1.2	1.7	2.6
SS A1	1.2	4.5	---
SS A2	1.0	1.9	2.9
SS A3	1.3	---	---
SS A4	1.0	2.0	3.4
SS A5	1.2	---	---
SS A6	1.0	1.9	3.1
SS A7	1.4	---	---
SS A8	1.0	2.2	3.9
SS B1	1.4	---	---
SS B2	1.1	2.0	3.1
SS B3	1.5	---	---
SS B4	1.1	2.1	3.4
SS B5	1.5	---	---
SS B6	1.1	2.1	3.3
SS B7	1.7	---	---
SS B8	1.2	2.3	3.9
SS C1	1.5	---	---
SS C2	1.2	2.2	3.7
SS C3	1.7	---	---
SS C4	1.3	2.5	4.4
SS D1	1.1	3.3	---
SS D2	.9	1.6	2.4
SS D3	1.1	4.9	---
SS D4	.9	1.6	2.5
SS E1	1.3	---	---
SS E2	1.0	1.9	3.0
SS E3	1.4	---	---
SS E4	1.1	2.0	3.4

PORTLAND, OR
SOLAR SAVINGS

	10%	50%	80%
WW A1	1.5	---	---
WW A2	1.3	4.1	---
WW A3	1.3	3.0	---
WW A4	1.3	2.7	15.4
WW A5	1.2	2.5	7.2
WW A6	1.2	2.4	5.9
WW B1	1.8	---	---
WW B2	1.2	2.1	3.7
WW B3	1.1	1.8	2.9
WW B4	1.0	1.5	2.2
WW B5	1.0	1.5	2.1
WW C1	1.0	1.6	2.5
WW C2	1.0	1.6	2.4
WW C3	.9	1.3	1.8
WW C4	.9	1.3	1.8
TW A1	1.5	---	---
TW A2	1.4	4.2	---
TW A3	1.4	3.2	---
TW A4	1.4	2.9	9.5
TW B1	1.5	---	---
TW B2	1.5	3.4	---
TW B3	1.5	3.1	12.2
TW B4	1.5	3.0	7.9
TW C1	1.6	4.2	---
TW C2	1.6	3.3	11.3
TW C3	1.6	3.3	8.7
TW C4	1.7	3.5	8.0
TW D1	2.0	---	---
TW D2	1.3	2.3	4.1
TW D3	1.2	2.1	3.7
TW D4	1.1	1.7	2.5
TW D5	1.1	1.6	2.3
TW E1	1.1	1.8	2.9
TW E2	1.1	1.7	2.6
TW E3	.9	1.4	2.0
TW E4	1.0	1.4	2.0
TW F1	1.6	---	---
TW F2	1.5	5.1	---
TW F3	1.5	3.6	---
TW F4	1.6	3.3	12.5
TW G1	1.7	---	---
TW G2	1.7	4.2	---
TW G3	1.7	3.7	---
TW G4	1.8	3.7	9.9
TW H1	1.8	6.4	---
TW H2	1.9	4.2	---
TW H3	2.0	4.1	12.2
TW H4	2.3	4.5	10.4
TW I1	2.2	---	---
TW I2	1.4	2.4	4.4
TW I3	1.3	2.2	3.7
TW I4	1.2	1.8	2.6
TW I5	1.1	1.7	2.4
TW J1	1.1	1.9	2.9
TW J2	1.1	1.8	2.7
TW J3	1.0	1.4	2.0
TW J4	1.0	1.5	2.0
DG A1	1.8	---	---
DG A2	1.5	---	---
DG A3	1.2	2.0	4.8
DG B1	1.8	---	---
DG B2	1.4	3.6	---
DG B3	1.2	1.8	3.0
DG C1	1.5	---	---

PORTLAND, OR				REDMOND, OR				SALEM, OR				SALEM, OR			
	SOLAR SAVINGS				SOLAR SAVINGS				SOLAR SAVINGS				SOLAR SAVINGS		
	10%	50%	80%		10%	50%	80%		10%	50%	80%		10%	50%	80%
DG C2	1.3	2.4	---	TW D4	1.0	1.4	1.9	WW A1	1.5	---	---	DG C2	1.3	2.1	---
DG C3	1.1	1.6	2.4	TW D5	1.0	1.4	1.8	WW A2	1.3	2.9	---	DG C3	1.1	1.5	2.2
SS A1	1.1	3.1	---	TW E1	1.0	1.5	2.0	WW A3	1.3	2.5	8.3	SS A1	1.1	2.6	16.0
SS A2	.9	1.7	2.7	TW E2	1.0	1.4	1.9	WW A4	1.2	2.3	5.5	SS A2	.9	1.6	2.5
SS A3	1.2	---	---	TW E3	.9	1.2	1.6	WW A5	1.2	2.1	4.5	SS A3	1.1	5.1	---
SS A4	.9	1.8	3.1	TW E4	.9	1.2	1.6	WW A6	1.2	2.1	4.1	SS A4	.9	1.7	2.8
SS A5	1.1	9.6	---	TW F1	1.5	2.9	5.7	WW B1	1.7	---	---	SS A5	1.1	3.5	---
SS A6	.9	1.8	2.9	TW F2	1.4	2.3	3.6	WW B2	1.2	1.9	3.2	SS A6	.9	1.6	2.6
SS A7	1.2	---	---	TW F3	1.4	2.1	3.2	WW B3	1.0	1.6	2.5	SS A7	1.2	---	---
SS A8	1.0	1.9	3.5	TW F4	1.4	2.1	3.1	WW B4	1.0	1.4	2.1	SS A8	.9	1.8	3.1
SS B1	1.3	4.1	---	TW G1	1.6	2.7	4.7	WW B5	1.0	1.4	2.0	SS B1	1.3	3.1	---
SS B2	1.0	1.8	2.9	TW G2	1.5	2.4	3.6	WW C1	1.0	1.5	2.3	SS B2	1.0	1.7	2.6
SS B3	1.4	---	---	TW G3	1.5	2.3	3.5	WW C2	1.0	1.5	2.2	SS B3	1.3	6.4	---
SS B4	1.0	1.9	3.1	TW G4	1.7	2.5	3.6	WW C3	.9	1.2	1.7	SS B4	1.0	1.8	2.9
SS B5	1.4	---	---	TW H1	1.7	2.7	4.3	WW C4	.9	1.3	1.7	SS B5	1.4	5.7	---
SS B6	1.1	1.9	3.1	TW H2	1.7	2.7	3.9	TW A1	1.5	---	---	SS B6	1.0	1.8	2.8
SS B7	1.5	---	---	TW H3	1.8	2.8	4.1	TW A2	1.4	3.1	---	SS B7	1.5	---	---
SS B8	1.1	2.1	3.5	TW H4	2.1	3.2	4.6	TW A3	1.3	2.7	9.1	SS B8	1.1	1.9	3.1
SS C1	1.4	---	---	TW I1	1.9	5.2	---	TW A4	1.3	2.5	5.5	SS C1	1.4	3.4	---
SS C2	1.2	2.0	3.5	TW I2	1.2	1.8	2.6	TW B1	1.5	4.5	---	SS C2	1.2	1.9	3.0
SS C3	1.6	---	---	TW I3	1.2	1.7	2.4	TW B2	1.4	2.9	9.8	SS C3	1.6	4.5	---
SS C4	1.3	2.3	4.0	TW I4	1.1	1.5	2.0	TW B3	1.4	2.7	6.3	SS C4	1.2	2.1	3.5
SS D1	1.0	2.5	---	TW I5	1.0	1.4	1.9	TW B4	1.5	2.7	5.5	SS D1	1.0	2.1	6.2
SS D2	.9	1.5	2.2	TW J1	1.0	1.5	2.1	TW C1	1.5	3.4	---	SS D2	.8	1.4	2.1
SS D3	1.0	3.1	---	TW J2	1.1	1.5	2.0	TW C2	1.5	2.9	6.6	SS D3	1.0	2.5	---
SS D4	.9	1.5	2.4	TW J3	.9	1.2	1.6	TW C3	1.6	3.0	6.0	SS D4	.9	1.4	2.2
SS E1	1.2	---	---	TW J4	.9	1.3	1.7	TW C4	1.7	3.2	6.1	SS E1	1.1	3.1	---
SS E2	1.0	1.7	2.8	DG A1	1.6	---	---	TW D1	1.9	---	---	SS E2	.9	1.6	2.5
SS E3	1.3	---	---	DG A2	1.4	2.5	---	TW D2	1.2	2.1	3.5	SS E3	1.2	4.0	---
SS E4	1.0	1.9	3.1	DG A3	1.2	1.6	2.6	TW D3	1.2	1.9	3.2	SS E4	1.0	1.7	2.8
				DG B1	1.6	---	---	TW D4	1.1	1.6	2.3	ALLENTOWN, PA			
REDMOND, OR				DG B2	1.3	2.1	---	TW D5	1.1	1.5	2.2		SOLAR SAVINGS		
	SOLAR SAVINGS			DG B3	1.2	1.5	2.1	TW E1	1.1	1.7	2.6		10%	50%	80%
	10%	50%	80%	DG C1	1.4	3.0	---	TW E2	1.1	1.6	2.4	WW A1	2.1	---	---
WW A1	1.5	3.1	8.9	DG C2	1.2	1.7	3.2	TW E3	.9	1.3	1.9	WW A2	1.8	3.5	8.0
WW A2	1.2	2.0	3.2	DG C3	1.1	1.4	1.8	TW E4	1.0	1.4	1.9	WW A3	1.6	2.9	4.7
WW A3	1.2	1.8	2.7	SS A1	1.1	2.0	3.3	TW F1	1.6	---	---	WW A4	1.6	2.6	4.0
WW A4	1.1	1.7	2.6	SS A2	.9	1.4	2.0	TW F2	1.5	3.4	---	WW A5	1.5	2.4	3.6
WW A5	1.1	1.7	2.4	SS A3	1.1	2.4	5.5	TW F3	1.5	2.9	11.9	WW A6	1.5	2.3	3.4
WW A6	1.1	1.6	2.3	SS A4	.9	1.5	2.2	TW F4	1.5	2.8	6.4	WW B1	3.1	---	---
WW B1	1.5	4.5	---	SS A5	1.1	2.3	4.5	TW G1	1.7	5.1	---	WW B2	1.4	2.2	3.0
WW B2	1.1	1.6	2.2	SS A6	.9	1.4	2.1	TW G2	1.6	3.3	---	WW B3	1.3	1.8	2.5
WW B3	1.0	1.4	1.9	SS A7	1.2	3.3	---	TW G3	1.6	3.1	7.9	WW B4	1.2	1.6	2.1
WW B4	.9	1.3	1.7	SS A8	.9	1.5	2.3	TW G4	1.8	3.2	6.6	WW B5	1.2	1.6	2.0
WW B5	.9	1.3	1.7	SS B1	1.2	2.2	3.8	TW H1	1.8	4.1	---	WW C1	1.2	1.7	2.3
WW C1	.9	1.3	1.8	SS B2	1.0	1.5	2.2	TW H2	1.9	3.5	9.7	WW C2	1.2	1.7	2.2
WW C2	1.0	1.3	1.8	SS B3	1.3	2.6	5.6	TW H3	2.0	3.6	7.7	WW C3	1.0	1.3	1.7
WW C3	.8	1.1	1.5	SS B4	1.0	1.6	2.3	TW H4	2.3	4.0	7.6	WW C4	1.1	1.4	1.8
WW C4	.9	1.2	1.5	SS B5	1.3	2.8	6.0	TW I1	2.1	---	---	TW A1	2.1	---	---
TW A1	1.5	2.9	6.1	SS B6	1.0	1.6	2.2	TW I2	1.3	2.2	3.7	TW A2	1.9	3.8	8.7
TW A2	1.3	2.2	3.5	SS B7	1.5	3.7	---	TW I3	1.2	2.0	3.2	TW A3	1.8	3.1	5.2
TW A3	1.3	2.0	3.0	SS B8	1.0	1.6	2.4	TW I4	1.1	1.7	2.4	TW A4	1.7	2.9	4.4
TW A4	1.3	1.9	2.8	SS C1	1.3	2.1	3.4	TW I5	1.1	1.6	2.2	TW B1	2.1	5.8	---
TW B1	1.5	2.7	4.6	SS C2	1.1	1.6	2.2	TW J1	1.1	1.7	2.7	TW B2	1.9	3.4	5.7
TW B2	1.4	2.3	3.3	SS C3	1.5	2.6	4.4	TW J2	1.1	1.7	2.5	TW B3	1.8	3.1	4.8
TW B3	1.3	2.1	3.1	SS C4	1.2	1.8	2.5	TW J3	1.0	1.4	1.9	TW B4	1.9	3.1	4.6
TW B4	1.4	2.2	3.1	SS D1	.9	1.7	2.7	TW J4	1.0	1.4	1.9	TW C1	2.1	4.0	7.6
TW C1	1.5	2.5	3.9	SS D2	.8	1.2	1.7	DG A1	1.8	---	---	TW C2	2.0	3.4	5.2
TW C2	1.5	2.3	3.4	SS D3	1.0	1.9	3.2	DG A2	1.4	4.3	---	TW C3	2.0	3.4	5.1
TW C3	1.5	2.4	3.5	SS D4	.8	1.3	1.8	DG A3	1.2	1.9	3.9	TW C4	2.2	3.7	5.4
TW C4	1.6	2.6	3.8	SS E1	1.1	2.0	3.5	DG B1	1.8	---	---	TW D1	3.3	---	---
TW D1	1.7	4.2	---	SS E2	.9	1.4	2.0	DG B2	1.4	2.9	---	TW D2	1.5	2.4	3.3
TW D2	1.2	1.7	2.4	SS E3	1.2	2.4	4.6	DG B3	1.2	1.7	2.7	TW D3	1.4	2.2	3.0
TW D3	1.1	1.6	2.3	SS E4	.9	1.5	2.1	DG C1	1.5	---	---				

Conservation Factor Tables / 363

ALLENTOWN, PA				ERIE, PA				ERIE, PA				HARRISBURG, PA			
	SOLAR SAVINGS				SOLAR SAVINGS				SOLAR SAVINGS				SOLAR SAVINGS		
	10%	50%	80%		10%	50%	80%		10%	50%	80%		10%	50%	80%
WW A1	2.1	---	---	WW A1	3.0	---	---	DG C2	2.2	---	---	TW D4	1.2	1.7	2.2
WW A2	1.8	3.5	8.0	WW A2	3.1	---	---	DG C3	1.6	2.4	4.6	TW D5	1.2	1.6	2.1
WW A3	1.6	2.9	4.7	WW A3	2.9	---	---	SS A1	1.8	---	---	TW E1	1.2	1.8	2.4
WW A4	1.6	2.6	4.0	WW A4	2.8	---	---	SS A2	1.3	2.7	5.0	TW E2	1.2	1.7	2.3
WW A5	1.5	2.4	3.6	WW A5	2.7	---	---	SS A3	2.1	---	---	TW E3	1.0	1.4	1.8
WW A6	1.5	2.3	3.4	WW A6	2.6	---	---	SS A4	1.4	3.3	15.4	TW E4	1.1	1.4	1.9
WW B1	3.1	---	---	WW B1	---	---	---	SS A5	1.8	---	---	TW F1	2.1	5.9	---
WW B2	1.4	2.2	3.0	WW B2	2.0	---	---	SS A6	1.4	3.0	7.2	TW F2	1.9	3.4	5.6
WW B3	1.3	1.8	2.5	WW B3	1.7	4.4	---	SS A7	2.0	---	---	TW F3	1.8	3.0	5.1
WW B4	1.2	1.6	2.1	WW B4	1.5	2.3	3.4	SS A8	1.4	4.6	---	TW F4	1.8	2.9	4.2
WW B5	1.2	1.6	2.0	WW B5	1.4	2.1	2.9	SS B1	2.2	---	---	TW G1	2.2	4.5	10.4
WW C1	1.2	1.7	2.3	WW C1	1.6	3.0	7.0	SS B2	1.5	2.9	4.9	TW G2	2.0	3.4	5.2
WW C2	1.2	1.7	2.2	WW C2	1.5	2.5	4.0	SS B3	2.7	---	---	TW G3	2.0	3.2	4.7
WW C3	1.0	1.3	1.7	WW C3	1.2	1.7	2.3	SS B4	1.5	3.3	7.1	TW G4	2.2	3.3	4.7
WW C4	1.1	1.4	1.8	WW C4	1.3	1.8	2.4	SS B5	2.5	---	---	TW H1	2.3	4.0	6.6
TW A1	2.1	---	---	TW A1	2.9	---	---	SS B6	1.5	3.2	6.4	TW H2	2.3	3.6	5.3
TW A2	1.9	3.8	8.7	TW A2	3.0	---	---	SS B7	3.1	---	---	TW H3	2.4	3.7	5.4
TW A3	1.8	3.1	5.2	TW A3	2.9	---	---	SS B8	1.6	4.2	---	TW H4	2.7	4.2	5.9
TW A4	1.7	2.9	4.4	TW A4	2.8	---	---	SS C1	---	---	---	TW I1	3.2	---	---
TW B1	2.1	5.8	---	TW B1	3.1	---	---	SS C2	1.9	---	---	TW I2	1.6	2.3	3.2
TW B2	1.9	3.4	5.7	TW B2	2.9	---	---	SS C3	4.4	---	---	TW I3	1.5	2.1	2.9
TW B3	1.8	3.1	4.8	TW B3	2.8	---	---	SS C4	2.1	---	---	TW I4	1.3	1.8	2.3
TW B4	1.9	3.1	4.6	TW B4	2.7	---	---	SS D1	1.7	---	---	TW I5	1.3	1.7	2.2
TW C1	2.1	4.0	7.6	TW C1	3.0	---	---	SS D2	1.2	2.2	3.5	TW J1	1.3	1.8	2.5
TW C2	2.0	3.4	5.2	TW C2	2.8	---	---	SS D3	1.7	---	---	TW J2	1.3	1.8	2.4
TW C3	2.0	3.4	5.1	TW C3	2.8	---	---	SS D4	1.2	2.3	3.8	TW J3	1.1	1.4	1.9
TW C4	2.2	3.7	5.4	TW C4	2.8	---	---	SS E1	2.3	---	---	TW J4	1.1	1.5	1.9
TW D1	3.3	---	---	TW D1	---	---	---	SS E2	1.4	3.2	11.7	DG A1	2.7	---	---
TW D2	1.5	2.4	3.3	TW D2	2.1	---	---	SS E3	2.3	---	---	DG A2	1.8	---	---
TW D3	1.4	2.2	3.0	TW D3	2.0	---	---	SS E4	1.5	3.7	---	DG A3	1.4	2.0	3.6
DG B1	3.1	---	---	TW D4	1.6	2.7	4.1					DG B1	2.6	---	---
DG B2	1.8	4.2	---	TW D5	1.5	2.3	3.2	HARRISBURG, PA				DG B2	1.7	3.3	---
DG B3	1.5	1.9	2.8	TW E1	1.7	3.9	---		SOLAR SAVINGS			DG B3	1.4	1.8	2.6
DG C1	2.1	---	---	TW E2	1.6	2.9	5.1		10%	50%	80%	DG C1	2.0	---	---
DG C2	1.6	2.5	---	TW E3	1.3	1.9	2.7	WW A1	2.0	---	---	DG C2	1.5	2.3	---
DG C3	1.4	1.7	2.3	TW E4	1.3	1.9	2.7	WW A2	1.7	3.0	4.8	DG C3	1.3	1.6	2.2
SS A1	1.5	3.0	5.4	TW F1	4.3	---	---	WW A3	1.6	2.5	3.8	SS A1	1.4	2.7	4.2
SS A2	1.1	1.8	2.5	TW F2	3.9	---	---	WW A4	1.5	2.4	3.4	SS A2	1.1	1.7	2.3
SS A3	1.6	5.9	---	TW F3	3.6	---	---	WW A5	1.4	2.2	3.1	SS A3	1.5	3.8	---
SS A4	1.2	1.9	2.7	TW F4	3.4	---	---	WW A6	1.4	2.1	3.0	SS A4	1.1	1.8	2.5
SS A5	1.5	4.2	---	TW G1	5.1	---	---	WW B1	2.6	---	---	SS A5	1.5	3.4	7.3
SS A6	1.2	1.8	2.6	TW G2	4.1	---	---	WW B2	1.4	2.0	2.8	SS A6	1.1	1.7	2.4
SS A7	1.8	---	---	TW G3	3.8	---	---	WW B3	1.2	1.7	2.3	SS A7	1.7	---	---
SS A8	1.2	2.0	2.9	TW G4	3.7	---	---	WW B4	1.1	1.5	2.0	SS A8	1.2	1.9	2.6
SS B1	1.7	3.6	8.0	TW H1	5.2	---	---	WW B5	1.1	1.5	2.0	SS B1	1.6	3.1	5.3
SS B2	1.2	1.9	2.6	TW H2	4.3	---	---	WW C1	1.1	1.6	2.2	SS B2	1.2	1.8	2.5
SS B3	1.9	7.4	---	TW H3	4.1	---	---	WW C2	1.2	1.6	2.1	SS B3	1.8	4.3	---
SS B4	1.3	2.0	2.8	TW H4	4.3	---	---	WW C3	1.0	1.3	1.7	SS B4	1.2	1.9	2.6
SS B5	1.9	7.3	---	TW I1	---	---	---	WW C4	1.0	1.4	1.7	SS B5	1.8	4.6	---
SS B6	1.3	2.0	2.7	TW I2	2.3	---	---	TW A1	2.0	6.6	---	SS B6	1.2	1.9	2.6
SS B7	2.2	---	---	TW I3	2.1	---	---	TW A2	1.8	3.2	5.3	SS B7	2.1	---	---
SS B8	1.3	2.1	3.0	TW I4	1.7	2.7	4.2	TW A3	1.7	2.8	4.2	SS B8	1.3	2.0	2.8
SS C1	1.9	4.0	---	TW I5	1.6	2.4	3.4	TW A4	1.6	2.6	3.7	SS C1	1.8	3.2	5.6
SS C2	1.4	2.1	2.9	TW J1	1.8	3.7	17.9	TW B1	2.0	4.3	10.0	SS C2	1.4	2.0	2.7
SS C3	2.2	5.5	---	TW J2	1.7	2.9	4.8	TW B2	1.8	3.0	4.5	SS C3	2.1	4.1	8.6
SS C4	1.6	2.4	3.3	TW J3	1.3	1.9	2.7	TW B3	1.8	2.8	4.1	SS C4	1.5	2.2	3.0
SS D1	1.3	2.4	4.0	TW J4	1.4	2.0	2.7	TW B4	1.8	2.9	4.1	SS D1	1.2	2.2	3.4
SS D2	1.0	1.5	2.1	DG A1	---	---	---	TW C1	2.0	3.5	5.6	SS D2	1.0	1.4	2.0
SS D3	1.4	2.9	5.5	DG A2	2.7	---	---	TW C2	1.9	3.1	4.5	SS D3	1.3	2.6	4.2
SS D4	1.0	1.6	2.1	DG A3	1.8	3.4	---	TW C3	1.9	3.1	4.5	SS D4	1.0	1.5	2.0
SS E1	1.5	3.4	---	DG B1	---	---	---	TW C4	2.1	3.4	4.8	SS E1	1.5	2.9	5.2
SS E2	1.2	1.8	2.4	DG B2	2.6	---	---	TW D1	2.8	---	---	SS E2	1.1	1.7	2.3
SS E3	1.7	4.8	---	DG B3	1.7	2.8	---	TW D2	1.5	2.2	3.0	SS E3	1.6	3.7	8.6
SS E4	1.2	1.9	2.7	DG C1	---	---	---	TW D3	1.4	2.0	2.8	SS E4	1.2	1.8	2.5

PHILADELPHIA, PA				PHILADELPHIA, PA				PITTSBURGH, PA				WILKES-BARRE, PA			
	SOLAR SAVINGS				SOLAR SAVINGS				SOLAR SAVINGS				SOLAR SAVINGS		
	10%	50%	80%		10%	50%	80%		10%	50%	80%		10%	50%	80%
WW A1	1.8	4.3	10.7	DG C2	1.4	2.0	4.9	TW D4	1.5	2.2	3.1	WW A1	2.5	---	---
WW A2	1.5	2.5	3.6	DG C3	1.2	1.5	2.0	TW D5	1.4	2.0	2.7	WW A2	2.2	---	---
WW A3	1.4	2.2	3.0	SS A1	1.3	2.3	3.4	TW E1	1.5	2.6	3.9	WW A3	2.0	---	---
WW A4	1.4	2.0	2.8	SS A2	1.0	1.6	2.1	TW E2	1.5	2.3	3.3	WW A4	1.9	6.1	---
WW A5	1.3	1.9	2.6	SS A3	1.4	2.9	4.9	TW E3	1.2	1.7	2.3	WW A5	1.9	4.3	---
WW A6	1.3	1.9	2.5	SS A4	1.1	1.6	2.2	TW E4	1.2	1.7	2.3	WW A6	1.8	3.8	---
WW B1	2.1	---	---	SS A5	1.4	2.8	4.5	TW F1	3.1	---	---	WW B1	---	---	---
WW B2	1.3	1.8	2.4	SS A6	1.1	1.6	2.2	TW F2	2.7	---	---	WW B2	1.7	2.8	4.4
WW B3	1.1	1.6	2.1	SS A7	1.6	4.3	---	TW F3	2.6	---	---	WW B3	1.4	2.3	3.3
WW B4	1.1	1.4	1.9	SS A8	1.1	1.7	2.4	TW F4	2.5	---	---	WW B4	1.3	1.8	2.5
WW B5	1.1	1.4	1.8	SS B1	1.5	2.6	3.9	TW G1	3.2	---	---	WW B5	1.3	1.8	2.3
WW C1	1.1	1.5	2.0	SS B2	1.1	1.7	2.3	TW G2	2.8	---	---	WW C1	1.3	2.1	2.9
WW C2	1.1	1.5	1.9	SS B3	1.6	3.1	5.3	TW G3	2.8	---	---	WW C2	1.3	2.0	2.7
WW C3	.9	1.2	1.6	SS B4	1.2	1.8	2.4	TW G4	2.9	---	---	WW C3	1.1	1.5	1.9
WW C4	1.0	1.3	1.6	SS B5	1.7	3.4	5.8	TW H1	3.3	---	---	WW C4	1.2	1.6	2.0
TW A1	1.8	3.9	7.1	SS B6	1.2	1.7	2.4	TW H2	3.1	---	---	TW A1	2.5	---	---
TW A2	1.6	2.7	3.9	SS B7	1.9	5.1	---	TW H3	3.2	---	---	TW A2	2.3	---	---
TW A3	1.5	2.4	3.4	SS B8	1.2	1.8	2.5	TW H4	3.6	11.0	---	TW A3	2.1	---	---
TW A4	1.5	2.3	3.1	SS C1	1.6	2.6	3.8	TW I1	---	---	---	TW A4	2.1	5.1	---
TW B1	1.8	3.3	5.2	SS C2	1.3	1.8	2.4	TW I2	2.0	3.9	7.9	TW B1	2.5	---	---
TW B2	1.6	2.6	3.7	SS C3	1.9	3.2	4.9	TW I3	1.8	3.2	5.3	TW B2	2.2	---	---
TW B3	1.6	2.5	3.4	SS C4	1.4	2.0	2.7	TW I4	1.5	2.3	3.2	TW B3	2.2	5.8	---
TW B4	1.7	2.5	3.5	SS D1	1.1	1.9	2.8	TW I5	1.5	2.1	2.8	TW B4	2.2	4.8	---
TW C1	1.8	3.0	4.3	SS D2	.9	1.3	1.8	TW J1	1.6	2.6	3.8	TW C1	2.4	---	---
TW C2	1.7	2.7	3.8	SS D3	1.2	2.2	3.3	TW J2	1.5	2.3	3.3	TW C2	2.3	5.8	---
TW C3	1.8	2.8	3.8	SS D4	1.0	1.4	1.9	TW J3	1.2	1.7	2.3	TW C3	2.4	5.3	---
TW C4	1.9	3.0	4.2	SS E1	1.3	2.3	3.5	TW J4	1.3	1.8	2.3	TW C4	2.5	5.2	14.3
TW D1	2.3	---	---	SS E2	1.0	1.5	2.1	DG A1	---	---	---	TW D1	---	---	---
TW D2	1.4	2.0	2.7	SS E3	1.5	2.9	4.6	DG A2	2.3	---	---	TW D2	1.8	3.1	5.1
TW D3	1.3	1.8	2.5	SS E4	1.1	1.7	2.3	DG A3	1.7	2.7	---	TW D3	1.7	2.8	4.5
TW D4	1.2	1.6	2.1					DG B1	---	---	---	TW D4	1.4	2.1	2.8
TW D5	1.1	1.5	2.0	PITTSBURGH, PA				DG B2	2.2	---	---	TW D5	1.4	1.9	2.6
TW E1	1.2	1.6	2.2		SOLAR SAVINGS			DG B3	1.6	2.3	4.3	TW E1	1.5	2.3	3.4
TW E2	1.1	1.6	2.1		10%	50%	80%	DG C1	4.1	---	---	TW E2	1.4	2.1	3.0
TW E3	1.0	1.3	1.7	WW A1	2.6	---	---	DG C2	1.9	---	---	TW E3	1.2	1.6	2.2
TW E4	1.0	1.4	1.7	WW A2	2.3	---	---	DG C3	1.5	2.1	3.1	TW E4	1.2	1.7	2.2
TW F1	1.9	3.8	6.5	WW A3	2.2	---	---	SS A1	1.7	---	---	TW F1	2.9	---	---
TW F2	1.7	2.8	4.1	WW A4	2.1	---	---	SS A2	1.3	2.2	3.2	TW F2	2.5	---	---
TW F3	1.7	2.6	3.6	WW A5	2.0	---	---	SS A3	1.9	---	---	TW F3	2.4	---	---
TW F4	1.7	2.5	3.5	WW A6	1.9	---	---	SS A4	1.3	2.4	3.8	TW F4	2.4	6.4	---
TW G1	2.0	3.4	5.3	WW B1	---	---	---	SS A5	1.7	---	---	TW G1	2.9	---	---
TW G2	1.8	2.8	4.0	WW B2	1.7	3.3	6.0	SS A6	1.3	2.3	3.5	TW G2	2.6	---	---
TW G3	1.8	2.8	3.8	WW B3	1.5	2.5	3.9	SS A7	2.0	---	---	TW G3	2.6	11.2	---
TW G4	2.0	2.9	4.0	WW B4	1.3	1.9	2.6	SS A8	1.3	2.7	4.5	TW G4	2.7	6.0	---
TW H1	2.0	3.3	4.8	WW B5	1.3	1.8	2.5	SS B1	2.0	---	---	TW H1	3.0	---	---
TW H2	2.1	3.1	4.4	WW C1	1.4	2.2	3.2	SS B2	1.4	2.3	3.4	TW H2	2.9	---	---
TW H3	2.2	3.3	4.5	WW C2	1.4	2.1	2.9	SS B3	2.3	---	---	TW H3	3.0	7.1	---
TW H4	2.5	3.7	5.0	WW C3	1.1	1.5	2.0	SS B4	1.4	2.5	3.8	TW H4	3.4	6.8	---
TW I1	2.6	---	---	WW C4	1.2	1.6	2.1	SS B5	2.3	---	---	TW I1	---	---	---
TW I2	1.5	2.1	2.8	TW A1	2.6	---	---	SS B6	1.4	2.5	3.7	TW I2	1.9	3.3	5.3
TW I3	1.4	1.9	2.6	TW A2	2.4	---	---	SS B7	2.7	---	---	TW I3	1.7	2.8	4.3
TW I4	1.2	1.7	2.2	TW A3	2.3	---	---	SS B8	1.5	2.8	4.5	TW I4	1.5	2.2	2.9
TW I5	1.2	1.6	2.1	TW A4	2.2	---	---	SS C1	2.7	---	---	TW I5	1.4	2.0	2.7
TW J1	1.2	1.7	2.2	TW B1	2.6	---	---	SS C2	1.7	3.1	5.1	TW J1	1.5	2.4	3.4
TW J2	1.2	1.6	2.2	TW B2	2.4	---	---	SS C3	3.0	---	---	TW J2	1.5	2.2	3.0
TW J3	1.0	1.4	1.7	TW B3	2.3	---	---	SS C4	1.8	3.5	6.4	TW J3	1.2	1.7	2.2
TW J4	1.1	1.4	1.8	TW B4	2.3	12.1	---	SS D1	1.5	---	---	TW J4	1.3	1.7	2.2
DG A1	2.2	---	---	TW C1	2.6	---	---	SS D2	1.1	1.8	2.6	DG A1	---	---	---
DG A2	1.6	3.8	---	TW C2	2.4	---	---	SS D3	1.6	---	---	DG A2	2.2	---	---
DG A3	1.3	1.8	3.0	TW C3	2.5	10.2	---	SS D4	1.2	1.9	2.7	DG A3	1.6	2.5	---
DG B1	2.1	---	---	TW C4	2.6	7.0	---	SS E1	1.9	---	---	DG B1	---	---	---
DG B2	1.6	2.6	---	TW D1	---	---	---	SS E2	1.3	2.3	3.5	DG B2	2.1	---	---
DG B3	1.3	1.7	2.3	TW D2	1.8	3.6	7.5	SS E3	2.1	---	---	DG B3	1.6	2.2	3.7
DG C1	1.7	---	---	TW D3	1.7	3.3	6.1	SS E4	1.4	2.5	3.9	DG C1	3.3	---	---

WILKES-BARRE, PA				PROVIDENCE, RI				CHARLESTON, SC				CHARLESTON, SC			
	SOLAR SAVINGS				SOLAR SAVINGS				SOLAR SAVINGS				SOLAR SAVINGS		
	10%	50%	80%		10%	50%	80%		10%	50%	80%		10%	50%	80%
DG C2	1.8	4.2	---	TW D4	1.3	1.8	2.3	WW A1	1.0	1.6	2.1	DG C2	.9	1.1	1.4
DG C3	1.5	1.9	2.8	TW D5	1.2	1.7	2.2	WW A2	.9	1.2	1.6	DG C3	.8	1.0	1.2
SS A1	1.7	9.5	---	TW E1	1.3	1.9	2.5	WW A3	.8	1.1	1.5	SS A1	.8	1.2	1.5
SS A2	1.2	2.1	3.0	TW E2	1.3	1.8	2.4	WW A4	.8	1.1	1.4	SS A2	.7	.9	1.2
SS A3	1.9	---	---	TW E3	1.1	1.4	1.9	WW A5	.8	1.0	1.3	SS A3	.8	1.2	1.7
SS A4	1.3	2.3	3.4	TW E4	1.1	1.5	1.9	WW A6	.8	1.0	1.3	SS A4	.7	1.0	1.2
SS A5	1.7	---	---	TW F1	2.3	---	---	WW B1	1.0	1.4	1.9	SS A5	.8	1.3	1.7
SS A6	1.3	2.2	3.2	TW F2	2.0	3.7	7.3	WW B2	.8	1.1	1.4	SS A6	.7	.9	1.2
SS A7	2.0	---	---	TW F3	1.9	3.2	5.1	WW B3	.7	.9	1.2	SS A7	.9	1.4	2.0
SS A8	1.3	2.5	3.9	TW F4	1.9	3.1	4.5	WW B4	.7	.9	1.2	SS A8	.7	1.0	1.3
SS B1	2.0	---	---	TW G1	2.3	5.3	---	WW B5	.7	.9	1.2	SS B1	.9	1.3	1.7
SS B2	1.4	2.2	3.2	TW G2	2.1	3.6	6.0	WW C1	.7	.9	1.2	SS B2	.7	1.0	1.3
SS B3	2.2	---	---	TW G3	2.1	3.4	5.2	WW C2	.7	.9	1.2	SS B3	.9	1.4	1.8
SS B4	1.4	2.4	3.5	TW G4	2.3	3.5	5.1	WW C3	.6	.8	1.0	SS B4	.7	1.0	1.4
SS B5	2.2	---	---	TW H1	2.4	4.5	8.6	WW C4	.7	.9	1.1	SS B5	1.0	1.5	2.0
SS B6	1.4	2.3	3.4	TW H2	2.4	3.9	5.9	TW A1	1.0	1.6	2.1	SS B6	.7	1.0	1.4
SS B7	2.6	---	---	TW H3	2.5	4.0	5.8	TW A2	.9	1.3	1.7	SS B7	1.0	1.6	2.2
SS B8	1.5	2.6	4.0	TW H4	2.9	4.4	6.3	TW A3	.9	1.2	1.6	SS B8	.8	1.1	1.4
SS C1	2.4	---	---	TW I1	3.8	---	---	TW A4	.9	1.2	1.6	SS C1	.9	1.2	1.6
SS C2	1.6	2.7	4.1	TW I2	1.6	2.4	3.4	TW B1	1.0	1.5	2.0	SS C2	.8	1.1	1.4
SS C3	2.8	---	---	TW I3	1.5	2.2	3.0	TW B2	.9	1.3	1.8	SS C3	1.0	1.5	2.0
SS C4	1.8	3.1	4.8	TW I4	1.3	1.8	2.4	TW B3	.9	1.3	1.7	SS C4	.8	1.2	1.5
SS D1	1.5	4.4	---	TW I5	1.3	1.7	2.3	TW B4	1.0	1.4	1.8	SS D1	.7	1.0	1.3
SS D2	1.1	1.7	2.4	TW J1	1.3	1.9	2.6	TW C1	1.0	1.5	2.0	SS D2	.6	.8	1.1
SS D3	1.6	---	---	TW J2	1.3	1.8	2.4	TW C2	1.0	1.5	1.9	SS D3	.7	1.1	1.5
SS D4	1.1	1.8	2.6	TW J3	1.1	1.5	1.9	TW C3	1.1	1.5	2.0	SS D4	.6	.8	1.1
SS E1	1.8	---	---	TW J4	1.1	1.5	2.0	TW C4	1.2	1.7	2.2	SS E1	.7	1.1	1.4
SS E2	1.3	2.1	3.1	DG A1	3.3	---	---	TW D1	1.1	1.6	2.1	SS E2	.7	.9	1.2
SS E3	2.0	---	---	DG A2	1.9	---	---	TW D2	.8	1.1	1.5	SS E3	.9	1.3	1.7
SS E4	1.4	2.3	3.5	DG A3	1.5	2.0	3.9	TW D3	.8	1.1	1.4	SS E4	.7	1.0	1.3
				DG B1	3.0	---	---	TW D4	.7	1.0	1.3				
PROVIDENCE, RI				DG B2	1.8	3.8	---	TW D5	.8	1.0	1.3	COLUMBIA, SC			
	SOLAR SAVINGS			DG B3	1.5	1.9	2.7	TW E1	.7	1.0	1.3		SOLAR SAVINGS		
	10%	50%	80%	DG C1	2.1	---	---	TW E2	.7	1.0	1.3		10%	50%	80%
WW A1	2.1	---	---	DG C2	1.6	2.5	---	TW E3	.7	.9	1.1	WW A1	1.1	1.7	2.3
WW A2	1.8	3.3	6.0	DG C3	1.4	1.7	2.2	TW E4	.7	.9	1.1	WW A2	.9	1.3	1.7
WW A3	1.6	2.7	4.3	SS A1	1.5	2.9	5.0	TW F1	1.1	1.6	2.1	WW A3	.9	1.2	1.6
WW A4	1.6	2.5	3.8	SS A2	1.1	1.7	2.4	TW F2	1.0	1.4	1.8	WW A4	.8	1.2	1.5
WW A5	1.5	2.3	3.4	SS A3	1.6	5.0	---	TW F3	.9	1.3	1.7	WW A5	.8	1.1	1.4
WW A6	1.5	2.3	3.2	SS A4	1.2	1.8	2.6	TW F4	1.0	1.3	1.7	WW A6	.8	1.1	1.4
WW B1	3.0	---	---	SS A5	1.5	3.9	---	TW G1	1.1	1.6	2.0	WW B1	1.1	1.6	2.1
WW B2	1.4	2.1	2.9	SS A6	1.1	1.8	2.5	TW G2	1.0	1.4	1.9	WW B2	.8	1.1	1.4
WW B3	1.2	1.8	2.4	SS A7	1.7	---	---	TW G3	1.1	1.5	1.9	WW B3	.7	1.0	1.3
WW B4	1.2	1.6	2.1	SS A8	1.2	1.9	2.8	TW G4	1.2	1.6	2.0	WW B4	.7	1.0	1.2
WW B5	1.2	1.5	2.0	SS B1	1.7	3.4	6.8	TW H1	1.1	1.6	2.1	WW B5	.7	1.0	1.2
WW C1	1.2	1.7	2.2	SS B2	1.2	1.9	2.6	TW H2	1.2	1.5	2.1	WW C1	.7	1.0	1.2
WW C2	1.2	1.6	2.2	SS B3	1.8	5.7	---	TW H3	1.3	1.8	2.3	WW C2	.7	1.0	1.3
WW C3	1.0	1.3	1.7	SS B4	1.3	2.0	2.7	TW H4	1.5	2.0	2.6	WW C3	.7	.9	1.1
WW C4	1.1	1.4	1.8	SS B5	1.9	5.9	---	TW I1	1.2	1.7	2.2	WW C4	.7	.9	1.1
TW A1	2.1	---	---	SS B6	1.3	1.9	2.7	TW I2	.9	1.2	1.6	TW A1	1.1	1.7	2.3
TW A2	1.9	3.5	6.6	SS B7	2.2	---	---	TW I3	.8	1.1	1.5	TW A2	1.0	1.4	1.9
TW A3	1.8	3.0	4.7	SS B8	1.3	2.1	3.0	TW I4	.8	1.0	1.3	TW A3	.9	1.3	1.7
TW A4	1.7	2.8	4.1	SS C1	1.9	3.6	8.9	TW I5	.8	1.0	1.3	TW A4	.9	1.3	1.7
TW B1	2.1	5.0	---	SS C2	1.4	2.1	2.8	TW J1	.8	1.0	1.3	TW B1	1.1	1.6	2.2
TW B2	1.9	3.2	5.1	SS C3	2.2	4.8	---	TW J2	.8	1.0	1.3	TW B2	1.0	1.4	1.9
TW B3	1.8	3.0	4.5	SS C4	1.5	2.3	3.2	TW J3	.7	.9	1.1	TW B3	1.0	1.4	1.8
TW B4	1.9	3.0	4.4	SS D1	1.3	2.3	3.8	TW J4	.7	.9	1.2	TW B4	1.1	1.5	1.9
TW C1	2.1	3.8	6.5	SS D2	1.0	1.5	2.0	DG A1	1.0	1.7	---	TW C1	1.1	1.6	2.1
TW C2	2.0	3.3	4.9	SS D3	1.4	2.8	5.0	DG A2	1.0	1.3	2.1	TW C2	1.1	1.6	2.0
TW C3	2.0	3.3	4.9	SS D4	1.0	1.5	2.1	DG A3	.9	1.1	1.5	TW C3	1.1	1.6	2.1
TW C4	2.2	3.6	5.2	SS E1	1.5	3.2	10.0	DG B1	1.0	1.4	2.9	TW C4	1.3	1.8	2.3
TW D1	3.2	---	---	SS E2	1.2	1.7	2.4	DG B2	.9	1.2	1.7	TW D1	1.2	1.7	2.3
TW D2	1.5	2.3	3.2	SS E3	1.7	4.3	---	DG B3	.9	1.0	1.3	TW D2	.9	1.2	1.6
TW D3	1.4	2.1	2.9	SS E4	1.2	1.9	2.6	DG C1	.9	1.2	1.8	TW D3	.8	1.1	1.5

COLUMBIA, SC	10%	50%	80%
WW A1	1.1	1.7	2.3
WW A2	.9	1.3	1.7
WW A3	.9	1.2	1.6
WW A4	.8	1.2	1.5
WW A5	.8	1.1	1.4
WW A6	.8	1.1	1.4
WW B1	1.1	1.6	2.1
WW B2	.8	1.1	1.4
WW B3	.7	1.0	1.3
WW B4	.7	1.0	1.2
WW B5	.7	1.0	1.2
WW C1	.7	1.0	1.2
WW C2	.7	1.0	1.3
WW C3	.7	.9	1.1
WW C4	.7	.9	1.1
TW A1	1.1	1.7	2.3
TW A2	1.0	1.4	1.9
TW A3	.9	1.3	1.7
TW A4	.9	1.3	1.7
TW B1	1.1	1.6	2.2
TW B2	1.0	1.4	1.9
TW B3	1.0	1.4	1.8
TW B4	1.1	1.5	1.9
TW C1	1.1	1.6	2.1
TW C2	1.1	1.6	2.0
TW C3	1.1	1.6	2.1
TW C4	1.3	1.8	2.3
TW D1	1.2	1.7	2.3
TW D2	.9	1.2	1.6
TW D3	.8	1.1	1.5
DG B1	1.1	1.5	---
DG B2	1.0	1.3	1.8
DG B3	.9	1.1	1.4
DG C1	1.0	1.3	2.1
DG C2	.9	1.1	1.5
DG C3	.9	1.0	1.2
SS A1	.8	1.3	1.7
SS A2	.7	1.0	1.3
SS A3	.9	1.4	1.8
SS A4	.7	1.0	1.3
SS A5	.9	1.4	1.9
SS A6	.7	1.0	1.3
SS A7	1.0	1.6	2.2
SS A8	.7	1.0	1.4
SS B1	.9	1.4	1.9
SS B2	.8	1.1	1.4
SS B3	1.0	1.5	2.0
SS B4	.8	1.1	1.5
SS B5	1.0	1.6	2.2
SS B6	.8	1.1	1.5
SS B7	1.1	1.8	2.4
SS B8	.8	1.1	1.5
SS C1	.9	1.3	1.7
SS C2	.8	1.1	1.4
SS C3	1.1	1.6	2.1
SS C4	.9	1.2	1.6
SS D1	.7	1.1	1.4
SS D2	.6	.9	1.1
SS D3	.8	1.2	1.6
SS D4	.7	.9	1.2
SS E1	.8	1.2	1.6
SS E2	.7	1.0	1.2
SS E3	.9	1.4	1.9
SS E4	.7	1.0	1.4

GREENVILLE, SC	10%	50%	80%
WW A1	1.2	2.0	2.6
WW A2	1.0	1.5	1.9
WW A3	1.0	1.4	1.8
WW A4	.9	1.3	1.7
WW A5	.9	1.2	1.6
WW A6	.9	1.2	1.6
WW B1	1.2	1.9	2.5
WW B2	.9	1.2	1.6
WW B3	.8	1.1	1.4
WW B4	.8	1.1	1.4
WW B5	.8	1.1	1.4
WW C1	.8	1.1	1.4
WW C2	.8	1.1	1.4
WW C3	.7	.9	1.2
WW C4	.8	1.0	1.3
TW A1	1.2	2.0	2.6
TW A2	1.1	1.6	2.1
TW A3	1.1	1.5	2.0
TW A4	1.0	1.5	1.9
TW B1	1.2	1.9	2.5
TW B2	1.1	1.6	2.1
TW B3	1.1	1.6	2.1
TW B4	1.2	1.6	2.2
TW C1	1.2	1.8	2.4
TW C2	1.2	1.8	2.3
TW C3	1.3	1.8	2.4
TW C4	1.4	2.0	2.6
TW D1	1.3	2.0	2.8
TW D2	1.0	1.4	1.8
TW D3	.9	1.3	1.6
TW D4	.9	1.2	1.5
TW D5	.9	1.2	1.5
TW E1	.8	1.2	1.5
TW E2	.9	1.2	1.5
TW E3	.8	1.0	1.3
TW E4	.8	1.0	1.3
TW F1	1.3	1.9	2.6
TW F2	1.2	1.7	2.2
TW F3	1.1	1.6	2.1
TW F4	1.2	1.6	2.1
TW G1	1.3	1.9	2.5
TW G2	1.2	1.7	2.3
TW G3	1.3	1.8	2.3
TW G4	1.4	1.9	2.5
TW H1	1.4	2.0	2.6
TW H2	1.4	2.0	2.6
TW H3	1.5	2.1	2.8
TW H4	1.8	2.4	3.2
TW I1	1.4	2.2	2.9
TW I2	1.0	1.4	1.9
TW I3	1.0	1.3	1.7
TW I4	.9	1.2	1.6
TW I5	.9	1.2	1.5
TW J1	.9	1.2	1.5
TW J2	.9	1.2	1.5
TW J3	.8	1.0	1.3
TW J4	.8	1.1	1.4
DG A1	1.2	3.3	---
DG A2	1.1	1.7	3.2
DG A3	1.0	1.3	1.8
DG B1	1.2	1.9	---
DG B2	1.1	1.5	2.3
DG B3	1.0	1.2	1.6
DG C1	1.1	1.5	3.7

GREENVILLE, SC	10%	50%	80%
DG C2	1.0	1.3	1.8
DG C3	.9	1.1	1.4
SS A1	.9	1.4	1.9
SS A2	.8	1.1	1.5
SS A3	1.0	1.5	2.1
SS A4	.8	1.1	1.5
SS A5	1.0	1.6	2.1
SS A6	.8	1.1	1.5
SS A7	1.0	1.8	2.5
SS A8	.8	1.2	1.5
SS B1	1.0	1.6	2.1
SS B2	.9	1.2	1.6
SS B3	1.1	1.7	2.3
SS B4	.9	1.2	1.6
SS B5	1.2	1.8	2.5
SS B6	.9	1.2	1.6
SS B7	1.3	2.1	2.8
SS B8	.9	1.3	1.7
SS C1	1.1	1.5	2.0
SS C2	.9	1.2	1.6
SS C3	1.2	1.8	2.4
SS C4	1.0	1.4	1.8
SS D1	.8	1.2	1.6
SS D2	.7	1.0	1.3
SS D3	.9	1.3	1.8
SS D4	.7	1.0	1.3
SS E1	.9	1.3	1.8
SS E2	.8	1.1	1.4
SS E3	1.0	1.6	2.2
SS E4	.8	1.2	1.5

HURON, SD	10%	50%	80%
WW A1	2.4	---	---
WW A2	2.0	---	---
WW A3	1.9	---	---
WW A4	1.8	4.9	---
WW A5	1.7	3.9	16.0
WW A6	1.7	3.5	8.6
WW B1	---	---	---
WW B2	1.6	2.7	4.2
WW B3	1.4	2.2	3.2
WW B4	1.3	1.8	2.4
WW B5	1.2	1.7	2.3
WW C1	1.3	2.0	2.8
WW C2	1.3	1.9	2.6
WW C3	1.1	1.5	1.9
WW C4	1.1	1.5	2.0
TW A1	2.4	---	---
TW A2	2.1	---	---
TW A3	2.0	---	---
TW A4	1.9	4.7	---
TW B1	2.3	---	---
TW B2	2.1	---	---
TW B3	2.0	5.2	---
TW B4	2.1	4.5	11.5
TW C1	2.3	---	---
TW C2	2.2	5.3	---
TW C3	2.2	4.9	12.7
TW C4	2.3	5.0	10.3
TW D1	---	---	---
TW D2	1.7	3.0	4.8
TW D3	1.6	2.7	4.3
TW D4	1.4	2.0	2.8
TW D5	1.3	1.9	2.5
TW E1	1.4	2.3	3.3
TW E2	1.4	2.1	2.9
TW E3	1.1	1.6	2.2
TW E4	1.2	1.6	2.2
TW F1	2.6	---	---
TW F2	2.3	---	---
TW F3	2.2	---	---
TW F4	2.2	5.6	---
TW G1	2.7	---	---
TW G2	2.4	---	---
TW G3	2.4	7.4	---
TW G4	2.6	5.6	15.9
TW H1	2.8	---	---
TW H2	2.7	10.0	---
TW H3	2.9	6.5	26.5
TW H4	3.2	6.4	13.6
TW I1	---	---	---
TW I2	1.8	3.2	5.0
TW I3	1.7	2.8	4.2
TW I4	1.4	2.1	2.9
TW I5	1.4	2.0	2.6
TW J1	1.5	2.3	3.3
TW J2	1.4	2.1	3.0
TW J3	1.2	1.6	2.2
TW J4	1.2	1.7	2.2
DG A1	---	---	---
DG A2	2.1	---	---
DG A3	1.6	2.4	---
DG B1	---	---	---
DG B2	2.0	---	---
DG B3	1.5	2.2	3.6
DG C1	2.7	---	---
DG C2	1.7	3.8	---
DG C3	1.4	1.9	2.8
SS A1	1.6	---	---
SS A2	1.2	2.1	3.1
SS A3	1.9	---	---
SS A4	1.3	2.4	3.7
SS A5	1.7	---	---
SS A6	1.3	2.2	3.4
SS A7	2.0	---	---
SS A8	1.3	2.6	4.3
SS B1	1.9	---	---
SS B2	1.4	2.3	3.3
SS B3	2.2	---	---
SS B4	1.4	2.5	3.6
SS B5	2.2	---	---
SS B6	1.4	2.4	3.5
SS B7	2.6	---	---
SS B8	1.5	2.7	4.2
SS C1	2.2	---	---
SS C2	1.6	2.6	4.0
SS C3	2.5	---	---
SS C4	1.7	3.0	4.6
SS D1	1.5	7.2	---
SS D2	1.1	1.8	2.5
SS D3	1.5	---	---
SS D4	1.1	1.9	2.7
SS E1	1.8	---	---
SS E2	1.3	2.2	3.2
SS E3	2.0	---	---
SS E4	1.4	2.4	3.6

Conservation Factor Tables

PIERRE, SD / PIERRE, SD / RAPID CITY, SD / SIOUX FALLS, SD

System	\tSolar Savings Pierre			System	Solar Savings Pierre			System	Solar Savings Rapid City			System	Solar Savings Sioux Falls		
	10%	50%	80%		10%	50%	80%		10%	50%	80%		10%	50%	80%
WW A1	2.0	---	---	DG C2	1.5	2.5	---	TW D4	1.1	1.6	2.1	WW A1	2.2	---	---
WW A2	1.7	3.6	9.2	DG C3	1.3	1.7	2.3	TW D5	1.1	1.5	2.0	WW A2	1.8	5.3	---
WW A3	1.6	2.9	5.2	SS A1	1.4	3.4	8.2	TW E1	1.1	1.6	2.2	WW A3	1.7	3.6	9.0
WW A4	1.5	2.7	4.3	SS A2	1.1	1.9	2.6	TW E2	1.1	1.6	2.1	WW A4	1.7	3.1	5.9
WW A5	1.5	2.5	3.8	SS A3	1.6	---	---	TW E3	1.0	1.3	1.7	WW A5	1.6	2.8	4.9
WW A6	1.4	2.4	3.6	SS A4	1.2	2.0	2.9	TW E4	1.0	1.3	1.7	WW A6	1.6	2.7	4.4
WW B1	2.7	---	---	SS A5	1.5	7.4	---	TW F1	1.9	3.8	8.5	WW B1	4.0	---	---
WW B2	1.4	2.2	3.1	SS A6	1.1	1.9	2.7	TW F2	1.7	2.8	4.3	WW B2	1.5	2.4	3.4
WW B3	1.2	1.8	2.5	SS A7	1.7	---	---	TW F3	1.6	2.6	3.7	WW B3	1.3	2.0	2.8
WW B4	1.1	1.6	2.1	SS A8	1.2	2.1	3.2	TW F4	1.6	2.5	3.6	WW B4	1.2	1.7	2.2
WW B5	1.1	1.6	2.0	SS B1	1.7	4.3	---	TW G1	1.9	3.4	5.9	WW B5	1.2	1.6	2.1
WW C1	1.2	1.7	2.3	SS B2	1.2	2.0	2.8	TW G2	1.8	2.8	4.2	WW C1	1.2	1.8	2.5
WW C2	1.2	1.7	2.2	SS B3	1.8	---	---	TW G3	1.8	2.8	4.0	WW C2	1.2	1.8	2.4
WW C3	1.0	1.3	1.7	SS B4	1.3	2.1	3.0	TW G4	1.9	2.9	4.1	WW C3	1.0	1.4	1.8
WW C4	1.0	1.4	1.8	SS B5	1.9	---	---	TW H1	2.0	3.3	5.1	WW C4	1.1	1.5	1.9
TW A1	2.0	---	---	SS B6	1.3	2.1	2.9	TW H2	2.0	3.1	4.5	TW A1	2.2	---	---
TW A2	1.8	3.9	10.1	SS B7	2.1	---	---	TW H3	2.1	3.3	4.6	TW A2	1.9	5.6	---
TW A3	1.7	3.2	5.7	SS B8	1.3	2.3	3.3	TW H4	2.4	3.7	5.2	TW A3	1.8	3.9	9.9
TW A4	1.7	2.9	4.6	SS C1	1.8	4.3	---	TW I1	2.4	---	---	TW A4	1.8	3.3	6.0
TW B1	2.0	6.7	---	SS C2	1.4	2.1	3.0	TW I2	1.4	2.1	2.8	TW B1	2.2	---	---
TW B2	1.8	3.4	6.2	SS C3	2.1	6.2	---	TW I3	1.3	1.9	2.6	TW B2	1.9	4.2	10.7
TW B3	1.8	3.2	5.1	SS C4	1.5	2.4	3.4	TW I4	1.2	1.6	2.2	TW B3	1.9	3.7	6.8
TW B4	1.8	3.1	4.9	SS D1	1.3	2.7	5.3	TW I5	1.2	1.6	2.1	TW B4	2.0	3.6	5.9
TW C1	1.9	4.1	8.7	SS D2	1.0	1.6	2.2	TW J1	1.2	1.7	2.3	TW C1	2.1	5.3	---
TW C2	1.9	3.4	5.5	SS D3	1.4	3.4	9.5	TW J2	1.2	1.6	2.2	TW C2	2.0	3.9	7.1
TW C3	1.9	3.5	5.4	SS D4	1.0	1.6	2.3	TW J3	1.0	1.3	1.8	TW C3	2.1	3.9	6.6
TW C4	2.1	3.7	5.6	SS E1	1.5	5.0	---	TW J4	1.0	1.4	1.8	TW C4	2.2	4.1	6.6
TW D1	2.9	---	---	SS E2	1.2	1.8	2.6	DG A1	2.0	---	---	TW D1	4.0	---	---
TW D2	1.5	2.4	3.4	SS E3	1.7	---	---	DG A2	1.6	3.6	---	TW D2	1.6	2.6	3.8
TW D3	1.4	2.2	3.1	SS E4	1.2	2.0	2.9	DG A3	1.3	1.8	3.1	TW D3	1.5	2.4	3.5
TW D4	1.2	1.8	2.4					DG B1	2.0	---	---	TW D4	1.3	1.9	2.5
TW D5	1.2	1.7	2.2	RAPID CITY, SD				DG B2	1.5	2.5	---	TW D5	1.3	1.8	2.3
TW E1	1.2	1.9	2.7		10%	50%	80%	DG B3	1.3	1.7	2.4	TW E1	1.3	2.1	2.9
TW E2	1.2	1.8	2.5	WW A1	1.8	4.5	---	DG C1	1.6	---	---	TW E2	1.3	1.9	2.6
TW E3	1.0	1.5	1.9	WW A2	1.5	2.5	3.8	DG C2	1.4	2.0	5.1	TW E3	1.1	1.5	2.0
TW E4	1.1	1.5	1.9	WW A3	1.4	2.2	3.2	DG C3	1.2	1.5	2.0	TW E4	1.1	1.6	2.0
TW F1	2.1	---	---	WW A4	1.3	2.0	2.9	SS A1	1.3	2.4	3.9	TW F1	2.4	---	---
TW F2	1.9	4.2	12.3	WW A5	1.3	1.9	2.7	SS A2	1.0	1.6	2.2	TW F2	2.1	6.9	---
TW F3	1.8	3.5	6.3	WW A6	1.3	1.9	2.6	SS A3	1.4	3.3	9.7	TW F3	2.0	4.3	13.0
TW F4	1.9	3.2	5.2	WW B1	2.0	---	---	SS A4	1.0	1.7	2.4	TW F4	2.0	3.8	6.9
TW G1	2.2	7.8	---	WW B2	1.2	1.8	2.5	SS A5	1.4	3.0	6.0	TW G1	2.4	---	---
TW G2	2.0	3.9	7.7	WW B3	1.1	1.6	2.1	SS A6	1.0	1.6	2.3	TW G2	2.2	5.1	---
TW G3	2.0	3.6	6.0	WW B4	1.0	1.4	1.9	SS A7	1.5	---	---	TW G3	2.2	4.3	8.5
TW G4	2.2	3.7	5.7	WW B5	1.0	1.4	1.8	SS A8	1.1	1.8	2.5	TW G4	2.4	4.2	7.1
TW H1	2.3	5.0	14.5	WW C1	1.0	1.5	2.0	SS B1	1.5	2.8	4.6	TW H1	2.5	8.6	---
TW H2	2.3	4.1	7.0	WW C2	1.1	1.5	1.9	SS B2	1.1	1.7	2.4	TW H2	2.5	4.9	10.4
TW H3	2.4	4.2	6.5	WW C3	.9	1.2	1.6	SS B3	1.6	3.5	9.2	TW H3	2.6	4.8	8.3
TW H4	2.7	4.6	6.9	WW C4	1.0	1.3	1.6	SS B4	1.2	1.8	2.5	TW H4	3.0	5.2	8.3
TW I1	3.5	---	---	TW A1	1.8	3.9	10.2	SS B5	1.7	3.8	9.3	TW I1	8.1	---	---
TW I2	1.6	2.5	3.6	TW A2	1.6	2.7	4.1	SS B6	1.2	1.8	2.5	TW I2	1.7	2.7	4.0
TW I3	1.5	2.3	3.2	TW A3	1.5	2.4	3.5	SS B7	1.9	---	---	TW I3	1.6	2.5	3.5
TW I4	1.3	1.9	2.5	TW A4	1.5	2.3	3.2	SS B8	1.2	1.9	2.7	TW I4	1.4	2.0	2.6
TW I5	1.3	1.7	2.3	TW B1	1.8	3.3	5.7	SS C1	1.5	2.6	4.1	TW I5	1.3	1.8	2.4
TW J1	1.3	2.0	2.7	TW B2	1.6	2.6	3.8	SS C2	1.2	1.8	2.4	TW J1	1.4	2.1	2.9
TW J2	1.3	1.9	2.5	TW B3	1.6	2.5	3.5	SS C3	1.8	3.3	5.4	TW J2	1.4	2.0	2.7
TW J3	1.1	1.5	2.0	TW B4	1.6	2.5	3.5	SS C4	1.3	2.0	2.7	TW J3	1.1	1.6	2.0
TW J4	1.1	1.5	2.0	TW C1	1.8	3.0	4.5	SS D1	1.1	2.0	3.1	TW J4	1.2	1.6	2.1
DG A1	2.8	---	---	TW C2	1.7	2.7	3.9	SS D2	.9	1.4	1.9	DG A1	5.3	---	---
DG A2	1.8	---	---	TW C3	1.7	2.8	3.9	SS D3	1.2	2.3	3.8	DG A2	1.9	---	---
DG A3	1.4	2.1	4.3	TW C4	1.9	3.0	4.2	SS D4	.9	1.4	1.9	DG A3	1.5	2.2	5.4
DG B1	2.7	---	---	TW D1	2.2	---	---	SS E1	1.3	2.5	4.4	DG B1	4.0	---	---
DG B2	1.7	4.1	---	TW D2	1.3	2.0	2.7	SS E2	1.0	1.6	2.2	DG B2	1.9	---	---
DG B3	1.4	1.9	2.8	TW D3	1.2	1.8	2.5	SS E3	1.5	3.1	6.3	DG B3	1.5	2.0	3.1
DG C1	2.0	---	---					SS E4	1.1	1.7	2.4	DG C1	2.3	---	---

SIOUX FALLS, SD

	SOLAR SAVINGS		
	10%	50%	80%
DG C2	1.7	2.9	---
DG C3	1.4	1.8	2.5
SS A1	1.5	4.2	---
SS A2	1.2	2.0	2.8
SS A3	1.7	---	---
SS A4	1.2	2.1	3.1
SS A5	1.6	---	---
SS A6	1.2	2.0	2.9
SS A7	1.8	---	---
SS A8	1.3	2.3	3.5
SS B1	1.8	6.5	---
SS B2	1.3	2.1	2.9
SS B3	2.0	---	---
SS B4	1.3	2.2	3.2
SS B5	2.0	---	---
SS B6	1.3	2.2	3.1
SS B7	2.4	---	---
SS B8	1.4	2.4	3.6
SS C1	2.0	---	---
SS C2	1.5	2.3	3.3
SS C3	2.3	---	---
SS C4	1.6	2.6	3.8
SS D1	1.4	3.2	8.8
SS D2	1.1	1.6	2.3
SS D3	1.5	4.2	---
SS D4	1.1	1.7	2.4
SS E1	1.7	---	---
SS E2	1.2	2.0	2.8
SS E3	1.8	---	---
SS E4	1.3	2.1	3.1

CHATTANOOGA, TN

	SOLAR SAVINGS		
	10%	50%	80%
WW A1	1.5	2.6	3.8
WW A2	1.2	1.9	2.5
WW A3	1.1	1.7	2.3
WW A4	1.1	1.6	2.1
WW A5	1.1	1.5	2.0
WW A6	1.1	1.5	2.0
WW B1	1.5	2.7	4.1
WW B2	1.1	1.5	2.0
WW B3	.9	1.3	1.7
WW B4	.9	1.2	1.6
WW B5	.9	1.2	1.6
WW C1	.9	1.3	1.6
WW C2	.9	1.3	1.6
WW C3	.8	1.1	1.4
WW C4	.9	1.1	1.4
TW A1	1.5	2.6	3.7
TW A2	1.3	2.0	2.8
TW A3	1.3	1.9	2.5
TW A4	1.2	1.8	2.4
TW B1	1.5	2.4	3.3
TW B2	1.3	2.0	2.7
TW B3	1.3	2.0	2.6
TW B4	1.4	2.0	2.7
TW C1	1.5	2.3	3.1
TW C2	1.4	2.2	2.9
TW C3	1.5	2.2	3.0
TW C4	1.6	2.4	3.3
TW D1	1.7	2.9	4.3
TW D2	1.1	1.6	2.2
TW D3	1.1	1.5	2.0

CHATTANOOGA, TN

	SOLAR SAVINGS		
	10%	50%	80%
TW D4	1.0	1.4	1.8
TW D5	1.0	1.3	1.7
TW E1	1.0	1.4	1.8
TW E2	1.0	1.4	1.8
TW E3	.9	1.2	1.5
TW E4	.9	1.2	1.5
TW F1	1.5	2.6	3.6
TW F2	1.4	2.1	2.8
TW F3	1.3	2.0	2.7
TW F4	1.4	2.0	2.6
TW G1	1.6	2.5	3.4
TW G2	1.5	2.2	2.9
TW G3	1.5	2.2	2.9
TW G4	1.6	2.3	3.1
TW H1	1.6	2.5	3.4
TW H2	1.7	2.5	3.3
TW H3	1.8	2.6	3.5
TW H4	2.1	3.0	3.9
TW I1	1.8	3.2	4.7
TW I2	1.2	1.7	2.3
TW I3	1.1	1.6	2.1
TW I4	1.0	1.4	1.8
TW I5	1.0	1.4	1.8
TW J1	1.0	1.4	1.9
TW J2	1.0	1.4	1.8
TW J3	.9	1.2	1.5
TW J4	.9	1.2	1.6
DG A1	1.6	---	---
DG A2	1.4	2.3	---
DG A3	1.2	1.5	2.3
DG B1	1.5	---	---
DG B2	1.3	1.9	4.3
DG B3	1.2	1.4	1.9
DG C1	1.3	2.2	---
DG C2	1.2	1.6	2.5
DG C3	1.1	1.3	1.7
SS A1	1.1	1.7	2.4
SS A2	.9	1.3	1.7
SS A3	1.1	1.9	2.8
SS A4	.9	1.3	1.8
SS A5	1.1	2.0	2.8
SS A6	.9	1.3	1.8
SS A7	1.2	2.4	3.5
SS A8	.9	1.4	1.8
SS B1	1.2	2.0	2.7
SS B2	1.0	1.4	1.9
SS B3	1.3	2.1	3.0
SS B4	1.0	1.5	1.9
SS B5	1.3	2.3	3.3
SS B6	1.0	1.4	1.9
SS B7	1.5	2.7	4.0
SS B8	1.0	1.5	2.0
SS C1	1.3	1.9	2.6
SS C2	1.1	1.5	2.0
SS C3	1.5	2.4	3.2
SS C4	1.2	1.7	2.2
SS D1	.9	1.5	2.0
SS D2	.8	1.1	1.5
SS D3	1.0	1.6	2.3
SS D4	.8	1.2	1.5
SS E1	1.1	1.7	2.3
SS E2	.9	1.3	1.7
SS E3	1.2	2.0	2.8
SS E4	.9	1.4	1.8

KNOXVILLE, TN

	SOLAR SAVINGS		
	10%	50%	80%
WW A1	1.4	2.5	3.6
WW A2	1.2	1.8	2.4
WW A3	1.1	1.6	2.2
WW A4	1.1	1.6	2.1
WW A5	1.1	1.5	2.0
WW A6	1.0	1.5	1.9
WW B1	1.5	2.6	3.8
WW B2	1.1	1.5	1.9
WW B3	.9	1.3	1.7
WW B4	.9	1.2	1.6
WW B5	.9	1.2	1.6
WW C1	.9	1.2	1.6
WW C2	.9	1.2	1.6
WW C3	.8	1.1	1.4
WW C4	.9	1.1	1.4
TW A1	1.4	2.5	3.5
TW A2	1.3	2.0	2.7
TW A3	1.2	1.8	2.4
TW A4	1.2	1.8	2.3
TW B1	1.4	2.3	3.2
TW B2	1.3	2.0	2.6
TW B3	1.3	1.9	2.5
TW B4	1.3	2.0	2.6
TW C1	1.4	2.2	3.0
TW C2	1.4	2.1	2.8
TW C3	1.5	2.2	2.9
TW C4	1.6	2.4	3.2
TW D1	1.6	2.8	4.0
TW D2	1.1	1.6	2.1
TW D3	1.0	1.5	2.0
TW D4	1.0	1.3	1.7
TW D5	1.0	1.3	1.7
TW E1	1.0	1.4	1.8
TW E2	1.0	1.3	1.7
TW E3	.8	1.1	1.5
TW E4	.9	1.2	1.5
TW F1	1.5	2.5	3.4
TW F2	1.3	2.0	2.8
TW F3	1.3	1.9	2.6
TW F4	1.3	1.9	2.6
TW G1	1.5	2.4	3.3
TW G2	1.4	2.1	2.8
TW G3	1.5	2.1	2.8
TW G4	1.6	2.3	3.0
TW H1	1.6	2.4	3.3
TW H2	1.6	2.4	3.2
TW H3	1.8	2.5	3.4
TW H4	2.0	2.9	3.8
TW I1	1.8	3.0	4.4
TW I2	1.2	1.7	2.2
TW I3	1.1	1.6	2.1
TW I4	1.0	1.4	1.8
TW I5	1.0	1.4	1.7
TW J1	1.0	1.4	1.8
TW J2	1.0	1.4	1.8
TW J3	.9	1.2	1.5
TW J4	.9	1.2	1.6
DG A1	1.5	---	---
DG A2	1.3	2.2	---
DG A3	1.1	1.5	2.3
DG B1	1.5	9.6	---
DG B2	1.3	1.8	3.8
DG B3	1.1	1.4	1.9
DG C1	1.3	2.1	---

KNOXVILLE, TN

	SOLAR SAVINGS		
	10%	50%	80%
DG C2	1.2	1.6	2.4
DG C3	1.0	1.3	1.6
SS A1	1.0	1.7	2.3
SS A2	.9	1.3	1.7
SS A3	1.1	1.9	2.7
SS A4	.9	1.3	1.8
SS A5	1.1	1.9	2.7
SS A6	.9	1.3	1.7
SS A7	1.2	2.3	3.4
SS A8	.9	1.4	1.8
SS B1	1.2	1.9	2.6
SS B2	1.0	1.4	1.8
SS B3	1.3	2.1	2.9
SS B4	1.0	1.4	1.9
SS B5	1.3	2.3	3.2
SS B6	1.0	1.4	1.9
SS B7	1.4	2.7	3.9
SS B8	1.0	1.5	2.0
SS C1	1.2	1.9	2.5
SS C2	1.0	1.5	1.9
SS C3	1.5	2.3	3.1
SS C4	1.1	1.6	2.1
SS D1	.9	1.4	2.0
SS D2	.8	1.1	1.5
SS D3	1.0	1.6	2.2
SS D4	.8	1.1	1.5
SS E1	1.0	1.6	2.2
SS E2	.9	1.2	1.6
SS E3	1.2	2.0	2.8
SS E4	.9	1.3	1.8

MEMPHIS, TN

	SOLAR SAVINGS		
	10%	50%	80%
WW A1	1.3	2.2	3.1
WW A2	1.1	1.6	2.2
WW A3	1.0	1.5	2.0
WW A4	1.0	1.4	1.9
WW A5	1.0	1.4	1.8
WW A6	1.0	1.3	1.7
WW B1	1.3	2.1	3.0
WW B2	1.0	1.4	1.8
WW B3	.9	1.2	1.5
WW B4	.9	1.1	1.5
WW B5	.9	1.1	1.5
WW C1	.8	1.1	1.5
WW C2	.9	1.2	1.5
WW C3	.8	1.0	1.3
WW C4	.8	1.1	1.3
TW A1	1.3	2.2	3.0
TW A2	1.2	1.8	2.4
TW A3	1.1	1.7	2.2
TW A4	1.1	1.6	2.1
TW B1	1.3	2.1	2.8
TW B2	1.2	1.8	2.4
TW B3	1.2	1.7	2.3
TW B4	1.3	1.8	2.4
TW C1	1.3	2.0	2.7
TW C2	1.3	1.9	2.5
TW C3	1.4	2.0	2.6
TW C4	1.5	2.2	2.9
TW D1	1.5	2.3	3.3
TW D2	1.1	1.5	1.9
TW D3	1.0	1.4	1.8

Conservation Factor Tables / 369

MEMPHIS, TN				NASHVILLE, TN				NASHVILLE, TN				ABILENE, TX			
	SOLAR SAVINGS				SOLAR SAVINGS				SOLAR SAVINGS				SOLAR SAVINGS		
	10%	50%	80%		10%	50%	80%		10%	50%	80%		10%	50%	80%
WW A1	1.3	2.2	3.1	WW A1	1.6	3.2	5.1	DG C2	1.3	1.8	3.2	TW D4	.7	1.0	1.2
WW A2	1.1	1.6	2.2	WW A2	1.3	2.1	3.0	DG C3	1.1	1.4	1.8	TW D5	.7	1.0	1.2
WW A3	1.0	1.5	2.0	WW A3	1.3	1.9	2.6	SS A1	1.2	2.0	2.8	TW E1	.7	1.0	1.2
WW A4	1.0	1.4	1.9	WW A4	1.2	1.8	2.4	SS A2	.9	1.4	1.9	TW E2	.7	1.0	1.2
WW A5	1.0	1.4	1.8	WW A5	1.2	1.7	2.3	SS A3	1.2	2.3	3.4	TW E3	.6	.8	1.1
WW A6	1.0	1.3	1.7	WW A6	1.2	1.7	2.2	SS A4	1.0	1.5	2.0	TW E4	.7	.9	1.1
WW B1	1.3	2.1	3.0	WW B1	1.7	3.9	8.1	SS A5	1.2	2.3	3.3	TW F1	1.1	1.5	2.0
WW B2	1.0	1.4	1.8	WW B2	1.2	1.7	2.2	SS A6	.9	1.4	1.9	TW F2	1.0	1.3	1.7
WW B3	.9	1.2	1.5	WW B3	1.0	1.4	1.9	SS A7	1.3	2.9	4.7	TW F3	.9	1.3	1.6
WW B4	.9	1.1	1.5	WW B4	1.0	1.3	1.7	SS A8	1.0	1.5	2.0	TW F4	1.0	1.3	1.7
WW B5	.9	1.1	1.5	WW B5	1.0	1.3	1.7	SS B1	1.3	2.2	3.1	TW G1	1.1	1.5	2.0
WW C1	.8	1.1	1.5	WW C1	1.0	1.4	1.8	SS B2	1.0	1.5	2.0	TW G2	1.0	1.4	1.8
WW C2	.9	1.2	1.5	WW C2	1.0	1.4	1.8	SS B3	1.4	2.5	3.7	TW G3	1.0	1.4	1.8
WW C3	.8	1.0	1.3	WW C3	.9	1.2	1.5	SS B4	1.1	1.6	2.1	TW G4	1.1	1.5	2.0
WW C4	.8	1.1	1.3	WW C4	.9	1.2	1.5	SS B5	1.5	2.7	4.0	TW H1	1.1	1.6	2.0
TW A1	1.3	2.2	3.0	TW A1	1.6	3.1	4.7	SS B6	1.1	1.6	2.1	TW H2	1.2	1.6	2.0
TW A2	1.2	1.8	2.4	TW A2	1.4	2.3	3.2	SS B7	1.6	3.3	5.4	TW H3	1.3	1.7	2.2
TW A3	1.1	1.7	2.2	TW A3	1.4	2.1	2.9	SS B8	1.1	1.7	2.2	TW H4	1.5	2.0	2.5
TW A4	1.1	1.6	2.1	TW A4	1.3	2.0	2.7	SS C1	1.4	2.2	3.1	TW I1	1.2	1.6	2.1
TW B1	1.3	2.1	2.8	TW B1	1.6	2.8	4.0	SS C2	1.1	1.6	2.2	TW I2	.9	1.2	1.5
TW B2	1.2	1.8	2.4	TW B2	1.4	2.3	3.1	SS C3	1.6	2.7	3.9	TW I3	.8	1.1	1.4
TW B3	1.2	1.7	2.3	TW B3	1.4	2.2	3.0	SS C4	1.2	1.8	2.4	TW I4	.8	1.0	1.3
TW B4	1.3	1.8	2.4	TW B4	1.5	2.3	3.0	SS D1	1.0	1.6	2.3	TW I5	.8	1.0	1.3
TW C1	1.3	2.0	2.7	TW C1	1.6	2.6	3.6	SS D2	.8	1.2	1.6	TW J1	.7	1.0	1.3
TW C2	1.3	1.9	2.5	TW C2	1.6	2.4	3.3	SS D3	1.1	1.9	2.6	TW J2	.8	1.0	1.3
TW C3	1.4	2.0	2.6	TW C3	1.6	2.5	3.4	SS D4	.9	1.3	1.7	TW J3	.7	.9	1.1
TW C4	1.5	2.2	2.9	TW C4	1.7	2.7	3.7	SS E1	1.2	1.9	2.7	TW J4	.7	.9	1.1
TW D1	1.5	2.3	3.3	TW D1	1.9	3.9	7.1	SS E2	.9	1.4	1.8	DG A1	1.0	1.6	---
TW D2	1.1	1.5	1.9	TW D2	1.2	1.8	2.4	SS E3	1.3	2.3	3.4	DG A2	1.0	1.3	1.9
TW D3	1.0	1.4	1.8	TW D3	1.2	1.7	2.2	SS E4	1.0	1.5	2.0	DG A3	.9	1.1	1.5
DG B1	1.3	2.6	---	TW D4	1.1	1.5	1.9					DG B1	1.0	1.3	2.4
DG B2	1.2	1.6	2.8	TW D5	1.1	1.4	1.9	ABILENE, TX				DG B2	.9	1.1	1.6
DG B3	1.1	1.3	1.7	TW E1	1.1	1.5	2.0		SOLAR SAVINGS			DG B3	.9	1.0	1.3
DG C1	1.2	1.7	---	TW E2	1.1	1.5	1.9		10%	50%	80%	DG C1	.9	1.1	1.6
DG C2	1.1	1.4	2.0	TW E3	.9	1.2	1.6	WW A1	1.0	1.5	2.0	DG C2	.9	1.0	1.3
DG C3	1.0	1.2	1.5	TW E4	.9	1.3	1.6	WW A2	.9	1.2	1.5	DG C3	.8	.9	1.1
SS A1	1.0	1.6	2.1	TW F1	1.7	3.0	4.5	WW A3	.8	1.1	1.4	SS A1	.8	1.1	1.5
SS A2	.8	1.2	1.6	TW F2	1.5	2.4	3.4	WW A4	.8	1.0	1.3	SS A2	.7	.9	1.2
SS A3	1.0	1.7	2.4	TW F3	1.5	2.2	3.1	WW A5	.8	1.0	1.3	SS A3	.8	1.2	1.6
SS A4	.8	1.2	1.6	TW F4	1.5	2.2	3.0	WW A6	.7	1.0	1.3	SS A4	.7	.9	1.2
SS A5	1.0	1.7	2.4	TW G1	1.7	2.9	4.1	WW B1	1.0	1.4	1.8	SS A5	.8	1.2	1.6
SS A6	.8	1.2	1.6	TW G2	1.6	2.5	3.4	WW B2	.8	1.0	1.3	SS A6	.7	.9	1.2
SS A7	1.1	2.1	2.9	TW G3	1.6	2.4	3.3	WW B3	.7	.9	1.2	SS A7	.9	1.4	1.9
SS A8	.8	1.3	1.7	TW G4	1.8	2.6	3.5	WW B4	.7	.9	1.1	SS A8	.7	.9	1.2
SS B1	1.1	1.7	2.4	TW H1	1.8	2.9	4.0	WW B5	.7	.9	1.1	SS B1	.9	1.3	1.7
SS B2	.9	1.3	1.7	TW H2	1.8	2.8	3.8	WW C1	.7	.9	1.1	SS B2	.7	1.0	1.3
SS B3	1.2	1.9	2.6	TW H3	2.0	2.9	3.9	WW C2	.7	.9	1.1	SS B3	.9	1.3	1.7
SS B4	.9	1.3	1.8	TW H4	2.2	3.3	4.5	WW C3	.6	.8	1.0	SS B4	.7	1.0	1.3
SS B5	1.2	2.0	2.8	TW I1	2.1	4.5	8.7	WW C4	.7	.8	1.1	SS B5	1.0	1.4	1.9
SS B6	.9	1.3	1.7	TW I2	1.3	1.9	2.5	TW A1	1.0	1.5	2.0	SS B6	.7	1.0	1.3
SS B7	1.4	2.3	3.3	TW I3	1.2	1.8	2.3	TW A2	.9	1.3	1.7	SS B7	1.0	1.6	2.1
SS B8	.9	1.4	1.8	TW I4	1.1	1.5	2.0	TW A3	.9	1.2	1.6	SS B8	.8	1.0	1.4
SS C1	1.1	1.7	2.2	TW I5	1.1	1.5	1.9	TW A4	.9	1.2	1.5	SS C1	.9	1.2	1.5
SS C2	1.0	1.4	1.8	TW J1	1.1	1.6	2.0	TW B1	1.0	1.5	1.9	SS C2	.8	1.0	1.3
SS C3	1.3	2.0	2.7	TW J2	1.1	1.5	2.0	TW B2	.9	1.3	1.7	SS C3	1.0	1.4	1.9
SS C4	1.1	1.5	2.0	TW J3	1.0	1.3	1.6	TW B3	.9	1.3	1.6	SS C4	.8	1.1	1.5
SS D1	.9	1.3	1.8	TW J4	1.0	1.3	1.7	TW B4	1.0	1.3	1.7	SS D1	.7	1.0	1.3
SS D2	.7	1.0	1.4	DG A1	1.8	---	---	TW C1	1.0	1.5	1.9	SS D2	.6	.8	1.0
SS D3	.9	1.5	2.0	DG A2	1.5	2.8	---	TW C2	1.0	1.4	1.8	SS D3	.7	1.1	1.4
SS D4	.8	1.1	1.4	DG A3	1.2	1.7	2.6	TW C3	1.1	1.5	1.9	SS D4	.6	.8	1.1
SS E1	1.0	1.5	2.0	DG B1	1.7	---	---	TW C4	1.2	1.6	2.1	SS E1	.7	1.1	1.4
SS E2	.8	1.2	1.5	DG B2	1.4	2.2	---	TW D1	1.1	1.5	2.0	SS E2	.6	.9	1.1
SS E3	1.1	1.8	2.4	DG B3	1.2	1.6	2.1	TW D2	.8	1.1	1.4	SS E3	.9	1.3	1.7
SS E4	.9	1.2	1.6	DG C1	1.4	3.4	---	TW D3	.8	1.0	1.3	SS E4	.7	.9	1.2

370 / APPENDIX 17

AMARILLO, TX

	SOLAR SAVINGS		
	10%	50%	80%
WW A1	1.1	1.7	2.2
WW A2	1.0	1.3	1.7
WW A3	.9	1.2	1.6
WW A4	.9	1.2	1.5
WW A5	.8	1.1	1.4
WW A6	.8	1.1	1.4
WW B1	1.1	1.6	2.1
WW B2	.9	1.1	1.4
WW B3	.8	1.0	1.3
WW B4	.8	1.0	1.2
WW B5	.8	1.0	1.2
WW C1	.7	1.0	1.2
WW C2	.8	1.0	1.3
WW C3	.7	.9	1.1
WW C4	.7	.9	1.1
TW A1	1.2	1.7	2.3
TW A2	1.0	1.4	1.9
TW A3	1.0	1.3	1.7
TW A4	1.0	1.3	1.7
TW B1	1.2	1.7	2.2
TW B2	1.1	1.5	1.9
TW B3	1.1	1.4	1.8
TW B4	1.1	1.5	1.9
TW C1	1.2	1.6	2.1
TW C2	1.2	1.6	2.0
TW C3	1.2	1.6	2.1
TW C4	1.3	1.8	2.3
TW D1	1.2	1.7	2.3
TW D2	.9	1.2	1.6
TW D3	.9	1.1	1.5
TW D4	.8	1.1	1.3
TW D5	.8	1.1	1.3
TW E1	.8	1.1	1.3
TW E2	.8	1.1	1.3
TW E3	.7	.9	1.2
TW E4	.7	1.0	1.2
TW F1	1.2	1.7	2.2
TW F2	1.1	1.5	1.9
TW F3	1.1	1.4	1.8
TW F4	1.1	1.4	1.8
TW G1	1.2	1.7	2.2
TW G2	1.2	1.6	2.0
TW G3	1.2	1.6	2.0
TW G4	1.3	1.7	2.2
TW H1	1.3	1.7	2.3
TW H2	1.3	1.8	2.3
TW H3	1.4	1.9	2.4
TW H4	1.6	2.2	2.8
TW I1	1.3	1.9	2.4
TW I2	1.0	1.3	1.6
TW I3	.9	1.2	1.5
TW I4	.9	1.1	1.4
TW I5	.9	1.1	1.4
TW J1	.8	1.1	1.4
TW J2	.8	1.1	1.4
TW J3	.7	1.0	1.2
TW J4	.8	1.0	1.2
DG A1	1.1	2.1	---
DG A2	1.1	1.5	2.4
DG A3	1.0	1.2	1.6
DG B1	1.1	1.5	---
DG B2	1.1	1.3	1.8
DG B3	1.0	1.1	1.4
DG C1	1.0	1.3	2.0
DG C2	1.0	1.1	1.5
DG C3	.9	1.0	1.2
SS A1	.9	1.3	1.7
SS A2	.7	1.0	1.3
SS A3	.9	1.4	1.8
SS A4	.7	1.0	1.4
SS A5	.9	1.4	1.9
SS A6	.7	1.0	1.3
SS A7	1.0	1.6	2.2
SS A8	.7	1.1	1.4
SS B1	1.0	1.4	1.9
SS B2	.8	1.1	1.4
SS B3	1.0	1.5	2.0
SS B4	.8	1.1	1.5
SS B5	1.1	1.6	2.2
SS B6	.8	1.1	1.5
SS B7	1.2	1.8	2.4
SS B8	.8	1.2	1.5
SS C1	1.0	1.3	1.7
SS C2	.9	1.1	1.4
SS C3	1.2	1.6	2.1
SS C4	.9	1.3	1.6
SS D1	.8	1.1	1.4
SS D2	.7	.9	1.1
SS D3	.8	1.2	1.6
SS D4	.7	.9	1.2
SS E1	.8	1.2	1.6
SS E2	.7	1.0	1.3
SS E3	1.0	1.4	1.9
SS E4	.8	1.1	1.4

AUSTIN, TX

	SOLAR SAVINGS		
	10%	50%	80%
WW A1	.9	1.3	1.8
WW A2	.8	1.1	1.4
WW A3	.7	1.0	1.3
WW A4	.7	.9	1.2
WW A5	.7	.9	1.2
WW A6	.7	.9	1.1
WW B1	.9	1.2	1.6
WW B2	.7	.9	1.2
WW B3	.6	.8	1.1
WW B4	.6	.8	1.0
WW B5	.6	.8	1.1
WW C1	.6	.8	1.0
WW C2	.6	.8	1.1
WW C3	.6	.7	.9
WW C4	.6	.8	1.0
TW A1	.9	1.4	1.8
TW A2	.8	1.2	1.5
TW A3	.8	1.1	1.4
TW A4	.8	1.1	1.4
TW B1	.9	1.3	1.7
TW B2	.8	1.2	1.5
TW B3	.8	1.2	1.5
TW B4	.9	1.2	1.6
TW C1	.9	1.3	1.7
TW C2	.9	1.3	1.7
TW C3	1.0	1.3	1.7
TW C4	1.0	1.5	1.9
TW D1	1.0	1.4	1.8
TW D2	.7	1.0	1.3
TW D3	.7	.9	1.2
TW D4	.7	.9	1.1
TW D5	.7	.9	1.1
TW E1	.6	.9	1.1
TW E2	.7	.9	1.1
TW E3	.6	.8	1.0
TW E4	.6	.8	1.0
TW F1	.9	1.4	1.8
TW F2	.9	1.2	1.5
TW F3	.8	1.1	1.5
TW F4	.9	1.2	1.5
TW G1	1.0	1.4	1.8
TW G2	.9	1.3	1.6
TW G3	.9	1.3	1.6
TW G4	1.0	1.4	1.8
TW H1	1.0	1.4	1.8
TW H2	1.1	1.4	1.9
TW H3	1.1	1.5	2.0
TW H4	1.3	1.8	2.3
TW I1	1.0	1.4	1.9
TW I2	.8	1.1	1.4
TW I3	.7	1.0	1.3
TW I4	.7	.9	1.2
TW I5	.7	.9	1.2
TW J1	.7	.9	1.2
TW J2	.7	.9	1.2
TW J3	.6	.8	1.0
TW J4	.6	.8	1.1
DG A1	.9	1.4	2.8
DG A2	.9	1.2	1.7
DG A3	.8	1.0	1.3
DG B1	.9	1.1	1.8
DG B2	.8	1.0	1.4
DG B3	.8	.9	1.2
DG C1	.8	1.0	1.4
DG C2	.8	.9	1.2
DG C3	.7	.9	1.0
SS A1	.7	1.0	1.3
SS A2	.6	.8	1.1
SS A3	.7	1.1	1.4
SS A4	.6	.8	1.1
SS A5	.7	1.1	1.5
SS A6	.6	.8	1.1
SS A7	.8	1.2	1.6
SS A8	.6	.8	1.1
SS B1	.8	1.1	1.5
SS B2	.7	.9	1.2
SS B3	.8	1.2	1.5
SS B4	.7	.9	1.2
SS B5	.9	1.3	1.7
SS B6	.7	.9	1.2
SS B7	.9	1.4	1.9
SS B8	.7	.9	1.2
SS C1	.8	1.1	1.4
SS C2	.7	.9	1.2
SS C3	.9	1.3	1.7
SS C4	.8	1.0	1.3
SS D1	.6	.9	1.1
SS D2	.5	.7	.9
SS D3	.6	1.0	1.3
SS D4	.5	.7	1.0
SS E1	.7	.9	1.2
SS E2	.6	.8	1.0
SS E3	.8	1.1	1.5
SS E4	.6	.9	1.1

BROWNSVILLE, TX

	SOLAR SAVINGS		
	10%	50%	80%
WW A1	.6	.9	1.2
WW A2	.5	.7	.9
WW A3	.5	.7	.9
WW A4	.5	.6	.8
WW A5	.5	.6	.8
WW A6	.5	.6	.8
WW B1	.6	.8	1.0
WW B2	.5	.6	.8
WW B3	.4	.6	.7
WW B4	.4	.6	.7
WW B5	.5	.6	.7
WW C1	.4	.6	.7
WW C2	.4	.6	.7
WW C3	.4	.5	.7
WW C4	.4	.6	.7
TW A1	.6	.9	1.2
TW A2	.6	.8	1.0
TW A3	.5	.7	1.0
TW A4	.5	.7	.9
TW B1	.6	.9	1.2
TW B2	.6	.8	1.0
TW B3	.6	.8	1.0
TW B4	.6	.8	1.1
TW C1	.6	.9	1.2
TW C2	.6	.9	1.1
TW C3	.7	.9	1.2
TW C4	.7	1.0	1.3
TW D1	.6	.9	1.1
TW D2	.5	.7	.9
TW D3	.5	.6	.8
TW D4	.5	.6	.8
TW D5	.5	.6	.8
TW E1	.5	.6	.8
TW E2	.5	.6	.8
TW E3	.4	.5	.7
TW E4	.4	.6	.7
TW F1	.6	.9	1.2
TW F2	.6	.8	1.0
TW F3	.6	.8	1.0
TW F4	.6	.8	1.0
TW G1	.7	.9	1.2
TW G2	.6	.9	1.1
TW G3	.7	.9	1.1
TW G4	.7	1.0	1.2
TW H1	.7	1.0	1.2
TW H2	.7	1.0	1.3
TW H3	.8	1.1	1.4
TW H4	.9	1.2	1.6
TW I1	.7	.9	1.2
TW I2	.6	.7	.9
TW I3	.5	.7	.9
TW I4	.5	.7	.8
TW I5	.5	.7	.8
TW J1	.5	.6	.8
TW J2	.5	.6	.8
TW J3	.4	.6	.7
TW J4	.5	.6	.8
DG A1	.6	.8	1.1
DG A2	.6	.8	1.0
DG A3	.6	.7	.9
DG B1	.6	.7	.9
DG B2	.6	.7	.9
DG B3	.6	.7	.8
DG C1	.5	.6	.8

Conservation Factor Tables / 371

BROWNSVILLE, TX
SOLAR SAVINGS

	10%	50%	80%
DG C2	.6	.6	.8
DG C3	.5	.6	.7
SS A1	.5	.7	.9
SS A2	.4	.6	.7
SS A3	.5	.7	.9
SS A4	.4	.6	.7
SS A5	.5	.7	1.0
SS A6	.4	.6	.7
SS A7	.5	.8	1.0
SS A8	.4	.6	.7
SS B1	.5	.7	1.0
SS B2	.5	.6	.8
SS B3	.5	.8	1.0
SS B4	.5	.6	.8
SS B5	.6	.8	1.1
SS B6	.5	.6	.8
SS B7	.6	.9	1.2
SS B8	.5	.6	.8
SS C1	.5	.7	.9
SS C2	.5	.6	.8
SS C3	.6	.9	1.1
SS C4	.5	.7	.9
SS D1	.4	.6	.7
SS D2	.4	.5	.6
SS D3	.4	.6	.8
SS D4	.4	.5	.7
SS E1	.5	.6	.8
SS E2	.4	.5	.7
SS E3	.5	.7	1.0
SS E4	.4	.6	.8

CORPUS CHRISTI, TX
SOLAR SAVINGS

	10%	50%	80%
WW A1	.7	1.0	1.3
WW A2	.6	.8	1.1
WW A3	.6	.8	1.0
WW A4	.6	.7	.9
WW A5	.5	.7	.9
WW A6	.5	.7	.9
WW B1	.7	.9	1.2
WW B2	.6	.7	.9
WW B3	.5	.6	.8
WW B4	.5	.6	.8
WW B5	.5	.7	.8
WW C1	.5	.6	.8
WW C2	.5	.7	.8
WW C3	.5	.6	.8
WW C4	.5	.6	.8
TW A1	.7	1.0	1.4
TW A2	.7	.9	1.2
TW A3	.6	.8	1.1
TW A4	.6	.8	1.1
TW B1	.7	1.0	1.3
TW B2	.7	.9	1.2
TW B3	.7	.9	1.2
TW B4	.7	.9	1.2
TW C1	.8	1.0	1.3
TW C2	.7	1.0	1.3
TW C3	.8	1.0	1.4
TW C4	.9	1.2	1.5
TW D1	.8	1.0	1.3
TW D2	.6	.8	1.0
TW D3	.6	.7	.9

CORPUS CHRISTI, TX
SOLAR SAVINGS

	10%	50%	80%
TW D4	.5	.7	.9
TW D5	.6	.7	.9
TW E1	.5	.7	.9
TW E2	.5	.7	.9
TW E3	.5	.6	.8
TW E4	.5	.6	.8
TW F1	.7	1.0	1.4
TW F2	.7	.9	1.2
TW F3	.7	.9	1.2
TW F4	.7	.9	1.2
TW G1	.8	1.0	1.4
TW G2	.7	1.0	1.3
TW G3	.8	1.0	1.3
TW G4	.8	1.1	1.4
TW H1	.8	1.1	1.4
TW H2	.9	1.1	1.4
TW H3	.9	1.2	1.6
TW H4	1.1	1.4	1.8
TW I1	.8	1.1	1.4
TW I2	.6	.8	1.1
TW I3	.6	.8	1.0
TW I4	.6	.7	.9
TW I5	.6	.7	.9
TW J1	.6	.7	.9
TW J2	.6	.7	.9
TW J3	.5	.6	.8
TW J4	.5	.7	.9
DG A1	.7	.9	1.4
DG A2	.7	.9	1.2
DG A3	.7	.8	1.0
DG B1	.7	.8	1.1
DG B2	.7	.8	1.0
DG B3	.7	.7	.9
DG C1	.6	.7	1.0
DG C2	.7	.7	.9
DG C3	.6	.7	.8
SS A1	.6	.8	1.0
SS A2	.5	.6	.8
SS A3	.6	.8	1.1
SS A4	.5	.6	.8
SS A5	.6	.8	1.1
SS A6	.5	.6	.8
SS A7	.6	.9	1.2
SS A8	.5	.7	.9
SS B1	.6	.9	1.1
SS B2	.5	.7	.9
SS B3	.6	.9	1.2
SS B4	.5	.7	.9
SS B5	.7	1.0	1.3
SS B6	.5	.7	.9
SS B7	.7	1.0	1.4
SS B8	.5	.7	1.0
SS C1	.6	.8	1.1
SS C2	.6	.7	.9
SS C3	.7	1.0	1.3
SS C4	.6	.8	1.0
SS D1	.5	.7	.9
SS D2	.4	.6	.7
SS D3	.5	.7	1.0
SS D4	.4	.6	.8
SS E1	.5	.7	.9
SS E2	.5	.6	.8
SS E3	.6	.8	1.1
SS E4	.5	.7	.9

DALLAS, TX
SOLAR SAVINGS

	10%	50%	80%
WW A1	1.0	1.6	2.1
WW A2	.9	1.2	1.6
WW A3	.8	1.1	1.4
WW A4	.8	1.1	1.4
WW A5	.8	1.0	1.3
WW A6	.8	1.0	1.3
WW B1	1.0	1.4	1.9
WW B2	.8	1.1	1.4
WW B3	.7	.9	1.2
WW B4	.7	.9	1.2
WW B5	.7	.9	1.2
WW C1	.7	.9	1.2
WW C2	.7	.9	1.2
WW C3	.6	.8	1.0
WW C4	.7	.9	1.1
TW A1	1.1	1.6	2.1
TW A2	.9	1.3	1.7
TW A3	.9	1.2	1.6
TW A4	.9	1.2	1.6
TW B1	1.1	1.5	2.0
TW B2	1.0	1.3	1.7
TW B3	1.0	1.3	1.7
TW B4	1.0	1.4	1.8
TW C1	1.1	1.5	2.0
TW C2	1.0	1.5	1.9
TW C3	1.1	1.5	2.0
TW C4	1.2	1.7	2.2
TW D1	1.1	1.6	2.1
TW D2	.8	1.1	1.5
TW D3	.8	1.1	1.4
TW D4	.8	1.0	1.3
TW D5	.8	1.0	1.3
TW E1	.7	1.0	1.3
TW E2	.7	1.0	1.3
TW E3	.7	.9	1.1
TW E4	.7	.9	1.1
TW F1	1.1	1.6	2.1
TW F2	1.0	1.4	1.8
TW F3	1.0	1.3	1.7
TW F4	1.0	1.3	1.7
TW G1	1.1	1.6	2.0
TW G2	1.0	1.4	1.9
TW G3	1.1	1.5	1.9
TW G4	1.2	1.6	2.0
TW H1	1.2	1.6	2.1
TW H2	1.2	1.6	2.1
TW H3	1.3	1.8	2.3
TW H4	1.5	2.0	2.6
TW I1	1.2	1.7	2.2
TW I2	.9	1.2	1.5
TW I3	.8	1.1	1.4
TW I4	.8	1.0	1.3
TW I5	.8	1.0	1.3
TW J1	.8	1.0	1.3
TW J2	.8	1.0	1.3
TW J3	.7	.9	1.1
TW J4	.7	.9	1.2
DG A1	1.0	1.7	---
DG A2	1.0	1.3	2.1
DG A3	.9	1.1	1.5
DG B1	1.0	1.4	2.9
DG B2	1.0	1.2	1.7
DG B3	.9	1.0	1.3
DG C1	.9	1.1	1.7

DALLAS, TX
SOLAR SAVINGS

	10%	50%	80%
DG C2	.9	1.1	1.4
DG C3	.8	1.0	1.2
SS A1	.8	1.2	1.5
SS A2	.7	.9	1.2
SS A3	.8	1.2	1.7
SS A4	.7	1.0	1.3
SS A5	.8	1.3	1.7
SS A6	.7	.9	1.2
SS A7	.9	1.5	2.0
SS A8	.7	1.0	1.3
SS B1	.9	1.3	1.7
SS B2	.7	1.0	1.3
SS B3	.9	1.4	1.8
SS B4	.8	1.1	1.4
SS B5	1.0	1.5	2.0
SS B6	.8	1.0	1.4
SS B7	1.1	1.6	2.2
SS B8	.8	1.1	1.4
SS C1	.9	1.2	1.6
SS C2	.8	1.1	1.4
SS C3	1.1	1.5	2.0
SS C4	.9	1.2	1.5
SS D1	.7	1.0	1.3
SS D2	.6	.8	1.1
SS D3	.7	1.1	1.5
SS D4	.6	.8	1.1
SS E1	.8	1.1	1.4
SS E2	.7	.9	1.2
SS E3	.9	1.3	1.7
SS E4	.7	1.0	1.3

DEL RIO, TX
SOLAR SAVINGS

	10%	50%	80%
WW A1	.8	1.2	1.6
WW A2	.7	1.0	1.2
WW A3	.7	.9	1.2
WW A4	.6	.9	1.1
WW A5	.6	.8	1.1
WW A6	.6	.8	1.0
WW B1	.8	1.1	1.4
WW B2	.6	.8	1.1
WW B3	.6	.8	1.0
WW B4	.6	.7	.9
WW B5	.6	.8	1.0
WW C1	.6	.7	.9
WW C2	.6	.8	1.0
WW C3	.5	.7	.9
WW C4	.6	.7	.9
TW A1	.8	1.2	1.6
TW A2	.7	1.0	1.4
TW A3	.7	1.0	1.3
TW A4	.7	1.0	1.3
TW B1	.8	1.2	1.6
TW B2	.8	1.1	1.4
TW B3	.8	1.0	1.4
TW B4	.8	1.1	1.4
TW C1	.8	1.2	1.6
TW C2	.8	1.2	1.5
TW C3	.9	1.2	1.6
TW C4	1.0	1.3	1.8
TW D1	.9	1.2	1.6
TW D2	.7	.9	1.2
TW D3	.6	.9	1.1

DEL RIO, TX				EL PASO, TX				EL PASO, TX				FORT WORTH, TX			
	SOLAR SAVINGS				SOLAR SAVINGS				SOLAR SAVINGS				SOLAR SAVINGS		
	10%	50%	80%		10%	50%	80%		10%	50%	80%		10%	50%	80%
WW A1	.8	1.2	1.6	WW A1	.9	1.3	1.7	DG C2	.8	.9	1.1	TW D4	.8	1.0	1.3
WW A2	.7	1.0	1.2	WW A2	.8	1.0	1.3	DG C3	.7	.8	1.0	TW D5	.8	1.0	1.3
WW A3	.7	.9	1.2	WW A3	.7	.9	1.2	SS A1	.7	1.0	1.3	TW E1	.8	1.0	1.3
WW A4	.6	.9	1.1	WW A4	.7	.9	1.2	SS A2	.6	.8	1.1	TW E2	.8	1.0	1.3
WW A5	.6	.8	1.1	WW A5	.7	.9	1.1	SS A3	.7	1.1	1.4	TW E3	.7	.9	1.1
WW A6	.6	.8	1.0	WW A6	.7	.9	1.1	SS A4	.6	.8	1.1	TW E4	.7	.9	1.2
WW B1	.8	1.1	1.4	WW B1	.8	1.2	1.5	SS A5	.7	1.1	1.4	TW F1	1.1	1.6	2.1
WW B2	.6	.8	1.1	WW B2	.7	.9	1.1	SS A6	.6	.8	1.1	TW F2	1.0	1.4	1.8
WW B3	.6	.8	1.0	WW B3	.6	.8	1.0	SS A7	.8	1.2	1.6	TW F3	1.0	1.4	1.8
WW B4	.6	.7	.9	WW B4	.6	.8	1.0	SS A8	.6	.8	1.1	TW F4	1.0	1.4	1.8
WW B5	.6	.8	1.0	WW B5	.6	.8	1.0	SS B1	.8	1.1	1.4	TW G1	1.1	1.6	2.1
WW C1	.6	.7	.9	WW C1	.6	.8	1.0	SS B2	.7	.9	1.2	TW G2	1.1	1.5	1.9
WW C2	.6	.8	1.0	WW C2	.6	.8	1.0	SS B3	.8	1.2	1.5	TW G3	1.1	1.5	1.9
WW C3	.5	.7	.9	WW C3	.6	.7	.9	SS B4	.7	.9	1.2	TW G4	1.2	1.6	2.1
WW C4	.6	.7	.9	WW C4	.6	.8	.9	SS B5	.9	1.3	1.7	TW H1	1.2	1.7	2.2
TW A1	.8	1.2	1.6	TW A1	.9	1.3	1.7	SS B6	.7	.9	1.2	TW H2	1.2	1.7	2.2
TW A2	.7	1.0	1.4	TW A2	.8	1.1	1.4	SS B7	.9	1.4	1.8	TW H3	1.3	1.8	2.3
TW A3	.7	1.0	1.3	TW A3	.8	1.0	1.3	SS B8	.7	.9	1.2	TW H4	1.5	2.1	2.7
TW A4	.7	1.0	1.3	TW A4	.8	1.0	1.3	SS C1	.8	1.0	1.3	TW I1	1.2	1.7	2.3
TW B1	.8	1.2	1.6	TW B1	.9	1.3	1.7	SS C2	.7	.9	1.2	TW I2	.9	1.2	1.6
TW B2	.8	1.1	1.4	TW B2	.8	1.1	1.5	SS C3	.9	1.3	1.6	TW I3	.9	1.2	1.5
TW B3	.8	1.0	1.4	TW B3	.8	1.1	1.4	SS C4	.7	1.0	1.3	TW I4	.8	1.1	1.4
TW B4	.8	1.1	1.4	TW B4	.9	1.2	1.5	SS D1	.6	.8	1.1	TW I5	.8	1.1	1.3
TW C1	.8	1.2	1.6	TW C1	.9	1.3	1.6	SS D2	.5	.7	.9	TW J1	.8	1.0	1.3
TW C2	.8	1.2	1.5	TW C2	.9	1.2	1.6	SS D3	.7	.9	1.2	TW J2	.8	1.1	1.3
TW C3	.9	1.2	1.6	TW C3	1.0	1.3	1.7	SS D4	.5	.7	1.0	TW J3	.7	.9	1.2
TW C4	1.0	1.3	1.8	TW C4	1.0	1.4	1.8	SS E1	.7	.9	1.2	TW J4	.7	1.0	1.2
TW D1	.9	1.2	1.6	TW D1	.9	1.3	1.7	SS E2	.6	.8	1.0	DG A1	1.1	1.8	---
TW D2	.7	.9	1.2	TW D2	.7	1.0	1.3	SS E3	.8	1.1	1.4	DG A2	1.0	1.4	2.2
TW D3	.6	.9	1.1	TW D3	.7	.9	1.2	SS E4	.6	.8	1.1	DG A3	.9	1.1	1.6
DG B1	.8	1.0	1.5	TW D4	.7	.9	1.1					DG B1	1.1	1.4	3.7
DG B2	.8	.9	1.2	TW D5	.7	.9	1.1	FORT WORTH, TX				DG B2	1.0	1.2	1.7
DG B3	.7	.9	1.1	TW E1	.6	.8	1.1		SOLAR SAVINGS			DG B3	.9	1.1	1.4
DG C1	.7	.9	1.2	TW E2	.6	.8	1.1		10%	50%	80%	DG C1	1.0	1.2	1.9
DG C2	.7	.9	1.1	TW E3	.6	.7	.9	WW A1	1.1	1.6	2.1	DG C2	.9	1.1	1.5
DG C3	.7	.8	1.0	TW E4	.6	.8	1.0	WW A2	.9	1.2	1.6	DG C3	.8	1.0	1.2
SS A1	.6	.9	1.2	TW F1	.9	1.3	1.7	WW A3	.8	1.2	1.5	SS A1	.8	1.2	1.6
SS A2	.5	.8	1.0	TW F2	.8	1.1	1.5	WW A4	.8	1.1	1.4	SS A2	.7	1.0	1.3
SS A3	.7	1.0	1.3	TW F3	.8	1.1	1.4	WW A5	.8	1.1	1.4	SS A3	.9	1.3	1.7
SS A4	.5	.8	1.0	TW F4	.9	1.1	1.4	WW A6	.8	1.0	1.3	SS A4	.7	1.0	1.3
SS A5	.7	1.0	1.3	TW G1	1.0	1.3	1.7	WW B1	1.0	1.5	2.0	SS A5	.9	1.3	1.8
SS A6	.5	.8	1.0	TW G2	.9	1.2	1.5	WW B2	.8	1.1	1.4	SS A6	.7	1.0	1.3
SS A7	.7	1.1	1.5	TW G3	.9	1.2	1.6	WW B3	.7	1.0	1.2	SS A7	.9	1.5	2.0
SS A8	.6	.8	1.0	TW G4	1.0	1.3	1.7	WW B4	.7	.9	1.2	SS A8	.7	1.0	1.3
SS B1	.7	1.0	1.3	TW H1	1.0	1.4	1.7	WW B5	.7	.9	1.2	SS B1	.9	1.3	1.8
SS B2	.6	.8	1.1	TW H2	1.0	1.4	1.8	WW C1	.7	.9	1.2	SS B2	.8	1.1	1.4
SS B3	.7	1.1	1.4	TW H3	1.1	1.5	1.9	WW C2	.7	.9	1.2	SS B3	.9	1.4	1.9
SS B4	.6	.8	1.1	TW H4	1.3	1.7	2.2	WW C3	.7	.8	1.1	SS B4	.8	1.1	1.4
SS B5	.8	1.2	1.5	TW I1	1.0	1.4	1.8	WW C4	.7	.9	1.1	SS B5	1.0	1.5	2.1
SS B6	.6	.8	1.1	TW I2	.8	1.0	1.3	TW A1	1.1	1.6	2.2	SS B6	.8	1.1	1.4
SS B7	.8	1.3	1.7	TW I3	.7	1.0	1.2	TW A2	1.0	1.4	1.8	SS B7	1.1	1.7	2.3
SS B8	.6	.9	1.1	TW I4	.7	.9	1.1	TW A3	.9	1.3	1.7	SS B8	.8	1.1	1.4
SS C1	.7	1.0	1.3	TW I5	.7	.9	1.1	TW A4	.9	1.3	1.6	SS C1	.9	1.3	1.7
SS C2	.6	.9	1.1	TW J1	.7	.9	1.1	TW B1	1.1	1.6	2.1	SS C2	.8	1.1	1.4
SS C3	.8	1.2	1.5	TW J2	.7	.9	1.1	TW B2	1.0	1.4	1.8	SS C3	1.1	1.5	2.0
SS C4	.7	.9	1.2	TW J3	.6	.8	1.0	TW B3	1.0	1.4	1.8	SS C4	.9	1.2	1.5
SS D1	.5	.8	1.0	TW J4	.6	.8	1.0	TW B4	1.0	1.4	1.8	SS D1	.7	1.0	1.3
SS D2	.5	.7	.9	DG A1	.9	1.3	2.3	TW C1	1.1	1.6	2.0	SS D2	.6	.8	1.1
SS D3	.6	.9	1.1	DG A2	.9	1.1	1.6	TW C2	1.1	1.5	2.0	SS D3	.8	1.1	1.5
SS D4	.5	.7	.9	DG A3	.8	.9	1.3	TW C3	1.1	1.6	2.0	SS D4	.6	.9	1.1
SS E1	.6	.8	1.1	DG B1	.9	1.1	1.6	TW C4	1.2	1.7	2.3	SS E1	.8	1.1	1.5
SS E2	.5	.7	.9	DG B2	.8	1.0	1.3	TW D1	1.1	1.6	2.2	SS E2	.7	.9	1.2
SS E3	.7	1.0	1.3	DG B3	.8	.9	1.1	TW D2	.9	1.2	1.5	SS E3	.9	1.3	1.8
SS E4	.6	.8	1.0	DG C1	.8	.9	1.3	TW D3	.8	1.1	1.4	SS E4	.7	1.0	1.3

HOUSTON, TX

	SOLAR SAVINGS		
	10%	50%	80%
WW A1	.9	1.3	1.8
WW A2	.8	1.1	1.4
WW A3	.7	1.0	1.3
WW A4	.7	.9	1.2
WW A5	.7	.9	1.2
WW A6	.7	.9	1.2
WW B1	.8	1.2	1.6
WW B2	.7	.9	1.2
WW B3	.6	.8	1.1
WW B4	.6	.8	1.0
WW B5	.6	.8	1.1
WW C1	.6	.8	1.0
WW C2	.6	.8	1.1
WW C3	.6	.7	.9
WW C4	.6	.8	1.0
TW A1	.9	1.4	1.8
TW A2	.8	1.2	1.5
TW A3	.8	1.1	1.4
TW A4	.8	1.1	1.4
TW B1	.9	1.3	1.8
TW B2	.8	1.2	1.5
TW B3	.8	1.2	1.5
TW B4	.9	1.2	1.6
TW C1	.9	1.3	1.7
TW C2	.9	1.3	1.7
TW C3	1.0	1.3	1.8
TW C4	1.0	1.5	1.9
TW D1	1.0	1.4	1.8
TW D2	.7	1.0	1.3
TW D3	.7	.9	1.2
TW D4	.7	.9	1.1
TW D5	.7	.9	1.1
TW E1	.7	.9	1.1
TW E2	.7	.9	1.1
TW E3	.6	.8	1.0
TW E4	.6	.8	1.0
TW F1	.9	1.4	1.8
TW F2	.9	1.2	1.6
TW F3	.8	1.2	1.5
TW F4	.9	1.2	1.5
TW G1	1.0	1.4	1.8
TW G2	.9	1.3	1.6
TW G3	.9	1.3	1.7
TW G4	1.0	1.4	1.8
TW H1	1.0	1.4	1.8
TW H2	1.1	1.4	1.9
TW H3	1.2	1.6	2.0
TW H4	1.3	1.8	2.3
TW I1	1.0	1.4	1.9
TW I2	.8	1.1	1.4
TW I3	.8	1.0	1.3
TW I4	.7	.9	1.2
TW I5	.7	.9	1.2
TW J1	.7	.9	1.2
TW J2	.7	.9	1.2
TW J3	.6	.8	1.0
TW J4	.7	.8	1.1
DG A1	.9	1.4	3.0
DG A2	.9	1.2	1.7
DG A3	.8	1.0	1.3
DG B1	.9	1.1	1.9
DG B2	.9	1.0	1.4
DG B3	.8	.9	1.2
DG C1	.8	1.0	1.4

HOUSTON, TX

	SOLAR SAVINGS		
	10%	50%	80%
DG C2	.8	.9	1.2
DG C3	.7	.9	1.1
SS A1	.7	1.0	1.3
SS A2	.6	.8	1.1
SS A3	.7	1.1	1.4
SS A4	.6	.8	1.1
SS A5	.7	1.1	1.5
SS A6	.6	.8	1.1
SS A7	.8	1.2	1.7
SS A8	.6	.8	1.1
SS B1	.8	1.1	1.5
SS B2	.7	.9	1.2
SS B3	.8	1.2	1.5
SS B4	.7	.9	1.2
SS B5	.8	1.3	1.7
SS B6	.7	.9	1.2
SS B7	.9	1.4	1.9
SS B8	.7	.9	1.2
SS C1	.8	1.1	1.4
SS C2	.7	.9	1.2
SS C3	.9	1.3	1.7
SS C4	.8	1.0	1.3
SS D1	.6	.8	1.1
SS D2	.5	.7	.9
SS D3	.6	.9	1.3
SS D4	.5	.7	1.0
SS E1	.7	.9	1.2
SS E2	.6	.8	1.0
SS E3	.8	1.1	1.5
SS E4	.6	.9	1.1

LAREDO, TX

	SOLAR SAVINGS		
	10%	50%	80%
WW A1	.7	1.0	1.3
WW A2	.6	.8	1.0
WW A3	.5	.7	.9
WW A4	.5	.7	.9
WW A5	.5	.7	.9
WW A6	.5	.7	.9
WW B1	.6	.9	1.1
WW B2	.5	.7	.9
WW B3	.5	.6	.8
WW B4	.5	.6	.8
WW B5	.5	.6	.8
WW C1	.5	.6	.8
WW C2	.5	.6	.8
WW C3	.4	.6	.7
WW C4	.5	.6	.8
TW A1	.7	1.0	1.3
TW A2	.6	.8	1.1
TW A3	.6	.8	1.0
TW A4	.6	.8	1.0
TW B1	.7	1.0	1.3
TW B2	.6	.9	1.1
TW B3	.6	.9	1.1
TW B4	.7	.9	1.2
TW C1	.7	1.0	1.3
TW C2	.7	1.0	1.2
TW C3	.7	1.0	1.3
TW C4	.8	1.1	1.4
TW D1	.7	1.0	1.3
TW D2	.6	.8	1.0
TW D3	.5	.7	.9

LAREDO, TX

	SOLAR SAVINGS		
	10%	50%	80%
TW D4	.5	.7	.9
TW D5	.5	.7	.9
TW E1	.5	.7	.9
TW E2	.5	.7	.9
TW E3	.5	.6	.8
TW E4	.5	.6	.8
TW F1	.7	1.0	1.3
TW F2	.6	.9	1.1
TW F3	.6	.8	1.1
TW F4	.7	.9	1.1
TW G1	.7	1.0	1.3
TW G2	.7	.9	1.2
TW G3	.7	.9	1.2
TW G4	.8	1.0	1.3
TW H1	.8	1.0	1.4
TW H2	.8	1.1	1.4
TW H3	.9	1.2	1.5
TW H4	1.0	1.3	1.7
TW I1	.8	1.0	1.3
TW I2	.6	.8	1.0
TW I3	.6	.8	1.0
TW I4	.5	.7	.9
TW I5	.5	.7	.9
TW J1	.5	.7	.9
TW J2	.5	.7	.9
TW J3	.5	.6	.8
TW J4	.5	.6	.8
DG A1	.6	.9	1.3
DG A2	.7	.8	1.1
DG A3	.6	.7	1.0
DG B1	.7	.8	1.1
DG B2	.7	.8	1.0
DG B3	.6	.7	.9
DG C1	.6	.7	.9
DG C2	.6	.7	.9
DG C3	.6	.7	.8
SS A1	.5	.7	1.0
SS A2	.5	.6	.8
SS A3	.5	.8	1.0
SS A4	.5	.6	.8
SS A5	.5	.8	1.1
SS A6	.5	.6	.8
SS A7	.6	.9	1.2
SS A8	.5	.6	.8
SS B1	.6	.8	1.1
SS B2	.5	.7	.9
SS B3	.6	.8	1.1
SS B4	.5	.7	.9
SS B5	.6	.9	1.2
SS B6	.5	.7	.9
SS B7	.7	1.0	1.3
SS B8	.5	.7	.9
SS C1	.6	.8	1.0
SS C2	.5	.7	.9
SS C3	.7	1.0	1.3
SS C4	.6	.8	1.0
SS D1	.5	.6	.8
SS D2	.4	.5	.7
SS D3	.5	.7	.9
SS D4	.4	.6	.7
SS E1	.5	.7	.9
SS E2	.4	.6	.8
SS E3	.6	.8	1.1
SS E4	.5	.6	.8

LUBBOCK, TX

	SOLAR SAVINGS		
	10%	50%	80%
WW A1	1.0	1.5	2.0
WW A2	.9	1.2	1.5
WW A3	.8	1.1	1.4
WW A4	.8	1.1	1.4
WW A5	.8	1.0	1.3
WW A6	.8	1.0	1.3
WW B1	1.0	1.4	1.8
WW B2	.8	1.0	1.3
WW B3	.7	.9	1.2
WW B4	.7	.9	1.1
WW B5	.7	.9	1.1
WW C1	.7	.9	1.1
WW C2	.7	.9	1.2
WW C3	.6	.8	1.0
WW C4	.7	.9	1.1
TW A1	1.1	1.6	2.0
TW A2	1.0	1.3	1.7
TW A3	.9	1.2	1.6
TW A4	.9	1.2	1.5
TW B1	1.1	1.5	2.0
TW B2	1.0	1.3	1.7
TW B3	1.0	1.3	1.7
TW B4	1.0	1.4	1.7
TW C1	1.1	1.5	1.9
TW C2	1.1	1.4	1.8
TW C3	1.1	1.5	1.9
TW C4	1.2	1.7	2.1
TW D1	1.1	1.6	2.0
TW D2	.9	1.1	1.4
TW D3	.8	1.0	1.3
TW D4	.8	1.0	1.2
TW D5	.8	1.0	1.2
TW E1	.7	1.0	1.2
TW E2	.7	1.0	1.2
TW E3	.7	.9	1.1
TW E4	.7	.9	1.1
TW F1	1.1	1.5	2.0
TW F2	1.0	1.3	1.7
TW F3	1.0	1.3	1.7
TW F4	1.0	1.3	1.7
TW G1	1.1	1.5	2.0
TW G2	1.1	1.4	1.8
TW G3	1.1	1.4	1.8
TW G4	1.2	1.6	2.0
TW H1	1.2	1.6	2.0
TW H2	1.2	1.6	2.1
TW H3	1.3	1.7	2.2
TW H4	1.5	2.0	2.6
TW I1	1.2	1.6	2.1
TW I2	.9	1.2	1.5
TW I3	.9	1.1	1.4
TW I4	.8	1.0	1.3
TW I5	.8	1.0	1.3
TW J1	.8	1.0	1.3
TW J2	.8	1.0	1.3
TW J3	.7	.9	1.1
TW J4	.7	.9	1.2
DG A1	1.0	1.7	---
DG A2	1.0	1.3	2.0
DG A3	.9	1.1	1.5
DG B1	1.0	1.3	2.5
DG B2	1.0	1.2	1.6
DG B3	.9	1.0	1.3
DG C1	.9	1.1	1.6

LUBBOCK, TX				LUFKIN, TX				MIDLAND-ODESSA, TX				MIDLAND-ODESSA, TX			
	SOLAR SAVINGS				SOLAR SAVINGS				SOLAR SAVINGS				SOLAR SAVINGS		
	10%	50%	80%		10%	50%	80%		10%	50%	80%		10%	50%	80%
DG C2	.9	1.0	1.4	TW D4	.7	1.0	1.2	WW A1	.9	1.3	1.7	DG C2	.8	.9	1.2
DG C3	.8	.9	1.1	TW D5	.7	1.0	1.2	WW A2	.8	1.0	1.3	DG C3	.7	.8	1.0
SS A1	.8	1.2	1.5	TW E1	.7	.9	1.2	WW A3	.7	1.0	1.2	SS A1	.7	1.0	1.3
SS A2	.7	.9	1.2	TW E2	.7	.9	1.2	WW A4	.7	.9	1.2	SS A2	.6	.8	1.1
SS A3	.8	1.2	1.6	TW E3	.6	.8	1.0	WW A5	.7	.9	1.1	SS A3	.7	1.1	1.4
SS A4	.7	1.0	1.2	TW E4	.7	.9	1.1	WW A6	.7	.9	1.1	SS A4	.6	.8	1.1
SS A5	.8	1.3	1.7	TW F1	1.0	1.5	1.9	WW B1	.9	1.2	1.6	SS A5	.8	1.1	1.5
SS A6	.7	.9	1.2	TW F2	.9	1.3	1.7	WW B2	.7	.9	1.2	SS A6	.6	.8	1.1
SS A7	.9	1.4	1.9	TW F3	.9	1.2	1.6	WW B3	.6	.8	1.0	SS A7	.8	1.2	1.7
SS A8	.7	1.0	1.3	TW F4	.9	1.3	1.6	WW B4	.6	.8	1.0	SS A8	.6	.9	1.1
SS B1	.9	1.3	1.7	TW G1	1.0	1.5	1.9	WW B5	.6	.8	1.0	SS B1	.8	1.1	1.5
SS B2	.8	1.0	1.3	TW G2	1.0	1.4	1.8	WW C1	.6	.8	1.0	SS B2	.7	.9	1.2
SS B3	.9	1.4	1.8	TW G3	1.0	1.4	1.8	WW C2	.6	.8	1.0	SS B3	.8	1.2	1.5
SS B4	.8	1.0	1.4	TW G4	1.1	1.5	1.9	WW C3	.6	.7	.9	SS B4	.7	.9	1.2
SS B5	1.0	1.5	2.0	TW H1	1.1	1.5	2.0	WW C4	.6	.8	1.0	SS B5	.9	1.3	1.7
SS B6	.8	1.0	1.3	TW H2	1.1	1.6	2.0	TW A1	1.0	1.3	1.8	SS B6	.7	.9	1.2
SS B7	1.1	1.6	2.2	TW H3	1.2	1.7	2.2	TW A2	.8	1.1	1.5	SS B7	1.0	1.4	1.9
SS B8	.8	1.1	1.4	TW H4	1.4	1.9	2.5	TW A3	.8	1.1	1.4	SS B8	.7	1.0	1.3
SS C1	.9	1.2	1.6	TW I1	1.1	1.6	2.1	TW A4	.8	1.1	1.3	SS C1	.8	1.1	1.4
SS C2	.8	1.0	1.3	TW I2	.8	1.1	1.5	TW B1	.9	1.3	1.7	SS C2	.7	.9	1.2
SS C3	1.1	1.5	1.9	TW I3	.8	1.1	1.4	TW B2	.9	1.2	1.5	SS C3	.9	1.3	1.7
SS C4	.9	1.2	1.5	TW I4	.8	1.0	1.3	TW B3	.9	1.1	1.5	SS C4	.8	1.0	1.3
SS D1	.7	1.0	1.3	TW I5	.8	1.0	1.3	TW B4	.9	1.2	1.5	SS D1	.6	.9	1.1
SS D2	.6	.8	1.0	TW J1	.7	1.0	1.2	TW C1	1.0	1.3	1.7	SS D2	.5	.7	.9
SS D3	.8	1.1	1.4	TW J2	.7	1.0	1.3	TW C2	.9	1.3	1.6	SS D3	.7	1.0	1.3
SS D4	.6	.8	1.1	TW J3	.7	.9	1.1	TW C3	1.0	1.3	1.7	SS D4	.6	.7	1.0
SS E1	.8	1.1	1.4	TW J4	.7	.9	1.1	TW C4	1.1	1.5	1.9	SS E1	.7	.9	1.2
SS E2	.7	.9	1.2	DG A1	1.0	1.5	---	TW D1	1.0	1.3	1.7	SS E2	.6	.8	1.0
SS E3	.9	1.3	1.7	DG A2	.9	1.3	1.9	TW D2	.8	1.0	1.3	SS E3	.8	1.1	1.5
SS E4	.7	1.0	1.3	DG A3	.8	1.0	1.4	TW D3	.7	.9	1.2	SS E4	.6	.9	1.1
				DG B1	.9	1.3	2.3	TW D4	.7	.9	1.1				
LUFKIN, TX				DG B2	.9	1.1	1.6	TW D5	.7	.9	1.1	PORT ARTHUR, TX			
	SOLAR SAVINGS			DG B3	.8	1.0	1.3	TW E1	.7	.9	1.1		SOLAR SAVINGS		
	10%	50%	80%	DG C1	.9	1.1	1.6	TW E2	.7	.9	1.1		10%	50%	80%
WW A1	1.0	1.5	1.9	DG C2	.8	1.0	1.3	TW E3	.6	.8	1.0	WW A1	.9	1.3	1.7
WW A2	.8	1.1	1.5	DG C3	.8	.9	1.1	TW E4	.6	.8	1.0	WW A2	.7	1.0	1.4
WW A3	.8	1.1	1.4	SS A1	.7	1.1	1.4	TW F1	1.0	1.3	1.7	WW A3	.7	1.0	1.3
WW A4	.8	1.0	1.3	SS A2	.6	.9	1.2	TW F2	.9	1.2	1.5	WW A4	.7	.9	1.2
WW A5	.7	1.0	1.3	SS A3	.8	1.2	1.5	TW F3	.9	1.1	1.4	WW A5	.7	.9	1.2
WW A6	.7	1.0	1.2	SS A4	.6	.9	1.2	TW F4	.9	1.2	1.5	WW A6	.7	.9	1.1
WW B1	.9	1.3	1.8	SS A5	.8	1.2	1.6	TW G1	1.0	1.3	1.7	WW B1	.8	1.2	1.6
WW B2	.7	1.0	1.3	SS A6	.6	.9	1.2	TW G2	.9	1.2	1.6	WW B2	.7	.9	1.2
WW B3	.7	.9	1.1	SS A7	.8	1.3	1.8	TW G3	.9	1.3	1.6	WW B3	.6	.8	1.0
WW B4	.7	.9	1.1	SS A8	.6	.9	1.2	TW G4	1.0	1.4	1.7	WW B4	.6	.8	1.0
WW B5	.7	.9	1.1	SS B1	.8	1.2	1.6	TW H1	1.0	1.4	1.8	WW B5	.6	.8	1.0
WW C1	.6	.9	1.1	SS B2	.7	1.0	1.3	TW H2	1.1	1.4	1.8	WW C1	.6	.8	1.0
WW C2	.7	.9	1.1	SS B3	.9	1.3	1.7	TW H3	1.2	1.5	1.9	WW C2	.6	.8	1.0
WW C3	.6	.8	1.0	SS B4	.7	1.0	1.3	TW H4	1.3	1.8	2.3	WW C3	.6	.7	.9
WW C4	.6	.8	1.0	SS B5	.9	1.4	1.9	TW I1	1.0	1.4	1.8	WW C4	.6	.8	1.0
TW A1	1.0	1.5	2.0	SS B6	.7	1.0	1.3	TW I2	.8	1.1	1.3	TW A1	.9	1.3	1.8
TW A2	.9	1.3	1.6	SS B7	1.0	1.5	2.0	TW I3	.8	1.0	1.3	TW A2	.8	1.1	1.5
TW A3	.8	1.2	1.5	SS B8	.7	1.0	1.3	TW I4	.7	.9	1.2	TW A3	.8	1.1	1.4
TW A4	.8	1.2	1.5	SS C1	.8	1.2	1.5	TW I5	.7	.9	1.1	TW A4	.8	1.1	1.4
TW B1	1.0	1.4	1.9	SS C2	.7	1.0	1.3	TW J1	.7	.9	1.1	TW B1	.9	1.3	1.7
TW B2	.9	1.3	1.7	SS C3	1.0	1.4	1.9	TW J2	.7	.9	1.1	TW B2	.8	1.2	1.5
TW B3	.9	1.3	1.6	SS C4	.8	1.1	1.4	TW J3	.6	.8	1.0	TW B3	.8	1.1	1.5
TW B4	.9	1.3	1.7	SS D1	.6	.9	1.2	TW J4	.6	.8	1.0	TW B4	.9	1.2	1.5
TW C1	1.0	1.4	1.9	SS D2	.6	.8	1.0	DG A1	.9	1.3	2.5	TW C1	.9	1.3	1.7
TW C2	1.0	1.4	1.8	SS D3	.7	1.0	1.4	DG A2	.9	1.1	1.6	TW C2	.9	1.3	1.6
TW C3	1.0	1.4	1.9	SS D4	.6	.8	1.0	DG A3	.8	1.0	1.3	TW C3	.9	1.3	1.7
TW C4	1.1	1.6	2.1	SS E1	.7	1.0	1.3	DG B1	.9	1.1	1.7	TW C4	1.0	1.5	1.9
TW D1	1.0	1.5	2.0	SS E2	.6	.9	1.1	DG B2	.9	1.0	1.4	TW D1	.9	1.3	1.7
TW D2	.8	1.1	1.4	SS E3	.8	1.2	1.6	DG B3	.8	.9	1.1	TW D2	.7	1.0	1.3
TW D3	.7	1.0	1.3	SS E4	.7	.9	1.2	DG C1	.8	1.0	1.3	TW D3	.7	.9	1.2

Conservation Factor Tables / 375

PORT ARTHUR, TX
SOLAR SAVINGS

	10%	50%	80%
WW A1	.9	1.3	1.7
WW A2	.7	1.0	1.4
WW A3	.7	1.0	1.3
WW A4	.7	.9	1.2
WW A5	.7	.9	1.2
WW A6	.7	.9	1.1
WW B1	.8	1.2	1.6
WW B2	.7	.9	1.2
WW B3	.6	.8	1.0
WW B4	.6	.8	1.0
WW B5	.6	.8	1.0
WW C1	.6	.8	1.0
WW C2	.6	.8	1.0
WW C3	.6	.7	.9
WW C4	.6	.8	1.0
TW A1	.9	1.3	1.8
TW A2	.8	1.1	1.5
TW A3	.8	1.1	1.4
TW A4	.8	1.1	1.4
TW B1	.9	1.3	1.7
TW B2	.8	1.2	1.5
TW B3	.8	1.1	1.5
TW B4	.9	1.2	1.5
TW C1	.9	1.3	1.7
TW C2	.9	1.3	1.6
TW C3	.9	1.3	1.7
TW C4	1.0	1.5	1.9
TW D1	.9	1.3	1.7
TW D2	.7	1.0	1.3
TW D3	.7	.9	1.2
DG B1	.9	1.1	1.8
DG B2	.8	1.0	1.4
DG B3	.8	.9	1.2
DG C1	.8	1.0	1.3
DG C2	.8	.9	1.2
DG C3	.7	.8	1.0
SS A1	.7	1.0	1.3
SS A2	.6	.8	1.1
SS A3	.7	1.0	1.4
SS A4	.6	.8	1.1
SS A5	.7	1.1	1.4
SS A6	.6	.8	1.1
SS A7	.8	1.2	1.6
SS A8	.6	.8	1.1
SS B1	.7	1.1	1.4
SS B2	.6	.9	1.2
SS B3	.8	1.1	1.5
SS B4	.6	.9	1.2
SS B5	.8	1.2	1.7
SS B6	.6	.9	1.2
SS B7	.9	1.3	1.8
SS B8	.7	.9	1.2
SS C1	.8	1.1	1.4
SS C2	.7	.9	1.2
SS C3	.9	1.3	1.7
SS C4	.7	1.0	1.3
SS D1	.6	.8	1.1
SS D2	.5	.7	.9
SS D3	.6	.9	1.2
SS D4	.5	.7	.9
SS E1	.6	.9	1.2
SS E2	.6	.8	1.0
SS E3	.7	1.1	1.4
SS E4	.6	.8	1.1

SAN ANGELO, TX
SOLAR SAVINGS

	10%	50%	80%
WW A1	1.0	1.4	1.8
WW A2	.8	1.1	1.4
WW A3	.7	1.0	1.3
WW A4	.7	1.0	1.2
WW A5	.7	.9	1.2
WW A6	.7	.9	1.2
WW B1	.9	1.2	1.6
WW B2	.7	.9	1.2
WW B3	.6	.8	1.1
WW B4	.6	.8	1.0
WW B5	.7	.8	1.1
WW C1	.6	.8	1.0
WW C2	.6	.8	1.1
WW C3	.6	.7	.9
WW C4	.6	.8	1.0
TW A1	1.0	1.4	1.8
TW A2	.9	1.2	1.5
TW A3	.8	1.1	1.4
TW A4	.8	1.1	1.4
TW B1	1.0	1.4	1.8
TW B2	.9	1.2	1.5
TW B3	.9	1.2	1.5
TW B4	.9	1.2	1.6
TW C1	1.0	1.3	1.7
TW C2	1.0	1.3	1.7
TW C3	1.0	1.4	1.8
TW C4	1.1	1.5	1.9
TW D1	1.0	1.4	1.8
TW D2	.8	1.0	1.3
TW D3	.7	1.0	1.2
TW D4	.7	.9	1.1
TW D5	.7	.9	1.2
TW E1	.7	.9	1.1
TW E2	.7	.9	1.1
TW E3	.6	.8	1.0
TW E4	.6	.8	1.0
TW F1	1.0	1.4	1.8
TW F2	.9	1.2	1.6
TW F3	.9	1.2	1.5
TW F4	.9	1.2	1.5
TW G1	1.0	1.4	1.8
TW G2	.9	1.3	1.6
TW G3	1.0	1.3	1.7
TW G4	1.1	1.4	1.8
TW H1	1.1	1.4	1.8
TW H2	1.1	1.5	1.9
TW H3	1.2	1.6	2.0
TW H4	1.4	1.8	2.3
TW I1	1.1	1.5	1.9
TW I2	.8	1.1	1.4
TW I3	.8	1.0	1.3
TW I4	.7	.9	1.2
TW I5	.7	.9	1.2
TW J1	.7	.9	1.2
TW J2	.7	.9	1.2
TW J3	.6	.8	1.0
TW J4	.7	.9	1.1
DG A1	.9	1.4	3.0
DG A2	.9	1.2	1.7
DG A3	8	1.0	1.3
DG B1	.9	1.1	1.9
DG B2	.9	1.1	1.4
DG B3	.8	.9	1.2
DG C1	.8	1.0	1.4

SAN ANGELO, TX
SOLAR SAVINGS

	10%	50%	80%
DG C2	.8	1.0	1.2
DG C3	.8	.9	1.1
SS A1	.7	1.0	1.4
SS A2	.6	.9	1.1
SS A3	.8	1.1	1.5
SS A4	.6	.9	1.1
SS A5	.8	1.1	1.5
SS A6	.6	.9	1.1
SS A7	.8	1.3	1.7
SS A8	.6	.9	1.1
SS B1	.8	1.2	1.5
SS B2	.7	.9	1.2
SS B3	.8	1.2	1.6
SS B4	.7	.9	1.2
SS B5	.9	1.3	1.7
SS B6	.7	.9	1.2
SS B7	1.0	1.4	1.9
SS B8	.7	1.0	1.3
SS C1	.8	1.1	1.4
SS C2	.7	1.0	1.2
SS C3	1.0	1.3	1.7
SS C4	.8	1.1	1.3
SS D1	.6	.9	1.1
SS D2	.5	.7	1.0
SS D3	.7	1.0	1.3
SS D4	.6	.8	1.0
SS E1	.7	1.0	1.3
SS E2	.6	.8	1.0
SS E3	.8	1.2	1.5
SS E4	.6	.9	1.1

SAN ANTONIO, TX
SOLAR SAVINGS

	10%	50%	80%
WW A1	.9	1.3	1.7
WW A2	.7	1.0	1.3
WW A3	.7	.9	1.2
WW A4	.7	.9	1.2
WW A5	.7	.9	1.1
WW A6	.6	.9	1.1
WW B1	.8	1.2	1.5
WW B2	.7	.9	1.1
WW B3	.6	.8	1.0
WW B4	.6	.8	1.0
WW B5	.6	.8	1.0
WW C1	.6	.8	1.0
WW C2	.6	.8	1.0
WW C3	.6	.7	.9
WW C4	.6	.7	.9
TW A1	.9	1.3	1.7
TW A2	.8	1.1	1.4
TW A3	.8	1.0	1.3
TW A4	.8	1.0	1.3
TW B1	.9	1.3	1.7
TW B2	.8	1.1	1.5
TW B3	.8	1.1	1.4
TW B4	.8	1.2	1.5
TW C1	.9	1.3	1.6
TW C2	.9	1.2	1.6
TW C3	.9	1.3	1.7
TW C4	1.0	1.4	1.8
TW D1	.9	1.3	1.7
TW D2	.7	1.0	1.2
TW D3	.7	.9	1.2

SAN ANTONIO, TX
SOLAR SAVINGS

	10%	50%	80%
TW D4	.6	.9	1.1
TW D5	.7	.9	1.1
TW E1	.6	.8	1.1
TW E2	.6	.8	1.1
TW E3	.6	.7	.9
TW E4	.6	.8	1.0
TW F1	.9	1.3	1.7
TW F2	.8	1.1	1.5
TW F3	.8	1.1	1.4
TW F4	.8	1.1	1.4
TW G1	.9	1.3	1.7
TW G2	.9	1.2	1.5
TW G3	.9	1.2	1.6
TW G4	1.0	1.3	1.7
TW H1	1.0	1.3	1.7
TW H2	1.0	1.4	1.8
TW H3	1.1	1.5	1.9
TW H4	1.3	1.7	2.2
TW I1	1.0	1.4	1.8
TW I2	.8	1.0	1.3
TW I3	.7	1.0	1.2
TW I4	.7	.9	1.1
TW I5	.7	.9	1.1
TW J1	.7	.9	1.1
TW J2	.7	.9	1.1
TW J3	.6	.8	1.0
TW J4	.6	.8	1.0
DG A1	.8	1.2	2.3
DG A2	.8	1.1	1.6
DG A3	.8	.9	1.3
DG B1	.8	1.1	1.6
DG B2	.8	1.0	1.3
DG B3	.8	.9	1.1
DG C1	.8	.9	1.3
DG C2	.8	.9	1.1
DG C3	.7	.8	1.0
SS A1	.7	1.0	1.3
SS A2	.6	.8	1.0
SS A3	.7	1.0	1.3
SS A4	.6	.8	1.0
SS A5	.7	1.0	1.4
SS A6	.6	.8	1.0
SS A7	.7	1.2	1.6
SS A8	.6	.8	1.1
SS B1	.7	1.1	1.4
SS B2	.6	.9	1.1
SS B3	.8	1.1	1.5
SS B4	.6	.9	1.1
SS B5	.8	1.2	1.6
SS B6	.6	.9	1.1
SS B7	.9	1.3	1.7
SS B8	.6	.9	1.2
SS C1	.7	1.0	1.3
SS C2	.7	.9	1.1
SS C3	.9	1.2	1.6
SS C4	.7	1.0	1.3
SS D1	.6	.8	1.1
SS D2	.5	.7	.9
SS D3	.6	.9	1.2
SS D4	.5	.7	.9
SS E1	.6	.9	1.2
SS E2	.6	.8	1.0
SS E3	.7	1.1	1.4
SS E4	.6	.8	1.1

SHERMAN, TX

	SOLAR SAVINGS		
	10%	50%	80%
WW A1	1.2	1.8	2.4
WW A2	1.0	1.4	1.8
WW A3	.9	1.3	1.7
WW A4	.9	1.2	1.6
WW A5	.9	1.2	1.5
WW A6	.9	1.2	1.5
WW B1	1.1	1.7	2.3
WW B2	.9	1.2	1.5
WW B3	.8	1.0	1.3
WW B4	.8	1.0	1.3
WW B5	.8	1.0	1.3
WW C1	.8	1.0	1.3
WW C2	.8	1.0	1.3
WW C3	.7	.9	1.1
WW C4	.7	1.0	1.2
TW A1	1.2	1.8	2.4
TW A2	1.1	1.5	2.0
TW A3	1.0	1.4	1.8
TW A4	1.0	1.4	1.8
TW B1	1.2	1.8	2.3
TW B2	1.1	1.5	2.0
TW B3	1.1	1.5	1.9
TW B4	1.1	1.6	2.0
TW C1	1.2	1.7	2.3
TW C2	1.2	1.7	2.2
TW C3	1.2	1.7	2.2
TW C4	1.3	1.9	2.5
TW D1	1.3	1.9	2.5
TW D2	.9	1.3	1.7
TW D3	.9	1.2	1.5
TW D4	.8	1.1	1.4
TW D5	.8	1.1	1.4
TW E1	.8	1.1	1.4
TW E2	.8	1.1	1.4
TW E3	.7	1.0	1.2
TW E4	.8	1.0	1.3
TW F1	1.2	1.8	2.4
TW F2	1.1	1.6	2.0
TW F3	1.1	1.5	1.9
TW F4	1.1	1.5	2.0
TW G1	1.2	1.8	2.4
TW G2	1.2	1.6	2.1
TW G3	1.2	1.7	2.1
TW G4	1.3	1.8	2.3
TW H1	1.3	1.8	2.4
TW H2	1.4	1.9	2.4
TW H3	1.5	2.0	2.6
TW H4	1.7	2.3	3.0
TW I1	1.4	2.0	2.6
TW I2	1.0	1.4	1.7
TW I3	.9	1.3	1.6
TW I4	.9	1.2	1.5
TW I5	.9	1.1	1.4
TW J1	.9	1.1	1.5
TW J2	.9	1.1	1.5
TW J3	.8	1.0	1.3
TW J4	.8	1.0	1.3
DG A1	1.2	2.4	---
DG A2	1.1	1.6	2.7
DG A3	1.0	1.2	1.7
DG B1	1.2	1.7	---
DG B2	1.1	1.4	2.0
DG B3	1.0	1.2	1.5
DG C1	1.0	1.4	2.5

SHERMAN, TX

	SOLAR SAVINGS		
	10%	50%	80%
DG C2	1.0	1.2	1.6
DG C3	.9	1.1	1.3
SS A1	.9	1.3	1.8
SS A2	.7	1.1	1.4
SS A3	.9	1.4	1.9
SS A4	.8	1.1	1.4
SS A5	.9	1.5	2.0
SS A6	.7	1.1	1.4
SS A7	1.0	1.7	2.3
SS A8	.8	1.1	1.4
SS B1	1.0	1.5	2.0
SS B2	.8	1.2	1.5
SS B3	1.0	1.6	2.1
SS B4	.8	1.2	1.5
SS B5	1.1	1.7	2.3
SS B6	.8	1.2	1.5
SS B7	1.2	1.9	2.6
SS B8	.9	1.2	1.6
SS C1	1.0	1.4	1.8
SS C2	.9	1.2	1.5
SS C3	1.2	1.7	2.3
SS C4	1.0	1.3	1.7
SS D1	.8	1.1	1.5
SS D2	.7	.9	1.2
SS D3	.8	1.3	1.7
SS D4	.7	.9	1.2
SS E1	.9	1.2	1.7
SS E2	.7	1.0	1.3
SS E3	1.0	1.5	2.0
SS E4	.8	1.1	1.4

WACO, TX

	SOLAR SAVINGS		
	10%	50%	80%
WW A1	1.0	1.5	2.0
WW A2	.8	1.2	1.5
WW A3	.8	1.1	1.4
WW A4	.8	1.0	1.3
WW A5	.7	1.0	1.3
WW A6	.7	1.0	1.2
WW B1	.9	1.4	1.8
WW B2	.8	1.0	1.3
WW B3	.7	.9	1.1
WW B4	.7	.9	1.1
WW B5	.7	.9	1.1
WW C1	.7	.9	1.1
WW C2	.7	.9	1.1
WW C3	.6	.8	1.0
WW C4	.6	.8	1.1
TW A1	1.0	1.5	2.0
TW A2	.9	1.3	1.6
TW A3	.9	1.2	1.5
TW A4	.9	1.2	1.5
TW B1	1.0	1.5	1.9
TW B2	.9	1.3	1.7
TW B3	.9	1.3	1.6
TW B4	1.0	1.3	1.7
TW C1	1.0	1.4	1.9
TW C2	1.0	1.4	1.8
TW C3	1.0	1.5	1.9
TW C4	1.1	1.6	2.1
TW D1	1.1	1.5	2.0
TW D2	.8	1.1	1.4
TW D3	.8	1.0	1.3

WACO, TX

	SOLAR SAVINGS		
	10%	50%	80%
TW D4	.7	1.0	1.2
TW D5	.7	1.0	1.2
TW E1	.7	.9	1.2
TW E2	.7	1.0	1.2
TW E3	.6	.8	1.1
TW E4	.7	.9	1.1
TW F1	1.0	1.5	2.0
TW F2	.9	1.3	1.7
TW F3	.9	1.3	1.6
TW F4	.9	1.3	1.6
TW G1	1.0	1.5	1.9
TW G2	1.0	1.4	1.8
TW G3	1.0	1.4	1.8
TW G4	1.1	1.5	1.9
TW H1	1.1	1.5	2.0
TW H2	1.2	1.6	2.0
TW H3	1.2	1.7	2.2
TW H4	1.4	2.0	2.5
TW I1	1.1	1.6	2.1
TW I2	.9	1.2	1.5
TW I3	.8	1.1	1.4
TW I4	.8	1.0	1.3
TW I5	.8	1.0	1.3
TW J1	.7	1.0	1.3
TW J2	.7	1.0	1.3
TW J3	.7	.9	1.1
TW J4	.7	.9	1.1
DG A1	1.0	1.6	---
DG A2	.9	1.3	1.9
DG A3	.9	1.1	1.4
DG B1	1.0	1.3	2.4
DG B2	.9	1.1	1.6
DG B3	.8	1.0	1.3
DG C1	.9	1.1	1.6
DG C2	.8	1.0	1.3
DG C3	.8	.9	1.1
SS A1	.8	1.1	1.5
SS A2	.6	.9	1.2
SS A3	.8	1.2	1.6
SS A4	.6	.9	1.2
SS A5	.8	1.2	1.6
SS A6	.6	.9	1.2
SS A7	.9	1.4	1.8
SS A8	.6	.9	1.2
SS B1	.8	1.2	1.6
SS B2	.7	1.0	1.3
SS B3	.9	1.3	1.7
SS B4	.7	1.0	1.3
SS B5	.9	1.4	1.9
SS B6	.7	1.0	1.3
SS B7	1.0	1.5	2.1
SS B8	.7	1.0	1.3
SS C1	.8	1.2	1.5
SS C2	.8	1.0	1.3
SS C3	1.0	1.4	1.9
SS C4	.8	1.1	1.4
SS D1	.7	.9	1.2
SS D2	.6	.8	1.0
SS D3	.7	1.0	1.4
SS D4	.6	.8	1.1
SS E1	.7	1.0	1.4
SS E2	.6	.9	1.1
SS E3	.8	1.2	1.6
SS E4	.7	.9	1.2

WICHITA FALLS, TX

	SOLAR SAVINGS		
	10%	50%	80%
WW A1	1.1	1.7	2.2
WW A2	.9	1.3	1.7
WW A3	.9	1.2	1.5
WW A4	.8	1.1	1.5
WW A5	.8	1.1	1.4
WW A6	.8	1.1	1.4
WW B1	1.1	1.5	2.0
WW B2	.8	1.1	1.4
WW B3	.7	1.0	1.3
WW B4	.7	1.0	1.2
WW B5	.7	1.0	1.2
WW C1	.7	1.0	1.2
WW C2	.7	1.0	1.2
WW C3	.7	.9	1.1
WW C4	.7	.9	1.1
TW A1	1.1	1.7	2.2
TW A2	1.0	1.4	1.8
TW A3	1.0	1.3	1.7
TW A4	.9	1.3	1.7
TW B1	1.1	1.6	2.1
TW B2	1.0	1.4	1.8
TW B3	1.0	1.4	1.8
TW B4	1.1	1.5	1.9
TW C1	1.1	1.6	2.1
TW C2	1.1	1.5	2.0
TW C3	1.2	1.6	2.1
TW C4	1.3	1.8	2.3
TW D1	1.2	1.7	2.2
TW D2	.9	1.2	1.6
TW D3	.8	1.1	1.4
TW D4	.8	1.0	1.3
TW D5	.8	1.0	1.3
TW E1	.8	1.0	1.3
TW E2	.8	1.0	1.3
TW E3	.7	.9	1.1
TW E4	.7	.9	1.2
TW F1	1.1	1.7	2.2
TW F2	1.0	1.4	1.9
TW F3	1.0	1.4	1.8
TW F4	1.0	1.4	1.8
TW G1	1.2	1.7	2.2
TW G2	1.1	1.5	2.0
TW G3	1.1	1.5	2.0
TW G4	1.2	1.7	2.1
TW H1	1.2	1.7	2.2
TW H2	1.3	1.7	2.2
TW H3	1.4	1.9	2.4
TW H4	1.6	2.1	2.8
TW I1	1.3	1.8	2.4
TW I2	.9	1.3	1.6
TW I3	.9	1.2	1.5
TW I4	.8	1.1	1.4
TW I5	.8	1.1	1.4
TW J1	.8	1.1	1.4
TW J2	.8	1.1	1.4
TW J3	.7	.9	1.2
TW J4	.8	1.0	1.2
DG A1	1.1	2.0	---
DG A2	1.0	1.4	2.3
DG A3	.9	1.1	1.6
DG B1	1.1	1.5	5.1
DG B2	1.0	1.3	1.8
DG B3	.9	1.1	1.4
DG C1	1.0	1.2	2.0

WICHITA FALLS, TX				BRYCE CANYON, UT				CEDAR CITY, UT				CEDAR CITY, UT			
	SOLAR SAVINGS				SOLAR SAVINGS				SOLAR SAVINGS				SOLAR SAVINGS		
	10%	50%	80%		10%	50%	80%		10%	50%	80%		10%	50%	80%
DG C2	.9	1.1	1.5	TW D4	1.0	1.3	1.7	WW A1	1.3	2.0	2.7	DG C2	1.1	1.3	1.8
DG C3	.9	1.0	1.2	TW D5	1.0	1.3	1.7	WW A2	1.1	1.5	2.0	DG C3	1.0	1.1	1.4
SS A1	.8	1.2	1.6	TW E1	1.0	1.3	1.7	WW A3	1.0	1.4	1.8	SS A1	1.0	1.5	2.0
SS A2	.7	1.0	1.3	TW E2	1.0	1.3	1.7	WW A4	1.0	1.3	1.7	SS A2	.8	1.1	1.5
SS A3	.9	1.3	1.8	TW E3	.9	1.1	1.4	WW A5	1.0	1.3	1.6	SS A3	1.0	1.6	2.2
SS A4	.7	1.0	1.3	TW E4	.9	1.2	1.5	WW A6	.9	1.3	1.6	SS A4	.8	1.2	1.6
SS A5	.9	1.4	1.8	TW F1	1.7	2.4	3.2	WW B1	1.3	1.9	2.6	SS A5	1.0	1.6	2.3
SS A6	.7	1.0	1.3	TW F2	1.4	2.0	2.6	WW B2	1.0	1.3	1.6	SS A6	.8	1.1	1.5
SS A7	1.0	1.6	2.1	TW F3	1.4	1.9	2.5	WW B3	.9	1.1	1.4	SS A7	1.1	1.9	2.7
SS A8	.7	1.0	1.4	TW F4	1.4	1.9	2.4	WW B4	.8	1.1	1.4	SS A8	.8	1.2	1.6
SS B1	1.0	1.4	1.8	TW G1	1.7	2.3	3.1	WW B5	.8	1.1	1.4	SS B1	1.1	1.6	2.2
SS B2	.8	1.1	1.4	TW G2	1.5	2.1	2.7	WW C1	.8	1.1	1.4	SS B2	.9	1.2	1.6
SS B3	1.0	1.5	1.9	TW G3	1.5	2.1	2.7	WW C2	.8	1.1	1.4	SS B3	1.1	1.8	2.4
SS B4	.8	1.1	1.4	TW G4	1.7	2.2	2.9	WW C3	.8	1.0	1.2	SS B4	.9	1.3	1.7
SS B5	1.1	1.6	2.1	TW H1	1.7	2.4	3.1	WW C4	.8	1.0	1.3	SS B5	1.2	1.9	2.6
SS B6	.8	1.1	1.4	TW H2	1.7	2.3	3.0	TW A1	1.4	2.0	2.7	SS B6	.9	1.3	1.7
SS B7	1.1	1.8	2.4	TW H3	1.9	2.5	3.2	TW A2	1.2	1.7	2.2	SS B7	1.3	2.2	3.0
SS B8	.8	1.1	1.5	TW H4	2.1	2.9	3.7	TW A3	1.1	1.5	2.0	SS B8	.9	1.3	1.7
SS C1	.9	1.3	1.7	TW I1	1.9	2.8	3.8	TW A4	1.1	1.5	2.0	SS C1	1.1	1.5	2.0
SS C2	.8	1.1	1.4	TW I2	1.3	1.7	2.1	TW B1	1.3	1.9	2.6	SS C2	1.0	1.3	1.6
SS C3	1.1	1.6	2.1	TW I3	1.2	1.5	2.0	TW B2	1.2	1.7	2.2	SS C3	1.3	1.9	2.5
SS C4	.9	1.2	1.6	TW I4	1.1	1.4	1.8	TW B3	1.2	1.6	2.1	SS C4	1.1	1.4	1.8
SS D1	.7	1.1	1.4	TW I5	1.1	1.3	1.7	TW B4	1.2	1.7	2.2	SS D1	.8	1.2	1.7
SS D2	.6	.9	1.1	TW J1	1.0	1.4	1.8	TW C1	1.3	1.9	2.5	SS D2	.7	1.0	1.3
SS D3	.8	1.2	1.6	TW J2	1.1	1.4	1.7	TW C2	1.3	1.8	2.3	SS D3	.9	1.4	1.9
SS D4	.7	.9	1.2	TW J3	.9	1.2	1.5	TW C3	1.4	1.9	2.4	SS D4	.7	1.0	1.3
SS E1	.8	1.2	1.5	TW J4	.9	1.2	1.5	TW C4	1.5	2.1	2.7	SS E1	.9	1.4	1.9
SS E2	.7	1.0	1.2	DG A1	1.6	---	---	TW D1	1.4	2.1	2.8	SS E2	.8	1.1	1.4
SS E3	.9	1.4	1.8	DG A2	1.4	2.2	---	TW D2	1.0	1.4	1.8	SS E3	1.1	1.7	2.3
SS E4	.7	1.0	1.3	DG A3	1.2	1.5	2.2	TW D3	1.0	1.3	1.7	SS E4	.8	1.2	1.6
				DG B1	1.6	4.4	---	TW D4	.9	1.2	1.5				
BRYCE CANYON, UT				DG B2	1.4	1.8	3.3	TW D5	.9	1.2	1.5	SALT LAKE CITY, UT			
	SOLAR SAVINGS			DG B3	1.2	1.4	1.8	TW E1	.9	1.2	1.5		SOLAR SAVINGS		
	10%	50%	80%	DG C1	1.4	2.0	---	TW E2	.9	1.2	1.5		10%	50%	80%
WW A1	1.6	2.4	3.3	DG C2	1.3	1.5	2.2	TW E3	.8	1.0	1.3	WW A1	1.4	2.6	4.3
WW A2	1.3	1.8	2.3	DG C3	1.1	1.3	1.6	TW E4	.8	1.1	1.3	WW A2	1.2	1.8	2.7
WW A3	1.2	1.6	2.1	SS A1	1.1	1.6	2.2	TW F1	1.4	2.0	2.7	WW A3	1.1	1.7	2.4
WW A4	1.1	1.5	2.0	SS A2	.9	1.2	1.6	TW F2	1.2	1.7	2.2	WW A4	1.1	1.6	2.2
WW A5	1.1	1.5	1.9	SS A3	1.1	1.8	2.6	TW F3	1.2	1.6	2.1	WW A5	1.0	1.5	2.1
WW A6	1.1	1.4	1.8	SS A4	.9	1.3	1.7	TW F4	1.2	1.6	2.1	WW A6	1.0	1.5	2.0
WW B1	1.6	2.4	3.2	SS A5	1.1	1.8	2.6	TW G1	1.4	2.0	2.6	WW B1	1.4	2.8	6.2
WW B2	1.1	1.5	1.9	SS A6	.9	1.2	1.7	TW G2	1.3	1.8	2.3	WW B2	1.0	1.5	2.0
WW B3	1.0	1.3	1.6	SS A7	1.3	2.2	3.3	TW G3	1.3	1.8	2.3	WW B3	.9	1.3	1.8
WW B4	.9	1.2	1.5	SS A8	.9	1.3	1.8	TW G4	1.4	1.9	2.5	WW B4	.9	1.2	1.6
WW B5	.9	1.2	1.5	SS B1	1.2	1.8	2.5	TW H1	1.5	2.0	2.6	WW B5	.9	1.2	1.6
WW C1	.9	1.2	1.6	SS B2	1.0	1.3	1.8	TW H2	1.5	2.0	2.6	WW C1	.9	1.2	1.7
WW C2	1.0	1.2	1.6	SS B3	1.3	2.0	2.8	TW H3	1.6	2.2	2.8	WW C2	.9	1.2	1.7
WW C3	.8	1.1	1.3	SS B4	1.0	1.4	1.8	TW H4	1.9	2.5	3.2	WW C3	.8	1.1	1.4
WW C4	.9	1.1	1.4	SS B5	1.4	2.2	3.0	TW I1	1.5	2.2	3.0	WW C4	.8	1.1	1.4
TW A1	1.7	2.4	3.2	SS B6	1.0	1.4	1.8	TW I2	1.1	1.5	1.9	TW A1	1.4	2.5	4.0
TW A2	1.4	1.9	2.5	SS B7	1.5	2.5	3.6	TW I3	1.0	1.4	1.8	TW A2	1.3	2.0	2.9
TW A3	1.3	1.8	2.3	SS B8	1.0	1.4	1.9	TW I4	1.0	1.2	1.6	TW A3	1.2	1.8	2.6
TW A4	1.3	1.7	2.2	SS C1	1.3	1.8	2.4	TW I5	.9	1.2	1.5	TW A4	1.2	1.8	2.5
TW B1	1.6	2.3	3.0	SS C2	1.1	1.4	1.9	TW J1	.9	1.2	1.6	TW B1	1.4	2.4	3.6
TW B2	1.4	1.9	2.5	SS C3	1.6	2.2	2.9	TW J2	.9	1.2	1.6	TW B2	1.3	2.0	2.8
TW B3	1.4	1.9	2.4	SS C4	1.2	1.6	2.1	TW J3	.8	1.1	1.3	TW B3	1.3	1.9	2.7
TW B4	1.4	1.9	2.5	SS D1	.9	1.4	1.9	TW J4	.9	1.1	1.4	TW B4	1.3	2.0	2.8
TW C1	1.6	2.2	2.9	SS D2	.8	1.1	1.4	DG A1	1.3	4.6	---	TW C1	1.4	2.3	3.3
TW C2	1.5	2.1	2.7	SS D3	1.0	1.5	2.1	DG A2	1.2	1.9	3.5	TW C2	1.4	2.1	3.0
TW C3	1.6	2.1	2.8	SS D4	.8	1.1	1.5	DG A3	1.1	1.3	1.9	TW C3	1.4	2.2	3.1
TW C4	1.7	2.4	3.1	SS E1	1.0	1.5	2.1	DG B1	1.3	2.1	---	TW C4	1.6	2.4	3.4
TW D1	1.7	2.6	3.5	SS E2	.9	1.2	1.6	DG B2	1.2	1.5	2.4	TW D1	1.6	3.0	5.9
TW D2	1.2	1.6	2.0	SS E3	1.2	1.9	2.6	DG B3	1.1	1.2	1.6	TW D2	1.1	1.6	2.2
TW D3	1.1	1.5	1.9	SS E4	.9	1.3	1.7	DG C1	1.2	1.5	4.9	TW D3	1.0	1.5	2.0

SALT LAKE CITY, UT				BURLINGTON, VT				BURLINGTON, VT				NORFOLK, VA			
	SOLAR SAVINGS				SOLAR SAVINGS				SOLAR SAVINGS				SOLAR SAVINGS		
	10%	50%	80%		10%	50%	80%		10%	50%	80%		10%	50%	80%
WW A1	1.4	2.6	4.3	WW A1	4.4	---	---	DG C2	2.3	---	---	TW D4	.9	1.2	1.6
WW A2	1.2	1.8	2.7	WW A2	4.4	---	---	DG C3	1.7	2.5	4.6	TW D5	.9	1.2	1.5
WW A3	1.1	1.7	2.4	WW A3	3.7	---	---	SS A1	2.2	---	---	TW E1	.9	1.2	1.6
WW A4	1.1	1.6	2.2	WW A4	3.4	---	---	SS A2	1.5	2.9	5.5	TW E2	.9	1.2	1.6
WW A5	1.0	1.5	2.1	WW A5	3.2	---	---	SS A3	2.8	---	---	TW E3	.8	1.0	1.3
WW A6	1.0	1.5	2.0	WW A6	3.0	---	---	SS A4	1.5	3.8	---	TW E4	.8	1.1	1.4
WW B1	1.4	2.8	6.2	WW B1	---	---	---	SS A5	2.2	---	---	TW F1	1.3	2.1	2.8
WW B2	1.0	1.5	2.0	WW B2	2.2	---	---	SS A6	1.5	3.3	19.3	TW F2	1.2	1.8	2.3
WW B3	.9	1.3	1.8	WW B3	1.8	4.5	---	SS A7	2.7	---	---	TW F3	1.2	1.7	2.2
WW B4	.9	1.2	1.6	WW B4	1.5	2.3	3.4	SS A8	1.6	9.2	---	TW F4	1.2	1.7	2.2
WW B5	.9	1.2	1.6	WW B5	1.5	2.1	2.9	SS B1	2.8	---	---	TW G1	1.4	2.0	2.7
WW C1	.9	1.2	1.7	WW C1	1.7	3.1	8.1	SS B2	1.6	3.0	5.1	TW G2	1.3	1.8	2.4
WW C2	.9	1.2	1.7	WW C2	1.6	2.6	4.0	SS B3	4.0	---	---	TW G3	1.3	1.8	2.4
WW C3	.8	1.1	1.4	WW C3	1.3	1.7	2.3	SS B4	1.7	3.6	9.6	TW G4	1.4	2.0	2.6
WW C4	.8	1.1	1.4	WW C4	1.3	1.8	2.4	SS B5	3.2	---	---	TW H1	1.5	2.1	2.8
TW A1	1.4	2.5	4.0	TW A1	3.9	---	---	SS B6	1.7	3.5	7.5	TW H2	1.5	2.1	2.7
TW A2	1.3	2.0	2.9	TW A2	3.8	---	---	SS B7	5.6	---	---	TW H3	1.6	2.2	2.9
TW A3	1.2	1.8	2.6	TW A3	3.5	---	---	SS B8	1.8	5.0	---	TW H4	1.8	2.6	3.3
TW A4	1.2	1.8	2.5	TW A4	3.2	---	---	SS C1	---	---	---	TW I1	1.5	2.4	3.2
TW B1	1.4	2.4	3.6	TW B1	4.0	---	---	SS C2	2.1	---	---	TW I2	1.1	1.5	2.0
TW B2	1.3	2.0	2.8	TW B2	3.4	---	---	SS C3	---	---	---	TW I3	1.0	1.4	1.8
TW B3	1.3	1.9	2.7	TW B3	3.2	---	---	SS C4	2.3	---	---	TW I4	.9	1.3	1.6
TW B4	1.3	2.0	2.8	TW B4	3.1	---	---	SS D1	2.0	---	---	TW I5	.9	1.2	1.6
TW C1	1.4	2.3	3.3	TW C1	3.6	---	---	SS D2	1.3	2.3	3.7	TW J1	.9	1.3	1.6
TW C2	1.4	2.1	3.0	TW C2	3.2	---	---	SS D3	2.1	---	---	TW J2	.9	1.3	1.6
TW C3	1.4	2.2	3.1	TW C3	3.2	---	---	SS D4	1.4	2.5	4.0	TW J3	.8	1.1	1.4
TW C4	1.6	2.4	3.4	TW C4	3.2	---	---	SS E1	3.3	---	---	TW J4	.9	1.1	1.4
TW D1	1.6	3.0	5.9	TW D1	---	---	---	SS E2	1.6	3.6	---	DG A1	1.3	---	---
TW D2	1.1	1.6	2.2	TW D2	2.3	---	---	SS E3	3.1	---	---	DG A2	1.2	1.8	4.0
TW D3	1.0	1.5	2.0	TW D3	2.2	---	---	SS E4	1.7	4.1	---	DG A3	1.1	1.3	1.9
DG B1	1.5	---	---	TW D4	1.7	2.7	4.1					DG B1	1.3	2.2	---
DG B2	1.3	1.9	4.7	TW D5	1.6	2.3	3.2	NORFOLK, VA				DG B2	1.2	1.5	2.5
DG B3	1.1	1.4	1.9	TW E1	1.8	4.0	---		SOLAR SAVINGS			DG B3	1.0	1.3	1.7
DG C1	1.3	2.2	---	TW E2	1.7	3.0	5.1		10%	50%	80%	DG C1	1.2	1.6	---
DG C2	1.2	1.6	2.5	TW E3	1.4	2.0	2.7	WW A1	1.3	2.1	2.9	DG C2	1.1	1.4	1.9
DG C3	1.0	1.3	1.7	TW E4	1.4	2.0	2.7	WW A2	1.1	1.6	2.1	DG C3	1.0	1.2	1.5
SS A1	1.0	1.8	2.7	TW F1	---	---	---	WW A3	1.0	1.4	1.9	SS A1	1.0	1.5	2.0
SS A2	.9	1.3	1.8	TW F2	7.0	---	---	WW A4	1.0	1.4	1.8	SS A2	.8	1.2	1.5
SS A3	1.1	2.1	3.5	TW F3	4.8	---	---	WW A5	1.0	1.3	1.7	SS A3	1.0	1.7	2.3
SS A4	.9	1.4	1.9	TW F4	4.1	---	---	WW A6	.9	1.3	1.7	SS A4	.8	1.2	1.6
SS A5	1.1	2.1	3.3	TW G1	---	---	---	WW B1	1.3	2.0	2.8	SS A5	1.0	1.7	2.3
SS A6	.9	1.3	1.9	TW G2	6.0	---	---	WW B2	1.0	1.3	1.7	SS A6	.8	1.2	1.6
SS A7	1.2	2.7	5.3	TW G3	4.6	---	---	WW B3	.9	1.2	1.5	SS A7	1.1	2.0	2.8
SS A8	.9	1.4	2.0	TW G4	4.3	---	---	WW B4	.8	1.1	1.4	SS A8	.8	1.2	1.6
SS B1	1.2	2.0	3.1	TW H1	---	---	---	WW B5	.8	1.1	1.4	SS B1	1.1	1.7	2.3
SS B2	.9	1.4	2.0	TW H2	5.3	---	---	WW C1	.8	1.1	1.4	SS B2	.9	1.3	1.7
SS B3	1.2	2.3	3.7	TW H3	4.8	---	---	WW C2	.8	1.1	1.4	SS B3	1.2	1.8	2.5
SS B4	1.0	1.5	2.1	TW H4	4.9	---	---	WW C3	.8	1.0	1.2	SS B4	.9	1.3	1.7
SS B5	1.3	2.5	4.0	TW I1	---	---	---	WW C4	.8	1.0	1.3	SS B5	1.2	2.0	2.7
SS B6	1.0	1.5	2.0	TW I2	2.5	---	---	TW A1	1.3	2.1	2.9	SS B6	.9	1.3	1.7
SS B7	1.4	3.0	5.8	TW I3	2.2	---	---	TW A2	1.2	1.7	2.3	SS B7	1.3	2.2	3.1
SS B8	1.0	1.5	2.2	TW I4	1.8	2.8	4.2	TW A3	1.1	1.6	2.1	SS B8	.9	1.3	1.8
SS C1	1.2	1.9	2.8	TW I5	1.7	2.4	3.4	TW A4	1.1	1.5	2.0	SS C1	1.1	1.6	2.1
SS C2	1.0	1.5	2.0	TW J1	1.9	3.8	---	TW B1	1.3	2.0	2.7	SS C2	1.0	1.3	1.7
SS C3	1.4	2.3	3.4	TW J2	1.8	3.0	4.7	TW B2	1.2	1.7	2.3	SS C3	1.3	2.0	2.6
SS C4	1.1	1.6	2.2	TW J3	1.4	2.0	2.7	TW B3	1.2	1.7	2.2	SS C4	1.0	1.5	1.9
SS D1	.9	1.5	2.2	TW J4	1.4	2.0	2.7	TW B4	1.2	1.7	2.3	SS D1	.8	1.3	1.7
SS D2	.8	1.1	1.6	DG A1	---	---	---	TW C1	1.3	1.9	2.6	SS D2	.7	1.0	1.3
SS D3	1.0	1.7	2.6	DG A2	3.1	---	---	TW C2	1.3	1.8	2.4	SS D3	.9	1.4	1.9
SS D4	.8	1.2	1.6	DG A3	1.9	3.5	---	TW C3	1.3	1.9	2.5	SS D4	.7	1.0	1.4
SS E1	1.0	1.7	2.7	DG B1	---	---	---	TW C4	1.5	2.1	2.8	SS E1	.9	1.4	1.9
SS E2	.9	1.3	1.8	DG B2	3.0	---	---	TW D1	1.4	2.2	3.0	SS E2	.8	1.1	1.5
SS E3	1.2	2.1	3.3	DG B3	1.8	2.9	---	TW D2	1.0	1.4	1.9	SS E3	1.1	1.7	2.3
SS E4	.9	1.4	1.9	DG C1	---	---	---	TW D3	1.0	1.3	1.7	SS E4	.9	1.2	1.6

Conservation Factor Tables

RICHMOND, VA

	SOLAR SAVINGS		
	10%	50%	80%
WW A1	1.5	2.6	3.7
WW A2	1.2	1.8	2.5
WW A3	1.2	1.7	2.2
WW A4	1.1	1.6	2.1
WW A5	1.1	1.5	2.0
WW A6	1.1	1.5	2.0
WW B1	1.5	2.7	4.0
WW B2	1.1	1.5	2.0
WW B3	1.0	1.3	1.7
WW B4	.9	1.2	1.6
WW B5	.9	1.2	1.6
WW C1	.9	1.3	1.6
WW C2	.9	1.3	1.6
WW C3	.8	1.1	1.4
WW C4	.9	1.1	1.4
TW A1	1.5	2.6	3.6
TW A2	1.3	2.0	2.7
TW A3	1.3	1.9	2.5
TW A4	1.2	1.8	2.4
TW B1	1.5	2.4	3.3
TW B2	1.3	2.0	2.7
TW B3	1.3	1.9	2.6
TW B4	1.4	2.0	2.7
TW C1	1.5	2.3	3.1
TW C2	1.4	2.1	2.9
TW C3	1.5	2.2	3.0
TW C4	1.6	2.4	3.3
TW D1	1.7	2.9	4.2
TW D2	1.2	1.6	2.2
TW D3	1.1	1.5	2.0
TW D4	1.0	1.4	1.8
TW D5	1.0	1.3	1.7
TW E1	1.0	1.4	1.8
TW E2	1.0	1.4	1.8
TW E3	.9	1.1	1.5
TW E4	.9	1.2	1.5
TW F1	1.5	2.5	3.5
TW F2	1.4	2.1	2.8
TW F3	1.4	2.0	2.6
TW F4	1.4	2.0	2.6
TW G1	1.6	2.4	3.4
TW G2	1.5	2.2	2.9
TW G3	1.5	2.2	2.9
TW G4	1.6	2.3	3.1
TW H1	1.7	2.5	3.3
TW H2	1.7	2.4	3.3
TW H3	1.8	2.6	3.4
TW H4	2.1	3.0	3.9
TW I1	1.8	3.1	4.6
TW I2	1.2	1.7	2.3
TW I3	1.1	1.6	2.1
TW I4	1.1	1.4	1.8
TW I5	1.0	1.4	1.8
TW J1	1.0	1.4	1.9
TW J2	1.0	1.4	1.8
TW J3	.9	1.2	1.5
TW J4	.9	1.2	1.6
DG A1	1.6	---	---
DG A2	1.4	2.3	---
DG A3	1.2	1.5	2.3
DG B1	1.5	---	---
DG B2	1.3	1.8	4.1
DG B3	1.2	1.4	1.9
DG C1	1.3	2.2	---

RICHMOND, VA

	SOLAR SAVINGS		
	10%	50%	80%
DG C2	1.2	1.6	2.5
DG C3	1.1	1.3	1.7
SS A1	1.1	1.8	2.4
SS A2	.9	1.3	1.8
SS A3	1.2	2.0	2.9
SS A4	.9	1.4	1.8
SS A5	1.1	2.0	2.9
SS A6	.9	1.3	1.8
SS A7	1.3	2.5	3.7
SS A8	.9	1.4	1.9
SS B1	1.2	2.0	2.8
SS B2	1.0	1.4	1.9
SS B3	1.3	2.2	3.1
SS B4	1.0	1.5	2.0
SS B5	1.4	2.4	3.4
SS B6	1.0	1.5	1.9
SS B7	1.5	2.8	4.2
SS B8	1.0	1.5	2.0
SS C1	1.3	1.9	2.6
SS C2	1.1	1.5	2.0
SS C3	1.5	2.3	3.2
SS C4	1.2	1.7	2.2
SS D1	1.0	1.5	2.0
SS D2	.8	1.1	1.5
SS D3	1.0	1.7	2.3
SS D4	.8	1.2	1.6
SS E1	1.1	1.7	2.4
SS E2	.9	1.3	1.7
SS E3	1.2	2.1	2.9
SS E4	1.0	1.4	1.8

ROANOKE, VA

	SOLAR SAVINGS		
	10%	50%	80%
WW A1	1.5	2.6	3.7
WW A2	1.2	1.8	2.5
WW A3	1.2	1.7	2.2
WW A4	1.1	1.6	2.1
WW A5	1.1	1.5	2.0
WW A6	1.1	1.5	2.0
WW B1	1.5	2.7	4.0
WW B2	1.1	1.5	2.0
WW B3	1.0	1.3	1.7
WW B4	.9	1.2	1.6
WW B5	.9	1.2	1.6
WW C1	.9	1.3	1.6
WW C2	.9	1.3	1.6
WW C3	.8	1.1	1.4
WW C4	.9	1.1	1.4
TW A1	1.5	2.6	3.6
TW A2	1.3	2.0	2.7
TW A3	1.3	1.9	2.5
TW A4	1.3	1.8	2.4
TW B1	1.5	2.4	3.3
TW B2	1.4	2.0	2.7
TW B3	1.3	1.9	2.6
TW B4	1.4	2.0	2.7
TW C1	1.5	2.3	3.1
TW C2	1.5	2.1	2.9
TW C3	1.5	2.2	3.0
TW C4	1.6	2.4	3.2
TW D1	1.7	2.9	4.2
TW D2	1.2	1.6	2.2
TW D3	1.1	1.5	2.0

ROANOKE, VA

	SOLAR SAVINGS		
	10%	50%	80%
TW D4	1.0	1.4	1.8
TW D5	1.0	1.3	1.7
TW E1	1.0	1.4	1.8
TW E2	1.0	1.4	1.8
TW E3	.9	1.1	1.5
TW E4	.9	1.2	1.5
TW F1	1.6	2.5	3.5
TW F2	1.4	2.1	2.8
TW F3	1.4	2.0	2.6
TW F4	1.4	2.0	2.6
TW G1	1.6	2.4	3.3
TW G2	1.5	2.2	2.9
TW G3	1.5	2.2	2.9
TW G4	1.6	2.3	3.1
TW H1	1.7	2.5	3.3
TW H2	1.7	2.4	3.2
TW H3	1.8	2.6	3.4
TW H4	2.1	3.0	3.9
TW I1	1.9	3.1	4.6
TW I2	1.2	1.7	2.2
TW I3	1.2	1.6	2.1
TW I4	1.1	1.4	1.8
TW I5	1.0	1.4	1.8
TW J1	1.0	1.4	1.8
TW J2	1.0	1.4	1.8
TW J3	.9	1.2	1.5
TW J4	.9	1.2	1.6
DG A1	1.6	---	---
DG A2	1.4	2.3	---
DG A3	1.2	1.5	2.3
DG B1	1.6	---	---
DG B2	1.3	1.8	4.1
DG B3	1.2	1.4	1.9
DG C1	1.4	2.2	---
DG C2	1.2	1.6	2.4
DG C3	1.1	1.3	1.6
SS A1	1.1	1.8	2.4
SS A2	.9	1.3	1.8
SS A3	1.2	2.0	2.8
SS A4	.9	1.3	1.8
SS A5	1.2	2.0	2.8
SS A6	.9	1.3	1.8
SS A7	1.3	2.5	3.7
SS A8	.9	1.4	1.9
SS B1	1.2	2.0	2.7
SS B2	1.0	1.4	1.9
SS B3	1.3	2.2	3.1
SS B4	1.0	1.5	2.0
SS B5	1.4	2.4	3.4
SS B6	1.0	1.5	1.9
SS B7	1.5	2.8	4.1
SS B8	1.0	1.5	2.0
SS C1	1.3	1.9	2.6
SS C2	1.1	1.5	2.0
SS C3	1.5	2.3	3.2
SS C4	1.2	1.7	2.2
SS D1	1.0	1.5	2.0
SS D2	.8	1.1	1.5
SS D3	1.0	1.7	2.3
SS D4	.8	1.2	1.6
SS E1	1.1	1.7	2.3
SS E2	.9	1.3	1.7
SS E3	1.2	2.0	2.9
SS E4	1.0	1.4	1.8

OLYMPIA, WA

	SOLAR SAVINGS		
	10%	50%	80%
WW A1	1.7	---	---
WW A2	1.5	---	---
WW A3	1.4	---	---
WW A4	1.4	---	---
WW A5	1.4	4.4	---
WW A6	1.3	3.7	---
WW B1	2.1	---	---
WW B2	1.3	2.5	7.3
WW B3	1.1	2.0	3.9
WW B4	1.1	1.7	2.5
WW B5	1.1	1.6	2.3
WW C1	1.1	1.8	3.2
WW C2	1.1	1.8	2.8
WW C3	.9	1.4	1.9
WW C4	1.0	1.4	2.0
TW A1	1.8	---	---
TW A2	1.6	---	---
TW A3	1.5	---	---
TW A4	1.5	4.7	---
TW B1	1.7	---	---
TW B2	1.6	---	---
TW B3	1.6	5.1	---
TW B4	1.6	4.2	---
TW C1	1.7	---	---
TW C2	1.7	4.9	---
TW C3	1.8	4.4	---
TW C4	1.9	4.4	---
TW D1	2.3	---	---
TW D2	1.4	2.7	10.6
TW D3	1.3	2.5	7.4
TW D4	1.2	1.8	2.9
TW D5	1.1	1.7	2.5
TW E1	1.2	2.1	3.9
TW E2	1.1	1.9	3.1
TW E3	1.0	1.5	2.2
TW E4	1.0	1.5	2.2
TW F1	1.8	---	---
TW F2	1.7	---	---
TW F3	1.7	---	---
TW F4	1.7	7.4	---
TW G1	1.9	---	---
TW G2	1.8	---	---
TW G3	1.9	---	---
TW G4	2.0	5.6	---
TW H1	2.0	---	---
TW H2	2.1	---	---
TW H3	2.2	6.6	---
TW H4	2.5	6.0	---
TW I1	2.7	---	---
TW I2	1.5	2.9	11.3
TW I3	1.4	2.5	5.6
TW I4	1.2	1.9	3.0
TW I5	1.2	1.8	2.7
TW J1	1.2	2.1	3.8
TW J2	1.2	2.0	3.1
TW J3	1.0	1.5	2.2
TW J4	1.1	1.6	2.2
DG A1	2.2	---	---
DG A2	1.6	---	---
DG A3	1.3	2.2	---
DG B1	2.1	---	---
DG B2	1.5	---	---
DG B3	1.3	2.0	3.7
DG C1	1.7	---	---

OLYMPIA, WA
SOLAR SAVINGS
	10%	50%	80%
DG C2	1.4	3.0	---
DG C3	1.2	1.7	2.8
SS A1	1.2	---	---
SS A2	1.0	1.9	3.3
SS A3	1.3	---	---
SS A4	1.0	2.0	4.1
SS A5	1.3	---	---
SS A6	1.0	2.0	3.7
SS A7	1.4	---	---
SS A8	1.0	2.2	5.7
SS B1	1.4	---	---
SS B2	1.1	2.0	3.4
SS B3	1.5	---	---
SS B4	1.1	2.1	3.9
SS B5	1.5	---	---
SS B6	1.1	2.1	3.8
SS B7	1.7	---	---
SS B8	1.1	2.3	5.0
SS C1	1.6	---	---
SS C2	1.3	2.4	5.6
SS C3	1.8	---	---
SS C4	1.4	2.7	7.5
SS D1	1.1	4.9	---
SS D2	.9	1.6	2.6
SS D3	1.1	---	---
SS D4	.9	1.7	2.8
SS E1	1.3	---	---
SS E2	1.0	1.9	3.7
SS E3	1.4	---	---
SS E4	1.1	2.1	4.2

SEATTLE-TACOMA, WA
SOLAR SAVINGS
	10%	50%	80%
WW A1	1.6	---	---
WW A2	1.4	---	---
WW A3	1.3	---	---
WW A4	1.3	5.8	---
WW A5	1.3	3.9	---
WW A6	1.3	3.4	---
WW B1	1.8	---	---
WW B2	1.2	2.4	7.5
WW B3	1.1	2.0	4.0
WW B4	1.0	1.6	2.5
WW B5	1.0	1.6	2.3
WW C1	1.0	1.8	3.2
WW C2	1.1	1.7	2.7
WW C3	.9	1.3	1.9
WW C4	1.0	1.4	2.0
TW A1	1.6	---	---
TW A2	1.5	---	---
TW A3	1.4	---	---
TW A4	1.4	4.2	---
TW B1	1.6	---	---
TW B2	1.5	---	---
TW B3	1.5	4.6	---
TW B4	1.5	3.9	---
TW C1	1.6	---	---
TW C2	1.6	4.5	---
TW C3	1.6	4.2	---
TW C4	1.8	4.2	---
TW D1	2.1	---	---
TW D2	1.3	2.6	11.3
TW D3	1.2	2.4	7.5

SEATTLE-TACOMA, WA
SOLAR SAVINGS
	10%	50%	80%
TW D4	1.1	1.8	2.9
TW D5	1.1	1.7	2.5
TW E1	1.1	2.0	3.9
TW E2	1.1	1.9	3.1
TW E3	1.0	1.5	2.2
TW E4	1.0	1.5	2.1
TW F1	1.7	---	---
TW F2	1.6	---	---
TW F3	1.6	---	---
TW F4	1.6	5.8	---
TW G1	1.8	---	---
TW G2	1.7	---	---
TW G3	1.7	17.7	---
TW G4	1.9	5.2	---
TW H1	1.9	---	---
TW H2	2.0	---	---
TW H3	2.1	6.0	---
TW H4	2.4	5.7	---
TW I1	2.4	---	---
TW I2	1.4	2.8	12.3
TW I3	1.3	2.5	5.6
TW I4	1.2	1.9	3.0
TW I5	1.2	1.8	2.6
TW J1	1.2	2.0	3.8
TW J2	1.2	1.9	3.1
TW J3	1.0	1.5	2.2
TW J4	1.0	1.5	2.2
DG A1	1.9	---	---
DG A2	1.5	---	---
DG A3	1.2	2.1	---
DG B1	1.9	---	---
DG B2	1.5	---	---
DG B3	1.2	1.9	3.6
DG C1	1.5	---	---
DG C2	1.3	2.8	---
DG C3	1.1	1.7	2.8
SS A1	1.1	20.5	---
SS A2	.9	1.8	3.3
SS A3	1.2	---	---
SS A4	1.0	2.0	4.1
SS A5	1.2	---	---
SS A6	1.0	1.9	3.7
SS A7	1.3	---	---
SS A8	1.0	2.2	5.8
SS B1	1.3	---	---
SS B2	1.0	2.0	3.4
SS B3	1.4	---	---
SS B4	1.1	2.1	3.9
SS B5	1.4	---	---
SS B6	1.1	2.1	3.8
SS B7	1.6	---	---
SS B8	1.1	2.3	5.0
SS C1	1.5	---	---
SS C2	1.2	2.3	5.7
SS C3	1.7	---	---
SS C4	1.3	2.6	7.6
SS D1	1.0	4.3	---
SS D2	.9	1.6	2.6
SS D3	1.1	---	---
SS D4	.9	1.6	2.8
SS E1	1.2	---	---
SS E2	1.0	1.9	3.7
SS E3	1.3	---	---
SS E4	1.0	2.1	4.3

SPOKANE, WA
SOLAR SAVINGS
	10%	50%	80%
WW A1	1.8	---	---
WW A2	1.7	---	---
WW A3	1.6	---	---
WW A4	1.6	---	---
WW A5	1.5	---	---
WW A6	1.5	---	---
WW B1	2.8	---	---
WW B2	1.4	3.4	---
WW B3	1.2	2.5	6.3
WW B4	1.2	1.8	2.9
WW B5	1.1	1.7	2.6
WW C1	1.2	2.2	4.1
WW C2	1.2	2.0	3.2
WW C3	1.0	1.5	2.1
WW C4	1.1	1.5	2.1
TW A1	1.8	---	---
TW A2	1.7	---	---
TW A3	1.7	---	---
TW A4	1.7	---	---
TW B1	1.9	---	---
TW B2	1.8	---	---
TW B3	1.8	---	---
TW B4	1.8	---	---
TW C1	1.9	---	---
TW C2	1.9	---	---
TW C3	1.9	---	---
TW C4	2.0	12.6	---
TW D1	3.0	---	---
TW D2	1.5	3.8	---
TW D3	1.4	3.4	---
TW D4	1.2	2.1	3.4
TW D5	1.2	1.9	2.8
TW E1	1.3	2.5	5.7
TW E2	1.2	2.2	3.8
TW E3	1.1	1.6	2.4
TW E4	1.1	1.6	2.4
TW F1	2.0	---	---
TW F2	1.9	---	---
TW F3	1.9	---	---
TW F4	1.9	---	---
TW G1	2.1	---	---
TW G2	2.0	---	---
TW G3	2.1	---	---
TW G4	2.2	---	---
TW H1	2.3	---	---
TW H2	2.3	---	---
TW H3	2.5	---	---
TW H4	2.8	---	---
TW I1	4.1	---	---
TW I2	1.6	4.1	---
TW I3	1.5	3.2	37.1
TW I4	1.3	2.2	3.5
TW I5	1.3	2.0	3.0
TW J1	1.3	2.5	5.1
TW J2	1.3	2.2	3.7
TW J3	1.1	1.6	2.4
TW J4	1.1	1.7	2.4
DG A1	2.9	---	---
DG A2	1.7	---	---
DG A3	1.4	2.5	---
DG B1	2.6	---	---
DG B2	1.7	---	---
DG B3	1.4	2.2	4.9
DG C1	2.0	---	---

SPOKANE, WA
SOLAR SAVINGS
	10%	50%	80%
DG C2	1.5	6.2	---
DG C3	1.2	1.9	3.4
SS A1	1.4	---	---
SS A2	1.1	2.2	4.3
SS A3	1.5	---	---
SS A4	1.1	2.6	7.0
SS A5	1.4	---	---
SS A6	1.1	2.4	5.3
SS A7	1.5	---	---
SS A8	1.1	3.1	---
SS B1	1.6	---	---
SS B2	1.2	2.3	4.2
SS B3	1.8	---	---
SS B4	1.2	2.6	5.4
SS B5	1.7	---	---
SS B6	1.2	2.5	5.0
SS B7	2.0	---	---
SS B8	1.3	3.0	10.4
SS C1	1.8	---	---
SS C2	1.4	3.1	---
SS C3	2.0	---	---
SS C4	1.5	3.5	---
SS D1	1.2	---	---
SS D2	1.0	1.9	3.2
SS D3	1.3	---	---
SS D4	1.0	1.9	3.4
SS E1	1.5	---	---
SS E2	1.1	2.4	5.6
SS E3	1.6	---	---
SS E4	1.2	2.7	7.1

YAKIMA, WA
SOLAR SAVINGS
	10%	50%	80%
WW A1	1.5	---	---
WW A2	1.4	---	---
WW A3	1.3	6.0	---
WW A4	1.3	3.9	---
WW A5	1.3	3.3	---
WW A6	1.3	3.0	---
WW B1	1.9	---	---
WW B2	1.2	2.3	4.5
WW B3	1.1	1.9	3.3
WW B4	1.0	1.6	2.4
WW B5	1.0	1.6	2.2
WW C1	1.0	1.8	2.9
WW C2	1.1	1.7	2.6
WW C3	.9	1.3	1.9
WW C4	1.0	1.4	1.9
TW A1	1.6	---	---
TW A2	1.5	---	---
TW A3	1.4	5.6	---
TW A4	1.4	3.8	---
TW B1	1.6	---	---
TW B2	1.5	5.5	---
TW B3	1.5	4.1	---
TW B4	1.6	3.7	---
TW C1	1.6	---	---
TW C2	1.6	4.2	---
TW C3	1.7	4.0	---
TW C4	1.8	4.1	13.8
TW D1	2.1	---	---
TW D2	1.3	2.6	5.2
TW D3	1.2	2.4	4.6

YAKIMA, WA				CHARLESTON, WV				CHARLESTON, WV				HUNTINGTON, WV			
	SOLAR SAVINGS				SOLAR SAVINGS				SOLAR SAVINGS				SOLAR SAVINGS		
	10%	50%	80%		10%	50%	80%		10%	50%	80%		10%	50%	80%
WW A1	1.5	---	---	WW A1	1.9	---	---	DG C2	1.5	2.4	---	TW D4	1.2	1.6	2.2
WW A2	1.4	---	---	WW A2	1.6	3.1	5.2	DG C3	1.3	1.7	2.2	TW D5	1.1	1.6	2.1
WW A3	1.3	6.0	---	WW A3	1.5	2.6	4.0	SS A1	1.3	2.7	4.2	TW E1	1.2	1.7	2.3
WW A4	1.3	3.9	---	WW A4	1.5	2.4	3.6	SS A2	1.1	1.7	2.3	TW E2	1.2	1.7	2.2
WW A5	1.3	3.3	---	WW A5	1.4	2.3	3.3	SS A3	1.5	3.7	9.1	TW E3	1.0	1.4	1.8
WW A6	1.3	3.0	---	WW A6	1.4	2.2	3.1	SS A4	1.1	1.8	2.5	TW E4	1.0	1.4	1.8
WW B1	1.9	---	---	WW B1	2.5	---	---	SS A5	1.4	3.4	6.7	TW F1	1.9	4.6	10.3
WW B2	1.2	2.3	4.5	WW B2	1.4	2.1	2.8	SS A6	1.1	1.7	2.4	TW F2	1.7	3.1	4.7
WW B3	1.1	1.9	3.3	WW B3	1.2	1.8	2.4	SS A7	1.6	---	---	TW F3	1.7	2.8	4.3
WW B4	1.0	1.6	2.4	WW B4	1.1	1.6	2.0	SS A8	1.1	1.9	2.6	TW F4	1.7	2.7	3.8
WW B5	1.0	1.6	2.2	WW B5	1.1	1.5	2.0	SS B1	1.6	3.1	5.2	TW G1	2.0	3.9	6.6
WW C1	1.0	1.8	2.9	WW C1	1.1	1.6	2.2	SS B2	1.2	1.8	2.5	TW G2	1.8	3.1	4.5
WW C2	1.1	1.7	2.6	WW C2	1.1	1.6	2.1	SS B3	1.7	4.2	10.8	TW G3	1.9	3.0	4.2
WW C3	.9	1.3	1.9	WW C3	1.0	1.3	1.7	SS B4	1.2	1.9	2.6	TW G4	2.0	3.1	4.4
WW C4	1.0	1.4	1.9	WW C4	1.0	1.4	1.8	SS B5	1.7	4.6	12.8	TW H1	2.1	3.7	5.6
TW A1	1.6	---	---	TW A1	1.9	14.6	---	SS B6	1.2	1.9	2.6	TW H2	2.1	3.4	4.8
TW A2	1.5	---	---	TW A2	1.7	3.4	5.6	SS B7	1.9	---	---	TW H3	2.2	3.5	4.9
TW A3	1.4	5.6	---	TW A3	1.6	2.9	4.4	SS B8	1.2	2.0	2.8	TW H4	2.5	3.9	5.5
TW A4	1.4	3.8	---	TW A4	1.6	2.7	3.9	SS C1	1.7	3.4	6.1	TW I1	2.7	---	---
TW B1	1.6	---	---	TW B1	1.9	4.7	10.8	SS C2	1.3	2.0	2.8	TW I2	1.5	2.2	3.0
TW B2	1.5	5.5	---	TW B2	1.7	3.1	4.8	SS C3	2.0	4.5	9.3	TW I3	1.4	2.0	2.7
TW B3	1.5	4.1	---	TW B3	1.7	2.9	4.3	SS C4	1.5	2.3	3.1	TW I4	1.2	1.7	2.3
TW B4	1.6	3.7	---	TW B4	1.8	2.9	4.2	SS D1	1.2	2.2	3.3	TW I5	1.2	1.6	2.1
TW C1	1.6	---	---	TW C1	1.9	3.7	5.9	SS D2	1.0	1.4	1.9	TW J1	1.2	1.8	2.4
TW C2	1.6	4.2	---	TW C2	1.8	3.2	4.7	SS D3	1.3	2.6	4.1	TW J2	1.2	1.7	2.3
TW C3	1.7	4.0	---	TW C3	1.9	3.2	4.7	SS D4	1.0	1.5	2.0	TW J3	1.0	1.4	1.8
TW C4	1.8	4.1	13.8	TW C4	2.0	3.5	5.0	SS E1	1.4	2.9	5.0	TW J4	1.1	1.4	1.9
TW D1	2.1	---	---	TW D1	2.7	---	---	SS E2	1.1	1.7	2.3	DG A1	2.2	---	---
TW D2	1.3	2.6	5.2	TW D2	1.5	2.3	3.1	SS E3	1.6	3.7	7.6	DG A2	1.6	5.0	---
TW D3	1.2	2.4	4.6	TW D3	1.4	2.1	2.9	SS E4	1.2	1.8	2.5	DG A3	1.4	1.9	3.3
DG B1	1.9	---	---	TW D4	1.2	1.7	2.3					DG B1	2.2	---	---
DG B2	1.5	6.8	---	TW D5	1.2	1.7	2.2	HUNTINGTON, WV				DG B2	1.6	2.9	---
DG B3	1.2	1.9	3.4	TW E1	1.2	1.8	2.5		SOLAR SAVINGS			DG B3	1.3	1.7	2.5
DG C1	1.6	---	---	TW E2	1.2	1.8	2.3		10%	50%	80%	DG C1	1.7	---	---
DG C2	1.3	2.8	---	TW E3	1.0	1.4	1.9	WW A1	1.8	6.3	---	DG C2	1.4	2.2	---
DG C3	1.1	1.7	2.7	TW E4	1.1	1.4	1.9	WW A2	1.5	2.7	4.1	DG C3	1.2	1.6	2.1
SS A1	1.2	---	---	TW F1	2.1	8.1	---	WW A3	1.4	2.3	3.4	SS A1	1.3	2.4	3.7
SS A2	1.0	1.9	3.1	TW F2	1.8	3.6	6.0	WW A4	1.4	2.2	3.1	SS A2	1.0	1.6	2.2
SS A3	1.3	---	---	TW F3	1.8	3.1	4.8	WW A5	1.3	2.1	2.9	SS A3	1.4	3.1	5.7
SS A4	1.0	2.1	3.7	TW F4	1.8	3.0	4.3	WW A6	1.3	2.0	2.8	SS A4	1.0	1.7	2.3
SS A5	1.2	---	---	TW G1	2.1	4.9	11.1	WW B1	2.1	---	---	SS A5	1.3	3.0	5.0
SS A6	1.0	2.0	3.4	TW G2	2.0	3.5	5.5	WW B2	1.3	1.9	2.6	SS A6	1.0	1.6	2.3
SS A7	1.3	---	---	TW G3	2.0	3.3	4.9	WW B3	1.1	1.6	2.2	SS A7	1.5	5.3	---
SS A8	1.0	2.3	4.6	TW G4	2.1	3.4	4.9	WW B4	1.1	1.5	1.9	SS A8	1.1	1.7	2.4
SS B1	1.4	---	---	TW H1	2.2	4.3	7.1	WW B5	1.1	1.5	1.9	SS B1	1.5	2.8	4.3
SS B2	1.1	2.0	3.2	TW H2	2.2	3.8	5.6	WW C1	1.1	1.6	2.1	SS B2	1.1	1.7	2.4
SS B3	1.5	---	---	TW H3	2.4	3.9	5.6	WW C2	1.1	1.5	2.0	SS B3	1.6	3.4	6.3
SS B4	1.1	2.2	3.7	TW H4	2.7	4.3	6.1	WW C3	.9	1.3	1.6	SS B4	1.2	1.8	2.5
SS B5	1.5	---	---	TW I1	3.1	---	---	WW C4	1.0	1.3	1.7	SS B5	1.6	3.7	7.0
SS B6	1.1	2.1	3.5	TW I2	1.6	2.4	3.3	TW A1	1.8	4.8	13.3	SS B6	1.2	1.8	2.4
SS B7	1.7	---	---	TW I3	1.4	2.2	3.0	TW A2	1.6	2.9	4.5	SS B7	1.8	7.0	---
SS B8	1.2	2.4	4.3	TW I4	1.3	1.8	2.4	TW A3	1.5	2.6	3.8	SS B8	1.2	1.9	2.6
SS C1	1.5	---	---	TW I5	1.2	1.7	2.2	TW A4	1.5	2.4	3.5	SS C1	1.6	2.9	4.5
SS C2	1.2	2.3	4.1	TW J1	1.3	1.9	2.5	TW B1	1.8	3.8	6.5	SS C2	1.3	1.9	2.6
SS C3	1.7	---	---	TW J2	1.3	1.8	2.4	TW B2	1.6	2.8	4.1	SS C3	1.9	3.7	6.1
SS C4	1.3	2.5	4.9	TW J3	1.1	1.5	1.9	TW B3	1.6	2.6	3.8	SS C4	1.4	2.1	2.9
SS D1	1.1	4.4	---	TW J4	1.1	1.5	1.9	TW B4	1.7	2.7	3.8	SS D1	1.1	2.0	2.9
SS D2	.9	1.6	2.5	DG A1	2.6	---	---	TW C1	1.8	3.2	4.9	SS D2	.9	1.4	1.8
SS D3	1.1	---	---	DG A2	1.7	---	---	TW C2	1.7	2.9	4.1	SS D3	1.2	2.3	3.5
SS D4	.9	1.7	2.7	DG A3	1.4	2.0	3.8	TW C3	1.8	3.0	4.2	SS D4	.9	1.4	1.9
SS E1	1.3	---	---	DG B1	2.5	---	---	TW C4	1.9	3.2	4.5	SS E1	1.3	2.5	4.0
SS E2	1.0	1.9	3.3	DG B2	1.7	3.5	---	TW D1	2.4	---	---	SS E2	1.0	1.6	2.1
SS E3	1.4	---	---	DG B3	1.4	1.8	2.7	TW D2	1.4	2.1	2.9	SS E3	1.5	3.1	5.3
SS E4	1.1	2.1	3.7	DG C1	1.9	---	---	TW D3	1.3	1.9	2.6	SS E4	1.1	1.7	2.3

RALEIGH-DURHAM, NORTH CAROLINA ELEV 440 LAT 35.9										AKRON-CANTON, OHIO ELEV 1237 LAT 40.9									
	HS	VS	TA	D50	D55	D60	D65	D70	KT	LD	HS	VS	TA	D50	D55	D60	D65	D70	KT LD
JAN	694	924	41	300	451	605	760	915	.46	57	428	570	26	735	890	1045	1200	1355	.34 62
FEB	943	1046	42	227	360	499	638	778	.48	50	649	750	28	625	764	904	1044	1184	.38 55
MAR	1276	1040	49	95	198	338	502	645	.50	38	964	856	36	430	583	738	893	1048	.41 43
APR	1644	959	60	5	26	87	180	319	.53	26	1357	896	49	108	212	349	495	645	.45 31
MAY	1808	866	67	0	2	11	48	126	.52	17	1668	889	59	9	38	108	231	354	.48 22
JUN	1864	846	74	0	0	0	0	26	.51	13	1839	897	68	0	2	10	33	111	.50 18
JUL	1776	832	78	0	0	0	0	10	.50	15	1787	903	72	0	1	4	9	62	.50 20
AUG	1611	865	77	0	0	0	0	14	.50	22	1596	964	70	0	1	5	16	83	.50 27
SEP	1377	996	71	0	1	4	12	71	.51	34	1272	1058	64	2	9	37	101	207	.50 39
OCT	1105	1134	60	4	21	79	186	309	.52	46	908	1041	53	42	115	224	369	518	.48 51
NOV	812	1072	50	79	172	304	450	600	.50	55	505	666	41	286	431	579	729	879	.37 60
DEC	636	893	41	280	429	583	738	893	.45	59	353	476	29	639	794	949	1104	1259	.31 64
YR	1297	955	59	990	1659	2509	3514	4706	.50		1113	831	50	2876	3839	4951	6224	7705	.46

BISMARCK, NORTH DAKOTA ELEV 1647 LAT 46.8										CINCINNATI, OHIO ELEV 889 LAT 39.1									
	HS	VS	TA	D50	D55	D60	D65	D70	KT	LD	HS	VS	TA	D50	D55	D60	D65	D70	KT LD
JAN	467	857	8	1296	1451	1606	1761	1916	.50	68	500	659	31	587	741	896	1051	1206	.37 60
FEB	776	1237	14	1022	1162	1302	1442	1582	.55	60	738	839	33	469	608	748	888	1028	.41 53
MAR	1168	1358	25	772	927	1082	1237	1392	.56	49	1027	878	42	272	416	568	722	877	.42 42
APR	1459	1150	43	230	365	511	660	810	.51	37	1398	883	54	44	112	209	341	485	.46 29
MAY	1848	1127	54	40	108	204	339	486	.54	28	1672	861	63	4	15	53	138	233	.48 20
JUN	2060	1116	64	3	13	45	122	211	.57	23	1837	872	72	0	1	5	9	65	.50 16
JUL	2184	1241	71	0	2	8	18	86	.62	25	1771	869	76	0	0	0	0	30	.50 18
AUG	1877	1369	69	1	3	12	35	111	.62	33	1634	943	74	0	0	0	0	40	.51 26
SEP	1354	1408	58	18	60	135	252	380	.58	45	1312	1038	68	1	4	17	44	130	.50 37
OCT	908	1321	47	152	270	413	564	719	.56	57	990	1098	57	22	72	152	271	413	.50 49
NOV	507	877	29	633	783	933	1083	1233	.49	66	588	769	44	211	343	488	636	786	.41 58
DEC	373	688	16	1066	1221	1376	1531	1686	.46	70	432	587	34	507	661	815	970	1125	.35 62
YR	1251	1145	41	5235	6364	7627	9044	10612	.56		1160	858	54	2117	2973	3951	5070	6419	.47

FARGO, NORTH DAKOTA ELEV 899 LAT 46.9										CLEVELAND, OHIO ELEV 804 LAT 41.4									
	HS	VS	TA	D50	D55	D60	D65	D70	KT	LD	HS	VS	TA	D50	D55	D60	D65	D70	KT LD
JAN	415	720	6	1367	1522	1677	1832	1987	.44	68	388	507	27	716	871	1026	1181	1336	.32 63
FEB	706	1078	11	1100	1240	1380	1520	1660	.51	61	601	687	28	619	759	899	1039	1179	.36 55
MAR	1098	1249	24	800	955	1110	1265	1420	.52	49	922	822	36	433	586	741	896	1051	.40 44
APR	1476	1170	42	243	384	532	681	831	.52	37	1349	901	48	112	217	355	501	651	.45 32
MAY	1835	1122	55	31	95	192	334	479	.54	28	1681	904	58	11	43	116	244	366	.49 22
JUN	1994	1087	65	1	7	29	97	184	.55	24	1843	906	68	1	3	11	40	119	.51 18
JUL	2120	1210	71	0	1	5	13	78	.60	26	1828	929	71	0	1	4	9	68	.52 20
AUG	1825	1330	69	0	2	8	33	102	.60	33	1583	969	70	0	1	6	17	88	.50 28
SEP	1304	1342	58	12	47	120	234	366	.56	45	1239	1041	64	2	9	36	95	202	.49 40
OCT	874	1253	47	140	260	406	558	713	.54	57	867	995	54	38	108	212	354	503	.46 51
NOV	457	753	29	642	792	942	1092	1242	.44	66	466	607	42	262	404	552	702	852	.35 61
DEC	337	594	13	1147	1302	1457	1612	1767	.42	70	318	420	30	611	766	921	1076	1231	.29 65
YR	1206	1075	41	5485	6607	7858	9271	10829	.54		1093	808	50	2804	3768	4879	6154	7646	.45

MINOT, NORTH DAKOTA ELEV 1713 LAT 48.3										COLUMBUS, OHIO ELEV 833 LAT 40.0									
	HS	VS	TA	D50	D55	D60	D65	D70	KT	LD	HS	VS	TA	D50	D55	D60	D65	D70	KT LD
JAN	384	691	8	1305	1460	1615	1770	1925	.45	70	459	606	28	670	825	980	1135	1290	.35 61
FEB	656	1029	13	1042	1182	1322	1462	1602	.50	62	677	769	30	552	692	832	972	1112	.39 54
MAR	1044	1227	24	819	973	1128	1283	1438	.51	51	980	851	39	339	491	645	800	955	.41 42
APR	1461	1208	41	280	420	568	717	867	.52	39	1353	873	51	67	150	272	418	564	.45 30
MAY	1846	1174	53	60	136	244	384	535	.55	29	1647	864	61	4	19	71	176	284	.48 21
JUN	1975	1116	62	6	21	68	150	257	.54	25	1813	875	70	0	1	5	13	78	.50 17
JUL	2098	1249	69	1	4	14	27	119	.60	27	1755	876	74	0	0	0	0	39	.50 19
AUG	1800	1371	67	1	6	21	70	147	.60	35	1641	969	72	0	1	3	8	59	.52 27
SEP	1277	1373	56	26	79	160	286	418	.56	46	1282	1038	65	1	5	24	76	171	.50 38
OCT	850	1277	46	168	289	434	586	741	.55	58	945	1064	54	33	100	201	342	491	.49 50
NOV	438	759	28	663	813	963	1113	1263	.45	68	538	703	42	259	401	549	699	849	.38 59
DEC	310	567	15	1094	1249	1404	1559	1714	.42	72	387	521	31	599	753	908	1063	1218	.33 63
YR	1181	1087	40	5465	6633	7943	9407	11025	.54		1120	834	52	2524	3438	4491	5702	7111	.46

EAU CLAIRE, WI				EAU CLAIRE, WI				GREEN BAY, WI				LA CROSSE, WI			
	SOLAR SAVINGS				SOLAR SAVINGS				SOLAR SAVINGS				SOLAR SAVINGS		
	10%	50%	80%		10%	50%	80%		10%	50%	80%		10%	50%	80%
WW A1	3.0	---	---	DG C2	2.0	---	---	TW D4	1.4	2.2	3.1	WW A1	2.5	---	---
WW A2	2.6	---	---	DG C3	1.5	2.1	3.4	TW D5	1.4	2.0	2.7	WW A2	2.2	---	---
WW A3	2.4	---	---	SS A1	1.9	---	---	TW E1	1.5	2.5	4.0	WW A3	2.0	---	---
WW A4	2.2	---	---	SS A2	1.4	2.5	3.9	TW E2	1.5	2.3	3.3	WW A4	2.0	---	---
WW A5	2.2	---	---	SS A3	2.3	---	---	TW E3	1.2	1.7	2.3	WW A5	1.9	5.7	---
WW A6	2.1	---	---	SS A4	1.4	2.9	5.6	TW E4	1.2	1.7	2.3	WW A6	1.8	4.4	---
WW B1	---	---	---	SS A5	2.0	---	---	TW F1	3.0	---	---	WW B1	---	---	---
WW B2	1.8	3.8	---	SS A6	1.4	2.7	4.6	TW F2	2.6	---	---	WW B2	1.7	2.9	5.0
WW B3	1.6	2.8	5.1	SS A7	2.3	---	---	TW F3	2.5	---	---	WW B3	1.4	2.3	3.5
WW B4	1.4	2.0	2.8	SS A8	1.5	3.5	---	TW F4	2.5	---	---	WW B4	1.3	1.9	2.5
WW B5	1.4	1.9	2.6	SS B1	2.3	---	---	TW G1	3.1	---	---	WW B5	1.3	1.8	2.4
WW C1	1.5	2.4	3.7	SS B2	1.5	2.6	4.0	TW G2	2.8	---	---	WW C1	1.4	2.1	3.0
WW C2	1.4	2.2	3.1	SS B3	2.8	---	---	TW G3	2.7	---	---	WW C2	1.3	2.0	2.7
WW C3	1.2	1.6	2.1	SS B4	1.6	2.9	4.9	TW G4	2.8	---	---	WW C3	1.1	1.5	2.0
WW C4	1.2	1.6	2.2	SS B5	2.6	---	---	TW H1	3.2	---	---	WW C4	1.2	1.6	2.0
TW A1	2.9	---	---	SS B6	1.6	2.8	4.6	TW H2	3.1	---	---	TW A1	2.5	---	---
TW A2	2.6	---	---	SS B7	3.5	---	---	TW H3	3.2	---	---	TW A2	2.3	---	---
TW A3	2.5	---	---	SS B8	1.7	3.4	7.4	TW H4	3.5	11.1	---	TW A3	2.2	---	---
TW A4	2.4	---	---	SS C1	3.2	---	---	TW I1	---	---	---	TW A4	2.1	7.0	---
TW B1	2.9	---	---	SS C2	1.8	3.4	14.0	TW I2	2.0	3.8	10.6	TW B1	2.5	---	---
TW B2	2.6	---	---	SS C3	3.4	---	---	TW I3	1.8	3.2	5.6	TW B2	2.3	---	---
TW B3	2.5	---	---	SS C4	1.9	4.0	---	TW I4	1.5	2.2	3.2	TW B3	2.2	8.9	---
TW B4	2.5	---	---	SS D1	1.7	---	---	TW I5	1.4	2.1	2.8	TW B4	2.2	5.4	---
TW C1	2.8	---	---	SS D2	1.2	2.0	3.0	TW J1	1.6	2.5	3.9	TW C1	2.5	---	---
TW C2	2.6	---	---	SS D3	1.8	---	---	TW J2	1.5	2.3	3.3	TW C2	2.3	7.1	---
TW C3	2.6	---	---	SS D4	1.3	2.1	3.2	TW J3	1.2	1.7	2.3	TW C3	2.4	5.8	---
TW C4	2.7	---	---	SS E1	2.3	---	---	TW J4	1.3	1.8	2.3	TW C4	2.5	5.5	---
TW D1	---	---	---	SS E2	1.4	2.7	4.7	DG A1	---	---	---	TW D1	---	---	---
TW D2	1.9	4.3	---	SS E3	2.4	---	---	DG A2	2.3	---	---	TW D2	1.8	3.2	5.9
TW D3	1.8	3.9	---	SS E4	1.5	3.0	5.7	DG A3	1.7	2.6	---	TW D3	1.7	2.9	5.2
TW D4	1.5	2.3	3.3					DG B1	---	---	---	TW D4	1.4	2.1	2.9
TW D5	1.4	2.1	2.8	GREEN BAY, WI				DG B2	2.2	---	---	TW D5	1.4	1.9	2.6
TW E1	1.6	2.8	4.8		SOLAR SAVINGS			DG B3	1.6	2.3	4.2	TW E1	1.5	2.4	3.6
TW E2	1.5	2.4	3.6		10%	50%	80%	DG C1	3.7	---	---	TW E2	1.4	2.2	3.1
TW E3	1.2	1.8	2.4	WW A1	2.6	---	---	DG C2	1.9	7.9	---	TW E3	1.2	1.7	2.2
TW E4	1.3	1.8	2.4	WW A2	2.3	---	---	DG C3	1.5	2.0	3.1	TW E4	1.2	1.7	2.2
TW F1	3.6	---	---	WW A3	2.1	---	---	SS A1	1.7	---	---	TW F1	2.9	---	---
TW F2	3.0	---	---	WW A4	2.0	---	---	SS A2	1.3	2.3	3.5	TW F2	2.5	---	---
TW F3	2.8	---	---	WW A5	1.9	---	---	SS A3	2.0	---	---	TW F3	2.4	---	---
TW F4	2.7	---	---	WW A6	1.9	---	---	SS A4	1.3	2.6	4.4	TW F4	2.4	---	---
TW G1	3.6	---	---	WW B1	---	---	---	SS A5	1.8	---	---	TW G1	3.0	---	---
TW G2	3.1	---	---	WW B2	1.7	3.2	7.0	SS A6	1.3	2.4	3.9	TW G2	2.7	---	---
TW G3	3.0	---	---	WW B3	1.5	2.5	4.0	SS A7	2.0	---	---	TW G3	2.6	---	---
TW G4	3.1	---	---	WW B4	1.3	1.9	2.6	SS A8	1.4	2.9	6.3	TW G4	2.8	7.1	---
TW H1	3.7	---	---	WW B5	1.3	1.8	2.5	SS B1	2.1	---	---	TW H1	3.1	---	---
TW H2	3.4	---	---	WW C1	1.4	2.2	3.3	SS B2	1.4	2.4	3.6	TW H2	3.0	---	---
TW H3	3.5	---	---	WW C2	1.4	2.0	2.9	SS B3	2.4	---	---	TW H3	3.1	8.8	---
TW H4	3.8	---	---	WW C3	1.1	1.5	2.0	SS B4	1.5	2.6	4.2	TW H4	3.4	7.4	---
TW I1	---	---	---	WW C4	1.2	1.6	2.1	SS B5	2.3	---	---	TW I1	---	---	---
TW I2	2.1	4.6	---	TW A1	2.5	---	---	SS B6	1.5	2.6	4.0	TW I2	1.9	3.4	6.2
TW I3	1.9	3.6	9.5	TW A2	2.3	---	---	SS B7	2.8	---	---	TW I3	1.8	2.9	4.7
TW I4	1.6	2.4	3.4	TW A3	2.2	---	---	SS B8	1.5	3.0	5.3	TW I4	1.5	2.2	3.0
TW I5	1.5	2.2	3.0	TW A4	2.2	---	---	SS C1	2.6	---	---	TW I5	1.4	2.0	2.7
TW J1	1.6	2.8	4.5	TW B1	2.6	---	---	SS C2	1.7	3.0	5.5	TW J1	1.5	2.4	3.5
TW J2	1.6	2.5	3.6	TW B2	2.3	---	---	SS C3	2.9	---	---	TW J2	1.5	2.2	3.1
TW J3	1.3	1.8	2.4	TW B3	2.3	---	---	SS C4	1.8	3.4	7.2	TW J3	1.2	1.7	2.2
TW J4	1.3	1.8	2.4	TW B4	2.3	16.4	---	SS D1	1.6	---	---	TW J4	1.3	1.7	2.3
DG A1	---	---	---	TW C1	2.5	---	---	SS D2	1.2	1.9	2.8	DG A1	---	---	---
DG A2	2.5	---	---	TW C2	2.4	---	---	SS D3	1.6	---	---	DG A2	2.2	---	---
DG A3	1.7	2.8	---	TW C3	2.4	10.2	---	SS D4	1.2	2.0	2.9	DG A3	1.6	2.5	---
DG B1	---	---	---	TW C4	2.5	6.8	---	SS E1	2.0	---	---	DG B1	---	---	---
DG B2	2.4	---	---	TW D1	---	---	---	SS E2	1.4	2.4	3.9	DG B2	2.1	---	---
DG B3	1.7	2.4	5.0	TW D2	1.8	3.5	9.9	SS E3	2.1	---	---	DG B3	1.6	2.2	3.8
DG C1	---	---	---	TW D3	1.7	3.2	7.3	SS E4	1.4	2.6	4.4	DG C1	3.3	---	---

LA CROSSE, WI

	SOLAR SAVINGS		
	10%	50%	80%
DG C2	1.8	4.6	---
DG C3	1.5	2.0	2.9
SS A1	1.7	---	---
SS A2	1.3	2.2	3.3
SS A3	2.0	---	---
SS A4	1.3	2.4	3.9
SS A5	1.8	---	---
SS A6	1.3	2.3	3.5
SS A7	2.1	---	---
SS A8	1.4	2.7	4.8
SS B1	2.1	---	---
SS B2	1.4	2.3	3.4
SS B3	2.4	---	---
SS B4	1.5	2.5	3.8
SS B5	2.3	---	---
SS B6	1.5	2.5	3.7
SS B7	2.9	---	---
SS B8	1.5	2.8	4.6
SS C1	2.5	---	---
SS C2	1.6	2.8	4.5
SS C3	2.8	---	---
SS C4	1.8	3.1	5.4
SS D1	1.6	---	---
SS D2	1.2	1.8	2.6
SS D3	1.6	---	---
SS D4	1.2	1.9	2.8
SS E1	2.0	---	---
SS E2	1.3	2.3	3.5
SS E3	2.1	---	---
SS E4	1.4	2.5	3.9

MADISON, WI

	SOLAR SAVINGS		
	10%	50%	80%
WW A1	2.4	---	---
WW A2	2.0	---	---
WW A3	1.9	11.2	---
WW A4	1.8	4.4	---
WW A5	1.8	3.6	---
WW A6	1.7	3.3	---
WW B1	---	---	---
WW B2	1.6	2.6	4.2
WW B3	1.4	2.2	3.2
WW B4	1.3	1.8	2.4
WW B5	1.3	1.7	2.3
WW C1	1.3	2.0	2.8
WW C2	1.3	1.9	2.6
WW C3	1.1	1.5	1.9
WW C4	1.1	1.5	2.0
TW A1	2.4	---	---
TW A2	2.1	---	---
TW A3	2.0	7.9	---
TW A4	2.0	4.3	---
TW B1	2.4	---	---
TW B2	2.1	7.6	---
TW B3	2.1	4.8	---
TW B4	2.1	4.3	---
TW C1	2.3	---	---
TW C2	2.2	4.9	---
TW C3	2.3	4.7	---
TW C4	2.4	4.8	14.5
TW D1	---	---	---
TW D2	1.7	2.9	4.8
TW D3	1.6	2.7	4.3

MADISON, WI

	SOLAR SAVINGS		
	10%	50%	80%
TW D4	1.4	2.0	2.7
TW D5	1.3	1.9	2.5
TW E1	1.4	2.2	3.3
TW E2	1.4	2.1	2.9
TW E3	1.2	1.6	2.1
TW E4	1.2	1.6	2.1
TW F1	2.7	---	---
TW F2	2.4	---	---
TW F3	2.3	---	---
TW F4	2.3	5.1	---
TW G1	2.7	---	---
TW G2	2.5	---	---
TW G3	2.5	6.3	---
TW G4	2.6	5.2	---
TW H1	2.9	---	---
TW H2	2.8	7.8	---
TW H3	2.9	6.0	---
TW H4	3.2	6.1	---
TW I1	---	---	---
TW I2	1.8	3.1	5.0
TW I3	1.7	2.7	4.1
TW I4	1.5	2.1	2.8
TW I5	1.4	1.9	2.6
TW J1	1.5	2.3	3.3
TW J2	1.4	2.1	2.9
TW J3	1.2	1.6	2.1
TW J4	1.2	1.7	2.2
DG A1	---	---	---
DG A2	2.1	---	---
DG A3	1.6	2.4	---
DG B1	---	---	---
DG B2	2.1	---	---
DG B3	1.6	2.1	3.5
DG C1	2.8	---	---
DG C2	1.8	3.6	---
DG C3	1.4	1.9	2.7
SS A1	1.6	9.9	---
SS A2	1.2	2.1	3.0
SS A3	1.8	---	---
SS A4	1.3	2.2	3.5
SS A5	1.7	---	---
SS A6	1.3	2.1	3.2
SS A7	1.9	---	---
SS A8	1.3	2.5	4.1
SS B1	1.9	---	---
SS B2	1.4	2.2	3.2
SS B3	2.2	---	---
SS B4	1.4	2.4	3.5
SS B5	2.2	---	---
SS B6	1.4	2.3	3.4
SS B7	2.6	---	---
SS B8	1.5	2.6	4.0
SS C1	2.3	---	---
SS C2	1.6	2.5	3.9
SS C3	2.6	---	---
SS C4	1.7	2.9	4.6
SS D1	1.5	4.4	---
SS D2	1.1	1.7	2.4
SS D3	1.5	---	---
SS D4	1.1	1.8	2.6
SS E1	1.8	---	---
SS E2	1.3	2.1	3.1
SS E3	2.0	---	---
SS E4	1.4	2.3	3.5

MILWAUKEE, WI

	SOLAR SAVINGS		
	10%	50%	80%
WW A1	2.3	---	---
WW A2	2.0	---	---
WW A3	1.9	---	---
WW A4	1.8	5.2	---
WW A5	1.8	4.0	---
WW A6	1.7	3.6	---
WW B1	---	---	---
WW B2	1.6	2.7	4.4
WW B3	1.4	2.2	3.3
WW B4	1.3	1.8	2.4
WW B5	1.3	1.7	2.3
WW C1	1.3	2.0	2.9
WW C2	1.3	1.9	2.6
WW C3	1.1	1.5	1.9
WW C4	1.1	1.5	2.0
TW A1	2.3	---	---
TW A2	2.1	---	---
TW A3	2.0	---	---
TW A4	2.0	4.7	---
TW B1	2.4	---	---
TW B2	2.1	---	---
TW B3	2.1	5.3	---
TW B4	2.1	4.5	---
TW C1	2.3	---	---
TW C2	2.2	5.3	---
TW C3	2.3	4.9	---
TW C4	2.4	4.9	16.8
TW D1	---	---	---
TW D2	1.7	3.0	5.1
TW D3	1.6	2.7	4.5
TW D4	1.4	2.0	2.8
TW D5	1.3	1.9	2.5
TW E1	1.4	2.3	3.4
TW E2	1.4	2.1	2.9
TW E3	1.2	1.6	2.2
TW E4	1.2	1.6	2.2
TW F1	2.7	---	---
TW F2	2.4	---	---
TW F3	2.3	---	---
TW F4	2.3	5.8	---
TW G1	2.7	---	---
TW G2	2.5	---	---
TW G3	2.5	8.1	---
TW G4	2.6	5.6	---
TW H1	2.9	---	---
TW H2	2.8	14.2	---
TW H3	2.9	6.6	---
TW H4	3.3	6.4	---
TW I1	---	---	---
TW I2	1.8	3.1	5.3
TW I3	1.7	2.8	4.3
TW I4	1.5	2.1	2.9
TW I5	1.4	2.0	2.6
TW J1	1.5	2.3	3.3
TW J2	1.4	2.1	3.0
TW J3	1.2	1.6	2.2
TW J4	1.2	1.7	2.2
DG A1	---	---	---
DG A2	2.1	---	---
DG A3	1.6	2.4	---
DG B1	---	---	---
DG B2	2.1	---	---
DG B3	1.6	2.1	3.6
DG C1	2.8	---	---

MILWAUKEE, WI

	SOLAR SAVINGS		
	10%	50%	80%
DG C2	1.8	3.8	---
DG C3	1.4	1.9	2.8
SS A1	1.6	---	---
SS A2	1.2	2.1	3.1
SS A3	1.8	---	---
SS A4	1.2	2.3	3.5
SS A5	1.6	---	---
SS A6	1.2	2.2	3.3
SS A7	1.9	---	---
SS A8	1.3	2.5	4.2
SS B1	1.9	---	---
SS B2	1.3	2.2	3.2
SS B3	2.1	---	---
SS B4	1.4	2.4	3.6
SS B5	2.1	---	---
SS B6	1.4	2.3	3.5
SS B7	2.5	---	---
SS B8	1.5	2.6	4.1
SS C1	2.3	---	---
SS C2	1.6	2.6	4.1
SS C3	2.6	---	---
SS C4	1.7	2.9	4.8
SS D1	1.4	4.8	---
SS D2	1.1	1.7	2.5
SS D3	1.5	---	---
SS D4	1.1	1.8	2.6
SS E1	1.8	---	---
SS E2	1.3	2.1	3.2
SS E3	1.9	---	---
SS E4	1.3	2.3	3.6

CASPER, WY

	SOLAR SAVINGS		
	10%	50%	80%
WW A1	1.5	2.5	3.6
WW A2	1.2	1.8	2.4
WW A3	1.2	1.6	2.2
WW A4	1.1	1.6	2.1
WW A5	1.1	1.5	2.0
WW A6	1.1	1.5	1.9
WW B1	1.6	2.6	4.0
WW B2	1.1	1.5	1.9
WW B3	1.0	1.3	1.7
WW B4	.9	1.2	1.6
WW B5	.9	1.2	1.6
WW C1	.9	1.2	1.6
WW C2	.9	1.2	1.6
WW C3	.8	1.1	1.4
WW C4	.9	1.1	1.4
TW A1	1.5	2.5	3.5
TW A2	1.3	2.0	2.7
TW A3	1.3	1.8	2.4
TW A4	1.3	1.8	2.3
TW B1	1.5	2.3	3.2
TW B2	1.4	2.0	2.6
TW B3	1.4	1.9	2.6
TW B4	1.4	2.0	2.6
TW C1	1.5	2.2	3.0
TW C2	1.5	2.1	2.8
TW C3	1.5	2.2	2.9
TW C4	1.7	2.4	3.2
TW D1	1.7	2.8	4.2
TW D2	1.2	1.6	2.1
TW D3	1.1	1.5	2.0

CASPER, WY				CHEYENNE, WY				CHEYENNE, WY				ROCK SPRINGS, WY			
	\multicolumn{3}{c}{SOLAR SAVINGS}			\multicolumn{3}{c}{SOLAR SAVINGS}			\multicolumn{3}{c}{SOLAR SAVINGS}			\multicolumn{3}{c}{SOLAR SAVINGS}	
	10%	50%	80%		10%	50%	80%		10%	50%	80%		10%	50%	80%
WW A1	1.5	2.5	3.6	WW A1	1.5	2.3	3.1	DG C2	1.2	1.5	2.1	TW D4	1.0	1.4	1.8
WW A2	1.2	1.8	2.4	WW A2	1.2	1.7	2.2	DG C3	1.1	1.2	1.5	TW D5	1.0	1.3	1.7
WW A3	1.2	1.6	2.2	WW A3	1.1	1.5	2.0	SS A1	1.1	1.6	2.2	TW E1	1.0	1.4	1.8
WW A4	1.1	1.6	2.1	WW A4	1.1	1.5	1.9	SS A2	.9	1.3	1.7	TW E2	1.0	1.4	1.8
WW A5	1.1	1.5	2.0	WW A5	1.1	1.4	1.8	SS A3	1.2	1.9	2.6	TW E3	.9	1.2	1.5
WW A6	1.1	1.5	1.9	WW A6	1.0	1.4	1.8	SS A4	.9	1.3	1.7	TW E4	.9	1.2	1.5
WW B1	1.6	2.6	4.0	WW B1	1.5	2.3	3.0	SS A5	1.1	1.9	2.6	TW F1	1.6	2.6	3.6
WW B2	1.1	1.5	1.9	WW B2	1.1	1.4	1.8	SS A6	.9	1.3	1.7	TW F2	1.5	2.1	2.9
WW B3	1.0	1.3	1.7	WW B3	.9	1.2	1.6	SS A7	1.3	2.2	3.2	TW F3	1.4	2.0	2.7
WW B4	.9	1.2	1.6	WW B4	.9	1.2	1.5	SS A8	.9	1.3	1.8	TW F4	1.4	2.0	2.6
WW B5	.9	1.2	1.6	WW B5	.9	1.2	1.5	SS B1	1.2	1.8	2.5	TW G1	1.7	2.5	3.4
WW C1	.9	1.2	1.6	WW C1	.9	1.2	1.5	SS B2	1.0	1.4	1.8	TW G2	1.6	2.2	2.9
WW C2	.9	1.2	1.6	WW C2	.9	1.2	1.5	SS B3	1.3	2.0	2.8	TW G3	1.6	2.2	2.9
WW C3	.8	1.1	1.4	WW C3	.8	1.0	1.3	SS B4	1.0	1.4	1.8	TW G4	1.7	2.3	3.1
WW C4	.9	1.1	1.4	WW C4	.9	1.1	1.4	SS B5	1.4	2.2	3.0	TW H1	1.8	2.5	3.4
TW A1	1.5	2.5	3.5	TW A1	1.5	2.3	3.1	SS B6	1.0	1.4	1.8	TW H2	1.8	2.5	3.3
TW A2	1.3	2.0	2.7	TW A2	1.3	1.8	2.4	SS B7	1.5	2.5	3.6	TW H3	1.9	2.6	3.5
TW A3	1.3	1.8	2.4	TW A3	1.2	1.7	2.2	SS B8	1.0	1.5	1.9	TW H4	2.2	3.0	4.0
TW A4	1.3	1.8	2.3	TW A4	1.2	1.7	2.1	SS C1	1.3	1.7	2.3	TW I1	2.0	3.2	5.0
TW B1	1.5	2.3	3.2	TW B1	1.5	2.2	2.9	SS C2	1.1	1.4	1.8	TW I2	1.3	1.7	2.3
TW B2	1.4	2.0	2.6	TW B2	1.3	1.9	2.4	SS C3	1.5	2.1	2.8	TW I3	1.2	1.6	2.1
TW B3	1.4	1.9	2.6	TW B3	1.3	1.8	2.3	SS C4	1.2	1.6	2.0	TW I4	1.1	1.4	1.8
TW B4	1.4	2.0	2.6	TW B4	1.4	1.9	2.4	SS D1	.9	1.4	1.9	TW I5	1.1	1.4	1.8
TW C1	1.5	2.2	3.0	TW C1	1.5	2.1	2.8	SS D2	.8	1.1	1.4	TW J1	1.1	1.4	1.9
TW C2	1.5	2.1	2.8	TW C2	1.4	2.0	2.6	SS D3	1.0	1.6	2.1	TW J2	1.1	1.4	1.8
TW C3	1.5	2.2	2.9	TW C3	1.5	2.1	2.7	SS D4	.8	1.1	1.5	TW J3	.9	1.2	1.5
TW C4	1.7	2.4	3.2	TW C4	1.6	2.3	3.0	SS E1	1.1	1.6	2.1	TW J4	1.0	1.2	1.6
TW D1	1.7	2.8	4.2	TW D1	1.7	2.5	3.3	SS E2	.9	1.2	1.6	DG A1	1.7	---	---
TW D2	1.2	1.6	2.1	TW D2	1.1	1.5	2.0	SS E3	1.2	1.9	2.6	DG A2	1.4	2.3	---
TW D3	1.1	1.5	2.0	TW D3	1.1	1.4	1.8	SS E4	.9	1.3	1.7	DG A3	1.2	1.5	2.3
DG B1	1.6	---	---	TW D4	1.0	1.3	1.6					DG B1	1.7	---	---
DG B2	1.3	1.8	3.7	TW D5	1.0	1.3	1.6	ROCK SPRINGS, WY				DG B2	1.4	1.9	4.3
DG B3	1.2	1.4	1.9	TW E1	1.0	1.3	1.7		\multicolumn{3}{c}{SOLAR SAVINGS}		DG B3	1.2	1.4	1.9
DG C1	1.3	2.1	---	TW E2	1.0	1.3	1.6		10%	50%	80%	DG C1	1.4	2.3	---
DG C2	1.2	1.5	2.4	TW E3	.8	1.1	1.4	WW A1	1.6	2.7	3.8	DG C2	1.3	1.6	2.5
DG C3	1.1	1.3	1.6	TW E4	.9	1.1	1.4	WW A2	1.3	1.9	2.5	DG C3	1.1	1.3	1.7
SS A1	1.1	1.8	2.6	TW F1	1.5	2.3	3.0	WW A3	1.2	1.7	2.3	SS A1	1.1	1.8	2.6
SS A2	.9	1.3	1.8	TW F2	1.4	1.9	2.5	WW A4	1.2	1.6	2.1	SS A2	.9	1.3	1.8
SS A3	1.2	2.1	3.2	TW F3	1.3	1.8	2.3	WW A5	1.1	1.5	2.0	SS A3	1.2	2.1	3.3
SS A4	.9	1.4	1.9	TW F4	1.4	1.8	2.3	WW A6	1.1	1.5	2.0	SS A4	.9	1.4	1.9
SS A5	1.2	2.1	3.0	TW G1	1.6	2.2	2.9	WW B1	1.6	2.8	4.4	SS A5	1.2	2.1	3.1
SS A6	.9	1.3	1.8	TW G2	1.5	2.0	2.6	WW B2	1.1	1.5	2.0	SS A6	.9	1.4	1.9
SS A7	1.3	2.6	4.4	TW G3	1.5	2.0	2.6	WW B3	1.0	1.3	1.7	SS A7	1.3	2.7	4.6
SS A8	.9	1.4	2.0	TW G4	1.6	2.2	2.8	WW B4	1.0	1.2	1.6	SS A8	1.0	1.4	2.0
SS B1	1.3	2.0	2.9	TW H1	1.6	2.3	2.9	WW B5	1.0	1.2	1.6	SS B1	1.3	2.1	2.9
SS B2	1.0	1.4	1.9	TW H2	1.7	2.3	2.9	WW C1	1.0	1.3	1.7	SS B2	1.0	1.5	2.0
SS B3	1.3	2.3	3.3	TW H3	1.8	2.4	3.1	WW C2	1.0	1.3	1.6	SS B3	1.4	2.3	3.4
SS B4	1.0	1.5	2.0	TW H4	2.1	2.8	3.5	WW C3	.8	1.1	1.4	SS B4	1.0	1.5	2.1
SS B5	1.4	2.4	3.6	TW I1	1.8	2.6	3.5	WW C4	.9	1.1	1.4	SS B5	1.5	2.5	3.7
SS B6	1.0	1.5	2.0	TW I2	1.2	1.6	2.1	TW A1	1.6	2.6	3.7	SS B6	1.0	1.5	2.0
SS B7	1.6	2.9	4.7	TW I3	1.1	1.5	1.9	TW A2	1.4	2.0	2.8	SS B7	1.6	3.0	5.0
SS B8	1.1	1.6	2.1	TW I4	1.0	1.3	1.7	TW A3	1.3	1.9	2.5	SS B8	1.1	1.6	2.1
SS C1	1.3	1.9	2.5	TW I5	1.0	1.3	1.7	TW A4	1.3	1.8	2.4	SS C1	1.3	1.9	2.6
SS C2	1.1	1.5	1.9	TW J1	1.0	1.3	1.7	TW B1	1.6	2.4	3.4	SS C2	1.1	1.5	2.0
SS C3	1.5	2.3	3.1	TW J2	1.0	1.3	1.7	TW B2	1.4	2.0	2.7	SS C3	1.6	2.4	3.3
SS C4	1.2	1.6	2.1	TW J3	.9	1.1	1.4	TW B3	1.4	2.0	2.6	SS C4	1.2	1.7	2.2
SS D1	1.0	1.5	2.1	TW J4	.9	1.2	1.5	TW B4	1.5	2.0	2.7	SS D1	1.0	1.5	2.2
SS D2	.8	1.1	1.5	DG A1	1.5	---	---	TW C1	1.6	2.3	3.1	SS D2	.8	1.2	1.6
SS D3	1.1	1.7	2.5	DG A2	1.4	2.0	8.1	TW C2	1.5	2.2	2.9	SS D3	1.1	1.7	2.5
SS D4	.8	1.2	1.6	DG A3	1.2	1.4	2.1	TW C3	1.6	2.2	3.0	SS D4	.8	1.2	1.6
SS E1	1.1	1.7	2.5	DG B1	1.5	3.1	---	TW C4	1.7	2.5	3.3	SS E1	1.1	1.8	2.5
SS E2	.9	1.3	1.7	DG B2	1.3	1.7	2.9	TW D1	1.8	3.0	4.5	SS E2	.9	1.3	1.8
SS E3	1.3	2.1	3.0	DG B3	1.2	1.3	1.8	TW D2	1.2	1.7	2.2	SS E3	1.3	2.1	3.1
SS E4	1.0	1.4	1.9	DG C1	1.3	1.8	---	TW D3	1.1	1.5	2.0	SS E4	1.0	1.4	1.9

SHERIDAN, WY

	SOLAR SAVINGS		
	10%	50%	80%
WW A1	1.8	7.6	---
WW A2	1.5	2.7	4.5
WW A3	1.4	2.3	3.6
WW A4	1.4	2.2	3.2
WW A5	1.3	2.1	3.0
WW A6	1.3	2.0	2.9
WW B1	2.1	---	---
WW B2	1.3	1.9	2.6
WW B3	1.1	1.6	2.2
WW B4	1.1	1.5	1.9
WW B5	1.1	1.4	1.9
WW C1	1.1	1.5	2.1
WW C2	1.1	1.5	2.0
WW C3	.9	1.2	1.6
WW C4	1.0	1.3	1.7
TW A1	1.8	4.9	---
TW A2	1.6	2.9	4.9
TW A3	1.5	2.6	4.0
TW A4	1.5	2.4	3.6
TW B1	1.8	3.7	8.5
TW B2	1.6	2.8	4.3
TW B3	1.6	2.6	3.9
TW B4	1.7	2.7	3.9
TW C1	1.8	3.2	5.2
TW C2	1.8	2.9	4.3
TW C3	1.8	2.9	4.3
TW C4	1.9	3.2	4.6
TW D1	2.3	---	---
TW D2	1.4	2.1	2.9
TW D3	1.3	1.9	2.7
TW D4	1.2	1.6	2.2
TW D5	1.1	1.6	2.1
TW E1	1.2	1.7	2.3
TW E2	1.2	1.6	2.2
TW E3	1.0	1.3	1.8
TW E4	1.0	1.4	1.8
TW F1	1.9	4.6	---
TW F2	1.7	3.1	5.2
TW F3	1.7	2.8	4.3
TW F4	1.7	2.7	3.9
TW G1	2.0	3.9	8.8
TW G2	1.8	3.1	4.8
TW G3	1.8	3.0	4.4
TW G4	2.0	3.1	4.5
TW H1	2.1	3.6	6.2
TW H2	2.1	3.4	5.1
TW H3	2.2	3.5	5.1
TW H4	2.5	3.9	5.6
TW I1	2.6	---	---
TW I2	1.5	2.2	3.0
TW I3	1.4	2.0	2.8
TW I4	1.2	1.7	2.3
TW I5	1.2	1.6	2.1
TW J1	1.2	1.8	2.4
TW J2	1.2	1.7	2.3
TW J3	1.0	1.4	1.8
TW J4	1.1	1.4	1.9
DG A1	2.2	---	---
DG A2	1.6	4.6	---
DG A3	1.4	1.9	3.3
DG B1	2.1	---	---
DG B2	1.6	2.8	---
DG B3	1.3	1.7	2.5
DG C1	1.7	---	---

SHERIDAN, WY

	SOLAR SAVINGS		
	10%	50%	80%
DG C2	1.4	2.1	---
DG C3	1.2	1.6	2.1
SS A1	1.3	2.6	4.7
SS A2	1.1	1.7	2.3
SS A3	1.4	4.2	---
SS A4	1.1	1.8	2.6
SS A5	1.4	3.4	12.7
SS A6	1.1	1.7	2.4
SS A7	1.5	---	---
SS A8	1.1	1.9	2.7
SS B1	1.5	3.1	5.9
SS B2	1.2	1.8	2.5
SS B3	1.6	4.4	---
SS B4	1.2	1.9	2.7
SS B5	1.7	4.6	---
SS B6	1.2	1.8	2.6
SS B7	1.9	---	---
SS B8	1.2	2.0	2.9
SS C1	1.6	2.9	5.2
SS C2	1.3	1.9	2.6
SS C3	1.9	3.6	7.6
SS C4	1.4	2.1	2.9
SS D1	1.2	2.2	3.6
SS D2	.9	1.4	2.0
SS D3	1.2	2.5	4.6
SS D4	1.0	1.5	2.1
SS E1	1.3	2.8	6.5
SS E2	1.1	1.6	2.3
SS E3	1.5	3.6	16.8
SS E4	1.1	1.8	2.5

EDMONTON, ALTA

	SOLAR SAVINGS		
	10%	50%	80%
WW A1	2.1	---	---
WW A2	1.8	---	---
WW A3	1.7	---	---
WW A4	1.6	---	---
WW A5	1.6	---	---
WW A6	1.6	---	---
WW B1	4.2	---	---
WW B2	1.5	3.3	---
WW B3	1.3	2.5	---
WW B4	1.2	1.9	2.9
WW B5	1.2	1.8	2.5
WW C1	1.2	2.2	4.3
WW C2	1.2	2.0	3.2
WW C3	1.0	1.5	2.1
WW C4	1.1	1.5	2.1
TW A1	2.1	---	---
TW A2	1.9	---	---
TW A3	1.8	---	---
TW A4	1.8	---	---
TW B1	2.1	---	---
TW B2	1.9	---	---
TW B3	1.9	---	---
TW B4	1.9	---	---
TW C1	2.1	---	---
TW C2	2.0	---	---
TW C3	2.0	---	---
TW C4	2.2	14.2	---
TW D1	4.2	---	---
TW D2	1.6	3.7	---
TW D3	1.5	3.5	---

EDMONTON, ALTA

	SOLAR SAVINGS		
	10%	50%	80%
TW D4	1.3	2.1	3.4
TW D5	1.3	1.9	2.8
TW E1	1.3	2.5	6.7
TW E2	1.3	2.2	3.8
TW E3	1.1	1.7	2.4
TW E4	1.1	1.7	2.4
TW F1	2.3	---	---
TW F2	2.1	---	---
TW F3	2.0	---	---
TW F4	2.0	---	---
TW G1	2.3	---	---
TW G2	2.2	---	---
TW G3	2.2	---	---
TW G4	2.4	---	---
TW H1	2.5	---	---
TW H2	2.5	---	---
TW H3	2.6	---	---
TW H4	2.9	---	---
TW I1	---	---	---
TW I2	1.7	4.0	---
TW I3	1.6	3.3	---
TW I4	1.4	2.2	3.5
TW I5	1.3	2.0	2.9
TW J1	1.4	2.5	5.5
TW J2	1.4	2.2	3.7
TW J3	1.1	1.7	2.4
TW J4	1.2	1.7	2.4
DG A1	4.7	---	---
DG A2	1.9	---	---
DG A3	1.5	2.5	---
DG B1	3.3	---	---
DG B2	1.8	---	---
DG B3	1.4	2.2	4.9
DG C1	2.1	---	---
DG C2	1.6	6.0	---
DG C3	1.3	2.0	3.4
SS A1	1.6	---	---
SS A2	1.2	2.5	5.3
SS A3	1.8	---	---
SS A4	1.3	3.2	---
SS A5	1.7	---	---
SS A6	1.2	2.7	32.3
SS A7	2.0	---	---
SS A8	1.3	4.6	---
SS B1	1.9	---	---
SS B2	1.3	2.5	4.8
SS B3	2.1	---	---
SS B4	1.4	3.0	9.1
SS B5	2.1	---	---
SS B6	1.4	2.8	6.5
SS B7	2.5	---	---
SS B8	1.5	3.6	---
SS C1	2.0	---	---
SS C2	1.5	3.1	---
SS C3	2.2	---	---
SS C4	1.6	3.5	---
SS D1	1.4	---	---
SS D2	1.1	2.0	3.6
SS D3	1.5	---	---
SS D4	1.1	2.1	3.9
SS E1	1.8	---	---
SS E2	1.3	2.7	---
SS E3	1.9	---	---
SS E4	1.3	3.0	---

SUFFIELD, ALTA

	SOLAR SAVINGS		
	10%	50%	80%
WW A1	1.7	10.9	---
WW A2	1.4	2.5	6.1
WW A3	1.3	2.2	4.2
WW A4	1.3	2.1	3.6
WW A5	1.2	2.0	3.3
WW A6	1.2	1.9	3.1
WW B1	1.9	---	---
WW B2	1.2	1.8	2.7
WW B3	1.1	1.6	2.3
WW B4	1.0	1.4	1.9
WW B5	1.0	1.4	1.9
WW C1	1.0	1.5	2.1
WW C2	1.0	1.5	2.0
WW C3	.9	1.2	1.6
WW C4	.9	1.3	1.7
TW A1	1.7	4.6	---
TW A2	1.5	2.7	6.6
TW A3	1.4	2.4	4.6
TW A4	1.4	2.3	3.9
TW B1	1.7	3.5	---
TW B2	1.5	2.6	5.0
TW B3	1.5	2.5	4.3
TW B4	1.6	2.5	4.2
TW C1	1.7	3.0	6.5
TW C2	1.6	2.7	4.7
TW C3	1.7	2.8	4.6
TW C4	1.8	3.0	4.9
TW D1	2.1	---	---
TW D2	1.3	2.0	3.0
TW D3	1.2	1.8	2.8
TW D4	1.1	1.6	2.2
TW D5	1.1	1.5	2.0
TW E1	1.1	1.6	2.4
TW E2	1.1	1.6	2.2
TW E3	.9	1.3	1.8
TW E4	1.0	1.3	1.8
TW F1	1.8	4.4	---
TW F2	1.6	2.9	7.4
TW F3	1.5	2.6	5.1
TW F4	1.6	2.5	4.4
TW G1	1.8	3.7	---
TW G2	1.7	2.9	6.0
TW G3	1.7	2.8	5.0
TW G4	1.9	2.9	4.9
TW H1	1.9	3.4	8.7
TW H2	1.9	3.2	5.8
TW H3	2.1	3.3	5.6
TW H4	2.4	3.7	6.0
TW I1	2.3	---	---
TW I2	1.4	2.1	3.1
TW I3	1.3	1.9	2.9
TW I4	1.2	1.6	2.3
TW I5	1.1	1.6	2.1
TW J1	1.1	1.7	2.4
TW J2	1.1	1.6	2.3
TW J3	1.0	1.3	1.8
TW J4	1.0	1.4	1.8
DG A1	1.9	---	---
DG A2	1.5	3.5	---
DG A3	1.3	1.8	3.3
DG B1	1.9	---	---
DG B2	1.5	2.6	---
DG B3	1.3	1.6	2.5
DG C1	1.6	---	---

SUFFIELD, ALTA				NANAIMO, BC				VANCOUVER, BC				VANCOUVER, BC			
	SOLAR SAVINGS				SOLAR SAVINGS				SOLAR SAVINGS				SOLAR SAVINGS		
	10%	50%	80%		10%	50%	80%		10%	50%	80%		10%	50%	80%
DG C2	1.4	2.0	---	TW D4	1.0	1.7	2.5	WW A1	1.8	---	---	DG C2	1.4	3.4	---
DG C3	1.2	1.5	2.1	TW D5	1.0	1.6	2.3	WW A2	1.5	---	---	DG C3	1.2	1.9	2.8
SS A1	1.3	2.8	9.3	TW E1	1.0	1.8	3.1	WW A3	1.5	---	---	SS A1	1.3	---	---
SS A2	1.0	1.7	2.5	TW E2	1.0	1.7	2.7	WW A4	1.5	---	---	SS A2	1.0	2.0	3.3
SS A3	1.4	---	---	TW E3	.9	1.4	2.0	WW A5	1.4	4.9	---	SS A3	1.4	---	---
SS A4	1.1	1.8	2.8	TW E4	.9	1.4	2.0	WW A6	1.4	4.0	---	SS A4	1.0	2.3	4.1
SS A5	1.3	3.9	---	TW F1	1.5	---	---	WW B1	2.3	---	---	SS A5	1.3	---	---
SS A6	1.0	1.7	2.6	TW F2	1.4	5.5	---	WW B2	1.3	2.7	5.6	SS A6	1.0	2.1	3.7
SS A7	1.5	---	---	TW F3	1.4	3.7	---	WW B3	1.2	2.1	3.6	SS A7	1.4	---	---
SS A8	1.1	1.9	3.1	TW F4	1.5	3.3	---	WW B4	1.1	1.7	2.5	SS A8	1.1	2.5	5.5
SS B1	1.5	3.2	37.5	TW G1	1.6	---	---	WW B5	1.1	1.7	2.3	SS B1	1.5	---	---
SS B2	1.1	1.8	2.6	TW G2	1.5	4.3	---	WW C1	1.1	1.9	3.0	SS B2	1.1	2.2	3.4
SS B3	1.6	---	---	TW G3	1.6	3.7	---	WW C2	1.1	1.8	2.7	SS B3	1.6	---	---
SS B4	1.2	1.9	2.9	TW G4	1.7	3.7	---	WW C3	1.0	1.4	1.9	SS B4	1.2	2.3	3.9
SS B5	1.6	5.9	---	TW H1	1.7	7.1	---	WW C4	1.0	1.5	2.0	SS B5	1.6	---	---
SS B6	1.2	1.8	2.7	TW H2	1.8	4.3	---	TW A1	1.8	---	---	SS B6	1.2	2.3	3.7
SS B7	1.8	---	---	TW H3	1.9	4.1	---	TW A2	1.6	---	---	SS B7	1.8	---	---
SS B8	1.2	2.0	3.1	TW H4	2.2	4.4	---	TW A3	1.6	---	---	SS B8	1.2	2.6	4.8
SS C1	1.5	2.7	8.3	TW I1	2.0	---	---	TW A4	1.6	5.6	---	SS C1	1.7	---	---
SS C2	1.2	1.8	2.7	TW I2	1.3	2.4	5.2	TW B1	1.8	---	---	SS C2	1.3	2.5	4.8
SS C3	1.7	3.4	27.3	TW I3	1.2	2.2	4.0	TW B2	1.7	---	---	SS C3	1.9	---	---
SS C4	1.3	2.0	3.0	TW I4	1.1	1.7	2.6	TW B3	1.6	6.3	---	SS C4	1.4	2.9	5.9
SS D1	1.1	2.3	5.6	TW I5	1.1	1.6	2.4	TW B4	1.7	4.7	---	SS D1	1.1	---	---
SS D2	.9	1.4	2.1	TW J1	1.1	1.8	3.1	TW C1	1.8	---	---	SS D2	1.0	1.7	2.6
SS D3	1.2	2.7	11.7	TW J2	1.1	1.8	2.7	TW C2	1.8	5.8	---	SS D3	1.2	---	---
SS D4	1.0	1.5	2.2	TW J3	.9	1.4	2.0	TW C3	1.8	5.0	---	SS D4	1.0	1.8	2.8
SS E1	1.3	3.1	---	TW J4	1.0	1.4	2.0	TW C4	1.9	4.9	---	SS E1	1.3	---	---
SS E2	1.0	1.6	2.5	DG A1	1.6	---	---	TW D1	2.6	---	---	SS E2	1.1	2.1	3.6
SS E3	1.5	4.1	---	DG A2	1.4	9.5	---	TW D2	1.4	2.9	7.0	SS E3	1.5	---	---
SS E4	1.1	1.8	2.7	DG A3	1.2	1.9	5.2	TW D3	1.3	2.7	5.7	SS E4	1.1	2.3	4.1
				DG B1	1.6	---	---	TW D4	1.2	1.9	2.9				
NANAIMO, BC				DG B2	1.3	3.4	---	TW D5	1.2	1.8	2.6	WINNIPEG, MAN			
	SOLAR SAVINGS			DG B3	1.2	1.8	3.0	TW E1	1.2	2.2	3.6		SOLAR SAVINGS		
	10%	50%	80%	DG C1	1.4	---	---	TW E2	1.2	2.0	3.1		10%	50%	80%
WW A1	1.5	---	---	DG C2	1.2	2.3	---	TW E3	1.0	1.6	2.2	WW A1	2.3	---	---
WW A2	1.3	4.2	---	DG C3	1.1	1.6	2.5	TW E4	1.1	1.6	2.2	WW A2	2.0	---	---
WW A3	1.2	3.0	---	SS A1	1.1	3.5	---	TW F1	1.9	---	---	WW A3	1.8	---	---
WW A4	1.2	2.7	---	SS A2	.9	1.7	2.9	TW F2	1.8	---	---	WW A4	1.8	---	---
WW A5	1.2	2.5	---	SS A3	1.1	---	---	TW F3	1.7	---	---	WW A5	1.7	---	---
WW A6	1.1	2.4	---	SS A4	.9	1.9	3.4	TW F4	1.8	9.8	---	WW A6	1.7	5.1	---
WW B1	1.6	---	---	SS A5	1.1	---	---	TW G1	2.0	---	---	WW B1	---	---	---
WW B2	1.1	2.1	4.2	SS A6	.9	1.8	3.2	TW G2	1.9	---	---	WW B2	1.6	2.7	6.2
WW B3	1.0	1.7	3.0	SS A7	1.2	---	---	TW G3	1.9	---	---	WW B3	1.4	2.2	3.8
WW B4	1.0	1.5	2.2	SS A8	.9	2.0	4.2	TW G4	2.1	6.3	---	WW B4	1.2	1.8	2.5
WW B5	1.0	1.5	2.1	SS B1	1.2	5.0	---	TW H1	2.1	---	---	WW B5	1.2	1.7	2.4
WW C1	1.0	1.6	2.6	SS B2	1.0	1.8	3.0	TW H2	2.2	---	---	WW C1	1.3	2.0	3.2
WW C2	1.0	1.6	2.4	SS B3	1.3	---	---	TW H3	2.3	7.5	---	WW C2	1.3	1.9	2.8
WW C3	.9	1.3	1.8	SS B4	1.0	2.0	3.4	TW H4	2.6	6.7	---	WW C3	1.1	1.5	2.0
WW C4	.9	1.3	1.8	SS B5	1.4	---	---	TW I1	3.1	---	---	WW C4	1.1	1.5	2.0
TW A1	1.5	---	---	SS B6	1.0	1.9	3.3	TW I2	1.5	3.1	7.5	TW A1	2.3	---	---
TW A2	1.3	4.3	---	SS B7	1.5	---	---	TW I3	1.4	2.7	4.9	TW A2	2.1	---	---
TW A3	1.3	3.3	---	SS B8	1.1	2.1	4.0	TW I4	1.3	2.0	3.0	TW A3	2.0	---	---
TW A4	1.3	2.9	---	SS C1	1.3	---	---	TW I5	1.2	1.9	2.7	TW A4	1.9	---	---
TW B1	1.5	---	---	SS C2	1.1	2.0	3.8	TW J1	1.3	2.2	3.6	TW B1	2.3	---	---
TW B2	1.4	3.4	---	SS C3	1.5	---	---	TW J2	1.2	2.1	3.1	TW B2	2.1	---	---
TW B3	1.4	3.1	---	SS C4	1.2	2.2	4.5	TW J3	1.1	1.6	2.2	TW B3	2.0	---	---
TW B4	1.4	3.0	---	SS D1	1.0	2.8	---	TW J4	1.1	1.6	2.2	TW B4	2.1	5.3	---
TW C1	1.5	4.2	---	SS D2	.8	1.5	2.4	DG A1	2.5	---	---	TW C1	2.3	---	---
TW C2	1.5	3.3	---	SS D3	1.0	3.5	---	DG A2	1.6	---	---	TW C2	2.2	7.6	---
TW C3	1.5	3.3	---	SS D4	.8	1.5	2.5	DG A3	1.3	2.3	---	TW C3	2.2	5.6	---
TW C4	1.6	3.4	---	SS E1	1.1	---	---	DG B1	2.3	---	---	TW C4	2.3	5.2	---
TW D1	1.8	---	---	SS E2	.9	1.7	3.1	DG B2	1.6	---	---	TW D1	17.0	---	---
TW D2	1.2	2.2	4.9	SS E3	1.2	---	---	DG B3	1.3	2.1	3.7	TW D2	1.7	3.0	8.5
TW D3	1.1	2.1	4.3	SS E4	1.0	1.9	3.4	DG C1	1.8	---	---	TW D3	1.6	2.8	7.1

WINNIPEG, MAN			
	\multicolumn{3}{c}{SOLAR SAVINGS}
	10%	50%	80%
WW A1	2.3	---	---
WW A2	2.0	---	---
WW A3	1.8	---	---
WW A4	1.8	---	---
WW A5	1.7	---	---
WW A6	1.7	5.1	---
WW B1	---	---	---
WW B2	1.6	2.7	6.2
WW B3	1.4	2.2	3.8
WW B4	1.2	1.8	2.5
WW B5	1.2	1.7	2.4
WW C1	1.3	2.0	3.2
WW C2	1.3	1.9	2.8
WW C3	1.1	1.5	2.0
WW C4	1.1	1.5	2.0
TW A1	2.3	---	---
TW A2	2.1	---	---
TW A3	2.0	---	---
TW A4	1.9	---	---
TW B1	2.3	---	---
TW B2	2.1	---	---
TW B3	2.0	---	---
TW B4	2.1	5.3	---
TW C1	2.3	---	---
TW C2	2.2	7.6	---
TW C3	2.2	5.6	---
TW C4	2.3	5.2	---
TW D1	17.0	---	---
TW D2	1.7	3.0	8.5
TW D3	1.6	2.8	7.1
DG B1	5.3	---	---
DG B2	2.0	---	---
DG B3	1.5	2.1	3.8
DG C1	2.5	---	---
DG C2	1.7	3.8	---
DG C3	1.4	1.9	2.9
SS A1	1.7	---	---
SS A2	1.3	2.3	3.7
SS A3	2.0	---	---
SS A4	1.3	2.6	5.3
SS A5	1.8	---	---
SS A6	1.3	2.4	4.3
SS A7	2.1	---	---
SS A8	1.4	3.1	---
SS B1	2.0	---	---
SS B2	1.4	2.3	3.7
SS B3	2.4	---	---
SS B4	1.5	2.6	4.5
SS B5	2.3	---	---
SS B6	1.4	2.5	4.2
SS B7	2.8	---	---
SS B8	1.5	3.0	6.3
SS C1	2.2	---	---
SS C2	1.5	2.6	5.1
SS C3	2.5	---	---
SS C4	1.7	3.0	6.6
SS D1	1.5	---	---
SS D2	1.1	1.9	2.9
SS D3	1.6	---	---
SS D4	1.2	2.0	3.1
SS E1	1.9	---	---
SS E2	1.3	2.3	4.2
SS E3	2.1	---	---
SS E4	1.4	2.6	4.9

DARTMOUTH, NS			
	\multicolumn{3}{c}{SOLAR SAVINGS}
	10%	50%	80%
WW A1	2.2	---	---
WW A2	1.8	3.6	---
WW A3	1.7	2.9	5.8
WW A4	1.6	2.7	4.5
WW A5	1.6	2.5	3.9
WW A6	1.5	2.4	3.6
WW B1	3.3	---	---
WW B2	1.5	2.2	3.1
WW B3	1.3	1.8	2.5
WW B4	1.2	1.6	2.1
WW B5	1.2	1.6	2.0
WW C1	1.2	1.7	2.3
WW C2	1.2	1.7	2.2
WW C3	1.0	1.3	1.7
WW C4	1.1	1.4	1.8
TW A1	2.1	---	---
TW A2	1.9	3.9	---
TW A3	1.8	3.2	6.3
TW A4	1.8	2.9	4.7
TW B1	2.1	6.5	---
TW B2	1.9	3.4	6.8
TW B3	1.9	3.2	5.2
TW B4	1.9	3.1	4.9
TW C1	2.1	4.1	23.6
TW C2	2.0	3.4	5.6
TW C3	2.1	3.5	5.4
TW C4	2.2	3.7	5.6
TW D1	3.4	---	---
TW D2	1.6	2.4	3.4
TW D3	1.5	2.2	3.1
TW D4	1.3	1.8	2.4
TW D5	1.3	1.7	2.2
TW E1	1.3	1.9	2.6
TW E2	1.3	1.8	2.5
TW E3	1.1	1.5	1.9
TW E4	1.1	1.5	1.9
TW F1	2.3	---	---
TW F2	2.1	4.2	---
TW F3	2.0	3.5	7.5
TW F4	2.0	3.2	5.3
TW G1	2.4	7.5	---
TW G2	2.2	3.9	12.1
TW G3	2.2	3.6	6.3
TW G4	2.3	3.7	5.8
TW H1	2.5	5.0	---
TW H2	2.4	4.1	7.4
TW H3	2.6	4.2	6.6
TW H4	2.9	4.6	7.0
TW I1	4.3	---	---
TW I2	1.7	2.5	3.6
TW I3	1.5	2.3	3.2
TW I4	1.4	1.9	2.5
TW I5	1.3	1.8	2.3
TW J1	1.4	2.0	2.7
TW J2	1.3	1.9	2.5
TW J3	1.1	1.5	1.9
TW J4	1.2	1.5	2.0
DG A1	3.7	---	---
DG A2	1.9	---	---
DG A3	1.5	2.1	4.2
DG B1	3.3	---	---
DG B2	1.9	4.4	---
DG B3	1.5	1.9	2.8
DG C1	2.3	---	---

DARTMOUTH, NS			
	\multicolumn{3}{c}{SOLAR SAVINGS}
	10%	50%	80%
DG C2	1.6	2.6	---
DG C3	1.4	1.7	2.3
SS A1	1.5	3.1	10.3
SS A2	1.1	1.8	2.5
SS A3	1.6	---	---
SS A4	1.2	1.9	2.8
SS A5	1.5	5.0	---
SS A6	1.2	1.8	2.7
SS A7	1.7	---	---
SS A8	1.2	2.0	3.1
SS B1	1.7	3.8	---
SS B2	1.2	1.9	2.7
SS B3	1.9	---	---
SS B4	1.3	2.0	2.9
SS B5	1.9	---	---
SS B6	1.3	2.0	2.8
SS B7	2.2	---	---
SS B8	1.3	2.2	3.2
SS C1	1.9	4.2	---
SS C2	1.4	2.1	3.0
SS C3	2.2	6.1	---
SS C4	1.6	2.4	3.4
SS D1	1.3	2.5	5.3
SS D2	1.0	1.5	2.1
SS D3	1.4	3.0	---
SS D4	1.0	1.6	2.2
SS E1	1.5	4.0	---
SS E2	1.2	1.8	2.5
SS E3	1.7	6.6	---
SS E4	1.2	1.9	2.8

MOOSONEE, ONT			
	\multicolumn{3}{c}{SOLAR SAVINGS}
	10%	50%	80%
WW A1	4.4	---	---
WW A2	3.3	---	---
WW A3	2.9	---	---
WW A4	2.7	---	---
WW A5	2.6	---	---
WW A6	2.5	---	---
WW B1	---	---	---
WW B2	2.0	---	---
WW B3	1.7	---	---
WW B4	1.5	2.3	3.7
WW B5	1.4	2.1	3.1
WW C1	1.6	3.3	---
WW C2	1.5	2.5	4.6
WW C3	1.2	1.7	2.4
WW C4	1.3	1.8	2.4
TW A1	3.9	---	---
TW A2	3.2	---	---
TW A3	2.9	---	---
TW A4	2.8	---	---
TW B1	3.6	---	---
TW B2	3.0	---	---
TW B3	2.9	---	---
TW B4	2.8	---	---
TW C1	3.2	---	---
TW C2	2.9	---	---
TW C3	2.9	---	---
TW C4	3.0	---	---
TW D1	---	---	---
TW D2	2.1	---	---
TW D3	2.0	---	---
TW D4	1.6	2.6	4.7
TW D5	1.5	2.3	3.4
TW E1	1.7	4.8	---
TW E2	1.6	2.9	6.4
TW E3	1.3	1.9	2.9
TW E4	1.3	1.9	2.7
TW F1	6.1	---	---
TW F2	4.0	---	---
TW F3	3.6	---	---
TW F4	3.3	---	---
TW G1	5.9	---	---
TW G2	4.0	---	---
TW G3	3.7	---	---
TW G4	3.6	---	---
TW H1	5.3	---	---
TW H2	4.2	---	---
TW H3	4.1	---	---
TW H4	4.3	---	---
TW I1	---	---	---
TW I2	2.3	---	---
TW I3	2.1	---	---
TW I4	1.7	2.7	4.7
TW I5	1.6	2.4	3.6
TW J1	1.8	4.1	---
TW J2	1.7	2.9	5.6
TW J3	1.4	1.9	2.8
TW J4	1.4	1.9	2.8
DG A1	---	---	---
DG A2	2.8	---	---
DG A3	1.8	3.4	---
DG B1	---	---	---
DG B2	2.7	---	---
DG B3	1.8	2.8	---
DG C1	---	---	---
DG C2	2.2	---	---
DG C3	1.6	2.4	5.4
SS A1	2.3	---	---
SS A2	1.5	3.3	---
SS A3	3.2	---	---
SS A4	1.6	---	---
SS A5	2.4	---	---
SS A6	1.5	4.5	---
SS A7	3.3	---	---
SS A8	1.7	---	---
SS B1	2.9	---	---
SS B2	1.6	3.3	7.9
SS B3	4.9	---	---
SS B4	1.7	4.6	---
SS B5	3.5	---	---
SS B6	1.7	4.0	---
SS B7	---	---	---
SS B8	1.8	---	---
SS C1	---	---	---
SS C2	2.0	---	---
SS C3	5.0	---	---
SS C4	2.1	---	---
SS D1	2.1	---	---
SS D2	1.3	2.5	5.1
SS D3	2.2	---	---
SS D4	1.4	2.7	5.8
SS E1	3.6	---	---
SS E2	1.6	11.7	---
SS E3	3.2	---	---
SS E4	1.7	---	---

OTTAWA, ONT				OTTAWA, ONT				TORONTO, ONT				NORMANDIN, QUE			
	SOLAR SAVINGS				SOLAR SAVINGS				SOLAR SAVINGS				SOLAR SAVINGS		
	10%	50%	80%		10%	50%	80%		10%	50%	80%		10%	50%	80%
WW A1	2.2	---	---	DG C2	1.7	3.0	---	TW D4	1.3	1.8	2.5	WW A1	2.8	---	---
WW A2	1.9	---	---	DG C3	1.4	1.8	2.5	TW D5	1.3	1.7	2.3	WW A2	2.2	---	---
WW A3	1.8	3.9	---	SS A1	1.6	5.2	---	TW E1	1.3	2.0	2.8	WW A3	2.0	---	---
WW A4	1.7	3.3	---	SS A2	1.2	2.0	2.9	TW E2	1.3	1.9	2.5	WW A4	2.0	---	---
WW A5	1.6	3.0	---	SS A3	1.8	---	---	TW E3	1.1	1.5	2.0	WW A5	1.9	---	---
WW A6	1.6	2.8	6.4	SS A4	1.2	2.2	3.4	TW E4	1.1	1.5	2.0	WW A6	1.8	---	---
WW B1	4.6	---	---	SS A5	1.6	---	---	TW F1	2.4	---	---	WW B1	---	---	---
WW B2	1.5	2.4	3.6	SS A6	1.2	2.1	3.1	TW F2	2.1	5.0	---	WW B2	1.7	3.2	10.2
WW B3	1.3	2.0	2.9	SS A7	1.8	---	---	TW F3	2.0	3.8	---	WW B3	1.5	2.5	4.5
WW B4	1.2	1.7	2.3	SS A8	1.3	2.4	3.8	TW F4	2.0	3.5	6.7	WW B4	1.3	1.9	2.7
WW B5	1.2	1.6	2.2	SS B1	1.8	---	---	TW G1	2.4	---	---	WW B5	1.3	1.8	2.5
WW C1	1.2	1.8	2.6	SS B2	1.3	2.1	3.0	TW G2	2.2	4.4	---	WW C1	1.4	2.2	3.5
WW C2	1.3	1.8	2.4	SS B3	2.1	---	---	TW G3	2.2	3.9	9.3	WW C2	1.4	2.0	3.0
WW C3	1.0	1.4	1.8	SS B4	1.4	2.3	3.4	TW G4	2.4	3.9	6.6	WW C3	1.1	1.5	2.0
WW C4	1.1	1.5	1.9	SS B5	2.0	---	---	TW H1	2.5	6.0	---	WW C4	1.2	1.6	2.1
TW A1	2.2	---	---	SS B6	1.4	2.2	3.3	TW H2	2.5	4.5	15.3	TW A1	2.8	---	---
TW A2	2.0	9.8	---	SS B7	2.4	---	---	TW H3	2.6	4.5	7.7	TW A2	2.3	---	---
TW A3	1.9	4.2	---	SS B8	1.4	2.5	3.8	TW H4	3.0	4.9	7.7	TW A3	2.2	---	---
TW A4	1.8	3.5	---	SS C1	2.1	---	---	TW I1	5.4	---	---	TW A4	2.1	---	---
TW B1	2.2	---	---	SS C2	1.5	2.3	3.5	TW I2	1.7	2.6	3.8	TW B1	2.6	---	---
TW B2	2.0	4.5	---	SS C3	2.4	---	---	TW I3	1.6	2.4	3.3	TW B2	2.3	---	---
TW B3	1.9	3.8	---	SS C4	1.6	2.6	4.0	TW I4	1.4	1.9	2.6	TW B3	2.2	---	---
TW B4	2.0	3.6	8.0	SS D1	1.4	3.6	---	TW I5	1.3	1.8	2.4	TW B4	2.3	---	---
TW C1	2.2	6.0	---	SS D2	1.1	1.7	2.4	TW J1	1.4	2.0	2.8	TW C1	2.5	---	---
TW C2	2.1	4.1	---	SS D3	1.5	6.2	---	TW J2	1.4	1.9	2.6	TW C2	2.4	---	---
TW C3	2.1	4.0	8.7	SS D4	1.1	1.7	2.5	TW J3	1.1	1.5	2.0	TW C3	2.4	---	---
TW C4	2.2	4.2	7.7	SS E1	1.7	---	---	TW J4	1.2	1.6	2.0	TW C4	2.5	7.5	---
TW D1	4.5	---	---	SS E2	1.2	2.0	3.0	DG A1	4.5	---	---	TW D1	---	---	---
TW D2	1.6	2.6	4.1	SS E3	1.9	---	---	DG A2	1.9	---	---	TW D2	1.8	3.5	---
TW D3	1.5	2.4	3.7	SS E4	1.3	2.2	3.3	DG A3	1.5	2.1	4.8	TW D3	1.7	3.3	---
TW D4	1.3	1.9	2.6					DG B1	3.7	---	---	TW D4	1.4	2.1	3.1
TW D5	1.3	1.8	2.4	TORONTO, ONT				DG B2	1.9	6.0	---	TW D5	1.4	2.0	2.7
TW E1	1.4	2.1	3.0		SOLAR SAVINGS			DG B3	1.5	1.9	3.0	TW E1	1.5	2.5	4.4
TW E2	1.3	1.9	2.7		10%	50%	80%	DG C1	2.3	---	---	TW E2	1.4	2.2	3.4
TW E3	1.1	1.5	2.0	WW A1	2.2	---	---	DG C2	1.7	2.7	---	TW E3	1.2	1.7	2.3
TW E4	1.1	1.6	2.1	WW A2	1.8	4.2	---	DG C3	1.4	1.8	2.4	TW E4	1.2	1.7	2.3
TW F1	2.4	---	---	WW A3	1.7	3.2	---	SS A1	1.5	3.5	---	TW F1	3.0	---	---
TW F2	2.2	---	---	WW A4	1.6	2.9	5.9	SS A2	1.2	1.9	2.6	TW F2	2.6	---	---
TW F3	2.1	4.8	---	WW A5	1.6	2.6	4.6	SS A3	1.7	---	---	TW F3	2.4	---	---
TW F4	2.1	3.9	---	WW A6	1.6	2.5	4.1	SS A4	1.2	2.0	2.9	TW F4	2.4	---	---
TW G1	2.5	---	---	WW B1	3.7	---	---	SS A5	1.6	---	---	TW G1	3.0	---	---
TW G2	2.3	5.9	---	WW B2	1.5	2.3	3.2	SS A6	1.2	1.9	2.8	TW G2	2.7	---	---
TW G3	2.3	4.5	---	WW B3	1.3	1.9	2.7	SS A7	1.8	---	---	TW G3	2.6	---	---
TW G4	2.4	4.4	11.7	WW B4	1.2	1.6	2.2	SS A8	1.2	2.1	3.2	TW G4	2.8	---	---
TW H1	2.6	---	---	WW B5	1.2	1.6	2.1	SS B1	1.8	4.6	---	TW H1	3.1	---	---
TW H2	2.6	5.3	---	WW C1	1.2	1.8	2.4	SS B2	1.3	2.0	2.8	TW H2	3.0	---	---
TW H3	2.7	5.0	---	WW C2	1.2	1.7	2.3	SS B3	1.9	---	---	TW H3	3.1	---	---
TW H4	3.0	5.3	10.1	WW C3	1.0	1.4	1.8	SS B4	1.3	2.1	3.0	TW H4	3.4	---	---
TW I1	---	---	---	WW C4	1.1	1.4	1.9	SS B5	2.0	---	---	TW I1	---	---	---
TW I2	1.7	2.8	4.3	TW A1	2.2	---	---	SS B6	1.3	2.1	3.0	TW I2	1.9	3.8	---
TW I3	1.6	2.5	3.7	TW A2	1.9	4.5	---	SS B7	2.3	---	---	TW I3	1.8	3.2	6.7
TW I4	1.4	2.0	2.7	TW A3	1.8	3.5	---	SS B8	1.4	2.3	3.3	TW I4	1.5	2.2	3.2
TW I5	1.3	1.8	2.5	TW A4	1.8	3.1	5.6	SS C1	2.0	6.0	---	TW I5	1.4	2.0	2.8
TW J1	1.4	2.1	3.0	TW B1	2.2	---	---	SS C2	1.5	2.2	3.2	TW J1	1.5	2.5	4.2
TW J2	1.4	2.0	2.7	TW B2	1.9	3.8	---	SS C3	2.3	---	---	TW J2	1.5	2.3	3.4
TW J3	1.2	1.6	2.1	TW B3	1.9	3.4	6.4	SS C4	1.6	2.5	3.6	TW J3	1.2	1.7	2.3
TW J4	1.2	1.6	2.1	TW B4	2.0	3.3	5.5	SS D1	1.3	2.8	---	TW J4	1.3	1.7	2.3
DG A1	---	---	---	TW C1	2.1	4.6	---	SS D2	1.0	1.6	2.2	DG A1	---	---	---
DG A2	2.0	---	---	TW C2	2.0	3.7	6.6	SS D3	1.4	3.5	---	DG A2	2.2	---	---
DG A3	1.5	2.2	6.1	TW C3	2.1	3.7	6.1	SS D4	1.1	1.6	2.3	DG A3	1.6	2.6	---
DG B1	4.2	---	---	TW C4	2.2	3.9	6.2	SS E1	1.6	6.3	---	DG B1	---	---	---
DG B2	1.9	---	---	TW D1	3.8	---	---	SS E2	1.2	1.9	2.7	DG B2	2.2	---	---
DG B3	1.5	2.0	3.2	TW D2	1.6	2.5	3.6	SS E3	1.8	---	---	DG B3	1.6	2.3	4.4
DG C1	2.5	---	---	TW D3	1.5	2.3	3.3	SS E4	1.3	2.0	2.9	DG C1	3.4	---	---

Conservation Factor Tables / 389

CONSERVATION FACTORS

NORMANDIN, QUE

	SOLAR SAVINGS		
	10%	50%	80%
DG C2	1.8	6.7	---
DG C3	1.5	2.0	3.2
SS A1	1.8	---	---
SS A2	1.3	2.4	4.0
SS A3	2.1	---	---
SS A4	1.4	2.8	6.1
SS A5	1.9	---	---
SS A6	1.3	2.5	4.7
SS A7	2.3	---	---
SS A8	1.4	3.3	---
SS B1	2.1	---	---
SS B2	1.4	2.5	4.0
SS B3	2.5	---	---
SS B4	1.5	2.8	5.0
SS B5	2.4	---	---
SS B6	1.5	2.7	4.6
SS B7	3.1	---	---
SS B8	1.6	3.2	7.6
SS C1	2.5	---	---
SS C2	1.7	3.0	6.5
SS C3	2.9	---	---
SS C4	1.8	3.4	9.9
SS D1	1.6	---	---
SS D2	1.2	1.9	3.1
SS D3	1.7	---	---
SS D4	1.2	2.0	3.3
SS E1	2.0	---	---
SS E2	1.4	2.5	4.7
SS E3	2.2	---	---
SS E4	1.4	2.8	5.7

Appendix 18

Solar Load Ratio (*SLR*) Correlations

Masonry properties	
thermal conductivity (k)	
sunspace floor	0.5 Btu/hr/ft/°F
all other masonry	1.0 Btu/hr/ft/°F
density (ϱ)	150 lb/ft³
specific heat (c)	0.2 Btu/lb/°F
infrared emittance of normal surface	0.9
infrared emittance of selective surface	0.1
Solar absorptances	
waterwall	1.0
masonry, Trombe wall	1.0
direct gain and sunspace	0.8
sunspace: water containers	0.9
lightweight common wall	0.7
other lightweight surfaces	0.3
Glazing properties	
transmission characteristics	diffuse
orientation	due south
index of refraction	1.526
extinction coefficient	0.5 inch^{-1}
thickness of each pane	one-eighth inch
air gap between panes	one-half inch
infrared emittance	0.9
Control range	
room temperature	65° to 75°F
sunspace temperature	45° to 95°F
internal heat generation	0
Thermocirculation vents	
(when used)	
vent area/projected area (sum of both	
upper and lower vents)	0.06
height between vents	8 ft
reverse flow	none
Nighttime insulation	
(when used)	
thermal resistance	R9
in place, solar time	5:30 P.M. to 7:30 A.M.
Solar radiation assumptions	
shading	none
ground diffuse reflectance	0.3

18-1: Reference design characteristics for *SLR* correlations.

This appendix summarizes the solar load ratio (*SLR*) correlations for 94 passive systems. Figure 18-1 lists selected characteristics of the reference designs. Refer to Chapters O and P for more details on the direct-gain and sunspace systems. Figures 18-2 through 18-6 summarize the system types and provide short designations for them. Figure 18-7 lists the *SLR* correlation coefficients for each system. The correlations are in the form:

$$SSF = 1 - K[1 - F(X)]$$

where:
$$F(X) = \begin{cases} AX, & X < R \\ B - C \exp(-DX), & X \geq R \end{cases}$$

$$K = 1 + G/LCR$$

$$X = \left(\frac{S/DD - (LCR_s \times H)}{LCR \times K}\right)$$

Designation	Thermal Storage Capacity* (in Btu/ft²/°F)	Wall Thickness (inches)	No. of Glazings	Wall Surface	Nighttime Insulation
A1	15.6	3	2	normal	no
A2	31.2	6	2	normal	no
A3	46.8	9	2	normal	no
A4	62.4	12	2	normal	no
A5	93.6	18	2	normal	no
A6	124.8	24	2	normal	no
B1	46.8	9	1	normal	no
B2	46.8	9	3	normal	no
B3	46.8	9	1	normal	yes
B4	46.8	9	2	normal	yes
B5	46.8	9	3	normal	yes
C1	46.8	9	1	selective	no
C2	46.8	9	2	selective	no
C3	46.8	9	1	selective	yes
C4	46.8	9	2	selective	yes

*per unit of projected area

18-2: Short designations and characteristics of 15 waterwall system types.

Designation	Thermal Storage Capacity* (Btu/ft²/°F)	Wall Thickness* (inches)	ρck (Btu²/hr/ft⁴/°F²)	No. of Glazings	Wall Surface	Nighttime Insulation
A1	15	6	30	2	normal	no
A2	22.5	9	30	2	normal	no
A3	30	12	30	2	normal	no
A4	45	18	30	2	normal	no
B1	15	6	15	2	normal	no
B2	22.5	9	15	2	normal	no
B3	30	12	15	2	normal	no
B4	45	18	15	2	normal	no
C1	15	6	7.5	2	normal	no
C2	22.5	9	7.5	2	normal	no
C3	30	12	7.5	2	normal	no
C4	45	18	7.5	2	normal	no
D1	30	12	30	1	normal	no
D2	30	12	30	3	normal	no
D3	30	12	30	1	normal	yes
D4	30	12	30	2	normal	yes
D5	30	12	30	3	normal	yes
E1	30	12	30	1	selective	no
E2	30	12	30	2	selective	no
E3	30	12	30	1	selective	yes
E4	30	12	30	2	selective	yes

*The thermal storage capacity is per unit of projected area, or, equivalently, the quantity ρct. The wall thickness is listed only as an appropriate guide by assuming ρc = 30 Btu/ft³/°F.

18-3: Short designations and characteristics of 21 vented Trombe-wall system types.

Designation	Thermal Storage Capacity* (Btu/ft²/°F)	Wall Thickness* (inches)	ρck (Btu²/hr/ft⁴/°F²)	No. of Glazings	Wall Surface	Nighttime Insulation
F1	15	6	30	2	normal	no
F2	22.5	9	30	2	normal	no
F3	30	12	30	2	normal	no
F4	45	18	30	2	normal	no
G1	15	6	15	2	normal	no
G2	22.5	9	15	2	normal	no
G3	30	12	15	2	normal	no
G4	45	18	15	2	normal	no
H1	15	6	7.5	2	normal	no
H2	22.5	9	7.5	2	normal	no
H3	30	12	7.5	2	normal	no
H4	45	18	7.5	2	normal	no
I1	30	12	30	1	normal	no
I2	30	12	30	3	normal	no
I3	30	12	30	1	normal	yes
I4	30	12	30	2	normal	yes
I5	30	12	30	3	normal	yes
J1	30	12	30	1	selective	no
J2	30	12	30	2	selective	no
J3	30	12	30	1	selective	yes
J4	30	12	30	2	selective	yes

*The thermal storage capacity is per unit of projected area, or, equivalently, the quantity ρct. The wall thickness is listed only as an appropriate guide by assuming ρc = 30 Btu/ft³/°F.

18-4: Short designations and characteristics of 21 unvented Trombe-wall system types.

The quantity X is a generalized SLR, defined so as to be applicable to all of the system types, with suitable choices of the constants that appear in it. Two forms occur. For thermal storage walls and sunspaces, $G = 0$; therefore;

$$X = SLR' = \frac{S/DD - (LCR_s \times H)}{LCR}$$

For direct gain, $LCR_s = H = 0$; therefore:

$$X = SLR = (S/DD)/(LCR + G)$$

The quantity S is the monthly solar radiation absorbed in the building per unit of projected area. (The projected area is the principal net glazing area projected on a vertical plane. See the definition in Chapter M). The procedure recommended to estimate S for the reference designs is as follows:

$$S = N \times HS \times \left(\frac{T}{Q_h}\right) \times \left(\frac{S}{T}\right)$$

where:
- N = number of days in the month
- HS = average daily value of total radiation incident on a horizontal surface (in Btu/ft²/day)
- T/Q_h = ratio of monthly total solar radiation transmitted through a unit area of the glazing plane to that on a unit area of horizontal surface
- S/T = ratio of monthly total solar radiation absorbed in the building per unit of projected area to that transmitted through a unit area of glazing

The data required are reproduced in later tables and appendixes: values of HS are reproduced in Appendix 19 for 219 locations in the United States and Canada; the ratio T/Q_h is expressed as a least-squares correlation with the parameters $LAT - DEC$ (latitude minus mid-month solar declination) and K_T (the average monthly clearness ratio); both are tabulated in Appendix 19. Refer to Appendix 19 for a definition of K_T. The form of the correlation is:

$$T/Q_h = B_1 + B_2 Y + B_3 Y^2 + K_T(B_4 + B_5 Y + B_6 Y^2)$$

Designation	Thermal Storage Capacity* (in Btu/ft²/°F)	Mass Thickness* (inches)	Mass-Area-to-Glazing-Area Ratio	No. of Glazings	Nighttime Insulation
A1	30	2	6	2	no
A2	30	2	6	3	no
A3	30	2	6	2	yes
B1	45	6	3	2	no
B2	45	6	3	3	no
B3	45	6	3	2	yes
C1	60	4	6	2	no
C2	60	4	6	3	no
C3	60	4	6	2	yes

*The thermal storage capacity is per unit of projected area, or, equivalently, the quantity ρct. The wall thickness is listed only as an appropriate guide by assuming $\rho c = 30$ Btu/ft³/°F.

18-5: Short designations and characterstics of 9 direct-gain system types.

Designation	Type	Tilt (degrees)	Common Wall	End Walls	Nighttime Insulation
A1	attached	50	masonry	opaque	no
A2	attached	50	masonry	opaque	yes
A3	attached	50	masonry	glazed	no
A4	attached	50	masonry	glazed	yes
A5	attached	50	insulated	opaque	no
A6	attached	50	insulated	opaque	yes
A7	attached	50	insulated	glazed	no
A8	attached	50	insulated	glazed	yes
B1	attached	90/30	masonry	opaque	no
B2	attached	90/30	masonry	opaque	yes
B3	attached	90/30	masonry	glazed	no
B4	attached	90/30	masonry	glazed	yes
B5	attached	90/30	insulated	opaque	no
B6	attached	90/30	insulated	opaque	yes
B7	attached	90/30	insulated	glazed	no
B8	attached	90/30	insulated	glazed	yes
C1	semienclosed	90	masonry	common	no
C2	semienclosed	90	masonry	common	yes
C3	semienclosed	90	insulated	common	no
C4	semienclosed	90	insulated	common	yes
D1	semienclosed	50	masonry	common	no
D2	semienclosed	50	masonry	common	yes
D3	semienclosed	50	insulated	common	no
D4	semienclosed	50	insulated	common	yes
E1	semienclosed	90/30	masonry	common	no
E2	semienclosed	90/30	masonry	common	yes
E3	semienclosed	90/30	insulated	common	no
E4	semienclosed	90/30	insulated	common	yes

18-6: Short designations and characteristics of 28 sunspace system types.

TYPE	A	B	C	D	R	G	H	LCRS	STDV
WW A1	0.0000	1.0000	.9172	.4841	-9.0000	0.00	1.17	13.0	.053
WW A2	0.0000	1.0000	.9833	.7603	-9.0000	0.00	.92	13.0	.046
WW A3	0.0000	1.0000	1.0171	.8852	-9.0000	0.00	.85	13.0	.040
WW A4	0.0000	1.0000	1.0395	.9569	-9.0000	0.00	.81	13.0	.037
WW A5	0.0000	1.0000	1.0604	1.0387	-9.0000	0.00	.78	13.0	.034
WW A6	0.0000	1.0000	1.0735	1.0827	-9.0000	0.00	.76	13.0	.033
WW B1	0.0000	1.0000	.9754	.5518	-9.0000	0.00	.92	22.0	.051
WW B2	0.0000	1.0000	1.0487	1.0851	-9.0000	0.00	.78	9.2	.036
WW B3	0.0000	1.0000	1.0673	1.0087	-9.0000	0.00	.95	8.9	.038
WW B4	0.0000	1.0000	1.1028	1.1811	-9.0000	0.00	.74	5.8	.034
WW B5	0.0000	1.0000	1.1146	1.2771	-9.0000	0.00	.56	4.5	.032
WW C1	0.0000	1.0000	1.0667	1.0437	-9.0000	0.00	.62	12.0	.038
WW C2	0.0000	1.0000	1.0846	1.1482	-9.0000	0.00	.59	8.7	.035
WW C3	0.0000	1.0000	1.1419	1.1756	-9.0000	0.00	.28	5.5	.033
WW C4	0.0000	1.0000	1.1401	1.2378	-9.0000	0.00	.23	4.3	.032
TW A1	0.0000	1.0000	.9194	.4601	-9.0000	0.00	1.11	13.0	.048
TW A2	0.0000	1.0000	.9680	.6318	-9.0000	0.00	.92	13.0	.043
TW A3	0.0000	1.0000	.9964	.7123	-9.0000	0.00	.85	13.0	.038
TW A4	0.0000	1.0000	1.0190	.7332	-9.0000	0.00	.79	13.0	.032
TW B1	0.0000	1.0000	.9364	.4777	-9.0000	0.00	1.01	13.0	.045
TW B2	0.0000	1.0000	.9821	.6020	-9.0000	0.00	.85	13.0	.038
TW B3	0.0000	1.0000	.9980	.6191	-9.0000	0.00	.80	13.0	.033
TW B4	0.0000	1.0000	.9981	.5615	-9.0000	0.00	.76	13.0	.028
TW C1	0.0000	1.0000	.9558	.4709	-9.0000	0.00	.89	13.0	.039
TW C2	0.0000	1.0000	.9788	.4964	-9.0000	0.00	.79	13.0	.033
TW C3	0.0000	1.0000	.9760	.4519	-9.0000	0.00	.76	13.0	.029
TW C4	0.0000	1.0000	.9588	.3612	-9.0000	0.00	.73	13.0	.026
TW D1	0.0000	1.0000	.9842	.4418	-9.0000	0.00	.89	22.0	.040
TW D2	0.0000	1.0000	1.0150	.8994	-9.0000	0.00	.80	9.2	.036
TW D3	0.0000	1.0000	1.0346	.7810	-9.0000	0.00	1.08	8.9	.036
TW D4	0.0000	1.0000	1.0606	.9770	-9.0000	0.00	.85	5.8	.035
TW D5	0.0000	1.0000	1.0721	1.0718	-9.0000	0.00	.61	4.5	.033
TW E1	0.0000	1.0000	1.0345	.8753	-9.0000	0.00	.68	12.0	.037
TW E2	0.0000	1.0000	1.0476	1.0050	-9.0000	0.00	.66	8.7	.035
TW E3	0.0000	1.0000	1.0919	1.0739	-9.0000	0.00	.61	5.5	.034
TW E4	0.0000	1.0000	1.0971	1.1429	-9.0000	0.00	.47	4.3	.033
TW F1	0.0000	1.0000	.9430	.4744	-9.0000	0.00	1.09	13.0	.047
TW F2	0.0000	1.0000	.9900	.6053	-9.0000	0.00	.93	13.0	.041
TW F3	0.0000	1.0000	1.0189	.6502	-9.0000	0.00	.86	13.0	.036
TW F4	0.0000	1.0000	1.0419	.6258	-9.0000	0.00	.80	13.0	.032
TW G1	0.0000	1.0000	.9693	.4714	-9.0000	0.00	1.01	13.0	.042
TW G2	0.0000	1.0000	1.0133	.5462	-9.0000	0.00	.88	13.0	.035
TW G3	0.0000	1.0000	1.0325	.5269	-9.0000	0.00	.82	13.0	.031
TW G4	0.0000	1.0000	1.0401	.4400	-9.0000	0.00	.77	13.0	.030
TW H1	0.0000	1.0000	1.0002	.4356	-9.0000	0.00	.93	13.0	.034
TW H2	0.0000	1.0000	1.0280	.4151	-9.0000	0.00	.83	13.0	.030

18-7: Solar load ratio (*SLR*) correlations for 94 passive system types.

TYPE	A	B	C	D	R	G	H	LCRS	STDV
TW H3	0.0000	1.0000	1.0327	.3522	-9.0000	0.00	.78	13.0	.029
TW H4	0.0000	1.0000	1.0287	.2600	-9.0000	0.00	.74	13.0	.024
TW I1	0.0000	1.0000	.9974	.4036	-9.0000	0.00	.91	22.0	.038
TW I2	0.0000	1.0000	1.0386	.8313	-9.0000	0.00	.80	9.2	.034
TW I3	0.0000	1.0000	1.0514	.6886	-9.0000	0.00	1.01	8.9	.034
TW I4	0.0000	1.0000	1.0781	.8952	-9.0000	0.00	.82	5.8	.032
TW I5	0.0000	1.0000	1.0902	1.0284	-9.0000	0.00	.65	4.5	.032
TW J1	0.0000	1.0000	1.0537	.8227	-9.0000	0.00	.65	12.0	.037
TW J2	0.0000	1.0000	1.0677	.9312	-9.0000	0.00	.62	8.7	.035
TW J3	0.0000	1.0000	1.1153	.9831	-9.0000	0.00	.44	5.5	.034
TW J4	0.0000	1.0000	1.1154	1.0607	-9.0000	0.00	.38	4.3	.033
DG A1	.5650	1.0090	1.0440	.7175	.3931	9.36	0.00	0.0	.046
DG A2	.5906	1.0060	1.0650	.8099	.4681	5.28	0.00	0.0	.039
DG A3	.5442	.9715	1.1300	.9273	.7068	2.64	0.00	0.0	.036
DG B1	.5739	.9948	1.2510	1.0610	.7905	9.60	0.00	0.0	.042
DG B2	.6180	1.0000	1.2760	1.1560	.7528	5.52	0.00	0.0	.035
DG B3	.5601	.9839	1.3520	1.1510	.8879	2.38	0.00	0.0	.032
DG C1	.6344	.9887	1.5270	1.4380	.8632	9.60	0.00	0.0	.039
DG C2	.6763	.9994	1.4000	1.3940	.7604	5.28	0.00	0.0	.033
DG C3	.6182	.9859	1.5660	1.4370	.8990	2.40	0.00	0.0	.031
SS A1	0.0000	1.0000	.9587	.4770	-9.0000	0.00	.83	18.6	.027
SS A2	0.0000	1.0000	.9982	.6614	-9.0000	0.00	.77	10.4	.026
SS A3	0.0000	1.0000	.9552	.4230	-9.0000	0.00	.83	23.6	.030
SS A4	0.0000	1.0000	.9956	.6277	-9.0000	0.00	.80	12.4	.026
SS A5	0.0000	1.0000	.9300	.4041	-9.0000	0.00	.96	18.6	.031
SS A6	0.0000	1.0000	.9981	.6660	-9.0000	0.00	.86	10.4	.028
SS A7	0.0000	1.0000	.9219	.3225	-9.0000	0.00	.96	23.6	.035
SS A8	0.0000	1.0000	.9922	.6173	-9.0000	0.00	.90	12.4	.028
SS B1	0.0000	1.0000	.9683	.4954	-9.0000	0.00	.84	16.3	.028
SS B2	0.0000	1.0000	1.0029	.6802	-9.0000	0.00	.74	8.5	.026
SS B3	0.0000	1.0000	.9689	.4685	-9.0000	0.00	.82	19.3	.029
SS B4	0.0000	1.0000	1.0029	.6641	-9.0000	0.00	.76	9.7	.026
SS B5	0.0000	1.0000	.9408	.3866	-9.0000	0.00	.97	16.3	.030
SS B6	0.0000	1.0000	1.0068	.6778	-9.0000	0.00	.84	8.5	.028
SS B7	0.0000	1.0000	.9395	.3363	-9.0000	0.00	.95	19.3	.032
SS B8	0.0000	1.0000	1.0047	.6469	-9.0000	0.00	.87	9.7	.027
SS C1	0.0000	1.0000	1.0087	.7683	-9.0000	0.00	.76	16.3	.025
SS C2	0.0000	1.0000	1.0412	.9281	-9.0000	0.00	.78	10.0	.027
SS C3	0.0000	1.0000	.9699	.5106	-9.0000	0.00	.79	16.3	.024
SS C4	0.0000	1.0000	1.0152	.7523	-9.0000	0.00	.81	10.0	.025
SS D1	0.0000	1.0000	.9889	.6643	-9.0000	0.00	.84	17.8	.028
SS D2	0.0000	1.0000	1.0493	.8753	-9.0000	0.00	.70	9.9	.028
SS D3	0.0000	1.0000	.9570	.5285	-9.0000	0.00	.90	17.8	.029
SS D4	0.0000	1.0000	1.0356	.8142	-9.0000	0.00	.73	9.9	.028
SS E1	0.0000	1.0000	.9968	.7004	-9.0000	0.00	.77	19.6	.027
SS E2	0.0000	1.0000	1.0468	.9054	-9.0000	0.00	.76	10.8	.027
SS E3	0.0000	1.0000	.9565	.4827	-9.0000	0.00	.81	19.6	.028
SS E4	0.0000	1.0000	1.0214	.7694	-9.0000	0.00	.79	10.8	.027

18-7: Continued.

Number of Glazings	$T/Q_h = B_1 + B_2Y + B_3Y^2 + K_T(B_4 + B_5Y + B_6Y^2)$						
	B_1	B_2	B_3	B_4	B_5	B_6	STDV
1	0.5136	−0.4020	0.9059	−0.3306	−0.3787	5.1344	0.056
2	0.4146	−0.2847	0.7160	−0.2817	−0.4251	4.5913	0.049
3	0.3484	−0.1610	0.4980	−0.2049	−0.6715	4.4288	0.043

18-8: Solar radiation correlation coefficients for vertical south-facing surfaces.

In this equation, $Y = (LAT - DEC)/100$; the coefficients B_1 through B_6 are given in figure 18-8.

Absorption factors S/T are given in figure 18-9, assuming mass surface absorptances of 0.8 for direct-gain and 0.95 for waterwall and Trombe-wall systems.

The ratios T/Q_h and S/T are combined in figure 18-10 for the sunspace reference designs into the single ratio S/Q_h.

The solar radiation approximations expressed in figures 18-8 through 18-10 were used in generating the LCR tables of Appendix 21.

More general solar radiation approximations that are capable of treating off-reference glazing configurations —such as off-south azimuths, off-vertical tilts, and sunspace geometries other than the reference designs—are described in Appendix 20.

Notice that, for a given reference design, the monthly SSF depends only on the parameter LCR and the monthly variable S/DD. Plots of SSF versus S/DD with LCR as a parameter are given in figure 18-11 for all of the direct-gain reference designs and in figure 18-12 for all of the sunspace reference designs.

Regarding the mass properties in figures 18-2 through 18-5, the relevant mass properties may be expressed in terms of just two parameter groupings: ϱcL, where ϱ is the density (in lb/ft^3), c is the specific heat (in Btu/°F/lb), and L is the thickness (in ft); and ϱck, where k is the thermal conductivity (in Btu/hr/ft/°F). The quantity ϱcL is exactly the "thermal storage capacity" (per square foot of mass surface) listed in figures 18-2 through 18-5. The thicknesses listed in the next columns are not primary properties, but are the thicknesses of mass having the given values of ϱcL. The thicknesses were computed using $\varrho c = 62.4$ Btu/ft^3/°F for water (figure 18-2) and $\varrho c = 30$ Btu/ft^3/°F for masonry (figures 18-2 through 18-5). Masonry properties vary from material to material, so the value of thickness to be associated with a given value of ϱcL must be computed by knowledge of the actual value of ϱc in the application at hand.

Design	Characteristic	S/T
Waterwall and Trombe wall	1 glazing	0.957
	2 glazings	0.962
	3 glazings	0.964
Direct gain	$A_m/A_g = 3$	0.948
	$A_m/A_g = 6$	0.976

A_m/A_g = mass-area-to-glazing-area ratio.

18-9: Absorption factors, S/T, for different reference designs.

In applying the correlations, a reference design is best chosen on the basis of the two mass properties, ϱcL and ϱck (as well as the other relevant reference design characteristics). Therefore, the procedure is to evaluate the quantities ϱcL and ϱck for the mass in question, and then to choose a reference design that best matches the values. Even, better is to choose a pair of bounding reference designs as the basis of an interpolation.

Caution should be exercised in applying the SLR correlations to sunspace and thermal storage wall systems at very small values of the SLR—say, below 0.5. In particular, the correlations may overpredict SSF by a few percent for some cases without nighttime insulation at large values of LCR (greater than about 100). The accuracy of the correlations for these high-LCR cases can be tested by calculating SSF for a comparable case *with* nighttime insulation: if SSF for the system without nighttime insulation is greater, then reduce it to the value for the system with nighttime insulation.

18-10: Solar radiation correlation coefficients for sunspace reference designs.

	$S/Q_h = B_1 + B_2Y + B_3Y^2 + K_T(B_4 + B_5Y + B_6Y^2)$						
TYPE	B_1	B_2	B_3	B_4	B_5	B_6	STDV
A1, 2, 5, 6	0.72008	−0.15181	0.49973	−0.15039	0.14384	3.6374	0.054
A3, 4, 7, 8	0.81554	−0.23988	0.60252	−0.16445	0.33730	3.1695	0.048
B1, 2, 5, 6	0.58932	−0.09693	0.38955	−0.14699	−0.39149	3.9171	0.044
B3, 4, 7, 8	0.62569	−0.13941	0.43331	−0.14982	−0.26401	3.5685	0.040
C1, 2, 3, 4	0.39436	−0.21103	0.58815	−0.24083	−0.60746	4.6546	0.046
D1, 2, 3, 4	0.73147	−0.15418	0.50763	−0.15276	0.14608	3.6950	0.055
E1, 2, 3, 4	0.61661	−0.10127	0.40733	−0.15367	−0.40940	4.0969	0.046

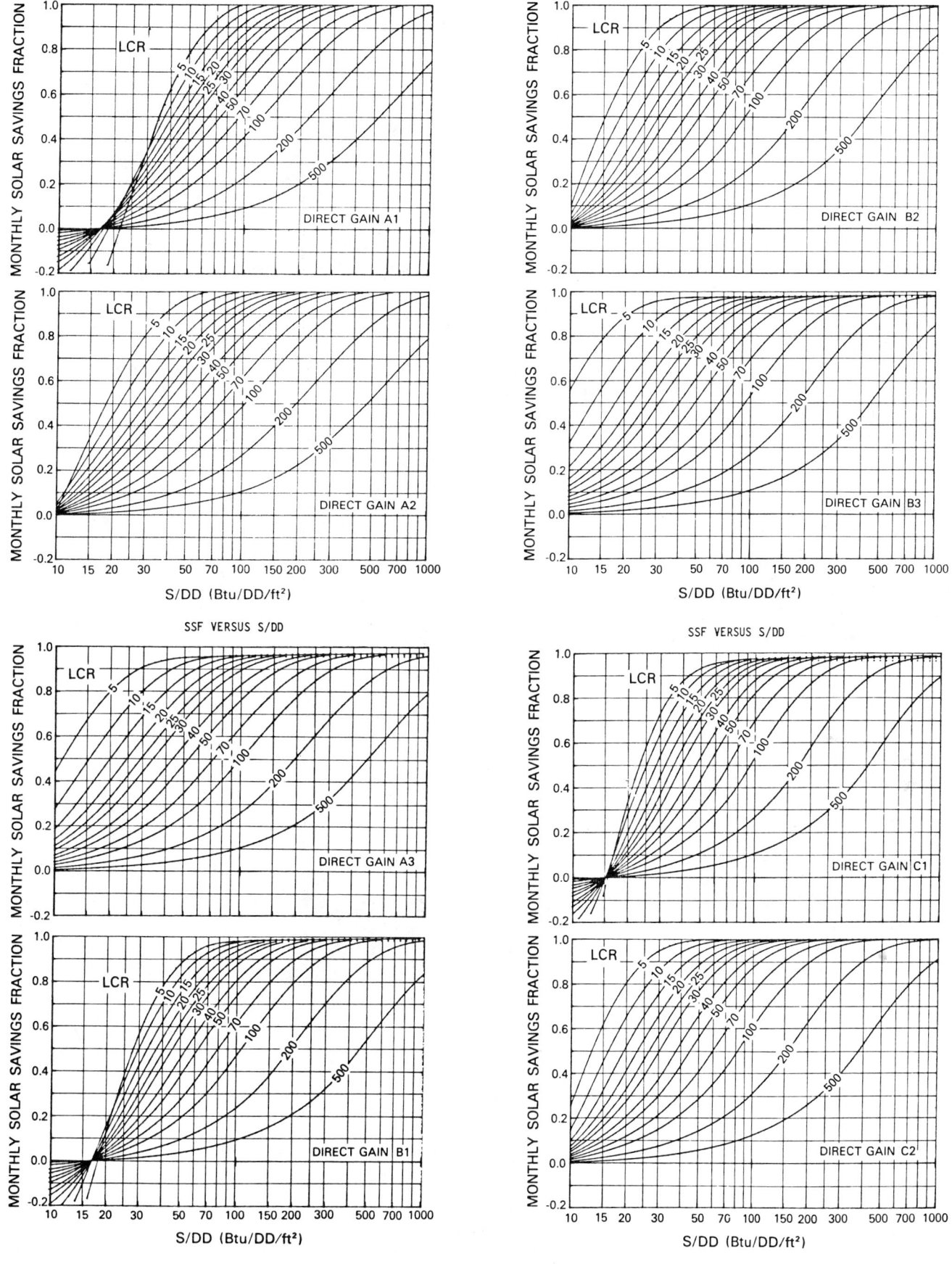

18-11: Plots of monthly *SSF* versus monthly *S/DD* at 12 different values of *LCR* for 9 direct-gain system types.

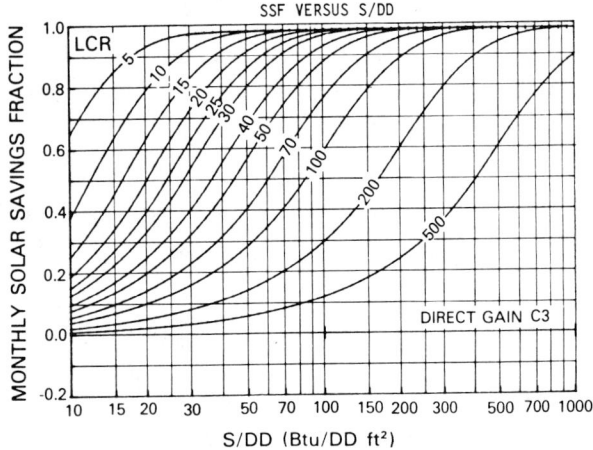
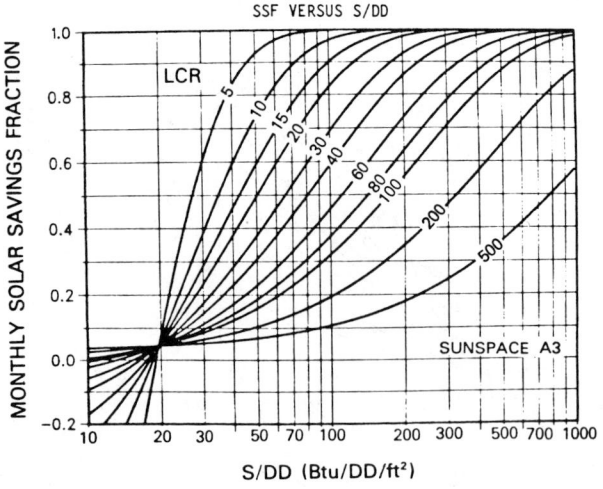

18-11: Continued.

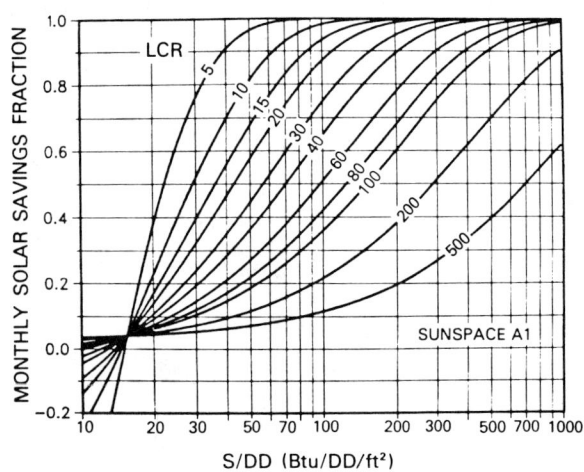

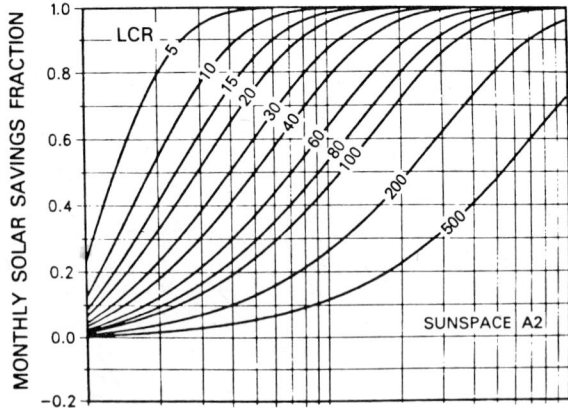

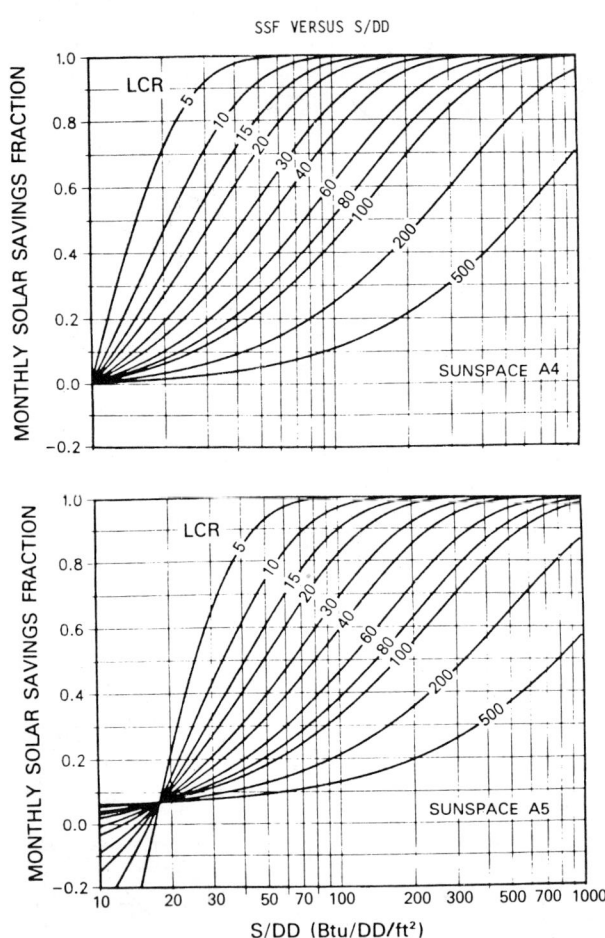

18-12: Plots of monthly *SSF* versus monthly *S/DD* at 11 different values of *LCR* for 28 sunspace system types.

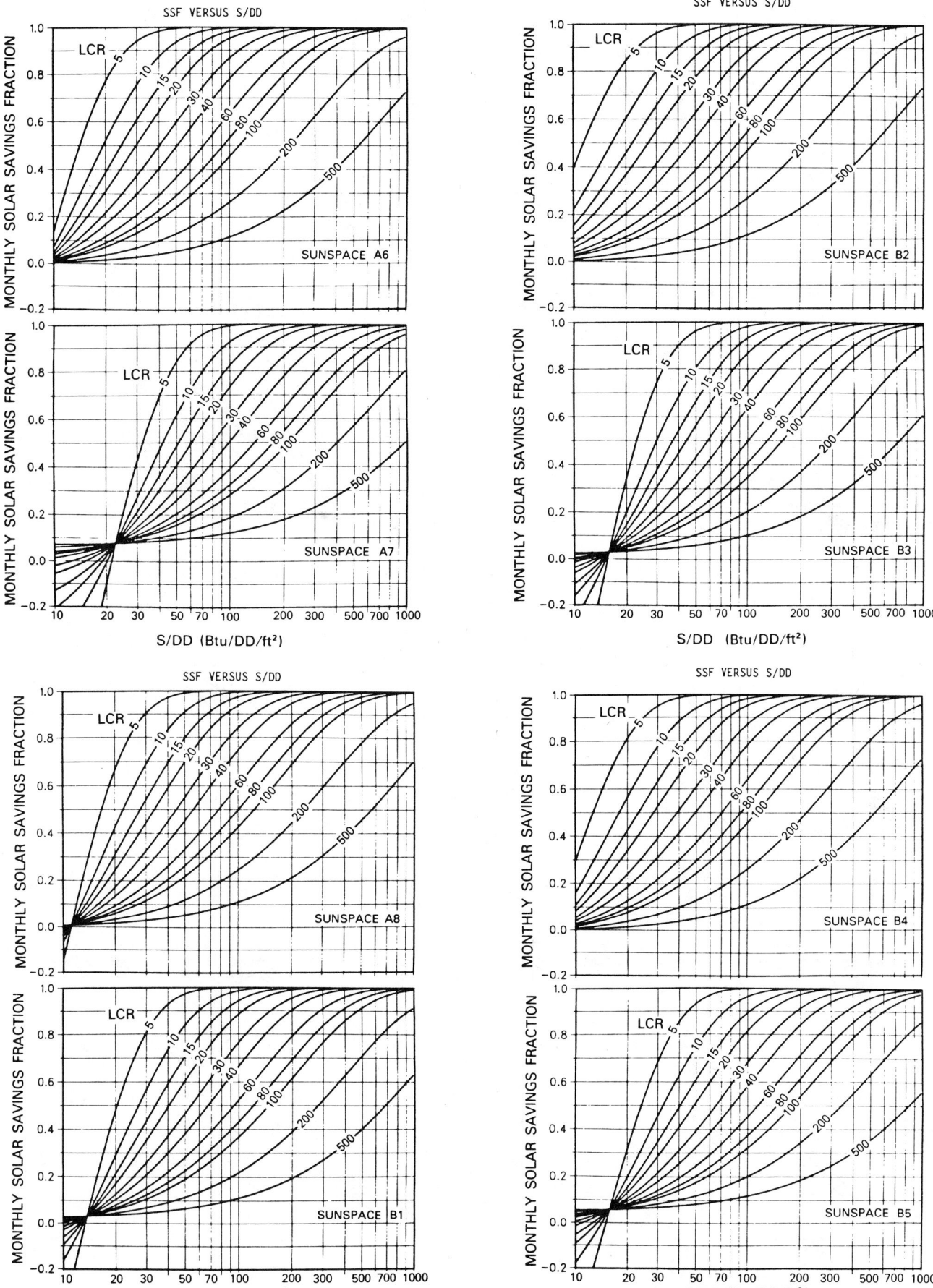

18-12: Continued.

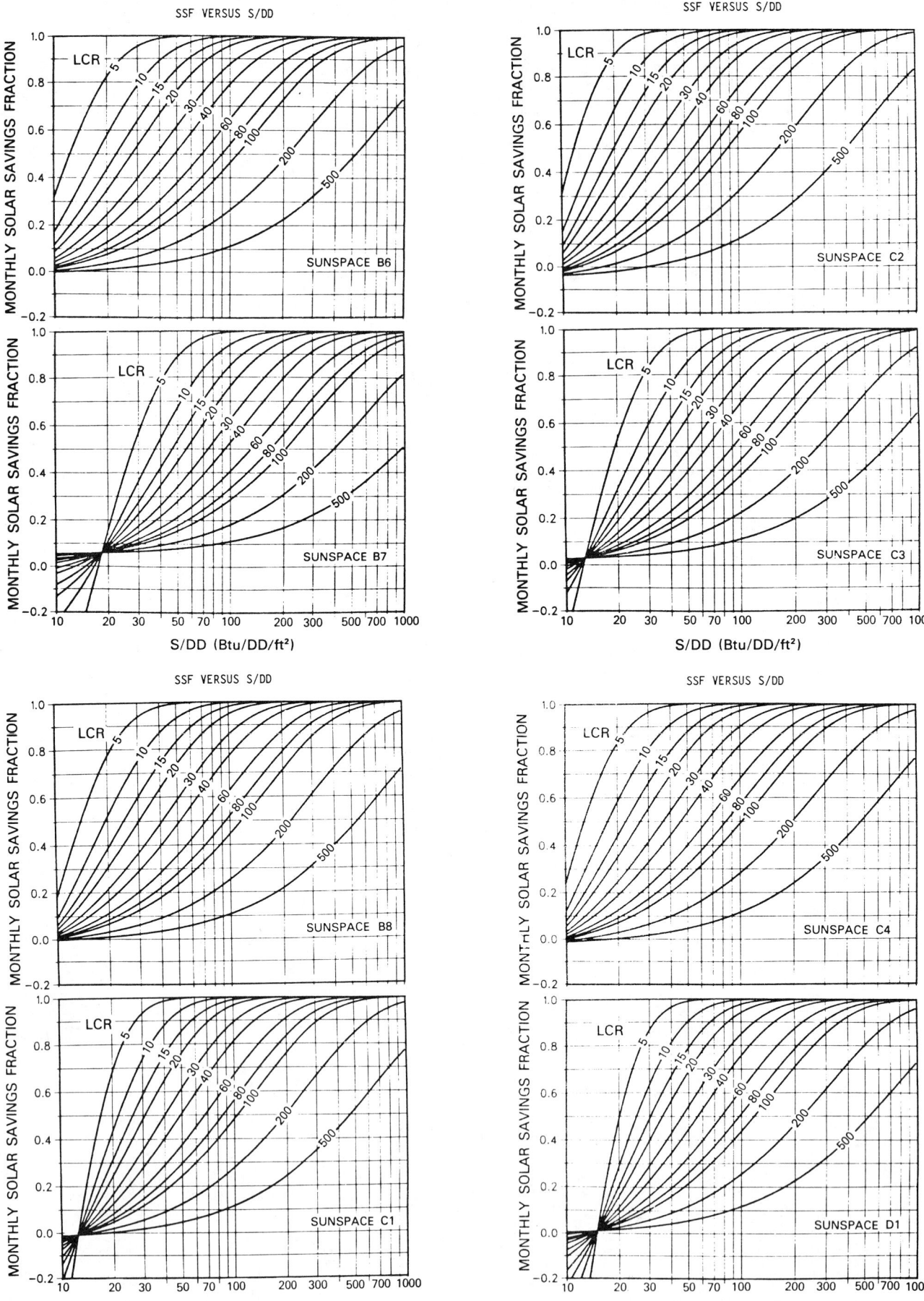

18-12: Continued.

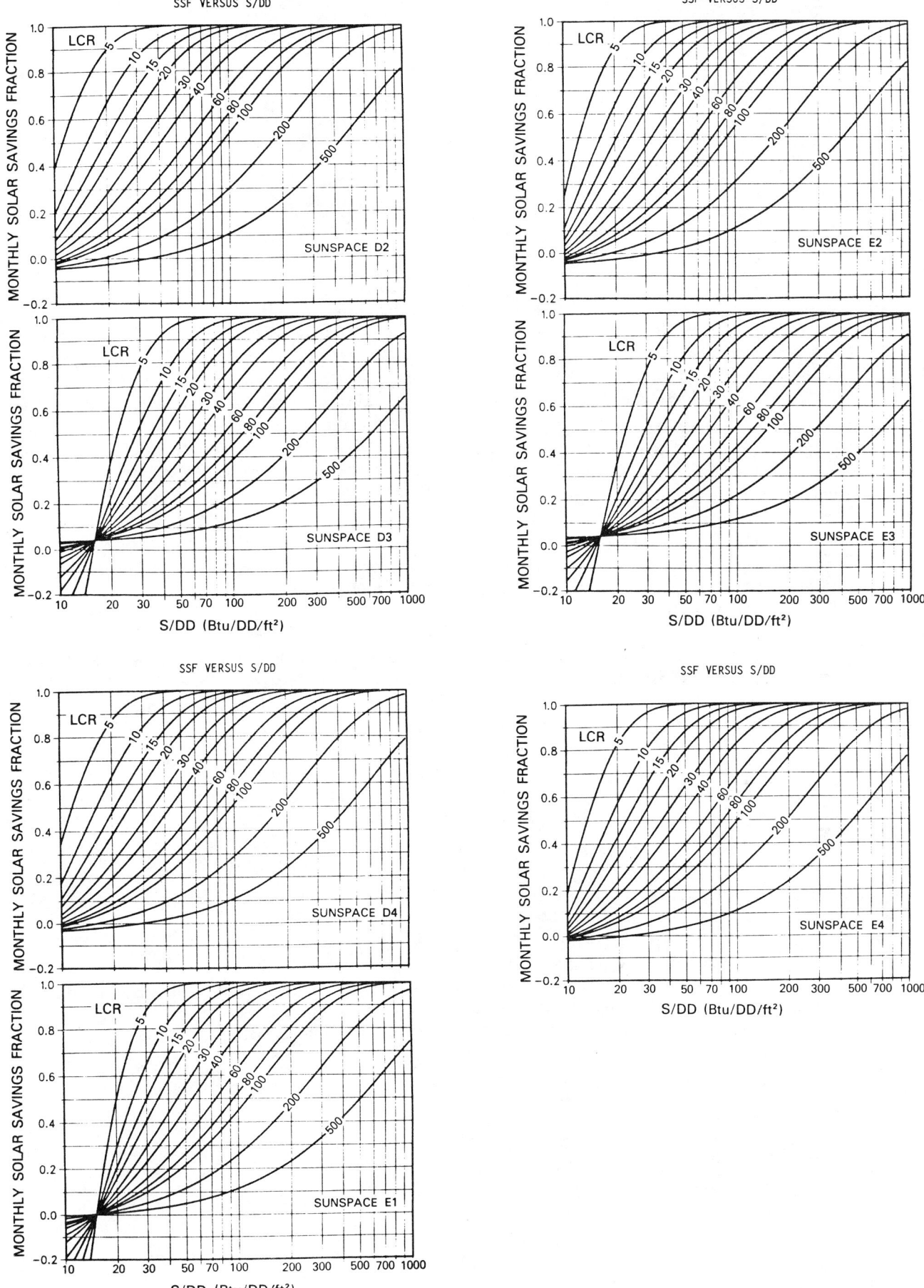

18-12: Continued.

Appendix 19 Weather Data

This appendix contains tabulated monthly average solar radiation, temperature, heating degree-days, and solar position data useful in solar design analysis. Much more detailed information on solar radiation is available in Appendix 20. The quantities tabulated are the following:

- HS = normal daily value of total hemispheric radiation incident on a horizontal surface (in Btu/ft^2/day).
- VS = normal daily value of total solar radiation incident on a vertical south-facing surface (in Btu/ft^2/day).
- TA = $(T_{min} + T_{max})/2$ where T_{min} and T_{max} are monthly (or annual) normals of daily minimum and maximum ambient temperatures (in °F).
- Dxx = monthly (or annual) normals of heating degree-days below the base temperature xx (in F degree-days).
- K_T = average monthly (or annual) clearness ratio—i.e., the ratio of total hemispheric radiation incident on a horizontal surface to the extraterrestrial radiation incident on a horizontal surface.
- LD = $LAT - DEC$, latitude minus midmonth solar declination (in degrees).

For the United States cities tabulated here, values of HS, TA, and $D65$ were taken from reference [CIN]. "Normals" are mean values for the period 1941 to 1970. Degree-days below base temperatures other than 65°F were calculated by the method described in [THO-1], [THO-2], based on the values of TA and $D65$ from [CIN].

For the Canadian cities, values of HS and TA were taken from [AES], [PHI]. Degree-days below various base temperatures were calculated by the method described in [THO-1], [THO-2], based on the values of TA from [AES] and the degree-days below 18°C from [AST-1], [AST-2], [AST-3], [AST-4], [AST-5], [AST-6].

For all cities, monthly values of VS were computed using the correlation:

$$\frac{VS}{HS} = 0.6866 - 0.6623Y + 1.3269Y^2 + K_T(-0.4458 + 0.3090Y + 4.7776Y^2)$$

In the above equation, $Y = (LAT - DEC)/100$, LAT is the latitude (in degrees), DEC is the midmonth solar declination (in degrees), and K_T is the average monthly clearness ratio.

The average monthly clearness ratio (K_T) is the ratio of the normal monthly value of total hemispheric radiation incident on a horizontal surface (Q_h) to the monthly extraterrestrial radiation incident on a horizontal surface (Q_{he}):

$$K_T = Q_h/Q_{he}$$

The monthly horizontal surface radiation is determined from the normal daily value:

$$Q_h = N \times HS$$

where N is the number of days in the month. The monthly extraterrestrial horizontal surface radiation is determined from the daily value at midmonth:

$$Q_{he} = NIY$$

where I is the extraterrestrial solar flux at normal incidence at midmonth (in Btu/hr/ft^2), and Y is the ratio of the daily total normal incidence radiation to horizontal surface radiation (the integral of the sine of the solar altitude from sunrise to sunset):

$$Y = (24/\pi)[\cos L \cos D \sin H_s + H_s \sin L \sin D]$$

where:

$$H_s = \cos^{-1}(-\tan L \tan D)$$

H_s is the sunrise hour angle in radians, L is latitude, and D is solar declination at midmonth. The quantities I and D are approximated by the following equations:

$$I = I_{sc}[1 + 0.033 \cos(360n/365)]$$
$$D = 23.45 \sin[360(284 + n)/365]$$

where D and the arguments of the cosine and sine are in degrees, n is the day of the year, taken to be the midday of each month, and I_{sc} is the solar constant taken to equal 428 Btu/hr/ft^2.

BIRMINGHAM, ALABAMA — ELEV 630 LAT 33.6

	HS	VS	TA	D50	D55	D60	D65	D70	KT	LD
JAN	707	864	44	216	346	493	654	800	.43	55
FEB	967	995	47	144	246	372	517	647	.47	47
MAR	1296	986	53	61	135	237	389	520	.49	36
APR	1673	921	63	6	20	60	116	232	.53	24
MAY	1857	852	71	1	3	12	20	100	.53	15
JUN	1918	843	77	0	0	0	0	24	.53	10
JUL	1810	821	80	0	0	0	0	14	.51	12
AUG	1724	874	79	0	0	0	0	16	.53	20
SEP	1455	990	74	0	1	5	6	52	.52	32
OCT	1211	1179	63	6	20	61	137	237	.54	44
NOV	858	1056	52	75	151	258	391	539	.49	53
DEC	661	857	45	193	318	463	614	769	.43	57
YR	1346	936	62	702	1240	1961	2844	3950	.51	

PRESCOTT, ARIZONA — ELEV 5023 LAT 34.6

	HS	VS	TA	D50	D55	D60	D65	D70	KT	LD
JAN	1016	1533	37	401	555	710	865	1020	.64	56
FEB	1335	1623	41	270	407	546	686	826	.66	48
MAR	1777	1502	44	193	335	487	642	797	.69	37
APR	2275	1251	52	49	127	248	394	540	.72	25
MAY	2629	1060	60	3	18	73	165	302	.75	16
JUN	2762	998	69	0	1	5	33	91	.76	11
JUL	2309	953	76	0	0	0	0	18	.65	13
AUG	2092	1043	73	0	0	0	0	38	.64	21
SEP	1954	1430	68	0	1	7	23	107	.71	33
OCT	1543	1709	57	10	48	127	254	398	.71	45
NOV	1140	1674	46	152	281	427	576	726	.68	54
DEC	927	1477	39	356	509	663	818	973	.63	58
YR	1815	1352	55	1436	2282	3294	4456	5837	.69	

MOBILE, ALABAMA — ELEV 220 LAT 30.7

	HS	VS	TA	D50	D55	D60	D65	D70	KT	LD
JAN	828	964	51	97	180	294	451	585	.46	52
FEB	1100	1066	54	54	119	204	337	452	.50	44
MAR	1407	993	59	19	55	125	221	343	.52	33
APR	1722	884	68	3	8	26	40	145	.54	21
MAY	1872	828	75	0	0	0	0	51	.54	12
JUN	1868	819	80	0	0	0	0	15	.52	7
JUL	1715	778	82	0	0	0	0	11	.49	9
AUG	1641	799	82	0	0	0	0	12	.50	17
SEP	1449	909	78	0	0	0	0	28	.50	29
OCT	1299	1164	69	2	7	21	39	133	.55	41
NOV	955	1096	59	22	63	135	211	356	.51	50
DEC	759	928	53	73	149	252	385	533	.45	54
YR	1386	935	67	270	581	1056	1684	2664	.51	

TUCSON, ARIZONA — ELEV 2556 LAT 32.1

	HS	VS	TA	D50	D55	D60	D65	D70	KT	LD
JAN	1099	1539	51	80	166	292	442	593	.64	53
FEB	1432	1615	54	41	106	201	333	463	.67	46
MAR	1864	1447	58	15	56	133	243	388	.70	35
APR	2363	1186	66	2	7	26	81	169	.74	22
MAY	2671	1001	74	0	0	0	0	44	.77	13
JUN	2730	953	82	0	0	0	0	4	.76	9
JUL	2341	922	86	0	0	0	0	1	.66	11
AUG	2183	1007	84	0	0	0	0	3	.67	19
SEP	1979	1322	80	0	0	0	0	7	.70	30
OCT	1602	1623	70	0	2	8	29	91	.70	42
NOV	1208	1631	59	12	44	114	221	350	.67	51
DEC	996	1464	52	64	144	262	403	559	.62	55
YR	1874	1307	68	214	525	1036	1752	2673	.70	

MONTGOMERY, ALABAMA — ELEV 203 LAT 32.3

	HS	VS	TA	D50	D55	D60	D65	D70	KT	LD
JAN	752	896	48	148	256	394	556	698	.44	54
FEB	1013	1011	51	88	166	277	419	544	.48	46
MAR	1341	987	57	30	85	166	299	424	.50	35
APR	1729	920	65	3	12	39	76	186	.54	23
MAY	1897	849	72	1	2	7	8	72	.54	13
JUN	1972	848	79	0	0	0	0	16	.55	9
JUL	1841	820	81	0	0	0	0	10	.52	11
AUG	1746	861	81	0	0	0	0	11	.53	19
SEP	1468	963	76	0	0	0	0	32	.52	30
OCT	1262	1188	66	3	11	35	93	180	.55	42
NOV	915	1099	55	41	103	191	306	454	.51	52
DEC	719	916	49	131	231	365	512	667	.45	56
YR	1390	946	65	445	866	1474	2269	3295	.52	

WINSLOW, ARIZONA — ELEV 4882 LAT 35.0

	HS	VS	TA	D50	D55	D60	D65	D70	KT	LD
JAN	985	1490	33	540	694	849	1004	1159	.63	56
FEB	1327	1637	39	309	446	585	725	865	.67	49
MAR	1780	1528	45	184	321	472	626	781	.69	37
APR	2283	1272	54	33	101	205	348	490	.73	25
MAY	2595	1064	63	2	10	45	124	238	.74	16
JUN	2712	1001	72	0	1	3	14	55	.75	12
JUL	2347	968	78	0	0	0	0	9	.66	14
AUG	2141	1074	76	0	0	0	0	17	.66	22
SEP	1928	1429	70	0	1	6	19	88	.70	33
OCT	1513	1691	57	12	50	129	252	396	.70	45
NOV	1119	1658	43	217	357	505	654	804	.67	54
DEC	894	1424	34	503	657	812	967	1122	.62	58
YR	1804	1351	55	1800	2639	3611	4733	6025	.69	

PHOENIX, ARIZONA — ELEV 1112 LAT 33.4

	HS	VS	TA	D50	D55	D60	D65	D70	KT	LD
JAN	1021	1462	51	78	162	285	428	584	.62	55
FEB	1374	1608	55	29	85	168	292	419	.66	47
MAR	1814	1471	60	9	36	100	185	327	.69	36
APR	2355	1236	68	1	4	16	60	130	.74	24
MAY	2676	1034	76	0	0	0	0	24	.77	14
JUN	2739	973	85	0	0	0	0	2	.76	10
JUL	2486	964	91	0	0	0	0	0	.70	12
AUG	2293	1080	89	0	0	0	0	1	.70	20
SEP	2015	1415	84	0	0	0	0	3	.72	32
OCT	1576	1674	72	0	1	5	17	64	.70	43
NOV	1150	1606	60	9	34	96	182	314	.66	53
DEC	932	1407	53	60	137	250	388	544	.61	57
YR	1371	1326	70	187	459	919	1552	2411	.71	

YUMA, ARIZONA — ELEV 207 LAT 32.7

	HS	VS	TA	D50	D55	D60	D65	D70	KT	LD
JAN	1096	1575	55	28	87	177	308	455	.65	54
FEB	1443	1675	59	8	33	93	192	303	.69	46
MAR	1919	1534	64	3	11	41	97	212	.72	35
APR	2413	1231	71	0	1	5	24	73	.76	23
MAY	2728	1023	79	0	0	0	0	11	.78	14
JUN	2814	966	86	0	0	0	0	1	.78	9
JUL	2453	948	94	0	0	0	0	0	.69	11
AUG	2329	1070	93	0	0	0	0	0	.71	19
SEP	2051	1406	87	0	0	0	0	1	.73	31
OCT	1623	1694	76	0	0	1	5	25	.72	43
NOV	1215	1690	64	3	12	44	108	215	.68	52
DEC	1000	1515	56	22	74	158	276	427	.64	56
YR	1925	1358	74	64	218	520	1010	1724	.72	

FORT SMITH, ARKANSAS — ELEV 463 LAT 35.3

	HS	VS	TA	D50	D55	D60	D65	D70	KT	LD
JAN	744	996	39	346	497	651	806	961	.48	57
FEB	999	1108	43	204	331	468	608	748	.50	49
MAR	1312	1056	50	85	177	308	471	611	.51	38
APR	1616	929	62	3	15	56	132	246	.52	26
MAY	1912	892	70	0	1	7	17	88	.55	16
JUN	2089	898	78	0	0	0	0	12	.58	12
JUL	2065	908	82	0	0	0	0	3	.58	14
AUG	1877	973	81	0	0	0	0	4	.58	22
SEP	1501	1080	74	0	0	0	0	36	.55	33
OCT	1201	1242	63	3	11	46	135	228	.56	45
NOV	851	1119	50	81	169	295	438	588	.52	55
DEC	682	963	42	274	421	574	729	884	.48	59
YR	1406	1013	61	996	1622	2405	3336	4409	.54	

FRESNO, CALIFORNIA — ELEV 328 LAT 36.8

	HS	VS	TA	D50	D55	D60	D65	D70	KT	LD
JAN	657	886	45	176	308	457	611	766	.45	58
FEB	1012	1195	50	81	167	288	423	563	.53	50
MAR	1566	1389	54	37	107	209	344	500	.62	39
APR	2093	1246	60	6	25	81	182	298	.68	27
MAY	2484	1092	67	1	3	14	51	132	.71	18
JUN	2733	1045	74	0	0	2	9	36	.75	13
JUL	2685	1076	81	0	0	0	0	5	.76	15
AUG	2423	1258	78	0	0	0	0	11	.75	23
SEP	1985	1587	74	0	0	0	0	37	.74	35
OCT	1429	1681	64	2	8	34	90	202	.68	47
NOV	888	1271	54	40	110	213	345	496	.57	56
DEC	574	800	46	165	293	442	595	750	.42	60
YR	1714	1210	62	507	1021	1741	2650	3797	.67	

LITTLE ROCK, ARKANSAS — ELEV 266 LAT 34.7

	HS	VS	TA	D50	D55	D60	D65	D70	KT	LD
JAN	731	947	40	331	482	636	791	946	.46	56
FEB	1003	1089	43	212	342	479	619	759	.50	48
MAR	1313	1037	50	83	175	308	470	611	.51	37
APR	1611	913	62	4	16	61	139	259	.51	25
MAY	1929	888	70	0	1	7	21	91	.55	16
JUN	2106	896	78	0	0	0	0	11	.58	11
JUL	2032	891	81	0	0	0	0	4	.57	13
AUG	1860	952	81	0	0	0	0	5	.57	21
SEP	1518	1074	73	0	0	2	5	41	.55	33
OCT	1228	1251	62	3	14	54	143	248	.56	45
NOV	847	1084	50	81	169	298	441	591	.50	54
DEC	674	922	42	270	418	571	725	880	.46	58
YR	1406	995	61	984	1617	2414	3354	4447	.54	

LONG BEACH, CALIFORNIA — ELEV 56 LAT 33.8

	HS	VS	TA	D50	D55	D60	D65	D70	KT	LD
JAN	928	1291	54	17	79	189	339	490	.57	55
FEB	1215	1373	56	9	53	140	273	406	.59	47
MAR	1610	1289	57	5	35	116	247	397	.61	36
APR	1938	1056	61	1	10	55	148	284	.61	24
MAY	2064	913	64	0	2	18	71	192	.59	15
JUN	2140	894	67	0	1	5	23	110	.59	10
JUL	2300	939	72	0	0	0	0	35	.65	12
AUG	2100	1024	73	0	0	0	0	24	.64	20
SEP	1701	1186	72	0	0	1	7	39	.61	32
OCT	1326	1342	67	0	1	6	48	122	.60	44
NOV	1003	1335	61	1	10	55	155	284	.58	53
DEC	847	1241	56	10	59	155	295	450	.56	57
YR	1600	1156	63	46	250	740	1606	2834	.61	

BAKERSFIELD, CALIFORNIA — ELEV 492 LAT 35.4

	HS	VS	TA	D50	D55	D60	D65	D70	KT	LD
JAN	766	1044	48	131	246	391	543	698	.49	57
FEB	1102	1276	52	50	120	225	353	493	.56	49
MAR	1595	1350	57	18	66	149	266	418	.62	38
APR	2095	1191	63	3	13	50	140	233	.67	26
MAY	2509	1057	70	0	2	7	22	92	.72	16
JUN	2749	1013	77	0	0	0	0	16	.76	12
JUL	2683	1037	84	0	0	0	0	2	.76	14
AUG	2421	1198	82	0	0	0	0	4	.75	22
SEP	1992	1508	77	0	0	0	0	18	.73	34
OCT	1458	1632	67	1	4	17	55	143	.68	45
NOV	942	1305	56	21	72	156	276	422	.57	55
DEC	677	957	48	124	236	379	530	685	.47	59
YR	1752	1213	65	348	759	1372	2185	3224	.68	

LOS ANGELES, CALIFORNIA — ELEV 105 LAT 33.9

	HS	VS	TA	D50	D55	D60	D65	D70	KT	LD
JAN	926	1293	55	21	83	186	331	481	.57	55
FEB	1214	1377	56	13	59	143	270	404	.59	48
MAR	1619	1303	57	11	53	138	267	419	.62	36
APR	1951	1066	59	5	26	91	195	338	.62	24
MAY	2060	914	62	2	9	47	114	257	.59	15
JUN	2119	891	65	1	4	20	71	180	.59	11
JUL	2307	942	69	0	1	5	19	99	.65	13
AUG	2079	1019	70	0	1	4	15	83	.64	20
SEP	1681	1174	69	0	1	5	23	93	.60	32
OCT	1317	1335	65	0	3	17	77	168	.59	44
NOV	1004	1343	61	2	15	65	158	289	.58	53
DEC	848	1249	57	9	47	129	279	407	.56	57
YR	1596	1157	62	64	299	849	1819	3216	.61	

DAGGETT, CALIFORNIA — ELEV 1929 LAT 34.9

	HS	VS	TA	D50	D55	D60	D65	D70	KT	LD
JAN	958	1422	47	144	257	398	549	704	.61	56
FEB	1281	1549	52	63	135	239	371	505	.64	49
MAR	1772	1514	57	23	74	155	271	416	.69	37
APR	2274	1263	64	3	11	41	118	200	.73	25
MAY	2591	1061	72	0	1	5	14	65	.74	16
JUN	2766	1004	80	0	0	0	0	9	.76	12
JUL	2603	1012	87	0	0	0	0	1	.73	14
AUG	2383	1164	86	0	0	0	0	2	.73	21
SEP	2008	1493	79	0	0	0	0	12	.73	33
OCT	1516	1688	68	1	4	16	57	130	.70	45
NOV	1085	1575	56	30	88	174	296	438	.65	54
DEC	876	1374	48	132	238	378	527	682	.60	58
YR	1845	1342	66	396	809	1406	2203	3165	.71	

MOUNT SHASTA, CALIFORNIA — ELEV 3586 LAT 41.3

	HS	VS	TA	D50	D55	D60	D65	D70	KT	LD
JAN	561	857	34	509	663	818	973	1128	.46	63
FEB	857	1128	38	344	482	622	762	902	.51	55
MAR	1250	1217	40	303	454	608	763	918	.54	44
APR	1756	1200	46	143	268	412	561	711	.59	32
MAY	2186	1129	53	37	109	221	371	518	.63	22
JUN	2436	1102	60	4	22	78	178	304	.67	18
JUL	2577	1201	68	0	2	9	37	118	.73	20
AUG	2213	1356	66	1	3	17	64	154	.70	28
SEP	1735	1595	61	3	15	62	145	271	.68	39
OCT	1155	1492	51	61	146	274	422	577	.62	51
NOV	659	998	42	257	400	549	699	849	.49	61
DEC	505	814	36	451	605	760	915	1070	.46	65
YR	1494	1174	50	2112	3169	4430	5890	7520	.62	

NEEDLES, CALIFORNIA								ELEV 886 LAT 34.8	
	HS	VS	TA	D50	D55	D60	D65	D70	KT LD
JAN	985	1477	52	80	161	278	421	572	.62 56
FEB	1353	1669	57	25	74	147	261	382	.68 48
MAR	1825	1565	62	8	27	80	150	278	.71 37
APR	2317	1280	70	1	3	10	42	94	.74 25
MAY	2652	1070	80	0	0	0	0	13	.76 16
JUN	2791	1005	88	0	0	0	0	1	.77 11
JUL	2541	1000	95	0	0	0	0	0	.72 13
AUG	2278	1121	93	0	0	0	0	0	.70 21
SEP	2015	1494	87	0	0	0	0	2	.73 33
OCT	1537	1714	74	0	1	4	10	46	.71 45
NOV	1124	1654	61	9	32	90	163	292	.67 54
DEC	913	1457	53	65	141	249	381	538	.63 58
YR	1863	1373	73	188	439	858	1428	2219	.71

SAN DIEGO, CALIFORNIA								ELEV 30 LAT 32.7	
	HS	VS	TA	D50	D55	D60	D65	D70	KT LD
JAN	976	1325	55	9	58	160	314	459	.58 54
FEB	1266	1392	57	4	33	111	237	373	.60 46
MAR	1632	1261	58	3	23	95	219	372	.61 35
APR	1937	1024	61	1	7	49	144	281	.61 23
MAY	2003	882	63	0	2	20	79	213	.57 14
JUN	2062	870	66	0	1	8	52	147	.57 9
JUL	2186	902	70	0	0	1	6	68	.62 11
AUG	2057	981	71	0	0	0	0	41	.63 19
SEP	1717	1155	70	0	0	1	16	61	.61 31
OCT	1373	1349	66	0	1	6	43	137	.61 43
NOV	1063	1387	61	1	7	48	140	278	.60 52
DEC	904	1302	57	5	37	123	257	413	.58 56
YR	1600	1151	63	23	170	623	1507	2841	.60

OAKLAND, CALIFORNIA								ELEV 7 LAT 37.7	
	HS	VS	TA	D50	D55	D60	D65	D70	KT LD
JAN	708	1028	49	77	202	354	508	663	.50 59
FEB	1017	1247	52	25	101	228	367	507	.55 51
MAR	1456	1307	54	12	75	199	350	505	.59 40
APR	1922	1178	56	4	36	128	270	417	.62 28
MAY	2211	1035	59	1	11	72	193	344	.64 19
JUN	2350	994	62	0	2	26	114	244	.65 14
JUL	2322	1018	63	0	1	16	80	216	.65 16
AUG	2053	1122	64	0	1	13	74	205	.64 24
SEP	1701	1359	65	0	1	8	59	170	.64 36
OCT	1212	1380	61	0	4	37	135	277	.59 48
NOV	822	1180	55	5	47	148	291	441	.54 57
DEC	647	993	50	55	164	314	468	623	.50 61
YR	1538	1152	57	179	646	1543	2909	4613	.61

SAN FRANCISCO, CALIFORNIA								ELEV 16 LAT 37.6	
	HS	VS	TA	D50	D55	D60	D65	D70	KT LD
JAN	708	1023	48	82	210	363	518	673	.49 59
FEB	1009	1229	51	32	117	247	386	526	.54 51
MAR	1455	1301	53	17	88	219	372	527	.59 40
APR	1920	1173	55	5	47	148	291	441	.62 28
MAY	2226	1038	58	1	15	82	210	363	.64 19
JUN	2377	998	62	0	3	30	120	253	.65 14
JUL	2392	1034	63	0	2	21	93	234	.67 16
AUG	2116	1149	63	0	1	17	84	219	.66 24
SEP	1742	1394	64	0	1	10	66	181	.65 36
OCT	1226	1397	61	0	4	39	137	280	.60 48
NOV	821	1172	55	5	47	148	291	441	.54 57
DEC	642	977	50	58	170	320	474	629	.49 61
YR	1556	1156	57	202	705	1643	3042	4768	.62

RED BLUFF, CALIFORNIA								ELEV 354 LAT 40.1	
	HS	VS	TA	D50	D55	D60	D65	D70	KT LD
JAN	570	831	45	189	316	462	614	769	.44 61
FEB	892	1138	50	92	175	290	420	561	.51 54
MAR	1354	1296	53	58	132	236	366	523	.57 43
APR	1910	1266	60	12	42	107	218	325	.63 30
MAY	2375	1164	67	2	6	22	64	146	.69 21
JUN	2600	1111	76	0	1	3	8	33	.71 17
JUL	2672	1184	82	0	0	0	0	6	.75 19
AUG	2311	1356	80	0	0	0	0	12	.73 27
SEP	1845	1649	75	0	0	0	0	35	.71 38
OCT	1227	1549	65	3	11	39	82	194	.63 50
NOV	706	1047	54	50	120	217	339	491	.50 59
DEC	511	782	46	165	283	426	577	732	.44 63
YR	1585	1198	63	572	1086	1802	2688	3827	.65

SANTA MARIA, CALIFORNIA								ELEV 236 LAT 34.9	
	HS	VS	TA	D50	D55	D60	D65	D70	KT LD
JAN	854	1198	51	51	150	296	450	605	.54 56
FEB	1141	1313	52	28	103	226	364	504	.57 49
MAR	1582	1312	53	22	97	226	378	533	.61 37
APR	1921	1082	55	9	58	161	303	453	.61 25
MAY	2141	953	57	3	30	112	245	400	.61 16
JUN	2349	947	60	1	10	63	167	313	.65 12
JUL	2341	965	62	0	3	30	112	247	.66 14
AUG	2106	1057	62	0	3	27	102	241	.65 21
SEP	1730	1255	63	0	3	24	94	225	.63 33
OCT	1353	1440	60	1	7	53	159	299	.62 45
NOV	974	1341	56	5	41	131	270	417	.58 54
DEC	804	1207	52	33	118	256	409	564	.55 58
YR	1610	1172	57	155	624	1604	3053	4801	.62

SACRAMENTO, CALIFORNIA								ELEV 26 LAT 38.5	
	HS	VS	TA	D50	D55	D60	D65	D70	KT LD
JAN	597	829	45	183	315	464	617	772	.43 60
FEB	939	1149	50	86	172	292	426	566	.52 52
MAR	1458	1348	53	50	125	235	372	528	.60 41
APR	2004	1263	58	12	45	116	227	355	.65 29
MAY	2435	1131	64	2	9	37	120	202	.70 19
JUN	2684	1082	71	0	1	6	20	82	.74 15
JUL	2688	1131	75	0	0	0	0	29	.76 17
AUG	2368	1310	74	0	0	0	0	38	.74 25
SEP	1907	1615	72	0	1	5	5	68	.72 37
OCT	1315	1602	63	3	12	47	101	227	.65 48
NOV	782	1135	53	49	121	227	360	511	.53 58
DEC	538	784	46	168	295	442	595	750	.43 62
YR	1646	1198	60	554	1097	1871	2843	4129	.66

COLORADO SPRINGS, COLORADO								ELEV 6171 LAT 38.8	
	HS	VS	TA	D50	D55	D60	D65	D70	KT LD
JAN	891	1534	29	663	818	973	1128	1283	.65 60
FEB	1178	1619	31	524	664	804	944	1084	.65 52
MAR	1550	1477	35	457	611	766	921	1076	.64 41
APR	1931	1227	46	145	271	415	564	714	.63 29
MAY	2129	1035	56	19	74	166	301	451	.61 20
JUN	2369	1021	65	1	5	25	103	181	.65 15
JUL	2212	1012	71	0	1	4	9	71	.62 17
AUG	2025	1145	69	0	1	6	13	96	.63 25
SEP	1759	1475	61	3	17	66	155	279	.67 37
OCT	1359	1702	51	74	166	300	456	605	.68 49
NOV	944	1529	38	377	525	675	825	975	.64 58
DEC	782	1399	31	589	744	899	1054	1209	.63 62
YR	1596	1346	48	2853	3896	5098	6473	8023	.64

```
DENVER, COLORADO                    ELEV 5331  LAT 39.7      HARTFORD, CONNECTICUT              ELEV  180  LAT 41.9
      HS   VS  TA   D50   D55   D60   D65   D70  KT LD              HS   VS  TA   D50   D55   D60   D65   D70  KT LD
JAN  840 1465  30   623   778   933  1088  1243 .64 61       JAN  477  694  25   781   936  1091  1246  1401 .40 63
FEB 1127 1577  33   482   622   762   902  1042 .64 53       FEB  715  891  27   650   790   930  1070  1210 .43 56
MAR 1530 1503  37   406   559   713   868  1023 .64 42       MAR  978  900  36   447   601   756   911  1066 .42 44
APR 1879 1227  48   130   240   379   525   675 .62 30       APR 1315  888  48   109   226   370   519   669 .44 32
MAY 2135 1061  57    18    63   143   253   406 .62 21       MAY 1568  859  58     5    30   101   226   364 .45 23
JUN 2351 1037  66     1     5    23    80   158 .65 16       JUN 1686  853  68     0     1     6    24   107 .46 19
JUL 2273 1053  73     0     0     0     0    50 .64 18       JUL 1649  861  73     0     0     0     0    36 .47 21
AUG 2044 1188  72     0     0     0     0    69 .64 26       AUG 1422  880  70     0     0     2    12    67 .45 28
SEP 1727 1491  63     3    14    51   120   232 .66 38       SEP 1154  967  63     1     6    34   106   224 .46 40
OCT 1300 1657  52    63   143   261   408   559 .67 50       OCT  853  991  53    37   114   237   384   540 .46 52
NOV  883 1441  39   324   469   618   768   918 .62 59       NOV  497  679  41   265   412   561   711   861 .38 61
DEC  732 1323  33   540   695   849  1004  1159 .61 63       DEC  385  562  28   676   831   986  1141  1296 .36 65
YR  1570 1334  50  2592  3588  4733  6016  7535 .64          YR  1060  835  49  2971  3948  5075  6350  7841 .44

EAGLE, COLORADO                     ELEV 6512  LAT 39.6      WILMINGTON, DELAWARE               ELEV   79  LAT 39.7
      HS   VS  TA   D50   D55   D60   D65   D70  KT LD              HS   VS  TA   D50   D55   D60   D65   D70  KT LD
JAN  754 1234  18   992  1147  1302  1457  1612 .57 61       JAN  571  819  32   558   713   868  1023  1178 .43 61
FEB 1078 1470  23   748   888  1028  1168  1308 .61 53       FEB  827 1006  34   460   599   739   879  1019 .47 53
MAR 1502 1461  31   586   741   896  1051  1206 .63 42       MAR 1149 1032  42   268   417   571   725   880 .48 42
APR 1933 1261  42   245   393   543   693   843 .64 30       APR 1480  952  52    47   123   240   381   531 .49 30
MAY 2255 1103  51    38   129   271   425   580 .65 21       MAY 1710  887  62     2    11    47   128   246 .49 21
JUN 2509 1075  59     1    13    72   190   333 .69 16       JUN 1883  895  71     0     0     0     0    59 .52 16
JUL 2384 1085  66     0     0     5    43   139 .67 18       JUL 1823  898  76     0     0     0     0    18 .51 18
AUG 2084 1207  64     0     1    14    79   200 .65 26       AUG 1615  947  74     0     0     0     0    29 .51 26
SEP 1767 1530  56     6    46   142   285   432 .68 38       SEP 1318 1064  68     0     2     9    32   113 .51 38
OCT 1307 1663  45   168   317   471   626   781 .67 50       OCT  984 1113  57    11    50   129   254   399 .50 50
NOV  869 1400  31   573   723   873  1023  1173 .61 59       NOV  645  901  46   156   285   430   579   729 .45 59
DEC  691 1203  20   921  1076  1231  1386  1541 .58 63       DEC  489  720  35   475   629   784   939  1094 .41 63
YR  1597 1306  42  4278  5474  6849  8426 10147 .65          YR  1210  935  54  1978  2829  3818  4940  6295 .49

GRAND JUNCTION, COLORADO            ELEV 4839  LAT 39.1      WASHINGTON, DC                     ELEV  289  LAT 38.9
      HS   VS  TA   D50   D55   D60   D65   D70  KT LD              HS   VS  TA   D50   D55   D60   D65   D70  KT LD
JAN  791 1296  27   726   880  1035  1190  1345 .59 60       JAN  572  793  32   555   710   865  1020  1175 .42 60
FEB 1119 1520  34   460   599   739   879  1019 .62 53       FEB  815  956  34   454   594   734   874  1014 .45 53
MAR 1553 1498  41   283   430   583   738   893 .64 42       MAR 1125  979  42   262   411   564   719   874 .46 41
APR 1986 1276  52    63   142   260   404   550 .65 29       APR 1459  919  53    38   109   219   357   507 .48 29
MAY 2380 1132  62     4    16    58   133   255 .69 20       MAY 1718  877  63     2    10    45   131   240 .50 20
JUN 2598 1081  71     0     1     5    20    69 .71 16       JUN 1901  890  71     0     1     3     5    63 .52 16
JUL 2465 1094  79     0     0     0     0    10 .70 18       JUL 1817  883  75     0     0     0     0    20 .51 18
AUG 2182 1240  75     0     0     0     0    26 .68 26       AUG 1617  929  74     0     0     0     0    34 .51 25
SEP 1834 1573  67     1     3    15    60   133 .70 37       SEP 1340 1058  67     0     2    12    43   131 .51 37
OCT 1345 1698  55    30    91   186   324   470 .68 49       OCT 1004 1111  56    17    68   156   291   438 .50 49
NOV  918 1486  40   312   457   606   756   906 .63 58       NOV  651  883  45   179   313   460   609   759 .45 58
DEC  731 1280  30   636   791   946  1101  1256 .60 62       DEC  481  678  34   497   651   806   961  1116 .39 62
YR  1661 1346  53  2514  3412  4434  5605  6931 .67          YR  1210  912  54  2004  2869  3864  5010  6372 .49

PUEBLO, COLORADO                    ELEV 4721  LAT 38.3      APALACHICOLA, FLORIDA              ELEV   20  LAT 29.7
      HS   VS  TA   D50   D55   D60   D65   D70  KT LD              HS   VS  TA   D50   D55   D60   D65   D70  KT LD
JAN  894 1504  30   617   772   927  1082  1237 .64 60       JAN  853  967  54    53   125   225   368   508 .46 51
FEB 1172 1572  35   429   569   708   848   988 .64 52       FEB 1126 1061  56    30    84   161   290   401 .50 43
MAR 1564 1467  40   315   466   620   775   930 .64 41       MAR 1474 1016  61    10    34    94   175   302 .53 32
APR 1956 1223  52    56   136   257   405   549 .64 29       APR 1879  928  68     1     5    18    30   128 .58 20
MAY 2162 1034  61     3    17    67   148   283 .62 19       MAY 2091  870  75     0     0     0     0    41 .60 11
JUN 2434 1025  71     0     1     4    28    70 .67 15       JUN 1998  844  80     0     0     0     0    12 .56  6
JUL 2312 1030  76     0     0     0     0    15 .65 17       JUL 1814  801  81     0     0     0     0     8 .52  8
AUG 2102 1167  75     0     0     0     0    27 .66 25       AUG 1688  804  82     0     0     0     0     8 .51 16
SEP 1779 1468  66     1     3    16    55   146 .67 36       SEP 1535  938  79     0     0     0     0    16 .53 28
OCT 1361 1669  55    27    91   191   335   481 .67 48       OCT 1371 1204  71     1     3    10    22    92 .57 40
NOV  954 1518  41   282   427   576   726   876 .64 58       NOV 1040 1188  61     9    30    85   158   282 .54 49
DEC  782 1364  33   527   682   837   992  1147 .62 62       DEC  818  991  55    38   101   191   318   462 .47 53
YR  1625 1335  53  2258  3163  4203  5394  6751 .65          YR  1475  967  69   142   382   783  1361  2261 .54
```

DAYTONA BEACH, FLORIDA										ELEV 39 LAT 29.2		TALLAHASSEE, FLORIDA										ELEV 69 LAT 30.4	
	HS	VS	TA	D50	D55	D60	D65	D70	KT	LD			HS	VS	TA	D50	D55	D60	D65	D70	KT	LD	
JAN	958	1116	58	19	60	133	241	368	.51	50		JAN	877	1033	53	73	150	256	408	542	.48	52	
FEB	1213	1150	60	13	42	103	210	302	.54	43		FEB	1138	1104	55	42	102	185	323	429	.52	44	
MAR	1548	1056	64	5	17	54	120	222	.56	32		MAR	1479	1042	60	13	42	106	187	316	.54	33	
APR	1884	919	70	1	4	14	17	108	.58	19		APR	1823	920	68	2	7	23	34	140	.57	21	
MAY	1968	840	75	0	0	0	0	42	.56	10		MAY	1936	842	75	0	0	0	0	47	.56	11	
JUN	1826	807	79	0	0	0	0	15	.51	6		JUN	1883	821	80	0	0	0	0	14	.53	7	
JUL	1784	792	81	0	0	0	0	10	.51	8		JUL	1748	786	81	0	0	0	0	11	.50	9	
AUG	1682	796	81	0	0	0	0	10	.51	16		AUG	1675	808	81	0	0	0	0	11	.51	17	
SEP	1478	892	80	0	0	0	0	14	.51	27		SEP	1493	930	78	0	0	0	0	22	.52	29	
OCT	1251	1054	73	0	2	6	5	61	.52	39		OCT	1318	1174	69	1	5	17	31	122	.56	40	
NOV	1035	1157	65	4	12	40	97	189	.53	49		NOV	1008	1168	59	18	55	124	204	344	.53	50	
DEC	870	1060	60	14	46	114	212	334	.50	53		DEC	813	1011	53	65	140	241	376	524	.48	54	
YR	1459	969	71	57	183	463	902	1677	.53			YR	1434	969	68	215	501	951	1563	2521	.53		

JACKSONVILLE, FLORIDA										ELEV 30 LAT 30.5		TAMPA, FLORIDA										ELEV 10 LAT 28.0	
	HS	VS	TA	D50	D55	D60	D65	D70	KT	LD			HS	VS	TA	D50	D55	D60	D65	D70	KT	LD	
JAN	900	1076	55	47	114	207	348	481	.50	52		JAN	1011	1148	60	16	46	110	203	316	.52	49	
FEB	1164	1141	56	29	80	155	282	389	.53	44		FEB	1259	1156	62	10	31	81	176	253	.54	42	
MAR	1522	1079	61	10	33	91	176	291	.56	33		MAR	1594	1051	66	4	14	41	90	185	.57	30	
APR	1856	936	68	2	6	21	24	135	.58	21		APR	1908	904	72	1	3	11	9	84	.59	18	
MAY	1956	847	74	0	0	0	0	50	.56	11		MAY	1998	839	77	0	0	0	0	32	.57	9	
JUN	1885	822	79	0	0	0	0	16	.53	7		JUN	1847	813	81	0	0	0	0	13	.52	5	
JUL	1802	800	81	0	0	0	0	11	.51	9		JUL	1753	783	82	0	0	0	0	11	.50	7	
AUG	1694	816	81	0	0	0	0	11	.51	17		AUG	1653	775	82	0	0	0	0	10	.50	15	
SEP	1442	900	78	0	0	0	0	20	.50	29		SEP	1492	873	81	0	0	0	0	14	.51	26	
OCT	1223	1070	71	1	3	12	19	101	.52	40		OCT	1346	1108	75	0	0	0	0	53	.55	38	
NOV	996	1153	61	10	32	88	161	281	.53	50		NOV	1108	1213	67	3	11	33	71	164	.55	47	
DEC	818	1024	55	40	102	190	317	457	.48	54		DEC	935	1119	62	12	36	93	169	285	.51	51	
YR	1439	971	68	138	372	762	1327	2241	.53			YR	1493	981	72	47	141	369	718	1421	.54		

MIAMI, FLORIDA										ELEV 7 LAT 25.8		WEST PALM BEACH, FLORIDA										ELEV 20 LAT 26.7	
	HS	VS	TA	D50	D55	D60	D65	D70	KT	LD			HS	VS	TA	D50	D55	D60	D65	D70	KT	LD	
JAN	1057	1121	67	1	4	18	53	142	.52	47		JAN	1000	1075	66	2	8	30	83	178	.50	48	
FEB	1314	1130	68	1	3	14	67	118	.55	39		FEB	1233	1075	66	2	6	23	91	150	.52	40	
MAR	1603	990	71	0	1	6	17	76	.56	28		MAR	1556	985	70	1	2	10	25	100	.55	29	
APR	1859	853	75	0	0	0	0	31	.57	16		APR	1814	851	74	0	0	0	0	43	.56	17	
MAY	1844	800	78	0	0	0	0	15	.53	7		MAY	1845	801	78	0	0	0	0	18	.53	8	
JUN	1708	787	81	0	0	0	0	6	.48	2		JUN	1706	783	81	0	0	0	0	8	.48	3	
JUL	1763	787	82	0	0	0	0	4	.51	4		JUL	1779	789	82	0	0	0	0	5	.51	5	
AUG	1630	751	83	0	0	0	0	4	.49	12		AUG	1663	767	82	0	0	0	0	5	.50	13	
SEP	1456	811	82	0	0	0	0	5	.49	24		SEP	1419	807	82	0	0	0	0	6	.48	25	
OCT	1303	991	78	0	0	0	0	15	.51	36		OCT	1224	946	77	0	0	0	0	20	.49	37	
NOV	1119	1131	72	0	1	4	13	61	.53	45		NOV	1060	1087	71	0	2	7	22	80	.51	46	
DEC	1019	1157	68	1	3	13	56	122	.53	49		DEC	958	1100	67	1	6	22	78	152	.51	50	
YR	1474	941	76	3	14	55	206	599	.52			YR	1439	922	75	6	24	92	299	765	.51		

ORLANDO, FLORIDA										ELEV 118 LAT 28.5		ATLANTA, GEORGIA										ELEV 1033 LAT 33.6	
	HS	VS	TA	D50	D55	D60	D65	D70	KT	LD			HS	VS	TA	D50	D55	D60	D65	D70	KT	LD	
JAN	999	1151	60	13	42	105	197	316	.52	50		JAN	718	884	42	246	393	546	701	856	.44	55	
FEB	1243	1158	62	9	29	79	184	256	.54	42		FEB	969	998	45	161	284	421	560	700	.47	47	
MAR	1582	1058	66	3	11	36	94	181	.57	31		MAR	1304	993	51	67	153	283	443	586	.50	36	
APR	1898	910	71	1	3	10	13	88	.59	19		APR	1686	927	61	3	16	65	144	274	.53	24	
MAY	1989	840	76	0	0	0	0	33	.57	9		MAY	1854	851	69	0	1	7	27	98	.53	15	
JUN	1831	809	80	0	0	0	0	13	.51	5		JUN	1914	842	76	0	0	0	0	19	.53	10	
JUL	1801	795	81	0	0	0	0	10	.51	7		JUL	1812	821	78	0	0	0	0	9	.51	12	
AUG	1673	786	82	0	0	0	0	9	.51	15		AUG	1708	867	78	0	0	0	0	11	.52	20	
SEP	1497	887	80	0	0	0	0	13	.51	27		SEP	1422	966	72	0	0	2	8	49	.51	32	
OCT	1304	1084	74	0	0	0	0	51	.53	38		OCT	1200	1165	62	2	11	49	137	246	.54	44	
NOV	1096	1219	67	3	9	30	75	162	.55	48		NOV	883	1101	51	61	142	266	408	558	.51	53	
DEC	926	1126	62	10	32	87	170	284	.52	52		DEC	674	881	44	217	360	512	667	822	.44	57	
YR	1488	984	72	39	126	348	733	1415	.54			YR	1347	941	61	758	1362	2150	3095	4228	.51		

AUGUSTA, GEORGIA — ELEV 148 LAT 33.4

	HS	VS	TA	D50	D55	D60	D65	D70	KT	LD
JAN	751	934	46	173	297	443	601	750	.45	55
FEB	1015	1055	48	114	208	333	475	608	.49	47
MAR	1338	1018	55	39	105	199	346	480	.51	36
APR	1728	944	64	3	13	45	90	211	.55	24
MAY	1865	852	72	0	2	6	10	73	.53	14
JUN	1904	838	78	0	0	0	0	15	.53	10
JUL	1803	817	80	0	0	0	0	9	.51	12
AUG	1667	847	80	0	0	0	0	11	.51	20
SEP	1410	951	74	0	0	0	0	41	.50	32
OCT	1220	1182	64	3	12	44	104	211	.55	43
NOV	916	1151	54	46	116	214	344	491	.52	53
DEC	721	962	46	161	281	425	577	732	.47	57
YR	1363	962	63	539	1033	1709	2547	3631	.51	

MACON, GEORGIA — ELEV 361 LAT 32.7

	HS	VS	TA	D50	D55	D60	D65	D70	KT	LD
JAN	769	940	48	138	245	384	543	689	.45	54
FEB	1020	1035	50	86	166	280	423	550	.49	46
MAR	1363	1018	57	26	80	162	298	423	.51	35
APR	1736	932	66	2	9	30	66	170	.55	23
MAY	1885	850	74	0	1	4	6	53	.54	14
JUN	1919	838	80	0	0	0	0	12	.53	9
JUL	1785	807	81	0	0	0	0	8	.50	11
AUG	1718	856	81	0	0	0	0	9	.52	19
SEP	1439	953	76	0	0	0	0	30	.51	31
OCT	1247	1186	66	2	9	32	82	178	.55	43
NOV	940	1161	55	35	95	182	304	447	.53	52
DEC	729	950	48	129	233	370	518	673	.46	56
YR	1381	960	65	420	838	1444	2240	3240	.52	

SAVANNAH, GEORGIA — ELEV 52 LAT 32.1

	HS	VS	TA	D50	D55	D60	D65	D70	KT	LD
JAN	795	962	50	100	192	323	483	624	.46	53
FEB	1044	1046	52	61	133	236	379	502	.49	46
MAR	1398	1029	58	16	56	131	256	378	.52	35
APR	1761	930	66	2	7	26	63	161	.55	22
MAY	1852	835	73	0	0	0	0	53	.53	13
JUN	1844	817	79	0	0	0	0	12	.51	9
JUL	1783	803	81	0	0	0	0	7	.50	11
AUG	1621	810	81	0	0	0	0	8	.49	19
SEP	1364	885	76	0	0	0	0	25	.48	30
OCT	1217	1125	67	1	5	21	60	147	.53	42
NOV	941	1134	57	20	66	143	253	391	.52	51
DEC	754	972	50	92	181	308	458	608	.47	55
YR	1366	945	66	293	641	1188	1952	2917	.51	

BOISE, IDAHO — ELEV 2867 LAT 43.6

	HS	VS	TA	D50	D55	D60	D65	D70	KT	LD
JAN	485	770	29	651	806	961	1116	1271	.44	65
FEB	840	1207	36	408	547	686	826	966	.54	57
MAR	1304	1402	41	288	434	587	741	896	.58	46
APR	1827	1355	49	110	206	337	480	630	.62	34
MAY	2277	1252	57	19	63	141	252	395	.66	25
JUN	2463	1183	65	2	9	35	97	188	.68	20
JUL	2613	1309	75	0	0	0	0	39	.74	22
AUG	2196	1460	72	0	1	5	12	65	.71	30
SEP	1737	1749	63	4	15	52	127	227	.71	42
OCT	1138	1610	52	67	146	261	406	556	.64	54
NOV	628	1034	40	314	458	607	756	906	.52	63
DEC	437	735	32	556	710	865	1020	1175	.44	67
YR	1499	1255	51	2420	3395	4536	5833	7315	.64	

LEWISTON, IDAHO — ELEV 1437 LAT 46.4

	HS	VS	TA	D50	D55	D60	D65	D70	KT	LD
JAN	340	523	31	583	738	893	1048	1203	.35	68
FEB	609	849	38	337	474	613	753	893	.43	60
MAR	1020	1108	43	237	379	531	685	840	.48	49
APR	1435	1114	50	84	172	299	441	591	.50	37
MAY	1842	1111	58	13	49	123	232	373	.54	27
JUN	2015	1083	65	2	7	29	84	179	.55	23
JUL	2336	1304	73	0	0	0	0	45	.66	25
AUG	1931	1396	72	0	1	5	17	70	.63	33
SEP	1435	1502	63	3	12	45	124	219	.61	45
OCT	860	1198	52	65	147	267	409	565	.53	56
NOV	413	633	41	293	437	585	735	885	.39	66
DEC	286	454	35	473	627	781	936	1091	.34	70
YR	1214	1024	52	2090	3042	4171	5464	6955	.54	

POCATELLO, IDAHO — ELEV 4478 LAT 42.9

	HS	VS	TA	D50	D55	D60	D65	D70	KT	LD
JAN	539	871	23	831	986	1141	1296	1451	.47	64
FEB	882	1259	29	577	717	857	997	1137	.55	57
MAR	1371	1463	35	454	608	763	918	1073	.60	45
APR	1820	1317	45	166	296	442	591	741	.61	33
MAY	2280	1227	54	29	93	194	336	484	.66	24
JUN	2480	1165	62	3	14	56	138	255	.68	20
JUL	2600	1273	72	0	0	0	0	62	.74	22
AUG	2239	1454	70	0	1	6	20	92	.72	29
SEP	1769	1744	59	6	28	90	192	322	.71	41
OCT	1203	1697	48	110	219	363	515	670	.67	53
NOV	689	1151	36	430	579	729	879	1029	.55	62
DEC	477	809	27	716	871	1026	1181	1336	.47	66
YR	1533	1285	47	3322	4413	5666	7063	8651	.65	

CHICAGO, ILLINOIS — ELEV 623 LAT 41.8

	HS	VS	TA	D50	D55	D60	D65	D70	KT	LD
JAN	507	756	24	797	952	1107	1262	1417	.42	63
FEB	759	967	27	633	773	913	1053	1193	.46	55
MAR	1107	1054	37	414	565	719	874	1029	.48	44
APR	1459	993	50	98	187	313	453	604	.49	32
MAY	1789	963	60	10	36	99	208	320	.52	23
JUN	2007	973	71	1	2	9	26	90	.55	18
JUL	1944	984	75	0	0	0	0	39	.55	20
AUG	1719	1065	74	0	1	4	8	49	.55	28
SEP	1354	1179	66	2	8	28	57	167	.54	40
OCT	969	1181	55	33	94	183	316	456	.52	52
NOV	566	815	40	299	441	589	738	888	.43	61
DEC	401	593	29	667	822	977	1132	1287	.37	65
YR	1217	960	51	2954	3881	4940	6127	7537	.51	

MOLINE, ILLINOIS — ELEV 594 LAT 41.4

	HS	VS	TA	D50	D55	D60	D65	D70	KT	LD
JAN	535	803	22	884	1039	1194	1349	1504	.44	63
FEB	812	1048	26	681	820	960	1100	1240	.48	55
MAR	1119	1055	36	446	599	753	908	1063	.48	44
APR	1459	982	51	81	167	291	436	582	.49	32
MAY	1754	938	61	6	23	77	184	287	.51	22
JUN	1969	952	71	0	1	6	20	79	.54	18
JUL	1939	974	75	0	0	0	0	35	.55	20
AUG	1715	1051	73	0	1	3	11	52	.54	28
SEP	1357	1166	65	2	8	33	79	189	.53	40
OCT	996	1209	54	36	103	200	344	485	.53	51
NOV	595	862	39	330	475	624	774	924	.45	61
DEC	433	651	27	726	880	1035	1190	1345	.39	65
YR	1226	973	50	3191	4117	5178	6395	7786	.51	

SPRINGFIELD, ILLINOIS — ELEV 614 LAT 39.8

	HS	VS	TA	D50	D55	D60	D65	D70	KT	LD
JAN	585	852	27	723	877	1032	1187	1342	.35	61
FEB	861	1069	30	540	689	829	969	1109	.49	53
MAR	1143	1028	39	337	486	639	794	949	.48	42
APR	1515	979	53	54	126	229	363	508	.50	30
MAY	1865	954	63	4	15	52	132	229	.54	21
JUN	2097	966	73	0	1	4	12	56	.58	16
JUL	2058	984	76	0	0	0	0	28	.58	18
AUG	1806	1058	74	0	1	3	8	42	.57	26
SEP	1454	1205	67	1	5	20	48	142	.56	38
OCT	1068	1254	57	24	76	158	282	419	.55	50
NOV	677	971	42	259	397	544	693	843	.48	59
DEC	490	725	31	605	760	915	1070	1225	.41	63
YR	1304	1003	53	2558	3434	4425	5558	6891	.53	

EVANSVILLE, INDIANA — ELEV 387 LAT 38.0

	HS	VS	TA	D50	D55	D60	D65	D70	KT	LD
JAN	574	767	33	541	695	849	1004	1159	.41	59
FEB	823	937	36	328	536	675	815	955	.45	52
MAR	1151	979	44	206	340	489	652	797	.47	40
APR	1501	925	57	22	71	150	263	403	.49	28
MAY	1783	888	66	2	8	30	95	175	.53	19
JUN	1983	903	75	0	1	3	5	37	.57	15
JUL	1920	905	78	0	0	0	0	17	.54	17
AUG	1735	971	76	0	0	0	0	26	.54	25
SEP	1403	1087	69	1	3	12	34	109	.53	36
OCT	1087	1199	58	16	54	128	230	372	.54	48
NOV	682	909	45	186	312	455	603	753	.45	57
DEC	499	688	35	458	611	766	921	1076	.39	61
YR	1264	929	56	1830	2630	3556	4629	5879	.50	

FORT WAYNE, INDIANA — ELEV 827 LAT 41.0

	HS	VS	TA	D50	D55	D60	D65	D70	KT	LD
JAN	455	624	25	766	921	1076	1231	1386	.36	62
FEB	698	833	28	627	767	907	1047	1187	.41	55
MAR	982	880	37	421	574	729	884	1039	.42	43
APR	1361	901	49	98	194	326	471	621	.45	31
MAY	1672	893	60	8	32	96	216	328	.48	22
JUN	1842	900	70	0	2	8	23	95	.51	18
JUL	1787	905	73	0	0	0	0	48	.50	20
AUG	1594	966	71	0	1	5	12	71	.50	28
SEP	1274	1064	65	2	8	31	90	189	.50	39
OCT	924	1071	54	41	112	218	363	509	.49	51
NOV	516	690	40	301	446	594	744	894	.38	60
DEC	369	509	29	664	818	973	1128	1283	.33	64
YR	1125	853	50	2927	3874	4963	6209	7650	.46	

INDIANAPOLIS, INDIANA — ELEV 807 LAT 39.7

	HS	VS	TA	D50	D55	D60	D65	D70	KT	LD
JAN	496	668	28	685	840	995	1150	1305	.38	61
FEB	747	872	31	541	681	820	960	1100	.42	53
MAR	1037	905	40	328	477	630	784	939	.43	42
APR	1398	897	52	63	139	248	387	532	.46	30
MAY	1688	877	62	5	20	66	159	259	.49	22
JUN	1868	890	72	0	1	6	11	71	.51	16
JUL	1806	891	75	0	0	0	0	33	.51	18
AUG	1643	962	73	0	1	4	5	54	.52	27
SEP	1324	1070	66	2	7	25	63	157	.51	38
OCT	977	1102	56	30	88	175	302	446	.50	50
NOV	579	771	42	264	403	550	699	849	.41	59
DEC	417	572	31	593	747	902	1057	1212	.35	63
YR	1157	873	52	2511	3403	4421	5577	6960	.47	

SOUTH BEND, INDIANA — ELEV 774 LAT 41.7

	HS	VS	TA	D50	D55	D60	D65	D70	KT	LD
JAN	416	566	24	806	961	1116	1271	1426	.34	63
FEB	660	791	26	664	804	944	1084	1224	.40	55
MAR	992	911	35	458	611	766	921	1076	.43	44
APR	1387	926	48	122	226	362	507	657	.46	32
MAY	1722	929	59	13	48	122	245	365	.50	23
JUN	1922	940	69	1	3	12	35	114	.53	18
JUL	1852	944	72	0	1	4	6	62	.52	20
AUG	1666	1029	71	0	2	6	24	81	.53	28
SEP	1291	1106	64	3	12	43	98	209	.51	40
OCT	909	1075	53	48	122	226	368	516	.49	52
NOV	497	673	40	319	464	612	762	912	.38	61
DEC	340	467	28	676	831	986	1141	1296	.31	65
YR	1140	864	49	3112	4084	5199	6462	7938	.48	

BURLINGTON, IOWA — ELEV 702 LAT 40.8

	HS	VS	TA	D50	D55	D60	D65	D70	KT	LD
JAN	579	878	23	840	995	1150	1305	1460	.46	62
FEB	853	1109	27	636	776	916	1056	1196	.50	54
MAR	1165	1090	37	411	562	716	871	1026	.49	43
APR	1538	1023	51	78	159	275	416	562	.51	31
MAY	1876	981	62	6	23	74	172	271	.54	22
JUN	2121	992	71	0	2	7	16	77	.58	17
JUL	2035	1016	75	0	0	0	0	33	.59	19
AUG	1829	1101	74	0	1	4	8	47	.58	27
SEP	1416	1206	65	2	9	32	70	177	.55	39
OCT	1061	1293	55	34	95	185	320	459	.56	51
NOV	664	987	40	315	458	607	756	906	.49	60
DEC	481	739	28	695	849	1004	1159	1314	.42	64
YR	1308	1034	51	3018	3930	4969	6149	7527	.54	

DES MOINES, IOWA — ELEV 965 LAT 41.5

	HS	VS	TA	D50	D55	D60	D65	D70	KT	LD
JAN	581	912	19	949	1104	1259	1414	1569	.48	63
FEB	861	1145	24	723	862	1002	1142	1282	.51	55
MAR	1180	1135	34	502	655	809	964	1119	.51	44
APR	1557	1057	50	109	200	325	465	615	.52	32
MAY	1867	993	61	10	33	91	186	297	.54	22
JUN	2125	1008	71	1	3	10	26	94	.58	18
JUL	2097	1037	75	0	0	0	0	39	.59	20
AUG	1828	1124	73	0	1	5	13	58	.58	28
SEP	1434	1256	64	4	14	45	94	204	.57	40
OCT	1068	1342	54	47	116	211	350	489	.57	51
NOV	658	1005	38	373	518	666	816	966	.50	61
DEC	487	778	25	775	930	1085	1240	1395	.44	65
YR	1314	1065	49	3491	4435	5510	6710	8129	.55	

MASON CITY, IOWA — ELEV 1224 LAT 43.1

	HS	VS	TA	D50	D55	D60	D65	D70	KT	LD
JAN	554	917	14	1110	1265	1420	1575	1730	.49	64
FEB	836	1173	19	832	1022	1162	1302	1442	.53	57
MAR	1168	1184	29	651	806	961	1116	1271	.52	46
APR	1519	1078	46	155	288	431	579	729	.51	33
MAY	1895	1046	57	16	58	136	265	394	.55	24
JUN	2114	1041	67	1	4	16	64	135	.53	20
JUL	2084	1074	71	0	1	5	13	74	.59	22
AUG	1833	1183	70	0	2	8	31	94	.59	30
SEP	1405	1294	60	7	28	87	165	302	.57	41
OCT	1010	1322	51	85	174	303	457	605	.56	53
NOV	600	943	34	493	642	792	942	1092	.48	62
DEC	443	731	20	927	1082	1237	1392	1547	.44	66
YR	1291	1081	45	4338	5373	6559	7901	9415	.53	

```
SIOUX CITY, IOWA                    ELEV 1102  LAT 42.4    WICHITA, KANSAS                     ELEV 1339  LAT 37.6
     HS   VS  TA   D50  D55   D60   D65   D70  KT LD            HS   VS  TA   D50  D55   D60   D65   D70  KT LD
JAN  569  922 18   992 1147  1302  1457  1612 .48 64       JAN  784 1191 31   581  735   890  1045  1200 .55 59
FEB  842 1151 23   745  885  1025  1165  1305 .52 56       FEB 1058 1315 36   389  525   664   804   944 .57 51
MAR 1170 1158 33   523  676   831   986  1141 .51 45       MAR 1405 1243 44   231  363   511   671   819 .57 40
APR 1578 1102 49   109  201   328   474   619 .53 33       APR 1782 1088 57    30   83   161   275   408 .58 28
MAY 1901 1031 61     9   32    90   189   297 .55 23       MAY 2036  974 66     3   11    35    90   176 .59 19
JUN 2124 1028 70     1    3    10    33    95 .58 19       JUN 2264  972 76     0    1     3     7    36 .62 14
JUL 2122 1071 75     0    0     0     0    36 .60 21       JUL 2238  994 81     0    0     0     0    12 .63 16
AUG 1845 1166 74     0    1     5    10    55 .59 29       AUG 2032 1108 80     0    0     0     0    15 .63 24
SEP 1421 1281 63     5   17    54   113   224 .57 41       SEP 1616 1272 71     1    3    12    32    97 .61 36
OCT 1038 1336 53    59  134   239   378   526 .57 52       OCT 1250 1436 60    15   48   115   211   335 .61 48
NOV  643 1012 36   415  562   711   861  1011 .50 62       NOV  871 1280 45   197  319   460   606   756 .57 57
DEC  469  769 24   822  977  1132  1287  1442 .45 66       DEC  690 1088 35   484  637   791   946  1101 .53 61
YR  1313 1085 48  3680 4634  5726  6953  8362 .55          YR  1504 1162 57  1931 2725  3642  4687  5897 .60

DODGE CITY, KANSAS                  ELEV 2582  LAT 37.8    LEXINGTON, KENTUCKY                 ELEV  988  LAT 38.0
     HS   VS  TA   D50  D55   D60   D65   D70  KT LD            HS   VS  TA   D50  D55   D60   D65   D70  KT LD
JAN  827 1303 31   596  750   905  1060  1215 .58 59       JAN  546  714 33   531  685   840   995  1150 .39 59
FEB 1122 1444 35   417  555   695  834   974 .60 51        FEB  779  868 35   414  552   692   832   972 .42 52
MAR 1476 1335 41   289  432   584  738   893 .60 40        MAR 1099  924 44   223  360   510   673   818 .45 40
APR 1886 1160 54    48  116   210  344   482 .61 28        APR 1479  911 55    31   90   177   302   444 .48 28
MAY 2090  997 64     4   15    50  115   218 .60 19        MAY 1747  874 65     3   10    37   106   197 .50 19
JUN 2358  998 74     0    1     4   21    51 .65 14        JUN 1897  877 73     0    1     4     8    54 .52 15
JUL 2295 1013 79     0    0     0    0    14 .65 16        JUL 1850  881 76     0    0     0     0    26 .52 17
AUG 2055 1126 78     0    0     0    0    19 .64 24        AUG 1685  945 75     0    0     0     0    35 .53 25
SEP 1687 1350 69     1    4    15   41   118 .63 36        SEP 1362 1049 69     1    3    13    40   117 .51 36
OCT 1301 1532 58    20   63   139  247   382 .64 48        OCT 1044 1134 58    17   58   134   246   383 .51 48
NOV  894 1343 43   239  373   518  666   816 .59 57        NOV  657  862 45   192  320   464   612   762 .44 57
DEC  732 1202 33   517  670   825  980  1135 .56 61        DEC  485  660 36   452  605   760   915  1070 .38 61
YR  1562 1232 55  2132 2980  3945 5046  6318 .62           YR  1221  892 55  1865 2686  3632  4729  6026 .49

GOODLAND, KANSAS                    ELEV 3688  LAT 39.4    LOUISVILLE, KENTUCKY                ELEV  489  LAT 38.2
     HS   VS  TA   D50  D55   D60   D65   D70  KT LD            HS   VS  TA   D50  D55   D60   D65   D70  KT LD
JAN  789 1310 28   695  850  1004  1159  1314 .59 61       JAN  545  718 33   519  673   828   983  1138 .39 59
FEB 1056 1414 32   519  658   798   938  1078 .59 53       FEB  789  890 36   400  538   678   818   958 .43 52
MAR 1424 1350 36   429  581   735   890  1045 .59 42       MAR 1102  933 44   214  349   498   661   806 .45 41
APR 1829 1181 49   121  217   347   489   640 .60 30       APR 1467  908 56    28   83   166   286   426 .48 28
MAY 2062 1025 59    16   51   122   216   353 .59 20       MAY 1720  867 65     3   11    38   105   195 .50 19
JUN 2357 1031 69     1    4    14    55   115 .65 16       JUN 1903  882 73     0    1     4     5    52 .52 15
JUL 2319 1060 76     0    0     0     0    33 .65 18       JUL 1837  880 77     0    0     0     0    22 .52 17
AUG 2046 1178 74     0    0     0     0    49 .64 26       AUG 1680  947 76     0    0     0     0    29 .52 25
SEP 1642 1383 64     4   14    45   108   204 .63 38       SEP 1361 1055 69     1    3    12    35   110 .51 36
OCT 1268 1578 53    64  140   247   387   535 .64 49       OCT 1042 1139 58    16   56   130   241   375 .52 48
NOV  857 1357 39   353  497   646   795   945 .60 59       NOV  653  861 45   185  310   452   600   750 .44 58
DEC  695 1202 30   618  772   927  1082  1237 .57 63       DEC  488  672 36   449  602   757   911  1066 .38 62
YR  1531 1255 51  2819 3784  4885  6119  7547 .62          YR  1218  896 56  1816 2625  3563  4645  5927 .49

TOPEKA, KANSAS                      ELEV  886  LAT 39.1    BATON ROUGE, LOUISIANA              ELEV   75  LAT 30.5
     HS   VS  TA   D50  D55   D60   D65   D70  KT LD            HS   VS  TA   D50  D55   D60   D65   D70  KT LD
JAN  681 1033 28   683  837   992  1147  1302 .50 60       JAN  785  889 51    90  174   294   451   590 .43 52
FEB  941 1181 33   467  605   745   885  1025 .52 53       FEB 1054 1001 54    46  111   199   335   453 .48 44
MAR 1257 1135 41   291  433   584   745   893 .52 42       MAR 1379  965 60    13   43   109   208   330 .51 33
APR 1642 1045 55    47  113   203   329   468 .54 29       APR 1681  864 68     1    5    17    33   127 .52 21
MAY 1915  960 65     5   16    49   118   210 .55 20       MAY 1871  826 75     0    0     0     0    42 .54 11
JUN 2126  962 74     0    2     6    13    58 .58 16       JUN 1926  831 80     0    0     0     0    11 .54  7
JUL 2128  993 78     0    0     0     0    21 .60 18       JUL 1746  786 82     0    0     0     0     7 .50  9
AUG 1910 1094 77     0    0     0     0    27 .60 26       AUG 1677  810 82     0    0     0     0     8 .51 17
SEP 1516 1239 68     2    6    21    55   134 .58 37       SEP 1464  914 78     0    0     0     0    22 .51 29
OCT 1147 1350 58    24   71   148   259   392 .58 49       OCT 1301 1159 69     1    5    17    54   129 .55 40
NOV  772 1144 43   240  371   515   663   813 .53 58       NOV  920 1032 59    17   53   123   208   350 .49 50
DEC  583  907 32   566  720   874  1029  1184 .47 62       DEC  737  883 53    63  139   245   381   532 .44 54
YR  1387 1036 54  2325 3175  4137  5243  6527 .56          YR  1380  913 67   232  530  1006  1670  2601 .51
```

LAKE CHARLES, LOUISIANA — ELEV 10 LAT 30.1

	HS	VS	TA	D50	D55	D60	D65	D70	KT	LD
JAN	728	790	52	79	158	265	415	551	.40	51
FEB	1010	934	55	42	100	181	306	422	.45	44
MAR	1313	903	60	15	44	108	200	317	.48	33
APR	1570	810	69	2	6	19	26	127	.49	20
MAY	1849	818	75	0	0	0	0	45	.53	11
JUN	1970	840	81	0	0	0	0	13	.55	7
JUL	1788	795	82	0	0	0	0	9	.51	9
AUG	1657	798	82	0	0	0	0	9	.50	17
SEP	1485	917	78	0	0	0	0	22	.51	28
OCT	1381	1233	70	1	5	16	36	112	.58	40
NOV	917	1012	60	14	44	106	177	310	.48	49
DEC	706	819	54	54	124	217	338	491	.41	53
YR	1366	889	68	207	481	912	1498	2427	.50	

NEW ORLEANS, LOUISIANA — ELEV 10 LAT 30.0

	HS	VS	TA	D50	D55	D60	D65	D70	KT	LD
JAN	835	950	53	73	150	252	403	533	.46	51
FEB	1112	1055	56	39	96	173	299	409	.50	44
MAR	1415	979	61	14	42	105	188	308	.52	32
APR	1780	895	69	2	7	22	29	133	.55	20
MAY	1968	846	75	0	0	0	0	48	.56	11
JUN	2004	846	80	0	0	0	0	15	.56	7
JUL	1813	801	82	0	0	0	0	11	.51	9
AUG	1717	817	82	0	0	0	0	11	.52	17
SEP	1514	933	78	0	0	0	0	24	.52	28
OCT	1335	1176	70	2	5	17	40	118	.56	40
NOV	973	1095	60	16	46	110	179	313	.51	49
DEC	779	936	55	51	118	208	327	476	.45	53
YR	1438	943	68	197	465	887	1465	2399	.53	

SHREVEPORT, LOUISIANA — ELEV 259 LAT 32.5

	HS	VS	TA	D50	D55	D60	D65	D70	KT	LD
JAN	762	920	47	154	264	403	552	707	.45	54
FEB	1038	1052	51	90	169	279	416	547	.49	46
MAR	1341	993	57	28	81	161	291	415	.50	35
APR	1613	870	66	3	9	30	65	163	.51	23
MAY	1886	848	73	0	2	6	5	59	.54	13
JUN	2065	869	80	0	0	0	0	12	.57	9
JUL	2014	864	83	0	0	0	0	6	.57	11
AUG	1877	913	83	0	0	0	0	6	.57	19
SEP	1554	1030	77	0	0	0	0	24	.55	31
OCT	1303	1248	68	2	7	24	70	148	.57	42
NOV	929	1132	56	31	87	167	278	419	.52	52
DEC	731	946	49	120	215	346	490	646	.46	56
YR	1428	973	66	428	832	1415	2167	3152	.53	

CARIBOU, MAINE — ELEV 623 LAT 46.9

	HS	VS	TA	D50	D55	D60	D65	D70	KT	LD
JAN	419	730	11	1218	1373	1528	1683	1838	.45	68
FEB	724	1120	13	1039	1179	1319	1459	1599	.52	61
MAR	1133	1306	24	818	973	1128	1283	1438	.54	49
APR	1414	1111	37	400	549	699	849	999	.50	37
MAY	1578	963	50	84	183	323	474	629	.46	28
JUN	1757	970	60	4	23	82	170	315	.48	24
JUL	1762	1016	65	1	4	21	84	178	.50	26
AUG	1501	1063	62	2	10	46	122	247	.49	33
SEP	1103	1072	54	26	90	192	327	477	.47	45
OCT	688	885	44	206	350	503	657	812	.43	57
NOV	366	544	31	558	708	858	1008	1158	.35	66
DEC	310	524	16	1051	1206	1361	1516	1671	.38	70
YR	1065	941	39	5408	6648	8061	9632	11363	.48	

PORTLAND, MAINE — ELEV 62 LAT 43.6

	HS	VS	TA	D50	D55	D60	D65	D70	KT	LD
JAN	450	689	22	884	1039	1194	1349	1504	.41	65
FEB	682	891	23	759	899	1039	1179	1319	.44	57
MAR	970	941	32	564	719	874	1029	1184	.43	46
APR	1304	920	43	225	370	519	669	819	.44	34
MAY	1567	888	53	32	108	232	381	536	.46	25
JUN	1712	888	62	1	6	37	106	239	.47	20
JUL	1659	894	68	0	1	5	27	102	.47	22
AUG	1461	943	66	0	1	8	55	135	.47	30
SEP	1158	1025	59	3	22	87	200	340	.47	42
OCT	822	1003	49	83	193	339	493	648	.47	54
NOV	459	651	39	343	492	642	792	942	.38	63
DEC	363	559	26	753	908	1063	1218	1373	.37	67
YR	1052	857	45	3648	4758	6039	7498	9142	.45	

BALTIMORE, MARYLAND — ELEV 154 LAT 39.2

	HS	VS	TA	D50	D55	D60	D65	D70	KT	LD
JAN	587	834	33	515	670	825	980	1135	.44	60
FEB	840	1008	35	426	566	706	846	986	.47	53
MAR	1162	1030	43	236	381	534	688	843	.48	42
APR	1488	945	54	33	101	203	340	487	.49	29
MAY	1714	880	64	2	8	36	110	212	.49	20
JUN	1879	887	72	0	0	0	0	49	.52	16
JUL	1823	890	77	0	0	0	0	15	.51	18
AUG	1599	926	75	0	0	0	0	25	.50	26
SEP	1330	1059	69	0	2	8	27	105	.51	37
OCT	998	1114	57	12	50	128	250	393	.50	49
NOV	660	912	46	150	274	418	567	717	.46	59
DEC	499	725	35	457	611	766	921	1076	.41	63
YR	1217	933	55	1831	2662	3623	4729	6042	.49	

BOSTON, MASSACHUSETTS — ELEV 16 LAT 42.4

	HS	VS	TA	D50	D55	D60	D65	D70	KT	LD
JAN	475	706	29	645	800	955	1110	1265	.40	64
FEB	710	900	30	549	689	829	969	1109	.44	56
MAR	1016	961	38	371	524	679	834	989	.44	45
APR	1326	908	49	96	203	344	492	642	.45	33
MAY	1620	893	59	5	29	99	218	355	.47	23
JUN	1817	912	68	0	1	7	27	106	.50	19
JUL	1749	914	73	0	0	0	0	32	.49	21
AUG	1486	932	71	0	0	2	8	57	.47	29
SEP	1260	1098	65	1	4	21	76	180	.50	41
OCT	890	1072	55	16	70	164	301	453	.49	52
NOV	503	705	45	163	297	445	594	744	.39	62
DEC	403	615	33	527	682	837	992	1147	.38	66
YR	1106	884	51	2374	3300	4381	5621	7080	.47	

ALPENA, MICHIGAN — ELEV 689 LAT 45.1

	HS	VS	TA	D50	D55	D60	D65	D70	KT	LD
JAN	362	538	18	998	1153	1308	1463	1618	.35	66
FEB	617	820	18	888	1028	1168	1308	1448	.41	59
MAR	1028	1069	26	738	893	1048	1203	1358	.47	48
APR	1407	1047	40	303	448	597	747	897	.48	35
MAY	1720	1004	51	79	169	301	455	605	.50	26
JUN	1879	989	61	4	19	71	150	280	.52	22
JUL	1885	1035	66	1	5	22	75	169	.53	24
AUG	1583	1071	64	2	7	32	110	200	.51	32
SEP	1156	1074	56	17	65	147	265	413	.48	43
OCT	743	921	47	132	250	396	549	704	.44	55
NOV	382	534	35	454	603	753	903	1053	.34	64
DEC	270	392	23	825	980	1135	1290	1445	.30	68
YR	1089	875	42	4440	5621	6978	8518	10190	.48	

```
DETROIT, MICHIGAN                  ELEV  627  LAT 42.4     TRAVERSE CITY, MICHIGAN            ELEV  630  LAT 44.7
       HS   VS  TA  D50  D55  D60  D65  D70  KT LD                HS   VS  TA  D50  D55  D60  D65  D70  KT LD
JAN   417  585  26  760  915 1070 1225 1380 .36 64         JAN   311  426  21  905 1060 1215 1370 1525 .30 66
FEB   680  847  27  647  787  927 1067 1207 .42 56         FEB   567  716  21  820  960 1100 1240 1380 .38 58
MAR  1000  941  35  454  608  763  918 1073 .44 45         MAR  1001 1018  29  660  815  970 1125 1280 .46 47
APR  1399  963  48  116  223  360  507  657 .47 33         APR  1405 1033  43  231  371  519  669  819 .48 35
MAY  1716  941  58   10   42  115  238  364 .50 23         MAY  1729  999  53   45  121  236  387  534 .50 26
JUN  1866  932  69    0    2    8   26  100 .51 19         JUN  1912  995  64    2    8   34  104  205 .52 21
JUL  1835  951  73    0    0    0    0   43 .52 21         JUL  1910 1037  69    0    2    8   33  105 .54 23
AUG  1575  989  72    0    1    4   11   61 .50 29         AUG  1609 1078  68    0    2   11   66  126 .52 31
SEP  1253 1090  65    2    7   31   80  188 .50 41         SEP  1165 1071  59    6   28   91  178  322 .48 43
OCT   876 1048  54   33  100  199  342  488 .48 52         OCT   754  926  50   87  184  321  471  626 .44 55
NOV   478  655  41  276  419  567  717  867 .37 62         NOV   377  515  37  395  543  693  843  993 .33 64
DEC   343  487  30  633  787  942 1097 1252 .33 66         DEC   257  360  26  747  902 1057 1212 1367 .28 68
YR   1122  869  50 2931 3890 4986 6228 7679 .47            YR   1086  848  45 3899 4997 6257 7698 9284 .47

FLINT, MICHIGAN                    ELEV  764  LAT 43.0     DULUTH, MINNESOTA                  ELEV 1417  LAT 46.8
       HS   VS  TA  D50  D55  D60  D65  D70  KT LD                HS   VS  TA  D50  D55  D60  D65  D70  KT LD
JAN   383  532  22  859 1014 1169 1324 1479 .34 64         JAN   389  650   9 1287 1442 1597 1751 1907 .41 68
FEB   636  788  24  734  874 1014 1154 1294 .40 57         FEB   673 1000  12 1061 1201 1341 1481 1621 .48 60
MAR   957  906  33  540  694  849 1004 1159 .42 45         MAR  1034 1145  24  822  977 1132 1287 1442 .49 49
APR  1339  932  46  153  280  424  573  723 .45 33         APR  1373 1069  39  345  492  642  792  942 .48 37
MAY  1658  924  56   19   72  161  306  442 .48 24         MAY  1643 1001  49   91  192  332  484  639 .48 28
JUN  1813  921  66    1    4   18   65  155 .50 20         JUN  1767  973  59    6   29   94  194  333 .49 23
JUL  1797  946  70    0    1    6   14   89 .51 22         JUL  1854 1063  66    1    4   18   67  163 .53 25
AUG  1555  992  68    0    2    9   36  113 .50 30         AUG  1547 1097  64    1    6   29  104  200 .51 33
SEP  1195 1047  61    4   17   67  147  277 .48 41         SEP  1095 1058  54   26   88  186  318  469 .47 45
OCT   829  992  51   66  152  280  433  583 .46 53         OCT   725  950  45  170  306  457  611  766 .45 57
NOV   429  576  38  354  502  651  801  951 .34 62         NOV   381  574  28  648  798  948 1098 1248 .37 66
DEC   309  432  27  719  874 1029 1184 1339 .30 66         DEC   292  477  14 1104 1259 1414 1569 1724 .36 70
YR   1077  832  47 3449 4485 5677 7041 8604 .46            YR   1067  921  39 5560 6792 8189 9756 11452 .48

GRAND RAPIDS, MICHIGAN             ELEV  804  LAT 42.9     INTERNATIONAL FALLS, MINNESOTA ELEV 1184 LAT 48.6
       HS   VS  TA  D50  D55  D60  D65  D70  KT LD                HS   VS  TA  D50  D55  D60  D65  D70  KT LD
JAN   370  504  23  831  986 1141 1296 1451 .32 64         JAN   356  626   2 1491 1646 1801 1956 2111 .42 70
FEB   648  806  25  714  854  994 1134 1274 .40 57         FEB   663 1061   7 1204 1344 1484 1624 1764 .51 62
MAR  1014  974  33  525  679  834  989 1144 .45 45         MAR  1046 1244  21  911 1066 1221 1376 1531 .52 51
APR  1412  987  47  145  265  407  555  705 .48 33         APR  1444 1202  38  357  505  654  804  954 .52 39
MAY  1755  971  57   15   58  138  270  403 .51 24         MAY  1716 1096  50   81  176  312  462  617 .51 30
JUN  1956  976  67    1    3   13   44  128 .54 20         JUN  1853 1059  60    4   20   74  168  293 .51 25
JUL  1914  996  72    0    1    4    8   66 .54 22         JUL  1921 1154  66    1    4   18   66  160 .55 27
AUG  1676 1070  70    0    1    6   27   88 .54 29         AUG  1618 1220  63    2    9   39  112  224 .54 35
SEP  1262 1118  62    3   13   52  114  241 .51 41         SEP  1121 1161  53   40  112  222  364  510 .49 47
OCT   858 1037  52   59  140  260  409  559 .48 53         OCT   704  979  44  216  360  512  667  822 .46 59
NOV   446  606  39  343  490  639  789  939 .36 62         NOV   345  541  25  753  903 1053 1203 1353 .36 68
DEC   311  434  27  701  856 1011 1166 1321 .30 66         DEC   272  475   9 1280 1435 1590 1745 1900 .38 72
YR   1138  873  48 3337 4345 5499 6801 8317 .48            YR   1091  984  37 6341 7580 8983 10547 12241 .50

SAULT STE. MARIE, MICHIGAN         ELEV  725  LAT 46.5     MINNEAPOLIS, MINNESOTA             ELEV  837  LAT 44.9
       HS   VS  TA  D50  D55  D60  D65  D70  KT LD                HS   VS  TA  D50  D55  D60  D65  D70  KT LD
JAN   325  492  14 1110 1265 1420 1575 1730 .34 68         JAN   464  768  12 1172 1327 1482 1637 1792 .45 66
FEB   603  840  15  974 1114 1254 1394 1534 .43 60         FEB   764 1110  17  938 1078 1218 1358 1498 .51 59
MAR  1029 1125  24  806  961 1116 1271 1426 .49 49         MAR  1103 1169  28  673  828  983 1138 1293 .51 47
APR  1383 1069  38  357  504  654  804  954 .48 37         APR  1442 1071  45  178  305  449  597  747 .50 35
MAY  1688 1021  49   98  203  344  496  651 .50 27         MAY  1737 1009  57   18   63  143  271  403 .51 26
JUN  1811  988  59    7   33  100  200  342 .50 23         JUN  1927 1006  67    1    4   18   65  142 .53 22
JUL  1835 1045  64    1    7   33   96  208 .52 25         JUL  1970 1071  72    0    1    5   11   66 .56 24
AUG  1523 1068  63    2    8   39  125  224 .50 33         AUG  1687 1142  70    0    2    8   21   90 .55 31
SEP  1049  990  55   20   75  165  291  442 .45 45         SEP  1255 1188  60    8   30   90  173  308 .52 43
OCT   673  844  46  150  280  429  583  738 .41 56         OCT   860 1126  50   93  186  318  472  620 .50 55
NOV   332  465  33  516  666  816  966 1116 .31 66         NOV   480  736  32  529  678  828  978 1128 .42 64
DEC   253  382  20  927 1082 1237 1392 1547 .30 70         DEC   353  572  19  973 1128 1283 1438 1593 .39 68
YR   1044  861  40 4969 6198 7607 9193 10912 .47           YR   1172  996  44 4584 5631 6824 8159 9680 .51
```

ROCHESTER, MINNESOTA — ELEV 1319 LAT 43.9

	HS	VS	TA	D50	D55	D60	D65	D70	KT	LD
JAN	477	762	13	1150	1305	1460	1615	1770	.44	65
FEB	753	1041	17	927	1067	1207	1347	1487	.49	58
MAR	1082	1099	28	688	843	998	1153	1308	.49	46
APR	1410	1013	45	188	320	466	615	765	.48	34
MAY	1696	963	56	20	71	156	292	430	.49	25
JUN	1902	974	66	1	5	20	78	155	.52	21
JUL	1909	1017	70	0	1	6	21	87	.54	23
AUG	1662	1091	69	0	2	10	35	112	.53	30
SEP	1250	1142	59	8	33	97	185	326	.51	42
OCT	870	1100	50	95	192	328	485	633	.50	54
NOV	494	734	33	523	672	822	972	1122	.41	63
DEC	370	583	19	964	1119	1274	1429	1584	.38	67
YR	1158	959	44	4565	5631	6845	8227	9779	.50	

JACKSON, MISSISSIPPI — ELEV 331 LAT 32.3

	HS	VS	TA	D50	D55	D60	D65	D70	KT	LD
JAN	753	898	47	158	268	407	569	710	.44	54
FEB	1026	1029	50	102	184	298	442	567	.48	46
MAR	1369	1011	56	35	94	177	313	436	.51	35
APR	1708	910	66	3	12	37	74	179	.54	23
MAY	1941	861	73	1	2	7	6	71	.56	13
JUN	2024	859	79	0	0	0	0	16	.56	9
JUL	1909	837	82	0	0	0	0	9	.54	11
AUG	1780	873	81	0	0	0	0	11	.54	19
SEP	1509	992	76	0	0	0	0	35	.53	30
OCT	1271	1199	66	3	12	38	91	183	.56	42
NOV	902	1077	55	40	102	187	301	445	.50	52
DEC	709	898	49	128	224	355	504	655	.45	56
YR	1410	953	65	471	898	1506	2300	3318	.53	

MERIDIAN, MISSISSIPPI — ELEV 308 LAT 32.3

	HS	VS	TA	D50	D55	D60	D65	D70	KT	LD
JAN	744	883	47	163	274	413	575	717	.43	54
FEB	1012	1010	50	103	185	298	443	567	.48	46
MAR	1328	976	56	36	95	178	312	437	.50	35
APR	1662	889	65	4	13	41	79	186	.52	23
MAY	1860	839	72	1	2	8	7	76	.53	13
JUN	1963	846	79	0	0	0	0	17	.54	9
JUL	1823	815	81	0	0	0	0	11	.52	11
AUG	1739	858	81	0	0	0	0	13	.53	19
SEP	1454	953	75	0	0	0	0	42	.51	30
OCT	1258	1183	65	5	16	48	111	205	.55	42
NOV	897	1068	54	52	120	211	331	478	.50	52
DEC	699	880	48	146	249	384	530	686	.44	56
YR	1371	933	65	510	955	1582	2388	3434	.51	

COLUMBIA, MISSOURI — ELEV 886 LAT 38.8

	HS	VS	TA	D50	D55	D60	D65	D70	KT	LD
JAN	611	869	29	642	797	952	1107	1262	.45	60
FEB	875	1052	34	461	600	739	879	1019	.48	52
MAR	1179	1035	42	275	417	569	730	877	.48	41
APR	1526	960	55	37	99	187	314	453	.50	29
MAY	1880	940	64	4	13	44	117	207	.54	20
JUN	2089	946	73	0	1	5	11	58	.57	15
JUL	2116	983	77	0	0	0	0	22	.60	17
AUG	1878	1067	76	0	1	2	5	30	.59	25
SEP	1450	1161	68	1	5	17	42	126	.55	37
OCT	1101	1259	58	19	60	135	247	379	.55	49
NOV	703	983	44	212	341	485	633	783	.48	58
DEC	522	760	33	535	689	843	998	1153	.42	62
YR	1330	1001	54	2186	3022	3979	5083	6370	.53	

KANSAS CITY, MISSOURI — ELEV 1033 LAT 39.3

	HS	VS	TA	D50	D55	D60	D65	D70	KT	LD
JAN	648	969	27	710	865	1020	1175	1330	.48	61
FEB	895	1108	32	497	636	776	916	1056	.50	53
MAR	1203	1080	41	303	447	599	753	908	.50	42
APR	1575	1006	54	46	114	206	336	477	.52	30
MAY	1873	948	64	4	15	49	127	216	.54	20
JUN	2080	952	73	0	1	5	15	60	.57	16
JUL	2102	989	78	0	0	0	0	22	.59	18
AUG	1862	1074	77	0	0	0	0	28	.58	26
SEP	1452	1183	68	1	5	19	50	133	.56	37
OCT	1092	1269	58	22	67	144	259	391	.55	49
NOV	737	1077	42	252	387	533	681	831	.51	59
DEC	561	865	31	581	735	890	1045	1200	.46	63
YR	1342	1042	54	2417	3273	4242	5357	6651	.54	

SAINT LOUIS, MISSOURI — ELEV 564 LAT 38.7

	HS	VS	TA	D50	D55	D60	D65	D70	KT	LD
JAN	627	898	31	581	735	890	1045	1200	.46	60
FEB	886	1066	35	421	558	697	837	977	.49	52
MAR	1205	1061	43	237	371	520	682	828	.49	41
APR	1564	982	57	29	83	162	272	410	.51	29
MAY	1871	935	66	3	11	36	103	181	.54	20
JUN	2092	945	75	0	1	4	10	42	.58	15
JUL	2049	959	79	0	0	0	0	19	.58	17
AUG	1816	1032	77	0	0	0	0	26	.57	25
SEP	1459	1166	70	1	4	14	35	110	.55	37
OCT	1100	1252	59	16	52	122	224	349	.55	49
NOV	718	1009	45	191	313	454	600	750	.49	58
DEC	531	776	35	481	633	788	942	1097	.43	62
YR	1329	1006	56	1961	2762	3686	4750	5989	.53	

SPRINGFIELD, MISSOURI — ELEV 1270 LAT 37.2

	HS	VS	TA	D50	D55	D60	D65	D70	KT	LD
JAN	684	956	33	532	686	840	995	1150	.47	58
FEB	926	1071	37	369	505	644	784	924	.49	51
MAR	1235	1042	44	217	350	498	660	806	.50	40
APR	1604	968	57	27	79	158	275	410	.52	27
MAY	1882	912	65	3	11	38	94	192	.54	18
JUN	2075	918	74	0	1	4	10	52	.57	14
JUL	2063	936	78	0	0	0	0	20	.58	16
AUG	1873	1019	77	0	0	2	6	24	.58	24
SEP	1481	1130	69	1	4	13	35	111	.55	35
OCT	1144	1249	59	15	49	120	227	350	.55	47
NOV	775	1058	46	178	298	438	585	735	.50	57
DEC	603	874	36	438	590	744	899	1054	.45	61
YR	1364	1011	56	1779	2573	3501	4570	5828	.54	

BILLINGS, MONTANA — ELEV 3570 LAT 45.8

	HS	VS	TA	D50	D55	D60	D65	D70	KT	LD
JAN	486	863	22	871	1026	1181	1336	1491	.49	67
FEB	763	1153	27	633	773	913	1053	1193	.53	59
MAR	1189	1340	33	540	695	849	1004	1159	.55	48
APR	1526	1177	45	190	319	464	612	762	.53	36
MAY	1913	1134	55	36	102	198	333	482	.56	27
JUN	2174	1139	63	4	15	55	131	238	.60	22
JUL	2384	1303	72	0	1	5	10	68	.68	24
AUG	2022	1441	70	0	2	8	15	92	.66	32
SEP	1470	1517	59	11	41	108	221	339	.62	44
OCT	987	1432	49	105	203	338	487	642	.60	56
NOV	561	972	36	431	580	729	879	1029	.51	65
DEC	421	782	27	719	874	1029	1184	1339	.48	69
YR	1328	1188	46	3541	4630	5878	7265	8835	.59	

Weather Data / 413

```
CUT BANK, MONTANA                    ELEV 3839 LAT 48.6      HELENA, MONTANA                      ELEV 3898 LAT 46.6
       HS   VS TA  D50  D55  D60  D65   D70 KT LD                   HS   VS TA  D50  D55  D60  D65   D70 KT LD
JAN   402  753 16 1048 1203 1358 1513  1668 .48 70          JAN    419  719 18  989 1144 1299 1454  1609 .44 68
FEB   688 1122 22  773  913 1053 1193  1333 .53 62          FEB    709 1071 25  689  829  969 1109  1249 .50 60
MAR  1128 1385 27  719  874 1029 1184  1339 .56 51          MAR   1145 1310 31  602  756  911 1066  1221 .54 49
APR  1485 1245 40  318  466  615  765   915 .53 39          APR   1487 1170 43  232  372  520  669   819 .52 37
MAY  1883 1209 50   85  185  326  477   633 .56 30          MAY   1860 1128 52   56  135  254  401   552 .55 28
JUN  2045 1162 57   12   55  136  267   406 .56 25          JUN   2040 1101 59    8   32   97  194   329 .56 23
JUL  2287 1372 64    1    5   25   82   190 .65 27          JUL   2334 1312 68    1    3   12   33   122 .66 25
AUG  1897 1475 63    2    9   42  125   239 .63 35          AUG   1930 1404 66    1    4   19   57   155 .63 33
SEP  1352 1503 53   34  105  215  368   504 .60 47          SEP   1412 1480 56   23   78  165  304   436 .60 45
OCT   871 1343 44  198  341  493  648   803 .57 59          OCT    926 1348 45  175  307  457  611   766 .57 57
NOV   480  887 30  609  759  909 1059  1209 .51 68          NOV    521  904 32  549  699  849  999  1149 .49 66
DEC   334  647 21  887 1042 1197 1352  1507 .46 72          DEC    364  656 23  828  983 1138 1293  1448 .44 70
YR   1241 1175 41 4686 5955 7398 9033 10745 .57             YR    1266 1134 43 4151 5342 6689 8190  9855 .57

     DILLON, MONTANA                 ELEV 5210 LAT 45.2          LEWISTOWN, MONTANA              ELEV 4147 LAT 47.0
       HS   VS TA  D50  D55  D60  D65   D70 KT LD                   HS   VS TA  D50  D55  D60  D65   D70 KT LD
JAN   526  943 20  924 1079 1234 1389  1544 .51 66          JAN    420  737 19  958 1113 1268 1423  1578 .45 68
FEB   846 1310 26  686  826  966 1106  1246 .57 59          FEB    692 1051 24  734  874 1014 1154  1294 .50 61
MAR  1279 1451 30  633  787  942 1097  1252 .59 48          MAR   1128 1303 28  698  853 1008 1163  1318 .54 49
APR  1639 1257 41  276  419  567  717   867 .57 35          APR   1444 1143 40  306  449  598  747   897 .51 37
MAY  1989 1158 50   84  174  305  453   608 .58 26          MAY   1807 1107 50  102  198  330  477   633 .53 28
JUN  2143 1108 58   14   52  127  238   378 .59 22          JUN   2059 1121 57   21   70  150  265   405 .57 24
JUL  2392 1282 66    1    4   19   54   153 .68 24          JUL   2288 1304 66    2    8   29   70   177 .65 26
AUG  2023 1412 65    2    8   32   85   193 .66 32          AUG   1901 1399 64    3   10   39   94   202 .62 34
SEP  1521 1554 55   30   91  184  325   460 .63 43          SEP   1372 1444 54   41  109  205  348   482 .59 45
OCT  1023 1472 45  183  317  466  620   775 .61 55          OCT    905 1326 46  177  305  452  605   760 .57 57
NOV   602 1050 32  547  696  846  996  1146 .53 65          NOV    502  872 32  535  684  834  984  1134 .49 66
DEC   450  838 24  809  964 1119 1274  1429 .50 69          DEC    363  668 25  791  946 1101 1256  1411 .45 70
YR   1373 1236 43 4188 5418 6809 8354 10051 .60             YR    1243 1123 42 4367 5617 7026 8586 10289 .56

     GLASGOW, MONTANA                ELEV 2297 LAT 48.2          MILES CITY, MONTANA             ELEV 2634 LAT 46.4
       HS   VS TA  D50  D55  D60  D65   D70 KT LD                   HS   VS TA  D50  D55  D60  D65   D70 KT LD
JAN   388  698  9 1265 1420 1575 1730  1885 .45 69          JAN    457  811 15 1073 1228 1383 1538  1693 .48 68
FEB   671 1060 15  974 1114 1254 1394  1534 .51 62          FEB    745 1143 22  795  935 1075 1215  1355 .52 60
MAR  1105 1324 25  769  924 1079 1234  1389 .54 51          MAR   1185 1365 30  615  769  924 1079  1234 .56 49
APR  1488 1232 43  234  370  517  666   816 .53 38          APR   1542 1215 45  190  308  446  591   741 .54 37
MAY  1828 1159 54   40  108  206  344   491 .54 29          MAY   1896 1143 56   37   95  177  288   432 .56 27
JUN  2047 1151 62    5   19   64  151   254 .56 25          JUN   2146 1144 65    5   16   47  117   199 .59 23
JUL  2193 1300 71    0    2    7   15    88 .63 27          JUL   2293 1282 74    1    2    6    9    54 .65 25
AUG  1863 1423 69    1    3   11   30   112 .62 35          AUG   1977 1434 73    1    3    9   16    78 .65 33
SEP  1340 1462 57   18   61  138  263   388 .59 46          SEP   1444 1515 60   16   47  111  217   318 .61 45
OCT   877 1333 46  158  279  425  577   732 .57 58          OCT    961 1413 49  134  230  359  508   658 .59 56
NOV   479  865 29  630  780  930 1080  1230 .49 68          NOV    551  975 32  531  679  828  978  1128 .52 66
DEC   334  632 17 1020 1175 1330 1485  1640 .45 72          DEC    399  746 22  868 1023 1178 1333  1488 .48 70
YR   1221 1137 42 5114 6256 7537 8969 10559 .56             YR    1303 1182 45 4265 5334 6544 7889  9378 .58

     GREAT FALLS, MONTANA            ELEV 3661 LAT 47.5          MISSOULA, MONTANA               ELEV 3189 LAT 46.9
       HS   VS TA  D50  D55  D60  D65   D70 KT LD                   HS   VS TA  D50  D55  D60  D65   D70 KT LD
JAN   420  757 21  915 1070 1225 1380  1535 .47 69          JAN    312  473 21  905 1060 1215 1370  1525 .33 68
FEB   720 1141 27  655  795  935 1075  1215 .53 61          FEB    574  796 27  638  778  918 1058  1198 .41 61
MAR  1170 1398 31  605  760  915 1070  1225 .57 50          MAR    981 1069 33  518  673  828  983  1138 .47 49
APR  1489 1205 43  215  352  499  648   798 .53 38          APR   1382 1081 44  194  335  483  633   783 .48 37
MAY  1848 1149 53   44  117  225  367   519 .54 28          MAY   1782 1089 52   42  122  248  397   552 .52 28
JUN  2101 1157 61    5   21   74  162   284 .58 24          JUN   1933 1057 59    4   24   88  201   335 .53 24
JUL  2329 1348 69    0    2    8   18   100 .66 26          JUL   2327 1321 67    0    2   10   39   134 .66 26
AUG  1933 1451 67    1    3   14   42   132 .64 34          AUG   1881 1377 65    0    3   17   71   172 .62 33
SEP  1378 1480 57   14   54  130  260   384 .60 46          SEP   1358 1418 55   15   66  159  301   441 .58 45
OCT   925 1400 48  117  225  367  524   673 .59 57          OCT    813 1126 44  195  340  493  648   803 .51 57
NOV   498  885 35  463  612  762  912  1062 .49 67          NOV    410  641 32  531  681  831  981  1131 .40 66
DEC   336  612 27  729  884 1039 1194  1349 .43 71          DEC    267  421 25  784  939 1094 1249  1404 .33 70
YR   1266 1165 45 3761 4893 6191 7652  9274 .57             YR    1172  990 44 3828 5023 6385 7931  9617 .53
```

GRAND ISLAND, NEBRASKA									ELEV 1857 LAT 41.0	
	HS	VS	TA	D50	D55	D60	D65	D70	KT	LD
JAN	661	1081	22	859	1014	1169	1324	1479	.53	62
FEB	917	1230	28	625	765	904	1044	1184	.54	55
MAR	1265	1223	36	453	606	760	915	1070	.54	43
APR	1692	1141	50	103	192	315	461	604	.56	31
MAY	1972	1028	61	10	35	95	184	303	.57	22
JUN	2242	1035	71	1	3	10	35	92	.62	18
JUL	2216	1068	76	0	1	2	6	30	.63	20
AUG	1939	1175	75	0	1	3	5	41	.61	28
SEP	1509	1317	64	4	14	45	107	203	.59	39
OCT	1138	1442	54	54	127	226	362	508	.60	51
NOV	738	1166	38	362	506	655	804	954	.55	60
DEC	569	960	27	713	868	1023	1178	1333	.51	64
YR	1407	1155	50	3186	4130	5208	6425	7801	.58	

ELKO, NEVADA									ELEV 5075 LAT 40.8	
	HS	VS	TA	D50	D55	D60	D65	D70	KT	LD
JAN	689	1140	23	831	986	1141	1296	1451	.55	62
FEB	1034	1456	29	583	722	862	1002	1142	.61	54
MAR	1463	1476	35	467	620	775	930	1085	.62	43
APR	1900	1288	44	216	350	496	645	795	.63	31
MAY	2303	1161	52	67	147	265	406	562	.67	22
JUN	2534	1115	60	9	34	97	190	319	.70	17
JUL	2623	1197	70	1	2	9	27	102	.74	19
AUG	2316	1394	67	1	5	19	60	145	.73	27
SEP	1893	1754	58	16	55	130	248	376	.74	39
OCT	1322	1782	47	147	265	409	561	716	.70	51
NOV	812	1336	35	458	606	756	906	1056	.60	60
DEC	617	1075	26	747	902	1057	1212	1367	.54	64
YR	1629	1346	45	3541	4696	6018	7483	9117	.67	

NORTH OMAHA, NEBRASKA									ELEV 1325 LAT 41.4	
	HS	VS	TA	D50	D55	D60	D65	D70	KT	LD
JAN	634	1034	20	924	1079	1234	1389	1544	.52	63
FEB	892	1201	26	686	826	966	1106	1246	.53	55
MAR	1222	1185	35	480	633	788	942	1097	.52	44
APR	1558	1055	50	99	187	311	456	601	.52	32
MAY	1873	993	61	9	31	89	186	296	.54	22
JUN	2122	1005	70	1	3	10	33	96	.58	18
JUL	2106	1038	75	0	1	3	7	38	.59	20
AUG	1858	1139	74	0	1	4	10	52	.59	28
SEP	1373	1184	64	3	13	43	99	200	.54	40
OCT	1050	1304	54	44	112	207	342	486	.56	51
NOV	644	969	38	369	515	663	813	963	.48	61
DEC	511	832	26	754	908	1063	1218	1373	.46	65
YR	1323	1078	49	3369	4309	5381	6601	7992	.55	

ELY, NEVADA									ELEV 6253 LAT 39.3	
	HS	VS	TA	D50	D55	D60	D65	D70	KT	LD
JAN	819	1380	24	818	973	1128	1283	1438	.61	61
FEB	1141	1578	28	619	759	899	1039	1179	.64	53
MAR	1606	1580	33	534	688	843	998	1153	.67	42
APR	2009	1301	41	268	412	561	711	861	.66	30
MAY	2311	1115	50	81	178	315	470	620	.67	20
JUN	2513	1068	58	9	42	116	241	371	.69	16
JUL	2447	1094	67	0	2	11	23	130	.69	18
AUG	2230	1274	66	1	4	19	62	166	.70	26
SEP	1935	1698	57	13	55	135	265	400	.74	37
OCT	1408	1832	46	154	285	435	589	744	.71	49
NOV	926	1521	34	481	630	780	930	1080	.64	59
DEC	723	1271	26	738	893	1048	1203	1358	.59	63
YR	1675	1391	44	3716	4922	6291	7814	9500	.68	

NORTH PLATTE, NEBRASKA									ELEV 2785 LAT 41.1	
	HS	VS	TA	D50	D55	D60	D65	D70	KT	LD
JAN	692	1165	23	825	980	1135	1290	1445	.56	62
FEB	958	1317	28	614	753	893	1033	1173	.57	55
MAR	1333	1316	34	490	643	797	952	1107	.57	44
APR	1724	1169	48	138	241	373	522	667	.57	31
MAY	1988	1037	58	20	60	134	238	371	.58	22
JUN	2266	1045	68	2	6	20	65	135	.62	18
JUL	2277	1093	74	0	1	4	7	49	.64	20
AUG	1989	1209	73	0	2	6	8	64	.63	28
SEP	1565	1385	62	7	24	71	141	253	.61	39
OCT	1177	1521	51	92	176	295	439	590	.62	51
NOV	759	1223	36	419	565	714	864	1014	.56	60
DEC	605	1059	27	720	874	1029	1184	1339	.54	64
YR	1447	1211	49	3326	4325	5473	6743	8206	.60	

LAS VEGAS, NEVADA									ELEV 2178 LAT 36.1	
	HS	VS	TA	D50	D55	D60	D65	D70	KT	LD
JAN	978	1553	44	216	346	493	645	800	.65	57
FEB	1339	1739	49	110	197	315	451	586	.69	50
MAR	1823	1646	55	45	111	202	324	475	.72	39
APR	2319	1343	64	5	17	53	126	218	.75	26
MAY	2646	1109	73	0	2	6	10	61	.76	17
JUN	2778	1033	82	0	0	0	0	7	.77	13
JUL	2588	1039	90	0	0	0	0	1	.73	15
AUG	2355	1200	87	0	0	0	0	2	.73	23
SEP	2037	1593	80	0	0	0	0	13	.75	34
OCT	1540	1817	67	2	8	27	74	156	.73	46
NOV	1085	1665	53	59	131	229	357	503	.67	55
DEC	880	1467	45	193	318	463	614	769	.63	59
YR	1866	1431	66	631	1129	1788	2601	3591	.73	

SCOTTSBLUFF, NEBRASKA									ELEV 3957 LAT 41.9	
	HS	VS	TA	D50	D55	D60	D65	D70	KT	LD
JAN	676	1170	25	778	933	1088	1243	1398	.56	63
FEB	950	1346	30	575	714	854	994	1134	.58	56
MAR	1307	1320	34	489	642	797	952	1107	.57	44
APR	1668	1155	46	163	279	418	564	714	.56	32
MAY	1933	1033	57	27	81	163	280	423	.56	23
JUN	2237	1055	66	2	9	30	91	169	.61	19
JUL	2284	1119	74	0	0	0	0	52	.65	21
AUG	1999	1247	72	1	2	8	8	80	.64	28
SEP	1599	1466	61	8	28	83	160	279	.63	40
OCT	1145	1510	50	100	190	316	459	615	.62	52
NOV	723	1180	36	418	565	714	864	1014	.55	61
DEC	575	1021	28	695	850	1004	1159	1314	.53	65
YR	1427	1218	48	3256	4293	5475	6774	8299	.60	

LOVELOCK, NEVADA									ELEV 3904 LAT 40.1	
	HS	VS	TA	D50	D55	D60	D65	D70	KT	LD
JAN	804	1396	29	654	809	964	1119	1274	.62	61
FEB	1165	1687	35	416	555	695	834	974	.67	54
MAR	1656	1700	40	316	464	617	772	927	.70	43
APR	2165	1451	49	116	216	350	495	645	.72	30
MAY	2555	1230	58	17	59	137	255	392	.74	21
JUN	2749	1145	66	2	7	27	86	169	.75	17
JUL	2784	1215	74	0	0	0	0	1727	.79	19
AUG	2484	1454	71	0	1	6	17	76	.78	27
SEP	2027	1865	63	4	15	55	126	236	.78	38
OCT	1451	1984	51	77	162	285	428	583	.75	50
NOV	929	1589	38	353	499	648	798	948	.66	59
DEC	714	1299	31	596	750	905	1060	1215	.61	63
YR	1793	1500	50	2551	3538	4688	5990	9166	.73	

```
RENO, NEVADA                         ELEV 4400  LAT 39.5    NEWARK, NEW JERSEY                   ELEV   30  LAT 40.7
      HS   VS TA  D50  D55  D60  D65  D70  KT LD               HS   VS TA  D50  D55  D60  D65  D70  KT LD
JAN   800 1345 32  561  716  871 1026 1181 .60 61     JAN   552  814 31  577  732  887 1042 1197 .44 62
FEB  1150 1611 37  363  501  641  781  921 .65 53     FEB   793  985 33  488  627  767  907 1047 .46 54
MAR  1649 1650 40  305  456  611  766  921 .69 42     MAR  1109 1019 41  297  447  602  756  911 .47 43
APR  2159 1415 47  131  253  397  546  696 .71 30     APR  1449  956 52   55  135  257  399  549 .48 31
MAY  2523 1196 55   24   85  186  328  478 .73 20     MAY  1687  894 62    2   13   54  143  260 .49 22
JUN  2701 1115 62    2   12   55  145  261 .74 16     JUN  1795  878 71    0    0    0    0   59 .49 17
JUL  2692 1167 69    0    1    5   17   90 .76 18     JUL  1760  889 76    0    0    0    0   15 .50 19
AUG  2406 1379 67    0    2   11   50  133 .76 26     AUG  1565  941 75    0    0    0    0   26 .49 27
SEP  1998 1785 60    4   19   73  168  298 .77 38     SEP  1273 1052 68    0    2    9   34  115 .50 39
OCT  1431 1893 50   74  169  305  456  611 .73 49     OCT   951 1102 58   10   46  124  243  389 .50 51
NOV   912 1500 40  301  448  597  747  897 .64 59     NOV   596  838 46  145  271  415  564  714 .44 60
DEC   705 1235 33  527  682  837  992 1147 .58 63     DEC   454  676 35  481  636  791  946 1101 .40 64
YR   1764 1439 49 2292 3345 4590 6022 7635 .71        YR   1167  920 54 2056 2908 3905 5034 6382 .48

TONOPAH, NEVADA                      ELEV 5423  LAT 38.1    ALBUQUERQUE, NEW MEXICO              ELEV 5312  LAT 35.0
      HS   VS TA  D50  D55  D60  D65  D70  KT LD               HS   VS TA  D50  D55  D60  D65  D70  KT LD
JAN   918 1552 30  614  769  924 1079 1234 .65 59     JAN  1016 1562 35  459  614  769  924 1079 .65 56
FEB  1274 1765 35  432  571  711  851  991 .69 52     FEB  1342 1664 40  281  420  560  700  840 .67 49
MAR  1777 1725 40  329  479  633  787  942 .72 41     MAR  1768 1515 46  145  287  440  595  750 .69 37
APR  2251 1404 48  118  224  361  512  657 .73 28     APR  2228 1244 56    7   49  140  282  426 .71 25
MAY  2577 1161 57   17   63  143  269  409 .74 19     MAY  2538 1052 65    0    1    9   58  158 .73 16
JUN  2788 1089 65    1    6   26   92  171 .77 15     JUN  2679  997 75    0    0    0    0   12 .74 12
JUL  2703 1121 73    0    0    0    0   48 .76 17     JUL  2489  995 79    0    0    0    0    2 .70 14
AUG  2438 1325 71    0    1    6   13   80 .76 25     AUG  2290 1132 77    0    0    0    0    5 .70 22
SEP  2043 1734 64    2   10   41  108  214 .77 36     SEP  1972 1468 70    0    0    1    7   59 .72 33
OCT  1520 1945 52   60  140  258  407  556 .75 48     OCT  1547 1745 58    3   22   93  218  366 .71 45
NOV  1031 1694 40  312  457  606  756  906 .69 57     NOV  1134 1692 45  173  316  465  615  765 .68 54
DEC   827 1473 32  562  716  871 1026 1181 .65 61     DEC   928 1508 36  428  583  738  893 1048 .64 58
YR   1848 1497 51 2448 3436 4580 5900 7389 .74        YR   1830 1379 57 1497 2292 3216 4292 5511 .70

WINNEMUCCA, NEVADA                   ELEV 4340  LAT 40.9    CLAYTON, NEW MEXICO                  ELEV 4970  LAT 36.4
      HS   VS TA  D50  D55  D60  D65  D70  KT LD               HS   VS TA  D50  D55  D60  D65  D70  KT LD
JAN   690 1148 28  676  831  986 1141 1296 .55 62     JAN   962 1536 33  524  679  834  989 1144 .64 58
FEB  1028 1449 34  446  585  725  865 1005 .60 55     FEB  1241 1576 36  390  529  669  809  949 .64 50
MAR  1472 1494 38  388  540  695  849 1004 .63 43     MAR  1652 1466 40  303  454  608  763  918 .65 39
APR  1967 1343 45  176  304  449  597  747 .65 31     APR  2039 1199 51   67  154  282  431  576 .66 27
MAY  2362 1188 54   40  110  213  359  503 .68 22     MAY  2222 1007 60    4   22   81  172  314 .64 17
JUN  2569 1127 62    4   18   64  149  260 .71 18     JUN  2418  984 69    0    1    6   38   91 .67 13
JUL  2678 1217 71    0    1    5    6   76 .76 20     JUL  2284  981 74    0    0    0    0   34 .64 15
AUG  2348 1418 68    1    3   13   42  127 .74 27     AUG  2097 1100 72    0    0    2    5   47 .65 23
SEP  1907 1778 59    9   35  101  199  329 .75 39     SEP  1802 1390 65    1    4   22   73  172 .67 35
OCT  1322 1789 48  119  226  367  518  673 .70 51     OCT  1433 1660 55   24   84  182  324  472 .68 46
NOV   809 1335 37  384  532  681  831  981 .60 60     NOV  1028 1556 42  240  383  531  681  831 .65 56
DEC   618 1083 30  608  763  918 1073 1228 .55 64     DEC   861 1439 35  463  617  772  927 1082 .63 60
YR   1651 1363 48 2851 3949 5216 6629 8230 .68        YR   1672 1323 53 2017 2928 3988 5212 6630 .65

CONCORD, NEW HAMPSHIRE               ELEV  344  LAT 43.2    FARMINGTON, NEW MEXICO               ELEV 5502  LAT 36.7
      HS   VS TA  D50  D55  D60  D65  D70  KT LD               HS   VS TA  D50  D55  D60  D65  D70  KT LD
JAN   459  696 21  911 1066 1221 1376 1531 .41 64     JAN   944 1513 29  663  818  973 1128 1283 .64 58
FEB   686  884 23  767  907 1047 1187 1327 .43 57     FEB  1281 1673 35  420  560  700  840  980 .67 50
MAR   974  933 32  549  704  859 1014 1169 .43 46     MAR  1693 1531 41  294  447  601  756  911 .67 39
APR  1317  920 44  185  326  474  624  774 .45 33     APR  2133 1265 50   73  172  311  465  609 .69 27
MAY  1582  889 55   14   68  167  315  462 .46 24     MAY  2452 1080 60    3   18   79  184  327 .70 18
JUN  1705  879 65    0    2   16   58  172 .47 20     JUN  2665 1032 68    0    1    5   36  102 .73 13
JUL  1675  894 70    0    0    3   16   75 .47 22     JUL  2478 1032 75    0    0    0    0   15 .70 15
AUG  1455  930 67    0    1    7   45  119 .47 30     AUG  2252 1179 73    0    0    1    6   35 .70 23
SEP  1140  992 60    3   17   75  182  317 .46 41     SEP  1934 1531 65    0    3   17   67  175 .72 35
OCT   817  979 49   81  188  333  487  642 .46 53     OCT  1479 1759 53   32  106  227  375  530 .71 47
NOV   463  648 38  361  510  660  810  960 .38 63     NOV  1047 1622 39  326  474  624  774  924 .66 56
DEC   362  546 25  781  936 1091 1246 1401 .36 67     DEC   837 1398 30  617  772  927 1082 1237 .62 60
YR   1055  849 46 3653 4726 5954 7360 8949 .45        YR   1768 1382 51 2428 3371 4465 5713 7130 .69
```

LOS ALAMOS, NEW MEXICO — ELEV 7380 LAT 35.9

	HS	VS	TA	D50	D55	D60	D65	D70	KT	LD
JAN	952	1476	29	651	806	961	1117	1271	.63	57
FEB	1279	1612	32	501	641	781	929	1061	.66	50
MAR	1568	1346	38	379	533	688	856	998	.62	38
APR	1929	1119	46	142	278	426	560	726	.62	26
MAY	2071	951	55	11	62	162	304	459	.59	17
JUN	2132	915	65	0	1	11	87	159	.59	13
JUL	1889	866	68	0	0	3	18	95	.53	15
AUG	1759	934	66	0	1	9	44	145	.54	22
SEP	1656	1235	60	1	10	57	147	290	.61	34
OCT	1267	1368	50	61	160	303	448	611	.60	46
NOV	1037	1540	38	355	504	654	796	954	.64	55
DEC	880	1452	30	608	763	918	1053	1228	.63	59
YR	1535	1232	48	2710	3760	4974	6359	7997	.60	

ROSWELL, NEW MEXICO — ELEV 3619 LAT 33.4

	HS	VS	TA	D50	D55	D60	D65	D70	KT	LD
JAN	1046	1516	38	371	524	679	834	989	.63	55
FEB	1373	1607	43	209	341	479	619	759	.66	47
MAR	1807	1464	49	92	195	335	487	642	.69	36
APR	2218	1175	60	5	24	83	185	313	.70	24
MAY	2459	997	69	0	1	7	20	106	.71	14
JUN	2610	962	77	0	0	0	0	11	.72	10
JUL	2441	957	79	0	0	0	0	6	.69	12
AUG	2242	1063	78	0	0	0	0	9	.69	20
SEP	1913	1335	70	0	1	4	17	73	.68	32
OCT	1527	1603	60	5	25	87	195	326	.68	43
NOV	1131	1567	47	131	251	395	543	693	.65	53
DEC	952	1453	39	335	487	642	797	952	.62	57
YR	1812	1306	59	1149	1849	2711	3697	4878	.68	

TRUTH OR CONSEQ., NEW MEXICO — ELEV 4859 LAT 33.2

	HS	VS	TA	D50	D55	D60	D65	D70	KT	LD
JAN	1118	1661	40	315	466	620	775	930	.67	54
FEB	1451	1724	45	163	287	424	563	703	.70	47
MAR	1886	1532	50	79	173	309	459	614	.71	36
APR	2338	1220	60	5	26	88	188	319	.74	23
MAY	2557	1010	68	6	27	91	19	330	.73	14
JUN	2650	963	77	0	0	0	0	13	.73	10
JUL	2365	942	79	0	0	0	0	6	.67	12
AUG	2216	1048	77	0	0	0	0	11	.68	20
SEP	1940	1346	72	0	1	3	5	57	.69	31
OCT	1579	1665	61	3	16	63	144	277	.70	43
NOV	1217	1733	49	100	204	342	489	639	.69	53
DEC	1003	1558	41	291	441	595	750	905	.65	57
YR	1861	1364	60	963	1641	2535	3392	4804	.70	

TUCUMCARI, NEW MEXICO — ELEV 4039 LAT 35.2

	HS	VS	TA	D50	D55	D60	D65	D70	KT	LD
JAN	1009	1560	37	405	558	713	868	1023	.65	56
FEB	1297	1597	41	256	390	529	669	809	.65	49
MAR	1712	1466	47	141	266	414	567	722	.67	38
APR	2098	1185	57	13	54	133	260	395	.67	25
MAY	2314	1005	66	1	4	20	57	165	.66	16
JUN	2484	974	75	0	0	1	7	23	.69	12
JUL	2349	972	78	0	0	0	0	8	.66	14
AUG	2164	1090	77	0	0	0	0	14	.67	22
SEP	1829	1354	70	0	1	6	20	87	.67	33
OCT	1443	1595	59	8	35	105	217	354	.67	45
NOV	1073	1571	46	147	271	415	564	714	.65	55
DEC	910	1478	39	357	509	664	818	973	.63	59
YR	1725	1319	58	1328	2090	3000	4047	5288	.66	

ZUNI, NEW MEXICO — ELEV 6447 LAT 35.1

	HS	VS	TA	D50	D55	D60	D65	D70	KT	LD
JAN	986	1500	30	611	766	921	1076	1231	.63	56
FEB	1297	1590	35	431	571	711	851	991	.65	49
MAR	1688	1434	40	324	478	632	787	942	.66	38
APR	2167	1217	48	96	213	358	507	657	.69	25
MAY	2473	1041	57	7	45	130	264	416	.71	16
JUN	2602	989	65	0	2	11	68	153	.72	12
JUL	2264	952	71	0	0	0	0	47	.64	14
AUG	2078	1052	69	0	0	2	13	77	.64	22
SEP	1895	1406	63	0	4	25	91	208	.69	33
OCT	1496	1671	53	33	111	238	388	543	.69	45
NOV	1088	1596	40	299	447	597	747	897	.66	54
DEC	893	1429	32	558	713	868	1023	1178	.62	58
YR	1745	1321	50	2361	3349	4493	5815	7341	.67	

ALBANY, NEW YORK — ELEV 292 LAT 42.7

	HS	VS	TA	D50	D55	D60	D65	D70	KT	LD
JAN	456	674	22	884	1039	1194	1349	1504	.39	64
FEB	688	871	24	742	882	1022	1162	1302	.43	58
MAR	986	933	33	515	670	825	980	1135	.43	45
APR	1335	922	47	136	253	395	543	693	.45	33
MAY	1570	873	58	12	49	125	253	384	.46	24
JUN	1730	882	68	1	3	12	39	125	.47	19
JUL	1725	908	72	0	1	3	9	58	.49	21
AUG	1499	947	70	0	1	7	22	93	.48	29
SEP	1170	1009	62	3	15	58	135	253	.47	41
OCT	817	961	51	66	150	276	422	577	.45	53
NOV	457	623	40	317	463	612	762	912	.36	62
DEC	356	521	26	747	902	1057	1212	1367	.34	66
YR	1068	843	48	3424	4428	5586	6888	8403	.45	

BINGHAMTON, NEW YORK — ELEV 1637 LAT 42.2

	HS	VS	TA	D50	D55	D60	D65	D70	KT	LD
JAN	386	520	22	868	1023	1178	1333	1488	.33	63
FEB	576	667	23	762	902	1042	1182	1322	.35	56
MAR	861	772	31	580	735	890	1045	1200	.37	45
APR	1242	840	45	176	312	459	609	759	.42	32
MAY	1496	828	55	18	75	172	320	462	.43	23
JUN	1681	855	65	1	4	20	75	173	.46	19
JUL	1659	871	69	0	1	5	21	91	.47	21
AUG	1425	888	67	0	1	9	40	123	.45	29
SEP	1131	952	60	3	17	71	172	298	.45	40
OCT	779	883	50	72	167	304	456	611	.42	52
NOV	414	531	38	356	504	654	804	954	.32	62
DEC	297	395	25	763	918	1073	1228	1383	.28	66
YR	998	750	46	3598	4657	5875	7285	8863	.42	

BUFFALO, NEW YORK — ELEV 705 LAT 42.9

	HS	VS	TA	D50	D55	D60	D65	D70	KT	LD
JAN	349	465	24	815	970	1125	1280	1435	.30	64
FEB	546	636	24	717	857	997	1137	1277	.34	57
MAR	888	820	32	555	710	865	1020	1175	.39	45
APR	1315	911	45	170	306	453	603	753	.44	33
MAY	1596	890	55	17	72	170	321	462	.46	24
JUN	1804	915	66	0	2	13	58	151	.49	20
JUL	1776	935	70	0	0	3	12	72	.50	22
AUG	1513	961	68	0	1	5	33	101	.48	29
SEP	1152	996	62	2	10	49	138	257	.46	41
OCT	784	914	52	52	137	269	419	574	.44	53
NOV	403	526	40	309	456	606	756	906	.32	62
DEC	283	380	28	685	840	995	1150	1305	.28	66
YR	1037	780	47	3322	4363	5551	6927	8468	.44	

```
MASSENA, NEW YORK              ELEV  207  LAT 44.9    ASHEVILLE, NORTH CAROLINA       ELEV 2169  LAT 35.4
      HS   VS TA  D50  D55  D60  D65  D70  KT LD            HS   VS TA  D50  D55  D60  D65  D70  KT LD
JAN  391  596 15 1101 1256 1411 1566 1721 .38 66     JAN   722  958 38  377  531  685  840  995 .47 57
FEB  620  819 17  932 1072 1212 1352 1492 .41 59     FEB   971 1069 39  300  437  577  717  857 .49 49
MAR  977  992 28  694  849 1004 1159 1314 .45 47     MAR  1306 1053 46  157  288  438  592  747 .51 38
APR 1343  986 42  242  385  534  684  834 .46 35     APR  1668  960 56   17   66  152  279  424 .53 26
MAY 1613  939 54   27   93  198  350  493 .47 26     MAY  1804  858 64    1    7   33  100  210 .52 16
JUN 1779  940 64    1    5   24   78  187 .49 22     JUN  1854  839 71    0    1    4   14   70 .51 12
JUL 1751  964 64    1    5   25   22  193 .50 24     JUL  1776  827 74    0    0    0    0   35 .50 14
AUG 1484  993 67    0    2   12   57  136 .48 31     AUG  1627  864 73    0    0    0    0   43 .50 22
SEP 1124 1029 59    5   25   88  192  327 .47 43     SEP  1361  969 67    0    3   13   50  135 .50 34
OCT  736  901 49  103  214  359  512  667 .43 55     OCT  1147 1171 57   13   56  138  269  411 .53 45
NOV  388  541 36  424  573  723  873 1023 .34 64     NOV   849 1120 46  143  268  412  561  711 .52 55
DEC  294  439 20  927 1082 1237 1392 1547 .32 68     DEC   658  918 39  353  506  660  815  970 .46 59
YR  1044  845 43 4456 5561 6827 8237 9934 .46        YR   1313  966 56 1362 2162 3112 4237 5609 .51

NEW YORK, NEW YORK             ELEV  187  LAT 40.8    CAPE HATTERAS, NORTH CAROLINA   ELEV    7  LAT 35.3
      HS   VS TA  D50  D55  D60  D65  D70  KT LD            HS   VS TA  D50  D55  D60  D65  D70  KT LD
JAN  500  708 32  552  707  862 1017 1172 .40 62     JAN   686  886 45  167  305  456  611  766 .44 57
FEB  721  865 33  465  605  745  885 1025 .42 54     FEB   952 1036 46  140  262  398  538  678 .48 49
MAR 1037  937 41  282  432  586  741  896 .44 43     MAR  1326 1070 51   68  160  296  458  602 .52 38
APR 1364  898 52   50  127  246  387  537 .45 31     APR  1774 1015 59    5   27   92  188  335 .57 26
MAY 1636  873 62    2   11   49  137  248 .47 22     MAY  1962  907 67    0    2   10   47  130 .56 16
JUN 1710  848 72    0    0    0    0   56 .47 17     JUN  2036  886 74    0    0    0    0   24 .56 12
JUL 1688  861 77    0    0    0    0   14 .48 19     JUL  1921  869 78    0    0    0    0    7 .54 14
AUG 1483  894 75    0    0    0    0   23 .47 27     AUG  1705  897 78    0    0    0    0    9 .53 22
SEP 1214  996 68    0    1    7   29  104 .47 39     SEP  1470 1055 74    0    0    0    0   29 .54 33
OCT  895 1017 59    7   33  103  209  353 .47 51     OCT  1137 1153 65    1    3   18   76  170 .53 45
NOV  533  716 47  122  238  380  528  678 .39 60     NOV   873 1162 56   14   60  146  277  421 .53 55
DEC  404  574 36  451  605  760  915 1070 .36 64     DEC   659  916 48  116  235  383  536  691 .46 59
YR  1101  849 55 1931 2759 3737 4848 6177 .45        YR   1377  987 62  512 1055 1800 2731 3860 .53

ROCHESTER, NEW YORK            ELEV  554  LAT 43.1    CHARLOTTE, NORTH CAROLINA       ELEV  768  LAT 35.2
      HS   VS TA  D50  D55  D60  D65  D70  KT LD            HS   VS TA  D50  D55  D60  D65  D70  KT LD
JAN  364  497 24  806  961 1116 1271 1426 .32 64     JAN   719  944 42  255  402  555  710  865 .46 56
FEB  559  661 25  706  846  986 1126 1266 .35 57     FEB   971 1061 44  184  311  449  588  728 .49 49
MAR  903  843 33  528  682  837  992 1147 .40 46     MAR  1317 1057 51   74  164  297  461  602 .51 38
APR 1339  935 46  152  275  419  567  717 .45 33     APR  1695  970 61    4   18   69  145  282 .54 25
MAY 1606  899 57   18   66  149  285  421 .47 24     MAY  1856  872 69    0    1    7   34  102 .53 16
JUN 1817  924 67    1    3   15   46  136 .50 20     JUN  1921  855 76    0    0    0    0   17 .53 12
JUL 1781  941 71    0    1    4    9   70 .50 24     JUL  1831  841 79    0    0    0    0    8 .52 14
AUG 1519  970 69    0    2    8   26   99 .49 30     AUG  1695  891 78    0    0    0    0   10 .52 22
SEP 1160 1011 62    3   13   53  126  243 .47 41     SEP  1416 1007 72    0    0    2   10   52 .52 33
OCT  782  917 52   55  133  251  398  549 .44 53     OCT  1173 1198 62    3   14   58  152  266 .54 45
NOV  404  532 41  292  436  585  735  885 .33 62     NOV   865 1141 51   66  150  277  420  570 .52 55
DEC  281  380 28  673  828  983 1138 1293 .28 66     DEC   672  938 43  243  390  543  698  853 .47 59
YR  1045  793 48 3233 4247 5405 6719 8252 .44        YR   1346  981 61  828 1451 2257 3218 4355 .52

SYRACUSE, NEW YORK             ELEV  407  LAT 43.1    GREENSBORO, NORTH CAROLINA      ELEV  886  LAT 36.1
      HS   VS TA  D50  D55  D60  D65  D70  KT LD            HS   VS TA  D50  D55  D60  D65  D70  KT LD
JAN  385  538 24  818  973 1128 1283 1438 .34 64     JAN   715  973 39  354  506  660  815  970 .47 57
FEB  571  681 25  711  851  991 1131 1271 .36 57     FEB   970 1096 41  270  405  543  683  823 .50 50
MAR  890  827 33  521  676  831  986 1141 .39 46     MAR  1313 1085 48  123  237  381  544  688 .52 39
APR 1324  923 47  141  263  407  555  705 .45 33     APR  1683  986 59    9   37  106  203  346 .54 26
MAY 1578  885 57   14   58  140  272  411 .46 24     MAY  1868  889 67    1    3   14   59  136 .54 17
JUN 1778  908 67    1    3   13   46  133 .49 20     JUN  1953  872 74    0    0    0    0   30 .54 13
JUL 1758  931 72    0    1    3   11   62 .50 22     JUL  1864  861 77    0    0    0    0   14 .53 15
AUG 1504  960 70    0    1    6   18   89 .48 30     AUG  1697  910 76    0    0    0    0   19 .52 23
SEP 1165 1017 63    2   10   44  120  228 .47 41     SEP  1418 1037 70    0    1    6   24   88 .52 34
OCT  777  909 53   49  126  244  392  543 .43 53     OCT  1141 1193 59    7   33   99  209  339 .54 46
NOV  399  523 41  277  421  570  720  870 .32 62     NOV   839 1134 48  111  216  354  501  651 .52 55
DEC  285  387 28  679  834  989 1144 1299 .28 66     DEC   659  949 40  328  478  633  787  942 .47 59
YR  1037  791 48 3215 4218 5366 6678 8192 .44        YR   1345  998 58 1202 1916 2797 3825 5047 .52
```

```
DAYTON, OHIO                      ELEV 1004  LAT 39.9     TULSA, OKLAHOMA                    ELEV  676  LAT 36.2
     HS   VS  TA  D50  D55  D60  D65   D70 KT LD              HS   VS  TA  D50  D55  D60  D65   D70 KT LD
JAN  489  660 28  679  834  989 1144 1299 .37 61        JAN  732 1011 37  419  571  726  880 1035 .49 57
FEB  725  843 30  549  689  829  969 1109 .41 54        FEB  978 1113 41  258  389  527  666  806 .51 50
MAR 1025  898 39  346  497  651  806  961 .43 42        MAR 1305 1080 48  126  230  369  528  673 .52 39
APR 1403  905 51   68  149  268  413  559 .46 30        APR 1603  943 61    7   28   83  176  287 .52 26
MAY 1699  886 62    4   19   67  166  272 .49 21        MAY 1822  875 69    1    3   13   28  117 .52 17
JUN 1874  895 71    0    1    5   13   69 .51 17        JUN 2021  891 77    0    0    0    0   19 .56 13
JUL 1810  896 75    0    0    0    0   32 .51 19        JUL 2030  911 82    0    0    0    0    5 .57 15
AUG 1645  968 73    0    1    3    7   49 .52 26        AUG 1865  990 81    0    0    0    0    6 .58 23
SEP 1318 1071 66    1    5   20   63  150 .51 38        SEP 1473 1087 73    0    1    4   10   50 .54 34
OCT  969 1098 56   25   83  173  307  451 .50 50        OCT 1164 1231 63    4   17   57  143  241 .55 46
NOV  564  749 42  257  399  547  696  846 .40 59        NOV  827 1115 49  104  197  326  468  618 .52 56
DEC  407  558 31  593  747  902 1057 1212 .34 63        DEC  659  953 40  325  473  627  781  936 .48 60
YR  1163  869 52 2523 3422 4453 5641 7009 .47            YR  1375 1016 60 1245 1910 2731 3680 4796 .54

TOLEDO, OHIO                      ELEV  692  LAT 41.6     ASTORIA, OREGON                    ELEV   23  LAT 46.1
     HS   VS  TA  D50  D55  D60  D65   D70 KT LD              HS   VS  TA  D50  D55  D60  D65   D70 KT LD
JAN  435  600 25  781  936 1091 1246 1401 .36 63        JAN  315  462 41  292  446  601  756  911 .32 67
FEB  680  821 27  641  781  921 1061 1201 .41 55        FEB  545  715 44  183  320  459  599  739 .38 60
MAR  997  914 36  442  596  750  905 1060 .43 44        MAR  866  879 44  180  329  484  639  794 .41 49
APR 1384  932 48  113  216  352  498  648 .46 32        APR 1253  940 48   92  219  366  516  666 .44 36
MAY 1717  925 59   10   40  110  229  352 .50 23        MAY 1608  963 52   26  106  241  394  549 .47 27
JUN 1878  922 69    0    2    9   32  105 .52 18        JUN 1626  890 57    4   34  120  255  405 .45 23
JUL 1849  941 72    0    1    4    5   57 .52 20        JUL 1746  988 60    1    8   57  163  311 .50 25
AUG 1616  994 71    0    1    6   18   79 .51 28        AUG 1499 1038 60    1    7   52  151  302 .49 33
SEP 1276 1087 64    2   10   39   99  206 .50 40        SEP 1183 1145 58    2   16   81  201  348 .50 44
OCT  911 1074 53   49  124  234  379  528 .49 52        OCT  713  902 53   21   95  226  378  533 .43 56
NOV  498  672 40  318  463  612  762  912 .38 61        NOV  387  568 47  120  256  405  555  705 .36 65
DEC  355  494 28  682  837  992 1147 1302 .32 65        DEC  261  392 43  226  378  533  688  843 .31 69
YR  1135  865 49 3040 4007 5120 6381 7852 .47            YR  1003  824 51 1147 2215 3626 5295 7106 .44

YOUNGSTOWN, OHIO                  ELEV 1184  LAT 41.3     BURNS, OREGON                      ELEV 4170  LAT 43.6
     HS   VS  TA  D50  D55  D60  D65   D70 KT LD              HS   VS  TA  D50  D55  D60  D65   D70 KT LD
JAN  385  499 26  753  908 1063 1218 1373 .31 63        JAN  490  782 25  769  924 1079 1234 1389 .44 65
FEB  586  662 27  652  792  932 1072 1212 .35 55        FEB  792 1107 31  532  672  812  952 1092 .51 57
MAR  890  784 35  457  611  766  921 1076 .38 44        MAR 1187 1232 36  433  586  741  896 1051 .53 46
APR 1278  848 48  120  231  372  519  669 .43 32        APR 1649 1203 44  195  329  475  624  774 .56 34
MAY 1586  857 58   12   49  126  258  387 .46 22        MAY 2052 1139 52   57  136  254  402  552 .60 25
JUN 1759  873 67    1    3   14   42  133 .48 18        JUN 2280 1116 59    8   35  101  205  335 .63 20
JUL 1734  888 71    0    1    4    9   75 .49 20        JUL 2460 1247 68    0    2   10   30  115 .70 22
AUG 1506  919 69    0    1    7   22   99 .48 28        AUG 2083 1379 66    1    5   20   68  158 .67 30
SEP 1194  991 63    2   11   46  118  231 .47 39        SEP 1620 1592 58   11   43  114  226  358 .66 42
OCT  851  966 53   49  126  242  384  540 .45 51        OCT 1043 1415 47  134  251  396  549  704 .59 54
NOV  456  586 40  297  442  591  741  891 .34 61        NOV  593  948 36  428  576  726  876 1026 .49 63
DEC  315  413 29  657  812  967 1122 1277 .28 65        DEC  430  718 28  685  840  995 1150 1305 .44 67
YR  1047  774 49 3001 3988 5130 6426 7963 .43            YR  1393 1156 46 3253 4399 5725 7212 8858 .60

OKLAHOMA CITY, OKLAHOMA           ELEV 1302  LAT 35.4     MEDFORD, OREGON                    ELEV 1299  LAT 42.4
     HS   VS  TA  D50  D55  D60  D65   D70 KT LD              HS   VS  TA  D50  D55  D60  D65   D70 KT LD
JAN  801 1114 37  413  565  719  874 1029 .52 57        JAN  407  565 37  417  571  725  880 1035 .35 64
FEB 1055 1200 41  255  386  524  664  804 .53 49        FEB  737  949 41  251  385  524  664  804 .45 56
MAR 1400 1147 48  126  232  371  532  676 .55 38        MAR 1133 1109 45  184  321  472  626  781 .50 45
APR 1725  991 60    8   29   87  180  298 .55 26        APR 1639 1150 50   78  169  299  444  594 .55 33
MAY 1918  895 68    1    4   14   36  124 .55 16        MAY 2034 1094 57   12   51  130  250  396 .59 23
JUN 2144  912 77    0    0    0    0   20 .59 12        JUN 2278 1082 64    1    6   29   94  190 .63 19
JUL 2128  925 82    0    0    0    0    6 .60 14        JUL 2475 1207 72    0    1    3   11   59 .70 21
AUG 1950 1007 81    0    0    0    0    7 .60 22        AUG 2121 1348 70    0    1    4   21   78 .68 29
SEP 1554 1128 73    0    1    4   12   52 .57 34        SEP 1589 1482 64    1    6   28   89  188 .63 41
OCT 1233 1292 62    5   18   62  148  253 .57 45        OCT  982 1233 53   38  110  219  360  515 .54 52
NOV  901 1222 49  106  201  331  474  624 .55 55        NOV  504  707 44  210  348  496  645  795 .39 62
DEC  725 1059 40  319  467  621  775  930 .51 59        DEC  337  475 38  384  537  691  846 1001 .32 66
YR  1463 1073 60 1232 1903 2734 3695 4823 .56            YR  1356 1033 53 1576 2505 3621 4930 6436 .57
```

NORTH BEND, OREGON — ELEV 16 LAT 43.4

	HS	VS	TA	D50	D55	D60	D65	D70	KT	LD
JAN	438	656	45	202	332	480	632	788	.39	65
FEB	704	925	47	133	244	377	515	655	.45	57
MAR	1058	1048	47	119	254	406	561	716	.47	46
APR	1510	1080	49	77	185	328	477	627	.51	34
MAY	1857	1035	53	26	98	220	369	524	.54	24
JUN	1994	1001	57	4	34	115	243	393	.55	20
JUL	2108	1093	59	2	16	79	188	342	.60	22
AUG	1786	1161	60	2	14	71	168	321	.57	30
SEP	1377	1273	58	3	22	88	201	349	.56	42
OCT	893	1120	55	13	67	170	313	468	.50	53
NOV	525	784	50	61	159	299	447	597	.43	63
DEC	381	594	47	142	270	420	574	729	.38	67
YR	1222	981	52	784	1696	3053	4688	6508	.52	

PENDLETON, OREGON — ELEV 1496 LAT 45.7

	HS	VS	TA	D50	D55	D60	D65	D70	KT	LD
JAN	348	523	32	558	713	868	1023	1178	.35	67
FEB	614	834	39	314	451	591	731	871	.42	59
MAR	1044	1115	44	209	351	503	657	812	.49	48
APR	1503	1152	51	68	153	279	423	573	.52	36
MAY	1925	1138	59	8	37	108	220	360	.56	27
JUN	2144	1122	66	1	4	19	70	160	.59	22
JUL	2396	1305	74	0	0	2	6	37	.68	24
AUG	1994	1413	72	0	1	3	13	62	.65	32
SEP	1502	1557	64	1	7	31	97	197	.63	44
OCT	908	1260	53	47	124	241	384	540	.55	56
NOV	438	668	41	266	410	558	708	858	.40	65
DEC	293	454	36	445	599	753	908	1063	.34	69
YR	1263	1046	52	1917	2848	3957	5240	6710	.56	

PORTLAND, OREGON — ELEV 39 LAT 45.6

	HS	VS	TA	D50	D55	D60	D65	D70	KT	LD
JAN	310	441	38	371	524	679	834	989	.31	67
FEB	554	718	43	210	343	482	622	762	.38	59
MAR	895	903	46	158	293	444	598	753	.42	48
APR	1308	975	51	67	156	286	432	582	.45	36
MAY	1663	984	57	12	53	136	264	413	.49	27
JUN	1772	951	62	2	10	48	128	247	.49	22
JUL	2037	1124	67	0	2	10	48	128	.58	24
AUG	1674	1157	67	0	2	12	56	139	.54	32
SEP	1217	1168	62	2	9	45	119	242	.51	44
OCT	724	905	54	29	98	206	347	503	.43	56
NOV	388	558	45	162	295	442	591	741	.35	65
DEC	260	381	41	293	444	598	753	908	.30	69
YR	1070	856	53	1307	2230	3389	4792	6407	.47	

REDMOND, OREGON — ELEV 3084 LAT 44.3

	HS	VS	TA	D50	D55	D60	D65	D70	KT	LD
JAN	491	811	30	614	769	924	1079	1234	.46	66
FEB	775	1105	36	399	538	678	818	958	.51	58
MAR	1190	1269	39	356	509	663	818	973	.54	47
APR	1683	1260	44	345	493	642	618	942	.58	35
MAY	2079	1177	51	62	148	277	425	580	.61	25
JUN	2287	1140	58	8	37	107	220	356	.63	21
JUL	2446	1270	66	1	4	18	55	161	.69	23
AUG	2069	1403	64	1	7	32	102	208	.67	31
SEP	1584	1586	58	9	42	116	233	371	.65	42
OCT	999	1367	48	107	218	362	515	670	.58	54
NOV	572	929	39	333	481	630	780	930	.49	64
DEC	424	729	33	515	670	825	980	1135	.45	68
YR	1387	1170	47	2750	3913	5274	6643	8517	.60	

SALEM, OREGON — ELEV 200 LAT 44.9

	HS	VS	TA	D50	D55	D60	D65	D70	KT	LD
JAN	332	471	39	348	502	657	812	967	.32	66
FEB	588	760	43	204	340	479	619	759	.39	59
MAR	947	951	45	163	306	459	614	769	.43	47
APR	1370	1009	50	68	168	308	456	606	.47	35
MAY	1738	1009	56	10	57	151	295	444	.51	26
JUN	1849	972	61	1	8	48	133	267	.51	22
JUL	2142	1154	67	0	1	7	43	130	.61	24
AUG	1775	1208	66	0	1	9	53	141	.58	31
SEP	1328	1280	62	1	6	39	120	247	.55	43
OCT	769	959	53	26	97	217	366	521	.45	55
NOV	410	585	45	158	296	444	594	744	.36	64
DEC	277	403	41	285	437	592	747	902	.30	68
YR	1130	897	52	1265	2220	3411	4852	6496	.49	

ALLENTOWN, PENNSYLVANIA — ELEV 384 LAT 40.6

	HS	VS	TA	D50	D55	D60	D65	D70	KT	LD
JAN	527	758	28	688	843	998	1153	1308	.41	62
FEB	763	929	29	577	717	857	997	1137	.44	54
MAR	1078	979	38	372	525	679	834	989	.46	43
APR	1410	926	50	85	178	308	453	603	.47	31
MAY	1637	870	60	6	25	85	190	313	.47	22
JUN	1777	870	70	0	1	7	21	91	.49	17
JUL	1765	890	74	0	0	0	0	33	.50	19
AUG	1546	927	72	0	1	3	6	62	.49	27
SEP	1238	1014	65	1	6	27	85	182	.48	39
OCT	926	1058	54	33	101	203	344	494	.48	51
NOV	568	778	42	242	383	531	681	831	.41	60
DEC	430	622	31	599	753	908	1063	1218	.38	64
YR	1141	885	51	2604	3534	4607	5827	7262	.47	

ERIE, PENNSYLVANIA — ELEV 738 LAT 42.1

	HS	VS	TA	D50	D55	D60	D65	D70	KT	LD
JAN	346	445	25	772	927	1082	1237	1392	.29	63
FEB	577	666	25	694	834	974	1114	1254	.35	56
MAR	920	837	33	530	685	840	995	1150	.40	45
APR	1359	925	45	173	309	457	606	756	.46	32
MAY	1646	900	55	22	83	184	336	478	.48	23
JUN	1847	919	65	1	4	21	80	178	.51	19
JUL	1833	944	69	0	1	5	24	98	.52	21
AUG	1455	905	68	0	1	8	43	119	.46	29
SEP	1201	1023	61	2	12	54	141	263	.48	40
OCT	827	956	52	53	137	266	415	571	.45	52
NOV	416	532	40	300	448	597	747	897	.32	61
DEC	278	359	29	648	803	958	1113	1268	.26	65
YR	1061	785	47	3197	4244	5448	6851	8424	.44	

HARRISBURG, PENNSYLVANIA — ELEV 348 LAT 40.2

	HS	VS	TA	D50	D55	D60	D65	D70	KT	LD
JAN	536	763	30	617	772	927	1082	1237	.41	61
FEB	771	929	32	496	636	776	916	1056	.44	54
MAR	1083	972	41	288	436	589	744	899	.46	43
APR	1410	916	53	48	122	231	370	517	.47	30
MAY	1652	870	63	3	12	46	128	230	.48	21
JUN	1805	875	72	0	0	0	0	58	.50	17
JUL	1764	883	76	0	0	0	0	21	.50	19
AUG	1550	921	74	0	0	0	0	38	.49	27
SEP	1267	1030	67	1	4	16	51	136	.49	38
OCT	934	1055	56	23	77	165	293	442	.48	50
NOV	579	787	44	205	340	487	636	786	.42	60
DEC	447	646	33	540	695	849	1004	1159	.38	64
YR	1152	887	53	2221	3093	4086	5224	6579	.47	

```
PHILADELPHIA, PENNSYLVANIA        ELEV   30  LAT 39.9    CHARLESTON, SOUTH CAROLINA        ELEV   39  LAT 32.9
      HS   VS  TA   D50   D55    D60   D65    D70 KT LD         HS   VS  TA   D50   D55    D60   D65    D70 KT LD
JAN  555  792  32   549   704    859  1014   1169 .42 61  JAN  744  904  49   120   222    360   521    664 .44 54
FEB  794  957  34   452   591    731   871   1011 .45 54  FEB  995 1009  51    81   161    275   419    547 .48 47
MAR 1108  991  42   262   408    562   716    871 .46 42  MAR 1339 1003  57    23    75    157   300    422 .51 35
APR 1434  926  53    46   119    227   367    514 .47 30  APR 1732  934  65     2    10     36    69    192 .55 23
MAY 1660  868  63     2    11     44   122    227 .48 21  MAY 1860  845  72     0     1      5     5     66 .53 14
JUN 1811  873  72     0     0      0     0     53 .50 17  JUN 1844  820  78     0     0      0     0     16 .51 10
JUL 1758  876  77     0     0      0     0     16 .50 19  JUL 1799  813  80     0     0      0     0      9 .51 12
AUG 1574  928  75     0     0      0     0     29 .50 26  AUG 1585  806  80     0     0      0     0     10 .48 19
SEP 1281 1034  68     0     2     11    38    115 .49 38  SEP 1394  926  75     0     0      0     0     31 .50 31
OCT  958 1081  57    14    53    132   249    393 .49 50  OCT 1193 1127  66     2     7     26    74    165 .53 43
NOV  619  856  46   150   273    416   564    714 .44 59  NOV  934 1159  56    24    75    156   271    414 .53 52
DEC  470  685  35   460   614    769   924   1079 .40 63  DEC  721  943  49   108   205    339   487    642 .46 56
YR  1170  905  55  1935  2775   3749  4865   6190 .48     YR  1346  940  65   360   756   1355  2146   3178 .51

    PITTSBURGH, PENNSYLVANIA      ELEV 1224  LAT 40.5    COLUMBIA, SOUTH CAROLINA         ELEV  226  LAT 33.9
      HS   VS  TA   D50   D55    D60   D65    D70 KT LD         HS   VS  TA   D50   D55    D60   D65    D70 KT LD
JAN  424  553  28   679   834    989  1144   1299 .33 62  JAN  762  973  45   185   310    456   608    763 .47 55
FEB  625  702  29   580   720    860  1000   1140 .36 54  FEB 1020 1082  48   129   227    353   493    628 .50 48
MAR  943  823  38   372   525    679   834    989 .40 43  MAR 1355 1050  54    47   116    212   360    492 .52 36
APR 1317  859  50    80   170    300   444    594 .44 31  APR 1747  965  64     4    14     46    83    207 .55 24
MAY 1602  852  60     6    27     89   208    321 .46 21  MAY 1895  867  72     0     2      7    12     72 .54 15
JUN 1762  863  69     0     2      8    26    105 .48 17  JUN 1947  852  79     0     0      0     0     15 .54 11
JUL 1689  857  72     0     1      3     7     58 .48 19  JUL 1842  832  81     0     0      0     0      8 .52 13
AUG 1510  904  70     0     1      5    16     83 .48 27  AUG 1703  871  80     0     0      0     0     11 .52 20
SEP 1209  981  64     2     8     35    98    203 .47 39  SEP 1439  987  75     0     0      0     0     42 .52 32
OCT  895 1005  53    42   115    226   372    521 .47 50  OCT 1211 1192  64     4    14     47   112    212 .55 44
NOV  505  656  41   269   413    561   711    861 .37 60  NOV  921 1184  54    49   118    214   341    488 .54 53
DEC  347  457  31   605   760    915  1070   1225 .30 64  DEC  722  985  46   173   294    438   589    744 .48 57
YR  1071  793  50  2635  3574   4669  5930   7399 .44     YR  1382  986  64   590  1094   1772  2598   3683 .52

     SCRANTON, PENNSYLVANIA       ELEV  948  LAT 41.3    GREENVILLE, SOUTH CAROLINA        ELEV  971  LAT 34.9
      HS   VS  TA   D50   D55    D60   D65    D70 KT LD         HS   VS  TA   D50   D55    D60   D65    D70 KT LD
JAN  455  632  26   744   899   1054  1209   1364 .37 63  JAN  730  953  42   248   395    549   704    859 .46 56
FEB  689  827  27   636   776    916  1056   1196 .41 55  FEB  982 1066  44   173   300    437   577    717 .49 49
MAR  991  898  36   435   589    744   899   1054 .42 44  MAR 1328 1058  51    68   156    288   450    592 .51 37
APR 1339  892  49   104   209    348   495    645 .45 32  APR 1697  964  61     3    16     64   144    276 .54 25
MAY 1591  860  59     7    32    100   219    347 .46 22  MAY 1839  863  69     0     1      6    29     96 .53 16
JUN 1760  873  68     0     2      9    28    114 .48 18  JUN 1918  852  76     0     0      0     0     16 .53 12
JUL 1746  893  72     0     0      2     7     51 .49 20  JUL 1830  838  78     0     0      0     0      8 .52 14
AUG 1513  923  70     0     1      5    18     82 .48 28  AUG 1699  887  78     0     0      0     0     10 .52 21
SEP 1199  996  63     2     9     41   116    225 .47 39  SEP 1406  990  72     0     0      2     9     55 .51 33
OCT  897 1039  53    46   123    241   391    540 .48 51  OCT 1180 1195  62     3    13     56   145    265 .54 45
NOV  490  649  41   282   427    576   726    876 .37 61  NOV  880 1156  51    64   149    276   420    570 .53 54
DEC  368  513  29   648   803    958  1113   1268 .33 65  DEC  670  922  43   232   377    531   685    840 .46 58
YR  1089  833  49  2904  3871   4994  6277   7762 .45     YR  1348  978  61   791  1409   2210  3163   4304 .52

    PROVIDENCE, RHODE ISLAND      ELEV   62  LAT 41.7    HURON, SOUTH DAKOTA              ELEV 1289  LAT 44.4
      HS   VS  TA   D50   D55    D60   D65    D70 KT LD         HS   VS  TA   D50   D55    D60   D65    D70 KT LD
JAN  506  750  28   670   825    980  1135   1290 .42 63  JAN  488  808  13  1163  1318   1473  1627   1783 .46 66
FEB  738  925  29   577   717    857   997   1137 .44 55  FEB  745 1047  18   899  1039   1179  1319   1459 .49 58
MAR 1032  959  37   406   561    716   871   1026 .45 44  MAR 1114 1163  29   652   806    961  1116   1271 .51 47
APR 1374  927  47   108   235    381   531    681 .46 32  APR 1530 1130  46   169   288    429   576    726 .52 35
MAY 1655  897  57     5    37    120   259    406 .48 23  MAY 1871 1069  57    22    71    151   273    408 .55 25
JUN 1775  885  66     0     1      6    36    127 .49 18  JUN 2101 1068  67     1     6     21    72    144 .58 21
JUL 1695  878  72     0     0      0     0     35 .48 20  JUL 2183 1156  74     0     1      4     9     50 .62 23
AUG 1499  924  70     0     0      1    10     58 .48 28  AUG 1892 1276  72     0     2      6    13     70 .61 31
SEP 1209 1019  63     0     3     21    93    204 .48 40  SEP 1418 1372  61     8    30     87   169    291 .58 43
OCT  907 1072  54    19    85    202   350    505 .49 52  OCT  988 1351  50   107   201    332   482    633 .57 54
NOV  538  754  43   207   352    501   651    801 .41 61  NOV  577  946  32   529   678    828   978   1128 .49 64
DEC  418  627  32   574   729    884  1039   1194 .38 65  DEC  405  684  19   955  1110   1265  1420   1575 .43 68
YR  1114  884  50  2566  3543   4669  5972   7464 .46     YR  1279 1089  45  4506  5549   6735  8054   9537 .55
```

PIERRE, SOUTH DAKOTA — ELEV 1726 LAT 44.4

	HS	VS	TA	D50	D55	D60	D65	D70	KT	LD
JAN	530	915	16	1068	1222	1377	1531	1686	.50	66
FEB	795	1152	20	832	970	1110	1249	1389	.52	58
MAR	1206	1297	30	641	789	940	1091	1247	.55	47
APR	1614	1203	46	239	332	447	561	720	.55	35
MAY	1966	1120	57	86	151	228	267	434	.57	25
JUN	2195	1106	67	22	44	83	74	221	.60	21
JUL	2278	1199	75	7	15	31	6	112	.65	23
AUG	1992	1350	74	9	18	37	10	129	.64	31
SEP	1496	1474	62	45	86	150	152	315	.62	43
OCT	1052	1483	51	47	89	155	451	326	.61	54
NOV	623	1063	34	512	649	792	936	1087	.53	64
DEC	442	780	22	888	1041	1195	1349	1504	.47	68
YR	1352	1178	46	4396	5405	6544	7677	9171	.59	

RAPID CITY, SOUTH DAKOTA — ELEV 3169 LAT 44.0

	HS	VS	TA	D50	D55	D60	D65	D70	KT	LD
JAN	542	928	22	871	1026	1181	1336	1491	.50	65
FEB	826	1198	26	678	818	958	1098	1238	.53	58
MAR	1229	1312	31	585	738	893	1048	1203	.55	46
APR	1589	1167	45	202	325	466	612	762	.54	34
MAY	1887	1066	55	43	107	196	319	463	.55	25
JUN	2131	1070	64	5	17	52	134	211	.58	21
JUL	2223	1161	73	1	2	8	13	73	.63	23
AUG	1963	1311	72	1	3	10	17	87	.63	31
SEP	1518	1481	61	13	39	100	191	301	.62	42
OCT	1064	1482	50	110	200	325	474	621	.61	54
NOV	647	1104	35	443	589	738	888	1038	.54	63
DEC	476	855	27	729	884	1039	1194	1349	.49	67
YR	1344	1177	47	3681	4749	5965	7324	8837	.58	

SIOUX FALLS, SOUTH DAKOTA — ELEV 1427 LAT 43.6

	HS	VS	TA	D50	D55	D60	D65	D70	KT	LD
JAN	533	887	14	1110	1265	1420	1575	1730	.48	65
FEB	802	1128	19	857	997	1137	1277	1417	.51	57
MAR	1152	1183	30	621	775	930	1085	1240	.51	46
APR	1543	1114	46	165	282	421	567	717	.52	34
MAY	1894	1059	58	20	65	141	259	388	.55	25
JUN	2100	1048	68	2	6	20	65	138	.58	20
JUL	2150	1117	73	0	1	5	10	57	.61	22
AUG	1844	1209	72	1	2	7	18	77	.59	30
SEP	1410	1323	61	9	31	87	165	287	.57	42
OCT	1005	1340	50	100	190	316	465	615	.57	54
NOV	607	982	33	509	657	807	957	1107	.50	63
DEC	441	745	20	930	1085	1240	1395	1550	.45	67
YR	1293	1094	45	4323	5355	6531	7838	9322	.55	

CHATTANOOGA, TENNESSEE — ELEV 689 LAT 35.0

	HS	VS	TA	D50	D55	D60	D65	D70	KT	LD
JAN	630	776	40	314	461	614	769	924	.40	56
FEB	859	891	43	218	344	480	625	759	.43	49
MAR	1176	916	50	102	195	325	483	627	.46	37
APR	1550	887	61	8	30	88	165	296	.49	25
MAY	1732	828	69	1	4	15	51	123	.50	16
JUN	1831	829	76	0	0	0	0	27	.51	12
JUL	1735	810	79	0	0	0	0	13	.49	14
AUG	1630	858	78	0	0	0	0	17	.50	22
SEP	1335	937	72	0	1	6	9	69	.49	33
OCT	1108	1102	61	8	29	87	182	297	.51	45
NOV	773	963	49	113	209	340	483	633	.46	54
DEC	580	753	41	286	431	584	738	893	.40	58
YR	1247	879	60	1050	1705	2539	3505	4677	.48	

KNOXVILLE, TENNESSEE — ELEV 981 LAT 35.8

	HS	VS	TA	D50	D55	D60	D65	D70	KT	LD
JAN	621	785	41	302	449	602	756	911	.41	57
FEB	863	923	43	219	346	483	630	762	.44	49
MAR	1191	953	50	98	191	322	484	624	.47	38
APR	1599	932	60	8	30	89	173	300	.51	26
MAY	1803	863	68	1	3	14	47	123	.52	17
JUN	1902	856	76	0	0	0	0	29	.52	12
JUL	1804	839	78	0	0	0	0	15	.51	14
AUG	1666	889	77	0	0	0	0	19	.51	22
SEP	1383	998	72	0	1	6	10	71	.51	34
OCT	1121	1152	61	7	27	83	175	294	.53	46
NOV	759	969	49	106	201	331	474	624	.47	55
DEC	569	757	42	277	422	574	729	884	.40	59
YR	1275	909	60	1018	1671	2504	3478	4654	.49	

MEMPHIS, TENNESSEE — ELEV 285 LAT 35.0

	HS	VS	TA	D50	D55	D60	D65	D70	KT	LD
JAN	683	870	41	312	455	606	760	915	.44	56
FEB	945	1015	44	207	324	457	594	734	.47	49
MAR	1278	1013	51	99	182	299	457	591	.50	37
APR	1639	935	63	9	27	75	131	252	.52	25
MAY	1885	879	71	1	4	13	22	100	.54	16
JUN	2045	885	79	0	0	0	0	21	.56	12
JUL	1972	879	82	0	0	0	0	11	.56	14
AUG	1824	943	80	0	0	0	0	14	.56	22
SEP	1471	1046	74	1	2	7	7	61	.54	33
OCT	1204	1232	63	8	25	70	142	248	.56	45
NOV	817	1042	51	97	178	292	423	575	.49	54
DEC	629	845	43	255	391	539	691	847	.43	58
YR	1368	965	62	988	1588	2357	3227	4368	.53	

NASHVILLE, TENNESSEE — ELEV 591 LAT 36.1

	HS	VS	TA	D50	D55	D60	D65	D70	KT	LD
JAN	580	721	38	369	519	673	828	983	.38	57
FEB	824	876	41	263	395	533	672	812	.42	50
MAR	1130	902	49	119	220	357	524	661	.45	39
APR	1544	907	60	9	33	93	176	306	.50	26
MAY	1825	874	69	1	4	14	45	122	.52	17
JUN	1963	875	77	0	0	0	0	22	.54	13
JUL	1891	869	80	0	0	0	0	10	.53	15
AUG	1737	928	79	0	0	0	0	14	.54	23
SEP	1398	1020	72	0	1	5	10	66	.52	34
OCT	1114	1155	61	7	27	84	180	294	.53	46
NOV	711	893	48	120	220	354	498	648	.44	55
DEC	521	676	40	308	455	608	763	918	.38	59
YR	1272	891	59	1195	1874	2720	3696	4856	.49	

ABILENE, TEXAS — ELEV 1752 LAT 32.4

	HS	VS	TA	D50	D55	D60	D65	D70	KT	LD
JAN	924	1208	44	236	364	510	660	816	.54	54
FEB	1183	1253	48	138	230	349	479	620	.56	46
MAR	1576	1197	55	58	127	218	354	486	.59	35
APR	1843	974	65	6	17	49	104	197	.58	23
MAY	2037	887	72	1	4	11	11	84	.58	13
JUN	2209	895	80	0	0	0	0	17	.61	9
JUL	2139	890	84	0	0	0	0	8	.60	11
AUG	1956	939	84	0	0	0	0	8	.60	19
SEP	1598	1058	76	0	0	0	0	41	.57	31
OCT	1315	1259	66	5	15	42	89	185	.58	42
NOV	1008	1267	54	60	129	219	336	482	.56	52
DEC	863	1199	46	180	292	430	577	732	.54	56
YR	1556	1084	65	683	1177	1828	2610	3675	.58	

AMARILLO, TEXAS — ELEV 3602 LAT 35.2

	HS	VS	TA	D50	D55	D60	D65	D70	KT	LD
JAN	960	1447	36	437	590	744	899	1054	.62	56
FEB	1243	1502	40	296	430	569	708	848	.63	49
MAR	1631	1379	46	174	302	449	601	757	.63	38
APR	2019	1143	57	22	71	151	275	408	.65	25
MAY	2212	978	66	2	7	28	81	174	.63	16
JUN	2393	959	75	0	1	2	10	35	.66	12
JUL	2280	958	79	0	0	0	0	12	.64	14
AUG	2103	1065	78	0	0	0	0	17	.65	22
SEP	1760	1294	70	1	2	9	20	95	.64	33
OCT	1403	1533	60	10	37	103	206	333	.65	45
NOV	1033	1483	46	155	273	414	561	711	.62	55
DEC	872	1384	39	362	513	667	822	977	.61	59
YR	1661	1259	57	1457	2225	3136	4183	5421	.64	

DALLAS, TEXAS — ELEV 489 LAT 32.8

	HS	VS	TA	D50	D55	D60	D65	D70	KT	LD
JAN	821	1035	45	189	312	457	608	763	.49	54
FEB	1071	1110	49	106	190	307	437	578	.51	46
MAR	1422	1073	56	35	95	181	314	445	.54	35
APR	1627	882	66	3	9	30	71	163	.51	23
MAY	1888	852	74	0	0	0	0	55	.54	14
JUN	2135	885	82	0	0	0	0	9	.59	9
JUL	2122	890	86	0	0	0	0	3	.60	11
AUG	1950	945	86	0	0	0	0	3	.60	19
SEP	1587	1063	78	0	0	0	0	20	.56	31
OCT	1276	1228	68	2	6	21	55	140	.56	43
NOV	936	1158	56	34	91	173	284	428	.53	52
DEC	780	1052	48	137	239	374	521	676	.50	56
YR	1470	1014	66	505	943	1543	2290	3282	.55	

AUSTIN, TEXAS — ELEV 620 LAT 30.3

	HS	VS	TA	D50	D55	D60	D65	D70	KT	LD
JAN	864	1008	50	116	207	333	483	631	.48	52
FEB	1125	1083	53	58	125	216	344	470	.51	44
MAR	1429	999	60	17	51	119	223	339	.52	33
APR	1605	828	69	2	6	20	44	130	.50	21
MAY	1834	815	75	0	0	0	0	44	.53	11
JUN	2072	860	82	0	0	0	0	10	.58	7
JUL	2105	865	85	0	0	0	0	5	.60	9
AUG	1931	891	85	0	0	0	0	5	.59	17
SEP	1606	999	79	0	0	0	0	18	.56	28
OCT	1333	1187	70	1	4	14	39	109	.56	40
NOV	987	1130	59	18	53	122	205	339	.52	50
DEC	825	1028	52	78	156	264	399	551	.49	54
YR	1478	974	68	289	602	1088	1737	2649	.54	

DEL RIO, TEXAS — ELEV 1027 LAT 29.4

	HS	VS	TA	D50	D55	D60	D65	D70	KT	LD
JAN	958	1125	51	95	180	301	449	596	.51	51
FEB	1206	1149	56	33	87	165	283	404	.54	43
MAR	1580	1087	63	7	23	69	163	254	.57	32
APR	1699	852	72	1	2	8	16	75	.53	20
MAY	1827	808	78	0	0	0	0	20	.52	10
JUN	2024	849	84	0	0	0	0	4	.57	6
JUL	2054	851	87	0	0	0	0	2	.58	8
AUG	1936	878	86	0	0	0	0	3	.59	16
SEP	1584	961	80	0	0	0	0	12	.55	28
OCT	1360	1179	71	1	3	10	34	89	.57	39
NOV	1059	1204	60	14	45	110	184	323	.54	49
DEC	903	1127	52	74	152	261	394	551	.52	53
YR	1518	1005	70	223	492	924	1523	2335	.55	

BROWNSVILLE, TEXAS — ELEV 20 LAT 25.9

	HS	VS	TA	D50	D55	D60	D65	D70	KT	LD
JAN	913	923	60	18	51	116	225	321	.45	47
FEB	1135	944	63	8	24	65	151	220	.47	40
MAR	1458	901	68	3	11	31	89	158	.51	28
APR	1737	816	75	0	0	0	0	53	.53	16
MAY	1927	817	79	0	0	0	0	22	.56	7
JUN	2115	868	83	0	0	0	0	10	.60	3
JUL	2212	867	84	0	0	0	0	7	.63	5
AUG	2027	854	84	0	0	0	0	7	.61	12
SEP	1694	933	82	0	0	0	0	13	.57	24
OCT	1439	1120	76	1	2	5	5	47	.57	36
NOV	1054	1047	68	3	9	28	35	146	.50	45
DEC	862	922	63	10	31	80	145	258	.45	49
YR	1550	917	74	44	127	325	650	1263	.55	

EL PASO, TEXAS — ELEV 3917 LAT 31.8

	HS	VS	TA	D50	D55	D60	D65	D70	KT	LD
JAN	1125	1572	44	210	355	509	663	818	.65	53
FEB	1480	1671	48	91	194	326	465	605	.69	45
MAR	1909	1472	55	21	81	183	328	478	.71	34
APR	2363	1174	64	1	4	25	89	195	.74	22
MAY	2601	985	72	0	0	0	0	44	.75	13
JUN	2682	946	80	0	0	0	0	3	.75	8
JUL	2450	934	82	0	0	0	0	1	.69	10
AUG	2284	1030	81	0	0	0	0	3	.70	18
SEP	1987	1313	74	0	0	0	0	23	.70	30
OCT	1639	1654	64	1	4	25	92	199	.71	42
NOV	1244	1681	52	50	131	257	402	552	.68	51
DEC	1031	1521	44	188	331	484	639	794	.64	55
YR	1901	1327	63	561	1102	1810	2678	3714	.71	

CORPUS CHRISTI, TEXAS — ELEV 43 LAT 27.8

	HS	VS	TA	D50	D55	D60	D65	D70	KT	LD
JAN	898	967	56	36	94	176	304	431	.46	49
FEB	1147	1018	60	16	47	108	199	304	.49	41
MAR	1430	930	65	5	16	49	120	205	.51	30
APR	1642	807	73	0	0	0	0	71	.51	18
MAY	1866	810	78	0	0	0	0	26	.54	9
JUN	2094	861	82	0	0	0	0	9	.59	4
JUL	2186	866	85	0	0	0	0	5	.62	6
AUG	1991	870	85	0	0	0	0	5	.60	14
SEP	1687	978	81	0	0	0	0	12	.57	26
OCT	1416	1172	74	1	2	6	7	59	.57	38
NOV	1043	1108	65	5	16	47	81	198	.51	47
DEC	845	963	59	19	57	128	219	351	.46	51
YR	1522	945	72	82	232	515	930	1676	.55	

FORT WORTH, TEXAS — ELEV 538 LAT 32.8

	HS	VS	TA	D50	D55	D60	D65	D70	KT	LD
JAN	805	1007	45	202	329	475	626	781	.48	54
FEB	1069	1107	49	116	205	325	456	597	.51	46
MAR	1409	1062	55	43	107	198	335	469	.53	35
APR	1616	877	65	3	12	39	88	187	.51	23
MAY	1890	853	73	0	0	0	0	71	.54	14
JUN	2153	888	81	0	0	0	0	11	.60	9
JUL	2155	897	85	0	0	0	0	4	.61	11
AUG	1983	957	85	0	0	0	0	4	.61	19
SEP	1621	1088	78	0	0	0	0	22	.58	31
OCT	1293	1249	68	2	7	23	60	147	.57	43
NOV	938	1162	56	34	93	175	287	430	.53	52
DEC	766	1024	48	142	246	383	530	686	.49	56
YR	1477	1014	66	542	999	1618	2382	3410	.55	

HOUSTON, TEXAS										
	HS	VS	TA	D50	D55	D60	D65	D70	KT	LD
						ELEV 108	LAT 30.0			
JAN	772	852	52	71	150	263	416	556	.42	51
FEB	1034	960	55	31	87	168	294	414	.46	44
MAR	1297	889	61	9	31	89	189	298	.47	32
APR	1522	788	69	1	3	12	23	107	.47	20
MAY	1775	798	76	0	0	0	0	31	.51	11
JUN	1898	824	81	0	0	0	0	8	.53	7
JUL	1828	805	83	0	0	0	0	4	.52	9
AUG	1686	807	83	0	0	0	0	4	.51	17
SEP	1471	906	79	0	0	0	0	13	.51	28
OCT	1276	1110	71	1	2	8	24	88	.54	40
NOV	924	1019	61	8	28	82	155	281	.48	49
DEC	730	855	55	41	108	201	333	480	.43	53
YR	1353	884	69	161	409	825	1434	2285	.49	

MIDLAND-ODESSA, TEXAS										
	HS	VS	TA	D50	D55	D60	D65	D70	KT	LD
						ELEV 2858	LAT 31.9			
JAN	1081	1488	44	225	361	510	663	819	.62	53
FEB	1383	1525	48	124	222	347	482	622	.65	46
MAR	1839	1413	54	44	113	209	349	489	.69	34
APR	2192	1109	64	3	13	43	98	202	.69	22
MAY	2430	961	72	0	0	0	0	68	.70	13
JUN	2562	938	80	0	0	0	0	11	.71	9
JUL	2389	927	82	0	0	0	0	6	.68	11
AUG	2210	1011	82	0	0	0	0	7	.67	18
SEP	1844	1216	75	0	0	0	0	33	.65	30
OCT	1522	1502	66	2	9	31	81	176	.66	42
NOV	1176	1553	53	54	125	225	356	503	.65	51
DEC	1000	1460	46	173	296	441	592	747	.62	55
YR	1804	1257	64	627	1137	1806	2621	3682	.67	

LAREDO, TEXAS										
	HS	VS	TA	D50	D55	D60	D65	D70	KT	LD
						ELEV 518	LAT 27.5			
JAN	959	1046	57	36	93	174	299	426	.49	49
FEB	1195	1062	61	12	37	92	177	273	.51	41
MAR	1516	981	68	3	9	28	87	154	.54	30
APR	1727	833	76	0	0	0	0	36	.53	18
MAY	1952	827	81	0	0	0	0	12	.56	8
JUN	2073	857	86	0	0	0	0	4	.58	4
JUL	2131	857	88	0	0	0	0	3	.61	6
AUG	2009	870	88	0	0	0	0	3	.61	14
SEP	1705	979	83	0	0	0	0	8	.58	26
OCT	1408	1152	76	0	1	5	8	44	.57	37
NOV	1041	1093	65	5	15	45	74	193	.51	47
DEC	889	1022	59	23	64	138	231	365	.48	51
YR	1552	965	74	79	219	482	876	1522	.56	

PORT ARTHUR, TEXAS										
	HS	VS	TA	D50	D55	D60	D65	D70	KT	LD
						ELEV 23	LAT 29.9			
JAN	800	892	52	78	157	269	420	560	.44	51
FEB	1071	1001	55	37	95	176	302	421	.48	44
MAR	1353	929	60	13	41	105	202	320	.49	32
APR	1609	824	69	1	5	17	33	121	.50	20
MAY	1871	822	75	0	0	0	0	42	.54	11
JUN	2011	847	81	0	0	0	0	10	.56	7
JUL	1846	809	83	0	0	0	0	6	.52	9
AUG	1736	822	83	0	0	0	0	6	.53	16
SEP	1527	939	79	0	0	0	0	17	.53	28
OCT	1321	1157	70	1	4	13	35	108	.55	40
NOV	953	1060	60	12	39	101	184	307	.50	49
DEC	754	891	54	50	120	215	342	493	.44	53
YR	1406	915	69	193	462	897	1518	2412	.51	

LUBBOCK, TEXAS										
	HS	VS	TA	D50	D55	D60	D65	D70	KT	LD
						ELEV 3241	LAT 33.6			
JAN	1031	1497	39	343	494	648	803	958	.63	55
FEB	1332	1551	43	218	348	485	624	764	.65	47
MAR	1762	1430	49	108	211	349	508	654	.67	36
APR	2168	1159	60	7	28	87	190	307	.69	24
MAY	2396	989	69	1	2	11	29	115	.69	15
JUN	2544	957	77	0	0	0	0	16	.70	10
JUL	2412	956	80	0	0	0	0	8	.68	12
AUG	2208	1057	78	0	0	0	0	11	.68	20
SEP	1820	1271	71	0	1	5	8	73	.65	32
OCT	1468	1530	61	5	22	76	162	288	.66	44
NOV	1116	1550	49	106	206	341	486	636	.64	53
DEC	934	1425	41	280	427	580	735	890	.61	57
YR	1768	1279	60	1069	1739	2582	3545	4720	.67	

SAN ANGELO, TEXAS										
	HS	VS	TA	D50	D55	D60	D65	D70	KT	LD
						ELEV 1909	LAT 31.4			
JAN	962	1228	46	178	291	429	577	732	.55	53
FEB	1208	1241	50	100	178	286	413	551	.56	45
MAR	1606	1183	57	33	86	165	287	409	.59	34
APR	1851	954	67	3	10	31	74	158	.58	22
MAY	2031	874	75	0	0	0	0	56	.58	12
JUN	2186	885	82	0	0	0	0	12	.61	8
JUL	2123	877	85	0	0	0	0	6	.60	10
AUG	1966	922	85	0	0	0	0	6	.60	18
SEP	1607	1033	77	0	0	0	0	34	.56	30
OCT	1337	1240	67	3	11	32	73	164	.58	41
NOV	1044	1279	56	45	106	189	298	441	.56	51
DEC	895	1211	48	145	244	375	518	674	.55	55
YR	1570	1076	66	508	927	1509	2240	3244	.58	

LUFKIN, TEXAS										
	HS	VS	TA	D50	D55	D60	D65	D70	KT	LD
						ELEV 315	LAT 31.2			
JAN	794	927	49	125	223	356	509	658	.45	52
FEB	1069	1045	52	66	137	237	371	500	.49	45
MAR	1376	983	58	20	62	138	256	379	.51	34
APR	1624	851	67	2	7	23	56	145	.51	21
MAY	1867	831	74	0	0	0	0	50	.54	12
JUN	2055	860	80	0	0	0	0	11	.57	8
JUL	2006	852	83	0	0	0	0	6	.57	10
AUG	1864	885	83	0	0	0	0	6	.57	18
SEP	1531	976	78	0	0	0	0	22	.54	29
OCT	1349	1245	68	1	5	19	52	134	.58	41
NOV	963	1131	57	24	71	147	256	390	.52	51
DEC	768	962	51	93	179	300	440	596	.47	55
YR	1441	962	67	331	684	1220	1940	2897	.53	

SAN ANTONIO, TEXAS										
	HS	VS	TA	D50	D55	D60	D65	D70	KT	LD
						ELEV 794	LAT 29.5			
JAN	895	1026	51	93	179	302	451	599	.48	51
FEB	1154	1089	55	39	100	185	310	436	.51	43
MAR	1450	991	61	9	32	91	194	299	.52	32
APR	1612	819	70	1	3	12	31	106	.50	20
MAY	1894	825	76	0	0	0	0	31	.54	10
JUN	2069	857	82	0	0	0	0	6	.58	6
JUL	2121	863	85	0	0	0	0	3	.60	8
AUG	1947	883	85	0	0	0	0	3	.59	16
SEP	1638	996	79	0	0	0	0	13	.56	28
OCT	1350	1172	71	1	3	10	32	96	.56	39
NOV	1009	1129	60	12	40	104	179	319	.52	49
DEC	847	1033	53	58	132	236	373	523	.49	53
YR	1501	973	69	213	490	941	1570	2435	.55	

```
SHERMAN, TEXAS                   ELEV  764  LAT 33.7    CEDAR CITY, UTAH                 ELEV 5617  LAT 37.7
       HS   VS TA  D50  D55   D60   D65   D70  KT LD           HS   VS TA  D50  D55   D60   D65   D70  KT LD
JAN   794 1024 42  269  415   568   722   877 .48 55    JAN   882 1431 29  660  815   970  1125  1280 .62 59
FEB  1037 1098 46  149  263   397   535   675 .50 47    FEB  1180 1548 33  474  613   753   893  1033 .63 51
MAR  1366 1053 52   59  137   253   411   549 .52 36    MAR  1636 1521 38  364  515   670   825   980 .66 40
APR  1610  892 64    2   10    41   114   209 .51 24    APR  2092 1284 47  134  249   390   537   687 .68 28
MAY  1852  852 71    0    1     5    13    74 .53 15    MAY  2467 1115 56   20   71   156   281   430 .71 19
JUN  2114  888 79    0    0     0     0     9 .59 10    JUN  2706 1064 65    1    6    27    86   177 .74 14
JUL  2077  890 84    0    0     0     0     3 .59 12    JUL  2503 1064 73    0    0     0     0    45 .71 16
AUG  1932  958 84    0    0     0     0     2 .59 20    AUG  2241 1213 71    0    1     4     6    70 .70 24
SEP  1580 1088 76    0    0     0     0    22 .57 32    SEP  1968 1627 63    2   11    43   114   220 .74 36
OCT  1268 1259 66    1    6    24    90   167 .57 44    OCT  1460 1802 52   67  150   274   424   574 .71 48
NOV   919 1171 53   44  114   217   353   499 .53 53    NOV   992 1566 39  340  487   636   786   936 .65 57
DEC   744 1020 45  190  323   473   626   781 .49 57    DEC   785 1331 31  596  750   905  1060  1215 .60 61
YR   1443 1015 64  715 1271  1977  2864  3869 .55       YR   1745 1379 50 2658 3669  4829  6137  7647 .69

WACO, TEXAS                      ELEV  509  LAT 31.6    SALT LAKE CITY, UTAH             ELEV 4226  LAT 40.8
       HS   VS TA  D50  D55   D60   D65   D70  KT LD           HS   VS TA  D50  D55   D60   D65   D70  KT LD
JAN   833 1007 47  156  269   409   558   713 .48 53    JAN   639 1017 28  683  837   992  1147  1302 .51 62
FEB  1096 1096 51   84  160   269   401   536 .51 45    FEB   989 1363 33  467  605   745   885  1025 .58 54
MAR  1427 1038 57   25   74   153   280   403 .53 34    MAR  1454 1463 40  334  481   633   787   942 .62 43
APR  1612  853 67    2    7    23    56   146 .51 22    APR  1894 1284 49  116  208   334   474   625 .63 31
MAY  1774  808 75    0    0     0     0    46 .51 13    MAY  2362 1184 58   20   61   135   237   371 .68 22
JUN  2112  873 82    0    0     0     0     8 .59  8    JUN  2561 1122 66    3    9    31    88   167 .70 17
JUL  2130  880 86    0    0     0     0     3 .60 10    JUL  2590 1186 77    0    0     0     0    29 .73 19
AUG  1958  923 86    0    0     0     0     3 .60 18    AUG  2254 1356 75    0    1     4     5    47 .71 27
SEP  1601 1035 79    0    0     0     0    16 .56 30    SEP  1843 1693 65    4   13    43   105   195 .72 39
OCT  1301 1205 69    1    4    16    51   121 .56 42    OCT  1293 1724 52   72  150   259   402   548 .68 51
NOV   957 1139 58   22   68   143   241   381 .52 51    NOV   788 1276 39  337  480   628   777   927 .58 60
DEC   803 1043 50  109  201   328   471   627 .49 55    DEC   570  953 30  612  766   921  1076  1231 .50 64
YR   1469  991 67  399  782  1340  2058  3003 .54       YR   1606 1301 51 2648 3612  4725  5983  7409 .66

WICHITA FALLS, TEXAS             ELEV 1030  LAT 34.0    BURLINGTON, VERMONT              ELEV  341  LAT 44.5
       HS   VS TA  D50  D55   D60   D65   D70  KT LD           HS   VS TA  D50  D55   D60   D65   D70  KT LD
JAN   862 1168 42  284  425   575   729   884 .53 55    JAN   385  572 17 1029 1184  1339  1494  1649 .36 66
FEB  1123 1240 46  164  271   400   535   675 .55 48    FEB   607  782 19  879 1019  1159  1299  1439 .40 58
MAR  1472 1164 53   74  152   259   409   545 .56 36    MAR   940  929 29  648  803   958  1113  1268 .43 47
APR  1763  975 64    5   17    51   112   209 .56 24    APR  1296  936 43  220  362   510   660   810 .44 35
MAY  2017  904 72    1    2     8    13    77 .58 15    MAY  1574  909 55   22   81   180   331   472 .46 25
JUN  2221  912 81    0    0     0     0    10 .61 11    JUN  1729  911 65    1    3    18    63   165 .47 21
JUL  2166  914 86    0    0     0     0     4 .61 13    JUL  1721  941 70    0    1     4    20    81 .49 23
AUG  1969  979 86    0    0     0     0     4 .60 21    AUG  1475  976 67    0    2     9    49   122 .48 31
SEP  1602 1116 77    0    0     0     0    28 .58 32    SEP  1122 1013 59    5   24    87   191   324 .46 43
OCT  1291 1304 66    3   11    36    92   180 .58 44    OCT   740  895 49   98  206   350   502   657 .43 54
NOV   957 1257 53   67  140   241   369   516 .56 53    NOV   375  507 37  391  540   690   840   990 .32 64
DEC   799 1148 44  218  347   493   645   800 .53 57    DEC   283  408 23  849 1004  1159  1314  1469 .30 68
YR   1522 1089 64  816 1365  2063  2904  3931 .58       YR   1023  815 44 4142 5230  6464  7876  9447 .44

BRYCE CANYON, UTAH               ELEV 7588  LAT 37.7    NORFOLK, VIRGINIA                ELEV   30  LAT 36.9
       HS   VS TA  D50  D55   D60   D65   D70  KT LD           HS   VS TA  D50  D55   D60   D65   D70  KT LD
JAN   914 1511 20  936 1091  1246  1401  1556 .64 59    JAN   678  932 41  300  450   605   760   915 .46 58
FEB  1236 1657 23  750  890  1030  1170  1310 .66 51    FEB   932 1068 41  247  382   521   661   801 .49 51
MAR  1685 1582 29  660  815   970  1125  1280 .68 40    MAR  1281 1080 48  112  226   371   532   679 .51 39
APR  2133 1310 38  370  519   669   819   969 .69 28    APR  1677 1004 58    9   40   114   226   368 .54 27
MAY  2454 1111 46  141  276   428   583   738 .71 19    MAY  1887  909 67    0    2    13    53   139 .54 18
JUN  2655 1055 54  137  268   415   330   714 .73 14    JUN  2000  894 75    0    0     0     0    25 .55 14
JUL  2424 1044 62    1    8    47   128   265 .68 16    JUL  1853  868 78    0    0     0     0     8 .52 16
AUG  2157 1172 60    2   16    72   176   315 .67 24    AUG  1680  918 77    0    0     0     0    12 .52 23
SEP  1920 1578 53   31  102   220   363   513 .72 36    SEP  1396 1044 72    0    0     2     9    53 .52 35
OCT  1465 1811 43  230  379   533   688   843 .72 48    OCT  1083 1145 62    2   13    56   141   265 .52 47
NOV  1015 1622 31  579  729   879  1029  1179 .67 57    NOV   811 1117 52   55  136   259   402   552 .52 56
DEC   818 1419 22  856 1011  1166  1321  1476 .63 61    DEC   624  907 42  248  395   549   704   859 .46 60
YR   1742 1404 40 4693 6106  7675  9133 11159 .69       YR   1327  990 59  974 1646  2489  3488  4674 .52
```

426 / APPENDIX 19

RICHMOND, VIRGINIA — ELEV 164 LAT 37.5

	HS	VS	TA	D50	D55	D60	D65	D70	KT	LD
JAN	632	863	38	390	543	698	853	1008	.44	59
FEB	877	1004	39	301	438	577	717	857	.47	51
MAR	1210	1025	47	140	262	408	569	716	.49	40
APR	1566	953	58	11	46	119	226	369	.51	28
MAY	1762	872	67	1	4	17	64	148	.51	18
JUN	1872	864	74	0	0	0	0	32	.52	14
JUL	1774	849	78	0	0	0	0	11	.50	16
AUG	1601	891	76	0	0	0	0	18	.50	24
SEP	1348	1020	70	0	1	6	21	84	.50	36
OCT	1033	1097	59	7	32	98	203	336	.50	47
NOV	733	988	49	100	199	334	480	630	.48	57
DEC	567	810	39	345	497	651	806	961	.43	61
YR	1250	936	58	1296	2021	2909	3939	5170	.49	

ROANOKE, VIRGINIA — ELEV 1175 LAT 37.3

	HS	VS	TA	D50	D55	D60	D65	D70	KT	LD
JAN	660	911	36	423	577	732	887	1042	.46	59
FEB	899	1032	38	336	474	613	753	893	.48	51
MAR	1236	1046	45	171	306	457	611	766	.50	40
APR	1581	957	56	17	67	152	283	424	.51	28
MAY	1764	870	64	1	6	28	101	193	.51	18
JUN	1882	865	72	0	0	0	0	56	.52	14
JUL	1796	854	75	0	0	0	0	22	.51	16
AUG	1620	897	74	0	0	0	0	30	.50	24
SEP	1358	1023	68	0	2	9	32	112	.51	35
OCT	1080	1157	58	10	43	119	235	381	.52	47
NOV	765	1043	47	136	257	401	549	699	.50	57
DEC	591	853	37	393	546	701	856	1011	.45	61
YR	1271	958	56	1486	2277	3211	4307	5627	.50	

OLYMPIA, WASHINGTON — ELEV 200 LAT 47.0

	HS	VS	TA	D50	D55	D60	D65	D70	KT	LD
JAN	269	385	37	398	552	707	862	1017	.29	68
FEB	503	662	41	255	392	532	672	812	.36	61
MAR	845	876	43	219	367	521	676	831	.40	49
APR	1255	966	48	97	211	355	504	654	.44	37
MAY	1632	999	54	21	86	196	341	496	.48	28
JUN	1693	940	59	3	22	85	197	334	.47	24
JUL	1913	1101	64	1	4	25	89	207	.54	26
AUG	1549	1105	63	1	5	32	103	230	.51	34
SEP	1157	1146	59	4	24	90	198	343	.50	45
OCT	636	795	51	61	155	294	446	601	.40	57
NOV	339	489	43	209	352	501	651	801	.33	66
DEC	221	324	40	327	481	636	791	946	.27	70
YR	1004	816	50	1595	2652	3974	5530	7273	.45	

SEATTLE-TACOMA, WASHINGTON — ELEV 400 LAT 47.4

	HS	VS	TA	D50	D55	D60	D65	D70	KT	LD
JAN	262	378	38	367	521	676	831	986	.29	69
FEB	495	657	42	220	356	496	636	776	.36	61
MAR	849	893	44	192	339	493	648	803	.41	50
APR	1293	1013	49	85	197	340	489	639	.46	38
MAY	1714	1061	55	13	68	171	313	468	.50	28
JUN	1802	1004	60	2	14	67	167	308	.50	24
JUL	2248	1299	65	0	2	16	80	181	.64	26
AUG	1616	1174	64	0	3	20	82	200	.53	34
SEP	1148	1150	60	2	15	71	170	313	.50	46
OCT	656	843	52	36	116	246	397	552	.42	57
NOV	337	494	45	173	313	462	612	762	.33	67
DEC	211	309	41	297	450	605	760	915	.27	71
YR	1056	857	51	1386	2393	3662	5185	6903	.48	

SPOKANE, WASHINGTON — ELEV 2365 LAT 47.6

	HS	VS	TA	D50	D55	D60	D65	D70	KT	LD
JAN	315	496	25	763	918	1073	1228	1383	.35	69
FEB	606	887	32	499	639	778	918	1058	.45	61
MAR	1041	1190	38	391	543	698	853	1008	.50	50
APR	1495	1215	46	153	276	419	567	717	.53	38
MAY	1918	1197	55	30	94	190	327	476	.57	29
JUN	2083	1151	62	4	18	65	144	265	.57	24
JUL	2357	1368	70	0	2	7	21	94	.67	26
AUG	1942	1465	68	1	3	12	47	122	.64	34
SEP	1435	1572	60	7	30	92	196	318	.62	46
OCT	841	1220	48	125	238	382	533	688	.54	58
NOV	398	635	36	437	585	735	885	1035	.40	67
DEC	255	409	29	651	806	961	1116	1271	.33	71
YR	1227	1068	47	3061	4150	5411	6835	8433	.56	

YAKIMA, WASHINGTON — ELEV 1066 LAT 46.6

	HS	VS	TA	D50	D55	D60	D65	D70	KT	LD
JAN	365	585	28	698	853	1008	1163	1318	.38	68
FEB	666	976	36	402	541	680	820	960	.47	60
MAR	1122	1273	42	265	412	565	719	874	.53	49
APR	1598	1277	50	93	188	320	465	615	.56	37
MAY	2008	1218	58	12	48	123	239	378	.59	28
JUN	2169	1161	65	2	7	30	94	188	.60	23
JUL	2358	1324	71	0	1	5	20	78	.67	25
AUG	1975	1442	69	0	2	10	37	111	.65	33
SEP	1483	1583	61	4	18	67	147	270	.63	45
OCT	891	1273	50	86	180	313	462	617	.55	57
NOV	444	711	38	352	499	648	798	948	.42	66
DEC	295	480	31	580	735	890	1045	1200	.36	70
YR	1285	1109	50	2494	3482	4658	6009	7557	.57	

CHARLESTON, WEST VIRGINIA — ELEV 951 LAT 38.4

	HS	VS	TA	D50	D55	D60	D65	D70	KT	LD
JAN	498	638	35	483	636	791	946	1101	.36	60
FEB	706	770	37	381	519	658	798	938	.39	52
MAR	1009	841	45	202	334	483	642	791	.41	41
APR	1356	842	56	27	83	165	287	426	.44	29
MAY	1639	837	65	3	11	40	113	202	.47	19
JUN	1776	843	72	0	1	5	10	67	.49	15
JUL	1682	827	75	0	0	0	0	35	.47	17
AUG	1514	863	74	0	0	0	0	49	.47	25
SEP	1272	979	68	1	5	18	46	136	.48	37
OCT	972	1043	57	21	69	149	267	407	.48	48
NOV	613	793	45	176	298	441	588	738	.41	58
DEC	440	585	36	431	584	738	893	1048	.35	62
YR	1125	822	55	1726	2540	3488	4590	5938	.45	

HUNTINGTON, WEST VIRGINIA — ELEV 837 LAT 38.4

	HS	VS	TA	D50	D55	D60	D65	D70	KT	LD
JAN	526	689	34	489	642	797	952	1107	.38	60
FEB	757	847	36	393	530	669	809	949	.41	52
MAR	1067	901	44	208	340	489	649	797	.44	41
APR	1448	900	56	30	87	171	293	432	.47	29
MAY	1710	866	65	3	12	41	115	203	.49	19
JUN	1844	865	72	0	1	5	11	64	.51	15
JUL	1769	859	75	0	0	0	0	34	.50	17
AUG	1580	899	74	0	0	0	0	48	.49	25
SEP	1306	1010	68	1	5	18	46	134	.49	37
OCT	1004	1091	57	22	70	149	265	405	.50	48
NOV	638	840	46	175	296	438	585	735	.43	58
DEC	467	636	36	438	590	744	899	1054	.37	62
YR	1178	867	55	1759	2574	3522	4624	5961	.47	

```
EAU CLAIRE, WISCONSIN            ELEV  896  LAT 44.9    MILWAUKEE, WISCONSIN             ELEV  692  LAT 42.9
     HS   VS  TA   D50  D55   D60   D65   D70  KT  LD        HS   VS  TA   D50  D55   D60   D65   D70  KT  LD
JAN  452  738  12 1187 1342  1497  1652  1807 .43 66   JAN  479  731  19  949 1104  1259  1414  1569 .42 64
FEB  746 1072  15  969 1109  1249  1389  1529 .50 59   FEB  736  967  23  770  910  1050  1190  1330 .46 57
MAR 1090 1150  27  704  859  1014  1169  1324 .50 47   MAR 1089 1071  31  577  732   887  1042  1197 .48 45
APR 1426 1057  45  187  320   466   615   765 .49 35   APR 1443 1011  45  177  313   460   609   759 .49 33
MAY 1681  977  56   18   68   154   293   430 .49 26   MAY 1768  977  54   27   92   196   348   490 .51 24
JUN 1872  982  66    1    4    18    65   151 .51 22   JUN 1977  984  65    1    5    24    90   182 .54 20
JUL 1886 1030  71    0    1     5    14    78 .53 24   JUL 1962 1017  70    0    1     4    15    81 .55 22
AUG 1621 1093  68    0    2     9    37   112 .53 31   AUG 1719 1099  69    0    1     5    36    92 .55 29
SEP 1196 1115  59    8   36   104   202   343 .50 43   SEP 1310 1173  61    3   14    61   140   273 .53 41
OCT  826 1062  49  107  213   354   505   660 .48 55   OCT  908 1124  51   65  153   285   440   589 .50 53
NOV  450  670  32  540  690   840   990  1140 .39 64   NOV  525  767  37  406  555   705   855  1005 .42 62
DEC  341  544  18  992 1147  1302  1457  1612 .37 68   DEC  378  573  24  800  955  1110  1265  1420 .37 66
YR  1134  957  43 4714 5790  7011  8388  9951 .49      YR  1194  957  46 3774 4833  6045  7444  8986 .51

GREEN BAY, WISCONSIN             ELEV  702  LAT 44.5    CASPER, WYOMING                  ELEV 5289  LAT 42.9
     HS   VS  TA   D50  D55   D60   D65   D70  KT  LD        HS   VS  TA   D50  D55   D60   D65   D70  KT  LD
JAN  451  722  15 1073 1228  1383  1538  1693 .42 66   JAN  683 1251  23  831  986  1141  1296  1451 .60 64
FEB  725 1010  18  896 1036  1176  1316  1456 .48 58   FEB 1013 1550  27  650  790   930  1070  1210 .63 57
MAR 1104 1153  29  664  818   973  1128  1283 .50 47   MAR 1441 1568  31  589  744   899  1054  1209 .64 45
APR 1439 1056  44  202  339   487   636   786 .49 35   APR 1847 1340  43  234  373   520   669   819 .62 33
MAY 1719  989  55   28   92   191   338   481 .50 25   MAY 2204 1191  53   53  129   242   388   537 .64 24
JUN 1908  989  65    1    6    27    91   185 .52 21   JUN 2501 1172  62    4   17    61   147   255 .69 20
JUL 1888 1022  69    0    1     7    22    97 .54 23   JUL 2535 1249  71    0    1     5    13    76 .72 22
AUG 1622 1082  68    0    2    11    54   122 .52 31   AUG 2225 1444  70    0    2     8    17    98 .71 29
SEP 1218 1127  59    7   32    98   191   336 .50 43   SEP 1749 1717  59   10   40   109   229   344 .70 41
OCT  821 1037  49   97  199   339   490   645 .48 54   OCT 1219 1731  48  129  242   385   536   691 .68 53
NOV  465  690  34  478  627   777   927  1077 .40 64   NOV  765 1352  34  484  633   783   933  1083 .61 62
DEC  350  554  21  902 1057  1212  1367  1522 .37 68   DEC  594 1134  26  738  893  1048  1203  1358 .58 66
YR  1145  952  44 4348 5437  6680  8098  9684 .50      YR  1568 1390  45 3723 4850  6131  7555  9131 .66

LA CROSSE, WISCONSIN             ELEV  673  LAT 43.9    CHEYENNE, WYOMING                ELEV 6142  LAT 41.1
     HS   VS  TA   D50  D55   D60   D65   D70  KT  LD        HS   VS  TA   D50  D55   D60   D65   D70  KT  LD
JAN  481  771  16 1051 1206  1361  1516  1671 .44 65   JAN  766 1362  27  725  880  1035  1190  1345 .62 62
FEB  765 1065  20  840  980  1120  1260  1400 .49 58   FEB 1068 1548  29  588  728   868  1008  1148 .63 55
MAR 1101 1125  31  586  741   896  1051  1206 .49 46   MAR 1433 1451  32  571  725   880  1035  1190 .61 44
APR 1426 1027  48  127  237   375   522   672 .49 34   APR 1770 1203  43  230  371   519   669   819 .59 31
MAY 1713  972  59   10   38   107   224   346 .50 25   MAY 1995 1040  52   49  127   246   394   546 .58 22
JUN 1905  976  69    1    2    11    39   112 .52 21   JUN 2258 1043  61    3   15    62   156   268 .62 18
JUL 1900 1013  73    0    1     3    10    52 .54 23   JUL 2230 1076  69    0    1     7    22    98 .63 20
AUG 1666 1094  71    0    1     5    17    71 .54 30   AUG 1966 1195  68    0    2    11    31   123 .62 28
SEP 1242 1133  62    4   17    63   130   258 .51 42   SEP 1667 1502  58    9   38   109   225   357 .65 39
OCT  863 1087  52   65  146   266   421   565 .49 54   OCT 1242 1645  48  118  233   378   530   685 .66 51
NOV  494  734  35  440  588   738   888  1038 .41 63   NOV  823 1384  36  436  585   735   885  1035 .61 60
DEC  369  581  22  874 1029  1184  1339  1494 .38 67   DEC  671 1244  29  645  800   955  1110  1265 .60 64
YR  1163  964  46 3998 4987  6130  7417  8884 .50      YR  1493 1306  46 3375 4507  5805  7255  8880 .62

MADISON, WISCONSIN               ELEV  860  LAT 43.1    ROCK SPRINGS, WYOMING            ELEV 6745  LAT 41.6
     HS   VS  TA   D50  D55   D60   D65   D70  KT  LD        HS   VS  TA   D50  D55   D60   D65   D70  KT  LD
JAN  515  822  17 1029 1184  1339  1494  1649 .45 64   JAN  735 1311  19  955 1110  1265  1420  1575 .60 63
FEB  804 1108  20  832  972  1112  1252  1392 .51 57   FEB 1089 1631  23  745  885  1025  1165  1305 .65 55
MAR 1136 1141  30  614  769   924  1079  1234 .50 46   MAR 1530 1619  29  654  809   964  1119  1274 .66 44
APR 1398  981  45  167  297   442   591   741 .47 33   APR 1944 1358  40  299  447   597   747   897 .65 32
MAY 1743  969  56   19   70   157   297   436 .51 24   MAY 2344 1206  50   62  158   300   453   608 .68 23
JUN 1948  977  66    1    4    19    72   156 .53 20   JUN 2574 1151  59    3   20    83   198   334 .71 18
JUL 1934 1009  70    0    1     5    14    83 .55 22   JUL 2547 1202  68    0    1     4    18    98 .72 20
AUG 1708 1098  69    0    2     8    39   106 .55 30   AUG 2240 1387  66    0    1     9    49   141 .71 28
SEP 1299 1168  60    6   26    87   173   314 .52 41   SEP 1832 1734  56    8   46   130   269   408 .72 40
OCT  911 1139  50   86  182   318   474   623 .51 53   OCT 1306 1812  45  176  321   475   629   784 .70 52
NOV  504  729  35  460  609   759   909  1059 .41 62   NOV  826 1427  31  579  729   879  1029  1179 .63 61
DEC  389  603  22  871 1026  1181  1336  1491 .38 66   DEC  651 1219  23  849 1004  1159  1314  1469 .60 65
YR  1193  978  45 4086 5143  6352  7730  9284 .51      YR  1638 1420  43 4330 5531  6889  8410 10073 .68
```

```
SHERIDAN, WYOMING              ELEV 3966 LAT 44.8
      HS   VS TA  D50  D55   D60   D65   D70  KT LD
JAN  517  900 21  899 1054  1209  1364  1519 .49 66
FEB  788 1157 26  675  815   955  1095  1235 .52 58
MAR 1205 1315 31  590  744   899  1054  1209 .55 47
APR 1537 1150 44  217  349   494   642   792 .53 35
MAY 1883 1087 53   57  131   237   375   526 .55 26
JUN 2156 1102 61    7   27    81   168   280 .59 21
JUL 2329 1237 70    1    2     9    28    94 .66 23
AUG 2006 1379 69    1    3    13    31   113 .65 31
SEP 1502 1504 58   17   57   131   245   369 .62 43
OCT 1005 1409 48  137  245   384   533   689 .59 55
NOV  591 1000 33  500  648   798   948  1098 .51 64
DEC  441  794 26  760  915  1070  1225  1380 .48 68
YR  1333 1170 45 3860 4991  6279  7708  9303 .58
```

```
EDMONTON, ALBERTA              ELEV 2220 LAT 53.6
      HS   VS TA   D50  D55   D60   D65   D70  KT LD
JAN  324  746  4 1421 1574  1728  1883  2038 .55 75
FEB  622 1238 12 1076 1215  1354  1494  1634 .60 67
MAR 1104 1645 21  916 1068  1221  1375  1530 .62 56
APR 1554 1566 38  374  510   655   802   952 .59 44
MAY 1818 1352 51   97  184   307   452   604 .55 35
JUN 1945 1271 57   10   46   121   238   380 .54 30
JUL 1977 1378 62    1    7    38   117   241 .57 32
AUG 1598 1412 60    4   21    83   185   327 .55 40
SEP 1113 1378 51   92  177   297   437   585 .54 52
OCT  689 1185 41  322  458   605   757   911 .55 64
NOV  359  762 23  806  952  1101  1250  1400 .52 73
DEC  235  535 11 1198 1351  1505  1660  1815 .50 77
YR  1114 1205 36 6317 7563  9016 10650 12416 .56
```

```
VANCOUVER, BRITISH COLUMBIA    ELEV  310 LAT 49.3
      HS   VS TA  D50  D55   D60   D65   D70  KT LD
JAN  254  395 37  425  572   724   878  1033 .31 71
FEB  495  710 40  287  417   553   693   832 .39 63
MAR  877  997 43  256  392   541   695   849 .44 52
APR 1331 1111 48  127  233   369   516   665 .48 40
MAY 1790 1169 55   24   87   188   329   481 .53 30
JUN 1922 1116 60    4   21    80   179   316 .53 26
JUL 2021 1239 63    0    1    13    82   215 .58 28
AUG 1648 1275 63    0    1    16    92   231 .55 36
SEP 1185 1284 58    4   28   100   218   363 .53 47
OCT  638  874 51   81  171   301   450   604 .43 59
NOV  315  490 43  233  363   508   656   805 .35 69
DEC  201  317 39  351  495   647   800   955 .29 73
YR  1060  916 50 1791 2781  4041  5588  7349 .49
```

```
SUFFIELD, ALBERTA              ELEV 2549 LAT 50.3
      HS   VS TA   D50  D55   D60   D65   D70  KT LD
JAN  433  937  7 1333 1486  1640  1794  1949 .57 72
FEB  748 1388 17  943 1080  1219  1358  1498 .62 64
MAR 1232 1688 24  815  965  1117  1270  1424 .64 53
APR 1583 1429 41  321  446   584   729   877 .58 41
MAY 1903 1287 53   84  161   262   394   541 .57 31
JUN 2074 1235 61    6   24    78   164   289 .57 27
JUL 2173 1377 67    0    2    11    49   128 .62 29
AUG 1801 1468 65    1    5    23    81   173 .61 37
SEP 1274 1475 55   48  111   195   312   449 .58 48
OCT  822 1329 45  242  355   491   638   790 .57 60
NOV  456  903 28  678  821   968  1116  1265 .53 70
DEC  333  722 17 1029 1181  1334  1487  1642 .52 74
YR  1239 1269 40 5500 6637  7923  9393 11025 .59
```

```
WINNIPEG, MANITOBA             ELEV  820 LAT 49.9
      HS   VS TA   D50  D55   D60   D65   D70  KT LD
JAN  461 1011  0 1588 1740  1893  2047  2201 .59 71
FEB  799 1508  4 1303 1440  1578  1717  1857 .65 64
MAR 1240 1676 17 1026 1175  1326  1479  1632 .63 52
APR 1548 1370 38  409  536   674   818   965 .56 40
MAY 1847 1231 51  125  207   314   447   592 .55 31
JUN 2000 1179 62    5   20    68   147   262 .55 27
JUL 2025 1264 67    0    2     9    45   122 .58 29
AUG 1687 1337 66    1    5    21    76   166 .57 36
SEP 1180 1305 55   61  127   214   330   467 .53 48
OCT  726 1082 44  275  387   519   664   814 .50 60
NOV  406  737 24  797  939  1084  1231  1380 .46 69
DEC  336  712  7 1334 1485  1637  1791  1944 .51 73
YR  1190 1199 36 6925 8062  9338 10790 12401 .56
```

```
NANAIMO, BRITISH COLUMBIA      ELEV   60 LAT 49.2
      HS   VS TA  D50  D55   D60   D65   D70  KT LD
JAN  256  398 37  432  579   730   884  1039 .32 70
FEB  538  799 40  304  433   569   708   847 .42 63
MAR  921 1063 42  280  416   564   717   871 .46 52
APR 1409 1189 47  140  246   381   527   676 .51 39
MAY 1860 1215 54   33  100   202   341   493 .55 30
JUN 1956 1132 60    4   20    79   179   316 .54 26
JUL 2092 1279 64    0    2    17    81   197 .60 28
AUG 1727 1344 63    0    2    17    85   211 .58 36
SEP 1234 1354 58    5   28    95   206   348 .55 47
OCT  680  954 50   97  190   319   467   621 .46 59
NOV  333  530 42  256  386   530   678   827 .36 69
DEC  206  327 39  370  513   664   817   972 .30 73
YR  1104  966 50 1921 2915  4169  5691  7417 .51
```

```
HALIFAX, NOVA SCOTIA           ELEV  136 LAT 44.6
      HS   VS TA  D50  D55   D60   D65   D70  KT LD
JAN  456  737 26  752  900  1051  1204  1358 .43 66
FEB  695  954 26  682  816   953  1092  1231 .46 58
MAR 1085 1130 32  575  718   868  1020  1174 .49 47
APR 1298  940 41  321  448   588   734   882 .45 35
MAY 1522  882 49  128  219   343   487   638 .44 26
JUN 1738  917 58   10   42   115   225   365 .48 21
JUL 1694  929 65    0    1     8    57   164 .48 23
AUG 1582 1056 65    0    0     5    47   149 .51 31
SEP 1211 1122 60    4   21    76   169   302 .50 43
OCT  812 1024 51   93  176   291   432   583 .47 55
NOV  463  688 42  280  402   540   685   833 .40 64
DEC  332  514 31  612  757   907  1059  1213 .36 68
YR  1076  907 46 3457 4500  5746  7211  8892 .47
```

MOOSONEE, ONTARIO   ELEV   34 LAT 51.3
      HS   VS TA   D50 D55   D60   D65   D70 KT LD
JAN 375 800 -4 1696 1848 2000 2154 2308 .53 73
FEB 721 1380 0 1424 1562 1700 1839 1978 .62 65
MAR 1122 1531 11 1237 1387 1538 1691 1845 .59 54
APR 1466 1338 28   692 830   973 1119 1267 .54 42
MAY 1603 1096 42   333 451   586   731   881 .48 32
JUN 1780 1089 53   94 168   258   376   512 .49 28
JUL 1625 1044 60   15   47   113   203   328 .47 30
AUG 1339 1055 58   24   69   144   248   383 .46 38
SEP 956 1026 50   143 227   334   464   606 .44 49
OCT 543 752 40   364 486   624   770   920 .39 61
NOV 297 498 24   795 936 1080 1227 1375 .37 71
DEC 274 571 4 1425 1575 1727 1881 2034 .47 75
YR 1010 1012 30 8243 9584 11078 12702 14436 .49

TORONTO, ONTARIO   ELEV 443 LAT 43.7
      HS   VS TA   D50 D55   D60   D65   D70 KT LD
JAN 487 777 22   891 1041 1194 1348 1502 .44 65
FEB 747 1021 23   773 910 1048 1187 1327 .48 57
MAR 1100 1116 31   615 761   911 1064 1218 .49 46
APR 1468 1055 44   248 362   495   638   785 .50 34
MAY 1764 994 54   61 132   223   348   492 .51 25
JUN 1950 990 65   2   7   28   86   176 .53 20
JUL 1958 1035 70   0   0   3   18   76 .55 22
AUG 1686 1102 69   0   1   6   32   102 .54 30
SEP 1248 1132 61   10   34   91   172   287 .51 42
OCT 833 1025 50   127 212   326   464   612 .47 54
NOV 428 590 39   371 501   642   788   936 .35 63
DEC 354 542 26   744 893 1044 1198 1352 .36 67
YR 1171 948 46 3842 4853 6013 7343 8865 .50

OTTAWA, ONTARIO   ELEV 377 LAT 45.5
      HS   VS TA   D50 D55   D60   D65   D70 KT LD
JAN 510 914 13 1169 1320 1473 1627 1782 .51 67
FEB 819 1264 15   979 1116 1255 1394 1533 .56 59
MAR 1203 1347 27   742 889 1040 1193 1347 .56 48
APR 1490 1133 42   287 404   539   682   829 .52 36
MAY 1754 1033 55   61 130   221   344   487 .51 26
JUN 1884 1000 65   1   6   27   87   181 .52 22
JUL 1875 1040 69   0   1   4   23   88 .53 24
AUG 1602 1098 67   0   3   13   57   139 .52 32
SEP 1167 1102 58   23   65   137   231   358 .49 44
OCT 771 986 48   179 275   399   541   690 .46 55
NOV 414 609 35   483 618   760   907 1055 .37 65
DEC 383 667 19   987 1138 1290 1443 1598 .43 69
YR 1158 1015 43 4912 5965 7158 8529 10086 .51

NORMANDIN, QUEBEC   ELEV 450 LAT 48.8
      HS   VS TA   D50 D55   D60   D65   D70 KT LD
JAN 454 921 0 1564 1719 1873 2028 2183 .55 70
FEB 785 1386 4 1302 1441 1581 1721 1860 .61 62
MAR 1275 1668 17 1042 1195 1349 1504 1658 .63 51
APR 1578 1353 33   532 674   821   970 1119 .56 39
MAY 1652 1059 47   172 277   411   558   711 .49 30
JUN 1831 1052 58   15   54   127   232   367 .50 25
JUL 1707 1031 62   1   7   40   118   242 .49 27
AUG 1471 1097 60   7   29   91   186   320 .49 35
SEP 1013 1026 51   86 165   276   413   558 .45 47
OCT 595 775 41   307 438   584   735   889 .39 59
NOV 375 615 27   689 834   982 1131 1281 .40 68
DEC 343 582 8 1320 1473 1628 1782 1937 .48 72
YR 1092 1053 34 7037 8308 9762 11376 13125 .51

Appendix 20: Solar Radiation Approximations

To use the present *SLR* correlations, it is necessary to have an estimate of *S,* the monthly solar radiation absorbed in the building per unit of projected area. (Recall that the projected area is the principal net glazing area projected on a vertical plane. See the definition in Chapter M.) The most usual solar radiation parameter available to the designer is the monthly total solar radiation incident on a horizontal surface. From this parameter, reasonable estimates can be obtained of the solar radiation incident on the glazed surfaces of a particular configuration, the radiation transmitted through the glazings, and the radiation absorbed in the building.

The approximations presented here are least-squares correlations of the monthly radiation quantities as functions of a solar-position parameter (latitude minus midmonth solar declination) and a clearness parameter (fraction of extraterrestrial radiation). The typical meterological year (*TMY*) hourly weather files for 26 United States cities are the starting point for the correlation development. Using total horizontal and direct normal solar radiation from these weather files, and assuming isotropic sky diffuse and ground-reflected radiations, the radiation incident on the glazing surface is found hourly. The transmitted radiation is found hourly by considering absorption in the glazings and reflection from the glazing surfaces. Reflection loss from the glazing surfaces is calculated using the Fresnel equation with an index of refraction of 1.526 for the glazing. Transmission of diffuse radiation incident on the glazings is calculated assuming an effective angle of incidence of 60°.

The absorbed radiation is then calculated from an internal reflection matrix, assuming completely diffuse transmission and reflections, including radiation absorbed on all internal surfaces (excluding the glazings) and the space air. The calculation depends on the absorptances of the surfaces, as well as on the solar energy system's geometry.

Assuming that only the monthly total horizontal and extraterrestrial horizontal* radiation data are available to the designer, correlations were made of the ratios incident-to-total horizontal and transmitted-to-incident.

---

*Radiation incident on a horizontal surface outside the atmosphere.

The primary independent parameter for these correlations is the solar-noon zenith angle at midmonth, *LAT − DEC,* where *LAT* is the latitude and *DEC* is the midmonth solar declination. A second correlating parameter indicating the degree of clearness is also necessary. The most readily available such parameter is the ratio of monthly total horizontal to monthly extraterrestrial horizontal radiation (K_T). Equations for the extraterrestrial horizontal solar radiation and the solar declination are given in Appendix 19. Values of K_T and *LAT − DEC* are given in Appendix 19 for 219 sites in the United States and Canada.

20.1 INCIDENT RADIATION

Figures 20-1, 20-2, and 20-3 show the ratio of the monthly total solar radiation incident on a unit area of a plane surface of different orientations to that on a horizontal plane for surface tilts of 30°, 60°, and 90°, respectively from the horizontal, and for azimuths of 0°, 30°, 60°, and 90° from true south. The curves give the incident-to-horizontal ratio as functions of the latitude minus midmonth declination (*LAT − DEC*) and the clearness ratio (K_T). Figure 20-4 gives the correlating equations used to plot figures 20-1 through 20-3.

20.2 TRANSMITTED RADIATION

Figures 20-5 through 20-8 give the ratio of the monthly total transmitted radiation to incident radiation (transmittance) of plane glazings at different surface orientations for 1, 2 and 3 glazing layers. Figures 20-9 and 20-10 give the correlating equations used to plot figures 20-5 through 20-8.

20.3 TRANSMITTANCE CORRECTIONS

20.3.a EXTINCTION COEFFICIENT–THICKNESS PRODUCT

The transmittances given in figures 20-5 through 20-8 are for a glazing material extinction coefficient–thickness product (*E*) of 0.0625 per layer. This value of *E* corresponds roughly to standard glass that is ⅛ inch thick.

INCIDENT-TO-HORIZONTAL
TILT = 90

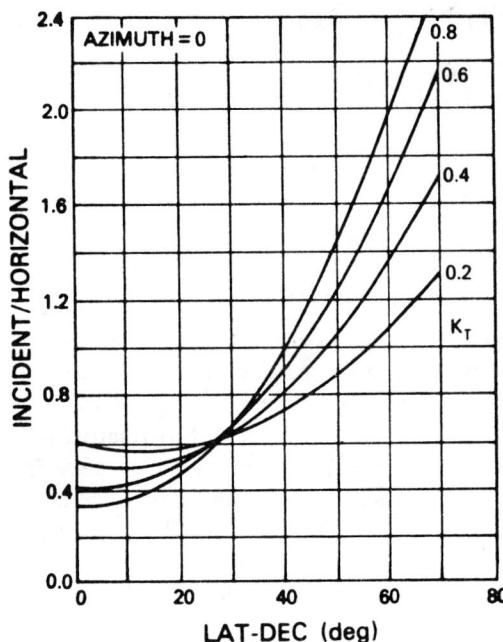
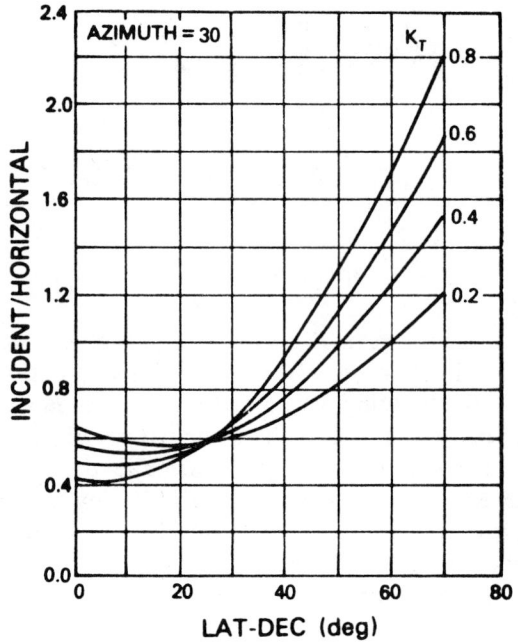
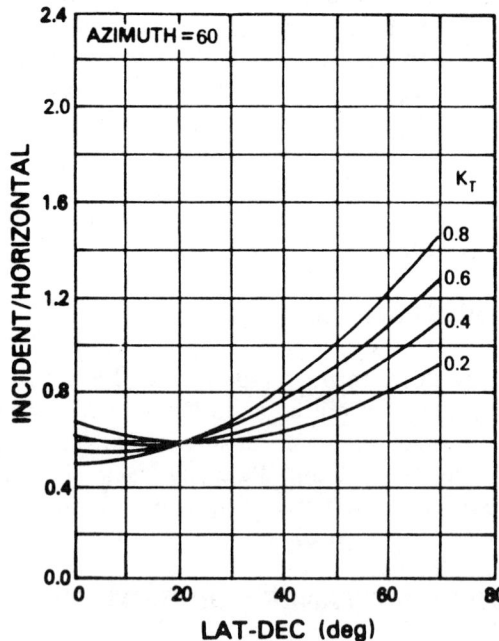
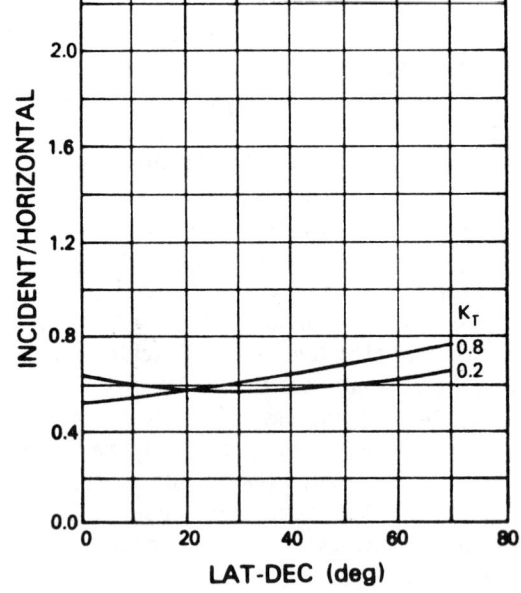

20-1: Monthly average tilted-to-incident radiation—at 90° tilt, four different azimuths, and two or four clearness factors—plotted against different values of $LAT - DEC$.

INCIDENT-TO-HORIZONTAL
TILT = 60

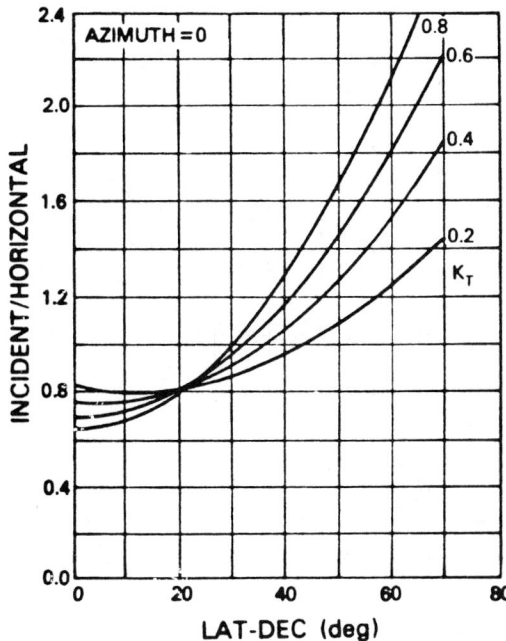
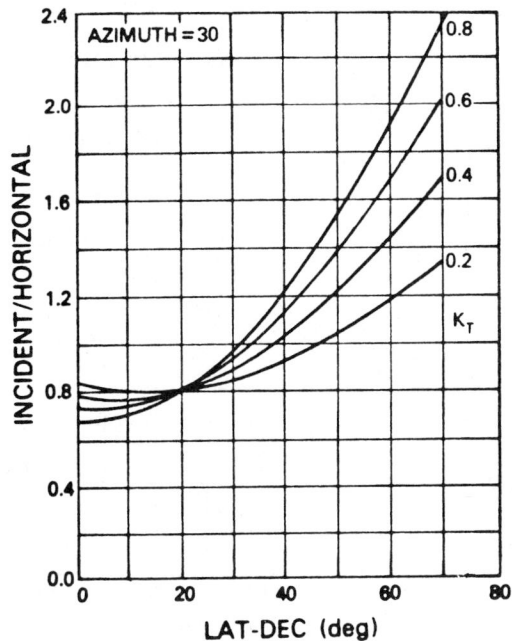
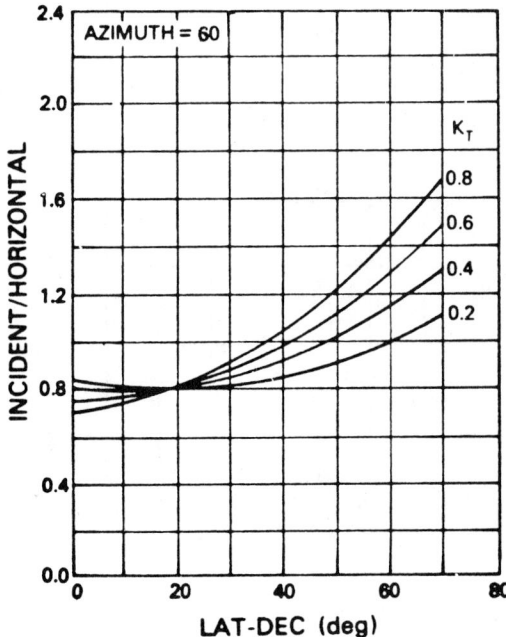
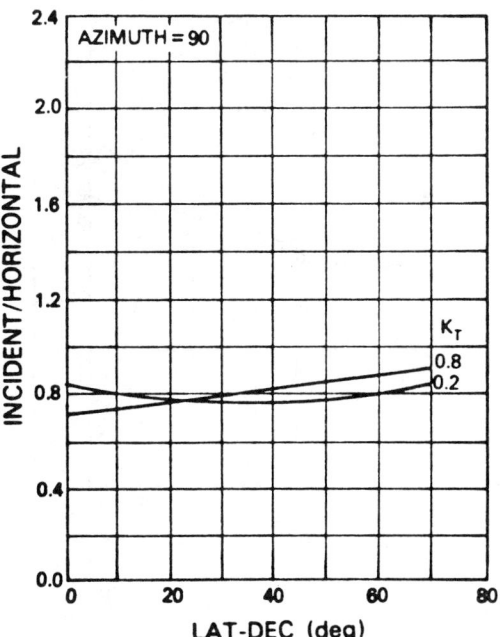

20-2: Monthly average tilted-to-incident radiation— at 60°tilt, four different azimuths, and two or four clearness factors—plotted against different values of $LAT - DEC$.

Solar Radiation Approximations / 433

INCIDENT-TO-HORIZONTAL
TILT = 30

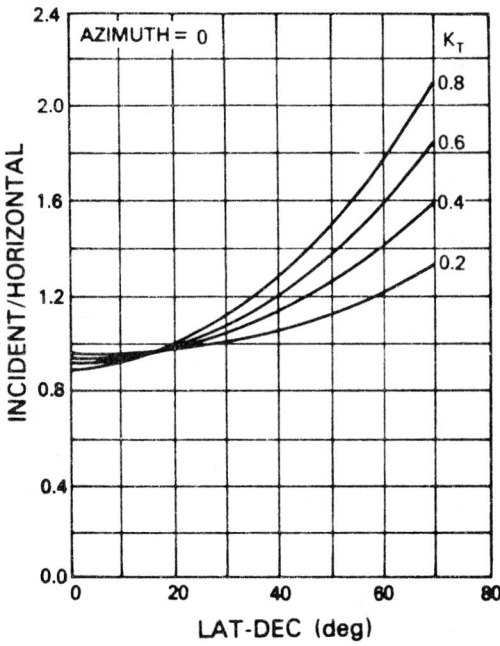
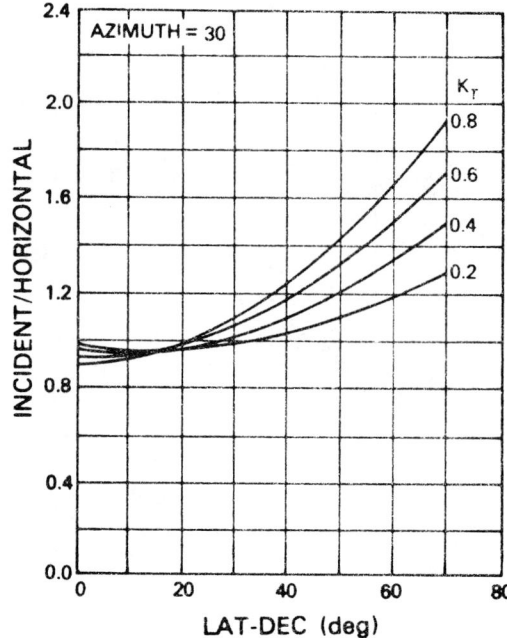
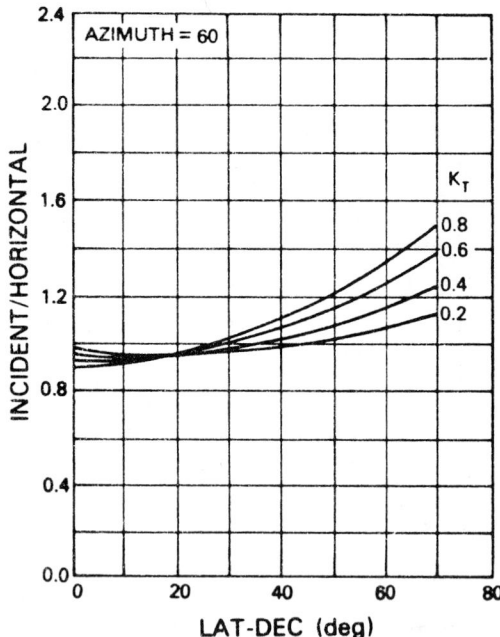
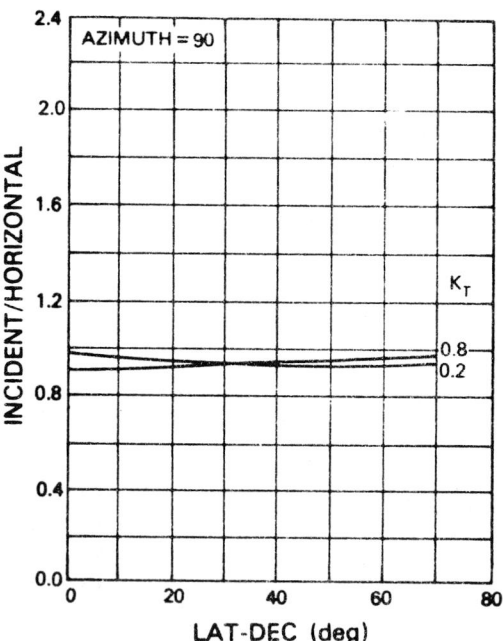

20-3: Monthly average tilted-to-incident radiation—at 30° tilt, four different azimuths, and two or four clearness factors—plotted against different values of $LAT - DEC$.

COEFFICIENTS FOR INCIDENT/HORIZONTAL EQUATIONS OF FORM:

$$I/Q_h = B_1 + B_2Y + B_3Y^2 + K_T(B_4 + B_5Y + B_6Y^2)$$

WHERE $Y = (LAT - DEC)/100$

TILT	AZIM	B_1	B_2	B_3	B_4	B_5	B_6
30	0	0.9786	-0.1895	0.4573	-0.0918	0.1596	2.5034
30	30	0.9935	-0.2581	0.4834	-0.1134	0.3354	1.9329
30	60	0.9937	-0.2691	0.3902	-0.1155	0.4634	0.8290
30	90	0.9959	-0.3156	0.3208	-0.1217	0.5485	-0.4429
60	0	0.8682	-0.4587	0.9788	-0.2817	0.4051	4.0989
60	30	0.8850	-0.5729	1.0557	-0.2689	0.6560	3.0255
60	60	0.8749	-0.5617	0.8552	-0.2151	0.9028	1.0646
60	90	0.8710	-0.5943	0.7018	-0.1966	1.0178	-0.7940
90	0	0.6866	-0.6623	1.3269	-0.4458	0.3090	4.7776
90	30	0.7183	-0.8937	1.5666	-0.3680	0.6415	3.2484
90	60	0.7203	-0.9037	1.3357	-0.2920	1.2595	0.5969
90	90	0.6749	-0.6051	0.7595	-0.1959	1.0720	-0.7838

20-4: Coefficients for the incident-to-horizontal equation at different tilts and azimuths.

TRANSMITTED-TO-INCIDENT
TILT = 90

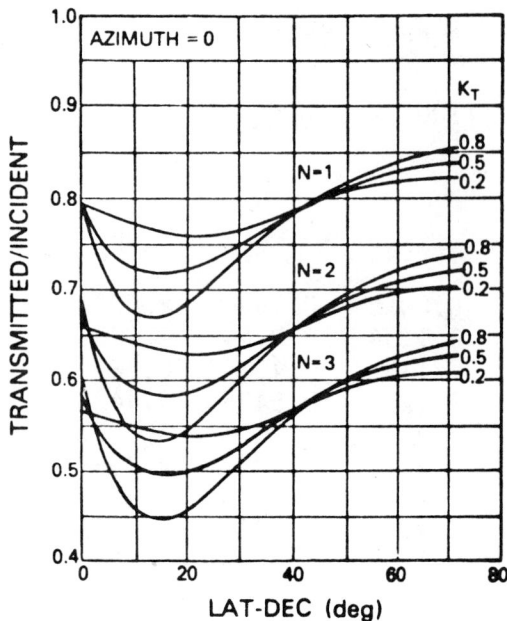
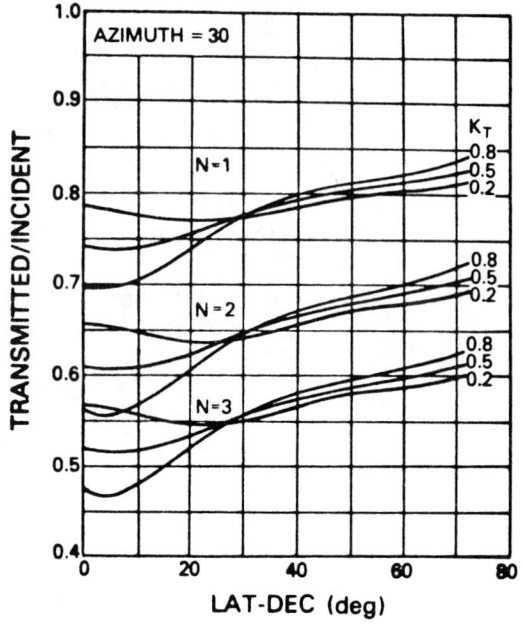
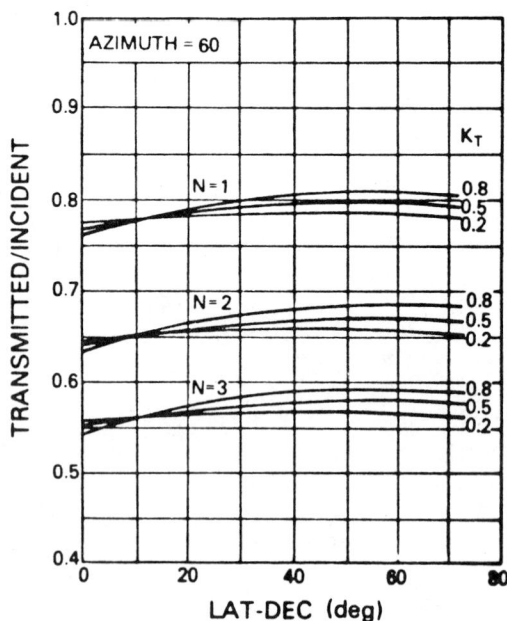
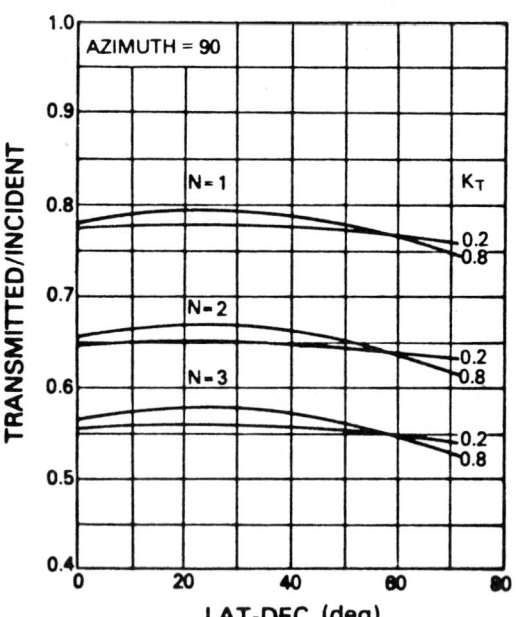

20-5: Monthly average transmitted-to-incident radiation—at 90° tilt, four different azimuths, and two or three clearness factors, for N layers of glazing—plotted against different values of $LAT - DEC$.

TRANSMITTED-TO-INCIDENT
TILT = 60

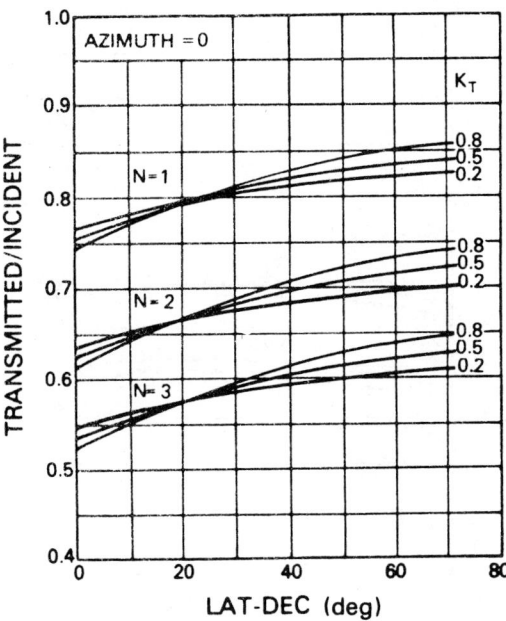
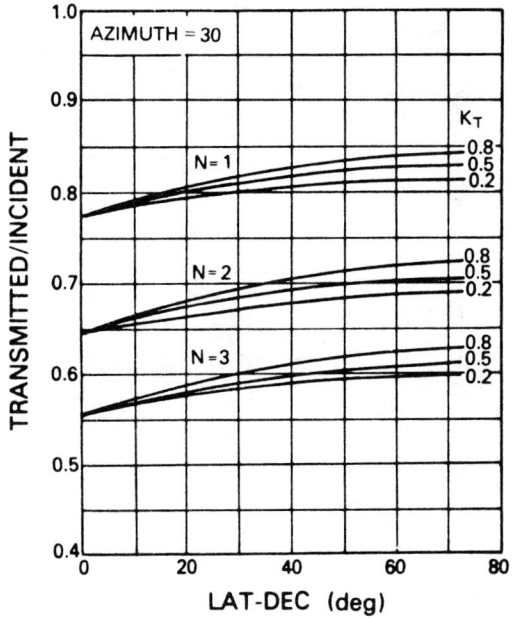
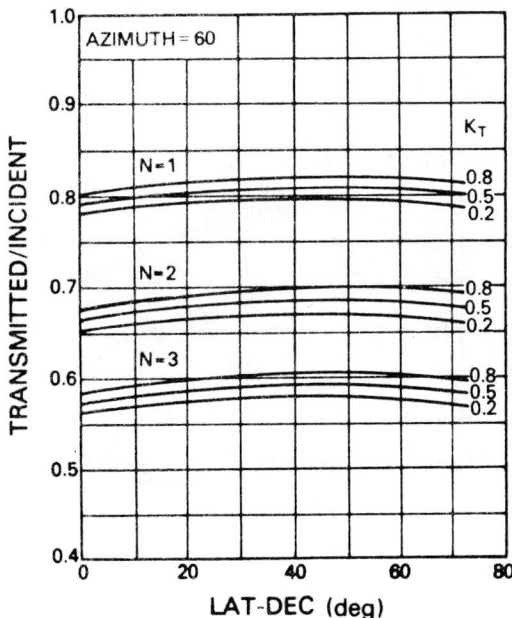
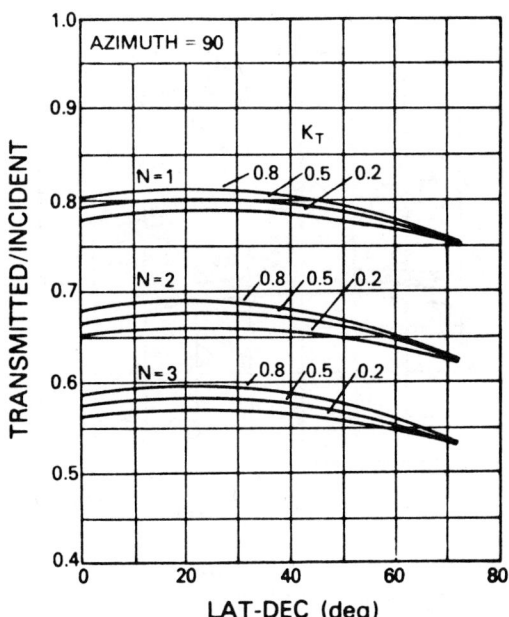

20-6: Monthly average transmitted-to-incident radiation—at 60° tilt, four different azimuths, and two or three clearness factors, for N layers of glazing—plotted against different values of $LAT - DEC$.

Solar Radiation Approximations / 437

TRANSMITTED-TO-INCIDENT
TILT = 30

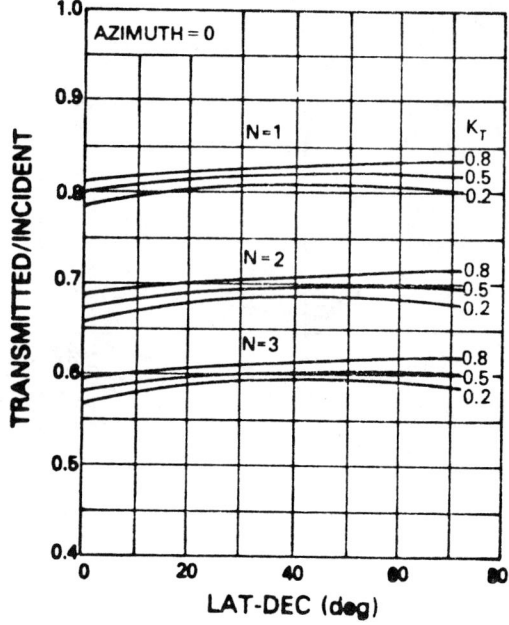
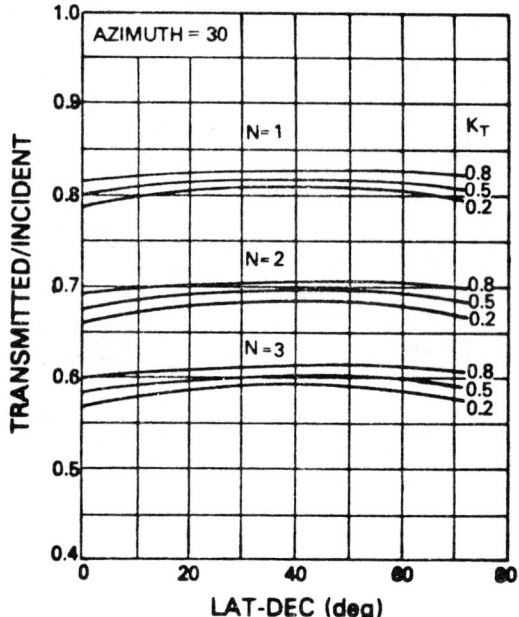
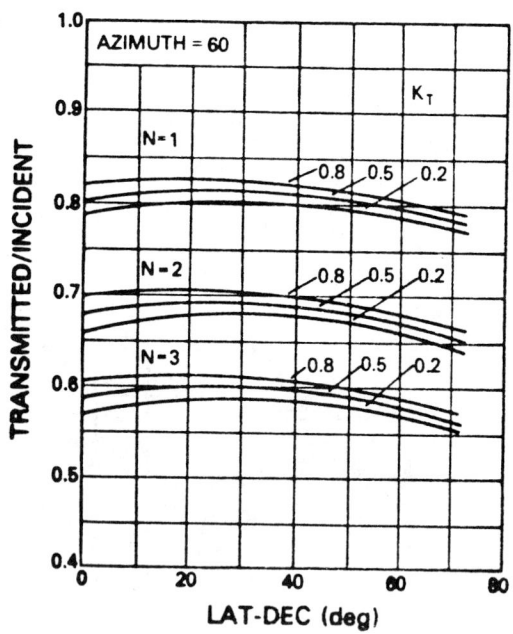
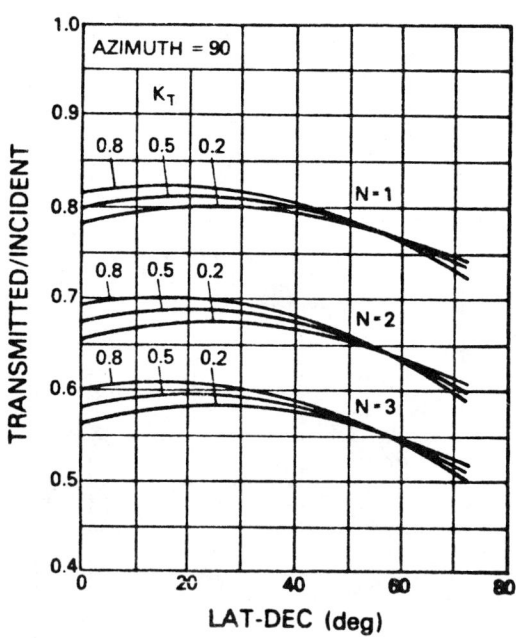

20-7: Monthly average transmitted-to-incident radiation—at 30° tilt, four different azimuths, and two or three clearness factors, for N layers of glazing—plotted against different values of $LAT - DEC$.

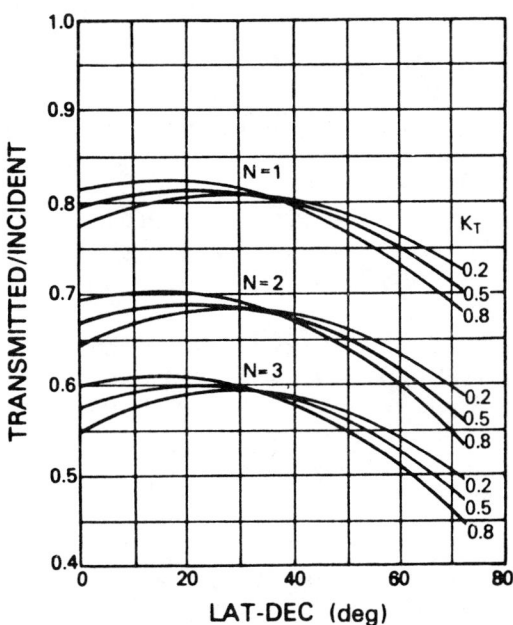

20-8: Monthly average transmitted-to-incident radiation—at 0° tilt, (horizontal position) and three different clearness factors, for *N* layers of glazing—plotted against different values of $LAT - DEC$.

COEFFICIENTS FOR TRANSMITTED/INCIDENT EQUATIONS OF FORM:

$$T/I = B_1 + B_2 Y + B_3 Y^2 + K_T(B_4 + B_5 Y + B_6 Y^2)$$

WHERE $Y = (LAT - DEC)/100$

1 GLAZING

TILT	AZIM	B_1	B_2	B_3	B_4	B_5	B_6
0	0	0.7646	0.2637	-0.4133	0.0640	-0.1718	-0.0349
30	0	0.7766	0.1372	-0.1614	0.0442	-0.1075	0.1715
30	30	0.7766	0.1334	-0.1689	0.0473	-0.0989	0.1389
30	60	0.7731	0.1413	-0.2102	0.0577	-0.1078	0.1027
30	90	0.7729	0.1342	-0.2378	0.0525	-0.0396	-0.0952
60	0	0.7735	0.0970	-0.0568	-0.0354	0.2104	-0.1189
60	30	0.7761	0.0713	-0.0414	-0.0032	0.1268	-0.0821
60	60	0.7754	0.0620	-0.0796	0.0343	0.0210	-0.0121
60	90	0.7722	0.0742	-0.1474	0.0394	0.0218	-0.0850
90	0						
90	30						
90	60	0.7728	0.0154	-0.0251	-0.0136	0.1710	-0.1321
90	90	0.7728	0.0087	-0.0350	0.0117	0.1094	-0.2094

20-9: Coefficients for the simpler transmitted-to-incident equation at different tilts and azimuths.

Solar Radiation Approximations / 439

2 GLAZINGS

TILT	AZIM	B_1	B_2	B_3	B_4	B_5	B_6
0	0	0.6331	0.3094	-0.4862	0.0759	-0.2092	-0.0221
30	0	0.6471	0.1596	-0.1862	0.0521	-0.1214	0.1884
30	30	0.6476	0.1537	-0.1945	0.0551	-0.1094	0.1525
30	60	0.6434	0.1643	-0.2449	0.0678	-0.1264	0.1220
30	90	0.6432	0.1565	-0.2766	0.0624	-0.0531	-0.0966
60	0	0.6429	0.1186	-0.0727	-0.0375	0.2257	-0.1159
60	30	0.6467	0.0844	-0.0496	-0.0026	0.1454	-0.0947
60	60	0.6456	0.0763	-0.0971	0.0404	0.0263	-0.0153
60	90	0.6450	0.0672	-0.1409	0.0436	0.0516	-0.1418
90	0						
90	30						
90	60	0.6436	0.0174	-0.0275	-0.0148	0.1946	-0.1459
90	90	0.6431	0.0115	-0.0392	0.0156	0.1215	-0.2342

3 GLAZINGS

TILT	AZIM	B_1	B_2	B_3	B_4	B_5	B_6
0	0	0.5397	0.3355	-0.5234	0.0766	-0.2504	0.0516
30	0	0.5594	0.1422	-0.1649	0.0454	-0.0998	0.1580
30	30	0.5600	0.1357	-0.1718	0.0482	-0.0876	0.1230
30	60	0.5553	0.1503	-0.2254	0.0617	-0.1137	0.1081
30	90	0.5520	0.1665	-0.2900	0.0610	-0.0796	-0.0454
60	0	0.5542	0.1125	-0.0709	-0.0366	0.2092	-0.1014
60	30	0.5582	0.0776	-0.0455	-0.0038	0.1365	-0.0882
60	60	0.5571	0.0699	-0.0903	0.0356	0.0319	-0.0234
60	90	0.5568	0.0590	-0.1280	0.0389	0.0547	-0.1440
90	0						
90	30						
90	60	0.5553	0.0134	-0.0238	-0.0155	0.1853	-0.1405
90	90	0.5549	0.0084	-0.0344	0.0126	0.1205	-0.2304

20-9: Continued.

COEFFICIENTS FOR TRANSMITTED/INCIDENT EQUATIONS OF FORM:

$$T/I = B_1 + B_2Y + B_3Y^2 + B_4Y^3 + B_5Y^4 + B_6Y^5$$
$$+ K_T(B_7 + B_8Y + B_9Y^2 + B_{10}Y^3 + B_{11}Y^4 + B_{12}Y^5)$$

WHERE $Y = (LAT - DEC)/100$

1 GLAZING

TILT	AZIM	B_1 / B_7	B_2 / B_8	B_3 / B_9	B_4 / B_{10}	B_5 / B_{11}	B_6 / B_{12}
90	0	0.7899	0.5388	-5.868	19.912	-26.876	12.777
		0.0098	-3.5791	24.622	-63.976	74.729	-32.865
90	30	0.8130	0.0561	-3.151	13.473	-20.389	10.529
		-0.1433	-0.3334	8.718	-30.567	41.434	-19.727

2 GLAZINGS

TILT	AZIM	B_1 / B_7	B_2 / B_8	B_3 / B_9	B_4 / B_{10}	B_5 / B_{11}	B_6 / B_{12}
90	0	0.6465	0.7766	-7.157	22.771	-29.310	13.294
		0.0523	-4.2457	26.952	-65.486	71.106	-28.793
90	30	0.6863	0.1325	-4.078	16.838	-25.266	13.027
		-0.1539	-0.5816	11.321	-38.883	52.567	-25.112

3 GLAZINGS

TILT	AZIM	B_1 / B_7	B_2 / B_8	B_3 / B_9	B_4 / B_{10}	B_5 / B_{11}	B_6 / B_{12}
90	0	0.5497	0.8103	-6.904	21.070	-26.118	11.345
		0.0730	-4.2675	25.534	-58.122	57.904	-20.849
90	30	0.5967	0.1515	-4.225	17.334	-26.076	13.513
		-0.1504	-0.7208	12.499	-42.924	58.527	-28.285

20-10: Coefficients for the more complex transmitted-to-incident equation at different tilts and azimuths.

The transmittance for other glazing types or thicknesses can be estimated using the following equation:

$$\tau(E,N) = \tau(.0625,N) \exp[-CN(E-.0625)]$$

where $\tau(E,N)$ is the transmittance through N glazing layers and C is an empirical constant: $C = 1.15$ for $0 \le E \le 0.125$, and $1 \le N \le 3$ with a standard deviation of 0.026. Therefore, values of transmittance obtained from figures 20-5 through 20-8 or from figures 20-9 and 20-10 can be corrected for an off-reference value of E by multiplying them by the ratio $\tau(E,N)/\tau(0.0625,N)$. This ratio, the quantity $\exp[CN(E-0.0625)]$, is plotted in figure 20-11 for $N = 1, 2,$ and 3.

20.3.b GROUND REFLECTANCE

The incident solar radiation given in figures 20-1 through 20-3 (and in figure 20-4) and the transmitted solar radiation given in figures 20-5 through 20-8 (and in figures 20-9 and 20-10) assume a diffuse ground reflectance (ϱ_G) of 0.3. These quantities may be estimated for other ground reflectances as explained below.

Since the solar radiation reflected from the ground is assumed to be isotropically diffuse, the transmitted radiation due to ground reflection can be given by:

$$T_{GR} = I_{GR}\tau_D = Q_h \varrho_G F_{CG} \tau_D$$

where:
I_{GR} = incident radiation due to ground reflection
τ_D = transmittance for isotropically diffuse radiation
Q_h = total horizontal radiation
ϱ_G = ground reflectance (assumed to be uniform)
F_{CG} = view factor from the collector plane to the ground.

In the simplest case of a uniform, horiztontal ground plane extending from the glazing plane to infinity:

$$F_{CG} = \tfrac{1}{2}[1 - \cos(TILT)]$$

where $TILT$ is the tilt of the glazing plane relative to horizontal. The transmittance for isotropically diffuse radiation (τ_D) is assumed to be a constant for any given glazing system. The change in total transmitted radiation due to a change in ground reflectance is:

$$T(\varrho_G) - T(0.3) - Q_h F_{CG} \tau_D (\varrho_G - 0.3)$$

or:

$$\left(\frac{T(\varrho_G)}{Q_h}\right) = \left(\frac{T(0.3)}{Q_h}\right) + F_{CG}\tau_D(\varrho_G - 0.3)$$

Thus, the ratio of transmitted to horizontal surface radiation may be corrected for an off-reference value of ground reflectance by adding the quantity $F_{CG}\tau_D(\varrho_G - 0.3)$.

The diffuse transmittance (τ_D) is given in figure 20-12 as a function of E for 1, 2, and 3 glazing layers and indexes of refraction of 1.3, 1.526, and 1.7. These were calculated assuming that the diffuse radiation seen by the collector would be uniform over a segment of a hemisphere. This results in average angles of incidence for diffuse radiation close to 60°—the value often assumed for calculation of diffuse transmittance. The effective angle of incidence can be significantly higher than 60° for ground-reflected radiation incident on surfaces at very small tilt angles (where the ground-reflected radiation is a small part of the total radiation striking the surface).

20.4 ABSORBED RADIATION

Once the solar radiation transmitted through each glazing surface has been determined, it remains to determine the "absorption factor," the fraction of transmitted radiation that is absorbed in the building. The simplest case is that of a thermal storage wall whose absorption factor is approximately equal to its absorptance. More precisely, reflections from the glazing system back to the wall increase the absorption factor slightly relative to the wall absorptance. Absorption factors for various wall absorptances, glazing extinction coefficient–thickness products (E), and numbers of glazings are given in figure 20-13.

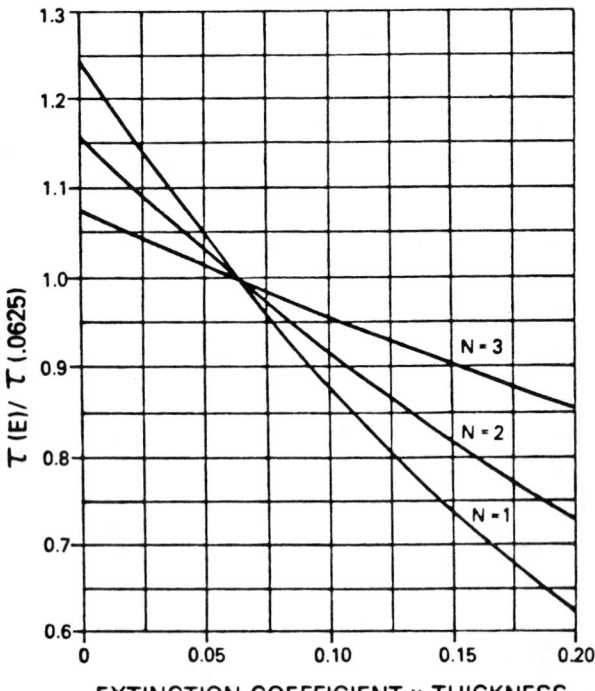

20-11: Transmittance correction ratio plotted against the extinction coefficient–thickness product.

TRANSMITTANCE CORRECTION

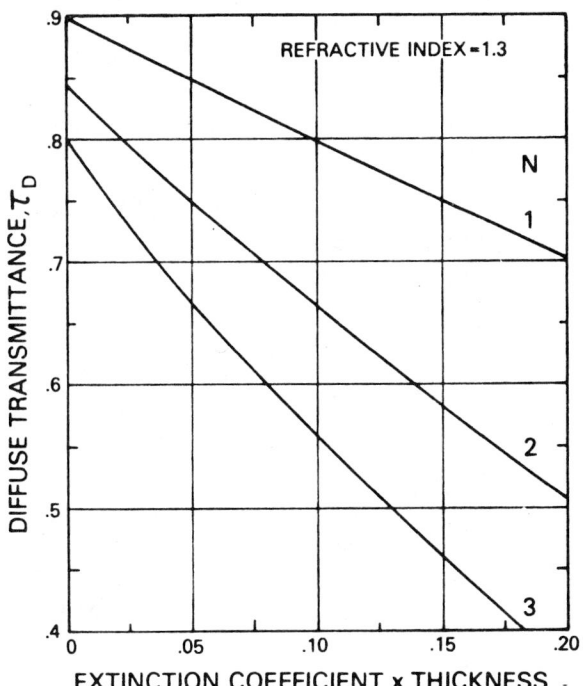

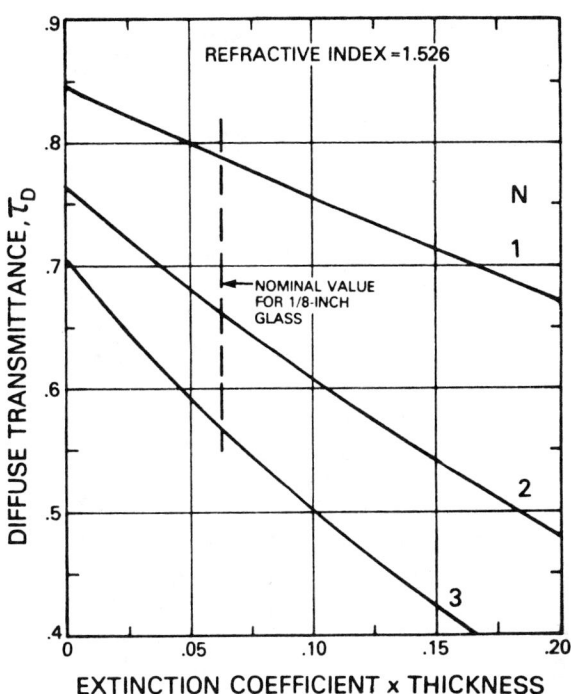

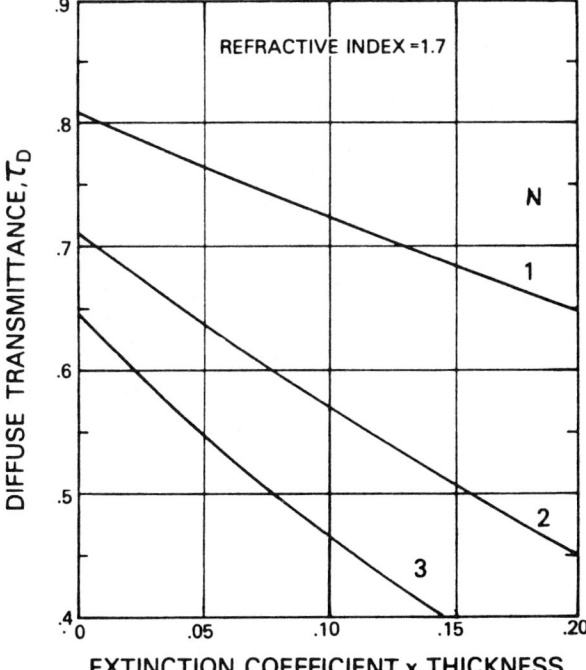

20-12: Diffuse transmittance (τ_D) plotted against extinction coefficient-thickness product for three different refractive indexes and N layers of glazing; this relationship is used to correct the transmitted solar radiation for off-reference values of ground reflectance.

For direct-gain and sunspace systems, the absorption factor depends on the interior surface absorptances and the interior geometry in a complex way. In these cases, the absorption factor may be determined from the tables at the end of this appendix. Because of the diffuse transmission assumption, this factor is not dependent on sun angles, clearness ratios, or ground reflection.

The absorption factors in the tables at the end of this appendix are given for various surface absorptances and for several values of the geometric variables that are shown in figure 20-14. The various dimensions shown are ratios to the north-wall height. In the following discussion and tables, "ALFSM" refers to the surface absorptances of the primary mass surfaces—the floor and sunspace north wall (surfaces 1 and 5 in figure 20-14)—while "ALFSO" refers to all other surfaces.

Solar Radiation Approximations / 443

THERMAL STORAGE WALL ABSORPTION FACTORS

WALL ABS	ABSORPTION IN WALL								
	E=.0125			E=.0625			E=.2500		
	1 GL	2 GL	3 GL	1 GL	2 GL	3 GL	1 GL	2 GL	3 GL
1.00	1.000	1.000	1.000	1.000	1.000	1.000	1.000	1.000	1.000
.95	.957	.963	.967	.957	.962	.964	.956	.959	.959
.90	.914	.925	.932	.914	.922	.928	.912	.916	.918
.85	.870	.885	.897	.869	.882	.890	.867	.873	.876
.80	.826	.845	.860	.824	.840	.851	.821	.829	.833
.75	.780	.803	.821	.779	.798	.811	.775	.785	.789
.70	.734	.761	.781	.733	.754	.769	.728	.739	.744
.65	.688	.717	.740	.686	.710	.726	.680	.693	.698
.60	.640	.671	.697	.638	.664	.682	.632	.646	.651
.55	.591	.625	.652	.589	.617	.636	.584	.598	.604
.50	.542	.577	.605	.540	.568	.588	.534	.549	.555
.45	.492	.527	.556	.490	.519	.539	.484	.499	.505
.40	.441	.476	.505	.439	.468	.488	.433	.448	.454
.35	.389	.423	.452	.387	.415	.435	.382	.396	.402
.30	.337	.369	.396	.335	.361	.380	.330	.343	.348

20-13: Absorption factors for various wall absorptances, extinction coefficient-thickness products (E), and numbers of glazings.

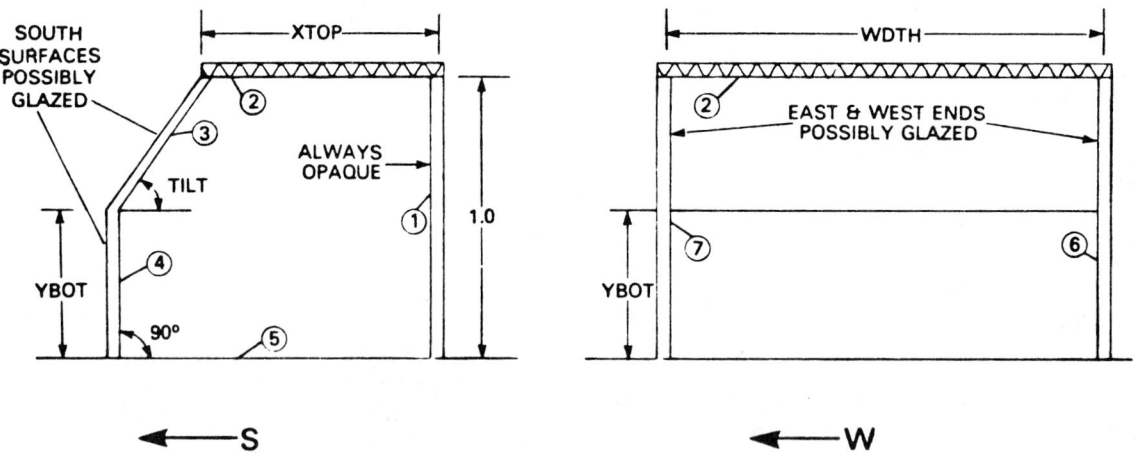

20-14: Surface types having different surface absorption factors. The circled numbers— ① through ⑦—refer to surface numbers.

Two different assumptions about internal surface absorptances are made:

1. All opaque surfaces have the same solar absorptance ($ALFSO = ALFSM$), which is varied from 0.3 to 1.0.
2. The primary mass surfaces have the same solar absorptance ($ALFSM$), which is varied; and the other opaque surfaces, if any, have a solar absorptance of $ALFSO = 0.3$.

Absorption factors for other combinations of absorptances may be approximated by calculating area-weighted averages.

In calculating the absorption factors, the following assumptions were made:

1. Transmission of solar radiation through glazed surfaces is isotropically diffuse.
2. Reflection of solar radiation from interior surfaces is isotropically diffuse.
3. Reflection of radiation from the interior of glazed surfaces is—for two layers of standard glass (index of refraction = 1.526)—about 24%. Radiation not reflected is assumed to pass through to the outside and be lost from the system.
4. Twenty percent of the transmitted radiation is absorbed by lightweight objects before striking any interior surface. Twenty percent of the radiation in each reflection is likewise absorbed by lightweight objects.

The tables at the end of this appendix give absorption factors for three different classes of geometries. Class I geometries, shown in figure 20–15, have a negligibly small bottom section (surface 4 in figure 20–14); surface 3 is glazed, and the tilt angle of this glazing is varied. Class II geometries, shown in figure 20–16, have a surface 4 height that is two-thirds of the north-wall height; surfaces 3 and 4 are glazed, and the tilt of surface 3 is varied. Class III geometries, shown in figure 20–17, have an opaque bottom section (surface 4); the length of glazing (Z) is fixed at two different values, and the glazing tilt is varied.

For all three classes, the factors are given as a function of north-south depth ($XTOP$) and east-west width ($WDTH$), for opaque end walls (surfaces 6 and 7) and glazed end walls, and for both surface absorptance assumptions. Absorption factors are given for each glazed surface of the particular geometry being considered. When the end walls are glazed, the factors are the same for surface 6 and surface 7 because of symmetry.

Caution should be used in applying the tables to cases with substantially lower absorptances than those in the reference designs. In such cases a high fraction of the total absorbed radiation goes directly to the air—an occurrence due to the assumed lightweight absorption fraction (20%). The S calculated in such a case will generally lead to an overprediction of the monthly SSF because solar radiation absorbed in the air is usually less effective than solar radiation absorbed on mass surfaces.

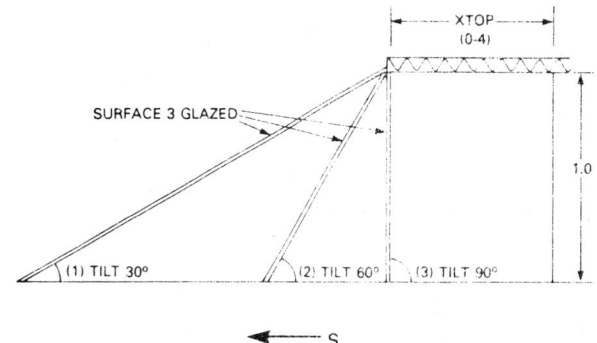

20-15: Class I geometries for internal absorption, $YBOT = 0$.

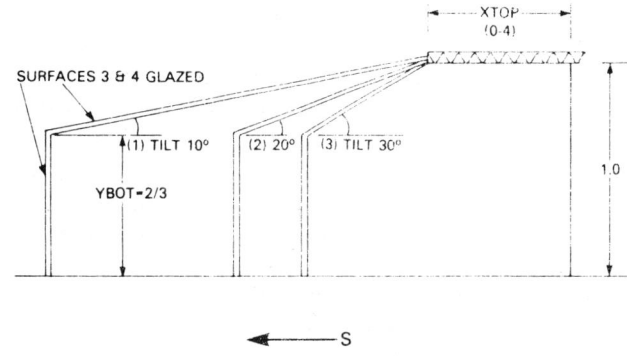

20-16: Class II geometries for internal absorption, $YBOT = \frac{2}{3}$.

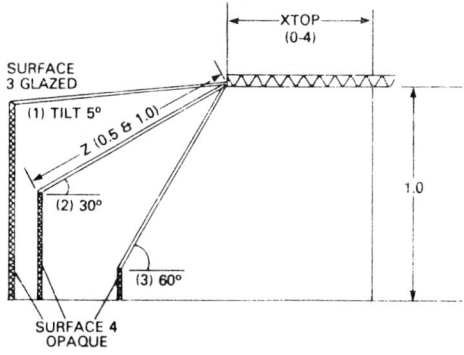

20-17: Class III geometries for internal absorption, opaque bottom section (surface 4).

Solar Radiation Approximations / 445

20.5 CALCULATION PROCEDURES

A step-by-step procedure follows for calculating S, the monthly solar radiation absorbed in the building per unit of projected area. A simple form of the procedure is given first, in which the glazing extinction coefficient–thickness product and the ground reflectance are assumed to have their reference values. As a preliminary step, divide the solar aperture into its distinct components—that is, into components with different orientations or numbers of glazings. Each of the following steps is performed monthly.

1. Determine the total monthly average solar radiation incident on a unit area of horizontal surface (Appendix 19: multiply entries by the number of days in the month).
2. For each component of the solar aperture,
   a. Determine the net glazing area of the component.
   b. Determine the monthly ratio of incident to horizontal surface radiation (figures 20–1 through 20–4).
   c. Determine the monthly ratio of transmitted to incident radiation (figures 20–5 through 20–10).
   d. Determine the monthly ratio of absorbed to transmitted radiation (figure 20–13 and tables at back of this appendix).
   e. Multiply all five of the above-determined numbers together. The product equals the total solar radiation absorbed due to the radiation incident upon each aperture component.
3. Add up each result of Step 2e. The sum equals the total radiation absorbed due to all of the aperture components combined.
4. Divide the result of Step 3 by the projected area. (The projected area is the principal net glazing area projected on a vertical plane. See the definition in Chapter M. The result is S, the monthly solar radiation absorbed in the building per unit of projected area.

If it is desired to account for an off-reference value of the glazing extinction coefficient–thickness product, then the ratios determined in Step 2c must be multiplied by correction factors drawn from figure 20–11 or from the equation for $\tau(E,N)$ in section 20.3.a.

If it is desired to account for off-reference values of the ground reflectance, then Steps 2c and 2e must be modified as follows:

2c. Determine the monthly ratio of transmitted to incident radiation in three steps:
   i. Determine the monthly ratio of transmitted to incident radiation for a ground reflectance of 0.3 (figures 20–5 through 20–10).
   ii. Multiply the result of Step i by the result of Step b. This product is the total monthly ratio of transmitted to horizontal surface radiation for the reference ground reflectance of 0.3.
   iii. Add to the product of Step ii the quantity $F_{CG}\tau_D (\varrho_G - 0.3)$, where F_{CG} is the view factor from the glazing plane to the ground (see the F_{CG} equation in section 20.3.b) τ_D is the diffuse transmittance from figure 20–12, and ϱ_G is the off-reference ground reflectance.
2d. Multiply together the results of Steps 1, 2a, 2c, and 2d. The product is the total monthly solar radiation absorbed due to the radiation incident on each aperture component.

The purpose of Step 4 is to convert the total solar radiation absorbed in the building to S, the total radiation absorbed per unit of projected area. The definition of the projected area as the primary glazing area parameter is an arbitrary convention chosen because most systems of interest are associated with vertical wall areas. It is possible, however, to use any other convention, if it is consistently applied throughout the analysis. Indeed, the vertical projection convention is inappropriate to some situations, especially in the case of a horizontal skylight that has no vertical projection.

ABSORPTION FACTORS - CLASS I
OPAQUE END WALLS - YBOT= .00
ALFSO=ALFSM

WDTH	TILT	XTOP	Z	\multicolumn{8}{c}{ABSORPTION FOR RADIATION THRU SURFACE 3 ALFSM}							
				0.3	0.4	0.5	0.6	0.7	0.8	0.9	1.0
1.0	30.	.0	2.00	.76	.81	.85	.89	.92	.95	.98	1.00
1.0	30.	.5	2.00	.81	.85	.88	.91	.94	.96	.98	1.00
1.0	30.	1.5	2.00	.86	.89	.92	.94	.96	.97	.99	1.00
1.0	30.	3.0	2.00	.90	.92	.94	.96	.97	.98	.99	1.00
1.0	60.	.0	1.15	.73	.78	.83	.87	.91	.94	.97	1.00
1.0	60.	.5	1.15	.83	.87	.90	.93	.95	.97	.98	1.00
1.0	60.	1.5	1.15	.90	.92	.94	.96	.97	.98	.99	1.00
1.0	60.	3.0	1.15	.94	.95	.97	.97	.98	.99	.99	1.00
1.0	90.	.0	1.00	.62	.68	.74	.79	.85	.90	.95	1.00
1.0	90.	.5	1.00	.81	.85	.88	.91	.94	.96	.98	1.00
1.0	90.	1.5	1.00	.90	.93	.95	.96	.97	.98	.99	1.00
1.0	90.	3.0	1.00	.94	.96	.97	.98	.99	.99	1.00	1.00
3.0	30.	.0	2.00	.71	.76	.81	.86	.90	.93	.97	1.00
3.0	30.	.5	2.00	.76	.80	.85	.88	.92	.95	.97	1.00
3.0	30.	1.5	2.00	.81	.85	.88	.91	.94	.96	.98	1.00
3.0	30.	3.0	2.00	.87	.89	.92	.94	.96	.97	.98	1.00
3.0	60.	.0	1.15	.70	.75	.80	.85	.89	.93	.97	1.00
3.0	60.	.5	1.15	.79	.83	.87	.90	.93	.96	.98	1.00
3.0	60.	1.5	1.15	.87	.90	.92	.94	.96	.97	.99	1.00
3.0	60.	3.0	1.15	.91	.93	.95	.96	.97	.98	.99	1.00
3.0	90.	.0	1.00	.62	.68	.74	.79	.85	.90	.95	1.00
3.0	90.	.5	1.00	.77	.81	.86	.89	.92	.95	.98	1.00
3.0	90.	1.5	1.00	.87	.90	.93	.95	.96	.98	.99	1.00
3.0	90.	3.0	1.00	.92	.94	.96	.97	.98	.99	.99	1.00
5.0	30.	.0	2.00	.70	.75	.80	.85	.89	.93	.97	1.00
5.0	30.	.5	2.00	.74	.79	.84	.87	.91	.94	.97	1.00
5.0	30.	1.5	2.00	.80	.84	.88	.91	.93	.96	.98	1.00
5.0	30.	3.0	2.00	.86	.89	.91	.93	.95	.97	.99	1.00
5.0	60.	.0	1.15	.69	.74	.80	.84	.89	.93	.96	1.00
5.0	60.	.5	1.15	.78	.82	.86	.90	.93	.95	.98	1.00
5.0	60.	1.5	1.15	.86	.89	.92	.94	.96	.98	.99	1.00
5.0	60.	3.0	1.15	.91	.93	.95	.96	.97	.98	.99	1.00
5.0	90.	.0	1.00	.62	.68	.74	.79	.85	.90	.95	1.00
5.0	90.	.5	1.00	.76	.81	.85	.89	.92	.95	.98	1.00
5.0	90.	1.5	1.00	.86	.89	.92	.94	.96	.98	.99	1.00
5.0	90.	3.0	1.00	.92	.94	.95	.97	.98	.99	.99	1.00

ABSORPTION FACTORS - CLASS I
GLAZED END WALLS - YBOT= .00
ALFSO=ALFSM

WDTH	TILT	XTOP	Z	\multicolumn{8}{c}{ABSORPTION FOR RADIATION THRU SURFACE 3 ALFSM}							
				0.3	0.4	0.5	0.6	0.7	0.8	0.9	1.0
1.0	30.	.0	2.00	.54	.58	.62	.66	.70	.74	.77	.80
1.0	30.	.5	2.00	.52	.56	.60	.64	.67	.70	.73	.76
1.0	30.	1.5	2.00	.52	.56	.59	.63	.66	.69	.72	.74
1.0	30.	3.0	2.00	.53	.56	.60	.63	.66	.69	.71	.74
1.0	60.	.0	1.15	.58	.63	.68	.72	.77	.80	.84	.88
1.0	60.	.5	1.15	.55	.59	.63	.66	.69	.72	.75	.77
1.0	60.	1.5	1.15	.52	.56	.59	.62	.64	.67	.69	.71
1.0	60.	3.0	1.15	.52	.55	.58	.61	.63	.66	.68	.70
1.0	90.	.0	1.00	.62	.68	.74	.79	.85	.90	.95	1.00
1.0	90.	.5	1.00	.57	.62	.65	.69	.72	.75	.78	.81
1.0	90.	1.5	1.00	.52	.56	.59	.62	.64	.67	.69	.71
1.0	90.	3.0	1.00	.51	.54	.57	.60	.62	.64	.66	.68
3.0	30.	.0	2.00	.62	.67	.72	.77	.81	.85	.89	.93
3.0	30.	.5	2.00	.63	.68	.73	.77	.81	.84	.88	.91
3.0	30.	1.5	2.00	.65	.70	.74	.77	.81	.84	.86	.89
3.0	30.	3.0	2.00	.67	.71	.75	.78	.81	.84	.86	.88
3.0	60.	.0	1.15	.64	.70	.75	.80	.84	.88	.92	.96
3.0	60.	.5	1.15	.67	.72	.76	.80	.83	.86	.89	.91
3.0	60.	1.5	1.15	.68	.72	.76	.79	.81	.84	.86	.88
3.0	60.	3.0	1.15	.69	.72	.75	.78	.80	.82	.84	.86
3.0	90.	.0	1.00	.62	.68	.74	.79	.85	.90	.95	1.00
3.0	90.	.5	1.00	.67	.72	.77	.81	.84	.87	.90	.93
3.0	90.	1.5	1.00	.69	.73	.76	.79	.82	.84	.86	.88
3.0	90.	3.0	1.00	.69	.72	.75	.78	.80	.82	.83	.85
5.0	30.	.0	2.00	.64	.70	.75	.79	.84	.88	.92	.96
5.0	30.	.5	2.00	.67	.72	.76	.80	.84	.88	.91	.94
5.0	30.	1.5	2.00	.70	.74	.78	.82	.85	.88	.90	.93
5.0	30.	3.0	2.00	.73	.77	.80	.83	.86	.88	.90	.93
5.0	60.	.0	1.15	.66	.71	.76	.81	.86	.90	.94	.97
5.0	60.	.5	1.15	.70	.75	.79	.83	.86	.89	.92	.95
5.0	60.	1.5	1.15	.74	.78	.81	.84	.86	.89	.91	.92
5.0	60.	3.0	1.15	.75	.79	.82	.84	.86	.88	.89	.91
5.0	90.	.0	1.00	.62	.68	.74	.79	.85	.90	.95	1.00
5.0	90.	.5	1.00	.70	.75	.79	.83	.87	.90	.93	.96
5.0	90.	1.5	1.00	.75	.78	.82	.84	.87	.89	.91	.92
5.0	90.	3.0	1.00	.76	.79	.82	.84	.86	.88	.89	.90

Solar Radiation Approximations / 447

ABSORPTION FACTORS - CLASS I
OPAQUE END WALLS - YBOT= .00
ALFSO= .3

ABSORPTION FOR RADIATION THRU SURFACE 3
ALFSM

WDTH	TILT	XTOP	Z	0.3	0.4	0.5	0.6	0.7	0.8	0.9	1.0
1.0	30.	.0	2.00	.76	.79	.83	.86	.89	.91	.94	.96
1.0	30.	.5	2.00	.81	.83	.86	.88	.90	.92	.94	.96
1.0	30.	1.5	2.00	.86	.88	.90	.91	.93	.94	.96	.97
1.0	30.	3.0	2.00	.90	.92	.93	.94	.95	.96	.97	.98
1.0	60.	.0	1.15	.73	.77	.81	.85	.88	.92	.95	.98
1.0	60.	.5	1.15	.83	.85	.87	.89	.91	.93	.95	.96
1.0	60.	1.5	1.15	.90	.91	.92	.93	.94	.95	.96	.97
1.0	60.	3.0	1.15	.94	.94	.95	.96	.96	.97	.98	.98
1.0	90.	.0	1.00	.62	.68	.74	.79	.85	.90	.95	1.00
1.0	90.	.5	1.00	.81	.83	.85	.87	.89	.91	.93	.95
1.0	90.	1.5	1.00	.90	.91	.92	.93	.94	.95	.96	.95
1.0	90.	3.0	1.00	.94	.95	.95	.96	.96	.97	.97	.97
3.0	30.	.0	2.00	.71	.76	.80	.84	.88	.92	.95	.98
3.0	30.	.5	2.00	.76	.80	.83	.87	.90	.93	.96	.98
3.0	30.	1.5	2.00	.81	.84	.87	.90	.92	.95	.97	.99
3.0	30.	3.0	2.00	.87	.89	.91	.93	.94	.96	.98	.99
3.0	60.	.0	1.15	.70	.75	.80	.84	.88	.92	.96	.99
3.0	60.	.5	1.15	.79	.82	.85	.88	.91	.94	.96	.98
3.0	60.	1.5	1.15	.87	.89	.91	.92	.94	.95	.97	.98
3.0	60.	3.0	1.15	.91	.93	.94	.95	.96	.97	.98	.99
3.0	90.	.0	1.00	.62	.68	.74	.79	.85	.90	.95	1.00
3.0	90.	.5	1.00	.77	.80	.83	.86	.89	.92	.94	.96
3.0	90.	1.5	1.00	.87	.89	.90	.92	.93	.94	.95	.96
3.0	90.	3.0	1.00	.92	.93	.94	.95	.95	.96	.97	.97
5.0	30.	.0	2.00	.70	.75	.80	.84	.88	.92	.96	.99
5.0	30.	.5	2.00	.74	.79	.83	.86	.90	.93	.96	.99
5.0	30.	1.5	2.00	.80	.84	.87	.89	.92	.94	.97	.99
5.0	30.	3.0	2.00	.86	.88	.90	.92	.94	.96	.98	.99
5.0	60.	.0	1.15	.69	.74	.79	.84	.88	.92	.96	.99
5.0	60.	.5	1.15	.78	.81	.85	.88	.91	.94	.96	.99
5.0	60.	1.5	1.15	.86	.88	.90	.92	.94	.96	.97	.99
5.0	60.	3.0	1.15	.91	.92	.94	.95	.96	.97	.98	.99
5.0	90.	.0	1.00	.62	.68	.74	.79	.85	.90	.95	1.00
5.0	90.	.5	1.00	.76	.79	.83	.86	.89	.92	.94	.96
5.0	90.	1.5	1.00	.86	.88	.90	.91	.93	.94	.95	.97
5.0	90.	3.0	1.00	.92	.93	.94	.95	.95	.96	.97	.97

GLAZED END WALLS - YBOT= .00
ALFSO=ALFSM

ABSORPTION FOR RADIATION THRU SURFACE 6
ALFSM

WDTH	TILT	XTOP	Z	0.3	0.4	0.5	0.6	0.7	0.8	0.9	1.0
1.0	30.	.0	2.00	.47	.50	.53	.56	.59	.62	.65	.67
1.0	30.	.5	2.00	.49	.52	.55	.58	.61	.64	.66	.68
1.0	30.	1.5	2.00	.51	.54	.57	.60	.63	.66	.68	.70
1.0	30.	3.0	2.00	.52	.55	.59	.61	.64	.67	.69	.71
1.0	60.	.0	1.15	.50	.54	.57	.60	.63	.66	.69	.72
1.0	60.	.5	1.15	.53	.56	.59	.62	.65	.68	.70	.72
1.0	60.	1.5	1.15	.54	.57	.60	.63	.66	.68	.71	.73
1.0	60.	3.0	1.15	.54	.57	.60	.63	.66	.68	.71	.73
1.0	90.	.0	1.00	.47	.51	.54	.57	.61	.64	.67	.70
1.0	90.	.5	1.00	.53	.57	.60	.63	.66	.69	.71	.73
1.0	90.	1.5	1.00	.54	.58	.61	.64	.67	.69	.71	.74
1.0	90.	3.0	1.00	.54	.58	.61	.64	.66	.69	.71	.73
3.0	30.	.0	2.00	.52	.56	.59	.63	.66	.69	.72	.74
3.0	30.	.5	2.00	.57	.61	.64	.68	.70	.73	.75	.78
3.0	30.	1.5	2.00	.63	.67	.70	.73	.76	.78	.80	.82
3.0	30.	3.0	2.00	.67	.71	.74	.77	.79	.82	.83	.85
3.0	60.	.0	1.15	.53	.57	.60	.64	.67	.70	.73	.75
3.0	60.	.5	1.15	.60	.64	.68	.71	.73	.76	.78	.80
3.0	60.	1.5	1.15	.67	.71	.74	.77	.79	.81	.83	.85
3.0	60.	3.0	1.15	.70	.74	.77	.80	.82	.84	.86	.87
3.0	90.	.0	1.00	.47	.51	.54	.57	.61	.64	.67	.70
3.0	90.	.5	1.00	.59	.63	.66	.69	.72	.75	.77	.79
3.0	90.	1.5	1.00	.67	.71	.74	.77	.79	.81	.83	.85
3.0	90.	3.0	1.00	.71	.74	.77	.80	.82	.84	.86	.87
5.0	30.	.0	2.00	.53	.57	.60	.64	.67	.70	.73	.75
5.0	30.	.5	2.00	.59	.63	.66	.69	.72	.75	.77	.79
5.0	30.	1.5	2.00	.66	.70	.73	.76	.78	.81	.83	.84
5.0	30.	3.0	2.00	.71	.75	.78	.81	.83	.85	.86	.88
5.0	60.	.0	1.15	.53	.57	.60	.64	.67	.70	.73	.76
5.0	60.	.5	1.15	.62	.66	.69	.72	.75	.77	.79	.81
5.0	60.	1.5	1.15	.70	.74	.77	.79	.81	.83	.85	.86
5.0	60.	3.0	1.15	.75	.78	.81	.83	.85	.87	.88	.90
5.0	90.	.0	1.00	.47	.51	.54	.57	.61	.64	.67	.70
5.0	90.	.5	1.00	.60	.64	.67	.70	.73	.76	.78	.80
5.0	90.	1.5	1.00	.70	.74	.77	.79	.81	.83	.85	.86
5.0	90.	3.0	1.00	.76	.79	.81	.84	.86	.87	.89	.90

ABSORPTION FACTORS - CLASS I
GLAZED END WALLS - YBOT= .00
ALFSO= .3

WDTH	TILT	XTOP	Z	0.3	0.4	0.5	0.6	0.7	0.8	0.9	1.0
1.0	30.	.0	2.00	.54	.58	.62	.66	.70	.74	.77	.80
1.0	30.	.5	2.00	.52	.56	.60	.63	.67	.70	.73	.76
1.0	30.	1.5	2.00	.52	.56	.59	.62	.65	.68	.71	.74
1.0	30.	3.0	2.00	.53	.56	.59	.62	.65	.68	.70	.73
1.0	60.	.0	1.15	.58	.63	.68	.72	.77	.80	.84	.88
1.0	60.	.5	1.15	.55	.58	.62	.65	.68	.71	.73	.76
1.0	60.	1.5	1.15	.52	.55	.58	.60	.63	.65	.67	.69
1.0	60.	3.0	1.15	.52	.54	.56	.59	.61	.63	.65	.67
1.0	90.	.0	1.00	.62	.68	.74	.79	.85	.90	.95	1.00
1.0	90.	.5	1.00	.57	.61	.64	.67	.70	.72	.75	.77
1.0	90.	1.5	1.00	.52	.55	.57	.59	.60	.62	.64	.65
1.0	90.	3.0	1.00	.51	.53	.54	.56	.58	.59	.61	.62
3.0	30.	.0	2.00	.62	.67	.72	.77	.81	.85	.89	.93
3.0	30.	.5	2.00	.63	.68	.72	.77	.80	.84	.87	.91
3.0	30.	1.5	2.00	.65	.69	.73	.77	.80	.83	.86	.89
3.0	30.	3.0	2.00	.67	.71	.74	.77	.80	.83	.85	.88
3.0	60.	.0	1.15	.64	.70	.75	.80	.84	.88	.92	.96
3.0	60.	.5	1.15	.67	.71	.75	.79	.82	.85	.88	.91
3.0	60.	1.5	1.15	.68	.71	.74	.77	.80	.82	.84	.86
3.0	60.	3.0	1.15	.69	.71	.74	.76	.78	.80	.82	.84
3.0	90.	.0	1.00	.62	.68	.74	.79	.85	.90	.95	1.00
3.0	90.	.5	1.00	.67	.71	.75	.78	.82	.85	.88	.90
3.0	90.	1.5	1.00	.69	.72	.74	.76	.78	.80	.82	.83
3.0	90.	3.0	1.00	.69	.71	.73	.75	.76	.78	.79	.80
5.0	30.	.0	2.00	.64	.70	.75	.79	.84	.88	.92	.96
5.0	30.	.5	2.00	.67	.71	.76	.80	.84	.88	.91	.94
5.0	30.	1.5	2.00	.70	.74	.78	.81	.84	.87	.90	.93
5.0	30.	3.0	2.00	.73	.76	.79	.82	.85	.88	.90	.92
5.0	60.	.0	1.15	.66	.71	.76	.81	.86	.90	.94	.97
5.0	60.	.5	1.15	.70	.75	.79	.82	.85	.89	.91	.94
5.0	60.	1.5	1.15	.74	.77	.80	.82	.85	.87	.89	.91
5.0	60.	3.0	1.15	.75	.78	.80	.82	.84	.86	.88	.89
5.0	90.	.0	1.00	.62	.68	.74	.79	.85	.90	.95	1.00
5.0	90.	.5	1.00	.70	.74	.78	.81	.84	.88	.90	.93
5.0	90.	1.5	1.00	.75	.77	.79	.82	.84	.85	.87	.89
5.0	90.	3.0	1.00	.76	.78	.80	.81	.83	.84	.85	.87

ABSORPTION FACTORS - CLASS I
GLAZED END WALLS - YBOT= .00
ALFSO= .3

ABSORPTION FOR RADIATION THRU SURFACE 6

WDTH	TILT	XTOP	Z	0.3	0.4	0.5	0.6	0.7	0.8	0.9	1.0
1.0	30.	.0	2.00	.47	.50	.53	.56	.59	.62	.65	.67
1.0	30.	.5	2.00	.49	.52	.55	.57	.60	.62	.65	.67
1.0	30.	1.5	2.00	.51	.53	.56	.58	.61	.63	.65	.67
1.0	30.	3.0	2.00	.52	.54	.56	.59	.61	.63	.65	.66
1.0	60.	.0	1.15	.50	.54	.57	.60	.63	.66	.69	.72
1.0	60.	.5	1.15	.53	.56	.58	.61	.63	.66	.68	.70
1.0	60.	1.5	1.15	.54	.56	.58	.60	.63	.65	.67	.68
1.0	60.	3.0	1.15	.54	.56	.58	.60	.62	.64	.66	.67
1.0	90.	.0	1.00	.47	.51	.54	.57	.61	.64	.67	.70
1.0	90.	.5	1.00	.53	.56	.59	.61	.64	.66	.68	.70
1.0	90.	1.5	1.00	.54	.57	.59	.61	.63	.65	.66	.68
1.0	90.	3.0	1.00	.54	.56	.58	.60	.62	.64	.65	.67
3.0	30.	.0	2.00	.52	.56	.59	.63	.66	.69	.72	.74
3.0	30.	.5	2.00	.57	.61	.64	.67	.70	.72	.75	.77
3.0	30.	1.5	2.00	.63	.66	.69	.71	.74	.76	.78	.80
3.0	30.	3.0	2.00	.67	.70	.72	.74	.77	.79	.81	.82
3.0	60.	.0	1.15	.53	.57	.60	.64	.67	.70	.73	.75
3.0	60.	.5	1.15	.60	.64	.67	.69	.72	.74	.76	.78
3.0	60.	1.5	1.15	.67	.69	.72	.74	.76	.78	.80	.82
3.0	60.	3.0	1.15	.70	.73	.75	.77	.79	.80	.82	.84
3.0	90.	.0	1.00	.47	.51	.54	.57	.61	.64	.67	.70
3.0	90.	.5	1.00	.59	.62	.65	.67	.70	.72	.74	.76
3.0	90.	1.5	1.00	.67	.69	.72	.74	.76	.77	.79	.81
3.0	90.	3.0	1.00	.71	.73	.75	.77	.78	.80	.82	.83
5.0	30.	.0	2.00	.53	.57	.60	.64	.67	.70	.73	.75
5.0	30.	.5	2.00	.59	.62	.66	.69	.71	.74	.76	.79
5.0	30.	1.5	2.00	.66	.69	.72	.75	.77	.79	.81	.83
5.0	30.	3.0	2.00	.71	.74	.76	.79	.81	.83	.84	.86
5.0	60.	.0	1.15	.53	.57	.61	.64	.67	.70	.73	.76
5.0	60.	.5	1.15	.62	.65	.68	.71	.73	.76	.78	.80
5.0	60.	1.5	1.15	.70	.73	.75	.77	.79	.81	.83	.84
5.0	60.	3.0	1.15	.75	.77	.79	.81	.83	.84	.86	.87
5.0	90.	.0	1.00	.47	.51	.54	.57	.61	.64	.67	.70
5.0	90.	.5	1.00	.60	.63	.66	.68	.71	.73	.75	.77
5.0	90.	1.5	1.00	.70	.72	.75	.77	.78	.80	.81	.83
5.0	90.	3.0	1.00	.76	.78	.79	.81	.82	.84	.85	.86

ABSORPTION FACTORS - CLASS II
OPAQUE END WALLS - YBOT= .67
ALFSO=ALFSM

ABSORPTION FOR RADIATION THRU SURFACE 4

WDTH	TILT	XTOP	Z	0.3	0.4	0.5	0.6	0.7	0.8	0.9	1.0
1.0	10.	.0	1.92	.69	.72	.76	.78	.81	.83	.85	.86
1.0	10.	.5	1.92	.72	.75	.78	.80	.82	.83	.85	.86
1.0	10.	1.5	1.92	.75	.78	.80	.82	.83	.84	.85	.86
1.0	10.	3.0	1.92	.79	.80	.82	.83	.84	.85	.86	.86
1.0	20.	.0	.97	.71	.75	.78	.81	.84	.87	.89	.91
1.0	20.	.5	.97	.76	.79	.82	.84	.86	.88	.89	.91
1.0	20.	1.5	.97	.81	.83	.85	.87	.88	.89	.90	.91
1.0	20.	3.0	.97	.84	.86	.87	.88	.89	.90	.90	.91
1.0	30.	.0	.67	.71	.76	.80	.83	.86	.89	.92	.94
1.0	30.	.5	.67	.79	.82	.85	.87	.89	.91	.93	.94
1.0	30.	1.5	.67	.85	.87	.89	.90	.91	.92	.93	.94
1.0	30.	3.0	.67	.88	.90	.91	.92	.92	.93	.93	.94
3.0	10.	.0	1.92	.61	.65	.69	.72	.75	.77	.80	.82
3.0	10.	.5	1.92	.64	.68	.71	.74	.76	.78	.80	.82
3.0	10.	1.5	1.92	.68	.71	.74	.76	.78	.79	.81	.82
3.0	10.	3.0	1.92	.72	.74	.76	.78	.79	.80	.81	.82
3.0	20.	.0	.97	.64	.69	.73	.77	.80	.83	.86	.88
3.0	20.	.5	.97	.70	.74	.77	.80	.82	.84	.86	.88
3.0	20.	1.5	.97	.76	.79	.81	.83	.84	.86	.87	.88
3.0	20.	3.0	.97	.80	.82	.83	.85	.86	.87	.87	.88
3.0	30.	.0	.67	.66	.71	.75	.79	.83	.86	.90	.92
3.0	30.	.5	.67	.74	.78	.81	.84	.86	.89	.91	.92
3.0	30.	1.5	.67	.81	.83	.86	.87	.89	.90	.91	.92
3.0	30.	3.0	.67	.85	.87	.88	.89	.90	.91	.92	.92
5.0	10.	.0	1.92	.59	.63	.67	.70	.73	.76	.79	.81
5.0	10.	.5	1.92	.62	.66	.69	.72	.75	.77	.79	.81
5.0	10.	1.5	1.92	.67	.70	.72	.74	.76	.78	.80	.81
5.0	10.	3.0	1.92	.70	.73	.75	.76	.78	.79	.80	.81
5.0	20.	.0	.97	.63	.67	.71	.75	.79	.82	.85	.88
5.0	20.	.5	.97	.69	.72	.76	.79	.81	.84	.86	.88
5.0	20.	1.5	.97	.75	.78	.80	.82	.84	.85	.86	.88
5.0	20.	3.0	.97	.79	.81	.82	.84	.85	.86	.87	.88
5.0	30.	.0	.67	.64	.70	.74	.78	.82	.86	.89	.92
5.0	30.	.5	.67	.72	.76	.80	.83	.86	.88	.90	.92
5.0	30.	1.5	.67	.79	.82	.85	.87	.88	.90	.91	.92
5.0	30.	3.0	.67	.84	.86	.87	.89	.90	.91	.91	.92

ABSORPTION FACTORS - CLASS II
OPAQUE END WALLS - YBOT= .67
ALFSO=ALFSM

ABSORPTION FOR RADIATION THRU SURFACE 3

WDTH	TILT	XTOP	Z	0.3	0.4	0.5	0.6	0.7	0.8	0.9	1.0
1.0	10.	.0	1.92	.75	.79	.83	.86	.89	.91	.93	.95
1.0	10.	.5	1.92	.78	.82	.85	.88	.90	.92	.94	.95
1.0	10.	1.5	1.92	.82	.85	.88	.90	.91	.93	.94	.95
1.0	10.	3.0	1.92	.86	.88	.90	.91	.92	.93	.94	.95
1.0	20.	.0	.97	.72	.77	.81	.84	.87	.89	.91	.94
1.0	20.	.5	.97	.78	.81	.84	.87	.89	.91	.92	.94
1.0	20.	1.5	.97	.83	.86	.88	.89	.90	.92	.93	.94
1.0	20.	3.0	.97	.87	.88	.90	.91	.92	.92	.93	.94
1.0	30.	.0	.67	.71	.75	.79	.83	.86	.89	.92	.94
1.0	30.	.5	.67	.79	.82	.85	.87	.89	.91	.93	.94
1.0	30.	1.5	.67	.84	.87	.88	.90	.91	.92	.93	.94
1.0	30.	3.0	.67	.88	.89	.91	.91	.92	.93	.93	.94
3.0	10.	.0	1.92	.68	.73	.77	.81	.85	.88	.91	.94
3.0	10.	.5	1.92	.71	.76	.80	.83	.86	.89	.91	.94
3.0	10.	1.5	1.92	.76	.80	.83	.85	.88	.90	.92	.94
3.0	10.	3.0	1.92	.81	.83	.86	.88	.89	.91	.92	.94
3.0	20.	.0	.97	.66	.71	.76	.79	.83	.86	.89	.92
3.0	20.	.5	.97	.72	.76	.80	.83	.85	.88	.90	.92
3.0	20.	1.5	.97	.78	.81	.84	.86	.87	.89	.90	.92
3.0	20.	3.0	.97	.82	.85	.86	.88	.89	.90	.91	.92
3.0	30.	.0	.67	.66	.71	.75	.79	.83	.86	.90	.92
3.0	30.	.5	.67	.73	.77	.81	.84	.86	.89	.91	.92
3.0	30.	1.5	.67	.80	.83	.85	.87	.89	.90	.91	.92
3.0	30.	3.0	.67	.84	.86	.88	.89	.90	.91	.92	.92
5.0	10.	.0	1.92	.66	.71	.75	.80	.83	.87	.90	.93
5.0	10.	.5	1.92	.69	.74	.78	.82	.85	.88	.91	.93
5.0	10.	1.5	1.92	.74	.78	.81	.84	.87	.89	.91	.93
5.0	10.	3.0	1.92	.79	.82	.85	.87	.88	.90	.92	.93
5.0	20.	.0	.97	.65	.70	.74	.78	.82	.85	.88	.91
5.0	20.	.5	.97	.71	.75	.78	.82	.84	.87	.89	.91
5.0	20.	1.5	.97	.77	.80	.82	.85	.87	.88	.90	.91
5.0	20.	3.0	.97	.81	.84	.85	.87	.88	.89	.90	.91
5.0	30.	.0	.67	.64	.69	.74	.78	.82	.86	.89	.92
5.0	30.	.5	.67	.72	.76	.80	.83	.85	.88	.90	.92
5.0	30.	1.5	.67	.79	.82	.84	.86	.88	.90	.91	.92
5.0	30.	3.0	.67	.83	.85	.87	.88	.89	.90	.91	.92

ABSORPTION FACTORS - CLASS II
GLAZED END WALLS - YBOT= .67
ALFSO=ALFSM

ABSORPTION FOR RADIATION THRU SURFACE 4
ALFSM

WDTH	TILT	XTOP	Z	0.3	0.4	0.5	0.6	0.7	0.8	0.9	1.0
1.0	10.	.0	1.92	.41	.43	.45	.48	.50	.52	.54	.56
1.0	10.	.5	1.92	.41	.43	.45	.48	.50	.52	.53	.55
1.0	10.	1.5	1.92	.41	.43	.46	.48	.50	.51	.53	.55
1.0	10.	3.0	1.92	.42	.44	.46	.48	.50	.51	.53	.55
1.0	20.	.0	.97	.46	.49	.52	.55	.57	.60	.62	.65
1.0	20.	.5	.97	.45	.48	.51	.53	.55	.57	.59	.61
1.0	20.	1.5	.97	.45	.47	.49	.52	.54	.56	.57	.59
1.0	20.	3.0	.97	.44	.47	.49	.51	.53	.55	.57	.58
1.0	30.	.0	.67	.50	.54	.57	.61	.64	.67	.70	.73
1.0	30.	.5	.67	.49	.52	.55	.57	.60	.62	.64	.66
1.0	30.	1.5	.67	.47	.50	.52	.55	.57	.59	.61	.62
1.0	30.	3.0	.67	.46	.49	.52	.54	.56	.58	.60	.61
3.0	10.	.0	1.92	.48	.52	.55	.58	.61	.64	.66	.69
3.0	10.	.5	1.92	.50	.53	.56	.58	.61	.63	.66	.68
3.0	10.	1.5	1.92	.51	.54	.57	.59	.61	.63	.65	.67
3.0	10.	3.0	1.92	.52	.55	.57	.59	.61	.63	.64	.66
3.0	20.	.0	.97	.54	.58	.62	.66	.69	.72	.75	.78
3.0	20.	.5	.97	.56	.60	.63	.66	.69	.71	.74	.76
3.0	20.	1.5	.97	.58	.61	.64	.66	.68	.70	.72	.73
3.0	20.	3.0	.97	.58	.61	.64	.66	.68	.69	.71	.72
3.0	30.	.0	.67	.58	.62	.67	.71	.75	.78	.82	.85
3.0	30.	.5	.67	.61	.65	.68	.71	.74	.77	.79	.81
3.0	30.	1.5	.67	.62	.65	.68	.71	.73	.75	.77	.78
3.0	30.	3.0	.67	.62	.65	.68	.70	.72	.74	.75	.76
5.0	10.	.0	1.92	.51	.55	.58	.61	.64	.67	.70	.73
5.0	10.	.5	1.92	.53	.56	.59	.62	.65	.67	.70	.72
5.0	10.	1.5	1.92	.55	.58	.61	.63	.65	.67	.69	.71
5.0	10.	3.0	1.92	.57	.59	.62	.64	.66	.67	.69	.70
5.0	20.	.0	.97	.56	.61	.65	.69	.72	.75	.79	.81
5.0	20.	.5	.97	.60	.64	.67	.70	.73	.75	.78	.80
5.0	20.	1.5	.97	.63	.66	.69	.71	.73	.75	.77	.78
5.0	20.	3.0	.97	.64	.67	.69	.71	.73	.74	.76	.77
5.0	30.	.0	.67	.60	.64	.69	.73	.77	.81	.84	.87
5.0	30.	.5	.67	.64	.68	.72	.75	.78	.81	.83	.85
5.0	30.	1.5	.67	.67	.71	.73	.76	.78	.80	.82	.83
5.0	30.	3.0	.67	.69	.71	.74	.76	.78	.79	.80	.82

ABSORPTION FACTORS - CLASS II
GLAZED END WALLS - YBOT= .67
ALFSO=ALFSM

ABSORPTION FOR RADIATION THRU SURFACE 3
ALFSM

WDTH	TILT	XTOP	Z	0.3	0.4	0.5	0.6	0.7	0.8	0.9	1.0
1.0	10.	.0	1.92	.45	.48	.51	.54	.57	.60	.62	.65
1.0	10.	.5	1.92	.44	.47	.50	.52	.55	.57	.60	.62
1.0	10.	1.5	1.92	.44	.47	.50	.52	.54	.57	.59	.61
1.0	10.	3.0	1.92	.45	.48	.50	.53	.55	.57	.59	.61
1.0	20.	.0	.97	.48	.51	.55	.58	.61	.64	.66	.69
1.0	20.	.5	.97	.46	.49	.51	.54	.56	.58	.61	.63
1.0	20.	1.5	.97	.45	.47	.50	.52	.54	.56	.58	.60
1.0	20.	3.0	.97	.45	.47	.50	.52	.54	.56	.58	.59
1.0	30.	.0	.67	.51	.55	.58	.62	.65	.68	.71	.74
1.0	30.	.5	.67	.48	.51	.53	.56	.58	.61	.63	.65
1.0	30.	1.5	.67	.46	.48	.51	.53	.55	.57	.59	.60
1.0	30.	3.0	.67	.45	.48	.50	.52	.54	.56	.58	.59
3.0	10.	.0	1.92	.55	.59	.64	.68	.71	.75	.78	.81
3.0	10.	.5	1.92	.56	.60	.64	.68	.71	.74	.77	.80
3.0	10.	1.5	1.92	.58	.62	.65	.68	.71	.74	.77	.79
3.0	10.	3.0	1.92	.60	.64	.67	.70	.72	.74	.77	.79
3.0	20.	.0	.97	.57	.61	.65	.69	.73	.76	.79	.82
3.0	20.	.5	.97	.58	.62	.66	.69	.72	.75	.77	.80
3.0	20.	1.5	.97	.60	.63	.66	.70	.72	.75	.76	.78
3.0	20.	3.0	.97	.61	.64	.67	.69	.71	.73	.75	.77
3.0	30.	.0	.67	.58	.63	.67	.71	.75	.79	.82	.85
3.0	30.	.5	.67	.60	.64	.68	.71	.74	.76	.79	.81
3.0	30.	1.5	.67	.63	.65	.68	.70	.72	.75	.76	.78
3.0	30.	3.0	.67	.66	.69	.72	.73	.77	.80	.82	.84
5.0	10.	.0	1.92	.58	.63	.67	.71	.75	.79	.82	.86
5.0	10.	.5	1.92	.60	.64	.68	.72	.75	.79	.82	.85
5.0	10.	1.5	1.92	.63	.67	.70	.73	.76	.79	.82	.84
5.0	10.	3.0	1.92	.66	.69	.72	.75	.77	.80	.82	.84
5.0	20.	.0	.97	.59	.63	.68	.72	.76	.79	.83	.86
5.0	20.	.5	.97	.62	.66	.70	.73	.76	.79	.81	.84
5.0	20.	1.5	.97	.65	.68	.71	.74	.76	.79	.81	.83
5.0	20.	3.0	.97	.67	.70	.73	.75	.77	.79	.80	.82
5.0	30.	.0	.67	.60	.65	.69	.73	.77	.81	.84	.88
5.0	30.	.5	.67	.64	.68	.71	.75	.78	.80	.83	.85
5.0	30.	1.5	.67	.67	.70	.73	.76	.78	.80	.82	.83
5.0	30.	3.0	.67	.68	.71	.74	.76	.78	.79	.81	.82

ABSORPTION FACTORS - CLASS II
OPAQUE END WALLS - YBOT= .67
ALFSO= .3

ABSORPTION FOR RADIATION THRU SURFACE 3
ALFSM

WDTH	TILT	XTOP	Z	0.3	0.4	0.5	0.6	0.7	0.8	0.9	1.0
1.0	10.	.0	1.92	.75	.77	.80	.82	.83	.85	.87	.89
1.0	10.	.5	1.92	.78	.80	.82	.83	.85	.86	.88	.89
1.0	10.	1.5	1.92	.82	.84	.85	.86	.87	.88	.90	.91
1.0	10.	3.0	1.92	.86	.87	.88	.89	.90	.90	.91	.92
1.0	20.	.0	.97	.72	.75	.78	.80	.82	.84	.86	.88
1.0	20.	.5	.97	.78	.80	.81	.83	.84	.86	.87	.88
1.0	20.	1.5	.97	.83	.84	.85	.86	.87	.88	.89	.90
1.0	20.	3.0	.97	.87	.87	.88	.89	.89	.90	.90	.91
1.0	30.	.0	.67	.71	.74	.77	.80	.82	.85	.87	.89
1.0	30.	.5	.67	.79	.80	.82	.84	.85	.86	.88	.89
1.0	30.	1.5	.67	.84	.85	.86	.87	.88	.89	.89	.90
1.0	30.	3.0	.67	.88	.89	.89	.90	.90	.91	.91	.91
3.0	10.	.0	1.92	.68	.72	.75	.79	.82	.85	.88	.91
3.0	10.	.5	1.92	.71	.75	.78	.81	.83	.86	.88	.91
3.0	10.	1.5	1.92	.76	.79	.81	.83	.86	.88	.89	.91
3.0	10.	3.0	1.92	.81	.83	.84	.86	.87	.89	.90	.92
3.0	20.	.0	.97	.66	.70	.74	.78	.81	.84	.87	.90
3.0	20.	.5	.97	.72	.75	.78	.81	.83	.85	.87	.89
3.0	20.	1.5	.97	.78	.80	.82	.84	.85	.87	.88	.89
3.0	20.	3.0	.97	.82	.84	.85	.86	.87	.88	.89	.90
3.0	30.	.0	.67	.66	.70	.74	.78	.81	.85	.88	.91
3.0	30.	.5	.67	.73	.76	.79	.82	.84	.86	.88	.90
3.0	30.	1.5	.67	.80	.82	.84	.85	.87	.88	.89	.90
3.0	30.	3.0	.67	.84	.85	.87	.88	.89	.89	.90	.91
5.0	10.	.0	1.92	.66	.69	.73	.77	.81	.85	.88	.91
5.0	10.	.5	1.92	.69	.74	.77	.80	.83	.86	.89	.91
5.0	10.	1.5	1.92	.74	.79	.81	.83	.85	.87	.90	.92
5.0	10.	3.0	1.92	.79	.81	.83	.85	.87	.89	.90	.92
5.0	20.	.0	.97	.65	.69	.73	.77	.81	.84	.87	.90
5.0	20.	.5	.97	.71	.74	.77	.80	.83	.85	.88	.90
5.0	20.	1.5	.97	.77	.79	.81	.83	.85	.87	.88	.90
5.0	20.	3.0	.97	.81	.83	.84	.86	.87	.89	.89	.90
5.0	30.	.0	.67	.64	.69	.73	.77	.81	.85	.88	.91
5.0	30.	.5	.67	.72	.75	.78	.81	.84	.86	.89	.91
5.0	30.	1.5	.67	.79	.81	.83	.85	.86	.88	.89	.91
5.0	30.	3.0	.67	.83	.85	.86	.87	.88	.89	.90	.91

ABSORPTION FACTORS - CLASS II
GLAZED END WALLS - YBOT= .67
ALFSO=ALFSM

ABSORPTION FOR RADIATION THRU SURFACE 6
ALFSM

WDTH	TILT	XTOP	Z	0.3	0.4	0.5	0.6	0.7	0.8	0.9	1.0
1.0	10.	.0	1.92	.42	.45	.47	.50	.52	.55	.57	.59
1.0	10.	.5	1.92	.45	.47	.50	.52	.55	.57	.59	.61
1.0	10.	1.5	1.92	.47	.50	.53	.56	.58	.60	.63	.65
1.0	10.	3.0	1.92	.49	.53	.55	.58	.61	.63	.65	.67
1.0	20.	.0	.97	.45	.48	.51	.53	.56	.58	.61	.63
1.0	20.	.5	.97	.48	.51	.54	.56	.59	.61	.63	.65
1.0	20.	1.5	.97	.50	.54	.57	.59	.62	.64	.66	.68
1.0	20.	3.0	.97	.52	.55	.58	.61	.63	.66	.68	.70
1.0	30.	.0	.67	.46	.49	.52	.55	.58	.60	.63	.65
1.0	30.	.5	.67	.50	.53	.56	.59	.61	.64	.66	.68
1.0	30.	1.5	.67	.52	.55	.58	.61	.64	.66	.68	.70
1.0	30.	3.0	.67	.53	.56	.59	.62	.65	.67	.69	.71
3.0	10.	.0	1.92	.48	.51	.54	.57	.60	.63	.65	.68
3.0	10.	.5	1.92	.52	.56	.59	.62	.64	.67	.69	.71
3.0	10.	1.5	1.92	.58	.62	.65	.68	.70	.73	.75	.77
3.0	10.	3.0	1.92	.63	.67	.70	.73	.75	.77	.79	.81
3.0	20.	.0	.97	.49	.53	.56	.59	.62	.64	.67	.69
3.0	20.	.5	.97	.56	.59	.62	.65	.68	.70	.72	.74
3.0	20.	1.5	.97	.63	.66	.69	.72	.75	.77	.79	.80
3.0	20.	3.0	.97	.67	.71	.74	.76	.79	.81	.82	.84
3.0	30.	.0	.67	.50	.53	.56	.59	.62	.65	.67	.70
3.0	30.	.5	.67	.57	.61	.64	.67	.70	.72	.74	.76
3.0	30.	1.5	.67	.65	.68	.71	.74	.76	.78	.80	.82
3.0	30.	3.0	.67	.69	.72	.75	.78	.80	.82	.84	.85
5.0	10.	.0	1.92	.49	.52	.56	.59	.62	.65	.67	.69
5.0	10.	.5	1.92	.54	.58	.61	.64	.66	.69	.71	.73
5.0	10.	1.5	1.92	.62	.65	.68	.71	.73	.75	.77	.79
5.0	10.	3.0	1.92	.67	.71	.74	.76	.78	.80	.82	.84
5.0	20.	.0	.97	.50	.54	.57	.60	.63	.66	.68	.70
5.0	20.	.5	.97	.58	.61	.64	.67	.69	.72	.74	.76
5.0	20.	1.5	.97	.66	.70	.72	.75	.77	.79	.81	.82
5.0	20.	3.0	.97	.72	.75	.78	.80	.82	.84	.85	.87
5.0	30.	.0	.67	.50	.54	.57	.60	.63	.66	.68	.70
5.0	30.	.5	.67	.59	.63	.66	.68	.71	.73	.75	.77
5.0	30.	1.5	.67	.68	.71	.74	.77	.79	.81	.82	.84
5.0	30.	3.0	.67	.74	.77	.79	.82	.84	.85	.87	.88

ABSORPTION FACTORS - CLASS II
OPAQUE END WALLS - YBOT= .67
ALFSO= .3

WDTH	TILT	XTOP	Z	\multicolumn{8}{c}{ABSORPTION FOR RADIATION THRU SURFACE 4}
				0.3	0.4	0.5	0.6 ALFSM	0.7	0.8	0.9	1.0
1.0	10.	.0	1.92	.69	.71	.72	.74	.76	.77	.79	.80
1.0	10.	.5	1.92	.72	.73	.75	.76	.77	.78	.80	.81
1.0	10.	1.5	1.92	.75	.77	.78	.79	.80	.80	.81	.82
1.0	10.	3.0	1.92	.79	.79	.80	.81	.81	.82	.83	.83
1.0	20.	.0	.97	.71	.73	.75	.77	.79	.81	.83	.85
1.0	20.	.5	.97	.76	.78	.79	.80	.82	.83	.84	.85
1.0	20.	1.5	.97	.81	.82	.83	.84	.85	.85	.86	.87
1.0	20.	3.0	.97	.84	.85	.86	.86	.87	.87	.88	.88
1.0	30.	.0	.67	.71	.74	.77	.80	.82	.85	.87	.89
1.0	30.	.5	.67	.79	.80	.82	.84	.85	.86	.88	.89
1.0	30.	1.5	.67	.85	.86	.87	.87	.88	.89	.89	.90
1.0	30.	3.0	.67	.88	.89	.89	.90	.90	.91	.91	.91
3.0	10.	.0	1.92	.61	.64	.67	.70	.72	.74	.77	.79
3.0	10.	.5	1.92	.64	.67	.69	.71	.73	.75	.77	.79
3.0	10.	1.5	1.92	.68	.70	.72	.74	.75	.77	.78	.80
3.0	10.	3.0	1.92	.72	.73	.75	.76	.77	.78	.79	.80
3.0	20.	.0	.97	.64	.68	.72	.75	.78	.81	.83	.86
3.0	20.	.5	.97	.70	.73	.75	.78	.80	.82	.84	.85
3.0	20.	1.5	.97	.76	.78	.79	.81	.82	.83	.85	.86
3.0	20.	3.0	.97	.80	.81	.82	.83	.84	.85	.86	.86
3.0	30.	.0	.67	.66	.70	.74	.78	.81	.85	.88	.91
3.0	30.	.5	.67	.74	.77	.79	.82	.84	.86	.88	.90
3.0	30.	1.5	.67	.81	.82	.84	.85	.87	.88	.89	.90
3.0	30.	3.0	.67	.85	.86	.87	.88	.88	.89	.90	.91
5.0	10.	.0	1.92	.59	.62	.66	.69	.71	.74	.77	.79
5.0	10.	.5	1.92	.62	.65	.68	.70	.73	.75	.77	.79
5.0	10.	1.5	1.92	.67	.69	.71	.73	.75	.76	.78	.79
5.0	10.	3.0	1.92	.70	.72	.73	.75	.76	.77	.78	.80
5.0	20.	.0	.97	.63	.67	.71	.74	.77	.81	.83	.86
5.0	20.	.5	.97	.69	.72	.75	.77	.80	.82	.84	.86
5.0	20.	1.5	.97	.75	.77	.79	.80	.82	.83	.85	.86
5.0	20.	3.0	.97	.79	.80	.81	.82	.84	.84	.85	.86
5.0	30.	.0	.67	.64	.69	.73	.77	.81	.85	.88	.91
5.0	30.	.5	.67	.72	.76	.79	.81	.84	.86	.88	.90
5.0	30.	1.5	.67	.79	.81	.83	.85	.86	.88	.89	.90
5.0	30.	3.0	.67	.84	.85	.86	.87	.88	.89	.90	.91

ABSORPTION FACTORS - CLASS II
GLAZED END WALLS - YBOT= .67
ALFSO= .3

WDTH	TILT	XTOP	Z	\multicolumn{8}{c}{ABSORPTION FOR RADIATION THRU SURFACE 3}
				0.3	0.4	0.5	0.6 ALFSM	0.7	0.8	0.9	1.0
1.0	10.	.0	1.92	.45	.48	.51	.54	.57	.60	.62	.65
1.0	10.	.5	1.92	.44	.47	.50	.52	.55	.57	.59	.62
1.0	10.	1.5	1.92	.44	.47	.49	.52	.54	.56	.58	.60
1.0	10.	3.0	1.92	.45	.47	.50	.52	.54	.56	.58	.60
1.0	20.	.0	.97	.48	.51	.55	.58	.61	.64	.66	.69
1.0	20.	.5	.97	.46	.48	.51	.54	.56	.58	.60	.62
1.0	20.	1.5	.97	.45	.47	.49	.51	.54	.56	.57	.59
1.0	20.	3.0	.97	.45	.47	.49	.51	.53	.55	.57	.58
1.0	30.	.0	.67	.51	.55	.58	.62	.65	.68	.71	.74
1.0	30.	.5	.67	.48	.50	.53	.55	.58	.60	.62	.64
1.0	30.	1.5	.67	.46	.48	.50	.52	.54	.56	.58	.59
1.0	30.	3.0	.67	.45	.47	.49	.51	.53	.55	.56	.58
3.0	10.	.0	1.92	.55	.59	.64	.68	.71	.75	.78	.81
3.0	10.	.5	1.92	.56	.60	.64	.67	.71	.74	.77	.80
3.0	10.	1.5	1.92	.58	.62	.65	.68	.71	.74	.76	.79
3.0	10.	3.0	1.92	.60	.63	.66	.69	.71	.74	.76	.78
3.0	20.	.0	.97	.57	.61	.65	.69	.73	.76	.79	.82
3.0	20.	.5	.97	.58	.62	.65	.68	.71	.74	.77	.79
3.0	20.	1.5	.97	.60	.63	.66	.68	.71	.73	.75	.77
3.0	20.	3.0	.97	.61	.64	.66	.68	.71	.73	.74	.76
3.0	30.	.0	.67	.58	.63	.67	.71	.75	.79	.82	.85
3.0	30.	.5	.67	.60	.64	.67	.70	.73	.76	.78	.81
3.0	30.	1.5	.67	.61	.64	.67	.69	.72	.74	.76	.78
3.0	30.	3.0	.67	.62	.64	.67	.69	.71	.73	.74	.76
5.0	10.	.0	1.92	.58	.63	.67	.71	.75	.79	.82	.86
5.0	10.	.5	1.92	.60	.64	.68	.72	.75	.79	.82	.85
5.0	10.	1.5	1.92	.63	.66	.70	.73	.76	.79	.81	.84
5.0	10.	3.0	1.92	.66	.69	.72	.74	.77	.79	.82	.84
5.0	20.	.0	.97	.59	.63	.68	.72	.76	.79	.83	.86
5.0	20.	.5	.97	.62	.66	.69	.73	.76	.79	.82	.84
5.0	20.	1.5	.97	.65	.68	.71	.73	.76	.78	.81	.83
5.0	20.	3.0	.97	.67	.70	.72	.74	.76	.78	.80	.82
5.0	30.	.0	.67	.60	.65	.69	.73	.77	.81	.84	.88
5.0	30.	.5	.67	.64	.67	.71	.74	.77	.80	.83	.85
5.0	30.	1.5	.67	.67	.70	.72	.75	.77	.79	.81	.83
5.0	30.	3.0	.67	.68	.71	.73	.75	.77	.79	.80	.82

ABSORPTION FACTORS - CLASS II
GLAZED END WALLS - YBOT= .67
ALFSO= .3

ABSORPTION FOR RADIATION THRU SURFACE 6
ALFSM

WDTH	TILT	XTOP	Z	0.3	0.4	0.5	0.6	0.7	0.8	0.9	1.0
1.0	10.	.0	1.92	.42	.45	.47	.50	.52	.55	.57	.59
1.0	10.	.5	1.92	.45	.47	.49	.52	.54	.56	.58	.60
1.0	10.	1.5	1.92	.47	.50	.52	.54	.56	.58	.60	.62
1.0	10.	3.0	1.92	.49	.51	.54	.56	.58	.60	.61	.63
1.0	20.	.0	.97	.45	.48	.51	.53	.56	.58	.61	.63
1.0	20.	.5	.97	.48	.50	.53	.55	.58	.60	.62	.64
1.0	20.	1.5	.97	.50	.53	.55	.57	.59	.61	.63	.65
1.0	20.	3.0	.97	.52	.54	.56	.58	.60	.62	.63	.65
1.0	30.	.0	.67	.46	.49	.52	.55	.58	.60	.63	.65
1.0	30.	.5	.67	.50	.52	.55	.57	.60	.62	.64	.66
1.0	30.	1.5	.67	.52	.54	.56	.58	.60	.62	.64	.66
1.0	30.	3.0	.67	.53	.55	.57	.59	.61	.62	.64	.66
3.0	10.	.0	1.92	.48	.51	.54	.57	.60	.63	.65	.68
3.0	10.	.5	1.92	.52	.56	.58	.61	.64	.66	.68	.71
3.0	10.	1.5	1.92	.58	.61	.64	.66	.69	.71	.73	.75
3.0	10.	3.0	1.92	.63	.66	.68	.71	.73	.75	.77	.78
3.0	20.	.0	.97	.49	.53	.56	.59	.62	.64	.67	.69
3.0	20.	.5	.97	.56	.59	.62	.64	.67	.69	.71	.73
3.0	20.	1.5	.97	.63	.65	.68	.70	.72	.74	.76	.78
3.0	20.	3.0	.97	.67	.70	.72	.74	.76	.78	.79	.81
3.0	30.	.0	.67	.50	.53	.56	.59	.62	.65	.67	.70
3.0	30.	.5	.67	.57	.60	.63	.66	.68	.70	.73	.74
3.0	30.	1.5	.67	.65	.67	.70	.72	.74	.76	.78	.79
3.0	30.	3.0	.67	.69	.71	.73	.75	.77	.79	.80	.82
5.0	10.	.0	1.92	.49	.52	.56	.59	.62	.64	.67	.69
5.0	10.	.5	1.92	.54	.57	.60	.63	.66	.68	.70	.73
5.0	10.	1.5	1.92	.62	.64	.67	.69	.72	.74	.76	.78
5.0	10.	3.0	1.92	.67	.70	.72	.75	.77	.78	.80	.82
5.0	20.	.0	.97	.50	.54	.57	.60	.63	.65	.68	.70
5.0	20.	.5	.97	.58	.61	.63	.66	.68	.71	.73	.75
5.0	20.	1.5	.97	.66	.69	.71	.73	.75	.77	.79	.81
5.0	20.	3.0	.97	.72	.74	.76	.78	.80	.82	.83	.84
5.0	30.	.0	.67	.50	.54	.57	.60	.63	.66	.68	.70
5.0	30.	.5	.67	.59	.62	.65	.67	.70	.72	.74	.76
5.0	30.	1.5	.67	.68	.70	.73	.75	.77	.79	.80	.82
5.0	30.	3.0	.67	.74	.76	.78	.79	.81	.83	.84	.85

ABSORPTION FACTORS - CLASS II
GLAZED END WALLS - YBOT= .67
ALFSO= .3

ABSORPTION FOR RADIATION THRU SURFACE 4
ALFSM

WDTH	TILT	XTOP	Z	0.3	0.4	0.5	0.6	0.7	0.8	0.9	1.0
1.0	10.	.0	1.92	.41	.43	.45	.48	.50	.52	.54	.56
1.0	10.	.5	1.92	.41	.43	.45	.47	.49	.51	.53	.55
1.0	10.	1.5	1.92	.41	.43	.45	.47	.49	.51	.52	.54
1.0	10.	3.0	1.92	.42	.43	.45	.47	.49	.50	.52	.54
1.0	20.	.0	.97	.46	.49	.52	.55	.57	.60	.62	.65
1.0	20.	.5	.97	.45	.48	.50	.52	.55	.57	.59	.61
1.0	20.	1.5	.97	.45	.47	.49	.51	.52	.54	.56	.58
1.0	20.	3.0	.97	.44	.46	.48	.50	.52	.53	.55	.56
1.0	30.	.0	.67	.50	.54	.57	.61	.64	.67	.70	.73
1.0	30.	.5	.67	.49	.51	.54	.56	.59	.61	.63	.65
1.0	30.	1.5	.67	.47	.49	.51	.53	.55	.57	.59	.60
1.0	30.	3.0	.67	.46	.48	.50	.52	.54	.55	.57	.58
3.0	10.	.0	1.92	.48	.52	.55	.58	.61	.64	.66	.69
3.0	10.	.5	1.92	.50	.53	.55	.58	.61	.63	.65	.67
3.0	10.	1.5	1.92	.51	.54	.56	.58	.60	.62	.64	.66
3.0	10.	3.0	1.92	.52	.54	.56	.58	.60	.62	.64	.65
3.0	20.	.0	.97	.54	.58	.62	.66	.69	.72	.75	.78
3.0	20.	.5	.97	.56	.60	.63	.66	.68	.71	.73	.75
3.0	20.	1.5	.97	.58	.60	.63	.65	.67	.69	.71	.72
3.0	20.	3.0	.97	.58	.60	.62	.64	.66	.68	.69	.71
3.0	30.	.0	.67	.58	.62	.67	.71	.75	.78	.82	.85
3.0	30.	.5	.67	.61	.64	.67	.70	.73	.76	.78	.80
3.0	30.	1.5	.67	.62	.64	.67	.69	.71	.73	.75	.76
3.0	30.	3.0	.67	.62	.64	.66	.68	.70	.71	.73	.74
5.0	10.	.0	1.92	.51	.55	.58	.61	.64	.67	.70	.73
5.0	10.	.5	1.92	.53	.56	.59	.62	.65	.67	.69	.72
5.0	10.	1.5	1.92	.55	.58	.60	.63	.65	.67	.69	.70
5.0	10.	3.0	1.92	.57	.59	.61	.63	.65	.66	.68	.70
5.0	20.	.0	.97	.56	.60	.63	.67	.70	.73	.75	.78
5.0	20.	.5	.97	.60	.63	.66	.69	.72	.75	.77	.79
5.0	20.	1.5	.97	.63	.65	.68	.70	.72	.74	.76	.77
5.0	20.	3.0	.97	.64	.66	.68	.70	.71	.73	.74	.76
5.0	30.	.0	.67	.60	.64	.69	.73	.77	.81	.84	.87
5.0	30.	.5	.67	.64	.68	.71	.74	.77	.80	.82	.84
5.0	30.	1.5	.67	.67	.70	.72	.74	.76	.78	.80	.82
5.0	30.	3.0	.67	.69	.71	.72	.74	.76	.77	.79	.80

ABSORPTION FACTORS - CLASS III
OPAQUE BOTTOM SECTION, Z=1.0
OPAQUE END WALLS
ALFSO=ALFSM

WDTH	TILT	XTOP	YBOT	ABSORPTION FOR RADIATION THRU SURFACE 3 ALFSM							
				0.3	0.4	0.5	0.6	0.7	0.8	0.9	1.0
1.0	5.	.0	.913	.87	.90	.92	.94	.96	.98	.99	1.00
1.0	5.	.5	.913	.90	.92	.94	.96	.97	.98	.99	1.00
1.0	5.	1.5	.913	.93	.94	.96	.97	.98	.99	.99	1.00
1.0	5.	3.0	.913	.95	.96	.97	.98	.98	.99	1.00	1.00
1.0	30.	.0	.500	.83	.87	.90	.93	.95	.97	.98	1.00
1.0	30.	.5	.500	.88	.91	.93	.95	.97	.98	.99	1.00
1.0	30.	1.5	.500	.92	.94	.96	.96	.98	.99	.99	1.00
1.0	30.	3.0	.500	.95	.96	.97	.98	.99	.99	1.00	1.00
1.0	60.	.0	.135	.75	.80	.84	.88	.92	.95	.97	1.00
1.0	60.	.5	.135	.85	.88	.91	.94	.96	.97	.99	1.00
1.0	60.	1.5	.135	.91	.93	.95	.96	.98	.98	.99	1.00
1.0	60.	3.0	.135	.95	.96	.97	.98	.99	.99	1.00	1.00
3.0	5.	.0	.913	.83	.87	.90	.93	.95	.97	.99	1.00
3.0	5.	.5	.913	.86	.89	.92	.94	.95	.97	.99	1.00
3.0	5.	1.5	.913	.90	.92	.94	.95	.97	.98	.99	1.00
3.0	5.	3.0	.913	.92	.94	.95	.97	.98	.98	.99	1.00
3.0	30.	.0	.500	.79	.84	.87	.91	.93	.96	.98	1.00
3.0	30.	.5	.500	.84	.88	.91	.93	.95	.97	.99	1.00
3.0	30.	1.5	.500	.89	.92	.94	.95	.97	.98	.99	1.00
3.0	30.	3.0	.500	.93	.94	.96	.97	.98	.99	.99	1.00
3.0	60.	.0	.135	.72	.77	.82	.86	.90	.94	.97	1.00
3.0	60.	.5	.135	.81	.85	.89	.91	.94	.96	.98	1.00
3.0	60.	1.5	.135	.88	.91	.93	.95	.97	.98	.99	1.00
3.0	60.	3.0	.135	.92	.94	.96	.97	.98	.99	.99	1.00
5.0	5.	.0	.913	.82	.86	.89	.92	.95	.97	.98	1.00
5.0	5.	.5	.913	.85	.88	.91	.93	.95	.97	.99	1.00
5.0	5.	1.5	.913	.89	.91	.93	.95	.96	.98	.99	1.00
5.0	5.	3.0	.913	.92	.93	.95	.96	.97	.98	.99	1.00
5.0	30.	.0	.500	.78	.83	.87	.90	.93	.96	.98	1.00
5.0	30.	.5	.500	.83	.87	.90	.93	.95	.97	.99	1.00
5.0	30.	1.5	.500	.88	.91	.93	.95	.96	.98	.99	1.00
5.0	30.	3.0	.500	.92	.94	.95	.96	.98	.98	.99	1.00
5.0	60.	.0	.135	.71	.77	.81	.86	.90	.93	.97	1.00
5.0	60.	.5	.135	.80	.84	.88	.91	.94	.96	.98	1.00
5.0	60.	1.5	.135	.87	.90	.93	.95	.96	.98	.99	1.00
5.0	60.	3.0	.135	.92	.94	.95	.97	.98	.99	.99	1.00

ABSORPTION FACTORS - CLASS III
OPAQUE BOTTOM SECTION, Z=1.0
GLAZED END WALLS
ALFSO=ALFSM

WDTH	TILT	XTOP	YBOT	ABSORPTION FOR RADIATION THRU SURFACE 3 ALFSM							
				0.3	0.4	0.5	0.6	0.7	0.8	0.9	1.0
1.0	5.	.0	.913	.54	.58	.61	.64	.67	.70	.72	.74
1.0	5.	.5	.913	.52	.55	.58	.60	.63	.65	.67	.69
1.0	5.	1.5	.913	.50	.53	.56	.59	.61	.64	.66	.68
1.0	5.	3.0	.913	.50	.53	.56	.59	.61	.63	.65	.67
1.0	30.	.0	.500	.57	.61	.64	.68	.71	.74	.76	.79
1.0	30.	.5	.500	.53	.56	.59	.62	.65	.67	.69	.71
1.0	30.	1.5	.500	.51	.54	.57	.59	.62	.64	.66	.68
1.0	30.	3.0	.500	.50	.53	.56	.59	.61	.63	.65	.67
1.0	60.	.0	.135	.60	.65	.69	.73	.77	.81	.84	.87
1.0	60.	.5	.135	.55	.59	.62	.66	.68	.71	.74	.76
1.0	60.	1.5	.135	.52	.55	.58	.61	.63	.66	.68	.70
1.0	60.	3.0	.135	.51	.54	.57	.59	.62	.64	.66	.68
3.0	5.	.0	.913	.69	.73	.77	.80	.83	.86	.88	.90
3.0	5.	.5	.913	.68	.72	.76	.79	.81	.84	.86	.88
3.0	5.	1.5	.913	.68	.72	.75	.78	.80	.83	.85	.86
3.0	5.	3.0	.913	.69	.72	.75	.78	.80	.82	.84	.86
3.0	30.	.0	.500	.68	.73	.77	.81	.84	.87	.90	.92
3.0	30.	.5	.500	.69	.73	.76	.79	.82	.85	.87	.89
3.0	30.	1.5	.500	.69	.72	.75	.78	.81	.83	.85	.86
3.0	30.	3.0	.500	.69	.72	.75	.78	.80	.82	.84	.85
3.0	60.	.0	.135	.66	.72	.76	.81	.85	.89	.92	.96
3.0	60.	.5	.135	.69	.73	.77	.80	.83	.86	.88	.91
3.0	60.	1.5	.135	.69	.73	.76	.79	.81	.83	.85	.87
3.0	60.	3.0	.135	.69	.72	.75	.78	.80	.82	.83	.85
5.0	5.	.0	.913	.73	.77	.81	.84	.87	.90	.92	.94
5.0	5.	.5	.913	.74	.78	.81	.84	.86	.89	.91	.93
5.0	5.	1.5	.913	.75	.78	.81	.84	.86	.88	.90	.92
5.0	5.	3.0	.913	.76	.79	.82	.84	.86	.88	.90	.91
5.0	30.	.0	.500	.72	.76	.80	.84	.87	.90	.93	.95
5.0	30.	.5	.500	.73	.77	.81	.84	.87	.89	.91	.93
5.0	30.	1.5	.500	.75	.78	.81	.84	.86	.88	.90	.92
5.0	30.	3.0	.500	.76	.79	.82	.84	.86	.88	.89	.91
5.0	60.	.0	.135	.68	.73	.78	.82	.87	.90	.94	.97
5.0	60.	.5	.135	.72	.77	.81	.84	.87	.90	.92	.95
5.0	60.	1.5	.135	.75	.79	.82	.84	.87	.89	.90	.92
5.0	60.	3.0	.135	.76	.79	.82	.84	.86	.88	.89	.90

ABSORPTION FACTORS - CLASS III
OPAQUE BOTTOM SECTION, Z= .5
OPAQUE END WALLS
ALFSO=ALFSM

				ABSORPTION FOR RADIATION THRU SURFACE 3 ALFSM							
WDTH	TILT	XTOP	YBOT	0.3	0.4	0.5	0.6	0.7	0.8	0.9	1.0
1.0	5.	.0	.956	.90	.93	.95	.96	.97	.98	.99	1.00
1.0	5.	.5	.956	.93	.95	.96	.97	.98	.99	.99	1.00
1.0	5.	1.5	.956	.95	.96	.97	.98	.99	.99	1.00	1.00
1.0	5.	3.0	.956	.97	.97	.98	.99	.99	1.00	1.00	1.00
1.0	45.	.0	.646	.86	.89	.92	.94	.96	.97	.99	1.00
1.0	45.	.5	.646	.92	.94	.96	.97	.98	.99	.99	1.00
1.0	45.	1.5	.646	.96	.96	.97	.98	.99	1.00	1.00	1.00
1.0	45.	3.0	.646	.97	.98	.98	.99	.99	1.00	1.00	1.00
1.0	90.	.0	.500	.78	.83	.86	.89	.92	.95	.98	1.00
1.0	90.	.5	.500	.89	.92	.94	.95	.97	.98	.99	1.00
1.0	90.	1.5	.500	.95	.96	.97	.98	.98	.99	1.00	1.00
1.0	90.	3.0	.500	.97	.98	.98	.99	.99	1.00	1.00	1.00
3.0	5.	.0	.956	.88	.91	.93	.95	.95	.98	.99	1.00
3.0	5.	.5	.956	.91	.93	.95	.96	.97	.98	.99	1.00
3.0	5.	1.5	.956	.94	.95	.96	.97	.98	.99	.99	1.00
3.0	5.	3.0	.956	.95	.96	.97	.98	.99	.99	1.00	1.00
3.0	45.	.0	.646	.84	.88	.91	.93	.95	.97	.98	1.00
3.0	45.	.5	.646	.90	.92	.94	.96	.97	.98	.99	1.00
3.0	45.	1.5	.646	.94	.95	.96	.97	.98	.99	1.00	1.00
3.0	45.	3.0	.646	.96	.97	.97	.98	.99	.99	1.00	1.00
3.0	90.	.0	.500	.78	.83	.86	.89	.92	.95	.98	1.00
3.0	90.	.5	.500	.87	.90	.92	.94	.96	.97	.99	1.00
3.0	90.	1.5	.500	.93	.95	.96	.97	.98	.99	1.00	1.00
3.0	90.	3.0	.500	.96	.97	.97	.98	.99	.99	1.00	1.00
5.0	5.	.0	.956	.88	.91	.93	.95	.95	.98	.99	1.00
5.0	5.	.5	.956	.91	.93	.95	.96	.97	.98	.99	1.00
5.0	5.	1.5	.956	.93	.95	.96	.97	.98	.99	1.00	1.00
5.0	5.	3.0	.956	.95	.96	.97	.98	.98	.99	1.00	1.00
5.0	45.	.0	.646	.84	.87	.90	.93	.95	.97	.98	1.00
5.0	45.	.5	.646	.89	.92	.94	.96	.97	.98	.99	1.00
5.0	45.	1.5	.646	.93	.95	.96	.97	.98	.99	1.00	1.00
5.0	45.	3.0	.646	.96	.97	.97	.98	.98	.99	1.00	1.00
5.0	90.	.0	.500	.78	.83	.86	.89	.92	.95	.98	1.00
5.0	90.	.5	.500	.86	.89	.92	.94	.96	.97	.99	1.00
5.0	90.	1.5	.500	.93	.95	.96	.97	.98	.99	.99	1.00
5.0	90.	3.0	.500	.96	.97	.98	.98	.99	1.00	1.00	1.00

ABSORPTION FACTORS - CLASS III
OPAQUE BOTTOM SECTION, Z=1.0
GLAZED END WALLS
ALFSO=ALFSM

				ABSORPTION FOR RADIATION THRU SURFACE 6 ALFSM							
WDTH	TILT	XTOP	YBOT	0.3	0.4	0.5	0.6	0.7	0.8	0.9	1.0
1.0	5.	.0	.913	.54	.58	.61	.64	.67	.69	.72	.74
1.0	5.	.5	.913	.54	.58	.61	.64	.66	.69	.71	.73
1.0	5.	1.5	.913	.54	.58	.61	.64	.66	.69	.71	.73
1.0	5.	3.0	.913	.54	.58	.61	.64	.66	.69	.71	.73
1.0	30.	.0	.500	.54	.58	.61	.64	.67	.70	.72	.74
1.0	30.	.5	.500	.54	.58	.61	.64	.67	.69	.71	.73
1.0	30.	1.5	.500	.54	.58	.61	.64	.67	.69	.71	.73
1.0	30.	3.0	.500	.54	.58	.61	.64	.66	.69	.71	.73
1.0	60.	.0	.135	.53	.57	.60	.63	.66	.69	.72	.74
1.0	60.	.5	.135	.54	.58	.61	.64	.67	.69	.72	.74
1.0	60.	1.5	.135	.54	.58	.61	.64	.67	.69	.72	.74
1.0	60.	3.0	.135	.54	.58	.61	.64	.67	.69	.71	.73
3.0	5.	.0	.913	.64	.68	.71	.74	.76	.79	.81	.83
3.0	5.	.5	.913	.67	.71	.74	.77	.79	.81	.83	.85
3.0	5.	1.5	.913	.70	.74	.77	.80	.82	.84	.86	.87
3.0	5.	3.0	.913	.72	.76	.79	.81	.84	.86	.87	.89
3.0	30.	.0	.500	.62	.66	.69	.72	.75	.77	.79	.81
3.0	30.	.5	.500	.66	.69	.72	.75	.78	.80	.82	.83
3.0	30.	1.5	.500	.70	.73	.76	.79	.81	.83	.85	.87
3.0	30.	3.0	.500	.72	.75	.78	.81	.83	.85	.87	.88
3.0	60.	.0	.135	.57	.61	.64	.67	.70	.73	.75	.78
3.0	60.	.5	.135	.63	.67	.70	.73	.75	.78	.80	.82
3.0	60.	1.5	.135	.68	.72	.75	.78	.80	.82	.84	.86
3.0	60.	3.0	.135	.71	.75	.78	.80	.83	.85	.86	.88
5.0	5.	.0	.913	.66	.70	.73	.76	.78	.80	.82	.84
5.0	5.	.5	.913	.70	.74	.77	.79	.81	.83	.85	.86
5.0	5.	1.5	.913	.75	.78	.81	.83	.85	.87	.88	.90
5.0	5.	3.0	.913	.78	.81	.84	.86	.88	.89	.91	.92
5.0	30.	.0	.500	.63	.67	.70	.73	.76	.78	.80	.82
5.0	30.	.5	.500	.68	.72	.75	.77	.80	.82	.83	.85
5.0	30.	1.5	.500	.74	.77	.80	.82	.84	.86	.87	.89
5.0	30.	3.0	.500	.77	.80	.83	.85	.87	.89	.90	.91
5.0	60.	.0	.135	.57	.61	.65	.68	.71	.73	.76	.78
5.0	60.	.5	.135	.65	.68	.71	.74	.77	.79	.81	.83
5.0	60.	1.5	.135	.72	.75	.78	.81	.83	.85	.86	.88
5.0	60.	3.0	.135	.76	.80	.82	.84	.86	.88	.89	.90

ABSORPTION FACTORS - CLASS III
OPAQUE BOTTOM SECTION, Z= .5
GLAZED END WALLS
ALFSO=ALFSM

ABSORPTION FOR RADIATION THRU SURFACE 6
ALFSM

WDTH	TILT	XTOP	YBOT	0.3	0.4	0.5	0.6	0.7	0.8	0.9	1.0
1.0	5.	.0	.956	.64	.68	.71	.74	.77	.79	.81	.83
1.0	5.	.5	.956	.61	.65	.68	.71	.74	.76	.78	.80
1.0	5.	1.5	.956	.58	.62	.65	.68	.71	.73	.76	.78
1.0	5.	3.0	.956	.56	.60	.63	.66	.69	.72	.74	.76
1.0	45.	.0	.646	.66	.70	.73	.76	.78	.81	.83	.84
1.0	45.	.5	.646	.62	.66	.69	.72	.74	.77	.79	.81
1.0	45.	1.5	.646	.58	.62	.65	.68	.71	.73	.76	.78
1.0	45.	3.0	.646	.56	.60	.63	.66	.69	.71	.74	.76
1.0	90.	.0	.500	.70	.73	.76	.78	.80	.82	.83	.85
1.0	90.	.5	.500	.64	.68	.72	.74	.77	.79	.81	.83
1.0	90.	1.5	.500	.59	.63	.66	.69	.72	.75	.77	.79
1.0	90.	3.0	.500	.57	.61	.64	.67	.69	.72	.74	.76
3.0	5.	.0	.956	.73	.77	.79	.82	.84	.85	.87	.88
3.0	5.	.5	.956	.75	.79	.81	.84	.86	.87	.89	.90
3.0	5.	1.5	.956	.76	.80	.83	.85	.87	.89	.90	.92
3.0	5.	3.0	.956	.76	.80	.83	.85	.87	.89	.90	.92
3.0	45.	.0	.646	.73	.76	.79	.81	.83	.85	.86	.88
3.0	45.	.5	.646	.74	.78	.80	.83	.85	.86	.88	.89
3.0	45.	1.5	.646	.76	.79	.82	.84	.86	.88	.89	.91
3.0	45.	3.0	.646	.76	.79	.82	.84	.86	.88	.90	.91
3.0	90.	.0	.500	.70	.73	.76	.78	.80	.82	.83	.85
3.0	90.	.5	.500	.74	.78	.80	.83	.85	.86	.88	.89
3.0	90.	1.5	.500	.76	.79	.82	.84	.86	.88	.90	.91
3.0	90.	3.0	.500	.76	.79	.82	.85	.87	.88	.90	.91
5.0	5.	.0	.956	.75	.78	.81	.83	.85	.85	.86	.89
5.0	5.	.5	.956	.79	.82	.84	.86	.88	.86	.88	.90
5.0	5.	1.5	.956	.81	.84	.87	.89	.90	.92	.93	.94
5.0	5.	3.0	.956	.83	.85	.88	.90	.91	.93	.94	.95
5.0	45.	.0	.646	.74	.77	.80	.82	.84	.86	.87	.88
5.0	45.	.5	.646	.77	.80	.83	.85	.86	.88	.89	.90
5.0	45.	1.5	.646	.80	.83	.86	.87	.89	.90	.92	.93
5.0	45.	3.0	.646	.82	.85	.87	.89	.90	.92	.93	.94
5.0	90.	.0	.500	.70	.73	.76	.78	.80	.82	.83	.85
5.0	90.	.5	.500	.76	.79	.82	.84	.86	.87	.89	.90
5.0	90.	1.5	.500	.80	.83	.85	.87	.89	.90	.92	.93
5.0	90.	3.0	.500	.82	.85	.87	.89	.90	.92	.93	.94

ABSORPTION FACTORS - CLASS III
OPAQUE BOTTOM SECTION, Z= .5
GLAZED END WALLS
ALFSO=ALFSM

ABSORPTION FOR RADIATION THRU SURFACE 3
ALFSM

WDTH	TILT	XTOP	YBOT	0.3	0.4	0.5	0.6	0.7	0.8	0.9	1.0
1.0	5.	.0	.956	.63	.67	.70	.73	.75	.77	.79	.81
1.0	5.	.5	.956	.56	.60	.63	.65	.68	.70	.72	.74
1.0	5.	1.5	.956	.53	.57	.60	.62	.65	.67	.69	.71
1.0	5.	3.0	.956	.52	.56	.59	.61	.64	.66	.68	.70
1.0	45.	.0	.646	.67	.71	.74	.77	.80	.83	.85	.87
1.0	45.	.5	.646	.57	.61	.64	.66	.68	.71	.72	.74
1.0	45.	1.5	.646	.52	.55	.58	.60	.62	.64	.66	.68
1.0	45.	3.0	.646	.50	.53	.56	.58	.60	.62	.64	.66
1.0	90.	.0	.500	.78	.83	.86	.89	.92	.95	.97	1.00
1.0	90.	.5	.500	.62	.66	.69	.72	.75	.77	.79	.81
1.0	90.	1.5	.500	.54	.58	.60	.63	.65	.67	.69	.71
1.0	90.	3.0	.500	.52	.55	.58	.60	.63	.65	.67	.68
3.0	5.	.0	.956	.77	.81	.83	.86	.88	.90	.91	.93
3.0	5.	.5	.956	.74	.78	.81	.83	.85	.87	.88	.90
3.0	5.	1.5	.956	.72	.76	.79	.81	.83	.85	.86	.88
3.0	5.	3.0	.956	.72	.75	.78	.80	.82	.84	.86	.87
3.0	45.	.0	.646	.77	.81	.83	.86	.88	.90	.91	.93
3.0	45.	.5	.646	.75	.78	.81	.83	.85	.87	.88	.90
3.0	45.	1.5	.646	.72	.76	.78	.80	.82	.83	.85	.86
3.0	45.	3.0	.646	.70	.74	.76	.78	.80	.82	.83	.84
3.0	90.	.0	.500	.78	.83	.86	.89	.92	.95	.98	1.00
3.0	90.	.5	.500	.76	.80	.83	.85	.88	.90	.92	.93
3.0	90.	1.5	.500	.73	.76	.79	.81	.83	.85	.87	.88
3.0	90.	3.0	.500	.71	.74	.77	.79	.81	.82	.84	.85
5.0	5.	.0	.956	.81	.84	.87	.89	.91	.93	.94	.96
5.0	5.	.5	.956	.80	.83	.86	.88	.90	.91	.92	.94
5.0	5.	1.5	.956	.79	.82	.85	.87	.88	.90	.91	.93
5.0	5.	3.0	.956	.79	.82	.84	.86	.88	.90	.91	.92
5.0	45.	.0	.646	.79	.83	.86	.89	.91	.94	.95	.97
5.0	45.	.5	.646	.80	.83	.86	.88	.90	.91	.93	.94
5.0	45.	1.5	.646	.79	.82	.84	.86	.88	.89	.90	.91
5.0	45.	3.0	.646	.78	.81	.83	.85	.87	.88	.89	.90
5.0	90.	.0	.500	.78	.83	.86	.89	.92	.95	.98	1.00
5.0	90.	.5	.500	.79	.83	.86	.88	.91	.93	.94	.96
5.0	90.	1.5	.500	.80	.83	.85	.87	.89	.90	.91	.92
5.0	90.	3.0	.500	.79	.82	.84	.86	.87	.88	.89	.90

ABSORPTION FACTORS - CLASS III
OPAQUE BOTTOM SECTION, Z=1.0
GLAZED END WALLS
ALFSO= .3

WDTH	TILT	XTOP	YBOT	\multicolumn{8}{c}{ABSORPTION FOR RADIATION THRU SURFACE 3 ALFSM}
				0.3	0.4	0.5	0.6	0.7	0.8	0.9	1.0
1.0	5.	.0	.913	.54	.57	.59	.62	.64	.66	.68	.70
1.0	5.	.5	.913	.52	.54	.56	.58	.60	.61	.63	.65
1.0	5.	1.5	.913	.50	.52	.54	.56	.58	.59	.61	.63
1.0	5.	3.0	.913	.50	.52	.54	.55	.57	.59	.60	.62
1.0	30.	.0	.500	.57	.60	.63	.67	.69	.72	.75	.77
1.0	30.	.5	.500	.53	.56	.58	.61	.63	.65	.67	.70
1.0	30.	1.5	.500	.51	.53	.55	.58	.60	.62	.64	.65
1.0	30.	3.0	.500	.50	.52	.54	.57	.59	.61	.62	.64
1.0	60.	.0	.135	.60	.64	.69	.73	.77	.81	.84	.87
1.0	60.	.5	.135	.55	.58	.61	.64	.67	.70	.72	.74
1.0	60.	1.5	.135	.52	.54	.57	.59	.61	.63	.65	.67
1.0	60.	3.0	.135	.51	.53	.55	.57	.59	.61	.63	.65
3.0	5.	.0	.913	.69	.72	.75	.78	.80	.82	.84	.87
3.0	5.	.5	.913	.68	.71	.74	.76	.78	.80	.82	.84
3.0	5.	1.5	.913	.68	.71	.73	.75	.77	.79	.81	.83
3.0	5.	3.0	.913	.69	.71	.73	.75	.77	.79	.80	.82
3.0	30.	.0	.500	.68	.72	.76	.80	.83	.86	.89	.91
3.0	30.	.5	.500	.69	.72	.75	.78	.81	.83	.86	.88
3.0	30.	1.5	.500	.69	.72	.74	.77	.79	.81	.83	.85
3.0	30.	3.0	.500	.69	.71	.74	.76	.78	.80	.82	.84
3.0	60.	.0	.135	.66	.71	.76	.81	.85	.89	.92	.96
3.0	60.	.5	.135	.69	.72	.76	.79	.82	.85	.88	.90
3.0	60.	1.5	.135	.69	.72	.75	.77	.79	.82	.84	.85
3.0	60.	3.0	.135	.69	.71	.74	.76	.78	.80	.81	.83
5.0	5.	.0	.913	.73	.76	.79	.81	.84	.87	.89	.91
5.0	5.	.5	.913	.74	.76	.79	.81	.84	.86	.88	.89
5.0	5.	1.5	.913	.75	.77	.79	.81	.83	.85	.87	.88
5.0	5.	3.0	.913	.76	.78	.80	.82	.83	.85	.86	.88
5.0	30.	.0	.500	.72	.76	.80	.83	.86	.89	.92	.94
5.0	30.	.5	.500	.73	.77	.80	.83	.85	.88	.90	.92
5.0	30.	1.5	.500	.75	.78	.80	.83	.85	.87	.89	.90
5.0	30.	3.0	.500	.76	.78	.81	.83	.84	.86	.88	.89
5.0	60.	.0	.135	.68	.73	.78	.82	.86	.90	.94	.97
5.0	60.	.5	.135	.72	.76	.80	.83	.85	.89	.91	.94
5.0	60.	1.5	.135	.75	.78	.80	.83	.85	.87	.89	.91
5.0	60.	3.0	.135	.76	.78	.80	.82	.84	.86	.87	.89

ABSORPTION FACTORS - CLASS III
OPAQUE BOTTOM SECTION, Z=1.0
OPAQUE END WALLS
ALFSO= .3

WDTH	TILT	XTOP	YBOT	\multicolumn{8}{c}{ABSORPTION FOR RADIATION THRU SURFACE 3 ALFSM}
				0.3	0.4	0.5	0.6	0.7	0.8	0.9	1.0
1.0	5.	.0	.913	.87	.88	.90	.91	.92	.93	.94	.95
1.0	5.	.5	.913	.90	.91	.92	.92	.93	.94	.95	.95
1.0	5.	1.5	.913	.93	.93	.94	.94	.95	.95	.96	.96
1.0	5.	3.0	.913	.95	.95	.96	.96	.96	.97	.97	.97
1.0	30.	.0	.500	.83	.86	.88	.90	.92	.93	.95	.96
1.0	30.	.5	.500	.88	.90	.91	.92	.93	.94	.95	.96
1.0	30.	1.5	.500	.92	.93	.94	.95	.95	.96	.97	.97
1.0	30.	3.0	.500	.95	.95	.96	.96	.97	.97	.98	.98
1.0	60.	.0	.135	.75	.79	.83	.86	.89	.92	.95	.98
1.0	60.	.5	.135	.85	.87	.89	.90	.92	.93	.95	.96
1.0	60.	1.5	.135	.91	.92	.93	.94	.95	.96	.96	.97
1.0	60.	3.0	.135	.95	.95	.96	.96	.97	.97	.98	.98
3.0	5.	.0	.913	.83	.85	.88	.90	.92	.93	.95	.96
3.0	5.	.5	.913	.86	.88	.90	.91	.93	.94	.95	.96
3.0	5.	1.5	.913	.90	.91	.92	.93	.94	.95	.96	.97
3.0	5.	3.0	.913	.92	.93	.94	.95	.95	.96	.97	.97
3.0	30.	.0	.500	.79	.83	.86	.89	.91	.93	.95	.98
3.0	30.	.5	.500	.84	.87	.89	.91	.93	.94	.95	.98
3.0	30.	1.5	.500	.89	.91	.92	.93	.95	.96	.97	.98
3.0	30.	3.0	.500	.93	.94	.95	.96	.96	.97	.98	.99
3.0	60.	.0	.135	.72	.77	.81	.85	.89	.93	.96	.99
3.0	60.	.5	.135	.81	.84	.87	.90	.92	.94	.96	.98
3.0	60.	1.5	.135	.88	.90	.92	.93	.95	.96	.97	.98
3.0	60.	3.0	.135	.92	.94	.95	.96	.96	.97	.98	.99
5.0	5.	.0	.913	.82	.85	.87	.90	.92	.94	.95	.97
5.0	5.	.5	.913	.85	.87	.89	.91	.93	.94	.96	.97
5.0	5.	1.5	.913	.89	.90	.92	.93	.94	.95	.96	.97
5.0	5.	3.0	.913	.92	.93	.94	.95	.95	.96	.97	.97
5.0	30.	.0	.500	.78	.82	.86	.89	.91	.94	.96	.99
5.0	30.	.5	.500	.83	.86	.89	.91	.93	.95	.97	.99
5.0	30.	1.5	.500	.88	.90	.92	.93	.95	.96	.97	.99
5.0	30.	3.0	.500	.92	.93	.94	.95	.96	.97	.98	.99
5.0	60.	.0	.135	.71	.76	.81	.85	.89	.93	.96	.99
5.0	60.	.5	.135	.80	.83	.87	.89	.92	.94	.97	.99
5.0	60.	1.5	.135	.87	.89	.91	.93	.95	.96	.97	.99
5.0	60.	3.0	.135	.92	.93	.94	.95	.96	.97	.98	.99

ABSORPTION FACTORS - CLASS III
OPAQUE BOTTOM SECTION, Z= .5
OPAQUE END WALLS
ALFSO= .3

ABSORPTION FOR RADIATION THRU SURFACE 3
ALFSM

WDTH	TILT	XTOP	YBOT	0.3	0.4	0.5	0.6	0.7	0.8	0.9	1.0
1.0	5.	.0	.956	.90	.91	.92	.93	.94	.95	.96	.97
1.0	5.	.5	.956	.93	.94	.94	.95	.95	.96	.96	.97
1.0	5.	1.5	.956	.95	.96	.96	.96	.97	.97	.97	.97
1.0	5.	3.0	.956	.97	.97	.97	.97	.97	.98	.98	.98
1.0	45.	.0	.646	.86	.88	.90	.92	.94	.96	.97	.99
1.0	45.	.5	.646	.92	.93	.94	.95	.96	.96	.97	.97
1.0	45.	1.5	.646	.96	.96	.96	.96	.97	.97	.98	.98
1.0	45.	3.0	.646	.97	.98	.98	.98	.98	.98	.99	.99
1.0	90.	.0	.500	.78	.82	.86	.89	.92	.95	.97	1.00
1.0	90.	.5	.500	.89	.90	.91	.92	.93	.94	.95	.96
1.0	90.	1.5	.500	.95	.95	.96	.96	.96	.96	.97	.97
1.0	90.	3.0	.500	.97	.97	.97	.98	.98	.98	.98	.98
3.0	5.	.0	.956	.88	.90	.92	.93	.94	.95	.96	.97
3.0	5.	.5	.956	.91	.92	.93	.94	.95	.96	.96	.97
3.0	5.	1.5	.956	.94	.94	.95	.96	.96	.97	.97	.98
3.0	5.	3.0	.956	.95	.96	.96	.97	.97	.97	.97	.98
3.0	45.	.0	.646	.84	.87	.90	.92	.94	.96	.98	.99
3.0	45.	.5	.646	.90	.92	.93	.94	.96	.97	.98	.99
3.0	45.	1.5	.646	.94	.95	.96	.96	.97	.98	.98	.99
3.0	45.	3.0	.646	.96	.97	.97	.97	.98	.98	.99	.99
3.0	90.	.0	.500	.78	.82	.86	.89	.92	.95	.97	1.00
3.0	90.	.5	.500	.87	.89	.90	.92	.93	.94	.95	.97
3.0	90.	1.5	.500	.93	.94	.95	.95	.96	.96	.97	.97
3.0	90.	3.0	.500	.96	.96	.97	.97	.97	.98	.98	.98
5.0	5.	.0	.956	.88	.90	.91	.93	.94	.95	.97	.98
5.0	5.	.5	.956	.91	.92	.93	.94	.95	.96	.97	.98
5.0	5.	1.5	.956	.93	.94	.95	.95	.96	.97	.98	.98
5.0	5.	3.0	.956	.95	.95	.96	.96	.97	.98	.98	.98
5.0	45.	.0	.646	.84	.87	.89	.92	.94	.96	.98	1.00
5.0	45.	.5	.646	.89	.91	.93	.94	.96	.97	.98	.99
5.0	45.	1.5	.646	.93	.94	.95	.96	.97	.98	.99	.99
5.0	45.	3.0	.646	.96	.96	.96	.97	.98	.98	.99	.99
5.0	90.	.0	.500	.78	.82	.86	.89	.92	.95	.97	1.00
5.0	90.	.5	.500	.86	.88	.90	.92	.93	.94	.96	.97
5.0	90.	1.5	.500	.93	.94	.95	.95	.96	.97	.97	.98
5.0	90.	3.0	.500	.96	.96	.97	.97	.97	.98	.98	.98

ABSORPTION FACTORS - CLASS III
OPAQUE BOTTOM SECTION, Z=1.0
GLAZED END WALLS
ALFSO= .3

ABSORPTION FOR RADIATION THRU SURFACE 6
ALFSM

WDTH	TILT	XTOP	YBOT	0.3	0.4	0.5	0.6	0.7	0.8	0.9	1.0
1.0	5.	.0	.913	.54	.57	.59	.61	.63	.66	.67	.69
1.0	5.	.5	.913	.54	.57	.59	.61	.63	.65	.66	.68
1.0	5.	1.5	.913	.54	.56	.58	.60	.62	.64	.66	.67
1.0	5.	3.0	.913	.54	.56	.58	.60	.62	.64	.65	.67
1.0	30.	.0	.500	.54	.57	.60	.63	.65	.67	.70	.72
1.0	30.	.5	.500	.54	.57	.60	.62	.64	.66	.68	.70
1.0	30.	1.5	.500	.54	.57	.59	.61	.63	.65	.67	.68
1.0	30.	3.0	.500	.54	.56	.58	.60	.62	.64	.66	.67
1.0	60.	.0	.135	.53	.57	.60	.63	.66	.69	.71	.74
1.0	60.	.5	.135	.54	.57	.60	.62	.65	.67	.69	.71
1.0	60.	1.5	.135	.54	.57	.59	.61	.63	.65	.67	.69
1.0	60.	3.0	.135	.54	.56	.58	.60	.62	.64	.66	.68
3.0	5.	.0	.913	.64	.67	.69	.71	.74	.76	.77	.79
3.0	5.	.5	.913	.67	.69	.72	.74	.76	.78	.79	.81
3.0	5.	1.5	.913	.70	.73	.75	.77	.78	.80	.82	.83
3.0	5.	3.0	.913	.72	.74	.76	.78	.80	.82	.83	.84
3.0	30.	.0	.500	.62	.65	.68	.70	.73	.75	.77	.79
3.0	30.	.5	.500	.66	.68	.71	.73	.75	.77	.79	.81
3.0	30.	1.5	.500	.70	.72	.74	.76	.78	.80	.82	.83
3.0	30.	3.0	.500	.72	.74	.76	.78	.80	.82	.83	.85
3.0	60.	.0	.135	.57	.60	.64	.67	.70	.72	.75	.77
3.0	60.	.5	.135	.63	.66	.69	.71	.74	.76	.78	.80
3.0	60.	1.5	.135	.68	.71	.73	.75	.77	.79	.81	.83
3.0	60.	3.0	.135	.71	.74	.76	.78	.79	.81	.83	.84
5.0	5.	.0	.913	.66	.69	.71	.73	.75	.77	.79	.81
5.0	5.	.5	.913	.70	.72	.75	.77	.78	.80	.82	.83
5.0	5.	1.5	.913	.75	.77	.79	.81	.82	.84	.85	.87
5.0	5.	3.0	.913	.78	.80	.82	.83	.85	.86	.87	.89
5.0	30.	.0	.500	.63	.66	.69	.72	.74	.76	.79	.80
5.0	30.	.5	.500	.68	.71	.73	.76	.78	.80	.81	.83
5.0	30.	1.5	.500	.74	.76	.78	.80	.82	.84	.85	.86
5.0	30.	3.0	.500	.77	.79	.81	.83	.84	.86	.87	.89
5.0	60.	.0	.135	.57	.61	.64	.67	.70	.73	.75	.78
5.0	60.	.5	.135	.65	.68	.70	.73	.75	.77	.79	.81
5.0	60.	1.5	.135	.72	.74	.77	.79	.81	.82	.84	.85
5.0	60.	3.0	.135	.76	.78	.80	.82	.84	.85	.86	.88

ABSORPTION FACTORS - CLASS III
OPAQUE BOTTOM SECTION, Z= .5
GLAZED END WALLS
ALFSO= .3

WDTH	TILT	XTOP	YBOT	\multicolumn{8}{c}{ABSORPTION FOR RADIATION THRU SURFACE 3 ALFSM}
				0.3	0.4	0.5	0.6	0.7	0.8	0.9	1.0
1.0	5.	.0	.956	.63	.65	.68	.70	.72	.73	.75	.77
1.0	5.	.5	.956	.56	.58	.60	.62	.64	.65	.67	.68
1.0	5.	1.5	.956	.53	.55	.57	.58	.60	.61	.63	.64
1.0	5.	3.0	.956	.52	.54	.56	.57	.59	.60	.62	.63
1.0	45.	.0	.646	.67	.70	.73	.76	.79	.82	.84	.86
1.0	45.	.5	.646	.57	.60	.62	.64	.66	.68	.70	.72
1.0	45.	1.5	.646	.52	.54	.56	.58	.59	.61	.63	.64
1.0	45.	3.0	.646	.50	.52	.54	.56	.57	.59	.60	.62
1.0	90.	.0	.500	.78	.82	.86	.89	.92	.95	.97	1.00
1.0	90.	.5	.500	.62	.65	.67	.69	.71	.73	.75	.76
1.0	90.	1.5	.500	.54	.56	.57	.59	.60	.62	.63	.64
1.0	90.	3.0	.500	.52	.53	.55	.56	.57	.59	.60	.61
3.0	5.	.0	.956	.77	.79	.82	.83	.85	.87	.88	.90
3.0	5.	.5	.956	.74	.76	.78	.80	.82	.83	.85	.86
3.0	5.	1.5	.956	.72	.74	.76	.78	.79	.81	.82	.83
3.0	5.	3.0	.956	.72	.74	.75	.77	.78	.80	.81	.82
3.0	45.	.0	.646	.77	.80	.83	.86	.88	.91	.93	.95
3.0	45.	.5	.646	.75	.77	.80	.82	.84	.86	.87	.89
3.0	45.	1.5	.646	.72	.74	.76	.78	.80	.81	.83	.84
3.0	45.	3.0	.646	.70	.72	.74	.76	.78	.79	.81	.82
3.0	90.	.0	.500	.78	.82	.86	.89	.92	.95	.97	1.00
3.0	90.	.5	.500	.76	.78	.81	.83	.84	.86	.88	.89
3.0	90.	1.5	.500	.73	.75	.77	.78	.80	.81	.82	.83
3.0	90.	3.0	.500	.71	.73	.74	.76	.77	.78	.79	.81
5.0	5.	.0	.956	.81	.83	.85	.87	.89	.90	.92	.93
5.0	5.	.5	.956	.80	.82	.84	.85	.87	.88	.89	.91
5.0	5.	1.5	.956	.79	.81	.83	.84	.85	.87	.88	.89
5.0	5.	3.0	.956	.79	.81	.82	.84	.85	.86	.87	.88
5.0	45.	.0	.646	.79	.83	.86	.88	.91	.93	.95	.97
5.0	45.	.5	.646	.80	.82	.84	.87	.88	.90	.92	.93
5.0	45.	1.5	.646	.79	.81	.83	.85	.86	.88	.89	.90
5.0	45.	3.0	.646	.78	.80	.82	.83	.85	.86	.87	.88
5.0	90.	.0	.500	.78	.82	.86	.89	.92	.95	.97	1.00
5.0	90.	.5	.500	.79	.82	.84	.86	.88	.89	.91	.92
5.0	90.	1.5	.500	.80	.81	.83	.84	.86	.87	.88	.89
5.0	90	3.0	.500	.79	.80	.82	.83	.84	.85	.86	.87

Appendix 21 *LCR* Tables

This appendix contains tabulated annual *SSF* versus *LCR* values for all 94 reference designs for 219 locations in the United States and Canada. In their normal application, the tables are entered with a value of the *LCR*, and an *SSF* value is then determined by interpolation. This is the annual calculation or *LCR* method. See section M.3 for more details and a step-by-step procedure.

There is a table for each city, arranged with United States locations first, in alphabetical order by state and then by city. The Canadian locations follow, in alphabetical order by province and then by city. The table entries are values of *LCR*—each value corresponding to the annual *SSF* listed at the head of the column. The system type is designated at the left of the row. When a dash is entered instead of a value, the indicated *SSF* cannot be achieved. The system type is designated first by the abbreviations:

    WW = Waterwall
    TW = Trombe wall
    DG = Direct gain
    SS = Sunspace

The system type is designated second by a reference design designation consisting of a letter and a number. Refer to figures 18-2 through 18-6 for definitions of the reference design designations. More details of the reference design characteristics are listed in figure 21-1. These are identical to the characteristics of figure 18-1, except that the waterwall and Trombe-wall absorptances, which were taken to be 1.0 for the *SLR* correlations, and taken to be 0.95 for the *LCR* tables here. An absorptance of 0.95 is typical of both flat back paint and selective surfaces. The corresponding absorption factors are from figure 20-13: 0.957, 0.962, and 0.964 for one, two, and three glazings, respectively. The absorption factors for the direct-gain systems are 0.948 and 0.976 for mass-area-to-glazing-area ratios of 3 and 6, respectively.

Caution should be used in applying the *LCR* tables to very large *LCR* values—say, above 100. In particular, the *SSF* values for some sunspaces and thermal storage walls without nighttime insulation are overpredicted for *LCR* values above about 100. The accuracy in such a case may be tested by looking up *SSF* for a comparable case *with* nighttime insulation. If *SSF* for the uninsulated case is greater, reduce it to the value for the system with nighttime insulation.

Masonry properties - direct gain and sunspace	
thermal conductivity (k)	
sunspace floor	0.5 Btu/hr/ft/°F
all other masonry	1.0 Btu/hr/ft/°F
density (ϱ)	150 lb/ft^3
specific heat (c)	0.2 Btu/lb/°F
Solar absorptances	
waterwall	0.95
masonry, Trombe wall	0.95
direct gain and sunspace	0.8
sunspace. water containers	0.9
lightweight common wall	0.7
other lightweight surfaces	0.3
Infrared emittances	
normal surface	0.9
selective surface	0.1
Glazing properties	
transmission characteristics	diffuse
orientation	due south
index of refraction	1.526
extinction coefficient	0.5 inch^{-1}
thickness of each pane	1/8 inch
air gap between panes	1/2 inch
Control range	
room temperature	65° to 75°F
sunspace temperature	45° to 95°F
internal heat generation	0
Thermocirculation vents (when used)	
vent area/projected area (sum of both upper and lower vents)	0.06
height between vents	8 feet
reverse flow	none
Nighttime insulation (when used)	
thermal resistance	R9
in place, solar time	5:30 P.M. to 7:30 A.M.
Solar radiation assumptions	
shading	none
ground diffuse reflectance	0.3

21-1: Reference design characteristics for *LCR* tables.

BIRMINGHAM, ALABAMA								2844 DD	
SSF =	.10	.20	.30	.40	.50	.60	.70	.80	.90
WW A1	991	111	51	29	19	13	9	6	4
WW A2	317	123	69	45	31	22	16	11	7
WW A3	265	124	75	50	35	25	18	13	9
WW A4	242	124	77	52	37	27	20	14	9
WW A5	230	125	80	55	40	29	22	16	10
WW A6	223	125	81	57	41	31	23	16	11
WW B1	257	91	49	30	20	14	10	7	4
WW B2	239	126	80	55	40	29	22	16	10
WW B3	278	154	100	70	51	38	28	20	14
WW B4	244	147	99	71	53	40	30	22	15
WW B5	225	138	95	69	51	39	29	22	15
WW C1	295	164	106	74	54	40	30	22	15
WW C2	255	148	98	70	51	38	29	21	14
WW C3	261	167	117	86	65	49	38	28	19
WW C4	235	151	105	78	58	45	34	25	17
TW A1	847	107	49	29	19	13	9	6	4
TW A2	327	112	61	38	26	19	13	9	6
TW A3	261	110	64	42	29	21	15	11	7
TW A4	221	104	63	42	30	22	16	11	7
TW B1	465	101	50	30	20	14	10	7	4
TW B2	262	101	57	37	25	18	13	9	6
TW B3	228	97	57	37	26	19	14	10	6
TW B4	209	90	52	34	24	17	13	9	6
TW C1	304	91	48	30	20	14	10	7	5
TW C2	230	87	49	31	22	15	11	8	5
TW C3	220	81	45	29	20	14	10	7	5
TW C4	233	73	39	25	17	12	9	6	4
TW D1	186	71	39	25	17	12	8	6	4
TW D2	257	120	73	49	34	25	18	13	9
TW D3	262	132	82	56	40	29	21	15	10
TW D4	251	138	89	62	45	34	25	18	12
TW D5	233	132	87	61	45	34	25	18	12
TW E1	303	153	95	65	47	34	25	18	12
TW E2	275	146	93	64	46	34	25	18	12
TW E3	288	170	114	82	60	45	34	25	17
TW E4	257	154	104	75	55	42	31	23	16
TW F1	375	92	46	28	19	13	9	6	4
TW F2	231	94	53	35	24	17	12	9	6
TW F3	191	90	54	36	26	18	13	10	6
TW F4	156	80	50	34	24	18	13	9	6
TW G1	231	79	43	27	18	13	9	7	4
TW G2	167	77	46	30	21	15	11	8	5
TW G3	140	70	43	29	21	15	11	8	5
TW G4	113	58	36	25	18	13	9	7	4
TW H1	149	64	37	24	17	12	9	6	4
TW H2	114	56	34	23	16	12	9	6	4
TW H3	95	47	29	20	14	10	8	5	4
TW H4	74	36	22	15	11	8	6	4	3
TW I1	143	59	33	21	14	10	7	5	3
TW I2	196	100	63	43	31	22	17	12	8
TW I3	207	110	70	48	35	26	19	14	9
TW I4	208	119	79	56	41	30	23	17	11
TW I5	201	119	80	57	42	31	24	17	12
TW J1	251	135	86	60	43	32	24	17	11
TW J2	226	126	82	58	42	31	23	17	11
TW J3	240	148	101	73	55	41	31	23	16
TW J4	221	136	94	68	51	38	29	21	15
DG A1	185	78	43	24	12	–	–	–	–
DG A2	206	92	55	37	25	17	11	6	–
DG A3	252	116	72	50	36	27	20	13	7
DG B1	185	80	47	30	20	12	–	–	–
DG B2	210	96	58	40	29	21	15	10	4
DG B3	255	120	74	52	39	30	23	16	10
DG C1	225	101	60	40	29	21	14	7	–

BIRMINGHAM, ALABAMA								CONTINUED	
SSF =	.10	.20	.30	.40	.50	.60	.70	.80	.90
DG C2	247	114	70	48	35	27	20	14	8
DG C3	294	139	86	60	45	35	27	21	14
SS A1	551	165	86	53	36	25	18	12	8
SS A2	490	207	120	78	54	39	28	20	13
SS A3	533	151	77	47	31	21	15	10	6
SS A4	499	207	118	77	53	38	27	19	13
SS A5	941	166	78	46	30	21	14	10	6
SS A6	487	205	119	78	54	38	28	20	13
SS A7	1083	141	63	36	23	16	11	7	4
SS A8	502	203	115	74	51	36	26	18	12
SS B1	376	125	67	42	28	20	14	10	6
SS B2	383	167	98	65	45	32	24	17	11
SS B3	347	115	61	38	25	18	12	9	5
SS B4	376	163	95	63	44	31	23	16	11
SS B5	495	112	55	33	22	15	10	7	4
SS B6	361	160	95	62	44	31	23	16	11
SS B7	439	95	46	27	18	12	8	6	4
SS B8	353	154	90	59	41	30	21	15	10
SS C1	238	107	63	41	29	21	15	11	7
SS C2	251	129	81	56	40	29	22	16	10
SS C3	249	86	47	30	20	14	10	7	5
SS C4	249	117	70	47	33	24	18	13	8
SS D1	501	197	111	71	48	34	24	17	11
SS D2	430	225	141	97	69	51	37	27	18
SS D3	637	187	97	60	40	28	20	14	9
SS D4	441	220	136	92	65	47	35	25	16
SS E1	374	154	88	56	39	27	20	14	9
SS E2	356	185	115	79	56	41	30	22	14
SS E3	458	133	69	42	28	19	14	9	6
SS E4	367	174	105	70	49	36	26	19	12

MOBILE, ALABAMA								1684 DD	
SSF =	.10	.20	.30	.40	.50	.60	.70	.80	.90
WW A1	1637	204	95	57	38	26	19	13	9
WW A2	546	216	123	80	56	41	30	21	14
WW A3	455	216	131	89	63	47	35	25	17
WW A4	416	214	134	93	67	50	37	27	18
WW A5	394	215	139	97	71	53	40	29	20
WW A6	381	215	141	100	73	55	41	30	21
WW B1	478	176	97	63	43	31	23	16	11
WW B2	399	212	135	94	69	51	38	28	19
WW B3	459	256	167	117	86	65	49	36	24
WW B4	391	236	160	115	86	66	50	37	25
WW B5	355	219	150	109	82	63	48	35	24
WW C1	483	269	175	124	91	68	51	38	25
WW C2	412	239	159	114	85	64	48	36	24
WW C3	407	261	183	135	102	79	60	45	31
WW C4	366	234	164	121	92	71	54	40	28
TW A1	1396	193	91	54	36	26	18	13	8
TW A2	558	196	108	69	48	35	25	18	12
TW A3	447	192	112	75	53	39	28	20	14
TW A4	376	180	110	74	53	39	29	21	14
TW B1	782	180	90	55	38	27	19	14	9
TW B2	445	175	100	65	46	33	24	17	12
TW B3	386	168	99	66	46	34	25	18	12
TW B4	353	153	90	60	43	31	23	17	11
TW C1	512	159	85	54	37	27	19	14	9
TW C2	388	149	84	55	39	28	20	15	10
TW C3	369	139	78	51	36	26	19	14	9
TW C4	385	124	67	43	30	21	15	11	7
TW D1	345	135	76	50	35	25	18	13	9
TW D2	428	202	123	83	59	44	32	24	16
TW D3	436	221	138	95	69	51	38	28	19

MOBILE, ALABAMA						CONTINUED				MONTGOMERY, ALABAMA								2269 DD	
SSF =	.10	.20	.30	.40	.50	.60	.70	.80	.90		SSF =	.10	.20	.30	.40	.50	.60	.70	.80	.90
TW D4	404	223	144	102	75	56	42	31	21	WW A1	1236	149	69	41	27	18	13	9	6
TW D5	369	210	138	98	72	55	41	30	21	WW A2	406	161	91	59	41	29	21	15	10
TW E1	499	253	158	109	79	59	44	32	21	WW A3	340	162	98	66	47	34	25	18	12
TW E2	447	237	151	105	77	57	43	31	21	WW A4	311	161	101	69	49	36	27	19	13
TW E3	456	270	181	130	97	73	55	41	28	WW A5	296	162	104	72	53	39	29	21	14
TW E4	403	241	163	118	88	67	51	37	26	WW A6	287	162	106	74	54	40	30	22	15
TW F1	645	167	85	53	36	26	18	13	9	WW B1	344	126	69	44	30	21	15	10	7
TW F2	401	166	96	63	44	32	24	17	11	WW B2	303	161	103	71	51	38	28	20	14
TW F3	329	157	95	65	46	34	25	18	12	WW B3	350	196	127	89	65	48	36	26	18
TW F4	268	139	87	60	44	33	24	18	12	WW B4	303	183	124	89	66	50	38	28	19
TW G1	401	142	78	50	35	25	18	13	9	WW B5	277	171	118	85	64	48	37	27	18
TW G2	289	135	81	55	39	29	21	15	10	WW C1	370	207	135	95	69	51	38	28	19
TW G3	241	121	75	52	37	28	20	15	10	WW C2	318	185	123	88	65	48	36	27	18
TW G4	192	99	62	43	31	23	17	13	8	WW C3	320	206	144	106	80	61	46	35	24
TW H1	258	113	66	44	31	23	17	12	8	WW C4	288	185	130	95	72	55	42	31	21
TW H2	196	97	60	41	29	22	16	12	8	TW A1	1055	141	66	39	26	18	13	9	6
TW H3	162	82	51	35	25	19	14	10	7	TW A2	417	146	80	51	35	25	18	13	8
TW H4	125	62	38	26	19	14	10	8	5	TW A3	334	144	84	55	39	28	20	15	10
TW I1	269	114	66	44	31	22	16	12	8	TW A4	282	135	82	55	39	29	21	15	10
TW I2	327	168	106	73	53	39	29	21	14	TW B1	588	132	66	41	27	19	13	9	6
TW I3	344	184	117	82	60	45	33	24	16	TW B2	333	131	74	48	34	24	17	12	8
TW I4	335	192	127	90	67	50	38	28	19	TW B3	290	126	74	49	34	25	18	13	9
TW I5	320	189	127	91	68	51	39	29	20	TW B4	266	116	68	45	32	23	17	12	8
TW J1	411	222	142	99	73	54	41	30	20	TW C1	385	118	63	40	27	19	14	10	6
TW J2	366	205	134	95	70	52	39	29	20	TW C2	292	112	63	41	28	20	15	11	7
TW J3	377	232	159	116	87	66	51	38	26	TW C3	278	105	59	38	26	19	14	10	6
TW J4	345	213	146	107	80	61	47	35	24	TW C4	292	94	51	32	22	16	11	8	5
DG A1	340	151	89	58	39	26	16	8	-	TW D1	249	97	54	35	24	17	12	9	6
DG A2	347	158	96	66	47	34	24	16	9	TW D2	324	154	93	63	45	33	24	17	11
DG A3	407	185	115	81	60	46	34	24	14	TW D3	330	168	105	72	51	38	28	20	13
DG B1	345	152	93	64	47	34	24	15	6	TW D4	312	173	112	78	57	43	32	23	16
DG B2	360	162	100	71	53	40	30	21	13	TW D5	288	164	108	76	56	42	31	23	16
DG B3	414	189	117	84	63	49	38	29	19	TW E1	381	194	121	83	60	44	32	23	16
DG C1	412	181	112	79	59	45	34	24	13	TW E2	343	183	117	81	59	43	32	23	16
DG C2	418	188	117	83	62	48	37	27	18	TW E3	356	211	142	101	75	56	42	31	21
DG C3	476	217	135	96	73	58	46	35	24	TW E4	316	190	128	92	68	52	39	29	20
SS A1	911	285	150	94	65	46	33	23	15	TW F1	480	122	62	38	26	18	13	9	6
SS A2	792	339	197	130	92	67	49	35	23	TW F2	298	123	71	46	32	23	17	12	8
SS A3	896	266	138	86	59	42	30	21	13	TW F3	245	117	71	48	34	25	18	13	9
SS A4	815	342	197	129	91	66	48	34	23	TW F4	201	104	66	45	32	24	18	13	8
SS A5	1506	290	139	84	56	39	28	20	13	TW G1	297	105	58	37	25	18	13	9	6
SS A6	792	338	196	130	91	66	49	35	23	TW G2	215	101	60	40	29	21	15	11	7
SS A7	1721	254	117	69	46	32	22	16	10	TW G3	180	91	56	38	27	20	15	11	7
SS A8	823	338	193	126	89	64	47	33	22	TW G4	144	75	47	32	23	17	12	9	6
SS B1	632	218	118	75	52	37	27	19	12	TW H1	192	84	49	33	23	16	12	9	6
SS B2	618	272	160	106	75	55	40	29	19	TW H2	147	73	45	30	22	16	12	8	6
SS B3	595	205	110	70	48	35	25	18	11	TW H3	122	61	38	26	19	14	10	7	5
SS B4	611	268	157	104	74	54	40	28	19	TW H4	94	47	29	20	14	10	8	5	4
SS B5	819	199	99	61	41	29	21	15	9	TW I1	193	82	47	30	21	15	11	8	5
SS B6	587	263	156	104	74	54	40	29	19	TW I2	248	128	80	55	40	29	22	16	10
SS B7	739	174	86	52	35	25	18	12	8	TW I3	262	140	90	62	45	33	25	18	12
SS B8	578	255	150	100	71	52	38	27	18	TW I4	259	149	99	70	51	38	29	21	14
SS C1	416	189	113	76	54	40	29	21	14	TW I5	249	147	99	71	52	39	30	22	15
SS C2	416	216	136	94	69	51	38	28	19	TW J1	315	170	109	76	55	41	30	22	15
SS C3	431	154	85	54	38	27	20	14	9	TW J2	282	159	104	73	53	40	30	22	15
SS C4	413	195	118	80	57	42	31	23	15	TW J3	295	183	125	91	68	51	39	29	20
SS D1	843	339	192	125	88	63	46	33	21	TW J4	271	168	115	84	63	47	36	27	18
SS D2	697	365	230	159	116	86	64	46	31	DG A1	247	109	62	39	24	14	-	-	-
SS D3	1054	323	169	106	73	52	37	26	17	DG A2	261	119	72	49	34	24	16	10	5
SS D4	715	359	222	152	109	81	60	43	29	DG A3	312	145	90	63	46	34	25	18	10
SS E1	645	271	156	103	72	52	38	27	18	DG B1	248	111	66	44	31	22	14	6	-
SS E2	585	304	192	132	96	71	53	38	26	DG B2	268	124	76	52	38	29	21	14	8
SS E3	771	235	123	77	53	38	27	19	12	DG B3	315	149	92	65	48	37	29	21	14
SS E4	603	288	174	118	84	62	46	33	22	DG C1	297	136	81	56	41	30	22	14	-

LCR Tables / 463

MONTGOMERY, ALABAMA								CONTINUED	
SSF =.10	.20	.30	.40	.50	.60	.70	.80	.90	
DG C2	312	146	89	62	46	35	27	19	12
DG C3	361	173	106	75	56	44	34	26	18
SS A1	687	214	113	70	47	33	24	16	11
SS A2	606	260	152	99	69	50	36	26	17
SS A3	669	197	102	63	42	29	21	14	9
SS A4	620	261	151	98	68	49	35	25	16
SS A5	1155	215	103	62	40	28	20	14	9
SS A6	604	259	151	99	69	49	36	26	17
SS A7	1326	186	85	50	32	22	15	10	7
SS A8	625	258	147	95	66	47	34	24	16
SS B1	473	163	88	55	38	27	19	13	9
SS B2	473	210	123	82	57	41	30	22	14
SS B3	442	152	81	51	35	24	17	12	8
SS B4	467	206	121	80	56	40	29	21	14
SS B5	618	147	73	44	29	20	14	10	6
SS B6	448	202	120	79	56	40	29	21	14
SS B7	552	127	62	37	25	17	12	8	5
SS B8	440	195	115	76	53	38	28	20	13
SS C1	308	141	84	56	39	28	21	15	10
SS C2	316	165	104	72	52	38	28	20	14
SS C3	320	113	62	40	27	19	14	10	6
SS C4	313	149	90	61	43	31	23	17	11
SS D1	633	255	144	93	64	46	33	23	15
SS D2	535	282	178	122	87	64	47	34	23
SS D3	795	242	127	79	53	37	26	18	12
SS D4	548	277	171	116	83	60	44	32	21
SS E1	479	202	116	75	52	37	27	19	12
SS E2	446	234	147	100	72	53	39	28	19
SS E3	576	174	91	56	38	27	19	13	8
SS E4	459	220	133	89	63	46	34	24	16

PHOENIX, ARIZONA								1552 DD	
SSF =.10	.20	.30	.40	.50	.60	.70	.80	.90	
WW A1	2869	355	170	104	70	49	35	25	16
WW A2	923	370	216	143	102	74	54	39	26
WW A3	766	369	229	157	113	84	62	45	30
WW A4	697	366	234	163	119	89	66	48	32
WW A5	661	368	242	171	126	95	71	52	35
WW A6	639	367	245	175	130	98	73	54	36
WW B1	836	314	179	117	82	60	43	31	20
WW B2	659	356	231	163	119	89	67	49	33
WW B3	752	426	282	201	148	111	83	61	41
WW B4	633	387	266	194	145	111	84	62	42
WW B5	570	356	247	181	137	105	79	59	40
WW C1	787	446	295	210	155	116	87	64	43
WW C2	669	395	267	192	143	108	82	60	41
WW C3	647	420	297	221	168	129	99	73	50
WW C4	582	377	267	198	151	116	89	66	45
TW A1	2430	334	161	99	67	47	34	24	16
TW A2	945	335	189	124	87	63	46	33	22
TW A3	751	327	196	132	94	69	51	37	24
TW A4	629	306	190	131	95	70	52	37	25
TW B1	1343	308	158	100	68	49	35	25	16
TW B2	749	298	173	115	82	59	44	31	21
TW B3	646	284	171	115	82	60	44	32	21
TW B4	589	259	156	105	75	55	41	29	19
TW C1	866	270	147	95	66	48	35	25	16
TW C2	649	252	146	97	68	50	36	26	17
TW C3	615	234	134	89	63	46	33	24	16
TW C4	642	208	114	74	52	37	27	19	13
TW D1	600	241	140	93	66	48	35	25	16
TW D2	706	338	209	144	104	77	57	41	27
TW D3	716	369	235	163	119	88	66	48	32

PHOENIX, ARIZONA								CONTINUED	
SSF =.10	.20	.30	.40	.50	.60	.70	.80	.90	
TW D4	654	365	241	171	126	95	71	52	35
TW D5	592	341	227	163	121	91	69	50	34
TW E1	813	419	267	186	135	100	75	54	36
TW E2	726	391	253	178	131	98	73	53	36
TW E3	729	437	297	215	160	122	92	68	46
TW E4	644	390	267	194	145	110	84	62	42
TW F1	1113	288	151	96	66	47	34	24	16
TW F2	678	284	168	112	80	58	43	31	20
TW F3	553	269	167	115	83	61	45	33	22
TW F4	449	237	152	107	78	58	43	31	21
TW G1	683	245	138	91	64	46	34	24	16
TW G2	487	231	142	97	70	52	38	28	18
TW G3	404	207	131	91	66	49	37	27	18
TW G4	321	169	108	76	55	41	31	22	15
TW H1	437	194	117	79	57	41	31	22	15
TW H2	329	166	104	72	52	39	29	21	14
TW H3	271	139	88	61	45	33	25	18	12
TW H4	207	105	66	46	33	25	18	13	9
TW I1	470	205	122	82	58	43	31	22	15
TW I2	541	283	181	127	93	69	51	37	25
TW I3	564	307	199	140	103	77	57	42	28
TW I4	542	316	212	152	113	85	64	47	32
TW I5	514	307	209	151	113	86	65	48	33
TW J1	671	367	239	169	124	93	69	51	34
TW J2	595	338	224	160	119	89	67	49	33
TW J3	601	375	260	191	144	110	83	62	42
TW J4	550	344	239	175	132	101	77	57	39
DG A1	590	273	168	114	80	57	40	26	14
DG A2	571	268	168	117	85	63	46	32	20
DG A3	640	302	193	137	103	78	58	42	26
DG B1	583	274	175	124	92	70	52	36	21
DG B2	582	275	176	126	95	73	56	41	27
DG B3	645	305	196	142	107	83	65	49	33
DG C1	679	320	205	147	111	85	66	48	31
DG C2	667	315	202	146	110	86	67	50	34
DG C3	738	349	225	163	124	97	77	59	41
SS A1	1458	459	247	157	107	76	55	38	25
SS A2	1234	533	314	209	147	107	78	55	36
SS A3	1429	429	227	143	97	69	49	34	22
SS A4	1260	534	311	206	145	104	76	54	35
SS A5	2440	470	230	141	94	66	47	33	21
SS A6	1238	534	314	209	147	107	78	55	36
SS A7	2788	413	195	117	77	54	38	26	17
SS A8	1278	530	307	202	142	102	74	53	34
SS B1	1011	353	195	126	87	62	45	32	21
SS B2	958	426	254	170	121	88	64	46	30
SS B3	952	332	183	118	81	58	42	29	19
SS B4	945	419	249	167	118	86	63	45	29
SS B5	1326	325	166	103	70	49	35	25	16
SS B6	913	414	249	167	119	87	63	45	30
SS B7	1197	284	144	89	60	42	30	21	13
SS B8	897	401	239	160	113	83	60	43	28
SS C1	704	327	199	136	98	72	53	38	25
SS C2	684	361	232	162	119	88	66	48	32
SS C3	733	264	149	98	69	50	36	26	17
SS C4	679	325	201	138	100	74	54	39	26
SS D1	1344	547	315	207	145	104	76	54	35
SS D2	1084	575	367	254	184	136	101	73	48
SS D3	1689	522	279	177	121	86	62	43	28
SS D4	1111	564	353	243	175	128	95	68	45
SS E1	1035	443	260	173	122	88	64	46	30
SS E2	915	484	308	214	155	115	85	61	41
SS E3	1242	383	205	130	89	64	46	32	21
SS E4	943	456	280	191	137	100	74	53	35

PRESCOTT, ARIZONA								4456 DD		PRESCOTT, ARIZONA								CONTINUED	
SSF =	.10	.20	.30	.40	.50	.60	.70	.80	.90	SSF =	.10	.20	.30	.40	.50	.60	.70	.80	.90
WW A1	1105	136	65	39	27	19	13	10	6	DG C2	282	133	85	61	46	36	28	20	13
WW A2	371	149	87	58	41	30	22	16	11	DG C3	331	157	101	73	56	45	36	27	19
WW A3	313	151	94	64	46	34	26	19	13	SS A1	672	208	112	71	48	34	25	17	11
WW A4	287	151	97	67	49	37	28	20	14	SS A2	592	255	151	100	71	51	38	27	18
WW A5	274	153	100	71	53	40	30	22	15	SS A3	654	191	101	63	43	30	21	15	10
WW A6	266	153	102	73	54	41	31	23	16	SS A4	605	255	149	99	69	50	36	26	17
WW B1	313	117	66	43	30	22	16	11	8	SS A5	1125	209	102	62	42	29	20	14	9
WW B2	281	152	99	70	51	38	29	21	15	SS A6	590	254	150	100	70	51	37	27	18
WW B3	327	186	123	88	65	49	37	27	19	SS A7	1284	180	84	50	33	23	16	11	7
WW B4	284	174	120	88	66	50	39	29	20	SS A8	609	251	146	96	67	48	35	25	16
WW B5	260	163	113	84	63	49	37	28	19	SS B1	453	156	86	55	38	27	20	14	9
WW C1	347	197	130	93	69	52	39	29	20	SS B2	455	202	121	81	58	42	31	22	15
WW C2	296	176	119	86	64	49	37	28	19	SS B3	423	145	79	51	35	25	18	13	8
WW C3	303	197	140	105	80	62	48	36	25	SS B4	448	198	118	79	56	41	30	21	14
WW C4	271	176	125	94	72	55	43	32	22	SS B5	590	140	71	44	30	21	15	10	7
TW A1	946	129	62	38	26	18	13	9	6	SS B6	430	195	117	79	56	41	30	22	14
TW A2	380	135	76	50	35	25	19	13	9	SS B7	528	121	61	37	25	17	12	9	5
TW A3	306	134	80	54	39	28	21	15	10	SS B8	422	188	112	75	53	39	28	20	13
TW A4	260	127	79	54	39	29	22	16	11	SS C1	283	131	80	54	39	29	21	16	10
TW B1	531	122	63	39	27	19	14	10	7	SS C2	293	155	100	70	51	38	29	21	14
TW B2	305	122	71	47	33	24	18	13	9	SS C3	292	105	59	39	27	20	14	10	7
TW B3	266	117	70	48	34	25	19	14	9	SS C4	290	139	86	59	43	32	24	17	12
TW B4	244	108	65	44	31	23	17	12	8	SS D1	617	249	144	94	66	47	34	24	16
TW C1	350	110	60	39	27	19	14	10	7	SS D2	522	278	178	123	90	66	49	36	24
TW C2	267	104	60	40	28	21	15	11	7	SS D3	778	236	126	80	54	39	28	19	13
TW C3	255	97	56	37	26	19	14	10	7	SS D4	535	272	171	117	85	62	46	33	22
TW C4	267	87	48	31	22	16	12	8	6	SS E1	458	194	114	75	53	38	28	20	13
TW D1	227	90	52	35	24	18	13	9	6	SS E2	428	226	144	100	73	54	40	29	19
TW D2	299	144	89	61	44	33	24	18	12	SS E3	552	166	89	56	38	27	20	14	9
TW D3	308	159	101	71	52	38	29	21	14	SS E4	440	212	131	89	64	47	35	25	17
TW D4	290	163	108	77	57	43	32	24	16										
TW D5	269	155	104	75	56	42	32	24	16	TUCSON, ARIZONA								1752 DD	
TW E1	355	184	117	82	60	45	33	24	17	SSF =	.10	.20	.30	.40	.50	.60	.70	.80	.90
TW E2	319	173	112	79	58	44	33	24	17	WW A1	2522	323	155	95	65	46	33	24	16
TW E3	335	201	137	100	75	57	43	32	22	WW A2	835	339	198	133	95	70	51	37	25
TW E4	296	180	124	90	68	52	40	30	20	WW A3	697	339	211	145	106	79	59	43	29
TW F1	433	112	59	37	26	18	13	9	6	WW A4	637	336	216	151	111	84	63	46	31
TW F2	272	114	67	45	32	24	17	13	8	WW A5	605	338	223	159	118	89	67	50	34
TW F3	225	110	68	47	34	25	19	14	9	WW A6	585	338	226	163	121	92	70	52	35
TW F4	185	98	63	44	32	24	18	13	9	WW B1	752	287	164	108	77	56	41	30	20
TW G1	270	97	55	36	25	18	13	10	6	WW B2	603	328	213	151	111	84	63	47	32
TW G2	197	94	58	40	29	21	16	11	8	WW B3	691	394	262	187	139	105	80	59	40
TW G3	166	85	54	38	27	20	15	11	8	WW B4	583	358	247	180	136	104	80	60	41
TW G4	133	70	45	31	23	17	13	9	6	WW B5	526	329	229	169	128	99	76	57	39
TW H1	176	78	47	32	23	17	12	9	6	WW C1	725	412	274	196	146	110	83	62	42
TW H2	135	68	43	30	22	16	12	9	6	WW C2	615	365	247	179	134	102	78	58	40
TW H3	112	58	37	25	19	14	10	8	5	WW C3	600	390	278	207	159	123	94	71	49
TW H4	86	44	28	19	14	10	8	6	4	WW C4	538	350	249	185	142	110	85	64	44
TW I1	176	76	45	30	21	16	11	8	6	TW A1	2141	304	148	91	62	44	32	23	15
TW I2	229	120	77	54	40	30	22	16	11	TW A2	851	307	173	114	81	59	43	31	21
TW I3	244	133	87	61	45	34	25	19	13	TW A3	682	300	180	122	88	65	48	35	24
TW I4	241	141	95	68	51	39	29	22	15	TW A4	573	281	175	121	88	66	49	36	24
TW I5	233	140	95	69	52	40	30	22	16	TW B1	1195	281	145	92	63	46	33	24	16
TW J1	294	161	105	75	55	41	31	23	16	TW B2	677	273	159	106	76	56	41	30	20
TW J2	263	150	100	71	53	40	30	22	15	TW B3	587	261	157	107	77	57	42	31	21
TW J3	279	175	122	90	68	52	40	30	21	TW B4	535	238	143	97	70	52	39	28	19
TW J4	254	160	111	82	62	48	37	27	19	TW C1	778	247	135	88	62	45	33	24	16
DG A1	221	100	59	38	24	15	7	-	-	TW C2	587	231	134	89	64	47	34	25	17
DG A2	237	110	69	47	34	24	17	11	5	TW C3	556	215	124	82	58	43	32	23	15
DG A3	283	135	86	61	46	35	26	18	11	TW C4	578	191	105	69	48	35	26	19	13
DG B1	217	101	63	44	32	23	15	8	-	TW D1	542	220	129	86	61	45	33	24	16
DG B2	241	114	72	51	39	29	22	15	9	TW D2	644	311	193	133	97	72	54	39	27
DG B3	287	137	88	63	48	38	30	22	15	TW D3	657	341	217	152	111	83	63	46	31
DG C1	262	123	78	55	41	32	23	15	6										

TUCSON, ARIZONA						CONTINUED			WINSLOW, ARIZONA								4733 DD	
SSF =	.10	.20	.30	.40	.50	.60	.70	.80	.90	SSF =	.10	.20	.30	.40	.50	.60	.70	.80	.90
TW D4	601	337	223	159	118	89	68	50	34	WW A1	993	117	55	34	22	16	11	8	5
TW D5	545	315	211	152	113	86	65	49	33	WW A2	327	130	76	50	35	25	19	13	9
TW E1	746	387	247	173	127	95	71	52	36	WW A3	275	132	82	56	40	29	22	16	10
TW E2	666	361	235	166	123	92	70	51	35	WW A4	252	132	84	59	43	32	23	17	11
TW E3	674	405	276	201	151	115	88	65	45	WW A5	241	134	88	62	45	34	25	18	12
TW E4	594	361	248	181	136	104	80	59	41	WW A6	234	134	89	64	47	35	26	19	13
TW F1	992	263	139	88	61	44	32	23	15	WW B1	268	99	56	36	25	18	13	9	6
TW F2	614	260	154	104	74	55	41	30	20	WW B2	249	134	87	61	45	33	25	18	12
TW F3	504	247	154	106	77	58	43	32	21	WW B3	290	164	109	77	57	43	32	23	16
TW F4	410	218	140	99	73	55	41	30	21	WW B4	254	156	107	78	58	44	34	25	17
TW G1	615	224	127	84	59	43	32	23	15	WW B5	234	146	102	75	56	43	33	24	17
TW G2	443	212	131	90	65	49	36	26	18	WW C1	308	174	115	82	60	45	34	25	17
TW G3	369	190	121	84	62	46	35	25	17	WW C2	265	157	106	76	57	43	32	24	16
TW G4	293	155	100	70	51	39	29	21	15	WW C3	272	177	126	93	71	55	42	31	22
TW H1	397	178	108	73	53	39	29	21	14	WW C4	245	159	113	84	64	49	38	28	20
TW H2	300	152	96	67	49	37	27	20	14	TW A1	850	112	54	33	22	15	11	8	5
TW H3	247	128	81	57	42	31	23	17	12	TW A2	335	118	66	43	30	22	16	11	7
TW H4	189	96	61	42	31	23	17	13	9	TW A3	270	117	70	47	33	24	18	13	8
TW I1	425	187	112	76	54	40	30	21	14	TW A4	229	111	69	47	34	25	18	13	9
TW I2	494	260	167	117	86	65	49	36	24	TW B1	473	106	54	34	23	16	12	8	5
TW I3	518	283	185	131	97	73	55	40	28	TW B2	269	107	62	41	29	21	15	11	7
TW I4	499	292	196	142	106	81	61	46	31	TW B3	235	103	62	41	29	22	16	11	8
TW I5	473	284	194	141	106	81	62	46	32	TW B4	216	95	57	38	27	20	15	11	7
TW J1	617	339	222	157	116	88	66	49	33	TW C1	311	96	52	33	23	17	12	9	6
TW J2	547	312	208	149	111	84	64	47	32	TW C2	237	92	53	35	25	18	13	9	6
TW J3	557	349	243	179	136	104	80	60	41	TW C3	226	86	49	32	23	16	12	9	6
TW J4	508	319	222	164	124	96	73	55	38	TW C4	239	77	42	27	19	14	10	7	5
DG A1	533	248	153	104	74	53	37	25	14	TW D1	195	77	44	29	20	14	10	7	5
DG A2	518	244	154	108	79	59	43	31	19	TW D2	265	127	78	54	39	28	21	15	10
DG A3	586	277	177	127	96	74	56	40	26	TW D3	272	140	89	62	45	33	25	18	12
DG B1	530	248	158	113	85	65	49	35	21	TW D4	260	146	96	68	50	38	28	21	14
DG B2	532	250	160	116	88	69	53	39	26	TW D5	241	139	93	67	50	37	28	21	14
DG B3	594	279	180	131	100	79	62	47	32	TW E1	315	163	103	72	52	39	29	21	14
DG C1	621	289	186	135	103	81	63	47	30	TW E2	285	154	100	70	51	38	29	21	14
DG C2	611	286	184	134	103	81	64	48	33	TW E3	304	180	123	89	66	50	38	28	19
DG C3	680	319	206	150	116	92	74	57	40	TW E4	266	162	111	81	61	46	35	26	18
SS A1	1341	433	235	150	104	75	54	38	25	TW F1	383	97	51	32	22	15	11	8	5
SS A2	1151	506	300	201	143	104	77	55	37	TW F2	239	100	59	39	28	20	15	10	7
SS A3	1314	406	217	138	94	67	49	35	23	TW F3	198	96	59	41	29	21	16	11	8
SS A4	1177	508	299	199	141	102	75	54	36	TW F4	163	86	55	38	28	21	15	11	7
SS A5	2213	442	219	135	91	64	46	33	21	TW G1	237	84	47	31	21	15	11	8	5
SS A6	1154	506	300	201	143	104	77	55	37	TW G2	173	82	50	34	24	18	13	10	6
SS A7	2518	388	185	112	75	53	37	26	17	TW G3	146	75	47	33	24	17	13	9	6
SS A8	1193	504	294	195	138	100	73	53	35	TW G4	117	62	39	27	20	15	11	8	5
SS B1	928	331	184	120	84	60	44	31	21	TW H1	154	68	41	27	19	14	10	7	5
SS B2	890	402	241	163	116	85	63	46	31	TW H2	119	60	37	26	19	14	10	7	5
SS B3	875	312	174	113	78	56	41	29	19	TW H3	99	51	32	22	16	12	9	6	4
SS B4	880	396	237	160	114	83	62	44	30	TW H4	76	39	24	17	12	9	7	5	3
SS B5	1203	303	156	98	67	48	34	24	16	TW I1	151	65	38	25	18	13	9	6	4
SS B6	849	391	236	160	114	84	62	45	30	TW I2	203	106	68	47	34	26	19	14	9
SS B7	1086	266	136	85	58	41	30	21	14	TW I3	216	117	76	54	39	29	22	16	11
SS B8	836	379	228	154	110	80	59	43	29	TW I4	216	126	85	61	45	34	26	19	13
SS C1	640	299	183	126	91	67	50	37	25	TW I5	209	125	85	62	46	35	27	20	13
SS C2	626	332	214	151	111	83	63	46	31	TW J1	261	143	93	66	48	36	27	20	13
SS C3	660	242	137	90	64	47	34	25	17	TW J2	234	134	89	63	47	35	26	19	13
SS C4	620	299	186	128	93	69	52	38	26	TW J3	250	157	109	80	60	46	35	26	18
SS D1	1248	517	301	199	140	102	75	54	36	TW J4	229	144	100	73	56	42	32	24	17
SS D2	1021	547	351	245	179	133	99	73	49	DG A1	191	86	49	30	18	9	-	-	-
SS D3	1551	492	265	169	117	84	61	43	28	DG A2	209	97	60	41	29	20	13	8	3
SS D4	1045	537	338	234	169	126	93	68	46	DG A3	254	121	77	55	41	30	22	16	9
SS E1	958	417	246	165	117	85	63	45	30	DG B1	187	87	54	37	26	17	10	-	-
SS E2	857	457	293	205	149	111	83	61	41	DG B2	212	101	63	45	33	25	18	12	7
SS E3	1135	358	194	124	86	62	45	32	21	DG B3	257	124	79	56	43	33	26	19	12
SS E4	879	430	266	182	131	97	72	52	35	DG C1	227	108	67	47	34	25	18	11	-

466 / APPENDIX 21

WINSLOW, ARIZONA (CONTINUED)

SSF =	.10	.20	.30	.40	.50	.60	.70	.80	.90
DG C2	248	119	75	53	40	31	23	17	10
DG C3	295	142	91	65	50	39	31	24	16
SS A1	579	178	95	59	40	28	20	14	9
SS A2	514	221	130	86	60	43	31	22	14
SS A3	557	161	84	52	35	24	17	11	7
SS A4	522	220	128	84	58	42	30	21	14
SS A5	983	178	86	52	34	24	16	11	7
SS A6	512	220	129	85	60	43	31	22	14
SS A7	1121	151	70	41	26	18	12	8	5
SS A8	525	216	125	81	56	40	29	20	13
SS B1	389	133	73	46	32	22	16	11	7
SS B2	397	176	105	70	49	36	26	19	12
SS B3	360	123	67	42	29	20	14	10	6
SS B4	389	172	102	68	48	34	25	18	12
SS B5	510	119	60	37	25	17	12	8	5
SS B6	375	169	102	68	48	35	25	18	12
SS B7	452	101	50	30	20	14	10	6	4
SS B8	365	163	97	64	45	33	24	17	11
SS C1	247	114	69	47	33	24	18	13	8
SS C2	260	137	88	61	45	33	25	18	12
SS C3	256	91	51	33	23	17	12	9	6
SS C4	257	123	76	52	37	28	20	15	10
SS D1	529	213	122	79	55	39	28	19	12
SS D2	455	241	154	106	76	56	41	30	20
SS D3	669	201	107	67	45	32	22	15	10
SS D4	465	236	148	101	72	53	39	28	18
SS E1	391	165	96	63	44	31	22	16	10
SS E2	372	196	125	86	62	45	33	24	16
SS E3	473	142	75	47	32	22	16	11	7
SS E4	382	184	113	76	54	40	29	21	14

YUMA, ARIZONA 1010 DD

SSF =	.10	.20	.30	.40	.50	.60	.70	.80	.90
WW A1	4704	570	274	167	113	79	56	39	26
WW A2	1459	589	343	228	161	117	85	61	40
WW A3	1208	586	363	248	178	131	97	70	46
WW A4	1099	579	370	258	187	139	103	74	50
WW A5	1041	582	382	270	198	148	110	80	54
WW A6	1006	580	386	275	203	152	114	83	56
WW B1	1346	509	291	191	133	96	70	49	33
WW B2	1031	559	362	254	186	138	103	75	50
WW B3	1170	664	439	311	229	171	128	93	63
WW B4	977	598	410	297	222	168	127	93	63
WW B5	877	547	379	277	208	158	120	88	60
WW C1	1222	693	457	324	239	178	133	97	65
WW C2	1035	612	412	296	220	165	124	91	61
WW C3	989	641	452	335	253	194	147	109	74
WW C4	889	576	407	301	228	174	133	98	67
TW A1	3964	535	259	159	107	75	54	38	25
TW A2	1495	532	300	196	137	99	72	51	34
TW A3	1184	519	310	209	148	108	79	57	38
TW A4	989	484	300	206	148	109	80	58	39
TW B1	2153	490	253	159	109	77	55	39	26
TW B2	1179	472	275	182	128	93	68	49	32
TW B3	1016	449	270	182	129	94	69	50	33
TW B4	924	409	245	165	118	86	63	45	30
TW C1	1372	428	234	151	105	75	54	39	25
TW C2	1019	398	230	152	107	78	57	40	27
TW C3	966	369	212	140	98	71	52	37	24
TW C4	1009	327	180	117	81	58	42	30	20
TW D1	965	390	227	150	106	76	56	40	26
TW D2	1103	531	328	225	161	119	88	63	42
TW D3	1118	578	367	254	184	136	101	73	49

YUMA, ARIZONA (CONTINUED)

SSF =	.10	.20	.30	.40	.50	.60	.70	.80	.90
TW D4	1011	566	372	263	194	145	108	79	53
TW D5	910	525	349	249	184	138	104	76	51
TW E1	1263	654	415	288	209	154	114	83	55
TW E2	1125	608	393	276	202	150	112	81	54
TW E3	1119	670	454	327	244	184	138	101	69
TW E4	986	598	408	295	220	167	125	92	63
TW F1	1785	462	243	154	106	75	54	38	25
TW F2	1072	452	267	178	126	92	67	48	32
TW F3	873	426	265	181	131	96	71	51	34
TW F4	707	375	241	168	122	91	67	49	33
TW G1	1086	391	221	145	101	73	53	38	25
TW G2	768	367	226	154	110	81	60	43	29
TW G3	636	327	207	144	104	77	57	41	27
TW G4	505	266	170	119	86	64	47	34	23
TW H1	691	309	186	126	89	65	48	34	23
TW H2	519	263	165	114	82	61	45	32	22
TW H3	426	219	139	96	70	52	38	28	18
TW H4	326	165	104	72	52	38	28	21	14
TW I1	757	332	198	133	94	68	50	36	24
TW I2	846	444	284	198	144	107	79	57	38
TW I3	879	479	311	218	159	118	88	64	43
TW I4	838	489	327	234	173	130	98	71	48
TW I5	791	473	321	231	172	130	98	72	49
TW J1	1042	571	372	261	191	142	106	77	52
TW J2	921	525	347	247	182	136	102	74	50
TW J3	921	574	397	290	218	165	125	92	63
TW J4	842	526	365	267	200	152	115	85	58
DG A1	951	444	274	187	133	96	69	47	29
DG A2	899	425	266	186	135	100	73	52	34
DG A3	990	471	300	211	157	119	89	64	40
DG B1	944	448	286	201	149	114	85	61	39
DG B2	917	437	281	198	148	114	88	65	43
DG B3	994	476	306	218	163	126	98	74	50
DG C1	1088	520	334	237	177	136	106	79	53
DG C2	1043	499	321	229	171	133	103	78	54
DG C3	1133	544	350	251	189	147	116	89	62
SS A1	2280	712	382	242	166	118	84	59	38
SS A2	1887	812	477	317	223	161	117	83	55
SS A3	2250	668	353	222	152	107	76	53	34
SS A4	1927	813	474	313	220	158	115	81	53
SS A5	3902	733	359	219	147	103	73	51	33
SS A6	1896	815	479	318	224	162	117	83	55
SS A7	4523	646	305	183	122	84	59	41	26
SS A8	1961	810	469	308	216	155	112	79	52
SS B1	1583	550	304	196	135	97	69	49	32
SS B2	1465	650	387	259	183	133	97	69	46
SS B3	1495	520	287	184	127	91	65	46	30
SS B4	1446	640	380	254	179	130	95	67	44
SS B5	2110	509	260	162	110	77	55	38	25
SS B6	1398	633	380	255	181	131	96	68	45
SS B7	1915	448	227	141	95	67	47	33	21
SS B8	1376	614	366	245	173	125	91	65	43
SS C1	1116	521	318	216	154	113	83	60	40
SS C2	1067	565	362	253	184	137	101	73	49
SS C3	1162	421	238	156	109	79	57	41	27
SS C4	1059	510	315	215	155	114	84	60	40
SS D1	2085	846	487	320	224	161	116	82	54
SS D2	1649	873	556	385	278	205	151	109	72
SS D3	2645	809	432	274	187	133	95	67	43
SS D4	1692	858	536	367	264	194	142	102	67
SS E1	1617	692	406	269	190	137	99	70	46
SS E2	1403	741	471	326	236	174	128	92	61
SS E3	1958	599	321	204	140	99	71	50	32
SS E4	1447	699	429	291	208	152	111	80	53

FORT SMITH, ARKANSAS							3336 DD		FORT SMITH, ARKANSAS							CONTINUED	
SSF = .10	.20	.30	.40	.50	.60	.70	.80	.90	SSF = .10	.20	.30	.40	.50	.60	.70	.80	.90
WW A1 803	95	43	25	16	11	8	5	3	DG C2 217	101	62	43	32	24	18	12	7
WW A2 273	107	60	39	27	19	14	10	6	DG C3 261	124	77	54	41	32	25	19	13
WW A3 230	109	66	44	31	23	17	12	8	SS A1 453	139	73	45	30	21	15	11	7
WW A4 212	109	68	46	33	24	18	13	9	SS A2 416	179	104	68	48	34	25	18	12
WW A5 202	110	71	49	36	26	20	14	9	SS A3 429	124	64	39	26	18	13	9	5
WW A6 196	110	72	50	37	27	20	15	10	SS A4 419	176	102	66	46	33	24	17	11
WW B1 214	76	41	26	17	12	9	6	4	SS A5 753	138	66	39	25	17	12	8	5
WW B2 211	112	71	49	36	26	20	14	10	SS A6 412	177	103	67	47	34	25	17	11
WW B3 245	137	89	62	46	34	25	18	12	SS A7 839	115	52	30	19	13	9	6	4
WW B4 218	132	90	65	48	36	27	20	14	SS A8 420	173	99	64	44	32	23	16	11
WW B5 202	125	86	62	47	36	27	20	14	SS B1 311	106	57	36	24	17	12	8	5
WW C1 261	146	95	67	49	36	27	20	13	SS B2 327	145	86	57	40	29	21	15	10
WW C2 227	133	88	63	46	35	26	19	13	SS B3 284	96	51	32	21	15	11	7	5
WW C3 235	151	106	78	59	45	35	26	18	SS B4 319	141	83	54	38	28	20	14	9
WW C4 213	137	96	71	54	41	31	23	16	SS B5 400	94	46	28	18	13	9	6	4
TW A1 691	91	42	25	16	11	8	5	3	SS B6 308	139	82	55	38	28	20	14	10
TW A2 280	97	53	34	23	16	12	8	5	SS B7 349	78	38	22	15	10	7	5	3
TW A3 227	97	56	37	26	19	14	10	6	SS B8 298	132	78	51	36	26	19	13	9
TW A4 193	92	56	37	27	19	14	10	7	SS C1 205	93	55	36	25	18	13	10	6
TW B1 391	87	43	26	18	12	9	6	4	SS C2 221	115	72	50	36	27	20	14	10
TW B2 226	88	50	32	22	16	12	8	5	SS C3 213	75	41	26	18	13	9	6	4
TW B3 198	85	50	33	23	17	12	9	6	SS C4 218	103	63	42	30	22	16	12	8
TW B4 182	79	46	31	21	16	11	8	5	SS D1 418	167	94	61	42	30	21	15	10
TW C1 260	79	42	26	18	13	9	6	4	SS D2 369	195	123	85	61	45	33	24	16
TW C2 200	76	43	28	19	14	10	7	5	SS D3 523	158	82	51	34	24	17	12	7
TW C3 191	72	40	26	18	13	9	7	4	SS D4 378	191	118	80	57	42	31	22	15
TW C4 202	64	35	22	15	11	8	5	4	SS E1 311	130	74	48	33	23	17	12	8
TW D1 156	60	33	21	14	10	7	5	3	SS E2 305	160	100	69	49	36	27	19	13
TW D2 226	107	65	43	31	23	17	12	8	SS E3 374	111	58	36	24	17	12	8	5
TW D3 229	117	73	50	36	26	19	14	9	SS E4 313	150	91	61	43	31	23	16	11
TW D4 224	124	80	56	41	31	23	17	11									
TW D5 209	120	79	56	41	31	23	17	12	LITTLE ROCK, ARKANSAS							3354 DD	
TW E1 267	136	85	58	42	31	23	17	11	SSF = .10	.20	.30	.40	.50	.60	.70	.80	.90
TW E2 244	130	83	58	42	31	23	17	11	WW A1 832	93	42	24	16	11	7	5	3
TW E3 258	153	103	74	55	41	31	23	16	WW A2 270	104	58	38	26	18	13	9	6
TW E4 232	139	94	68	51	38	29	21	15	WW A3 226	106	63	42	30	22	16	11	7
TW F1 315	79	40	24	16	11	8	6	4	WW A4 208	106	66	45	32	23	17	12	8
TW F2 200	82	47	30	21	15	11	8	5	WW A5 198	107	68	47	34	25	19	13	9
TW F3 166	79	48	32	23	16	12	9	6	WW A6 192	107	70	49	35	26	19	14	9
TW F4 137	71	44	30	22	16	12	9	6	WW B1 211	74	39	24	16	11	8	5	3
TW G1 197	69	37	24	16	11	8	6	4	WW B2 207	109	69	48	34	25	19	14	9
TW G2 145	67	40	27	19	14	10	7	5	WW B3 241	134	87	61	44	33	24	18	12
TW G3 122	61	38	26	18	13	10	7	5	WW B4 214	129	87	63	47	35	27	20	13
TW G4 99	51	32	22	16	11	8	6	4	WW B5 199	122	84	61	45	34	26	19	13
TW H1 128	56	33	21	15	11	8	6	4	WW C1 257	143	93	65	47	35	26	19	13
TW H2 99	49	30	20	14	11	8	6	4	WW C2 223	130	86	61	45	34	25	19	13
TW H3 83	42	26	18	13	9	7	5	3	WW C3 231	148	104	77	58	44	34	25	17
TW H4 64	32	20	13	10	7	5	4	2	WW C4 209	134	94	69	52	40	30	23	16
TW I1 120	50	28	18	12	9	6	4	3	TW A1 713	89	41	24	16	11	7	5	3
TW I2 172	89	56	38	28	20	15	11	7	TW A2 278	95	51	32	22	16	11	8	5
TW I3 182	98	62	43	31	23	17	12	8	TW A3 223	94	55	36	25	18	13	9	6
TW I4 186	107	71	50	37	28	21	15	10	TW A4 189	89	54	36	26	19	14	10	6
TW I5 181	107	72	52	38	29	22	16	11	TW B1 393	85	42	25	17	12	8	6	4
TW J1 222	120	77	53	39	29	21	16	10	TW B2 224	86	48	31	22	15	11	8	5
TW J2 201	113	74	52	38	29	21	16	11	TW B3 195	83	48	32	22	16	12	8	5
TW J3 216	134	92	67	50	38	29	21	15	TW B4 180	77	45	30	21	15	11	8	5
TW J4 199	124	85	62	46	35	27	20	14	TW C1 259	77	41	25	17	12	9	6	4
DG A1 156	66	35	19	–	–	–	–	–	TW C2 198	74	42	27	19	13	10	7	4
DG A2 179	81	49	32	22	14	9	5	–	TW C3 189	70	39	25	17	12	9	6	4
DG A3 224	104	64	45	33	24	18	12	6	TW C4 201	63	34	21	14	10	7	5	3
DG B1 156	67	39	26	17	9	–	–	–	TW D1 154	58	31	20	13	9	7	5	3
DG B2 183	84	51	35	26	19	13	8	3	TW D2 222	104	63	42	30	22	16	11	8
DG B3 227	107	66	47	35	27	21	15	10	TW D3 226	114	71	48	34	25	19	13	9
DG C1 193	86	51	35	25	17	11	–	–									

LITTLE ROCK, ARKANSAS							CONTINUED		BAKERSFIELD, CALIFORNIA								2185 DD	
SSF =	.10	.20	.30	.40	.50	.60	.70	.80	.90	SSF =	.10	.20	.30	.40	.50	.60	.70	.80	.90
TW D4	220	121	78	55	40	30	22	16	11	WW A1	1754	213	97	57	38	26	18	12	7
TW D5	206	117	77	54	40	30	22	16	11	WW A2	574	223	125	81	56	40	29	20	12
TW E1	264	133	83	57	41	30	22	16	11	WW A3	476	222	133	89	63	46	33	23	15
TW E2	241	127	81	56	41	30	22	16	11	WW A4	432	220	137	93	67	49	35	25	16
TW E3	254	150	101	72	53	40	30	22	15	WW A5	409	221	141	98	71	52	38	27	17
TW E4	228	136	92	66	49	37	28	21	14	WW A6	395	220	143	100	73	54	39	28	18
TW F1	315	77	38	23	15	11	8	5	3	WW B1	505	181	99	63	43	30	21	14	9
TW F2	197	79	45	29	20	14	10	7	5	WW B2	414	217	137	95	68	50	37	26	17
TW F3	163	77	46	31	22	16	11	8	5	WW B3	475	261	169	118	86	63	46	33	21
TW F4	134	69	43	29	21	15	11	8	5	WW B4	403	241	162	116	86	64	48	35	23
TW G1	195	67	36	23	15	11	8	5	4	WW B5	366	223	152	110	82	62	46	34	22
TW G2	143	65	39	26	18	13	10	7	4	WW C1	500	275	178	124	90	67	49	35	23
TW G3	120	60	37	25	18	13	9	7	4	WW C2	425	245	162	115	84	63	46	33	22
TW G4	97	49	31	21	15	11	8	6	4	WW C3	418	265	185	135	102	77	58	42	28
TW H1	126	54	31	21	14	10	7	5	3	WW C4	376	239	166	122	92	70	52	38	26
TW H2	98	48	29	20	14	10	7	5	4	TW A1	1493	202	93	55	36	25	17	12	7
TW H3	82	41	25	17	12	9	6	5	3	TW A2	589	203	110	70	48	34	24	17	10
TW H4	63	31	19	13	9	7	5	4	2	TW A3	469	198	115	75	53	38	27	19	12
TW I1	118	48	27	17	12	8	6	4	2	TW A4	393	185	112	75	53	38	28	19	12
TW I2	169	87	54	37	27	20	14	10	7	TW B1	832	187	92	56	38	26	18	12	8
TW I3	180	95	61	42	30	22	17	12	8	TW B2	468	181	102	66	46	32	23	16	10
TW I4	183	105	69	49	36	27	20	15	10	TW B3	405	173	100	66	46	33	24	17	11
TW I5	178	105	70	50	37	28	21	15	11	TW B4	370	158	92	61	42	31	22	15	10
TW J1	218	117	75	52	38	28	21	15	10	TW C1	542	165	87	54	37	26	18	13	8
TW J2	198	111	72	51	37	28	21	15	10	TW C2	408	154	86	56	38	27	20	14	9
TW J3	212	131	90	65	49	37	28	21	14	TW C3	388	144	80	51	36	25	18	13	8
TW J4	196	121	83	60	45	34	26	19	13	TW C4	406	129	68	43	30	21	15	10	7
DG A1	152	63	33	17	-	-	-	-	-	TW D1	363	140	78	50	34	24	17	11	7
DG A2	177	79	47	31	21	14	8	4	-	TW D2	446	208	125	84	59	43	31	22	14
DG A3	222	102	63	44	32	24	17	12	6	TW D3	453	227	140	95	68	50	36	26	16
DG B1	153	65	38	24	15	7	-	-	-	TW D4	419	228	147	102	74	55	41	29	19
DG B2	181	82	49	34	25	18	12	8	3	TW D5	382	215	140	99	72	54	40	29	19
DG B3	225	105	65	46	34	26	20	15	9	TW E1	518	260	161	110	78	57	42	30	19
DG C1	190	84	49	33	23	16	10	-	-	TW E2	464	243	154	106	77	56	41	30	19
DG C2	214	98	60	42	31	23	17	12	6	TW E3	471	275	183	131	96	72	53	39	25
DG C3	259	122	75	53	40	31	24	18	12	TW E4	416	246	165	118	88	66	49	36	24
SS A1	461	139	72	45	30	21	15	10	6	TW F1	685	173	87	54	36	25	17	12	7
SS A2	419	178	103	67	47	34	24	17	11	TW F2	421	171	97	63	44	31	22	16	10
SS A3	439	124	63	38	25	17	12	8	5	TW F3	343	161	97	65	46	33	24	17	11
SS A4	424	176	101	66	45	32	23	17	11	TW F4	278	142	89	61	44	32	23	16	10
SS A5	782	138	65	38	25	17	12	8	5	TW G1	423	147	80	51	35	25	17	12	7
SS A6	415	176	102	67	46	33	24	17	11	TW G2	302	139	83	55	39	28	20	14	9
SS A7	887	115	51	29	19	12	9	6	3	TW G3	251	124	76	52	37	27	20	14	9
SS A8	424	172	98	63	44	31	22	16	10	TW G4	200	102	63	43	31	23	16	12	7
SS B1	314	105	56	35	24	17	12	8	5	TW H1	271	116	68	45	31	22	16	11	7
SS B2	329	144	84	56	39	28	20	15	10	TW H2	204	100	61	41	29	21	15	11	7
SS B3	287	95	50	31	21	15	10	7	4	TW H3	169	84	52	35	25	18	13	9	6
SS B4	321	140	82	54	37	27	20	14	9	TW H4	130	64	39	26	19	14	10	7	4
SS B5	410	93	46	27	18	12	9	6	4	TW I1	283	118	67	44	30	21	15	10	6
SS B6	309	138	81	54	38	27	20	14	9	TW I2	340	173	108	74	53	39	28	20	13
SS B7	359	78	37	22	14	10	7	4	3	TW I3	357	188	119	82	59	43	32	23	15
SS B8	300	131	77	51	35	25	18	13	9	TW I4	346	197	129	91	67	50	37	26	17
SS C1	202	90	53	35	24	18	13	9	6	TW I5	330	193	128	91	67	50	38	27	18
SS C2	217	112	70	48	35	26	19	14	9	TW J1	426	227	144	100	72	53	39	28	18
SS C3	211	73	39	25	17	12	9	6	4	TW J2	379	210	136	95	70	52	38	27	18
SS C4	215	101	61	41	29	21	15	11	7	TW J3	388	237	161	117	87	65	49	35	23
SS D1	421	166	93	60	41	29	21	14	9	TW J4	355	217	148	107	80	60	45	33	22
SS D2	370	194	122	83	60	44	32	23	15	DG A1	359	157	91	58	38	25	14	4	-
SS D3	533	157	81	50	33	23	16	11	7	DG A2	364	164	99	66	47	33	23	15	7
SS D4	378	190	117	79	56	41	30	22	14	DG A3	425	192	119	83	60	45	32	22	13
SS E1	312	129	73	47	32	23	16	11	7	DG B1	362	158	97	66	46	33	22	12	-
SS E2	304	158	99	68	48	35	26	19	12	DG B2	378	168	104	73	53	39	28	20	11
SS E3	380	111	57	35	23	16	11	8	5	DG B3	435	194	121	86	64	48	36	26	17
SS E4	314	149	90	60	42	31	22	16	11	DG C1	433	188	117	82	59	43	31	20	9

BAKERSFIELD, CALIFORNIA									CONTINUED		DAGGETT, CALIFORNIA									CONTINUED
SSF	=.10	.20	.30	.40	.50	.60	.70	.80	.90		SSF	=.10	.20	.30	.40	.50	.60	.70	.80	.90
DG C2	439	195	121	86	63	47	35	25	16		TW D4	504	279	182	129	95	71	53	39	26
DG C3	501	223	140	100	75	57	43	32	21		TW D5	459	262	173	123	91	69	52	38	26
SS A1	959	289	149	92	61	42	29	20	12		TW E1	624	319	201	139	101	75	55	40	27
SS A2	819	342	196	127	88	62	44	30	19		TW E2	558	298	192	134	98	73	55	40	27
SS A3	940	268	136	83	54	37	25	17	10		TW E3	565	335	226	163	121	92	69	51	35
SS A4	837	342	194	125	86	60	43	29	18		TW E4	499	300	204	147	110	84	63	47	32
SS A5	1615	296	139	82	53	36	24	16	10		TW F1	828	215	111	70	48	34	25	17	11
SS A6	819	341	195	126	87	62	44	30	19		TW F2	511	213	124	82	58	42	31	22	15
SS A7	1854	258	116	67	42	28	19	12	7		TW F3	418	201	124	84	61	45	33	24	16
SS A8	847	339	190	122	83	59	41	28	18		TW F4	340	178	113	78	57	42	32	23	15
SS B1	659	220	117	73	49	34	24	16	10		TW G1	513	183	102	66	46	33	24	17	11
SS B2	636	274	159	104	72	52	37	26	16		TW G2	368	173	105	71	51	38	28	20	13
SS B3	618	206	109	68	45	31	22	15	9		TW G3	306	155	97	67	49	36	27	19	13
SS B4	627	269	156	102	70	50	36	25	16		TW G4	244	127	80	56	41	30	22	16	11
SS B5	867	202	99	60	39	27	18	12	7		TW H1	330	145	86	58	41	30	22	16	11
SS B6	603	265	155	102	71	50	36	25	16		TW H2	249	124	77	53	38	28	21	15	10
SS B7	781	176	85	51	33	22	15	10	6		TW H3	206	104	65	45	33	24	18	13	9
SS B8	592	256	148	97	67	48	34	23	15		TW H4	158	79	49	34	25	18	13	10	7
SS C1	435	195	115	76	54	38	28	19	12		TW I1	349	150	88	59	41	30	22	16	10
SS C2	432	222	139	95	68	50	36	26	17		TW I2	413	214	136	94	68	51	38	28	18
SS C3	455	159	86	55	38	27	19	13	8		TW I3	431	232	149	104	76	57	42	31	21
SS C4	430	200	120	81	57	42	30	21	14		TW I4	418	241	160	115	85	64	48	35	24
SS D1	875	342	190	122	83	58	40	28	17		TW I5	398	236	159	114	85	65	49	36	25
SS D2	711	366	228	155	110	79	57	40	26		TW J1	514	279	180	126	92	69	52	38	25
SS D3	1111	328	169	104	69	47	33	22	14		TW J2	457	258	170	120	89	67	50	37	25
SS D4	731	360	220	148	104	75	54	37	24		TW J3	466	288	199	145	109	83	63	47	32
SS E1	665	273	155	100	68	48	34	23	14		TW J4	427	265	183	133	101	77	58	43	30
SS E2	596	305	190	129	91	66	48	33	21		DG A1	441	200	120	80	55	38	26	16	7
SS E3	809	238	122	75	50	34	24	16	10		DG A2	437	201	125	86	62	45	32	22	13
SS E4	618	289	173	115	80	58	41	29	18		DG A3	502	232	146	103	77	58	43	31	19
											DG B1	443	201	125	88	65	48	35	23	12
											DG B2	452	207	130	93	69	53	40	29	18
DAGGETT, CALIFORNIA								2203 DD			DG B3	510	235	148	106	81	62	49	36	24
SSF	=.10	.20	.30	.40	.50	.60	.70	.80	.90		DG C1	523	236	148	106	79	60	46	33	20
WW A1	2112	265	125	75	50	35	25	18	11		DG C2	522	238	150	108	81	63	49	36	24
WW A2	696	278	160	105	74	54	39	28	19		DG C3	585	269	170	122	94	73	58	44	30
WW A3	580	277	170	115	83	61	45	33	22		SS A1	1115	346	184	116	79	55	40	28	18
WW A4	528	274	174	120	87	65	48	35	23		SS A2	951	407	238	157	110	79	57	41	27
WW A5	501	276	179	126	93	69	52	38	25		SS A3	1089	321	168	104	70	49	35	24	15
WW A6	484	275	182	129	95	72	54	39	26		SS A4	967	406	235	154	107	77	56	39	26
WW B1	619	230	129	84	58	42	30	22	14		SS A5	1876	354	171	103	69	48	34	23	15
WW B2	503	269	173	121	88	66	49	36	24		SS A6	952	407	238	157	110	79	57	41	27
WW B3	575	323	212	150	110	83	62	45	30		SS A7	2148	308	143	85	55	38	27	18	11
WW B4	487	295	201	146	109	83	63	46	32		SS A8	980	403	231	151	105	75	54	38	25
WW B5	441	273	188	137	104	79	60	44	30		SS B1	770	265	145	93	64	45	32	23	15
WW C1	604	339	222	157	116	87	65	48	32		SS B2	739	326	193	128	91	66	48	34	22
WW C2	514	301	202	144	107	81	61	45	30		SS B3	722	248	135	86	59	42	30	21	13
WW C3	502	323	227	168	128	98	75	56	38		SS B4	727	320	189	125	88	64	46	33	22
WW C4	452	290	204	151	115	88	67	50	34		SS B5	1011	243	123	76	51	36	25	18	11
TW A1	1795	250	119	72	48	34	24	17	11		SS B6	702	316	188	126	89	64	47	33	22
TW A2	711	252	140	91	63	46	33	24	16		SS B7	910	212	106	65	43	30	21	15	9
TW A3	568	246	145	97	69	50	37	27	18		SS B8	688	305	181	120	84	61	44	32	21
TW A4	477	230	141	96	69	51	38	27	18		SS C1	531	244	147	99	71	52	38	28	18
TW B1	1001	231	117	73	50	35	25	18	12		SS C2	523	273	174	121	88	65	49	35	24
TW B2	566	224	129	85	60	43	32	23	15		SS C3	550	198	110	71	50	36	26	19	12
TW B3	490	214	127	85	60	44	33	23	15		SS C4	519	247	151	103	74	54	40	29	19
TW B4	447	196	116	78	55	40	30	21	14		SS D1	1024	412	235	153	106	76	55	39	25
TW C1	652	204	110	70	48	35	25	18	12		SS D2	834	439	278	191	138	101	75	54	35
TW C2	491	190	109	71	50	36	27	19	13		SS D3	1291	393	208	130	88	62	44	31	20
TW C3	466	177	100	66	46	33	24	17	12		SS D4	855	431	268	182	131	95	70	50	33
TW C4	486	158	86	55	38	28	20	14	9		SS E1	784	332	193	127	88	64	46	33	21
TW D1	446	177	101	66	47	34	24	17	11		SS E2	701	367	232	160	115	85	63	45	30
TW D2	539	256	157	107	77	57	42	30	20		SS E3	945	287	152	95	65	46	33	23	15
TW D3	547	280	176	121	88	65	48	35	23		SS E4	723	347	211	143	102	74	54	39	26

FRESNO, CALIFORNIA   2650 DD
   SSF =.10   .20   .30   .40   .50   .60   .70   .80   .90
WW A1 1464   172   77   44   28   19   12   8   5
WW A2 473   182   100   64   43   30   21   14   8
WW A3 392   181   107   70   49   34   24   16   10
WW A4 356   179   109   73   51   37   26   18   11
WW A5 337   180   113   77   55   39   28   19   12
WW A6 325   179   115   79   56   41   29   20   13
WW B1 409   144   76   47   31   21   14   9   5
WW B2 344   179   111   76   54   39   28   19   12
WW B3 396   216   138   95   68   49   35   25   16
WW B4 338   201   133   94   69   51   38   27   17
WW B5 308   187   126   90   66   49   36   26   17
WW C1 418   229   146   100   72   52   38   26   17
WW C2 356   204   133   93   67   49   36   26   16
WW C3 354   223   154   111   83   62   46   33   22
WW C4 318   201   139   100   75   56   42   30   20
TW A1 1247   164   74   43   27   18   12   8   5
TW A2 487   166   88   55   37   25   18   12   7
TW A3 387   162   92   59   41   28   20   13   8
TW A4 324   151   90   59   41   29   21   14   9
TW B1 692   152   74   44   29   19   13   9   5
TW B2 387   148   82   52   35   25   17   12   7
TW B3 335   142   81   52   36   25   18   12   7
TW B4 307   130   74   48   33   23   16   11   7
TW C1 450   135   70   43   29   20   14   9   5
TW C2 338   127   70   44   30   21   15   10   6
TW C3 322   118   65   41   28   19   14   9   6
TW C4 339   106   56   35   23   16   11   8   5
TW D1 295   111   60   37   25   17   11   7   4
TW D2 371   171   102   67   47   33   24   16   10
TW D3 377   187   114   76   54   38   27   19   12
TW D4 351   190   121   83   60   44   32   22   14
TW D5 322   180   116   81   59   43   32   22   14
TW E1 434   216   132   88   62   45   32   22   14
TW E2 388   202   127   86   61   44   32   22   14
TW E3 398   231   152   107   78   58   42   30   19
TW E4 352   207   138   97   71   53   39   28   18
TW F1 566   140   69   41   27   18   12   8   5
TW F2 346   139   78   50   34   24   16   11   7
TW F3 283   131   78   51   35   25   18   12   7
TW F4 229   116   71   48   34   24   17   12   7
TW G1 348   119   64   40   26   18   12   8   5
TW G2 248   113   66   43   30   21   15   10   6
TW G3 207   101   61   41   29   20   14   10   6
TW G4 165   83   51   34   24   17   12   8   5
TW H1 222   95   54   35   24   17   12   8   5
TW H2 168   81   49   32   23   16   11   8   5
TW H3 139   68   42   28   19   14   10   7   4
TW H4 107   52   31   21   15   10   7   5   3
TW I1 228   93   52   33   22   15   10   6   4
TW I2 282   142   87   59   41   30   21   15   9
TW I3 297   156   97   66   47   34   24   17   11
TW I4 290   164   106   74   54   39   29   20   13
TW I5 278   161   106   75   55   40   30   21   14
TW J1 357   188   118   80   57   41   30   21   13
TW J2 317   175   112   77   56   41   30   21   13
TW J3 329   199   134   96   70   52   39   28   18
TW J4 301   183   124   88   65   49   36   26   17
DG A1 292   125   69   42   25   14   -   -   -
DG A2 304   135   80   52   35   24   16   9   3
DG A3 361   162   99   67   48   35   25   17   9
DG B1 297   128   75   49   32   21   12   -   -
DG B2 316   140   84   57   41   29   20   13   6
DG B3 368   166   102   71   51   38   28   20   12
DG C1 357   154   92   62   43   30   20   11   -

FRESNO, CALIFORNIA   CONTINUED
   SSF =.10   .20   .30   .40   .50   .60   .70   .80   .90
DG C2 368   164   99   69   49   36   26   18   10
DG C3 422   192   117   82   60   45   34   25   16
SS A1 820   242   122   73   47   32   21   14   8
SS A2 703   290   163   104   70   49   34   23   14
SS A3 803   223   110   65   41   27   18   12   7
SS A4 719   290   161   101   68   47   33   22   13
SS A5 1403   248   113   65   41   27   18   11   7
SS A6 701   289   162   103   69   48   34   23   14
SS A7 1626   215   93   52   32   20   13   8   5
SS A8 727   286   157   99   66   45   31   21   13
SS B1 558   183   95   58   38   26   17   11   7
SS B2 544   232   132   85   58   41   29   20   12
SS B3 523   170   88   53   34   23   15   10   6
SS B4 536   227   129   83   56   39   27   19   12
SS B5 741   168   80   47   30   20   13   8   5
SS B6 515   224   128   83   56   40   28   19   12
SS B7 666   145   68   39   25   16   10   7   4
SS B8 506   216   123   78   53   37   26   18   11
SS C1 357   158   92   59   41   29   20   13   8
SS C2 360   183   113   76   54   39   28   19   12
SS C3 375   129   69   43   29   20   14   9   5
SS C4 359   166   98   65   45   32   23   16   10
SS D1 743   286   155   96   64   44   30   20   12
SS D2 609   309   189   126   88   63   44   31   19
SS D3 950   275   138   82   53   36   24   16   9
SS D4 626   304   182   120   83   59   42   29   18
SS E1 559   225   125   78   52   36   24   16   10
SS E2 506   256   156   104   72   51   36   25   16
SS E3 686   197   99   59   38   26   17   11   7
SS E4 526   243   142   93   64   45   32   22   13

LONG BEACH, CALIFORNIA   1606 DD
   SSF =.10   .20   .30   .40   .50   .60   .70   .80   .90
WW A1 3088   387   184   112   75   53   38   27   18
WW A2 998   402   233   154   109   80   59   42   29
WW A3 830   401   247   169   122   90   67   49   33
WW A4 756   396   252   176   128   95   71   52   36
WW A5 716   399   261   184   135   102   77   56   39
WW A6 693   398   264   188   139   105   79   59   40
WW B1 917   346   196   128   90   65   48   34   23
WW B2 715   385   249   174   128   96   72   53   36
WW B3 821   464   306   217   160   121   91   67   46
WW B4 683   417   285   207   155   118   90   67   46
WW B5 616   383   265   194   146   112   85   63   44
WW C1 859   486   320   227   167   126   95   70   48
WW C2 722   426   286   206   153   116   88   65   45
WW C3 704   455   321   238   181   139   107   80   55
WW C4 627   405   286   212   161   124   95   71   50
TW A1 2615   364   175   107   72   51   37   26   17
TW A2 1020   364   204   133   93   68   49   36   24
TW A3 813   355   212   142   101   74   55   40   27
TW A4 681   332   205   141   101   75   56   41   28
TW B1 1446   335   172   108   74   53   38   27   18
TW B2 809   323   187   124   88   64   47   34   23
TW B3 699   308   184   124   88   65   48   35   24
TW B4 636   281   168   113   81   59   44   32   22
TW C1 934   293   160   103   71   51   37   27   18
TW C2 700   274   157   104   73   53   39   28   19
TW C3 664   254   145   96   67   49   36   26   18
TW C4 691   225   124   80   56   40   29   21   14
TW D1 659   265   154   101   71   52   38   28   18
TW D2 765   367   226   154   111   82   61   45   30
TW D3 783   403   255   177   129   96   71   52   36

LONG BEACH, CALIFORNIA							CONTINUED			LOS ANGELES, CALIFORNIA								1819 DD	
SSF =.10	.20	.30	.40	.50	.60	.70	.80	.90		SSF =.10	.20	.30	.40	.50	.60	.70	.80	.90
TW D4	706	394	258	183	135	101	76	56	39	WW A1	3168	406	192	116	78	55	39	28	18
TW D5	640	367	244	174	129	97	74	54	37	WW A2	1044	421	242	160	112	82	60	43	29
TW E1	888	458	290	201	146	109	81	59	40	WW A3	869	418	256	174	125	92	69	50	34
TW E2	784	422	272	191	140	105	79	58	40	WW A4	792	413	261	181	132	98	73	54	36
TW E3	794	474	321	231	173	131	99	74	51	WW A5	750	415	270	190	139	104	78	58	39
TW E4	694	419	286	207	155	118	90	67	46	WW A6	725	414	273	194	143	108	81	60	41
TW F1	1200	314	164	104	71	51	37	27	18	WW B1	968	365	205	134	93	68	49	35	24
TW F2	733	309	181	121	86	63	46	34	23	WW B2	748	401	258	180	131	98	73	54	37
TW F3	599	292	180	123	89	66	49	36	24	WW B3	864	486	318	225	166	124	93	69	47
TW F4	487	257	164	114	84	62	47	34	23	WW B4	713	432	294	213	159	121	92	68	47
TW G1	739	266	150	98	68	50	36	26	18	WW B5	642	397	273	199	150	114	87	65	45
TW G2	527	250	154	105	75	56	41	30	20	WW C1	904	508	333	235	173	130	98	72	49
TW G3	438	224	141	98	71	53	40	29	20	WW C2	754	442	296	212	157	119	90	66	46
TW G4	348	183	116	81	59	44	33	24	17	WW C3	737	473	333	246	186	143	110	82	57
TW H1	473	211	126	85	61	45	33	24	16	WW C4	652	419	294	217	165	127	97	73	50
TW H2	357	180	113	78	56	42	31	23	16	TW A1	2685	382	183	111	74	52	38	27	18
TW H3	294	150	95	66	48	36	27	19	13	TW A2	1065	381	212	137	96	69	51	36	24
TW H4	225	114	71	49	36	27	20	14	10	TW A3	851	371	219	147	104	76	56	41	27
TW I1	517	226	134	90	64	47	34	25	17	TW A4	713	346	213	145	104	77	57	42	28
TW I2	586	307	195	136	99	74	55	41	28	TW B1	1496	351	179	111	76	54	39	28	18
TW I3	616	335	216	152	111	83	62	46	31	TW B2	845	338	194	128	90	66	48	35	23
TW I4	585	340	227	163	121	91	69	51	35	TW B3	731	322	191	128	91	67	49	36	24
TW I5	555	331	224	161	121	92	69	52	36	TW B4	666	293	174	117	83	61	45	33	22
TW J1	732	400	260	183	134	100	75	55	38	TW C1	972	307	166	106	73	53	38	28	18
TW J2	642	365	241	171	127	95	72	53	37	TW C2	731	286	163	107	75	55	40	29	19
TW J3	654	407	281	206	155	118	90	67	46	TW C3	693	265	151	99	69	50	37	27	18
TW J4	593	369	256	187	141	108	82	61	43	TW C4	719	236	129	82	57	41	30	22	14
DG A1	644	297	182	123	87	62	44	30	17	TW D1	696	280	161	106	74	54	39	28	19
DG A2	623	290	181	126	91	67	50	35	23	TW D2	801	383	234	159	114	84	62	46	31
DG A3	698	327	206	146	110	84	63	46	29	TW D3	824	423	266	183	133	99	74	54	37
DG B1	645	299	187	132	99	76	57	41	25	TW D4	738	410	267	188	138	104	78	58	39
DG B2	640	299	188	133	101	78	60	45	30	TW D5	668	382	252	179	132	100	75	56	38
DG B3	702	332	209	149	113	89	70	53	36	TW E1	935	479	302	208	151	112	84	61	42
DG C1	748	351	219	156	118	93	72	54	36	TW E2	820	439	282	197	144	107	80	59	40
DG C2	731	344	216	154	117	92	72	55	38	TW E3	833	495	333	239	178	135	102	76	52
DG C3	802	381	239	172	131	104	83	64	45	TW E4	723	435	295	213	159	121	92	68	47
SS A1	1722	545	293	185	127	90	65	46	30	TW F1	1246	329	171	107	73	52	38	27	18
SS A2	1450	628	369	245	172	125	91	65	43	TW F2	768	323	188	125	88	64	47	34	23
SS A3	1723	519	275	172	117	83	59	41	27	TW F3	628	304	187	127	92	68	50	37	25
SS A4	1506	639	373	245	172	124	90	64	42	TW F4	510	268	170	118	86	64	48	35	24
SS A5	2893	559	274	167	112	78	55	39	25	TW G1	771	279	156	101	70	51	37	27	18
SS A6	1455	630	370	245	172	125	91	65	43	TW G2	553	261	159	108	77	57	42	31	21
SS A7	3382	500	236	141	93	65	46	32	20	TW G3	459	234	146	101	73	54	40	30	20
SS A8	1529	636	368	241	169	121	88	63	41	TW G4	364	190	121	84	61	45	34	25	17
SS B1	1191	417	230	147	102	73	53	37	25	TW H1	496	221	131	88	63	46	34	25	17
SS B2	1120	499	297	198	140	102	75	54	36	TW H2	374	187	117	80	58	43	32	23	16
SS B3	1137	398	219	140	96	69	50	35	23	TW H3	308	157	98	68	49	36	27	20	13
SS B4	1118	496	294	196	138	100	73	53	35	TW H4	235	118	74	51	37	27	20	15	10
SS B5	1566	384	196	121	82	58	41	29	19	TW I1	546	238	140	93	66	48	35	26	17
SS B6	1068	485	290	194	138	100	74	53	35	TW I2	614	320	202	140	102	76	57	41	28
SS B7	1434	341	173	106	72	50	36	25	16	TW I3	649	350	225	157	115	86	64	47	32
SS B8	1063	476	283	188	133	97	71	51	34	TW I4	611	353	235	167	124	94	71	52	36
SS C1	767	356	217	147	105	77	57	42	28	TW I5	580	344	231	166	124	94	71	53	36
SS C2	743	392	250	175	127	95	71	52	36	TW J1	771	419	270	189	138	104	78	57	39
SS C3	796	288	162	106	74	54	39	28	19	TW J2	671	379	249	176	130	98	74	54	37
SS C4	737	354	218	149	107	79	59	43	29	TW J3	686	424	292	213	160	122	93	69	48
SS D1	1588	649	373	245	171	123	89	63	42	TW J4	618	383	264	192	145	110	84	63	43
SS D2	1275	677	430	298	215	159	118	85	57	DG A1	679	310	189	127	90	64	45	31	18
SS D3	1995	619	331	209	143	101	73	51	33	DG A2	655	303	187	130	94	69	51	36	23
SS D4	1306	664	415	284	204	150	110	80	53	DG A3	734	341	213	150	113	86	64	47	30
SS E1	1221	524	307	203	142	103	75	54	36	DG B1	684	314	194	136	102	78	58	42	26
SS E2	1073	566	359	249	180	133	99	71	48	DG B2	678	314	195	138	104	80	62	46	31
SS E3	1465	453	242	153	105	74	54	38	25	DG B3	745	346	217	154	116	91	72	54	37
SS E4	1105	534	327	222	158	116	85	62	41	DG C1	800	367	228	161	121	95	74	55	36

472 / APPENDIX 21

LOS ANGELES, CALIFORNIA   CONTINUED

SSF	=.10	.20	.30	.40	.50	.60	.70	.80	.90
DG C2	777	360	224	159	120	94	74	56	39
DG C3	853	397	248	177	134	106	85	66	46
SS A1	1909	601	319	202	138	98	70	49	32
SS A2	1604	688	401	265	187	135	98	70	46
SS A3	1927	578	303	189	129	91	65	45	29
SS A4	1679	706	408	268	188	135	98	70	46
SS A5	3188	619	300	182	122	85	60	42	27
SS A6	1610	690	402	266	187	135	98	70	46
SS A7	3751	559	261	156	103	71	50	35	22
SS A8	1706	703	403	264	184	133	96	68	44
SS B1	1312	457	250	160	110	79	57	40	26
SS B2	1229	543	321	213	151	110	80	57	38
SS B3	1261	439	239	153	105	75	54	38	25
SS B4	1235	543	320	212	150	109	79	57	37
SS B5	1719	423	213	132	89	63	45	31	20
SS B6	1173	528	314	210	148	108	79	57	38
SS B7	1586	378	189	116	78	55	39	27	18
SS B8	1175	521	308	204	144	105	76	54	36
SS C1	806	373	225	152	109	80	59	43	29
SS C2	780	409	260	180	131	98	73	53	36
SS C3	833	303	169	110	76	55	40	29	19
SS C4	773	370	226	154	110	81	60	44	30
SS D1	1762	713	407	266	186	134	97	69	45
SS D2	1404	738	467	323	233	172	127	92	61
SS D3	2211	683	361	228	155	110	79	56	36
SS D4	1441	725	450	308	221	162	119	86	57
SS E1	1347	573	333	220	154	111	81	58	38
SS E2	1176	615	388	268	194	143	106	76	51
SS E3	1613	497	263	166	113	81	58	41	27
SS E4	1214	582	354	239	171	125	92	66	44

MOUNT SHASTA, CALIFORNIA   5890 DD

SSF	=.10	.20	.30	.40	.50	.60	.70	.80	.90
WW A1	721	80	34	19	12	8	5	3	2
WW A2	243	91	49	31	21	14	10	7	4
WW A3	203	92	54	35	24	17	12	8	5
WW A4	186	92	56	37	26	19	13	9	6
WW A5	176	93	58	39	28	20	14	10	6
WW A6	171	93	59	41	29	21	15	11	7
WW B1	188	61	31	18	11	7	4	2	-
WW B2	187	96	60	41	29	21	15	11	7
WW B3	220	119	76	52	37	27	20	14	9
WW B4	195	116	77	55	40	30	22	16	11
WW B5	181	110	75	54	40	30	22	16	11
WW C1	236	128	82	56	40	30	22	15	10
WW C2	203	116	76	53	39	29	21	15	10
WW C3	213	135	94	69	51	39	29	21	14
WW C4	192	121	84	61	46	35	26	19	13
TW A1	623	78	34	19	12	8	5	3	2
TW A2	251	83	43	27	18	12	8	6	3
TW A3	202	82	46	30	20	14	10	7	4
TW A4	171	78	46	30	21	15	11	7	5
TW B1	353	74	35	21	13	9	6	4	2
TW B2	203	75	41	26	17	12	8	6	3
TW B3	177	73	41	27	18	13	9	6	4
TW B4	163	68	39	25	17	12	8	6	4
TW C1	235	68	35	21	14	10	7	4	3
TW C2	180	66	36	23	15	11	7	5	3
TW C3	172	62	34	21	14	10	7	5	3
TW C4	183	56	29	18	12	8	6	4	2
TW D1	137	48	25	15	9	6	4	2	-
TW D2	202	92	54	36	25	18	13	9	6
TW D3	207	102	62	41	29	21	15	10	7

MOUNT SHASTA, CALIFORNIA   CONTINUED

SSF	=.10	.20	.30	.40	.50	.60	.70	.80	.90
TW D4	201	109	69	48	34	25	19	13	9
TW D5	189	106	69	48	35	26	19	14	9
TW E1	243	120	73	49	35	25	18	13	8
TW E2	220	114	71	49	35	25	19	13	9
TW E3	235	137	90	64	47	35	26	19	12
TW E4	209	123	82	59	43	32	24	17	12
TW F1	282	66	32	19	12	8	5	3	2
TW F2	177	69	38	24	16	11	8	5	3
TW F3	146	66	39	25	18	12	9	6	4
TW F4	120	60	36	24	17	12	9	6	4
TW G1	175	58	30	18	12	8	6	4	2
TW G2	128	57	33	21	15	10	7	5	3
TW G3	108	52	31	21	14	10	7	5	3
TW G4	87	43	26	18	12	9	6	4	3
TW H1	113	47	26	17	11	8	6	4	2
TW H2	88	41	25	16	11	8	6	4	2
TW H3	73	35	21	14	10	7	5	3	2
TW H4	57	27	16	11	8	5	4	3	2
TW I1	104	39	21	12	8	5	3	1	-
TW I2	153	76	47	31	22	16	12	8	5
TW I3	164	85	53	36	26	19	13	9	6
TW I4	167	94	61	43	31	23	17	12	8
TW I5	163	94	62	44	32	24	18	13	9
TW J1	200	105	66	45	32	23	17	12	8
TW J2	180	99	64	44	32	23	17	12	8
TW J3	196	119	81	58	43	32	24	17	12
TW J4	180	109	74	54	40	30	22	16	11
DG A1	134	52	24	-	-	-	-	-	-
DG A2	161	69	40	25	16	10	5	-	-
DG A3	207	93	56	38	27	20	14	9	4
DG B1	134	54	29	16	-	-	-	-	-
DG B2	166	72	43	28	20	14	9	4	-
DG B3	212	95	58	40	30	22	17	12	7
DG C1	171	71	40	26	17	10	-	-	-
DG C2	199	87	52	36	26	18	13	8	3
DG C3	246	111	68	47	35	27	20	15	9
SS A1	470	135	67	39	25	16	11	7	4
SS A2	423	173	97	61	41	29	20	14	8
SS A3	453	121	58	33	20	13	8	5	3
SS A4	432	172	95	59	40	27	19	12	8
SS A5	789	135	60	33	20	13	8	5	3
SS A6	420	171	95	60	41	28	19	13	8
SS A7	894	113	47	25	14	8	5	3	1
SS A8	433	168	92	57	38	26	18	12	7
SS B1	313	99	51	30	19	13	8	5	3
SS B2	326	138	78	50	34	24	17	12	7
SS B3	288	90	45	26	16	11	7	4	2
SS B4	321	134	76	48	33	23	16	11	7
SS B5	407	89	41	23	14	9	6	3	2
SS B6	306	132	75	48	33	23	16	11	7
SS B7	360	74	33	18	11	6	4	2	-
SS B8	299	126	71	45	31	21	15	10	6
SS C1	181	78	44	28	19	13	9	6	4
SS C2	198	100	61	41	29	21	15	11	7
SS C3	190	63	33	20	13	9	6	4	2
SS C4	197	90	53	35	24	17	12	9	6
SS D1	426	159	85	52	34	23	15	10	6
SS D2	368	186	113	76	53	37	27	18	11
SS D3	543	152	75	44	28	18	12	8	4
SS D4	378	183	109	72	50	35	25	17	11
SS E1	308	120	65	40	26	17	11	7	4
SS E2	298	149	91	60	42	30	21	14	9
SS E3	379	105	51	30	19	12	8	5	3
SS E4	309	141	82	54	37	26	18	12	8

LCR *Tables* / 473

NEEDLES, CALIFORNIA								1428 DD		NEEDLES, CALIFORNIA								CONTINUED	
SSF =	.10	.20	.30	.40	.50	.60	.70	.80	.90	SSF =	.10	.20	.30	.40	.50	.60	.70	.80	.90
WW A1	3252	394	188	114	77	54	38	27	17	DG C2	735	350	225	160	120	92	71	53	36
WW A2	1021	409	238	157	111	80	58	41	27	DG C3	808	386	248	179	134	104	81	62	43
WW A3	845	408	252	172	123	91	67	48	31	SS A1	1567	487	261	165	113	80	57	40	25
WW A4	769	403	257	179	130	96	71	51	34	SS A2	1313	563	330	219	154	111	80	57	37
WW A5	729	406	266	187	137	102	76	55	37	SS A3	1532	453	238	150	101	71	51	35	22
WW A6	704	405	269	191	141	105	79	57	38	SS A4	1334	561	326	215	150	108	78	55	36
WW B1	927	347	197	129	90	65	47	33	21	SS A5	2658	500	244	148	99	69	49	34	21
WW B2	726	392	254	178	130	96	72	52	35	SS A6	1317	564	331	219	154	111	80	57	37
WW B3	824	466	308	218	160	120	89	65	43	SS A7	3044	437	205	122	81	56	39	27	17
WW B4	694	424	290	210	157	119	89	66	44	SS A8	1355	557	322	211	147	106	76	54	35
WW B5	625	390	270	197	148	112	85	62	42	SS B1	1091	377	208	133	92	65	47	33	21
WW C1	862	488	321	228	167	125	93	68	45	SS B2	1024	453	270	180	127	92	67	48	31
WW C2	734	433	291	209	155	117	87	64	43	SS B3	1024	354	194	124	86	61	43	30	20
WW C3	706	457	323	239	181	138	105	77	53	SS B4	1007	444	263	175	124	90	65	46	30
WW C4	636	411	291	215	163	124	95	70	48	SS B5	1445	348	177	110	74	52	37	26	16
TW A1	2749	370	178	109	73	51	37	26	16	SS B6	976	441	264	177	125	91	66	47	31
TW A2	1048	370	208	136	95	68	49	35	23	SS B7	1302	303	153	94	63	44	31	22	14
TW A3	830	361	216	145	103	75	55	39	26	SS B8	956	425	253	169	119	86	63	44	29
TW A4	694	337	209	143	103	76	56	40	26	SS C1	778	361	220	149	106	78	57	41	27
TW B1	1504	340	175	110	75	53	38	27	17	SS C2	752	397	254	177	129	95	71	51	34
TW B2	828	329	191	126	89	64	47	33	22	SS C3	812	292	165	107	75	54	39	28	18
TW B3	713	314	188	126	90	65	48	34	22	SS C4	747	358	221	151	108	79	58	42	28
TW B4	649	286	171	115	82	60	44	31	21	SS D1	1434	579	333	218	152	109	78	55	36
TW C1	963	298	163	105	72	52	37	26	17	SS D2	1148	607	385	266	192	141	104	74	49
TW C2	717	278	160	106	74	54	39	28	18	SS D3	1816	553	294	186	127	90	64	44	29
TW C3	680	258	148	97	68	49	36	25	17	SS D4	1178	596	372	254	182	133	98	70	46
TW C4	712	229	126	81	56	41	29	21	14	SS E1	1112	473	277	183	128	92	67	47	31
TW D1	665	266	154	102	71	52	37	26	17	SS E2	977	514	327	226	163	120	88	63	42
TW D2	779	373	230	157	113	83	61	44	29	SS E3	1345	409	218	138	94	67	48	33	21
TW D3	786	405	257	178	129	95	70	51	34	SS E4	1008	485	297	201	144	105	77	55	36
TW D4	718	401	263	186	137	102	76	55	37										
TW D5	650	373	249	177	131	98	74	54	36										
TW E1	892	459	291	202	146	108	80	58	38	OAKLAND, CALIFORNIA								2909 DD	
TW E2	797	429	277	194	142	106	78	57	38	SSF =	.10	.20	.30	.40	.50	.60	.70	.80	.90
TW E3	796	476	323	232	173	130	98	71	48	WW A1	1847	243	116	70	46	32	22	15	10
TW E4	704	426	291	210	157	119	89	65	44	WW A2	630	258	150	98	68	49	35	25	16
TW F1	1244	319	167	106	72	51	37	26	17	WW A3	531	259	159	108	76	55	40	29	19
TW F2	750	314	185	123	87	63	46	33	22	WW A4	487	257	163	112	80	59	43	31	20
TW F3	611	297	184	126	90	66	49	35	23	WW A5	463	259	168	118	85	63	46	33	22
TW F4	495	261	167	117	85	63	46	33	22	WW A6	449	258	171	120	88	65	48	35	23
TW G1	758	271	153	100	69	50	36	26	17	WW B1	567	218	123	79	54	38	27	19	12
TW G2	537	255	156	107	76	56	41	29	19	WW B2	465	253	163	113	82	60	44	32	21
TW G3	446	228	144	100	72	53	39	28	19	WW B3	543	309	203	142	103	76	56	41	27
TW G4	354	186	119	83	60	44	33	24	16	WW B4	456	279	190	137	102	76	57	42	28
TW H1	483	215	129	87	62	45	33	23	15	WW B5	413	258	178	129	96	73	55	40	27
TW H2	363	183	115	79	57	42	31	22	15	WW C1	571	324	213	149	109	80	59	43	28
TW H3	299	153	97	67	48	36	26	19	13	WW C2	479	284	190	135	100	74	55	40	27
TW H4	229	115	73	50	36	27	20	14	9	WW C3	480	311	218	160	121	91	69	51	34
TW I1	521	226	135	90	64	46	33	24	16	WW C4	425	275	193	142	107	81	62	45	31
TW I2	596	312	199	139	101	75	55	40	26	TW A1	1574	229	111	67	44	31	22	15	9
TW I3	619	336	218	152	111	83	61	44	30	TW A2	641	234	131	85	58	41	30	21	13
TW I4	595	346	231	165	122	92	69	50	34	TW A3	518	230	136	91	64	46	33	23	15
TW I5	564	337	228	164	122	92	69	51	34	TW A4	438	215	133	90	64	47	34	24	16
TW J1	735	402	261	183	134	100	74	54	36	TW B1	888	213	110	68	46	32	23	16	10
TW J2	653	371	245	174	128	96	72	52	35	TW B2	513	209	121	79	55	39	28	20	13
TW J3	656	409	283	206	155	117	89	65	44	TW B3	447	200	119	79	56	40	29	21	13
TW J4	602	376	260	190	143	108	82	60	41	TW B4	408	183	109	73	51	37	27	19	12
DG A1	657	303	186	126	88	63	44	29	16	TW C1	584	189	103	65	45	32	23	16	10
DG A2	632	297	185	129	93	68	49	34	21	TW C2	445	177	102	67	46	33	24	17	11
DG A3	704	334	212	150	111	84	62	44	27	TW C3	422	165	94	62	43	31	22	15	10
DG B1	650	307	195	137	100	76	56	39	23	TW C4	438	146	81	52	35	25	18	13	8
DG B2	643	305	196	138	103	79	60	44	28	TW D1	410	168	96	63	43	31	22	15	10
DG B3	708	337	217	155	116	89	69	51	34	TW D2	495	240	148	100	71	52	38	27	18
DG C1	753	358	229	163	121	92	71	52	33	TW D3	513	267	168	115	83	60	44	31	21

474 / APPENDIX 21

OAKLAND, CALIFORNIA						CONTINUED				RED BLUFF, CALIFORNIA									2688 DD
SSF =	.10	.20	.30	.40	.50	.60	.70	.80	.90	SSF =	.10	.20	.30	.40	.50	.60	.70	.80	.90
TW D4	468	263	172	121	88	65	48	35	23	WW A1	1373	159	71	41	26	18	12	8	4
TW D5	427	247	164	116	85	63	47	34	23	WW A2	440	169	93	59	40	28	20	13	8
TW E1	585	305	192	132	95	69	51	36	24	WW A3	364	168	100	66	46	33	23	16	10
TW E2	517	281	181	126	91	67	50	36	24	WW A4	331	166	102	69	48	35	25	17	11
TW E3	536	322	217	155	115	86	64	46	31	WW A5	313	167	106	72	52	37	27	19	12
TW E4	467	283	193	139	103	77	58	42	28	WW A6	303	167	107	74	53	39	28	20	12
TW F1	736	199	104	65	44	31	22	15	10	WW B1	375	132	70	43	28	19	13	8	5
TW F2	464	198	116	77	54	38	28	19	13	WW B2	321	167	104	71	51	37	27	19	12
TW F3	383	188	116	79	56	41	29	21	14	WW B3	369	202	129	89	64	47	34	24	15
TW F4	314	167	106	73	53	39	28	20	13	WW B4	318	188	126	89	66	49	36	26	17
TW G1	461	170	95	62	42	30	22	15	10	WW B5	290	176	119	85	63	47	35	25	17
TW G2	336	162	99	66	47	34	25	18	11	WW C1	390	213	136	94	68	50	36	26	16
TW G3	281	145	91	62	45	33	24	17	11	WW C2	334	191	125	88	64	47	35	25	16
TW G4	225	119	75	52	37	27	20	14	9	WW C3	333	210	145	105	79	59	45	32	21
TW H1	300	136	81	54	38	27	20	14	9	WW C4	300	189	131	95	71	54	40	29	20
TW H2	229	116	73	50	35	26	19	13	9	TW A1	1170	151	68	40	25	17	12	8	4
TW H3	189	98	61	42	30	22	16	11	8	TW A2	453	154	82	51	35	24	17	11	7
TW H4	145	74	46	32	23	17	12	9	6	TW A3	360	150	86	55	38	27	19	13	8
TW I1	322	142	84	55	38	27	19	14	9	TW A4	302	141	84	55	39	28	20	14	9
TW I2	381	201	128	88	63	46	34	24	16	TW B1	646	141	68	41	27	18	13	8	5
TW I3	406	222	143	99	72	53	39	28	18	TW B2	361	138	76	48	33	23	16	11	7
TW I4	389	227	151	107	79	59	44	32	21	TW B3	312	132	75	49	34	24	17	12	7
TW I5	371	222	150	107	79	59	44	32	22	TW B4	286	121	69	45	31	22	16	11	7
TW J1	485	267	172	120	87	64	47	34	22	TW C1	419	125	65	40	27	19	13	9	5
TW J2	425	243	160	113	82	61	45	33	22	TW C2	316	118	65	41	28	20	14	10	6
TW J3	444	277	191	138	103	78	58	42	28	TW C3	301	110	60	38	26	18	13	9	5
TW J4	400	250	173	125	94	71	53	39	26	TW C4	317	99	52	32	22	15	11	7	5
DG A1	403	187	112	73	49	33	21	12	-	TW D1	270	102	55	34	23	16	11	7	4
DG A2	401	190	118	80	57	41	29	19	11	TW D2	347	160	95	63	44	32	23	16	10
DG A3	459	222	139	97	71	53	39	27	16	TW D3	351	174	106	71	50	36	26	18	12
DG B1	401	192	118	81	58	42	30	19	8	TW D4	329	178	114	79	57	42	30	22	14
DG B2	409	199	124	86	64	48	35	25	15	TW D5	303	170	110	77	56	41	30	22	14
DG B3	462	226	143	100	75	57	44	32	21	TW E1	404	201	123	83	59	42	31	21	14
DG C1	471	228	142	98	72	54	40	28	15	TW E2	364	190	119	81	58	42	31	22	14
DG C2	471	230	145	101	75	57	43	31	20	TW E3	373	217	143	101	74	55	41	29	19
DG C3	528	259	165	116	87	67	52	39	26	TW E4	331	195	130	92	68	51	38	27	18
SS A1	1173	383	204	126	84	57	39	26	16	TW F1	527	129	64	38	25	17	12	8	5
SS A2	1020	449	262	170	116	81	57	39	24	TW F2	322	129	72	46	32	22	16	11	6
SS A3	1174	366	192	117	76	51	35	23	14	TW F3	263	122	72	48	33	24	17	12	7
SS A4	1064	460	265	171	116	80	56	38	23	TW F4	214	108	67	45	32	23	16	11	7
SS A5	1919	389	191	114	73	49	33	22	13	TW G1	323	110	59	37	25	17	12	8	5
SS A6	1022	449	262	170	116	81	57	39	24	TW G2	231	105	61	40	28	20	14	10	6
SS A7	2225	349	164	96	61	40	26	17	10	TW G3	192	94	57	38	27	19	14	10	6
SS A8	1077	456	261	168	113	78	54	37	23	TW G4	154	77	48	32	23	16	12	8	5
SS B1	804	289	158	99	66	46	32	21	13	TW H1	207	88	50	32	22	16	11	8	5
SS B2	780	353	209	137	95	67	47	33	21	TW H2	157	75	45	30	21	15	11	7	5
SS B3	769	276	151	94	62	43	29	20	12	TW H3	130	64	39	26	18	13	9	7	4
SS B4	783	353	208	136	93	65	46	32	20	TW H4	100	49	29	20	14	10	7	5	3
SS B5	1032	264	134	81	53	36	25	16	10	TW I1	209	85	47	30	20	14	9	6	4
SS B6	744	343	204	134	93	65	46	32	20	TW I2	264	133	82	55	39	28	21	14	9
SS B7	945	235	118	71	46	31	21	14	8	TW I3	277	145	91	62	44	32	23	16	10
SS B8	743	337	199	130	89	63	44	30	19	TW I4	272	154	100	70	51	37	28	20	13
SS C1	488	229	139	93	65	47	34	24	16	TW I5	262	152	100	71	52	39	29	20	13
SS C2	486	258	165	114	82	60	44	32	21	TW J1	333	176	110	76	54	39	29	20	13
SS C3	499	185	104	67	46	33	23	16	10	TW J2	297	164	105	73	53	39	28	20	13
SS C4	479	233	143	97	69	50	36	26	17	TW J3	308	187	126	91	67	50	37	27	18
SS D1	1099	457	260	166	112	78	54	36	22	TW J4	283	172	117	84	62	46	35	25	17
SS D2	908	484	304	206	145	104	74	51	32	DG A1	269	114	63	38	23	12	-	-	-
SS D3	1357	434	231	142	94	64	44	29	18	DG A2	283	125	74	49	33	23	15	9	3
SS D4	929	475	294	197	137	98	69	48	30	DG A3	338	152	93	63	45	33	24	16	8
SS E1	832	362	210	135	92	64	45	30	19	DG B1	273	116	69	45	30	20	11	-	-
SS E2	752	399	251	170	120	86	61	43	27	DG B2	295	130	78	54	38	27	19	13	6
SS E3	978	312	166	103	68	46	32	21	13	DG B3	345	155	95	66	49	36	27	19	12
SS E4	771	377	229	152	106	75	53	37	23	DG C1	330	141	84	57	40	28	19	10	-

RED BLUFF, CALIFORNIA   CONTINUED

	SSF =	.10	.20	.30	.40	.50	.60	.70	.80	.90
DG	C2	344	152	92	64	46	34	25	17	10
DG	C3	396	179	109	77	57	43	32	24	15
SS	A1	731	216	109	66	43	29	20	13	8
SS	A2	633	262	147	94	64	45	31	21	13
SS	A3	710	198	97	57	37	24	16	11	6
SS	A4	644	260	145	92	62	43	30	20	13
SS	A5	1249	221	101	58	36	24	16	10	6
SS	A6	631	260	146	93	63	44	31	21	13
SS	A7	1438	189	82	45	28	18	12	7	4
SS	A8	650	257	141	89	60	41	29	19	12
SS	B1	502	165	86	52	34	24	16	11	6
SS	B2	494	211	121	78	53	38	27	18	12
SS	B3	467	152	78	47	31	21	14	9	5
SS	B4	486	206	117	75	51	36	25	17	11
SS	B5	663	151	72	42	27	18	12	8	5
SS	B6	468	203	117	76	52	37	26	18	11
SS	B7	592	129	60	35	22	14	9	6	3
SS	B8	457	195	111	71	49	34	24	16	10
SS	C1	331	146	85	55	38	27	19	13	8
SS	C2	336	171	106	71	51	37	27	19	12
SS	C3	348	119	64	40	27	19	13	9	5
SS	C4	335	155	92	61	43	31	22	15	10
SS	D1	664	255	139	87	58	40	27	18	11
SS	D2	550	280	171	115	80	58	41	29	18
SS	D3	846	245	123	74	48	32	22	14	9
SS	D4	565	275	165	109	76	54	39	27	17
SS	E1	502	203	112	70	47	33	22	15	9
SS	E2	460	233	142	95	67	48	34	24	15
SS	E3	615	177	89	53	35	23	16	10	6
SS	E4	477	221	130	85	59	42	29	20	13

SACRAMENTO, CALIFORNIA   2843 DD

	SSF =	.10	.20	.30	.40	.50	.60	.70	.80	.90
WW	A1	1440	164	74	43	27	18	12	8	4
WW	A2	450	174	97	62	42	29	20	14	8
WW	A3	373	173	103	68	47	33	24	16	10
WW	A4	340	172	106	71	50	36	25	17	11
WW	A5	322	173	109	75	53	38	27	19	12
WW	A6	311	173	111	77	55	40	29	20	12
WW	B1	388	137	74	45	30	20	13	8	5
WW	B2	329	172	108	74	52	38	27	19	12
WW	B3	380	209	134	92	66	48	35	24	15
WW	B4	325	194	130	92	67	50	37	26	17
WW	B5	297	181	123	88	65	48	36	26	17
WW	C1	402	221	142	98	70	51	37	26	16
WW	C2	342	197	129	91	66	48	35	25	16
WW	C3	342	217	150	109	81	61	45	33	21
WW	C4	307	194	135	98	73	55	41	30	20
TW	A1	1222	155	71	41	27	18	12	8	4
TW	A2	464	158	85	53	36	25	17	11	7
TW	A3	369	155	89	58	39	28	19	13	8
TW	A4	309	145	87	57	40	28	20	14	8
TW	B1	664	145	71	42	28	19	13	8	5
TW	B2	369	142	79	50	34	24	17	11	7
TW	B3	319	136	78	51	35	25	17	12	7
TW	B4	292	124	72	47	32	23	16	11	7
TW	C1	429	129	67	42	28	19	13	9	5
TW	C2	323	121	67	43	29	20	14	10	6
TW	C3	308	113	62	40	27	19	13	9	5
TW	C4	324	101	54	34	23	16	11	7	4
TW	D1	280	106	58	36	24	16	11	7	4
TW	D2	355	165	98	65	45	32	23	16	10
TW	D3	361	180	111	74	52	37	27	18	12

SACRAMENTO, CALIFORNIA   CONTINUED

	SSF =	.10	.20	.30	.40	.50	.60	.70	.80	.90
TW	D4	337	183	117	81	58	43	31	22	14
TW	D5	310	174	113	79	57	42	31	22	14
TW	E1	416	208	128	86	61	44	31	22	14
TW	E2	372	195	123	84	60	43	31	22	14
TW	E3	383	224	148	105	76	56	41	29	19
TW	E4	338	200	134	95	70	52	38	27	18
TW	F1	541	133	66	40	26	18	12	8	5
TW	F2	329	133	75	48	33	23	16	11	6
TW	F3	269	126	75	50	34	24	17	12	7
TW	F4	219	112	69	46	33	23	17	11	7
TW	G1	331	114	61	38	26	18	12	8	5
TW	G2	237	108	64	42	29	20	14	10	6
TW	G3	197	97	59	40	28	20	14	10	6
TW	G4	157	80	49	33	23	17	12	8	5
TW	H1	212	90	52	34	23	16	11	8	5
TW	H2	160	78	47	31	22	16	11	8	5
TW	H3	133	66	40	27	19	13	10	7	4
TW	H4	102	50	30	20	14	10	7	5	3
TW	I1	216	89	50	32	21	14	10	6	3
TW	I2	270	137	85	57	40	29	21	14	9
TW	I3	285	150	94	64	46	33	24	16	10
TW	I4	279	158	103	72	52	38	28	20	13
TW	I5	267	156	103	73	53	39	29	21	13
TW	J1	342	182	115	78	56	40	29	20	13
TW	J2	304	168	109	75	54	40	29	20	13
TW	J3	317	193	131	94	69	51	38	27	18
TW	J4	289	177	120	86	64	48	35	25	17
DG	A1	276	119	66	40	24	13	-	-	-
DG	A2	289	129	77	50	34	23	15	9	2
DG	A3	343	156	96	66	47	34	24	16	8
DG	B1	278	122	72	47	31	20	11	-	-
DG	B2	298	134	82	56	39	28	20	13	6
DG	B3	348	160	99	69	50	37	28	20	12
DG	C1	334	148	89	60	42	29	19	10	-
DG	C2	347	157	96	67	48	35	25	17	10
DG	C3	400	184	114	80	59	44	33	24	15
SS	A1	798	237	121	72	46	31	21	13	8
SS	A2	684	284	161	102	69	48	33	22	14
SS	A3	783	219	109	64	40	26	17	11	6
SS	A4	702	285	159	100	67	46	32	21	13
SS	A5	1375	241	112	64	40	26	17	11	6
SS	A6	683	283	160	101	68	47	32	22	13
SS	A7	1607	210	92	51	31	20	12	8	4
SS	A8	709	282	156	97	65	44	30	20	12
SS	B1	542	179	94	57	37	25	17	11	6
SS	B2	529	227	131	84	57	40	28	19	12
SS	B3	508	167	87	52	34	22	15	9	6
SS	B4	522	223	128	82	55	38	27	18	11
SS	B5	721	163	79	46	29	19	13	8	5
SS	B6	501	219	126	81	55	39	27	18	11
SS	B7	649	141	67	38	24	15	10	6	3
SS	B8	492	212	121	77	52	36	25	17	10
SS	C1	340	151	88	58	40	28	19	13	8
SS	C2	344	176	109	74	52	38	27	19	12
SS	C3	357	123	66	42	28	19	13	9	5
SS	C4	343	159	95	63	44	31	22	15	10
SS	D1	722	280	153	95	63	42	29	19	11
SS	D2	594	305	187	124	86	61	43	30	18
SS	D3	924	268	136	81	52	35	23	15	9
SS	D4	610	300	180	119	82	57	40	28	17
SS	E1	542	220	123	77	51	34	23	15	9
SS	E2	493	251	154	102	71	50	35	24	15
SS	E3	665	192	97	58	37	25	16	11	6
SS	E4	511	238	141	92	63	44	31	21	13

476 / APPENDIX 21

```
SAN DIEGO, CALIFORNIA                              1507 DD    SAN DIEGO, CALIFORNIA                              CONTINUED
  SSF =.10   .20   .30   .40   .50   .60   .70   .80   .90      SSF =.10   .20   .30   .40   .50   .60   .70   .80   .90
WW A1 3585   454   216   131    88    62    45    32    21    DG C2   854   397   250   179   135   107    84    64    44
WW A2 1165   468   271   179   127    93    68    49    33    DG C3   934   436   275   197   150   119    95    74    52
WW A3  968   466   287   196   141   105    78    57    38    SS A1  2065   655   353   224   153   109    78    55    36
WW A4  881   460   292   204   148   111    83    61    41    SS A2  1730   750   442   293   207   149   109    78    51
WW A5  834   462   302   213   157   118    89    65    45    SS A3  2079   629   334   210   143   101    72    50    33
WW A6  805   461   305   218   161   122    92    68    46    SS A4  1806   769   449   297   208   150   108    77    50
WW B1 1082   408   231   151   106    77    56    41    27    SS A5  3469   673   331   202   136    95    67    47    30
WW B2  830   446   287   202   148   111    83    61    42    SS A6  1737   753   443   294   207   150   109    78    51
WW B3  955   539   355   252   186   140   105    78    53    SS A7  4082   607   288   173   115    79    56    39    25
WW B4  788   479   327   238   178   136   103    77    53    SS A8  1836   766   444   292   204   147   106    75    49
WW B5  709   440   304   222   168   128    98    73    50    SS B1  1425   500   276   178   123    88    63    45    29
WW C1  998   563   370   263   194   146   110    81    55    SS B2  1331   593   353   236   167   121    89    64    42
WW C2  835   490   329   236   176   134   101    75    52    SS B3  1367   480   265   170   117    83    60    42    28
WW C3  811   523   369   273   208   160   123    92    64    SS B4  1334   593   353   235   166   120    88    63    41
WW C4  719   463   327   242   184   142   109    82    57    SS B5  1875   462   236   147    99    70    50    35    23
TW A1 3035   426   204   125    84    60    43    31    20    SS B6  1271   577   346   232   164   120    88    63    42
TW A2 1190   424   237   155   108    79    58    42    28    SS B7  1727   413   210   130    87    61    44    30    20
TW A3  948   413   245   165   117    86    64    46    31    SS B8  1270   570   339   226   160   116    85    60    40
TW A4  793   385   238   163   118    87    65    47    32    SS C1   898   416   252   171   123    90    67    49    33
TW B1 1683   391   200   125    86    61    44    32    21    SS C2   864   454   290   202   147   110    82    61    41
TW B2  943   376   217   144   102    74    55    40    27    SS C3   931   337   189   123    86    63    46    33    22
TW B3  815   358   213   144   102    75    56    40    27    SS C4   858   410   252   172   124    92    68    50    34
TW B4  741   326   194   131    93    69    51    37    25    SS D1  1904   779   450   295   206   149   108    76    50
TW C1 1088   342   185   119    83    60    44    31    21    SS D2  1520   808   515   357   258   190   140   101    67
TW C2  816   318   182   120    85    62    46    33    22    SS D3  2392   744   399   253   173   123    88    62    40
TW C3  773   295   168   111    78    57    42    30    20    SS D4  1558   794   497   340   244   179   132    95    63
TW C4  804   262   143    92    64    47    34    25    16    SS E1  1463   629   369   245   172   124    90    65    43
TW D1  777   313   180   119    84    61    45    32    22    SS E2  1278   675   429   297   215   159   117    85    57
TW D2  889   425   261   178   128    95    71    52    35    SS E3  1755   544   291   185   127    90    65    46    30
TW D3  912   469   296   205   149   111    83    61    41    SS E4  1316   637   391   265   190   139   102    73    49
TW D4  817   454   297   210   155   117    88    65    44
TW D5  738   422   280   200   148   112    85    62    43
TW E1 1033   531   336   233   169   126    94    69    47    SAN FRANCISCO, CALIFORNIA                           3042 DD
TW E2  908   487   314   220   161   121    91    67    46      SSF =.10   .20   .30   .40   .50   .60   .70   .80   .90
TW E3  918   547   370   267   199   151   115    85    58    WW A1 1752   230   110    67    44    30    21    15     9
TW E4  798   481   327   237   178   135   103    76    53    WW A2  600   246   143    93    65    47    33    24    15
TW F1 1400   367   191   121    83    60    43    31    21    WW A3  506   247   152   103    73    53    38    27    18
TW F2  857   360   211   141   100    73    54    39    26    WW A4  464   245   155   107    77    56    41    29    19
TW F3  699   339   209   143   103    77    57    42    28    WW A5  442   247   161   112    81    60    44    32    21
TW F4  567   298   190   133    97    72    54    40    27    WW A6  428   247   163   115    84    62    46    33    22
TW G1  863   311   174   114    80    58    42    31    20    WW B1  538   207   117    75    51    36    26    18    11
TW G2  615   291   178   121    87    65    48    35    24    WW B2  444   241   155   108    78    58    42    30    20
TW G3  510   260   164   113    82    61    46    34    23    WW B3  519   296   194   136    99    73    54    39    26
TW G4  405   212   135    94    69    51    38    28    19    WW B4  436   267   182   131    97    73    55    40    27
TW H1  553   246   147    99    71    52    38    28    19    WW B5  396   247   171   124    92    70    52    38    26
TW H2  416   209   131    90    65    49    36    27    18    WW C1  546   310   204   143   104    77    57    41    27
TW H3  342   175   110    76    55    41    31    23    15    WW C2  458   271   182   130    96    71    53    39    26
TW H4  261   132    83    57    41    31    23    17    11    WW C3  461   298   210   154   116    88    66    49    33
TW I1  610   266   157   105    75    55    40    29    20    WW C4  407   264   186   137   103    78    59    44    30
TW I2  681   355   226   157   115    85    64    47    32    TW A1 1493   218   105    64    42    29    20    14     9
TW I3  718   388   251   176   129    97    73    53    36    TW A2  609   223   125    81    56    39    28    20    13
TW I4  676   392   261   187   139   105    80    59    40    TW A3  493   219   130    86    61    44    32    22    15
TW I5  641   381   257   185   139   105    80    59    41    TW A4  417   205   127    86    61    44    32    23    15
TW J1  852   464   301   211   155   116    87    64    44    TW B1  843   203   104    65    44    30    22    15    10
TW J2  743   420   277   197   146   110    83    61    42    TW B2  488   199   115    76    53    38    27    19    12
TW J3  755   469   324   237   178   136   104    77    53    TW B3  425   191   114    76    53    38    28    20    13
TW J4  681   423   293   214   162   124    94    70    49    TW B4  388   174   104    69    49    35    26    18    12
DG A1  757   348   213   145   103    74    53    37    22    TW C1  556   180    98    62    43    30    22    15    10
DG A2  727   337   210   146   107    79    58    42    27    TW C2  424   169    97    64    44    32    23    16    10
DG A3  809   376   237   168   126    96    73    53    34    TW C3  402   157    90    59    41    29    21    15    10
DG B1  759   350   219   155   117    90    68    49    31    TW C4  417   140    77    49    34    24    17    12     8
DG B2  747   347   218   155   117    91    70    52    35    TW D1  389   159    92    60    41    29    21    14     9
DG B3  817   381   240   172   131   103    81    61    42    TW D2  472   229   141    96    68    49    36    26    17
DG C1  882   408   256   183   138   109    85    64    43    TW D3  491   255   161   111    79    58    42    30    20
```

LCR *Tables* / 477

SAN FRANCISCO, CALIFORNIA						CONTINUED				SANTA MARIA, CALIFORNIA								3053 DD	
SSF	=.10	.20	.30	.40	.50	.60	.70	.80	.90		SSF	=.10	.20	.30	.40	.50	.60	.70	.80	.90
TW D4	448	252	165	116	84	63	46	34	22	WW A1	1776	240	119	73	50	35	25	18	12
TW D5	409	237	157	111	82	61	45	33	22	WW A2	617	259	154	103	74	54	39	28	19
TW E1	560	292	184	127	91	66	49	35	23	WW A3	523	261	164	114	82	61	45	33	22
TW E2	494	268	173	120	87	64	47	34	23	WW A4	482	260	169	119	87	65	48	35	24
TW E3	514	309	208	149	110	82	61	45	30	WW A5	461	263	175	125	92	69	52	38	26
TW E4	447	272	185	133	99	74	56	41	27	WW A6	447	263	177	128	95	72	54	40	27
TW F1	699	189	99	62	42	29	21	14	9	WW B1	556	220	128	85	60	43	32	23	15
TW F2	441	189	111	73	51	37	26	19	12	WW B2	462	256	168	119	88	66	49	36	25
TW F3	365	179	111	75	53	39	28	20	13	WW B3	542	315	211	151	112	85	64	47	32
TW F4	299	159	101	70	50	37	27	19	13	WW B4	455	283	197	144	109	83	63	47	32
TW G1	439	162	91	59	40	29	20	14	9	WW B5	414	263	184	136	103	79	60	45	31
TW G2	320	154	94	63	45	33	24	17	11	WW C1	569	330	221	159	118	89	67	49	33
TW G3	268	138	87	60	43	31	23	16	11	WW C2	478	288	197	143	107	81	61	45	31
TW G4	214	113	72	50	36	26	19	14	9	WW C3	483	318	228	170	130	100	77	57	40
TW H1	286	129	77	51	36	26	19	13	9	WW C4	426	280	200	149	114	88	68	51	35
TW H2	218	111	69	47	34	25	18	13	8	TW A1	1515	227	113	70	48	34	24	17	11
TW H3	180	93	59	40	29	21	15	11	7	TW A2	625	234	134	89	63	46	33	24	16
TW H4	138	71	44	30	22	16	12	8	5	TW A3	508	231	140	95	68	50	37	27	18
TW I1	306	135	80	52	36	26	18	13	8	TW A4	431	217	137	95	69	51	38	28	19
TW I2	363	192	122	84	61	44	33	23	15	TW B1	859	212	112	71	49	35	25	18	12
TW I3	388	212	137	95	69	51	37	27	18	TW B2	502	209	124	83	59	43	32	23	15
TW I4	372	217	145	103	76	56	42	30	20	TW B3	438	201	123	84	60	44	33	24	16
TW I5	355	213	144	103	76	57	43	31	21	TW B4	400	184	112	76	55	40	30	22	14
TW J1	464	255	165	115	83	61	45	32	21	TW C1	568	188	105	69	48	35	25	18	12
TW J2	407	232	153	108	79	59	44	31	21	TW C2	435	178	105	70	50	36	27	19	13
TW J3	426	266	183	133	99	75	56	41	27	TW C3	413	165	97	64	46	33	25	18	12
TW J4	384	240	166	121	90	68	51	37	25	TW C4	426	146	82	54	38	27	20	14	10
DG A1	382	177	106	69	46	31	19	10	-	TW D1	403	170	101	67	48	35	25	18	12
DG A2	382	181	112	76	54	39	27	18	10	TW D2	488	242	152	105	76	57	42	31	21
DG A3	438	212	133	93	68	51	37	26	15	TW D3	509	271	175	123	90	67	50	36	25
DG B1	380	182	112	77	55	40	28	17	7	TW D4	464	266	177	127	94	71	53	39	27
DG B2	390	189	118	82	61	45	34	24	14	TW D5	425	250	169	122	91	69	52	38	26
DG B3	442	216	137	96	72	55	42	31	20	TW E1	581	309	199	140	102	76	57	42	28
DG C1	447	217	135	93	68	51	38	26	14	TW E2	512	283	186	132	97	73	55	40	27
DG C2	449	219	138	96	72	54	41	30	19	TW E3	537	328	225	164	123	94	71	53	36
DG C3	505	248	158	111	83	64	50	38	25	TW E4	466	287	199	145	109	83	63	47	32
SS A1	1129	369	198	122	81	55	38	25	15	TW F1	713	198	107	68	47	34	25	18	12
SS A2	984	434	254	165	113	79	56	38	24	TW F2	455	199	120	81	58	42	31	22	15
SS A3	1130	353	186	113	74	50	34	22	13	TW F3	378	190	120	83	60	45	33	24	16
SS A4	1028	445	258	166	113	78	55	37	23	TW F4	311	169	110	77	57	42	32	23	16
SS A5	1843	375	184	110	71	47	32	21	13	TW G1	450	170	98	65	46	33	24	17	12
SS A6	986	434	254	165	113	79	55	38	24	TW G2	331	163	102	70	51	38	28	20	14
SS A7	2137	336	159	93	59	39	26	17	10	TW G3	278	147	94	66	48	36	27	20	13
SS A8	1040	442	254	163	110	76	53	36	22	TW G4	222	120	78	55	40	30	22	16	11
SS B1	772	278	153	96	64	44	31	21	13	TW H1	295	137	84	57	41	30	22	16	11
SS B2	752	341	202	133	92	65	46	32	20	TW H2	226	118	75	52	38	28	21	15	10
SS B3	739	266	145	91	60	41	28	19	12	TW H3	187	99	64	44	33	24	18	13	9
SS B4	754	341	201	132	90	64	45	31	19	TW H4	143	75	48	33	24	18	14	10	7
SS B5	990	254	129	79	51	35	24	16	10	TW I1	318	144	88	59	42	31	23	16	11
SS B6	717	331	197	130	90	63	45	31	20	TW I2	377	203	132	93	68	51	38	28	19
SS B7	906	226	114	69	44	30	20	13	8	TW I3	404	226	149	106	78	58	44	32	22
SS B8	716	326	193	126	87	61	43	29	18	TW I4	387	230	156	113	84	64	48	36	24
SS C1	464	218	132	89	62	45	33	23	15	TW I5	370	226	155	113	85	65	49	36	25
SS C2	464	247	157	109	78	58	42	30	20	TW J1	483	271	179	127	94	71	53	39	26
SS C3	474	176	99	64	44	31	22	16	10	TW J2	422	246	165	119	88	67	50	37	25
SS C4	457	222	137	93	66	48	35	25	16	TW J3	446	283	199	146	111	85	65	48	33
SS D1	1058	441	252	161	109	75	52	35	22	TW J4	400	254	178	132	100	77	58	43	30
SS D2	878	469	296	200	141	101	72	50	32	DG A1	392	188	117	79	55	38	26	16	7
SS D3	1306	419	223	138	91	62	43	28	17	DG A2	389	190	121	85	61	45	32	22	14
SS D4	897	461	285	191	134	95	68	47	30	DG A3	443	220	142	102	77	58	44	31	19
SS E1	799	349	203	131	89	62	43	29	18	DG B1	384	191	122	86	64	48	35	24	13
SS E2	725	386	243	165	116	83	60	42	26	DG B2	394	196	127	91	69	53	40	29	19
SS E3	939	300	160	99	66	45	31	21	13	DG B3	445	222	145	105	80	62	49	37	25
SS E4	742	364	221	147	103	73	52	36	23	DG C1	451	225	146	104	78	61	47	34	21

SANTA MARIA, CALIFORNIA   CONTINUED

	SSF=.10	.20	.30	.40	.50	.60	.70	.80	.90
DG C2	453	226	148	106	80	63	49	37	25
DG C3	509	254	167	121	92	73	58	45	31
SS A1	1171	396	220	142	98	69	49	34	22
SS A2	1028	468	283	190	135	98	71	50	33
SS A3	1174	380	209	133	91	64	45	31	20
SS A4	1077	481	289	193	136	98	71	50	32
SS A5	1896	400	204	127	86	60	42	29	18
SS A6	1030	468	283	190	135	97	71	50	32
SS A7	2199	359	178	109	72	50	35	24	15
SS A8	1089	478	285	190	133	96	69	48	31
SS B1	802	298	170	111	77	55	40	28	18
SS B2	785	366	224	152	108	79	57	41	27
SS B3	770	287	163	106	74	52	37	26	17
SS B4	790	368	224	152	108	78	57	40	26
SS B5	1022	271	144	91	62	44	31	22	14
SS B6	750	356	219	149	106	77	56	40	26
SS B7	937	242	127	80	54	38	27	19	12
SS B8	750	352	215	146	103	75	55	39	25
SS C1	481	232	144	99	71	52	39	28	19
SS C2	482	262	170	120	88	66	49	36	24
SS C3	487	185	107	71	50	36	27	19	13
SS C4	473	235	147	102	74	55	41	30	20
SS D1	1107	477	282	188	132	95	68	48	31
SS D2	928	511	332	232	169	125	92	66	43
SS D3	1353	449	248	160	110	78	56	39	25
SS D4	946	500	319	222	160	117	86	61	40
SS E1	838	378	227	153	108	78	56	40	26
SS E2	766	419	272	190	138	102	75	54	36
SS E3	973	322	178	115	79	56	40	28	18
SS E4	780	393	247	170	122	89	65	47	31

COLORADO SPRINGS, COLORADO   6473 DD

	SSF=.10	.20	.30	.40	.50	.60	.70	.80	.90
WW A1	760	90	42	25	17	12	8	6	4
WW A2	256	103	59	39	27	20	15	10	7
WW A3	218	105	65	44	31	23	17	12	8
WW A4	201	105	67	46	34	25	19	14	9
WW A5	193	107	70	49	36	27	20	15	10
WW A6	188	108	71	51	37	28	21	16	11
WW B1	200	74	41	26	18	13	9	7	4
WW B2	201	109	70	49	36	27	20	15	10
WW B3	236	134	88	63	46	35	26	19	13
WW B4	210	129	89	64	48	37	28	21	14
WW B5	195	122	85	62	47	36	28	21	14
WW C1	252	143	94	67	49	37	28	21	14
WW C2	218	129	87	63	47	35	27	20	14
WW C3	229	149	106	78	60	46	35	27	18
WW C4	206	134	95	71	54	42	32	24	17
TW A1	655	87	41	25	17	12	8	6	4
TW A2	263	93	52	34	23	17	12	9	6
TW A3	214	93	55	37	26	19	14	10	7
TW A4	183	89	55	37	27	20	15	11	7
TW B1	367	83	42	26	18	13	9	6	4
TW B2	213	85	49	32	23	16	12	9	6
TW B3	187	82	49	33	23	17	13	9	6
TW B4	172	76	45	30	22	16	12	8	6
TW C1	244	76	41	26	18	13	9	7	4
TW C2	188	73	42	28	19	14	10	7	5
TW C3	180	69	39	26	18	13	10	7	5
TW C4	190	62	34	22	15	11	8	6	4
TW D1	146	58	33	21	15	11	8	5	4
TW D2	214	103	63	43	31	23	17	12	8
TW D3	220	113	72	50	36	27	20	15	10

COLORADO SPRINGS, COLORADO   CONTINUED

	SSF=.10	.20	.30	.40	.50	.60	.70	.80	.90
TW D4	214	120	79	56	41	31	23	17	12
TW D5	201	116	78	56	41	31	24	18	12
TW E1	256	133	84	58	42	31	24	17	12
TW E2	234	127	82	58	42	32	24	17	12
TW E3	250	150	102	74	55	42	32	24	16
TW E4	223	136	93	68	51	39	29	22	15
TW F1	294	75	39	24	17	12	8	6	4
TW F2	188	79	46	30	21	16	11	8	6
TW F3	157	76	47	32	23	17	13	9	6
TW F4	130	69	44	30	22	16	12	9	6
TW G1	184	66	37	24	16	12	9	6	4
TW G2	137	65	40	27	19	14	10	8	5
TW G3	116	59	37	26	19	14	10	7	5
TW G4	94	49	31	22	16	12	9	6	4
TW H1	121	54	32	21	15	11	8	6	4
TW H2	94	47	30	20	15	11	8	6	4
TW H3	79	40	25	18	13	9	7	5	3
TW H4	61	31	19	13	10	7	5	4	3
TW I1	113	48	28	18	13	9	7	5	3
TW I2	164	86	55	38	28	21	15	11	8
TW I3	175	95	62	43	32	24	18	13	9
TW I4	178	104	70	50	37	28	21	16	11
TW I5	174	105	71	51	38	29	22	17	11
TW J1	213	117	76	53	39	29	22	16	11
TW J2	193	110	73	52	38	29	22	16	11
TW J3	209	131	91	67	50	39	29	22	15
TW J4	192	121	84	62	47	36	27	20	14
DG A1	145	63	35	19	8	-	-	-	-
DG A2	168	78	48	32	22	15	10	5	-
DG A3	210	100	63	45	33	25	18	13	7
DG B1	143	65	39	26	17	10	-	-	-
DG B2	171	80	50	35	26	19	14	9	4
DG B3	213	102	65	46	35	28	21	16	10
DG C1	178	82	50	35	25	19	12	6	-
DG C2	202	96	60	43	32	25	19	13	8
DG C3	245	118	75	54	41	32	26	20	13
SS A1	448	140	74	46	31	22	16	11	7
SS A2	413	179	105	70	49	35	26	18	12
SS A3	425	124	64	40	26	18	13	9	6
SS A4	417	177	102	67	47	34	24	17	11
SS A5	747	138	67	40	26	18	13	9	6
SS A6	410	177	104	69	48	35	25	18	12
SS A7	838	115	52	30	20	13	9	6	4
SS A8	417	173	99	65	45	32	23	16	11
SS B1	303	104	57	36	25	17	12	9	6
SS B2	321	143	85	57	40	29	21	15	10
SS B3	277	95	51	32	22	15	11	8	5
SS B4	313	139	82	55	39	28	20	15	10
SS B5	389	92	46	28	19	13	9	6	4
SS B6	302	138	82	55	39	28	21	15	10
SS B7	339	77	38	23	15	10	7	5	3
SS B8	293	131	78	52	36	26	19	14	9
SS C1	194	90	54	36	26	19	14	10	7
SS C2	211	112	71	50	36	27	20	15	10
SS C3	199	72	40	26	18	13	9	7	4
SS C4	208	100	62	42	30	22	17	12	8
SS D1	414	168	95	62	43	31	22	15	10
SS D2	368	196	125	86	62	46	34	24	16
SS D3	517	158	83	52	35	25	17	12	8
SS D4	376	192	120	82	59	43	32	23	15
SS E1	304	129	74	49	34	24	17	12	8
SS E2	300	159	101	69	50	37	27	20	13
SS E3	364	110	58	36	24	17	12	8	5
SS E4	308	149	91	62	44	32	24	17	11

```
DENVER, COLORADO                                    6016 DD      DENVER, COLORADO                                    CONTINUED
   SSF =.10    .20   .30   .40   .50   .60   .70   .80   .90        SSF =.10    .20   .30   .40   .50   .60   .70   .80   .90
   WW A1  729    93    44    26    17    12     9     6     4       DG C2  209    99    62    44    33    25    19    13     8
   WW A2  262   106    61    40    28    20    15    11     7       DG C3  253   121    77    55    42    33    26    20    14
   WW A3  224   108    66    45    32    24    18    13     9       SS A1  444   140    74    46    31    22    16    11     7
   WW A4  207   109    69    48    35    26    19    14     9       SS A2  413   180   105    70    49    35    26    18    12
   WW A5  198   110    72    51    37    28    21    15    10       SS A3  419   125    64    39    26    18    13     9     6
   WW A6  193   111    73    52    38    29    22    16    11       SS A4  416   177   102    67    47    34    24    17    11
   WW B1  206    77    42    27    19    13     9     7     4       SS A5  714   139    67    40    26    18    13     9     5
   WW B2  207   112    72    51    37    28    21    15    10       SS A6  410   178   104    69    48    35    25    18    12
   WW B3  242   137    90    64    47    35    27    20    13       SS A7  782   115    52    30    19    13     9     6     4
   WW B4  216   132    91    66    49    38    29    21    15       SS A8  416   173    99    65    45    32    23    16    11
   WW B5  200   125    87    64    48    37    28    21    14       SS B1  303   106    57    36    25    18    13     9     6
   WW C1  258   146    96    68    50    38    28    21    14       SS B2  322   145    86    57    40    29    21    15    10
   WW C2  224   133    89    64    48    36    27    20    14       SS B3  276    96    51    32    22    15    11     8     5
   WW C3  233   152   108    80    61    47    36    27    19       SS B4  314   140    83    55    39    28    20    15    10
   WW C4  211   137    97    72    55    42    33    24    17       SS B5  380    93    47    28    19    13     9     6     4
   TW A1  631    90    43    26    17    12     8     6     4       SS B6  304   139    83    55    39    28    21    15    10
   TW A2  267    96    54    35    24    17    13     9     6       SS B7  331    78    38    23    15    10     7     5     3
   TW A3  219    96    57    38    27    20    14    10     7       SS B8  294   132    78    52    36    26    19    14     9
   TW A4  188    92    56    38    28    20    15    11     7       SS C1  199    93    56    37    27    19    14    10     7
   TW B1  364    86    44    27    18    13     9     7     4       SS C2  216   115    73    51    37    28    21    15    10
   TW B2  217    88    50    33    23    17    12     9     6       SS C3  203    74    41    27    18    13    10     7     5
   TW B3  191    85    51    34    24    18    13     9     6       SS C4  213   103    63    43    31    23    17    12     8
   TW B4  176    78    47    31    22    16    12     9     6       SS D1  415   168    96    62    43    30    22    15    10
   TW C1  246    78    42    27    19    13    10     7     5       SS D2  370   197   125    86    62    46    34    24    16
   TW C2  192    76    43    28    20    14    11     8     5       SS D3  511   159    83    52    35    25    17    12     8
   TW C3  183    71    40    26    19    13    10     7     5       SS D4  378   192   120    82    59    43    32    23    15
   TW C4  191    63    35    22    16    11     8     6     4       SS E1  307   130    75    49    34    24    17    12     8
   TW D1  151    60    34    22    15    11     8     6     4       SS E2  303   160   101    70    50    37    27    20    13
   TW D2  220   106    65    44    32    24    17    13     9       SS E3  362   111    58    36    24    17    12     8     5
   TW D3  225   116    74    51    37    27    20    15    10       SS E4  310   150    92    62    44    32    24    17    11
   TW D4  219   123    81    57    42    32    24    18    12
   TW D5  206   119    80    57    42    32    24    18    12
   TW E1  262   136    86    60    43    32    24    17    12       EAGLE, COLORADO                                     8426 DD
   TW E2  239   130    84    59    43    32    24    18    12          SSF =.10    .20   .30   .40   .50   .60   .70   .80   .90
   TW E3  255   154   104    75    56    43    32    24    17       WW A1  487    56    25    14     9     6     4     2     1
   TW E4  228   139    95    69    52    39    30    22    15       WW A2  176    68    38    24    16    12     8     6     3
   TW F1  295    78    40    25    17    12     9     6     4       WW A3  151    70    42    28    19    14    10     7     4
   TW F2  192    81    47    31    22    16    12     8     6       WW A4  140    71    44    30    21    15    11     8     5
   TW F3  161    79    48    33    24    17    13     9     6       WW A5  134    73    46    32    23    17    12     9     6
   TW F4  134    71    45    31    23    17    13     9     6       WW A6  131    73    47    33    24    17    13     9     6
   TW G1  188    68    38    24    17    12     9     6     4       WW B1  124    41    21    12     7     4     2     1     -
   TW G2  141    67    41    28    20    14    11     8     5       WW B2  145    77    49    33    24    18    13     9     6
   TW G3  120    61    38    26    19    14    11     8     5       WW B3  172    96    62    43    31    23    17    12     8
   TW G4   97    51    32    22    16    12     9     7     4       WW B4  157    95    65    47    35    26    20    14    10
   TW H1  124    55    33    22    16    11     8     6     4       WW B5  148    92    63    46    34    26    20    14    10
   TW H2   97    49    31    21    15    11     8     6     4       WW C1  186   103    67    47    34    25    19    13     9
   TW H3   81    42    26    18    13    10     7     5     4       WW C2  162    95    63    45    33    25    18    13     9
   TW H4   63    32    20    14    10     7     5     4     3       WW C3  176   113    80    59    44    34    26    19    13
   TW I1  117    50    29    19    13     9     7     5     3       WW C4  159   102    72    53    40    31    23    17    12
   TW I2  169    89    56    39    29    21    16    12     8       TW A1  427    55    25    14     9     6     4     3     1
   TW I3  180    98    63    44    32    24    18    13     9       TW A2  181    62    33    21    14    10     7     5     3
   TW I4  183   107    72    51    38    29    22    16    11       TW A3  149    63    36    24    16    11     8     6     4
   TW I5  178   107    73    53    39    30    23    17    12       TW A4  128    60    36    24    17    12     9     6     4
   TW J1  218   120    78    55    40    30    22    16    11       TW B1  248    54    26    16    10     7     5     3     2
   TW J2  198   113    75    53    39    30    22    16    11       TW B2  149    57    32    21    14    10     7     5     3
   TW J3  214   134    93    68    51    39    30    22    15       TW B3  131    56    33    21    15    10     7     5     3
   TW J4  197   123    86    63    48    36    28    21    14       TW B4  122    52    31    20    14    10     7     5     3
   DG A1  150    66    37    21    10     -     -     -     -       TW C1  169    51    27    17    11     8     5     4     2
   DG A2  172    80    49    33    23    15    10     6     -       TW C2  133    50    28    18    12     9     6     4     3
   DG A3  216   103    65    46    34    25    19    13     7       TW C3  128    48    26    17    12     8     6     4     3
   DG B1  149    68    40    27    18    11     -     -     -       TW C4  136    43    23    15    10     7     5     3     2
   DG B2  177    83    52    36    27    20    14     9     4       TW D1   91    33    17    10     6     4     2     1     -
   DG B3  220   105    67    47    36    28    22    16    10       TW D2  154    73    44    29    21    15    11     8     5
   DG C1  186    85    52    36    26    19    13     6     -       TW D3  159    80    50    34    24    17    13     9     6
```

480 / APPENDIX 21

EAGLE, COLORADO							CONTINUED			GRAND JUNCTION, COLORADO								5605 DD	
SSF =	.10	.20	.30	.40	.50	.60	.70	.80	.90	SSF =	.10	.20	.30	.40	.50	.60	.70	.80	.90
TW D4	160	88	57	40	29	22	16	12	8	WW A1	781	90	40	24	15	10	7	5	3
TW D5	153	87	58	41	30	22	17	12	8	WW A2	262	101	57	37	25	18	13	9	6
TW E1	189	96	60	41	29	21	15	11	7	WW A3	220	103	62	41	29	21	15	11	7
TW E2	173	92	59	41	30	22	16	12	8	WW A4	202	103	64	44	31	23	17	12	8
TW E3	191	113	76	55	40	30	23	17	11	WW A5	192	104	67	46	33	25	18	13	9
TW E4	171	103	70	50	37	28	21	16	11	WW A6	187	105	68	48	35	26	19	14	9
TW F1	195	48	23	14	9	6	4	3	2	WW B1	204	71	38	23	16	11	7	5	3
TW F2	129	52	29	19	13	9	6	4	3	WW B2	202	107	68	47	34	25	18	13	9
TW F3	108	51	30	20	14	10	7	5	3	WW B3	235	130	85	59	43	32	24	17	11
TW F4	90	46	29	20	14	10	7	5	3	WW B4	210	126	86	62	46	35	26	19	13
TW G1	125	43	23	14	9	6	4	3	2	WW B5	194	120	82	60	45	34	26	19	13
TW G2	94	43	26	17	12	8	6	4	3	WW C1	251	139	91	64	46	34	26	18	12
TW G3	81	40	25	16	12	8	6	4	3	WW C2	218	127	84	60	44	33	25	18	12
TW G4	66	34	21	14	10	7	5	4	2	WW C3	226	145	102	75	57	43	33	24	17
TW H1	83	35	20	13	9	6	4	3	2	WW C4	205	131	92	68	52	39	30	22	15
TW H2	65	32	19	13	9	7	5	3	2	TW A1	673	86	40	23	15	10	7	5	3
TW H3	55	28	17	11	8	6	4	3	2	TW A2	270	92	50	32	22	15	11	8	5
TW H4	43	21	13	9	6	4	3	2	1	TW A3	218	92	53	35	24	17	13	9	6
TW I1	69	27	14	8	5	3	2	-	-	TW A4	184	87	53	35	25	18	13	9	6
TW I2	118	60	38	26	18	13	10	7	5	TW B1	379	82	41	25	16	11	8	5	3
TW I3	127	68	43	30	21	16	11	8	5	TW B2	218	84	47	30	21	15	11	8	5
TW I4	133	77	51	36	26	20	15	11	7	TW B3	190	81	47	31	22	16	11	8	5
TW I5	132	78	53	38	28	21	16	12	8	TW B4	175	75	44	29	20	15	11	7	5
TW J1	157	84	54	37	27	20	15	10	7	TW C1	252	75	40	25	17	12	8	6	4
TW J2	143	81	53	37	27	20	15	11	7	TW C2	193	72	41	26	18	13	9	7	4
TW J3	160	99	68	50	37	28	21	16	11	TW C3	185	68	38	24	17	12	9	6	4
TW J4	147	92	63	46	35	26	20	15	10	TW C4	195	61	33	21	14	10	7	5	3
DG A1	90	34	-	-	-	-	-	-	-	TW D1	148	55	30	19	13	9	6	4	2
DG A2	119	53	31	20	12	7	-	-	-	TW D2	217	101	61	41	29	21	16	11	7
DG A3	160	75	46	32	23	17	12	8	3	TW D3	220	111	69	47	34	25	18	13	9
DG B1	89	37	19	7	-	-	-	-	-	TW D4	215	118	77	54	39	29	22	16	11
DG B2	122	55	34	23	16	11	6	-	-	TW D5	202	115	76	54	39	30	22	16	11
DG B3	164	77	48	34	25	19	14	10	6	TW E1	257	130	81	55	40	29	21	15	10
DG C1	117	51	29	19	11	-	-	-	-	TW E2	235	124	79	55	40	30	22	16	11
DG C2	147	68	42	29	21	15	11	6	-	TW E3	249	147	99	71	52	39	30	22	15
DG C3	190	90	56	40	30	23	18	13	8	TW E4	223	134	90	65	49	37	28	20	14
SS A1	343	102	52	31	20	13	9	6	3	TW F1	305	74	37	23	15	10	7	5	3
SS A2	326	137	78	51	35	24	17	12	7	TW F2	191	77	44	29	20	14	10	7	5
SS A3	322	88	43	25	15	10	6	4	2	TW F3	158	74	45	30	21	15	11	8	5
SS A4	329	134	76	48	33	23	16	11	7	TW F4	130	67	42	29	20	15	11	8	5
SS A5	562	100	45	26	16	10	7	4	2	TW G1	189	65	35	22	15	11	8	5	3
SS A6	322	135	77	50	34	24	17	11	7	TW G2	139	64	38	25	18	13	9	7	4
SS A7	618	80	34	18	10	6	3	2	-	TW G3	117	58	36	24	17	13	9	7	4
SS A8	327	130	73	46	31	21	15	10	6	TW G4	94	48	30	21	15	11	8	6	4
SS B1	226	74	39	23	15	10	7	4	3	TW H1	123	53	31	20	14	10	7	5	3
SS B2	251	109	64	42	29	20	15	10	6	TW H2	95	46	28	19	14	10	7	5	3
SS B3	205	66	33	20	12	8	5	3	2	TW H3	79	40	24	17	12	9	6	4	3
SS B4	245	106	61	40	27	19	14	9	6	TW H4	62	30	19	13	9	7	5	3	2
SS B5	287	64	30	17	11	7	4	3	2	TW I1	113	46	26	16	11	8	5	4	2
SS B6	235	104	61	40	27	19	14	10	6	TW I2	165	85	53	36	26	19	14	10	7
SS B7	247	52	23	13	8	5	3	1	-	TW I3	175	93	59	41	30	22	16	12	8
SS B8	227	98	57	37	25	18	12	9	5	TW I4	178	102	68	48	35	26	20	14	10
SS C1	131	58	34	22	15	11	8	5	3	TW I5	174	103	69	49	37	28	21	15	10
SS C2	153	79	50	34	24	18	13	9	6	TW J1	213	114	73	51	37	27	20	14	10
SS C3	135	47	25	16	10	7	5	3	2	TW J2	193	108	71	50	36	27	20	15	10
SS C4	150	71	43	29	20	15	11	8	5	TW J3	208	128	88	64	48	36	27	20	14
SS D1	315	122	67	41	27	19	13	8	5	TW J4	192	119	82	59	45	34	26	19	13
SS D2	289	150	93	63	45	32	23	16	10	DG A1	147	61	32	16	-	-	-	-	-
SS D3	396	115	58	34	22	15	10	6	4	DG A2	172	77	46	30	20	13	8	3	-
SS D4	295	146	89	60	42	30	22	15	10	DG A3	216	99	62	43	31	23	17	11	6
SS E1	223	89	49	31	20	14	9	6	3	DG B1	146	63	37	23	14	-	-	-	-
SS E2	230	118	73	49	35	25	18	12	8	DG B2	175	80	49	33	24	17	12	7	2
SS E3	270	77	38	23	14	9	6	4	2	DG B3	220	102	64	45	34	26	20	14	9
SS E4	237	111	66	44	30	22	15	11	7	DG C1	181	81	48	32	23	15	9	-	-

GRAND JUNCTION, COLORADO — CONTINUED

SSF =	.10	.20	.30	.40	.50	.60	.70	.80	.90
DG C2	208	96	59	41	30	23	17	11	6
DG C3	254	118	74	52	39	31	24	18	12
SS A1	445	131	68	41	27	19	13	9	5
SS A2	405	169	97	63	44	31	22	16	10
SS A3	419	115	57	34	22	15	10	7	4
SS A4	406	166	94	61	42	30	21	15	9
SS A5	749	131	61	35	23	15	10	7	4
SS A6	401	167	96	62	43	31	22	15	10
SS A7	833	107	46	26	16	10	7	4	3
SS A8	406	162	91	58	40	28	20	14	9
SS B1	300	99	52	32	22	15	10	7	4
SS B2	316	137	80	53	37	26	19	13	9
SS B3	272	88	46	28	19	13	9	6	4
SS B4	307	132	76	50	35	25	18	13	8
SS B5	390	87	42	25	16	11	7	5	3
SS B6	297	131	77	51	35	25	18	13	8
SS B7	339	72	34	20	12	8	5	4	2
SS B8	286	124	72	47	33	23	17	12	7
SS C1	196	87	52	34	24	17	12	9	6
SS C2	212	109	69	47	34	25	18	13	9
SS C3	204	70	38	24	16	12	8	6	4
SS C4	210	98	59	40	28	21	15	11	7
SS D1	405	157	87	55	37	26	18	13	8
SS D2	355	184	115	79	56	41	30	21	14
SS D3	514	148	76	46	31	21	15	10	6
SS D4	364	180	111	75	53	38	28	20	13
SS E1	297	120	68	43	29	20	14	10	6
SS E2	291	149	93	63	45	33	24	17	11
SS E3	363	104	53	32	21	14	10	7	4
SS E4	300	141	84	56	40	28	21	14	9

PUEBLO, COLORADO — 5394 DD

SSF =	.10	.20	.30	.40	.50	.60	.70	.80	.90
WW A1	841	101	47	28	19	13	9	6	4
WW A2	288	113	65	43	30	22	16	11	8
WW A3	243	115	70	48	34	25	19	13	9
WW A4	223	115	73	50	37	27	20	15	10
WW A5	213	117	76	53	39	29	22	16	11
WW A6	206	117	77	55	41	30	23	17	11
WW B1	229	83	46	29	20	14	10	7	5
WW B2	222	118	76	53	39	29	22	16	11
WW B3	257	144	95	67	50	37	28	20	14
WW B4	228	139	95	69	52	39	30	22	15
WW B5	211	131	91	66	50	38	29	22	15
WW C1	274	154	101	72	53	40	30	22	15
WW C2	238	139	94	67	50	38	29	21	15
WW C3	245	158	112	83	63	49	37	28	19
WW C4	221	143	101	75	57	44	34	25	18
TW A1	725	97	46	27	18	13	9	6	4
TW A2	295	103	57	37	26	18	13	10	6
TW A3	239	102	60	40	29	21	15	11	7
TW A4	203	97	60	41	29	22	16	12	8
TW B1	411	92	46	29	19	14	10	7	5
TW B2	238	93	53	35	25	18	13	9	6
TW B3	208	90	53	36	25	19	14	10	7
TW B4	192	83	49	33	24	17	13	9	6
TW C1	273	84	45	29	20	14	10	7	5
TW C2	210	81	46	30	21	15	11	8	5
TW C3	201	75	43	28	20	14	10	7	5
TW C4	211	68	37	24	16	12	9	6	4
TW D1	167	64	36	24	16	12	8	6	4
TW D2	237	112	69	47	34	25	18	13	9
TW D3	241	123	77	54	39	29	21	16	11

PUEBLO, COLORADO — CONTINUED

SSF =	.10	.20	.30	.40	.50	.60	.70	.80	.90
TW D4	233	130	85	60	44	33	25	18	13
TW D5	218	125	83	59	44	33	25	19	13
TW E1	280	143	90	63	45	34	25	18	12
TW E2	256	137	88	62	45	34	25	19	13
TW E3	269	161	109	79	59	45	34	25	17
TW E4	241	146	99	72	54	41	31	23	16
TW F1	332	84	43	27	18	13	9	7	4
TW F2	211	87	50	33	23	17	12	9	6
TW F3	175	84	51	35	25	18	14	10	7
TW F4	144	75	48	33	24	18	13	10	7
TW G1	208	73	40	26	18	13	9	7	4
TW G2	153	71	43	29	21	15	11	8	6
TW G3	129	65	41	28	20	15	11	8	5
TW G4	104	54	34	24	17	13	9	7	5
TW H1	136	59	35	23	17	12	9	6	4
TW H2	105	52	32	22	16	12	9	6	4
TW H3	87	44	28	19	14	10	8	6	4
TW H4	68	34	21	15	10	8	6	4	3
TW I1	128	54	31	20	14	10	7	5	3
TW I2	181	94	60	41	30	22	17	12	8
TW I3	192	103	66	47	34	25	19	14	9
TW I4	194	112	75	54	40	30	23	17	12
TW I5	189	112	76	55	41	31	24	18	12
TW J1	232	126	82	57	42	31	23	17	12
TW J2	210	119	78	56	41	31	23	17	12
TW J3	225	140	97	71	53	41	31	23	16
TW J4	207	129	90	66	50	38	29	22	15
DG A1	165	72	40	23	12	—	—	—	—
DG A2	187	85	52	35	24	17	11	6	1
DG A3	231	108	68	48	36	27	20	14	8
DG B1	162	73	44	30	20	13	5	—	—
DG B2	190	88	55	38	28	21	15	10	5
DG B3	235	110	70	50	38	29	23	17	11
DG C1	200	92	56	39	28	21	14	8	—
DG C2	225	105	65	46	35	27	20	14	9
DG C3	272	127	81	58	44	35	27	21	14
SS A1	484	147	77	48	33	23	16	11	7
SS A2	439	187	109	72	51	37	27	19	13
SS A3	459	131	67	41	28	19	13	9	6
SS A4	442	183	106	70	49	35	25	18	12
SS A5	805	146	70	42	28	19	13	9	6
SS A6	436	185	108	71	50	36	26	19	12
SS A7	895	121	55	32	21	14	10	6	4
SS A8	443	180	103	67	47	33	24	17	11
SS B1	329	111	60	38	26	18	13	9	6
SS B2	343	151	89	59	42	31	22	16	11
SS B3	300	100	54	34	23	16	11	8	5
SS B4	334	146	86	57	40	29	21	15	10
SS B5	424	98	49	30	20	14	10	7	4
SS B6	323	144	86	57	41	30	22	15	10
SS B7	371	82	40	24	16	11	8	5	3
SS B8	312	137	81	54	38	27	20	14	9
SS C1	217	99	59	40	28	21	15	11	7
SS C2	232	121	77	54	39	29	22	16	11
SS C3	224	79	44	28	20	14	10	7	5
SS C4	229	109	67	45	33	24	18	13	9
SS D1	444	176	100	65	45	32	23	16	11
SS D2	388	204	129	89	64	48	35	25	17
SS D3	559	166	87	54	37	26	18	13	8
SS D4	397	199	124	85	61	45	33	24	16
SS E1	329	136	79	51	36	26	18	13	8
SS E2	318	166	105	73	52	39	29	21	14
SS E3	397	117	61	38	26	18	13	9	6
SS E4	328	156	95	64	46	33	25	18	12

HARTFORD, CONNECTICUT								6350 DD	
SSF =	.10	.20	.30	.40	.50	.60	.70	.80	.90
WW A1	237	-	-	-	-	-	-	-	-
WW A2	82	26	11	5	-	-	-	-	-
WW A3	71	29	15	8	4	-	-	-	-
WW A4	67	30	17	10	6	3	1	-	-
WW A5	65	32	18	11	7	4	2	1	-
WW A6	63	32	19	12	8	5	3	2	-
WW B1	-	-	-	-	-	-	-	-	-
WW B2	79	39	23	15	10	7	5	3	2
WW B3	96	51	32	21	15	10	7	5	3
WW B4	97	57	38	27	19	14	10	7	5
WW B5	94	57	39	28	20	15	11	8	5
WW C1	107	57	36	24	17	12	9	6	4
WW C2	98	55	36	25	18	13	9	7	4
WW C3	115	73	50	37	27	21	16	11	8
WW C4	105	66	46	34	25	19	14	10	7
TW A1	215	14	-	-	-	-	-	-	-
TW A2	87	24	10	4	-	-	-	-	-
TW A3	72	26	13	7	4	-	-	-	-
TW A4	64	27	14	9	5	3	2	-	-
TW B1	122	19	5	-	-	-	-	-	-
TW B2	74	24	12	6	3	1	-	-	-
TW B3	67	25	13	7	4	3	1	-	-
TW B4	64	24	13	8	5	3	2	-	-
TW C1	86	21	9	4	2	-	-	-	-
TW C2	70	23	12	7	4	2	1	-	-
TW C3	69	23	11	7	4	2	1	-	-
TW C4	76	21	10	6	4	2	1	-	-
TW D1	-	-	-	-	-	-	-	-	-
TW D2	85	37	21	13	9	6	4	3	1
TW D3	87	41	24	15	10	7	5	3	2
TW D4	98	52	33	22	16	12	8	6	4
TW D5	98	54	35	24	18	13	10	7	4
TW E1	108	52	31	20	14	10	7	5	3
TW E2	104	53	33	22	15	11	8	5	3
TW E3	121	70	46	33	24	18	13	9	6
TW E4	111	66	44	31	23	17	13	9	6
TW F1	85	12	-	-	-	-	-	-	-
TW F2	58	19	8	3	-	-	-	-	-
TW F3	50	20	10	6	3	-	-	-	-
TW F4	43	20	11	7	4	2	1	-	-
TW G1	54	13	-	-	-	-	-	-	-
TW G2	43	17	8	4	2	-	-	-	-
TW G3	38	17	9	5	3	2	-	-	-
TW G4	33	15	8	5	3	2	1	-	-
TW H1	37	13	6	2	-	-	-	-	-
TW H2	31	13	7	4	2	1	-	-	-
TW H3	27	12	7	4	3	2	-	-	-
TW H4	22	10	6	3	2	1	-	-	-
TW I1	-	-	-	-	-	-	-	-	-
TW I2	64	31	18	12	8	5	4	2	1
TW I3	70	35	21	14	10	7	5	3	2
TW I4	81	45	29	20	15	11	8	5	4
TW I5	83	48	32	22	16	12	9	6	4
TW J1	90	46	28	19	13	9	7	5	3
TW J2	86	47	30	20	14	10	7	5	3
TW J3	103	63	43	30	22	17	12	9	6
TW J4	96	59	40	29	21	16	12	9	6
DG A1	-	-	-	-	-	-	-	-	-
DG A2	63	24	10	-	-	-	-	-	-
DG A3	102	45	27	18	12	8	5	2	-
DG B1	-	-	-	-	-	-	-	-	-
DG B2	64	25	13	6	-	-	-	-	-
DG B3	106	48	29	19	14	10	7	5	-
DG C1	41	-	-	-	-	-	-	-	-

HARTFORD, CONNECTICUT								CONTINUED	
SSF =	.10	.20	.30	.40	.50	.60	.70	.80	.90
DG C2	84	35	20	12	7	-	-	-	-
DG C3	124	56	34	23	17	13	9	6	3
SS A1	174	45	20	10	5	-	-	-	-
SS A2	188	76	42	26	18	12	8	5	3
SS A3	148	31	9	-	-	-	-	-	-
SS A4	185	72	39	24	16	10	7	5	3
SS A5	288	40	14	-	-	-	-	-	-
SS A6	182	73	40	25	16	11	8	5	3
SS A7	298	22	-	-	-	-	-	-	-
SS A8	180	68	36	22	14	9	6	4	2
SS B1	111	31	13	6	-	-	-	-	-
SS B2	149	62	35	22	15	10	7	5	3
SS B3	91	22	-	-	-	-	-	-	-
SS B4	142	59	33	20	14	9	6	4	3
SS B5	138	23	6	-	-	-	-	-	-
SS B6	137	58	33	21	14	9	7	4	3
SS B7	105	-	-	-	-	-	-	-	-
SS B8	128	53	29	18	12	8	5	3	2
SS C1	55	19	8	-	-	-	-	-	-
SS C2	85	41	25	16	11	8	5	3	2
SS C3	60	15	5	-	-	-	-	-	-
SS C4	84	37	21	13	9	6	4	3	2
SS D1	156	54	26	14	8	4	1	-	-
SS D2	167	84	51	34	23	16	12	8	5
SS D3	199	50	22	11	5	-	-	-	-
SS D4	170	82	49	32	22	15	11	7	5
SS E1	101	33	14	-	-	-	-	-	-
SS E2	129	63	38	25	17	12	8	5	3
SS E3	127	28	10	-	-	-	-	-	-
SS E4	133	60	34	22	14	10	7	4	3

WILMINGTON, DELAWARE								4940 DD	
SSF =	.10	.20	.30	.40	.50	.60	.70	.80	.90
WW A1	428	45	19	10	6	3	2	1	-
WW A2	151	57	31	19	13	9	6	4	2
WW A3	129	59	35	22	15	11	8	5	3
WW A4	120	60	36	24	17	12	9	6	4
WW A5	115	61	38	26	18	13	9	7	4
WW A6	112	62	39	27	19	14	10	7	4
WW B1	96	29	13	6	-	-	-	-	-
WW B2	126	66	41	28	20	15	11	7	5
WW B3	150	83	53	37	26	19	14	10	7
WW B4	140	84	57	41	30	23	17	12	8
WW B5	133	82	56	41	30	23	17	13	8
WW C1	163	90	58	40	29	21	16	11	7
WW C2	144	84	55	39	29	21	16	11	8
WW C3	158	101	71	52	39	30	23	17	11
WW C4	143	92	65	47	36	27	21	15	10
TW A1	376	45	19	10	6	4	2	1	-
TW A2	156	52	27	17	11	7	5	3	2
TW A3	128	53	30	19	13	9	6	4	3
TW A4	110	51	30	20	14	10	7	5	3
TW B1	216	45	21	12	8	5	3	2	1
TW B2	129	48	27	17	11	8	5	4	2
TW B3	114	48	27	17	12	8	6	4	2
TW B4	106	45	26	17	11	8	6	4	2
TW C1	148	43	22	13	9	6	4	3	2
TW C2	116	43	24	15	10	7	5	3	2
TW C3	112	41	22	14	9	7	5	3	2
TW C4	120	37	20	12	8	6	4	3	2
TW D1	71	24	11	6	3	-	-	-	-
TW D2	135	63	37	25	17	12	9	6	4
TW D3	138	69	42	28	20	14	10	7	5

WILMINGTON, DELAWARE							CONTINUED			WASHINGTON, DC								5010 DD	
SSF =.10	.20	.30	.40	.50	.60	.70	.80	.90		SSF =.10	.20	.30	.40	.50	.60	.70	.80	.90	
TW D4	143	78	51	35	25	19	14	10	7	WW A1	419	41	16	8	4	2	-	-	-
TW D5	137	78	51	36	26	20	15	11	7	WW A2	144	53	28	17	11	8	5	3	2
TW E1	165	83	51	35	25	18	13	9	6	WW A3	123	55	32	20	14	10	7	4	3
TW E2	154	81	51	35	25	19	14	10	6	WW A4	113	56	34	22	15	11	7	5	3
TW E3	170	101	67	48	35	26	20	14	10	WW A5	109	57	36	24	17	12	8	6	4
TW E4	154	92	62	45	33	25	19	14	9	WW A6	106	58	37	25	17	13	9	6	4
TW F1	167	39	18	10	6	4	3	2	-	WW B1	88	25	9	-	-	-	-	-	-
TW F2	110	43	24	15	10	7	5	3	2	WW B2	121	63	39	26	19	13	10	7	4
TW F3	93	43	25	16	11	8	5	4	2	WW B3	144	79	50	34	25	18	13	9	6
TW F4	78	39	24	16	11	8	6	4	2	WW B4	135	81	54	39	28	21	16	11	8
TW G1	106	35	18	11	7	5	3	2	1	WW B5	128	79	54	39	29	22	16	12	8
TW G2	81	36	21	13	9	6	4	3	2	WW C1	156	86	55	38	27	20	15	10	7
TW G3	69	34	20	13	9	7	5	3	2	WW C2	139	80	53	37	27	20	15	11	7
TW G4	57	28	17	12	8	6	4	3	2	WW C3	153	98	68	50	38	28	22	16	11
TW H1	70	29	16	10	7	5	3	2	1	WW C4	139	89	62	45	34	26	20	14	10
TW H2	56	27	16	11	7	5	4	2	2	TW A1	367	41	17	9	5	3	2	-	-
TW H3	48	23	14	9	7	5	3	2	1	TW A2	149	48	25	15	10	6	4	3	2
TW H4	37	18	11	7	5	4	3	2	1	TW A3	122	49	28	17	12	8	5	4	2
TW I1	53	19	9	4	-	-	-	-	-	TW A4	105	48	28	18	12	9	6	4	3
TW I2	103	52	32	22	15	11	8	6	4	TW B1	208	42	19	11	7	4	3	2	-
TW I3	111	58	37	25	18	13	9	7	4	TW B2	123	45	25	15	10	7	5	3	2
TW I4	119	68	45	31	23	17	13	9	6	TW B3	109	45	25	16	11	7	5	3	2
TW I5	118	70	47	33	24	18	14	10	7	TW B4	102	42	24	15	10	7	5	3	2
TW J1	137	73	46	32	23	17	12	9	6	TW C1	142	40	20	12	8	5	3	2	1
TW J2	127	71	46	32	23	17	13	9	6	TW C2	111	40	22	14	9	6	4	3	2
TW J3	144	89	61	44	33	25	19	14	9	TW C3	108	38	21	13	9	6	4	3	2
TW J4	133	82	57	41	31	23	17	13	9	TW C4	116	35	18	11	7	5	4	2	1
DG A1	71	21	-	-	-	-	-	-	-	TW D1	66	21	9	-	-	-	-	-	-
DG A2	104	45	26	15	9	-	-	-	-	TW D2	130	59	35	23	16	11	8	6	4
DG A3	144	67	41	28	20	14	10	6	-	TW D3	132	65	40	27	19	13	9	7	4
DG B1	71	26	-	-	-	-	-	-	-	TW D4	137	75	48	33	24	18	13	9	6
DG B2	106	47	28	18	12	8	3	-	-	TW D5	133	75	49	34	25	19	14	10	7
DG B3	147	69	43	30	22	16	12	9	5	TW E1	159	79	49	33	23	17	12	8	5
DG C1	98	40	22	12	-	-	-	-	-	TW E2	148	78	49	33	24	17	13	9	6
DG C2	130	59	36	24	17	12	8	4	-	TW E3	164	97	65	46	34	25	19	14	9
DG C3	170	81	50	35	26	20	15	11	7	TW E4	149	89	60	43	32	24	18	13	9
SS A1	272	80	40	24	15	10	7	4	3	TW F1	160	36	16	9	5	3	2	-	-
SS A2	269	113	65	42	29	20	14	10	6	TW F2	104	40	22	13	9	6	4	2	1
SS A3	249	67	32	18	11	7	4	3	1	TW F3	88	40	23	15	10	7	5	3	2
SS A4	268	110	62	40	27	19	13	9	6	TW F4	74	37	22	15	10	7	5	3	2
SS A5	443	77	34	19	12	7	5	3	2	TW G1	101	32	16	10	6	4	2	1	-
SS A6	264	111	63	41	28	20	14	10	6	TW G2	76	34	19	12	8	6	4	3	2
SS A7	474	59	24	12	6	3	2	-	-	TW G3	65	31	19	12	8	6	4	3	2
SS A8	266	106	59	38	25	18	12	8	5	TW G4	54	27	16	11	7	5	4	3	2
SS B1	183	59	30	18	12	8	5	3	2	TW H1	67	27	15	9	6	4	3	2	1
SS B2	212	92	54	35	24	17	12	9	5	TW H2	53	25	15	10	7	5	3	2	1
SS B3	161	51	25	15	9	6	4	2	1	TW H3	45	22	13	9	6	4	3	2	1
SS B4	205	88	51	33	23	16	11	8	5	TW H4	36	17	10	7	5	3	2	2	1
SS B5	229	50	23	13	8	5	3	2	1	TW I1	48	16	6	-	-	-	-	-	-
SS B6	198	87	51	33	23	16	12	8	5	TW I2	98	49	30	20	14	10	7	5	3
SS B7	192	39	17	9	5	3	1	-	-	TW I3	106	56	35	24	17	12	9	6	4
SS B8	189	82	47	30	21	15	10	7	4	TW I4	114	65	43	30	22	16	12	9	6
SS C1	111	48	27	17	11	8	5	4	2	TW I5	114	67	45	32	23	17	13	9	6
SS C2	134	69	43	29	20	15	11	8	5	TW J1	132	70	44	30	21	16	11	8	5
SS C3	115	38	20	12	8	5	3	2	1	TW J2	122	68	44	30	22	16	12	8	6
SS C4	132	61	37	24	17	12	9	6	4	TW J3	139	85	58	42	31	23	18	13	9
SS D1	250	96	52	32	21	14	10	6	4	TW J4	129	79	54	39	29	22	17	12	8
SS D2	239	125	78	53	37	27	19	14	9	DG A1	65	-	-	-	-	-	-	-	-
SS D3	313	90	45	27	17	11	7	5	3	DG A2	100	43	24	14	7	-	-	-	-
SS D4	245	122	74	50	35	25	18	13	8	DG A3	140	64	39	27	19	13	9	6	-
SS E1	178	70	38	23	15	10	7	4	2	DG B1	65	22	-	-	-	-	-	-	-
SS E2	193	99	61	41	29	21	15	10	7	DG B2	102	45	26	17	11	6	-	-	-
SS E3	215	60	29	17	11	7	4	3	1	DG B3	143	67	41	28	21	15	11	8	4
SS E4	198	93	55	36	25	18	13	9	6	DG C1	92	37	19	9	-	-	-	-	-

WASHINGTON, DC								CONTINUED		APALACHICOLA, FLORIDA								CONTINUED	
SSF =	.10	.20	.30	.40	.50	.60	.70	.80	.90	SSF =	.10	.20	.30	.40	.50	.60	.70	.80	.90
DG C2	125	57	34	22	16	11	7	3	-	TW D4	507	278	181	127	93	70	52	38	26
DG C3	166	78	48	33	25	19	14	10	6	TW D5	461	261	172	122	90	68	51	37	26
SS A1	269	77	38	22	14	9	6	4	2	TW E1	632	319	200	138	100	74	55	40	27
SS A2	264	110	63	40	27	19	14	9	6	TW E2	563	297	190	132	97	72	54	39	27
SS A3	246	64	30	17	10	6	4	2	-	TW E3	569	335	225	161	120	91	69	51	35
SS A4	265	107	60	38	26	18	12	9	5	TW E4	501	299	202	146	109	82	62	46	32
SS A5	442	74	32	18	11	7	4	2	1	TW F1	850	216	111	69	47	33	24	17	11
SS A6	260	108	61	39	27	19	13	9	6	TW F2	519	213	123	81	57	42	31	22	15
SS A7	478	56	22	11	5	2	-	-	-	TW F3	423	201	122	83	60	44	32	23	16
SS A8	262	103	57	36	24	17	11	8	5	TW F4	343	177	112	77	56	42	31	23	15
SS B1	179	57	29	17	11	7	5	3	2	TW G1	523	183	101	65	45	33	24	17	11
SS B2	209	90	52	34	23	16	12	8	5	TW G2	372	173	104	70	50	37	27	20	13
SS B3	158	48	24	13	8	5	3	2	-	TW G3	309	155	96	66	48	35	26	19	13
SS B4	202	86	49	32	22	15	11	7	5	TW G4	245	126	79	55	40	30	22	16	11
SS B5	226	48	21	12	7	4	2	1	-	TW H1	334	145	86	57	41	30	22	16	10
SS B6	194	85	49	32	22	15	11	8	5	TW H2	252	124	77	52	38	28	21	15	10
SS B7	189	37	15	7	4	1	-	-	-	TW H3	207	104	65	45	32	24	18	13	9
SS B8	186	79	45	29	20	14	10	7	4	TW H4	159	79	49	33	24	18	13	10	6
SS C1	104	44	25	15	10	7	4	3	2	TW I1	355	150	87	58	41	30	22	16	10
SS C2	128	65	40	27	19	14	10	7	4	TW I2	416	213	134	93	67	50	37	27	18
SS C3	110	35	18	11	7	4	3	2	-	TW I3	436	232	148	104	76	56	42	31	21
SS C4	127	58	34	23	16	11	8	6	4	TW I4	419	240	159	113	84	63	47	35	24
SS D1	245	92	50	30	20	13	9	6	3	TW I5	399	234	157	113	84	64	48	35	24
SS D2	235	121	75	51	36	26	18	13	8	TW J1	520	279	179	125	92	68	51	37	25
SS D3	309	87	43	25	16	10	7	4	2	TW J2	460	257	168	119	87	66	49	36	25
SS D4	240	118	72	48	33	24	17	12	8	TW J3	469	288	198	144	108	82	63	46	32
SS E1	173	67	36	21	14	9	6	4	2	TW J4	428	263	181	132	99	76	57	43	29
SS E2	188	95	59	39	27	20	14	10	6	DG A1	447	199	119	78	54	37	25	15	6
SS E3	211	57	27	16	9	6	4	2	1	DG A2	444	201	123	85	61	44	32	22	13
SS E4	193	90	53	35	24	17	12	8	5	DG A3	511	231	145	101	76	57	43	30	19
										DG B1	451	200	124	86	63	47	34	23	12
										DG B2	460	206	129	91	68	52	39	28	18
APALACHICOLA, FLORIDA								1361	DD	DG B3	519	235	147	105	79	61	48	36	24
SSF =	.10	.20	.30	.40	.50	.60	.70	.80	.90	DG C1	533	236	147	104	77	59	45	33	20
WW A1	2181	267	124	74	50	35	25	17	11	DG C2	531	238	149	106	79	62	48	35	24
WW A2	708	278	158	104	73	53	39	28	18	DG C3	596	269	169	121	91	72	57	44	30
WW A3	586	276	168	114	82	60	44	32	21	SS A1	1193	367	193	122	84	60	43	30	20
WW A4	532	273	172	119	86	64	47	34	23	SS A2	1013	428	249	165	116	85	62	44	29
WW A5	503	275	177	125	91	68	51	37	25	SS A3	1183	345	179	112	77	54	39	27	18
WW A6	486	274	180	127	94	70	53	39	26	SS A4	1046	433	249	165	116	84	61	44	29
WW B1	634	231	128	83	58	42	30	21	14	SS A5	1998	376	180	109	73	51	37	26	17
WW B2	506	268	171	119	87	65	48	35	24	SS A6	1015	428	248	164	116	84	62	44	29
WW B3	580	322	210	148	109	82	61	45	30	SS A7	2309	333	153	91	60	42	30	21	13
WW B4	488	294	199	144	108	82	62	46	31	SS A8	1060	430	246	161	113	82	60	43	28
WW B5	441	271	186	136	102	78	59	44	30	SS B1	826	281	152	97	67	48	35	25	16
WW C1	609	338	221	156	115	86	64	47	32	SS B2	787	343	201	134	95	70	51	37	24
WW C2	516	299	200	143	106	80	60	44	30	SS B3	782	265	143	92	63	45	33	23	15
WW C3	504	322	226	167	127	97	74	55	38	SS B4	781	339	199	132	94	68	50	36	24
WW C4	452	289	202	149	113	87	66	50	34	SS B5	1083	258	129	79	54	38	27	19	12
TW A1	1852	252	118	71	48	33	24	17	11	SS B6	748	332	196	131	93	68	50	36	24
TW A2	726	252	139	89	62	45	33	23	15	SS B7	984	227	112	69	46	33	23	16	11
TW A3	577	246	144	96	68	50	36	26	17	SS B8	740	323	190	127	90	66	48	34	23
TW A4	482	230	140	95	68	50	37	27	18	SS C1	538	244	146	98	70	51	38	27	18
TW B1	1029	232	116	72	49	35	25	18	12	SS C2	528	273	172	119	87	65	48	35	24
TW B2	576	224	128	84	59	43	31	22	15	SS C3	562	198	109	71	49	35	26	18	12
TW B3	497	214	126	84	59	43	32	23	15	SS C4	526	247	150	102	73	54	40	29	19
TW B4	453	195	115	77	54	40	29	21	14	SS D1	1095	434	246	161	113	82	59	42	28
TW C1	666	204	109	69	48	34	25	18	12	SS D2	883	460	290	201	146	108	80	58	39
TW C2	500	191	108	70	49	36	26	19	12	SS D3	1382	416	218	137	94	67	48	34	22
TW C3	475	177	100	65	45	33	24	17	11	SS D4	907	452	280	192	138	102	76	55	36
TW C4	497	158	85	54	38	27	20	14	9	SS E1	839	350	202	133	94	68	50	35	23
TW D1	455	178	101	66	46	33	24	17	11	SS E2	744	385	242	168	122	90	67	49	32
TW D2	545	256	155	105	75	56	41	30	20	SS E3	1014	304	159	100	69	49	35	25	16
TW D3	554	280	175	120	87	65	48	35	23	SS E4	770	364	221	149	107	79	58	42	28

DAYTONA BEACH, FLORIDA								902 DD		DAYTONA BEACH, FLORIDA								CONTINUED	
SSF =	.10	.20	.30	.40	.50	.60	.70	.80	.90	SSF =	.10	.20	.30	.40	.50	.60	.70	.80	.90
WW A1	3501	425	203	124	84	60	43	31	21	DG C2	793	370	237	171	130	103	81	62	43
WW A2	1098	440	257	171	122	89	66	48	32	DG C3	872	407	261	190	146	115	93	72	51
WW A3	909	439	272	187	136	101	75	55	37	SS A1	1823	567	307	198	138	100	73	52	34
WW A4	827	434	278	195	143	107	80	59	40	SS A2	1517	654	388	261	187	137	101	73	49
WW A5	783	437	287	204	151	114	86	64	44	SS A3	1819	539	288	184	128	92	67	48	32
WW A6	757	436	291	208	155	118	89	66	45	SS A4	1569	663	391	262	187	137	101	73	49
WW B1	1008	378	216	142	101	73	54	39	26	SS A5	3107	583	287	178	121	86	62	44	29
WW B2	778	421	273	193	142	107	80	59	41	SS A6	1523	655	389	262	187	138	102	74	49
WW B3	890	504	334	239	177	134	101	75	51	SS A7	3637	519	247	151	102	72	52	37	24
WW B4	743	454	312	228	172	132	100	75	52	SS A8	1594	659	386	258	184	134	99	71	48
WW B5	668	417	290	213	161	124	95	71	49	SS B1	1264	437	243	159	111	81	59	43	28
WW C1	931	527	349	249	185	140	106	78	54	SS B2	1178	523	314	212	152	112	83	60	41
WW C2	786	464	314	227	170	129	98	73	50	SS B3	1202	416	231	151	106	77	56	40	27
WW C3	760	493	350	260	199	154	118	89	62	SS B4	1172	518	311	210	151	111	82	60	40
WW C4	680	441	313	233	178	137	106	79	55	SS B5	1682	403	207	131	90	64	47	33	22
TW A1	2957	399	193	118	81	57	42	30	20	SS B6	1123	508	307	209	150	111	82	60	40
TW A2	1126	398	224	147	104	76	56	40	27	SS B7	1537	357	182	114	79	56	41	29	19
TW A3	892	389	233	157	113	83	62	45	30	SS B8	1114	497	299	202	145	107	79	57	39
TW A4	745	363	226	156	113	84	63	46	31	SS C1	839	390	238	163	117	87	65	47	32
TW B1	1617	367	189	119	82	59	43	31	21	SS C2	809	427	274	192	141	106	80	59	40
TW B2	889	354	206	137	98	72	53	38	26	SS C3	876	315	178	117	83	60	44	32	22
TW B3	767	337	202	137	98	73	54	39	27	SS C4	804	385	238	164	119	88	66	48	33
TW B4	698	307	184	125	90	66	49	36	24	SS D1	1666	676	393	262	186	136	100	72	48
TW C1	1034	321	175	113	79	58	42	30	20	SS D2	1326	706	454	319	234	175	131	96	65
TW C2	770	299	173	115	81	60	44	32	22	SS D3	2114	645	347	223	155	112	82	58	39
TW C3	730	277	159	106	75	55	40	29	20	SS D4	1361	693	438	304	222	165	123	90	61
TW C4	764	246	136	88	62	45	33	24	16	SS E1	1288	550	326	219	156	115	85	61	41
TW D1	723	290	169	112	80	59	43	31	21	SS E2	1124	595	382	268	197	147	110	81	55
TW D2	835	400	247	170	123	92	68	50	34	SS E3	1562	475	256	165	115	83	60	43	29
TW D3	849	438	279	194	142	106	80	59	40	SS E4	1160	561	347	239	173	128	96	70	47
TW D4	769	430	283	202	149	113	85	63	43										
TW D5	694	399	267	191	143	108	82	61	42	JACKSONVILLE, FLORIDA								1327 DD	
TW E1	963	496	316	220	161	121	91	67	45	SSF =	.10	.20	.30	.40	.50	.60	.70	.80	.90
TW E2	855	461	299	211	155	117	88	65	44	WW A1	2307	286	134	81	55	39	28	20	13
TW E3	858	514	349	254	190	145	111	82	57	WW A2	752	297	171	113	80	59	43	31	21
TW E4	753	456	312	227	171	131	100	74	51	WW A3	623	296	182	124	90	66	50	36	24
TW F1	1338	344	181	115	80	57	42	30	20	WW A4	567	293	186	129	94	71	53	39	26
TW F2	806	338	200	134	96	71	52	38	26	WW A5	537	295	192	136	100	75	57	42	29
TW F3	656	319	198	137	99	74	55	40	27	WW A6	518	294	195	139	103	78	59	43	30
TW F4	532	281	181	127	93	70	53	39	26	WW B1	674	249	140	91	64	47	34	25	16
TW G1	816	292	165	108	76	56	41	30	20	WW B2	539	287	184	129	95	71	54	39	27
TW G2	578	274	169	116	84	62	46	34	23	WW B3	616	345	226	161	119	90	68	50	34
TW G3	479	245	156	108	79	59	44	33	22	WW B4	519	314	214	156	117	90	68	51	35
TW G4	380	200	128	90	66	49	37	27	19	WW B5	469	290	200	146	111	85	65	48	34
TW H1	520	231	139	94	68	50	37	27	18	WW C1	646	361	237	169	125	94	71	53	36
TW H2	390	197	124	86	63	47	35	26	18	WW C2	549	320	215	154	115	88	67	49	34
TW H3	321	165	105	73	53	40	30	22	15	WW C3	534	343	242	180	137	106	81	61	42
TW H4	246	124	78	54	40	30	22	16	11	WW C4	480	308	217	161	123	95	73	55	38
TW I1	566	247	147	99	71	52	39	28	19	TW A1	1959	269	127	77	52	37	27	19	13
TW I2	639	334	214	150	110	82	62	46	31	TW A2	769	270	150	97	68	50	36	26	18
TW I3	668	363	236	167	123	93	70	51	35	TW A3	613	263	156	104	74	55	41	29	20
TW I4	637	371	249	179	134	102	77	57	39	TW A4	513	246	151	104	75	56	41	30	20
TW I5	603	360	245	178	133	102	77	58	40	TW B1	1089	248	125	78	54	38	28	20	13
TW J1	793	434	283	201	148	112	84	62	42	TW B2	611	240	138	91	64	47	35	25	17
TW J2	700	398	264	189	140	106	81	60	41	TW B3	528	229	136	91	65	48	36	26	17
TW J3	706	441	306	225	171	131	100	75	52	TW B4	481	209	124	83	60	44	33	24	16
TW J4	644	402	280	206	156	120	92	69	47	TW C1	706	218	117	75	52	38	28	20	13
DG A1	705	326	202	138	98	71	51	35	21	TW C2	530	204	116	76	54	40	29	21	14
DG A2	676	316	199	140	102	76	56	40	26	TW C3	503	189	107	70	50	36	27	19	13
DG A3	755	353	225	161	122	93	71	51	33	TW C4	526	169	92	59	41	30	22	16	11
DG B1	702	326	208	148	112	86	65	47	30	TW D1	485	191	109	72	51	37	27	20	13
DG B2	693	323	207	148	113	88	68	51	34	TW D2	579	273	167	114	83	61	46	33	23
DG B3	763	356	229	165	126	100	78	60	41	TW D3	588	299	188	130	95	71	53	39	26
DG C1	818	379	243	175	133	105	83	62	41										

JACKSONVILLE, FLORIDA (CONTINUED)

	SSF =.10	.20	.30	.40	.50	.60	.70	.80	.90
TW D4	538	297	194	137	102	77	58	43	29
TW D5	489	278	184	131	98	74	56	42	29
TW E1	669	341	214	149	109	81	61	45	30
TW E2	597	318	204	143	105	79	59	44	30
TW E3	602	357	241	174	130	99	76	56	39
TW E4	531	318	216	157	118	90	69	51	35
TW F1	901	232	119	75	52	37	27	19	13
TW F2	552	228	133	88	63	46	34	25	17
TW F3	450	216	132	91	66	49	36	26	18
TW F4	365	190	121	84	62	46	35	25	17
TW G1	555	197	109	71	50	36	27	19	13
TW G2	396	185	113	77	55	41	30	22	15
TW G3	329	166	104	72	52	39	29	21	15
TW G4	261	135	86	60	44	33	25	18	12
TW H1	355	156	93	62	45	33	24	18	12
TW H2	268	133	83	57	42	31	23	17	11
TW H3	221	112	70	48	35	26	20	14	10
TW H4	169	84	53	36	26	20	15	11	7
TW I1	379	162	95	64	45	33	25	18	12
TW I2	442	228	145	101	74	55	41	30	21
TW I3	463	248	160	112	83	62	47	34	23
TW I4	445	256	171	122	91	69	52	39	27
TW I5	423	250	169	122	91	70	53	39	27
TW J1	551	298	192	135	100	75	56	42	28
TW J2	488	274	181	129	95	72	54	40	28
TW J3	496	307	212	155	117	90	69	51	35
TW J4	454	281	194	142	108	82	63	47	33
DG A1	476	214	130	87	61	43	29	19	10
DG A2	469	214	133	92	67	49	36	25	16
DG A3	537	245	154	110	83	63	47	34	21
DG B1	479	213	134	95	71	54	40	27	15
DG B2	486	219	138	99	75	58	44	32	21
DG B3	547	247	156	113	86	68	53	40	27
DG C1	565	251	158	113	86	67	52	38	24
DG C2	561	252	159	115	87	68	54	40	27
DG C3	629	284	179	130	99	79	63	49	34
SS A1	1237	383	204	130	90	65	47	34	22
SS A2	1050	447	262	175	125	91	67	49	32
SS A3	1223	360	189	119	82	59	43	30	20
SS A4	1081	451	262	175	124	91	67	48	32
SS A5	2071	392	190	116	79	56	40	28	18
SS A6	1052	447	262	175	124	91	67	49	32
SS A7	2388	346	161	97	65	46	33	23	15
SS A8	1096	447	258	171	121	89	65	47	31
SS B1	859	294	161	104	73	52	38	27	18
SS B2	818	358	212	143	102	75	56	40	27
SS B3	812	278	152	98	68	49	36	26	17
SS B4	810	354	209	141	100	74	54	39	26
SS B5	1127	271	136	85	58	41	30	21	14
SS B6	778	347	207	140	100	74	55	40	26
SS B7	1021	238	119	74	50	36	26	18	12
SS B8	768	338	201	135	96	71	52	38	25
SS C1	573	262	158	107	77	57	42	31	21
SS C2	561	292	186	130	95	71	53	39	27
SS C3	596	212	118	77	54	39	29	21	14
SS C4	557	264	161	110	80	59	44	32	22
SS D1	1136	455	261	172	122	89	65	47	31
SS D2	917	481	307	214	157	117	87	64	43
SS D3	1433	435	230	146	101	73	53	38	25
SS D4	942	473	295	204	148	110	82	60	40
SS E1	874	367	214	143	101	74	55	39	26
SS E2	775	404	257	179	131	98	73	53	36
SS E3	1054	319	169	107	74	53	39	28	18
SS E4	801	382	233	159	115	85	63	46	31

MIAMI, FLORIDA

	SSF =.10	.20	.30	.40	.50	.60	.70	.80	.90
WW A1	12631	1698	832	513	346	244	175	124	81
WW A2	4197	1750	1031	688	486	354	259	187	125
WW A3	3514	1741	1086	746	538	396	293	213	143
WW A4	3211	1719	1106	773	563	418	311	227	154
WW A5	3050	1728	1140	807	593	444	332	243	165
WW A6	2951	1722	1152	822	608	457	343	252	172
WW B1	3955	1558	902	596	419	304	222	159	106
WW B2	2983	1640	1068	751	549	410	306	223	152
WW B3	3407	1961	1303	927	683	512	384	282	192
WW B4	2809	1733	1191	865	647	491	371	274	188
WW B5	2504	1574	1093	799	600	457	346	256	176
WW C1	3544	2037	1353	962	709	531	398	292	199
WW C2	2980	1780	1204	866	643	485	365	269	184
WW C3	2835	1849	1310	969	735	562	429	319	220
WW C4	2529	1647	1165	862	654	500	381	284	196
TW A1	10668	1588	785	485	328	232	166	118	78
TW A2	4255	1580	901	592	415	300	219	157	105
TW A3	3416	1539	929	627	447	327	241	174	117
TW A4	2870	1433	897	617	445	328	243	176	119
TW B1	5946	1454	763	483	331	236	170	121	80
TW B2	3377	1397	821	547	387	281	206	148	99
TW B3	2926	1330	805	544	388	284	209	151	101
TW B4	2654	1207	731	494	352	258	190	137	92
TW C1	3856	1266	702	455	316	227	165	118	79
TW C2	2903	1174	686	455	321	233	171	123	82
TW C3	2741	1086	631	418	294	213	156	112	75
TW C4	2821	958	535	348	242	175	127	91	60
TW D1	2845	1192	703	469	331	241	176	127	85
TW D2	3168	1557	969	665	478	353	261	189	127
TW D3	3244	1710	1094	761	553	410	305	222	150
TW D4	2898	1643	1085	769	566	423	317	232	158
TW D5	2590	1508	1008	720	532	400	300	221	150
TW E1	3648	1923	1229	855	622	461	343	250	169
TW E2	3226	1769	1150	808	591	441	329	240	163
TW E3	3212	1944	1323	955	711	537	405	299	205
TW E4	2806	1715	1174	850	635	480	363	268	183
TW F1	4997	1377	737	470	324	232	167	119	79
TW F2	3097	1346	803	539	383	279	205	148	99
TW F3	2542	1268	794	546	394	290	215	156	105
TW F4	2066	1114	719	503	367	273	203	148	100
TW G1	3109	1167	668	439	308	223	162	117	78
TW G2	2237	1091	677	464	333	245	181	132	89
TW G3	1857	973	620	431	313	232	172	126	85
TW G4	1472	789	508	355	259	192	143	105	71
TW H1	2007	922	560	379	271	198	146	106	71
TW H2	1512	780	495	343	248	184	136	99	67
TW H3	1241	650	415	288	209	155	115	84	57
TW H4	945	489	310	215	156	115	86	62	42
TW I1	2253	1019	616	415	296	216	159	115	77
TW I2	2444	1305	839	586	427	317	236	172	117
TW I3	2557	1416	924	651	476	355	265	194	132
TW I4	2406	1418	953	683	506	381	286	211	144
TW I5	2258	1363	927	669	498	376	284	209	143
TW J1	3018	1681	1100	776	569	424	317	232	158
TW J2	2647	1526	1015	723	533	400	300	220	150
TW J3	2641	1662	1154	843	633	481	365	270	186
TW J4	2397	1509	1048	766	576	438	332	246	169
DG A1	2824	1351	848	588	425	314	231	167	111
DG A2	2597	1254	793	557	409	305	227	165	112
DG A3	2799	1372	869	615	461	351	265	194	126
DG B1	2798	1380	877	619	468	361	278	208	141
DG B2	2641	1306	833	589	446	346	269	203	141
DG B3	2801	1389	888	627	474	371	290	220	152
DG C1	3197	1590	1018	718	541	426	336	259	181

```
MIAMI, FLORIDA                              CONTINUED    ORLANDO, FLORIDA                            CONTINUED
   SSF =.10   .20   .30   .40   .50   .60  .70  .80  .90    SSF =.10   .20   .30   .40   .50   .60  .70  .80  .90
DG C2 2985  1482   949   671   507   398  313  240  170  TW D4   936   530   351   250   186   140  106   78   54
DG C3 3185  1586  1019   720   543   429  340  264  187  TW D5   843   491   329   237   177   134  101   75   52
SS A1 6675  2252  1249   806   558   399  289  206  138  TW E1  1169   614   393   275   202   151  113   83   57
SS A2 5630  2548  1534  1031   731   532  390  282  190  TW E2  1038   568   371   262   193   145  110   81   55
SS A3 6690  2167  1190   764   526   376  271  194  129  TW E3  1045   632   432   314   236   180  137  102   70
SS A4 5842  2605  1559  1044   738   536  393  283  191  TW E4   917   561   386   281   212   162  123   92   63
SS A510965  2307  1177   733   497   351  251  178  118  TW F1  1560   426   227   145   101    73   53   38   25
SS A6 5668  2564  1543  1037   735   535  392  283  191  TW F2   972   420   250   169   121    89   66   48   32
SS A712724  2081  1030   634   426   299  213  151  100  TW F3   799   397   249   172   125    93   69   51   34
SS A8 5945  2602  1548  1033   729   529  387  279  188  TW F4   651   350   226   159   117    88   66   48   33
SS B1 4704  1744   990   647   451   324  236  169  113  TW G1   972   362   207   137    97    70   52   37   25
SS B2 4369  2027  1231   832   592   432  317  230  155  TW G2   703   341   212   146   106    78   59   43   29
SS B3 4512  1680   954   623   434   312  226  162  109  TW G3   585   305   195   136   100    75   56   41   28
SS B4 4368  2025  1229   829   590   430  316  228  154  TW G4   465   248   160   113    83    62   47   34   23
SS B5 6088  1610   851   539   368   262  188  134   89  TW H1   629   287   175   119    86    63   47   34   23
SS B6 4193  1981  1211   820   585   428  315  228  154  TW H2   476   245   155   108    79    59   44   32   22
SS B7 5594  1444   760   479   327   232  167  118   79  TW H3   392   205   131    91    67    50   37   28   19
SS B8 4177  1952  1189   803   571   417  307  222  150  TW H4   299   154    98    68    50    37   28   20   14
SS C1 3260  1559   960   655   470   345  255  184  124  TW I1   691   310   187   127    91    67   50   36   24
SS C2 3088  1662  1072   750   547   406  303  221  150  TW I2   778   414   267   188   138   103   78   57   39
SS C3 3332  1257   720   474   332   240  175  126   84  TW I3   816   450   295   209   154   116   87   64   44
SS C4 3045  1498   932   639   460   339  251  182  123  TW I4   777   458   309   223   167   126   96   71   49
SS D1 6282  2699  1597  1063   749   543  397  286  193  TW I5   735   444   303   220   165   126   96   71   49
SS D2 5041  2762  1788  1250   909   674  502  367  250  TW J1   967   537   352   250   185   139  105   77   53
SS D3 7732  2560  1414   911   629   450  326  233  155  TW J2   853   491   328   235   175   132  100   74   51
SS D4 5146  2708  1724  1194   863   637  472  344  234  TW J3   862   543   379   279   211   162  124   92   64
SS E1 4936  2219  1333   895   634   461  338  244  164  TW J4   785   495   345   254   193   148  113   84   58
SS E2 4298  2340  1512  1056   768   570  424  309  211  DG A1   863   410   257   177   127    93   67   47   29
SS E3 5770  1902  1050   677   468   335  242  173  115  DG A2   817   392   249   176   129    96   71   51   34
SS E4 4389  2202  1376   943   678   498  368  267  181  DG A3   905   436   280   200   152   116   88   64   41
                                                         DG B1   859   412   264   189   143   111   84   62   40
                                                         DG B2   838   403   259   187   142   111   86   65   44
ORLANDO, FLORIDA                             733 DD      DG B3   913   439   284   205   157   124   97   74   51
   SSF =.10   .20   .30   .40   .50   .60  .70  .80  .90 DG C1   996   477   308   222   169   133  105   80   54
WW A1 3941   523   255   157   107    76   55   39   26  DG C2   956   459   297   214   164   129  102   78   54
WW A2 1318   546   322   215   154   113   83   60   41  DG C3  1043   501   325   236   181   143  115   89   63
WW A3 1105   545   341   235   171   127   95   69   47  SS A1  2147   701   385   250   174   126   92   66   44
WW A4 1011   540   348   244   179   135  101   74   50  SS A2  1819   806   484   327   234   172  127   92   62
WW A5  961   543   359   256   190   143  108   80   54  SS A3  2140   668   362   233   162   117   85   61   40
WW A6  931   542   364   261   195   148  112   83   57  SS A4  1881   819   488   329   235   172  127   92   61
WW B1 1215   473   273   181   128    94   69   50   33  SS A5  3554   717   361   225   154   110   79   56   37
WW B2  950   521   340   241   178   134  101   74   51  SS A6  1828   809   485   328   235   173  128   92   62
WW B3 1088   625   416   298   222   168  127   94   64  SS A7  4121   641   311   192   130    92   66   47   31
WW B4  909   561   387   283   214   163  125   93   64  SS A8  1911   815   482   324   231   169  125   90   60
WW B5  816   514   358   264   200   153  117   88   61  SS B1  1501   541   305   200   141   102   75   54   36
WW C1 1136   652   434   311   231   175  132   98   67  SS B2  1413   644   390   265   190   141  104   76   51
WW C2  961   574   389   282   211   161  122   91   62  SS B3  1431   517   291   191   134    97   71   51   34
WW C3  928   606   432   322   246   190  146  110   76  SS B4  1407   639   387   263   189   139  103   75   50
WW C4  830   542   385   287   220   170  130   98   68  SS B5  1951   498   260   165   114    82   59   42   28
TW A1 3339   491   242   150   102    73   53   38   25  SS B6  1351   627   382   261   188   139  103   75   50
TW A2 1338   493   281   185   131    95   70   51   34  SS B7  1784   443   230   145   100    72   52   37   24
TW A3 1076   482   291   198   142   105   78   57   38  SS B8  1341   614   373   253   182   134  100   72   48
TW A4  906   450   282   195   142   106   79   58   39  SS C1  1020   485   299   205   148   110   82   60   40
TW B1 1867   452   236   150   104    75   54   39   26  SS C2   985   529   342   240   177   133  100   73   50
TW B2 1065   438   257   172   123    90   67   48   32  SS C3  1044   391   223   148   104    76   56   40   27
TW B3  924   418   253   172   124    91   68   49   33  SS C4   972   476   297   205   148   110   82   60   41
TW B4  840   380   230   157   113    83   62   45   30  SS D1  2001   839   494   330   235   172  127   91   61
TW C1 1217   396   219   143   100    73   53   38   26  SS D2  1615   874   566   399   293   220  164  120   82
TW C2  919   369   216   144   102    75   55   40   27  SS D3  2486   797   436   282   197   142  103   74   49
TW C3  870   342   199   132    94    69   51   37   25  SS D4  1652   857   545   380   278   207  155  113   76
TW C4  898   303   169   110    77    56   41   30   20  SS E1  1557   685   410   276   198   145  107   78   52
TW D1  875   363   214   143   102    75   55   40   27  SS E2  1370   738   477   335   246   185  138  101   69
TW D2 1010   494   308   212   154   115   86   63   42  SS E3  1840   588   322   208   145   105   77   55   36
TW D3 1034   543   347   243   178   133  100   73   50  SS E4  1405   694   433   299   217   161  120   87   59
```

TALLAHASSEE, FLORIDA — 1563 DD

	SSF=.10	.20	.30	.40	.50	.60	.70	.80	.90
WW A1	1899	237	110	66	44	31	22	16	10
WW A2	627	249	142	93	66	48	35	25	17
WW A3	522	248	151	102	74	54	40	29	20
WW A4	476	245	155	107	78	58	43	31	21
WW A5	451	247	160	112	82	62	46	34	23
WW A6	436	246	162	115	85	64	48	35	24
WW B1	556	205	114	74	51	37	27	19	13
WW B2	454	242	154	108	79	59	44	32	22
WW B3	521	291	190	134	99	74	56	41	28
WW B4	442	267	181	131	98	75	57	42	29
WW B5	400	247	170	124	93	71	54	40	28
WW C1	548	306	200	141	104	78	59	43	29
WW C2	466	271	181	130	96	73	55	41	28
WW C3	458	293	206	152	116	89	68	51	35
WW C4	411	263	185	137	104	80	61	46	32
TW A1	1615	224	105	63	43	30	21	15	10
TW A2	642	226	124	80	56	40	30	21	14
TW A3	513	221	129	86	61	45	33	24	16
TW A4	431	206	126	86	62	46	34	25	17
TW B1	903	207	104	64	44	31	23	16	11
TW B2	511	201	115	75	53	39	28	20	14
TW B3	443	192	113	76	54	39	29	21	14
TW B4	404	176	104	69	49	36	27	19	13
TW C1	589	183	98	62	43	31	22	16	11
TW C2	444	171	97	63	45	32	24	17	11
TW C3	422	159	90	59	41	30	22	16	10
TW C4	440	142	77	49	34	25	18	13	9
TW D1	400	158	89	58	41	30	22	16	10
TW D2	488	230	140	95	68	50	37	27	18
TW D3	496	252	158	109	79	59	44	32	22
TW D4	458	252	164	116	85	64	48	35	24
TW D5	417	237	156	111	82	62	47	35	24
TW E1	567	288	180	125	90	67	50	37	25
TW E2	507	269	172	120	88	66	49	36	24
TW E3	515	304	205	147	110	83	63	47	32
TW E4	454	272	184	133	99	76	57	43	29
TW F1	746	192	99	62	42	30	22	15	10
TW F2	461	190	110	73	52	38	28	20	13
TW F3	377	180	110	75	54	40	29	21	14
TW F4	306	159	101	70	51	38	28	21	14
TW G1	462	164	90	58	41	29	21	15	10
TW G2	331	155	94	63	45	33	25	18	12
TW G3	276	139	87	60	43	32	24	17	12
TW G4	220	114	72	50	36	27	20	15	10
TW H1	297	130	77	51	36	27	20	14	10
TW H2	225	111	69	47	34	25	19	14	9
TW H3	185	94	58	40	29	22	16	12	8
TW H4	142	71	44	30	22	16	12	9	6
TW I1	313	133	78	51	36	26	19	14	9
TW I2	373	192	121	84	61	45	34	25	17
TW I3	391	209	134	94	69	51	38	28	19
TW I4	379	218	144	103	76	58	44	32	22
TW I5	361	213	143	103	77	58	44	33	22
TW J1	467	252	162	113	83	62	47	34	23
TW J2	415	233	152	108	80	60	45	33	23
TW J3	425	262	180	131	99	75	57	43	29
TW J4	388	240	165	120	91	69	53	39	27
DG A1	395	177	105	69	47	32	21	12	4
DG A2	396	180	111	76	55	40	28	19	12
DG A3	459	209	131	92	69	52	39	28	17
DG B1	399	177	110	76	56	42	30	20	10
DG B2	411	185	115	82	61	47	35	26	16
DG B3	468	212	133	95	72	56	44	33	22
DG C1	474	209	130	92	69	53	41	29	17

TALLAHASSEE, FLORIDA — CONTINUED

	SSF=.10	.20	.30	.40	.50	.60	.70	.80	.90
DG C2	476	214	134	95	72	56	43	32	21
DG C3	537	244	152	110	83	66	52	40	28
SS A1	1043	325	171	108	74	53	38	27	18
SS A2	897	383	223	147	104	76	56	40	27
SS A3	1028	304	158	99	68	48	35	24	16
SS A4	923	386	222	147	103	75	55	39	26
SS A5	1733	332	159	96	65	45	32	23	15
SS A6	898	382	222	147	104	76	56	40	26
SS A7	1988	292	135	80	53	37	26	18	12
SS A8	934	382	219	144	101	73	54	38	25
SS B1	723	249	135	86	60	43	31	22	15
SS B2	698	307	181	120	85	63	46	33	22
SS B3	682	234	127	81	56	40	29	21	13
SS B4	692	303	178	118	84	61	45	32	22
SS B5	942	228	114	70	48	34	24	17	11
SS B6	664	297	176	118	84	61	45	33	22
SS B7	851	200	99	60	41	29	21	14	9
SS B8	655	289	170	113	80	59	43	31	21
SS C1	478	218	131	88	63	46	34	25	17
SS C2	474	246	156	108	79	59	44	32	22
SS C3	496	177	98	63	44	32	23	17	11
SS C4	470	222	135	92	66	49	36	26	18
SS D1	963	386	219	143	100	73	53	38	25
SS D2	787	412	260	180	131	97	73	53	35
SS D3	1207	369	193	122	84	60	43	30	20
SS D4	807	404	251	172	124	92	68	49	33
SS E1	738	310	179	118	83	61	44	32	21
SS E2	662	344	217	150	109	81	60	44	29
SS E3	885	269	141	89	61	43	31	22	15
SS E4	683	325	197	134	96	71	52	38	25

TAMPA, FLORIDA — 718 DD

	SSF=.10	.20	.30	.40	.50	.60	.70	.80	.90
WW A1	3985	522	253	156	107	76	55	39	26
WW A2	1320	543	320	214	153	112	83	60	41
WW A3	1104	542	338	234	170	126	94	69	47
WW A4	1009	536	346	243	179	134	101	74	50
WW A5	958	540	357	255	189	143	108	79	54
WW A6	927	539	362	260	194	147	111	82	56
WW B1	1218	471	272	180	128	93	69	50	33
WW B2	947	518	338	239	177	133	100	74	50
WW B3	1084	621	414	297	221	167	126	93	64
WW B4	905	558	385	282	213	163	124	92	64
WW B5	812	510	356	262	199	153	117	87	60
WW C1	1132	648	432	309	230	174	132	97	67
WW C2	957	570	387	280	210	160	122	90	62
WW C3	923	603	429	320	245	189	146	109	76
WW C4	826	538	383	286	219	169	130	97	68
TW A1	3375	490	240	149	101	72	52	37	25
TW A2	1343	491	279	184	130	95	70	51	34
TW A3	1077	480	289	196	141	104	77	56	38
TW A4	905	448	280	194	141	105	79	57	39
TW B1	1882	450	235	149	103	74	54	39	26
TW B2	1067	436	256	171	122	90	66	48	32
TW B3	925	416	251	171	123	91	68	49	33
TW B4	841	378	229	156	112	83	61	45	30
TW C1	1223	395	218	142	99	72	53	38	26
TW C2	921	368	214	143	102	75	55	40	27
TW C3	872	341	197	131	93	68	51	37	25
TW C4	902	302	168	110	77	56	41	30	20
TW D1	876	361	212	142	101	74	55	40	27
TW D2	1009	491	306	211	153	114	85	62	42
TW D3	1031	539	345	242	177	133	100	73	50

TAMPA, FLORIDA							CONTINUED			WEST PALM BEACH, FLORIDA								
SSF =	.10	.20	.30	.40	.50	.60	.70	.80	.90	SSF =	.10	.20	.30	.40	.50	.60	.70	.80	.90
TW D4	933	527	349	249	185	140	106	78	53	WW A1	8289	1119	550	340	230	163	117	82	54
TW D5	840	488	327	235	176	133	101	75	51	WW A2	2771	1160	686	458	325	237	173	125	84
TW E1	1166	610	391	274	201	150	113	83	56	WW A3	2325	1156	724	498	360	265	196	143	96
TW E2	1036	565	368	261	192	145	109	80	55	WW A4	2127	1143	738	516	377	280	209	152	103
TW E3	1041	629	430	313	235	179	136	101	70	WW A5	2022	1150	760	540	398	298	223	163	111
TW E4	913	558	383	280	211	161	123	91	63	WW A6	1958	1146	769	550	407	306	230	169	115
TW F1	1571	424	226	144	100	72	53	38	25	WW B1	2593	1025	596	395	278	201	147	106	71
TW F2	973	418	249	168	120	89	66	48	32	WW B2	1983	1094	714	504	369	275	206	150	102
TW F3	798	395	247	171	124	93	69	51	34	WW B3	2264	1308	871	621	459	344	258	189	129
TW F4	649	348	225	158	117	87	66	48	33	WW B4	1877	1162	801	583	437	331	251	185	127
TW G1	976	361	206	136	96	70	52	37	25	WW B5	1677	1058	737	540	406	309	235	174	120
TW G2	702	339	210	145	105	78	58	43	29	WW C1	2357	1360	906	646	477	357	268	197	134
TW G3	584	303	194	135	99	74	56	41	28	WW C2	1988	1191	808	583	433	327	247	182	124
TW G4	464	247	159	112	82	62	46	34	23	WW C3	1900	1243	883	655	497	381	291	216	150
TW H1	629	286	174	118	85	63	47	34	23	WW C4	1697	1109	787	583	443	339	259	193	133
TW H2	475	243	154	108	79	59	44	32	22	TW A1	7007	1047	520	322	219	155	111	78	52
TW H3	391	203	130	91	67	50	37	27	19	TW A2	2807	1046	599	395	277	201	146	105	70
TW H4	298	153	97	68	50	37	28	20	14	TW A3	2258	1021	619	419	299	219	161	116	78
TW I1	691	308	186	126	90	67	49	36	24	TW A4	1900	952	598	412	298	220	163	118	80
TW I2	776	412	265	186	137	103	77	57	39	TW B1	3913	961	506	321	221	158	114	81	54
TW I3	813	447	293	208	153	115	87	64	44	TW B2	2231	927	547	365	259	188	138	99	67
TW I4	774	455	307	222	166	126	96	71	49	TW B3	1936	883	536	364	260	190	140	101	68
TW I5	731	441	301	219	164	125	95	71	49	TW B4	1757	802	487	330	236	173	127	92	62
TW J1	964	534	350	249	184	139	105	77	53	TW C1	2544	839	466	303	211	152	111	79	53
TW J2	850	488	326	234	174	132	100	74	51	TW C2	1920	780	457	304	215	156	115	82	55
TW J3	858	539	376	277	210	161	123	92	64	TW C3	1813	721	420	279	197	143	105	75	50
TW J4	781	492	343	253	192	147	113	84	58	TW C4	1865	636	357	233	162	117	85	61	41
DG A1	862	408	255	176	126	92	67	47	29	TW D1	1867	785	464	310	220	160	117	84	56
DG A2	816	390	247	175	128	96	71	51	34	TW D2	2102	1037	648	445	321	237	175	127	86
DG A3	902	433	278	199	151	116	88	64	41	TW D3	2152	1138	730	509	371	275	205	149	101
DG B1	856	409	263	188	143	110	84	61	40	TW D4	1932	1100	728	517	381	286	214	157	107
DG B2	835	400	258	186	141	111	86	64	44	TW D5	1732	1012	679	486	360	271	203	150	102
DG B3	910	436	283	204	156	123	97	74	51	TW E1	2423	1282	822	574	417	310	231	168	114
DG C1	992	473	307	221	168	133	105	80	54	TW E2	2149	1182	771	543	398	297	222	162	110
DG C2	953	455	295	213	163	129	102	78	54	TW E3	2146	1303	890	644	480	363	274	202	139
DG C3	1041	497	324	235	180	143	115	89	63	TW E4	1879	1152	791	574	430	325	246	182	125
SS A1	2154	697	384	250	175	126	92	66	44	TW F1	3286	909	488	312	216	154	112	80	53
SS A2	1816	802	482	327	234	172	127	92	62	TW F2	2045	892	534	359	256	187	137	99	66
SS A3	2148	664	361	234	163	117	85	61	40	TW F3	1682	842	529	364	263	194	144	105	71
SS A4	1877	815	487	329	235	173	127	92	61	TW F4	1369	741	480	336	246	183	136	99	67
SS A5	3593	714	359	225	154	110	79	56	37	TW G1	2049	772	443	292	205	149	109	78	52
SS A6	1824	805	484	328	235	173	128	92	62	TW G2	1479	724	451	309	223	164	121	88	59
SS A7	4178	637	310	192	130	92	66	47	31	TW G3	1230	646	413	288	209	155	116	84	57
SS A8	1907	811	482	324	231	170	125	90	60	TW G4	975	525	339	238	173	129	96	70	48
SS B1	1504	538	304	200	141	102	75	54	36	TW H1	1326	611	373	253	181	133	98	71	48
SS B2	1410	640	389	265	190	141	104	75	51	TW H2	1001	518	330	229	166	123	91	66	45
SS B3	1433	514	290	191	134	97	71	51	34	TW H3	822	432	277	193	140	104	77	56	38
SS B4	1404	636	386	263	189	139	103	75	50	TW H4	626	325	207	144	104	77	57	42	28
SS B5	1966	496	259	165	114	82	59	42	28	TW I1	1478	671	407	275	196	143	105	76	51
SS B6	1347	624	381	260	188	139	103	75	50	TW I2	1624	870	561	393	287	213	159	116	79
SS B7	1798	440	229	145	100	72	52	37	24	TW I3	1698	943	618	436	320	239	178	130	89
SS B8	1337	611	372	253	182	134	100	72	48	TW I4	1606	950	640	460	341	257	193	142	97
SS C1	1020	482	297	204	147	109	81	59	40	TW I5	1511	915	624	452	337	255	192	142	97
SS C2	983	525	340	239	176	132	99	73	50	TW J1	2006	1121	736	520	382	285	213	156	106
SS C3	1048	389	222	147	104	76	56	40	27	TW J2	1765	1021	681	486	359	270	202	149	101
SS C4	971	473	295	203	148	110	82	60	41	TW J3	1767	1115	777	569	428	326	247	183	126
SS D1	1998	834	492	330	235	172	127	91	61	TW J4	1606	1014	707	518	390	297	225	167	115
SS D2	1607	870	566	399	294	220	165	120	82	DG A1	1855	891	561	388	281	206	152	108	71
SS D3	2496	792	434	282	197	142	104	74	49	DG A2	1716	834	528	372	273	203	151	110	74
SS D4	1644	853	544	380	278	207	155	113	76	DG A3	1855	918	584	414	310	237	179	130	85
SS E1	1553	681	408	276	198	145	107	78	52	DG B1	1831	911	582	411	310	239	184	136	92
SS E2	1364	734	476	335	246	185	138	101	69	DG B2	1742	867	557	394	298	232	180	135	94
SS E3	1847	585	320	208	145	105	77	55	36	DG B3	1857	927	598	423	319	250	196	149	103
SS E4	1400	690	432	299	217	161	120	87	59	DG C1	2097	1048	678	478	359	282	223	171	119

WEST PALM BEACH, FLORIDA								CONTINUED	
SSF =	.10	.20	.30	.40	.50	.60	.70	.80	.90
DG C2	1972	985	636	450	340	267	210	161	113
DG C3	2113	1057	687	487	366	289	230	179	127
SS A1	4363	1477	822	532	369	264	191	137	91
SS A2	3701	1681	1015	684	486	354	260	187	126
SS A3	4361	1417	780	502	347	248	179	128	85
SS A4	3834	1715	1030	691	490	356	261	188	127
SS A5	7150	1509	773	483	328	232	166	118	78
SS A6	3724	1690	1021	687	488	356	261	188	127
SS A7	8271	1357	674	416	280	197	140	99	66
SS A8	3898	1711	1021	683	483	351	257	185	125
SS B1	3083	1146	653	428	299	215	156	112	75
SS B2	2882	1342	818	554	395	289	212	153	104
SS B3	2950	1102	628	411	287	206	150	107	72
SS B4	2877	1338	815	551	393	287	211	152	103
SS B5	3980	1056	560	355	244	173	125	89	59
SS B6	2765	1310	804	546	390	286	210	152	103
SS B7	3647	945	499	315	216	153	110	78	52
SS B8	2749	1289	787	534	380	278	204	148	100
SS C1	2153	1033	639	437	314	231	170	123	83
SS C2	2052	1108	717	503	367	273	204	149	101
SS C3	2197	832	478	315	222	161	117	84	56
SS C4	2021	997	623	428	309	228	169	122	83
SS D1	4112	1772	1052	702	496	360	263	189	127
SS D2	3321	1825	1186	831	605	449	334	244	167
SS D3	5052	1679	930	601	416	298	215	154	103
SS D4	3388	1789	1143	793	574	424	315	229	156
SS E1	3236	1459	879	592	420	306	224	162	109
SS E2	2836	1549	1004	703	512	380	283	206	141
SS E3	3775	1248	691	447	310	222	160	115	76
SS E4	2893	1456	913	628	452	332	246	178	121

ATLANTA, GEORGIA								3095	DD
SSF =	.10	.20	.30	.40	.50	.60	.70	.80	.90
WW A1	905	102	46	27	17	12	8	6	3
WW A2	293	113	63	41	28	20	14	10	7
WW A3	245	114	68	46	32	23	17	12	8
WW A4	224	114	71	48	34	25	18	13	9
WW A5	213	115	74	51	37	27	20	14	10
WW A6	207	116	75	52	38	28	21	15	10
WW B1	234	82	44	27	18	13	9	6	4
WW B2	222	117	74	51	37	27	20	14	10
WW B3	259	143	93	65	47	35	26	19	13
WW B4	228	137	93	67	49	37	28	21	14
WW B5	211	130	89	64	48	36	27	20	14
WW C1	276	153	99	69	50	37	28	20	14
WW C2	238	138	92	65	48	36	27	20	13
WW C3	246	157	110	81	61	47	35	26	18
WW C4	221	142	99	73	55	42	32	24	16
TW A1	775	98	45	26	17	12	8	6	3
TW A2	302	103	56	35	24	17	12	8	5
TW A3	242	102	59	38	27	19	14	10	6
TW A4	205	96	58	39	27	20	15	10	7
TW B1	428	92	45	27	18	13	9	6	4
TW B2	242	93	52	34	23	17	12	8	5
TW B3	211	90	52	34	24	17	12	9	6
TW B4	194	83	48	32	22	16	12	8	5
TW C1	281	84	44	28	19	13	9	7	4
TW C2	214	80	45	29	20	14	10	7	5
TW C3	204	75	42	27	19	13	10	7	4
TW C4	217	68	36	23	16	11	8	6	4
TW D1	170	64	35	22	15	10	7	5	3
TW D2	239	112	67	45	32	23	17	12	8
TW D3	244	122	76	52	37	27	20	14	9

ATLANTA, GEORGIA								CONTINUED	
SSF =	.10	.20	.30	.40	.50	.60	.70	.80	.90
TW D4	235	129	83	58	42	31	23	17	11
TW D5	219	124	82	58	42	32	24	17	12
TW E1	283	143	89	61	43	32	23	17	11
TW E2	257	136	86	60	43	32	24	17	11
TW E3	271	160	107	76	57	42	32	23	16
TW E4	242	145	98	70	52	39	29	22	15
TW F1	344	84	42	25	17	12	8	6	4
TW F2	214	86	49	32	22	16	11	8	5
TW F3	176	83	50	33	23	17	12	9	6
TW F4	145	74	46	31	22	16	12	9	6
TW G1	212	73	39	25	17	12	8	6	4
TW G2	154	71	42	28	20	14	10	7	5
TW G3	130	64	40	27	19	14	10	7	5
TW G4	104	53	33	23	16	12	9	6	4
TW H1	137	59	34	22	15	11	8	6	4
TW H2	105	51	31	21	15	11	8	6	4
TW H3	88	44	27	18	13	9	7	5	3
TW H4	68	34	21	14	10	7	5	4	2
TW I1	130	53	30	19	13	9	6	4	3
TW I2	182	93	58	40	28	21	15	11	7
TW I3	193	103	65	45	32	24	18	13	8
TW I4	195	111	73	52	38	28	21	16	10
TW I5	189	111	74	53	39	30	22	16	11
TW J1	234	125	80	55	40	30	22	16	11
TW J2	211	118	77	54	39	29	22	16	11
TW J3	226	139	95	69	52	39	29	22	15
TW J4	208	128	88	64	48	36	27	20	14
DG A1	168	70	38	21	8	-	-	-	-
DG A2	191	85	51	33	23	15	9	5	-
DG A3	236	109	67	46	34	25	18	12	7
DG B1	168	72	42	27	17	10	-	-	-
DG B2	195	89	54	37	26	19	13	8	4
DG B3	240	112	69	48	36	28	21	15	10
DG C1	207	92	54	36	26	18	12	-	-
DG C2	230	106	64	45	33	25	18	13	7
DG C3	277	130	80	56	42	33	25	19	13
SS A1	516	154	80	49	33	23	16	11	7
SS A2	462	195	113	74	51	37	26	19	12
SS A3	497	140	71	43	28	20	14	9	6
SS A4	471	194	111	72	50	35	26	18	12
SS A5	877	154	73	43	28	19	13	9	6
SS A6	459	193	111	73	50	36	26	18	12
SS A7	1003	130	58	33	21	14	10	6	4
SS A8	472	190	108	70	48	34	24	17	11
SS B1	351	116	62	39	26	18	13	9	6
SS B2	361	157	92	61	42	30	22	16	10
SS B3	324	107	56	35	23	16	11	8	5
SS B4	354	153	89	59	41	29	21	15	10
SS B5	460	104	51	30	20	14	9	6	4
SS B6	340	150	89	59	41	29	21	15	10
SS B7	407	88	42	25	16	11	8	5	3
SS B8	332	144	84	55	38	28	20	14	9
SS C1	219	98	57	38	26	19	14	10	6
SS C2	234	120	75	52	37	27	20	14	10
SS C3	229	79	43	27	18	13	9	6	4
SS C4	232	108	65	44	31	22	16	12	8
SS D1	470	184	103	66	45	32	23	16	10
SS D2	405	212	133	91	65	48	35	25	17
SS D3	597	174	90	55	37	26	18	12	8
SS D4	415	207	128	86	61	45	33	23	15
SS E1	349	143	81	52	36	25	18	13	8
SS E2	334	173	108	74	53	38	28	20	13
SS E3	426	123	63	39	26	18	13	9	5
SS E4	345	163	98	65	46	33	24	17	11

AUGUSTA, GEORGIA   2547 DD

SSF =	.10	.20	.30	.40	.50	.60	.70	.80	.90
WW A1	1116	135	63	37	24	17	12	8	5
WW A2	371	147	84	54	38	27	20	14	9
WW A3	311	148	90	60	43	31	23	16	11
WW A4	285	147	93	63	46	33	25	18	12
WW A5	271	149	96	67	49	36	27	19	13
WW A6	263	149	98	69	50	37	28	20	14
WW B1	309	113	62	39	27	19	14	9	6
WW B2	279	149	95	66	48	35	26	19	13
WW B3	322	181	118	83	60	45	34	24	17
WW B4	281	170	116	83	62	47	35	26	18
WW B5	258	160	110	80	60	45	34	25	17
WW C1	342	192	125	88	64	48	36	26	18
WW C2	294	172	115	82	60	45	34	25	17
WW C3	298	192	135	99	75	57	44	33	22
WW C4	269	173	121	89	68	52	39	29	20
TW A1	955	128	60	36	24	16	12	8	5
TW A2	380	133	73	47	32	23	17	12	8
TW A3	306	132	77	51	36	26	19	13	9
TW A4	259	124	76	51	36	27	19	14	9
TW B1	534	121	61	37	25	17	12	9	6
TW B2	305	120	68	44	31	22	16	11	8
TW B3	266	116	68	45	32	23	17	12	8
TW B4	244	106	63	41	29	21	15	11	7
TW C1	351	108	58	37	25	18	13	9	6
TW C2	267	103	58	38	26	19	14	10	6
TW C3	255	96	54	35	24	17	13	9	6
TW C4	268	86	47	30	20	14	10	7	5
TW D1	224	88	49	31	22	15	11	8	5
TW D2	298	142	86	58	41	30	22	16	11
TW D3	304	155	97	66	48	35	26	19	13
TW D4	289	160	104	73	53	40	30	22	15
TW D5	267	153	101	71	53	39	30	22	15
TW E1	351	179	112	77	55	41	30	22	15
TW E2	317	170	109	75	55	41	30	22	15
TW E3	330	196	132	95	70	53	40	29	20
TW E4	294	177	120	86	64	48	37	27	18
TW F1	435	111	57	35	23	16	12	8	5
TW F2	272	112	65	42	30	21	15	11	7
TW F3	224	108	65	44	31	23	17	12	8
TW F4	184	96	60	41	30	22	16	12	8
TW G1	270	96	53	34	23	16	12	8	5
TW G2	197	92	55	37	26	19	14	10	7
TW G3	165	83	52	35	25	19	14	10	7
TW G4	132	69	43	30	21	16	12	8	6
TW H1	175	77	45	30	21	15	11	8	5
TW H2	134	67	41	28	20	15	11	8	5
TW H3	112	56	35	24	17	13	9	7	4
TW H4	86	43	27	18	13	10	7	5	3
TW I1	174	74	42	27	19	14	10	7	4
TW I2	228	118	74	51	37	27	20	15	10
TW I3	241	129	83	57	42	31	23	17	11
TW I4	240	138	92	65	48	36	27	20	13
TW I5	231	137	92	66	49	37	28	20	14
TW J1	290	157	101	70	51	38	28	21	14
TW J2	261	147	96	68	50	37	28	20	14
TW J3	275	170	117	85	64	48	37	27	19
TW J4	252	157	108	78	59	45	34	25	17
DG A1	223	98	56	34	21	11	-	-	-
DG A2	239	109	66	45	31	21	14	9	4
DG A3	288	134	84	58	43	32	24	16	9
DG B1	223	100	60	40	28	19	11	-	-
DG B2	245	114	70	48	35	26	19	13	7
DG B3	291	138	86	60	45	35	27	20	13
DG C1	269	123	74	51	37	27	19	12	-

AUGUSTA, GEORGIA   CONTINUED

SSF =	.10	.20	.30	.40	.50	.60	.70	.80	.90
DG C2	286	134	82	58	43	33	25	18	11
DG C3	334	160	99	70	53	41	32	25	17
SS A1	619	193	102	63	43	30	21	15	10
SS A2	551	237	139	91	64	46	33	24	16
SS A3	598	176	91	56	38	26	19	13	8
SS A4	561	237	137	89	62	45	32	23	15
SS A5	1036	193	93	55	36	25	18	12	8
SS A6	548	236	137	90	63	45	33	24	15
SS A7	1179	165	76	44	29	20	14	9	6
SS A8	564	233	134	87	60	43	31	22	14
SS B1	426	147	80	50	34	24	17	12	8
SS B2	431	192	113	75	53	38	28	20	13
SS B3	396	136	73	46	31	22	16	11	7
SS B4	424	187	110	73	51	37	27	19	13
SS B5	554	132	66	40	27	18	13	9	6
SS B6	408	184	110	73	51	37	27	19	13
SS B7	492	113	56	33	22	15	11	7	5
SS B8	399	178	105	69	48	35	25	18	12
SS C1	281	128	77	51	36	26	19	14	9
SS C2	291	152	96	66	48	35	26	19	13
SS C3	291	103	57	36	25	18	13	9	6
SS C4	288	137	83	56	40	29	21	15	10
SS D1	570	230	130	84	58	41	30	21	14
SS D2	488	258	163	112	80	59	44	32	21
SS D3	715	218	115	71	48	34	24	17	11
SS D4	499	253	157	106	76	56	41	30	20
SS E1	431	182	105	68	47	34	24	17	11
SS E2	406	213	134	92	66	49	36	26	17
SS E3	518	157	82	51	34	24	17	12	8
SS E4	418	201	122	82	58	42	31	22	15

MACON, GEORGIA   2240 DD

SSF =	.10	.20	.30	.40	.50	.60	.70	.80	.90
WW A1	1295	155	72	43	28	19	14	10	6
WW A2	423	167	95	62	43	31	22	16	11
WW A3	354	168	102	68	49	35	26	19	12
WW A4	324	167	105	72	52	38	28	20	14
WW A5	308	169	108	75	55	41	30	22	15
WW A6	298	169	110	77	57	42	31	23	15
WW B1	359	131	72	46	31	22	16	11	7
WW B2	315	168	107	74	53	40	29	21	14
WW B3	363	203	132	92	68	50	38	27	19
WW B4	314	190	128	92	69	52	39	29	20
WW B5	287	177	121	88	66	50	38	28	19
WW C1	383	214	140	98	72	53	40	29	20
WW C2	329	192	128	91	67	50	38	28	19
WW C3	330	212	149	109	83	63	48	36	25
WW C4	297	191	134	98	74	57	43	32	22
TW A1	1104	148	69	41	27	19	13	9	6
TW A2	434	152	83	53	37	26	19	13	9
TW A3	348	150	87	58	40	29	21	15	10
TW A4	294	141	86	58	41	30	22	16	11
TW B1	613	138	69	42	28	20	14	10	7
TW B2	347	136	78	50	35	25	18	13	9
TW B3	302	131	77	51	36	26	19	14	9
TW B4	276	120	71	47	33	24	17	13	8
TW C1	400	123	66	41	28	20	14	10	7
TW C2	303	117	66	43	30	21	16	11	7
TW C3	289	109	61	40	27	20	14	10	7
TW C4	304	97	53	33	23	16	12	8	6
TW D1	260	102	57	37	25	18	13	9	6
TW D2	337	160	97	65	46	34	25	18	12
TW D3	343	175	109	74	53	39	29	21	14

MACON, GEORGIA (CONTINUED)

SSF =	.10	.20	.30	.40	.50	.60	.70	.80	.90
TW D4	323	179	116	81	59	44	33	24	16
TW D5	298	170	112	79	58	44	33	24	16
TW E1	394	201	126	86	62	46	34	25	16
TW E2	356	190	121	84	61	45	34	25	17
TW E3	368	218	146	105	77	58	44	32	22
TW E4	327	196	132	95	71	53	40	30	20
TW F1	500	127	65	40	27	19	14	10	6
TW F2	310	128	74	48	34	24	18	13	8
TW F3	255	122	74	50	35	26	19	14	9
TW F4	209	109	68	47	34	25	18	13	9
TW G1	310	109	60	38	26	19	14	10	6
TW G2	224	105	63	42	30	22	16	11	8
TW G3	187	94	59	40	29	21	15	11	7
TW G4	150	78	49	33	24	18	13	10	6
TW H1	200	88	51	34	24	17	13	9	6
TW H2	153	76	47	32	23	17	12	9	6
TW H3	127	64	40	27	19	14	11	8	5
TW H4	97	49	30	20	15	11	8	6	4
TW I1	202	86	49	32	22	16	12	8	5
TW I2	258	133	84	57	41	31	23	16	11
TW I3	271	145	93	64	47	35	26	19	13
TW I4	268	154	102	72	53	40	30	22	15
TW I5	258	152	102	73	54	41	31	23	15
TW J1	326	176	113	78	57	42	32	23	15
TW J2	292	164	107	75	55	41	31	23	15
TW J3	305	188	129	94	70	53	40	30	20
TW J4	280	173	119	86	65	49	37	28	19
DG A1	258	114	66	41	26	15	7	-	-
DG A2	272	124	75	51	35	25	17	11	5
DG A3	324	150	93	65	48	36	26	18	11
DG B1	261	115	69	47	33	23	15	7	-
DG B2	280	128	79	55	40	30	22	15	9
DG B3	328	154	95	67	50	39	30	22	14
DG C1	312	141	85	59	43	32	23	15	5
DG C2	326	151	92	65	48	37	28	20	13
DG C3	376	178	109	77	58	46	36	27	19
SS A1	703	219	115	72	49	34	25	17	11
SS A2	619	266	155	102	71	51	37	27	18
SS A3	684	202	104	64	43	30	21	15	10
SS A4	633	267	153	100	70	50	37	26	17
SS A5	1182	221	106	63	42	29	20	14	9
SS A6	618	265	154	101	71	51	37	26	17
SS A7	1355	191	87	51	33	23	16	11	7
SS A8	638	263	150	97	68	49	35	25	16
SS B1	486	167	90	57	39	28	20	14	9
SS B2	485	215	126	83	59	43	31	22	15
SS B3	453	155	84	52	36	25	18	13	8
SS B4	477	210	123	81	57	41	30	22	14
SS B5	633	151	75	46	30	21	15	11	7
SS B6	459	207	122	81	57	42	30	22	14
SS B7	565	130	64	39	26	18	13	9	6
SS B8	450	200	117	77	54	39	29	21	14
SS C1	321	146	87	58	41	30	22	16	10
SS C2	328	171	108	74	54	40	29	21	14
SS C3	333	118	65	42	29	20	15	10	7
SS C4	325	154	94	63	45	33	24	17	12
SS D1	648	261	148	95	66	47	34	24	16
SS D2	547	288	181	125	90	66	49	35	24
SS D3	813	248	130	81	55	38	27	19	12
SS D4	560	283	175	119	85	62	46	33	22
SS E1	492	207	119	77	54	39	28	20	13
SS E2	457	239	150	103	74	55	40	29	19
SS E3	591	179	94	58	39	28	20	14	9
SS E4	471	226	136	92	65	47	35	25	17

SAVANNAH, GEORGIA 1952 DD

SSF =	.10	.20	.30	.40	.50	.60	.70	.80	.90
WW A1	1405	181	85	51	34	23	17	12	8
WW A2	480	194	111	73	51	37	27	19	13
WW A3	404	195	119	80	57	42	31	22	15
WW A4	371	193	122	84	61	45	33	24	16
WW A5	353	195	126	88	64	48	36	26	18
WW A6	342	195	128	90	66	50	37	27	18
WW B1	414	156	87	56	38	27	20	14	9
WW B2	358	192	123	86	62	46	34	25	17
WW B3	412	233	152	107	78	59	44	32	22
WW B4	355	216	147	106	79	60	45	33	23
WW B5	323	201	138	100	75	57	44	32	22
WW C1	435	245	160	113	83	62	46	34	23
WW C2	372	219	146	104	77	58	44	32	22
WW C3	372	240	169	124	94	72	55	41	28
WW C4	334	215	151	112	85	65	50	37	25
TW A1	1200	171	82	49	33	23	16	11	7
TW A2	489	176	98	63	43	31	23	16	11
TW A3	395	173	102	68	48	35	25	18	12
TW A4	334	163	100	67	48	35	26	19	13
TW B1	678	160	81	50	34	24	17	12	8
TW B2	392	157	90	59	41	30	22	16	10
TW B3	342	151	89	59	42	31	22	16	11
TW B4	313	138	82	55	39	28	21	15	10
TW C1	448	142	77	49	33	24	17	12	8
TW C2	342	134	77	50	35	25	18	13	9
TW C3	325	125	71	46	32	23	17	12	8
TW C4	338	112	61	39	27	19	14	10	7
TW D1	300	120	68	44	31	22	16	11	7
TW D2	381	183	112	76	54	40	29	21	14
TW D3	389	200	126	86	62	46	34	25	17
TW D4	365	203	132	93	68	51	38	28	19
TW D5	335	192	127	90	66	50	38	28	19
TW E1	446	230	145	99	72	53	39	29	19
TW E2	402	216	139	97	70	52	39	28	19
TW E3	414	247	166	119	89	67	51	37	25
TW E4	367	221	150	108	81	61	46	34	23
TW F1	559	148	77	48	32	23	16	12	8
TW F2	353	149	86	57	40	29	21	15	10
TW F3	292	141	87	59	42	31	23	16	11
TW F4	239	126	79	55	40	29	22	16	11
TW G1	351	127	71	45	31	22	16	12	8
TW G2	256	121	74	50	35	26	19	14	9
TW G3	214	109	68	47	34	25	18	13	9
TW G4	171	90	57	39	28	21	15	11	8
TW H1	228	102	60	40	28	21	15	11	7
TW H2	174	88	54	37	27	20	14	10	7
TW H3	144	74	46	32	23	17	12	9	6
TW H4	111	56	35	24	17	13	9	7	5
TW I1	234	101	59	39	27	20	14	10	7
TW I2	293	153	97	67	48	36	27	19	13
TW I3	308	167	107	75	54	40	30	22	15
TW I4	303	176	117	83	61	46	35	25	17
TW I5	290	173	116	83	62	47	35	26	18
TW J1	370	202	130	91	66	49	37	27	18
TW J2	330	187	123	87	64	48	36	26	18
TW J3	344	213	147	107	80	61	46	34	23
TW J4	315	196	135	98	74	56	43	32	22
DG A1	299	135	80	51	33	22	13	5	-
DG A2	308	143	88	60	42	30	21	14	8
DG A3	362	170	106	74	55	41	31	22	13
DG B1	301	137	84	57	41	30	20	12	-
DG B2	317	148	91	64	47	36	27	19	11
DG B3	367	174	108	77	58	45	35	26	17
DG C1	359	165	101	71	52	39	29	20	10

SAVANNAH, GEORGIA   CONTINUED

	SSF =.10	.20	.30	.40	.50	.60	.70	.80	.90
DG C2	368	173	107	76	56	44	33	24	16
DG C3	420	201	125	89	67	52	41	32	22
SS A1	789	253	135	84	57	41	29	21	13
SS A2	699	305	178	118	82	60	44	31	21
SS A3	770	235	123	76	51	36	26	18	12
SS A4	716	307	178	116	81	59	43	30	20
SS A5	1288	256	124	75	50	35	24	17	11
SS A6	698	304	178	117	82	59	43	31	20
SS A7	1457	223	104	61	40	28	19	13	9
SS A8	722	303	174	114	79	57	41	29	19
SS B1	549	194	106	67	46	33	24	17	11
SS B2	547	245	145	96	68	49	36	26	17
SS B3	515	182	99	62	43	30	22	15	10
SS B4	540	241	142	94	66	48	35	25	17
SS B5	703	176	89	54	36	26	18	13	8
SS B6	519	237	141	94	66	48	35	25	17
SS B7	630	153	76	46	31	22	15	11	7
SS B8	510	230	136	90	63	46	34	24	16
SS C1	368	170	102	68	49	35	26	19	12
SS C2	373	196	124	86	62	46	34	25	17
SS C3	377	137	76	49	34	24	18	13	8
SS C4	368	177	108	73	52	38	28	20	14
SS D1	737	303	173	112	78	56	40	29	19
SS D2	622	330	209	144	104	77	57	41	28
SS D3	912	287	152	95	65	46	33	23	15
SS D4	636	324	201	137	98	72	53	38	26
SS E1	564	242	140	92	64	46	33	24	16
SS E2	521	275	174	119	86	64	47	34	23
SS E3	666	208	110	69	47	33	24	17	11
SS E4	535	259	158	106	76	55	41	29	19

BOISE, IDAHO   5833 DD

	SSF =.10	.20	.30	.40	.50	.60	.70	.80	.90
WW A1	688	74	30	16	8	4	-	-	-
WW A2	226	84	44	27	17	11	7	4	2
WW A3	190	85	49	30	20	13	8	5	3
WW A4	174	85	50	32	22	14	9	6	3
WW A5	165	86	52	34	23	16	11	7	4
WW A6	160	86	54	36	24	17	11	7	4
WW B1	170	54	24	11	-	-	-	-	-
WW B2	176	90	55	37	25	17	12	8	5
WW B3	206	111	70	47	33	23	16	11	6
WW B4	185	109	72	50	36	26	19	13	8
WW B5	173	104	70	50	36	27	19	14	9
WW C1	221	120	75	51	36	25	18	12	7
WW C2	192	109	71	49	35	25	18	12	8
WW C3	202	127	88	63	46	35	25	18	12
WW C4	183	115	79	57	42	32	23	17	11
TW A1	593	72	30	16	9	5	2	-	-
TW A2	234	77	39	23	14	9	6	3	2
TW A3	188	76	42	26	17	11	7	4	2
TW A4	160	73	42	27	18	12	8	5	3
TW B1	330	69	32	18	10	6	3	2	-
TW B2	189	70	37	23	15	9	6	4	2
TW B3	165	68	38	23	15	10	7	4	2
TW B4	153	63	35	22	14	10	6	4	2
TW C1	219	64	32	19	12	7	5	3	1
TW C2	168	61	33	20	13	9	6	3	2
TW C3	162	58	31	19	12	8	5	3	2
TW C4	172	53	27	16	10	7	5	3	2
TW D1	124	42	20	10	-	-	-	-	-
TW D2	190	86	50	32	22	15	10	7	4
TW D3	193	94	56	37	25	17	12	8	4

BOISE, IDAHO   CONTINUED

	SSF =.10	.20	.30	.40	.50	.60	.70	.80	.90
TW D4	191	103	65	44	31	22	16	11	7
TW D5	180	100	65	45	32	23	17	12	7
TW E1	227	112	67	44	30	21	15	10	6
TW E2	208	108	67	45	31	22	16	11	6
TW E3	222	129	84	59	42	31	22	16	10
TW E4	199	117	78	55	40	29	21	15	10
TW F1	263	61	28	15	9	5	2	-	-
TW F2	165	64	34	21	13	8	5	3	1
TW F3	136	62	35	22	14	9	6	4	2
TW F4	112	55	33	21	14	10	6	4	2
TW G1	163	53	27	16	9	6	3	2	-
TW G2	119	52	30	18	12	8	5	3	1
TW G3	101	48	28	18	12	8	5	3	2
TW G4	81	40	24	15	10	7	5	3	2
TW H1	106	43	24	14	9	6	4	2	1
TW H2	82	38	22	14	9	6	4	2	1
TW H3	69	33	19	12	8	6	4	2	1
TW H4	54	25	15	10	6	4	3	2	1
TW I1	94	34	16	8	-	-	-	-	-
TW I2	144	72	43	28	19	13	9	6	3
TW I3	154	79	49	32	22	15	11	7	4
TW I4	158	89	57	39	28	20	14	10	6
TW I5	155	90	59	41	29	21	16	11	7
TW J1	188	98	61	40	28	20	14	9	6
TW J2	171	93	59	40	28	20	14	10	6
TW J3	186	112	75	53	39	29	21	15	9
TW J4	171	104	70	50	36	27	20	14	9
DG A1	124	46	16	-	-	-	-	-	-
DG A2	152	66	37	22	13	6	-	-	-
DG A3	197	89	53	36	25	17	12	7	-
DG B1	125	50	24	-	-	-	-	-	-
DG B2	157	69	40	26	16	10	5	-	-
DG B3	202	92	56	38	27	19	14	9	5
DG C1	161	67	36	21	9	-	-	-	-
DG C2	189	84	50	33	22	15	9	5	-
DG C3	233	107	66	45	32	24	17	12	7
SS A1	403	115	55	31	19	11	7	4	2
SS A2	371	151	83	52	34	23	15	10	6
SS A3	379	100	45	24	13	7	3	-	-
SS A4	373	148	80	49	32	21	14	9	5
SS A5	675	115	49	26	14	8	4	2	-
SS A6	367	149	82	50	33	22	15	9	5
SS A7	752	93	36	17	7	-	-	-	-
SS A8	373	144	77	46	30	19	13	8	4
SS B1	270	85	42	24	14	9	5	3	1
SS B2	289	122	69	43	29	20	13	9	5
SS B3	244	75	36	19	11	6	2	-	-
SS B4	281	117	65	41	27	18	12	8	5
SS B5	349	75	33	18	10	5	2	-	-
SS B6	271	116	65	41	27	18	12	8	5
SS B7	302	61	25	12	5	-	-	-	-
SS B8	261	110	61	38	25	16	11	7	4
SS C1	168	71	39	24	15	10	6	3	1
SS C2	186	93	56	37	25	18	12	8	5
SS C3	176	58	29	17	10	6	4	2	-
SS C4	184	84	49	31	21	15	10	6	4
SS D1	366	136	70	41	25	16	10	6	3
SS D2	324	163	98	64	43	30	21	14	8
SS D3	465	130	62	34	20	12	7	4	2
SS D4	333	160	94	61	41	28	19	13	8
SS E1	265	102	53	31	18	11	6	3	1
SS E2	263	131	78	50	34	23	16	10	6
SS E3	325	89	42	23	13	7	4	2	-
SS E4	273	124	71	45	30	20	14	9	5

LEWISTON, IDAHO 5464 DD

SSF =	.10	.20	.30	.40	.50	.60	.70	.80	.90
WW A1	521	45	11	-	-	-	-	-	-
WW A2	165	54	25	11	-	-	-	-	-
WW A3	136	55	28	15	7	-	-	-	-
WW A4	124	56	30	17	9	2	-	-	-
WW A5	117	56	31	18	10	5	1	-	-
WW A6	113	57	32	19	11	6	2	-	-
WW B1	105	-	-	-	-	-	-	-	-
WW B2	131	63	37	23	14	9	5	3	2
WW B3	154	79	47	30	20	13	8	5	3
WW B4	143	82	53	36	25	18	12	8	5
WW B5	135	80	52	36	26	18	13	9	6
WW C1	168	87	52	34	23	15	10	6	4
WW C2	148	81	51	34	23	16	11	7	4
WW C3	160	99	67	47	34	25	18	12	8
WW C4	146	90	61	43	31	23	17	12	7
TW A1	452	45	14	-	-	-	-	-	-
TW A2	173	50	22	10	-	-	-	-	-
TW A3	138	51	25	13	6	-	-	-	-
TW A4	117	49	26	14	8	3	-	-	-
TW B1	249	45	17	-	-	-	-	-	-
TW B2	140	47	22	12	5	-	-	-	-
TW B3	122	46	23	13	7	3	-	-	-
TW B4	114	44	22	13	7	4	1	-	-
TW C1	165	43	19	9	3	-	-	-	-
TW C2	127	42	21	11	6	3	-	-	-
TW C3	123	40	20	11	6	3	1	-	-
TW C4	133	38	18	10	6	3	1	-	-
TW D1	77	-	-	-	-	-	-	-	-
TW D2	144	61	33	20	12	8	4	2	1
TW D3	145	66	37	23	14	9	5	3	2
TW D4	148	77	47	31	21	14	10	6	4
TW D5	142	77	48	33	23	16	11	8	5
TW E1	173	81	46	29	19	12	8	5	3
TW E2	161	80	47	31	20	14	9	6	3
TW E3	175	99	63	43	30	21	15	10	6
TW E4	158	91	59	41	29	21	15	10	6
TW F1	193	37	13	-	-	-	-	-	-
TW F2	119	41	19	8	-	-	-	-	-
TW F3	97	40	20	10	4	-	-	-	-
TW F4	80	36	20	11	6	2	-	-	-
TW G1	117	33	13	-	-	-	-	-	-
TW G2	85	34	17	8	-	-	-	-	-
TW G3	72	31	17	9	5	-	-	-	-
TW G4	59	27	15	8	5	2	-	-	-
TW H1	75	27	13	6	-	-	-	-	-
TW H2	59	25	13	7	3	-	-	-	-
TW H3	50	22	12	7	4	2	-	-	-
TW H4	39	17	9	5	3	2	-	-	-
TW I1	55	-	-	-	-	-	-	-	-
TW I2	107	50	28	17	11	7	4	2	1
TW I3	115	56	33	20	13	8	5	3	2
TW I4	122	66	41	28	19	13	9	6	4
TW I5	121	68	44	30	21	15	10	7	4
TW J1	143	71	42	27	18	12	8	5	3
TW J2	132	69	42	28	19	13	9	6	3
TW J3	147	87	57	40	28	20	14	10	6
TW J4	136	81	53	37	27	19	14	10	6
DG A1	78	-	-	-	-	-	-	-	-
DG A2	115	45	22	10	-	-	-	-	-
DG A3	159	69	40	25	16	11	6	2	-
DG B1	80	-	-	-	-	-	-	-	-
DG B2	119	48	25	13	5	-	-	-	-
DG B3	164	72	42	27	19	13	8	5	-
DG C1	111	39	-	-	-	-	-	-	-

LEWISTON, IDAHO CONTINUED

SSF =	.10	.20	.30	.40	.50	.60	.70	.80	.90
DG C2	147	61	34	20	12	6	-	-	-
DG C3	192	84	50	33	23	16	11	7	3
SS A1	302	78	33	16	6	-	-	-	-
SS A2	287	111	59	35	22	14	9	5	3
SS A3	278	63	22	-	-	-	-	-	-
SS A4	286	107	55	32	19	12	7	4	2
SS A5	515	76	27	9	-	-	-	-	-
SS A6	282	108	57	33	21	13	8	5	3
SS A7	568	56	-	-	-	-	-	-	-
SS A8	284	103	52	29	17	10	6	3	2
SS B1	200	56	24	11	-	-	-	-	-
SS B2	226	91	49	30	19	12	8	5	3
SS B3	177	46	17	-	-	-	-	-	-
SS B4	218	86	46	27	17	11	7	4	2
SS B5	262	48	17	-	-	-	-	-	-
SS B6	210	85	46	28	17	11	7	4	2
SS B7	221	35	-	-	-	-	-	-	-
SS B8	200	79	42	24	15	9	5	3	2
SS C1	117	44	20	7	-	-	-	-	-
SS C2	140	66	38	24	15	9	6	3	2
SS C3	127	36	15	-	-	-	-	-	-
SS C4	140	60	33	20	12	8	5	3	1
SS D1	269	91	42	21	9	-	-	-	-
SS D2	247	119	69	44	28	19	12	8	5
SS D3	349	87	37	17	-	-	-	-	-
SS D4	254	117	66	41	27	17	11	7	4
SS E1	189	64	27	-	-	-	-	-	-
SS E2	198	94	53	33	21	13	8	5	3
SS E3	240	57	22	-	-	-	-	-	-
SS E4	207	89	49	29	18	11	7	4	2

POCATELLO, IDAHO 7063 DD

SSF =	.10	.20	.30	.40	.50	.60	.70	.80	.90
WW A1	540	58	24	12	6	-	-	-	-
WW A2	185	70	37	23	14	9	6	3	2
WW A3	157	71	41	26	17	11	7	4	2
WW A4	145	72	43	28	19	13	8	5	3
WW A5	138	73	45	30	20	14	9	6	3
WW A6	134	73	46	31	21	15	10	6	4
WW B1	130	41	18	-	-	-	-	-	-
WW B2	149	78	48	32	22	16	11	7	4
WW B3	176	96	61	41	29	21	15	10	6
WW B4	161	96	64	45	33	24	18	12	8
WW B5	151	93	63	45	33	24	18	13	8
WW C1	190	104	66	45	32	23	16	11	7
WW C2	167	96	63	44	31	23	16	11	7
WW C3	179	114	79	57	43	32	24	17	11
WW C4	162	103	72	52	39	29	22	16	10
TW A1	469	57	24	13	7	4	-	-	-
TW A2	191	64	33	20	12	8	5	3	1
TW A3	155	64	36	22	14	10	6	4	2
TW A4	133	61	36	23	15	10	7	4	2
TW B1	266	56	26	15	9	5	3	1	-
TW B2	156	59	32	19	13	8	5	3	2
TW B3	137	57	32	20	13	9	6	4	2
TW B4	127	54	30	19	13	9	6	4	2
TW C1	179	53	27	16	10	6	4	2	1
TW C2	139	52	28	17	11	7	5	3	2
TW C3	134	49	26	16	11	7	5	3	2
TW C4	143	44	23	14	9	6	4	3	1
TW D1	96	33	15	7	-	-	-	-	-
TW D2	160	74	43	28	19	13	9	6	4
TW D3	164	81	49	32	22	15	11	7	4

POCATELLO, IDAHO									CONTINUED	CHICAGO, ILLINOIS									6127 DD
SSF =	.10	.20	.30	.40	.50	.60	.70	.80	.90	SSF =	.10	.20	.30	.40	.50	.60	.70	.80	.90
TW D4	165	90	57	39	28	20	15	10	6	WW A1	346	23	-	-	-	-	-	-	-
TW D5	157	89	58	40	29	21	15	11	7	WW A2	109	36	17	9	5	-	-	-	-
TW E1	194	97	59	39	27	19	13	9	5	WW A3	92	38	20	12	7	4	2	-	-
TW E2	179	94	59	40	28	20	14	10	6	WW A4	85	39	22	13	8	5	3	2	-
TW E3	195	114	76	53	39	28	21	15	9	WW A5	82	40	24	15	10	6	4	2	1
TW E4	175	104	70	49	36	27	20	14	9	WW A6	80	41	25	16	10	7	4	3	1
TW F1	209	49	23	13	7	4	1	-	-	WW B1	45	-	-	-	-	-	-	-	-
TW F2	135	53	29	17	11	7	4	2	1	WW B2	95	47	28	19	13	9	6	4	2
TW F3	113	52	30	19	12	8	5	3	2	WW B3	115	61	38	25	17	12	9	6	4
TW F4	94	47	28	18	12	8	6	4	2	WW B4	112	65	43	30	22	16	12	8	5
TW G1	131	44	22	13	8	5	3	1	-	WW B5	108	65	44	31	23	17	13	9	6
TW G2	98	44	25	16	10	7	4	3	1	WW C1	127	67	42	28	20	14	10	7	4
TW G3	84	40	24	15	10	7	5	3	2	WW C2	114	64	41	29	20	15	11	8	5
TW G4	68	34	20	13	9	6	4	3	1	WW C3	129	82	56	41	30	23	17	12	8
TW H1	86	36	20	12	8	5	3	2	-	WW C4	118	75	52	37	28	21	16	11	8
TW H2	68	32	19	12	8	5	4	2	1	TW A1	305	25	6	-	-	-	-	-	-
TW H3	57	28	17	11	7	5	3	2	1	TW A2	116	33	15	8	4	-	-	-	-
TW H4	45	22	13	8	6	4	3	2	-	TW A3	93	35	18	10	6	3	2	-	-
TW I1	72	26	12	-	-	-	-	-	-	TW A4	81	34	19	11	7	5	3	2	-
TW I2	122	61	37	25	17	12	8	5	3	TW B1	167	27	10	4	-	-	-	-	-
TW I3	131	68	42	28	20	14	10	7	4	TW B2	96	32	16	9	5	3	1	-	-
TW I4	137	78	51	35	25	18	13	9	6	TW B3	84	32	17	10	6	4	2	1	-
TW I5	135	79	52	37	27	20	14	10	7	TW B4	80	31	17	10	6	4	3	1	-
TW J1	161	85	53	36	25	18	13	9	5	TW C1	113	28	13	7	4	2	-	-	-
TW J2	147	82	52	36	26	19	13	9	6	TW C2	88	30	15	9	5	3	2	1	-
TW J3	164	100	68	49	36	26	19	14	9	TW C3	87	29	15	9	5	3	2	1	-
TW J4	151	93	63	45	33	25	18	13	9	TW C4	96	27	13	8	5	3	2	1	-
DG A1	96	34	-	-	-	-	-	-	-	TW D1	35	-	-	-	-	-	-	-	-
DG A2	126	55	31	19	11	4	-	-	-	TW D2	104	45	26	16	11	7	5	3	2
DG A3	168	78	47	32	22	16	10	6	-	TW D3	105	50	29	19	13	9	6	4	2
DG B1	97	38	17	-	-	-	-	-	-	TW D4	114	60	38	26	18	13	10	7	4
DG B2	130	58	34	22	14	9	3	-	-	TW D5	112	62	40	28	20	15	11	8	5
DG B3	171	80	49	34	24	18	13	9	4	TW E1	129	62	37	24	16	12	8	5	3
DG C1	128	54	29	16	-	-	-	-	-	TW E2	122	62	38	26	18	13	9	6	4
DG C2	157	72	43	28	19	13	8	4	-	TW E3	138	80	53	37	27	20	14	10	7
DG C3	198	93	58	40	29	22	16	11	6	TW E4	127	74	49	35	25	19	14	10	7
SS A1	338	99	48	27	16	10	6	3	2	TW F1	122	21	6	-	-	-	-	-	-
SS A2	321	134	75	47	31	21	14	9	6	TW F2	78	27	13	7	3	-	-	-	-
SS A3	314	84	39	20	11	5	1	-	-	TW F3	66	27	14	8	5	3	1	-	-
SS A4	322	130	72	44	29	19	13	8	5	TW F4	55	26	15	9	6	4	2	1	-
SS A5	559	97	42	22	12	7	3	1	-	TW G1	75	20	8	3	-	-	-	-	-
SS A6	317	131	73	46	30	20	14	9	5	TW G2	57	23	12	7	4	2	-	-	-
SS A7	612	77	30	14	5	-	-	-	-	TW G3	49	22	12	7	5	3	2	-	-
SS A8	320	126	68	42	27	18	12	7	4	TW G4	41	19	11	7	4	3	2	1	-
SS B1	225	72	36	21	12	8	4	2	1	TW H1	49	18	9	5	2	-	-	-	-
SS B2	250	108	61	39	26	18	12	8	5	TW H2	40	17	10	6	3	2	1	-	-
SS B3	201	63	30	16	9	5	1	-	-	TW H3	35	16	9	5	3	2	1	-	-
SS B4	242	103	58	37	24	17	11	7	4	TW H4	28	12	7	4	3	2	1	-	-
SS B5	286	63	28	15	8	4	2	-	-	TW I1	20	-	-	-	-	-	-	-	-
SS B6	234	102	58	37	25	17	12	8	5	TW I2	78	37	22	14	10	7	4	3	2
SS B7	245	50	21	10	3	-	-	-	-	TW I3	85	43	26	17	12	8	6	4	2
SS B8	225	96	54	34	22	15	10	7	4	TW I4	94	52	34	23	17	12	9	6	4
SS C1	137	59	33	20	13	8	5	3	1	TW I5	96	55	36	25	18	13	10	7	5
SS C2	158	80	49	33	23	16	11	7	4	TW J1	107	55	34	22	15	11	8	5	3
SS C3	142	48	24	14	9	5	3	1	-	TW J2	101	54	34	23	17	12	9	6	4
SS C4	156	72	42	28	19	13	9	6	4	TW J3	117	71	48	34	25	19	14	10	7
SS D1	310	117	62	37	23	14	9	5	3	TW J4	109	66	45	32	24	18	13	9	6
SS D2	284	145	88	58	40	28	19	13	8	DG A1	31	-	-	-	-	-	-	-	-
SS D3	389	111	54	30	18	11	6	3	2	DG A2	79	31	15	-	-	-	-	-	-
SS D4	291	142	85	55	38	26	18	12	7	DG A3	119	53	31	21	14	10	6	3	-
SS E1	221	86	45	26	16	9	5	2	1	DG B1	35	-	-	-	-	-	-	-	-
SS E2	228	115	70	46	31	21	15	10	6	DG B2	81	33	18	10	-	-	-	-	-
SS E3	268	75	35	19	11	6	3	1	-	DG B3	123	55	33	22	16	12	8	5	2
SS E4	235	109	63	40	27	19	13	8	5	DG C1	64	17	-	-	-	-	-	-	-

CHICAGO, ILLINOIS CONTINUED

	SSF =.10	.20	.30	.40	.50	.60	.70	.80	.90
DG C2	102	44	24	15	10	6	-	-	-
DG C3	143	65	39	27	19	14	11	7	4
SS A1	217	56	25	13	7	4	2	-	-
SS A2	220	88	48	30	20	14	9	6	4
SS A3	192	42	16	-	-	-	-	-	-
SS A4	218	84	45	27	18	12	8	5	3
SS A5	374	52	19	8	-	-	-	-	-
SS A6	215	85	46	29	19	13	9	6	3
SS A7	403	34	-	-	-	-	-	-	-
SS A8	214	79	42	25	16	11	7	4	3
SS B1	140	39	18	9	4	-	-	-	-
SS B2	174	72	40	25	17	12	8	5	3
SS B3	119	31	11	-	-	-	-	-	-
SS B4	166	68	37	23	15	10	7	5	3
SS B5	181	31	11	-	-	-	-	-	-
SS B6	160	67	38	24	16	11	7	5	3
SS B7	146	20	-	-	-	-	-	-	-
SS B8	151	61	34	21	14	9	6	4	2
SS C1	75	28	13	7	-	-	-	-	-
SS C2	102	50	30	19	13	9	6	4	2
SS C3	82	23	10	4	-	-	-	-	-
SS C4	102	45	25	16	11	7	5	3	2
SS D1	193	67	33	18	11	6	3	1	-
SS D2	193	96	58	38	26	18	13	9	6
SS D3	250	62	28	14	8	4	-	-	-
SS D4	197	94	56	36	25	17	12	8	5
SS E1	128	44	20	10	-	-	-	-	-
SS E2	151	74	44	28	19	13	9	6	4
SS E3	165	38	15	6	-	-	-	-	-
SS E4	157	70	40	25	17	11	8	5	3

MOLINE, ILLINOIS 6395 DD

	SSF =.10	.20	.30	.40	.50	.60	.70	.80	.90
WW A1	338	23	-	-	-	-	-	-	-
WW A2	109	36	17	9	5	2	-	-	-
WW A3	92	38	20	12	7	4	2	1	-
WW A4	85	39	22	14	9	6	3	2	1
WW A5	82	40	24	15	10	7	4	3	1
WW A6	80	41	25	16	11	7	5	3	2
WW B1	45	-	-	-	-	-	-	-	-
WW B2	96	47	28	19	13	9	6	4	2
WW B3	115	61	38	25	18	12	9	6	4
WW B4	112	65	43	30	22	16	12	9	6
WW B5	108	65	44	31	23	17	13	9	6
WW C1	127	67	42	28	20	14	10	7	4
WW C2	114	64	41	29	20	15	11	8	5
WW C3	129	82	56	41	30	23	17	13	8
WW C4	118	75	52	37	28	21	16	12	8
TW A1	299	25	7	-	-	-	-	-	-
TW A2	115	33	15	8	4	2	-	-	-
TW A3	93	35	18	10	6	4	2	1	-
TW A4	81	34	19	12	7	5	3	2	-
TW B1	165	28	10	4	-	-	-	-	-
TW B2	95	32	16	9	5	3	2	-	-
TW B3	84	32	17	10	6	4	3	1	-
TW B4	80	31	17	10	6	4	3	2	-
TW C1	112	28	13	7	4	2	-	-	-
TW C2	88	30	15	9	5	3	2	1	-
TW C3	86	29	15	9	5	3	2	1	-
TW C4	95	27	13	8	5	3	2	1	-
TW D1	35	-	-	-	-	-	-	-	-
TW D2	104	45	26	16	11	7	5	3	2
TW D3	105	50	29	19	13	9	6	4	2

MOLINE, ILLINOIS CONTINUED

	SSF =.10	.20	.30	.40	.50	.60	.70	.80	.90
TW D4	114	61	38	26	18	13	10	7	4
TW D5	112	62	40	28	20	15	11	8	5
TW E1	129	62	37	24	17	12	8	6	3
TW E2	122	62	38	26	18	13	9	6	4
TW E3	138	80	52	37	27	20	15	10	7
TW E4	126	74	49	35	26	19	14	10	7
TW F1	121	21	6	-	-	-	-	-	-
TW F2	78	27	13	7	4	-	-	-	-
TW F3	66	27	15	9	5	3	2	-	-
TW F4	55	26	15	9	6	4	2	1	-
TW G1	75	20	8	3	-	-	-	-	-
TW G2	57	23	12	7	4	2	1	-	-
TW G3	49	22	12	7	5	3	2	1	-
TW G4	41	19	11	7	4	3	2	1	-
TW H1	49	18	9	5	2	-	-	-	-
TW H2	40	17	10	6	4	2	1	-	-
TW H3	35	16	9	5	4	2	1	-	-
TW H4	28	12	7	4	3	2	1	-	-
TW I1	20	-	-	-	-	-	-	-	-
TW I2	78	37	22	14	10	7	5	3	2
TW I3	85	43	26	17	12	8	6	4	2
TW I4	94	52	34	23	17	12	9	6	4
TW I5	96	55	36	25	18	14	10	7	5
TW J1	107	55	34	22	16	11	8	5	3
TW J2	101	54	34	23	17	12	9	6	4
TW J3	117	71	48	34	25	19	14	10	7
TW J4	109	66	45	32	24	18	13	10	6
DG A1	31	-	-	-	-	-	-	-	-
DG A2	79	31	15	5	-	-	-	-	-
DG A3	119	53	31	21	14	10	6	3	-
DG B1	34	-	-	-	-	-	-	-	-
DG B2	81	33	17	10	4	-	-	-	-
DG B3	123	55	33	22	16	12	8	5	2
DG C1	63	17	-	-	-	-	-	-	-
DG C2	102	43	24	15	10	6	-	-	-
DG C3	144	65	39	27	19	14	11	7	4
SS A1	213	55	25	13	7	4	2	-	-
SS A2	218	87	48	30	20	14	9	6	4
SS A3	187	41	15	5	-	-	-	-	-
SS A4	215	83	45	27	18	12	8	5	3
SS A5	363	51	19	8	-	-	-	-	-
SS A6	212	84	46	28	19	13	9	6	4
SS A7	388	33	-	-	-	-	-	-	-
SS A8	211	78	41	25	16	11	7	5	3
SS B1	138	39	18	9	5	2	-	-	-
SS B2	172	71	40	25	17	12	8	6	3
SS B3	116	30	11	-	-	-	-	-	-
SS B4	164	67	37	23	15	11	7	5	3
SS B5	177	31	11	3	-	-	-	-	-
SS B6	158	66	37	23	16	11	8	5	3
SS B7	141	19	-	-	-	-	-	-	-
SS B8	149	61	33	21	14	9	6	4	2
SS C1	75	28	14	7	3	-	-	-	-
SS C2	102	50	30	19	13	9	6	4	3
SS C3	82	23	10	4	-	-	-	-	-
SS C4	102	45	25	16	11	7	5	3	2
SS D1	190	66	33	18	11	6	4	2	-
SS D2	191	95	58	38	26	19	13	9	6
SS D3	245	61	27	14	8	4	2	-	-
SS D4	196	93	55	36	25	17	12	8	5
SS E1	127	43	20	10	4	-	-	-	-
SS E2	150	73	43	28	19	13	9	6	4
SS E3	161	37	15	6	-	-	-	-	-
SS E4	155	69	39	25	17	11	8	5	3

```
SPRINGFIELD, ILLINOIS                      5558 DD   SPRINGFIELD, ILLINOIS                     CONTINUED
  SSF =.10  .20  .30  .40  .50  .60  .70  .80  .90     SSF =.10  .20  .30  .40  .50  .60  .70  .80  .90
WW A1  463   39   14    6    3    -    -    -    -   DG C2  126   55   32   21   14   10    6    -    -
WW A2  143   50   26   16   10    7    4    3    1   DG C3  168   77   46   32   23   18   13   10    6
WW A3  121   52   29   19   12    8    6    4    2   SS A1  269   72   35   20   12    8    5    3    2
WW A4  111   53   31   20   14   10    7    4    3   SS A2  261  105   59   37   25   18   12    9    5
WW A5  106   54   33   22   15   11    7    5    3   SS A3  245   59   26   14    8    4    2    1    -
WW A6  103   55   34   23   16   11    8    5    3   SS A4  260  102   56   35   23   16   11    8    5
WW B1   85   20    -    -    -    -    -    -    -   SS A5  465   69   29   15    9    5    3    2    -
WW B2  119   60   37   25   17   12    9    6    4   SS A6  256  103   57   36   24   17   12    8    5
WW B3  141   76   48   32   23   17   12    8    5   SS A7  514   51   18    8    -    -    -    -    -
WW B4  133   78   52   37   27   20   15   11    7   SS A8  257   97   53   33   22   15   10    7    4
WW B5  126   76   52   37   27   20   15   11    7   SS B1  177   53   26   15    9    6    4    2    1
WW C1  154   83   52   36   25   19   13   10    6   SS B2  206   86   49   31   21   15   11    7    5
WW C2  136   77   50   35   25   19   14   10    7   SS B3  155   44   20   11    6    4    2    -    -
WW C3  150   95   66   48   36   27   20   15   10   SS B4  198   82   46   29   20   14   10    7    4
WW C4  137   87   60   44   33   25   19   14    9   SS B5  231   44   19   10    6    3    2    -    -
TW A1  402   39   15    8    4    2    -    -    -   SS B6  191   81   46   29   20   14   10    7    4
TW A2  151   46   23   13    9    6    4    2    1   SS B7  192   33   12    5    -    -    -    -    -
TW A3  121   47   26   16   10    7    5    3    2   SS B8  182   75   42   27   18   12    9    6    4
TW A4  104   46   26   17   11    8    5    4    2   SS C1  102   41   22   14    9    6    4    2    1
TW B1  218   40   17    9    6    3    2    1    -   SS C2  126   63   38   25   18   13    9    6    4
TW B2  123   43   23   14    9    6    4    3    2   SS C3  110   33   16    9    6    4    2    1    -
TW B3  108   43   24   15   10    7    5    3    2   SS C4  126   56   33   21   15   10    7    5    3
TW B4  101   41   22   14    9    6    4    3    2   SS D1  240   86   45   27   17   11    8    5    3
TW C1  145   39   19   11    7    5    3    2    1   SS D2  228  115   70   47   33   24   17   12    8
TW C2  112   39   21   13    8    6    4    3    2   SS D3  310   81   38   22   13    9    6    3    2
TW C3  109   37   20   12    8    5    4    2    1   SS D4  234  112   67   44   31   22   16   11    7
TW C4  119   34   17   10    7    5    3    2    1   SS E1  167   61   31   18   11    7    4    3    1
TW D1   63   17    5    -    -    -    -    -    -   SS E2  182   90   55   36   25   18   13    9    6
TW D2  129   57   33   22   15   10    7    5    3   SS E3  211   53   24   13    8    5    3    1    -
TW D3  131   63   38   25   17   12    9    6    4   SS E4  189   85   49   32   22   15   11    7    5
TW D4  136   73   46   32   23   17   12    9    6
TW D5  131   73   47   33   24   18   13    9    6
TW E1  157   76   46   31   21   15   11    8    5   EVANSVILLE, INDIANA                       4629 DD
TW E2  147   75   47   32   23   16   12    8    5     SSF =.10  .20  .30  .40  .50  .60  .70  .80  .90
TW E3  162   94   62   44   32   24   18   13    9   WW A1  493   47   18    9    5    2    1    -    -
TW E4  147   87   58   41   30   23   17   12    8   WW A2  161   58   30   18   12    8    5    3    2
TW F1  165   33   14    7    4    2    -    -    -   WW A3  136   60   34   21   14   10    7    5    3
TW F2  103   38   20   12    8    5    3    2    1   WW A4  125   60   36   23   16   11    8    5    3
TW F3   86   38   21   13    9    6    4    3    2   WW A5  119   61   37   25   17   12    9    6    4
TW F4   72   35   21   13    9    6    4    3    2   WW A6  116   62   38   26   18   13    9    6    4
TW G1  101   30   15    8    5    3    2    1    -   WW B1  104   28   10    -    -    -    -    -    -
TW G2   75   32   18   11    7    5    3    2    1   WW B2  131   67   41   27   19   14   10    7    4
TW G3   64   30   17   11    8    5    4    2    1   WW B3  155   84   53   36   25   18   13    9    6
TW G4   53   25   15   10    7    5    3    2    1   WW B4  144   85   57   40   29   22   16   12    8
TW H1   66   26   14    8    5    4    2    1    -   WW B5  136   83   56   40   29   22   16   12    8
TW H2   52   24   14    9    6    4    3    2    1   WW C1  169   91   58   39   28   20   15   11    7
TW H3   44   21   12    8    5    4    3    2    1   WW C2  149   84   55   38   28   20   15   11    7
TW H4   35   16   10    6    4    3    2    1    -   WW C3  162  102   71   52   38   29   22   16   11
TW I1   45   12    -    -    -    -    -    -    -   WW C4  147   93   64   47   35   26   20   15   10
TW I2   97   47   29   19   13    9    7    5    3   TW A1  429   47   19   10    5    3    2    -    -
TW I3  104   53   33   22   15   11    8    6    4   TW A2  168   53   27   16   10    7    4    3    2
TW I4  113   63   41   28   21   15   11    8    5   TW A3  136   54   29   18   12    8    6    4    2
TW I5  113   65   43   30   22   16   12    9    6   TW A4  116   52   30   19   13    9    6    4    3
TW J1  130   67   42   28   20   14   10    7    5   TW B1  239   46   21   11    7    4    3    2    -
TW J2  121   66   42   29   21   15   11    8    5   TW B2  137   49   26   16   10    7    5    3    2
TW J3  137   83   56   40   30   22   17   12    8   TW B3  120   49   27   17   11    8    5    4    2
TW J4  127   77   53   38   28   21   16   12    8   TW B4  112   46   25   16   11    7    5    4    2
DG A1   62    -    -    -    -    -    -    -    -   TW C1  160   44   22   13    8    5    4    2    1
DG A2  100   41   22   12    5    -    -    -    -   TW C2  124   44   23   14    9    6    4    3    2
DG A3  141   63   37   25   18   13    9    5    -   TW C3  120   42   22   14    9    6    4    3    2
DG B1   63   17    -    -    -    -    -    -    -   TW C4  129   38   20   12    8    5    4    2    2
DG B2  102   43   24   15   10    5    -    -    -   TW D1   77   23   10    4    -    -    -    -    -
DG B3  145   65   39   27   20   15   11    7    4   TW D2  142   64   37   24   17   12    8    6    4
DG C1   91   33   15    -    -    -    -    -    -   TW D3  144   70   42   28   19   14   10    7    4
```

EVANSVILLE, INDIANA							CONTINUED			FORT WAYNE, INDIANA								6209 DD	
SSF =	.10	.20	.30	.40	.50	.60	.70	.80	.90	SSF =	.10	.20	.30	.40	.50	.60	.70	.80	.90
TW D4	148	79	50	35	25	18	13	9	6	WW A1	247	-	-	-	-	-	-	-	-
TW D5	142	79	51	36	26	19	14	10	7	WW A2	80	22	6	-	-	-	-	-	-
TW E1	173	84	51	34	24	17	12	9	6	WW A3	68	25	11	-	-	-	-	-	-
TW E2	160	83	51	35	25	18	13	9	6	WW A4	64	27	13	7	-	-	-	-	-
TW E3	175	102	67	47	35	26	19	14	9	WW A5	61	28	15	8	4	-	-	-	-
TW E4	159	94	62	44	32	24	18	13	9	WW A6	60	29	16	9	5	2	-	-	-
TW F1	185	40	18	9	6	3	2	1	-	WW B1	-	-	-	-	-	-	-	-	-
TW F2	117	44	23	14	9	6	4	3	2	WW B2	76	37	21	13	9	6	4	2	1
TW F3	97	43	24	15	10	7	5	3	2	WW B3	94	48	29	19	13	9	6	4	2
TW F4	81	39	23	15	10	7	5	4	2	WW B4	95	55	36	25	18	13	9	7	4
TW G1	114	36	18	10	6	4	3	2	-	WW B5	92	55	37	26	19	14	10	7	5
TW G2	85	36	20	13	8	6	4	3	2	WW C1	105	55	34	22	15	11	7	5	3
TW G3	72	34	20	13	9	6	4	3	2	WW C2	96	53	34	23	16	12	8	6	4
TW G4	59	29	17	11	8	5	4	3	2	WW C3	113	71	49	35	26	19	14	10	7
TW H1	74	30	16	10	6	4	3	2	1	WW C4	103	65	45	32	24	18	13	10	6
TW H2	59	27	16	10	7	5	3	2	1	TW A1	222	-	-	-	-	-	-	-	-
TW H3	50	24	14	9	6	4	3	2	1	TW A2	86	21	6	-	-	-	-	-	-
TW H4	39	18	11	7	5	3	2	2	1	TW A3	70	24	10	4	-	-	-	-	-
TW I1	56	18	7	-	-	-	-	-	-	TW A4	62	24	12	6	3	-	-	-	-
TW I2	107	53	32	21	15	10	7	5	3	TW B1	123	16	-	-	-	-	-	-	-
TW I3	115	59	36	24	17	12	9	6	4	TW B2	73	22	9	4	-	-	-	-	-
TW I4	123	69	44	31	22	16	12	9	6	TW B3	65	23	11	6	2	-	-	-	-
TW I5	122	71	46	33	24	18	13	9	6	TW B4	62	23	11	6	3	2	-	-	-
TW J1	143	74	46	31	22	16	12	8	5	TW C1	86	19	7	-	-	-	-	-	-
TW J2	132	72	46	32	23	17	12	9	6	TW C2	69	22	10	5	2	-	-	-	-
TW J3	148	90	61	43	32	24	18	13	9	TW C3	68	21	10	5	3	1	-	-	-
TW J4	137	83	56	41	30	22	17	12	8	TW C4	77	20	9	5	3	2	-	-	-
DG A1	76	20	-	-	-	-	-	-	-	TW D1	-	-	-	-	-	-	-	-	-
DG A2	111	46	25	15	8	-	-	-	-	TW D2	83	35	19	12	7	5	3	2	-
DG A3	153	68	41	28	19	14	9	6	-	TW D3	85	39	22	13	9	6	4	2	1
DG B1	78	25	-	-	-	-	-	-	-	TW D4	96	50	31	21	15	10	7	5	3
DG B2	114	49	27	17	11	7	-	-	-	TW D5	96	53	34	23	17	12	9	6	4
DG B3	157	71	43	29	21	16	12	8	4	TW E1	106	50	29	18	12	8	6	4	2
DG C1	107	40	20	10	-	-	-	-	-	TW E2	102	51	31	20	14	10	7	5	3
DG C2	140	61	35	23	16	11	7	3	-	TW E3	119	68	45	31	22	16	12	8	6
DG C3	182	84	50	34	25	19	14	10	6	TW E4	110	64	42	30	22	16	12	8	6
SS A1	302	84	40	23	15	10	6	4	2	TW F1	85	-	-	-	-	-	-	-	-
SS A2	289	117	66	42	28	20	14	10	6	TW F2	56	16	-	-	-	-	-	-	-
SS A3	279	70	32	17	10	6	4	2	1	TW F3	48	18	8	-	-	-	-	-	-
SS A4	290	114	63	39	26	18	13	9	5	TW F4	41	18	9	5	-	-	-	-	-
SS A5	507	81	35	19	11	7	4	3	1	TW G1	52	10	-	-	-	-	-	-	-
SS A6	284	115	64	40	27	19	13	9	6	TW G2	41	15	6	-	-	-	-	-	-
SS A7	559	63	24	11	6	3	1	-	-	TW G3	37	15	7	3	-	-	-	-	-
SS A8	288	110	60	37	25	17	12	8	5	TW G4	31	14	7	4	2	-	-	-	-
SS B1	201	61	30	18	11	7	5	3	2	TW H1	35	10	-	-	-	-	-	-	-
SS B2	227	96	54	35	24	17	12	8	5	TW H2	30	12	6	3	-	-	-	-	-
SS B3	178	53	25	14	8	5	3	2	1	TW H3	26	11	6	3	2	-	-	-	-
SS B4	220	92	52	33	22	16	11	8	5	TW H4	21	9	5	3	2	-	-	-	-
SS B5	259	53	23	12	7	4	3	2	-	TW I1	-	-	-	-	-	-	-	-	-
SS B6	211	90	51	33	22	16	11	8	5	TW I2	62	29	16	10	6	4	3	2	-
SS B7	220	41	16	8	4	2	-	-	-	TW I3	69	34	20	13	8	6	4	2	1
SS B8	203	85	48	30	20	14	10	7	4	TW I4	80	44	28	19	13	10	7	5	3
SS C1	117	48	26	16	10	7	5	3	2	TW I5	82	47	30	21	15	11	8	6	4
SS C2	139	69	42	28	20	14	10	7	5	TW J1	88	44	27	17	12	8	6	4	2
SS C3	124	39	19	11	7	5	3	2	1	TW J2	84	45	28	19	13	9	7	5	3
SS C4	139	62	36	24	16	11	8	6	4	TW J3	102	61	41	29	21	16	12	8	5
SS D1	272	100	52	32	20	14	9	6	4	TW J4	95	57	39	27	20	15	11	8	5
SS D2	253	128	78	52	36	26	19	13	9	DG A1	-	-	-	-	-	-	-	-	-
SS D3	348	94	45	26	16	11	7	4	3	DG A2	62	22	6	-	-	-	-	-	-
SS D4	259	125	75	49	34	24	18	12	8	DG A3	102	44	26	17	11	7	4	-	-
SS E1	193	72	38	22	14	9	6	4	2	DG B1	-	-	-	-	-	-	-	-	-
SS E2	203	101	61	41	28	20	14	10	6	DG B2	64	24	11	-	-	-	-	-	-
SS E3	240	63	29	16	10	6	4	2	1	DG B3	106	47	28	18	13	9	6	4	-
SS E4	210	96	55	36	25	17	12	8	5	DG C1	38	-	-	-	-	-	-	-	-

```
FORT WAYNE, INDIANA                        CONTINUED      INDIANAPOLIS, INDIANA                      CONTINUED
   SSF =.10  .20  .30  .40  .50  .60  .70  .80  .90         SSF =.10  .20  .30  .40  .50  .60  .70  .80  .90
DG C2    83   34   18   10    5    -    -    -    -      TW D4   111   59   37   25   18   13    9    6    4
DG C3   124   56   33   22   16   12    8    6    3      TW D5   109   60   39   27   19   14   10    7    5
SS A1   180   44   18    8    -    -    -    -    -      TW E1   125   60   35   23   16   11    8    5    3
SS A2   191   75   41   25   16   11    7    5    3      TW E2   118   60   37   25   17   12    9    6    4
SS A3   155   29    -    -    -    -    -    -    -      TW E3   135   78   51   36   26   19   14   10    7
SS A4   189   71   38   22   14    9    6    4    2      TW E4   124   72   48   34   25   18   14   10    6
SS A5   306   39   11    -    -    -    -    -    -      TW F1   114   19    -    -    -    -    -    -    -
SS A6   186   72   39   23   15   10    7    4    2      TW F2    74   25   11    6    3    -    -    -    -
SS A7   323   18    -    -    -    -    -    -    -      TW F3    63   26   13    8    4    2    1    -    -
SS A8   184   67   34   20   12    8    5    3    2      TW F4    53   24   14    8    5    3    2    1    -
SS B1   114   29   11    -    -    -    -    -    -      TW G1    71   19    7    -    -    -    -    -    -
SS B2   151   62   34   21   14    9    6    4    3      TW G2    54   21   11    6    3    2    -    -    -
SS B3    93   19    -    -    -    -    -    -    -      TW G3    47   21   11    7    4    3    1    -    -
SS B4   144   58   31   19   12    8    6    4    2      TW G4    39   18   10    6    4    3    2    1    -
SS B5   144   21    -    -    -    -    -    -    -      TW H1    47   17    8    4    2    -    -    -    -
SS B6   138   57   31   19   13    9    6    4    2      TW H2    38   16    9    5    3    2    1    -    -
SS B7   111    -    -    -    -    -    -    -    -      TW H3    33   15    8    5    3    2    1    -    -
SS B8   130   52   28   17   11    7    5    3    2      TW H4    27   12    7    4    3    2    1    -    -
SS C1    52   15    -    -    -    -    -    -    -      TW I1     -    -    -    -    -    -    -    -    -
SS C2    83   39   23   14    9    6    4    3    1      TW I2    75   36   21   13    9    6    4    3    2
SS C3    58   12    -    -    -    -    -    -    -      TW I3    82   41   25   16   11    8    5    4    2
SS C4    83   35   19   12    7    5    3    2    1      TW I4    92   51   33   22   16   12    8    6    4
SS D1   159   52   24   11    5    -    -    -    -      TW I5    93   54   35   24   18   13   10    7    5
SS D2   168   83   49   32   22   15   11    7    4      TW J1   104   53   32   21   15   11    7    5    3
SS D3   206   48   19    8    -    -    -    -    -      TW J2    98   53   33   23   16   11    8    6    4
SS D4   172   81   47   30   20   14   10    6    4      TW J3   115   69   47   33   24   18   13   10    6
SS E1   101   30    -    -    -    -    -    -    -      TW J4   107   65   44   31   23   17   13    9    6
SS E2   129   62   36   23   15   10    7    4    3      DG A1     -    -    -    -    -    -    -    -    -
SS E3   131   26    -    -    -    -    -    -    -      DG A2    76   29   14    -    -    -    -    -    -
SS E4   134   58   32   20   13    9    6    4    2      DG A3   116   51   30   20   14    9    6    3    -
                                                         DG B1    29    -    -    -    -    -    -    -    -
                                                         DG B2    78   31   16    9    -    -    -    -    -
INDIANAPOLIS, INDIANA                      5577 DD       DG B3   120   54   32   22   15   11    8    5    2
   SSF =.10  .20  .30  .40  .50  .60  .70  .80  .90      DG C1    60   13    -    -    -    -    -    -    -
WW A1   316   20    -    -    -    -    -    -    -      DG C2    99   42   23   14    9    5    -    -    -
WW A2   103   33   15    8    4    -    -    -    -      DG C3   140   64   38   26   19   14   10    7    4
WW A3    88   36   19   11    6    4    2    -    -      SS A1   214   56   25   13    7    4    2    -    -
WW A4    82   37   20   12    8    5    3    2    -      SS A2   218   87   48   30   20   14    9    6    4
WW A5    78   38   22   14    9    6    4    2    1      SS A3   190   42   16    5    -    -    -    -    -
WW A6    76   39   23   15   10    6    4    3    1      SS A4   217   84   45   28   18   12    8    5    3
WW B1    38    -    -    -    -    -    -    -    -      SS A5   358   51   19    8    -    -    -    -    -
WW B2    92   45   27   18   12    8    6    4    2      SS A6   213   84   46   28   19   13    9    6    4
WW B3   111   59   36   24   17   12    8    6    3      SS A7   385   34    -    -    -    -    -    -    -
WW B4   109   64   42   29   21   16   11    8    5      SS A8   213   80   42   25   16   11    7    5    3
WW B5   105   63   42   30   22   16   12    9    6      SS B1   138   39   17    9    5    2    -    -    -
WW C1   123   65   41   27   19   14   10    7    4      SS B2   172   71   40   25   17   12    8    6    3
WW C2   111   62   40   27   20   14   10    7    5      SS B3   118   30   11    -    -    -    -    -    -
WW C3   127   80   55   40   30   22   17   12    8      SS B4   165   68   37   23   15   11    7    5    3
WW C4   115   73   50   36   27   20   15   11    8      SS B5   175   31   11    -    -    -    -    -    -
TW A1   280   23    -    -    -    -    -    -    -      SS B6   159   67   37   23   16   11    7    5    3
TW A2   110   31   14    7    3    -    -    -    -      SS B7   142   20    -    -    -    -    -    -    -
TW A3    89   33   16    9    5    3    2    -    -      SS B8   151   61   34   21   14    9    6    4    2
TW A4    77   33   18   11    7    4    3    1    -      SS C1    71   26   12    5    -    -    -    -    -
TW B1   156   26    9    -    -    -    -    -    -      SS C2    99   48   28   18   12    9    6    4    2
TW B2    91   30   15    8    5    3    1    -    -      SS C3    78   21    8    2    -    -    -    -    -
TW B3    81   31   16    9    6    4    2    1    -      SS C4    98   43   24   15   10    7    5    3    2
TW B4    77   30   16    9    6    4    2    1    -      SS D1   191   66   33   18   11    6    4    2    -
TW C1   107   27   12    6    3    2    -    -    -      SS D2   192   96   58   38   26   18   13    9    6
TW C2    84   28   14    8    5    3    2    1    -      SS D3   245   62   27   14    8    4    2    -    -
TW C3    83   27   14    8    5    3    2    1    -      SS D4   197   94   55   36   24   17   12    8    5
TW C4    91   26   12    7    4    3    2    1    -      SS E1   128   43   20   10    4    -    -    -    -
TW D1    31    -    -    -    -    -    -    -    -      SS E2   150   73   43   28   19   13    9    6    4
TW D2   100   43   24   15   10    7    5    3    2      SS E3   161   38   15    6    -    -    -    -    -
TW D3   102   48   28   18   12    8    6    4    2      SS E4   156   69   39   25   16   11    8    5    3
```

SOUTH BEND, INDIANA — 6462 DD

	SSF =.10	.20	.30	.40	.50	.60	.70	.80	.90
WW A1	261	-	-	-	-	-	-	-	-
WW A2	76	17	-	-	-	-	-	-	-
WW A3	64	21	-	-	-	-	-	-	-
WW A4	59	22	9	-	-	-	-	-	-
WW A5	56	24	11	-	-	-	-	-	-
WW A6	55	25	12	-	-	-	-	-	-
WW B1	-	-	-	-	-	-	-	-	-
WW B2	73	33	19	11	7	4	2	-	-
WW B3	90	45	26	17	11	7	4	3	1
WW B4	92	52	34	23	16	12	8	6	4
WW B5	90	53	35	24	18	13	9	7	4
WW C1	101	51	31	20	13	9	6	4	2
WW C2	92	50	31	21	15	10	7	5	3
WW C3	110	68	46	33	24	18	13	10	6
WW C4	100	62	43	31	22	17	12	9	6
TW A1	234	-	-	-	-	-	-	-	-
TW A2	83	17	-	-	-	-	-	-	-
TW A3	67	20	-	-	-	-	-	-	-
TW A4	58	21	9	-	-	-	-	-	-
TW B1	125	-	-	-	-	-	-	-	-
TW B2	70	19	5	-	-	-	-	-	-
TW B3	62	20	8	-	-	-	-	-	-
TW B4	60	20	9	4	-	-	-	-	-
TW C1	85	17	-	-	-	-	-	-	-
TW C2	67	19	8	-	-	-	-	-	-
TW C3	67	19	8	4	-	-	-	-	-
TW C4	77	19	8	4	2	-	-	-	-
TW D1	-	-	-	-	-	-	-	-	-
TW D2	81	32	17	10	6	3	1	-	-
TW D3	82	36	19	11	7	4	2	-	-
TW D4	94	48	29	19	13	9	7	4	3
TW D5	94	51	32	22	15	11	8	6	3
TW E1	103	47	26	16	11	7	4	3	1
TW E2	99	48	29	18	12	9	6	4	2
TW E3	116	66	42	29	21	15	11	8	5
TW E4	107	62	40	28	20	15	11	8	5
TW F1	83	-	-	-	-	-	-	-	-
TW F2	53	11	-	-	-	-	-	-	-
TW F3	45	14	-	-	-	-	-	-	-
TW F4	38	15	6	-	-	-	-	-	-
TW G1	49	-	-	-	-	-	-	-	-
TW G2	38	11	-	-	-	-	-	-	-
TW G3	34	12	4	-	-	-	-	-	-
TW G4	29	12	5	-	-	-	-	-	-
TW H1	32	6	-	-	-	-	-	-	-
TW H2	28	10	-	-	-	-	-	-	-
TW H3	25	10	4	-	-	-	-	-	-
TW H4	20	8	4	2	-	-	-	-	-
TW I1	-	-	-	-	-	-	-	-	-
TW I2	59	26	14	8	5	3	1	-	-
TW I3	66	31	18	11	7	4	3	1	-
TW I4	77	42	26	17	12	9	6	4	3
TW I5	80	45	29	20	14	10	7	5	3
TW J1	85	42	24	15	10	7	4	3	2
TW J2	82	42	26	17	12	8	6	4	2
TW J3	99	59	39	27	20	15	11	8	5
TW J4	93	55	37	26	19	14	10	7	5
DG A1	-	-	-	-	-	-	-	-	-
DG A2	60	19	-	-	-	-	-	-	-
DG A3	102	43	24	15	10	6	3	-	-
DG B1	-	-	-	-	-	-	-	-	-
DG B2	62	21	8	-	-	-	-	-	-
DG B3	106	46	26	17	12	8	5	3	-
DG C1	29	-	-	-	-	-	-	-	-

SOUTH BEND, INDIANA CONTINUED

	SSF =.10	.20	.30	.40	.50	.60	.70	.80	.90
DG C2	82	32	16	8	-	-	-	-	-
DG C3	124	54	32	21	15	10	7	5	2
SS A1	183	40	14	-	-	-	-	-	-
SS A2	191	73	38	23	15	9	6	4	2
SS A3	157	24	-	-	-	-	-	-	-
SS A4	189	69	35	20	12	8	5	3	1
SS A5	324	35	-	-	-	-	-	-	-
SS A6	185	69	36	21	13	8	5	3	2
SS A7	351	-	-	-	-	-	-	-	-
SS A8	184	64	32	18	10	6	4	2	-
SS B1	113	26	-	-	-	-	-	-	-
SS B2	151	59	32	20	13	8	6	4	2
SS B3	92	12	-	-	-	-	-	-	-
SS B4	143	55	29	17	11	7	5	3	2
SS B5	148	17	-	-	-	-	-	-	-
SS B6	137	55	29	18	11	7	5	3	2
SS B7	114	-	-	-	-	-	-	-	-
SS B8	129	49	26	15	9	6	3	2	1
SS C1	47	-	-	-	-	-	-	-	-
SS C2	80	36	20	12	8	5	3	1	-
SS C3	55	-	-	-	-	-	-	-	-
SS C4	80	32	17	10	6	4	2	-	-
SS D1	157	48	19	-	-	-	-	-	-
SS D2	165	80	46	30	20	13	9	6	4
SS D3	210	45	15	-	-	-	-	-	-
SS D4	170	78	44	28	18	12	8	5	3
SS E1	97	24	-	-	-	-	-	-	-
SS E2	126	59	33	20	13	8	5	3	2
SS E3	131	22	-	-	-	-	-	-	-
SS E4	132	55	30	18	11	7	4	3	1

BURLINGTON, IOWA — 6149 DD

	SSF =.10	.20	.30	.40	.50	.60	.70	.80	.90
WW A1	412	34	12	4	-	-	-	-	-
WW A2	132	46	24	14	9	5	3	2	1
WW A3	112	48	27	17	11	7	5	3	2
WW A4	103	49	29	18	12	8	6	4	2
WW A5	99	50	30	20	14	10	7	4	3
WW A6	96	51	31	21	14	10	7	5	3
WW B1	74	13	-	-	-	-	-	-	-
WW B2	112	56	34	23	16	11	8	6	3
WW B3	133	71	45	30	21	15	11	8	5
WW B4	127	75	50	35	26	19	14	10	7
WW B5	121	73	49	35	26	19	15	11	7
WW C1	145	78	49	34	24	17	13	9	6
WW C2	130	74	48	33	24	18	13	9	6
WW C3	144	91	63	46	34	26	19	14	10
WW C4	131	83	58	42	31	24	18	13	9
TW A1	360	35	13	6	3	-	-	-	-
TW A2	139	42	21	12	7	5	3	2	-
TW A3	113	44	23	14	9	6	4	3	1
TW A4	97	43	24	15	10	7	5	3	2
TW B1	199	36	16	8	5	3	1	-	-
TW B2	114	40	21	13	8	5	3	2	1
TW B3	101	40	22	13	9	6	4	3	2
TW B4	94	38	21	13	9	6	4	3	2
TW C1	134	36	17	10	6	4	2	1	-
TW C2	104	36	19	12	7	5	3	2	1
TW C3	101	35	18	11	7	5	3	2	1
TW C4	110	32	16	10	6	4	3	2	1
TW D1	55	13	-	-	-	-	-	-	-
TW D2	121	54	31	20	14	10	7	5	3
TW D3	123	59	35	23	16	11	8	5	3

LCR *Tables* / 501

BURLINGTON, IOWA (CONTINUED)

	SSF =.10	.20	.30	.40	.50	.60	.70	.80	.90
TW D4	129	69	44	30	21	16	11	8	5
TW D5	126	70	45	32	23	17	12	9	6
TW E1	148	72	43	29	20	14	10	7	4
TW E2	139	71	44	30	21	15	11	8	5
TW E3	155	90	59	42	31	23	17	12	8
TW E4	141	83	55	39	29	22	16	12	8
TW F1	150	30	12	6	3	-	-	-	-
TW F2	95	35	18	11	7	4	3	1	-
TW F3	80	35	19	12	8	5	3	2	1
TW F4	67	32	19	12	8	6	4	3	1
TW G1	93	28	13	7	4	2	1	-	-
TW G2	70	29	16	10	6	4	3	2	-
TW G3	60	28	16	10	7	5	3	2	1
TW G4	49	24	14	9	6	4	3	2	1
TW H1	61	23	12	7	5	3	2	1	-
TW H2	48	22	13	8	5	4	2	2	-
TW H3	41	19	11	7	5	3	2	2	-
TW H4	33	15	9	6	4	3	2	1	-
TW I1	39	7	-	-	-	-	-	-	-
TW I2	91	45	27	18	12	9	6	4	3
TW I3	98	50	31	21	14	10	7	5	3
TW I4	107	60	39	27	19	14	10	7	5
TW I5	108	62	41	29	21	16	12	8	6
TW J1	123	64	39	27	19	13	10	7	4
TW J2	115	62	40	27	20	14	10	7	5
TW J3	131	79	54	39	28	21	16	12	8
TW J4	122	74	50	36	27	20	15	11	7
DG A1	54	-	-	-	-	-	-	-	-
DG A2	93	38	20	11	-	-	-	-	-
DG A3	134	60	36	24	17	12	8	4	-
DG B1	55	-	-	-	-	-	-	-	-
DG B2	96	40	22	14	8	-	-	-	-
DG B3	138	62	38	26	19	14	10	7	3
DG C1	83	29	12	-	-	-	-	-	-
DG C2	119	52	30	19	13	9	5	-	-
DG C3	160	73	44	30	22	17	13	9	5
SS A1	248	67	32	18	11	7	4	2	1
SS A2	246	99	55	35	23	16	11	8	5
SS A3	222	53	23	11	6	2	-	-	-
SS A4	244	95	52	32	21	15	10	7	4
SS A5	421	63	26	13	7	4	2	-	-
SS A6	241	97	53	34	22	16	11	7	5
SS A7	457	45	15	4	-	-	-	-	-
SS A8	240	91	49	30	20	13	9	6	4
SS B1	163	48	23	13	8	5	3	2	-
SS B2	194	81	46	29	20	14	10	7	4
SS B3	141	40	18	9	5	2	-	-	-
SS B4	186	77	43	27	18	13	9	6	4
SS B5	209	40	16	8	4	2	-	-	-
SS B6	179	76	43	28	19	13	9	6	4
SS B7	172	29	10	-	-	-	-	-	-
SS B8	170	70	39	25	17	11	8	5	3
SS C1	94	38	20	12	7	5	3	1	-
SS C2	119	59	36	24	16	12	8	6	4
SS C3	101	30	15	8	5	3	1	-	-
SS C4	118	53	31	20	14	9	7	5	3
SS D1	223	80	41	24	15	10	6	4	2
SS D2	216	109	66	44	31	22	16	11	7
SS D3	285	75	35	19	12	7	4	3	1
SS D4	221	106	63	42	29	20	14	10	6
SS E1	153	56	28	16	9	6	3	2	-
SS E2	171	85	51	34	23	16	12	8	5
SS E3	192	48	21	11	6	3	1	-	-
SS E4	177	80	46	30	20	14	10	7	4

DES MOINES, IOWA   6710 DD

	SSF =.10	.20	.30	.40	.50	.60	.70	.80	.90
WW A1	391	31	10	-	-	-	-	-	-
WW A2	126	43	22	12	8	5	3	2	-
WW A3	107	45	25	15	10	7	4	3	2
WW A4	98	46	27	17	11	8	5	3	2
WW A5	94	47	28	18	13	9	6	4	2
WW A6	92	48	29	19	13	9	7	4	3
WW B1	66	-	-	-	-	-	-	-	-
WW B2	107	54	33	22	15	11	8	5	3
WW B3	128	68	42	29	20	15	10	7	5
WW B4	122	72	48	34	25	18	13	10	6
WW B5	117	71	48	34	25	19	14	10	7
WW C1	140	75	47	32	23	16	12	8	5
WW C2	125	71	46	32	23	17	12	9	6
WW C3	140	88	61	44	33	25	19	14	9
WW C4	128	80	56	41	30	23	17	13	9
TW A1	343	33	11	5	-	-	-	-	-
TW A2	133	40	19	11	7	4	3	1	-
TW A3	108	41	22	13	8	6	4	2	1
TW A4	92	40	23	14	9	6	4	3	2
TW B1	190	34	14	7	4	2	-	-	-
TW B2	109	38	19	12	7	5	3	2	1
TW B3	96	38	20	12	8	5	4	2	1
TW B4	91	36	19	12	8	5	4	2	1
TW C1	128	34	16	9	6	4	2	1	-
TW C2	100	34	18	11	7	5	3	2	1
TW C3	98	33	17	10	7	5	3	2	1
TW C4	107	31	15	9	6	4	3	2	1
TW D1	50	8	-	-	-	-	-	-	-
TW D2	117	51	30	19	13	9	6	4	3
TW D3	118	56	33	22	15	11	7	5	3
TW D4	125	67	42	29	21	15	11	8	5
TW D5	122	68	44	30	22	16	12	9	6
TW E1	143	69	41	27	19	14	10	7	4
TW E2	135	69	42	29	20	15	11	7	5
TW E3	150	87	57	40	29	22	16	12	8
TW E4	137	80	54	38	28	21	15	11	8
TW F1	143	27	11	5	-	-	-	-	-
TW F2	91	32	16	9	6	4	2	1	-
TW F3	76	32	18	11	7	5	3	2	1
TW F4	64	30	18	11	8	5	4	2	1
TW G1	88	25	12	6	4	2	-	-	-
TW G2	66	27	15	9	6	4	2	1	-
TW G3	57	26	15	9	6	4	3	2	1
TW G4	47	22	13	8	6	4	3	2	1
TW H1	58	22	11	7	4	3	2	-	-
TW H2	46	21	12	7	5	3	2	1	-
TW H3	40	18	11	7	4	3	2	1	-
TW H4	31	14	8	5	4	2	2	1	-
TW I1	34	-	-	-	-	-	-	-	-
TW I2	88	42	25	17	11	8	6	4	2
TW I3	95	48	29	19	14	10	7	5	3
TW I4	104	58	37	26	19	14	10	7	5
TW I5	104	60	40	28	20	15	11	8	5
TW J1	118	61	38	25	18	13	9	6	4
TW J2	111	60	38	26	19	14	10	7	5
TW J3	127	77	52	37	27	20	15	11	7
TW J4	118	72	49	35	26	19	14	11	7
DG A1	48	-	-	-	-	-	-	-	-
DG A2	90	36	19	10	-	-	-	-	-
DG A3	131	58	34	23	16	11	7	4	-
DG B1	50	-	-	-	-	-	-	-	-
DG B2	92	38	21	13	7	-	-	-	-
DG B3	135	60	36	25	18	13	10	6	3
DG C1	78	25	-	-	-	-	-	-	-

DES MOINES, IOWA — CONTINUED

SSF =	.10	.20	.30	.40	.50	.60	.70	.80	.90
DG C2	115	49	28	18	12	8	4	-	-
DG C3	157	71	43	29	21	16	12	9	5
SS A1	238	63	29	16	10	6	4	2	1
SS A2	237	95	52	33	22	15	11	7	5
SS A3	211	49	20	9	4	-	-	-	-
SS A4	234	91	49	30	20	14	9	6	4
SS A5	405	59	23	11	6	3	1	-	-
SS A6	231	92	51	32	21	15	10	7	4
SS A7	437	41	12	-	-	-	-	-	-
SS A8	230	86	46	28	18	12	8	6	3
SS B1	155	45	21	12	7	4	2	1	-
SS B2	187	78	44	28	19	13	9	6	4
SS B3	133	36	15	7	3	-	-	-	-
SS B4	179	73	41	26	17	12	8	6	4
SS B5	200	37	15	7	3	-	-	-	-
SS B6	173	73	41	26	18	12	9	6	4
SS B7	163	25	7	-	-	-	-	-	-
SS B8	163	67	37	23	15	11	7	5	3
SS C1	89	35	18	11	6	4	2	1	-
SS C2	115	56	34	22	15	11	8	5	3
SS C3	96	28	13	7	4	2	1	-	-
SS C4	114	50	29	19	13	9	6	4	3
SS D1	212	75	38	22	14	9	6	3	2
SS D2	207	104	63	42	29	21	15	10	7
SS D3	273	70	32	18	10	6	4	2	1
SS D4	212	102	60	39	27	19	14	9	6
SS E1	144	51	25	14	8	5	2	1	-
SS E2	164	81	48	32	22	15	11	7	5
SS E3	183	44	19	10	5	2	-	-	-
SS E4	170	76	44	28	19	13	9	6	4

MASON CITY, IOWA — 7901 DD

SSF =	.10	.20	.30	.40	.50	.60	.70	.80	.90
WW A1	335	21	-	-	-	-	-	-	-
WW A2	108	34	16	8	4	-	-	-	-
WW A3	91	37	19	11	7	4	2	-	-
WW A4	84	38	21	13	8	5	3	2	-
WW A5	80	39	23	14	9	6	4	2	1
WW A6	78	40	24	15	10	7	4	3	1
WW B1	41	-	-	-	-	-	-	-	-
WW B2	95	46	28	18	12	9	6	4	2
WW B3	114	59	37	24	17	12	8	6	4
WW B4	111	65	42	30	22	16	12	8	5
WW B5	107	64	43	31	22	17	12	9	6
WW C1	125	66	41	28	19	14	10	7	4
WW C2	113	63	41	28	20	15	11	7	5
WW C3	128	81	56	40	30	22	17	12	8
WW C4	117	74	51	37	27	21	16	11	8
TW A1	296	24	5	-	-	-	-	-	-
TW A2	114	32	14	7	4	-	-	-	-
TW A3	93	34	17	10	6	3	2	-	-
TW A4	80	34	18	11	7	4	3	2	-
TW B1	164	27	10	4	-	-	-	-	-
TW B2	95	31	15	9	5	3	2	-	-
TW B3	84	31	16	10	6	4	2	1	-
TW B4	79	30	16	10	6	4	3	2	-
TW C1	112	28	12	6	4	2	-	-	-
TW C2	88	29	14	8	5	3	2	1	-
TW C3	86	28	14	8	5	3	2	1	-
TW C4	95	26	13	7	5	3	2	1	-
TW D1	33	-	-	-	-	-	-	-	-
TW D2	103	44	25	16	11	7	5	3	2
TW D3	104	49	28	18	12	8	6	4	2

MASON CITY, IOWA — CONTINUED

SSF =	.10	.20	.30	.40	.50	.60	.70	.80	.90
TW D4	113	60	37	25	18	13	9	7	4
TW D5	111	61	39	27	20	14	11	8	5
TW E1	128	61	36	23	16	11	8	5	3
TW E2	121	61	37	25	17	12	9	6	4
TW E3	137	79	52	36	26	19	14	10	7
TW E4	126	73	49	34	25	19	14	10	7
TW F1	120	20	5	-	-	-	-	-	-
TW F2	77	26	12	6	3	-	-	-	-
TW F3	65	26	14	8	5	3	1	-	-
TW F4	55	25	14	9	5	4	2	1	-
TW G1	74	19	8	2	-	-	-	-	-
TW G2	56	22	11	6	4	2	-	-	-
TW G3	49	21	12	7	4	3	2	-	-
TW G4	40	19	11	7	4	3	2	1	-
TW H1	49	17	8	4	2	-	-	-	-
TW H2	39	17	9	5	3	2	1	-	-
TW H3	34	15	8	5	3	2	1	-	-
TW H4	27	12	7	4	3	2	1	-	-
TW I1	14	-	-	-	-	-	-	-	-
TW I2	77	37	21	14	9	6	4	3	2
TW I3	84	42	25	16	11	8	5	4	2
TW I4	94	52	33	23	16	12	9	6	4
TW I5	95	54	36	25	18	13	10	7	5
TW J1	106	54	33	22	15	11	8	5	3
TW J2	100	54	34	23	16	12	8	6	4
TW J3	116	70	47	33	25	18	14	10	7
TW J4	109	66	44	32	23	17	13	9	6
DG A1	27	-	-	-	-	-	-	-	-
DG A2	79	30	15	-	-	-	-	-	-
DG A3	120	52	30	20	14	10	6	3	-
DG B1	32	-	-	-	-	-	-	-	-
DG B2	81	32	17	9	-	-	-	-	-
DG B3	124	55	32	22	16	11	8	5	2
DG C1	62	14	-	-	-	-	-	-	-
DG C2	103	43	24	15	9	5	-	-	-
DG C3	145	65	39	26	19	14	10	7	4
SS A1	213	53	23	12	6	3	-	-	-
SS A2	216	85	46	29	19	13	9	6	4
SS A3	186	38	13	-	-	-	-	-	-
SS A4	212	80	42	26	17	11	7	5	3
SS A5	368	49	17	7	-	-	-	-	-
SS A6	211	82	44	27	18	12	8	5	3
SS A7	394	30	-	-	-	-	-	-	-
SS A8	208	76	39	23	15	10	6	4	2
SS B1	137	37	16	8	4	-	-	-	-
SS B2	171	69	39	24	16	11	8	5	3
SS B3	114	27	9	-	-	-	-	-	-
SS B4	162	65	36	22	15	10	7	5	3
SS B5	177	29	9	-	-	-	-	-	-
SS B6	157	65	36	22	15	10	7	5	3
SS B7	141	17	-	-	-	-	-	-	-
SS B8	147	59	32	20	13	9	6	4	2
SS C1	74	27	13	6	-	-	-	-	-
SS C2	102	49	29	19	13	9	6	4	3
SS C3	81	22	9	4	-	-	-	-	-
SS C4	101	44	25	16	10	7	5	3	2
SS D1	188	63	30	17	9	5	3	1	-
SS D2	188	93	56	36	25	18	12	8	5
SS D3	245	59	25	13	7	3	-	-	-
SS D4	193	91	53	34	23	16	11	8	5
SS E1	124	40	18	8	-	-	-	-	-
SS E2	147	71	42	27	18	13	9	6	4
SS E3	160	35	13	4	-	-	-	-	-
SS E4	153	67	38	24	16	11	7	5	3

SIOUX CITY, IOWA								6953 DD		SIOUX CITY, IOWA								CONTINUED	
SSF =.10	.20	.30	.40	.50	.60	.70	.80	.90		SSF =.10	.20	.30	.40	.50	.60	.70	.80	.90	
WW A1	373	30	9	-	-	-	-	-	-	DG C2	114	49	27	18	12	8	4	-	-
WW A2	124	42	21	12	7	4	2	1	-	DG C3	156	70	42	29	21	16	12	8	5
WW A3	105	44	24	15	9	6	4	2	1	SS A1	230	60	28	15	9	5	3	2	-
WW A4	97	45	26	16	11	7	5	3	2	SS A2	231	92	51	32	21	15	10	7	4
WW A5	92	46	28	18	12	8	6	4	2	SS A3	203	46	18	8	-	-	-	-	-
WW A6	90	47	28	19	13	9	6	4	2	SS A4	228	88	47	29	19	13	9	6	4
WW B1	63	-	-	-	-	-	-	-	-	SS A5	388	57	22	10	5	2	-	-	-
WW B2	106	53	32	21	15	10	7	5	3	SS A6	226	90	49	30	20	14	10	6	4
WW B3	126	67	42	28	20	14	10	7	4	SS A7	414	38	9	-	-	-	-	-	-
WW B4	121	71	47	33	24	18	13	9	6	SS A8	224	84	44	27	17	12	8	5	3
WW B5	116	70	47	34	25	18	14	10	7	SS B1	150	43	20	11	6	3	2	-	-
WW C1	138	74	46	31	22	16	11	8	5	SS B2	183	76	43	27	18	13	9	6	4
WW C2	124	70	45	31	22	16	12	9	6	SS B3	128	34	14	6	-	-	-	-	-
WW C3	138	87	60	44	33	25	18	13	9	SS B4	174	71	39	25	17	11	8	5	3
WW C4	126	80	55	40	30	23	17	12	8	SS B5	192	35	13	6	-	-	-	-	-
TW A1	329	32	11	4	-	-	-	-	-	SS B6	169	71	40	25	17	12	8	6	3
TW A2	130	39	18	10	6	4	2	1	-	SS B7	156	23	-	-	-	-	-	-	-
TW A3	106	40	21	13	8	5	3	2	1	SS B8	159	65	36	22	15	10	7	5	3
TW A4	91	39	22	14	9	6	4	3	1	SS C1	87	34	17	10	6	3	2	-	-
TW B1	184	33	13	7	3	1	-	-	-	SS C2	113	55	33	22	15	11	7	5	3
TW B2	107	37	19	11	7	4	3	2	-	SS C3	94	27	12	7	4	2	-	-	-
TW B3	95	37	20	12	8	5	3	2	1	SS C4	112	50	28	18	12	9	6	4	3
TW B4	89	35	19	12	8	5	3	2	1	SS D1	206	72	36	21	13	8	5	3	1
TW C1	126	33	15	9	5	3	2	1	-	SS D2	203	101	61	40	28	20	14	10	6
TW C2	98	34	17	10	7	4	3	2	1	SS D3	265	67	30	16	9	6	3	2	-
TW C3	96	32	17	10	6	4	3	2	1	SS D4	208	99	58	38	26	18	13	9	6
TW C4	104	30	15	9	6	4	3	2	-	SS E1	140	49	23	12	7	3	-	-	-
TW D1	48	-	-	-	-	-	-	-	-	SS E2	160	78	47	30	21	15	10	7	4
TW D2	115	51	29	18	12	9	6	4	2	SS E3	177	42	18	8	4	-	-	-	-
TW D3	116	55	32	21	14	10	7	5	3	SS E4	166	74	42	27	18	12	9	6	4
TW D4	124	66	41	28	20	15	11	8	5										
TW D5	121	67	43	30	22	16	12	8	6										
TW E1	141	68	41	27	18	13	9	6	4	DODGE CITY, KANSAS								5046 DD	
TW E2	133	68	42	28	20	14	10	7	5	SSF =.10	.20	.30	.40	.50	.60	.70	.80	.90	
TW E3	149	86	56	40	29	21	16	11	8	WW A1	796	91	41	24	16	11	8	5	3
TW E4	136	80	53	37	27	20	15	11	7	WW A2	266	102	58	37	26	19	13	10	6
TW F1	139	27	10	4	-	-	-	-	-	WW A3	223	104	63	42	30	22	16	11	8
TW F2	89	31	16	9	5	3	2	-	-	WW A4	204	104	65	44	32	23	17	13	8
TW F3	75	32	17	10	7	4	3	2	-	WW A5	194	105	68	47	34	25	19	14	9
TW F4	63	30	17	11	7	5	3	2	1	WW A6	189	106	69	48	35	26	20	14	10
TW G1	86	25	11	6	3	1	-	-	-	WW B1	207	72	39	24	16	11	8	6	4
TW G2	65	27	14	9	5	3	2	1	-	WW B2	204	108	69	48	34	26	19	14	9
TW G3	56	25	14	9	6	4	3	2	-	WW B3	237	132	86	60	44	33	24	18	12
TW G4	46	22	13	8	5	4	2	2	-	WW B4	211	127	87	62	47	35	27	20	14
TW H1	57	21	11	6	4	2	1	-	-	WW B5	196	121	83	61	45	35	26	19	13
TW H2	45	20	11	7	5	3	2	1	-	WW C1	253	141	92	64	47	35	26	19	13
TW H3	39	18	10	6	4	3	2	1	-	WW C2	220	128	85	61	45	34	25	19	13
TW H4	31	14	8	5	3	2	2	1	-	WW C3	228	147	103	76	58	44	34	25	17
TW I1	32	-	-	-	-	-	-	-	-	WW C4	206	133	93	69	52	40	31	23	16
TW I2	86	42	25	16	11	8	5	4	2	TW A1	685	87	40	24	16	11	8	5	3
TW I3	93	47	29	19	13	9	7	5	3	TW A2	274	93	51	32	22	16	11	8	5
TW I4	103	57	37	25	18	13	10	7	5	TW A3	220	93	54	35	25	18	13	9	6
TW I5	103	59	39	27	20	15	11	8	5	TW A4	187	88	53	36	26	19	14	10	7
TW J1	117	60	37	25	17	12	9	6	4	TW B1	385	83	41	25	17	12	8	6	4
TW J2	110	59	38	26	18	13	10	7	4	TW B2	221	85	48	31	21	15	11	8	5
TW J3	126	76	51	37	27	20	15	11	7	TW B3	192	82	48	32	22	16	12	8	6
TW J4	117	71	48	34	25	19	14	10	7	TW B4	177	76	44	29	21	15	11	8	5
DG A1	47	-	-	-	-	-	-	-	-	TW C1	255	76	40	25	17	12	9	6	4
DG A2	88	35	18	9	-	-	-	-	-	TW C2	195	73	41	27	18	13	10	7	5
DG A3	129	57	34	23	16	11	7	4	-	TW C3	187	69	38	25	17	12	9	6	4
DG B1	48	-	-	-	-	-	-	-	-	TW C4	198	62	33	21	14	10	7	5	3
DG B2	91	37	20	12	7	-	-	-	-	TW D1	151	56	31	20	13	9	7	5	3
DG B3	133	60	36	24	17	13	9	6	3	TW D2	219	102	62	42	30	22	16	12	8
DG C1	76	24	-	-	-	-	-	-	-	TW D3	223	112	70	48	34	25	19	14	9

DODGE CITY, KANSAS							CONTINUED		GOODLAND, KANSAS								6119 DD
SSF =.10	.20	.30	.40	.50	.60	.70	.80	.90	SSF =.10	.20	.30	.40	.50	.60	.70	.80	.90
TW D4 217	119	77	54	40	30	22	16	11	WW A1 625	76	35	20	13	9	6	4	3
TW D5 204	116	76	54	40	30	23	17	11	WW A2 224	88	50	32	22	16	12	8	5
TW E1 260	131	82	56	40	30	22	16	11	WW A3 191	90	54	37	26	19	14	10	7
TW E2 237	126	80	56	41	30	22	16	11	WW A4 176	91	57	39	28	20	15	11	7
TW E3 251	148	100	72	53	40	30	22	15	WW A5 169	92	59	41	30	22	16	12	8
TW E4 225	135	91	66	49	37	28	21	14	WW A6 164	93	60	42	31	23	17	13	8
TW F1 310	75	38	23	15	11	8	5	3	WW B1 169	59	32	20	13	9	6	4	3
TW F2 194	78	45	29	20	15	11	8	5	WW B2 178	95	61	42	30	23	17	12	8
TW F3 160	75	46	31	22	16	12	8	6	WW B3 209	117	76	54	39	29	22	16	11
TW F4 132	68	42	29	21	15	11	8	6	WW B4 188	114	78	56	42	32	24	18	12
TW G1 192	66	36	23	15	11	8	6	4	WW B5 176	109	75	55	41	31	24	18	12
TW G2 140	64	38	26	18	13	10	7	5	WW C1 224	125	82	58	42	31	23	17	12
TW G3 118	59	36	25	18	13	10	7	5	WW C2 195	114	76	55	40	30	23	17	11
TW G4 95	49	30	21	15	11	8	6	4	WW C3 206	133	94	69	52	40	31	23	16
TW H1 124	53	31	20	14	10	8	5	4	WW C4 186	120	85	63	48	36	28	21	14
TW H2 96	47	29	19	14	10	7	5	4	TW A1 543	74	34	20	13	9	6	4	3
TW H3 80	40	25	17	12	9	7	5	3	TW A2 229	80	44	28	19	14	10	7	5
TW H4 62	31	19	13	9	7	5	4	2	TW A3 187	80	47	31	22	16	11	8	5
TW I1 115	47	26	17	12	8	6	4	3	TW A4 160	77	47	31	22	16	12	9	6
TW I2 167	85	54	37	27	20	15	11	7	TW B1 313	71	35	21	14	10	7	5	3
TW I3 177	94	60	42	30	22	17	12	8	TW B2 187	73	41	27	19	13	10	7	5
TW I4 180	103	68	49	36	27	20	15	10	TW B3 164	71	42	28	19	14	10	7	5
TW I5 176	104	70	50	37	28	21	16	11	TW B4 152	66	39	26	18	13	10	7	5
TW J1 215	115	74	51	37	28	21	15	10	TW C1 212	66	35	22	15	11	8	5	3
TW J2 195	109	71	50	37	28	21	15	10	TW C2 165	64	36	23	16	12	8	6	4
TW J3 210	129	89	65	49	37	28	21	14	TW C3 159	60	34	22	15	11	8	6	4
TW J4 193	120	82	60	45	34	26	19	13	TW C4 167	54	29	18	13	9	7	5	3
DG A1 149	62	33	17	-	-	-	-	-	TW D1 124	47	26	16	11	8	5	4	2
DG A2 174	77	47	31	21	14	9	4	-	TW D2 190	90	55	37	26	19	14	10	7
DG A3 218	100	62	43	32	24	17	12	6	TW D3 194	99	62	42	30	22	17	12	8
DG B1 148	63	37	24	15	8	-	-	-	TW D4 192	107	69	49	36	27	20	15	10
DG B2 177	80	49	34	25	18	13	8	3	TW D5 181	104	69	49	36	27	20	15	10
DG B3 222	103	64	45	34	26	20	15	9	TW E1 228	116	73	50	36	27	20	14	10
DG C1 183	81	48	33	23	17	10	-	-	TW E2 209	112	72	50	36	27	20	15	10
DG C2 210	96	59	41	31	23	17	12	7	TW E3 225	134	90	65	48	36	27	20	14
DG C3 256	119	74	52	40	31	24	18	12	TW E4 202	122	83	60	45	34	26	19	13
SS A1 457	135	70	43	29	20	14	10	6	TW F1 251	64	32	20	13	9	6	4	3
SS A2 414	173	100	65	46	33	24	17	11	TW F2 164	67	38	25	17	13	9	6	4
SS A3 433	119	60	36	24	16	11	8	5	TW F3 137	65	40	27	19	14	10	7	5
SS A4 417	170	97	63	43	31	22	16	10	TW F4 114	59	37	25	18	13	10	7	5
SS A5 770	134	63	37	24	16	11	8	5	TW G1 160	56	30	19	13	9	7	5	3
SS A6 410	171	99	64	45	32	23	16	11	TW G2 120	56	33	22	16	11	8	6	4
SS A7 862	110	49	28	17	12	8	5	3	TW G3 102	51	32	21	15	11	8	6	4
SS A8 417	166	94	61	42	30	21	15	10	TW G4 82	43	27	18	13	10	7	5	3
SS B1 309	101	54	34	23	16	11	8	5	TW H1 106	46	27	18	12	9	6	5	3
SS B2 323	140	82	54	38	27	20	14	9	TW H2 83	41	25	17	12	9	7	5	3
SS B3 281	91	48	30	20	14	10	7	4	TW H3 69	35	22	15	11	8	6	4	3
SS B4 315	135	79	52	36	26	19	13	9	TW H4 54	27	17	11	8	6	4	3	2
SS B5 402	90	44	26	17	12	8	6	4	TW I1 95	39	22	14	9	7	5	3	2
SS B6 304	134	79	52	36	26	19	14	9	TW I2 146	75	47	33	23	17	13	9	6
SS B7 351	74	35	21	13	9	6	4	3	TW I3 155	83	53	37	27	20	15	11	7
SS B8 294	127	74	49	34	24	18	13	8	TW I4 160	92	61	44	32	24	18	13	9
SS C1 198	88	52	35	24	18	13	9	6	TW I5 157	93	63	45	34	25	19	14	10
SS C2 214	110	70	48	35	26	19	14	9	TW J1 189	102	66	46	33	25	18	13	9
SS C3 207	71	39	25	17	12	9	6	4	TW J2 173	97	64	45	33	25	19	14	9
SS C4 212	99	60	41	29	21	16	11	8	TW J3 188	117	81	59	44	34	25	19	13
SS D1 415	161	90	57	39	28	20	14	9	TW J4 174	108	75	55	41	31	24	18	12
SS D2 363	188	118	81	58	43	31	23	15	DG A1 123	51	25	10	-	-	-	-	-
SS D3 528	152	78	48	32	22	16	11	7	DG A2 149	67	41	26	18	11	7	-	-
SS D4 372	185	114	77	55	40	29	21	14	DG A3 191	90	56	39	29	21	15	10	5
SS E1 305	124	70	45	31	22	16	11	7	DG B1 122	53	30	19	11	-	-	-	-
SS E2 297	153	96	65	47	34	25	18	12	DG B2 153	70	43	29	21	15	10	6	-
SS E3 374	107	55	33	22	15	11	7	5	DG B3 196	92	58	41	31	24	18	13	8
SS E4 307	144	87	58	41	30	22	15	10	DG C1 155	69	41	27	19	13	7	-	-

GOODLAND, KANSAS (CONTINUED)

	SSF =.10	.20	.30	.40	.50	.60	.70	.80	.90
DG C2	183	85	52	36	27	20	15	10	5
DG C3	227	106	67	47	36	28	22	17	11
SS A1	388	118	61	38	25	17	12	8	5
SS A2	364	155	90	59	41	29	21	15	10
SS A3	363	103	52	31	20	14	10	6	4
SS A4	365	152	87	56	39	28	20	14	9
SS A5	630	116	54	32	21	14	10	6	4
SS A6	360	153	88	58	40	29	21	15	10
SS A7	688	95	41	23	15	10	6	4	3
SS A8	365	148	84	54	37	26	19	13	9
SS B1	262	88	47	29	20	14	10	7	4
SS B2	284	125	74	49	34	25	18	13	8
SS B3	238	79	41	25	17	12	8	6	3
SS B4	276	121	71	46	32	23	17	12	8
SS B5	332	78	38	22	15	10	7	5	3
SS B6	267	119	71	47	33	24	17	12	8
SS B7	286	63	30	17	11	7	5	3	2
SS B8	258	113	66	43	30	22	16	11	7
SS C1	168	76	45	30	21	15	11	8	5
SS C2	187	98	62	43	31	23	17	12	8
SS C3	173	61	33	21	14	10	7	5	3
SS C4	184	87	53	36	26	19	14	10	7
SS D1	359	141	79	50	34	24	17	12	8
SS D2	324	169	106	73	52	38	28	20	13
SS D3	447	133	69	42	28	19	14	9	6
SS D4	331	166	102	69	49	36	26	19	12
SS E1	262	108	61	39	27	19	13	9	6
SS E2	263	137	86	58	42	31	22	16	11
SS E3	313	92	47	29	19	13	9	6	4
SS E4	271	129	78	52	37	26	19	14	9

TOPEKA, KANSAS 5243 DD

	SSF =.10	.20	.30	.40	.50	.60	.70	.80	.90
WW A1	583	61	26	14	9	6	4	2	1
WW A2	194	72	39	24	16	11	8	5	3
WW A3	163	74	43	28	19	14	10	7	4
WW A4	150	74	45	30	21	15	11	8	5
WW A5	143	75	47	32	23	17	12	8	5
WW A6	139	76	48	33	24	17	13	9	6
WW B1	137	43	21	12	7	4	2	1	-
WW B2	154	80	50	34	24	18	13	9	6
WW B3	181	99	63	44	31	23	17	12	8
WW B4	165	99	66	47	35	26	19	14	10
WW B5	155	95	64	46	35	26	20	14	10
WW C1	195	107	68	47	34	25	18	13	9
WW C2	171	98	65	45	33	25	18	13	9
WW C3	182	116	81	59	44	34	26	19	13
WW C4	165	105	73	54	40	31	23	17	12
TW A1	506	60	26	14	9	6	4	2	1
TW A2	201	66	34	21	14	10	7	5	3
TW A3	162	66	37	24	16	11	8	6	4
TW A4	138	63	37	25	17	12	9	6	4
TW B1	284	58	27	16	10	7	5	3	2
TW B2	163	61	33	21	14	10	7	5	3
TW B3	143	59	34	22	15	10	7	5	3
TW B4	133	55	31	20	14	10	7	5	3
TW C1	190	54	28	17	11	8	5	4	2
TW C2	146	53	29	18	12	9	6	4	3
TW C3	141	50	27	17	12	8	6	4	3
TW C4	151	46	24	15	10	7	5	3	2
TW D1	100	34	17	10	6	4	2	1	-
TW D2	166	76	45	30	21	15	11	8	5
TW D3	169	83	51	34	24	17	13	9	6

TOPEKA, KANSAS (CONTINUED)

	SSF =.10	.20	.30	.40	.50	.60	.70	.80	.90
TW D4	169	92	59	41	30	22	16	12	8
TW D5	161	91	59	42	30	23	17	12	8
TW E1	200	99	61	41	29	21	15	11	7
TW E2	184	96	60	41	30	22	16	11	8
TW E3	199	116	77	55	40	30	23	16	11
TW E4	179	106	71	51	38	28	21	16	11
TW F1	224	51	24	14	9	6	4	3	2
TW F2	141	55	30	19	13	9	6	4	3
TW F3	117	53	31	20	14	10	7	5	3
TW F4	97	48	30	20	14	10	7	5	3
TW G1	139	45	23	14	9	6	4	3	2
TW G2	102	45	26	17	12	8	6	4	3
TW G3	87	42	25	17	12	8	6	4	3
TW G4	70	35	21	14	10	7	5	4	2
TW H1	90	37	21	13	9	6	4	3	2
TW H2	70	33	20	13	9	7	5	3	2
TW H3	59	29	17	12	8	5	4	3	2
TW H4	46	22	13	9	6	4	3	2	1
TW I1	75	28	14	8	5	3	2	1	-
TW I2	126	63	39	26	19	13	10	7	5
TW I3	134	70	44	30	21	16	11	8	5
TW I4	141	80	52	37	27	20	15	11	7
TW I5	139	81	54	38	28	21	16	11	8
TW J1	165	87	55	38	27	20	14	10	7
TW J2	152	84	54	38	27	20	15	11	7
TW J3	167	102	69	50	37	28	21	15	10
TW J4	154	95	65	47	35	26	20	15	10
DG A1	100	36	-	-	-	-	-	-	-
DG A2	131	56	32	20	12	7	-	-	-
DG A3	173	78	48	33	23	17	12	8	3
DG B1	100	38	19	-	-	-	-	-	-
DG B2	134	59	34	23	16	11	6	-	-
DG B3	177	81	50	34	25	19	14	10	6
DG C1	131	54	30	18	11	-	-	-	-
DG C2	162	72	43	29	21	15	11	6	-
DG C3	205	95	58	40	30	23	18	13	8
SS A1	341	98	49	29	19	13	9	6	3
SS A2	321	132	75	48	33	23	17	12	7
SS A3	316	83	40	23	14	9	6	4	2
SS A4	321	129	72	46	31	22	15	11	7
SS A5	572	96	42	24	15	10	6	4	2
SS A6	317	130	73	47	32	23	16	11	7
SS A7	631	75	31	16	9	6	3	2	1
SS A8	320	125	69	43	29	20	14	10	6
SS B1	229	73	37	22	14	10	7	4	3
SS B2	252	108	62	40	28	20	14	10	6
SS B3	205	63	32	18	12	8	5	3	2
SS B4	244	103	59	38	26	18	13	9	6
SS B5	296	63	29	16	10	7	4	3	2
SS B6	236	102	59	38	26	19	13	9	6
SS B7	252	50	22	12	7	4	3	2	-
SS B8	226	96	55	35	24	17	12	8	5
SS C1	143	61	35	22	15	11	7	5	3
SS C2	163	82	51	34	24	18	13	9	6
SS C3	150	49	26	16	10	7	5	3	2
SS C4	162	74	44	29	20	15	11	7	5
SS D1	309	116	63	39	26	18	12	8	5
SS D2	282	144	89	60	42	31	22	16	10
SS D3	393	110	54	32	21	14	9	6	4
SS D4	289	141	85	57	40	29	21	15	10
SS E1	223	87	47	29	19	13	9	6	4
SS E2	229	116	71	48	33	24	17	12	8
SS E3	274	75	37	21	14	9	6	4	2
SS E4	236	109	64	42	29	21	15	10	7

WICHITA, KANSAS									4687 DD		WICHITA, KANSAS									CONTINUED
SSF =	.10	.20	.30	.40	.50	.60	.70	.80	.90		SSF =	.10	.20	.30	.40	.50	.60	.70	.80	.90
WW A1	774	85	38	22	14	10	7	5	3		DG C2	202	92	56	39	28	21	16	11	6
WW A2	252	97	54	35	24	17	12	9	6		DG C3	246	115	71	50	38	29	23	17	11
WW A3	212	98	59	39	28	20	15	10	7		SS A1	432	127	65	40	26	18	13	9	6
WW A4	194	98	61	41	30	22	16	11	8		SS A2	394	165	95	62	43	31	22	16	10
WW A5	185	100	64	44	32	23	17	12	8		SS A3	408	112	56	33	22	15	10	7	4
WW A6	179	100	65	45	33	24	18	13	9		SS A4	397	162	92	59	41	29	21	15	9
WW B1	194	67	35	22	14	10	7	5	3		SS A5	729	126	58	34	22	15	10	7	4
WW B2	195	103	65	45	32	24	18	13	9		SS A6	391	163	93	61	42	30	22	15	10
WW B3	227	126	81	57	41	31	23	17	11		SS A7	816	104	45	25	16	10	7	5	3
WW B4	203	122	83	59	44	33	25	18	13		SS A8	397	158	89	57	39	28	20	14	9
WW B5	189	116	80	58	43	33	25	18	13		SS B1	292	96	50	31	21	15	10	7	4
WW C1	243	134	87	61	44	33	24	18	12		SS B2	309	134	78	51	36	26	19	13	9
WW C2	211	122	81	58	42	32	24	17	12		SS B3	265	86	45	27	18	12	9	6	4
WW C3	220	141	99	73	55	42	32	24	16		SS B4	300	129	75	49	34	24	18	12	8
WW C4	199	127	89	66	50	38	29	22	15		SS B5	380	85	41	24	16	11	7	5	3
TW A1	665	82	37	22	14	10	7	5	3		SS B6	290	128	75	49	34	25	18	13	8
TW A2	261	88	48	30	20	14	10	7	5		SS B7	331	70	33	19	12	8	5	4	2
TW A3	209	88	51	33	23	16	12	8	6		SS B8	280	121	70	46	32	23	16	12	8
TW A4	178	83	50	34	24	17	13	9	6		SS C1	188	83	49	32	22	16	12	8	5
TW B1	369	79	39	23	15	11	7	5	3		SS C2	205	105	66	45	32	24	18	13	9
TW B2	210	80	45	29	20	14	10	7	5		SS C3	197	67	36	23	15	11	8	5	4
TW B3	183	78	45	30	21	15	11	8	5		SS C4	203	95	57	38	27	20	14	10	7
TW B4	169	72	42	27	19	14	10	7	5		SS D1	393	152	84	53	36	25	18	13	8
TW C1	243	72	38	24	16	11	8	6	4		SS D2	347	179	112	76	55	40	29	21	14
TW C2	186	70	39	25	17	12	9	6	4		SS D3	499	144	73	45	30	20	14	10	6
TW C3	178	65	36	23	16	11	8	6	4		SS D4	355	176	108	73	51	37	27	19	13
TW C4	190	59	31	20	13	10	7	5	3		SS E1	289	117	65	42	28	20	14	10	6
TW D1	141	52	28	18	12	8	6	4	2		SS E2	284	146	91	62	44	32	23	17	11
TW D2	210	98	59	39	28	20	15	11	7		SS E3	353	101	51	31	20	14	10	7	4
TW D3	213	107	66	45	32	23	17	12	8		SS E4	293	137	82	55	38	28	20	14	9
TW D4	208	114	74	52	38	28	21	15	10											
TW D5	196	111	73	52	38	28	21	16	11		LEXINGTON, KENTUCKY									4729 DD
TW E1	249	125	78	53	38	28	21	15	10		SSF =	.10	.20	.30	.40	.50	.60	.70	.80	.90
TW E2	227	120	76	53	38	28	21	15	10		WW A1	446	40	15	7	3	-	-	-	-
TW E3	241	142	95	68	50	38	29	21	14		WW A2	146	51	27	16	10	7	4	3	2
TW E4	216	129	87	63	47	35	27	20	13		WW A3	123	53	30	19	13	9	6	4	2
TW F1	295	71	35	21	14	10	7	5	3		WW A4	113	54	32	20	14	10	7	5	3
TW F2	184	74	42	27	19	13	10	7	4		WW A5	108	55	34	22	15	11	8	5	3
TW F3	152	71	43	28	20	15	11	8	5		WW A6	105	56	35	23	16	11	8	6	4
TW F4	125	64	40	27	19	14	10	7	5		WW B1	89	21	-	-	-	-	-	-	-
TW G1	182	62	33	21	14	10	7	5	3		WW B2	121	61	37	25	17	12	9	6	4
TW G2	133	61	36	24	17	12	9	6	4		WW B3	144	77	48	33	23	17	12	9	6
TW G3	112	56	34	23	16	12	9	6	4		WW B4	135	79	53	37	27	20	15	11	7
TW G4	91	46	29	20	14	10	8	5	4		WW B5	128	77	52	37	28	21	15	11	8
TW H1	118	50	29	19	13	9	7	5	3		WW C1	157	84	53	36	26	19	14	10	6
TW H2	91	44	27	18	13	9	7	5	3		WW C2	139	79	51	35	26	19	14	10	7
TW H3	76	38	23	16	11	8	6	4	3		WW C3	152	96	67	49	36	27	21	15	10
TW H4	59	29	18	12	9	6	5	3	2		WW C4	138	88	61	44	33	25	19	14	9
TW I1	108	43	24	15	10	7	5	3	2		TW A1	389	41	16	8	4	2	1	-	-
TW I2	159	81	51	35	25	18	13	10	6		TW A2	153	47	24	14	9	6	4	2	1
TW I3	169	90	57	39	28	21	15	11	7		TW A3	123	48	26	16	10	7	5	3	2
TW I4	173	99	65	46	34	25	19	14	9		TW A4	106	47	27	17	11	8	5	4	2
TW I5	169	99	67	48	35	26	20	15	10		TW B1	217	41	18	10	6	4	2	1	-
TW J1	206	110	70	49	35	26	19	14	9		TW B2	125	44	23	14	9	6	4	3	2
TW J2	187	104	68	48	35	26	19	14	10		TW B3	110	44	24	15	10	7	5	3	2
TW J3	202	124	85	62	46	35	26	20	13		TW B4	103	41	23	14	10	7	5	3	2
TW J4	186	115	79	57	43	33	25	18	13		TW C1	146	40	19	11	7	5	3	2	1
DG A1	140	57	29	14	-	-	-	-	-		TW C2	113	40	21	13	8	6	4	3	2
DG A2	166	74	44	29	19	12	7	3	-		TW C3	110	38	20	12	8	5	4	3	2
DG A3	210	96	59	41	30	22	16	11	6		TW C4	119	35	18	11	7	5	3	2	1
DG B1	140	59	34	21	13	-	-	-	-		TW D1	66	18	6	-	-	-	-	-	-
DG B2	170	76	46	32	23	16	11	7	-		TW D2	131	58	34	22	15	11	7	5	3
DG B3	214	99	61	43	32	25	19	14	8		TW D3	133	64	38	25	17	12	9	6	4
DG C1	176	77	45	30	21	14	8	-	-											

LEXINGTON, KENTUCKY — CONTINUED

	SSF =.10	.20	.30	.40	.50	.60	.70	.80	.90
TW D4	138	74	47	32	23	17	12	9	6
TW D5	133	74	48	33	24	18	13	9	6
TW E1	160	78	47	31	22	16	11	8	5
TW E2	149	77	47	32	23	16	12	8	6
TW E3	165	96	63	44	32	24	18	13	9
TW E4	149	88	59	42	30	23	17	12	8
TW F1	166	35	15	8	4	2	1	-	-
TW F2	105	39	20	12	8	5	3	2	1
TW F3	88	38	22	14	9	6	4	3	2
TW F4	73	35	21	14	9	6	4	3	2
TW G1	103	31	15	9	5	3	2	1	-
TW G2	77	33	18	11	7	5	3	2	1
TW G3	66	30	18	11	8	5	4	2	2
TW G4	54	26	15	10	7	5	3	2	1
TW H1	67	26	14	8	5	4	2	2	-
TW H2	53	24	14	9	6	4	3	2	1
TW H3	45	21	12	8	5	4	3	2	1
TW H4	36	17	10	6	4	3	2	1	-
TW I1	47	13	-	-	-	-	-	-	-
TW I2	99	48	29	19	13	9	7	5	3
TW I3	107	54	33	22	16	11	8	6	4
TW I4	114	64	41	29	21	15	11	8	5
TW I5	114	66	43	31	22	17	12	9	6
TW J1	132	69	43	29	20	15	11	7	5
TW J2	123	67	43	29	21	15	11	8	5
TW J3	139	84	57	41	30	23	17	12	8
TW J4	129	78	53	38	28	21	16	12	8
DG A1	65	-	-	-	-	-	-	-	-
DG A2	102	42	23	13	6	-	-	-	-
DG A3	143	64	38	26	18	13	9	5	-
DG B1	66	18	-	-	-	-	-	-	-
DG B2	105	44	25	15	10	5	-	-	-
DG B3	147	67	40	27	20	15	11	7	4
DG C1	94	34	16	-	-	-	-	-	-
DG C2	129	56	32	21	14	10	6	-	-
DG C3	171	78	47	32	24	18	13	10	6
SS A1	282	77	37	21	13	9	6	4	2
SS A2	272	110	62	39	26	18	13	9	6
SS A3	260	64	29	15	9	5	3	2	-
SS A4	274	108	59	37	25	17	12	8	5
SS A5	474	74	31	16	10	6	4	2	1
SS A6	268	108	60	38	25	18	12	9	5
SS A7	522	57	21	9	4	-	-	-	-
SS A8	271	104	56	35	23	16	11	7	5
SS B1	186	56	28	16	10	6	4	3	1
SS B2	214	90	51	33	22	16	11	8	5
SS B3	165	48	22	12	7	4	3	1	-
SS B4	207	86	48	31	21	15	10	7	4
SS B5	240	48	20	11	6	4	2	1	-
SS B6	199	85	48	31	21	15	10	7	5
SS B7	203	36	14	6	3	-	-	-	-
SS B8	191	79	44	28	19	13	9	6	4
SS C1	105	43	23	14	9	6	4	2	1
SS C2	129	64	39	26	18	13	9	6	4
SS C3	112	34	17	10	6	4	2	1	-
SS C4	128	57	33	21	15	10	7	5	3
SS D1	254	92	48	29	18	12	8	5	3
SS D2	239	120	74	49	34	25	18	12	8
SS D3	324	87	41	23	15	9	6	4	2
SS D4	245	118	70	46	32	23	16	11	7
SS E1	177	66	34	20	12	8	5	3	2
SS E2	191	95	57	38	26	19	13	9	6
SS E3	222	57	26	14	9	5	3	2	1
SS E4	198	89	52	33	23	16	11	8	5

LOUISVILLE, KENTUCKY 4645 DD

	SSF =.10	.20	.30	.40	.50	.60	.70	.80	.90
WW A1	454	42	16	7	4	2	-	-	-
WW A2	150	54	28	17	11	7	5	3	2
WW A3	127	56	31	20	13	9	6	4	3
WW A4	117	56	33	22	15	10	7	5	3
WW A5	112	58	35	23	16	11	8	6	3
WW A6	109	58	36	24	17	12	9	6	4
WW B1	93	24	7	-	-	-	-	-	-
WW B2	124	63	39	26	18	13	9	6	4
WW B3	147	79	50	34	24	17	13	9	6
WW B4	138	81	54	38	28	21	15	11	7
WW B5	130	79	54	38	28	21	16	12	8
WW C1	160	87	55	37	27	19	14	10	7
WW C2	142	81	52	37	26	19	14	10	7
WW C3	155	98	68	50	37	28	21	16	11
WW C4	141	89	62	45	34	26	19	14	10
TW A1	396	43	17	8	5	3	1	-	-
TW A2	157	49	25	15	9	6	4	3	1
TW A3	127	50	27	17	11	8	5	3	2
TW A4	109	49	28	18	12	8	6	4	2
TW B1	221	43	19	10	6	4	2	1	-
TW B2	128	46	24	15	10	7	4	3	2
TW B3	113	45	25	16	10	7	5	3	2
TW B4	105	43	24	15	10	7	5	3	2
TW C1	149	41	20	12	7	5	3	2	1
TW C2	116	41	22	13	9	6	4	3	2
TW C3	112	39	21	13	8	6	4	3	2
TW C4	121	36	18	11	7	5	3	2	1
TW D1	69	20	7	-	-	-	-	-	-
TW D2	134	60	35	23	16	11	8	5	3
TW D3	136	66	40	26	18	13	9	6	4
TW D4	141	76	48	33	24	17	13	9	6
TW D5	136	76	49	34	25	18	14	10	6
TW E1	163	80	48	32	22	16	12	8	5
TW E2	152	79	49	33	23	17	12	9	6
TW E3	168	98	65	46	33	25	18	13	9
TW E4	152	90	60	43	31	23	17	13	8
TW F1	170	36	16	8	5	3	2	-	-
TW F2	108	40	21	13	8	5	4	2	1
TW F3	91	40	23	14	10	7	4	3	2
TW F4	76	37	22	14	10	7	5	3	2
TW G1	106	33	16	9	6	4	2	1	-
TW G2	79	34	19	12	8	5	4	2	1
TW G3	67	32	18	12	8	6	4	3	2
TW G4	55	27	16	10	7	5	4	2	2
TW H1	69	27	15	9	6	4	3	2	1
TW H2	55	25	15	9	6	4	3	2	1
TW H3	47	22	13	8	6	4	3	2	1
TW H4	37	17	10	7	4	3	2	2	-
TW I1	50	15	5	-	-	-	-	-	-
TW I2	101	50	30	20	14	10	7	5	3
TW I3	109	56	35	23	16	12	8	6	4
TW I4	117	66	42	30	21	16	12	8	5
TW I5	116	68	45	31	23	17	13	9	6
TW J1	135	71	44	30	21	15	11	8	5
TW J2	125	69	44	30	22	16	12	8	5
TW J3	141	86	58	42	31	23	17	13	8
TW J4	131	80	54	39	29	22	16	12	8
DG A1	68	-	-	-	-	-	-	-	-
DG A2	104	43	24	13	7	-	-	-	-
DG A3	145	65	39	26	19	13	9	5	-
DG B1	70	21	-	-	-	-	-	-	-
DG B2	107	46	26	16	10	6	-	-	-
DG B3	149	68	41	28	20	15	11	8	4
DG C1	98	36	18	6	-	-	-	-	-

LOUISVILLE, KENTUCKY									CONTINUED
SSF =	.10	.20	.30	.40	.50	.60	.70	.80	.90
DG C2	132	58	33	22	15	10	6	2	-
DG C3	173	80	48	33	24	18	14	10	6
SS A1	284	79	38	22	14	9	6	4	2
SS A2	275	112	63	40	27	19	13	9	6
SS A3	262	66	30	16	9	6	3	2	1
SS A4	276	110	60	38	25	18	12	8	5
SS A5	475	76	32	17	10	6	4	2	1
SS A6	271	110	61	39	26	18	13	9	6
SS A7	521	58	22	10	5	2	-	-	-
SS A8	274	105	57	36	24	16	11	8	5
SS B1	189	58	28	16	10	7	4	3	2
SS B2	217	92	52	33	23	16	11	8	5
SS B3	167	49	23	13	8	5	3	2	-
SS B4	210	88	49	32	21	15	11	7	5
SS B5	242	49	21	11	7	4	2	1	-
SS B6	202	86	49	32	22	15	11	7	5
SS B7	205	38	15	7	3	1	-	-	-
SS B8	193	81	45	29	19	14	10	7	4
SS C1	108	45	24	15	10	6	4	3	2
SS C2	132	66	40	27	19	13	10	7	4
SS C3	115	36	18	10	6	4	3	2	-
SS C4	131	59	34	22	15	11	8	5	3
SS D1	257	94	49	30	19	13	9	6	3
SS D2	242	123	75	50	35	25	18	13	8
SS D3	327	89	43	24	15	10	6	4	2
SS D4	248	120	72	47	33	23	17	12	8
SS E1	181	68	35	21	13	9	6	3	2
SS E2	194	97	59	39	27	19	14	10	6
SS E3	224	59	27	15	9	6	3	2	1
SS E4	201	91	53	34	23	16	12	8	5

BATON ROUGE, LOUISIANA									1670 DD
SSF =	.10	.20	.30	.40	.50	.60	.70	.80	.90
WW A1	1567	197	91	54	36	25	17	12	8
WW A2	525	208	118	77	53	39	28	20	13
WW A3	439	208	126	84	60	44	32	23	15
WW A4	401	206	129	88	64	47	35	25	17
WW A5	380	207	133	93	67	50	37	27	18
WW A6	368	207	135	95	69	52	39	28	19
WW B1	458	168	92	59	41	29	21	15	10
WW B2	385	204	130	90	65	48	36	26	18
WW B3	443	247	160	112	82	61	46	33	22
WW B4	378	228	154	111	82	62	47	35	24
WW B5	344	211	144	105	78	60	45	33	23
WW C1	467	260	169	118	87	65	48	35	24
WW C2	398	231	153	109	81	61	46	33	23
WW C3	395	252	176	130	98	75	57	42	29
WW C4	355	226	158	116	88	67	51	38	26
TW A1	1336	186	87	52	34	24	17	12	8
TW A2	537	189	103	66	46	33	24	17	11
TW A3	431	185	108	71	50	36	27	19	13
TW A4	363	173	105	71	51	37	27	20	13
TW B1	751	173	86	53	36	25	18	13	8
TW B2	429	169	96	62	43	31	23	16	11
TW B3	372	162	95	62	44	32	23	17	11
TW B4	340	148	87	57	40	29	22	15	10
TW C1	493	153	81	51	35	25	18	13	8
TW C2	374	144	81	53	37	26	19	14	9
TW C3	355	134	75	49	34	24	18	13	8
TW C4	371	120	64	41	28	20	15	10	7
TW D1	331	130	73	47	33	23	17	12	8
TW D2	413	195	118	79	56	41	31	22	15
TW D3	421	213	133	91	65	48	36	26	17

BATON ROUGE, LOUISIANA									CONTINUED
SSF =	.10	.20	.30	.40	.50	.60	.70	.80	.90
TW D4	391	215	139	97	71	53	40	29	20
TW D5	358	203	133	94	69	52	39	29	19
TW E1	482	245	152	104	75	55	41	30	20
TW E2	432	229	146	101	73	54	41	30	20
TW E3	442	261	174	125	92	70	53	39	26
TW E4	391	233	157	113	84	63	48	35	24
TW F1	619	160	81	50	34	24	17	12	8
TW F2	386	159	91	60	42	30	22	16	10
TW F3	317	151	91	62	44	32	24	17	11
TW F4	258	134	84	58	42	31	23	16	11
TW G1	385	137	75	48	33	24	17	12	8
TW G2	278	130	78	52	37	27	20	14	9
TW G3	232	116	72	49	35	26	19	14	9
TW G4	185	95	60	41	30	22	16	12	8
TW H1	249	109	64	42	30	22	16	11	7
TW H2	189	93	57	39	28	20	15	11	7
TW H3	156	79	49	33	24	18	13	9	6
TW H4	120	60	37	25	18	13	10	7	5
TW I1	258	109	63	41	29	21	15	11	7
TW I2	316	162	102	70	50	37	28	20	13
TW I3	332	177	113	78	57	42	31	23	15
TW I4	324	186	122	87	64	48	36	26	18
TW I5	310	182	122	87	65	49	37	27	18
TW J1	398	214	137	95	69	51	38	28	19
TW J2	354	198	129	91	66	50	37	27	18
TW J3	366	225	154	111	83	63	48	36	24
TW J4	335	206	141	102	77	58	44	33	22
DG A1	328	145	85	54	36	23	14	6	-
DG A2	336	152	92	63	44	31	22	15	8
DG A3	395	180	111	78	57	43	32	22	13
DG B1	334	146	89	61	44	32	22	13	-
DG B2	350	157	96	67	50	38	28	20	12
DG B3	403	184	113	80	60	47	36	27	18
DG C1	400	175	107	75	55	42	31	21	11
DG C2	406	183	112	79	59	45	35	25	16
DG C3	462	212	130	93	70	55	43	33	22
SS A1	881	276	145	91	62	44	31	22	14
SS A2	768	328	190	125	88	64	46	33	22
SS A3	866	257	133	82	56	39	28	19	12
SS A4	790	331	190	125	87	63	46	32	21
SS A5	1454	281	134	80	53	37	26	18	12
SS A6	767	328	190	125	87	63	46	33	22
SS A7	1662	246	113	66	44	30	21	14	9
SS A8	798	328	187	122	85	61	44	31	21
SS B1	611	211	114	72	49	35	25	18	12
SS B2	599	264	155	102	72	53	38	27	18
SS B3	575	198	106	67	46	33	23	16	11
SS B4	593	260	152	101	71	51	38	27	18
SS B5	791	193	96	58	39	27	19	14	9
SS B6	569	255	150	100	71	51	38	27	18
SS B7	713	168	83	50	33	23	16	11	7
SS B8	561	248	145	96	68	49	36	26	17
SS C1	401	182	108	72	51	37	27	19	13
SS C2	402	209	131	90	65	48	36	26	17
SS C3	415	148	81	52	36	26	19	13	9
SS C4	399	188	114	77	55	40	29	21	14
SS D1	816	328	185	120	83	60	43	31	20
SS D2	676	354	223	153	111	82	60	44	29
SS D3	1019	313	163	102	69	49	35	25	16
SS D4	693	348	214	146	105	77	57	41	27
SS E1	624	262	150	98	69	49	36	25	17
SS E2	567	295	185	127	92	68	50	36	24
SS E3	745	227	119	74	50	35	25	18	11
SS E4	585	279	168	113	81	59	43	31	21

LCR *Tables* / 509

```
LAKE CHARLES, LOUISIANA                        1498 DD    LAKE CHARLES, LOUISIANA                      CONTINUED
   SSF =.10  .20  .30  .40  .50  .60  .70  .80  .90         SSF =.10  .20  .30  .40  .50  .60  .70  .80  .90
WW A1  1680  202   91   54   36   25   17   12    8      DG C2   422  183  113   79   59   45   35   25   16
WW A2   547  210  118   77   53   39   28   20   13      DG C3   483  211  130   93   70   55   43   33   22
WW A3   452  209  126   84   60   44   32   23   15      SS A1   927  281  146   91   62   44   32   22   14
WW A4   410  207  129   88   64   47   35   25   17      SS A2   796  333  192  126   89   65   47   33   22
WW A5   387  208  133   93   68   50   37   27   18      SS A3   915  263  134   84   57   40   28   20   13
WW A6   374  208  135   95   70   52   39   28   19      SS A4   821  336  192  126   89   64   46   33   21
WW B1   479  170   92   59   41   29   21   15    9      SS A5  1550  287  135   81   54   38   27   18   12
WW B2   394  205  130   90   65   48   36   26   17      SS A6   796  332  191  126   89   64   47   33   22
WW B3   452  248  160  112   82   61   46   33   22      SS A7  1783  253  114   67   44   31   21   15    9
WW B4   384  228  154  111   82   62   47   35   23      SS A8   830  333  189  123   86   62   45   32   21
WW B5   348  212  144  105   78   60   45   33   23      SS B1   642  215  115   73   50   36   26   18   12
WW C1   476  261  169  118   87   65   48   35   24      SS B2   621  268  156  103   73   53   39   28   18
WW C2   405  232  153  109   81   61   46   33   22      SS B3   605  202  108   68   47   33   24   17   11
WW C3   399  253  176  130   98   75   57   42   29      SS B4   615  264  154  102   72   52   38   27   18
WW C4   358  227  158  116   88   67   51   38   26      SS B5   840  197   97   59   40   28   20   14    9
TW A1  1430  191   87   52   34   24   17   12    8      SS B6   589  259  152  101   71   52   38   27   18
TW A2   562  192  104   66   46   33   24   17   11      SS B7   759  172   83   51   34   24   17   12    7
TW A3   446  187  108   71   50   36   26   19   12      SS B8   582  251  147   97   69   50   36   26   17
TW A4   373  175  105   71   51   37   27   20   13      SS C1   414  184  108   72   51   37   27   19   13
TW B1   796  176   87   53   36   25   18   13    8      SS C2   412  210  131   90   65   48   36   26   17
TW B2   446  171   96   62   43   31   23   16   11      SS C3   434  150   81   52   36   26   19   13    9
TW B3   386  163   95   62   44   32   23   17   11      SS C4   411  190  114   77   55   40   29   21   14
TW B4   352  149   87   57   40   29   21   15   10      SS D1   849  332  187  121   85   61   44   31   20
TW C1   518  156   82   51   35   25   18   13    8      SS D2   692  357  224  155  112   83   61   44   29
TW C2   389  146   81   53   37   26   19   14    9      SS D3  1073  318  165  103   70   50   36   25   16
TW C3   370  136   75   49   34   24   18   13    8      SS D4   712  351  216  147  106   78   57   41   27
TW C4   389  122   65   41   28   20   15   10    7      SS E1   648  266  152   99   70   50   36   26   17
TW D1   344  131   73   47   33   23   17   12    8      SS E2   582  298  187  128   93   69   51   36   24
TW D2   425  197  118   79   57   41   30   22   15      SS E3   786  232  120   75   51   36   26   18   12
TW D3   432  215  133   91   65   48   36   26   17      SS E4   603  282  170  114   82   60   44   31   21
TW D4   399  216  139   97   71   53   40   29   19
TW D5   365  204  133   94   69   52   39   29   19
TW E1   494  246  152  104   75   56   41   30   20      NEW ORLEANS, LOUISIANA                         1465 DD
TW E2   442  231  146  101   73   54   41   29   20         SSF =.10  .20  .30  .40  .50  .60  .70  .80  .90
TW E3   449  262  174  125   93   70   53   39   26      WW A1  1930  243  113   67   45   31   22   16   10
TW E4   397  234  157  113   84   63   48   35   24      WW A2   641  254  144   94   66   48   35   25   17
TW F1   655  163   82   50   34   24   17   12    8      WW A3   534  253  153  104   74   54   40   29   19
TW F2   400  161   92   60   42   30   22   16   10      WW A4   487  250  157  108   78   58   43   31   21
TW F3   326  152   91   62   44   32   24   17   11      WW A5   461  252  162  113   83   62   46   34   23
TW F4   264  134   84   58   42   31   23   16   11      WW A6   445  251  164  116   85   64   48   35   24
TW G1   403  138   75   48   33   24   17   12    8      WW B1   569  210  116   75   52   37   27   19   13
TW G2   287  131   78   52   37   27   20   14    9      WW B2   464  247  157  109   79   59   44   32   22
TW G3   238  117   72   49   35   26   19   14    9      WW B3   533  297  193  136  100   75   56   41   28
TW G4   190   96   60   41   30   22   16   12    8      WW B4   451  271  184  132   99   75   57   42   29
TW H1   257  110   64   42   30   22   16   11    7      WW B5   408  251  172  125   94   71   54   40   28
TW H2   194   94   57   39   28   20   15   11    7      WW C1   560  312  203  143  105   79   59   43   29
TW H3   160   79   49   33   24   18   13    9    6      WW C2   476  276  184  131   97   73   55   41   28
TW H4   123   60   37   25   18   13   10    7    5      WW C3   467  298  209  154  117   89   68   51   35
TW I1   267  110   63   41   29   21   15   11    7      WW C4   419  268  187  138  104   80   61   46   31
TW I2   323  163  102   70   50   37   28   20   13      TW A1  1642  229  108   64   43   30   21   15   10
TW I3   339  178  113   78   57   42   31   23   15      TW A2   655  231  127   81   56   41   30   21   14
TW I4   330  186  122   87   64   48   36   26   18      TW A3   524  225  132   87   62   45   33   24   16
TW I5   315  183  122   87   65   49   37   27   18      TW A4   440  211  128   87   62   46   34   24   16
TW J1   406  215  137   95   69   51   38   28   19      TW B1   919  212  106   65   44   31   23   16   10
TW J2   361  199  129   91   67   50   37   27   18      TW B2   522  206  117   76   53   39   28   20   13
TW J3   370  225  154  111   83   63   48   35   24      TW B3   452  196  115   76   54   39   29   21   14
TW J4   339  207  141  102   77   58   44   33   22      TW B4   413  179  105   70   50   36   27   19   13
DG A1   339  146   85   54   36   23   14    6    -      TW C1   600  187  100   63   43   31   22   16   11
DG A2   347  153   92   63   44   31   22   15    8      TW C2   453  175   99   64   45   33   24   17   11
DG A3   408  181  111   78   58   43   32   22   13      TW C3   430  163   91   59   41   30   22   16   10
DG B1   344  146   89   61   44   32   22   13    -      TW C4   449  145   78   50   34   25   18   13    8
DG B2   362  157   97   67   50   38   28   20   12      TW D1   410  161   91   59   41   30   22   15   10
DG B3   419  184  114   81   60   47   36   27   17      TW D2   498  235  142   96   69   51   37   27   18
DG C1   414  174  107   75   55   42   31   21   10      TW D3   507  257  160  110   79   59   44   32   21
```

NEW ORLEANS, LOUISIANA								CONTINUED		SHREVEPORT, LOUISIANA									2167 DD
SSF =	.10	.20	.30	.40	.50	.60	.70	.80	.90	SSF =	.10	.20	.30	.40	.50	.60	.70	.80	.90
TW D4	467	257	166	117	86	64	48	35	24	WW A1	1329	158	73	43	28	20	14	10	6
TW D5	425	241	158	112	83	62	47	34	23	WW A2	431	170	96	62	43	31	23	16	11
TW E1	579	294	183	126	91	67	50	37	25	WW A3	360	170	103	69	49	36	26	19	13
TW E2	517	275	175	121	88	66	49	36	24	WW A4	329	169	105	72	52	38	28	20	14
TW E3	525	310	207	149	110	84	63	47	32	WW A5	312	170	109	76	55	41	30	22	15
TW E4	463	277	187	134	100	76	57	42	29	WW A6	302	170	111	78	57	42	32	23	16
TW F1	760	197	101	62	42	30	22	15	10	WW B1	367	133	73	46	32	22	16	11	7
TW F2	470	195	112	74	52	38	28	20	13	WW B2	319	169	107	74	54	40	30	22	14
TW F3	385	184	112	76	54	40	29	21	14	WW B3	368	205	133	93	68	51	38	28	19
TW F4	313	163	102	70	51	38	28	20	14	WW B4	317	191	129	93	69	52	39	29	20
TW G1	471	168	92	59	41	30	21	15	10	WW B5	290	179	122	89	66	50	38	28	19
TW G2	339	158	95	64	46	34	25	18	12	WW C1	389	216	141	99	72	54	40	29	20
TW G3	282	142	88	60	43	32	24	17	12	WW C2	333	194	129	91	67	51	38	28	19
TW G4	224	116	73	50	36	27	20	15	10	WW C3	334	214	150	110	83	64	48	36	25
TW H1	303	133	78	52	37	27	20	14	9	WW C4	301	192	135	99	75	57	44	32	22
TW H2	230	114	70	48	34	25	19	14	9	TW A1	1133	150	70	42	27	19	13	9	6
TW H3	190	95	59	41	29	22	16	12	8	TW A2	442	154	84	54	37	26	19	14	9
TW H4	145	72	45	30	22	16	12	9	6	TW A3	354	151	88	58	41	30	22	15	10
TW I1	320	136	79	52	37	27	19	14	9	TW A4	298	142	86	58	41	30	22	16	11
TW I2	381	196	123	85	61	46	34	25	17	TW B1	627	140	70	43	29	20	14	10	7
TW I3	400	214	136	95	69	51	38	28	19	TW B2	353	138	78	51	35	25	18	13	9
TW I4	387	222	146	104	77	58	44	32	22	TW B3	307	133	78	51	36	26	19	14	9
TW I5	368	217	145	104	77	58	44	33	22	TW B4	281	122	71	47	33	24	18	13	8
TW J1	477	257	165	115	84	62	47	34	23	TW C1	408	125	66	42	29	20	15	10	7
TW J2	423	237	155	109	80	60	45	33	22	TW C2	309	118	67	43	30	22	16	11	7
TW J3	434	266	182	133	99	76	58	43	29	TW C3	294	110	62	40	28	20	14	10	7
TW J4	396	244	167	122	91	70	53	39	27	TW C4	309	99	53	34	23	16	12	8	6
DG A1	405	181	107	70	48	32	21	12	4	TW D1	265	103	58	37	25	18	13	9	6
DG A2	406	185	113	77	55	40	28	19	11	TW D2	342	161	98	66	47	34	25	18	12
DG A3	471	214	133	93	69	52	39	28	17	TW D3	348	176	110	75	54	40	29	21	14
DG B1	412	182	111	77	57	42	30	20	9	TW D4	327	180	117	82	60	44	33	24	16
DG B2	423	190	117	82	62	47	35	25	16	TW D5	301	171	113	80	58	44	33	24	16
DG B3	480	218	135	96	72	56	44	33	22	TW E1	400	203	127	87	62	46	34	25	17
DG C1	490	216	132	93	69	53	40	29	17	TW E2	361	192	122	84	61	45	34	25	17
DG C2	490	220	136	96	72	56	43	32	21	TW E3	372	220	147	105	78	59	44	33	22
DG C3	552	251	155	111	84	66	52	40	27	TW E4	330	198	133	96	71	54	41	30	20
SS A1	1072	335	176	111	76	54	39	27	18	TW F1	511	129	66	40	27	19	14	10	6
SS A2	922	394	229	151	106	77	57	41	27	TW F2	316	130	74	49	34	24	18	13	8
SS A3	1059	315	163	102	69	49	35	25	16	TW F3	260	124	75	50	36	26	19	14	9
SS A4	950	398	229	151	106	77	56	40	26	TW F4	212	110	69	47	34	25	19	13	9
SS A5	1777	342	164	99	66	46	33	23	15	TW G1	316	111	61	39	27	19	14	10	6
SS A6	923	393	228	151	106	77	56	40	27	TW G2	228	106	63	42	30	22	16	12	8
SS A7	2042	302	139	82	55	38	27	19	12	TW G3	190	95	59	40	29	21	16	11	7
SS A8	962	394	225	148	103	75	55	39	26	TW G4	152	78	49	34	24	18	13	10	6
SS B1	744	257	139	88	61	44	32	22	15	TW H1	203	89	52	34	24	17	13	9	6
SS B2	718	316	185	123	87	64	47	34	22	TW H2	155	76	47	32	23	17	12	9	6
SS B3	703	242	131	83	57	41	29	21	14	TW H3	129	65	40	27	20	14	11	8	5
SS B4	712	312	183	121	86	62	46	33	22	TW H4	99	49	30	21	15	11	8	6	4
SS B5	967	236	118	72	49	34	25	17	11	TW I1	206	87	50	32	22	16	12	8	5
SS B6	683	306	181	120	85	62	46	33	22	TW I2	261	134	84	58	42	31	23	17	11
SS B7	876	207	102	62	42	29	21	15	9	TW I3	275	147	94	65	47	35	26	19	13
SS B8	674	298	175	116	82	60	44	31	21	TW I4	271	156	103	73	54	40	30	22	15
SS C1	489	223	133	89	63	46	34	25	16	TW I5	261	154	103	74	54	41	31	23	16
SS C2	484	251	158	109	79	59	44	32	22	TW J1	330	178	114	79	57	43	32	23	16
SS C3	507	181	100	64	44	32	23	17	11	TW J2	296	166	108	76	56	42	31	23	15
SS C4	480	227	138	93	66	49	36	26	18	TW J3	308	190	130	94	71	54	41	30	21
SS D1	990	398	225	147	103	74	54	38	25	TW J4	283	175	120	87	65	49	38	28	19
SS D2	809	423	267	184	134	99	74	53	36	DG A1	263	115	66	41	26	16	7	-	-
SS D3	1240	380	199	125	85	61	44	31	20	DG A2	277	125	76	51	36	25	17	11	5
SS D4	830	416	257	176	127	93	69	50	33	DG A3	330	151	94	65	48	36	27	19	11
SS E1	760	320	184	121	85	62	45	32	21	DG B1	266	117	70	47	34	24	15	8	-
SS E2	681	354	223	154	111	82	61	44	30	DG B2	286	129	79	55	41	30	22	15	9
SS E3	909	277	145	91	62	44	32	22	15	DG B3	335	155	96	67	51	39	30	22	14
SS E4	703	335	203	137	98	72	53	38	25	DG C1	320	142	86	59	43	32	24	15	5

SHREVEPORT, LOUISIANA   CONTINUED

SSF =	.10	.20	.30	.40	.50	.60	.70	.80	.90
DG C2	333	152	93	65	49	37	28	20	13
DG C3	384	179	110	78	59	46	36	27	19
SS A1	715	222	117	73	49	35	25	17	11
SS A2	629	270	156	103	72	52	38	27	18
SS A3	696	205	106	65	44	31	22	15	10
SS A4	643	270	155	101	71	51	37	26	17
SS A5	1202	225	107	64	42	29	21	14	9
SS A6	627	268	156	102	71	52	37	27	18
SS A7	1379	194	89	52	34	23	16	11	7
SS A8	648	266	152	99	69	49	36	25	17
SS B1	494	170	91	58	39	28	20	14	9
SS B2	492	217	128	84	59	43	31	23	15
SS B3	461	158	85	53	36	26	18	13	8
SS B4	485	213	125	82	58	42	31	22	14
SS B5	645	154	76	46	31	22	15	11	7
SS B6	466	210	124	82	58	42	31	22	14
SS B7	576	133	65	39	26	18	13	9	6
SS B8	457	202	119	78	55	40	29	21	14
SS C1	327	148	88	58	41	30	22	16	10
SS C2	333	173	109	75	54	40	30	21	14
SS C3	340	120	66	42	29	21	15	11	7
SS C4	330	156	94	63	45	33	24	18	12
SS D1	659	265	149	96	67	48	35	24	16
SS D2	555	292	183	126	91	67	49	36	24
SS D3	827	252	131	82	55	39	28	19	13
SS D4	569	286	177	120	86	63	46	33	22
SS E1	501	210	120	78	54	39	28	20	13
SS E2	464	242	152	104	75	55	41	29	20
SS E3	602	182	95	59	40	28	20	14	9
SS E4	478	228	138	93	66	48	35	25	17

CARIBOU, MAINE   9632 DD

SSF =	.10	.20	.30	.40	.50	.60	.70	.80	.90
WW A1	160	-	-	-	-	-	-	-	-
WW A2	63	13	-	-	-	-	-	-	-
WW A3	55	18	-	-	-	-	-	-	-
WW A4	52	20	-	-	-	-	-	-	-
WW A5	50	22	9	-	-	-	-	-	-
WW A6	50	23	11	-	-	-	-	-	-
WW B1	-	-	-	-	-	-	-	-	-
WW B2	66	31	18	10	6	3	-	-	-
WW B3	82	42	25	16	10	6	3	2	-
WW B4	86	50	32	22	16	11	8	5	3
WW B5	84	50	34	24	17	13	9	6	4
WW C1	93	48	29	19	13	8	5	3	2
WW C2	86	47	30	20	14	10	7	4	3
WW C3	104	65	45	32	24	18	13	9	6
WW C4	95	60	41	30	22	16	12	9	6
TW A1	152	-	-	-	-	-	-	-	-
TW A2	68	14	-	-	-	-	-	-	-
TW A3	57	17	-	-	-	-	-	-	-
TW A4	51	19	8	-	-	-	-	-	-
TW B1	93	-	-	-	-	-	-	-	-
TW B2	59	17	-	-	-	-	-	-	-
TW B3	54	18	7	-	-	-	-	-	-
TW B4	53	18	8	-	-	-	-	-	-
TW C1	69	14	-	-	-	-	-	-	-
TW C2	57	17	7	-	-	-	-	-	-
TW C3	57	17	8	-	-	-	-	-	-
TW C4	64	17	7	4	-	-	-	-	-
TW D1	-	-	-	-	-	-	-	-	-
TW D2	72	30	16	9	5	-	-	-	-
TW D3	73	33	18	11	6	2	-	-	-

CARIBOU, MAINE   CONTINUED

SSF =	.10	.20	.30	.40	.50	.60	.70	.80	.90
TW D4	86	45	28	19	13	9	6	4	2
TW D5	88	48	31	21	15	11	8	5	3
TW E1	94	44	25	16	10	6	4	2	-
TW E2	91	45	27	18	12	8	5	3	2
TW E3	109	62	40	28	20	15	10	7	4
TW E4	101	59	39	27	20	14	10	7	5
TW F1	61	-	-	-	-	-	-	-	-
TW F2	44	-	-	-	-	-	-	-	-
TW F3	39	12	-	-	-	-	-	-	-
TW F4	34	14	-	-	-	-	-	-	-
TW G1	39	-	-	-	-	-	-	-	-
TW G2	33	10	-	-	-	-	-	-	-
TW G3	30	11	-	-	-	-	-	-	-
TW G4	26	11	5	-	-	-	-	-	-
TW H1	27	-	-	-	-	-	-	-	-
TW H2	24	9	-	-	-	-	-	-	-
TW H3	22	9	4	-	-	-	-	-	-
TW H4	18	7	4	-	-	-	-	-	-
TW I1	-	-	-	-	-	-	-	-	-
TW I2	54	24	13	8	4	-	-	-	-
TW I3	60	29	17	10	6	4	2	-	-
TW I4	72	39	25	17	12	8	6	4	2
TW I5	75	43	28	19	14	10	7	5	3
TW J1	78	39	23	15	10	6	4	2	1
TW J2	76	40	25	17	11	8	5	3	2
TW J3	93	56	37	26	19	14	10	7	5
TW J4	87	53	35	25	18	14	10	7	5
DG A1	-	-	-	-	-	-	-	-	-
DG A2	52	17	-	-	-	-	-	-	-
DG A3	92	40	23	15	10	6	3	-	-
DG B1	-	-	-	-	-	-	-	-	-
DG B2	54	20	7	-	-	-	-	-	-
DG B3	96	42	25	17	12	8	5	3	-
DG C1	-	-	-	-	-	-	-	-	-
DG C2	73	29	15	8	-	-	-	-	-
DG C3	113	50	30	21	14	10	7	4	-
SS A1	147	33	8	-	-	-	-	-	-
SS A2	167	65	34	21	13	8	5	3	1
SS A3	120	-	-	-	-	-	-	-	-
SS A4	162	60	31	17	10	6	3	1	-
SS A5	235	26	-	-	-	-	-	-	-
SS A6	161	61	32	19	11	7	4	2	-
SS A7	229	-	-	-	-	-	-	-	-
SS A8	157	55	27	15	8	3	-	-	-
SS B1	91	20	-	-	-	-	-	-	-
SS B2	132	53	29	18	11	7	5	3	1
SS B3	70	-	-	-	-	-	-	-	-
SS B4	124	49	26	15	10	6	3	2	-
SS B5	110	-	-	-	-	-	-	-	-
SS B6	120	49	26	16	10	6	4	2	1
SS B7	79	-	-	-	-	-	-	-	-
SS B8	111	43	22	13	7	4	2	-	-
SS C1	39	-	-	-	-	-	-	-	-
SS C2	72	34	19	12	7	4	1	-	-
SS C3	44	-	-	-	-	-	-	-	-
SS C4	72	30	16	9	5	3	-	-	-
SS D1	131	39	13	-	-	-	-	-	-
SS D2	147	72	42	27	18	12	8	5	3
SS D3	168	36	-	-	-	-	-	-	-
SS D4	151	70	40	25	16	11	7	4	2
SS E1	77	-	-	-	-	-	-	-	-
SS E2	111	52	29	18	11	7	4	2	-
SS E3	101	-	-	-	-	-	-	-	-
SS E4	115	49	26	16	9	5	3	1	-

PORTLAND, MAINE							7498 DD		
SSF =	.10	.20	.30	.40	.50	.60	.70	.80	.90
WW A1	233	-	-	-	-	-	-	-	-
WW A2	80	23	9	-	-	-	-	-	-
WW A3	69	26	13	6	-	-	-	-	-
WW A4	64	28	15	8	4	-	-	-	-
WW A5	62	29	16	9	5	3	-	-	-
WW A6	61	30	17	10	6	4	2	-	-
WW B1	-	-	-	-	-	-	-	-	-
WW B2	77	37	22	14	9	6	4	3	1
WW B3	94	49	30	20	14	9	6	4	3
WW B4	95	55	36	25	18	13	10	7	4
WW B5	93	56	37	27	19	14	11	8	5
WW C1	106	55	34	23	16	11	8	5	3
WW C2	96	54	34	24	17	12	9	6	4
WW C3	113	71	49	36	26	20	15	11	7
WW C4	103	65	45	33	24	18	14	10	7
TW A1	212	9	-	-	-	-	-	-	-
TW A2	86	22	8	-	-	-	-	-	-
TW A3	71	24	11	5	-	-	-	-	-
TW A4	62	25	13	7	4	2	-	-	-
TW B1	122	17	-	-	-	-	-	-	-
TW B2	73	23	10	5	-	-	-	-	-
TW B3	65	24	12	6	3	-	-	-	-
TW B4	63	23	12	7	4	2	1	-	-
TW C1	86	20	8	3	-	-	-	-	-
TW C2	69	22	10	6	3	1	-	-	-
TW C3	68	22	10	6	3	2	-	-	-
TW C4	76	21	10	5	3	2	1	-	-
TW D1	-	-	-	-	-	-	-	-	-
TW D2	84	36	20	12	8	5	3	2	1
TW D3	86	39	22	14	9	6	4	3	1
TW D4	97	51	32	21	15	11	8	5	3
TW D5	97	53	34	24	17	12	9	6	4
TW E1	107	50	30	19	13	9	6	4	2
TW E2	103	51	31	21	15	10	7	5	3
TW E3	120	69	45	32	23	17	12	9	6
TW E4	110	64	43	30	22	16	12	9	6
TW F1	85	-	-	-	-	-	-	-	-
TW F2	57	17	6	-	-	-	-	-	-
TW F3	49	19	9	4	-	-	-	-	-
TW F4	42	18	10	5	3	-	-	-	-
TW G1	53	11	-	-	-	-	-	-	-
TW G2	42	15	7	-	-	-	-	-	-
TW G3	37	16	8	4	2	-	-	-	-
TW G4	32	14	8	4	3	1	-	-	-
TW H1	36	11	4	-	-	-	-	-	-
TW H2	30	12	6	3	1	-	-	-	-
TW H3	27	11	6	3	2	1	-	-	-
TW H4	22	9	5	3	2	1	-	-	-
TW I1	-	-	-	-	-	-	-	-	-
TW I2	63	29	17	11	7	5	3	2	1
TW I3	69	34	20	13	9	6	4	3	2
TW I4	80	44	28	19	14	10	7	5	3
TW I5	82	47	31	21	16	11	8	6	4
TW J1	89	45	27	18	12	9	6	4	2
TW J2	85	45	28	19	14	10	7	5	3
TW J3	102	62	41	29	22	16	12	9	6
TW J4	95	58	39	28	20	15	11	8	5
DG A1	-	-	-	-	-	-	-	-	-
DG A2	62	22	8	-	-	-	-	-	-
DG A3	102	44	26	17	12	8	5	-	-
DG B1	-	-	-	-	-	-	-	-	-
DG B2	64	24	12	-	-	-	-	-	-
DG B3	107	47	28	19	13	10	7	4	-
DG C1	37	-	-	-	-	-	-	-	-

PORTLAND, MAINE								CONTINUED	
SSF =	.10	.20	.30	.40	.50	.60	.70	.80	.90
DG C2	84	34	19	11	6	-	-	-	-
DG C3	125	55	33	23	16	12	9	6	3
SS A1	180	44	18	8	-	-	-	-	-
SS A2	192	75	41	25	17	11	7	5	3
SS A3	155	30	-	-	-	-	-	-	-
SS A4	189	71	38	23	14	9	6	4	2
SS A5	302	39	12	-	-	-	-	-	-
SS A6	186	72	39	24	15	10	7	4	2
SS A7	316	19	-	-	-	-	-	-	-
SS A8	184	67	35	20	13	8	5	3	2
SS B1	114	29	12	-	-	-	-	-	-
SS B2	151	62	34	21	14	10	7	4	3
SS B3	93	20	-	-	-	-	-	-	-
SS B4	144	58	32	19	13	8	6	4	2
SS B5	144	21	-	-	-	-	-	-	-
SS B6	139	57	32	20	13	9	6	4	2
SS B7	112	-	-	-	-	-	-	-	-
SS B8	130	52	28	17	11	7	5	3	2
SS C1	53	16	-	-	-	-	-	-	-
SS C2	83	40	23	15	10	7	5	3	2
SS C3	59	13	-	-	-	-	-	-	-
SS C4	83	35	20	12	8	5	4	2	1
SS D1	160	52	24	12	5	-	-	-	-
SS D2	168	83	50	32	22	15	11	7	4
SS D3	207	48	20	8	-	-	-	-	-
SS D4	172	81	47	31	21	14	10	7	4
SS E1	101	31	10	-	-	-	-	-	-
SS E2	129	62	36	23	15	10	7	5	3
SS E3	131	26	-	-	-	-	-	-	-
SS E4	134	58	33	20	13	9	6	4	2

BALTIMORE, MARYLAND								4729 DD	
SSF =	.10	.20	.30	.40	.50	.60	.70	.80	.90
WW A1	463	49	21	11	7	4	2	1	-
WW A2	161	61	33	21	14	10	7	4	3
WW A3	138	63	37	24	17	12	8	6	4
WW A4	127	64	39	26	18	13	9	6	4
WW A5	122	65	41	28	20	14	10	7	5
WW A6	119	66	42	29	21	15	11	8	5
WW B1	106	33	15	8	4	-	-	-	-
WW B2	133	70	44	30	21	15	11	8	5
WW B3	158	87	56	39	28	20	15	11	7
WW B4	147	88	60	43	32	24	18	13	9
WW B5	139	85	59	42	32	24	18	13	9
WW C1	171	95	61	42	31	23	17	12	8
WW C2	151	88	58	41	30	22	17	12	8
WW C3	164	106	74	54	41	31	24	17	12
WW C4	149	96	67	49	37	28	21	16	11
TW A1	405	49	21	12	7	4	3	2	-
TW A2	167	56	29	18	12	8	6	4	2
TW A3	136	57	32	20	14	10	7	5	3
TW A4	117	55	32	21	15	10	7	5	3
TW B1	231	48	23	13	8	6	4	2	1
TW B2	137	52	29	18	12	8	6	4	2
TW B3	121	51	29	19	13	9	6	4	3
TW B4	112	48	27	18	12	9	6	4	3
TW C1	157	46	24	14	9	6	4	3	2
TW C2	123	46	25	16	11	7	5	4	2
TW C3	119	43	24	15	10	7	5	3	2
TW C4	127	39	21	13	9	6	4	3	2
TW D1	79	27	13	7	4	2	-	-	-
TW D2	143	66	40	26	18	13	9	7	4
TW D3	146	73	45	30	21	15	11	8	5

LCR *Tables* / 513

BALTIMORE, MARYLAND					CONTINUED				BOSTON, MASSACHUSETTS							5621 DD	
SSF =.10	.20	.30	.40	.50	.60	.70	.80	.90	SSF =.10	.20	.30	.40	.50	.60	.70	.80	.90
TW D4 149	82	53	37	27	20	15	10	7	WW A1 368	28	9	-	-	-	-	-	-
TW D5 143	81	54	38	28	21	15	11	7	WW A2 119	41	20	12	7	5	3	2	-
TW E1 174	87	54	37	26	19	14	10	6	WW A3 101	43	24	15	10	6	4	3	1
TW E2 161	85	54	37	27	20	14	10	7	WW A4 93	44	26	16	11	7	5	3	2
TW E3 177	105	70	50	37	28	21	15	10	WW A5 89	45	27	18	12	8	6	4	2
TW E4 161	96	65	47	35	26	19	14	10	WW A6 87	46	28	19	13	9	6	4	3
TW F1 180	42	20	12	7	5	3	2	1	WW B1 59	-	-	-	-	-	-	-	-
TW F2 117	46	26	16	11	7	5	3	2	WW B2 103	52	31	21	15	10	7	5	3
TW F3 99	46	27	18	12	8	6	4	3	WW B3 123	66	41	28	20	14	10	7	5
TW F4 82	42	26	17	12	9	6	4	3	WW B4 118	70	46	33	24	18	13	9	6
TW G1 114	38	20	12	8	5	3	2	1	WW B5 113	69	46	33	25	18	14	10	7
TW G2 86	39	23	15	10	7	5	3	2	WW C1 135	72	46	31	22	16	12	8	5
TW G3 73	36	22	14	10	7	5	3	2	WW C2 121	68	44	31	22	16	12	9	6
TW G4 60	30	19	12	9	6	4	3	2	WW C3 136	86	60	44	33	25	19	14	9
TW H1 75	31	18	11	8	5	4	2	1	WW C4 124	78	54	40	30	23	17	12	8
TW H2 60	29	17	11	8	6	4	3	2	TW A1 324	30	11	4	-	-	-	-	-
TW H3 51	25	15	10	7	5	4	2	2	TW A2 126	37	18	10	6	4	2	1	-
TW H4 40	19	12	8	5	4	3	2	1	TW A3 102	39	21	13	8	5	3	2	1
TW I1 59	22	11	6	3	-	-	-	-	TW A4 88	38	22	14	9	6	4	3	2
TW I2 109	55	34	23	16	12	9	6	4	TW B1 180	32	13	7	4	2	-	-	-
TW I3 117	62	39	27	19	14	10	7	5	TW B2 104	36	19	11	7	5	3	2	1
TW I4 124	71	47	33	24	18	13	10	6	TW B3 92	36	19	12	8	5	3	2	1
TW I5 123	73	49	35	26	19	14	10	7	TW B4 86	34	19	12	8	5	4	2	1
TW J1 144	77	49	34	24	18	13	9	6	TW C1 122	32	15	9	5	3	2	1	-
TW J2 133	75	48	34	25	18	13	10	6	TW C2 95	33	17	10	7	4	3	2	1
TW J3 150	92	63	46	34	26	19	14	10	TW C3 93	31	16	10	6	4	3	2	1
TW J4 139	86	59	43	32	24	18	13	9	TW C4 102	29	15	9	6	4	3	2	1
DG A1 78	26	-	-	-	-	-	-	-	TW D1 45	-	-	-	-	-	-	-	-
DG A2 110	48	28	17	10	4	-	-	-	TW D2 112	49	28	18	12	9	6	4	3
DG A3 151	70	43	29	21	15	11	7	-	TW D3 113	54	32	21	15	10	7	5	3
DG B1 79	30	13	-	-	-	-	-	-	TW D4 121	64	41	28	20	15	11	8	5
DG B2 113	51	30	20	13	9	4	-	-	TW D5 118	66	42	30	21	16	12	8	6
DG B3 154	72	45	31	23	17	13	9	5	TW E1 138	67	40	27	18	13	9	7	4
DG C1 106	44	24	14	6	-	-	-	-	TW E2 130	66	41	28	20	14	10	7	5
DG C2 137	63	38	26	18	13	9	5	-	TW E3 146	84	56	39	29	21	16	11	8
DG C3 178	84	52	37	27	21	16	12	7	TW E4 133	78	52	37	27	20	15	11	7
SS A1 289	85	43	26	17	11	7	5	3	TW F1 134	25	10	4	-	-	-	-	-
SS A2 282	119	68	44	30	21	15	11	7	TW F2 86	30	16	9	5	3	2	1	-
SS A3 265	72	35	20	12	8	5	3	2	TW F3 72	31	17	11	7	4	3	2	1
SS A4 282	116	65	42	29	20	14	10	6	TW F4 61	29	17	11	7	5	3	2	1
SS A5 473	82	37	21	13	8	5	3	2	TW G1 83	24	11	6	3	2	-	-	-
SS A6 277	116	67	43	29	21	15	10	6	TW G2 63	26	14	9	5	4	2	1	-
SS A7 510	64	26	14	8	4	2	1	-	TW G3 54	25	14	9	6	4	3	2	1
SS A8 279	112	63	40	27	19	13	9	6	TW G4 45	21	12	8	5	4	3	2	1
SS B1 194	63	33	20	13	8	6	4	2	TW H1 54	21	11	6	4	2	1	-	-
SS B2 223	97	56	37	26	18	13	9	6	TW H2 44	20	11	7	5	3	2	1	-
SS B3 172	55	27	16	10	7	4	3	1	TW H3 38	17	10	6	4	3	2	1	-
SS B4 215	93	54	35	24	17	12	8	5	TW H4 30	14	8	5	3	2	2	1	-
SS B5 245	54	25	14	9	6	4	2	1	TW I1 30	-	-	-	-	-	-	-	-
SS B6 207	92	54	35	24	17	12	9	5	TW I2 84	41	24	16	11	8	6	4	2
SS B7 207	42	19	10	6	3	2	-	-	TW I3 91	46	28	19	13	9	7	5	3
SS B8 199	86	50	32	22	16	11	8	5	TW I4 100	56	36	25	18	13	10	7	5
SS C1 119	52	29	19	13	9	6	4	2	TW I5 101	58	38	27	20	15	11	8	5
SS C2 141	72	45	31	22	16	11	8	5	TW J1 114	59	37	25	17	12	9	6	4
SS C3 123	41	22	13	9	6	4	3	1	TW J2 107	58	37	25	18	13	10	7	4
SS C4 139	65	39	26	18	13	9	7	4	TW J3 123	75	51	36	27	20	15	11	7
SS D1 265	102	56	35	23	16	11	7	4	TW J4 115	70	47	34	25	19	14	10	7
SS D2 251	130	81	55	39	28	20	14	9	DG A1 43	-	-	-	-	-	-	-	-
SS D3 332	96	48	29	18	12	8	5	3	DG A2 85	34	18	9	-	-	-	-	-
SS D4 256	127	78	52	37	26	19	13	9	DG A3 125	56	33	22	16	11	7	4	-
SS E1 189	75	41	25	17	11	7	5	3	DG B1 44	-	-	-	-	-	-	-	-
SS E2 202	104	64	43	31	22	16	11	7	DG B2 87	36	20	12	7	-	-	-	-
SS E3 230	65	32	19	12	8	5	3	2	DG B3 129	58	35	24	17	13	9	6	3
SS E4 208	98	58	38	27	19	14	9	6	DG C1 71	23	-	-	-	-	-	-	-

BOSTON, MASSACHUSETTS								CONTINUED	
SSF =	.10	.20	.30	.40	.50	.60	.70	.80	.90
DG C2	109	47	27	17	12	8	4	-	-
DG C3	151	68	41	28	21	16	12	8	5
SS A1	230	61	29	16	10	6	4	2	1
SS A2	231	93	52	33	22	15	11	7	5
SS A3	205	48	20	10	4	-	-	-	-
SS A4	229	90	49	31	20	14	9	6	4
SS A5	389	58	23	11	6	3	-	-	-
SS A6	226	91	50	32	21	15	10	7	4
SS A7	420	40	12	-	-	-	-	-	-
SS A8	226	86	46	28	19	12	8	6	3
SS B1	151	44	21	12	7	4	2	1	-
SS B2	183	77	43	28	19	13	9	6	4
SS B3	129	36	16	8	3	-	-	-	-
SS B4	176	73	41	26	17	12	8	6	4
SS B5	193	36	15	7	3	-	-	-	-
SS B6	169	72	41	26	18	12	9	6	4
SS B7	157	25	7	-	-	-	-	-	-
SS B8	160	66	37	23	16	11	7	5	3
SS C1	84	33	17	10	6	4	2	1	-
SS C2	110	54	33	22	15	11	8	5	3
SS C3	91	26	12	7	4	2	-	-	-
SS C4	109	48	28	18	12	9	6	4	3
SS D1	206	73	38	22	14	9	5	3	2
SS D2	203	103	63	42	29	21	15	10	6
SS D3	264	69	32	18	10	6	4	2	1
SS D4	208	100	60	39	27	19	14	9	6
SS E1	140	51	25	14	8	4	2	-	-
SS E2	161	80	48	32	22	15	11	7	5
SS E3	177	44	19	10	5	2	-	-	-
SS E4	166	75	43	28	19	13	9	6	4

ALPENA, MICHIGAN								8518 DD	
SSF =	.10	.20	.30	.40	.50	.60	.70	.80	.90
WW A1	172	-	-	-	-	-	-	-	-
WW A2	61	-	-	-	-	-	-	-	-
WW A3	51	-	-	-	-	-	-	-	-
WW A4	48	-	-	-	-	-	-	-	-
WW A5	46	14	-	-	-	-	-	-	-
WW A6	45	15	-	-	-	-	-	-	-
WW B1	-	-	-	-	-	-	-	-	-
WW B2	64	28	14	6	-	-	-	-	-
WW B3	80	38	21	12	7	-	-	-	-
WW B4	84	47	30	20	14	9	6	4	2
WW B5	83	48	31	22	15	11	8	5	3
WW C1	91	45	26	16	10	6	3	1	-
WW C2	84	45	27	18	12	8	5	3	2
WW C3	102	63	43	30	22	16	12	8	5
WW C4	94	58	39	28	20	15	11	8	5
TW A1	162	-	-	-	-	-	-	-	-
TW A2	68	-	-	-	-	-	-	-	-
TW A3	55	-	-	-	-	-	-	-	-
TW A4	49	14	-	-	-	-	-	-	-
TW B1	96	-	-	-	-	-	-	-	-
TW B2	58	11	-	-	-	-	-	-	-
TW B3	53	14	-	-	-	-	-	-	-
TW B4	52	16	-	-	-	-	-	-	-
TW C1	70	-	-	-	-	-	-	-	-
TW C2	57	14	-	-	-	-	-	-	-
TW C3	57	15	-	-	-	-	-	-	-
TW C4	65	16	5	-	-	-	-	-	-
TW D1	-	-	-	-	-	-	-	-	-
TW D2	71	27	13	5	-	-	-	-	-
TW D3	72	30	15	7	-	-	-	-	-

ALPENA, MICHIGAN								CONTINUED	
SSF =	.10	.20	.30	.40	.50	.60	.70	.80	.90
TW D4	85	43	26	17	11	7	5	3	2
TW D5	87	46	29	19	13	9	7	4	3
TW E1	93	41	22	13	7	4	-	-	-
TW E2	90	43	25	15	10	6	4	2	-
TW E3	108	60	38	26	18	13	9	6	4
TW E4	99	57	37	25	18	13	9	6	4
TW F1	61	-	-	-	-	-	-	-	-
TW F2	41	-	-	-	-	-	-	-	-
TW F3	36	-	-	-	-	-	-	-	-
TW F4	31	8	-	-	-	-	-	-	-
TW G1	37	-	-	-	-	-	-	-	-
TW G2	30	-	-	-	-	-	-	-	-
TW G3	28	-	-	-	-	-	-	-	-
TW G4	24	8	-	-	-	-	-	-	-
TW H1	25	-	-	-	-	-	-	-	-
TW H2	22	-	-	-	-	-	-	-	-
TW H3	21	6	-	-	-	-	-	-	-
TW H4	17	6	-	-	-	-	-	-	-
TW I1	-	-	-	-	-	-	-	-	-
TW I2	52	21	10	4	-	-	-	-	-
TW I3	59	26	14	7	3	-	-	-	-
TW I4	71	37	23	15	10	7	5	3	2
TW I5	73	41	26	17	12	9	6	4	2
TW J1	77	36	20	12	7	4	2	-	-
TW J2	74	38	23	14	9	6	4	2	1
TW J3	92	54	35	25	18	13	9	6	4
TW J4	86	51	34	23	17	12	9	6	4
DG A1	-	-	-	-	-	-	-	-	-
DG A2	52	13	-	-	-	-	-	-	-
DG A3	94	39	22	13	8	4	-	-	-
DG B1	-	-	-	-	-	-	-	-	-
DG B2	54	16	-	-	-	-	-	-	-
DG B3	99	42	24	15	10	6	4	-	-
DG C1	-	-	-	-	-	-	-	-	-
DG C2	74	27	12	-	-	-	-	-	-
DG C3	116	50	29	19	13	9	6	3	-
SS A1	164	33	-	-	-	-	-	-	-
SS A2	179	66	34	19	11	7	4	2	-
SS A3	139	-	-	-	-	-	-	-	-
SS A4	176	62	30	16	9	4	-	-	-
SS A5	271	26	-	-	-	-	-	-	-
SS A6	173	63	31	17	10	5	2	-	-
SS A7	278	-	-	-	-	-	-	-	-
SS A8	171	57	26	13	5	-	-	-	-
SS B1	100	18	-	-	-	-	-	-	-
SS B2	140	54	28	17	10	6	4	2	-
SS B3	79	-	-	-	-	-	-	-	-
SS B4	133	50	25	14	8	5	2	-	-
SS B5	126	-	-	-	-	-	-	-	-
SS B6	127	49	25	15	9	5	3	1	-
SS B7	95	-	-	-	-	-	-	-	-
SS B8	119	44	21	11	6	-	-	-	-
SS C1	32	-	-	-	-	-	-	-	-
SS C2	70	31	16	8	-	-	-	-	-
SS C3	42	-	-	-	-	-	-	-	-
SS C4	71	27	13	6	-	-	-	-	-
SS D1	143	38	-	-	-	-	-	-	-
SS D2	155	73	41	25	16	10	6	4	2
SS D3	188	36	-	-	-	-	-	-	-
SS D4	159	71	39	24	15	9	6	3	2
SS E1	82	-	-	-	-	-	-	-	-
SS E2	116	52	28	16	9	5	-	-	-
SS E3	114	-	-	-	-	-	-	-	-
SS E4	121	49	25	14	7	3	-	-	-

```
DETROIT, MICHIGAN                          6228 DD      DETROIT, MICHIGAN                           CONTINUED
   SSF =.10  .20  .30  .40  .50  .60  .70  .80  .90        SSF =.10  .20  .30  .40  .50  .60  .70  .80  .90
WW A1   254    -    -    -    -    -    -    -    -     DG C2    84   34   18   10    5    -    -    -    -
WW A2    81   22    -    -    -    -    -    -    -     DG C3   125   56   33   22   16   11    8    5    2
WW A3    69   25   11    -    -    -    -    -    -     SS A1   180   43   17    7    -    -    -    -    -
WW A4    64   27   13    6    -    -    -    -    -     SS A2   191   75   40   25   16   11    7    5    3
WW A5    62   28   15    8    -    -    -    -    -     SS A3   154   28    -    -    -    -    -    -    -
WW A6    60   29   16    9    4    -    -    -    -     SS A4   188   71   37   22   14    9    6    3    2
WW B1     -    -    -    -    -    -    -    -    -     SS A5   307   38    9    -    -    -    -    -    -
WW B2    77   37   21   13    8    5    3    2    1     SS A6   185   72   38   23   15   10    6    4    2
WW B3    94   48   29   19   13    9    6    4    2     SS A7   324   16    -    -    -    -    -    -    -
WW B4    95   55   36   25   18   13    9    6    4     SS A8   183   66   34   20   12    7    5    3    1
WW B5    93   55   37   26   19   14   10    7    5     SS B1   113   29   10    -    -    -    -    -    -
WW C1   105   55   34   22   15   10    7    5    3     SS B2   151   61   34   21   14    9    6    4    2
WW C2    96   53   34   23   16   12    8    6    3     SS B3    92   19    -    -    -    -    -    -    -
WW C3   113   71   49   35   26   19   14   10    7     SS B4   143   57   31   19   12    8    5    3    2
WW C4   103   65   45   32   24   18   13   10    6     SS B5   144   21    -    -    -    -    -    -    -
TW A1   229    -    -    -    -    -    -    -    -     SS B6   138   57   31   19   12    8    5    4    2
TW A2    87   21    6    -    -    -    -    -    -     SS B7   111    -    -    -    -    -    -    -    -
TW A3    71   24   10    -    -    -    -    -    -     SS B8   129   51   27   16   10    7    4    3    1
TW A4    62   25   12    6    -    -    -    -    -     SS C1    53   14    -    -    -    -    -    -    -
TW B1   126   16    -    -    -    -    -    -    -     SS C2    83   39   23   14    9    6    4    2    1
TW B2    73   22    9    -    -    -    -    -    -     SS C3    59   12    -    -    -    -    -    -    -
TW B3    65   23   11    5    -    -    -    -    -     SS C4    83   35   19   12    7    5    3    2    -
TW B4    63   23   11    6    3    -    -    -    -     SS D1   158   51   23   10    -    -    -    -    -
TW C1    87   19    7    -    -    -    -    -    -     SS D2   167   82   49   32   21   15   10    7    4
TW C2    70   22   10    5    2    -    -    -    -     SS D3   206   48   18    6    -    -    -    -    -
TW C3    69   21   10    5    3    -    -    -    -     SS D4   171   80   47   30   20   14    9    6    4
TW C4    78   20    9    5    3    1    -    -    -     SS E1   100   29    -    -    -    -    -    -    -
TW D1     -    -    -    -    -    -    -    -    -     SS E2   128   61   36   22   15   10    6    4    2
TW D2    84   35   19   11    7    4    3    1    -     SS E3   130   26    -    -    -    -    -    -    -
TW D3    85   39   22   13    9    5    3    2    1     SS E4   134   58   32   20   13    8    5    3    2
TW D4    97   51   31   21   15   10    7    5    3
TW D5    97   53   34   23   17   12    9    6    4
TW E1   107   50   29   18   12    8    6    4    2     FLINT, MICHIGAN                              7041 DD
TW E2   103   51   31   20   14   10    7    5    3        SSF =.10  .20  .30  .40  .50  .60  .70  .80  .90
TW E3   120   68   45   31   22   16   12    8    5     WW A1   207    -    -    -    -    -    -    -    -
TW E4   110   64   42   30   22   16   12    8    5     WW A2    62    -    -    -    -    -    -    -    -
TW F1    87    -    -    -    -    -    -    -    -     WW A3    52   11    -    -    -    -    -    -    -
TW F2    57   16    -    -    -    -    -    -    -     WW A4    49   15    -    -    -    -    -    -    -
TW F3    49   18    7    -    -    -    -    -    -     WW A5    47   17    -    -    -    -    -    -    -
TW F4    42   18    9    4    -    -    -    -    -     WW A6    46   18    -    -    -    -    -    -    -
TW G1    53    9    -    -    -    -    -    -    -     WW B1     -    -    -    -    -    -    -    -    -
TW G2    42   15    5    -    -    -    -    -    -     WW B2    64   29   15    8    4    -    -    -    -
TW G3    37   15    7    3    -    -    -    -    -     WW B3    81   39   22   14    8    5    3    1    -
TW G4    31   14    7    4    2    -    -    -    -     WW B4    84   48   30   21   15   10    7    5    3
TW H1    35   11    -    -    -    -    -    -    -     WW B5    83   49   32   22   16   12    8    6    4
TW H2    30   12    5    2    -    -    -    -    -     WW C1    92   46   27   17   11    7    5    3    1
TW H3    26   11    6    3    1    -    -    -    -     WW C2    85   45   28   19   13    9    6    4    2
TW H4    22    9    5    3    2    -    -    -    -     WW C3   103   64   43   31   23   17   12    9    6
TW I1     -    -    -    -    -    -    -    -    -     WW C4    94   58   40   28   21   15   11    8    5
TW I2    62   29   16   10    6    4    2    1    -     TW A1   190    -    -    -    -    -    -    -    -
TW I3    69   34   20   12    8    5    4    2    1     TW A2    69    -    -    -    -    -    -    -    -
TW I4    80   44   28   19   13   10    7    5    3     TW A3    56   13    -    -    -    -    -    -    -
TW I5    82   47   30   21   15   11    8    6    4     TW A4    49   16    -    -    -    -    -    -    -
TW J1    89   44   27   17   12    8    5    4    2     TW B1   103    -    -    -    -    -    -    -    -
TW J2    85   45   28   19   13    9    6    4    3     TW B2    59   13    -    -    -    -    -    -    -
TW J3   102   61   41   29   21   16   11    8    5     TW B3    53   15    -    -    -    -    -    -    -
TW J4    95   57   39   27   20   15   11    8    5     TW B4    52   16    6    -    -    -    -    -    -
DG A1     -    -    -    -    -    -    -    -    -     TW C1    73   11    -    -    -    -    -    -    -
DG A2    62   22    -    -    -    -    -    -    -     TW C2    58   15    -    -    -    -    -    -    -
DG A3   103   45   26   17   11    7    4    -    -     TW C3    59   16    6    -    -    -    -    -    -
DG B1     -    -    -    -    -    -    -    -    -     TW C4    68   16    6    -    -    -    -    -    -
DG B2    64   24   11    -    -    -    -    -    -     TW D1     -    -    -    -    -    -    -    -    -
DG B3   106   47   28   18   13    9    6    4    -     TW D2    72   28   14    7    3    -    -    -    -
DG C1    39    -    -    -    -    -    -    -    -     TW D3    73   31   16    9    4    -    -    -    -
```

516 / APPENDIX 21

FLINT, MICHIGAN (CONTINUED)

SSF =	.10	.20	.30	.40	.50	.60	.70	.80	.90
TW D4	86	44	26	17	12	8	6	4	2
TW D5	87	47	29	20	14	10	7	5	3
TW E1	93	42	23	14	9	5	3	2	-
TW E2	91	44	25	16	11	7	5	3	2
TW E3	108	61	39	27	19	14	10	7	4
TW E4	100	57	37	26	19	13	10	7	4
TW F1	65	-	-	-	-	-	-	-	-
TW F2	42	-	-	-	-	-	-	-	-
TW F3	36	-	-	-	-	-	-	-	-
TW F4	32	11	-	-	-	-	-	-	-
TW G1	38	-	-	-	-	-	-	-	-
TW G2	31	-	-	-	-	-	-	-	-
TW G3	28	8	-	-	-	-	-	-	-
TW G4	25	9	-	-	-	-	-	-	-
TW H1	25	-	-	-	-	-	-	-	-
TW H2	23	6	-	-	-	-	-	-	-
TW H3	21	7	-	-	-	-	-	-	-
TW H4	18	7	3	-	-	-	-	-	-
TW I1	-	-	-	-	-	-	-	-	-
TW I2	52	22	11	6	3	-	-	-	-
TW I3	59	27	15	9	5	3	1	-	-
TW I4	71	38	23	16	11	7	5	3	2
TW I5	74	41	26	18	13	9	6	4	3
TW J1	77	37	21	13	8	5	3	2	-
TW J2	75	38	23	15	10	7	5	3	2
TW J3	93	55	36	25	18	13	10	7	4
TW J4	87	51	34	24	17	13	9	7	4
DG A1	-	-	-	-	-	-	-	-	-
DG A2	52	14	-	-	-	-	-	-	-
DG A3	94	39	22	14	9	5	-	-	-
DG B1	-	-	-	-	-	-	-	-	-
DG B2	54	17	-	-	-	-	-	-	-
DG B3	99	42	24	16	11	7	5	2	-
DG C1	-	-	-	-	-	-	-	-	-
DG C2	73	28	13	4	-	-	-	-	-
DG C3	117	50	29	19	13	9	6	4	-
SS A1	163	33	-	-	-	-	-	-	-
SS A2	177	66	34	20	13	8	5	3	2
SS A3	137	-	-	-	-	-	-	-	-
SS A4	174	62	31	17	10	6	3	2	-
SS A5	286	27	-	-	-	-	-	-	-
SS A6	171	63	32	18	11	7	4	2	1
SS A7	302	-	-	-	-	-	-	-	-
SS A8	168	57	27	15	8	4	1	-	-
SS B1	99	20	-	-	-	-	-	-	-
SS B2	139	54	29	17	11	7	5	3	2
SS B3	78	-	-	-	-	-	-	-	-
SS B4	132	50	26	15	9	6	4	2	1
SS B5	129	-	-	-	-	-	-	-	-
SS B6	126	50	26	15	10	6	4	2	1
SS B7	95	16	-	-	-	-	-	-	-
SS B8	118	44	22	12	7	4	2	1	-
SS C1	34	-	-	-	-	-	-	-	-
SS C2	71	31	17	10	6	3	-	-	-
SS C3	43	-	-	-	-	-	-	-	-
SS C4	71	28	14	8	4	-	-	-	-
SS D1	140	39	11	-	-	-	-	-	-
SS D2	153	73	42	26	17	12	8	5	3
SS D3	187	37	-	-	-	-	-	-	-
SS D4	157	71	40	25	16	11	7	4	2
SS E1	81	-	-	-	-	-	-	-	-
SS E2	115	52	29	17	11	7	4	2	1
SS E3	114	-	-	-	-	-	-	-	-
SS E4	120	50	26	15	9	5	3	2	-

GRAND RAPIDS, MICHIGAN 6801 DD

SSF =	.10	.20	.30	.40	.50	.60	.70	.80	.90
WW A1	235	-	-	-	-	-	-	-	-
WW A2	67	-	-	-	-	-	-	-	-
WW A3	57	15	-	-	-	-	-	-	-
WW A4	53	18	-	-	-	-	-	-	-
WW A5	50	19	-	-	-	-	-	-	-
WW A6	49	20	-	-	-	-	-	-	-
WW B1	-	-	-	-	-	-	-	-	-
WW B2	67	30	16	9	4	-	-	-	-
WW B3	84	41	24	14	9	5	3	1	-
WW B4	87	49	31	21	15	11	7	5	3
WW B5	86	50	33	23	16	12	9	6	4
WW C1	95	48	28	18	12	8	5	3	1
WW C2	88	47	29	19	13	9	6	4	2
WW C3	105	65	44	32	23	17	12	9	6
WW C4	96	60	41	29	21	16	12	8	5
TW A1	212	-	-	-	-	-	-	-	-
TW A2	75	-	-	-	-	-	-	-	-
TW A3	60	16	-	-	-	-	-	-	-
TW A4	53	18	-	-	-	-	-	-	-
TW B1	113	-	-	-	-	-	-	-	-
TW B2	64	15	-	-	-	-	-	-	-
TW B3	57	17	-	-	-	-	-	-	-
TW B4	55	18	7	-	-	-	-	-	-
TW C1	78	13	-	-	-	-	-	-	-
TW C2	62	17	5	-	-	-	-	-	-
TW C3	62	17	6	-	-	-	-	-	-
TW C4	72	17	7	2	-	-	-	-	-
TW D1	-	-	-	-	-	-	-	-	-
TW D2	75	29	15	8	4	-	-	-	-
TW D3	76	32	17	9	5	-	-	-	-
TW D4	89	45	27	18	12	8	6	4	2
TW D5	90	48	30	20	14	10	7	5	3
TW E1	97	43	24	14	9	5	3	2	-
TW E2	94	45	27	17	11	7	5	3	2
TW E3	111	62	40	27	19	14	10	7	4
TW E4	103	59	38	26	19	14	10	7	4
TW F1	73	-	-	-	-	-	-	-	-
TW F2	46	-	-	-	-	-	-	-	-
TW F3	40	9	-	-	-	-	-	-	-
TW F4	34	12	-	-	-	-	-	-	-
TW G1	43	-	-	-	-	-	-	-	-
TW G2	34	-	-	-	-	-	-	-	-
TW G3	30	10	-	-	-	-	-	-	-
TW G4	26	10	-	-	-	-	-	-	-
TW H1	28	-	-	-	-	-	-	-	-
TW H2	25	7	-	-	-	-	-	-	-
TW H3	22	8	-	-	-	-	-	-	-
TW H4	19	7	3	-	-	-	-	-	-
TW I1	-	-	-	-	-	-	-	-	-
TW I2	55	24	12	6	3	-	-	-	-
TW I3	62	28	16	9	5	3	-	-	-
TW I4	73	39	24	16	11	8	5	3	2
TW I5	76	42	27	18	13	9	7	5	3
TW J1	80	39	22	14	9	6	3	2	-
TW J2	77	40	24	16	11	7	5	3	2
TW J3	95	56	37	26	19	14	10	7	4
TW J4	89	53	35	25	18	13	9	7	4
DG A1	-	-	-	-	-	-	-	-	-
DG A2	55	16	-	-	-	-	-	-	-
DG A3	97	41	23	14	9	5	-	-	-
DG B1	-	-	-	-	-	-	-	-	-
DG B2	57	19	-	-	-	-	-	-	-
DG B3	101	43	25	16	11	7	5	2	-
DG C1	-	-	-	-	-	-	-	-	-

GRAND RAPIDS, MICHIGAN   CONTINUED

SSF =	.10	.20	.30	.40	.50	.60	.70	.80	.90
DG C2	77	29	14	6	-	-	-	-	-
DG C3	120	52	30	20	14	9	6	4	-
SS A1	171	36	8	-	-	-	-	-	-
SS A2	182	68	35	21	13	8	5	3	2
SS A3	144	-	-	-	-	-	-	-	-
SS A4	179	64	32	18	10	6	3	2	-
SS A5	303	29	-	-	-	-	-	-	-
SS A6	176	65	33	19	11	7	4	2	1
SS A7	324	-	-	-	-	-	-	-	-
SS A8	174	59	29	15	8	4	1	-	-
SS B1	104	22	-	-	-	-	-	-	-
SS B2	143	56	30	18	11	7	5	3	2
SS B3	82	-	-	-	-	-	-	-	-
SS B4	136	52	27	16	10	6	4	2	1
SS B5	136	-	-	-	-	-	-	-	-
SS B6	130	51	27	16	10	6	4	2	1
SS B7	102	-	-	-	-	-	-	-	-
SS B8	121	46	23	13	8	4	2	1	-
SS C1	39	-	-	-	-	-	-	-	-
SS C2	74	33	18	10	6	3	-	-	-
SS C3	48	-	-	-	-	-	-	-	-
SS C4	75	30	15	8	4	-	-	-	-
SS D1	146	42	13	-	-	-	-	-	-
SS D2	157	75	43	27	18	12	8	5	3
SS D3	196	39	-	-	-	-	-	-	-
SS D4	161	73	41	25	16	11	7	4	2
SS E1	86	-	-	-	-	-	-	-	-
SS E2	118	54	30	18	11	7	4	2	1
SS E3	120	-	-	-	-	-	-	-	-
SS E4	124	51	27	16	9	5	3	1	-

SAULT STE. MARIE, MICHIGAN   9193 DD

SSF =	.10	.20	.30	.40	.50	.60	.70	.80	.90
WW A1	120	-	-	-	-	-	-	-	-
WW A2	44	-	-	-	-	-	-	-	-
WW A3	38	-	-	-	-	-	-	-	-
WW A4	36	-	-	-	-	-	-	-	-
WW A5	35	-	-	-	-	-	-	-	-
WW A6	34	-	-	-	-	-	-	-	-
WW B1	-	-	-	-	-	-	-	-	-
WW B2	54	22	9	-	-	-	-	-	-
WW B3	70	33	17	8	-	-	-	-	-
WW B4	76	42	26	17	12	8	5	3	2
WW B5	76	44	29	20	14	10	7	4	3
WW C1	81	39	22	13	7	-	-	-	-
WW C2	75	40	24	15	10	6	4	2	-
WW C3	95	58	39	28	20	14	10	7	4
WW C4	87	54	36	26	19	14	10	7	4
TW A1	119	-	-	-	-	-	-	-	-
TW A2	52	-	-	-	-	-	-	-	-
TW A3	43	-	-	-	-	-	-	-	-
TW A4	39	-	-	-	-	-	-	-	-
TW B1	73	-	-	-	-	-	-	-	-
TW B2	46	-	-	-	-	-	-	-	-
TW B3	43	-	-	-	-	-	-	-	-
TW B4	43	10	-	-	-	-	-	-	-
TW C1	56	-	-	-	-	-	-	-	-
TW C2	47	8	-	-	-	-	-	-	-
TW C3	48	11	-	-	-	-	-	-	-
TW C4	55	12	-	-	-	-	-	-	-
TW D1	-	-	-	-	-	-	-	-	-
TW D2	61	22	8	-	-	-	-	-	-
TW D3	62	24	10	-	-	-	-	-	-

SAULT STE. MARIE, MICHIGAN   CONTINUED

SSF =	.10	.20	.30	.40	.50	.60	.70	.80	.90
TW D4	77	39	23	14	9	6	4	2	1
TW D5	79	42	26	17	12	8	6	4	2
TW E1	82	35	18	9	-	-	-	-	-
TW E2	80	38	21	13	8	4	2	-	-
TW E3	99	55	35	23	16	11	8	5	3
TW E4	91	52	34	23	16	12	8	5	3
TW F1	37	-	-	-	-	-	-	-	-
TW F2	28	-	-	-	-	-	-	-	-
TW F3	26	-	-	-	-	-	-	-	-
TW F4	24	-	-	-	-	-	-	-	-
TW G1	19	-	-	-	-	-	-	-	-
TW G2	21	-	-	-	-	-	-	-	-
TW G3	21	-	-	-	-	-	-	-	-
TW G4	19	-	-	-	-	-	-	-	-
TW H1	14	-	-	-	-	-	-	-	-
TW H2	17	-	-	-	-	-	-	-	-
TW H3	16	-	-	-	-	-	-	-	-
TW H4	14	4	-	-	-	-	-	-	-
TW I1	-	-	-	-	-	-	-	-	-
TW I2	44	17	-	-	-	-	-	-	-
TW I3	51	22	10	-	-	-	-	-	-
TW I4	64	34	20	13	8	5	3	2	1
TW I5	67	37	23	16	11	7	5	3	2
TW J1	68	31	17	9	4	-	-	-	-
TW J2	67	34	20	12	8	5	3	1	-
TW J3	85	50	33	22	16	11	8	5	3
TW J4	80	47	31	21	15	11	8	5	3
DG A1	-	-	-	-	-	-	-	-	-
DG A2	43	-	-	-	-	-	-	-	-
DG A3	85	36	20	12	7	3	-	-	-
DG B1	-	-	-	-	-	-	-	-	-
DG B2	45	10	-	-	-	-	-	-	-
DG B3	89	38	22	14	9	5	3	-	-
DG C1	-	-	-	-	-	-	-	-	-
DG C2	65	23	8	-	-	-	-	-	-
DG C3	105	46	27	17	11	7	4	2	-
SS A1	138	23	-	-	-	-	-	-	-
SS A2	159	59	29	16	9	4	2	-	-
SS A3	112	-	-	-	-	-	-	-	-
SS A4	156	54	25	12	4	-	-	-	-
SS A5	224	-	-	-	-	-	-	-	-
SS A6	153	55	27	14	7	-	-	-	-
SS A7	220	-	-	-	-	-	-	-	-
SS A8	150	49	21	7	-	-	-	-	-
SS B1	82	-	-	-	-	-	-	-	-
SS B2	125	48	25	14	8	4	2	-	-
SS B3	60	-	-	-	-	-	-	-	-
SS B4	118	44	22	11	6	-	-	-	-
SS B5	100	-	-	-	-	-	-	-	-
SS B6	113	44	22	12	6	3	-	-	-
SS B7	69	-	-	-	-	-	-	-	-
SS B8	105	38	18	8	-	-	-	-	-
SS C1	-	-	-	-	-	-	-	-	-
SS C2	61	25	12	-	-	-	-	-	-
SS C3	27	-	-	-	-	-	-	-	-
SS C4	61	23	9	-	-	-	-	-	-
SS D1	120	26	-	-	-	-	-	-	-
SS D2	139	65	36	21	13	8	4	2	1
SS D3	158	24	-	-	-	-	-	-	-
SS D4	143	63	34	20	12	7	4	2	-
SS E1	63	-	-	-	-	-	-	-	-
SS E2	102	45	23	12	4	-	-	-	-
SS E3	91	-	-	-	-	-	-	-	-
SS E4	107	42	21	10	-	-	-	-	-

518 / APPENDIX 21

TRAVERSE CITY, MICHIGAN — 7698 DD

SSF =	.10	.20	.30	.40	.50	.60	.70	.80	.90
WW A1	183	-	-	-	-	-	-	-	-
WW A2	51	-	-	-	-	-	-	-	-
WW A3	42	-	-	-	-	-	-	-	-
WW A4	39	-	-	-	-	-	-	-	-
WW A5	37	-	-	-	-	-	-	-	-
WW A6	37	-	-	-	-	-	-	-	-
WW B1	-	-	-	-	-	-	-	-	-
WW B2	58	23	9	-	-	-	-	-	-
WW B3	74	34	17	8	-	-	-	-	-
WW B4	79	43	27	18	12	8	5	3	2
WW B5	79	45	29	20	14	10	7	5	3
WW C1	85	41	22	13	7	3	-	-	-
WW C2	79	41	24	15	10	6	4	2	1
WW C3	98	60	40	28	20	15	10	7	5
WW C4	90	55	37	26	19	14	10	7	4
TW A1	170	-	-	-	-	-	-	-	-
TW A2	61	-	-	-	-	-	-	-	-
TW A3	48	-	-	-	-	-	-	-	-
TW A4	42	-	-	-	-	-	-	-	-
TW B1	94	-	-	-	-	-	-	-	-
TW B2	53	-	-	-	-	-	-	-	-
TW B3	47	-	-	-	-	-	-	-	-
TW B4	47	11	-	-	-	-	-	-	-
TW C1	66	-	-	-	-	-	-	-	-
TW C2	53	10	-	-	-	-	-	-	-
TW C3	54	12	-	-	-	-	-	-	-
TW C4	64	13	-	-	-	-	-	-	-
TW D1	-	-	-	-	-	-	-	-	-
TW D2	66	23	8	-	-	-	-	-	-
TW D3	66	25	10	-	-	-	-	-	-
TW D4	81	40	23	15	9	6	4	2	1
TW D5	83	43	27	18	12	8	6	4	2
TW E1	87	36	19	10	4	-	-	-	-
TW E2	85	39	22	13	8	5	2	1	-
TW E3	103	56	35	24	16	11	8	5	3
TW E4	95	54	34	23	16	12	8	6	3
TW F1	53	-	-	-	-	-	-	-	-
TW F2	33	-	-	-	-	-	-	-	-
TW F3	28	-	-	-	-	-	-	-	-
TW F4	26	-	-	-	-	-	-	-	-
TW G1	28	-	-	-	-	-	-	-	-
TW G2	24	-	-	-	-	-	-	-	-
TW G3	23	-	-	-	-	-	-	-	-
TW G4	21	-	-	-	-	-	-	-	-
TW H1	18	-	-	-	-	-	-	-	-
TW H2	18	-	-	-	-	-	-	-	-
TW H3	18	-	-	-	-	-	-	-	-
TW H4	15	4	-	-	-	-	-	-	-
TW I1	-	-	-	-	-	-	-	-	-
TW I2	47	18	6	-	-	-	-	-	-
TW I3	54	23	11	-	-	-	-	-	-
TW I4	67	34	21	13	9	6	4	2	1
TW I5	70	38	24	16	11	7	5	3	2
TW J1	72	32	17	10	5	-	-	-	-
TW J2	70	35	20	12	8	5	3	1	-
TW J3	88	51	33	23	16	11	8	5	3
TW J4	83	48	31	22	15	11	8	5	3
DG A1	-	-	-	-	-	-	-	-	-
DG A2	48	-	-	-	-	-	-	-	-
DG A3	91	37	20	12	7	3	-	-	-
DG B1	-	-	-	-	-	-	-	-	-
DG B2	49	11	-	-	-	-	-	-	-
DG B3	96	40	22	14	9	5	3	-	-
DG C1	-	-	-	-	-	-	-	-	-

TRAVERSE CITY, MICHIGAN — CONTINUED

SSF =	.10	.20	.30	.40	.50	.60	.70	.80	.90
DG C2	69	24	8	-	-	-	-	-	-
DG C3	114	47	27	17	11	7	5	2	-
SS A1	159	27	-	-	-	-	-	-	-
SS A2	172	62	31	17	9	5	3	1	-
SS A3	131	-	-	-	-	-	-	-	-
SS A4	169	57	27	13	6	-	-	-	-
SS A5	278	-	-	-	-	-	-	-	-
SS A6	166	58	28	15	7	3	-	-	-
SS A7	292	-	-	-	-	-	-	-	-
SS A8	164	52	23	9	-	-	-	-	-
SS B1	94	-	-	-	-	-	-	-	-
SS B2	135	50	26	15	9	5	3	1	-
SS B3	71	-	-	-	-	-	-	-	-
SS B4	128	46	23	12	6	3	-	-	-
SS B5	123	-	-	-	-	-	-	-	-
SS B6	122	46	23	12	7	3	1	-	-
SS B7	90	-	-	-	-	-	-	-	-
SS B8	113	40	19	9	-	-	-	-	-
SS C1	-	-	-	-	-	-	-	-	-
SS C2	65	26	12	-	-	-	-	-	-
SS C3	34	-	-	-	-	-	-	-	-
SS C4	66	24	10	-	-	-	-	-	-
SS D1	133	30	-	-	-	-	-	-	-
SS D2	147	67	37	22	14	8	5	3	1
SS D3	182	28	-	-	-	-	-	-	-
SS D4	151	66	36	21	13	7	4	2	1
SS E1	71	-	-	-	-	-	-	-	-
SS E2	108	47	24	13	6	-	-	-	-
SS E3	107	-	-	-	-	-	-	-	-
SS E4	115	45	22	11	4	-	-	-	-

DULUTH, MINNESOTA — 9756 DD

SSF =	.10	.20	.30	.40	.50	.60	.70	.80	.90
WW A1	124	-	-	-	-	-	-	-	-
WW A2	47	-	-	-	-	-	-	-	-
WW A3	41	-	-	-	-	-	-	-	-
WW A4	39	-	-	-	-	-	-	-	-
WW A5	38	-	-	-	-	-	-	-	-
WW A6	38	10	-	-	-	-	-	-	-
WW B1	-	-	-	-	-	-	-	-	-
WW B2	56	25	12	5	-	-	-	-	-
WW B3	71	35	19	11	6	-	-	-	-
WW B4	77	44	28	19	13	9	6	4	2
WW B5	77	45	30	21	15	11	8	5	3
WW C1	82	41	24	15	9	6	3	1	-
WW C2	77	41	26	17	11	8	5	3	2
WW C3	95	59	40	29	21	15	11	8	5
WW C4	88	55	37	27	20	14	11	7	5
TW A1	123	-	-	-	-	-	-	-	-
TW A2	53	-	-	-	-	-	-	-	-
TW A3	45	-	-	-	-	-	-	-	-
TW A4	41	11	-	-	-	-	-	-	-
TW B1	74	-	-	-	-	-	-	-	-
TW B2	48	-	-	-	-	-	-	-	-
TW B3	44	11	-	-	-	-	-	-	-
TW B4	44	13	-	-	-	-	-	-	-
TW C1	57	-	-	-	-	-	-	-	-
TW C2	48	12	-	-	-	-	-	-	-
TW C3	49	13	-	-	-	-	-	-	-
TW C4	56	13	4	-	-	-	-	-	-
TW D1	-	-	-	-	-	-	-	-	-
TW D2	62	24	11	4	-	-	-	-	-
TW D3	63	26	13	5	-	-	-	-	-

DULUTH, MINNESOTA — CONTINUED

	SSF =.10	.20	.30	.40	.50	.60	.70	.80	.90
TW D4	78	40	24	16	10	7	5	3	2
TW D5	80	43	27	18	13	9	6	4	3
TW E1	82	37	20	12	7	3	-	-	-
TW E2	81	39	23	14	9	6	4	2	1
TW E3	99	56	36	25	17	12	9	6	4
TW E4	92	53	35	24	17	12	9	6	4
TW F1	40	-	-	-	-	-	-	-	-
TW F2	31	-	-	-	-	-	-	-	-
TW F3	28	-	-	-	-	-	-	-	-
TW F4	26	-	-	-	-	-	-	-	-
TW G1	25	-	-	-	-	-	-	-	-
TW G2	24	-	-	-	-	-	-	-	-
TW G3	23	-	-	-	-	-	-	-	-
TW G4	20	6	-	-	-	-	-	-	-
TW H1	18	-	-	-	-	-	-	-	-
TW H2	18	-	-	-	-	-	-	-	-
TW H3	17	5	-	-	-	-	-	-	-
TW H4	15	5	-	-	-	-	-	-	-
TW I1	-	-	-	-	-	-	-	-	-
TW I2	45	19	9	-	-	-	-	-	-
TW I3	52	24	13	7	-	-	-	-	-
TW I4	64	35	21	14	10	7	4	3	2
TW I5	68	38	24	17	12	8	6	4	2
TW J1	68	33	19	11	7	4	2	-	-
TW J2	67	35	21	14	9	6	4	2	1
TW J3	85	51	34	24	17	12	9	6	4
TW J4	80	48	32	22	16	12	9	6	4
DG A1	-	-	-	-	-	-	-	-	-
DG A2	44	-	-	-	-	-	-	-	-
DG A3	84	36	20	12	8	4	-	-	-
DG B1	-	-	-	-	-	-	-	-	-
DG B2	45	14	-	-	-	-	-	-	-
DG B3	88	39	23	14	10	6	4	-	-
DG C1	-	-	-	-	-	-	-	-	-
DG C2	64	24	11	-	-	-	-	-	-
DG C3	104	46	27	18	12	8	5	3	-
SS A1	130	24	-	-	-	-	-	-	-
SS A2	153	57	30	17	10	6	3	2	-
SS A3	100	-	-	-	-	-	-	-	-
SS A4	148	52	25	13	7	2	-	-	-
SS A5	212	-	-	-	-	-	-	-	-
SS A6	146	54	27	15	8	4	2	-	-
SS A7	202	-	-	-	-	-	-	-	-
SS A8	142	48	22	10	-	-	-	-	-
SS B1	77	-	-	-	-	-	-	-	-
SS B2	121	48	25	15	9	5	3	2	-
SS B3	54	-	-	-	-	-	-	-	-
SS B4	113	43	22	12	7	4	2	-	-
SS B5	93	-	-	-	-	-	-	-	-
SS B6	109	43	22	13	7	4	2	-	-
SS B7	60	-	-	-	-	-	-	-	-
SS B8	100	37	18	9	4	-	-	-	-
SS C1	-	-	-	-	-	-	-	-	-
SS C2	62	27	14	7	-	-	-	-	-
SS C3	30	-	-	-	-	-	-	-	-
SS C4	62	24	12	6	-	-	-	-	-
SS D1	112	28	-	-	-	-	-	-	-
SS D2	134	64	36	22	14	9	6	3	2
SS D3	148	25	-	-	-	-	-	-	-
SS D4	137	62	35	21	13	8	5	3	1
SS E1	59	-	-	-	-	-	-	-	-
SS E2	99	45	24	14	7	3	-	-	-
SS E3	84	-	-	-	-	-	-	-	-
SS E4	103	42	22	12	6	-	-	-	-

INTERNATIONAL FALLS, MINNESOTA   10547 DD

	SSF =.10	.20	.30	.40	.50	.60	.70	.80	.90
WW A1	111	-	-	-	-	-	-	-	-
WW A2	36	-	-	-	-	-	-	-	-
WW A3	32	-	-	-	-	-	-	-	-
WW A4	31	-	-	-	-	-	-	-	-
WW A5	30	-	-	-	-	-	-	-	-
WW A6	30	-	-	-	-	-	-	-	-
WW B1	-	-	-	-	-	-	-	-	-
WW B2	51	20	-	-	-	-	-	-	-
WW B3	65	30	15	5	-	-	-	-	-
WW B4	73	41	25	17	11	7	5	3	2
WW B5	73	43	28	19	13	9	6	4	3
WW C1	76	37	20	11	6	-	-	-	-
WW C2	72	38	23	14	9	6	3	2	-
WW C3	91	56	38	27	19	14	10	7	4
WW C4	84	52	35	25	18	13	9	7	4
TW A1	113	-	-	-	-	-	-	-	-
TW A2	45	-	-	-	-	-	-	-	-
TW A3	38	-	-	-	-	-	-	-	-
TW A4	35	-	-	-	-	-	-	-	-
TW B1	67	-	-	-	-	-	-	-	-
TW B2	42	-	-	-	-	-	-	-	-
TW B3	39	-	-	-	-	-	-	-	-
TW B4	39	-	-	-	-	-	-	-	-
TW C1	52	-	-	-	-	-	-	-	-
TW C2	44	-	-	-	-	-	-	-	-
TW C3	45	9	-	-	-	-	-	-	-
TW C4	53	11	-	-	-	-	-	-	-
TW D1	-	-	-	-	-	-	-	-	-
TW D2	57	20	-	-	-	-	-	-	-
TW D3	57	22	-	-	-	-	-	-	-
TW D4	74	37	22	14	9	5	3	2	1
TW D5	77	41	25	17	11	8	5	4	2
TW E1	77	33	16	8	-	-	-	-	-
TW E2	77	36	20	12	7	4	2	-	-
TW E3	95	53	33	22	16	11	7	5	3
TW E4	89	51	33	22	16	11	8	5	3
TW F1	-	-	-	-	-	-	-	-	-
TW F2	19	-	-	-	-	-	-	-	-
TW F3	21	-	-	-	-	-	-	-	-
TW F4	21	-	-	-	-	-	-	-	-
TW G1	-	-	-	-	-	-	-	-	-
TW G2	16	-	-	-	-	-	-	-	-
TW G3	18	-	-	-	-	-	-	-	-
TW G4	17	-	-	-	-	-	-	-	-
TW H1	-	-	-	-	-	-	-	-	-
TW H2	14	-	-	-	-	-	-	-	-
TW H3	14	-	-	-	-	-	-	-	-
TW H4	13	-	-	-	-	-	-	-	-
TW I1	-	-	-	-	-	-	-	-	-
TW I2	41	16	-	-	-	-	-	-	-
TW I3	47	20	9	-	-	-	-	-	-
TW I4	61	32	19	12	8	5	3	2	1
TW I5	65	36	23	15	10	7	5	3	2
TW J1	64	29	16	8	3	-	-	-	-
TW J2	64	32	19	12	7	4	2	1	-
TW J3	82	48	31	22	15	11	8	5	3
TW J4	77	46	30	21	15	11	8	5	3
DG A1	-	-	-	-	-	-	-	-	-
DG A2	40	-	-	-	-	-	-	-	-
DG A3	82	34	19	11	6	-	-	-	-
DG B1	-	-	-	-	-	-	-	-	-
DG B2	42	-	-	-	-	-	-	-	-
DG B3	86	37	21	13	8	5	2	-	-
DG C1	-	-	-	-	-	-	-	-	-

INTERNATIONAL FALLS, MINNESOTA — CONTINUED

SSF =	.10	.20	.30	.40	.50	.60	.70	.80	.90
DG C2	60	22	-	-	-	-	-	-	-
DG C3	102	44	26	16	11	7	4	2	-
SS A1	115	-	-	-	-	-	-	-	-
SS A2	141	51	25	13	7	3	-	-	-
SS A3	81	-	-	-	-	-	-	-	-
SS A4	135	45	20	8	-	-	-	-	-
SS A5	193	-	-	-	-	-	-	-	-
SS A6	135	47	22	11	4	-	-	-	-
SS A7	177	-	-	-	-	-	-	-	-
SS A8	128	40	15	-	-	-	-	-	-
SS B1	65	-	-	-	-	-	-	-	-
SS B2	112	43	22	12	7	4	2	-	-
SS B3	36	-	-	-	-	-	-	-	-
SS B4	104	38	18	9	4	-	-	-	-
SS B5	80	-	-	-	-	-	-	-	-
SS B6	101	38	19	10	5	-	-	-	-
SS B7	-	-	-	-	-	-	-	-	-
SS B8	91	32	14	-	-	-	-	-	-
SS C1	-	-	-	-	-	-	-	-	-
SS C2	57	23	10	-	-	-	-	-	-
SS C3	-	-	-	-	-	-	-	-	-
SS C4	57	21	8	-	-	-	-	-	-
SS D1	97	-	-	-	-	-	-	-	-
SS D2	123	57	31	18	11	6	4	2	-
SS D3	131	-	-	-	-	-	-	-	-
SS D4	126	55	30	17	10	5	3	1	-
SS E1	37	-	-	-	-	-	-	-	-
SS E2	89	38	19	8	-	-	-	-	-
SS E3	69	-	-	-	-	-	-	-	-
SS E4	94	36	17	7	-	-	-	-	-

MINNEAPOLIS-ST. PAUL, MINNESOTA   8159 DD

SSF =	.10	.20	.30	.40	.50	.60	.70	.80	.90
WW A1	225	-	-	-	-	-	-	-	-
WW A2	74	19	-	-	-	-	-	-	-
WW A3	64	22	8	-	-	-	-	-	-
WW A4	59	24	11	-	-	-	-	-	-
WW A5	57	25	12	5	-	-	-	-	-
WW A6	56	26	14	7	-	-	-	-	-
WW B1	-	-	-	-	-	-	-	-	-
WW B2	73	34	20	12	8	5	3	1	-
WW B3	89	45	27	17	11	8	5	3	2
WW B4	91	53	34	24	17	12	9	6	4
WW B5	90	53	36	25	18	13	10	7	4
WW C1	100	52	31	21	14	10	6	4	2
WW C2	92	51	32	22	15	11	8	5	3
WW C3	109	68	47	34	25	19	14	10	6
WW C4	100	63	43	31	23	17	13	9	6
TW A1	205	-	-	-	-	-	-	-	-
TW A2	80	18	-	-	-	-	-	-	-
TW A3	66	21	8	-	-	-	-	-	-
TW A4	58	22	10	4	-	-	-	-	-
TW B1	114	12	-	-	-	-	-	-	-
TW B2	68	20	7	-	-	-	-	-	-
TW B3	61	21	9	3	-	-	-	-	-
TW B4	59	21	10	5	-	-	-	-	-
TW C1	81	17	5	-	-	-	-	-	-
TW C2	65	20	9	4	-	-	-	-	-
TW C3	65	20	9	4	-	-	-	-	-
TW C4	73	19	9	4	2	-	-	-	-
TW D1	-	-	-	-	-	-	-	-	-
TW D2	80	33	18	10	6	4	2	-	-
TW D3	80	36	20	12	7	5	3	1	-

MINNEAPOLIS-ST. PAUL, MINNESOTA — CONTINUED

SSF =	.10	.20	.30	.40	.50	.60	.70	.80	.90
TW D4	93	48	30	20	14	10	7	5	3
TW D5	93	51	32	22	16	11	8	6	4
TW E1	101	47	27	17	11	7	5	3	2
TW E2	98	49	29	19	13	9	6	4	2
TW E3	115	66	43	30	21	15	11	8	5
TW E4	106	62	41	28	21	15	11	8	5
TW F1	77	-	-	-	-	-	-	-	-
TW F2	52	13	-	-	-	-	-	-	-
TW F3	45	15	-	-	-	-	-	-	-
TW F4	39	16	7	-	-	-	-	-	-
TW G1	48	-	-	-	-	-	-	-	-
TW G2	38	12	-	-	-	-	-	-	-
TW G3	34	13	6	-	-	-	-	-	-
TW G4	29	12	6	3	-	-	-	-	-
TW H1	32	8	-	-	-	-	-	-	-
TW H2	28	10	4	-	-	-	-	-	-
TW H3	25	10	5	2	-	-	-	-	-
TW H4	20	8	4	2	1	-	-	-	-
TW I1	-	-	-	-	-	-	-	-	-
TW I2	59	27	15	9	6	3	2	-	-
TW I3	65	31	18	11	7	5	3	2	-
TW I4	77	42	26	18	13	9	6	4	3
TW I5	79	45	29	20	14	10	8	5	3
TW J1	84	42	25	16	11	7	5	3	2
TW J2	81	43	27	18	12	9	6	4	2
TW J3	98	59	39	28	20	15	11	8	5
TW J4	92	56	37	26	19	14	11	8	5
DG A1	-	-	-	-	-	-	-	-	-
DG A2	59	20	-	-	-	-	-	-	-
DG A3	99	43	25	16	10	7	4	-	-
DG B1	-	-	-	-	-	-	-	-	-
DG B2	61	22	9	-	-	-	-	-	-
DG B3	103	45	27	18	12	9	6	3	-
DG C1	31	-	-	-	-	-	-	-	-
DG C2	80	32	17	9	-	-	-	-	-
DG C3	122	54	32	21	15	11	8	5	2
SS A1	158	35	12	-	-	-	-	-	-
SS A2	174	67	36	22	14	9	6	4	2
SS A3	129	14	-	-	-	-	-	-	-
SS A4	169	62	32	18	11	7	4	2	1
SS A5	269	29	-	-	-	-	-	-	-
SS A6	168	64	34	20	12	8	5	3	2
SS A7	273	-	-	-	-	-	-	-	-
SS A8	163	58	29	16	9	5	3	1	-
SS B1	98	23	-	-	-	-	-	-	-
SS B2	138	56	30	19	12	8	5	3	2
SS B3	76	-	-	-	-	-	-	-	-
SS B4	130	51	27	16	10	7	4	3	1
SS B5	124	12	-	-	-	-	-	-	-
SS B6	126	51	28	17	11	7	5	3	2
SS B7	89	-	-	-	-	-	-	-	-
SS B8	116	45	24	14	8	5	3	2	-
SS C1	47	-	-	-	-	-	-	-	-
SS C2	79	37	21	13	8	5	3	2	-
SS C3	53	-	-	-	-	-	-	-	-
SS C4	79	33	18	11	7	4	2	1	-
SS D1	139	42	16	-	-	-	-	-	-
SS D2	153	74	44	28	19	13	9	6	3
SS D3	181	39	12	-	-	-	-	-	-
SS D4	156	72	42	26	17	12	8	5	3
SS E1	84	19	-	-	-	-	-	-	-
SS E2	116	55	31	19	12	8	5	3	2
SS E3	112	17	-	-	-	-	-	-	-
SS E4	121	51	28	17	11	7	4	2	1

ROCHESTER, MINNESOTA								8227 DD		ROCHESTER, MINNESOTA								CONTINUED	
SSF =	.10	.20	.30	.40	.50	.60	.70	.80	.90	SSF =	.10	.20	.30	.40	.50	.60	.70	.80	.90
WW A1	220	-	-	-	-	-	-	-	-	DG C2	78	31	16	9	-	-	-	-	-
WW A2	71	17	-	-	-	-	-	-	-	DG C3	120	53	31	21	15	11	8	5	2
WW A3	61	21	7	-	-	-	-	-	-	SS A1	162	35	12	-	-	-	-	-	-
WW A4	57	23	10	-	-	-	-	-	-	SS A2	176	67	36	22	14	9	6	4	2
WW A5	55	24	12	5	-	-	-	-	-	SS A3	133	16	-	-	-	-	-	-	-
WW A6	54	25	13	7	-	-	-	-	-	SS A4	171	63	32	19	12	7	5	3	1
WW B1	-	-	-	-	-	-	-	-	-	SS A5	279	29	-	-	-	-	-	-	-
WW B2	71	33	19	12	7	5	3	1	-	SS A6	170	64	34	20	13	8	5	3	2
WW B3	87	44	26	17	11	8	5	3	2	SS A7	289	-	-	-	-	-	-	-	-
WW B4	90	52	33	23	17	12	9	6	4	SS A8	166	58	29	16	10	6	3	2	-
WW B5	88	52	35	25	18	13	10	7	4	SS B1	100	22	-	-	-	-	-	-	-
WW C1	98	51	31	20	14	9	6	4	2	SS B2	139	56	30	19	12	8	5	4	2
WW C2	90	50	31	21	15	11	8	5	3	SS B3	78	-	-	-	-	-	-	-	-
WW C3	108	67	46	33	24	18	14	10	7	SS B4	131	51	27	16	10	7	4	3	2
WW C4	99	62	42	31	23	17	13	9	6	SS B5	127	12	-	-	-	-	-	-	-
TW A1	201	-	-	-	-	-	-	-	-	SS B6	127	51	28	17	11	7	5	3	2
TW A2	77	17	-	-	-	-	-	-	-	SS B7	93	-	-	-	-	-	-	-	-
TW A3	63	20	7	-	-	-	-	-	-	SS B8	118	45	24	14	9	5	3	2	1
TW A4	56	21	10	4	-	-	-	-	-	SS C1	44	-	-	-	-	-	-	-	-
TW B1	112	10	-	-	-	-	-	-	-	SS C2	77	36	20	13	8	5	3	2	-
TW B2	66	19	7	-	-	-	-	-	-	SS C3	51	-	-	-	-	-	-	-	-
TW B3	59	20	9	3	-	-	-	-	-	SS C4	77	32	17	10	6	4	2	1	-
TW B4	57	20	9	5	-	-	-	-	-	SS D1	140	42	17	-	-	-	-	-	-
TW C1	79	16	-	-	-	-	-	-	-	SS D2	154	74	44	28	19	13	9	6	4
TW C2	63	19	8	4	-	-	-	-	-	SS D3	185	39	12	-	-	-	-	-	-
TW C3	63	19	8	4	2	-	-	-	-	SS D4	157	73	42	26	18	12	8	5	3
TW C4	72	18	8	4	2	-	-	-	-	SS E1	84	19	-	-	-	-	-	-	-
TW D1	-	-	-	-	-	-	-	-	-	SS E2	116	55	31	19	13	8	5	3	2
TW D2	78	32	17	10	6	4	2	1	-	SS E3	114	17	-	-	-	-	-	-	-
TW D3	78	35	19	12	7	5	3	1	-	SS E4	121	51	28	17	11	7	4	3	1
TW D4	91	47	29	19	14	10	7	5	3										
TW D5	92	50	32	22	16	11	8	6	4										
TW E1	99	46	26	17	11	7	5	3	2	JACKSON, MISSISSIPPI								2300 DD	
TW E2	96	48	29	19	13	9	6	4	2	SSF =	.10	.20	.30	.40	.50	.60	.70	.80	.90
TW E3	113	64	42	29	21	15	11	8	5	WW A1	1202	144	67	39	26	18	12	9	6
TW E4	105	61	40	28	20	15	11	8	5	WW A2	394	156	89	57	40	29	21	15	10
TW F1	74	-	-	-	-	-	-	-	-	WW A3	330	157	95	64	45	33	24	17	11
TW F2	50	12	-	-	-	-	-	-	-	WW A4	302	156	98	67	48	35	26	19	12
TW F3	43	14	-	-	-	-	-	-	-	WW A5	288	158	101	70	51	38	28	20	14
TW F4	37	15	7	-	-	-	-	-	-	WW A6	279	158	103	72	53	39	29	21	14
TW G1	46	-	-	-	-	-	-	-	-	WW B1	332	121	66	42	29	20	14	10	6
TW G2	37	11	-	-	-	-	-	-	-	WW B2	294	157	100	69	50	37	27	20	13
TW G3	33	13	5	-	-	-	-	-	-	WW B3	341	191	124	87	63	47	35	26	17
TW G4	28	12	6	3	-	-	-	-	-	WW B4	296	179	121	87	65	49	37	27	18
TW H1	31	7	-	-	-	-	-	-	-	WW B5	271	167	115	83	62	47	36	26	18
TW H2	26	10	4	-	-	-	-	-	-	WW C1	361	202	131	92	67	50	37	27	18
TW H3	24	10	5	2	-	-	-	-	-	WW C2	310	181	120	85	63	47	35	26	18
TW H4	20	8	4	2	1	-	-	-	-	WW C3	313	201	141	103	78	60	45	34	23
TW I1	-	-	-	-	-	-	-	-	-	WW C4	281	181	127	93	70	54	41	30	21
TW I2	57	26	14	9	5	3	2	-	-	TW A1	1026	137	64	38	25	17	12	8	5
TW I3	64	30	18	11	7	5	3	2	-	TW A2	404	142	78	50	34	24	17	12	8
TW I4	75	41	26	18	12	9	6	4	3	TW A3	324	140	82	54	38	27	20	14	9
TW I5	78	44	29	20	14	10	8	5	3	TW A4	274	132	80	54	38	28	20	15	10
TW J1	82	41	24	16	10	7	5	3	2	TW B1	570	128	64	39	26	18	13	9	6
TW J2	80	42	26	17	12	8	6	4	2	TW B2	323	127	72	47	33	23	17	12	8
TW J3	97	58	39	27	20	15	11	8	5	TW B3	281	122	72	47	33	24	18	13	8
TW J4	91	55	37	26	19	14	10	7	5	TW B4	258	112	66	44	31	22	16	12	8
DG A1	-	-	-	-	-	-	-	-	-	TW C1	373	115	61	39	26	19	13	9	6
DG A2	57	19	-	-	-	-	-	-	-	TW C2	283	109	62	40	28	20	14	10	7
DG A3	98	42	24	15	10	7	4	-	-	TW C3	270	102	57	37	26	18	13	9	6
DG B1	-	-	-	-	-	-	-	-	-	TW C4	284	91	49	31	21	15	11	8	5
DG B2	59	21	8	-	-	-	-	-	-	TW D1	240	94	53	34	23	16	12	8	5
DG B3	102	44	26	17	12	8	6	3	-	TW D2	315	149	91	61	43	32	23	17	11
DG C1	24	-	-	-	-	-	-	-	-	TW D3	321	164	102	70	50	37	27	20	13

JACKSON, MISSISSIPPI — CONTINUED

SSF =	.10	.20	.30	.40	.50	.60	.70	.80	.90
TW D4	304	168	109	76	56	41	31	23	15
TW D5	281	160	105	75	55	41	31	23	15
TW E1	370	189	118	81	58	43	32	23	15
TW E2	334	178	114	79	57	42	31	23	15
TW E3	347	206	138	99	73	55	41	30	21
TW E4	308	185	125	90	67	50	38	28	19
TW F1	464	118	60	37	25	17	12	9	6
TW F2	289	119	69	45	31	22	16	12	8
TW F3	238	114	69	46	33	24	18	13	8
TW F4	195	101	64	44	31	23	17	12	8
TW G1	288	102	56	36	24	17	12	9	6
TW G2	209	98	59	39	28	20	15	11	7
TW G3	175	88	55	37	27	19	14	10	7
TW G4	140	73	45	31	22	16	12	9	6
TW H1	186	81	48	32	22	16	12	8	5
TW H2	142	71	43	29	21	15	11	8	5
TW H3	118	60	37	25	18	13	10	7	5
TW H4	91	45	28	19	14	10	7	5	4
TW I1	186	79	45	29	20	14	10	7	5
TW I2	241	125	78	54	39	28	21	15	10
TW I3	255	137	87	60	44	32	24	17	12
TW I4	252	145	96	68	50	37	28	21	14
TW I5	243	144	96	69	51	38	29	21	14
TW J1	306	166	106	74	53	40	29	21	14
TW J2	275	155	101	71	52	39	29	21	14
TW J3	288	178	122	89	66	50	38	28	19
TW J4	264	164	113	82	61	46	35	26	18
DG A1	238	105	60	37	23	13	-	-	-
DG A2	254	116	70	47	33	23	15	10	4
DG A3	305	141	88	61	45	33	25	17	10
DG B1	241	107	64	43	30	21	13	4	-
DG B2	261	120	73	51	37	28	20	14	8
DG B3	308	145	90	63	47	36	28	21	13
DG C1	289	131	78	54	39	29	21	13	-
DG C2	304	142	87	61	45	34	26	19	11
DG C3	353	168	104	73	55	43	34	26	17
SS A1	665	208	109	68	46	32	23	16	10
SS A2	589	254	148	97	68	49	35	25	17
SS A3	647	191	99	61	41	28	20	14	9
SS A4	603	255	147	96	66	48	35	25	16
SS A5	1116	209	100	60	39	27	19	13	8
SS A6	587	253	147	96	67	48	35	25	16
SS A7	1279	180	83	48	31	21	15	10	6
SS A8	607	251	143	93	64	46	33	24	15
SS B1	458	158	86	54	37	26	19	13	8
SS B2	461	204	120	80	56	40	29	21	14
SS B3	427	147	79	50	34	24	17	12	8
SS B4	454	201	118	78	54	39	29	20	13
SS B5	596	143	71	43	29	20	14	10	6
SS B6	436	197	117	77	54	39	29	21	14
SS B7	532	123	60	36	24	17	12	8	5
SS B8	428	190	112	74	52	37	27	19	13
SS C1	299	136	81	54	38	27	20	14	9
SS C2	308	161	101	70	50	37	27	20	13
SS C3	310	110	60	39	26	19	14	10	6
SS C4	305	145	88	59	42	31	22	16	11
SS D1	614	248	140	90	62	44	32	23	15
SS D2	522	275	173	119	85	63	46	33	22
SS D3	769	235	123	76	52	36	26	18	12
SS D4	534	270	167	113	81	59	43	31	21
SS E1	465	196	112	73	51	36	26	18	12
SS E2	435	228	143	98	70	52	38	27	18
SS E3	557	169	88	55	37	26	18	13	8
SS E4	447	215	130	87	62	45	33	24	16

MERIDIAN, MISSISSIPPI 2388 DD

SSF =	.10	.20	.30	.40	.50	.60	.70	.80	.90
WW A1	1121	135	63	37	24	17	12	8	5
WW A2	371	148	84	55	38	27	20	14	9
WW A3	312	149	90	61	43	31	23	16	11
WW A4	286	148	93	64	46	34	25	18	12
WW A5	272	150	97	67	49	36	27	19	13
WW A6	264	150	98	69	50	37	28	20	14
WW B1	310	114	63	40	27	19	14	9	6
WW B2	279	150	95	66	48	35	26	19	13
WW B3	324	182	119	83	61	45	34	24	16
WW B4	282	171	116	84	62	47	35	26	18
WW B5	258	160	110	80	60	45	34	25	17
WW C1	343	193	126	88	64	48	36	26	18
WW C2	295	173	115	82	61	45	34	25	17
WW C3	299	193	135	100	75	57	44	33	22
WW C4	269	174	122	90	68	52	39	29	20
TW A1	958	129	61	36	24	16	12	8	5
TW A2	380	134	74	47	32	23	17	12	8
TW A3	306	132	78	51	36	26	19	13	9
TW A4	259	125	76	51	36	27	19	14	9
TW B1	534	121	61	37	25	17	12	9	6
TW B2	305	121	69	45	31	22	16	11	8
TW B3	266	116	68	45	32	23	17	12	8
TW B4	244	107	63	42	29	21	15	11	7
TW C1	351	109	58	37	25	18	13	9	6
TW C2	267	103	59	38	26	19	14	10	6
TW C3	255	97	55	35	24	18	13	9	6
TW C4	268	86	47	30	20	15	10	7	5
TW D1	224	88	50	32	22	15	11	8	5
TW D2	298	142	87	58	41	30	22	16	11
TW D3	305	156	98	67	48	35	26	19	12
TW D4	289	161	105	73	54	40	30	22	15
TW D5	268	153	101	72	53	39	30	22	15
TW E1	352	180	113	77	56	41	30	22	15
TW E2	318	170	109	76	55	41	30	22	15
TW E3	331	197	133	95	70	53	40	29	20
TW E4	295	178	120	87	64	49	37	27	18
TW F1	435	111	57	35	24	16	12	8	5
TW F2	272	113	65	43	30	21	15	11	7
TW F3	225	108	66	44	31	23	17	12	8
TW F4	184	96	61	42	30	22	16	12	8
TW G1	270	96	53	34	23	16	12	8	5
TW G2	197	92	56	37	26	19	14	10	7
TW G3	165	84	52	35	25	19	14	10	7
TW G4	133	69	43	30	21	16	12	8	6
TW H1	175	77	45	30	21	15	11	8	5
TW H2	134	67	41	28	20	15	11	8	5
TW H3	112	57	35	24	17	13	9	7	4
TW H4	86	43	27	18	13	10	7	5	3
TW I1	174	74	43	28	19	14	10	7	4
TW I2	228	119	75	51	37	27	20	15	10
TW I3	242	130	83	58	42	31	23	17	11
TW I4	240	139	92	65	48	36	27	20	13
TW I5	232	138	93	66	49	37	28	20	14
TW J1	291	158	102	71	51	38	28	20	14
TW J2	261	148	97	68	50	37	28	20	14
TW J3	276	171	118	85	64	48	37	27	18
TW J4	253	157	108	79	59	45	34	25	17
DG A1	223	99	56	34	21	11	-	-	-
DG A2	240	110	67	45	31	21	14	9	4
DG A3	288	135	84	59	43	32	24	16	9
DG B1	224	101	60	40	28	19	11	-	-
DG B2	245	114	70	49	35	26	19	13	7
DG B3	291	139	86	61	45	35	27	20	13
DG C1	269	124	74	51	37	27	19	12	-

```
MERIDIAN, MISSISSIPPI                      CONTINUED    COLUMBIA, MISSOURI                          CONTINUED
    SSF =.10   .20   .30   .40   .50   .60   .70   .80   .90        SSF =.10   .20   .30   .40   .50   .60   .70   .80   .90
DG C2   285   135    83    58    43    33    25    18    11    TW D4   150    81    51    35    25    19    14    10     6
DG C3   333   161   100    70    53    41    32    24    16    TW D5   144    80    52    36    26    20    14    10     7
SS A1   628   197   104    65    44    31    22    15    10    TW E1   176    86    52    35    24    18    13     9     6
SS A2   560   242   142    93    65    47    34    24    16    TW E2   163    84    52    36    25    18    13    10     6
SS A3   609   181    94    58    39    27    19    13     8    TW E3   178   103    68    48    35    26    20    14    10
SS A4   572   243   141    92    64    46    33    23    15    TW E4   161    95    63    45    33    25    19    13     9
SS A5  1050   198    96    57    37    26    18    12     8    TW F1   193    41    18    10     6     4     2     1     -
SS A6   558   241   141    92    64    46    34    24    16    TW F2   120    45    24    15    10     7     4     3     2
SS A7  1200   170    79    46    30    20    14    10     6    TW F3   100    44    25    16    11     8     5     4     2
SS A8   576   239   137    89    62    44    32    23    15    TW F4    83    40    24    16    11     8     5     4     2
SS B1   432   150    81    51    35    25    18    12     8    TW G1   118    37    18    11     7     4     3     2     1
SS B2   438   195   115    76    54    39    28    20    13    TW G2    87    37    21    13     9     6     4     3     2
SS B3   403   139    75    47    32    22    16    11     7    TW G3    74    35    20    13     9     6     5     3     2
SS B4   431   192   113    74    52    38    27    20    13    TW G4    60    29    18    12     8     6     4     3     2
SS B5   561   135    68    41    27    19    13     9     6    TW H1    76    30    17    10     7     5     3     2     1
SS B6   414   188   112    74    52    38    27    20    13    TW H2    60    28    16    10     7     5     4     2     1
SS B7   500   116    57    34    23    16    11     7     5    TW H3    51    24    14     9     6     5     3     2     1
SS B8   406   182   107    71    49    36    26    18    12    TW H4    40    19    11     7     5     4     3     2     1
SS C1   281   129    77    51    36    26    19    14     9    TW I1    58    19     8     4     -     -     -     -     -
SS C2   292   153    97    67    48    35    26    19    13    TW I2   110    54    33    22    15    11     8     6     4
SS C3   291   104    57    37    25    18    13     9     6    TW I3   118    60    37    25    18    13     9     7     4
SS C4   289   138    84    56    40    29    21    15    10    TW I4   125    70    45    32    23    17    12     9     6
SS D1   580   236   134    86    59    42    30    21    14    TW I5   124    72    47    33    24    18    14    10     7
SS D2   497   263   167   114    82    60    45    32    21    TW J1   145    76    47    32    23    16    12     9     6
SS D3   726   223   118    73    49    34    24    17    11    TW J2   134    73    47    32    23    17    13     9     6
SS D4   508   258   160   109    78    57    42    30    20    TW J3   150    91    62    44    33    25    18    13     9
SS E1   439   186   107    70    48    34    25    17    11    TW J4   139    85    57    41    31    23    17    13     9
SS E2   414   218   137    94    67    49    36    26    17    DG A1    79    22     -     -     -     -     -     -     -
SS E3   525   160    84    52    35    25    17    12     8    DG A2   114    47    26    15     9     -     -     -     -
SS E4   425   205   124    84    59    43    31    22    15    DG A3   156    69    42    28    20    14    10     6     -
                                                                DG B1    80    27     -     -     -     -     -     -     -
COLUMBIA, MISSOURI                         5083 DD              DG B2   117    50    28    18    12     7     -     -     -
    SSF =.10   .20   .30   .40   .50   .60   .70   .80   .90    DG B3   159    72    44    30    22    16    12     8     5
WW A1   531    48    19    10     5     3     2     -     -    DG C1   109    42    21    11     -     -     -     -     -
WW A2   166    59    31    19    12     8     6     4     2    DG C2   142    62    36    24    17    12     8     4     -
WW A3   139    61    35    22    15    11     7     5     3    DG C3   185    85    51    35    26    20    15    11     7
WW A4   128    62    37    24    17    12     8     6     4    SS A1   307    84    41    24    15    10     7     4     2
WW A5   122    63    39    26    18    13     9     6     4    SS A2   292   118    66    42    29    20    14    10     6
WW A6   118    64    40    27    19    14    10     7     4    SS A3   284    71    32    18    11     7     4     2     1
WW B1   108    30    12     5     -     -     -     -     -    SS A4   292   115    63    40    27    19    13     9     6
WW B2   134    68    42    28    20    14    10     7     5    SS A5   528    82    35    19    11     7     5     3     1
WW B3   158    85    54    37    26    19    14    10     6    SS A6   287   116    64    41    28    19    14     9     6
WW B4   147    87    58    41    30    22    17    12     8    SS A7   590    63    24    12     6     3     1     -     -
WW B5   139    84    57    41    30    23    17    12     8    SS A8   290   111    60    38    25    17    12     8     5
WW C1   172    93    59    40    29    21    15    11     7    SS B1   204    62    31    18    11     8     5     3     2
WW C2   151    86    56    39    28    21    16    11     7    SS B2   229    96    55    35    24    17    12     8     5
WW C3   164   104    72    52    39    30    22    17    11    SS B3   181    53    25    14     9     6     3     2     1
WW C4   149    94    65    48    36    27    20    15    10    SS B4   222    92    52    33    23    16    11     8     5
TW A1   460    48    20    10     6     4     2     1     -    SS B5   267    53    23    13     8     5     3     2     -
TW A2   174    54    28    17    11     7     5     3     2    SS B6   213    91    52    33    23    16    11     8     5
TW A3   139    55    30    19    13     9     6     4     2    SS B7   226    41    17     8     4     2     -     -     -
TW A4   119    53    31    20    14     9     7     5     3    SS B8   204    85    48    30    21    14    10     7     4
TW B1   250    48    22    12     7     5     3     2     1    SS C1   120    50    27    17    11     8     5     3     2
TW B2   141    51    27    17    11     8     5     4     2    SS C2   142    71    43    29    20    15    11     7     5
TW B3   123    50    28    17    12     8     6     4     2    SS C3   128    40    20    12     8     5     3     2     1
TW B4   115    47    26    17    11     8     5     4     2    SS C4   142    64    37    24    17    12     9     6     4
TW C1   166    45    23    13     9     6     4     3     2    SS D1   275   100    53    32    21    14    10     6     4
TW C2   127    45    24    15    10     7     5     3     2    SS D2   255   129    79    53    37    27    19    14     9
TW C3   123    43    23    14     9     6     5     3     2    SS D3   354    95    46    26    17    11     7     5     3
TW C4   134    39    20    12     8     6     4     3     2    SS D4   261   126    76    50    35    25    18    13     8
TW D1    80    25    11     5     -     -     -     -     -    SS E1   195    73    38    23    15    10     6     4     2
TW D2   145    65    38    25    17    12     9     6     4    SS E2   205   102    62    41    29    20    15    10     7
TW D3   147    71    43    28    20    14    10     7     5    SS E3   244    63    30    17    10     7     4     3     1
                                                                SS E4   212    96    56    36    25    18    13     9     6
```

KANSAS CITY, MISSOURI								5357 DD	
SSF =	.10	.20	.30	.40	.50	.60	.70	.80	.90
WW A1	515	51	21	11	7	4	3	1	-
WW A2	171	63	34	21	14	10	7	4	3
WW A3	144	65	37	24	17	12	8	6	4
WW A4	133	65	39	26	18	13	9	6	4
WW A5	127	66	41	28	20	14	10	7	5
WW A6	123	67	42	29	21	15	11	8	5
WW B1	114	34	15	8	4	2	-	-	-
WW B2	139	71	44	30	21	15	11	8	5
WW B3	163	89	56	39	28	20	15	11	7
WW B4	151	90	60	43	32	24	18	13	9
WW B5	142	87	59	42	32	24	18	13	9
WW C1	176	96	61	42	31	22	16	12	8
WW C2	156	89	58	41	30	22	16	12	8
WW C3	168	107	74	54	41	31	23	17	12
WW C4	153	97	68	50	37	28	21	16	11
TW A1	448	51	21	12	7	5	3	2	1
TW A2	178	57	30	18	12	8	6	4	2
TW A3	144	58	32	20	14	10	7	5	3
TW A4	123	56	33	21	15	10	7	5	3
TW B1	252	50	23	13	8	6	4	2	1
TW B2	145	53	29	18	12	8	6	4	3
TW B3	127	52	29	19	13	9	6	4	3
TW B4	118	49	28	18	12	9	6	4	3
TW C1	169	48	24	14	9	6	4	3	2
TW C2	130	47	25	16	11	7	5	4	2
TW C3	126	45	24	15	10	7	5	3	2
TW C4	136	41	21	13	9	6	4	3	2
TW D1	84	27	13	7	4	2	1	-	-
TW D2	150	68	40	26	18	13	9	7	4
TW D3	152	74	45	30	21	15	11	8	5
TW D4	154	84	53	37	27	20	15	10	7
TW D5	148	83	54	38	28	21	15	11	7
TW E1	180	89	54	37	26	19	14	10	6
TW E2	167	87	55	37	27	20	14	10	7
TW E3	182	107	71	50	37	28	21	15	10
TW E4	165	98	66	47	35	26	19	14	10
TW F1	196	44	20	11	7	5	3	2	1
TW F2	124	47	26	16	11	7	5	3	2
TW F3	103	47	27	17	12	8	6	4	3
TW F4	86	42	26	17	12	9	6	4	3
TW G1	122	39	20	12	8	5	4	2	1
TW G2	90	40	23	15	10	7	5	3	2
TW G3	77	37	22	14	10	7	5	4	2
TW G4	63	31	19	12	9	6	4	3	2
TW H1	79	32	18	11	8	5	4	2	2
TW H2	62	29	17	11	8	6	4	3	2
TW H3	53	25	15	10	7	5	4	2	2
TW H4	41	20	12	8	5	4	3	2	1
TW I1	62	22	10	6	3	1	-	-	-
TW I2	113	56	34	23	16	12	9	6	4
TW I3	121	63	39	27	19	14	10	7	5
TW I4	128	72	47	33	24	18	13	10	6
TW I5	127	74	49	35	26	19	14	10	7
TW J1	149	79	49	34	24	18	13	9	6
TW J2	138	76	49	34	25	18	13	10	6
TW J3	153	94	64	46	34	26	19	14	10
TW J4	142	87	59	43	32	24	18	13	9
DG A1	83	27	-	-	-	-	-	-	-
DG A2	117	50	28	17	10	4	-	-	-
DG A3	158	71	43	29	21	15	11	7	-
DG B1	84	30	13	-	-	-	-	-	-
DG B2	120	52	30	20	13	9	4	-	-
DG B3	162	74	45	31	23	17	13	9	5
DG C1	113	45	24	14	6	-	-	-	-

KANSAS CITY, MISSOURI								CONTINUED	
SSF =	.10	.20	.30	.40	.50	.60	.70	.80	.90
DG C2	145	65	38	26	18	13	9	5	-
DG C3	188	87	53	37	27	21	16	12	7
SS A1	307	86	43	25	16	11	7	5	3
SS A2	293	120	68	44	30	21	15	10	7
SS A3	283	73	34	19	12	7	5	3	2
SS A4	293	117	65	41	28	19	14	9	6
SS A5	517	84	37	20	12	8	5	3	2
SS A6	289	118	66	42	29	20	14	10	6
SS A7	567	65	25	13	7	4	2	1	-
SS A8	291	113	62	39	26	18	13	9	5
SS B1	205	64	32	19	12	8	6	4	2
SS B2	231	98	56	36	25	18	13	9	6
SS B3	182	55	27	15	10	6	4	3	1
SS B4	223	94	53	34	23	17	12	8	5
SS B5	265	55	25	14	8	5	3	2	1
SS B6	215	93	53	35	24	17	12	8	5
SS B7	224	43	18	9	5	3	2	-	-
SS B8	206	87	49	32	22	15	11	7	5
SS C1	125	53	30	19	13	9	6	4	3
SS C2	147	74	45	31	22	16	11	8	5
SS C3	132	42	22	13	9	6	4	3	2
SS C4	146	66	39	26	18	13	9	7	4
SS D1	278	103	55	34	22	15	10	7	4
SS D2	258	131	81	54	38	28	20	14	9
SS D3	354	97	48	28	18	12	8	5	3
SS D4	264	128	78	52	36	26	19	13	9
SS E1	198	76	40	25	16	11	7	5	3
SS E2	208	104	64	43	30	22	16	11	7
SS E3	245	65	31	18	11	7	5	3	2
SS E4	215	98	58	38	26	19	13	9	6

SAINT LOUIS, MISSOURI								4750 DD	
SSF =	.10	.20	.30	.40	.50	.60	.70	.80	.90
WW A1	577	57	23	12	7	5	3	2	-
WW A2	186	68	36	22	15	10	7	5	3
WW A3	156	70	40	26	18	12	9	6	4
WW A4	143	70	42	28	19	14	10	7	4
WW A5	136	71	44	30	21	15	11	8	5
WW A6	132	72	45	31	22	16	11	8	5
WW B1	129	39	18	9	5	2	-	-	-
WW B2	148	76	47	32	22	16	12	8	5
WW B3	174	94	60	41	29	21	16	11	7
WW B4	159	95	63	45	33	25	18	13	9
WW B5	150	91	62	44	33	25	18	14	9
WW C1	188	102	65	45	32	23	17	12	8
WW C2	165	94	62	43	31	23	17	12	8
WW C3	177	112	78	57	42	32	24	18	12
WW C4	160	102	71	52	39	29	22	16	11
TW A1	499	56	24	13	8	5	3	2	1
TW A2	194	62	32	19	13	9	6	4	2
TW A3	155	62	35	22	15	10	7	5	3
TW A4	132	60	35	23	16	11	8	5	3
TW B1	277	55	25	14	9	6	4	3	2
TW B2	157	57	31	19	13	9	6	4	3
TW B3	137	56	31	20	14	9	7	5	3
TW B4	127	52	30	19	13	9	6	4	3
TW C1	183	52	26	15	10	7	5	3	2
TW C2	141	51	27	17	11	8	6	4	2
TW C3	136	48	26	16	11	7	5	4	2
TW C4	146	44	23	14	9	6	4	3	2
TW D1	94	31	15	8	5	3	1	-	-
TW D2	160	72	43	28	19	14	10	7	4
TW D3	162	79	48	32	22	16	12	8	5

```
SAINT LOUIS, MISSOURI                    CONTINUED    SPRINGFIELD, MISSOURI                         4570 DD
   SSF =.10  .20  .30  .40  .50  .60  .70  .80  .90      SSF =.10  .20  .30  .40  .50  .60  .70  .80  .90
TW D4    164   88   56   39   28   20   15   11    7   WW A1   674   66   28   16   10    6    4    3    2
TW D5    156   87   57   40   29   21   16   11    8   WW A2   209   77   42   27   18   13    9    6    4
TW E1    192   95   58   39   27   20   14   10    7   WW A3   175   79   46   30   21   15   11    8    5
TW E2    178   92   58   39   28   20   15   11    7   WW A4   160   79   48   32   23   16   12    8    6
TW E3    192  112   74   52   38   29   21   16   10   WW A5   152   81   51   35   25   18   13    9    6
TW E4    174  103   69   49   36   27   20   15   10   WW A6   148   81   52   36   26   19   14   10    7
TW F1    216   48   22   13    8    5    3    2    1   WW B1   152   48   24   14    9    6    4    2    1
TW F2    135   51   28   17   12    8    5    4    2   WW B2   163   85   53   36   26   19   14   10    7
TW F3    112   50   29   19   13    9    6    4    3   WW B3   192  105   67   46   33   25   18   13    9
TW F4     92   46   28   18   13    9    6    4    3   WW B4   174  103   70   50   37   28   21   15   10
TW G1    133   42   22   13    8    6    4    2    1   WW B5   162   99   68   49   36   27   21   15   10
TW G2     98   43   24   16   11    7    5    4    2   WW C1   206  113   72   50   36   27   20   14   10
TW G3     83   39   23   15   11    8    5    4    2   WW C2   180  103   68   48   35   26   19   14   10
TW G4     67   33   20   13    9    7    5    3    2   WW C3   191  122   85   62   47   35   27   20   14
TW H1     86   35   19   12    8    6    4    3    2   WW C4   173  110   77   56   42   32   24   18   12
TW H2     67   31   19   12    8    6    4    3    2   TW A1   579   65   28   16   10    7    5    3    2
TW H3     57   27   16   11    7    5    4    3    2   TW A2   218   71   37   23   15   11    8    5    3
TW H4     44   21   13    8    6    4    3    2    1   TW A3   174   71   40   26   18   12    9    6    4
TW I1     70   25   12    7    4    2    -    -    -   TW A4   147   68   40   26   18   13   10    7    4
TW I2    121   60   37   24   17   12    9    6    4   TW B1   314   63   30   17   11    8    5    4    2
TW I3    129   67   42   28   20   14   10    7    5   TW B2   175   65   36   23   15   11    8    5    3
TW I4    136   76   50   35   25   19   14   10    7   TW B3   153   63   36   23   16   11    8    6    4
TW I5    134   78   52   36   27   20   15   11    7   TW B4   142   59   34   22   15   11    8    5    4
TW J1    159   83   52   35   25   18   13   10    6   TW C1   206   58   30   18   12    8    6    4    3
TW J2    146   80   51   36   26   19   14   10    7   TW C2   157   57   31   20   13    9    7    5    3
TW J3    161   98   67   48   35   27   20   15   10   TW C3   151   54   29   18   13    9    6    4    3
TW J4    149   91   62   45   33   25   19   14    9   TW C4   162   49   25   16   11    7    5    4    2
DG A1     93   32    -    -    -    -    -    -    -   TW D1   111   38   20   12    8    5    3    2    1
DG A2    126   53   30   18   11    5    -    -    -   TW D2   177   81   48   32   22   16   12    8    5
DG A3    168   76   46   31   22   16   11    7    -   TW D3   179   88   54   36   26   19   14   10    6
DG B1     95   34   16    -    -    -    -    -    -   TW D4   178   97   62   43   31   23   17   12    8
DG B2    129   56   32   21   14    9    5    -    -   TW D5   169   95   62   44   32   24   18   13    9
DG B3    172   79   48   33   24   18   13    9    5   TW E1   212  105   64   44   31   23   16   12    8
DG C1    125   50   27   16    8    -    -    -    -   TW E2   194  101   64   44   31   23   17   12    8
DG C2    156   69   41   27   19   14    9    5    -   TW E3   209  122   81   58   43   32   24   17   12
DG C3    199   92   56   38   28   22   17   12    8   TW E4   188  111   75   53   40   30   22   16   11
SS A1    335   94   46   27   17   12    8    5    3   TW F1   247   55   27   16   10    7    5    3    2
SS A2    314  128   72   46   31   22   16   11    7   TW F2   152   59   33   21   14   10    7    5    3
SS A3    311   80   37   21   13    8    5    3    2   TW F3   125   57   34   22   15   11    8    6    4
SS A4    315  125   69   44   30   21   15   10    6   TW F4   103   52   32   21   15   11    8    6    4
SS A5    569   92   40   22   13    9    6    4    2   TW G1   150   49   26   16   10    7    5    3    2
SS A6    310  126   71   45   30   21   15   11    7   TW G2   110   49   28   18   13    9    7    5    3
SS A7    633   72   29   15    8    5    3    1    -   TW G3    93   45   27   18   13    9    7    5    3
SS A8    314  121   66   41   28   19   14    9    6   TW G4    75   37   23   15   11    8    6    4    3
SS B1    224   70   35   21   13    9    6    4    2   TW H1    97   40   23   14   10    7    5    3    2
SS B2    247  104   60   39   26   19   13    9    6   TW H2    75   36   21   14   10    7    5    4    2
SS B3    200   61   30   17   11    7    5    3    2   TW H3    63   31   19   12    9    6    5    3    2
SS B4    239  100   57   36   25   18   12    9    6   TW H4    49   24   14   10    7    5    4    2    2
SS B5    292   60   27   15    9    6    4    2    1   TW I1    83   31   16   10    6    4    3    2    1
SS B6    231   99   57   37   25   18   13    9    6   TW I2   134   67   41   28   20   14   11    8    5
SS B7    249   48   20   11    6    4    2    1    -   TW I3   143   74   47   32   23   17   12    9    6
SS B8    222   93   53   34   23   16   11    8    5   TW I4   148   84   55   38   28   21   16   11    8
SS C1    136   57   32   20   14    9    6    4    3   TW I5   145   85   56   40   30   22   17   12    8
SS C2    156   79   48   32   23   16   12    8    5   TW J1   175   92   58   40   29   21   15   11    7
SS C3    144   46   24   14    9    6    4    3    2   TW J2   160   88   57   40   29   21   16   11    8
SS C4    156   71   42   27   19   13   10    7    4   TW J3   175  107   73   53   39   29   22   16   11
SS D1    302  112   59   36   24   16   11    8    5   TW J4   161   99   68   49   36   28   21   15   10
SS D2    275  140   86   57   40   29   21   15   10   DG A1   109   41   16    -    -    -    -    -    -
SS D3    386  106   52   30   19   13    9    6    3   DG A2   139   60   35   22   14    8    3    -    -
SS D4    282  137   82   55   38   27   20   14    9   DG A3   182   82   50   34   25   18   13    8    4
SS E1    216   83   44   27   18   12    8    5    3   DG B1   109   43   22   12    -    -    -    -    -
SS E2    222  112   68   45   32   23   16   12    7   DG B2   142   62   37   25   17   12    7    3    -
SS E3    269   72   34   20   12    8    5    3    2   DG B3   186   85   52   36   27   20   15   11    7
SS E4    231  105   62   40   28   20   14   10    6   DG C1   141   59   33   21   13    7    -    -    -
```

SPRINGFIELD, MISSOURI — CONTINUED

SSF =	.10	.20	.30	.40	.50	.60	.70	.80	.90
DG C2	171	76	46	31	22	16	12	7	3
DG C3	215	99	61	42	31	24	19	14	9
SS A1	379	107	54	32	21	14	10	7	4
SS A2	349	143	81	52	36	26	18	13	8
SS A3	358	93	45	26	17	11	8	5	3
SS A4	352	140	79	50	34	24	17	12	8
SS A5	657	106	47	27	17	11	8	5	3
SS A6	344	141	80	51	35	25	18	13	8
SS A7	748	85	36	19	12	7	5	3	2
SS A8	351	136	76	48	33	23	16	11	7
SS B1	254	80	41	25	16	11	8	5	3
SS B2	273	116	67	43	30	22	15	11	7
SS B3	229	71	36	21	14	9	6	4	3
SS B4	266	112	64	41	29	20	15	10	7
SS B5	335	70	33	19	12	8	5	4	2
SS B6	255	110	64	42	29	21	15	10	7
SS B7	290	57	25	14	9	6	4	2	1
SS B8	247	104	60	39	27	19	13	9	6
SS C1	153	66	38	24	17	12	8	6	4
SS C2	173	87	54	37	26	19	14	10	7
SS C3	163	53	28	17	11	8	6	4	2
SS C4	172	78	47	31	22	16	11	8	5
SS D1	341	128	69	43	29	20	14	10	6
SS D2	304	155	96	65	46	34	24	17	11
SS D3	438	121	60	36	23	16	11	7	5
SS D4	312	152	92	62	44	31	23	16	11
SS E1	246	96	53	33	22	15	11	7	4
SS E2	247	125	77	52	37	26	19	14	9
SS E3	307	83	41	24	16	11	7	5	3
SS E4	256	118	70	46	32	23	17	12	8

BILLINGS, MONTANA — 7265 DD

SSF =	.10	.20	.30	.40	.50	.60	.70	.80	.90
WW A1	467	46	18	8	3	-	-	-	-
WW A2	155	57	30	18	11	7	4	3	1
WW A3	132	59	34	21	14	9	6	4	2
WW A4	122	60	35	23	15	10	7	4	3
WW A5	117	61	37	25	17	12	8	5	3
WW A6	114	62	38	25	18	12	8	6	3
WW B1	99	28	-	-	-	-	-	-	-
WW B2	129	67	41	27	19	13	9	6	4
WW B3	152	83	52	35	25	18	13	9	5
WW B4	142	85	57	40	29	21	16	11	7
WW B5	135	82	56	40	29	22	16	12	8
WW C1	165	90	57	39	28	20	14	10	6
WW C2	146	84	55	38	27	20	15	10	7
WW C3	160	102	70	51	38	29	21	16	10
WW C4	145	93	64	47	35	26	20	14	10
TW A1	406	46	19	9	5	2	-	-	-
TW A2	161	52	27	16	10	6	4	2	1
TW A3	131	53	29	18	12	8	5	3	2
TW A4	113	52	30	19	13	9	6	4	2
TW B1	225	46	21	11	6	4	2	-	-
TW B2	132	49	26	16	10	7	4	3	1
TW B3	117	48	27	17	11	7	5	3	2
TW B4	109	45	25	16	10	7	5	3	2
TW C1	152	44	22	13	8	5	3	2	1
TW C2	119	43	23	14	9	6	4	3	1
TW C3	115	41	22	13	9	6	4	3	1
TW C4	123	38	19	12	8	5	3	2	1
TW D1	74	23	8	-	-	-	-	-	-
TW D2	138	63	37	24	16	11	8	5	3
TW D3	141	69	42	27	19	13	9	6	4

BILLINGS, MONTANA — CONTINUED

SSF =	.10	.20	.30	.40	.50	.60	.70	.80	.90
TW D4	145	79	50	34	25	18	13	9	6
TW D5	140	79	51	36	26	19	14	10	6
TW E1	168	83	51	34	23	16	12	8	5
TW E2	157	82	51	35	24	18	13	9	6
TW E3	172	101	67	47	34	25	19	13	9
TW E4	156	93	62	44	32	24	18	13	8
TW F1	174	39	17	9	5	2	-	-	-
TW F2	113	43	23	14	9	5	3	2	-
TW F3	95	43	24	15	10	7	4	3	1
TW F4	79	39	23	15	10	7	5	3	2
TW G1	109	35	17	10	6	3	2	-	-
TW G2	82	36	20	13	8	5	3	2	1
TW G3	71	34	20	13	8	6	4	2	1
TW G4	58	29	17	11	7	5	3	2	1
TW H1	72	29	16	10	6	4	2	1	-
TW H2	57	27	16	10	7	4	3	2	1
TW H3	49	23	14	9	6	4	3	2	1
TW H4	38	18	11	7	5	3	2	1	-
TW I1	54	17	-	-	-	-	-	-	-
TW I2	105	53	32	21	14	10	7	5	3
TW I3	113	59	36	24	17	12	8	6	4
TW I4	121	68	44	31	22	16	12	8	5
TW I5	120	70	46	33	24	18	13	9	6
TW J1	139	74	46	31	22	15	11	8	5
TW J2	129	71	46	31	22	16	12	8	5
TW J3	145	89	60	43	32	24	17	13	8
TW J4	135	83	56	40	30	22	17	12	8
DG A1	74	19	-	-	-	-	-	-	-
DG A2	108	46	25	15	7	-	-	-	-
DG A3	148	68	41	28	19	14	9	5	-
DG B1	76	26	-	-	-	-	-	-	-
DG B2	111	49	28	18	11	6	-	-	-
DG B3	152	71	43	29	21	16	11	8	4
DG C1	104	41	20	8	-	-	-	-	-
DG C2	136	61	36	24	16	11	7	-	-
DG C3	176	83	51	35	25	19	14	10	6
SS A1	272	77	37	21	12	7	4	2	1
SS A2	267	111	62	39	26	18	12	8	5
SS A3	245	63	28	14	6	-	-	-	-
SS A4	265	106	58	36	23	16	11	7	4
SS A5	453	74	31	16	8	4	2	-	-
SS A6	262	108	60	37	25	17	11	8	5
SS A7	490	55	19	6	-	-	-	-	-
SS A8	262	102	55	33	22	14	9	6	4
SS B1	181	57	28	16	9	5	3	2	-
SS B2	211	90	51	33	22	15	11	7	4
SS B3	159	48	22	11	6	-	-	-	-
SS B4	203	86	48	30	20	14	9	6	4
SS B5	230	48	21	10	5	2	-	-	-
SS B6	196	85	49	31	21	14	10	7	4
SS B7	191	36	14	4	-	-	-	-	-
SS B8	186	79	44	28	18	12	8	6	3
SS C1	113	48	26	16	10	6	4	2	1
SS C2	136	69	42	28	19	14	10	7	4
SS C3	118	39	19	11	7	4	2	1	-
SS C4	135	62	36	23	16	11	8	5	3
SS D1	248	92	48	28	17	11	7	4	2
SS D2	237	121	74	49	33	24	16	11	7
SS D3	313	87	41	23	13	8	5	2	1
SS D4	242	118	71	46	31	22	15	10	6
SS E1	175	67	34	19	11	6	3	1	-
SS E2	190	96	58	38	26	18	12	8	5
SS E3	214	58	26	14	8	4	1	-	-
SS E4	196	90	52	33	22	15	11	7	4

CUT BANK, MONTANA 9033 DD

	SSF=.10	.20	.30	.40	.50	.60	.70	.80	.90
WW A1	360	33	-	-	-	-	-	-	-
WW A2	129	44	22	11	-	-	-	-	-
WW A3	110	47	25	14	8	-	-	-	-
WW A4	102	47	27	16	9	5	-	-	-
WW A5	97	48	28	17	11	6	3	-	-
WW A6	94	49	29	18	11	7	4	1	-
WW B1	71	-	-	-	-	-	-	-	-
WW B2	110	55	33	21	14	9	6	4	2
WW B3	131	69	43	28	19	13	9	5	3
WW B4	125	73	48	34	24	17	12	8	5
WW B5	119	72	48	34	25	18	13	9	6
WW C1	143	77	48	32	22	15	10	7	4
WW C2	128	72	46	32	22	16	11	8	5
WW C3	142	90	62	44	33	24	18	13	8
WW C4	130	82	56	41	30	22	17	12	8
TW A1	320	34	11	-	-	-	-	-	-
TW A2	135	41	19	10	4	-	-	-	-
TW A3	110	42	22	12	7	2	-	-	-
TW A4	95	41	23	13	8	4	2	-	-
TW B1	186	35	14	6	-	-	-	-	-
TW B2	112	39	20	11	6	2	-	-	-
TW B3	99	39	20	12	7	4	1	-	-
TW B4	93	37	20	12	7	4	2	-	-
TW C1	129	35	16	8	4	-	-	-	-
TW C2	102	35	18	10	6	3	1	-	-
TW C3	99	34	17	10	6	4	2	-	-
TW C4	106	31	16	9	5	3	2	-	-
TW D1	53	-	-	-	-	-	-	-	-
TW D2	119	53	30	19	12	8	5	3	1
TW D3	121	58	34	21	14	9	6	3	2
TW D4	128	68	43	29	20	14	10	7	4
TW D5	124	69	44	31	22	16	11	8	5
TW E1	146	71	42	27	18	12	8	5	3
TW E2	137	70	43	29	20	14	9	6	4
TW E3	153	88	58	40	29	21	15	10	7
TW E4	140	82	54	38	28	20	15	10	7
TW F1	142	29	10	-	-	-	-	-	-
TW F2	93	33	16	8	-	-	-	-	-
TW F3	79	33	18	10	5	-	-	-	-
TW F4	66	31	18	11	6	3	1	-	-
TW G1	90	26	11	-	-	-	-	-	-
TW G2	68	28	15	8	4	-	-	-	-
TW G3	59	27	15	9	5	2	-	-	-
TW G4	48	23	13	8	5	3	1	-	-
TW H1	59	22	11	6	-	-	-	-	-
TW H2	48	21	12	7	4	2	-	-	-
TW H3	41	19	11	6	4	2	1	-	-
TW H4	32	15	8	5	3	2	1	-	-
TW I1	37	-	-	-	-	-	-	-	-
TW I2	90	43	26	16	11	7	4	3	1
TW I3	97	49	29	19	13	9	6	3	2
TW I4	106	59	38	26	18	13	9	6	4
TW I5	106	61	40	28	20	14	10	7	5
TW J1	121	62	38	25	17	12	8	5	3
TW J2	113	61	39	26	18	13	9	6	4
TW J3	129	78	52	37	27	20	14	10	6
TW J4	120	73	49	35	26	19	14	10	6
DG A1	53	-	-	-	-	-	-	-	-
DG A2	92	38	19	9	-	-	-	-	-
DG A3	133	60	35	23	16	11	7	3	-
DG B1	55	-	-	-	-	-	-	-	-
DG B2	95	40	22	12	6	-	-	-	-
DG B3	137	62	38	25	18	12	9	5	2
DG C1	82	29	-	-	-	-	-	-	-

CUT BANK, MONTANA CONTINUED

	SSF=.10	.20	.30	.40	.50	.60	.70	.80	.90
DG C2	118	52	30	18	11	6	-	-	-
DG C3	160	73	45	30	21	15	11	7	4
SS A1	239	64	28	14	6	-	-	-	-
SS A2	240	96	52	32	21	13	9	5	3
SS A3	213	49	17	-	-	-	-	-	-
SS A4	237	92	49	29	18	11	7	4	2
SS A5	391	60	22	7	-	-	-	-	-
SS A6	235	93	50	31	19	12	8	5	3
SS A7	409	42	-	-	-	-	-	-	-
SS A8	233	87	45	26	16	10	6	3	1
SS B1	157	46	20	9	-	-	-	-	-
SS B2	189	79	44	27	18	12	8	5	3
SS B3	135	36	13	-	-	-	-	-	-
SS B4	181	74	41	25	16	10	7	4	2
SS B5	197	38	13	-	-	-	-	-	-
SS B6	175	74	41	25	16	11	7	4	2
SS B7	161	26	-	-	-	-	-	-	-
SS B8	165	68	37	22	14	9	5	3	2
SS C1	92	36	18	8	-	-	-	-	-
SS C2	117	58	34	22	15	10	6	4	2
SS C3	98	29	13	5	-	-	-	-	-
SS C4	116	52	29	18	12	8	5	3	2
SS D1	216	76	37	19	9	-	-	-	-
SS D2	211	105	63	40	27	18	12	8	5
SS D3	275	71	31	15	4	-	-	-	-
SS D4	216	103	60	38	25	17	11	7	4
SS E1	148	52	23	-	-	-	-	-	-
SS E2	167	81	48	30	20	13	8	5	3
SS E3	184	45	18	-	-	-	-	-	-
SS E4	173	77	43	27	17	11	7	4	2

DILLON, MONTANA 8354 DD

	SSF=.10	.20	.30	.40	.50	.60	.70	.80	.90
WW A1	456	48	20	10	5	-	-	-	-
WW A2	161	60	32	20	13	8	5	3	2
WW A3	137	62	36	23	15	10	7	4	2
WW A4	127	63	38	25	17	12	8	5	3
WW A5	121	64	40	27	18	13	9	6	3
WW A6	118	64	41	28	19	14	10	6	4
WW B1	106	32	13	-	-	-	-	-	-
WW B2	133	69	43	29	20	15	10	7	4
WW B3	157	86	55	38	27	19	14	10	6
WW B4	146	87	59	42	31	23	17	12	8
WW B5	138	85	58	42	31	23	17	12	8
WW C1	170	93	60	41	29	21	15	11	7
WW C2	150	87	57	40	29	21	16	11	7
WW C3	164	104	73	53	40	30	23	16	11
WW C4	148	95	66	48	36	28	21	15	10
TW A1	400	48	20	11	6	3	-	-	-
TW A2	167	55	28	17	11	7	5	3	1
TW A3	136	56	31	20	13	9	6	4	2
TW A4	117	54	32	20	14	10	6	4	2
TW B1	231	48	22	13	8	5	3	1	-
TW B2	137	51	28	17	11	8	5	3	2
TW B3	121	50	28	18	12	8	6	4	2
TW B4	113	47	27	17	11	8	5	4	2
TW C1	157	46	23	14	9	6	4	2	1
TW C2	123	45	25	15	10	7	5	3	2
TW C3	119	43	23	15	10	7	4	3	2
TW C4	127	39	20	13	8	6	4	2	1
TW D1	78	26	12	4	-	-	-	-	-
TW D2	143	66	39	26	18	12	9	6	4
TW D3	146	72	44	29	20	14	10	7	4

DILLON, MONTANA CONTINUED

SSF =	.10	.20	.30	.40	.50	.60	.70	.80	.90
TW D4	149	81	52	36	26	19	14	10	6
TW D5	143	81	53	37	27	20	15	10	7
TW E1	174	86	53	36	25	18	13	9	5
TW E2	161	84	53	36	26	19	14	9	6
TW E3	177	104	69	49	36	27	20	14	9
TW E4	160	95	64	46	34	25	19	13	9
TW F1	180	41	19	11	6	3	2	-	-
TW F2	117	45	25	15	10	6	4	2	1
TW F3	98	45	26	17	11	8	5	3	2
TW F4	82	41	25	16	11	8	5	3	2
TW G1	114	37	19	11	7	4	3	1	-
TW G2	86	38	22	14	9	6	4	3	1
TW G3	73	35	21	14	9	6	4	3	2
TW G4	60	30	18	12	8	6	4	3	2
TW H1	75	31	17	11	7	5	3	2	-
TW H2	59	28	17	11	7	5	3	2	1
TW H3	50	24	15	10	7	5	3	2	1
TW H4	40	19	11	7	5	4	2	2	1
TW I1	58	20	9	-	-	-	-	-	-
TW I2	108	54	33	22	16	11	8	5	3
TW I3	116	61	38	26	18	13	9	6	4
TW I4	124	70	46	32	23	17	13	9	6
TW I5	123	72	48	34	25	19	14	10	6
TW J1	144	76	48	33	23	17	12	8	5
TW J2	133	74	48	33	24	17	13	9	6
TW J3	149	91	62	45	33	25	19	13	9
TW J4	138	85	58	42	31	23	17	13	8
DG A1	78	24	-	-	-	-	-	-	-
DG A2	111	48	27	16	9	-	-	-	-
DG A3	151	69	42	29	20	15	10	6	-
DG B1	78	29	9	-	-	-	-	-	-
DG B2	113	50	30	19	12	8	-	-	-
DG B3	155	72	45	31	22	17	12	8	4
DG C1	105	44	23	12	-	-	-	-	-
DG C2	138	62	38	25	18	12	8	3	-
DG C3	180	84	52	36	27	20	15	11	6
SS A1	296	84	41	24	14	9	5	3	1
SS A2	286	118	67	42	28	19	13	9	5
SS A3	271	70	32	17	9	4	-	-	-
SS A4	285	114	63	40	26	18	12	8	4
SS A5	487	81	35	19	11	6	3	-	-
SS A6	281	115	65	41	27	19	13	8	5
SS A7	526	62	24	10	-	-	-	-	-
SS A8	283	110	60	37	24	16	11	7	4
SS B1	196	62	31	18	11	7	4	2	-
SS B2	224	96	55	35	24	17	12	8	5
SS B3	173	53	25	14	8	4	-	-	-
SS B4	216	91	52	33	22	15	10	7	4
SS B5	249	53	24	13	7	4	-	-	-
SS B6	208	90	52	33	23	16	11	7	4
SS B7	210	41	17	8	-	-	-	-	-
SS B8	199	84	48	30	20	14	9	6	4
SS C1	118	50	28	18	11	7	5	3	1
SS C2	141	71	44	30	21	15	10	7	4
SS C3	124	40	21	12	8	5	3	2	-
SS C4	139	64	38	25	17	12	8	6	4
SS D1	269	101	53	32	20	13	8	5	2
SS D2	252	129	79	53	37	26	18	12	8
SS D3	341	95	46	26	16	10	6	3	1
SS D4	258	126	76	50	34	24	17	11	7
SS E1	190	73	39	23	14	8	5	2	-
SS E2	201	102	62	41	28	20	14	9	6
SS E3	233	63	30	17	9	5	3	-	-
SS E4	208	96	56	36	25	17	12	8	5

GLASGOW, MONTANA 8969 DD

SSF =	.10	.20	.30	.40	.50	.60	.70	.80	.90
WW A1	271	-	-	-	-	-	-	-	-
WW A2	90	26	-	-	-	-	-	-	-
WW A3	77	28	11	-	-	-	-	-	-
WW A4	71	30	13	-	-	-	-	-	-
WW A5	68	31	15	-	-	-	-	-	-
WW A6	67	31	16	7	-	-	-	-	-
WW B1	-	-	-	-	-	-	-	-	-
WW B2	84	40	22	13	8	5	2	1	-
WW B3	101	51	30	19	12	8	5	3	1
WW B4	101	58	37	26	18	13	9	6	4
WW B5	98	58	39	27	19	14	10	7	4
WW C1	112	58	35	22	15	10	6	4	2
WW C2	102	56	35	24	16	11	8	5	3
WW C3	118	74	50	36	26	19	14	10	6
WW C4	108	68	46	33	24	18	13	9	6
TW A1	243	12	-	-	-	-	-	-	-
TW A2	96	24	-	-	-	-	-	-	-
TW A3	79	27	10	-	-	-	-	-	-
TW A4	69	27	12	5	-	-	-	-	-
TW B1	136	19	-	-	-	-	-	-	-
TW B2	81	25	10	-	-	-	-	-	-
TW B3	72	26	11	4	-	-	-	-	-
TW B4	69	25	12	6	-	-	-	-	-
TW C1	95	22	7	-	-	-	-	-	-
TW C2	75	24	11	4	-	-	-	-	-
TW C3	74	24	11	5	-	-	-	-	-
TW C4	82	23	10	5	2	-	-	-	-
TW D1	-	-	-	-	-	-	-	-	-
TW D2	91	38	20	12	7	4	2	-	-
TW D3	92	42	23	13	8	4	2	-	-
TW D4	103	54	33	22	15	10	7	5	3
TW D5	102	56	35	24	17	12	8	6	4
TW E1	114	53	30	19	12	8	5	3	1
TW E2	109	55	32	21	14	9	6	4	2
TW E3	126	72	46	32	22	16	11	8	5
TW E4	116	67	44	31	22	16	11	8	5
TW F1	97	-	-	-	-	-	-	-	-
TW F2	64	18	-	-	-	-	-	-	-
TW F3	55	20	7	-	-	-	-	-	-
TW F4	46	20	9	-	-	-	-	-	-
TW G1	60	12	-	-	-	-	-	-	-
TW G2	47	16	-	-	-	-	-	-	-
TW G3	41	17	7	-	-	-	-	-	-
TW G4	35	15	7	3	-	-	-	-	-
TW H1	40	12	-	-	-	-	-	-	-
TW H2	33	13	5	-	-	-	-	-	-
TW H3	29	12	6	2	-	-	-	-	-
TW H4	24	10	5	3	-	-	-	-	-
TW I1	-	-	-	-	-	-	-	-	-
TW I2	68	31	17	10	6	3	1	-	-
TW I3	74	36	20	12	8	5	3	1	-
TW I4	85	47	29	19	13	9	6	4	3
TW I5	87	50	32	22	15	11	8	5	3
TW J1	94	47	28	17	11	7	5	3	2
TW J2	90	48	29	19	13	9	6	4	2
TW J3	107	64	42	30	21	15	11	8	5
TW J4	100	60	40	28	20	15	11	8	5
DG A1	-	-	-	-	-	-	-	-	-
DG A2	69	25	8	-	-	-	-	-	-
DG A3	109	48	28	17	11	7	4	-	-
DG B1	-	-	-	-	-	-	-	-	-
DG B2	72	28	12	-	-	-	-	-	-
DG B3	113	51	30	19	13	9	6	3	-
DG C1	51	-	-	-	-	-	-	-	-

GLASGOW, MONTANA — CONTINUED

SSF	=.10	.20	.30	.40	.50	.60	.70	.80	.90
DG C2	93	39	20	11	4	-	-	-	-
DG C3	132	60	36	23	16	11	8	5	2
SS A1	174	41	13	-	-	-	-	-	-
SS A2	187	73	38	22	14	9	5	3	2
SS A3	144	20	-	-	-	-	-	-	-
SS A4	181	67	34	19	11	6	4	2	-
SS A5	291	35	-	-	-	-	-	-	-
SS A6	181	70	36	21	12	8	5	3	1
SS A7	295	-	-	-	-	-	-	-	-
SS A8	176	62	30	16	9	4	2	-	-
SS B1	111	27	-	-	-	-	-	-	-
SS B2	149	60	32	19	12	8	5	3	2
SS B3	88	-	-	-	-	-	-	-	-
SS B4	140	55	29	17	10	6	4	2	1
SS B5	138	18	-	-	-	-	-	-	-
SS B6	136	56	30	18	11	7	4	3	1
SS B7	104	-	-	-	-	-	-	-	-
SS B8	126	49	25	14	8	5	3	1	-
SS C1	61	16	-	-	-	-	-	-	-
SS C2	90	42	24	14	9	5	3	2	-
SS C3	66	14	-	-	-	-	-	-	-
SS C4	89	38	20	12	7	4	2	-	-
SS D1	154	48	18	-	-	-	-	-	-
SS D2	164	80	46	29	19	13	8	5	3
SS D3	199	45	13	-	-	-	-	-	-
SS D4	168	78	44	27	18	12	8	5	3
SS E1	98	24	-	-	-	-	-	-	-
SS E2	127	60	33	20	12	8	5	3	1
SS E3	127	22	-	-	-	-	-	-	-
SS E4	132	56	30	18	11	6	4	2	1

GREAT FALLS, MONTANA — 7652 DD

SSF	=.10	.20	.30	.40	.50	.60	.70	.80	.90
WW A1	444	40	13	-	-	-	-	-	-
WW A2	145	51	26	14	8	-	-	-	-
WW A3	122	53	29	17	10	6	2	-	-
WW A4	113	54	31	19	12	7	4	1	-
WW A5	108	55	32	20	13	8	5	2	1
WW A6	105	55	33	21	14	9	5	3	1
WW B1	87	17	-	-	-	-	-	-	-
WW B2	121	61	37	24	16	11	7	4	2
WW B3	142	76	47	31	21	15	10	6	4
WW B4	134	79	52	36	26	19	14	9	6
WW B5	128	77	52	37	27	20	14	10	6
WW C1	155	83	52	35	24	17	12	8	5
WW C2	138	78	50	35	24	17	12	8	5
WW C3	151	96	66	48	35	26	19	14	9
WW C4	138	87	60	44	32	24	18	13	8
TW A1	388	40	15	5	-	-	-	-	-
TW A2	152	47	23	12	7	-	-	-	-
TW A3	123	48	25	15	9	5	2	-	-
TW A4	105	47	26	16	10	6	3	2	-
TW B1	216	41	17	8	-	-	-	-	-
TW B2	124	44	23	13	8	4	2	-	-
TW B3	109	44	23	14	9	5	3	1	-
TW B4	102	41	22	13	8	5	3	2	-
TW C1	145	40	19	10	6	3	-	-	-
TW C2	112	40	21	12	7	4	2	1	-
TW C3	109	38	20	12	7	4	3	1	-
TW C4	118	35	17	10	6	4	2	1	-
TW D1	64	16	-	-	-	-	-	-	-
TW D2	130	58	33	21	14	9	6	4	2
TW D3	132	63	37	24	16	11	7	4	2

GREAT FALLS, MONTANA — CONTINUED

SSF	=.10	.20	.30	.40	.50	.60	.70	.80	.90
TW D4	137	74	46	31	22	16	11	7	5
TW D5	133	74	48	33	23	17	12	8	5
TW E1	158	77	46	30	20	14	9	6	3
TW E2	148	76	47	31	22	15	11	7	4
TW E3	163	95	62	43	31	23	16	11	7
TW E4	149	88	58	41	30	22	16	11	7
TW F1	165	34	14	5	-	-	-	-	-
TW F2	105	39	19	11	6	-	-	-	-
TW F3	88	38	21	12	7	4	1	-	-
TW F4	73	35	20	12	8	5	3	1	-
TW G1	102	31	14	7	-	-	-	-	-
TW G2	76	32	17	10	6	3	-	-	-
TW G3	65	30	17	10	6	4	2	-	-
TW G4	54	26	15	9	6	4	2	1	-
TW H1	67	26	13	7	4	-	-	-	-
TW H2	53	24	13	8	5	3	1	-	-
TW H3	45	21	12	7	5	3	2	-	-
TW H4	36	17	10	6	4	2	1	-	-
TW I1	46	-	-	-	-	-	-	-	-
TW I2	98	48	29	18	12	8	5	3	2
TW I3	105	54	33	21	14	10	7	4	2
TW I4	114	64	41	28	20	14	10	7	4
TW I5	114	66	43	30	22	16	11	8	5
TW J1	131	68	42	28	19	13	9	6	3
TW J2	122	67	42	29	20	14	10	7	4
TW J3	138	84	56	40	29	21	16	11	7
TW J4	128	78	53	38	27	20	15	11	7
DG A1	65	-	-	-	-	-	-	-	-
DG A2	102	42	22	12	-	-	-	-	-
DG A3	142	65	38	25	17	12	7	4	-
DG B1	67	18	-	-	-	-	-	-	-
DG B2	104	45	25	15	8	-	-	-	-
DG B3	145	67	41	27	19	14	10	6	2
DG C1	95	36	14	-	-	-	-	-	-
DG C2	128	57	33	21	13	8	3	-	-
DG C3	169	79	48	33	23	17	12	8	4
SS A1	258	70	32	17	9	3	-	-	-
SS A2	254	103	57	35	23	15	10	6	4
SS A3	231	56	22	-	-	-	-	-	-
SS A4	251	98	53	32	20	13	8	5	3
SS A5	438	67	26	12	-	-	-	-	-
SS A6	249	100	55	33	22	14	9	6	3
SS A7	472	48	13	-	-	-	-	-	-
SS A8	247	94	50	29	18	11	7	4	2
SS B1	171	51	24	12	6	-	-	-	-
SS B2	201	84	47	30	20	13	9	6	3
SS B3	148	42	17	-	-	-	-	-	-
SS B4	192	80	44	27	18	12	8	5	3
SS B5	220	43	17	6	-	-	-	-	-
SS B6	186	79	44	28	18	12	8	5	3
SS B7	181	31	7	-	-	-	-	-	-
SS B8	176	73	40	25	16	10	6	4	2
SS C1	104	42	22	12	5	-	-	-	-
SS C2	128	63	38	25	17	11	7	5	3
SS C3	111	34	16	8	-	-	-	-	-
SS C4	127	57	33	21	14	9	6	4	2
SS D1	232	83	42	23	13	6	-	-	-
SS D2	223	112	67	44	30	20	14	9	5
SS D3	297	79	36	18	9	-	-	-	-
SS D4	229	110	65	42	28	19	13	8	5
SS E1	162	59	28	14	-	-	-	-	-
SS E2	178	88	52	33	22	15	10	6	3
SS E3	202	51	22	9	-	-	-	-	-
SS E4	184	83	47	30	19	13	8	5	3

HELENA, MONTANA								8190 DD	
SSF =	.10	.20	.30	.40	.50	.60	.70	.80	.90
WW A1	379	32	-	-	-	-	-	-	-
WW A2	127	44	21	11	4	-	-	-	-
WW A3	108	46	25	14	8	3	-	-	-
WW A4	99	47	26	16	9	5	1	-	-
WW A5	95	48	28	17	11	6	3	1	-
WW A6	93	48	29	18	11	7	4	2	-
WW B1	68	-	-	-	-	-	-	-	-
WW B2	108	54	33	21	14	9	6	4	2
WW B3	129	69	43	28	19	13	9	6	3
WW B4	123	73	48	33	24	17	12	8	5
WW B5	118	71	48	34	25	18	13	9	6
WW C1	141	76	47	32	22	15	10	7	4
WW C2	126	71	46	32	22	16	11	8	5
WW C3	141	89	61	44	33	24	18	13	8
WW C4	128	81	56	41	30	22	17	12	8
TW A1	334	33	11	-	-	-	-	-	-
TW A2	133	40	19	10	4	-	-	-	-
TW A3	108	42	22	12	7	3	-	-	-
TW A4	93	41	22	13	8	5	2	-	-
TW B1	189	34	14	5	-	-	-	-	-
TW B2	110	38	19	11	6	2	-	-	-
TW B3	97	38	20	12	7	4	2	-	-
TW B4	91	36	20	12	7	4	2	-	-
TW C1	128	34	16	8	4	-	-	-	-
TW C2	100	35	18	10	6	3	2	-	-
TW C3	98	33	17	10	6	4	2	-	-
TW C4	106	31	15	9	5	3	2	-	-
TW D1	51	-	-	-	-	-	-	-	-
TW D2	117	52	30	19	12	8	5	3	1
TW D3	119	57	33	21	14	9	6	3	2
TW D4	126	67	42	29	20	14	10	7	4
TW D5	122	68	44	30	22	16	11	8	5
TW E1	144	70	42	27	18	12	8	5	3
TW E2	135	69	43	28	20	14	9	6	4
TW E3	151	88	58	40	29	21	15	11	7
TW E4	138	81	54	38	27	20	15	10	7
TW F1	143	28	10	-	-	-	-	-	-
TW F2	91	33	16	8	-	-	-	-	-
TW F3	77	33	18	10	5	-	-	-	-
TW F4	64	31	17	10	6	3	1	-	-
TW G1	88	26	11	-	-	-	-	-	-
TW G2	67	28	15	8	4	-	-	-	-
TW G3	57	26	15	9	5	3	-	-	-
TW G4	47	23	13	8	5	3	1	-	-
TW H1	58	22	11	6	-	-	-	-	-
TW H2	47	21	11	7	4	2	-	-	-
TW H3	40	18	10	6	4	2	1	-	-
TW H4	32	15	8	5	3	2	1	-	-
TW I1	35	-	-	-	-	-	-	-	-
TW I2	88	43	25	16	11	7	4	3	1
TW I3	95	48	29	19	13	9	6	3	2
TW I4	104	58	37	26	18	13	9	6	4
TW I5	105	61	40	28	20	14	10	7	5
TW J1	119	62	38	25	17	12	8	5	3
TW J2	112	61	38	26	18	13	9	6	4
TW J3	128	77	52	37	27	20	14	10	7
TW J4	119	72	49	35	26	19	14	10	6
DG A1	51	-	-	-	-	-	-	-	-
DG A2	90	37	19	9	-	-	-	-	-
DG A3	131	59	35	23	16	11	7	3	-
DG B1	53	-	-	-	-	-	-	-	-
DG B2	92	40	22	12	6	-	-	-	-
DG B3	134	61	37	25	18	12	9	5	2
DG C1	80	29	-	-	-	-	-	-	-

HELENA, MONTANA								CONTINUED	
SSF =	.10	.20	.30	.40	.50	.60	.70	.80	.90
DG C2	115	51	29	18	11	6	-	-	-
DG C3	155	72	44	30	21	15	11	7	4
SS A1	240	64	29	14	7	-	-	-	-
SS A2	238	96	53	32	21	14	9	6	3
SS A3	213	49	18	-	-	-	-	-	-
SS A4	236	92	49	29	18	12	7	4	2
SS A5	405	60	23	8	-	-	-	-	-
SS A6	233	93	51	31	20	13	8	5	3
SS A7	433	42	-	-	-	-	-	-	-
SS A8	232	87	46	27	16	10	6	3	2
SS B1	156	46	21	10	-	-	-	-	-
SS B2	188	78	44	27	18	12	8	5	3
SS B3	134	36	14	-	-	-	-	-	-
SS B4	180	74	41	25	16	11	7	4	2
SS B5	200	37	14	-	-	-	-	-	-
SS B6	174	73	41	26	17	11	7	4	3
SS B7	164	26	-	-	-	-	-	-	-
SS B8	164	68	37	22	14	9	6	3	2
SS C1	90	35	17	8	-	-	-	-	-
SS C2	115	57	34	22	15	10	6	4	2
SS C3	96	29	13	5	-	-	-	-	-
SS C4	115	51	29	18	12	8	5	3	2
SS D1	214	76	37	20	10	-	-	-	-
SS D2	209	105	63	41	27	19	13	8	5
SS D3	276	71	32	15	6	-	-	-	-
SS D4	214	103	60	39	26	17	12	7	4
SS E1	146	52	23	9	-	-	-	-	-
SS E2	165	81	48	31	20	13	9	5	3
SS E3	184	45	18	-	-	-	-	-	-
SS E4	171	77	43	27	17	11	7	4	2

LEWISTOWN, MONTANA								8586 DD	
SSF =	.10	.20	.30	.40	.50	.60	.70	.80	.90
WW A1	356	32	9	-	-	-	-	-	-
WW A2	126	44	22	12	6	-	-	-	-
WW A3	107	46	25	15	9	5	2	-	-
WW A4	99	47	27	16	10	6	3	1	-
WW A5	95	48	28	18	11	7	4	2	1
WW A6	93	49	29	19	12	8	5	3	1
WW B1	68	-	-	-	-	-	-	-	-
WW B2	108	54	33	21	15	10	7	4	2
WW B3	129	69	43	29	20	14	9	6	4
WW B4	123	73	48	34	24	18	13	9	6
WW B5	118	71	48	34	25	18	13	10	6
WW C1	141	76	48	32	22	16	11	7	4
WW C2	126	71	46	32	23	16	12	8	5
WW C3	141	89	62	45	33	25	18	13	9
WW C4	128	81	56	41	30	23	17	12	8
TW A1	316	33	11	-	-	-	-	-	-
TW A2	132	40	19	10	5	-	-	-	-
TW A3	108	42	22	13	7	4	2	-	-
TW A4	93	41	23	14	9	5	3	1	-
TW B1	183	34	14	6	-	-	-	-	-
TW B2	109	38	20	11	7	4	2	-	-
TW B3	97	38	20	12	7	5	3	1	-
TW B4	91	36	20	12	7	5	3	2	-
TW C1	126	34	16	9	5	2	-	-	-
TW C2	100	35	18	11	6	4	2	1	-
TW C3	97	33	17	10	6	4	2	1	-
TW C4	105	31	15	9	6	4	2	1	-
TW D1	51	-	-	-	-	-	-	-	-
TW D2	117	52	30	19	12	8	6	3	2
TW D3	119	57	34	22	14	10	6	4	2

LEWISTOWN, MONTANA (CONTINUED)

	SSF=.10	.20	.30	.40	.50	.60	.70	.80	.90
TW D4	126	67	42	29	20	15	10	7	4
TW D5	122	68	44	30	22	16	12	8	5
TW E1	144	70	42	27	19	13	9	6	3
TW E2	135	69	43	29	20	14	10	7	4
TW E3	151	88	58	40	29	21	16	11	7
TW E4	138	81	54	38	28	20	15	11	7
TW F1	139	28	11	-	-	-	-	-	-
TW F2	91	33	16	9	4	-	-	-	-
TW F3	77	33	18	11	6	3	-	-	-
TW F4	64	31	18	11	7	4	2	1	-
TW G1	88	26	12	5	-	-	-	-	-
TW G2	67	28	15	9	5	2	-	-	-
TW G3	57	26	15	9	6	3	2	-	-
TW G4	47	23	13	8	5	3	2	1	-
TW H1	58	22	11	6	3	-	-	-	-
TW H2	46	21	12	7	4	3	1	-	-
TW H3	40	18	11	7	4	3	2	-	-
TW H4	32	15	8	5	3	2	1	-	-
TW I1	35	-	-	-	-	-	-	-	-
TW I2	88	43	25	16	11	7	5	3	2
TW I3	95	48	29	19	13	9	6	4	2
TW I4	104	58	38	26	18	13	9	7	4
TW I5	105	61	40	28	20	15	11	8	5
TW J1	119	62	38	25	17	12	8	6	3
TW J2	112	61	38	26	18	13	9	6	4
TW J3	128	78	52	37	27	20	15	11	7
TW J4	119	72	49	35	26	19	14	10	7
DG A1	50	-	-	-	-	-	-	-	-
DG A2	90	37	19	9	-	-	-	-	-
DG A3	130	59	35	23	16	11	7	3	-
DG B1	52	-	-	-	-	-	-	-	-
DG B2	92	39	22	13	6	-	-	-	-
DG B3	134	61	37	25	18	13	9	6	2
DG C1	78	28	-	-	-	-	-	-	-
DG C2	115	50	29	18	12	7	-	-	-
DG C3	156	72	44	30	22	16	11	8	4
SS A1	242	65	30	15	8	3	-	-	-
SS A2	241	97	54	33	22	14	10	6	3
SS A3	217	51	20	-	-	-	-	-	-
SS A4	239	93	50	30	19	12	8	5	3
SS A5	398	61	24	10	-	-	-	-	-
SS A6	236	94	52	32	20	13	9	6	3
SS A7	421	43	9	-	-	-	-	-	-
SS A8	236	89	47	28	17	11	7	4	2
SS B1	158	46	21	11	5	-	-	-	-
SS B2	190	79	45	28	19	13	8	5	3
SS B3	136	37	15	-	-	-	-	-	-
SS B4	182	75	42	26	17	11	7	5	3
SS B5	200	38	15	-	-	-	-	-	-
SS B6	176	74	42	26	17	12	8	5	3
SS B7	164	27	-	-	-	-	-	-	-
SS B8	166	69	38	23	15	10	6	4	2
SS C1	90	35	18	9	3	-	-	-	-
SS C2	115	57	34	22	15	10	7	4	3
SS C3	96	29	13	6	-	-	-	-	-
SS C4	114	51	29	19	12	8	6	3	2
SS D1	218	77	38	21	12	6	-	-	-
SS D2	212	106	64	42	28	19	13	9	5
SS D3	278	73	33	17	8	-	-	-	-
SS D4	217	104	61	40	27	18	12	8	5
SS E1	149	53	25	12	-	-	-	-	-
SS E2	167	82	49	31	21	14	9	6	3
SS E3	186	46	19	8	-	-	-	-	-
SS E4	173	78	44	28	18	12	8	5	3

MILES CITY, MONTANA 7889 DD

	SSF=.10	.20	.30	.40	.50	.60	.70	.80	.90
WW A1	384	33	7	-	-	-	-	-	-
WW A2	128	44	21	11	6	-	-	-	-
WW A3	109	46	24	14	8	5	2	-	-
WW A4	101	47	26	16	10	6	3	2	-
WW A5	96	48	28	17	11	7	4	2	1
WW A6	94	49	29	18	12	8	5	3	1
WW B1	69	-	-	-	-	-	-	-	-
WW B2	110	55	33	21	14	10	6	4	2
WW B3	130	69	42	28	19	13	9	6	4
WW B4	125	73	48	33	24	17	13	9	6
WW B5	119	72	48	34	25	18	13	9	6
WW C1	142	76	47	31	22	15	11	7	4
WW C2	128	72	46	31	22	16	11	8	5
WW C3	142	89	61	44	33	24	18	13	9
WW C4	129	82	56	40	30	22	17	12	8
TW A1	337	34	11	-	-	-	-	-	-
TW A2	134	41	19	10	5	-	-	-	-
TW A3	109	42	21	12	7	4	2	-	-
TW A4	94	41	22	13	8	5	3	2	-
TW B1	188	35	14	6	-	-	-	-	-
TW B2	111	39	19	11	6	3	2	-	-
TW B3	98	38	20	12	7	4	2	1	-
TW B4	92	37	19	12	7	5	3	2	-
TW C1	128	35	16	8	5	2	-	-	-
TW C2	101	35	18	10	6	4	2	1	-
TW C3	98	34	17	10	6	4	2	1	-
TW C4	106	31	15	9	5	3	2	1	-
TW D1	52	-	-	-	-	-	-	-	-
TW D2	119	53	30	19	12	8	5	3	2
TW D3	120	57	33	21	14	9	6	4	2
TW D4	127	68	42	29	20	14	10	7	4
TW D5	124	69	44	30	22	16	11	8	5
TW E1	145	70	41	27	18	12	9	6	3
TW E2	137	70	43	28	20	14	10	7	4
TW E3	152	88	57	40	29	21	15	11	7
TW E4	139	82	54	38	27	20	15	10	7
TW F1	142	28	10	-	-	-	-	-	-
TW F2	93	33	16	8	4	-	-	-	-
TW F3	78	33	18	10	6	3	1	-	-
TW F4	65	31	17	10	7	4	2	1	-
TW G1	89	26	11	5	-	-	-	-	-
TW G2	68	28	14	8	5	2	-	-	-
TW G3	58	26	15	9	5	3	2	-	-
TW G4	48	23	13	8	5	3	2	1	-
TW H1	59	22	11	6	3	-	-	-	-
TW H2	47	21	11	7	4	2	1	-	-
TW H3	40	19	10	6	4	2	1	-	-
TW H4	32	15	8	5	3	2	1	-	-
TW I1	36	-	-	-	-	-	-	-	-
TW I2	89	43	25	16	11	7	5	3	2
TW I3	96	49	29	19	13	9	6	4	2
TW I4	105	59	37	26	18	13	9	6	4
TW I5	106	61	40	28	20	14	11	7	5
TW J1	120	62	38	25	17	12	8	5	3
TW J2	113	61	38	26	18	13	9	6	4
TW J3	129	78	52	37	27	20	15	10	7
TW J4	120	73	49	35	25	19	14	10	6
DG A1	52	-	-	-	-	-	-	-	-
DG A2	92	37	19	9	-	-	-	-	-
DG A3	133	60	35	23	16	11	7	3	-
DG B1	55	-	-	-	-	-	-	-	-
DG B2	95	40	21	12	6	-	-	-	-
DG B3	137	62	37	25	17	12	9	6	2
DG C1	83	27	-	-	-	-	-	-	-

MILES CITY, MONTANA							CONTINUED		
SSF =	.10	.20	.30	.40	.50	.60	.70	.80	.90
DG C2	119	52	29	18	11	7	-	-	-
DG C3	159	73	44	30	21	15	11	8	4
SS A1	229	61	27	13	7	2	-	-	-
SS A2	232	93	51	31	20	13	9	6	3
SS A3	202	46	16	-	-	-	-	-	-
SS A4	228	88	46	28	18	11	7	5	3
SS A5	383	57	21	8	-	-	-	-	-
SS A6	227	90	49	29	19	13	8	5	3
SS A7	404	38	-	-	-	-	-	-	-
SS A8	224	84	43	25	16	10	6	4	2
SS B1	151	44	19	9	4	-	-	-	-
SS B2	184	77	43	26	17	12	8	5	3
SS B3	128	34	12	-	-	-	-	-	-
SS B4	175	72	39	24	16	10	7	4	3
SS B5	191	36	12	-	-	-	-	-	-
SS B6	170	72	40	25	16	11	7	5	3
SS B7	154	23	-	-	-	-	-	-	-
SS B8	160	65	35	21	14	9	6	4	2
SS C1	92	35	17	9	-	-	-	-	-
SS C2	117	57	34	22	15	10	7	4	3
SS C3	97	29	13	6	-	-	-	-	-
SS C4	116	51	29	18	12	8	5	3	2
SS D1	207	72	35	18	10	5	2	-	-
SS D2	204	102	60	39	26	18	13	8	5
SS D3	264	68	29	14	7	-	-	-	-
SS D4	209	99	58	37	25	17	12	8	5
SS E1	142	49	22	9	-	-	-	-	-
SS E2	162	79	46	29	19	13	9	6	3
SS E3	177	43	16	5	-	-	-	-	-
SS E4	168	75	42	26	17	11	7	5	3

MISSOULA, MONTANA								7931 DD	
SSF =	.10	.20	.30	.40	.50	.60	.70	.80	.90
WW A1	286	-	-	-	-	-	-	-	-
WW A2	91	23	-	-	-	-	-	-	-
WW A3	77	26	-	-	-	-	-	-	-
WW A4	71	27	-	-	-	-	-	-	-
WW A5	67	28	-	-	-	-	-	-	-
WW A6	65	29	10	-	-	-	-	-	-
WW B1	-	-	-	-	-	-	-	-	-
WW B2	83	38	20	11	-	-	-	-	-
WW B3	101	50	29	17	9	4	-	-	-
WW B4	101	57	36	24	16	11	7	5	3
WW B5	98	57	37	26	18	13	9	6	3
WW C1	113	57	34	21	13	8	4	2	-
WW C2	102	55	34	22	15	10	6	4	2
WW C3	119	73	49	35	25	18	13	9	6
WW C4	108	67	45	32	23	17	12	8	5
TW A1	256	-	-	-	-	-	-	-	-
TW A2	99	23	-	-	-	-	-	-	-
TW A3	79	25	-	-	-	-	-	-	-
TW A4	68	26	9	-	-	-	-	-	-
TW B1	143	17	-	-	-	-	-	-	-
TW B2	82	24	-	-	-	-	-	-	-
TW B3	73	24	9	-	-	-	-	-	-
TW B4	70	24	10	-	-	-	-	-	-
TW C1	98	21	-	-	-	-	-	-	-
TW C2	77	23	9	-	-	-	-	-	-
TW C3	76	23	9	-	-	-	-	-	-
TW C4	85	22	9	3	-	-	-	-	-
TW D1	-	-	-	-	-	-	-	-	-
TW D2	91	37	19	10	-	-	-	-	-
TW D3	93	41	21	11	-	-	-	-	-

MISSOULA, MONTANA							CONTINUED		
SSF =	.10	.20	.30	.40	.50	.60	.70	.80	.90
TW D4	103	53	32	20	13	9	6	3	2
TW D5	102	55	34	23	16	11	7	5	3
TW E1	115	53	29	17	10	5	2	-	-
TW E2	110	54	31	20	12	8	5	2	1
TW E3	126	71	45	31	21	15	10	7	4
TW E4	116	67	43	30	21	15	10	7	4
TW F1	102	-	-	-	-	-	-	-	-
TW F2	65	16	-	-	-	-	-	-	-
TW F3	54	18	-	-	-	-	-	-	-
TW F4	46	18	-	-	-	-	-	-	-
TW G1	62	-	-	-	-	-	-	-	-
TW G2	47	15	-	-	-	-	-	-	-
TW G3	41	15	-	-	-	-	-	-	-
TW G4	34	14	5	-	-	-	-	-	-
TW H1	40	10	-	-	-	-	-	-	-
TW H2	33	12	-	-	-	-	-	-	-
TW H3	29	11	4	-	-	-	-	-	-
TW H4	24	10	4	-	-	-	-	-	-
TW I1	-	-	-	-	-	-	-	-	-
TW I2	68	30	16	8	-	-	-	-	-
TW I3	75	35	19	11	5	-	-	-	-
TW I4	85	46	28	18	12	8	5	3	2
TW I5	87	49	31	21	14	10	7	4	3
TW J1	95	46	27	16	10	5	3	-	-
TW J2	90	47	28	18	12	8	5	3	1
TW J3	107	63	42	29	20	14	10	7	4
TW J4	100	60	39	27	19	14	10	7	4
DG A1	-	-	-	-	-	-	-	-	-
DG A2	70	24	-	-	-	-	-	-	-
DG A3	112	48	27	17	10	6	-	-	-
DG B1	-	-	-	-	-	-	-	-	-
DG B2	73	27	10	-	-	-	-	-	-
DG B3	115	51	30	19	12	8	5	2	-
DG C1	50	-	-	-	-	-	-	-	-
DG C2	94	38	19	9	-	-	-	-	-
DG C3	134	60	36	23	15	10	6	4	-
SS A1	201	46	13	-	-	-	-	-	-
SS A2	206	79	41	23	14	8	4	2	-
SS A3	176	29	-	-	-	-	-	-	-
SS A4	204	74	37	20	10	5	-	-	-
SS A5	345	42	-	-	-	-	-	-	-
SS A6	201	75	38	21	12	6	3	1	-
SS A7	367	-	-	-	-	-	-	-	-
SS A8	200	70	33	17	7	-	-	-	-
SS B1	127	30	-	-	-	-	-	-	-
SS B2	162	64	34	20	12	7	4	2	1
SS B3	105	15	-	-	-	-	-	-	-
SS B4	155	60	31	17	10	5	3	1	-
SS B5	164	22	-	-	-	-	-	-	-
SS B6	149	59	31	18	10	6	3	1	-
SS B7	130	-	-	-	-	-	-	-	-
SS B8	140	54	27	14	7	3	-	-	-
SS C1	60	-	-	-	-	-	-	-	-
SS C2	90	41	22	12	6	-	-	-	-
SS C3	68	-	-	-	-	-	-	-	-
SS C4	90	37	19	10	4	-	-	-	-
SS D1	176	54	18	-	-	-	-	-	-
SS D2	179	85	49	30	19	12	7	4	2
SS D3	231	51	-	-	-	-	-	-	-
SS D4	184	84	47	28	17	11	6	3	2
SS E1	112	28	-	-	-	-	-	-	-
SS E2	137	63	35	20	11	6	2	-	-
SS E3	148	26	-	-	-	-	-	-	-
SS E4	144	60	31	17	9	4	-	-	-

```
GRAND ISLAND, NEBRASKA                    6425 DD    GRAND ISLAND, NEBRASKA                    CONTINUED
   SSF = .10   .20   .30   .40   .50   .60   .70   .80   .90       SSF = .10   .20   .30   .40   .50   .60   .70   .80   .90
WW A1    524    52    21    11     7     4     3     2     -    DG C2    147    65    38    26    18    13     9     5     -
WW A2    172    63    34    21    14    10     7     5     3    DG C3    190    87    53    37    27    21    16    12     7
WW A3    145    65    38    24    17    12     8     6     4    SS A1    305    86    42    25    16    10     7     5     3
WW A4    133    65    39    26    18    13     9     6     4    SS A2    292   120    67    43    29    21    15    10     6
WW A5    127    67    42    28    20    14    10     7     5    SS A3    280    71    33    18    11     7     4     3     1
WW A6    124    67    43    29    21    15    11     8     5    SS A4    290   116    64    40    27    19    13     9     6
WW B1    115    34    15     8     4     2     -     -     -    SS A5    517    83    36    20    12     8     5     3     2
WW B2    139    72    45    30    21    16    11     8     5    SS A6    287   117    66    42    28    20    14    10     6
WW B3    164    89    57    39    28    20    15    11     7    SS A7    568    64    25    12     7     4     2     -     -
WW B4    152    90    60    43    32    24    18    13     9    SS A8    288   111    61    38    25    18    12     8     5
WW B5    143    87    59    43    32    24    18    13     9    SS B1    203    63    32    19    12     8     5     4     2
WW C1    177    97    62    43    31    22    17    12     8    SS B2    230    97    56    36    25    18    13     9     6
WW C2    156    90    59    41    30    22    17    12     8    SS B3    179    54    26    15     9     6     4     2     1
WW C3    169   107    75    55    41    31    24    17    12    SS B4    221    93    53    34    23    16    12     8     5
WW C4    153    98    68    50    37    28    22    16    11    SS B5    263    54    24    13     8     5     3     2     1
TW A1    454    51    22    12     7     5     3     2     1    SS B6    214    92    53    34    23    17    12     8     5
TW A2    179    58    30    18    12     8     6     4     2    SS B7    221    42    18     9     5     3     1     -     -
TW A3    144    58    33    21    14    10     7     5     3    SS B8    204    86    49    31    21    15    10     7     5
TW A4    123    56    33    21    15    10     7     5     3    SS C1    125    53    30    19    13     9     6     4     3
TW B1    252    51    23    13     8     6     4     2     1    SS C2    147    74    46    31    22    16    11     8     5
TW B2    146    53    29    18    12     8     6     4     3    SS C3    132    43    22    13     9     6     4     3     2
TW B3    128    52    30    19    13     9     6     4     3    SS C4    146    67    39    26    18    13     9     7     4
TW B4    119    49    28    18    12     9     6     4     3    SS D1    276   102    54    33    22    15    10     7     4
TW C1    169    48    24    15     9     6     4     3     2    SS D2    256   130    80    54    38    27    20    14     9
TW C2    131    47    26    16    11     8     5     4     2    SS D3    352    96    47    27    17    11     8     5     3
TW C3    126    45    24    15    10     7     5     3     2    SS D4    262   128    77    51    36    26    18    13     8
TW C4    136    41    21    13     9     6     4     3     2    SS E1    196    75    40    24    16    10     7     5     3
TW D1     85    28    13     7     4     2     1     -     -    SS E2    206   104    63    42    30    21    15    11     7
TW D2    150    68    40    27    18    13    10     7     4    SS E3    243    65    31    18    11     7     5     3     2
TW D3    152    75    45    30    21    15    11     8     5    SS E4    213    98    57    37    26    18    13     9     6
TW D4    155    84    54    37    27    20    15    11     7
TW D5    149    83    54    38    28    21    15    11     7
TW E1    181    89    55    37    26    19    14    10     6    NORTH OMAHA, NEBRASKA                     6601 DD
TW E2    168    87    55    37    27    20    14    10     7       SSF = .10   .20   .30   .40   .50   .60   .70   .80   .90
TW E3    183   107    71    50    37    28    21    15    10    WW A1    444    38    14     7     3     1     -     -     -
TW E4    166    98    66    47    35    26    19    14    10    WW A2    142    50    26    16    10     7     5     3     2
TW F1    196    44    20    12     7     5     3     2     1    WW A3    119    52    29    19    13     9     6     4     2
TW F2    125    48    26    16    11     7     5     3     2    WW A4    110    53    31    20    14    10     7     5     3
TW F3    104    47    27    18    12     8     6     4     3    WW A5    105    54    33    22    15    11     8     5     3
TW F4     86    43    26    17    12     9     6     4     3    WW A6    102    55    34    23    16    12     8     6     4
TW G1    122    39    20    12     8     5     4     2     1    WW B1     83    19     -     -     -     -     -     -     -
TW G2     91    40    23    15    10     7     5     3     2    WW B2    118    60    37    25    17    13     9     6     4
TW G3     77    37    22    14    10     7     5     4     2    WW B3    140    75    47    32    23    17    12     9     6
TW G4     63    31    19    13     9     6     5     3     2    WW B4    132    78    52    37    27    20    15    11     7
TW H1     80    32    18    11     8     5     4     2     2    WW B5    126    76    52    37    27    21    15    11     8
TW H2     63    29    17    11     8     6     4     3     2    WW C1    152    82    52    36    26    19    14    10     6
TW H3     53    26    15    10     7     5     4     3     2    WW C2    136    77    50    35    25    19    14    10     7
TW H4     42    20    12     8     5     4     3     2     1    WW C3    149    95    66    48    36    27    21    15    10
TW I1     63    22    10     6     3     1     -     -     -    WW C4    136    86    60    44    33    25    19    14    10
TW I2    114    57    35    23    16    12     9     6     4    TW A1    387    39    15     8     4     2     1     -     -
TW I3    122    63    39    27    19    14    10     7     5    TW A2    149    46    23    14     9     6     4     3     1
TW I4    129    73    47    33    24    18    13    10     6    TW A3    120    47    26    16    11     7     5     3     2
TW I5    128    74    49    35    26    19    14    10     7    TW A4    103    46    26    17    11     8     6     4     2
TW J1    150    79    50    34    24    18    13     9     6    TW B1    212    40    17    10     6     4     2     1     -
TW J2    139    76    49    34    25    18    13    10     6    TW B2    122    43    23    14     9     6     4     3     2
TW J3    154    94    64    46    34    26    19    14    10    TW B3    107    43    24    15    10     7     5     3     2
TW J4    143    87    60    43    32    24    18    13     9    TW B4    100    40    22    14    10     7     5     3     2
DG A1     84    27     -     -     -     -     -     -     -    TW C1    142    39    19    11     7     5     3     2     1
DG A2    117    50    28    17    10     4     -     -     -    TW C2    110    39    20    13     8     6     4     3     2
DG A3    159    72    43    30    21    15    11     7     -    TW C3    107    37    20    12     8     6     4     3     2
DG B1     85    31    13     -     -     -     -     -     -    TW C4    117    34    17    10     7     5     3     2     1
DG B2    120    52    30    20    13     9     5     -     -    TW D1     62    17     6     -     -     -     -     -     -
DG B3    163    75    45    31    23    17    13     9     5    TW D2    128    57    33    22    15    11     8     5     3
DG C1    113    46    24    14     6     -     -     -     -    TW D3    129    62    37    25    17    12     9     6     4
```

NORTH OMAHA, NEBRASKA								CONTINUED	
SSF =.10	.20	.30	.40	.50	.60	.70	.80	.90	
TW D4	135	72	46	32	23	17	12	9	6
TW D5	131	73	47	33	24	18	13	10	6
TW E1	156	76	46	31	21	15	11	8	5
TW E2	146	75	47	32	23	16	12	9	6
TW E3	161	94	62	44	32	24	18	13	9
TW E4	147	86	58	41	30	23	17	12	8
TW F1	161	33	14	8	4	3	1	-	-
TW F2	102	38	20	12	8	5	4	2	1
TW F3	85	37	21	13	9	6	4	3	2
TW F4	71	34	21	13	9	7	5	3	2
TW G1	100	30	15	8	5	3	2	1	-
TW G2	74	32	18	11	7	5	4	2	1
TW G3	64	30	17	11	8	5	4	3	2
TW G4	52	25	15	10	7	5	3	2	1
TW H1	65	25	14	8	6	4	3	2	-
TW H2	52	24	14	9	6	4	3	2	1
TW H3	44	21	12	8	5	4	3	2	1
TW H4	35	16	10	6	4	3	2	1	-
TW I1	44	12	-	-	-	-	-	-	-
TW I2	96	47	29	19	13	10	7	5	3
TW I3	104	53	33	22	16	11	8	6	4
TW I4	112	63	41	28	21	15	11	8	5
TW I5	112	65	43	30	22	17	12	9	6
TW J1	129	67	42	28	20	15	11	8	5
TW J2	120	65	42	29	21	15	11	8	5
TW J3	136	83	56	40	30	22	17	12	8
TW J4	126	77	52	38	28	21	16	12	8
DG A1	61	-	-	-	-	-	-	-	-
DG A2	99	41	22	12	6	-	-	-	-
DG A3	140	62	37	25	18	13	9	5	-
DG B1	62	17	-	-	-	-	-	-	-
DG B2	101	42	24	15	10	5	-	-	-
DG B3	144	65	39	27	20	15	11	7	4
DG C1	89	32	15	-	-	-	-	-	-
DG C2	125	54	31	21	14	10	6	-	-
DG C3	168	76	46	32	23	18	14	10	6
SS A1	261	70	33	19	12	8	5	3	2
SS A2	255	103	57	36	25	17	12	8	5
SS A3	234	56	25	13	7	4	2	1	-
SS A4	252	99	54	34	23	16	11	7	5
SS A5	447	67	28	14	8	5	3	2	-
SS A6	250	100	56	35	24	16	12	8	5
SS A7	490	48	17	7	-	-	-	-	-
SS A8	249	94	51	31	21	14	10	7	4
SS B1	171	51	25	14	9	6	4	2	1
SS B2	201	84	48	31	21	15	10	7	5
SS B3	148	42	19	10	6	3	2	-	-
SS B4	193	80	45	28	19	13	10	7	4
SS B5	222	43	18	9	5	3	2	-	-
SS B6	186	79	45	29	20	14	10	7	4
SS B7	183	31	11	5	-	-	-	-	-
SS B8	177	73	41	26	17	12	9	6	4
SS C1	101	41	22	14	9	6	4	3	1
SS C2	125	62	38	25	18	13	9	7	4
SS C3	108	33	16	9	6	4	2	2	-
SS C4	125	56	33	21	15	10	7	5	3
SS D1	233	84	43	26	17	11	7	5	3
SS D2	223	112	69	46	32	23	17	12	8
SS D3	300	79	37	21	13	8	5	3	2
SS D4	229	110	66	43	30	21	15	11	7
SS E1	161	59	30	18	11	7	5	3	1
SS E2	178	88	53	35	24	17	12	9	6
SS E3	203	51	23	13	7	5	3	1	-
SS E4	184	83	48	31	21	15	11	7	5

NORTH PLATTE, NEBRASKA									6743 DD
SSF =.10	.20	.30	.40	.50	.60	.70	.80	.90	
WW A1	553	59	25	14	9	6	4	2	1
WW A2	185	71	39	25	17	12	8	6	4
WW A3	158	73	43	28	20	14	10	7	5
WW A4	145	73	45	30	21	15	11	8	5
WW A5	139	74	47	32	23	17	12	9	6
WW A6	135	75	48	33	24	18	13	9	6
WW B1	130	42	21	12	7	5	3	2	-
WW B2	150	79	50	34	24	18	13	9	6
WW B3	176	97	63	44	32	23	17	12	8
WW B4	162	97	66	47	35	26	20	14	10
WW B5	152	94	64	46	35	26	20	15	10
WW C1	190	105	68	47	34	25	19	14	9
WW C2	167	97	64	45	33	25	19	14	9
WW C3	179	115	80	59	45	34	26	19	13
WW C4	163	104	73	54	41	31	24	17	12
TW A1	479	58	26	14	9	6	4	3	2
TW A2	191	64	34	21	14	10	7	5	3
TW A3	156	65	37	24	16	12	8	6	4
TW A4	133	62	37	25	17	12	9	6	4
TW B1	267	57	27	16	10	7	5	3	2
TW B2	156	59	33	21	14	10	7	5	3
TW B3	137	58	33	22	15	11	8	5	3
TW B4	128	54	31	20	14	10	7	5	3
TW C1	180	53	27	17	11	8	5	4	2
TW C2	140	52	29	18	13	9	6	4	3
TW C3	134	49	27	17	12	8	6	4	3
TW C4	143	45	24	15	10	7	5	3	2
TW D1	96	34	17	10	7	4	3	2	-
TW D2	161	75	45	30	21	15	11	8	5
TW D3	164	82	50	34	24	18	13	9	6
TW D4	165	91	58	41	30	22	16	12	8
TW D5	158	89	59	41	30	23	17	12	8
TW E1	194	97	60	41	29	21	16	11	7
TW E2	179	95	60	41	30	22	16	12	8
TW E3	195	115	77	55	41	30	23	17	11
TW E4	176	105	71	51	38	28	21	16	11
TW F1	210	50	24	14	9	6	4	3	2
TW F2	135	54	30	19	13	9	6	4	3
TW F3	113	52	31	20	14	10	7	5	3
TW F4	94	48	29	20	14	10	7	5	3
TW G1	132	44	23	14	10	7	5	3	2
TW G2	99	45	26	17	12	8	6	4	3
TW G3	84	41	25	17	12	8	6	4	3
TW G4	68	35	21	14	10	7	5	4	2
TW H1	87	36	21	13	9	6	5	3	2
TW H2	68	33	20	13	9	7	5	3	2
TW H3	58	28	17	12	8	6	4	3	2
TW H4	45	22	13	9	6	5	3	2	2
TW I1	72	27	14	9	5	3	2	1	-
TW I2	123	62	39	26	19	14	10	7	5
TW I3	131	69	44	30	21	16	12	8	5
TW I4	137	79	52	36	27	20	15	11	7
TW I5	136	80	53	38	28	21	16	12	8
TW J1	161	86	55	38	27	20	15	11	7
TW J2	148	82	54	38	27	20	15	11	7
TW J3	164	101	69	50	37	28	21	16	11
TW J4	151	94	64	47	35	26	20	15	10
DG A1	95	35	-	-	-	-	-	-	-
DG A2	126	55	32	20	13	7	-	-	-
DG A3	167	77	47	32	24	17	12	8	3
DG B1	95	38	20	8	-	-	-	-	-
DG B2	129	57	34	23	16	11	7	-	-
DG B3	171	79	49	34	25	19	15	10	6
DG C1	125	53	30	19	12	-	-	-	-

```
NORTH PLATTE, NEBRASKA                    CONTINUED     SCOTTSBLUFF, NEBRASKA                     CONTINUED
   SSF =.10   .20   .30   .40   .50   .60   .70   .80   .90      SSF =.10   .20   .30   .40   .50   .60   .70   .80   .90
DG C2    156    71    43    29    21    15    11     7     2   TW D4    167    92    60    42    30    23    17    12     8
DG C3    199    92    57    40    30    23    18    13     8   TW D5    159    91    60    42    31    23    17    13     9
SS A1    327    95    48    29    19    13     9     6     3   TW E1    195    99    62    42    30    22    16    12     8
SS A2    311   130    74    48    33    23    17    12     7   TW E2    181    96    61    42    31    23    17    12     8
SS A3    302    81    39    22    14     9     6     4     2   TW E3    196   117    78    56    41    31    23    17    12
SS A4    311   126    71    45    31    22    15    11     7   TW E4    177   107    72    52    39    29    22    16    11
SS A5    546    93    42    24    15    10     6     4     2   TW F1    207    51    25    15    10     7     5     3     2
SS A6    307   128    73    47    32    23    16    11     7   TW F2    136    55    31    20    14    10     7     5     3
SS A7    601    73    30    16     9     6     3     2     1   TW F3    114    54    32    21    15    11     8     5     3
SS A8    309   122    68    43    29    20    14    10     6   TW F4     95    49    30    21    15    11     8     5     4
SS B1    219    71    36    22    14    10     7     4     3   TW G1    132    45    24    15    10     7     5     3     2
SS B2    244   106    61    40    28    20    14    10     6   TW G2    100    46    27    18    12     9     6     4     3
SS B3    195    62    31    18    12     8     5     3     2   TW G3     85    42    26    17    12     9     6     5     3
SS B4    236   101    58    38    26    18    13     9     6   TW G4     69    35    22    15    11     8     6     4     3
SS B5    280    61    28    16    10     7     4     3     2   TW H1     87    37    22    14    10     7     5     3     2
SS B6    228   100    58    38    26    19    13     9     6   TW H2     69    34    20    14    10     7     5     4     2
SS B7    238    48    21    12     7     4     3     2     -   TW H3     58    29    18    12     8     6     4     3     2
SS B8    219    94    54    35    24    17    12     8     5   TW H4     45    22    14     9     7     5     3     2     2
SS C1    137    60    35    22    15    11     8     5     3   TW I1     73    29    15     9     6     4     3     2     -
SS C2    158    81    51    34    25    18    13     9     6   TW I2    124    64    40    27    19    14    10     7     5
SS C3    143    48    26    16    11     7     5     3     2   TW I3    132    70    45    31    22    16    12     9     6
SS C4    156    73    44    29    20    15    11     8     5   TW I4    139    80    53    37    27    20    15    11     8
SS D1    299   114    62    39    26    18    12     8     5   TW I5    137    81    55    39    29    22    16    12     8
SS D2    275   142    88    60    42    31    22    16    10   TW J1    162    87    56    39    28    21    15    11     7
SS D3    377   107    54    32    21    14     9     6     4   TW J2    149    84    55    38    28    21    16    11     8
SS D4    282   139    85    57    40    29    21    15    10   TW J3    165   102    70    51    38    29    22    16    11
SS E1    214    85    46    29    19    13     9     6     4   TW J4    153    95    65    48    36    27    20    15    10
SS E2    222   114    70    47    33    24    17    12     8   DG A1     97    37    13     -     -     -     -     -     -
SS E3    261    73    36    21    14     9     6     4     2   DG A2    126    56    33    21    13     8     -     -     -
SS E4    229   107    63    42    29    21    15    10     7   DG A3    167    78    48    33    24    18    13     8     4
                                                                DG B1     97    39    21    11     -     -     -     -     -
                                                                DG B2    129    59    35    24    17    12     7     3     -
   SCOTTSBLUFF, NEBRASKA                         6774 DD        DG B3    171    80    50    35    26    20    15    11     6
   SSF =.10   .20   .30   .40   .50   .60   .70   .80   .90    DG C1    127    55    31    20    13     6     -     -     -
WW A1    526    60    27    15     9     6     4     3     2   DG C2    157    72    44    30    22    16    11     7     3
WW A2    186    72    40    26    17    12     9     6     4   DG C3    198    93    59    41    31    24    18    14     9
WW A3    159    74    44    29    20    15    11     7     5   SS A1    323    97    49    30    19    13     9     6     4
WW A4    147    75    46    31    22    16    12     8     5   SS A2    311   132    76    49    34    24    17    12     8
WW A5    141    76    49    33    24    18    13     9     6   SS A3    298    83    40    23    15    10     6     4     2
WW A6    137    77    50    35    25    18    13    10     6   SS A4    310   128    73    46    32    22    16    11     7
WW B1    131    44    22    13     8     5     3     2     1   SS A5    525    95    43    25    15    10     7     4     3
WW B2    152    80    51    35    25    18    14    10     6   SS A6    307   130    74    48    33    23    17    12     7
WW B3    178    99    64    45    32    24    18    13     9   SS A7    568    75    31    17    10     6     4     2     1
WW B4    164    99    67    48    36    27    20    15    10   SS A8    308   124    69    44    30    21    15    10     6
WW B5    154    95    65    47    35    27    20    15    10   SS B1    218    72    38    23    15    10     7     5     3
WW C1    192   107    69    49    35    26    19    14     9   SS B2    244   107    62    41    28    20    15    10     7
WW C2    169    99    66    47    34    26    19    14     9   SS B3    194    63    32    19    12     8     5     4     2
WW C3    181   117    82    60    46    35    26    20    13   SS B4    236   103    59    39    27    19    14     9     6
WW C4    164   106    74    55    41    32    24    18    12   SS B5    273    62    30    17    11     7     5     3     2
TW A1    458    59    27    15    10     6     4     3     2   SS B6    229   102    60    39    27    19    14    10     6
TW A2    191    66    35    22    15    10     7     5     3   SS B7    232    50    22    12     8     5     3     2     -
TW A3    157    66    38    25    17    12     9     6     4   SS B8    219    96    55    36    25    17    12     9     6
TW A4    135    64    38    25    18    13     9     7     4   SS C1    139    62    36    23    16    11     8     6     4
TW B1    262    58    28    17    11     7     5     3     2   SS C2    160    83    52    35    25    19    14    10     7
TW B2    156    61    34    22    15    10     7     5     3   SS C3    143    49    26    16    11     8     5     4     2
TW B3    138    59    34    22    16    11     8     6     4   SS C4    157    74    45    30    21    15    11     8     5
TW B4    128    55    32    21    15    10     8     5     3   SS D1    299   116    64    40    27    18    13     9     5
TW C1    178    54    28    17    12     8     6     4     3   SS D2    277   144    90    61    43    32    23    16    11
TW C2    139    53    30    19    13     9     7     5     3   SS D3    372   109    55    33    21    14    10     7     4
TW C3    134    50    28    18    12     9     6     4     3   SS D4    283   141    86    58    41    29    21    15    10
TW C4    142    45    24    15    10     7     5     4     2   SS E1    215    87    48    30    20    14     9     6     4
TW D1     97    35    18    11     7     5     3     2     1   SS E2    224   116    72    49    34    25    18    13     8
TW D2    162    76    46    31    22    16    11     8     5   SS E3    258    75    37    22    14     9     6     4     2
TW D3    165    84    52    35    25    18    13    10     6   SS E4    230   109    65    43    30    21    15    11     7
```

ELKO, NEVADA							7483 DD		
SSF =.10	.20	.30	.40	.50	.60	.70	.80	.90	
WW A1	619	70	31	18	11	7	5	3	2
WW A2	213	82	46	29	20	14	10	6	4
WW A3	180	84	50	33	23	16	12	8	5
WW A4	166	84	52	35	25	18	13	9	6
WW A5	159	86	54	37	27	19	14	10	6
WW A6	154	86	56	39	28	20	15	10	7
WW B1	159	54	28	16	10	6	4	2	-
WW B2	169	89	56	39	28	20	15	10	7
WW B3	199	110	71	50	36	26	19	14	9
WW B4	180	108	73	53	39	29	22	16	11
WW B5	168	103	71	51	38	29	22	16	11
WW C1	214	119	77	54	39	29	21	15	10
WW C2	186	108	72	51	37	28	21	15	10
WW C3	198	127	89	65	49	37	28	21	14
WW C4	178	114	80	59	45	34	26	19	13
TW A1	537	68	31	18	11	7	5	3	2
TW A2	220	75	40	25	17	12	8	5	3
TW A3	178	75	43	28	19	14	10	7	4
TW A4	152	72	43	29	20	14	10	7	4
TW B1	306	66	32	19	12	8	6	4	2
TW B2	178	68	38	24	17	12	8	6	3
TW B3	157	67	39	25	17	12	9	6	4
TW B4	145	62	36	23	16	12	8	6	4
TW C1	205	61	32	20	13	9	6	4	3
TW C2	159	60	33	21	15	10	7	5	3
TW C3	152	56	31	20	14	10	7	5	3
TW C4	162	51	27	17	11	8	6	4	2
TW D1	116	43	23	14	9	6	3	2	-
TW D2	181	85	51	34	24	17	12	9	6
TW D3	186	93	58	39	28	20	14	10	7
TW D4	183	101	65	46	33	24	18	13	9
TW D5	174	99	65	46	34	25	19	13	9
TW E1	219	110	68	47	33	24	17	12	8
TW E2	200	106	67	46	33	25	18	13	8
TW E3	216	128	86	61	45	34	25	18	12
TW E4	193	116	78	56	42	31	23	17	11
TW F1	243	59	29	17	11	7	5	3	2
TW F2	156	62	35	23	15	11	7	5	3
TW F3	130	61	36	24	17	12	8	6	4
TW F4	107	55	34	23	16	12	8	6	4
TW G1	153	52	28	17	11	8	5	4	2
TW G2	113	52	31	20	14	10	7	5	3
TW G3	96	47	29	19	14	10	7	5	3
TW G4	78	40	25	17	12	8	6	4	3
TW H1	100	43	24	16	11	8	5	4	2
TW H2	78	38	23	15	11	8	5	4	2
TW H3	66	33	20	13	9	7	5	3	2
TW H4	51	25	15	10	7	5	4	3	2
TW I1	88	35	19	11	7	5	3	1	-
TW I2	138	70	44	30	21	15	11	8	5
TW I3	148	79	50	34	24	18	13	9	6
TW I4	152	87	58	41	30	22	16	12	8
TW I5	150	88	59	42	31	23	17	13	8
TW J1	181	97	62	43	31	22	16	12	8
TW J2	165	92	60	42	31	23	17	12	8
TW J3	181	112	77	55	41	31	23	17	11
TW J4	166	103	71	51	38	29	22	16	11
DG A1	115	45	21	-	-	-	-	-	-
DG A2	142	63	37	24	15	9	4	-	-
DG A3	184	85	53	37	26	19	14	9	4
DG B1	114	48	26	15	-	-	-	-	-
DG B2	145	66	40	27	19	13	8	4	-
DG B3	188	88	55	38	29	21	16	11	7
DG C1	145	64	37	24	15	9	-	-	-

ELKO, NEVADA							CONTINUED		
SSF =.10	.20	.30	.40	.50	.60	.70	.80	.90	
DG C2	173	80	49	34	24	18	13	8	3
DG C3	217	102	64	45	34	26	20	14	9
SS A1	401	118	60	36	23	15	10	7	4
SS A2	370	155	89	57	39	27	19	13	8
SS A3	379	104	51	29	18	12	7	4	2
SS A4	374	153	86	55	37	26	18	12	7
SS A5	669	117	54	30	19	12	8	5	3
SS A6	366	153	87	56	38	27	19	13	8
SS A7	746	96	41	22	13	7	4	2	1
SS A8	374	149	83	52	35	24	17	11	7
SS B1	266	87	46	28	18	12	8	5	3
SS B2	286	124	72	47	32	23	16	11	7
SS B3	242	78	40	24	15	10	6	4	2
SS B4	279	120	69	45	31	21	15	10	6
SS B5	343	77	36	21	13	8	5	3	2
SS B6	268	118	69	45	31	22	15	11	7
SS B7	298	63	29	16	10	6	3	2	-
SS B8	260	112	65	42	28	20	14	9	6
SS C1	159	70	41	27	18	13	9	6	4
SS C2	178	92	57	39	28	20	15	10	7
SS C3	165	57	30	19	13	9	6	4	2
SS C4	176	82	50	33	23	17	12	8	5
SS D1	366	141	77	48	32	21	14	9	6
SS D2	326	169	105	71	50	36	25	18	11
SS D3	463	134	68	40	26	17	11	7	4
SS D4	334	165	101	67	47	33	24	16	10
SS E1	262	106	59	36	24	16	11	7	4
SS E2	262	135	83	56	39	28	20	14	9
SS E3	319	91	46	27	17	11	7	5	3
SS E4	270	127	76	50	34	24	17	12	7

ELY, NEVADA							7814 DD		
SSF =.10	.20	.30	.40	.50	.60	.70	.80	.90	
WW A1	615	73	34	20	13	9	6	4	3
WW A2	217	86	49	32	22	16	12	8	5
WW A3	185	88	54	36	26	19	14	10	7
WW A4	171	89	56	39	28	20	15	11	7
WW A5	164	90	59	41	30	22	16	12	8
WW A6	159	91	60	42	31	23	17	13	8
WW B1	163	58	32	20	13	9	6	4	3
WW B2	173	93	60	42	30	23	17	12	8
WW B3	205	115	76	54	39	29	22	16	11
WW B4	184	112	77	56	42	32	24	18	12
WW B5	172	107	74	54	41	31	24	18	12
WW C1	219	124	81	58	42	32	24	17	12
WW C2	190	112	76	54	40	30	23	17	11
WW C3	203	132	93	69	53	40	31	23	16
WW C4	182	118	84	62	47	36	28	21	14
TW A1	534	71	33	20	13	9	6	4	3
TW A2	222	78	43	28	19	14	10	7	4
TW A3	181	78	46	31	22	16	11	8	5
TW A4	155	75	46	31	22	16	12	9	6
TW B1	306	69	35	21	14	10	7	5	3
TW B2	181	71	41	27	19	13	10	7	5
TW B3	159	69	41	27	19	14	10	7	5
TW B4	147	64	38	26	18	13	10	7	5
TW C1	207	64	34	22	15	11	8	5	3
TW C2	161	62	35	23	16	12	8	6	4
TW C3	154	58	33	22	15	11	8	6	4
TW C4	163	52	29	18	13	9	7	5	3
TW D1	120	46	26	16	11	8	5	4	2
TW D2	184	88	54	37	26	19	14	10	7
TW D3	190	97	61	42	31	23	17	12	8

ELY, NEVADA (CONTINUED) / LAS VEGAS, NEVADA — 2601 DD

ID	SSF=.10	.20	.30	.40	.50	.60	.70	.80	.90	ID	SSF=.10	.20	.30	.40	.50	.60	.70	.80	.90
TW D4	187	105	69	49	36	27	20	15	10	WW A1	1901	237	112	68	45	32	23	16	10
TW D5	177	102	68	49	36	27	20	15	10	WW A2	625	250	144	95	67	49	36	26	17
TW E1	223	115	72	50	36	27	20	14	10	WW A3	522	250	154	105	75	55	41	30	20
TW E2	204	110	71	50	36	27	20	15	10	WW A4	476	248	157	109	79	59	44	32	21
TW E3	221	132	90	65	48	37	28	20	14	WW A5	452	250	163	115	84	63	47	34	23
TW E4	197	120	82	59	44	34	26	19	13	WW A6	437	250	165	117	87	65	49	36	24
TW F1	244	62	32	19	13	9	6	4	3	WW B1	550	205	115	75	52	38	27	19	13
TW F2	158	65	38	25	17	13	9	6	4	WW B2	455	245	158	110	81	60	45	33	22
TW F3	133	64	39	26	19	14	10	7	5	WW B3	519	293	192	136	100	75	56	41	28
TW F4	110	58	37	25	18	13	10	7	5	WW B4	444	270	184	134	100	76	58	43	29
TW G1	155	54	30	19	13	9	7	5	3	WW B5	403	250	173	126	95	73	55	41	28
TW G2	116	54	33	22	16	11	8	6	4	WW C1	546	308	202	143	105	79	59	43	29
TW G3	99	50	31	21	15	11	8	6	4	WW C2	468	275	184	132	98	74	56	41	28
TW G4	80	42	26	18	13	10	7	5	3	WW C3	458	296	209	154	117	90	69	51	35
TW H1	102	45	26	18	12	9	6	5	3	WW C4	413	267	188	139	106	81	62	46	32
TW H2	80	40	25	17	12	9	7	5	3	TW A1	1616	224	107	65	44	31	22	15	10
TW H3	67	34	21	15	11	8	6	4	3	TW A2	639	227	126	82	57	41	30	21	14
TW H4	52	26	16	11	8	6	4	3	2	TW A3	511	222	132	88	63	46	34	24	16
TW I1	92	38	22	14	10	7	5	3	2	TW A4	430	208	128	88	63	46	34	25	17
TW I2	141	73	47	32	23	17	13	9	6	TW B1	901	207	106	66	45	32	23	16	11
TW I3	152	82	53	37	27	20	15	11	7	TW B2	509	202	117	77	54	39	29	21	14
TW I4	156	91	61	43	32	24	18	13	9	TW B3	441	194	115	77	55	40	30	21	14
TW I5	153	91	62	45	33	25	19	14	10	TW B4	403	177	105	71	50	37	27	19	13
TW J1	185	101	66	46	34	25	19	14	9	TW C1	586	183	99	63	44	32	23	16	11
TW J2	168	96	63	45	33	25	19	14	9	TW C2	443	172	98	65	46	33	24	17	11
TW J3	185	116	80	59	44	34	26	19	13	TW C3	420	160	91	60	42	30	22	16	11
TW J4	170	106	74	54	41	31	24	18	12	TW C4	439	142	78	50	35	25	18	13	9
DG A1	117	49	25	10	-	-	-	-	-	TW D1	396	158	91	60	42	30	22	16	10
DG A2	143	66	40	26	18	11	7	-	-	TW D2	487	233	143	97	70	52	38	28	19
DG A3	184	87	55	39	29	21	15	10	5	TW D3	493	253	160	110	80	59	44	32	21
DG B1	115	51	30	19	11	-	-	-	-	TW D4	458	255	167	118	87	65	49	36	24
DG B2	145	68	42	29	21	15	10	6	-	TW D5	418	240	159	113	84	63	48	35	24
DG B3	188	89	57	40	31	24	18	13	8	TW E1	563	289	183	126	92	68	51	37	25
DG C1	146	67	41	28	19	13	7	-	-	TW E2	507	272	175	123	90	67	50	37	25
DG C2	174	82	51	36	27	20	15	10	5	TW E3	513	306	207	149	111	84	64	47	32
DG C3	217	103	66	47	36	28	22	17	11	TW E4	456	275	187	135	101	77	58	43	29
SS A1	406	125	66	41	27	19	13	9	5	TW F1	743	193	100	63	43	31	22	16	10
SS A2	378	163	95	63	44	31	22	16	10	TW F2	459	192	112	75	53	38	28	20	13
SS A3	384	111	57	35	23	15	10	7	4	TW F3	376	182	112	76	55	41	30	22	14
SS A4	382	161	93	61	42	30	21	15	9	TW F4	307	161	103	71	52	39	29	21	14
SS A5	668	123	59	35	23	15	10	7	4	TW G1	460	165	92	60	42	30	22	16	10
SS A6	374	161	94	62	43	31	22	15	10	TW G2	331	156	95	65	46	34	25	18	12
SS A7	742	102	46	26	16	11	7	4	2	TW G3	276	140	88	61	44	33	24	18	12
SS A8	382	157	90	58	40	28	20	14	9	TW G4	220	115	73	51	37	27	20	15	10
SS B1	270	92	50	31	21	15	10	7	4	TW H1	296	131	78	52	37	27	20	14	10
SS B2	291	129	77	51	36	26	19	13	9	TW H2	224	112	70	48	35	26	19	14	9
SS B3	246	83	45	28	18	13	9	6	3	TW H3	185	94	59	41	30	22	16	12	8
SS B4	285	126	74	49	34	25	18	12	8	TW H4	142	72	45	31	22	17	12	9	6
SS B5	344	81	40	24	16	11	7	5	3	TW I1	310	134	79	52	37	27	20	14	9
SS B6	274	124	74	49	35	25	18	13	8	TW I2	373	194	124	86	62	46	35	25	17
SS B7	300	67	33	19	12	8	5	3	2	TW I3	389	211	136	95	70	52	39	28	19
SS B8	265	118	70	46	32	23	16	12	7	TW I4	380	220	147	105	78	59	44	33	22
SS C1	163	74	45	30	21	15	11	8	5	TW I5	363	216	146	105	78	59	45	33	23
SS C2	182	96	61	42	31	23	17	12	8	TW J1	465	253	164	115	84	63	47	34	23
SS C3	168	59	33	21	14	10	7	5	3	TW J2	415	235	155	110	81	61	46	34	23
SS C4	179	86	52	36	26	19	14	10	7	TW J3	424	264	182	133	100	76	58	43	29
SS D1	375	150	85	55	37	26	18	12	8	TW J4	390	243	168	123	92	71	54	40	27
SS D2	337	178	113	78	56	41	30	21	14	DG A1	393	179	108	71	49	33	22	13	4
SS D3	468	141	74	46	30	21	14	10	6	DG A2	394	182	113	78	56	41	29	20	12
SS D4	344	174	109	74	53	38	28	20	13	DG A3	454	211	133	94	70	53	40	28	17
SS E1	270	113	65	42	29	20	14	10	6	DG B1	393	180	112	79	58	43	31	20	10
SS E2	271	142	90	62	44	32	24	17	11	DG B2	405	188	118	84	63	48	36	26	16
SS E3	323	96	50	31	21	14	10	7	4	DG B3	460	215	136	97	74	57	45	33	22
SS E4	278	134	81	55	39	28	20	14	9	DG C1	465	213	133	95	71	54	41	29	17

LAS VEGAS, NEVADA — CONTINUED

SSF =	.10	.20	.30	.40	.50	.60	.70	.80	.90
DG C2	468	217	137	98	74	57	44	33	22
DG C3	527	246	156	112	86	67	53	41	28
SS A1	967	304	162	102	69	49	35	24	16
SS A2	833	361	212	140	98	71	51	37	24
SS A3	934	279	147	91	61	43	31	21	14
SS A4	842	358	208	136	95	68	50	35	23
SS A5	1628	309	151	91	60	42	30	20	13
SS A6	833	360	211	139	98	70	51	36	24
SS A7	1850	266	124	74	48	33	23	16	10
SS A8	851	354	204	133	93	67	48	34	22
SS B1	670	233	128	82	56	40	29	20	13
SS B2	650	290	172	115	81	59	43	31	20
SS B3	623	217	119	75	52	37	26	18	12
SS B4	637	283	168	111	78	57	41	29	19
SS B5	877	213	108	67	45	31	22	15	10
SS B6	618	281	168	112	79	58	42	30	20
SS B7	783	184	92	56	38	26	19	13	8
SS B8	602	270	160	106	75	54	40	28	19
SS C1	476	220	133	90	64	47	35	25	17
SS C2	473	248	158	110	80	60	45	32	22
SS C3	493	178	99	65	45	33	24	17	11
SS C4	468	224	137	94	67	50	37	27	18
SS D1	890	362	208	135	94	67	48	34	22
SS D2	736	390	248	171	123	91	67	48	32
SS D3	1119	345	183	115	78	55	39	27	18
SS D4	753	383	239	163	117	85	63	45	30
SS E1	682	292	170	112	78	56	41	29	19
SS E2	619	327	207	143	103	76	56	40	27
SS E3	819	251	134	84	57	40	29	20	13
SS E4	636	308	188	127	91	66	49	35	23

LOVELOCK, NEVADA — 5990 DD

SSF =	.10	.20	.30	.40	.50	.60	.70	.80	.90
WW A1	929	111	51	30	20	14	9	6	4
WW A2	313	124	70	46	32	23	16	11	7
WW A3	263	125	76	51	36	26	19	14	9
WW A4	242	125	78	54	39	28	21	15	10
WW A5	231	127	82	57	41	30	22	16	11
WW A6	224	127	83	58	43	32	23	17	11
WW B1	254	92	50	32	21	15	10	7	4
WW B2	239	128	82	57	41	30	22	16	11
WW B3	278	156	102	71	52	39	29	21	14
WW B4	244	148	101	73	54	41	31	22	15
WW B5	225	140	96	70	52	40	30	22	15
WW C1	296	166	108	76	55	41	31	22	15
WW C2	255	149	100	71	52	39	29	21	14
WW C3	262	169	119	87	66	50	38	28	19
WW C4	236	152	107	79	60	46	35	26	18
TW A1	798	107	50	30	19	13	9	6	4
TW A2	321	112	62	39	27	19	14	10	6
TW A3	259	111	65	43	30	22	16	11	7
TW A4	220	105	64	43	31	22	16	12	8
TW B1	448	101	51	31	21	14	10	7	4
TW B2	258	102	58	38	26	19	13	9	6
TW B3	225	98	58	38	27	19	14	10	6
TW B4	207	90	53	35	25	18	13	9	6
TW C1	297	91	49	31	21	15	11	7	5
TW C2	227	87	49	32	22	16	11	8	5
TW C3	217	82	46	30	21	15	11	8	5
TW C4	228	73	40	25	17	12	9	6	4
TW D1	185	72	40	25	17	12	8	6	4
TW D2	255	121	74	50	35	26	19	14	9
TW D3	261	133	83	57	41	30	22	16	10

LOVELOCK, NEVADA — CONTINUED

SSF =	.10	.20	.30	.40	.50	.60	.70	.80	.90
TW D4	250	139	90	64	46	35	26	19	12
TW D5	233	133	88	63	46	35	26	19	13
TW E1	303	155	97	66	48	35	26	18	12
TW E2	274	147	94	65	47	35	26	19	13
TW E3	289	172	116	83	61	46	35	25	17
TW E4	257	155	105	76	56	43	32	24	16
TW F1	363	92	47	29	19	13	9	6	4
TW F2	229	95	54	36	25	18	13	9	6
TW F3	190	91	55	37	26	19	14	10	6
TW F4	156	81	51	35	25	18	14	10	6
TW G1	227	80	44	28	19	14	10	7	4
TW G2	166	78	47	31	22	16	12	8	5
TW G3	140	71	44	30	21	16	11	8	5
TW G4	112	58	37	25	18	13	10	7	5
TW H1	147	65	38	25	18	13	9	6	4
TW H2	114	56	35	24	17	12	9	6	4
TW H3	95	48	30	20	15	11	8	6	4
TW H4	73	37	23	15	11	8	6	4	3
TW I1	143	60	34	22	15	11	7	5	3
TW I2	195	101	64	44	32	23	17	12	8
TW I3	207	111	71	49	36	26	19	14	9
TW I4	208	120	80	57	42	31	23	17	11
TW I5	202	120	81	58	43	32	24	18	12
TW J1	251	136	87	61	44	33	24	17	12
TW J2	226	127	84	59	43	32	24	17	12
TW J3	241	149	103	75	56	42	32	24	16
TW J4	221	137	95	69	52	39	30	22	15
DG A1	183	80	44	26	14	-	-	-	-
DG A2	203	93	56	38	26	17	11	6	-
DG A3	249	117	73	51	37	26	20	14	7
DG B1	182	82	49	32	21	13	-	-	-
DG B2	208	96	59	41	30	22	16	10	5
DG B3	253	119	75	53	40	30	23	17	11
DG C1	224	102	62	42	30	21	14	7	-
DG C2	245	114	71	50	37	28	21	14	8
DG C3	292	138	87	62	46	36	28	21	14
SS A1	543	165	86	52	35	24	16	11	7
SS A2	488	207	119	77	53	38	27	19	12
SS A3	519	148	75	45	29	20	13	9	5
SS A4	493	204	116	75	51	36	25	17	11
SS A5	907	166	78	45	29	20	13	9	5
SS A6	485	205	118	76	53	37	26	18	12
SS A7	1019	139	62	35	22	14	9	6	3
SS A8	495	200	113	72	49	34	24	17	10
SS B1	366	124	66	41	27	19	13	9	6
SS B2	377	165	97	64	44	32	23	16	10
SS B3	337	113	60	37	24	17	11	8	5
SS B4	369	161	93	61	42	30	21	15	10
SS B5	474	111	54	32	21	14	10	6	4
SS B6	356	159	93	61	43	31	22	15	10
SS B7	418	94	45	26	17	11	7	5	3
SS B8	346	152	88	58	40	28	20	14	9
SS C1	236	108	64	43	30	22	16	11	7
SS C2	250	131	83	57	41	30	22	16	11
SS C3	244	87	48	30	21	15	10	7	5
SS C4	247	117	71	48	34	25	18	13	9
SS D1	499	197	110	70	47	33	23	16	10
SS D2	430	224	140	95	68	49	35	25	16
SS D3	628	187	96	59	39	27	18	12	8
SS D4	440	220	135	91	64	46	33	23	15
SS E1	368	152	86	55	38	26	18	12	8
SS E2	352	182	114	77	55	40	29	20	13
SS E3	444	132	68	41	27	19	13	9	5
SS E4	362	172	103	69	48	35	25	17	11

LCR *Tables* / 539

RENO, NEVADA   6022 DD
SSF	=.10	.20	.30	.40	.50	.60	.70	.80	.90
WW A1	950	115	53	32	21	14	10	7	4
WW A2	322	127	73	47	33	24	17	12	8
WW A3	271	129	78	53	38	27	20	14	9
WW A4	248	128	81	56	40	29	22	15	10
WW A5	237	130	84	59	43	32	23	17	11
WW A6	230	130	85	60	44	33	24	18	12
WW B1	265	96	53	33	23	16	11	8	5
WW B2	245	131	84	58	42	31	23	17	11
WW B3	286	160	105	74	54	40	30	22	14
WW B4	250	152	103	75	56	42	32	23	16
WW B5	230	143	98	72	54	41	31	23	16
WW C1	304	170	111	79	57	43	32	23	15
WW C2	261	153	102	73	54	41	30	22	15
WW C3	268	173	122	90	68	52	40	29	20
WW C4	241	155	109	81	61	47	36	27	18
TW A1	816	110	52	31	20	14	10	7	4
TW A2	330	115	64	41	28	20	14	10	6
TW A3	266	114	67	44	31	23	16	12	8
TW A4	226	108	66	45	32	23	17	12	8
TW B1	460	104	52	32	22	15	11	7	5
TW B2	265	104	59	39	27	19	14	10	6
TW B3	232	101	59	39	28	20	15	10	7
TW B4	213	93	55	36	26	19	14	10	6
TW C1	305	94	50	32	22	15	11	8	5
TW C2	233	90	51	33	23	17	12	8	6
TW C3	223	84	47	31	21	15	11	8	5
TW C4	234	75	41	26	18	13	9	6	4
TW D1	192	75	42	27	18	13	9	6	4
TW D2	262	124	76	51	37	27	20	14	9
TW D3	269	137	86	59	42	31	23	16	11
TW D4	256	142	93	65	48	36	27	19	13
TW D5	238	136	90	64	47	36	27	20	13
TW E1	311	159	100	69	50	36	27	19	13
TW E2	281	150	96	67	49	36	27	20	13
TW E3	296	176	119	85	63	48	36	26	18
TW E4	263	158	108	78	58	44	33	24	17
TW F1	374	95	48	30	20	14	10	7	4
TW F2	235	97	56	37	26	19	13	9	6
TW F3	195	93	57	38	27	20	15	10	7
TW F4	160	83	53	36	26	19	14	10	7
TW G1	233	82	45	29	20	14	10	7	5
TW G2	171	80	48	32	23	17	12	9	6
TW G3	144	72	45	31	22	16	12	9	6
TW G4	115	60	38	26	19	14	10	7	5
TW H1	152	66	39	26	18	13	10	7	4
TW H2	117	58	36	24	17	13	9	7	4
TW H3	97	49	31	21	15	11	8	6	4
TW H4	75	38	23	16	11	8	6	4	3
TW I1	149	62	36	23	16	11	8	6	3
TW I2	200	104	65	45	33	24	18	13	9
TW I3	213	115	74	51	37	27	20	15	10
TW I4	213	123	82	58	43	32	24	18	12
TW I5	206	122	83	59	44	33	25	18	12
TW J1	258	140	90	63	46	34	25	18	12
TW J2	231	130	86	61	44	33	25	18	12
TW J3	247	153	106	77	58	44	33	24	17
TW J4	226	140	97	71	53	40	31	23	15
DG A1	189	82	46	27	15	-	-	-	-
DG A2	208	95	58	39	27	18	12	7	2
DG A3	254	119	74	52	38	29	21	14	8
DG B1	187	84	50	34	23	15	7	-	-
DG B2	213	98	61	42	31	23	16	11	5
DG B3	258	122	77	54	41	31	24	18	11
DG C1	229	104	64	44	32	23	16	8	-

RENO, NEVADA   CONTINUED
SSF	=.10	.20	.30	.40	.50	.60	.70	.80	.90
DG C2	250	117	73	51	38	29	22	15	9
DG C3	297	140	89	63	48	37	29	22	15
SS A1	587	179	93	57	38	26	18	12	7
SS A2	522	222	128	84	58	41	29	20	13
SS A3	567	163	83	50	33	22	15	10	6
SS A4	532	221	126	81	56	39	28	19	12
SS A5	982	180	85	50	32	22	15	10	6
SS A6	520	220	127	83	57	40	29	20	13
SS A7	1116	153	69	39	25	16	11	7	4
SS A8	535	217	123	79	54	38	27	18	11
SS B1	393	133	71	45	30	21	14	10	6
SS B2	401	176	103	68	47	34	24	17	11
SS B3	365	123	65	40	27	18	13	8	5
SS B4	394	172	100	66	46	33	23	16	10
SS B5	511	120	59	35	23	16	11	7	4
SS B6	378	169	100	66	46	33	24	17	11
SS B7	454	102	49	29	19	13	8	5	3
SS B8	370	163	95	62	43	31	22	15	10
SS C1	243	111	66	44	31	23	16	12	8
SS C2	256	134	85	59	42	31	23	17	11
SS C3	252	89	49	32	22	15	11	8	5
SS C4	253	120	73	50	35	26	19	14	9
SS D1	538	213	119	76	52	36	25	17	11
SS D2	460	240	151	103	73	53	38	27	17
SS D3	678	202	105	64	43	29	20	14	8
SS D4	471	236	145	98	69	50	36	25	16
SS E1	395	164	94	60	41	29	20	14	9
SS E2	375	195	122	83	59	43	31	22	14
SS E3	478	142	73	45	30	20	14	9	6
SS E4	386	184	111	74	52	37	27	19	12

TONOPAH, NEVADA   5900 DD
SSF	=.10	.20	.30	.40	.50	.60	.70	.80	.90
WW A1	947	116	55	33	22	15	11	8	5
WW A2	322	129	74	49	34	25	18	13	9
WW A3	272	131	80	55	39	29	21	15	10
WW A4	250	131	83	58	42	31	23	17	11
WW A5	239	133	86	61	45	33	25	18	12
WW A6	232	133	88	62	46	35	26	19	13
WW B1	264	98	55	35	24	17	12	9	6
WW B2	247	133	86	60	44	33	24	18	12
WW B3	288	163	107	76	56	42	31	23	16
WW B4	252	154	106	77	58	44	33	25	17
WW B5	232	145	101	74	56	42	32	24	17
WW C1	306	173	114	81	60	45	33	25	17
WW C2	263	155	105	75	56	42	32	24	16
WW C3	271	176	124	92	70	54	41	31	21
WW C4	243	158	112	83	63	49	37	28	19
TW A1	813	111	53	32	21	15	11	7	5
TW A2	329	117	65	42	29	21	15	11	7
TW A3	266	116	69	46	33	24	17	13	8
TW A4	227	110	68	46	33	25	18	13	9
TW B1	458	105	53	33	23	16	11	8	5
TW B2	265	106	61	40	28	20	15	11	7
TW B3	232	102	61	41	29	21	16	11	7
TW B4	213	94	56	38	27	20	14	10	7
TW C1	304	95	51	33	23	16	12	8	6
TW C2	233	91	52	34	24	17	13	9	6
TW C3	222	85	49	32	22	16	12	8	6
TW C4	233	76	42	27	19	13	10	7	5
TW D1	192	76	43	28	20	14	10	7	5
TW D2	263	126	78	53	38	28	21	15	10
TW D3	270	139	88	61	44	33	24	18	12

TONOPAH, NEVADA						CONTINUED				WINNEMUCCA, NEVADA									6629 DD
SSF =	.10	.20	.30	.40	.50	.60	.70	.80	.90	SSF =	.10	.20	.30	.40	.50	.60	.70	.80	.90
TW D4	258	144	95	67	49	37	28	21	14	WW A1	730	85	38	22	14	10	6	4	2
TW D5	240	138	92	66	49	37	28	21	14	WW A2	247	97	55	35	24	17	12	8	5
TW E1	312	161	102	71	51	38	28	21	14	WW A3	209	99	59	40	28	20	14	10	6
TW E2	282	153	99	69	51	38	28	21	14	WW A4	192	99	62	42	30	22	16	11	7
TW E3	298	179	121	88	65	50	38	28	19	WW A5	184	100	64	44	32	23	17	12	8
TW E4	265	161	110	80	60	45	35	26	18	WW A6	179	101	66	46	33	24	18	13	8
TW F1	372	96	50	31	21	15	11	8	5	WW B1	192	68	36	22	14	10	6	4	2
TW F2	236	99	58	38	27	20	14	10	7	WW B2	193	103	65	45	33	24	18	13	8
TW F3	196	95	59	40	29	21	16	11	7	WW B3	227	127	82	58	42	31	23	16	11
TW F4	161	85	54	38	27	20	15	11	7	WW B4	202	122	83	60	44	33	25	18	12
TW G1	233	83	47	30	21	15	11	8	5	WW B5	188	116	80	58	43	33	25	18	12
TW G2	172	81	50	34	24	18	13	9	6	WW C1	242	135	88	62	45	33	25	18	12
TW G3	145	74	46	32	23	17	13	9	6	WW C2	210	123	82	58	43	32	24	17	12
TW G4	116	61	39	27	20	14	11	8	5	WW C3	220	142	100	73	55	42	32	24	16
TW H1	152	68	40	27	19	14	10	7	5	WW C4	198	128	90	66	50	38	29	21	15
TW H2	117	59	37	25	18	13	10	7	5	TW A1	630	82	38	22	14	10	7	4	3
TW H3	98	50	32	22	16	12	9	6	4	TW A2	254	88	48	30	21	14	10	7	4
TW H4	76	38	24	16	12	9	7	5	3	TW A3	206	88	51	33	23	17	12	8	5
TW I1	149	64	37	25	17	12	9	6	4	TW A4	175	84	51	34	24	17	12	9	6
TW I2	201	106	67	47	34	25	19	14	9	TW B1	354	78	39	23	15	11	7	5	3
TW I3	214	117	75	53	39	29	21	16	11	TW B2	206	80	45	29	20	14	10	7	4
TW I4	214	125	84	60	44	34	25	19	13	TW B3	180	78	45	30	21	15	11	7	5
TW I5	208	124	84	61	46	35	26	19	13	TW B4	166	72	42	28	19	14	10	7	4
TW J1	259	142	92	65	47	35	26	19	13	TW C1	236	72	38	24	16	11	8	5	3
TW J2	233	133	88	62	46	35	26	19	13	TW C2	182	69	39	25	17	12	9	6	4
TW J3	249	155	108	79	59	45	35	26	18	TW C3	174	65	37	24	16	11	8	6	4
TW J4	228	143	99	73	55	42	32	24	16	TW C4	184	59	32	20	14	10	7	5	3
DG A1	189	84	48	29	17	8	-	-	-	TW D1	140	53	29	18	12	8	5	4	2
DG A2	208	96	60	40	28	20	13	8	3	TW D2	206	97	59	40	28	20	15	10	7
DG A3	254	120	76	54	40	30	22	15	9	TW D3	212	107	67	46	33	24	17	12	8
DG B1	188	86	52	36	25	17	9	-	-	TW D4	207	114	74	52	38	28	21	15	10
DG B2	212	100	62	44	33	24	18	12	6	TW D5	194	111	73	52	38	29	21	15	10
DG B3	258	122	78	55	42	33	25	19	12	TW E1	247	126	79	54	39	28	21	15	10
DG C1	229	106	65	46	34	25	18	10	-	TW E2	225	120	77	53	39	28	21	15	10
DG C2	250	118	74	52	40	30	23	16	10	TW E3	241	143	96	69	51	38	29	21	14
DG C3	296	141	90	64	49	39	31	23	16	TW E4	215	130	88	63	47	35	27	20	13
SS A1	570	178	94	58	39	27	19	13	8	TW F1	284	71	36	22	14	10	7	4	3
SS A2	512	221	129	85	59	42	31	22	14	TW F2	181	74	42	27	19	13	9	7	4
SS A3	548	161	83	51	34	23	16	11	7	TW F3	150	71	43	29	20	15	10	7	5
SS A4	520	219	126	82	57	41	29	20	13	TW F4	124	64	40	27	20	14	10	7	5
SS A5	946	178	85	51	33	23	16	11	7	TW G1	178	62	34	21	14	10	7	5	3
SS A6	510	219	128	84	58	42	30	21	14	TW G2	131	61	36	24	17	12	9	6	4
SS A7	1067	151	69	40	26	17	12	8	5	TW G3	111	56	34	23	16	12	9	6	4
SS A8	522	215	123	80	55	39	28	20	13	TW G4	90	46	29	20	14	10	7	5	3
SS B1	384	133	72	46	31	22	15	11	7	TW H1	116	50	29	19	13	9	7	5	3
SS B2	394	175	104	69	49	35	25	18	12	TW H2	90	44	27	18	13	9	7	5	3
SS B3	355	122	66	41	28	19	14	9	6	TW H3	76	38	24	16	11	8	6	4	3
SS B4	387	171	101	67	47	34	24	17	11	TW H4	59	29	18	12	9	6	5	3	2
SS B5	495	119	59	36	24	17	12	8	5	TW I1	107	44	25	15	10	7	5	3	2
SS B6	373	169	100	67	47	34	25	18	12	TW I2	158	81	51	35	25	18	13	10	6
SS B7	438	101	50	30	20	13	9	6	4	TW I3	169	90	58	40	29	21	15	11	7
SS B8	363	162	95	63	44	32	23	16	11	TW I4	172	99	66	47	34	26	19	14	9
SS C1	244	113	68	46	33	24	18	13	8	TW I5	168	99	67	48	35	27	20	15	10
SS C2	258	136	87	61	44	33	24	18	12	TW J1	205	111	71	49	36	26	19	14	9
SS C3	251	91	51	33	23	16	12	8	6	TW J2	186	105	68	48	35	26	19	14	9
SS C4	254	122	75	51	37	27	20	15	10	TW J3	202	125	86	62	47	35	27	19	13
SS D1	527	212	120	78	53	38	27	19	12	TW J4	185	115	79	58	43	33	25	18	12
SS D2	455	240	152	104	75	55	40	29	19	DG A1	138	58	30	14	-	-	-	-	-
SS D3	658	201	105	65	44	31	22	15	9	DG A2	163	74	44	29	19	12	7	-	-
SS D4	465	235	146	99	71	52	38	27	18	DG A3	206	96	60	42	30	22	16	11	5
SS E1	389	164	95	62	43	30	22	15	10	DG B1	137	60	35	22	13	-	-	-	-
SS E2	371	195	123	84	61	45	33	23	15	DG B2	166	77	47	32	23	16	11	6	-
SS E3	465	141	74	46	31	22	15	10	7	DG B3	209	99	62	44	33	25	19	13	8
SS E4	381	183	112	75	53	39	28	20	13	DG C1	172	77	46	31	21	14	7	-	-

WINNEMUCCA, NEVADA — CONTINUED

	SSF =.10	.20	.30	.40	.50	.60	.70	.80	.90
DG C2	197	92	57	39	29	21	16	10	5
DG C3	241	115	72	51	38	30	23	17	11
SS A1	448	136	70	42	28	19	13	8	5
SS A2	411	174	100	65	45	31	22	15	10
SS A3	426	121	60	36	23	15	10	6	3
SS A4	416	172	98	63	42	30	21	14	9
SS A5	745	135	63	36	23	15	10	6	4
SS A6	407	172	99	64	44	31	22	15	9
SS A7	835	112	49	27	16	10	6	4	2
SS A8	416	168	95	60	41	28	20	13	8
SS B1	300	101	53	33	22	15	10	7	4
SS B2	317	139	81	53	37	26	19	13	8
SS B3	275	91	48	29	19	12	8	5	3
SS B4	310	135	79	51	35	25	18	12	8
SS B5	386	89	43	25	16	11	7	5	3
SS B6	299	133	78	51	36	25	18	12	8
SS B7	338	74	35	20	12	8	5	3	2
SS B8	290	127	74	48	33	23	16	11	7
SS C1	186	84	49	33	23	16	11	8	5
SS C2	203	106	67	46	33	24	18	13	8
SS C3	192	67	37	23	16	11	8	5	3
SS C4	200	95	57	39	27	20	14	10	7
SS D1	412	162	90	56	38	26	18	12	7
SS D2	363	189	118	80	57	41	29	21	13
SS D3	517	153	78	47	31	21	14	9	5
SS D4	372	186	114	76	53	38	27	19	12
SS E1	300	123	69	44	29	20	14	9	5
SS E2	294	153	95	64	45	33	23	16	10
SS E3	361	106	54	33	21	14	9	6	4
SS E4	303	144	86	57	40	28	20	14	9

CONCORD, NEW HAMPSHIRE — 7360 DD

	SSF =.10	.20	.30	.40	.50	.60	.70	.80	.90
WW A1	217	-	-	-	-	-	-	-	-
WW A2	72	20	-	-	-	-	-	-	-
WW A3	62	23	10	-	-	-	-	-	-
WW A4	58	25	12	6	-	-	-	-	-
WW A5	56	26	14	8	4	-	-	-	-
WW A6	55	27	15	9	5	2	-	-	-
WW B1	-	-	-	-	-	-	-	-	-
WW B2	72	34	20	13	8	6	4	2	1
WW B3	88	46	28	18	12	9	6	4	2
WW B4	90	53	35	24	17	13	9	7	4
WW B5	89	53	36	25	19	14	10	7	5
WW C1	99	52	32	21	15	10	7	5	3
WW C2	91	51	32	22	16	11	8	6	4
WW C3	109	68	47	34	25	19	14	10	7
WW C4	99	63	43	31	23	18	13	10	6
TW A1	198	-	-	-	-	-	-	-	-
TW A2	78	19	5	-	-	-	-	-	-
TW A3	64	21	9	3	-	-	-	-	-
TW A4	57	23	11	6	3	-	-	-	-
TW B1	112	13	-	-	-	-	-	-	-
TW B2	67	20	9	3	-	-	-	-	-
TW B3	60	21	10	5	2	-	-	-	-
TW B4	58	21	11	6	3	2	-	-	-
TW C1	79	17	6	-	-	-	-	-	-
TW C2	64	20	9	5	2	-	-	-	-
TW C3	63	20	9	5	3	1	-	-	-
TW C4	72	19	9	5	3	2	-	-	-
TW D1	-	-	-	-	-	-	-	-	-
TW D2	78	33	18	11	7	5	3	2	-
TW D3	80	36	21	13	8	6	4	2	1

CONCORD, NEW HAMPSHIRE — CONTINUED

	SSF =.10	.20	.30	.40	.50	.60	.70	.80	.90
TW D4	92	48	30	20	14	10	7	5	3
TW D5	92	51	33	22	16	12	9	6	4
TW E1	100	47	28	18	12	8	6	4	2
TW E2	97	49	30	20	14	10	7	5	3
TW E3	114	66	43	30	22	16	12	8	5
TW E4	105	62	41	29	21	16	12	8	5
TW F1	75	-	-	-	-	-	-	-	-
TW F2	51	14	-	-	-	-	-	-	-
TW F3	44	16	7	-	-	-	-	-	-
TW F4	38	16	8	4	-	-	-	-	-
TW G1	47	7	-	-	-	-	-	-	-
TW G2	38	13	5	-	-	-	-	-	-
TW G3	34	14	7	3	-	-	-	-	-
TW G4	29	13	7	4	2	-	-	-	-
TW H1	31	9	-	-	-	-	-	-	-
TW H2	27	11	5	2	-	-	-	-	-
TW H3	24	10	5	3	2	-	-	-	-
TW H4	20	9	5	3	2	-	-	-	-
TW I1	-	-	-	-	-	-	-	-	-
TW I2	58	27	15	10	6	4	3	2	-
TW I3	65	32	19	12	8	5	4	2	1
TW I4	76	42	27	18	13	9	7	5	3
TW I5	78	45	29	20	15	11	8	6	4
TW J1	84	42	25	17	11	8	5	4	2
TW J2	80	43	27	18	13	9	7	5	3
TW J3	98	59	40	28	21	15	11	8	5
TW J4	91	55	37	27	20	15	11	8	5
DG A1	-	-	-	-	-	-	-	-	-
DG A2	57	20	-	-	-	-	-	-	-
DG A3	97	42	25	16	11	7	4	-	-
DG B1	-	-	-	-	-	-	-	-	-
DG B2	58	22	10	-	-	-	-	-	-
DG B3	101	45	27	18	13	9	6	4	-
DG C1	29	-	-	-	-	-	-	-	-
DG C2	78	32	17	10	5	-	-	-	-
DG C3	119	53	32	21	15	11	8	6	3
SS A1	166	39	16	6	-	-	-	-	-
SS A2	180	70	38	24	15	10	7	5	3
SS A3	139	24	-	-	-	-	-	-	-
SS A4	176	66	35	21	13	9	6	3	2
SS A5	282	33	7	-	-	-	-	-	-
SS A6	174	67	36	22	14	9	6	4	2
SS A7	294	-	-	-	-	-	-	-	-
SS A8	171	62	32	19	11	7	4	3	1
SS B1	103	26	9	-	-	-	-	-	-
SS B2	143	58	32	20	13	9	6	4	2
SS B3	82	15	-	-	-	-	-	-	-
SS B4	135	54	29	18	12	8	5	3	2
SS B5	131	17	-	-	-	-	-	-	-
SS B6	130	53	30	18	12	8	5	4	2
SS B7	98	-	-	-	-	-	-	-	-
SS B8	121	48	26	16	10	6	4	3	1
SS C1	46	12	-	-	-	-	-	-	-
SS C2	78	37	21	14	9	6	4	3	1
SS C3	52	10	-	-	-	-	-	-	-
SS C4	78	33	18	11	7	5	3	2	1
SS D1	145	47	21	10	-	-	-	-	-
SS D2	158	78	47	30	21	14	10	7	4
SS D3	190	43	17	5	-	-	-	-	-
SS D4	162	76	45	29	19	13	9	6	4
SS E1	89	26	-	-	-	-	-	-	-
SS E2	120	58	34	22	14	10	6	4	2
SS E3	118	22	-	-	-	-	-	-	-
SS E4	125	54	30	19	12	8	5	3	2

NEWARK, NEW JERSEY								5034 DD	
SSF =	.10	.20	.30	.40	.50	.60	.70	.80	.90
WW A1	400	41	16	8	4	2	1	-	-
WW A2	142	53	28	17	11	8	5	3	2
WW A3	121	55	32	20	14	10	7	5	3
WW A4	113	56	34	22	15	11	8	5	3
WW A5	108	57	36	24	17	12	9	6	4
WW A6	105	58	37	25	18	13	9	6	4
WW B1	86	24	9	-	-	-	-	-	-
WW B2	120	62	39	26	19	13	10	7	4
WW B3	142	78	50	34	25	18	13	9	6
WW B4	134	81	54	39	29	21	16	12	8
WW B5	128	78	54	39	29	22	16	12	8
WW C1	155	85	55	38	27	20	15	10	7
WW C2	138	80	52	37	27	20	15	11	7
WW C3	152	97	68	50	38	28	22	16	11
WW C4	138	88	62	45	34	26	20	15	10
TW A1	353	41	17	9	5	3	2	-	-
TW A2	147	48	25	15	10	7	4	3	2
TW A3	121	49	28	17	12	8	6	4	2
TW A4	104	48	28	18	13	9	6	4	3
TW B1	203	41	19	11	7	4	3	2	-
TW B2	121	45	25	15	10	7	5	3	2
TW B3	107	45	25	16	11	8	5	4	2
TW B4	100	42	24	15	10	7	5	3	2
TW C1	139	40	20	12	8	5	4	2	1
TW C2	110	40	22	14	9	6	4	3	2
TW C3	106	38	21	13	9	6	4	3	2
TW C4	114	35	18	11	7	5	4	2	2
TW D1	64	20	9	3	-	-	-	-	-
TW D2	129	59	35	23	16	11	8	6	4
TW D3	131	65	40	26	19	13	10	7	4
TW D4	136	75	48	33	24	18	13	9	6
TW D5	132	75	49	34	25	19	14	10	7
TW E1	157	79	48	33	23	17	12	8	5
TW E2	147	77	49	33	24	17	13	9	6
TW E3	163	96	64	46	34	25	19	14	9
TW E4	148	89	60	43	32	24	18	13	9
TW F1	156	35	16	9	5	3	2	1	-
TW F2	103	40	22	13	9	6	4	3	2
TW F3	87	40	23	15	10	7	5	3	2
TW F4	73	36	22	15	10	7	5	3	2
TW G1	99	32	16	10	6	4	3	2	-
TW G2	76	33	19	12	8	6	4	3	2
TW G3	65	31	19	12	8	6	4	3	2
TW G4	53	27	16	11	7	5	4	3	2
TW H1	66	27	15	9	6	4	3	2	1
TW H2	53	25	15	10	7	5	3	2	1
TW H3	45	22	13	9	6	4	3	2	1
TW H4	35	17	10	7	5	3	2	2	1
TW I1	47	16	6	-	-	-	-	-	-
TW I2	98	49	30	20	14	10	7	5	3
TW I3	105	55	35	23	17	12	9	6	4
TW I4	113	65	42	30	22	16	12	9	6
TW I5	113	67	45	32	23	17	13	9	6
TW J1	131	69	44	30	21	16	11	8	5
TW J2	122	68	44	30	22	16	12	9	6
TW J3	138	85	58	42	31	23	18	13	9
TW J4	128	79	54	39	29	22	17	12	8
DG A1	64	-	-	-	-	-	-	-	-
DG A2	99	43	24	14	7	-	-	-	-
DG A3	138	64	39	27	19	14	9	6	-
DG B1	65	22	-	-	-	-	-	-	-
DG B2	101	44	26	17	11	7	-	-	-
DG B3	141	66	41	28	21	16	12	8	4
DG C1	92	36	19	9	-	-	-	-	-

NEWARK, NEW JERSEY								CONTINUED	
SSF =	.10	.20	.30	.40	.50	.60	.70	.80	.90
DG C2	124	56	33	22	16	11	7	3	-
DG C3	164	78	48	33	25	19	14	10	6
SS A1	256	74	37	22	14	9	6	4	2
SS A2	255	107	61	39	27	19	13	9	6
SS A3	232	61	29	16	9	6	3	2	-
SS A4	254	104	58	37	25	17	12	8	5
SS A5	415	71	31	17	10	6	4	2	1
SS A6	250	105	59	38	26	18	13	9	6
SS A7	440	53	21	10	5	2	-	-	-
SS A8	251	100	55	35	23	16	11	8	5
SS B1	171	55	28	16	10	7	4	3	2
SS B2	202	88	51	33	23	16	11	8	5
SS B3	150	46	23	13	8	5	3	2	-
SS B4	195	84	48	31	21	15	11	7	5
SS B5	214	46	21	11	7	4	2	1	-
SS B6	188	83	48	31	21	15	11	7	5
SS B7	178	35	14	7	3	-	-	-	-
SS B8	179	77	44	28	19	14	10	7	4
SS C1	103	44	25	15	10	7	5	3	2
SS C2	127	65	40	27	19	14	10	7	5
SS C3	108	35	18	11	7	4	3	2	1
SS C4	125	58	34	23	16	11	8	6	4
SS D1	235	89	48	29	19	13	9	6	3
SS D2	228	118	73	49	35	25	18	13	8
SS D3	294	83	41	24	15	10	6	4	2
SS D4	232	115	70	47	33	23	17	12	8
SS E1	166	65	35	21	13	9	6	3	2
SS E2	182	93	57	38	27	19	14	10	6
SS E3	201	55	27	15	9	6	4	2	1
SS E4	188	88	52	34	23	17	12	8	5

ALBUQUERQUE, NEW MEXICO								4292 DD	
SSF =	.10	.20	.30	.40	.50	.60	.70	.80	.90
WW A1	1052	130	62	38	25	18	13	9	6
WW A2	354	144	84	56	39	29	21	15	10
WW A3	300	146	90	62	45	33	24	18	12
WW A4	276	146	93	65	47	35	26	19	13
WW A5	264	148	97	69	50	38	28	21	14
WW A6	256	148	99	70	52	39	30	22	15
WW B1	293	111	63	41	28	20	15	11	7
WW B2	270	147	96	67	49	37	28	20	14
WW B3	314	179	119	84	62	47	35	26	18
WW B4	275	169	116	85	64	49	37	28	19
WW B5	252	159	110	81	61	47	36	27	19
WW C1	333	190	126	89	66	50	38	28	19
WW C2	287	171	115	83	62	47	36	27	18
WW C3	293	191	136	101	77	59	46	34	24
WW C4	264	172	122	91	69	54	41	31	22
TW A1	900	124	60	37	25	17	12	9	6
TW A2	361	130	73	48	33	24	18	13	8
TW A3	293	129	77	52	37	27	20	15	10
TW A4	249	123	76	52	38	28	21	15	10
TW B1	502	117	60	38	26	18	13	9	6
TW B2	291	118	68	45	32	23	17	12	8
TW B3	254	114	68	46	33	24	18	13	9
TW B4	233	104	63	42	30	22	16	12	8
TW C1	332	106	58	37	26	19	14	10	6
TW C2	255	101	58	39	27	20	15	11	7
TW C3	243	94	54	36	25	18	13	10	7
TW C4	254	84	46	30	21	15	11	8	5
TW D1	213	86	50	33	23	17	12	9	6
TW D2	287	139	86	59	43	32	24	17	12
TW D3	294	153	97	68	49	37	27	20	14

ALBUQUERQUE, NEW MEXICO						CONTINUED			CLAYTON, NEW MEXICO								5212 DD
SSF =.10	.20	.30	.40	.50	.60	.70	.80	.90	SSF =.10	.20	.30	.40	.50	.60	.70	.80	.90
TW D4 281	158	104	74	55	41	31	23	16	WW A1 924	112	52	31	21	15	10	7	5
TW D5 260	151	101	73	54	41	31	23	16	WW A2 312	124	72	47	33	24	18	13	9
TW E1 339	177	113	78	57	43	32	23	16	WW A3 264	126	77	53	38	28	21	15	10
TW E2 308	168	109	77	56	42	32	23	16	WW A4 243	126	80	55	40	30	22	16	11
TW E3 323	195	133	96	72	55	42	31	21	WW A5 231	128	83	59	43	32	24	18	12
TW E4 287	175	120	88	66	50	38	28	20	WW A6 225	128	85	60	45	34	25	19	13
TW F1 409	108	57	36	24	17	13	9	6	WW B1 254	93	52	33	23	17	12	9	6
TW F2 260	110	65	43	31	22	17	12	8	WW B2 239	129	83	58	43	32	24	18	12
TW F3 216	106	66	45	33	24	18	13	9	WW B3 279	157	104	73	54	41	31	23	16
TW F4 178	95	61	42	31	23	17	13	9	WW B4 245	149	102	74	56	43	32	24	17
TW G1 256	93	53	34	24	17	13	9	6	WW B5 226	141	98	71	54	41	32	24	16
TW G2 189	91	56	38	27	20	15	11	7	WW C1 296	167	110	78	58	43	33	24	17
TW G3 159	82	52	36	26	20	15	11	7	WW C2 256	151	101	73	54	41	31	23	16
TW G4 128	68	43	30	22	16	12	9	6	WW C3 263	170	121	90	68	53	40	30	21
TW H1 168	76	45	31	22	16	12	9	6	WW C4 237	153	108	81	61	47	36	27	19
TW H2 130	66	41	29	21	15	11	8	6	TW A1 793	107	51	31	21	14	10	7	5
TW H3 108	56	35	25	18	13	10	7	5	TW A2 320	113	63	41	28	20	15	11	7
TW H4 83	42	27	19	13	10	7	5	4	TW A3 259	112	66	44	32	23	17	12	8
TW I1 166	73	43	29	20	15	11	8	5	TW A4 220	106	65	45	32	24	18	13	9
TW I2 221	117	75	52	38	28	21	16	11	TW B1 445	101	51	32	22	15	11	8	5
TW I3 234	128	83	59	43	32	24	18	12	TW B2 258	102	59	39	27	20	15	11	7
TW I4 234	137	92	66	49	37	28	21	14	TW B3 225	99	59	39	28	20	15	11	7
TW I5 226	136	93	67	50	38	29	22	15	TW B4 207	91	54	36	26	19	14	10	7
TW J1 282	156	102	72	53	40	30	22	15	TW C1 295	92	50	32	22	16	11	8	5
TW J2 254	146	97	69	51	39	29	22	15	TW C2 226	88	50	33	23	17	12	9	6
TW J3 269	169	118	86	65	50	38	29	20	TW C3 216	82	47	31	22	16	11	8	6
TW J4 247	155	108	80	60	46	35	26	18	TW C4 227	74	40	26	18	13	9	7	5
DG A1 211	97	57	36	22	13	5	-	-	TW D1 185	73	41	27	19	14	10	7	5
DG A2 227	107	67	46	32	23	16	10	5	TW D2 255	122	75	51	37	27	20	15	10
DG A3 274	131	83	59	44	34	25	18	10	TW D3 262	134	85	59	43	32	24	17	12
DG B1 210	97	60	42	30	21	13	6	-	TW D4 251	140	92	65	48	36	27	20	14
DG B2 232	110	69	49	37	28	21	14	8	TW D5 234	134	89	64	47	36	27	20	14
DG B3 277	134	85	61	47	37	28	21	14	TW E1 303	156	98	68	50	37	28	20	14
DG C1 253	120	74	53	39	30	22	14	-	TW E2 275	148	95	67	49	37	28	20	14
DG C2 271	130	82	59	45	35	26	19	12	TW E3 290	173	117	85	64	48	37	27	19
DG C3 318	155	98	71	54	43	34	26	18	TW E4 258	156	107	77	58	44	34	25	17
SS A1 591	187	101	64	44	31	22	16	10	TW F1 361	93	48	30	20	15	10	7	5
SS A2 531	232	137	92	65	47	34	25	16	TW F2 229	95	55	37	26	19	14	10	7
SS A3 566	170	90	56	38	27	19	13	8	TW F3 190	92	56	38	28	20	15	11	7
SS A4 537	230	135	89	63	45	33	23	15	TW F4 156	82	52	36	26	20	15	11	7
SS A5 980	187	92	56	37	26	18	13	8	TW G1 226	80	45	29	20	15	11	8	5
SS A6 529	231	136	91	64	47	34	24	16	TW G2 166	78	48	32	23	17	13	9	6
SS A7 1103	158	74	44	29	20	14	10	6	TW G3 140	71	45	31	22	17	12	9	6
SS A8 540	226	131	87	61	44	32	23	15	TW G4 113	59	37	26	19	14	11	8	5
SS B1 403	141	78	50	35	25	18	13	8	TW H1 147	65	39	26	18	13	10	7	5
SS B2 412	186	111	75	53	39	28	20	14	TW H2 114	57	35	24	18	13	10	7	5
SS B3 372	130	71	46	31	22	16	11	7	TW H3 95	48	30	21	15	11	8	6	4
SS B4 403	181	108	72	51	37	27	20	13	TW H4 73	37	23	16	12	9	6	5	3
SS B5 518	127	65	40	27	19	13	9	6	TW I1 143	61	35	23	16	12	9	6	4
SS B6 390	179	108	73	52	38	28	20	13	TW I2 196	102	65	45	33	24	18	13	9
SS B7 457	108	54	33	22	16	11	8	5	TW I3 208	112	73	51	37	28	21	15	10
SS B8 379	171	102	69	49	35	26	19	12	TW I4 208	121	81	58	43	33	25	18	13
SS C1 270	126	77	52	37	27	20	15	10	TW I5 202	121	82	59	44	34	26	19	13
SS C2 282	150	97	68	49	37	28	20	14	TW J1 251	137	89	62	46	34	26	19	13
SS C3 276	101	57	37	26	19	14	10	7	TW J2 226	128	85	60	45	34	25	19	13
SS C4 277	135	83	57	41	31	23	17	11	TW J3 242	151	104	76	58	44	34	25	17
SS D1 548	225	130	85	59	43	31	22	14	TW J4 222	138	96	70	53	41	31	23	16
SS D2 474	253	162	113	82	61	45	33	22	DG A1 182	80	46	28	16	7	-	-	-
SS D3 683	212	113	72	49	35	25	17	11	DG A2 202	93	57	39	27	19	13	8	3
SS D4 484	248	156	107	77	57	42	30	20	DG A3 248	116	73	52	39	29	22	15	9
SS E1 410	176	103	68	48	35	25	18	12	DG B1 180	81	49	34	24	16	9	-	-
SS E2 390	208	133	92	67	50	37	27	18	DG B2 207	96	60	42	31	24	17	12	6
SS E3 487	151	80	51	35	25	18	12	8	DG B3 253	118	75	53	41	32	25	18	12
SS E4 400	195	120	82	59	43	32	23	15	DG C1 222	101	62	43	32	24	17	10	-

CLAYTON, NEW MEXICO								CONTINUED	
SSF =	.10	.20	.30	.40	.50	.60	.70	.80	.90
DG C2	244	113	71	50	38	30	22	16	10
DG C3	292	136	87	62	47	38	30	23	16
SS A1	546	168	89	56	38	27	19	14	9
SS A2	491	210	123	82	57	42	30	22	14
SS A3	524	152	79	49	33	23	16	11	7
SS A4	498	209	121	79	56	40	29	21	14
SS A5	907	168	81	49	32	22	16	11	7
SS A6	488	209	122	81	57	41	30	21	14
SS A7	1024	142	65	38	25	17	12	8	5
SS A8	500	205	118	77	54	39	28	20	13
SS B1	370	126	69	44	30	21	15	11	7
SS B2	380	168	100	67	47	34	25	18	12
SS B3	341	116	63	40	27	19	14	10	6
SS B4	373	164	97	64	45	33	24	17	12
SS B5	477	113	56	35	23	16	12	8	5
SS B6	359	162	96	65	46	33	24	18	12
SS B7	421	96	47	29	19	13	9	6	4
SS B8	350	155	92	61	43	31	23	16	11
SS C1	237	109	65	44	32	23	17	12	8
SS C2	251	132	84	58	43	32	24	17	12
SS C3	244	87	49	31	22	16	12	8	6
SS C4	247	118	72	49	36	26	20	14	10
SS D1	503	201	115	75	52	37	27	19	13
SS D2	434	229	146	101	73	54	40	29	19
SS D3	630	190	100	63	43	30	22	15	10
SS D4	445	224	140	96	69	51	37	27	18
SS E1	372	156	90	59	41	30	22	15	10
SS E2	356	187	118	82	59	44	32	24	16
SS E3	447	134	70	44	30	21	15	11	7
SS E4	366	175	107	73	52	38	28	20	13

FARMINGTON, NEW MEXICO								5713 DD	
SSF =	.10	.20	.30	.40	.50	.60	.70	.80	.90
WW A1	852	100	46	28	18	13	9	6	4
WW A2	286	113	64	42	30	22	16	11	7
WW A3	241	114	70	47	34	25	18	13	9
WW A4	221	114	72	50	36	27	20	14	10
WW A5	211	116	75	53	39	29	22	16	11
WW A6	205	116	77	55	40	30	23	16	11
WW B1	229	82	45	29	20	14	10	7	4
WW B2	220	118	76	53	39	29	22	16	11
WW B3	257	144	95	67	49	37	28	20	14
WW B4	227	138	94	69	51	39	30	22	15
WW B5	210	130	90	66	50	38	29	22	15
WW C1	273	153	101	71	53	40	30	22	15
WW C2	236	139	93	67	50	38	28	21	14
WW C3	245	158	112	83	63	49	37	28	19
WW C4	220	142	101	75	57	44	34	25	17
TW A1	732	96	45	27	18	13	9	6	4
TW A2	293	102	56	37	25	18	13	9	6
TW A3	237	102	60	40	28	21	15	11	7
TW A4	201	97	59	40	29	21	16	11	8
TW B1	410	91	46	29	19	14	10	7	4
TW B2	236	93	53	35	24	18	13	9	6
TW B3	207	90	53	35	25	18	13	10	6
TW B4	190	83	49	33	23	17	13	9	6
TW C1	272	83	45	28	20	14	10	7	5
TW C2	208	80	46	30	21	15	11	8	5
TW C3	199	75	42	28	19	14	10	7	5
TW C4	210	67	37	23	16	12	8	6	4
TW D1	166	64	36	23	16	12	8	6	4
TW D2	235	111	68	46	33	25	18	13	9
TW D3	241	123	77	53	39	29	21	15	10

FARMINGTON, NEW MEXICO								CONTINUED	
SSF =	.10	.20	.30	.40	.50	.60	.70	.80	.90
TW D4	232	129	84	60	44	33	25	18	12
TW D5	217	124	83	59	44	33	25	18	13
TW E1	280	143	90	62	45	34	25	18	12
TW E2	254	136	88	61	45	34	25	18	12
TW E3	269	160	109	78	59	44	34	25	17
TW E4	240	145	99	72	54	41	31	23	16
TW F1	331	83	43	27	18	13	9	6	4
TW F2	209	86	50	33	23	17	12	9	6
TW F3	173	83	51	35	25	18	13	10	6
TW F4	143	74	47	33	24	18	13	10	6
TW G1	206	72	40	26	18	13	9	7	4
TW G2	152	71	43	29	21	15	11	8	5
TW G3	128	65	40	28	20	15	11	8	5
TW G4	103	53	34	23	17	13	9	7	5
TW H1	134	59	35	23	16	12	9	6	4
TW H2	104	52	32	22	16	12	9	6	4
TW H3	87	44	28	19	14	10	8	5	4
TW H4	67	34	21	14	10	8	6	4	3
TW I1	128	54	31	20	14	10	7	5	3
TW I2	180	93	59	41	30	22	16	12	8
TW I3	191	103	66	46	34	25	19	14	9
TW I4	193	112	75	53	40	30	23	17	11
TW I5	188	112	76	55	41	31	23	17	12
TW J1	232	126	81	57	42	31	23	17	12
TW J2	209	118	78	55	41	31	23	17	12
TW J3	225	140	97	71	53	41	31	23	16
TW J4	206	128	89	65	49	38	29	21	15
DG A1	163	71	39	23	12	-	-	-	-
DG A2	185	84	52	35	24	16	11	6	-
DG A3	230	107	67	48	35	27	19	13	7
DG B1	161	72	44	29	20	12	-	-	-
DG B2	189	87	54	38	28	21	15	10	5
DG B3	235	109	69	49	38	29	23	17	11
DG C1	199	90	56	39	28	20	14	7	-
DG C2	224	103	65	46	35	26	20	14	8
DG C3	271	126	80	57	44	35	27	21	14
SS A1	508	153	81	50	34	24	17	12	7
SS A2	457	194	113	75	52	38	27	19	13
SS A3	485	137	70	43	29	20	14	9	6
SS A4	462	191	110	72	50	36	26	18	12
SS A5	852	153	73	43	29	20	14	9	6
SS A6	454	192	112	74	52	37	27	19	12
SS A7	962	128	58	34	22	14	10	7	4
SS A8	464	188	107	70	48	35	25	17	11
SS B1	341	114	62	39	27	19	13	9	6
SS B2	353	155	91	61	43	31	23	16	11
SS B3	313	104	56	35	24	17	12	8	5
SS B4	345	150	88	59	41	30	22	15	10
SS B5	443	102	51	31	20	14	10	7	4
SS B6	333	148	88	59	42	30	22	16	10
SS B7	389	86	42	25	16	11	8	5	3
SS B8	323	142	83	55	39	28	20	14	9
SS C1	215	98	59	40	28	21	15	11	7
SS C2	230	120	77	53	39	29	22	16	11
SS C3	223	79	44	28	20	14	10	7	5
SS C4	228	108	66	45	32	24	18	13	9
SS D1	464	183	104	67	46	33	23	16	10
SS D2	403	211	134	92	66	49	36	26	17
SS D3	586	173	91	56	38	27	19	13	8
SS D4	412	207	128	88	63	46	34	24	16
SS E1	340	141	81	53	37	26	19	13	8
SS E2	328	171	108	74	54	39	29	21	14
SS E3	413	121	63	39	26	18	13	9	6
SS E4	338	161	98	66	47	34	25	18	12

```
LOS ALAMOS, NEW MEXICO                            6359 DD    LOS ALAMOS, NEW MEXICO                         CONTINUED
   SSF =.10    .20    .30    .40    .50    .60    .70    .80    .90       SSF =.10    .20    .30    .40    .50    .60    .70    .80    .90
WW A1   707    86     40     24     16     11     8      6      4    DG C2   197    91     57     41     31     24     18     13     7
WW A2   247    98     57     37     26     19     14     10     7    DG C3   241   114     72     52     40     32     25     19    13
WW A3   210   101     62     42     30     22     17     12     8    SS A1   458   141     75     47     32     23     16     11     7
WW A4   194   101     64     44     32     24     18     13     9    SS A2   420   180    106     70     49     36     26     19    13
WW A5   185   103     67     47     35     26     20     14    10    SS A3   439   127     65     41     27     19     14      9     6
WW A6   180   103     68     49     36     27     20     15    10    SS A4   427   179    104     68     48     35     25     18    12
WW B1   193    70     39     25     17     12     9      6     4    SS A5   754   140     67     40     27     19     13      9     6
WW B2   194   105     67     47     35     26     20     14    10    SS A6   417   178    104     69     49     35     26     19    12
WW B3   229   129     85     61     45     34     25     19    13    SS A7   847   117     53     31     20     14     10      7     4
WW B4   204   124     85     62     47     36     27     20    14    SS A8   428   175    101     66     46     33     24     17    11
WW B5   189   118     82     60     46     35     27     20    14    SS B1   308   105     57     36     25     18     13      9     6
WW C1   244   138     91     65     48     36     27     20    14    SS B2   325   144     85     57     40     30     22     16    10
WW C2   211   125     84     61     45     34     26     19    13    SS B3   283    96     51     33     22     16     11      8     5
WW C3   223   144    102     76     58     45     35     26    18    SS B4   319   140     83     55     39     28     21     15    10
WW C4   200   130     92     68     52     40     31     23    16    SS B5   394    93     46     28     19     13      9      7     4
TW A1   611    83     39     24     16     11     8      6     4    SS B6   306   138     82     55     39     29     21     15    10
TW A2   253    89     50     32     22     16     12     9      6    SS B7   347    78     38     23     15     11      7      5     3
TW A3   206    89     53     35     25     18     14     10     7    SS B8   298   132     78     52     37     27     20     14     9
TW A4   176    85     52     36     26     19     14     10     7    SS C1   187    86     52     35     25     18     14     10     7
TW B1   349    79     40     25     17     12     9      6     4    SS C2   204   107     69     48     35     26     20     14    10
TW B2   205    81     47     31     22     16     12     8      6    SS C3   192    68     38     25     17     12      9      7     4
TW B3   180    79     47     31     22     16     12     9      6    SS C4   201    96     59     40     29     22     16     12     8
TW B4   166    73     43     29     21     15     11     8      6    SS D1   422   169     96     63     44     31     23     16    11
TW C1   235    73     39     25     17     13     9      7     4    SS D2   373   197    125     87     63     47     35     25    17
TW C2   182    70     40     26     19     14     10     7      5    SS D3   528   159     84     53     36     25     18     13     8
TW C3   174    66     38     25     17     13     9      7     5    SS D4   381   193    120     83     59     44     32     24    16
TW C4   183    59     32     21     15     11     8      6     4    SS E1   308   129     74     49     34     25     18     13     8
TW D1   141    55     31     20     14     10     7      5     3    SS E2   303   159    101     70     50     37     28     20    14
TW D2   207    99     61     42     30     22     17     12     8    SS E3   370   110     58     36     25     17     12      9     6
TW D3   213   110     69     48     35     26     19     14    10    SS E4   311   149     91     62     44     32     24     17    12
TW D4   208   116     76     54     40     30     23     17    12
TW D5   195   112     75     54     40     30     23     17    12    ROSWELL, NEW MEXICO                            3697 DD
TW E1   249   128     81     56     41     31     23     17    11       SSF =.10    .20    .30    .40    .50    .60    .70    .80    .90
TW E2   226   122     79     55     41     31     23     17    12    WW A1  1223   147     69     42     28     20     14     10     7
TW E3   243   146     99     72     54     41     31     23    16    WW A2   401   160     93     61     43     32     23     17    11
TW E4   217   131     90     65     49     38     29     21    15    WW A3   336   161    100     68     49     36     27     20    13
TW F1   281    72     37     23     16     11     8      6     4    WW A4   307   161    103     71     52     39     29     21    14
TW F2   181    75     44     29     21     15     11     8      5    WW A5   293   163    107     75     56     42     31     23    16
TW F3   151    73     45     31     22     16     12     9      6    WW A6   284   163    108     77     57     43     32     24    16
TW F4   125    66     42     29     21     16     12     9      6    WW B1   338   125     70     46     32     23     17     12     8
TW G1   177    63     35     23     16     11     8      6     4    WW B2   299   161    104     73     54     40     30     22    15
TW G2   132    62     38     26     18     14     10     7      5    WW B3   346   196    130     92     68     51     39     28    19
TW G3   112    57     36     25     18     13     10     7      5    WW B4   300   184    126     92     69     53     40     30    21
TW G4    90    47     30     21     15     11     8      6     4    WW B5   275   172    119     88     66     51     39     29    20
TW H1   117    51     31     20     15     11     8      6     4    WW C1   366   207    137     98     72     54     41     30    21
TW H2    91    45     28     19     14     10     8      6     4    WW C2   315   186    125     90     68     51     39     29    20
TW H3    76    39     24     17     12     9      7      5     3    WW C3   318   206    146    109     83     64     49     37    26
TW H4    59    30     19     13     9      7      5      4     3    WW C4   286   186    132     98     75     58     44     33    23
TW I1   108    46     27     17     12     9      6      5     3    TW A1  1045   140     67     41     27     19     14     10     6
TW I2   159    83     53     37     27     20     15     11     8    TW A2   412   145     81     53     37     27     20     14     9
TW I3   170    92     60     42     31     23     17     13     9    TW A3   330   143     85     57     41     30     22     16    11
TW I4   173   101     67     48     36     27     21     15    11    TW A4   279   135     84     57     41     31     23     17    11
TW I5   169   101     69     50     37     28     22     16    11    TW B1   582   131     67     42     29     20     15     10     7
TW J1   207   113     73     52     38     29     21     16    11    TW B2   329   130     75     50     35     26     19     14     9
TW J2   187   106     70     50     37     28     21     16    11    TW B3   286   125     75     51     36     26     20     14     9
TW J3   204   127     88     65     49     38     29     22    15    TW B4   262   115     69     46     33     24     18     13     9
TW J4   187   117     81     60     45     35     27     20    14    TW C1   380   117     64     41     29     21     15     11     7
DG A1   138    60     32     18      -      -      -      -     -    TW C2   288   111     64     42     30     22     16     12     8
DG A2   162    74     46     31     21     14     9      5      -    TW C3   275   104     59     39     28     20     15     11     7
DG A3   205    96     61     43     32     24     18     12     7    TW C4   289    93     51     33     23     17     12      9     6
DG B1   136    60     36     25     16     9       -      -     -    TW D1   245    97     56     37     26     19     14     10     6
DG B2   165    76     48     34     25     19     13     9      4    TW D2   320   153     94     65     47     35     26     19    13
DG B3   209    98     62     45     34     27     21     15    10    TW D3   326   168    107     74     54     40     30     22    15
DG C1   171    77     47     33     24     18     12     5       -
```

546 / APPENDIX 21

ROSWELL, NEW MEXICO							CONTINUED			TRUTH OR CONSEQUENCES, NEW MEXICO								3392 DD	
SSF =.10	.20	.30	.40	.50	.60	.70	.80	.90	SSF =.10	.20	.30	.40	.50	.60	.70	.80	.90		
TW D4	308	172	113	81	60	45	34	25	17	WW A1	1423	174	83	51	34	24	18	12	8
TW D5	284	164	109	78	58	44	33	25	17	WW A2	466	189	110	73	52	38	28	20	14
TW E1	376	193	123	86	62	46	35	25	17	WW A3	391	190	118	81	59	43	32	24	16
TW E2	339	183	118	83	61	46	34	25	17	WW A4	359	189	121	85	62	46	35	26	17
TW E3	352	211	144	104	78	59	45	33	23	WW A5	342	191	126	89	66	50	37	28	19
TW E4	313	190	130	95	71	54	41	31	21	WW A6	332	192	128	92	68	51	39	29	20
TW F1	474	121	63	40	27	19	14	10	7	WW B1	400	151	86	56	40	29	21	15	10
TW F2	294	122	72	48	34	25	18	13	9	WW B2	347	188	122	86	64	48	36	26	18
TW F3	242	117	73	50	36	26	20	14	10	WW B3	400	228	151	108	80	60	46	34	23
TW F4	198	104	67	47	34	25	19	14	9	WW B4	345	212	146	107	80	62	47	35	24
TW G1	293	104	58	38	27	19	14	10	7	WW B5	315	197	138	101	77	59	45	34	23
TW G2	212	100	61	42	30	22	16	12	8	WW C1	422	241	160	114	84	64	48	36	24
TW G3	178	91	57	40	29	21	16	12	8	WW C2	362	215	146	105	79	60	45	34	23
TW G4	142	75	48	33	24	18	14	10	7	WW C3	363	237	168	125	96	74	57	43	30
TW H1	189	84	50	34	24	18	13	9	6	WW C4	326	212	151	113	86	66	51	38	27
TW H2	145	73	46	31	23	17	13	9	6	TW A1	1211	165	80	49	33	24	17	12	8
TW H3	120	61	39	27	20	15	11	8	5	TW A2	476	171	96	63	44	32	24	17	11
TW H4	93	47	29	20	15	11	8	6	4	TW A3	383	168	101	68	49	36	27	19	13
TW I1	190	82	48	32	23	16	12	9	6	TW A4	324	159	99	68	49	37	27	20	13
TW I2	245	128	82	57	42	31	23	17	12	TW B1	670	155	80	50	34	25	18	13	8
TW I3	258	140	91	64	47	35	26	19	13	TW B2	381	153	89	59	42	31	23	16	11
TW I4	256	149	100	72	54	41	31	23	16	TW B3	331	147	89	60	43	32	23	17	11
TW I5	246	147	100	73	54	41	31	23	16	TW B4	303	135	81	55	39	29	21	16	11
TW J1	311	170	111	78	57	43	32	24	16	TW C1	437	138	76	49	34	25	18	13	9
TW J2	279	159	105	75	56	42	32	23	16	TW C2	332	131	76	50	36	26	19	14	9
TW J3	293	183	127	93	71	54	41	31	21	TW C3	316	122	70	46	33	24	18	13	9
TW J4	269	168	117	86	65	50	38	28	20	TW C4	331	108	60	39	27	20	14	10	7
DG A1	240	108	64	41	27	16	8	-	-	TW D1	289	117	68	45	32	23	17	12	8
DG A2	254	118	73	50	36	25	18	11	6	TW D2	369	178	111	76	55	41	30	22	15
DG A3	302	143	90	64	48	36	27	19	11	TW D3	377	196	125	87	64	47	36	25	18
DG B1	236	110	68	47	34	24	16	9	-	TW D4	354	199	132	94	69	52	40	29	20
DG B2	258	121	77	54	41	31	23	16	10	TW D5	325	188	126	91	68	51	39	29	20
DG B3	306	145	93	66	51	40	31	23	15	TW E1	433	225	143	100	73	55	41	30	20
DG C1	283	133	83	59	44	33	25	16	7	TW E2	390	212	138	97	71	54	40	30	20
DG C2	301	142	90	64	49	38	29	21	14	TW E3	403	243	166	120	90	69	52	39	27
DG C3	351	167	107	77	59	47	37	28	19	TW E4	357	218	149	109	82	63	48	36	25
SS A1	677	208	112	71	49	34	25	17	11	TW F1	548	143	75	48	33	24	17	12	8
SS A2	592	255	151	101	71	52	38	27	18	TW F2	342	145	86	57	41	30	22	16	11
SS A3	653	190	100	63	43	30	21	15	10	TW F3	282	138	86	59	43	32	24	17	12
SS A4	600	253	149	98	69	50	36	26	17	TW F4	231	123	79	55	41	30	23	17	11
SS A5	1149	209	103	62	42	29	21	14	9	TW G1	340	123	70	46	32	23	17	12	8
SS A6	590	254	150	100	71	51	37	27	18	TW G2	248	118	73	50	36	27	20	14	10
SS A7	1310	178	84	50	33	23	16	11	7	TW G3	207	107	68	47	34	26	19	14	10
SS A8	605	250	145	96	67	48	35	25	16	TW G4	166	88	56	39	29	22	16	12	8
SS B1	459	157	87	56	39	28	20	14	9	TW H1	221	99	60	40	29	21	16	11	8
SS B2	458	204	122	82	58	42	31	22	15	TW H2	169	86	54	37	27	20	15	11	8
SS B3	425	145	80	51	35	25	18	13	8	TW H3	140	72	46	32	23	17	13	10	6
SS B4	449	199	119	80	56	41	30	21	14	TW H4	108	55	35	24	17	13	10	7	5
SS B5	604	142	72	45	30	21	15	11	7	TW I1	226	99	59	39	28	21	15	11	7
SS B6	434	196	118	80	57	41	30	22	15	TW I2	284	149	96	67	49	37	28	20	14
SS B7	537	121	61	38	25	18	12	9	6	TW I3	299	164	107	75	55	42	31	23	16
SS B8	423	189	113	76	54	39	28	20	14	TW I4	294	172	116	84	62	47	36	27	18
SS C1	303	140	85	58	41	30	22	16	11	TW I5	282	170	116	84	63	48	37	27	19
SS C2	312	164	105	74	54	40	30	22	15	TW J1	359	198	129	91	67	51	38	28	19
SS C3	315	112	63	41	29	21	15	11	7	TW J2	321	184	122	87	65	49	37	28	19
SS C4	309	148	91	62	45	33	25	18	12	TW J3	335	210	146	108	81	62	48	36	25
SS D1	618	249	144	95	66	47	34	24	16	TW J4	307	193	134	99	75	58	44	33	23
SS D2	522	278	178	124	90	66	49	36	24	DG A1	285	131	79	52	35	23	14	7	-
SS D3	783	236	126	80	55	39	28	19	13	DG A2	294	138	87	60	43	31	22	15	8
SS D4	534	272	171	118	85	62	46	33	22	DG A3	346	164	105	75	56	43	32	23	14
SS E1	462	196	115	76	53	39	28	20	13	DG B1	282	132	83	58	43	32	22	14	4
SS E2	431	228	146	101	73	54	40	29	20	DG B2	300	142	90	65	49	37	28	20	12
SS E3	561	168	90	57	39	28	20	14	9	DG B3	349	167	107	77	59	46	36	27	18
SS E4	443	214	132	90	64	47	35	25	17	DG C1	336	159	100	71	54	42	31	22	12

LCR *Tables* / 547

TRUTH OR CONSEQUENCES, NEW MEXICO CONTINUED

	SSF =.10	.20	.30	.40	.50	.60	.70	.80	.90
DG C2	348	166	105	76	58	45	35	26	17
DG C3	400	192	123	89	68	54	43	33	23
SS A1	771	244	132	84	58	41	30	21	14
SS A2	674	295	176	117	83	60	44	32	21
SS A3	745	224	119	75	51	36	26	18	12
SS A4	684	294	173	115	81	59	43	31	20
SS A5	1302	245	122	75	50	35	25	17	11
SS A6	673	294	175	117	83	60	44	32	21
SS A7	1489	211	100	60	40	28	19	13	9
SS A8	690	290	170	112	79	57	42	30	20
SS B1	526	185	103	67	46	33	24	17	11
SS B2	522	235	141	95	68	50	36	26	18
SS B3	489	172	95	61	42	30	22	15	10
SS B4	512	230	138	93	66	48	35	25	17
SS B5	688	167	86	54	36	26	18	13	8
SS B6	495	227	138	93	66	49	36	26	17
SS B7	613	144	73	45	31	22	15	11	7
SS B8	483	219	132	89	63	46	34	24	16
SS C1	355	166	101	69	50	37	27	20	13
SS C2	361	192	123	87	63	48	36	26	18
SS C3	366	133	75	49	35	25	19	13	9
SS C4	356	172	107	73	53	39	29	21	15
SS D1	710	292	170	112	79	57	41	29	19
SS D2	599	322	207	144	105	78	58	42	28
SS D3	892	276	149	95	65	46	33	24	15
SS D4	612	315	199	137	99	73	54	39	26
SS E1	535	231	137	91	64	46	34	24	16
SS E2	496	265	170	118	86	64	48	35	23
SS E3	642	198	107	68	47	33	24	17	11
SS E4	509	249	154	105	76	56	41	30	20

TUCUMCARI, NEW MEXICO 4047 DD

	SSF =.10	.20	.30	.40	.50	.60	.70	.80	.90
WW A1	1189	141	66	40	27	19	14	10	6
WW A2	386	154	89	59	42	30	22	16	11
WW A3	323	155	96	66	47	35	26	19	13
WW A4	296	155	99	69	50	37	28	21	14
WW A5	282	157	103	73	53	40	30	22	15
WW A6	274	157	104	74	55	42	31	23	16
WW B1	324	119	67	44	30	22	16	11	8
WW B2	289	156	101	71	52	39	29	22	15
WW B3	334	189	125	89	66	50	37	28	19
WW B4	291	178	122	89	67	51	39	29	20
WW B5	267	167	116	85	64	49	38	28	20
WW C1	354	200	132	94	70	52	40	29	20
WW C2	305	180	121	87	65	50	38	28	19
WW C3	308	200	142	105	81	62	48	36	25
WW C4	278	180	128	95	73	56	43	32	23
TW A1	1015	134	64	39	26	19	13	9	6
TW A2	397	139	78	51	36	26	19	14	9
TW A3	318	138	82	55	39	29	21	15	10
TW A4	269	130	81	55	40	30	22	16	11
TW B1	562	126	64	40	27	20	14	10	7
TW B2	317	125	73	48	34	25	18	13	9
TW B3	276	121	72	49	35	25	19	14	9
TW B4	253	111	66	45	32	23	17	13	9
TW C1	367	113	61	39	27	20	14	10	7
TW C2	278	107	62	41	29	21	15	11	8
TW C3	265	100	57	38	27	19	14	10	7
TW C4	279	90	49	32	22	16	12	8	6
TW D1	234	93	53	35	25	18	13	9	6
TW D2	309	148	91	62	45	33	25	18	12
TW D3	315	162	103	71	52	39	29	21	14

TUCUMCARI, NEW MEXICO CONTINUED

	SSF =.10	.20	.30	.40	.50	.60	.70	.80	.90
TW D4	299	167	110	78	58	43	33	24	17
TW D5	276	159	106	76	57	43	32	24	17
TW E1	363	187	119	82	60	45	34	25	17
TW E2	328	177	114	81	59	44	33	25	17
TW E3	341	204	139	101	75	57	44	32	22
TW E4	304	184	126	92	69	53	40	30	21
TW F1	457	116	60	38	26	19	14	10	6
TW F2	283	118	69	46	33	24	18	13	9
TW F3	233	113	70	48	34	26	19	14	9
TW F4	191	101	64	45	33	24	18	13	9
TW G1	282	100	56	37	26	18	14	10	7
TW G2	204	97	59	40	29	21	16	12	8
TW G3	171	87	55	38	28	21	15	11	8
TW G4	137	72	46	32	23	17	13	10	7
TW H1	182	80	48	32	23	17	13	9	6
TW H2	139	70	44	30	22	16	12	9	6
TW H3	116	59	37	26	19	14	10	8	5
TW H4	89	45	28	20	14	11	8	6	4
TW I1	182	78	46	31	22	16	12	8	6
TW I2	237	124	79	55	40	30	23	17	11
TW I3	250	135	88	62	45	34	26	19	13
TW I4	248	144	97	70	52	39	30	22	15
TW I5	239	143	97	70	53	40	31	23	16
TW J1	300	164	107	75	55	42	31	23	16
TW J2	270	153	102	73	54	41	31	23	16
TW J3	284	177	123	90	68	52	40	30	21
TW J4	261	163	113	83	63	48	37	28	19
DG A1	230	104	61	39	25	15	7	-	-
DG A2	246	113	71	48	34	24	17	11	6
DG A3	293	138	87	62	47	35	26	19	11
DG B1	227	104	65	45	33	23	15	8	-
DG B2	250	117	74	52	39	30	22	16	9
DG B3	297	141	89	64	49	38	30	22	15
DG C1	274	128	79	56	42	32	24	16	7
DG C2	292	138	87	62	47	37	28	21	13
DG C3	341	162	103	74	57	45	36	28	19
SS A1	649	199	106	67	46	33	24	17	11
SS A2	570	245	144	96	68	49	36	26	17
SS A3	625	181	95	59	40	28	20	14	9
SS A4	578	242	141	94	66	48	35	25	16
SS A5	1104	200	97	59	39	28	20	14	9
SS A6	568	243	143	95	67	49	36	26	17
SS A7	1261	170	79	47	31	21	15	10	7
SS A8	582	239	138	91	64	46	33	24	16
SS B1	442	151	83	53	37	26	19	13	9
SS B2	443	196	117	78	56	41	30	22	14
SS B3	409	139	76	49	33	24	17	12	8
SS B4	434	191	113	76	54	39	29	21	14
SS B5	581	136	69	42	29	20	14	10	7
SS B6	419	189	113	76	54	40	29	21	14
SS B7	515	116	58	35	24	17	12	8	5
SS B8	408	181	108	72	51	37	27	20	13
SS C1	292	135	82	55	40	29	22	16	11
SS C2	302	159	102	71	52	39	29	21	15
SS C3	304	108	61	40	28	20	15	11	7
SS C4	299	143	88	60	43	32	24	18	12
SS D1	593	238	137	90	63	45	33	23	15
SS D2	502	266	170	118	86	64	47	34	23
SS D3	751	225	120	76	52	37	26	19	12
SS D4	514	261	163	112	81	60	44	32	21
SS E1	444	187	109	72	51	37	27	19	13
SS E2	416	219	139	97	70	52	39	28	19
SS E3	539	161	85	54	37	26	19	13	9
SS E4	428	206	126	86	62	45	33	24	16

```
ZUNI, NEW MEXICO                          5815 DD    ZUNI, NEW MEXICO                        CONTINUED
   SSF =.10  .20  .30  .40  .50  .60  .70  .80  .90    SSF =.10  .20  .30  .40  .50  .60  .70  .80  .90
WW A1   801   97   46   27   18   13    9    6    4  DG C2   218  101   64   45   34   26   20   14    8
WW A2   275  110   63   42   29   21   16   11    8  DG C3   263  124   79   57   43   34   27   21   14
WW A3   234  112   69   47   34   25   18   13    9  SS A1   516  160   84   53   36   25   18   12    8
WW A4   215  112   71   49   36   27   20   15   10  SS A2   467  201  118   78   55   39   29   20   13
WW A5   206  114   74   52   38   29   22   16   11  SS A3   497  145   75   46   31   21   15   10    7
WW A6   200  114   76   54   40   30   23   17   11  SS A4   477  200  116   76   53   38   28   19   13
WW B1   221   81   45   29   20   14   10    7    5  SS A5   855  159   76   46   30   21   15   10    6
WW B2   214  115   74   52   38   29   21   16   11  SS A6   465  199  116   77   54   39   28   20   13
WW B3   252  142   94   67   49   37   28   21   14  SS A7   968  135   62   36   23   16   11    7    5
WW B4   222  136   93   68   51   39   30   22   15  SS A8   478  196  113   74   51   37   26   19   12
WW B5   205  128   89   65   49   38   29   22   15  SS B1   346  118   64   41   28   20   14   10    6
WW C1   268  152  100   71   52   40   30   22   15  SS B2   359  159   94   63   45   32   24   17   11
WW C2   231  136   92   66   49   37   28   21   14  SS B3   320  109   59   37   25   18   13    9    6
WW C3   241  157  111   82   63   49   37   28   19  SS B4   353  156   92   61   43   31   23   16   11
WW C4   216  140   99   74   56   44   34   25   18  SS B5   445  105   53   32   21   15   11    7    5
TW A1   690   93   44   27   18   13    9    6    4  SS B6   339  153   91   61   43   31   23   16   11
TW A2   282  100   55   36   25   18   13    9    6  SS B7   394   89   44   26   18   12    8    6    4
TW A3   229  100   59   39   28   20   15   11    7  SS B8   331  147   87   58   41   29   21   15   10
TW A4   195   95   58   40   29   21   16   11    8  SS C1   209   96   58   39   28   20   15   11    7
TW B1   391   89   45   28   19   14   10    7    5  SS C2   225  118   76   53   38   29   22   16   11
TW B2   228   91   52   34   24   18   13    9    6  SS C3   215   77   43   28   19   14   10    7    5
TW B3   200   88   52   35   25   18   13   10    7  SS C4   222  106   65   45   32   24   18   13    9
TW B4   184   81   48   32   23   17   12    9    6  SS D1   476  191  109   71   49   35   25   18   11
TW C1   261   81   44   28   19   14   10    7    5  SS D2   415  219  139   96   69   51   38   27   18
TW C2   201   78   45   29   21   15   11    8    5  SS D3   596  180   95   59   40   28   20   14    9
TW C3   192   73   42   27   19   14   10    7    5  SS D4   424  214  133   91   66   48   35   25   17
TW C4   202   66   36   23   16   12    8    6    4  SS E1   348  146   84   55   38   28   20   14    9
TW D1   161   63   36   23   16   12    8    6    4  SS E2   336  176  111   77   56   41   30   22   15
TW D2   228  109   67   46   33   24   18   13    9  SS E3   417  125   66   41   28   20   14   10    6
TW D3   235  121   76   53   39   29   21   16   11  SS E4   345  166  101   68   49   36   26   19   12
TW D4   227  127   83   59   44   33   25   18   13
TW D5   212  122   81   58   43   33   25   18   13
TW E1   274  141   89   62   45   34   25   18   12  ALBANY, NEW YORK                           6888 DD
TW E2   248  134   86   61   44   33   25   18   13    SSF =.10  .20  .30  .40  .50  .60  .70  .80  .90
TW E3   265  159  108   78   58   44   34   25   17  WW A1   219    -    -    -    -    -    -    -    -
TW E4   235  143   97   71   53   41   31   23   16  WW A2    70   18    -    -    -    -    -    -    -
TW F1   316   81   42   26   18   13    9    6    4  WW A3    61   22    9    -    -    -    -    -    -
TW F2   202   84   49   33   23   17   12    9    6  WW A4    57   24   11    5    -    -    -    -    -
TW F3   168   81   50   34   25   18   13   10    7  WW A5    55   25   13    7    -    -    -    -    -
TW F4   139   73   46   32   24   18   13   10    6  WW A6    54   26   14    8    4    -    -    -    -
TW G1   198   71   39   25   18   13    9    7    4  WW B1     -    -    -    -    -    -    -    -    -
TW G2   147   69   42   29   21   15   11    8    5  WW B2    70   34   19   12    8    5    3    2    -
TW G3   124   63   40   27   20   15   11    8    5  WW B3    87   45   27   18   12    8    6    4    2
TW G4   100   52   33   23   17   13    9    7    5  WW B4    89   52   34   24   17   12    9    6    4
TW H1   130   57   34   23   16   12    9    6    4  WW B5    88   52   35   25   18   13   10    7    5
TW H2   101   51   32   22   16   12    9    6    4  WW C1    98   51   31   21   14   10    7    5    3
TW H3    84   43   27   19   14   10    8    5    4  WW C2    90   50   32   22   15   11    8    5    3
TW H4    65   33   21   14   10    8    6    4    3  WW C3   107   68   46   34   25   19   14   10    7
TW I1   124   53   31   20   14   10    7    5    3  WW C4    98   62   43   31   23   17   13    9    6
TW I2   175   91   58   40   29   22   16   12    8  TW A1   199    -    -    -    -    -    -    -    -
TW I3   188  102   66   46   34   25   19   14    9  TW A2    76   18    -    -    -    -    -    -    -
TW I4   189  110   74   53   39   30   23   17   11  TW A3    63   20    8    -    -    -    -    -    -
TW I5   184  110   75   54   40   31   23   17   12  TW A4    55   22   11    5    -    -    -    -    -
TW J1   227  124   81   57   42   31   23   17   12  TW B1   110   11    -    -    -    -    -    -    -
TW J2   204  116   77   55   41   31   23   17   12  TW B2    65   19    8    -    -    -    -    -    -
TW J3   221  138   96   70   53   41   31   23   16  TW B3    59   20    9    4    -    -    -    -    -
TW J4   202  127   88   65   49   37   29   21   15  TW B4    57   20   10    5    3    -    -    -    -
DG A1   157   69   39   22   12    -    -    -    -  TW C1    77   17    5    -    -    -    -    -    -
DG A2   179   83   51   34   24   16   11    6    -  TW C2    63   19    9    4    2    -    -    -    -
DG A3   223  105   66   47   35   26   19   13    8  TW C3    62   19    9    5    2    -    -    -    -
DG B1   156   70   42   29   20   12    -    -    -  TW C4    71   18    8    4    2    1    -    -    -
DG B2   183   85   53   37   28   21   15   10    5  TW D1     -    -    -    -    -    -    -    -    -
DG B3   228  107   68   49   37   29   23   17   11  TW D2    77   32   17   11    7    4    3    1    -
DG C1   194   88   54   38   28   20   14    7    -  TW D3    78   35   20   12    8    5    3    2    -
```

ALBANY, NEW YORK — CONTINUED

	SSF =.10	.20	.30	.40	.50	.60	.70	.80	.90
TW D4	90	47	29	20	14	10	7	5	3
TW D5	91	50	32	22	16	12	8	6	4
TW E1	99	46	27	17	11	8	5	3	2
TW E2	96	48	29	19	13	9	7	4	3
TW E3	113	65	42	30	21	16	11	8	5
TW E4	104	61	40	28	21	15	11	8	5
TW F1	72	-	-	-	-	-	-	-	-
TW F2	49	13	-	-	-	-	-	-	-
TW F3	43	15	6	-	-	-	-	-	-
TW F4	37	16	8	4	-	-	-	-	-
TW G1	45	-	-	-	-	-	-	-	-
TW G2	37	12	4	-	-	-	-	-	-
TW G3	33	13	6	2	-	-	-	-	-
TW G4	28	12	6	3	2	-	-	-	-
TW H1	30	8	-	-	-	-	-	-	-
TW H2	26	10	5	-	-	-	-	-	-
TW H3	24	10	5	3	-	-	-	-	-
TW H4	19	8	4	2	1	-	-	-	-
TW I1	-	-	-	-	-	-	-	-	-
TW I2	57	26	15	9	6	4	2	1	-
TW I3	64	31	18	12	8	5	3	2	1
TW I4	75	41	26	18	13	9	7	5	3
TW I5	78	44	29	20	14	11	8	5	4
TW J1	82	41	25	16	11	8	5	3	2
TW J2	79	42	26	18	12	9	6	4	3
TW J3	97	58	39	28	20	15	11	8	5
TW J4	90	55	37	26	19	14	11	8	5
DG A1	-	-	-	-	-	-	-	-	-
DG A2	56	19	-	-	-	-	-	-	-
DG A3	96	42	24	16	11	7	4	-	-
DG B1	-	-	-	-	-	-	-	-	-
DG B2	57	21	9	-	-	-	-	-	-
DG B3	100	44	26	17	12	9	6	4	-
DG C1	26	-	-	-	-	-	-	-	-
DG C2	77	31	16	9	4	-	-	-	-
DG C3	118	52	31	21	15	11	8	5	2
SS A1	162	38	15	5	-	-	-	-	-
SS A2	177	69	37	23	15	10	7	4	3
SS A3	135	22	-	-	-	-	-	-	-
SS A4	173	65	34	20	13	8	5	3	2
SS A5	278	32	-	-	-	-	-	-	-
SS A6	171	66	35	21	14	9	6	4	2
SS A7	290	-	-	-	-	-	-	-	-
SS A8	168	61	31	18	11	7	4	2	1
SS B1	101	25	8	-	-	-	-	-	-
SS B2	140	57	31	20	13	9	6	4	2
SS B3	80	13	-	-	-	-	-	-	-
SS B4	133	53	29	18	11	8	5	3	2
SS B5	127	16	-	-	-	-	-	-	-
SS B6	128	52	29	18	12	8	5	3	2
SS B7	94	-	-	-	-	-	-	-	-
SS B8	119	47	25	15	10	6	4	2	1
SS C1	44	9	-	-	-	-	-	-	-
SS C2	77	36	21	13	9	6	4	2	1
SS C3	50	-	-	-	-	-	-	-	-
SS C4	76	32	18	11	7	4	3	2	-
SS D1	142	45	20	9	-	-	-	-	-
SS D2	155	77	46	30	20	14	10	7	4
SS D3	185	42	15	-	-	-	-	-	-
SS D4	159	75	43	28	19	13	9	6	4
SS E1	87	24	-	-	-	-	-	-	-
SS E2	118	57	33	21	14	9	6	4	2
SS E3	114	20	-	-	-	-	-	-	-
SS E4	123	53	29	18	12	8	5	3	2

BINGHAMTON, NEW YORK 7285 DD

	SSF =.10	.20	.30	.40	.50	.60	.70	.80	.90
WW A1	113	-	-	-	-	-	-	-	-
WW A2	30	-	-	-	-	-	-	-	-
WW A3	28	-	-	-	-	-	-	-	-
WW A4	28	-	-	-	-	-	-	-	-
WW A5	28	-	-	-	-	-	-	-	-
WW A6	28	-	-	-	-	-	-	-	-
WW B1	-	-	-	-	-	-	-	-	-
WW B2	48	20	9	-	-	-	-	-	-
WW B3	63	30	16	9	5	2	-	-	-
WW B4	71	40	25	17	12	8	6	4	2
WW B5	71	42	27	19	14	10	7	5	3
WW C1	74	36	21	13	8	5	3	2	-
WW C2	70	37	23	15	10	7	5	3	2
WW C3	90	55	38	27	20	15	11	8	5
WW C4	82	51	35	25	18	14	10	7	5
TW A1	115	-	-	-	-	-	-	-	-
TW A2	41	-	-	-	-	-	-	-	-
TW A3	34	-	-	-	-	-	-	-	-
TW A4	32	-	-	-	-	-	-	-	-
TW B1	63	-	-	-	-	-	-	-	-
TW B2	38	-	-	-	-	-	-	-	-
TW B3	36	-	-	-	-	-	-	-	-
TW B4	37	8	-	-	-	-	-	-	-
TW C1	49	-	-	-	-	-	-	-	-
TW C2	41	-	-	-	-	-	-	-	-
TW C3	43	9	-	-	-	-	-	-	-
TW C4	52	10	-	-	-	-	-	-	-
TW D1	-	-	-	-	-	-	-	-	-
TW D2	54	19	8	-	-	-	-	-	-
TW D3	56	22	10	4	-	-	-	-	-
TW D4	72	36	21	14	9	6	4	3	2
TW D5	75	40	25	17	12	8	6	4	3
TW E1	75	32	17	10	6	3	1	-	-
TW E2	74	35	20	13	8	5	4	2	1
TW E3	93	52	33	23	16	12	8	6	4
TW E4	87	49	32	22	16	12	8	6	4
TW F1	-	-	-	-	-	-	-	-	-
TW F2	-	-	-	-	-	-	-	-	-
TW F3	18	-	-	-	-	-	-	-	-
TW F4	19	-	-	-	-	-	-	-	-
TW G1	-	-	-	-	-	-	-	-	-
TW G2	12	-	-	-	-	-	-	-	-
TW G3	16	-	-	-	-	-	-	-	-
TW G4	16	-	-	-	-	-	-	-	-
TW H1	-	-	-	-	-	-	-	-	-
TW H2	13	-	-	-	-	-	-	-	-
TW H3	13	-	-	-	-	-	-	-	-
TW H4	12	3	-	-	-	-	-	-	-
TW I1	-	-	-	-	-	-	-	-	-
TW I2	39	15	6	-	-	-	-	-	-
TW I3	46	20	10	5	2	-	-	-	-
TW I4	59	31	19	13	9	6	4	3	2
TW I5	63	35	22	15	11	8	5	4	2
TW J1	62	29	16	10	6	4	2	-	-
TW J2	62	31	19	12	8	5	4	2	1
TW J3	80	47	31	22	16	11	8	6	4
TW J4	75	45	30	21	15	11	8	6	4
DG A1	-	-	-	-	-	-	-	-	-
DG A2	37	-	-	-	-	-	-	-	-
DG A3	80	33	18	11	7	3	-	-	-
DG B1	-	-	-	-	-	-	-	-	-
DG B2	38	-	-	-	-	-	-	-	-
DG B3	84	35	20	13	8	6	3	-	-
DG C1	-	-	-	-	-	-	-	-	-

BINGHAMTON, NEW YORK — CONTINUED

SSF =	.10	.20	.30	.40	.50	.60	.70	.80	.90
DG C2	57	19	6	-	-	-	-	-	-
DG C3	100	42	24	16	11	8	5	3	-
SS A1	129	21	-	-	-	-	-	-	-
SS A2	151	56	28	16	10	6	4	2	1
SS A3	101	-	-	-	-	-	-	-	-
SS A4	148	51	25	14	8	4	2	-	-
SS A5	226	-	-	-	-	-	-	-	-
SS A6	145	52	26	15	9	5	3	1	-
SS A7	229	-	-	-	-	-	-	-	-
SS A8	142	47	21	11	5	-	-	-	-
SS B1	74	-	-	-	-	-	-	-	-
SS B2	119	46	24	14	9	6	4	2	1
SS B3	51	-	-	-	-	-	-	-	-
SS B4	112	42	21	12	7	4	3	1	-
SS B5	94	-	-	-	-	-	-	-	-
SS B6	107	41	21	12	7	5	3	2	-
SS B7	58	-	-	-	-	-	-	-	-
SS B8	99	36	18	9	5	3	-	-	-
SS C1	-	-	-	-	-	-	-	-	-
SS C2	55	23	11	6	-	-	-	-	-
SS C3	-	-	-	-	-	-	-	-	-
SS C4	55	20	9	4	-	-	-	-	-
SS D1	109	25	-	-	-	-	-	-	-
SS D2	131	62	35	22	14	10	6	4	2
SS D3	147	21	-	-	-	-	-	-	-
SS D4	135	60	34	21	13	9	6	4	2
SS E1	53	-	-	-	-	-	-	-	-
SS E2	96	43	23	13	8	5	2	-	-
SS E3	81	-	-	-	-	-	-	-	-
SS E4	100	40	20	11	6	4	2	-	-

BUFFALO, NEW YORK — 6927 DD

SSF =	.10	.20	.30	.40	.50	.60	.70	.80	.90
WW A1	148	-	-	-	-	-	-	-	-
WW A2	39	-	-	-	-	-	-	-	-
WW A3	34	-	-	-	-	-	-	-	-
WW A4	33	-	-	-	-	-	-	-	-
WW A5	32	-	-	-	-	-	-	-	-
WW A6	32	-	-	-	-	-	-	-	-
WW B1	-	-	-	-	-	-	-	-	-
WW B2	52	22	10	-	-	-	-	-	-
WW B3	67	32	17	10	5	2	-	-	-
WW B4	74	41	26	18	12	8	6	4	2
WW B5	74	43	28	19	14	10	7	5	3
WW C1	78	38	22	13	8	5	3	2	-
WW C2	74	39	24	15	10	7	5	3	2
WW C3	93	57	39	28	20	15	11	8	5
WW C4	85	53	36	25	19	14	10	7	5
TW A1	142	-	-	-	-	-	-	-	-
TW A2	48	-	-	-	-	-	-	-	-
TW A3	40	-	-	-	-	-	-	-	-
TW A4	36	-	-	-	-	-	-	-	-
TW B1	75	-	-	-	-	-	-	-	-
TW B2	44	-	-	-	-	-	-	-	-
TW B3	40	-	-	-	-	-	-	-	-
TW B4	41	10	-	-	-	-	-	-	-
TW C1	55	-	-	-	-	-	-	-	-
TW C2	46	8	-	-	-	-	-	-	-
TW C3	47	10	-	-	-	-	-	-	-
TW C4	56	12	-	-	-	-	-	-	-
TW D1	-	-	-	-	-	-	-	-	-
TW D2	59	21	9	-	-	-	-	-	-
TW D3	60	24	11	-	-	-	-	-	-

BUFFALO, NEW YORK — CONTINUED

SSF =	.10	.20	.30	.40	.50	.60	.70	.80	.90
TW D4	75	38	22	14	10	6	4	3	2
TW D5	78	41	26	17	12	8	6	4	3
TW E1	79	34	18	10	6	3	1	-	-
TW E2	78	37	21	13	8	5	3	2	1
TW E3	97	54	34	23	16	12	8	6	4
TW E4	90	51	33	23	16	12	8	6	4
TW F1	-	-	-	-	-	-	-	-	-
TW F2	24	-	-	-	-	-	-	-	-
TW F3	23	-	-	-	-	-	-	-	-
TW F4	22	-	-	-	-	-	-	-	-
TW G1	-	-	-	-	-	-	-	-	-
TW G2	18	-	-	-	-	-	-	-	-
TW G3	19	-	-	-	-	-	-	-	-
TW G4	18	-	-	-	-	-	-	-	-
TW H1	-	-	-	-	-	-	-	-	-
TW H2	15	-	-	-	-	-	-	-	-
TW H3	15	-	-	-	-	-	-	-	-
TW H4	13	4	-	-	-	-	-	-	-
TW I1	-	-	-	-	-	-	-	-	-
TW I2	42	17	7	-	-	-	-	-	-
TW I3	49	21	11	5	-	-	-	-	-
TW I4	62	33	20	13	9	6	4	3	2
TW I5	66	36	23	16	11	8	5	4	2
TW J1	66	31	17	10	6	3	2	-	-
TW J2	65	33	19	12	8	5	4	2	1
TW J3	83	49	32	22	16	12	8	6	4
TW J4	78	46	30	21	15	11	8	6	4
DG A1	-	-	-	-	-	-	-	-	-
DG A2	41	-	-	-	-	-	-	-	-
DG A3	84	34	19	11	7	4	-	-	-
DG B1	-	-	-	-	-	-	-	-	-
DG B2	43	-	-	-	-	-	-	-	-
DG B3	88	37	21	13	9	6	3	-	-
DG C1	-	-	-	-	-	-	-	-	-
DG C2	61	21	7	-	-	-	-	-	-
DG C3	104	44	26	16	11	8	5	3	-
SS A1	139	24	-	-	-	-	-	-	-
SS A2	158	58	30	17	10	6	4	2	1
SS A3	111	-	-	-	-	-	-	-	-
SS A4	155	54	26	14	8	4	2	-	-
SS A5	246	-	-	-	-	-	-	-	-
SS A6	152	55	27	15	9	5	3	1	-
SS A7	255	-	-	-	-	-	-	-	-
SS A8	149	49	22	11	5	-	-	-	-
SS B1	81	-	-	-	-	-	-	-	-
SS B2	125	48	25	15	9	6	4	2	1
SS B3	58	-	-	-	-	-	-	-	-
SS B4	118	44	22	12	7	4	2	1	-
SS B5	104	-	-	-	-	-	-	-	-
SS B6	112	43	22	13	8	5	3	1	-
SS B7	70	-	-	-	-	-	-	-	-
SS B8	104	38	18	10	5	2	-	-	-
SS C1	-	-	-	-	-	-	-	-	-
SS C2	59	25	12	6	-	-	-	-	-
SS C3	-	-	-	-	-	-	-	-	-
SS C4	59	22	10	4	-	-	-	-	-
SS D1	117	28	-	-	-	-	-	-	-
SS D2	137	64	37	23	15	9	6	4	2
SS D3	159	25	-	-	-	-	-	-	-
SS D4	140	63	35	21	13	9	6	3	2
SS E1	60	-	-	-	-	-	-	-	-
SS E2	101	45	24	14	8	4	2	-	-
SS E3	90	-	-	-	-	-	-	-	-
SS E4	106	42	21	12	6	3	1	-	-

MASSENA, NEW YORK   8237 DD

SSF =	.10	.20	.30	.40	.50	.60	.70	.80	.90
WW A1	127	-	-	-	-	-	-	-	-
WW A2	38	-	-	-	-	-	-	-	-
WW A3	34	-	-	-	-	-	-	-	-
WW A4	32	-	-	-	-	-	-	-	-
WW A5	32	-	-	-	-	-	-	-	-
WW A6	32	-	-	-	-	-	-	-	-
WW B1	-	-	-	-	-	-	-	-	-
WW B2	51	22	11	4	-	-	-	-	-
WW B3	66	31	17	10	5	2	-	-	-
WW B4	73	41	26	18	12	9	6	4	2
WW B5	73	43	28	20	14	10	7	5	3
WW C1	77	38	22	14	9	5	3	2	-
WW C2	73	39	24	16	11	7	5	3	2
WW C3	92	57	39	28	20	15	11	8	5
WW C4	84	52	36	25	19	14	10	7	5
TW A1	126	-	-	-	-	-	-	-	-
TW A2	46	-	-	-	-	-	-	-	-
TW A3	38	-	-	-	-	-	-	-	-
TW A4	35	-	-	-	-	-	-	-	-
TW B1	70	-	-	-	-	-	-	-	-
TW B2	42	-	-	-	-	-	-	-	-
TW B3	39	-	-	-	-	-	-	-	-
TW B4	40	10	-	-	-	-	-	-	-
TW C1	53	-	-	-	-	-	-	-	-
TW C2	44	9	-	-	-	-	-	-	-
TW C3	46	11	-	-	-	-	-	-	-
TW C4	54	12	-	-	-	-	-	-	-
TW D1	-	-	-	-	-	-	-	-	-
TW D2	57	21	9	-	-	-	-	-	-
TW D3	58	23	11	4	-	-	-	-	-
TW D4	74	37	22	15	10	7	4	3	2
TW D5	77	41	26	17	12	9	6	4	3
TW E1	78	34	18	10	6	3	-	-	-
TW E2	77	37	21	13	9	6	4	2	1
TW E3	95	53	34	23	17	12	8	6	4
TW E4	89	51	33	23	16	12	9	6	4
TW F1	-	-	-	-	-	-	-	-	-
TW F2	23	-	-	-	-	-	-	-	-
TW F3	22	-	-	-	-	-	-	-	-
TW F4	22	-	-	-	-	-	-	-	-
TW G1	-	-	-	-	-	-	-	-	-
TW G2	18	-	-	-	-	-	-	-	-
TW G3	19	-	-	-	-	-	-	-	-
TW G4	18	-	-	-	-	-	-	-	-
TW H1	-	-	-	-	-	-	-	-	-
TW H2	15	-	-	-	-	-	-	-	-
TW H3	15	-	-	-	-	-	-	-	-
TW H4	13	4	-	-	-	-	-	-	-
TW I1	-	-	-	-	-	-	-	-	-
TW I2	42	17	8	-	-	-	-	-	-
TW I3	48	21	11	6	2	-	-	-	-
TW I4	61	32	20	13	9	6	4	3	2
TW I5	65	36	23	16	11	8	6	4	2
TW J1	64	30	17	10	6	4	2	-	-
TW J2	64	33	20	13	8	6	4	2	1
TW J3	82	48	32	22	16	12	9	6	4
TW J4	77	46	30	21	16	11	8	6	4
DG A1	-	-	-	-	-	-	-	-	-
DG A2	40	-	-	-	-	-	-	-	-
DG A3	82	34	19	12	7	4	-	-	-
DG B1	-	-	-	-	-	-	-	-	-
DG B2	41	8	-	-	-	-	-	-	-
DG B3	86	36	21	13	9	6	4	-	-
DG C1	-	-	-	-	-	-	-	-	-

MASSENA, NEW YORK   CONTINUED

SSF =	.10	.20	.30	.40	.50	.60	.70	.80	.90
DG C2	59	21	8	-	-	-	-	-	-
DG C3	102	43	26	17	11	8	5	3	-
SS A1	125	20	-	-	-	-	-	-	-
SS A2	148	55	28	16	10	6	3	2	-
SS A3	94	-	-	-	-	-	-	-	-
SS A4	143	50	24	13	7	3	-	-	-
SS A5	217	-	-	-	-	-	-	-	-
SS A6	142	51	25	14	8	5	2	-	-
SS A7	214	-	-	-	-	-	-	-	-
SS A8	137	45	20	10	3	-	-	-	-
SS B1	72	-	-	-	-	-	-	-	-
SS B2	118	45	24	14	9	5	3	2	1
SS B3	47	-	-	-	-	-	-	-	-
SS B4	110	41	21	12	7	4	2	-	-
SS B5	90	-	-	-	-	-	-	-	-
SS B6	106	41	21	12	7	4	2	1	-
SS B7	52	-	-	-	-	-	-	-	-
SS B8	96	35	17	9	4	-	-	-	-
SS C1	-	-	-	-	-	-	-	-	-
SS C2	58	25	12	6	-	-	-	-	-
SS C3	-	-	-	-	-	-	-	-	-
SS C4	58	22	10	5	-	-	-	-	-
SS D1	105	23	-	-	-	-	-	-	-
SS D2	129	61	35	22	14	9	6	4	2
SS D3	143	19	-	-	-	-	-	-	-
SS D4	132	59	33	20	13	8	5	3	2
SS E1	50	-	-	-	-	-	-	-	-
SS E2	94	42	23	13	7	4	1	-	-
SS E3	78	-	-	-	-	-	-	-	-
SS E4	99	40	20	11	6	3	-	-	-

NEW YORK (CENTRAL PARK), NEW YORK   4848 DD

SSF =	.10	.20	.30	.40	.50	.60	.70	.80	.90
WW A1	345	31	10	3	-	-	-	-	-
WW A2	121	43	22	13	8	5	3	2	1
WW A3	104	45	25	16	10	7	5	3	2
WW A4	96	46	27	17	12	8	6	4	2
WW A5	93	48	29	19	13	9	6	4	3
WW A6	90	48	30	20	14	10	7	5	3
WW B1	64	-	-	-	-	-	-	-	-
WW B2	105	54	33	22	15	11	8	5	3
WW B3	126	68	43	29	21	15	11	8	5
WW B4	121	72	48	34	25	18	14	10	7
WW B5	115	70	48	34	25	19	14	10	7
WW C1	138	75	48	32	23	17	12	9	6
WW C2	123	71	46	32	23	17	13	9	6
WW C3	139	88	61	45	33	25	19	14	10
WW C4	126	80	56	41	31	23	18	13	9
TW A1	306	32	12	5	2	-	-	-	-
TW A2	127	39	19	11	7	4	3	2	-
TW A3	104	41	22	13	9	6	4	2	1
TW A4	90	40	23	15	10	7	5	3	2
TW B1	176	33	14	8	4	2	1	-	-
TW B2	105	38	20	12	8	5	3	2	1
TW B3	93	37	21	13	8	6	4	3	1
TW B4	88	36	20	12	8	6	4	3	2
TW C1	121	33	16	9	6	4	2	1	-
TW C2	96	34	18	11	7	5	3	2	1
TW C3	93	33	17	10	7	5	3	2	1
TW C4	101	30	15	9	6	4	3	2	1
TW D1	48	10	-	-	-	-	-	-	-
TW D2	113	51	30	19	13	9	7	4	3
TW D3	115	56	34	22	15	11	8	5	3

NEW YORK (CENTRAL PARK), NEW YORK — CONTINUED

	SSF=.10	.20	.30	.40	.50	.60	.70	.80	.90
TW D4	123	66	42	29	21	15	11	8	5
TW D5	120	67	44	31	22	16	12	9	6
TW E1	140	69	42	28	19	14	10	7	4
TW E2	132	68	43	29	21	15	11	8	5
TW E3	148	87	57	41	30	22	16	12	8
TW E4	135	80	54	38	28	21	16	11	8
TW F1	132	27	11	5	2	-	-	-	-
TW F2	87	32	17	10	6	4	2	1	-
TW F3	74	33	18	11	7	5	3	2	1
TW F4	63	30	18	12	8	5	4	2	1
TW G1	84	25	12	7	4	2	1	-	-
TW G2	64	27	15	9	6	4	3	2	-
TW G3	56	26	15	10	6	4	3	2	1
TW G4	46	22	13	9	6	4	3	2	1
TW H1	56	22	12	7	4	3	2	1	-
TW H2	45	21	12	7	5	3	2	1	-
TW H3	39	18	11	7	5	3	2	1	-
TW H4	31	14	8	5	4	3	2	1	-
TW I1	34	-	-	-	-	-	-	-	-
TW I2	86	42	25	17	12	8	6	4	3
TW I3	93	48	29	20	14	10	7	5	3
TW I4	102	58	37	26	19	14	10	7	5
TW I5	103	60	40	28	20	15	11	8	5
TW J1	116	61	38	26	18	13	9	7	4
TW J2	109	60	38	26	19	14	10	7	5
TW J3	126	77	52	37	28	21	16	11	8
TW J4	117	72	49	35	26	20	15	11	7
DG A1	47	-	-	-	-	-	-	-	-
DG A2	86	36	19	10	-	-	-	-	-
DG A3	127	57	34	23	16	11	8	4	-
DG B1	48	-	-	-	-	-	-	-	-
DG B2	89	37	21	13	8	-	-	-	-
DG B3	130	60	36	25	18	13	10	7	3
DG C1	76	26	10	-	-	-	-	-	-
DG C2	112	48	28	18	12	8	5	-	-
DG C3	151	70	43	29	22	16	12	9	5
SS A1	229	64	31	17	11	7	4	3	1
SS A2	233	96	54	34	23	16	11	8	5
SS A3	206	52	22	11	6	3	-	-	-
SS A4	232	93	51	32	21	15	10	7	4
SS A5	374	60	25	13	7	4	2	-	-
SS A6	228	94	52	33	22	15	11	7	5
SS A7	395	44	15	5	-	-	-	-	-
SS A8	228	89	48	30	20	13	9	6	4
SS B1	152	47	23	13	8	5	3	2	-
SS B2	185	79	45	29	20	14	10	7	4
SS B3	131	38	17	9	5	2	-	-	-
SS B4	178	75	42	27	18	13	9	6	4
SS B5	190	38	16	8	4	2	-	-	-
SS B6	171	74	42	27	19	13	9	6	4
SS B7	156	28	10	-	-	-	-	-	-
SS B8	163	69	39	25	16	11	8	5	3
SS C1	87	35	19	11	7	4	3	1	-
SS C2	112	56	34	23	16	11	8	6	4
SS C3	91	28	13	7	4	3	1	-	-
SS C4	111	50	29	19	13	9	6	4	3
SS D1	209	77	40	24	15	10	6	4	2
SS D2	207	106	65	43	30	22	16	11	7
SS D3	263	72	34	19	12	7	5	3	1
SS D4	212	104	62	41	29	20	14	10	6
SS E1	145	54	27	16	9	6	3	2	-
SS E2	165	83	50	33	23	16	12	8	5
SS E3	178	46	21	11	6	3	2	-	-
SS E4	170	78	45	29	20	14	10	7	4

ROCHESTER, NEW YORK — 6719 DD

	SSF=.10	.20	.30	.40	.50	.60	.70	.80	.90
WW A1	169	-	-	-	-	-	-	-	-
WW A2	47	-	-	-	-	-	-	-	-
WW A3	40	-	-	-	-	-	-	-	-
WW A4	38	-	-	-	-	-	-	-	-
WW A5	37	-	-	-	-	-	-	-	-
WW A6	37	-	-	-	-	-	-	-	-
WW B1	-	-	-	-	-	-	-	-	-
WW B2	56	24	12	5	-	-	-	-	-
WW B3	71	34	19	11	6	3	-	-	-
WW B4	77	43	27	19	13	9	6	4	3
WW B5	77	45	29	20	15	10	8	5	3
WW C1	83	41	23	14	9	6	4	2	-
WW C2	77	41	25	16	11	8	5	3	2
WW C3	96	59	40	28	21	15	11	8	5
WW C4	88	54	37	26	19	14	10	7	5
TW A1	159	-	-	-	-	-	-	-	-
TW A2	55	-	-	-	-	-	-	-	-
TW A3	45	-	-	-	-	-	-	-	-
TW A4	41	8	-	-	-	-	-	-	-
TW B1	84	-	-	-	-	-	-	-	-
TW B2	49	-	-	-	-	-	-	-	-
TW B3	44	9	-	-	-	-	-	-	-
TW B4	44	12	-	-	-	-	-	-	-
TW C1	61	-	-	-	-	-	-	-	-
TW C2	50	11	-	-	-	-	-	-	-
TW C3	51	12	-	-	-	-	-	-	-
TW C4	60	13	4	-	-	-	-	-	-
TW D1	-	-	-	-	-	-	-	-	-
TW D2	63	23	10	4	-	-	-	-	-
TW D3	64	26	12	6	-	-	-	-	-
TW D4	78	39	24	15	10	7	5	3	2
TW D5	81	43	27	18	13	9	6	4	3
TW E1	84	36	20	11	7	4	2	-	-
TW E2	82	39	22	14	9	6	4	2	1
TW E3	100	56	36	24	17	12	9	6	4
TW E4	93	53	34	24	17	12	9	6	4
TW F1	43	-	-	-	-	-	-	-	-
TW F2	31	-	-	-	-	-	-	-	-
TW F3	28	-	-	-	-	-	-	-	-
TW F4	25	-	-	-	-	-	-	-	-
TW G1	24	-	-	-	-	-	-	-	-
TW G2	23	-	-	-	-	-	-	-	-
TW G3	22	-	-	-	-	-	-	-	-
TW G4	20	5	-	-	-	-	-	-	-
TW H1	17	-	-	-	-	-	-	-	-
TW H2	18	-	-	-	-	-	-	-	-
TW H3	17	-	-	-	-	-	-	-	-
TW H4	15	5	-	-	-	-	-	-	-
TW I1	-	-	-	-	-	-	-	-	-
TW I2	45	18	9	3	-	-	-	-	-
TW I3	52	23	12	6	3	-	-	-	-
TW I4	65	34	21	14	9	6	4	3	2
TW I5	68	38	24	16	11	8	6	4	2
TW J1	69	33	18	11	7	4	2	1	-
TW J2	68	35	21	13	9	6	4	3	1
TW J3	86	51	33	23	17	12	9	6	4
TW J4	81	48	32	22	16	12	8	6	4
DG A1	-	-	-	-	-	-	-	-	-
DG A2	44	-	-	-	-	-	-	-	-
DG A3	87	36	20	12	7	4	-	-	-
DG B1	-	-	-	-	-	-	-	-	-
DG B2	46	12	-	-	-	-	-	-	-
DG B3	91	39	22	14	9	6	4	-	-
DG C1	-	-	-	-	-	-	-	-	-

ROCHESTER, NEW YORK — CONTINUED

SSF =	.10	.20	.30	.40	.50	.60	.70	.80	.90
DG C2	65	23	9	-	-	-	-	-	-
DG C3	108	46	27	17	12	8	5	3	-
SS A1	144	27	-	-	-	-	-	-	-
SS A2	162	60	31	18	11	7	4	2	1
SS A3	116	-	-	-	-	-	-	-	-
SS A4	159	56	27	15	8	5	2	1	-
SS A5	255	-	-	-	-	-	-	-	-
SS A6	156	57	28	16	9	6	3	2	-
SS A7	265	-	-	-	-	-	-	-	-
SS A8	153	51	24	12	6	-	-	-	-
SS B1	85	-	-	-	-	-	-	-	-
SS B2	128	50	26	15	10	6	4	2	1
SS B3	63	-	-	-	-	-	-	-	-
SS B4	121	46	23	13	8	5	3	2	-
SS B5	110	-	-	-	-	-	-	-	-
SS B6	116	45	23	13	8	5	3	2	-
SS B7	75	-	-	-	-	-	-	-	-
SS B8	107	40	19	10	6	3	1	-	-
SS C1	-	-	-	-	-	-	-	-	-
SS C2	63	27	14	7	3	-	-	-	-
SS C3	30	-	-	-	-	-	-	-	-
SS C4	63	24	11	5	-	-	-	-	-
SS D1	122	31	-	-	-	-	-	-	-
SS D2	140	66	38	24	15	10	7	4	2
SS D3	165	28	-	-	-	-	-	-	-
SS D4	144	65	36	22	14	9	6	4	2
SS E1	66	-	-	-	-	-	-	-	-
SS E2	104	47	25	15	9	5	3	1	-
SS E3	96	-	-	-	-	-	-	-	-
SS E4	109	44	23	13	7	4	2	-	-

SYRACUSE, NEW YORK — 6678 DD

SSF =	.10	.20	.30	.40	.50	.60	.70	.80	.90
WW A1	167	-	-	-	-	-	-	-	-
WW A2	48	-	-	-	-	-	-	-	-
WW A3	41	-	-	-	-	-	-	-	-
WW A4	39	-	-	-	-	-	-	-	-
WW A5	38	-	-	-	-	-	-	-	-
WW A6	38	12	-	-	-	-	-	-	-
WW B1	-	-	-	-	-	-	-	-	-
WW B2	56	24	12	6	-	-	-	-	-
WW B3	72	35	19	11	7	4	2	-	-
WW B4	77	44	28	19	13	9	7	5	3
WW B5	77	45	30	21	15	11	8	5	3
WW C1	83	41	24	15	10	6	4	2	1
WW C2	77	41	25	17	11	8	5	4	2
WW C3	96	59	40	29	21	16	11	8	5
WW C4	88	55	37	27	20	14	11	8	5
TW A1	157	-	-	-	-	-	-	-	-
TW A2	55	-	-	-	-	-	-	-	-
TW A3	45	-	-	-	-	-	-	-	-
TW A4	41	11	-	-	-	-	-	-	-
TW B1	84	-	-	-	-	-	-	-	-
TW B2	49	-	-	-	-	-	-	-	-
TW B3	45	11	-	-	-	-	-	-	-
TW B4	45	13	-	-	-	-	-	-	-
TW C1	61	-	-	-	-	-	-	-	-
TW C2	50	12	-	-	-	-	-	-	-
TW C3	51	13	-	-	-	-	-	-	-
TW C4	60	13	4	-	-	-	-	-	-
TW D1	-	-	-	-	-	-	-	-	-
TW D2	63	24	11	5	-	-	-	-	-
TW D3	64	26	13	7	-	-	-	-	-

SYRACUSE, NEW YORK — CONTINUED

SSF =	.10	.20	.30	.40	.50	.60	.70	.80	.90
TW D4	79	40	24	16	11	7	5	3	2
TW D5	81	43	27	18	13	9	7	5	3
TW E1	84	37	20	12	7	4	2	1	-
TW E2	82	39	23	14	9	6	4	3	1
TW E3	100	56	36	25	17	13	9	6	4
TW E4	93	53	35	24	17	13	9	6	4
TW F1	44	-	-	-	-	-	-	-	-
TW F2	31	-	-	-	-	-	-	-	-
TW F3	28	-	-	-	-	-	-	-	-
TW F4	26	-	-	-	-	-	-	-	-
TW G1	25	-	-	-	-	-	-	-	-
TW G2	23	-	-	-	-	-	-	-	-
TW G3	22	-	-	-	-	-	-	-	-
TW G4	20	6	-	-	-	-	-	-	-
TW H1	18	-	-	-	-	-	-	-	-
TW H2	18	-	-	-	-	-	-	-	-
TW H3	17	5	-	-	-	-	-	-	-
TW H4	15	5	-	-	-	-	-	-	-
TW I1	-	-	-	-	-	-	-	-	-
TW I2	46	19	9	4	-	-	-	-	-
TW I3	52	24	13	7	4	2	-	-	-
TW I4	65	35	21	14	10	7	5	3	2
TW I5	68	38	24	17	12	8	6	4	3
TW J1	69	33	19	11	7	5	3	2	-
TW J2	68	35	21	14	9	6	4	3	2
TW J3	86	51	34	23	17	12	9	6	4
TW J4	81	48	32	22	16	12	9	6	4
DG A1	-	-	-	-	-	-	-	-	-
DG A2	45	-	-	-	-	-	-	-	-
DG A3	86	36	20	12	8	4	-	-	-
DG B1	-	-	-	-	-	-	-	-	-
DG B2	46	12	-	-	-	-	-	-	-
DG B3	91	39	22	14	9	6	4	2	-
DG C1	-	-	-	-	-	-	-	-	-
DG C2	65	24	10	-	-	-	-	-	-
DG C3	107	46	27	17	12	8	6	3	-
SS A1	142	27	-	-	-	-	-	-	-
SS A2	161	60	31	18	11	7	4	3	1
SS A3	114	-	-	-	-	-	-	-	-
SS A4	157	56	27	15	9	5	3	1	-
SS A5	250	15	-	-	-	-	-	-	-
SS A6	155	57	29	16	10	6	4	2	-
SS A7	259	-	-	-	-	-	-	-	-
SS A8	151	51	24	12	6	3	-	-	-
SS B1	85	11	-	-	-	-	-	-	-
SS B2	127	49	26	16	10	6	4	3	1
SS B3	62	-	-	-	-	-	-	-	-
SS B4	120	45	23	13	8	5	3	2	-
SS B5	108	-	-	-	-	-	-	-	-
SS B6	115	45	23	14	8	5	3	2	-
SS B7	74	-	-	-	-	-	-	-	-
SS B8	106	40	20	11	6	3	2	-	-
SS C1	-	-	-	-	-	-	-	-	-
SS C2	63	27	14	8	4	-	-	-	-
SS C3	30	-	-	-	-	-	-	-	-
SS C4	63	24	12	6	3	-	-	-	-
SS D1	121	31	-	-	-	-	-	-	-
SS D2	140	66	38	24	16	10	7	4	3
SS D3	162	28	-	-	-	-	-	-	-
SS D4	143	65	36	22	14	9	6	4	2
SS E1	66	-	-	-	-	-	-	-	-
SS E2	104	47	26	15	9	6	3	2	-
SS E3	94	-	-	-	-	-	-	-	-
SS E4	109	44	23	13	8	4	2	1	-

ASHEVILLE, NORTH CAROLINA								4237 DD	
SSF =	.10	.20	.30	.40	.50	.60	.70	.80	.90
WW A1	704	81	37	21	14	9	6	4	3
WW A2	240	93	52	34	23	16	12	8	5
WW A3	202	95	57	38	27	19	14	10	7
WW A4	186	95	59	40	29	21	15	11	7
WW A5	177	96	62	43	31	23	17	12	8
WW A6	172	97	63	44	32	24	18	13	8
WW B1	184	64	34	21	14	9	7	4	3
WW B2	187	99	63	44	31	23	17	12	8
WW B3	220	122	79	56	41	30	22	16	11
WW B4	196	119	81	58	43	33	25	18	12
WW B5	183	113	78	56	42	32	24	18	12
WW C1	235	131	85	60	44	32	24	17	12
WW C2	204	119	79	56	42	31	23	17	12
WW C3	214	138	97	71	54	41	31	23	16
WW C4	193	124	87	64	49	37	28	21	15
TW A1	609	78	36	21	14	9	6	4	3
TW A2	247	84	46	29	20	14	10	7	4
TW A3	199	84	49	32	22	16	12	8	5
TW A4	170	80	49	33	23	17	12	9	6
TW B1	344	75	37	22	15	10	7	5	3
TW B2	200	77	43	28	19	14	10	7	5
TW B3	175	75	44	29	20	14	10	7	5
TW B4	161	69	41	27	19	13	10	7	5
TW C1	230	69	36	23	15	11	8	5	3
TW C2	177	67	38	24	17	12	9	6	4
TW C3	170	63	35	23	16	11	8	6	4
TW C4	180	57	30	19	13	9	7	5	3
TW D1	134	50	27	17	11	8	5	4	2
TW D2	201	94	57	38	27	20	14	10	7
TW D3	205	104	64	44	32	23	17	12	8
TW D4	201	111	72	51	37	27	20	15	10
TW D5	189	108	71	50	37	28	21	15	10
TW E1	240	122	76	52	37	27	20	15	10
TW E2	219	116	74	52	37	28	21	15	10
TW E3	235	139	94	67	50	37	28	21	14
TW E4	210	126	85	62	46	35	26	19	13
TW F1	276	68	34	21	14	9	7	5	3
TW F2	175	71	40	26	18	13	9	6	4
TW F3	145	69	41	28	20	14	10	7	5
TW F4	120	62	39	26	19	14	10	7	5
TW G1	172	59	32	20	14	10	7	5	3
TW G2	127	58	35	23	16	12	9	6	4
TW G3	107	54	33	22	16	12	8	6	4
TW G4	87	45	28	19	14	10	7	5	3
TW H1	112	48	28	18	13	9	7	5	3
TW H2	87	43	26	18	13	9	7	5	3
TW H3	73	37	23	15	11	8	6	4	3
TW H4	57	28	17	12	8	6	4	3	2
TW I1	103	42	23	15	10	7	5	3	2
TW I2	153	78	49	34	24	18	13	9	6
TW I3	163	87	56	38	28	20	15	11	7
TW I4	167	96	64	45	33	25	19	14	9
TW I5	163	97	65	47	34	26	20	14	10
TW J1	199	107	69	48	35	26	19	14	9
TW J2	181	101	66	47	34	25	19	14	9
TW J3	196	121	84	61	45	34	26	19	13
TW J4	180	112	77	56	42	32	24	18	12
DG A1	132	55	28	12	-	-	-	-	-
DG A2	158	71	42	28	18	12	7	-	-
DG A3	201	93	58	40	30	22	16	10	5
DG B1	131	57	32	20	12	-	-	-	-
DG B2	161	74	45	31	22	16	11	6	-
DG B3	205	96	60	42	31	24	18	13	8
DG C1	165	74	44	29	20	13	7	-	-

ASHEVILLE, NORTH CAROLINA								CONTINUED	
SSF =	.10	.20	.30	.40	.50	.60	.70	.80	.90
DG C2	192	89	55	38	28	21	15	10	5
DG C3	237	111	70	49	37	29	22	17	11
SS A1	434	131	68	42	28	19	13	9	6
SS A2	399	169	98	64	44	32	23	16	10
SS A3	415	117	59	36	23	16	11	7	5
SS A4	406	168	96	62	43	30	22	15	10
SS A5	722	130	61	36	23	15	11	7	4
SS A6	395	167	96	63	44	31	22	16	10
SS A7	813	108	48	27	17	11	7	5	3
SS A8	406	164	93	60	41	29	21	14	9
SS B1	294	98	52	32	22	15	11	7	5
SS B2	311	136	80	52	37	26	19	14	9
SS B3	269	89	47	29	19	13	9	6	4
SS B4	305	132	77	51	35	25	18	13	8
SS B5	379	86	42	25	16	11	8	5	3
SS B6	293	130	76	51	35	25	18	13	9
SS B7	332	72	34	20	13	9	6	4	2
SS B8	285	124	72	47	33	24	17	12	8
SS C1	179	80	47	31	22	16	11	8	5
SS C2	197	102	64	44	32	23	17	12	8
SS C3	187	64	35	22	15	11	7	5	3
SS C4	195	92	55	37	26	19	14	10	7
SS D1	398	156	87	56	38	27	19	13	8
SS D2	352	184	115	79	56	41	30	22	14
SS D3	501	148	76	47	31	21	15	10	6
SS D4	361	180	111	75	53	39	28	20	13
SS E1	292	119	67	43	29	21	15	10	6
SS E2	288	149	93	63	45	33	24	17	11
SS E3	354	103	53	32	21	15	10	7	4
SS E4	296	140	84	56	40	29	21	15	10

CAPE HATTERAS, NORTH CAROLINA								2731 DD	
SSF =	.10	.20	.30	.40	.50	.60	.70	.80	.90
WW A1	1050	131	60	35	23	16	11	8	5
WW A2	361	142	80	52	36	26	18	13	9
WW A3	303	143	86	58	41	30	22	15	10
WW A4	278	143	89	60	43	32	23	17	11
WW A5	264	144	92	64	46	34	25	18	12
WW A6	256	144	94	65	48	35	26	19	13
WW B1	301	109	59	37	25	18	12	9	6
WW B2	272	144	91	63	45	34	25	18	12
WW B3	316	176	114	79	58	43	32	23	16
WW B4	275	165	112	80	59	45	34	25	17
WW B5	252	155	106	77	57	43	33	24	16
WW C1	335	186	121	84	61	46	34	25	17
WW C2	287	167	111	79	58	43	32	24	16
WW C3	292	187	131	96	72	55	42	31	21
WW C4	263	168	118	86	65	50	38	28	19
TW A1	901	125	58	34	22	15	11	7	5
TW A2	369	129	70	45	31	22	16	11	7
TW A3	298	128	74	49	34	24	18	13	8
TW A4	252	120	73	49	35	25	18	13	9
TW B1	511	117	58	35	24	16	12	8	5
TW B2	296	117	66	42	29	21	15	11	7
TW B3	259	112	65	43	30	22	16	11	7
TW B4	237	103	60	40	28	20	15	10	7
TW C1	339	105	56	35	24	17	12	8	5
TW C2	260	100	56	36	25	18	13	9	6
TW C3	248	94	52	34	23	17	12	8	6
TW C4	259	84	45	28	19	14	10	7	5
TW D1	218	84	47	30	20	14	10	7	5
TW D2	291	138	83	56	39	29	21	15	10
TW D3	297	151	93	64	46	33	25	18	12

CAPE HATTERAS, NORTH CAROLINA							CONTINUED			CHARLOTTE, NORTH CAROLINA								3218 DD	
SSF =	.10	.20	.30	.40	.50	.60	.70	.80	.90	SSF =	.10	.20	.30	.40	.50	.60	.70	.80	.90
TW D4	282	156	101	70	51	38	28	21	14	WW A1	939	107	49	28	18	13	9	6	4
TW D5	262	149	98	69	51	38	28	21	14	WW A2	306	119	67	43	30	21	15	11	7
TW E1	344	175	109	74	53	39	29	21	14	WW A3	256	120	72	48	34	25	18	13	8
TW E2	310	165	105	72	52	39	29	21	14	WW A4	235	120	75	51	36	26	19	14	9
TW E3	324	191	128	91	68	51	38	28	19	WW A5	223	121	78	54	39	29	21	15	10
TW E4	288	172	116	83	62	47	35	26	18	WW A6	217	121	79	55	40	30	22	16	11
TW F1	418	107	54	33	22	15	11	8	5	WW B1	247	87	47	29	20	14	10	7	4
TW F2	264	109	62	40	28	20	14	10	7	WW B2	232	123	78	54	39	28	21	15	10
TW F3	219	104	63	42	30	22	16	11	7	WW B3	269	150	97	68	49	37	27	20	13
TW F4	179	93	58	40	28	21	15	11	7	WW B4	238	143	97	70	52	39	29	21	15
TW G1	262	93	50	32	22	15	11	8	5	WW B5	219	135	93	67	50	38	29	21	14
TW G2	192	89	53	35	25	18	13	9	6	WW C1	287	159	103	72	53	39	29	21	14
TW G3	161	81	50	34	24	17	13	9	6	WW C2	248	144	96	68	50	37	28	20	14
TW G4	129	66	41	28	20	15	11	8	5	WW C3	254	163	114	84	63	48	37	27	19
TW H1	170	74	43	28	20	14	10	7	5	WW C4	230	147	103	76	57	44	33	25	17
TW H2	131	64	39	27	19	14	10	7	5	TW A1	804	103	47	28	18	12	9	6	4
TW H3	109	55	34	23	16	12	9	6	4	TW A2	315	108	59	37	25	18	13	9	6
TW H4	84	42	26	17	12	9	7	5	3	TW A3	253	107	62	41	28	20	15	10	7
TW I1	169	71	40	26	18	13	9	6	4	TW A4	214	101	61	41	29	21	15	11	7
TW I2	223	115	72	49	35	26	19	14	9	TW B1	446	97	48	29	19	13	9	7	4
TW I3	236	126	80	55	40	29	22	16	11	TW B2	253	98	55	36	25	17	13	9	6
TW I4	234	134	89	63	46	34	26	19	13	TW B3	220	94	55	36	25	18	13	9	6
TW I5	226	133	89	64	47	35	27	19	13	TW B4	203	87	51	33	23	17	12	9	6
TW J1	284	153	98	68	49	36	27	19	13	TW C1	293	88	47	29	20	14	10	7	5
TW J2	255	143	93	65	48	35	26	19	13	TW C2	223	84	47	30	21	15	11	8	5
TW J3	269	166	113	82	61	46	35	26	18	TW C3	213	79	44	28	20	14	10	7	5
TW J4	247	152	104	76	57	43	32	24	16	TW C4	225	71	38	24	16	12	8	6	4
DG A1	217	94	53	32	18	9	-	-	-	TW D1	179	68	37	24	16	11	8	5	3
DG A2	234	106	64	43	29	20	13	8	3	TW D2	249	117	71	47	34	24	18	13	8
DG A3	284	131	81	56	41	31	22	15	9	TW D3	254	128	79	54	39	28	21	15	10
DG B1	219	96	57	37	26	17	10	-	-	TW D4	244	134	87	61	44	33	24	18	12
DG B2	243	110	67	46	34	25	18	12	6	TW D5	228	129	85	60	44	33	25	18	12
DG B3	290	134	84	58	43	33	26	19	12	TW E1	294	149	93	63	45	33	24	18	12
DG C1	267	118	71	48	34	25	18	10	-	TW E2	268	142	90	62	45	33	25	18	12
DG C2	285	130	80	55	41	31	23	16	10	TW E3	281	166	111	79	59	44	33	24	17
DG C3	334	155	96	68	51	39	31	23	16	TW E4	251	150	101	73	54	41	31	23	15
SS A1	616	191	100	62	41	29	20	14	9	TW F1	360	89	45	27	18	12	9	6	4
SS A2	551	236	136	89	62	44	32	23	15	TW F2	224	91	52	34	23	17	12	8	6
SS A3	598	176	90	55	36	25	17	12	8	TW F3	185	87	52	35	25	18	13	9	6
SS A4	564	236	135	87	60	43	31	22	14	TW F4	151	78	49	33	24	17	13	9	6
SS A5	1009	193	91	54	35	24	17	11	7	TW G1	222	77	42	26	18	13	9	6	4
SS A6	549	234	135	88	61	44	32	22	15	TW G2	162	74	44	30	21	15	11	8	5
SS A7	1139	165	75	43	28	19	13	9	5	TW G3	136	68	42	28	20	15	11	8	5
SS A8	567	232	131	85	58	42	30	21	14	TW G4	109	56	35	24	17	12	9	7	4
SS B1	424	145	78	49	33	23	16	11	7	TW H1	144	62	36	24	16	12	9	6	4
SS B2	431	190	111	73	51	37	27	19	12	TW H2	110	54	33	22	16	12	8	6	4
SS B3	395	135	71	44	30	21	15	10	6	TW H3	92	46	28	19	14	10	7	5	3
SS B4	424	186	108	71	50	36	26	18	12	TW H4	71	35	22	15	10	8	6	4	3
SS B5	544	131	64	39	26	18	12	8	5	TW I1	138	57	32	20	14	10	7	5	3
SS B6	408	183	107	71	50	36	26	18	12	TW I2	190	97	61	42	30	22	16	12	8
SS B7	485	112	54	32	21	14	10	7	4	TW I3	201	107	68	47	34	25	18	13	9
SS B8	399	176	103	67	47	34	24	17	11	TW I4	202	116	77	54	40	30	22	16	11
SS C1	274	124	73	48	34	24	18	13	8	TW I5	197	116	78	55	41	31	23	17	12
SS C2	285	148	93	64	46	34	25	18	12	TW J1	243	131	84	58	42	31	23	17	11
SS C3	283	100	55	35	24	17	12	8	6	TW J2	220	123	80	56	41	31	23	17	11
SS C4	282	133	80	54	38	28	20	15	10	TW J3	234	144	99	72	53	40	31	23	15
SS D1	571	228	128	82	56	40	28	20	13	TW J4	215	133	91	66	50	38	28	21	14
SS D2	488	255	160	109	78	57	42	30	20	DG A1	178	75	41	23	11	-	-	-	-
SS D3	712	217	112	69	46	32	23	16	10	DG A2	199	89	54	35	24	16	10	6	-
SS D4	499	250	154	104	74	54	39	28	19	DG A3	245	113	70	48	35	26	19	13	7
SS E1	430	179	102	66	45	32	23	16	10	DG B1	177	77	45	29	19	11	-	-	-
SS E2	405	210	131	89	64	47	34	25	16	DG B2	203	93	56	39	28	20	14	9	4
SS E3	514	155	80	49	33	23	16	11	7	DG B3	249	117	72	50	38	29	22	16	10
SS E4	417	198	119	80	56	41	30	21	14	DG C1	217	98	58	39	27	20	13	6	-

CHARLOTTE, NORTH CAROLINA — CONTINUED

SSF =	.10	.20	.30	.40	.50	.60	.70	.80	.90
DG C2	240	111	68	47	34	26	19	13	8
DG C3	287	135	84	59	44	34	27	20	13
SS A1	524	157	82	50	34	23	17	11	7
SS A2	469	198	114	75	52	37	27	19	13
SS A3	504	142	72	44	29	20	14	9	6
SS A4	476	196	112	73	50	36	26	18	12
SS A5	886	157	74	44	28	19	13	9	6
SS A6	466	196	113	74	51	37	26	19	12
SS A7	1007	132	59	34	22	14	10	7	4
SS A8	478	192	109	70	48	34	25	17	11
SS B1	358	119	64	40	27	19	13	9	6
SS B2	367	160	94	62	43	31	23	16	11
SS B3	330	109	58	36	24	17	12	8	5
SS B4	360	156	91	60	42	30	22	15	10
SS B5	468	106	52	31	20	14	10	7	4
SS B6	346	153	90	60	42	30	22	16	10
SS B7	413	90	43	25	16	11	8	5	3
SS B8	337	147	86	56	39	28	20	14	9
SS C1	230	103	61	40	28	20	15	10	7
SS C2	244	126	79	54	39	29	21	15	10
SS C3	240	83	45	29	19	14	10	7	4
SS C4	241	113	68	46	32	24	17	12	8
SS D1	478	187	105	67	46	32	23	16	10
SS D2	412	215	135	92	66	48	35	26	17
SS D3	606	178	92	56	38	26	18	13	8
SS D4	422	210	130	88	62	45	33	24	16
SS E1	356	146	83	53	37	26	18	13	8
SS E2	341	176	110	75	54	39	29	21	14
SS E3	435	126	65	40	26	18	13	9	6
SS E4	352	166	100	67	47	34	25	18	12

GREENSBORO, NORTH CAROLINA — 3825 DD

SSF =	.10	.20	.30	.40	.50	.60	.70	.80	.90
WW A1	773	90	41	24	15	10	7	5	3
WW A2	262	102	57	37	25	18	13	9	6
WW A3	220	103	62	41	29	21	15	11	7
WW A4	202	103	64	44	31	23	17	12	8
WW A5	193	105	67	46	34	25	18	13	9
WW A6	187	105	68	48	35	26	19	14	9
WW B1	204	72	38	24	16	11	7	5	3
WW B2	202	107	68	47	34	25	18	13	9
WW B3	236	131	85	60	43	32	24	17	12
WW B4	210	127	86	62	46	35	26	19	13
WW B5	195	120	83	60	45	34	26	19	13
WW C1	252	140	91	64	47	35	26	19	12
WW C2	219	127	85	60	44	33	25	18	12
WW C3	227	146	103	75	57	44	33	25	17
WW C4	205	132	93	68	52	39	30	22	15
TW A1	666	87	40	23	15	10	7	5	3
TW A2	269	92	50	32	22	15	11	8	5
TW A3	217	92	53	35	24	17	13	9	6
TW A4	185	87	53	35	25	18	13	9	6
TW B1	376	83	41	25	16	11	8	5	4
TW B2	217	84	47	31	21	15	11	8	5
TW B3	190	81	47	31	22	16	11	8	5
TW B4	175	75	44	29	20	15	11	7	5
TW C1	250	76	40	25	17	12	8	6	4
TW C2	192	73	41	26	18	13	9	7	4
TW C3	184	68	38	25	17	12	9	6	4
TW C4	195	61	33	21	14	10	7	5	3
TW D1	149	56	31	19	13	9	6	4	3
TW D2	217	102	62	41	29	21	16	11	7
TW D3	221	112	69	47	34	25	18	13	9

GREENSBORO, NORTH CAROLINA — CONTINUED

SSF =	.10	.20	.30	.40	.50	.60	.70	.80	.90
TW D4	215	119	77	54	39	29	22	16	11
TW D5	202	115	76	54	39	30	22	16	11
TW E1	258	131	81	56	40	29	22	16	10
TW E2	235	125	80	55	40	30	22	16	11
TW E3	250	148	99	71	53	40	30	22	15
TW E4	224	134	91	65	49	37	28	20	14
TW F1	303	75	37	23	15	10	7	5	3
TW F2	191	78	44	29	20	14	10	7	5
TW F3	159	75	45	30	21	15	11	8	5
TW F4	131	67	42	29	20	15	11	8	5
TW G1	189	65	35	22	15	11	8	5	3
TW G2	139	64	38	25	18	13	9	7	4
TW G3	117	58	36	24	17	13	9	7	4
TW G4	94	48	30	21	15	11	8	6	4
TW H1	123	53	31	20	14	10	7	5	3
TW H2	95	47	29	19	14	10	7	5	3
TW H3	80	40	25	17	12	9	6	5	3
TW H4	62	30	19	13	9	7	5	2	2
TW I1	114	47	26	17	11	8	5	4	2
TW I2	166	85	53	36	26	19	14	10	7
TW I3	176	94	60	41	30	22	16	12	8
TW I4	179	103	68	48	35	26	20	14	10
TW I5	174	103	69	50	37	28	21	15	10
TW J1	214	115	74	51	37	27	20	15	10
TW J2	194	109	71	50	36	27	20	15	10
TW J3	209	129	89	64	48	36	27	20	14
TW J4	192	119	82	60	45	34	26	19	13
DG A1	148	62	32	16	–	–	–	–	–
DG A2	172	77	46	30	20	13	8	4	–
DG A3	216	100	62	43	32	23	17	11	6
DG B1	147	64	37	23	14	5	–	–	–
DG B2	175	80	49	33	24	17	12	7	2
DG B3	220	103	64	45	33	26	20	14	9
DG C1	183	82	48	32	22	16	9	–	–
DG C2	208	96	59	41	30	22	17	11	6
DG C3	254	119	75	52	39	30	24	18	12
SS A1	454	136	71	43	29	20	14	10	6
SS A2	414	175	101	66	46	33	24	17	11
SS A3	433	121	61	37	24	17	11	8	5
SS A4	420	173	99	64	44	31	23	16	10
SS A5	753	135	63	37	24	16	11	8	5
SS A6	411	173	100	65	45	32	23	16	11
SS A7	842	112	50	28	18	12	8	5	3
SS A8	420	169	96	62	42	30	21	15	10
SS B1	309	103	55	34	23	16	11	8	5
SS B2	325	141	83	55	38	27	20	14	9
SS B3	283	93	49	30	20	14	10	7	4
SS B4	317	137	80	53	37	26	19	13	9
SS B5	399	91	44	26	17	12	8	6	3
SS B6	305	135	80	53	37	26	19	14	9
SS B7	349	76	36	21	14	9	6	4	3
SS B8	296	129	75	49	34	25	18	13	8
SS C1	196	88	52	34	24	17	12	9	6
SS C2	213	110	69	47	34	25	18	13	9
SS C3	204	71	38	24	16	12	8	6	4
SS C4	210	99	60	40	28	21	15	11	7
SS D1	416	162	91	58	40	28	20	14	9
SS D2	365	190	119	82	58	43	31	22	15
SS D3	524	154	79	49	32	22	16	11	7
SS D4	374	186	115	78	55	40	29	21	14
SS E1	307	126	71	46	31	22	16	11	7
SS E2	300	155	97	66	47	34	25	18	12
SS E3	373	108	55	34	22	15	11	7	5
SS E4	309	146	88	59	41	30	22	15	10

```
RALEIGH-DURHAM, NORTH CAROLINA               3514 DD    RALEIGH-DURHAM, NORTH CAROLINA              CONTINUED
   SSF =.10  .20  .30  .40  .50  .60  .70  .80  .90       SSF =.10  .20  .30  .40  .50  .60  .70  .80  .90
   WW A1  798   92   42   24   16   11    7    5    3    DG C2  210   98   60   42   30   23   17   11    6
   WW A2  265  103   58   38   26   18   13    9    6    DG C3  256  121   76   53   40   31   24   18   12
   WW A3  223  105   63   42   30   21   16   11    7    SS A1  458  138   72   44   30   21   14   10    6
   WW A4  205  105   66   45   32   23   17   12    8    SS A2  418  177  103   67   47   33   24   17   11
   WW A5  196  107   68   47   34   25   18   13    9    SS A3  437  124   63   38   25   17   12    8    5
   WW A6  190  107   70   49   35   26   19   14    9    SS A4  423  175  101   65   45   32   23   16   11
   WW B1  208   73   39   24   16   11    8    5    3    SS A5  767  137   65   38   25   17   12    8    5
   WW B2  205  109   69   48   34   25   19   13    9    SS A6  414  175  101   66   46   33   24   17   11
   WW B3  239  133   87   61   44   33   24   18   12    SS A7  864  114   51   29   18   12    8    5    3
   WW B4  213  129   87   63   47   35   26   19   13    SS A8  424  171   97   63   43   31   22   15   10
   WW B5  197  122   84   61   45   34   26   19   13    SS B1  312  104   56   35   23   16   11    8    5
   WW C1  255  142   93   65   47   35   26   19   13    SS B2  328  143   84   56   39   28   20   14    9
   WW C2  222  129   86   61   45   34   25   18   12    SS B3  286   95   50   31   21   14   10    7    4
   WW C3  230  148  104   77   58   44   34   25   17    SS B4  320  139   81   54   37   27   19   14    9
   WW C4  208  134   94   69   52   40   30   23   15    SS B5  405   93   45   27   18   12    8    6    4
   TW A1  686   88   41   24   15   11    7    5    3    SS B6  308  137   81   54   38   27   20   14    9
   TW A2  273   94   51   32   22   16   11    8    5    SS B7  355   77   37   22   14    9    6    4    3
   TW A3  220   94   54   36   25   18   13    9    6    SS B8  299  131   77   50   35   25   18   13    8
   TW A4  187   89   54   36   26   18   13   10    6    SS C1  199   90   53   35   24   17   13    9    6
   TW B1  384   84   42   25   17   12    8    6    4    SS C2  215  112   70   48   35   25   19   13    9
   TW B2  220   85   48   31   21   15   11    8    5    SS C3  207   72   39   25   17   12    8    6    4
   TW B3  192   83   48   32   22   16   12    8    5    SS C4  213  100   61   41   29   21   15   11    7
   TW B4  177   77   45   30   21   15   11    8    5    SS D1  419  165   93   59   40   28   20   14    9
   TW C1  254   77   41   25   17   12    9    6    4    SS D2  369  193  121   83   59   43   32   23   15
   TW C2  194   74   42   27   19   13    9    7    4    SS D3  529  156   81   50   33   23   16   11    7
   TW C3  186   69   39   25   17   12    9    6    4    SS D4  377  189  117   79   56   41   30   21   14
   TW C4  197   62   34   21   14   10    7    5    3    SS E1  311  128   73   47   32   22   16   11    7
   TW D1  151   57   31   20   13    9    6    4    3    SS E2  304  158   99   67   48   35   26   18   12
   TW D2  220  103   63   42   30   22   16   11    7    SS E3  377  110   57   35   23   16   11    8    5
   TW D3  224  113   71   48   34   25   18   13    9    SS E4  313  148   89   60   42   30   22   16   10
   TW D4  218  120   78   55   40   30   22   16   11
   TW D5  204  117   77   54   40   30   22   16   11    BISMARCK, NORTH DAKOTA                       9044 DD
   TW E1  261  133   83   57   41   30   22   16   10       SSF =.10  .20  .30  .40  .50  .60  .70  .80  .90
   TW E2  238  127   81   56   41   30   22   16   11    WW A1  285   11    -    -    -    -    -    -    -
   TW E3  253  150  101   72   53   40   30   22   15    WW A2   95   29   12    -    -    -    -    -    -
   TW E4  226  136   92   66   49   37   28   21   14    WW A3   81   32   15    7    -    -    -    -    -
   TW F1  308   76   38   23   15   11    7    5    3    WW A4   75   33   17    9    4    -    -    -    -
   TW F2  194   79   45   29   20   14   10    7    5    WW A5   72   34   19   11    6    -    -    -    -
   TW F3  161   76   46   31   22   16   11    8    5    WW A6   70   35   20   12    7    3    -    -    -
   TW F4  133   68   43   29   21   15   11    8    5    WW B1    -    -    -    -    -    -    -    -    -
   TW G1  191   66   36   23   15   11    8    5    3    WW B2   87   42   25   16   10    7    4    3    1
   TW G2  141   65   39   26   18   13    9    7    4    WW B3  104   54   33   21   14   10    7    4    2
   TW G3  119   59   37   25   18   13    9    7    4    WW B4  104   60   39   27   20   14   10    7    4
   TW G4   96   49   31   21   15   11    8    6    4    WW B5  100   60   40   28   21   15   11    8    5
   TW H1  125   54   31   21   14   10    7    5    3    WW C1  116   61   37   25   17   12    8    5    3
   TW H2   96   47   29   20   14   10    7    5    3    WW C2  105   59   37   25   18   13    9    6    4
   TW H3   81   40   25   17   12    9    6    5    3    WW C3  121   76   52   38   28   21   15   11    7
   TW H4   63   31   19   13    9    7    5    3    2    WW C4  111   70   48   35   26   19   14   10    7
   TW I1  116   48   27   17   11    8    6    4    2    TW A1  254   18    -    -    -    -    -    -    -
   TW I2  168   86   54   37   27   19   14   10    7    TW A2  100   27   11    -    -    -    -    -    -
   TW I3  178   95   61   42   30   22   16   12    8    TW A3   82   29   14    6    -    -    -    -    -
   TW I4  181  104   69   49   36   27   20   15   10    TW A4   72   29   15    8    4    -    -    -    -
   TW I5  177  105   70   50   37   28   21   15   10    TW B1  142   22    -    -    -    -    -    -    -
   TW J1  216  117   75   52   38   28   20   15   10    TW B2   84   27   12    6    -    -    -    -    -
   TW J2  196  110   72   51   37   28   21   15   10    TW B3   75   28   14    7    4    -    -    -    -
   TW J3  211  131   90   65   49   37   28   21   14    TW B4   71   27   14    8    4    2    -    -    -
   TW J4  194  121   83   60   45   34   26   19   13    TW C1   99   24   10    4    -    -    -    -    -
   DG A1  150   63   33   17    -    -    -    -    -    TW C2   78   26   12    7    3    -    -    -    -
   DG A2  174   79   47   31   21   14    8    4    -    TW C3   77   25   12    7    4    2    -    -    -
   DG A3  218  102   63   44   32   24   17   11    6    TW C4   85   24   11    6    4    2    1    -    -
   DG B1  149   65   38   24   15    7    -    -    -    TW D1    -    -    -    6    -    -    -    -    -
   DG B2  177   82   50   34   24   18   12    7    3    TW D2   94   40   22   14    9    6    3    2    1
   DG B3  222  105   65   46   34   26   20   14    9    TW D3   95   44   25   15   10    6    4    2    1
   DG C1  185   84   50   33   23   16   10    -    -
```

558 / APPENDIX 21

BISMARCK, NORTH DAKOTA — CONTINUED

SSF =	.10	.20	.30	.40	.50	.60	.70	.80	.90
TW D4	105	56	35	23	16	11	8	6	3
TW D5	104	58	37	25	18	13	9	7	4
TW E1	117	56	32	21	14	9	6	4	2
TW E2	112	57	34	23	16	11	8	5	3
TW E3	129	74	48	33	24	17	13	9	6
TW E4	118	69	46	32	23	17	12	9	6
TW F1	102	15	-	-	-	-	-	-	-
TW F2	67	21	8	-	-	-	-	-	-
TW F3	57	23	11	5	-	-	-	-	-
TW F4	49	22	11	6	3	-	-	-	-
TW G1	64	15	-	-	-	-	-	-	-
TW G2	50	19	8	3	-	-	-	-	-
TW G3	43	18	9	5	2	-	-	-	-
TW G4	36	16	9	5	3	1	-	-	-
TW H1	42	14	5	-	-	-	-	-	-
TW H2	35	15	7	4	-	-	-	-	-
TW H3	31	13	7	4	2	-	-	-	-
TW H4	25	11	6	3	2	1	-	-	-
TW I1	-	-	-	-	-	-	-	-	-
TW I2	71	33	19	12	8	5	3	2	-
TW I3	77	38	22	14	9	6	4	3	1
TW I4	87	48	31	21	15	10	7	5	3
TW I5	89	51	33	23	17	12	9	6	4
TW J1	97	49	30	19	13	9	6	4	2
TW J2	93	50	31	21	14	10	7	5	3
TW J3	109	66	44	31	23	17	12	9	6
TW J4	102	62	42	30	22	16	12	8	6
DG A1	-	-	-	-	-	-	-	-	-
DG A2	71	27	12	-	-	-	-	-	-
DG A3	111	49	29	19	12	8	5	-	-
DG B1	-	-	-	-	-	-	-	-	-
DG B2	74	29	14	6	-	-	-	-	-
DG B3	115	52	31	20	14	10	7	4	-
DG C1	53	-	-	-	-	-	-	-	-
DG C2	95	40	21	13	7	-	-	-	-
DG C3	135	61	37	24	17	13	9	6	3
SS A1	181	44	17	6	-	-	-	-	-
SS A2	192	75	41	25	16	10	7	4	2
SS A3	152	27	-	-	-	-	-	-	-
SS A4	187	70	36	21	13	8	5	3	2
SS A5	306	39	9	-	-	-	-	-	-
SS A6	187	72	38	23	15	9	6	4	2
SS A7	315	-	-	-	-	-	-	-	-
SS A8	182	66	33	19	11	7	4	2	1
SS B1	116	30	11	-	-	-	-	-	-
SS B2	153	62	34	21	14	9	6	4	2
SS B3	93	19	-	-	-	-	-	-	-
SS B4	144	58	31	19	12	8	5	3	2
SS B5	146	22	-	-	-	-	-	-	-
SS B6	140	58	32	19	12	8	5	3	2
SS B7	111	-	-	-	-	-	-	-	-
SS B8	130	52	27	16	10	6	4	2	1
SS C1	64	21	-	-	-	-	-	-	-
SS C2	93	45	26	16	11	7	5	3	2
SS C3	70	17	-	-	-	-	-	-	-
SS C4	92	40	22	14	9	6	4	2	1
SS D1	161	52	23	10	-	-	-	-	-
SS D2	169	83	49	32	21	15	10	7	4
SS D3	208	49	18	-	-	-	-	-	-
SS D4	173	81	47	30	20	13	9	6	4
SS E1	103	30	-	-	-	-	-	-	-
SS E2	131	63	36	23	15	10	6	4	2
SS E3	133	27	-	-	-	-	-	-	-
SS E4	136	59	33	20	13	8	5	3	2

FARGO, NORTH DAKOTA 9271 DD

SSF =	.10	.20	.30	.40	.50	.60	.70	.80	.90
WW A1	196	-	-	-	-	-	-	-	-
WW A2	64	-	-	-	-	-	-	-	-
WW A3	55	14	-	-	-	-	-	-	-
WW A4	51	17	-	-	-	-	-	-	-
WW A5	50	19	-	-	-	-	-	-	-
WW A6	49	20	-	-	-	-	-	-	-
WW B1	-	-	-	-	-	-	-	-	-
WW B2	66	30	16	9	5	-	-	-	-
WW B3	82	40	23	14	9	5	3	2	-
WW B4	86	49	31	21	15	10	7	5	3
WW B5	85	50	33	23	16	12	9	6	4
WW C1	93	47	28	17	11	7	5	3	2
WW C2	86	47	29	19	13	9	6	4	2
WW C3	104	64	44	31	23	17	12	9	6
WW C4	95	59	40	29	21	16	12	8	5
TW A1	181	-	-	-	-	-	-	-	-
TW A2	71	-	-	-	-	-	-	-	-
TW A3	58	15	-	-	-	-	-	-	-
TW A4	51	17	-	-	-	-	-	-	-
TW B1	101	-	-	-	-	-	-	-	-
TW B2	61	15	-	-	-	-	-	-	-
TW B3	55	17	-	-	-	-	-	-	-
TW B4	54	18	7	-	-	-	-	-	-
TW C1	73	13	-	-	-	-	-	-	-
TW C2	59	16	5	-	-	-	-	-	-
TW C3	59	17	6	-	-	-	-	-	-
TW C4	67	17	7	2	-	-	-	-	-
TW D1	-	-	-	-	-	-	-	-	-
TW D2	73	29	15	8	4	-	-	-	-
TW D3	74	32	16	9	5	-	-	-	-
TW D4	87	45	27	18	12	8	6	4	2
TW D5	88	48	30	20	14	10	7	5	3
TW E1	94	42	23	14	9	5	3	2	-
TW E2	92	45	26	17	11	7	5	3	2
TW E3	109	62	39	27	19	14	10	7	4
TW E4	101	58	38	26	19	14	10	7	4
TW F1	65	-	-	-	-	-	-	-	-
TW F2	45	-	-	-	-	-	-	-	-
TW F3	39	8	-	-	-	-	-	-	-
TW F4	34	12	-	-	-	-	-	-	-
TW G1	40	-	-	-	-	-	-	-	-
TW G2	33	-	-	-	-	-	-	-	-
TW G3	30	9	-	-	-	-	-	-	-
TW G4	26	10	-	-	-	-	-	-	-
TW H1	27	-	-	-	-	-	-	-	-
TW H2	24	7	-	-	-	-	-	-	-
TW H3	22	8	-	-	-	-	-	-	-
TW H4	18	7	3	-	-	-	-	-	-
TW I1	-	-	-	-	-	-	-	-	-
TW I2	54	23	12	6	3	-	-	-	-
TW I3	60	28	15	9	5	3	1	-	-
TW I4	72	39	24	16	11	8	5	4	2
TW I5	75	42	27	18	13	9	7	5	3
TW J1	78	38	22	13	9	5	3	2	1
TW J2	76	40	24	16	10	7	5	3	2
TW J3	93	55	37	26	18	13	10	7	4
TW J4	88	52	35	24	18	13	9	7	4
DG A1	-	-	-	-	-	-	-	-	-
DG A2	54	16	-	-	-	-	-	-	-
DG A3	95	40	23	14	9	5	-	-	-
DG B1	-	-	-	-	-	-	-	-	-
DG B2	56	18	-	-	-	-	-	-	-
DG B3	99	43	25	16	11	7	5	2	-
DG C1	-	-	-	-	-	-	-	-	-

FARGO, NORTH DAKOTA — CONTINUED

SSF =	.10	.20	.30	.40	.50	.60	.70	.80	.90
DG C2	76	29	14	5	-	-	-	-	-
DG C3	116	51	30	19	13	9	6	4	-
SS A1	142	28	-	-	-	-	-	-	-
SS A2	162	61	31	18	11	7	4	3	1
SS A3	111	-	-	-	-	-	-	-	-
SS A4	155	55	27	15	8	5	2	1	-
SS A5	241	18	-	-	-	-	-	-	-
SS A6	156	58	29	16	10	6	3	2	-
SS A7	237	-	-	-	-	-	-	-	-
SS A8	149	51	23	12	6	-	-	-	-
SS B1	87	14	-	-	-	-	-	-	-
SS B2	129	51	27	16	10	6	4	3	1
SS B3	63	-	-	-	-	-	-	-	-
SS B4	120	46	24	13	8	5	3	2	-
SS B5	108	-	-	-	-	-	-	-	-
SS B6	117	46	24	14	9	5	3	2	-
SS B7	73	-	-	-	-	-	-	-	-
SS B8	107	40	20	11	6	3	1	-	-
SS C1	38	-	-	-	-	-	-	-	-
SS C2	73	33	18	10	6	3	1	-	-
SS C3	45	-	-	-	-	-	-	-	-
SS C4	73	29	15	8	4	2	-	-	-
SS D1	123	33	-	-	-	-	-	-	-
SS D2	141	67	39	24	16	10	7	4	2
SS D3	162	30	-	-	-	-	-	-	-
SS D4	145	66	37	23	14	9	6	4	2
SS E1	70	-	-	-	-	-	-	-	-
SS E2	106	48	26	15	9	6	3	2	-
SS E3	97	-	-	-	-	-	-	-	-
SS E4	111	46	24	13	8	4	2	1	-

MINOT, NORTH DAKOTA — 9407 DD

SSF =	.10	.20	.30	.40	.50	.60	.70	.80	.90
WW A1	234	-	-	-	-	-	-	-	-
WW A2	76	14	-	-	-	-	-	-	-
WW A3	64	19	-	-	-	-	-	-	-
WW A4	60	21	-	-	-	-	-	-	-
WW A5	57	23	-	-	-	-	-	-	-
WW A6	56	23	8	-	-	-	-	-	-
WW B1	-	-	-	-	-	-	-	-	-
WW B2	74	33	18	10	5	2	-	-	-
WW B3	90	44	25	15	9	6	3	2	-
WW B4	92	52	33	22	16	11	8	5	3
WW B5	90	53	35	24	17	12	9	6	4
WW C1	101	51	30	19	12	8	5	3	2
WW C2	93	50	31	20	14	9	6	4	2
WW C3	110	68	46	33	24	17	13	9	6
WW C4	101	63	42	30	22	16	12	8	5
TW A1	213	-	-	-	-	-	-	-	-
TW A2	83	16	-	-	-	-	-	-	-
TW A3	67	19	-	-	-	-	-	-	-
TW A4	59	21	6	-	-	-	-	-	-
TW B1	120	-	-	-	-	-	-	-	-
TW B2	70	19	-	-	-	-	-	-	-
TW B3	63	20	6	-	-	-	-	-	-
TW B4	60	20	8	-	-	-	-	-	-
TW C1	84	16	-	-	-	-	-	-	-
TW C2	67	19	7	-	-	-	-	-	-
TW C3	67	19	8	-	-	-	-	-	-
TW C4	75	19	8	3	-	-	-	-	-
TW D1	-	-	-	-	-	-	-	-	-
TW D2	81	32	16	9	4	-	-	-	-
TW D3	81	35	18	10	5	-	-	-	-

MINOT, NORTH DAKOTA — CONTINUED

SSF =	.10	.20	.30	.40	.50	.60	.70	.80	.90
TW D4	94	48	29	19	13	9	6	4	2
TW D5	94	51	32	21	15	11	7	5	3
TW E1	103	46	26	15	9	6	3	2	-
TW E2	100	48	28	18	12	8	5	3	2
TW E3	116	65	42	28	20	14	10	7	4
TW E4	108	62	40	28	20	14	10	7	5
TW F1	82	-	-	-	-	-	-	-	-
TW F2	53	-	-	-	-	-	-	-	-
TW F3	45	13	-	-	-	-	-	-	-
TW F4	39	14	-	-	-	-	-	-	-
TW G1	50	-	-	-	-	-	-	-	-
TW G2	39	10	-	-	-	-	-	-	-
TW G3	34	12	-	-	-	-	-	-	-
TW G4	30	11	4	-	-	-	-	-	-
TW H1	33	-	-	-	-	-	-	-	-
TW H2	28	9	-	-	-	-	-	-	-
TW H3	25	9	3	-	-	-	-	-	-
TW H4	21	8	3	-	-	-	-	-	-
TW I1	-	-	-	-	-	-	-	-	-
TW I2	60	26	14	7	4	-	-	-	-
TW I3	66	31	17	10	6	3	1	-	-
TW I4	78	42	26	17	12	8	5	4	2
TW I5	80	45	29	19	14	10	7	5	3
TW J1	85	41	23	14	9	6	4	2	1
TW J2	82	43	26	17	11	7	5	3	2
TW J3	99	59	39	27	19	14	10	7	4
TW J4	93	55	37	26	18	13	10	7	4
DG A1	-	-	-	-	-	-	-	-	-
DG A2	61	19	-	-	-	-	-	-	-
DG A3	102	44	24	15	9	6	2	-	-
DG B1	-	-	-	-	-	-	-	-	-
DG B2	64	22	-	-	-	-	-	-	-
DG B3	106	47	27	17	11	8	5	2	-
DG C1	34	-	-	-	-	-	-	-	-
DG C2	84	33	16	7	-	-	-	-	-
DG C3	125	55	32	21	14	10	7	4	-
SS A1	161	33	-	-	-	-	-	-	-
SS A2	176	66	34	19	12	7	4	2	1
SS A3	130	-	-	-	-	-	-	-	-
SS A4	170	60	29	16	9	5	2	-	-
SS A5	274	26	-	-	-	-	-	-	-
SS A6	170	63	31	17	10	6	3	2	-
SS A7	276	-	-	-	-	-	-	-	-
SS A8	164	55	25	13	6	-	-	-	-
SS B1	100	19	-	-	-	-	-	-	-
SS B2	140	55	29	17	10	7	4	3	1
SS B3	77	-	-	-	-	-	-	-	-
SS B4	131	50	25	14	8	5	3	2	-
SS B5	127	-	-	-	-	-	-	-	-
SS B6	127	50	26	15	9	5	3	2	-
SS B7	92	-	-	-	-	-	-	-	-
SS B8	117	44	22	12	6	3	1	-	-
SS C1	47	-	-	-	-	-	-	-	-
SS C2	80	36	19	11	6	3	1	-	-
SS C3	55	-	-	-	-	-	-	-	-
SS C4	80	32	16	9	5	2	-	-	-
SS D1	139	38	-	-	-	-	-	-	-
SS D2	153	72	41	25	16	11	7	4	2
SS D3	184	36	-	-	-	-	-	-	-
SS D4	157	71	39	24	15	10	6	4	2
SS E1	83	-	-	-	-	-	-	-	-
SS E2	116	53	28	16	10	6	3	2	-
SS E3	114	-	-	-	-	-	-	-	-
SS E4	121	50	26	14	8	5	2	-	-

AKRON-CANTON, OHIO								6224 DD	
SSF =	.10	.20	.30	.40	.50	.60	.70	.80	.90
WW A1	260	-	-	-	-	-	-	-	-
WW A2	77	19	-	-	-	-	-	-	-
WW A3	65	22	8	-	-	-	-	-	-
WW A4	60	24	11	-	-	-	-	-	-
WW A5	58	25	12	6	-	-	-	-	-
WW A6	57	26	13	7	-	-	-	-	-
WW B1	-	-	-	-	-	-	-	-	-
WW B2	74	34	19	12	7	5	3	2	-
WW B3	91	46	27	17	12	8	5	3	2
WW B4	92	53	34	24	17	12	9	6	4
WW B5	90	54	35	25	18	13	10	7	4
WW C1	102	53	32	21	14	10	7	4	3
WW C2	93	51	32	22	15	11	8	5	3
WW C3	111	69	47	34	25	19	14	10	7
WW C4	101	63	43	31	23	17	13	9	6
TW A1	233	-	-	-	-	-	-	-	-
TW A2	84	18	-	-	-	-	-	-	-
TW A3	68	21	8	-	-	-	-	-	-
TW A4	59	22	10	4	-	-	-	-	-
TW B1	124	12	-	-	-	-	-	-	-
TW B2	71	20	7	-	-	-	-	-	-
TW B3	63	21	9	4	-	-	-	-	-
TW B4	61	21	10	5	2	-	-	-	-
TW C1	85	17	4	-	-	-	-	-	-
TW C2	68	20	9	4	-	-	-	-	-
TW C3	67	20	9	4	2	-	-	-	-
TW C4	76	19	9	4	2	-	-	-	-
TW D1	-	-	-	-	-	-	-	-	-
TW D2	81	33	18	10	6	4	2	1	-
TW D3	83	37	20	12	8	5	3	2	-
TW D4	94	49	30	20	14	10	7	5	3
TW D5	95	51	32	22	16	11	8	6	4
TW E1	104	48	27	17	11	8	5	3	2
TW E2	100	49	29	19	13	9	6	4	3
TW E3	117	66	43	30	21	15	11	8	5
TW E4	108	62	41	28	21	15	11	8	5
TW F1	84	-	-	-	-	-	-	-	-
TW F2	54	13	-	-	-	-	-	-	-
TW F3	46	15	4	-	-	-	-	-	-
TW F4	39	16	7	-	-	-	-	-	-
TW G1	50	-	-	-	-	-	-	-	-
TW G2	39	12	-	-	-	-	-	-	-
TW G3	35	13	6	-	-	-	-	-	-
TW G4	30	12	6	3	-	-	-	-	-
TW H1	33	8	-	-	-	-	-	-	-
TW H2	28	10	4	-	-	-	-	-	-
TW H3	25	10	5	2	-	-	-	-	-
TW H4	21	8	4	2	1	-	-	-	-
TW I1	-	-	-	-	-	-	-	-	-
TW I2	60	27	15	9	5	3	2	1	-
TW I3	67	32	18	11	7	5	3	2	1
TW I4	78	42	26	18	13	9	6	4	3
TW I5	80	45	29	20	14	10	8	5	3
TW J1	86	42	25	16	11	7	5	3	2
TW J2	83	43	27	18	12	9	6	4	3
TW J3	100	59	40	28	20	15	11	8	5
TW J4	93	56	37	26	19	14	10	8	5
DG A1	-	-	-	-	-	-	-	-	-
DG A2	60	20	-	-	-	-	-	-	-
DG A3	102	43	25	16	10	7	4	-	-
DG B1	-	-	-	-	-	-	-	-	-
DG B2	62	22	9	-	-	-	-	-	-
DG B3	106	46	27	18	12	9	6	3	-
DG C1	32	-	-	-	-	-	-	-	-

AKRON-CANTON, OHIO								CONTINUED	
SSF =	.10	.20	.30	.40	.50	.60	.70	.80	.90
DG C2	82	32	17	9	-	-	-	-	-
DG C3	124	55	32	21	15	11	8	5	2
SS A1	184	42	16	5	-	-	-	-	-
SS A2	193	74	40	24	15	10	7	4	3
SS A3	159	27	-	-	-	-	-	-	-
SS A4	191	71	37	21	13	9	6	3	2
SS A5	325	37	-	-	-	-	-	-	-
SS A6	187	71	37	22	14	9	6	4	2
SS A7	354	-	-	-	-	-	-	-	-
SS A8	187	66	33	19	11	7	4	2	1
SS B1	115	28	9	-	-	-	-	-	-
SS B2	152	61	33	20	13	9	6	4	2
SS B3	94	17	-	-	-	-	-	-	-
SS B4	145	57	30	18	12	8	5	3	2
SS B5	150	19	-	-	-	-	-	-	-
SS B6	139	56	30	18	12	8	5	3	2
SS B7	116	-	-	-	-	-	-	-	-
SS B8	131	51	27	16	10	6	4	2	1
SS C1	48	-	-	-	-	-	-	-	-
SS C2	81	37	21	13	8	5	3	2	1
SS C3	56	-	-	-	-	-	-	-	-
SS C4	81	33	18	11	7	4	2	1	-
SS D1	160	50	22	9	-	-	-	-	-
SS D2	168	81	48	31	21	14	10	7	4
SS D3	212	47	17	-	-	-	-	-	-
SS D4	172	80	46	29	19	13	9	6	4
SS E1	100	27	-	-	-	-	-	-	-
SS E2	128	60	34	22	14	9	6	4	2
SS E3	133	24	-	-	-	-	-	-	-
SS E4	134	57	31	19	12	8	5	3	2

CINCINNATI, OHIO								5070 DD	
SSF =	.10	.20	.30	.40	.50	.60	.70	.80	.90
WW A1	350	27	6	-	-	-	-	-	-
WW A2	116	39	19	11	6	4	2	1	-
WW A3	99	42	23	14	9	6	4	2	1
WW A4	91	43	24	15	10	7	4	3	2
WW A5	88	44	26	17	11	8	5	3	2
WW A6	85	44	27	17	12	8	6	4	2
WW B1	56	-	-	-	-	-	-	-	-
WW B2	101	50	30	20	14	10	7	5	3
WW B3	121	64	40	27	19	13	10	7	4
WW B4	117	69	45	32	23	17	13	9	6
WW B5	112	68	46	32	24	18	13	10	6
WW C1	133	71	45	30	21	15	11	8	5
WW C2	119	67	43	30	22	16	12	8	5
WW C3	135	85	59	43	32	24	18	13	9
WW C4	122	77	54	39	29	22	16	12	8
TW A1	309	29	9	-	-	-	-	-	-
TW A2	122	36	17	9	5	3	2	-	-
TW A3	100	38	20	12	7	5	3	2	-
TW A4	86	37	21	13	8	6	4	2	1
TW B1	173	31	12	6	3	-	-	-	-
TW B2	101	35	18	10	6	4	3	1	-
TW B3	90	35	19	11	7	5	3	2	1
TW B4	85	33	18	11	7	5	3	2	1
TW C1	118	31	14	8	5	3	2	-	-
TW C2	93	32	16	10	6	4	3	2	-
TW C3	91	31	16	9	6	4	3	2	1
TW C4	99	29	14	8	5	3	2	1	-
TW D1	43	-	-	-	-	-	-	-	-
TW D2	109	48	27	17	12	8	6	4	2
TW D3	111	53	31	20	14	10	7	5	3

LCR Tables / 561

CINCINNATI, OHIO — CONTINUED

	SSF = .10	.20	.30	.40	.50	.60	.70	.80	.90
TW D4	119	64	40	27	19	14	10	7	5
TW D5	116	65	42	29	21	15	11	8	5
TW E1	136	66	39	26	18	13	9	6	4
TW E2	128	65	40	27	19	14	10	7	4
TW E3	144	84	55	39	28	21	15	11	7
TW E4	131	77	51	36	27	20	15	11	7
TW F1	129	24	9	-	-	-	-	-	-
TW F2	83	29	14	8	5	3	1	-	-
TW F3	70	30	16	10	6	4	2	1	-
TW F4	59	28	16	10	7	4	3	2	1
TW G1	80	23	10	5	2	-	-	-	-
TW G2	61	25	13	8	5	3	2	1	-
TW G3	53	24	13	8	5	4	2	1	-
TW G4	44	21	12	8	5	3	2	1	-
TW H1	53	20	10	6	3	2	1	-	-
TW H2	43	19	11	6	4	3	2	1	-
TW H3	37	17	10	6	4	3	2	1	-
TW H4	29	13	8	5	3	2	1	1	-
TW I1	28	-	-	-	-	-	-	-	-
TW I2	82	40	24	15	10	7	5	3	2
TW I3	90	45	27	18	13	9	6	4	3
TW I4	99	55	35	24	18	13	9	7	4
TW I5	100	57	38	26	19	14	10	8	5
TW J1	113	58	36	24	17	12	8	6	4
TW J2	105	57	36	25	18	13	9	7	4
TW J3	122	74	50	36	26	20	15	11	7
TW J4	113	69	47	33	25	18	14	10	7
DG A1	40	-	-	-	-	-	-	-	-
DG A2	84	33	17	8	-	-	-	-	-
DG A3	124	55	33	22	15	10	7	4	-
DG B1	43	-	-	-	-	-	-	-	-
DG B2	86	35	19	11	6	-	-	-	-
DG B3	128	58	35	23	17	12	9	6	3
DG C1	71	21	-	-	-	-	-	-	-
DG C2	108	46	26	17	11	7	-	-	-
DG C3	148	68	41	28	20	15	11	8	5
SS A1	233	63	29	16	9	6	3	2	1
SS A2	234	95	52	33	22	15	11	7	4
SS A3	210	50	20	10	4	-	-	-	-
SS A4	234	92	50	31	20	14	9	6	4
SS A5	390	59	23	11	6	3	-	-	-
SS A6	229	92	51	32	21	14	10	7	4
SS A7	421	42	12	-	-	-	-	-	-
SS A8	231	87	47	28	18	12	8	6	3
SS B1	153	45	21	11	7	4	2	1	-
SS B2	185	77	44	28	19	13	9	6	4
SS B3	132	36	15	7	3	-	-	-	-
SS B4	178	74	41	26	17	12	8	6	3
SS B5	194	37	14	6	3	-	-	-	-
SS B6	171	72	41	26	17	12	8	6	4
SS B7	160	26	7	-	-	-	-	-	-
SS B8	163	67	37	23	15	11	7	5	3
SS C1	82	32	16	9	5	3	1	-	-
SS C2	108	53	32	21	14	10	7	5	3
SS C3	88	25	11	6	3	-	-	-	-
SS C4	107	47	27	17	12	8	6	4	2
SS D1	210	75	38	22	13	9	5	3	2
SS D2	206	104	63	42	29	20	15	10	6
SS D3	268	70	32	17	10	6	4	2	1
SS D4	211	101	60	39	27	19	14	9	6
SS E1	143	51	25	13	8	4	2	-	-
SS E2	163	80	48	31	21	15	11	7	5
SS E3	179	44	19	9	5	2	-	-	-
SS E4	169	76	43	28	19	13	9	6	4

CLEVELAND, OHIO   6154 DD

	SSF = .10	.20	.30	.40	.50	.60	.70	.80	.90
WW A1	240	-	-	-	-	-	-	-	-
WW A2	68	-	-	-	-	-	-	-	-
WW A3	57	16	-	-	-	-	-	-	-
WW A4	53	18	-	-	-	-	-	-	-
WW A5	51	20	-	-	-	-	-	-	-
WW A6	50	21	8	-	-	-	-	-	-
WW B1	-	-	-	-	-	-	-	-	-
WW B2	68	30	16	9	5	3	-	-	-
WW B3	84	41	24	15	9	6	4	2	1
WW B4	87	49	32	21	15	11	8	5	3
WW B5	86	50	33	23	17	12	9	6	4
WW C1	96	48	28	18	12	8	5	3	2
WW C2	88	47	29	19	13	9	7	4	3
WW C3	106	65	44	32	23	17	13	9	6
WW C4	97	60	41	29	21	16	12	8	6
TW A1	216	-	-	-	-	-	-	-	-
TW A2	76	10	-	-	-	-	-	-	-
TW A3	61	16	-	-	-	-	-	-	-
TW A4	53	18	5	-	-	-	-	-	-
TW B1	114	-	-	-	-	-	-	-	-
TW B2	64	16	-	-	-	-	-	-	-
TW B3	57	17	5	-	-	-	-	-	-
TW B4	56	18	7	-	-	-	-	-	-
TW C1	79	13	-	-	-	-	-	-	-
TW C2	62	17	6	-	-	-	-	-	-
TW C3	62	17	7	-	-	-	-	-	-
TW C4	72	17	7	3	-	-	-	-	-
TW D1	-	-	-	-	-	-	-	-	-
TW D2	75	29	15	8	4	2	-	-	-
TW D3	77	33	17	10	5	3	1	-	-
TW D4	89	45	27	18	12	9	6	4	2
TW D5	90	48	30	20	14	10	7	5	3
TW E1	98	44	24	15	9	6	4	2	1
TW E2	94	45	27	17	11	8	5	3	2
TW E3	112	63	40	28	20	14	10	7	5
TW E4	103	59	38	27	19	14	10	7	5
TW F1	73	-	-	-	-	-	-	-	-
TW F2	47	-	-	-	-	-	-	-	-
TW F3	40	10	-	-	-	-	-	-	-
TW F4	35	12	-	-	-	-	-	-	-
TW G1	43	-	-	-	-	-	-	-	-
TW G2	34	-	-	-	-	-	-	-	-
TW G3	31	10	-	-	-	-	-	-	-
TW G4	27	10	4	-	-	-	-	-	-
TW H1	28	-	-	-	-	-	-	-	-
TW H2	25	8	-	-	-	-	-	-	-
TW H3	22	8	3	-	-	-	-	-	-
TW H4	19	7	3	-	-	-	-	-	-
TW I1	-	-	-	-	-	-	-	-	-
TW I2	55	24	12	7	4	2	-	-	-
TW I3	62	29	16	9	6	3	2	1	-
TW I4	74	39	24	16	11	8	5	4	2
TW I5	76	42	27	19	13	9	7	5	3
TW J1	81	39	22	14	9	6	4	2	1
TW J2	78	40	24	16	11	7	5	3	2
TW J3	95	56	37	26	19	14	10	7	5
TW J4	89	53	35	25	18	13	10	7	4
DG A1	-	-	-	-	-	-	-	-	-
DG A2	56	16	-	-	-	-	-	-	-
DG A3	98	41	23	14	9	6	2	-	-
DG B1	-	-	-	-	-	-	-	-	-
DG B2	58	18	-	-	-	-	-	-	-
DG B3	102	44	25	16	11	7	5	2	-
DG C1	-	-	-	-	-	-	-	-	-

CLEVELAND, OHIO — CONTINUED

	SSF = .10	.20	.30	.40	.50	.60	.70	.80	.90
DG C2	77	29	14	6	-	-	-	-	-
DG C3	120	52	30	20	14	10	7	4	-
SS A1	177	38	11	-	-	-	-	-	-
SS A2	186	70	37	22	14	9	6	4	2
SS A3	151	19	-	-	-	-	-	-	-
SS A4	184	66	33	19	11	7	4	2	1
SS A5	314	32	-	-	-	-	-	-	-
SS A6	180	67	34	20	12	8	5	3	2
SS A7	342	-	-	-	-	-	-	-	-
SS A8	179	62	30	16	9	5	3	1	-
SS B1	108	23	-	-	-	-	-	-	-
SS B2	147	57	31	18	12	8	5	3	2
SS B3	87	-	-	-	-	-	-	-	-
SS B4	140	53	28	16	10	7	4	3	1
SS B5	142	11	-	-	-	-	-	-	-
SS B6	133	52	28	17	10	7	4	3	1
SS B7	109	-	-	-	-	-	-	-	-
SS B8	125	47	24	14	8	5	3	2	-
SS C1	39	-	-	-	-	-	-	-	-
SS C2	75	33	18	11	6	4	2	-	-
SS C3	49	-	-	-	-	-	-	-	-
SS C4	75	30	15	8	5	3	1	-	-
SS D1	151	44	16	-	-	-	-	-	-
SS D2	161	77	44	28	19	13	8	6	3
SS D3	203	41	11	-	-	-	-	-	-
SS D4	165	75	42	26	17	12	8	5	3
SS E1	91	17	-	-	-	-	-	-	-
SS E2	122	56	31	19	12	8	5	3	2
SS E3	125	17	-	-	-	-	-	-	-
SS E4	128	53	28	16	10	6	4	2	1

COLUMBUS, OHIO — 5702 DD

	SSF = .10	.20	.30	.40	.50	.60	.70	.80	.90
WW A1	269	-	-	-	-	-	-	-	-
WW A2	88	26	10	-	-	-	-	-	-
WW A3	75	29	14	7	2	-	-	-	-
WW A4	70	30	16	9	5	2	-	-	-
WW A5	67	32	17	10	6	3	2	-	-
WW A6	66	32	18	11	7	4	2	1	-
WW B1	-	-	-	-	-	-	-	-	-
WW B2	82	40	23	15	10	7	4	3	2
WW B3	100	52	31	21	14	10	7	5	3
WW B4	99	58	38	26	19	14	10	7	5
WW B5	97	58	39	27	20	15	11	8	5
WW C1	111	58	36	24	16	12	8	6	3
WW C2	101	56	36	24	17	12	9	6	4
WW C3	118	74	51	37	27	20	15	11	7
WW C4	107	67	46	34	25	19	14	10	7
TW A1	241	14	-	-	-	-	-	-	-
TW A2	94	25	9	-	-	-	-	-	-
TW A3	77	27	12	6	2	-	-	-	-
TW A4	67	27	14	8	4	2	1	-	-
TW B1	133	19	-	-	-	-	-	-	-
TW B2	79	25	11	5	2	-	-	-	-
TW B3	70	26	13	7	4	2	-	-	-
TW B4	67	25	13	7	4	2	1	-	-
TW C1	93	22	9	4	-	-	-	-	-
TW C2	74	24	11	6	3	2	-	-	-
TW C3	73	23	11	6	4	2	1	-	-
TW C4	81	22	10	6	3	2	-	-	-
TW D1	-	-	-	-	-	-	-	-	-
TW D2	89	38	21	13	8	5	4	2	1
TW D3	91	42	24	15	10	7	4	3	2

COLUMBUS, OHIO — CONTINUED

	SSF = .10	.20	.30	.40	.50	.60	.70	.80	.90
TW D4	101	53	33	22	16	11	8	6	4
TW D5	100	55	35	24	17	13	9	7	4
TW E1	113	53	31	20	13	9	6	4	3
TW E2	108	54	33	22	15	11	7	5	3
TW E3	125	72	47	33	23	17	13	9	6
TW E4	114	67	44	31	23	17	12	9	6
TW F1	94	11	-	-	-	-	-	-	-
TW F2	62	19	7	-	-	-	-	-	-
TW F3	53	21	10	5	-	-	-	-	-
TW F4	45	20	11	6	3	2	-	-	-
TW G1	59	13	-	-	-	-	-	-	-
TW G2	46	17	8	3	-	-	-	-	-
TW G3	40	17	9	5	2	-	-	-	-
TW G4	34	15	8	5	3	2	-	-	-
TW H1	39	13	5	-	-	-	-	-	-
TW H2	33	13	7	4	2	-	-	-	-
TW H3	29	12	7	4	2	1	-	-	-
TW H4	23	10	5	3	2	1	-	-	-
TW I1	-	-	-	-	-	-	-	-	-
TW I2	66	31	18	11	7	5	3	2	1
TW I3	73	36	21	14	9	6	4	3	2
TW I4	84	46	29	20	14	10	7	5	3
TW I5	86	49	32	22	16	12	9	6	4
TW J1	94	47	28	19	13	9	6	4	3
TW J2	89	47	30	20	14	10	7	5	3
TW J3	106	64	43	30	22	16	12	9	6
TW J4	99	60	40	29	21	16	12	8	6
DG A1	-	-	-	-	-	-	-	-	-
DG A2	67	24	10	-	-	-	-	-	-
DG A3	107	47	27	18	12	8	5	-	-
DG B1	-	-	-	-	-	-	-	-	-
DG B2	69	26	13	5	-	-	-	-	-
DG B3	111	49	29	19	14	10	7	4	-
DG C1	46	-	-	-	-	-	-	-	-
DG C2	89	37	20	12	7	-	-	-	-
DG C3	130	59	35	23	17	12	9	6	3
SS A1	192	48	21	10	5	-	-	-	-
SS A2	202	80	44	27	18	12	8	5	3
SS A3	168	35	10	-	-	-	-	-	-
SS A4	200	76	41	24	16	10	7	4	3
SS A5	324	44	14	-	-	-	-	-	-
SS A6	196	77	41	25	16	11	7	5	3
SS A7	345	26	-	-	-	-	-	-	-
SS A8	196	72	37	22	14	9	6	4	2
SS B1	123	33	14	6	-	-	-	-	-
SS B2	159	65	36	23	15	10	7	5	3
SS B3	103	24	-	-	-	-	-	-	-
SS B4	152	62	34	21	14	9	6	4	2
SS B5	155	25	5	-	-	-	-	-	-
SS B6	146	61	34	21	14	9	6	4	3
SS B7	123	11	-	-	-	-	-	-	-
SS B8	138	56	30	18	12	8	5	3	2
SS C1	59	19	-	-	-	-	-	-	-
SS C2	88	42	24	16	10	7	5	3	2
SS C3	65	15	-	-	-	-	-	-	-
SS C4	88	38	21	13	8	6	4	2	1
SS D1	172	58	27	14	8	4	1	-	-
SS D2	177	88	53	34	23	16	11	8	5
SS D3	221	54	22	10	5	-	-	-	-
SS D4	182	86	50	32	22	15	11	7	4
SS E1	111	35	14	-	-	-	-	-	-
SS E2	137	66	39	25	17	11	8	5	3
SS E3	142	31	10	-	-	-	-	-	-
SS E4	142	62	35	22	14	10	6	4	2

DAYTON, OHIO — 5641 DD

	SSF =.10	.20	.30	.40	.50	.60	.70	.80	.90
WW A1	310	17	-	-	-	-	-	-	-
WW A2	99	31	14	7	3	-	-	-	-
WW A3	84	34	18	10	6	3	1	-	-
WW A4	78	35	19	12	7	4	3	1	-
WW A5	75	36	21	13	8	5	3	2	1
WW A6	73	37	22	14	9	6	4	2	1
WW B1	28	-	-	-	-	-	-	-	-
WW B2	89	44	26	17	11	8	5	4	2
WW B3	108	57	35	23	16	11	8	5	3
WW B4	106	62	41	29	21	15	11	8	5
WW B5	103	62	41	29	22	16	12	9	6
WW C1	120	63	39	26	18	13	9	6	4
WW C2	108	60	39	27	19	14	10	7	5
WW C3	124	78	54	39	29	22	16	12	8
WW C4	113	71	49	36	27	20	15	11	7
TW A1	275	21	-	-	-	-	-	-	-
TW A2	106	29	13	6	3	-	-	-	-
TW A3	86	31	15	9	5	3	1	-	-
TW A4	74	31	17	10	6	4	2	1	-
TW B1	151	24	8	-	-	-	-	-	-
TW B2	88	29	14	8	4	2	1	-	-
TW B3	78	29	15	9	5	3	2	1	-
TW B4	74	28	15	9	5	4	2	1	-
TW C1	104	26	11	6	3	1	-	-	-
TW C2	82	27	13	8	5	3	2	-	-
TW C3	80	26	13	8	5	3	2	1	-
TW C4	89	25	12	7	4	3	2	1	-
TW D1	26	-	-	-	-	-	-	-	-
TW D2	97	42	23	15	10	7	4	3	2
TW D3	99	46	27	17	11	8	5	4	2
TW D4	108	57	36	24	17	12	9	6	4
TW D5	107	59	38	26	19	14	10	7	5
TW E1	122	58	34	22	15	11	7	5	3
TW E2	116	58	36	24	17	12	8	6	4
TW E3	132	76	50	35	25	19	14	10	6
TW E4	121	71	47	33	24	18	13	10	6
TW F1	109	17	-	-	-	-	-	-	-
TW F2	71	23	10	5	-	-	-	-	-
TW F3	60	24	12	7	4	2	-	-	-
TW F4	51	23	13	8	5	3	2	-	-
TW G1	68	17	6	-	-	-	-	-	-
TW G2	52	20	10	5	3	1	-	-	-
TW G3	45	20	11	6	4	2	1	-	-
TW G4	38	17	10	6	4	2	2	-	-
TW H1	45	16	7	3	-	-	-	-	-
TW H2	37	16	8	5	3	2	-	-	-
TW H3	32	14	8	5	3	2	1	-	-
TW H4	26	11	6	4	3	2	1	-	-
TW I1	-	-	-	-	-	-	-	-	-
TW I2	73	34	20	13	9	6	4	3	2
TW I3	80	40	24	15	11	7	5	3	2
TW I4	90	50	32	22	16	11	8	6	4
TW I5	91	52	34	24	17	13	9	7	4
TW J1	101	51	31	21	14	10	7	5	3
TW J2	95	51	32	22	15	11	8	6	4
TW J3	112	68	46	32	24	18	13	10	6
TW J4	105	63	43	31	22	17	12	9	6
DG A1	-	-	-	-	-	-	-	-	-
DG A2	73	28	13	-	-	-	-	-	-
DG A3	114	50	29	19	13	9	6	3	-
DG B1	18	-	-	-	-	-	-	-	-
DG B2	76	30	15	8	-	-	-	-	-
DG B3	118	53	31	21	15	11	8	5	2
DG C1	56	-	-	-	-	-	-	-	-

DAYTON, OHIO — CONTINUED

	SSF =.10	.20	.30	.40	.50	.60	.70	.80	.90
DG C2	97	40	22	14	9	4	-	-	-
DG C3	137	62	37	25	18	13	10	7	4
SS A1	209	54	24	12	7	4	2	-	-
SS A2	215	85	47	29	19	13	9	6	4
SS A3	185	40	14	-	-	-	-	-	-
SS A4	213	82	44	27	17	12	8	5	3
SS A5	354	49	18	7	-	-	-	-	-
SS A6	209	82	45	28	18	12	8	6	3
SS A7	382	32	-	-	-	-	-	-	-
SS A8	209	78	41	24	16	10	7	4	3
SS B1	135	37	16	8	4	-	-	-	-
SS B2	169	70	39	25	16	11	8	5	3
SS B3	114	29	10	-	-	-	-	-	-
SS B4	162	66	36	23	15	10	7	5	3
SS B5	172	29	10	-	-	-	-	-	-
SS B6	156	65	36	23	15	10	7	5	3
SS B7	138	18	-	-	-	-	-	-	-
SS B8	147	60	33	20	13	9	6	4	2
SS C1	68	24	11	4	-	-	-	-	-
SS C2	96	46	27	18	12	8	6	4	2
SS C3	74	19	7	-	-	-	-	-	-
SS C4	96	41	23	15	10	7	5	3	2
SS D1	187	64	31	17	10	6	3	2	-
SS D2	188	94	56	37	26	18	13	9	5
SS D3	240	60	26	13	7	4	1	-	-
SS D4	193	92	54	35	24	17	12	8	5
SS E1	123	41	18	8	-	-	-	-	-
SS E2	147	71	42	27	18	13	9	6	4
SS E3	157	36	14	5	-	-	-	-	-
SS E4	152	67	38	24	16	11	7	5	3

TOLEDO, OHIO — 6381 DD

	SSF =.10	.20	.30	.40	.50	.60	.70	.80	.90
WW A1	265	-	-	-	-	-	-	-	-
WW A2	80	21	-	-	-	-	-	-	-
WW A3	68	24	10	-	-	-	-	-	-
WW A4	63	26	12	5	-	-	-	-	-
WW A5	60	27	14	7	-	-	-	-	-
WW A6	59	28	15	8	3	-	-	-	-
WW B1	-	-	-	-	-	-	-	-	-
WW B2	76	36	20	13	8	5	3	2	-
WW B3	93	47	28	18	12	8	5	3	2
WW B4	94	54	35	24	17	13	9	6	4
WW B5	92	55	36	26	19	14	10	7	5
WW C1	105	54	33	21	15	10	7	5	3
WW C2	96	52	33	22	16	11	8	5	3
WW C3	113	70	48	35	25	19	14	10	7
WW C4	103	64	44	32	23	18	13	9	6
TW A1	237	-	-	-	-	-	-	-	-
TW A2	87	20	-	-	-	-	-	-	-
TW A3	70	23	9	-	-	-	-	-	-
TW A4	61	24	11	5	-	-	-	-	-
TW B1	128	14	-	-	-	-	-	-	-
TW B2	73	21	8	-	-	-	-	-	-
TW B3	65	22	10	5	-	-	-	-	-
TW B4	63	22	11	5	3	-	-	-	-
TW C1	88	19	6	-	-	-	-	-	-
TW C2	70	21	9	4	-	-	-	-	-
TW C3	69	21	10	5	2	-	-	-	-
TW C4	78	20	9	5	3	1	-	-	-
TW D1	-	-	-	-	-	-	-	-	-
TW D2	84	34	18	11	7	4	2	1	-
TW D3	85	38	21	13	8	5	3	2	-

TOLEDO, OHIO							CONTINUED			YOUNGSTOWN, OHIO								6426 DD	
SSF =	.10	.20	.30	.40	.50	.60	.70	.80	.90	SSF =	.10	.20	.30	.40	.50	.60	.70	.80	.90
TW D4	96	50	31	20	14	10	7	5	3	WW A1	185	-	-	-	-	-	-	-	-
TW D5	96	52	33	23	16	12	8	6	4	WW A2	53	-	-	-	-	-	-	-	-
TW E1	107	49	28	18	12	8	5	3	2	WW A3	46	-	-	-	-	-	-	-	-
TW E2	102	50	30	20	13	9	7	4	3	WW A4	43	-	-	-	-	-	-	-	-
TW E3	119	68	44	30	22	16	12	8	5	WW A5	41	13	-	-	-	-	-	-	-
TW E4	110	63	42	29	21	15	11	8	5	WW A6	41	15	-	-	-	-	-	-	-
TW F1	88	-	-	-	-	-	-	-	-	WW B1	-	-	-	-	-	-	-	-	-
TW F2	56	15	-	-	-	-	-	-	-	WW B2	59	26	13	7	4	-	-	-	-
TW F3	48	17	6	-	-	-	-	-	-	WW B3	75	36	21	12	8	5	3	1	-
TW F4	41	17	8	4	-	-	-	-	-	WW B4	80	45	29	20	14	10	7	5	3
TW G1	53	-	-	-	-	-	-	-	-	WW B5	80	47	31	21	15	11	8	6	4
TW G2	41	14	4	-	-	-	-	-	-	WW C1	86	43	25	16	10	7	4	3	1
TW G3	36	14	6	-	-	-	-	-	-	WW C2	80	43	26	17	12	8	6	4	2
TW G4	31	13	7	3	-	-	-	-	-	WW C3	99	61	41	30	22	16	12	8	6
TW H1	35	10	-	-	-	-	-	-	-	WW C4	90	56	38	27	20	15	11	8	5
TW H2	29	11	5	-	-	-	-	-	-	TW A1	172	-	-	-	-	-	-	-	-
TW H3	26	11	5	3	-	-	-	-	-	TW A2	61	-	-	-	-	-	-	-	-
TW H4	21	9	5	3	1	-	-	-	-	TW A3	50	-	-	-	-	-	-	-	-
TW I1	-	-	-	-	-	-	-	-	-	TW A4	44	13	-	-	-	-	-	-	-
TW I2	62	28	16	9	6	4	2	1	-	TW B1	92	-	-	-	-	-	-	-	-
TW I3	69	33	19	12	8	5	3	2	1	TW B2	53	9	-	-	-	-	-	-	-
TW I4	80	43	27	18	13	9	7	5	3	TW B3	48	13	-	-	-	-	-	-	-
TW I5	82	46	30	21	15	11	8	5	4	TW B4	48	14	-	-	-	-	-	-	-
TW J1	88	44	26	17	11	8	5	3	2	TW C1	65	-	-	-	-	-	-	-	-
TW J2	84	44	27	18	13	9	6	4	3	TW C2	53	13	-	-	-	-	-	-	-
TW J3	102	61	40	29	21	15	11	8	5	TW C3	54	14	4	-	-	-	-	-	-
TW J4	95	57	38	27	20	15	11	8	5	TW C4	63	14	5	-	-	-	-	-	-
DG A1	-	-	-	-	-	-	-	-	-	TW D1	-	-	-	-	-	-	-	-	-
DG A2	62	21	-	-	-	-	-	-	-	TW D2	66	25	12	6	3	-	-	-	-
DG A3	103	44	25	16	11	7	4	-	-	TW D3	67	28	14	8	4	-	-	-	-
DG B1	-	-	-	-	-	-	-	-	-	TW D4	81	41	25	16	11	8	5	4	2
DG B2	64	23	10	-	-	-	-	-	-	TW D5	83	45	28	19	13	9	7	5	3
DG B3	108	47	27	18	13	9	6	3	-	TW E1	88	39	21	13	8	5	3	2	-
DG C1	37	-	-	-	-	-	-	-	-	TW E2	85	41	24	15	10	7	5	3	2
DG C2	84	34	17	10	4	-	-	-	-	TW E3	104	58	37	25	18	13	9	7	4
DG C3	126	56	33	22	15	11	8	5	2	TW E4	96	55	36	25	18	13	9	7	4
SS A1	185	43	17	6	-	-	-	-	-	TW F1	53	-	-	-	-	-	-	-	-
SS A2	194	75	40	24	16	10	7	4	3	TW F2	36	-	-	-	-	-	-	-	-
SS A3	159	28	-	-	-	-	-	-	-	TW F3	31	-	-	-	-	-	-	-	-
SS A4	191	71	37	22	14	9	6	3	2	TW F4	28	7	-	-	-	-	-	-	-
SS A5	323	38	7	-	-	-	-	-	-	TW G1	31	-	-	-	-	-	-	-	-
SS A6	188	72	38	23	14	9	6	4	2	TW G2	26	-	-	-	-	-	-	-	-
SS A7	348	-	-	-	-	-	-	-	-	TW G3	25	-	-	-	-	-	-	-	-
SS A8	187	66	33	19	12	7	4	2	1	TW G4	22	7	-	-	-	-	-	-	-
SS B1	116	28	10	-	-	-	-	-	-	TW H1	21	-	-	-	-	-	-	-	-
SS B2	153	61	33	21	13	9	6	4	2	TW H2	20	-	-	-	-	-	-	-	-
SS B3	95	18	-	-	-	-	-	-	-	TW H3	19	6	-	-	-	-	-	-	-
SS B4	146	57	31	19	12	8	5	3	2	TW H4	16	6	2	-	-	-	-	-	-
SS B5	150	20	-	-	-	-	-	-	-	TW I1	-	-	-	-	-	-	-	-	-
SS B6	140	56	31	19	12	8	5	3	2	TW I2	48	20	10	5	2	-	-	-	-
SS B7	116	-	-	-	-	-	-	-	-	TW I3	55	25	13	8	5	2	1	-	-
SS B8	131	51	27	16	10	6	4	2	1	TW I4	67	36	22	15	10	7	5	3	2
SS C1	51	12	-	-	-	-	-	-	-	TW I5	70	39	25	17	12	9	6	4	3
SS C2	83	38	22	14	9	6	4	2	1	TW J1	73	35	20	12	8	5	3	2	1
SS C3	59	10	-	-	-	-	-	-	-	TW J2	71	36	22	14	10	7	4	3	2
SS C4	83	34	19	11	7	4	3	2	-	TW J3	89	52	35	24	17	13	9	7	4
SS D1	161	51	22	10	-	-	-	-	-	TW J4	83	49	33	23	17	12	9	6	4
SS D2	169	82	48	31	21	14	10	7	4	DG A1	-	-	-	-	-	-	-	-	-
SS D3	212	47	18	-	-	-	-	-	-	DG A2	47	11	-	-	-	-	-	-	-
SS D4	173	80	46	29	20	13	9	6	4	DG A3	90	37	21	13	8	5	-	-	-
SS E1	101	28	-	-	-	-	-	-	-	DG B1	-	-	-	-	-	-	-	-	-
SS E2	129	61	35	22	14	10	6	4	2	DG B2	50	14	-	-	-	-	-	-	-
SS E3	134	25	-	-	-	-	-	-	-	DG B3	94	40	23	15	10	7	4	2	-
SS E4	135	58	32	19	12	8	5	3	2	DG C1	-	-	-	-	-	-	-	-	-

YOUNGSTOWN, OHIO — CONTINUED

SSF =	.10	.20	.30	.40	.50	.60	.70	.80	.90
DG C2	68	25	11	-	-	-	-	-	-
DG C3	111	48	28	18	12	9	6	4	-
SS A1	155	31	-	-	-	-	-	-	-
SS A2	170	64	33	19	12	8	5	3	2
SS A3	128	-	-	-	-	-	-	-	-
SS A4	168	60	30	17	10	6	4	2	1
SS A5	272	24	-	-	-	-	-	-	-
SS A6	164	61	31	18	11	7	4	2	1
SS A7	289	-	-	-	-	-	-	-	-
SS A8	162	55	26	14	8	4	2	-	-
SS B1	93	17	-	-	-	-	-	-	-
SS B2	134	52	28	17	11	7	5	3	2
SS B3	72	-	-	-	-	-	-	-	-
SS B4	127	49	25	15	9	6	4	2	1
SS B5	120	-	-	-	-	-	-	-	-
SS B6	122	48	25	15	9	6	4	2	1
SS B7	87	-	-	-	-	-	-	-	-
SS B8	113	43	21	12	7	4	2	1	-
SS C1	25	-	-	-	-	-	-	-	-
SS C2	66	29	15	9	5	2	-	-	-
SS C3	36	-	-	-	-	-	-	-	-
SS C4	66	26	13	7	3	-	-	-	-
SS D1	132	37	-	-	-	-	-	-	-
SS D2	148	70	41	26	17	11	8	5	3
SS D3	177	34	-	-	-	-	-	-	-
SS D4	152	69	39	24	16	10	7	4	3
SS E1	75	-	-	-	-	-	-	-	-
SS E2	110	50	28	17	10	6	4	2	1
SS E3	106	-	-	-	-	-	-	-	-
SS E4	116	48	25	14	9	5	3	2	-

OKLAHOMA CITY, OKLAHOMA — 3695 DD

SSF =	.10	.20	.30	.40	.50	.60	.70	.80	.90
WW A1	879	100	45	27	17	12	8	6	4
WW A2	287	111	63	41	28	20	15	10	7
WW A3	241	113	68	46	32	24	17	12	8
WW A4	221	113	70	48	35	25	19	14	9
WW A5	210	114	73	51	37	27	20	15	10
WW A6	204	114	75	52	38	28	21	15	10
WW B1	228	80	43	27	18	13	9	6	4
WW B2	219	116	74	51	37	27	20	15	10
WW B3	255	142	92	65	47	35	26	19	13
WW B4	226	136	92	67	50	37	28	21	14
WW B5	209	129	88	64	48	37	28	21	14
WW C1	271	151	98	69	50	38	28	21	14
WW C2	235	137	91	65	48	36	27	20	14
WW C3	242	156	109	81	61	47	36	27	18
WW C4	219	141	99	73	55	42	32	24	17
TW A1	754	96	44	26	17	12	8	6	4
TW A2	296	101	55	35	24	17	12	9	6
TW A3	238	101	58	38	27	19	14	10	7
TW A4	201	95	58	39	28	20	15	11	7
TW B1	419	91	45	27	18	13	9	6	4
TW B2	238	92	52	34	23	17	12	9	6
TW B3	207	89	52	34	24	17	13	9	6
TW B4	191	82	48	32	22	16	12	8	6
TW C1	275	83	44	27	19	13	9	7	4
TW C2	210	79	45	29	20	14	10	7	5
TW C3	201	74	42	27	19	13	10	7	5
TW C4	213	67	36	23	16	11	8	6	4
TW D1	166	63	35	22	15	11	8	5	3
TW D2	236	110	67	45	32	23	17	12	8
TW D3	239	121	75	51	37	27	20	14	10

OKLAHOMA CITY, OKLAHOMA — CONTINUED

SSF =	.10	.20	.30	.40	.50	.60	.70	.80	.90
TW D4	232	128	83	58	42	32	24	17	12
TW D5	217	123	81	57	42	32	24	18	12
TW E1	278	141	88	60	43	32	24	17	11
TW E2	254	135	86	60	43	32	24	17	12
TW E3	267	158	106	76	56	43	32	24	16
TW E4	239	143	97	70	52	39	30	22	15
TW F1	337	83	42	25	17	12	8	6	4
TW F2	210	85	49	32	22	16	11	8	5
TW F3	173	82	49	33	24	17	13	9	6
TW F4	142	73	46	31	23	17	12	9	6
TW G1	208	72	39	25	17	12	9	6	4
TW G2	152	70	42	28	20	14	10	7	5
TW G3	128	64	39	27	19	14	10	7	5
TW G4	103	53	33	23	16	12	9	6	4
TW H1	135	58	34	22	16	11	8	6	4
TW H2	104	51	31	21	15	11	8	6	4
TW H3	87	43	27	18	13	10	7	5	3
TW H4	67	33	20	14	10	7	5	4	3
TW I1	127	52	30	19	13	9	7	5	3
TW I2	180	92	58	40	29	21	16	11	8
TW I3	190	101	64	45	32	24	18	13	9
TW I4	192	110	73	52	38	29	22	16	11
TW I5	187	110	74	53	39	30	22	17	11
TW J1	230	124	79	55	40	30	22	16	11
TW J2	209	117	76	54	39	29	22	16	11
TW J3	223	138	94	69	51	39	30	22	15
TW J4	205	127	87	64	48	36	28	20	14
DG A1	164	69	38	21	9	-	-	-	-
DG A2	187	84	50	33	23	15	10	5	-
DG A3	233	107	66	46	34	25	18	13	7
DG B1	164	71	41	27	18	10	-	-	-
DG B2	192	87	53	37	27	20	14	9	4
DG B3	237	110	68	48	36	28	21	16	10
DG C1	203	90	53	36	26	19	12	5	-
DG C2	227	104	64	44	33	25	19	13	7
DG C3	273	127	79	56	42	33	26	20	13
SS A1	488	146	76	47	32	22	16	11	7
SS A2	440	186	108	71	49	36	26	18	12
SS A3	465	131	67	41	27	19	13	9	6
SS A4	445	184	106	69	48	34	25	17	11
SS A5	824	146	69	41	27	18	13	9	6
SS A6	437	184	107	70	49	35	25	18	12
SS A7	930	122	55	31	20	13	9	6	4
SS A8	446	180	102	66	46	33	24	17	11
SS B1	332	111	59	37	25	18	13	9	6
SS B2	345	150	88	58	41	30	22	15	10
SS B3	304	100	53	33	22	16	11	8	5
SS B4	337	146	85	56	39	28	21	15	10
SS B5	434	98	48	29	19	13	9	6	4
SS B6	325	144	85	56	40	29	21	15	10
SS B7	381	82	40	24	15	10	7	5	3
SS B8	315	137	80	53	37	27	19	14	9
SS C1	215	96	57	38	27	19	14	10	7
SS C2	230	119	75	52	37	27	20	15	10
SS C3	225	78	42	27	18	13	9	7	4
SS C4	228	107	65	44	31	23	17	12	8
SS D1	445	175	98	63	43	31	22	16	10
SS D2	387	202	128	88	63	46	34	25	16
SS D3	564	165	86	53	36	25	18	12	8
SS D4	397	198	123	83	59	43	32	23	15
SS E1	330	136	77	50	34	24	18	12	8
SS E2	319	165	104	71	51	37	28	20	13
SS E3	403	117	60	37	25	17	12	8	5
SS E4	329	156	94	63	45	32	24	17	11

TULSA, OKLAHOMA   3680 DD

	SSF=.10	.20	.30	.40	.50	.60	.70	.80	.90
WW A1	789	86	39	22	14	10	7	5	3
WW A2	255	98	55	35	24	17	12	9	6
WW A3	214	99	60	40	28	20	15	11	7
WW A4	196	99	62	42	30	22	16	12	8
WW A5	187	101	64	45	32	24	18	13	8
WW A6	181	101	66	46	33	25	18	13	9
WW B1	197	68	36	22	15	10	7	5	3
WW B2	197	104	66	45	33	24	18	13	9
WW B3	229	127	82	57	42	31	23	17	11
WW B4	205	123	83	60	45	34	25	19	13
WW B5	190	117	80	58	44	33	25	19	13
WW C1	244	135	88	62	45	33	25	18	12
WW C2	213	123	82	58	43	32	24	18	12
WW C3	221	142	99	73	55	42	32	24	17
WW C4	200	128	90	66	50	38	29	22	15
TW A1	678	83	38	22	14	10	7	5	3
TW A2	263	89	48	30	21	15	11	7	5
TW A3	212	89	51	33	23	17	12	9	6
TW A4	179	84	51	34	24	18	13	9	6
TW B1	374	80	39	24	16	11	8	5	3
TW B2	212	81	45	29	20	14	10	7	5
TW B3	185	78	46	30	21	15	11	8	5
TW B4	171	73	42	28	19	14	10	7	5
TW C1	246	73	38	24	16	11	8	6	4
TW C2	188	70	39	25	17	12	9	6	4
TW C3	180	66	37	24	16	12	8	6	4
TW C4	192	60	32	20	14	10	7	5	3
TW D1	143	53	29	18	12	8	6	4	3
TW D2	212	98	59	40	28	21	15	11	7
TW D3	214	108	67	45	33	24	18	13	8
TW D4	210	115	75	52	38	28	21	15	10
TW D5	197	112	74	52	38	29	22	16	11
TW E1	251	126	79	54	38	28	21	15	10
TW E2	229	121	77	53	39	29	21	15	10
TW E3	243	143	96	69	51	38	29	21	14
TW E4	218	130	88	63	47	36	27	20	14
TW F1	299	72	36	22	14	10	7	5	3
TW F2	186	75	42	27	19	14	10	7	4
TW F3	154	72	43	29	20	15	11	8	5
TW F4	127	65	40	28	20	14	11	8	5
TW G1	184	63	34	21	14	10	7	5	3
TW G2	135	61	36	24	17	12	9	6	4
TW G3	113	56	34	23	17	12	9	6	4
TW G4	92	47	29	20	14	10	8	5	4
TW H1	119	51	29	19	13	10	7	5	3
TW H2	92	45	27	18	13	10	7	5	3
TW H3	77	38	24	16	11	8	6	4	3
TW H4	60	29	18	12	9	6	5	3	2
TW I1	109	44	24	16	11	7	5	4	2
TW I2	161	82	51	35	25	18	14	10	7
TW I3	170	90	57	40	29	21	16	11	8
TW I4	174	100	66	47	34	26	19	14	10
TW I5	170	100	67	48	36	27	20	15	10
TW J1	207	111	71	49	36	26	20	14	10
TW J2	189	105	69	48	35	26	20	14	10
TW J3	203	125	86	62	47	35	27	20	14
TW J4	188	116	80	58	43	33	25	18	13
DG A1	142	58	30	14	-	-	-	-	-
DG A2	167	74	44	29	19	13	8	3	-
DG A3	212	97	60	42	30	23	16	11	6
DG B1	142	60	34	22	13	-	-	-	-
DG B2	172	77	47	32	23	17	11	7	2
DG B3	216	100	62	43	32	25	19	14	9
DG C1	178	77	46	31	21	15	8	-	-

TULSA, OKLAHOMA   CONTINUED

	SSF=.10	.20	.30	.40	.50	.60	.70	.80	.90
DG C2	204	93	56	39	29	22	16	11	6
DG C3	249	116	72	51	38	30	23	17	11
SS A1	431	128	67	41	27	19	13	9	6
SS A2	394	166	96	63	44	31	23	16	11
SS A3	408	113	57	35	23	16	11	7	5
SS A4	396	163	94	61	42	30	22	15	10
SS A5	733	127	60	35	23	15	11	7	5
SS A6	390	164	95	62	43	31	22	16	10
SS A7	824	104	46	26	17	11	8	5	3
SS A8	397	160	91	59	40	29	21	15	9
SS B1	294	97	52	32	22	15	11	7	5
SS B2	310	135	79	52	37	26	19	14	9
SS B3	267	87	46	28	19	13	9	6	4
SS B4	301	130	76	50	35	25	18	13	8
SS B5	383	86	42	25	16	11	8	5	3
SS B6	291	129	76	50	35	25	18	13	9
SS B7	334	71	34	20	13	9	6	4	2
SS B8	281	122	72	47	33	24	17	12	8
SS C1	190	84	49	33	23	16	12	8	5
SS C2	207	106	67	46	33	24	18	13	9
SS C3	199	68	37	23	16	11	8	6	4
SS C4	205	95	58	39	27	20	15	11	7
SS D1	392	153	86	55	37	27	19	13	8
SS D2	347	181	114	78	56	41	30	22	14
SS D3	498	145	75	46	31	21	15	10	7
SS D4	355	177	109	74	53	38	28	20	13
SS E1	290	118	67	43	29	21	15	10	7
SS E2	285	148	92	63	45	33	24	17	12
SS E3	355	102	52	32	21	15	10	7	4
SS E4	294	139	84	56	39	29	21	15	10

ASTORIA, OREGON   5295 DD

	SSF=.10	.20	.30	.40	.50	.60	.70	.80	.90
WW A1	510	62	26	13	6	-	-	-	-
WW A2	189	73	40	24	15	9	5	1	-
WW A3	162	75	43	27	17	11	7	3	1
WW A4	150	76	45	29	19	12	8	4	2
WW A5	144	77	47	31	21	14	9	5	2
WW A6	140	77	48	32	22	14	9	5	3
WW B1	137	45	20	-	-	-	-	-	-
WW B2	154	81	50	33	23	16	10	7	4
WW B3	183	101	64	43	30	21	14	9	5
WW B4	166	100	67	47	34	24	17	12	7
WW B5	156	96	65	46	34	25	18	12	8
WW C1	197	109	70	47	33	23	16	11	6
WW C2	172	100	65	45	32	23	16	11	7
WW C3	185	119	82	59	44	33	24	17	11
WW C4	166	107	74	54	40	30	22	15	10
TW A1	446	60	26	14	7	-	-	-	-
TW A2	193	67	35	21	13	7	4	1	-
TW A3	159	67	38	23	15	9	5	3	1
TW A4	137	65	38	24	16	10	6	4	2
TW B1	261	59	28	15	9	5	-	-	-
TW B2	159	62	34	20	13	8	5	2	-
TW B3	140	60	34	21	14	9	5	3	1
TW B4	130	56	32	20	13	8	5	3	1
TW C1	179	55	28	16	10	6	3	1	-
TW C2	141	54	29	18	12	7	5	3	1
TW C3	136	51	28	17	11	7	4	3	1
TW C4	143	46	24	15	9	6	4	2	1
TW D1	101	36	17	-	-	-	-	-	-
TW D2	164	77	46	29	20	13	9	5	3
TW D3	169	85	52	34	23	16	10	6	3

ASTORIA, OREGON						CONTINUED				BURNS, OREGON								7212 DD	
SSF =	.10	.20	.30	.40	.50	.60	.70	.80	.90	SSF =	.10	.20	.30	.40	.50	.60	.70	.80	.90
TW D4	169	93	60	41	29	20	14	10	6	WW A1	502	50	19	9	-	-	-	-	-
TW D5	161	92	60	41	30	21	15	11	7	WW A2	169	61	32	19	12	7	4	2	-
TW E1	200	101	62	41	28	19	13	9	5	WW A3	142	63	36	22	14	9	6	3	2
TW E2	183	97	61	41	29	20	14	9	6	WW A4	131	63	37	24	16	11	7	4	2
TW E3	201	119	79	55	40	29	21	14	9	WW A5	125	64	39	26	17	12	8	5	3
TW E4	180	108	72	51	37	27	20	14	9	WW A6	121	65	40	27	18	13	8	5	3
TW F1	208	52	25	13	7	-	-	-	-	WW B1	113	32	11	-	-	-	-	-	-
TW F2	138	56	30	18	11	7	3	-	-	WW B2	137	70	43	28	20	14	10	6	4
TW F3	116	54	31	20	13	8	5	2	-	WW B3	162	87	55	37	26	18	13	9	5
TW F4	97	49	30	19	13	8	5	3	1	WW B4	149	88	59	41	30	22	16	11	7
TW G1	133	46	24	14	8	4	-	-	-	WW B5	141	85	58	41	30	22	16	12	8
TW G2	101	46	26	16	10	6	4	2	-	WW C1	176	95	60	41	29	20	15	10	6
TW G3	87	43	25	16	10	7	4	2	-	WW C2	154	87	57	39	28	21	15	10	7
TW G4	70	36	22	14	9	6	4	2	1	WW C3	167	106	73	53	39	29	22	16	10
TW H1	89	38	21	13	8	5	2	-	-	WW C4	151	96	66	48	36	27	20	14	10
TW H2	70	34	20	13	8	5	3	2	-	TW A1	438	50	20	10	5	-	-	-	-
TW H3	59	29	17	11	7	5	3	2	-	TW A2	176	56	28	17	10	6	4	2	-
TW H4	46	23	14	9	6	4	2	1	-	TW A3	142	56	31	19	12	8	5	3	1
TW I1	77	29	13	-	-	-	-	-	-	TW A4	121	54	31	20	13	9	6	4	2
TW I2	125	64	39	26	17	12	8	5	3	TW B1	248	49	22	12	7	4	2	-	-
TW I3	136	72	45	30	20	14	9	6	3	TW B2	143	52	28	17	11	7	4	2	1
TW I4	141	81	53	36	26	19	13	9	5	TW B3	126	51	28	17	11	8	5	3	2
TW I5	139	82	54	38	27	20	14	10	6	TW B4	117	48	27	17	11	7	5	3	2
TW J1	166	89	56	38	26	18	12	8	5	TW C1	167	47	23	13	8	5	3	2	-
TW J2	152	85	55	37	26	19	13	9	5	TW C2	129	46	24	15	10	6	4	2	1
TW J3	169	104	71	50	37	27	20	14	9	TW C3	125	44	23	14	9	6	4	2	1
TW J4	155	96	65	47	34	25	18	13	8	TW C4	134	40	20	12	8	5	3	2	1
DG A1	102	38	-	-	-	-	-	-	-	TW D1	83	26	10	-	-	-	-	-	-
DG A2	130	58	33	20	11	-	-	-	-	TW D2	148	67	39	25	17	12	8	5	3
DG A3	170	81	50	33	23	16	10	6	-	TW D3	151	73	44	29	20	14	9	6	4
DG B1	104	43	19	-	-	-	-	-	-	TW D4	153	82	52	36	25	18	13	9	6
DG B2	132	62	36	23	14	8	-	-	-	TW D5	146	82	53	37	26	19	14	10	6
DG B3	172	84	52	35	25	18	13	8	4	TW E1	180	88	53	35	24	17	12	8	5
DG C1	133	59	32	17	-	-	-	-	-	TW E2	166	86	53	36	25	18	13	9	5
DG C2	159	76	46	30	20	13	8	-	-	TW E3	182	105	70	49	35	26	19	13	9
DG C3	199	98	61	42	30	22	16	11	6	TW E4	164	96	64	45	33	25	18	13	8
SS A1	356	111	56	32	19	11	6	2	-	TW F1	194	42	19	10	5	-	-	-	-
SS A2	342	148	84	53	34	23	15	9	5	TW F2	123	46	24	15	9	6	3	2	-
SS A3	339	100	48	25	13	5	-	-	-	TW F3	102	45	26	16	10	7	4	2	1
SS A4	349	148	83	51	33	21	14	8	4	TW F4	85	41	25	16	11	7	5	3	1
SS A5	566	109	50	26	14	7	-	-	-	TW G1	120	38	19	11	6	3	1	-	-
SS A6	338	145	82	51	33	22	14	9	5	TW G2	89	38	21	13	9	5	3	2	-
SS A7	621	91	38	18	7	-	-	-	-	TW G3	76	36	21	13	9	6	4	2	1
SS A8	348	144	79	49	31	20	12	7	4	TW G4	62	30	18	12	8	5	3	2	1
SS B1	242	83	42	24	14	8	4	-	-	TW H1	78	31	17	10	6	4	2	1	-
SS B2	267	119	69	44	29	19	13	8	5	TW H2	61	28	16	10	7	5	3	2	-
SS B3	221	75	37	20	11	4	-	-	-	TW H3	52	25	15	9	6	4	3	2	-
SS B4	263	116	67	42	27	18	12	7	4	TW H4	41	19	11	7	5	3	2	1	-
SS B5	300	72	34	18	9	4	-	-	-	TW I1	61	20	7	-	-	-	-	-	-
SS B6	251	113	66	42	27	18	12	8	4	TW I2	112	55	33	22	15	10	7	5	3
SS B7	261	60	27	13	-	-	-	-	-	TW I3	120	62	38	25	18	12	9	6	3
SS B8	245	109	62	39	25	16	11	6	3	TW I4	127	71	46	32	23	17	12	8	5
SS C1	142	63	35	21	13	7	3	-	-	TW I5	126	73	48	34	24	18	13	9	6
SS C2	163	84	52	34	23	16	11	7	4	TW J1	149	77	48	32	23	16	11	8	5
SS C3	145	50	26	15	9	5	-	-	-	TW J2	136	75	48	33	23	17	12	8	5
SS C4	160	75	45	29	19	13	9	5	3	TW J3	153	93	63	45	33	24	18	13	8
SS D1	334	133	72	42	26	15	8	4	1	TW J4	141	86	58	42	31	23	17	12	8
SS D2	307	161	99	65	44	30	20	13	7	DG A1	83	24	-	-	-	-	-	-	-
SS D3	410	126	63	35	21	12	6	2	-	DG A2	116	49	27	16	8	-	-	-	-
SS D4	313	158	95	62	41	28	19	12	7	DG A3	158	71	43	29	20	14	9	5	-
SS E1	243	100	54	31	18	10	-	-	-	DG B1	83	29	-	-	-	-	-	-	-
SS E2	249	129	79	51	34	23	15	10	5	DG B2	119	51	30	19	12	6	-	-	-
SS E3	286	86	42	23	13	6	-	-	-	DG B3	162	74	45	31	22	16	11	8	4
SS E4	255	122	71	46	30	20	13	8	4	DG C1	112	45	23	10	-	-	-	-	-

BURNS, OREGON								CONTINUED		MEDFORD, OREGON								CONTINUED	
SSF =	.10	.20	.30	.40	.50	.60	.70	.80	.90	SSF =	.10	.20	.30	.40	.50	.60	.70	.80	.90
DG C2	145	64	38	25	17	11	7	-	-	TW D4	177	93	58	39	27	19	13	9	5
DG C3	188	86	53	36	26	20	14	10	6	TW D5	168	92	58	40	28	20	14	10	6
SS A1	327	91	43	24	14	8	5	2	1	TW E1	213	101	60	39	26	18	12	8	4
SS A2	308	125	69	43	29	19	13	8	5	TW E2	194	98	59	39	27	19	13	9	5
SS A3	305	77	34	17	9	-	-	-	-	TW E3	208	118	77	53	38	27	20	13	8
SS A4	310	122	67	41	26	17	12	7	4	TW E4	186	108	71	49	36	26	19	13	8
SS A5	550	89	38	19	10	5	1	-	-	TW F1	256	53	23	12	5	-	-	-	-
SS A6	304	123	68	42	28	18	12	8	5	TW F2	153	56	29	17	10	5	2	-	-
SS A7	609	70	26	11	-	-	-	-	-	TW F3	125	54	30	18	11	7	3	1	-
SS A8	308	118	63	38	25	16	10	6	4	TW F4	102	48	28	18	11	7	4	2	1
SS B1	215	66	32	18	11	6	3	1	-	TW G1	153	46	22	12	7	-	-	-	-
SS B2	240	101	57	36	24	17	11	7	4	TW G2	109	46	25	15	9	5	3	1	-
SS B3	192	57	27	14	7	-	-	-	-	TW G3	92	42	24	15	9	6	3	2	-
SS B4	233	97	54	34	22	15	10	7	4	TW G4	74	35	20	13	8	5	3	2	-
SS B5	278	57	25	13	6	2	-	-	-	TW H1	97	38	20	12	7	4	1	-	-
SS B6	223	95	54	34	23	15	10	7	4	TW H2	75	34	19	12	7	5	3	1	-
SS B7	238	45	18	7	-	-	-	-	-	TW H3	63	29	17	10	7	4	3	1	-
SS B8	215	90	50	31	20	14	9	6	3	TW H4	49	23	13	8	5	3	2	1	-
SS C1	123	51	28	17	10	6	3	1	-	TW I1	83	27	10	-	-	-	-	-	-
SS C2	145	72	44	29	20	14	10	6	4	TW I2	133	64	38	24	16	11	7	4	2
SS C3	131	41	21	12	7	4	2	-	-	TW I3	142	71	43	28	19	13	9	5	3
SS C4	144	65	38	24	17	11	8	5	3	TW I4	146	80	51	35	25	17	12	8	5
SS D1	294	108	56	33	20	12	7	4	2	TW I5	144	82	53	37	26	19	13	9	6
SS D2	269	136	82	54	37	26	18	12	7	TW J1	175	89	54	36	24	17	11	7	4
SS D3	377	102	49	27	16	9	5	2	1	TW J2	158	85	53	36	25	18	12	8	5
SS D4	276	133	79	51	35	24	16	11	7	TW J3	173	103	69	48	35	26	18	13	8
SS E1	207	78	40	23	13	7	3	1	-	TW J4	160	96	64	45	33	24	17	12	8
SS E2	215	107	64	42	28	19	13	9	5	DG A1	110	35	-	-	-	-	-	-	-
SS E3	257	68	31	17	9	4	-	-	-	DG A2	142	58	32	18	9	-	-	-	-
SS E4	223	101	58	37	25	17	11	7	4	DG A3	187	82	48	32	22	15	9	5	-
										DG B1	110	40	15	-	-	-	-	-	-
										DG B2	146	61	35	21	13	7	-	-	-
MEDFORD, OREGON								4930 DD		DG B3	193	84	51	34	24	17	12	7	3
SSF =	.10	.20	.30	.40	.50	.60	.70	.80	.90	DG C1	144	57	29	13	-	-	-	-	-
WW A1	708	64	24	11	-	-	-	-	-	DG C2	177	75	44	28	19	12	6	-	-
WW A2	212	73	38	22	13	7	3	-	-	DG C3	224	98	60	41	29	21	14	10	5
WW A3	174	75	41	25	16	9	5	2	-	SS A1	415	110	51	28	16	9	4	2	-
WW A4	158	74	43	27	17	11	6	3	1	SS A2	372	146	79	48	31	21	14	8	5
WW A5	149	75	45	29	19	12	7	4	2	SS A3	397	96	42	21	10	-	-	-	-
WW A6	144	75	46	30	20	13	8	4	2	SS A4	379	144	76	46	29	19	12	7	4
WW B1	154	43	16	-	-	-	-	-	-	SS A5	732	111	45	23	12	5	-	-	-
WW B2	162	80	48	31	21	14	9	6	3	SS A6	368	143	77	47	30	20	13	8	4
WW B3	190	100	62	41	28	19	13	8	5	SS A7	846	90	33	14	-	-	-	-	-
WW B4	171	99	65	45	32	23	16	11	7	SS A8	379	140	73	44	27	17	11	6	3
WW B5	160	95	63	45	32	23	17	12	7	SS B1	274	81	38	21	12	6	3	-	-
WW C1	205	108	67	45	31	21	15	10	6	SS B2	288	117	65	40	26	18	12	7	4
WW C2	178	99	63	43	30	22	15	10	6	SS B3	249	71	33	17	8	-	-	-	-
WW C3	189	117	80	57	42	31	23	16	10	SS B4	282	113	62	38	25	16	11	7	4
WW C4	170	106	72	52	38	28	21	15	9	SS B5	368	72	30	15	7	-	-	-	-
TW A1	607	63	25	12	5	-	-	-	-	SS B6	269	111	62	38	25	17	11	7	4
TW A2	222	68	33	19	11	6	2	-	-	SS B7	323	58	23	10	-	-	-	-	-
TW A3	175	67	36	21	13	8	4	2	-	SS B8	262	106	57	35	23	15	9	6	3
TW A4	147	64	36	22	14	9	5	3	1	SS C1	153	62	33	19	11	5	-	-	-
TW B1	327	61	27	14	7	3	-	-	-	SS C2	172	83	50	32	22	15	10	6	3
TW B2	178	62	32	19	12	7	4	2	-	SS C3	166	51	24	13	7	3	-	-	-
TW B3	154	60	33	20	12	8	4	2	1	SS C4	173	76	43	27	18	12	8	5	3
TW B4	143	56	31	19	12	8	5	2	1	SS D1	367	129	65	37	22	13	7	3	1
TW C1	212	56	27	15	9	5	2	-	-	SS D2	318	156	92	60	40	27	18	12	7
TW C2	159	55	28	17	11	7	4	2	-	SS D3	480	124	57	31	18	10	5	2	-
TW C3	154	52	27	16	10	6	4	2	1	SS D4	328	153	89	57	38	26	17	11	6
TW C4	167	48	24	14	9	5	3	2	-	SS E1	262	95	48	27	15	7	-	-	-
TW D1	112	34	14	-	-	-	-	-	-	SS E2	257	124	73	47	31	21	14	9	5
TW D2	177	77	44	28	18	12	8	5	3	SS E3	334	84	38	20	10	4	-	-	-
TW D3	180	85	50	32	21	14	9	6	3	SS E4	269	118	67	42	27	18	12	7	4

```
NORTH BEND, OREGON                           4688 DD    NORTH BEND, OREGON                           CONTINUED
   SSF =.10  .20  .30  .40  .50  .60  .70  .80  .90       SSF =.10  .20  .30  .40  .50  .60  .70  .80  .90
  WW A1  815  108   51   31   20   13    9    6    3     DG C2  229  114   73   51   37   27   20   13    7
  WW A2  295  122   71   46   32   22   16   11    6     DG C3  273  136   89   63   47   36   27   20   13
  WW A3  253  124   76   51   36   26   18   13    8     SS A1  544  180   97   60   39   26   17   11    6
  WW A4  234  124   79   54   39   28   20   14    9     SS A2  501  224  133   87   60   42   29   19   12
  WW A5  224  126   82   57   41   30   22   15    9     SS A3  531  167   88   54   34   22   14    9    5
  WW A6  218  126   83   59   42   31   22   16   10     SS A4  517  227  133   86   59   41   28   18   11
  WW B1  241   92   51   32   21   14    9    6    3     SS A5  870  179   88   53   34   22   14    9    5
  WW B2  231  126   82   57   41   30   22   15   10     SS A6  499  223  132   86   59   41   28   19   11
  WW B3  273  156  103   73   53   39   28   20   13     SS A7  979  155   73   42   26   16   10    6    3
  WW B4  239  147  101   73   54   41   30   22   14     SS A8  519  223  130   84   57   39   26   17   10
  WW B5  220  139   96   70   52   40   30   21   14     SS B1  372  135   75   47   31   21   14    9    5
  WW C1  290  166  110   77   56   41   30   21   14     SS B2  388  178  107   71   49   35   24   16   10
  WW C2  249  148  100   71   53   39   29   21   13     SS B3  349  126   69   43   28   19   12    8    4
  WW C3  260  170  120   89   67   51   38   28   18     SS B4  385  176  105   69   48   33   23   16    9
  WW C4  232  151  107   79   60   46   34   25   17     SS B5  466  120   62   37   24   16   10    6    3
  TW A1  704  103   50   30   19   13    9    6    3     SS B6  368  172  104   69   48   33   23   16   10
  TW A2  299  110   62   40   27   19   13    9    5     SS B7  416  104   53   31   20   13    8    5    2
  TW A3  246  110   65   43   30   21   15   10    6     SS B8  362  167  100   66   45   32   22   15    9
  TW A4  211  104   65   44   31   22   16   11    7     SS C1  226  107   65   43   30   21   15   10    6
  TW B1  406   98   51   31   21   14   10    6    4     SS C2  242  130   83   58   41   30   22   15   10
  TW B2  243  100   58   38   26   18   13    9    5     SS C3  229   85   48   31   21   15   10    7    4
  TW B3  214   97   58   38   27   19   14    9    6     SS C4  237  116   72   49   35   25   18   12    8
  TW B4  197   89   53   36   25   18   13    9    5     SS D1  514  216  124   80   53   36   24   16    9
  TW C1  275   89   49   31   21   15   10    7    4     SS D2  453  245  156  107   75   54   38   26   16
  TW C2  214   86   50   32   22   16   11    8    5     SS D3  628  203  109   68   44   29   19   12    7
  TW C3  204   80   46   30   21   15   10    7    4     SS D4  462  240  150  102   71   50   35   24   15
  TW C4  212   72   40   25   17   12    9    6    4     SS E1  384  168   98   63   42   29   19   12    7
  TW D1  176   72   41   26   17   12    8    5    3     SS E2  371  200  127   87   61   43   31   21   13
  TW D2  244  119   74   50   35   26   18   13    8     SS E3  446  143   77   47   31   21   14    9    5
  TW D3  254  133   84   58   41   30   22   15    9     SS E4  378  187  115   77   54   38   26   18   11
  TW D4  243  138   91   64   47   34   25   18   12
  TW D5  226  132   88   63   46   34   25   18   12
  TW E1  294  154   98   68   48   35   25   18   11    PENDLETON, OREGON                              5240 DD
  TW E2  265  145   94   66   48   35   26   18   12       SSF =.10  .20  .30  .40  .50  .60  .70  .80  .90
  TW E3  284  172  117   84   62   47   34   25   16     WW A1  608   49   14    -    -    -    -    -    -
  TW E4  251  154  105   76   57   43   32   23   15     WW A2  179   58   27   13    -    -    -    -    -
  TW F1  331   90   47   29   19   13    9    6    3     WW A3  146   59   30   16    8    -    -    -    -
  TW F2  217   93   55   36   25   17   12    8    5     WW A4  132   59   32   18   10    4    -    -    -
  TW F3  182   90   55   37   26   19   13    9    6     WW A5  125   60   34   20   11    6    2    -    -
  TW F4  151   81   51   35   25   18   13    9    6     WW A6  120   60   34   21   12    7    3    1    -
  TW G1  212   78   44   28   19   13    9    6    4     WW B1  118   20    -    -    -    -    -    -    -
  TW G2  159   77   47   32   22   16   11    8    5     WW B2  139   67   39   24   15   10    6    3    2
  TW G3  135   70   44   30   21   15   11    8    5     WW B3  163   83   50   32   21   14    9    5    3
  TW G4  109   58   37   25   18   13    9    7    4     WW B4  150   85   55   37   26   18   13    8    5
  TW H1  140   64   38   25   18   12    9    6    4     WW B5  141   83   54   38   27   19   14    9    6
  TW H2  109   56   35   24   17   12    9    6    4     WW C1  177   91   55   36   24   16   10    7    4
  TW H3   91   47   30   20   15   10    8    5    3     WW C2  155   85   53   35   24   17   11    7    4
  TW H4   71   36   23   16   11    8    6    4    2     WW C3  166  102   69   49   35   26   18   13    8
  TW I1  137   60   35   23   15   10    7    4    2     WW C4  151   93   63   45   32   24   17   12    8
  TW I2  188  100   64   44   32   23   17   12    7     TW A1  522   49   16    -    -    -    -    -    -
  TW I3  203  112   72   50   36   26   19   13    9     TW A2  189   54   24   12    -    -    -    -    -
  TW I4  203  119   80   57   42   31   23   16   11     TW A3  148   54   27   14    7    -    -    -    -
  TW I5  197  119   81   58   43   32   24   17   11     TW A4  125   52   27   16    9    4    1    -    -
  TW J1  245  136   88   62   45   33   24   17   11     TW B1  278   48   19    7    -    -    -    -    -
  TW J2  220  126   84   59   43   32   24   17   11     TW B2  152   50   24   13    6    -    -    -    -
  TW J3  238  150  104   76   57   43   32   23   15     TW B3  131   49   25   14    8    4    -    -    -
  TW J4  217  137   95   69   52   39   29   21   14     TW B4  122   46   24   13    8    4    2    -    -
  DG A1  175   79   45   26   13    -    -    -    -     TW C1  181   46   20   10    4    -    -    -    -
  DG A2  193   92   57   38   26   17   11    5    -     TW C2  137   45   22   12    7    3    1    -    -
  DG A3  235  116   74   52   38   27   20   13    7     TW C3  132   43   21   12    7    4    2    -    -
  DG B1  172   83   50   33   21   12    -    -    -     TW C4  145   40   19   10    6    3    2    -    -
  DG B2  194   96   61   42   30   22   15    9    3     TW D1   86   19    -    -    -    -    -    -    -
  DG B3  237  119   77   54   40   30   23   16   10     TW D2  153   65   35   21   13    8    5    3    1
  DG C1  209  103   64   43   30   21   13    -    -     TW D3  154   70   39   24   15    9    6    3    2
```

570 / APPENDIX 21

PENDLETON, OREGON								CONTINUED		PORTLAND, OREGON									4792 DD
SSF =.10	.20	.30	.40	.50	.60	.70	.80	.90		SSF =.10	.20	.30	.40	.50	.60	.70	.80	.90	
TW D4 156	80	49	32	22	15	10	7	4	WW A1 609	57	19	-	-	-	-	-	-	-	
TW D5 149	80	50	34	23	17	12	8	5	WW A2 198	66	31	17	8	-	-	-	-	-	
TW E1 184	85	49	31	20	13	8	5	3	WW A3 162	66	35	20	11	5	-	-	-	-	
TW E2 169	83	50	32	21	14	10	6	4	WW A4 147	66	36	21	13	7	2	-	-	-	
TW E3 183	102	65	45	31	22	16	11	7	WW A5 138	67	38	23	14	8	4	1	-	-	
TW E4 165	94	61	42	30	21	15	10	6	WW A6 133	67	39	24	15	9	4	2	-	-	
TW F1 215	41	15	-	-	-	-	-	-	WW B1 140	32	-	-	-	-	-	-	-	-	
TW F2 129	44	20	10	-	-	-	-	-	WW B2 152	73	42	27	17	11	7	4	2		
TW F3 105	42	22	12	5	-	-	-	-	WW B3 178	91	55	35	23	16	10	6	3		
TW F4 86	39	21	12	7	3	-	-	-	WW B4 161	92	59	40	28	20	14	9	6		
TW G1 128	36	15	5	-	-	-	-	-	WW B5 151	89	58	40	29	21	15	10	6		
TW G2 92	36	18	9	4	-	-	-	-	WW C1 193	99	60	39	26	18	12	8	4		
TW G3 77	33	18	10	5	2	-	-	-	WW C2 168	91	57	38	26	18	13	8	5		
TW G4 63	28	16	9	5	3	1	-	-	WW C3 178	109	74	52	38	28	20	14	9		
TW H1 81	29	14	7	-	-	-	-	-	WW C4 161	99	67	47	34	25	18	13	8		
TW H2 63	27	14	8	4	1	-	-	-	TW A1 528	57	20	7	-	-	-	-	-	-	
TW H3 53	23	13	7	4	2	-	-	-	TW A2 207	61	28	15	7	-	-	-	-	-	
TW H4 42	18	10	6	3	2	1	-	-	TW A3 164	60	30	17	9	4	-	-	-	-	
TW I1 61	-	-	-	-	-	-	-	-	TW A4 138	58	31	18	11	6	3	-	-	-	
TW I2 114	53	30	18	12	7	4	2	1	TW B1 297	55	22	10	-	-	-	-	-	-	
TW I3 122	59	34	22	14	9	6	3	2	TW B2 167	56	27	15	8	4	-	-	-	-	
TW I4 128	69	43	29	20	14	9	6	4	TW B3 145	54	28	16	9	5	2	-	-	-	
TW I5 127	71	45	31	22	15	11	7	4	TW B4 134	51	27	15	9	5	3	1	-	-	
TW J1 151	74	44	28	19	12	8	5	3	TW C1 196	51	23	12	6	-	-	-	-	-	
TW J2 138	72	44	29	20	14	9	6	3	TW C2 149	50	25	14	8	5	2	-	-	-	
TW J3 153	90	59	41	29	21	15	10	6	TW C3 144	47	23	13	8	5	2	-	-	-	
TW J4 142	84	55	39	28	20	14	10	6	TW C4 156	44	21	12	7	4	2	1	-	-	
DG A1 86	-	-	-	-	-	-	-	-	TW D1 101	26	-	-	-	-	-	-	-	-	
DG A2 123	48	24	11	-	-	-	-	-	TW D2 167	71	39	24	15	10	6	3	2		
DG A3 168	72	41	26	17	11	7	3	-	TW D3 169	77	44	27	17	11	7	4	2		
DG B1 88	23	-	-	-	-	-	-	-	TW D4 168	86	53	35	24	17	11	7	4		
DG B2 127	51	27	15	6	-	-	-	-	TW D5 160	86	54	36	25	18	13	8	5		
DG B3 174	75	44	29	19	13	9	5	1	TW E1 201	93	54	34	22	15	10	6	3		
DG C1 119	43	13	-	-	-	-	-	-	TW E2 183	90	54	35	24	16	11	7	4		
DG C2 156	65	36	21	13	7	-	-	-	TW E3 197	110	70	48	34	24	17	12	7		
DG C3 202	87	52	34	23	16	11	7	4	TW E4 177	101	65	45	32	23	16	11	7		
SS A1 332	83	36	17	7	-	-	-	-	TW F1 234	47	18	7	-	-	-	-	-	-	
SS A2 307	117	62	37	23	15	9	6	3	TW F2 143	49	24	12	6	-	-	-	-	-	
SS A3 309	68	25	-	-	-	-	-	-	TW F3 116	48	25	14	8	3	-	-	-	-	
SS A4 308	113	58	34	21	13	8	5	2	TW F4 95	43	24	14	8	5	2	-	-	-	
SS A5 583	82	30	11	-	-	-	-	-	TW G1 143	41	18	8	-	-	-	-	-	-	
SS A6 302	114	60	35	22	14	8	5	3	TW G2 102	41	21	12	6	-	-	-	-	-	
SS A7 660	62	16	-	-	-	-	-	-	TW G3 85	37	20	12	7	4	-	-	-	-	
SS A8 306	109	55	31	18	11	6	4	2	TW G4 69	31	18	11	6	4	2	-	-	-	
SS B1 219	61	26	12	-	-	-	-	-	TW H1 91	33	16	9	4	-	-	-	-	-	
SS B2 241	96	51	31	20	13	8	5	3	TW H2 69	30	16	9	5	3	-	-	-	-	
SS B3 195	51	19	-	-	-	-	-	-	TW H3 59	26	14	8	5	3	1	-	-	-	
SS B4 233	91	48	29	18	11	7	4	2	TW H4 46	20	11	7	4	2	1	-	-	-	
SS B5 293	52	19	-	-	-	-	-	-	TW I1 74	19	-	-	-	-	-	-	-	-	
SS B6 224	90	48	29	18	12	7	5	3	TW I2 125	58	33	21	13	8	5	3	1		
SS B7 250	39	-	-	-	-	-	-	-	TW I3 133	65	38	24	16	10	7	4	2		
SS B8 214	84	44	26	16	10	6	3	2	TW I4 138	74	46	31	21	15	10	7	4		
SS C1 127	47	22	9	-	-	-	-	-	TW I5 136	76	48	33	23	17	12	8	5		
SS C2 148	70	40	25	16	10	6	4	2	TW J1 164	81	48	31	21	14	9	6	3		
SS C3 139	39	17	6	-	-	-	-	-	TW J2 150	78	48	32	22	15	10	7	4		
SS C4 149	63	35	21	13	8	5	3	1	TW J3 164	96	63	44	31	23	16	11	7		
SS D1 292	97	46	23	11	3	-	-	-	TW J4 152	89	59	41	29	21	15	11	7		
SS D2 262	125	72	46	30	20	13	8	5	DG A1 101	23	-	-	-	-	-	-	-	-	
SS D3 384	94	40	19	7	-	-	-	-	DG A2 135	53	27	14	-	-	-	-	-	-	
SS D4 270	123	70	43	28	18	12	8	4	DG A3 182	77	44	28	19	12	8	4	-		
SS E1 205	69	31	12	-	-	-	-	-	DG B1 103	31	-	-	-	-	-	-	-	-	
SS E2 210	99	56	35	22	14	9	5	3	DG B2 141	56	30	17	9	-	-	-	-	-	
SS E3 265	62	24	8	-	-	-	-	-	DG B3 188	80	47	31	21	15	10	6	2		
SS E4 221	94	51	31	19	12	7	4	2	DG C1 138	49	21	-	-	-	-	-	-	-	

LCR Tables / 571

PORTLAND, OREGON — CONTINUED

	SSF =	.10	.20	.30	.40	.50	.60	.70	.80	.90
DG	C2	171	70	39	24	15	8	-	-	-
DG	C3	218	94	55	37	25	18	12	8	4
SS	A1	384	99	44	22	11	5	-	-	-
SS	A2	350	134	71	42	27	17	11	7	4
SS	A3	366	86	34	14	-	-	-	-	-
SS	A4	356	132	68	40	24	15	9	5	3
SS	A5	649	100	38	17	6	-	-	-	-
SS	A6	345	131	69	41	25	16	10	6	3
SS	A7	731	80	25	-	-	-	-	-	-
SS	A8	356	128	65	37	22	14	8	4	2
SS	B1	256	73	32	16	8	-	-	-	-
SS	B2	273	109	58	35	23	15	10	6	3
SS	B3	232	63	26	11	-	-	-	-	-
SS	B4	267	105	56	33	21	13	8	5	3
SS	B5	336	64	25	10	-	-	-	-	-
SS	B6	255	103	55	33	21	14	9	5	3
SS	B7	293	51	17	-	-	-	-	-	-
SS	B8	247	97	51	30	19	12	7	4	2
SS	C1	142	54	26	14	-	-	-	-	-
SS	C2	162	76	44	28	18	12	7	4	2
SS	C3	155	45	20	10	-	-	-	-	-
SS	C4	163	69	38	23	15	10	6	3	2
SS	D1	341	116	55	30	16	8	-	-	-
SS	D2	298	142	82	52	34	23	15	10	5
SS	D3	443	112	49	24	12	4	-	-	-
SS	D4	308	140	79	49	32	21	14	9	5
SS	E1	243	84	39	19	-	-	-	-	-
SS	E2	241	113	64	40	26	17	11	6	3
SS	E3	309	75	31	14	-	-	-	-	-
SS	E4	253	108	59	36	23	14	9	5	3

REDMOND, OREGON — 6643 DD

	SSF =	.10	.20	.30	.40	.50	.60	.70	.80	.90
WW	A1	612	69	29	16	9	5	3	1	-
WW	A2	213	80	43	27	18	12	8	5	3
WW	A3	180	82	47	31	21	14	10	6	4
WW	A4	165	82	49	33	22	16	11	7	4
WW	A5	157	83	52	35	24	17	12	8	5
WW	A6	153	83	53	36	25	18	13	8	5
WW	B1	158	51	25	13	6	-	-	-	-
WW	B2	168	87	54	36	26	18	13	9	5
WW	B3	198	108	69	47	33	24	17	12	7
WW	B4	178	106	71	50	37	27	20	14	9
WW	B5	166	101	69	49	36	27	20	14	9
WW	C1	213	116	74	51	36	26	19	13	8
WW	C2	185	106	69	48	35	26	19	13	8
WW	C3	196	125	86	63	47	35	26	19	13
WW	C4	177	112	78	57	42	32	24	17	12
TW	A1	531	68	29	16	10	6	3	2	-
TW	A2	220	73	38	23	15	10	7	4	2
TW	A3	178	73	41	26	17	12	8	5	3
TW	A4	152	70	41	27	18	13	9	6	3
TW	B1	304	65	31	18	11	7	4	3	1
TW	B2	178	67	36	23	15	10	7	4	2
TW	B3	157	65	37	23	16	11	7	5	3
TW	B4	145	61	34	22	15	10	7	5	3
TW	C1	205	60	31	18	12	8	5	3	2
TW	C2	159	59	32	20	13	9	6	4	2
TW	C3	152	55	30	19	12	9	6	4	2
TW	C4	162	50	26	16	11	7	5	3	2
TW	D1	116	40	20	11	6	-	-	-	-
TW	D2	181	83	49	32	22	16	11	7	5
TW	D3	185	91	55	37	26	18	13	9	5

REDMOND, OREGON — CONTINUED

	SSF =	.10	.20	.30	.40	.50	.60	.70	.80	.90
TW	D4	183	99	63	44	31	23	16	12	7
TW	D5	173	97	63	44	32	23	17	12	8
TW	E1	218	108	66	44	31	22	16	11	7
TW	E2	199	104	65	44	31	23	16	11	7
TW	E3	215	125	83	59	43	32	23	17	11
TW	E4	192	114	76	54	40	29	22	16	10
TW	F1	243	58	27	16	10	6	4	2	-
TW	F2	155	61	33	21	14	9	6	4	2
TW	F3	129	59	34	22	15	10	7	4	3
TW	F4	107	53	32	21	15	10	7	5	3
TW	G1	153	51	26	16	10	7	4	2	1
TW	G2	113	50	29	19	12	8	6	4	2
TW	G3	96	46	28	18	12	9	6	4	2
TW	G4	77	39	23	15	11	7	5	3	2
TW	H1	100	41	23	15	10	6	4	3	1
TW	H2	78	37	22	14	10	7	5	3	2
TW	H3	65	32	19	12	9	6	4	3	2
TW	H4	51	24	15	10	7	5	3	2	1
TW	I1	88	33	17	9	5	-	-	-	-
TW	I2	137	69	42	28	20	14	10	7	4
TW	I3	147	77	48	32	23	16	11	8	5
TW	I4	152	86	56	39	28	21	15	11	7
TW	I5	149	87	57	41	30	22	16	11	7
TW	J1	180	95	60	40	29	21	15	10	6
TW	J2	164	90	58	40	29	21	15	11	7
TW	J3	180	110	74	53	39	29	22	16	10
TW	J4	165	101	69	49	37	27	20	15	10
DG	A1	115	43	16	-	-	-	-	-	-
DG	A2	143	62	36	22	14	7	-	-	-
DG	A3	187	85	52	35	25	18	12	8	-
DG	B1	116	46	24	10	-	-	-	-	-
DG	B2	148	65	38	25	17	11	6	-	-
DG	B3	192	88	54	37	27	20	15	10	6
DG	C1	149	62	35	21	12	-	-	-	-
DG	C2	177	80	48	32	23	16	11	6	-
DG	C3	222	102	63	44	32	24	18	13	8
SS	A1	399	116	57	33	20	13	8	5	2
SS	A2	369	152	85	54	36	25	17	11	7
SS	A3	378	102	48	26	15	9	5	2	-
SS	A4	374	150	83	51	34	23	15	10	6
SS	A5	657	115	51	28	16	10	6	3	1
SS	A6	365	150	84	53	35	24	16	11	6
SS	A7	730	94	38	19	10	4	-	-	-
SS	A8	374	146	80	49	32	21	14	9	5
SS	B1	266	86	43	25	16	10	6	4	2
SS	B2	286	122	70	45	30	21	14	10	6
SS	B3	243	76	38	21	13	7	4	2	-
SS	B4	280	118	67	42	28	19	13	9	5
SS	B5	341	75	34	19	11	7	4	2	-
SS	B6	268	116	67	43	29	20	14	9	5
SS	B7	297	62	27	14	7	3	-	-	-
SS	B8	260	111	62	39	26	18	12	8	5
SS	C1	158	68	39	24	16	11	7	4	2
SS	C2	178	90	55	37	26	19	13	9	6
SS	C3	165	55	29	17	11	7	5	3	1
SS	C4	176	81	48	31	22	15	11	7	4
SS	D1	364	138	73	44	28	18	12	7	4
SS	D2	324	165	100	67	46	32	22	15	9
SS	D3	460	131	64	37	23	14	9	5	3
SS	D4	332	162	97	63	43	30	21	14	9
SS	E1	262	103	55	33	21	13	8	4	2
SS	E2	262	132	80	53	36	25	17	12	7
SS	E3	319	89	43	24	15	9	5	3	1
SS	E4	271	124	73	47	32	22	15	10	6

SALEM, OREGON									4852 DD
SSF =	.10	.20	.30	.40	.50	.60	.70	.80	.90
WW A1	640	61	22	9	-	-	-	-	-
WW A2	207	70	35	20	11	5	-	-	-
WW A3	170	71	38	23	14	8	4	1	-
WW A4	154	71	40	25	15	9	5	2	-
WW A5	145	71	42	26	17	11	6	3	1
WW A6	140	71	43	27	18	11	7	4	2
WW B1	149	38	-	-	-	-	-	-	-
WW B2	158	77	46	29	20	13	8	5	3
WW B3	186	96	58	38	26	18	12	8	4
WW B4	167	96	62	43	30	22	15	10	6
WW B5	156	92	61	43	31	22	16	11	7
WW C1	200	104	64	42	29	20	14	9	5
WW C2	174	95	60	41	29	20	14	9	6
WW C3	184	114	77	55	40	30	22	15	10
WW C4	167	103	70	50	37	27	20	14	9
TW A1	554	60	23	10	-	-	-	-	-
TW A2	216	65	31	17	10	5	-	-	-
TW A3	171	64	33	20	12	7	3	1	-
TW A4	144	61	34	20	13	8	4	2	-
TW B1	310	58	24	12	6	-	-	-	-
TW B2	174	59	30	17	10	6	3	-	-
TW B3	151	58	30	18	11	7	4	2	-
TW B4	140	54	29	17	11	7	4	2	-
TW C1	205	54	25	14	8	4	2	-	-
TW C2	156	53	27	16	10	6	3	2	-
TW C3	150	50	25	15	9	6	3	2	-
TW C4	162	46	22	13	8	5	3	2	-
TW D1	108	31	10	-	-	-	-	-	-
TW D2	173	74	42	26	17	11	7	4	2
TW D3	176	81	47	30	20	13	8	5	3
TW D4	174	90	56	37	26	18	13	8	5
TW D5	165	89	56	38	27	19	14	9	6
TW E1	208	98	57	37	25	17	11	7	4
TW E2	190	94	57	37	26	18	12	8	5
TW E3	204	115	74	51	36	26	19	13	8
TW E4	182	105	68	47	34	25	18	12	8
TW F1	245	50	21	10	-	-	-	-	-
TW F2	149	53	27	15	8	4	-	-	-
TW F3	122	51	28	16	10	6	3	-	-
TW F4	99	46	26	16	10	6	3	2	-
TW G1	149	44	20	11	5	-	-	-	-
TW G2	107	43	23	14	8	4	2	-	-
TW G3	89	40	22	14	8	5	3	1	-
TW G4	72	34	19	12	8	5	3	1	-
TW H1	95	36	18	10	6	3	-	-	-
TW H2	73	32	18	11	7	4	2	-	-
TW H3	61	28	16	10	6	4	2	1	-
TW H4	48	21	12	8	5	3	2	1	-
TW I1	79	24	-	-	-	-	-	-	-
TW I2	130	61	36	23	15	10	6	4	2
TW I3	139	68	41	26	18	12	8	5	3
TW I4	143	78	49	33	23	16	12	8	5
TW I5	141	79	51	35	25	18	13	9	5
TW J1	171	85	51	34	23	16	11	7	4
TW J2	155	82	51	34	24	17	11	8	4
TW J3	170	100	66	46	33	24	18	12	8
TW J4	156	93	62	43	31	23	17	12	7
DG A1	107	30	-	-	-	-	-	-	-
DG A2	140	56	29	16	8	-	-	-	-
DG A3	186	79	46	30	20	14	9	5	-
DG B1	108	36	-	-	-	-	-	-	-
DG B2	145	59	33	20	11	5	-	-	-
DG B3	193	82	49	33	23	16	11	7	3
DG C1	143	53	26	-	-	-	-	-	-

SALEM, OREGON									CONTINUED
SSF =	.10	.20	.30	.40	.50	.60	.70	.80	.90
DG C2	176	72	42	26	17	10	5	-	-
DG C3	224	96	57	39	27	20	14	9	5
SS A1	402	106	48	26	14	7	3	-	-
SS A2	364	141	75	46	29	19	13	8	4
SS A3	385	92	39	19	8	-	-	-	-
SS A4	371	139	73	43	27	18	11	7	4
SS A5	683	106	42	20	10	-	-	-	-
SS A6	360	138	73	44	28	18	12	7	4
SS A7	773	87	30	11	-	-	-	-	-
SS A8	371	135	70	41	25	16	10	6	3
SS B1	268	78	36	19	10	5	-	-	-
SS B2	283	114	62	38	25	17	11	7	4
SS B3	244	68	30	15	6	-	-	-	-
SS B4	278	110	59	36	23	15	10	6	3
SS B5	352	69	28	13	6	-	-	-	-
SS B6	265	108	59	36	23	15	10	6	3
SS B7	309	55	21	7	-	-	-	-	-
SS B8	257	102	55	33	21	14	9	5	3
SS C1	149	58	30	17	9	-	-	-	-
SS C2	168	80	47	30	20	14	9	5	3
SS C3	162	48	22	12	6	-	-	-	-
SS C4	169	73	41	26	17	11	7	4	2
SS D1	357	124	61	34	20	11	5	2	-
SS D2	311	150	88	57	38	26	17	11	6
SS D3	465	119	54	28	16	8	3	-	-
SS D4	321	148	85	54	36	24	16	10	6
SS E1	256	91	44	24	12	-	-	-	-
SS E2	251	120	69	44	29	19	13	8	4
SS E3	324	81	35	18	8	-	-	-	-
SS E4	264	114	63	39	25	17	11	6	3

ALLENTOWN, PENNSYLVANIA									5827 DD
SSF =	.10	.20	.30	.40	.50	.60	.70	.80	.90
WW A1	331	26	8	-	-	-	-	-	-
WW A2	113	39	20	11	7	4	2	1	-
WW A3	96	41	23	14	9	6	4	2	1
WW A4	89	42	25	16	10	7	5	3	2
WW A5	86	44	26	17	12	8	5	4	2
WW A6	84	44	27	18	12	9	6	4	2
WW B1	53	-	-	-	-	-	-	-	-
WW B2	99	50	31	20	14	10	7	5	3
WW B3	119	64	40	27	19	14	10	7	4
WW B4	115	68	45	32	24	17	13	9	6
WW B5	110	67	46	33	24	18	13	10	7
WW C1	131	71	45	31	22	16	11	8	5
WW C2	117	67	44	30	22	16	12	8	6
WW C3	133	85	59	43	32	24	18	13	9
WW C4	121	77	54	39	29	22	17	12	8
TW A1	293	28	10	3	-	-	-	-	-
TW A2	118	36	17	10	6	4	2	1	-
TW A3	97	37	20	12	8	5	3	2	1
TW A4	84	37	21	13	9	6	4	3	1
TW B1	166	30	13	6	3	-	-	-	-
TW B2	98	35	18	11	7	4	3	2	-
TW B3	87	35	19	12	7	5	3	2	1
TW B4	82	33	18	11	7	5	3	2	1
TW C1	114	31	15	8	5	3	2	1	-
TW C2	90	31	16	10	6	4	3	2	1
TW C3	88	30	16	10	6	4	3	2	1
TW C4	96	28	14	8	5	4	2	2	-
TW D1	41	-	-	-	-	-	-	-	-
TW D2	107	48	28	18	12	8	6	4	2
TW D3	109	53	31	21	14	10	7	5	3

ALLENTOWN, PENNSYLVANIA — CONTINUED

SSF	=.10	.20	.30	.40	.50	.60	.70	.80	.90
TW D4	117	63	40	28	20	14	10	7	5
TW D5	115	64	42	29	21	16	11	8	5
TW E1	133	65	39	26	18	13	9	6	4
TW E2	125	65	40	27	19	14	10	7	5
TW E3	142	83	55	39	28	21	16	11	7
TW E4	129	77	51	37	27	20	15	11	7
TW F1	123	24	9	4	-	-	-	-	-
TW F2	81	29	15	9	5	3	2	-	-
TW F3	69	30	16	10	6	4	3	2	-
TW F4	58	28	16	10	7	5	3	2	1
TW G1	77	23	10	5	3	-	-	-	-
TW G2	60	25	14	8	5	3	2	1	-
TW G3	52	24	14	9	6	4	3	2	-
TW G4	43	21	12	8	5	4	2	2	-
TW H1	51	20	10	6	4	2	1	-	-
TW H2	42	19	11	7	4	3	2	1	-
TW H3	36	17	10	6	4	3	2	1	-
TW H4	29	13	8	5	3	2	2	1	-
TW I1	27	-	-	-	-	-	-	-	-
TW I2	81	39	24	16	11	8	5	4	2
TW I3	88	45	28	18	13	9	6	4	3
TW I4	97	55	35	25	18	13	10	7	4
TW I5	98	57	38	27	19	14	11	8	5
TW J1	110	57	36	24	17	12	9	6	4
TW J2	104	57	36	25	18	13	9	7	4
TW J3	120	73	50	36	26	20	15	11	7
TW J4	112	69	47	34	25	19	14	10	7
DG A1	39	-	-	-	-	-	-	-	-
DG A2	81	33	17	8	-	-	-	-	-
DG A3	121	54	33	22	15	11	7	4	-
DG B1	40	-	-	-	-	-	-	-	-
DG B2	83	35	19	12	6	-	-	-	-
DG B3	124	57	35	24	17	13	9	6	3
DG C1	67	22	-	-	-	-	-	-	-
DG C2	104	46	26	17	11	7	3	-	-
DG C3	145	67	41	28	21	15	12	8	5
SS A1	221	60	29	16	9	6	3	2	-
SS A2	226	92	52	33	22	15	11	7	4
SS A3	197	48	20	10	4	-	-	-	-
SS A4	224	89	49	30	20	14	9	6	4
SS A5	369	56	23	11	6	3	-	-	-
SS A6	221	90	50	31	21	14	10	7	4
SS A7	394	40	12	-	-	-	-	-	-
SS A8	220	85	46	28	18	12	8	6	3
SS B1	145	43	21	11	7	4	2	1	-
SS B2	178	75	43	28	19	13	9	6	4
SS B3	124	35	15	7	3	-	-	-	-
SS B4	171	72	40	26	17	12	8	6	3
SS B5	183	35	14	7	3	-	-	-	-
SS B6	165	71	40	26	17	12	9	6	4
SS B7	148	24	7	-	-	-	-	-	-
SS B8	156	65	37	23	15	11	7	5	3
SS C1	80	32	16	9	6	3	1	-	-
SS C2	106	53	32	21	15	10	7	5	3
SS C3	85	25	12	6	3	1	-	-	-
SS C4	105	47	27	18	12	8	6	4	2
SS D1	200	73	37	22	14	9	5	3	2
SS D2	200	102	62	41	29	21	15	10	6
SS D3	254	68	32	17	10	6	4	2	-
SS D4	204	99	60	39	27	19	14	9	6
SS E1	136	50	25	14	8	4	2	-	-
SS E2	158	79	48	31	22	15	11	7	5
SS E3	169	43	19	9	5	2	-	-	-
SS E4	163	74	43	28	19	13	9	6	4

ERIE, PENNSYLVANIA 6851 DD

SSF	=.10	.20	.30	.40	.50	.60	.70	.80	.90
WW A1	195	-	-	-	-	-	-	-	-
WW A2	54	-	-	-	-	-	-	-	-
WW A3	45	-	-	-	-	-	-	-	-
WW A4	42	-	-	-	-	-	-	-	-
WW A5	40	-	-	-	-	-	-	-	-
WW A6	39	-	-	-	-	-	-	-	-
WW B1	-	-	-	-	-	-	-	-	-
WW B2	59	25	12	-	-	-	-	-	-
WW B3	76	35	19	11	6	-	-	-	-
WW B4	80	44	28	19	13	9	6	4	2
WW B5	80	46	30	21	15	10	7	5	3
WW C1	87	42	24	15	9	5	3	1	-
WW C2	80	42	26	17	11	7	5	3	2
WW C3	99	61	41	29	21	15	11	8	5
WW C4	91	56	37	27	19	14	10	7	5
TW A1	180	-	-	-	-	-	-	-	-
TW A2	63	-	-	-	-	-	-	-	-
TW A3	50	-	-	-	-	-	-	-	-
TW A4	44	-	-	-	-	-	-	-	-
TW B1	98	-	-	-	-	-	-	-	-
TW B2	54	-	-	-	-	-	-	-	-
TW B3	49	10	-	-	-	-	-	-	-
TW B4	48	13	-	-	-	-	-	-	-
TW C1	68	-	-	-	-	-	-	-	-
TW C2	55	12	-	-	-	-	-	-	-
TW C3	55	13	-	-	-	-	-	-	-
TW C4	65	14	4	-	-	-	-	-	-
TW D1	-	-	-	-	-	-	-	-	-
TW D2	67	24	11	-	-	-	-	-	-
TW D3	68	27	13	5	-	-	-	-	-
TW D4	82	41	24	15	10	7	5	3	2
TW D5	84	44	27	18	13	9	6	4	3
TW E1	89	38	20	11	6	3	-	-	-
TW E2	86	40	23	14	9	6	4	2	1
TW E3	105	58	36	25	17	12	9	6	4
TW E4	97	54	35	24	17	12	9	6	4
TW F1	57	-	-	-	-	-	-	-	-
TW F2	35	-	-	-	-	-	-	-	-
TW F3	31	-	-	-	-	-	-	-	-
TW F4	27	-	-	-	-	-	-	-	-
TW G1	31	-	-	-	-	-	-	-	-
TW G2	26	-	-	-	-	-	-	-	-
TW G3	24	-	-	-	-	-	-	-	-
TW G4	22	5	-	-	-	-	-	-	-
TW H1	20	-	-	-	-	-	-	-	-
TW H2	19	-	-	-	-	-	-	-	-
TW H3	18	-	-	-	-	-	-	-	-
TW H4	16	5	-	-	-	-	-	-	-
TW I1	-	-	-	-	-	-	-	-	-
TW I2	48	19	9	-	-	-	-	-	-
TW I3	55	24	12	6	-	-	-	-	-
TW I4	68	35	21	14	9	6	4	3	2
TW I5	71	39	24	16	11	8	6	4	2
TW J1	73	34	19	11	6	4	2	-	-
TW J2	71	36	21	13	9	6	4	2	1
TW J3	89	52	34	24	17	12	9	6	4
TW J4	84	49	32	22	16	12	8	6	4
DG A1	-	-	-	-	-	-	-	-	-
DG A2	49	-	-	-	-	-	-	-	-
DG A3	92	37	20	12	7	4	-	-	-
DG B1	-	-	-	-	-	-	-	-	-
DG B2	50	13	-	-	-	-	-	-	-
DG B3	97	40	23	14	9	6	4	-	-
DG C1	-	-	-	-	-	-	-	-	-

ERIE, PENNSYLVANIA — CONTINUED

SSF =	.10	.20	.30	.40	.50	.60	.70	.80	.90
DG C2	70	25	10	-	-	-	-	-	-
DG C3	115	48	27	18	12	8	5	3	-
SS A1	168	32	-	-	-	-	-	-	-
SS A2	179	65	33	19	11	7	4	2	1
SS A3	143	-	-	-	-	-	-	-	-
SS A4	178	61	30	16	9	5	2	-	-
SS A5	297	24	-	-	-	-	-	-	-
SS A6	173	62	31	17	10	6	3	1	-
SS A7	319	-	-	-	-	-	-	-	-
SS A8	172	57	26	13	6	-	-	-	-
SS B1	101	16	-	-	-	-	-	-	-
SS B2	140	53	28	16	10	6	4	2	1
SS B3	79	-	-	-	-	-	-	-	-
SS B4	134	49	25	14	8	5	3	1	-
SS B5	133	-	-	-	-	-	-	-	-
SS B6	127	48	25	14	8	5	3	1	-
SS B7	100	-	-	-	-	-	-	-	-
SS B8	119	43	21	11	6	3	-	-	-
SS C1	-	-	-	-	-	-	-	-	-
SS C2	66	28	14	7	-	-	-	-	-
SS C3	37	-	-	-	-	-	-	-	-
SS C4	67	25	12	5	-	-	-	-	-
SS D1	142	37	-	-	-	-	-	-	-
SS D2	153	71	40	25	16	10	7	4	2
SS D3	193	34	-	-	-	-	-	-	-
SS D4	158	70	38	23	15	9	6	4	2
SS E1	80	-	-	-	-	-	-	-	-
SS E2	114	50	27	15	9	5	2	-	-
SS E3	116	-	-	-	-	-	-	-	-
SS E4	120	48	24	13	7	4	1	-	-

HARRISBURG, PENNSYLVANIA — 5224 DD

SSF =	.10	.20	.30	.40	.50	.60	.70	.80	.90
WW A1	340	32	11	4	-	-	-	-	-
WW A2	123	44	23	14	9	6	4	2	1
WW A3	106	47	26	17	11	7	5	3	2
WW A4	98	48	28	18	12	9	6	4	2
WW A5	94	49	30	20	14	10	7	5	3
WW A6	92	50	31	21	14	10	7	5	3
WW B1	66	12	-	-	-	-	-	-	-
WW B2	107	55	34	23	16	11	8	6	4
WW B3	128	69	44	30	21	15	11	8	5
WW B4	122	73	49	35	25	19	14	10	7
WW B5	117	71	49	35	26	19	15	11	7
WW C1	140	76	49	33	24	17	13	9	6
WW C2	125	72	47	33	24	18	13	9	6
WW C3	140	89	62	45	34	26	20	14	10
WW C4	127	81	57	42	31	24	18	13	9
TW A1	302	33	13	6	3	-	-	-	-
TW A2	129	41	20	12	7	5	3	2	1
TW A3	106	42	23	14	9	6	4	3	2
TW A4	92	41	24	15	10	7	5	3	2
TW B1	176	34	15	8	5	3	1	-	-
TW B2	107	39	20	12	8	5	4	2	1
TW B3	95	38	21	13	9	6	4	3	2
TW B4	89	36	20	13	9	6	4	3	2
TW C1	122	34	17	10	6	4	3	2	-
TW C2	97	35	19	11	7	5	3	2	1
TW C3	94	33	18	11	7	5	3	2	1
TW C4	102	31	16	10	6	4	3	2	1
TW D1	50	12	-	-	-	-	-	-	-
TW D2	115	52	30	20	14	10	7	5	3
TW D3	117	57	35	23	16	11	8	6	3

HARRISBURG, PENNSYLVANIA — CONTINUED

SSF =	.10	.20	.30	.40	.50	.60	.70	.80	.90
TW D4	124	67	43	30	21	16	11	8	5
TW D5	121	68	44	31	23	17	12	9	6
TW E1	142	70	43	28	20	14	10	7	5
TW E2	133	70	43	30	21	15	11	8	5
TW E3	150	88	58	41	30	23	17	12	8
TW E4	137	81	54	39	29	21	16	12	8
TW F1	134	28	12	6	3	-	-	-	-
TW F2	89	33	18	10	7	4	3	2	-
TW F3	76	34	19	12	8	5	4	2	1
TW F4	64	31	19	12	8	6	4	3	2
TW G1	85	26	13	7	4	2	1	-	-
TW G2	66	28	16	10	6	4	3	2	1
TW G3	57	27	16	10	7	5	3	2	1
TW G4	47	23	14	9	6	4	3	2	1
TW H1	57	23	12	7	5	3	2	1	-
TW H2	46	21	12	8	5	4	2	2	-
TW H3	39	19	11	7	5	3	2	2	-
TW H4	31	15	9	6	4	3	2	1	-
TW I1	35	7	-	-	-	-	-	-	-
TW I2	87	43	26	17	12	9	6	4	3
TW I3	94	49	30	20	14	10	7	5	3
TW I4	103	58	38	27	19	14	10	8	5
TW I5	104	61	40	29	21	16	12	8	6
TW J1	118	62	39	26	19	13	10	7	4
TW J2	110	61	39	27	19	14	10	7	5
TW J3	127	78	53	38	28	21	16	12	8
TW J4	118	73	50	36	27	20	15	11	7
DG A1	49	-	-	-	-	-	-	-	-
DG A2	88	37	20	11	-	-	-	-	-
DG A3	127	58	35	24	17	12	8	5	-
DG B1	50	-	-	-	-	-	-	-	-
DG B2	90	39	22	14	8	3	-	-	-
DG B3	131	60	37	25	18	14	10	7	3
DG C1	76	28	12	-	-	-	-	-	-
DG C2	112	50	29	19	13	9	5	-	-
DG C3	152	71	44	30	22	17	13	9	5
SS A1	231	65	31	18	11	7	5	3	1
SS A2	235	97	55	35	24	17	12	8	5
SS A3	208	52	23	12	7	3	1	-	-
SS A4	234	94	52	33	22	15	10	7	4
SS A5	373	61	26	13	8	4	2	1	-
SS A6	230	95	53	34	23	16	11	8	5
SS A7	391	44	15	6	-	-	-	-	-
SS A8	230	90	49	31	20	14	9	6	4
SS B1	153	47	23	13	8	5	3	2	1
SS B2	186	80	46	29	20	14	10	7	4
SS B3	133	39	18	10	5	3	1	-	-
SS B4	179	76	43	28	19	13	9	6	4
SS B5	191	39	17	8	5	2	1	-	-
SS B6	173	75	43	28	19	13	9	6	4
SS B7	156	28	10	3	-	-	-	-	-
SS B8	164	69	39	25	17	12	8	6	3
SS C1	89	36	20	12	7	5	3	2	-
SS C2	114	57	35	23	16	12	8	6	4
SS C3	93	29	14	8	5	3	2	-	-
SS C4	113	51	30	20	13	9	7	5	3
SS D1	212	78	41	25	16	10	7	4	2
SS D2	209	107	66	44	31	22	16	11	7
SS D3	266	73	35	20	12	8	5	3	2
SS D4	214	105	63	42	29	21	15	10	7
SS E1	146	55	28	16	10	6	4	2	1
SS E2	166	84	51	34	24	17	12	8	5
SS E3	179	47	22	12	7	4	2	-	-
SS E4	171	79	46	30	20	14	10	7	4

PITTSBURGH, PENNSYLVANIA — CONTINUED

SSF =	.10	.20	.30	.40	.50	.60	.70	.80	.90
TW D4	94	49	30	20	14	10	7	5	3
TW D5	95	51	33	22	16	11	8	6	4
TW E1	104	48	27	17	11	7	5	3	2
TW E2	100	49	29	19	13	9	6	4	3
TW E3	117	67	43	30	21	15	11	8	5
TW E4	108	62	41	29	21	15	11	8	5
TW F1	83	-	-	-	-	-	-	-	-
TW F2	54	13	-	-	-	-	-	-	-
TW F3	46	16	-	-	-	-	-	-	-
TW F4	39	16	7	-	-	-	-	-	-
TW G1	50	-	-	-	-	-	-	-	-
TW G2	39	13	-	-	-	-	-	-	-
TW G3	35	13	6	-	-	-	-	-	-
TW G4	30	12	6	3	-	-	-	-	-
TW H1	33	8	-	-	-	-	-	-	-
TW H2	28	10	4	-	-	-	-	-	-
TW H3	25	10	5	2	-	-	-	-	-
TW H4	21	9	4	2	1	-	-	-	-
TW I1	-	-	-	-	-	-	-	-	-
TW I2	60	27	15	9	5	3	2	1	-
TW I3	67	32	18	11	7	5	3	2	1
TW I4	78	42	27	18	13	9	6	4	3
TW I5	80	45	29	20	14	10	8	5	3
TW J1	86	43	25	16	11	7	5	3	2
TW J2	83	43	27	18	12	9	6	4	3
TW J3	100	60	40	28	20	15	11	8	5
TW J4	93	56	37	26	19	14	10	7	5
DG A1	-	-	-	-	-	-	-	-	-
DG A2	60	20	-	-	-	-	-	-	-
DG A3	101	43	25	16	10	7	4	-	-
DG B1	-	-	-	-	-	-	-	-	-
DG B2	62	22	9	-	-	-	-	-	-
DG B3	105	46	27	18	12	8	6	3	-
DG C1	33	-	-	-	-	-	-	-	-
DG C2	82	33	17	9	-	-	-	-	-
DG C3	124	55	32	21	15	11	8	5	2
SS A1	183	43	16	5	-	-	-	-	-
SS A2	193	74	40	24	15	10	7	4	3
SS A3	159	28	-	-	-	-	-	-	-
SS A4	191	71	37	22	13	9	6	3	2
SS A5	321	38	-	-	-	-	-	-	-
SS A6	187	71	38	22	14	9	6	4	2
SS A7	349	-	-	-	-	-	-	-	-
SS A8	187	67	33	19	12	7	4	3	1
SS B1	115	28	9	-	-	-	-	-	-
SS B2	152	61	33	20	13	9	6	4	2
SS B3	94	17	-	-	-	-	-	-	-
SS B4	145	57	31	18	12	8	5	3	2
SS B5	149	20	-	-	-	-	-	-	-
SS B6	139	56	31	19	12	8	5	3	2
SS B7	115	-	-	-	-	-	-	-	-
SS B8	131	51	27	16	10	6	4	2	1
SS C1	49	-	-	-	-	-	-	-	-
SS C2	81	37	21	13	8	5	3	2	1
SS C3	56	-	-	-	-	-	-	-	-
SS C4	81	33	18	11	6	4	2	1	-
SS D1	160	51	22	9	-	-	-	-	-
SS D2	168	82	48	31	21	14	10	7	4
SS D3	210	47	17	-	-	-	-	-	-
SS D4	172	80	46	29	19	13	9	6	4
SS E1	100	28	-	-	-	-	-	-	-
SS E2	129	61	35	22	14	9	6	4	2
SS E3	133	24	-	-	-	-	-	-	-
SS E4	134	57	31	19	12	8	5	3	2

WILKES-BARRE-SCRANTON, PENNSYLVANIA   6277 DD

SSF =	.10	.20	.30	.40	.50	.60	.70	.80	.90
WW A1	264	-	-	-	-	-	-	-	-
WW A2	83	24	8	-	-	-	-	-	-
WW A3	71	27	12	6	-	-	-	-	-
WW A4	66	28	14	8	4	-	-	-	-
WW A5	63	30	16	9	5	2	-	-	-
WW A6	62	30	17	10	6	3	1	-	-
WW B1	-	-	-	-	-	-	-	-	-
WW B2	78	38	22	14	9	6	4	3	1
WW B3	96	50	30	20	13	9	6	4	2
WW B4	96	56	37	25	18	13	10	7	4
WW B5	94	56	38	27	19	14	11	8	5
WW C1	107	56	34	23	16	11	8	5	3
WW C2	98	54	35	24	17	12	9	6	4
WW C3	115	72	49	36	26	20	15	11	7
WW C4	105	66	45	33	24	18	14	10	7
TW A1	237	10	-	-	-	-	-	-	-
TW A2	89	22	8	-	-	-	-	-	-
TW A3	73	25	11	5	-	-	-	-	-
TW A4	64	25	13	7	4	1	-	-	-
TW B1	129	17	-	-	-	-	-	-	-
TW B2	75	23	10	5	-	-	-	-	-
TW B3	67	24	12	6	3	-	-	-	-
TW B4	64	24	12	7	4	2	-	-	-
TW C1	89	20	8	2	-	-	-	-	-
TW C2	71	22	11	6	3	1	-	-	-
TW C3	70	22	11	6	3	2	-	-	-
TW C4	79	21	10	5	3	2	1	-	-
TW D1	-	-	-	-	-	-	-	-	-
TW D2	86	36	20	12	8	5	3	2	1
TW D3	87	40	23	14	9	6	4	2	1
TW D4	98	51	32	22	15	11	8	5	3
TW D5	98	54	34	24	17	12	9	6	4
TW E1	109	51	30	19	13	9	6	4	2
TW E2	104	52	32	21	14	10	7	5	3
TW E3	121	70	45	32	23	17	12	9	6
TW E4	111	65	43	30	22	16	12	9	6
TW F1	89	-	-	-	-	-	-	-	-
TW F2	58	17	6	-	-	-	-	-	-
TW F3	50	19	9	4	-	-	-	-	-
TW F4	43	19	10	5	3	-	-	-	-
TW G1	55	11	-	-	-	-	-	-	-
TW G2	43	15	7	-	-	-	-	-	-
TW G3	38	16	8	4	2	-	-	-	-
TW G4	32	14	8	4	3	1	-	-	-
TW H1	37	11	3	-	-	-	-	-	-
TW H2	31	12	6	3	-	-	-	-	-
TW H3	27	12	6	3	2	-	-	-	-
TW H4	22	10	5	3	2	1	-	-	-
TW I1	-	-	-	-	-	-	-	-	-
TW I2	64	30	17	10	7	5	3	2	-
TW I3	70	35	20	13	9	6	4	3	1
TW I4	81	45	28	19	14	10	7	5	3
TW I5	83	48	31	22	15	11	8	6	4
TW J1	90	45	27	18	12	8	6	4	2
TW J2	86	46	29	19	13	10	7	5	3
TW J3	103	62	42	30	22	16	12	9	6
TW J4	96	58	39	28	20	15	11	8	5
DG A1	-	-	-	-	-	-	-	-	-
DG A2	64	23	8	-	-	-	-	-	-
DG A3	104	45	26	17	12	8	5	-	-
DG B1	-	-	-	-	-	-	-	-	-
DG B2	65	25	12	-	-	-	-	-	-
DG B3	108	48	28	19	13	9	7	4	-
DG C1	41	-	-	-	-	-	-	-	-

576 / APPENDIX 21

PHILADELPHIA, PENNSYLVANIA — 4865 DD

SSF =	.10	.20	.30	.40	.50	.60	.70	.80	.90
WW A1	415	43	17	9	5	3	1	-	-
WW A2	147	54	29	18	12	8	5	4	2
WW A3	125	57	33	21	14	10	7	5	3
WW A4	116	57	35	23	16	11	8	5	3
WW A5	111	59	37	25	17	12	9	6	4
WW A6	109	59	38	26	18	13	9	7	4
WW B1	91	26	10	-	-	-	-	-	-
WW B2	123	64	40	27	19	14	10	7	5
WW B3	146	80	51	35	25	18	13	10	6
WW B4	137	82	55	39	29	22	16	12	8
WW B5	130	80	55	39	29	22	16	12	8
WW C1	159	87	56	39	28	20	15	11	7
WW C2	141	81	54	38	27	20	15	11	7
WW C3	155	99	69	51	38	29	22	16	11
WW C4	141	90	63	46	35	26	20	15	10
TW A1	365	43	18	10	6	3	2	1	-
TW A2	152	50	26	16	10	7	5	3	2
TW A3	125	51	28	18	12	8	6	4	2
TW A4	107	49	29	19	13	9	6	4	3
TW B1	210	43	20	11	7	4	3	2	1
TW B2	125	47	25	16	10	7	5	3	2
TW B3	111	46	26	17	11	8	5	4	2
TW B4	104	43	25	16	11	7	5	4	2
TW C1	144	41	21	13	8	5	4	2	1
TW C2	113	41	23	14	9	7	5	3	2
TW C3	110	39	21	13	9	6	4	3	2
TW C4	118	36	19	12	8	5	4	3	2
TW D1	68	22	10	4	-	-	-	-	-
TW D2	132	61	36	24	16	12	8	6	4
TW D3	135	67	41	27	19	14	10	7	4
TW D4	140	76	49	34	25	18	13	10	6
TW D5	135	76	50	35	26	19	14	10	7
TW E1	161	80	49	33	23	17	12	9	6
TW E2	151	79	50	34	24	18	13	9	6
TW E3	167	98	66	47	34	26	19	14	9
TW E4	151	90	61	43	32	24	18	13	9
TW F1	163	37	17	9	6	4	2	1	-
TW F2	106	41	22	14	9	6	4	3	2
TW F3	90	41	24	15	10	7	5	3	2
TW F4	75	38	23	15	10	7	5	4	2
TW G1	103	33	17	10	6	4	3	2	1
TW G2	78	35	20	13	9	6	4	3	2
TW G3	67	32	19	13	9	6	4	3	2
TW G4	55	27	17	11	8	5	4	3	2
TW H1	68	28	16	10	6	4	3	2	1
TW H2	54	26	15	10	7	5	3	2	1
TW H3	46	22	13	9	6	4	3	2	1
TW H4	36	18	10	7	5	3	2	2	1
TW I1	50	17	7	-	-	-	-	-	-
TW I2	100	50	31	21	15	11	8	5	3
TW I3	108	57	35	24	17	12	9	6	4
TW I4	116	66	43	30	22	16	12	9	6
TW I5	116	68	45	32	24	18	13	10	6
TW J1	134	71	45	31	22	16	12	8	5
TW J2	125	69	45	31	22	17	12	9	6
TW J3	141	87	59	43	32	24	18	13	9
TW J4	131	81	55	40	30	22	17	12	8
DG A1	67	16	-	-	-	-	-	-	-
DG A2	102	44	25	14	8	-	-	-	-
DG A3	142	65	40	27	19	14	10	6	-
DG B1	68	23	-	-	-	-	-	-	-
DG B2	104	46	27	17	11	7	-	-	-
DG B3	145	68	42	29	21	16	12	8	4
DG C1	95	38	20	10	-	-	-	-	-

PHILADELPHIA, PENNSYLVANIA — CONTINUED

SSF =	.10	.20	.30	.40	.50	.60	.70	.80	.90
DG C2	128	58	34	23	16	11	7	3	-
DG C3	168	79	49	34	25	19	15	11	7
SS A1	266	77	38	23	14	9	6	4	2
SS A2	263	110	63	40	28	19	14	10	6
SS A3	243	64	30	17	10	6	4	2	1
SS A4	263	107	60	38	26	18	13	9	5
SS A5	432	74	33	18	11	7	4	3	1
SS A6	259	108	61	39	27	19	13	9	6
SS A7	461	56	22	11	6	3	-	-	-
SS A8	260	103	57	36	24	17	12	8	5
SS B1	178	57	29	17	11	7	5	3	2
SS B2	209	90	52	34	23	17	12	8	5
SS B3	157	48	24	14	8	5	3	2	1
SS B4	201	86	49	32	22	15	11	8	5
SS B5	224	48	22	12	7	4	3	2	-
SS B6	194	85	49	32	22	16	11	8	5
SS B7	187	37	15	8	4	2	-	-	-
SS B8	185	79	45	29	20	14	10	7	4
SS C1	107	46	26	16	11	7	5	3	2
SS C2	131	66	41	28	20	14	10	7	5
SS C3	112	36	19	11	7	5	3	2	1
SS C4	129	59	35	23	16	11	8	6	4
SS D1	244	93	50	31	20	13	9	6	4
SS D2	234	121	75	51	36	26	19	13	8
SS D3	306	87	43	25	16	10	7	4	3
SS D4	239	118	72	48	34	24	17	12	8
SS E1	173	67	36	22	14	9	6	4	2
SS E2	188	96	59	40	28	20	14	10	6
SS E3	210	58	28	16	10	6	4	2	1
SS E4	194	90	53	35	24	17	12	8	5

PITTSBURGH, PENNSYLVANIA — 5930 DD

SSF =	.10	.20	.30	.40	.50	.60	.70	.80	.90
WW A1	256	-	-	-	-	-	-	-	-
WW A2	77	19	-	-	-	-	-	-	-
WW A3	65	22	7	-	-	-	-	-	-
WW A4	61	24	10	-	-	-	-	-	-
WW A5	58	25	12	5	-	-	-	-	-
WW A6	57	26	13	7	-	-	-	-	-
WW B1	-	-	-	-	-	-	-	-	-
WW B2	74	35	19	12	7	5	3	2	-
WW B3	91	46	27	17	11	8	5	3	2
WW B4	93	53	34	24	17	12	9	6	4
WW B5	91	54	36	25	18	13	10	7	4
WW C1	103	53	32	21	14	10	7	4	3
WW C2	93	51	32	22	15	11	8	5	3
WW C3	111	69	47	34	25	19	14	10	7
WW C4	101	63	43	31	23	17	13	9	6
TW A1	230	-	-	-	-	-	-	-	-
TW A2	84	18	-	-	-	-	-	-	-
TW A3	68	21	7	-	-	-	-	-	-
TW A4	59	22	10	4	-	-	-	-	-
TW B1	123	12	-	-	-	-	-	-	-
TW B2	71	20	7	-	-	-	-	-	-
TW B3	63	21	9	3	-	-	-	-	-
TW B4	61	21	10	5	2	-	-	-	-
TW C1	85	18	4	-	-	-	-	-	-
TW C2	67	20	9	4	-	-	-	-	-
TW C3	67	20	9	4	2	-	-	-	-
TW C4	76	19	9	4	2	-	-	-	-
TW D1	-	-	-	-	-	-	-	-	-
TW D2	81	33	18	10	6	4	2	1	-
TW D3	83	37	20	12	7	5	3	2	-

WILKES-BARRE-SCRANTON, PENNSYLVANIA   CONTINUED

	SSF =	.10	.20	.30	.40	.50	.60	.70	.80	.90
DG	C2	85	35	19	11	6	-	-	-	-
DG	C3	127	57	34	23	16	12	9	6	3
SS	A1	185	45	19	9	3	-	-	-	-
SS	A2	195	77	42	26	17	11	8	5	3
SS	A3	160	31	-	-	-	-	-	-	-
SS	A4	192	73	39	23	15	10	6	4	2
SS	A5	319	40	12	-	-	-	-	-	-
SS	A6	189	74	40	24	16	10	7	4	3
SS	A7	343	21	-	-	-	-	-	-	-
SS	A8	188	69	35	21	13	8	5	3	2
SS	B1	117	31	12	4	-	-	-	-	-
SS	B2	154	63	35	22	14	10	7	4	3
SS	B3	96	21	-	-	-	-	-	-	-
SS	B4	147	59	32	20	13	9	6	4	2
SS	B5	150	22	-	-	-	-	-	-	-
SS	B6	141	58	32	20	13	9	6	4	2
SS	B7	116	-	-	-	-	-	-	-	-
SS	B8	132	53	29	17	11	7	5	3	2
SS	C1	55	16	-	-	-	-	-	-	-
SS	C2	85	40	23	15	10	7	4	3	2
SS	C3	61	13	-	-	-	-	-	-	-
SS	C4	85	36	20	12	8	5	3	2	1
SS	D1	163	54	25	13	6	-	-	-	-
SS	D2	171	84	50	33	22	16	11	7	5
SS	D3	212	50	20	9	-	-	-	-	-
SS	D4	175	82	48	31	21	14	10	7	4
SS	E1	104	32	11	-	-	-	-	-	-
SS	E2	132	63	37	23	16	11	7	5	3
SS	E3	135	28	-	-	-	-	-	-	-
SS	E4	137	60	33	21	13	9	6	4	2

PROVIDENCE, RHODE ISLAND   5972 DD

	SSF =	.10	.20	.30	.40	.50	.60	.70	.80	.90
WW	A1	337	27	9	-	-	-	-	-	-
WW	A2	114	40	20	12	7	5	3	2	-
WW	A3	98	42	24	15	10	6	4	3	1
WW	A4	91	43	25	16	11	8	5	3	2
WW	A5	87	45	27	18	12	8	6	4	2
WW	A6	85	45	28	19	13	9	6	4	3
WW	B1	55	-	-	-	-	-	-	-	-
WW	B2	100	51	31	21	15	10	7	5	3
WW	B3	120	65	41	28	20	14	10	7	5
WW	B4	116	69	46	33	24	18	13	10	6
WW	B5	111	68	46	33	25	18	14	10	7
WW	C1	132	72	45	31	22	16	12	8	5
WW	C2	119	67	44	31	22	17	12	9	6
WW	C3	134	85	59	43	33	25	19	14	9
WW	C4	122	78	54	40	30	23	17	13	9
TW	A1	298	29	10	4	-	-	-	-	-
TW	A2	120	37	18	10	6	4	2	1	-
TW	A3	98	38	21	13	8	5	4	2	1
TW	A4	85	38	22	14	9	6	4	3	2
TW	B1	169	31	13	7	4	2	-	-	-
TW	B2	100	35	18	11	7	5	3	2	1
TW	B3	88	35	19	12	8	5	4	2	1
TW	B4	83	33	19	12	8	5	4	2	1
TW	C1	116	31	15	9	5	3	2	1	-
TW	C2	92	32	17	10	7	4	3	2	1
TW	C3	89	31	16	10	6	4	3	2	1
TW	C4	98	28	14	9	6	4	3	2	1
TW	D1	42	-	-	-	-	-	-	-	-
TW	D2	108	48	28	18	13	9	6	4	3
TW	D3	110	53	32	21	15	10	7	5	3

PROVIDENCE, RHODE ISLAND   CONTINUED

	SSF =	.10	.20	.30	.40	.50	.60	.70	.80	.90
TW	D4	118	64	40	28	20	15	11	8	5
TW	D5	116	65	42	29	21	16	12	8	6
TW	E1	134	66	40	27	19	13	9	7	4
TW	E2	127	65	41	28	20	14	10	7	5
TW	E3	143	83	55	39	29	21	16	12	8
TW	E4	131	77	52	37	27	20	15	11	7
TW	F1	126	25	10	5	-	-	-	-	-
TW	F2	82	30	15	9	6	3	2	1	-
TW	F3	70	30	17	11	7	5	3	2	1
TW	F4	59	28	17	11	7	5	3	2	1
TW	G1	79	23	11	6	3	2	-	-	-
TW	G2	60	25	14	9	6	4	2	1	-
TW	G3	52	24	14	9	6	4	3	2	1
TW	G4	43	21	12	8	5	4	3	2	1
TW	H1	52	20	11	6	4	2	1	-	-
TW	H2	42	19	11	7	5	3	2	1	-
TW	H3	37	17	10	6	4	3	2	1	-
TW	H4	29	14	8	5	3	2	2	1	-
TW	I1	28	-	-	-	-	-	-	-	-
TW	I2	82	40	24	16	11	8	6	4	2
TW	I3	89	46	28	19	13	9	7	5	3
TW	I4	98	55	36	25	18	13	10	7	5
TW	I5	99	58	38	27	20	15	11	8	5
TW	J1	112	58	36	25	17	13	9	6	4
TW	J2	105	57	37	25	18	13	10	7	5
TW	J3	121	74	50	36	27	20	15	11	7
TW	J4	113	69	47	34	25	19	14	10	7
DG	A1	40	-	-	-	-	-	-	-	-
DG	A2	82	33	18	9	-	-	-	-	-
DG	A3	122	55	33	22	16	11	7	4	-
DG	B1	41	-	-	-	-	-	-	-	-
DG	B2	83	35	20	12	7	-	-	-	-
DG	B3	126	57	35	24	17	13	9	6	3
DG	C1	67	23	-	-	-	-	-	-	-
DG	C2	105	46	27	17	12	8	4	-	-
DG	C3	147	67	41	28	21	16	12	8	5
SS	A1	224	61	29	17	10	6	4	2	1
SS	A2	228	93	52	33	23	16	11	7	5
SS	A3	200	49	21	10	5	-	-	-	-
SS	A4	226	90	50	31	21	14	10	7	4
SS	A5	374	57	24	12	6	3	1	-	-
SS	A6	223	90	51	32	22	15	10	7	4
SS	A7	401	41	13	-	-	-	-	-	-
SS	A8	223	86	47	29	19	13	9	6	3
SS	B1	147	44	21	12	7	4	2	1	-
SS	B2	180	76	44	28	19	13	9	6	4
SS	B3	126	36	16	8	4	-	-	-	-
SS	B4	173	72	41	26	18	12	9	6	4
SS	B5	186	36	15	7	3	-	-	-	-
SS	B6	166	71	41	26	18	12	9	6	4
SS	B7	151	25	8	-	-	-	-	-	-
SS	B8	158	66	37	24	16	11	8	5	3
SS	C1	81	32	17	10	6	4	2	1	-
SS	C2	107	53	32	22	15	11	8	5	3
SS	C3	86	26	12	7	4	2	-	-	-
SS	C4	106	48	28	18	12	9	6	4	3
SS	D1	202	74	38	23	14	9	6	3	2
SS	D2	201	103	63	42	30	21	15	10	7
SS	D3	258	69	32	18	11	7	4	2	1
SS	D4	206	100	60	40	28	20	14	10	6
SS	E1	138	51	26	14	8	5	2	1	-
SS	E2	159	80	48	32	22	16	11	8	5
SS	E3	172	43	19	10	5	3	-	-	-
SS	E4	164	75	44	28	19	13	9	6	4

CHARLESTON, SOUTH CAROLINA								2146 DD		CHARLESTON, SOUTH CAROLINA								CONTINUED	
SSF =	.10	.20	.30	.40	.50	.60	.70	.80	.90	SSF =	.10	.20	.30	.40	.50	.60	.70	.80	.90
WW A1	1330	160	74	44	29	20	14	10	6	DG C2	333	155	95	67	49	38	29	21	13
WW A2	434	172	98	64	44	32	23	16	11	DG C3	384	182	113	79	60	46	37	28	19
WW A3	363	173	105	70	50	36	27	19	13	SS A1	718	225	119	74	50	35	25	18	11
WW A4	333	172	108	74	53	39	29	21	14	SS A2	633	273	159	104	73	53	38	27	18
WW A5	316	173	111	78	56	42	31	23	15	SS A3	699	208	107	66	44	31	22	15	10
WW A6	306	173	113	79	58	43	32	23	16	SS A4	647	273	158	103	72	52	37	27	17
WW B1	370	136	75	47	32	23	16	12	8	SS A5	1205	227	109	65	43	30	21	15	9
WW B2	322	172	110	76	55	41	30	22	15	SS A6	631	272	158	104	72	52	38	27	18
WW B3	372	208	135	95	69	52	38	28	19	SS A7	1381	196	90	53	34	24	16	11	7
WW B4	321	194	132	95	70	53	40	30	20	SS A8	652	270	154	100	70	50	36	26	17
WW B5	293	181	124	90	67	51	39	29	20	SS B1	497	172	93	59	40	28	20	14	9
WW C1	393	220	143	100	73	55	41	30	20	SS B2	496	220	130	86	60	44	32	23	15
WW C2	337	197	131	93	69	51	39	28	19	SS B3	464	160	86	54	37	26	19	13	8
WW C3	338	217	152	112	84	64	49	37	25	SS B4	488	216	127	84	59	42	31	22	15
WW C4	304	195	137	101	76	58	44	33	23	SS B5	647	156	78	47	31	22	16	11	7
TW A1	1133	152	71	43	28	19	14	10	6	SS B6	470	212	126	83	59	43	31	22	15
TW A2	445	156	86	55	38	27	19	14	9	SS B7	578	134	66	40	26	18	13	9	6
TW A3	357	154	90	59	42	30	22	16	10	SS B8	460	205	121	80	56	40	29	21	14
TW A4	301	145	88	59	42	31	23	16	11	SS C1	330	151	90	60	42	30	22	16	11
TW B1	628	142	71	44	29	21	15	10	7	SS C2	337	176	111	76	55	41	30	22	15
TW B2	356	140	80	52	36	26	19	13	9	SS C3	342	122	67	43	29	21	15	11	7
TW B3	309	135	79	52	37	27	19	14	9	SS C4	333	158	96	65	46	34	25	18	12
TW B4	283	124	73	48	34	24	18	13	8	SS D1	663	268	152	98	68	49	35	25	16
TW C1	410	127	68	43	29	21	15	10	7	SS D2	560	295	186	128	92	68	50	36	24
TW C2	311	120	68	44	31	22	16	11	7	SS D3	831	255	134	83	56	40	28	20	13
TW C3	296	112	63	41	28	20	15	10	7	SS D4	573	290	179	122	87	64	47	34	22
TW C4	311	100	54	34	24	17	12	9	6	SS E1	505	214	123	80	55	40	29	20	13
TW D1	267	105	59	38	26	19	13	9	6	SS E2	469	245	154	106	76	56	41	30	20
TW D2	345	164	99	67	48	35	26	18	12	SS E3	605	184	96	60	40	28	20	14	9
TW D3	351	179	112	76	55	40	30	22	14	SS E4	482	231	140	94	67	49	36	26	17
TW D4	331	183	119	83	61	45	34	25	17										
TW D5	304	174	114	81	59	44	33	24	17	COLUMBIA, SOUTH CAROLINA								2598 DD	
TW E1	404	206	129	88	63	47	35	25	17	SSF =	.10	.20	.30	.40	.50	.60	.70	.80	.90
TW E2	364	194	124	86	62	46	34	25	17	WW A1	1132	137	64	38	25	17	12	8	5
TW E3	376	223	150	107	79	60	45	33	23	WW A2	378	149	85	55	38	28	20	14	9
TW E4	334	200	135	97	72	55	41	30	21	WW A3	317	151	91	61	44	32	23	17	11
TW F1	513	131	67	41	28	19	14	10	6	WW A4	290	150	94	64	46	34	25	18	12
TW F2	318	132	76	50	35	25	18	13	9	WW A5	276	151	98	68	49	36	27	20	13
TW F3	262	126	76	51	36	27	19	14	9	WW A6	268	152	99	70	51	38	28	21	14
TW F4	214	112	70	48	35	25	19	14	9	WW B1	315	115	63	40	27	19	14	10	6
TW G1	317	113	62	39	27	19	14	10	6	WW B2	283	151	96	67	48	36	27	19	13
TW G2	230	108	65	43	31	22	16	12	8	WW B3	327	183	119	84	61	46	34	25	17
TW G3	192	97	60	41	29	22	16	11	8	WW B4	285	173	117	84	63	47	36	26	18
TW G4	154	80	50	34	25	18	13	10	6	WW B5	261	162	111	81	60	46	35	26	18
TW H1	205	90	53	35	24	18	13	9	6	WW C1	347	194	127	89	65	48	36	26	18
TW H2	157	78	48	32	23	17	12	9	6	WW C2	299	175	116	83	61	46	34	25	17
TW H3	130	66	41	28	20	15	11	8	5	WW C3	302	194	136	100	76	58	44	33	23
TW H4	100	50	31	21	15	11	8	6	4	WW C4	272	175	123	90	68	52	40	30	20
TW I1	208	88	51	33	23	16	12	8	5	TW A1	968	131	61	37	24	17	12	8	5
TW I2	264	137	86	59	42	31	23	17	11	TW A2	387	136	75	48	33	23	17	12	8
TW I3	278	149	95	66	48	35	26	19	13	TW A3	311	134	78	52	36	26	19	14	9
TW I4	274	158	105	74	54	41	31	22	15	TW A4	263	126	77	52	37	27	20	14	9
TW I5	264	156	105	75	55	42	31	23	16	TW B1	543	123	62	38	25	18	13	9	6
TW J1	334	181	116	80	58	43	32	23	16	TW B2	310	122	69	45	31	23	16	12	8
TW J2	299	168	110	77	57	42	32	23	16	TW B3	270	118	69	46	32	23	17	12	8
TW J3	312	193	132	96	72	54	41	30	21	TW B4	248	108	64	42	30	21	16	11	7
TW J4	286	177	122	88	66	50	38	28	19	TW C1	357	110	59	37	25	18	13	9	6
DG A1	266	118	68	43	27	16	8	-	-	TW C2	272	105	59	38	27	19	14	10	7
DG A2	279	127	77	52	36	26	18	11	6	TW C3	259	98	55	36	25	18	13	9	6
DG A3	331	154	95	66	49	37	27	19	11	TW C4	272	88	47	30	21	15	11	8	5
DG B1	268	120	72	48	34	24	16	8	-	TW D1	228	89	50	32	22	16	11	8	5
DG B2	287	132	81	56	41	31	23	16	9	TW D2	303	144	87	59	42	31	23	16	11
DG B3	335	158	98	69	51	40	31	23	15	TW D3	308	157	98	67	48	35	26	19	13
DG C1	321	146	88	61	44	33	24	16	6										

COLUMBIA, SOUTH CAROLINA — CONTINUED

	SSF =	.10	.20	.30	.40	.50	.60	.70	.80	.90
TW	D4	293	162	106	74	54	40	30	22	15
TW	D5	271	155	102	72	53	40	30	22	15
TW	E1	356	182	114	78	56	41	31	22	15
TW	E2	322	172	110	76	55	41	31	22	15
TW	E3	335	199	134	96	71	53	40	30	20
TW	E4	298	179	121	87	65	49	37	27	19
TW	F1	443	113	58	36	24	17	12	8	5
TW	F2	277	114	66	43	30	22	16	11	7
TW	F3	228	109	66	45	32	23	17	12	8
TW	F4	187	97	61	42	30	22	16	12	8
TW	G1	275	97	53	34	23	17	12	8	6
TW	G2	200	94	56	38	27	19	14	10	7
TW	G3	168	85	53	36	26	19	14	10	7
TW	G4	135	70	44	30	22	16	12	8	6
TW	H1	178	78	46	30	21	15	11	8	5
TW	H2	137	68	42	28	20	15	11	8	5
TW	H3	113	57	36	24	17	13	9	7	5
TW	H4	88	44	27	18	13	10	7	5	3
TW	I1	177	75	43	28	19	14	10	7	5
TW	I2	232	120	76	52	37	28	20	15	10
TW	I3	244	131	84	58	42	31	23	17	11
TW	I4	243	140	93	66	49	36	27	20	14
TW	I5	235	139	93	67	50	37	28	21	14
TW	J1	294	159	102	71	52	38	29	21	14
TW	J2	265	149	98	69	50	38	28	21	14
TW	J3	278	172	118	86	64	49	37	27	19
TW	J4	256	159	109	79	59	45	34	25	17
DG	A1	227	100	57	35	21	11	-	-	-
DG	A2	244	111	67	45	31	22	15	9	4
DG	A3	293	136	85	59	44	32	24	17	9
DG	B1	228	102	61	41	29	19	12	-	-
DG	B2	249	116	71	49	36	27	20	13	7
DG	B3	296	140	87	61	46	35	27	20	13
DG	C1	274	126	75	52	38	28	20	12	-
DG	C2	291	137	84	58	43	33	25	18	11
DG	C3	339	163	100	71	53	42	33	25	17
SS	A1	625	193	102	64	43	30	22	15	10
SS	A2	554	238	139	91	64	46	34	24	16
SS	A3	603	177	91	56	38	26	19	13	8
SS	A4	564	237	137	90	62	45	33	23	15
SS	A5	1046	194	93	56	37	25	18	12	8
SS	A6	552	237	138	90	63	46	33	24	16
SS	A7	1188	166	76	44	29	20	14	9	6
SS	A8	567	234	134	87	60	43	31	22	15
SS	B1	430	148	80	50	34	24	17	12	8
SS	B2	435	192	114	75	53	38	28	20	13
SS	B3	399	136	73	46	31	22	16	11	7
SS	B4	427	188	111	73	51	37	27	19	13
SS	B5	560	133	66	40	27	19	13	9	6
SS	B6	411	185	110	73	51	37	27	19	13
SS	B7	497	114	56	34	22	15	11	7	5
SS	B8	401	178	105	69	49	35	26	18	12
SS	C1	286	130	78	52	36	26	19	14	9
SS	C2	296	154	98	67	48	36	26	19	13
SS	C3	297	105	58	37	25	18	13	9	6
SS	C4	293	139	85	57	41	30	22	16	10
SS	D1	574	231	131	84	58	42	30	21	14
SS	D2	490	259	163	112	81	59	44	32	21
SS	D3	722	219	115	71	48	34	24	17	11
SS	D4	502	253	157	107	76	56	41	30	20
SS	E1	435	183	105	68	47	34	25	17	11
SS	E2	409	214	135	92	66	49	36	26	17
SS	E3	523	157	82	51	34	24	17	12	8
SS	E4	420	202	122	82	58	42	31	22	15

GREENVILLE-SPARTANBURG, SOUTH CAROLINA 3163 DD

	SSF =	.10	.20	.30	.40	.50	.60	.70	.80	.90
WW	A1	954	109	49	29	19	13	9	6	4
WW	A2	310	120	68	44	30	22	15	11	7
WW	A3	259	122	73	49	35	25	18	13	9
WW	A4	237	121	76	52	37	27	20	14	9
WW	A5	226	123	79	54	39	29	21	15	10
WW	A6	219	123	80	56	41	30	22	16	11
WW	B1	250	88	48	30	20	14	10	7	4
WW	B2	235	124	79	54	39	29	21	15	10
WW	B3	272	151	98	69	50	37	28	20	13
WW	B4	240	145	98	70	52	39	30	22	15
WW	B5	221	136	93	68	51	38	29	21	15
WW	C1	290	161	105	73	53	40	29	21	14
WW	C2	250	146	97	69	51	38	28	21	14
WW	C3	257	165	115	85	64	49	37	28	19
WW	C4	232	149	104	77	58	44	34	25	17
TW	A1	816	104	48	28	18	13	9	6	4
TW	A2	319	109	60	38	26	18	13	9	6
TW	A3	256	108	63	41	29	21	15	11	7
TW	A4	216	103	62	42	29	21	16	11	7
TW	B1	452	99	49	30	20	14	10	7	4
TW	B2	256	99	56	36	25	18	13	9	6
TW	B3	223	95	56	37	26	18	13	10	6
TW	B4	205	88	51	34	24	17	12	9	6
TW	C1	296	89	47	30	20	14	10	7	5
TW	C2	225	85	48	31	21	15	11	8	5
TW	C3	215	80	45	29	20	14	10	7	5
TW	C4	228	72	38	24	17	12	8	6	4
TW	D1	181	69	38	24	16	11	8	6	4
TW	D2	252	118	71	48	34	25	18	13	9
TW	D3	256	129	80	55	39	29	21	15	10
TW	D4	246	136	88	62	45	33	25	18	12
TW	D5	230	130	86	61	45	33	25	18	12
TW	E1	297	151	94	64	46	34	25	18	12
TW	E2	270	143	91	63	46	34	25	18	12
TW	E3	283	168	112	8C	59	45	34	25	17
TW	E4	253	152	102	74	55	41	31	23	16
TW	F1	365	90	45	28	18	13	9	6	4
TW	F2	226	92	52	34	24	17	12	9	6
TW	F3	187	88	53	36	25	18	13	9	6
TW	F4	153	79	49	34	24	18	13	9	6
TW	G1	225	78	42	27	18	13	9	6	4
TW	G2	164	75	45	30	21	15	11	8	5
TW	G3	137	68	42	29	20	15	11	8	5
TW	G4	110	57	35	24	17	13	9	7	4
TW	H1	145	63	37	24	17	12	9	6	4
TW	H2	112	55	34	23	16	12	9	6	4
TW	H3	93	47	29	20	14	10	7	5	4
TW	H4	72	36	22	15	11	8	6	4	3
TW	I1	140	58	33	21	14	10	7	5	3
TW	I2	192	98	62	42	30	22	16	12	8
TW	I3	203	108	69	48	34	25	19	14	9
TW	I4	204	117	78	55	40	30	23	16	11
TW	I5	198	117	78	56	41	31	23	17	12
TW	J1	246	132	85	59	42	31	23	17	11
TW	J2	222	124	81	57	42	31	23	17	11
TW	J3	236	146	100	72	54	41	31	23	16
TW	J4	217	134	92	67	50	38	29	21	15
DG	A1	180	76	42	24	12	-	-	-	-
DG	A2	201	90	54	36	24	16	11	6	-
DG	A3	247	114	71	49	36	27	19	13	7
DG	B1	180	78	46	30	20	12	-	-	-
DG	B2	205	94	57	39	28	21	15	10	4
DG	B3	251	118	73	51	38	29	22	16	10
DG	C1	219	99	59	40	28	20	13	6	-

GREENVILLE-SPARTANBURG, SOUTH CAROLINA   CONTINUED
SSF =	.10	.20	.30	.40	.50	.60	.70	.80	.90
DG C2	242	112	68	48	35	26	20	14	8
DG C3	289	136	85	59	45	34	27	20	14
SS A1	530	159	83	51	34	24	17	12	7
SS A2	473	200	116	76	53	38	27	19	13
SS A3	509	144	73	45	29	20	14	10	6
SS A4	481	199	114	74	51	37	26	19	12
SS A5	898	159	75	44	29	20	14	9	6
SS A6	470	198	115	75	52	37	27	19	12
SS A7	1022	134	60	35	22	15	10	7	4
SS A8	483	195	111	72	49	35	25	18	12
SS B1	362	121	65	40	27	19	14	9	6
SS B2	371	162	95	63	44	32	23	16	11
SS B3	333	110	59	36	24	17	12	8	5
SS B4	363	158	92	61	42	30	22	16	10
SS B5	474	108	53	32	21	14	10	7	4
SS B6	350	155	92	61	42	31	22	16	10
SS B7	419	91	44	26	17	12	8	5	3
SS B8	340	149	87	57	40	29	21	15	10
SS C1	233	104	62	41	28	21	15	11	7
SS C2	246	127	80	55	39	29	21	15	10
SS C3	243	84	46	29	20	14	10	7	5
SS C4	244	114	69	46	33	24	18	13	8
SS D1	483	190	107	68	47	33	24	17	11
SS D2	416	217	137	94	67	49	36	26	17
SS D3	613	180	93	58	38	27	19	13	8
SS D4	426	213	132	89	63	46	34	24	16
SS E1	360	148	84	54	37	26	19	13	9
SS E2	344	178	112	76	55	40	29	21	14
SS E3	440	128	66	41	27	19	13	9	6
SS E4	355	168	101	68	48	35	25	18	12

HURON, SOUTH DAKOTA   8054 DD
SSF =	.10	.20	.30	.40	.50	.60	.70	.80	.90
WW A1	285	-	-	-	-	-	-	-	-
WW A2	95	28	11	-	-	-	-	-	-
WW A3	80	31	15	7	-	-	-	-	-
WW A4	74	32	16	9	5	-	-	-	-
WW A5	71	33	18	10	6	3	1	-	-
WW A6	70	34	19	11	7	4	2	-	-
WW B1	-	-	-	-	-	-	-	-	-
WW B2	86	41	24	15	10	7	4	3	2
WW B3	104	53	32	21	14	10	7	4	3
WW B4	103	60	39	27	19	14	10	7	5
WW B5	100	60	40	28	20	15	11	8	5
WW C1	115	60	37	24	17	12	8	5	3
WW C2	105	58	37	25	18	13	9	6	4
WW C3	121	76	52	37	27	20	15	11	7
WW C4	111	69	47	34	25	19	14	10	7
TW A1	255	17	-	-	-	-	-	-	-
TW A2	101	27	10	-	-	-	-	-	-
TW A3	82	28	13	6	-	-	-	-	-
TW A4	71	29	15	8	4	2	-	-	-
TW B1	142	21	-	-	-	-	-	-	-
TW B2	84	27	12	6	-	-	-	-	-
TW B3	75	27	13	7	4	2	-	-	-
TW B4	71	26	13	7	4	2	1	-	-
TW C1	99	24	9	4	-	-	-	-	-
TW C2	79	25	12	6	3	2	-	-	-
TW C3	77	25	12	6	4	2	1	-	-
TW C4	85	23	11	6	3	2	1	-	-
TW D1	-	-	-	-	-	-	-	-	-
TW D2	94	40	22	13	8	6	4	2	1
TW D3	95	43	24	15	10	6	4	3	1

HURON, SOUTH DAKOTA   CONTINUED
SSF =	.10	.20	.30	.40	.50	.60	.70	.80	.90
TW D4	105	55	34	23	16	11	8	6	4
TW D5	104	57	36	25	18	13	9	7	4
TW E1	118	55	32	20	14	9	6	4	2
TW E2	112	56	34	22	15	11	8	5	3
TW E3	129	73	48	33	24	17	13	9	6
TW E4	118	69	45	32	23	17	12	9	6
TW F1	103	13	-	-	-	-	-	-	-
TW F2	67	21	8	-	-	-	-	-	-
TW F3	57	22	10	5	-	-	-	-	-
TW F4	48	21	11	6	3	1	-	-	-
TW G1	64	15	-	-	-	-	-	-	-
TW G2	49	18	8	3	-	-	-	-	-
TW G3	43	18	9	5	2	-	-	-	-
TW G4	36	16	9	5	3	2	-	-	-
TW H1	42	14	5	-	-	-	-	-	-
TW H2	35	14	7	4	2	-	-	-	-
TW H3	30	13	7	4	2	1	-	-	-
TW H4	25	11	6	3	2	1	-	-	-
TW I1	-	-	-	-	-	-	-	-	-
TW I2	70	33	18	11	7	5	3	2	1
TW I3	76	37	22	14	9	6	4	3	2
TW I4	87	48	30	20	14	10	7	5	3
TW I5	89	51	33	23	16	12	9	6	4
TW J1	97	49	29	19	13	9	6	4	2
TW J2	93	49	31	20	14	10	7	5	3
TW J3	109	65	44	31	22	17	12	9	6
TW J4	102	62	41	29	21	16	12	8	6
DG A1	-	-	-	-	-	-	-	-	-
DG A2	71	26	11	-	-	-	-	-	-
DG A3	112	49	28	18	12	8	5	-	-
DG B1	-	-	-	-	-	-	-	-	-
DG B2	74	28	14	6	-	-	-	-	-
DG B3	116	51	30	20	14	10	7	4	-
DG C1	52	-	-	-	-	-	-	-	-
DG C2	95	39	21	12	7	-	-	-	-
DG C3	136	61	36	24	17	12	9	6	3
SS A1	188	45	18	7	-	-	-	-	-
SS A2	198	77	41	25	16	11	7	5	3
SS A3	160	29	-	-	-	-	-	-	-
SS A4	193	72	37	22	14	9	6	4	2
SS A5	317	40	10	-	-	-	-	-	-
SS A6	192	74	39	23	15	10	6	4	2
SS A7	329	17	-	-	-	-	-	-	-
SS A8	188	68	34	20	12	7	5	3	1
SS B1	120	31	11	-	-	-	-	-	-
SS B2	157	63	35	21	14	10	6	4	3
SS B3	98	20	-	-	-	-	-	-	-
SS B4	148	59	32	19	12	8	5	3	2
SS B5	152	22	-	-	-	-	-	-	-
SS B6	143	59	32	20	13	9	6	4	2
SS B7	117	-	-	-	-	-	-	-	-
SS B8	134	53	28	17	10	7	4	3	2
SS C1	64	20	-	-	-	-	-	-	-
SS C2	93	44	25	16	11	7	5	3	2
SS C3	70	17	-	-	-	-	-	-	-
SS C4	92	39	22	13	9	6	4	2	1
SS D1	167	54	24	11	4	-	-	-	-
SS D2	173	84	50	32	22	15	10	7	4
SS D3	216	50	19	7	-	-	-	-	-
SS D4	177	83	48	30	20	14	10	6	4
SS E1	107	31	-	-	-	-	-	-	-
SS E2	134	64	37	23	15	10	7	4	3
SS E3	139	28	-	-	-	-	-	-	-
SS E4	139	60	33	20	13	9	6	4	2

PIERRE, SOUTH DAKOTA   7677 DD
SSF	=.10	.20	.30	.40	.50	.60	.70	.80	.90
WW A1	384	32	9	-	-	-	-	-	-
WW A2	128	43	21	12	7	4	2	-	-
WW A3	109	46	25	15	9	6	4	2	1
WW A4	100	46	26	16	11	7	4	3	1
WW A5	95	48	28	18	12	8	5	3	2
WW A6	93	48	29	19	13	8	6	4	2
WW B1	68	-	-	-	-	-	-	-	-
WW B2	109	54	33	21	15	10	7	5	3
WW B3	129	68	42	28	20	14	10	7	4
WW B4	124	72	48	33	24	18	13	9	6
WW B5	118	71	48	34	25	18	14	10	6
WW C1	142	75	47	32	22	16	11	8	5
WW C2	127	71	46	32	23	16	12	8	5
WW C3	141	89	61	44	33	25	18	13	9
WW C4	129	81	56	40	30	23	17	12	8
TW A1	337	33	11	-	-	-	-	-	-
TW A2	134	40	19	10	6	3	2	-	-
TW A3	109	41	21	13	8	5	3	2	-
TW A4	94	41	22	14	9	6	4	2	1
TW B1	189	34	14	7	3	-	-	-	-
TW B2	111	38	19	11	7	4	3	1	-
TW B3	98	38	20	12	8	5	3	2	1
TW B4	92	36	19	12	8	5	3	2	1
TW C1	129	34	16	9	5	3	2	-	-
TW C2	101	35	18	10	7	4	3	2	-
TW C3	98	33	17	10	6	4	3	2	-
TW C4	107	31	15	9	6	4	2	2	-
TW D1	51	-	-	-	-	-	-	-	-
TW D2	118	52	29	19	12	9	6	4	2
TW D3	119	57	33	21	14	10	7	5	3
TW D4	127	67	42	29	20	15	10	7	5
TW D5	123	68	44	30	22	16	12	8	5
TW E1	145	70	41	27	18	13	9	6	4
TW E2	136	69	42	28	20	14	10	7	4
TW E3	152	87	57	40	29	21	16	11	7
TW E4	138	81	54	38	28	20	15	11	7
TW F1	143	28	10	3	-	-	-	-	-
TW F2	92	33	16	9	5	3	1	-	-
TW F3	77	33	18	10	6	4	2	1	-
TW F4	65	30	17	11	7	5	3	2	1
TW G1	89	26	11	6	2	-	-	-	-
TW G2	67	28	15	9	5	3	2	1	-
TW G3	58	26	15	9	6	4	2	1	-
TW G4	48	22	13	8	5	4	2	1	-
TW H1	59	22	11	6	4	2	-	-	-
TW H2	47	21	11	7	4	3	2	1	-
TW H3	40	18	10	6	4	3	2	1	-
TW H4	32	15	8	5	3	2	2	1	-
TW I1	35	-	-	-	-	-	-	-	-
TW I2	89	43	25	16	11	8	5	4	2
TW I3	96	48	29	19	13	9	6	4	3
TW I4	105	58	37	26	18	13	10	7	4
TW I5	106	61	40	28	20	15	11	8	5
TW J1	120	61	37	25	17	12	9	6	4
TW J2	112	61	38	26	18	13	10	7	4
TW J3	128	77	52	37	27	20	15	11	7
TW J4	120	72	49	35	26	19	14	10	7
DG A1	50	-	-	-	-	-	-	-	-
DG A2	92	37	19	9	-	-	-	-	-
DG A3	132	59	34	23	16	11	7	4	-
DG B1	53	-	-	-	-	-	-	-	-
DG B2	94	39	21	12	7	-	-	-	-
DG B3	137	62	37	25	18	13	9	6	3
DG C1	80	26	-	-	-	-	-	-	-

PIERRE, SOUTH DAKOTA   CONTINUED
SSF	=.10	.20	.30	.40	.50	.60	.70	.80	.90
DG C2	118	50	28	18	12	7	3	-	-
DG C3	160	72	43	29	21	16	12	8	5
SS A1	234	61	27	14	8	4	2	1	-
SS A2	235	94	51	32	21	14	10	6	4
SS A3	207	46	17	6	-	-	-	-	-
SS A4	231	89	47	29	18	12	8	5	3
SS A5	392	58	22	9	4	-	-	-	-
SS A6	230	91	49	30	20	13	9	6	4
SS A7	416	39	-	-	-	-	-	-	-
SS A8	227	84	44	26	17	11	7	5	3
SS B1	153	44	20	10	5	3	-	-	-
SS B2	186	77	43	27	18	12	9	6	4
SS B3	131	34	13	4	-	-	-	-	-
SS B4	177	72	40	25	16	11	7	5	3
SS B5	195	36	13	5	-	-	-	-	-
SS B6	171	72	40	25	17	11	8	5	3
SS B7	158	24	-	-	-	-	-	-	-
SS B8	162	66	36	22	14	10	6	4	3
SS C1	91	35	18	10	5	2	-	-	-
SS C2	116	57	34	22	15	10	7	5	3
SS C3	97	28	13	6	3	-	-	-	-
SS C4	115	51	29	18	12	8	6	4	2
SS D1	210	73	36	20	12	7	4	2	1
SS D2	206	102	61	40	27	19	13	9	6
SS D3	269	69	30	16	8	5	2	-	-
SS D4	211	100	59	38	26	18	12	8	5
SS E1	143	49	23	11	5	-	-	-	-
SS E2	163	79	47	30	20	14	10	7	4
SS E3	180	43	17	8	-	-	-	-	-
SS E4	169	75	42	27	18	12	8	5	3

RAPID CITY, SOUTH DAKOTA   7324 DD
SSF	=.10	.20	.30	.40	.50	.60	.70	.80	.90
WW A1	474	47	19	9	5	2	-	-	-
WW A2	157	58	31	19	12	8	5	3	2
WW A3	134	60	35	22	15	10	7	5	3
WW A4	124	61	36	24	16	11	8	5	3
WW A5	119	62	38	26	18	12	9	6	4
WW A6	116	63	39	27	19	13	9	6	4
WW B1	102	29	11	-	-	-	-	-	-
WW B2	130	68	42	28	20	14	10	7	4
WW B3	154	84	53	36	26	19	13	9	6
WW B4	144	86	57	41	30	22	16	12	8
WW B5	136	83	57	41	30	22	17	12	8
WW C1	167	91	58	40	28	21	15	11	7
WW C2	148	85	56	39	28	21	15	11	7
WW C3	161	103	71	52	39	29	22	16	11
WW C4	146	93	65	47	36	27	20	15	10
TW A1	413	47	19	10	6	3	2	-	-
TW A2	163	53	27	16	10	7	5	3	2
TW A3	133	54	30	19	12	8	6	4	2
TW A4	114	52	30	20	13	9	6	4	3
TW B1	230	46	21	12	7	4	3	2	-
TW B2	134	50	27	16	11	7	5	3	2
TW B3	118	49	27	17	11	8	5	4	2
TW B4	110	46	26	16	11	8	5	4	2
TW C1	155	44	22	13	8	6	4	2	1
TW C2	121	44	24	15	10	7	5	3	2
TW C3	117	42	23	14	9	6	4	3	2
TW C4	125	38	20	12	8	5	4	2	1
TW D1	75	24	10	4	-	-	-	-	-
TW D2	140	64	38	25	17	12	8	6	4
TW D3	142	70	42	28	19	14	10	7	4

RAPID CITY, SOUTH DAKOTA							CONTINUED			SIOUX FALLS, SOUTH DAKOTA								7838 DD	
SSF =	.10	.20	.30	.40	.50	.60	.70	.80	.90	SSF =	.10	.20	.30	.40	.50	.60	.70	.80	.90
TW D4	146	80	51	35	25	18	13	10	6	WW A1	331	22	-	-	-	-	-	-	-
TW D5	141	79	52	36	26	19	14	10	7	WW A2	108	35	16	8	4	-	-	-	-
TW E1	170	85	52	34	24	17	12	9	5	WW A3	92	37	20	11	7	4	2	-	-
TW E2	158	83	52	35	25	18	13	9	6	WW A4	85	38	21	13	8	5	3	2	-
TW E3	174	102	68	48	35	26	19	14	9	WW A5	81	40	23	14	9	6	4	2	1
TW E4	158	94	63	45	33	25	18	13	9	WW A6	79	40	24	15	10	7	4	3	1
TW F1	177	40	18	10	6	3	2	-	-	WW B1	43	-	-	-	-	-	-	-	-
TW F2	114	44	24	14	9	6	4	3	1	WW B2	95	47	28	18	12	9	6	4	2
TW F3	96	44	25	16	11	7	5	3	2	WW B3	114	60	37	24	17	12	8	6	4
TW F4	80	40	24	16	11	7	5	4	2	WW B4	111	65	43	30	22	16	12	8	5
TW G1	111	36	18	11	7	4	3	2	-	WW B5	107	65	43	31	22	17	12	9	6
TW G2	84	37	21	13	9	6	4	3	2	WW C1	126	67	41	28	19	14	10	7	4
TW G3	72	34	20	13	9	6	4	3	2	WW C2	114	64	41	28	20	15	11	7	5
TW G4	59	29	17	11	8	6	4	3	2	WW C3	129	81	56	40	30	22	17	12	8
TW H1	73	30	16	10	7	4	3	2	1	WW C4	118	74	51	37	28	21	16	11	8
TW H2	58	27	16	10	7	5	3	2	1	TW A1	292	25	-	-	-	-	-	-	-
TW H3	49	24	14	9	6	4	3	2	1	TW A2	114	33	15	7	4	-	-	-	-
TW H4	39	19	11	7	5	3	2	2	1	TW A3	93	34	17	10	6	3	2	-	-
TW I1	56	19	8	-	-	-	-	-	-	TW A4	80	34	18	11	7	4	3	2	-
TW I2	106	53	32	22	15	11	8	5	3	TW B1	162	27	10	3	-	-	-	-	-
TW I3	114	60	37	25	17	13	9	6	4	TW B2	95	32	15	9	5	3	1	-	-
TW I4	122	69	45	31	23	17	12	9	6	TW B3	84	32	16	10	6	4	2	1	-
TW I5	121	71	47	33	24	18	13	10	6	TW B4	79	31	16	10	6	4	2	1	-
TW J1	141	75	47	32	22	16	12	8	5	TW C1	111	28	13	6	3	2	-	-	-
TW J2	131	72	46	32	23	17	12	9	6	TW C2	88	29	15	8	5	3	2	1	-
TW J3	147	90	61	44	32	24	18	13	9	TW C3	86	28	14	8	5	3	2	1	-
TW J4	136	84	57	41	30	23	17	12	8	TW C4	94	27	13	7	5	3	2	1	-
DG A1	75	22	-	-	-	-	-	-	-	TW D1	34	-	-	-	-	-	-	-	-
DG A2	109	47	26	15	8	-	-	-	-	TW D2	104	45	25	16	11	7	5	3	2
DG A3	150	69	41	28	20	14	10	6	-	TW D3	105	49	28	18	12	8	6	4	2
DG B1	76	27	-	-	-	-	-	-	-	TW D4	114	60	38	25	18	13	9	7	4
DG B2	112	49	28	18	12	7	-	-	-	TW D5	112	62	40	27	20	14	11	7	5
DG B3	153	71	44	30	22	16	12	8	4	TW E1	128	61	36	23	16	11	8	5	3
DG C1	105	42	21	11	-	-	-	-	-	TW E2	122	62	38	25	18	12	9	6	4
DG C2	137	62	36	24	17	12	7	3	-	TW E3	138	79	52	36	26	19	14	10	7
DG C3	177	83	51	35	26	19	15	11	6	TW E4	126	74	49	34	25	19	14	10	7
SS A1	282	81	39	22	14	9	5	3	2	TW F1	119	20	4	-	-	-	-	-	-
SS A2	275	114	64	40	27	19	13	9	5	TW F2	77	26	12	6	3	-	-	-	-
SS A3	256	66	30	16	9	5	2	-	-	TW F3	65	27	14	8	5	3	1	-	-
SS A4	273	110	61	38	25	17	12	8	5	TW F4	55	25	14	9	5	3	2	1	-
SS A5	473	78	33	17	10	6	3	2	-	TW G1	74	20	8	-	-	-	-	-	-
SS A6	270	111	62	39	26	18	12	8	5	TW G2	57	22	11	6	4	2	-	-	-
SS A7	515	59	22	9	-	-	-	-	-	TW G3	49	22	12	7	4	3	2	-	-
SS A8	270	106	57	35	23	16	11	7	4	TW G4	41	19	11	7	4	3	2	1	-
SS B1	187	59	29	17	10	6	4	2	1	TW H1	49	17	8	4	2	-	-	-	-
SS B2	216	93	53	34	23	16	11	8	5	TW H2	40	17	9	5	3	2	1	-	-
SS B3	165	50	24	13	7	4	2	-	-	TW H3	34	15	9	5	3	2	1	-	-
SS B4	208	88	50	32	21	15	10	7	4	TW H4	28	12	7	4	3	2	1	-	-
SS B5	239	50	22	12	6	4	2	-	-	TW I1	17	-	-	-	-	-	-	-	-
SS B6	201	88	50	32	22	15	11	7	5	TW I2	78	37	22	14	9	6	4	3	2
SS B7	200	38	15	7	-	-	-	-	-	TW I3	84	42	25	16	11	8	5	4	2
SS B8	192	82	46	29	19	13	9	6	4	TW I4	94	52	33	23	16	12	9	6	4
SS C1	115	49	27	17	11	7	5	3	2	TW I5	95	55	36	25	18	13	10	7	5
SS C2	138	70	43	29	20	14	10	7	5	TW J1	106	54	33	22	15	11	8	5	3
SS C3	120	39	20	12	7	5	3	2	-	TW J2	100	54	34	23	16	12	8	6	4
SS C4	136	63	37	24	17	12	8	6	4	TW J3	117	70	47	34	25	18	14	10	6
SS D1	257	96	50	30	19	12	8	5	3	TW J4	109	66	44	32	23	17	13	9	6
SS D2	243	124	76	51	35	25	18	12	8	DG A1	29	-	-	-	-	-	-	-	-
SS D3	324	91	43	25	15	9	6	4	2	DG A2	79	31	15	-	-	-	-	-	-
SS D4	249	122	73	48	33	23	17	11	7	DG A3	120	53	31	20	14	10	6	3	-
SS E1	181	70	36	21	13	8	5	3	1	DG B1	34	-	-	-	-	-	-	-	-
SS E2	195	98	60	39	27	19	14	9	6	DG B2	81	32	17	9	-	-	-	-	-
SS E3	222	60	28	15	9	5	3	2	-	DG B3	124	55	33	22	16	11	8	5	2
SS E4	201	93	54	35	24	16	12	8	5	DG C1	63	15	-	-	-	-	-	-	-

SIOUX FALLS, SOUTH DAKOTA   CONTINUED
SSF =	.10	.20	.30	.40	.50	.60	.70	.80	.90
DG C2	103	43	24	15	9	5	-	-	-
DG C3	145	65	39	26	19	14	10	7	4
SS A1	209	53	23	11	6	3	-	-	-
SS A2	214	84	46	28	19	13	9	6	4
SS A3	181	38	12	-	-	-	-	-	-
SS A4	210	80	42	26	16	11	7	5	3
SS A5	355	48	17	6	-	-	-	-	-
SS A6	208	81	44	27	18	12	8	5	3
SS A7	378	29	-	-	-	-	-	-	-
SS A8	205	75	39	23	15	10	6	4	2
SS B1	135	37	16	8	3	-	-	-	-
SS B2	169	69	39	24	16	11	8	5	3
SS B3	112	27	9	-	-	-	-	-	-
SS B4	161	65	35	22	14	10	7	4	3
SS B5	172	29	9	-	-	-	-	-	-
SS B6	155	65	36	22	15	10	7	5	3
SS B7	136	16	-	-	-	-	-	-	-
SS B8	146	59	32	19	13	8	6	4	2
SS C1	75	27	13	6	-	-	-	-	-
SS C2	102	49	29	19	13	9	6	4	2
SS C3	81	22	9	3	-	-	-	-	-
SS C4	101	44	25	16	10	7	5	3	2
SS D1	185	63	30	16	9	5	2	-	-
SS D2	187	93	55	36	25	17	12	8	5
SS D3	240	59	25	12	6	2	-	-	-
SS D4	192	91	53	34	23	16	11	8	5
SS E1	123	40	17	7	-	-	-	-	-
SS E2	146	71	42	27	18	12	9	6	3
SS E3	157	35	13	-	-	-	-	-	-
SS E4	152	67	38	23	16	11	7	5	3

CHATTANOOGA, TENNESSEE   3505 DD
SSF =	.10	.20	.30	.40	.50	.60	.70	.80	.90
WW A1	659	72	32	18	11	7	5	3	2
WW A2	219	84	46	29	20	14	10	7	4
WW A3	185	86	50	33	23	16	12	8	5
WW A4	170	86	52	35	25	18	13	9	6
WW A5	163	87	55	37	27	19	14	10	7
WW A6	158	87	56	39	28	20	15	11	7
WW B1	164	55	28	16	10	7	4	3	2
WW B2	173	91	57	39	28	20	15	11	7
WW B3	203	112	72	50	36	26	19	14	9
WW B4	183	110	74	53	39	29	22	16	11
WW B5	171	105	71	51	38	29	22	16	11
WW C1	218	120	77	54	39	28	21	15	10
WW C2	190	110	72	51	37	28	21	15	10
WW C3	201	128	89	65	49	37	28	21	14
WW C4	181	116	81	59	44	34	26	19	13
TW A1	569	71	31	18	11	7	5	3	2
TW A2	226	76	41	25	17	12	8	6	4
TW A3	183	76	43	28	19	13	10	7	4
TW A4	156	73	43	29	20	14	10	7	5
TW B1	318	68	33	19	12	8	6	4	2
TW B2	183	70	39	24	17	12	8	6	4
TW B3	161	68	39	25	17	12	9	6	4
TW B4	149	63	36	24	16	11	8	6	4
TW C1	212	63	32	20	13	9	6	4	3
TW C2	163	61	34	21	14	10	7	5	3
TW C3	157	57	32	20	14	10	7	5	3
TW C4	166	52	27	17	11	8	6	4	3
TW D1	120	43	23	13	9	6	4	3	1
TW D2	186	86	51	34	24	17	12	9	6
TW D3	190	95	58	39	28	20	15	10	7

CHATTANOOGA, TENNESSEE   CONTINUED
SSF =	.10	.20	.30	.40	.50	.60	.70	.80	.90
TW D4	187	103	66	46	33	24	18	13	9
TW D5	177	100	66	46	34	25	19	14	9
TW E1	223	112	69	47	33	24	18	13	8
TW E2	204	108	68	47	33	24	18	13	9
TW E3	219	129	86	61	45	34	25	18	12
TW E4	197	117	79	56	42	31	23	17	12
TW F1	252	61	29	17	11	8	5	3	2
TW F2	160	64	36	23	15	11	8	5	3
TW F3	133	62	37	24	17	12	8	6	4
TW F4	110	56	34	23	16	12	8	6	4
TW G1	157	53	28	17	11	8	5	4	2
TW G2	116	53	31	20	14	10	7	5	3
TW G3	98	48	29	19	14	10	7	5	3
TW G4	80	40	25	17	12	8	6	4	3
TW H1	103	43	25	16	11	8	5	4	2
TW H2	80	39	23	15	11	8	6	4	3
TW H3	67	33	20	13	9	7	5	3	2
TW H4	52	26	15	10	7	5	4	3	2
TW I1	91	35	19	11	7	5	3	2	1
TW I2	141	72	44	30	21	15	11	8	5
TW I3	151	80	50	34	24	18	13	9	6
TW I4	156	89	58	41	30	22	16	12	8
TW I5	153	90	60	42	31	23	17	13	9
TW J1	185	98	62	43	31	22	16	12	8
TW J2	168	94	61	42	31	23	17	12	8
TW J3	184	113	77	55	41	31	23	17	12
TW J4	169	104	71	52	38	29	22	16	11
DG A1	119	46	21	-	-	-	-	-	-
DG A2	147	65	38	24	15	9	4	-	-
DG A3	190	87	53	36	26	19	14	9	4
DG B1	120	49	26	15	-	-	-	-	-
DG B2	150	68	40	27	19	13	8	4	-
DG B3	194	91	55	38	28	21	16	12	7
DG C1	152	66	37	23	15	9	-	-	-
DG C2	180	83	49	33	24	18	13	8	3
DG C3	224	105	65	45	33	25	20	15	9
SS A1	397	117	60	36	24	16	11	8	5
SS A2	367	154	88	57	39	28	20	14	9
SS A3	377	104	51	30	19	13	9	6	4
SS A4	373	153	86	55	38	27	19	13	9
SS A5	664	116	53	30	19	13	9	6	4
SS A6	364	152	87	56	38	27	20	14	9
SS A7	747	96	41	22	14	9	6	4	2
SS A8	373	149	83	53	36	25	18	13	8
SS B1	269	88	46	28	18	13	9	6	4
SS B2	288	125	72	47	33	23	17	12	8
SS B3	245	79	41	24	16	11	7	5	3
SS B4	282	121	70	45	31	22	16	11	7
SS B5	347	77	37	21	14	9	6	4	3
SS B6	271	119	69	45	31	22	16	11	7
SS B7	302	64	29	17	10	7	4	3	2
SS B8	263	113	65	42	29	21	15	10	7
SS C1	164	72	41	27	18	13	9	6	4
SS C2	182	93	58	39	28	20	15	11	7
SS C3	170	58	31	19	13	9	6	4	3
SS C4	180	84	50	33	23	17	12	9	6
SS D1	362	140	77	48	32	23	16	11	7
SS D2	324	168	104	71	50	36	27	19	12
SS D3	458	133	67	40	26	18	12	8	5
SS D4	332	164	100	67	47	34	25	18	12
SS E1	265	107	59	37	25	17	12	8	5
SS E2	265	136	84	57	40	29	21	15	10
SS E3	323	92	46	27	18	12	8	6	3
SS E4	273	128	76	50	35	25	18	13	8

KNOXVILLE, TENNESSEE									3478 DD
SSF =	.10	.20	.30	.40	.50	.60	.70	.80	.90
WW A1	707	77	34	19	12	8	5	3	2
WW A2	232	88	49	31	21	15	10	7	5
WW A3	195	90	53	35	24	17	12	9	6
WW A4	179	90	55	37	26	19	14	10	6
WW A5	171	91	57	39	28	20	15	11	7
WW A6	166	92	59	40	29	21	16	11	7
WW B1	176	59	30	18	11	7	5	3	2
WW B2	181	95	59	40	29	21	15	11	7
WW B3	212	117	75	52	37	27	20	14	10
WW B4	190	114	77	55	40	30	23	16	11
WW B5	177	109	74	53	40	30	22	16	11
WW C1	227	125	80	56	40	30	22	16	10
WW C2	198	114	75	53	39	29	21	16	10
WW C3	208	132	92	68	51	38	29	22	15
WW C4	188	120	84	61	46	35	26	20	13
TW A1	609	75	33	19	12	8	5	4	2
TW A2	240	81	43	27	18	12	9	6	4
TW A3	193	80	46	29	20	14	10	7	5
TW A4	164	77	46	30	21	15	11	8	5
TW B1	339	72	35	20	13	9	6	4	3
TW B2	194	74	41	26	17	12	9	6	4
TW B3	169	71	41	26	18	13	9	6	4
TW B4	157	66	38	25	17	12	9	6	4
TW C1	224	66	34	21	14	10	7	5	3
TW C2	172	64	35	22	15	11	8	5	3
TW C3	165	60	33	21	14	10	7	5	3
TW C4	176	55	29	18	12	8	6	4	3
TW D1	128	46	24	15	9	6	4	3	2
TW D2	195	90	54	36	25	18	13	9	6
TW D3	198	99	61	41	29	21	15	11	7
TW D4	195	107	68	47	34	25	19	14	9
TW D5	184	104	68	48	35	26	19	14	9
TW E1	233	117	72	48	34	25	18	13	9
TW E2	213	112	71	48	35	25	19	13	9
TW E3	228	134	89	63	46	35	26	19	13
TW E4	204	122	82	58	43	32	24	18	12
TW F1	269	64	31	18	12	8	6	4	2
TW F2	169	67	38	24	16	11	8	6	4
TW F3	140	65	38	25	18	12	9	6	4
TW F4	116	59	36	24	17	12	9	6	4
TW G1	167	56	30	18	12	8	6	4	3
TW G2	123	55	32	21	15	10	7	5	3
TW G3	104	51	31	20	14	10	7	5	3
TW G4	84	42	26	17	12	9	6	5	3
TW H1	109	46	26	17	11	8	6	4	3
TW H2	84	41	24	16	11	8	6	4	3
TW H3	71	35	21	14	10	7	5	4	2
TW H4	55	27	16	11	8	5	4	3	2
TW I1	97	38	20	12	8	5	4	2	1
TW I2	148	75	46	31	22	16	12	8	6
TW I3	158	83	52	36	25	19	14	10	6
TW I4	162	92	60	42	31	23	17	12	8
TW I5	159	93	62	44	32	24	18	13	9
TW J1	193	102	65	44	32	23	17	12	8
TW J2	175	97	63	44	32	23	17	13	8
TW J3	190	117	80	57	43	32	24	18	12
TW J4	176	108	74	53	40	30	23	17	11
DG A1	127	50	23	-	-	-	-	-	-
DG A2	155	68	40	25	16	10	5	-	-
DG A3	198	91	55	38	27	20	14	9	4
DG B1	129	52	28	16	-	-	-	-	-
DG B2	159	71	42	28	20	14	9	5	-
DG B3	202	94	57	40	29	22	17	12	7
DG C1	162	70	39	25	17	10	-	-	-

KNOXVILLE, TENNESSEE									CONTINUED
SSF =	.10	.20	.30	.40	.50	.60	.70	.80	.90
DG C2	189	86	51	35	25	19	13	9	4
DG C3	234	109	67	46	34	26	20	15	10
SS A1	414	122	62	37	24	17	12	8	5
SS A2	380	159	91	59	40	29	21	14	9
SS A3	394	108	53	31	20	14	9	6	4
SS A4	386	157	89	57	39	27	20	14	9
SS A5	697	121	55	32	20	13	9	6	4
SS A6	376	157	89	58	39	28	20	14	9
SS A7	787	99	43	23	14	9	6	4	2
SS A8	385	153	86	54	37	26	18	13	8
SS B1	281	91	48	29	19	13	9	6	4
SS B2	298	129	75	49	34	24	17	12	8
SS B3	256	82	42	25	17	11	8	5	3
SS B4	292	125	72	47	32	23	16	12	7
SS B5	364	81	38	22	14	9	6	4	3
SS B6	280	123	71	47	32	23	17	12	8
SS B7	318	67	31	17	11	7	5	3	2
SS B8	272	117	67	44	30	21	15	11	7
SS C1	173	76	44	28	19	14	10	7	4
SS C2	191	98	60	41	29	21	15	11	7
SS C3	181	61	32	20	13	9	6	4	3
SS C4	189	88	52	35	24	17	13	9	6
SS D1	377	145	80	50	34	23	16	11	7
SS D2	334	173	107	73	51	37	27	19	13
SS D3	478	137	69	42	27	19	13	9	5
SS D4	342	169	103	69	49	35	25	18	12
SS E1	277	111	61	38	26	18	13	9	5
SS E2	274	140	86	58	41	30	22	15	10
SS E3	338	96	48	29	19	13	9	6	4
SS E4	283	132	78	52	36	26	19	13	9

MEMPHIS, TENNESSEE									3227 DD
SSF =	.10	.20	.30	.40	.50	.60	.70	.80	.90
WW A1	851	92	41	23	15	10	7	5	3
WW A2	271	103	57	37	25	18	13	9	6
WW A3	227	105	62	41	29	21	15	11	7
WW A4	207	104	64	43	31	22	16	12	8
WW A5	197	106	67	46	33	24	18	13	8
WW A6	191	106	68	47	34	25	19	13	9
WW B1	212	73	38	23	15	11	7	5	3
WW B2	207	108	68	47	33	25	18	13	9
WW B3	241	132	85	59	43	32	24	17	11
WW B4	214	128	86	62	46	34	26	19	13
WW B5	198	121	83	60	45	34	25	19	13
WW C1	257	142	91	64	46	34	25	18	12
WW C2	223	129	85	60	44	33	25	18	12
WW C3	231	147	103	75	57	43	33	24	17
WW C4	208	133	93	68	51	39	30	22	15
TW A1	729	89	40	23	15	10	7	5	3
TW A2	280	94	50	32	21	15	11	7	5
TW A3	224	93	54	35	24	17	12	9	6
TW A4	190	89	53	35	25	18	13	9	6
TW B1	400	85	41	25	16	11	8	5	3
TW B2	225	85	48	30	21	15	11	7	5
TW B3	196	83	48	31	22	15	11	8	5
TW B4	181	76	44	29	20	14	10	7	5
TW C1	262	77	40	25	17	12	8	6	4
TW C2	199	74	41	26	18	13	9	6	4
TW C3	191	69	38	24	17	12	9	6	4
TW C4	203	63	33	21	14	10	7	5	3
TW D1	154	57	31	19	13	9	6	4	3
TW D2	223	103	62	41	29	21	15	11	7
TW D3	226	113	70	47	33	24	18	13	8

MEMPHIS, TENNESSEE							CONTINUED			NASHVILLE, TENNESSEE								3696 DD	
SSF =	.10	.20	.30	.40	.50	.60	.70	.80	.90	SSF =	.10	.20	.30	.40	.50	.60	.70	.80	.90
TW D4	220	120	77	54	39	29	21	16	10	WW A1	588	60	24	13	8	5	3	2	1
TW D5	206	116	76	54	39	29	22	16	11	WW A2	192	70	38	23	15	11	7	5	3
TW E1	264	132	82	55	40	29	21	15	10	WW A3	161	72	42	27	18	13	9	6	4
TW E2	241	126	80	55	40	29	22	16	10	WW A4	148	72	43	29	20	14	10	7	5
TW E3	254	149	100	71	52	39	29	22	15	WW A5	141	74	46	31	22	16	11	8	5
TW E4	227	135	91	65	48	36	27	20	14	WW A6	137	74	47	32	22	16	12	8	5
TW F1	320	76	38	23	15	10	7	5	3	WW B1	135	41	19	10	6	3	2	-	-
TW F2	198	79	44	28	20	14	10	7	4	WW B2	152	78	48	33	23	17	12	9	6
TW F3	163	76	45	30	21	15	11	8	5	WW B3	179	97	61	42	30	22	16	12	8
TW F4	134	68	42	28	20	15	11	8	5	WW B4	164	97	65	46	34	25	19	14	9
TW G1	196	66	35	22	15	10	7	5	3	WW B5	153	93	63	45	33	25	19	14	9
TW G2	143	65	38	25	18	13	9	6	4	WW C1	193	105	67	46	33	24	18	13	8
TW G3	120	59	36	24	17	12	9	6	4	WW C2	169	97	63	44	32	24	18	13	8
TW G4	97	49	30	20	15	11	8	6	4	WW C3	181	115	79	58	43	33	25	18	12
TW H1	127	54	31	20	14	10	7	5	3	WW C4	164	104	72	53	39	30	23	17	11
TW H2	98	47	29	19	13	10	7	5	3	TW A1	509	59	25	13	8	5	3	2	1
TW H3	82	40	25	17	12	9	6	4	3	TW A2	199	64	33	20	13	9	6	4	3
TW H4	63	31	19	13	9	6	5	3	2	TW A3	160	65	36	23	15	11	8	5	3
TW I1	118	47	26	16	11	8	5	4	2	TW A4	136	62	36	23	16	11	8	6	4
TW I2	169	86	53	36	26	19	14	10	7	TW B1	282	57	26	15	9	6	4	3	2
TW I3	179	95	60	41	29	22	16	11	8	TW B2	161	59	32	20	13	9	6	4	3
TW I4	182	104	68	48	35	26	20	14	10	TW B3	141	58	32	21	14	10	7	5	3
TW I5	178	104	69	49	36	27	20	15	10	TW B4	131	54	30	19	13	9	7	5	3
TW J1	218	116	74	51	37	27	20	14	10	TW C1	188	53	27	16	10	7	5	3	2
TW J2	198	110	71	50	36	27	20	15	10	TW C2	144	52	28	18	12	8	6	4	2
TW J3	212	130	89	64	48	36	27	20	14	TW C3	139	49	27	17	11	8	5	4	2
TW J4	195	120	82	59	44	34	25	19	13	TW C4	149	45	23	14	9	7	5	3	2
DG A1	153	62	32	16	-	-	-	-	-	TW D1	99	33	16	9	5	3	2	1	-
DG A2	178	78	46	30	20	13	8	3	-	TW D2	164	75	44	29	20	14	10	7	5
DG A3	223	102	62	43	31	23	17	11	6	TW D3	167	82	49	33	23	17	12	8	5
DG B1	154	64	36	23	14	-	-	-	-	TW D4	168	91	58	40	29	21	15	11	7
DG B2	182	81	49	33	24	17	12	7	2	TW D5	160	89	58	40	29	22	16	12	8
DG B3	227	105	64	45	33	25	19	14	9	TW E1	198	98	59	40	28	20	15	10	7
DG C1	191	83	48	32	22	15	9	-	-	TW E2	182	95	59	40	29	21	15	11	7
DG C2	216	98	59	41	30	22	16	11	6	TW E3	197	115	76	54	39	29	22	16	11
DG C3	261	122	75	52	39	30	23	18	12	TW E4	178	105	70	50	37	27	20	15	10
SS A1	469	138	71	44	29	20	14	10	6	TW F1	221	50	23	13	8	5	4	2	1
SS A2	422	177	102	66	46	33	24	17	11	TW F2	139	53	29	18	12	8	6	4	2
SS A3	447	124	62	37	24	17	12	8	5	TW F3	116	52	30	19	13	9	7	5	3
SS A4	428	175	100	64	44	32	23	16	10	TW F4	96	47	28	19	13	9	7	5	3
SS A5	802	138	64	37	24	16	11	8	5	TW G1	137	44	22	13	9	6	4	3	2
SS A6	419	175	101	65	45	32	23	16	11	TW G2	101	44	25	16	11	8	5	4	2
SS A7	914	115	51	28	18	12	8	5	3	TW G3	86	41	24	16	11	8	6	4	2
SS A8	429	172	97	62	42	30	22	15	10	TW G4	69	34	21	14	10	7	5	3	2
SS B1	318	104	55	34	23	16	11	8	5	TW H1	89	36	20	13	8	6	4	3	2
SS B2	331	143	84	55	38	27	20	14	9	TW H2	69	33	19	12	9	6	4	3	2
SS B3	291	95	49	30	20	14	10	7	4	TW H3	59	28	17	11	8	5	4	3	2
SS B4	324	139	81	53	37	26	19	13	9	TW H4	46	22	13	9	6	4	3	2	1
SS B5	419	93	45	26	17	12	8	6	3	TW I1	74	26	13	7	4	2	1	-	-
SS B6	312	137	80	53	37	26	19	14	9	TW I2	125	62	38	25	18	13	9	7	4
SS B7	367	78	37	21	14	9	6	4	3	TW I3	133	69	43	29	20	15	11	8	5
SS B8	302	131	76	50	34	25	18	13	8	TW I4	139	78	51	35	26	19	14	10	7
SS C1	202	89	52	34	23	17	12	8	5	TW I5	137	80	53	37	27	20	15	11	7
SS C2	217	111	69	47	34	25	18	13	9	TW J1	164	86	54	36	26	19	14	10	6
SS C3	212	72	39	24	16	11	8	6	4	TW J2	150	82	53	36	26	19	14	10	7
SS C4	216	100	60	40	28	20	15	11	7	TW J3	165	101	68	49	36	27	20	15	10
SS D1	425	165	92	58	39	28	20	14	9	TW J4	153	93	63	46	34	25	19	14	10
SS D2	371	192	120	82	58	43	31	22	15	DG A1	98	34	-	-	-	-	-	-	-
SS D3	542	156	80	49	32	22	16	11	7	DG A2	130	55	31	19	11	6	-	-	-
SS D4	380	189	116	78	55	40	29	21	14	DG A3	173	78	47	32	23	16	11	7	2
SS E1	315	128	72	46	31	22	16	11	7	DG B1	100	36	17	-	-	-	-	-	-
SS E2	305	157	98	66	47	34	25	18	12	DG B2	134	58	33	22	15	10	6	-	-
SS E3	386	110	56	34	22	15	11	7	5	DG B3	177	81	49	33	24	18	14	10	6
SS E4	315	148	89	59	41	30	22	15	10	DG C1	131	52	28	17	9	-	-	-	-

NASHVILLE, TENNESSEE — CONTINUED

SSF =	.10	.20	.30	.40	.50	.60	.70	.80	.90
DG C2	161	71	42	28	20	14	10	6	-
DG C3	205	94	57	39	29	22	17	12	8
SS A1	351	100	50	29	19	13	9	6	4
SS A2	328	135	76	49	33	24	17	12	8
SS A3	330	87	41	24	15	10	6	4	2
SS A4	331	133	74	47	32	22	16	11	7
SS A5	595	98	43	24	15	10	7	4	2
SS A6	324	132	75	48	32	23	16	11	7
SS A7	668	79	32	17	10	6	4	2	1
SS A8	330	129	71	45	30	21	15	10	6
SS B1	236	74	38	23	15	10	7	5	3
SS B2	258	110	63	41	28	20	14	10	6
SS B3	212	65	32	19	12	8	5	3	2
SS B4	251	105	60	39	27	19	13	9	6
SS B5	307	65	30	17	10	7	4	3	2
SS B6	241	104	60	39	27	19	14	10	6
SS B7	264	52	23	12	7	5	3	2	-
SS B8	233	98	56	36	25	17	12	9	5
SS C1	141	60	33	21	14	10	7	5	3
SS C2	161	81	50	33	23	17	12	9	6
SS C3	149	48	25	15	10	7	4	3	2
SS C4	160	73	43	28	19	14	10	7	5
SS D1	317	119	64	39	26	18	13	8	5
SS D2	287	147	90	61	43	31	23	16	10
SS D3	405	113	55	33	21	14	10	6	4
SS D4	295	144	87	58	40	29	21	15	10
SS E1	229	89	48	29	19	13	9	6	4
SS E2	233	118	72	48	34	24	18	12	8
SS E3	283	77	37	22	14	9	6	4	2
SS E4	242	111	65	43	29	21	15	11	7

ABILENE, TEXAS — 2610 DD

SSF =	.10	.20	.30	.40	.50	.60	.70	.80	.90
WW A1	1296	161	76	46	30	21	15	11	7
WW A2	434	174	100	66	46	34	25	18	12
WW A3	364	175	108	73	52	39	29	21	14
WW A4	334	174	111	77	56	41	31	22	15
WW A5	318	176	115	81	59	44	33	24	16
WW A6	308	176	116	83	61	46	34	25	17
WW B1	370	138	77	50	35	25	18	13	8
WW B2	323	174	112	78	57	43	32	23	16
WW B3	374	211	139	98	72	54	41	30	20
WW B4	323	197	135	98	73	56	42	31	21
WW B5	295	184	127	93	70	53	41	30	21
WW C1	395	223	147	104	76	57	43	32	22
WW C2	339	199	134	96	71	54	41	30	21
WW C3	340	220	156	115	88	67	52	39	27
WW C4	306	198	140	104	79	61	46	35	24
TW A1	1108	153	73	44	29	21	15	10	7
TW A2	443	158	88	57	40	29	21	15	10
TW A3	357	155	92	62	44	32	23	17	11
TW A4	302	146	90	62	44	33	24	17	12
TW B1	621	143	73	45	31	22	16	11	7
TW B2	355	142	82	54	38	27	20	14	10
TW B3	309	136	81	54	38	28	21	15	10
TW B4	283	125	74	50	35	26	19	14	9
TW C1	408	128	69	44	30	22	16	11	7
TW C2	310	121	69	46	32	23	17	12	8
TW C3	295	113	64	42	30	21	16	11	7
TW C4	309	101	55	35	25	18	13	9	6
TW D1	267	107	61	40	28	20	15	10	7
TW D2	345	165	102	69	50	37	27	20	13
TW D3	352	181	115	79	57	42	32	23	16

ABILENE, TEXAS — CONTINUED

SSF =	.10	.20	.30	.40	.50	.60	.70	.80	.90
TW D4	332	185	121	86	63	47	36	26	18
TW D5	305	176	117	83	62	46	35	26	18
TW E1	405	209	132	91	66	49	37	27	18
TW E2	365	197	127	89	65	48	36	27	18
TW E3	378	226	153	110	82	62	47	35	24
TW E4	335	203	138	100	75	57	43	32	22
TW F1	509	132	69	43	29	21	15	11	7
TW F2	318	133	78	52	36	26	19	14	9
TW F3	263	127	78	53	38	28	21	15	10
TW F4	215	113	72	50	36	27	20	15	10
TW G1	317	114	63	41	29	21	15	11	7
TW G2	230	109	66	45	32	24	17	13	8
TW G3	193	98	62	43	31	23	17	12	8
TW G4	154	81	51	36	26	19	14	10	7
TW H1	205	91	54	36	26	19	14	10	7
TW H2	157	79	49	34	24	18	13	10	7
TW H3	130	67	42	29	21	15	11	8	6
TW H4	100	51	32	22	16	12	9	6	4
TW I1	208	90	53	35	24	18	13	9	6
TW I2	265	138	88	61	44	33	25	18	12
TW I3	279	151	98	68	50	37	28	20	14
TW I4	275	160	107	76	57	43	32	24	16
TW I5	265	158	107	77	57	44	33	24	17
TW J1	335	183	119	83	61	46	34	25	17
TW J2	300	170	113	80	59	44	33	25	17
TW J3	314	196	135	99	74	57	43	32	22
TW J4	288	180	124	91	69	52	40	30	20
DG A1	265	119	71	45	29	18	10	-	-
DG A2	277	128	79	54	38	27	19	12	7
DG A3	328	154	97	68	51	38	29	20	12
DG B1	264	121	74	51	37	27	18	10	-
DG B2	283	133	83	58	43	33	25	17	10
DG B3	331	158	99	71	54	42	32	24	16
DG C1	315	147	90	64	47	36	27	18	8
DG C2	329	155	97	69	52	40	31	23	14
DG C3	379	181	114	82	62	49	39	30	20
SS A1	721	228	122	77	52	37	27	19	12
SS A2	636	277	163	108	76	55	40	29	19
SS A3	700	210	110	69	47	33	23	16	11
SS A4	648	277	162	106	74	54	39	28	18
SS A5	1204	229	112	68	45	31	22	16	10
SS A6	634	276	162	107	75	55	40	29	19
SS A7	1372	198	93	55	36	25	17	12	8
SS A8	654	273	158	104	72	52	38	27	18
SS B1	496	173	95	61	42	30	22	15	10
SS B2	495	222	132	88	62	45	33	24	16
SS B3	463	161	88	56	38	27	20	14	9
SS B4	488	218	129	86	61	44	32	23	15
SS B5	645	157	80	49	33	23	16	12	7
SS B6	470	214	128	86	61	44	33	23	16
SS B7	576	135	68	41	28	19	14	10	6
SS B8	460	207	123	82	58	42	31	22	15
SS C1	330	153	92	62	44	32	24	17	11
SS C2	337	178	113	79	57	43	32	23	16
SS C3	341	123	69	44	31	22	16	12	8
SS C4	333	160	98	67	48	35	26	19	13
SS D1	667	272	157	102	71	51	37	26	17
SS D2	564	301	192	133	96	71	53	38	26
SS D3	834	258	138	87	59	42	30	21	14
SS D4	577	295	184	126	91	67	49	36	24
SS E1	505	216	126	83	58	42	30	22	14
SS E2	470	249	158	109	79	58	43	31	21
SS E3	604	186	99	62	42	30	21	15	10
SS E4	482	234	143	97	69	51	37	27	18

AMARILLO, TEXAS								4183 DD		AMARILLO, TEXAS								CONTINUED	
SSF =.10	.20	.30	.40	.50	.60	.70	.80	.90		SSF =.10	.20	.30	.40	.50	.60	.70	.80	.90	
WW A1 1065	125	59	35	24	16	12	8	5		DG C2 264	125	78	55	42	32	25	18	11	
WW A2 347	138	79	52	37	27	20	14	9		DG C3 312	149	94	67	51	41	32	25	17	
WW A3 291	139	86	58	42	31	23	17	11		SS A1 595	182	97	61	41	29	21	15	10	
WW A4 267	139	88	61	44	33	25	18	12		SS A2 527	226	132	88	62	45	33	23	15	
WW A5 254	141	92	65	47	36	27	20	13		SS A3 573	165	86	53	36	25	18	12	8	
WW A6 247	141	93	66	49	37	28	20	14		SS A4 534	224	130	85	60	43	31	22	15	
WW B1 287	105	59	38	26	19	14	10	6		SS A5 1011	182	88	53	35	24	17	12	8	
WW B2 262	141	91	64	47	35	26	19	13		SS A6 524	224	131	87	61	44	32	23	15	
WW B3 304	171	113	80	59	44	33	25	17		SS A7 1153	154	71	42	27	19	13	9	6	
WW B4 266	162	111	81	61	46	35	26	18		SS A8 537	220	127	83	58	42	30	21	14	
WW B5 244	152	106	77	58	45	34	25	18		SS B1 404	137	75	48	33	23	17	12	8	
WW C1 322	182	120	85	63	47	35	26	18		SS B2 409	181	107	72	51	37	27	19	13	
WW C2 278	164	110	79	59	45	34	25	17		SS B3 373	126	68	43	30	21	15	11	7	
WW C3 283	184	130	96	73	57	43	33	23		SS B4 401	176	104	69	49	36	26	19	12	
WW C4 255	165	117	87	66	51	39	29	20		SS B5 530	123	62	38	25	18	13	9	6	
TW A1 911	119	57	34	23	16	12	8	5		SS B6 386	174	104	70	49	36	26	19	13	
TW A2 357	125	69	45	31	23	17	12	8		SS B7 470	105	52	31	21	15	10	7	5	
TW A3 286	124	73	49	35	25	19	14	9		SS B8 376	167	99	66	46	34	25	18	12	
TW A4 242	117	72	49	35	26	19	14	10		SS C1 262	120	73	49	35	26	19	14	9	
TW B1 506	112	57	36	24	17	12	9	6		SS C2 274	144	92	64	47	35	26	19	13	
TW B2 286	113	65	43	30	22	16	12	8		SS C3 273	97	54	35	24	18	13	9	6	
TW B3 249	109	65	43	31	23	17	12	8		SS C4 271	129	79	54	39	29	21	16	11	
TW B4 229	100	60	40	28	21	15	11	7		SS D1 543	217	124	81	56	40	29	21	14	
TW C1 331	101	55	35	24	17	13	9	6		SS D2 464	246	156	108	78	58	43	31	21	
TW C2 251	97	55	36	26	19	14	10	7		SS D3 688	206	109	68	46	33	23	16	11	
TW C3 240	90	51	34	24	17	13	9	6		SS D4 475	240	150	103	74	54	40	29	19	
TW C4 253	81	44	28	20	14	10	7	5		SS E1 405	170	99	65	45	33	24	17	11	
TW D1 208	81	46	30	21	15	11	8	5		SS E2 383	201	128	88	64	47	35	25	17	
TW D2 280	133	82	56	40	30	22	16	11		SS E3 492	146	77	48	33	23	17	12	8	
TW D3 286	147	93	64	46	35	26	19	13		SS E4 394	189	116	78	56	41	30	22	15	
TW D4 272	152	100	71	52	39	29	22	15											
TW D5 253	145	97	69	51	39	29	22	15		AUSTIN, TEXAS								1737 DD	
TW E1 330	170	107	74	54	40	30	22	15		SSF =.10	.20	.30	.40	.50	.60	.70	.80	.90	
TW E2 299	161	104	73	53	40	30	22	15		WW A1 1627	204	95	57	38	27	19	13	9	
TW E3 313	187	127	92	69	52	39	29	20		WW A2 544	217	124	81	57	41	30	22	14	
TW E4 278	169	115	84	63	48	36	27	19		WW A3 455	217	132	89	64	47	35	25	17	
TW F1 410	103	53	33	23	16	12	8	5		WW A4 415	215	135	93	68	50	37	27	18	
TW F2 254	105	62	41	29	21	15	11	7		WW A5 394	216	140	98	72	54	40	29	20	
TW F3 210	101	62	42	30	22	17	12	8		WW A6 381	216	142	100	74	55	42	30	21	
TW F4 172	90	58	40	29	22	16	12	8		WW B1 476	176	98	63	44	31	23	16	11	
TW G1 253	89	50	32	22	16	12	8	6		WW B2 399	213	136	95	69	51	38	28	19	
TW G2 184	86	53	36	26	19	14	10	7		WW B3 459	257	168	118	87	65	49	36	24	
TW G3 154	78	49	34	25	18	14	10	7		WW B4 391	237	161	116	87	66	50	37	25	
TW G4 124	65	41	29	21	15	12	8	6		WW B5 355	219	151	110	83	63	48	35	24	
TW H1 164	72	43	29	20	15	11	8	5		WW C1 483	270	177	125	92	69	51	38	26	
TW H2 125	63	39	27	19	14	11	8	5		WW C2 412	240	160	115	85	64	48	36	24	
TW H3 104	53	33	23	17	12	9	7	5		WW C3 408	262	184	136	103	79	60	45	31	
TW H4 81	40	25	17	13	9	7	5	3		WW C4 366	235	165	122	92	71	54	40	28	
TW I1 161	69	40	26	19	13	10	7	5		TW A1 1387	193	91	55	37	26	18	13	8	
TW I2 214	112	71	49	36	27	20	15	10		TW A2 556	197	108	70	49	35	25	18	12	
TW I3 227	123	79	56	41	30	23	17	11		TW A3 446	193	113	75	53	39	28	20	14	
TW I4 226	132	88	63	47	35	27	20	14		TW A4 376	181	110	75	54	40	29	21	14	
TW I5 219	131	89	64	48	36	28	20	14		TW B1 778	180	90	56	38	27	19	14	9	
TW J1 273	149	97	68	50	37	28	21	14		TW B2 444	176	100	66	46	33	24	17	12	
TW J2 246	140	92	66	49	37	28	20	14		TW B3 386	168	99	66	47	34	25	18	12	
TW J3 261	163	113	82	62	48	36	27	19		TW B4 352	154	91	60	43	31	23	17	11	
TW J4 239	149	104	76	57	44	34	25	17		TW C1 510	159	85	54	37	27	19	14	9	
DG A1 204	91	53	32	20	11	-	-	-		TW C2 387	150	85	55	39	28	21	15	10	
DG A2 222	102	63	43	30	21	14	9	4		TW C3 367	139	79	51	36	26	19	14	9	
DG A3 269	126	79	56	42	32	23	16	10		TW C4 383	124	67	43	30	21	16	11	7	
DG B1 202	92	56	39	27	19	12	-	-		TW D1 344	136	77	50	35	25	18	13	9	
DG B2 226	106	66	46	35	26	19	13	7		TW D2 427	203	123	84	60	44	33	24	16	
DG B3 272	129	82	58	44	35	27	20	13		TW D3 435	222	139	96	69	51	38	28	19	
DG C1 244	114	70	49	36	27	20	12	-											

AUSTIN, TEXAS						CONTINUED				BROWNSVILLE, TEXAS								650 DD	
SSF =	.10	.20	.30	.40	.50	.60	.70	.80	.90	SSF =	.10	.20	.30	.40	.50	.60	.70	.80	.90
TW D4	404	223	145	102	75	56	42	31	21	WW A1	3797	446	210	127	85	60	43	31	20
TW D5	369	210	139	99	73	55	41	30	21	WW A2	1162	458	264	174	123	89	66	47	32
TW E1	498	254	159	110	79	59	44	32	21	WW A3	955	455	279	190	137	101	75	54	37
TW E2	446	238	152	106	77	58	43	32	21	WW A4	866	449	285	198	144	107	80	58	39
TW E3	457	271	182	131	97	74	56	41	28	WW A5	818	452	294	207	152	114	85	63	43
TW E4	404	242	164	118	88	67	51	38	26	WW A6	790	450	297	211	156	117	88	65	44
TW F1	642	167	86	53	36	26	19	13	9	WW B1	1070	396	223	146	102	74	54	39	26
TW F2	400	166	96	63	45	32	24	17	11	WW B2	814	435	280	195	143	107	80	59	40
TW F3	328	157	96	65	47	34	25	18	12	WW B3	931	523	344	243	179	135	101	74	51
TW F4	268	139	88	61	44	33	24	18	12	WW B4	773	468	319	231	173	131	100	74	51
TW G1	399	142	79	51	35	25	18	13	9	WW B5	694	429	296	216	162	124	94	70	48
TW G2	288	135	82	55	39	29	21	15	10	WW C1	973	547	359	254	187	141	106	78	53
TW G3	240	121	76	52	37	28	21	15	10	WW C2	819	479	321	230	171	129	98	72	49
TW G4	192	99	63	43	31	23	17	13	8	WW C3	790	509	358	265	201	154	118	88	61
TW H1	258	113	67	45	32	23	17	12	8	WW C4	705	453	319	236	179	137	105	78	54
TW H2	196	97	60	41	30	22	16	12	8	TW A1	3202	419	199	121	81	57	41	29	19
TW H3	162	82	51	35	25	19	14	10	7	TW A2	1196	415	231	150	105	76	55	40	26
TW H4	124	62	39	26	19	14	10	8	5	TW A3	941	404	239	160	114	83	61	44	30
TW I1	268	115	67	44	31	22	16	12	8	TW A4	783	376	232	158	114	84	62	45	31
TW I2	327	169	107	74	53	40	29	21	14	TW B1	1736	384	195	122	83	59	43	30	20
TW I3	344	185	118	82	60	45	33	24	16	TW B2	941	368	212	140	98	72	53	38	25
TW I4	335	193	128	91	67	51	38	28	19	TW B3	808	351	208	140	99	73	54	39	26
TW I5	320	189	127	91	68	51	39	29	20	TW B4	735	319	190	127	90	66	49	35	24
TW J1	411	223	143	100	73	55	41	30	20	TW C1	1102	335	181	116	80	58	42	30	20
TW J2	366	206	135	95	70	53	40	29	20	TW C2	815	311	178	117	82	60	44	32	21
TW J3	378	233	160	117	88	67	51	38	26	TW C3	773	289	164	108	76	55	40	29	19
TW J4	345	214	147	107	81	62	47	35	24	TW C4	813	257	140	90	62	45	33	24	16
DG A1	340	152	90	58	39	26	16	8	-	TW D1	766	303	175	115	81	59	43	31	21
DG A2	346	158	97	66	47	34	24	16	9	TW D2	876	414	254	173	124	92	68	49	33
DG A3	404	186	116	82	61	46	34	24	14	TW D3	891	455	287	199	144	107	80	58	39
DG B1	343	153	94	65	47	35	24	15	6	TW D4	803	444	290	205	150	113	85	62	42
DG B2	358	163	101	71	53	40	30	22	13	TW D5	723	412	273	194	143	108	81	60	41
DG B3	411	189	118	84	64	50	38	29	19	TW E1	1010	516	326	225	163	121	91	66	45
DG C1	408	183	113	80	59	45	34	24	13	TW E2	894	477	306	214	156	117	87	64	44
DG C2	415	189	118	84	63	49	37	28	18	TW E3	895	532	359	258	192	146	110	82	56
DG C3	472	218	136	97	74	58	46	35	24	TW E4	783	470	319	231	172	131	99	74	50
SS A1	906	286	152	95	65	47	33	24	15	TW F1	1434	360	187	118	80	57	42	30	20
SS A2	790	341	199	131	93	67	49	35	23	TW F2	852	352	206	137	97	71	52	37	25
SS A3	890	267	139	87	59	42	30	21	14	TW F3	690	331	204	139	100	74	55	40	27
SS A4	812	343	199	131	92	67	49	35	23	TW F4	557	291	185	129	94	70	52	38	26
SS A5	1493	291	140	85	57	40	28	20	13	TW G1	866	304	170	110	77	56	41	29	19
SS A6	790	340	198	131	92	67	49	35	23	TW G2	608	285	174	118	85	62	46	34	23
SS A7	1702	255	118	70	46	32	23	16	10	TW G3	502	254	160	110	80	59	44	32	22
SS A8	820	340	195	128	90	65	47	34	22	TW G4	398	207	131	91	66	49	37	27	18
SS B1	629	219	119	76	52	37	27	19	12	TW H1	548	240	143	96	68	50	37	27	18
SS B2	616	273	161	107	76	55	41	29	19	TW H2	410	204	127	88	63	47	35	25	17
SS B3	592	206	111	71	49	35	25	18	11	TW H3	337	171	107	74	54	40	30	22	15
SS B4	610	269	159	105	74	54	40	29	19	TW H4	258	129	80	55	40	30	22	16	11
SS B5	812	200	100	62	42	29	21	15	9	TW I1	599	258	152	102	72	53	39	28	19
SS B6	585	264	157	105	74	54	40	29	19	TW I2	669	346	220	152	111	82	62	45	30
SS B7	732	174	86	53	35	25	18	12	8	TW I3	700	377	243	170	125	93	70	51	35
SS B8	577	257	151	101	71	52	38	27	18	TW I4	664	383	255	182	135	102	77	57	39
SS C1	415	190	114	76	54	40	29	21	14	TW I5	627	371	250	180	134	102	77	57	39
SS C2	416	217	137	95	69	51	38	28	19	TW J1	831	451	292	205	150	112	84	62	42
SS C3	429	154	85	55	38	27	20	14	9	TW J2	730	411	270	192	141	106	80	59	40
SS C4	412	196	119	81	58	43	31	23	15	TW J3	736	456	314	229	172	131	100	74	51
SS D1	841	341	194	127	88	64	46	33	22	TW J4	668	414	286	208	157	120	91	68	47
SS D2	698	368	233	161	117	87	64	47	31	DG A1	749	340	208	140	99	71	51	34	21
SS D3	1048	324	171	107	73	52	38	26	17	DG A2	718	328	204	142	103	76	56	40	25
SS D4	715	361	224	153	110	81	60	44	29	DG A3	799	369	231	164	122	93	70	51	32
SS E1	643	273	158	104	73	53	38	27	18	DG B1	747	343	213	151	113	86	65	46	29
SS E2	585	306	193	133	97	72	53	38	26	DG B2	733	340	212	150	114	88	68	50	33
SS E3	766	236	124	78	53	38	27	19	12	DG B3	803	375	234	167	127	99	78	59	40
SS E4	602	289	176	119	85	62	46	33	22	DG C1	864	402	249	177	134	105	81	61	40

LCR Tables / 589

BROWNSVILLE, TEXAS   CONTINUED   CORPUS CHRISTI, TEXAS   CONTINUED
   SSF =.10   .20   .30   .40   .50   .60   .70   .80   .90   SSF =.10   .20   .30   .40   .50   .60   .70   .80   .90

	SSF=.10	.20	.30	.40	.50	.60	.70	.80	.90		SSF=.10	.20	.30	.40	.50	.60	.70	.80	.90
DG C2	836	390	243	173	132	103	80	61	42	TW D4	602	338	223	159	117	88	66	49	33
DG C3	916	431	269	192	146	116	92	71	50	TW D5	546	315	211	151	112	65	64	47	32
SS A1	1954	615	334	214	148	106	76	54	35	TW E1	748	388	248	173	126	94	70	51	34
SS A2	1626	707	421	281	199	145	106	76	51	TW E2	668	361	235	166	122	91	68	50	34
SS A3	1958	589	317	202	138	99	71	50	33	TW E3	675	406	277	201	150	114	86	64	44
SS A4	1690	724	428	285	201	146	107	77	51	TW E4	594	361	248	180	135	103	78	58	40
SS A5	3342	630	313	193	130	92	66	46	30	TW F1	1031	263	139	88	61	44	32	22	15
SS A6	1632	709	422	282	200	145	106	76	51	TW F2	620	261	154	104	74	54	40	29	19
SS A7	3930	566	272	166	111	77	55	39	25	TW F3	506	247	154	106	77	57	42	30	20
SS A8	1718	720	423	281	198	144	105	75	50	TW F4	411	218	141	98	72	54	40	29	20
SS B1	1357	473	263	171	119	85	62	44	29	TW G1	627	224	127	84	59	43	31	22	15
SS B2	1262	564	338	227	162	118	87	62	42	TW G2	445	212	131	90	65	48	35	26	17
SS B3	1295	453	253	164	114	81	59	42	28	TW G3	370	190	121	84	61	46	34	25	17
SS B4	1260	563	337	227	161	118	86	62	41	TW G4	294	155	100	70	51	38	28	21	14
SS B5	1810	436	225	141	96	68	49	34	23	TW H1	400	178	108	73	52	38	28	20	14
SS B6	1204	549	331	223	159	116	86	62	41	TW H2	301	152	96	67	49	36	27	19	13
SS B7	1658	389	199	124	85	60	43	30	20	TW H3	248	128	81	57	41	31	23	17	11
SS B8	1199	540	325	218	155	113	83	60	40	TW H4	190	96	61	42	31	23	17	12	8
SS C1	884	405	245	166	118	87	64	47	31	TW I1	428	188	112	76	54	39	29	21	14
SS C2	848	443	282	196	143	106	79	58	39	TW I2	496	261	167	117	86	64	48	35	23
SS C3	930	329	184	120	84	60	44	32	21	TW I3	519	284	185	131	96	72	54	39	27
SS C4	845	400	245	167	120	88	66	48	32	TW I4	499	292	197	141	105	80	60	44	30
SS D1	1788	733	427	283	199	144	105	75	50	TW I5	474	284	194	140	105	80	60	45	30
SS D2	1427	765	491	343	249	185	137	100	67	TW J1	618	340	223	158	116	87	65	48	32
SS D3	2266	698	378	242	166	119	86	61	40	TW J2	547	313	208	149	110	83	62	46	31
SS D4	1463	750	473	327	236	174	129	93	63	TW J3	557	350	243	179	135	103	78	58	40
SS E1	1385	596	353	235	166	121	88	63	42	TW J4	508	319	222	163	123	94	72	53	37
SS E2	1209	643	412	287	208	155	115	83	56	DG A1	537	249	154	104	73	52	36	23	13
SS E3	1679	514	278	177	122	87	63	45	29	DG A2	522	246	155	108	78	58	42	29	18
SS E4	1246	606	374	256	184	135	100	72	48	DG A3	585	280	179	127	95	73	54	39	24
										DG B1	527	252	160	113	84	64	47	33	19
										DG B2	529	253	162	116	87	67	51	38	25
										DG B3	589	283	182	131	99	78	61	45	31
CORPUS CHRISTI, TEXAS							930 DD			DG C1	614	295	189	135	101	79	61	45	28
WW A1	2719	324	155	95	64	46	33	23	15	DG C2	606	291	187	134	102	79	62	46	31
WW A2	845	339	198	132	94	68	50	36	24	DG C3	672	323	209	151	115	91	72	55	38
WW A3	700	339	211	145	105	77	57	42	28	SS A1	1382	441	243	158	110	79	57	40	26
WW A4	638	336	216	151	110	82	61	45	30	SS A2	1170	516	311	210	150	110	80	57	38
WW A5	605	339	223	158	117	88	66	48	32	SS A3	1369	418	228	147	102	73	52	37	24
WW A6	586	338	226	162	120	90	68	50	34	SS A4	1207	524	314	211	150	110	80	57	38
WW B1	763	287	165	109	76	55	40	29	19	SS A5	2347	449	226	141	96	68	49	34	22
WW B2	604	328	213	150	110	83	62	45	30	SS A6	1173	517	311	210	150	110	80	57	38
WW B3	692	395	262	187	139	104	78	57	39	SS A7	2730	398	194	120	81	57	40	28	18
WW B4	583	358	247	180	135	103	78	58	39	SS A8	1224	521	310	208	148	107	78	56	37
WW B5	526	330	229	169	127	97	74	55	38	SS B1	962	340	192	126	88	64	46	33	21
WW C1	726	413	275	196	145	109	82	60	41	SS B2	913	414	251	171	122	90	66	47	31
WW C2	616	365	247	178	133	101	76	56	38	SS B3	911	324	183	120	84	60	44	31	20
WW C3	601	392	278	207	158	121	93	69	48	SS B4	906	410	249	169	121	89	65	47	31
WW C4	538	350	249	185	141	108	83	62	43	SS B5	1273	311	163	103	71	50	36	25	17
TW A1	2298	305	148	91	62	44	31	22	14	SS B6	870	402	246	167	120	88	65	47	31
TW A2	868	307	174	114	80	58	42	30	20	SS B7	1156	275	143	90	62	44	31	22	14
TW A3	688	301	180	122	87	64	47	34	23	SS B8	860	393	239	163	116	85	63	45	30
TW A4	576	281	175	121	87	65	48	35	23	SS C1	644	300	184	126	90	66	49	35	24
TW B1	1251	281	145	92	63	45	33	23	15	SS C2	629	333	215	151	110	82	61	45	30
TW B2	686	273	160	106	75	55	40	29	19	SS C3	673	242	137	90	64	46	34	24	16
TW B3	592	261	157	106	76	56	41	30	20	SS C4	624	300	186	128	93	68	51	37	25
TW B4	539	238	143	97	70	51	38	27	18	SS D1	1270	529	312	209	148	107	78	56	37
TW C1	800	247	136	88	61	44	32	23	15	SS D2	1036	562	365	257	188	140	104	75	50
TW C2	596	231	134	89	63	46	34	24	16	SS D3	1601	501	275	178	123	88	64	45	29
TW C3	565	215	124	82	58	42	31	22	15	SS D4	1059	551	352	245	178	132	98	71	47
TW C4	593	191	106	69	48	35	25	18	12	SS E1	983	430	258	173	123	90	66	47	31
TW D1	548	221	129	86	61	44	32	23	15	SS E2	877	472	306	215	157	117	87	63	42
TW D2	647	311	193	133	96	71	53	38	26	SS E3	1184	368	202	130	90	65	47	33	22
TW D3	659	342	218	152	111	83	61	45	30	SS E4	900	444	278	191	138	102	75	54	36

590 / APPENDIX 21

DALLAS, TEXAS — 2290 DD

	SSF =	.10	.20	.30	.40	.50	.60	.70	.80	.90
WW	A1	1218	153	71	43	28	20	14	10	6
WW	A2	415	165	95	62	43	31	23	16	11
WW	A3	349	166	101	69	49	36	27	19	13
WW	A4	319	165	104	72	52	39	29	21	14
WW	A5	304	167	108	76	55	41	31	23	15
WW	A6	294	167	110	78	57	43	32	24	16
WW	B1	351	130	72	46	32	23	16	12	8
WW	B2	310	166	106	74	54	40	30	22	15
WW	B3	358	201	131	93	68	51	38	28	19
WW	B4	310	188	128	93	69	53	40	30	20
WW	B5	284	176	121	88	66	51	39	29	20
WW	C1	378	212	139	98	72	54	41	30	20
WW	C2	325	190	127	91	68	51	39	28	19
WW	C3	327	210	148	109	83	64	49	36	25
WW	C4	294	189	133	99	75	58	44	33	23
TW	A1	1042	145	69	41	27	19	14	10	6
TW	A2	424	150	83	53	37	27	19	14	9
TW	A3	342	148	87	58	41	30	22	16	10
TW	A4	289	139	85	58	41	31	23	16	11
TW	B1	589	136	69	42	29	20	15	10	7
TW	B2	340	135	77	50	35	26	19	13	9
TW	B3	296	130	76	51	36	26	19	14	9
TW	B4	271	119	70	47	33	24	18	13	9
TW	C1	389	122	65	41	29	20	15	11	7
TW	C2	297	115	65	43	30	22	16	11	8
TW	C3	283	108	61	40	28	20	15	10	7
TW	C4	295	96	52	33	23	17	12	9	6
TW	D1	254	100	57	37	26	18	13	9	6
TW	D2	332	158	96	65	47	34	25	18	12
TW	D3	338	173	108	75	54	40	30	22	14
TW	D4	319	177	115	81	60	45	34	25	17
TW	D5	294	168	111	79	58	44	33	24	17
TW	E1	388	199	125	86	62	46	34	25	17
TW	E2	351	188	120	84	61	46	34	25	17
TW	E3	363	216	146	105	78	59	45	33	23
TW	E4	323	194	132	95	71	54	41	30	21
TW	F1	483	126	64	40	27	19	14	10	6
TW	F2	304	127	73	48	34	25	18	13	9
TW	F3	251	121	74	50	36	26	19	14	9
TW	F4	206	107	68	47	34	25	19	14	9
TW	G1	302	108	60	38	27	19	14	10	7
TW	G2	220	103	63	42	30	22	16	12	8
TW	G3	185	93	58	40	29	21	16	11	8
TW	G4	148	77	48	33	24	18	13	10	7
TW	H1	196	86	51	34	24	18	13	9	6
TW	H2	150	75	46	32	23	17	12	9	6
TW	H3	125	63	39	27	20	14	11	8	5
TW	H4	96	48	30	20	15	11	8	6	4
TW	I1	198	84	49	32	23	16	12	8	6
TW	I2	254	132	83	58	42	31	23	17	11
TW	I3	267	144	92	64	47	35	26	19	13
TW	I4	265	153	102	72	54	40	30	22	15
TW	I5	255	151	102	73	55	41	31	23	16
TW	J1	321	174	112	79	57	43	32	23	16
TW	J2	288	163	107	76	56	42	31	23	16
TW	J3	302	187	129	94	71	54	41	30	21
TW	J4	277	172	119	87	65	50	38	28	19
DG	A1	252	112	65	41	26	16	8	–	–
DG	A2	266	122	75	51	36	25	17	11	6
DG	A3	318	147	92	65	48	36	27	19	11
DG	B1	254	113	69	47	34	24	16	8	–
DG	B2	274	126	78	55	41	31	23	16	9
DG	B3	324	150	94	67	51	39	31	23	15
DG	C1	306	137	84	59	44	33	24	16	6

DALLAS, TEXAS — CONTINUED

	SSF =	.10	.20	.30	.40	.50	.60	.70	.80	.90
DG	C2	321	147	92	65	49	38	29	21	13
DG	C3	373	174	108	78	59	46	37	28	19
SS	A1	675	214	114	71	49	35	25	18	11
SS	A2	603	261	153	101	71	52	38	27	18
SS	A3	653	197	102	64	43	30	22	15	10
SS	A4	614	261	151	99	70	51	37	26	17
SS	A5	1103	216	104	63	42	29	21	14	9
SS	A6	601	260	152	100	71	51	38	27	18
SS	A7	1237	186	86	51	33	23	16	11	7
SS	A8	619	257	148	97	68	49	36	25	17
SS	B1	469	164	89	57	39	28	20	14	9
SS	B2	472	211	125	83	59	43	31	23	15
SS	B3	436	152	82	52	36	26	18	13	8
SS	B4	465	206	122	81	57	42	30	22	14
SS	B5	600	148	74	45	31	21	15	11	7
SS	B6	448	203	121	81	57	42	31	22	15
SS	B7	534	127	63	38	26	18	13	9	6
SS	B8	438	196	116	77	54	40	29	21	14
SS	C1	316	145	87	58	41	30	22	16	11
SS	C2	324	169	107	74	54	40	30	22	15
SS	C3	325	117	65	42	29	21	15	11	7
SS	C4	320	152	93	63	45	33	25	18	12
SS	D1	629	256	146	95	66	48	35	25	16
SS	D2	536	284	180	124	90	67	50	36	24
SS	D3	780	243	128	80	55	39	28	20	13
SS	D4	548	278	173	118	85	63	46	34	22
SS	E1	479	203	118	77	54	39	28	20	13
SS	E2	448	235	149	103	74	55	41	30	20
SS	E3	567	175	92	58	39	28	20	14	9
SS	E4	460	222	135	91	65	48	35	25	17

DEL RIO, TEXAS — 1523 DD

	SSF =	.10	.20	.30	.40	.50	.60	.70	.80	.90
WW	A1	2039	252	117	70	47	33	24	17	11
WW	A2	668	264	151	99	70	51	37	26	17
WW	A3	555	263	160	109	78	57	42	31	20
WW	A4	505	260	164	113	82	61	45	33	22
WW	A5	478	261	169	119	87	65	49	36	24
WW	A6	462	261	171	122	90	67	51	37	25
WW	B1	594	218	121	79	55	39	29	20	13
WW	B2	481	255	163	114	83	62	46	34	23
WW	B3	551	307	201	142	104	78	59	43	29
WW	B4	466	281	191	138	103	79	60	44	30
WW	B5	422	260	178	130	98	75	57	42	29
WW	C1	579	322	211	149	110	82	62	45	30
WW	C2	492	286	191	137	102	77	58	43	29
WW	C3	481	308	217	160	122	93	71	53	36
WW	C4	432	277	194	144	109	84	64	48	33
TW	A1	1733	238	112	68	45	32	23	16	10
TW	A2	684	239	132	85	59	43	31	22	15
TW	A3	546	234	137	91	65	47	35	25	17
TW	A4	457	218	133	91	65	48	36	26	17
TW	B1	966	220	110	68	47	33	24	17	11
TW	B2	544	213	121	80	56	41	30	21	14
TW	B3	470	204	120	80	57	42	31	22	15
TW	B4	429	186	110	73	52	38	28	20	13
TW	C1	628	194	103	66	46	33	24	17	11
TW	C2	473	181	103	67	47	34	25	18	12
TW	C3	449	169	95	62	43	31	23	16	11
TW	C4	469	150	81	52	36	26	19	13	9
TW	D1	429	168	95	62	44	32	23	16	11
TW	D2	517	244	148	101	72	53	39	28	19
TW	D3	525	266	167	115	83	62	46	33	22

DEL RIO, TEXAS										EL PASO, TEXAS									2678 DD
SSF =	.10	.20	.30	.40	.50	.60	.70	.80	.90	SSF =	.10	.20	.30	.40	.50	.60	.70	.80	.90
TW D4	483	266	173	122	90	67	50	37	25	WW A1	1571	207	100	61	41	29	21	15	10
TW D5	440	249	164	117	86	65	49	36	24	WW A2	541	223	130	86	61	44	33	24	16
TW E1	599	304	190	131	95	71	53	38	26	WW A3	457	224	139	95	68	51	38	27	18
TW E2	535	284	182	127	93	69	52	38	25	WW A4	420	222	142	99	72	54	40	30	20
TW E3	542	320	215	155	115	87	66	49	33	WW A5	400	225	147	104	77	58	43	32	22
TW E4	478	286	193	140	104	79	60	44	30	WW A6	388	225	150	107	79	60	45	33	23
TW F1	798	205	105	66	45	32	23	16	11	WW B1	472	182	103	68	47	34	25	18	12
TW F2	490	202	117	77	55	40	29	21	14	WW B2	403	220	143	100	74	55	41	30	21
TW F3	401	191	117	79	57	42	31	22	15	WW B3	465	266	176	125	92	69	52	39	26
TW F4	325	169	107	74	54	40	30	22	14	WW B4	399	245	168	123	92	70	53	40	27
TW G1	493	174	96	62	43	31	23	16	11	WW B5	362	227	158	116	87	67	51	38	26
TW G2	352	164	99	67	48	35	26	19	12	WW C1	489	279	185	131	97	73	55	41	28
TW G3	293	147	92	63	46	34	25	18	12	WW C2	418	249	168	121	90	68	52	38	26
TW G4	233	120	76	52	38	28	21	15	10	WW C3	416	271	192	143	109	84	64	48	33
TW H1	316	138	81	54	39	28	21	15	10	WW C4	374	243	173	128	98	75	58	43	30
TW H2	238	118	73	50	36	27	20	14	10	TW A1	1341	196	95	58	39	28	20	14	9
TW H3	197	99	62	43	31	23	17	12	8	TW A2	549	201	114	74	52	38	28	20	13
TW H4	151	75	47	32	23	17	13	9	6	TW A3	445	198	119	80	57	42	31	22	15
TW I1	334	142	83	55	39	28	21	15	10	TW A4	377	186	116	80	57	42	32	23	16
TW I2	395	203	128	89	64	48	36	26	17	TW B1	759	183	95	59	41	29	21	15	10
TW I3	414	221	142	99	72	54	40	29	20	TW B2	441	180	105	70	49	36	26	19	13
TW I4	400	229	152	108	80	61	46	34	23	TW B3	385	173	104	70	50	37	27	20	13
TW I5	380	224	151	108	81	61	46	34	23	TW B4	351	158	95	64	46	34	25	18	12
TW J1	494	266	171	120	88	65	49	36	24	TW C1	501	162	89	57	40	29	21	15	10
TW J2	438	245	161	114	84	63	47	35	23	TW C2	383	153	89	59	41	30	22	16	11
TW J3	447	275	189	138	104	79	60	45	30	TW C3	364	143	82	54	38	28	20	15	10
TW J4	409	252	173	127	95	73	55	41	28	TW C4	377	127	70	45	32	23	17	12	8
DG A1	421	188	113	74	51	35	23	14	5	TW D1	342	140	81	54	38	27	20	14	10
DG A2	420	191	117	81	58	42	30	21	12	TW D2	427	208	129	88	64	47	35	26	17
DG A3	485	220	138	97	73	55	41	29	18	TW D3	438	229	145	101	73	55	41	30	20
DG B1	425	188	117	82	60	45	32	21	11	TW D4	408	231	152	108	80	60	45	33	23
DG B2	436	195	123	87	65	50	37	27	17	TW D5	374	217	145	104	77	58	44	33	22
DG B3	495	223	141	101	76	59	46	34	23	TW E1	501	262	167	116	84	63	47	34	23
DG C1	504	222	140	99	74	56	43	31	18	TW E2	450	246	159	112	82	62	46	34	23
DG C2	504	226	142	101	76	59	46	34	22	TW E3	463	279	190	137	103	78	59	44	30
DG C3	569	256	161	116	88	69	54	42	29	TW E4	410	250	171	124	93	71	54	40	28
SS A1	1103	341	181	115	79	56	40	28	18	TW F1	628	170	90	57	39	28	20	14	10
SS A2	942	401	235	156	110	80	58	42	28	TW F2	398	171	101	67	48	35	26	19	12
SS A3	1086	320	167	105	72	51	36	25	16	TW F3	330	163	101	69	50	37	27	20	14
SS A4	967	404	235	155	109	79	58	41	27	TW F4	270	144	93	65	47	35	26	19	13
SS A5	1844	349	169	103	69	48	34	24	15	TW G1	395	146	83	54	38	27	20	14	10
SS A6	943	401	234	156	110	80	58	42	27	TW G2	289	139	86	59	42	31	23	17	11
SS A7	2120	307	143	85	57	39	28	19	12	TW G3	242	126	80	55	40	30	22	16	11
SS A8	980	401	231	152	107	77	56	40	26	TW G4	194	103	66	46	34	25	19	14	9
SS B1	763	261	143	92	63	45	33	23	15	TW H1	258	117	70	47	34	25	18	13	9
SS B2	732	321	190	127	90	66	48	35	23	TW H2	197	101	63	44	32	24	18	13	9
SS B3	720	246	134	86	59	42	30	21	14	TW H3	163	85	54	37	27	20	15	11	7
SS B4	725	317	187	125	89	64	47	34	22	TW H4	125	64	40	28	20	15	11	8	6
SS B5	1000	240	121	75	51	36	25	18	11	TW I1	268	119	71	47	33	24	18	13	9
SS B6	696	311	185	124	88	65	47	34	22	TW I2	329	174	112	78	57	42	32	23	16
SS B7	904	210	105	64	43	31	22	15	10	TW I3	347	191	124	87	64	48	36	26	18
SS B8	686	302	179	120	85	62	45	32	21	TW I4	340	199	134	96	71	54	41	30	21
SS C1	509	231	139	93	67	49	36	26	17	TW I5	325	196	133	96	72	55	42	31	21
SS C2	501	260	165	114	83	62	46	33	22	TW J1	415	230	150	106	78	58	44	32	22
SS C3	529	188	104	67	47	34	25	17	12	TW J2	371	213	141	101	75	56	42	31	21
SS C4	498	235	143	97	70	51	38	27	18	TW J3	385	241	168	123	93	71	54	40	28
SS D1	1013	406	232	152	107	77	56	40	26	TW J4	352	221	154	113	85	65	50	37	26
SS D2	824	433	275	191	138	103	76	55	37	DG A1	339	158	96	63	43	29	19	11	-
SS D3	1277	388	205	129	89	63	45	32	21	DG A2	343	162	102	71	51	37	26	18	11
SS D4	846	425	265	182	131	97	71	51	34	DG A3	399	191	121	86	65	49	37	26	16
SS E1	776	326	190	125	88	64	46	33	22	DG B1	339	159	99	70	52	38	27	18	8
SS E2	693	361	229	159	115	85	63	46	30	DG B2	351	168	106	76	57	44	33	24	15
SS E3	936	283	149	94	65	46	33	23	15	DG B3	402	195	123	88	68	53	41	31	21
SS E4	716	341	208	141	101	74	55	39	26	DG C1	401	191	119	85	64	49	38	27	15

EL PASO, TEXAS									CONTINUED
SSF =	.10	.20	.30	.40	.50	.60	.70	.80	.90
DG C2	406	195	123	88	67	52	41	30	20
DG C3	460	224	142	102	78	62	49	38	26
SS A1	876	286	155	98	67	48	34	24	16
SS A2	774	343	203	135	95	69	51	36	24
SS A3	849	265	141	88	60	42	30	21	14
SS A4	787	342	200	132	93	67	49	35	23
SS A5	1425	289	143	87	58	41	29	20	13
SS A6	774	342	202	134	95	69	50	36	24
SS A7	1599	250	119	71	47	33	23	16	10
SS A8	794	338	197	129	91	65	48	34	23
SS B1	604	218	121	78	54	38	28	20	13
SS B2	599	273	163	109	78	57	42	30	20
SS B3	565	203	112	72	50	35	26	18	12
SS B4	589	267	160	107	75	55	40	29	19
SS B5	771	197	101	63	42	30	22	15	10
SS B6	569	264	159	107	76	55	41	29	20
SS B7	689	171	87	54	36	25	18	13	8
SS B8	557	255	152	102	72	53	39	28	18
SS C1	415	196	120	81	58	43	32	23	15
SS C2	418	224	144	100	73	55	41	30	20
SS C3	424	157	89	58	41	30	22	16	10
SS C4	412	201	124	85	61	45	34	25	17
SS D1	822	343	198	130	91	65	48	34	22
SS D2	692	372	238	165	119	89	66	48	32
SS D3	1012	325	175	110	75	54	39	27	18
SS D4	707	365	229	157	113	83	62	45	30
SS E1	624	273	160	106	75	54	39	28	19
SS E2	576	308	196	136	99	73	54	40	27
SS E3	732	234	126	80	54	39	28	20	13
SS E4	589	290	178	121	87	64	47	34	23

FORT WORTH, TEXAS									2382 DD
SSF =	.10	.20	.30	.40	.50	.60	.70	.80	.90
WW A1	1145	144	67	40	27	19	13	9	6
WW A2	392	157	90	59	41	30	22	15	10
WW A3	330	158	96	65	47	34	25	18	12
WW A4	303	157	99	68	49	36	27	20	13
WW A5	288	159	103	72	53	39	29	21	14
WW A6	279	159	104	74	54	41	30	22	15
WW B1	329	122	67	43	30	21	15	11	7
WW B2	295	158	101	70	51	38	28	21	14
WW B3	341	192	125	88	65	49	36	27	18
WW B4	296	180	122	89	66	50	38	28	19
WW B5	271	168	116	85	64	48	37	27	19
WW C1	361	203	133	94	69	52	39	28	19
WW C2	310	182	122	87	65	49	37	27	18
WW C3	313	202	142	105	80	61	47	35	24
WW C4	282	182	128	95	72	55	42	31	22
TW A1	981	137	65	39	26	18	13	9	6
TW A2	401	142	79	51	35	25	18	13	8
TW A3	324	140	82	55	39	28	21	15	10
TW A4	274	132	81	55	39	29	21	15	10
TW B1	556	129	65	40	27	19	14	10	6
TW B2	322	128	73	48	33	24	18	13	8
TW B3	281	123	73	48	34	25	18	13	9
TW B4	257	113	67	44	31	23	17	12	8
TW C1	368	115	62	39	27	19	14	10	6
TW C2	282	109	62	41	28	21	15	11	7
TW C3	268	102	58	38	26	19	14	10	7
TW C4	280	91	50	32	22	16	11	8	5
TW D1	239	94	53	35	24	17	12	9	6
TW D2	315	150	92	62	44	33	24	17	12
TW D3	321	164	103	71	51	38	28	20	14

FORT WORTH, TEXAS									CONTINUED
SSF =	.10	.20	.30	.40	.50	.60	.70	.80	.90
TW D4	304	169	110	78	57	43	32	23	16
TW D5	281	161	107	76	56	42	32	23	16
TW E1	370	190	119	82	59	44	33	24	16
TW E2	335	179	115	80	59	44	33	24	16
TW E3	347	207	140	100	75	57	43	31	21
TW E4	309	186	126	91	68	52	39	29	20
TW F1	455	118	61	38	26	18	13	9	6
TW F2	288	120	70	46	32	23	17	12	8
TW F3	238	115	70	47	34	25	18	13	9
TW F4	195	102	65	45	32	24	18	13	9
TW G1	285	102	56	36	25	18	13	9	6
TW G2	209	98	59	40	29	21	15	11	7
TW G3	175	89	55	38	27	20	15	11	7
TW G4	140	73	46	32	23	17	13	9	6
TW H1	186	82	48	32	23	17	12	9	6
TW H2	142	71	44	30	22	16	12	8	6
TW H3	118	60	38	26	19	14	10	7	5
TW H4	91	46	28	19	14	10	8	6	4
TW I1	186	79	46	30	21	15	11	8	5
TW I2	241	125	79	55	40	29	22	16	11
TW I3	255	137	88	61	45	33	25	18	12
TW I4	253	146	97	69	51	39	29	21	14
TW I5	243	145	97	70	52	39	30	22	15
TW J1	306	167	107	75	55	41	30	22	15
TW J2	275	155	102	72	53	40	30	22	15
TW J3	289	179	124	90	68	52	39	29	20
TW J4	265	165	114	83	62	48	36	27	18
DG A1	237	106	61	38	24	14	5	-	-
DG A2	252	116	71	48	34	24	16	10	5
DG A3	302	141	88	62	46	35	25	18	10
DG B1	238	107	65	44	32	22	14	6	-
DG B2	260	120	74	52	39	29	21	15	8
DG B3	308	144	90	64	48	38	29	21	14
DG C1	288	130	80	56	41	31	22	14	-
DG C2	304	141	88	62	46	36	27	20	12
DG C3	355	166	104	74	56	44	35	27	18
SS A1	643	205	109	68	47	33	24	17	11
SS A2	577	251	147	97	68	50	36	26	17
SS A3	621	188	98	61	41	29	21	14	9
SS A4	588	251	146	96	67	48	35	25	16
SS A5	1046	206	100	60	40	28	20	14	9
SS A6	575	250	146	97	68	49	36	26	17
SS A7	1170	177	82	48	32	22	15	10	7
SS A8	592	248	143	93	65	47	34	24	16
SS B1	446	156	85	54	37	27	19	13	9
SS B2	452	202	120	80	56	41	30	22	14
SS B3	415	145	79	50	34	24	17	12	8
SS B4	445	198	117	78	55	40	29	21	14
SS B5	569	141	71	43	29	20	14	10	6
SS B6	428	195	116	78	55	40	29	21	14
SS B7	505	121	60	36	24	17	12	8	5
SS B8	419	188	111	74	52	38	28	20	13
SS C1	298	137	82	55	39	28	21	15	10
SS C2	308	161	102	71	51	38	28	21	14
SS C3	307	110	61	39	27	20	14	10	7
SS C4	304	145	89	60	43	32	23	17	11
SS D1	600	245	140	91	63	45	33	23	15
SS D2	514	273	173	120	87	64	47	34	23
SS D3	742	232	123	77	52	37	26	19	12
SS D4	526	268	167	114	82	60	44	32	21
SS E1	456	194	112	74	51	37	27	19	12
SS E2	429	226	143	99	71	53	39	28	19
SS E3	539	167	88	55	37	26	19	13	9
SS E4	440	213	130	88	62	46	34	24	16

HOUSTON, TEXAS — 1434 DD

	SSF=.10	.20	.30	.40	.50	.60	.70	.80	.90
WW A1	1758	209	95	57	38	26	19	13	8
WW A2	566	218	123	81	57	41	30	21	14
WW A3	468	217	131	89	64	47	34	25	16
WW A4	424	215	135	93	67	50	37	27	18
WW A5	401	216	139	98	71	53	40	29	19
WW A6	387	216	141	100	73	55	41	30	20
WW B1	497	177	97	63	43	31	22	16	10
WW B2	407	213	135	94	69	51	38	28	19
WW B3	466	257	167	118	87	65	48	35	24
WW B4	395	236	160	116	86	66	50	37	25
WW B5	359	219	150	109	82	63	47	35	24
WW C1	491	270	176	124	91	68	51	37	25
WW C2	417	240	160	114	85	64	48	35	24
WW C3	410	261	183	135	103	79	60	45	31
WW C4	369	235	164	121	92	71	54	40	28
TW A1	1495	198	91	55	36	25	18	13	8
TW A2	583	199	108	70	48	35	25	18	12
TW A3	462	194	113	75	53	39	28	20	13
TW A4	386	181	110	75	53	39	29	21	14
TW B1	830	183	90	56	38	27	19	13	9
TW B2	462	177	100	65	46	33	24	17	11
TW B3	399	169	99	66	47	34	25	18	12
TW B4	364	155	90	60	43	31	23	16	11
TW C1	537	161	85	54	37	27	19	14	9
TW C2	403	151	85	55	39	28	20	15	10
TW C3	383	141	78	51	36	26	19	13	9
TW C4	403	126	67	43	30	21	15	11	7
TW D1	357	136	77	50	35	25	18	13	8
TW D2	439	204	123	83	60	44	32	23	16
TW D3	445	222	138	95	69	51	38	27	18
TW D4	411	224	145	102	75	56	42	31	21
TW D5	376	211	138	98	73	55	41	30	20
TW E1	509	255	159	109	79	59	43	32	21
TW E2	455	239	152	106	77	57	43	31	21
TW E3	462	271	181	130	97	73	55	41	28
TW E4	409	242	163	118	88	67	50	37	25
TW F1	682	169	85	53	36	26	18	13	8
TW F2	415	167	96	63	44	32	23	17	11
TW F3	337	158	96	65	46	34	25	18	12
TW F4	273	140	88	61	44	33	24	17	12
TW G1	418	144	78	50	35	25	18	13	8
TW G2	297	136	81	55	39	29	21	15	10
TW G3	246	122	75	52	37	28	20	15	10
TW G4	196	100	62	43	31	23	17	12	8
TW H1	266	114	67	44	31	23	17	12	8
TW H2	201	98	60	41	29	22	16	12	8
TW H3	166	82	51	35	25	19	14	10	7
TW H4	127	62	38	26	19	14	10	7	5
TW I1	277	115	66	44	31	22	16	11	7
TW I2	334	169	106	73	53	39	29	21	14
TW I3	350	185	118	82	60	45	33	24	16
TW I4	340	193	127	91	67	50	38	28	19
TW I5	324	189	127	91	68	51	39	28	19
TW J1	419	223	143	100	73	54	40	29	20
TW J2	372	206	134	95	70	52	39	29	19
TW J3	381	233	160	116	87	67	51	37	26
TW J4	349	214	147	107	80	61	46	34	24
DG A1	351	152	90	58	39	26	16	8	-
DG A2	357	159	97	66	47	34	24	16	9
DG A3	417	186	116	81	60	46	34	24	14
DG B1	352	153	94	65	47	34	24	15	5
DG B2	370	162	101	71	53	40	30	21	13
DG B3	426	189	118	84	63	49	38	28	19
DG C1	421	182	113	79	59	45	33	23	12

HOUSTON, TEXAS — CONTINUED

	SSF=.10	.20	.30	.40	.50	.60	.70	.80	.90
DG C2	430	189	118	83	63	48	37	27	18
DG C3	491	217	136	97	74	58	45	35	24
SS A1	954	288	151	96	66	47	34	24	15
SS A2	815	341	198	132	93	68	49	35	23
SS A3	941	270	140	88	60	42	30	21	14
SS A4	840	345	199	131	93	67	49	35	23
SS A5	1602	295	140	85	57	40	28	20	13
SS A6	815	341	198	131	93	67	49	35	23
SS A7	1848	259	118	70	47	32	23	16	10
SS A8	850	341	195	129	90	65	48	34	22
SS B1	661	221	119	76	53	38	27	19	12
SS B2	637	275	161	108	76	56	41	29	19
SS B3	622	208	112	71	49	35	25	18	11
SS B4	631	271	159	106	75	55	40	29	19
SS B5	869	203	100	62	42	29	21	15	9
SS B6	604	266	157	105	75	55	40	29	19
SS B7	785	177	87	53	36	25	18	12	8
SS B8	596	258	152	101	72	52	38	27	18
SS C1	428	191	113	76	54	40	29	21	14
SS C2	425	218	137	95	69	51	38	28	18
SS C3	450	155	85	55	38	27	20	14	9
SS C4	424	197	119	80	58	42	31	23	15
SS D1	871	341	194	127	89	64	46	33	21
SS D2	708	367	233	161	117	87	64	46	31
SS D3	1104	327	171	108	74	53	38	26	17
SS D4	728	361	224	154	111	82	60	43	29
SS E1	666	274	158	104	73	53	39	27	18
SS E2	597	307	194	134	97	72	53	38	26
SS E3	811	238	124	78	54	38	27	19	12
SS E4	618	290	176	119	86	63	46	33	22

LAREDO, TEXAS — 876 DD

	SSF=.10	.20	.30	.40	.50	.60	.70	.80	.90
WW A1	3049	368	176	107	72	51	36	25	16
WW A2	959	383	223	148	104	76	55	39	26
WW A3	793	382	236	162	116	85	63	45	30
WW A4	722	378	241	168	122	91	67	49	33
WW A5	683	380	249	176	129	96	72	52	35
WW A6	660	380	253	180	133	99	74	54	37
WW B1	872	326	186	122	85	61	44	32	21
WW B2	681	367	238	167	122	91	68	49	33
WW B3	778	441	292	207	153	114	85	62	42
WW B4	653	399	273	199	149	113	85	63	43
WW B5	588	366	254	186	140	106	80	59	41
WW C1	815	461	305	217	160	119	89	65	44
WW C2	690	407	274	197	147	110	83	61	41
WW C3	669	434	307	228	173	132	101	75	51
WW C4	600	388	275	203	154	118	90	67	46
TW A1	2579	346	167	102	69	48	35	24	16
TW A2	984	347	195	127	89	64	47	33	22
TW A3	779	339	202	136	97	71	52	37	25
TW A4	651	316	196	135	97	71	53	38	25
TW B1	1413	319	164	103	70	50	36	25	17
TW B2	778	308	179	119	84	61	44	32	21
TW B3	670	294	176	119	84	62	45	32	22
TW B4	610	268	161	108	77	56	41	30	20
TW C1	905	279	152	98	68	49	35	25	16
TW C2	674	261	150	99	70	51	37	26	18
TW C3	640	242	139	91	64	47	34	24	16
TW C4	670	215	118	77	53	38	28	20	13
TW D1	626	250	145	96	68	49	36	25	17
TW D2	731	349	215	147	106	78	58	42	28
TW D3	743	383	243	169	122	91	67	49	33

LAREDO, TEXAS									CONTINUED	LUBBOCK, TEXAS								3545 DD
SSF =	.10	.20	.30	.40	.50	.60	.70	.80	.90	SSF =	.10	.20	.30	.40	.50	.60	.70	.80	.90
TW D4	675	377	248	176	129	97	72	53	36	WW A1	1232	149	71	43	29	20	15	10	7
TW D5	611	351	234	167	124	93	70	51	35	WW A2	405	163	95	63	44	32	24	17	11
TW E1	843	434	276	192	139	103	76	55	37	WW A3	340	164	102	69	50	37	27	20	14
TW E2	750	404	261	183	134	100	74	54	36	WW A4	312	164	105	73	53	40	30	22	15
TW E3	754	452	307	222	165	125	94	69	47	WW A5	298	166	109	77	57	42	32	23	16
TW E4	664	402	274	199	149	112	85	62	43	WW A6	289	166	110	79	58	44	33	24	17
TW F1	1168	299	156	99	68	48	35	25	16	WW B1	342	128	72	47	33	24	17	12	8
TW F2	704	294	173	116	82	60	44	31	21	WW B2	304	164	106	75	55	41	31	23	15
TW F3	573	278	172	118	85	63	46	33	22	WW B3	351	199	132	94	69	52	39	29	20
TW F4	464	245	157	110	80	59	44	32	21	WW B4	305	187	128	94	70	54	41	30	21
TW G1	712	254	143	93	65	47	34	24	16	WW B5	279	175	121	89	67	52	39	29	20
TW G2	504	239	147	100	72	53	39	28	19	WW C1	372	211	139	99	73	55	42	31	21
TW G3	418	214	135	94	68	50	37	27	18	WW C2	319	189	128	92	69	52	39	29	20
TW G4	332	174	111	78	56	42	31	23	15	WW C3	323	210	149	111	84	65	50	37	26
TW H1	453	201	121	81	58	42	31	22	15	WW C4	290	189	134	99	76	59	45	34	24
TW H2	341	171	108	74	54	40	29	21	14	TW A1	1051	142	68	42	28	20	14	10	7
TW H3	280	143	91	63	46	34	25	18	12	TW A2	414	147	83	54	38	27	20	14	10
TW H4	215	108	68	47	34	25	19	13	9	TW A3	333	146	87	58	42	31	23	16	11
TW I1	490	213	127	85	60	44	32	23	15	TW A4	282	138	85	59	42	31	23	17	11
TW I2	559	292	187	130	95	70	52	38	25	TW B1	583	133	68	43	29	21	15	11	7
TW I3	585	318	206	145	106	79	59	43	29	TW B2	332	133	77	51	36	26	19	14	9
TW I4	559	326	218	156	116	87	65	48	32	TW B3	289	128	76	51	37	27	20	14	10
TW I5	530	316	215	155	115	87	66	48	33	TW B4	265	117	70	47	34	25	18	13	9
TW J1	695	380	248	174	128	95	71	52	35	TW C1	382	119	65	42	29	21	15	11	7
TW J2	614	349	231	164	121	91	68	50	34	TW C2	290	113	65	43	30	22	16	12	8
TW J3	622	388	269	197	148	112	85	63	43	TW C3	277	106	61	40	28	21	15	11	7
TW J4	567	354	246	180	135	103	78	58	39	TW C4	290	94	52	34	23	17	12	9	6
DG A1	614	283	173	117	82	58	41	27	15	TW D1	248	99	57	38	26	19	14	10	7
DG A2	592	277	173	121	87	64	46	32	20	TW D2	324	156	96	66	48	35	26	19	13
DG A3	661	314	199	141	105	79	59	42	26	TW D3	331	171	108	75	55	41	31	22	15
DG B1	604	287	181	127	94	71	52	37	22	TW D4	313	175	115	82	61	46	34	25	17
DG B2	601	286	183	129	96	74	56	41	27	TW D5	289	167	111	80	59	45	34	25	17
DG B3	665	317	204	145	109	85	66	49	33	TW E1	381	197	125	87	63	47	35	26	18
DG C1	702	335	215	151	113	87	67	49	31	TW E2	344	186	120	85	62	47	35	26	18
DG C2	688	328	211	149	112	87	67	51	34	TW E3	357	215	146	106	79	60	46	34	23
DG C3	759	362	235	168	126	98	78	60	41	TW E4	317	193	132	96	72	55	42	31	22
SS A1	1567	497	271	174	120	85	61	43	28	TW F1	475	123	64	41	28	20	14	10	7
SS A2	1320	577	344	231	163	118	86	61	41	TW F2	297	125	73	49	35	25	19	14	9
SS A3	1556	471	254	162	111	79	56	39	26	TW F3	245	119	74	51	37	27	20	15	10
SS A4	1363	586	347	232	164	118	86	61	40	TW F4	201	106	68	48	35	26	19	14	10
SS A5	2650	507	253	156	105	74	52	36	24	TW G1	295	106	60	39	27	20	14	10	7
SS A6	1324	578	345	231	163	118	86	61	40	TW G2	215	102	63	43	31	23	17	12	8
SS A7	3078	451	218	133	88	62	43	30	19	TW G3	180	93	58	41	29	22	16	12	8
SS A8	1383	582	343	228	160	116	84	60	39	TW G4	145	76	49	34	25	18	14	10	7
SS B1	1091	383	214	139	97	69	50	35	23	TW H1	191	85	51	34	25	18	13	10	6
SS B2	1027	461	278	187	133	97	71	51	33	TW H2	147	74	46	32	23	17	13	9	6
SS B3	1034	364	204	132	92	65	47	33	22	TW H3	122	63	40	27	20	15	11	8	5
SS B4	1021	458	276	186	132	96	70	50	33	TW H4	94	48	30	21	15	11	8	6	4
SS B5	1441	352	182	114	78	55	39	27	18	TW I1	193	84	49	33	23	17	12	9	6
SS B6	980	449	272	184	131	95	70	50	33	TW I2	248	130	83	58	42	32	24	17	12
SS B7	1311	311	160	100	68	48	34	24	15	TW I3	262	143	93	65	48	36	27	20	13
SS B8	970	439	265	179	127	92	67	48	32	TW I4	260	152	102	73	54	41	31	23	16
SS C1	731	338	206	140	100	73	54	39	26	TW I5	250	150	102	74	55	42	32	24	16
SS C2	708	373	239	167	122	90	67	49	33	TW J1	315	173	113	79	58	44	33	24	17
SS C3	764	274	154	101	71	51	37	26	17	TW J2	283	161	107	76	57	43	32	24	16
SS C4	704	336	208	142	102	75	55	40	27	TW J3	297	186	129	95	72	55	42	31	22
SS D1	1441	594	347	230	162	116	84	60	39	TW J4	273	171	119	87	66	51	39	29	20
SS D2	1163	626	403	281	204	151	111	80	54	DG A1	244	111	66	42	27	17	9	-	-
SS D3	1816	564	306	196	135	96	69	48	31	DG A2	258	120	75	51	36	26	18	12	6
SS D4	1191	613	388	268	194	142	105	75	50	DG A3	306	145	92	65	49	37	28	20	12
SS E1	1115	482	287	191	135	98	71	50	33	DG B1	242	112	69	48	35	25	17	9	-
SS E2	985	526	338	236	171	126	93	67	45	DG B2	262	124	78	55	42	32	24	17	10
SS E3	1343	415	225	144	99	71	51	35	23	DG B3	310	148	94	67	51	40	31	23	16
SS E4	1013	495	307	210	151	110	81	58	39	DG C1	290	136	85	60	45	34	25	17	8

LUBBOCK, TEXAS									CONTINUED		LUFKIN, TEXAS									CONTINUED
SSF	=.10	.20	.30	.40	.50	.60	.70	.80	.90		SSF	=.10	.20	.30	.40	.50	.60	.70	.80	.90
DG C2	306	145	92	65	50	39	30	22	14		TW D4	355	196	127	90	66	49	37	27	18
DG C3	356	170	109	78	60	47	38	29	20		TW D5	326	186	122	87	64	48	36	27	18
SS A1	681	212	114	73	50	35	25	18	12		TW E1	436	222	139	96	69	51	38	28	19
SS A2	600	260	154	103	72	53	39	28	18		TW E2	392	209	133	93	68	50	38	27	18
SS A3	658	194	103	64	44	31	22	15	10		TW E3	403	239	160	115	86	65	49	36	25
SS A4	609	258	152	100	71	51	37	27	18		TW E4	357	214	145	104	78	59	45	33	23
SS A5	1149	214	105	64	43	30	21	15	10		TW F1	551	143	73	45	31	22	16	11	7
SS A6	598	259	153	102	72	52	38	27	18		TW F2	345	143	83	54	38	28	20	14	10
SS A7	1310	182	86	51	34	23	16	11	7		TW F3	284	136	83	56	40	29	22	16	10
SS A8	613	255	148	98	69	49	36	26	17		TW F4	232	121	76	52	38	28	21	15	10
SS B1	464	161	89	57	39	28	20	14	10		TW G1	344	122	67	43	30	21	16	11	7
SS B2	465	208	124	83	59	43	32	23	15		TW G2	249	117	70	47	34	25	18	13	9
SS B3	431	149	82	52	36	26	19	13	9		TW G3	208	105	65	45	32	24	18	13	9
SS B4	456	203	121	81	58	42	31	22	15		TW G4	167	86	54	37	27	20	15	11	7
SS B5	606	145	74	46	31	22	16	11	7		TW H1	223	98	57	38	27	20	14	10	7
SS B6	440	200	121	81	58	42	31	22	15		TW H2	170	84	52	35	25	19	14	10	7
SS B7	539	124	63	39	26	18	13	9	6		TW H3	141	71	44	30	22	16	12	9	6
SS B8	429	192	115	77	55	40	29	21	14		TW H4	108	54	33	23	16	12	9	6	4
SS C1	308	143	87	59	42	31	23	17	11		TW I1	228	97	56	37	26	19	14	10	6
SS C2	317	168	107	75	55	41	31	23	15		TW I2	285	147	93	64	46	34	25	19	12
SS C3	318	115	64	42	29	21	16	11	7		TW I3	300	161	103	72	52	39	29	21	14
SS C4	313	150	93	64	46	34	25	18	12		TW I4	294	170	112	80	59	44	33	25	17
SS D1	626	254	147	97	67	49	35	25	17		TW I5	282	167	112	80	60	45	34	25	17
SS D2	531	283	181	126	92	68	50	37	25		TW J1	360	195	125	87	64	47	35	26	17
SS D3	788	241	129	82	56	40	28	20	13		TW J2	321	181	118	83	61	46	35	25	17
SS D4	543	277	174	120	87	64	47	34	23		TW J3	334	206	142	103	77	59	45	33	23
SS E1	470	200	118	78	55	40	29	21	14		TW J4	306	189	130	95	71	54	41	31	21
SS E2	439	233	149	103	75	56	41	30	20		DG A1	290	129	75	48	31	20	11	-	-
SS E3	565	172	92	58	40	28	20	14	9		DG A2	301	137	84	57	40	29	20	13	7
SS E4	451	219	135	92	66	48	36	26	17		DG A3	356	164	102	72	53	40	30	21	12
											DG B1	293	130	79	54	39	28	19	11	-
											DG B2	311	141	87	61	46	34	26	18	11
LUFKIN, TEXAS									1940 DD		DG B3	362	167	104	74	56	43	33	25	16
SSF	=.10	.20	.30	.40	.50	.60	.70	.80	.90		DG C1	351	157	96	67	50	38	28	19	9
WW A1	1393	175	81	48	32	22	16	11	7		DG C2	362	165	102	72	54	42	32	23	15
WW A2	470	187	106	70	49	35	26	18	12		DG C3	417	193	119	85	65	51	40	31	21
WW A3	394	187	114	77	55	40	30	21	14		SS A1	783	247	131	82	56	40	28	20	13
WW A4	360	186	117	80	58	43	32	23	15		SS A2	690	298	173	114	80	58	43	30	20
WW A5	342	187	121	85	62	46	34	25	17		SS A3	765	230	119	74	50	35	25	18	11
WW A6	331	187	123	87	64	48	36	26	18		SS A4	708	299	173	113	79	57	42	30	20
WW B1	405	149	82	53	36	26	19	13	9		SS A5	1284	251	121	72	48	34	24	17	11
WW B2	348	185	118	82	60	44	33	24	16		SS A6	689	297	173	114	80	58	42	30	20
WW B3	401	224	146	103	75	56	42	31	21		SS A7	1456	218	101	59	39	27	19	13	8
WW B4	344	208	141	102	76	58	44	32	22		SS A8	714	296	169	111	77	56	41	29	19
WW B5	314	194	133	97	73	55	42	31	21		SS B1	543	189	102	65	45	32	23	16	11
WW C1	423	237	154	109	80	60	45	33	22		SS B2	539	239	141	94	66	48	35	25	17
WW C2	362	211	141	100	74	56	42	31	21		SS B3	509	177	95	60	41	29	21	15	10
WW C3	361	232	163	120	91	70	53	40	27		SS B4	532	235	138	92	65	47	34	25	16
WW C4	325	208	146	108	82	63	48	36	25		SS B5	698	172	86	52	35	25	18	12	8
TW A1	1190	166	78	47	31	22	15	11	7		SS B6	511	231	137	91	65	47	35	25	16
TW A2	480	170	93	60	41	30	22	15	10		SS B7	626	149	74	45	30	21	15	10	7
TW A3	386	167	98	65	46	33	24	17	12		SS B8	503	224	132	87	62	45	33	23	15
TW A4	326	156	95	65	46	34	25	18	12		SS C1	358	164	98	65	46	34	25	18	12
TW B1	670	155	78	48	32	23	16	12	7		SS C2	363	189	119	83	60	44	33	24	16
TW B2	384	152	86	56	39	29	21	15	10		SS C3	370	132	73	47	32	23	17	12	8
TW B3	334	146	86	57	40	29	21	15	10		SS C4	359	171	104	70	50	37	27	20	13
TW B4	306	134	79	52	37	27	20	14	9		SS D1	728	295	167	109	76	54	39	28	18
TW C1	441	138	73	47	32	23	16	12	8		SS D2	611	322	203	140	101	75	56	40	27
TW C2	335	130	73	48	33	24	18	13	8		SS D3	905	280	147	92	63	44	32	22	14
TW C3	319	121	68	44	31	22	16	12	8		SS D4	625	316	196	133	96	71	52	38	25
TW C4	333	108	58	37	26	18	13	10	6		SS E1	555	235	135	89	62	45	33	23	15
TW D1	293	115	65	42	29	21	15	11	7		SS E2	511	267	168	116	84	62	46	33	22
TW D2	372	176	107	72	52	38	28	20	14		SS E3	659	203	106	66	45	32	23	16	10
TW D3	380	193	121	83	60	44	33	24	16		SS E4	526	252	153	103	74	54	40	29	19

MIDLAND-ODESSA, TEXAS								2621 DD	MIDLAND-ODESSA, TEXAS								CONTINUED
SSF =.10	.20	.30	.40	.50	.60	.70	.80	.90	SSF =.10	.20	.30	.40	.50	.60	.70	.80	.90
WW A1 1487	195	94	57	39	27	20	14	9	DG C2 386	185	117	84	64	50	39	29	19
WW A2 513	211	123	82	58	42	31	22	15	DG C3 439	213	135	97	75	59	47	36	25
WW A3 433	212	132	90	65	48	36	26	18	SS A1 831	273	148	94	64	46	33	24	15
WW A4 398	211	135	94	69	51	38	28	19	SS A2 737	327	194	129	91	67	49	35	23
WW A5 380	213	140	99	73	55	41	30	21	SS A3 806	252	134	84	58	41	29	21	13
WW A6 368	213	142	101	75	57	43	32	22	SS A4 751	327	192	127	90	65	48	34	23
WW B1 445	171	97	64	45	32	24	17	11	SS A5 1350	275	136	83	56	39	28	20	13
WW B2 383	209	136	95	70	53	39	29	20	SS A6 737	327	194	129	91	66	49	35	23
WW B3 442	253	167	119	88	66	50	37	25	SS A7 1514	238	113	68	45	31	22	15	10
WW B4 380	234	161	117	88	67	51	38	26	SS A8 757	323	189	124	87	63	46	33	22
WW B5 346	217	151	111	84	64	49	37	25	SS B1 574	207	115	74	51	37	27	19	13
WW C1 466	266	176	125	93	70	53	39	27	SS B2 572	261	157	105	75	55	40	29	19
WW C2 399	237	160	116	86	65	50	37	25	SS B3 537	194	107	69	48	34	25	18	11
WW C3 398	259	184	137	104	80	62	46	32	SS B4 563	256	153	103	73	53	39	28	19
WW C4 358	233	165	123	94	72	56	42	29	SS B5 732	188	97	60	41	29	21	15	10
TW A1 1270	185	90	55	37	26	19	13	9	SS B6 544	253	152	103	73	53	39	29	19
TW A2 521	191	108	70	49	36	26	19	13	SS B7 654	163	83	51	34	24	17	12	8
TW A3 422	188	113	76	54	40	29	21	14	SS B8 532	244	146	98	70	51	37	27	18
TW A4 358	177	110	76	55	40	30	22	15	SS C1 393	186	113	77	55	41	30	22	15
TW B1 719	173	89	56	38	27	20	14	9	SS C2 398	213	137	96	70	52	39	29	20
TW B2 418	171	100	66	47	34	25	18	12	SS C3 401	149	84	55	39	28	21	15	10
TW B3 365	164	99	66	47	35	26	19	13	SS C4 392	191	118	81	59	43	32	24	16
TW B4 334	150	90	61	44	32	24	17	12	SS D1 781	327	189	124	87	63	46	33	22
TW C1 476	154	84	54	38	27	20	14	10	SS D2 661	356	228	158	115	85	64	46	31
TW C2 364	145	84	56	39	29	21	15	10	SS D3 960	309	167	106	72	52	37	26	17
TW C3 346	135	78	51	36	26	19	14	9	SS D4 675	349	220	151	109	80	60	43	29
TW C4 359	120	67	43	30	22	16	11	8	SS E1 593	260	153	101	71	52	38	27	18
TW D1 322	132	77	51	36	26	19	14	9	SS E2 550	295	188	131	95	71	53	38	26
TW D2 406	198	123	84	61	45	34	24	17	SS E3 695	223	120	76	52	37	27	19	13
TW D3 416	217	138	96	70	52	39	29	19	SS E4 562	277	171	116	83	61	45	33	22
TW D4 389	220	145	103	76	57	43	32	22									
TW D5 357	207	138	99	74	56	42	31	22	PORT ARTHUR, TEXAS								1518 DD
TW E1 476	249	159	110	80	60	45	33	22	SSF =.10	.20	.30	.40	.50	.60	.70	.80	.90
TW E2 429	234	152	107	78	59	44	33	22	WW A1 1741	216	100	59	39	28	20	14	9
TW E3 442	267	182	132	98	75	57	42	29	WW A2 576	227	129	84	59	43	31	22	15
TW E4 392	239	164	119	89	68	52	39	27	WW A3 479	226	137	92	66	49	36	26	17
TW F1 594	161	85	54	37	26	19	14	9	WW A4 437	224	140	96	70	52	38	28	19
TW F2 377	162	96	64	45	33	25	18	12	WW A5 413	225	145	101	74	55	41	30	20
TW F3 313	154	96	66	48	35	26	19	13	WW A6 399	225	147	104	76	57	43	31	21
TW F4 257	137	88	61	45	34	25	18	13	WW B1 507	185	102	66	45	33	24	17	11
TW G1 374	138	78	51	36	26	19	14	9	WW B2 418	221	141	98	71	53	40	29	19
TW G2 274	132	81	56	40	30	22	16	11	WW B3 480	267	173	122	90	67	50	37	25
TW G3 230	119	76	52	38	28	21	16	11	WW B4 408	245	166	120	89	68	51	38	26
TW G4 184	98	63	44	32	24	18	13	9	WW B5 370	227	156	113	85	65	49	36	25
TW H1 244	111	67	45	32	24	17	13	9	WW C1 506	281	183	129	94	71	53	39	26
TW H2 187	95	60	42	30	22	17	12	8	WW C2 430	249	166	118	88	66	50	37	25
TW H3 155	80	51	35	26	19	14	10	7	WW C3 424	271	190	140	106	81	62	46	32
TW H4 119	61	38	27	19	14	11	8	5	WW C4 381	243	170	125	95	73	56	42	29
TW I1 253	112	67	45	32	23	17	12	8	TW A1 1483	204	95	57	38	27	19	13	9
TW I2 313	166	106	74	54	40	30	22	15	TW A2 590	206	113	72	50	36	26	19	12
TW I3 330	181	118	83	61	46	34	25	17	TW A3 471	201	117	78	55	40	29	21	14
TW I4 324	190	128	92	68	52	39	29	20	TW A4 396	188	114	77	55	41	30	22	15
TW I5 310	187	127	92	69	52	40	30	20	TW B1 829	189	94	58	39	28	20	14	9
TW J1 395	219	143	101	74	56	42	31	21	TW B2 470	184	104	68	48	34	25	18	12
TW J2 353	203	135	96	71	54	41	30	21	TW B3 407	176	103	68	48	35	26	19	12
TW J3 367	231	160	118	89	68	52	39	27	TW B4 372	161	94	63	44	32	24	17	11
TW J4 336	211	147	108	82	63	48	36	25	TW C1 541	167	89	56	39	28	20	14	9
DG A1 320	149	90	59	40	27	17	9	-	TW C2 409	156	88	57	40	29	21	15	10
DG A2 325	154	97	67	48	35	25	17	10	TW C3 389	146	82	53	37	27	20	14	9
DG A3 380	182	115	82	62	47	35	25	15	TW C4 406	130	70	45	31	22	16	11	8
DG B1 319	150	94	66	49	36	26	17	7	TW D1 365	143	80	52	36	26	19	14	9
DG B2 334	159	100	72	54	42	31	23	14	TW D2 449	211	128	86	62	45	34	24	16
DG B3 384	185	118	85	64	51	40	30	20	TW D3 457	231	144	99	71	53	39	28	19
DG C1 379	180	112	80	60	47	36	25	14									

PORT ARTHUR, TEXAS							CONTINUED		SAN ANGELO, TEXAS								2240 DD
SSF =.10	.20	.30	.40	.50	.60	.70	.80	.90	SSF =.10	.20	.30	.40	.50	.60	.70	.80	.90
TW D4 423	232	150	105	77	58	43	32	21	WW A1 1427	187	90	54	36	26	18	13	8
TW D5 386	218	143	102	75	56	42	31	21	WW A2 492	202	117	77	55	40	29	21	14
TW E1 523	265	165	113	82	61	45	33	22	WW A3 416	203	125	85	61	45	34	24	16
TW E2 468	247	157	109	80	59	44	32	22	WW A4 382	202	129	89	65	48	36	26	18
TW E3 476	280	188	135	100	76	57	42	29	WW A5 364	204	133	94	69	52	39	28	19
TW E4 421	251	169	122	91	69	52	39	26	WW A6 353	204	135	96	71	53	40	30	20
TW F1 684	175	89	55	38	27	19	14	9	WW B1 426	163	92	60	42	30	22	16	10
TW F2 423	174	100	66	46	34	25	18	12	WW B2 368	200	130	91	66	50	37	27	19
TW F3 346	164	100	67	48	35	26	19	13	WW B3 425	242	160	113	84	63	47	35	24
TW F4 281	145	91	63	45	34	25	18	12	WW B4 366	225	154	112	84	64	48	36	25
TW G1 424	149	82	53	36	26	19	14	9	WW B5 333	209	145	106	80	61	46	35	24
TW G2 304	141	85	57	41	30	22	16	11	WW C1 448	255	169	120	88	66	50	37	25
TW G3 253	127	78	54	39	29	21	15	10	WW C2 384	228	153	110	82	62	47	35	24
TW G4 202	104	65	45	32	24	18	13	9	WW C3 384	250	177	131	100	77	59	44	30
TW H1 272	118	69	46	33	24	17	13	8	WW C4 345	224	158	118	89	69	53	40	27
TW H2 206	101	62	43	31	23	17	12	8	TW A1 1219	177	86	52	35	25	18	12	8
TW H3 170	85	53	36	26	19	14	10	7	TW A2 500	182	103	67	47	34	25	18	12
TW H4 131	65	40	27	20	15	11	8	5	TW A3 406	180	107	72	51	37	28	20	13
TW I1 285	120	70	46	32	23	17	12	8	TW A4 344	169	105	72	52	38	28	21	14
TW I2 343	176	110	76	55	41	30	22	15	TW B1 691	165	85	53	36	26	19	13	9
TW I3 361	192	122	85	62	46	34	25	17	TW B2 402	164	95	63	44	32	24	17	11
TW I4 350	200	132	94	69	52	39	29	20	TW B3 351	157	94	63	45	33	24	18	12
TW I5 334	196	131	94	70	53	40	29	20	TW B4 321	144	86	58	41	30	22	16	11
TW J1 431	231	148	103	75	56	42	31	21	TW C1 457	147	81	52	36	26	19	13	9
TW J2 383	214	139	98	72	54	41	30	20	TW C2 350	139	80	53	37	27	20	14	10
TW J3 393	242	165	120	90	69	52	39	26	TW C3 332	130	74	49	34	25	18	13	9
TW J4 359	221	152	110	83	63	48	36	24	TW C4 345	115	64	41	29	21	15	11	7
DG A1 360	159	94	61	41	27	17	9	-	TW D1 308	126	73	48	33	24	18	13	8
DG A2 365	165	101	69	49	35	25	17	10	TW D2 391	190	117	80	58	43	32	23	15
DG A3 427	193	120	84	63	47	35	25	15	TW D3 400	209	132	92	66	49	37	27	18
DG B1 365	160	98	68	49	36	25	16	6	TW D4 375	211	139	98	72	54	41	30	21
DG B2 380	169	105	74	55	42	31	22	14	TW D5 344	199	133	95	70	53	40	30	20
DG B3 436	196	122	87	66	51	40	29	19	TW E1 459	239	152	105	76	57	42	31	21
DG C1 436	190	117	83	61	47	35	25	13	TW E2 413	225	145	102	75	56	42	31	21
DG C2 441	197	122	86	65	50	39	28	19	TW E3 426	257	174	126	94	71	54	40	27
DG C3 501	226	140	100	76	60	47	36	25	TW E4 378	230	157	114	85	65	49	37	25
SS A1 970	302	159	100	68	49	35	25	16	TW F1 570	154	81	51	35	25	18	13	8
SS A2 838	357	207	137	96	70	51	37	24	TW F2 362	155	91	61	43	31	23	17	11
SS A3 957	283	146	91	62	44	32	22	14	TW F3 300	148	91	62	45	33	25	18	12
SS A4 864	361	208	137	96	70	51	36	24	TW F4 246	131	84	58	42	32	24	17	12
SS A5 1608	308	147	89	59	42	30	21	13	TW G1 359	132	74	48	34	24	18	13	8
SS A6 838	356	207	136	96	70	51	37	24	TW G2 263	126	78	53	38	28	21	15	10
SS A7 1846	271	124	74	49	34	24	17	11	TW G3 221	114	72	50	36	27	20	15	10
SS A8 874	357	204	134	94	68	49	35	23	TW G4 177	94	60	41	30	22	17	12	8
SS B1 673	231	125	79	55	39	28	20	13	TW H1 234	106	63	43	30	22	16	12	8
SS B2 653	286	168	112	79	58	42	30	20	TW H2 179	91	57	39	28	21	16	11	8
SS B3 635	217	117	74	51	37	26	19	12	TW H3 149	77	49	34	24	18	13	10	7
SS B4 648	283	166	110	78	57	42	30	20	TW H4 114	58	37	25	18	14	10	7	5
SS B5 875	211	105	64	44	31	22	15	10	TW I1 242	107	63	42	30	21	16	11	7
SS B6 621	277	164	109	77	57	42	30	20	TW I2 301	159	101	71	51	38	29	21	14
SS B7 791	185	91	56	37	26	19	13	8	TW I3 318	174	113	79	58	43	32	24	16
SS B8 613	270	158	105	74	54	40	28	19	TW I4 312	183	122	88	65	49	37	27	19
SS C1 439	199	118	79	56	41	30	22	14	TW I5 299	180	122	88	66	50	38	28	19
SS C2 437	226	142	98	71	53	39	29	19	TW J1 380	210	136	96	70	53	39	29	20
SS C3 456	161	89	57	39	28	21	15	10	TW J2 340	195	129	92	68	51	38	28	19
SS C4 434	204	123	83	60	44	32	23	16	TW J3 354	222	154	112	85	65	49	37	25
SS D1 895	358	203	132	92	67	48	34	22	TW J4 324	203	141	103	78	60	45	34	23
SS D2 735	384	242	167	121	90	67	48	32	DG A1 307	142	85	55	37	25	15	8	-
SS D3 1122	342	179	112	77	55	39	28	18	DG A2 314	148	92	63	45	33	23	15	9
SS D4 754	377	233	159	115	85	63	45	30	DG A3 367	175	111	78	59	44	33	24	14
SS E1 686	287	165	109	76	55	40	29	19	DG B1 307	143	89	62	45	33	23	15	5
SS E2 618	321	202	139	101	75	55	40	27	DG B2 322	153	96	68	51	39	29	21	13
SS E3 822	249	130	82	56	40	28	20	13	DG B3 371	179	113	81	61	48	37	28	19
SS E4 638	303	183	124	89	65	48	34	23	DG C1 364	172	107	76	56	43	33	23	12

SAN ANGELO, TEXAS								CONTINUED	
SSF =	.10	.20	.30	.40	.50	.60	.70	.80	.90
DG C2	372	179	112	80	60	47	36	27	18
DG C3	425	206	130	93	71	56	45	34	24
SS A1	805	263	142	90	62	44	32	22	15
SS A2	716	317	187	124	88	64	47	34	22
SS A3	784	244	130	81	55	39	28	20	13
SS A4	731	318	186	123	86	63	46	33	22
SS A5	1309	265	131	80	53	37	27	19	12
SS A6	715	316	187	124	87	63	46	33	22
SS A7	1474	230	109	65	43	30	21	15	9
SS A8	738	314	183	120	84	61	44	32	21
SS B1	559	201	111	71	49	35	25	18	12
SS B2	557	253	152	101	72	52	39	28	19
SS B3	523	188	104	66	46	33	23	17	11
SS B4	550	249	149	99	70	51	38	27	18
SS B5	713	182	93	58	39	27	20	14	9
SS B6	530	245	148	99	70	51	38	27	18
SS B7	638	158	80	49	33	23	17	12	7
SS B8	520	237	142	95	67	49	36	26	17
SS C1	378	178	108	73	52	38	28	20	14
SS C2	383	204	131	91	66	49	37	27	18
SS C3	386	143	80	52	37	26	19	14	9
SS C4	377	183	113	77	56	41	30	22	15
SS D1	756	315	182	119	83	60	44	31	21
SS D2	641	344	220	152	111	82	61	44	30
SS D3	931	298	160	101	69	49	35	25	16
SS D4	654	337	212	145	105	77	57	41	28
SS E1	577	252	148	97	68	50	36	26	17
SS E2	535	286	182	126	91	68	50	37	25
SS E3	676	216	116	73	50	36	26	18	12
SS E4	547	269	165	112	80	59	44	31	21

SAN ANTONIO, TEXAS								1570 DD	
SSF =	.10	.20	.30	.40	.50	.60	.70	.80	.90
WW A1	1829	228	106	63	42	30	21	15	10
WW A2	605	239	136	89	63	46	33	24	16
WW A3	503	239	145	98	70	52	38	28	18
WW A4	459	236	148	103	74	55	41	30	20
WW A5	434	238	154	108	79	59	44	32	22
WW A6	420	237	156	110	81	61	46	33	23
WW B1	534	196	109	70	49	35	25	18	12
WW B2	438	233	149	104	76	56	42	31	21
WW B3	504	281	183	129	95	71	53	39	26
WW B4	427	257	175	126	94	72	54	40	27
WW B5	387	238	164	119	90	68	52	39	26
WW C1	530	295	193	136	100	75	56	41	28
WW C2	450	262	175	125	93	70	53	39	26
WW C3	443	284	199	147	112	86	66	49	34
WW C4	398	255	179	132	100	77	59	44	30
TW A1	1556	215	101	61	41	29	20	14	9
TW A2	619	217	119	77	54	39	28	20	13
TW A3	494	212	124	83	59	43	31	23	15
TW A4	415	199	121	82	59	43	32	23	16
TW B1	870	199	100	62	42	30	21	15	10
TW B2	493	194	110	72	51	37	27	19	13
TW B3	427	185	109	72	51	38	28	20	13
TW B4	390	169	100	66	47	34	25	18	12
TW C1	568	176	94	60	41	29	21	15	10
TW C2	429	165	93	61	43	31	23	16	11
TW C3	407	153	86	56	39	28	21	15	10
TW C4	425	137	74	47	33	23	17	12	8
TW D1	385	151	86	56	39	28	21	15	10
TW D2	471	222	135	91	66	48	36	26	17
TW D3	479	243	152	105	76	56	42	30	20

SAN ANTONIO, TEXAS								CONTINUED	
SSF =	.10	.20	.30	.40	.50	.60	.70	.80	.90
TW D4	442	243	158	111	82	61	46	34	23
TW D5	403	229	151	107	79	60	45	33	22
TW E1	548	278	174	120	87	64	48	35	23
TW E2	489	260	166	116	84	63	47	34	23
TW E3	498	294	198	142	106	80	61	45	30
TW E4	440	263	178	128	96	73	55	41	28
TW F1	718	185	95	59	40	29	21	15	10
TW F2	444	183	106	70	49	36	26	19	12
TW F3	363	173	106	72	51	38	28	20	14
TW F4	295	153	97	67	49	36	27	20	13
TW G1	445	157	87	56	39	28	20	15	10
TW G2	319	149	90	61	43	32	24	17	11
TW G3	266	134	83	57	41	31	23	16	11
TW G4	212	109	69	48	35	26	19	14	9
TW H1	286	125	74	49	35	25	19	13	9
TW H2	216	107	66	45	33	24	18	13	9
TW H3	179	90	56	39	28	21	15	11	7
TW H4	137	68	42	29	21	15	11	8	6
TW I1	300	128	74	49	35	25	18	13	9
TW I2	360	185	117	81	58	43	32	24	16
TW I3	378	202	129	90	66	49	37	27	18
TW I4	366	210	139	99	73	55	42	31	21
TW I5	349	206	138	99	74	56	42	31	21
TW J1	451	243	156	109	80	60	45	33	22
TW J2	401	224	147	104	77	57	43	32	21
TW J3	411	253	174	127	95	73	55	41	28
TW J4	376	232	159	116	87	67	51	38	26
DG A1	380	169	101	66	45	30	19	11	-
DG A2	382	174	107	73	52	38	27	18	11
DG A3	444	202	126	89	66	50	37	26	16
DG B1	383	170	105	73	53	39	28	18	8
DG B2	397	178	111	78	58	45	34	24	15
DG B3	453	206	129	92	69	54	42	31	21
DG C1	456	201	125	88	66	50	38	27	15
DG C2	459	207	129	92	69	53	41	30	20
DG C3	520	236	148	106	80	63	50	38	26
SS A1	1012	316	167	105	72	52	37	26	17
SS A2	872	373	217	144	102	74	54	39	26
SS A3	997	296	154	97	66	47	33	23	15
SS A4	898	377	218	144	101	73	54	38	25
SS A5	1678	322	155	94	63	44	31	22	14
SS A6	873	373	217	144	101	74	54	39	25
SS A7	1924	284	131	78	52	36	25	18	11
SS A8	909	373	214	140	99	71	52	37	24
SS B1	701	242	131	84	58	42	30	21	14
SS B2	679	299	176	117	83	61	45	32	21
SS B3	662	228	123	79	54	39	28	20	13
SS B4	673	295	173	115	82	60	44	31	21
SS B5	912	221	111	68	46	33	23	16	11
SS B6	645	289	171	115	82	60	44	31	21
SS B7	825	194	96	59	40	28	20	14	9
SS B8	637	281	166	110	78	57	42	30	20
SS C1	461	210	125	84	60	44	32	23	16
SS C2	457	237	150	104	76	56	42	30	21
SS C3	478	170	94	61	42	30	22	16	10
SS C4	454	214	130	88	63	47	34	25	17
SS D1	935	375	214	140	98	71	51	37	24
SS D2	766	402	254	176	128	95	70	51	34
SS D3	1171	358	189	119	81	58	42	29	19
SS D4	786	395	245	168	121	89	66	48	32
SS E1	715	301	174	115	81	59	43	30	20
SS E2	644	335	211	146	106	79	58	42	28
SS E3	857	261	137	86	59	42	30	21	14
SS E4	664	317	192	130	93	69	51	36	24

LCR Tables / 599

SHERMAN, TEXAS									2864 DD
SSF =	.10	.20	.30	.40	.50	.60	.70	.80	.90
WW A1	1003	119	55	32	21	15	10	7	5
WW A2	335	131	74	48	34	24	18	12	8
WW A3	280	132	80	54	38	28	21	15	10
WW A4	257	132	82	57	41	30	22	16	11
WW A5	244	133	86	60	43	32	24	17	12
WW A6	237	133	87	61	45	33	25	18	12
WW B1	274	98	54	34	23	16	12	8	5
WW B2	253	134	85	59	43	32	24	17	12
WW B3	292	163	106	75	55	41	30	22	15
WW B4	256	155	105	76	57	43	32	24	16
WW B5	236	146	100	73	55	41	32	23	16
WW C1	310	173	113	79	58	43	32	24	16
WW C2	268	156	104	74	55	41	31	23	16
WW C3	273	175	123	91	69	53	40	30	21
WW C4	246	158	111	82	62	48	36	27	19
TW A1	860	114	53	31	21	14	10	7	5
TW A2	344	119	65	42	29	20	15	10	7
TW A3	276	118	69	45	32	23	17	12	8
TW A4	233	111	68	46	33	24	17	13	8
TW B1	483	107	53	33	22	15	11	8	5
TW B2	276	107	61	40	28	20	14	10	7
TW B3	240	103	61	40	28	21	15	11	7
TW B4	221	95	56	37	26	19	14	10	7
TW C1	318	97	51	32	22	16	11	8	5
TW C2	242	92	52	34	24	17	12	9	6
TW C3	231	86	48	31	22	16	11	8	5
TW C4	244	78	42	27	18	13	9	7	4
TW D1	198	76	43	27	19	13	10	7	4
TW D2	271	128	77	52	37	27	20	14	10
TW D3	275	140	87	60	43	32	23	17	11
TW D4	264	145	94	66	49	36	27	20	13
TW D5	245	139	92	65	48	36	27	20	14
TW E1	319	162	101	70	50	37	27	20	13
TW E2	289	154	98	68	50	37	27	20	14
TW E3	302	179	120	86	64	48	37	27	18
TW E4	269	162	109	79	59	45	34	25	17
TW F1	392	98	50	31	21	14	10	7	5
TW F2	245	100	57	38	26	19	14	10	6
TW F3	202	96	58	39	28	20	15	11	7
TW F4	166	86	54	37	27	20	15	11	7
TW G1	243	85	46	30	20	15	10	7	5
TW G2	177	82	49	33	23	17	12	9	6
TW G3	149	74	46	31	23	17	12	9	6
TW G4	119	61	39	26	19	14	10	8	5
TW H1	158	68	40	26	19	13	10	7	5
TW H2	121	60	37	25	18	13	10	7	5
TW H3	101	50	31	21	15	11	8	6	4
TW H4	78	39	24	16	12	9	6	5	3
TW I1	153	64	37	24	16	12	8	6	4
TW I2	207	106	67	46	33	25	18	13	9
TW I3	218	117	75	52	38	28	21	15	10
TW I4	219	126	83	59	44	33	25	18	12
TW I5	212	125	84	60	45	34	25	19	13
TW J1	264	142	91	64	46	34	26	19	13
TW J2	238	133	87	62	45	34	25	19	13
TW J3	251	155	107	78	58	44	34	25	17
TW J4	231	143	99	72	54	41	31	23	16
DG A1	197	85	48	28	16	6	-	-	-
DG A2	216	98	59	40	27	19	12	7	3
DG A3	264	122	76	53	39	29	21	15	8
DG B1	197	86	51	34	24	16	8	-	-
DG B2	222	101	62	43	32	24	17	11	6
DG B3	269	125	78	55	41	32	25	18	12
DG C1	241	107	65	44	32	24	16	9	-

SHERMAN, TEXAS									CONTINUED
SSF =	.10	.20	.30	.40	.50	.60	.70	.80	.90
DG C2	262	120	74	52	38	29	22	16	9
DG C3	310	144	90	64	48	38	30	22	15
SS A1	558	172	91	56	38	27	19	13	9
SS A2	500	215	125	82	58	42	30	22	14
SS A3	536	156	81	50	33	23	16	11	7
SS A4	508	214	123	81	56	40	29	21	14
SS A5	932	172	82	49	32	22	16	11	7
SS A6	498	213	124	81	57	41	30	21	14
SS A7	1053	146	67	39	25	17	12	8	5
SS A8	510	210	120	78	54	39	28	20	13
SS B1	384	131	71	45	30	22	15	11	7
SS B2	392	173	102	68	48	35	25	18	12
SS B3	354	120	64	40	27	19	14	10	6
SS B4	384	169	99	66	46	33	24	17	12
SS B5	498	117	58	35	23	16	12	8	5
SS B6	370	167	99	66	46	34	25	18	12
SS B7	440	100	49	29	19	13	9	6	4
SS B8	361	160	94	62	44	32	23	16	11
SS C1	252	114	68	45	32	23	17	12	8
SS C2	264	137	87	60	43	32	24	17	12
SS C3	262	92	50	32	22	16	11	8	5
SS C4	262	123	75	51	36	26	19	14	9
SS D1	513	206	116	75	52	37	27	19	12
SS D2	443	233	147	101	73	54	40	29	19
SS D3	645	195	102	63	43	30	21	15	10
SS D4	453	229	142	96	69	51	37	27	18
SS E1	386	162	93	60	42	30	22	15	10
SS E2	367	192	121	83	60	44	33	23	16
SS E3	465	139	73	45	30	21	15	11	7
SS E4	378	181	110	74	52	38	28	20	13

WACO, TEXAS									2058 DD
SSF =	.10	.20	.30	.40	.50	.60	.70	.80	.90
WW A1	1352	170	79	47	31	22	16	11	7
WW A2	458	182	104	68	48	34	25	18	12
WW A3	384	183	111	75	54	39	29	21	14
WW A4	351	182	114	79	57	42	31	23	15
WW A5	334	183	118	83	61	45	34	25	17
WW A6	323	183	120	85	62	47	35	26	17
WW B1	393	145	80	52	36	25	18	13	8
WW B2	340	181	116	81	59	44	33	24	16
WW B3	392	219	143	101	74	55	42	30	21
WW B4	337	204	139	100	75	57	43	32	22
WW B5	308	190	131	95	71	54	41	31	21
WW C1	413	232	151	107	78	59	44	32	22
WW C2	354	207	138	99	73	55	42	31	21
WW C3	354	227	160	118	89	69	52	39	27
WW C4	318	204	144	106	80	62	47	35	24
TW A1	1156	161	76	46	30	21	15	11	7
TW A2	468	165	91	59	41	29	21	15	10
TW A3	376	163	95	63	45	33	24	17	11
TW A4	318	153	93	63	45	33	25	18	12
TW B1	651	151	76	47	32	22	16	11	7
TW B2	375	148	85	55	39	28	20	15	10
TW B3	326	142	84	56	39	29	21	15	10
TW B4	298	130	77	51	36	26	19	14	9
TW C1	429	134	72	46	31	22	16	11	8
TW C2	327	127	72	47	33	24	17	12	8
TW C3	311	118	67	43	30	22	16	11	8
TW C4	325	106	57	36	25	18	13	9	6
TW D1	284	112	63	41	29	21	15	11	7
TW D2	363	173	105	71	51	37	28	20	13
TW D3	370	189	118	81	59	43	32	23	16

600 / APPENDIX 21

WACO, TEXAS							CONTINUED			WICHITA FALLS, TEXAS								2904 DD	
SSF =	.10	.20	.30	.40	.50	.60	.70	.80	.90	SSF =	.10	.20	.30	.40	.50	.60	.70	.80	.90
TW D4	347	192	125	88	65	48	36	27	18	WW A1	1120	136	63	38	25	17	12	9	6
TW D5	319	182	120	85	63	47	36	26	18	WW A2	375	148	85	55	39	28	20	15	10
TW E1	425	217	136	94	68	50	37	27	18	WW A3	314	150	91	62	44	32	24	17	12
TW E2	383	204	131	91	66	49	37	27	18	WW A4	288	149	94	65	47	35	26	19	13
TW E3	394	234	157	113	84	64	48	35	24	WW A5	274	150	97	68	50	37	28	20	14
TW E4	350	210	142	103	77	58	44	33	22	WW A6	265	151	99	70	51	39	29	21	14
TW F1	536	139	71	44	30	21	15	11	7	WW B1	312	114	63	40	28	20	14	10	7
TW F2	336	140	81	53	37	27	20	14	9	WW B2	281	150	96	67	49	36	27	20	13
TW F3	277	133	81	55	39	29	21	15	10	WW B3	324	182	119	84	62	46	35	25	17
TW F4	226	118	74	51	37	28	20	15	10	WW B4	283	172	117	85	63	48	36	27	19
TW G1	334	119	66	42	29	21	15	11	7	WW B5	260	161	111	81	61	46	35	26	18
TW G2	243	114	69	46	33	24	18	13	9	WW C1	344	193	126	89	65	49	37	27	18
TW G3	203	103	64	44	32	23	17	12	8	WW C2	296	174	116	83	62	46	35	26	18
TW G4	163	84	53	37	26	20	15	11	7	WW C3	300	193	136	101	76	59	45	34	23
TW H1	217	95	56	37	26	19	14	10	7	WW C4	270	174	123	91	69	53	41	30	21
TW H2	165	82	51	35	25	18	14	10	7	TW A1	959	130	61	37	24	17	12	8	6
TW H3	137	69	43	30	21	16	12	8	6	TW A2	384	134	74	48	33	24	17	12	8
TW H4	105	53	33	22	16	12	9	6	4	TW A3	309	133	78	52	37	27	20	14	9
TW I1	221	94	55	36	25	18	13	9	6	TW A4	261	125	77	52	37	27	20	15	10
TW I2	278	144	91	63	45	34	25	18	12	TW B1	538	122	61	38	26	18	13	9	6
TW I3	293	157	101	70	51	38	28	21	14	TW B2	308	121	69	45	32	23	17	12	8
TW I4	288	166	110	78	58	44	33	24	16	TW B3	268	117	69	46	32	24	17	13	8
TW I5	276	164	110	79	59	44	34	25	17	TW B4	246	107	63	42	30	22	16	12	8
TW J1	351	190	122	85	62	46	35	25	17	TW C1	354	109	59	37	26	18	13	9	6
TW J2	314	177	116	82	60	45	34	25	17	TW C2	270	104	59	39	27	19	14	10	7
TW J3	327	202	139	101	76	58	44	33	22	TW C3	257	97	55	36	25	18	13	9	6
TW J4	300	186	128	93	70	53	41	30	21	TW C4	270	87	47	30	21	15	11	8	5
DG A1	282	125	73	47	30	19	11	-	-	TW D1	226	88	50	32	22	16	12	8	5
DG A2	293	134	82	56	39	28	19	13	7	TW D2	301	143	87	59	42	31	23	17	11
DG A3	347	160	100	70	52	39	29	20	12	TW D3	306	156	98	67	49	36	27	19	13
DG B1	284	126	77	53	38	27	18	10	-	TW D4	291	161	105	74	54	41	31	22	15
DG B2	303	138	86	60	45	34	25	17	10	TW D5	269	154	102	72	54	40	30	22	15
DG B3	353	164	102	73	55	43	33	24	16	TW E1	353	181	113	78	57	42	31	23	15
DG C1	341	153	94	66	49	37	27	18	8	TW E2	320	171	110	77	56	42	31	23	15
DG C2	353	162	100	71	53	41	31	23	15	TW E3	332	198	133	96	71	54	41	30	21
DG C3	406	189	117	84	64	50	39	30	21	TW E4	296	178	121	87	65	50	38	28	19
SS A1	754	239	126	79	54	39	28	19	13	TW F1	439	112	57	36	24	17	12	9	6
SS A2	667	289	168	111	78	57	42	30	20	TW F2	274	113	66	43	30	22	16	12	8
SS A3	735	221	115	72	49	34	24	17	11	TW F3	226	109	66	45	32	24	17	13	8
SS A4	683	290	167	110	77	56	41	29	19	TW F4	186	97	61	42	31	23	17	12	8
SS A5	1236	242	117	70	47	33	23	16	10	TW G1	273	96	53	34	24	17	12	9	6
SS A6	666	288	168	111	78	57	41	30	19	TW G2	199	93	56	38	27	20	15	11	7
SS A7	1396	209	97	57	38	26	18	13	8	TW G3	166	84	52	36	26	19	14	10	7
SS A8	688	286	164	107	75	54	39	28	18	TW G4	133	69	44	30	22	16	12	9	6
SS B1	523	183	99	63	43	31	22	16	10	TW H1	177	77	46	30	22	16	11	8	6
SS B2	522	232	137	91	64	47	34	25	16	TW H2	135	67	42	28	20	15	11	8	5
SS B3	490	170	92	58	40	29	21	14	9	TW H3	113	57	36	24	18	13	10	7	5
SS B4	514	228	134	89	63	46	34	24	16	TW H4	87	43	27	18	13	10	7	5	4
SS B5	672	166	83	51	34	24	17	12	8	TW I1	175	74	43	28	20	14	10	7	5
SS B6	495	224	133	89	63	46	34	24	16	TW I2	230	119	75	52	38	28	21	15	10
SS B7	601	143	71	43	29	20	14	10	6	TW I3	242	130	84	58	43	32	24	17	12
SS B8	486	217	128	85	60	44	32	23	15	TW I4	241	140	93	66	49	37	28	20	14
SS C1	349	160	95	64	45	33	24	17	12	TW I5	233	138	93	67	50	38	29	21	15
SS C2	354	185	117	81	59	44	32	24	16	TW J1	292	158	102	71	52	39	29	21	14
SS C3	360	129	71	46	32	23	17	12	8	TW J2	263	148	97	69	51	38	29	21	14
SS C4	351	167	101	69	49	36	27	19	13	TW J3	276	171	118	86	65	49	38	28	19
SS D1	702	285	162	105	74	53	38	27	18	TW J4	254	158	109	80	60	46	35	26	18
SS D2	592	312	198	136	99	73	54	39	26	DG A1	224	99	57	35	22	12	-	-	-
SS D3	872	271	143	89	61	43	31	22	14	DG A2	241	110	67	46	32	22	15	10	5
SS D4	605	306	190	130	93	69	51	37	24	DG A3	290	135	84	59	44	33	24	17	10
SS E1	535	227	131	86	60	43	32	22	15	DG B1	224	100	61	41	29	20	13	5	-
SS E2	495	259	164	113	82	60	45	32	22	DG B2	247	114	70	49	36	27	20	14	8
SS E3	635	196	103	64	44	31	22	16	10	DG B3	294	138	86	61	46	36	28	21	13
SS E4	509	245	149	100	72	53	39	28	18	DG C1	272	123	75	52	38	29	21	13	-

```
WICHITA FALLS, TEXAS                       CONTINUED    BRYCE CANYON, UTAH                         CONTINUED
   SSF =.10  .20  .30  .40  .50  .60  .70  .80  .90       SSF =.10  .20  .30  .40  .50  .60  .70  .80  .90
DG C2    289  134   83   59   44   34   26   19   12    TW D4   170   96   64   45   34   25   19   14    9
DG C3    338  159   99   71   54   42   33   26   17    TW D5   161   94   63   46   34   26   19   14   10
SS A1    614  191  101   63   43   30   22   15   10    TW E1   201  106   68   47   34   25   19   14    9
SS A2    546  236  138   91   64   46   34   24   16    TW E2   184  101   66   46   34   25   19   14    9
SS A3    590  174   90   56   38   26   19   13    8    TW E3   203  124   85   61   46   35   26   20   13
SS A4    554  234  136   89   62   45   33   23   15    TW E4   181  111   77   56   42   32   24   18   12
SS A5   1025  192   92   56   37   26   18   13    8    TW F1   202   54   28   18   12    8    6    4    3
SS A6    544  234  137   90   63   46   34   24   16    TW F2   137   59   34   23   16   12    8    6    4
SS A7   1159  163   75   44   29   20   14   10    6    TW F3   117   58   36   24   17   13    9    7    4
SS A8    557  230  132   86   60   43   32   22   15    TW F4    98   52   34   23   17   12    9    7    4
SS B1    423  145   79   50   34   24   18   12    8    TW G1   131   48   27   17   12    9    6    4    3
SS B2    428  190  113   75   53   38   28   20   13    TW G2   102   49   30   20   14   11    8    6    4
SS B3    391  134   73   46   31   22   16   11    7    TW G3    87   45   29   20   14   10    8    6    4
SS B4    419  185  109   73   51   37   27   20   13    TW G4    71   38   24   17   12    9    7    5    3
SS B5    549  131   66   40   27   19   13    9    6    TW H1    89   40   24   16   11    8    6    4    3
SS B6    405  183  109   73   51   37   27   20   13    TW H2    71   36   23   16   11    8    6    4    3
SS B7    486  111   55   33   22   15   11    8    5    TW H3    60   31   20   14   10    7    5    4    3
SS B8    394  176  104   69   49   35   26   18   12    TW H4    46   24   15   10    7    6    4    3    2
SS C1    283  129   78   52   37   27   20   14    9    TW I1    78   34   19   13    9    6    4    3    2
SS C2    294  153   97   67   49   36   27   20   13    TW I2   126   67   43   30   22   16   12    9    6
SS C3    294  104   58   37   26   18   13   10    6    TW I3   138   76   49   35   25   19   14   10    7
SS C4    290  138   84   57   41   30   22   16   11    TW I4   142   84   57   41   30   23   17   13    9
SS D1    565  228  130   84   59   42   30   22   14    TW I5   140   85   58   42   31   24   18   13    9
SS D2    484  256  162  112   81   60   44   32   22    TW J1   168   94   61   43   32   24   18   13    9
SS D3    710  216  114   71   48   34   24   17   11    TW J2   153   88   59   42   31   23   18   13    9
SS D4    495  251  156  107   76   56   42   30   20    TW J3   172  108   76   56   42   32   25   18   13
SS E1    427  180  104   68   47   34   25   18   12    TW J4   156   99   69   51   39   30   23   17   12
SS E2    403  212  134   92   66   49   36   26   18    DG A1   100   43   21    -    -    -    -    -    -
SS E3    513  155   82   51   35   24   17   12    8    DG A2   125   59   36   24   16   10    6    -    -
SS E4    414  199  121   82   58   43   31   23   15    DG A3   164   80   51   36   27   20   14   10    5
                                                        DG B1    98   45   26   16    8    -    -    -    -
                                                        DG B2   127   61   39   27   20   14    9    5    -
BRYCE CANYON, UTAH                          9133 DD     DG B3   167   82   53   38   29   22   17   12    8
   SSF =.10  .20  .30  .40  .50  .60  .70  .80  .90     DG C1   126   60   36   24   17   11    -    -    -
WW A1    489   63   30   18   12    8    6    4    2    DG C2   153   75   47   33   25   19   14    9    5
WW A2    187   77   45   29   20   15   11    8    5    DG C3   193   95   62   44   34   26   21   16   10
WW A3    162   80   49   33   24   17   13    9    6    SS A1   376  122   65   40   27   19   13    9    5
WW A4    152   80   51   36   26   19   14   10    7    SS A2   360  159   94   62   43   31   22   16   10
WW A5    146   82   54   38   28   21   15   11    8    SS A3   358  109   57   34   22   15   10    7    4
WW A6    143   83   55   39   29   21   16   12    8    SS A4   367  159   92   60   42   30   21   15    9
WW B1    138   51   28   18   12    8    6    4    3    SS A5   597  119   58   34   22   15   10    7    4
WW B2    155   85   55   39   28   21   16   11    8    SS A6   357  157   93   61   43   30   22   15   10
WW B3    186  107   71   50   37   28   21   15   10    SS A7   654   99   46   26   16   11    7    4    3
WW B4    168  104   72   52   39   30   23   17   12    SS A8   366  155   89   58   40   28   20   14    9
WW B5    157  100   69   51   39   30   23   17   12    SS B1   249   89   48   31   21   14   10    7    4
WW C1    200  115   76   54   40   30   22   16   11    SS B2   275  125   75   50   35   25   18   13    9
WW C2    173  104   70   51   38   29   22   16   11    SS B3   229   81   44   27   18   12    9    6    4
WW C3    189  124   88   66   50   39   30   22   15    SS B4   271  123   73   48   34   24   18   12    8
WW C4    168  110   79   59   45   35   27   20   14    SS B5   308   77   39   24   15   10    7    5    3
TW A1    429   62   30   18   12    8    6    4    3    SS B6   259  120   72   48   34   24   18   13    8
TW A2    190   69   39   25   17   13    9    6    4    SS B7   269   65   32   19   12    8    5    3    2
TW A3    158   70   42   28   20   14   11    8    5    SS B8   252  115   68   45   32   23   16   11    7
TW A4    137   68   42   29   21   15   11    8    5    SS C1   142   67   41   27   19   14   10    7    5
TW B1    253   61   31   19   13    9    6    5    3    SS C2   163   88   56   39   29   21   16   12    8
TW B2    157   64   37   25   17   12    9    6    4    SS C3   143   53   30   19   13    9    7    5    3
TW B3    139   63   38   25   18   13   10    7    5    SS C4   160   78   48   33   24   18   13    9    6
TW B4    129   58   35   24   17   12    9    6    4    SS D1   355  147   84   54   37   26   18   13    8
TW C1    176   57   31   20   14   10    7    5    3    SS D2   326  175  112   77   55   40   30   21   14
TW C2    139   56   32   21   15   11    8    6    4    SS D3   433  137   73   45   30   21   14   10    6
TW C3    134   53   30   20   14   10    7    5    3    SS D4   332  171  107   73   52   38   28   20   13
TW C4    140   47   26   17   12    8    6    4    3    SS E1   253  109   63   41   28   20   14   10    6
TW D1    102   41   23   15   10    7    5    3    2    SS E2   259  138   88   61   43   32   23   17   11
TW D2    163   80   50   34   24   18   13   10    6    SS E3   295   93   49   30   20   14   10    6    4
TW D3    171   90   57   40   29   21   16   11    8    SS E4   264  130   80   54   38   28   20   14    9
```

602 / APPENDIX 21

CEDAR CITY, UTAH — 6137 DD

	SSF = .10	.20	.30	.40	.50	.60	.70	.80	.90
WW A1	765	93	43	26	17	12	8	6	4
WW A2	265	106	61	40	28	20	15	10	7
WW A3	225	108	66	45	32	23	17	12	8
WW A4	208	108	68	47	34	25	19	14	9
WW A5	199	110	71	50	36	27	20	15	10
WW A6	193	110	73	51	38	28	21	15	10
WW B1	210	77	42	27	18	13	9	6	4
WW B2	207	112	72	50	36	27	20	15	10
WW B3	243	137	90	64	47	35	26	19	13
WW B4	216	132	90	65	49	37	28	21	14
WW B5	200	125	86	63	48	36	28	21	14
WW C1	259	147	96	68	50	37	28	21	14
WW C2	224	132	89	64	47	36	27	20	14
WW C3	234	152	107	80	60	46	36	27	18
WW C4	211	136	97	72	54	42	32	24	17
TW A1	659	90	42	25	17	12	8	6	4
TW A2	270	96	53	34	24	17	12	9	6
TW A3	221	96	57	38	27	19	14	10	7
TW A4	188	91	56	38	27	20	15	11	7
TW B1	373	86	43	27	18	13	9	6	4
TW B2	219	87	50	33	23	17	12	9	6
TW B3	193	85	50	33	24	17	13	9	6
TW B4	178	78	46	31	22	16	12	8	6
TW C1	250	78	42	27	18	13	9	7	4
TW C2	193	75	43	28	20	14	10	7	5
TW C3	185	71	40	26	18	13	10	7	5
TW C4	194	63	35	22	15	11	8	6	4
TW D1	153	60	34	22	15	11	8	5	3
TW D2	221	106	65	44	32	23	17	12	8
TW D3	227	117	73	51	37	27	20	15	10
TW D4	220	123	81	57	42	31	24	17	12
TW D5	206	119	79	56	42	31	24	18	12
TW E1	264	136	86	59	43	32	24	17	12
TW E2	240	130	83	58	43	32	24	17	12
TW E3	257	154	104	75	56	42	32	24	16
TW E4	228	139	95	69	51	39	30	22	15
TW F1	301	78	40	25	17	12	8	6	4
TW F2	194	81	47	31	22	16	11	8	5
TW F3	162	78	48	32	23	17	13	9	6
TW F4	134	70	45	31	22	17	12	9	6
TW G1	190	68	38	24	17	12	9	6	4
TW G2	142	67	41	27	19	14	10	8	5
TW G3	120	61	38	26	19	14	10	7	5
TW G4	97	51	32	22	16	12	9	6	4
TW H1	125	55	33	22	15	11	8	6	4
TW H2	97	49	30	21	15	11	8	6	4
TW H3	81	42	26	18	13	10	7	5	3
TW H4	63	32	20	14	10	7	5	4	3
TW I1	118	50	29	19	13	9	7	5	3
TW I2	169	88	56	39	28	21	16	11	8
TW I3	181	98	63	44	32	24	18	13	9
TW I4	183	107	71	51	38	28	21	16	11
TW I5	178	107	72	52	39	29	22	17	11
TW J1	220	120	78	54	40	30	22	16	11
TW J2	198	113	74	53	39	29	22	16	11
TW J3	215	134	93	68	51	39	30	22	15
TW J4	197	123	85	62	47	36	27	20	14
DG A1	151	66	36	20	9	-	-	-	-
DG A2	173	80	49	33	22	15	10	5	-
DG A3	217	103	64	45	34	25	18	13	7
DG B1	150	67	40	27	18	10	-	-	-
DG B2	178	83	51	36	26	19	14	9	4
DG B3	221	105	66	47	36	28	21	16	10
DG C1	187	85	52	36	26	19	12	5	-

CEDAR CITY, UTAH — CONTINUED

	SSF = .10	.20	.30	.40	.50	.60	.70	.80	.90
DG C2	211	99	62	43	33	25	19	13	7
DG C3	255	121	77	55	42	33	26	20	13
SS A1	479	149	78	48	32	23	16	11	7
SS A2	439	189	110	72	50	36	26	18	12
SS A3	457	134	68	42	27	19	13	9	5
SS A4	445	187	107	70	48	35	25	17	11
SS A5	786	148	70	42	27	19	13	9	5
SS A6	436	187	109	71	50	36	26	18	12
SS A7	879	124	56	32	20	14	9	6	4
SS A8	446	183	104	68	47	33	24	16	11
SS B1	322	111	60	38	25	18	13	9	6
SS B2	339	150	89	59	41	30	22	15	10
SS B3	296	101	54	34	23	16	11	7	5
SS B4	332	146	86	57	40	29	21	15	10
SS B5	410	98	49	29	19	13	9	6	4
SS B6	319	144	86	57	40	29	21	15	10
SS B7	361	83	40	24	15	10	7	5	3
SS B8	311	138	81	53	37	27	19	14	9
SS C1	201	92	55	37	26	19	14	10	7
SS C2	217	114	73	51	37	27	20	15	10
SS C3	206	74	41	26	18	13	9	7	4
SS C4	214	103	63	43	31	22	17	12	8
SS D1	444	178	100	64	44	31	22	15	10
SS D2	391	206	130	89	64	47	34	25	16
SS D3	553	168	88	54	36	25	18	12	8
SS D4	399	202	125	85	60	44	32	23	15
SS E1	325	136	78	50	35	25	18	12	8
SS E2	317	166	104	72	51	38	28	20	13
SS E3	387	117	61	38	25	17	12	8	5
SS E4	326	156	95	64	45	33	24	17	11

SALT LAKE CITY, UTAH — 5983 DD

	SSF = .10	.20	.30	.40	.50	.60	.70	.80	.90
WW A1	690	79	34	19	12	7	5	3	1
WW A2	235	90	50	31	21	14	10	6	4
WW A3	198	92	54	35	24	17	12	8	5
WW A4	182	92	56	37	26	18	13	9	5
WW A5	174	93	58	40	28	20	14	10	6
WW A6	169	93	60	41	29	21	15	10	6
WW B1	179	61	31	18	11	6	3	2	-
WW B2	184	96	60	41	29	21	15	11	7
WW B3	216	118	76	52	37	27	20	14	9
WW B4	193	116	78	55	41	30	22	16	11
WW B5	180	110	75	54	40	30	22	16	11
WW C1	231	127	82	56	40	29	21	15	10
WW C2	201	116	76	54	39	29	21	15	10
WW C3	211	134	93	68	51	39	29	21	14
WW C4	190	121	85	62	46	35	26	19	13
TW A1	596	77	34	19	12	8	5	3	2
TW A2	242	82	44	27	18	12	8	5	3
TW A3	196	82	47	30	20	14	10	7	4
TW A4	167	78	46	30	21	15	10	7	4
TW B1	336	73	35	21	13	9	6	4	2
TW B2	196	75	41	26	18	12	8	6	3
TW B3	172	73	42	27	18	13	9	6	4
TW B4	159	67	39	25	17	12	8	6	4
TW C1	225	67	35	21	14	10	6	4	3
TW C2	174	65	36	23	15	11	7	5	3
TW C3	166	61	34	21	14	10	7	5	3
TW C4	176	55	29	18	12	8	6	4	2
TW D1	131	48	25	15	9	6	3	2	-
TW D2	197	92	55	36	25	18	13	9	6
TW D3	202	101	62	41	29	21	15	10	6

LCR Tables / 603

SALT LAKE CITY, UTAH						CONTINUED			BURLINGTON, VERMONT								7876 DD
SSF =.10	.20	.30	.40	.50	.60	.70	.80	.90	SSF =.10	.20	.30	.40	.50	.60	.70	.80	.90
TW D4 198	108	69	48	35	25	19	13	9	WW A1 121	-	-	-	-	-	-	-	-
TW D5 186	105	69	48	35	26	19	14	9	WW A2 34	-	-	-	-	-	-	-	-
TW E1 236	119	73	49	35	25	18	12	8	WW A3 31	-	-	-	-	-	-	-	-
TW E2 216	114	72	49	35	25	19	13	8	WW A4 31	-	-	-	-	-	-	-	-
TW E3 231	136	90	64	47	35	26	18	12	WW A5 30	-	-	-	-	-	-	-	-
TW E4 207	123	83	59	43	32	24	17	12	WW A6 30	-	-	-	-	-	-	-	-
TW F1 269	66	32	19	12	8	5	3	2	WW B1 -	-	-	-	-	-	-	-	-
TW F2 172	69	38	24	16	11	8	5	3	WW B2 50	21	10	-	-	-	-	-	-
TW F3 143	66	39	26	18	12	9	6	4	WW B3 65	31	17	10	5	-	-	-	-
TW F4 118	60	37	24	17	12	9	6	4	WW B4 72	41	26	17	12	9	6	4	2
TW G1 169	57	30	19	12	8	6	4	2	WW B5 72	42	28	19	14	10	7	5	3
TW G2 125	57	33	21	15	10	7	5	3	WW C1 75	37	21	13	8	5	3	1	-
TW G3 105	52	31	21	14	10	7	5	3	WW C2 71	38	23	15	10	7	5	3	2
TW G4 85	43	26	18	12	9	6	4	3	WW C3 91	56	38	27	20	15	11	8	5
TW H1 110	47	27	17	11	8	5	4	2	WW C4 83	52	35	25	19	14	10	7	5
TW H2 86	41	25	16	11	8	6	4	2	TW A1 121	-	-	-	-	-	-	-	-
TW H3 72	35	21	14	10	7	5	3	2	TW A2 43	-	-	-	-	-	-	-	-
TW H4 56	27	16	11	8	5	4	3	2	TW A3 36	-	-	-	-	-	-	-	-
TW I1 100	39	21	12	8	5	3	1	-	TW A4 34	-	-	-	-	-	-	-	-
TW I2 150	76	47	32	22	16	12	8	5	TW B1 66	-	-	-	-	-	-	-	-
TW I3 160	84	53	36	26	18	13	9	6	TW B2 40	-	-	-	-	-	-	-	-
TW I4 164	94	61	43	31	23	17	12	8	TW B3 37	-	-	-	-	-	-	-	-
TW I5 161	94	63	44	33	24	18	13	9	TW B4 38	9	-	-	-	-	-	-	-
TW J1 196	104	66	45	32	23	17	12	8	TW C1 50	-	-	-	-	-	-	-	-
TW J2 178	99	64	44	32	23	17	12	8	TW C2 43	8	-	-	-	-	-	-	-
TW J3 193	118	81	58	43	32	24	17	11	TW C3 44	10	-	-	-	-	-	-	-
TW J4 178	109	75	54	40	30	22	16	11	TW C4 53	11	-	-	-	-	-	-	-
DG A1 131	52	24	-	-	-	-	-	-	TW D1 -	-	-	-	-	-	-	-	-
DG A2 157	69	40	26	16	10	5	-	-	TW D2 56	20	9	-	-	-	-	-	-
DG A3 200	92	56	39	28	20	14	9	4	TW D3 57	23	11	-	-	-	-	-	-
DG B1 131	55	30	16	-	-	-	-	-	TW D4 73	37	22	14	10	7	4	3	2
DG B2 161	73	43	29	20	14	9	4	-	TW D5 76	40	25	17	12	9	6	4	3
DG B3 205	95	59	41	30	22	16	12	7	TW E1 76	33	18	10	6	3	-	-	-
DG C1 166	72	41	26	17	9	-	-	-	TW E2 76	36	21	13	8	6	4	2	1
DG C2 193	88	53	36	26	18	13	8	3	TW E3 94	53	34	23	16	12	8	6	4
DG C3 237	110	69	48	35	27	20	15	9	TW E4 88	50	33	23	16	12	9	6	4
SS A1 421	124	62	37	23	15	10	6	4	TW F1 -	-	-	-	-	-	-	-	-
SS A2 388	161	91	58	39	27	19	13	8	TW F2 19	-	-	-	-	-	-	-	-
SS A3 398	109	52	30	18	11	7	4	2	TW F3 21	-	-	-	-	-	-	-	-
SS A4 392	158	88	55	37	25	18	12	7	TW F4 20	-	-	-	-	-	-	-	-
SS A5 695	124	55	31	19	12	8	5	2	TW G1 -	-	-	-	-	-	-	-	-
SS A6 385	159	89	57	38	27	18	12	8	TW G2 16	-	-	-	-	-	-	-	-
SS A7 772	101	42	22	13	7	4	2	1	TW G3 18	-	-	-	-	-	-	-	-
SS A8 392	154	85	53	35	24	16	11	7	TW G4 17	-	-	-	-	-	-	-	-
SS B1 283	92	47	28	18	12	8	5	3	TW H1 -	-	-	-	-	-	-	-	-
SS B2 301	130	74	48	33	23	16	11	7	TW H2 14	-	-	-	-	-	-	-	-
SS B3 257	82	41	24	15	10	6	4	2	TW H3 14	-	-	-	-	-	-	-	-
SS B4 294	125	71	46	31	22	15	10	6	TW H4 13	4	-	-	-	-	-	-	-
SS B5 362	82	38	21	13	8	5	3	2	TW I1 -	-	-	-	-	-	-	-	-
SS B6 283	124	71	46	31	22	15	11	7	TW I2 40	16	7	-	-	-	-	-	-
SS B7 315	67	30	16	10	6	3	2	-	TW I3 47	21	11	5	-	-	-	-	-
SS B8 274	117	67	43	29	20	14	9	6	TW I4 60	32	20	13	9	6	4	3	2
SS C1 176	77	44	29	19	13	9	6	4	TW I5 64	36	23	15	11	8	5	4	2
SS C2 194	99	61	42	29	21	15	11	7	TW J1 63	30	17	10	6	3	2	-	-
SS C3 182	62	33	20	13	9	6	4	2	TW J2 63	32	19	12	8	5	4	2	1
SS C4 191	89	53	35	24	17	12	9	5	TW J3 81	48	32	22	16	12	8	6	4
SS D1 386	147	79	49	32	21	14	9	5	TW J4 76	45	30	21	15	11	8	6	4
SS D2 342	174	107	71	50	36	25	17	11	DG A1 -	-	-	-	-	-	-	-	-
SS D3 486	140	69	41	26	17	11	7	4	DG A2 38	-	-	-	-	-	-	-	-
SS D4 350	171	103	68	47	33	24	16	10	DG A3 80	33	18	11	7	4	-	-	-
SS E1 281	111	60	37	24	16	11	7	4	DG B1 -	-	-	-	-	-	-	-	-
SS E2 277	140	86	57	40	28	20	14	9	DG B2 39	-	-	-	-	-	-	-	-
SS E3 339	97	47	28	17	11	7	4	2	DG B3 85	36	21	13	9	6	3	-	-
SS E4 286	133	78	51	35	24	17	12	7	DG C1 -	-	-	-	-	-	-	-	-

BURLINGTON, VERMONT								CONTINUED	
SSF =	.10	.20	.30	.40	.50	.60	.70	.80	.90
DG C2	58	20	8	-	-	-	-	-	-
DG C3	100	43	25	16	11	8	5	3	-
SS A1	123	19	-	-	-	-	-	-	-
SS A2	146	54	28	16	10	6	3	2	-
SS A3	91	-	-	-	-	-	-	-	-
SS A4	141	49	24	13	7	3	-	-	-
SS A5	214	-	-	-	-	-	-	-	-
SS A6	140	51	25	14	8	5	2	-	-
SS A7	211	-	-	-	-	-	-	-	-
SS A8	135	44	20	10	3	-	-	-	-
SS B1	70	-	-	-	-	-	-	-	-
SS B2	116	45	24	14	9	5	3	2	1
SS B3	45	-	-	-	-	-	-	-	-
SS B4	108	41	21	12	7	4	2	-	-
SS B5	88	-	-	-	-	-	-	-	-
SS B6	104	40	21	12	7	4	2	1	-
SS B7	49	-	-	-	-	-	-	-	-
SS B8	95	35	17	9	4	-	-	-	-
SS C1	-	-	-	-	-	-	-	-	-
SS C2	56	24	12	6	-	-	-	-	-
SS C3	-	-	-	-	-	-	-	-	-
SS C4	56	21	10	4	-	-	-	-	-
SS D1	103	23	-	-	-	-	-	-	-
SS D2	127	60	34	21	14	9	6	4	2
SS D3	140	18	-	-	-	-	-	-	-
SS D4	131	59	33	20	13	8	5	3	2
SS E1	48	-	-	-	-	-	-	-	-
SS E2	93	42	22	13	7	4	1	-	-
SS E3	75	-	-	-	-	-	-	-	-
SS E4	97	39	20	11	6	3	-	-	-

NORFOLK, VIRGINIA								3488 DD	
SSF =	.10	.20	.30	.40	.50	.60	.70	.80	.90
WW A1	835	96	44	25	16	11	8	5	3
WW A2	276	108	61	39	27	19	14	10	6
WW A3	233	109	66	44	31	22	16	12	8
WW A4	214	109	68	46	33	24	18	13	8
WW A5	204	111	71	49	35	26	19	14	9
WW A6	198	111	72	50	37	27	20	14	10
WW B1	218	77	41	26	17	12	8	6	4
WW B2	213	113	72	49	36	26	19	14	9
WW B3	248	138	89	63	45	34	25	18	12
WW B4	220	133	90	65	48	36	27	20	14
WW B5	204	126	86	62	47	35	27	20	13
WW C1	264	147	96	67	49	36	27	19	13
WW C2	229	133	89	63	46	35	26	19	13
WW C3	237	152	107	79	59	45	34	26	17
WW C4	214	137	96	71	54	41	31	23	16
TW A1	717	92	43	25	16	11	8	5	3
TW A2	284	98	53	34	23	16	12	8	5
TW A3	229	97	57	37	26	18	13	9	6
TW A4	195	93	56	37	26	19	14	10	7
TW B1	399	88	43	26	17	12	8	6	4
TW B2	229	89	50	32	22	16	11	8	5
TW B3	200	86	50	33	23	17	12	8	6
TW B4	184	80	46	31	21	15	11	8	5
TW C1	264	80	42	26	18	13	9	6	4
TW C2	202	77	43	28	19	14	10	7	5
TW C3	193	72	40	26	18	13	9	6	4
TW C4	204	65	35	22	15	11	8	5	3
TW D1	159	60	33	21	14	10	7	5	3
TW D2	228	107	65	44	31	22	16	12	8
TW D3	232	117	73	50	36	26	19	14	9

NORFOLK, VIRGINIA								CONTINUED	
SSF =	.10	.20	.30	.40	.50	.60	.70	.80	.90
TW D4	225	124	81	56	41	31	23	17	11
TW D5	211	120	79	56	41	31	23	17	11
TW E1	270	137	86	58	42	31	23	16	11
TW E2	246	131	83	58	42	31	23	17	11
TW E3	261	154	104	74	55	41	31	23	15
TW E4	233	140	95	68	51	38	29	21	14
TW F1	321	80	40	24	16	11	8	5	3
TW F2	202	82	47	30	21	15	11	8	5
TW F3	167	79	48	32	22	16	12	8	6
TW F4	138	71	44	30	22	16	12	8	6
TW G1	199	69	38	24	16	11	8	6	4
TW G2	147	68	40	27	19	14	10	7	5
TW G3	123	62	38	26	18	13	10	7	5
TW G4	100	51	32	22	16	11	8	6	4
TW H1	130	56	33	21	15	11	8	5	4
TW H2	100	49	30	20	14	10	8	5	4
TW H3	84	42	26	18	13	9	7	5	3
TW H4	65	32	20	13	10	7	5	4	2
TW I1	122	50	28	18	12	8	6	4	3
TW I2	174	89	56	38	27	20	15	11	7
TW I3	184	98	63	43	31	23	17	12	8
TW I4	187	108	71	50	37	28	21	15	10
TW I5	182	108	72	52	38	29	22	16	11
TW J1	224	121	77	53	39	29	21	15	10
TW J2	203	114	74	52	38	28	21	15	10
TW J3	218	134	92	67	50	38	29	21	14
TW J4	200	124	85	62	46	35	27	20	13
DG A1	158	67	36	19	-	-	-	-	-
DG A2	181	82	49	32	22	14	9	4	-
DG A3	226	105	65	45	33	24	18	12	6
DG B1	158	69	40	25	16	8	-	-	-
DG B2	186	85	52	35	25	19	13	8	3
DG B3	230	108	67	47	35	27	20	15	9
DG C1	196	87	52	35	24	17	11	-	-
DG C2	220	101	62	43	32	24	18	12	7
DG C3	265	125	78	55	41	32	25	19	12
SS A1	470	143	74	46	31	21	15	10	6
SS A2	428	182	105	69	48	34	25	17	11
SS A3	448	128	65	39	26	18	12	8	5
SS A4	434	180	103	67	46	33	24	17	11
SS A5	786	142	67	39	25	17	12	8	5
SS A6	425	180	104	68	47	34	24	17	11
SS A7	885	118	53	30	19	13	9	6	3
SS A8	434	176	100	65	44	32	23	16	10
SS B1	321	108	58	36	24	17	12	8	5
SS B2	336	147	87	57	40	29	21	15	10
SS B3	295	98	52	32	21	15	10	7	4
SS B4	329	143	84	55	38	27	20	14	9
SS B5	416	96	47	28	18	13	9	6	4
SS B6	317	141	83	55	39	28	20	14	9
SS B7	365	80	39	23	15	10	7	4	3
SS B8	307	135	79	52	36	26	19	13	9
SS C1	208	93	55	36	25	18	13	9	6
SS C2	223	116	73	50	36	26	19	14	9
SS C3	215	75	41	26	18	12	9	6	4
SS C4	221	104	63	42	30	22	16	11	8
SS D1	431	170	95	61	42	29	21	15	9
SS D2	378	198	125	85	61	45	33	23	15
SS D3	543	161	83	51	34	24	17	11	7
SS D4	387	194	120	81	58	42	31	22	14
SS E1	321	132	75	48	33	23	16	11	7
SS E2	312	162	101	69	49	36	26	19	12
SS E3	388	114	59	36	24	16	11	8	5
SS E4	322	153	92	61	43	31	23	16	11

RICHMOND, VIRGINIA								3939 DD		RICHMOND, VIRGINIA								CONTINUED	
SSF =	.10	.20	.30	.40	.50	.60	.70	.80	.90	SSF =	.10	.20	.30	.40	.50	.60	.70	.80	.90
WW A1	647	72	32	18	11	7	5	3	2	DG C2	178	81	49	34	24	18	13	8	4
WW A2	218	83	46	29	20	14	10	7	4	DG C3	222	104	65	45	34	26	20	15	10
WW A3	184	85	50	33	23	17	12	8	5	SS A1	381	112	57	35	23	16	11	7	5
WW A4	169	85	52	35	25	18	13	9	6	SS A2	354	149	85	55	38	27	19	14	9
WW A5	161	87	55	38	27	20	14	10	7	SS A3	359	98	49	29	18	12	8	5	3
WW A6	157	87	56	39	28	20	15	11	7	SS A4	358	146	83	53	37	26	18	13	8
WW B1	161	54	28	16	10	7	4	3	2	SS A5	634	111	51	29	19	12	8	6	3
WW B2	172	90	57	39	28	20	15	11	7	SS A6	351	146	84	54	37	27	19	13	9
WW B3	201	111	72	50	36	26	19	14	9	SS A7	703	90	39	21	13	8	5	3	2
WW B4	182	109	74	53	39	29	22	16	11	SS A8	357	142	80	51	35	24	17	12	8
WW B5	170	104	71	52	38	29	22	16	11	SS B1	258	84	44	27	18	12	9	6	4
WW C1	216	119	77	54	39	29	21	15	10	SS B2	279	121	70	46	32	23	16	12	8
WW C2	188	109	72	51	37	28	21	15	10	SS B3	234	75	39	23	15	10	7	5	3
WW C3	199	127	89	65	49	37	28	21	14	SS B4	272	117	68	44	30	22	16	11	7
WW C4	180	115	81	59	45	34	26	19	13	SS B5	333	74	35	21	13	9	6	4	2
TW A1	560	70	31	18	11	8	5	3	2	SS B6	262	115	67	44	31	22	16	11	7
TW A2	225	76	41	25	17	12	8	6	4	SS B7	288	60	28	16	10	6	4	3	2
TW A3	182	76	43	28	19	14	10	7	4	SS B8	253	109	63	41	28	20	14	10	6
TW A4	155	73	43	29	20	14	10	7	5	SS C1	162	71	41	27	18	13	9	6	4
TW B1	315	67	33	19	13	9	6	4	2	SS C2	181	93	58	39	28	20	15	11	7
TW B2	182	69	39	25	17	12	8	6	4	SS C3	169	57	31	19	13	9	6	4	3
TW B3	160	67	39	25	17	12	9	6	4	SS C4	179	83	50	33	23	17	12	9	6
TW B4	148	63	36	24	16	12	8	6	4	SS D1	348	134	74	47	31	22	15	10	7
TW C1	210	62	32	20	13	9	6	4	3	SS D2	313	162	101	69	49	36	26	18	12
TW C2	162	61	34	21	15	10	7	5	3	SS D3	440	127	64	39	25	17	12	8	5
TW C3	156	57	31	20	14	10	7	5	3	SS D4	320	159	97	65	46	33	24	17	11
TW C4	166	52	27	17	12	8	6	4	3	SS E1	254	102	57	36	24	17	12	8	5
TW D1	118	43	23	14	9	6	4	3	2	SS E2	256	131	81	55	39	28	21	15	10
TW D2	184	86	51	34	24	17	13	9	6	SS E3	310	88	44	27	17	12	8	5	3
TW D3	188	94	58	39	28	20	15	10	7	SS E4	264	123	74	49	34	24	18	12	8
TW D4	186	102	66	46	33	25	18	13	9										
TW D5	176	100	65	46	34	25	19	14	9										
TW E1	220	111	69	47	33	24	18	13	8	ROANOKE, VIRGINIA								4307 DD	
TW E2	203	107	68	47	34	25	18	13	9	SSF =	.10	.20	.30	.40	.50	.60	.70	.80	.90
TW E3	217	128	86	61	45	34	25	18	12	WW A1	599	69	31	18	11	7	5	3	2
TW E4	195	117	79	56	42	31	24	17	12	WW A2	209	81	45	29	20	14	10	7	4
TW F1	251	60	29	17	11	8	5	3	2	WW A3	177	83	50	33	23	16	12	8	5
TW F2	159	63	36	23	15	11	8	5	3	WW A4	164	83	52	35	25	18	13	9	6
TW F3	132	61	37	24	17	12	9	6	4	WW A5	156	85	54	37	27	19	14	10	7
TW F4	109	56	34	23	16	12	9	6	4	WW A6	152	85	55	38	28	20	15	11	7
TW G1	156	53	28	17	12	8	6	4	2	WW B1	153	53	27	16	10	7	5	3	2
TW G2	115	52	31	20	14	10	7	5	3	WW B2	167	88	56	38	28	20	15	11	7
TW G3	98	48	29	20	14	10	7	5	3	WW B3	196	109	71	49	36	26	19	14	9
TW G4	79	40	25	17	12	9	6	4	3	WW B4	177	107	73	52	39	29	22	16	11
TW H1	102	43	25	16	11	8	5	4	2	WW B5	166	103	70	51	38	29	22	16	11
TW H2	79	38	23	15	11	8	6	4	3	WW C1	210	117	76	53	39	29	21	15	10
TW H3	67	33	20	14	10	7	5	4	2	WW C2	184	107	71	51	37	28	21	15	10
TW H4	52	25	15	10	7	5	4	3	2	WW C3	195	126	88	65	49	37	28	21	14
TW I1	89	35	19	12	7	5	3	2	1	WW C4	176	114	80	59	44	34	26	19	13
TW I2	140	71	44	30	21	16	11	8	5	TW A1	521	67	31	18	11	8	5	3	2
TW I3	149	79	50	34	25	18	13	9	6	TW A2	215	74	40	25	17	12	8	6	4
TW I4	154	88	58	41	30	22	17	12	8	TW A3	175	74	43	28	19	14	10	7	4
TW I5	152	89	60	43	31	23	18	13	9	TW A4	149	71	43	28	20	14	10	7	5
TW J1	183	98	62	43	31	23	17	12	8	TW B1	297	65	32	19	12	9	6	4	2
TW J2	167	93	60	42	31	23	17	12	8	TW B2	175	67	38	24	17	12	8	6	4
TW J3	182	112	77	55	41	31	23	17	12	TW B3	154	66	38	25	17	12	9	6	4
TW J4	168	104	71	52	39	29	22	16	11	TW B4	142	61	36	23	16	12	8	6	4
DG A1	117	46	21	-	-	-	-	-	-	TW C1	200	60	32	20	13	9	6	4	3
DG A2	145	64	38	24	15	9	5	-	-	TW C2	156	59	33	21	14	10	7	5	3
DG A3	188	87	53	37	27	19	14	9	4	TW C3	149	55	31	20	14	10	7	5	3
DG B1	117	49	26	15	-	-	-	-	-	TW C4	158	50	27	17	11	8	6	4	3
DG B2	148	67	40	27	19	13	8	4	-	TW D1	113	42	22	14	9	6	4	3	2
DG B3	192	89	55	38	28	22	16	12	7	TW D2	178	84	51	34	24	17	13	9	6
DG C1	149	65	37	24	16	9	-	-	-	TW D3	182	92	57	39	28	20	15	10	7

ROANOKE, VIRGINIA   CONTINUED   OLYMPIA, WASHINGTON   5530 DD

SSF	=.10	.20	.30	.40	.50	.60	.70	.80	.90		SSF	=.10	.20	.30	.40	.50	.60	.70	.80	.90
TW D4	181	100	65	45	33	24	18	13	9		WW A1	448	43	-	-	-	-	-	-	-
TW D5	171	98	65	46	34	25	19	14	9		WW A2	158	53	24	10	-	-	-	-	-
TW E1	214	109	68	46	33	24	18	13	8		WW A3	132	54	27	14	-	-	-	-	-
TW E2	197	105	67	46	33	25	18	13	9		WW A4	121	54	29	16	5	-	-	-	-
TW E3	212	126	85	61	45	34	25	18	12		WW A5	115	55	30	17	8	-	-	-	-
TW E4	191	115	78	56	42	31	24	17	12		WW A6	111	56	31	18	9	-	-	-	-
TW F1	237	58	29	17	11	8	5	4	2		WW B1	101	-	-	-	-	-	-	-	-
TW F2	152	62	35	22	15	11	8	5	3		WW B2	128	62	36	22	14	8	4	1	-
TW F3	127	60	36	24	17	12	9	6	4		WW B3	152	79	47	30	19	12	7	4	2
TW F4	106	54	34	23	16	12	9	6	4		WW B4	141	81	52	35	24	17	12	8	4
TW G1	149	51	27	17	12	8	6	4	2		WW B5	133	79	52	36	25	18	13	8	5
TW G2	111	51	30	20	14	10	7	5	3		WW C1	166	86	52	33	22	14	9	5	3
TW G3	94	47	29	19	14	10	7	5	3		WW C2	145	80	50	33	23	15	10	6	4
TW G4	77	39	24	17	12	9	6	4	3		WW C3	159	98	66	47	34	24	18	12	7
TW H1	98	42	24	16	11	8	5	4	2		WW C4	144	89	60	43	31	22	16	11	7
TW H2	77	37	23	15	11	8	6	4	3		TW A1	394	44	13	-	-	-	-	-	-
TW H3	64	32	20	13	9	7	5	4	2		TW A2	164	49	22	9	-	-	-	-	-
TW H4	50	25	15	10	7	5	4	3	2		TW A3	133	50	24	12	-	-	-	-	-
TW I1	86	34	19	11	7	5	3	2	1		TW A4	113	48	25	14	7	-	-	-	-
TW I2	136	70	44	30	21	15	11	8	5		TW B1	227	44	16	-	-	-	-	-	-
TW I3	145	77	49	34	24	18	13	9	6		TW B2	134	46	22	11	-	-	-	-	-
TW I4	150	87	57	41	30	22	17	12	8		TW B3	118	45	23	12	6	-	-	-	-
TW I5	148	88	59	42	31	23	18	13	9		TW B4	110	43	22	12	6	-	-	-	-
TW J1	178	96	61	42	31	22	17	12	8		TW C1	155	42	18	8	-	-	-	-	-
TW J2	162	91	60	42	31	23	17	12	8		TW C2	121	41	20	11	5	-	-	-	-
TW J3	178	110	76	55	41	31	23	17	12		TW C3	117	40	19	11	6	-	-	-	-
TW J4	164	102	70	51	38	29	22	16	11		TW C4	126	37	17	9	5	2	-	-	-
DG A1	112	45	20	-	-	-	-	-	-		TW D1	75	-	-	-	-	-	-	-	-
DG A2	140	62	37	24	15	9	5	-	-		TW D2	139	60	33	19	12	7	3	-	-
DG A3	181	85	52	36	26	19	14	9	4		TW D3	142	66	37	22	14	8	4	2	-
DG B1	112	47	26	15	-	-	-	-	-		TW D4	145	76	46	30	20	14	9	6	3
DG B2	142	65	39	27	19	13	9	4	-		TW D5	139	76	48	32	22	16	11	7	4
DG B3	184	87	54	38	28	21	16	12	7		TW E1	170	80	46	29	18	12	7	4	2
DG C1	142	63	36	24	16	9	-	-	-		TW E2	157	79	47	30	20	13	9	5	3
DG C2	170	79	48	33	24	18	13	8	4		TW E3	173	98	63	43	30	21	15	10	6
DG C3	213	101	63	45	33	26	20	15	10		TW E4	156	90	58	40	28	20	14	10	6
SS A1	368	111	57	35	23	16	11	7	5		TW F1	177	36	11	-	-	-	-	-	-
SS A2	347	147	85	56	38	27	20	14	9		TW F2	114	40	18	7	-	-	-	-	-
SS A3	347	98	49	29	19	12	8	6	3		TW F3	95	39	19	10	-	-	-	-	-
SS A4	351	145	83	54	37	26	18	13	8		TW F4	78	36	19	10	5	-	-	-	-
SS A5	602	109	51	29	19	12	8	6	3		TW G1	111	32	13	-	-	-	-	-	-
SS A6	343	145	84	54	38	27	19	13	9		TW G2	83	33	16	8	-	-	-	-	-
SS A7	661	89	39	21	13	8	5	3	2		TW G3	70	31	16	9	-	-	-	-	-
SS A8	350	141	80	51	35	25	17	12	8		TW G4	57	26	14	8	4	-	-	-	-
SS B1	250	83	44	27	18	12	9	6	4		TW H1	73	27	12	-	-	-	-	-	-
SS B2	273	120	70	46	32	23	17	12	8		TW H2	57	25	13	7	-	-	-	-	-
SS B3	227	74	39	24	15	10	7	5	3		TW H3	48	22	11	6	3	-	-	-	-
SS B4	266	116	67	44	31	22	16	11	7		TW H4	38	17	9	5	3	-	-	-	-
SS B5	317	73	35	21	13	9	6	4	2		TW I1	53	-	-	-	-	-	-	-	-
SS B6	256	114	67	44	31	22	16	11	7		TW I2	105	49	28	17	10	6	3	-	-
SS B7	274	60	28	16	10	6	4	3	2		TW I3	113	56	32	20	13	8	4	2	-
SS B8	247	108	63	41	28	20	14	10	7		TW I4	120	65	41	27	18	13	9	5	3
SS C1	156	70	41	27	18	13	9	6	4		TW I5	119	67	43	29	20	14	10	7	4
SS C2	175	91	57	39	28	20	15	11	7		TW J1	140	71	42	26	17	11	7	4	2
SS C3	162	56	30	19	13	9	6	4	3		TW J2	129	68	42	27	18	12	8	5	3
SS C4	173	81	49	33	23	17	12	9	6		TW J3	145	86	57	39	28	20	14	10	6
SS D1	340	133	74	47	32	22	15	11	7		TW J4	134	80	53	37	26	19	13	9	6
SS D2	308	161	101	69	49	36	26	19	12		DG A1	75	-	-	-	-	-	-	-	-
SS D3	425	126	64	39	26	17	12	8	5		DG A2	111	44	22	9	-	-	-	-	-
SS D4	315	158	97	65	46	33	24	17	11		DG A3	154	68	39	25	16	10	6	-	-
SS E1	248	101	57	36	24	17	12	8	5		DG B1	78	-	-	-	-	-	-	-	-
SS E2	252	130	81	55	39	28	21	15	10		DG B2	116	47	25	13	-	-	-	-	-
SS E3	299	87	44	27	17	12	8	5	3		DG B3	159	71	42	27	18	12	8	4	-
SS E4	259	122	74	49	34	25	18	12	8		DG C1	109	37	-	-	-	-	-	-	-

OLYMPIA, WASHINGTON   CONTINUED

SSF =	.10	.20	.30	.40	.50	.60	.70	.80	.90
DG C2	142	60	33	20	11	4	-	-	-
DG C3	185	84	49	33	22	15	10	6	3
SS A1	312	85	37	17	-	-	-	-	-
SS A2	299	119	63	37	23	14	9	5	2
SS A3	294	72	27	-	-	-	-	-	-
SS A4	304	117	60	35	21	12	7	4	2
SS A5	510	84	31	11	-	-	-	-	-
SS A6	295	116	61	35	22	13	8	4	2
SS A7	560	66	18	-	-	-	-	-	-
SS A8	302	113	57	32	18	10	5	2	-
SS B1	208	62	27	12	-	-	-	-	-
SS B2	234	96	52	31	20	12	8	5	2
SS B3	187	53	20	-	-	-	-	-	-
SS B4	228	93	49	29	18	11	7	4	2
SS B5	264	53	19	-	-	-	-	-	-
SS B6	218	91	49	29	18	11	7	4	2
SS B7	227	41	-	-	-	-	-	-	-
SS B8	211	86	45	26	16	9	5	3	1
SS C1	114	43	19	-	-	-	-	-	-
SS C2	136	65	37	23	14	9	5	2	-
SS C3	121	36	14	-	-	-	-	-	-
SS C4	136	59	32	19	12	7	4	1	-
SS D1	284	100	47	23	9	-	-	-	-
SS D2	261	128	74	46	30	19	12	7	4
SS D3	360	96	41	19	-	-	-	-	-
SS D4	268	125	71	44	28	18	11	7	4
SS E1	201	70	30	-	-	-	-	-	-
SS E2	209	100	57	34	22	13	8	4	2
SS E3	247	63	24	-	-	-	-	-	-
SS E4	217	95	52	31	19	11	7	3	1

SEATTLE-TACOMA, WASHINGTON   5185 DD

SSF =	.10	.20	.30	.40	.50	.60	.70	.80	.90
WW A1	542	52	14	-	-	-	-	-	-
WW A2	183	61	27	12	-	-	-	-	-
WW A3	151	61	31	16	-	-	-	-	-
WW A4	137	61	32	17	7	-	-	-	-
WW A5	129	62	34	19	9	-	-	-	-
WW A6	125	62	35	20	10	-	-	-	-
WW B1	126	22	-	-	-	-	-	-	-
WW B2	143	69	39	24	15	9	4	1	-
WW B3	169	86	51	32	20	13	7	4	2
WW B4	154	87	55	37	26	18	12	8	4
WW B5	145	85	55	38	26	19	13	9	5
WW C1	183	94	56	36	23	15	9	5	3
WW C2	160	87	54	35	24	16	11	7	4
WW C3	171	105	70	49	35	25	18	12	8
WW C4	155	95	64	45	32	23	17	11	7
TW A1	472	52	16	-	-	-	-	-	-
TW A2	191	56	25	11	-	-	-	-	-
TW A3	152	56	27	14	-	-	-	-	-
TW A4	129	54	28	15	8	-	-	-	-
TW B1	269	51	19	-	-	-	-	-	-
TW B2	154	52	25	12	-	-	-	-	-
TW B3	135	51	25	14	7	-	-	-	-
TW B4	125	48	24	13	7	-	-	-	-
TW C1	180	48	21	10	-	-	-	-	-
TW C2	139	47	22	12	6	-	-	-	-
TW C3	134	44	21	12	6	-	-	-	-
TW C4	144	41	19	10	6	3	-	-	-
TW D1	91	20	-	-	-	-	-	-	-
TW D2	156	67	36	21	13	7	3	-	-
TW D3	159	73	40	24	15	9	4	2	-

SEATTLE-TACOMA, WASHINGTON   CONTINUED

SSF =	.10	.20	.30	.40	.50	.60	.70	.80	.90
TW D4	160	82	50	32	22	15	10	6	3
TW D5	152	82	51	34	23	16	11	7	4
TW E1	189	88	50	31	20	12	7	4	2
TW E2	174	86	50	32	21	14	9	5	3
TW E3	188	105	67	45	31	22	15	10	6
TW E4	169	96	62	42	30	21	15	10	6
TW F1	211	43	15	-	-	-	-	-	-
TW F2	132	45	21	9	-	-	-	-	-
TW F3	108	44	22	11	-	-	-	-	-
TW F4	89	40	21	12	5	-	-	-	-
TW G1	131	37	15	-	-	-	-	-	-
TW G2	95	37	18	9	-	-	-	-	-
TW G3	80	35	18	10	4	-	-	-	-
TW G4	65	29	16	9	5	-	-	-	-
TW H1	84	30	14	6	-	-	-	-	-
TW H2	65	28	14	7	-	-	-	-	-
TW H3	55	24	13	7	4	-	-	-	-
TW H4	43	19	10	6	3	1	-	-	-
TW I1	66	-	-	-	-	-	-	-	-
TW I2	117	54	30	18	11	6	3	-	-
TW I3	126	61	35	22	14	8	5	2	-
TW I4	132	71	44	29	20	13	9	6	3
TW I5	130	73	46	31	21	15	10	7	4
TW J1	156	77	45	28	18	12	7	4	2
TW J2	142	74	45	29	20	13	9	5	3
TW J3	157	92	60	41	29	21	15	10	6
TW J4	145	86	56	39	27	20	14	9	6
DG A1	92	-	-	-	-	-	-	-	-
DG A2	127	50	24	11	-	-	-	-	-
DG A3	172	74	42	26	17	11	6	-	-
DG B1	95	25	-	-	-	-	-	-	-
DG B2	133	53	28	15	-	-	-	-	-
DG B3	178	78	45	29	19	13	8	5	-
DG C1	129	45	14	-	-	-	-	-	-
DG C2	161	67	37	22	12	5	-	-	-
DG C3	207	91	53	35	24	16	11	7	3
SS A1	358	95	40	19	6	-	-	-	-
SS A2	332	129	67	39	24	15	9	5	2
SS A3	340	81	30	-	-	-	-	-	-
SS A4	338	127	64	36	22	13	7	4	2
SS A5	595	94	34	13	-	-	-	-	-
SS A6	328	126	65	37	23	14	8	4	2
SS A7	662	75	21	-	-	-	-	-	-
SS A8	337	123	61	34	19	11	5	2	-
SS B1	238	69	29	13	-	-	-	-	-
SS B2	259	104	55	33	21	13	8	5	2
SS B3	216	59	23	-	-	-	-	-	-
SS B4	253	100	53	31	19	11	7	4	2
SS B5	307	60	22	-	-	-	-	-	-
SS B6	242	98	52	31	19	12	7	4	2
SS B7	267	47	13	-	-	-	-	-	-
SS B8	234	93	48	28	16	10	5	3	1
SS C1	131	49	22	-	-	-	-	-	-
SS C2	152	72	41	25	15	9	5	2	-
SS C3	142	41	17	-	-	-	-	-	-
SS C4	153	65	35	21	13	7	4	1	-
SS D1	322	110	51	25	10	-	-	-	-
SS D2	286	137	78	48	31	20	13	7	4
SS D3	413	106	45	20	-	-	-	-	-
SS D4	295	135	75	46	29	19	12	7	4
SS E1	228	78	34	12	-	-	-	-	-
SS E2	230	108	60	37	23	14	8	4	2
SS E3	286	71	28	-	-	-	-	-	-
SS E4	241	103	55	33	20	12	7	3	1

SPOKANE, WASHINGTON — 6835 DD

SSF =	.10	.20	.30	.40	.50	.60	.70	.80	.90
WW A1	451	30	-	-	-	-	-	-	-
WW A2	137	41	15	-	-	-	-	-	-
WW A3	112	43	19	-	-	-	-	-	-
WW A4	102	43	20	-	-	-	-	-	-
WW A5	97	44	22	-	-	-	-	-	-
WW A6	93	44	23	9	-	-	-	-	-
WW B1	73	-	-	-	-	-	-	-	-
WW B2	112	52	29	17	9	4	-	-	-
WW B3	132	66	38	23	14	8	4	2	-
WW B4	126	71	45	30	20	14	9	6	3
WW B5	120	70	45	31	22	15	11	7	4
WW C1	145	74	44	27	17	11	6	3	2
WW C2	129	70	43	28	19	12	8	5	3
WW C3	143	88	59	41	29	21	15	10	6
WW C4	130	80	54	38	27	20	14	10	6
TW A1	393	33	-	-	-	-	-	-	-
TW A2	146	39	14	-	-	-	-	-	-
TW A3	115	40	17	-	-	-	-	-	-
TW A4	97	39	18	7	-	-	-	-	-
TW B1	214	34	-	-	-	-	-	-	-
TW B2	118	37	16	-	-	-	-	-	-
TW B3	103	37	17	7	-	-	-	-	-
TW B4	97	35	17	8	-	-	-	-	-
TW C1	142	34	13	-	-	-	-	-	-
TW C2	108	34	15	7	-	-	-	-	-
TW C3	105	33	15	7	-	-	-	-	-
TW C4	116	31	14	7	3	-	-	-	-
TW D1	54	-	-	-	-	-	-	-	-
TW D2	123	51	27	15	8	2	-	-	-
TW D3	124	55	30	17	9	4	-	-	-
TW D4	130	66	40	26	17	11	7	4	2
TW D5	126	68	42	28	19	13	9	6	3
TW E1	150	69	38	23	14	8	4	2	-
TW E2	140	68	40	25	16	10	6	4	2
TW E3	155	87	55	37	26	18	12	8	5
TW E4	141	80	52	35	25	17	12	8	5
TW F1	162	26	-	-	-	-	-	-	-
TW F2	98	31	10	-	-	-	-	-	-
TW F3	80	30	13	-	-	-	-	-	-
TW F4	66	28	14	-	-	-	-	-	-
TW G1	97	24	-	-	-	-	-	-	-
TW G2	70	26	10	-	-	-	-	-	-
TW G3	59	24	11	-	-	-	-	-	-
TW G4	49	21	11	4	-	-	-	-	-
TW H1	61	20	-	-	-	-	-	-	-
TW H2	48	19	9	-	-	-	-	-	-
TW H3	41	17	9	3	-	-	-	-	-
TW H4	33	14	7	3	-	-	-	-	-
TW I1	34	-	-	-	-	-	-	-	-
TW I2	91	41	22	13	7	-	-	-	-
TW I3	99	47	26	16	9	5	2	-	-
TW I4	107	57	35	23	15	10	7	4	2
TW I5	107	60	38	25	17	12	8	5	3
TW J1	123	60	35	21	13	8	5	2	1
TW J2	115	60	36	23	15	10	6	4	2
TW J3	130	77	50	34	24	17	12	8	5
TW J4	121	72	47	32	23	16	12	8	5
DG A1	54	-	-	-	-	-	-	-	-
DG A2	98	37	16	-	-	-	-	-	-
DG A3	141	60	34	21	13	8	4	-	-
DG B1	58	-	-	-	-	-	-	-	-
DG B2	102	40	19	-	-	-	-	-	-
DG B3	146	63	37	24	15	10	6	3	-
DG C1	88	25	-	-	-	-	-	-	-

SPOKANE, WASHINGTON — CONTINUED

SSF =	.10	.20	.30	.40	.50	.60	.70	.80	.90
DG C2	127	52	28	15	6	-	-	-	-
DG C3	170	74	44	29	19	13	8	5	-
SS A1	267	64	24	-	-	-	-	-	-
SS A2	256	97	50	29	17	10	6	3	2
SS A3	242	49	-	-	-	-	-	-	-
SS A4	255	93	46	25	14	8	4	1	-
SS A5	465	62	17	-	-	-	-	-	-
SS A6	251	94	48	27	16	9	5	2	1
SS A7	513	42	-	-	-	-	-	-	-
SS A8	252	88	43	23	12	5	-	-	-
SS B1	173	45	16	-	-	-	-	-	-
SS B2	201	79	42	25	15	9	5	3	2
SS B3	150	35	-	-	-	-	-	-	-
SS B4	193	75	39	22	13	8	4	2	-
SS B5	229	37	-	-	-	-	-	-	-
SS B6	186	74	39	23	13	8	4	2	1
SS B7	190	24	-	-	-	-	-	-	-
SS B8	176	68	34	19	11	6	2	-	-
SS C1	94	31	-	-	-	-	-	-	-
SS C2	120	55	31	18	10	5	2	-	-
SS C3	105	27	-	-	-	-	-	-	-
SS C4	120	50	26	15	8	4	-	-	-
SS D1	233	75	31	-	-	-	-	-	-
SS D2	220	104	59	36	23	14	9	5	3
SS D3	308	72	27	-	-	-	-	-	-
SS D4	226	103	57	34	21	13	8	5	2
SS E1	158	48	-	-	-	-	-	-	-
SS E2	173	80	44	26	15	9	4	2	-
SS E3	207	44	-	-	-	-	-	-	-
SS E4	182	76	40	23	13	7	3	1	-

YAKIMA, WASHINGTON — 6009 DD

SSF =	.10	.20	.30	.40	.50	.60	.70	.80	.90
WW A1	615	51	14	-	-	-	-	-	-
WW A2	185	60	27	13	-	-	-	-	-
WW A3	151	61	31	16	7	-	-	-	-
WW A4	137	61	32	18	9	-	-	-	-
WW A5	128	61	34	19	10	4	-	-	-
WW A6	124	61	35	20	11	5	1	-	-
WW B1	124	20	-	-	-	-	-	-	-
WW B2	143	68	39	24	15	9	5	3	1
WW B3	167	85	50	32	20	13	8	5	3
WW B4	153	87	55	37	26	18	12	8	5
WW B5	144	84	55	38	27	19	13	9	5
WW C1	181	93	55	36	23	15	10	6	3
WW C2	159	86	53	35	24	16	11	7	4
WW C3	169	104	70	49	35	25	18	13	8
WW C4	154	95	64	45	32	23	17	12	7
TW A1	531	51	16	-	-	-	-	-	-
TW A2	196	56	25	11	-	-	-	-	-
TW A3	153	55	27	14	6	-	-	-	-
TW A4	129	53	28	15	8	3	-	-	-
TW B1	289	50	19	6	-	-	-	-	-
TW B2	157	52	25	13	6	-	-	-	-
TW B3	136	50	25	14	7	2	-	-	-
TW B4	126	47	24	13	7	4	-	-	-
TW C1	188	47	21	10	-	-	-	-	-
TW C2	141	46	22	12	6	3	-	-	-
TW C3	137	44	21	12	6	3	-	-	-
TW C4	150	41	19	10	6	3	1	-	-
TW D1	90	19	-	-	-	-	-	-	-
TW D2	157	66	36	21	13	8	4	2	1
TW D3	158	72	40	24	15	9	5	3	1

YAKIMA, WASHINGTON — CONTINUED

SSF =	.10	.20	.30	.40	.50	.60	.70	.80	.90
TW D4	159	82	49	32	22	15	10	6	4
TW D5	152	81	51	34	24	16	11	8	5
TW E1	188	87	49	31	20	13	8	5	2
TW E2	174	85	50	32	21	14	9	5	3
TW E3	186	104	66	45	31	22	15	10	6
TW E4	169	96	62	42	30	21	15	10	6
TW F1	224	42	15	-	-	-	-	-	-
TW F2	133	45	21	9	-	-	-	-	-
TW F3	108	44	22	11	4	-	-	-	-
TW F4	88	40	21	12	6	-	-	-	-
TW G1	134	37	15	-	-	-	-	-	-
TW G2	95	37	18	9	-	-	-	-	-
TW G3	79	34	18	10	5	-	-	-	-
TW G4	64	29	16	9	5	2	-	-	-
TW H1	84	30	14	6	-	-	-	-	-
TW H2	65	27	14	8	4	-	-	-	-
TW H3	55	24	13	7	4	2	-	-	-
TW H4	43	19	10	6	3	2	-	-	-
TW I1	64	-	-	-	-	-	-	-	-
TW I2	117	54	30	18	11	7	4	2	-
TW I3	125	60	35	22	14	9	5	3	1
TW I4	131	70	44	29	20	13	9	6	3
TW I5	130	72	46	31	22	15	10	7	4
TW J1	154	76	45	28	18	12	8	5	2
TW J2	142	74	45	29	20	13	9	6	3
TW J3	156	91	60	41	29	21	15	10	6
TW J4	145	85	56	39	28	20	14	10	6
DG A1	91	-	-	-	-	-	-	-	-
DG A2	127	49	24	11	-	-	-	-	-
DG A3	173	74	42	26	17	11	6	2	-
DG B1	93	25	-	-	-	-	-	-	-
DG B2	132	53	28	15	6	-	-	-	-
DG B3	179	77	45	29	19	13	8	5	-
DG C1	126	45	13	-	-	-	-	-	-
DG C2	161	66	37	22	12	6	-	-	-
DG C3	207	89	53	35	24	16	11	7	3
SS A1	343	85	35	16	5	-	-	-	-
SS A2	314	118	62	36	22	14	9	5	3
SS A3	319	69	24	-	-	-	-	-	-
SS A4	314	114	58	33	20	12	7	4	2
SS A5	605	84	30	10	-	-	-	-	-
SS A6	309	115	59	35	21	13	8	4	2
SS A7	681	63	14	-	-	-	-	-	-
SS A8	312	110	54	30	17	10	5	3	1
SS B1	226	61	26	11	-	-	-	-	-
SS B2	245	97	51	31	19	12	8	5	3
SS B3	200	51	18	-	-	-	-	-	-
SS B4	237	92	48	28	17	11	7	4	2
SS B5	304	53	19	-	-	-	-	-	-
SS B6	228	91	48	29	18	11	7	4	2
SS B7	260	40	-	-	-	-	-	-	-
SS B8	218	84	44	25	15	9	5	3	1
SS C1	131	48	22	8	-	-	-	-	-
SS C2	152	71	40	25	16	10	6	3	2
SS C3	145	41	17	-	-	-	-	-	-
SS C4	153	65	35	21	13	8	4	2	1
SS D1	299	98	45	22	9	-	-	-	-
SS D2	266	126	72	45	29	19	12	8	4
SS D3	397	95	39	18	-	-	-	-	-
SS D4	275	124	69	43	27	18	11	7	4
SS E1	210	69	30	-	-	-	-	-	-
SS E2	214	99	56	34	21	13	8	5	3
SS E3	274	63	24	-	-	-	-	-	-
SS E4	225	95	51	30	19	11	7	4	2

CHARLESTON, WEST VIRGINIA 4590 DD

SSF =	.10	.20	.30	.40	.50	.60	.70	.80	.90
WW A1	408	35	11	-	-	-	-	-	-
WW A2	134	46	23	13	8	5	3	2	1
WW A3	113	48	27	16	11	7	5	3	2
WW A4	105	49	28	18	12	8	5	4	2
WW A5	100	50	30	20	13	9	6	4	3
WW A6	97	51	31	20	14	10	7	5	3
WW B1	76	-	-	-	-	-	-	-	-
WW B2	113	57	34	23	16	11	8	5	3
WW B3	135	72	45	30	21	15	11	8	5
WW B4	128	75	50	35	25	19	14	10	7
WW B5	121	73	49	35	26	19	14	10	7
WW C1	147	79	49	34	24	17	12	9	6
WW C2	131	74	48	33	24	17	13	9	6
WW C3	145	92	63	46	34	26	19	14	9
WW C4	132	83	58	42	31	23	18	13	9
TW A1	358	36	13	5	-	-	-	-	-
TW A2	141	43	21	12	7	4	3	2	-
TW A3	114	44	23	14	9	6	4	2	1
TW A4	98	43	24	15	10	7	5	3	2
TW B1	200	37	15	8	4	2	1	-	-
TW B2	116	40	21	12	8	5	3	2	1
TW B3	102	40	22	13	9	6	4	2	1
TW B4	96	38	21	13	8	6	4	3	2
TW C1	135	36	17	10	6	4	2	1	-
TW C2	105	37	19	11	7	5	3	2	1
TW C3	102	35	18	11	7	5	3	2	1
TW C4	112	32	16	10	6	4	3	2	1
TW D1	57	12	-	-	-	-	-	-	-
TW D2	122	54	31	20	13	9	6	4	3
TW D3	125	60	35	23	16	11	8	5	3
TW D4	130	70	44	30	21	15	11	8	5
TW D5	126	70	45	31	23	17	12	9	6
TW E1	150	73	44	29	20	14	10	7	4
TW E2	140	72	44	30	21	15	11	8	5
TW E3	157	91	60	42	30	22	17	12	8
TW E4	142	83	55	39	29	21	16	11	8
TW F1	152	30	12	6	2	-	-	-	-
TW F2	97	35	18	10	6	4	2	1	-
TW F3	81	35	19	12	8	5	3	2	1
TW F4	68	32	19	12	8	5	4	2	1
TW G1	94	28	13	7	4	2	1	-	-
TW G2	70	29	16	10	6	4	3	2	-
TW G3	60	28	16	10	6	4	3	2	1
TW G4	50	24	14	9	6	4	3	2	1
TW H1	61	24	12	7	4	3	2	1	-
TW H2	49	22	12	8	5	3	2	1	-
TW H3	42	19	11	7	5	3	2	1	-
TW H4	33	15	9	6	4	3	2	1	-
TW I1	40	-	-	-	-	-	-	-	-
TW I2	92	45	27	17	12	8	6	4	3
TW I3	100	51	31	20	14	10	7	5	3
TW I4	108	60	39	27	19	14	10	7	5
TW I5	108	62	41	29	21	15	11	8	5
TW J1	125	64	40	26	19	13	9	7	4
TW J2	116	63	40	27	19	14	10	7	5
TW J3	132	80	54	38	28	21	16	11	8
TW J4	123	74	50	36	27	20	15	11	7
DG A1	55	-	-	-	-	-	-	-	-
DG A2	95	38	20	10	-	-	-	-	-
DG A3	136	60	36	24	17	12	8	4	-
DG B1	57	-	-	-	-	-	-	-	-
DG B2	97	41	22	13	8	-	-	-	-
DG B3	140	63	38	26	18	13	10	7	3
DG C1	85	29	11	-	-	-	-	-	-

CHARLESTON, WEST VIRGINIA — CONTINUED

SSF =	.10	.20	.30	.40	.50	.60	.70	.80	.90
DG C2	121	52	30	19	13	8	4	-	-
DG C3	162	74	45	30	22	16	12	9	5
SS A1	264	71	34	19	11	7	5	3	2
SS A2	258	104	58	36	24	17	12	8	5
SS A3	242	59	25	13	7	4	2	-	-
SS A4	260	101	55	34	23	16	11	7	4
SS A5	443	68	28	14	8	4	2	1	-
SS A6	253	102	56	35	23	16	11	8	5
SS A7	487	51	17	7	-	-	-	-	-
SS A8	257	97	52	32	21	14	10	7	4
SS B1	174	52	25	14	8	5	3	2	1
SS B2	204	85	48	31	21	14	10	7	4
SS B3	153	43	20	10	5	3	1	-	-
SS B4	197	81	45	29	19	13	9	6	4
SS B5	224	43	18	9	5	2	1	-	-
SS B6	189	80	45	29	19	13	9	6	4
SS B7	188	32	11	4	-	-	-	-	-
SS B8	181	75	42	26	17	12	8	6	3
SS C1	96	38	20	11	7	4	2	1	-
SS C2	121	59	36	23	16	11	8	6	3
SS C3	103	31	14	8	4	2	1	-	-
SS C4	120	53	31	20	13	9	6	4	3
SS D1	237	85	44	26	16	11	7	4	2
SS D2	226	114	69	46	32	23	16	11	7
SS D3	304	80	37	21	13	8	5	3	2
SS D4	232	111	66	43	30	21	15	10	7
SS E1	165	60	30	17	10	6	4	2	1
SS E2	180	89	53	35	24	17	12	8	5
SS E3	207	52	23	12	7	4	2	1	-
SS E4	187	84	48	31	21	15	10	7	4

HUNTINGTON, WEST VIRGINIA — 4624 DD

SSF =	.10	.20	.30	.40	.50	.60	.70	.80	.90
WW A1	457	42	16	7	3	-	-	-	-
WW A2	151	53	28	16	10	7	4	3	2
WW A3	127	55	31	19	13	9	6	4	2
WW A4	117	56	33	21	14	10	7	5	3
WW A5	111	57	35	23	16	11	8	5	3
WW A6	108	58	36	24	16	12	8	6	3
WW B1	94	23	-	-	-	-	-	-	-
WW B2	124	63	38	26	18	13	9	6	4
WW B3	147	79	50	34	24	17	12	9	6
WW B4	138	81	54	38	28	21	15	11	7
WW B5	130	79	53	38	28	21	16	11	8
WW C1	160	86	55	37	26	19	14	10	6
WW C2	142	80	52	36	26	19	14	10	7
WW C3	155	98	68	49	37	28	21	15	10
WW C4	141	89	62	45	34	25	19	14	9
TW A1	399	42	17	8	4	2	-	-	-
TW A2	157	49	24	14	9	6	4	2	1
TW A3	127	50	27	17	11	7	5	3	2
TW A4	109	48	28	18	12	8	6	4	2
TW B1	223	43	19	10	6	4	2	1	-
TW B2	129	46	24	15	9	6	4	3	2
TW B3	113	45	25	15	10	7	5	3	2
TW B4	106	43	24	15	10	7	5	3	2
TW C1	150	41	20	12	7	5	3	2	1
TW C2	116	41	22	13	9	6	4	3	2
TW C3	113	39	21	13	8	6	4	3	2
TW C4	122	36	18	11	7	5	3	2	1
TW D1	69	20	6	-	-	-	-	-	-
TW D2	134	60	35	22	15	11	8	5	3
TW D3	137	66	39	26	18	13	9	6	4

HUNTINGTON, WEST VIRGINIA — CONTINUED

SSF =	.10	.20	.30	.40	.50	.60	.70	.80	.90
TW D4	141	76	48	33	23	17	12	9	6
TW D5	136	76	49	34	25	18	13	10	6
TW E1	164	80	48	32	22	16	11	8	5
TW E2	152	78	49	33	23	17	12	9	6
TW E3	168	98	64	45	33	24	18	13	9
TW E4	152	90	60	42	31	23	17	12	8
TW F1	172	36	16	8	4	2	1	-	-
TW F2	109	40	21	13	8	5	3	2	1
TW F3	91	40	22	14	9	6	4	3	2
TW F4	76	37	22	14	9	7	5	3	2
TW G1	106	33	16	9	5	3	2	1	-
TW G2	79	34	19	12	8	5	3	2	1
TW G3	67	32	18	12	8	5	4	2	1
TW G4	55	27	16	10	7	5	3	2	1
TW H1	69	27	15	9	6	4	2	2	-
TW H2	55	25	14	9	6	4	3	2	1
TW H3	47	22	13	8	6	4	3	2	1
TW H4	37	17	10	6	4	3	2	1	-
TW I1	50	15	-	-	-	-	-	-	-
TW I2	101	50	30	20	14	10	7	5	3
TW I3	109	56	34	23	16	11	8	6	4
TW I4	117	65	42	29	21	15	11	8	5
TW I5	116	67	44	31	23	17	12	9	6
TW J1	136	70	44	29	21	15	11	8	5
TW J2	125	68	44	30	21	16	11	8	5
TW J3	142	86	58	42	31	23	17	12	8
TW J4	131	80	54	39	29	21	16	12	8
DG A1	68	-	-	-	-	-	-	-	-
DG A2	105	43	23	13	6	-	-	-	-
DG A3	146	65	39	26	18	13	9	5	-
DG B1	70	21	-	-	-	-	-	-	-
DG B2	107	46	26	16	10	5	-	-	-
DG B3	150	68	41	28	20	15	11	7	4
DG C1	97	37	17	-	-	-	-	-	-
DG C2	132	58	33	22	15	10	6	-	-
DG C3	174	80	48	33	24	18	14	10	6
SS A1	288	80	38	22	13	9	6	4	2
SS A2	278	113	63	40	27	19	13	9	6
SS A3	267	67	30	16	9	5	3	2	-
SS A4	280	110	60	38	25	17	12	8	5
SS A5	485	77	32	17	10	6	4	2	1
SS A6	273	110	61	39	26	18	13	9	5
SS A7	536	59	22	10	4	-	-	-	-
SS A8	277	106	57	35	23	16	11	7	5
SS B1	191	58	28	16	10	6	4	3	1
SS B2	219	92	52	33	23	16	11	8	5
SS B3	170	50	23	13	7	4	3	1	-
SS B4	212	88	50	31	21	15	10	7	4
SS B5	247	49	21	11	6	4	2	1	-
SS B6	203	87	49	32	21	15	11	7	5
SS B7	209	38	15	7	3	-	-	-	-
SS B8	195	81	46	29	19	13	9	6	4
SS C1	109	44	24	14	9	6	4	2	1
SS C2	132	65	40	26	18	13	9	6	4
SS C3	116	36	17	10	6	4	2	1	-
SS C4	131	59	34	22	15	11	7	5	3
SS D1	260	95	49	30	19	12	8	5	3
SS D2	244	123	75	50	35	25	18	12	8
SS D3	332	89	43	24	15	10	6	4	2
SS D4	250	120	72	47	33	23	17	12	7
SS E1	183	68	35	20	13	8	5	3	2
SS E2	195	97	59	39	27	19	13	9	6
SS E3	228	59	27	15	9	5	3	2	1
SS E4	202	92	53	34	23	16	11	8	5

LCR *Tables* / 611

EAU CLAIRE, WISCONSIN — 8388 DD

SSF =	.10	.20	.30	.40	.50	.60	.70	.80	.90
WW A1	194	-	-	-	-	-	-	-	-
WW A2	64	12	-	-	-	-	-	-	-
WW A3	55	17	-	-	-	-	-	-	-
WW A4	52	19	-	-	-	-	-	-	-
WW A5	50	21	9	-	-	-	-	-	-
WW A6	49	22	10	-	-	-	-	-	-
WW B1	-	-	-	-	-	-	-	-	-
WW B2	66	31	17	10	6	3	1	-	-
WW B3	82	41	24	15	10	6	4	2	1
WW B4	86	49	32	22	16	11	8	5	3
WW B5	85	50	33	23	17	12	9	6	4
WW C1	93	48	29	19	12	8	6	4	2
WW C2	86	47	30	20	14	10	7	5	3
WW C3	104	65	44	32	24	17	13	9	6
WW C4	95	60	41	29	22	16	12	9	6
TW A1	180	-	-	-	-	-	-	-	-
TW A2	70	13	-	-	-	-	-	-	-
TW A3	58	17	-	-	-	-	-	-	-
TW A4	51	19	7	-	-	-	-	-	-
TW B1	101	-	-	-	-	-	-	-	-
TW B2	60	16	-	-	-	-	-	-	-
TW B3	55	18	7	-	-	-	-	-	-
TW B4	53	18	8	-	-	-	-	-	-
TW C1	72	14	-	-	-	-	-	-	-
TW C2	59	17	7	-	-	-	-	-	-
TW C3	59	17	7	3	-	-	-	-	-
TW C4	67	17	7	3	-	-	-	-	-
TW D1	-	-	-	-	-	-	-	-	-
TW D2	73	30	16	9	5	3	-	-	-
TW D3	74	33	18	10	6	3	-	-	-
TW D4	87	45	28	18	13	9	6	4	3
TW D5	88	48	31	21	15	11	8	5	3
TW E1	94	43	25	15	10	6	4	2	1
TW E2	92	45	27	18	12	8	6	4	2
TW E3	109	62	40	28	20	14	10	7	5
TW E4	101	59	39	27	19	14	10	7	5
TW F1	64	-	-	-	-	-	-	-	-
TW F2	44	-	-	-	-	-	-	-	-
TW F3	39	12	-	-	-	-	-	-	-
TW F4	34	13	4	-	-	-	-	-	-
TW G1	40	-	-	-	-	-	-	-	-
TW G2	33	9	-	-	-	-	-	-	-
TW G3	30	11	-	-	-	-	-	-	-
TW G4	26	11	5	-	-	-	-	-	-
TW H1	27	-	-	-	-	-	-	-	-
TW H2	24	8	-	-	-	-	-	-	-
TW H3	22	9	4	-	-	-	-	-	-
TW H4	18	7	3	1	-	-	-	-	-
TW I1	-	-	-	-	-	-	-	-	-
TW I2	54	24	13	8	4	2	-	-	-
TW I3	60	29	16	10	6	4	2	1	-
TW I4	72	39	25	17	12	8	6	4	2
TW I5	75	42	27	19	14	10	7	5	3
TW J1	78	38	23	14	9	6	4	2	1
TW J2	76	40	25	16	11	8	5	4	2
TW J3	93	56	37	26	19	14	10	7	5
TW J4	88	53	35	25	18	13	10	7	5
DG A1	-	-	-	-	-	-	-	-	-
DG A2	53	17	-	-	-	-	-	-	-
DG A3	93	40	23	15	10	6	3	-	-
DG B1	-	-	-	-	-	-	-	-	-
DG B2	55	19	-	-	-	-	-	-	-
DG B3	97	43	25	16	11	8	5	3	-
DG C1	-	-	-	-	-	-	-	-	-

EAU CLAIRE, WISCONSIN — CONTINUED

SSF =	.10	.20	.30	.40	.50	.60	.70	.80	.90
DG C2	74	29	15	7	-	-	-	-	-
DG C3	114	51	30	20	14	10	7	4	1
SS A1	146	31	-	-	-	-	-	-	-
SS A2	165	63	33	20	13	8	5	3	2
SS A3	116	-	-	-	-	-	-	-	-
SS A4	159	58	30	17	10	6	4	2	-
SS A5	251	24	-	-	-	-	-	-	-
SS A6	159	60	31	18	11	7	4	3	1
SS A7	252	-	-	-	-	-	-	-	-
SS A8	154	54	26	14	8	4	2	-	-
SS B1	89	19	-	-	-	-	-	-	-
SS B2	131	52	28	17	11	7	5	3	2
SS B3	67	-	-	-	-	-	-	-	-
SS B4	123	48	25	15	9	6	4	2	1
SS B5	112	-	-	-	-	-	-	-	-
SS B6	119	48	26	15	10	6	4	2	1
SS B7	78	-	-	-	-	-	-	-	-
SS B8	109	42	22	12	7	4	3	1	-
SS C1	38	-	-	-	-	-	-	-	-
SS C2	72	33	19	11	7	4	2	-	-
SS C3	45	-	-	-	-	-	-	-	-
SS C4	72	30	16	9	5	3	1	-	-
SS D1	127	37	12	-	-	-	-	-	-
SS D2	144	70	41	26	17	12	8	5	3
SS D3	167	34	-	-	-	-	-	-	-
SS D4	148	68	39	25	16	11	7	5	3
SS E1	74	-	-	-	-	-	-	-	-
SS E2	109	51	29	17	11	7	4	2	1
SS E3	100	-	-	-	-	-	-	-	-
SS E4	113	48	26	15	9	6	3	2	-

GREEN BAY, WISCONSIN — 8098 DD

SSF =	.10	.20	.30	.40	.50	.60	.70	.80	.90
WW A1	239	-	-	-	-	-	-	-	-
WW A2	78	20	-	-	-	-	-	-	-
WW A3	67	24	9	-	-	-	-	-	-
WW A4	62	25	12	-	-	-	-	-	-
WW A5	59	27	13	6	-	-	-	-	-
WW A6	58	27	14	7	-	-	-	-	-
WW B1	-	-	-	-	-	-	-	-	-
WW B2	75	35	20	12	8	5	3	1	-
WW B3	92	47	28	18	12	8	5	3	2
WW B4	94	54	35	24	17	12	9	6	4
WW B5	91	54	36	25	18	14	10	7	5
WW C1	103	53	32	21	14	10	7	4	2
WW C2	95	52	33	22	16	11	8	5	3
WW C3	112	70	48	34	25	19	14	10	7
WW C4	102	64	44	32	23	17	13	9	6
TW A1	216	-	-	-	-	-	-	-	-
TW A2	85	20	-	-	-	-	-	-	-
TW A3	69	22	9	-	-	-	-	-	-
TW A4	60	23	11	5	-	-	-	-	-
TW B1	122	14	-	-	-	-	-	-	-
TW B2	72	21	8	-	-	-	-	-	-
TW B3	64	22	10	4	-	-	-	-	-
TW B4	62	22	10	5	2	-	-	-	-
TW C1	85	18	6	-	-	-	-	-	-
TW C2	68	21	9	4	-	-	-	-	-
TW C3	68	21	9	5	2	-	-	-	-
TW C4	76	20	9	5	2	-	-	-	-
TW D1	-	-	-	-	-	-	-	-	-
TW D2	83	34	18	11	7	4	2	-	-
TW D3	84	37	21	13	8	5	3	1	-

GREEN BAY, WISCONSIN								CONTINUED		LA CROSSE, WISCONSIN								7417 DD	
SSF =	.10	.20	.30	.40	.50	.60	.70	.80	.90	SSF =	.10	.20	.30	.40	.50	.60	.70	.80	.90
TW D4	95	49	30	20	14	10	7	5	3	WW A1	260	-	-	-	-	-	-	-	-
TW D5	95	52	33	23	16	12	8	6	4	WW A2	82	23	7	-	-	-	-	-	-
TW E1	105	48	28	18	12	8	5	3	2	WW A3	70	26	12	4	-	-	-	-	-
TW E2	101	50	30	20	13	9	6	4	3	WW A4	65	27	14	7	-	-	-	-	-
TW E3	118	67	44	30	22	16	11	8	5	WW A5	62	29	15	9	4	-	-	-	-
TW E4	109	63	41	29	21	15	11	8	5	WW A6	61	30	16	10	5	-	-	-	-
TW F1	84	-	-	-	-	-	-	-	-	WW B1	-	-	-	-	-	-	-	-	-
TW F2	55	14	-	-	-	-	-	-	-	WW B2	78	37	22	14	9	6	4	2	1
TW F3	47	17	6	-	-	-	-	-	-	WW B3	94	49	29	19	13	9	6	4	2
TW F4	40	17	8	-	-	-	-	-	-	WW B4	96	55	36	25	18	13	10	7	4
TW G1	51	-	-	-	-	-	-	-	-	WW B5	93	56	37	26	19	14	10	7	5
TW G2	40	13	-	-	-	-	-	-	-	WW C1	106	55	34	22	15	11	7	5	3
TW G3	36	14	6	-	-	-	-	-	-	WW C2	97	54	34	23	16	12	8	6	4
TW G4	31	13	7	3	-	-	-	-	-	WW C3	113	71	49	35	26	20	15	11	7
TW H1	34	9	-	-	-	-	-	-	-	WW C4	104	65	45	32	24	18	13	10	7
TW H2	29	11	5	-	-	-	-	-	-	TW A1	233	-	-	-	-	-	-	-	-
TW H3	26	11	5	2	-	-	-	-	-	TW A2	88	22	7	-	-	-	-	-	-
TW H4	21	9	5	2	1	-	-	-	-	TW A3	72	24	11	4	-	-	-	-	-
TW I1	-	-	-	-	-	-	-	-	-	TW A4	63	25	13	7	3	-	-	-	-
TW I2	61	28	15	9	6	3	2	-	-	TW B1	127	16	-	-	-	-	-	-	-
TW I3	68	32	19	12	8	5	3	2	-	TW B2	74	23	10	4	-	-	-	-	-
TW I4	79	43	27	18	13	9	6	4	3	TW B3	66	23	11	6	2	-	-	-	-
TW I5	81	46	30	21	15	11	8	5	3	TW B4	64	23	12	6	3	1	-	-	-
TW J1	87	43	26	16	11	7	5	3	2	TW C1	88	20	7	-	-	-	-	-	-
TW J2	84	44	27	18	13	9	6	4	2	TW C2	70	22	10	5	3	-	-	-	-
TW J3	101	60	40	28	21	15	11	8	5	TW C3	70	22	10	6	3	1	-	-	-
TW J4	94	57	38	27	20	15	11	8	5	TW C4	78	21	10	5	3	2	-	-	-
DG A1	-	-	-	-	-	-	-	-	-	TW D1	-	-	-	-	-	-	-	-	-
DG A2	61	21	-	-	-	-	-	-	-	TW D2	85	36	19	12	8	5	3	2	-
DG A3	102	44	25	16	11	7	4	-	-	TW D3	86	39	22	14	9	6	4	2	1
DG B1	-	-	-	-	-	-	-	-	-	TW D4	97	51	32	21	15	11	8	5	3
DG B2	63	23	10	-	-	-	-	-	-	TW D5	97	53	34	23	17	12	9	6	4
DG B3	107	46	27	18	13	9	6	3	-	TW E1	107	50	29	19	12	8	6	4	2
DG C1	35	-	-	-	-	-	-	-	-	TW E2	103	52	31	21	14	10	7	5	3
DG C2	83	33	17	10	4	-	-	-	-	TW E3	120	69	45	31	22	16	12	9	6
DG C3	125	55	33	22	15	11	8	5	2	TW E4	111	64	43	30	22	16	12	8	6
SS A1	175	39	14	-	-	-	-	-	-	TW F1	88	-	-	-	-	-	-	-	-
SS A2	187	71	38	23	15	10	6	4	2	TW F2	58	17	-	-	-	-	-	-	-
SS A3	148	22	-	-	-	-	-	-	-	TW F3	49	18	8	-	-	-	-	-	-
SS A4	183	67	34	20	12	8	5	3	1	TW F4	42	18	9	5	-	-	-	-	-
SS A5	299	34	-	-	-	-	-	-	-	TW G1	54	10	-	-	-	-	-	-	-
SS A6	181	68	36	21	13	8	5	3	2	TW G2	42	15	6	-	-	-	-	-	-
SS A7	312	-	-	-	-	-	-	-	-	TW G3	37	15	8	4	-	-	-	-	-
SS A8	178	62	31	17	10	6	3	2	-	TW G4	32	14	7	4	2	-	-	-	-
SS B1	109	26	-	-	-	-	-	-	-	TW H1	36	11	-	-	-	-	-	-	-
SS B2	147	59	32	20	13	8	6	4	2	TW H2	30	12	6	3	-	-	-	-	-
SS B3	87	12	-	-	-	-	-	-	-	TW H3	27	11	6	3	2	-	-	-	-
SS B4	139	54	29	17	11	7	5	3	2	TW H4	22	9	5	3	2	-	-	-	-
SS B5	140	17	-	-	-	-	-	-	-	TW I1	-	-	-	-	-	-	-	-	-
SS B6	134	54	29	18	11	7	5	3	2	TW I2	63	29	17	10	7	4	3	2	-
SS B7	106	-	-	-	-	-	-	-	-	TW I3	69	34	20	13	8	6	4	2	1
SS B8	125	48	25	15	9	6	3	2	-	TW I4	81	44	28	19	14	10	7	5	3
SS C1	50	11	-	-	-	-	-	-	-	TW I5	83	47	31	21	15	11	8	6	4
SS C2	82	38	22	13	9	5	3	2	-	TW J1	89	45	27	17	12	8	6	4	2
SS C3	57	9	-	-	-	-	-	-	-	TW J2	85	45	28	19	13	9	7	5	3
SS C4	82	34	18	11	7	4	2	1	-	TW J3	102	61	41	29	21	16	12	8	5
SS D1	152	47	19	-	-	-	-	-	-	TW J4	96	58	39	28	20	15	11	8	5
SS D2	162	78	46	30	20	13	9	6	4	DG A1	-	-	-	-	-	-	-	-	-
SS D3	201	44	15	-	-	-	-	-	-	DG A2	63	23	8	-	-	-	-	-	-
SS D4	166	77	44	28	18	12	8	5	3	DG A3	104	45	26	17	11	7	4	-	-
SS E1	94	24	-	-	-	-	-	-	-	DG B1	-	-	-	-	-	-	-	-	-
SS E2	124	58	33	21	13	9	6	3	2	DG B2	65	24	11	-	-	-	-	-	-
SS E3	126	21	-	-	-	-	-	-	-	DG B3	108	47	28	19	13	9	6	4	-
SS E4	129	55	30	18	11	7	4	3	1	DG C1	39	-	-	-	-	-	-	-	-

LA CROSSE, WISCONSIN CONTINUED

	SSF =.10	.20	.30	.40	.50	.60	.70	.80	.90
DG C2	85	35	18	11	6	-	-	-	-
DG C3	126	56	33	22	16	12	8	6	3
SS A1	171	40	15	5	-	-	-	-	-
SS A2	183	71	38	23	15	10	7	4	3
SS A3	142	23	-	-	-	-	-	-	-
SS A4	179	66	35	21	13	8	5	3	2
SS A5	294	34	-	-	-	-	-	-	-
SS A6	178	68	36	22	14	9	6	4	2
SS A7	307	-	-	-	-	-	-	-	-
SS A8	173	62	31	18	11	7	4	2	1
SS B1	107	26	9	-	-	-	-	-	-
SS B2	146	59	32	20	13	9	6	4	2
SS B3	85	15	-	-	-	-	-	-	-
SS B4	137	54	29	18	12	8	5	3	2
SS B5	137	18	-	-	-	-	-	-	-
SS B6	133	54	30	18	12	8	5	3	2
SS B7	102	-	-	-	-	-	-	-	-
SS B8	123	49	26	15	10	6	4	2	1
SS C1	53	15	-	-	-	-	-	-	-
SS C2	84	40	23	14	10	6	4	3	1
SS C3	60	12	-	-	-	-	-	-	-
SS C4	84	35	19	12	8	5	3	2	-
SS D1	150	47	21	9	-	-	-	-	-
SS D2	161	79	47	30	21	14	10	7	4
SS D3	195	44	16	-	-	-	-	-	-
SS D4	165	77	44	28	19	13	9	6	4
SS E1	93	26	-	-	-	-	-	-	-
SS E2	123	59	34	21	14	9	6	4	2
SS E3	123	22	-	-	-	-	-	-	-
SS E4	128	55	30	19	12	8	5	3	2

MADISON, WISCONSIN 7730 DD

	SSF =.10	.20	.30	.40	.50	.60	.70	.80	.90
WW A1	278	-	-	-	-	-	-	-	-
WW A2	91	27	12	-	-	-	-	-	-
WW A3	77	30	15	8	3	-	-	-	-
WW A4	72	32	17	10	5	-	-	-	-
WW A5	69	33	19	11	7	4	-	-	-
WW A6	67	34	19	12	7	4	2	-	-
WW B1	-	-	-	-	-	-	-	-	-
WW B2	84	41	24	15	10	7	5	3	2
WW B3	102	53	32	21	15	10	7	5	3
WW B4	101	59	39	27	19	14	10	7	5
WW B5	98	59	39	28	20	15	11	8	5
WW C1	113	59	37	25	17	12	8	6	3
WW C2	103	57	37	25	18	13	9	6	4
WW C3	119	75	51	37	28	21	15	11	7
WW C4	109	68	47	34	25	19	14	10	7
TW A1	249	16	-	-	-	-	-	-	-
TW A2	97	26	11	4	-	-	-	-	-
TW A3	79	28	13	7	3	-	-	-	-
TW A4	69	28	15	9	5	3	-	-	-
TW B1	139	20	5	-	-	-	-	-	-
TW B2	81	26	12	6	3	-	-	-	-
TW B3	72	27	13	7	4	2	-	-	-
TW B4	69	26	13	8	5	3	1	-	-
TW C1	96	23	10	4	-	-	-	-	-
TW C2	76	25	12	7	4	2	-	-	-
TW C3	75	24	12	7	4	2	1	-	-
TW C4	84	23	11	6	4	2	1	-	-
TW D1	-	-	-	-	-	-	-	-	-
TW D2	91	39	22	13	9	6	4	2	1
TW D3	93	43	25	16	10	7	5	3	1

MADISON, WISCONSIN CONTINUED

	SSF =.10	.20	.30	.40	.50	.60	.70	.80	.90
TW D4	103	54	34	23	16	12	8	6	4
TW D5	102	56	36	25	18	13	10	7	4
TW E1	115	54	32	21	14	10	7	4	3
TW E2	110	55	34	22	16	11	8	5	3
TW E3	126	72	47	33	24	18	13	9	6
TW E4	116	68	45	32	23	17	13	9	6
TW F1	99	13	-	-	-	-	-	-	-
TW F2	65	20	8	-	-	-	-	-	-
TW F3	55	22	11	5	-	-	-	-	-
TW F4	47	21	11	7	4	2	-	-	-
TW G1	61	14	-	-	-	-	-	-	-
TW G2	47	18	8	4	-	-	-	-	-
TW G3	42	18	9	5	3	-	-	-	-
TW G4	35	16	9	5	3	2	-	-	-
TW H1	41	13	6	-	-	-	-	-	-
TW H2	34	14	7	4	2	-	-	-	-
TW H3	29	13	7	4	2	1	-	-	-
TW H4	24	10	6	3	2	1	-	-	-
TW I1	-	-	-	-	-	-	-	-	-
TW I2	68	32	18	12	8	5	3	2	1
TW I3	75	37	22	14	10	7	4	3	2
TW I4	85	47	30	21	15	11	8	5	3
TW I5	87	50	33	23	16	12	9	6	4
TW J1	95	48	29	19	13	9	6	4	3
TW J2	91	48	30	21	14	10	7	5	3
TW J3	108	65	43	31	23	17	12	9	6
TW J4	100	61	41	29	21	16	12	9	6
DG A1	-	-	-	-	-	-	-	-	-
DG A2	68	25	11	-	-	-	-	-	-
DG A3	109	47	28	18	12	8	5	-	-
DG B1	-	-	-	-	-	-	-	-	-
DG B2	70	27	14	6	-	-	-	-	-
DG B3	114	50	30	20	14	10	7	4	-
DG C1	47	-	-	-	-	-	-	-	-
DG C2	91	37	21	13	7	-	-	-	-
DG C3	133	59	35	24	17	13	9	6	3
SS A1	192	47	20	9	3	-	-	-	-
SS A2	200	78	42	26	17	12	8	5	3
SS A3	166	32	-	-	-	-	-	-	-
SS A4	197	74	39	23	15	10	6	4	2
SS A5	329	42	13	-	-	-	-	-	-
SS A6	195	75	40	25	16	11	7	5	3
SS A7	349	22	-	-	-	-	-	-	-
SS A8	192	69	36	21	13	8	5	3	2
SS B1	122	32	13	5	-	-	-	-	-
SS B2	158	64	36	22	15	10	7	5	3
SS B3	100	22	-	-	-	-	-	-	-
SS B4	150	60	33	20	13	9	6	4	2
SS B5	156	24	-	-	-	-	-	-	-
SS B6	145	59	33	20	13	9	6	4	2
SS B7	122	-	-	-	-	-	-	-	-
SS B8	136	54	29	18	11	7	5	3	2
SS C1	61	20	7	-	-	-	-	-	-
SS C2	90	43	25	16	11	7	5	3	2
SS C3	67	16	-	-	-	-	-	-	-
SS C4	90	38	22	13	9	6	4	2	1
SS D1	169	56	26	13	6	-	-	-	-
SS D2	175	86	51	34	23	16	11	7	5
SS D3	221	52	21	10	-	-	-	-	-
SS D4	179	84	49	32	21	15	10	7	4
SS E1	108	34	12	-	-	-	-	-	-
SS E2	135	65	38	24	16	11	7	5	3
SS E3	141	29	8	-	-	-	-	-	-
SS E4	140	61	34	21	14	9	6	4	2

MILWAUKEE, WISCONSIN   7444 DD
SSF =	.10	.20	.30	.40	.50	.60	.70	.80	.90
WW A1	286	-	-	-	-	-	-	-	-
WW A2	92	27	11	-	-	-	-	-	-
WW A3	78	30	15	7	-	-	-	-	-
WW A4	72	31	16	9	5	-	-	-	-
WW A5	69	32	18	11	6	3	-	-	-
WW A6	67	33	19	11	7	4	-	-	-
WW B1	-	-	-	-	-	-	-	-	-
WW B2	84	40	24	15	10	7	4	3	1
WW B3	102	53	32	21	14	10	7	4	3
WW B4	101	59	38	27	19	14	10	7	5
WW B5	98	59	39	28	20	15	11	8	5
WW C1	114	59	36	24	17	12	8	5	3
WW C2	103	57	36	25	18	13	9	6	4
WW C3	119	75	51	37	27	21	15	11	7
WW C4	109	68	47	34	25	19	14	10	7
TW A1	255	16	-	-	-	-	-	-	-
TW A2	98	25	10	-	-	-	-	-	-
TW A3	80	27	13	6	-	-	-	-	-
TW A4	69	28	14	8	4	2	-	-	-
TW B1	142	20	-	-	-	-	-	-	-
TW B2	82	26	12	6	-	-	-	-	-
TW B3	73	26	13	7	4	-	-	-	-
TW B4	70	26	13	7	4	2	-	-	-
TW C1	97	23	9	4	-	-	-	-	-
TW C2	77	24	12	6	3	2	-	-	-
TW C3	76	24	12	6	4	2	-	-	-
TW C4	85	23	11	6	3	2	1	-	-
TW D1	-	-	-	-	-	-	-	-	-
TW D2	92	39	21	13	8	6	4	2	1
TW D3	93	43	24	15	10	7	4	3	1
TW D4	103	54	34	23	16	11	8	6	3
TW D5	102	56	36	25	18	13	9	7	4
TW E1	116	54	32	20	14	9	6	4	2
TW E2	110	55	33	22	15	11	8	5	3
TW E3	127	72	47	33	24	17	13	9	6
TW E4	116	68	45	31	23	17	12	9	6
TW F1	101	12	-	-	-	-	-	-	-
TW F2	65	20	8	-	-	-	-	-	-
TW F3	55	21	10	5	-	-	-	-	-
TW F4	47	21	11	6	3	-	-	-	-
TW G1	62	14	-	-	-	-	-	-	-
TW G2	47	17	8	3	-	-	-	-	-
TW G3	42	17	9	5	2	-	-	-	-
TW G4	35	16	8	5	3	1	-	-	-
TW H1	41	13	5	-	-	-	-	-	-
TW H2	34	14	7	4	1	-	-	-	-
TW H3	29	13	7	4	2	1	-	-	-
TW H4	24	10	6	3	2	1	-	-	-
TW I1	-	-	-	-	-	-	-	-	-
TW I2	68	32	18	11	7	5	3	2	-
TW I3	75	37	22	14	9	6	4	3	1
TW I4	86	47	30	20	14	10	7	5	3
TW I5	87	50	32	22	16	12	9	6	4
TW J1	96	48	29	19	13	9	6	4	2
TW J2	91	48	30	20	14	10	7	5	3
TW J3	108	65	43	31	22	17	12	9	6
TW J4	101	61	41	29	21	16	12	8	6
DG A1	-	-	-	-	-	-	-	-	-
DG A2	69	25	11	-	-	-	-	-	-
DG A3	110	47	28	18	12	8	5	-	-
DG B1	-	-	-	-	-	-	-	-	-
DG B2	70	27	13	6	-	-	-	-	-
DG B3	115	50	30	20	14	10	7	4	-
DG C1	47	-	-	-	-	-	-	-	-

MILWAUKEE, WISCONSIN   CONTINUED
SSF =	.10	.20	.30	.40	.50	.60	.70	.80	.90
DG C2	92	37	20	12	7	-	-	-	-
DG C3	135	59	35	24	17	12	9	6	3
SS A1	201	48	20	9	-	-	-	-	-
SS A2	206	80	43	27	17	12	8	5	3
SS A3	175	34	-	-	-	-	-	-	-
SS A4	204	76	40	24	15	10	6	4	2
SS A5	345	44	14	-	-	-	-	-	-
SS A6	201	77	41	25	16	11	7	4	2
SS A7	370	25	-	-	-	-	-	-	-
SS A8	200	72	37	21	13	8	5	3	2
SS B1	127	33	13	4	-	-	-	-	-
SS B2	162	65	36	22	15	10	7	4	3
SS B3	106	23	-	-	-	-	-	-	-
SS B4	155	61	33	20	13	9	6	4	2
SS B5	164	25	-	-	-	-	-	-	-
SS B6	149	61	33	21	14	9	6	4	2
SS B7	130	-	-	-	-	-	-	-	-
SS B8	140	55	30	18	11	7	5	3	2
SS C1	61	20	-	-	-	-	-	-	-
SS C2	91	43	25	16	11	7	5	3	2
SS C3	68	16	-	-	-	-	-	-	-
SS C4	90	38	21	13	9	6	4	2	1
SS D1	176	57	27	13	6	-	-	-	-
SS D2	179	88	52	34	23	16	11	7	4
SS D3	230	54	22	9	-	-	-	-	-
SS D4	184	86	50	32	21	15	10	7	4
SS E1	113	35	12	-	-	-	-	-	-
SS E2	138	66	38	24	16	11	7	5	3
SS E3	148	30	8	-	-	-	-	-	-
SS E4	144	62	35	21	14	9	6	4	2

CASPER, WYOMING   7555 DD
SSF =	.10	.20	.30	.40	.50	.60	.70	.80	.90
WW A1	617	69	31	18	11	8	5	3	2
WW A2	207	81	46	29	20	14	10	7	5
WW A3	176	83	50	34	24	17	12	9	6
WW A4	163	84	52	36	25	18	13	10	6
WW A5	156	85	55	38	27	20	15	11	7
WW A6	152	86	56	39	28	21	15	11	7
WW B1	151	53	28	17	11	7	5	3	2
WW B2	166	89	57	39	28	21	15	11	7
WW B3	195	109	71	50	36	27	20	14	10
WW B4	177	108	73	53	39	30	22	16	11
WW B5	166	103	71	52	39	29	22	16	11
WW C1	209	117	77	54	39	29	22	16	10
WW C2	183	108	72	51	38	28	21	15	10
WW C3	195	126	89	65	50	38	29	21	15
WW C4	176	114	80	59	45	34	26	19	13
TW A1	533	67	31	18	12	8	5	4	2
TW A2	213	74	40	25	17	12	9	6	4
TW A3	173	74	43	28	20	14	10	7	5
TW A4	148	71	43	29	20	15	11	8	5
TW B1	298	65	32	19	13	9	6	4	3
TW B2	173	68	38	25	17	12	9	6	4
TW B3	152	66	39	25	18	13	9	6	4
TW B4	141	61	36	24	17	12	9	6	4
TW C1	199	60	32	20	14	9	7	5	3
TW C2	154	59	33	21	15	11	8	5	3
TW C3	148	55	31	20	14	10	7	5	3
TW C4	157	50	27	17	12	8	6	4	3
TW D1	111	42	23	14	9	6	4	3	2
TW D2	177	84	51	34	24	18	13	9	6
TW D3	181	92	58	39	28	20	15	11	7

```
CASPER, WYOMING                                CONTINUED    CHEYENNE, WYOMING                                         7255 DD
     SSF =.10   .20   .30   .40   .50   .60   .70   .80   .90         SSF =.10   .20   .30   .40   .50   .60   .70   .80   .90
  TW D4    180   100    65    46    34    25    19    13     9    WW A1    631    73    34    20    13     9     6     4     3
  TW D5    171    98    65    46    34    25    19    14     9    WW A2    217    86    49    32    22    16    12     8     6
  TW E1    213   109    68    47    33    25    18    13     9    WW A3    185    88    54    36    26    19    14    10     7
  TW E2    196   105    67    47    34    25    19    13     9    WW A4    171    88    56    39    28    21    15    11     7
  TW E3    212   126    85    61    45    34    26    19    13    WW A5    163    90    58    41    30    22    17    12     8
  TW E4    190   115    78    56    42    32    24    18    12    WW A6    159    91    60    42    31    23    17    13     9
  TW F1    235    58    29    18    12     8     5     4     2    WW B1    162    58    31    20    13     9     7     5     3
  TW F2    151    62    35    23    16    11     8     6     4    WW B2    173    93    60    42    30    23    17    12     8
  TW F3    127    60    36    24    17    12     9     6     4    WW B3    203   115    75    53    39    29    22    16    11
  TW F4    105    55    34    23    17    12     9     6     4    WW B4    184   112    77    56    42    32    24    18    12
  TW G1    147    51    28    18    12     8     6     4     3    WW B5    172   107    74    54    41    31    24    18    12
  TW G2    110    51    31    20    14    10     7     5     3    WW C1    218   123    81    57    42    32    24    17    12
  TW G3     94    47    29    20    14    10     7     5     3    WW C2    190   112    75    54    40    30    23    17    12
  TW G4     76    39    25    17    12     9     6     5     3    WW C3    202   131    93    69    52    40    31    23    16
  TW H1     97    42    25    16    11     8     6     4     3    WW C4    182   118    84    62    47    36    28    21    15
  TW H2     76    38    23    16    11     8     6     4     3    TW A1    547    71    33    20    13     9     6     4     3
  TW H3     64    32    20    14    10     7     5     4     2    TW A2    223    78    43    28    19    14    10     7     5
  TW H4     50    25    15    10     7     5     4     3     2    TW A3    182    78    46    31    22    16    11     8     5
  TW I1     85    35    19    12     8     5     4     2     1    TW A4    155    75    46    31    22    16    12     9     6
  TW I2    135    70    44    30    22    16    12     8     6    TW B1    311    69    34    21    14    10     7     5     3
  TW I3    144    78    50    34    25    18    13    10     6    TW B2    182    71    41    27    19    13    10     7     5
  TW I4    150    87    58    41    30    23    17    12     8    TW B3    160    69    41    27    19    14    10     7     5
  TW I5    148    88    59    43    32    24    18    13     9    TW B4    148    64    38    25    18    13    10     7     5
  TW J1    177    96    62    43    31    23    17    12     8    TW C1    208    64    34    22    15    11     8     5     4
  TW J2    162    92    60    42    31    23    17    12     8    TW C2    161    62    35    23    16    12     8     6     4
  TW J3    178   111    76    56    42    32    24    18    12    TW C3    155    58    33    22    15    11     8     6     4
  TW J4    164   102    71    52    39    29    22    16    11    TW C4    164    52    29    18    13     9     7     5     3
  DG A1    111    45    21     -     -     -     -     -     -    TW D1    119    45    25    16    11     8     6     4     3
  DG A2    138    63    37    24    16    10     5     -     -    TW D2    185    88    54    37    26    19    14    10     7
  DG A3    179    84    53    37    27    20    14     9     4    TW D3    189    97    61    42    30    22    17    12     8
  DG B1    110    48    27    16     -     -     -     -     -    TW D4    187   104    68    48    36    27    20    15    10
  DG B2    141    65    40    27    19    14     9     5     -    TW D5    177   102    68    49    36    27    20    15    10
  DG B3    182    87    55    38    29    22    17    12     7    TW E1    222   114    72    50    36    27    20    14    10
  DG C1    141    63    37    24    16    10     -     -     -    TW E2    204   109    71    49    36    27    20    15    10
  DG C2    168    79    49    34    25    18    13     9     4    TW E3    219   131    89    64    48    36    28    20    14
  DG C3    210   101    64    45    34    26    20    15    10    TW E4    197   119    82    59    44    34    26    19    13
  SS A1    358   109    56    34    22    15    10     7     4    TW F1    247    62    31    19    13     9     7     5     3
  SS A2    338   145    84    54    37    27    19    13     8    TW F2    159    65    38    25    17    13     9     7     4
  SS A3    332    94    47    27    17    11     7     5     3    TW F3    133    64    39    26    19    14    10     7     5
  SS A4    338   141    80    52    35    25    17    12     8    TW F4    110    58    37    25    18    13    10     7     5
  SS A5    599   106    50    28    18    12     8     5     3    TW G1    155    54    30    19    13     9     7     5     3
  SS A6    334   142    82    53    37    26    18    13     8    TW G2    116    54    33    22    16    11     8     6     4
  SS A7    662    85    37    20    12     7     5     3     1    TW G3     98    50    31    21    15    11     8     6     4
  SS A8    337   137    77    49    33    23    16    11     7    TW G4     80    41    26    18    13    10     7     5     4
  SS B1    240    81    43    26    17    12     8     5     3    TW H1    102    45    26    18    12     9     7     5     3
  SS B2    264   117    69    45    31    22    16    11     7    TW H2     80    40    25    17    12     9     7     5     3
  SS B3    216    72    37    22    15    10     6     4     2    TW H3     67    34    21    15    11     8     6     4     3
  SS B4    256   112    65    43    30    21    15    10     7    TW H4     52    26    16    11     8     6     4     3     2
  SS B5    308    70    34    20    13     8     6     4     2    TW I1     91    38    22    14    10     7     5     3     2
  SS B6    248   111    66    43    30    21    15    11     7    TW I2    141    73    47    32    23    17    13     9     6
  SS B7    263    57    27    15     9     6     4     2     1    TW I3    151    81    53    37    27    20    15    11     7
  SS B8    238   105    61    40    27    19    14    10     6    TW I4    156    91    61    43    32    24    18    14     9
  SS C1    154    70    41    27    19    13    10     7     4    TW I5    153    91    62    45    33    25    19    14    10
  SS C2    174    91    58    40    28    21    15    11     7    TW J1    184   100    65    46    33    25    19    14     9
  SS C3    159    56    30    19    13     9     6     4     3    TW J2    168    95    63    45    33    25    19    14     9
  SS C4    172    82    50    33    24    17    13     9     6    TW J3    184   115    80    58    44    34    26    19    13
  SS D1    330   130    72    46    31    21    15    10     6    TW J4    170   106    74    54    41    31    24    18    12
  SS D2    302   158    99    68    48    35    25    18    11    DG A1    117    49    25    10     -     -     -     -     -
  SS D3    413   122    63    38    25    17    11     8     5    DG A2    144    66    40    26    18    11     7     2     -
  SS D4    308   155    95    64    45    33    23    17    11    DG A3    184    87    55    39    29    21    15    10     5
  SS E1    239    98    55    35    23    16    11     7     4    DG B1    115    51    30    19    11     -     -     -     -
  SS E2    244   127    79    54    38    28    20    14     9    DG B2    145    68    42    29    21    15    11     6     -
  SS E3    287    84    43    26    17    11     8     5     3    DG B3    188    89    57    40    30    24    18    13     8
  SS E4    250   120    72    48    33    24    17    12     8    DG C1    146    67    40    27    19    13     7     -     -
```

CHEYENNE, WYOMING									CONTINUED	ROCK SPRINGS, WYOMING									CONTINUED
SSF =	.10	.20	.30	.40	.50	.60	.70	.80	.90	SSF =	.10	.20	.30	.40	.50	.60	.70	.80	.90
DG C2	174	82	51	36	27	20	15	10	5	TW D4	171	95	62	44	32	24	18	13	9
DG C3	217	104	66	47	36	28	22	17	11	TW D5	162	93	62	44	32	24	18	13	9
SS A1	382	116	61	38	25	18	12	9	5	TW E1	201	103	64	44	32	23	17	12	8
SS A2	357	153	90	59	41	30	22	15	10	TW E2	185	99	64	44	32	24	18	13	9
SS A3	358	102	52	32	21	14	10	7	4	TW E3	202	120	81	58	43	33	25	18	12
SS A4	358	150	87	57	39	28	20	14	9	TW E4	181	109	75	54	40	30	23	17	12
SS A5	636	114	54	32	21	14	10	7	4	TW F1	217	53	27	16	11	7	5	3	2
SS A6	353	151	88	58	41	29	21	15	10	TW F2	141	57	33	21	15	10	7	5	3
SS A7	703	93	42	24	15	10	7	4	3	TW F3	118	56	34	23	16	12	8	6	4
SS A8	358	146	84	54	38	27	19	13	9	TW F4	98	51	32	22	16	11	8	6	4
SS B1	257	87	47	30	20	14	10	7	4	TW G1	137	47	26	16	11	8	5	4	2
SS B2	279	123	73	49	34	25	18	13	9	TW G2	103	48	29	19	13	10	7	5	3
SS B3	232	78	41	26	17	12	8	6	4	TW G3	88	44	27	18	13	10	7	5	3
SS B4	271	119	70	47	33	24	17	12	8	TW G4	71	37	23	16	11	8	6	4	3
SS B5	329	76	38	23	15	10	7	5	3	TW H1	90	39	23	15	10	7	5	4	2
SS B6	261	118	70	47	33	24	17	12	8	TW H2	71	35	22	15	10	7	5	4	2
SS B7	284	62	30	18	11	8	5	4	2	TW H3	60	30	19	13	9	7	5	3	2
SS B8	252	112	66	44	31	22	16	11	7	TW H4	47	23	14	10	7	5	4	3	2
SS C1	163	74	44	30	21	15	11	8	5	TW I1	77	31	17	11	7	5	3	2	1
SS C2	182	95	61	42	31	23	17	12	8	TW I2	127	66	41	28	20	15	11	8	5
SS C3	168	59	33	21	14	10	7	5	3	TW I3	136	73	47	33	23	17	13	9	6
SS C4	179	85	52	36	26	19	14	10	7	TW I4	142	82	55	39	29	22	16	12	8
SS D1	351	140	79	51	35	25	18	12	8	TW I5	140	83	56	41	30	23	17	13	9
SS D2	318	168	107	73	53	39	29	21	14	TW J1	167	91	58	41	29	22	16	12	8
SS D3	441	131	69	43	28	20	14	10	6	TW J2	153	86	57	40	29	22	16	12	8
SS D4	324	164	102	70	50	36	27	19	13	TW J3	170	105	73	53	40	30	23	17	12
SS E1	255	106	61	39	27	19	14	10	6	TW J4	156	97	67	49	37	28	21	16	11
SS E2	258	135	86	59	42	31	23	16	11	DG A1	101	40	17	-	-	-	-	-	-
SS E3	308	91	47	29	19	13	9	6	4	DG A2	129	58	35	22	14	9	4	-	-
SS E4	265	127	77	52	37	27	20	14	9	DG A3	169	80	50	35	25	19	13	9	4
										DG B1	99	43	24	13	-	-	-	-	-
										DG B2	131	60	37	25	18	13	8	4	-
ROCK SPRINGS, WYOMING									8410 DD	DG B3	172	82	52	36	27	21	16	11	7
SSF =	.10	.20	.30	.40	.50	.60	.70	.80	.90	DG C1	128	58	34	22	15	8	-	-	-
WW A1	556	63	28	16	10	7	5	3	2	DG C2	157	74	46	32	23	17	12	8	3
WW A2	193	75	42	27	19	13	9	7	4	DG C3	199	95	60	43	32	25	19	14	9
WW A3	164	77	47	31	22	16	11	8	5	SS A1	351	105	54	33	22	15	10	7	4
WW A4	152	78	49	33	24	17	13	9	6	SS A2	332	141	82	53	37	26	18	13	8
WW A5	146	80	51	35	26	19	14	10	6	SS A3	327	91	45	27	17	11	7	5	3
WW A6	142	80	52	37	27	20	15	10	7	SS A4	333	138	79	51	35	24	17	12	7
WW B1	138	48	25	15	10	6	4	3	1	SS A5	583	103	48	28	17	11	8	5	3
WW B2	156	83	53	37	27	20	14	10	7	SS A6	328	139	80	52	36	25	18	12	8
WW B3	184	103	67	47	34	26	19	14	9	SS A7	641	83	36	19	12	7	4	2	1
WW B4	168	102	70	50	37	28	21	16	11	SS A8	331	134	76	48	33	23	16	11	7
WW B5	157	98	68	49	37	28	21	16	11	SS B1	233	78	41	25	17	11	8	5	3
WW C1	198	111	73	51	37	28	21	15	10	SS B2	257	113	67	44	31	22	16	11	7
WW C2	173	102	68	49	36	27	20	15	10	SS B3	209	69	36	22	14	9	6	4	2
WW C3	186	120	85	63	48	36	28	21	14	SS B4	250	109	64	42	29	20	15	10	6
WW C4	168	109	77	57	43	33	25	19	13	SS B5	298	67	33	19	12	8	5	3	2
TW A1	484	61	28	16	11	7	5	3	2	SS B6	241	108	64	42	29	21	15	10	7
TW A2	198	68	37	24	16	11	8	6	4	SS B7	255	55	25	14	9	6	3	2	1
TW A3	162	69	40	26	18	13	9	7	4	SS B8	232	102	59	39	27	19	13	9	6
TW A4	139	66	40	27	19	14	10	7	5	SS C1	143	65	38	25	17	12	9	6	4
TW B1	275	60	30	18	12	8	6	4	2	SS C2	164	86	54	37	27	20	15	10	7
TW B2	162	63	36	23	16	11	8	6	4	SS C3	148	52	28	18	12	8	6	4	3
TW B3	143	61	36	24	17	12	9	6	4	SS C4	162	76	47	31	22	16	12	8	6
TW B4	132	57	34	22	16	11	8	6	4	SS D1	322	127	70	44	30	20	14	9	6
TW C1	186	56	30	19	13	9	6	4	3	SS D2	295	155	97	66	47	34	25	17	11
TW C2	144	55	31	20	14	10	7	5	3	SS D3	405	119	61	37	24	16	11	7	4
TW C3	139	52	29	19	13	9	7	5	3	SS D4	301	151	93	63	44	32	23	16	10
TW C4	148	47	25	16	11	8	6	4	3	SS E1	230	94	53	33	22	15	11	7	4
TW D1	102	38	21	13	8	6	4	2	1	SS E2	237	123	77	52	37	27	19	14	9
TW D2	166	79	48	32	23	17	12	9	6	SS E3	278	81	41	25	16	11	7	5	3
TW D3	170	87	54	37	27	19	14	10	7	SS E4	243	116	70	46	32	23	17	12	7

```
SHERIDAN, WYOMING                         7708 DD    SHERIDAN, WYOMING                          CONTINUED
   SSF =.10  .20  .30  .40  .50  .60  .70  .80  .90     SSF =.10  .20  .30  .40  .50  .60  .70  .80  .90
WW A1   446   42   16    7    3    -    -    -    -  DG C2   129   58   34   22   15   10    6    -    -
WW A2   148   54   28   17   11    7    4    3    1  DG C3   170   79   49   34   25   18   14   10    6
WW A3   126   56   32   20   13    9    6    4    2  SS A1   273   76   36   20   12    7    4    2    1
WW A4   116   57   34   22   15   10    7    4    3  SS A2   265  109   61   38   26   18   12    8    5
WW A5   111   58   36   23   16   11    8    5    3  SS A3   247   62   27   14    7    2    -    -    -
WW A6   108   58   37   24   17   12    8    6    3  SS A4   264  105   58   36   23   16   11    7    4
WW B1    91   24    -    -    -    -    -    -    -  SS A5   463   73   31   16    8    4    2    -    -
WW B2   123   63   39   26   18   13    9    6    4  SS A6   260  106   59   37   25   17   11    8    5
WW B3   146   79   50   34   24   17   12    9    5  SS A7   504   54   19    7    -    -    -    -    -
WW B4   137   82   55   39   28   21   15   11    7  SS A8   261  101   54   33   22   14   10    6    4
WW B5   130   79   54   39   29   21   16   11    8  SS B1   180   55   27   15    9    5    3    2    -
WW C1   159   87   55   38   27   19   14   10    6  SS B2   209   89   51   32   22   15   11    7    4
WW C2   141   81   53   37   27   19   14   10    7  SS B3   157   46   21   11    6    2    -    -    -
WW C3   155   99   68   50   37   28   21   15   10  SS B4   201   84   48   30   20   14    9    6    4
WW C4   141   90   62   45   34   26   19   14    9  SS B5   232   47   20   10    5    2    -    -    -
TW A1   390   43   17    9    4    2    -    -    -  SS B6   194   84   48   30   20   14   10    7    4
TW A2   154   49   25   15    9    6    4    2    1  SS B7   193   35   13    4    -    -    -    -    -
TW A3   125   50   28   17   11    7    5    3    2  SS B8   184   78   44   27   18   12    8    6    3
TW A4   108   49   28   18   12    8    6    4    2  SS C1   107   45   25   15    9    6    4    2    1
TW B1   219   43   19   10    6    4    2    -    -  SS C2   131   66   40   27   19   13    9    6    4
TW B2   126   46   25   15   10    6    4    3    1  SS C3   113   36   18   10    6    4    2    1    -
TW B3   111   46   25   16   10    7    5    3    2  SS C4   129   59   35   23   15   11    8    5    3
TW B4   104   43   24   15   10    7    5    3    2  SS D1   246   91   47   28   17   11    7    4    2
TW C1   147   41   20   12    7    5    3    2    1  SS D2   234  119   73   48   33   23   16   11    7
TW C2   114   41   22   14    9    6    4    3    1  SS D3   314   85   40   22   13    8    5    3    1
TW C3   111   39   21   13    8    6    4    3    1  SS D4   239  116   70   46   31   22   15   10    6
TW C4   120   36   19   11    7    5    3    2    1  SS E1   172   65   33   19   11    6    3    2    -
TW D1    68   20    7    -    -    -    -    -    -  SS E2   187   94   57   37   25   18   12    8    5
TW D2   133   60   35   23   16   11    8    5    3  SS E3   213   56   26   14    7    4    2    -    -
TW D3   135   66   40   26   18   13    9    6    4  SS E4   193   88   51   33   22   15   11    7    4
TW D4   140   76   48   33   24   17   13    9    6
TW D5   135   76   49   35   25   18   14   10    6
TW E1   162   80   49   32   23   16   11    8    5
TW E2   151   79   49   33   24   17   12    9    6
TW E3   167   98   65   46   33   25   18   13    9
TW E4   151   90   60   43   31   23   17   13    8
TW F1   168   36   16    8    5    2    -    -    -
TW F2   107   41   22   13    8    5    3    2    1
TW F3    90   40   23   14   10    6    4    3    1
TW F4    75   37   22   14   10    7    5    3    2
TW G1   104   33   16    9    6    3    2    -    -
TW G2    78   34   19   12    8    5    3    2    1
TW G3    67   32   19   12    8    5    4    2    1
TW G4    55   27   16   11    7    5    3    2    1
TW H1    68   28   15    9    6    4    2    1    -
TW H2    54   25   15    9    6    4    3    2    1
TW H3    46   22   13    8    6    4    3    2    1
TW H4    36   17   10    7    5    3    2    1    -
TW I1    49   15    -    -    -    -    -    -    -
TW I2   101   50   30   20   14   10    7    5    3
TW I3   108   56   35   23   16   12    8    6    4
TW I4   116   66   43   30   21   16   11    8    5
TW I5   116   68   45   32   23   17   13    9    6
TW J1   134   71   44   30   21   15   11    8    5
TW J2   125   69   44   30   22   16   11    8    5
TW J3   141   86   58   42   31   23   17   12    8
TW J4   131   80   55   39   29   22   16   12    8
DG A1    68    -    -    -    -    -    -    -    -
DG A2   103   44   24   14    7    -    -    -    -
DG A3   143   65   39   27   19   13    9    5    -
DG B1    69   22    -    -    -    -    -    -    -
DG B2   105   46   26   17   10    6    -    -    -
DG B3   146   68   42   28   21   15   11    7    4
DG C1    96   38   18    -    -    -    -    -    -
```

EDMONTON, ALBERTA — 10645 DD

	SSF=.10	.20	.30	.40	.50	.60	.70	.80	.90
WW A1	278	-	-	-	-	-	-	-	-
WW A2	103	31	-	-	-	-	-	-	-
WW A3	88	33	12	-	-	-	-	-	-
WW A4	81	34	15	-	-	-	-	-	-
WW A5	78	35	16	-	-	-	-	-	-
WW A6	76	36	18	-	-	-	-	-	-
WW B1	-	-	-	-	-	-	-	-	-
WW B2	93	44	25	14	7	-	-	-	-
WW B3	110	56	33	20	12	6	3	-	-
WW B4	109	63	40	27	18	13	8	5	3
WW B5	105	63	41	28	20	14	10	7	4
WW C1	122	63	38	24	15	9	5	3	1
WW C2	111	61	38	25	17	11	7	4	2
WW C3	126	78	53	37	27	20	14	10	6
WW C4	115	72	49	35	25	18	13	9	6
TW A1	251	21	-	-	-	-	-	-	-
TW A2	109	30	-	-	-	-	-	-	-
TW A3	89	31	12	-	-	-	-	-	-
TW A4	77	31	14	-	-	-	-	-	-
TW B1	149	24	-	-	-	-	-	-	-
TW B2	91	29	11	-	-	-	-	-	-
TW B3	81	29	13	-	-	-	-	-	-
TW B4	77	29	13	5	-	-	-	-	-
TW C1	105	26	9	-	-	-	-	-	-
TW C2	84	28	12	-	-	-	-	-	-
TW C3	82	27	12	5	-	-	-	-	-
TW C4	89	25	11	5	-	-	-	-	-
TW D1	-	-	-	-	-	-	-	-	-
TW D2	101	43	23	13	6	-	-	-	-
TW D3	101	46	25	14	7	-	-	-	-
TW D4	111	58	35	23	15	10	7	4	2
TW D5	110	60	38	25	18	12	8	5	3
TW E1	124	58	33	20	12	7	3	1	-
TW E2	119	59	35	22	14	9	6	3	2
TW E3	134	76	49	33	23	16	11	7	4
TW E4	124	72	47	32	23	16	11	8	5
TW F1	111	16	-	-	-	-	-	-	-
TW F2	74	23	-	-	-	-	-	-	-
TW F3	63	24	-	-	-	-	-	-	-
TW F4	53	23	10	-	-	-	-	-	-
TW G1	70	16	-	-	-	-	-	-	-
TW G2	54	20	-	-	-	-	-	-	-
TW G3	47	19	8	-	-	-	-	-	-
TW G4	39	17	8	-	-	-	-	-	-
TW H1	47	15	-	-	-	-	-	-	-
TW H2	38	15	6	-	-	-	-	-	-
TW H3	33	14	7	-	-	-	-	-	-
TW H4	27	11	6	2	-	-	-	-	-
TW I1	-	-	-	-	-	-	-	-	-
TW I2	76	35	19	11	5	-	-	-	-
TW I3	81	39	22	13	7	3	-	-	-
TW I4	92	50	31	21	14	9	6	4	2
TW I5	94	53	34	23	16	11	8	5	3
TW J1	103	51	30	18	11	7	4	2	-
TW J2	98	52	32	20	13	9	6	3	2
TW J3	114	68	45	31	22	16	11	7	5
TW J4	107	64	43	30	21	15	11	7	4
DG A1	-	-	-	-	-	-	-	-	-
DG A2	78	29	11	-	-	-	-	-	-
DG A3	118	52	30	19	12	7	3	-	-
DG B1	34	-	-	-	-	-	-	-	-
DG B2	81	33	15	-	-	-	-	-	-
DG B3	122	55	33	21	14	9	5	3	-
DG C1	63	-	-	-	-	-	-	-	-

EDMONTON, ALBERTA — CONTINUED

	SSF=.10	.20	.30	.40	.50	.60	.70	.80	.90
DG C2	103	44	23	12	-	-	-	-	-
DG C3	143	65	39	26	17	12	7	4	-
SS A1	180	43	-	-	-	-	-	-	-
SS A2	193	75	39	22	13	7	4	2	-
SS A3	150	-	-	-	-	-	-	-	-
SS A4	186	69	34	18	9	3	-	-	-
SS A5	288	38	-	-	-	-	-	-	-
SS A6	188	72	36	20	11	6	2	-	-
SS A7	283	-	-	-	-	-	-	-	-
SS A8	181	64	30	14	-	-	-	-	-
SS B1	117	29	-	-	-	-	-	-	-
SS B2	155	63	33	20	12	7	4	2	1
SS B3	95	-	-	-	-	-	-	-	-
SS B4	145	57	30	17	9	5	2	-	-
SS B5	144	20	-	-	-	-	-	-	-
SS B6	142	58	30	17	10	6	3	1	-
SS B7	110	-	-	-	-	-	-	-	-
SS B8	131	51	26	14	7	-	-	-	-
SS C1	71	22	-	-	-	-	-	-	-
SS C2	99	47	26	15	8	4	-	-	-
SS C3	77	19	-	-	-	-	-	-	-
SS C4	98	42	22	13	7	-	-	-	-
SS D1	162	50	13	-	-	-	-	-	-
SS D2	170	82	47	28	18	11	7	4	2
SS D3	206	47	-	-	-	-	-	-	-
SS D4	174	80	45	27	16	10	6	3	2
SS E1	106	25	-	-	-	-	-	-	-
SS E2	133	62	34	19	11	5	-	-	-
SS E3	135	25	-	-	-	-	-	-	-
SS E4	138	59	31	17	9	4	-	-	-

SUFFIELD, ALBERTA — 9391 DD

	SSF=.10	.20	.30	.40	.50	.60	.70	.80	.90
WW A1	389	33	8	-	-	-	-	-	-
WW A2	132	45	22	11	5	-	-	-	-
WW A3	112	47	25	14	8	4	-	-	-
WW A4	103	48	27	16	9	5	2	-	-
WW A5	98	49	28	17	11	6	3	1	-
WW A6	95	49	29	18	12	7	4	2	-
WW B1	71	-	-	-	-	-	-	-	-
WW B2	112	56	33	21	14	10	6	4	2
WW B3	131	69	43	28	19	13	9	6	3
WW B4	126	74	48	34	24	17	12	9	6
WW B5	121	73	49	34	25	18	13	9	6
WW C1	144	76	47	32	22	15	10	7	4
WW C2	129	73	47	32	22	16	11	8	5
WW C3	143	90	62	44	33	24	18	13	8
WW C4	131	82	57	41	30	22	17	12	8
TW A1	343	35	11	-	-	-	-	-	-
TW A2	139	41	19	10	4	-	-	-	-
TW A3	112	43	22	12	7	3	-	-	-
TW A4	96	42	23	14	8	5	2	-	-
TW B1	196	36	14	6	-	-	-	-	-
TW B2	114	39	20	11	6	3	-	-	-
TW B3	101	39	21	12	7	4	2	-	-
TW B4	94	37	20	12	7	4	2	1	-
TW C1	133	35	16	9	4	-	-	-	-
TW C2	104	36	18	10	6	3	2	-	-
TW C3	101	34	17	10	6	4	2	-	-
TW C4	110	32	16	9	5	3	2	1	-
TW D1	53	-	-	-	-	-	-	-	-
TW D2	121	53	30	19	12	8	5	3	2
TW D3	121	57	33	21	14	9	6	4	2

SUFFIELD, ALBERTA							CONTINUED		NANAIMO, BRITISH COLUMBIA								5692 DD	
SSF =.10	.20	.30	.40	.50	.60	.70	.80	.90	SSF =.10	.20	.30	.40	.50	.60	.70	.80	.90
TW D4 129	68	43	29	20	14	10	7	4	WW A1 579	58	18	-	-	-	-	-	-
TW D5 126	69	45	31	22	16	11	8	5	WW A2 195	66	31	15	-	-	-	-	-
TW E1 147	71	42	27	18	12	8	5	3	WW A3 162	67	34	18	-	-	-	-	-
TW E2 139	71	43	29	20	14	10	6	4	WW A4 147	67	36	20	9	-	-	-	-
TW E3 154	88	58	40	29	21	15	11	7	WW A5 139	67	38	22	11	-	-	-	-
TW E4 141	82	54	38	28	20	15	10	7	WW A6 134	67	38	22	12	-	-	-	-
TW F1 149	29	10	-	-	-	-	-	-	WW B1 138	32	-	-	-	-	-	-	-
TW F2 95	34	16	8	-	-	-	-	-	WW B2 152	74	42	26	16	9	5	2	-
TW F3 80	34	18	10	6	2	-	-	-	WW B3 178	92	54	34	22	14	8	4	2
TW F4 67	31	18	11	6	4	2	-	-	WW B4 162	92	59	40	27	19	13	8	4
TW G1 93	27	12	4	-	-	-	-	-	WW B5 152	89	58	40	28	20	14	9	5
TW G2 69	28	15	8	4	-	-	-	-	WW C1 193	100	60	38	25	16	10	6	3
TW G3 59	27	15	9	5	3	1	-	-	WW C2 168	92	57	38	26	17	11	7	4
TW G4 49	23	13	8	5	3	2	-	-	WW C3 179	110	74	52	37	27	19	13	8
TW H1 60	23	11	6	-	-	-	-	-	WW C4 162	100	67	47	34	24	17	12	7
TW H2 48	21	12	7	4	2	-	-	-	TW A1 503	57	20	-	-	-	-	-	-
TW H3 41	19	11	6	4	2	1	-	-	TW A2 203	62	28	13	-	-	-	-	-
TW H4 33	15	8	5	3	2	1	-	-	TW A3 163	61	30	16	6	-	-	-	-
TW I1 37	-	-	-	-	-	-	-	-	TW A4 137	58	31	17	9	-	-	-	-
TW I2 91	44	26	16	11	7	5	3	1	TW B1 286	55	22	9	-	-	-	-	-
TW I3 97	49	29	19	13	9	6	4	2	TW B2 165	57	27	14	6	-	-	-	-
TW I4 107	59	38	26	18	13	9	6	4	TW B3 143	55	28	15	8	-	-	-	-
TW I5 108	62	40	28	20	15	11	7	5	TW B4 133	52	27	15	8	-	-	-	-
TW J1 122	62	38	25	17	12	8	5	3	TW C1 192	52	23	11	-	-	-	-	-
TW J2 114	62	39	26	18	13	9	6	4	TW C2 147	50	25	13	7	-	-	-	-
TW J3 130	78	52	37	27	20	14	10	7	TW C3 142	48	24	13	7	3	-	-	-
TW J4 121	73	49	35	26	19	14	10	6	TW C4 152	44	21	12	6	3	-	-	-
DG A1 54	-	-	-	-	-	-	-	-	TW D1 100	26	-	-	-	-	-	-	-
DG A2 94	38	19	9	-	-	-	-	-	TW D2 166	71	39	23	14	8	4	-	-
DG A3 135	60	36	23	16	11	7	3	-	TW D3 168	78	44	26	16	9	4	1	-
DG B1 56	-	-	-	-	-	-	-	-	TW D4 168	87	53	34	23	16	10	6	3
DG B2 97	41	22	13	6	-	-	-	-	TW D5 160	86	54	36	25	17	12	8	4
DG B3 139	63	38	25	18	13	9	5	2	TW E1 200	94	53	33	21	13	8	4	2
DG C1 83	29	-	-	-	-	-	-	-	TW E2 183	91	54	34	23	15	9	5	3
DG C2 120	52	30	19	12	7	-	-	-	TW E3 197	111	70	48	33	23	16	10	6
DG C3 162	73	45	30	22	16	11	7	4	TW E4 177	101	65	45	31	22	15	10	6
SS A1 227	58	25	12	-	-	-	-	-	TW F1 226	48	18	-	-	-	-	-	-
SS A2 228	90	49	30	19	12	8	5	3	TW F2 142	50	23	11	-	-	-	-	-
SS A3 196	42	11	-	-	-	-	-	-	TW F3 116	48	25	13	-	-	-	-	-
SS A4 221	84	44	26	16	10	6	4	2	TW F4 95	44	24	13	7	-	-	-	-
SS A5 379	54	19	-	-	-	-	-	-	TW G1 140	41	18	-	-	-	-	-	-
SS A6 223	87	47	28	18	12	7	4	2	TW G2 102	41	21	11	-	-	-	-	-
SS A7 393	33	-	-	-	-	-	-	-	TW G3 85	38	20	11	5	-	-	-	-
SS A8 218	79	41	24	14	9	5	3	1	TW G4 69	32	17	10	5	-	-	-	-
SS B1 149	42	18	8	-	-	-	-	-	TW H1 90	34	16	7	-	-	-	-	-
SS B2 182	74	41	26	17	11	7	5	3	TW H2 70	30	16	9	4	-	-	-	-
SS B3 125	31	9	-	-	-	-	-	-	TW H3 59	26	14	8	4	-	-	-	-
SS B4 172	69	38	23	15	10	6	4	2	TW H4 46	21	11	6	4	1	-	-	-
SS B5 190	34	11	-	-	-	-	-	-	TW I1 74	18	-	-	-	-	-	-	-
SS B6 167	69	38	24	15	10	7	4	2	TW I2 125	59	33	20	12	7	3	-	-
SS B7 152	21	-	-	-	-	-	-	-	TW I3 133	65	38	23	15	9	5	2	-
SS B8 156	63	34	20	13	8	5	3	2	TW I4 139	75	46	31	21	14	9	6	3
SS C1 94	36	18	8	-	-	-	-	-	TW I5 137	77	49	33	23	16	11	7	4
SS C2 119	58	34	22	15	10	6	4	2	TW J1 164	82	48	30	20	13	8	4	2
SS C3 101	29	13	5	-	-	-	-	-	TW J2 150	79	48	31	21	14	9	5	3
SS C4 118	52	30	19	12	8	5	3	2	TW J3 165	97	63	43	31	22	15	10	6
SS D1 202	68	32	16	7	-	-	-	-	TW J4 152	90	59	41	29	21	14	10	6
SS D2 199	98	58	38	25	17	12	7	4	DG A1 102	24	-	-	-	-	-	-	-
SS D3 261	64	27	12	-	-	-	-	-	DG A2 135	54	27	13	-	-	-	-	-
SS D4 204	96	56	36	24	16	11	7	4	DG A3 180	78	45	28	18	12	7	-	-
SS E1 137	46	20	-	-	-	-	-	-	DG B1 105	33	-	-	-	-	-	-	-
SS E2 158	76	45	28	18	12	8	5	3	DG B2 141	58	31	17	7	-	-	-	-
SS E3 175	40	15	-	-	-	-	-	-	DG B3 185	82	48	31	21	14	9	5	-
SS E4 164	72	40	25	16	10	7	4	2	DG C1 139	51	21	-	-	-	-	-	-

NANAIMO, BRITISH COLUMBIA — CONTINUED

SSF =	.10	.20	.30	.40	.50	.60	.70	.80	.90
DG C2	170	72	40	24	14	6	-	-	-
DG C3	215	96	56	37	25	17	11	7	3
SS A1	362	97	42	20	6	-	-	-	-
SS A2	337	132	69	40	25	15	9	5	2
SS A3	342	83	31	-	-	-	-	-	-
SS A4	341	129	66	37	22	13	7	3	1
SS A5	600	97	36	14	-	-	-	-	-
SS A6	333	129	67	39	23	14	8	4	2
SS A7	663	77	23	-	-	-	-	-	-
SS A8	340	125	62	35	20	11	5	2	-
SS B1	242	71	31	14	-	-	-	-	-
SS B2	263	107	57	34	21	13	8	5	2
SS B3	219	61	24	-	-	-	-	-	-
SS B4	256	103	54	32	19	12	7	4	2
SS B5	312	63	23	-	-	-	-	-	-
SS B6	246	101	54	32	20	12	7	4	2
SS B7	270	49	14	-	-	-	-	-	-
SS B8	237	95	49	29	17	10	5	2	-
SS C1	142	54	26	10	-	-	-	-	-
SS C2	161	77	44	27	17	10	5	2	-
SS C3	152	45	20	7	-	-	-	-	-
SS C4	162	70	38	23	14	8	4	1	-
SS D1	327	113	53	26	11	-	-	-	-
SS D2	291	140	80	50	32	20	13	7	4
SS D3	418	110	47	22	-	-	-	-	-
SS D4	300	138	78	47	30	19	12	7	3
SS E1	234	82	36	14	-	-	-	-	-
SS E2	235	111	63	38	23	14	8	4	2
SS E3	291	74	29	10	-	-	-	-	-
SS E4	245	106	57	34	21	12	7	3	1

VANCOUVER, BRITISH COLUMBIA — 5592 DD

SSF =	.10	.20	.30	.40	.50	.60	.70	.80	.90
WW A1	545	54	15	-	-	-	-	-	-
WW A2	186	63	29	13	-	-	-	-	-
WW A3	154	63	32	16	-	-	-	-	-
WW A4	140	63	33	18	7	-	-	-	-
WW A5	132	64	35	20	9	-	-	-	-
WW A6	128	64	36	20	11	-	-	-	-
WW B1	128	25	-	-	-	-	-	-	-
WW B2	146	70	40	24	15	9	4	1	-
WW B3	171	88	52	32	21	13	7	4	2
WW B4	157	89	57	38	26	18	12	8	4
WW B5	147	86	56	38	27	19	13	9	5
WW C1	186	96	57	36	24	15	9	5	3
WW C2	162	89	55	36	24	16	11	7	4
WW C3	173	107	71	50	36	26	18	12	7
WW C4	157	97	65	46	33	24	17	11	7
TW A1	475	54	17	-	-	-	-	-	-
TW A2	193	58	26	12	-	-	-	-	-
TW A3	155	58	28	14	-	-	-	-	-
TW A4	131	55	29	16	8	-	-	-	-
TW B1	271	52	20	-	-	-	-	-	-
TW B2	157	54	25	13	-	-	-	-	-
TW B3	137	52	26	14	7	-	-	-	-
TW B4	127	49	25	14	7	-	-	-	-
TW C1	182	49	22	10	-	-	-	-	-
TW C2	140	48	23	12	6	-	-	-	-
TW C3	135	46	22	12	6	-	-	-	-
TW C4	145	42	20	11	6	3	-	-	-
TW D1	94	22	-	-	-	-	-	-	-
TW D2	159	68	37	22	13	7	3	-	-
TW D3	161	74	41	25	15	9	4	1	-

VANCOUVER, BRITISH COLUMBIA — CONTINUED

SSF =	.10	.20	.30	.40	.50	.60	.70	.80	.90
TW D4	162	84	51	33	22	15	10	6	3
TW D5	154	83	52	35	24	16	11	7	4
TW E1	192	90	51	31	20	12	7	4	2
TW E2	176	88	52	33	22	14	9	5	3
TW E3	191	107	68	46	32	22	15	10	6
TW E4	172	98	63	43	30	21	15	10	6
TW F1	214	45	16	-	-	-	-	-	-
TW F2	134	47	22	9	-	-	-	-	-
TW F3	110	46	23	11	-	-	-	-	-
TW F4	91	41	22	12	5	-	-	-	-
TW G1	133	39	16	-	-	-	-	-	-
TW G2	97	39	19	9	-	-	-	-	-
TW G3	81	36	19	10	-	-	-	-	-
TW G4	66	30	16	9	5	-	-	-	-
TW H1	86	32	15	6	-	-	-	-	-
TW H2	66	29	15	8	-	-	-	-	-
TW H3	56	25	13	7	4	-	-	-	-
TW H4	44	19	10	6	3	-	-	-	-
TW I1	68	-	-	-	-	-	-	-	-
TW I2	119	56	31	19	11	6	3	-	-
TW I3	128	63	36	22	14	8	4	2	-
TW I4	134	72	45	29	20	13	9	5	3
TW I5	132	74	47	31	22	15	10	7	4
TW J1	158	79	46	29	19	12	7	4	2
TW J2	144	76	46	30	20	13	9	5	3
TW J3	159	94	61	42	30	21	15	10	6
TW J4	147	87	57	39	28	20	14	9	6
DG A1	95	-	-	-	-	-	-	-	-
DG A2	129	52	25	12	-	-	-	-	-
DG A3	174	76	43	27	17	11	6	-	-
DG B1	99	28	-	-	-	-	-	-	-
DG B2	134	55	29	15	-	-	-	-	-
DG B3	179	79	46	30	20	13	8	5	-
DG C1	132	47	16	-	-	-	-	-	-
DG C2	163	69	38	22	13	5	-	-	-
DG C3	208	93	54	36	24	16	11	6	3
SS A1	348	93	40	18	-	-	-	-	-
SS A2	326	127	66	39	23	14	8	5	2
SS A3	327	79	29	-	-	-	-	-	-
SS A4	330	125	63	36	21	12	6	3	1
SS A5	574	93	34	12	-	-	-	-	-
SS A6	321	125	64	37	22	13	7	4	2
SS A7	633	73	20	-	-	-	-	-	-
SS A8	328	120	60	33	19	10	5	2	-
SS B1	232	68	29	12	-	-	-	-	-
SS B2	255	103	55	33	20	13	8	4	2
SS B3	210	58	22	-	-	-	-	-	-
SS B4	248	99	52	30	18	11	6	3	2
SS B5	298	60	21	-	-	-	-	-	-
SS B6	238	98	52	30	19	11	7	4	2
SS B7	257	46	11	-	-	-	-	-	-
SS B8	230	92	47	27	16	9	5	2	-
SS C1	134	51	23	-	-	-	-	-	-
SS C2	155	74	42	25	16	9	5	2	-
SS C3	144	43	18	-	-	-	-	-	-
SS C4	155	67	36	21	13	7	4	1	-
SS D1	314	108	50	24	8	-	-	-	-
SS D2	282	136	77	48	30	19	12	7	4
SS D3	401	105	44	20	-	-	-	-	-
SS D4	290	134	75	45	29	18	11	6	3
SS E1	224	78	33	-	-	-	-	-	-
SS E2	227	107	60	36	22	13	8	4	2
SS E3	278	70	27	-	-	-	-	-	-
SS E4	237	102	55	32	19	11	6	3	1

WINNIPEG, MANITOBA — 10789 DD

SSF =	.10	.20	.30	.40	.50	.60	.70	.80	.90
WW A1	236	-	-	-	-	-	-	-	-
WW A2	83	25	8	-	-	-	-	-	-
WW A3	71	28	13	-	-	-	-	-	-
WW A4	66	29	15	7	-	-	-	-	-
WW A5	64	30	16	9	-	-	-	-	-
WW A6	63	31	17	10	4	-	-	-	-
WW B1	-	-	-	-	-	-	-	-	-
WW B2	79	39	23	14	9	6	3	2	-
WW B3	95	49	30	20	13	9	6	3	2
WW B4	97	56	37	26	18	13	10	7	4
WW B5	94	57	38	27	20	14	11	7	5
WW C1	106	56	35	23	16	11	7	5	3
WW C2	98	55	35	24	17	12	8	6	3
WW C3	114	72	49	36	26	20	15	10	7
WW C4	105	66	46	33	25	18	14	10	6
TW A1	214	12	-	-	-	-	-	-	-
TW A2	88	23	8	-	-	-	-	-	-
TW A3	73	25	12	-	-	-	-	-	-
TW A4	64	26	13	7	-	-	-	-	-
TW B1	124	18	-	-	-	-	-	-	-
TW B2	75	24	11	-	-	-	-	-	-
TW B3	67	25	12	6	-	-	-	-	-
TW B4	64	24	12	7	3	-	-	-	-
TW C1	88	21	8	-	-	-	-	-	-
TW C2	70	23	11	6	-	-	-	-	-
TW C3	69	22	11	6	3	-	-	-	-
TW C4	77	21	10	5	3	1	-	-	-
TW D1	-	-	-	-	-	-	-	-	-
TW D2	86	37	20	12	8	5	3	1	-
TW D3	86	40	23	14	9	5	3	2	-
TW D4	98	52	32	22	15	11	8	5	3
TW D5	98	54	35	24	17	12	9	6	4
TW E1	107	51	30	19	13	8	5	3	2
TW E2	104	53	32	21	15	10	7	5	3
TW E3	120	69	45	32	23	17	12	8	5
TW E4	111	65	43	31	22	16	12	8	5
TW F1	87	-	-	-	-	-	-	-	-
TW F2	59	18	-	-	-	-	-	-	-
TW F3	50	20	9	-	-	-	-	-	-
TW F4	43	19	10	5	-	-	-	-	-
TW G1	55	12	-	-	-	-	-	-	-
TW G2	43	16	7	-	-	-	-	-	-
TW G3	38	16	8	3	-	-	-	-	-
TW G4	32	15	8	4	2	-	-	-	-
TW H1	37	12	-	-	-	-	-	-	-
TW H2	31	13	6	2	-	-	-	-	-
TW H3	27	12	6	3	-	-	-	-	-
TW H4	22	10	5	3	2	-	-	-	-
TW I1	-	-	-	-	-	-	-	-	-
TW I2	64	30	17	11	7	4	2	1	-
TW I3	70	34	20	13	8	5	3	2	1
TW I4	81	45	29	20	14	10	7	5	3
TW I5	84	48	31	22	16	11	8	6	4
TW J1	89	45	27	18	12	8	5	3	2
TW J2	86	46	29	20	14	10	7	4	3
TW J3	103	62	42	30	22	16	12	8	5
TW J4	97	59	40	28	21	15	11	8	5
DG A1	-	-	-	-	-	-	-	-	-
DG A2	64	24	9	-	-	-	-	-	-
DG A3	103	45	27	17	12	8	4	-	-
DG B1	-	-	-	-	-	-	-	-	-
DG B2	65	26	13	-	-	-	-	-	-
DG B3	107	48	29	19	13	9	6	4	-
DG C1	41	-	-	-	-	-	-	-	-

WINNIPEG, MANITOBA — CONTINUED

SSF =	.10	.20	.30	.40	.50	.60	.70	.80	.90
DG C2	85	36	20	12	6	-	-	-	-
DG C3	125	56	34	23	17	12	8	6	2
SS A1	154	35	12	-	-	-	-	-	-
SS A2	172	67	36	22	14	9	6	3	2
SS A3	122	-	-	-	-	-	-	-	-
SS A4	164	61	31	18	11	7	4	2	-
SS A5	257	29	-	-	-	-	-	-	-
SS A6	166	64	34	20	13	8	5	3	1
SS A7	250	-	-	-	-	-	-	-	-
SS A8	158	56	28	16	9	5	2	-	-
SS B1	97	23	-	-	-	-	-	-	-
SS B2	137	56	31	19	12	8	5	3	2
SS B3	74	-	-	-	-	-	-	-	-
SS B4	128	51	27	16	10	7	4	2	1
SS B5	121	14	-	-	-	-	-	-	-
SS B6	125	51	28	17	11	7	5	3	2
SS B7	85	-	-	-	-	-	-	-	-
SS B8	114	45	24	14	8	5	3	2	-
SS C1	55	17	-	-	-	-	-	-	-
SS C2	85	41	24	15	10	6	4	2	1
SS C3	60	14	-	-	-	-	-	-	-
SS C4	84	36	20	12	8	5	3	2	-
SS D1	136	42	17	-	-	-	-	-	-
SS D2	151	74	44	28	19	13	9	5	3
SS D3	177	39	12	-	-	-	-	-	-
SS D4	154	72	42	26	17	12	8	5	3
SS E1	83	21	-	-	-	-	-	-	-
SS E2	116	55	32	20	13	8	5	3	2
SS E3	110	18	-	-	-	-	-	-	-
SS E4	120	52	29	17	11	7	4	2	1

HALIFAX, NOVA SCOTIA — 7210 DD

SSF =	.10	.20	.30	.40	.50	.60	.70	.80	.90
WW A1	317	25	-	-	-	-	-	-	-
WW A2	111	38	19	10	6	-	-	-	-
WW A3	95	40	22	13	8	5	2	-	-
WW A4	88	41	24	15	10	6	4	-	-
WW A5	84	42	25	16	11	7	4	2	-
WW A6	82	43	26	17	12	8	5	3	-
WW B1	51	-	-	-	-	-	-	-	-
WW B2	98	49	30	20	14	10	7	4	2
WW B3	118	63	39	27	19	13	9	6	4
WW B4	114	67	45	31	23	17	12	9	6
WW B5	109	66	45	32	24	18	13	9	6
WW C1	130	70	44	30	21	15	11	7	5
WW C2	116	66	43	30	21	16	11	8	5
WW C3	132	84	58	42	32	24	18	13	9
WW C4	120	76	53	38	29	22	16	12	8
TW A1	283	27	9	-	-	-	-	-	-
TW A2	117	35	17	9	5	-	-	-	-
TW A3	96	36	19	11	7	4	2	-	-
TW A4	83	36	20	13	8	5	3	2	-
TW B1	164	29	12	6	-	-	-	-	-
TW B2	97	34	17	10	6	4	2	-	-
TW B3	86	34	18	11	7	4	3	1	-
TW B4	82	32	18	11	7	5	3	2	-
TW C1	113	30	14	8	5	2	-	-	-
TW C2	90	31	16	10	6	4	2	1	-
TW C3	88	30	15	9	6	4	2	1	-
TW C4	96	28	14	8	5	3	2	1	-
TW D1	39	-	-	-	-	-	-	-	-
TW D2	106	47	27	17	12	8	5	4	2
TW D3	108	52	31	20	14	9	6	4	2

HALIFAX, NOVA SCOTIA — CONTINUED

SSF =	.10	.20	.30	.40	.50	.60	.70	.80	.90
TW D4	116	62	39	27	19	14	10	7	4
TW D5	114	63	41	29	21	15	11	8	5
TW E1	132	64	38	25	18	12	9	6	3
TW E2	124	64	39	27	19	14	10	7	4
TW E3	141	82	54	38	28	21	15	11	7
TW E4	128	76	50	36	26	20	15	11	7
TW F1	122	23	8	-	-	-	-	-	-
TW F2	80	28	14	8	4	-	-	-	-
TW F3	68	29	16	10	6	3	-	-	-
TW F4	57	27	16	10	6	4	3	1	-
TW G1	77	22	10	5	-	-	-	-	-
TW G2	59	24	13	8	5	3	-	-	-
TW G3	51	23	13	8	5	3	2	-	-
TW G4	42	20	12	7	5	3	2	1	-
TW H1	51	19	10	5	3	-	-	-	-
TW H2	41	18	10	6	4	3	1	-	-
TW H3	36	16	9	6	4	3	2	-	-
TW H4	28	13	8	5	3	2	1	-	-
TW I1	25	-	-	-	-	-	-	-	-
TW I2	80	39	23	15	10	7	5	3	2
TW I3	87	44	27	18	12	9	6	4	2
TW I4	96	54	35	24	17	13	9	7	4
TW I5	97	56	37	26	19	14	10	7	5
TW J1	110	57	35	24	17	12	8	6	3
TW J2	103	56	35	24	17	13	9	6	4
TW J3	120	73	49	35	26	19	14	10	7
TW J4	111	68	46	33	24	18	14	10	7
DG A1	36	-	-	-	-	-	-	-	-
DG A2	80	32	16	7	-	-	-	-	-
DG A3	121	53	32	22	15	10	7	3	-
DG B1	38	-	-	-	-	-	-	-	-
DG B2	82	34	19	11	6	-	-	-	-
DG B3	125	56	34	23	17	12	9	6	2
DG C1	64	21	-	-	-	-	-	-	-
DG C2	104	44	26	17	11	7	-	-	-
DG C3	146	66	40	28	20	15	11	8	4
SS A1	227	61	29	16	9	5	-	-	-
SS A2	230	93	52	33	22	15	10	7	4
SS A3	204	48	20	8	-	-	-	-	-
SS A4	229	90	49	30	20	13	9	6	3
SS A5	374	57	23	11	4	-	-	-	-
SS A6	225	90	50	31	21	14	9	6	3
SS A7	399	40	12	-	-	-	-	-	-
SS A8	226	86	46	28	18	12	8	5	2
SS B1	148	43	20	11	6	-	-	-	-
SS B2	181	76	43	27	19	13	9	6	3
SS B3	128	35	15	6	-	-	-	-	-
SS B4	174	72	40	25	17	12	8	5	3
SS B5	187	35	14	6	-	-	-	-	-
SS B6	167	71	40	26	17	12	8	5	3
SS B7	153	25	-	-	-	-	-	-	-
SS B8	159	66	37	23	15	10	7	4	2
SS C1	78	30	15	8	4	-	-	-	-
SS C2	105	52	31	20	14	10	7	5	3
SS C3	84	24	11	5	-	-	-	-	-
SS C4	104	46	27	17	12	8	5	4	2
SS D1	205	73	37	21	13	7	3	-	-
SS D2	202	102	62	41	29	20	14	9	6
SS D3	261	68	32	17	9	5	-	-	-
SS D4	207	100	60	39	27	19	13	9	5
SS E1	138	50	24	13	6	-	-	-	-
SS E2	159	79	47	31	21	15	10	7	4
SS E3	173	43	18	9	-	-	-	-	-
SS E4	165	74	43	27	18	13	8	5	3

MOOSONEE, ONTARIO — 11572 DD

SSF =	.10	.20	.30	.40	.50	.60	.70	.80	.90
WW A1	92	-	-	-	-	-	-	-	-
WW A2	42	-	-	-	-	-	-	-	-
WW A3	39	-	-	-	-	-	-	-	-
WW A4	38	-	-	-	-	-	-	-	-
WW A5	37	-	-	-	-	-	-	-	-
WW A6	37	-	-	-	-	-	-	-	-
WW B1	-	-	-	-	-	-	-	-	-
WW B2	54	25	13	-	-	-	-	-	-
WW B3	68	34	19	11	-	-	-	-	-
WW B4	75	43	28	19	13	9	6	4	2
WW B5	75	45	30	21	15	11	8	5	3
WW C1	78	40	24	15	9	5	2	-	-
WW C2	74	41	26	17	11	7	5	3	1
WW C3	93	58	40	29	21	15	11	8	5
WW C4	85	54	37	27	20	14	11	7	5
TW A1	97	-	-	-	-	-	-	-	-
TW A2	48	-	-	-	-	-	-	-	-
TW A3	42	-	-	-	-	-	-	-	-
TW A4	38	11	-	-	-	-	-	-	-
TW B1	63	-	-	-	-	-	-	-	-
TW B2	44	-	-	-	-	-	-	-	-
TW B3	41	11	-	-	-	-	-	-	-
TW B4	41	13	-	-	-	-	-	-	-
TW C1	51	-	-	-	-	-	-	-	-
TW C2	44	12	-	-	-	-	-	-	-
TW C3	45	13	-	-	-	-	-	-	-
TW C4	51	13	5	-	-	-	-	-	-
TW D1	-	-	-	-	-	-	-	-	-
TW D2	59	24	11	-	-	-	-	-	-
TW D3	59	26	13	-	-	-	-	-	-
TW D4	75	39	24	16	10	7	4	3	1
TW D5	77	43	27	18	13	9	6	4	2
TW E1	78	36	20	12	6	-	-	-	-
TW E2	78	39	23	15	9	6	3	2	-
TW E3	95	55	36	25	17	12	9	6	3
TW E4	89	52	35	24	17	13	9	6	4
TW F1	30	-	-	-	-	-	-	-	-
TW F2	28	-	-	-	-	-	-	-	-
TW F3	27	-	-	-	-	-	-	-	-
TW F4	25	-	-	-	-	-	-	-	-
TW G1	21	-	-	-	-	-	-	-	-
TW G2	22	-	-	-	-	-	-	-	-
TW G3	22	-	-	-	-	-	-	-	-
TW G4	20	6	-	-	-	-	-	-	-
TW H1	17	-	-	-	-	-	-	-	-
TW H2	17	-	-	-	-	-	-	-	-
TW H3	16	5	-	-	-	-	-	-	-
TW H4	14	5	-	-	-	-	-	-	-
TW I1	-	-	-	-	-	-	-	-	-
TW I2	44	19	9	-	-	-	-	-	-
TW I3	49	23	12	6	-	-	-	-	-
TW I4	62	34	21	14	10	6	4	2	1
TW I5	66	38	24	17	12	8	6	4	2
TW J1	65	32	19	11	6	2	-	-	-
TW J2	65	34	21	14	9	6	3	2	-
TW J3	82	50	33	23	17	12	9	6	4
TW J4	78	47	32	22	16	12	9	6	4
DG A1	-	-	-	-	-	-	-	-	-
DG A2	41	-	-	-	-	-	-	-	-
DG A3	79	35	20	13	8	4	-	-	-
DG B1	-	-	-	-	-	-	-	-	-
DG B2	42	14	-	-	-	-	-	-	-
DG B3	83	37	22	15	10	6	3	-	-
DG C1	-	-	-	-	-	-	-	-	-

```
MOOSONEE, ONTARIO                              CONTINUED      OTTAWA, ONTARIO                                CONTINUED
    SSF  =.10   .20   .30   .40   .50   .60   .70   .80   .90     SSF  =.10   .20   .30   .40   .50   .60   .70   .80   .90
DG C2    60    24    12     -     -     -     -     -     -    TW D4   111    59    37    25    18    13     9     6     4
DG C3    98    45    27    18    13     8     5     3     -    TW D5   109    60    39    27    19    14    10     7     5
SS A1   101    16     -     -     -     -     -     -     -    TW E1   125    59    35    23    16    11     8     5     3
SS A2   132    51    27    15     9     4     2     -     -    TW E2   119    60    37    25    17    12     9     6     4
SS A3    67     -     -     -     -     -     -     -     -    TW E3   135    78    51    36    26    19    14    10     7
SS A4   124    45    22    11     -     -     -     -     -    TW E4   124    72    48    34    25    18    14    10     6
SS A5   157     -     -     -     -     -     -     -     -    TW F1   114    19     -     -     -     -     -     -     -
SS A6   126    48    24    13     7     -     -     -     -    TW F2    74    25    12     6     -     -     -     -     -
SS A7   127     -     -     -     -     -     -     -     -    TW F3    63    26    14     8     4     2     -     -     -
SS A8   118    40    18     -     -     -     -     -     -    TW F4    53    24    14     8     5     3     2     -     -
SS B1    60     -     -     -     -     -     -     -     -    TW G1    71    19     7     -     -     -     -     -     -
SS B2   106    43    23    14     8     5     2     1     -    TW G2    54    21    11     6     3     -     -     -     -
SS B3    33     -     -     -     -     -     -     -     -    TW G3    47    21    11     7     4     2     1     -     -
SS B4    98    38    20    11     6     -     -     -     -    TW G4    39    18    10     6     4     3     2     -     -
SS B5    66     -     -     -     -     -     -     -     -    TW H1    47    17     8     4     -     -     -     -     -
SS B6    95    39    20    12     6     3     -     -     -    TW H2    38    16     9     5     3     2     -     -     -
SS B7     -     -     -     -     -     -     -     -     -    TW H3    33    15     8     5     3     2     1     -     -
SS B8    86    33    16     8     -     -     -     -     -    TW H4    27    12     7     4     3     2     1     -     -
SS C1     -     -     -     -     -     -     -     -     -    TW I1     -     -     -     -     -     -     -     -     -
SS C2    59    27    14     7     -     -     -     -     -    TW I2    75    36    21    14     9     6     4     3     1
SS C3    26     -     -     -     -     -     -     -     -    TW I3    82    41    25    16    11     8     5     3     2
SS C4    58    24    12     5     -     -     -     -     -    TW I4    92    51    33    23    16    12     8     6     4
SS D1    90    19     -     -     -     -     -     -     -    TW I5    93    54    35    25    18    13    10     7     4
SS D2   118    57    33    21    13     8     4     2     1    TW J1   104    53    32    21    15    11     7     5     3
SS D3   115     -     -     -     -     -     -     -     -    TW J2    98    53    33    23    16    12     8     6     4
SS D4   121    56    32    19    12     7     4     2     -    TW J3   114    69    46    33    24    18    13    10     6
SS E1    41     -     -     -     -     -     -     -     -    TW J4   107    65    44    31    23    17    13     9     6
SS E2    87    40    22    12     5     -     -     -     -    DG A1     -     -     -     -     -     -     -     -     -
SS E3    61     -     -     -     -     -     -     -     -    DG A2    76    29    14     -     -     -     -     -     -
SS E4    90    38    19    10     -     -     -     -     -    DG A3   117    51    30    20    14     9     6     3     -
                                                                DG B1    28     -     -     -     -     -     -     -     -
                                                                DG B2    78    31    17     9     -     -     -     -     -
OTTAWA, ONTARIO                                8735 DD          DG B3   122    53    32    22    16    11     8     5     2
    SSF  =.10   .20   .30   .40   .50   .60   .70   .80   .90  DG C1    57    15     -     -     -     -     -     -     -
WW A1   306    20     -     -     -     -     -     -     -    DG C2   100    41    24    15     9     5     -     -     -
WW A2   104    33    16     8     3     -     -     -     -    DG C3   142    63    38    26    19    14    10     7     4
WW A3    88    36    19    11     6     3     -     -     -    SS A1   204    50    22    11     5     -     -     -     -
WW A4    82    37    21    13     8     5     2     -     -    SS A2   209    82    45    28    18    12     8     5     3
WW A5    78    38    22    14     9     6     3     1     -    SS A3   176    35    10     -     -     -     -     -     -
WW A6    76    39    23    15    10     6     4     2     -    SS A4   205    77    41    25    16    10     7     4     2
WW B1    36     -     -     -     -     -     -     -     -    SS A5   343    46    16     -     -     -     -     -     -
WW B2    92    46    27    18    12     8     6     4     2    SS A6   204    79    43    26    17    11     8     5     3
WW B3   111    58    36    24    17    12     8     5     3    SS A7   360    26     -     -     -     -     -     -     -
WW B4   109    64    42    29    21    16    11     8     5    SS A8   200    72    38    22    14     9     6     3     2
WW B5   105    63    42    30    22    17    12     9     6    SS B1   130    35    15     7     -     -     -     -     -
WW C1   123    65    40    27    19    14    10     7     4    SS B2   165    67    37    24    16    11     7     5     3
WW C2   111    62    40    28    20    14    10     7     5    SS B3   108    25     6     -     -     -     -     -     -
WW C3   126    79    55    40    30    22    17    12     8    SS B4   157    63    34    21    14     9     6     4     2
WW C4   116    73    50    37    27    21    15    11     7    SS B5   166    27     8     -     -     -     -     -     -
TW A1   273    23     -     -     -     -     -     -     -    SS B6   152    62    35    22    14    10     7     4     3
TW A2   110    31    14     7     3     -     -     -     -    SS B7   131    14     -     -     -     -     -     -     -
TW A3    90    33    17     9     5     3     -     -     -    SS B8   142    56    31    19    12     8     5     3     2
TW A4    78    33    18    11     7     4     2     -     -    SS C1    71    26    12     5     -     -     -     -     -
TW B1   156    26     9     -     -     -     -     -     -    SS C2    99    48    28    19    13     9     6     4     2
TW B2    92    30    15     8     5     2     -     -     -    SS C3    78    21     9     -     -     -     -     -     -
TW B3    81    31    16     9     6     3     2     -     -    SS C4    99    43    24    15    10     7     5     3     2
TW B4    77    29    16     9     6     4     2     1     -    SS D1   180    60    29    15     8     4     -     -     -
TW C1   108    27    12     6     3     -     -     -     -    SS D2   182    90    54    35    24    17    12     8     5
TW C2    85    28    14     8     5     3     2     -     -    SS D3   234    56    24    11     5     -     -     -     -
TW C3    83    27    14     8     5     3     2     1     -    SS D4   187    88    51    33    23    16    11     7     4
TW C4    92    26    12     7     5     3     2     1     -    SS E1   117    38    16     -     -     -     -     -     -
TW D1    30     -     -     -     -     -     -     -     -    SS E2   142    68    40    26    17    12     8     5     3
TW D2   101    43    25    16    10     7     5     3     2    SS E3   152    33    12     -     -     -     -     -     -
TW D3   102    47    28    18    12     8     5     4     2    SS E4   148    64    36    23    15    10     7     4     2
```

624 / APPENDIX 21

TORONTO, ONTARIO 6827 DD

SSF =	.10	.20	.30	.40	.50	.60	.70	.80	.90
WW A1	325	23	-	-	-	-	-	-	-
WW A2	107	36	17	10	5	3	-	-	-
WW A3	91	38	21	12	8	5	3	-	-
WW A4	85	39	22	14	9	6	4	2	-
WW A5	81	40	24	15	10	7	4	3	1
WW A6	79	41	25	16	11	7	5	3	1
WW B1	44	-	-	-	-	-	-	-	-
WW B2	95	47	29	19	13	9	6	4	3
WW B3	114	61	38	25	18	13	9	6	4
WW B4	111	65	43	31	22	16	12	9	6
WW B5	107	65	44	31	23	17	13	9	6
WW C1	126	67	42	29	20	15	10	7	5
WW C2	113	64	41	29	21	15	11	8	5
WW C3	129	82	56	41	31	23	17	13	9
WW C4	118	74	52	38	28	21	16	12	8
TW A1	287	25	7	-	-	-	-	-	-
TW A2	113	33	15	8	5	2	-	-	-
TW A3	92	35	18	11	7	4	2	-	-
TW A4	80	34	19	12	8	5	3	2	-
TW B1	160	27	11	5	-	-	-	-	-
TW B2	94	32	16	9	6	4	2	-	-
TW B3	83	32	17	10	7	4	3	2	-
TW B4	79	31	17	10	7	4	3	2	-
TW C1	110	28	13	7	4	2	1	-	-
TW C2	87	29	15	9	6	4	2	1	-
TW C3	85	28	15	9	6	4	2	1	-
TW C4	93	27	13	8	5	3	2	1	-
TW D1	35	-	-	-	-	-	-	-	-
TW D2	103	45	26	16	11	8	5	4	2
TW D3	104	50	29	19	13	9	6	4	2
TW D4	113	60	38	26	19	14	10	7	4
TW D5	111	62	40	28	20	15	11	8	5
TW E1	128	62	37	24	17	12	8	6	4
TW E2	121	62	38	26	18	13	9	7	4
TW E3	138	80	53	37	27	20	15	11	7
TW E4	126	74	49	35	26	19	14	10	7
TW F1	118	21	7	-	-	-	-	-	-
TW F2	77	27	13	7	4	2	-	-	-
TW F3	65	27	15	9	5	3	2	-	-
TW F4	55	26	15	9	6	4	3	1	-
TW G1	73	20	9	4	-	-	-	-	-
TW G2	56	23	12	7	4	3	1	-	-
TW G3	49	22	12	8	5	3	2	1	-
TW G4	41	19	11	7	5	3	2	1	-
TW H1	48	18	9	5	3	-	-	-	-
TW H2	40	17	10	6	4	2	1	-	-
TW H3	34	16	9	6	4	2	2	-	-
TW H4	27	12	7	5	3	2	1	-	-
TW I1	20	-	-	-	-	-	-	-	-
TW I2	77	37	22	14	10	7	5	3	2
TW I3	84	42	26	17	12	8	6	4	2
TW I4	94	52	34	23	17	12	9	6	4
TW I5	95	55	36	25	19	14	10	7	5
TW J1	106	55	34	23	16	11	8	6	3
TW J2	100	54	34	24	17	12	9	6	4
TW J3	117	71	48	34	25	19	14	10	7
TW J4	109	66	45	32	24	18	13	10	6
DG A1	30	-	-	-	-	-	-	-	-
DG A2	78	31	15	6	-	-	-	-	-
DG A3	118	52	31	21	14	10	6	3	-
DG B1	33	-	-	-	-	-	-	-	-
DG B2	79	32	18	10	5	-	-	-	-
DG B3	123	55	33	22	16	12	8	5	2
DG C1	61	18	-	-	-	-	-	-	-

TORONTO, ONTARIO CONTINUED

SSF =	.10	.20	.30	.40	.50	.60	.70	.80	.90
DG C2	101	43	25	16	10	6	-	-	-
DG C3	143	64	39	27	19	15	11	8	4
SS A1	215	56	25	14	8	4	2	-	-
SS A2	219	88	48	30	20	14	10	6	4
SS A3	189	42	16	5	-	-	-	-	-
SS A4	217	84	45	28	18	12	8	5	3
SS A5	364	52	20	9	-	-	-	-	-
SS A6	214	85	46	29	19	13	9	6	4
SS A7	390	34	-	-	-	-	-	-	-
SS A8	213	79	42	25	16	11	7	5	3
SS B1	139	39	18	9	5	2	-	-	-
SS B2	173	72	40	26	17	12	8	6	3
SS B3	117	31	12	-	-	-	-	-	-
SS B4	165	68	38	24	16	11	7	5	3
SS B5	177	31	12	4	-	-	-	-	-
SS B6	159	67	38	24	16	11	8	5	3
SS B7	142	20	-	-	-	-	-	-	-
SS B8	150	61	34	21	14	9	6	4	2
SS C1	74	28	14	8	4	-	-	-	-
SS C2	102	50	30	20	14	9	7	4	3
SS C3	80	23	10	5	-	-	-	-	-
SS C4	101	44	25	16	11	8	5	4	2
SS D1	192	67	33	19	11	7	4	1	-
SS D2	192	96	58	38	27	19	13	9	6
SS D3	247	62	28	15	8	4	-	-	-
SS D4	197	94	56	36	25	17	12	8	5
SS E1	128	44	21	10	4	-	-	-	-
SS E2	150	74	44	29	20	14	9	6	4
SS E3	162	38	16	7	-	-	-	-	-
SS E4	156	69	40	25	17	12	8	5	3

NORMANDIN, QUEBEC 10528 DD

SSF =	.10	.20	.30	.40	.50	.60	.70	.80	.90
WW A1	175	-	-	-	-	-	-	-	-
WW A2	73	22	-	-	-	-	-	-	-
WW A3	65	25	11	-	-	-	-	-	-
WW A4	61	26	13	-	-	-	-	-	-
WW A5	59	28	15	6	-	-	-	-	-
WW A6	58	29	16	8	-	-	-	-	-
WW B1	-	-	-	-	-	-	-	-	-
WW B2	73	36	21	13	8	5	2	1	-
WW B3	90	47	29	19	12	8	5	3	1
WW B4	92	54	36	25	18	13	9	6	4
WW B5	90	55	37	26	19	14	10	7	4
WW C1	101	53	33	22	15	10	7	4	2
WW C2	92	52	33	23	16	11	8	5	3
WW C3	110	70	48	35	26	19	14	10	7
WW C4	100	64	44	32	24	18	13	9	6
TW A1	164	-	-	-	-	-	-	-	-
TW A2	77	21	-	-	-	-	-	-	-
TW A3	65	23	10	-	-	-	-	-	-
TW A4	58	24	12	5	-	-	-	-	-
TW B1	102	15	-	-	-	-	-	-	-
TW B2	67	22	9	-	-	-	-	-	-
TW B3	61	22	11	5	-	-	-	-	-
TW B4	58	22	11	6	-	-	-	-	-
TW C1	76	19	7	-	-	-	-	-	-
TW C2	63	21	10	5	-	-	-	-	-
TW C3	62	21	10	5	-	-	-	-	-
TW C4	68	20	9	5	2	-	-	-	-
TW D1	-	-	-	-	-	-	-	-	-
TW D2	79	34	19	11	7	4	2	-	-
TW D3	80	38	21	13	8	5	2	-	-

NORMANDIN, QUEBEC   CONTINUED
     SSF = .10   .20   .30   .40   .50   .60   .70   .80   .90
TW D4   92   49   31   21   15   10   7   5   3
TW D5   93   52   33   23   17   12   9   6   4
TW E1   101   48   28   18   12   8   5   3   1
TW E2   98   50   31   20   14   10   6   4   2
TW E3   115   67   44   31   22   16   12   8   5
TW E4   106   63   42   29   21   16   11   8   5
TW F1   71   -   -   -   -   -   -   -   -
TW F2   52   16   -   -   -   -   -   -   -
TW F3   46   18   7   -   -   -   -   -   -
TW F4   40   18   9   -   -   -   -   -   -
TW G1   47   9   -   -   -   -   -   -   -
TW G2   39   14   -   -   -   -   -   -   -
TW G3   35   15   7   -   -   -   -   -   -
TW G4   30   13   7   3   -   -   -   -   -
TW H1   33   10   -   -   -   -   -   -   -
TW H2   28   12   5   -   -   -   -   -   -
TW H3   25   11   6   3   -   -   -   -   -
TW H4   20   9   5   3   -   -   -   -   -
TW I1   -   -   -   -   -   -   -   -   -
TW I2   59   28   16   10   6   3   2   -   -
TW I3   65   33   19   12   8   5   3   1   -
TW I4   77   43   28   19   13   9   7   4   3
TW I5   79   46   30   21   15   11   8   5   3
TW J1   84   43   26   17   11   8   5   3   2
TW J2   81   44   28   19   13   9   6   4   2
TW J3   98   60   40   29   21   15   11   8   5
TW J4   92   56   38   27   20   15   11   8   5
DG A1   -   -   -   -   -   -   -   -   -
DG A2   58   22   -   -   -   -   -   -   -
DG A3   96   43   26   17   11   7   4   -   -
DG B1   -   -   -   -   -   -   -   -   -
DG B2   60   24   11   -   -   -   -   -   -
DG B3   100   46   28   19   13   9   6   3   -
DG C1   34   -   -   -   -   -   -   -   -
DG C2   79   33   19   11   5   -   -   -   -
DG C3   117   54   33   23   16   11   8   5   2
SS A1   148   36   13   -   -   -   -   -   -
SS A2   169   68   37   22   14   9   6   3   2
SS A3   120   17   -   -   -   -   -   -   -
SS A4   163   63   32   19   11   7   4   2   -
SS A5   231   30   -   -   -   -   -   -   -
SS A6   163   65   34   20   13   8   5   3   1
SS A7   219   -   -   -   -   -   -   -   -
SS A8   158   58   29   16   9   4   1   -   -
SS B1   94   24   -   -   -   -   -   -   -
SS B2   134   56   31   19   12   8   5   3   2
SS B3   74   -   -   -   -   -   -   -   -
SS B4   126   52   28   17   10   7   4   2   1
SS B5   111   15   -   -   -   -   -   -   -
SS B6   123   52   28   17   11   7   4   3   1
SS B7   80   -   -   -   -   -   -   -   -
SS B8   113   46   24   14   8   5   3   1   -
SS C1   49   14   -   -   -   -   -   -   -
SS C2   79   38   22   14   9   6   3   2   -
SS C3   52   11   -   -   -   -   -   -   -
SS C4   78   34   19   12   7   4   2   -   -
SS D1   135   44   18   -   -   -   -   -   -
SS D2   151   75   45   29   19   13   8   5   3
SS D3   169   40   13   -   -   -   -   -   -
SS D4   154   73   43   27   18   12   8   5   3
SS E1   83   22   -   -   -   -   -   -   -
SS E2   116   56   32   20   13   8   5   3   1
SS E3   105   19   -   -   -   -   -   -   -
SS E4   119   52   29   17   11   6   4   2   -

Appendix 22 Sensitivity Data

The data in this appendix express the sensitivity of the annual *SSF* for direct-gain and sunspace systems to selected designs parameters. In any given set of data, one or two parameters are varied while the others are held fixed at their reference values. Each set of data is presented for several different reference cities in the United States. For the direct-gain data, the cities are Albuquerque, New Mexico; Boston, Massachusetts; Madison, Wisconsin; Medford, Oregon; Nashville, Tennessee; and Santa Maria, California. For the sunspace data the cities are Albuquerque, New Mexico; Bismarck, North Dakota; Boston, Massachusetts; Ely, Nevada; Nashville, Tennessee; and Seattle, Washington. The cities were selected to represent wide geographical and climatological ranges.

To apply the data, choose one of the reference cities to represent the location of interest, on the basis of climate similarity. Two gross measures are of use in making the choice: the 65°F-base heating degree-days and the solar radiation incident on a vertical surface. These data are available as monthly averages for numerous locations in the United States and Canada in Appendix 19, where they are tabulated as *D65* and *VS*, respectively. Figure 22-1 shows the January values of *D65* and *VS* for each reference city.

Each parameter variation is usually represented by two different values of the load collector ratio (*LCR*). In applying the sensitivity data, the *SSF*—not the *LCR*—nearest that of the design being analyzed should be used.

Dots on the curves represent the parameter values corresponding to one of the reference designs.

Two applications of sensitivity data are important. One is quantitative, as a final step in a design analysis. In this application, an analysis is performed on the basis of the reference design that most nearly resembles the design of interest. The annual *SSF* is then corrected by use of sensitivity data.

Another application of sensitivity data is qualitative, wherein the data serve as a guide—perhaps very early in the design process—to the relative significance of various design parameters and their preferred values. Discussions of the sensitivity data in this qualitative vein are given in Chapters O and P. The discussions are illustrated by selected data from the full set provided here.

22.1 HIGH-MASS DIRECT-GAIN SYSTEMS

22.1.a HEATING PERFORMANCE

The graphs in this section plot the effect on solar savings fraction of changes in each of the following system parameters: mass thickness, ϱck product, mass absorptance, lightweight absorptance fraction, mass distribution, R-value of the mass covering, number of glazings, temperature bounds, and nighttime setback.

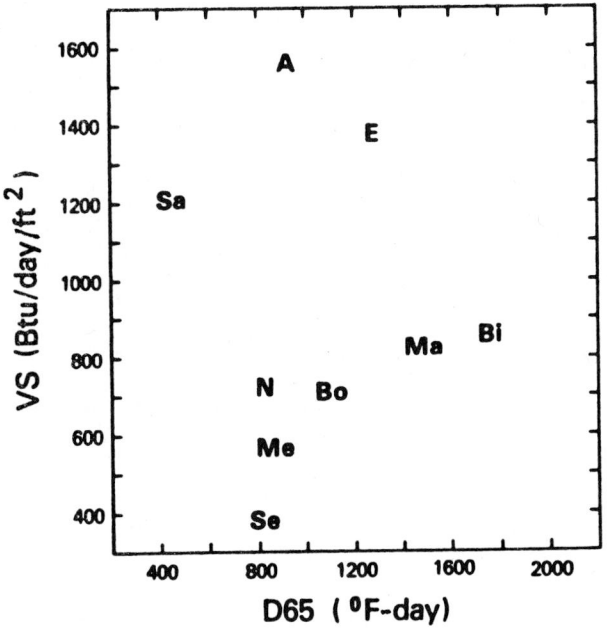

22-1: Climates of the reference cities mapped according to daily average solar radiation incident upon a vertical surface (*VS*) and the January 65°F-base heating degree-days (*D65*) in each city. The cities are Albuquerque (A), Bismarck (Bi), Boston (Bo), Ely (E), Madison (Ma), Medford (Me), Nashville, (N), Santa Maria (Sa), and Seattle (Se).

22.1.a.1 Mass Thickness

The graphs below plot the effect on solar savings fraction of changes in mass thickness—for the double-glazed configuration, with and without R9 nighttime insulation, at different LCRs, and at various values of mass-area-to-glazing-area ratio (A_m/A_g)—for each of six reference cities. Dots represent reference conditions.

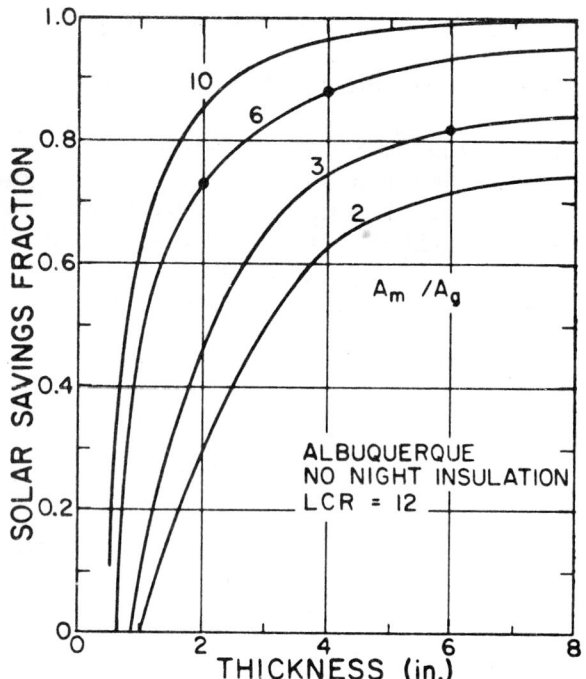

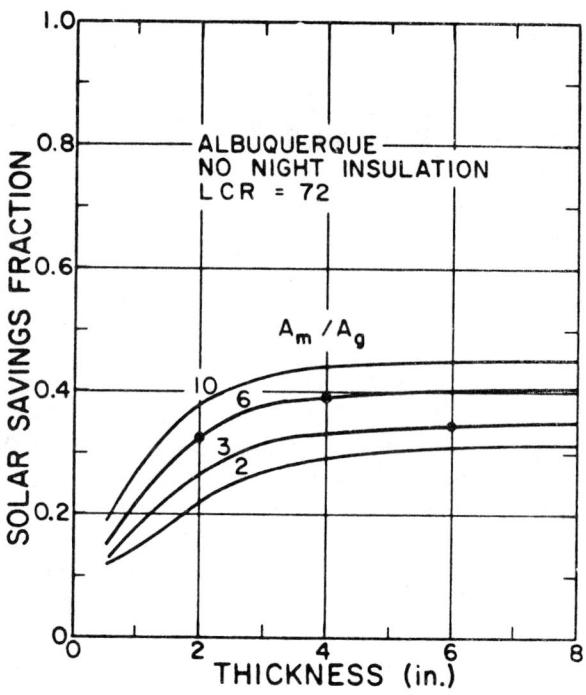

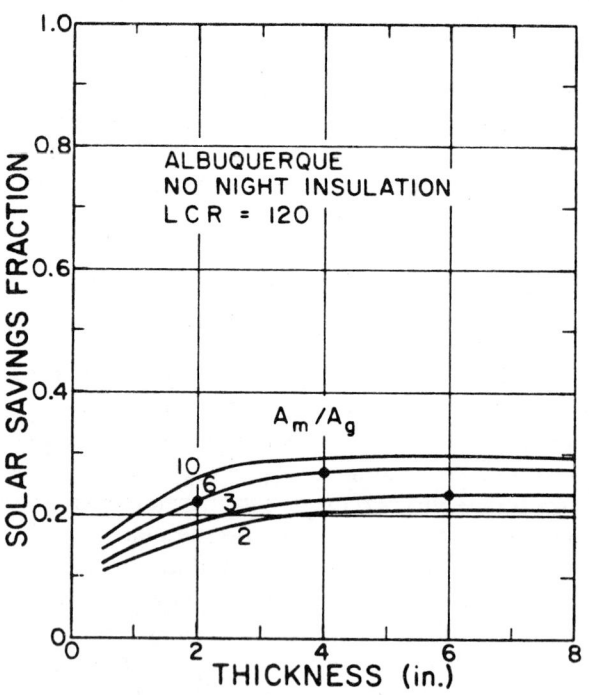

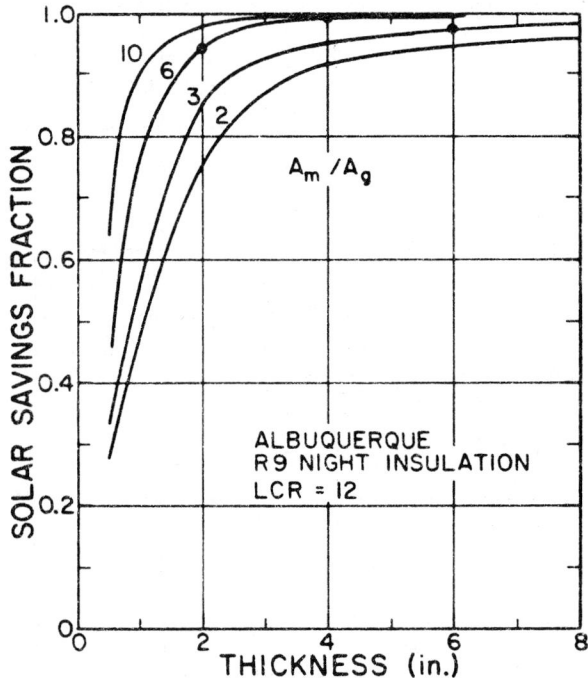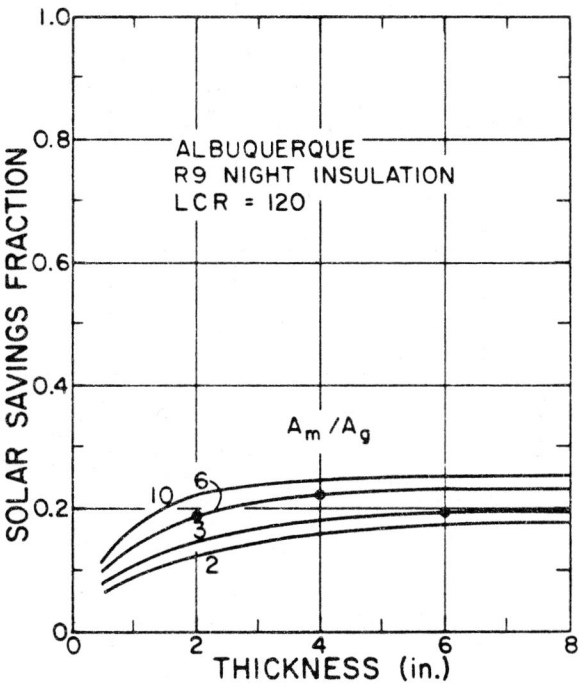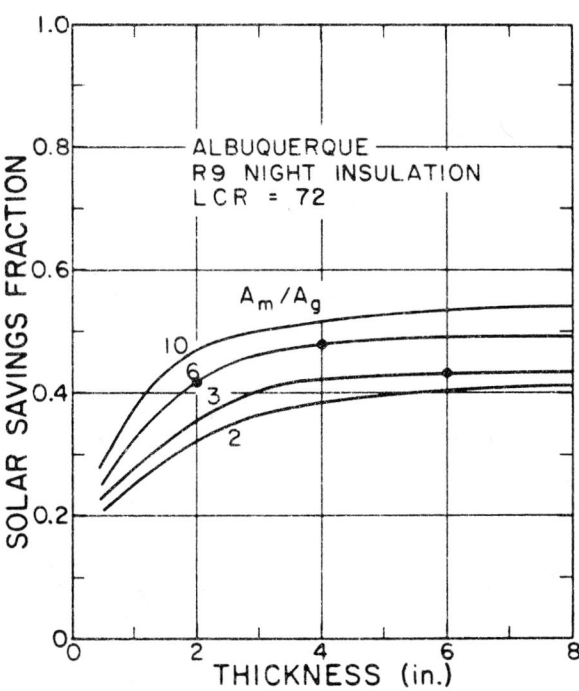

Sensitivity Data / 629

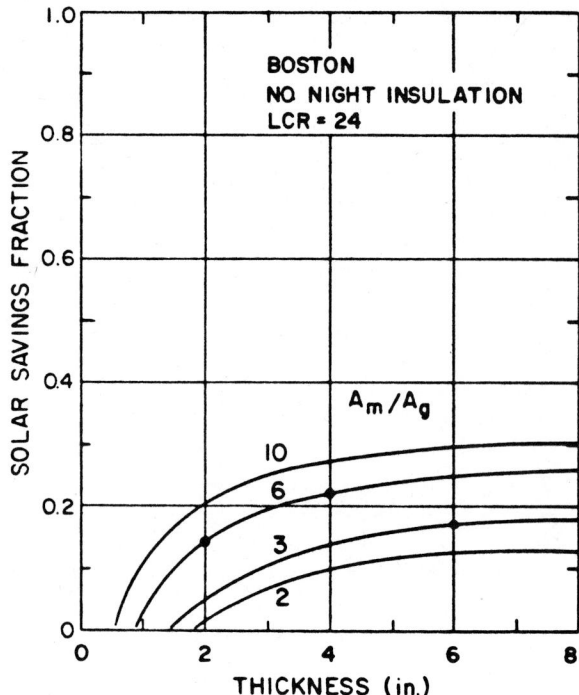
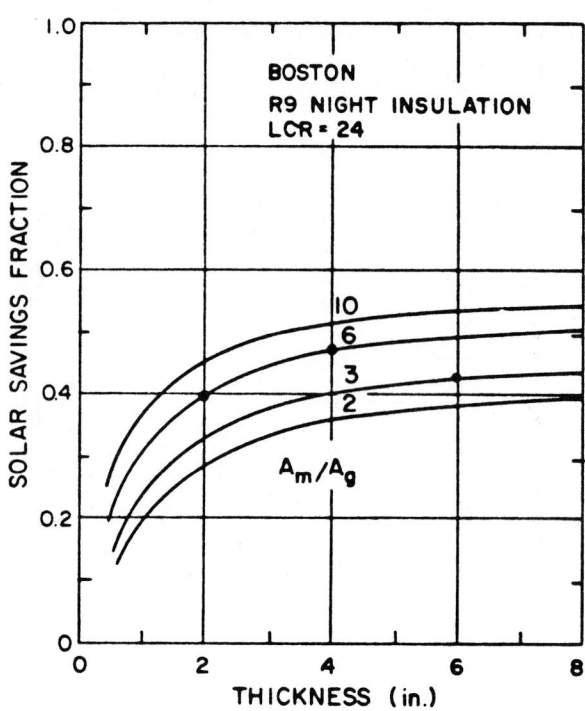
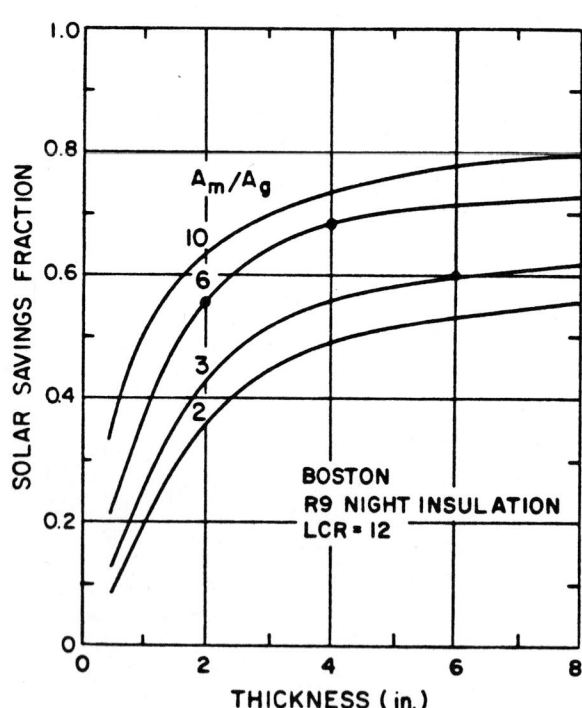

630 / APPENDIX 22

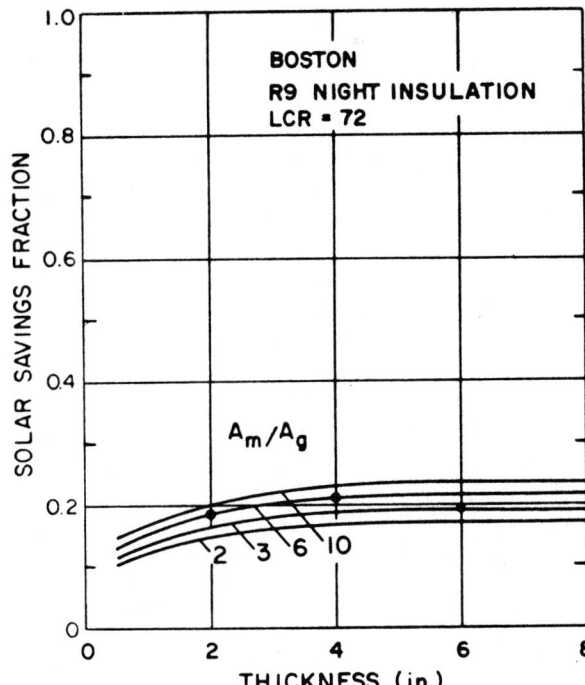
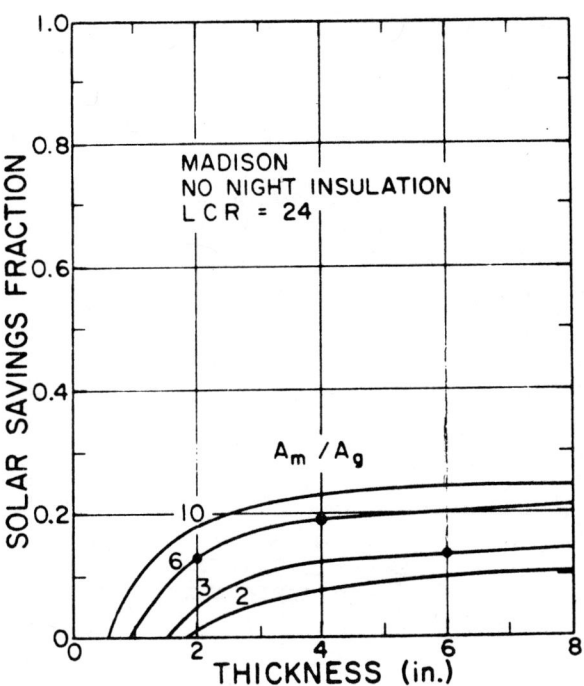
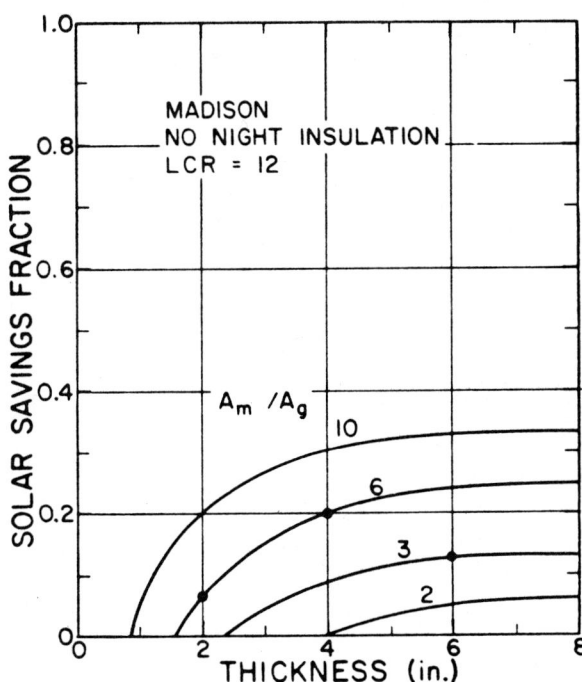

Sensitivity Data / 631

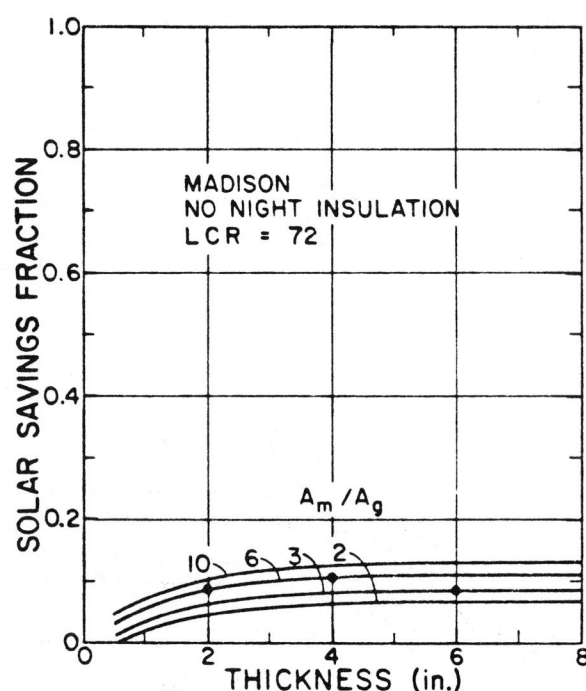
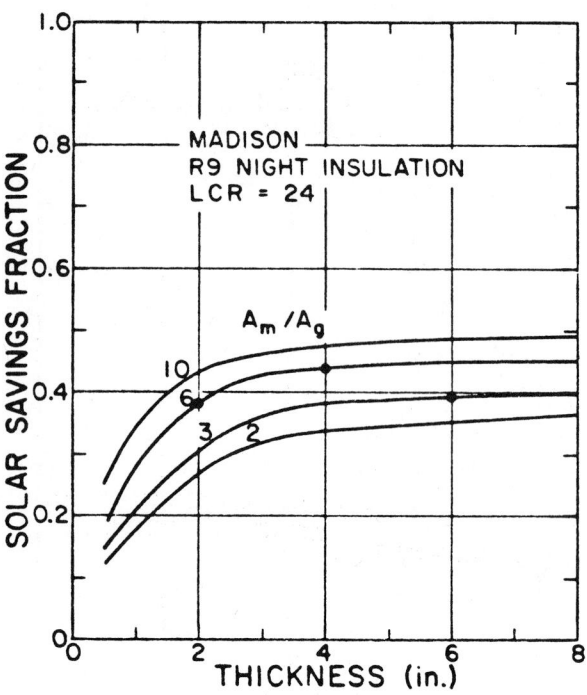
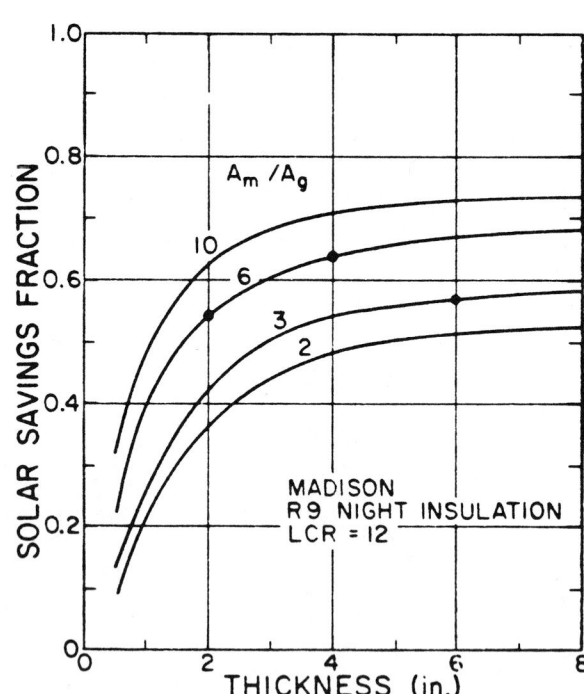

632 / APPENDIX 22

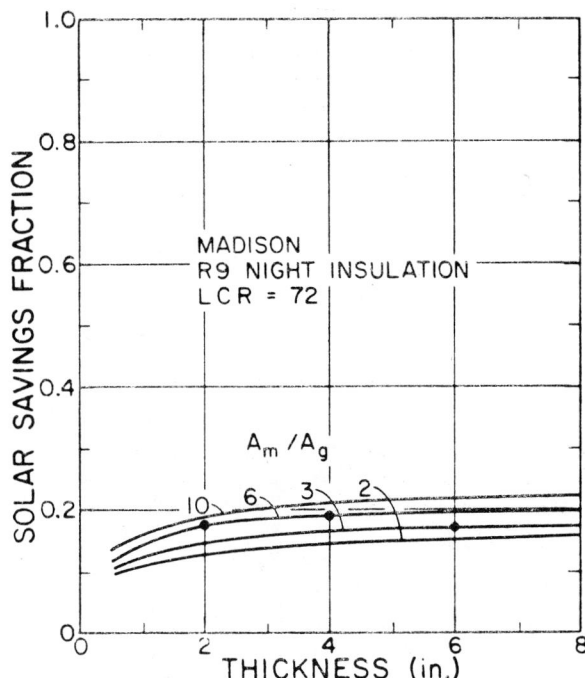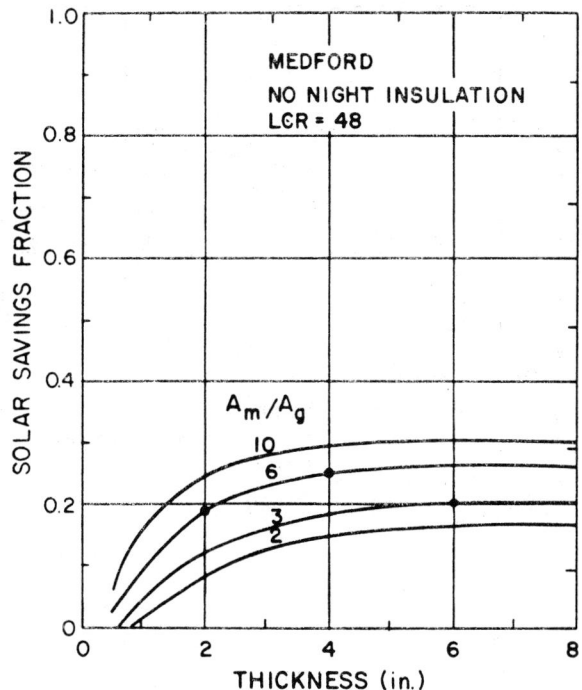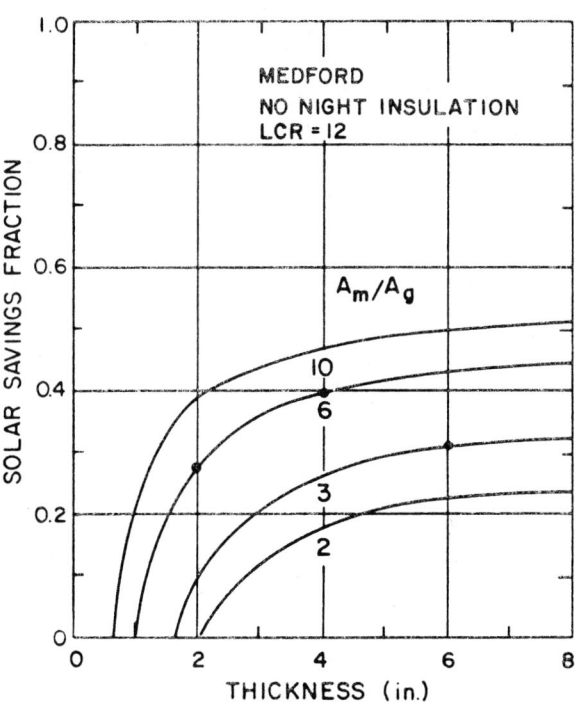

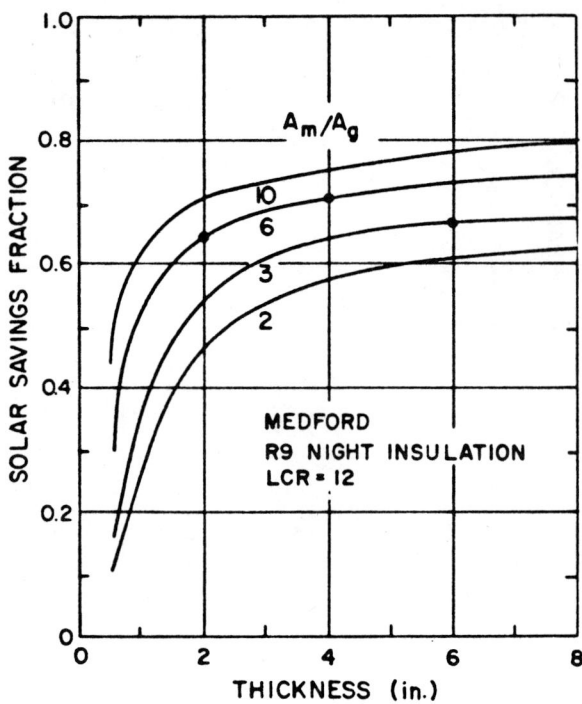
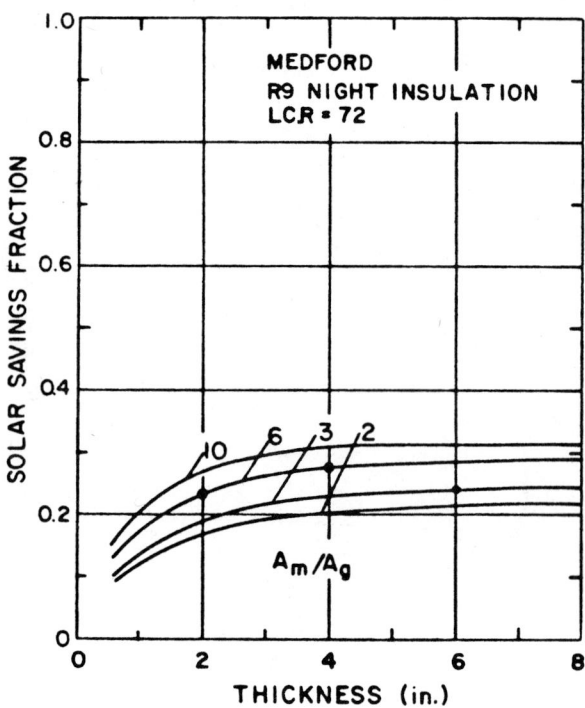
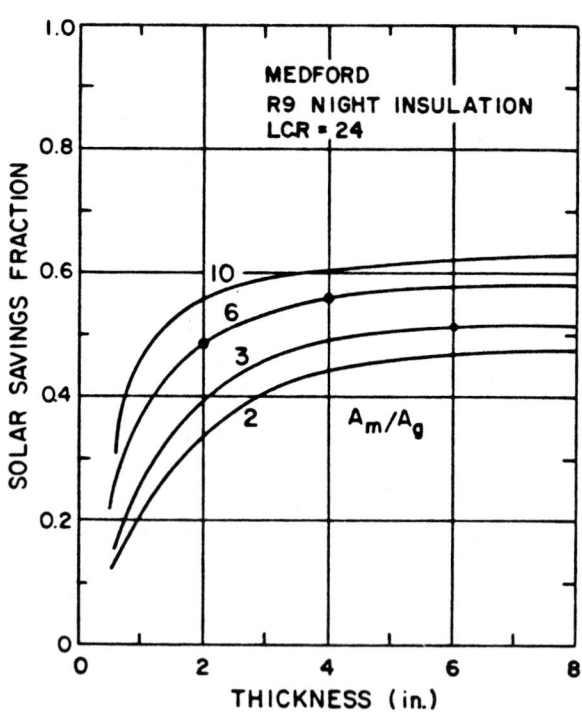

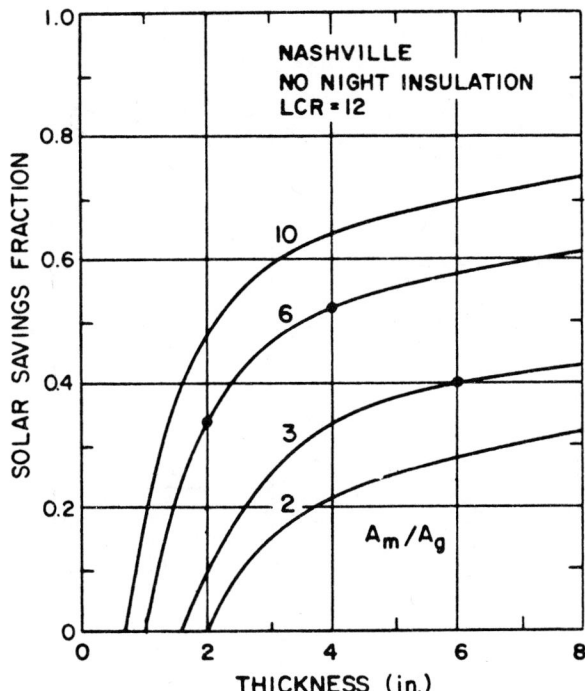
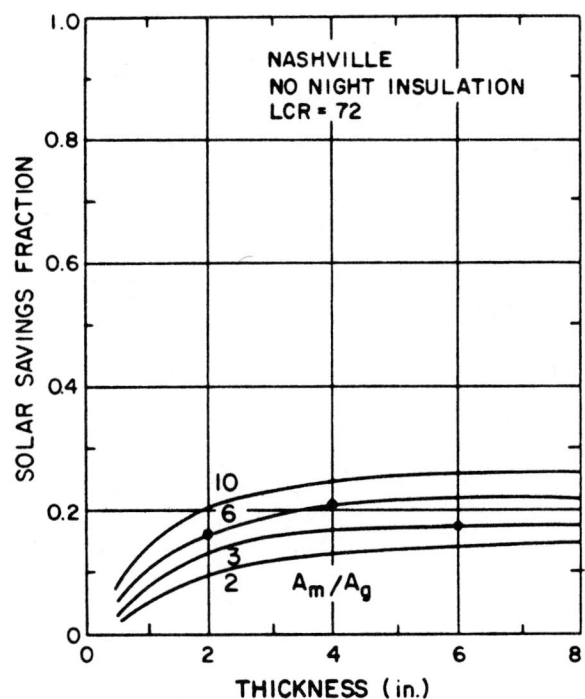
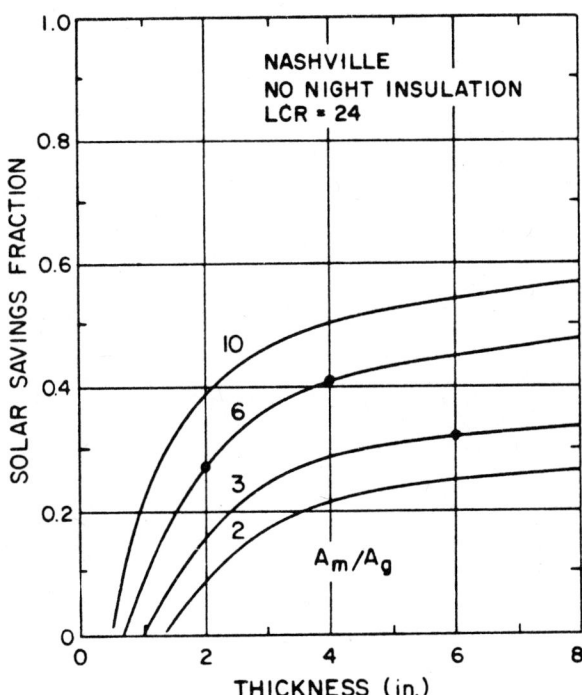

Sensitivity Data / 635

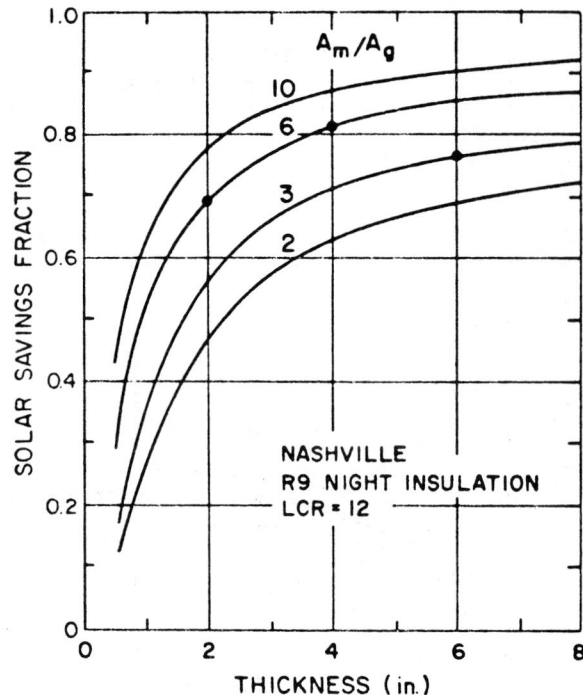
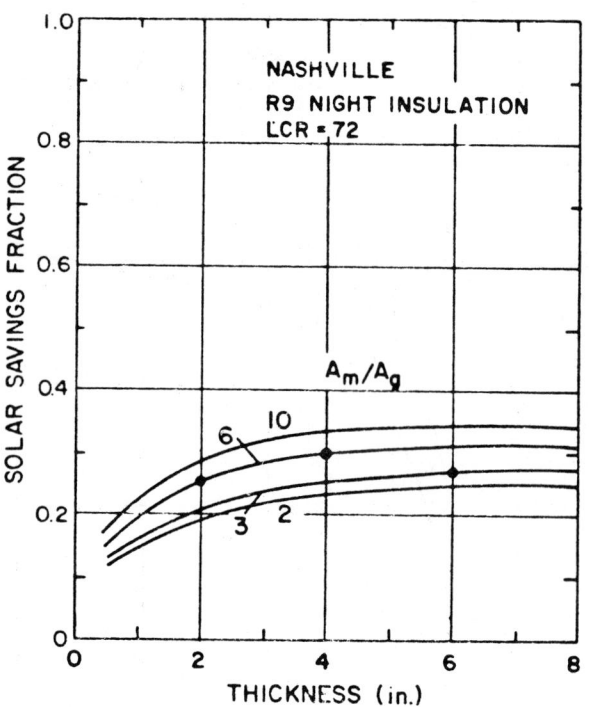
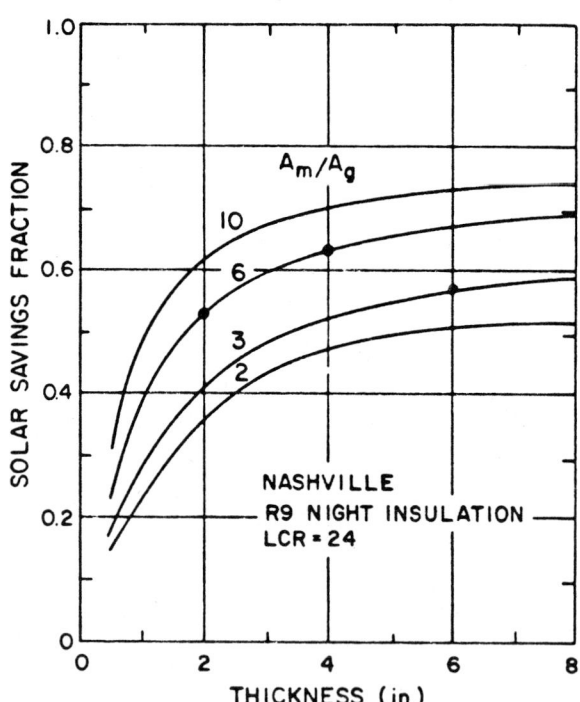

636 / APPENDIX 22

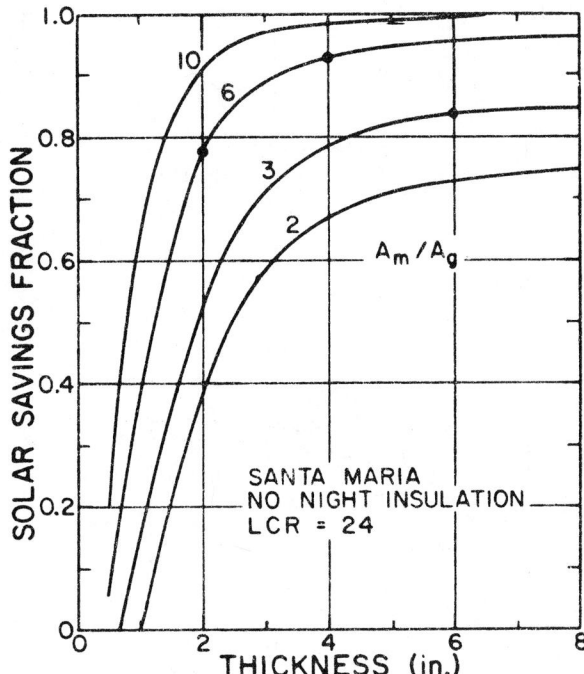
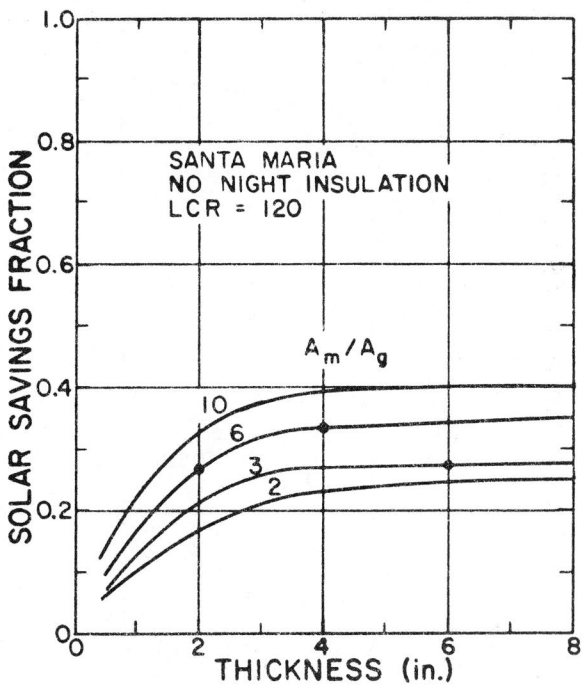
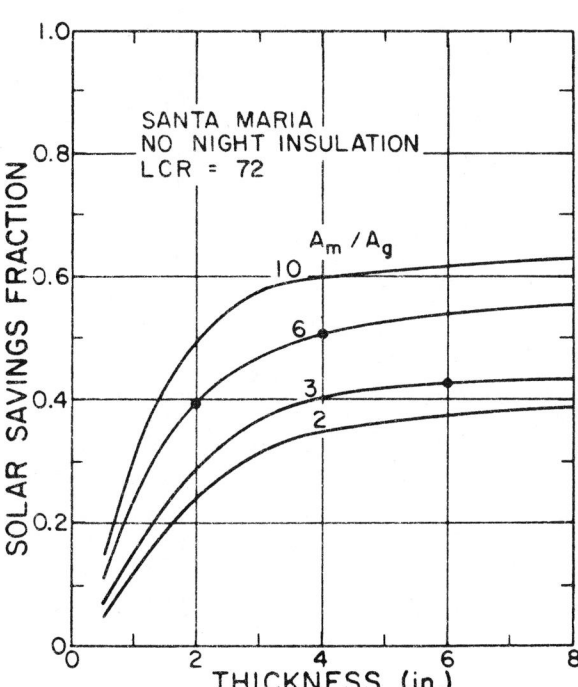

Sensitivity Data / 637

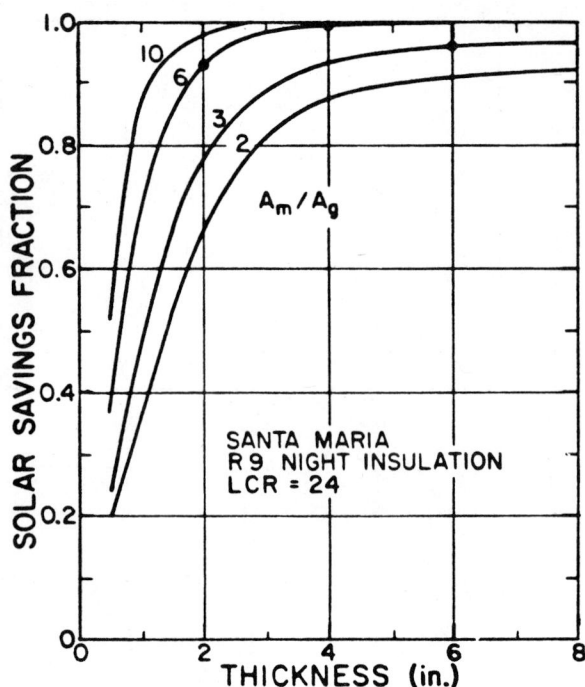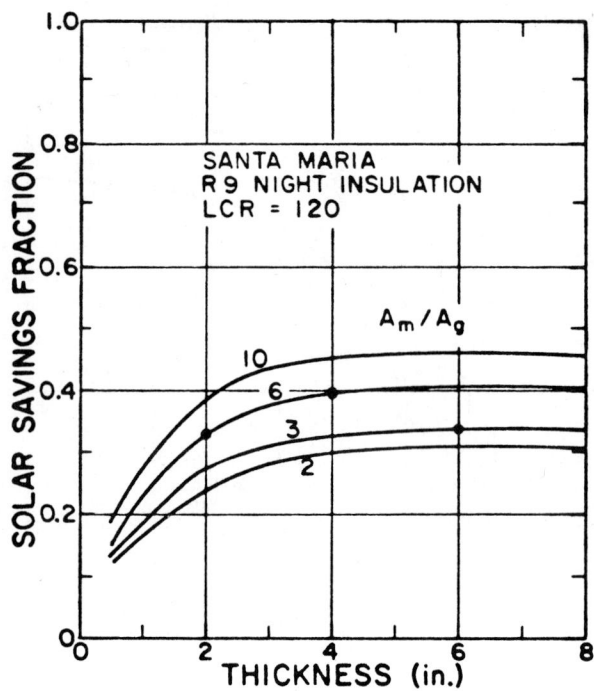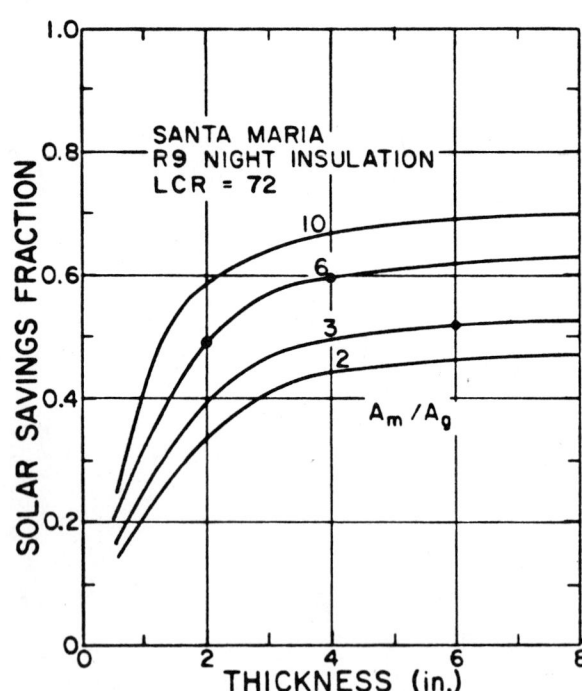

22.1.a.2 ϱck Product

Density (ϱ), specific heat (c), and thermal conductivity (k) are frequently grouped together to form the parameter ϱck. The graphs below plot the effect on solar savings fraction of changes in ϱck product—for the double-glazed configuration, with and without R9 nighttime insulation, at different *LCR*s, and at mass thickness of 6 inches and a mass-area-to-glazing-area ratio of 3—for each of three reference cities. Dots represent reference conditions.

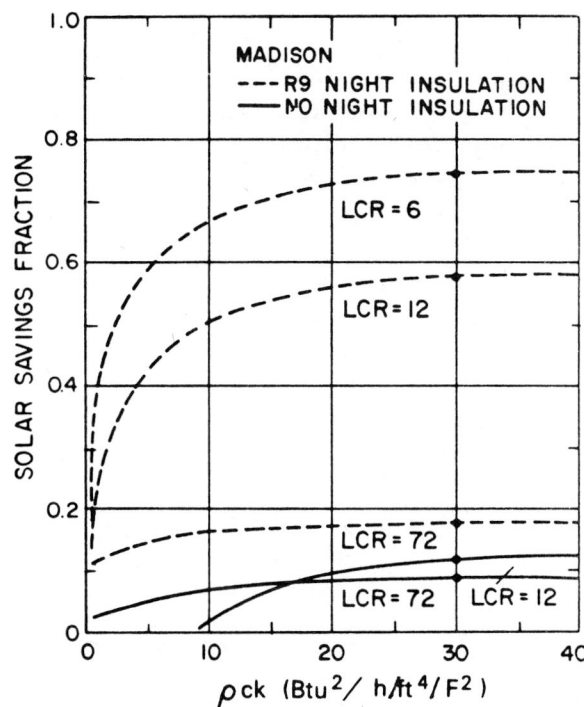

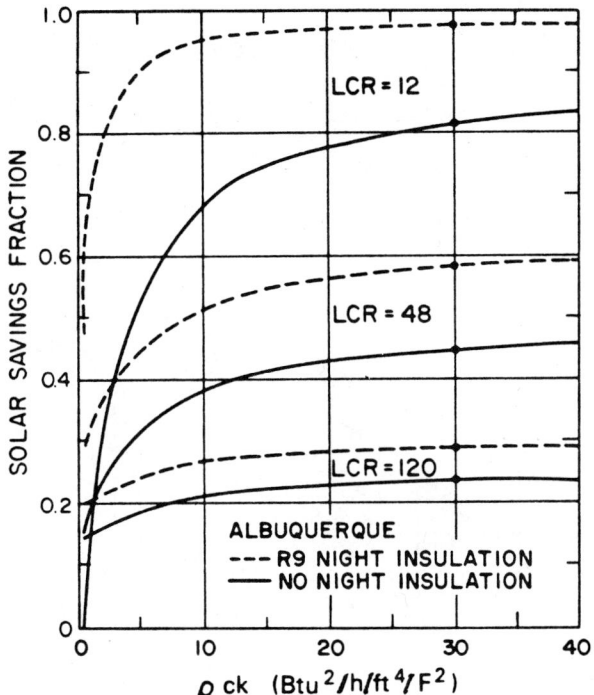

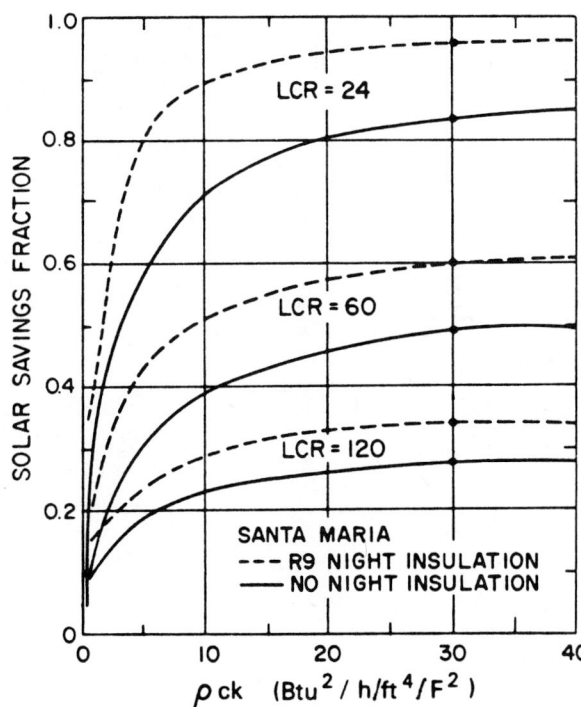

22.1.a.3 Mass Absorptance

The graphs below plot the effect on solar savings fraction of changes in mass absorptance—for the double-glazed configuration, with and without R9 nighttime insulation, at different *LCR*s, and at mass thickness of 6 inches and a mass-area-to-glazing-area ratio of 3—for each of three reference cities. Dots represent reference conditions.

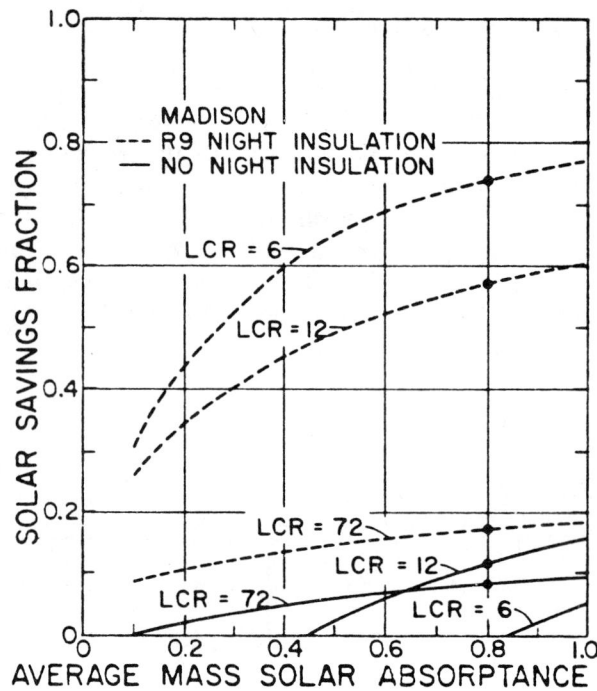

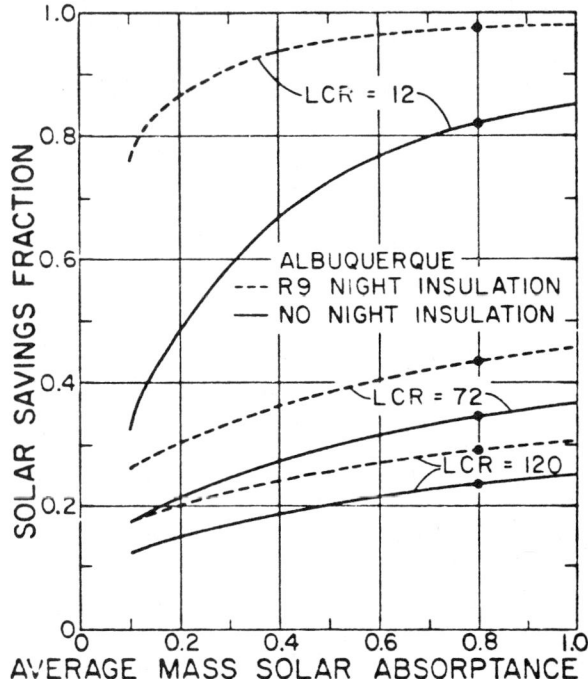

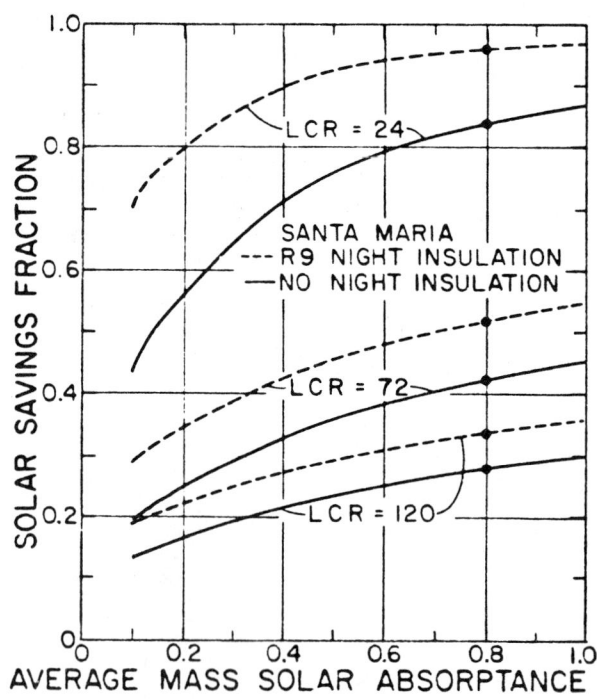

22.1.a.4 Lightweight Absorption Factor

The graphs below plot the effect on solar savings fraction of changes in lightweight absorption fraction—for the double-glazed configuration, with and without R9 nighttime insulation, at different LCRs, and at mass thickness of 6 inches and a mass-area-to-glazing-area ratio of 3—for each of three reference cities. Dots represent reference conditions.

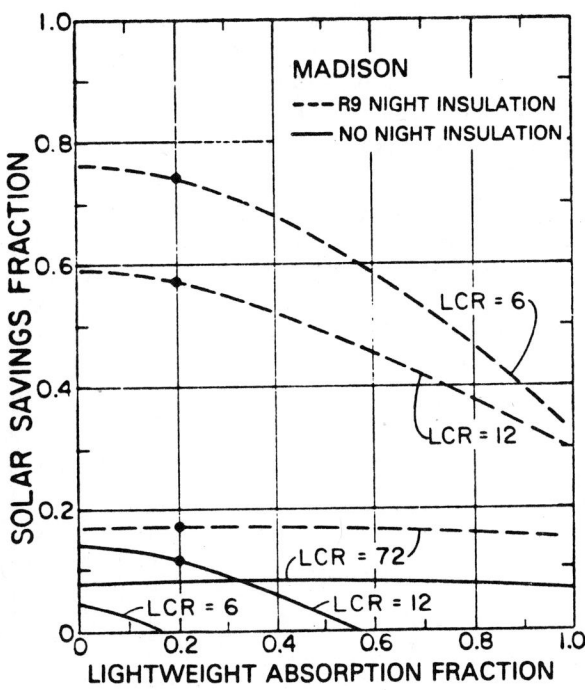

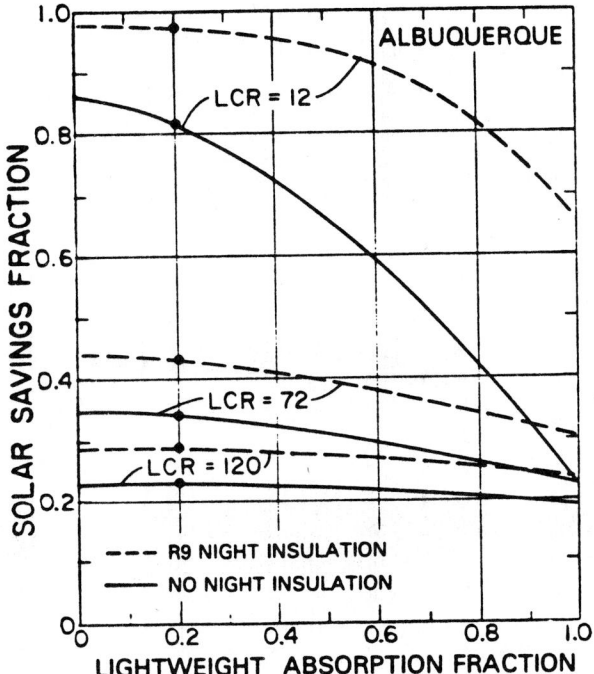

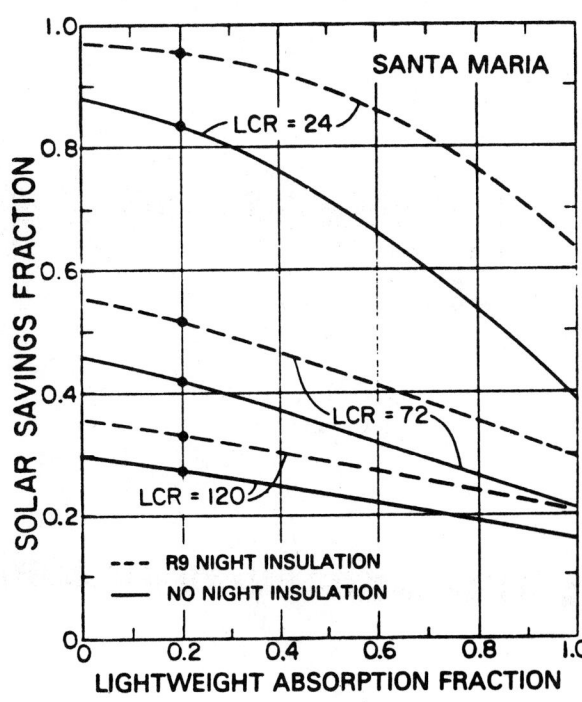

Sensitivity Data / 641

22.1.a.5 Mass Distribution

The chart below illustrates the effect on solar savings fraction of changes in the distribution of a building's thermal storage mass—for the double-glazed configuration, at $LCR = 12$, with no nighttime insulation, and at mass thickness of 6 inches and a mass-area-to-glazing-area ratio of 3—for Albuquerque. The hatching on each bar corresponds to an assumption of nondiffuse transmission by the glazing system; the solid portion above the hatching corresponds to the inclusion of diffuse transmission as assumed for the other sensitivity data and for the SLR correlations. The dashed horizontal line identifies the level of performance predicted by the two-dimensional model used for the other sensitivity data and for the SLR correlations.

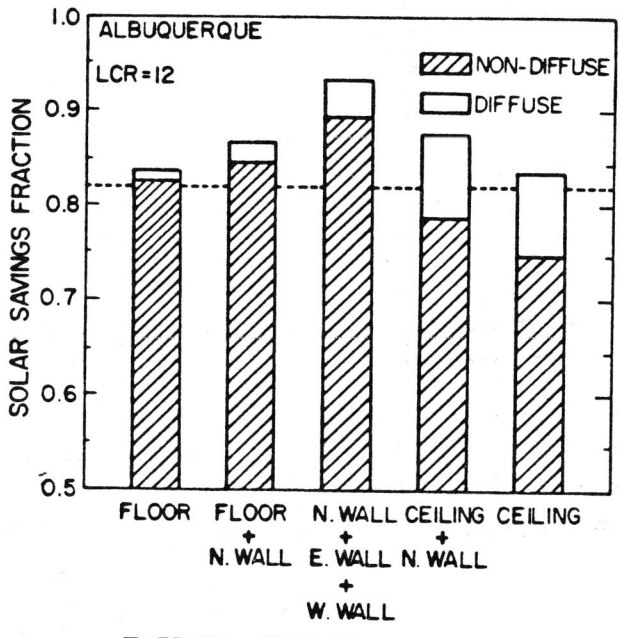

22.1.a.6 R-Value of the Mass Covering

The graphs below plot the effect on solar savings fraction of changes in R-values of the mass covering—for the double-glazed configuration, with and without R9 nighttime insulation, at different *LCR*s, and at mass thickness of 6 inches and a mass-area-to-glazing-area ratio of 3—for each of three reference cities. Two values of solar absorptance (*ALFC*) for the covering are also provided. Dots represent reference conditions.

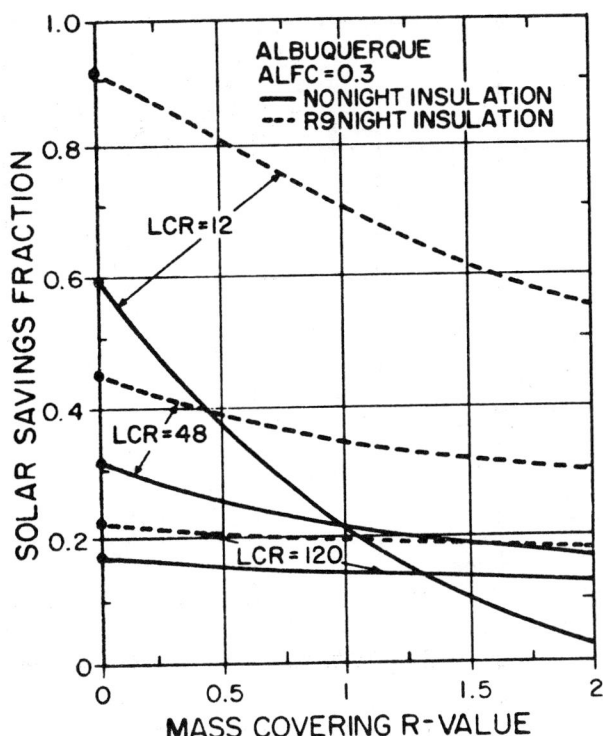

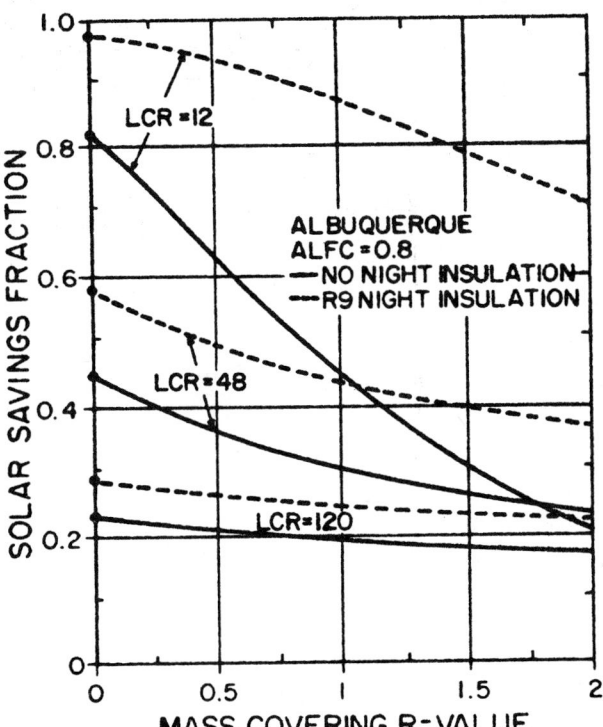

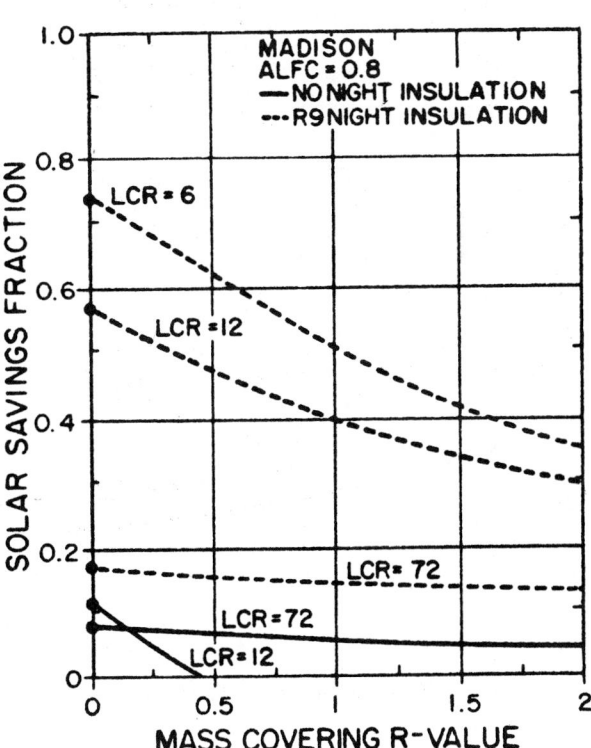

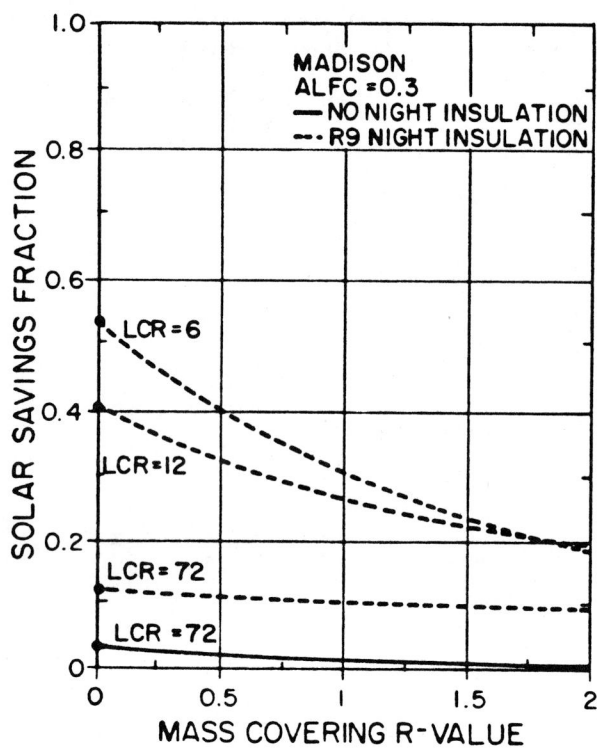
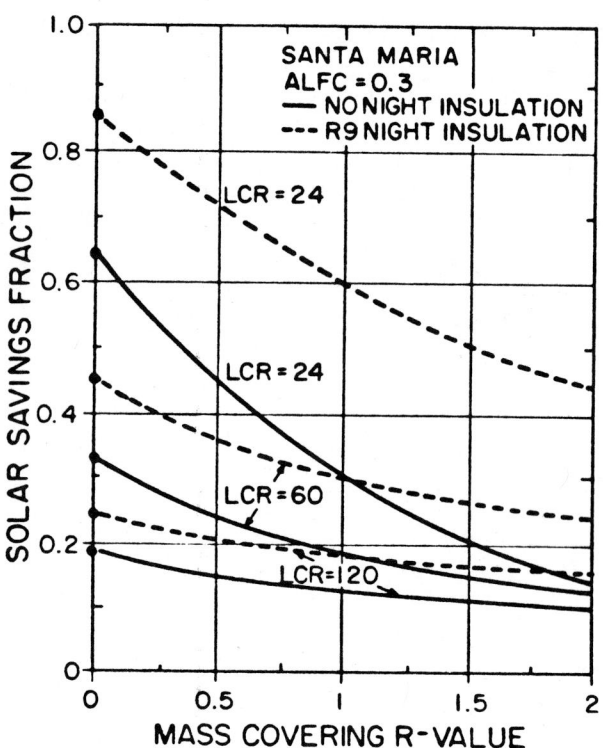
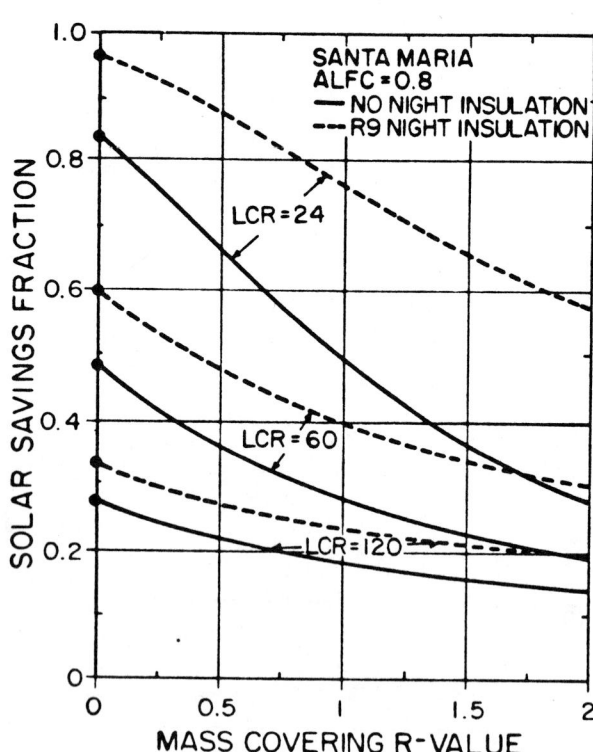

644 / APPENDIX 22

22.1.a.7 Number of Glazings

The graphs below plot the effect on solar savings fraction of changes in number of glazings—for a building with and without R9 nighttime insulation, at different *LCR*s, and at mass thickness of 6 inches and a mass-area-to-glazing-area ratio of 3—for each of six reference cities. Dots represent reference conditions.

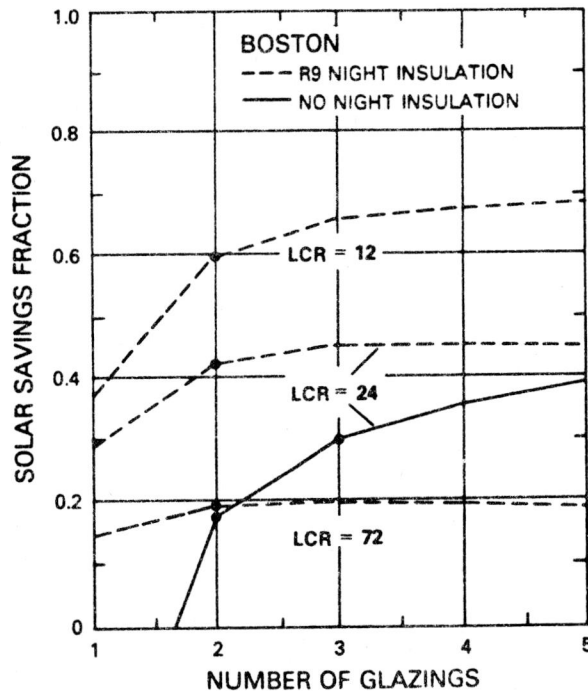

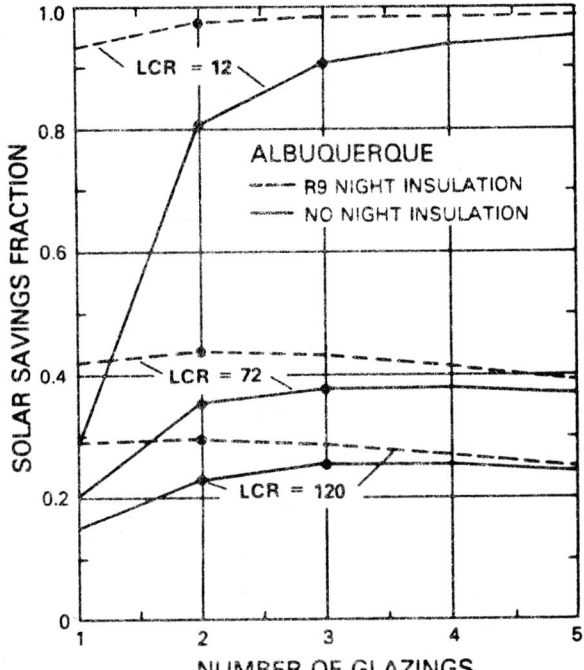

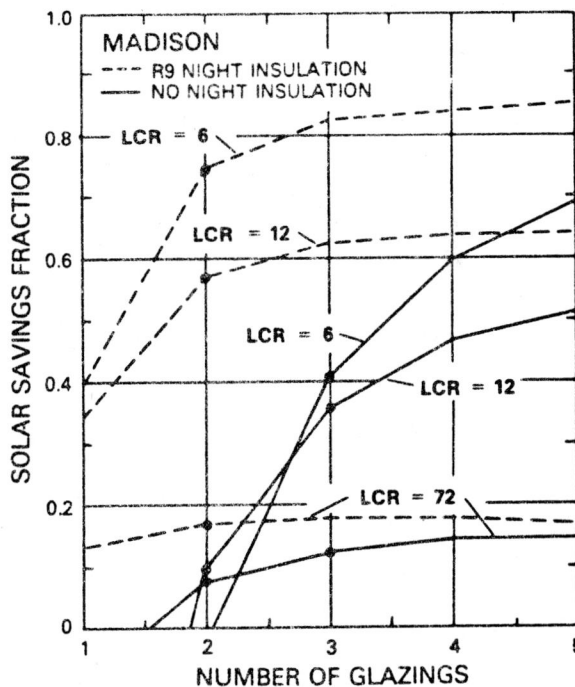

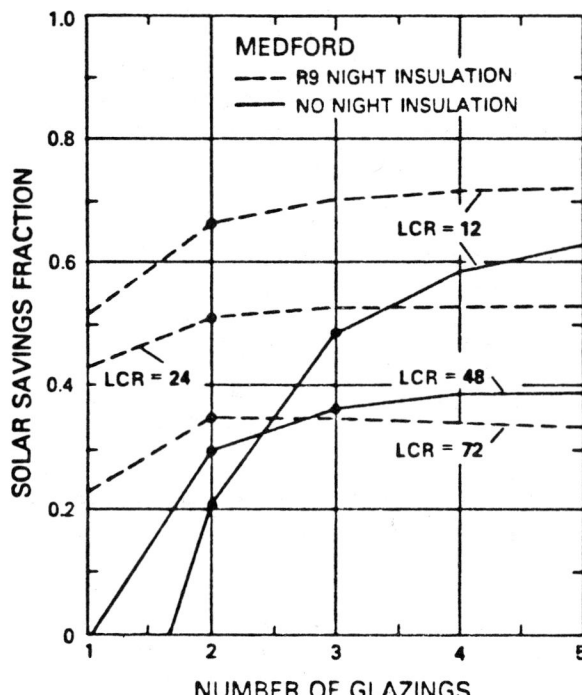
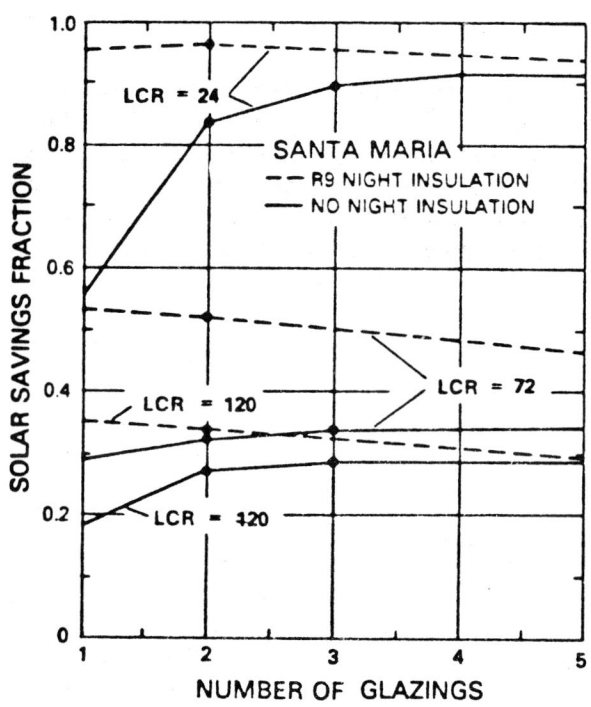
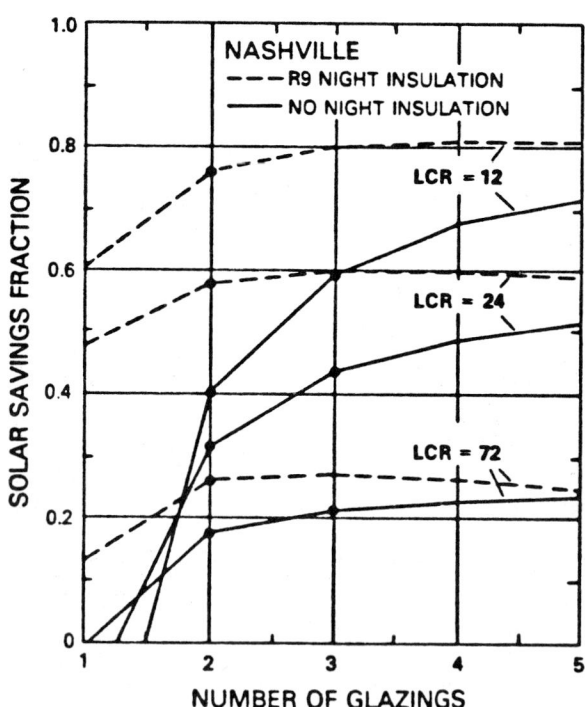

646 / APPENDIX 22

22.1.a.8 Temperature Bounds

The graphs below plot the effect on solar savings fraction of changes in thermostat setpoint of the auxiliary heating system—for the double-glazed configuration, with and without R9 nighttime insulation, at different *LCR*s, and at mass thickness of 6 inches and a mass-area-to-glazing-area ratio of 3—for each of three reference cities. The hatching indicates the sensitivity band for each value of *LCR:* the lower boundary of a sensitivity band delineates performance when the maximum room temperature is limited to 75°F; the upper boundary of the band delineates performance when the maximum room temperature is limited to 85°F. Dots represent reference conditions.

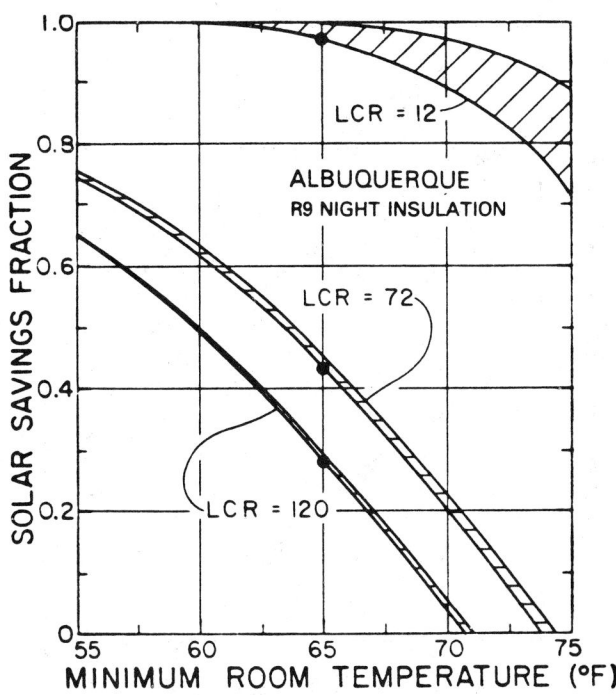

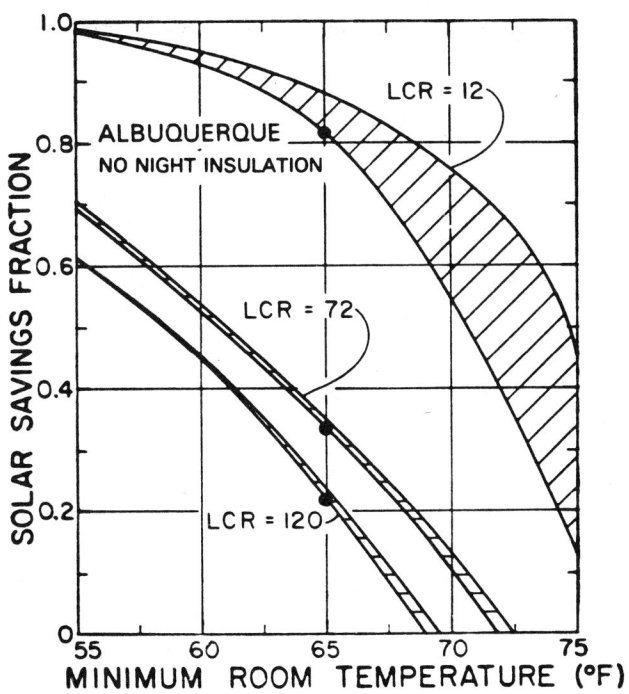

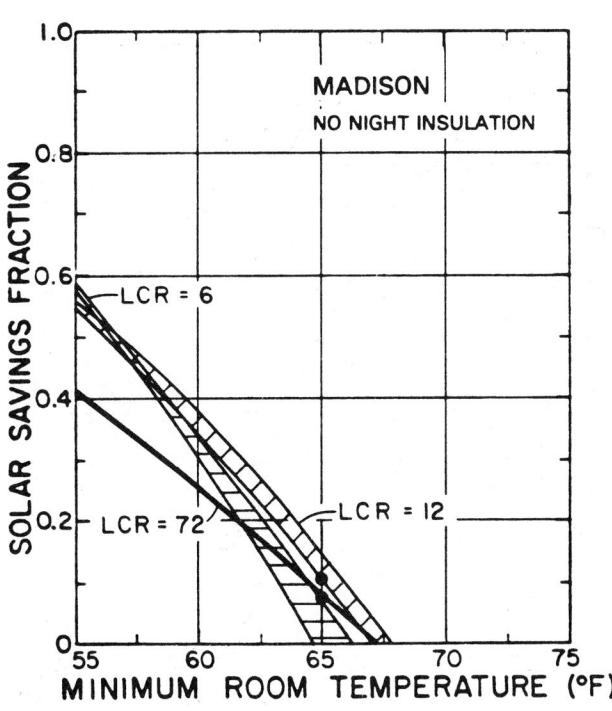

Sensitivity Data / 647

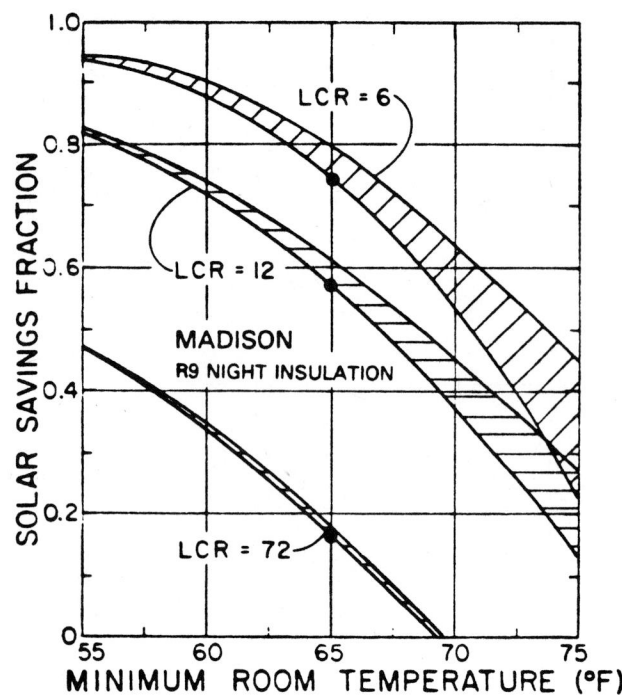
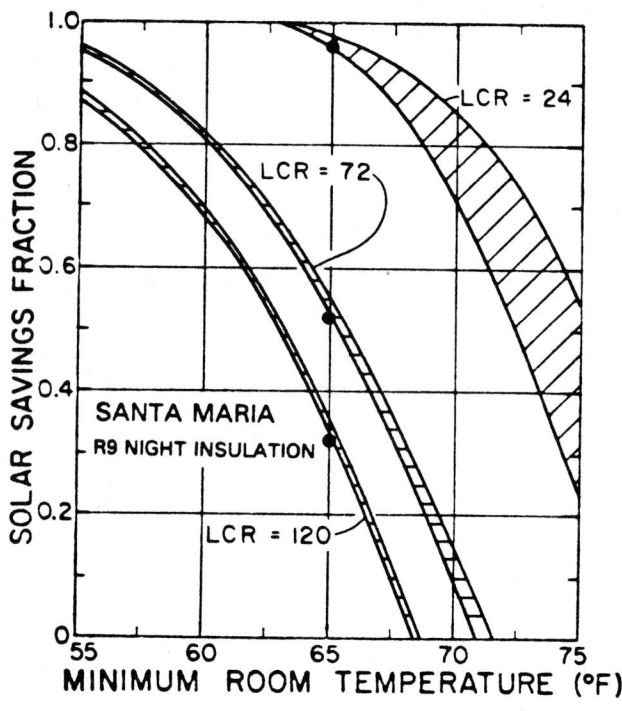
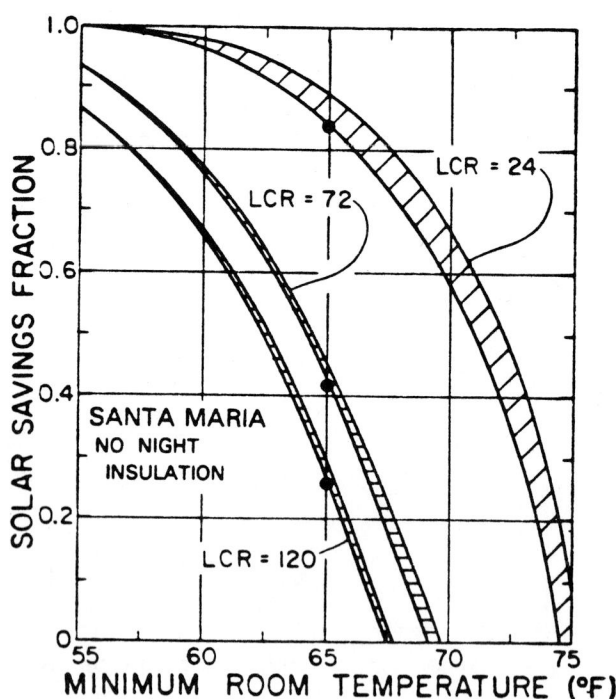

648 / APPENDIX 22

22.1.a.9 Nighttime Setback

The graphs below plot the effect on solar savings fraction of changes in the nighttime (11:00 P.M. to 6:00 A.M.) thermostat setpoint of the auxiliary heating system—for the double-glazed configuration, with and without R9 nighttime insulation, at different *LCR*s, and at mass thickness of 6 inches and a mass-area-to-glazing-area ratio of 3—for each of three reference cities. Dots represent reference conditions.

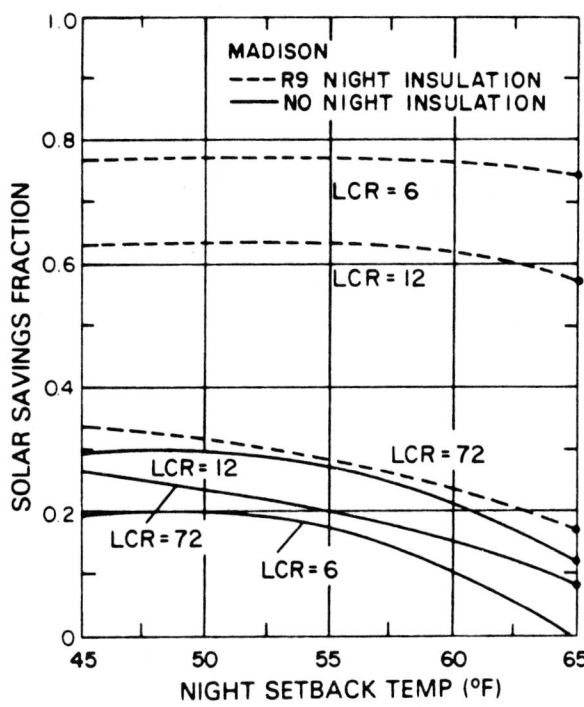

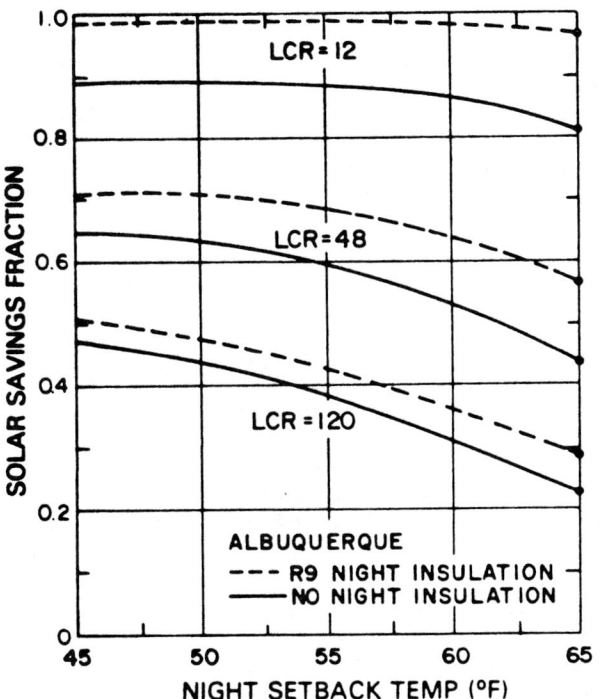

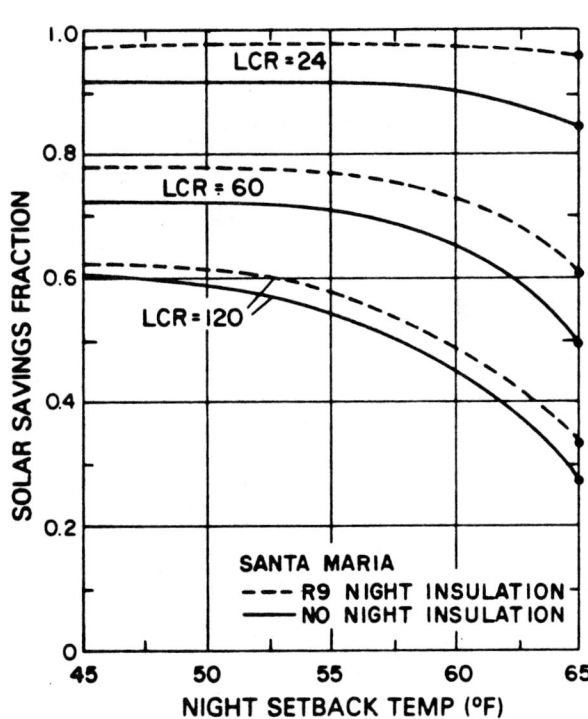

22.1.b OVERHEATING

The graphs in this section plot the effect on maximum room temperature of changes in each of the following system parameters: mass-area-to-glazing-area ratio and lightweight absorption fraction.

22.1.b.1 Mass-Area-to-Glazing-Area Ratio

The graphs below plot the effect on maximum room temperature of changes in mass-area-to-glazing-area ratio—for the double-glazed configuration, at different *LCR*s, with no nighttime insulation, and at various mass thicknesses—for each of three reference cities. Dots represent reference conditions.

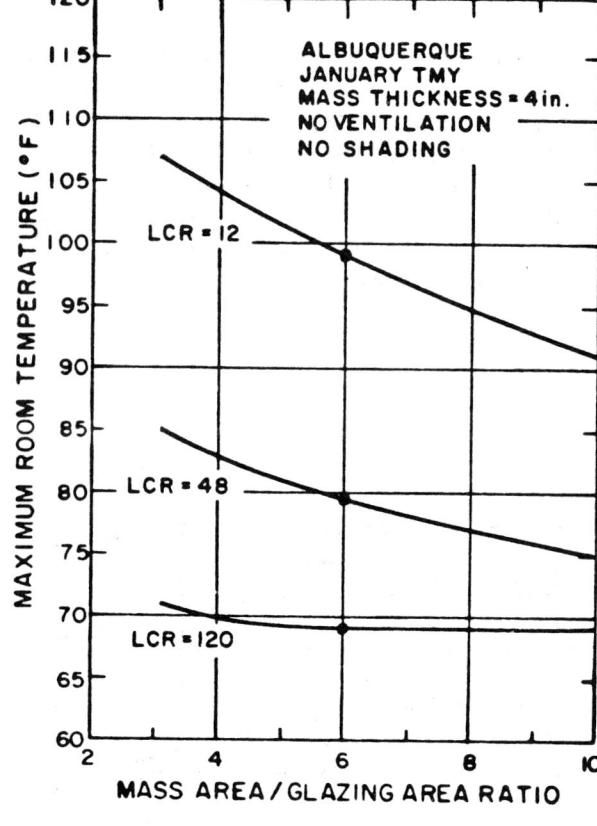

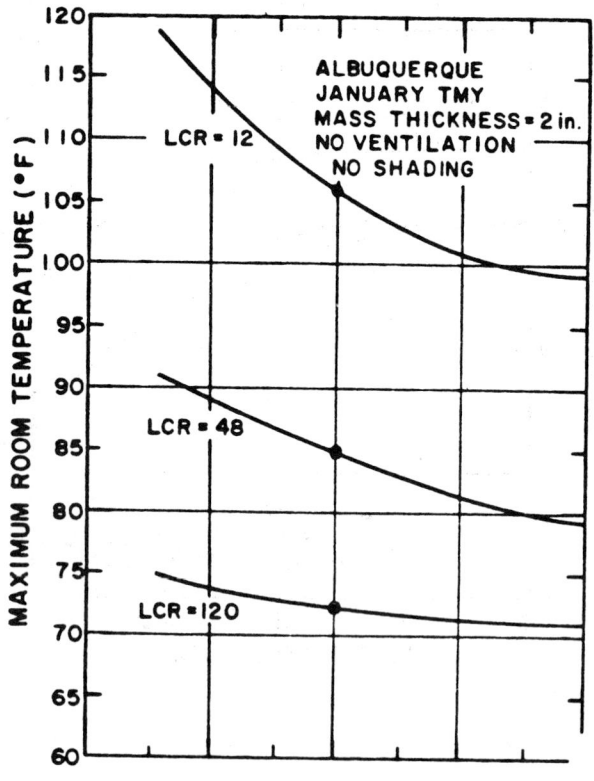

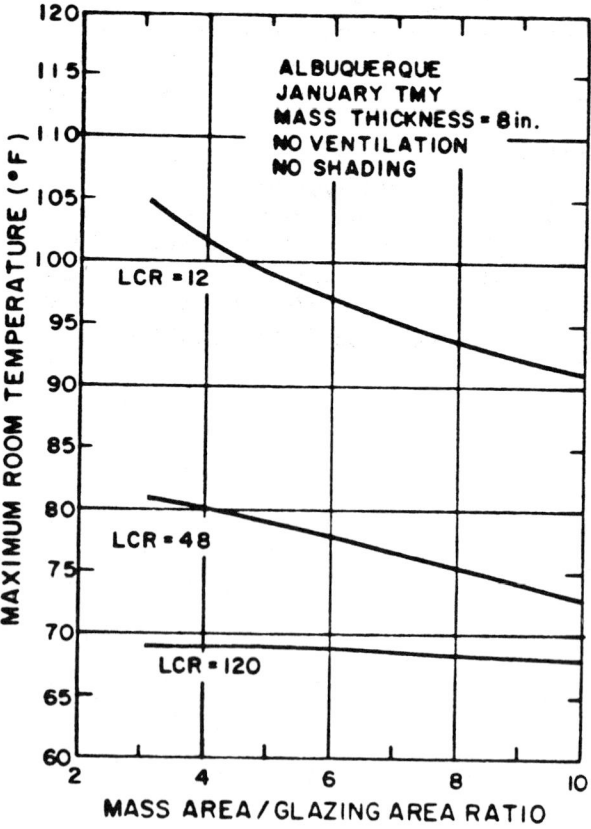

650 / APPENDIX 22

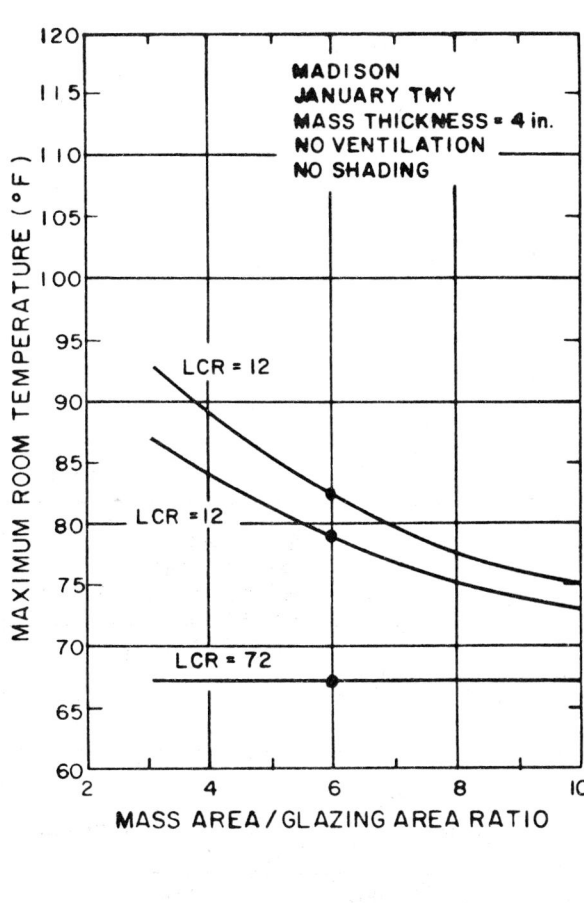

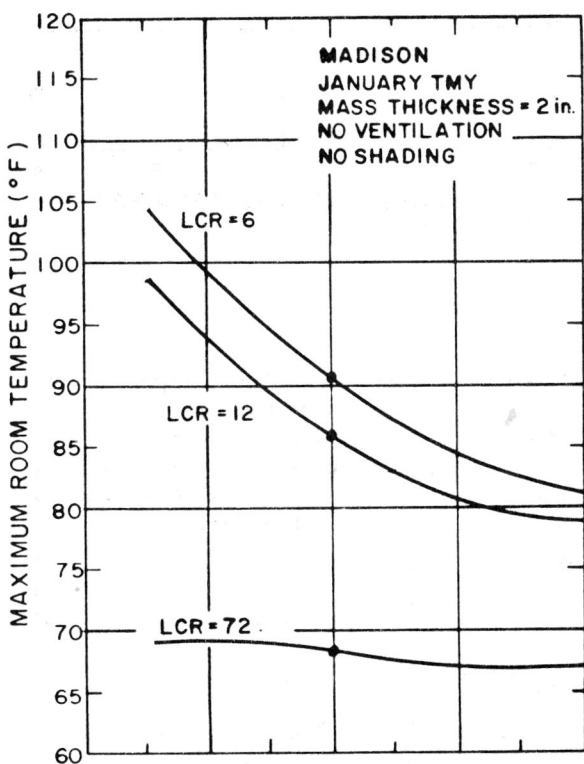

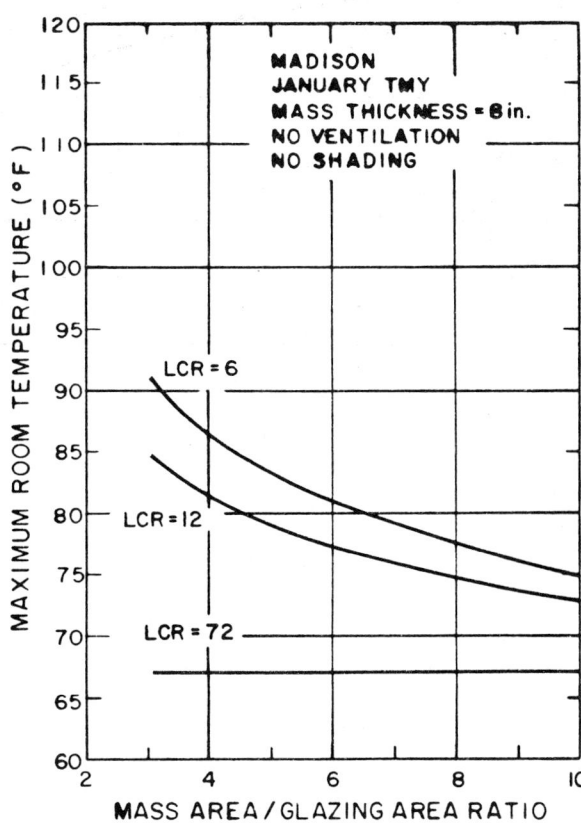

Sensitivity Data / 651

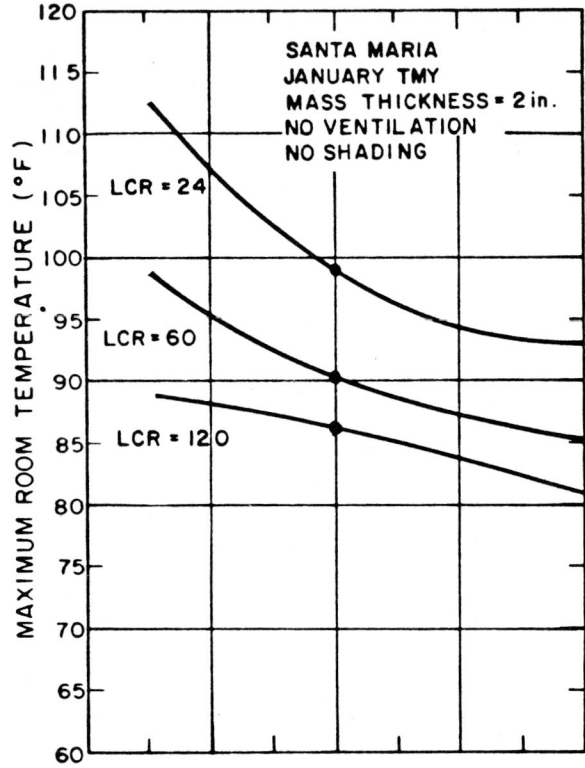
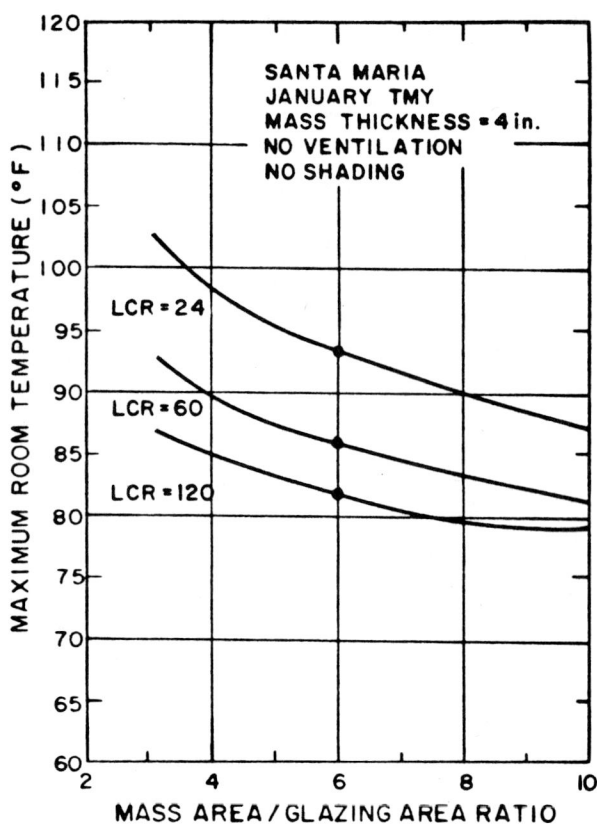
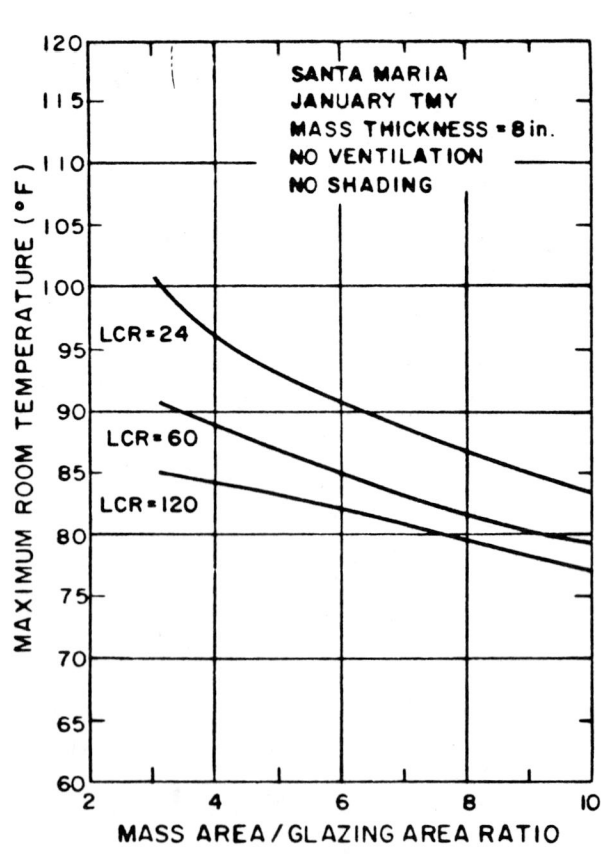

652 / APPENDIX 22

22.1.b.2 Lightweight Absorption Fraction

The graphs below plot the effect on maximum room temperature of changes in the lightweight absorption fraction—for the double-glazed configuration, at different *LCR*s, with no nighttime insulation, and at mass thickness of 6 inches and a mass-area-to-glazing-area ratio of 3—for each of three reference cities. Dots represent reference conditions.

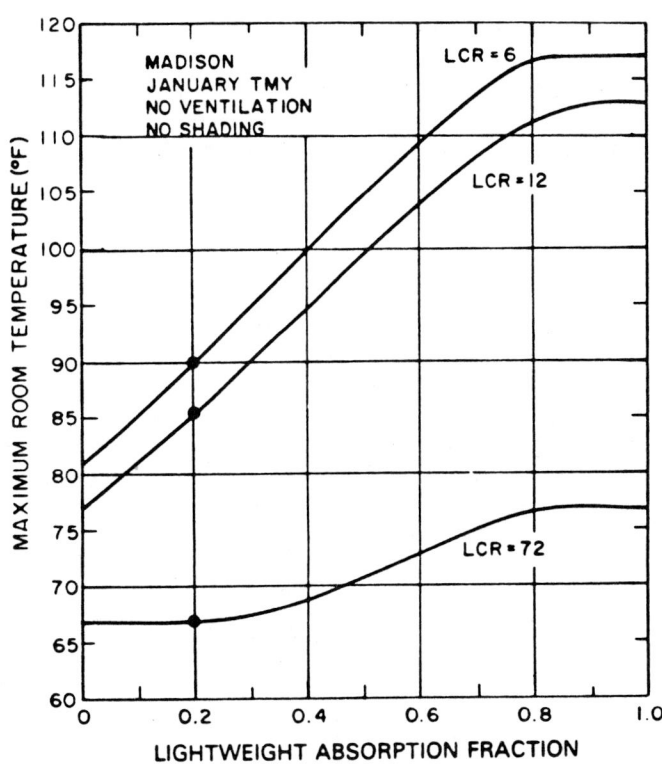

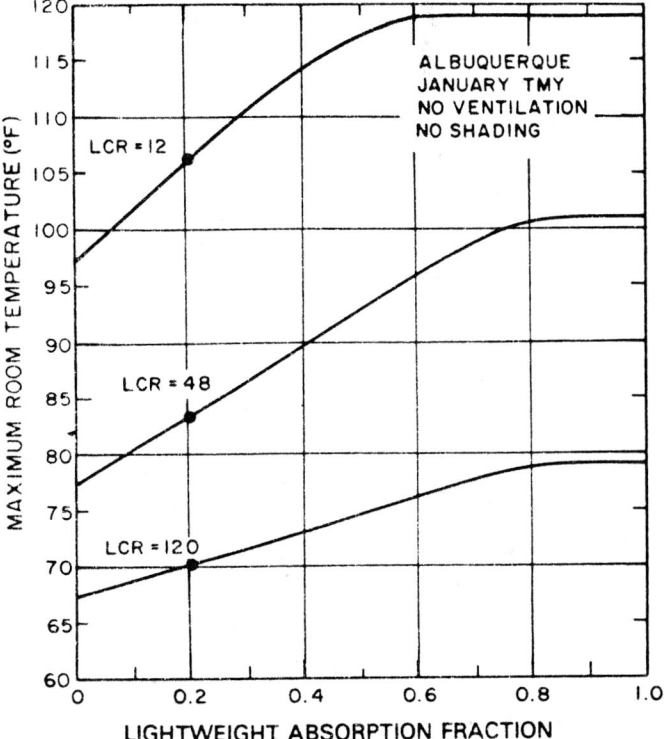

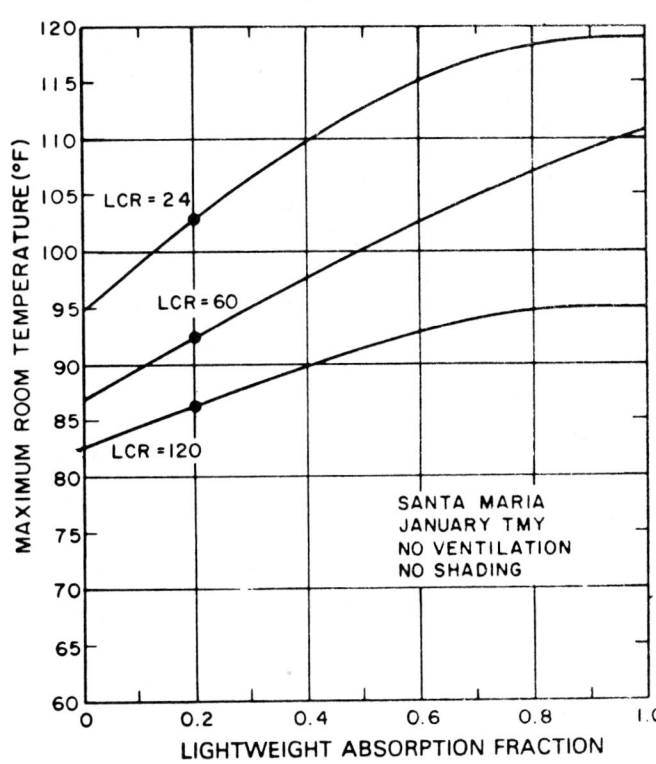

Sensitivity Data / 653

22.2 SUN-TEMPERED DIRECT-GAIN SYSTEMS

22.2.a HEATING PERFORMANCE

The graphs in this section plot the effect on solar savings fraction of changes in each of the following system parameters: mass-area-to-glazing-area ratio, number of glazings, and mass absorptance.

22.2.a.1 Mass-Area-to-Glazing-Area Ratio

The graphs below plot the effect on solar savings fraction of changes in mass-area-to-glazing-area ratio—for the sun-tempered reference design, with and without R9 nighttime insulation, and at different *LCRs*—for each of six reference cities. Dots represent reference conditions.

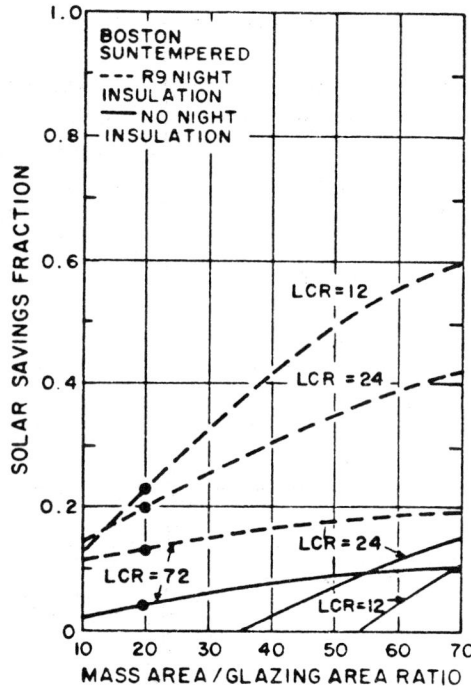

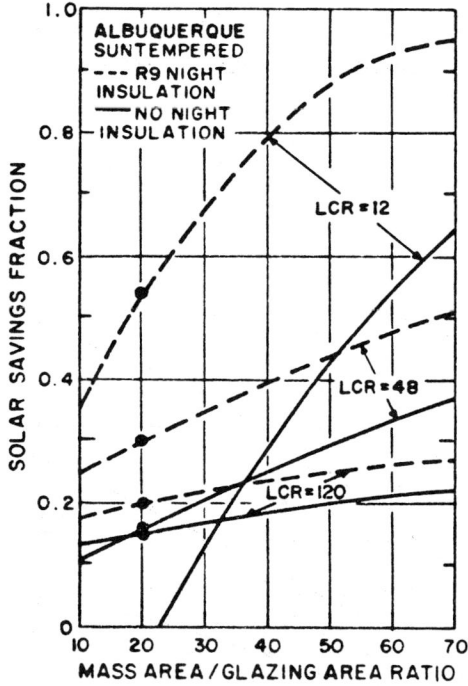

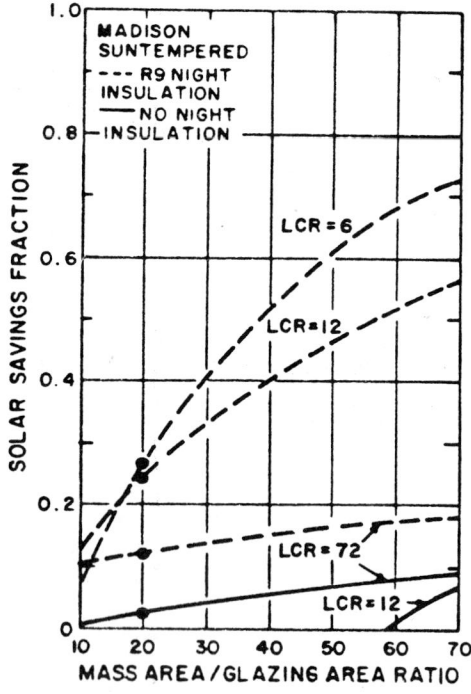

654 / APPENDIX 22

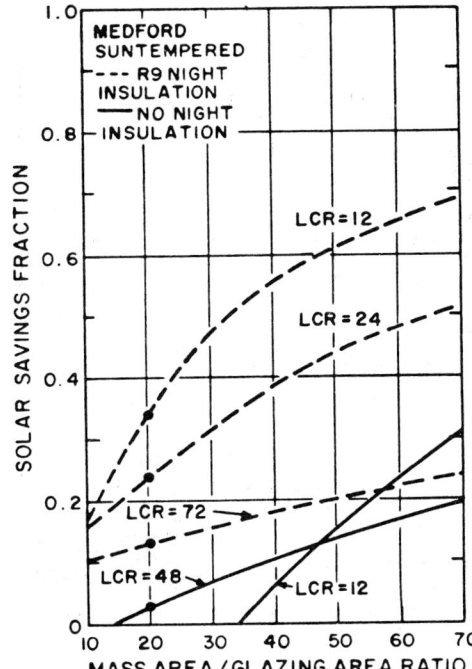
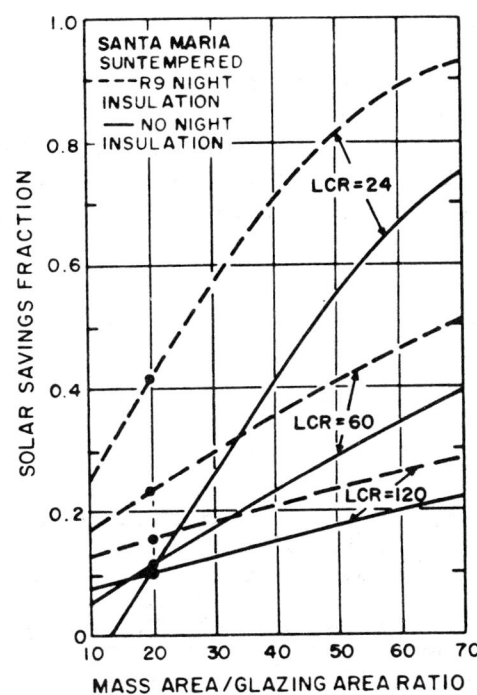
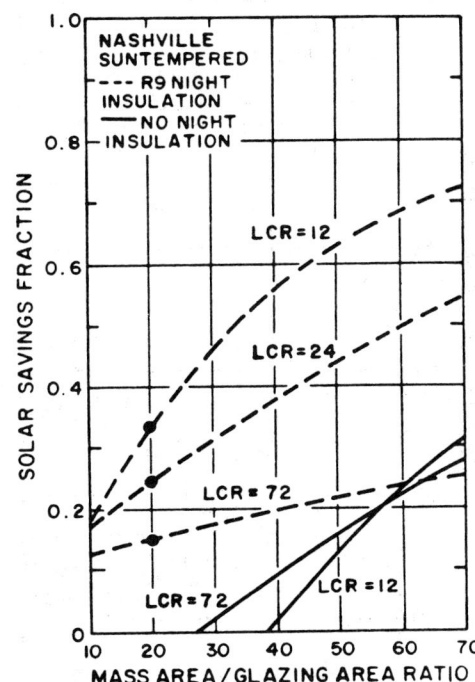

22.2.a.2 Number of Glazings

The graphs below plot the effect on solar savings fraction of changes in number of glazings—for the sun-tempered reference design, with and without R9 nighttime insulation, and at different *LCR*s—for each of six reference cities. Dots represent reference conditions.

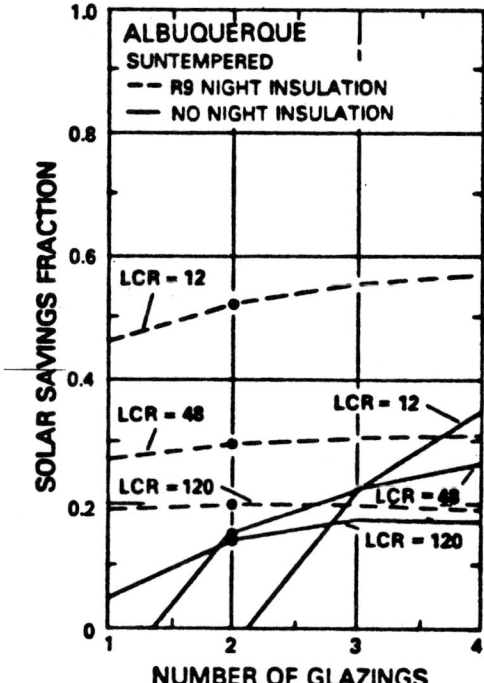

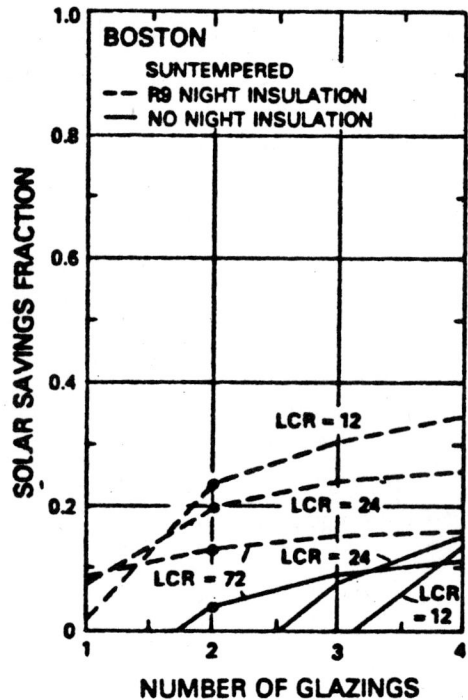

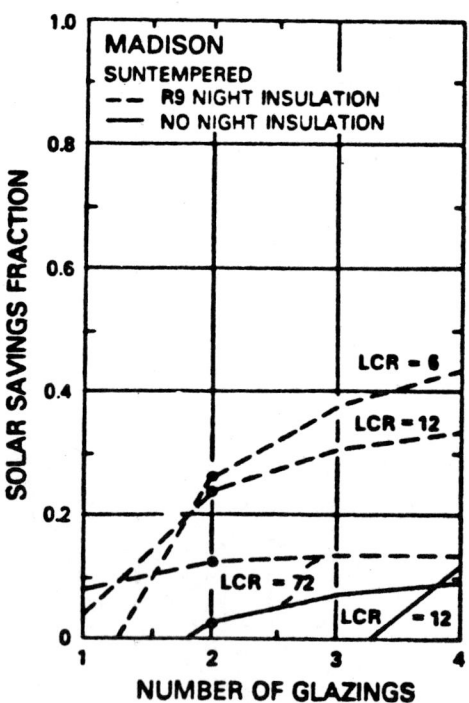

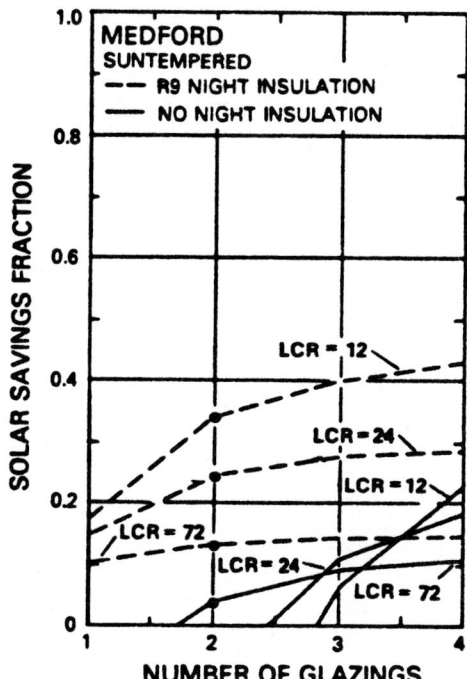
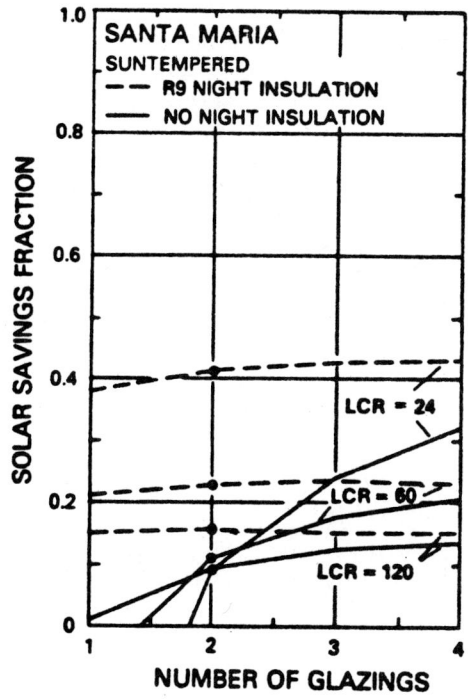
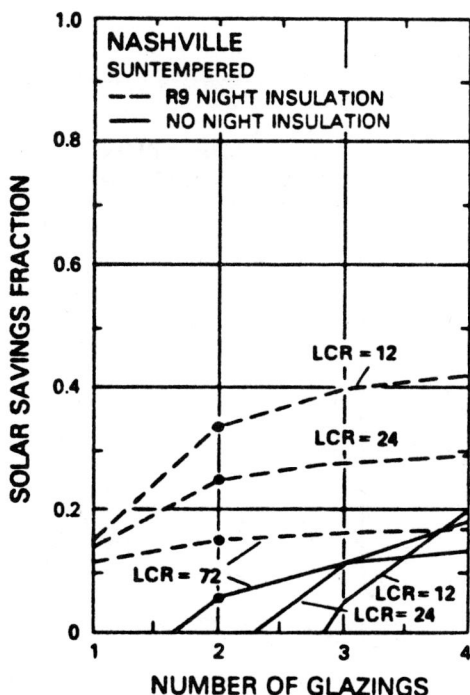

Sensitivity Data / 657

22.2.a.3 Mass Absorptance

The graphs below plot the effect on solar savings fraction of changes in solar absorptance—for the sun-tempered reference design, with and without R9 night-time insulation, and at different *LCR*s—for each of six reference cities. Dots represent reference conditions.

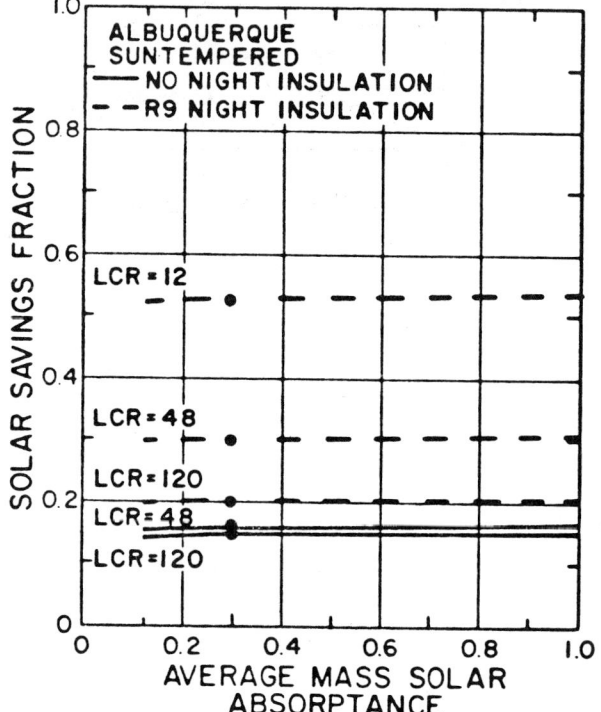

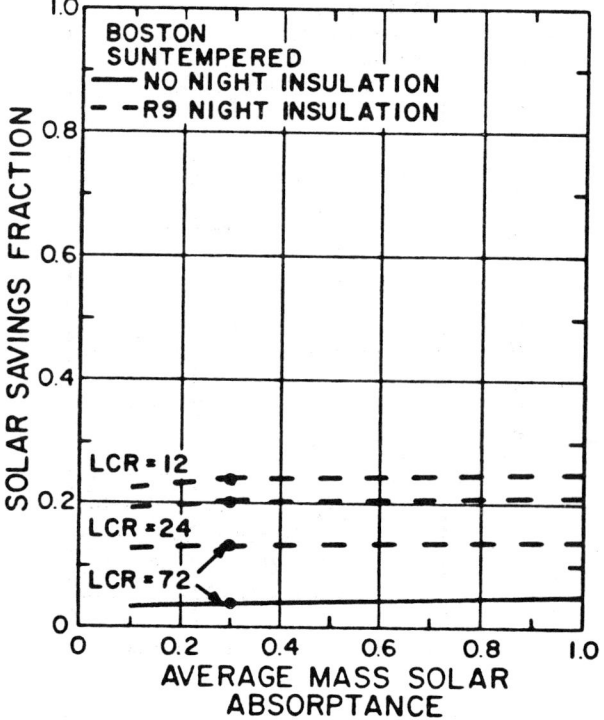

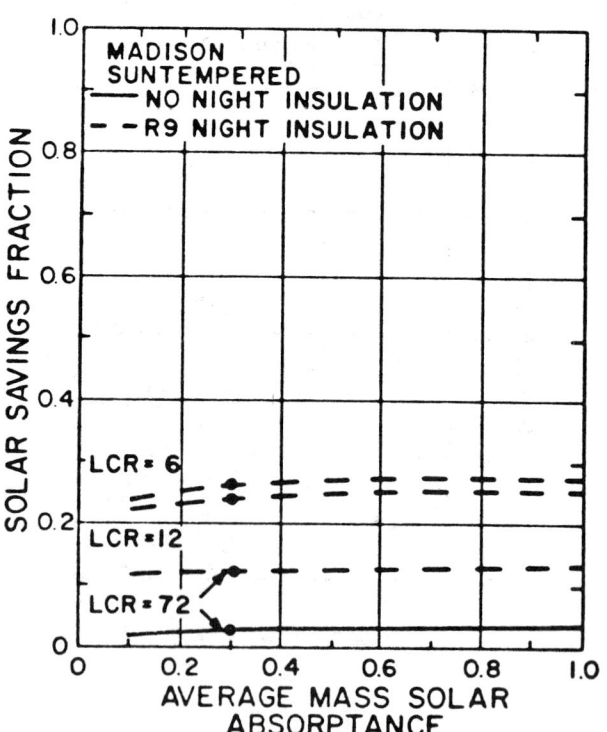

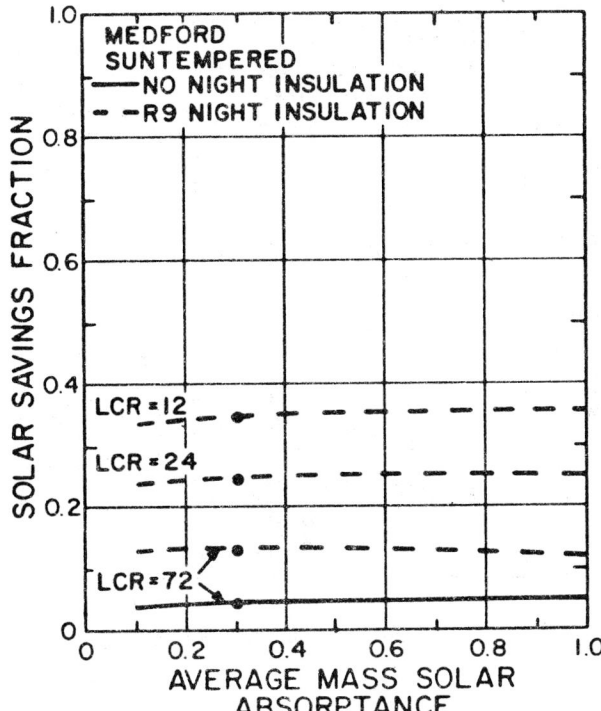
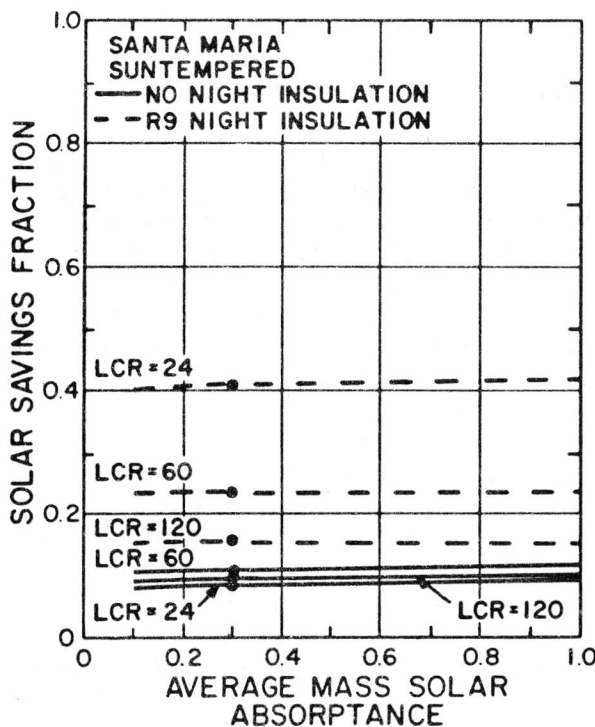
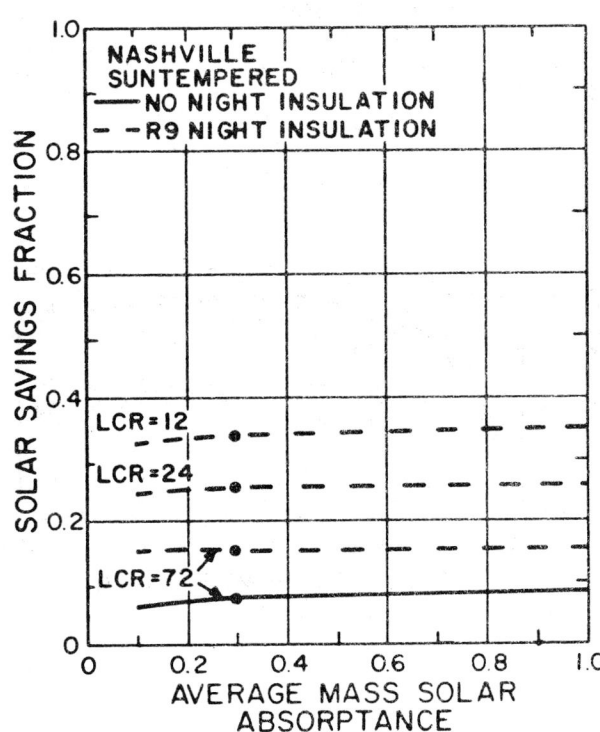

22.3 SUNSPACES

22.3.a HEATING PERFORMANCE

The graphs in this section plot the effect on solar savings fraction of changes in each of the following system parameters: storage-volume-to-projected-area ratio, wall orientation, glazing tilt, vertical glazing height, sunspace width, number of glazings, air-gap width between glazings, glazing thickness, mass absorptance, lightweight absorptance fraction, vent area fraction, sunspace infiltration, common-wall R-value, minimum room setpoint, and minimum sunspace setpoint.

22.3.a.1 Storage-Volume-to-Projected-Area Ratio

The graphs below plot the effect on solar savings fraction of changes in storage-volume-to-projected-area ratio—for geometry A with insulated end walls, no nighttime insulation, and both masonry and lightweight common wall (A1, A3, A5, and A7)—for each of six reference cities. In the masonry wall case, the storage volume ratio is equal to the wall thickness. Curves for different values of density (ρ) are included. Dots represent reference conditions.

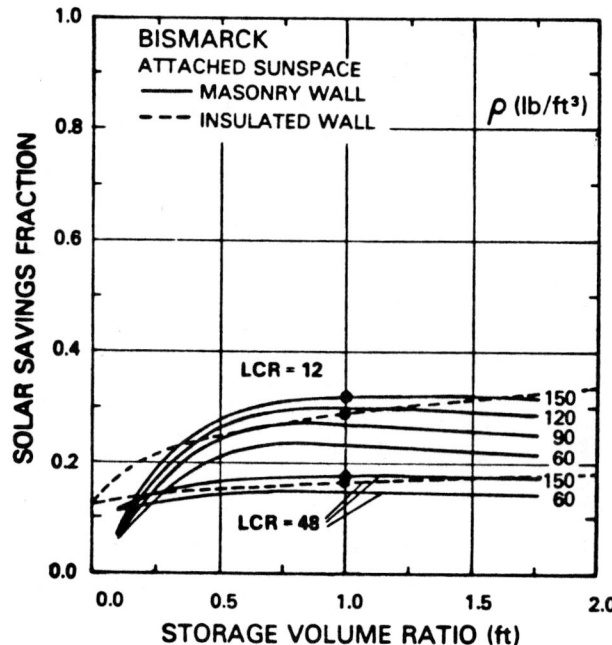

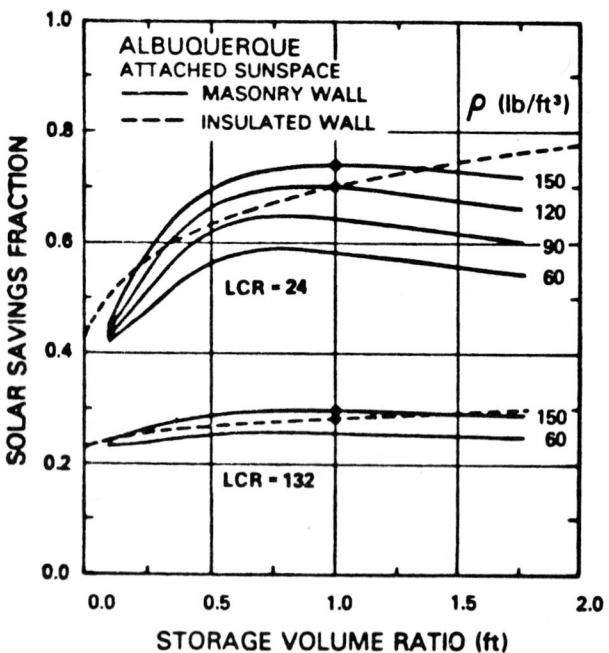

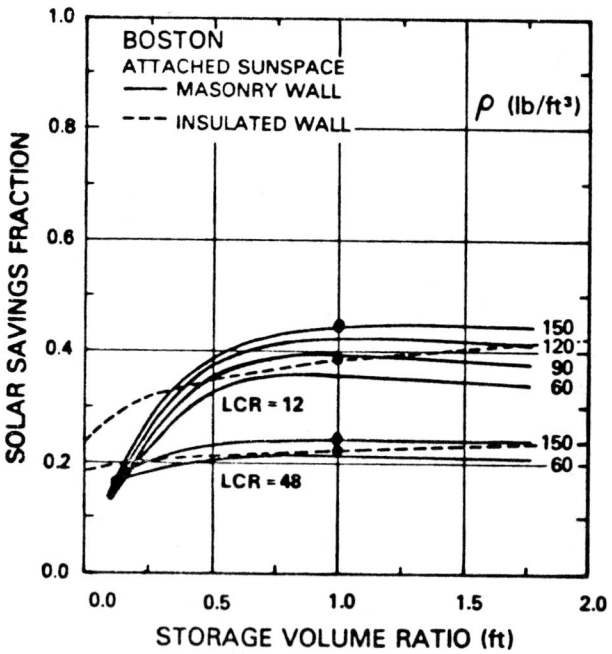

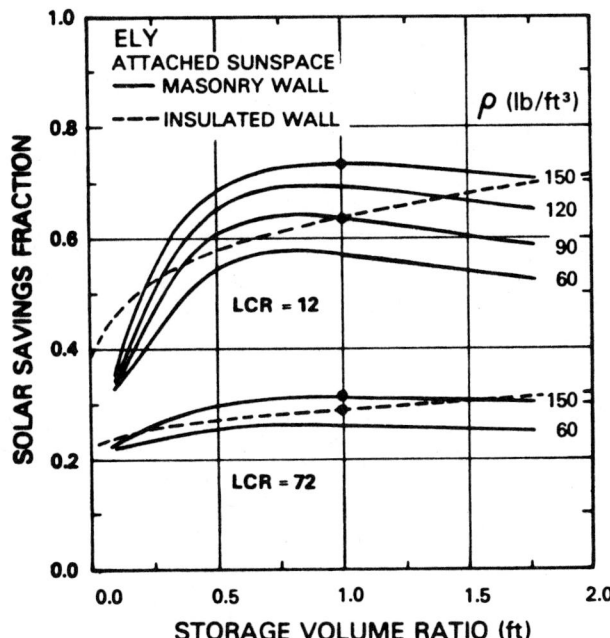
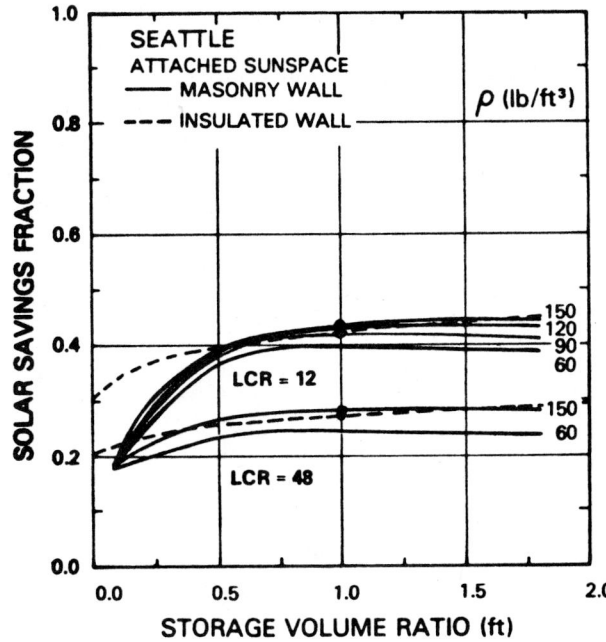
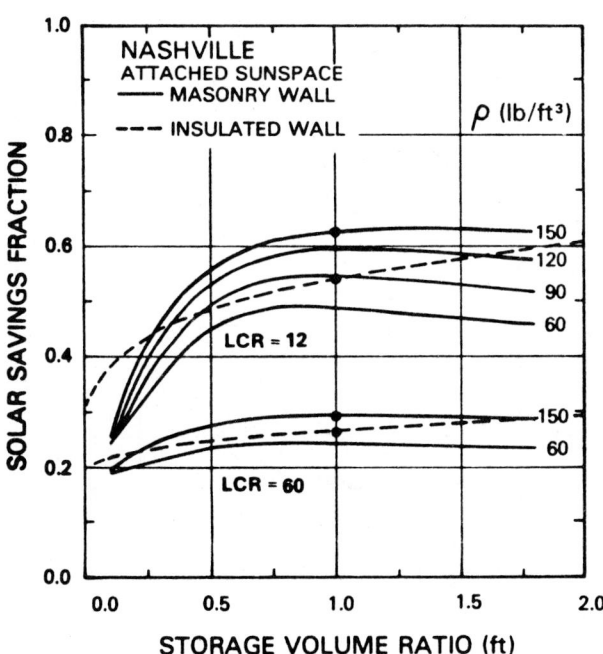

Sensitivity Data / 661

22.3.a.2 Common-Wall Thickness

The graphs below plot the effect on solar savings fraction of changes in masonry common-wall thickness—for geometry C with no nighttime insulation (C1)—for each of six reference cities. Curves for different values of density (ρ) are included. Dots represent reference conditions.

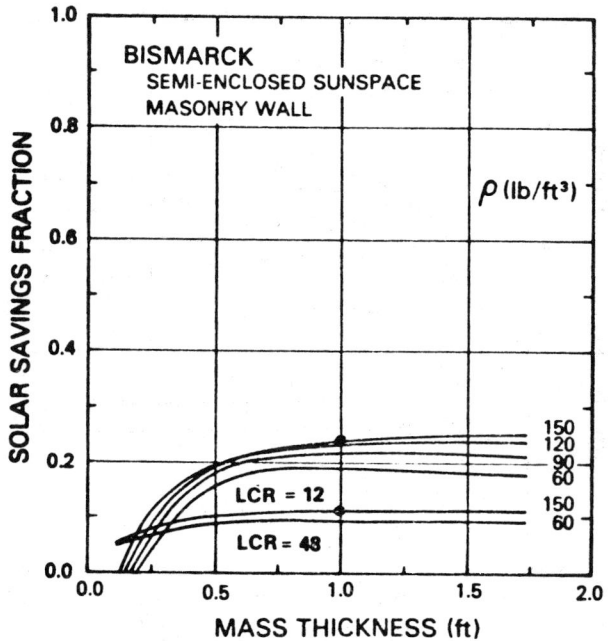

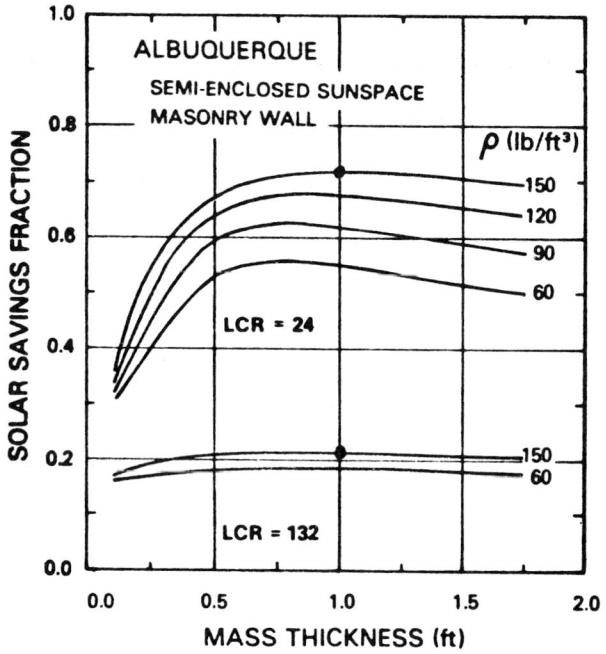

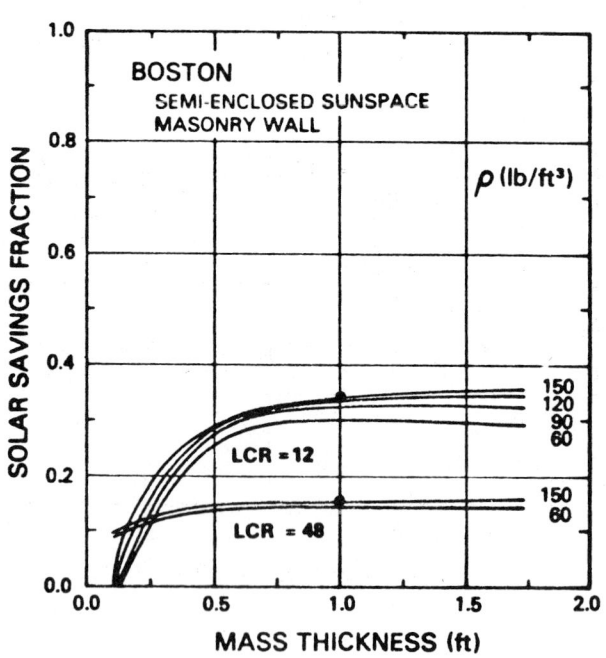

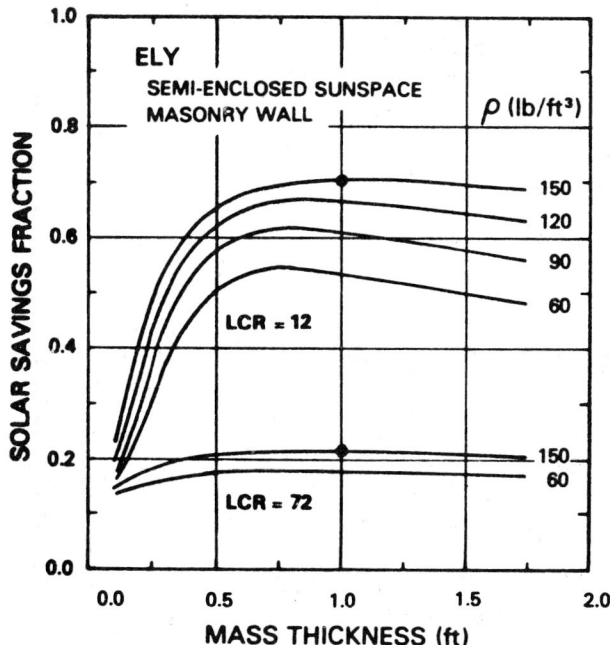
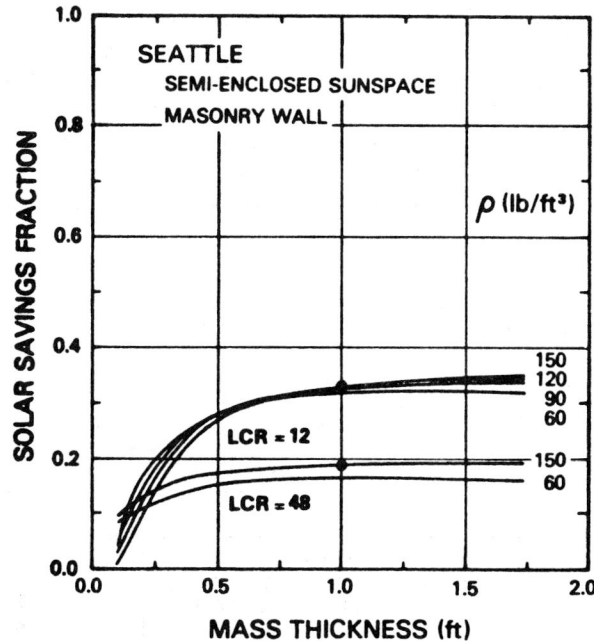
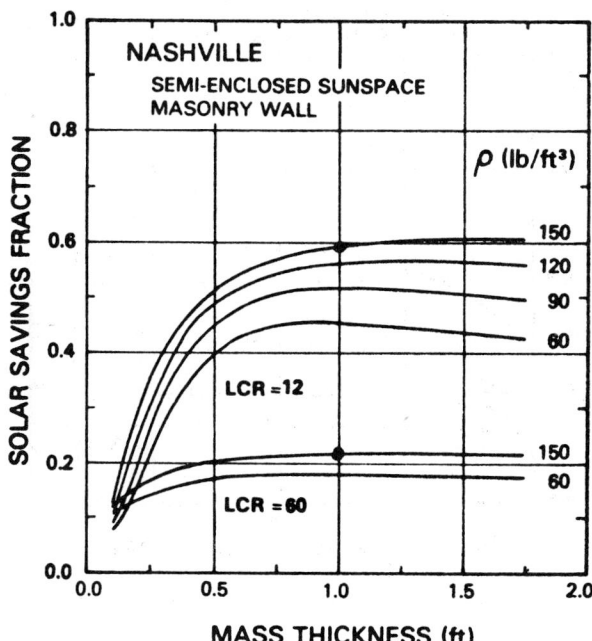

Sensitivity Data / 663

22.3.a.3 Water-Volume-to-Projected-Area Ratio

The graphs below plot the effect on solar savings fraction of changes in water-volume-to-projected-area ratio —for geometry C with water-storage containers, and with and without R9 nighttime insulation (C3 and C4)— for each of six reference cities. Dots represent reference conditions.

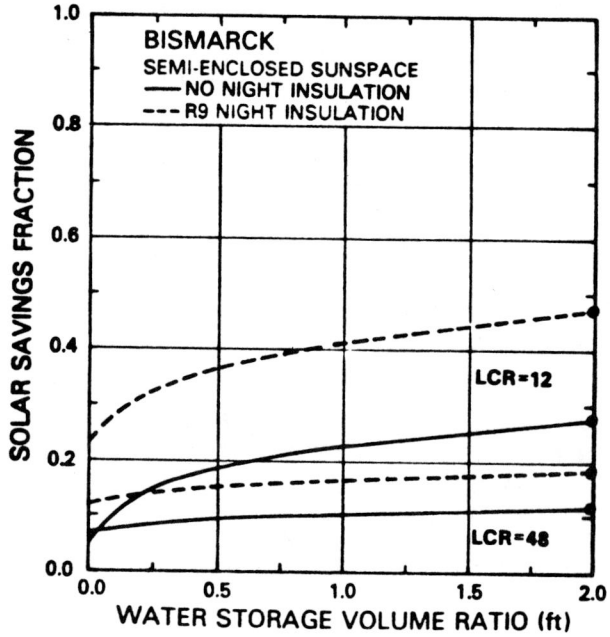

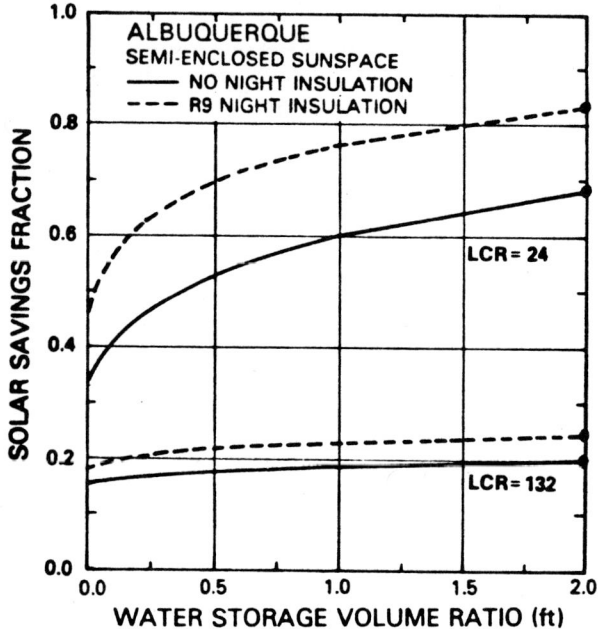

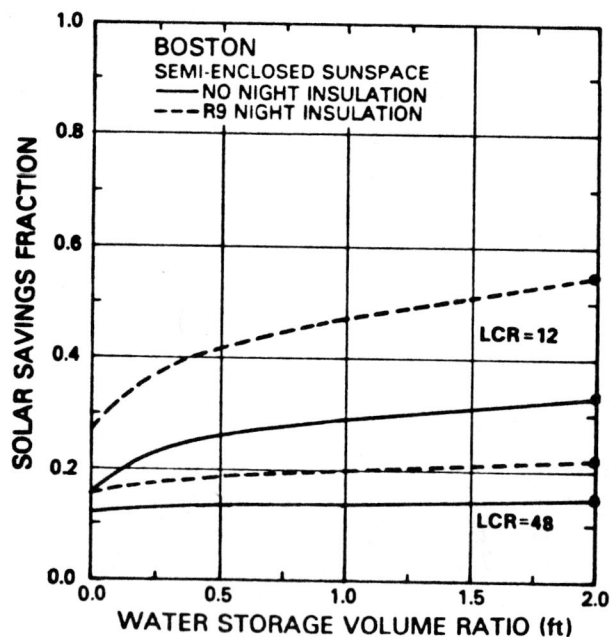

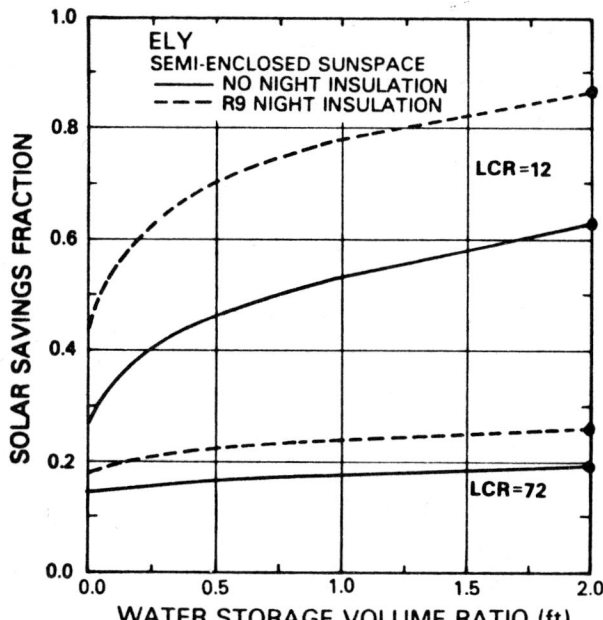
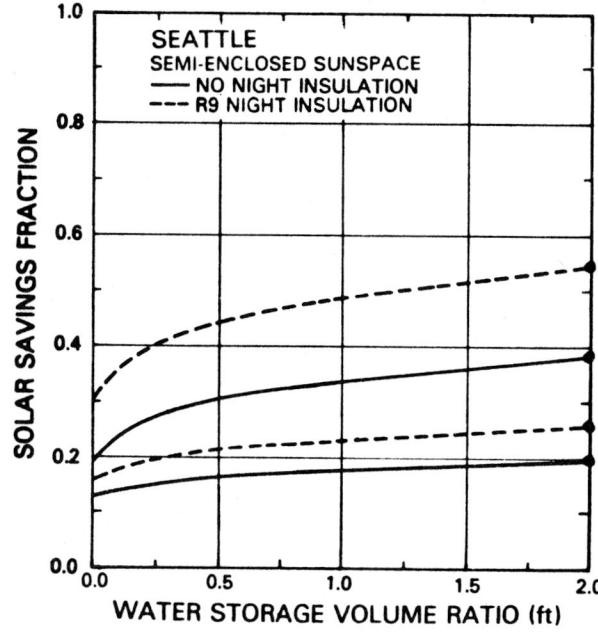
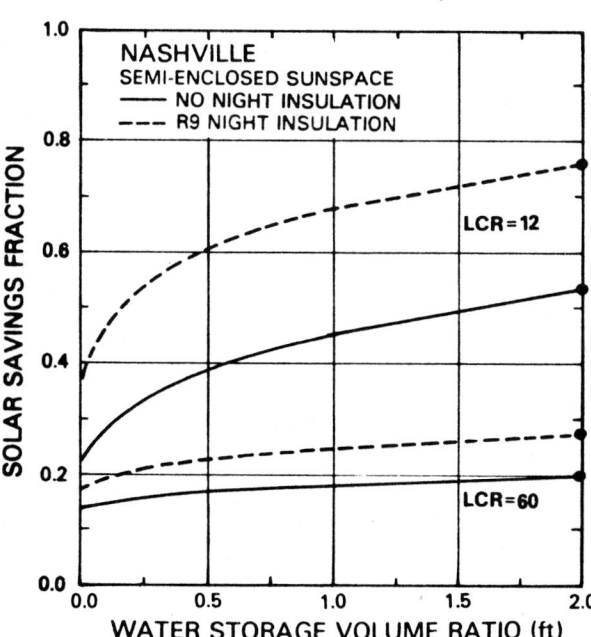

22.3.a.4 Sunspace Orientation

The graphs below plot the effect on solar savings fraction of changes in sunspace orientation with respect to true south—for geometry A with masonry common wall, no nighttime insulation, and both insulated and glazed end walls (A1 and A3)—for each of six reference cities. A wall azimuth of 0 indicates a true-south orientation; a negative wall azimuth indicates an orientation west of true south; a positive wall azimuth indicates an orientation east of true south. Dots represent reference conditions.

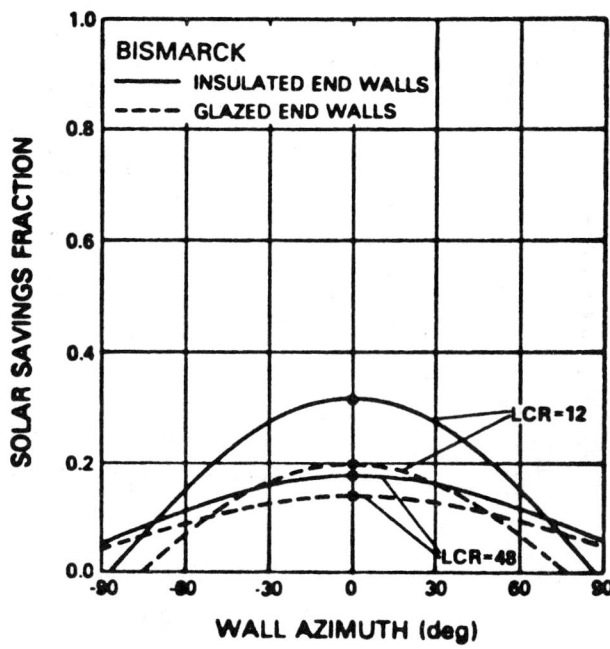

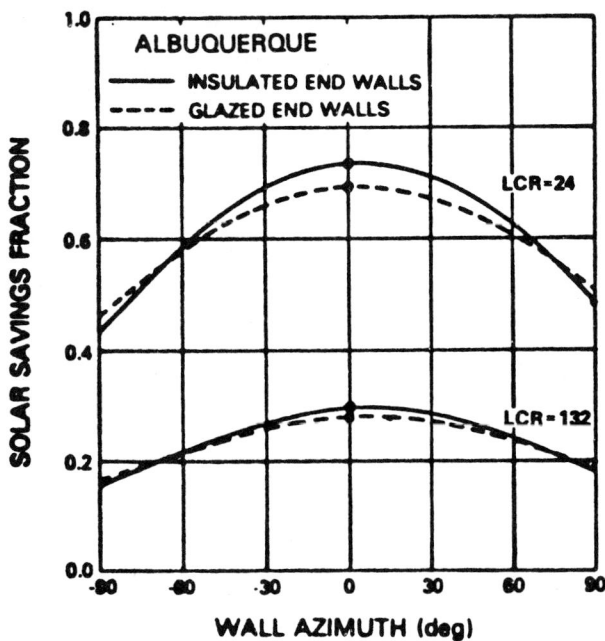

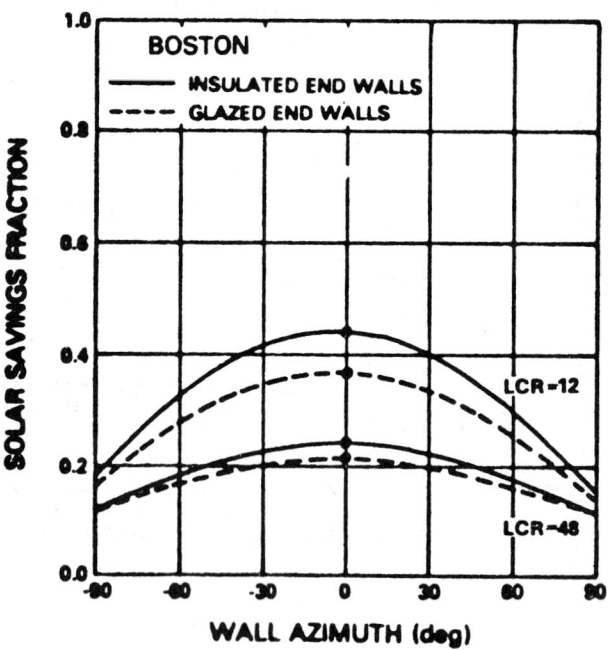

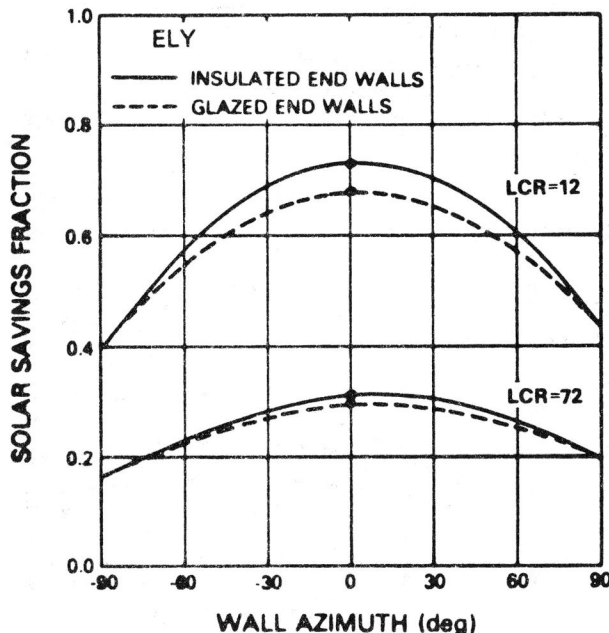
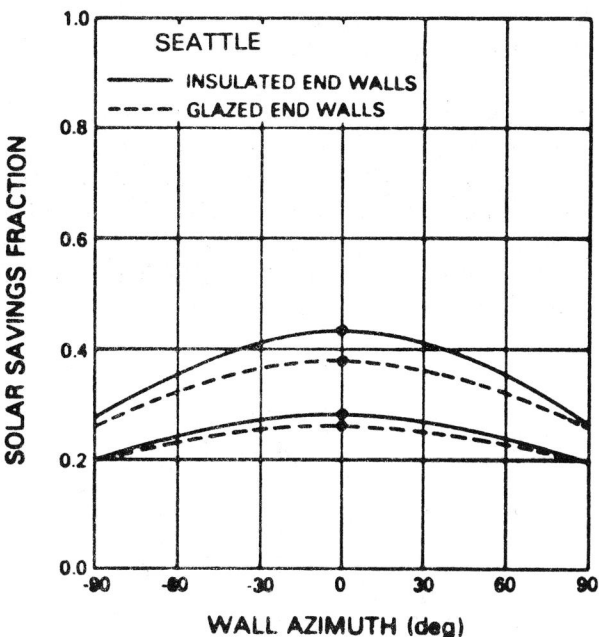
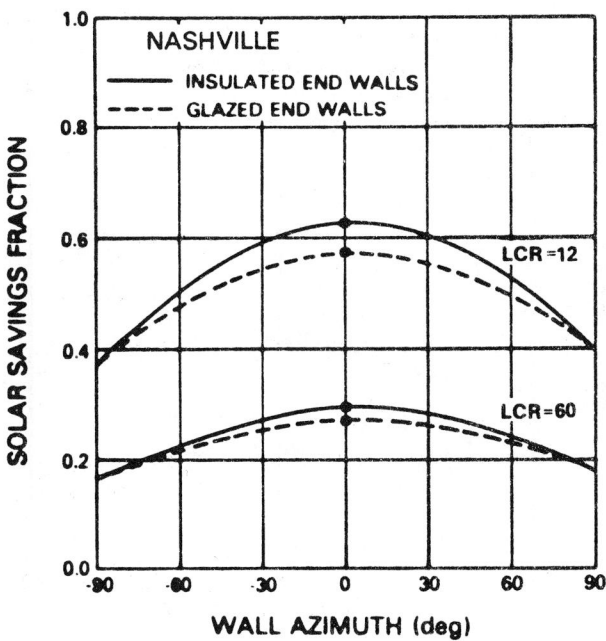

22.3.a.5 Glazing Tilt—Fixed-Length, Single-Plane Glazing

The graphs below plot the effect on solar savings fraction of changes in glazing tilt—for a sunspace with masonry common wall, with and without R9 nighttime insulation, and with insulated end walls—for each of six reference cities. The single, tilted glazing plane is held at a constant length by means of a variable-length knee wall, as shown in figure 22-2.

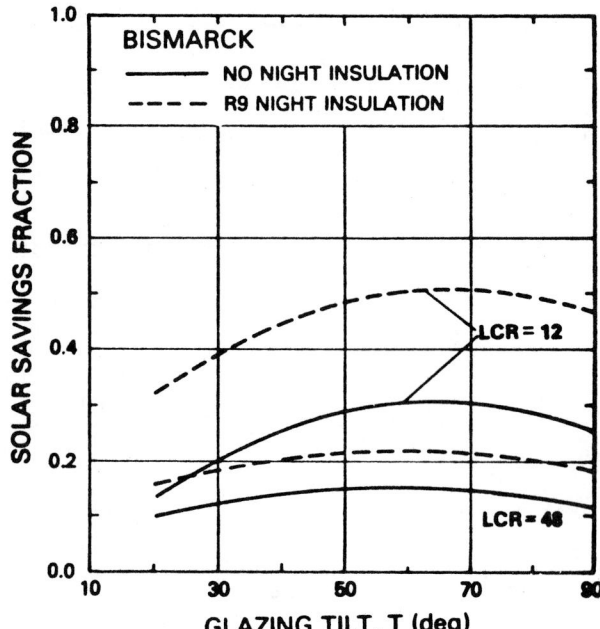

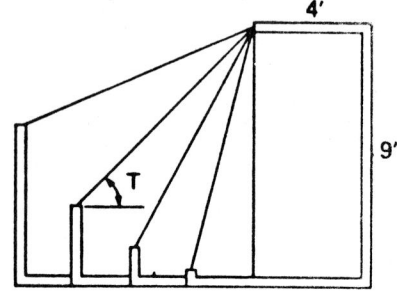

22-2: Geometry of the sunspace with fixed-length, single-plane glazing and variable-height knee wall.

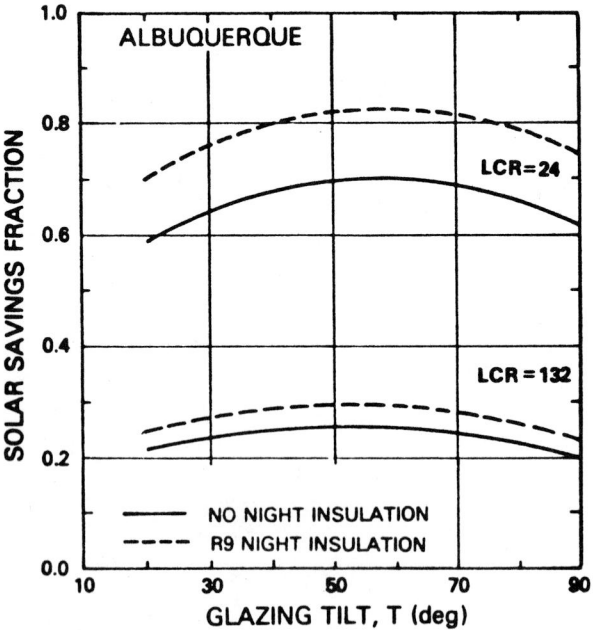

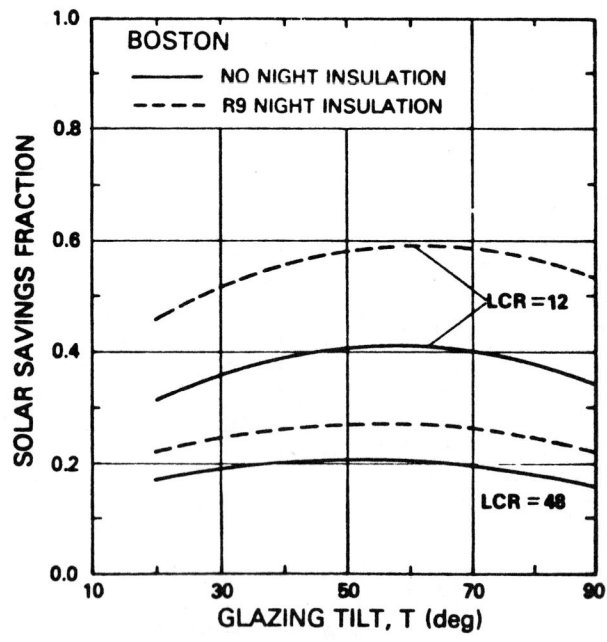

668 / APPENDIX 22

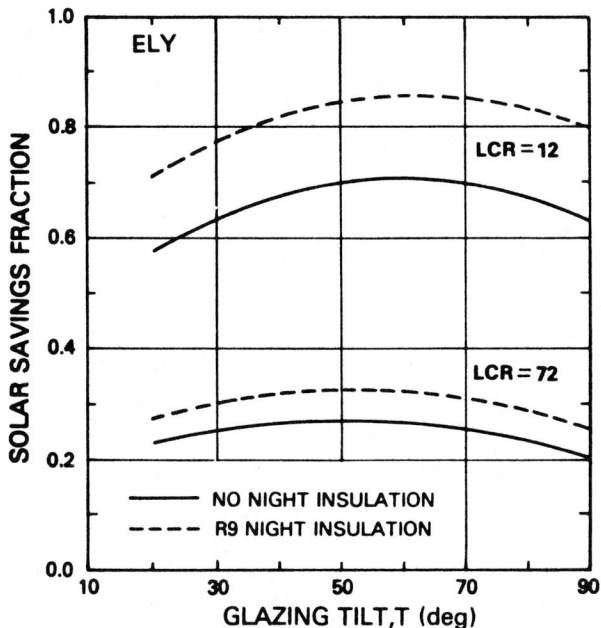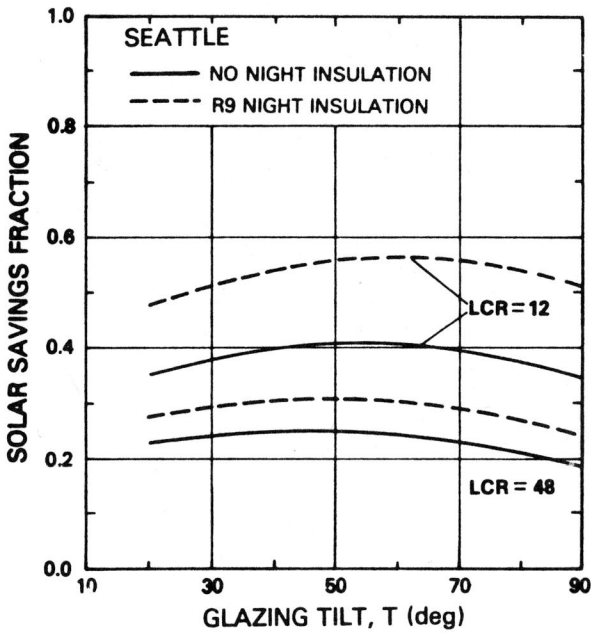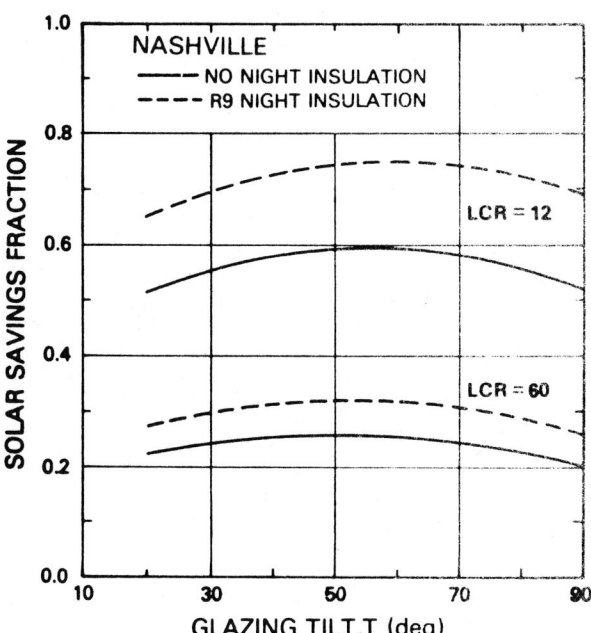

22.3.a.6 Glazing Tilt—Variable-Length, Single-Plane Glazing

The graphs below plot the effect on solar savings fraction of changes in glazing tilt—for a sunspace of geometry-A-like configuration, with masonry common wall, with and without R9 nighttime insulation, and both insulated and glazed end walls (A1, A2, A3, and A4)—for each of six reference cities. The single, tilted glazing plane is of variable length, as shown in figure 22-3. Dots represent reference conditions.

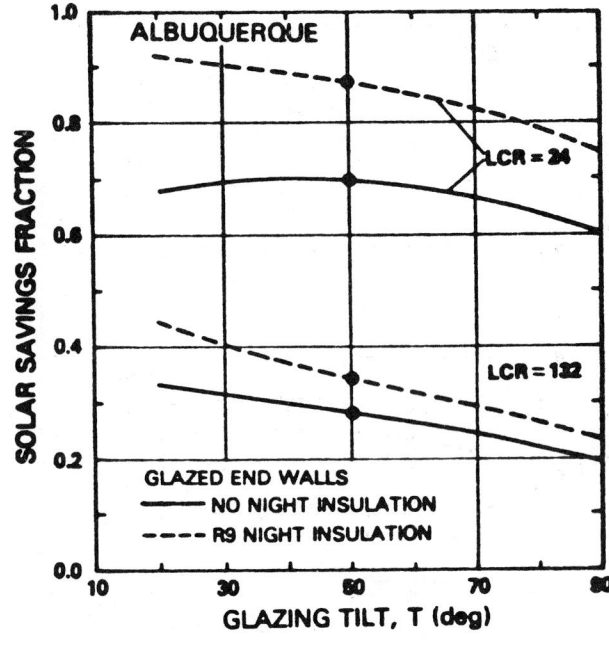

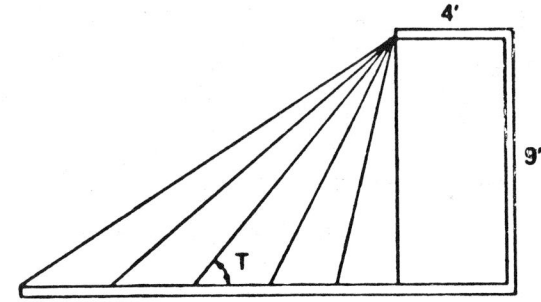

22-3: Geometry of the sunspace with variable-length, single-plane glazing.

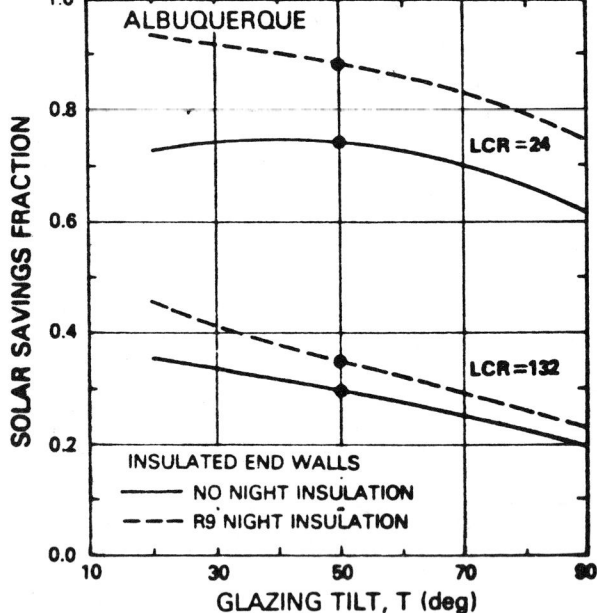

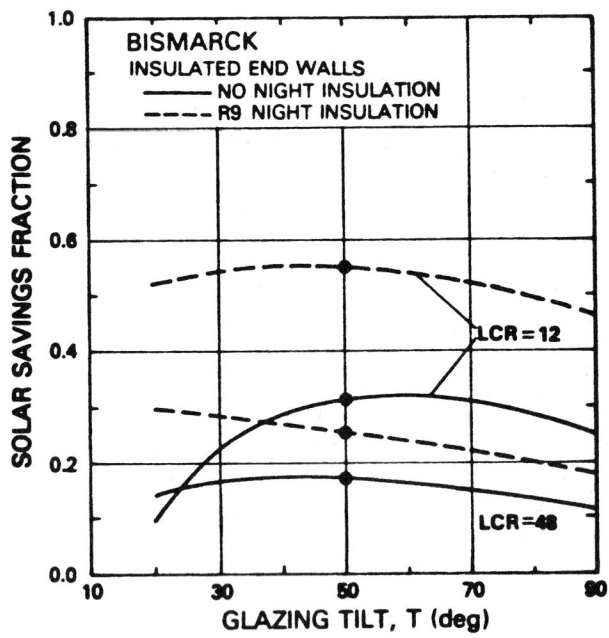

670 / APPENDIX 22

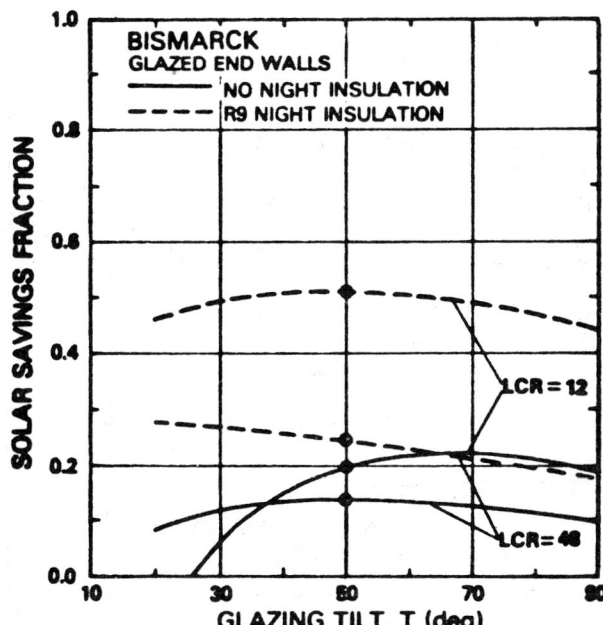
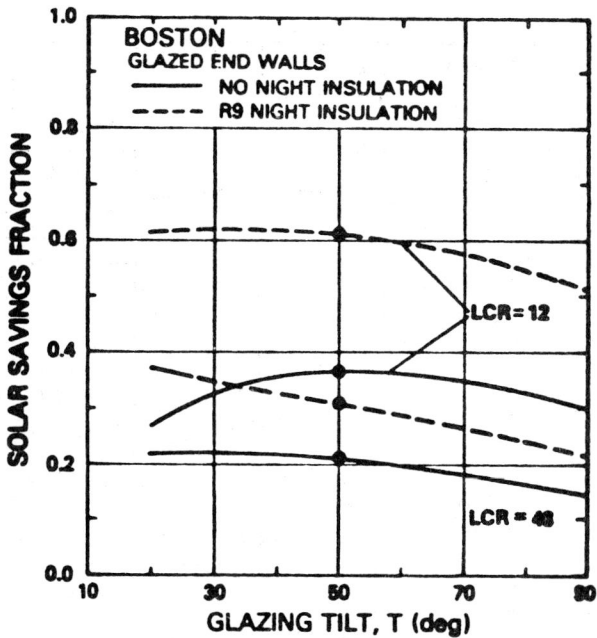
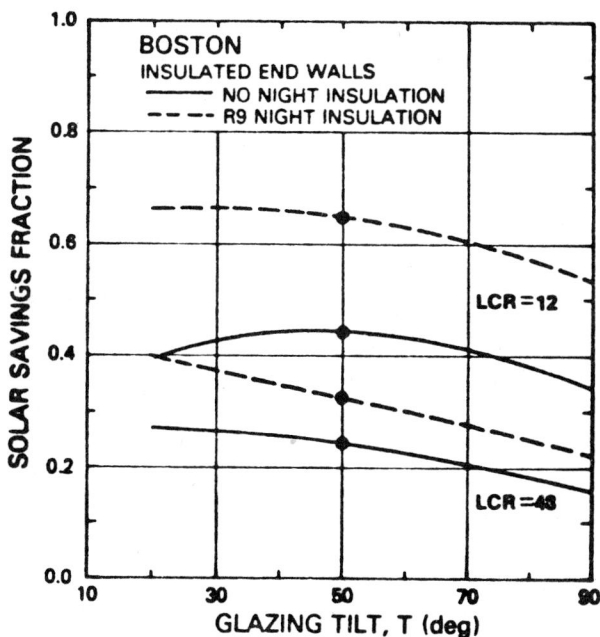

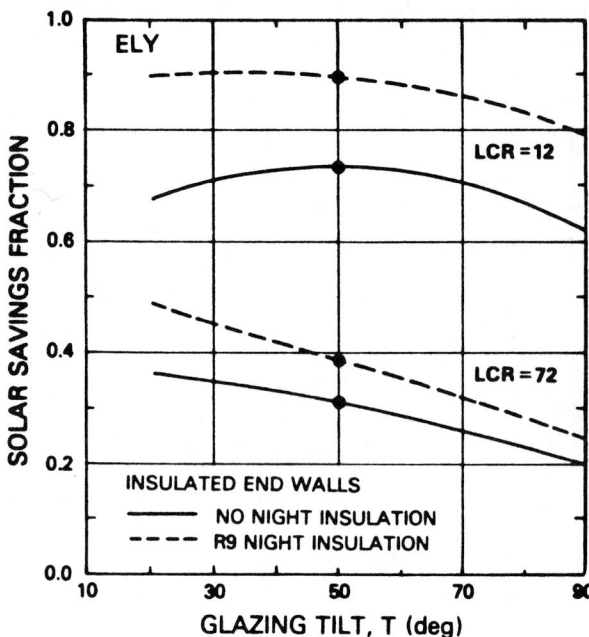
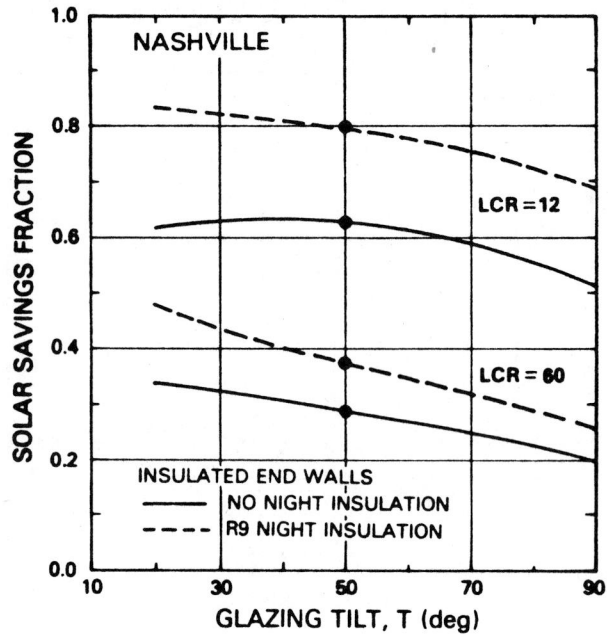
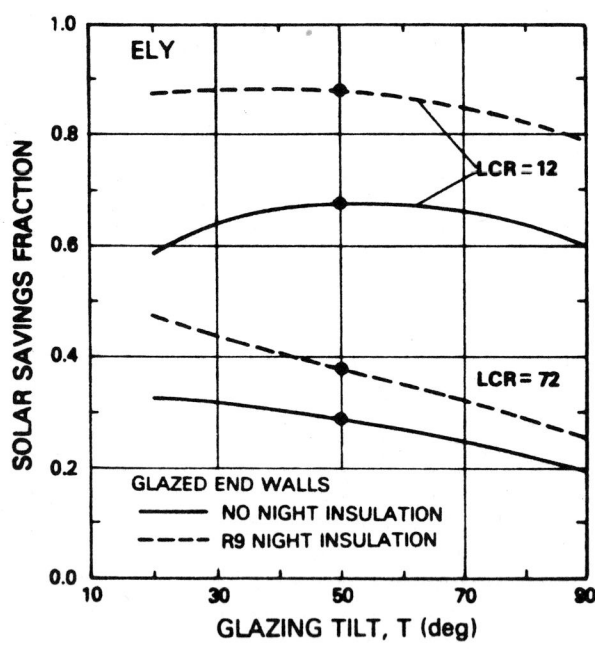

672 / APPENDIX 22

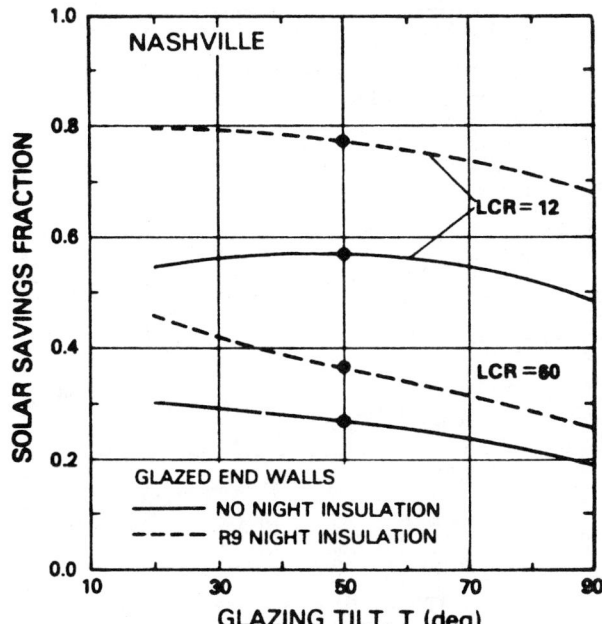
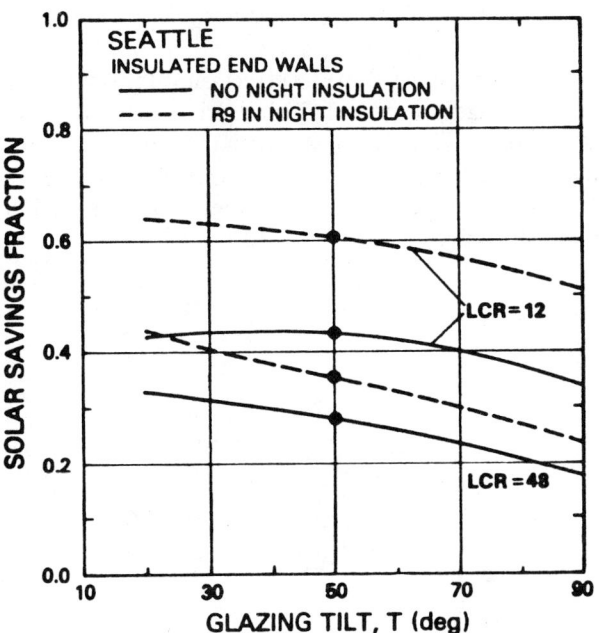
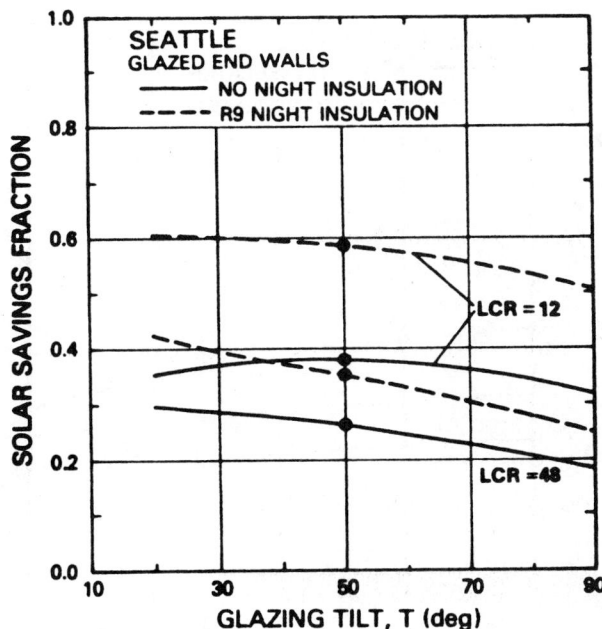

Sensitivity Data / 673

22.3.a.7 Vertical Glazing Height

The graphs below plot the effect on solar savings fraction of changes in vertical glazing height—for a sunspace with masonry common wall, with and without R9 nighttime insulation, and with insulated end walls—for each of six reference cities. The geometry of this design is shown in figure 22-4.

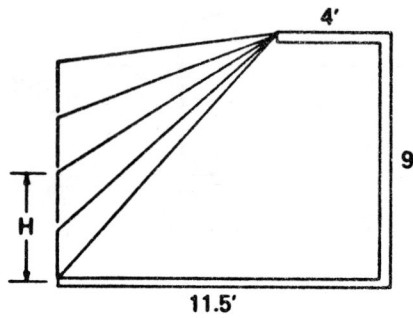

22-4: Geometry of the sunspace with variable-height vertical glazing and variable-length tilted glazing.

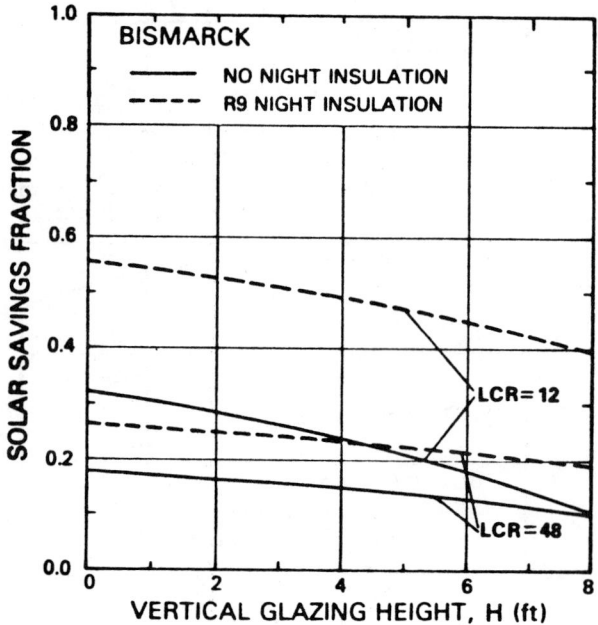

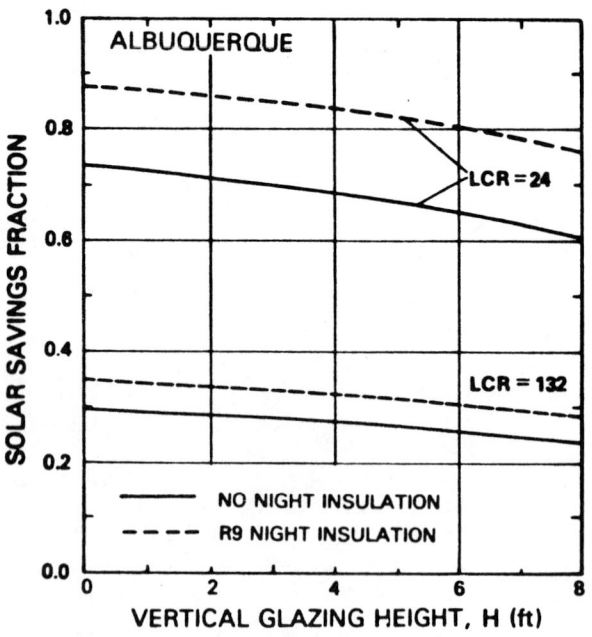

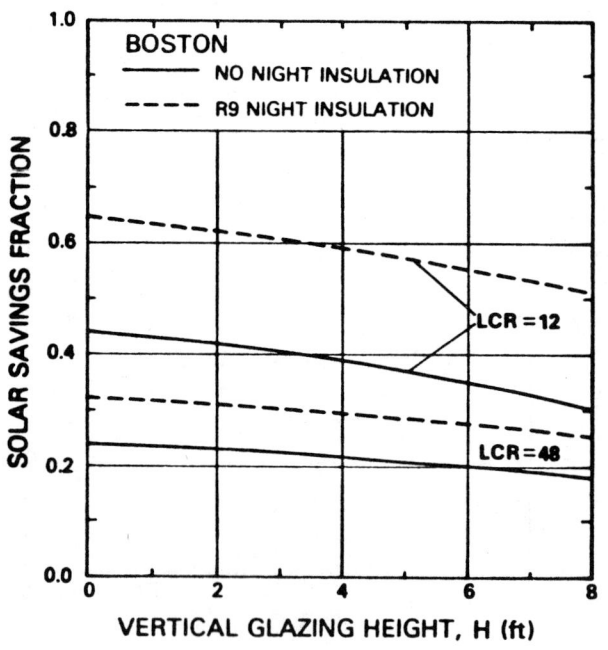

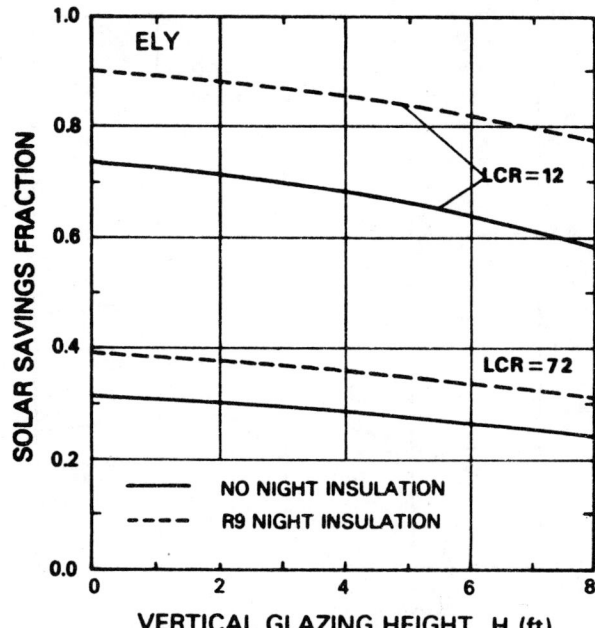
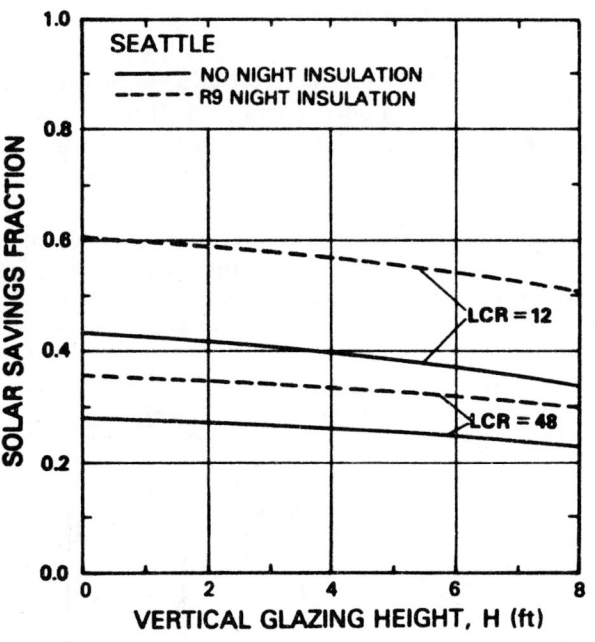
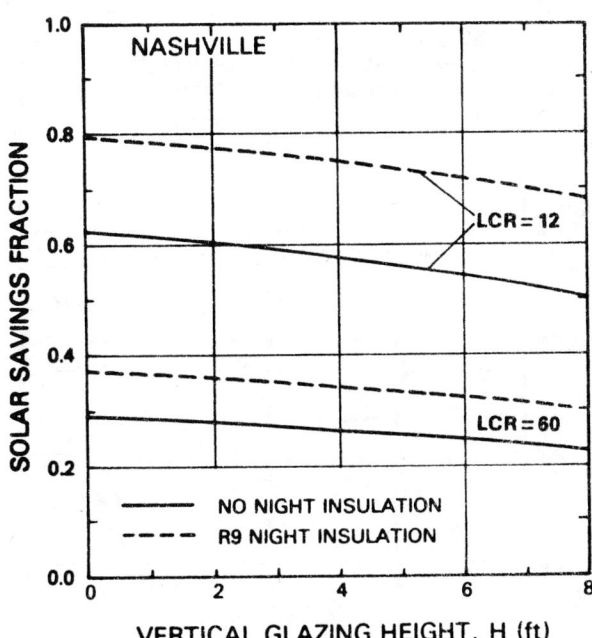

22.3.a.8 Upper Glazing Tilt

The graphs below plot the effect on solar savings fraction of changes in upper glazing tilt—for a sunspace of geometry-B-like configuration (recall, however, that geometry B does not include a 4-foot-long ceiling), with and without R9 nighttime insulation, with a 6-foot-high vertical glazing, and insulated end walls—for each of six reference cities. The geometry of this design is shown in figure 22–5.

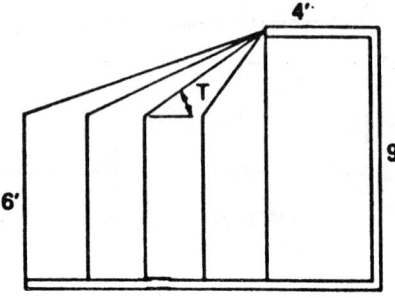

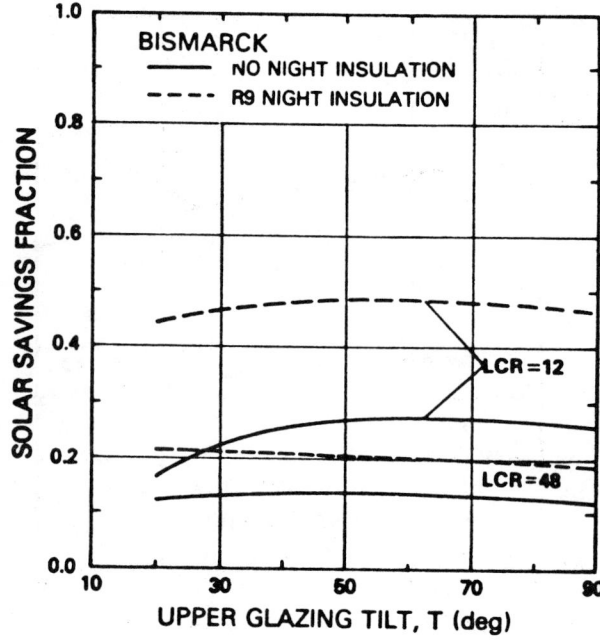

22-5: Geometry of the sunspace with fixed-height vertical glazing and variable-length tilted glazing.

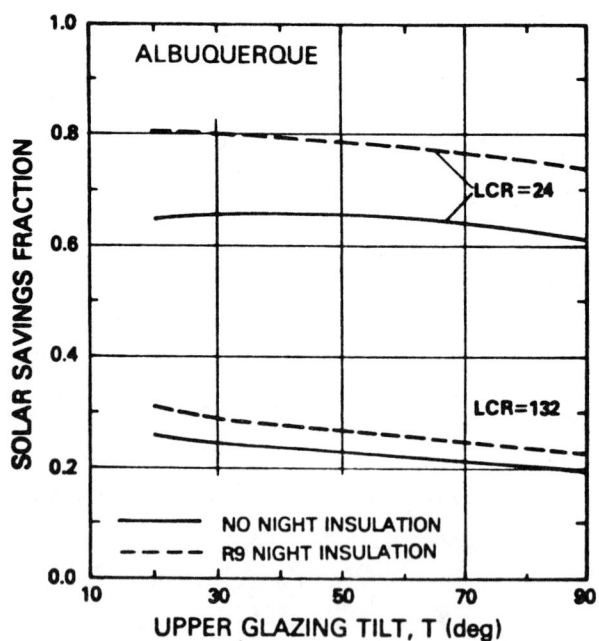

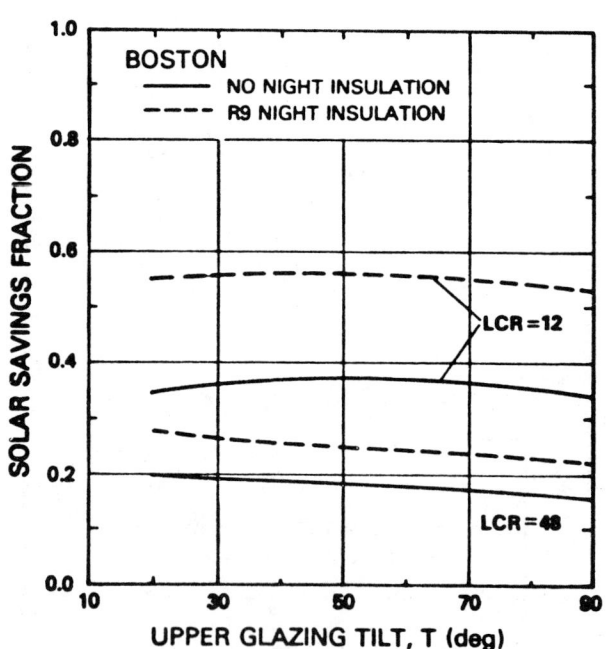

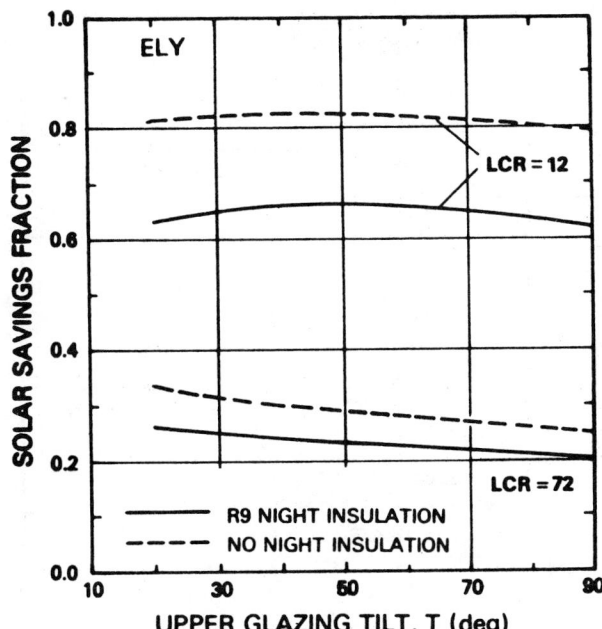
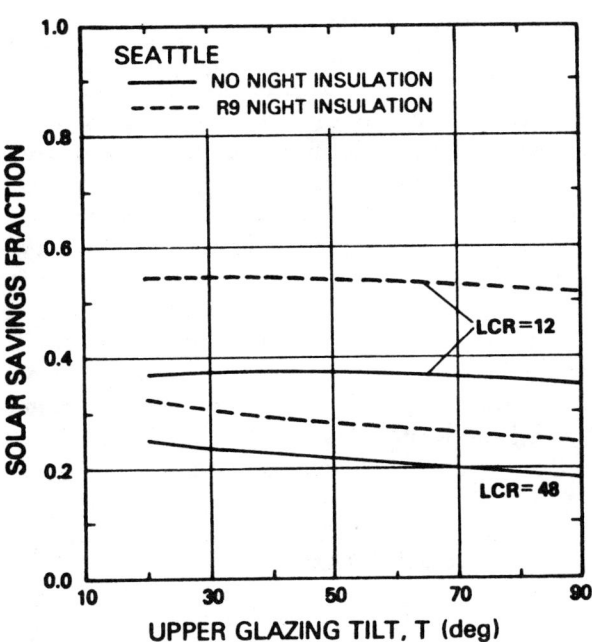
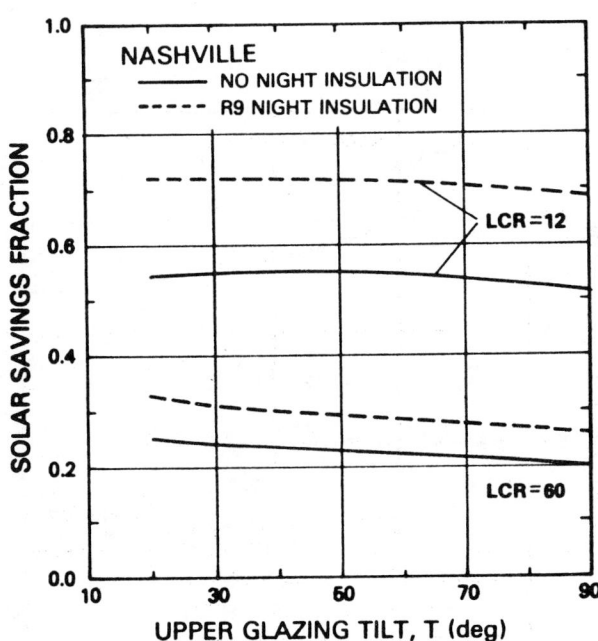

22.3.a.9 Sunspace Width

The graphs below plot the effect on solar savings fraction of changes in east-west width—for sunspaces of three basic configurations—for each of six reference cities. The first graph for each city plots performance for geometry A with masonry common wall, no nighttime insulation, and both insulated and glazed end walls (A1 and A3). The second graph for each city plots performance for geometry B with masonry common wall, no nighttime insulation, and both insulated and glazed end walls (B1 and B3). The third graph for each city plots performance for geometry C with masonry common wall, with and without R9 nighttime insulation (C1 and C2). Dots represent reference conditions in each case.

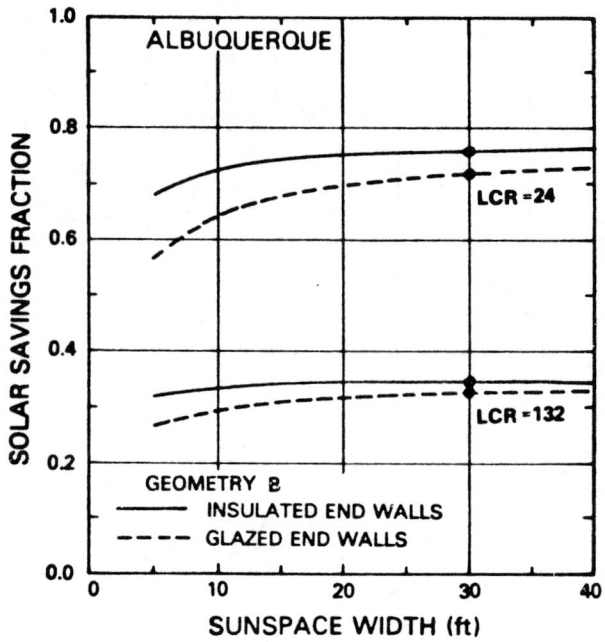

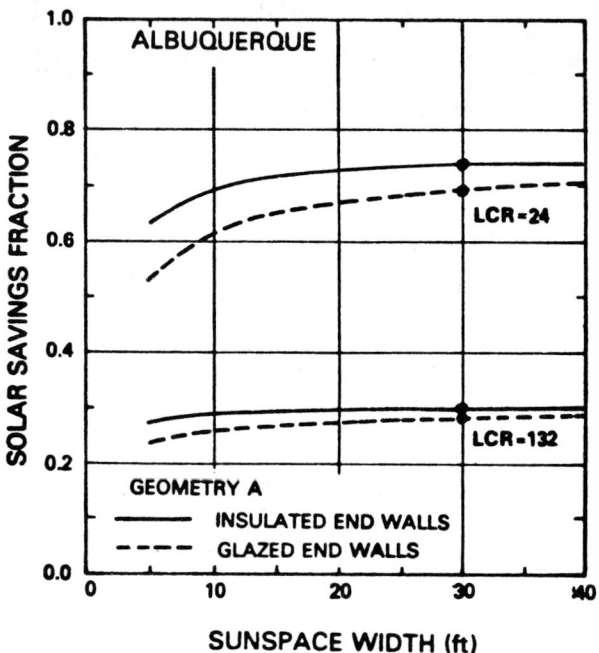

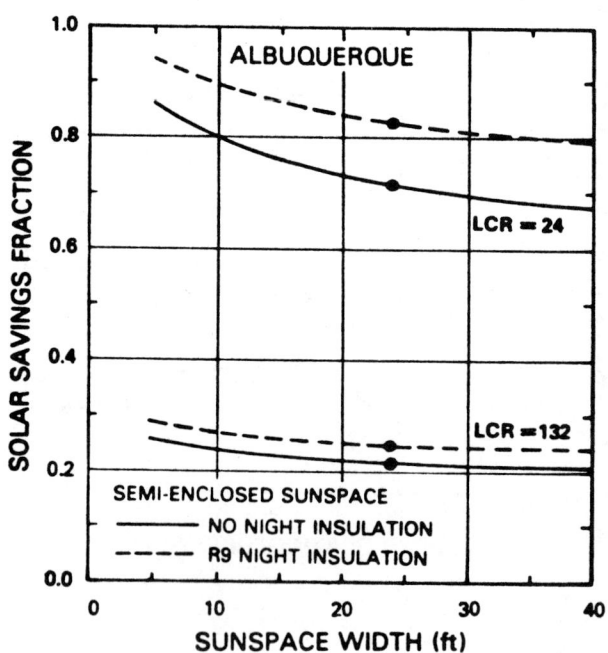

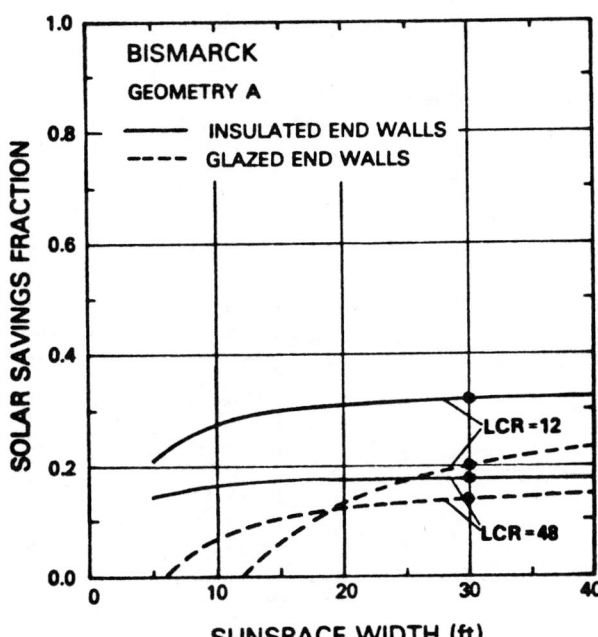
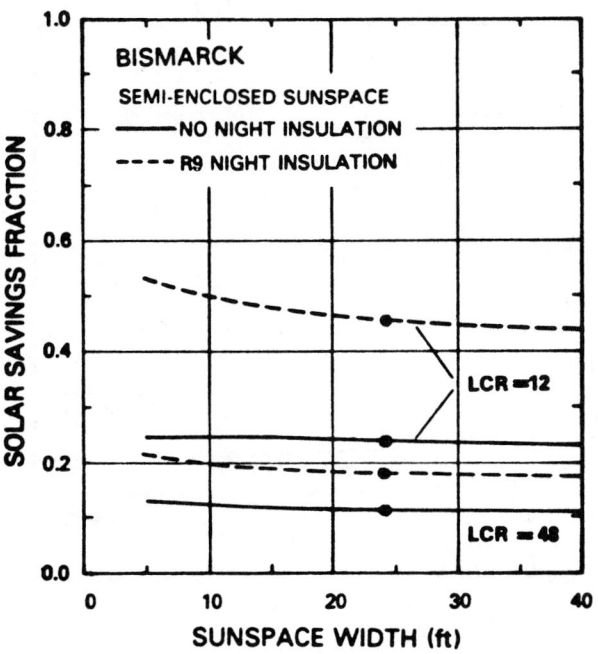
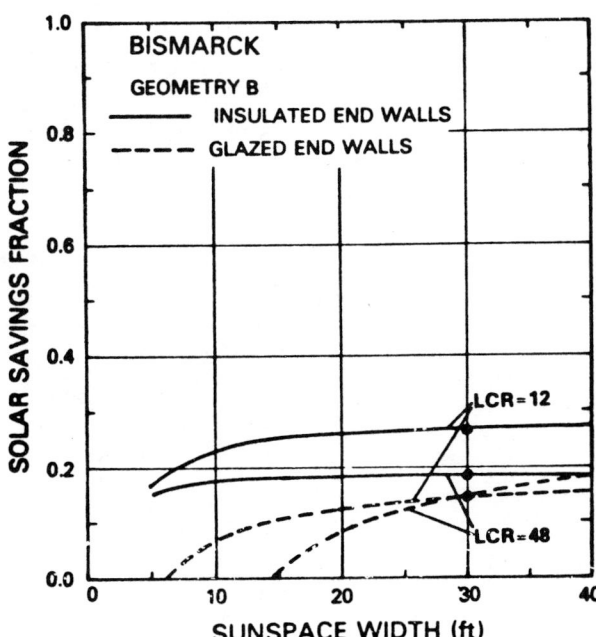

Sensitivity Data / 679

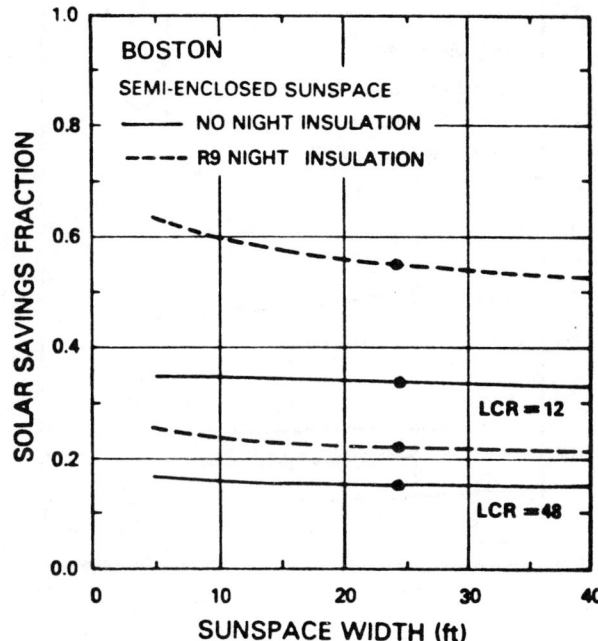
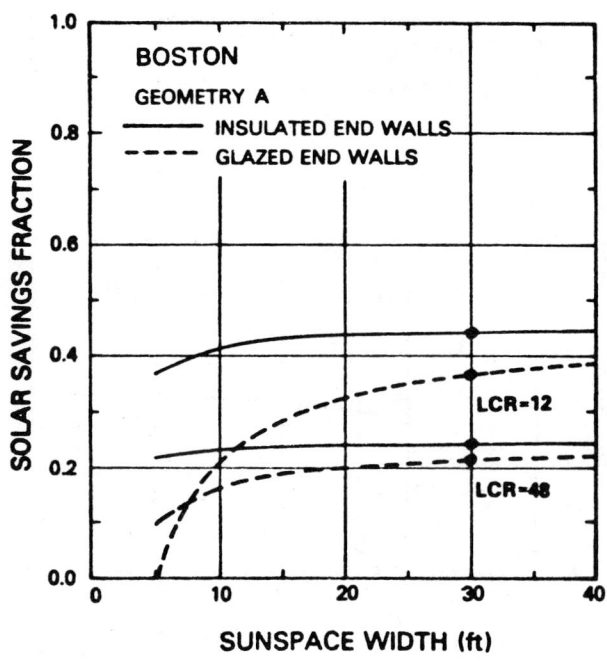
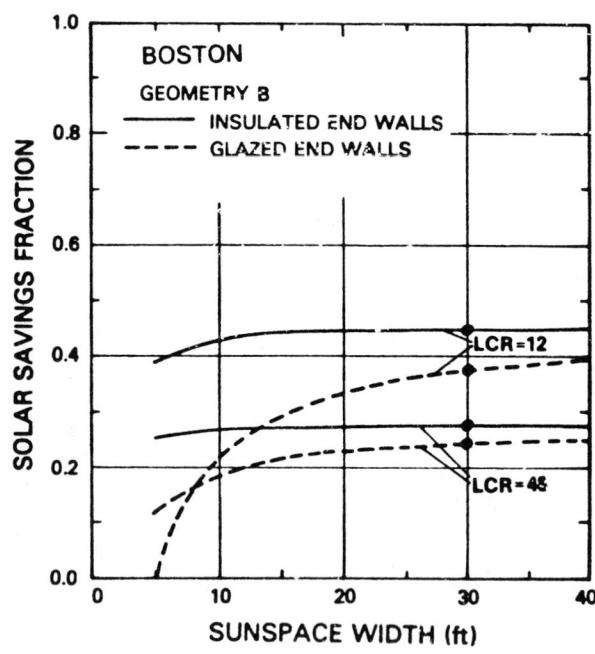

680 / APPENDIX 22

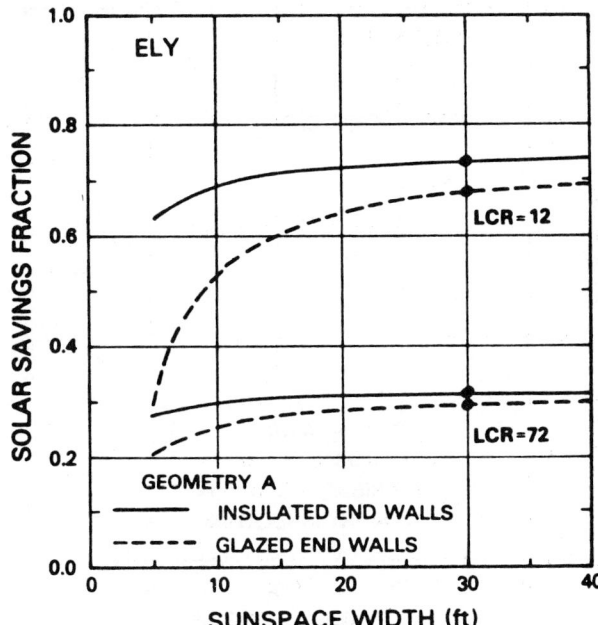
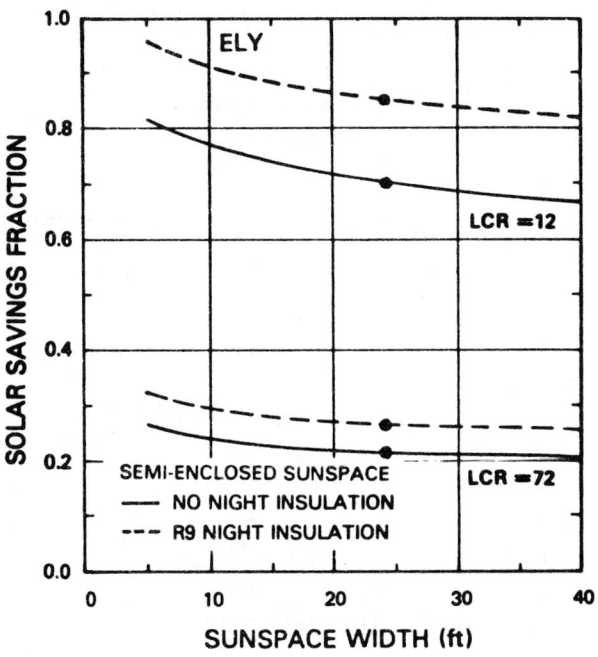
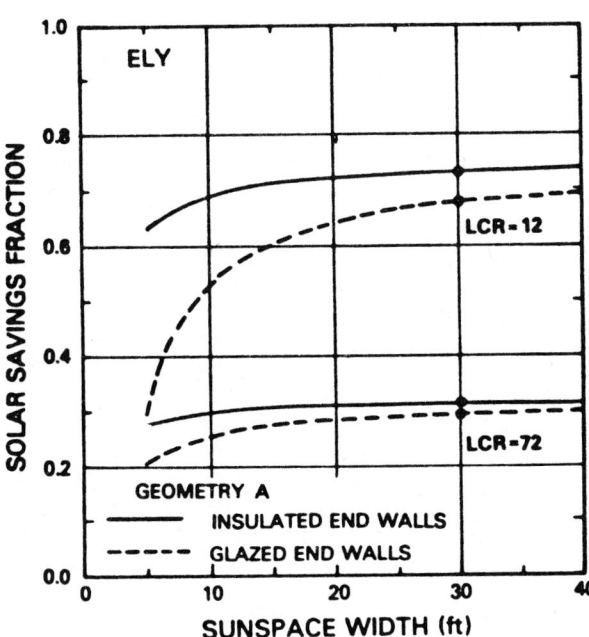

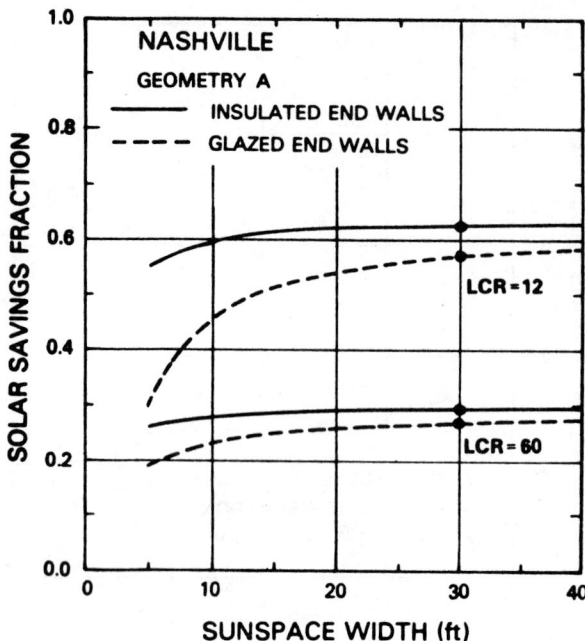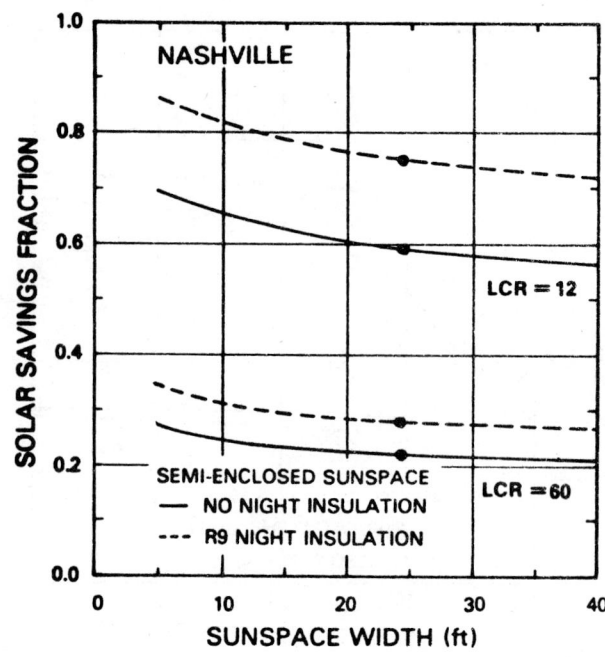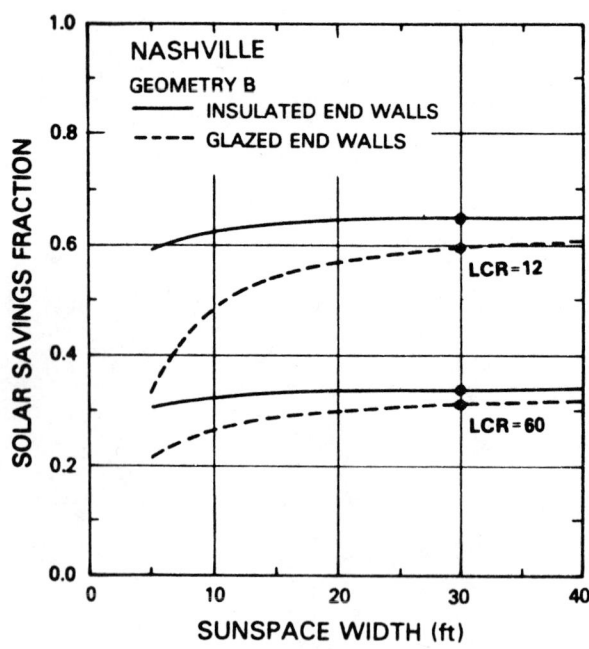

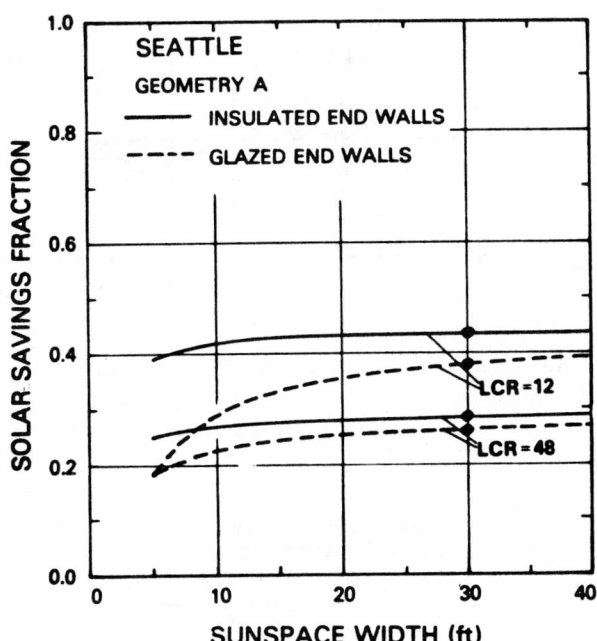
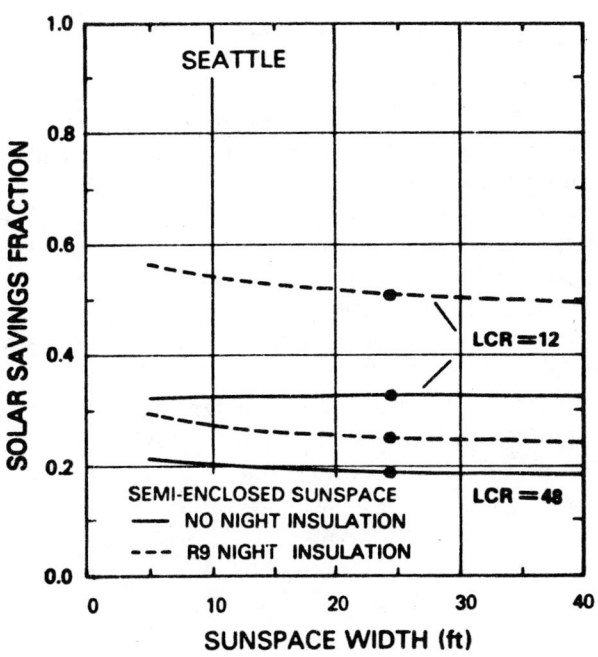
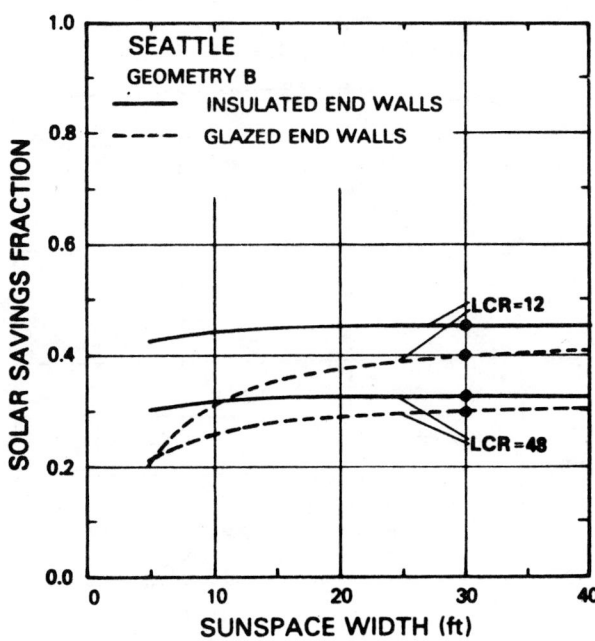

22.3.a.10 Number of Glazings

The graphs below plot the effect on solar savings fraction of changes in number of glazings—for sunspaces of geometry A with either masonry or lightweight insulated common wall, with and without R9 nighttime insulation (in some graphs, there are plots only for the "without insulation" case), and with insulated or both insulated and glazed end walls—for each of six reference cities. The four graphs for each city show one pair of geometry-A reference designs each, as follows: the first graph shows designs A1 and A2; the second graph shows designs A1 and A3; the third graph shows designs A5 and A6; and the fourth graph shows designs A5 and A7. Dots represent reference conditions.

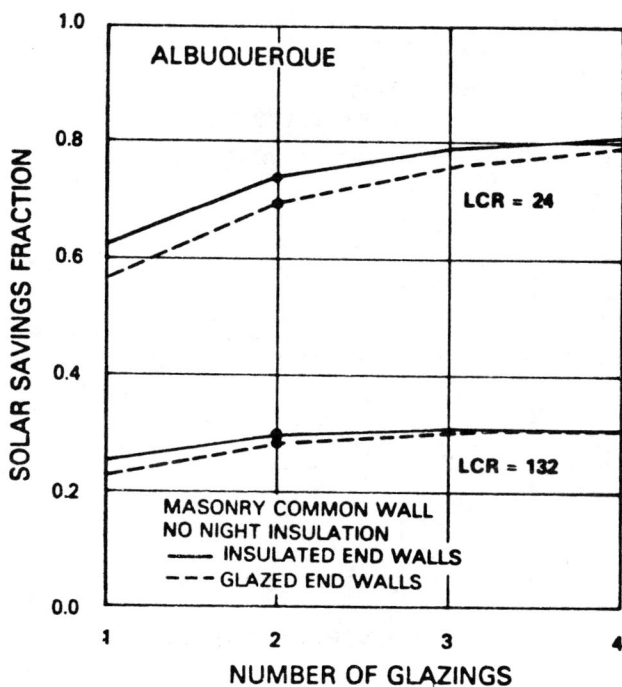

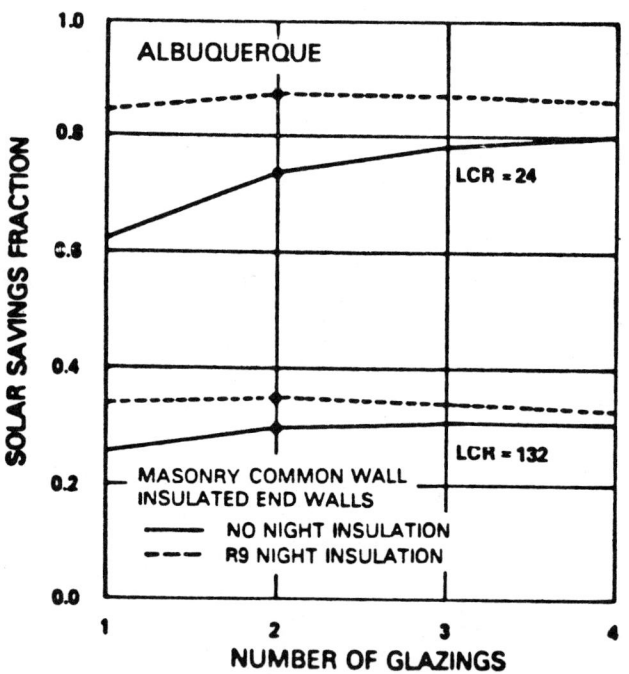

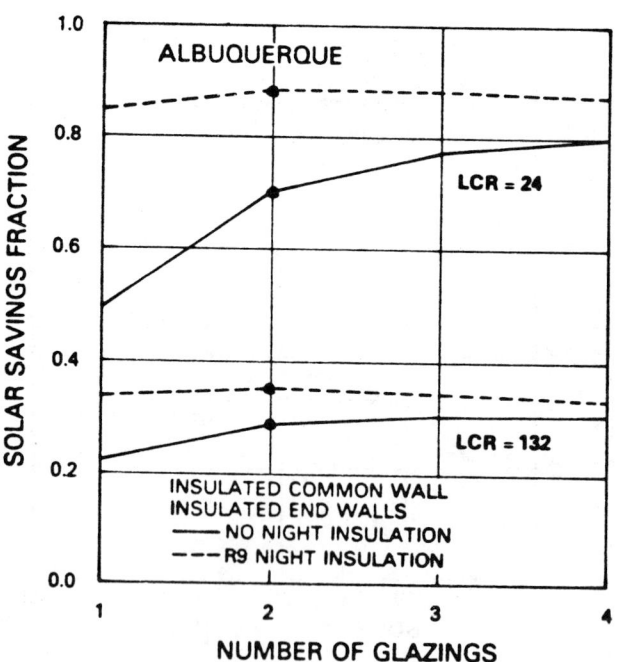

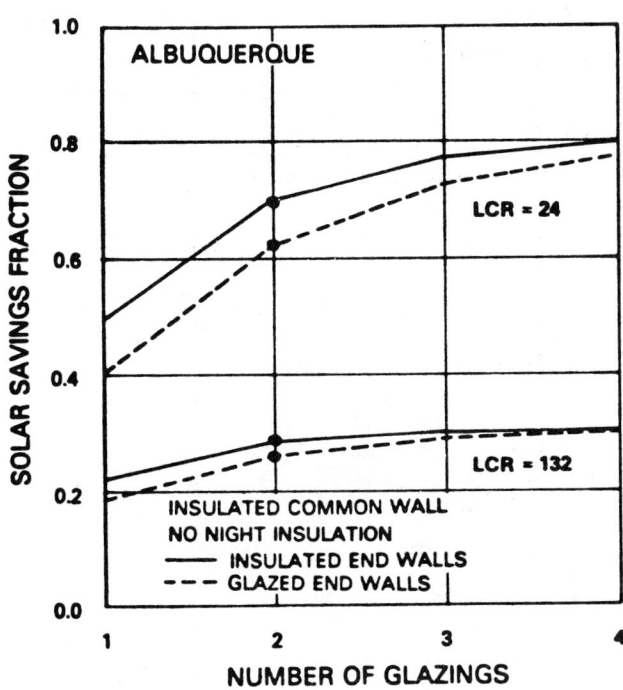
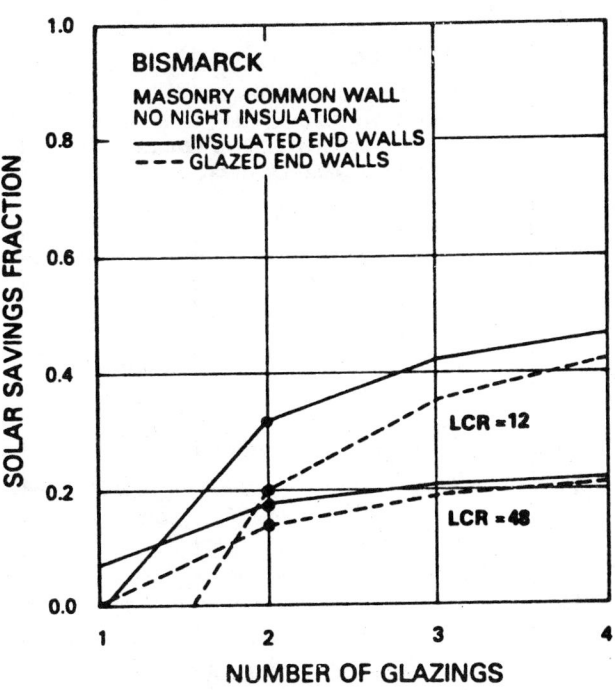
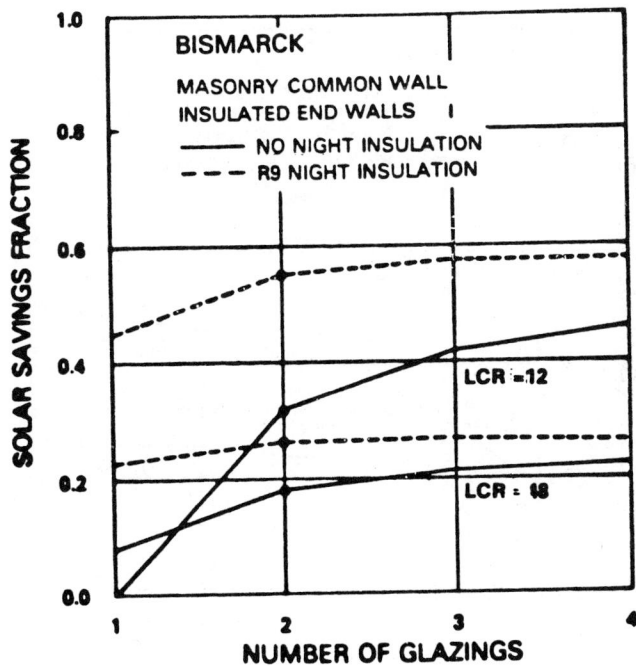
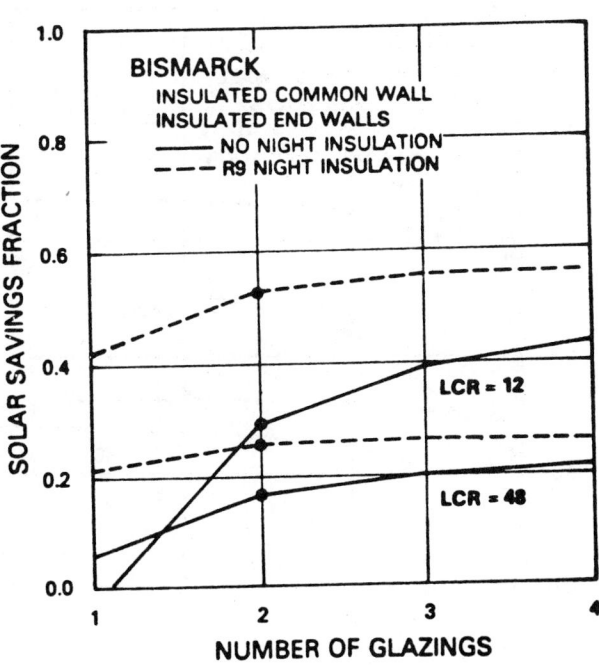

Sensitivity Data / 685

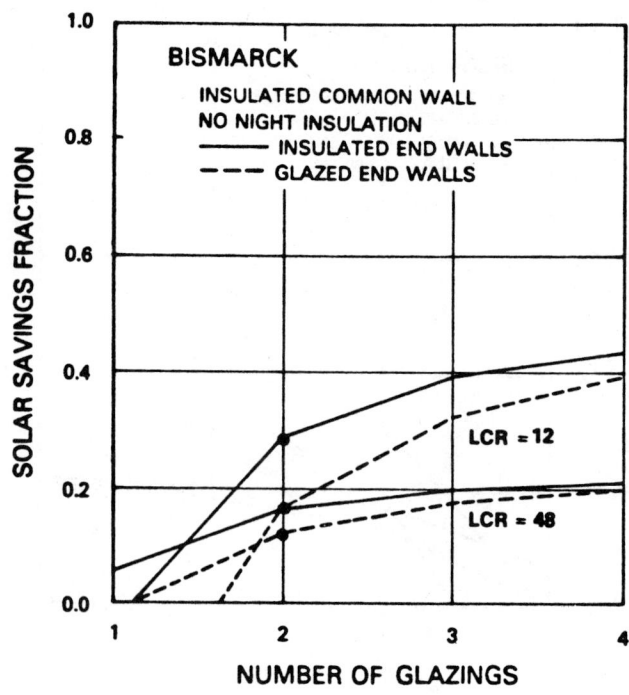
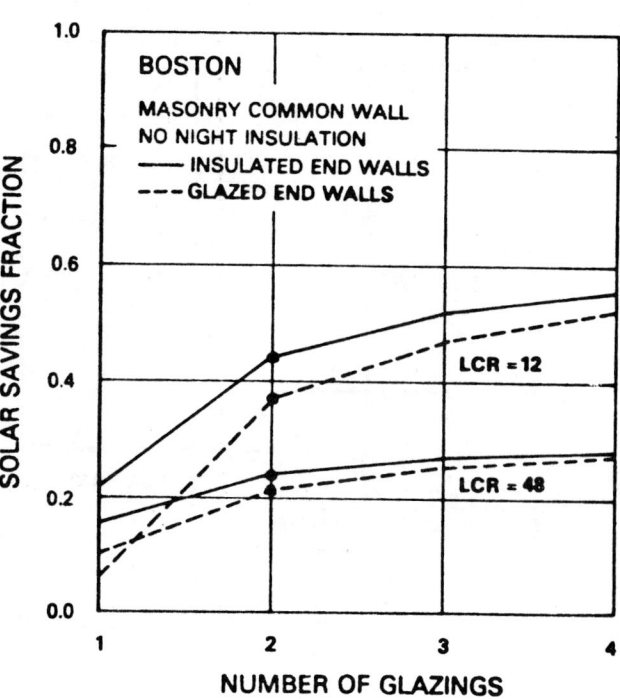
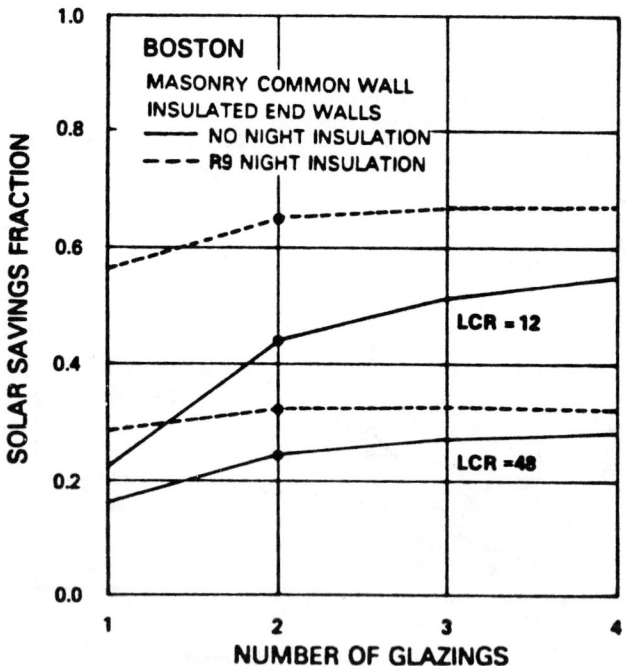
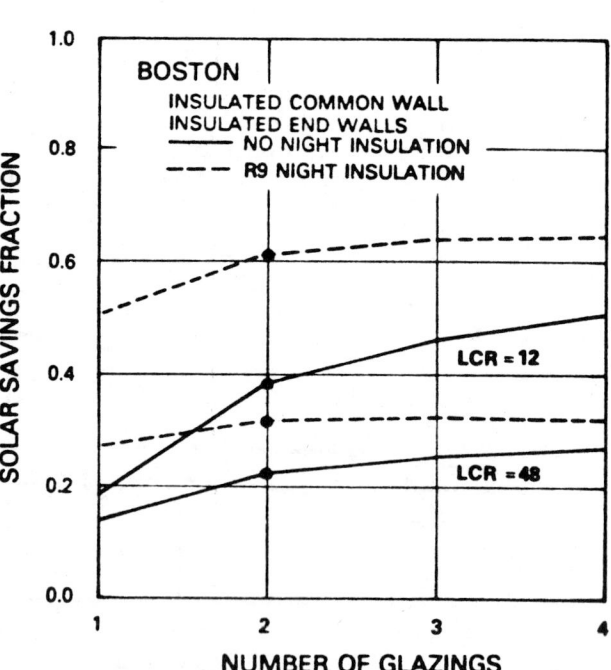

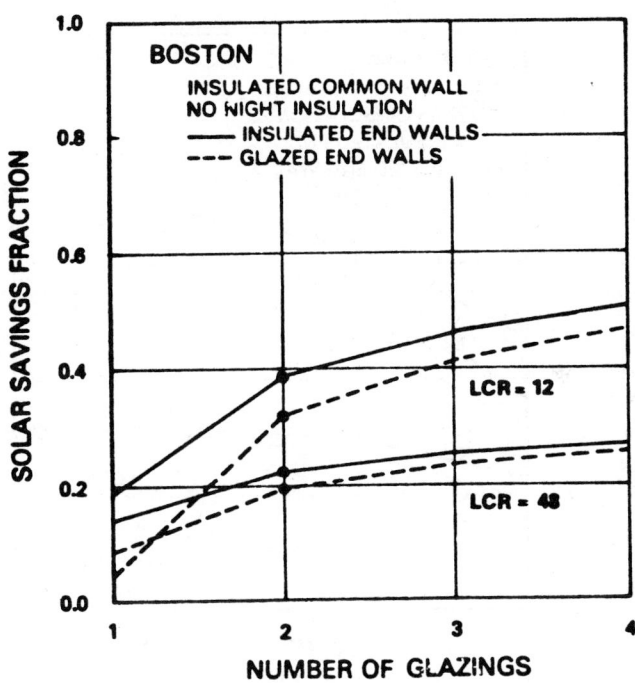
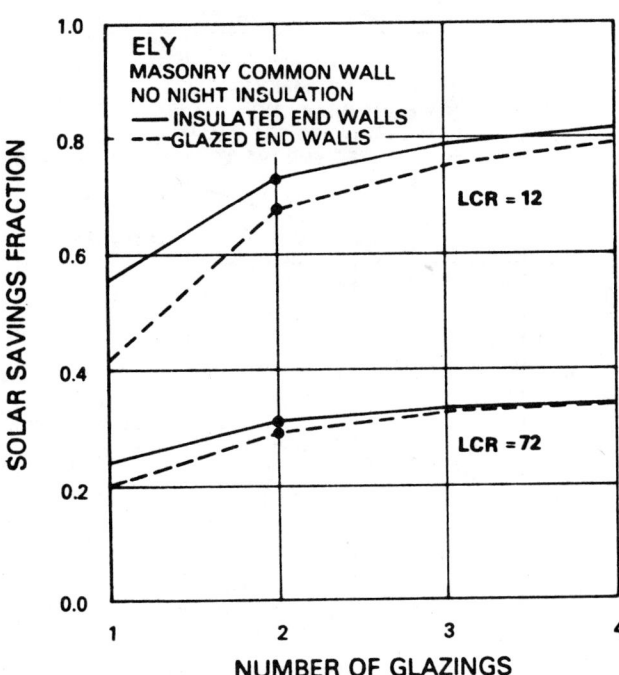
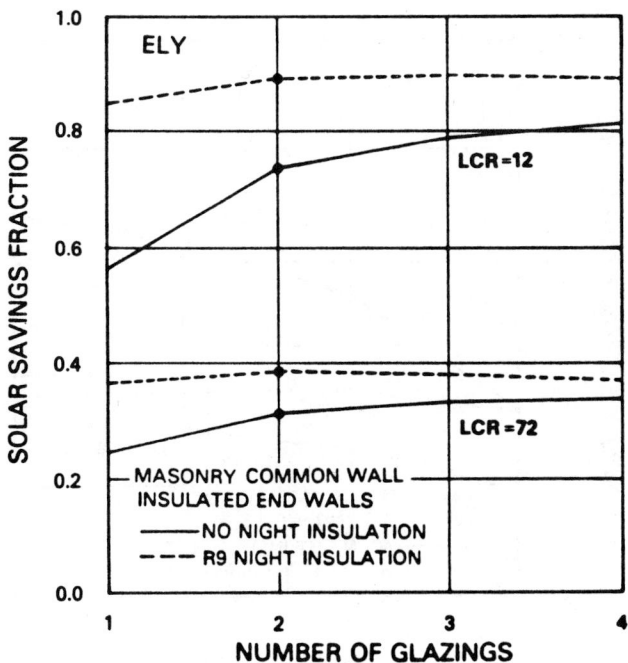
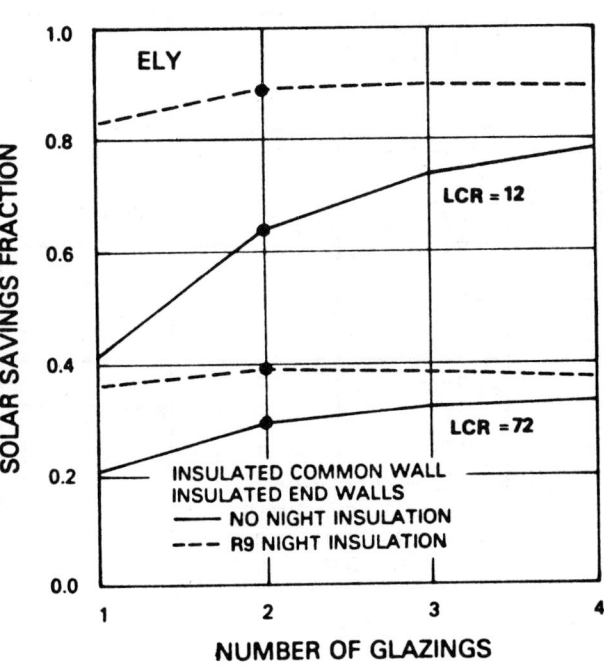

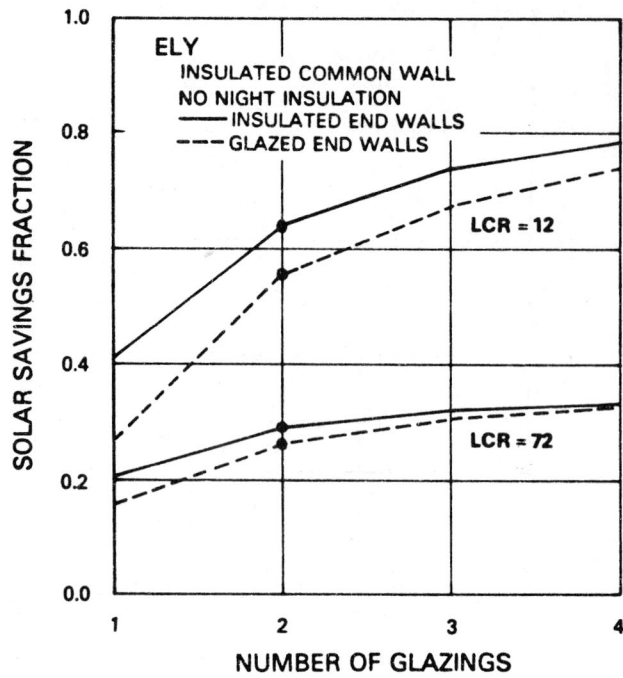
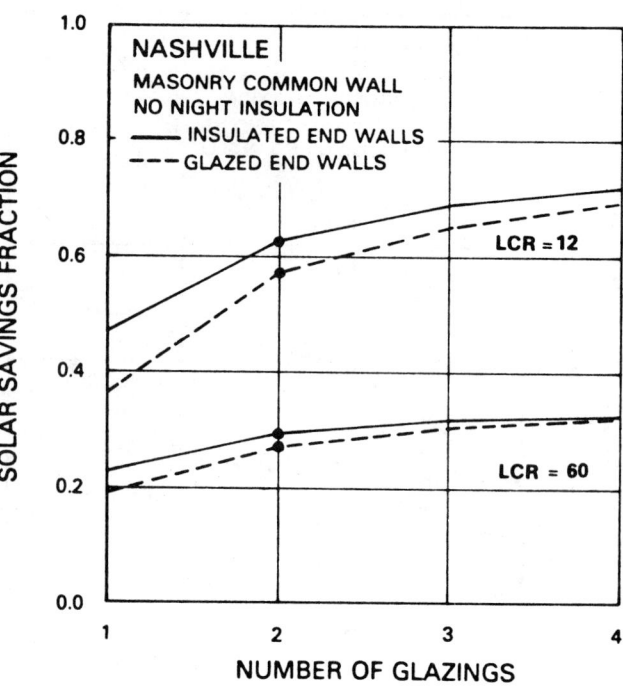
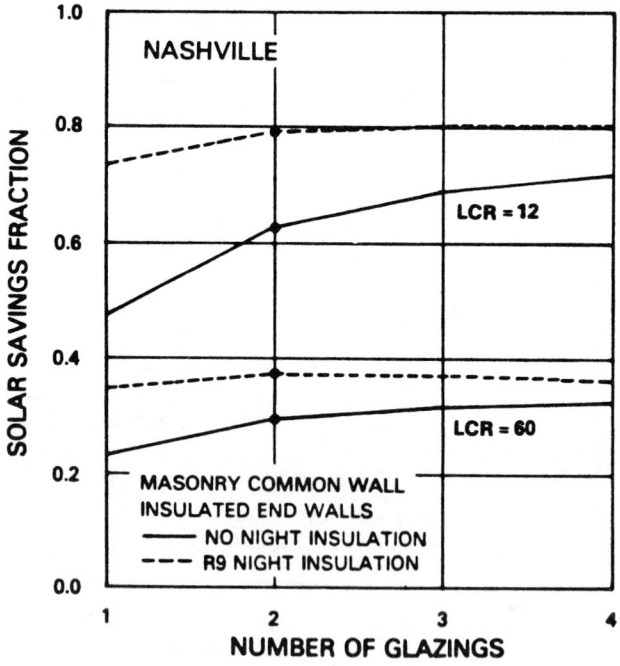
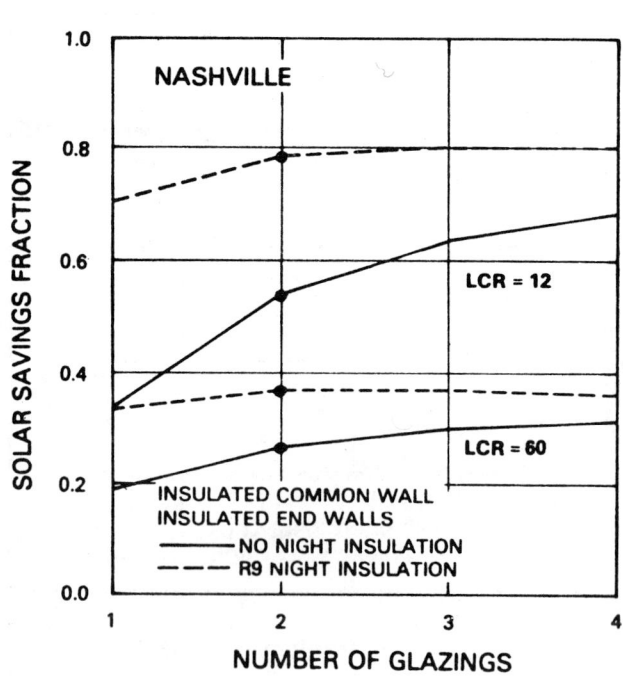

688 / *APPENDIX 22*

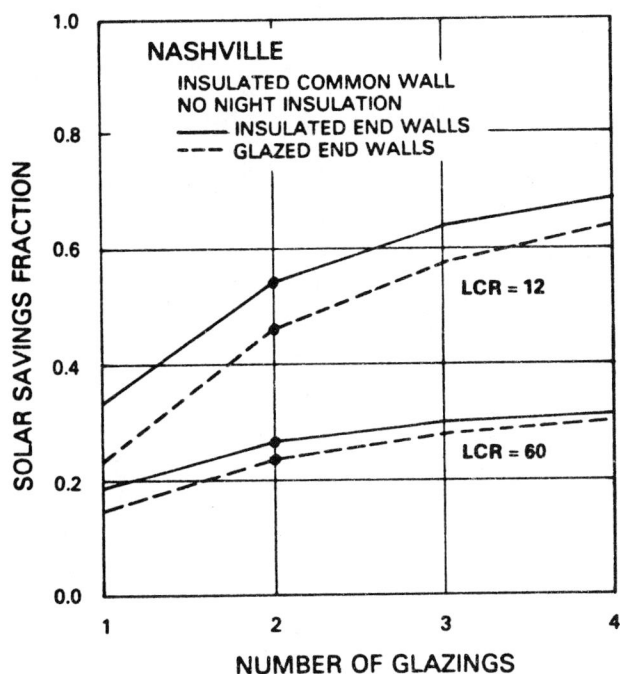
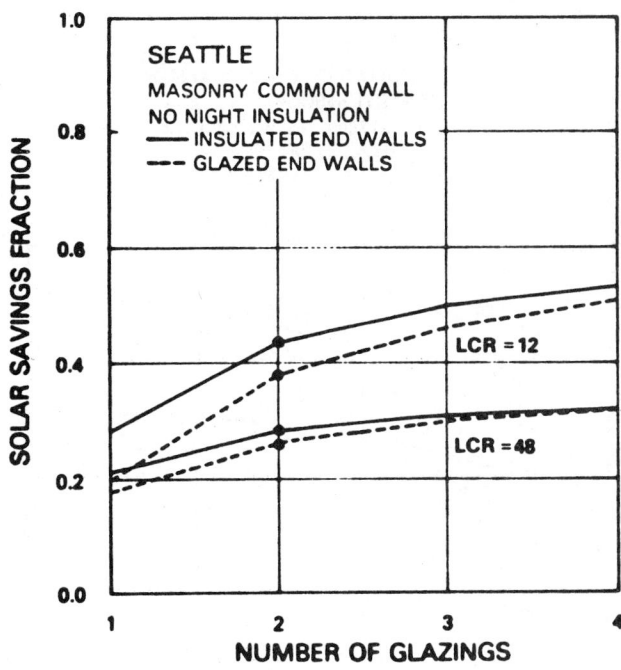
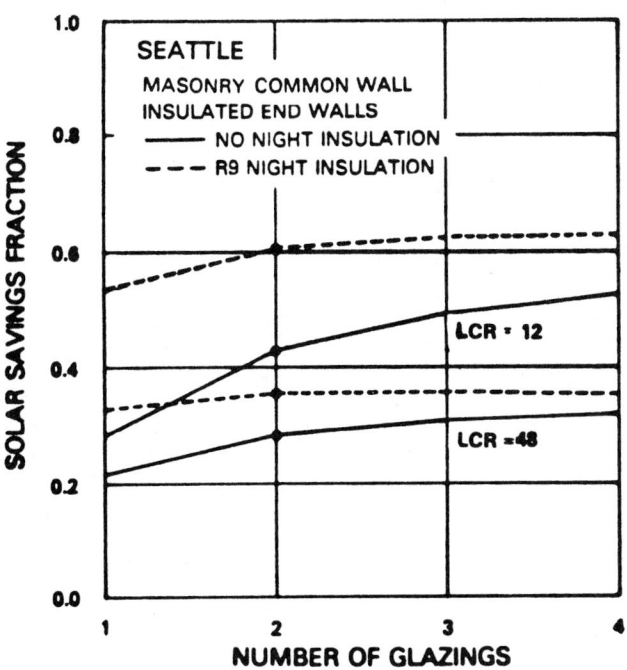
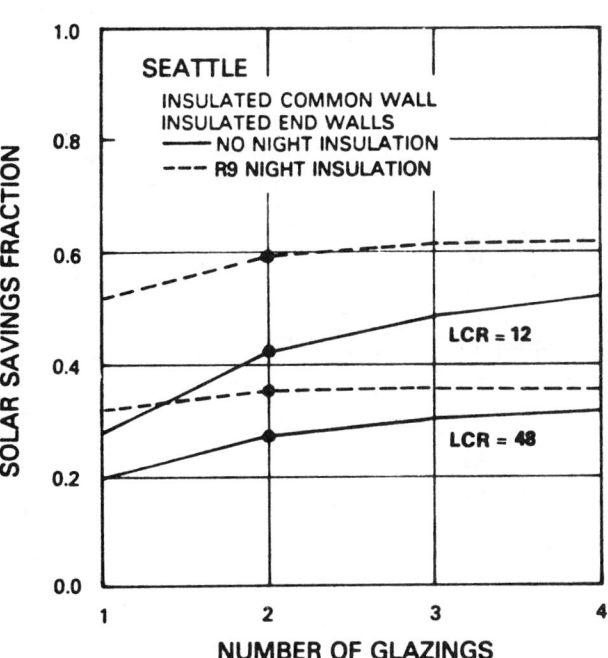

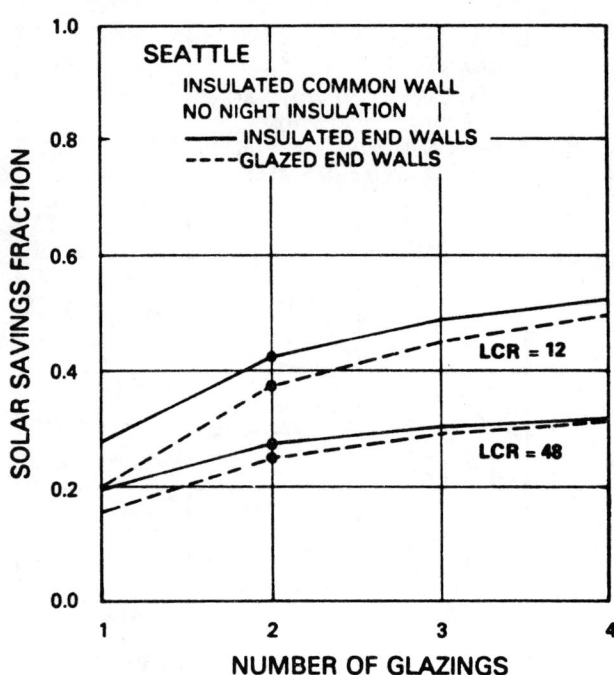

690 / *APPENDIX 22*

22.3.a.11 Air-Gap Width Between Glazings

The graphs below plot the effect on solar savings fraction of changes in the width of the air gap between glazings—for geometry A with masonry common wall, with and without R9 nighttime insulation, and both insulated end walls (A1 and A2)—for each of six reference cities. Dots represent reference conditions.

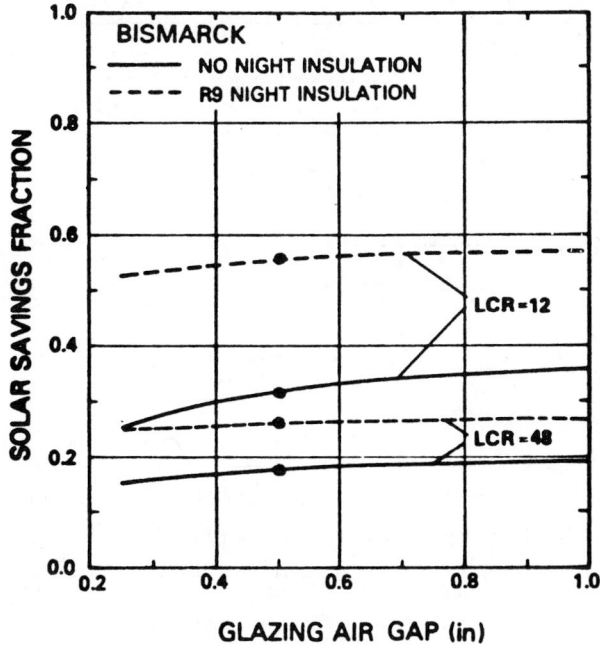

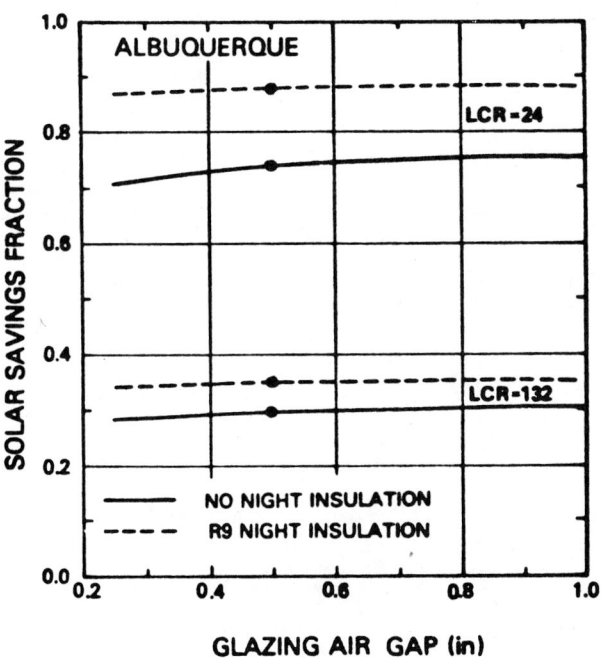

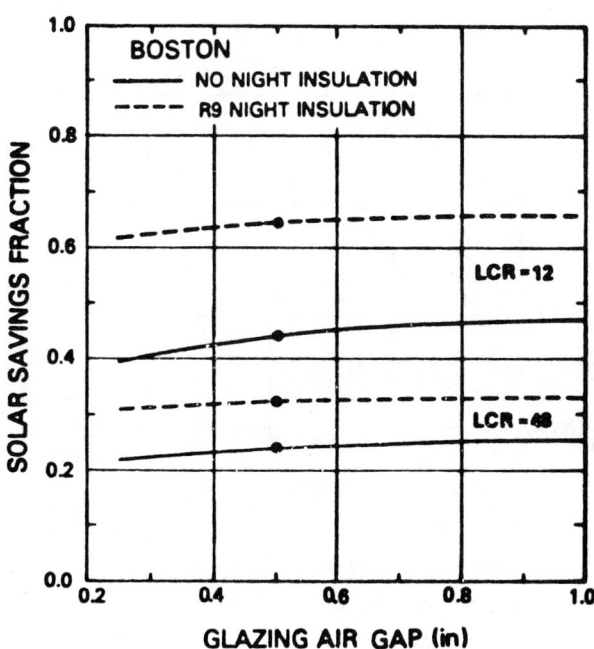

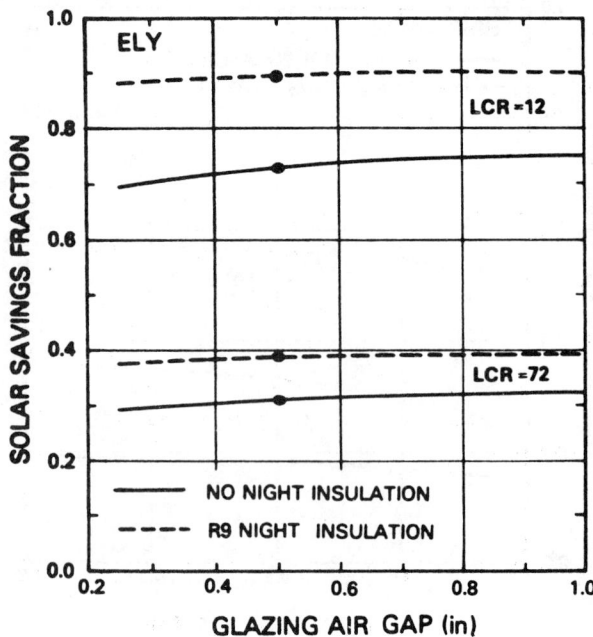
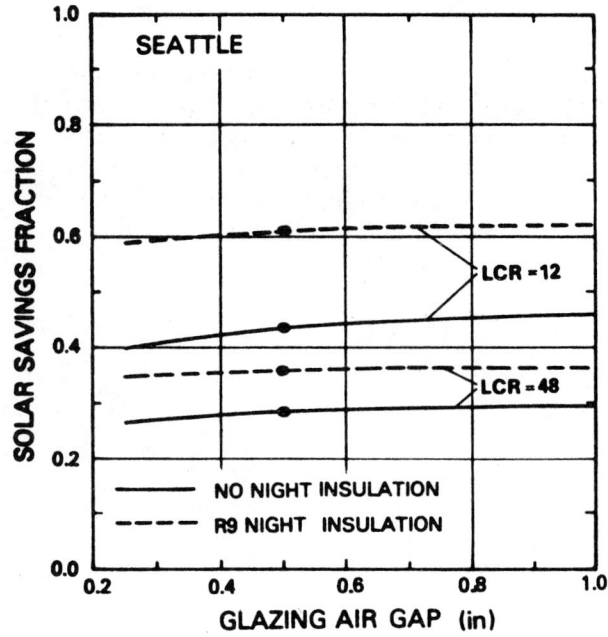
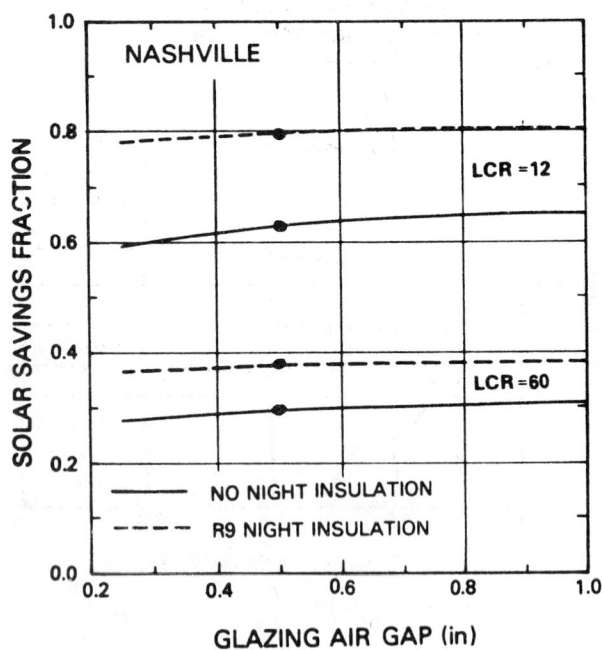

22.3.a.12 Glazing Thickness

The graphs below plot the effect on solar savings fraction of changes in glazing thickness—for geometry A with masonry common wall, with and without R9 nighttime insulation, and with insulated end walls (A1 and A2)—for each of six reference cities. Dots represent reference conditions.

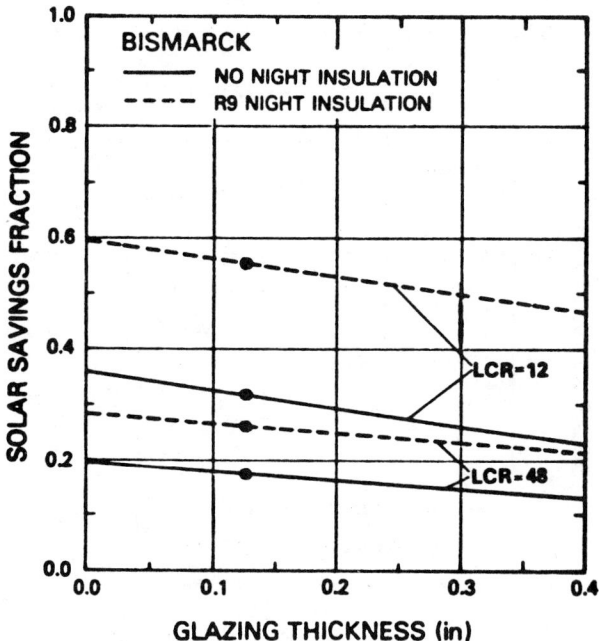

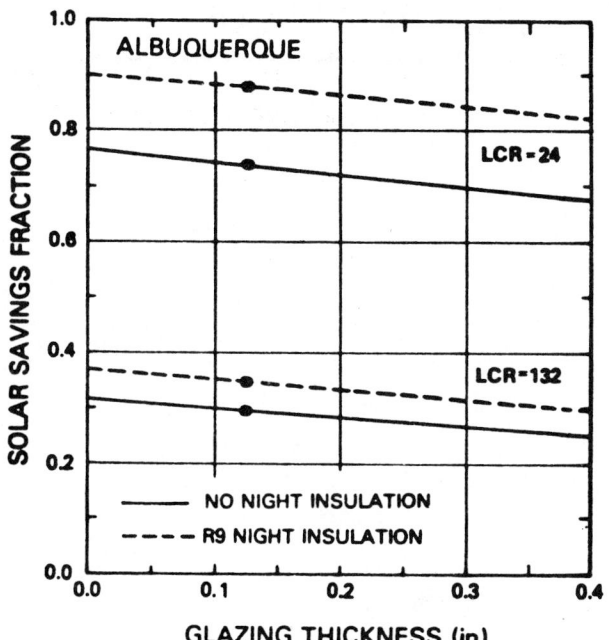

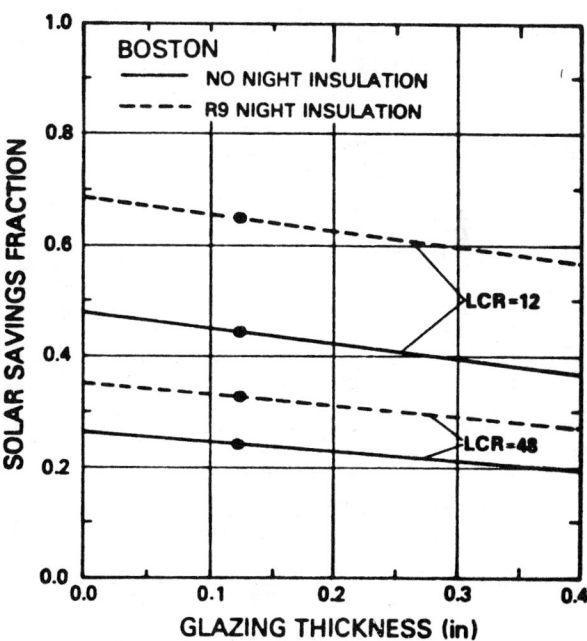

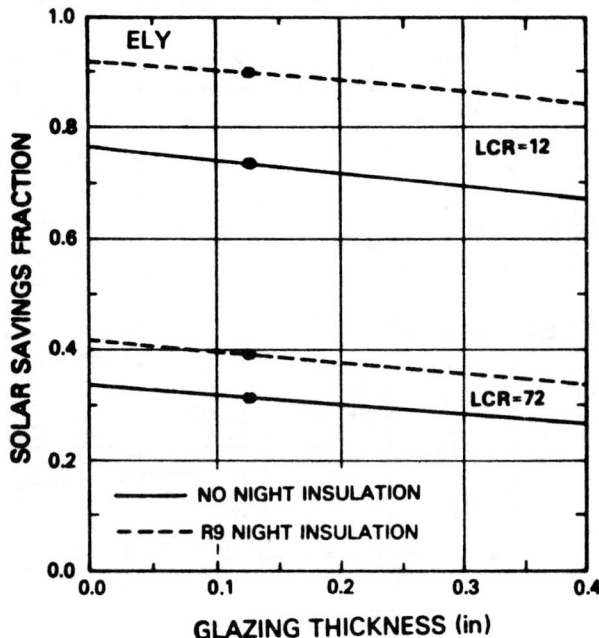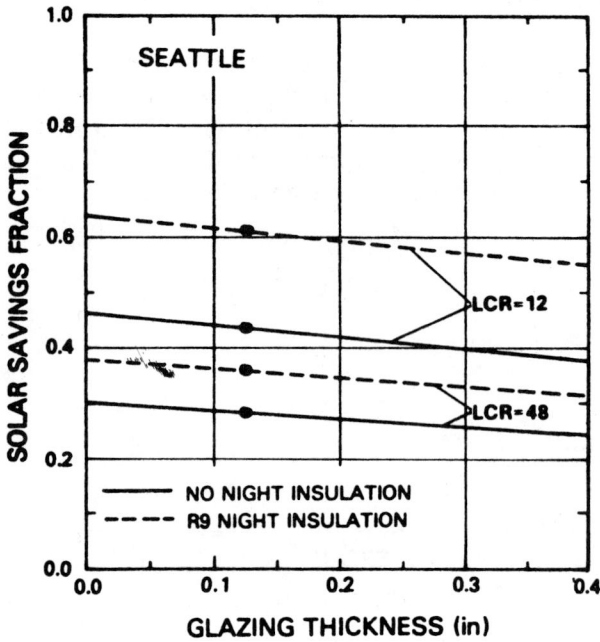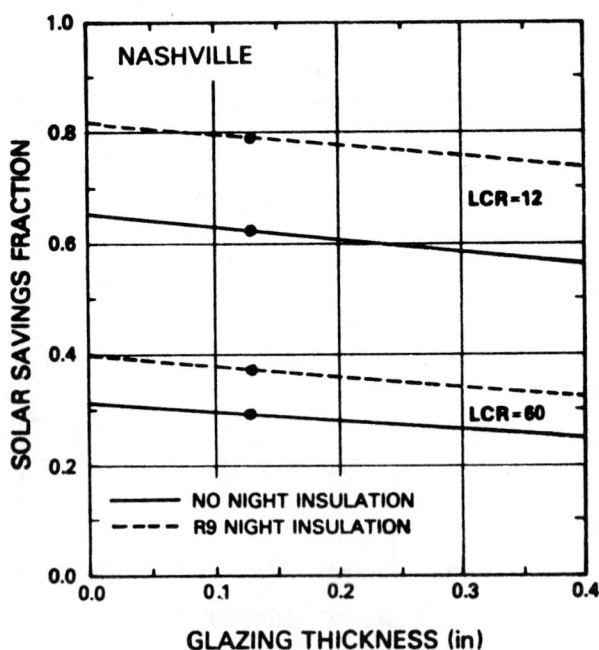

22.3.a.13 Common-Wall Absorptance

The graphs below plot the effect on solar savings fraction of changes in common-wall absorptance—for geometry A with masonry common wall, no nighttime insulation, and both insulated and glazed end walls (A1 and A3)—for each of six reference cities. Dots represent reference conditions.

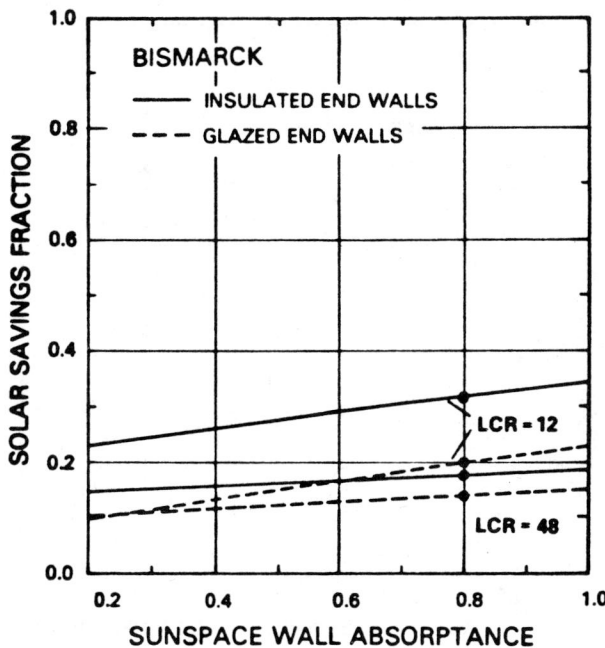

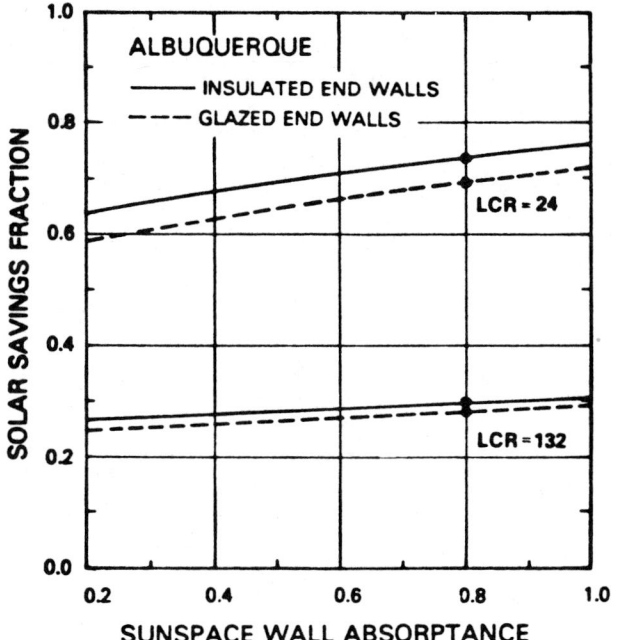

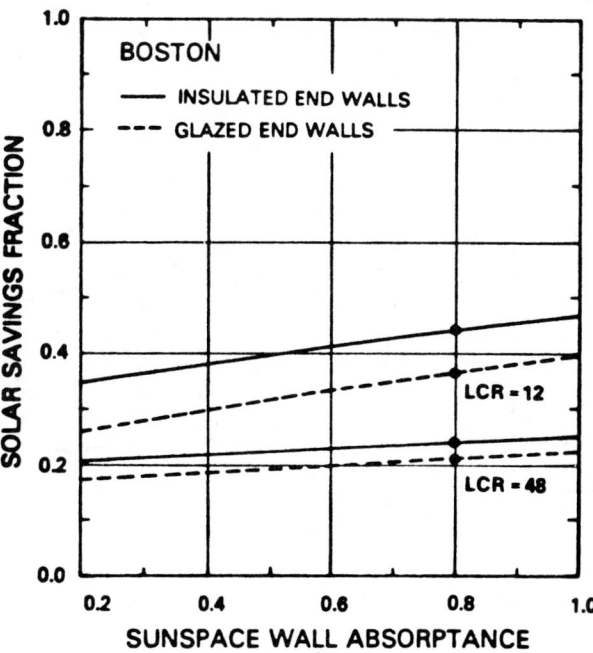

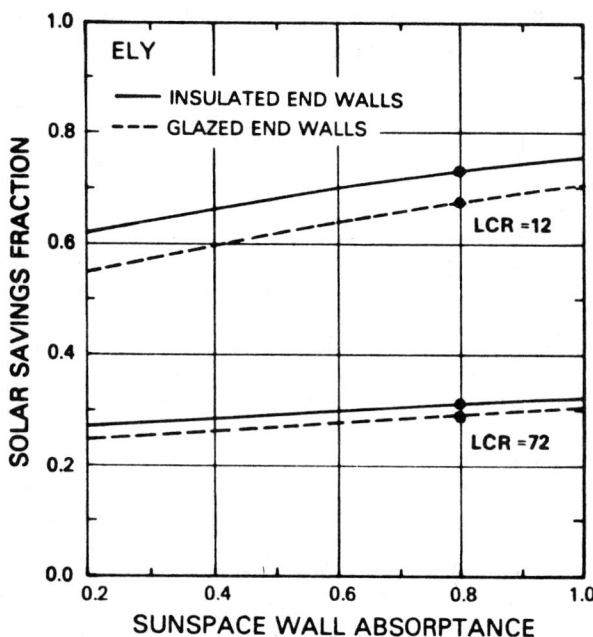
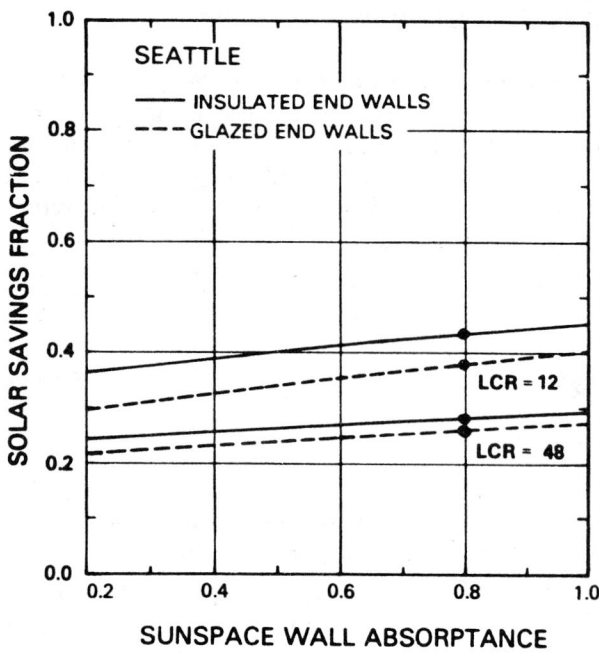
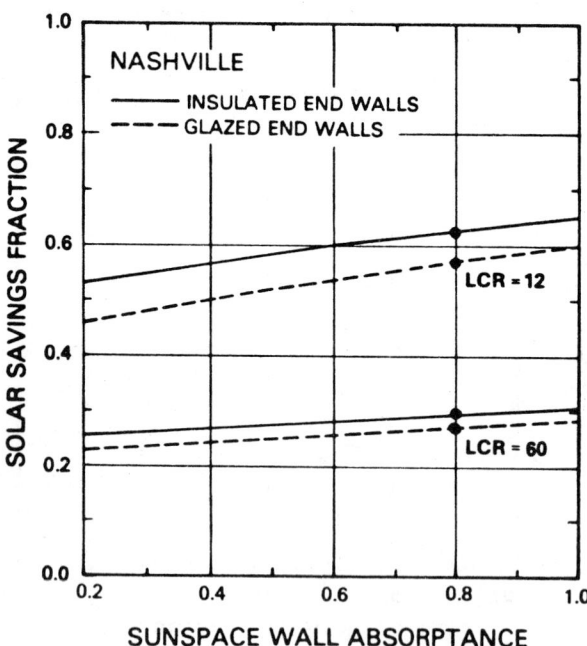

696 / *APPENDIX 22*

22.3.a.14 Floor Absorptance

The graphs below plot the effect on solar savings fraction of changes in floor absorptance— for geometry A with masonry common wall, no nighttime insulation, and both insulated and glazed end walls (A1 and A3)— for each of six reference cities. Dots represent reference conditions.

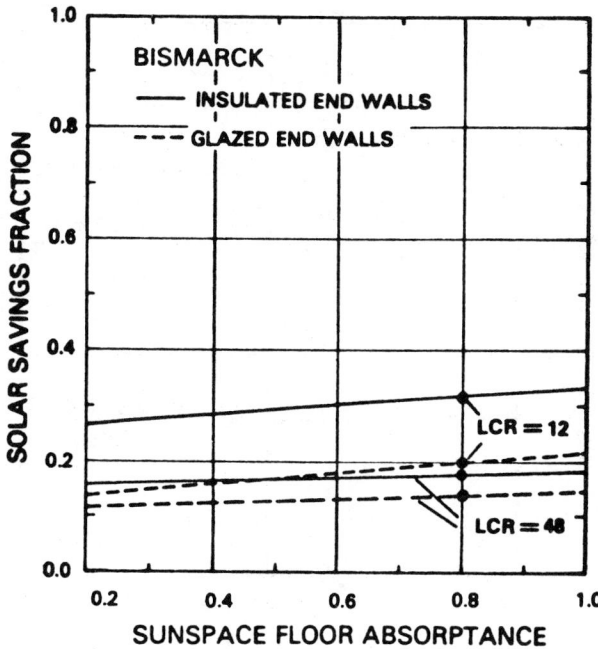

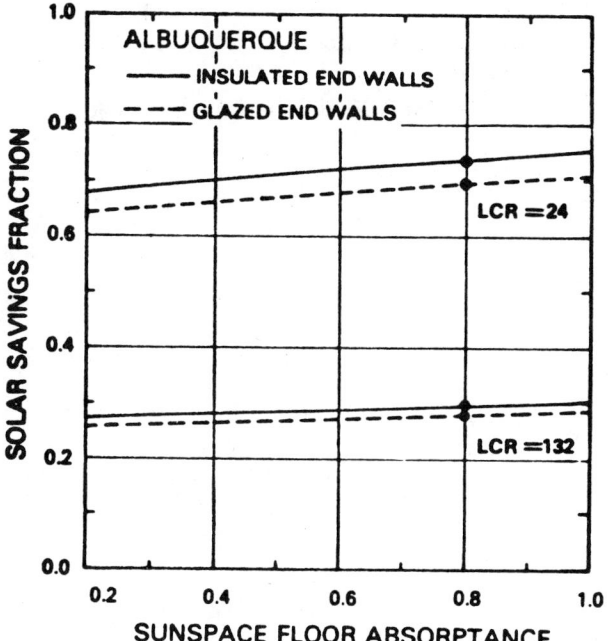

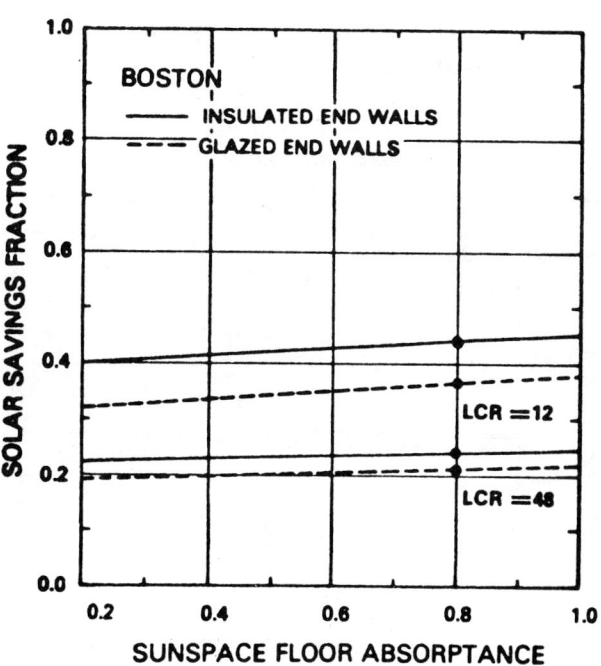

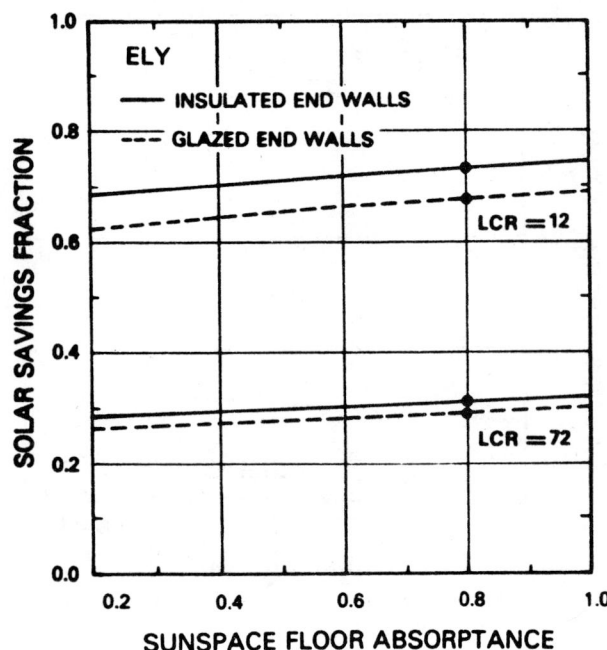
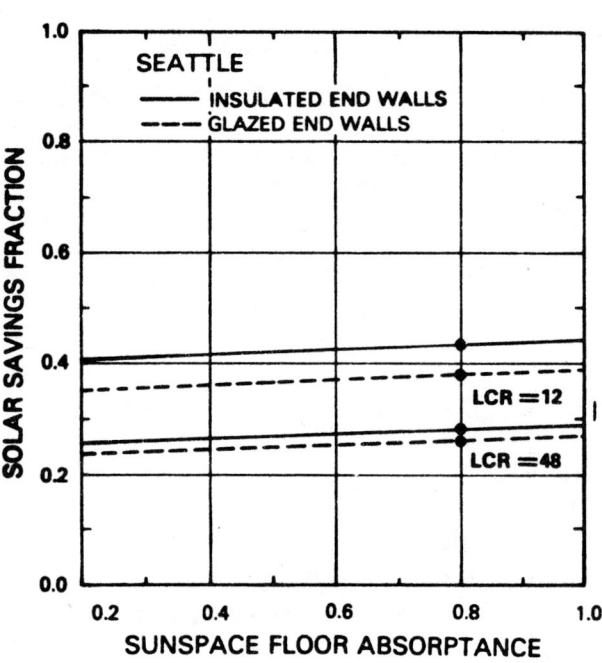
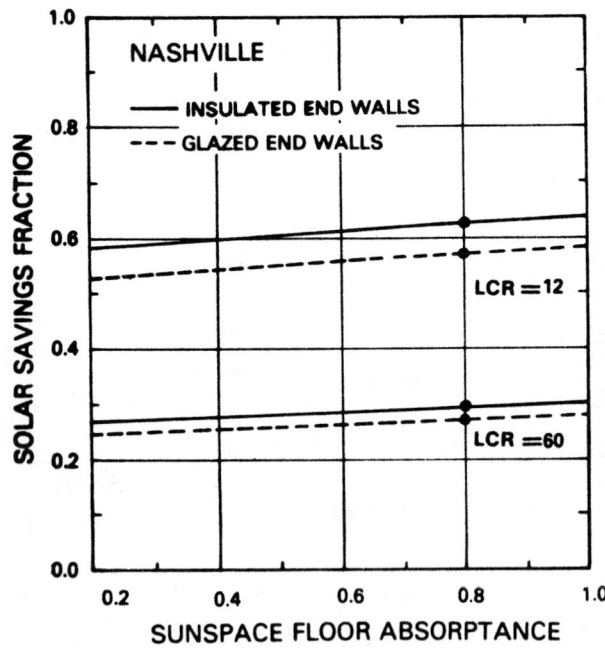

698 / *APPENDIX 22*

22.3.a.15 Floor and Wall Absorptance

The graphs below plot the effect on solar savings fraction of changes in floor and common-wall absorptance (varied simultaneously)—for sunspaces of two basic configurations—for each of six reference cities. The first graph for each city plots performance for geometry A with masonry common wall, no nighttime insulation, and both insulated and glazed end walls (A1 and A3). The second graph for each city plots performance for geometry C with masonry common wall, with and without R9 nighttime insulation (C1 and C2). Dots represent reference conditions.

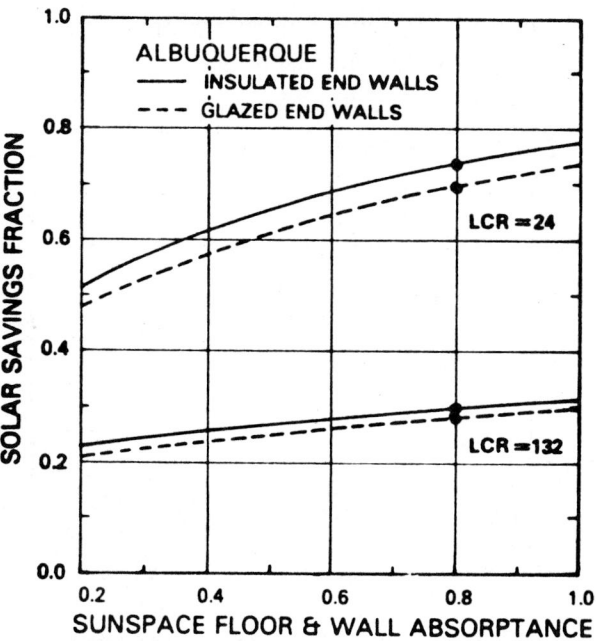

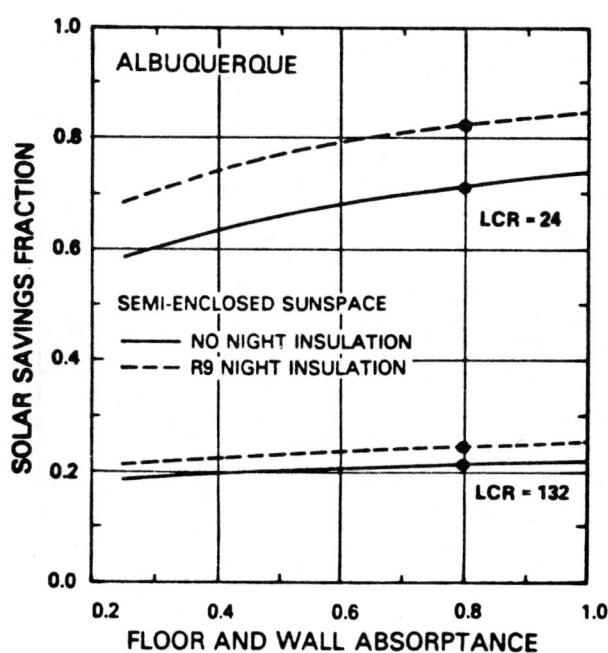

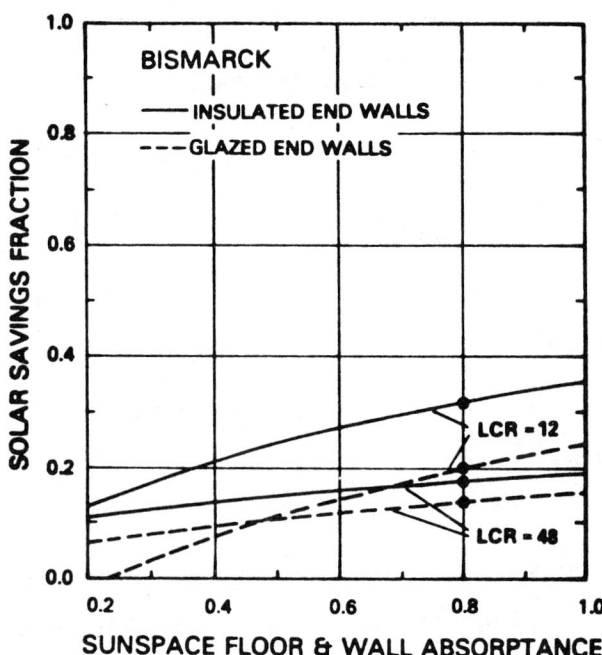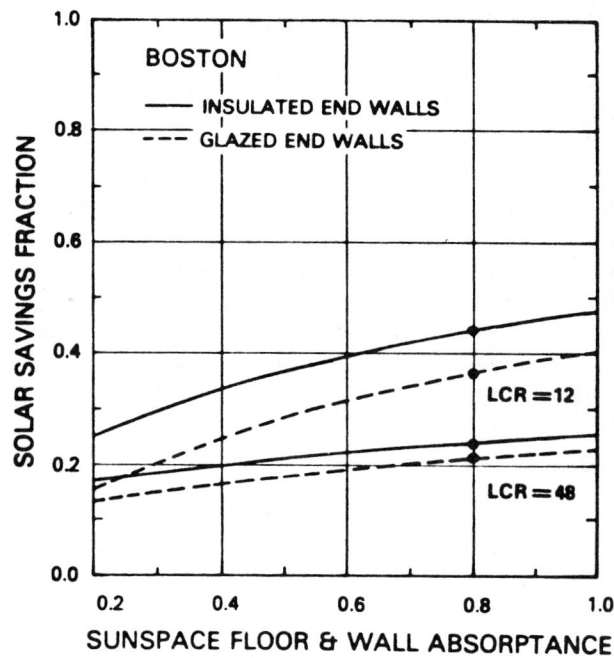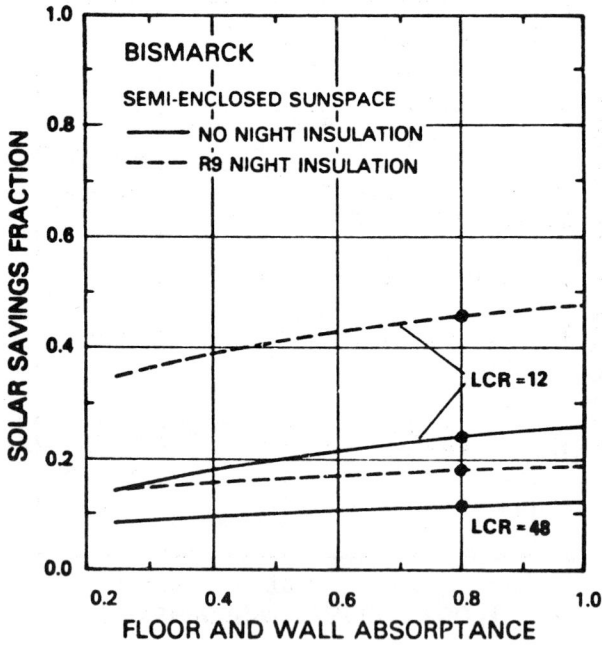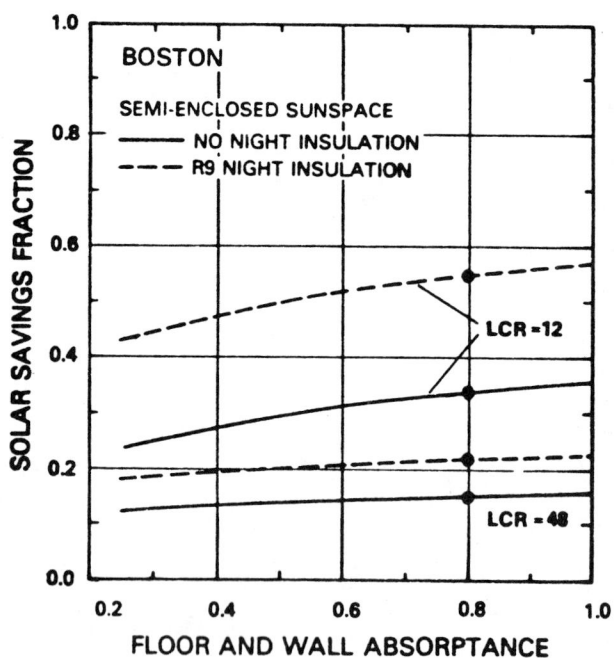

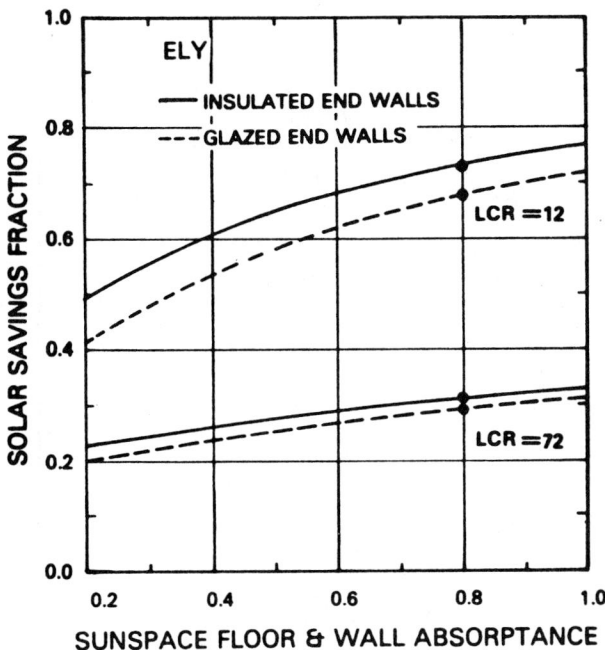
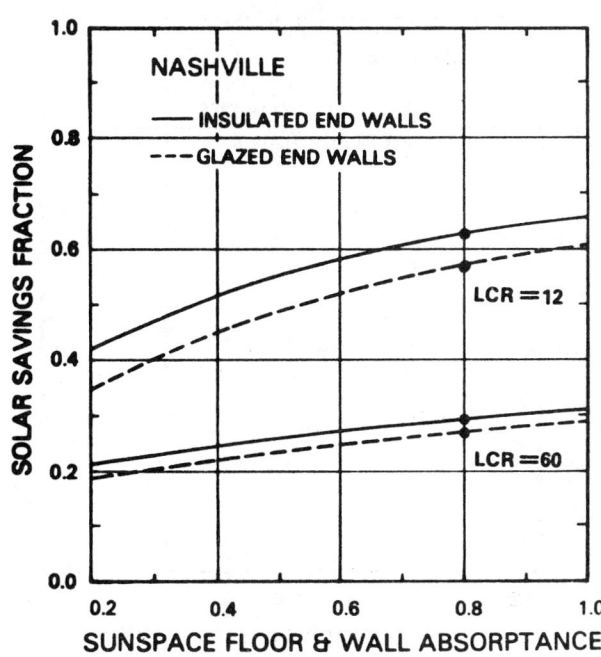
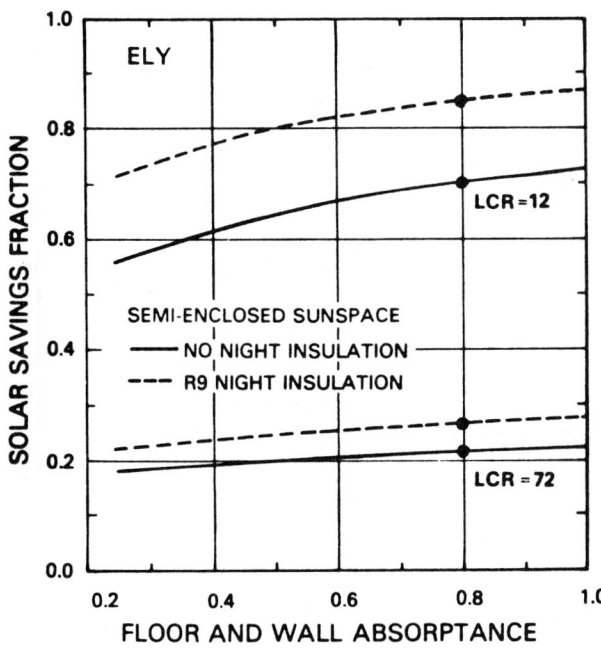
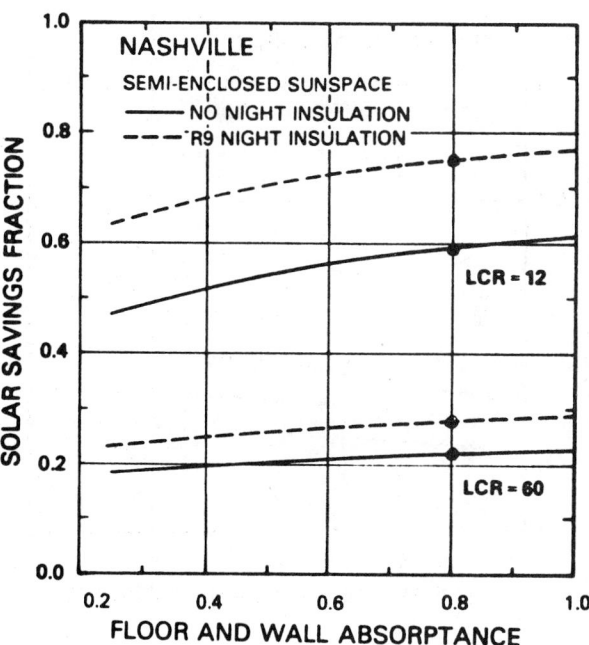

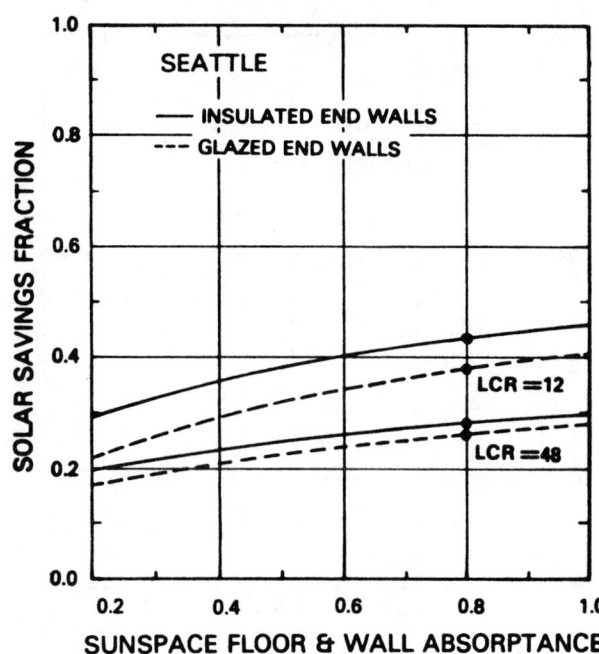
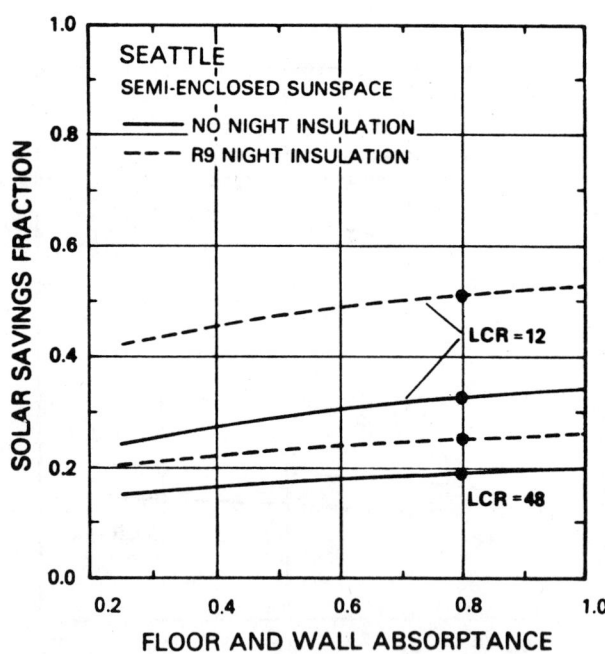

702 / APPENDIX 22

22.3.a.16 Water-Container Absorptance

The graphs below plot the effect on solar savings fraction of changes in water-container absorptance—for geometry A with lightweight insulated common wall, no nighttime insulation, and both insulated and glazed end walls (A5 and A7)—for each of six reference cities. Dots represent reference conditions.

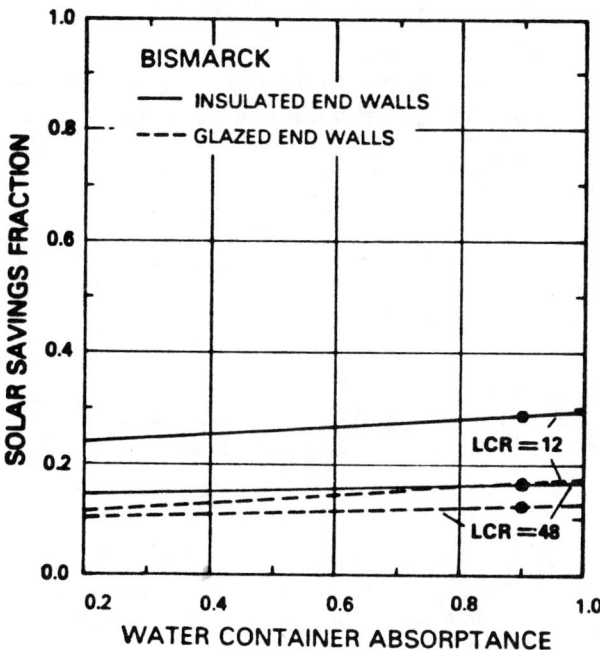

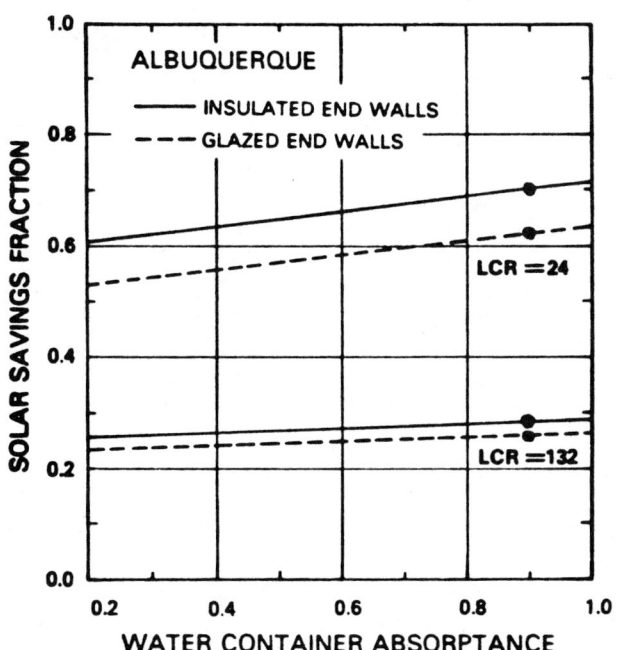

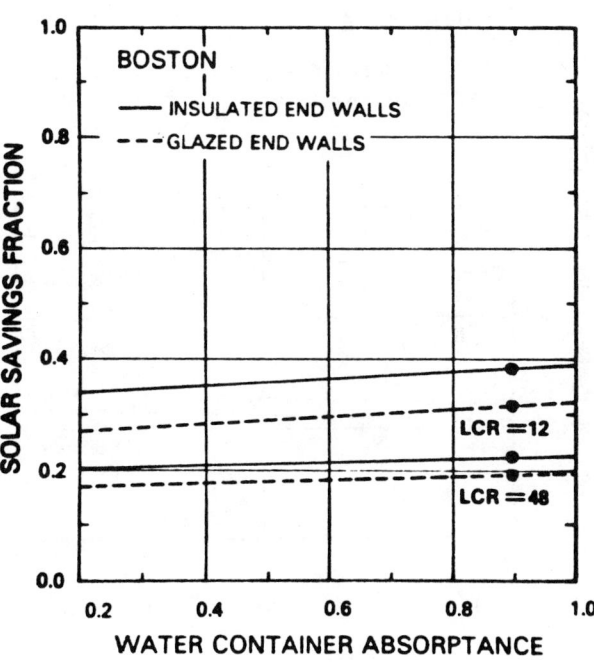

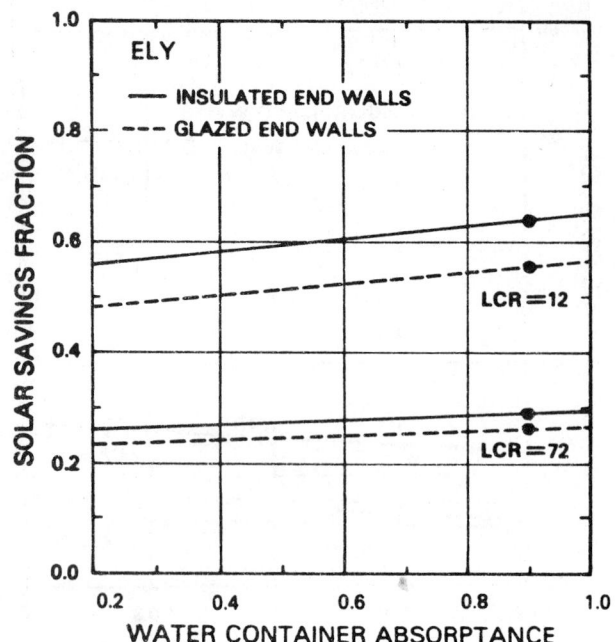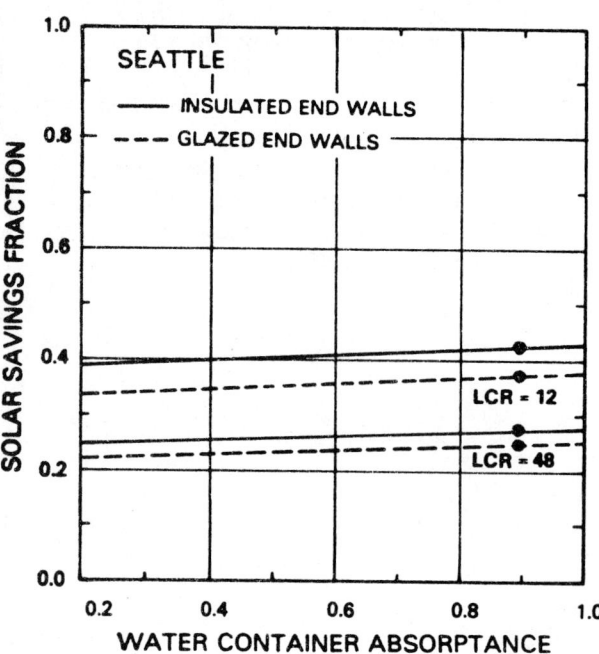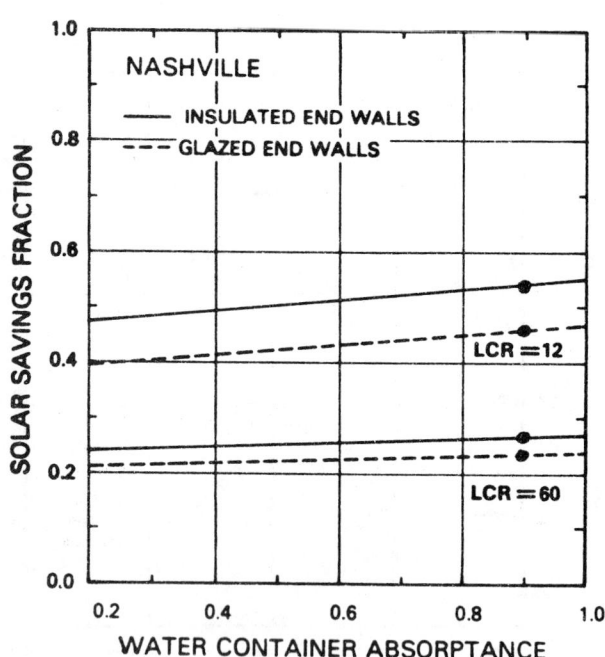

704 / APPENDIX 22

22.3.a.17 Lightweight Absorption Fraction

The graphs below plot the effect on solar savings fraction of changes in lightweight absorption fraction—for geometry A with either masonry or lightweight insulated common wall, with and without R9 nighttime insulation, and with insulated end walls (A1, A2, A5, and A6)—for each of six reference cities. Dots represent reference conditions.

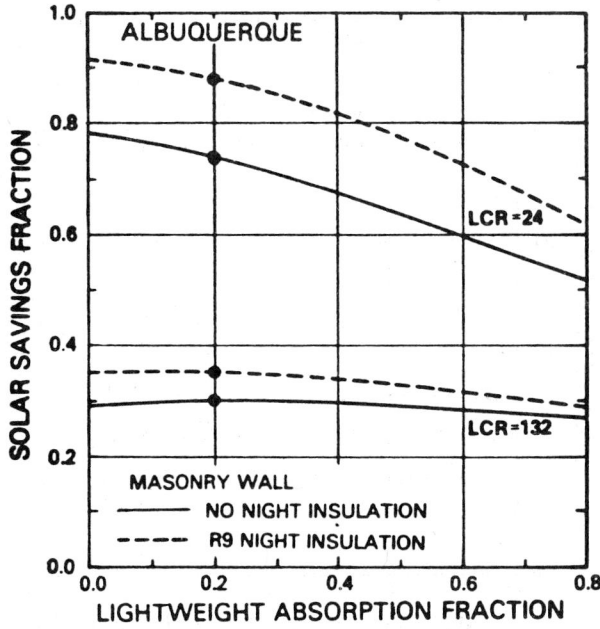

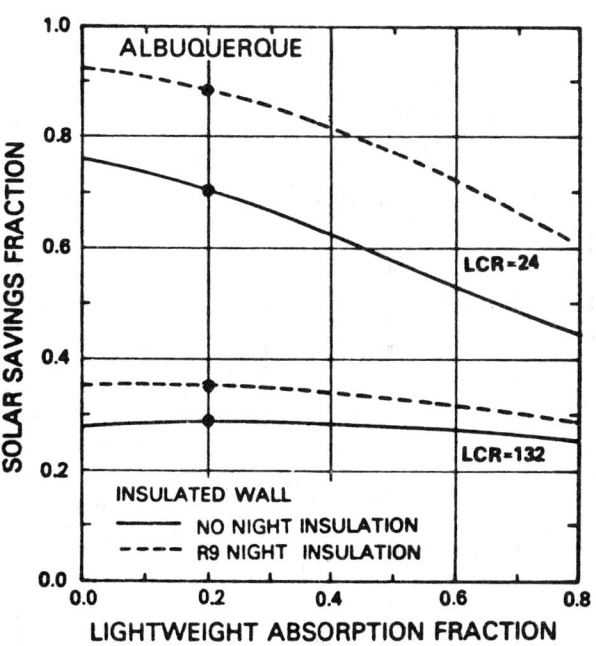

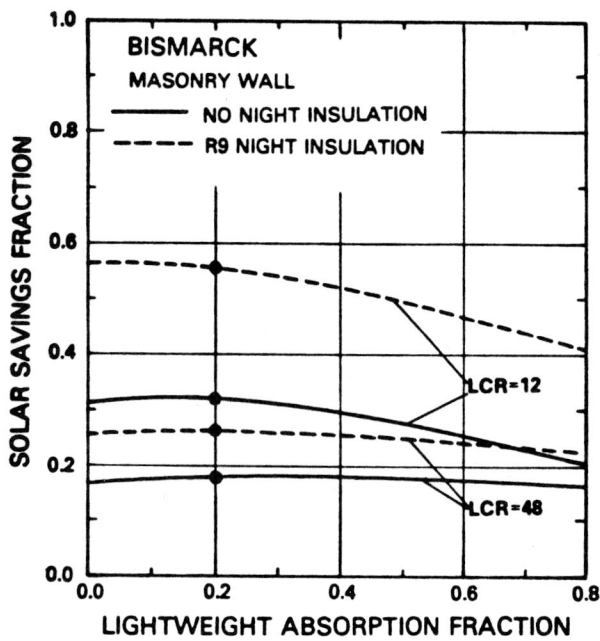
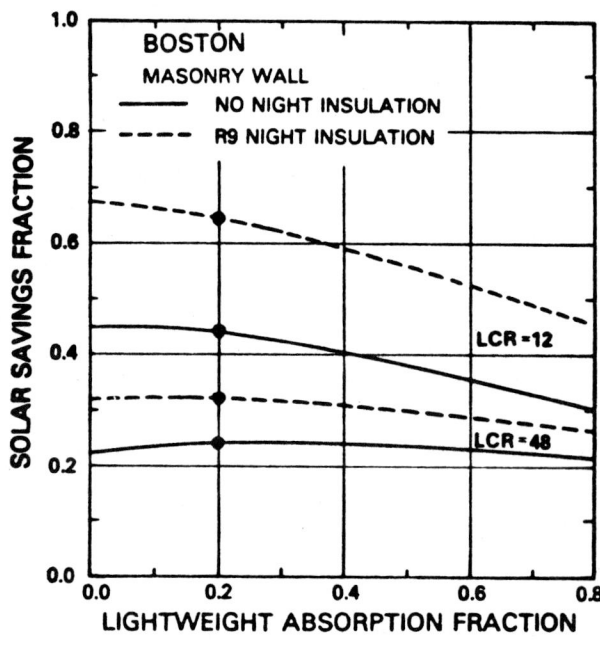
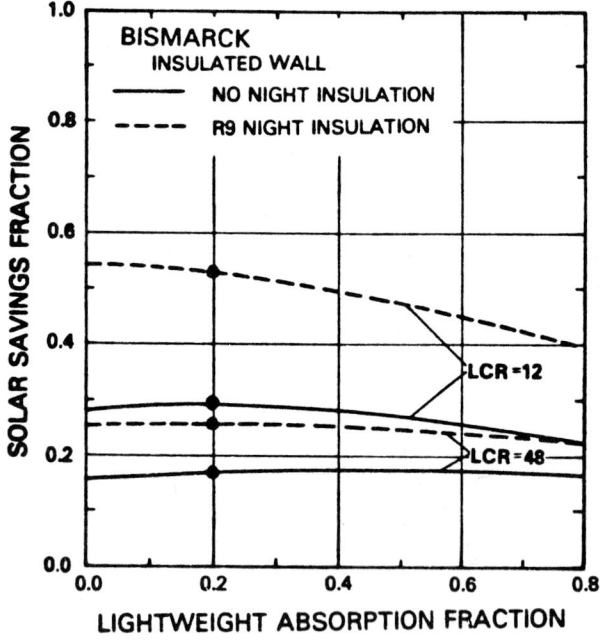
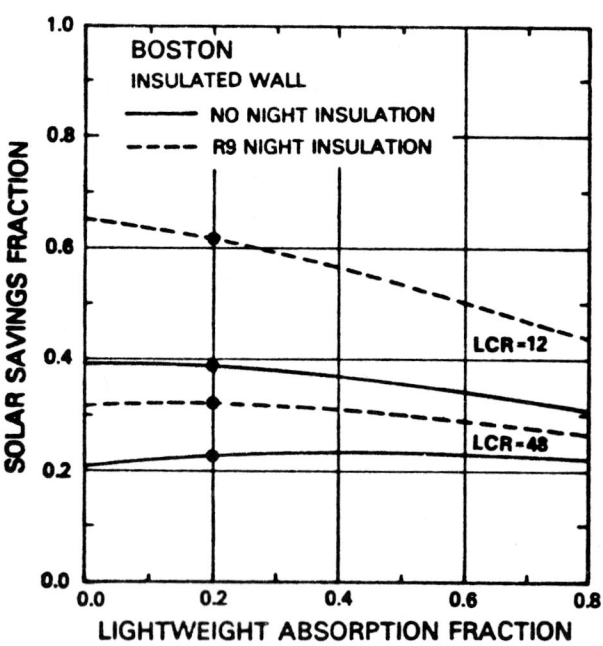

706 / APPENDIX 22

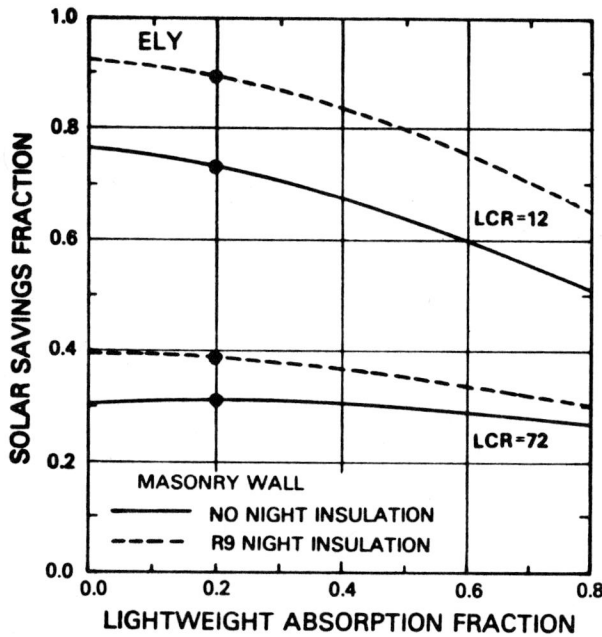
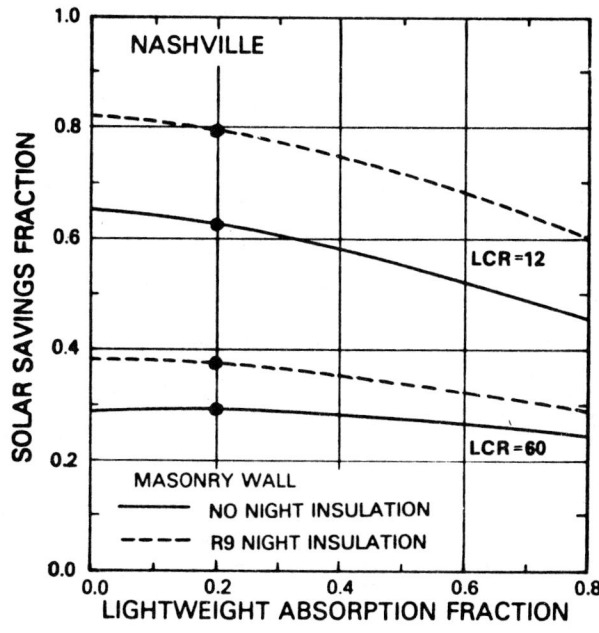
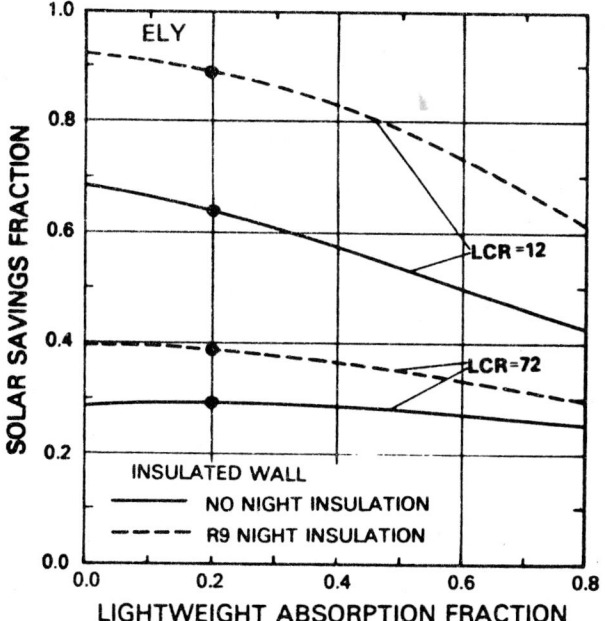
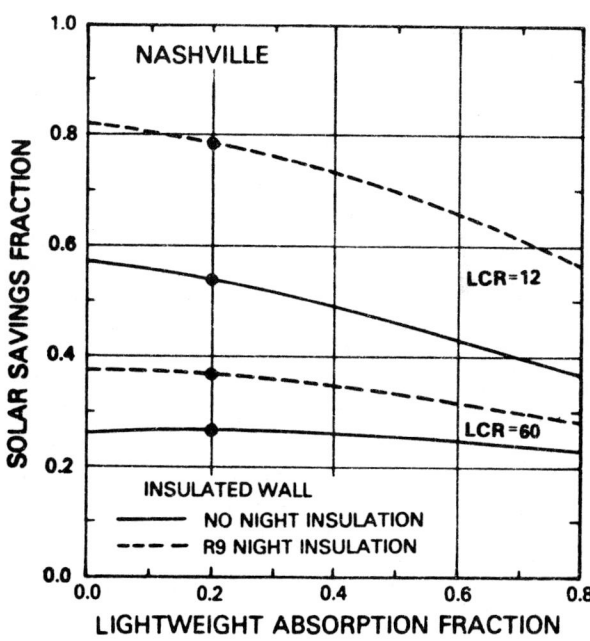

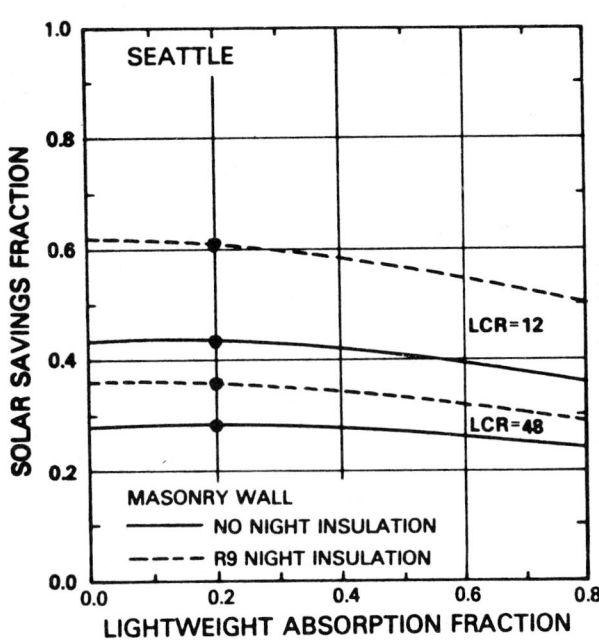

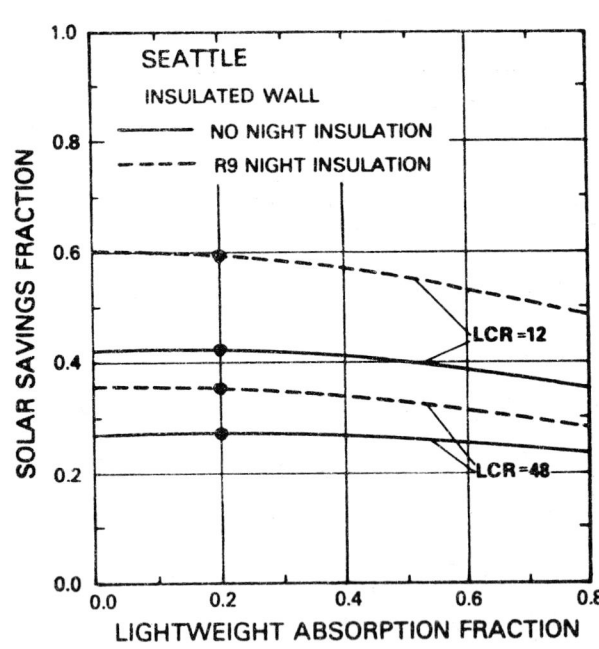

22.3.a.18 Vent Area Fraction

The graphs below plot the effect on solar savings fraction of changes in vent area fraction—for sunspaces of two basic configurations—for each of six reference cities. The first two graphs for each city plot performance for geometry A with insulated end walls, with and without R9 nighttime insulation, and with both masonry and lightweight insulated common wall (A1, A2, A5, and A6). The next two graphs for each city plot performance for geometry C with both masonry and lightweight insulated common wall, with and without R9 nighttime insulation (C1, C2, C3, and C4). Dots represent reference conditions in each case.

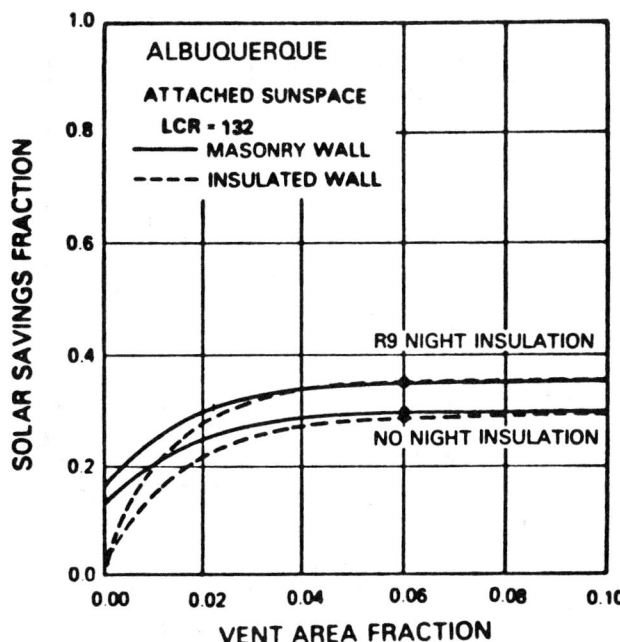

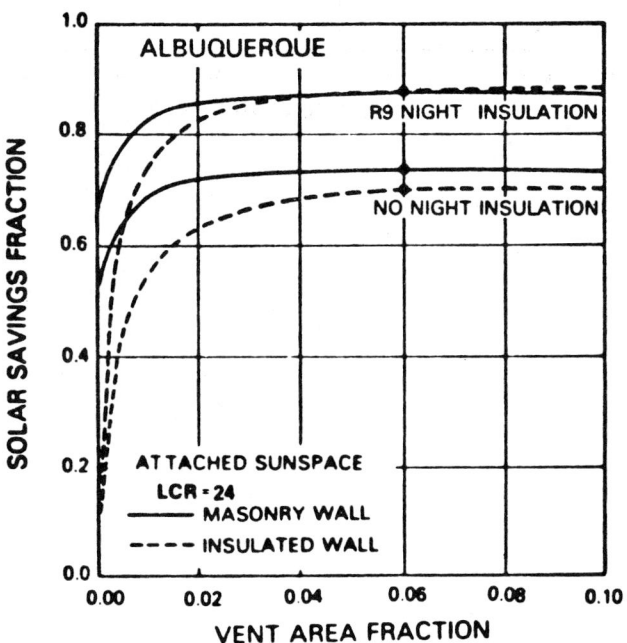

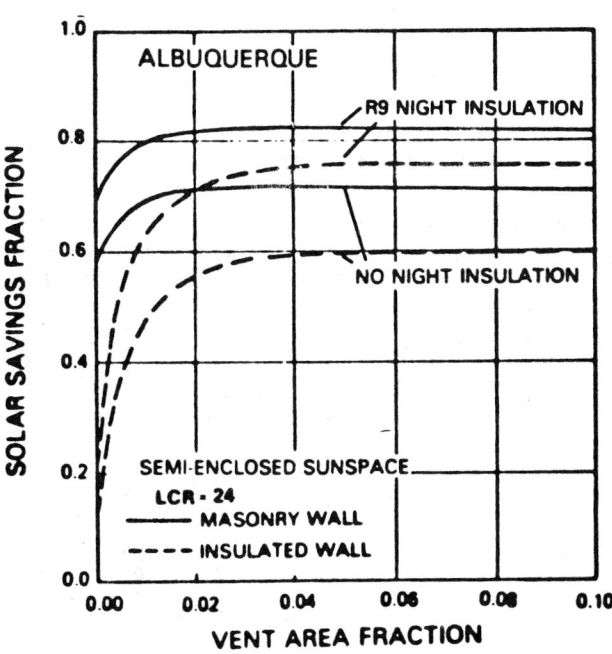

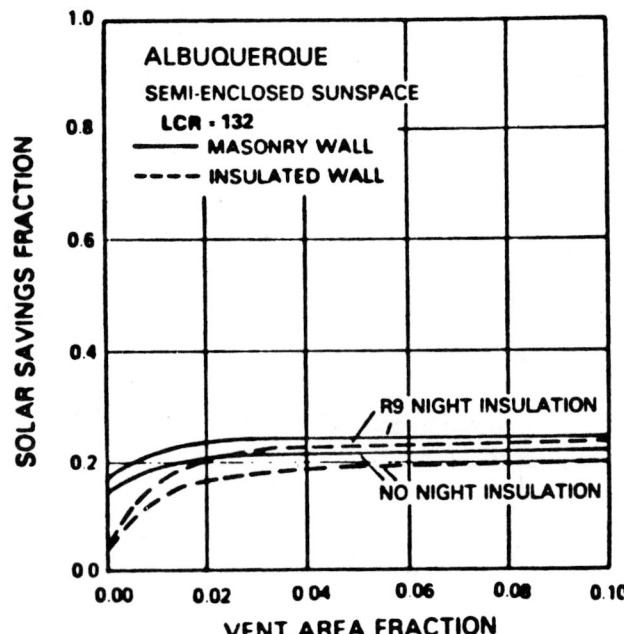
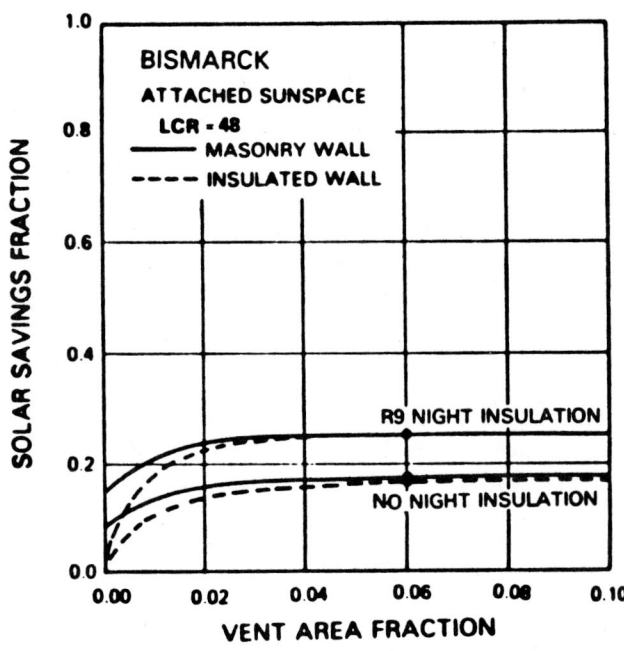
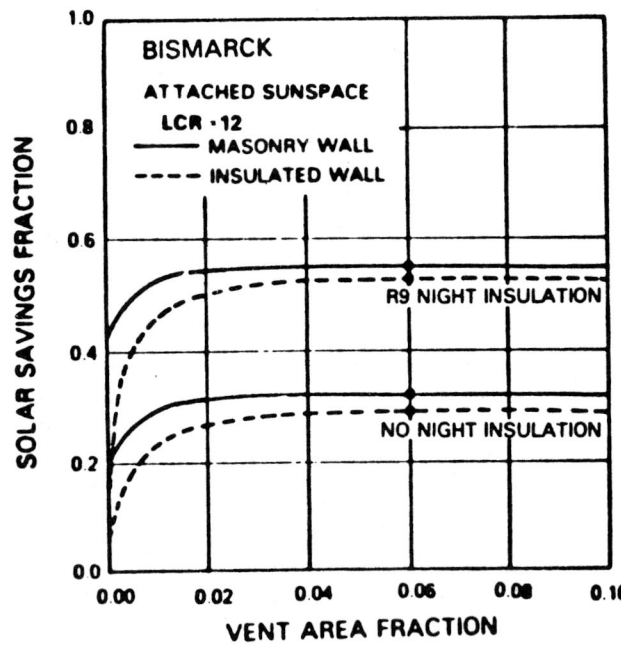

710 / APPENDIX 22

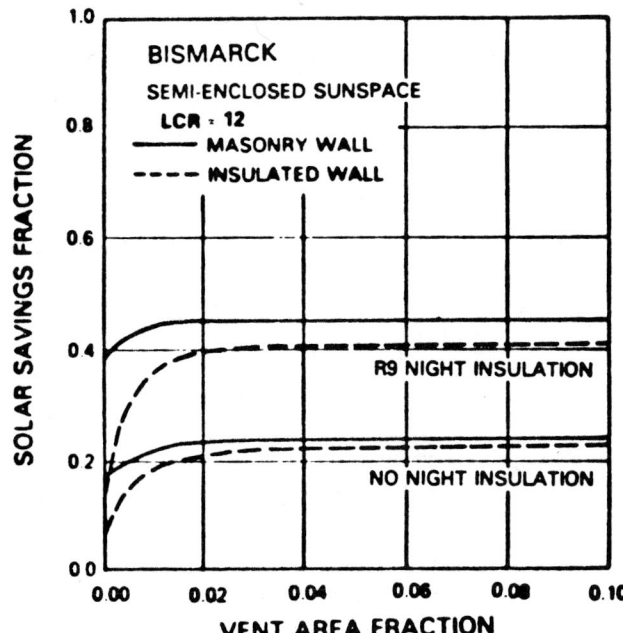
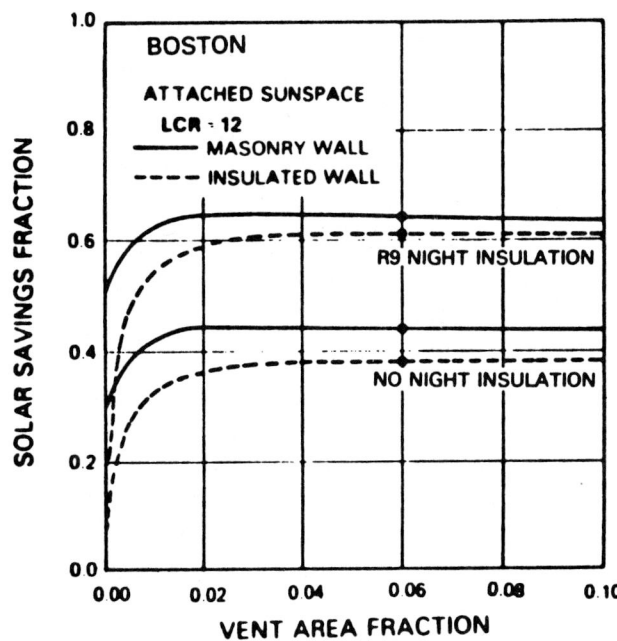
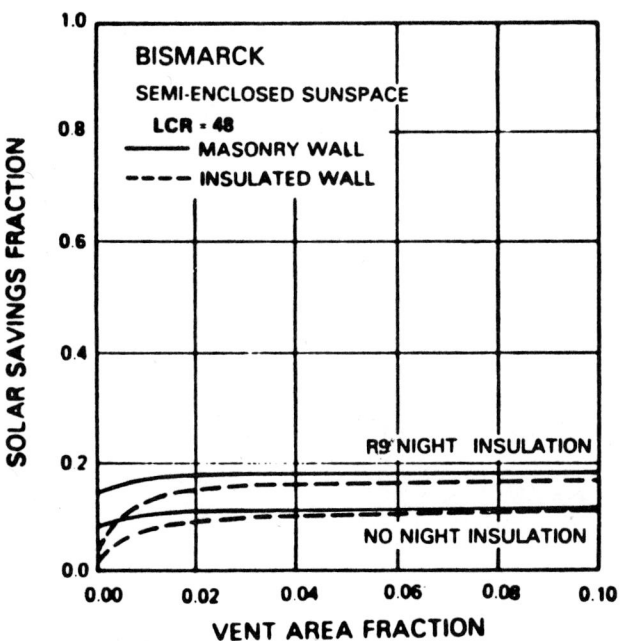

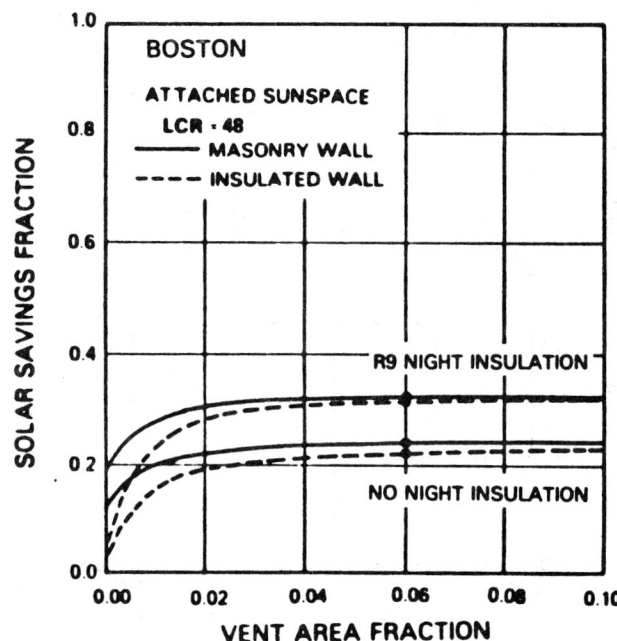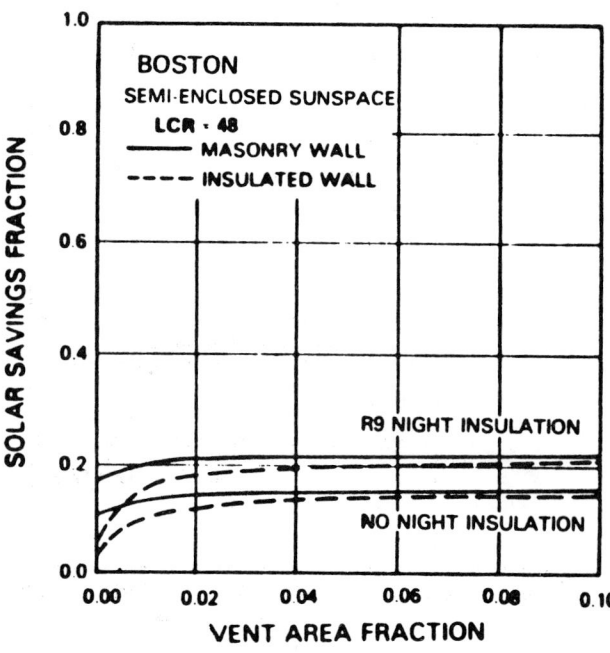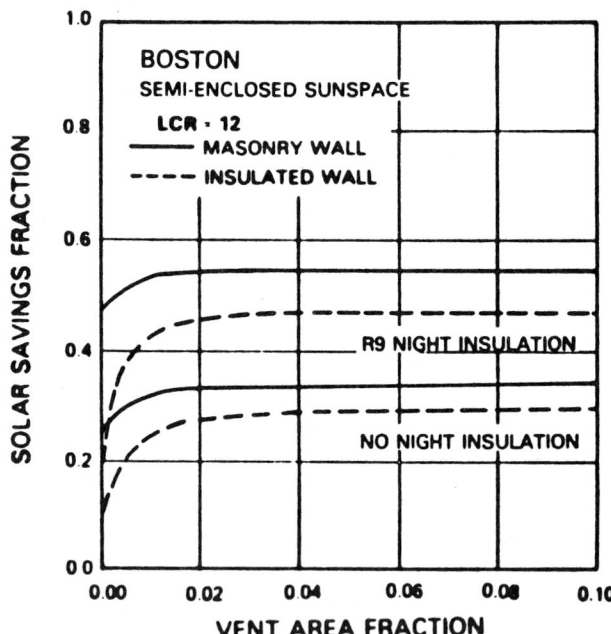

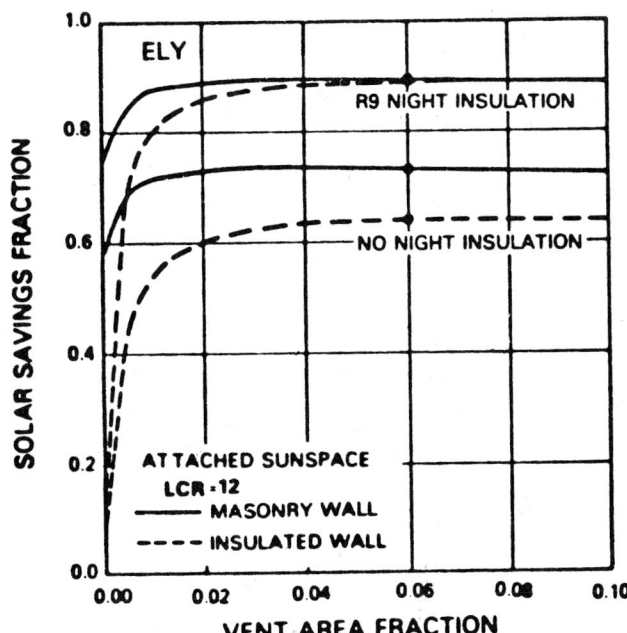
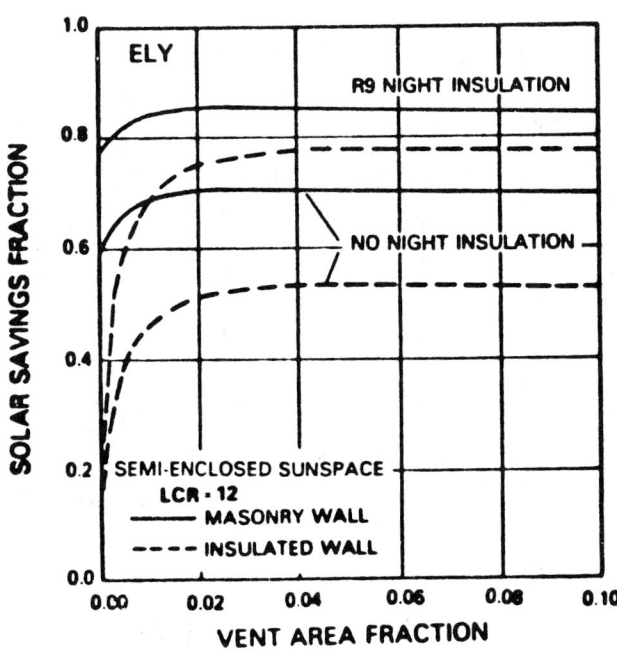
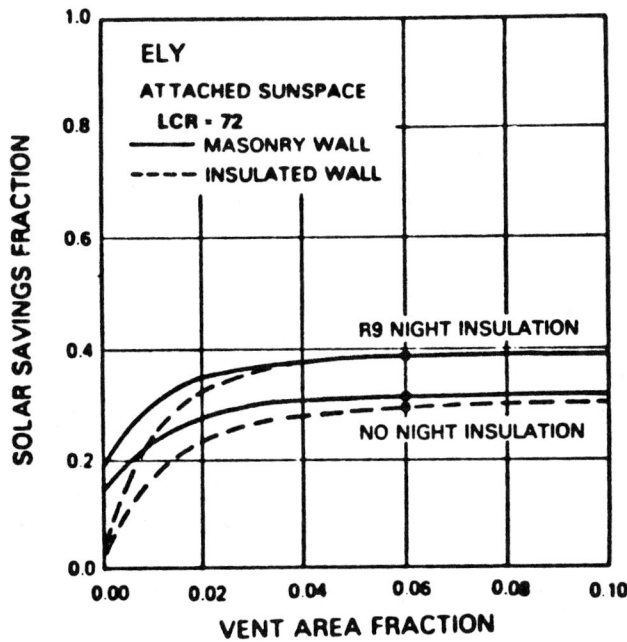

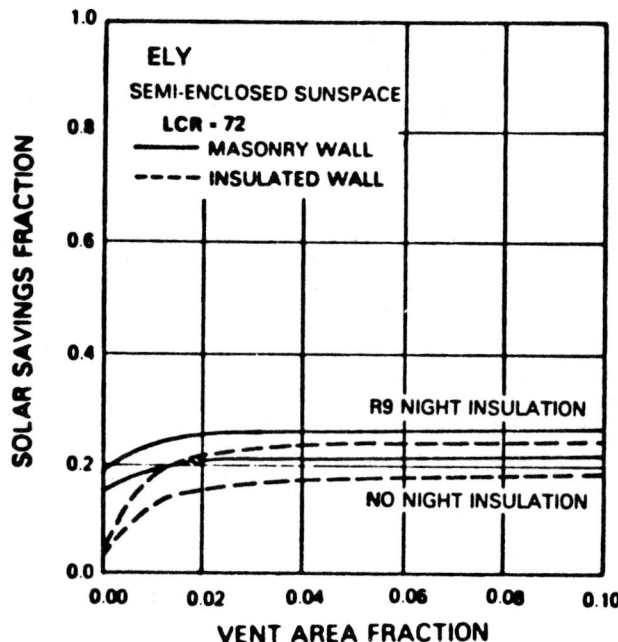
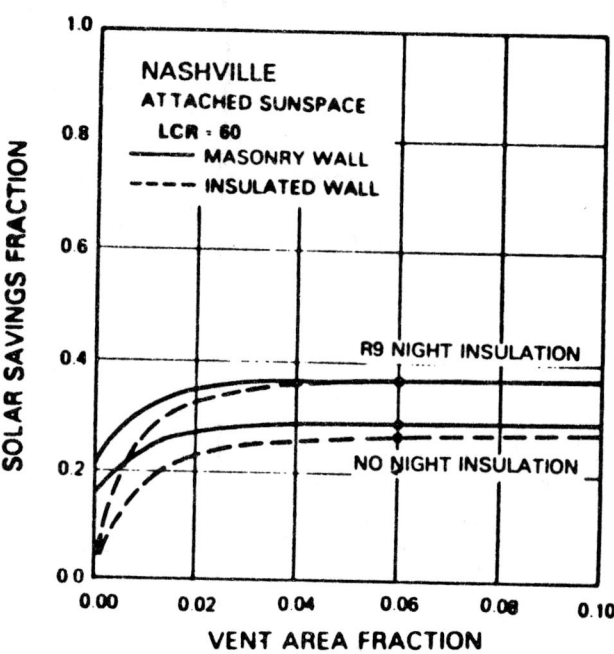
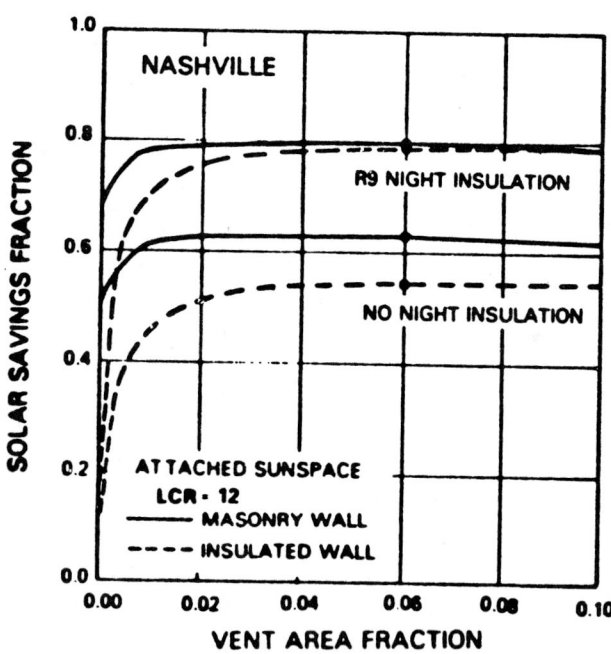

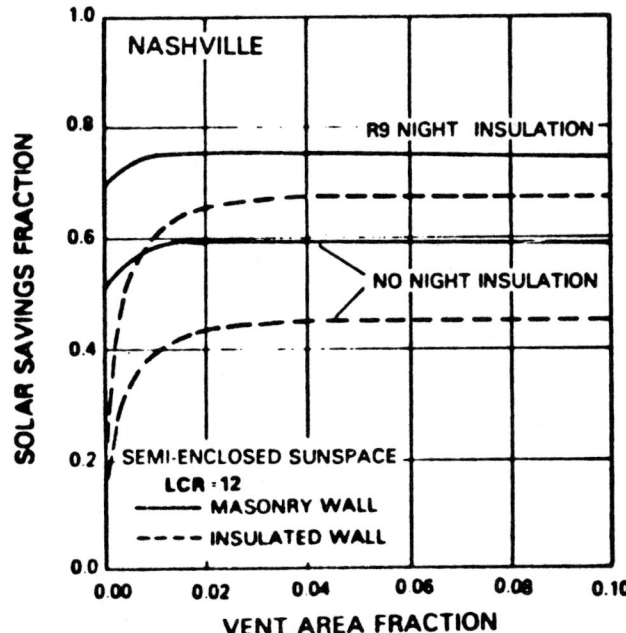
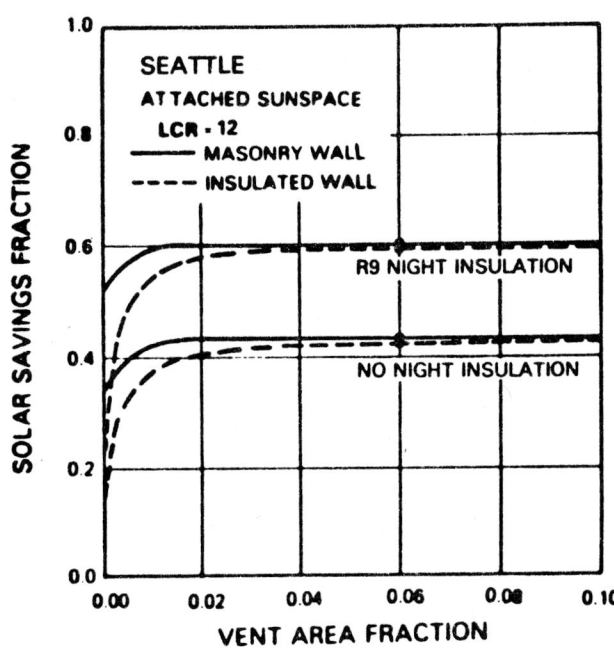
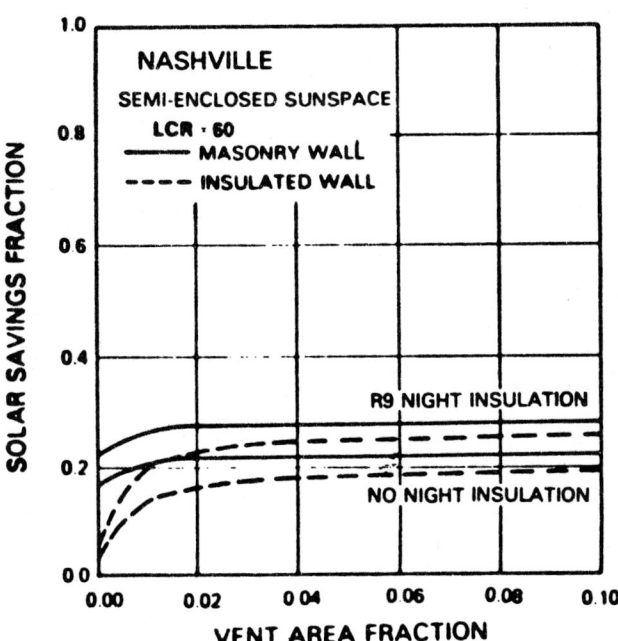

Sensitivity Data / 715

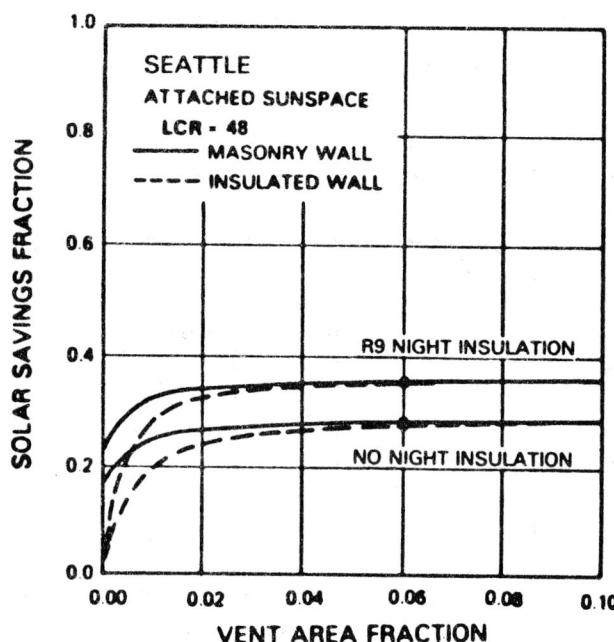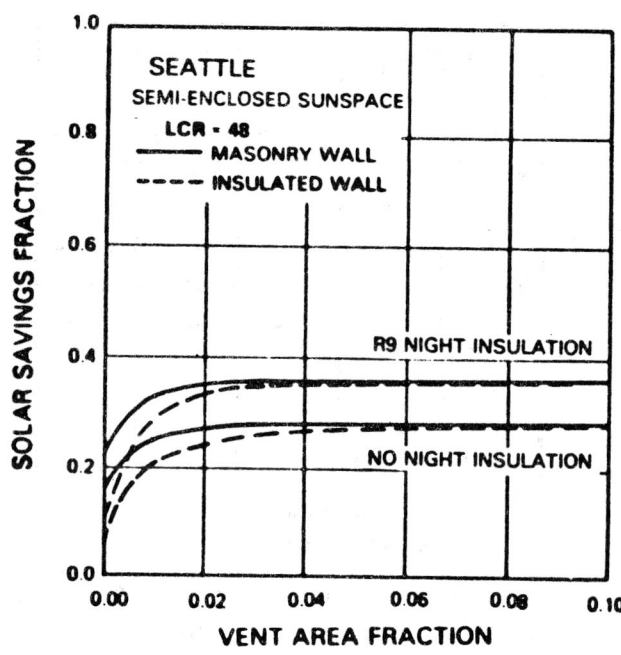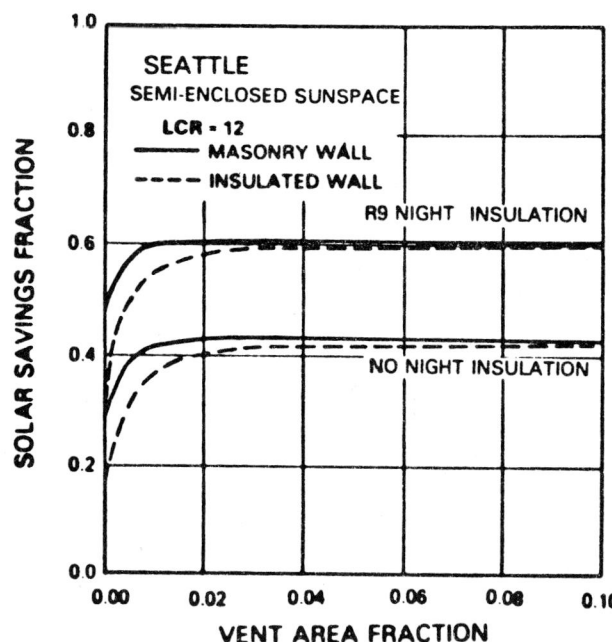

22.3.a.19 Sunspace Infiltration Rate

The graphs below plot the effect on solar savings fraction of changes in sunspace infiltration rate—for geometry A with masonry common wall, with and without R9 nighttime insulation, and with insulated end walls (A1 and A2)—for each of six reference cities. Dots represent reference conditions.

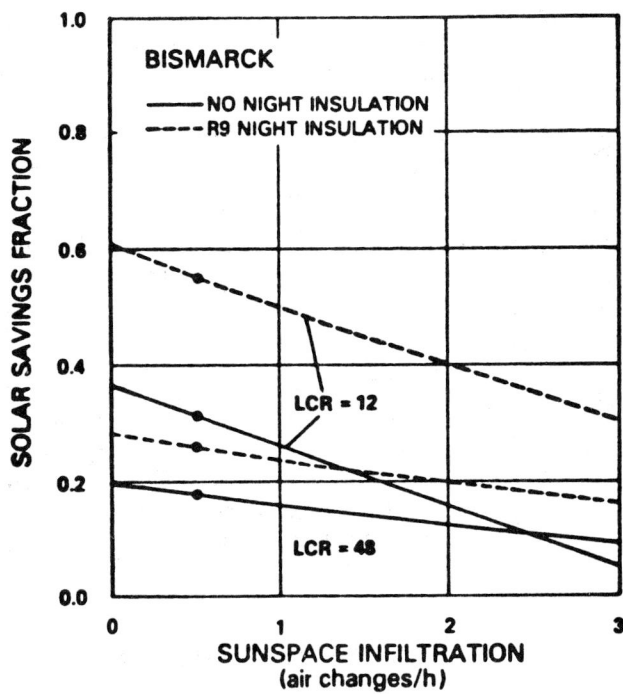

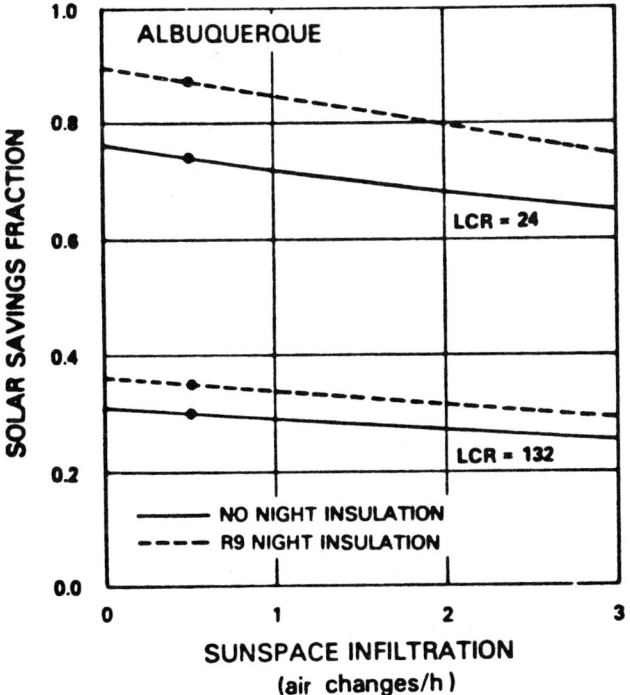

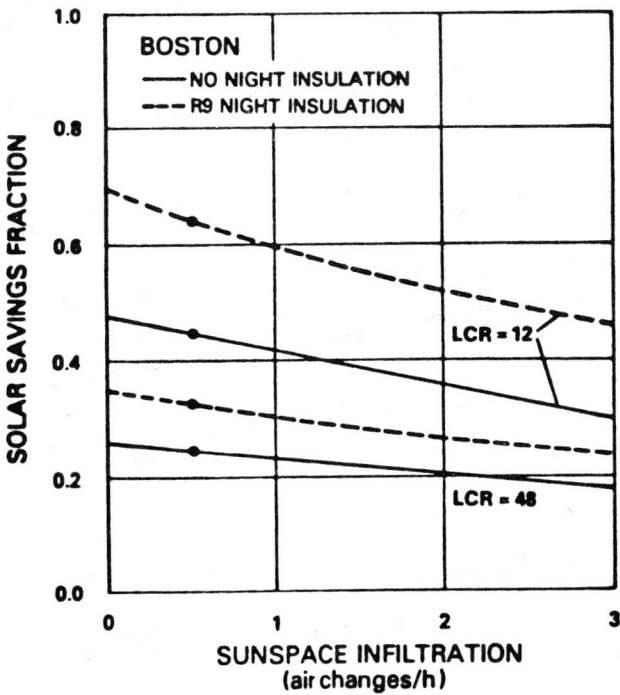

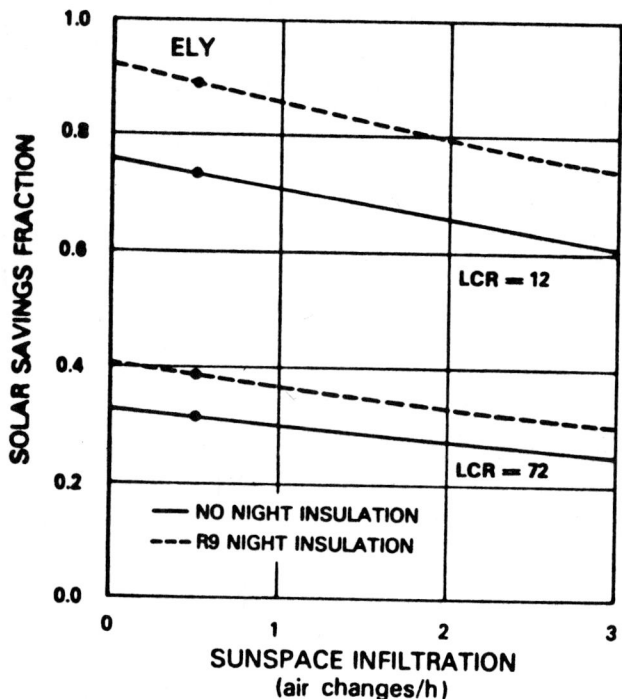
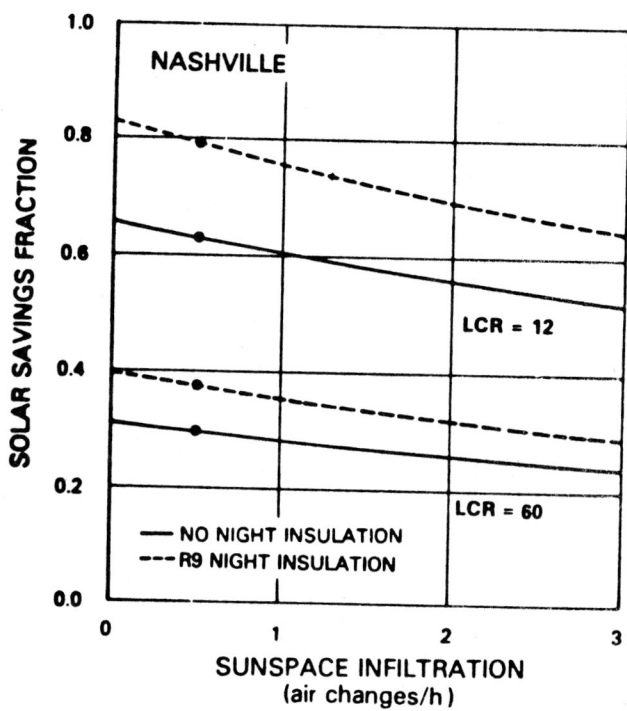
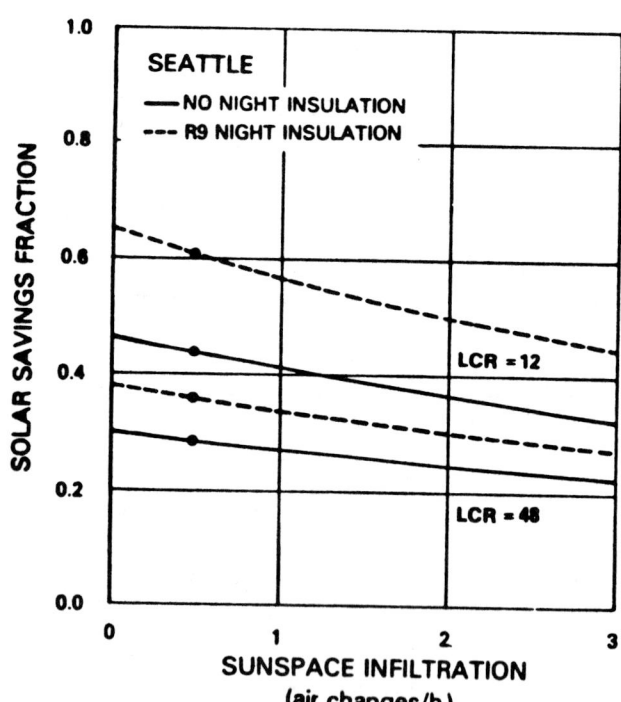

718 / APPENDIX 22

22.3.a.20 Common-Wall R-Value

The graphs below plot the effect on solar savings fraction of changes in common-wall R-value—for geometry A with lightweight insulated common wall, with and without R9 nighttime insulation, and with either insulated or glazed end walls (A1, A2, A3, and A4)—for each of six reference cities. Wall-surface-to-air film resistances are not included. Dots represent reference conditions.

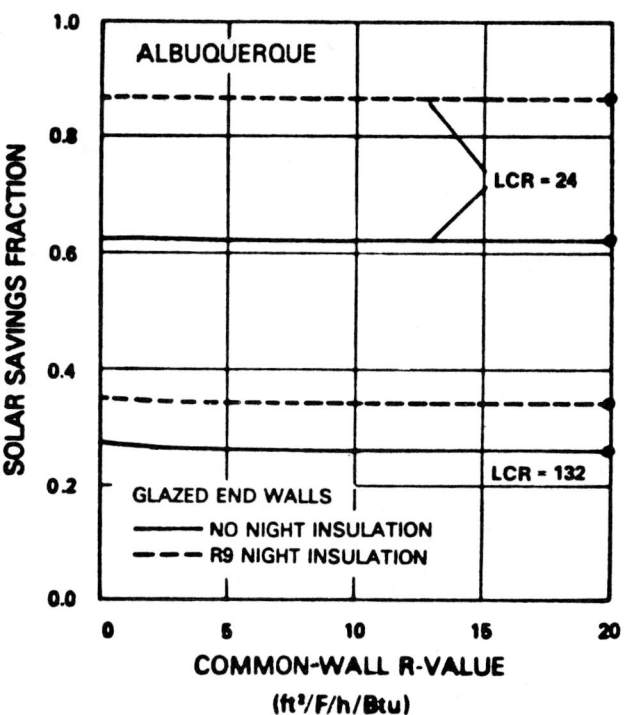

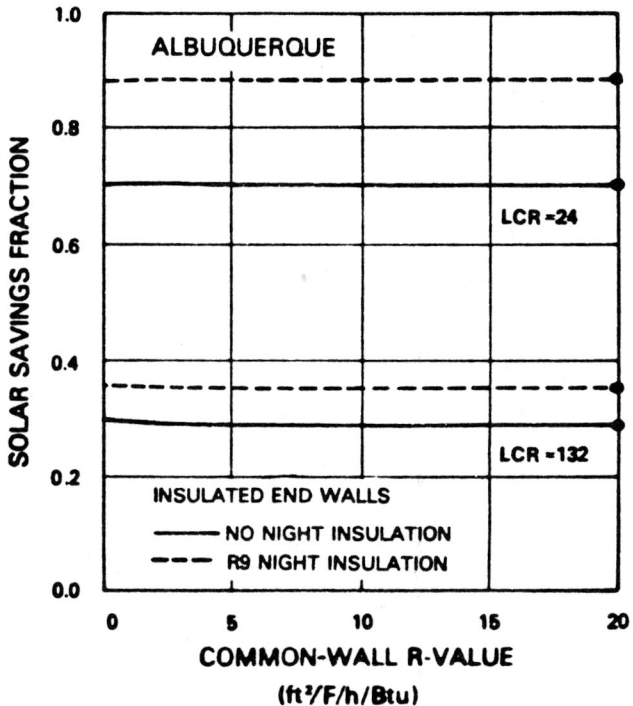

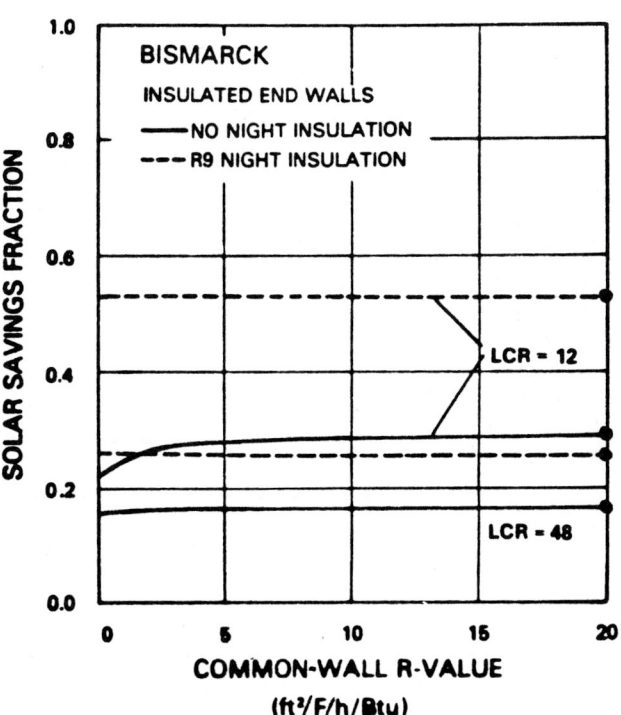

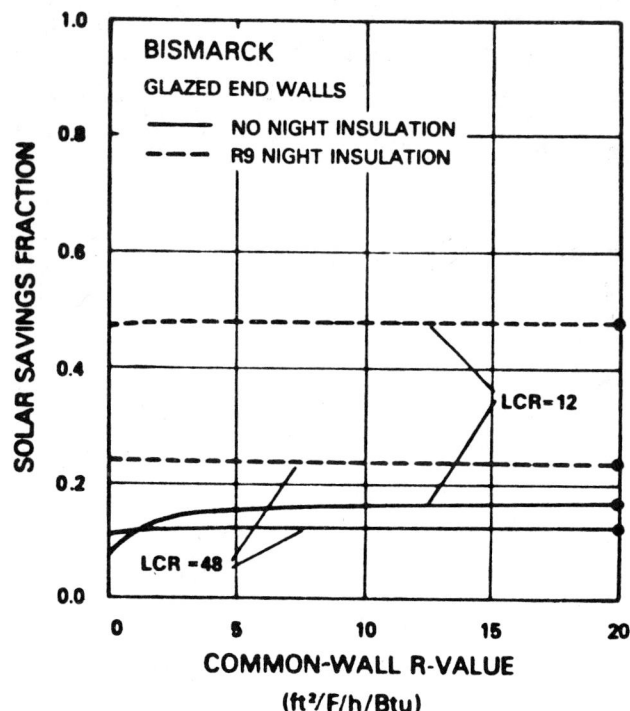
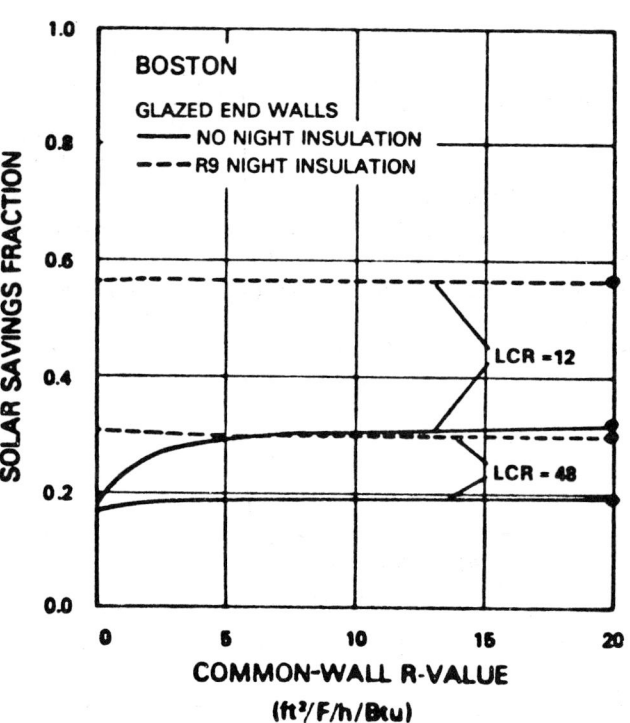
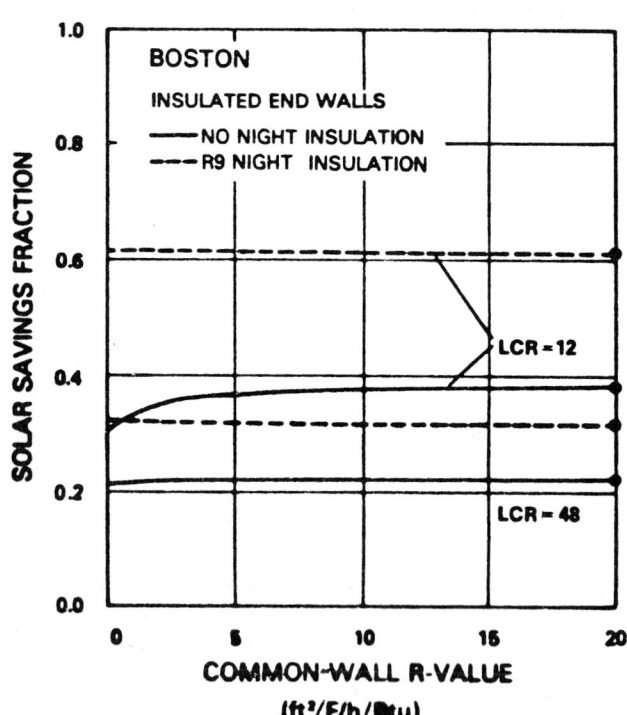

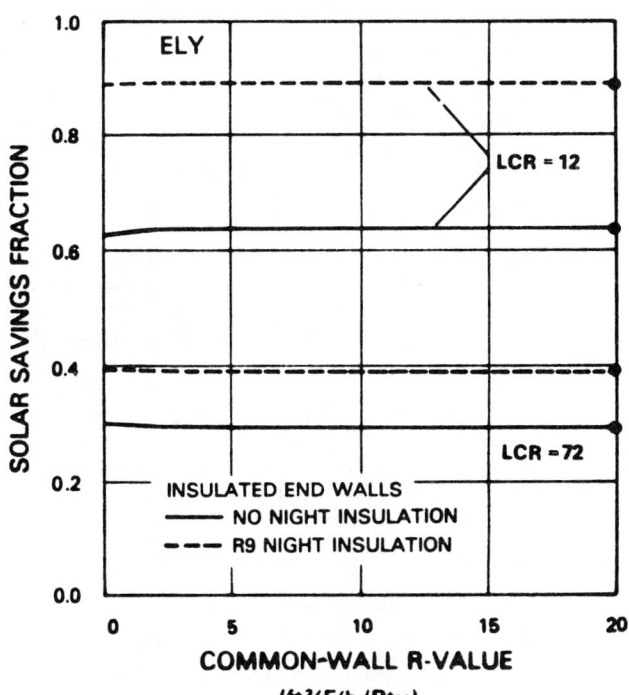
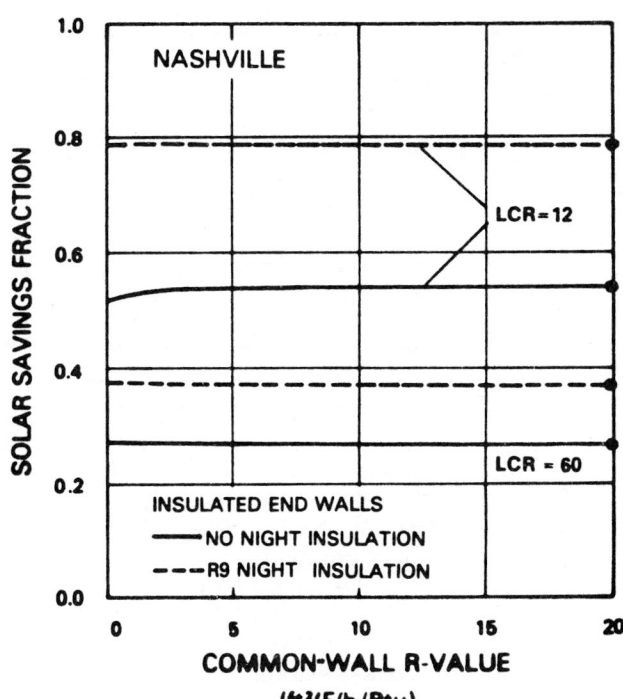
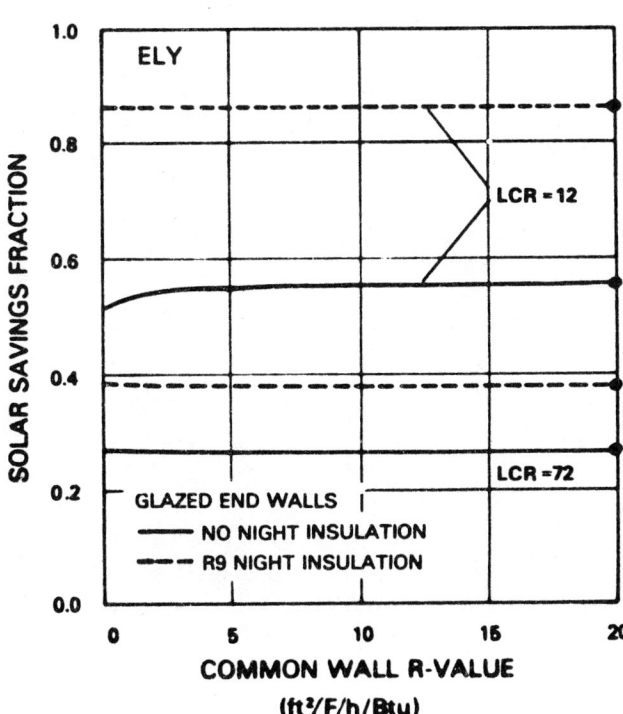

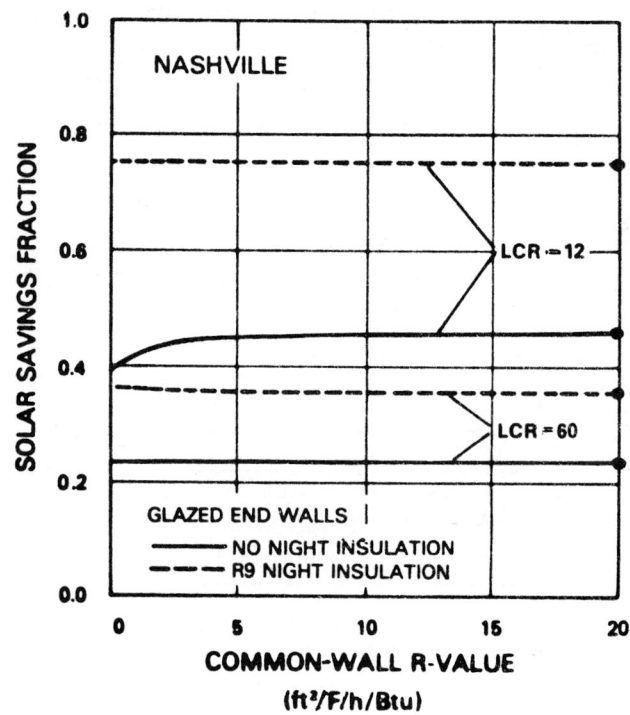
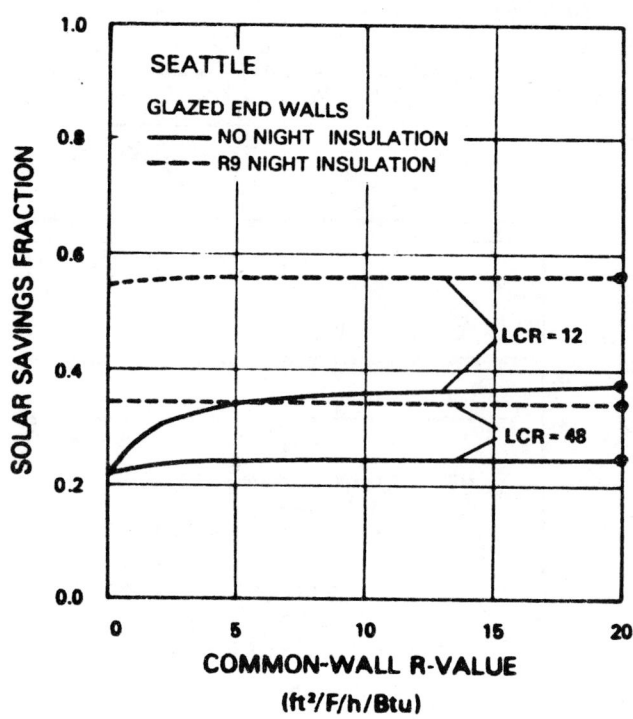
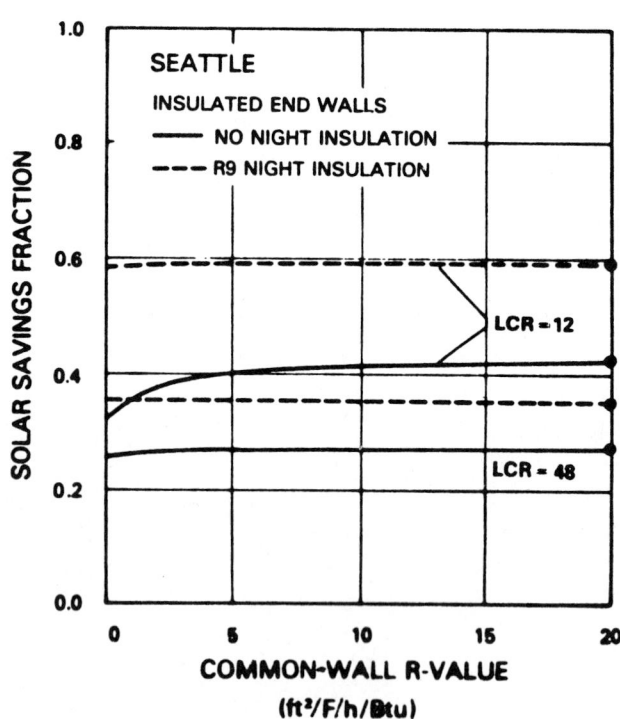

722 / APPENDIX 22

22.3.a.21 Minimum Room Setpoint

The graphs below plot the effect on solar savings fraction of changes in minimum room air temperature (as determined by the thermostat setpoint on the auxiliary heating system)—for geometry A with masonry common wall, no nighttime insulation, and both insulated and glazed end walls (A1 and A3)—for each of six reference cities. The first graph for each city plots *SSF* relative to a net reference load evaluated at the variable thermostat setpoint; the *SSF* in this case therefore reflects only the variation in solar savings, and not the variation in the building load, produced by the varying thermostat setpoint. The second graph for each city plots *SSF* relative to a 65°F reference load; it therefore reflects variation in both solar savings and in building load produced by the varying thermostat setpoint. Dots represent reference conditions in each case.

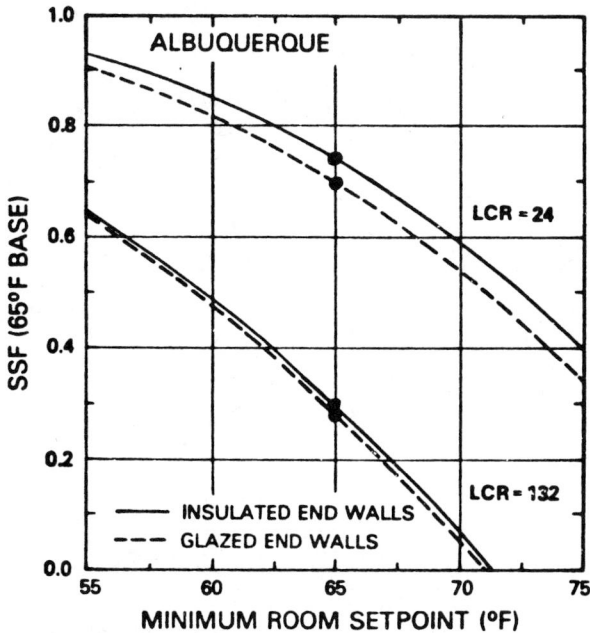

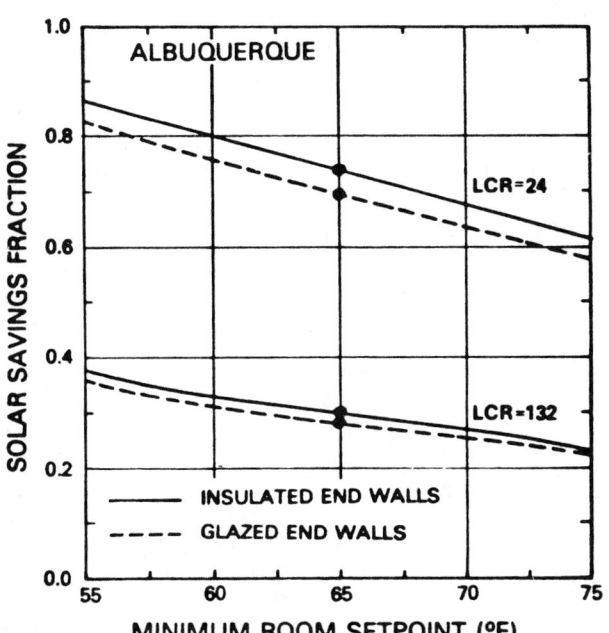

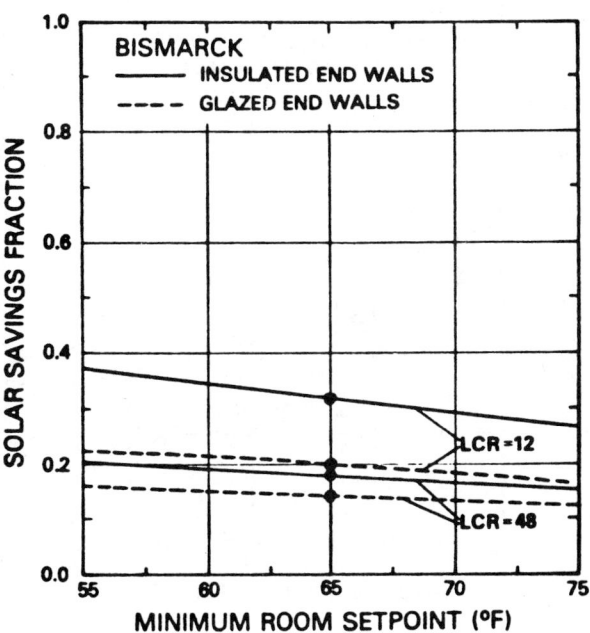

Sensitivity Data / 723

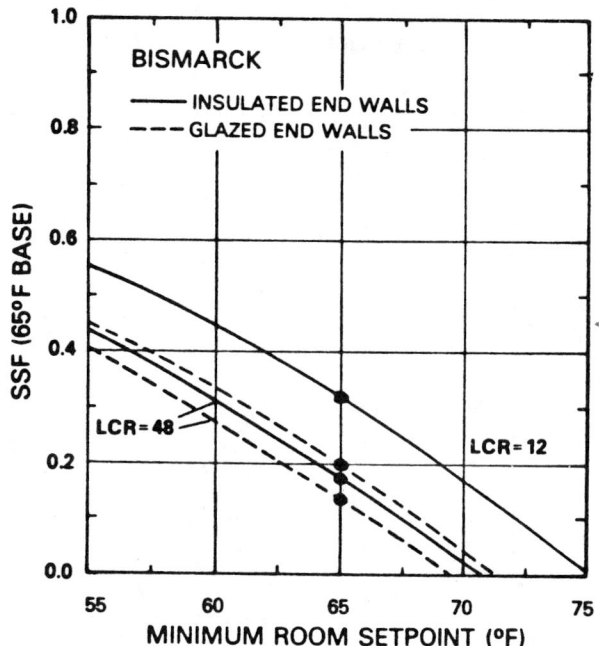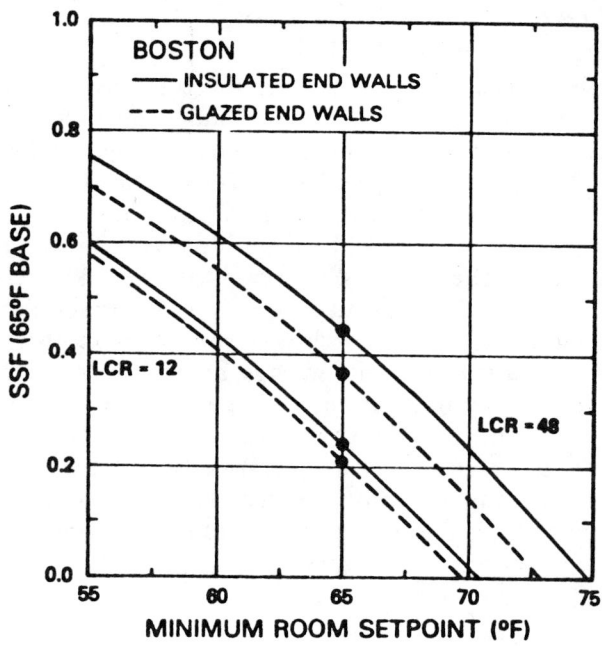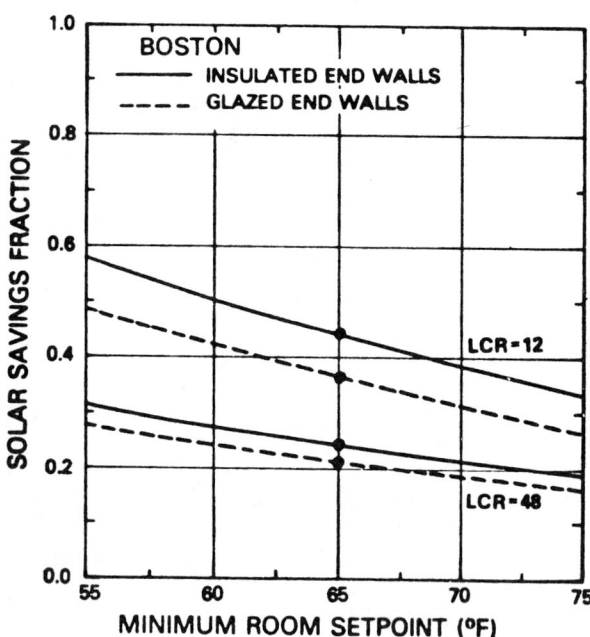

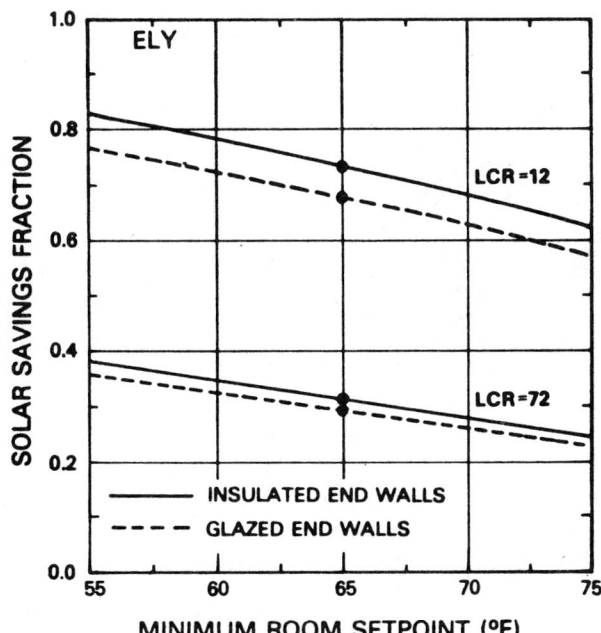
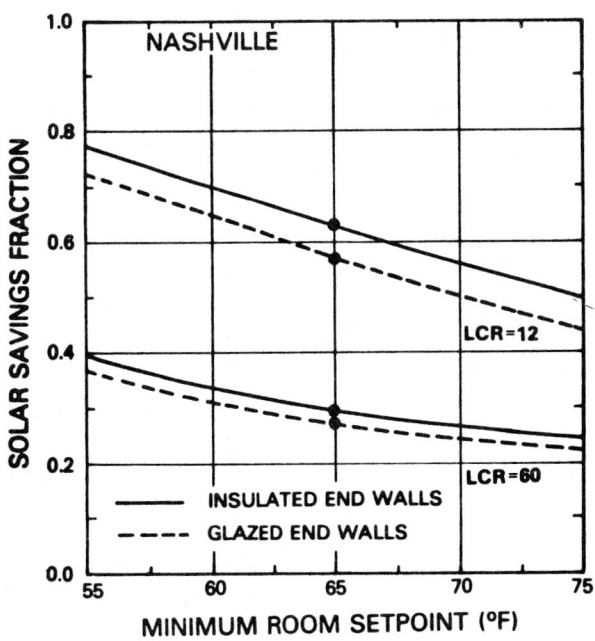
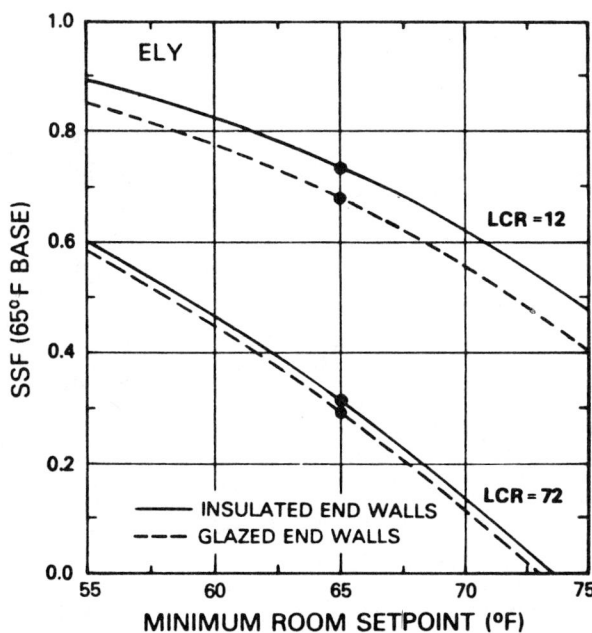

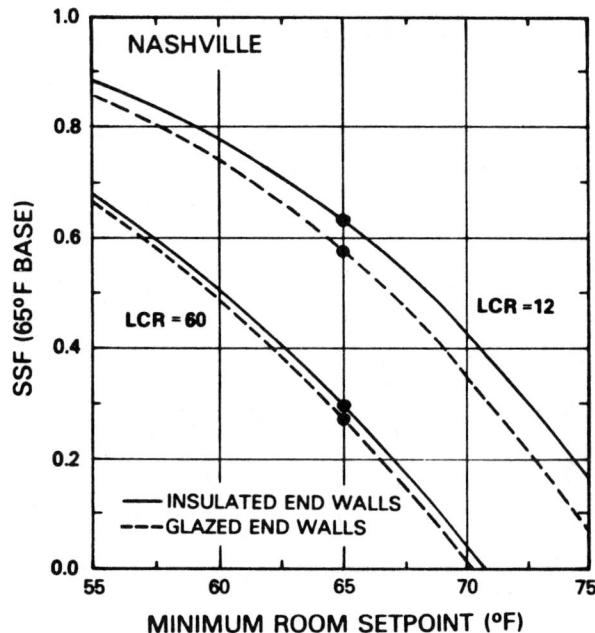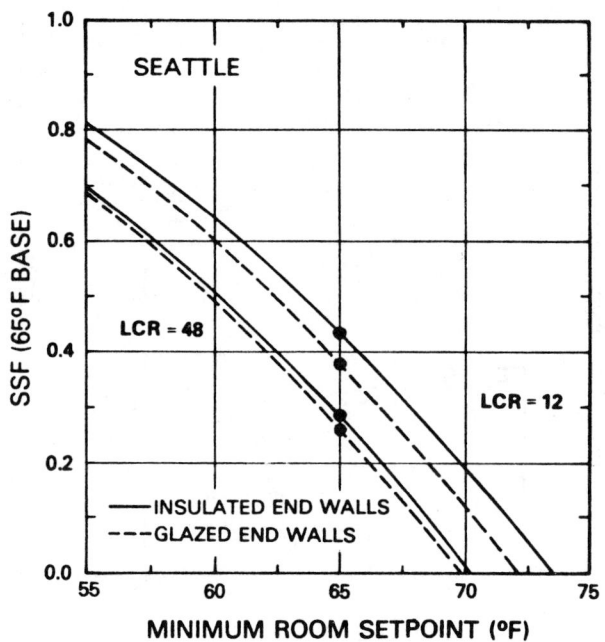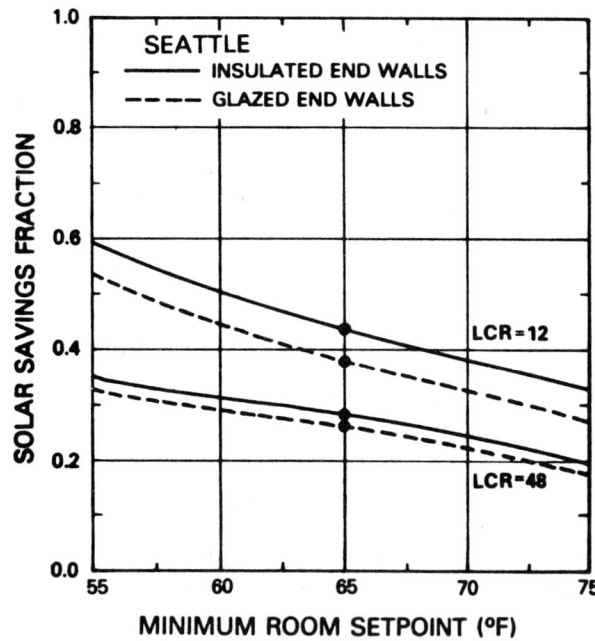

726 / APPENDIX 22

22.3.a.22 Minimum Sunspace Setpoint

The graphs below plot the effect on solar savings fraction of changes in minimum sunspace air temperature (as determined by the thermostat setpoint on the sunspace's auxiliary heating system)—for geometry A with masonry common wall, with and without R9 nighttime insulation in one case and with no nighttime insulation in the other case, with either insulated end walls only or both insulated and glazed end walls (designs A1 and A2 are plotted in the first graph for each city, and designs A1 and A3 in the second graph for each city)—for each of six reference cities. Dots represent reference conditions.

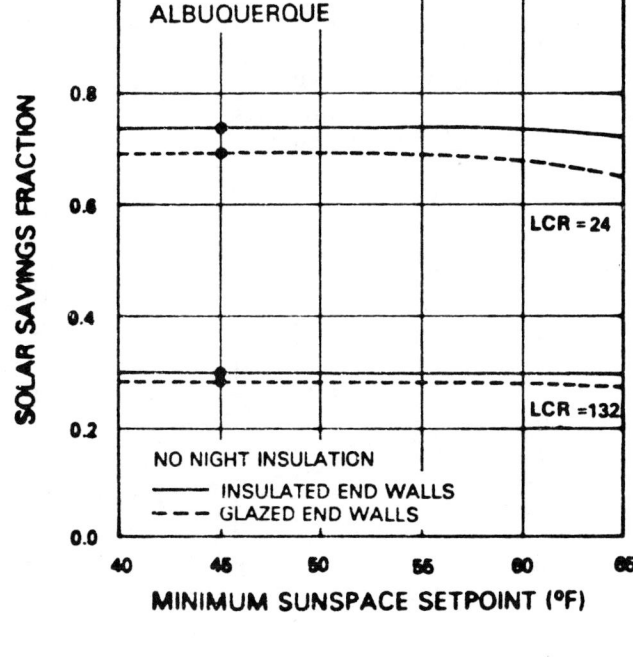

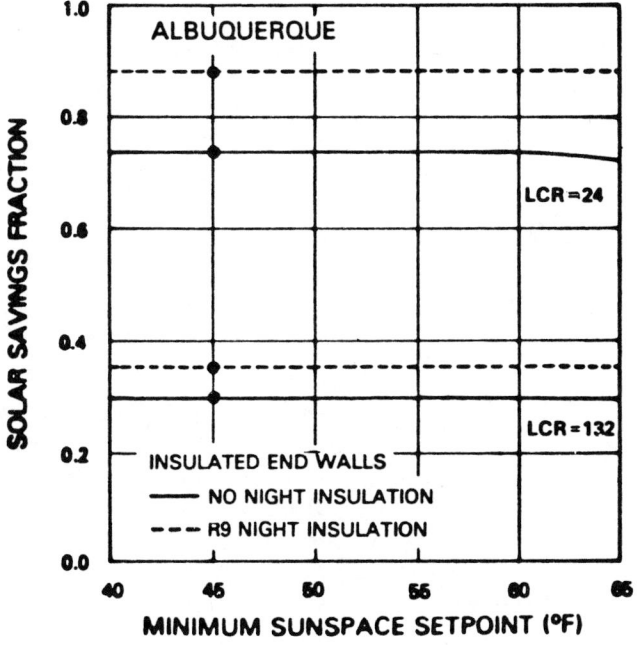

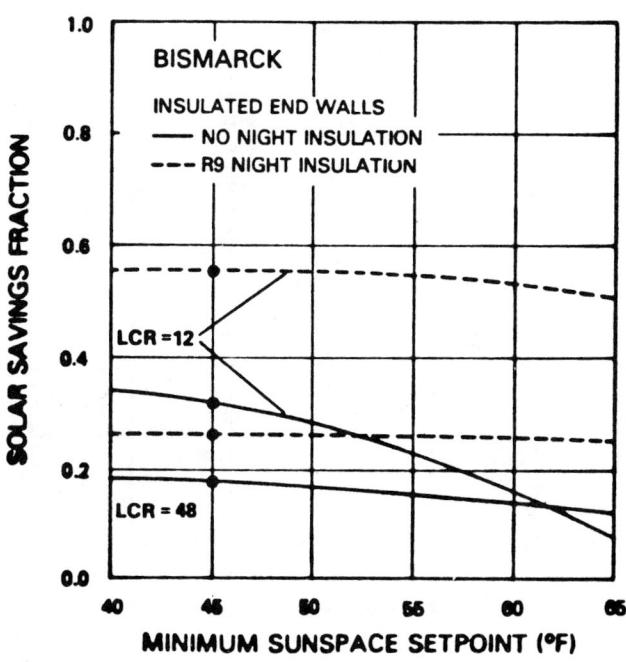

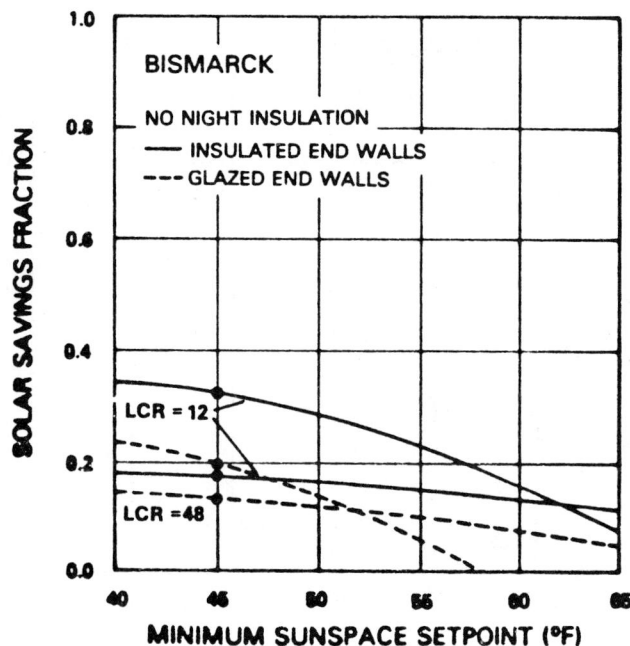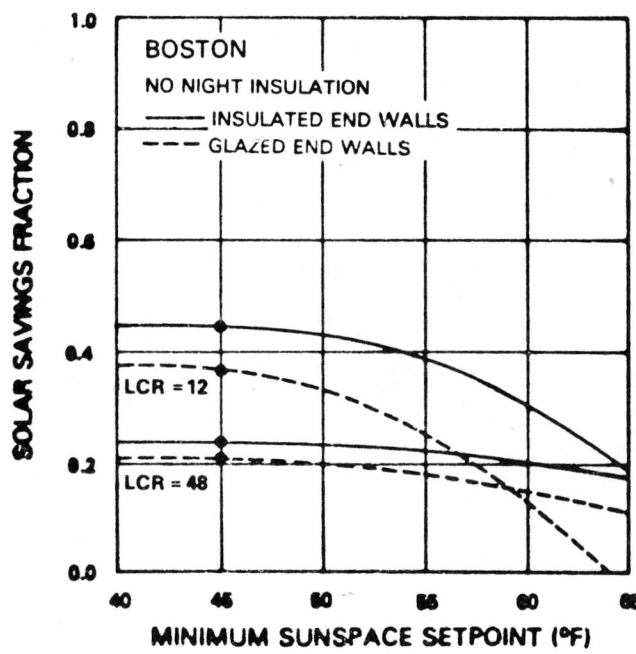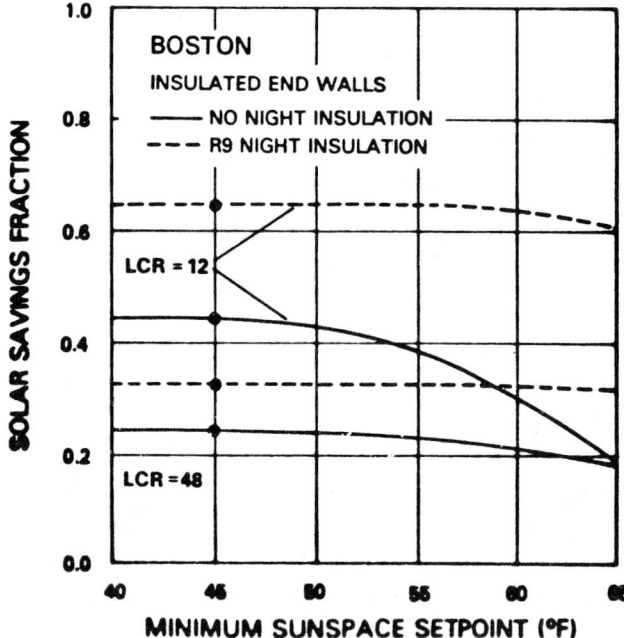

728 / APPENDIX 22

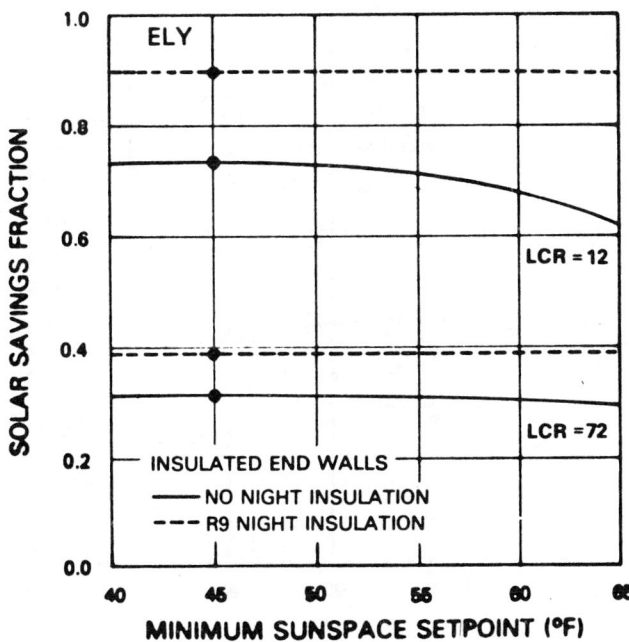
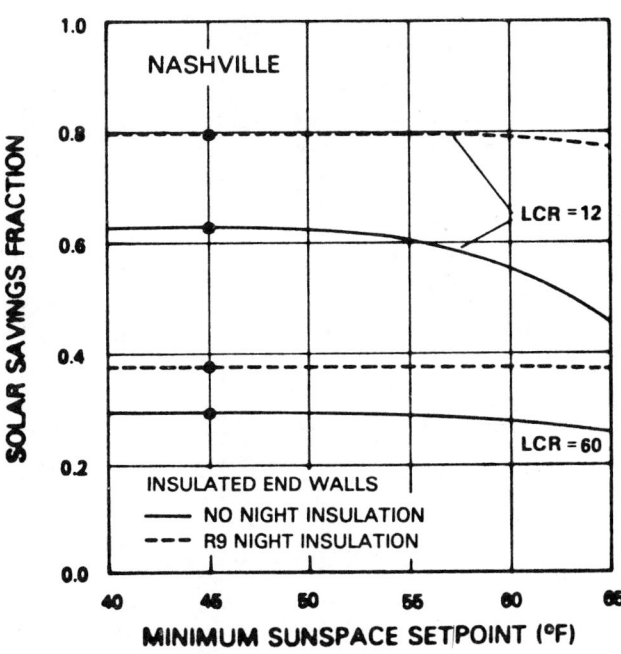
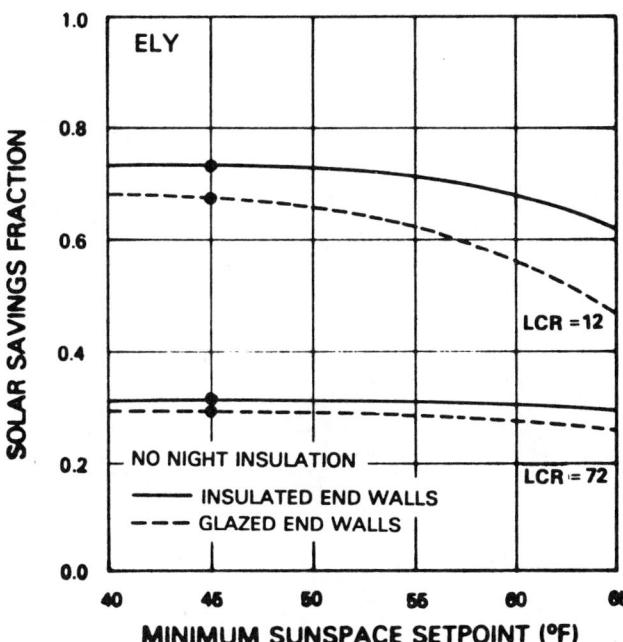

Sensitivity Data / 729

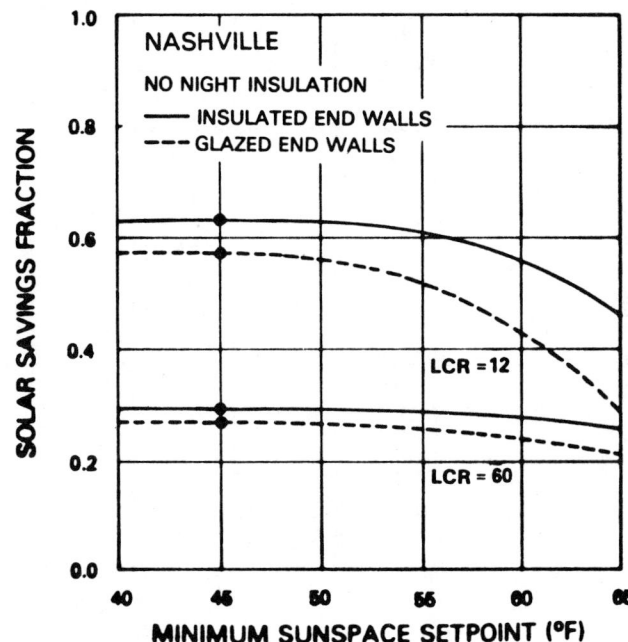
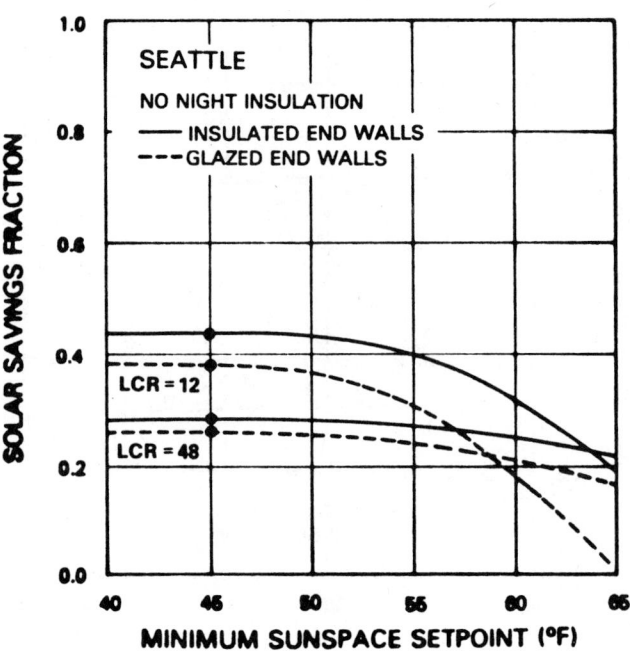
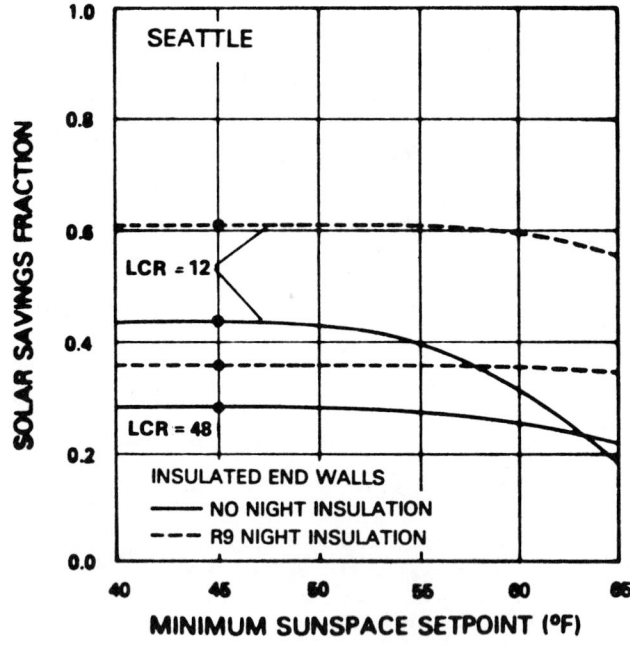

730 / APPENDIX 22

Glossary

The following terms make up a short list of commonly used expressions and concepts in the field of passive solar energy. While the listing is in no sense exhaustive, it may serve a helpful purpose on occasion to both experienced readers and newcomers to the field.

Absorptance – see *Solar absorptance*.

Absorption factor – the fraction of solar radiation transmitted through a glazing system that is absorbed inside the building.

Active system – a solar heating or cooling system that requires external mechanical power to transfer thermal energy.

Adobe – sun-dried block made of mud and straw, a traditional southwestern building material.

Air changes – a measure of the air exchange in a building due to infiltration. One air change is an exchange of a volume of air equal to the interior volume of a building.

Altitude – the angular distance from the horizon to the sun.

Annual calculation method – application of the tables of annual solar savings fraction versus load collector ratio (the "LCR tables") to a calculation of the annual auxiliary heat. Also called the *LCR method*.

Attached sunspace – a solar collector that doubles as useful building space; sometimes also called attached greenhouse or solarium. The term "attached" also more specifically implies a space that shares one common wall with the associated building. Compare with *Semienclosed sunspace*.

Auxiliary heat – the heat provided by a conventional heating system for periods of cloudiness or intense cold when a solar heating system is not sufficient.

Azimuth – the angular distance between the south and the point on the horizon directly below the sun. Also, the angle of a solar glazing or wall relative to an east-west vertical plane; an azimuth of zero defines a glazing or wall that faces due south, a positive azimuth defines an orientation east of south, and a negative azimuth defines an orientation west of south.

Base temperature – a fixed temperature in the definition of degree-days, usually 65°F.

Btu (British thermal unit) – the quantity of heat needed to raise the temperature of 1 pound of water 1 degree Fahrenheit; equivalent of 1,054 joules.

Building load coefficient (BLC) – the net reference load per degree of indoor-minus-outdoor temperature difference per day (in Btu/°F/day).

Calorie – the quantity of heat needed to raise the temperature of 1 gram of water 1°C.

Clerestory window – a window located above and behind (north of) the principal south windows to illuminate north building zones.

Coefficient of heat transmission (U-value) – the rate of heat-flow (in Btu per hour) through a square foot of wall or other building surface when the difference between the indoor and outdoor air temperatures is 1 Fahrenheit degree.

Collection area (A_c) – see *Projected area*.

Conductance – a measure of the ease of heat transfer between two points—namely, the heat flow rate per unit of area per degree of temperature difference (Btu/hr/ft²/°F).

Conduction – the transfer of heat through a static medium, usually a solid such as concrete.

Conductivity – see *Thermal conductivity*.

Convection – heat transfer *between* a surface and an adjacent fluid (usually air or water) and *by* the flow of fluid from one place to another. See also *Natural convection* and *Forced convection*.

Declination – the angular position of the sun at solar noon with respect to the plane of the equator. The decination varies between $-23.45°$ at the winter solstice and $+23.45°$ at the summer solstice.

Degree-days (DD) – see *Heating degree-days*.

Density (ϱ) – the mass of a unit of volume of material (in lb/ft³).

Diffuse radiation – the component of solar radiation that has been scattered by atmospheric molecules and particles. The diffuse radiation is assumed to be isotropic—that is, equally intense from all points of the sky. Also, solar radiation scattered by transmission through diffusing glazings. See *Diffuse transmission*.

Diffuse transmission – the type of solar radiation transmission through a diffusing or translucent glazing—namely, transmission that is scattered by interaction with the glazing material. The diffuse transmitted radiation is assumed to be isotropic, that is, equally intense in all directions.

Diffusing glazing – a translucent glazing, or a glazing that produces diffuse transmission. See *Diffuse transmission*.

Direct gain – the transmission of sunlight directly into the space to be heated, where it is converted to heat by absorption on the interior surfaces.

Double-glazed – covered by two panes of glass or other transparent material.

Emittance – see *Infrared emittance*.

Eutectic salts – a group of materials that melt at low temperatures, absorbing large quantities of heat. As they recrystallize, they release that heat. Used for storing energy as heat.

Extinction coefficient – a property of glazing material that characterizes the solar absorption in the material—namely, the fraction of radiation that is absorbed per unit of path length in the material.

Forced convection – heat transfer between a surface and an adjacent fluid (air in the present context) stream that is produced by external means such as wind or a fan. Compare with *Natural convection*.

Free convection – see *Natural convection*.

Glauber's salt – sodium sulfate ($Na_2SO_4 \cdot 10H_2O$), eutectic salt that melts at 90°F and absorbs about 104 Btu per pound as it does so.

Glazing – transparent or translucent material (glass or various plastics) used to cover the solar aperture.

Gravity convection – the natural movement of heat through a body of fluid that occurs when a warm fluid rises and cool fluid sinks under the influence of gravity. See also *Natural convection*.

Gross reference load – steady-state heat loss from a building, including the solar wall, assuming constant indoor temperature, normally 65°F.

Gypsum board – a common material used to finish interior walls and ceilings to give the appearance of plaster, also called plaster board, Sheetrock, and dry wall.

Heat capacity – a measure of the ability of an element of thermal storage mass to store heat—namely, the heat stored in an element of thermal storage mass per degree of temperature rise (in Btu/°F).

Heating degree-days (*DD*) – the summed differences between a fixed base temperature (usually, 65°F) and the daily mean outdoor temperatures. Only positive differences are counted—that is, when the outdoor mean is less than the base temperature. The daily mean is computed as the mean of the daily minimum and maximum temperatures. For example, if the mean outdoor temperature for one day is 40°F, and if 65°F is taken as the reference point, then twenty-five (65 minus 40) degree-days result. Degree-days are used to determine the demand of a heating season for different locales.

Heating load – the heat-loss (in Btu) from a building in a designated time period.

Heat transfer coefficient – see *Conductance*.

Hybrid system – a solar system that combines some passive and some active elements.

Indirect gain – the indirect transfer of solar heat into the space to be heated from a collector that is coupled to the space by an uninsulated, conductive or convective medium (such as a thermal storage wall or roof pond).

Infiltration – heat loss due to air exchange between heated spaces and the outdoors.

Infrared emittance – a measure of the ability of a surface to emit infrared radiation—namely, the ratio of the infrared radition emitted at a given temperature to the radiation emitted at that temperature by a perfect emitter.

Infrared radiation – the type of thermal radiation whose wave length is above the wave-length range of visible light. This is the preponderant form of radiation emitted by bodies with moderate temperatures, such as the elements of a passive solar building.

Internal heat – heat generated inside the building by sources other than the solar energy system or the space heating equipment—such as that produced by appliances, lights, and people.

Internal sources – the sources of internal heat other than the space heating equipment—such as appliances, lights, and people.

Irradiation – solar radiation—direct, diffuse, and reflected—that strikes a surface.

Index of refraction – a property of glazing material that determines the reflection characteristics of the glazing.

Isolated gain – the transfer of heat into the space to be heated from a collector that is thermally isolated from the space to be heated by physical separation or insulation (such as a convective loop collector or an attached sunspace with an insulated common wall).

Lightweight absorption fraction – the fraction of solar radiation that directly heats the air when it is transmitted through the glazing and when it is reflected from interior surfaces. It is intended to simulate the presence of lightweight objects that absorb solar radiation and rapidly convect heat to the air.

Load – see *Heat load*.

Load collector ratio (*LCR*) – the ratio of the building load coefficient (*BLC*) to the collection area.

Load collector ratio (*LCR*) *method* – see *Annual calculation method*.

Masonry – concrete, concrete block, brick, adobe, stone, and other similar materials.

Mass-area-to-glazing-area ratio – the ratio of the total surface area of massive elements in a direct gain building to the total solar collection area. Massive elements included in this definition are all floors, walls, ceilings, or other interior objects with densities comparable to high-density concrete, provided their surfaces are exposed and located in a room that is at least partially illuminated by direct solar gains.

Monthly calculation method – application of the monthly solar load ratio correlations to a calculation of the monthly auxiliary heat. Also called the monthly *SLR* method.

Monthly solar load ratio (*SLR*) *method* – see *Monthly calculation method*.

Natural convection – heat transfer *between* a surface and adjacent fluid (usually air or water), and *by* the flow of fluid from one place to another, both induced by temperature differences only rather than mechanical means, also called *Free convection*. Compare with *Forced convection*.

Net load coefficient (*NLC*) – see *Building load coefficient*.

Net reference load – steady-state heat-loss from a building, excluding the solar wall, assuming constant indoor temperature, normally 65°F. Compare with *Building load coefficient*.

Nighttime insulation – movable insulation that covers the solar aperture at night and is removed during the day.

Nocturnal cooling – the cooling of a building or heat-storage device by the radiation of heat to the night sky.

Opaque – not able to transmit light; for example, unglazed walls.

Projected area (A_p) – the principal net glazing area projected on a vertical plane.

Radiation – see *Thermal radiation*.

Reference city – a city whose solar radiation and weather data are used to generate sensitivity data.

Reference design – a detailed specification of the passive solar features of an hypothetical passive solar building used as the subject of performance analysis.

Reference nonsolar building – a building similar to the solar building but with an energy-neutral wall in place of the solar wall and with a constant indoor reference temperature.

Reference temperature – a fixed indoor temperature in the definition of the net reference load, usually 65°F. Also, the fixed indoor temperature in the reference nonsolar building, used in the definition of solar savings.

Relative solar savings fraction – a measure of comparison between two solar buildings.

Resistance (R-value) – a measure of the tendency of a material to retard the flow of heat (in hr/°F/ft^2/Btu); the reciprocal of conductance.

Retrofitting – the application of a solar heating or cooling system to an existing building.

Reverse thermocirculation – thermocirculation in the reverse direction—that is, from the heated space to the solar collector (sunspace or Trombe wall). This can occur at night when the heated space is warmer than the collector. It is assumed in the reference designs to be prevented by dampers.

R-value – see *Resistance*.

Selective surface – a surface with a high solar absorptance and a low infrared emittance (to reduce heat loss by infrared radiation).

Semienclosed sunspace – a sunspace that shares three common walls with the associated building.

Sensitivity data – data that express the dependence of the heating performance of a passive solar heating system on individual parameters of the system design.

Shading coefficient – the ratio of the solar heat gain through a specific glazing system under a given set of conditions, to the total solar heat gain through a single layer of clear, double-strength glass under the same conditions.

Shading mask – a section of a circle that is characteristic of a particular shading device. This mask is superimposed on a circular path diagram to determine the time of day and the months of the year when a window will be shaded by the device.

Solar absorptance – the fraction of incident solar radiation that is absorbed upon striking a surface. The radiation not absorbed is reflected.

Solar aperture – that portion of the solar wall covered by glazing.

Solar collection area – see *Projected area*.

Solar load ratio (SLR) – ratio of solar gain to building heat load used in *SLR* correlations.

Solar load ratio (SLR) correlations – correlations between monthly solar savings fractions and monthly solar load ratios.

Solar load ratio (SLR) method – see *Monthly calculation method*.

Solar radiation – thermal radiation emitted by the sun including infrared radiation, ultraviolet radiation and visible light.

Solar savings – the energy savings due to the solar energy system relative to the energy requirement of a reference nonsolar building that has an energy-neutral wall in place of a solar wall.

Solar savings fraction (SSF) – the ratio of the solar savings to the energy requirement of the reference nonsolar building.

Solar time – time of day adjusted so that the sun is due south at noon (due north in the southern hemisphere).

Solar transmittance – the fraction of solar radiation incident on a glazing that is transmitted through the glazing.

Solar wall – the portion or portions of glazed walls at which solar gains are accounted for.

Specific heat (c) – a measure of the ability of a unit of mass of material to store heat—namely, the heat stored in a unit of mass of material per degree of temperature rise (Btu/lb/°F).

Storage volume ratio – ratio of the volume of thermal storage material to the collection area (in feet).

Sun-path diagram – a circular projection of the sky vault, similar to a map, that can be used to determine solar positions and to calculate shading.

Sunspace – see *Attached sunspace*.

Sun-tempered building – a minimally solar building derived from a conventional building by orienting its long axis east-west and placing a substantial fraction of its window area on the south side.

Temperature swing – the range of indoor temperature in the building between day and night.

Thermal conductance – see *Conductance*.

thermal conductivity (k) – a measure of the ease with which heat flows in a material by conduction—namely the heat flow rate by conduction through a unit of distance in the material per unit of area per degree of temperature difference (Btu/hr/ft/°F).

Thermal radiation – energy transfer in the form of electromagnetic waves from a body by virtue of its temperature—including infrared radiation, ultraviolet radiation, and visible light.

Thermal resistance – see *Resistance*.

Thermal storage mass – building elements, usually masonry or water in containers, designed to absorb heat during daytime hours for release later when heat is needed.

Thermal storage volume ratio – see *Storage volume ratio*.

Thermal storage wall – a wall of massive material (masonry or water in containers) placed between the solar aperture and the heated space. Heat is transferred into the space by conduction through the masonry or conduction and convection through the water, and, if openings are provided, by natural convection. See also *Trombe wall* and *Waterwall*.

Thermocirculation – free convection from a warm zone (sunspace or Trombe-wall air space) to a cool zone through openings in a common wall.

Thermocirculation vents – openings in a common wall between cool and warm zones through which thermocirculation occurs. The vents are arranged in pairs, one of each pair near the floor and one near the ceiling.

Thermosiphoning – see *Gravity convection*.

Tilt – the angle of a solar glazing plane relative to a horizontal plane.

Total load coefficient (TLC) – gross reference load per degree of indoor-minus-outdoor temperature difference per day (Btu/°F/day).

Transmittance – see *Solar transmittance*.

Trombe wall – a thermal storage wall of masonry placed between the solar aperture and the heated space. Heat is transferred into the space by conduction through the masonry, and, if vents are provided, by natural convection.

U-value – see *Coefficient of heat transmission*.

Vapor barrier – a layer of material, impervious to water in the vapor state, used to prevent condensation of water within insulation.

Veiling reflection – a blinding or obscuring glare.

Volumetric heat capacity (ρc) – a measure of the ability of a unit of volume of material to store heat—namely, the heat stored in a unit of volume of material per degree of temperature rise (Btu/ft^3/°F).

Waterwall – a thermal storage wall of water in containers placed between the solar aperture and the heated space. Heat is transferred into the space by conduction and convection through the water.

References

CODED REFERENCES

[ABR] Abraham, F. F. "The Determination of Long Wave Atmospheric Radiation." *Journal of Meterology* 17 (1960).

[AES] Atmospheric Environment Service. *Canadian Normals, 1-SI, Temperature, 1941–1970.* Downsview, Ont.: AES, 1975.

[AND-1] Anderson, Bruce, and Riordan, Michael. *The Solar Home Book.* Harrisville, N.H.: Brick House Publishing, 1976.

[AND-2] Anderson, Bruce. *Solar Energy: Fundamentals in Building Design.* New York: McGraw-Hill, 1977.

[AND-3] Anderson, Bruce. "The Solar Heated and Cooled Tyrrell Residence." In *Proceedings of the Passive Solar Heating and Cooling Conference.* Los Alamos, N.M.: Los Alamos Scientific Laboratory, 1976.

[AND-4] Anderson, Bruce and Michal, C. J. "Passive Solar Design." In *Solar Heating and Cooling Workshop Notebook.* Washington, D.C.: Honeywell/DOE, 1978.

[ASH-1] American Society of Heating, Refrigerating, and Air-Conditioning Engineers, *ASHRAE Handbook of Fundamentals.* New York: ASHRAE, 1972.

[ASH-2] American Society of Heating, Refrigerating, and Air-Conditioning Engineers, *ASHRAE Handbook of Fundamentals.* New York: ASHRAE, 1977.

[ASH-3] American Society of Heating, Refrigerating, and Air-Conditioning Engineers, *ASHRAE Handbook of Fundamentals.* New York: ASHRAE, 1982.

[AST-1] Aston, D. "Degree Days, 1941–1970, Alberta." Toronto: Atmospheric Environment Service, August 1978.

[AST-2] Aston, D. "Degree Days, 1941–1970, Atlantic Provinces." Toronto: Atmospheric Environment Service, June 1977.

[AST-3] Aston, D. "Degree Days, 1941–1970, British Columbia." Toronto: Atmospheric Environment Service, December 1978.

[AST-4] Aston, D. "Degree Days, 1941–1970, Manitoba." Toronto: Atmospheric Environment Service, May 1978.

[AST-5] Aston, D. "Degree Days, 1941–1970, Ontario." Toronto: Atmospheric Environment Service, October 1977.

[AST-6] Aston, D. "Degree Days, 1941–1970, Quebec." Toronto: Atmospheric Environment Service, December 1977.

[BAE-1] Baer, Steve. *Sunspots.* Albuquerque: Zomeworks Corporation, 1975.

[BAE-2] Baer, Steve. "Untested Ideas on Spiking Natural Convection Air Heaters With Water Vapor." *New Mexico Solar Energy Association Bulletin,* Vol. 3, No. 2 (1978).

[BAH] Bahadori, Mehdi N. "Passive Cooling Systems in Iranian Architecture." *Scientific American,* February 1978.

[BAI] Bainbridge, David A. "Natural Cooling: Practical Use of Climate Resources for Space Conditioning in California." Working Draft, California: Energy Resources Conservation and Development Commission, 1978.

[BAK] Baker, M. S. of Mazria & Associates, "Passive Solar Technical Planning Study." Electric Power Research Institute Report EPRI EM-1591/TPS 79-750, October 1980.

[BAL-1] Balcomb, J. Douglas. "State of the Art in Passive Solar Heating and Cooling." In *Passive Solar State of the Art: Proceeding of the American Section of ISES,* vol. 2.1, 1978.

[BAL-2] Balcomb, J. Douglas; Hedstrom, J. D.; and McFarland, R. D. "Passive Solar Heating of Buildings." In *Solar Architecture.* Ann Arbor: Ann Arbor Science Publishers, 1977.

[BAL-3] Balcomb, J. Douglas; Hedstrom, J. D.; and Moore, S. W. "Performance Data Evaluation of the Balcomb Solar Home." Paper Presented at Second Annual Solar Heating and Cooling Systems Operational Results Conference, 1979, at Colorado Springs, Colo.

[BAL-4] Balcomb, J. Douglas; Jones, R. W.; McFarland, R. D.; and Wray, O. W. "Expanding the *SLR* Method." *Passive Solar Journal* 1:2 (1982): 67–90.

[BAL-5] Balcomb, J. Douglas. "Heat Storage and Distribution in Passive Solar Heated Buildings."

[BAL-6] Balcomb, J. Douglas; McFarland, R. D.; and Moore, S. W. "Passive Testing at Los Alamos." In *Proceedings of the Second National Passive Solar Conference of the American Section of ISES,* 1977.

[BAL-7] Balcomb, J. Douglas, "Conservation and Solar: Working Together." In *Proceedings of the Fifth National Solar Conference of the American Section of ISES,* 1980.

[BAR] Barret, E. C. *Climatology from Satellites.* London: Methuen & Co., 1974.

[BEN] Bennett, Robert. *Sun Angles for Design.* Bala Cynwyd, Pa.: Robert Bennett, 1978.

[BIE] Bier, Jim. "Vertical Solar Louvers: A System for Tempering and Storing Solar Energy." In *Passive Solar State of the Art: Proceedings of the American Section of ISES,* vol. 2.1, 1978.

[BLI] Bliss, Raymond W., Jr. "Atmospheric Radiation Near the Surface of the Ground: A Summary for Engineers." *Solar Energy,* July/September 1961.

[BRO] Brooks, F. A. *An Introduction to Phsyical Micrometeorology.* Davis, Calif.: University of California, 1959.

[BUD] Budyko, M. I. *The Heat Balance of the Earth's Surface.* Washington, D.C.: U. S. Department of Commerce, Office of Technical Services, 1958.

[CIN] Cinquemani, V.; Owenby, J. R.; and Baldwin, R. G.

Input Data for Solar Systems. Asheville, N.C.: National Climatic Center, 1978.

[COM] U.S. Department of Commerce, Environmental Science Services Administration. *Climatic Atlas of the United States.* Washington, D.C.: DOC, 1968.

[CRO] Crowther, Richard and Solar Group/Architects. *Sun/Earth.* Denver: A. B. Hirschfield Press, 1976.

[DAV] Davis, M. G. "The Thermal Admittance of Layered Walls." In *Building Science,* vol. 8. Oxford, U.K.: Pergamon Press, 1973.

[DUB] DuBois, P. "Matchtiche Effektive Ausstrahburg." *Gerlands Beitraege zur Geophysik* 22, 1929.

[DUF] Duffie, J. A. and Beckman, W. A. *Solar Engineering of Thermal Processes* New York: John Wiley & Sons, 1980.

[DUN] Dunkle, R. V. and Ellul, W. M. J., "Randomly-packed Particulate Bed Regenerators and Evaporative Coolers." *Mech. Chem.* 8, 1972.

[EEE] E^3 Education and Experience in Engineering. "Solar Collector Storage Panel." In *Passive Solar State of the Art: Proceedings of the American Section of ISES,* Vol. 2.2, 1978.

[ERD] Energy Research Development Group. "Energy Efficient Housing—A Prairie Approach." Regina: University of Sasketchewan Press, 1980.

[FAN] Fanger, P. O. "Proposed Nordic Standard for Ventilation and Thermal Comfort," In *Building Energy Management.* New York: Pergamon Press, 1981.

[GEI] Geiger, Rudolf. *The Climate Near the Ground.* Cambridge, Mass.: Harvard University Press, 1950.

[GRI-1] Griffith, J. W. "Analysis of Reflected Glare and Visual Effect from Windows." Paper presented at National Technical Conference of the Illuminating Engineering Society, New York City, September 1963.

[GRI-2] Griffith, J. W.; Balent, J. D.; and Hock, H. G. "Veiling Reflection Studies with Sidewall Lighting." Paper presented at National Technical Conference of the Illuminating Engineering Society, New York City, August 1965.

[GUB] Gubaroff, G. G. et al. *Thermal Radiation Properties Survey.* Minneapolis: Honeywell, 1960.

[HAG-1] Haggard, Kenneth. "First Cost Economic Evaluation for the Atascadero Skytherm House." In *Proceedings of the Passive Solar Heating and Cooling Conference.* Los Alamos, N.M.: Los Alamos Scientific Laboratory, 1976.

[HAG-2] Haggard, Keith; Francis, Barbara; and Palmiter, Larry. "Keeping a Cool Head and Warm Feet." *Solar Architecture.* Ann Arbor: Ann Arbor Science Publishers, 1977.

[HAM] Hammond, Jonathan. "Winters House." In *Proceedings of the Passive Solar Heating and Cooling Conference.* Los Alamos, N.M. Los Alamos Scientific Laboratory, 1976.

[HIL] Hill, John and Lau, Andrew S. Personal communication, 1982.

[HUN-1] Hunn, B. and Jones, M. M. "An Air Thermosiphon Solar Heating System: The Jones House." *New Mexico Solar Energy Association Bulletin,* Vol. 3, No. 7 (1978).

[HUN-2] Hunn, B. and Calafell, D. O. "Determination of Average Ground Reflectivity for Solar Collectors." *Solar Energy* 19: 87–89 (1977).

[IDC] International Daylighting Conference, *Workbook: Daylighting Design Tools Workshop,* 1983. Phoenix: Int'l Daylight Conf., 1983.

[IES] Illuminating Engineering Society. *IES Lighting Handbook.* 5th ed. New York: Illuminating Engineering Society.

[JON] Jones, R. W. and McFarland, R. D. "Attached Sunspace Heating Performance Estimates," *Proceedings of the Fifth National Passive Solar Conference of the American Section of ISES,* 1980.

[JOR] Jordan, R. C. and Threlkeld, J. L. "Solar Energy Availability for Heating in the United States." *Heating, Piping and Air Conditioning,* December 1953.

[KEL-1] Kelbaugh, Doug. "The Kelbaugh House." *Solar Age Magazine,* July 1976.

[KEL-2] Kelbaugh, Doug. "The Doug Kelbaugh House." In *Passive Solar Buildings: A Compilation of Data and Results.* Albuquerque: Sandia Laboratories, 1977.

[KEL-3] Kelbaugh, Doug. "Kelbaugh House: Recent Performance." In *Passive Solar State of the Art: Proceedings of the American Section of ISES,* Vol. 2.1, 1978.

[KNO] Knodrat'yez, K. Ya. *Radiative Heat Exchange in the Atmosphere.* New York: Pergamon Press, 1965.

[LAN-1] Lane, George A.; Hartwick, P. B., and Rossow, H. E. *Macro-Encapsulation of Heat Storage Phase-Change Materials for Use in Residential Buildings.* Midland, Mich. Dow Chemical Company, 1977.

[LAN-2] Lane, George A.; Warner, G. L.; Hartwick, P. B.; and Rossow, H. E. *Macro-Encapsulation of Heat Storage Phase-Change Materials for Use in Residential Buildings.* Midland, Mich.: Dow Chemical Company, 1978.

[LAN-3] Langdon, William K. "Thermal Curtains and Southwall Heating." *Alternative Sources of Energy Magazine,* December 1974.

[LAN-4] Langdon, William K. *Movable Insulation: A Guide to Reducing Heating and Cooling Losses through the Windows in Your Home.* Emmaus, Pa.: Rodale Press, 1980.

LAU] Lau, Andrew. "How to Design Fixed Overhangs." *Solar Age Magazine,* February 1983, 32–38.

[MAH] Mahone, Douglas, "Three Solutions for Persistent Passive Problems." *Solar Age Magazine,* September 1978, 20–23.

[MAL] Maloney, Tim. "Four Generations of Waterwall Design." In *Passive Solar State of the Art: Proceedings of the American Section of ISES,* Vol. 2.2, 1978.

[MAZ-1] Mazria, Edward; Baker, M. S.; and Wessling, F. C. "Predicting Performance of Passive Solar Heated Buildings." In *Solar Architecture.* Ann Arbor: Ann Arbor Science Publishers, Inc., 1977.

[MAZ-2] Mazria, Edward. *The Passive Solar Energy Book.* Emmaus, Pa.: Rodale Press, 1979.

[MCF-1] McFarland, R. D. "PASOLE: A General Simulation Program for Passive Solar Energy." Los Alamos National Laboratory Informal Report, LA-7433-MS, 1978.

[MCF-2] McFarland, R. D. and Jones, R. W. "Performance Estimates for Attached-Sunspace Passive Solar Heated Buildings," In *Proceedings of the 1980 Annual Meeting of the American Section of ISES,* 1980.

[MOR-1] Morris, Scott W. "Natural Convection Collectors." *Solar Age Magazine,* September 1978, 24–27.

[MOR-2] Morris, Scott W. "Storage for Convective Systems." *Solar Age Magazine,* January 1979, 38–41.

[MOR-3] Morris, Scott W. "Natural Convection Solar Collectors." In *Passive Solar State of the Art: Proceedings of the American Section of ISES,* Vol. 2.2, 1978.

[NEU] Neubauer, L. and Cramer, R. "Diurnal Radiant Exchange with the Skydome." *Solar Energy* 9 (1965).

[NIL] Niles, R. W. B. "Thermal Evaluation of a House Using

a Movable-Insulation Heating and Cooling System." *Solar Energy* 18 (1976).

[OLG-1] Olgyay, Victor. *Design with Climate*. Princeton: Princeton University Press, 1963.

[OLG-2] Olgyay, Aladar and Olgyay, V. *Solar Control and Shading Devices*. Princeton: Princeton University Press, 1967.

[PER] Perry, Joseph E., Jr. "The Wallasey School." In *Proceedings of the Passive Solar Heating and Cooling Conference*. Los Alamos, N.M.: Los Alamos Scientific Laboratory, 1976.

[PHI] Phillips, D. W. and Aston, D. *Canadian Solar Radiation Data*. Downsview, Ont.: Atmospheric Environment Service, 1980.

[PIN] Pinney, Neil; Fonda-Bonardi, Marie; and Yu Ying-Nien. "A New Nocturnal Air Cooling System." In *Passive Solar State of the Art: Proceedings of the American Section of ISES,* Vol. 2, 1978.

[PIT] Pittinger, A. L.; White, W. R.; and Yellott, J. I. *The Energy Roof*. Tempe: Solar Building Systems, 1978.

[PLE] Pleijel, G. "The Computation of Natural Radiation." In Statens: *Architecture and Town Planning,* 1954.

[PYD] Pyde, S. E. "Load Management and the La Vereda Passive Solar Community." In *Proceedings of the Sixth National Passive Solar Conference of the American Section of the ISES,* 1981.

[RAM] Ramsey, C. G. and Sleeper, B. R. *Architectural Graphic Standards*. New York: John Wiley & Sons, 1981.

[REI] Reitan, C. H. "Distribution of Precipitable Water Vapor over the Continental U.S." *Bulletin of the American Meterological Society,* 41 (1959).

[ROS] Roseme, G. D.; Berk, J. V.; Boegel, M. L; Halsey, H. I.; Hollowell, C. D.; Rosenfeld, A. H.; and Turiel, I. "Residential Ventilation with Heat Recovery: Improving Indoor Air Quality and Saving Energy." Lawrence Berkeley Laboratory Report No. LBL-9749, Berkeley, California, October 1980.

[SAN] Sandia Laboratories. *Passive Solar Buildings*. Albuquerque: Sandia Laboratories, 1979.

[SEA] Solar Energy Applications Laboratory, *Solar Heating and Cooling of Buildings, Design of Systems*. Fort Collins, Colorado State University.

[SEL] Sellers, W. D. *Physical Climatology*. Chicago: University of Chicago Press, 1965.

[SHU-1] Shurcliff, William A. *Solar Heated Buildings: A Brief Survey*. Cambridge, Mass.: Wm. A. Shurcliff, 1977.

[SHU-2] Shurcliff, William A. *Solar Heated Buildings of North America—120 Outstanding Examples*. Harrisville, N.H.: Brick House Publishing Co., 1978.

[STR] Stromberg, R. P. and Woodall, S. O. *Passive Solar Buildings: A Compilation of Data and Results*. Albuquerque: Sandia Laboratories, 1977.

[TEA-1] Total Environmental Action, Inc. *Design of Residential Buildings Utilizing Natural Thermal Storage: Final Report*. DOE: Brookhaven National Laboratories, 1979.

[TEA-2] Total Environmental Action, Inc. *The Brookhaven House*. Harrisville, N.H.: T.E.A. Inc., 1979.

[TEA-3] Total Environmental Action, Inc. *The Thermal Mass Pattern Book*. Harrisville, N.H.: T.E.A., Inc. 1980.

[TEM] Temple, Peter and Kohler, Joseph. "Glazing Choices." *Solar Age Magazine,* April 1979.

[THO-1] Thom, H. C. S. "The Rational Relationship Between Heating Degree Days and Temperature." *Monthly Weather Review,* January 1954, 1–6.

[THO-12 Thom, H. C. S. "Normal Degree-Days Below Any Base." *Monthly Weather Review,* May 1954, 111–115.

[WEB] Weber, D. D. and Kearney, R. J. "Natural Convection Heat Transfer Through an Aperture in Passive Solar Heated Buildings." In *Proceedings of the Fifth National Passive Solar Conference of the American Section of ISES,* 1980.

[WRA-1] Wray, W. O. and Balcomb, J. D. "Sensitivity of Direct Gain Space Heating Performance to Fundamental Parameter Variations." *Solar Energy* 23, 1979.

[WRA-2] Wray, W. O.; Schnurr, N. M.; and Moore, J. E. "Sensitivity of Direct Gain Performance to Detailed Characteristics of the Living Space." In *Proceedings of the Fifth National Passive Solar Conference of the American Section of ISES,* 1980.

[WRA-3] Wray, W. O. "Design and Analysis of Direct Gain Solar Heated Buildings." Los Alamos National Laboratory Informal Report, LA-8885-MS, 1981.

[WRA-4] Wray W. O. and Weber, D. D. "LASL Similarity Studies: Part I, Hot Zone/Cold Zone: A Quantitative Study of Natural Heat Distribution Mechanisms in Passive Solar Buildings." In *Proceedings of the Fourth National Passive Solar Conference of the American Section of ISES,* vol. 4, 1979.

[WRA-5] Wray, W. O.; Weber, D. D.; and Kearney, R. J. "LASL Sensitivity Studies: Part II, Similitude Modeling in Interzone Heat Transfer by Natural Convection." In *Proceedings of the Fourth National Passive Solar Conference of the American Section of ISES,* vol. 4, 1979.

[YEL-1] Yellott, John I. "When Sunshine Falls on Roofs and Walls." Paper presented at Heating, Piping, Air-Conditioning Conference on Controlling the Industrial Environment, November 2–4, 1970, at Chicago.

[YEL-2] Yellot, John I. "Passive Solar Heating and Cooling Systems." *ASHRAE Journal* (1978).

[YEL-3] Yellott, John I. "Early Tests of the Skytherm System." In *Proceedings of the Passive Solar Heating and Cooling Conference*. Los Alamos, N.M.: Los Alamos Scientific Laboratories, 1976.

ADDITIONAL REFERENCES

Adams, Anthony. *Your Energy Efficient House: Building and Remodeling Ideas*. Charlotte, Vt.: Garden Way Publishing, 1975.

American Institute of Architects. *House Beautiful: Climate Control Project*. Washington, D.C.: AIA Bulletin. 1950.

American Institute of Architects Research Corporation. *A Survey of Passive Solar Buildings*. Washington, D.C.: AIA, 1978.

American Institute of Architects Research Corporation. *Solar Oriented Architecture*. Tempe: Arizona State University, 1975.

American Society of Heating, Refrigerating, and Air-Conditioning Engineers, *ASHRAE Handbook & Product Directory, 1978 Applications*. New York: ASHRAE, 1978.

Aronin, Jeffrey Ellis. *Climate and Architecture*. New York: Van Nostrand Reinhold, 1953.

Arumi, Francisco N. *Thermal Inertia in Architectural Walls*.

McLean, Va.: National Concrete Masonry Association, 1978.

Askew, Gregory L. "Solar Heating Utilizing a Paraffin Phase-Change Material." In *Passive Solar State of the Art: Proceedings of the American Section of ISES,* vol. 2.2, 1978.

Beckman, William A.; Klein, S. A.; and Duffie, J. A. *Solar Heating Design by the F-Chart Method.* New York: John Wiley and Sons, 1977.

Bedrick, J. F.; Millet, M. S.; Spencer, G. S.; Heewagen, D. R.; and Varey, G. B. "The Development and Use of the Computer Program UWLIGHT for the Simulation of Natural and Artificial Illumination in Buildings." In *Passive Solar State of the Art: Proceedings of the American Section of ISES,* vol. 2.2, 1978.

Bitterice, M. G. and McKinley, R. W. *Use Solar Daylight and Heat From Windows to Save Fossil Fuel.* Pittsburgh: PPG Industries, 1978.

Boes, E. C. and Hall, I. J. *Estimating Monthly Means of Daily Totals of Direct Normal Solar Radiation and of Total Solar Radiation on a South-Facing, 45° Tilted Surface.* Albuquerque: Sandia Laboratories, 1977.

Buckley, Shawn. "Thermic Diode Solar Panels: Passive and Modular." In *Proceedings of the Passive Solar Heating and Cooling Conference.* Los Alamos, N.M.: Los Alamos Scientific Laboratories, 1976.

CHA-1 Chadroudi, Day. "Energy Processing Building Materials." In *Solar Architecture.* Ann Arbor: Ann Arbor Science Publishers, Inc., 1977.

CHA-2 Chahroudi, Day. "Buildings as Organisms." In *Passive Solar State of the Art: Proceedings of the American Section of ISES,* Vol. 2.2, 1978.

Clegg, Peter. *New Low-cost Sources of Energy for the Home.* Charlotte, Vt.: Garden Way Publishing, 1975.

Corliss, J. M.; Stickford, G. H.; Klausing, T. A.; Jakob, F. E., and Liu, C. Y. "An Analytical Evaluation of Heat Pipe Augmented Passive Solar Heating Systems." In *Passive Solar State of the Art: Proceedings of the American Section of ISES,* Vol. 2.1, 1978.

Daniels, Farrington. *Direct Use of the Sun's Energy.* Reprint. Westminster, Md.: Ballantine Books, 1974.

Danz, Ernst. *Architecture and the Sun.* London: Thames & Hudson.

Dietz, A. G. H. and Czapek, Edmund L. "Solar Heating of Houses by Vertical Wall Storage Panels." *Heating, Piping and Air Conditioning,* 1947.

Eagen, David M. *Concepts in Thermal Comfort.* Englewood Cliffs, N.J.: Prentice Hall, 1975.

Eccli, Eugene. *Low-Cost, Energy Efficient Shelter for the Owner and Builder.* Emmaus, Pa.: Rodale Press, 1975.

Elmer, Donald B. "Technical Note, Passive Solar Heating and Cooling." *Tracor Sciences & Systems Document,* June 1977.

Farber, E. A.; Smith, W. A.; Pennington, C. W.; and Reed, J. C. "Theoretical Analysis of Solar Heat Gain Through Insulating Glass with Inside Shading." *ASHRAE Journal* 5, 1963.

Faunce, Stuart F.; Guceri, S.; Meakin, J. D.; and Sliwkowski, J. J. "Application of Phase-Change Materials in a Passive Solar System." In *Passive Solar State of the Art: Proceedings of the American Section of ISES,* vol. 2.2, 1978.

Fisher, Rick and Yanda, W. *The Food and Heat Producing Solar Greenhouse: Design, Construction, Operation.* Santa Fe: John Muir Publications, 1976.

Fisk, Pliny, 3rd. "Spatial Distribution and Characteristics of Ten Highmass Earth Materials within the State of Texas." *Passive Solar State of the Art: Proceedings of the American Section of ISES,* Vol. 2.3, 1978.

Franta, Gregory, E. and Olson, K. R., editors. *Solar Architecture: Proceedings of the Aspen Energy Forum of 1977.* Ann Arbor: Ann Arbor Science Publishers, 1978.

Givoni, B. *Man, Climate, and Architecture.* Barking, Essex, U.K.: Applied Science Publishers, 1969.

Griffith, J. W. *Benefits of Daylighting: Cost and Energy Savings.* Reprint. Atlanta: ASHRAE.

Harrison, David D. "Review of Monte Vista Elementary School Greenhouse." In *Proceedings of the Passive Solar Heating and Cooling Conference.* Albuquerque: 1976.

Hastings and Crenshaw. *Window Design Strategies to Conserve Energy.* NBS Building Science Series 104. Washington, D.C.: U.S. Department of Commerce, 1977.

Hauer, Charles R.; Remillard, R. V.; and Nichols, L. "Passive Solar Collector Wall Incorporating Phase-Change." In *Passive Solar State of the Art: Proceedings of the American Section of ISES,* Vol. 2.2, 1978.

Hay, Harold. "Energy, Technology, and Solarchitecture." *Mechanical Engineering,* November 1973.

Henrikson, Hans. *Swedish Building Research Summaries.* Grant 740631. Stockholm, Sweden: Swedish Council for Building Research to the Royal Institute of Technology, Department of Town Planning.

Hunt, Marshall. "The Davis Experience." *Solar Age Magazine,* May 1978, 20–23.

Hutchinson, F. W. "The Solar House." *Heating and Ventilation,* March 1947.

Hutchinson, F. W. and Chapman, W. P. "A Rational Basis for Solar Heating Analysis." *Heating, Piping and Air Conditioning,* July 1946.

International Solar Energy Society, American Section. *Proceedings of the Annual Meeting, Denver, Colo. August 28–31, 1978.* 2 vols. Killeen, Tex.: American Section, International Solar Energy Society, Inc., 1978.

Johnson, Timothy E. "Lightweight Thermal Storage for Solar Heated Buildings." *Solar Energy,* January 1977.

Kassler, Helene. "The Thermic Diode." *Solar Age Magazine,* April 1978, 22–26.

Kreider, Jan F. and Kreith, F. *Solar Heating and Cooling.* New York: McGraw-Hill, 1975.

Kroner, Walter. "Passive Energy Technologies for Residential Construction." Research report, Rensselaer Polytechnic Institute, 1976.

Leckie, Masters, Whitehouse, and Young. *Other Homes and Garbage: Designs for Self-Sufficient Living.* San Francisco: Sierra Club Books, 1975.

Libbey-Owens-Ford Co. *Predicting Daylight as Interior Illumination.* Toledo, Ohio: Libbey-Owens-Ford Co.

Los Alamos Laboratories. *Passive Solar Heating and Cooling: Conference and Workshop Proceedings.* Albuquerque: Los Alamos Laboratories, May 1976.

Lunde, Peter J. *Solar Thermal Engineering Space and Heating and Hot Water Systems.* New York: John Wiley & Sons, 1980.

McClintock, Michael and Frantz, M. "Solar Space Heat and Domestic Hot Water by a System Operating Both Actively and Passively." In *Passive Solar State of the Art: Proceedings of the American Section of ISES,* Vol. 2.2, 1978.

Michal, Charles J. and Lewis, Daniel C. "Natural Thermal Storage: Performance Expectations and Design Techniques Using the M_e Factor." Paper Presented at Solar Energy Storage Options Conference, March 1979, at San Antonio, Tx.

National Solar Heating & Cooling Information Center. *Passive Design Ideas for the Energy Conscious Architect.* Rockville, Md.: NSHCIC, 1977.

National Solar Heating and Cooling Information Center. *Passive Design Ideas for the Conscious Builder.* Rockville, Md.: NSHCIC, 1977.

National Solar Heating and Cooling Information Center. *Passive Design Ideas for the Energy Conscious Consumer.* Rockville, Md.: NSHCIC, 1977.

Niles, Philip W. B. and Haggard, Kenneth L. *Passive Solar Book.* Sacramento: California Energy Commission, Publication Unit, 1981.

Palmiter, Larry; Wheeling, T.; and Corbett, B. *Performance of Passive Test Units in Butte, Montana.* Butte: National Center for Appropriate Technology, 1978.

Pennington, Clark W. "ASHRAE Solar Calorimeter and the Shading of Sunlit Glass." *ASHRAE Journal* 8 (1966).

Pennington, Clark W. and Moore, G. L. "Measurement of Solar-Optical Properties of Glazing Materials." *ASHRAE Journal* 13 (1971).

Pennington, C. W. and Smith, W. A. "Solar Heat Gain Through Double Glass with Between-Glass Shading." *ASHRAE Journal* 6 (1964).

Pennington, C. W.; Smith, W. A.; Farber, E. A.; and Reed, J. C. "Experimental Analysis of Solar Heat Gain Through Insulating Glass with Indoor Shading." *ASHRAE Journal* 6 (1964).

Sarcunanathan, Suppramanian, and Deonarine, S. "A Two-Pass Solar Air Heater." *Solar Energy,* 15, 1973.

Saunders, Norman. "The Overall Solution to Solar Heating." In *Proceedings of the Conference on Energy Conserving Solar Heated Greenhouses.* Marlboro, Vt.: Marlboro College, 1978.

Schade, John. "Insulated Shutters." *Alternative Sources of Energy Magazine,* July 1975.

Scully, Daniel V., "Knowing and Loving, and Never Knowing: Two Houses." In *Passive Solar State of the Art: Proceedings of the American Section of ISES,* Vol. 2.1, 1978.

Shippee, Paul. "The Sunearth Home." In *Passive Solar State of the Art: Proceedings of the American Section of ISES,* Vol. 2.1, 1978.

Shore, Ronald. "A Self-Inflated Movable Insulation System." In *Passive Solar State of the Art: Proceedings of the American Section of ISES,* Vol. 2.2, 1978.

Shurcliff, William A. *Thermal Shutters and Shades: A Systematic Survey of over 100 Schemes for Reducing Heat-Loss Through Large, Vertical, Double-Glazed, South Windows on Winter Nights.* Cambridge, Mass.: Wm. A. Shurcliff, 1977.

Shurcliff, William A. "An Amazing Furnace-Free House." *Solar Age Magazine,* November 1982, 33-35.

Smith, C.C.; Farrer, R. G.; Bedford, S.; and Hannon, J. J. "Solar Space and Soil Heating in a Combination Greenhouse/Residence Structure." In *Proceedings of the Third Annual Conference, Solar Energy for Heating Greenhouses and Greenhouse/Residence Combinations.* Fort Collins: Colorado State University, 1978.

Steadman, Philip. *Energy, Environment and Building.* Cambridge, Mass.: Cambridge University Press, 1975.

Taff, D. C.; Holdridge, R. B.; and Converse, A. O. "Passive vs. Active Collector Systems: A Comparative Study of Efficiency, Capacity and Economics." In *Passive Solar State of the Art: Proceedings of the American Section of ISES,* Vol. 2.3, 1978.

Telkes, Maria. "Trombe Wall with Phase-Change Storage Materials." In *Passive Solar State of the Art: Proceedings of the American Section of ISES,* Vol. 2.2, 1978.

Thomas, Wendell. "The Self-Heating, Self-Cooling House." *The Mother Earth News,* No. 10.

Trombe, Felix; Robert, J. F.; Cabanat, M.; and Sesolis, B. "Some Performance Characteristics of the CNRS House Collectors." In *Proceedings: Passive Solar Heating and Cooling Conference.* Los Alamos, N.M.: Los Alamos Scientific Laboratories, 1976.

U.S. Department of Housing and Urban Development. *Solar Dwelling Design Concepts.* Stock No. 023-000-00334-1. Washington, D.C.: HUD, 1976.

Vild, Donald J. *ASHRAE Research and Principles of Heat Transfer Through Glass Fenestration.* Publication No. 478. Washington, D.C.: Building Research Institute, National Academy of Sciences, 1957.

Wade, Alex and Ewenstein, Neal. *30 Energy Efficient Houses You Can Build.* Emmaus, Pa.: Rodale Press, 1977.

Wallis, Alva L., Jr. *Comparative Climatic Data through 1977.* Asheville, N.C.: National Climatic Center, 1978.

Water Information Center, Inc. *Climates of the States.* Washington, D.C.: Water Information Center, Inc., 1974.

Watson, Donald. *Designing and Building a Solar House: Your Place in the Sun.* Charlotte, Vt.: Garden Way Publishing, 1977.

Williams, Peter. "Annotated Bibliography: Passive Solar Systems." Internal Report. Harrisville, N.H.: Total Environmental Action, Inc., 1976.

Wright, David. *Natural Solar Architecture: A Passive Primer.* New York: Van Nostrand Reinhold Company, 1978.

Wright, David and Andrejko, Dennis A. *Passive Solar Architecture: Logic and Beauty. 35 Outstanding Houses Across the United States.* New York: Van Nostrand Reinhold, 1982.

Index

A. See adjustable direct-gain coefficients; direct-gain area; solar collection area; vent area
a. See cost per square foot of projected solar area
A_c. See solar collection area
A_m/A_g. See mass area to glazing area ratio
A_p. See solar projected area
AAC. See average annual cost
ABSC. See average annual capital, operation, and maintenance costs
absorptance (α). See also solar absorptance
  of covering material (ALFC), 161
  direct-gain
    effect on absorption factors, 442–445
    table of values, 447–460
  direct-gain mass
    effect on performance, 159–161, 640
    effect on solar radiation absorbed, 86–89, 442–445, 447–460
    estimating, 159–160
    reference values, 155, 391
  sunspace
    effect on absorbed solar radiation, 193, 442–445
    effect on performance, 186–188, 695–704
    reference values, 175, 177, 178, 393
    table of values, 447–460
  sun-tempered building mass, effect on performance, 171, 658–659
  thermal storage wall, effect on absorption factors, 443–445
  Trombe-wall
    effect on performance, 93
    reference value, 392
  of various materials (table), 159, 186
  waterwall, reference value, 391
absorption factors
  definition of, 393, 396, 731
  dependence on geometry, 445, 447–460
  direct-gain
    for load collection ratio tables, 461
    reference values, 396
    effect on solar radiation absorbed, 393, 396, 446
    reference designs, table of values, 396
  sunspace/direct gain
    description of, 443–445
    table of values, 447–460
  thermal storage wall, table of values, 444
  Trombe-wall
    for load collector ratio tables, 461
    reference values, 396
  waterwall
    for load collector ratio tables, 461
    reference values, 396
AC. See average cost
ACH. See air changes per hour
acrylics, 204
active system, definition of, 731
additional internal mass, effect on thermal storage wall performance, 103
adjustable direct-gain coefficients (*A, B, C, D, G,* and *R*), 167
  table of values, 167
adobe
  definition of, 731
  as direct-gain mass, 158–159
  properties of (table), 158
ADP. See equivalent down payment
air changes, definition of, 731
air changes per hour (ACH). See also infiltration
  effective, 146–147
  effect on infiltration heat loss, 146–147, 195, 196
  of optimum conservation level, 144
air films, R-values of (table), 221
airflow (Q)
  in thermosiphoning collectors, 33–36, 37
  in sunspaces, 58–59
air heat capacity, 117
air spaces, R-values of (table), 221

Albuquerque, New Mexico, 38, 40, 86, 95, 101, 156, 159, 161, 162, 164, 166, 168, 170, 171, 179, 181, 192, 627, 642
ALFC. See solar absorptance of covering material
ALFSM, definition of, 443
ALFSO, definition of, 443
altitude. See also elevation
  definition of, 731
  solar, 18
ambient temperature (TA)
  definition of, 402
  tables of monthly averages, 403–430
AMP. See equivalent annual payment
AMPB. See equivalent annual payment for backup system
analysis, total period of (*T*), 224
annual auxiliary heat required (Q_{aux}), 173
annual calculation method. See also load collector ratio method
  definition of, 731
annual heating load, total (HLOAD), 228
annualized operation, maintenance, and insurance charges (OMI), 226
  effect on fixed charge rate, 299
antireflective coatings, 94
area (*A*). See also collection area; projected area
  units of, 141, 216
assessment valuation factor (*V*), 230, 285
Atascadero, California, 52, 55
Atascadero House, 52–54, 82
attached greenhouse. See sunspace
attached sunspace, 56–67
  advantages and disadvantages of, 62
  compromise design for, 62, 63
  definition of, 731
  design fundamentals for, 62, 63
  description of, 15, 174, 175
  examples of, 64–67
  glazing systems for, 57
  heat transfer from, 60–62
  integration into building designs, 58
  rockbeds in, 59, 62
  thermal storage in, 59–60, 62

ventilation in, 58
auxiliary heat
  calculation of
    load collector ratio method, 140
    solar load ratio method, 200
  definition of, 731
  in definition of solar savings, 139
  in definition of solar savings fraction, 139
  in mixed system LCR method, 140
  role in design analysis, 139
  system choice, 150
average annual capital, operation, and maintenance costs ($ABSC$), 231
average annual costs (AAC), 231
average cost (AC), 228–229
azimuth. See also glazing azimuth
  definition of, 18, 731

B. See adjustable direct-gain coefficients; dollar benefits; solar radiation correlation coefficients
backdraft dampers
  in rockbeds, 121
  in thermocirculation systems, 37
  in Trombe Walls, 90
backup heat. See auxiliary heat
backup system capital cost (BSC), 230
backup system operation, maintenance and insurance charges (OMB), 230
Baer, Steve, 33, 38, 43
Bainbridge, David, 81
balancing conservation and solar, 142–152
  advantages, 152
  cooling implications, 143, 151–152
  design procedure for, 144–147
  example analysis, 148–151
Balcomb House, Doug, 115
Balcomb House, Kenneth, 115
Barkmann, Herman, 115
base heating degree-days, 65°F ($D65$), 156
base temperature
  definition of, 138–139, 731
  effect on sensitivity curves, 163
  in mixed systems, 200
  nonsolar building, 199
  solar building, 198, 200
Beadwall, 27, 42, 91, 92
Benedictine monastery, 20
benefits, dollar (B), 224
Bismarck, North Dakota, 179, 181, 192, 627
black chrome selective surface, 91
BLC. See building load coefficient
Blue Hill, Massachusetts, 20
Boston, Massachusetts, 36, 86, 99, 156, 168, 170, 171, 173, 179, 627
British thermal units (Btu)
  conversion to SI units, 141, 216
  definition of, 731
Brookhaven House, 47, 67
Brookhaven National Laboratories, 47, 67
BSC. See backup system capital cost
Btu. See British thermal units
building load coefficient (BLC or L)
  definition of, 138, 731
  in definition of load collector ratio, 139
  in definition of net reference load, 139
  effect on conservation-solar balancing, 144–145
  effect on degree-days base temperature, 198, 200
  effect on life-cycle cost, 122
  exclusion of solar glazing from, 138
  in mixed systems, 200
  in solar load ratio method, 198, 200
  units of, 141, 216
building materials, R-values of (table), 218–221

C. See adjustable direct-gain coefficients; empirical constant; sunspace correlation coefficients
c. See heat capacity
Calorie, definition of, 731
Cambridge, Massachusetts, 30
capital investments discount rate (r), 223
capital recovery factor (CRF),
  definition of, 224–225
  examples of use, 225
  tables of values, 241–246
cash-flow analysis, 232–237
  example, 232, 234
  worksheets for, 233, 235–237
caulking, 146, 190
CC. See conservation cost; construction cost
Celsius scale, 141, 216
Center for Advanced Computation, 123, 132
CF. See conservation factor
Champaign-Urbana, Illinois, 132
Charleston, South Carolina, 168, 170
cities. See reference cities
clearness ratio (K_T)
  definition of, 402
  in solar load ratio method, 198
  in solar radiation conversions, 194–195, 393, 402, 431, 432–434, 436–439
  tables of monthly averages, 403–430
clerestory windows
  definition of, 731
  in direct-gain buildings, 158
  effect of glazing tilt in, 97
coefficient of heat transmission (U-value). See also conductance
  definition of, 731
  description of, 27
coefficient of performance (COP), 130–131, 145
collection area (A_c). See also projected area; net glazing area
  compared with projected area, 138
  effect on life-cycle cost, 122
color. See absorptance
comfort sensation, 17–18
common wall, sunspace, reference designs, 174–177, 393, 461
compounding, 223
compounding or discounting periods, number of (t), 223
computers, role of, 139, 161
conductance (U-value). See also coefficient of heat transmission; conductance losses;
  definition of, 141, 731
  mass-to-air, storage wall performance, 98–99
  units of, 141, 216
conductance losses (L), 134
conduction, definition of, 731
conductivity. See thermal conductivity
conservation, 142–151
conservation cost (CC)
  determination of, 146–147
  effect on life-cycle cost, 122
  effect on optimum conservation level, 143–144
conservation factor (CF)
  determination of, 143–144
  effect on optimum conservation level, 143–144
  formula for, 307
  for mixed systems, 146
  table of values, 308–390
conservation formulas, 143–144
construction cost (CC), 124
control temperature. See thermostat setpoint
convection. See also natural convection; forced convection
  definition of, 731
  to remote mass, 158
conversion factors, metric, 216
cooling. See also overheating
  considerations in conservation-solar balancing, 143, 151–152
  general considerations, 141
  passive solar, 68–83
    convective, 74–78
    evaporative, 78
    for large buildings, 83
    ground, 82–83
    peak load reduction and, 83
    radiative, 78–82
    solar control and, 68–74
cooling load, reduction of, 94–97
Cool Pool, 81

COP. *See* coefficient of performance
correlation coefficients, sunspace (*C, D, H, LCR,* and *stdv*), 193
    table of values, 193
cost. *See names of particular types*
cost curve, 228–229
cost of energy, 217
cost evaluation, 123–127, 146–147
    add-on cost, 124, 145, 147
    categorization of costs, 123–124
    conservation cost (*CC*), 124, 146–147
    envelope cost estimation, 124
    fixed costs (*FC*), 124
    functional elements approach, 123
    replacement credits, 124, 127
    variable costs (*VC*), 124
    worksheet for, 125
cost optimization, 124, 128–136, 142–152
    concepts, 124, 128–129, 142
    example calculation, 132–134, 136, 148–151
    procedure, 129–136, 144–147
cost per square foot of projected solar area (*a*), 145
CRF. *See* capital recovery factor
CRFN. *See* nominal capital recovery factor
current cost of competing fuel alternative (P_o), 229

D. *See* adjustable direct-gain coefficients; declination; derivative of solar savings fraction; sunspace correlation coefficients
d. *See* undefined *D*-formula variable
δ. *See PTXR* expenditure escalation rate
dampers. *See* backdraft dampers
Davis, California, 75, 81
Davis House, Paul, 38–39
daylighting, 21
DCH. *See* delivered cost of heat
DD. *See* degree-days
DEC. *See* declination
declination (*DEC* or *D*). *See also* latitude minus declination
    definition of, 731
    formula for, 402
    in solar radiation conversions, 195, 393, 396, 402, 431–441
degree-days (*DD*)
    calculation of base temperature in, 199–200
    definition of, 138–139, 731
    effect of cost optimization on, 128, 130–131
    effect of life-cycle cost on, 122
    effect of life-cycle cost optimum on, 128, 145
    in mixed systems, 200
    in solar load ratio method, 198–199
    sunspace
        base temperature calculation, 147
        in definition of effective sunspace load, 194
        in definition of sunspace solar load ratio, 193, 194
        tables of monthly averages, 403–430
        units of, 141, 216
delivered cost of heat (*DCH*), 229–232
Denmark, 147
density (ρ)
    air, table of values, 117
    definition of, 731
    direct-gain mass
        effect on performance, 106, 107, 157–159,
        reference value, 156, 391
    masonry, reference value, 156, 391
    sunspace masonry
        effect on performance, 179–180, 660–663
        reference values, 175, 177, 393
    sun-tempered building mass, reference value, 168
    Trombe-wall
        effect on performance, 179–180, 660–663
        reference values, 392
    units of, 141, 216
    of various materials (table), 158, 205
Denver, Colorado, 28
derivative of solar savings fraction (*D*), 128–130
    in cost optimization, 145
    determination of, 145, 150
    formula for calculating, 145, 307
    tables of values, 257–284
design basics, 12–21
DG. *See* direct gain
DHC. *See* diurnal heat capacity
dhc. *See* diurnal heat capacity
differential series present worth factor (*DSPWF*)
    definition of, 225–226, 240
    tables of values, 247–252
diffuse ground reflectance (ρ_G), 442
diffuse radiation
    definition of, 731
    in solar radiation conversions, 431
diffuse transmission. *See also* diffusing glazing; diffuse transmittance
    definition of, 731
    in solar radiation conversions, 431
diffuse transmittance (τ_D)
    effect on ground reflectance correction, 442
    effect of extinction coefficient–thickness product on, 443
diffusing glazing
    definition of, 731
    direct-gain
        effect on performance, 162
        reference designs, 155–156
    sunspace, reference designs, 175–178
dimensionless thickness parameter (*x*), 303
    formula for, 304
direct-gain area, 28
direct-gain (DG) systems, 22–31, 153–173
    adjustable coefficients (*A, B, C, D, G,* and *R*), 161
    advantages and disadvantages of, 29
    definition of, 13, 153, 731
    design considerations, 22
    design fundamentals, 23–29
    examples of, 29–31
    movable insulation in, 27–28
    rockbeds in, 23
    solar gain in, 22–23
    thermal performance of, 28–29
direct heat storage, definition of, 107–108
discounted net annual cash flow (*DNCF*), 234
discounted payback (*DPBK*), 238–239
discounted present worth (*PW*), 224
discounting pay periods, number of (*t*), 223
discount rate
    on capital investments (*r*), 223
    in discounting calculations, 224, 226, 227
    effect on fixed charge rate, 301
diurnal heat capacity (*DHC, dhc,* or *s*)
    calculation of, 303–306
    definition of, 303
    dependence on thickness, 303–306
    direct-gain
        estimation procedure, 109
        example calculation, 109–113
        worksheet, 111
    finite thickness with insulation, 303
    formulas for, 303–306
    indirectly heated wall, 306
    infinite thickness, 303
    layered wall, 306
    of various materials (table), 108
DNCF. *See* discounted net annual cash flow
Dodge City, Kansas, 130, 136, 228, 230
dollar benefits (*B*), 224
doors, wood, R-values of (table), 221
double-glazed, definition of, 731
down payment (*DP*)
    in cash-flow analysis, 232
    equivalent time payment, 225
down payment ratio (*DPR*), 227
    effect on fixed charge rate, 300
DP. *See* down payment
ΔP. *See* pressure drop
DPBK. *See* discounted payback

DPR. See down payment ratio
Drumwall, 43, 49
D65. See base heating degree-days, 65°F
DSPWF. See differential series present worth factor
ΔT. See temperature swing

E. See annual energy savings; economic parameters; general escalation annual rate; glazing extinction coefficient–thickness product; real fuel escalation rate
E_1. See fixed charge rate (FCR)
E_2. See fuel cost leveling factor (FF)
e. See fuel cost escalation rate; nominal fuel escalation rate
ϵ. See emittance
EAC. See equivalent annual cost
east-west width (WDTH), 445
economic analysis, 122–136, 142–152, 228–239
   balancing conservation and solar, 142–152
   cash flow in, 232–237
   cost curves, 228–229
   cost evaluation, 123–124, 146–147
   cost optimization, 124–136
   in the design process, 142
   monetary versus physical, 122–123
   payback, 238–239
   terms and formulas, 223–227, 238–239
economic parameter corresponding to minimum heat supply cost (EP), 128
economic parameters (E), 122
effective base temperature. See base temperature
effective solar wall load coefficient (G), 166, 391
effective sunspace load (L), 193
efficiency of furnace, effect on fuel cost, 130–131, 145
elevation
   effect on air density, 117
   of various locations (tables), 403–430
Ely, Nevada, 179, 627
embrittlement, 203–204
emittance (ϵ). See also infrared emittance
   direct-gain glazing surface, reference value, 155
   direct-gain mass surface, reference value, 155
   mass surface, reference value, 391
   normal surface, reference value, 391
   selective surface, reference value, 391
   sunspace glazing, reference value, 177
   sunspace mass surface, reference value, 177
   in thermal storage walls, 93
end of period value (V), 131
end walls, sunspace, reference designs, 174, 175, 393
energy
   cost of (chart), 217
   units of, 141, 216
energy flux, units of, 141, 216
energy savings, annual (E), 28
enhancement factor $(1 + \alpha f)$, 113
EP. See economic parameter corresponding to minimum heat supply cost
equinoxes, 18
equivalent annual cost (EAC), 225
equivalent annual fuel price ($\bar{P}$), 226
equivalent annual payment (AMP), 225
equivalent annual payment for backup system (AMPB), 231
equivalent down payment (ADP), 225
equivalent glazing thickness, 185–186
equivalent thermocirculation area, 189–190
escalation rate (E or e). See fuel cost escalation rate; real fuel escalation rate
eutectic, definition of, 731
evaporation, sunspace reference designs, 175
evaporative cooling, 78
extinction coefficient. See also glazing extinction coefficient
   definition of, 731
extraterrestrial solar flux (I)
   in definition of clearness ratio, 402, 431
   formula for, 402

F. See solar savings fraction
f. See fraction of solar day affording direct sun-lighting
F_{CG}. See view factor
F_1. See function of x
F_2. See function of r and x
fan-forced distribution. See forced convection
fan-forced rockbeds. See rockbeds
FC. See fixed cost
FCR. See fixed charge rate
federal and state marginal income tax rate (FSLTX), 227
FF. See fuel cost leveling factor
fiberglass-reinforced polyester (FRP), 202–203
film coefficient. See conductance, mass-to-air
films
   air, R-values of (table), 221
   plastic, 203–204
finish materials, R-values of (table), 218
fins. See side shades
First Village, New Mexico, 64, 115
fixed charge rate (FCR)
   dependence on discount rate, 301
   dependence on down payment ratio, 300
   dependence on income tax rate, 300
   dependence on mortgage or loan interest rate, 300
   dependence on mortgage term, 301
   dependence on operation, maintenance, and insurance escalation rate, 300
   dependence on operation, maintenance, and insurance rate, 299
   dependence on ownership period, 301
   dependence on property tax rate, 299
   dependence on property value escalation rate, 301
   dependence on resale valuation factor, 302
   effect on cost optimization, 131, 145
   effect on life-cycle cost, 122–123
   tables of values, 285–293
fixed cost (FC)
   effect on cost curves, 228–229
   effect on life-cycle cost, 122
   evaluation of, 124
   in final passive add-on cost estimation,
fixed overhang. See overhang
floor, sunspace, reference designs, 174, 175
forced convection
   definition of, 731
   to remote mass, 157–158
   from waterwall, 99
FP. See fuel price
fraction of solar day affording direct sun-lighting (f), 109
free convection. See natural convection
FRP. See fiberglass-reinforced polyester
FSLTX. See federal and state marginal income tax rate
fuel cost. See fuel price
fuel cost escalation rate (E or e), 131, 226, 294
fuel cost leveling factor (FF)
   effect on cost optimization procedure, 131
   effect on life-cycle cost, 122–123
   tables of values, 294–298
fuel price (FP)
   effect on cost optimization, 128, 130, 145
   effect on life-cycle cost, 122
function of effective solar wall load coefficient (K), 391, 392
function of latitude and declination (Y), 195, 396, 402

function of r and x (F_2), 306
  formula for, 305
function of x (F_1), 303
  formula for, 304
furniture
  in direct-gain systems, 159–160
  in sunspaces, 175, 188
future worth (*FW*), 223
future worth factor (*FWF*)
  definition of, 223, 240
  tables of values, 247–252
FW. *See* future worth
FWF. *See* future worth factor

G. *See* adjustable direct-gain coefficients; effective solar wall load coefficient
γ. *See OMI* expenditure escalation rate
general annual escalation rate (*E*), 247
generalized solar load ratio (*X*), 392
  formula for, 391
geometry, sunspace, reference design description, 174–176
glass, properties of, 202
Glauber's salt, definition of, 732
glazed-vertical-surface-height-to-north-vertical-wall-height ratio (*YBOT*), 444, 445
glazing. *See also* number of glazings
  definition of, 732
glazing air gap
  direct-gain, reference value, 155, 391
  sunspace
    effect on performance, 185, 691–692
    reference value, 175, 177, 391
  thermal storage wall
    effect on performance, 97
    reference value, 391, 461
glazing area. *See* collection area; net glazing area; projected area
glazing azimuth
  direct-gain, reference value, 155, 391
  effect on solar radiation conversions, 431, 432–441
  reference value, 391
  sunspace
    effect on performance, 180–181, 666–667
    reference values, 174–177, 391
  thermal storage walls
    effect on performance, 97
    reference value, 391
glazing extinction coefficient
  direct-gain, reference value, 155, 391
  effect on diffuse transmittance, 442
  effect on equivalent thickness, 185–186
  effect on solar radiation transmitted, 431, 442
  reference value, 391
  sunspace, reference value, 177, 391

thermal storage wall
  effect on absorption factor, 442–443
  effect on performance, 93
  reference value, 391
  of various materials (table), 186
glazing extinction coefficient–thickness product (*E*), 431
glazing index of refraction, reference value
  all systems, 391
  direct-gain, 155
  sunspace, 177
glazing length (*Z*), 445
glazing materials, properties of, 202–204
glazing orientation. *See* glazing azimuth
glazing thickness
  direct-gain, reference value, 155, 391
  effect on diffuse transmittance, 442
  effect on solar radiation transmitted, 431, 442, 693–694
  reference value, 391
  sunspace
    effect on performance, 185–186
    reference value, 175, 177, 391
  thermal storage wall, effect on absorption factor, 442–443
glazing tilt (*TILT*)
  effect on solar radiation conversions, 431, 432–441
  sunspace
    effect on performance, 181–184, 668–677
    optimum value, 182
    problems with small values, 182
    reference values, 174–176, 393
  thermal storage wall, effect on performance, 97
Glenwood Springs, Colorado, 115
glossary, 731–733
gravity convection, definition of, 732
Griffith, J. W., 21
gross reference load, definition of, 732
ground cooling, 82–83
ground reflectance (*RHO*)
  direct-gain, reference value, 155, 391
  effect on solar radiation transmitted, 442
  reference value, 391
  snow and, 99, 100
  in solar radiation conversions, 431, 442
  sunspace, reference value, 177, 391
  typical values, 100
  thermal storage walls, effect on performance, 99–101
  Trombe-wall, reference value, 391
  waterwall
    effect on performance, 99–101
    reference value, 391

gypsum board
  definition of, 732
  properties of (table), 168
  in suntempered buildings, 171

H. *See* height; sunrise hour angle; sunspace correlation coefficients
h. *See* heat transfer coefficient; height; hour; leveled cost of backup heat
Hammond, John, 55, 81
Hay, Harold, 14, 52, 79
heat capacity (*c*). *See also* specific heat
  air, 117
  definition of, 732
  dependence on specific heat, 106
  dependence on temperature swing, 106
  direct-gain, reference values, 154–155, 293
  sunspace, reference values, 175, 177, 178
  sun-tempered buildings, reference value, 168
  Trombe-wall, reference values, 292
  units of, 141, 216
  of various materials (table), 205
  waterwall, reference values, 391
heating degree-days. *See* degree-days
heating load, definition of, 732
Heat Mirror, 30
heat recovery unit (*HRU*), 146
heat storage capacities of thermal storage walls, 43
heat storage effectiveness. *See* diurnal heat capacity
heat supply cost (*HSC*), 128
heat transfer coefficient (*HTC* or *L*), 33, 98, 99. *See also* conductance
heat transmission, coefficient of (*U*-value), 27
heat transmitted per unit time per unit area (*Qtran*), 109
height (*H* or *h*). *See* thermocirculation height
high-mass direct-gain buildings, 155–164
  definition of, 154
HLOAD. *See* total annual heating load
horizontal surface solar radiation (*HS*)
  definition of, 402
  in definition of clearness ratio, 402
  effect on solar radiation absorbed, 393, 446
  in solar load ratio method, 198
  tables of monthly averages, 403–430
horizontal total solar radiation (Q_h), 194, 442
hour angle. *See* sunrise hour angle
hours of sunshine
  mean monthly and hourly (chart), 206
  mean monthly (maps), 207–213

HRU. See heat recovery unit
HS. See horizontal surface solar radiation
HSC. See heat supply cost
HTC. See heat transfer coefficient
HVAC systems, 46
hybrid system. See also rockbeds,
    definition of, 174, 732

I. See extraterrestrial solar flux; incident radiation
i. See interest rate; mortgage interest
ICH. See incremental cost of heat supply
ID. See interest deductions
IES Lighting Handbook, 21
Illinois, University of, 123
Illuminating Engineering Society, 21
improvement factor (Y), 131, 197
incident radiation (I), 442
income tax rate,
    effect on property cost, 226–227
    versus fixed charge rate, 300
incremental cost of heat supply (ICH), 132
index pertaining to option (j), 224
index pertaining to year (t), 224
index of refraction. See also glazing index of refraction
    definition of, 732
Indio, California, 81
indirect gain
    definition of, 732
    in sunspaces, 174
indirect storage location, 107–108
infiltration
    definition of, 732
    estimation of air changes per hour due to, 146–147
    hazards from low rates of, 147
    strategies for reduction of, 146–147
    sunspace
        effect on performance, 190, 717–718
        effect on solar wall conductance, 197
        effect on sunspace load collector ratio, 195
        reference value, 175, 177
inflation rate (θ), 227
infrared emittance, definition of, 732
infrared radiation
    definition of, 732
    in direct-gain systems, 157
    effect on diurnal heat capacity, 107, 108
insulating materials, R-values of (table), 218
insulating panels, 204
insulation. See nighttime insulation
insurance cost
    in cash-flow analysis, 234

equivalent present value of, 226
interest deductions (ID)
    calculation of, 227
    in cash-flow analysis, 234
interest rate (i, R, or r), 131, 135
interest tables, 240–255
internal heat
    definition of, 732
    direct-gain
        effect on performance, 163–164
        reference value, 155, 391
    effect on degree-days base temperature, 198
    estimation of, 199
    in mixed systems, 200
    reference value, 391
    sunspace
        effect on equivalent thermostat setting, 191–192
        reference value, 177, 178, 391
    temperature increment due to, 191–192
    Trombe-wall, reference value, 391
    waterwall, reference value, 391
internal heat generation rate (Q_{int}), 197
irradiation, definition of, 732
isolated gain
    definition of, 732
    in sunspace designs, 174
isolated storage location, 108

j. See index pertaining to option
Jones House, Mark, 39–40

K. See function of effective solar wall load coefficient
K_T. See clearness ratio
k. See thermal conductivity
Kelbaugh House, Doug, 50
Kelvin scale, 141

L. See building load coefficient; conductance losses; effective sunspace load; latitude; rockbed length; wall thickness
landscaping, 21
LAT. See latitude
latitude (LAT or L). See also latitude minus declination
    in solar load ratio method, 198
    in solar radiation conversions, 195, 393, 396, 402, 431–441
    of various localities, 403, 430
latitude minus declination ($LAT - DEC$ or LD)
    tables of midmonth values, 403–430
LCC. See life-cycle cost, uniform annual
LCR. See load collector ratio
LCR_S. See sunspace load collector ratio
length, units of, 141, 216
leveled cost of backup heat (h), 145

leveled fuel cost. See fuel cost leveling factor
life-cycle cost, uniform annual (LCC), 122
    equation for, 122
    optimization procedure, 129–136, 145, 150–151
lightweight absorption fraction
    definition of, 732
    direct-gain
        effect on overheating, 166, 653
        effect on performance, 159–160, 641
        estimation of, 159–160
        reference value, 155
    effect on absorption factors, 445
    sunspace
        effect on performance, 188, 705–708
        estimation of, 188
        reference value, 175, 177, 178
lightweight objects. See lightweight absorption fraction
Living Systems, 81
load. See also building load coefficient; gross reference load; net load; total annual heating load
    in definition of solar load ratio, 139–140, 166, 193
load collector ratio (LCR). See also sunspace load collector ratio
    definition of, 139, 732
    in definition of solar load ratio, 166, 193
    effect on conservation-solar balancing, 145
    in solar load ratio method, 198–200
    starting value, 147
    units of, 141
load collector ratio (LCR) method
    application of, 140
    definition of, 732
load collector ratio (LCR) tables, 257–284, 462–626
    application to load collector ratio method, 140, 461
    caution for large load collector ratio values, 461
loan principal amount (LP), 225
loan term (T), 131, 225
    effect on fixed charge rate, 301
Los Alamos, New Mexico, 45, 93, 94, 99, 101, 115
Los Alamos National Laboratory, 154
Los Alamos Scientific Laboratory, 28, 39, 88, 91
lot orientation, 20
low-iron glass
    effect on performance, 93
    extinction coefficient of, 186
low-mass sun-tempered buildings, 167–173

definition of, 154
LP. See loan principal amount

M. See thermal storage capacity per unit area
Madison, Wisconsin, 86, 101, 156, 161, 162, 168, 170, 171, 173, 627
maintenance cost
   in cash-flow analysis, 234
   equivalent present value of, 226
Mason City, Iowa, 148
masonry, definition of, 732
masonry materials, R-values of (table), 219
masonry properties, reference designs
   all systems, 391
   direct-gain, 155
   sunspace, 177
mass. See also heat capacity; mass area; mass area; mass thickness; remote mass
   direct-gain
      effect on performance, 393, 396
      reference values, 155, 393
   Trombe-wall, effect on performance, 86–89
   waterwall, effect on performance, 86–89
   units of, 141, 216
mass absorptance. See absorptance
Massachusetts Institute of Technology, 30
mass area
   direct-gain
      effect on overheating, 164–166, 650–652
      effect on performance, 157–158, 628–638
      reference values, 155
   sun-tempered buildings
      effect on performance, 171, 654–655
      reference value, 168
mass-area-to-glazing-area ratio (A_m/A_g). See also mass area
   definition of, 157, 732
mass coverings
   direct-gain, effect on performance, 161–162, 643–644
   various, R-values of (table), 162
mass distribution. See mass location
mass location, direct-gain. See also remote mass
   direct, 107–108
   effect on performance, 157–158, 160–161, 642
   indirect, 107–108
   isolated, 108
   reference designs, 155
   rules of thumb, 161
mass thickness
   direct-gain
      effect on diurnal heat capacity, 107, 303–306
      effect on overheating, 164–166, 650–652
      effect on performance, 157–158, 628–638
      reference values, 155, 393
      rules of thumb, 157
   sunspace
      effect on performance, 179–180, 660–665
      reference value, 175, 177
   sun-tempered buildings, reference value, 168
   Trombe-wall
      effect on performance, 86, 88, 89
      reference values, 392
   waterwall, reference values, 391
maximum temperature. See overheating
McFarland, R. D., 154
mean radiant temperature (MRT), 17
Medford, Oregon, 86, 156, 168, 170, 627
metric conversion table, 216
Michel, Jacques, 41, 89
midmonth solar declination (DEC or D), 195
minimum room temperature. See thermostat setpoint
MIT Solar Building V, 30
mixed systems
   direct-gain and Trombe-wall, 105
   in load collector ratio method, 140
   in solar load ratio method, 200
model. See thermal network model
monthly building load (Q_{load}), 140
monthly calculation method. See also solar load ratio method
   definition of, 732
monthly horizontal solar radiation (Q_h), 194
monthly hours of sunshine
   chart of, 206
   maps of, 207–213
monthly solar load ratio method. See solar load ratio method
monthly solar radiation input to solar aperture (Q_s), 140
Morse, E. L., 41
mortgage interest (i)
   tax deductions for, 227
   versus fixed charge rate, 300
mortgage or loan payment factor (X), 230
mortgage payment, in cash-flow analysis, 232, 234
movable insulation. See nighttime insulation
movable shades, 95–97
MRT. See mean radiant temperature

multizone building, 200

N. See number of days in month; number of glazings; year of simple payback
Nashville, Tennessee, 86, 156, 168, 170, 179, 627
National Scientific Research Center, 35
natural convection
   definition of, 732
   in direct-gain zones, 161
   from waterwall, 99
NC. See net square-foot costs; net unit cost
NCF. See net annual cash flow
net annual cash flow (NCF), 232, 234
net glazing area, definition of, 138
net load, definition of, 136
net load coefficient (NLC). See building load coefficient
net present value (NPV), 136
net reference load
   definition of, 139, 732
   in definition of solar savings, 139
   in definition of solar savings fraction, 139
   as reference building heating requirement, 139
net square-foot costs (NC), 132
net unit cost (NC), 132
New York City, 19
Nichols, Susan and Wayne, 43, 64, 115
nighttime insulation
   in attached sunspaces, 57–58
   cost of, effect on optimum conservation level, 143
   definition of, 732
   in direct-gain systems, 27–28
   effectiveness compared to R9, 197
   interpolation for off-reference values, 197, 199
   reference designs
      all systems, 391
      direct-gain, 155
      sunspace, 175, 393
      Trombe-wall, 392
      waterwall, 391
   in thermal storage roofs, 53
   in thermal storage walls, 45–46
nighttime setback. See thermostat setback
nocturnal cooling, definition of, 732
nominal capital recovery factor ($CRFN$). See capital recovery factor
nominal discount rate (r), 227
nominal fuel escalation rate (e), 227
nominal versus real economic terms, 227
north-south depth ($XTOP$), 445
NPV. See net present value
number of compounding or discount-

ing periods (t), 223
number of days in month (N), 393
number of glazings (N)
    direct-gain
        effect on performance, 27–28, 162, 645–646
        reference values, 155, 393
    in equations, 196
    sunspace
        effect on performance, 57–58, 184–185, 684–690
        reference value, 175
    sun-tempered buildings
        effect on performance, 171, 656–657
        reference value, 168
    thermal storage wall
        effect on absorption factor, 444
        effect on performance, 45–46, 90–93
    Trombe-wall
        effect on absorption factor, 393, 396
        reference values, 392
    waterwall
        effect on absorption factor, 393
        effect on performance, 91–93
        reference values, 391
numerical model. *See* thermal network model

Odeillo, France, 13, 35, 89
OMB. *See* backup system operation, maintenance, and insurance charge
OMI. *See* annualized operation, maintenance, and insurance charges
OMI expenditure escalation rate (γ), 231
    effect on fixed charge rate, 300
opaque, definition of, 732
operation cost
    in cash-flow analysis, 232, 234
    equivalent present value of, 226
opportunity cost (r), 223
optimum conservation level, 143–144
optimum glazing air gap, 93
optimum glazing azimuth, 180–181
optimum glazing tilt
    sunspace, 182
    thermal storage wall, 97
optimum mix of conservation and solar, 142–152
optimum orientation. *See* optimum glazing azimuth
optimum reflector tilt, 101
optimum thickness
    sunspace masonry common wall, 179–180
    thermal storage wall, 86, 88–89
optimum vent size, 89–90
option, index pertaining to (j), 224
orientation. *See also* glazing azimuth;
    glazing tilt
    of direct-gain mass, 108
    of lot, 20
overhang
    definition of geometry, 94–95
    effect on heating and cooling loads, 94–97
    sizing procedure, 94
overheating
    direct-gain
        dependence on lightweight absorption fraction, 166
        dependence on mass area, 164–166
        dependence on mass thickness, 164–166
    reduction of by shading, 94–97
ownership period (PT), 131
    effect of fixed charge rate, 301

P. *See* fuel price; periodicity; pressure
$\bar{P}$. *See* equivalent annual fuel price
P_o. *See* current cost of competing fuel alternative
PASOLE, 154
passive solar energy systems
    basic types of, 12
    definition of, 12
payback analysis, 238–239. *See also* simple payback; discounted payback
peak load reduction, 83
Pecos, New Mexico, 49
performance curves
    direct-gain
        versus load collector ratio, 156
        versus S/DD, 167, 397–398
    sunspace
        versus load collector ratio, 178, 179
        versus S/DD, 194, 398–401
    sun-tempered buildings, 168–170
performance measures, 139
period of analysis, total (T), 224
periodicity (P), effect on diurnal heat capacity, 303
ϕ parameter, formula for, 305
plants
    in direct-gain systems, 159–160
    in sunspaces, 174, 175, 188
plastering materials, R-values of (table), 220
plastic films, 203–204
polycarbonates, 204
power, units of, 141, 216
present worth (PW), 223–224
present worth factor (PWF). *See also* uniform series present worth factor; differential series present worth factor
    definition of, 223–224
    tables of values, 247–252

pressure (P) driving airflow in thermosiphoning collector, 34
pressure drop (ΔP)
    equation for sites other than sea-level, 117
    table of values, 117
Princeton, New Jersey, 50
projected area (A_p). *See also* collection area
    definition of, 138, 732
    in definition of load collector ratio, 139
    effect on degree-days base temperature, 198, 200
    in mixed systems
        load collector ratio method, 140
        solar load ratio method, 198–200
    in solar load ratio method, 198–200
property tax
    in cash-flow analysis, 232, 234
    equivalent present value of, 226–227
property tax rate ($PTXR$), 226
    effect on fixed charge rate, 299
PT. *See* ownership period
$PTXR$. *See* property tax rate
$PTXR$ expenditure escalation rate (δ), 231
    effect on fixed charge rate, 301
PW. *See* discounted present worth; present worth
PWF. *See* present worth factor

Q. *See* airflow
Q_{aux}. *See* annual auxiliary heat required
Q_h. *See* horizontal solar radiation
Q_{int}. *See* internal heat generation rate
Q_{load}. *See* monthly building load
Q_s. *See* monthly solar radiation input
$Qtran$ (heat transmitted per unit time per unit area), 109

R. *See* adjustable direct-gain coefficients; radiative heat loss; R-value
r. *See* discount rate on capital investments; nominal rate of discount; opportunity cost; rate of interest; r parameter
ρ. *See* density
ρ_G. *See* diffuse ground reflectance
radiation. *See* infrared radiation; solar radiation; thermal radiation
radiative cooling, 78–82
radiative heat loss (R), 78
Rankine scale, 34
rate of interest (r), 223
$RATIO$, definition of, 104
RB. *See* remaining loan balance
RBF. *See* remaining loan balance factor
ρck. *See* thermal storage wall property

746